D0164116

Congratulations! You have purchased a 12-month, prepaid subscription to the Special Edition of *The Biology Place* for *BIOLOGY,* Fifth Edition, by Campbell, Reece, and Mitchell, a web learning environment for general biology that includes learning activities, study and testing aids, and a wide range of content to help you succeed in your biology course.

To activate your prepaid subscription:

1. Point your web browser to **www.biology.com/campbell**
2. Select *The Biology Place* Special Edition login
3. Enter your preassigned activation ID and password exactly as they appear below in the user ID and password fields:

Activation ID BIPPST12075935

Password temple

4. Select "Login"
5. Complete the online registration form to establish your personal user ID and password.
6. Once your personal ID and password are confirmed, go back to **www.biology.com/campbell** to enter the site with your new ID and password.

Your activation ID and password can be used only once to establish your subscription, which is not transferable.

If you did not purchase this book new and in a shrink-wrapped package, this activation ID and password may not be valid. However, if your instructor is recommending or requiring use of *The Biology Place,* you can find more information on purchasing a subscription directly online at **www.biology.com/campbell**, or you can purchase a subscription at your local college bookstore if your professor has specifically requested it.

BIOLOGY

FIFTH EDITION

Neil A. Campbell
University of California, Riverside

Jane B. Reece
Palo Alto, California

Lawrence G. Mitchell
University of Montana

An imprint of Addison Wesley Longman, Inc.

Menlo Park, California • Reading, Massachusetts • New York • Harlow, England
Don Mills, Ontario • Sydney • Mexico City • Madrid • Amsterdam

Publisher: Jim Green

Executive Editor: Erin Mulligan

Senior Developmental Editors: Pat Burner, Shelley Parlante

Managing Editor: Laura Kenney

Senior Production Editor: Angela Mann

Art and Design Manager: Don Kesner

Senior Art Supervisor: Donna Kalal

Principal Artists for the Fifth Edition: Karl Miyajima, Precision Graphics, Carla Simmons, Mary Ann Tenorio

Photo Researchers: Kathleen Cameron, Roberta Spieckerman

Permissions Editor: Ariane dePree-Kajfez

Text Designer/Layout Artist: Mark Ong, Side by Side Studios

Cover Designer: Yvo Riezebos

Copyeditor: Alan Titche

Proofreader: Anita Wagner

Indexer: Shane-Armstrong Information Systems

Prepress Supervisor: Vivian McDougal

Compositor: American Composition & Graphics, Inc.

Film House: H&S Graphics, Inc.

Market Development Manager: David Horwitz

Channel Marketing Manager: Gay Meixel

High School Binding distributed by
Scott Foresman-Addison Wesley
Glenview, Illinois
ISBN 0-8053-6566-4

On the cover:
Magnolia Blossom, Tower of Jewels, 1925.
Photograph by Imogen Cunningham, ©1974 The Imogen Cunningham Trust.

Library of Congress Cataloging-in-Publication Data
Campbell, Neil A., 1946–
Biology/Neil A. Campbell, Jane B. Reece, Lawrence G. Mitchell.—5th ed.
p. cm.
Includes index.
ISBN 0-8053-6573-7
1. Biology. I. Reece, Jane B. II. Mitchell, Lawrence G. III. Title.
QH308.2.C34 1999
570—dc21 98-45707
CIP

4 5 6 7 8 9 10—VH—02 01 00

Benjamin/Cummings, an imprint of Addison Wesley Longman, Inc.
2725 Sand Hill Road
Menlo Park, CA 94025

To Rochelle and Allison, with love
N.A.C.

To the memory of my parents and to Daniel, with love
J.B.R.

To Charlton, Paul, Rycho, and Wesley, with love and gratitude
L.G.M.

ABOUT THE AUTHORS

Neil A. Campbell combines the investigative nature of a research scientist with the heart of an experienced and caring teacher. He earned his M.A. in zoology from UCLA and his Ph.D. in plant biology from the University of California, Riverside. Dr. Campbell has published numerous research articles on how certain desert and coastal plants thrive in salty soil and how the sensitive plant (*Mimosa*) and other legumes move their leaves. His 30 years of teaching in diverse environments include general biology courses at Cornell University, Pomona College, and San Bernardino Valley College, where he received the college's first Outstanding Professor Award in 1986. Dr. Campbell is currently a visiting scholar in the Department of Botany and Plant Sciences at the University of California, Riverside. In addition to his authorship of this book, he coauthors *Biology: Concepts and Connections* with Drs. Mitchell and Reece.

Jane B. Reece has worked in biology publishing since 1978, when she joined the editorial staff of Benjamin/Cummings. Her education includes an A.B. from Harvard University, an M.S. from Rutgers University, and a Ph.D. in bacteriology from the University of California, Berkeley. At UC Berkeley and later as a postdoctoral fellow in genetics at Stanford University, her research focused on genetic recombination in bacteria. She has taught biology at Middlesex County College (New Jersey) and Queensborough Community College (New York). During her years at Benjamin/Cummings, Dr. Reece played major roles in a number of successful textbooks, including *Microbiology: An Introduction*, by G.J. Tortora, B.R. Funke, and C.L. Case; *Molecular Biology of the Gene*, Fourth Edition, by J.D. Watson et al.; and earlier editions of this book. She is a coauthor of *The World of the Cell*, Third Edition, with W.M. Becker and M.F. Poenie, and of *Biology: Concepts and Connections*.

Lawrence G. Mitchell has 21 years of experience teaching a broad range of life science courses at both undergraduate and graduate levels. He holds a B.S. in zoology from Pennsylvania State University and a Ph.D. in zoology and microbiology from the University of Montana. Following postdoctoral research with the National Institute of Allergy and Infectious Diseases, Dr. Mitchell joined the biology faculty at Iowa State University in 1971. He received the Outstanding Teacher Award at Iowa State in 1982. In addition to numerous research publications in aquatic parasitology, Dr. Mitchell has coauthored the textbooks *Zoology* and *Biology: Concepts and Connections*. He has also developed television courses in general biology and has written, produced, and narrated programs on wildlife biology for public television. Dr. Mitchell is currently an affiliate professor in the Division of Biological Sciences at the University of Montana. Since 1989, he has devoted most of his time to writing and environmental activism.

PREFACE

by Neil A. Campbell

Students of biology are in the right place at the right time. Biology has emerged as the central science, a junction of all the natural sciences and the busiest intersection between the natural sciences and the humanities and social sciences. Biology is daily news. Advances in biotechnology, health science, agriculture, and environmental monitoring are just a few examples of how biology weaves into the fabric of society as never before. Biology is as inspirational as it is important. We are moving ever closer to understanding how a single cell develops into a plant or animal, how the human mind works, how myriad interactions among organisms structure biological communities, and how the enormous diversity of life on Earth evolved from the first microbes. For biology students and their instructors, these are the best of times.

These are also the most challenging times to learn and teach biology. The same information explosion that makes modern biology so exhilarating also threatens to suffocate students under an avalanche of facts and terminology. Most students in an introductory course do not yet own a framework of biological concepts into which they can fit the many new things they learn. More than ever, a biology textbook must help students synthesize a coherent view of life that will serve them throughout their introductory course and during their continuing education. *BIOLOGY* has always reflected my belief that education is all about making connections.

This fifth edition of *BIOLOGY* builds upon the earlier versions' dual goals: to help students construct a conceptual appreciation of life within the context of integrating themes and to inspire students to develop more positive and realistic impressions of science as a human activity. Those teaching values evolved in the classroom, and it is gratifying that the book's conceptual approach and emphasis on science as a process have had such widespread appeal for biology courses all over the world. It is a privilege to help instructors share biology with so many students, but the book's success also brings a responsibility to serve the biology community even better. Thus, in 1996, as I began planning this new edition, I visited dozens of campuses to listen to what students and their professors had to say about their biology courses and textbooks. Those conversations with faculty and students helped shape this fifth edition of *BIOLOGY*, by far the most ambitious revision in the history of the book.

My New Coauthors

In planning this new edition, I was humbled by the task ahead and had to admit that it was finally time to seek the help of coauthors. Biology is a much bigger subject than it was when the first edition of the text was published in 1987, but in adapting to the growth of biology, I did not want the book to become bigger or encyclopedic. Adding content is not very hard, but deciding what to leave out is much tougher. Authoritative renewal across the whole book would require expert reevaluation of the current state of each field of biology and judicious decisions about which ideas are most important to develop in depth in an introductory textbook. Given the breathtaking progress of biology on its many fronts, I was concerned that if I continued as sole author, the text could no longer deliver the educational integrity and scientific accuracy that the biology community expects from the book.

Once I decided to recruit help, it was easy to choose coauthors: Jane Reece and Larry Mitchell. They are my longtime collaborators on our nonmajors textbook, *Biology: Concepts and Connections*, and they have contributed in various ways to earlier editions of *BIOLOGY*. Dr. Reece's knowledge of genetics and molecular biology and Dr. Mitchell's expertise in zoology and ecology complement my own research background in cell biology and botany. More importantly, Jane and Larry share my commitment to "getting it right" and to having a positive impact on biology education.

My new coauthors brought fresh ideas and reenergized me and the whole *BIOLOGY* team. Among Jane's many contributions are two new chapters: Cell Communication (Chapter 11) and The Genetic Basis of Development (Chapter 21). Larry's extensive work on this edition includes the book's new capstone chapter, Conservation Biology (Chapter 55).

Even with the help of my coauthors, I worked as hard as ever on this edition of *BIOLOGY*. In addition to writing chapters, I edited the entire book to ensure consistent voice and level. Because of my confidence in Jane and Larry as partners in the book, I also finally had more time to help plan the supplements, especially the new CD-ROM and web site. All that is new in this fifth edition of *BIOLOGY* and its supplements is firmly stamped with the hallmarks that have distinguished earlier editions.

BIOLOGY's Hallmark Features

Unifying Themes and Key Concepts. A thematic approach continues to distinguish *BIOLOGY* from an encyclopedia of life science. Chapter 1 introduces 11 themes that resurface throughout the text to help students make connections in their study of life. The core theme, the theme of all themes, is evolution, the thread that ties all of biology together.

On the scale of individual chapters, an emphasis on key concepts keeps students focused on each topic's most important ideas. For example, Chapter 9, Cellular Respiration, teaches about 10 key concepts, including "Feedback mechanisms control cellular respiration." Each chapter's key concepts appear in three places: in an overview of key concepts on the first page of the chapter; within the body of the chapter as the headings of the main subsections under broader topic headings; and in the review of key concepts at the end of the chapter. In placing such emphasis on a manageable number of big ideas in each chapter, we do not equate "conceptual" with "superficial." Oversimplifying a complex topic is a disservice to students, for without the support of sufficient detail, key concepts collapse into meaningless factual statements and lose their value as organizing principles. We avoid trying to cover everything, but if a concept is essential to include, we believe it should be developed in enough depth to leave a lasting impression on students.

BIOLOGY's key concepts and unifying themes work together to help students develop a coherent view of life. Notice, for example, that "Feedback mechanisms control cellular respiration" is a concept specific to the topic of a particular chapter. However, one of the book's overarching themes, the theme of regulation, helps students fit the concept of controlling respiration into a broader context that applies to many other biological processes. The key concepts give form to each chapter; the integrating themes connect the concepts and give form to the whole book.

Science as a Process. As one of the book's unifying themes, the nature of science is introduced at some length in Chapter 1, but *BIOLOGY*'s commitment to featuring science as a human activity does not end there. Case studies announced by the subtitle "Science as a Process" enrich many chapters throughout the book by balancing "what we know" with "how we know" and "what we do not yet know." *BIOLOGY* also features many Methods Boxes, which demystify science by walking students through laboratory and field methods in the context of experiments. For example, a new Methods Box in Chapter 21, "Using Mutation to Study the Genetic Basis of *Drosophila* Development," explains the rationale behind experiments that have revolutionized developmental biology. Another example is a new Methods Box in Chapter 55 that illustrates the use of computerized mapping techniques in conservation biology. Eight new interviews with influential scientists, which introduce the eight units of the book, help personalize science and portray it as a social activity of creative men and women. Many of the Challenge Questions at the ends of chapters also engage students in the process of science.

Another one of *BIOLOGY*'s themes, the relationship of science and technology to society, helps students connect biology to their other college courses and to their daily lives. It is important for students to realize that ethics has a place in science, even in basic research, and that the benefits of technology also bring the need to examine values and make choices. At the end of each chapter, Science, Technology, and Society questions encourage students to fit the biological concepts they have learned into their broader view of the world.

A Marriage of Text and Illustrations. Biology is a visual science, and many of our students are visual learners. We take as much care in creating new ways to illustrate the story of life as we do in writing the narrative. I have always authored the illustration program for *BIOLOGY* side by side with the text, and Drs. Reece and Mitchell take the same approach. Beginning with the first draft of this edition, we collaborated with the designer, artists, photo researchers, and developmental editors to embed the illustrations and their legends into the flow of each chapter. Compared to the more common publishing model of deferring work on the illustration program until the text manuscript is completed, our approach creates logistical challenges. There is a benefit, however: a union of words and pictures that makes each chapter an integrated lesson.

With this commitment to integrating text and graphics, earlier editions of *BIOLOGY* pioneered many innovations in how science textbooks are illustrated. For example, many chapters in *BIOLOGY* use a sequence of orientation diagrams as road signs to help students keep track of where they are in biological processes such as photosynthesis (Chapter 10) and gene expression (Chapter 17). In figures that walk students through stepwise processes, circled numbers in the illustration are keyed to circled numbers in the figure legend or text. We think this approach works better than cluttering the illustration itself with too much text. Throughout the book, color coding and icons help students connect concepts from chapter to chapter. For example, proteins are generally purple, and ATP always appears as a yellow sunburst. These and other features of *BIOLOGY*'s illustration tradition have had a noticeable impact on many other excellent textbooks, but we have continued to refine our model to raise the standard even higher. In addition to the many new diagrams that support the narrative in this fifth edition, we have found subtle ways to improve the teaching effectiveness of a great number of the other figures in the book.

An "Overview-Closer Look" Teaching Style. *BIOLOGY* begins the development of many complex topics, such as cellular respiration (Chapter 9) and gene expression (Chapter 17), with a panoramic view—an overview—of what the whole process accomplishes. Text and figures then invite the student inside the process for a closer look at how it works. The orientation diagrams, miniature versions of the overview illustration, with appropriate parts highlighted as "you are here" aids, keep the overall process in sight even as its steps unfold. This teaching approach complements *BIOLOGY*'s other hallmarks in helping students navigate through their biology course.

BIOLOGY's Versatile Organization

BIOLOGY makes no pretense that there is one "correct" sequence of topics for a general biology course. No two syllabi match, and the individual approaches of biology instructors are one of the strengths of biology education. Therefore, we have built *BIOLOGY* to be versatile enough to serve diverse courses, whether the instructor chooses to start with molecules or ecosystems or somewhere in between. The eight units of the book are largely self-contained, and most of the chapters can be assigned in a different sequence without substantial loss of continuity. For example, instructors who integrate plant and animal physiology into a systems approach can merge chapters from Units Six and Seven to fit their courses. As another example, instructors who begin their course with ecology and continue with a "top-down" approach can assign Unit Eight right after Chapter 1. It is the themes introduced in Chapter 1 that make the book so versatile in organization by providing students with a biological context no matter what the topic order of the syllabus.

An Overview of *BIOLOGY* and a Few Examples of What's New in the Fifth Edition

Unit One: The Chemistry of Life. Many students struggle in general biology courses because they are uncomfortable with basic chemistry. Chapters 2–4 help such students by developing just the chemical concepts that are essential to success in a beginning biology course. This edition features improved explanations of atomic and molecular electron orbitals and their influence on molecular shape and function. We designed the chapters of Unit One so that students of diverse educational backgrounds can use them for self-study, reducing the amount of valuable class time instructors need for reviewing chemistry before they move on to biology. However, Chapter 5 (The Structure and Function of Macromolecules) and Chapter 6 (An Introduction to Metabolism) provide important orientation even for students with solid chemistry backgrounds.

Unit Two: The Cell. Chapters 7–12 build the study of cells upon the theme of correlation between structure and function. For example, the role of membranes in ordering cell functions is emphasized throughout the unit. In the fifth edition, this unit includes a completely new chapter, called Cell Communication (Chapter 11). This chapter synthesizes our current understanding of the functions and mechanisms of cell communication and signal transduction, giving students a conceptual foundation that applies to many processes at the cellular and organismal levels.

Unit Three: Genetics. Chapters 13–21 trace the history of genetics, from Mendel to DNA technology, with the process of science as a major theme. In this edition, the chapters on molecular genetics (Chapters 16–20) have undergone a major revision, entailing both updating and organizational improvements. For instance, the explanations of transcription and translation in Chapter 17 (From Gene to Protein) have been reorganized into two distinct sections that should help students keep these two important stages of gene expression clear in their minds. The material in Chapter 19, The Organization and Control of Eukaryotic Genomes, has been extensively revised and expanded; it includes updated coverage of chromatin structure, repetitive DNA, transcriptional regulation, and cancer. Chapter 20, DNA Technology, features updated discussion of the Human Genome Project and introduces students to some exciting new techniques for studying gene function. As in previous editions, *BIOLOGY*'s extensive coverage of human genetics is integrated throughout the unit. Concluding the unit is a new chapter on one of the most important areas in biology, The Genetic Basis of Development (Chapter 21). This chapter builds on molecular, cellular, and genetic principles to introduce concepts of development that apply to both animals and plants. *Drosophila*, the nematode *C. elegans*, and the plant *Arabidopsis* provide the main case studies here. (More extensive coverage of vertebrate development and plant development appears in later units.)

Unit Four: Mechanisms of Evolution. As the core theme of *BIOLOGY*, evolution figures prominently in all parts of the book, but Chapters 22–25 focus specifically on how life evolves and how biologists study evolution and test evolutionary hypotheses. Chapter 22, Descent with Modification: A Darwinian View of Life, sets the stage for the unit by grounding evolutionary biology in the process of science. Students will then find many examples throughout the unit of research and debate about mechanisms of evolution. For instance, a new section in Chapter 24 summarizes the ongoing debate about how to define a species. An expanded discussion of systematics in Chapter 25 emphasizes current efforts to meld classification and evolutionary history. New and updated sections highlight molecular and cladistic methods and their use in formulating phylogenetic hypotheses.

Unit Five: The Evolutionary History of Biological Diversity. Chapters 26–34 view the diversity of life in the context of key evolutionary junctures, such as the origin of prokaryotes; the evolution of the eukaryotic cell; the genesis of multicellular life; the colonization of land; and the adaptive radiation of plants, fungi, and animals. This evolutionary theme contrasts with a "parade of phyla" approach. The scientific view of biological history and diversity is being transformed by discoveries of new fossils, increased application of molecular biology in systematics, and a growing consensus for cladistic classification. These trends demanded extensive revision of this unit. Chapter 26 sets the stage by comparing several classification schemes, including the traditional five-kingdom system, an eight-kingdom system, and a superkingdom, three-domain system. Chapters 27 and 28 reflect the view that the enormous diversity among prokaryotes and protists deserves recognition at the highest levels of classification. Unit Five centers on current thinking and ongoing debates about kingdoms and superkingdoms while retaining its applicability to courses with labs based on the five-kingdom system. Extensive emphasis on biodiversity is making a comeback in many courses, and we expanded our coverage of the evolution and diversity of plants (now two chapters instead of one) and animals (now three chapters instead of two.)

Unit Six: Plant Form and Function. Chapters 35–39 showcase plants in the evolutionary context of adaptation to terrestrial environments. Much of the new material emphasizes how plant cell biologists and molecular biologists are reshaping our understanding of plant morphology, physiology, and development. Just one example is the progress in our understanding of disease resistance in plants, now a major section in Chapter 39 (Control Systems in Plants). Throughout the unit in this edition, there is also more emphasis on symbiotic relationships between plants and bacteria and fungi.

Unit Seven: Animal Form and Function. The interaction between organisms and their environment is the focus of Chapters 40–49, which take a comparative approach to exploring adaptations that have evolved in diverse animal groups. The themes of regulation and bioenergetics are accented throughout, and many figures have been reconstructed to bring out these themes. Some highlights include new information on fat metabolism in Chapter 41, current findings about coreceptors and protease inhibitors relevant to HIV in Chapter 43, more about the role of cell adhesion in animal development in Chapter 47, an updated account of human brain development in Chapter 48, and an expanded section on locomotion in Chapter 49.

Unit Eight: Ecology. Chapters 50–55 emphasize the connections between ecology and evolution. The unit also reflects the urgent need for basic ecological research in an era when the growing human population and its technology are threatening the biosphere. Highlighting this issue, and serving as a capstone for the unit and the entire book, is a new chapter on conservation biology (Chapter 55). Firmly grounded in modern ecological science, this chapter examines the significance of biodiversity and discusses conservation strategies at the species, population, community, ecosystem, and landscape levels.

Learning Aids at the Ends of Chapters

At the end of each chapter is a **Review of Key Concepts**, which restates each concept with a short explanation. With each concept are page numbers and figure numbers telling students where to return if they are having trouble understanding that concept. A **Self-Quiz** helps students assess their comprehension of key concepts, and many of these questions also require students to apply concepts or solve problems. **Challenge Questions** give students opportunities to verbalize their interpretations of key concepts, to extrapolate from what they have learned to new situations, to think and write critically about controversies, to apply quantitative skill to biological problems, and to generate testable hypotheses of their own. The **Science, Technology, and Society** questions encourage students to think about biology's place in our culture and to confront the potential consequences of applied biology. A short, annotated **Further Reading** list directs students to articles and books that explore some of the chapter's topics in more depth. New to this edition is a list of **Web Links**, Internet resources selected and annotated by Dr. Iain Miller of the University of Cincinnati.

A Complete Teaching and Learning Package

This full range of supplements is highlighted by breakthrough media tools that have been developed specifically for use with *BIOLOGY*, Fifth Edition. They have been crafted with the same careful attention to scientific accuracy and teaching effectiveness found in the text. In each case, the value to instructor and student is based on integration with the text.

New! **Interactive Study Partner CD-ROM** by Richard Liebaert of Linn-Benton Community College. Packaged with the textbook at no additional charge, the Study Partner includes 120 interactive exercises, animations, and lab simulations. Icons (7.2) appear throughout the textbook to guide students to the applicable activity on the CD-ROM. Also featured are a glossary and 20 quiz questions per chapter, with feedback for all answers and page references to the relevant explanations in the textbook. The CD-ROM is also the student's gateway to the special edition of *The Biology Place*, the web site that now complements this textbook.

***New!* The Special Edition of *The Biology Place* for *BIOLOGY*, Fifth Edition.** For the past three years, I have been working with several other educators and Peregrine Publishers to develop *The Biology Place*, a web-based learning environment for general biology students. A subscription to the customized version of *The Biology Place*, specifically keyed to the fifth edition of *BIOLOGY*, is now available to all students (see www.biology.com/campbell for more details). *The Biology Place* includes BioCoach (interactive tutorials with self-assessment), investigative learning activities, LabBench (lab simulations), TestFlight (customizable practice exams), Biology News (summaries of newsworthy research breakthroughs), chapter-specific web links, and access to a web version of the Interactive Study Partner. An instructor's entrance to the special edition of *The Biology Place* provides access to fifth edition art, resources, and a new on-line Course Management System.

***New!* Course Management On-Line.** The Addison Wesley Longman Course Management System allows the instructor to offer on-line quizzing and to create a course-specific web site for posting a syllabus and/or lecture schedule. The instructor can also use the system to conduct discussion groups and administer the course.

***New!* Instructor's Presentation CD-ROM.** Over 100 animations and QuickTime movies and the complete art program from *BIOLOGY* are included. A presentation program enables instructors to design a slide show of images, import illustrations and photos from other sources, transfer figures into other software programs, and edit the figures for content.

***New!* More Transparency Acetates.** For the first time, all of the drawings in the text are available, more than 600 figures in full color.

***New!* Instructor's Guide to Media** by Iain Miller, University of Cincinnati. This useful guide to media supplements provides guidance on how best to utilize the electronic tools available with the text and how to integrate them into your course. It contains information on web sites, course management software, presentation software, the special edition of *The Biology Place*, and the Interactive Study Partner.

Instructor's Guide by Mark Sheridan, North Dakota State University. Tailored to match the content in *BIOLOGY*, Fifth Edition, the guide offers teaching assistance that is especially helpful for first-time instructors.

Student Study Guide by Martha Taylor, Cornell University. This printed learning aid provides a concept map of each chapter, chapter summaries, a variety of interactive questions, and chapter tests.

***Investigating Biology*, Third Edition,** by Judith Giles Morgan, Emory University, and M. Eloise Brown Carter, Oxford College of Emory University. This investigative laboratory manual asks students to pose their own hypotheses, design their own experiments, and then analyze their data. Also available: ***Investigating Biology*, Third Edition, Annotated Instructor's Edition** and **Preparation Guide**.

35-mm Slides. This package contains approximately 300 slides keyed to the text.

Printed Test Bank and **Computerized Testing Software for Macintosh and Windows**, edited by Dan Wivagg, Baylor University. Featuring more questions than ever, a new program allows the professor to view and edit electronic questions, transfer them to tests, and print them in a variety of formats.

▪ ▪ ▪

The real test of any textbook is how well it helps instructors teach and students learn. We welcome comments from the students and professors who use *BIOLOGY*. Please address your suggestions for improving the next edition directly to me:

Neil Campbell
Department of Botany and Plant Sciences
University of California
Riverside, California 92521

ACKNOWLEDGMENTS

Every edition of *BIOLOGY* has been an immense journey for many people, and the authors of this fifth edition wish to extend heartfelt thanks to the numerous instructors, researchers, students, editors, and artists who have contributed to this and previous editions. It is a privilege to be part of a global community dedicated to excellence in science education.

Special thanks go to several biologists who contributed in major ways to the revision of fourth edition chapters or helped with the creation of new chapters for the fifth edition. Chapter 21, The Genetic Basis of Development, greatly benefited from the advice of Ann Reynolds (University of Washington), Jeff Hardin (University of Wisconsin), and Nancy Hopkins (MIT) and from early drafts of new material that Ann and Jeff created. Jeff Hardin also helped reshape and refine Chapter 47, Animal Development, and was the main contributor of updated material for that chapter. Mary Jane Niles (University of San Francisco) did a superlative reorganization and updating of Chapter 43, The Body's Defenses. Steven Lebsack (Linn-Benton Community College) provided a number of new review questions for the ends of chapters. We thank these contributors for helping us make our fifth edition more accurate, up to date, coherent, and pedagogically effective. It was a pleasure to work with them.

Further helping us improve *BIOLOGY*'s scientific accuracy and pedagogy, 50 scholars and teachers, cited on page x, provided detailed reviews of one or more chapters for this edition. Numerous other professors and their students offered suggestions by writing directly to Neil Campbell. Those correspondents include: Anne Ashford (University of New South Wales), Peter Atsatt (University of California, Irvine), Anton Baudoin (Virginia Polytechnic Institute), Bruce Chorba (Mercer County Community College), Raymond Damian (University of Georgia), Marshall Darley (University of Georgia), Marianne Dauwalder (University of Texas), Earl Fleck (Whitman College), Joseph Frankel (University of Iowa), Warren Gallin (University of Alberta), Larry Giesman (Northern Kentucky University), Bruce Grant (Widener University), Albert Hendricks (Virginia Polytechnic Institute), Becky Houck (University of Portland), Roger Lloyd (Florida Community College of Jacksonville), Ernst Mayr (Harvard University), Matthias Ochs (Georg-August-Universitat Gottingen), Barry Palevitz (University of Georgia), Tom Rambo (Northern Kentucky University), Thomas Reimchen (University of Victoria, Canada), Cyril Thong (Simon Fraser University), Gordon Ultsch (University of Alabama), Itzick Vatnick (Widener University), and Cherie Wetzel (City College of San Francisco).

Of course, the authors alone bear the responsibility for any errors that remain in the text, but the dedication of our contributors, reviewers, and correspondents makes us especially confident in the accuracy of this edition.

Many colleagues have also helped shape this fifth edition by discussing their research fields and ideas about biology education. Neil Campbell thanks numerous UC Riverside colleagues, including Katharine Atkinson, Elizabeth Bray, Richard Cardullo, Mark Chappell, Timothy Close, Darleen DeMason, Norman Ellstrand, Robert Heath, Anthony Huang, Tracy Kahn, Elizabeth Lord, Carol Lovatt, Robert Neuman, Eugene Nothnagel, John Oross, Kathryn Platt, Mary Price, David Reznick, Rodolfo Ruibal, Clay Sassaman, Irwin Sherman, Vaughan Shoemaker, William Thomson, Linda Walling, Nickolas Waser, and John Moore (whose "Science as a Way of Knowing" essays have had such an important influence on the evolution of *BIOLOGY*). Neil also is grateful to Pius Horner, who taught him by example during their many years of team-teaching at San Bernardino Valley College. Among the scientists elsewhere who shared their expertise and ideas with us are Wayne Becker (University of Wisconsin), Deric Bownds (University of Wisconsin), Nicholas Davies (Cambridge University), Daniel Gagnon (Université du Quebec), David Glenn-Lewin (Wichita State University), Peter Grant (Princeton University), Ira Herskowitz (UC San Francisco), Nancy Hopkins (MIT), Richard Hutto (University of Montana), Robert Lambrecht (U.S. Forest Service), Christopher Murphy (James Madison University), Andrée Nault (Biodôme de Montréal), David Patterson (University of Sydney), Mitchell Sogin (Marine Biological Laboratory), and Kirwin Werner (Salish Kootenai College).

Interviews with prominent scientists have been a hallmark of *BIOLOGY* since its inception, and conducting these interviews was one of the great pleasures of revising the text. To open the eight units of this fifth edition and help put human faces on our science, we are proud to include interviews with Mario Molina, Bruce Alberts, Mary-Claire King, Richard Dawkins, Elisabeth Vrba, Gloria Coruzzi, Terence Dawson, and Michael Dombeck.

BIOLOGY, Fifth Edition, results from an unusually strong synergism between a team of scientists and a team of publishing professionals. A fresh design, ambitious revision of the art and photo program, and the nesting of key concepts into a pedagogical hierarchy are three examples of goals for this edition that created challenges for the publishing team. Heading Addison Wesley Longman's Biology Group, publisher Jim Green brought his high publishing standards and flair for team-building to our endeavor; we are grateful for his leadership. Linda Davis, general manager of Benjamin/Cummings Publishing, has shared the book team's commitment to excellence and provided strong support over the long haul. Our former editor Lisa Moller worked closely with the authors and book team in the planning and early draft stages. The authors are extremely grateful to Laura Kenney, production managing editor, who has been a key member of the book team since those early stages. Senior production editor Angela Mann brought her professional expertise and can-do attitude to the book's production; moreover, she brought calm and good cheer to the often-turbulent later stages of the production process. Two senior developmental editors, the incomparable Pat Burner and Shelley Parlante, worked closely with the authors through all phases of manuscript and art preparation; we are deeply grateful for their countless contributions to the book and for their professionalism, obsessive dedication, and exceptional patience. Our copyeditor and fellow biologist Alan Titche brought a rare mix of content knowledge and editing skills to late stages of the manuscript.

Art and design are key elements of *BIOLOGY*'s teaching effectiveness. Don Kesner, art and design manager of the AWL Biology Group, and book designer Mark Ong worked closely and patiently with us until we had the right design to make each chapter a better teaching tool. Senior art supervisor Donna Kalal shepherded all of our art through production. We are grateful to the artists who worked on the new and revised figures for this edition, including Mary Ann Tenorio, Karl Miyajima, Carla Simmons, and the artists of Precision Graphics. On the four previous editions Carla was a major shaper of the art program, which continues to serve students well. Photo researchers Kathleen Cameron and Roberta Spieckerman searched for just the right photos to reinforce key concepts. Yvo Riezebos designed a handsome cover that reflects the freshness of this new edition yet remains true to the elegant simplicity that distinguished the covers of earlier editions.

We are all indebted to our supreme leader for the project, executive editor Erin Mulligan. Erin brought her leadership ability, publishing expertise, native intelligence, and down-to-earth sense of humor to the *BIOLOGY* team in 1997. She immediately began mustering the forces that gave rise to the fifth edition's outstanding supplements package. Erin's fierce dedication to this project and to excellence in educational publishing has touched us all.

Working closely with Erin and with authors of the supplements, associate editor Thor Ekstrom and associate producer Claire Cameron coordinated the print and electronic supplements, respectively. Key players on the supplements included senior developmental editor Pat Burner, senior production editor Larry Olsen, project editor Kathy Yankton, and publishing assistant Maureen Kennedy. Important contributions to the electronic components of the package were made by Russell Chun (with his threefold expertise in art, computer animation, and biology), designer Peilin Nee, several freelancers, and our incomparable media lab manager, Guy Mills. We also thank Laura Maier, executive editor at Peregrine Publishers, and Claire Cameron for their collaborative leadership in developing the special edition of *The Biology Place* as a web site coordinated with our book. Guy Mills, Lee Stayton, Rachel Collett, Betsy Burr, and Todd Rodgers also played key roles in customizing *The Biology Place* for *BIOLOGY*, Fifth Edition. As a result of the efforts of all these people, students and professors have available a marvelous assemblage of support materials for their biology courses.

Dorothy Zinky is hands-down the most efficient and unflappable administrative assistant the authors have ever had the pleasure of depending upon. Publishing assistants Kelly Millon, Claire Cameron (before her promotion), Natalia Cortes and Maureen Kennedy provided essential support to the authors and the book team. Hilair Chism helped us by editing the transcripts of some of the interviews. Anita Wagner and Carol Lombardi were careful proofreaders for the fifth edition, and Charlotte Shane created a very useful index. Prepress manager Lillian Hom and prepress supervisor Vivian McDougal worked wonders to bring about the culmination of all our efforts—a bound book. The entire publishing group worked together to craft a book that teaches biological concepts even better than earlier editions.

Both before and after the publication of the book, we are fortunate to have the support of AWL's marketing professionals. We gratefully acknowledge the contributions of market development manager David Horwitz and members of Stacy Treco's marketing team, especially biology marketing manager Gay Meixel. We also thank Lillian Carr and Bob Leone for creating effective promotional materials.

The field staff that represents *BIOLOGY* on campus is our living link to the students and professors who use the text. The field representatives tell us what you like and don't like about the book, and they provide prompt service to biology departments. The field reps are good allies in science education, and we thank them for their professionalism in communicating the merits of our book without denigrating other publishers and their competing texts.

Finally, we wish to thank our families and friends for their encouragement and for enduring our obsession with *BIOLOGY*. In particular, we are grateful for the support of Rochelle and Allison Campbell (N.A.C.); Deborah Gale, Dan Gillen, Robin Heyden, Sharmon Hilfinger, Jeff Reece, Susan Weisberg, and Hugues d'Audiffret (J.B.R.); and Wesley, Paul, and Paula Mitchell (L.G.M.).

Neil Campbell
Jane Reece
Larry Mitchell

Reviewers of the Fifth Edition

Martin Adamson	University of British Columbia
Robert Atherton	University of Wyoming
Judy Bluemer	Morton College
Deric Bownds	University of Wisconsin, Madison
Charles H. Brenner	Berkeley, California
Mark Browning	Purdue University
Bruce Chase	University of Nebraska, Omaha
David Cone	Saint Mary's University
Marianne Dauwalder	University of Texas, Austin
Michael Dini	Texas Tech University
Andrew Dobson	Princeton University
Susan Dunford	University of Cincinnati
Carl Frankel	Pennsylvania State University, Hazleton
Simon Gilroy	Pennsylvania State University
David Glenn-Lewin	Wichita State University
Elliott Goldstein	Arizona State University
Linda Graham	University of Wisconsin, Madison
Mark Guyer	National Human Genome Research Institute
Ruth Levy Guyer	Bethesda, Maryland
Jeff Hardin	University of Wisconsin, Madison
Colin Henderson	University of Montana
Ira Herskowitz	University of California, San Francisco
R. James Hickey	Miami University
Charles Holliday	Lafayette College
Steven Hutcheson	University of Maryland, College Park
Robert Kitchin	University of Wyoming
Jacqueline McLaughlin	Pennsylvania State University, Lehigh Valley
Paul Melchior	North Hennepin Community College
Brian Metscher	University of California, Irvine
Michael Misamore	Louisiana State University
Deborah Mowshowitz	Columbia University
Gavin Naylor	Iowa State University
Raymond Neubauer	University of Texas, Austin
Caroline Niederman	Tomball College
Maria Nieto	California State University, Hayward
Patricia O'Hern	Emory University
Daniel Pavuk	Bowling Green State University
Debra Pearce	Northern Kentucky University
Martin Poenie	University of Texas, Austin
Deanna Raineri	University of Illinois, Champaign-Urbana
Gary Reiness	Lewis & Clark College
Walter Sakai	Santa Monica College
Gary Saunders	University of New Brunswick
Erik Scully	Towson State University
Edna Seaman	Northeastern University
Elaine Shea	Loyola College, Maryland
Susan Singer	Carleton College
John Smol	Queen's University
Mitchell Sogin	Woods Hole Marine Biological Laboratory
Kathryn VandenBosch	Texas A & M University

Reviewers of Previous Editions

Kenneth Able (State University of New York, Albany), John Alcock (Arizona State University), Richard Almon (State University of New York, Buffalo), Katherine Anderson (University of California, Berkeley), Richard J. Andren (Montgomery County Community College), J. David Archibald (Yale University), Leigh Auleb (San Francisco State University), P. Stephen Baenziger (University of Nebraska), Katherine Baker (Millersville University), William Barklow (Framingham State College), Steven Barnhart (Santa Rosa Junior College), Ron Basmajian (Merced College), Tom Beatty (University of British Columbia), Wayne Becker (University of Wisconsin, Madison), Jane Beiswenger (University of Wyoming), Anne Bekoff (University of Colorado, Boulder), Marc Bekoff (University of Colorado, Boulder), Tania Beliz (College of San Mateo), Adrianne Bendich (Hoffman-La Roche, Inc.), Barbara Bentley (State University of New York, Stony Brook), Darwin Berg (University of California, San Diego), Werner Bergen (Michigan State University), Gerald Bergstrom (University of Wisconsin, Milwaukee), Anna W. Berkovitz (Purdue University), Dorothy Berner (Temple University), Annalisa Berta (San Diego State University), Paulette Bierzychudek (Pomona College), Charles Biggers (Memphis State University), Robert Blystone (Trinity University), Robert Boley (University of Texas, Arlington), Eric Bonde (University of Colorado, Boulder), Richard Boohar (University of Nebraska, Omaha), Carey L. Booth (Reed College), James L. Botsford (New Mexico State University), J. Michael Bowes (Humboldt State University), Richard Bowker (Alma College), Barry Bowman (University of California, Santa Cruz), Jerry Brand (University of Texas, Austin), Theodore A. Bremner (Howard University), James Brenneman (University of Evansville), Donald P. Briskin (University of Illinois, Urbana), Danny Brower (University of Arizona), Carole Browne (Wake Forest University), Herbert Bruneau (Oklahoma State University), Gary Brusca (Humboldt State University), Alan H. Brush (University of Connecticut, Storrs), Meg Burke (University of North Dakota), Edwin Burling (De Anza College), William Busa (Johns Hopkins University), John Bushnell (University of Colorado), Linda Butler (University of Texas, Austin), Iain Campbell (University of Pittsburgh), Deborah Canington (University of California, Davis), Gregory Capelli (College of William and Mary), Richard Cardullo (University of California, Riverside), Nina Caris (Texas A & M University), Doug Cheeseman (De Anza College), Shepley Chen (University of Illinois, Chicago), Henry Claman (University of Colorado Health Science Center), Lynwood Clemens (Michigan State University), William P. Coffman (University of Pittsburgh), J. John Cohen (University of Colorado Health Science Center), John Corliss (University of Maryland), Stuart J. Coward (University of Georgia), Charles Creutz (University of Toledo), Bruce Criley (Illinois Wesleyan University), Norma Criley (Illinois Wesleyan University), Richard Cyr (Pennsylvania State University), Marianne Dauwalder (University of Texas, Austin), Bonnie J. Davis (San Francisco State University), Jerry Davis (University of Wisconsin, La Crosse), Thomas Davis (University of New Hampshire), John Dearn (University of Canberra), James Dekloe (University of California, Santa Cruz), T. Delevoryas (University of Texas, Austin), Diane C. DeNagel (Northwestern University), Jean DeSaix (University of North Carolina), John Drees (Temple University School of Medicine), Charles Drewes (Iowa State University), Marvin Druger (Syracuse University), Betsey Dyer (Wheaton College), Robert Eaton (University of Colorado), Robert S. Edgar (University of California, Santa Cruz), Betty J. Eidemiller (Lamar University), David Evans (University of Florida), Robert C. Evans (Rutgers University, Camden), Sharon Eversman (Montana State University), Lincoln Fairchild (Ohio State University), Bruce Fall (University of Minnesota), Lynn Fancher (College of DuPage), Larry Farrell (Idaho State University), Jerry F. Feldman (University of California, Santa Cruz), Russell Fernald (University of Oregon), Milton Fingerman (Tulane University), Barbara Finney (Regis College), David Fisher (University of Hawaii, Manoa), William Fixsen (Harvard University), Abraham Flexer (Manuscript Consultant, Boulder, Colorado), Kerry Foresman (University of Montana), Norma Fowler (University of Texas, Austin), David Fox (University of Tennessee, Knoxville), Otto Friesen (University of Virginia), Virginia Fry (Monterey Peninsula College), Alice Fulton (University of Iowa), Sara Fultz (Stanford University), Berdell Funke (North Dakota State University), Anne Funkhouser (University of the Pacific), Arthur W. Galston (Yale University), Carl Gans (University of Michigan), John Gapter (University of Northern Colorado), Reginald Garrett (University of Virginia), Patricia Gensel (University of North Carolina), Chris George (California Polytechnic State University, San Luis Obispo), Robert George (University of Wyoming), Frank Gilliam (Marshall University), Todd Gleeson (University of Colorado), William Glider (University of Nebraska), Elizabeth A. Godrick (Boston University), Lynda Goff (University of California, Santa Cruz), Paul Goldstein (University of Texas, El Paso), Anne Good (University of California, Berkeley), Judith Goodenough (University of Massachusetts, Amherst), Ester Goudsmit (Oakland University), Robert Grammer (Belmont University), Joseph Graves (Arizona State University), A. J. F. Griffiths (University of British Columbia), William Grimes (University of Arizona), Mark Gromko (Bowling Green State University), Serine Gropper (Auburn University), Katherine L. Gross (Ohio State University), Gary Gussin (University of Iowa), R. Wayne Habermehl (Montgomery County Community College), Mac Hadley (University of Arizona), Jack P. Hailman (University of Wisconsin), Leah Haimo (University of California, Riverside), Rebecca Halyard (Clayton State College), Penny Hanchey-Bauer (Colorado State University), Laszlo Hanzely (Northern Illinois University), Richard Harrison (Cornell University), H. D. Heath (California State University, Hayward), George Hechtel (State University of New York, Stony Brook), Jean Heitz-Johnson (University of Wisconsin, Madison), Caroll Henry (Chicago State University), Frank Heppner (University of Rhode Island), Paul E. Hertz (Barnard College), Ralph Hinegardner (University of California, Santa Cruz), William Hines (Foothill College), Helmut Hirsch (State University of New York, Albany), Tuan-hua David Ho (Washington University), Carl Hoagstrom (Ohio Northern University), James Hoffman (University of Vermont), James Holland (Indiana State University, Bloomington), Laura Hoopes (Occidental College), Nancy Hopkins (Massachusetts Institute of Technology), Kathy Hornberger (Widener University), Pius F. Horner (San Bernardino Valley College), Margaret Houk (Ripon College), Ronald R. Hoy (Cornell University), Donald Humphrey (Emory University School of Medicine), Robert J. Huskey (University of Virginia), Bradley Hyman (University of California, Riverside), Alice Jacklet (State University of New York, Albany), John Jackson (North Hennepin Community College), Dan Johnson (East Tennessee State University), Wayne Johnson (Ohio State University), Kenneth C. Jones (California State University, Northridge), Russell Jones (University of California, Berkeley), Alan Journet (Southeast Missouri State University), Thomas Kane (University of Cincinnati), E. L. Karlstrom (University of Puget Sound), George Khoury (National Cancer Institute), Robert Kitchin (University of Wyoming), Attila O. Klein (Brandeis University), Greg Kopf (University of Pennsylvania School of Medicine), Thomas Koppenheffer (Trinity University), Janis Kuby (San Francisco State University), J. A. Lackey (State University of New York, Oswego), Lynn Lamoreux (Texas A & M University), Carmine A. Lanciani (University of Florida), Kenneth Lang (Humboldt State University), Allan Larson (Washington University), Diane K. Lavett (State University of New York, Cortland, and Emory University), Charles Leavell (Fullerton College), C. S. Lee (University of Texas), Robert Leonard (University of California, Riverside), Joseph Levine (Boston College), Bill Lewis (Shoreline Community College), John Lewis (Loma Linda University), Lorraine Lica (California State University, Hayward), Harvey Lillywhite (University of Florida, Gainesville), Sam Loker (University of New Mexico), Jane Lubchenco (Oregon State University), James MacMahon (Utah State University), Charles Mallery (University of Miami), Lynn Margulis (Boston University), Edith Marsh (Angelo State University), Karl Mattox (Miami University of

Ohio), Joyce Maxwell (California State University, Northridge), Richard McCracken (Purdue University), John Merrill (University of Washington), Ralph Meyer (University of Cincinnati), Roger Milkman (University of Iowa), Helen Miller (Oklahoma State University), John Miller (University of California, Berkeley), Kenneth R. Miller (Brown University), John E. Minnich (University of Wisconsin, Milwaukee), Kenneth Mitchell (Tulane University School of Medicine), Russell Monson (University of Colorado, Boulder), Frank Moore (Oregon State University), Randy Moore (Wright State University), William Moore (Wayne State University), Carl Moos (Veterans Administration Hospital, Albany, New York), Michael Mote (Temple University), John Mutchmor (Iowa State University), Elliot Myerowitz (California Institute of Technology), John Neess (University of Wisconsin, Madison), Todd Newbury (University of California, Santa Cruz), Harvey Nichols (University of Colorado, Boulder), Deborah Nickerson (University of South Florida), Bette Nicotri (University of Washington), Charles R. Noback (College of Physicians and Surgeons, Columbia University), Mary C. Nolan (Irvine Valley College), David O. Norris (University of Colorado, Boulder), Cynthia Norton (University of Maine, Augusta), Bette H. Nybakken (Hartnell College), Brian O'Conner (University of Massachusetts, Amherst), Gerard O'Donovan (University of North Texas), Eugene Odum (University of Georgia), John Olsen (Rhodes College), Sharman O'Neill (University of California, Davis), Wan Ooi (Houston Community College), Gay Ostarello (Diablo Valley College), Barry Palevitz (University of Georgia), Peter Pappas (County College of Morris), Bulah Parker (North Carolina State University), Stanton Parmeter (Chemeketa Community College), Robert Patterson (San Francisco State University), Crellin Pauling (San Francisco State University), Kay Pauling (Foothill Community College), Patricia Pearson (Western Kentucky University), Bob Pittman (Michigan State University), James Platt (University of Denver), Scott Poethig (University of Pennsylvania), Jeffrey Pommerville (Texas A & M University), Warren Porter (University of Wisconsin), Donald Potts (University of California, Santa Cruz), David Pratt (University of California, Davis), Halina Presley (University of Illinois, Chicago), Rebecca Pyles (East Tennessee State University), Scott Quackenbush (Florida International University), Ralph Quatrano (Oregon State University), Charles Ralph (Colorado State University), Brian Reeder (Morehead State University), C. Gary Reiness (Pomona College), Charles Remington (Yale University), David Reznick (University of California, Riverside), Fred Rhoades (Western Washington State University), Christopher Riegle (Irvine Valley College), Donna Ritch (Pennsylvania State University), Thomas Rodella (Merced College), Rodney Rogers (Drake University), Wayne Rosing (Middle Tennessee State University), Thomas Rost (University of California, Davis), Stephen I. Rothstein (University of California, Santa Barbara), John Ruben (Oregon State University), Albert Ruesink (Indiana University), Don Sakaguchi (Iowa State University), Mark F. Sanders (University of California, Davis), Ted Sargent (University of Massachusetts, Amherst), Carl Schaefer (University of Connecticut), Lisa Shimeld (Crafton Hills College), David Schimpf (University of Minnesota, Duluth), William H. Schlesinger (Duke University), Erik P. Scully (Towson State University), Stephen Sheckler (Virginia Polytechnic Institute and State University), James Shinkle (Trinity University), Barbara Shipes (Hampton University), Peter Shugarman (University of Southern California), Alice Shuttey (DeKalb Community College), James Sidie (Ursinus College), Daniel Simberloff (Florida State University), John Smarrelli (Loyola University), Andrew T. Smith (Arizona State University), Andrew J. Snope (Essex Community College), Susan Sovonick-Dunford (University of Cincinnati), Karen Steudel (University of Wisconsin), Barbara Stewart (Swarthmore College), Cecil Still (Rutgers University, New Brunswick), John Stolz (California Institute of Technology), Richard D. Storey (Colorado College), Stephen Strand (University of California, Los Angeles), Eric Strauss (University of Massachusetts, Boston), Russell Stullken (Augusta College), John Sullivan (Southern Oregon State University), Gerald Summers (University of Missouri), Marshall Sundberg (Louisiana State University), Daryl Sweeney (University of Illinois, Champaign-Urbana), Samuel S. Sweet (University of California, Santa Barbara), Lincoln Taiz (University of California, Santa Cruz), Samuel Tarsitano (Southwest Texas State University), David Tauck (Santa Clara University), James Taylor (University of New Hampshire), Roger Thibault (Bowling Green State University), William Thomas (Colby-Sawyer College), John Thornton (Oklahoma State University), Robert Thornton (University of California, Davis), James Traniello (Boston University), Robert Tuveson (University of Illinois, Urbana), Maura G. Tyrrell (Stonehill College), Gordon Uno (University of Oklahoma), James W. Valentine (University of California, Santa Barbara), Joseph Vanable (Purdue University), Theodore Van Bruggen (University of South Dakota), Frank Visco (Orange Coast College), Laurie Vitt (University of California, Los Angeles), Susan D. Waaland (University of Washington), William Wade (Dartmouth Medical College), John Waggoner (Loyola Marymount University), Dan Walker (San Jose State University), Robert L. Wallace (Ripon College), Jeffrey Walters (North Carolina State University), Margaret Waterman (University of Pittsburgh), Charles Webber (Loyola University of Chicago), Peter Webster (University of Massachusetts, Amherst), Terry Webster (University of Connecticut, Storrs), Peter Wejksnora (University of Wisconsin, Milwaukee), Kentwood Wells (University of Connecticut), Stephen Williams (Glendale Community College), Christopher Wills (University of California, San Diego), Fred Wilt (University of California, Berkeley), Robert T. Woodland (University of Massachusetts Medical School), Joseph Woodring (Louisiana State University), Patrick Woolley (East Central College), Philip Yant (University of Michigan), Hideo Yonenaka (San Francisco State University), John Zimmerman (Kansas State University), Uko Zylstra (Calvin College).

THE INTERVIEWS

UNIT ONE

THE CHEMISTRY OF LIFE

Page 20
Mario Molina
Massachusetts Institute of Technology

UNIT TWO

THE CELL

Page 100
Bruce Alberts
University of California, San Francisco

UNIT THREE

GENETICS

Page 224
Mary-Claire King
University of Washington

UNIT FOUR

MECHANISMS OF EVOLUTION

Page 412
Richard Dawkins
Oxford University

UNIT FIVE

THE EVOLUTIONARY HISTORY OF BIOLOGICAL DIVERSITY

Page 488
Elisabeth Vrba
Yale University

UNIT SIX

PLANT FORM AND FUNCTION

Page 668
Gloria Coruzzi
New York University

UNIT SEVEN

ANIMAL FORM AND FUNCTION

Page 776
Terence Dawson
The University of New South Wales

UNIT EIGHT

ECOLOGY

Page 1024
Michael Dombeck
U.S. Forest Service

BRIEF CONTENTS

DETAILED CONTENTS

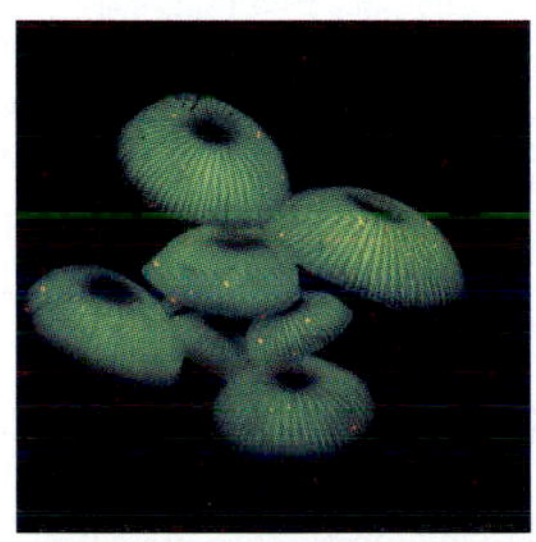

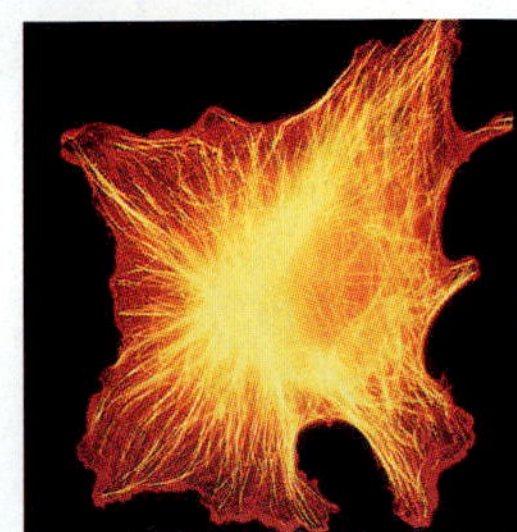

UNIT TWO

UNIT FIVE

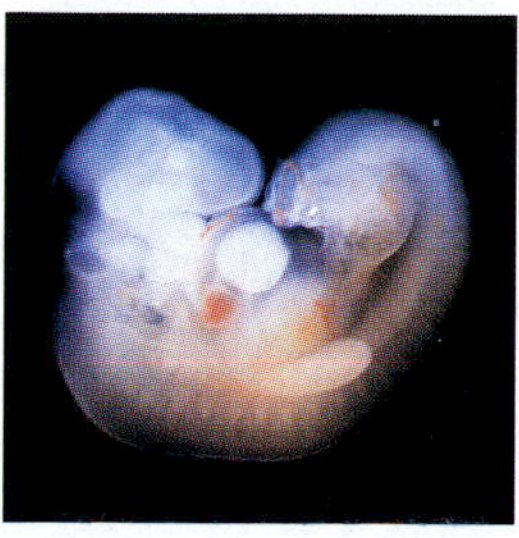

UNIT EIGHT

CHAPTER 1

INTRODUCTION: THEMES IN THE STUDY OF LIFE

Life's Hierarchical Order

- The living world is a hierarchy, with each level of biological structure building on the level below it
- Each level of biological structure has emergent properties
- Cells are an organism's basic units of structure and function
- The continuity of life is based on heritable information in the form of DNA
- Structure and function are correlated at all levels of biological organization
- Organisms are open systems that interact continuously with their environments
- Regulatory mechanisms ensure a dynamic balance in living systems

Evolution, Unity, and Diversity

- Diversity and unity are the dual faces of life on Earth
- Evolution is the core theme of biology

Science as a Process

- Testable hypotheses are the hallmarks of the scientific process
- Science and technology are functions of society
- Biology is a multidisciplinary adventure

Biology, the study of life, is rooted in the human spirit. People keep pets, nurture houseplants, invite avian visitors with backyard birdhouses, and visit zoos and nature preserves. Biology is the scientific extension of this human tendency to feel connected to and curious about all forms of life. It is a science for adventurous minds. It takes us, personally or vicariously, into jungles, deserts, seas, and other environments, where a variety of living forms and their physical surroundings are interwoven into complex webs called ecosystems. Studying life leads us into laboratories to examine more closely how living things, called organisms, work. Biology draws us into the microscopic world of the fundamental units of organisms known as cells, and into the submicroscopic realm of the molecules that make up those cells. Our intellectual journey also takes us back in time, for biology encompasses not only contemporary life, but also a history of ancestral forms stretching nearly four billion years into the past. The scope of biology is immense. The purpose of this book is to introduce you to this multifaceted science. It is unlikely that any biology course could or should cover all 55 chapters, but this textbook is designed to help you succeed in your general biology course and *to serve as a durable reference in your continuing education.*

You are becoming involved with biology during its most exciting era. Using fresh approaches and new research methods, biologists are unraveling some of life's most engaging mysteries. Though stimulating, the information explosion in biology is also intimidating. Most of the biologists who have ever lived are alive today, and they add about a half-million new research articles to the scientific literature annually. Each of biology's many subfields changes continuously, and it is very difficult for a professional biologist to remain current in more than one narrowly defined specialty. How, then, can beginning biology students hope to keep their heads above water in this deluge of data and discovery? The key is to recognize unifying themes that pervade all of biology—themes that will still apply decades from now, when much of the specific information presented in any textbook will be obsolete. This chapter introduces some broad, enduring themes in the study of life. The list on this page previews these themes.

LIFE'S HIERARCHICAL ORDER

In this very first section, we introduce one of life's most distinctive features, its order. Life is highly organized into a hierarchy of structural levels, with each level building on the levels below it. As we examine the hierarchy, we will see that special qualities, called emergent properties, result from the structure at each level. Life itself is associated with a set of properties that depend on structural order.

The cell is the lowest structural level in which all of life's properties, including reproduction, can occur. Within each

(a) Mary-Claire King, geneticist

(b) Elisabeth Vrba, paleontologist

(c) Gloria Coruzzi, plant molecular biologist

(d) Terry Dawson, physiologist

FIGURE 1.1 ▪ Biologists study life on many different scales of size and time. (a) Geneticist Mary-Claire King studies mutations in DNA that contribute to human disorders such as familial breast cancer and hereditary deafness. **(b)** Paleontologist Elisabeth Vrba investigates human evolution by studying relationships between the history of life as chronicled by the fossil record and environmental changes of the past. **(c)** Plant molecular biologist Gloria Coruzzi, photographed here with one of her student researchers, uses genetics and DNA technology to investigate key processes in plants. **(d)** Terry Dawson studies the structural and physiological adaptations of Australian organisms.

cell, life's order is coded at the molecular level. Molecules of DNA contain genes, units of inheritance, that enable organisms to reproduce, providing continuity from one generation to the next.

Biologists study life on many different levels (FIGURE 1.1). At the end of this section, we take a brief look at the regulatory mechanisms that keep living systems ordered and thereby functioning smoothly.

The living world is a hierarchy, with each level of biological structure building on the level below it

You can see order in the intricate pattern of veins throughout a leaf or in the colorful pattern of a bird's feathers. Biological order exists at all levels (FIGURE 1.2), even those invisible to the unaided eye. Starting at the lowest level, atoms, the chemical building blocks of all matter, are ordered into complex biological molecules. Many of the molecules of life are arranged into minute structures called organelles, which are in turn the components of cells.

Cells are subunits of organisms, and organisms are the units of life. Some organisms, such as amoebas, consist of single cells, but others are multicellular aggregates of many specialized types of cells. What an amoeba accomplishes with a single cell—the uptake and processing of nutrients, excretion of wastes, response to environmental stimuli, reproduction, and other functions—a human or other multicellular organism accomplishes with a division of labor among specialized cells. Unlike the amoeba, none of your cells could live for long on its own. The organism we recognize as an animal or plant is not a random collection of individual cells, but a multicellular cooperative.

(a) Molecule (chlorophyll)

(b) Organelle (chloroplast)

(c) Cells

(d) Tissues

(e) Organ

(f) An organism in a community

FIGURE 1.2 ▪ The hierarchy of biological organization. This sequence of images takes us all the way from atoms to a biological community of many interacting species. **(a)** Chlorophyll, represented here by a computer graphic model, is a molecule built from many atoms. This molecule in the leaves of plants absorbs sunlight as a source of energy for driving photosynthesis, the manufacture of food in the leaf. **(b)** The process of photosynthesis requires the participation of many other molecules organized within the cellular organelle called the chloroplast (the greenish structure in this micrograph, a photograph taken with a microscope). **(c)** Many organelles cooperate in the functioning of the living unit we call a cell. Chloroplasts are evident in these leaf cells. **(d)** In multicellular organisms, cells are usually organized into tissues, groups of similar cells forming a functional unit. The leaf in this micrograph has been cut obliquely, revealing two different specialized tissues. The honeycomblike tissue (upper half) consists of photosynthetic cells within the leaf. The dark green tissue with the small pores (lower half) is the epidermis, the "skin" of the plant. The pores in the epidermis allow carbon dioxide, a raw material that is converted to sugar by photosynthesis, to enter the leaf. **(e)** The aspen leaf, a plant organ, has a specific organization of many different tissues, including photosynthetic tissue, epidermis, and the vascular tissue that transports water from the roots to the leaves. **(f)** These aspens are members of a biological community that includes many other species of organisms.

Multicellular organisms exhibit three major structural levels above the cell. Similar cells are grouped into tissues, specific arrangements of different tissues form organs, and organs are grouped into organ systems. For example, the signals (nerve impulses) that coordinate your movements are transmitted along specialized cells called neurons. The nervous tissue within your brain has billions of neurons organized into a communications network of spectacular complexity. The brain, however, is not pure nervous tissue; it is an organ built of many different tissues, including a type called connective tissue that forms the protective covering of the brain. The brain is itself part of the nervous system, which also includes the spinal cord and the many nerves that transmit messages between the spinal cord and other parts of the body. The nervous system is only one of several organ systems characteristic of humans and other complex animals.

In the hierarchy of biological organization, there are tiers beyond the individual organism. A population is a localized group of organisms belonging to the same species; populations of different species living in the same area make up a biological community; and community interactions that include nonliving features of the environment, such as soil and water, form an ecosystem.

Identifying biological organization at its many levels is fundamental to the study of life. This text essentially follows such an organization, beginning by looking at the chemistry of life and ending with the study of ecosystems and the biosphere, the sum of all Earth's ecosystems. However, we will also see that biological processes often involve several levels of biological organization. For example, when a rattlesnake explodes from its coiled posture and strikes a mouse, the snake's coordinated movements result from complex interactions at the molecular, cellular, tissue, and organ levels within its body. This behavior also affects the biological community in which the snake and its prey live. Such episodes of predation can have an important cumulative impact on the sizes of both the mouse and the rattlesnake populations. Most biologists specialize in the study of life at a particular level, but they gain broader perspective when they integrate their discoveries with processes occurring at lower and higher levels.

Each level of biological structure has emergent properties

With each step upward in the hierarchy of biological order, novel properties emerge that were not present at the simpler levels of organization. These emergent properties result from interactions between components. A molecule such as a protein has attributes not exhibited by any of its component atoms, and a cell is certainly much more than a bag of molecules. If the intricate organization of the human brain is disrupted by a head injury, that organ will cease to function properly, even though all its parts may still be present. And an organism is a living whole greater than the sum of its parts.

The concept of emergent properties accents the importance of structural arrangement and applies to inanimate material as well as to life. Neither the head nor the handle of a hammer alone is very useful for driving nails, but put these parts together in a certain way, and the functional properties of a hammer emerge. Diamonds and graphite are both made of carbon, but they have different properties because their carbon atoms are arranged differently. The emergent properties of life reflect a hierarchy of structural organization without counterpart among inanimate objects.

Life resists a simple, one-sentence definition because it is associated with numerous emergent properties. Yet almost any child perceives that a dog or a bug or a tree is alive and a rock is not. We recognize life by what living things do. FIGURE 1.3 illustrates and describes some of the properties and processes we associate with the state of being alive.

Because the properties of life emerge from complex organization, scientists seeking to understand biological processes confront a dilemma. One horn of the dilemma is that we cannot fully explain a higher level of order by breaking it down into its parts. A dissected animal no longer functions; a cell reduced to its chemical ingredients is no longer a cell. Disrupting a living system interferes with the meaningful explanation of its processes. The other horn of the dilemma is the futility of trying to analyze something as complex as an organism or a cell without taking it apart. Reductionism—reducing complex systems to simpler components that are more manageable to study—is a powerful strategy in biology. For example, by studying the molecular structure of a substance called DNA that had been extracted from cells, James Watson and Francis Crick deduced, in 1953, how this molecule could serve as the chemical basis of inheritance. The central role of DNA was better understood, however, when it was possible to study its interactions with other substances in the cell. Biology balances the reductionist strategy with the longer-range objective of understanding how the parts of cells, organisms, and higher levels of order, such as ecosystems, are functionally integrated.

Cells are an organism's basic units of structure and function

The cell is the lowest level of structure capable of performing *all* the activities of life. All organisms are composed of cells, the basic units of structure and function.

Robert Hooke, an English scientist, first described and named cells in 1665, when he observed a slice of cork (bark from an oak tree) with a microscope that magnified 30 times (30×). Apparently believing that the tiny boxes, or "cells," that he saw were unique to cork, Hooke never realized the significance of his discovery. His contemporary, a Dutchman named Anton van Leeuwenhoek, discovered organisms we now know to be single-celled. Using grains of sand that he had polished into magnifying glasses as powerful as 300×, Leeuwenhoek discovered a microbial world in droplets of pond water and also observed the blood cells and sperm cells of animals. In 1839, nearly two centuries after the discoveries of Hooke and Leeuwenhoek, cells were finally acknowledged as the ubiquitous units of life by Matthias Schleiden and Theodor Schwann, two German biologists. In a classic case of inductive reasoning—reaching a generalization based on many concurring observations—Schleiden and Schwann summarized

(a) Order

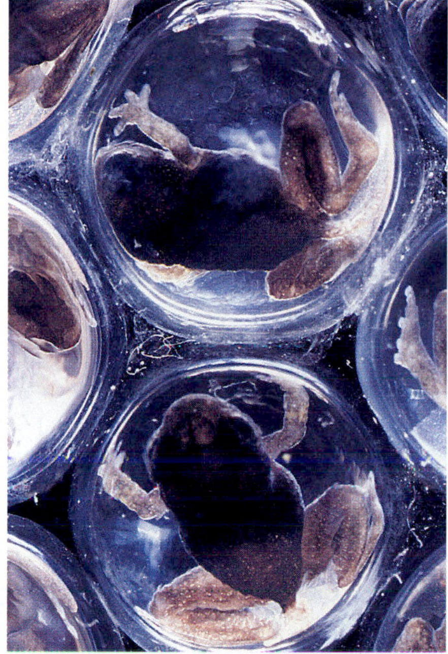

(c) Growth and development

(b) Reproduction

(d) Energy utilization

FIGURE 1.3 ▪ Some properties of life. **(a)** *Order:* All other characteristics of life emerge from an organism's complex organization, which is apparent in this closeup of a sunflower. **(b)** *Reproduction:* Organisms reproduce their own kind. Life comes only from life, an axiom known as biogenesis. Here, a Japanese macaque protects its offspring. **(c)** *Growth and development:* Heritable programs in the form of DNA direct the pattern of growth and development, producing an organism that is characteristic of its species. Shown here are embryos of a Costa Rican species of frog. **(d)** *Energy utilization:* Organisms take in energy and transform it to do many kinds of work. This bat obtains fuel in the form of nectar from the saguaro cactus. The bat will use energy stored in the molecules of its food to power flight and other work. **(e)** *Response to the environment:* This soon-to-be-digested cricket "tripped" the Venus flytrap when it stimulated hair cells on the surface of the modified leaves that make up the trap. The plant responded to this environmental stimulus with a rapid closure of the trap. **(f)** *Homeostasis:* Regulatory mechanisms maintain an organism's internal environment within tolerable limits, even though the external environment may fluctuate. This regulation is called homeostasis. In this example, regulation of the amount of blood flowing through the blood vessels of this blacktail jackrabbit's large ears constantly adjusts heat loss to its surroundings. This contributes to homeostasis of the animal's body temperature. **(g)** *Evolutionary adaptation:* Life evolves as a result of the interaction between organisms and their environments. One consequence of evolution is the adaptation of organisms to their environment. The white feathers of the white-tail ptarmigan in winter plumage make it nearly invisible against the animal's snowy surroundings.

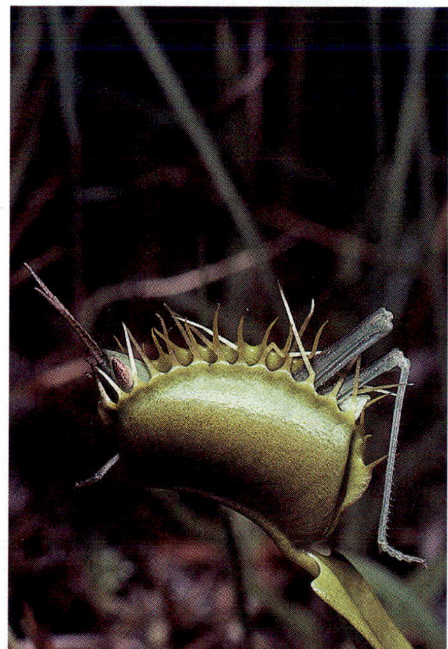

(e) Response to the environment

(f) Homeostasis

(g) Evolutionary adaptation

their own microscopic studies and those of others by concluding that all living things consist of cells. This generalization forms the basis of what is known as the cell theory. This theory was later expanded to include the idea that all cells come from other cells. The ability of cells to divide to form new cells is the basis for all reproduction and for the growth and repair of multicellular organisms, including humans.

All cells are enclosed by a membrane that regulates the passage of materials between the cell and its surroundings. Every cell, at some stage in its life, contains DNA, the heritable material that directs the cell's many activities.

Two major kinds of cells—prokaryotic cells and eukaryotic cells—can be distinguished by their structural organization (FIGURE 1.4). The cells of the microorganisms commonly called bacteria are prokaryotic. All other forms of life are composed of eukaryotic cells. Much more complex than the prokaryotic cell, the eukaryotic cell is subdivided by internal membranes into many different functional compartments, or organelles. In eukaryotic cells, the DNA is organized along with certain proteins into structures called chromosomes contained within a nucleus, the largest organelle of most eukaryotic cells. Surrounding the nucleus is the cytoplasm, a thick fluid in which are suspended the various organelles that perform most of the cell's functions. Some eukaryotic cells, including those of plants, have tough walls external to their membranes. Animal cells lack walls.

In the much simpler prokaryotic cell, the DNA is not separated from the rest of the cell in a nucleus. Prokaryotic cells also lack the cytoplasmic organelles typical of eukaryotic cells.

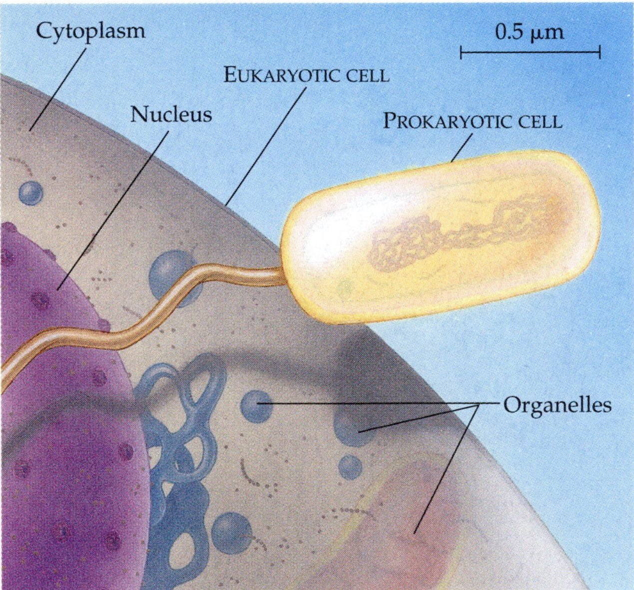

FIGURE 1.4 ▪ Structural organization of eukaryotic and prokaryotic cells. The eukaryotic cell, found in plants, animals, and all other organisms except bacteria, is characterized by an extensive subdivision into many different functional compartments called organelles. The prokaryotic cell, unique to bacteria, is much simpler, lacking most of the organelles found in eukaryotic cells. Compared to eukaryotic cells, most prokaryotic cells are also much smaller.

Almost all prokaryotic cells (bacteria) have tough external cell walls.

Although eukaryotic and prokaryotic cells contrast sharply in complexity, we will see that they have some key similarities. Cells vary widely in size, shape, and specific structural features, but all are highly ordered structures that carry out complicated processes necessary for maintaining life.

The continuity of life is based on heritable information in the form of DNA

Order implies information; instructions are required to arrange parts or processes in an organized way. Biological instructions are encoded in the molecule known as DNA (deoxyribonucleic acid). DNA is the substance of genes, the units of inheritance that transmit information from parents to offspring (FIGURE 1.5).

Each DNA molecule is made up of two long chains each composed of four kinds of chemical building blocks called

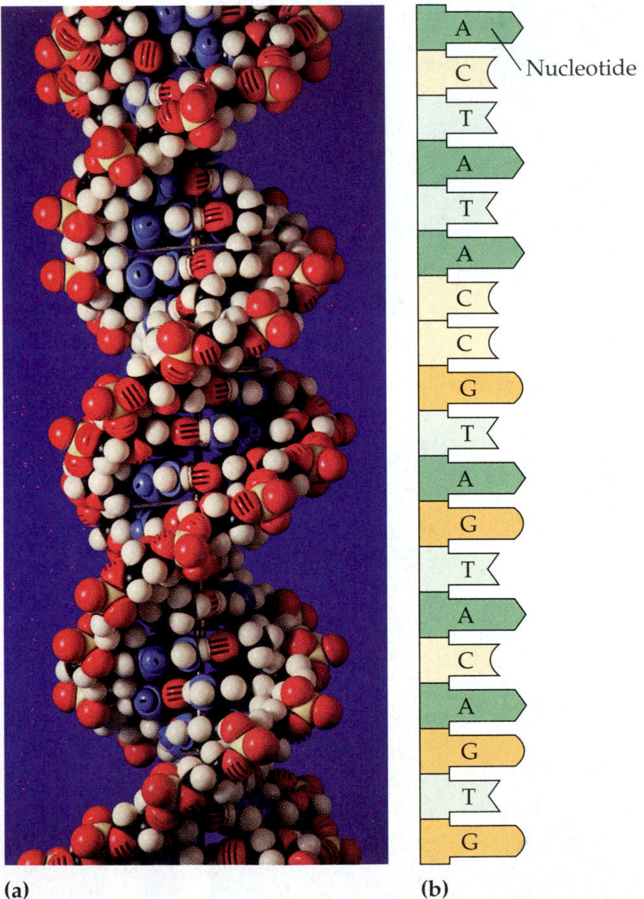

FIGURE 1.5 ▪ The genetic material: DNA. DNA molecules carry biological information from one generation to the next. **(a)** This model shows each atom in a segment of DNA. Made up of two long chains of building blocks called nucleotides, a DNA molecule takes the three-dimensional form of a double helix. **(b)** This diagram uses geometric shapes to represent the nucleotides in a small section of one of the two nucleotide chains in a DNA molecule. Each nucleotide is made up of many atoms. Genetic information is encoded in specific sequences of the nucleotides.

nucleotides. The way DNA conveys information is analogous to the way we arrange the letters of the alphabet into precise sequences with specific meanings. The word *rat,* for example, conjures up an image of a rodent; *tar* and *art,* which contain the same letters, mean quite different things. Libraries are filled with books containing information encoded in varying sequences of only 26 letters. We can think of nucleotides as the alphabet of inheritance. Specific sequential arrangements of these four chemical letters encode the precise information in a gene. If the entire library of genes stored within the microscopic nucleus of a single human cell were written in letters the size of those you are now reading, the information would fill more than a hundred books as large as this one. The complex structural organization of an organism is thus specified by an inherited script conveying an enormous amount of coded information.

All forms of life employ essentially the same genetic code. A particular sequence of nucleotides says the same thing to one organism as it does to another; differences between organisms reflect differences between their nucleotide sequences. The diverse forms of life are different expressions of a common language for programming biological order.

Inheritance itself depends on a mechanism for copying DNA and passing its sequence of chemical letters on to offspring. As a cell prepares to divide to form two cells, it copies its DNA. A mechanical system that moves chromosomes then distributes the DNA copies equally to the two "daughter" cells. In species that reproduce sexually, offspring inherit copies of DNA from the parents' sperm and egg cells. The continuity of life over the generations and over the eons has its molecular basis in the replication of DNA.

Structure and function are correlated at all levels of biological organization

Given a choice of tools, you would not loosen a screw with a hammer or pound a nail with a screwdriver. How a device works is correlated with its structure: Form fits function. Applied to biology, this theme is a guide to the anatomy of life at its many structural levels, from molecules to organisms. Analyzing a biological structure gives us clues about what it does and how it works. Conversely, knowing the function of a structure provides insight about its construction.

An example of this structure-function theme is the aerodynamically efficient shape of a bird's wing (FIGURE 1.6). The skeleton of the bird also has structural qualities that contribute to flight, with bones that have a strong but light

(a)

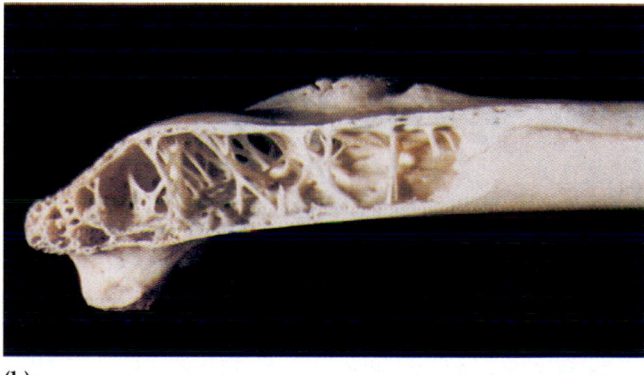

(b)

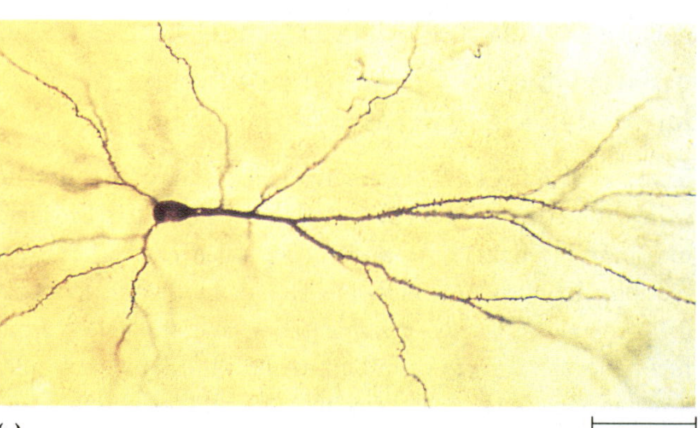

(c) 100 μm

Mitochondrion

Infoldings of membrane

(d) 0.5 μm

FIGURE 1.6 ▪ Form fits function. (a) A bird's build makes flight possible. The correlation between structure and function can apply to the shape of an entire organism, as you can see from this white tern in flight. **(b)** The structure-function theme also applies to organs and tissues. For example, the honeycombed construction of a bird's bones provides a lightweight skeleton of great strength. **(c)** The form of a cell fits its specialized function. Nerve cells, or neurons, have long extensions (processes) that transmit nervous impulses. **(d)** Functional beauty is also apparent at the subcellular level. This organelle, called a mitochondrion, has an inner membrane that is extensively folded, a structural solution to the problem of packing a relatively large amount of this membrane into a very small container.

honeycombed internal structure. The flight muscles of a bird are controlled by neurons (nerve cells), which transmit impulses. With long extensions, neurons are especially well structured for communication. The flight muscles need plenty of energy, which they obtain from organelles called mitochondria. These organelles are the sites of cellular respiration, the chemical process that powers the cell by using oxygen to help tap the energy stored in sugar and other food molecules. A mitochondrion is surrounded by an outer membrane, but it also has an inner membrane with many infoldings. Molecules embedded in the inner membrane carry out many of the steps in cellular respiration, and the infoldings pack a large amount of this membrane into a minute container (FIGURE 1.6d). In exploring life on its different structural levels, we will discover functional beauty at every turn.

Organisms are open systems that interact continuously with their environments

Life does not exist in a vacuum. An organism is an example of what scientists call an open system, an entity that exchanges materials and energy with its surroundings. Each organism interacts continuously with its environment, which includes other organisms as well as nonliving factors. The roots of a tree, for example, absorb water and minerals from the soil, and the leaves take in carbon dioxide from the air. Solar energy absorbed by chlorophyll, the green pigment of leaves, drives photosynthesis, which converts water and carbon dioxide to sugar and oxygen. The tree releases oxygen to the air, and its roots change the soil by breaking up rocks into smaller particles, secreting acid, and absorbing minerals. Both organism and environment are affected by the interaction between them. The tree also interacts with other life, including soil microorganisms associated with its roots and animals that eat its leaves and fruit.

The many interactions between organisms and their environment are interwoven to form the fabric of an ecosystem. The dynamics of any ecosystem include two major processes. One is the cycling of nutrients. For example, minerals acquired by plants will eventually be returned to the soil by microorganisms that decompose leaf litter, dead roots, and other organic debris. The second major process in an ecosystem is the flow of energy from sunlight to photosynthetic life (producers) to organisms that feed on plants (consumers) (FIGURE 1.7). The theme of organisms as open systems that exchange materials and energy with their surroundings is essential to understanding life on all levels of organization.

The exchange of energy between an organism and its surroundings involves the transformation of one form of energy to another. For example, when a leaf produces sugar, it converts solar energy to chemical energy in sugar molecules. When an animal's muscle cells use sugar as fuel to power movements, they convert chemical energy into kinetic energy, the energy of motion. All of the work of cells involves the transformation of chemical energy (which is ordered) into heat, which is the unordered energy of random molecular motion. Life requires continual uptake of ordered energy and the release of some unordered energy to the surroundings.

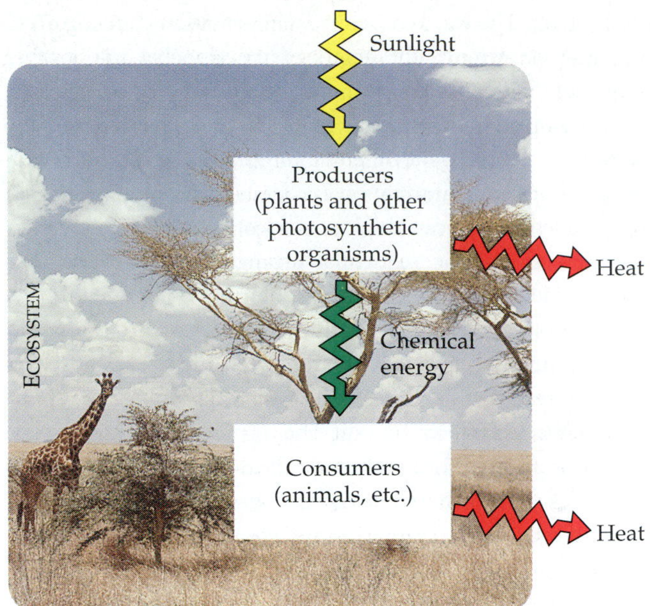

FIGURE 1.7 ▪ An introduction to energy flow and energy transformation in an ecosystem. Living is work, and work requires that organisms obtain and use energy. Most ecosystems are solar-powered. Plants and other photosynthetic organisms convert light energy to the chemical energy stored in sugar and other complex molecules. By breaking these fuel molecules down to simpler molecules, organisms can harvest the stored energy and put it to work. Photosynthetic organisms are called producers because the entire ecosystem (schematically represented here) depends on their photosynthetic products. Animals and other consumers acquire their energy in chemical form by eating plants, by eating animals that ate plants, or by decomposing organic refuse, such as leaf litter and dead animals. The energy that enters an ecosystem in the form of sunlight exits in the form of heat, which all organisms dissipate to their surroundings whenever they perform work.

Regulatory mechanisms ensure a dynamic balance in living systems

If you strike a match, it undergoes a chemical reaction in which the chemical energy in its molecules is transformed into heat and light. Burning is unregulated energy transformation, obviously not a suitable way for living cells to transform energy. Organisms are able to obtain useful energy from fuel molecules such as sugar because cells break the molecules down in a series of closely regulated chemical reactions.

Regulation of the chemical reactions within cells centers on protein molecules called enzymes. Produced by the cells in which they function, enzymes are catalysts, substances that speed up chemical reactions. When your muscle cells need a lot of energy during exercise, enzymes catalyze the rapid breakdown of sugar (glucose) molecules, releasing energy that can be put to work. In contrast, when you rest, other enzymes catalyze the formation of glucose, which may be added to the body's fuel reserves. With one group of enzymes catalyzing the breakdown of glucose and another group catalyzing its formation, how is order maintained in the cell? A big part of the answer is that regulatory mechanisms determine precisely when, where, and how fast certain reactions occur in a cell.

Many biological processes are self-regulating, operating by a mechanism called feedback, in which an output or product of a process regulates that process. Negative feedback, also called feedback inhibition, slows or stops processes; positive feedback speeds a process up (FIGURE 1.8).

Mammals and birds have a negative feedback system that keeps body temperature within a narrow range, despite wide fluctuations in their surroundings. A control center in the brain holds the temperature of the blood close to a set point (about 37°C in mammals). For instance, when the human body starts getting hot, signals from the control center increase the activity of sweat glands and the diameter of blood vessels in the skin. Evaporative cooling results as sweating increases, and heat radiates from the blood vessels as they fill with warm blood. Negative feedback occurs as soon as the blood cools back to the set point, causing the control center to stop sending signals to the skin. If the blood temperature drops below the set point, the brain's control center inactivates the sweat glands and constricts the skin's blood vessels. This shunts blood to deeper tissues, reducing heat loss. When the blood warms back to the set point, negative feedback occurs again, and signals from the control center cease.

The clotting of blood is one example of positive feedback. When a blood vessel is injured, structures in the blood called platelets start accumulating at the site. Positive feedback occurs as chemicals released by the platelets attract more platelets, and the platelet cluster initiates a complex sequence of chemical reactions (called a cascade) that seals the wound with a clot. Regulation by positive and negative feedback is a pervasive theme in biology, and we will see numerous examples throughout this text.

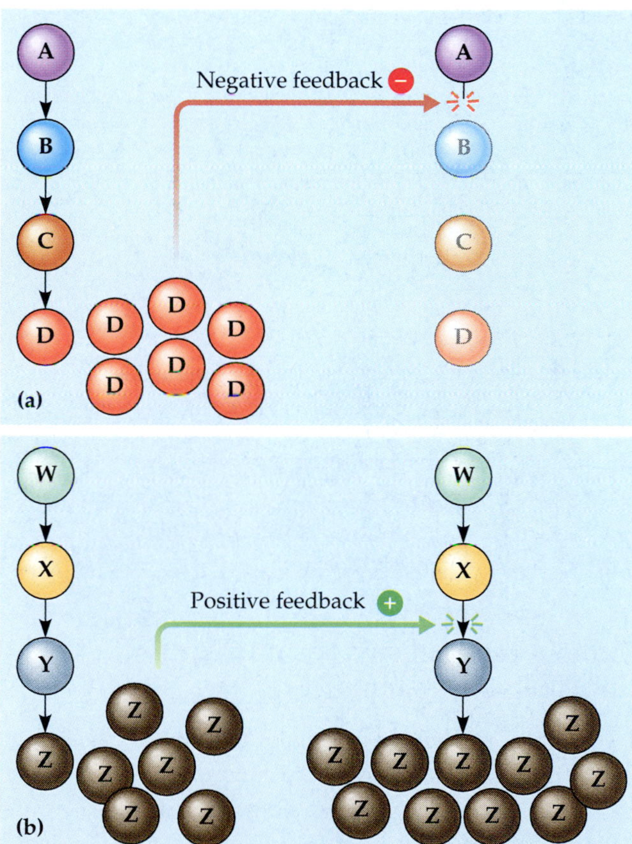

FIGURE 1.8 ▪ Regulation by feedback mechanisms. (a) This simple model illustrates the principle of negative feedback, or feedback inhibition, in the regulation of a chemical reaction sequence in a cell. The sequence involves four molecules (A–D). The black arrows represent three different enzymes catalyzing the conversion of one molecule to the next. The final product (D) inhibits the first enzyme in the sequence; when the concentration of D rises to a certain point, the reaction shuts itself down. Self-regulation by negative feedback is common in living systems. **(b)** In positive feedback, a product of the reaction sequence enhances the action of one of the enzymes, increasing the rate of production of the product.

EVOLUTION, UNITY, AND DIVERSITY

Evolutionary change has been a central feature of life since it arose about 4 billion years ago. The evolutionary connections among all organisms explain the unity and diversity of life.

Diversity and unity are the dual faces of life on Earth

Diversity is a hallmark of life. Biologists have identified and named about 1.5 million species, including over 260,000 plants, almost 50,000 vertebrates (animals with backbones), and more than 750,000 insects. Thousands of newly identified species are added to the list each year. Estimates of the total diversity of life range from about 5 million to over 100 million species.

If life is so diverse, how can biology have any unifying themes at all? What, for instance, can a mold, a tree, and a human possibly have in common? As it turns out, a great deal! Underlying the diversity of life is a striking unity, especially at the lower levels of organization. We can see it, for example, in certain similarities of cell structure (FIGURE 1.9 on p. 10). Unity is also evident in the universal genetic code shared by all organisms. Different expressions of this genetic code, however, result in the diversity of life.

Biological diversity is something to relish and preserve, but it can also be a bit overwhelming. To make diversity less daunting, people tend to group similar species together. For instance, we may talk about squirrels without distinguishing among the many different species of squirrels. Taxonomy, the branch of biology concerned with naming and classifying species, groups organisms according to a more formal scheme. The scheme consists of different levels of classification, each more comprehensive than those below it (FIGURE 1.10).

Traditionally, biologists divided the diversity of life into five kingdoms. All prokaryotic organisms, commonly called bacteria, were placed in the kingdom Monera. Eukaryotes were divided into the kingdom Protista (unicellular eukaryotes and their relatively simple multicellular relatives), the kingdom Plantae (plants), the kingdom Fungi (fungi), and the kingdom Animalia (animals).

Today, many biologists prefer classification schemes that recognize six, eight, or more kingdoms and that group kingdoms into a higher category, the domain. Reflecting strong evidence developed over the past two decades, prokaryotes are divided into two fundamentally different groups, the domains Bacteria (formerly called Eubacteria) and Archaea (formerly called Archaebacteria). These groups of prokaryotes seem to be at least as different from each other as either is from any eukaryote. The eukaryotes are grouped into a third domain, Eukarya (FIGURE 1.11).

Among the Eukarya, the enormous diversity among the protists (Kingdom Protista) has most researchers convinced that the group should be divided into several kingdoms. We will take a close look at these new developments in taxonomy at the kingdom and domain levels in Unit Five. Meanwhile, we follow the convention of referring to single-celled eukaryotes and closely related multicellular organisms as protists. The other eukaryotes, the kingdoms Plantae, Fungi, and Animalia, consist mainly of multicellular organisms. One way to distinguish them from one another is by their contrasting modes of nutrition (see FIGURE 1.11).

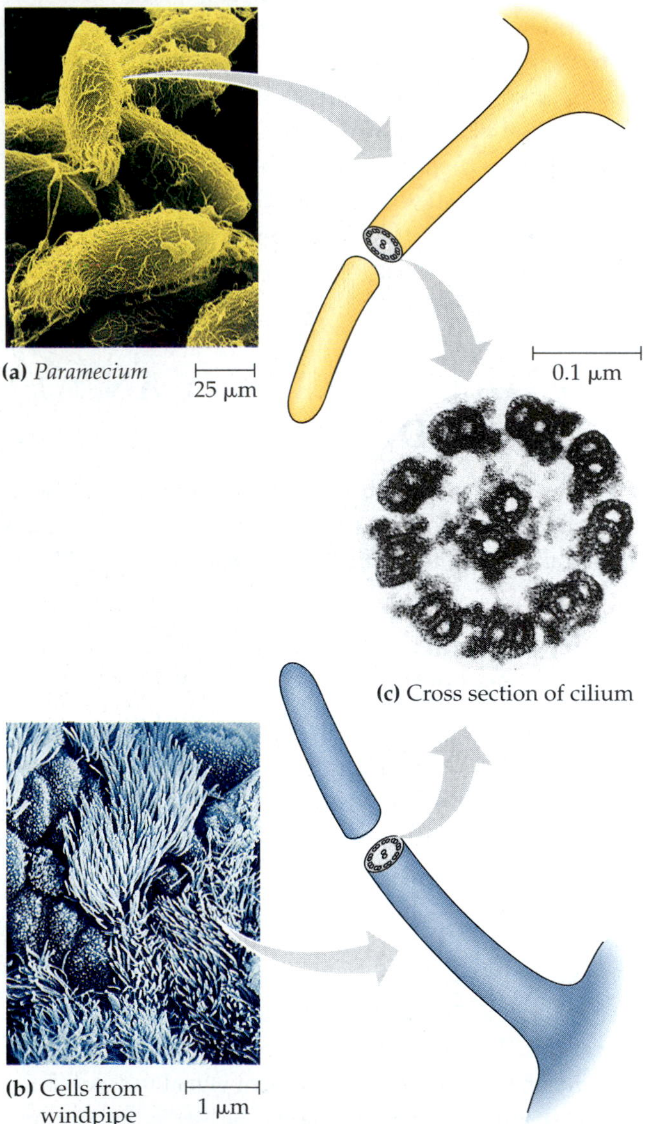

FIGURE 1.9 ▪ An example of unity underlying the diversity of life: the architecture of eukaryotic cilia. Eukaryotic organisms as diverse as the single-celled organism *Paramecium* and animals possess cilia, locomotory "hairs" that extend from cells. **(a)** The cilia of *Paramecium* propel the cell through pond water. **(b)** The cells that line the human windpipe are also equipped with cilia, which help keep the lungs clean by moving a film of debris-trapping mucus upward. **(c)** Comparing cross sections of cilia from diverse eukaryotes reveals a common structural organization. Such striking similarity in complex components contributes to the evidence that organisms as different as *Paramecium* and humans are, to some degree, related.

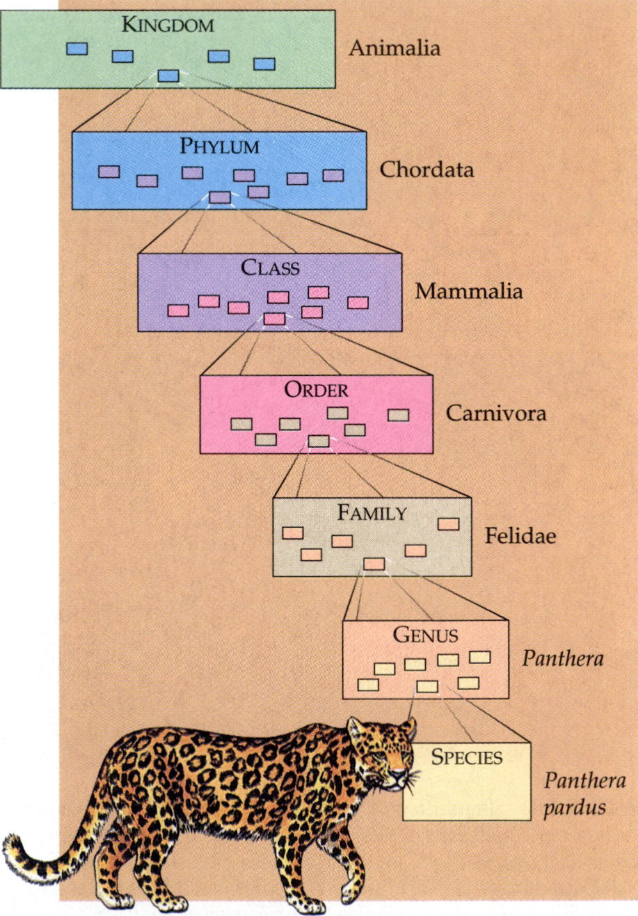

FIGURE 1.10 ▪ Classifying life. The taxonomic scheme classifies species into groups subordinate to more comprehensive groups. Species that are very similar are placed in the same genus, genera are grouped into families, and so on, each level of classification being more comprehensive than those it includes. This example classifies the species *Panthera pardus*, the leopard.

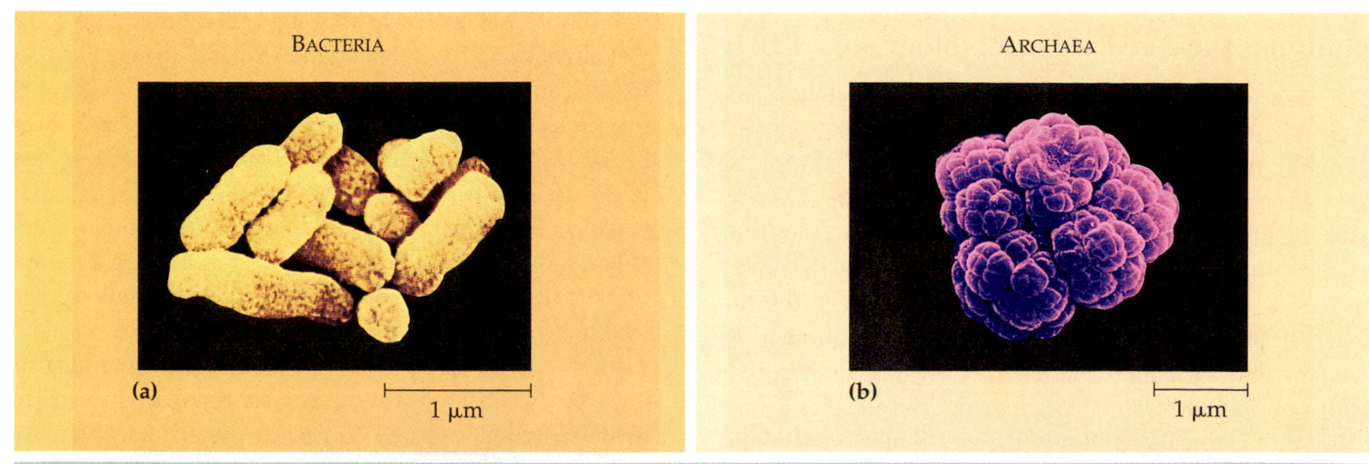

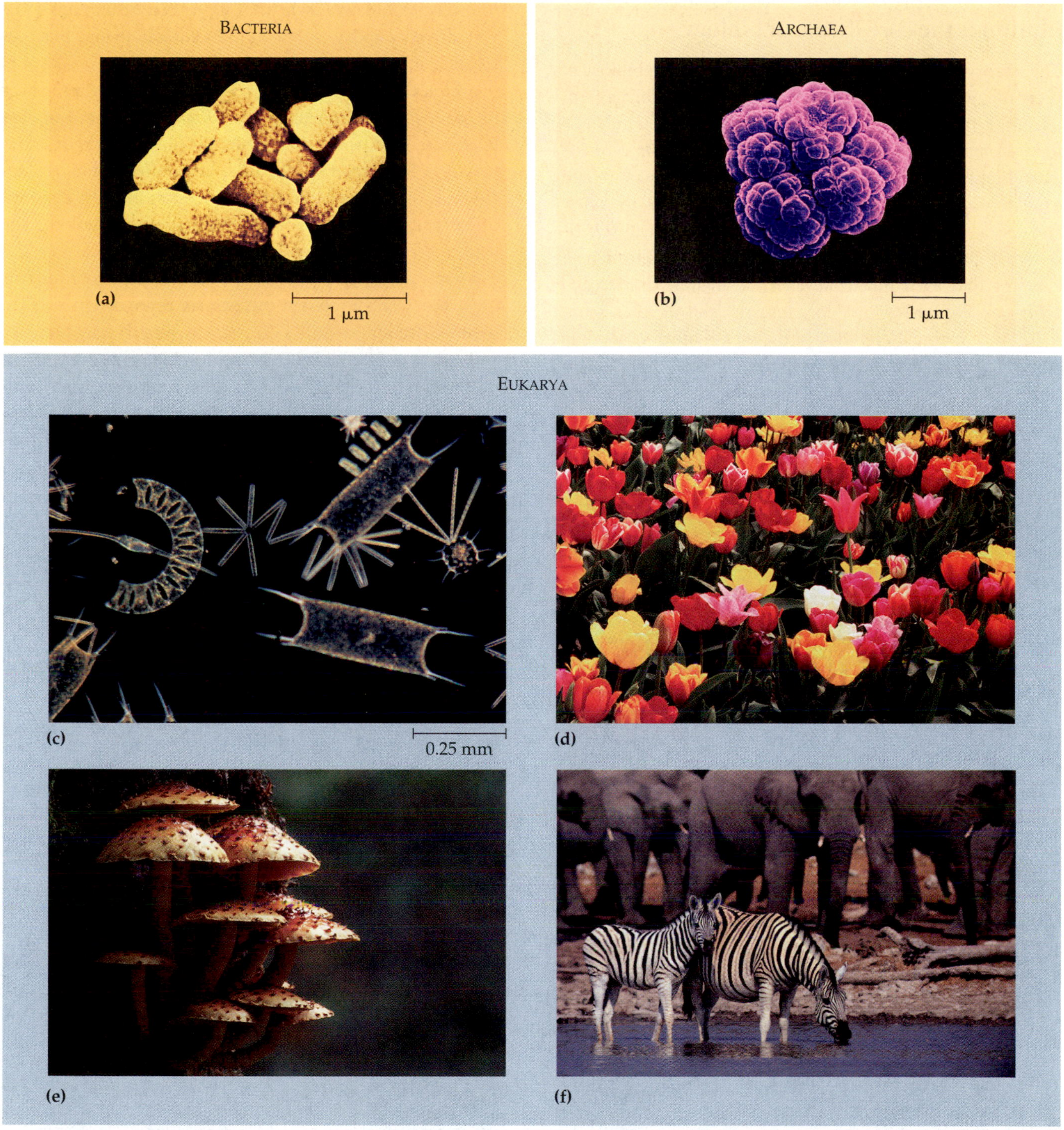

EUKARYA

(c)

0.25 mm

(d)

(e)

(f)

FIGURE 1.11 ▪ Domains of life. The domains Bacteria, Archaea, and Eukarya represent three fundamentally different kinds of organisms. Their most basic differences appear at the molecular level, in their nucleic acids. The domains Bacteria and Archaea are composed of prokaryotes. The more traditional five-kingdom system of classification recognized all prokaryotes as members of one kingdom, the Monera, plus the four kingdoms in the domain Eukarya shown here. **(a)** Members of Domain Bacteria (formerly called Eubacteria) are the most diverse and widespread prokaryotes. A unique set of molecular traits distinguishes them from other prokaryotes and eukaryotes. **(b)** Most Archaea (formerly called Archaebacteria) live in Earth's extreme environments, such as salty lakes and boiling hot springs. Molecular evidence indicates that archaeans have at least as much in common with eukaryotes as they do with members of Domain Bacteria. Within the domain Eukarya, **(c)** Kingdom Protista consists of unicellular eukaryotes and their relatively simple multicellular relatives. Pictured here is an assortment of protists inhabiting pond water. Scientists are currently debating how to split the protists into several kingdoms that better represent evolution and diversity. **(d)** Kingdom Plantae consists of multicellular eukaryotes, such as these tulips, that carry out photosynthesis. **(e)** Kingdom Fungi is defined, in part, by the nutritional mode of its members, organisms, such as these mushrooms, that absorb nutrients after decomposing organic material. **(f)** Kingdom Animalia consists of multicellular eukaryotes that ingest other organisms. You will learn more about the scientific debate over classification at the domain and kingdom levels in Unit Five.

Evolution is the core theme of biology

The history of life is a chronicle of a restless Earth billions of years old, inhabited by a changing cast of living forms (FIGURE 1.12). Life evolves. Just as an individual has a family history, each species is one twig on a branching tree of life extending back in time through ancestral species more and more remote. Species that are very similar, such as the horse and zebra, share a common ancestor that represents a relatively recent branch point on the tree of life. But through an ancestor that lived much farther back in time, horses and zebras are also related to rabbits, humans, and all other mammals. And mammals, reptiles, birds, and all other vertebrates share an even more ancient common ancestor. Trace evolution back far enough, and there are only the primeval prokaryotes that inhabited Earth more than three billion years ago. All of life is connected. Evolution, the processes that have transformed life on Earth from its earliest beginnings to its vast diversity today, is the one biological theme that ties together all others.

FIGURE 1.12 ▪ The fossil record is one type of historical documentation that chronicles evolution. *Archaeopteryx,* the animal documented in this fossil, lived about 150 million years ago. It had feathers (a hallmark of birds), but unlike any modern bird, *Archaeopteryx* had teeth, claws on its wings, and a bony tail. *Archaeopteryx* actually resembled certain dinosaurs more than it did any modern bird. The fossil record is compatible with other types of evidence supporting the hypothesis that birds evolved from dinosaurs.

Charles Darwin (FIGURE 1.13) brought biology into focus in 1859 when he published *The Origin of Species.* Darwin's book presented two main themes. First, Darwin argued convincingly from several lines of evidence that contemporary species arose from a succession of ancestors through a process of "descent with modification," his phrase for evolution. (The evidence for evolution is discussed in detail in Chapter 22.) The second theme of Darwin's book was his theory for *how* life evolves. This proposed mechanism of evolution is called natural selection.

Darwin synthesized the concept of natural selection from observations that by themselves were neither new nor profound. Others had the pieces of the puzzle, but Darwin saw how they fit together. He inferred natural selection by connecting two observations:

OBSERVATION #1: *Individual variation.* Individuals in a population of any species vary in many heritable traits.

OBSERVATION #2: *Struggle for existence.* Any population of a species has the potential to produce far more offspring than the environment can possibly support with food, space, and other resources. This overproduction makes a struggle for existence among the variant members of a population inevitable.

FIGURE 1.13 ▪ Charles Darwin (1809–1882). Darwin and his son William posed for this photograph in 1842. The author of numerous books and monographs on topics as diverse as barnacles, plant movements, and island geology, Darwin would be remembered as one of the greatest naturalists of the nineteenth century even if he had never published on the topic of evolution. But it was *The Origin of Species* that established Darwin's place as the most influential scientist in the development of modern biology. He is buried next to Isaac Newton in London's Westminster Abbey.

INFERENCE: *Differential reproductive success.* Those individuals with traits best suited to the local environment generally leave a disproportionately large number of surviving, fertile offspring. This differential reproductive success of some individuals over others means that certain heritable traits (those carried by the best-suited individuals) are more likely to appear in each new generation. Darwin called differential reproductive success natural selection, and he envisioned it as the cause of evolution.

We see the products of natural selection in the exquisite adaptations of organisms to the special problems posed by their environments (FIGURE 1.14). Notice, however, that natural selection does not *create* adaptations; rather, it screens the heritable variations in each generation, increasing the frequencies of some variations and decreasing the frequencies of others over the generations. Natural selection is an editing process, with heritable variations exposed to environmental factors that favor the reproductive success of some individuals over others. The camouflage of the sea horse in FIGURE 1.14 did not result from individuals changing during their lifetimes to look more like their backgrounds and then passing that improvement on to offspring. The adaptation evolved over many generations by the greater reproductive success in each generation of individuals who were innately better camouflaged than the average member of the sea horse population.

FIGURE 1.14 ▪ Evolutionary adaptation is a product of natural selection. This sea horse lives among kelp (seaweed). The fish looks so much like a seaweed that it lures prey into the seeming safety of the kelp forest and then eats them. In the Darwinian view of life, the best-camouflaged members of the sea horse population have the greatest probability of obtaining food and escaping predators—and thus the greatest probability of surviving and leaving offspring. By this mechanism of differential reproductive success, or natural selection, the interaction between environment and the heritable variation among members of the sea horse population gradually refined and maintained the camouflage over the generations.

Darwin proposed that natural selection, by its cumulative effects over vast spans of time, could produce new species from ancestral species. This would occur, for example, when a population fragments into several populations isolated in different environments. In these various arenas of natural selection, what begins as one species may gradually diversify into many as the geographically isolated populations adapt over many generations to different sets of environmental problems. Descent with modification accounts for both the unity and the diversity we observe in life. In many cases, features shared by two species are due to their descent from common ancestors, and differences between the species are due to natural selection modifying the ancestral equipment in different environmental contexts. Evolution is the core theme of biology—a unifying thread that ties every chapter of this text together.

SCIENCE AS A PROCESS

Biology is a natural science. Having identified unifying themes that apply specifically to the study of life, we now examine some general features of science as a process.

Like life, science is better understood by observing it than by trying to create a precise definition. The word *science* is derived from a Latin verb meaning "to know." Science is a way of knowing. It emerges from our curiosity about ourselves, the world, and the universe. Striving to understand seems to be one of our basic drives. At the heart of science are people asking questions about nature and believing that those questions are answerable. Scientists tend to be quite passionate in their quest for discovery. Max Perutz, a Nobel Prize–winning biochemist, put it this way: "A discovery is like falling in love and reaching the top of a mountain after a hard climb all in one, an ecstasy induced not by drugs but by the revelation of a face of nature that no one has seen before."

Testable hypotheses are the hallmarks of the scientific process

A process known as the scientific method outlines a series of steps for answering questions, but few scientists adhere rigidly to this prescription. Science is a less structured process than most people realize. Like other intellectual activities, the best science is a process of minds that are creative, intuitive, imaginative, and social. Perhaps science is distinguished by its conviction that natural phenomena, including the processes of life, have natural causes—and by its obsession with evidence. Scientists are generally skeptics. As you begin each unit of study in this text, you will meet such scientists through personal interviews and come to understand a bit more about how they think and why they enjoy their work.

1.1 Although it is counterproductive to reduce science to a stereotyped method, we *can* identify a common theme of the scientific process: *hypothetico-deductive thinking.* The first part of this term refers to *hypothesis,* which is a tentative answer to some question—an explanation on trial. Consider, for example, this imaginary scenario: Scott and Ian, identical twins, are sleepy every day in their 1:00 history class; they want to know why, so they can prevent the drowsiness and improve their history grades. Maybe eating a big lunch just before going to history every day makes them sleepy. Or maybe the classroom is too warm. Maybe Scott and Ian are listless because they sit in the back of their history class and are less involved than they are in their other classes, where they always sit in front. Or maybe it's just the time of day. These are all hypotheses, possible explanations for this daily behavior of sleeping through history class. It is possible to test these hypotheses.

The *deductive* in hypothetico-deductive thinking refers to the use of deductive reasoning to test hypotheses. Deduction contrasts with induction, which is reasoning from a set of specific observations to reach a general conclusion (as in "All organisms are composed of cells"). In deduction, the reasoning flows in the reverse direction, from the general to the specific. From general premises we extrapolate to specific results we should expect if the premises are true: If all organisms are made of cells (premise #1) and humans are organisms (premise #2), then humans are composed of cells (prediction about a specific case). In the scientific process, deduction usually takes the form of predictions about the results of experiments or observations we should expect *if* a particular hypothesis (premise) is correct. We then test the hypothesis by performing the experiments or making observations to see whether or not the predicted results occur. This deductive testing takes the form of "*If . . . then*" logic:

HYPOTHESIS #1: *If* the twins get sleepy because they eat lunch before their 1:00 class,

EXPERIMENT: and Scott postpones lunch until class ends at 2:00 (but Ian still eats at noon),

PREDICTED RESULT #1: *then* Scott should be less sleepy than Ian in history class.

PREDICTED RESULT #2: *then* Scott should get sleepy in his 3:00 class instead.

OR

HYPOTHESIS #2: *If* the twins get sleepy in their 1:00 class because they sit in the back of the room,

EXPERIMENT: and Scott (but not Ian) moves to the front of the class,

PREDICTED RESULT: *then* Scott should be more alert than Ian during the 1:00 class.

Now, suppose the result of the first experiment is that Scott is still as sleepy as Ian (but hungrier) in their 1:00 class. The twins try the second experiment, with the result that now Scott dozes in the front row of class instead of the back row. The twins continue to test and reject various hypotheses. And then they are the subjects of an experiment they did not even plan—daylight savings time begins. Scott and Ian (and most of their classmates) notice that for a day or two, they have more trouble than usual waking up in the morning. However, the twins find that they are more alert than usual in their 1:00 class and instead hit their low point during the 2:00–3:00 break in their schedule. Within a few days, patterns are back to normal, and Scott and Ian are drowsy during their 1:00 class again. The twins conclude that they are just naturally sleepy at this time of day and decide not to schedule 1:00 classes in future semesters.

Although this example is fanciful and there are flaws in the experiments, five important points about hypotheses are evident:

1. *Hypotheses are possible explanations.* A generalization based on inductive reasoning is not a hypothesis. The twins may conclude, "We are sleepy every day at about 1:00," but that conclusion merely summarizes a set of observations. A hypothesis is a possible *explanation* for what we have observed. "The warm temperature of the classroom puts Scott and Ian to sleep" is a hypothesis.
2. *Hypotheses reflect past experience.* Sometimes hypotheses are described as *educated* propositions about cause. For example, the hypothesis that a warm classroom causes drowsiness may be based on a general experience of sleepiness under such conditions. Although we should consider any hypothesis that is a possible explanation for what we have observed, the hypotheses we test first should be those that seem the most reasonable, based on what we already know.
3. *Multiple hypotheses should be proposed whenever possible.* Proposing alternative explanations that can answer a question is good science. If we operate with a single hypothesis, especially one we favor, we may direct our investigation toward a hunt for evidence in support of this hypothesis.

1.1 This symbol links topics in the text to interactive exercises in the CD-ROM that accompanies the book. The number indicates the appropriate activity in the CD.

4. *Hypotheses should be testable via the hypothetico-deductive approach.* Hypotheses should be phrased in a way that enables us to make predictions that can be tested by experiments or further observation. "Scott falls asleep in history class because the devil makes him do it" is not a testable hypothesis and therefore does not lend itself to the scientific process. Requiring that hypotheses be testable limits the scope of questions that science can answer.
5. *Hypotheses can be eliminated but not confirmed with absolute certainty.* If moving to the front of his 1:00 class does not reduce Ian's drowsiness, this casts doubt on one hypothesis. A hypothesis can be falsified by experimental tests, especially if the experiments are repeated with the same results. The onset of daylight savings time supported the "sleepy time of day" hypothesis, and our confidence in this explanation grows if the hypothesis continues to stand up to various kinds of experimental tests. But we can never *prove* that this hypothesis is the true explanation. It is impossible to repeat an experiment enough times to be absolutely certain that the results will *always* be the same. And some false hypotheses make accurate predictions. Consider this hypothesis: "Night and day are caused by the sun orbiting around Earth in an east-west direction." This hypothesis predicts that the sun will rise each morning in the east, move across the sky, and set in the west, which is exactly what we observe. However, the "geocentric universe" hypothesis makes many other predictions that enable us to falsify the hypothesis. Of the many hypotheses proposed to answer a particular question, the correct explanation may not even be included. Even the most thoroughly tested hypotheses are accepted only conditionally, pending further investigation.

The "sleepy twins" scenario introduces another important feature of the scientific process: the controlled experiment. In a *controlled experiment,* the subjects (Scott and Ian, in this case) are divided into two groups, an *experimental group* and a *control group.* Ideally, the two groups are treated exactly alike except for the one variable the experiment is designed to test. This provides a basis for comparison, enabling us to draw conclusions about the effects of our experimental manipulation. Scott and Ian attempted to control their experiments. When Scott moved to the front of the 1:00 class as the experimental, Ian remained in the back of the room as the control.

There are, of course, serious deficiencies in this experiment. If the twins were trying to test the effect of seating locale on sleepiness in class generally, then a single individual is an inadequate sample size for either a control group or an experimental group. Note also that this experiment included uncontrolled variables. Temperature and other unknown factors may have varied between the front and back of the classroom. Some experiments are easier to control than others, but setting up the best possible controls is characteristic of good experimental design.

Now that we have analyzed the hypothetico-deductive approach by using an imaginary example, you should be able to recognize the process in an elegant study reported in the scientific literature. For many years, David Reznick of the University of California, Riverside, and John Endler of the University of California, Santa Barbara, have been investigating differences between populations of guppies in Trinidad, a Caribbean island (FIGURE 1.15). Guppies *(Poecilia reticulata)* are small freshwater fishes you probably recognize as common aquarium pets. In the Aripo River system of Trinidad, guppies live in small pools as populations that are relatively isolated from one another. In some cases, two populations inhabiting

FIGURE 1.15 ▪ David Reznick conducting field experiments on guppy evolution in Trinidad.

the same stream live less than 100 meters (m) apart, but they are separated by a waterfall that impedes the migration of guppies between the two pools.

When Reznick and Endler compared guppy populations, they observed differences in what are called *life history characteristics*. These characteristics included the average age and size of guppies when they reach sexual maturity and begin to reproduce, as well as the average number of offspring per brood. The researchers were able to correlate variations in these life history characteristics with the types of predators present in different locations. In some pools, the main predator is a small fish called a killifish, which preys predominately on small, juvenile guppies. In other locations, a larger predator called a pike-cichlid preys more intensely on guppies and mainly eats relatively large, sexually mature individuals (FIGURE 1.16). Guppies in populations exposed to these pike-cichlids have larger broods, reproduce at a younger age, and are smaller at maturity, on average, than guppies that coexist with killifish.

What causes these life history differences between the guppy populations? Correlation with the type of predator present is suggestive, but a correlation does not necessarily imply a cause-and-effect relationship. The type of predator present and the life history characteristics of the guppy populations in a particular location may be independent consequences of some third factor. In fact, Reznick and Endler tested the hypothesis that the life history variations were due to differences in water temperature or other features of the physical environment. Notice the "*If* . . . *then*" logic characteristic of the hypothetico-deductive approach:

HYPOTHESIS #1: *If* differences in physical environments cause variations in the life histories of guppy populations,

EXPERIMENT: and samples from different wild guppy populations are collected and maintained for several generations in identical environments in predator-free aquaria,

PREDICTED RESULT: *then* the laboratory populations should become more similar in their life history characteristics.

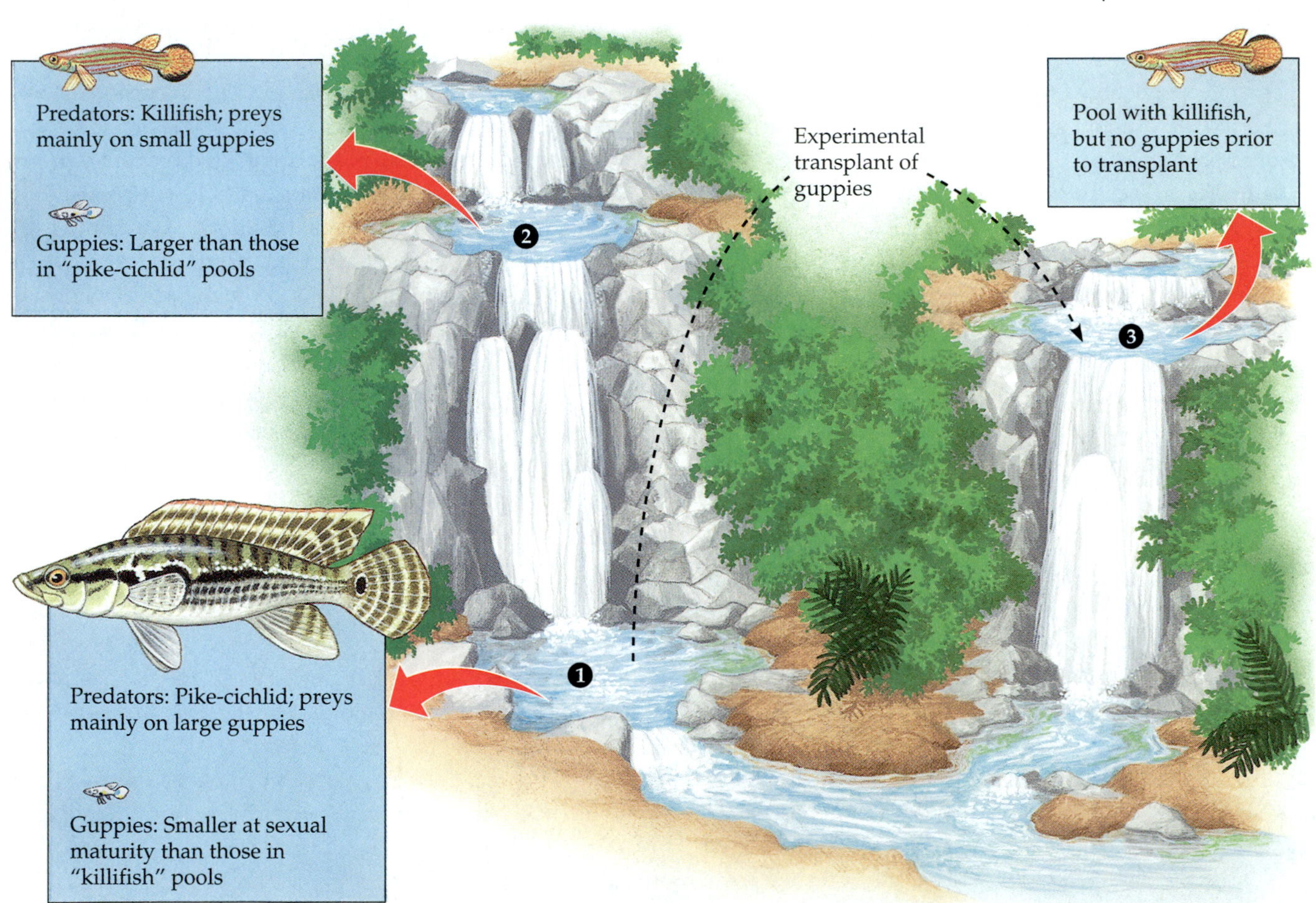

FIGURE 1.16 ▪ Testing the hypothesis that selective predation affects the evolution of guppy populations. This drawing represents three pools along streams of Trinidad's Aripo River system. ① In one pool, pike-cichlids prey intensively on guppies, mainly eating relatively large individuals. ② In another pool, the predators are killifish, which prey less intensively than pike-cichlids and feed mainly on relatively small guppies. In this and other killifish pools, guppies are larger and older at sexual maturity than guppies in pike-cichlid pools. This observation led to Reznick and Endler's hypothesis that selective predation was affecting the evolution of life history characteristics of the guppy populations. One way the researchers tested this hypothesis was to transplant guppies from pike-cichlid pools to pools that contained killifish but had no natural guppy populations ③. Reznick and Endler then tracked the evolution of life history in the experimental guppy populations for 11 years. They compared average age and size of mature guppies in the experimental pools to these life history characteristics of guppies in control pools inhabited by pike-cichlids. FIGURE 1.17 summarizes the results of these experiments.

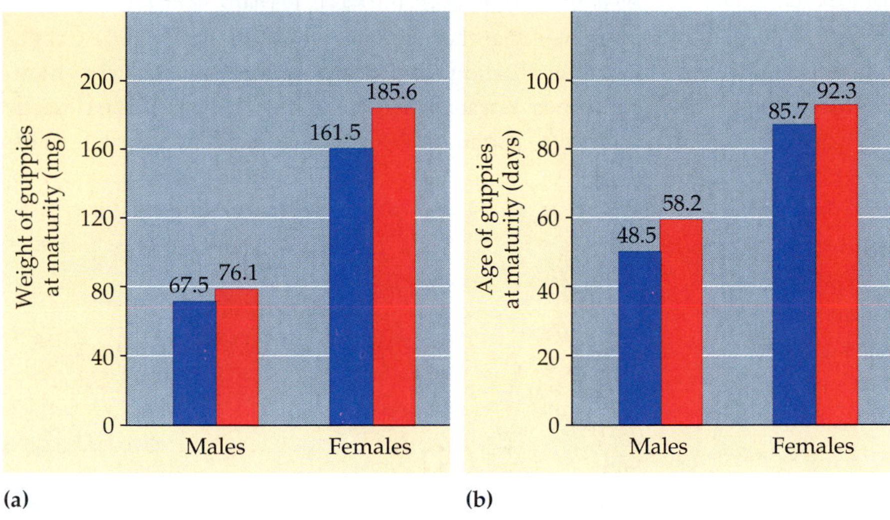

FIGURE 1.17 ▪ Experimental evidence for natural selection in action: results of the guppy transplant experiments. These data represent average measurements for two life history characteristics in guppy populations: **(a)** weight at sexual maturity and **(b)** age at sexual maturity. The vertical bars of the histograms compare control guppy populations to experimental populations. Controls were populations native to pools in which the main predator is the pike-cichlid, which feeds predominately on large, sexually mature guppies. The experimental populations consisted of guppies that were removed from pike-cichlid pools and transplanted to guppy-free pools inhabited by killifish (see FIGURE 1.16). Killifish prey mainly on small, immature guppies. After just 11 years in this study, the transplanted guppy populations had evolved measurably. More recent studies show that similar changes can occur in as few as 4 years in male guppies and 7.5 years in females.

When the researchers performed this experiment, the differences persisted for many generations. This result eliminates hypothesis #1 and also indicates that the life history differences in guppy populations are inherited. Based on the assumption that natural selection can lead to genetic differences in populations, Reznick and Endler tested the following explanation:

HYPOTHESIS #2: *If* the feeding preferences of different predators caused contrasting life histories in different guppy populations to evolve by natural selection,

EXPERIMENT: and guppies are transplanted from locations with pike-cichlids (predators of mature guppies) to guppy-free sites inhabited by killifish (predators of juvenile guppies),

PREDICTED RESULT: *then* the transplanted guppy populations should show a generation-to-generation trend toward later maturation, larger size, and smaller broods—life history characteristics typical of natural guppy populations that coexist with killifish.

In 1976, Reznick and Endler introduced guppies from locations with pike-cichlids to sites that had killifish but no guppies (see FIGURE 1.16). These transplanted populations were the researchers' experimental groups, and they studied them for 11 years, measuring age and size at maturity, brood size, and other life history characteristics. The scientists compared these measurements to data collected over the same period on control groups, guppies that remained in the original locations inhabited by pike-cichlids. To be certain that only heritable differences were counted, the measurements were made after samples from the experimental and control populations had been reared for two generations in identical aquarium environments. Over 11 years, or 30 to 60 generations, the average weight at maturity for guppies in the introduced (experimental) populations increased by about 14% compared to the control populations. Other life history characteristics also changed in the direction predicted by hypothesis #2 (FIGURE 1.17).

Without a control group for comparison, there would be no way to tell whether it was the killifish or some *other* factor that caused the transplanted guppy population to change. But because control sites and experimental sites were often nearby pools of the same stream, the main variable was probably the presence of different predators. And these careful researchers observed similar results when guppy populations were reared in artificial streams that were identical except for the type of predator.

Of the several hypotheses Reznick and Endler have tested (we examined only two), they are left with natural selection due to differential predation on larger versus smaller guppies as the most likely explanation for the observed differences between guppy populations. Apparently, when predators such as pike-cichlids prey mainly on reproductively mature adults, the chance that a guppy will survive to reproduce several times is relatively low. The guppies with greatest reproductive success should then be the individuals that mature at a young age and small size and produce at least one brood before growing to a size preferred by the local predator.

The popular press gave Reznick and Endler's research a lot of attention because it documents evolution in a natural setting over a relatively short time (within only 11 years in the study we have followed). In more recent studies, Reznick and others have documented similar evolutionary changes in guppies in as few as four years. We have examined the guppy experiments in some detail because they also provide a fine example of the key role of hypothetico-deductive thinking in science.

Another key feature of science is its progressive, self-correcting quality. A succession of scientists working on the same problem build on what has been learned earlier. It is also common for scientists to check on the conclusions of others by attempting to repeat observations and experiments. Both cooperation and competition exist among scientists working on the same question. Scientists share information through publications, seminars, meetings, and personal communication. They also subject one another's work to careful scrutiny.

Many people associate the word *discovery* with science. Often, what they have in mind is the discovery of new facts. But accumulating facts is not really what science is about; a telephone book is a catalog of facts, but it has little to do with science. It is true that facts, in the form of observations and experimental results, are the prerequisites of science. What really advances science, however, is a new idea that collectively explains a number of observations that previously seemed to be unrelated. The most exciting ideas in science are those that explain the greatest variety of phenomena. People like Newton, Darwin, and Albert Einstein stand out in the history of science not because they discovered a great many facts, but because they synthesized ideas with great explanatory power. Such ideas, much broader in scope than the hypotheses that pose possible causes for one set of observations, are known as theories. Because theories are comprehensive, they only become widely accepted if they are supported by a large body of evidence. Natural selection qualifies as a theory because of its broad application to so many situations. And the theory of natural selection is widely accepted because researchers working in the many fields of biology continue to validate the theory with new observations and experiments, including those of Reznick and Endler.

This book is only partly about the current state of biological knowledge. It is also important for you to learn, by example and by practice, how the process of science works. The power of hypothetico-deductive thinking is one of the themes of this text, and you will read many examples of how biologists have applied it in their research. However, it is mainly your own experience in the laboratory and in the field that will teach you how to do science. And your practice with the hypothetico-deductive approach will help you think more critically in general.

Science and technology are functions of society

Science and technology are associated. In many cases, technology results from scientific discoveries applied to the development of goods and services. Watson and Crick discovered the structure of DNA through the process of science. This breakthrough sparked an explosion of scientific activity that led to better understanding of DNA chemistry and the genetic code. These discoveries eventually made it possible to manipulate DNA, enabling genetic technologists to transplant foreign genes into microorganisms and produce such valuable products as human insulin. The new biotechnology is revolutionizing the pharmaceutical industry, and DNA technology has also had an enormous impact in other areas, including the legal profession (FIGURE 1.18). Perhaps Watson and Crick envisioned that their discovery would someday have technological applications, but that was probably not what motivated their research; nor could they have predicted exactly what the applications would be. Scientist/writer Lewis

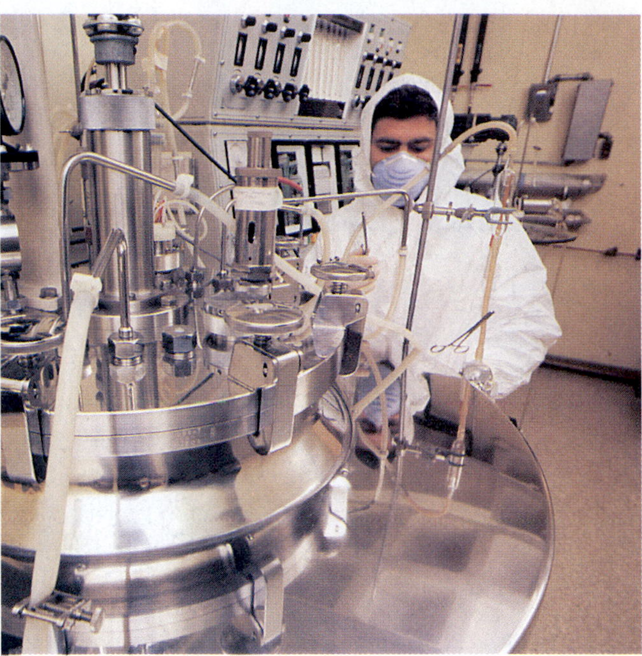

(a)

(b)

FIGURE 1.18 • Two examples of DNA technology. (a) This employee of a biotechnology company is monitoring a tank for growing yeast cells that have been engineered to carry genes of the virus that causes hepatitis B. The genetically engineered cells produce large amounts of a protein molecule that is found on the surface of the virus. The protein is used to make a vaccine against the virus. **(b)** Forensic technicians can use traces of DNA extracted from a blood sample or other body fluid to produce a molecular "fingerprint." The stained bands visible in this photograph represent fragments of DNA, and the pattern of bands varies from one person to another. The legal applications of DNA technology have become very public in recent years. You will learn more about DNA technology in Chapter 20.

Thomas put it this way in his essay "Making Science Work": "We cannot say to ourselves, we need this or that sort of technology, therefore we should be doing this or that sort of science. . . . Science is useful, indispensable sometimes, but whenever it moves forward it does so by producing a surprise; you cannot specify the surprise you'd like."*

Not all technology can be described as applied science. In fact, technology predates science, driven by inventive humans who built tools, crafted pots, mixed paints, designed musical instruments, and made clothing—all without necessarily understanding why their inventions worked. Science catalyzes certain technologies by complementing trial and error with more informed design. But the direction technology takes depends less on science than it does on the needs of humans and the values of society.

Technology has improved our standard of living in many ways, but it is a double-edged sword. Technology, especially technology that keeps people healthier, has enabled the human population to grow more than tenfold in the past three centuries. The environmental consequences are enormous. Acid rain, deforestation, global warming, nuclear accidents, ozone holes, toxic wastes, and extinction of species are just a few of the repercussions from more and more people wielding more and more technology. Science can help us identify problems and provide insight about what course of action may prevent further damage. But solutions to these problems have as much to do with politics, economics, culture, and values as with science and technology.

Now that both science and technology have become such powerful functions of society, it is more important than ever to distinguish "what we would like to understand" from "what we would like to build." Scientists should not distance themselves from technology but instead try to influence how society applies scientific discoveries. And scientists have a responsibility to help educate politicians, bureaucrats, corporate leaders, and voters about how science works and about the potential benefits and hazards of specific technologies. The crucial relationship among science, technology, and society is a theme that increases the significance of our study of life.

Biology is a multidisciplinary adventure

In some ways, biology is the most demanding of all sciences, partly because living systems are so complex and partly because biology is a multidisciplinary science that requires a knowledge of chemistry, physics, and mathematics. Modern biology is the decathlon of natural science. If you are a biology major or a preprofessional student, you have an opportunity to become a versatile scientist. If you are a physical science major or an engineering student, you will discover in the study of life many applications for what you have learned in your other science courses. If you are a nonscience student enrolled in biology as part of a liberal arts education, you have selected a course in which you can sample many scientific disciplines. And of all the sciences, biology is the most connected to the humanities and social sciences.

No matter what brings you to biology, you will find the study of life to be challenging and uplifting. Do not let the details of biology spoil a good time. The complexity of life is inspiring, but it can be overwhelming. To help you keep from "getting lost in the forest because of all the trees," each chapter of this book is constructed from a manageable number of key concepts. The concepts are listed at the beginning of the chapter, displayed throughout the chapter, and then reappear in the review at the end of the chapter. The details of a chapter enrich your understanding of the concepts and how they fit together. This introductory chapter is the exception; instead of presenting the key concepts of a particular area of biology, this chapter introduced themes that cut across all biological fields—ways of thinking about biology. These themes, along with the key concepts in each chapter, will provide you with a framework for fitting together the many things you will learn in your multidisciplinary exploration of life—and will encourage you to begin asking important questions of your own.

* Thomas, L. "Making Science Work." *Late Night Thoughts on Listening to Mahler's Ninth Symphony.* New York: Viking Press, 1983, p. 28.

UNIT ONE

THE CHEMISTRY OF LIFE

Mario Molina personifies two of this textbook's integrating themes: the multidisciplinary nature of biology and the relationship of science and technology to society. His research on the chemistry of the atmosphere's ozone layer and how certain pollutants are damaging that protective layer led to international awareness about a serious threat to life on Earth. In 1995, Dr. Molina was awarded the Nobel Prize in chemistry, along with Paul Crutzen and F. Sherwood Rowland. The Nobel citation clearly connected these chemists' research to biology and society: "By explaining the chemical mechanisms that affect the thickness of the ozone layer, the three researchers have contributed to our salvation from a global environmental problem that could have catastrophic consequences." I interviewed Dr. Molina in his office at the Massachusetts Institute of Technology, where he is a professor in the Department of Earth, Atmospheric, and Planetary Sciences.

Let's start with the atmosphere's ozone layer. Why is it important to life on Earth?

The ozone shield is important to biological systems because it shields the Earth's surface from powerful ultraviolet radiation that comes from the sun. This UV radiation is harmful to organisms, including humans. For example, UV radiation causes sunburn, and skin cancer can be a cumulative result of exposure. There is also evidence that UV can damage crops, such as soybeans. Certain developing animals, such as the larvae of fish, seem to be particularly sensitive. The atmosphere's ozone layer prevents most of this harmful UV radiation from reaching Earth. But it is the Achilles' heel of the atmosphere.

In what way?

The ozone shield, so crucial to life, is relatively thin and fragile. In fact, the whole atmosphere is thin, with most air located within about 30 miles of Earth. Compared to the size of the Earth, the atmosphere is like the skin of an apple. And within this thin atmosphere, ozone is very scarce, measured in parts per million. Most of that ozone is located about 15 miles from Earth, in the stratosphere. The ozone forms when high-energy solar radiation breaks apart oxygen molecules and frees oxygen atoms. These oxygen atoms then react with unbroken oxygen molecules, which consist of two bonded oxygen atoms. This forms ozone, which has three oxygen atoms. So ozone is continuously forming by the action of sunlight on the atmosphere, but this is balanced by continuous destruction of the unstable ozone molecules when they react with other chemical compounds that are naturally present in the atmosphere. We've disrupted that balance by releasing certain industrial chemicals that damage the fragile ozone layer.

That's a good segue to the story of your Nobel Prize. Tell us about that research.

In 1974 we published our first paper on the CFC-ozone depletion theory. We made a prediction based on laboratory experiments on atmospheric chemistry. We predicted that our continuous release of chlorofluorocarbons, or CFCs, such as Freon, which is a brand name,

INTERVIEW

MARIO MOLINA

would damage the protective ozone layer. These compounds were used as refrigerants, as propellants in spray cans, as solvents, and in the process for making plastic foams. The CFCs were actually developed as replacements for toxic refrigerants. In fact CFCs are very stable, or unreactive, which is why they are not toxic. You can even breathe them, and nothing happens. But it is also the stability of CFCs that allows them to reach the ozone layer. Most other pollutants released into the air are removed before they reach the stratosphere. Water-soluble pollutants come back to the Earth very fast. And most pollutants that are not soluble in water, such as hydrocarbons, are oxidized by other gases in the lower atmosphere and are converted to water-soluble compounds. So the atmosphere has a pretty efficient cleansing mechanism, but only at low altitudes. The problem is that CFCs make it up to the stratosphere, where solar radiation finally converts them to very reactive chemical species called free radicals. Those free radicals then destroy the ozone.

So, when you and your colleagues sounded this alarm in 1974, what was the reaction of the scientific community and the general public?

At first, there was very little reaction because not many people were aware of the importance of invisible things like the ozone layer and UV radiation. Experts in our research field quickly realized that the prediction of the CFC–ozone depletion theory was something to worry about, but many other scientists were skeptical, which is natural for scientists. Then as we and others in our field began doing experiments to test the idea, the case for the CFC–ozone depletion theory became strong and more and more people, including politicians, became concerned about the issue of ozone depletion. There was also growing awareness about the role of UV radiation in skin cancer and in other biological damage. Then, in 1985, scientists documented a drastic depletion of the ozone layer over the Antarctic, an ozone hole. This caused the United Nations to act in 1987.

And what was that action?

The UN Montreal Protocol of 1987 was an international agreement to reduce by one-half the amount of CFCs released into the atmosphere. Then, as evidence in support of the ozone depletion theory continued to grow, the Protocol was strengthened to require a complete phasing out of CFC production by developed countries by 1995. So that's already happened. Industrial countries are no longer manufacturing CFCs. Instead, these countries now use other refrigerants, which are destroyed in the atmosphere before they reach the ozone layer. The Montreal Protocol allowed a grace period for developing countries, a bit more time and some help from the industrial countries in the transition to CFC-free technology. That's in the works now, and CFCs will be phased out everywhere within the next decade.

Is the damage we've already inflicted on the ozone layer reversible?

Yes, because ozone is forming continuously in the atmosphere. But because CFCs are so stable and because they take so long to reach the ozone layer, the damage that we're seeing today is actually caused by CFCs released a decade ago. And CFCs will remain in the stratosphere for several more decades. So, we predict that the ozone layer won't recover until about the middle of the next century.

Your commitment to the relationship of science to society includes your service on the President's Committee of Advisors on Science and Technology. What have you learned from that experience about the interface between science and politics?

One lesson is that science is not always something that politicians care very much about. Often, supporting science is considered a luxury. But I think the scientific community has succeeded to some extent in putting science in the limelight. Supporting science is a good investment. Another challenge is that many issues related to science and technology are long-term issues that require patience and long-term commitment. An example is our excessive production of carbon dioxide and the related possibility of global warming. Let's say that ten to twenty years from now, we realize that we really are in trouble because of global warming. We will seem foolish if we did not invest in research now if global warming does turn out to be a serious problem, which we think is a reasonable possibility.

In terms of mounting an international effort, the global warming problem is an even tougher challenge than the ozone depletion problem, right?

Fortunately, we have the precedent of the Montreal Protocol, but, yes, the climate change issue is more complicated. First of all, the accumulation of gases that could warm the planet by what is known as the greenhouse effect is occurring mainly in the lower part of the atmosphere, which is a more complex system than the stratosphere. So, there are many more scientific uncertainties than we have for the ozone problem. But it is reasonable to predict that there will be consequences of our adding these greenhouse gases to the atmosphere. We can also tell that temperatures are indeed increasing, but it's not yet that clear that this is a consequence of human activities. One could argue that much of the change is natural, although there is more and more evidence that the planet's climate is being affected by human activities. But we don't have a smoking gun yet, like we did with the ozone hole over the Antarctic. The second important difference is that reducing the release of greenhouse gases requires much more social and economic change than does reducing the release of CFCs.

Tell us more about this challenge.

The most important greenhouse gas is the carbon dioxide we're releasing by burning fossil fuels and burning forests. This activity is related to energy sources, which are so important for society. Conserving energy and developing alternative energy requires much bigger change in our patterns than using other compounds in place of CFCs. What we're talking about is preserving a high standard of living and increasing the standard of living in developing countries. But we may be able to do that in clever ways, without consuming so much energy. We have some examples. Energy consumption per person in Europe is a lot less than it is in the United States. Clearly, there are some different habits. For one thing, gasoline for cars is very cheap here in the United States, so there is no incentive to be more efficient. With rapid, drastic change in energy use, we can predict that the economy would suffer. But with more gradual change and the right kind of planning and incentives, we can deal with the climate change issue. In fact, recent international agreements could change the way we use energy.

Will this seem like "changing rules" to the developing countries that are trying to improve their living standard?

That's an issue, but there is another important element that I've observed from my own experience. When I was a young person going to the university in Mexico, no one cared about pollution. Dealing with pollution seemed like a luxury. Today, everybody in Mexico City cares about pollution, so solving this problem no longer seems like something the industrialized countries want to impose upon them. The costs of pollution are enormous in Mexico City, where I was born. If only there had been insight about what would happen to the city, pollution could have been reduced at a much lower cost. It's much more difficult to repair the environment now, but I think there is a strong incentive to do that in developing countries. It's not easy because of the complications of social and economic issues. Science and technology are very important components, but not the only ones, so scientists will have to be humble and realize their limits.

You came to MIT in 1989 after several years as a researcher at the Jet Propulsion Lab in California. Why did you decide to return to the academic community?

The main reason was to have more interaction with students. I find it very stimulating both to teach undergraduate students in the classroom and also work with graduate students in research.

What general advice can you share with the undergraduate students reading this book?

First, students should be conscious about how fun it can be to study and learn, which is difficult to realize when there are so many hurdles to go over and the work sometimes seems overwhelming. But believe me, studying can be fun. To some extent, university education is changing. There are so many things to learn and so many new developments that there is no hope of covering everything. We are not just teaching facts, but teaching how to learn. Students should discuss ideas with their professors and become engaged. Try not to be just a passive student listening to lectures, but participate actively in the process of learning.

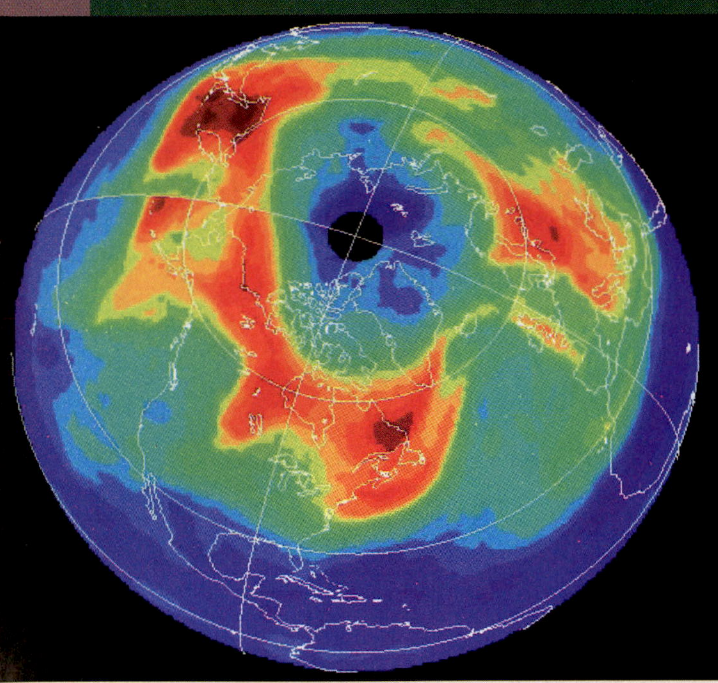

CHAPTER 2

THE CHEMICAL CONTEXT OF LIFE

Chemical Elements and Compounds

- Matter consists of chemical elements in pure form and in combinations called compounds
- Life requires about 25 chemical elements

Atoms and Molecules

- Atomic structure determines the behavior of an element
- Atoms combine by chemical bonding to form molecules
- Weak chemical bonds play important roles in the chemistry of life
- A molecule's biological function is related to its shape
- Chemical reactions make and break chemical bonds

Along with Mario Molina, whom you met in the preceding interview, many other scientists are studying the Earth's ozone layer and its importance to life on the planet. (In this image of the northern hemisphere at the left, the areas of thinnest ozone are blue and purple.) This research involves specialists in a number of different disciplines, for unlike a college catalog of courses, nature is not packaged into biology, chemistry, physics, and the other natural sciences. Biologists are scientists who specialize in the study of life, but organisms and the world they live in are natural systems to which basic concepts of chemistry and physics apply. Biology is a multidisciplinary science.

*This unit of chapters introduces key concepts of chemistry that will apply throughout our study of life. We will make many connections to the themes introduced in Chapter 1. One of those themes is the organization of life on a hierarchy of structural levels (*FIGURE 2.1*), with additional properties emerging at each successive level. In this unit, we will see how the theme of emergent properties applies to the lowest levels of biological organization—to the ordering of atoms into molecules and to the interactions of those molecules within cells. Somewhere in the transition from molecules to cells, we will cross the blurry boundary between nonlife and life.*

CHEMICAL ELEMENTS AND COMPOUNDS

Matter consists of chemical elements in pure form and in combinations called compounds

Organisms are composed of **matter**, which is anything that takes up space and has mass.* Matter exists in many diverse forms, each with its own characteristics. Rocks, metals, wood, glass, and you and I are just a few examples of what seems an endless assortment of matter.

Some of the ancient Greek philosophers believed that the great variety of matter arises from four basic ingredients, or elements. They imagined these elements to be air, water, fire, and earth—supposedly pure substances that could not be decomposed to other forms of matter. All other substances were thought to be formed by blending various proportions of two or more of the elements. Even though these classical philosophers proposed the wrong elements, their basic idea was correct.

* Sometimes we substitute the term *weight* for *mass*, although the two are not equivalent. We can think of mass as the amount of matter an object represents. The weight of an object measures how strongly that mass is pulled by gravity. An astronaut in orbit in a space shuttle is weightless, but the astronaut's mass is the same as it would be on Earth. However, as long as we are earthbound, the weight of an object is a measure of its quantity of matter; so for our purposes, we can use the terms interchangeably.

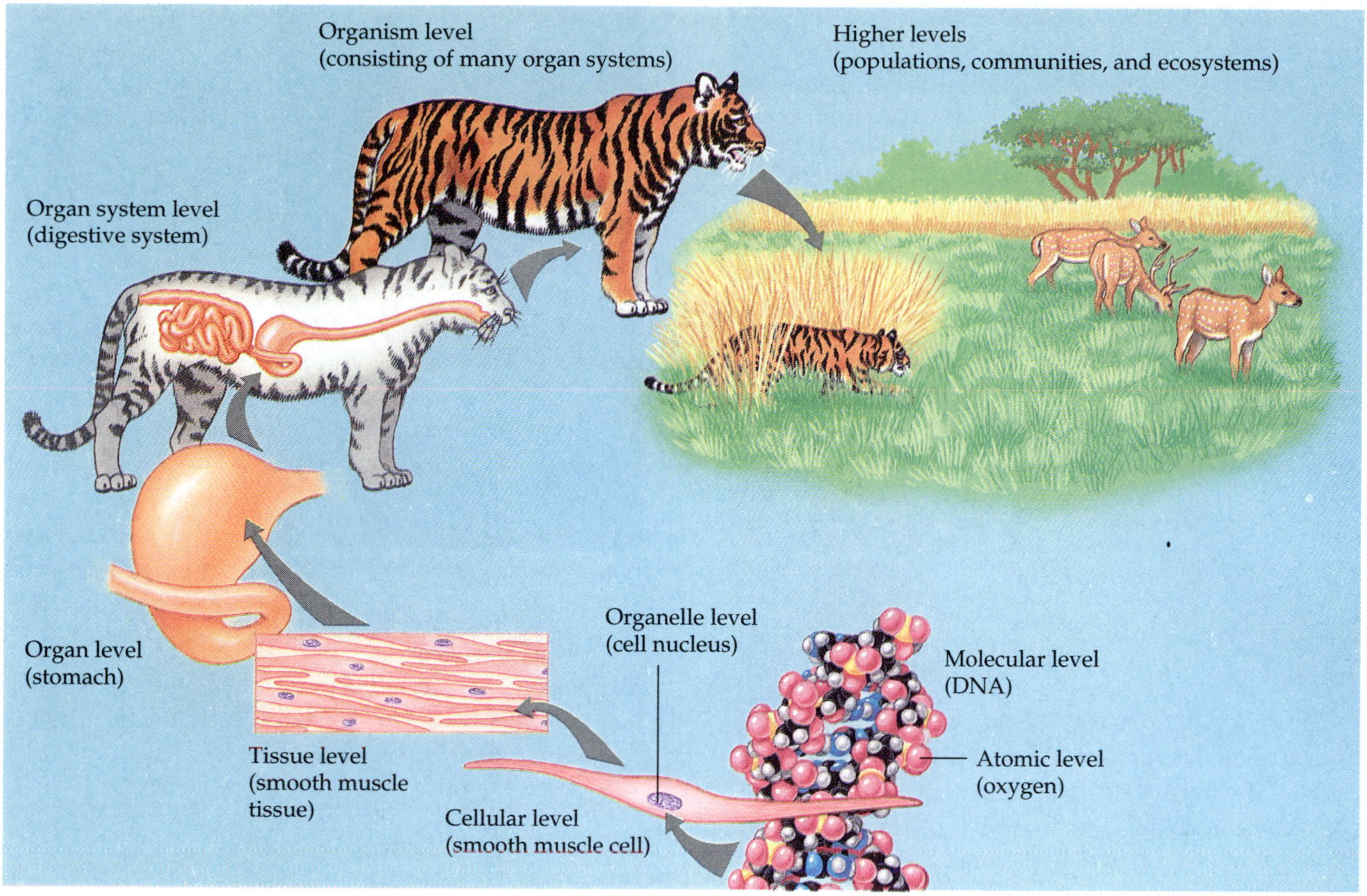

Organism level
(consisting of many organ systems)

Higher levels
(populations, communities, and ecosystems)

Organ system level
(digestive system)

Organ level
(stomach)

Organelle level
(cell nucleus)

Molecular level
(DNA)

Atomic level
(oxygen)

Tissue level
(smooth muscle tissue)

Cellular level
(smooth muscle cell)

FIGURE 2.1 ▪ The hierarchy of biological order.

An **element** is a substance that cannot be broken down to other substances by chemical reactions. Today, chemists recognize 92 elements occurring in nature; gold, copper, carbon, and oxygen are examples. Each element has a symbol, usually the first letter or two of its name. Some of the symbols are derived from Latin or German names; for instance, the symbol for sodium is Na, from the Latin word *natrium.*

A **compound** is a substance consisting of two or more elements combined in a fixed ratio. Table salt, for example, is sodium chloride (NaCl), a compound composed of the elements sodium (Na) and chlorine (Cl) in a 1:1 ratio. Pure sodium is a metal and pure chlorine is a poisonous gas. Chemically combined, however, sodium and chlorine form an edible compound. This is a simple example of organized matter having emergent properties: A compound has characteristics beyond those of its combined elements (FIGURE 2.2).

Life requires about 25 chemical elements

About 25 of the 92 natural elements are known to be essential to life. Just four of these—carbon (C), oxygen (O), hydrogen (H), and nitrogen (N)—make up 96% of living matter. Phosphorus (P), sulfur (S), calcium (Ca), potassium (K), and a few other elements account for most of the remaining 4% of an

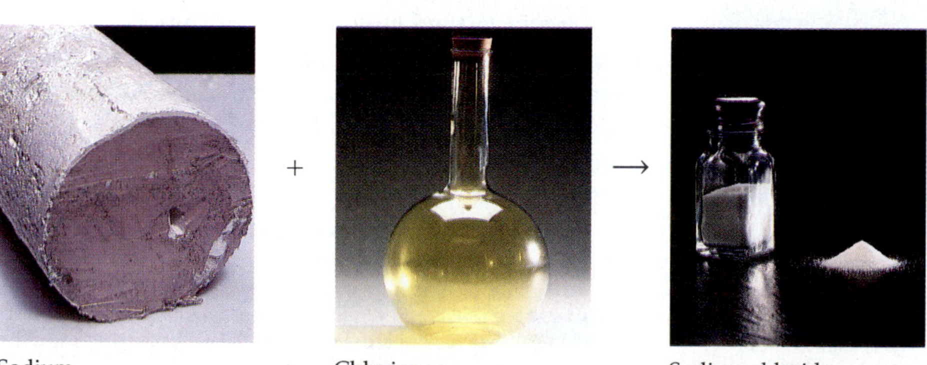

Sodium

Chlorine

Sodium chloride

FIGURE 2.2 ▪ The emergent properties of a compound. The metal sodium combines with the poisonous gas chlorine to form the edible compound sodium chloride, or table salt.

Table 2.1 ▪ Naturally Occurring Elements in the Human Body

SYMBOL	ELEMENT	ATOMIC NUMBER (See p. 25)	PERCENTAGE OF HUMAN BODY WEIGHT
O	Oxygen	8	65.0
C	Carbon	6	18.5
H	Hydrogen	1	9.5
N	Nitrogen	7	3.3
Ca	Calcium	20	1.5
P	Phosphorus	15	1.0
K	Potassium	19	0.4
S	Sulfur	16	0.3
Na	Sodium	11	0.2
Cl	Chlorine	17	0.2
Mg	Magnesium	12	0.1

Trace elements (less than 0.01%): boron (B), chromium (Cr), cobalt (Co), copper (Cu), fluorine (F), iodine (I), iron (Fe), manganese (Mn), molybdenum (Mo), selenium (Se), silicon (Si), tin (Sn), vanadium (V), and zinc (Zn).

organism's weight. TABLE 2.1 lists by percentage the elements that make up the human body. FIGURE 2.3 illustrates a deficiency of the essential element nitrogen in plants.

Trace elements are those required by an organism in only minute quantities. Some trace elements, such as iron (Fe), are needed by all forms of life; others are required only by certain species. For example, in vertebrates (animals with backbones), the element iodine (I) is an essential ingredient of a hormone produced by the thyroid gland. A daily intake of only 0.15 milligram (mg) of iodine is adequate for normal activity of the human thyroid. An iodine deficiency in the diet causes the thyroid gland to grow to abnormal size, a condition called goiter (FIGURE 2.4). Where it is available, iodized salt has reduced the incidence of goiter.

FIGURE 2.3 ▪ Nitrogen deficiency in corn. In this controlled experiment, the plants on the left are growing in soil that was fertilized with compounds containing nitrogen, an essential element. The soil on the right is deficient in nitrogen. If the poorly nourished crop growing in this soil is harvested, it will yield less food—and less nutritious food—than the crop on the left.

FIGURE 2.4 ▪ Goiter. The enlarged thyroid gland of this Malaysian woman is due to an iodine deficiency.

ATOMS AND MOLECULES

The properties of chemical elements and of the compounds they form, including the compounds crucial to life, ultimately result from the structure of atoms.

Atomic structure determines the behavior of an element

2.1 Each element consists of a certain kind of atom that is different from the atoms of any other element. An **atom** is the smallest unit of matter that still retains the properties of an element. Atoms are so small that it would take about a million of them to stretch across the period printed at the end of this sentence. We symbolize atoms with the same abbreviation used for the element made up of those atoms; thus, C stands for both the element carbon and a single carbon atom.

Subatomic Particles

Although the atom is the smallest unit having the properties of its element, these tiny bits of matter are composed of even smaller parts, called subatomic particles. Physicists have split the atom into more than a hundred types of particles, but only three kinds of particles are stable enough to be of relevance here: **neutrons**, **protons**, and **electrons**. Neutrons and protons are packed together tightly to form a dense core, or **atomic nucleus**, at the center of the atom. The electrons, moving at nearly the speed of light, form a cloud around the nucleus (FIGURE 2.5).

Electrons and protons are electrically charged. Each electron has one unit of negative charge, and each proton has one

2.1 This symbol links topics in the text to interactive exercises in the CD-ROM that accompanies the book. The number indicates the appropriate activity in the CD.

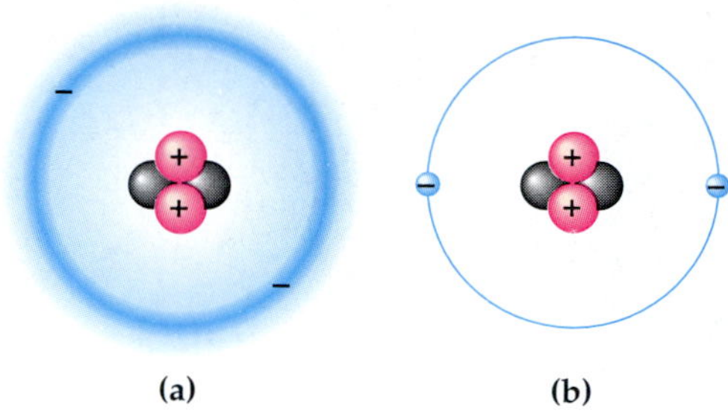

FIGURE 2.5 ▪ Two simplified models of a helium (He) atom. The helium nucleus consists of two neutrons (gray) and two protons (magenta). **(a)** The nucleus is surrounded by a cloud of negative charge (blue), owing to the rapid movement of two electrons. **(b)** The circle indicates the average distance of the electrons from the nucleus, although this distance is not drawn to scale compared to the size of the nucleus. (Our model of an atom will be refined as the chapter progresses.)

unit of positive charge. A neutron, as its name implies, is electrically neutral. Protons give the nucleus a positive charge, and it is the attraction between opposite charges that keeps the rapidly moving electrons in the vicinity of the nucleus.

The neutron and proton are almost identical in mass, each about 1.7×10^{-24} grams (g). Grams and other conventional units are not very useful for describing the mass of objects so minuscule. Thus, for atoms and subatomic particles, we use a unit of measurement called the **dalton**, in honor of John Dalton, the British scientist who helped develop atomic theory around 1800. Neutrons and protons have masses close to 1 dalton (1.009 and 1.007, respectively). Because the mass of an electron is only about 1/2000 that of a neutron or proton, we can ignore electrons when computing the total mass of an atom.

Atomic Number and Atomic Weight

Atoms of the various elements differ in their number of sub-
2.2 atomic particles. All atoms of a particular element have the same number of protons in their nuclei. This number of protons, which is unique to that element, is referred to as the **atomic number** and is written as a subscript to the left of the symbol for the element. The abbreviation ${}_{2}He$, for example, tells us that an atom of the element helium has two protons in its nucleus. Unless otherwise indicated, an atom is neutral in electrical charge, which means that its protons must be balanced by an equal number of electrons. Therefore, the atomic number tells us the number of protons and also the number of electrons in an electrically neutral atom.

We can deduce the number of neutrons from a second quantity, the **mass number**, which is the sum of protons plus neutrons in the nucleus of an atom. The mass number is written as a superscript to the left of an element's symbol. For example, we can use this shorthand to write an atom of helium as ${}^{4}_{2}He$. Because the atomic number indicates how many protons there are, we can determine the quantity of neutrons by subtracting the atomic number from the mass number: A ${}^{4}_{2}He$ atom has 2 neutrons. An atom of sodium, ${}^{23}_{11}Na$, has 11 protons, 11 electrons, and 12 neutrons. The simplest atom is hydrogen, ${}^{1}_{1}H$, which has no neutrons—a lone proton with a single electron moving around it constitutes a hydrogen atom.

Almost all of an atom's mass is concentrated in its nucleus, because the contribution of electrons to mass is negligible. Because neutrons and protons each have a mass very close to 1 dalton, the mass number is an approximation of the total mass of an atom, usually called its **atomic weight**. Thus, we might say that the atomic weight of helium (${}^{4}_{2}He$) is 4 daltons, although more precisely it is 4.003 daltons.

Isotopes

All atoms of a given element have the same number of protons, but some atoms have more neutrons than other atoms of the same element and therefore weigh more. These different atomic forms are referred to as **isotopes** of the element. In nature, an element occurs as a mixture of its isotopes. For example, consider the three isotopes of the element carbon, which has the atomic number 6. The most common isotope is carbon-12, ${}^{12}_{6}C$, which accounts for about 99% of the carbon in nature. It has 6 neutrons. Most of the remaining 1% of carbon consists of atoms of the isotope ${}^{13}_{6}C$, with 7 neutrons. A third, even rarer isotope, ${}^{14}_{6}C$, has 8 neutrons. Notice that all three isotopes of carbon have 6 protons—otherwise, they would not be carbon. (The number usually given as the atomic weight of an element is actually an average of the atomic weights of all the element's naturally occurring isotopes.)

Both ${}^{12}C$ and ${}^{13}C$ are stable isotopes, meaning that their nuclei do not have a tendency to lose particles. The isotope ${}^{14}C$, however, is unstable, or radioactive. A **radioactive isotope** is one in which the nucleus decays spontaneously, giving off particles and energy. When the decay leads to a change in the number of protons, it transforms the atom to an atom of a different element. For example, radioactive carbon decays to form nitrogen.

Radioactive isotopes have many useful applications in biology. In Chapter 25 you will learn how researchers use measurements of radioactivity in fossils to date those relics of past life. Radioactive isotopes are also useful as tracers to follow atoms through metabolism, the chemical processes of an organism (see the Methods Box, p. 30). Cells use the radioactive atoms as they would nonradioactive isotopes of the same element, but the radioactive tracers can be readily detected. Radioactive tracers have become important diagnostic tools in medicine. For example, certain kidney disorders can be diagnosed by injecting small doses of substances containing radioactive isotopes into the blood and then measuring the amount of tracer excreted in the urine.

Although radioactive isotopes are very useful in biological research and medicine, radiation from these decaying isotopes also poses a hazard to life by damaging cellular molecules. The

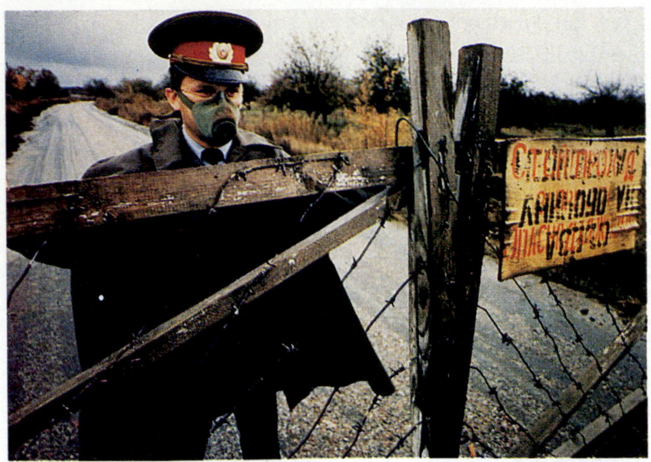

FIGURE 2.6 ▪ The Chernobyl accident. In 1986, an explosion and fire at a nuclear power plant in Chernobyl, Ukraine, released clouds of dangerously radioactive materials into the air, producing widespread contamination downwind. The militiaman in this photo controls access to an evacuated town in the region.

severity of this damage depends on the type and amount of radiation an organism absorbs. One of the most serious environmental threats is radioactive fallout from nuclear accidents (FIGURE 2.6).

The Energy Levels of Electrons

The simplified models of the atom in FIGURE 2.5 greatly exaggerate the size of the nucleus relative to the volume of the whole atom. If the nucleus were the size of a golf ball, the electrons would be moving about the nucleus at an average distance of approximately 1 kilometer (km). Atoms are mostly empty space.

When two atoms approach each other during a chemical reaction, their nuclei do not come close enough to interact. Of the three kinds of subatomic particles we have discussed, only electrons are directly involved in the chemical reactions between atoms.

An atom's electrons vary in the amount of energy they possess. **Energy** is defined as the ability to do work. **Potential energy** is the energy that matter stores because of its position or location. For example, because of its altitude, water in a reservoir on a hill has potential energy. When the gates of the reservoir's dam are opened and the water runs downhill, the energy is taken out of storage to do work, such as turning generators. Because potential energy has been expended, the water stores less energy at the bottom of the hill than it did in the reservoir. Matter has a natural tendency to move to the lowest possible state of potential energy; in this example, water runs downhill. To restore the potential energy of a reservoir, work must be done to elevate the water against gravity.

The electrons of an atom also have potential energy, because of their position in relation to the nucleus. The negatively charged electrons are attracted to the positively charged

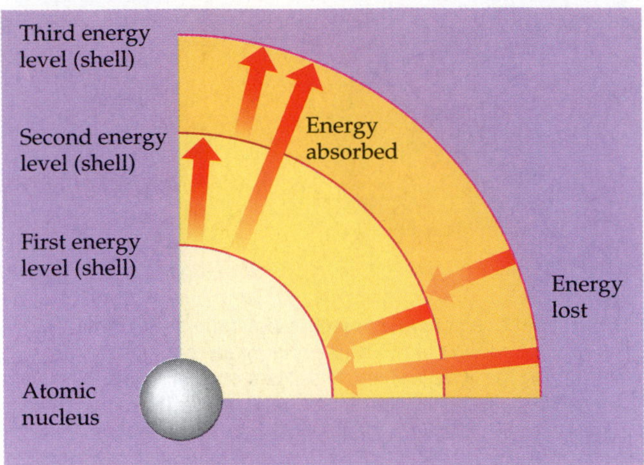

FIGURE 2.7 ▪ Energy levels of an atom's electrons. Electrons exist only at fixed levels of potential energy. An electron can move from one level to another only if the energy it gains or loses is exactly equal to the difference in energy between the two levels. Arrows indicate some of the stepwise changes in potential energy that are possible. These energy levels are also called electron shells.

nucleus; the more distant the electrons are from the nucleus, the greater their potential energy. Unlike the continuous flow of water downhill, changes in the potential energy of electrons can occur only in steps of fixed amounts. An electron having a certain discrete amount of energy is analogous to a ball on a staircase. The ball can have different amounts of potential energy, depending on which step it is on, but it cannot spend much time between the steps.

The different states of potential energy for electrons in an atom are called **energy levels**, or **electron shells** (FIGURE 2.7). The first shell is closest to the nucleus, and electrons in this shell have the lowest energy. Electrons in the second shell have more energy, electrons in the third shell more energy still, and so on. An electron can change its shell, but only by absorbing or losing an amount of energy equal to the difference in potential energy between the old shell and the new shell. To move to a shell farther out from the nucleus, the electron must absorb energy. For example, light energy can excite an electron to a higher energy level. (Indeed, this is the first step when plants harness light energy for photosynthesis, the process that produces food from carbon dioxide and water.) To move to a shell closer in, an electron must lose energy, which is usually released to the environment in the form of heat. For example, when sunlight excites electrons in the paint of a car roof to higher energy levels, the roof heats up as the electrons fall back to their original levels.

Electron Orbitals

Early in the twentieth century, the electron shells of an atom were visualized as concentric paths of electrons orbiting the nucleus, somewhat like planets orbiting the sun. It is still convenient to use concentric-circle diagrams to symbolize electron shells, as in FIGURE 2.7. The atom is not this simple, however. In fact, we can never know the exact path of an electron. What we can do instead is describe the volume of space in which an electron spends most of its time. The three-dimensional space where an electron is found 90% of the time is called an **orbital** (FIGURE 2.8).

No more than two electrons can occupy the same orbital. The first energy shell has a single orbital and can thereby accommodate a maximum of two electrons. This orbital, which is spherical in shape, is designated the 1*s* orbital. The lone electron of a hydrogen atom occupies the 1*s* orbital, as do the two electrons of a helium atom. Electrons, like all matter, tend to exist in the lowest available state of potential energy, which they have in the first shell. An atom with more than two electrons must use higher shells because the first shell is full.

The second electron shell can hold eight electrons, two in each of four orbitals. Electrons in the four different orbitals all have nearly the same energy, but they move in different volumes of space. There is a 2*s* orbital, spherical in shape like the 1*s* orbital, but with a greater diameter. The other three orbitals, called 2*p* orbitals, are dumbbell-shaped, each oriented at right angles to the other two. (At higher energy levels, the orbitals are referred to as 3*s*, 3*p*, and so on.)

Electron Configuration and Chemical Properties

The chemical behavior of an atom is determined by its electron configuration—that is, the distribution of electrons in the atom's electron shells. Beginning with hydrogen, the simplest atom, we can imagine building the atoms of other elements by adding one proton and one electron at a time (along with an appropriate number of neutrons). FIGURE 2.9, p. 28, an abbreviated version of what is called a periodic table, shows this distribution of electrons for the first 18 elements, from hydrogen ($_1$H) to argon ($_{18}$Ar). The elements are arranged in three rows, or periods, corresponding to the sequential addition of electrons to orbitals in the first three electron shells. The main point is that the chemical properties of an atom depend mostly on the number of electrons in its *outermost* shell. We refer to those outer electrons as **valence electrons**, and to the outermost electron shell as the **valence shell**.

An atom with a completed valence shell is unreactive; that is, it will not interact readily with other atoms it encounters. At the far right of the periodic table are helium, neon, and argon, the only three elements shown that have full valence shells. They are termed inert elements because they are chemically unreactive. All other atoms shown in FIGURE 2.9 are chemically reactive because they have unfilled valence shells. This reactivity actually arises from the presence of unpaired electrons in the valence shell. As we "build" the electronic

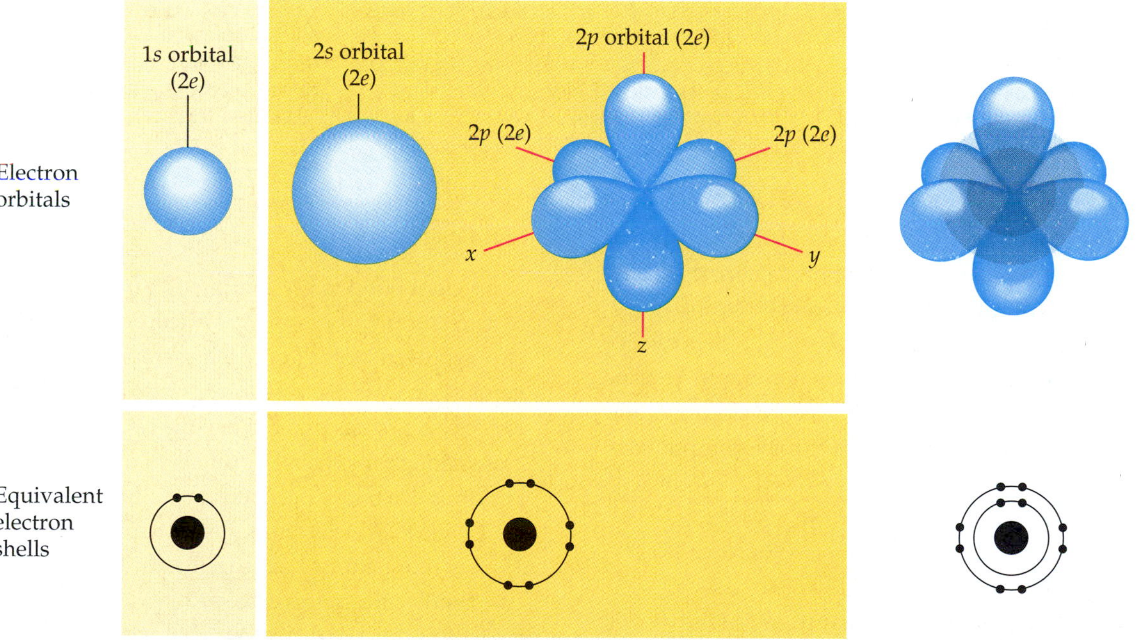

FIGURE 2.8 ▪ Electron orbitals. These three-dimensional shapes represent the volumes of space where the constantly moving electrons of an atom are most likely to be found. Each orbital holds a maximum of two electrons. **(a)** The first electron shell has one spherical *(s)* orbital, designated 1*s*. Only this orbital is occupied in hydrogen, which has one electron, and helium, which has two electrons. **(b)** The second and all higher shells each have one larger *s* orbital (designated 2*s*, in the case of the second shell) plus three dumbbell-shaped orbitals called *p* orbitals (2*p* for the second shell). The three 2*p* orbitals are arranged at right angles to one another along imaginary *x*, *y*, and *z* axes of the atom. The third and higher electron shells can hold additional electrons in orbitals of more complex shapes. **(c)** To symbolize the electron orbitals of the element neon, which has a total of ten electrons, we superimpose the 1*s* orbital of the first shell and the 2*s* and three 2*p* orbitals of the second shell. (Each orbital, remember, can hold two electrons.) The small dots in the electron-shell diagrams represent electrons.

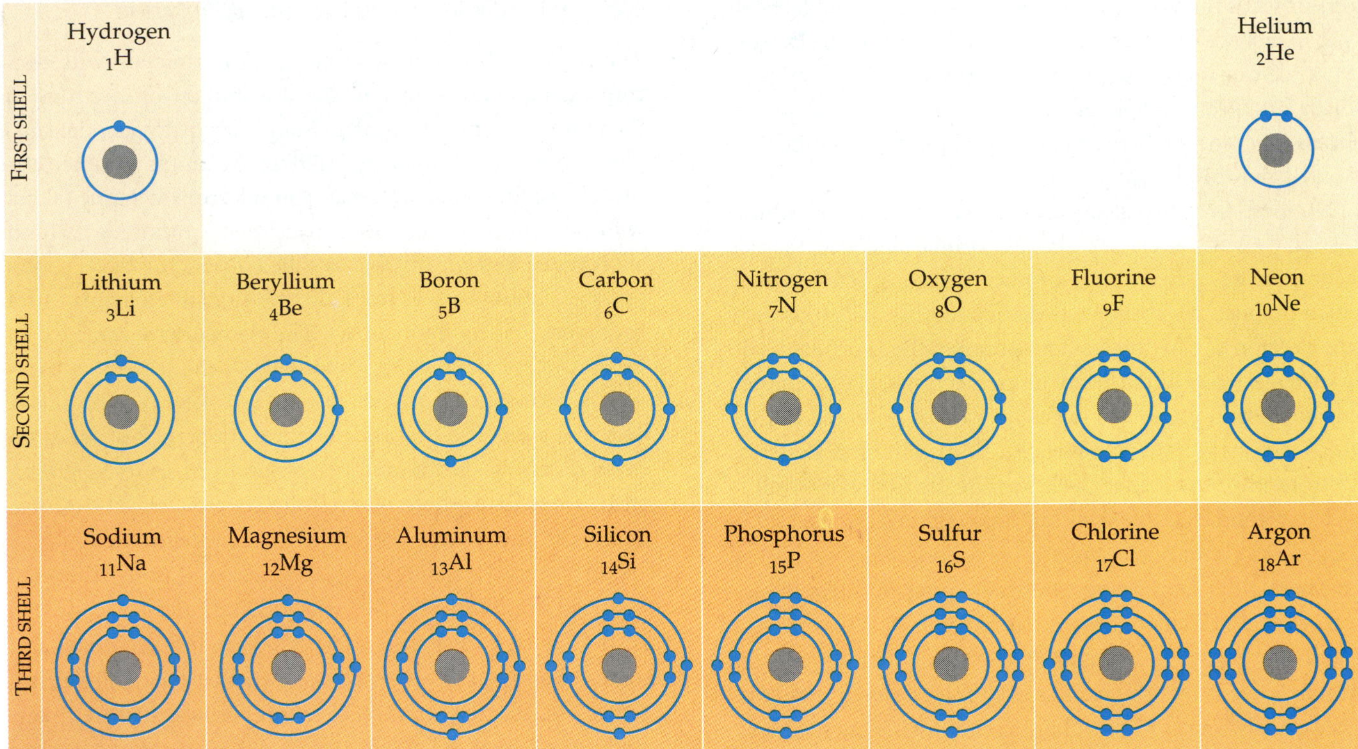

2.1 **FIGURE 2.9 ▪ Electron configurations of the first 18 elements.** In these diagrams, electrons are shown as blue dots and energy levels (electron shells) as concentric rings. (The orbitals of each shell are not shown, but two electrons occupying the same orbital are drawn as a pair.) The elements are arranged in rows, each representing the filling of an electron shell. As electrons are added, they occupy the lowest available shell. Hydrogen's one electron and helium's two electrons are located in the first shell. The next element, lithium, has three electrons. Two electrons fill the first shell while the third electron occupies the second shell (not some higher shell). This behavior of electrons illustrates the general tendency for matter to exist in its lowest state of potential energy. The outermost energy level occupied by electrons is called the valence shell. Of these first 18 elements, only helium, neon, and argon have atoms with full valence shells; such elements are called inert because they are unreactive. All other elements in this figure consist of atoms with incomplete valence shells; these atoms are chemically reactive. Note that, within a shell, one electron goes into each orbital before any orbital gets a second. The reactivity of an atom is due to unpaired electrons. Elements with the same number of valence electrons—fluorine and chlorine, for instance—have similar chemical properties.

configurations in FIGURE 2.9 by adding electrons, we first place just one electron in each orbital of the valence shell; only then do the orbitals begin to receive a second electron.

Atoms with the same number of electrons in their valence shells exhibit similar chemical behavior. For example, fluorine (F) and chlorine (Cl) both have seven valence electrons, and both combine with the element sodium to form compounds.

Atoms combine by chemical bonding to form molecules

Now that we have looked at the structure of atoms, we will move up in the hierarchy of organization and see how atoms combine to form molecules. Atoms with incomplete valence shells interact with certain other atoms in such a way that each partner completes its valence shell. Atoms do this by either sharing or transferring valence electrons. These interactions usually result in atoms staying close together, held by attractions called **chemical bonds.** The strongest kinds of chemical bonds are covalent bonds and ionic bonds.

Covalent Bonds

2.3 A **covalent bond** is the sharing of a pair of valence electrons by two atoms. For example, let's consider what happens when two hydrogen atoms approach each other. Recall that hydrogen has one valence electron in the first shell, but the shell's capacity is two electrons. When the two hydrogen atoms come close enough for their 1*s* orbitals to overlap, they share their electrons. Each hydrogen atom now has two electrons associated with it in what amounts to a completed valence shell. Two or more atoms held together by covalent bonds constitute a **molecule.** In this case, we have formed a hydrogen molecule (FIGURE 2.10a). We can abbreviate the structure of this molecule as H—H, where the line represents a covalent bond—that is, a pair of shared electrons. This notation, which represents both atoms and bonding, is called a **structural formula.** We can abbreviate even further by writing H_2, a **molecular formula** indicating simply that the molecule consists of two atoms of hydrogen.

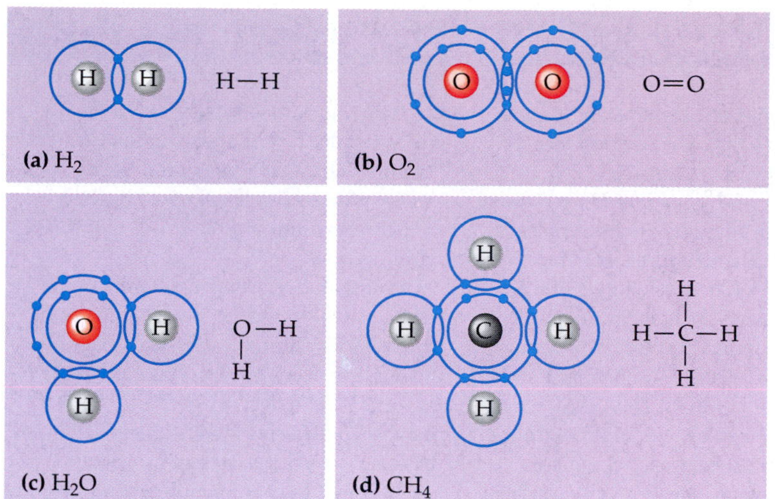

FIGURE 2.10 ▪ Covalent bonding in four molecules. Each covalent bond consists of a pair of shared electrons. The number of electrons required to complete an atom's valence shell determines how many bonds that atom will form. **(a)** If two unattached hydrogen atoms meet, they can form a single covalent bond. **(b)** Two oxygen atoms form a molecule by sharing two pairs of valence electrons; the atoms are joined by a double covalent bond. **(c)** Two hydrogen atoms can be joined to one oxygen atom by covalent bonds to produce a molecule of water. **(d)** Four hydrogen atoms can satisfy the valence of one carbon atom, forming methane.

2.3

With six electrons in its second electron shell, oxygen needs *two* more electrons to complete its valence shell. Two oxygen atoms form a molecule by sharing *two* pairs of valence electrons (FIGURE 2.10b). The atoms are thus joined by what is called a **double covalent bond.**

Each atom that can share valence electrons has a bonding capacity corresponding to the number of covalent bonds the atom can form. When the bonds form, they give the atom a full complement of valence electrons. This bonding capacity is called the atom's **valence** and equals the number of unpaired electrons in the atom's outermost (valence) shell. You can readily determine the valences of life's most abundant elements from the electron configurations in FIGURE 2.9: The valence of hydrogen is 1; oxygen, 2; nitrogen, 3; and carbon, 4. One slightly more complicated case is phosphorus (P), another element important to life. Phosphorus can have a valence of 3, as we would predict. In biologically important molecules, however, it generally has a valence of 5, forming three single bonds and one double bond.

The molecules H_2 and O_2 are pure elements, not compounds. (Recall that a compound is a combination of two or more *different* elements.) An example of a molecule that is a compound is water, with the molecular formula H_2O. It takes two atoms of hydrogen to satisfy the valence of one oxygen atom. FIGURE 2.10c shows the structure of a water molecule. This molecule is so important to life that the next chapter is devoted entirely to its structure and behavior.

Another molecule that is a compound is methane, the main component of natural gas, with the molecular formula CH_4 (FIGURE 2.10d). It takes four hydrogen atoms, each with a valence of 1, to complement one atom of carbon, with its valence of 4. We will look at many other compounds of carbon in Chapter 4.

Nonpolar and Polar Covalent Bonds. The attraction of an atom for the electrons of a covalent bond is called its **electronegativity.** The more electronegative an atom, the more strongly it pulls shared electrons toward itself. In a covalent bond between two atoms of the same element, the outcome of the tug-of-war for common electrons is a standoff; the two atoms are equally electronegative. Such a bond is a **nonpolar covalent bond**; the electrons are shared equally. The covalent bond of H_2 is nonpolar, as is the double bond of O_2. The bonds of methane (CH_4) are also nonpolar; although the partners are different elements, carbon and hydrogen do not differ substantially in electronegativity. This is not always the case in a compound where covalent bonds join atoms of different elements. If one atom is more electronegative than the other, electrons of the bond will not be shared equally. In such cases, the bond is called a **polar covalent bond.** In a water molecule, the bonds between oxygen and hydrogen are polar (FIGURE 2.11). Oxygen is one of the most electronegative of the 92 elements, attracting shared electrons much more strongly than hydrogen does. In a covalent bond between oxygen and hydrogen, the electrons spend more time near the oxygen atom than

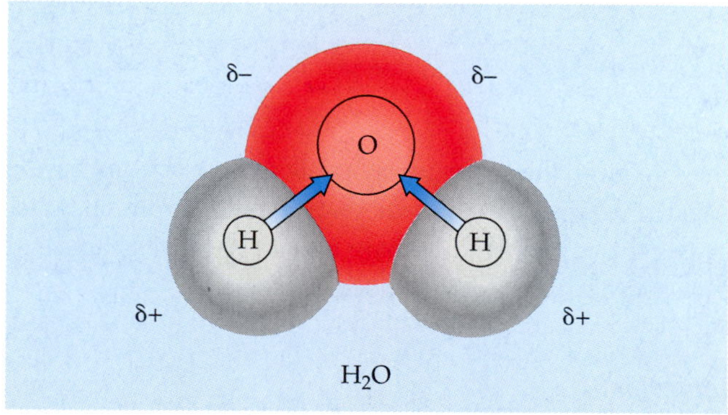

FIGURE 2.11 ▪ Polar covalent bonds in a water molecule. Oxygen, being much more electronegative than hydrogen, pulls the shared electrons of the bond toward itself, as indicated by the arrows. This unequal sharing of electrons gives the oxygen a slight negative charge and the hydrogens a small positive charge. The Greek symbol delta (δ) indicates that the charges are less than full units. This "space-filling" model approximates the true shape of H_2O.

METHODS: The Use of Radioactive Tracers in Biology

Radioactive isotopes are among the most versatile tools in biological research. Because organisms process radioactive and stable isotopes of the same element in exactly the same way, radioactive isotopes can serve as "spies" within an organism. Scientists use them to label certain chemical substances in order to follow a metabolic process or to locate the substance within an organism.

Here you see an experiment designed to determine how temperature affects the rate at which a population of dividing animal cells makes new copies of its DNA, the substance that functions as genetic information, and to locate the new DNA within the cells. First, the cells are grown in a medium that includes the ingredients for making DNA. One ingredient, thymidine, is labeled with ^{3}H, a radioactive isotope of hydrogen. Any new DNA the cells make will incorporate radioactive thymidine.

Cells are grown at various temperatures. After a certain amount of time, samples are taken from each culture, the cells are killed, and their DNA is bound to pieces of filter paper. The papers are then placed in vials containing scintillation fluid, which emits flashes of light whenever radiation from the decay of the tracer in the DNA excites certain chemicals in the fluid. The frequency of flashes, proportional to the amount of radioactive material present, is measured in counts per minute by placing the vials in a scintillation counter (FIGURE a). The higher the counts per minute, the more DNA the cells have made. When the counts for the various DNA samples are plotted against the temperature at which the cells were grown (FIGURE b), it is apparent that temperature dramatically affects the rate of DNA synthesis. This result might be useful to researchers studying the details of DNA synthesis.

A technique known as autoradiography can be used to locate the radioactively labeled DNA within the cells. Thin slices of the cells are placed on glass slides and kept for some time in the dark covered by a layer of photographic emulsion. Radiation from the radioactive tracer present in any new DNA exposes the photographic emulsion, creating a pattern of black dots. In FIGURE c, radioactive DNA in the cell on the left is clearly located in the nucleus.

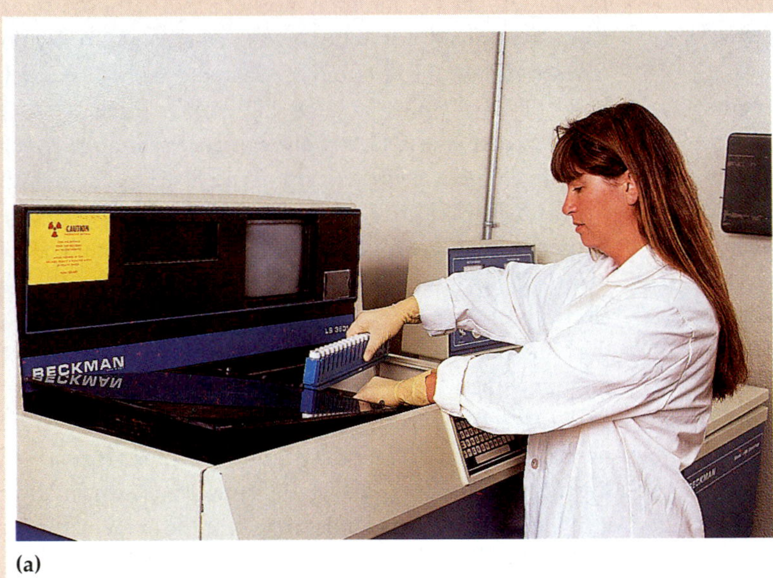

(a)

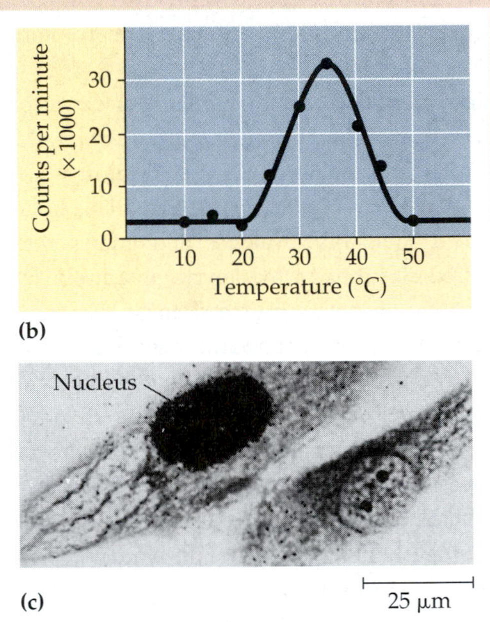

(b)

(c)

they do near the hydrogen atom. Because electrons have a negative charge, the unequal sharing of electrons in water causes the oxygen atom to have a slight negative charge and each hydrogen atom a slight positive charge.

Ionic Bonds

2.4 In some cases, two atoms are so unequal in their attraction for valence electrons that the more electronegative atom strips an electron completely away from its partner. This is what happens when an atom of sodium ($_{11}Na$) encounters an atom of chlorine ($_{17}Cl$) (FIGURE 2.12). A sodium atom has a total of 11 electrons, with its single valence electron in the third electron shell. A chlorine atom has a total of 17 electrons, with 7 electrons in its valence shell. When these two atoms meet, the lone valence electron of sodium is transferred to the chlorine atom, and both atoms end up with their valence shells complete. (Because sodium no longer has an electron in the third shell, the second shell is now outermost.)

The electron transfer between the two atoms moves one unit of negative charge from sodium to chlorine. Sodium, now with 11 protons but only 10 electrons, has a net electrical charge of +1. A charged atom (or molecule) is called an **ion**. When the charge is positive, the ion is specifically called a **cation**. Conversely, the chlorine atom, having gained an extra electron, now has 17 protons and 18 electrons, giving it a net

2.4

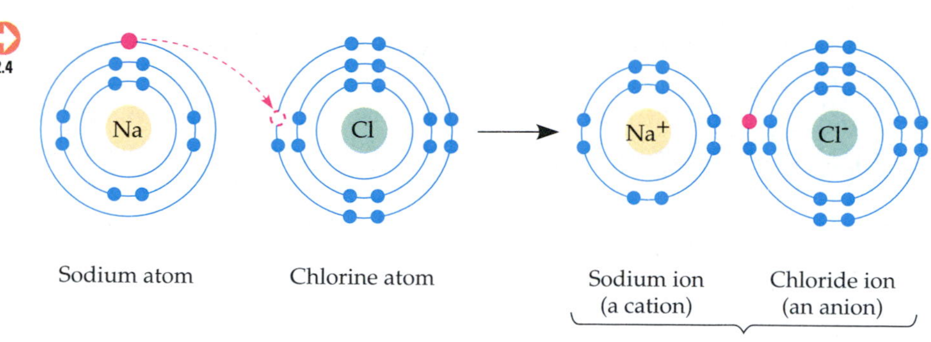

FIGURE 2.12 ▪ Electron transfer and ionic bonding. A valence electron can be transferred from sodium (Na) to chlorine (Cl), giving both atoms completed valence shells. The electron transfer leaves the sodium atom with a net charge of +1 (cation) and the chlorine atom with a net charge of −1 (anion). The attraction between the oppositely charged atoms, or ions, is an ionic bond. Ions can bond not only to atoms with which they have reacted, but to any other ions of opposite charge.

electrical charge of −1. It has become a chloride ion—an **anion**, or negatively charged ion. Because of their opposite charges, cations and anions attract each other in what is called an **ionic bond.** *Any* two ions of opposite charge can form an ionic bond. The ions need not have acquired their charge by an electron transfer with each other.

Ionic compounds are called salts. We know the compound sodium chloride (NaCl) as table salt (FIGURE 2.13). Salts are often found in nature as crystals of various sizes and shapes, each an aggregate of vast numbers of cations and anions bonded by their electrical attraction and arranged in a three-dimensional lattice. A salt crystal does not consist of molecules in the same sense that a covalent compound does, because a covalently bonded molecule has a definite size and number of atoms. The formula for an ionic compound, such as NaCl, indicates only the ratio of elements in a crystal of the salt.

Not all salts have equal numbers of cations and anions. For example, the ionic compound magnesium chloride ($MgCl_2$) has two chloride ions for each magnesium ion. Magnesium ($_{12}Mg$) must lose two outer electrons if the atom is to have a complete valence shell, so it tends to form a cation with a net charge of +2 (Mg^{2+}). One magnesium cation can then form ionic bonds with two chloride anions.

FIGURE 2.13 ▪ A sodium chloride crystal. The sodium ions (Na^+) and chloride ions (Cl^-) are held together by ionic bonds.

The term *ion* also applies to entire molecules that are electrically charged. In the salt ammonium chloride (NH_4Cl), for instance, the anion is a single chloride ion (Cl^-), but the cation is ammonium (NH_4^+), a nitrogen atom with four covalently bonded hydrogen atoms. The whole ammonium ion has an electrical charge of +1 because it is one electron short.

Environment affects the strength of ionic bonds. In a dry salt crystal, the bonds are so strong that it takes a hammer and chisel to break enough of them to crack the crystal in two. Place the same salt crystal in water, however, and the salt dissolves as the attractions between its ions decrease. In the next chapter you will learn how water dissolves salts.

Weak chemical bonds play important roles in the chemistry of life

Covalent bonds are essential to life because they link atoms to form a cell's molecules. But bonding *between* molecules is also indispensable in the cell, where the properties of life emerge from molecular interactions. When two molecules in the cell make contact, they may adhere temporarily by types of chemical bonds that are much weaker than covalent bonds. The advantage of weak bonding is that the contact between the molecules can be brief; the molecules come together, respond to one another in some way, and then separate.

The importance of weak bonding can be seen in the example of chemical signaling in the brain. One brain cell signals another by releasing molecules that use weak bonds to dock onto receptor molecules on the nearby surface of a receiving cell. The bonds last just long enough to trigger a momentary response by the receiving cell. If the signal molecules attached by stronger bonds, the receiving cell would continue to respond long after the transmitting cell ceased dispatching the message, with perhaps disastrous consequences. (Imagine what it would be like, for instance, if your brain continued to perceive the ringing sound of a bell for hours after nerve cells transmitted the information from the ears to the brain.)

Several types of weak chemical bonds are important in living organisms. One is the ionic bond, which is relatively weak in the presence of water. Another type of weak bond, crucial to life, is known as a hydrogen bond.

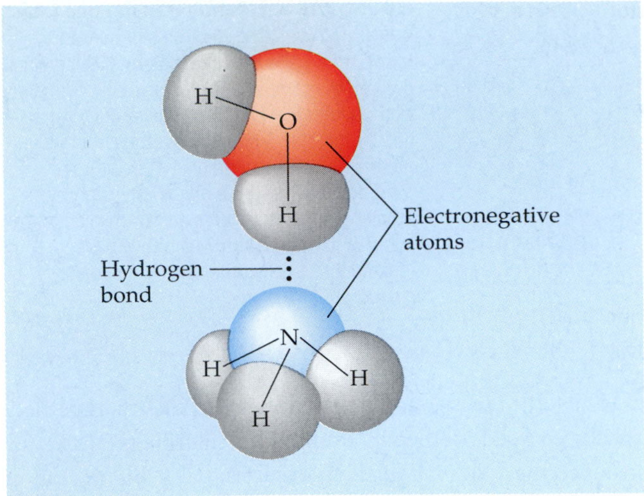

2.4 **FIGURE 2.14 ▪ A hydrogen bond.** Through a weak electrical attraction, one electronegative atom shares a hydrogen atom with another electronegative atom. In this diagram, a hydrogen bond joins a hydrogen atom of a water molecule (H_2O) with the nitrogen atom of an ammonia molecule (NH_3).

Hydrogen Bonds

2.5 Among the various kinds of weak chemical bonds, hydrogen bonds are so important in the chemistry of life that they deserve special attention. A **hydrogen bond** occurs when a hydrogen atom covalently bonded to one electronegative atom is also attracted to another electronegative atom. In living cells, the electronegative partners involved are usually oxygen or nitrogen atoms.

Let's examine the simple case of hydrogen bonding between water (H_2O) and ammonia (NH_3) (FIGURE 2.14). You have seen how the polar covalent bonds of water result in the oxygen atom having a slight negative charge and the hydrogen atoms having a slight positive charge. A similar situation arises in the ammonia molecule, where an electronegative nitrogen atom has a small amount of negative charge because of its pull on the electrons it shares covalently with hydrogen. If a water molecule and an ammonia molecule are close together, a weak attraction will occur between the negatively charged nitrogen atom and a positively charged hydrogen atom of the adjacent water molecule. This attraction is a hydrogen bond.

Van der Waals Interactions

Even a molecule with nonpolar covalent bonds may have positively and negatively charged regions. Because electrons are in constant motion, they are not always symmetrically distributed in the molecule; at any instant, they may accumulate by chance in one part of the molecule or another. The results are ever-changing "hot spots" of positive and negative charge that enable all atoms and molecules to stick to one another. These **van der Waals interactions** are weak and occur only when atoms and molecules are very close together.

Van der Waals interactions, hydrogen bonds, ionic bonds, and other weak bonds may form not only between molecules but also between different regions of a single large molecule, such as a protein. Although these bonds are individually weak, their cumulative effect is to reinforce the three-dimensional shape of a large molecule. You will learn more about the biological roles of weak bonds in Chapter 5.

A molecule's biological function is related to its shape

A molecule has a characteristic size and shape. The precise shape of a molecule is usually very important to its function in the living cell.

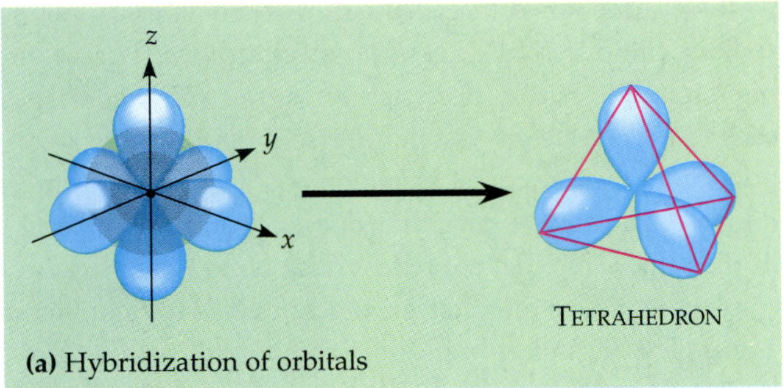

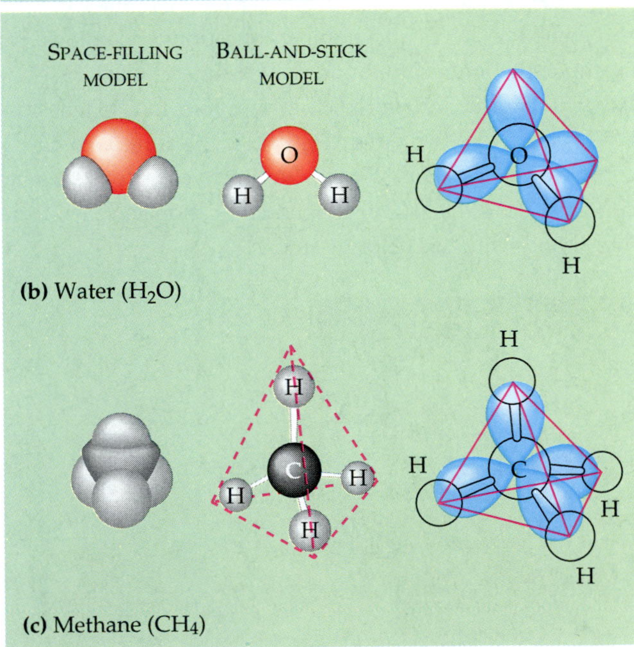

FIGURE 2.15 ▪ Molecular shapes due to hybrid orbitals. (a) The four teardrop-shaped orbitals of a valence shell involved in covalent bonding form by hybridization of the single *s* and three *p* orbitals. The orbitals extend to the four corners of an imaginary tetrahedron (outlined in magenta). **(b)** Because of the positions of the hybrid orbitals, the two covalent bonds of water are angled at 104.5°. This is seen most clearly in the ball-and-stick model. The space-filling model comes closer to the actual shape of the molecule. **(c)** The hydrogens of methane occupy all four corners of the tetrahedron, giving methane a tetrahedral shape.

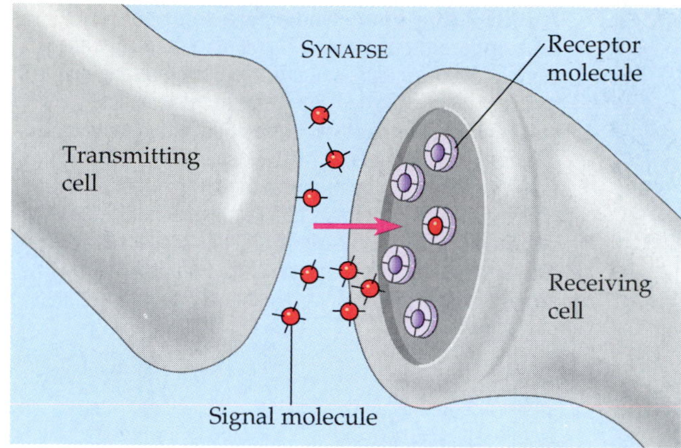

FIGURE 2.16 ▪ Molecular shape and brain chemistry. One nerve cell in the brain signals another by releasing molecules with a shape that is complementary to the shape of receptor molecules located on the surface of the receiving cell. The signal molecules pass across the tiny gap (called the synapse) between the two nerve cells and attach to the receptors by weak bonds, stimulating the receiving cell in the process. The actual molecules have much more complex shapes than represented here. The role of molecular shape in brain chemistry is an example of the correlation between structure and function, one of biology's unifying themes.

A molecule consisting of two atoms, such as H_2 or O_2, is always linear, but molecules with more than two atoms have more complicated shapes. These shapes are determined by the positions of the atoms' orbitals. When an atom forms covalent bonds, the orbitals in its valence shell rearrange. For atoms with valence electrons in both *s* and *p* orbitals (see FIGURE 2.8), the single *s* and three *p* orbitals hybridize to form four new orbitals shaped like identical teardrops extending from the region of the atomic nucleus (FIGURE 2.15a). If we connect the larger ends of the teardrops with lines, we have the outline of a pyramidal shape called a tetrahedron.

For the water molecule (H_2O), two of the hybrid orbitals in the oxygen atom's valence shell are shared with hydrogen atoms. The result is a molecule shaped roughly like a V, its two covalent bonds spread apart at an angle of 104.5° (FIGURE 2.15b).

The methane molecule (CH_4) has the shape of a completed tetrahedron, because all four hybrid orbitals are shared (FIGURE 2.15c). The nucleus of the carbon atom is at the center, with its four covalent bonds radiating to the hydrogen nuclei at the corners of the tetrahedron. Larger molecules containing multiple carbon atoms, including many of the molecules that make up living matter, have more complex overall shapes. However, the tetrahedral shape of a carbon atom bonded to four other atoms is often a repeating motif within such molecules.

Molecular shape is crucial in biology because it determines how most molecules of life recognize and respond to one another. For instance, in the brain-cell signaling example mentioned earlier, the signal molecules released by the transmitting cell have a unique shape that specifically fits together with the shape of the receptor molecules on the surface of the receiving cell, much as a key fits into a lock (FIGURE 2.16). (This complementarity of shape aids the formation of weak bonds.) Molecules that have shapes similar to the brain's signal molecules can affect mood and pain perception. Morphine, heroin, and other opiate drugs, for example, mimic natural signal molecules called endorphins (FIGURE 2.17). These drugs artificially produce euphoria and relieve pain by binding to endorphin receptors in the brain.

Chemical reactions make and break chemical bonds

The making and breaking of chemical bonds, leading to changes in the composition of matter, are called **chemical**

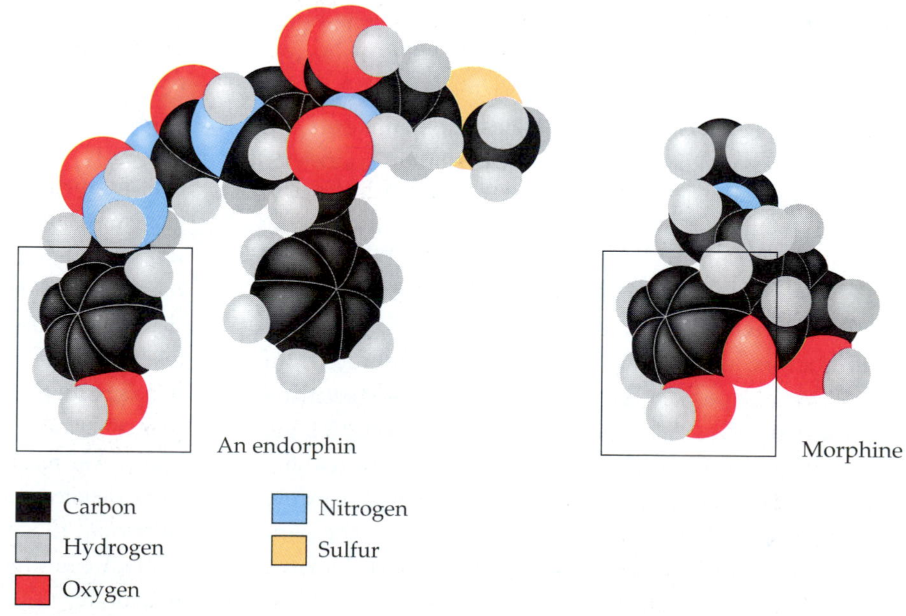

FIGURE 2.17 ▪ A molecular mimic. An endorphin is a natural brain signal molecule that helps us feel good. The boxed portion of the endorphin molecule is the shape that is recognized by receptor molecules on appropriate target cells in the brain. The boxed portion of the morphine molecule, an opiate drug, is a close match. Morphine, a chemical imposter, affects emotional state by mimicking the brain's natural endorphins. The discovery that opiates bind to receptors in the brain that normally recognize the brain's own endorphins was a major milestone in neurochemistry.

FIGURE 2.18 ▪ Photosynthesis: a solar-powered rearrangement of matter. This *Elodea*, a freshwater plant, produces sugars and other food by rearranging the atoms of carbon dioxide and water in the chemical process known as photosynthesis. Sunlight powers this chemical transformation. The oxygen bubbles escaping from the leaves are by-products of photosynthesis. Humans and other animals ultimately depend on photosynthesis for food and oxygen, and this process is at the foundation of almost all ecosystems. Photosynthesis is an important example of how chemistry applies to the study of life.

reactions. An example is the reaction between hydrogen and oxygen to form water:

$$2\,H_2 + O_2 \longrightarrow 2\,H_2O$$

Reactants — Products

This reaction breaks the covalent bonds of H_2 and O_2 and forms the new bonds of H_2O. When we write a chemical reaction, we use an arrow to indicate the conversion of the starting materials, called **reactants**, to the **products**. The coefficients indicate the number of molecules involved; for example, the 2 in front of the H_2 means that the reaction starts with two molecules of hydrogen. Notice that all atoms of the reactants must be accounted for in the products. Matter is conserved in a chemical reaction: Reactions cannot create or destroy matter but can only rearrange it.

Following is the chemical shorthand that summarizes the process of photosynthesis, a particularly important example of chemical reactions rearranging matter:

$$6\,CO_2 + 6\,H_2O \longrightarrow C_6H_{12}O_6 + 6\,O_2$$

Photosynthesis takes place within the cells of green plant tissues. The raw materials are carbon dioxide (CO_2), which is taken from the air, and water (H_2O), which is absorbed from the soil. Within the plant cells, sunlight powers the conversion of these ingredients to a sugar called glucose ($C_6H_{12}O_6$) and oxygen (O_2), which the plant releases into the surroundings (FIGURE 2.18). Although photosynthesis is actually a sequence of many chemical reactions, we still end up with the same number and kinds of atoms we had when we started. Matter has simply been rearranged.

Some chemical reactions go to completion; that is, all the reactants are converted to products. But most reactions are reversible, the products of the forward reaction becoming the reactants for the reverse reaction. For example, hydrogen and nitrogen molecules can combine to form ammonia, but ammonia can also decompose to regenerate hydrogen and nitrogen:

$$3\,H_2 + N_2 \rightleftharpoons 2\,NH_3$$

The opposite-headed arrows indicate that the reaction is reversible.

One of the factors affecting the rate of a reaction is the concentration of reactants. The greater the concentration of reactant molecules, the more frequently they collide with one another and have an opportunity to react to form products. The same holds true for the products. As products accumulate, collisions resulting in the reverse reaction become increasingly frequent. Eventually, the forward and reverse reactions occur at the same rate, and the relative concentrations of products and reactants stop changing. The point at which the reactions offset one another exactly is called **chemical equilibrium**. This is a dynamic equilibrium; reactions are still going on, but with no net effect on the concentrations of reactants and products. Equilibrium does *not* mean that the reactants and products are equal in concentration, but only that their concentrations have stabilized. For the reaction involving ammonia, equilibrium is reached when ammonia decomposes as rapidly as it forms. In this case, at equilibrium there is far more ammonia than hydrogen and nitrogen.

We will return to the subject of chemical reactions after more detailed study of the various types of molecules that are important to life. In the next chapter we focus on water, the substance in which all the chemical processes of living organisms occur.

CHAPTER REVIEW

REVIEW OF KEY CONCEPTS

(with page numbers and key figures)

CHEMICAL ELEMENTS AND COMPOUNDS

- **Matter consists of chemical elements in pure form and in combinations called compounds** (pp. 22–23, FIGURE 2.2) Elements cannot be broken down to other substances. A compound contains two or more elements in a fixed ratio.

- **Life requires about 25 chemical elements** (pp. 23–24, TABLE 2.1) Carbon, oxygen, hydrogen, and nitrogen make up approximately 96% of living matter.

ATOMS AND MOLECULES

Atomic structure determines the behavior of an element (pp. 24–
2.1 28, FIGURE 2.9) An atom is the smallest unit of an element. An atom
2.2 has a nucleus made up of positively charged protons and uncharged neutrons, and a surrounding cloud of negatively charged electrons. The number of electrons in an electrically neutral atom equals the number of protons. Most elements have two or more isotopes, different in neutron number and therefore mass. Some isotopes are unstable and give off particles and energy as radioactivity.

Electron configuration determines the chemical behavior of an atom—the way it reacts with other atoms. Electrons move within orbitals, three-dimensional spaces located within successive electron shells (energy levels). Chemical properties depend on the number of valence electrons, those in the outermost shell. An atom with an incomplete valence shell is reactive.

Atoms combine by chemical bonding to form molecules (pp. 28–
2.3 31, FIGURES 2.10 and 2.12) Chemical bonds form when atoms interact and complete their valence shells. A covalent bond is the sharing of a pair of valence electrons by two atoms. Molecules consist of two or more covalently bonded atoms. Electrons of a polar covalent bond are pulled closer to the more electronegative atom. A covalent bond is nonpolar if both atoms are equally electronegative.

Two atoms may differ so much in electronegativity that one or
2.4 more electrons are actually transferred from one atom to the other. The result is a negatively charged ion (anion) and a positively charged ion (cation). The attraction between two ions of opposite charge is called an ionic bond.

Weak chemical bonds play important roles in the chemistry of life
2.5 (pp. 31–32, FIGURE 2.14) Hydrogen bonds are relatively weak interactions between a partially positive hydrogen atom of one polar molecule and a partially negative atom of another polar molecule. Van der Waals interactions occur when transiently positive and negative regions of very close molecules are attracted to each other. Weak bonds reinforce the shapes of large biological molecules and also help molecules adhere to each other.

- **A molecule's biological function is related to its shape** (pp. 32–33, FIGURE 2.15) A molecule's shape is determined by the positions of its atoms' valence orbitals. When covalent bonds form, the *s* and *p* orbitals in the valence shell of an atom may combine to form four hybrid orbitals that extend to the corners of a tetrahedron; such orbitals are responsible for the shapes of H_2O, CH_4, and many more complex biological molecules. Shape is often the basis of the recognition of one biological molecule by another.

- **Chemical reactions make and break chemical bonds** (pp. 33–34) Chemical reactions change reactants into products while conserving matter. Most chemical reactions are reversible. Chemical equilibrium is reached when the forward and reverse reaction rates are equal.

SELF-QUIZ

1. An element is to a (an) ________ as a tissue is to a (an) ________.
   a. atom; organism
   b. compound; organ
   c. molecule; cell
   d. atom; organ
   e. compound; organelle

2. In the term *trace element,* the modifier *trace* means
   a. the element is required in very small amounts
   b. the element can be used as a label to trace atoms through an organism's metabolism
   c. the element is very rare on Earth
   d. the element enhances health but is not essential for the organism's long-term survival
   e. the element passes rapidly through the organism

3. Compared to ^{31}P, the radioactive isotope ^{32}P has
   a. a different atomic number
   b. one more neutron
   c. one more proton
   d. one more electron
   e. a different charge

4. Atoms can be represented by simply listing the number of protons, neutrons, and electrons—for example, $2p^+$; $2n^0$; $2e^-$ for helium. Which atom represents the ^{18}O isotope of oxygen?
   a. $6p^+$; $8n^0$; $6e^-$
   b. $8p^+$; $10n^0$; $8e^-$
   c. $9p^+$; $9n^0$; $9e^-$
   d. $7p^+$; $2n^0$; $9e^-$
   e. $10p^+$; $8n^0$; $10e^-$

5. The atomic number of sulfur is 16. Sulfur combines with hydrogen by covalent bonding to form a compound, hydrogen sulfide. Based on the electron configuration of sulfur, we can predict that the molecular formula of the compound will be (explain your answer)
   a. HS b. HS_2 c. H_2S d. H_3S_2 e. H_4S

6. Review the valences of carbon, oxygen, hydrogen, and nitrogen, and then determine which of the following molecules is most likely to exist.

```
a. O═C—H

b.          H   H
            |   |
     H—O—C—C═O
            |
            H

c.   H       H
     |       |
   H—C—H—C═O
     |
     H

d.   O
     |
   H—N═H
```

7. The reactivity of an atom arises from
   a. the average distance of the outermost electron shell from the nucleus
   b. the existence of unpaired electrons in the valence shell
   c. the sum of the potential energies of all the electron shells
   d. the potential energy of the valence shell
   e. the energy difference between the *s* and *p* orbitals

8. Which of these statements is true of all anionic atoms?
   a. The atom has more electrons than protons.
   b. The atom has more protons than electrons.
   c. The atom has fewer protons than does a neutral atom of the same element.
   d. The atom has more neutrons than protons.
   e. The net charge is −1.

9. What coefficients must be placed in the blanks to balance this chemical reaction?

$$C_6H_{12}O_6 \longrightarrow \_\_ C_2H_6O + \_\_ CO_2$$

   a. 1; 2 b. 2; 2 c. 1; 3 d. 1; 1 e. 3; 1

10. Which of the following statements correctly describes any chemical reaction that has reached equilibrium?
   a. The concentration of products equals the concentration of reactants.
   b. The rate of the forward reaction equals the rate of the reverse reaction.
   c. Both forward and reverse reactions have halted.
   d. The reaction is now irreversible.
   e. No reactants remain.

CHALLENGE QUESTIONS

1. Female silkworm moths *(Bombyx mori)* attract males by emitting chemical signals that spread through the air. A male hundreds of meters away can detect these molecules and fly toward their source. The sensory organs responsible for this behavior are the comblike antennae visible in the photograph below. Each filament of an antenna is equipped with thousands of receptor cells that detect the sex attractant. Based on what you learned in this chapter, propose a hypothesis to account for the ability of the male moth to detect a specific molecule in the presence of many other molecules in the air. What predictions does your hypothesis make? Design an experiment to test one of these predictions.

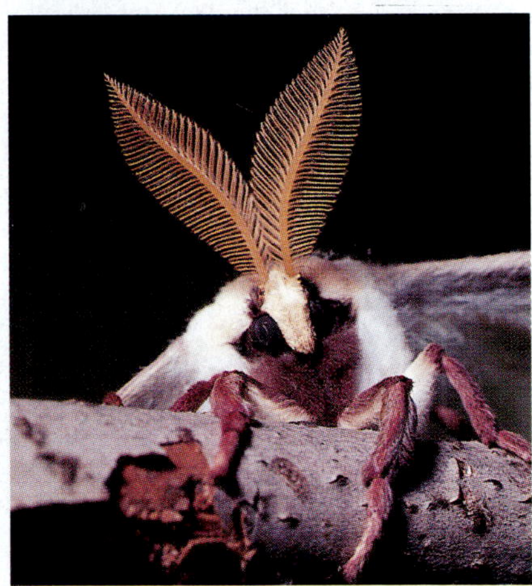

2. Vitalism is the belief that life possesses supernatural forces that cannot be explained by physical and chemical principles. Explain why the concept of emergent properties does not lend credibility to vitalism.

SCIENCE, TECHNOLOGY, AND SOCIETY

1. The use of radioactive isotopes as tracers in biochemical research is based on cells using the tracers in place of nonradioactive isotopes. Explain why this ability of radioactive isotopes to infiltrate the chemical processes of the cell also compounds the threat posed by radioactive contaminants in air, soil, and water.

2. While waiting at an airport, Neil Campbell once overhead this claim: "It's paranoid and ignorant to worry about industry or agriculture contaminating the environment with their chemical wastes. After all, this stuff is just made of the same atoms that were already present in our environment." How would you counter this argument?

FURTHER READING

Atkins, P. W. *Molecules.* New York: Scientific American Library, 1987. Beautifully illustrated tour of the world of molecules.

Kahn, P. "A Grisly Archive of Key Cancer Data." *Science,* January 22, 1993. Alarming health problems among uranium workers in Germany.

Pennisi, E. "Natureworks." *Science News,* May 16, 1992. How organisms make minerals.

Sackheim, G. *Introduction to Chemistry for Biology Students,* 5th ed. Menlo Park, CA: Benjamin/Cummings, 1996. A programmed review of the fundamentals of chemistry.

Shcherbak, Yuri M. "Ten Years of the Chernobyl Era." *Scientific American,* April 1996.

Thornton, R. M. *The Chemistry of Life.* Menlo Park, CA: Benjamin/Cummings, 1998. A CD-ROM that uses interactive exercises to help bridge the gap between chemistry and biology.

WEB LINKS

Visit the special edition of *The Biology Place* for BIOLOGY, Fifth Edition, at http://www.biology.com/campbell. Go to Chapter 2 for online resources, including learning activities, practice exams, and links to the following web sites:

"BioChemNet"
The main page of an excellent site that links chemistry and the life sciences. It contains links to sites ranging from general chemistry to biotechnology and has a good weekly news and review section.

"CHEMystery: An Interactive Guide to Chemistry"
Written by high school seniors for the ThinkQuest contest, this is a good review and reference site for students at all levels.

"ChemiCool"
Interactive periodic table. Click on any element in the table and a full page of information about that element is displayed.

"The American Chemical Society"
The education page of the ACS that provides links to many aspects of chemistry.

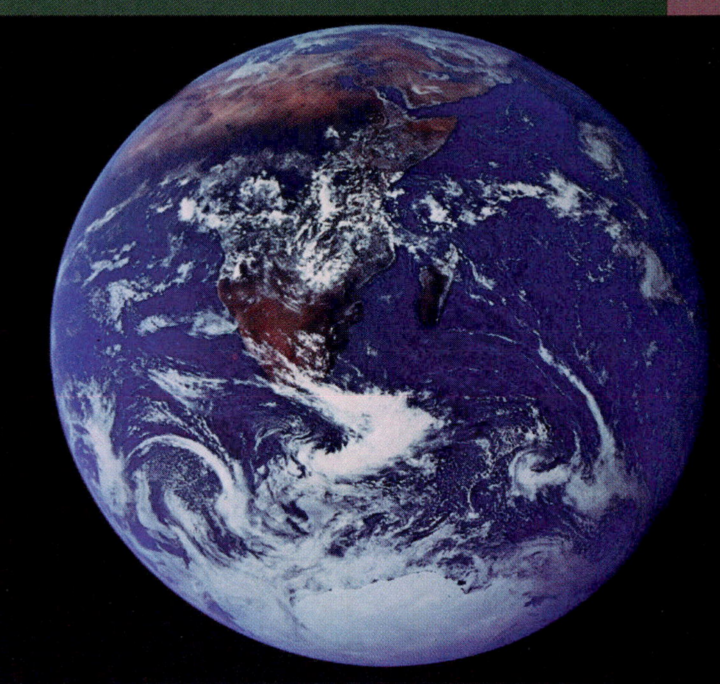

CHAPTER 3

WATER AND THE FITNESS OF THE ENVIRONMENT

Effects of Water's Polarity

- The polarity of water molecules results in hydrogen bonding
- Organisms depend on the cohesion of water molecules
- Water moderates temperatures on Earth
- Oceans and lakes don't freeze solid because ice floats
- Water is the solvent of life

Dissociation of Water Molecules

- Organisms are sensitive to changes in pH
- Acid precipitation threatens the fitness of the environment

As astronomers study newly discovered planets orbiting distant stars, they hope to find evidence for water on these far-off worlds, for water is the substance that makes possible life as we know it here on Earth. All organisms familiar to us are made mostly of water and live in an environment dominated by water. Water is the biological medium here on Earth, and possibly on other planets as well.

Life on Earth began in water and evolved there for three billion years before spreading onto land. Modern life, even terrestrial (land-dwelling) life, remains tied to water. Most cells are surrounded by water, and cells themselves are about 70% to 95% water. Three-quarters of Earth's surface is submerged in water. Although most of this water is in liquid form, water is also present on Earth as ice and vapor. Water is the only common substance to exist in the natural environment in all three physical states of matter: solid, liquid, and gas. These three states of water are visible in the photograph on this page, a view of Earth from space.

The abundance of water is a major reason Earth is habitable. In a classic book called The Fitness of the Environment, *Lawrence Henderson highlights the importance of water to life. While acknowledging that life adapts to its environment through natural selection, Henderson emphasizes that for life to exist at all in a particular location, the environment must first be a suitable abode. Your objective in this chapter is to develop a conceptual understanding of how water contributes to the fitness of Earth for life.*

EFFECTS OF WATER'S POLARITY

Water is so common that it is easy to overlook the fact that it is an exceptional substance with many extraordinary qualities. Following the theme of emergent properties, we can trace water's unique behavior to the structure and interactions of its molecules.

The polarity of water molecules results in hydrogen bonding

3.1 Studied in isolation, the water molecule is deceptively simple. Its two hydrogen atoms are joined to the oxygen atom by single covalent bonds. Because oxygen is more electronegative than hydrogen, the electrons of the polar bonds spend more time closer to the oxygen atom. This results in the oxygen region of the molecule having a slight negative charge and the hydrogens having a slight positive charge. The water molecule, shaped something like a wide V, is a **polar molecule**, meaning that opposite ends of the molecule have opposite charges.

The anomalous properties of water arise from attractions among these polar molecules. The attraction is electrical; a slightly positive hydrogen of one molecule is attracted to the slightly negative oxygen of a nearby molecule. The two molecules are thus held together by a hydrogen bond (FIGURE 3.1). Each water molecule can form hydrogen bonds to a maximum of four neighbors. The extraordinary qualities of water are emergent properties resulting from the hydrogen bonding that orders molecules into a higher level of structural organization.

We will examine four of water's properties that contribute to the fitness of Earth as an environment for life: water's cohesive behavior, its ability to stabilize temperature, its expansion upon freezing, and its versatility as a solvent.

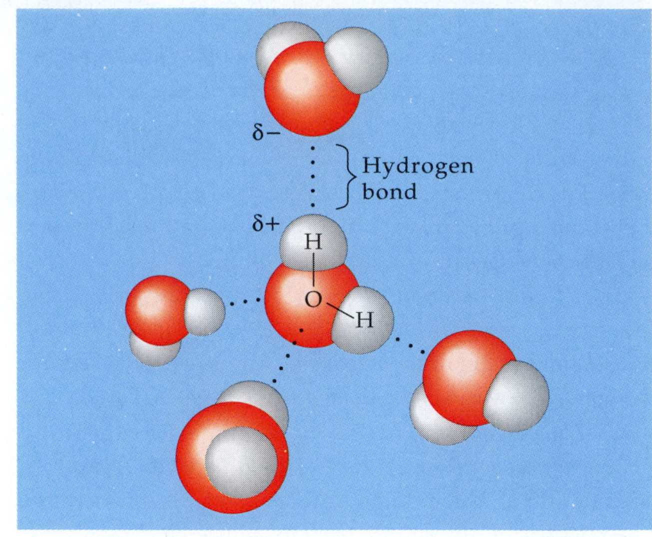

3.1 **FIGURE 3.1 ▪ Hydrogen bonds between water molecules.** The charged regions of a polar water molecule are attracted to oppositely charged parts of neighboring molecules. (The oxygen has a slight negative charge; the hydrogens have a slight positive charge.) Each molecule can hydrogen-bond to a maximum of four partners. At any instant in liquid water at 37°C (human body temperature), about 15% of the molecules are bonded to four partners in short-lived clusters.

Organisms depend on the cohesion of water molecules

Water molecules stick together as a result of hydrogen bonding. When water is in its liquid form, its hydrogen bonds are very fragile, about one-twentieth as strong as covalent bonds. They form, break, and re-form with great frequency. Each hydrogen bond lasts only a few trillionths of a second, but the molecules are constantly forming new bonds with a succession of partners. Thus, at any instant, a substantial percentage of all the water molecules are bonded to their neighbors, making water more structured than most other liquids. Collectively, the hydrogen bonds hold the substance together, a phenomenon called **cohesion**.

Cohesion due to hydrogen bonding contributes to the transport of water against gravity in plants (FIGURE 3.2). Water reaches the leaves through microscopic vessels that extend upward from the roots. Water that evaporates from a leaf is replaced by water from the vessels in the veins of the leaf. Hydrogen bonds cause water molecules leaving the veins to tug on molecules farther down in the vessel, and the upward pull is transmitted along the vessel all the way down to the roots. **Adhesion**, the clinging of one substance to another, also plays a role. Adhesion of water to the walls of the vessels helps counter the downward pull of gravity.

Related to cohesion is **surface tension**, a measure of how difficult it is to stretch or break the surface of a liquid. Water has a greater surface tension than most other liquids. At the interface between water and air is an ordered arrangement of water molecules, hydrogen-bonded to one another and to the water below. This makes the water behave as though coated with an invisible film. We can observe the surface tension of water by slightly overfilling a drinking glass; the water will stand above the rim. Water's surface tension also allows us to

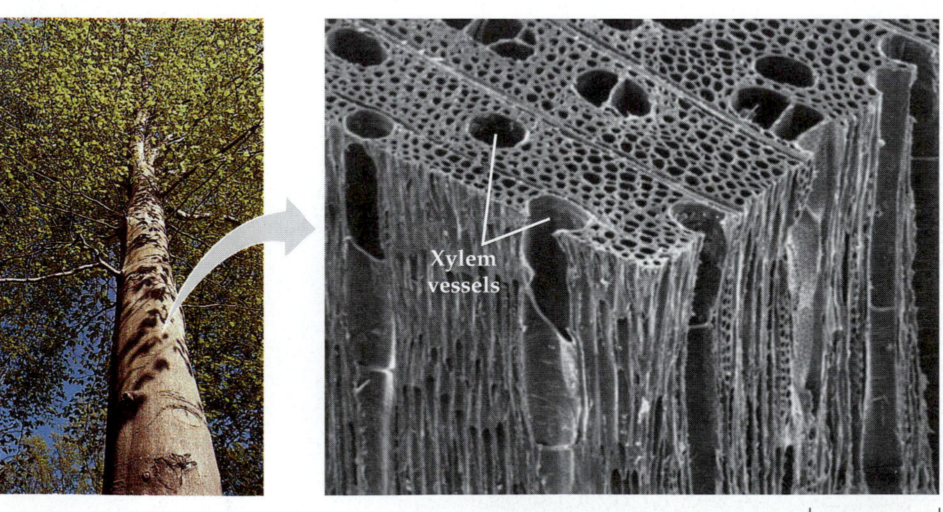

FIGURE 3.2 ▪ Water transport in plants. Evaporation from leaves pulls water upward from the roots through microscopic tubes called xylem vessels, in this case located in the trunk of a tree. Cohesion due to hydrogen bonding helps hold together the column of water within a vessel. Adhesion of the water to the vessel wall also helps in resisting the downward pull of gravity.

SEM courtesy of Drs. R. G. Kessel and C. Y. Shih,

skip rocks on a pond. In a more biological example, some animals can stand, walk, or run on water without breaking the surface (FIGURE 3.3).

Water moderates temperatures on Earth

Water stabilizes air temperatures by absorbing heat from air that is warmer and releasing the stored heat to air that is cooler. Water is effective as a heat bank because it can absorb or release a relatively large amount of heat with only a slight change in its own temperature. To understand this quality of water, we must first look briefly at heat and temperature.

Heat and Temperature

Anything that moves has **kinetic energy**, the energy of motion. Atoms and molecules have kinetic energy because they are always moving, although not necessarily in any particular direction. The faster a molecule moves, the greater its kinetic energy. **Heat** is a measure of the *total* quantity of kinetic energy due to molecular motion in a body of matter. **Temperature** measures the intensity of heat due to the *average* kinetic energy of the molecules. When the average speed of the molecules increases, a thermometer records this as a rise in temperature. Heat and temperature are related, but they are not the same. A swimmer crossing the English Channel has a higher temperature than the water, but the ocean contains far more heat because of its volume.

Whenever two objects of different temperature are brought together, heat passes from the warmer to the cooler body until the two are the same temperature. Molecules in the cooler object speed up at the expense of the kinetic energy of the warmer object. An ice cube cools a drink not by adding coldness to the liquid, but by absorbing heat as the ice melts.

Throughout this book, we will use the **Celsius scale** to indicate temperature (Celsius degrees are abbreviated as °C). At sea level, water freezes at 0°C and boils at 100°C. The temperature of the human body averages 37°C, and comfortable room temperature is about 20–25°C.

One convenient unit of heat used in this book is the **calorie (cal)**. A calorie is the amount of heat energy it takes to raise the temperature of 1 g of water by 1°C. Conversely, a calorie is also the amount of heat that 1 g of water releases when it cools by 1°C. A **kilocalorie (kcal)**, 1000 cal, is the quantity of heat required to raise the temperature of 1 kilogram (kg) of water by 1°C. (The "calories" on food packages are actually kilocalories.) Another energy unit used in this book is the **joule (J)**. One joule equals 0.239 cal; a calorie equals 4.184 J.

Water's High Specific Heat

The ability of water to stabilize temperature depends on its relatively high specific heat. The **specific heat** of a substance is defined as the amount of heat that must be absorbed or lost for 1 g of that substance to change its temperature by 1°C. We already know water's specific heat because we have defined a calorie as the amount of heat that causes 1 g of water to change its temperature by 1°C. Therefore, the specific heat of water is 1 calorie per gram per degree Celsius, abbreviated as 1 cal/g/°C. Compared with most other substances, water has an unusually high specific heat. For example, ethyl alcohol, the type of alcohol in alcoholic beverages, has a specific heat of 0.6 cal/g/°C.

Because of the high specific heat of water relative to other materials, water will change its temperature less when it absorbs or loses a given amount of heat. The reason you can burn your fingers by touching the metal handle of a pot on the stove when the water in the pot is still lukewarm is that the

FIGURE 3.3 ▪ Walking on water. The high surface tension of water, resulting from the collective strength of its hydrogen bonds, allows the water strider to walk on a pond without breaking the surface.

specific heat of water is ten times greater than that of iron. In other words, it will take only 0.1 cal to raise the temperature of 1 g of iron 1°C. Specific heat can be thought of as a measure of how well a substance resists changing its temperature when it absorbs or releases heat. Water resists changing its temperature; when it does change its temperature, it absorbs or loses a relatively large quantity of heat for each degree of change.

We can trace water's high specific heat, like many of its other properties, to hydrogen bonding. Heat must be absorbed in order to break hydrogen bonds, and heat is released when hydrogen bonds form. A calorie of heat causes a relatively small change in the temperature of water because much of the heat energy is used to disrupt hydrogen bonds before the water molecules can begin moving faster. And when the temperature of water drops slightly, many additional hydrogen bonds form, releasing a considerable amount of energy in the form of heat.

What is the relevance of water's high specific heat to life on Earth? A large body of water can absorb and store a huge amount of heat from the sun in the daytime and during summer, while warming up only a few degrees. At night and during winter, the gradually cooling water can warm the air. This is the reason coastal areas generally have milder climates than inland regions. The high specific heat of water also tends to stabilize ocean temperatures, creating a favorable environment for marine life. Thus, because of its high specific heat, the water that covers most of Earth keeps temperature fluctuations on land and in water within limits that permit life. Also, because organisms are made primarily of water, they are more able to resist changes in their own temperatures than if they were made of a liquid with a lower specific heat.

Evaporative Cooling

Molecules of any liquid stay close together because they are attracted to one another. Molecules moving fast enough to overcome these attractions can depart the liquid and enter the air as gas. This transformation from a liquid to a gas is called vaporization, or evaporation. Recall that the speed of molecular movement varies, and that temperature is the *average* kinetic energy of molecules. Even at low temperatures, the speediest molecules can escape into the air. Some evaporation occurs at any temperature; a glass of water, for example, will eventually evaporate at room temperature. If a liquid is heated, the average kinetic energy of molecules increases and the liquid evaporates more rapidly.

Heat of vaporization is the quantity of heat a liquid must absorb for 1 g of it to be converted from the liquid to the gaseous state. Compared with most other liquids, water has a high heat of vaporization. To evaporate each gram of water at room temperature, about 580 cal of heat are needed—nearly double the amount needed to vaporize a gram of alcohol or ammonia. Water's high heat of vaporization is another emergent property caused by hydrogen bonds, which must be broken before the molecules can make their exodus from the liquid.

FIGURE 3.4 ▪ Evaporative cooling. Because of water's high heat of vaporization, evaporation of sweat significantly cools the body surface.

Water's high heat of vaporization helps moderate Earth's climate. A considerable amount of solar heat absorbed by tropical seas is consumed during the evaporation of surface water. Then, as moist tropical air circulates poleward, it releases heat as it condenses to form rain.

As a liquid evaporates, the surface of the liquid that remains behind cools down. This **evaporative cooling** occurs because the "hottest" molecules, those with the greatest kinetic energy, are the most likely to leave as gas. It is as if the 100 fastest runners at a college transferred to another school; the average speed of the remaining students would decline.

Evaporative cooling of water contributes to the stability of temperature in lakes and ponds and also provides a mechanism that prevents terrestrial organisms from overheating. For example, evaporation of water from the leaves of a plant helps keep the tissues in the leaves from becoming too warm in the sunlight. Evaporation of sweat from human skin dissipates body heat and helps prevent overheating on a hot day or when excess heat is generated by strenuous activity (FIGURE 3.4). High humidity on a hot day increases discomfort because the high concentration of water vapor in the air inhibits the evaporation of sweat from the body.

Oceans and lakes don't freeze solid because ice floats

Water is one of the few substances that are less dense as a solid than as a liquid. In other words, ice floats. While other materials contract when they solidify, water expands. The cause of this exotic behavior is, once again, hydrogen bonding. At temperatures above 4°C, water behaves like other liquids, expand-

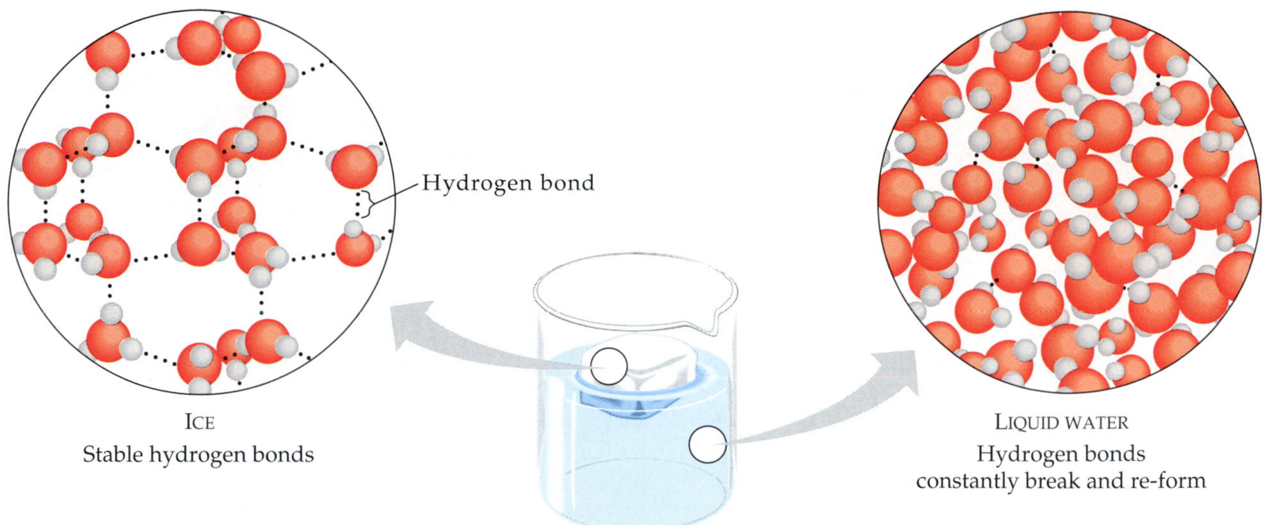

FIGURE 3.5 ▪ The structure of ice. Each molecule is hydrogen-bonded to four neighbors in a three-dimensional crystal with open channels. Because the hydrogen bonds make the crystal spacious, ice contains fewer molecules than an equal volume of liquid water. In other words, ice is less dense than liquid water.

ing as it warms and contracting as it cools. Water begins to freeze when its molecules are no longer moving vigorously enough to break their hydrogen bonds. As the temperature reaches 0°C, the water becomes locked into a crystalline lattice, each water molecule bonded to the maximum of four partners (FIGURE 3.5). The hydrogen bonds keep the molecules at "arm's length," far enough apart to make ice about 10% less dense (10% fewer molecules for the same volume) than liquid water at 4°C. When ice absorbs enough heat for its temperature to rise above 0°C, hydrogen bonds between molecules are disrupted. As the crystal collapses, the ice melts, and molecules are free to slip closer together. Water reaches its greatest density at 4°C and then begins to expand as the molecules move faster. Keep in mind, however, that even liquid water is semistructured because of transient hydrogen bonds.

The ability of ice to float because of the expansion of water as it solidifies is an important factor in the fitness of the environment. If ice sank, then eventually all ponds, lakes, and even oceans would freeze solid, making life as we know it impossible on Earth. During summer, only the upper few inches of the ocean would thaw. Instead, when a deep body of water cools, the floating ice insulates the liquid water below, preventing it from freezing and allowing life to exist under the frozen surface (FIGURE 3.6).

FIGURE 3.6 ▪ Floating ice and the fitness of the environment. Floating ice becomes a barrier that protects the liquid water below from the colder air. These are invertebrates called krill, photographed beneath Antarctic ice.

Water is the solvent of life

A sugar cube placed in a glass of water will dissolve. The glass will then contain a uniform mixture of sugar and water; the concentration of dissolved sugar will be the same everywhere in the mixture. A liquid that is a completely homogeneous mixture of two or more substances is called a **solution**. The dissolving agent of a solution is the **solvent**, and the substance that is dissolved is the **solute**. In this case, water is the solvent and sugar is the solute. An **aqueous solution** is one in which water is the solvent.

The medieval alchemists tried to find a universal solvent, one that would dissolve anything. They learned that nothing works better than water. However, water is not a universal solvent; if it were, it could not be stored in any container, including our cells. But water is a very versatile solvent, a quality we can trace to the polarity of the water molecule.

Suppose, for example, that a crystal of the ionic compound sodium chloride is placed in water (FIGURE 3.7, p. 42). At the surface of the crystal, the sodium and chloride ions are exposed to the solvent. The ions and water molecules have a mutual affinity through electrical attraction. The oxygen regions of the water molecules are negatively charged and cling to sodium cations. The hydrogen regions of the water molecules are positively charged and are attracted to chloride anions. Water surrounds the individual ions, separating the

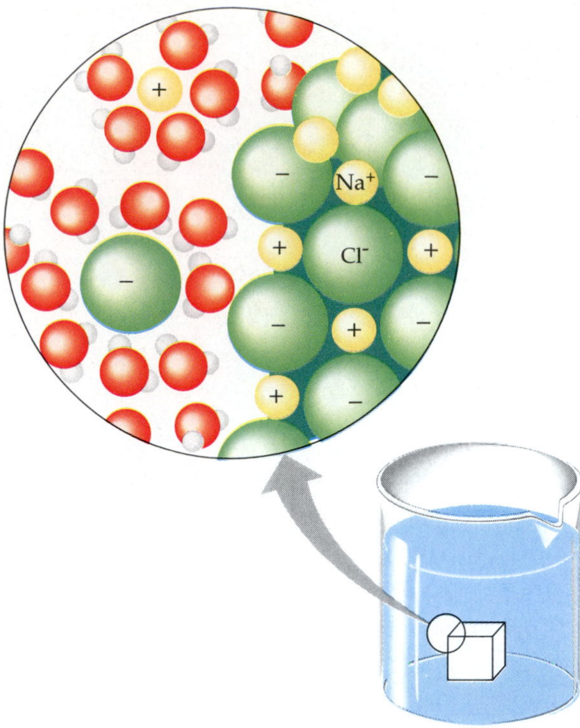

FIGURE 3.7 ▪ A crystal of table salt dissolving in water. The positive hydrogen regions of the polar water molecules are attracted to the chloride anions (green), whereas the negative oxygen regions cling to the sodium cations (yellow). The sphere of water molecules surrounding a solute ion (or molecule) is called a hydration shell.

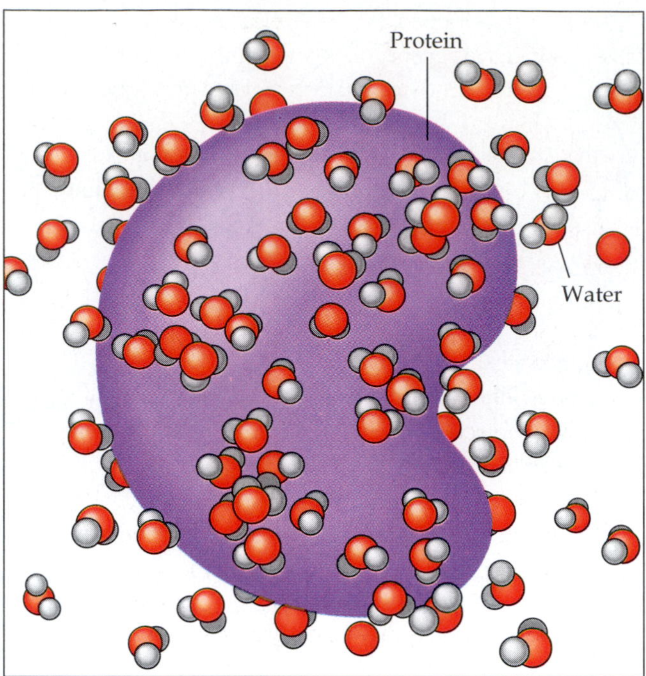

FIGURE 3.8 ▪ A water-soluble protein. Even a molecule as large as a protein can dissolve in water if it has enough ionic and polar regions on its surface. The mass of purple here represents a single such protein molecule, which water molecules are surrounding.

sodium from the chloride and shielding the ions from one another. Working inward from the surface of the salt crystal, water eventually dissolves all the ions. The result is a solution of two solutes, sodium and chloride, homogeneously mixed with water, the solvent. Other ionic compounds also dissolve in water. Seawater, for instance, contains a great variety of dissolved ions, as do living cells.

A compound does not need to be ionic in order to dissolve in water; polar compounds are also water-soluble. For example, sugars are soluble because water molecules can coat the polar sugar molecules. Even molecules as large as proteins can dissolve in water if they have ionic and polar regions on their surface (FIGURE 3.8). Many different kinds of polar compounds are dissolved (along with ions) in the water of such biological fluids as blood, the sap of plants, and the liquid within all cells. Water is the solvent of life.

Hydrophilic and Hydrophobic Substances

Whether ionic or polar, any substance that has an affinity for water is said to be **hydrophilic** (from the Greek *hydro,* "water," and *philios,* "loving"). This term is used even if the substance does not dissolve—because the molecules are too large, for instance. Cotton, a plant product, is an example of a hydrophilic substance that absorbs water without dissolving. Cotton consists of giant molecules of cellulose, a compound with numerous regions of partial positive and partial negative charges associated with polar bonds. Water adheres to the cellulose fibers. Thus, a cotton towel does a great job of drying the body, yet does not dissolve in the washing machine. Cellulose is also present in the walls of water-conducting vessels in a plant; you read earlier how the adhesion of water to these hydrophilic walls functions in water transport.

There are, of course, substances that do not have an affinity for water. Substances that are non-ionic and nonpolar actually seem to repel water; these substances are termed **hydrophobic** (Gr. *phobos,* "fearing"). An example from the kitchen is vegetable oil, which, as you know, does not mix stably with watery substances such as vinegar. The hydrophobic behavior of the oil molecules results from a prevalence of nonpolar bonds, in this case bonds between carbon and hydrogen, which share electrons almost equally. Hydrophobic molecules related to oils are major ingredients of cell membranes. (Imagine what would happen to a cell if its membrane dissolved.)

Solute Concentration in Aqueous Solutions

Biological chemistry is "wet" chemistry. Most of the chemical reactions that occur in organisms involve solutes dissolved in water. To understand the chemistry of life, it is important to learn how to calculate the concentrations of solutes dissolved in aqueous solutions.

Suppose we wanted to prepare an aqueous solution of table sugar having a specified concentration of sugar molecules (a certain number of solute molecules in a certain volume of solution). Because counting or weighing individual molecules is not practical, we instead usually measure substances in units called moles. A **mole (mol)** is equal in number to the molecular weight of a substance, but upscaled from daltons to units of grams. Imagine weighing out 1 mol of table sugar (sucrose), which has the molecular formula $C_{12}H_{22}O_{11}$. In round numbers, a carbon atom weighs 12 daltons, a hydrogen atom weighs 1 dalton, and an oxygen atom weighs 16 daltons. **Molecular weight** is the sum of the weights of all the atoms in a molecule; thus, the molecular weight of sucrose is 342 daltons. To obtain 1 mol of sucrose, we weigh out 342 g, the molecular weight of sucrose expressed in grams.

The practical advantage of measuring a quantity of chemicals in moles is that a mole of one substance has exactly the same number of molecules as a mole of any other substance. If substance A has a molecular weight of 10 daltons and substance B has a molecular weight of 100 daltons, then 10 g of A will have the same number of molecules as 100 g of B. The number of molecules in a mole, called Avogadro's number, is 6.02×10^{23}. A mole of table sugar contains 6.02×10^{23} sucrose molecules and weighs 342 g. A mole of ethyl alcohol (C_2H_6O) also contains 6.02×10^{23} molecules, but it weighs only 46 g because the molecules are smaller than those of sucrose. Measuring in moles makes it convenient for scientists working in the laboratory to combine substances in fixed ratios of molecules.

How would we make a liter (L) of solution consisting of 1 mol of sucrose dissolved in water? To obtain this concentration, we would weigh out 342 g of sucrose and then gradually add water, while stirring, until the sugar was completely dissolved. We would then add enough water to bring the total volume of the solution up to 1 L. At that point, we would have a one-molar (1 *M*) solution of sucrose. **Molarity**—the number of moles of solute per liter of solution—is the unit of concentration most often used by biologists for aqueous solutions.

DISSOCIATION OF WATER MOLECULES

3.2 Occasionally, a hydrogen atom shared between two water molecules in a hydrogen bond shifts from one molecule to the other. When this happens, the hydrogen atom leaves its electron behind, and what is actually transferred is a **hydrogen ion**, a single proton with a charge of +1. The water molecule that lost a proton is now a **hydroxide ion** (OH^-), which has a charge of −1. The proton binds to the other water molecule, making that molecule a hydronium ion (H_3O^+). We can picture the chemical reaction this way:

3.2
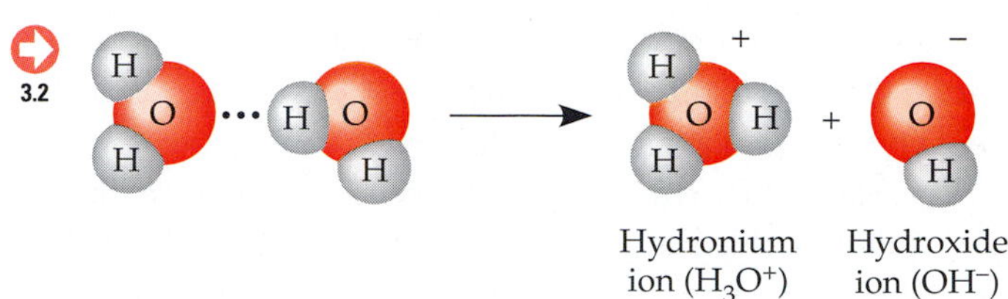

Although this is technically what happens, we will think of the process in a simplified way, as the dissociation (separation) of a water molecule into a hydrogen ion and a hydroxide ion:

$$H_2O \rightleftharpoons \underset{\text{Hydrogen ion}}{H^+} + \underset{\text{Hydroxide ion}}{OH^-}$$

As the double arrows indicate, this is a reversible reaction that will reach a state of dynamic equilibrium when water dissociates at the same rate that it is being re-formed from H^+ and OH^-. At this equilibrium point, the concentration of water molecules greatly exceeds the concentrations of H^+ and OH^-. In fact, in pure water, only one water molecule in every 554 million is dissociated. The concentration of each ion in pure water is 10^{-7} *M* (at 25°C). This means that there is only one ten-millionth of a mole of hydrogen ions per liter of pure water, and an equal number of hydroxide ions.

Although the dissociation of water is reversible and statistically rare, it is exceedingly important in the chemistry of life. Hydrogen and hydroxide ions are very reactive. Changes in their concentrations can drastically affect a cell's proteins and other complex molecules. As we have seen, the concentrations of H^+ and OH^- are equal in pure water, but adding certain kinds of solutes, called acids and bases, disrupts this balance. Biologists use something called the pH scale to describe how acidic or basic (the opposite of acidic) a solution is. In the remainder of the chapter you will learn about acids, bases, and pH, and why changes in pH can adversely affect organisms.

Organisms are sensitive to changes in pH

Before discussing the pH scale, let's see how acids and bases
3.3 actually work.

Acids and Bases

What would cause an aqueous solution to have an imbalance in its H^+ and OH^- concentrations? When the substances called acids dissolve in water, they donate additional hydrogen ions to the solution. An **acid**, according to the definition most often used by biologists, is a substance that increases the H^+ concentration of a solution. For example, when hydrochloric acid (HCl) is added to water, hydrogen ions dissociate from chloride ions:

$$HCl \longrightarrow H^+ + Cl^-$$

This additional source of H^+ in the solution (dissociation of water is the other source) results in more H^+ than OH^-. Such a solution is known as an acidic solution.

A substance that reduces the hydrogen ion concentration in a solution is called a **base**. Some bases reduce the H^+ concentration indirectly by dissociating to form hydroxide ions, which then combine with hydrogen ions to form water. One base that acts this way is sodium hydroxide (NaOH), which in water dissociates into its ions:

$$NaOH \longrightarrow Na^+ + OH^-$$

Other bases reduce H^+ concentration directly by accepting hydrogen ions. Ammonia (NH_3), for instance, acts as a base when the unshared electron pair in nitrogen's valence shell attracts a hydrogen ion from the solution, resulting in an ammonium ion (NH_4^+):

$$NH_3 + H^+ \rightleftharpoons NH_4^+$$

In either case, the base reduces the H^+ concentration. Solutions with a higher concentration of OH^- than H^+ are known as basic solutions. A solution in which the H^+ and OH^- concentrations are equal is said to be neutral.

Notice that single arrows were used in the reactions for HCl and NaOH. These compounds dissociate completely when mixed with water. Hydrochloric acid is called a strong acid and sodium hydroxide a strong base because they dissociate completely. In contrast, ammonia is a relatively weak base. The double arrows in the reaction for ammonia indicate that the binding and release of the hydrogen ion are reversible, although at equilibrium there will be a fixed ratio of NH_4^+ to NH_3.

There are also weak acids, which reversibly release and reaccept hydrogen ions. An example is carbonic acid:

$$\underset{\text{Carbonic acid}}{H_2CO_3} \rightleftharpoons \underset{\text{Bicarbonate ion}}{HCO_3^-} + \underset{\text{Hydrogen ion}}{H^+}$$

Here the equilibrium so favors the reaction in the left direction that when carbonic acid is added to water, only 1% of the molecules are dissociated at any particular time. Still, that is enough to shift the balance of H^+ and OH^- from neutrality.

The pH Scale

In any solution, the *product* of the H^+ and the OH^- concentrations is constant at 10^{-14} *M*. This can be written

$$[H^+][OH^-] = 10^{-14} M^2$$

In such an equation, brackets indicate molar concentration for the substance enclosed within them. In a neutral solution at room temperature (25°C), $[H^+] = 10^{-7}$ and $[OH^-] = 10^{-7}$, so the product is $10^{-14} M^2$ ($10^{-7} \times 10^{-7}$). If enough acid is added to a solution to increase $[H^+]$ to 10^{-5} *M*, then $[OH^-]$ will decline by an equivalent amount to 10^{-9} *M* ($10^{-5} \times 10^{-9} = 10^{-14}$). An acid not only adds hydrogen ions to a solution, but also removes hydroxide ions because of the tendency for H^+ to combine with OH^- to form water. A base has the opposite effect, increasing OH^- concentration but also reducing H^+ concentration by the formation of water. If enough of a base is added to raise the OH^- concentration to 10^{-4} *M*, the H^+ concentration will drop to 10^{-10} *M*. Whenever we know the concentration of either H^+ or OH^- in a solution, we can deduce the concentration of the other ion.

Because the H^+ and OH^- concentrations of solutions can vary by a factor of 100 trillion or more, scientists have developed a way to express this variation more conveniently, by use of the **pH scale**, which ranges from 0 to 14 (FIGURE 3.9). The pH scale compresses the range of H^+ and OH^- concentra-

3.3

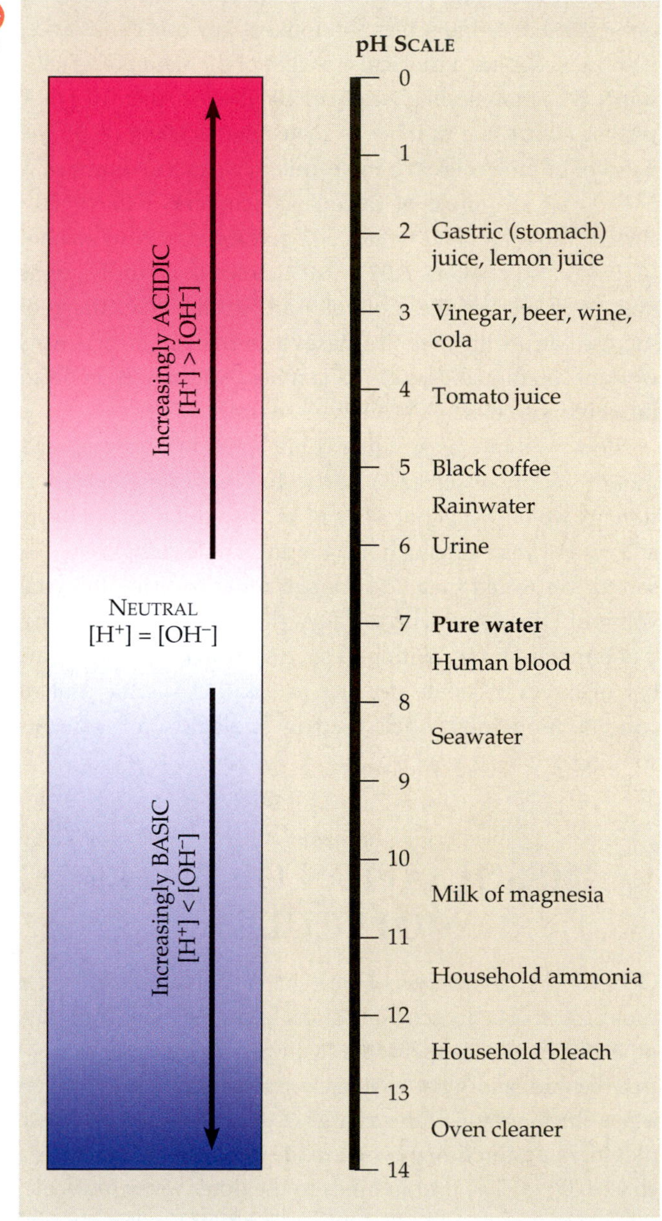

FIGURE 3.9 • The pH of some aqueous solutions.

tions by employing a common mathematical device: logarithms. The pH of a solution is defined as the negative logarithm (base 10) of the hydrogen ion concentration:

$$pH = -\log [H^+]$$

For a neutral solution, $[H^+]$ is 10^{-7} *M*, giving us

$$-\log 10^{-7} = -(-7) = 7$$

Notice that pH *declines* as H^+ concentration *increases.* Notice, too, that although the pH scale is based on H^+ concentration, it also implies OH^- concentration. A solution of pH 10 has a hydrogen ion concentration of 10^{-10} *M* and a hydroxide ion concentration of 10^{-4} *M.*

The pH of a neutral solution is 7, the midpoint of the scale. A pH value less than 7 denotes an acidic solution; the lower the number, the more acidic the solution. The pH for basic solutions is above 7. Most biological fluids are within the range pH 6 to pH 8. There are a few exceptions, however, including the strongly acidic digestive juice of the human stomach, which has a pH of about 2.

It is important to remember that each pH unit represents a tenfold difference in H^+ and OH^- concentrations. It is this mathematical feature that makes the pH scale so compact. A solution of pH 3 is not twice as acidic as a solution of pH 6, but a thousand times more acidic. When the pH of a solution changes slightly, the actual concentrations of H^+ and OH^- in the solution change substantially.

Buffers

The internal pH of most living cells is close to 7. Even a slight change in pH can be harmful, because the chemical processes of the cell are very sensitive to the concentrations of hydrogen and hydroxide ions.

Biological fluids resist changes to their own pH when acids or bases are introduced because of the presence of **buffers,** substances that minimize changes in the concentrations of H^+ and OH^- in a solution. Buffers in human blood, for example, normally maintain the blood pH very close to 7.4. A person cannot survive for more than a few minutes if the blood pH drops to 7 or rises to 7.8. Under normal circumstances, the buffering capacity of the blood prevents such swings in pH.

A buffer works by accepting hydrogen ions from the solution when they are in excess and donating hydrogen ions to the solution when they have been depleted. Most buffers are weak acids or weak bases that combine reversibly with hydrogen ions. One of the buffers that contributes to pH stability in human blood and many other biological solutions is carbonic acid (H_2CO_3), which, as already mentioned, dissociates to yield a bicarbonate ion (HCO_3^-) and a hydrogen ion (H^+):

$$\underset{\substack{H^+ \text{ donor} \\ \text{(acid)}}}{H_2CO_3} \; \underset{\substack{\text{Response to} \\ \text{a drop in pH}}}{\overset{\substack{\text{Response to} \\ \text{a rise in pH}}}{\rightleftharpoons}} \; \underset{\substack{H^+ \text{ acceptor} \\ \text{(base)}}}{HCO_3^-} + \underset{\substack{\text{Hydrogen} \\ \text{ion}}}{H^+}$$

The chemical equilibrium between carbonic acid and bicarbonate acts as a pH regulator, the reaction shifting left or right as other processes in the solution add or remove hydrogen ions. If the H^+ concentration in blood begins to fall (that is, if pH rises), more carbonic acid dissociates, replenishing hydrogen ions. But when H^+ concentration in blood begins to rise (pH drops), the bicarbonate ion acts as a base and removes the excess hydrogen ions from the solution. Thus, the carbonic acid–bicarbonate buffering system actually consists of an acid *and* a base in equilibrium with each other. Most other buffers are also acid-base pairs.

Acid precipitation threatens the fitness of the environment

Considering the dependence of all life on water, contamination of rivers, lakes, and seas is a dire environmental problem. One of the most serious assaults on water quality is acid precipitation.

Uncontaminated rain has a pH of about 5.6, slightly acidic, owing to the formation of carbonic acid from carbon dioxide and water. **Acid precipitation** refers to rain, snow, or fog more acidic than pH 5.6. What causes acid precipitation, and what are its effects on the fitness of the environment?

Acid precipitation is caused primarily by the presence in the atmosphere of sulfur oxides and nitrogen oxides, gaseous compounds that react with water in the air to form strong acids, which fall to earth with rain or snow. A major source of these oxides is the burning of fossil fuels (coal, oil, and gas) in factories and automobiles. Electrical power plants that burn coal produce more of these pollutants than any other single source. Ironically, the tall smokestacks built to reduce local pollution by dispersing factory exhaust help spread airborne pollutants. Winds carry the pollutants away, and acid rain may fall thousands of miles away from industrial centers. In the Adirondack Mountains of upstate New York, the pH of rainfall averages 4.2, about 25 times more acidic than normal rain. Acid precipitation falls on many other regions, including the Cascade Mountains of the Pacific Northwest and certain parts of Europe and Asia. One West Virginia storm dropped rain having a pH of 1.5, as acidic as the digestive juices in our stomachs!

The effect of acids in lakes and streams is most pronounced in the spring, as snow begins to melt. The surface snow melts first, drains down, and sends much of the acid that has accumulated over the winter into lakes and streams all at once. Early meltwater often has a pH as low as 3, and this acid surge hits when fish and other forms of aquatic life are producing eggs and young, which are especially vulnerable to acidic conditions. Strong acidity can alter the structure of biological molecules and prevent them from carrying out the essential chemical processes of life.

Although acid precipitation can clearly damage life in lakes and streams, its direct effects on forests and other terrestrial life are controversial. However, recent research indicates that acid rain and snow can bring about profound changes in soils by affecting the solubility of soil minerals. Acid precipitation falling on land washes away certain mineral ions, such as calcium and magnesium ions, that ordinarily help buffer the soil solution and are essential nutrients for plant growth. At the same time, other minerals, such as aluminum, reach toxic concentrations when acidification increases their solubility. The effects of acid precipitation on soil chemistry have contributed to the decline of European forests and are taking a toll on some North American forests (FIGURE 3.10). Nevertheless, recent studies indicate that currently the majority of North American forests are not suffering substantially from acid precipitation.

If there is reason for optimism about the future quality of water resources, it is our progress in reducing certain kinds of pollution. For example, in the United States, Canada, and Europe, emissions of sulfur oxides have declined markedly in recent decades, causing a decrease in acid precipitation. Continued progress can come only from the actions of individuals who are concerned about environmental quality. An essential part of their education is to understand the crucial role that water plays in the environmental fitness for continued life on Earth.

FIGURE 3.10 ▪ The effects of acid precipitation on a forest. Acid fog and acid rain are thought to be responsible for killing many of the spruce and fir trees on Mount Mitchell in North Carolina.

CHAPTER REVIEW

REVIEW OF KEY CONCEPTS

(with page numbers and key figures)

EFFECTS OF WATER'S POLARITY

- **The polarity of water molecules results in hydrogen bonding (pp.**
3.1 **37–38, FIGURE 3.1)** A hydrogen bond forms when the oxygen of one water molecule is electrically attracted to the hydrogen of a nearby molecule. Hydrogen bonding between water molecules is the basis for water's unusual properties.

- **Organisms depend on the cohesion of water molecules (pp. 38–39)** Hydrogen bonding makes water molecules stick together, and this cohesion helps pull water upward in the microscopic vessels of plants. Hydrogen bonding is also responsible for water's surface tension.

- **Water moderates temperatures on Earth (pp. 39–40)** Hydrogen bonding gives water a high specific heat. Heat is absorbed when hydrogen bonds break and is released when hydrogen bonds form, minimizing temperature fluctuations to within limits that permit life. Evaporative cooling is based on water's high heat of vaporization. Water molecules must have a relatively high kinetic energy to break hydrogen bonds. The evaporative loss of these energetic water molecules cools a surface.

- **Oceans and lakes don't freeze solid because ice floats (pp. 40–41, FIGURE 3.6)** Ice is less dense than liquid water because its more organized hydrogen bonding causes expansion into a crystal formation. Floating ice allows life to exist under the frozen surfaces of lakes and polar seas.

- **Water is the solvent of life (pp. 41–43, FIGURE 3.7)** Water is an unusually versatile solvent because its polarity attracts it to charged and polar substances. When ions or polar substances are surrounded by water molecules, they dissolve and are called solutes. Hydrophilic substances have an affinity for water. Hydrophobic substances do not; they seem to repel water. Biologists usually use molarity, the number of moles of solute per liter of solution, as a measure of solute concentration in solutions. A mole is the number of grams of a substance equal to its molecular weight.

DISSOCIATION OF WATER MOLECULES

- **Organisms are sensitive to changes in pH (pp. 43–45, FIGURE 3.9)**
3.2 Water can dissociate into H^+ and OH^-. The concentration of H^+
3.3 is expressed as pH, where $pH = -\log [H^+]$. Acids donate additional H^+ in aqueous solutions; bases donate OH^- or accept H^+. In a neutral solution, $[H^+] = [OH^-] = 10^{-7}$, and pH = 7. In an acidic solution, $[H^+]$ is greater than $[OH^-]$, and the pH is less than 7. In a basic solution, $[H^+]$ is less than $[OH^-]$, and the pH is greater than 7. Buffers in biological fluids resist changes in pH. A buffer consists of an acid-base pair that combines reversibly with hydrogen ions.

- **Acid precipitation threatens the fitness of the environment (pp. 45–46)** Acid precipitation is rain, snow, or fog with a pH below 5.6. It often results from a reaction in the air between water vapor and sulfur oxides and nitrogen oxides produced by the combustion of fossil fuels.

SELF-QUIZ

1. The main thesis of Lawrence Henderson's *The Fitness of the Environment* is
   a. Earth's environment is constant
   b. it is the physical environment, not life, that has evolved
   c. the environment of Earth has adapted to life
   d. life as we know it depends on certain environmental qualities on Earth
   e. water and other aspects of Earth's environment exist because they make the planet more suitable for life

2. Air temperature often increases slightly as clouds begin to drop rain or snow. Which behavior of water is *most directly* responsible for this phenomenon?
   a. water's change in density when it condenses
   b. water's reactions with other atmospheric compounds
   c. release of heat by the formation of hydrogen bonds
   d. release of heat by the breaking of hydrogen bonds
   e. water's high surface tension

3. For two bodies of matter in contact, heat always flows from
   a. the body with greater heat to the one with less heat
   b. the body of higher temperature to the one of lower temperature
   c. the denser body to the less dense body
   d. the body with more water to the one with less water
   e. the larger body to the smaller body

4. A slice of pizza has 500 kcal. If we could burn the pizza and use all the heat to warm a 50-L container of cold water, what would be the approximate increase in the temperature of the water? (*Note:* A liter of cold water weighs about 1 kg.)
   a. 50°C
   b. 5°C
   c. 10°C
   d. 100°C
   e. 1°C

5. The bonds that are broken when water vaporizes are
   a. ionic bonds
   b. bonds between water molecules
   c. bonds between atoms of individual water molecules
   d. polar covalent bonds
   e. nonpolar covalent bonds

6. Which of the following is an example of a hydrophobic material?
   a. paper
   b. table salt
   c. wax
   d. sugar
   e. pasta

7. We can be sure that a mole of table sugar and a mole of vitamin C are equal in their
   a. weight in daltons
   b. weight in grams
   c. number of molecules
   d. number of atoms
   e. volume

8. How many grams of acetic acid ($C_2H_4O_2$) would you use to make 10 L of a 0.1 *M* aqueous solution of acetic acid? (*Note:* The atomic weights, in daltons, are approximately 12 for carbon, 1 for hydrogen, and 16 for oxygen.)
   a. 10 g
   b. 0.1 g
   c. 6 g
   d. 60 g
   e. 0.6 g

9. Acid precipitation has lowered the pH of a particular lake to 4.0. What is the hydrogen ion concentration of the lake?
   a. 4.0 *M*
   b. 10^{-10} *M*
   c. 10^{-4} *M*
   d. 10^{4} *M*
   e. 4%

10. What is the *hydroxide* ion concentration of the lake described in question 9?
    a. 10^{-7} *M*
    b. 10^{-4} *M*
    c. 10^{-10} *M*
    d. 10^{-14} *M*
    e. 10 *M*

CHALLENGE QUESTIONS

1. Explain how panting helps regulate a dog's body temperature.
2. Design a controlled experiment to test the hypothesis that acid precipitation inhibits the growth of *Elodea,* a common freshwater plant.

SCIENCE, TECHNOLOGY, AND SOCIETY

1. Agriculture, industry, and the growing populations of cities all compete, through political influence, for water. If you were in charge of water resources in an arid region, what would your priorities be for allocating the limited water supply for various uses? How would you try to build consensus among the different special interest groups?
2. Discuss the special political obstacles to reducing acid precipitation. (Compare this issue with environmental issues confined to a more localized region.)

FURTHER READING

Henderson, L. J. *The Fitness of the Environment.* New York: Macmillan, 1913. A classic book highlighting the importance of water and carbon to life.

Kaiser, J. "Acid Rain's Dirty Business: Stealing Minerals from Soil." *Science,* April 12, 1996. The surprisingly long-term effects of soil acidification.

Lehninger, A. L., D. L. Nelson, and M. M. Cox. *Principles of Biochemistry,* 2nd ed. New York: Worth, 1993. A readable biochemistry text with an excellent discussion of water in Chapter 4.

Mohner, V. A. "The Challenge of Acid Rain." *Scientific American,* August 1988. Analysis of a complex environmental problem.

Matthews, R. "Wacky Water." *New Scientist,* June 21, 1997. More about hydrogen bonding in water.

Pennisi, E. "Water, Water Everywhere." *Science News,* February 20, 1993. How water stabilizes the structure of proteins and other large biological molecules.

Pennisi, E. "Water: The Power, Promise, and Turmoil of North America's Fresh Water." *National Geographic.* A special 1993 issue examining how we use and abuse our water resources.

WEB LINKS

Visit the special edition of *The Biology Place* for BIOLOGY, Fifth Edition, at http://www.biology.com/campbell. Go to Chapter 3 for online resources, including learning activities, practice exams, and links to the following web sites:

"The pH Playground"
An interactive site that allows you to see the relationship between pH and H^+ ion concentration at different levels of precision.

"Acid Rain: A Student's First Sourcebook"
All you need to know about acid rain and its effect on society and the environment from the Environmental Protection Agency.

"Water Resources of the United States"
Water is one of humanity's most precious resources. This link will take you to the water resources page of the United States Geological Survey, one of the government bodies that look after the nation's water resources.

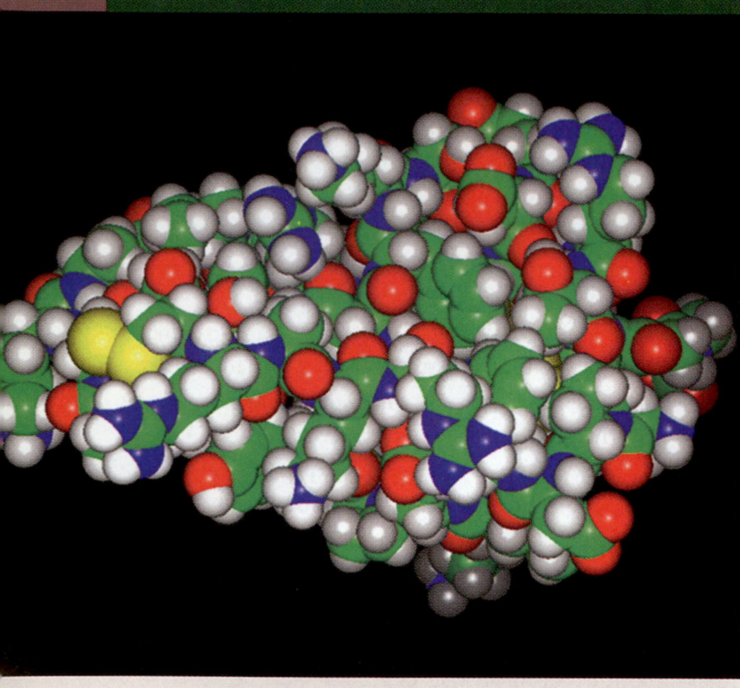

CHAPTER 4

CARBON AND THE MOLECULAR DIVERSITY OF LIFE

The Importance of Carbon

- Organic chemistry is the study of carbon compounds
- Carbon atoms are the most versatile building blocks of molecules
- Variation in carbon skeletons contributes to the diversity of organic molecules

Functional Groups

- Functional groups also contribute to the molecular diversity of life
- The chemical elements of life: *a review*

Although water is the universal medium for life on Earth, most of the chemicals that make up living organisms are based on the element carbon. Of all chemical elements, carbon is unparalleled in its ability to form molecules that are large, complex, and diverse, and this molecular diversity has made possible the diversity of organisms that have evolved on Earth. The protein shown in the adjacent computer graphic model is an example of a large, complex molecule based on carbon (the green atoms). Proteins are a major topic of Chapter 5. In this chapter we focus on smaller molecules, using them to illustrate a few concepts of molecular architecture that highlight carbon's importance to life and the theme that emergent properties arise from the organization of life's matter.

THE IMPORTANCE OF CARBON

Although a cell is composed of 70% to 95% water, most of the rest consists of carbon-based compounds. Proteins, DNA, carbohydrates, and other molecules that distinguish living matter from inanimate material are all composed of carbon atoms bonded to one another and to atoms of other elements. Hydrogen (H), oxygen (O), nitrogen (N), sulfur (S), and phosphorus (P) are other common ingredients of these compounds, but it is carbon (C) that accounts for the endless diversity of biological molecules.

Organic chemistry is the study of carbon compounds

Compounds containing carbon are said to be organic, and the branch of chemistry that specializes in the study of carbon compounds is called **organic chemistry**. Once thought to come only from living things, organic compounds range from simple molecules such as methane (CH_4) to colossal ones such as proteins with thousands of atoms and molecular weights in excess of 100,000 daltons. The percentages of the major elements of life—C, H, O, N, S, and P—are quite uniform from one organism to another. Because of carbon's versatility, however, this limited assortment of atomic building blocks, taken in roughly the same proportions, can be used to build an inexhaustible variety of organic molecules. Different species of organisms, and different individuals within a species, are distinguished by variations among their organic molecules.

Since the dawn of human history, people have used other organisms as sources of valued substances—from foods to medicines and fabrics. Organic chemistry originated in attempts to purify and improve the yield of such products. By the early nineteenth century, chemists had learned to make many simple compounds in the laboratory by combining elements under the right conditions. Artificial synthesis of the complex molecules extracted from living matter seemed impossible, however. It was at that time that the Swedish chemist Jöns Jakob Berzelius first made the distinction between organic compounds, those that seemingly could arise only within living organisms, and inorganic compounds, those that were found in the nonliving world. The new discipline of organic chemistry was first built on a foundation of vitalism, the belief in a life force outside the jurisdiction of physical and chemical laws.

Chemists began to chip away at the foundation of vitalism when they learned to synthesize organic compounds in their laboratories. In 1828, Friedrich Wöhler, a German chemist who had studied with Berzelius, attempted to make an inorganic salt, ammonium cyanate, by mixing solutions of ammonium (NH_4^+) and cyanate (CNO^-) ions. Wöhler was astonished to find that instead of the expected product, he had made urea, an organic compound present in the urine of animals. Wöhler challenged the vitalists when he wrote, "I must tell you that I can prepare urea without requiring a kidney or an animal, either man or dog." However, one of the ingredients used in the synthesis, the cyanate, had been extracted from animal blood, and the vitalists were not swayed by Wöhler's discovery. But then, a few years later, Hermann Kolbe, a student of Wöhler's, made the organic compound acetic acid from inorganic substances that could themselves be prepared directly from pure elements.

FIGURE 4.1 ▪ Abiotic synthesis of organic compounds under "early Earth" conditions. Stanley Miller re-creates his 1953 experiment, a laboratory simulation demonstrating that environmental conditions on the lifeless, primordial Earth favored the spontaneous synthesis of some organic molecules. Miller used electrical discharges (simulated lightning) to trigger reactions in a primitive "atmosphere" of H_2O, H_2, NH_3 (ammonia), and CH_4 (methane)—some of the gases released by volcanoes. From these ingredients, Miller's apparatus made a variety of organic compounds that play key roles in living cells. Similar chemical reactions may have set the stage for the origin of life on Earth, a hypothesis we will explore in more detail in Chapter 26.

The foundation of vitalism finally crumbled after several more decades of laboratory synthesis of increasingly complex organic compounds. In 1953, Stanley Miller, a graduate student at the University of Chicago, helped bring this abiotic (nonliving) synthesis of organic compounds into the context of evolution. Miller used a laboratory simulation of chemical conditions on the primitive Earth to demonstrate that the spontaneous synthesis of organic compounds could have been an early stage in the origin of life (FIGURE 4.1).

The pioneers of organic chemistry helped shift the mainstream of biological thought from vitalism to mechanism, the belief that all natural phenomena, including the processes of life, are governed by physical and chemical laws. Organic chemistry was redefined as the study of carbon compounds, regardless of their origin. Most naturally occurring organic compounds are produced by organisms, and these molecules represent a diversity and range of complexity unrivaled by inorganic compounds. However, the same rules of chemistry apply to inorganic and organic molecules alike. The foundation of organic chemistry is not some intangible life force, but the unique chemical versatility of the element carbon.

Carbon atoms are the most versatile building blocks of molecules

The key to the chemical characteristics of an atom, as you learned in Chapter 2, is in its configuration of electrons; electron configuration determines the kinds and number of bonds an atom will form with other atoms. Carbon has a total of six electrons, with two in the first electron shell and four in the second shell. Having four valence electrons in a shell that holds eight, carbon has little tendency to gain or lose electrons and form ionic bonds; it would have to donate or accept four electrons to do so. Instead, a carbon atom usually completes its valence shell by sharing electrons with other atoms in four covalent bonds. Each carbon atom thus acts as an intersection point from which a molecule can branch off in up to four directions. This *tetravalence* is one facet of carbon's versatility that makes large, complex molecules possible.

In Chapter 2 you also learned that when a carbon atom forms single covalent bonds, the arrangement of its four hybrid orbitals causes the bonds to angle toward the corners of an imaginary tetrahedron (see FIGURE 2.15c). The bond angles in methane (CH_4) are 109°, and they would be approximately the same in any molecule where carbon has four single bonds. For example, ethane (C_2H_6) is shaped like two tetrahedrons joined at their apexes (FIGURE 4.2, p. 50). In molecules with more carbons, every grouping of a carbon bonded to four other atoms has a tetrahedral shape. It is convenient to write structural formulas as though molecules were flat, but it is important to remember that molecules are three-dimensional and that the shape of a molecule often determines its function.

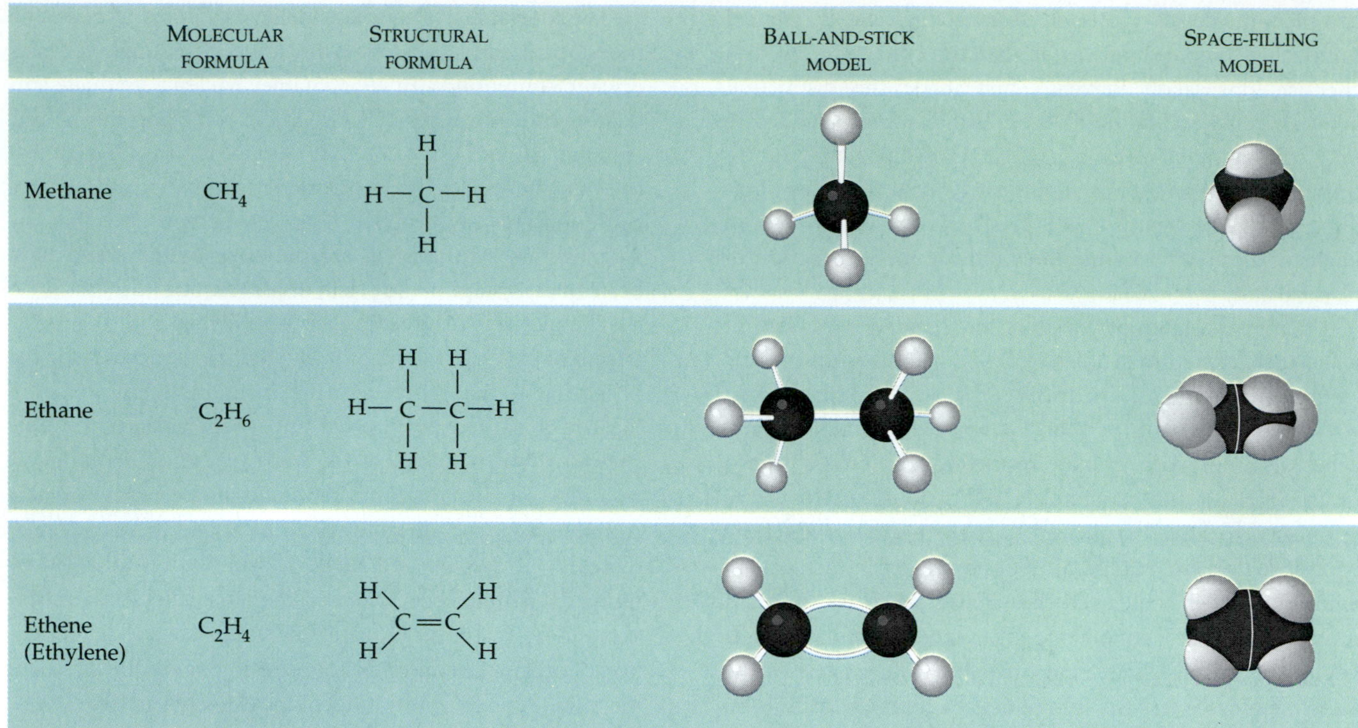

MOLECULAR FORMULA
STRUCTURAL FORMULA
BALL-AND-STICK MODEL
SPACE-FILLING MODEL
Methane
CH_4
Ethane
C_2H_6
Ethene (Ethylene)
C_2H_4

FIGURE 4.2 • The shapes of three simple organic molecules. Whenever a carbon atom has four single bonds, the bonds angle toward the corners of an imaginary tetrahedron. When two carbons are joined by a double bond, all bonds around those atoms are in the same plane. (Notice, for example, that ethene is a flat molecule; its atoms all lie in the same plane.)

The electron configuration of carbon gives it covalent compatibility with many different elements. FIGURE 4.3 reviews the valences of the four major atomic components of organic molecules: carbon and its most frequent partners, oxygen, hydrogen, and nitrogen. We can think of these valences as the rules of covalent bonding in organic chemistry—the building code that governs the architecture of organic molecules.

A couple of additional examples will show how the rules of covalent bonding apply to carbon atoms with partners other than hydrogen. In the carbon dioxide molecule (CO_2), a single carbon atom is joined to two atoms of oxygen by double covalent bonds. The structural formula for CO_2 is O=C=O. Each line (bond) in a structural formula represents a pair of shared electrons. Notice that the carbon atom in CO_2 is involved in four covalent bonds, two with each oxygen atom. The arrangement completes the valence shells of all atoms in the molecule. Carbon dioxide is such a simple molecule that it is often considered inorganic, even though it contains carbon. Whether we call CO_2 organic or inorganic is an arbitrary distinction, but there is no ambiguity about its importance to the living world. Taken from the air by plants and incorporated into sugar and other foods during photosynthesis, CO_2 is the source of carbon for all the organic molecules found in organisms.

Another relatively simple molecule is urea, $CO(NH_2)_2$. This is the organic compound found in urine that Wöhler learned to synthesize in the early nineteenth century. The structural formula for urea is:

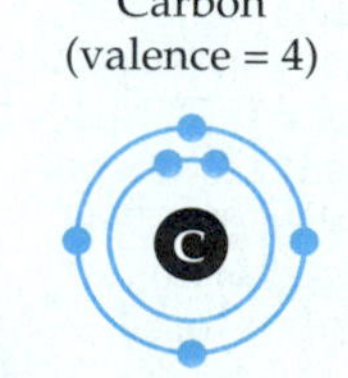

Again, each atom has the required number of covalent bonds. In this case, one carbon atom is involved in both single and double bonds.

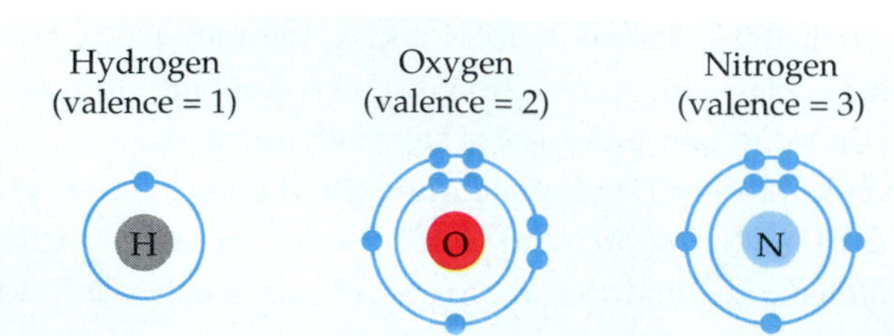

FIGURE 4.3 • Valences for the major elements of organic molecules. Valence is the number of covalent bonds an atom will usually form. It is equal to the number of electrons required to complete the atom's outermost (valence) electron shell. Forming bonds with other atoms gives each valence-shell orbital a pair of electrons.

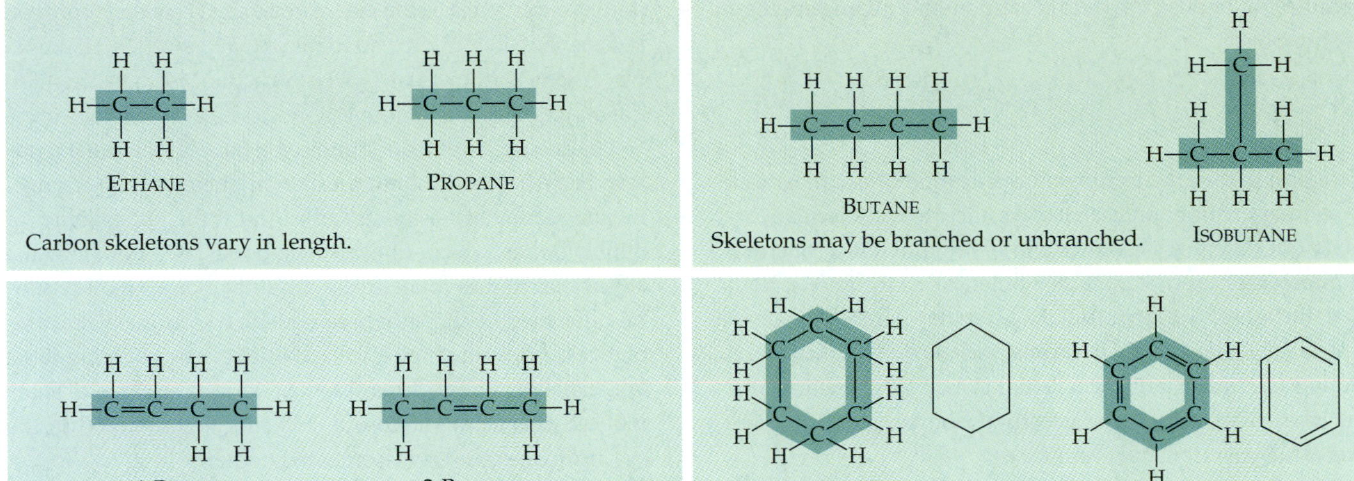
ETHANE
PROPANE
Carbon skeletons vary in length.
BUTANE
ISOBUTANE
Skeletons may be branched or unbranched.
1-BUTENE
2-BUTENE
The skeleton may have double bonds, which can vary in location.
CYCLOHEXANE
BENZENE
Some carbon skeletons are arranged in rings. (The abbreviated structural formulas omit the corner carbons and the hydrogens attached to them.)

FIGURE 4.4 ▪ Variations in carbon skeletons. Hydrocarbons, organic molecules consisting only of carbon and hydrogen, illustrate the diversity of the carbon skeletons of organic molecules.

Both urea and carbon dioxide are molecules with only one carbon atom. But as FIGURE 4.2 shows, a carbon atom can also use one or more of its valence electrons to form covalent bonds to other carbon atoms, making it possible to link the atoms into chains of seemingly infinite variety.

Variation in carbon skeletons contributes to the diversity of organic molecules

Carbon chains form the skeletons of most organic molecules. The skeletons vary in length and may be straight, branched, or arranged in closed rings (FIGURE 4.4). Some carbon skeletons have double bonds, which vary in number and location. Such variation in carbon skeletons is one important source of the molecular complexity and diversity that characterize living matter. In addition, atoms of other elements can be bonded to the skeletons at available sites.

All the molecules shown in FIGURES 4.2 and 4.4 are **hydrocarbons**, organic molecules consisting only of carbon and hydrogen. Atoms of hydrogen are attached to the carbon skeleton wherever electrons are available for covalent bonding. Hydrocarbons are the major components of petroleum, which is called a fossil fuel because it consists of the partially decomposed remains of organisms that lived millions of years ago. Although hydrocarbons are not prevalent in living organisms, many of a cell's organic molecules have regions consisting of only carbon and hydrogen. For example, the molecules known as fats have long hydrocarbon tails attached to a nonhydrocarbon component (FIGURE 4.5). Neither petroleum nor fat mixes with water; both are hydrophobic compounds

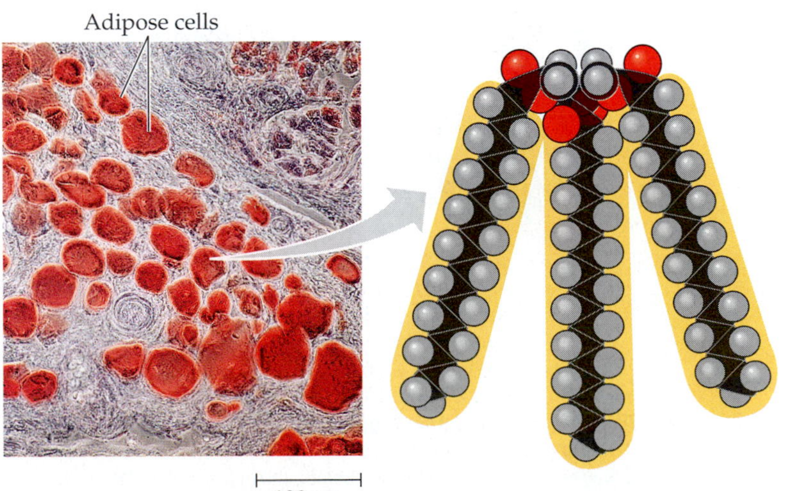
Adipose cells
100 µm

FIGURE 4.5 ▪ The role of hydrocarbons in the characteristics of fats. Humans and other mammals store fats in specialized cells called adipose cells. Each cell is almost completely filled with a large fat droplet (stained red in this light micrograph), which stockpiles an enormous number of fat molecules. The drawing is a space-filling model of one fat molecule (black = carbon; gray = hydrogen; red = oxygen). Three hydrocarbon tails are attached to a headpiece containing oxygen as well as carbon and hydrogen. Hydrocarbon bonds are nonpolar, which accounts for the hydrophobic behavior of fats. Another characteristic of hydrocarbons is that they store a relatively large amount of energy. The gasoline that fuels a car consists of hydrocarbons, and the hydrocarbon tails of fat molecules serve as stored fuel for your body.

because the bonds between the carbon and hydrogen atoms are nonpolar.

Isomers

Variation in the architecture of organic molecules can be seen
4.1 in **isomers**, compounds that have the same molecular formula but different structures and hence different properties. Compare, for example, the two butanes in FIGURE 4.4. Both have the molecular formula C_4H_{10}, but they differ in the covalent arrangement of their carbon skeletons. The skeleton is straight in butane, but branched in isobutane. We will examine three types of isomers: structural isomers, geometric isomers, and enantiomers (FIGURE 4.6).

Structural isomers differ in the covalent arrangements of their atoms. The number of possible isomers increases tremendously as carbon skeletons increase in size. There are only two butanes, but there are 18 variations of C_8H_{18} and 366,319 possible structural isomers of $C_{20}H_{42}$. Structural isomers may also differ in the location of double bonds.

Geometric isomers of a molecule have all the same covalent partnerships, but they differ in their spatial arrangements. Geometric isomers arise from the inflexibility of double bonds, which, unlike single bonds, will not allow the atoms they join to rotate freely about the bond axis. The subtle difference in shape between geometric isomers can dramatically affect the biological activities of organic molecules. For example, the biochemistry of vision involves a light-induced change of rhodopsin, a chemical compound in the eye, from one geometric isomer to another.

Enantiomers are molecules that are mirror images of each other. In the ball-and-stick models shown in FIGURE 4.6c, the middle carbon is called an *asymmetric carbon* because it is attached to four different atoms or groups of atoms. The four groups can be arranged in space about the asymmetric carbon in two different ways that are mirror images. They are, in a way, left-handed and right-handed versions of the molecule. A cell can distinguish these isomers based on their different shapes. Usually, one isomer is biologically active and the other is inactive.

The concept of enantiomers is important in the pharmaceutical industry, because the two enantiomers of a drug may not be equally effective (FIGURE 4.7). In some cases, one of the isomers may even produce harmful effects. This was the case with thalidomide, a drug prescribed for thousands of pregnant women in the early 1960s. The drug was a mixture of two enantiomers. One enantiomer produced the desired effect by acting as a sedative, but the other caused birth defects. Scientists are developing new techniques for synthesizing such drugs in pure isomeric form rather than as mixtures of enantiomers. The differing effects of enantiomers in the body demonstrate that organisms are sensitive to even the

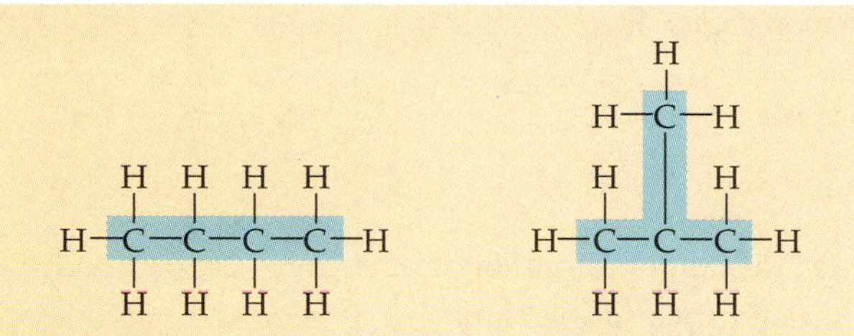

(a) Structural isomers: variation in covalent partners, as shown in the example of butane and isobutane.

(b) Geometric isomers: variation in arrangement about a double bond. (In these diagrams, "X" represents an atom or group of atoms attached to a double-bonded carbon.)

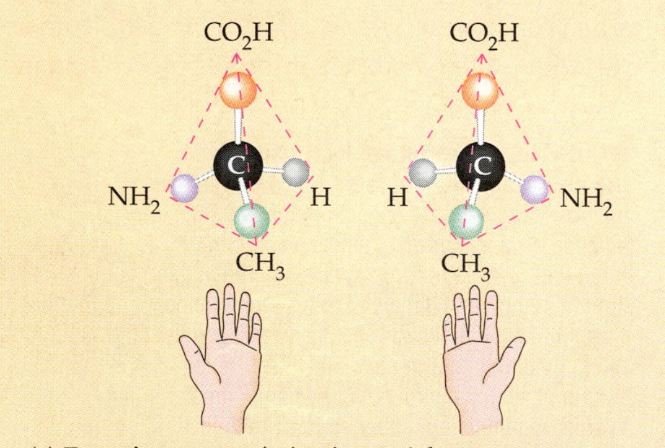

(c) Enantiomers: variation in spatial arrangement around an asymmetric carbon, resulting in molecules that are mirror images, like left and right hands. Enantiomers cannot be superimposed on each other.

FIGURE 4.6 ▪ Three types of isomers. Compounds with the same
4.1 molecular formula but different structures, isomers are a source of diversity in organic molecules.

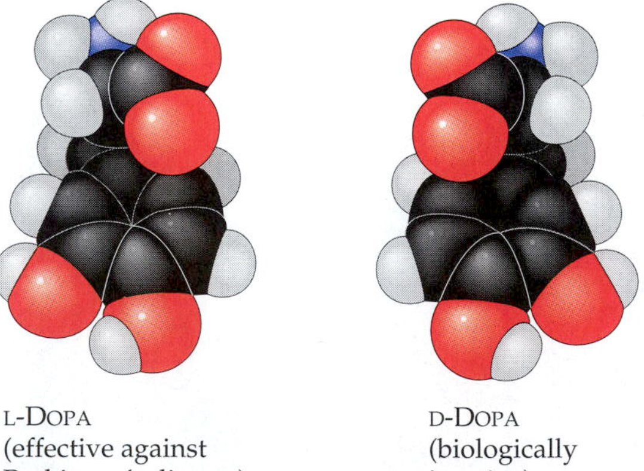

L-DOPA (effective against Parkinson's disease)

D-DOPA (biologically inactive)

FIGURE 4.7 ▪ The pharmacological importance of enantiomers. L-dopa is a drug used to treat Parkinson's disease, a disorder of the central nervous system. The drug's enantiomer, the mirror-image molecule designated D-dopa, has no effect on patients.

most subtle variations in molecular architecture. Once again, we see that molecules have emergent properties that depend on the specific arrangement of their atoms.

FUNCTIONAL GROUPS

The distinctive properties of an organic molecule depend not only on the arrangement of its carbon skeleton, but also on the molecular components attached to that skeleton. We will now examine certain groups of atoms that are frequently attached to the skeletons of organic molecules.

Functional groups also contribute to the molecular diversity of life

4.2 The components of organic molecules that are most commonly involved in chemical reactions are known as **functional groups**. If we think of hydrocarbons as the simplest organic molecules, we can, in general, view functional groups as attachments that replace one or more of the hydrogens bonded to the carbon skeleton of the hydrocarbon. (However, some functional groups include atoms of the carbon skeleton, as we will see.)

Each functional group behaves consistently from one organic molecule to another, and the number and arrangement of the groups help give each molecule its unique properties. Consider the differences between estrone and testosterone, female and male sex hormones, respectively, in humans and other vertebrates (FIGURE 4.8). Both are steroids, organic molecules with a common carbon skeleton in the form of four fused rings. These sex hormones differ only in the presence of certain functional groups on the four rings. The different actions of these two molecules on many targets throughout the body help produce the contrasting features of females and males. Thus, even our sexuality has its biological basis in variations of molecular architecture.

The six functional groups most important in the chemistry of life are the hydroxyl, carbonyl, carboxyl, amino, sulfhydryl, and phosphate groups (TABLE 4.1, p. 54). All are hydrophilic and thus increase the solubility of organic compounds in water.

The Hydroxyl Group

In a **hydroxyl group**, a hydrogen atom is bonded to an oxygen atom, which in turn is bonded to the carbon skeleton of the organic molecule. Organic compounds containing hydroxyl groups are called **alcohols**, and their specific names usually end in *-ol*, as in ethanol, the drug present in alcoholic beverages. In a structural formula, the hydroxyl group is usually abbreviated by omission of the covalent bond between the oxygen and hydrogen, and is written as —OH or HO—. (Do not confuse this functional group with the hydroxide ion, OH^-, formed by the dissociation of bases such as sodium hydroxide.) The hydroxyl group is polar as a result of the electronegative oxygen atom drawing electrons toward itself. Consequently, water molecules are attracted to the hydroxyl group, and this helps dissolve organic compounds containing such groups. Sugars, for example, owe their solubility in water to the presence of hydroxyl groups.

The Carbonyl Group

The **carbonyl group** ($\rangle$CO) consists of a carbon atom joined to an oxygen atom by a double bond. If the carbonyl group is on the end of a carbon skeleton, the organic compound is called an **aldehyde**; otherwise the compound is called a

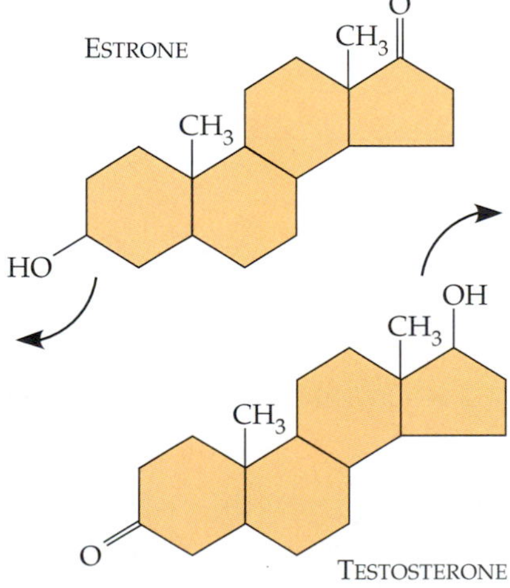

4.2 **FIGURE 4.8 • A comparison of functional groups of female (estrone) and male (testosterone) sex hormones.** The two molecules differ in the attachment of functional groups to a common carbon skeleton of four fused rings. The carbon skeleton has been simplified here by omitting the carbons in the rings, as well as their double bonds and hydrogens. The subtle variations in the molecular architecture of the molecules influence the development of the anatomical and physiological differences between female and male vertebrates. Compare, for example, the plumage of the female (left) and male (right) wood ducks.

Table 4.1 ■ Functional Groups of Organic Compounds

FUNCTIONAL GROUP	FORMULA	NAME OF COMPOUNDS	EXAMPLE
Hydroxyl	—OH	Alcohols	H—C(H)(H)—C(H)(H)—OH Ethanol (the drug of alcoholic beverages)
Carbonyl	—C(=O)—H	Aldehydes	H—C(H)(H)—C(H)(H)—C(=O)—H Propanal
	—C(=O)—	Ketones	H—C(H)(H)—C(=O)—C(H)(H)—H Acetone
Carboxyl	—C(=O)—OH (non-ionized) —C(=O)—O⁻ (ionized)	Carboxylic acids	H—C(H)(H)—C(=O)—OH Acetic acid* (the acid of vinegar)
Amino	—N(H)—H (non-ionized) —⁺N(H)(H)—H (ionized)	Amines	HO—C(=O)—C(H)(H)—N(H)—H Glycine* (an amino acid)
Sulfhydryl	—SH	Thiols	H—C(H)(H)—C(H)(H)—SH Ethanethiol
Phosphate	—O—P(=O)(O⁻)—O⁻	Organic phosphates	H—C(OH)(H)—C(OH)(H)—C(H)(H)—O—P(=O)(O⁻)—O⁻ Glycerol phosphate

*The ionized forms of the carboxyl and amino groups prevail in cells. However, acetic acid and glycine are represented here in their non-ionized forms.

ketone. The simplest ketone is acetone, which is three carbons long (see TABLE 4.1). Acetone has different properties from propanal, a three-carbon aldehyde (acetone and propanal are structural isomers). Thus, variation in locations of functional groups along carbon skeletons is a major source of molecular diversity.

The Carboxyl Group

When an oxygen atom is double-bonded to a carbon atom that is also bonded to a hydroxyl group, the entire assembly of atoms is called a **carboxyl group** (—COOH). Compounds containing carboxyl groups are known as **carboxylic acids**, or organic acids. The simplest is the one-carbon compound called formic acid (HCOOH), the substance some ants inject when they sting. Acetic acid, which has two carbons, gives vinegar its sour taste. (In general, acids, including carboxylic acids, taste sour.)

Why does a carboxyl group have acidic properties? A carboxyl group is a source of hydrogen ions. The covalent bond between the oxygen and the hydrogen is so polar that the hydrogen tends to dissociate reversibly from the molecule as an ion (H^+). In the case of acetic acid, we have:

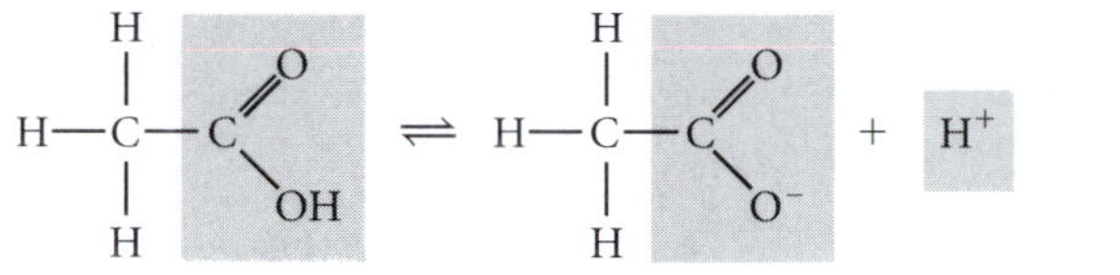

$$\mathrm{H{-}\underset{H}{\overset{H}{\underset{|}{\overset{|}{C}}}}{-}C{\genfrac{}{}{0pt}{}{\diagup\!\!O}{\diagdown OH}}} \rightleftharpoons \mathrm{H{-}\underset{H}{\overset{H}{\underset{|}{\overset{|}{C}}}}{-}C{\genfrac{}{}{0pt}{}{\diagup\!\!O}{\diagdown O^-}}} + \mathrm{H^+}$$

Acetic acid Acetate ion Hydrogen ion

Dissociation occurs as a result of the two electronegative oxygen atoms of the carboxyl group pulling shared electrons away from hydrogen. If the double-bonded oxygen and the hydroxyl group were attached to *separate* carbon atoms, there would be less tendency for the —OH group to dissociate because the second oxygen would be farther away. Here is another example of how emergent properties result from a specific arrangement of building components.

The Amino Group

The **amino group** (—NH_2) consists of a nitrogen atom bonded to two hydrogen atoms and to the carbon skeleton. Organic compounds with this functional group are called **amines**. An example is glycine, illustrated in TABLE 4.1. Because glycine *also* has a carboxyl group, it is both an amine and a carboxylic acid. Most of the cell's organic compounds have two or more different functional groups. Glycine belongs to a group of organic compounds named amino acids, which are the molecular building blocks of proteins.

The amino group acts as a base. You learned in Chapter 3 that ammonia (NH_3) can pick up a proton from the surrounding solution. Amino groups of organic compounds can do the same:

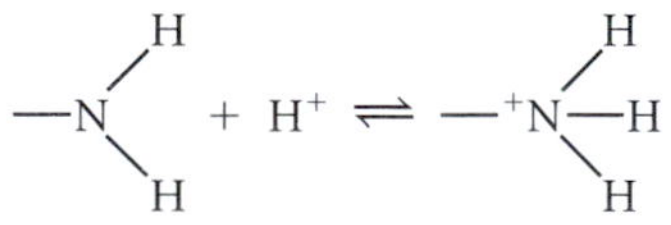

$$\mathrm{{-}N{\genfrac{}{}{0pt}{}{\diagup H}{\diagdown H}}} + \mathrm{H^+} \rightleftharpoons \mathrm{{-}{}^{+}N{\genfrac{}{}{0pt}{}{\diagup H}{\diagdown H}}{-}H}$$

This process gives the amino group a charge of +1, its most common state within the cell.

The Sulfhydryl Group

Sulfur is directly below oxygen in the periodic table; both have six valence electrons and form two covalent bonds. The organic functional group known as the **sulfhydryl group** (—SH), which consists of a sulfur atom bonded to an atom of hydrogen, resembles a hydroxyl group in shape (see TABLE 4.1). Organic compounds containing sulfhydryls are called **thiols**. In Chapter 5 you will learn how sulfhydryl groups can interact to help stabilize the intricate structure of many proteins.

The Phosphate Group

Phosphate is an anion formed by dissociation of an inorganic acid called phosphoric acid (H_3PO_4). The loss of hydrogen ions by dissociation leaves the phosphate with a negative charge. Organic compounds containing **phosphate groups** have a phosphate ion covalently attached by one of its oxygen atoms to the carbon skeleton (see TABLE 4.1). One function of phosphate groups is the transfer of energy between organic molecules. In Chapter 6 you will learn how cells harness the transfer of phosphate groups to perform work, such as the contraction of muscle cells.

The chemical elements of life: *a review*

Living matter, as you have learned, consists mainly of carbon, oxygen, hydrogen, and nitrogen, with smaller amounts of sulfur and phosphorus. These elements share the characteristic of forming strong covalent bonds, a quality that is essential in the architecture of complex organic molecules. Of all these elements, carbon is the virtuoso of the covalent bond. The chemical behavior of carbon makes it exceptionally versatile as a building block in molecular architecture: It can form four covalent bonds, link together into intricate molecular skeletons, and join with several other elements. The versatility of carbon makes possible the great diversity of organic molecules, each with special properties that emerge from the unique arrangement of its carbon skeleton and the functional groups appended to that skeleton. At the foundation of all biological diversity lies this variation at the molecular level.

Now that we have examined the basic architectural principles of organic compounds, we can move on to the next chapter, where we will explore the specific structures and functions of the large and complex molecules made by living cells: carbohydrates, lipids, proteins, and nucleic acids.

CHAPTER REVIEW

REVIEW OF KEY CONCEPTS

(with page numbers and key figures)

THE IMPORTANCE OF CARBON

- **Organic chemistry is the study of carbon compounds (pp. 48–49)** Organic compounds were once thought to arise only within living organisms, but this idea (vitalism) was disproved when chemists were able to synthesize organic compounds from inorganic ones.
- **Carbon atoms are the most versatile building blocks of molecules (pp. 49–51, FIGURE 4.2)** A covalent bonding capacity of four contributes to carbon's ability to form diverse molecules. Carbon can bond to a variety of atoms, including O, H, N, and S. Carbon atoms can also bond to other carbons, forming the carbon skeletons of organic compounds.
- 4.1 **Variation in carbon skeletons contributes to the diversity of organic molecules (pp. 51–53, FIGURE 4.4)** The carbon skeletons of organic molecules vary in length and shape and have bonding sites for atoms of other elements. Hydrocarbons consist only of carbon and hydrogen. Carbon's versatile bonding is the basis for isomers, molecules with the same molecular formula but different structures and thus different properties. Three types of isomers are structural isomers, geometric isomers, and enantiomers.

FUNCTIONAL GROUPS

- 4.2 **Functional groups also contribute to the molecular diversity of life (pp. 53–55, TABLE 4.1)** Functional groups are specific, chemically reactive groups of atoms within organic molecules that give the overall molecule distinctive chemical properties. The hydroxyl group (—OH), found in alcohols, has a polar covalent bond, which helps alcohols dissolve in water. The carbonyl group (>CO) can be either at the end of a carbon skeleton (aldehyde) or within the skeleton (ketone). The carboxyl group (—COOH) is found in carboxylic acids. The hydrogen of this group can dissociate, making the molecule a weak acid. The amino group ($—NH_2$) can accept an H^+, thereby acting as a base. The sulfhydryl group (—SH) helps stabilize the structure of some proteins. The phosphate group can bond to a carbon skeleton by one of its oxygen atoms and has an important role in the transfer of cellular energy.
- **The chemical elements of life: *a review* (p. 55)** Living matter is made mostly of carbon, oxygen, hydrogen, and nitrogen, with some sulfur and phosphorus. Biological diversity has its molecular basis in carbon's ability to form a huge number of molecules with particular shapes and chemical properties.

SELF-QUIZ

1. Organic chemistry is currently defined as
   a. the study of compounds that can be made only by living cells
   b. the study of carbon compounds
   c. the study of vital forces
   d. the study of natural (as opposed to synthetic) compounds
   e. the study of hydrocarbons

2. Choose the pair of terms that correctly completes this sentence: Hydroxyl is to ________ as ________ is to aldehyde.
   a. carbonyl; ketone
   b. oxygen; carbon
   c. alcohol; carbonyl
   d. amine; carboxyl
   e. alcohol; ketone

3. Which of the following hydrocarbons has a double bond in its carbon skeleton?
   a. C_3H_8
   b. C_2H_6
   c. CH_4
   d. C_2H_4
   e. C_2H_2

4. The gasoline consumed by an automobile is a fossil fuel consisting mostly of
   a. aldehydes
   b. amino acids
   c. alcohols
   d. hydrocarbons
   e. thiols

5. Choose the term that correctly describes the relationship between these two sugar molecules:

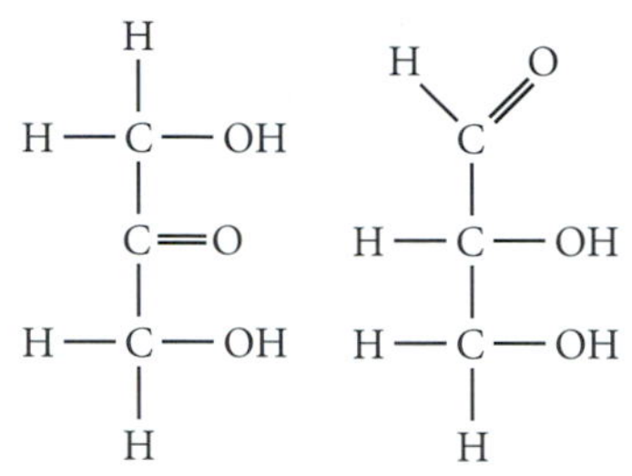

   a. structural isomers
   b. geometric isomers
   c. enantiomers
   d. isotopes

6. Identify the asymmetric carbon in this molecule:

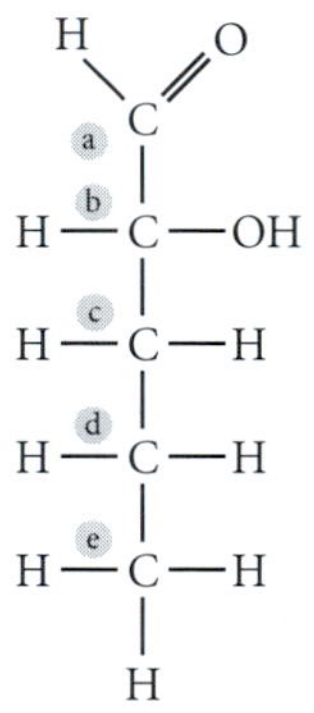

7. Which functional group is *not* present on this molecule?

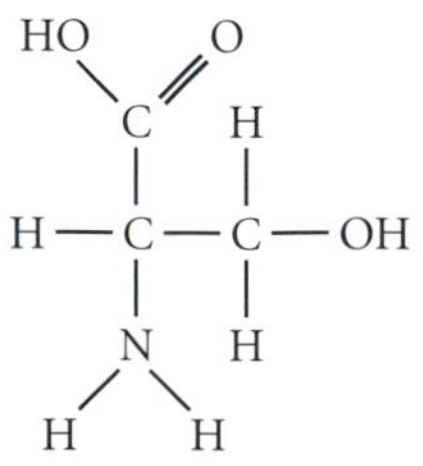

   a. carboxyl
   b. sulfhydryl
   c. hydroxyl
   d. amino

8. Which action could produce a carbonyl group?
   a. the removal of the hydroxyl from a carboxyl
   b. the addition of a thiol to a hydroxyl
   c. the addition of a hydroxyl to a phosphate
   d. the replacement of the nitrogen of an amine with oxygen
   e. the addition of a sulfhydryl to a carboxyl

9. Which functional group is most responsible for some organic molecules behaving as bases?
   a. hydroxyl
   b. carbonyl
   c. carboxyl
   d. amino
   e. phosphate

10. Which of the following molecules would be the strongest acid? Explain your answer.

a. 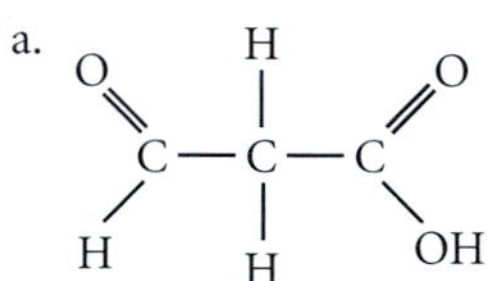

c. 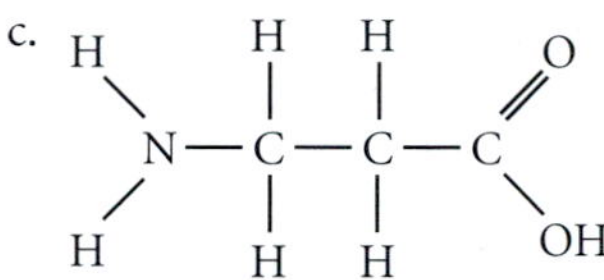

b. 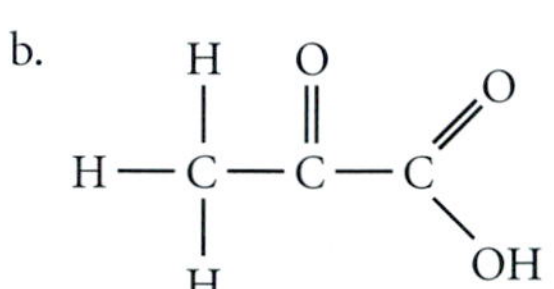

d. 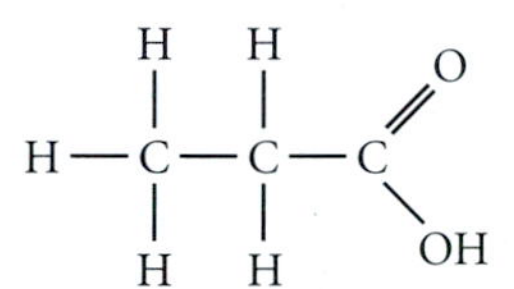

CHALLENGE QUESTIONS

1. Draw an organic molecule having all six functional groups described in this chapter.

2. Draw three structural isomers of the hydrocarbon pentane (C_5H_{12}).

3. Alice, in Lewis Carroll's classic *Alice in Wonderland,* poses this question: "Is looking-glass milk good to drink?" Respond, and justify your answer based on what you have learned about the structure of organic molecules, the importance of isomers, and the biological relevance of molecular shape.

SCIENCE, TECHNOLOGY, AND SOCIETY

1. The role of the Food and Drug Administration (FDA) in monitoring the testing of new drugs continues to be controversial. The thalidomide tragedy (p. 52) is often cited by those who argue that the FDA should not approve any drug until no doubt remains that the drug can be safely prescribed. Others would like to see FDA standards less rigid for drugs that may help people suffering from terminal diseases, such as AIDS. Where do you stand on this issue? Defend your position.

2. Each year, industrial chemists synthesize and test thousands of new organic compounds for use as insecticides and weed killers. In what ways are these chemicals useful and important to us? In what ways can they be harmful? What influences have shaped your opinions about these chemicals?

FURTHER READING

Asimov, I. *The World of Carbon,* 2nd ed. New York: Macmillan, 1962. A primer on the basics of organic chemistry by one of America's most popular science writers.

Bada, J. L. "Extraterrestrial Handedness?" *Science,* February 14, 1997. A possible explanation for why cells use certain enantiomers exclusively.

Bradley, D. "Frog Venom Cocktail Yields a One-Handed Painkiller." *Science,* August 27, 1993. Organic chemists synthesize a new drug based on the toxic compound secreted by the "poison arrow frog."

Bradley, D. "A New Twist in the Tale of Nature's Asymmetry." *Science,* May 13, 1994. The pharmaceutical importance of isomers.

Kessler, D. A., and K. L. Feiden. "Faster Evaluation of Vital Drugs." *Scientific American,* March 1995. When is it ethical to trade safety for speed in order to make new drugs available?

Ourisson, G., P. Albrecht, and M. Rohmer. "The Microbial Origin of Fossil Fuels." *Scientific American,* August 1984. The biology behind our important energy resources.

WEB LINKS

Visit the special edition of *The Biology Place* for BIOLOGY, Fifth Edition, at http://www.biology.com/campbell. Go to Chapter 4 for online resources, including learning activities, practice exams, and links to the following web sites:

"IUPAC Nomenclature of Organic Chemistry"
Exhaustive lists, descriptions, and explanations of correct IUPAC nomenclature for chemical compounds.

"Organic Chemistry from BioChemNet"
Comprehensive list of tutorial, reference, and other organic chemistry educational resources. Some sites offer free downloadable demo software.

"Basic Reactions in Organic Chemistry—Interactive"
An interesting virtual reality site that features a number of three-dimensional, animated, basic organic chemistry reactions.

"MathMol"
Good introductory site for those interested in molecular modeling. Numerous links to freely downloadable public domain software.

THE STRUCTURE AND FUNCTION OF MACROMOLECULES

We have applied the concept of emergent properties to our study of water and relatively simple organic molecules. These substances are central to life, each one having unique behavior arising from the orderly arrangement of its atoms. Another level in the hierarchy of biological organization is attained when cells join together small organic molecules to form larger molecules that belong to four classes: carbohydrates, lipids, proteins, and nucleic acids. Many of these cellular molecules are, on the molecular scale, huge. For example, a protein may consist of thousands of covalently connected atoms that form a molecular colossus weighing over 100,000 daltons. Biologists use the term ***macromolecule*** *for such giant molecules.*

*Considering the size and complexity of macromolecules, it is remarkable that biochemists have determined the detailed structures of so many of them (*FIGURE 5.1*). The architecture of a macromolecule helps explain how that molecule works. For example, the structure of the silk protein from which the orb spider in the photo on this page weaves its web gives the fibers their strength and springiness. Proteins and the other large molecules of life are the main subject of this chapter. For these molecules, as at all levels in the biological hierarchy, form and function are inseparable.*

POLYMER PRINCIPLES

In examining the relationship between the structure and function of life's macromolecules, we begin with a key generalization about how cells build these large molecules from smaller ones.

Most macromolecules are polymers

The large molecules in three of the four classes of life's organic compounds—carbohydrates, proteins, and nucleic acids—are chainlike molecules called polymers (from the Greek *polys*, "many," and *meris*, "part"). A **polymer** is a long molecule consisting of many identical or similar building blocks linked by covalent bonds, much as a train consists of a chain of cars. The repeating units that serve as the building blocks of a polymer are small molecules called **monomers**. Some of the molecules that serve as monomers also have other functions of their own.

The classes of polymeric macromolecules differ in the nature of their monomers, but the chemical mechanisms that cells use to make and break polymers are basically the same in all cases (FIGURE 5.2). Monomers are connected by a reaction in which two molecules are covalently bonded to each other through loss of a water molecule; this is called a **condensation reaction**, specifically a **dehydration reaction**, because the molecule lost is water (FIGURE 5.2a). When a bond forms between two monomers, each monomer contributes part of the water molecule that is lost: One molecule provides a

(a)

(b)

FIGURE 5.1 ▪ Building models to study the structure and function of macromolecules. (a) Linus Pauling (1901–1994), here with a model of part of a protein. In the 1950s, Pauling discovered several of the basic structural features of proteins. **(b)** Today, scientists use computers to help build molecular models. Though methods have improved, the goal remains to correlate the structure of macromolecules with their functions.

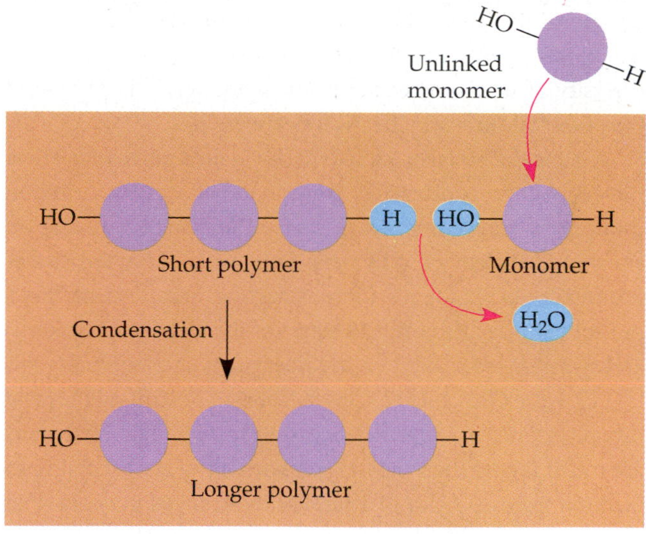

(a) Condensation (dehydration) synthesis of a polymer

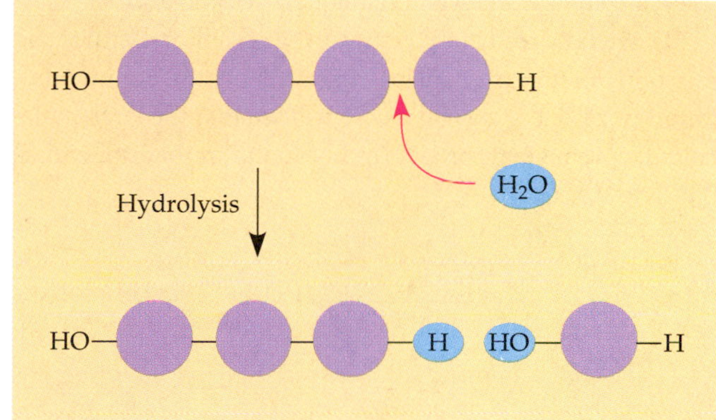

(b) Hydrolysis of a polymer

FIGURE 5.2 ▪ The synthesis and breakdown of polymers. (a) Monomers are joined by condensation reactions (dehydration reactions). The net effect is the removal of a water molecule. **(b)** The reverse of this process, hydrolysis, breaks bonds between monomers by adding water molecules.

hydroxyl group (—OH), while the other provides a hydrogen (—H). To make a polymer, this reaction is repeated as monomers are added to the chain one by one. The cell must expend energy to carry out these condensation reactions, and the process occurs only with the help of enzymes, specialized proteins that speed up chemical reactions in cells.

Polymers are disassembled to monomers by **hydrolysis**, a process that is essentially the reverse of the dehydration reaction (FIGURE 5.2b). Hydrolysis means to break with water (from the Greek *hydro*, "water," and *lysis*, "break"). Bonds between monomers are broken by the addition of water molecules, a hydrogen from the water attaching to one monomer, and a hydroxyl attaching to the adjacent monomer. An example of hydrolysis working in our bodies is the process of digestion. The bulk of the organic material in our food is in the form of polymers that are much too large to enter our cells. Within the digestive tract, various enzymes attack the polymers, speeding up hydrolysis. The released monomers are then absorbed into the bloodstream for distribution to all body cells. Those cells can then use dehydration reactions to assemble the monomers into new polymers that differ from the ones that were digested.

An immense variety of polymers can be built from a small set of monomers

Each cell has thousands of different kinds of macromolecules; the collection varies from one type of cell to another in the same organism. The inherent differences between human siblings reflect variations in polymers, particularly DNA and proteins. Molecular differences between unrelated individuals are more extensive, and between species greater still. The diversity of macromolecules in the living world is vast—and the potential variety is effectively limitless.

What is the basis for such diversity in life's polymers? These molecules are constructed from only 40 to 50 common monomers and some others that occur rarely. Building an enormous variety of polymers from such a limited list of monomers is analogous to constructing hundreds of thousands of words from only 26 letters of the alphabet. The key is arrangement—variation in the linear sequence the units follow. However, this analogy falls far short of describing the great diversity of macromolecules, because most biological polymers are much longer than the longest word. Proteins, for example, are built from 20 kinds of amino acids arranged in chains that are typically hundreds of amino acids long. The molecular logic of life is simple but elegant: Small molecules common to all organisms are ordered into unique macromolecules.

We are now ready to investigate the specific structures and functions of the four major classes of organic compounds found in cells. For each class, we will see that the large molecules have emergent properties not found in their individual building blocks.

CARBOHYDRATES—FUEL AND BUILDING MATERIAL

5.1 The term **carbohydrates** includes both sugars and their polymers. The simplest carbohydrates are the monosaccharides, single sugars also known as simple sugars (FIGURE 5.3). Disaccharides are double sugars, consisting of two monosaccharides joined by condensation. The carbohydrates that are macromolecules are polysaccharides, polymers of many sugars.

Sugars, the smallest carbohydrates, serve as fuel and carbon sources

Monosaccharides (from the Greek *monos,* "single," and *sacchar,* "sugar") generally have molecular formulas that are some multiple of CH_2O (see FIGURE 5.3). Glucose ($C_6H_{12}O_6$), the most common monosaccharide, is of central importance in the chemistry of life. In the structure of glucose, we can see all the trademarks of a sugar. A hydroxyl group is attached to each carbon except one, which is double-bonded to an oxygen to form a carbonyl group. Depending on the location of the carbonyl group, a sugar is either an aldose (aldehyde sugar) or a ketose (ketone sugar). Glucose, for example, is an aldose;

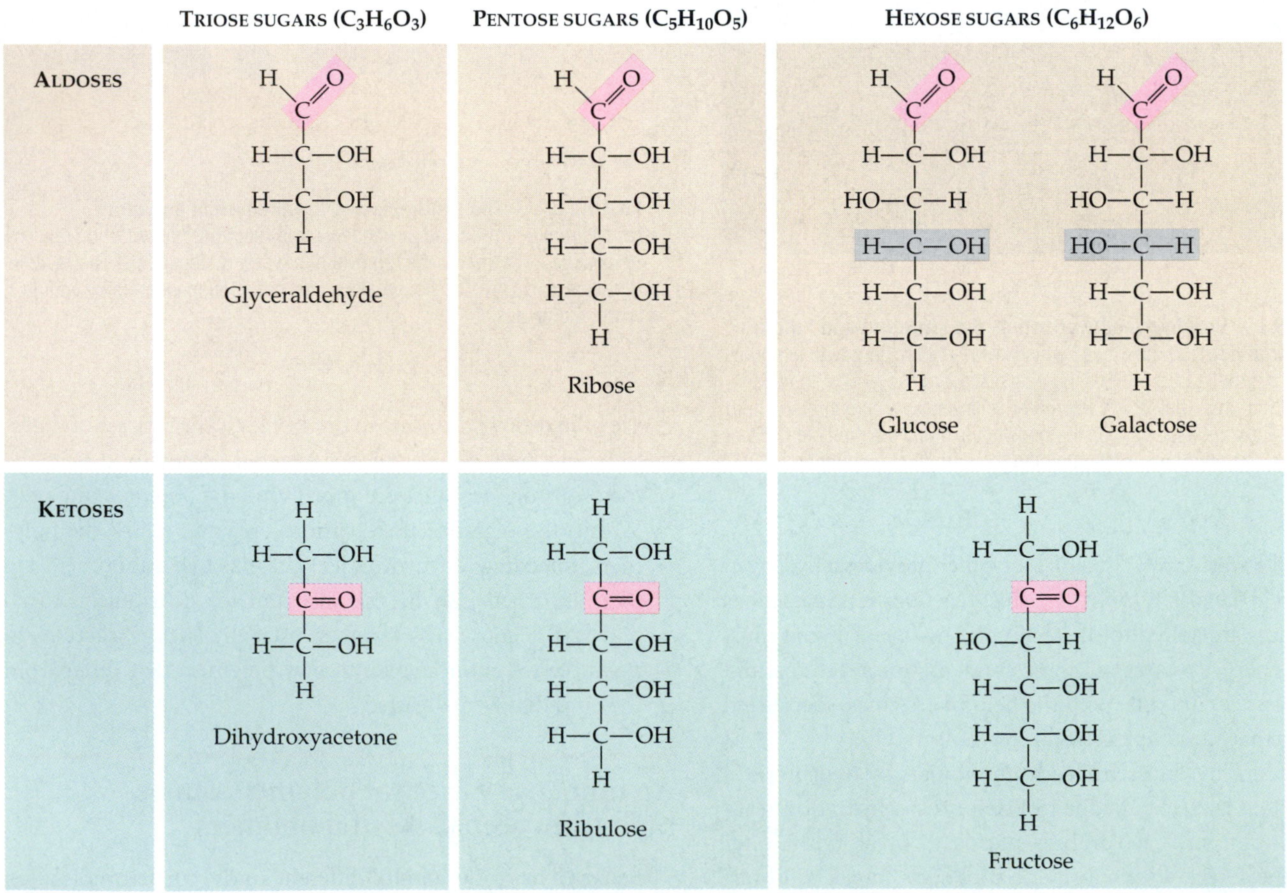

	TRIOSE SUGARS ($C_3H_6O_3$)	PENTOSE SUGARS ($C_5H_{10}O_5$)	HEXOSE SUGARS ($C_6H_{12}O_6$)	
ALDOSES	H–C=O H—C—OH H—C—OH H Glyceraldehyde	H–C=O H—C—OH H—C—OH H—C—OH H—C—OH H Ribose	H–C=O H—C—OH HO—C—H H—C—OH H—C—OH H—C—OH H Glucose	H–C=O H—C—OH HO—C—H HO—C—H H—C—OH H—C—OH H Galactose
KETOSES	H H—C—OH C=O H—C—OH H Dihydroxyacetone	H H—C—OH C=O H—C—OH H—C—OH H—C—OH H Ribulose	H H—C—OH C=O HO—C—H H—C—OH H—C—OH H—C—OH H Fructose	

FIGURE 5.3 ▪ The structure and classification of some monosaccharides. Sugars may be aldoses (aldehyde sugars) or ketoses (ketone sugars), depending on the location of the carbonyl group (pink). Sugars are also classified according to the length of their carbon skeletons. A third point of variation is in the spatial arrangement around asymmetric carbons (compare, for example, the gray portions of glucose and galactose).

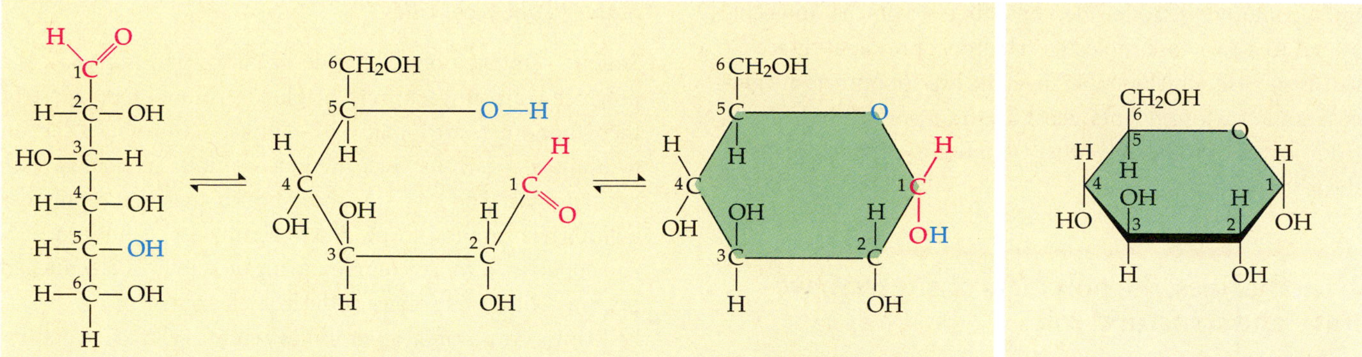

(a) Linear and ring forms

(b) Abbreviated ring structure

FIGURE 5.4 ▪ Linear and ring forms of glucose. **(a)** Chemical equilibrium between the linear and ring structures greatly favors the formation of rings. To form the glucose ring, carbon 1 bonds to the oxygen attached to carbon 5. **(b)** In this abbreviated ring formula, the carbons in the ring are omitted. The ring's thicker edge indicates that you are looking at the ring edge-on; the components attached to the ring lie above or below the plane of the ring.

fructose, a structural isomer of glucose, is a ketose. (Most names for sugars end in *-ose.*) Another criterion for classifying sugars is the size of the carbon skeleton, which ranges from three to seven carbons long. Glucose, fructose, and other sugars that have six carbons are called hexoses. Trioses and pentoses are also common.

Still another source of diversity for simple sugars is in the spatial arrangement of their parts around asymmetric carbons. (Recall from Chapter 4 that an asymmetric carbon is one attached to four different kinds of covalent partners.) Glucose and galactose, for example, differ only in the placement of parts around one asymmetric carbon (see the gray boxes in FIGURE 5.3). What may seem at first a small difference is significant enough to give the two sugars distinctive shapes and behaviors.

Although it is convenient to draw glucose with a linear carbon skeleton, this representation is not accurate. In aqueous solutions, glucose molecules, as well as most other sugars, form rings (FIGURE 5.4).

Monosaccharides, particularly glucose, are major nutrients for cells. In the process known as cellular respiration, cells extract the energy stored in glucose molecules. Not only are simple sugar molecules a major fuel for cellular work, but their carbon skeletons serve as raw material for the synthesis of other types of small organic molecules, including amino acids and fatty acids. Sugar molecules that are not immediately used in these ways are generally incorporated as monomers into disaccharides or polysaccharides.

A **disaccharide** (FIGURE 5.5) consists of two monosaccharides joined by a **glycosidic linkage**, a covalent bond formed between two monosaccharides by a dehydration reaction. For example, maltose is a disaccharide formed by linking two molecules of glucose. Also known as malt sugar, maltose is an ingredient for brewing beer. Lactose, the sugar present in

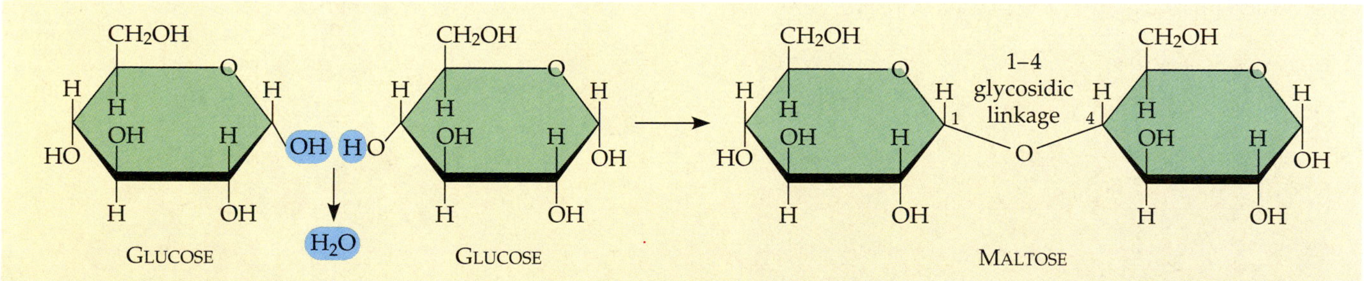

(a) Condensation synthesis of maltose

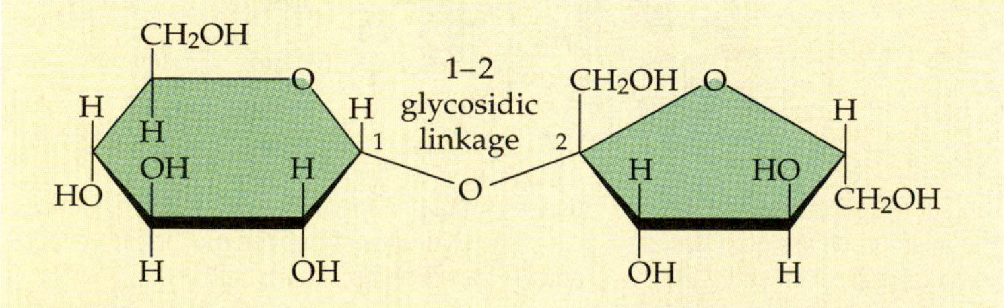

(b) Sucrose

FIGURE 5.5 ▪ Examples of disaccharides. **(a)** The bonding of two glucose units forms maltose. The glycosidic link joins the number 1 carbon of one glucose to the number 4 carbon of the second glucose. Joining the glucose monomers in a different way would result in a different disaccharide. **(b)** Sucrose is a disaccharide formed from glucose and fructose. Notice that fructose, though a hexose like glucose, forms a five-sided ring.

milk, is another disaccharide, consisting of a glucose molecule joined to a galactose molecule. The most prevalent disaccharide is sucrose, which is table sugar. Its two monomers are glucose and fructose. Plants generally transport carbohydrates from leaves to roots and other nonphotosynthetic organs in the form of sucrose.

Polysaccharides, the polymers of sugars, have storage and structural roles

Polysaccharides are macromolecules, polymers with a few hundred to a few thousand monosaccharides joined by glycosidic linkages. Some polysaccharides serve as storage material, hydrolyzed as needed to provide sugar for cells. Other polysaccharides serve as building material for structures that protect the cell or the whole organism. The architecture and function of a polysaccharide are determined by its sugar monomers and by the positions of its glycosidic linkages.

Storage Polysaccharides

Starch, a storage polysaccharide of plants (FIGURE 5.6a), is a polymer consisting entirely of glucose monomers. Most of these monomers are joined by 1–4 linkages (number 1 carbon to number 4 carbon), like the glucose units in maltose (see FIGURE 5.5a). The angle of these bonds makes the polymer helical. The simplest form of starch, amylose, is unbranched. Amylopectin, a more complex form of starch, is a branched polymer with 1–6 linkages at the branch points.

Plants store starch as granules within cellular structures called plastids, including chloroplasts (see FIGURE 5.6a). By synthesizing starch, the plant can stockpile surplus sugar. Because glucose is a major cellular fuel, starch represents stored energy. The sugar can later be withdrawn from this carbohydrate bank by hydrolysis, which breaks the bonds between the glucose monomers. Most animals, including humans, also have enzymes that can hydrolyze plant starch, making glucose available as a nutrient for cells. Potato tubers

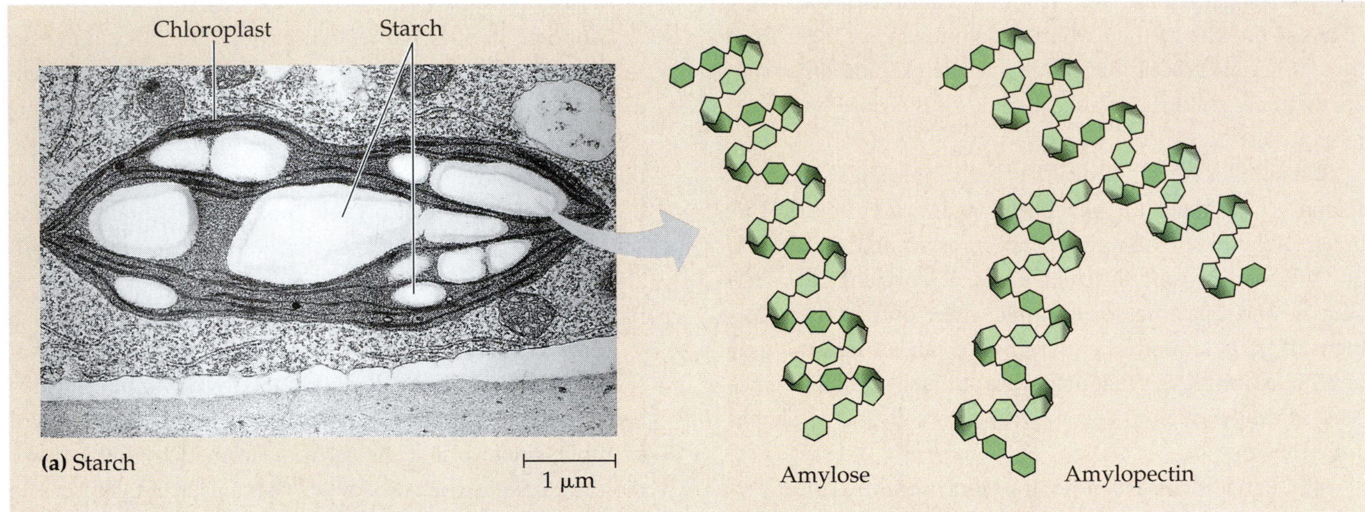

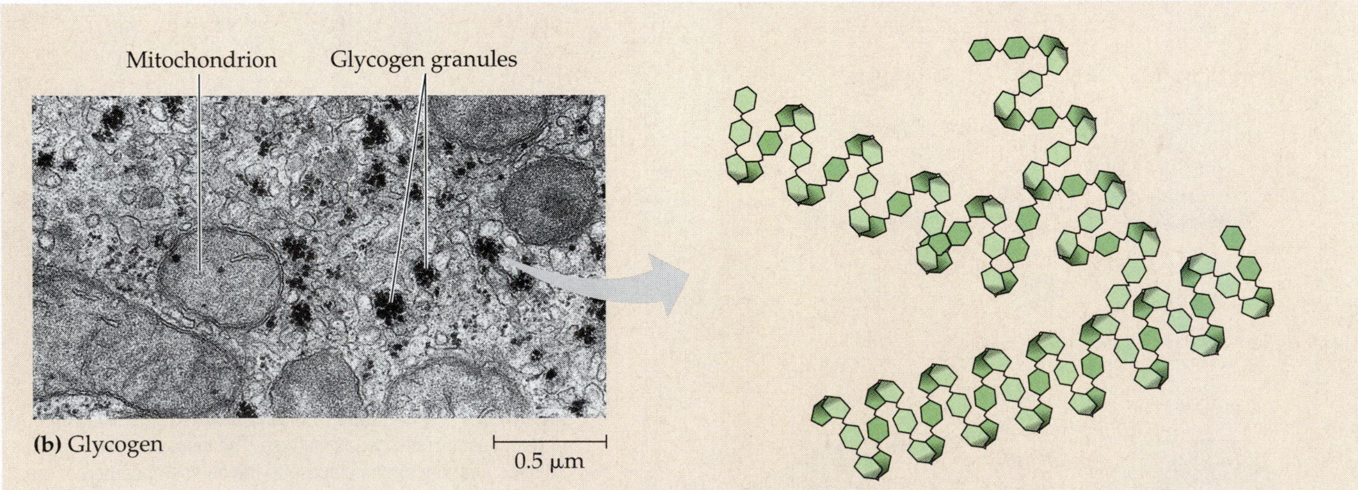

FIGURE 5.6 • Storage polysaccharides. These examples, starch and glycogen, are composed entirely of glucose monomers, abbreviated here as hexagons. The polymer chains spiral to form helices. **(a)** Two forms of starch are amylose (unbranched) and amylopectin (branched). The light ovals in the micrograph are granules of starch within a chloroplast of a plant cell (TEM). **(b)** Glycogen is more extensively branched than amylopectin. Animal cells stockpile glycogen as dense clusters of granules within liver and muscle cells. Hydrolysis frees the glucose from storage (TEM, portion of a liver cell).

and grains—the fruits of wheat, corn, rice, and other grasses—are the major sources of starch in the human diet.

Animals store a polysaccharide called **glycogen**, a polymer of glucose that is like amylopectin but more extensively branched (FIGURE 5.6b). Humans and other vertebrates store glycogen mainly in liver and muscle cells. Hydrolysis of glycogen in these cells releases glucose when the demand for sugar increases. This stored fuel cannot sustain an animal for long, however. In humans, for example, the glycogen bank is depleted in about a day unless it is replenished by consumption of food.

Structural Polysaccharides

Organisms build strong materials from structural polysaccharides. For example, the polysaccharide called **cellulose** is a major component of the tough walls that enclose plant cells. On a global scale, plants produce almost 10^{11} (100 billion) tons of cellulose per year; it is the most abundant organic compound on Earth. Like starch, cellulose is a polymer of glucose, but the glycosidic linkages in these two polymers differ. The difference is based on the fact that there are actually two, slightly different ring structures for glucose. When glucose forms a ring, the hydroxyl group attached to the number 1 carbon is locked into one of two alternative positions: lying either below or above the plane of the ring. These two ring forms for glucose are called alpha (α) and beta (β), respectively (FIGURE 5.7a). In starch, all the glucose monomers are in the α configuration (FIGURE 5.7b), the arrangement we saw in FIGURES 5.4 and 5.5. In contrast, the glucose monomers of cellulose are all in the β configuration, making every other glucose monomer upside down with respect to the others (FIGURE 5.7c).

The differing glycosidic links in starch and cellulose give the two molecules distinct three-dimensional shapes. Whereas a starch molecule is helical, a cellulose molecule is straight (and never branched), and its hydroxyl groups are free to hydrogen-bond with the hydroxyls of other cellulose molecules lying parallel to it. In plant cell walls, many parallel cellulose molecules, held together in this way, are grouped into units called microfibrils (FIGURE 5.8, p. 64). These cables are a strong building material for plants—as well as for humans, who use wood, which is rich in cellulose, for lumber.

Enzymes that digest starch by hydrolyzing its α linkages are unable to hydrolyze the β linkages of cellulose. In fact, few organisms possess enzymes that can digest cellulose. Humans do not; the cellulose fibrils in our food pass through the digestive tract and are eliminated with the feces. Along the way, the fibrils abrade the wall of the digestive tract and stimulate the lining to secrete mucus, which aids in the smooth passage of food through the tract. Thus, although cellulose is not a nutrient for humans, it is an important part of a healthful diet. Most fresh fruits, vegetables, and grains are rich in cellulose, or fiber.

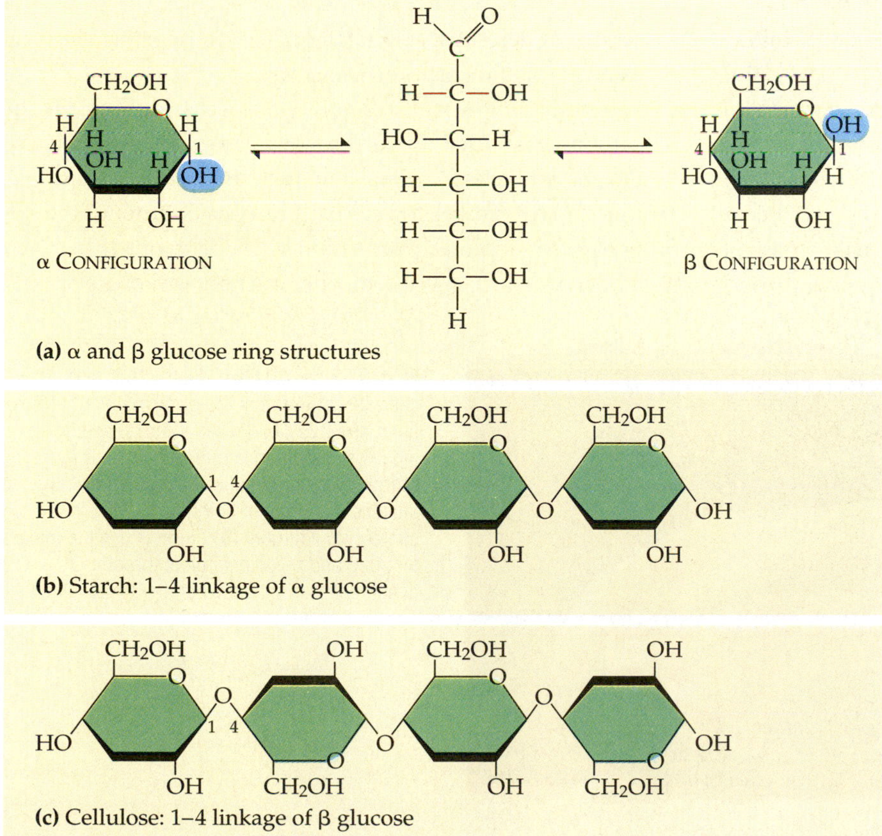

FIGURE 5.7 ▪ Starch and cellulose structures compared. **(a)** Glucose forms two interconvertible ring structures, designated α and β. These two forms differ in the placement of the hydroxyl group attached to the number 1 carbon. **(b)** The α ring form is the monomer for starch. **(c)** Cellulose consists of glucose monomers in the β configuration. The angles of the bonds that link the rings make every other glucose monomer "upside down."

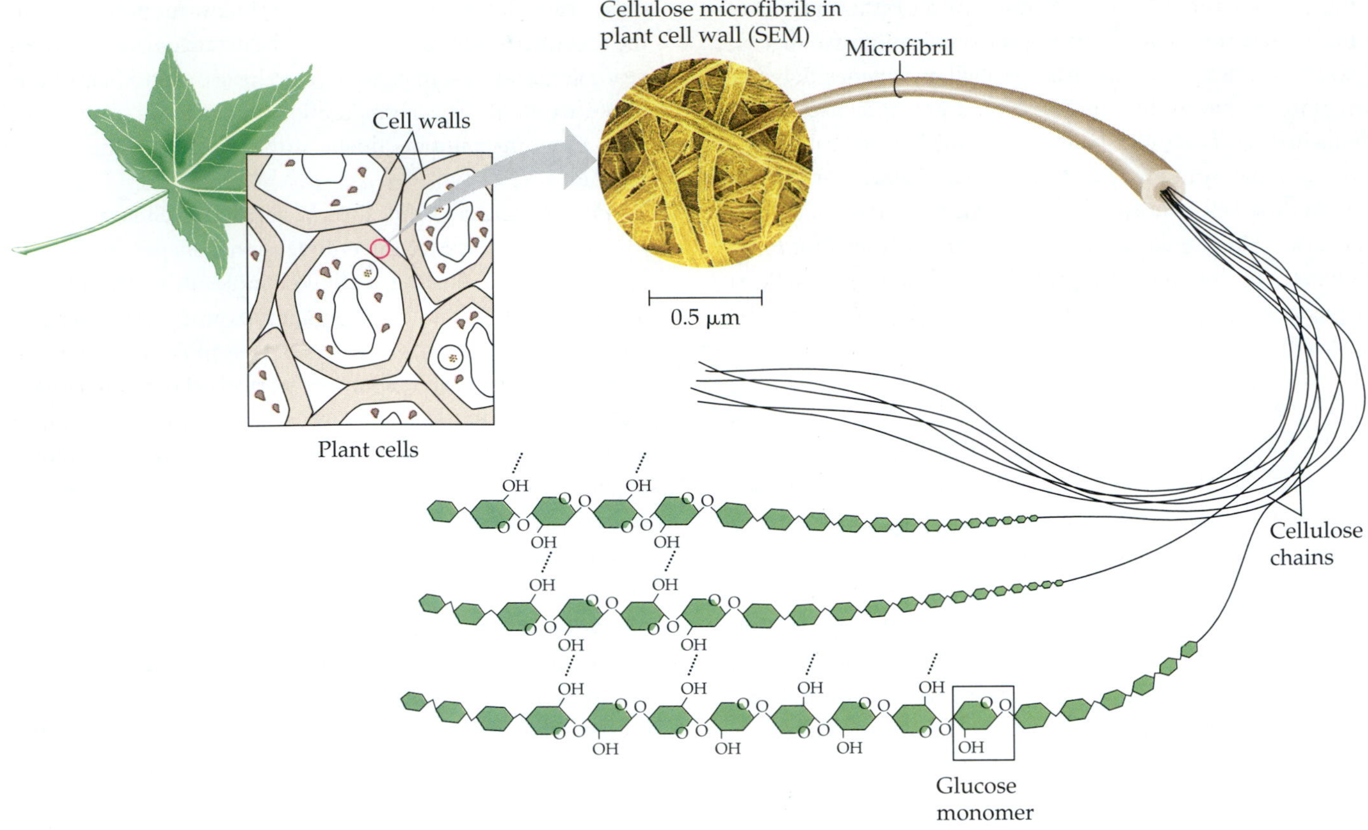

FIGURE 5.8 ▪ The arrangement of cellulose in plant cell walls. Cellulose is an unbranched polysaccharide. Parallel cellulose molecules are held together by hydrogen bonds (dotted lines) between hydroxyl groups projecting from both sides. About 80 cellulose molecules associate to form a microfibril, the main architectural unit of the plant cell wall.

Some bacteria and other microbes can digest cellulose, breaking it down to glucose monomers. A cow harbors cellulose-digesting bacteria in the rumen, the first compartment in its stomach. The bacteria hydrolyze the cellulose of hay and grass and convert the glucose to other nutrients that nourish the cow. Similarly, a termite, which is unable to digest cellulose for itself, has microbes living in its gut that can make a meal of wood. Some molds (fungi) can also digest cellulose, thereby serving as decomposers that are crucial in recycling chemical elements within Earth's ecosystems.

Another important structural polysaccharide is **chitin**, the carbohydrate used by arthropods (insects, spiders, crustaceans, and related animals) to build their exoskeletons (FIGURE 5.9). An exoskeleton is a hard case that surrounds the soft parts of the animal. Pure chitin is leathery, but it becomes hardened when encrusted with calcium carbonate, a salt.

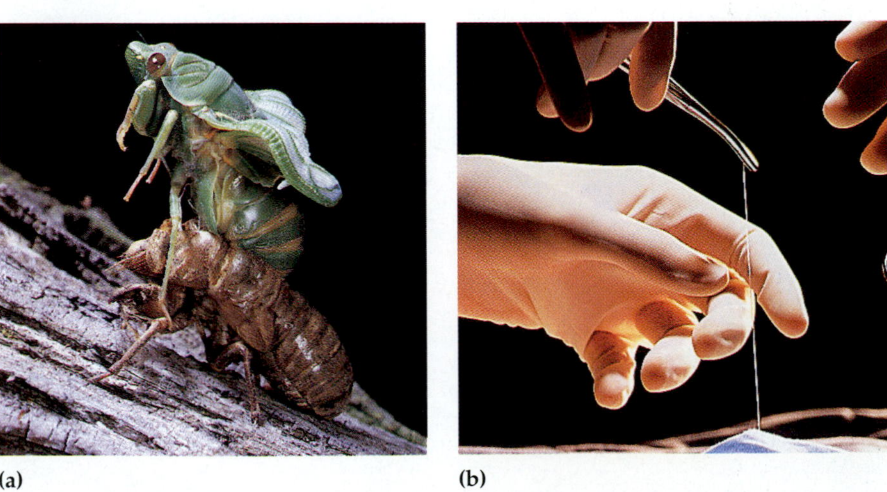

FIGURE 5.9 ▪ Chitin. A structural polysaccharide, chitin forms the exoskeleton of arthropods. **(a)** This cicada is molting, shedding its old exoskeleton and emerging in adult form. Chitin is similar to cellulose, except that the glucose monomer of chitin has a nitrogen-containing addition. **(b)** Chitin is used to make a strong and flexible surgical thread that decomposes after the wound or incision heals.

Chitin is also found in many fungi, which use this polysaccharide rather than cellulose as the building material for their cell walls. The monomer of chitin is a glucose molecule with a nitrogen-containing appendage:

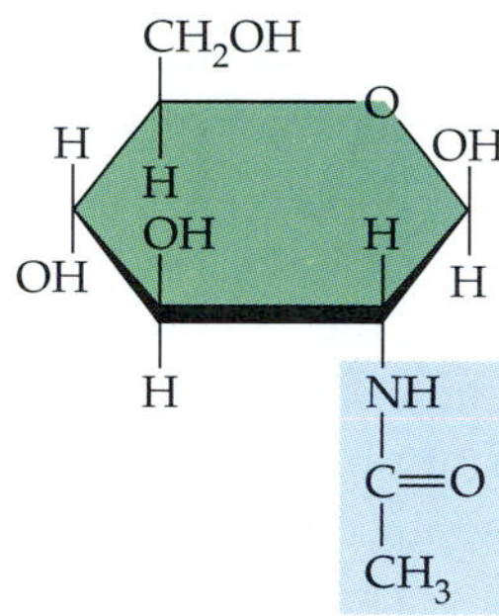

LIPIDS—DIVERSE HYDROPHOBIC MOLECULES

Lipids are the one class of large biological molecules that does
5.2 not include polymers. The compounds called **lipids** are grouped together because they share one important trait: They have little or no affinity for water. The hydrophobic behavior of lipids is based on their molecular structure. Although they may have some polar bonds associated with oxygen, lipids consist mostly of hydrocarbons. Smaller than true (polymeric) macromolecules, lipids are a highly varied group in both form and function. Lipids include waxes and certain pigments, but we will focus on the most important families of lipids: the fats, phospholipids, and steroids.

Fats store large amounts of energy

Although fats are not polymers, they are large molecules, and they are assembled from smaller molecules by dehydration reactions. A **fat** is constructed from two kinds of smaller molecules: glycerol and fatty acids (FIGURE 5.10). Glycerol is an alcohol with three carbons, each bearing a hydroxyl group. A **fatty acid** has a long carbon skeleton, usually 16 or 18 carbon atoms in length. At one end of the fatty acid is a "head" consisting of a carboxyl group, the functional group that gives these molecules the name fatty *acids*. Attached to the carboxyl group is a long hydrocarbon "tail." The nonpolar C—H bonds in the tails of fatty acids are the reason fats are hydrophobic. Fats separate from water because the water molecules hydrogen-bond to one another and exclude the fats. A common example of this phenomenon is the separation of

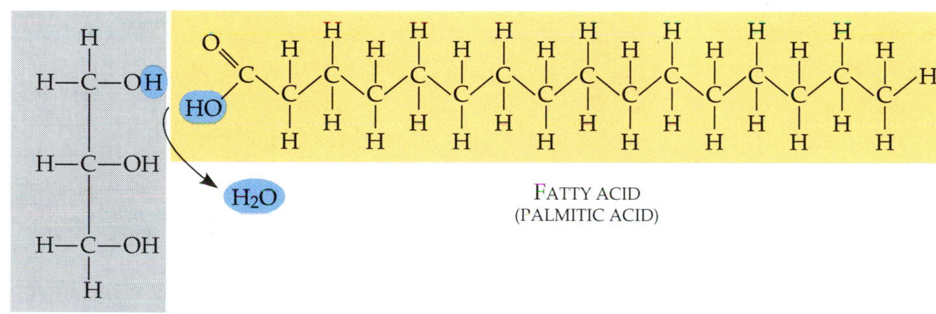

(a) Dehydration synthesis

Ester linkage

(b) Fat molecule

FIGURE 5.10 ▪ The synthesis and structure of a fat, or triacylglycerol. The molecular building blocks of a fat are one molecule of glycerol and three molecules of fatty acids. **(a)** One water molecule is removed for each fatty acid joined to the glycerol. **(b)** The result is a fat. Although the fat shown here has three identical fatty acid units, other fats have two or even three different kinds of fatty acids.

vegetable oil (a liquid fat) from the aqueous vinegar solution in a bottle of salad dressing.

In making a fat, three fatty acids each join to glycerol by an ester linkage, a bond between a hydroxyl group and a carboxyl group. The resulting fat, also called a **triacylglycerol**, thus consists of three fatty acids linked to one glycerol molecule. (Still another name for a fat is triglyceride, a word often found in the list of ingredients on packaged foods.) The fatty acids in a fat can be the same (see FIGURE 5.10b), or they can be of two or three different kinds.

Fatty acids vary in length and in the number and locations of double bonds. The terms *saturated fats* and *unsaturated fats* are commonly used in the context of nutrition. These terms refer to the structure of the hydrocarbon tails of the fatty acids. If there are no double bonds between the carbon atoms composing the tail, then as many hydrogen atoms as possible are bonded to the carbon skeleton, creating a **saturated fatty acid** (FIGURE 5.11a). An **unsaturated fatty acid** has one or more double bonds, formed by the removal of hydrogen atoms from the carbon skeleton. The fatty acid will have a kink in its shape wherever a double bond occurs (FIGURE 5.11b).

Most animal fats are saturated: The tails of their fatty acids lack double bonds. Saturated animal fats—such as lard and butter—are solid at room temperature. In contrast, the fats of plants and fishes are generally unsaturated. Usually liquid at room temperature, plant and fish fats are referred to as oils—for instance, corn oil and cod liver oil. The kinks where the double bonds are located prevent the molecules from packing together closely enough to solidify at room temperature. The phrase "hydrogenated vegetable oils," often found on food labels, means that unsaturated fats have been synthetically converted to saturated fats by adding hydrogen. Peanut butter, margarine, and many other products are hydrogenated to prevent lipids from separating out in liquid (oil) form.

A diet rich in saturated fats is one of several factors that may contribute to the human cardiovascular disease known as atherosclerosis. In this condition, deposits called plaques develop on the internal lining of blood vessels, impeding blood flow and reducing the resilience of the vessels.

Fat has come to have such a negative connotation in our culture that you might wonder whether fats serve any useful purpose. The major function of fats is energy storage. The hydrocarbon chains of fats are similar to gasoline molecules and are just as rich in energy. A gram of fat stores more than twice as much energy as a gram of a polysaccharide, such as starch. Because plants are relatively immobile, they can function with bulky energy storage in the form of starch. (Vegetable oils are generally obtained from seeds, where more compact storage is an asset to the plant.) Animals, on the other hand, must carry their energy stores with them, so there is an advantage to having a more compact reservoir of fuel—fat. Humans and other mammals stock their long-term food reserves in adipose cells (see FIGURE 4.5), which swell and shrink as fat is deposited and withdrawn from storage. In addition to storing energy, adipose tissue also cushions such vital organs as the kidneys, and a layer of fat beneath the skin insulates the body. This subcutaneous layer is especially thick in whales, seals, and most other marine mammals.

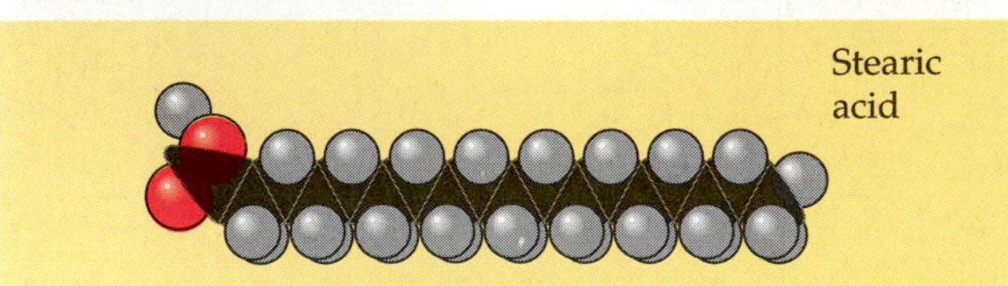

(a) Saturated fat and fatty acid

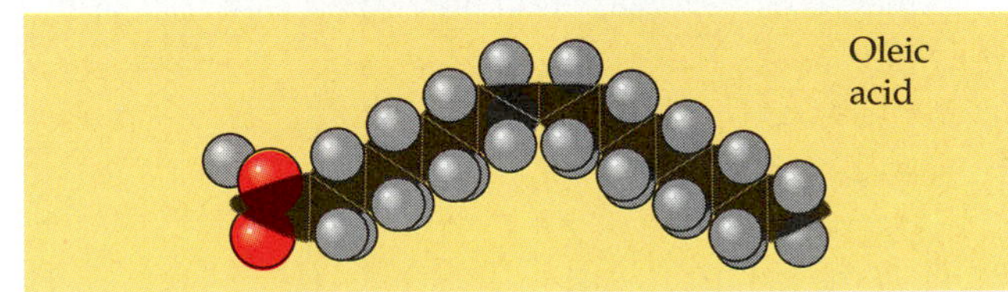

(b) Unsaturated fat and fatty acid

FIGURE 5.11 ▪ Saturated and unsaturated fats and fatty acids. (a) If its fatty acids are saturated (with hydrogen), a fat is also said to be saturated. Most animal fats, such as those in butter, are saturated. They are solids at room temperature. **(b)** Unsaturated fatty acids, such as oleic acid, have one or more double bonds between carbons. The fatty acid bends where double bonds are located. Unsaturated fats have one or more unsaturated fatty acids. Most vegetable fats are unsaturated and are called oils because they are liquids at room temperature. The kinks in the fatty acids prevent the fat molecules from packing together closely enough to solidify.

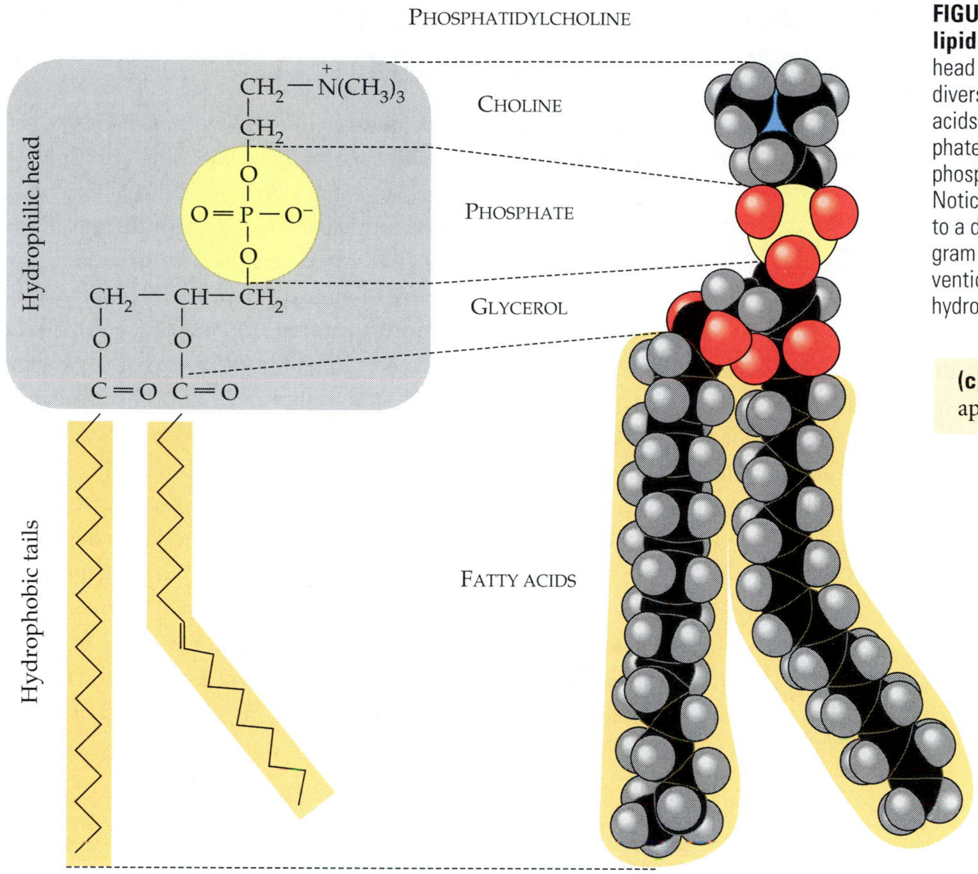

PHOSPHATIDYLCHOLINE

Hydrophilic head

CHOLINE

PHOSPHATE

GLYCEROL

Hydrophobic tails

FATTY ACIDS

Hydrophilic head

Hydrophobic tails

(a) Structural formula

(b) Space-filling model

(c) Phospholipid symbol

FIGURE 5.12 ■ The structure of a phospholipid. A phospholipid has a hydrophilic (polar) head and two hydrophobic tails. Phospholipid diversity is based on differences in the fatty acids and in the groups attached to the phosphate group of the head. **(a), (b)** This particular phospholipid is called a phosphatidylcholine. Notice that one of its fatty acids has a kink due to a double bond in its carbon chain. (The diagram in part **(a)** follows a common chemical convention of omitting the carbons and attached hydrogens of the hydrocarbon tails.)

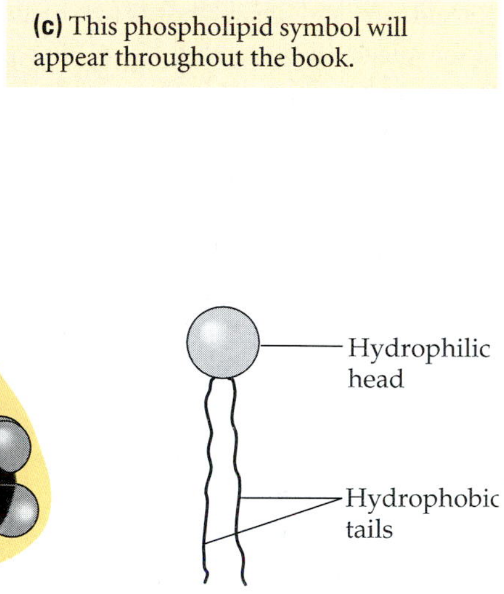

(c) This phospholipid symbol will appear throughout the book.

Phospholipids are major components of cell membranes

Phospholipids are similar to fats, but they have only two fatty acids rather than three. The third hydroxyl group of glycerol is joined to a phosphate group, which is negative in electrical charge. Additional small molecules, usually charged or polar, can be linked to the phosphate group to form a variety of phospholipids (FIGURE 5.12).

Phospholipids show ambivalent behavior toward water. Their tails, which consist of hydrocarbons, are hydrophobic and are excluded from water. However, the phosphate group and its attachments form a hydrophilic head that has an affinity for water.

When phospholipids are added to water, they self-assemble into aggregates that shield their hydrophobic portions from water. One such cluster is a micelle, a phospholipid droplet with the phosphate heads on the outside, in contact with water. The hydrocarbon tails are restricted to the water-free interior of the micelle (FIGURE 5.13a).

At the surface of a cell, phospholipids are arranged in a bilayer, or double layer (FIGURE 5.13b). The hydrophilic heads of the molecules are on the outside of the bilayer, in contact with the aqueous solutions inside and outside the cell. The

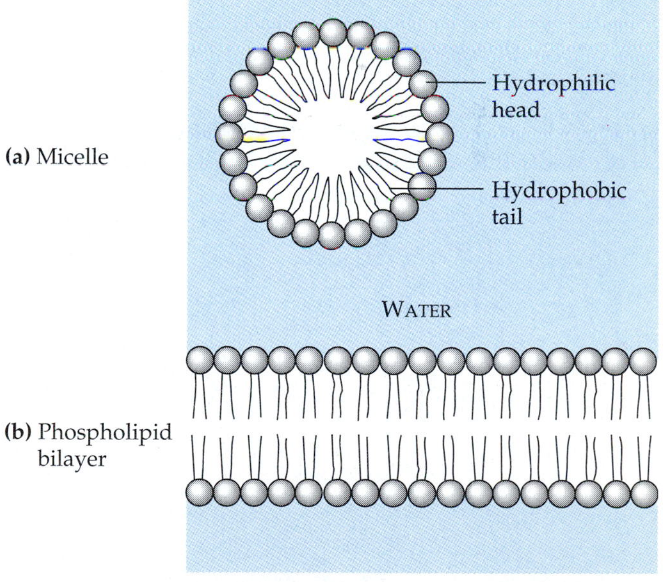

FIGURE 5.13 ■ Two structures formed by self-assembly of phospholipids in aqueous environments. (a) A micelle, in cross section. **(b)** A cross section of a phospholipid bilayer between two aqueous compartments. Such bilayers are the main fabric of biological membranes. The hydrophilic heads (spheres) of the phospholipids are in contact with water, whereas the hydrophobic tails are in contact with each other and remote from water.

hydrophobic tails point toward the interior of the membrane, away from the water. The phospholipid bilayer forms a boundary between the cell and its external environment; in fact, phospholipids are major components of cell membranes. This behavior provides another example of how form fits function at the molecular level.

Steroids include cholesterol and certain hormones

Steroids are lipids characterized by a carbon skeleton consisting of four fused rings (FIGURE 5.14). Different steroids vary in the functional groups attached to this ensemble of rings. One steroid, **cholesterol**, is a common component of animal cell membranes and is also the precursor from which other steroids are synthesized. Many hormones, including vertebrate sex hormones, are steroids produced from cholesterol (see FIGURE 4.8). Thus, cholesterol is a crucial molecule in animals, although a high level of it in the blood may contribute to atherosclerosis.

H_3C CH_3 CH_3 CH_3 CH_3 HO

FIGURE 5.14 ▪ Cholesterol: a steroid. Cholesterol is the molecule from which other steroids, including the sex hormones, are synthesized. Steroids vary in the functional groups attached to their four interconnected rings (shown in gold).

PROTEINS—THE MOLECULAR TOOLS OF THE CELL

5.3 The importance of proteins is implied by their name, which comes from the Greek word *proteios,* meaning "first place." **Proteins** account for more than 50% of the dry weight of most cells, and they are instrumental in almost everything organisms do (TABLE 5.1). Proteins are used for structural support, storage, transport of other substances, signaling from one part of the organism to another, movement, and defense against foreign substances. In addition, as enzymes, proteins regulate metabolism by selectively accelerating chemical reactions in the cell. A human has tens of thousands of different proteins, each with a specific structure and function.

Proteins are the most structurally sophisticated molecules known. Consistent with their diverse functions, they vary extensively in structure, each type of protein having a unique three-dimensional shape, or **conformation**. Diverse though proteins may be, they are all polymers constructed from the same set of 20 amino acids. Polymers of amino acids are called **polypeptides**. A protein consists of one or more polypeptides folded and coiled into specific conformations.

A polypeptide is a polymer of amino acids connected in a specific sequence

As mentioned in Chapter 4, **amino acids** are organic molecules possessing both carboxyl and amino groups. The figure at the right shows the general formula for an amino acid:

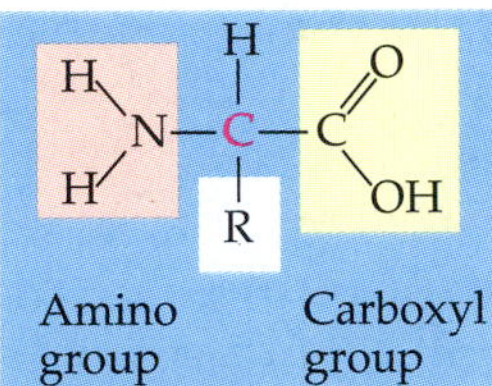

Table 5.1 ▪ An Overview of Protein Functions

TYPE OF PROTEIN	FUNCTION	EXAMPLES
Structural proteins	Support	Insects and spiders use silk fibers to make their cocoons and webs, respectively. Collagen and elastin provide a fibrous framework in animal connective tissues, such as tendons and ligaments. Keratin is the protein of hair, horns, feathers, and other skin appendages.
Storage proteins	Storage of amino acids	Ovalbumin is the protein of egg white, used as an amino acid source for the developing embryo. Casein, the protein of milk, is the major source of amino acids for baby mammals. Plants have storage proteins in their seeds.
Transport proteins	Transport of other substances	Hemoglobin, the iron-containing protein of vertebrate blood, transports oxygen from the lungs to other parts of the body. Other proteins transport molecules across cell membranes.
Hormonal proteins	Coordination of an organism's activities	Insulin, a hormone secreted by the pancreas, helps regulate the concentration of sugar in the blood of vertebrates.
Receptor proteins	Response of cell to chemical stimuli	Receptors built into the membrane of a nerve cell detect chemical signals released by other nerve cells.
Contractile proteins	Movement	Actin and myosin are responsible for the movement of muscles. Contractile proteins are responsible for the undulations of cilia and flagella, which propel many cells.
Defensive proteins	Protection against disease	Antibodies combat bacteria and viruses.
Enzymatic proteins	Selective acceleration of chemical reactions	Digestive enzymes hydrolyze the polymers in food.

At the center of the amino acid is an asymmetric carbon atom. Its four different partners are an amino group, a carboxyl group, a hydrogen atom, and a variable group symbolized by R. The R group, also called the side chain, differs with the amino acid. FIGURE 5.15 shows the 20 amino acids that cells use to build their thousands of proteins. (Here the amino and carboxyl groups are all depicted in ionized form.) The R group may be as simple as a hydrogen atom, as in the amino acid glycine (the one amino acid lacking an asymmetric carbon), or it may be a carbon skeleton with various functional

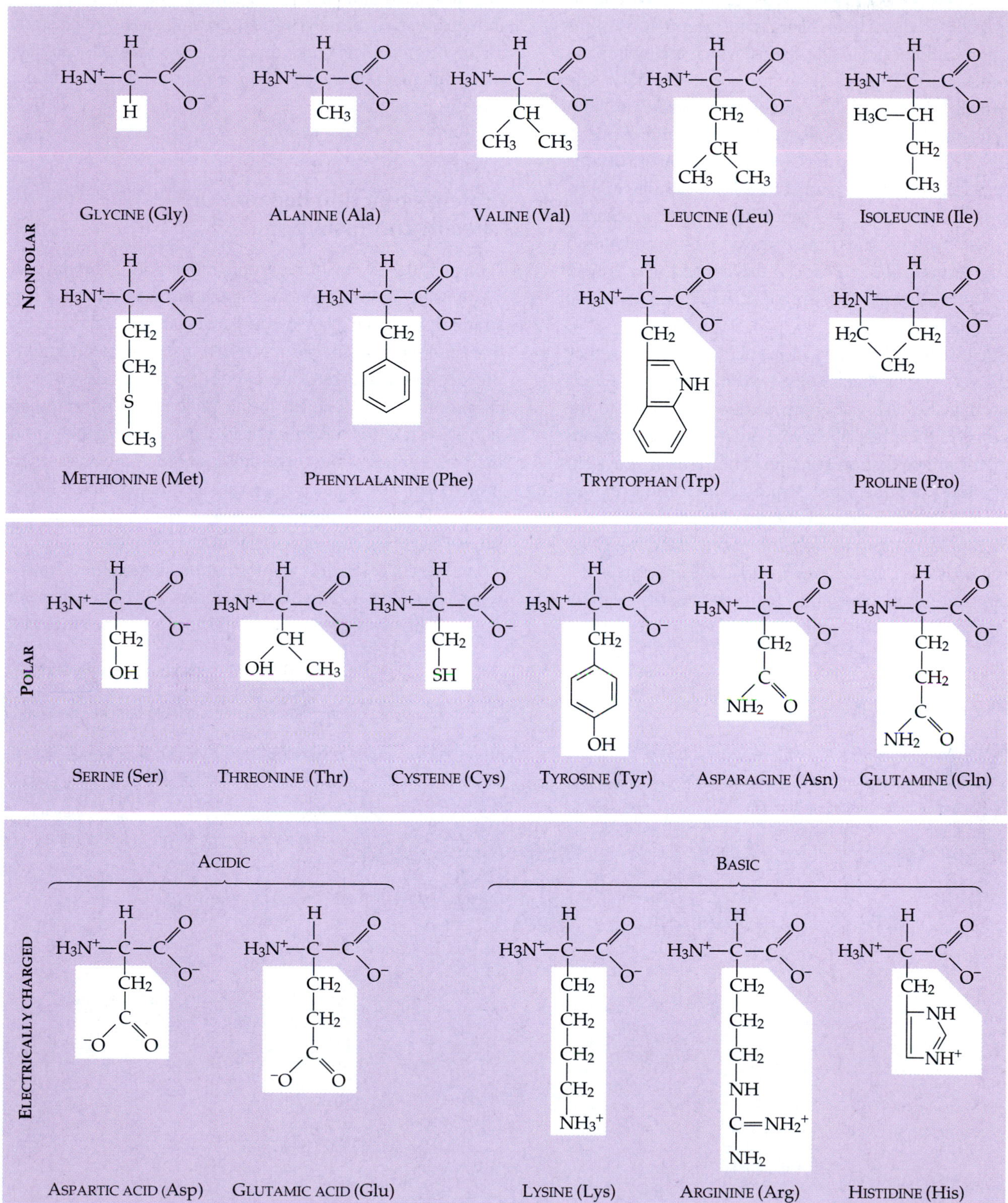

FIGURE 5.15 ▪ The 20 amino acids of proteins. The amino acids are grouped here according to the properties of their side chains (R groups), highlighted in white. The amino acids are shown in their prevailing ionic forms at pH 7, the approximate pH within a cell. In parentheses are the three-letter abbreviations for the amino acids.

groups attached, as in glutamine. (Other amino acids, not shown in FIGURE 5.15, have important functions in organisms, but these amino acids are not incorporated into proteins.)

The physical and chemical properties of the side chain determine the unique characteristics of a particular amino acid. In FIGURE 5.15 the amino acids are grouped according to the properties of their side chains. One group consists of amino acids with nonpolar side chains, which are hydrophobic. Another group consists of amino acids with polar side chains, which are hydrophilic. Acidic amino acids are those with side chains that are generally negative in charge owing to the presence of a carboxyl group, which is usually dissociated (ionized) at cellular pH. Basic amino acids have amino groups in their side chains that are generally positive in charge. (Notice that *all* amino acids have carboxyl groups and amino groups; the terms *acidic* and *basic* in this context refer only to the side chains.) Because they are ionic, acidic and basic side chains are also hydrophilic.

Now that we have examined amino acids, let's see how they are linked to form polymers. When two amino acids are positioned so that the carboxyl group of one is adjacent to the amino group of the other, an enzyme can join the amino acids by means of a dehydration reaction. The resulting covalent bond is called a **peptide bond**. Repeated over and over, this process yields a polypeptide, a polymer of many amino acids linked by peptide bonds (FIGURE 5.16). At one end of the polypeptide chain is a free amino group, and at the opposite end is a free carboxyl group. Thus, the chain has polarity, with an amino end (or N-terminus) and a carboxyl end (C-terminus).

The repeating sequence of atoms highlighted in purple in FIGURE 5.16b is called the polypeptide backbone. Attached to this repetitive backbone are different kinds of appendages, the side chains of the amino acids. Polypeptides range in length from a few monomers to a thousand or more. Each specific polypeptide has a unique linear sequence of amino acids. The immense variety of polypeptides existing in nature illustrates an important concept introduced earlier—that cells can make many different polymers by linking a limited set of monomers into diverse sequences.

A protein's function depends on its specific conformation

"Polypeptide" is not quite synonymous with "protein." The relationship is somewhat analogous to that between a long strand of yarn and a sweater of particular size and shape that one can knit from the yarn. A functional protein is not *just* a polypeptide chain, but one or more polypeptides precisely twisted, folded, and coiled into a molecule of unique shape (FIGURE 5.17). It is the amino acid sequence of a polypeptide that determines what three-dimensional conformation the protein will take. Many proteins are globular (roughly spherical), while others are fibrous in shape. However, within these broad categories, countless variations are possible.

A protein's specific conformation determines how it works. In almost every case, the function of a protein depends on its ability to recognize and bind to some other molecule.

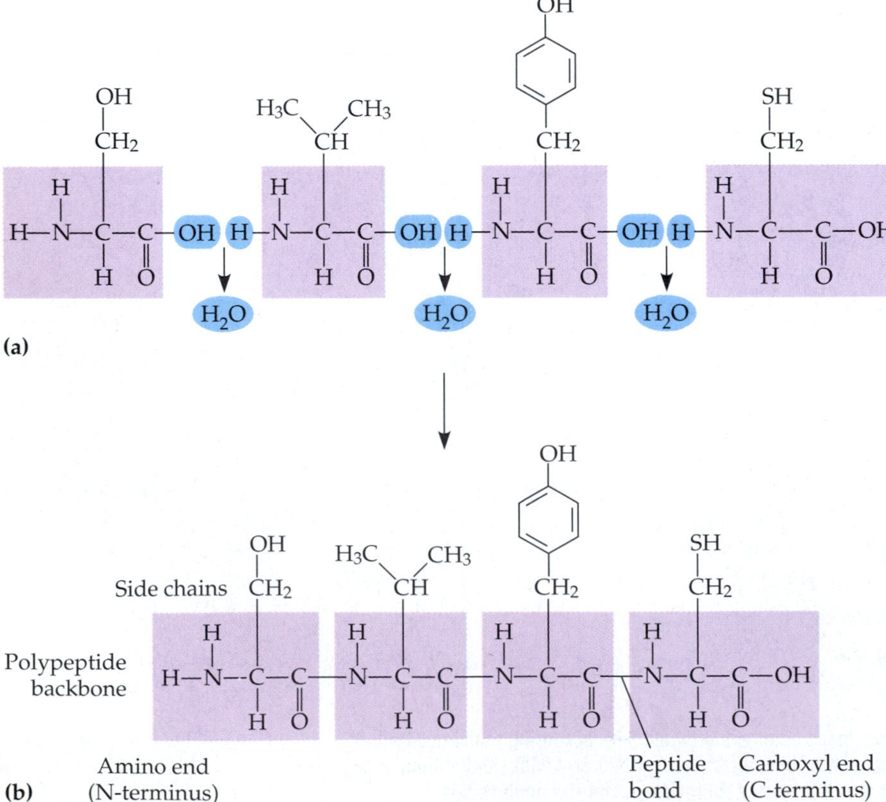

FIGURE 5.16 ▪ Making a polypeptide chain. (a) Peptide bonds formed by dehydration reactions link the carboxyl group of one amino acid to the amino group of the next. **(b)** The polypeptide has a repetitive backbone (purple) to which the amino acid side chains are attached. The peptide bonds are formed one at a time, starting with the amino acid at the amino end (N-terminus).

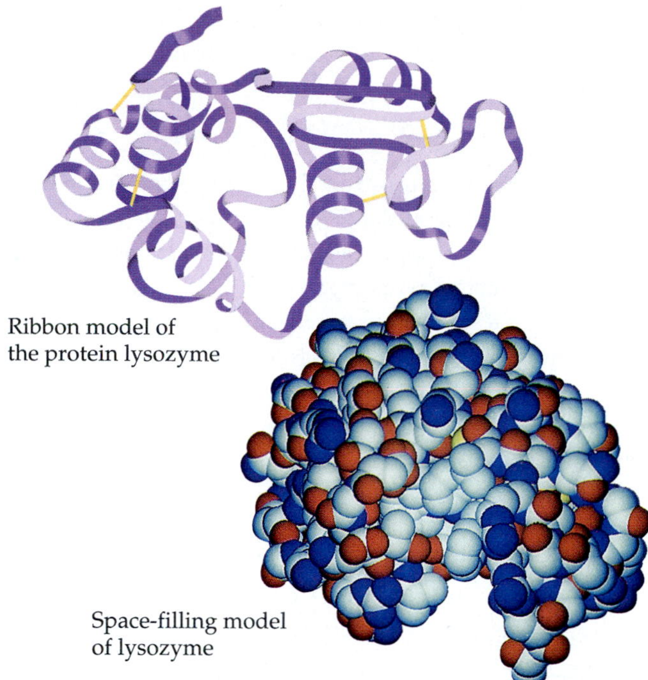

FIGURE 5.17 ▪ Functional conformation of a protein, the enzyme lysozyme. Present in our sweat, tears, and saliva, lysozyme is an enzyme that helps prevent infection by binding to and destroying specific molecules on the surface of many kinds of bacteria. In the ribbon model of lysozyme, you can see how the single polypeptide chain, represented by the ribbon, folds and coils to form the functional protein. (In this simplification, the yellow lines symbolize one type of chemical bond that stabilizes the protein's shape.) The space-filling model shows more clearly that lysozyme's overall shape is roughly spherical (globular), as are many other proteins. This *exact* conformation, however, is unique to lysozyme, and the protein's specialized function emerges from this shape. A groove on the surface of lysozyme is the part of the protein that recognizes and binds to the target molecules on bacterial walls.

For instance, an antibody binds to a particular foreign substance that has invaded the body, and an enzyme recognizes and binds to its substrate, the substance the enzyme works on. In Chapter 2 you learned that one nerve cell in the brain signals another by dispatching specific molecules that have a unique shape. The receptor molecules on the surface of the receiving cell are proteins that fit the signal molecules something like a lock and key (see FIGURE 2.16).

Four Levels of Protein Structure

5.4 When a cell synthesizes a polypeptide, the chain generally folds spontaneously to assume the functional conformation for that protein. This folding is driven and reinforced by the formation of a variety of bonds between parts of the chain. Thus, the function of a protein—the ability of a receptor protein to identify and associate with a particular chemical messenger, for instance—is an emergent property resulting from exquisite molecular order. In the complex architecture of a protein, we can recognize three superimposed levels of structure, known as primary, secondary, and tertiary structure. A fourth level, quaternary structure, occurs when a protein consists of two or more polypeptide chains.

Primary Structure. The **primary structure** of a protein is its unique sequence of amino acids. As an example, we will examine the primary structure of lysozyme, the antibacterial enzyme illustrated in its three-dimensional form in FIGURE 5.17. Lysozyme is a relatively small protein, its single polypeptide chain only 129 amino acids long. In FIGURE 5.18 the polypeptide chain is unraveled for a closer look at its primary structure. A specific one of the 20 amino acids occupies each of the 129 positions along the chain. The primary structure is

FIGURE 5.18 ▪ The primary structure of a protein. This is the unique amino acid sequence, or primary structure, of the enzyme lysozyme. The names of the amino acids are given as their three-letter abbreviations. (The chain was drawn in this serpentine fashion to make the entire sequence readily visible. The actual shape of lysozyme is shown in Figure 5.17.)

like the order of letters in a very long word. If left to chance, there would be 20^{129} different ways of arranging amino acids into a polypeptide chain of this length. However, the precise primary structure of a protein is determined not by the random linking of amino acids, but by inherited genetic information.

Even a slight change in primary structure can affect a protein's conformation and ability to function. For instance, the substitution of one amino acid for another at a particular position in the primary structure of hemoglobin, the protein that carries oxygen in red blood cells, causes sickle-cell anemia, an inherited blood disorder (FIGURE 5.19).

The pioneer in determining the primary structure of proteins was Frederick Sanger, who, with his colleagues at Cambridge University in England, worked out the amino acid sequence of the hormone insulin in the late 1940s and early 1950s. His approach was to use protein-digesting enzymes and other catalysts that break polypeptides at specific places rather than completely hydrolyzing the chain. Treatment with one of these agents would cleave the polypeptide into fragments that could be separated by a technique called chromatography. Hydrolysis with another agent would break the polypeptide at different sites, yielding a second group of fragments. Sanger used chemical methods to determine the sequence of amino acids in these small fragments. Then he searched for overlapping regions among the pieces obtained by hydrolyzing with the different agents. Consider, for instance, two fragments with the following sequences:

Cys-Ser-Leu-Tyr-Gln-Leu

Tyr-Gln-Leu-Glu-Asn

We can deduce from the overlapping regions that the intact polypeptide contains in its primary structure the following segment:

Cys-Ser-Leu-Tyr-Gln-Leu-Glu-Asn

Just as we could reconstruct this sentence from a collection of fragments with overlapping sequences of letters, Sanger and his co-workers were able, after years of effort, to reconstruct the complete primary structure of insulin. Since then, most of the steps involved in sequencing a polypeptide have been automated. However, it was Sanger's analysis of insulin that first demonstrated what is now a fundamental axiom of molecular biology: Each type of protein has a unique primary structure, a precise sequence of amino acids.

Secondary Structure. Most proteins have segments of their polypeptide chain repeatedly coiled or folded in patterns that contribute to the protein's overall conformation. These coils and folds, collectively referred to as **secondary structure**, are the result of hydrogen bonds at regular intervals along the polypeptide backbone (FIGURE 5.20). Because they are electronegative, both the oxygen and the nitrogen atoms of the backbone have weak negative charges (see Chapter 2). The weakly positive hydrogen atom attached to the nitrogen atom has an affinity for the oxygen atom of a nearby peptide bond. Individually, these hydrogen bonds are weak, but because they are repeated many times over a relatively long region of the polypeptide chain, they can support a particular shape for that part of the protein. One such secondary structure is the **alpha** (α) **helix**, a delicate coil held together by hydrogen bonding between every fourth amino acid. The regions of α

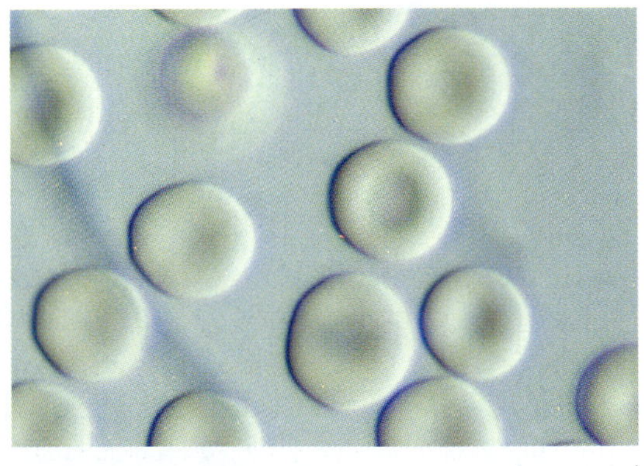

(a) Normal red blood cells and normal hemoglobin

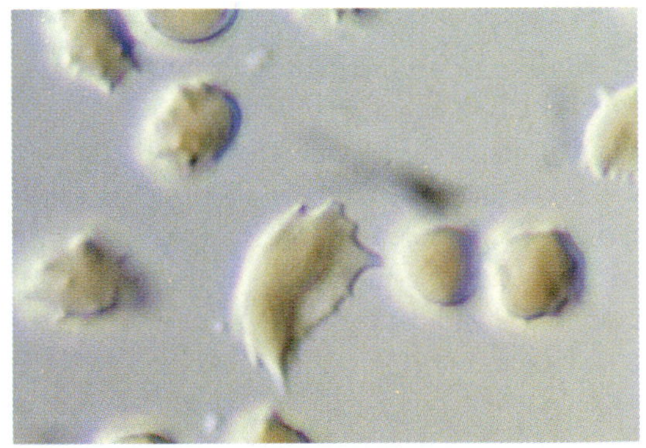

(b) Sickled red blood cells and sickle-cell hemoglobin

FIGURE 5.19 ▪ A single amino acid substitution in a protein causes sickle-cell disease. (a) Normal human red blood cells are disk-shaped, as seen in this light micrograph. Each cell contains millions of molecules of the protein hemoglobin, which transports oxygen molecules from the lungs to the other organs of the body. The first seven amino acids of one of the polypeptides of normal hemoglobin are shown below the micrograph; this polypeptide has a total of 146 amino acids. **(b)** A slight change in the primary structure of hemoglobin—an inherited substitution of one amino acid—causes sickle-cell disease. The substitution occurs in the number 6 position of the polypeptide. The abnormal hemoglobin molecules tend to crystallize, deforming some of the cells into a "sickle" shape. The life of someone with the disease is punctuated by "sickle-cell crises," which occur when the angular cells clog tiny blood vessels, impeding blood flow.

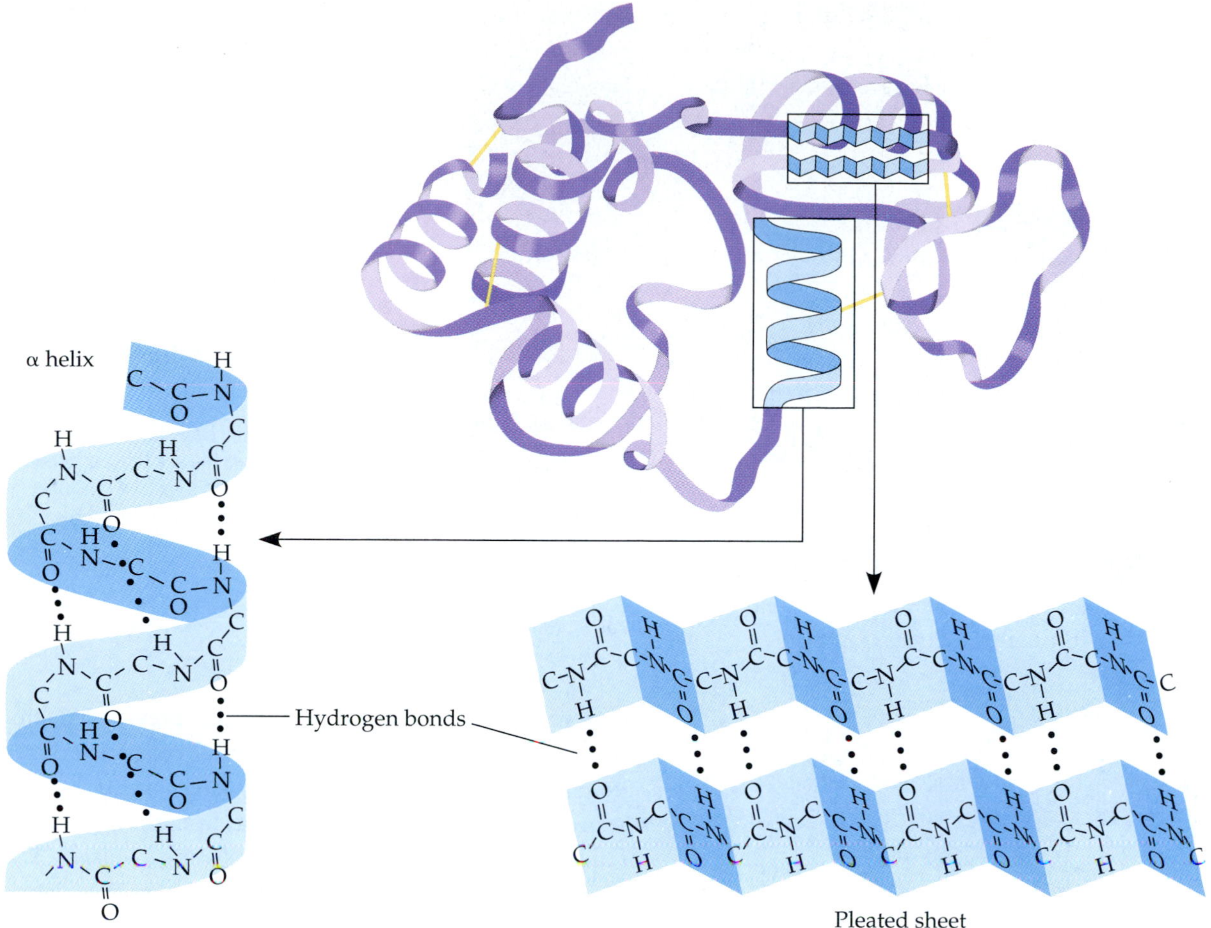

FIGURE 5.20 ▪ The secondary structure of a protein. Two types of secondary structure, the α helix and the pleated sheet, can both be found in the protein lysozyme. Both patterns depend on hydrogen bonding between $>$CO and $>$NH groups along the polypeptide backbone. The R groups of the amino acids are omitted in these diagrams, as are some H atoms.

helix in the enzyme lysozyme are evident in FIGURE 5.20, where one α helix is enlarged to show the hydrogen bonds. Lysozyme is fairly typical of a globular protein in that it has a few stretches of α helix separated by nonhelical regions. In contrast, some fibrous proteins, such as α-keratin, the structural protein of hair, have the α-helix formation over most of their length.

The other main type of secondary structure is the **pleated sheet**, in which two regions of the polypeptide chain lie parallel to each other. Hydrogen bonds between the parts of the backbone in the parallel regions hold the structure together. Pleated sheets make up the core of many globular proteins, and we can see one such region in lysozyme in FIGURE 5.20. Also, pleated sheets dominate some fibrous proteins, including the silk produced by many insects and spiders (FIGURE 5.21, p. 74).

Tertiary Structure. Superimposed on the patterns of secondary structure is a protein's **tertiary structure**, consisting of irregular contortions from bonding between side chains (R groups) of the various amino acids. One of the types of bonding that contributes to tertiary structure is called a **hydrophobic interaction.** As a polypeptide folds into its functional conformation, amino acids with hydrophobic (nonpolar) side chains usually congregate in clusters at the core of the protein, out of contact with water. Thus, what we call a hydrophobic interaction is actually initiated by the behavior of water molecules, which exclude nonpolar substances as the water molecules hydrogen-bond to one another and to hydrophilic parts of the protein. Once nonpolar amino acid side chains are close together, van der Waals attractions reinforce the hydrophobic interactions. Meanwhile, hydrogen bonds between polar side chains and ionic bonds between positively and negatively charged side chains also help stabilize tertiary structure. These are all weak interactions, but their cumulative effect helps give the protein a specific shape.

The conformation of a protein may be reinforced further by strong, covalent bonds called **disulfide bridges**. Disulfide bridges form where two cysteine monomers, amino acids with sulfhydryl groups (—SH) on their side chains, are brought close together by the folding of the protein. The

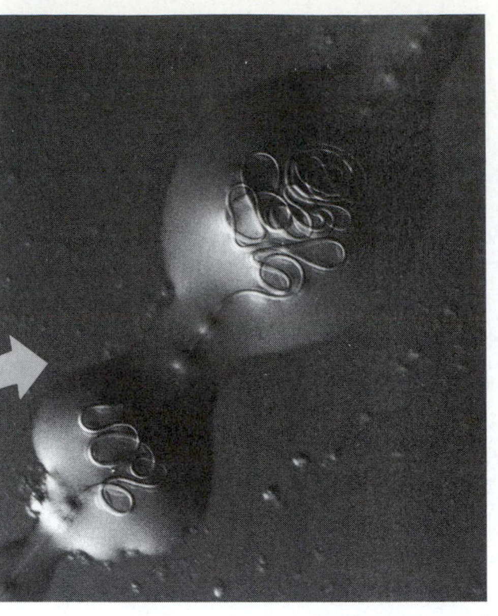

FIGURE 5.21 • Spider silk: a structural protein. Abdominal glands of this spider secrete silk fibers in liquid form; the protein solidifies upon contact with air. The silk protein owes its strength to its secondary structure, consisting of many pleated sheets. The teamwork of so many hydrogen bonds makes each silk fiber stronger than steel. The strands that radiate out from the web's center, made of dry silk protein, maintain the basic shape of the web. The strands forming the spiral, called capture strands, are elastic; they stretch and coil in response to wind, rain, and insects that are unfortunate enough to fly into the web. The light micrograph, right, helps us understand this resilience. A capture strand consists of a coiled silk fiber coated by a sticky fluid. Force on the strand unwinds the silk fiber, absorbing the shock. Then the fluid's surface tension—the tendency for the aqueous solution to "bead"—rewinds the fiber.

FIGURE 5.22 • Examples of bonds contributing to the tertiary structure of a protein. Hydrogen bonds, ionic bonds, hydrophobic interactions, and van der Waals interactions are weak bonds between side chains that collectively hold the protein in a specific conformation. Much stronger are the disulfide bridges, covalent bonds between the side chains of cysteine pairs.

sulfur of one cysteine bonds to the sulfur of the second, and the disulfide bridge (—S—S—) rivets parts of the protein together. (The yellow lines in FIGURES 5.17 and 5.20 represent disulfide bridges.) All of these different kinds of bonds can occur in one protein, as shown diagramatically in FIGURE 5.22.

Quaternary Structure. As mentioned previously, some proteins consist of two or more polypeptide chains aggregated into one functional macromolecule. **Quaternary structure** is the overall protein structure that results from the aggregation of these polypeptide subunits. For example, collagen is a fibrous protein that has helical subunits supercoiled into a larger triple helix (FIGURE 5.23a). This supercoiled organization of collagen, which is similar to the construction of a rope, gives the long fibers great strength. This suits collagen fibers to their function as the girders of connective tissue, such as tendons and ligaments. Hemoglobin, the oxygen-binding protein of red blood cells, is an example of a globular protein with quaternary structure (FIGURE 5.23b). It consists of two kinds of polypeptide chains, with two of each kind per hemoglobin molecule.

We have taken the reductionist approach in dissecting proteins to their four levels of structural organization. However, it is the overall product, a macromolecule with a unique shape, that works in a cell. The specific function of a protein is an emergent property that arises from the architecture of the molecule (FIGURE 5.24).

What Determines Protein Conformation?

You've learned that unique conformation endows each protein with a specific function, but what are the key factors determining conformation? You already know most of the answer: A polypeptide chain of a given amino acid sequence can spontaneously arrange itself into a three-dimensional shape maintained by the interactions responsible for secondary and tertiary structure. This normally occurs as the protein is being synthesized within the cell. However, protein conformation also depends on the physical and chemical con-

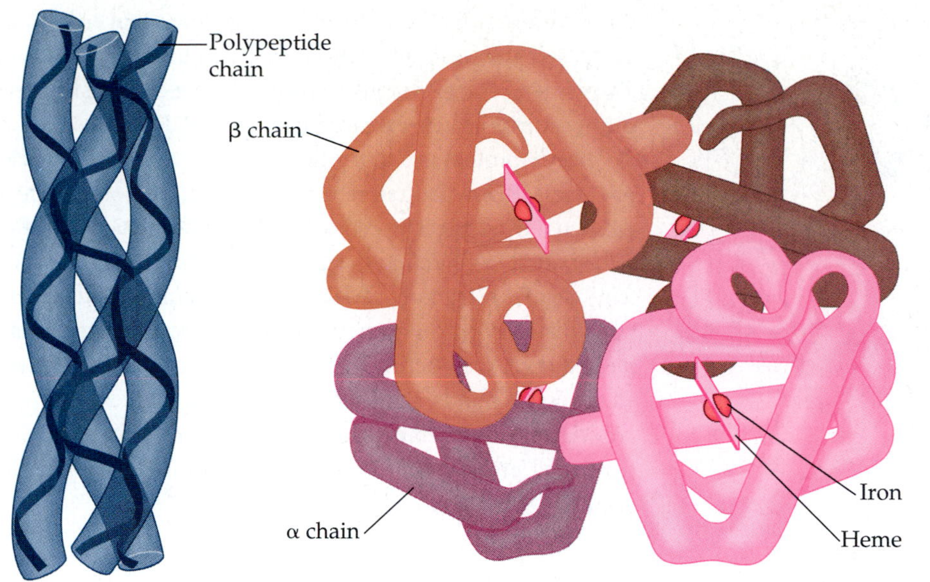

FIGURE 5.23 ▪ The quaternary structure of proteins. At this level of structure, two or more polypeptide subunits associate to form a functional protein. **(a)** Collagen is a fibrous protein consisting of three helical polypeptides that are supercoiled to form a ropelike structure of great strength. Accounting for 40% of the protein in the human body, collagen strengthens connective tissue in our skin, bones, ligaments, tendons, and other body parts. **(b)** Hemoglobin is a globular protein with four polypeptide subunits, two of one kind (α chains) and two of another kind (β chains). Both α and β subunits consist primarily of α-helical secondary structure, represented by the thicker cylindrical sections of the polypeptides in this model. (Each subunit has a nonpolypeptide component, called heme, with an iron atom that binds oxygen.)

ditions of the protein's environment. If the pH, salt concentration, temperature, or other aspects of its environment are altered, the protein may unravel and lose its native conformation, a change called **denaturation** (FIGURE 5.25, p. 76). Misshapen, the denatured protein is biologically inactive. Most proteins become denatured if they are transferred from an aqueous environment to an organic solvent, such as ether or chloroform; the protein turns inside out, its hydrophobic regions changing places with its hydrophilic portions. Other agents of denaturation include chemicals that disrupt the hydrogen bonds, ionic bonds, and disulfide bridges that maintain a protein's shape. Denaturation can also result from

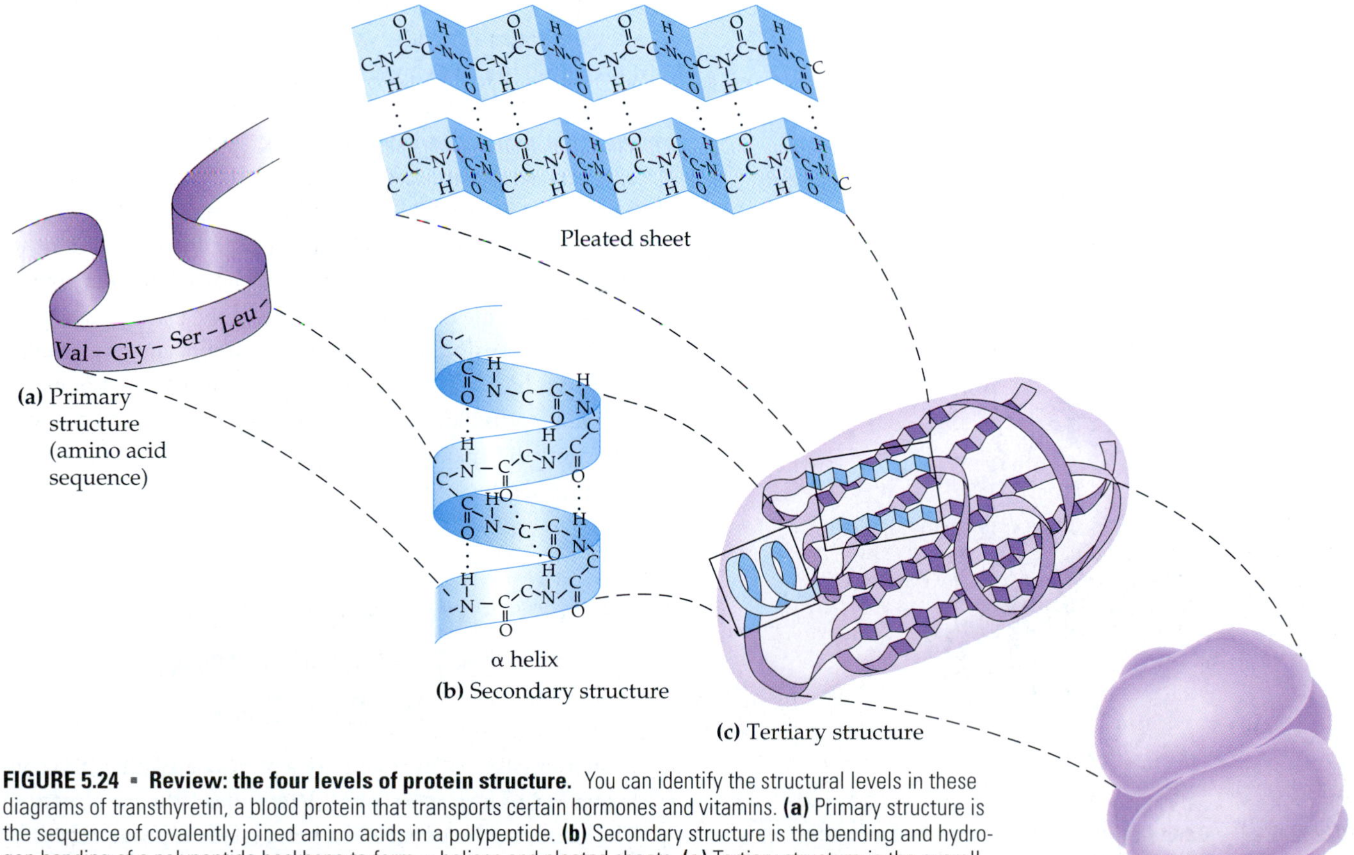

Val – Gly – Ser – Leu

(a) Primary structure (amino acid sequence)

Pleated sheet

α helix

(b) Secondary structure

(c) Tertiary structure

(d) Quaternary structure

5.4 **FIGURE 5.24 ▪ Review: the four levels of protein structure.** You can identify the structural levels in these diagrams of transthyretin, a blood protein that transports certain hormones and vitamins. **(a)** Primary structure is the sequence of covalently joined amino acids in a polypeptide. **(b)** Secondary structure is the bending and hydrogen bonding of a polypeptide backbone to form α helices and pleated sheets. **(c)** Tertiary structure is the overall conformation (shape) of a polypeptide, as reinforced by interactions between the side chains (R groups) of amino acids. **(d)** Quaternary structure is the association between two or more polypeptides that make up a protein. In the case of transthyretin, the whole protein consists of four identical polypeptide subunits.

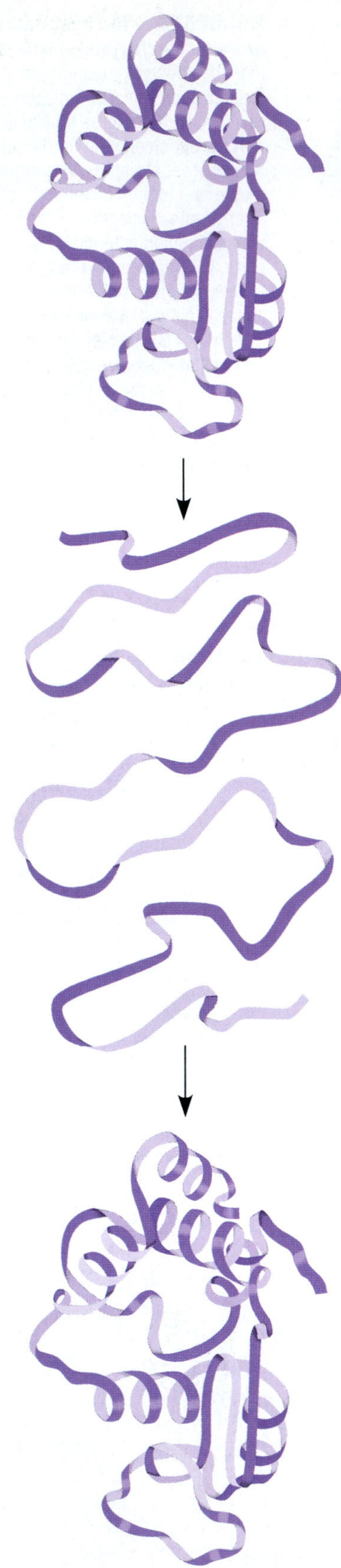

FIGURE 5.25 ▪ Denaturation and renaturation of a protein. High temperatures or various chemical treatments will denature a protein, causing it to lose its conformation and hence its ability to function. If the denatured protein remains dissolved, it can often renature when the chemical and physical aspects of its environment are restored to normal.

excessive heat, which agitates the polypeptide chain enough to overpower the weak interactions that stabilize conformation. The white of an egg becomes opaque during cooking because the denatured proteins are insoluble and solidify.

When a protein in a test-tube solution is denatured by heat or chemicals, it will often return to its functional shape when the denaturing agent is removed. We can conclude that the information for building specific shape is intrinsic in the protein's primary structure. The sequence of amino acids determines conformation—where an α helix can form, where pleated sheets can occur, where disulfide bridges are located, where ionic bonds can form, and so on. However, in the crowded environment inside a cell, correct refolding of a denatured protein—and even correct folding during protein synthesis—may be more of a problem.

The Protein-Folding Problem

Biochemists now know the amino acid sequences of more than 100,000 proteins and the three-dimensional shapes of about 10,000 (see the Methods Box on p. 80). One would think that by correlating the primary structures of many proteins with their conformations, it would be possible to discover the rules of protein folding, especially with the help of computers. Unfortunately, the protein-folding problem is not that simple. Most proteins probably go through several intermediate states on their way to a stable conformation, and looking at the "mature" conformation does not reveal the stages of folding that are required to achieve that form. However, biochemists have developed methods for tracking a protein through its intermediate stages of folding. Researchers have also discovered **chaperone proteins**, molecules that function as temporary braces in assisting the folding of other proteins. These breakthroughs will accelerate our understanding of protein folding.

The protein-folding problem is as important as it is challenging. Once the rules of protein folding are known, it should be possible to design proteins that will carry out specific tasks by making polypeptide chains with appropriate amino acid sequences.

NUCLEIC ACIDS—INFORMATIONAL POLYMERS

If the primary structure of polypeptides determines the con-
5.5 formation of a protein, what determines primary structure? The amino acid sequence of a polypeptide is programmed by a unit of inheritance known as a **gene**. Genes consist of DNA, which is a polymer belonging to the class of compounds known as **nucleic acids**.

Nucleic acids store and transmit hereditary information

There are two types of nucleic acids: **deoxyribonucleic acid (DNA)** and **ribonucleic acid (RNA)**. These are the molecules that enable living organisms to reproduce their complex components from one generation to the next. Unique among molecules, DNA provides directions for its own replication. DNA also directs RNA synthesis and, through RNA, controls protein synthesis.

DNA is the genetic material that organisms inherit from their parents. A DNA molecule is very long and usually consists of hundreds or thousands of genes. When a cell reproduces itself by dividing, its DNA is copied and passed along from one generation of cells to the next. Encoded in the structure of DNA is the information that programs all the cell's activities. The DNA, however, is not directly involved in running the operations of the cell, any more than computer software by itself can print a bank statement or read the bar code on a box of cereal. Just as a printer is needed to print out a statement and a scanner is needed to read a bar code, proteins are required to implement genetic programs. Proteins are the molecular hardware of the cell—the tools for most biological functions. For example, it is the protein hemoglobin that carries oxygen in the blood, not the DNA that specifies the structure of hemoglobin.

How does RNA, the other type of nucleic acid, fit into the flow of genetic information from DNA to proteins? Each gene along the length of the DNA molecule directs the synthesis of a type of RNA called messenger RNA (mRNA). The mRNA molecule then interacts with the cell's protein-synthesizing machinery to direct the production of a polypeptide. We can summarize the flow of genetic information as DNA⟶RNA⟶protein (FIGURE 5.26). The actual sites of protein synthesis are cellular structures called ribosomes. In a eukaryotic cell, ribosomes are located in the cytoplasm, but DNA resides in the nucleus. Messenger RNA conveys the genetic instructions for building proteins from the nucleus to the cytoplasm. Prokaryotic cells lack nuclei, but they still use RNA to send a message from the DNA to the ribosomes and other equipment of the cell that translate the coded information into amino acid sequences.

A nucleic acid strand is a polymer of nucleotides

Nucleic acids are polymers of monomers called **nucleotides**. Each nucleotide is itself composed of three parts: an organic molecule called a nitrogenous base, a pentose (five-carbon sugar), and a phosphate group (FIGURE 5.27, p. 78).

There are two families of nitrogenous bases: pyrimidines and purines. A **pyrimidine** has a six-membered ring of carbon and nitrogen atoms. (The nitrogen atoms tend to take up H^+ from solution, which explains the term *nitrogenous base.*) The members of the pyrimidine family are cytosine (C), thymine (T), and uracil (U). **Purines** are larger, with the six-membered ring fused to a five-membered ring. The purines are adenine (A) and guanine (G). The specific pyrimidines and purines differ in the functional groups attached to the rings. Adenine, guanine, and cytosine are found in both types of nucleic acid. Thymine is found only in DNA and uracil only in RNA.

The pentose connected to the nitrogenous base is **ribose** in the nucleotides of RNA and **deoxyribose** in DNA. The only difference between these two sugars is that deoxyribose lacks an oxygen atom on its number 2 carbon—hence its name. So far, we have built a nucleoside, which is a nitrogenous base joined to a sugar. To complete the construction of a nucleotide, we attach a phosphate group to the number 5 carbon of the sugar (FIGURE 5.27b). The molecule is now a nucleoside monophosphate, better known as a nucleotide.

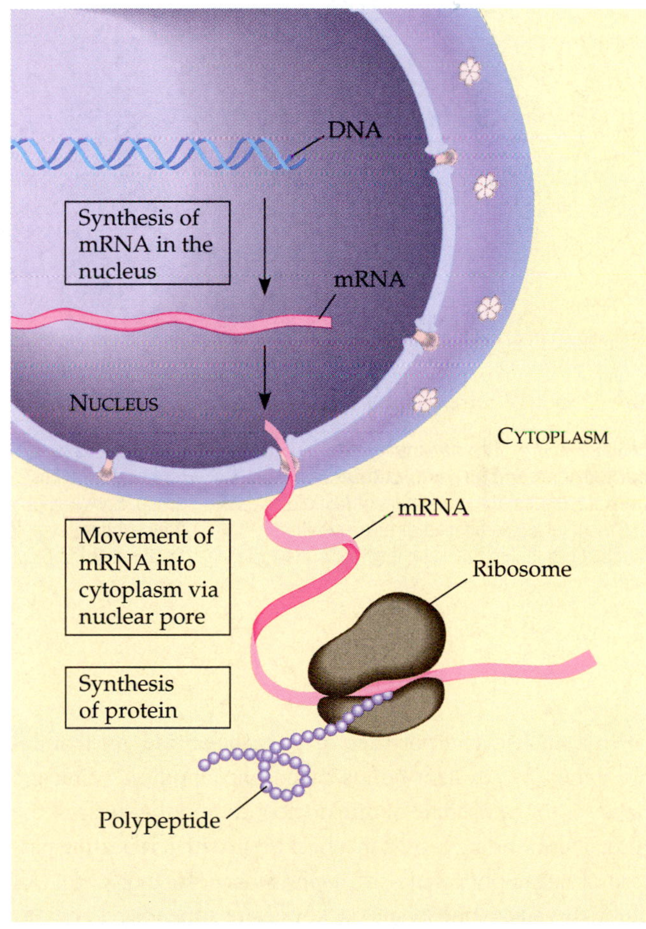

FIGURE 5.26 • DNA⟶RNA⟶protein: a diagrammatic overview of information flow in a cell. In a eukaryotic cell, DNA in the nucleus programs protein production in the cytoplasm by dictating the synthesis of messenger RNA (mRNA), which travels to the cytoplasm and binds to ribosomes. As a ribosome (greatly enlarged in this drawing) moves along the mRNA, the genetic message is translated into a polypeptide of specific amino acid sequence.

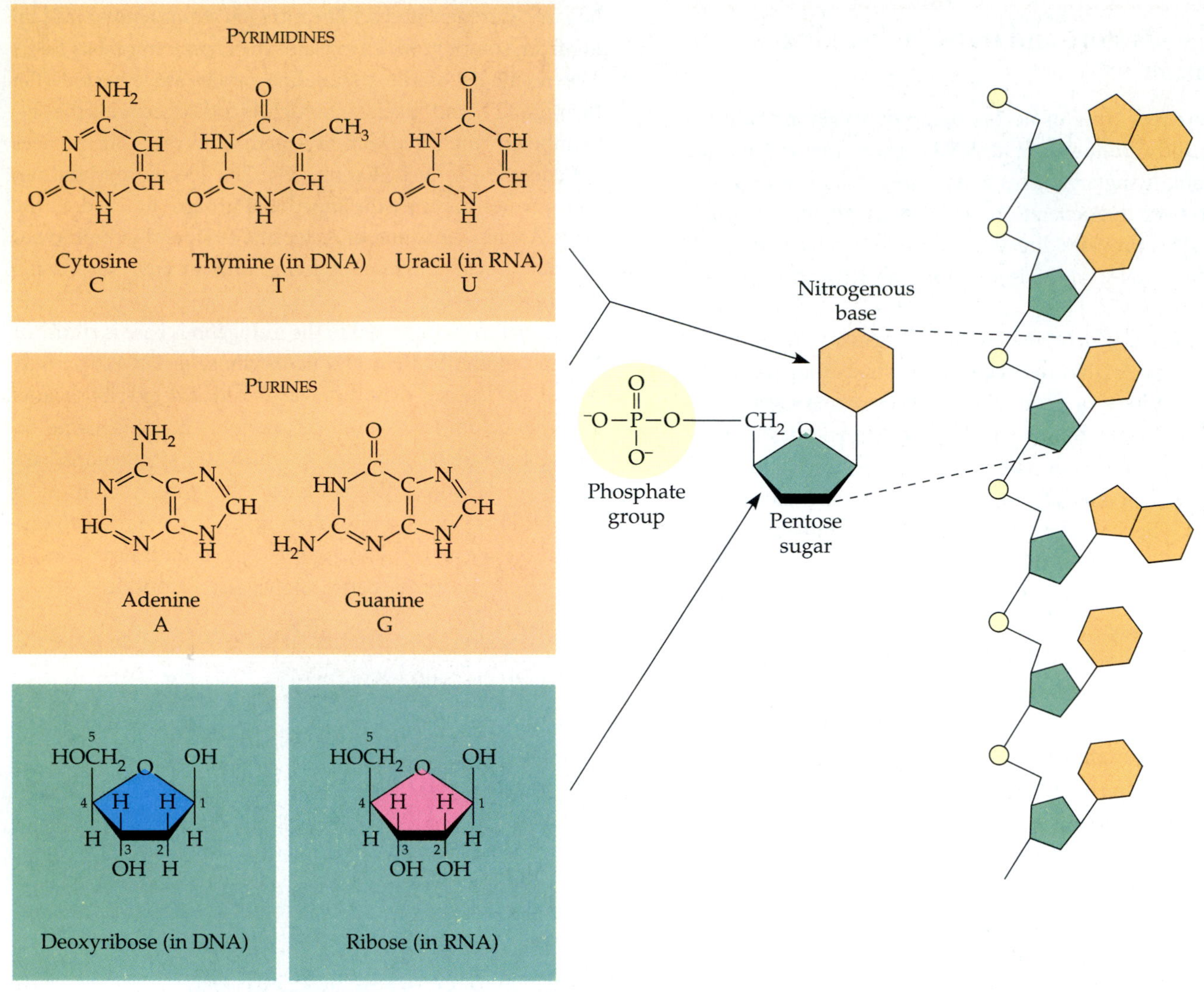

(a) Nucleotide components **(b)** Nucleotide **(c)** Polynucleotide

5.5 **FIGURE 5.27 ▪ The structures of nucleotides and polynucleotides.** **(a)** Nucleotides, the monomers of nucleic acids, are themselves composed of three smaller molecular building blocks: a nitrogenous base (either a purine or a pyrimidine), a pentose sugar, (either deoxyribose or ribose), and a phosphate group. RNA has ribose as its sugar, and DNA has deoxyribose. Also, RNA has uracil, and DNA has thymine. **(b)** In polynucleotides, each nucleotide monomer has its phosphate group bonded to the sugar of the next nucleotide. The polymer has a regular sugar-phosphate backbone with variable appendages, the four kinds of nitrogenous bases.

In a nucleic acid polymer, or **polynucleotide**, nucleotides are joined by covalent bonds called phosphodiester linkages between the phosphate of one nucleotide and the sugar of the next. This bonding results in a backbone with a repeating pattern of sugar-phosphate-sugar-phosphate (FIGURE 5.27c). All along this sugar-phosphate backbone are appendages consisting of the nitrogenous bases. Unlike the backbone, the sequence of bases along a DNA polymer is unique for each gene. Because genes are hundreds to thousands of nucleotides long, the number of possible base sequences is effectively limitless. A gene's meaning to the cell is encoded in its specific sequence of the four DNA bases. For example, the sequence AGGTAACTT means one thing, whereas the sequence CGCTTTAAC has a different translation. (Real genes, of course, are much longer.) The linear order of bases in a gene specifies the amino acid sequence—the primary structure—of a protein, which in turn specifies that protein's three-dimensional conformation and function in the cell.

Inheritance is based on replication of the DNA double helix

The DNA molecules of cells actually consist of two polynucleotides that spiral around an imaginary axis to form a **dou-**

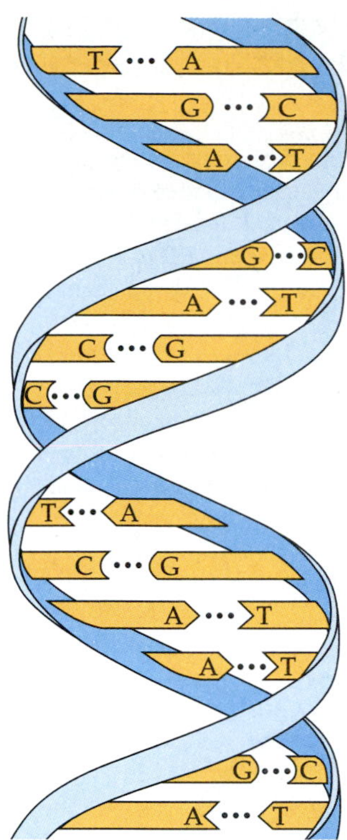

FIGURE 5.28 ▪ The double helix. The DNA molecule is usually double-stranded, with the sugar-phosphate backbone of the polynucleotides (abbreviated here by blue ribbons) on the outside of the helix. In the interior are pairs of nitrogenous bases, holding the two strands together by hydrogen bonds. Hydrogen bonding between the bases is specific. As illustrated here with symbolic shapes for the bases, adenine (A) can pair only with thymine (T), and guanine (G) can pair only with cytosine (C). As a cell prepares to divide, the two strands of the double helix separate, and each serves as a template for the precise ordering of nucleotides into new complementary strands.

ble helix. James Watson and Francis Crick, working at Cambridge University, first proposed the double helix as the three-dimensional structure of DNA in 1953 (FIGURE 5.28). The two sugar-phosphate backbones are on the outside of the helix, and the nitrogenous bases are paired in the interior of the helix. The two polynucleotides, or strands, as they are called, are held together by hydrogen bonds between the paired bases and by van der Waals attractions between the stacked bases. Most DNA molecules are very long, with thousands or even millions of base pairs connecting the two chains. One long DNA double helix includes many genes, each one a particular segment of the molecule.

Only certain bases in the double helix are compatible with each other. Adenine (A) always pairs with thymine (T), and guanine (G) always pairs with cytosine (C). If we were to read the sequence of bases along one strand as we traveled the length of the double helix, we would know the sequence of bases along the other strand. If a stretch of one strand has the base sequence AGGTCCG, then the base-pairing rules tell us that the same stretch of the other strand must have the sequence TCCAGGC. The two strands of the double helix are *complementary*, each the predictable counterpart of the other. It is this feature of DNA that makes possible the precise copying of genes that is responsible for inheritance. In preparation for cell division, the two strands of each DNA molecule separate, and each strand serves as a template to order nucleotides into a new complementary strand. The identical copies of the DNA molecule are then distributed to the two daughter cells.

Table 5.2 ▪ Polypeptide Sequence as Evidence for Evolutionary Relationships

SPECIES	NUMBER OF AMINO ACID DIFFERENCES IN THE β CHAIN OF HEMOGLOBIN, COMPARED TO HUMAN HEMOGLOBIN (TOTAL CHAIN LENGTH = 146 AMINO ACIDS)
Human	0
Gorilla	1
Gibbon	2
Rhesus monkey	8
Mouse	27
Frog	67

Thus, the structure of DNA accounts for its function in transmitting genetic information whenever a cell reproduces.

We can use DNA and proteins as tape measures of evolution

Genes (DNA) and their products (proteins) document the hereditary background of an organism. The linear sequences of nucleotides in DNA molecules are passed from parents to offspring, and these DNA sequences determine the amino acid sequences of proteins. Siblings have greater similarity in their DNA and proteins than do unrelated individuals of the same species. If the evolutionary view of life is valid, we should be able to extend this concept of "molecular genealogy" to relationships *between* species: We should expect two species that appear to be closely related based on fossil and anatomical evidence to also share a greater proportion of their DNA and protein sequences than do more distantly related species. That is the case. For example, TABLE 5.2 compares a polypeptide chain of human hemoglobin with the corresponding hemoglobin polypeptide in five other vertebrates. In this chain of 146 amino acids, humans and gorillas differ in just one amino acid. More distantly related species have chains that are less similar. Molecular biology has added a new tape measure to the toolkit biologists use to assess evolutionary kinship.

▪ ▪ ▪

We have concluded our survey of macromolecules, but not our study of the chemistry of life. Applying the reductionist strategy, we have examined the architecture of molecules, but we have yet to explore the dynamic interactions between molecules that result in the biochemical changes collectively called cellular metabolism. Chapter 6, the final chapter of this unit, will take us another step up the hierarchy of biological order by introducing the fundamental principles of metabolism.

METHODS: Molecular Models and Computer Graphics

Three-dimensional structures of biological macromolecules provide important insights into molecular function. Determining the structures of macromolecules as complex as proteins, each made up of thousands of atoms, is a formidable task. Pauling, Watson and Crick, and the other pioneers of molecular biology built models from wood, wire, and plastic model sets. Computers have made it possible to build models much more quickly.

In the illustrations in this box (from the Department of Biochemistry at the University of California, Riverside), we follow the development of a computer model for the structure of an enzymatic protein called ribonuclease, whose function involves binding to a nucleic acid molecule. The first step is to crystallize the protein. (In this case, the protein is combined with a short strand of nucleic acid.) Then, using a method called X-ray crystallography, an instrument aims an X-ray beam through the crystal. ① The regularly spaced atoms of the crystal diffract (deflect) the X-rays into an orderly array. ② The diffracted X-rays expose photographic film, producing a pattern of spots. ③ From such diffraction patterns, computers generate electron density maps of successive, cross-sectional slices through the protein.

By combining the information from electron density maps with the primary structure of the protein, as determined by chemical methods, it is possible to plot the three-dimensional (*x, y, z*) coordinates of each atom. ④ Finally, graphics software enables the computer to create a picture showing the position of each atom in the molecule. The scientist can use the software to view the molecule's appearance from various angles—even from *inside* the molecule. Thus, computers have expanded the ways we can view a molecule.

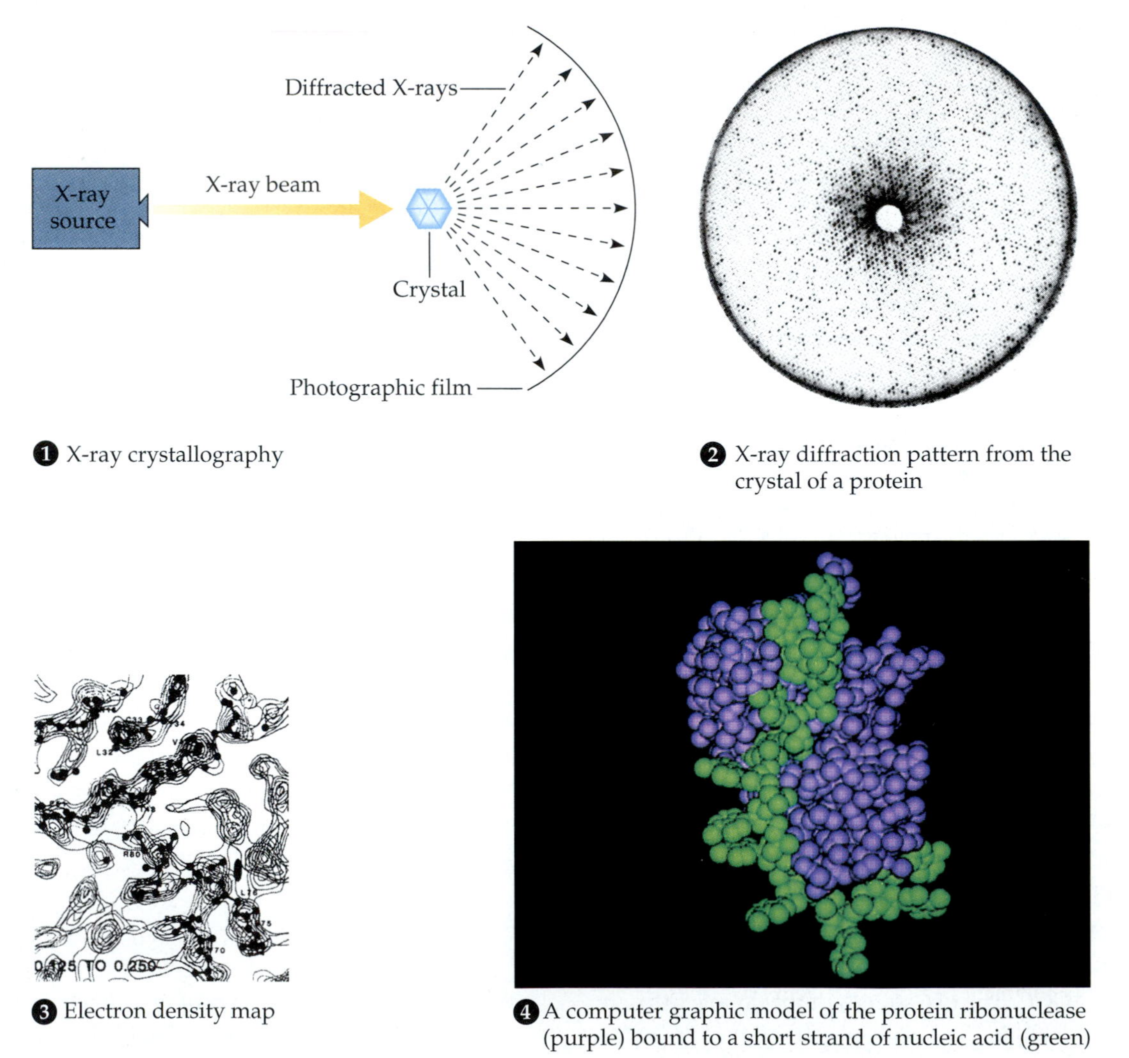

❶ X-ray crystallography

❷ X-ray diffraction pattern from the crystal of a protein

❸ Electron density map

❹ A computer graphic model of the protein ribonuclease (purple) bound to a short strand of nucleic acid (green)

CHAPTER REVIEW

REVIEW OF KEY CONCEPTS

(with page numbers and key figures)

POLYMER PRINCIPLES

- **Most macromolecules are polymers (pp. 58–59, FIGURE 5.2)** Carbohydrates, lipids, proteins, and nucleic acids are the four major classes of organic compounds in cells. Some of these compounds are *very* large and are called macromolecules. Most macromolecules are polymers, chains of identical or similar building blocks called monomers. Monomers form larger molecules by condensation reactions in which water molecules are released (dehydration). Polymers can disassemble by the reverse process, hydrolysis.
- **An immense variety of polymers can be built from a small set of monomers (pp. 59–60)** Each class of polymer is formed from a specific set of monomers. Although organisms share the same limited number of monomer types, each organism is unique because of the specific arrangement of monomers into polymers.

CARBOHYDRATES—FUEL AND BUILDING MATERIAL

5.1 **Sugars, the smallest carbohydrates, serve as fuel and carbon sources (pp. 60–61, FIGURES 5.3–5.5)** Monosaccharides are the simplest carbohydrates. They are used directly for fuel, converted to other types of organic molecules, or used as monomers for polymers. Disaccharides consist of two monosaccharides connected by a glycosidic linkage.

- **Polysaccharides, the polymers of sugars, have storage and structural roles (pp. 62–65, FIGURES 5.6–5.8)** The monosaccharide monomers of polysaccharides are connected by glycosidic linkages. Starch in plants and glycogen in animals are both storage polymers of glucose. Cellulose is an important structural polymer of glucose in plant cell walls. Starch, glycogen, and cellulose differ in the positions and orientations of their glycosidic linkages.

LIPIDS—DIVERSE HYDROPHOBIC MOLECULES

5.2 **Fats store large amounts of energy (p. 66, FIGURE 5.10)** Fats, also known as triacylglycerols, are constructed by joining a glycerol molecule to three fatty acids by dehydration reactions. Saturated fatty acids have the maximum number of hydrogen atoms. Unsaturated fatty acids (present in oils) have one or more double bonds between their carbons.

- **Phospholipids are major components of cell membranes (pp. 67–68, FIGURE 5.12)** Where fats have a third fatty acid linked to glycerol, phospholipids have a negatively charged phosphate group, which may be joined, in turn, to another small hydrophilic molecule. Thus, the "head" of a phospholipid is hydrophilic.
- **Steroids include cholesterol and certain hormones (p. 68, FIGURE 5.14)** Steroids have a basic structure of four fused rings of carbon atoms.

PROTEINS—THE MOLECULAR TOOLS OF THE CELL

5.3 A protein consists of one or more polypeptide chains folded into a specific three-dimensional conformation.

- **A polypeptide is a polymer of amino acids connected in a specific sequence (pp. 68–70, FIGURE 5.16)** Proteins are constructed from 20 different amino acids, each with a characteristic side chain (R group). The carboxyl and amino groups of adjacent amino acids link together in peptide bonds.

5.4 **A protein's function depends on its specific conformation (pp. 70–76, FIGURE 5.24)** The primary structure of a protein is its unique sequence of amino acids. Secondary structure is the folding or coiling of the polypeptide into repeating configurations, such as the α helix and the pleated sheet, which result from hydrogen bonding between parts of the polypeptide backbone. Tertiary structure is the overall three-dimensional shape of a polypeptide and results from interactions between amino acid side chains. Proteins made of more than one polypeptide chain (subunits) have a quaternary level of structure. The structure and function of a protein are sensitive to physical and chemical conditions. Protein shape is ultimately determined by its primary structure, but in the cell chaperone proteins may help the folding process.

NUCLEIC ACIDS—INFORMATIONAL POLYMERS

5.5 **Nucleic acids store and transmit hereditary information (pp. 76–77, FIGURE 5.26)** DNA stores information for the synthesis of specific proteins. RNA carries this genetic information to the protein-synthesizing machinery.

- **A nucleic acid strand is a polymer of nucleotides (pp. 77–78, FIGURE 5.27)** Each nucleotide monomer consists of a pentose covalently bonded to a phosphate group and to one of four different nitrogenous bases (A, G, C, and T or U). RNA has ribose as its pentose; DNA has deoxyribose. RNA has U and DNA T. In making a polynucleotide, nucleotides join to form a sugar-phosphate backbone from which the nitrogenous bases project. The sequence of bases along a gene specifies the amino acid sequence of a particular protein.
- **Inheritance is based on replication of the DNA double helix (pp. 78–79, FIGURE 5.28)** DNA is a helical, double-stranded macromolecule with bases projecting into the interior of the molecule. Because A always hydrogen-bonds to T, and C to G, the nucleotide sequences of the two strands are complementary. One strand can serve as a template for the formation of the other. This unique feature of DNA provides a mechanism for the continuity of life.
- **We can use DNA and proteins as tape measures of evolution (p. 79, TABLE 5.2)** Molecular comparisons help biologists sort out the evolutionary connections among species.

SELF-QUIZ

1. Which of the following terms includes all others in the list?

   a. monosaccharide
   b. disaccharide
   c. starch
   d. carbohydrate
   e. polysaccharide

2. The molecular formula for glucose is $C_6H_{12}O_6$. What would be the molecular formula for a polymer made by linking ten glucose molecules together by condensation reactions? Explain your answer.

   a. $C_{60}H_{120}O_{60}$
   b. $C_6H_{12}O_6$
   c. $C_{60}H_{102}O_{51}$
   d. $C_{60}H_{100}O_{50}$
   e. $C_{60}H_{111}O_{51}$

3. The two ring forms of glucose (α and β)

   a. are made from different structural isomers of glucose
   b. arise from different linear (nonring) glucose molecules
   c. arise when different carbons of the linear structure join to form the rings
   d. arise because the hydroxyl group at the point of ring closure can be trapped in either one of two possible positions
   e. include an aldose and a ketose

4. Choose the pair of terms that correctly completes this sentence: Nucleotides are to ____________ as ____________ are to proteins.

   a. nucleic acids; amino acids
   b. amino acids; polypeptides

c. glycosidic linkages; polypeptide linkages
d. genes; enzymes
e. polymers; polypeptides

5. Which of the following statements concerning *unsaturated* fats is correct?
   a. They are more common in animals than in plants.
   b. They have double bonds in the carbon chains of their fatty acids.
   c. They generally solidify at room temperature.
   d. They contain more hydrogen than saturated fats having the same number of carbon atoms.
   e. They have fewer fatty acid molecules per fat molecule.

6. The structural level of a protein least affected by a disruption in hydrogen bonding is the:
   a. primary level
   b. secondary level
   c. tertiary level
   d. quaternary level
   e. all structural levels are equally affected

7. To convert a nucleoside to a nucleotide, it would be necessary to:
   a. combine two nucleosides using dehydration synthesis
   b. remove the pentose from the nucleoside
   c. replace the purine with a pyrimidine
   d. add phosphate to the nucleoside
   e. replace ribose with deoxyribose

CHALLENGE QUESTIONS

1. A particular small polypeptide is nine amino acids long. Using three different enzymes to hydrolyze the polypeptide at various sites, we obtain the following five fragments (N denotes the amino end of the chain): Ala-Leu-Asp-Tyr-Val-Leu; Tyr-Val-Leu; N-Gly-Pro-Leu; Asp-Tyr-Val-Leu; N-Gly-Pro-Leu-Ala-Leu. Determine the primary structure of this polypeptide.
2. Most amino acids can exist in two forms:

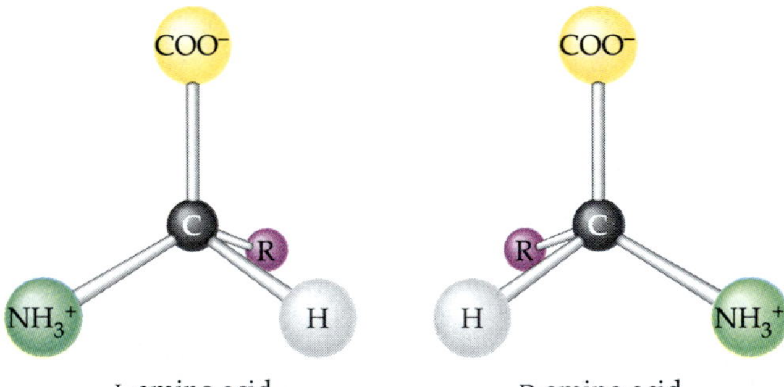

   a. Apply what you learned in Chapter 4 to explain the structural difference between these two molecules.
   b. When an organic chemist synthesizes amino acids, a mixture of these two forms is made; the result was probably the same when the first amino acids formed by abiotic synthesis in the "primordial soup" on the early Earth. However, with very few exceptions, only the L form of amino acids occurs in proteins. This is one example of molecular "handedness" in life. What are the possible advantages of life's becoming locked into the exclusive use of one of these two amino acid versions? Speculate on the possible evolutionary significance of this "handedness."

SCIENCE, TECHNOLOGY, AND SOCIETY

Some amateur and professional athletes take anabolic steroids to help them "bulk up" or build strength. The health risks of this practice are extensively documented. Apart from health considerations, how do you feel about the use of chemicals to enhance athletic performance? Is an athlete who takes anabolic steroids cheating, or is his or her use of such chemicals just part of the preparation required to succeed in a competitive sport? Defend your answer.

FURTHER READING

Dushesne, L. C., and D. W. Larson. "Cellulose and the Evolution of Plant Life." *Bioscience,* April 1989. The chemistry and natural history of the most abundant organic molecule in the biosphere.

Fackelmann, K. "Olestra: Too Good to Be True?" *Science News,* January 27, 1996. Researchers identify some health risks of fake fat.

Gibbons, A. "Geneticists Trace the DNA Trail of the First Americans." *Science,* January 15, 1993. Applying the "DNA tape measure" to a problem in anthropology.

Hoberman, J. M., and C. E. Yesalis. "The History of Synthetic Testosterone." *Scientific American,* February 1995. Anabolic steroids and sports.

Horgan, J. "Stubbornly Ahead of His Time." *Scientific American,* March 1993. A profile of Linus Pauling, a controversial giant of twentieth-century science.

Lewis, R. "Unraveling the Weave of Spider Silk." *BioScience,* October 1996. The structure of silk protein and its possible use in human inventions.

Mathews, C. K., and K. E. van Holde. *Biochemistry,* 2nd ed. Menlo Park, CA: Benjamin/Cummings, 1996. Excellent explanations and beautiful art.

Pennisi, E. "Twirling Ribbons, Billowing Bubbles." *Science News,* November 19, 1994. Computer visualization brings protein structure into view.

Strauss, E. "How Proteins Take Shape." *Science News,* September 6, 1997. How chaperone proteins help proteins fold correctly.

Taubes, G. "Misfolding the Way to Disease." *Science,* March 15, 1996. Alzheimer's disease and others may result from incorrect protein folding.

WEB LINKS

Visit the special edition of *The Biology Place* for BIOLOGY, Fifth Edition, at http://www.biology.com/campbell. Go to Chapter 5 for online resources, including learning activities, practice exams, and links to the following web sites:

"ChemCenter"
ChemCenter is a one-stop web site for chemists and also provides resources for educators, students, and individuals who want reliable, accurate information about the chemistry-related sciences.

"US Food & Drug Administration"
Advances in applied biology and chemistry have given rise to many new consumable items. The FDA ensures that the food we eat is safe and wholesome and that medicines are safe and effective.

"BioChemNet"
This is the main page of an excellent site that links chemistry and the life sciences. Contains links to sites from general chemistry to biotechnology and has a good weekly news section.

"ChemEd: Chemistry Education Resources"
A comprehensive list of links to chemistry education-related sites.

CHAPTER 6

AN INTRODUCTION TO METABOLISM

Metabolism, Energy, and Life

- The chemistry of life is organized into metabolic pathways
- Organisms transform energy
- The energy transformations of life are subject to two laws of thermodynamics
- Organisms live at the expense of free energy
- ATP powers cellular work by coupling exergonic reactions to endergonic reactions

Enzymes

- Enzymes speed up metabolic reactions by lowering energy barriers
- Enzymes are substrate-specific
- The active site is an enzyme's catalytic center
- A cell's physical and chemical environment affects enzyme activity

The Control of Metabolism

- Metabolic control often depends on allosteric regulation
- The localization of enzymes within a cell helps order metabolism
- The theme of emergent properties is manifest in the chemistry of life

The living cell is a chemical industry in miniature, where thousands of reactions occur within a microscopic space. Sugars are converted to amino acids, and vice versa. Small molecules are assembled into polymers, which may be hydrolyzed later as the needs of the cell change. In plants and animals, many cells export chemical products that are used in other parts of the organism. The chemical process known as cellular respiration drives the cellular economy by extracting the energy stored in sugars and other fuels. Cells apply this energy to perform various types of work, such as the synthesis of the macromolecules featured in Chapter 5. In a more exotic example, cells of the fungus from New Guinea in the photo at the right convert the energy stored in certain organic molecules to light, a process called bioluminescence. (The glow may attract insects that benefit the fungus by dispersing its spores.) Bioluminescence and all other metabolic activities carried out by a cell are precisely coordinated and controlled. In its complexity, its efficiency, its integration, and its responsiveness to subtle changes, the cell is peerless as a chemical institution. The concepts of metabolism you learn in this chapter will help you further understand the connections between chemistry and life.

METABOLISM, ENERGY, AND LIFE

The totality of an organism's chemical processes is called **metabolism** (from the Greek *metabole*, "change"). Metabolism is an emergent property of life that arises from specific interactions between molecules within the orderly environment of the cell.

The chemistry of life is organized into metabolic pathways

We can think of a cell's metabolism as an elaborate road map of the thousands of chemical reactions that occur in that cell (FIGURE 6.1, p. 84). These reactions are arranged in intricately branched metabolic pathways that alter molecules by a series of steps. Enzymes route matter through the metabolic pathways by selectively accelerating each step. Analogous to the red, yellow, and green lights that control the flow of traffic and prevent snarls, mechanisms that regulate enzymes balance metabolic supply and demand, averting deficits and surpluses of chemicals.

As a whole, metabolism is concerned with managing the material and energy resources of the cell. Some metabolic pathways release energy by breaking down complex molecules to simpler compounds. These degradative processes are called **catabolic pathways**. A major thoroughfare of catabolism is cellular respiration, in which the sugar glucose and other organic fuels are broken down to carbon dioxide and water. Energy that was stored in the organic molecules becomes available to do the work of the cell. **Anabolic pathways**, in contrast,

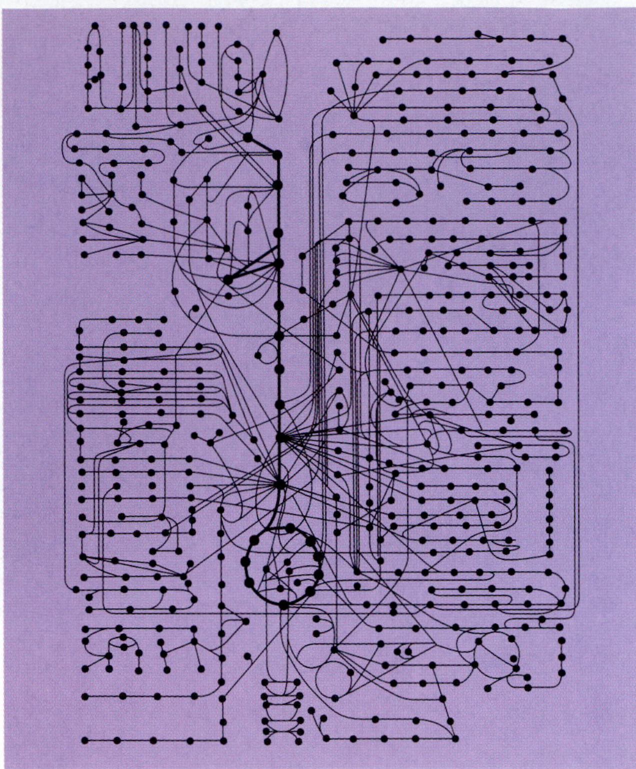

FIGURE 6.1 ▪ The complexity of metabolism. This schematic diagram traces only a few hundred of the thousands of metabolic reactions that occur in a cell. The dots represent molecules, and the lines represent the chemical reactions that transform them. The reactions proceed in stepwise sequences called metabolic pathways, each step catalyzed by a specific enzyme.

consume energy to build complicated molecules from simpler ones. An example of anabolism is the synthesis of a protein from amino acids. Catabolic and anabolic pathways are the downhill and uphill avenues of the metabolic map. The metabolic pathways intersect in such a way that energy released from the "downhill" reactions of catabolism can be used to drive the "uphill" reactions of the anabolic pathways. This transfer of energy from catabolism to anabolism is called energy coupling.

In this chapter we will focus on the mechanisms common to metabolic pathways. Since energy is fundamental to all metabolic processes, a basic knowledge of energy is necessary to understand how the living cell works. Although we will use some nonliving examples to study energy, keep in mind that the same concepts demonstrated by these examples also apply to **bioenergetics**, the study of how organisms manage their energy resources. An understanding of energy is as important for students of biology as it is for students of physics, chemistry, and engineering.

Organisms transform energy

Energy is the capacity to do work—that is, to move matter against opposing forces, such as gravity and friction. Put another way, energy is the ability to rearrange a collection of matter. For example, you expend energy to turn the pages of this book. Energy exists in various forms, and the work of life depends on the ability of cells to transform energy from one type into another.

Anything that moves possesses a form of energy called **kinetic energy**, the energy of motion. Moving objects perform work by imparting motion to other matter: A pool player uses the motion of the cue stick to push the cue ball, which in turn moves the other balls; water gushing through a dam turns turbines; electrons flowing along a wire run household appliances; the contraction of leg muscles pushes bicycle pedals. Light is also a type of kinetic energy that can be harnessed to perform work, such as powering photosynthesis in green plants. Heat, or thermal energy, is kinetic energy that results from the random movement of molecules.

A resting object not presently at work may still possess energy, which, remember, is the *capacity* to do work. Stored energy, or **potential energy**, is energy that matter possesses because of its location or structure. Water behind a dam, for instance, stores energy because of its altitude. Chemical energy, a form of potential energy especially important to biologists, is stored in molecules because of the structural arrangement of the atoms in those molecules.

How is energy converted from one form to another? Consider, for example, the playground scene in FIGURE 6.2. The girl at the bottom of the slide transformed kinetic energy to

FIGURE 6.2 ▪ Transformations between kinetic and potential energy. The children in this playground scene have more potential energy at the top of the slide (because of the effect of gravity) than they do at the bottom. They convert kinetic energy to potential (stored) energy when they climb up the slide, and convert potential energy back to kinetic energy during their descent.

potential energy when she climbed to the top of the slide. This stored energy was converted back to kinetic energy as she slid down. It was another source of potential energy, the chemical energy in the food she ate for breakfast, that enabled the girl to climb to the top in the first place.

Chemical energy can be tapped when chemical reactions rearrange the atoms of molecules in such a way that potential energy stored in the molecules is converted to kinetic energy. This transformation occurs, for example, in the engine of an automobile when the hydrocarbons of gasoline react explosively with oxygen, releasing the energy that pushes the pistons. Similarly, chemical energy fuels organisms. Cellular respiration and other catabolic pathways unleash energy stored in sugar and other complex molecules and make that energy available for cellular work. Each child who climbed the hill in FIGURE 6.2 transformed some of the chemical energy that was stored in the organic molecules of food to the kinetic energy of movements. The chemical energy stored in these fuel molecules had itself been derived from light energy by plants during photosynthesis. Organisms are energy transformers.

The energy transformations of life are subject to two laws of thermodynamics

The study of the energy transformations that occur in a collection of matter is called **thermodynamics**. Scientists use the word *system* to denote the matter under study and refer to the rest of the universe—everything outside the system—as the *surroundings*. A *closed system*, such as that approximated by liquid in a thermos bottle, is isolated from its surroundings. In an *open system*, energy (and often matter) can be transferred between the system and its surroundings. Organisms are open systems. They absorb energy—for instance, chemical energy in the form of organic molecules, or light energy—and release heat and metabolic waste products, such as carbon dioxide, to the surroundings. Two laws of thermodynamics govern energy transformations in organisms and all other collections of matter.

According to the **first law of thermodynamics**, the energy of the universe is constant. *Energy can be transferred and transformed, but it can be neither created nor destroyed.* The first law is also known as *conservation of energy*. The electric company does not manufacture energy, but merely converts it to a form that is convenient to use. By converting light to chemical energy, a green plant acts as an energy transformer, not an energy producer. What happens to energy after it has performed work in a machine or an organism? If energy cannot be destroyed, what prevents organisms from behaving like closed systems and recycling their energy? The second law answers these questions.

The **second law of thermodynamics** can be stated many ways. Let's begin with the following interpretation: Every energy transfer or transformation makes the universe more disordered. Scientists use a quantity called **entropy** as a measure of disorder, or randomness. The more random a collection of matter is, the greater its entropy. We can now restate the second law as follows: *Every energy transfer or transformation increases the entropy of the universe.* There is an unstoppable trend toward randomization. In many cases, increased entropy is evident in the physical disintegration of a system's organized structure. For example, you can observe this increasing entropy in the gradual decay of an unmaintained building. Much of the increasing entropy of the universe is less apparent, however, because it takes the form of an increasing amount of heat, which is the energy of random molecular motion.

In most energy transformations, ordered forms of energy are at least partly converted to heat. Only about 25% of the chemical energy stored in the fuel tank of an automobile is transformed into the motion of the car; the remaining 75% is lost from the engine as heat, which dissipates rapidly through the surroundings. Similarly, the children in FIGURE 6.2 convert only a fraction of the energy stored in their food to the kinetic energy of hill climbing and other play. In performing various kinds of work, living cells unavoidably convert organized forms of energy to heat. (This can make a room crowded with people uncomfortably warm.)

In machines and organisms, even energy that performs useful work is eventually converted to heat. The organized energy of an automobile's forward movement becomes heat when the friction of the brakes and tires stops the car. Conversion to heat is the fate of *all* the chemical energy a child uses to climb a slide: Metabolic breakdown of food generates heat during the climb, and the fraction of energy temporarily stored as gravitational potential energy is converted to heat on the way down, as friction between child and slide warms the surrounding air.

Conversion of other forms of energy to heat does not violate the first law of thermodynamics. Energy has been conserved, because heat is a form of energy, though energy in its most random state. By combining the first and second laws of thermodynamics, we can conclude that the *quantity* of energy in the universe is constant, but its *quality* is not. In a sense, heat is the lowest grade of energy. It is the uncoordinated movement of molecules, which many systems cannot harness in order to perform work. A system can only put heat to work when there is a temperature difference that results in the heat flowing from a warmer location to a cooler one. If temperature is uniform throughout a system, as it is in a living cell, then the only use for heat energy is to warm a body of matter, such as an organism.

Thus, an organism takes in organized forms of matter and energy from the surroundings and replaces them with less ordered forms. For example, an animal obtains starch, proteins, and other complex molecules from the food it eats. As catabolic pathways break these molecules down, the animal releases carbon dioxide and water, relatively small, simple molecules that store less chemical energy than the food. The depletion of chemical energy is accounted for by heat generated during metabolism. On a larger scale, energy flows into an ecosystem in the form of light and leaves in the form of heat. Living systems increase the entropy of their surroundings, as predicted by thermodynamic law.

How can we reconcile the second law of thermodynamics—the unstoppable increase in the entropy of the universe—with the orderliness of life, which is one of this book's themes? The key is to remember another theme: Organisms are open systems that exchange energy and materials with their surroundings. It is true that cells create ordered structures from less organized starting materials. For example, amino acids are ordered into the specific sequences of polypeptide chains. And during the early history of life, complex organisms evolved from simpler ancestors (FIGURE 6.3). However, this increase in organization in no way violates the second law. The entropy of a particular system, such as an organism, may actually decrease, so long as the total entropy of the *universe*—the system plus its surroundings—increases. Thus, organisms are islands of low entropy in an increasingly random universe. The evolution of biological order is perfectly consistent with the laws of thermodynamics.

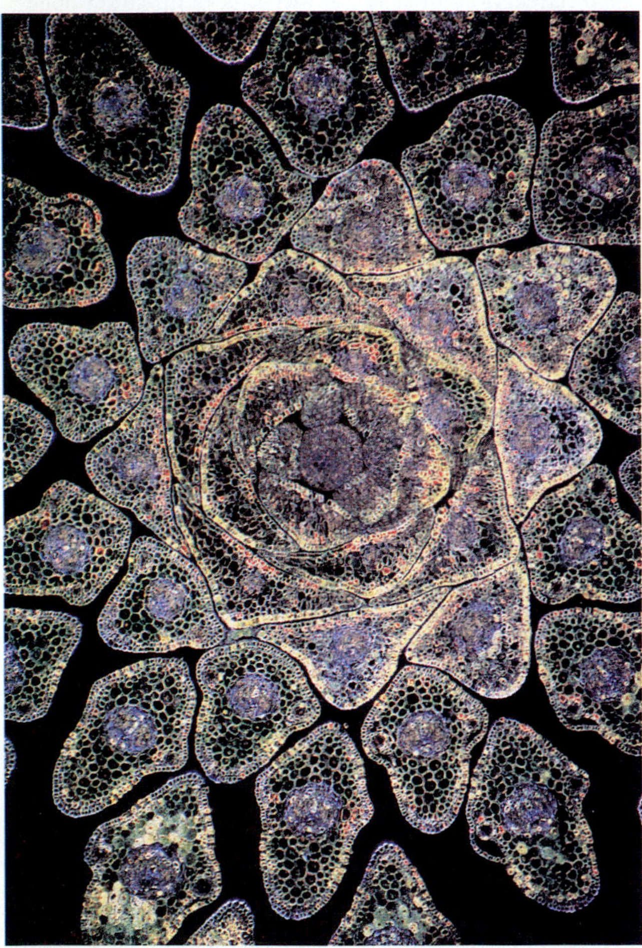

FIGURE 6.3 • The evolution of biological order is consistent with thermodynamic law. Order is a characteristic of life. It is evident, for example, in the detailed anatomy of this bud from a spruce tree (TEM of a cross section). Organisms decrease their entropy when they order raw materials, such as organic monomers, into macromolecules and then organize these macromolecules into cellular structures. Biological order has, at times, also increased over the grander scale of geological time. For example, we can trace the ancestry of the plant kingdom to much simpler organisms called green algae. An evolutionary interpretation of biological history does not violate thermodynamic law. The second law requires only that processes increase the entropy of the universe. Open systems can increase their order at the expense of the order of their surroundings.

Organisms live at the expense of free energy

How can we predict what can and cannot occur in nature? How can we distinguish the possible from the impossible? We know from experience that certain events occur spontaneously and others do not. For instance, we know that water flows downhill, that objects of opposite charge move toward each other, that an ice cube melts at room temperature, and that a sugar cube dissolves in water. Explaining *why* these processes occur spontaneously is tricky.

Let's begin by defining a spontaneous process as a change that can occur without outside help. A spontaneous change can be harnessed in order to perform work. The downhill flow of water can be used to turn a turbine in a power plant, for example. A process that cannot occur on its own is said to be nonspontaneous; it will happen only if energy is added to the system. Water moves uphill only when a windmill or some other machine pumps the water against gravity, and a cell must expend energy to synthesize a protein from amino acids.

When a spontaneous process occurs in a system, the stability of that system increases. Unstable systems tend to change in such a way that they become more stable. A body of elevated water, such as a reservoir, is less stable than the same water at sea level. A system of charged particles is less stable when opposite charges are apart than when they are together. In situations less familiar to us, how can we predict which changes lead to greater stability in a system—that is, which changes are spontaneous? You have already learned that a process can occur spontaneously only if it increases the disorder (entropy) of the universe. This principle is helpful in theory, but it does not give us a practical criterion to apply to biological systems because it requires that we measure changes in the surroundings. We need some standard for spontaneity that is based on the system alone. That criterion is called free energy.

Free Energy: A Criterion for Spontaneous Change

The concept of free energy is not easy to grasp, but the effort is worthwhile because we can apply the idea to many biological problems. ***Free energy*** *is the portion of a system's energy that can perform work when temperature is uniform throughout the system,* as in a living cell. It is called *free* energy because it is available for work, not because it can be spent without cost to the universe. In fact, you will soon understand that organisms can live only at the expense of free energy acquired from the surroundings.

A system's quantity of free energy is symbolized by the letter *G*. There are two components to *G*: the system's total energy (symbolized by *H*) and its entropy (symbolized by *S*). Free energy is related to these factors in the following way:

$$G = H - TS$$

T stands for absolute temperature (in Kelvin units, K, equal to °C + 273; see Appendix Two). Notice that temperature amplifies the entropy term of the equation. This makes sense if you remember that temperature measures the intensity of random molecular motion (heat), which tends to disrupt order. What does this equation tell us about free energy? Not all the energy stored in a system (*H*) is available for work. The system's disorder, the entropy factor, is subtracted from total energy in computing the maximum capacity of the system to perform useful work. We are then left with free energy, which is somewhat less than the system's total energy. The relationships between free energy, equilibrium, and work are illustrated in FIGURE 6.4.

How does the concept of free energy help us determine whether a particular process can occur spontaneously? Think of free energy, *G* in the previous equation, as a measure of a system's instability—its tendency to change to a more stable state. Systems that are rich in energy, such as stretched springs or separated charges, are unstable; so are highly ordered systems, such as complex molecules. Thus, those systems that tend to change spontaneously to a more stable state are those that have high energy, low entropy, or both. The free-energy equation weighs these two factors, which are consolidated in the system's *G* content. Now we can state a versatile criterion for spontaneous change: *In any spontaneous process, the free energy of a system decreases.*

The change in free energy as a system goes from a starting state to a different state is represented by ΔG:

$$\Delta G = G_{final\ state} - G_{starting\ state}$$

Or, put another way:

$$\Delta G = \Delta H - T\Delta S$$

For a process to occur spontaneously, the system must either give up energy (a decrease in *H*), give up order (an increase in *S*), or both. When these changes in *H* and *S* are tallied, ΔG must have a negative value. The greater this decrease in free energy, the greater the maximum amount of work the spontaneous process can perform. This is a formal, mathematical way of stating the obvious: Nature runs "downhill"—downhill in the metaphorical sense of a loss in useful energy, the capacity to perform work.

More free energy
Less stable
Greater work capacity

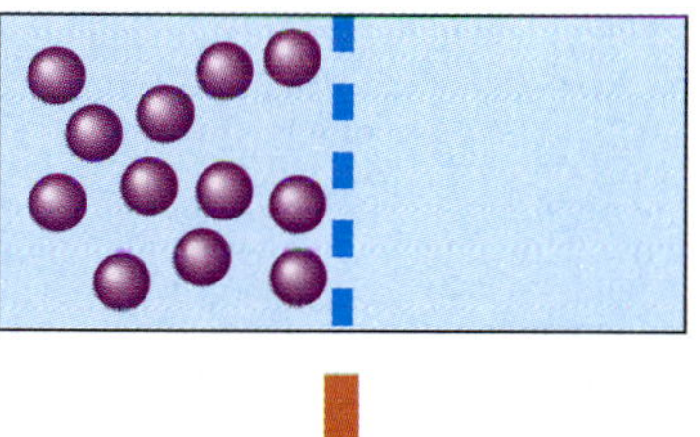

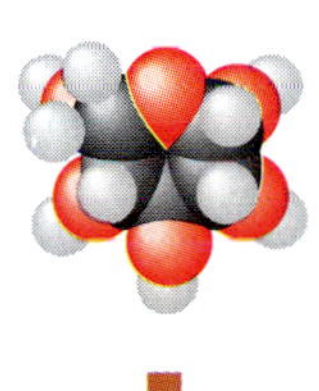

Free energy decreases; direction of spontaneous change; $\Delta G < 0$;
change in the direction of equilibrium; change can be harnessed to perform work

Less free energy
More stable
Less work capacity

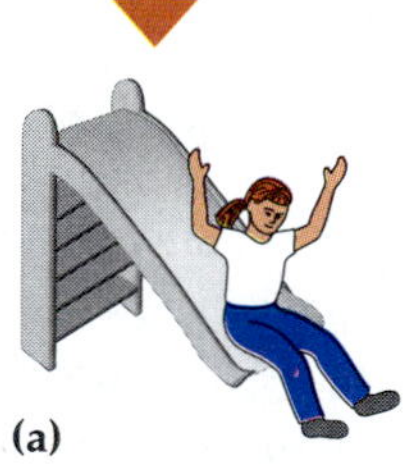

(a)

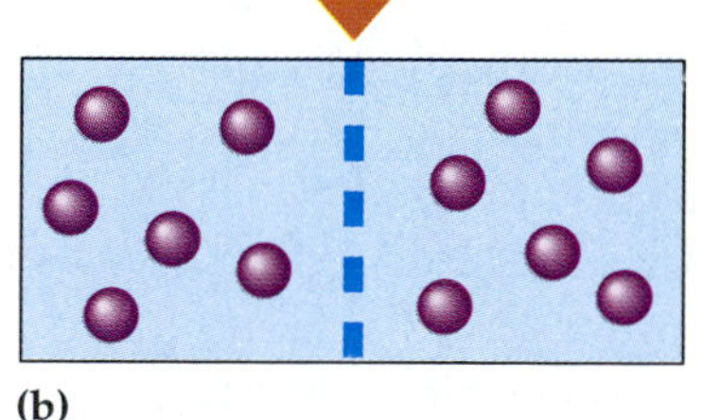

(b)

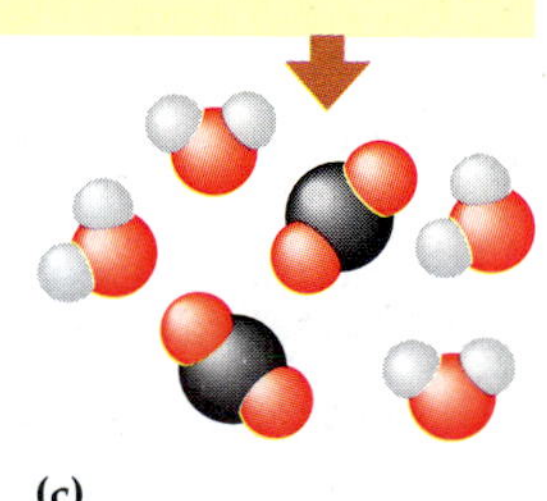

(c)

FIGURE 6.4 • The relationship of free energy to stability, spontaneous change, equilibrium, and work. An unstable system is rich in free energy. It has a tendency to change spontaneously to a more stable state, and it is possible to harness this "downhill" change in order to perform work. **(a)** In this case, free energy is proportional to the girl's altitude. **(b)** The free-energy concept also applies on the molecular scale, in this case to the physical movement of molecules known as diffusion. Here, molecules of a particular solute are distributed unequally across a membrane separating two aqueous compartments. This ordered state is unstable; it is rich in free energy. If the solute molecules can cross the membrane, there will be a net movement (diffusion) of the molecules until they are equally concentrated in both compartments. **(c)** Chemical reactions also involve free energy. The sugar molecule on top is less stable than the simpler molecules below. When catabolic pathways break down complex organic molecules, a cell can harness the free energy stored in the molecules to perform work.

Free Energy and Equilibrium

There is an important relationship between free energy and equilibrium, including chemical equilibrium. Recall from Chapter 2 that most chemical reactions are reversible and proceed until the forward and backward reactions occur at the same rate. The reaction is then said to be at chemical equilibrium, and there is no further change in the concentration of products or reactants. As a reaction proceeds toward equilibrium, the free energy of the mixture of reactants and products decreases. Free energy increases when a reaction is somehow pushed away from equilibrium. For a reaction at equilibrium, $\Delta G = 0$, because there is no net change in the system. We can think of equilibrium as an energy valley. A chemical reaction or physical process at equilibrium performs no work. A process is spontaneous and can perform work when sliding toward equilibrium. Movement away from equilibrium is nonspontaneous; it can occur only with the help of an outside energy source. We can now apply the free-energy concept more specifically to the chemistry of life.

Free Energy and Metabolism

Exergonic and Endergonic Reactions in Metabolism. Based on their free-energy changes, chemical reactions can be classified as either exergonic (meaning "energy outward") or endergonic ("energy inward"). An **exergonic reaction** proceeds with a net release of free energy. Since the chemical mixture loses free energy, ΔG is negative for an exergonic reaction. In other words, exergonic reactions are those that occur spontaneously. The magnitude of ΔG for an exergonic reaction is the maximum amount of work the reaction can perform. We can use the overall reaction for cellular respiration as an example:

$$C_6H_{12}O_6 + 6\,O_2 \longrightarrow 6\,CO_2 + 6\,H_2O$$

$$\Delta G = -686 \text{ kcal/mol } (-2870 \text{ kJ/mol})$$

For each mole (180 g) of glucose broken down by respiration, 686 kilocalories (or 2870 kilojoules) of energy are made available for work (under what scientists call standard conditions). Because energy must be conserved, the chemical products of respiration store 686 kcal less free energy than the reactants. The products are, in a sense, the spent exhaust of a process that tapped most of the free energy stored in the sugar molecules.

An **endergonic reaction** is one that absorbs free energy from its surroundings. Because this kind of reaction *stores* free energy in molecules, ΔG is positive. Such reactions are nonspontaneous, and the magnitude of ΔG is the quantity of energy required to drive the reaction. If a chemical process is exergonic (downhill) in one direction, then the reverse process must be endergonic (uphill). A reversible process cannot be downhill in both directions. If $\Delta G = -686$ kcal/mol for respiration, then in order for photosynthesis to produce sugar from carbon dioxide and water, $\Delta G = +686$ kcal/mol. Sugar production in the leaf cells of a plant is steeply endergonic, an uphill process powered by the absorption of light energy from the sun.

Metabolic Disequilibrium. The chemical reactions of metabolism are reversible and would reach equilibrium if they occurred in the isolation of a test tube. Because chemical systems at equilibrium have a ΔG of zero and can do no work, a cell that has reached equilibrium is dead! In fact, metabolic disequilibrium is one of the defining features of life. We'll use the catabolic pathway of respiration as an example. Some of the reversible reactions of respiration are pulled in one direction and kept out of equilibrium. The key to sustaining this disequilibrium is that the product of one reaction does not accumulate, but instead becomes a reactant in the next step along the metabolic pathway; FIGURE 6.5 shows an analogy. The overall sequence of reactions is pulled by the huge free-energy difference between glucose at the uphill end of respiration and carbon dioxide and water at the downhill end of the pathway. As long as the cell has a steady supply of glucose or other fuels and is able to expel the CO_2 waste to the surroundings, the cell does not reach equilibrium and continues to do the work of life.

We see once again how important it is to think of organisms as open systems. Sunlight provides a daily source of free energy for an ecosystem's plants and other photosynthetic organisms. Animals and other nonphotosynthetic organisms in an ecosystem depend on free-energy transfusions in the form of the organic products of photosynthesis.

Now that we have applied the free-energy concept to metabolism, we are ready to see how a cell actually performs the work of life. A key strategy in bioenergetics is **energy coupling**, the use of an exergonic process to drive an endergonic one. A molecule called ATP is responsible for mediating most energy coupling in cells.

ATP powers cellular work by coupling exergonic reactions to endergonic reactions

A cell does three main kinds of work:

1. *Mechanical work,* such as the beating of cilia, the contraction of muscle cells, and the movement of chromosomes during cellular reproduction.
2. *Transport work,* the pumping of substances across membranes against the direction of spontaneous movement.
3. *Chemical work,* the pushing of endergonic reactions that would not occur spontaneously, such as the synthesis of polymers from monomers.

In most cases, the immediate source of energy that powers cellular work is ATP.

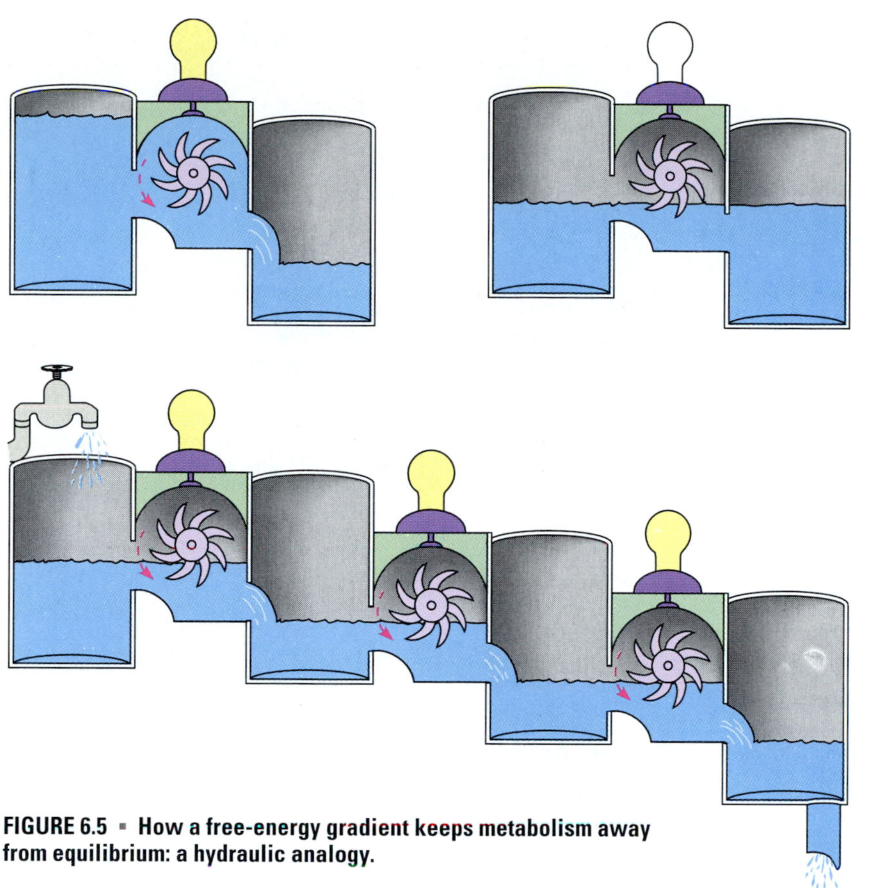

(a) This diagram portrays a system in which water generates electric energy only while it is falling. Once the levels in the two containers are equal, the turbine ceases to turn and the light goes out. The individual steps of respiration, in isolation, would also come to equilibrium, and cellular work would cease.

(b) If there is a series of drops in water level, electric energy can be generated at each drop. In respiration, there is a series of drops in free energy between glucose, the starting material, and the metabolic wastes at the end (carbon dioxide and water). The overall process never reaches equilibrium as long as the organism lives. A cell continues to acquire free energy in the form of glucose. The product of each reaction becomes the reactant for the next, and the metabolic wastes are expelled from the cell.

FIGURE 6.5 ▪ How a free-energy gradient keeps metabolism away from equilibrium: a hydraulic analogy.

The Structure and Hydrolysis of ATP

ATP (adenosine triphosphate) is closely related to one type of nucleotide found in nucleic acids. ATP has the nitrogenous base adenine bonded to ribose, as in an adenine nucleotide of RNA. However, in RNA, *one* phosphate group is attached to the ribose (see FIGURE 5.27a). Adenosine *tri*phosphate has a chain of *three* phosphate groups attached to the ribose (FIGURE 6.6a).

The bonds between the phosphate groups of ATP's tail can be broken by hydrolysis. When the terminal phosphate bond is broken, a molecule of inorganic phosphate (abbreviated Ⓟ$_i$ throughout this book) leaves the ATP, which becomes

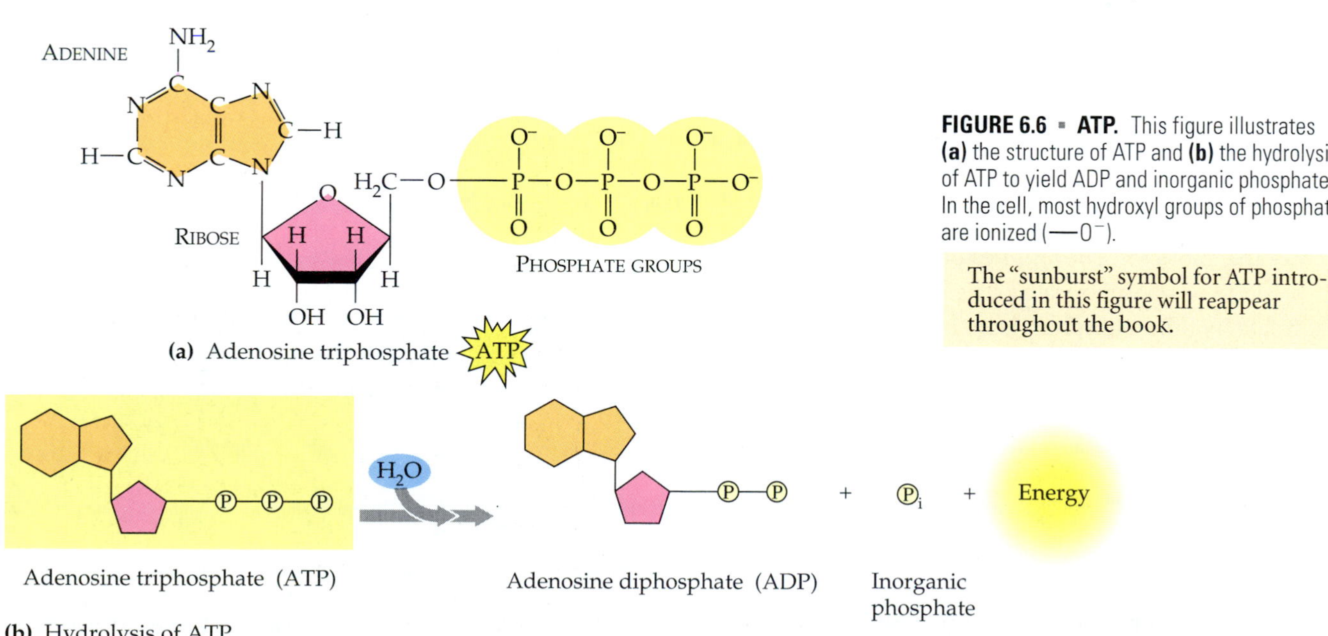

FIGURE 6.6 ▪ ATP. This figure illustrates **(a)** the structure of ATP and **(b)** the hydrolysis of ATP to yield ADP and inorganic phosphate. In the cell, most hydroxyl groups of phosphates are ionized (—O$^-$).

The "sunburst" symbol for ATP introduced in this figure will reappear throughout the book.

adenosine diphosphate, or ADP (FIGURE 6.6b). The reaction is exergonic and, under laboratory conditions, releases 7.3 kcal of energy per mole of ATP hydrolyzed:

$$ATP + H_2O \longrightarrow ADP + \text{Ⓟ}_i$$

$$\Delta G = -7.3 \text{ kcal/mol } (-31 \text{ kJ/mol})$$

This is the free-energy change measured under what are called standard conditions. However, the chemical and physical conditions in the cell do not conform to standard conditions. When the reaction occurs in the cellular environment rather than in a test tube, the actual ΔG is about -13 kcal/mole, 78% greater than the energy released by ATP hydrolysis under standard conditions.

Because their hydrolysis releases energy, the phosphate bonds of ATP are sometimes referred to as high-energy phosphate bonds, but the term is misleading. The phosphate bonds of ATP are not unusually strong bonds, as the words "high-energy" imply. In fact, compared to most bonds in organic molecules, these bonds are relatively weak, and it is *because* they are somewhat unstable that their hydrolysis yields energy. The products of hydrolysis (ADP and Ⓟ$_i$) are more stable than ATP. When a system changes in the direction of greater stability—as when a compressed spring relaxes, for instance—the change is exergonic. Thus, the release of energy during the hydrolysis of ATP comes from the chemical change to a more stable condition, not from the phosphate bonds themselves. Why are the phosphate bonds so fragile? If we reexamine the ATP molecule in FIGURE 6.6a, we can see that all three phosphate groups are negatively charged. These like charges are crowded together, and their repulsion contributes to the instability of this region of the ATP molecule. The triphosphate tail of ATP is the chemical equivalent of a loaded spring.

How ATP Performs Work

When ATP is hydrolyzed in a test tube, the release of free energy merely heats the surrounding water. In the cell, that would be an inefficient and dangerous use of a valuable energy resource. With the help of specific enzymes, the cell is able to couple the energy of ATP hydrolysis directly to endergonic processes by transferring a phosphate group from ATP to some other molecule. The recipient of the phosphate group is then said to be phosphorylated. The key to the coupling is the formation of this **phosphorylated intermediate**, which is more reactive (less stable) than the original molecule (FIGURE 6.7). Nearly all cellular work depends on ATP's energizing of other molecules by transferring phosphate groups. For instance, ATP powers the movement of muscles by transferring phosphate to contractile proteins.

The Regeneration of ATP

An organism at work uses ATP continuously, but ATP is a renewable resource that can be regenerated by the addition of phosphate to ADP (FIGURE 6.8). The ATP cycle moves at an astonishing pace. For example, a working muscle cell recycles its entire pool of ATP about once each minute. That turnover represents 10 million molecules of ATP consumed and regenerated per second per cell. If ATP could not be regenerated by the phosphorylation of ADP, humans would consume nearly their body weight in ATP each day.

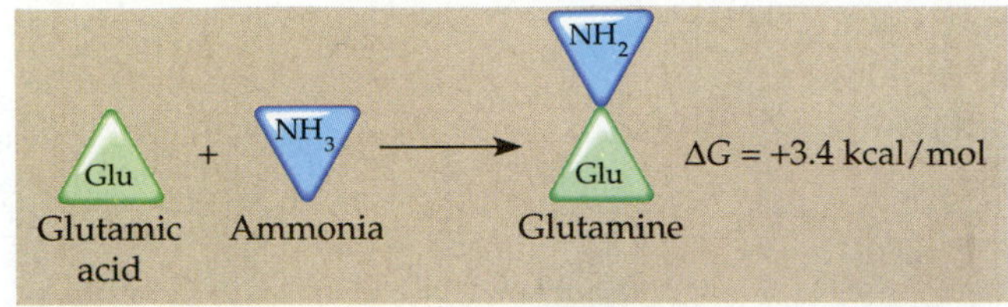

(a)

(b)

Glu + NH_3	⟶	Glu—NH_2	ΔG = +3.4 kcal/mol
ATP	⟶	ADP + Ⓟ$_i$	ΔG = −7.3 kcal/mol
			Net ΔG = −3.9 kcal/mol

(c)

FIGURE 6.7 • Energy coupling by phosphate transfer. In this example, ATP hydrolysis is used to drive an endergonic reaction, the conversion of the amino acid glutamic acid (Glu) to another amino acid, glutamine (Glu—NH_2). **(a)** Without the help of ATP, the conversion is nonspontaneous. **(b)** As it actually occurs in the cell, the synthesis of glutamine is a two-step reaction driven by ATP. The formation of a phosphorylated intermediate couples the two steps. ① ATP phosphorylates glutamic acid, making the amino acid less stable. ② Ammonia displaces the phosphate group, forming glutamine. **(c)** Adding the ΔG for the amino acid conversion to the ΔG for the ATP hydrolysis gives the free-energy change for the overall reaction. Because the overall process is exergonic (has a negative ΔG), it occurs spontaneously.

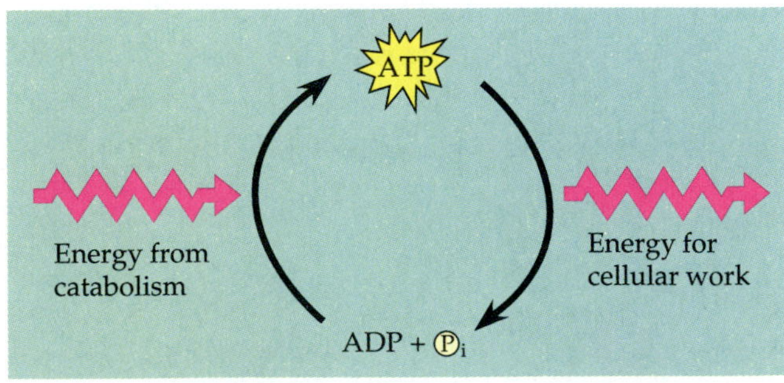

FIGURE 6.8 • The ATP cycle. Energy released by breakdown reactions (catabolism) in the cell is used to phosphorylate ADP, regenerating ATP. Energy stored in ATP drives most cellular work. Thus ATP couples the cell's energy-yielding processes to the energy-consuming ones.

Because a reversible process cannot go downhill both ways, the regeneration of ATP from ADP is necessarily endergonic:

$$\text{ADP} + Ⓟ_i \longrightarrow \text{ATP} + H_2O$$

$$\Delta G = +7.3 \text{ kcal/mol (standard conditions)}$$

Catabolic (exergonic) pathways, especially cellular respiration, provide the energy for the endergonic process of making ATP. Plants also use light energy to produce ATP.

Thus the ATP cycle is a turnstile through which energy passes during its transfer from catabolic to anabolic pathways.

ENZYMES

The laws of thermodynamics tell us what can and cannot hap-
6.1 pen but say nothing about the speed of these processes. A spontaneous chemical reaction may occur so slowly that it is imperceptible. For example, the hydrolysis of sucrose (table sugar) to glucose and fructose is exergonic, occurring spontaneously with a release of free energy ($\Delta G = -7$ kcal/mol). Yet a solution of sucrose dissolved in sterile water will sit for years at room temperature with no appreciable hydrolysis. However, if we add a small amount of the enzyme sucrase to the solution, then all the sucrose may be hydrolyzed within seconds. How does an enzyme do this?

Enzymes speed up metabolic reactions by lowering energy barriers

Enzymes are catalytic proteins. (Another class of biological catalysts, ribozymes, made of RNA, is discussed in Chapters 17 and 26.) A **catalyst** is a chemical agent that changes the rate of a reaction without being consumed by the reaction. In the absence of enzymes, chemical traffic through the pathways of metabolism would become hopelessly congested. What impedes a spontaneous reaction, and how does an enzyme change the situation?

Every chemical reaction involves both bond breaking and bond forming. For example, the hydrolysis of sucrose involves first breaking the bond between glucose and fructose and then forming new bonds with a hydrogen and a hydroxyl group from water. Whenever a reaction rearranges the atoms of molecules, existing bonds in the reactants must be broken and the new bonds of the products formed. The reactant molecules must absorb energy from their surroundings for their bonds to break, and energy is released when the new bonds of the product molecules are formed.

The initial investment of energy for starting a reaction—the energy required to break bonds in the reactant molecules—is known as the **free energy of activation**, or **activation energy**, abbreviated E_A in this book. It is usually provided in the form of heat that the reactant molecules absorb from the surroundings. If the reaction is exergonic, E_A will be repaid with dividends, as the formation of new bonds releases more energy than was invested in the breaking of old bonds. FIGURE 6.9 graphs these energy changes for a hypothetical reaction that swaps portions of two reactant molecules:

$$AB + CD \longrightarrow AC + BD$$

The bonds of the reactants break only when the molecules have absorbed enough energy to become unstable. (Recall that systems rich in free energy are intrinsically unstable, and unstable systems are reactive.) The activation energy is represented by the uphill portion of the graph, with the free-energy content of the reactants increasing. The absorption of thermal energy increases the speed of the reactants, so they are colliding more often and more forcefully. Also, thermal agitation of the atoms in the molecules makes the bonds more likely to break. At the summit, the reactants are in an unstable condition known as the *transition state;* they are primed, and the reaction can occur. As the molecules settle into their new bonding arrangements, energy is released to the surroundings. This phase of the reaction corresponds to the downhill part of the curve, which indicates a loss of free energy by the molecules. The difference in the free energy of the products and reactants is ΔG for the overall reaction, which is negative for an exergonic reaction.

As FIGURE 6.9 shows, even for an exergonic reaction, which is energetically downhill overall, the barrier of activation energy must be scaled before the reaction can occur. For some reactions, E_A is modest enough that even at room temperature there is sufficient thermal energy for many of the reactants to

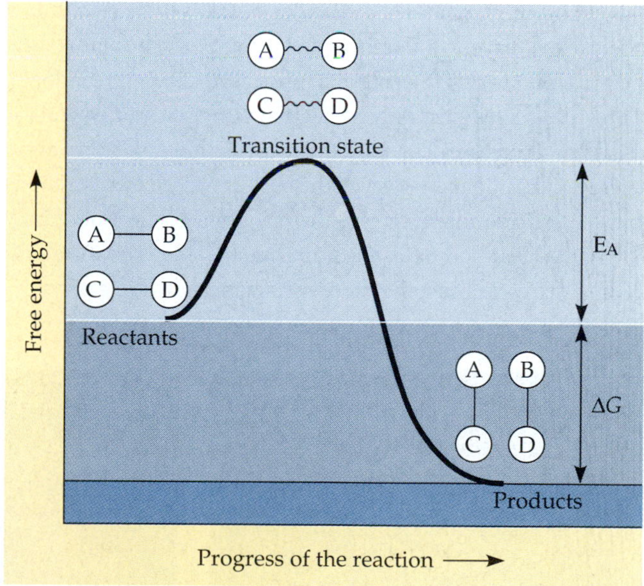

6.1 **FIGURE 6.9 • An energy profile of a reaction.** In this hypothetical reaction, the reactants AB and CD must absorb enough energy from the surroundings to surmount the hill of activation energy (E_A) and reach the unstable transition state. The bonds can then break, and as the reaction proceeds, energy is released to the surroundings during the formation of new bonds. This particular graph profiles the energy inputs and outputs of an exergonic reaction, which has a negative ΔG; the products have less free energy than the reactants.

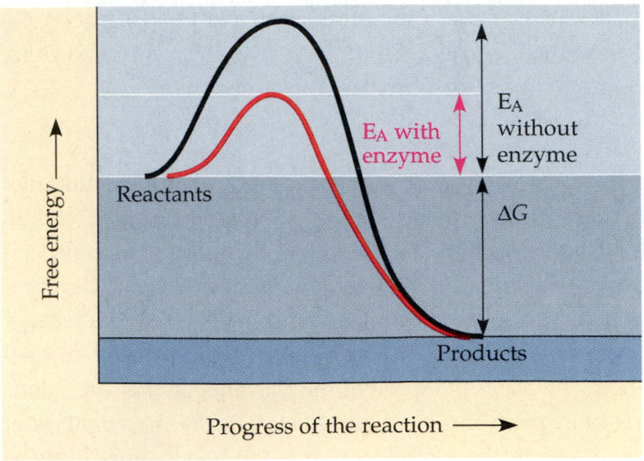

FIGURE 6.10 ▪ Enzymes lower the barrier of activation energy.
6.1 Without affecting the free-energy change (ΔG) for the reaction, an enzyme speeds the reaction by reducing the uphill climb to the transition state. The black curve shows the course of the reaction without an enzyme; the red curve shows the course of the reaction with an enzyme.

reach the transition state. In most cases, however, the E_A barrier is loftier, and the reaction will occur at a noticeable rate only if the reactants are heated. The spark plugs in an automobile engine energize the gasoline-oxygen mixture so that the molecules reach the transition state and react; only then can there be the explosive release of energy that pushes the pistons. Without a spark, the hydrocarbons of gasoline are too stable to react with oxygen.

The barrier of activation energy is essential to life. Proteins, DNA, and other complex molecules of the cell are rich in free energy and have the potential to decompose spontaneously; that is, the laws of thermodynamics favor their breakdown. These molecules exist only because at temperatures typical for cells, few molecules can make it over the hump of activation energy. Occasionally, however, the barrier for selected reactions must be surmounted, or else the cell would be metabolically stagnant. Heat speeds a reaction, but high temperature kills cells. Organisms must therefore use an alternative: a catalyst.

An enzyme speeds a reaction by lowering the E_A barrier (FIGURE 6.10), so that the precipice of the transition state is within reach even at moderate temperatures. An enzyme cannot change the ΔG for a reaction; it cannot make an endergonic reaction exergonic. Enzymes can only hasten reactions that would occur eventually anyway, but this function makes it possible for the cell to have a dynamic metabolism. Further, because enzymes are very selective in the reactions they catalyze, they determine which chemical processes will be going on in the cell at any particular time.

Enzymes are substrate-specific

The reactant an enzyme acts on is referred to as the enzyme's **substrate.** The enzyme binds to its substrate (or substrates, when there are two or more reactants). While enzyme and substrate are joined, the catalytic action of the enzyme converts the substrate to the product (or products) of the reaction. The overall process can be summarized as follows, with the name of the enzyme written above the reaction arrow:

$$\text{Substrate(s)} \xrightarrow{\text{Enzyme}} \text{Product(s)}$$

For example, the enzyme sucrase (most enzyme names end in *-ase*) breaks the disaccharide sucrose into its two monosaccharides, glucose and fructose:

$$\text{Sucrose} + H_2O \xrightarrow{\text{Sucrase}} \text{Glucose} + \text{Fructose}$$

An enzyme can distinguish its substrate from even closely related compounds, such as isomers, so that each type of enzyme catalyzes a particular reaction. For instance, sucrase will act only on sucrose and will reject other disaccharides, such as maltose. What accounts for this molecular recognition? Recall that enzymes are proteins, and proteins are macromolecules with unique three-dimensional conformations. The specificity of an enzyme results from its shape.

Only a restricted region of the enzyme molecule actually binds to the substrate. This region, called the **active site,** is

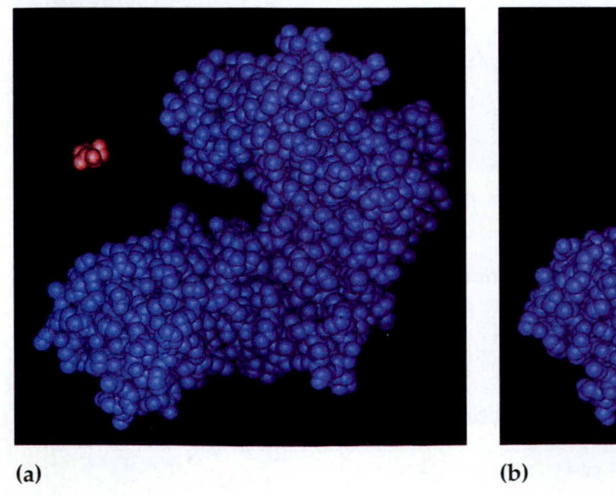

(a) (b)

FIGURE 6.11 ▪ The induced fit between an enzyme's active site and its substrate. **(a)** The active site of this enzyme, called hexokinase, can be seen in this computer graphic model as a groove on the surface of the blue protein. **(b)** On entering the active site, the substrate, which is glucose (red), induces a slight change in the shape of the protein, causing the active site to embrace the substrate.

typically a pocket or groove on the surface of the protein (FIGURE 6.11). Usually, the active site is formed by only a few of the enzyme's amino acids, with the rest of the protein molecule providing a framework that reinforces the configuration of the active site.

The specificity of an enzyme is attributed to a compatible fit between the shape of its active site and the shape of the substrate. The active site, however, is not a rigid receptacle for the substrate. As the substrate enters the active site, it induces the enzyme to change its shape slightly so that the active site fits even more snugly around the substrate. This **induced fit** is like a clasping handshake. Induced fit brings chemical groups of the active site into positions that enhance their ability to catalyze the chemical reaction.

The active site is an enzyme's catalytic center

In an enzymatic reaction, the substrate binds to the active site to form an enzyme-substrate complex (FIGURE 6.12). In most cases, the substrate is held in the active site by weak interactions, such as hydrogen bonds and ionic bonds. Side chains (R groups) of a few of the amino acids that make up the active site catalyze the conversion of substrate to product, and the product departs from the active site. The enzyme is then free to take another substrate molecule into its active site. The entire cycle happens so fast that a single enzyme molecule typically acts on about a thousand substrate molecules per second. Some enzymes are much faster. Enzymes, like other catalysts, emerge from the reaction in their original form. Therefore, very small amounts of enzyme can have a huge metabolic impact by functioning over and over again in catalytic cycles.

Enzymes use a variety of mechanisms that lower activation energy and speed up a reaction. In reactions involving two or more reactants, the active site provides a template for the substrates to come together in the proper orientation for a reaction to occur between them. As the active site clutches the substrates with an induced fit, the enzyme may stress the substrate molecules, stretching and bending critical chemical bonds that must be broken during the reaction. Since E_A is proportional to the difficulty of breaking bonds, distorting the substrate reduces the amount of thermal energy that must be absorbed to achieve a transition state.

The active site may also provide a microenvironment that is conducive to a particular type of reaction. For example, if the active site has amino acids with acidic side chains (R groups), the active site may be a pocket of low pH in an otherwise neutral cell. In such cases, an acidic amino acid may facilitate H^+ transfer to the substrate as a key step in catalyzing the reaction. Still another mechanism of catalysis is the direct participation of the active site in the chemical reaction. Sometimes this process even involves brief covalent bonding between the substrate and a side chain of an amino acid of the enzyme. Subsequent steps of the reaction restore the side chains to their original states, so the active site is the same after the reaction as it was before.

The rate at which a given amount of enzyme converts substrate to product is partly a function of the initial concentration

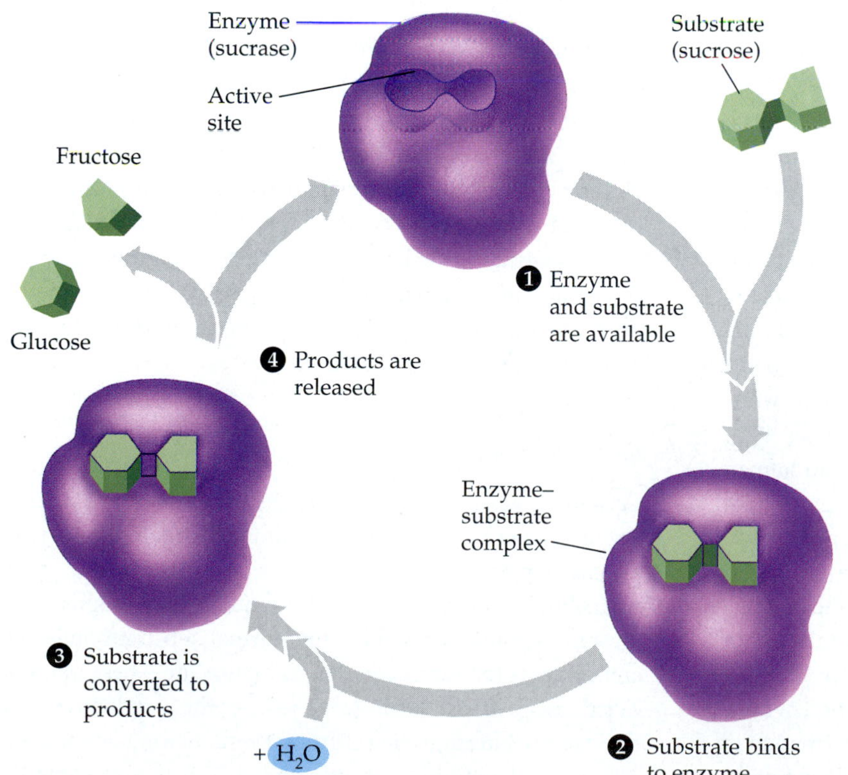

FIGURE 6.12 • The catalytic cycle of an enzyme. In this example, the enzyme sucrase catalyzes the hydrolysis of sucrose to glucose and fructose. ① When the active site of an enzyme is unoccupied and its substrate is available, the cycle begins. ② An enzyme-substrate complex forms when the substrate enters the active site and attaches by weak bonds. The active site changes shape to fit snugly around the substrate (induced fit). ③ The substrate is converted to products while in the active site. ④ The enzyme releases the products, and its active site is then available for another molecule of substrate. Most metabolic reactions are reversible, and an enzyme can catalyze both the forward and the reverse reactions. Which reaction prevails depends mainly on the relative concentrations of reactants and products; that is, the enzyme catalyzes the reaction in the direction of equilibrium.

6.1

of substrate: The more substrate molecules available, the more frequently they access the active sites of the enzyme molecules. However, there is a limit to how fast the reaction can be pushed by adding more substrate to a fixed concentration of enzyme. At some point, the concentration of substrate will be high enough that all enzyme molecules have their active sites engaged. As soon as the product exits an active site, another substrate molecule enters. At this substrate concentration, the enzyme is said to be saturated, and the rate of the reaction is determined by the speed at which the active site can convert substrate to product. When an enzyme population is saturated, the only way to increase productivity is to add more enzyme. Cells sometimes do this by making more enzyme molecules.

A cell's physical and chemical environment affects enzyme activity

The activity of an enzyme is affected by general environmental factors, such as temperature and pH, and also by particular chemicals that specifically influence that enzyme.

Effects of Temperature and pH

Recall from Chapter 5 that the three-dimensional structures of proteins are sensitive to their environment. As a protein, an enzyme has conditions under which it works optimally, because that environment favors the most active conformation for the enzyme molecule.

Temperature is one environmental factor important in the activity of an enzyme (FIGURE 6.13a). Up to a point, the velocity of an enzymatic reaction increases with increasing temperature, partly because substrates collide with active sites more frequently when the molecules move rapidly. Beyond that temperature, however, the speed of the enzymatic reaction drops sharply. The thermal agitation of the enzyme molecule disrupts the hydrogen bonds, ionic bonds, and other weak interactions that stabilize the active conformation, and the protein molecule denatures. Each enzyme has an optimal temperature at which its reaction rate is fastest. This temperature allows the greatest number of molecular collisions without denaturing the enzyme. Most human enzymes have optimal temperatures of about 35°C to 40°C (close to human body temperature). Bacteria that live in hot springs contain enzymes with optimal temperatures of 70°C or higher.

Just as each enzyme has an optimal temperature, it also has a pH at which it is most active (FIGURE 6.13b). The optimal pH values for most enzymes fall in the range of 6 to 8, but there are exceptions. For example, pepsin, a digestive enzyme in the stomach, works best at pH 2. Such an acidic environment denatures most enzymes, but the active conformation of pepsin is adapted to the acidic environment of the stomach.

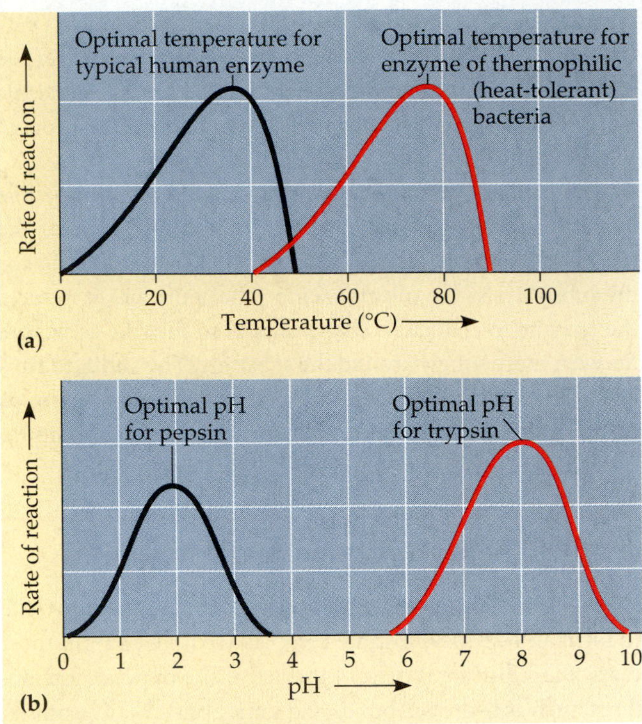

FIGURE 6.13 • Environmental factors affecting enzymes. Each enzyme has an optimal **(a)** temperature and **(b)** pH that favor the active conformation of the protein molecule.

In contrast, trypsin, a digestive enzyme residing in the alkaline environment of the intestine, has an optimal pH of 8.

Cofactors

Many enzymes require nonprotein helpers for catalytic activity. These adjuncts, called **cofactors**, may be bound tightly to the active site as permanent residents, or they may bind loosely and reversibly along with the substrate. The cofactors of some enzymes are inorganic, such as the metal atoms zinc, iron, and copper. If the cofactor is an organic molecule, it is more specifically called a **coenzyme**. Most vitamins are coenzymes or raw materials from which coenzymes are made. Cofactors function in various ways, but in all cases they are necessary for catalysis to take place.

Enzyme Inhibitors

Certain chemicals selectively inhibit the action of specific enzymes (FIGURE 6.14). If the inhibitor attaches to the enzyme by covalent bonds, inhibition is usually irreversible. It is reversible, however, if the inhibitor binds by weak bonds.

Some inhibitors resemble the normal substrate molecule and compete for admission into the active site. These mimics, called **competitive inhibitors**, reduce the productivity of enzymes by blocking the substrate from entering active sites. This kind of inhibition is reversible. It can be overcome by

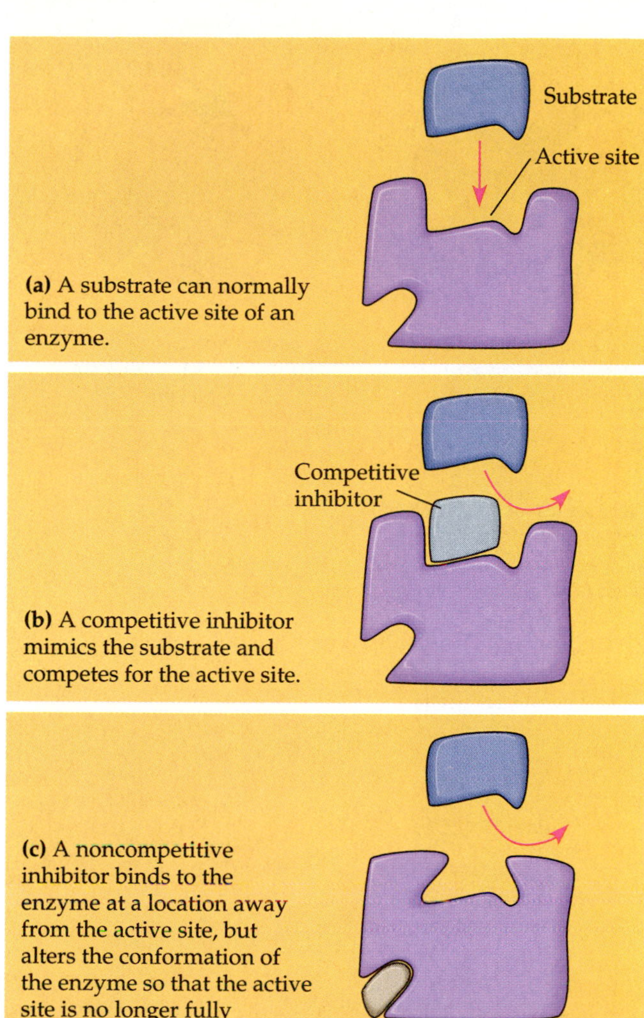

Substrate
Active site
(a) A substrate can normally bind to the active site of an enzyme.
Competitive inhibitor
(b) A competitive inhibitor mimics the substrate and competes for the active site.
(c) A noncompetitive inhibitor binds to the enzyme at a location away from the active site, but alters the conformation of the enzyme so that the active site is no longer fully functional.
Noncompetitive inhibitor

FIGURE 6.14 ▪ Enzyme inhibition.

increasing the concentration of substrate so that as active sites become available, more substrate molecules than inhibitor molecules are around to gain entry to the sites.

So-called **noncompetitive inhibitors** do not directly compete with the substrate at the active site. Instead, they impede enzymatic reactions by binding to another part of the enzyme. This interaction causes the enzyme molecule to change its shape, rendering the active site unreceptive to substrate, or leaving the enzyme less effective at catalyzing the conversion of substrate to product.

Some poisons absorbed from an organism's environment act by inhibiting enzymes. For example, the pesticides DDT and parathion are inhibitors of key enzymes in the nervous system. Many antibiotics are inhibitors of specific enzymes in bacteria. For instance, penicillin blocks the active site of an enzyme that many bacteria use to make their cell walls.

Mentioning enzyme inhibitors that are metabolic poisons may give the impression that enzyme inhibition is generally abnormal and harmful. In fact, selective inhibition and activation of enzymes by molecules naturally present in the cell are essential mechanisms in metabolic control, as we discuss next.

THE CONTROL OF METABOLISM

Chemical chaos would result if all of a cell's metabolic pathways were open simultaneously. Imagine, for example, a substance synthesized by one pathway being immediately broken down by another—the cell would be spinning its metabolic wheels. Actually, a cell tightly regulates its metabolic pathways by controlling when and where its various enzymes are active. Pathways are controlled either by switching on and off the genes that encode specific enzymes (as we will discuss in Unit Three) or, as we discuss here, by regulating the activity of enzymes once they are made.

Metabolic control often depends on allosteric regulation

In many cases, the molecules that naturally regulate enzyme activity in a cell behave something like reversible noncompetitive inhibitors (see FIGURE 6.14c): These regulatory molecules change an enzyme's shape and function by binding to an **allosteric site**, a specific receptor site on some part of the enzyme molecule remote from the active site. The effect may be either inhibition or stimulation of the enzyme's activity.

Allosteric Regulation

Most allosterically regulated enzymes are constructed from two or more polypeptide chains, or subunits. Each subunit has its own active site, and allosteric sites are usually located where subunits are joined (FIGURE 6.15a, p. 96). The entire complex oscillates between two conformational states, one catalytically active and the other inactive. The binding of an activator to an allosteric site stabilizes the conformation that has a functional active site, whereas the binding of an allosteric inhibitor stabilizes the inactive form of the enzyme (FIGURE 6.15b). The areas of contact between the subunits of an allosteric enzyme articulate in such a way that a conformational change in one subunit is transmitted to all others. Through this interaction of subunits, a single activator or inhibitor molecule that binds to one allosteric site will affect the active sites of all subunits.

Because allosteric regulators attach to an enzyme by weak bonds, the activity of the enzyme changes in response to fluctuating concentrations of the regulators. In some cases, an inhibitor and an activator are similar enough in shape to compete for the same allosteric site. For example, some enzymes of catabolic pathways, may have an allosteric site that fits both ATP and AMP (adenosine monophosphate), which the cell routinely derives from ADP. Such enzymes are inhibited by ATP and activated by AMP. This is logical because a major function of catabolism is to regenerate ATP. If ATP production lags behind its use, AMP accumulates and activates key enzymes that speed up catabolism. If the supply of ATP

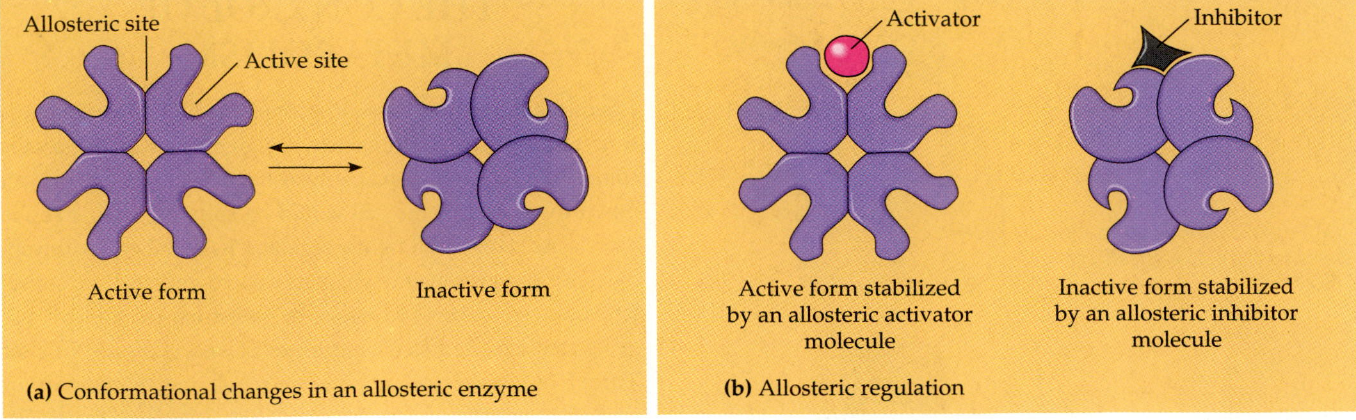

FIGURE 6.15 ▪ Allosteric regulation. (a) Most allosteric enzymes are constructed from two or more polypeptide subunits, each having its own active site. The enzyme oscillates between two conformational states, one active and the other inactive. Remote from the active sites are allosteric sites, specific receptors for regulators of the enzyme, which may be activators or inhibitors. **(b)** Here we see the opposing effects of an allosteric inhibitor and an allosteric activator on the conformation of all four subunits of an enzyme.

exceeds demand, then catabolism slows down as ATP molecules accumulate and compete for allosteric sites. In this way, allosteric enzymes control the rates of key reactions in metabolic pathways.

Feedback Inhibition

The inhibition of an ATP-generating catabolic pathway by the allosteric binding of ATP to an enzyme in the pathway is an example of feedback inhibition, one of the most common methods of metabolic control. **Feedback inhibition** is the switching off of a metabolic pathway by its end-product, which acts as an inhibitor of an enzyme within the pathway. FIGURE 6.16 shows an example of this control mechanism operating on an anabolic pathway. Some cells use this pathway of five steps to synthesize the amino acid isoleucine from threonine, another amino acid. As isoleucine, the end-product of the pathway, accumulates, it slows down its own synthesis by allosterically inhibiting the enzyme for the very first step of the pathway. Feedback inhibition thereby prevents the cell from wasting chemical resources to synthesize more isoleucine than is necessary.

Cooperativity

By a mechanism that resembles allosteric activation, substrate molecules may stimulate the catalytic powers of an enzyme (FIGURE 6.17). Recall that the binding of a substrate to an enzyme induces a favorable change in the shape of the active site (induced fit). If an enzyme has two or more subunits, this interaction with one substrate molecule triggers the same favorable conformational change in all other subunits of the enzyme. Called **cooperativity**, this mechanism amplifies the response of enzymes to substrates: One substrate molecule primes an enzyme to accept additional substrate molecules.

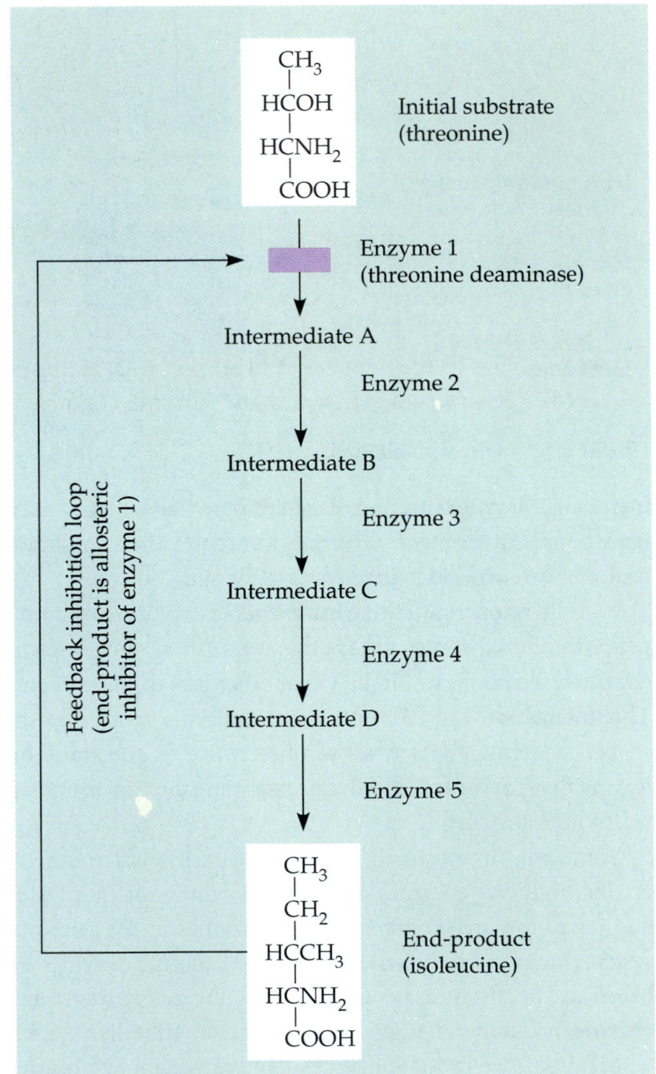

FIGURE 6.16 ▪ Feedback inhibition. Many metabolic pathways are switched off by an end-product, which acts as an allosteric inhibitor of an enzyme earlier in the pathway. This example is the pathway for synthesizing the amino acid isoleucine.

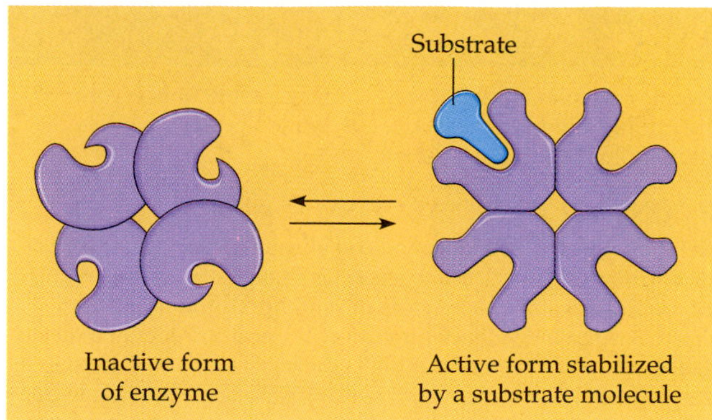

FIGURE 6.17 ▪ Cooperativity. In an enzyme molecule with multiple subunits, the binding of one substrate molecule to the active site of one subunit causes all the subunits to assume their active conformation, via the mechanism of induced fit.

The localization of enzymes within a cell helps order metabolism

The cell is not just a bag of chemicals with thousands of different kinds of enzymes and substrates wandering about randomly. Structures within the cell help bring order to metabolic pathways. In some cases, a team of enzymes for several steps of a metabolic pathway is assembled together as a multienzyme complex. The arrangement controls the sequence of reactions, as the product from the first enzyme becomes substrate for the adjacent enzyme in the complex, and so on, until the end-product is released. Some enzymes and enzyme complexes have fixed locations within the cell as structural components of particular membranes. Others are in solution within specific membrane-enclosed eukaryotic organelles, each with its own internal chemical environment. For example, in eukaryotic cells the enzymes for cellular respiration reside within mitochondria (FIGURE 6.18). If the cell had the same number of enzyme molecules for respiration but they were diluted throughout the entire volume of the cell, respiration would be very inefficient.

The structural basis of metabolic order brings us back to the theme with which this unit of chapters began.

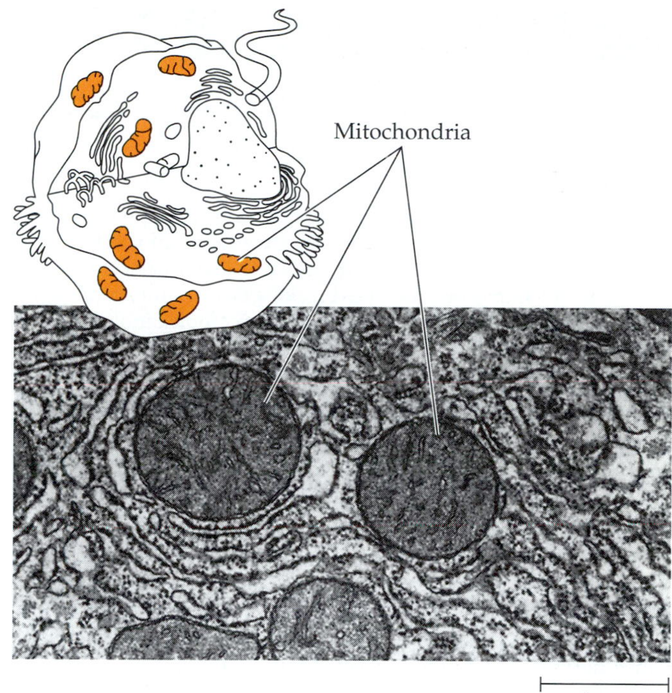

FIGURE 6.18 ▪ Organelles and structural order in metabolism. Membranes partition a eukaryotic cell into various metabolic compartments, or organelles, each with a corps of enzymes that carry out specific functions. The organelles shown here, called mitochondria, are the sites of cellular respiration (TEM).

The theme of emergent properties is manifest in the chemistry of life: *a review*

Recall that life is organized along a hierarchy of structural levels. With each increase in the level of order, new properties emerge in addition to those of the component parts. In Chapters 2–6 we have dissected the chemistry of life using the strategy of the reductionist. But we have also begun to develop a more integrated view of life as we have seen how properties emerge with increasing order.

We have seen that the peculiar behavior of water, so essential to life on Earth, results from interactions of the water molecules, themselves an ordered arrangement of hydrogen and oxygen atoms. We reduced the great complexity and diversity of organic compounds to the chemical characteristics of carbon, but we also saw that the unique properties of organic compounds are related to the specific structural arrangements of carbon skeletons and their appended functional groups. We learned that small organic molecules are often assembled into giant molecules, but we also discovered that a macromolecule does not behave like a simple composite of its monomers. For example, the unique form and function of a protein are consequences of a hierarchy of primary, secondary, and tertiary structures. And in this chapter we have seen that metabolism, that orderly chemistry that characterizes life, is a concerted interplay of thousands of different kinds of molecules in an organized cell.

By completing our overview of metabolism with an introduction to its structural basis in the compartmentalized cell, we have built a bridge to Unit Two, where we will study the cell's structure and function in more depth. We will maintain our balance between the need to reduce life to a conglomerate of simpler processes and the ultimate satisfaction of viewing those processes in their integrated context.

CHAPTER REVIEW

REVIEW OF KEY CONCEPTS

(with page numbers and key figures)

METABOLISM, ENERGY, AND LIFE

- **The chemistry of life is organized into metabolic pathways (pp. 83–84)** Metabolism is the collection of chemical reactions that occur in an organism. Aided by enzymes, it follows intersecting pathways, which may be catabolic (breaking down complex molecules, releasing energy) or anabolic (building complex molecules, consuming energy).
- **Organisms transform energy (pp. 84–85)** Energy is the capacity to do work by moving matter. A moving object has kinetic energy. Potential energy is stored in the location or structure of matter and includes chemical energy stored in molecular structure. Energy can change form, governed by the laws of thermodynamics.
- **The energy transformations of life are subject to two laws of thermodynamics (pp. 85–86, FIGURE 6.4)** The first, conservation of energy, states that energy cannot be created or destroyed. The second states that when energy changes form, entropy (S), or the disorder of the universe, increases. Matter can become more ordered only if the surroundings become more disordered.
- **Organisms live at the expense of free energy (pp. 86–88, FIGURE 6.5)** A living system's free energy is energy that can do work under cellular conditions. Free energy (G) is related directly to total energy (H) and to entropy (S): $\Delta G = \Delta H - T\Delta S$. Spontaneous changes involve a decrease in free energy ($-\Delta G$). In an exergonic (spontaneous) chemical reaction, the products have less free energy than the reactants ($-\Delta G$). Endergonic (nonspontaneous) reactions require an input of energy ($+\Delta G$). In cellular metabolism, exergonic reactions power endergonic reactions (energy coupling). The addition of starting materials and the removal of end-products prevent metabolism from reaching equilibrium.
- **ATP powers cellular work by coupling exergonic reactions to endergonic reactions (pp. 88–91, FIGURE 6.6)** ATP is the cell's energy shuttle. Hydrolysis of one of its phosphate bonds releases ADP, inorganic phosphate, and free energy. ATP drives endergonic reactions by transfer of the phosphate group to specific reactants, making them more reactive. In this way, cells can carry out work, such as movement and anabolism. Catabolic pathways, such as cellular respiration, drive the regeneration of ATP from ADP and phosphate.

ENZYMES

- 6.1 **Enzymes speed up metabolic reactions by lowering energy barriers (pp. 91–92, FIGURE 6.10)** Enzymes, which are proteins, are biological catalysts. They speed reactions by lowering activation energy (E_A), so bonds can break at the moderate temperatures of most organisms.
- **Enzymes are substrate-specific (pp. 92–93, FIGURE 6.11)** Each type of enzyme has a unique active site that combines specifically with its substrate, the reactant molecule on which it acts. The enzyme changes shape slightly when it binds the substrate (induced fit).
- **The active site is an enzyme's catalytic center (pp. 93–94, FIGURE 6.12)** The active site can lower activation energy by orienting substrates correctly, straining their bonds, and providing a suitable microenvironment.
- **A cell's physical and chemical environment affects enzyme activity (pp. 94–95, FIGURES 6.13–6.14)** As proteins, enzymes are sensitive to conditions that influence their three-dimensional structure. Each has an optimal temperature and pH. Cofactors are ions or molecules required for some enzymes to function. Coenzymes are organic cofactors. Inhibitors reduce enzyme function. A competitive inhibitor binds to the active site, while a noncompetitive inhibitor binds to a different site.

THE CONTROL OF METABOLISM

- **Metabolic control often depends on allosteric regulation (pp. 95–96, FIGURES 6.15–6.17)** Some enzymes change shape when regulatory molecules, either activators or inhibitors, bind to specific allosteric sites. In feedback inhibition, the end-product of a metabolic pathway allosterically inhibits the enzyme for an early step in the pathway. In cooperativity, a substrate molecule binding to one active site of a multi-subunit enzyme activates the other subunits.
- **The localization of enzymes within a cell helps order metabolism (p. 97, FIGURE 6.18)** Some enzymes are grouped into complexes, some are incorporated into membranes, and others are contained in organelles.
- **The theme of emergent properties is manifest in the chemistry of life: *a review* (p. 97)** Increasing levels of organization result in the emergence of new properties. Organization is the key to the chemistry of life.

SELF-QUIZ

1. Choose the pair of terms that correctly completes this sentence: Catabolism is to anabolism as ________________ is to ________________.
   a. exergonic; spontaneous
   b. exergonic; endergonic
   c. free energy; entropy
   d. work; energy
   e. entropy; order
2. Most cells cannot harness heat in order to perform work because
   a. heat is not a form of energy
   b. cells do not have much heat; they are relatively cool
   c. temperature is usually uniform throughout a cell
   d. there are no mechanisms in nature that can use heat to do work
   e. heat denatures enzymes
3. According to the first law of thermodynamics,
   a. matter can be neither created nor destroyed
   b. energy is conserved in all processes
   c. all processes increase the order of the universe
   d. systems rich in energy are intrinsically stable
   e. the universe constantly loses energy because of friction
4. Which of the following metabolic processes can occur without a net influx of energy from some other process?
   a. $ADP + Ⓟ_i \longrightarrow ATP + H_2O$
   b. $C_6H_{12}O_6 + 6\,O_2 \longrightarrow 6\,CO_2 + 6\,H_2O$
   c. $6\,CO_2 + 6\,H_2O \longrightarrow C_6H_{12}O_6 + 6\,O_2$
   d. amino acids $\longrightarrow$ protein
   e. glucose + fructose $\longrightarrow$ sucrose
5. If an enzyme has been noncompetitively inhibited
   a. the ΔG for the reaction it catalyzes will always be negative
   b. the active site will be occupied by the inhibitor molecule
   c. increasing the substrate concentration will increase the inhibition

d. a higher activation energy will be necessary to initiate the reaction
e. the inhibitor molecule may be chemically unrelated to the substrate

6. If an enzyme solution is saturated with substrate, the most effective way to obtain an even faster yield of products would be to
   a. add more of the enzyme
   b. heat the solution to 90°C
   c. add more substrate
   d. add an allosteric inhibitor
   e. add a noncompetitive inhibitor

7. An enzyme accelerates a metabolic reaction by
   a. altering the overall free-energy change for the reaction
   b. making an endergonic reaction occur spontaneously
   c. lowering the activation energy
   d. pushing the reaction away from equilibrium
   e. making the substrate molecule more stable

8. Some bacteria are metabolically active in hot springs because
   a. they are able to maintain an internal temperature much cooler than that of the surrounding water
   b. the high temperatures facilitate active metabolism without the need of catalysis
   c. their enzymes have high optimal temperatures
   d. their enzymes are insensitive to temperature
   e. they use molecules other than proteins as their main catalysts

9. Which of the following characteristics is not associated with allosteric regulation of an enzyme's activity?
   a. A molecule mimics the substrate and competes for the active site.
   b. A naturally occurring molecule stabilizes a catalytically active conformation.
   c. Regulatory molecules bind to a site remote from the active site.
   d. Inhibitor and activator molecules may compete with one another.
   e. The enzyme usually has a quaternary structure.

10. In the following branched metabolic pathway, a dotted arrow with a minus sign symbolizes inhibition of a metabolic step by an end-product:

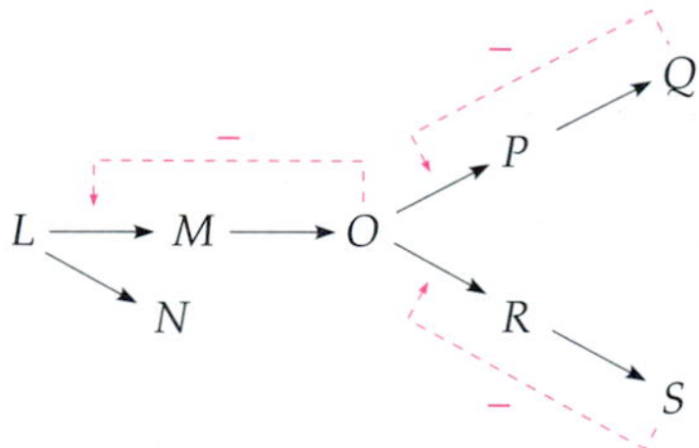

Which reaction would prevail if both Q and S are present in the cell in high concentrations?
   a. $L \longrightarrow M$
   b. $M \longrightarrow O$
   c. $L \longrightarrow N$
   d. $O \longrightarrow P$
   e. $R \longrightarrow S$

CHALLENGE QUESTION

A particular enzyme has an optimal temperature of 37°C and begins to denature at 45°C. During denaturation, entropy increases (the protein loses much of its organization). The protein also increases its energy content (energy must be absorbed from the surroundings to break the numerous weak bonds that reinforce the native conformation). Thus, for denaturation, ΔS and ΔH are both positive. Using the free-energy equation ($\Delta G = \Delta H - T\Delta S$), explain why denaturation becomes spontaneous at a certain temperature.

SCIENCE, TECHNOLOGY, AND SOCIETY

As mandated by the 1996 Food Quality Protection Act, the EPA has announced its intention to evaluate the safety of the most commonly used organophosphate insecticides. In agriculture, the most frequently used organophosphates account for half of the 58 million pounds applied annually nationwide. Organophosphates typically interfere with nerve transmission by inhibiting the enzymes that degrade the transmitter molecules that diffuse from one neuron to another. Noxious insects are not uniquely susceptible; humans and other vertebrates can be affected as well. Thus the use of pesticides creates some health risks. As a consumer, what levels of risk are you willing to accept in exchange for an abundant and affordable food supply? Is it prudent to expect "a reasonable certainty that no harm will result from pesticide exposure"? What other facts would you like to know about this situation before you could defend your opinion?

FURTHER READING

Adams, S. "No Way Back." *New Scientist*, October 22, 1994. An entertaining article about the second law of thermodynamics.
Becker, W. M., J. B. Reece, and M. F. Poenie. *The World of the Cell*, 3rd ed. Menlo Park, CA: Benjamin/Cummings, 1996. Provides a lucid explanation of cellular energetics and enzymes in Chapters 5 and 6.
Deeth, R. "Chemical Choreography." *New Scientist*, July 5, 1997. Computer modeling is being used to study how enzymes work in the cell.
Harold, F. M. *The Vital Force: A Study of Bioenergetics*. New York: W. H. Freeman, 1986. A challenging but clear introduction to energy and life in Chapters 1–3.
Kauffman, S. A. *The Origins of Order*. New York: Oxford University Press, 1993. An influential book applying thermodynamics and chaos theory to life.
Lewin, R. "A Simple Matter of Complexity." *New Scientist*, February 5, 1994. Our interpretation of life's history depends in part on how we define complexity.
Mathews, C. K., and K. E. van Holde. *Biochemistry*, 2nd ed. Menlo Park, CA: Benjamin/Cummings, 1996. An introduction to enzymes and metabolism in Chapters 10 and 12.

WEB LINKS

Visit the special edition of *The Biology Place* for BIOLOGY, Fifth Edition, at http://www.biology.com/campbell. Go to Chapter 6 for online resources, including learning activities, practice exams, and links to the following web sites:

"About Temperature"
This site clearly explains temperature and heat in an interesting manner.

"Elementary Thermodynamics and Biochemical Systems"
Helpful lecture notes from the University of Illinois.

"The Food Zone"
A lively site that students will find useful for reviewing macromolecules, energy transformations, enzymes, ATP, and related topics.

"Activation Energy and Enzymes"
Graphic explanation of the role of enzymes in lowering the activation energy of reactions.

THE CELL

Bruce Alberts first experienced scientific research by volunteering to work in a lab as an undergraduate. He went on to build a successful career investigating DNA chemistry, first at Princeton and then at the University of California, San Francisco, where he is a professor of biochemistry. Currently on leave from UCSF, Dr. Alberts is serving as president of the National Academy of Sciences in Washington, D.C. Under his leadership, the Academy has made the improvement of science education a top priority. Many biology students will have a chance to meet Bruce Alberts again if they use his classic text, Molecular Biology of the Cell.

How did you start in science?

I went to a large public high school in Illinois and really enjoyed my chemistry class. That's probably how I got so interested in science, but I don't think I really knew what science was. I do remember going to a career night at the high school where they invited parents and others working in various careers to come and talk about their occupations. I looked at all the occupations on the list that might use chemistry or science and there were two. One was medicine and one was chemical engineering, so I went to those two talks and decided I didn't want to be a chemical engineer, but thought maybe I would become a doctor. From my perspective as a 17-year-old, I didn't have any conception that there was an occupation of "scientist" for which you could get paid. I subsequently went to college at Harvard, where I took a lot of science courses because I was pre-med. But I still didn't get any hint from these courses about what doing science might be like.

So when did you begin to see that you could become a scientist?

During my junior year, Jacques Fresco, a postdoctoral researcher who was my tutor, asked me if I wanted to work in his lab during the summer. Since my girlfriend, now my wife, was going to Europe that summer, I had nothing else to do. I worked very hard all summer in his laboratory. The lab group worked on nucleic acids, and my project involved deciphering how errors in pairing between bases in DNA affect the helical structure of the DNA. I actually made a discovery that people were excited about, and the experience was completely different from the laboratory work that I had done in courses. We also used this discovery to analyze RNA— specifically, how helical structure forms due to base-pairing between certain regions of an RNA strand. So my research as an undergraduate that summer led to two papers, one published in *Nature*, the other in *Proceedings of the National Academy of Sciences*. Given the success of that one summer of work, and since it was so exciting to do this stuff, why go to medical school? I didn't even apply. I stayed at Harvard and did graduate work for a PhD degree.

How did your parents react to that?

I think parents like their children to choose safe careers. Being a doctor had a good, solid image with my father, while it seemed like being a scientist had a shakier future. But although my parents were a little surprised, they were supportive.

INTERVIEW

BRUCE ALBERTS

Did your graduate work go as well as your undergraduate research projects?

Actually, my research career has been a series of ups and downs. My success as an undergraduate researcher gave me a false view of science—that it is easy. So I was extremely naive when I went to graduate school. My major professor, Paul Doty, told me I could pick my own research project. Overconfident, I thought I might as well tackle something very important, so I decided that I would solve the genetic code, which had not been deciphered at that point. I got the idea when I took a discussion course in molecular biology and we had to write a term paper proposing experiments. My experiments were my idea of how you might solve the genetic code using a method that turned out to be very cumbersome and would never have worked. But I got an A on the term paper, and I got excited about my idea. This is a very dangerous thing in science, when you get too excited about your own ideas! So when I started my research for the PhD, I tried to solve the genetic code using my term paper as the outline for my experimental plan. I spent three years on experiments that really didn't have a prayer of working.

What was wrong with your plan?

Basically, the problem was that I did this long series of experiments to solve the genetic code without first doing the control experiment that would have told me that the plan couldn't work. About the time I realized the problem, Francis Crick visited our group at Harvard and I explained to him what I had been trying to do. I'm sure Francis Crick doesn't remember this at all, but it is burned in my memory. As soon as I started telling him what my plan was, before I even got to the punch line that it didn't work, Crick asked if I had done the very control experiment that I had left out! I was really impressed that he saw the problem immediately, when it had taken me three or four years to see it. So, I had a lot of trouble with my thesis research, but I was able to shift to a different project to complete my PhD. And, most importantly, I didn't make the same type of mistake again.

In the medical center environment of UCSF, you and your colleagues are at an intersection of basic research in cell biology and biotechnology. How do you see the science-technology relationship?

I've actually lived through the explosive revolution that led to the biotech industry. When I first arrived at UCSF in 1976, Herbert Boyer invited my wife and me over for dinner, and during a walk he told me how he was going to start this company, Genentech. He asked if I would like to be involved, and I said no. I was too busy, and like most academics at the time, I wasn't interested in starting companies. Herb was unusual in that he was comfortable with the idea of applying what we knew in commercial ventures. But most of us also had not seen the potential benefits from technology based on gene cloning and other methods in molecular biology—how the different techniques could be combined to produce useful things for society. But here it was, this ability to make commercial quantities of rare proteins, such as human growth factors. The technology has revolutionized the pharmaceutical industry and has improved human health on a scale that I think even surprises Herb Boyer and the other pioneers of biotechnology. All that knowledge came from the basic research of scientists trying to figure out how cells work. In turn, basic biological research is benefiting from new knowledge in chemistry combined with engineering breakthroughs, such as machines that can sequence DNA very rapidly. As this illustrates, the whole enterprise of science builds up units of knowledge that can be combined in many ways. The progress of science and technology is ever accelerating; the more things you know, the more ways you can combine that knowledge.

Can you give an example of how such a synthesis of existing knowledge could give us a new breakthrough in cell biology?

We know a lot of facts about what is called cell signaling. We know that various growth factors and hormones bind to the outside of a cell and send signals inside. Right now, all the steps and variations make cell signaling look incredibly complicated, like a complete jumble that doesn't make sense. There seem to be too many signal pathways, with the complexity compounded by cross-talk between the pathways. We have lots of facts, but we don't yet understand the network properties of cell signaling. We can understand every piece of the puzzle of cell signaling, but until we understand how it all fits together in a network of communication and control, the whole picture won't come together and seem simpler. Solving the problem demands different ways of thinking about cells as complex networks, and I think this provides a role for new kinds of people in this field—perhaps people who would otherwise have gone into physics. I'm sure when we understand these cellular networks, we'll say, "How were we so stupid not to see that this is what the cell is doing?"

Sometimes these signaling and response systems run amok, with cancer being an important example. What are some of your thoughts about the cell biology of cancer?

I think this is another case where we already have a lot of knowledge but need to start putting the facts together in new ways. We now know that for a normal cell to become a cancerous cell, it has to be disturbed in a large number of different ways. A lot of check points have evolved in our cells, safety mechanisms that allow large multicellular organisms to grow and develop without the individual cells being selfish and growing out of control. Cancer therefore requires that several different defects occur in the same line of cells. For that to happen, there must be some breakdown in the cell's DNA metabolism—in its ability to repair DNA damage, for example. Different tumors can be caused by different defects in DNA metabolism. And yet the chemotherapy used to treat cancer is usually a kitchen-sink approach, a mixture of drugs. As we learn more about these defects in DNA metabolism, we'll be able to screen a particular tumor cell for the type of defect that it contains, and then take a much more tailored approach to chemotherapy. We will soon see some major breakthroughs in both our understanding and treatment of cancer. If I wasn't in my present job, that's exactly what I'd be working on right now.

Speaking of your present job, tell us a little about the National Academy of Sciences.

The National Academy of Sciences was chartered by Abraham Lincoln in 1863 to be an honorary society of scientists. The charter we received is unusual, because it requires the members elected to this honorary society to advise the government, without compensation, on matters involving science and technology. For example, committees set up by the Academy now do a lot of work on environmental questions, such as, "What is a wetland, and why should we protect wetlands?" "What is the best strategy for cleaning up radioactive waste that has accumulated around the nuclear weapons sites that are now being dismantled all over the world?" As another example, in 1991 the federal government asked us, in response to an earlier request by the governors of all 50 states, to facilitate the development of the first-ever national standards for science education in our nation's elementary and secondary schools.

What advice do you have for students with aspirations for research or other careers that involve science?

First, I think it's important for science students to get a strong background in the fundamentals of chemistry, physics, and biology. But there is much more to being a successful scientist than that. Enthusiasm and persistence are so important. There are a lot of times when you're doing science that things go wrong, as you've heard from my failures. Of the 30 or so graduate students I've had during my career, the most successful were the ones who persevered and worked even harder when their research wasn't going well. Otherwise, you never get past the rough spots. To succeed in science, you also have to get satisfaction from making progress in small steps. We don't make big discoveries every day. Sometimes it's a victory just to get a particular technique to work a little better. The only way to find out if you're good at science and enjoy it is to work at it for a while. And I don't think we're very good at predicting who will succeed in science. It depends a lot more on creativity, imagination, and persistence than on how many facts you can recall on an exam.

What if a student likes biology or some other science as an undergraduate major, but doesn't want to go to medical school or become a research scientist?

I think far too many science professors still have the idea that their most important teaching goal is to produce more scientists. That's one reason to teach science, but a much bigger reason is that we need more people with science backgrounds in many fields, not just in research careers. For instance, we need more people with science backgrounds in industry, in journalism, in government, and certainly as teachers in elementary and secondary schools. There are so many ways to use a science background successfully without being a research scientist or professor. I encourage students to keep an open mind about how they can put their science abilities to work in different careers. And I urge science professors to become much more inclusive in how they define the science community—to welcome not just academic scientists and researchers as part of their world, but the many others who are using their science education to improve schools, businesses, government, and other institutions.

Spreading both science and its values throughout our society has become the central mission of the Academy. Please visit us at www.nas.edu

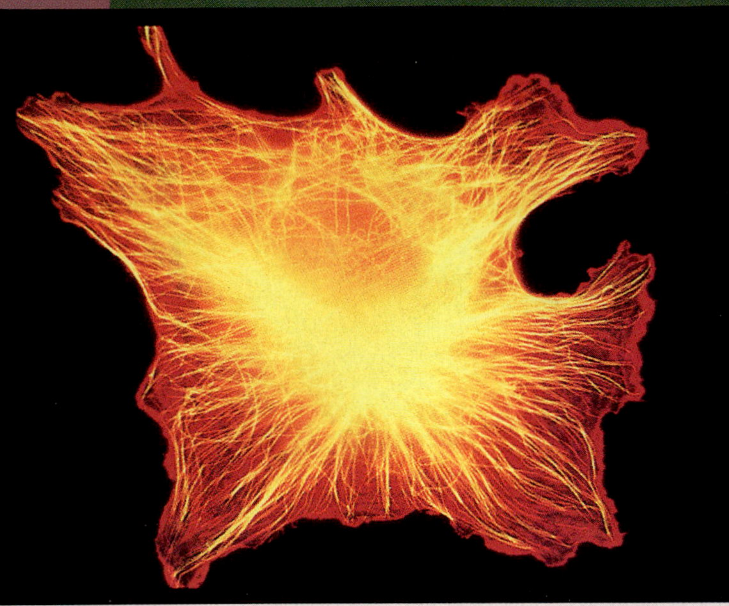

A TOUR OF THE CELL

The cell is as fundamental to biology as the atom is to chemistry: All organisms are made of cells. In the hierarchy of biological organization, the cell is the simplest collection of matter that can live. Indeed, there are diverse forms of life existing as single-celled organisms. More complex organisms, including plants and animals, are multicellular; their bodies are cooperatives of many kinds of specialized cells that could not survive for long on their own. However, even when they are arranged into higher levels of organization, such as tissues and organs, cells can be singled out as the organism's basic units of structure and function. The contraction of muscle cells moves your eyes as you read this sentence; when you decide to turn this page, nerve cells will transmit that decision from your brain to the muscle cells of your hand. Everything an organism does occurs fundamentally at the cellular level. This chapter introduces the microscopic world of the cell.

This book takes a thematic approach to the study of life, and the cell is a microcosmic model of many of the themes introduced in Chapter 1. We will see that life at the cellular level arises from structural order, reinforcing the themes of emergent properties and the correlation between structure and function. For example, the movement of an animal cell depends on an intricate interplay of structures that make up a cellular skeleton (visible in the light micrograph on this page of a cell called a fibroblast). Another recurring theme in biology is the interaction of organisms with their environment. Cells sense and respond to environmental fluctuations. As open systems, they continuously exchange both materials and energy with their surroundings. And keep in mind the one biological theme that unifies all others: evolution. All cells are related by their descent from earlier cells, but they have been modified in various ways during the long evolutionary history of life on Earth. For example, if one unicellular organism lives in fresh water and another inhabits the sea, we can expect these cells to be somewhat differently equipped as a result of their divergent adaptations to disparate environments.

HOW WE STUDY CELLS

Perhaps the greatest obstacle to becoming acquainted with the cell is imagining how something too small to be seen by the unaided eye can be so complex. How can cell biologists possibly dissect so small a package to investigate its inner workings? Before we actually tour the cell, it will be helpful to learn how cells are studied.

Microscopes provide windows to the world of the cell

The evolution of a science often parallels the invention of instruments that extend human senses to new limits. The discovery and early study of cells progressed with the invention and improvement of microscopes in the seventeenth century.

Microscopes of various types are still indispensable tools for the study of cells.

The microscopes first used by Renaissance scientists, as well as the microscopes you are likely to use in the laboratory, are all **light microscopes (LMs)**. Visible light is passed through the specimen and then through glass lenses. The lenses refract (bend) the light in such a way that the image of the specimen is magnified as it is projected into the eye. (See Appendix Four at the back of the book, which diagrams microscope structure.)

Two important values in microscopy are magnification and resolving power, or resolution. Magnification is how much larger the object appears compared to its real size. **Resolving power** is a measure of the clarity of the image; it is the minimum distance two points can be separated and still be distinguished as two separate points. For example, what appears to the unaided eye as one star in the sky may be resolved as twin stars with a telescope.

Just as the resolving power of the human eye is limited, the resolving power of telescopes and microscopes is limited. Microscopes can be designed to magnify objects as much as desired, but the light microscope can never resolve detail finer than about 0.2 μm, the size of a small bacterium (FIGURE 7.1). This resolution is limited by the wavelength of the visible light used to illuminate the specimen. Light microscopes can magnify effectively to about 1000 times the size of the actual specimen; greater magnifications increase blurriness. Most of the improvements in light microscopy since the beginning of the twentieth century have involved new methods for enhancing contrast, which makes the details that can be resolved stand out better to the eye (TABLE 7.1, p. 104).

Although cells were discovered by Robert Hooke in 1665, the geography of the cell was largely uncharted until the past few decades. Most subcellular structures, or **organelles**, are too small to be resolved by the light microscope. Cell biology advanced rapidly in the 1950s with the introduction of the electron microscope. Instead of using visible light, the **electron microscope (EM)** focuses a beam of electrons through the specimen. Resolving power is inversely related to the wavelength of radiation a microscope uses, and electron beams have wavelengths much shorter than the wavelengths of visible light. Modern electron microscopes can theoretically achieve a resolution of about 0.1 nanometer (nm), but the practical limit for biological structures is generally only about 2 nm—still a hundredfold improvement over the light microscope. Biologists use the term *cell ultrastructure* to refer to a cell's anatomy as resolved by an electron microscope.

There are two basic types of electron microscopes: the **transmission electron microscope (TEM)** and the **scanning electron microscope (SEM)**. The TEM aims an electron beam through a thin section of the specimen, similar to the way a light microscope transmits light through a slide. However, instead of using glass lenses, which are opaque to electrons, the TEM uses electromagnets as lenses to focus and magnify the image by bending the paths of the electrons. The image is ultimately focused onto a screen for viewing or onto photographic film. To enhance contrast in the image, very thin

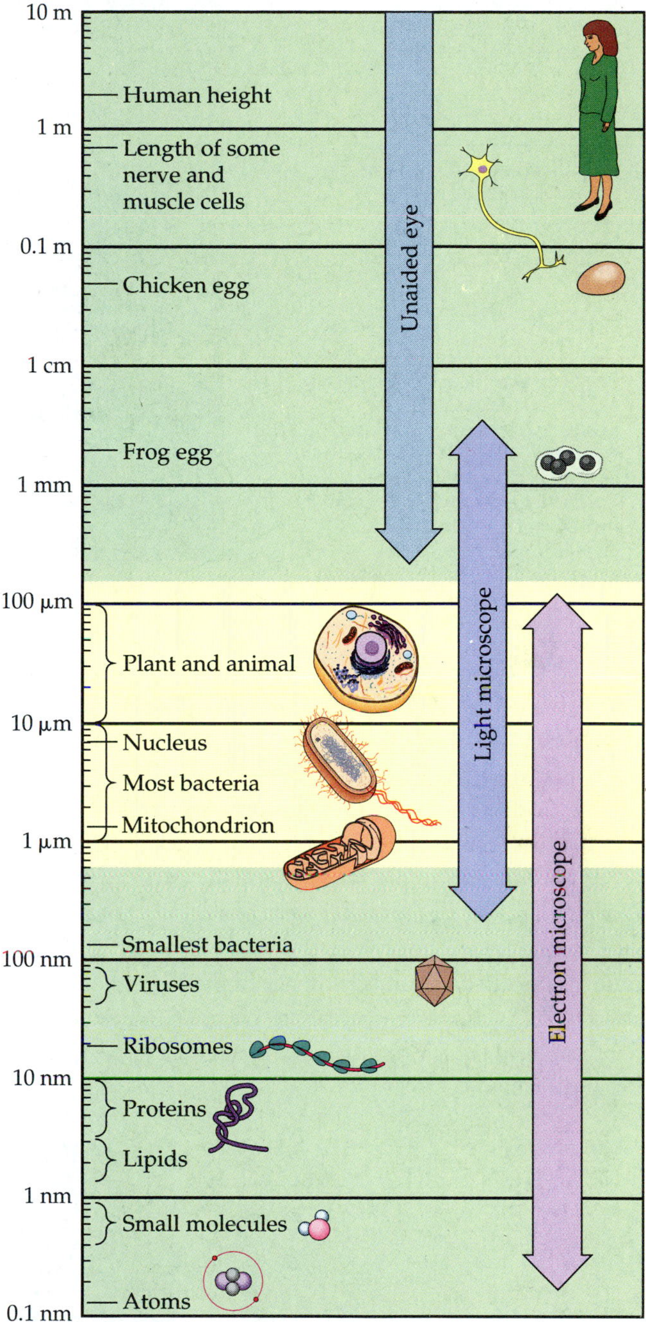

MEASUREMENTS
1 centimeter (cm) = 10^{-2} meter (m) = 0.4 inch
1 millimeter (mm) = 10^{-3} m
1 micrometer (μm) = 10^{-3} mm = 10^{-6} m
1 nanometer (nm) = 10^{-3} μm = 10^{-9} m

FIGURE 7.1 ▪ The size range of cells. Most cells are between 1 and 100 μm in diameter and are therefore visible only under a microscope. Notice that the scale is logarithmic to accommodate the range of sizes shown. Starting at the top of the scale with 10 meters and going down, each reference measurement along the left side marks a *tenfold* decrease in size.

Table 7.1 ▪ Different Types of Light Microscopy: A Comparison

TYPE OF MICROSCOPY	LIGHT MICROGRAPHS OF HUMAN CHEEK EPITHELIAL CELLS	TYPE OF MICROSCOPY
Brightfield (unstained specimen): Passes light directly through specimen; unless cell is naturally pigmented or artificially stained, image has little contrast.		**Phase-contrast:** Enhances contrast in unstained cells by amplifying variations in density within specimen; especially useful for examining living, unpigmented cells.
Brightfield (stained specimen): Staining with various dyes enhances contrast, but most staining procedures require that cells be fixed (preserved).		**Differential-interference-contrast (Nomarski):** Like phase-contrast microscopy, it uses optical modifications to exaggerate differences in density.
Fluorescence: Shows the locations of specific molecules in the cell. Fluorescent substances absorb short wavelength, ultraviolet radiation and emit longer wavelength, visible light. The fluorescing molecules may occur naturally in the specimen but more often are made by tagging the molecules of interest with fluorescent dyes or antibodies.	50 μm	**Confocal:** Uses lasers and special optics for "optical sectioning." Only those regions within a narrow depth of focus are imaged. Light at regions above and below the selected plane of view appear black rather than blurry.

sections of preserved cells are stained with atoms of heavy metals, which attach at certain places in the cells. Cell biologists use the TEM mainly to study the internal ultrastructure of cells (FIGURE 7.2a).

The SEM is especially useful for detailed study of the surface of the specimen (FIGURE 7.2b). The electron beam scans the surface of the sample, which is usually coated with a thin film of gold. The beam excites electrons on the sample's surface, and these secondary electrons are collected and focused onto a screen. The result is an image of the topography of the specimen. An important attribute of the SEM is its great depth of field, which results in an image that appears three-dimensional.

Electron microscopes reveal many organelles that are impossible to resolve with the light microscope. But the light microscope offers advantages, especially for the study of live cells. A disadvantage of electron microscopy is that the methods used to prepare the specimen kill cells. Also, they may introduce artifacts, structural features seen in micrographs that do not exist in the living cell. (Artifacts can occur in light microscopy, too.)

Microscopes are the most important tools of cytology, the study of cell structure. But simply describing the diverse organelles within the cell reveals little about their function. Modern cell biology developed from an integration of cytology with biochemistry, the study of the molecules and chemical

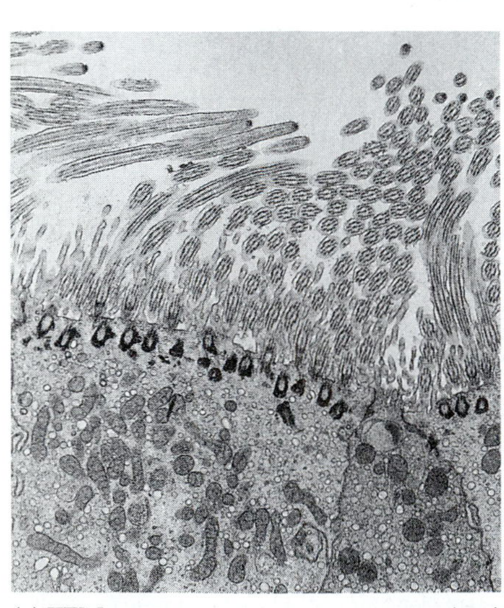

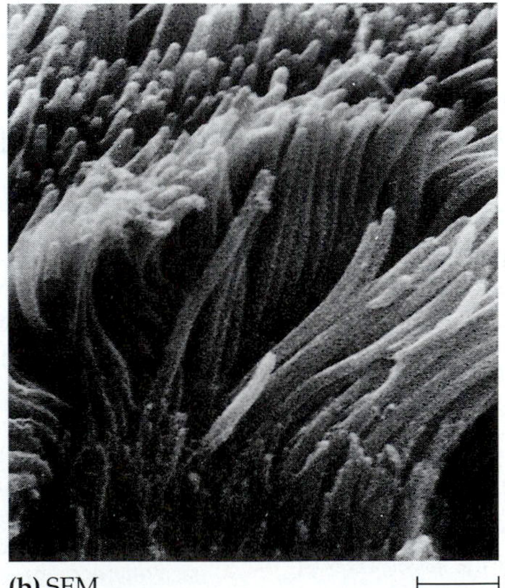

FIGURE 7.2 ▪ Electron micrographs: photographs taken with electron microscopes. (a) This micrograph, taken with a transmission electron microscope (TEM), profiles a thin section of part of a cell from a rabbit trachea (windpipe), revealing its ultrastructure. **(b)** The scanning electron microscope (SEM) produces a three-dimensional image of the surface of the same type of cell. Both micrographs show motile organelles called cilia. Beating of the cilia that line the windpipe helps move inhaled debris upward toward the pharynx (throat).

Throughout this book, micrographs are identified by the type of microscopy: LM for a light micrograph, TEM for a transmission electron micrograph, and SEM for a scanning electron micrograph.

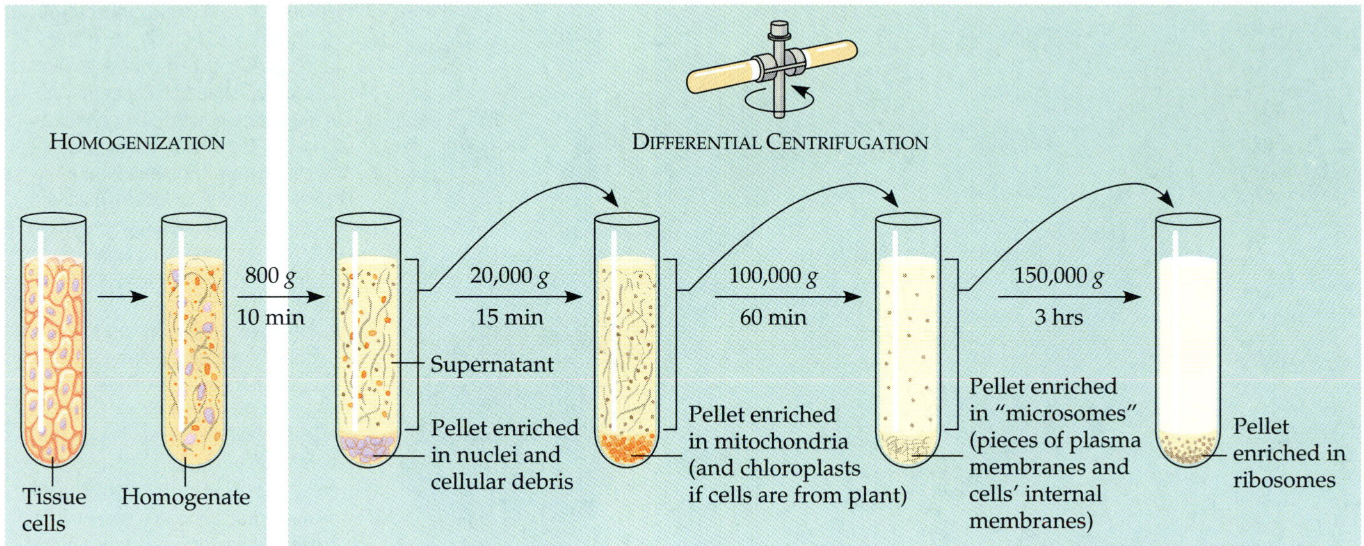

FIGURE 7.3 ▪ Cell fractionation. Disrupted cells are centrifuged at various speeds and durations to isolate components of different sizes. The process begins with homogenization, the disruption of a tissue and its cells with the help of such instruments as kitchen blenders or ultrasound devices. The homogenate, a soupy mixture of organelles, bits of membrane, and molecules from the broken cells, is then fractionated by a series of spins in a centrifuge. After each centrifugation, the unpelleted portion, or supernatant, is decanted and centrifuged again, at a higher speed. By determining which cell fractions are associated with particular metabolic processes, those functions can be tied to certain organelles.

processes of metabolism. A biochemical approach called cell fractionation has been particularly important in cell biology.

Cell biologists can isolate organelles to study their functions

The goal of **cell fractionation** is to take cells apart, separating the major organelles so that their individual functions can be studied (FIGURE 7.3). The instrument used to fractionate cells is the centrifuge, a merry-go-round for test tubes capable of spinning at various speeds. The most powerful machines, called **ultracentrifuges**, can spin as fast as 80,000 revolutions per minute (rpm) and apply forces on particles of up to 500,000 times the force of gravity (500,000 *g*).

Fractionation begins with homogenization, the disruption of cells. The objective is to break the cells apart without severely damaging their organelles. Spinning the soupy homogenate in a centrifuge separates the parts of the cell into two fractions: the pellet, consisting of the larger structures that become packed at the bottom of the test tube; and the supernatant, consisting of smaller parts of the cell suspended in the liquid above the pellet. The supernatant is decanted into another tube and centrifuged again. The process is repeated, increasing the speed with each step, collecting smaller and smaller components of the homogenized cells in successive pellets.

Cell fractionation enables the researcher to prepare specific components of cells in bulk quantity in order to study their composition and functions. By following this approach, biologists have been able to assign various functions of the cell to the different organelles, a task that would be far more difficult with intact cells. For example, one cellular fraction collected by centrifugation has enzymes that function in the metabolic process known as cellular respiration. The electron microscope reveals this fraction to be very rich in the organelles called mitochondria. This evidence helped cell biologists determine that mitochondria are the sites of cellular respiration. Cytology and biochemistry complement each other in correlating cellular structure and function.

A PANORAMIC VIEW OF THE CELL

Every organism is composed of one of two structurally different types of cells: prokaryotic cells or eukaryotic cells. Only the bacteria and archaea have prokaryotic cells. Protists, plants, fungi, and animals are all eukaryotes.

Prokaryotic and eukaryotic cells differ in size and complexity

7.1 A major difference between prokaryotic and eukaryotic cells is indicated by their names. The word *prokaryote* is from the Greek *pro*, "before," and *karyon*, "kernel," referring here to the nucleus. The **prokaryotic cell** has no nucleus (FIGURE 7.4, p. 106). Its genetic material (DNA) is concentrated in a region called the **nucleoid**, but no membrane separates this region from the rest of the cell. In contrast, the eukaryotic cell (Gr. *eu*,

This symbol links topics in the text to interactive exercises in the CD-ROM that accompanies the book. The number indicates the appropriate activity in the CD.

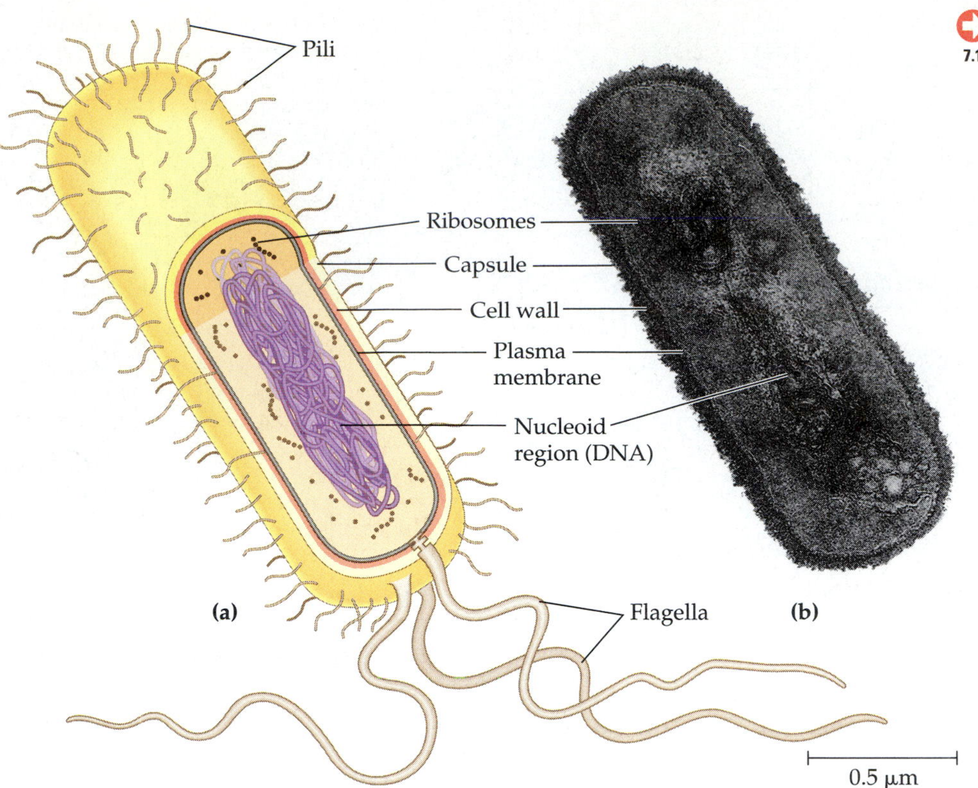

FIGURE 7.4 ▪ A prokaryotic cell.
7.1 Prokaryotes are bacteria, including cyanobacteria. **(a)** The drawing depicts a typical rod-shaped bacterium. Lacking the membrane-enclosed organelles of a eukaryote, the prokaryote is much simpler in structure. The area where the DNA is located is called the nucleoid region, and no membrane separates the DNA from the rest of the cell. A prokaryote has many ribosomes, where proteins are synthesized. The border of the cell is the plasma membrane. Outside the plasma membrane are a fairly rigid cell wall and often an outer capsule, usually jellylike. Some bacteria have flagella (locomotion organelles), pili (attachment structures), or both projecting from their surface. **(b)** This electron micrograph shows a thin section through the bacterium *Bacillus coagulans* (TEM).

"true," and *karyon*) has a true nucleus enclosed by a membranous nuclear envelope. The entire region between the nucleus and the membrane bounding the cell is called the **cytoplasm.** It consists of a semifluid medium called the **cytosol,** in which are located organelles of specialized form and function, most of them absent in prokaryotic cells. Thus the presence or absence of a true nucleus is just one example of the disparity in structural complexity between the two types of cells. Prokaryotic cells will be described in detail in Chapters 18 and 27, and the possible evolutionary relationships between the two types of cells will be discussed in Chapter 28. Most of the discussion of cell structure that follows in this chapter applies to eukaryotes.

Size is a general feature of cell structure that relates to function. The logistics of carrying out metabolism set limits

Surface area increases while total volume remains constant

	(a) 1	(b) 5	(c) 1
Total surface area (height × width × number of sides × number of boxes)	6	150	750
Total volume (height × width × length × number of boxes)	1	125	125
Surface-area-to-volume ratio (area ÷ volume)	6	1.2	6

FIGURE 7.5 ▪ Why are most cells microscopic? In this diagram, cells are represented as boxes. **(a)** In arbitrary units of length, this small cell measures 1 on each side. We can calculate the cell's surface area (in square units), volume (in cubic units), and surface-area-to-volume ratio. **(b)** As the size of the cell increases to 5 units of length per side, the ratio of surface area to volume decreases compared to the smaller cell. Rates of chemical exchange with the extracellular environment might be inadequate to maintain the cell because most of its cytoplasm is relatively far from the outer membrane. **(c)** By dividing the large cell into many smaller cells, we can restore a surface-area-to-volume ratio that can serve each cell's need for acquiring nutrients and expelling waste products. These geometric relationships explain why most cells are microscopic, and why larger organisms do not generally have *larger* cells than smaller organisms, but *more* cells.

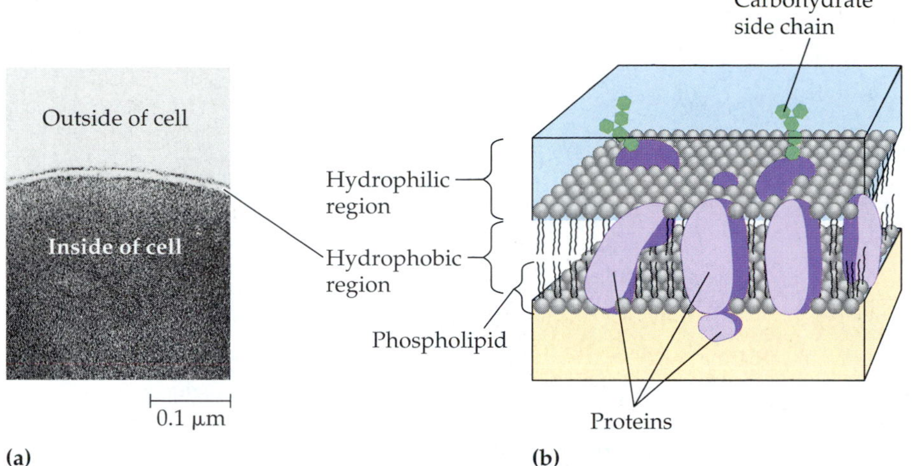

FIGURE 7.6 ▪ The plasma membrane. **(a)** In electron micrographs of sufficient magnification, the plasma membrane appears as a pair of dark bands separated by a light band. Shown here is the membrane of a red blood cell (TEM). **(b)** The plasma membrane and the various internal membranes of cells consist of a double layer (bilayer) of phospholipids with proteins of diverse functions attached to or embedded in it. (See Chapter 5 to review the dual hydrophilic/hydrophobic behavior of phospholipids.) The specific functions of the plasma membrane and of the various types of membranes within the cell depend on the kinds of phospholipids and proteins present. The plasma membrane also has carbohydrates attached to its outer surface.

on the size range of cells. The smallest cells known are bacteria called mycoplasmas, which have diameters between 0.1 and 1.0 micrometer (µm). These are perhaps the smallest packages with enough DNA to program metabolism and enough enzymes and other cellular equipment to carry out the activities necessary for a cell to sustain itself and reproduce. Most bacteria are 1 to 10 µm in diameter, about ten times larger than mycoplasmas (see FIGURE 7.1). Eukaryotic cells are typically 10 to 100 µm in diameter, ten times larger than bacteria.

Metabolic requirements also impose upper limits on the size that is practical for a single cell. As an object of a particular shape increases in size, its volume grows proportionately more than its surface area. (Area is proportional to a linear dimension squared, whereas volume is proportional to the linear dimension cubed.) Thus for objects of the same shape, the smaller the object, the greater its ratio of surface area to volume (FIGURE 7.5).

At the boundary of every cell, the **plasma membrane** functions as a selective barrier that allows sufficient passage of oxygen, nutrients, and wastes to service the entire volume of the cell (FIGURE 7.6). For each square micrometer of membrane, only so much of a particular substance can cross per second. The need for a surface sufficiently large to accommodate the volume helps explain the microscopic size of most cells.

Internal membranes compartmentalize the functions of a eukaryotic cell

In addition to the plasma membrane at its outer surface, a eukaryotic cell has extensive and elaborately arranged internal membranes, which partition the cell into compartments. These membranes also participate directly in the cell's metabolism; many enzymes are built right into the membranes. Because the cell's compartments provide different local environments that facilitate specific metabolic functions, incompatible processes can go on simultaneously inside the same cell.

Membranes of various kinds are fundamental to the organization of the cell. In general, biological membranes consist of a double layer of phospholipids and other lipids. Embedded in this lipid bilayer or attached to its surfaces are diverse proteins (see FIGURE 7.6). However, each membrane has a unique composition of lipids and proteins suited to that membrane's specific functions. For example, enzymes that function in cellular respiration are embedded in the membranes of the organelles called mitochondria.

Before continuing with this chapter, examine the over-
7.2 views of eukaryotic cells in FIGURES 7.7 and 7.8 on pages 108
7.3 and 109. These figures and their legends introduce the various organelles and provide a map of the cell for the detailed tour upon which we will soon embark. FIGURES 7.7 and 7.8 also contrast animal and plant cells. As eukaryotic cells, they have much more in common than either has with any prokaryote. As you will see, however, there are important differences between plant and animal cells.

The first stop on our detailed tour of the cell are two organelles involved in the genetic control of the cell.

THE NUCLEUS AND RIBOSOMES

The nucleus contains a eukaryotic cell's genetic library

The **nucleus** contains most of the genes that control the eukaryotic cell (some genes are located in mitochondria and chloroplasts). It is generally the most conspicuous organelle in a eukaryotic cell, averaging about 5 µm in diameter. The nuclear envelope encloses the nucleus (FIGURE 7.9 on page 110), separating its contents from the cytoplasm.

The nuclear envelope is a double membrane. The two membranes, each a lipid bilayer with associated proteins, are separated by a space of about 20–40 nm. The envelope is perforated by pores that are about 100 nm in diameter. At the lip of each pore, the inner and outer membranes of the nuclear

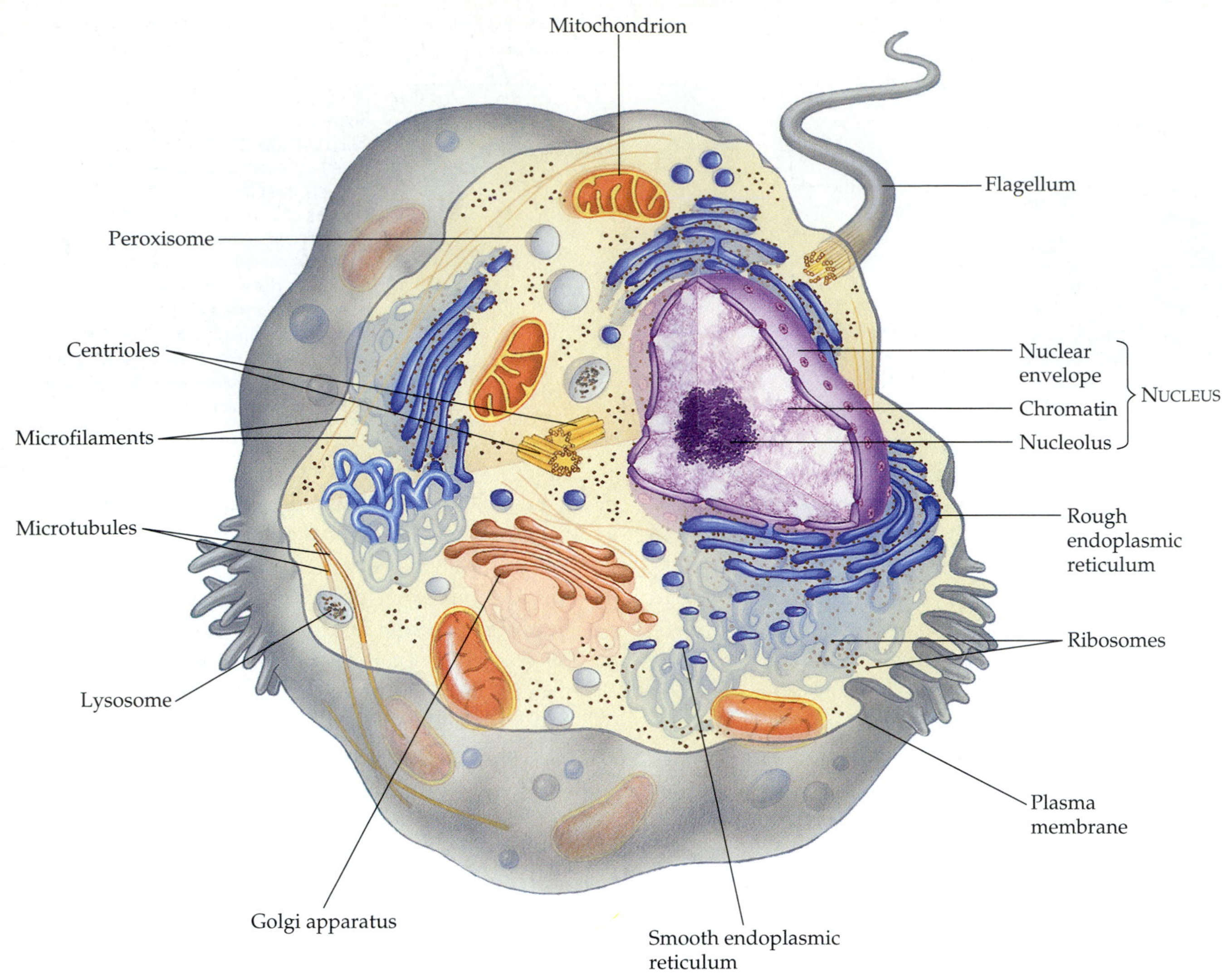

FIGURE 7.7 ▪ Overview of an animal cell.
7.2 This drawing of a generalized animal cell combines all the most common structures found in animal cells (no cell actually looks just like this). Within the cell are a variety of components called organelles ("little organs"), many of which are bounded by membranes. The most prominent organelle in an animal cell is usually the nucleus. The chromatin in the nucleus consists of DNA, which carries genes, along with proteins. Chromatin is actually a collection of separate structures called chromosomes, visible as separate units only in a dividing cell. Adjoining part of the chromatin in the nucleus are one or more nucleoli (singular, nucleolus). Nucleoli are involved in the production of particles called ribosomes, which synthesize proteins. The nucleus is bordered by a porous envelope consisting of two membranes.

Most of the cell's metabolic activities occur in the cytoplasm, the entire region between the nucleus and the plasma membrane surrounding the cell. The cytoplasm is full of specialized organelles suspended in a semifluid medium called the cytosol. Pervading much of the cytoplasm is the endoplasmic reticulum (ER), a labyrinth of membranes forming flattened sacs and tubes that segregate the contents of the ER from the cytosol. The ER takes two forms: rough (studded with ribosomes) and smooth. Many types of proteins are made by ribosomes attached to ER membranes, and the ER also plays a major role in assembling the cell's other membranes. The Golgi apparatus, another type of membranous organelle, consists of stacks of flattened sacs active in the synthesis, refinement, storage, sorting, and secretion of a variety of chemical products.

Other membrane-enclosed organelles are: lysosomes, which contain mixtures of digestive enzymes that hydrolyze macromolecules; peroxisomes, a diverse group of organelles containing enzymes that perform specialized metabolic processes; and vacuoles, which have a variety of storage and metabolic functions. The mitochondria (singular, mitochondrion) carry out cellular respiration, which generates ATP from organic fuels such as sugar.

Nonmembranous organelles within the cells include microtubules and microfilaments. They help form a framework called the cytoskeleton, which reinforces the cell's shape and functions in cell movement. The cell in the drawing has a flagellum, an organelle of locomotion, which is an assembly of microtubules. Also made of microtubules are centrioles, located near the nucleus. These play a role in cell division.

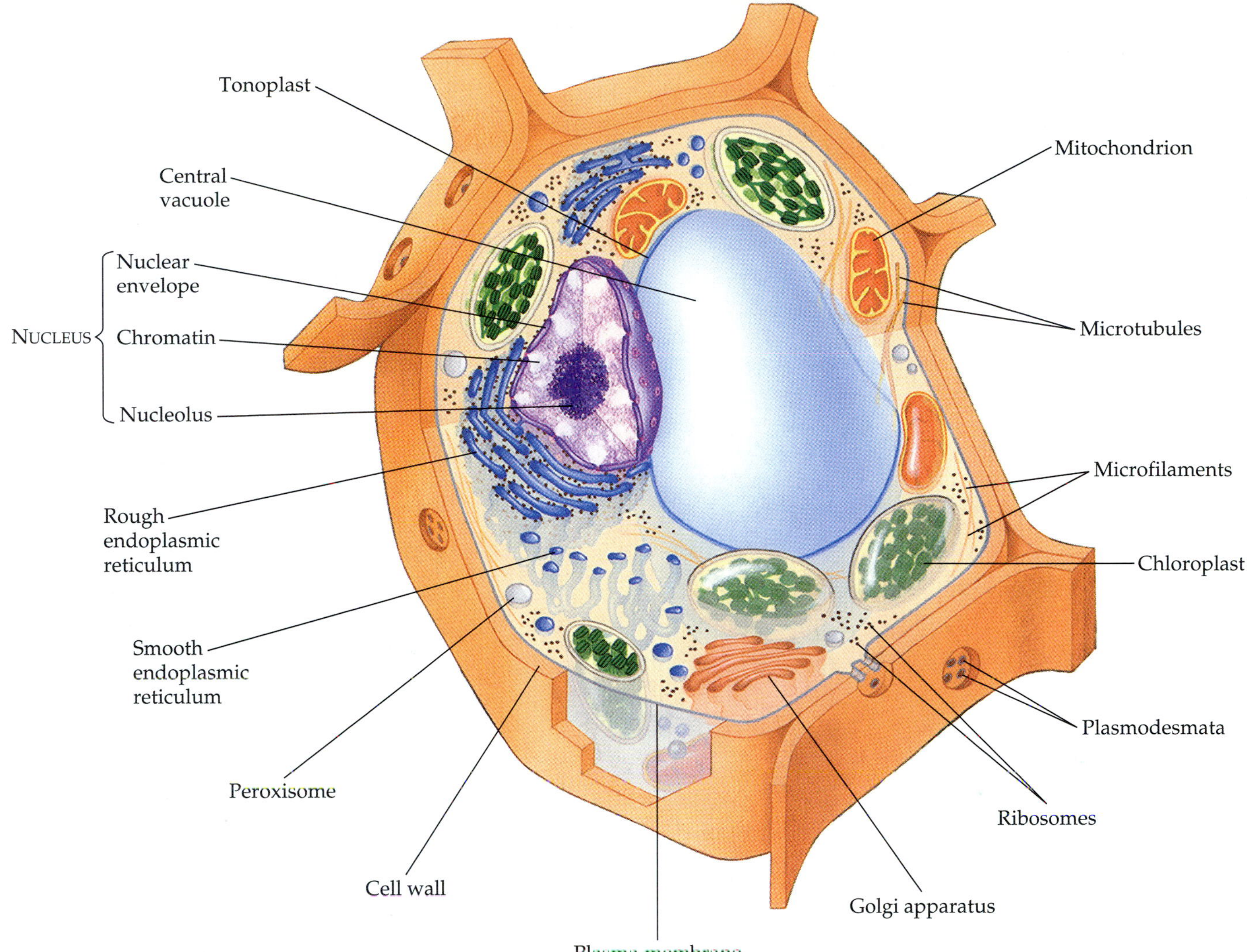

7.3 **FIGURE 7.8 • Overview of a plant cell.** This drawing of a generalized plant cell reveals the similarities and differences between an animal cell and a plant cell. Like the animal cell, the plant cell is surrounded by a plasma membrane and contains a nucleus, ribosomes, ER, Golgi apparatus, mitochondria, peroxisomes, and microfilaments and microtubules. However, a plant cell also contains a family of membrane-enclosed organelles called plastids. The most important type of plastid is the chloroplast, which carries out photosynthesis, converting sunlight to chemical energy stored in sugar and other organic molecules. Another prominent organelle in many plant cells, especially older ones, is a large central vacuole. The vacuole stores chemicals, breaks down macromolecules, and, by enlarging, plays a major role in plant growth. The vacuole membrane is called the tonoplast. Outside a plant cell's plasma membrane (and in fungi and some protists as well) is a thick cell wall, which helps maintain the cell's shape and protects the cell from mechanical damage. The cytosol of adjacent cells connects through trans-wall channels called plasmodesmata.

If you preview the rest of the chapter now, you'll see Figures 7.7 and 7.8 repeated in miniature as orientation diagrams. In each case, a particular organelle is highlighted, color-coded to its appearance in Figures 7.7 and 7.8. As we take a closer look at individual organelles, the orientation diagrams will help you place those structures in the context of the whole cell.

envelope are fused. An intricate protein structure called a pore complex lines each pore and regulates the entry and exit of certain large macromolecules and particles. Except at the pores, the nuclear side of the envelope is lined by the **nuclear lamina**, a netlike array of protein filaments that maintains the shape of the nucleus. There is also much evidence for a nuclear matrix, a framework of fibers extending throughout the nuclear interior. (We will examine possible functions of the nuclear lamina and matrix in Chapter 19.)

Within the nucleus, the DNA is organized along with proteins into material called **chromatin**. Stained chromatin appears through both light microscopes and electron microscopes as a diffuse mass. As a cell prepares to divide (reproduce), the stringy, entangled chromatin coils up (condenses), becoming thick enough to be discerned as separate structures called **chromosomes**. Each eukaryotic species has a characteristic number of chromosomes. A human cell, for example, has 46 chromosomes in its nucleus; the exceptions are the sex cells—eggs and sperm—which have only 23 chromosomes in humans.

A prominent structure within the nondividing nucleus is the **nucleolus**, where components of ribosomes are synthesized

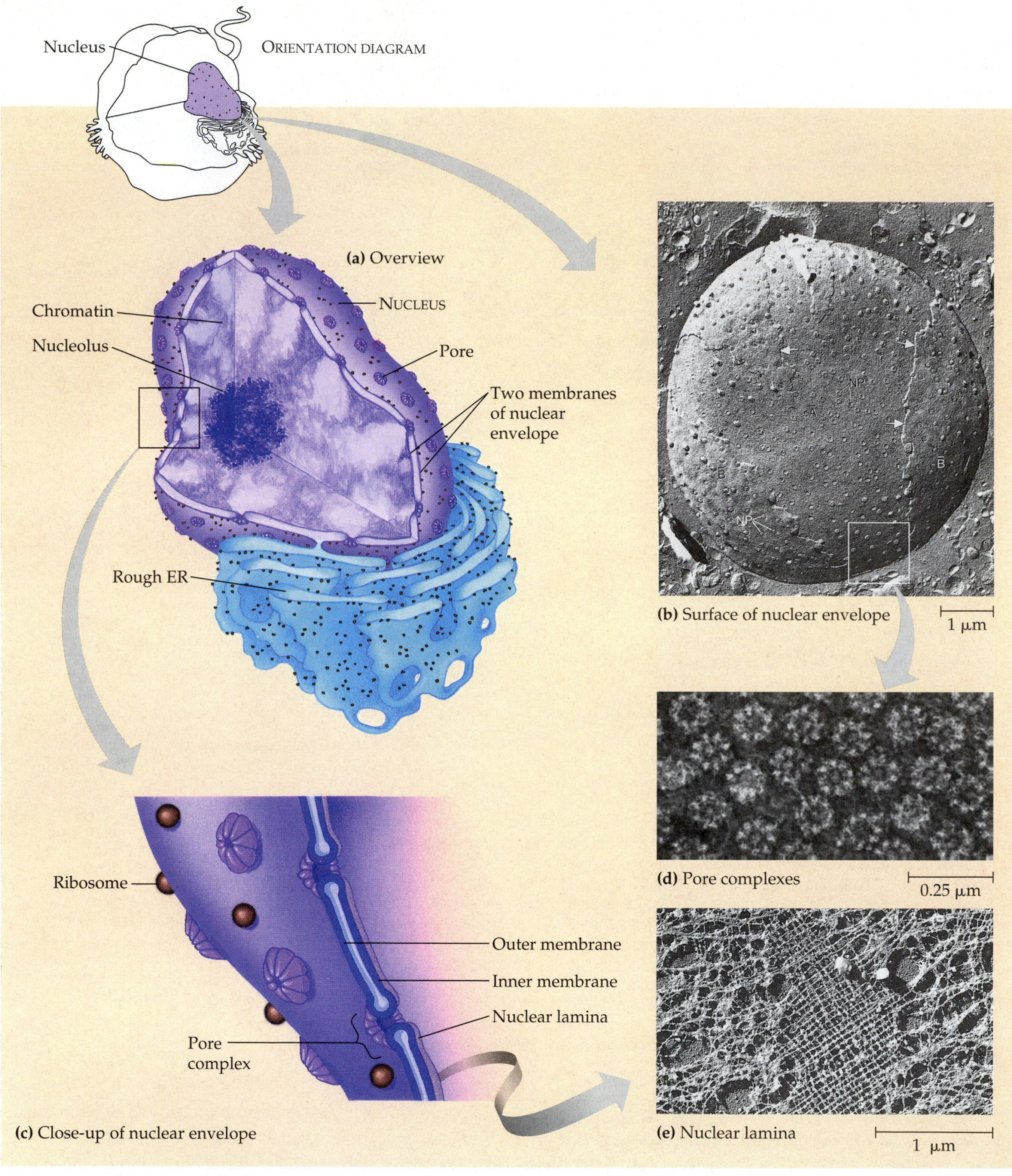

FIGURE 7.9 ▪ The nucleus and its envelope. **(a)** Within the nucleus is chromatin, consisting of DNA and proteins. When a cell prepares to divide, individual chromosomes become visible as the chromatin condenses. The nucleolus functions in ribosome synthesis. The nuclear envelope, which consists of two membranes separated by a narrow space, is perforated with pores. **(b)** Numerous nuclear pores (NP) through the envelope are evident in this electron micrograph, prepared by a method called freeze-fracture (TEM). (The A and B labels distinguish the two membranes of the envelope.) **(c)** The nuclear envelope. **(d)** An electron micrograph of the outer surface of the envelope reveals that each pore is bordered by a ring of protein particles (TEM). **(e)** The netlike nuclear lamina lines the inner surface of the envelope and reinforces the shape of the nucleus (TEM).

and assembled. These components then pass through the nuclear pores to the cytoplasm, where they combine to form ribosomes. Sometimes there are two or more nucleoli; the number depends on the species and the stage in the cell's reproductive cycle. A nucleolus is roughly spherical, and through the electron microscope it appears as a mass of densely stained granules and fibers adjoining part of the chromatin.

The nucleus controls protein synthesis in the cytoplasm by sending molecular messengers in the form of RNA. As we saw in FIGURE 5.26, this messenger RNA (mRNA) is synthesized in the nucleus according to instructions provided by the DNA. The mRNA then conveys the genetic messages to the cytoplasm via the nuclear pores. Once in the cytoplasm, an mRNA molecule attaches to ribosomes, where the genetic message is translated into the primary structure of a specific protein. This process of translating genetic information is described in detail in Chapter 17.

Ribosomes build a cell's proteins

Ribosomes are the sites where the cell makes proteins. Cells that have high rates of protein synthesis have a particularly large number of ribosomes. For example, a human liver cell has a few million ribosomes. Not surprisingly, cells active in protein synthesis also have prominent nucleoli.

Ribosomes build proteins in two cytoplasmic locales (FIGURE 7.10). *Free* ribosomes are suspended in the cytosol, while *bound* ribosomes are attached to the outside of the membranous network called the endoplasmic reticulum. Most of the proteins made by free ribosomes will function within the cytosol; examples are enzymes that catalyze metabolic processes localized in the cytosol. Bound ribosomes generally make proteins that are destined either for inclusion into membranes, for packaging within certain organelles such as lysosomes, or for export from the cell. Cells that specialize in protein secretion—for instance, the cells of the pancreas and other glands that secrete digestive enzymes—frequently have a high proportion of bound ribosomes. Bound and free ribosomes are structurally identical and interchangeable, and the cell can adjust the relative numbers of each as its metabolism changes. You will learn more about ribosome structure and function in Chapter 17.

THE ENDOMEMBRANE SYSTEM

Many of the different membranes of the eukaryotic cell are part of an **endomembrane system**. These membranes are related either through direct physical continuity or by the transfer of membrane segments as tiny vesicles (membrane-enclosed sacs). These relationships, however, do not mean that the various membranes are alike in structure and function. The thickness, molecular composition, and metabolic behavior of a membrane are not fixed, but may be modified several times during the membrane's life. The endomembrane system includes the nuclear envelope, endoplasmic reticulum, Golgi apparatus, lysosomes, various kinds of vacuoles, and the plasma membrane (not actually an *endo*membrane in physical location, but nevertheless related to the endoplasmic reticulum and other internal membranes). We have already discussed the nuclear envelope and will now focus on the endoplasmic reticulum and the other endomembranes to which it gives rise.

The endoplasmic reticulum manufactures membranes and performs many other biosynthetic functions

The **endoplasmic reticulum (ER)** is a membranous labyrinth so extensive that it accounts for more than half the total membrane in many eukaryotic cells. (The word *endoplasmic* means "within the cytoplasm," and *reticulum* is derived from the Latin for "network.") The ER consists of a network of membranous tubules and sacs called cisternae (L. *cisterna*, "box" or "chest"). The ER membrane separates its internal compartment, the cisternal space, from the cytosol. And because the ER membrane is continuous with the nuclear envelope, the space between the two membranes of the envelope is continuous with the cisternal space of the ER.

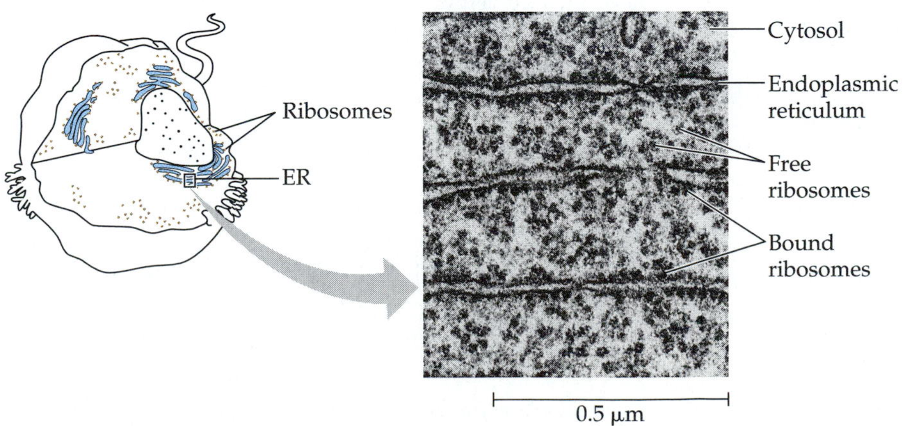

FIGURE 7.10 ▪ Ribosomes. Both free and bound ribosomes are abundant in this electron micrograph of part of a pancreas cell (TEM). The pancreas is a gland specialized for the secretion of proteins. It secretes hormones, including the protein insulin, into the bloodstream, and secretes digestive enzymes, which are also proteins, into the intestine. Bound ribosomes, those presently attached to the endoplasmic reticulum (ER), produce secretory proteins. Free ribosomes mainly make proteins that will remain dissolved in the cytosol. Bound and free ribosomes are identical and can alternate between these two roles.

There are two distinct, though connected, regions of ER that differ in structure and function: smooth ER and rough ER. **Smooth ER** is so named because its cytoplasmic surface lacks ribosomes. **Rough ER** appears rough through the electron microscope because ribosomes stud the cytoplasmic surface of the membrane. Ribosomes are also attached to the cytoplasmic side of the nuclear envelope's outer membrane, which is confluent with rough ER (FIGURE 7.11).

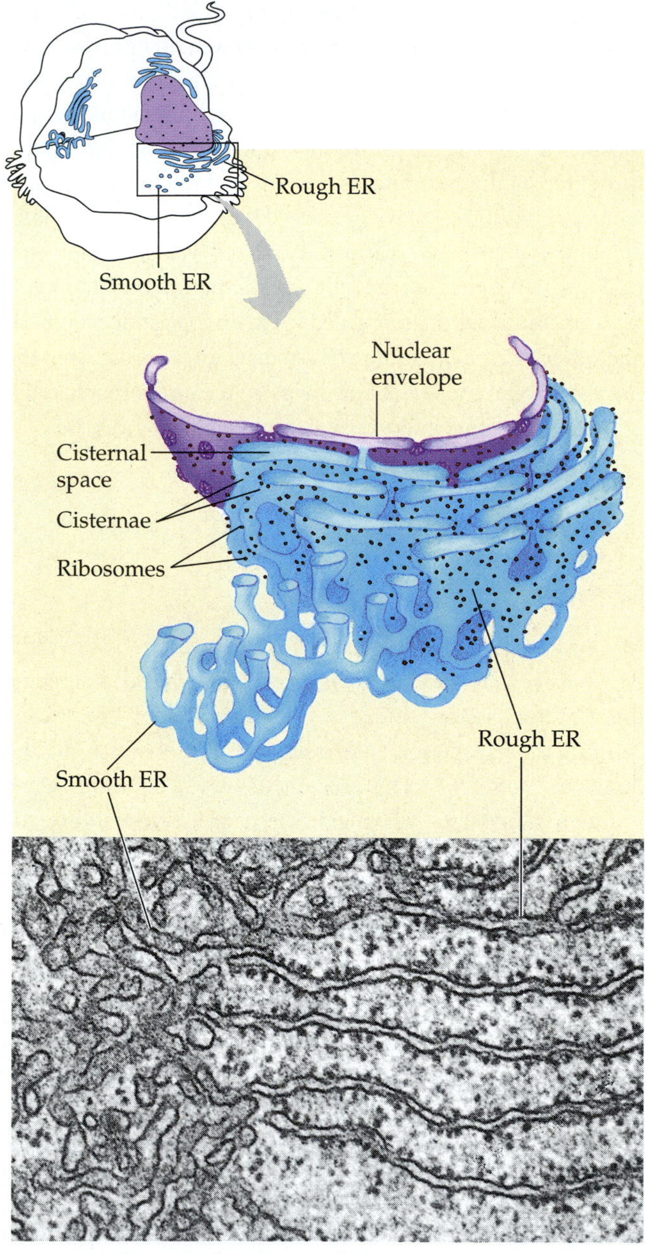

FIGURE 7.11 ▪ Endoplasmic reticulum (ER). A membranous system of interconnected tubules and flattened sacs called cisternae, the ER is also continuous with the nuclear envelope. (The drawing is a cutaway view.) The membrane of the ER encloses a compartment called the cisternal space. Rough ER, which is studded on its outer surface with ribosomes, can be distinguished from smooth ER in the electron micrograph (TEM).

Functions of Smooth ER

The smooth ER of various cell types functions in diverse metabolic processes, including synthesis of lipids, metabolism of carbohydrates, and detoxification of drugs and poisons.

Enzymes of the smooth ER are important to the synthesis of lipids, including phospholipids and steroids. Among the steroids produced by smooth ER are the sex hormones of vertebrates and the various steroid hormones secreted by the adrenal glands. The cells that actually synthesize and secrete these hormones—in the testes and ovaries, for example—are rich in smooth ER, a structural feature that fits the function of these cells.

Liver cells provide one example of the role of smooth ER in carbohydrate metabolism. Liver cells store carbohydrate in the form of glycogen, a polysaccharide. The hydrolysis of glycogen leads to the release of glucose from the liver cells, which is important in the regulation of sugar concentration in the blood. However, the first product of glycogen hydrolysis is glucose phosphate, an ionic form of the sugar that cannot exit the cell and enter the blood. An enzyme embedded in the membrane of the liver cell's smooth ER removes the phosphate from the glucose, which can then leave the cell.

Enzymes of the smooth ER help detoxify drugs and poisons, especially in liver cells. Detoxification usually involves adding hydroxyl groups to drugs, making them more soluble and easier to flush from the body. The sedative phenobarbital and other barbiturates are examples of drugs metabolized in this manner by smooth ER in liver cells. In fact, barbiturates, alcohol, and many other drugs induce the proliferation of smooth ER and its associated detoxification enzymes. This in turn increases tolerance to the drugs, meaning that higher doses are required to achieve a particular effect, such as sedation. Also, because some of the detoxification enzymes have relatively broad action, the proliferation of smooth ER in response to one drug can increase tolerance to other drugs as well. Barbiturate abuse, for example, may decrease the effectiveness of certain antibiotics and other useful drugs.

Muscle cells exhibit still another specialized function of smooth ER. The ER membrane pumps calcium ions from the cytosol into the cisternal space. When a muscle cell is stimulated by a nerve impulse, calcium rushes back across the ER membrane into the cytosol and triggers contraction of the muscle cell.

Rough ER and Synthesis of Secretory Proteins

Many types of specialized cells secrete proteins produced by ribosomes attached to rough ER. For example, certain cells in the pancreas secrete the protein insulin, a hormone, into the bloodstream. As a polypeptide chain grows from a bound ribosome, it is threaded into the cisternal space through a pore formed by a protein in the ER membrane. As it enters the cisternal space, the new protein folds into its native conformation. Most secretory proteins are **glycoproteins**, proteins that are covalently bonded to carbohydrates. The carbohydrate is attached to the protein in the ER by specialized molecules built into the ER membrane. The carbohydrate appendage of a glycoprotein is an oligosaccharide, the term for a relatively small polymer of sugar units.

Once secretory proteins are formed, the ER membrane keeps them separate from the proteins, produced by free ribosomes, that will remain in the cytosol. Secretory proteins depart from the ER wrapped in the membranes of vesicles that bud like bubbles from a specialized region called transitional ER. Such vesicles in transit from one part of the cell to another are called **transport vesicles**, and we will soon learn their fate.

Rough ER and Membrane Production

In addition to making secretory proteins, rough ER is a membrane factory that grows in place by adding proteins and phospholipids. As polypeptides destined to be membrane proteins grow from the ribosomes, they are inserted into the ER membrane itself and are anchored there by hydrophobic portions of the proteins. The rough ER also makes its own membrane phospholipids; enzymes built into the ER membrane assemble phospholipids from precursors in the cytosol. The ER membrane expands and can be transferred in the form of transport vesicles to other components of the endomembrane system.

The Golgi apparatus finishes, sorts, and ships cell products

After leaving the ER, many transport vesicles travel to the **Golgi apparatus.** We can think of the Golgi as a center of manufacturing, warehousing, sorting, and shipping. Here, products of the ER are modified and stored, and then sent to other destinations. Not surprisingly, the Golgi apparatus is especially extensive in cells specialized for secretion.

The Golgi apparatus consists of flattened membranous sacs—cisternae—looking like a stack of pita bread (FIGURE 7.12). A cell may have several of these stacks. The membrane of each cisterna in a stack separates its internal space from the cytosol. Vesicles concentrated in the vicinity of the Golgi apparatus are engaged in the transfer of material between the Golgi and other structures.

The Golgi apparatus has a distinct polarity, with the membranes of cisternae at opposite ends of a stack differing in thickness and molecular composition. The two poles of a

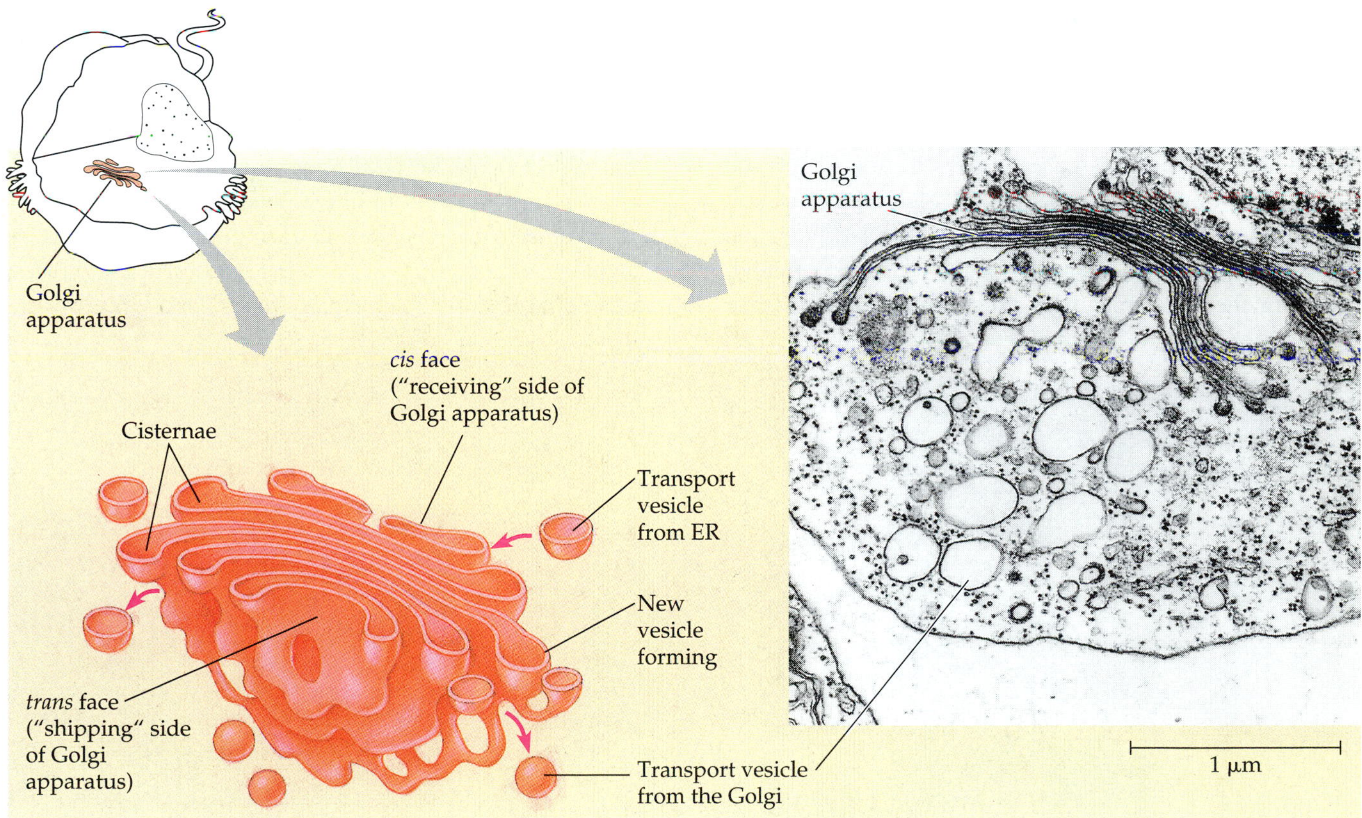

FIGURE 7.12 • The Golgi apparatus. A Golgi apparatus consists of stacks of flattened sacs, or cisternae, which unlike ER cisternae are not physically connected. (The drawing is a cutaway view.) The apparatus receives and dispatches transport vesicles and the products they contain. Materials received from the ER are modified and stored in the Golgi and eventually shipped to the cell surface or other destinations. Note the vesicles joining and leaving the cisternae. A stack has a structural and functional polarity, with a *cis* face that receives vesicles and a *trans* face that dispatches vesicles (at right, TEM).

Golgi stack are referred to as the *cis* face and the *trans* face; these act, respectively, as the receiving and shipping departments of the Golgi apparatus. The *cis* face is usually located near ER. Transport vesicles move material from the ER to the Golgi. A vesicle that buds from the ER will add its membrane and the contents of its lumen (cavity) to the *cis* face by fusing with a Golgi membrane. The *trans* face gives rise to vesicles, which pinch off and travel to other sites.

Products of the ER are usually modified during their transit from the *cis* pole to the *trans* pole of the Golgi. Proteins and phospholipids of membranes may be altered. For example, various Golgi enzymes modify the oligosaccharide portions of glycoproteins. When first added to proteins in the ER, the oligosaccharides of all glycoproteins are identical. The Golgi removes some sugar monomers and substitutes others, producing diverse oligosaccharides.

In addition to its finishing work, the Golgi apparatus manufactures certain macromolecules by itself. Many polysaccharides secreted by cells are Golgi products, including hyaluronic acid, a sticky substance that helps glue animal cells together. Golgi products that will be secreted depart from the *trans* faces of Golgi inside transport vesicles that eventually fuse with the plasma membrane.

The Golgi manufactures and refines its products in stages, with different cisternae between the *cis* and *trans* ends containing unique teams of enzymes. Products in various stages of processing appear to be transferred from one cisterna to the next by vesicles.

Before the Golgi apparatus dispatches its products by budding vesicles from the *trans* face, it sorts these products and targets them for various parts of the cell. Molecular identification tags, such as phosphate groups that have been added to the Golgi products, aid in sorting. And transport vesicles budded from the Golgi may have external molecules on their membranes that recognize "docking sites" on the surface of specific organelles.

Lysosomes are digestive compartments

A **lysosome** is a membrane-bounded sac of hydrolytic enzymes that the cell uses to digest macromolecules (FIGURE 7.13). There are lysosomal enzymes that can hydrolyze proteins, polysaccharides, fats, and nucleic acids—all the major classes of macromolecules. These enzymes work best in an acidic environment, at about pH 5. The lysosomal membrane maintains this low internal pH by pumping hydrogen ions from the cytosol into the lumen of the lysosome. If a lysosome breaks open or leaks its contents, the enzymes are not very active in the neutral environment of the cytosol. However, excessive leakage from a large number of lysosomes can destroy a cell by autodigestion. From this example we can see once again how important compartmental organization is to the functions of the cell: The lysosome provides a space where the cell can digest macromolecules safely, without the general destruction that would occur if hydrolytic enzymes roamed at large.

Hydrolytic enzymes and lysosomal membrane are made by rough ER and then transferred to a Golgi apparatus for further processing. At least some lysosomes probably arise by budding from the *trans* face of the Golgi apparatus (FIGURE 7.14). Proteins of the inner surface of the lysosomal membrane and the digestive enzymes themselves are probably spared from destruction by having three-dimensional conformations that protect vulnerable bonds from enzymatic attack.

Lysosomes function in intracellular digestion in a variety of circumstances. *Amoeba* and many other protists eat by engulfing smaller organisms or other food particles, a process called **phagocytosis** (Gr. *phagein*, "to eat," and *kytos*, "vessel,"

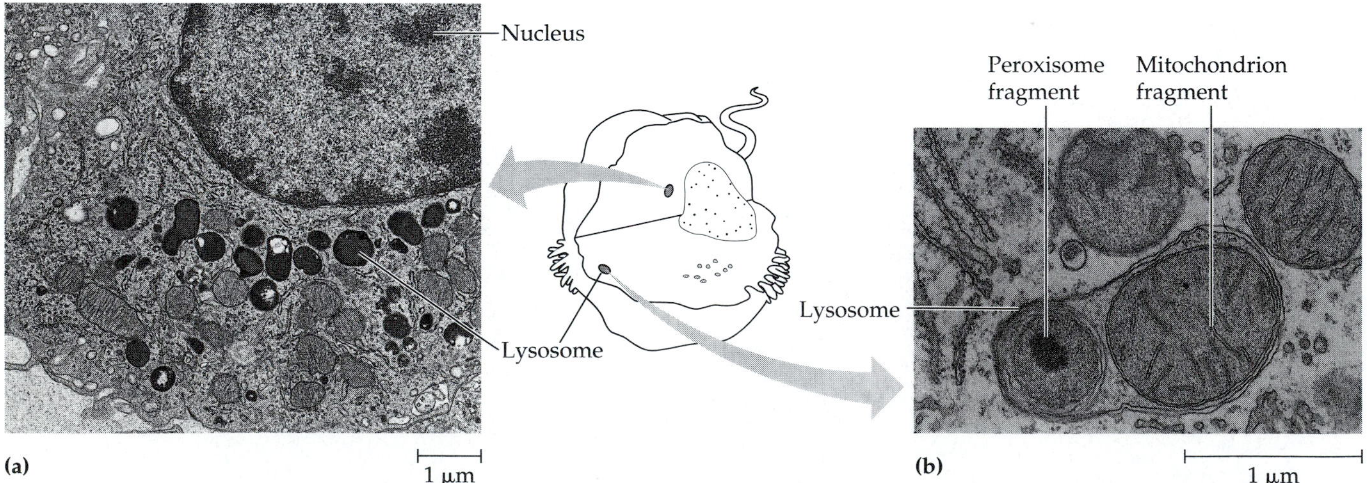

FIGURE 7.13 • Lysosomes. (a) In this white blood cell from a rat, the lysosomes are very dark because of a specific stain that reacts with one of the products of digestion within the lysosome. This type of white blood cell ingests bacteria and viruses and destroys them in the lysosomes (TEM). **(b)** In the cytoplasm of this rat liver cell, an autophagic lysosome has engulfed two disabled organelles, a mitochondrion and a peroxisome (TEM).

referring here to the cell). The food vacuole formed in this way then fuses with a lysosome, whose enzymes digest the food (see FIGURE 7.14). Some human cells also carry out phagocytosis. Among them are macrophages, cells that help defend the body by destroying bacteria and other invaders.

Lysosomes also use their hydrolytic enzymes to recycle the cell's own organic material, a process called autophagy. This occurs when a lysosome engulfs another organelle or a small parcel of cytosol (see FIGURE 7.13b). The lysosomal enzymes dismantle the ingested material, and the organic monomers are returned to the cytosol for reuse. With the help of lysosomes, the cell continually renews itself. A human liver cell, for example, recycles half of its macromolecules each week.

Programmed destruction of cells by their own lysosomal enzymes is important in the development of many organisms. During the transforming of a tadpole into a frog, for instance, lysosomes destroy the cells of the tail. And the hands of human embryos are webbed until lysosomes digest the tissue between the fingers.

A variety of inherited disorders called lysosomal storage diseases affects lysosomal metabolism. A person afflicted with a storage disease lacks one of the active hydrolytic enzymes normally present in lysosomes. The lysosomes become engorged with indigestible substrates, which begin to interfere with other cellular functions. In Pompe's disease, for example, the liver is damaged by an accumulation of glycogen due to the absence of a lysosomal enzyme needed to break down the polysaccharide. In Tay-Sachs disease, a lipid-digesting enzyme is missing or inactive, and the brain becomes impaired by an accumulation of lipids in the cells. Fortunately, storage diseases are rare in the general population. In the future, it might be possible to treat storage diseases by injecting the missing enzymes into the blood along with adaptor molecules that target the enzymes for engulfment by cells and fusion with lysosomes. It might also be possible someday to repair a disorder directly by inserting genes (DNA) for the missing enzyme into the appropriate cells (see Chapter 20).

Vacuoles have diverse functions in cell maintenance

Vacuoles and vesicles are both membrane-bounded sacs within the cell, but vacuoles are larger than vesicles. Vacuoles have various functions. **Food vacuoles**, formed by phagocytosis, have already been mentioned (see FIGURE 7.14). Many freshwater protists have **contractile vacuoles** that pump excess water out of the cell (see FIGURE 8.12). Mature plant cells generally contain a large **central vacuole** (FIGURE 7.15, p. 116) enclosed by a membrane called the **tonoplast**, which is part of their endomembrane system.

The plant cell vacuole is a versatile compartment. It is a place to store organic compounds, such as the proteins that are stockpiled in the vacuoles of storage cells in seeds. This vacuole is also the plant cell's main repository of inorganic ions, such as potassium and chloride. Many plant cells use their vacuoles as disposal sites for metabolic by-products that would endanger the cell if they accumulated in the cytosol. Some vacuoles are enriched in pigments that color the cells,

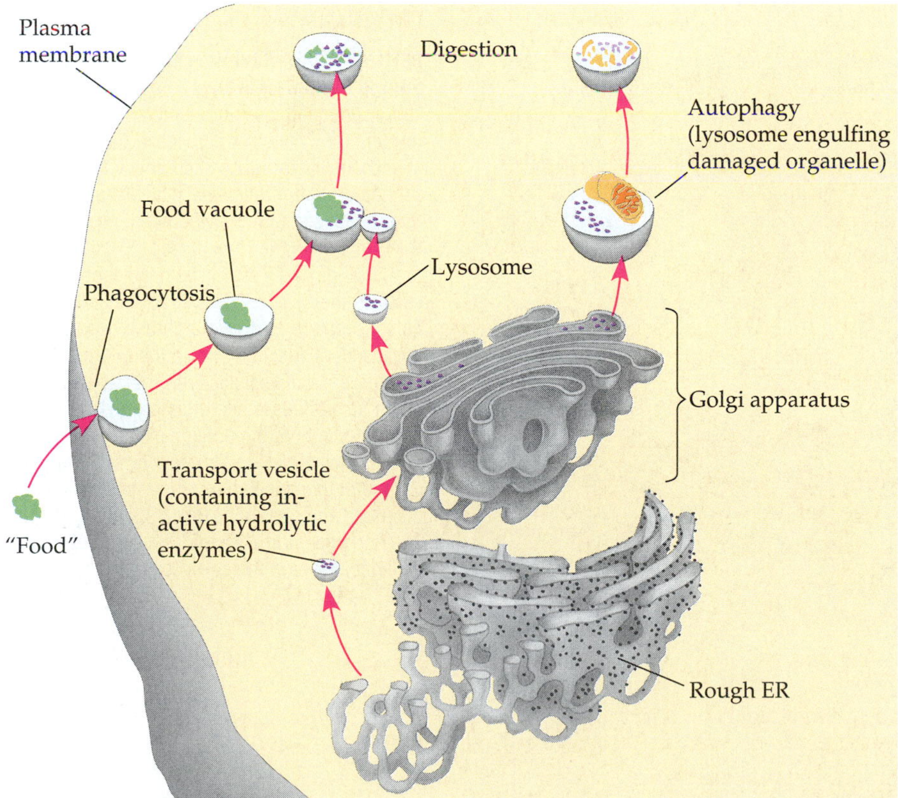

FIGURE 7.14 ▪ The formation and functions of lysosomes. Lysosomes digest materials taken into the cell and recycle materials from intracellular refuse. During phagocytosis, the cell encloses food in a vacuole with a membrane that pinches off internally from the plasma membrane. This food vacuole fuses with a lysosome, and hydrolytic enzymes digest the food. After hydrolysis, simple sugars, amino acids, and other monomers pass across the lysosomal membrane into the cytosol as nutrients for the cell. By the process of autophagy, lysosomes recycle the molecular ingredients of organelles. The ER and Golgi generally cooperate in the production of lysosomes containing active enzymes.

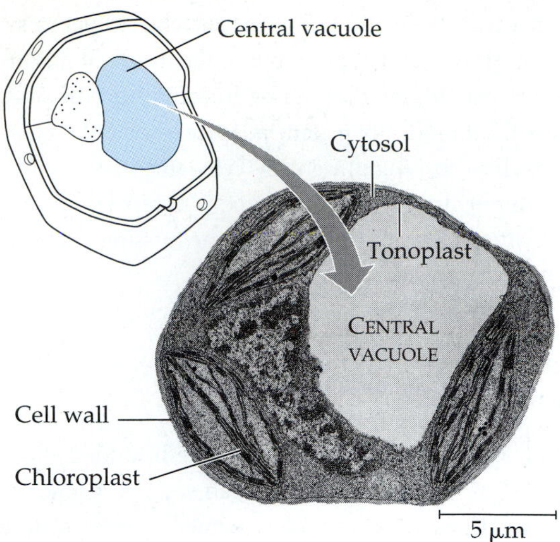

FIGURE 7.15 • The plant cell vacuole. The central vacuole is usually the largest compartment in a plant cell, comprising 80% or more of a mature cell. The rest of the cytoplasm is generally confined to a narrow zone between the vacuole and the plasma membrane. The membrane of the vacuole, the tonoplast, separates the cytosol from the solution inside the vacuole, which is called cell sap. Like all cellular membranes, the tonoplast is selective in transporting solutes; therefore, cell sap differs in composition from the cytosol. Functions of the vacuole include storage, waste disposal, protection, and growth (TEM).

such as the red and blue pigments of petals that help attract pollinating insects to flowers. Vacuoles may also help protect the plant against predators by containing compounds that are poisonous or unpalatable to animals. The vacuole has a major role in the growth of plant cells, which elongate as their vacuoles absorb water, enabling the cell to become larger with a minimal investment in new cytoplasm. And because the cytosol often occupies only a thin layer between the plasma membrane and the tonoplast, the ratio of membrane surface to cytosolic volume is great, even for a large plant cell.

The large vacuole of a plant cell develops by the coalescence of smaller vacuoles, themselves derived from the endoplasmic reticulum and Golgi apparatus. Through these relationships, the vacuole is an integral part of the endomembrane system. FIGURE 7.16 reviews the endomembrane system. We'll continue our tour of the cell with some membranous organelles that are *not* closely related to the endomembrane system.

OTHER MEMBRANOUS ORGANELLES

Mitochondria and chloroplasts are the main energy transformers of cells

One of this book's themes is that organisms are open systems that transform energy they acquire from their surroundings. In eukaryotic cells, mitochondria and chloroplasts are the organelles that convert energy to forms that cells can use for work. **Mitochondria** (singular, **mitochondrion**) are the sites of cellular respiration, the catabolic process that generates ATP by extracting energy from sugars, fats, and other fuels with the help of oxygen. **Chloroplasts**, found only in plants and eukaryotic algae, are the sites of photosynthesis. They convert solar energy to chemical energy by absorbing sunlight

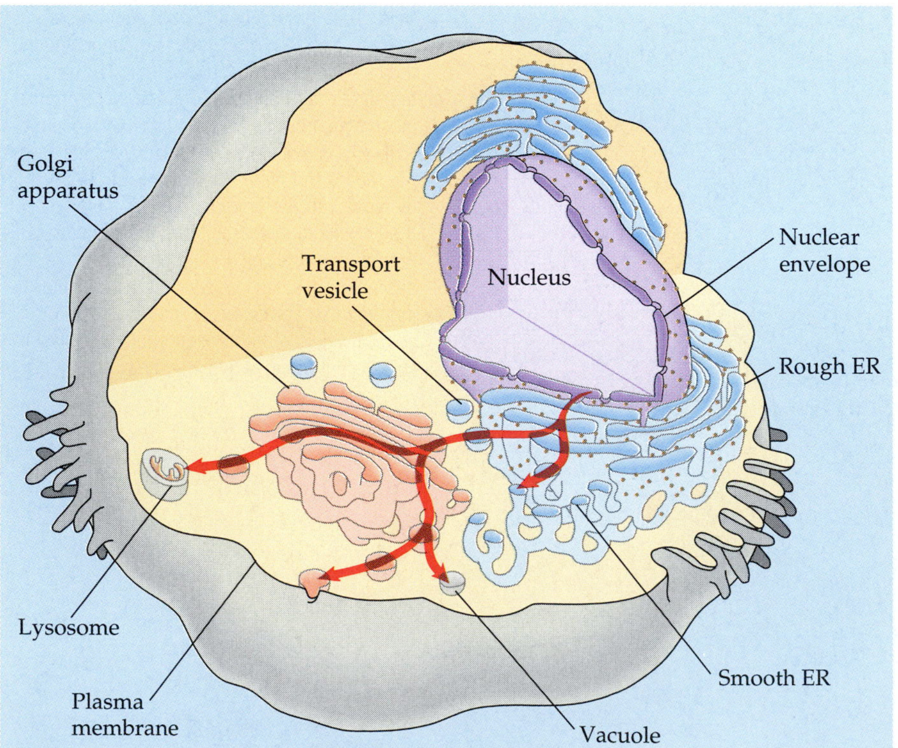

FIGURE 7.16 • Review: relationships among endomembranes. The red arrows show some pathways of membrane migration through the various organelles of the endomembrane system. The nuclear envelope is connected to the rough ER, which is also confluent with smooth ER. Membrane produced by the ER flows in the form of transport vesicles to the Golgi. The Golgi, in turn, pinches off vesicles that give rise to lysosomes and vacuoles. Even the plasma membrane expands by the fusion of vesicles born in the ER and Golgi. (Coalescence of vesicles with the plasma membrane also releases secretory proteins and other products to the outside of the cell.) As membranes of the system flow from ER to the Golgi and then elsewhere, their molecular compositions and metabolic functions are modified. The endomembrane system is a complex and dynamic player in the cell's compartmental organization.

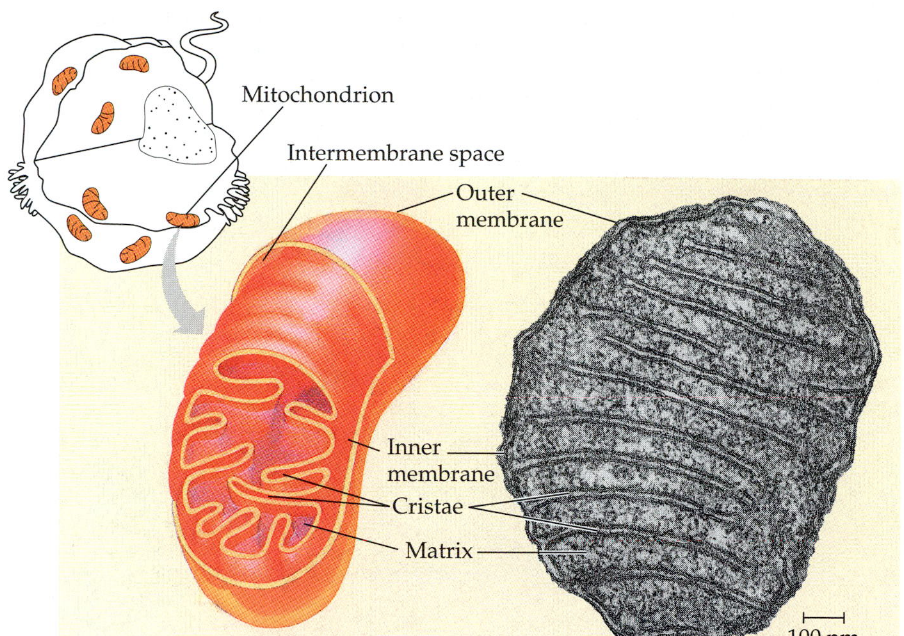

FIGURE 7.17 ▪ The mitochondrion, site of cellular respiration. The two membranes of the mitochondrion are evident in the drawing and micrograph (TEM). The cristae are infoldings of the inner membrane. The cutaway drawing shows the two compartments bounded by the membranes: the intermembrane space and the mitochondrial matrix.

and using it to drive the synthesis of organic compounds from carbon dioxide and water.

Although mitochondria and chloroplasts are enclosed by membranes, they are not part of the endomembrane system. Their membrane proteins are made not by the ER, but by free ribosomes in the cytosol and by ribosomes contained within the mitochondria and chloroplasts themselves. Not only do these organelles have ribosomes, but they also contain a small amount of DNA. It is this DNA that programs the synthesis of the proteins made on the organelle's own ribosomes. (Proteins imported from the cytosol—constituting most of the organelle's proteins—are programmed by nuclear DNA.) Mitochondria and chloroplasts are semiautonomous organelles that grow and reproduce within the cell. In Chapters 9 and 10 we will focus on how mitochondria and chloroplasts function. We will consider the evolution of these organelles in Chapter 28. Here we are concerned mainly with the structure of these energy transformers.

Mitochondria

Mitochondria are found in nearly all eukaryotic cells. In some cases, there is a single large mitochondrion, but more often a cell has hundreds or even thousands of mitochondria; the number is correlated with the cell's level of metabolic activity. Mitochondria are about 1 to 10 μm long. Time-lapse films of living cells reveal mitochondria moving around, changing their shapes, and dividing in two, unlike the static cylinders seen in electron micrographs of dead cells.

The mitochondrion is enclosed in an envelope of two membranes, each a phospholipid bilayer with a unique collection of embedded proteins (FIGURE 7.17). The outer membrane is smooth, but the inner membrane is convoluted, with infoldings called **cristae**. The inner membrane divides the mitochondrion into two internal compartments. The first is the intermembrane space, the narrow region between the inner and outer membranes. The second compartment, the **mitochondrial matrix**, is enclosed by the inner membrane. Some of the metabolic steps of cellular respiration occur in the matrix, where many different enzymes are concentrated. Other proteins that function in respiration, including the enzyme that makes ATP, are built into the inner membrane. The cristae give the inner mitochondrial membrane a large surface area that enhances the productivity of cellular respiration, another example of the correlation between structure and function.

Chloroplasts

The chloroplast is a specialized member of a family of closely related plant organelles called **plastids**. Amyloplasts are colorless plastids that store starch (amylose), particularly in roots and tubers. Chromoplasts are enriched in pigments that give fruits and flowers their orange and yellow hues. Chloroplasts contain the green pigment chlorophyll along with enzymes and other molecules that function in the photosynthetic production of food. These lens-shaped organelles, measuring about 2 μm by 5 μm, are found in leaves and other green organs of plants and in eukaryotic algae (FIGURE 7.18, p. 118).

The contents of a chloroplast are partitioned from the cytosol by an envelope consisting of two membranes separated by a very narrow intermembrane space. Inside the chloroplast is another membranous system, arranged into flattened sacs called **thylakoids**. In some regions, thylakoids are stacked like poker chips, forming structures called **grana** (singular, **granum**). The fluid outside the thylakoids is called

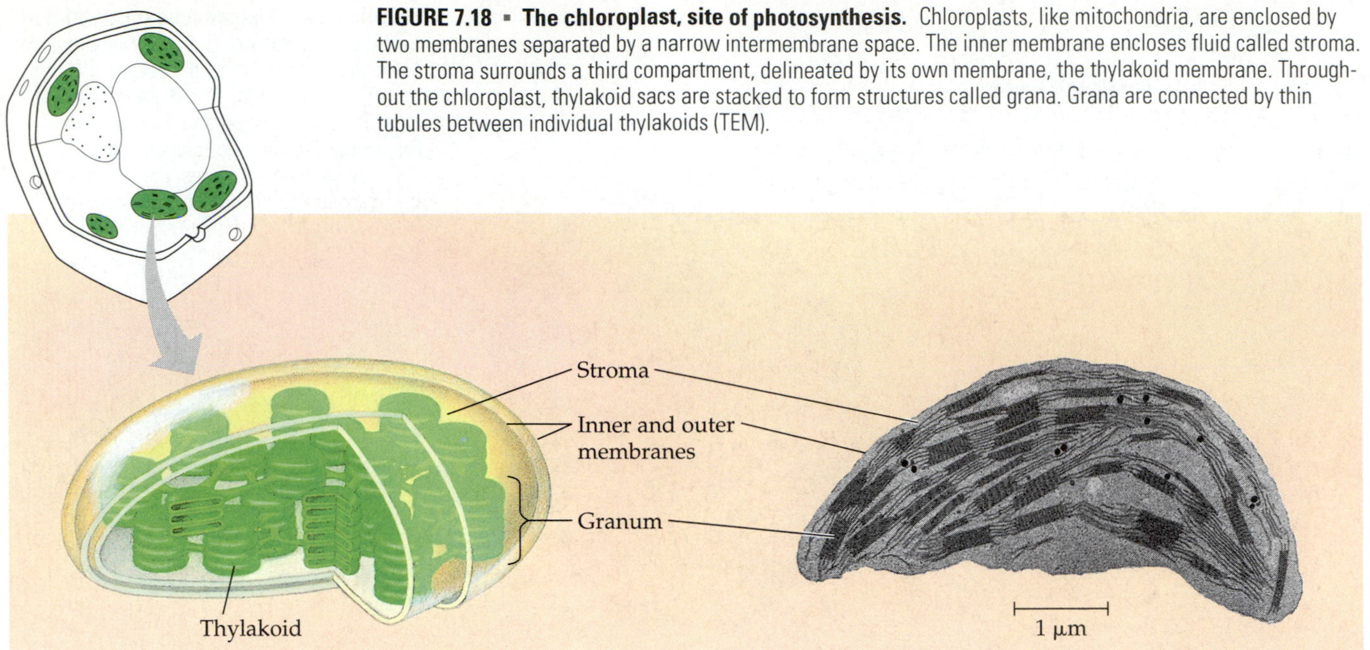

FIGURE 7.18 ▪ The chloroplast, site of photosynthesis. Chloroplasts, like mitochondria, are enclosed by two membranes separated by a narrow intermembrane space. The inner membrane encloses fluid called stroma. The stroma surrounds a third compartment, delineated by its own membrane, the thylakoid membrane. Throughout the chloroplast, thylakoid sacs are stacked to form structures called grana. Grana are connected by thin tubules between individual thylakoids (TEM).

the **stroma**. Thus, the thylakoid membrane divides the interior of the chloroplast into two compartments: the thylakoid space and the stroma. In Chapter 10 you will learn how this compartmental organization enables the chloroplast to convert light energy to chemical energy during photosynthesis.

As with mitochondria, the static and rigid appearance of chloroplasts in electron micrographs is not true to their dynamic behavior in the living cell. Their shapes are plastic, and they occasionally pinch in two. They are mobile and move around the cell with mitochondria and other organelles along tracks of the cytoskeleton.

Peroxisomes consume oxygen in various metabolic functions

The **peroxisome** is a specialized metabolic compartment bounded by a single membrane (FIGURE 7.19). Peroxisomes contain enzymes that transfer hydrogen from various substrates to oxygen, producing hydrogen peroxide (H_2O_2) as a by-product, from which the organelle derives its name. These reactions may have many different functions. Some peroxisomes use oxygen to break fatty acids down into smaller molecules that can then be transported to mitochondria as fuel for cellular respiration. Peroxisomes in the liver detoxify alcohol and other harmful compounds by transferring hydrogen from the poisons to oxygen. The H_2O_2 formed by peroxisome metabolism is itself toxic, but the organelle contains an enzyme that converts the H_2O_2 to water. Enclosing in the same space both the enzymes that produce hydrogen peroxide and those that dispose of this toxic compound is another example of how the cell's compartmental structure is crucial to its functions.

Specialized peroxisomes called glyoxysomes are found in the fat-storing tissues of plant seeds. These organelles contain enzymes that initiate the conversion of fatty acids to sugar, which the emerging seedling can use as an energy and carbon source until it is able to produce its own sugar by photosynthesis.

Unlike lysosomes, peroxisomes do not bud from the endomembrane system. They grow by incorporating proteins and lipids made in the cytosol, and increase in number by splitting in two when they reach a certain size.

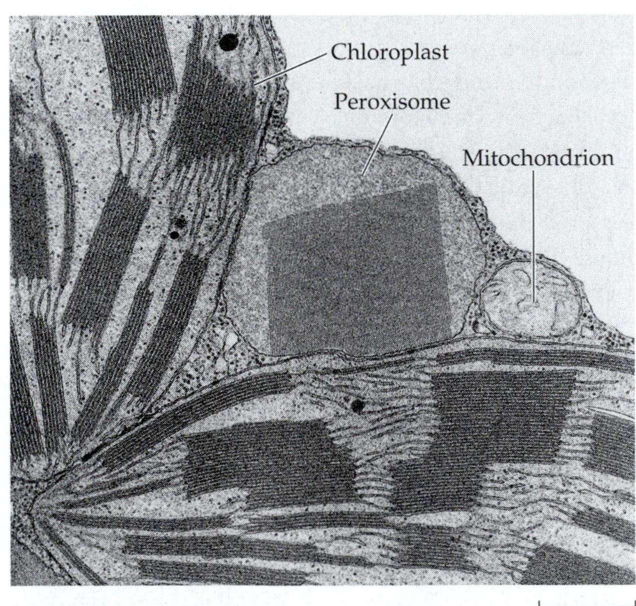

FIGURE 7.19 ▪ Peroxisomes. Peroxisomes are roughly spherical and often have a granular or crystalline core that is probably a dense collection of enzymes. This peroxisome is in a leaf cell. Notice its proximity to two chloroplasts and a mitochondrion. These organelles cooperate with peroxisomes in certain metabolic functions (TEM).

THE CYTOSKELETON

In the early days of electron microscopy, biologists thought that the organelles of a eukaryotic cell floated freely in the cytosol. But improvements in both light microscopy and electron microscopy have revealed the **cytoskeleton**, a network of fibers extending throughout the cytoplasm (FIGURE 7.20). The cytoskeleton plays a major role in organizing the structures and activities of the cell.

Providing structural support to the cell, the cytoskeleton also functions in cell motility and regulation

The most obvious function of the cytoskeleton is to give mechanical support to the cell and maintain its shape. This is especially important for animal cells, which lack walls. The remarkable strength and resilience of the cytoskeleton as a whole is based on its architecture. Like a geodesic dome, the cytoskeleton is stabilized by a balance between opposing forces exerted by its elements. And, just as the skeleton of an animal helps fix the positions of other body parts, the cytoskeleton provides anchorage for many organelles and even cytosolic enzyme molecules. The cytoskeleton is more dynamic than an animal skeleton, however. It can be quickly dismantled in one part of the cell and reassembled in a new location, changing the shape of the cell.

The cytoskeleton is also involved in several types of cell motility (movement). The term *cell motility* encompasses both changes in cell location and more limited movements of parts of the cell. Cell motility generally requires the interaction of the cytoskeleton with proteins called motor molecules (FIGURE 7.21). Examples of such cell motility abound. Motor molecules of the cytoskeleton wiggle cilia and flagella. They also cause muscle cells to contract. Vesicles may travel to their destinations in the cell along "monorails" provided by the cytoskeleton, and the cytoskeleton manipulates the plasma membrane to form food vacuoles during phagocytosis. The streaming of cytoplasm that circulates materials within many large plant cells is yet another kind of cellular movement brought about by components of the cytoskeleton.

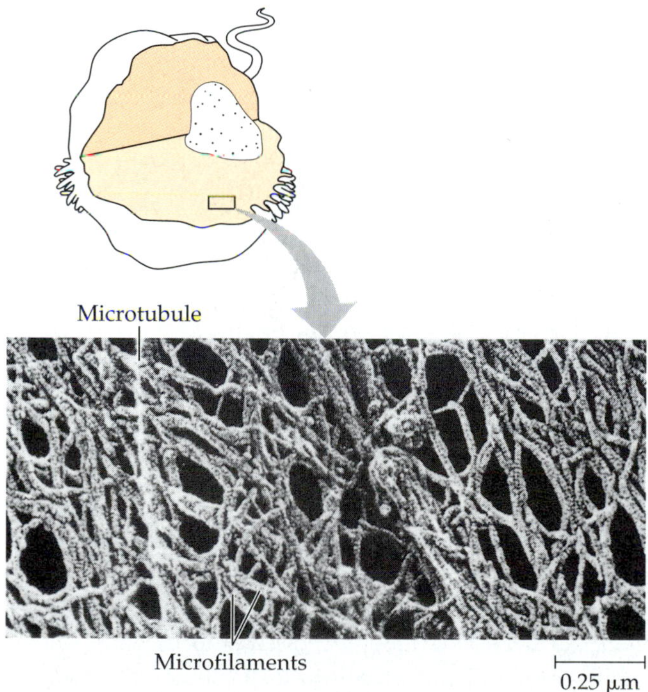

FIGURE 7.20 ▪ The cytoskeleton. The cytoskeleton gives the cell shape, anchors some organelles and directs the movement of others, and may enable the entire cell to change shape or move. It may even play a regulatory role, by mechanically transmitting signals from the cell's surface to its interior. In this electron micrograph, prepared by a method known as deep-etching, microtubules and microfilaments are visible. A third component of the cytoskeleton, intermediate filaments, is not evident here (TEM).

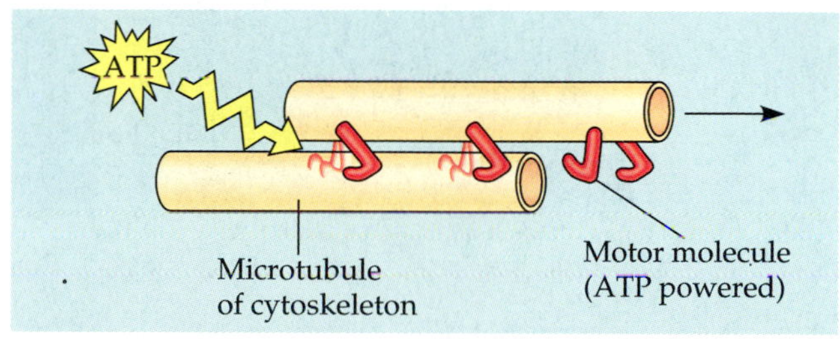

(a)

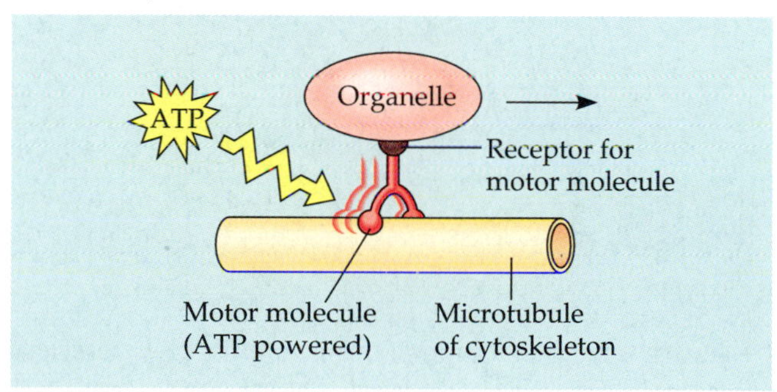

(b)

FIGURE 7.21 ▪ Motor molecules and the cytoskeleton. The microtubules and microfilaments of the cytoskeleton function in motility by interacting with proteins called motor molecules. Motor molecules work by changing their shapes, moving back and forth something like microscopic legs. ATP powers these conformational changes. With each cycle of shape change, the motor molecule releases at its free end and then grips at a site farther along a microtubule or microfilament. **(a)** In some types of cell motility, motor molecules attached to one element of the cytoskeleton cause it to slide past another cytoskeletal element. For example, a sliding of neighboring microtubules moves cilia and flagella. A similar mechanism causes muscle cells to contract, but in this case motor molecules slide microfilaments rather than microtubules. **(b)** Motor molecules can also attach to receptors on organelles such as vesicles and enable the organelles to "walk" along microtubules of the cytoskeleton. For example, this is how vesicles containing neurotransmitters migrate to the tips of axons, the long extensions of nerve cells that release transmitter molecules as chemical signals to adjacent nerve cells.

Table 7.2 ■ The Structure and Function of the Cytoskeleton

PROPERTY	MICROTUBULES	MICROFILAMENTS (ACTIN FILAMENTS)	INTERMEDIATE FILAMENTS
Structure	Hollow tubes; wall consists of 13 columns of tubulin molecules	Two intertwined strands of actin	Fibrous proteins supercoiled into thicker cables
Diameter	25 nm with 15-nm lumen	7 nm	8–12 nm
Protein subunits	Tubulin, consisting of α-tubulin and β-tubulin	Actin	One of several different proteins of the keratin family, depending on cell type
Functions	Maintenance of cell shape (compression-resisting "girders") Cell motility (as in cilia or flagella) Chromosome movements in cell division Organelle movements	Maintenance of cell shape (tension-bearing elements) Changes in cell shape Muscle contraction Cytoplasmic streaming Cell motility (as in pseudopodia) Cell division (cleavage furrow formation)	Maintenance of cell shape (tension-bearing elements) Anchorage of nucleus and certain other organelles Formation of nuclear lamina

SOURCE: Adapted from W. M. Becker, J .B. Reece, and M. F. Poenie, *The World of the Cell*, 3rd ed. Menlo Park, CA: Benjamin/Cummings, 1996, p. 555.

The most recent addition to the list of possible cytoskeletal functions is the regulation of biochemical activities in the cell. A growing body of evidence suggests that the cytoskeleton can transmit mechanical forces from the surface of the cell to its interior—and even, via other fibers, into the nucleus. In one experiment, investigators used a micromanipulation device to pull on certain plasma-membrane proteins attached to the cytoskeleton. A video microscope captured almost instantaneous rearrangements of nucleoli and other structures in the nucleus. It is easy to imagine that transmission of mechanical signals by the cytoskeleton may help regulate the functioning of the cell.

Let's now look more closely at the three main types of fibers that make up the cytoskeleton (TABLE 7.2). **Microtubules** are the thickest of the three types; **microfilaments** (also called actin filaments) are the thinnest. **Intermediate filaments** are fibers with diameters in a middle range.

Microtubules

Microtubules are found in the cytoplasm of all eukaryotic cells. They are straight, hollow rods measuring about 25 nm in diameter and from 200 nm to 25 μm in length. The wall of the hollow tube is constructed from a globular protein called tubulin. Each tubulin molecule consists of two similar polypeptide subunits, α-tubulin and β-tubulin. A microtubule elongates by adding tubulin molecules to its ends. Microtubules can be disassembled and their tubulin used to build microtubules elsewhere in the cell.

Microtubules shape and support the cell and also serve as tracks along which organelles equipped with motor molecules can move (see FIGURE 7.21). For example, microtubules seem to guide secretory vesicles from the Golgi apparatus to the plasma membrane. Microtubules are also involved in the separation of chromosomes during cell division, discussed in Chapter 12.

Centrosomes and Centrioles. In many cells, microtubules grow out from a **centrosome**, a region located near the nucleus. These microtubules function as compression-resisting girders of the cytoskeleton. Within the centrosome of an animal cell are a pair of **centrioles**, each composed of nine sets of triplet microtubules arranged in a ring (FIGURE 7.22). When a cell divides, the centrioles replicate. Although centrioles may help organize microtubule assembly, they are not essential for this function in all eukaryotes; centrosomes of most plants lack centrioles altogether.

Cilia and Flagella. In eukaryotes, a specialized arrangement of microtubules is responsible for the beating of **flagella** and **cilia**, locomotive appendages that protrude from some cells. Many unicellular eukaryotic organisms are propelled through water by cilia or flagella, and the sperm of animals, algae, and some plants are flagellated. If cilia or flagella extend from cells that are held in place as part of a tissue layer, they function to move fluid over the surface of the tissue. For example, the ciliated lining of the windpipe sweeps mucus with trapped debris out of the lungs (see FIGURE 7.2).

Cilia usually occur in large numbers on the cell surface. They are about 0.25 μm in diameter and about 2 to 20 μm in length. Flagella are the same diameter but longer than cilia, measuring 10 to 200 μm in length. Also, flagella are usually limited to just one or a few per cell.

Flagella and cilia also differ in their beating patterns. A flagellum has an undulating motion that generates force in the same direction as the flagellum's axis. In contrast, cilia work more like oars, with alternating power and recovery strokes generating force in a direction perpendicular to the cilium's axis (FIGURE 7.23).

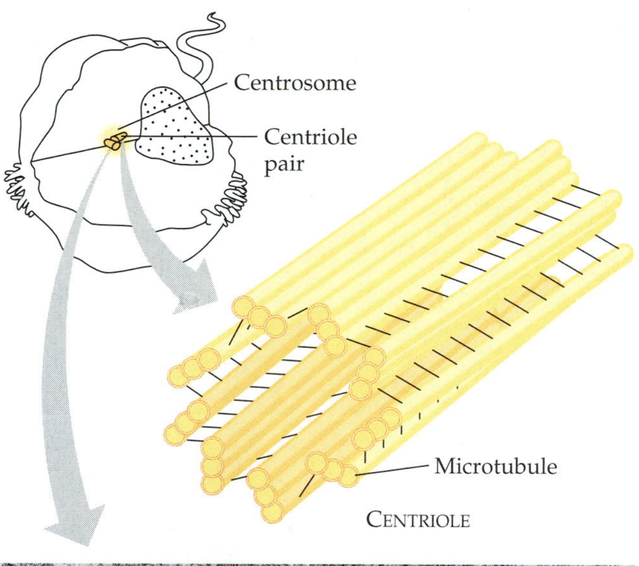

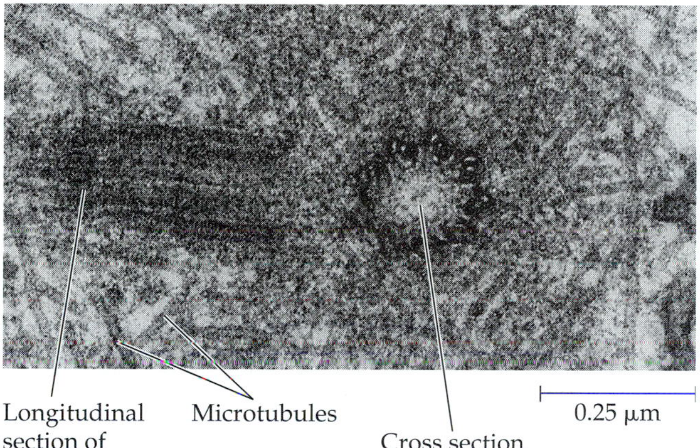

FIGURE 7.22 ▪ Centrosome containing a pair of centrioles. An animal cell has a pair of centrioles within its centrosome, the region near the nucleus where the cell's microtubules are initiated. The centrioles, each about 250 nm (0.25 μm) in diameter, are arranged at right angles to each other, and each is made up of nine sets of three microtubules (TEM).

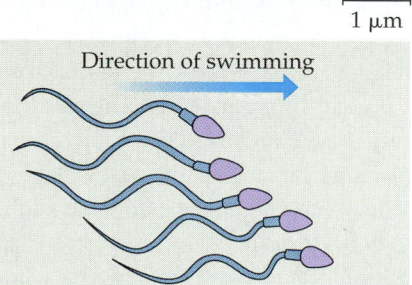

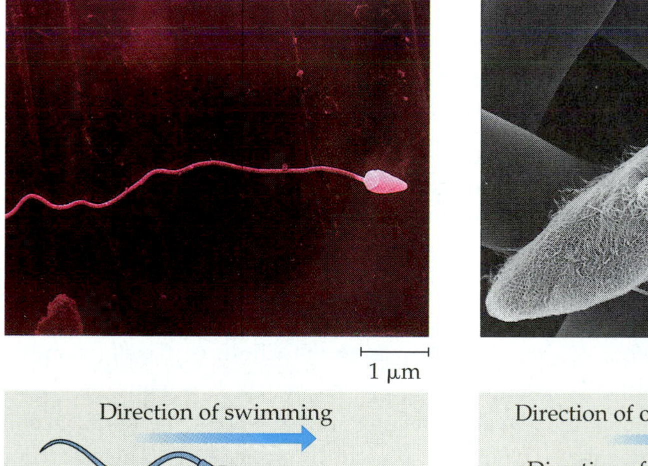

Direction of swimming

(a) Motion of flagella

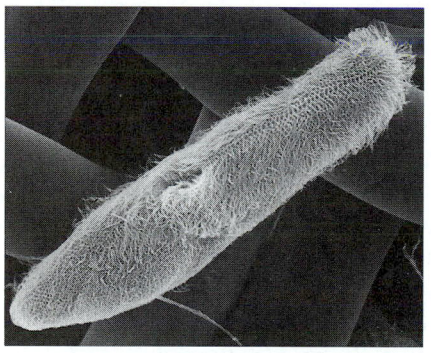

25 μm

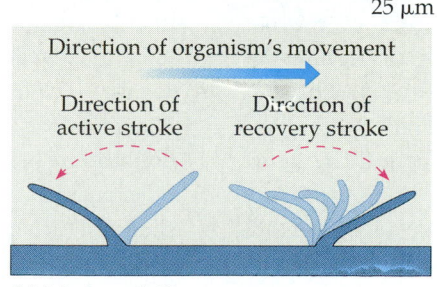

Direction of organism's movement

Direction of active stroke

Direction of recovery stroke

(b) Motion of cilia

FIGURE 7.23 ▪ A comparison of the beating of flagella and cilia. (a) A flagellum usually undulates, its snakelike motion driving a cell in the same direction as the axis of the flagellum. Propulsion of a sperm cell is an example of flagellate locomotion (SEM). **(b)** A dense nap of beating cilia covers this *Paramecium,* a motile protist (SEM). The cilia beat at a rate of about 40 to 60 strokes per second. Cilia have a back-and-forth motion, alternating active strokes with recovery strokes. This moves the cell, or moves a fluid over the surface of a stationary cell, in a direction perpendicular to the axis of the cilium.

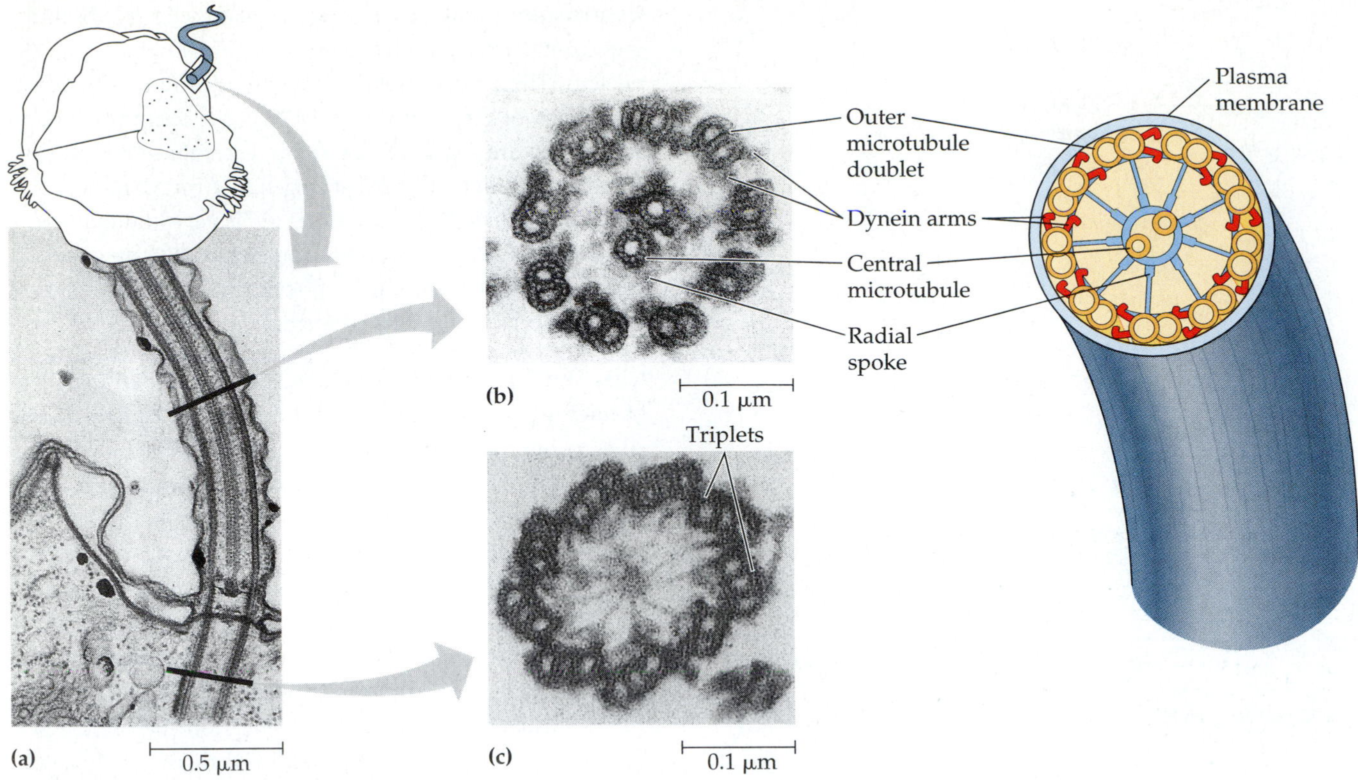

FIGURE 7.24 ▪ Ultrastructure of a eukaryotic flagellum or cilium. (a) In this electron micrograph of a longitudinal section of a cilium, microtubules can be seen running the length of the structure (TEM). **(b)** A cross section through the cilium shows the "9 + 2" arrangement of microtubules (TEM). **(c)** The basal body anchoring the cilium or flagellum to the cell has a ring of nine microtubule triplets (the basal body is structurally identical to a centriole). The nine doublets of the cilium extend into the basal body, where each doublet joins another microtubule to form the ring of nine triplets. The two central microtubules of the cilium terminate above the basal body (TEM).

Though different in length, number per cell, and beating pattern, cilia and flagella actually share a common ultrastructure. A cilium or flagellum has a core of microtubules sheathed in an extension of the plasma membrane (FIGURE 7.24). Nine doublets of microtubules, the members of each pair sharing part of their walls, are arranged in a ring. In the center of the ring are two single microtubules. This arrangement, referred to as the "9 + 2" pattern, is found in nearly all eukaryotic flagella and cilia. (The flagella of motile prokaryotes, which will be discussed in Chapter 27, are entirely different.) The doublets of the outer ring are connected to the center of the cilium or flagellum by radial spokes that terminate near the central pair of microtubules. Each doublet of the outer ring also has pairs of arms evenly spaced along its length and reaching toward the neighboring doublet of microtubules. The microtubule assembly of a cilium or flagellum is anchored in the cell by a **basal body**, which is structurally identical to a centriole. In fact, in many animals (including humans), the basal body of the fertilizing sperm's flagellum enters the egg and becomes a centriole.

The arms extending from each microtubule doublet to the next are the motors responsible for the bending movements of cilia and flagella. The motor molecule that makes up these arms is a very large protein called **dynein**. A dynein arm performs a complex cycle of movements caused by changes in the conformation of the protein, with ATP providing the energy for these changes.

The mechanics of dynein "walking" are reminiscent of a cat climbing a tree by attaching its claws, moving its legs, releasing its claws, and grabbing again farther up the tree. Similarly, the dynein arms of one doublet attach to an adjacent doublet and pull so that the doublets slide past each other in opposite directions. The arms then release from the other doublet and reattach a little farther along its length. Without any restraints on the movement of the microtubule doublets, one doublet would continue to "walk" along the surface of the other, elongating the cilium or flagellum rather than bending it. For lateral movement of a cilium or flagellum, the dynein "walking" must have something to pull against, as when the muscles in your leg pull against your bones to move your knee. In cilia and flagella, the microtubule doublets are held in place, presumably by the radial spokes or other structural elements. Thus neighboring doublets cannot slide past each other very far. Instead, the forces exerted by the dynein arms cause the doublets to curve, bending the cilium or flagellum (FIGURE 7.25).

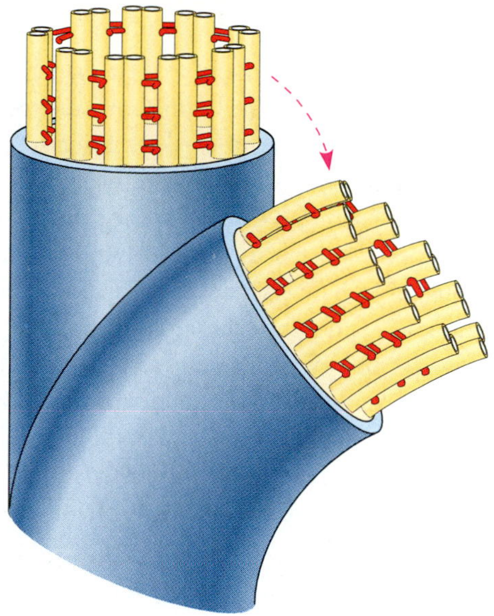

FIGURE 7.25 ▪ How dynein "walking" moves cilia and flagella. The dynein arms of one microtubule doublet grip the adjacent doublet, pull, release, and then grip again. This cycle of the dynein motors is powered by ATP. The doublets cannot slide far because they are physically restrained within the cilium. Instead, the action of the dynein arms causes the doublets to bend.

Microfilaments (Actin Filaments)

Microfilaments are solid rods about 7 nm in diameter. They are also called actin filaments, because they are built from molecules of **actin**, a globular protein. A microfilament is a twisted double chain of actin subunits (see TABLE 7.2). Microfilaments seem to be present in all eukaryotic cells.

In contrast to the compression-resisting role of microtubules, the structural role of microfilaments in the ctyoskeleton is to bear tension (pulling forces). In combination with other proteins, they often form a three-dimensional network just inside the plasma membrane, helping support the cell's shape. This network gives the cortex (outer cytoplasmic layer) of such a cell the semisolid consistency of a gel, in contrast with the more fluid (sol) state of the interior cytoplasm. In animal cells specialized for transporting materials across the plasma membrane, bundles of microfilaments make up the core of microvilli, delicate projections that increase the cell surface area (FIGURE 7.26).

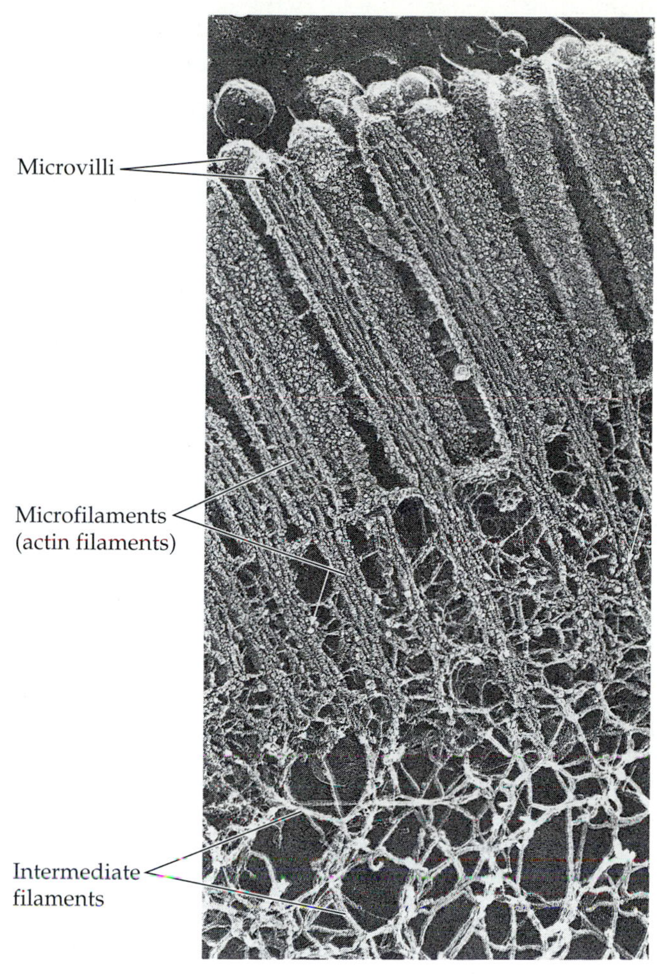

FIGURE 7.26 ▪ A structural role of microfilaments. The surface area of this nutrient-absorbing intestinal cell is increased by its many microvilli, cellular extensions reinforced by bundles of microfilaments. These actin filaments are anchored to a network of intermediate filaments (TEM, Hirokawa et al. 1982. *J. Cell Biol.* 94, pp. 425–443, Fig. 1).

Microfilaments are well known for their roles in cell motility, in particular as part of the contractile apparatus of muscle cells. Thousands of actin filaments are arranged parallel to one another along the length of a muscle cell, interdigitated with thicker filaments made of a protein called **myosin** (FIGURE 7.27a, p. 124). Contraction of the muscle cell results from the actin and myosin filaments sliding past one another, shortening the cell. In other kinds of cells, actin filaments are associated with myosin in miniature and less elaborate versions of the arrangement in muscle cells. These actin–myosin aggregates are responsible for localized contractions of cells. For example, a contracting belt of microfilaments forms a cleavage furrow that pinches a dividing animal cell into two daughter cells.

Localized contraction brought about by actin and myosin also plays a role in ameboid movement (FIGURE 7.27b), in which a cell crawls along a surface by extending and flowing into cellular extensions called **pseudopodia** (Gr. *pseudes,* "false," and *pod,* "foot"). Pseudopodia extend and contract through the *reversible* assembly of actin subunits into microfilaments and of microfilaments into networks that convert cytoplasm from sol to gel. Not only do amoebas move by crawling, but so do many cells in the animal body, such as white blood cells.

In plant cells, both actin–myosin interactions and sol-gel transformations brought about by actin may be involved in **cytoplasmic streaming**, a circular flow of cytoplasm within cells (FIGURE 7.27c). This movement, which is especially common in large plant cells, speeds the distribution of materials within the cell.

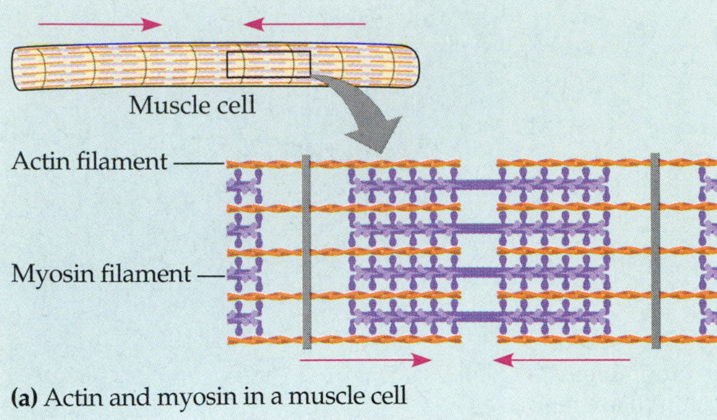

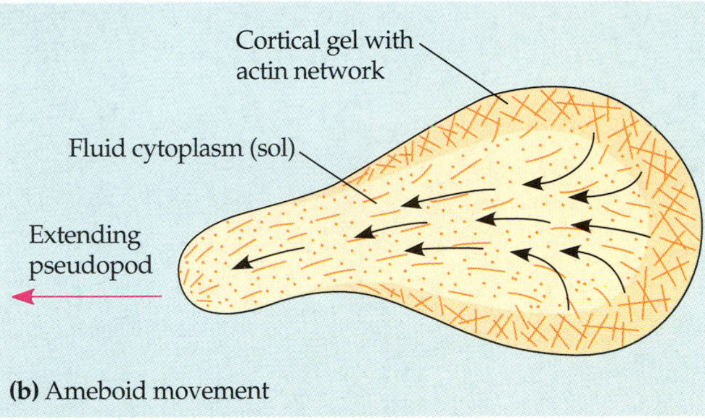

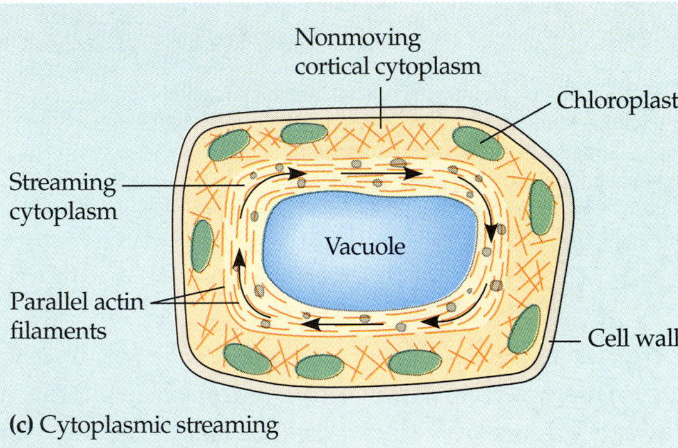

FIGURE 7.27 ▪ Microfilaments and motility. (a) In muscle cells, actin filaments (orange) lie parallel to thick myosin filaments (purple). Myosin acts as a motor molecule by means of arms that "walk" the two types of filaments past each other. The teamwork of many such sliding filaments enables the entire muscle cell to shorten. **(b)** In a crawling cell (ameboid movement), actin is organized into a network in the gel-like cortex (outer layer) but exists as subunits and linear filaments in the more fluid (sol) interior. According to one hypothesis, filaments at the cell's trailing end interact with myosin, causing contraction. Like squeezing on a toothpaste tube, this contraction forces the interior fluid into the pseudopod, where the actin network has been weakened. The pseudopod extends until the actin reassembles into a network. **(c)** In cytoplasmic streaming, a layer of cytoplasm cycles around the cell, moving over a carpet of parallel actin filaments. Myosin motors attached to organelles in the fluid cytosol may drive the streaming by interacting with the actin. Also, rapid changes between the gel and sol states may occur locally. (In this figure, cell nuclei and most other organelles have been omitted.)

Intermediate Filaments

Intermediate filaments are named for their diameter, which, at 8 to 12 nm, is larger than the diameter of microfilaments but smaller than that of microtubules (see TABLE 7.2 and FIGURE 7.26). Specialized for bearing tension (like microtubules), intermediate filaments are a diverse class of cytoskeletal elements. Each type is constructed from a different molecular subunit belonging to a diverse family of proteins called keratins. Microtubules and microfilaments, in contrast, are consistent in diameter and composition in all eukaryotic cells.

Intermediate filaments are more permanent fixtures of cells than are microfilaments and microtubules, which are often disassembled and reassembled in various parts of a cell. Chemical treatments that remove microfilaments and microtubules from the cytoplasm leave a web of intermediate filaments that retains its original shape. Such experiments suggest that intermediate filaments are especially important in reinforcing the shape of a cell and fixing the position of certain organelles. For example, the nucleus commonly sits within a cage made of intermediate filaments, fixed in location by branches of the filaments that extend into the cytoplasm. Other intermediate filaments make up the nuclear lamina that lines the interior of the nuclear envelope (see FIGURE 7.9c). In cases where the shape of the entire cell is correlated with function, intermediate filaments support that shape. For instance, the long extensions (axons) of nerve cells that transmit impulses are strengthened by one class of intermediate filament. The various kinds of intermediate filaments may function as the framework of the entire cytoskeleton.

CELL SURFACES AND JUNCTIONS

Having crisscrossed the interior of the cell to explore various organelles, we complete our tour of the cell by returning to the surface of this microscopic world, where there are additional structures with important functions. Although the plasma membrane is usually regarded as the boundary of the living cell, most cells synthesize and secrete coats of one kind or another that are external to the plasma membrane.

Plant cells are encased by cell walls

The **cell wall** is one of the features of plant cells that distinguishes them from animal cells. The wall protects the plant cell, maintains its shape, and prevents excessive uptake of water. On the level of the whole plant, the strong walls of specialized cells hold the plant up against the force of gravity. Prokaryotes, fungi, and some protists also have cell walls, but we will postpone discussion of them until Unit Five.

Plant cell walls are much thicker than the plasma membrane, ranging from 0.1 μm to several micrometers. The exact

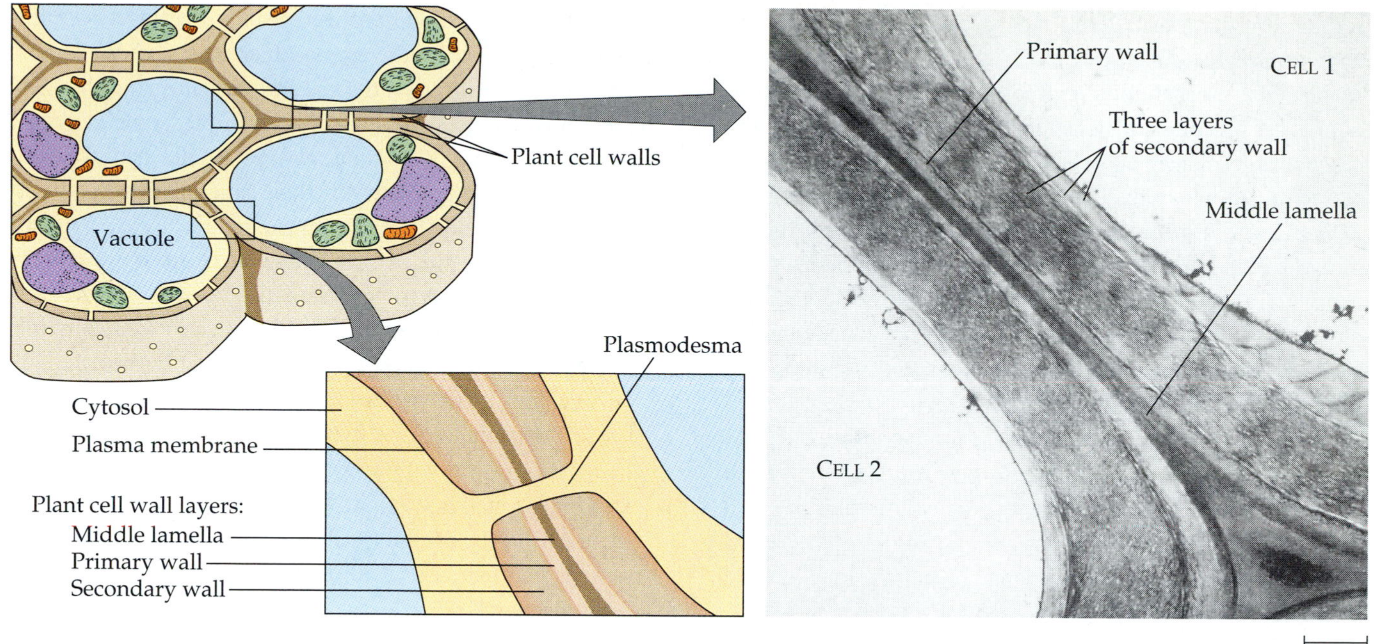

FIGURE 7.28 • Plant cell walls. Young cells first construct thin primary walls, often adding stronger secondary walls to the inside of the primary wall when growth ceases. A sticky middle lamella cements adjacent cells together. Thus, the multilayered partition between these cells consists of adjoining walls individually secreted by the cells. The walls do not isolate the cells: The cytoplasm of one cell is continuous with the cytoplasm of its neighbors via plasmodesmata, channels through the walls (TEM).

chemical composition of the wall varies from species to species and from one cell type to another in the same plant, but the basic design of the wall is consistent (see FIGURE 5.8). Microfibrils made of the polysaccharide cellulose are embedded in a matrix of other polysaccharides and protein. This combination of materials, strong fibers in a "ground substance" (matrix), is the same basic architectural design found in steel-reinforced concrete and in fiberglass.

A young plant cell first secretes a relatively thin and flexible wall called the **primary cell wall** (FIGURE 7.28). Between primary walls of adjacent cells is the **middle lamella**, a thin layer rich in sticky polysaccharides called pectins. The middle lamella glues the cells together (pectin is used as a thickening agent in jams and jellies). When the cell matures and stops growing, it strengthens its wall. Some plant cells do this simply by secreting hardening substances into the primary wall. Other cells add a **secondary cell wall** between the plasma membrane and the primary wall. The secondary wall, often deposited in several laminated layers, has a strong and durable matrix that affords the cell protection and support. Wood, for example, consists mainly of secondary walls.

The extracellular matrix (ECM) of animal cells functions in support, adhesion, movement, and regulation

Although animal cells lack walls akin to those of plant cells, they do have an elaborate **extracellular matrix (ECM)** (FIGURE 7.29). The main ingredients of the ECM are glycoproteins

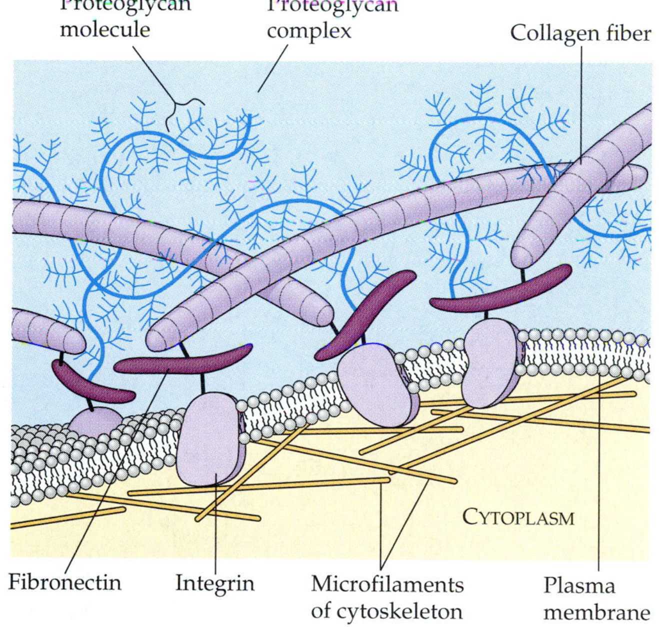

FIGURE 7.29 • Extracellular matrix (ECM) of an animal cell. The molecular composition and structure of the ECM varies from one cell type to another. In this example, three different types of glycoproteins are present. Fibers made of the glycoprotein collagen are embedded in a web of proteoglycans, which can be as much as 95% carbohydrate; the proteoglycan molecules have formed complexes by noncovalently attaching to long polysaccharide molecules. The third glycoprotein is fibronectin, the adhesive that attaches the ECM to the plasma membrane of the cell. Membrane proteins called integrins are bound to the ECM on one side and to the microfilaments of the cytoskeleton on the other. This linkage can transmit stimuli between the cell's external environment and its interior.

secreted by the cells. (Recall that glycoproteins are proteins with covalently bonded carbohydrate, usually short chains of sugars.) The most abundant glycoprotein in the ECM of most animal cells is **collagen**, which forms strong fibers outside the cells. In fact, collagen accounts for about half of the total protein in the human body. The collagen fibers are embedded in a network woven from **proteoglycans**, which are glycoproteins of another class. Proteoglycan molecules are especially rich in carbohydrate—up to 95%—and they can form large complexes, as shown in FIGURE 7.29. Some cells are attached to the ECM by still other kinds of glycoproteins, most commonly **fibronectins**. Fibronectins bind to receptor proteins called **integrins** that are built into the plasma membrane. Integrins span the membrane and bind on their cytoplasmic side to microfilaments of the cytoskeleton. Thus, integrins are in a position to transmit changes in the ECM to the cytoskeleton, and vice versa—to integrate changes occurring outside and inside the cell.

Current research on fibronectins and integrins is revealing the influential role of the extracellular matrix in the lives of cells. Communicating with a cell through integrins, the ECM can regulate a cell's behavior. For example, some cells in a developing embryo migrate along specific pathways by matching the orientation of their microfilaments to the "grain" of fibers in the extracellular matrix. Researchers are also learning that the extracellular matrix around a cell can influence the activity of genes in the nucleus. Information about the ECM probably reaches the nucleus by a combination of mechanical and chemical signaling pathways. Mechanical signaling involves fibronectins, integrins, and the cytoskeleton. Changes in the cytoskeleton may in turn trigger chemical signaling pathways inside the cell. In this way, the extracellular matrix of a particular tissue could help coordinate the behavior of all the cells within that tissue. Direct connections between cells also function in this coordination, as we discuss next.

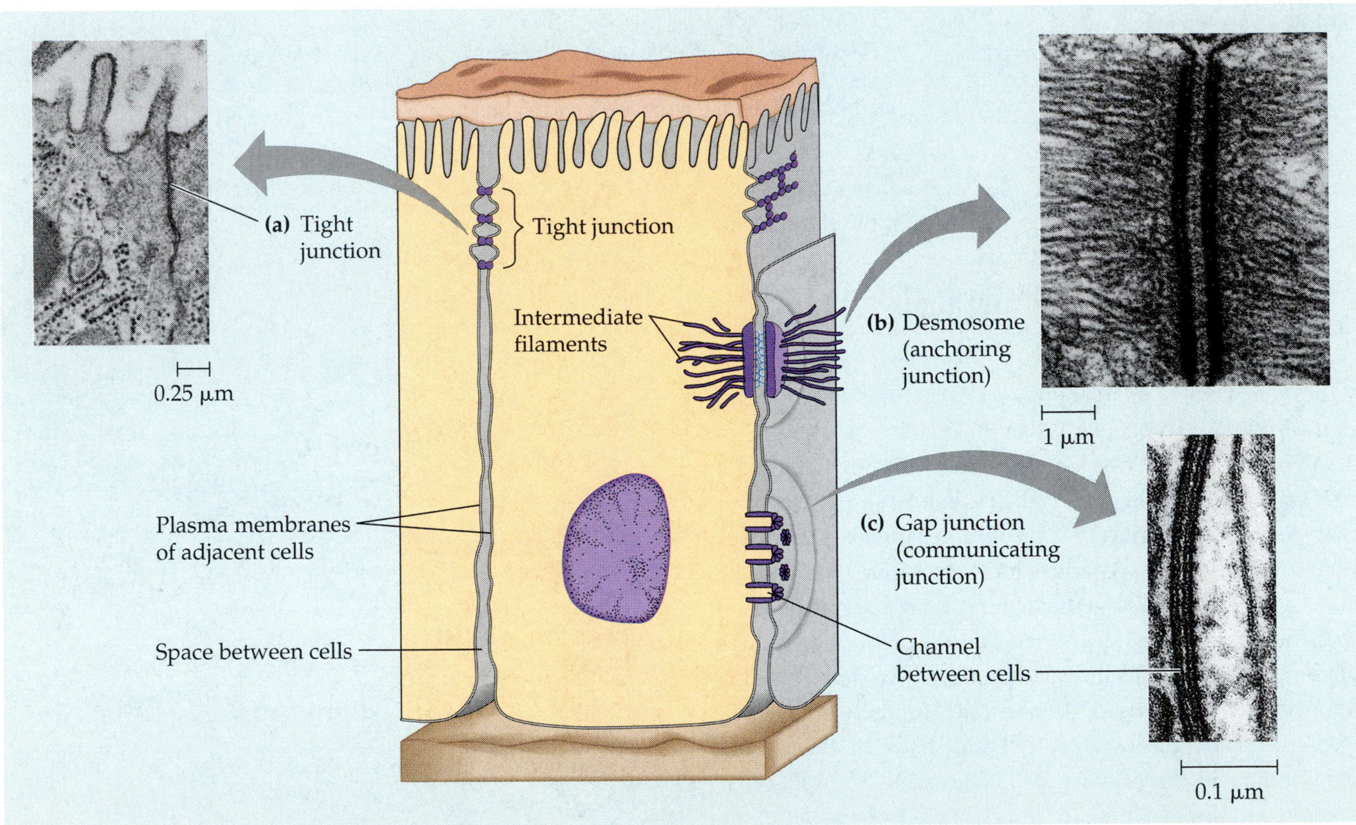

FIGURE 7.30 ▪ Intercellular junctions in animals. These specialized connections are especially common in epithelial tissue, which lines the internal surfaces of the body. Here, we use epithelial cells lining the intestine to describe three kinds of intercellular junctions, each with a structure well adapted for its function. **(a)** Tight junctions. These connections form continuous belts around the cell. The membranes of neighboring cells are actually fused at a tight junction, forming a seal that prevents leakage of extracellular fluid across a layer of epithelial cells. For example, the tight junctions of the intestinal epithelium keep the contents of the intestine separate from the body fluid on the opposite side of the epithelium (TEM).
(b) Desmosomes (anchoring junctions). These junctions function like rivets, fastening cells together into strong epithelial sheets. Intermediate filaments made of the sturdy protein keratin reinforce desmosomes (TEM).
(c) Gap junctions (communicating junctions). These connections provide cytoplasmic channels between adjacent cells. Special membrane proteins surround each pore, which is wide enough for salts, sugars, amino acids, and other small molecules to pass (TEM). In the muscle tissue of the heart, the flow of ions through gap junctions coordinates the contractions of the cells. Gap junctions are especially common in animal embryos, in which chemical communication between cells is essential for development.

7.30b from L. Orci and A. Perrelet, *Freeze-Hetch Histology*, (Heidelberg: Springer-Verlag, 1975). ©1975 Springer-Verlag.

Intercellular junctions help integrate cells into higher levels of structure and function

The many cells of an animal or plant are organized into tissues, organs, and organ systems. Neighboring cells often adhere, interact, and communicate through special patches of direct physical contact.

It might seem that the nonliving cell walls of plants would isolate cells from one another. In fact, as already mentioned, the walls are perforated with channels called **plasmodesmata** (singular, **plasmodesma**; Gr. *desmos,* "to bind"). Cytoplasm passes through the plasmodesmata and connects the living contents of adjacent cells (see FIGURES 7.8 and 7.28). This unifies most of the plant into one living continuum. The plasma membranes of adjacent cells are continuous through a plasmodesma; the membrane lines the channel. Water and small solutes can pass freely from cell to cell, and recent experiments have shown that, in certain circumstances, particular proteins and RNA molecules can also do so. The macromolecules to be transported seem to reach the plasmodesmata by moving along fibers of the cytoskeleton.

In animals, there are three main types of intercellular junctions: **tight junctions, desmosomes**, and **gap junctions**. These are illustrated and described in detail in FIGURE 7.30.

■ ■ ■

The cell is a living unit greater than the sum of its parts

From our panoramic view of the cell's overall compartmental organization to our closeup inspection of each organelle's architecture, this tour of the cell has provided many opportunities to correlate structure with function. (This would be a good time to review cell structure by returning to FIGURES 7.7 and 7.8.) But even as we dissect the cell, remember that none of its organelles works alone. As an example of cellular integration, consider the microscopic scene in FIGURE 7.31. The large cell is a macrophage. It helps defend the body against infections by ingesting bacteria (the smaller yellow cells). The macrophage crawls along a surface and reaches out to the bacteria with thin pseudopodia ("filopodia"). Actin filaments interact with other elements of the cytoskeleton in these movements. After the macrophage engulfs the bacteria, they are destroyed by lysosomes. The elaborate endomembrane system, which includes the ER and the Golgi apparatus, produces the lysosomes. The digestive enzymes of the lysosomes and the proteins of the cytoskeleton are all made on ribosomes. And the synthesis of these proteins is programmed by genetic messages dispatched from the DNA in the nucleus. All these processes require energy, which mitochondria supply in the form of ATP. Cellular functions arise from cellular order: The cell is a living unit greater than the sum of its parts.

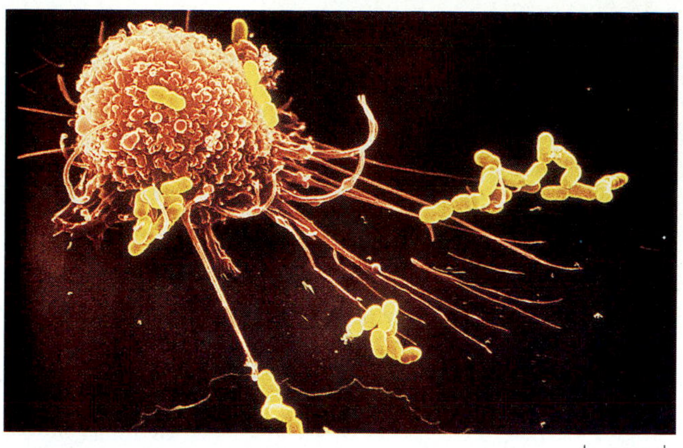

FIGURE 7.31 ▪ The emergence of cellular functions from the cooperation of many organelles. The ability of this macrophage (brown) to recognize, apprehend, and destroy bacteria (yellow) is a coordinated activity of the whole cell. The cytoskeleton, lysosomes, and plasma membrane are among the components that function in phagocytosis. Other cellular functions are also emergent properties that depend on interactions of the cell's parts (colorized SEM).

CHAPTER REVIEW

REVIEW OF KEY CONCEPTS

(with page numbers and key figures)

HOW WE STUDY CELLS

- **Microscopes provide windows to the world of the cell (pp. 102–105, FIGURE 7.1)** Improvements in microscopy have catalyzed progress in the study of cell structure.
- **Cell biologists can isolate organelles to study their functions (p. 105, FIGURE 7.3)** They use the ultracentrifuge to produce pellets enriched in specific organelles.

A PANORAMIC VIEW OF THE CELL

- **Prokaryotic and eukaryotic cells differ in size and complexity (pp.**
7.1 **105–107, FIGURE 7.5)** All cells are bounded by a plasma membrane. Bacteria and archaea have prokaryotic cells, without nuclei or other membrane-enclosed organelles. All other organisms have eukaryotic cells, with membrane-enclosed nuclei and other specialized organelles in their cytoplasm. The need for a high surface-to-volume ratio limits cell size.
- **Internal membranes compartmentalize the functions of a eukary-**
7.2 **otic cell (p. 107, FIGURES 7.7–7.8)** Plant and animal cells have most
7.3 of the same organelles.

THE NUCLEUS AND RIBOSOMES

- **The nucleus contains a eukaryotic cell's genetic library (pp. 107–111, FIGURE 7.9)** DNA is organized with proteins into chromosomes, which exist as chromatin in nondividing cells. Associated with the chromatin are one or more nucleoli, sites of ribosome synthesis. Macromolecules pass between nucleus and cytoplasm through pores in the nuclear envelope.

- **Ribosomes build a cell's proteins (p. 111, FIGURE 7.10)** Free ribosomes in the cytosol, and bound ribosomes on the outside of endoplasmic reticulum, synthesize proteins.

THE ENDOMEMBRANE SYSTEM

Many of the eukaryotic cell's membranes are connected either by physical continuity or through transport vesicles made of pinched-off pieces of membrane.

- **The endoplasmic reticulum manufactures membranes and performs many other biosynthetic functions (pp. 111–113, FIGURE 7.11)** Continuous with the nuclear envelope, the endoplasmic reticulum (ER) is a network of cisternae, membrane-enclosed compartments. Smooth ER lacks ribosomes; it synthesizes steroids, metabolizes carbohydrates, stores calcium in muscle, and detoxifies poisons in liver. Rough ER has bound ribosomes and produces cell membrane and secretory proteins. These products are distributed by transport vesicles budded from the ER.
- **The Golgi apparatus finishes, sorts, and ships cell products (pp. 113–114, FIGURE 7.12)** Stacks of separate cisternae make up the Golgi. Its *cis* face receives secretory proteins from the ER in transport vesicles. They are modified, sorted, and released in transport vesicles from the *trans* face.
- **Lysosomes are digestive compartments (pp. 114–115, FIGURE 7.14)** Lysosomes are membranous sacs of hydrolytic enzymes. They break down cell macromolecules for recycling as well as substances ingested by phagocytosis.
- **Vacuoles have diverse functions in cell maintenance (pp. 115–116, FIGURE 7.15)** A plant cell's central vacuole functions in storage, waste disposal, cell growth, and protection.

OTHER MEMBRANOUS ORGANELLES

- **Mitochondria and chloroplasts are the main energy transformers of cells (p. 118, FIGURE 7.17–7.18)** Mitochondria, the sites of cellular respiration in eukaryotes, have an outer membrane and an inner membrane that is folded into cristae. Some reactions of respiration occur in the mitochondrial matrix enclosed by the inner membrane, and others are catalyzed by enzymes built into the inner membrane. Chloroplasts, a type of plastid, contain chlorophyll and other pigments, which function in photosynthesis. In chloroplasts, two membranes surround the fluid stroma, which contains thylakoids, often stacked into grana.
- **Peroxisomes consume oxygen in various metabolic functions (p. 118, FIGURE 7.19)** Peroxisomes carry out processes that produce hydrogen peroxide as waste, and their enzymes convert the toxic peroxide to water.

THE CYTOSKELETON

- **Providing structural support to the cell, the cytoskeleton also functions in cell motility and regulation (pp. 119–124, TABLE 7.2, FIGURES 7.22–7.24)** The cytoskeleton is made of microtubules, microfilaments, and intermediate filaments. Microtubules are hollow cylinders; they radiate from the centrosome, a region near the nucleus that includes two centrioles in many animal cells. Microtubules shape the cell, guide movement of organelles, and aid chromosome separation in dividing cells. Cilia and flagella are motile appendages containing microtubule doublets that are moved past each other by the protein dynein. Microfilaments are thin rods built from the protein actin; they function in muscle contraction, ameboid movement, cytoplasmic streaming, and support for cellular projections. Intermediate filaments support cell shape and fix organelles in place.

CELL SURFACES AND JUNCTIONS

- **Plant cells are encased by cell walls (pp. 124–125, FIGURE 7.28)** Plant cell walls are composed of cellulose fibers embedded in other polysaccharides and protein.
- **The extracellular matrix (ECM) of animal cells functions in support, adhesion, movement, and regulation (pp. 125–126, FIGURE 7.29)** The cells secrete glycoproteins that form the ECM. Important components include collagen, proteoglycan complexes, and fibronectin attached to integrins in the plasma membrane.
- **Intercellular junctions help integrate cells into higher levels of structure and function (p. 127, FIGURE 7.30)** Plants have plasmodesmata, cytoplasmic channels that pass through adjoining cell walls. Animal cell contact is by tight junctions, desmosomes, and gap junctions.
- **The cell is a living unit greater than the sum of its parts (p. 127, FIGURE 7.31)**

SELF-QUIZ

1. The symptoms of a certain inherited disorder in humans include respiratory problems and, in males, sterility. Which of the following is a reasonable hypothesis for the molecular basis of this disorder? (Explain your answer.)
   a. a defective respiratory enzyme in the mitochondria
   b. defective actin molecules in cellular microfilaments
   c. defective dynein molecules in cilia and flagella
   d. abnormal hydrolytic enzymes in the lysosomes
   e. a defective secretory protein

2. From the following, choose the statement that correctly characterizes bound ribosomes.
   a. Bound ribosomes are enclosed in their own membrane.
   b. Bound ribosomes are structurally different from free ribosomes.
   c. Bound ribosomes generally synthesize membrane proteins and secretory proteins.
   d. The most common location for bound ribosomes is the cytoplasmic surface of the plasma membrane.
   e. Bound ribosomes are concentrated in the cisternal space of rough ER.

3. Which of the following organelles is least closely associated with the endomembrane system?
   a. nuclear envelope
   b. chloroplast
   c. Golgi apparatus
   d. plasma membrane
   e. ER

4. Cells of the pancreas will incorporate radioactively labeled amino acids into proteins. This "tagging" of newly synthesized proteins enables a researcher to track the location of these proteins in a cell. In this case, we are tracking an enzyme that is eventually secreted by pancreatic cells. Which of the following is the most likely pathway for movement of this protein in the cell?
   a. ER ⟶ Golgi ⟶ nucleus
   b. Golgi ⟶ ER ⟶ lysosome
   c. nucleus ⟶ ER ⟶ Golgi
   d. ER ⟶ Golgi ⟶ vesicles that fuse with plasma membrane
   e. ER ⟶ lysosomes ⟶ vesicles that fuse with plasma membrane

5. Which of the following organelles is common to plant *and* animal cells?
   a. chloroplasts
   b. wall made of cellulose
   c. tonoplast
   d. mitochondria
   e. centrioles

6. Which of the following components is present in a prokaryotic cell?
   a. mitochondria
   b. ribosomes
   c. nuclear envelope
   d. chloroplasts
   e. ER

7. Which type of cell would probably provide the best opportunity to study lysosomes? Explain your answer.
   a. muscle cell
   b. nerve cell
   c. phagocytic white blood cell
   d. leaf cell of a plant
   e. bacterial cell

8. Which of the following statements is a correct distinction between prokaryotic and eukaryotic cells attributable to the absence of a prokaryotic cytoskeleton?
   a. Compartmentalized organelles are found only in eukaryotic cells.
   b. Cytoplasmic streaming is not observed in prokaryotes.
   c. Only eukaryotic cells are capable of movement.
   d. Prokaryotic cells are usually 10 μm or less in diameter.
   e. Only the eukaryotic cell concentrates its genetic material in a region separate from the rest of the cell.

9. Which of the following structure-function pairs is *mismatched?*
   a. nucleolus; ribosome production
   b. lysosome; intracellular digestion
   c. ribosome; protein synthesis
   d. Golgi; secretion of cell products
   e. microtubules; muscle contraction

10. Cyanide binds with at least one of the molecules involved in the production of ATP. Following exposure of a cell to cyanide, most of the cyanide could be expected to be found within the
   a. mitochondria
   b. ribosomes
   c. peroxisomes
   d. lysosomes
   e. endoplasmic reticulum

CHALLENGE QUESTIONS

1. After very small viruses infect a plant cell by crossing its membrane, the viruses often spread rapidly throughout the entire plant without crossing additional membranes. Explain how this occurs.
2. Write a short essay describing similarities and differences between plant cells and animal cells.

SCIENCE, TECHNOLOGY, AND SOCIETY

1. Researchers working with breast-tumor cells in culture have made a startling discovery: The cell's transformation from a benign to a malignant state involves signaling across the membrane. When antibodies were used to block specific integrins, the cells lost their ability to become malignant. These observations suggested the possibility of a new treatment strategy for breast cancer. Current treatments seek to kill cancer cells through the accumulated uptake of toxic chemotherapeutic drugs that inhibit cell division. A treatment that inhibited integrins might spare the patient the unpleasant side effects of chemotherapy. Of course, it would also leave living tumor cells in the body. Would you accept a treatment that simply blocked the action of cancerous cells as opposed to destroying them? Explain your reasoning.
2. Doctors at a California university removed a man's spleen, standard treatment for a type of leukemia. The disease did not recur. Researchers kept some of the spleen cells alive in a nutrient medium. They found that some of the cells produced a blood protein called GM-CSF, which they are now testing to fight cancer and AIDS. The researchers patented the cells. The patient sued, claiming a share in profits from any products derived from his cells. In 1988, the California Supreme Court ruled against the plaintiff (patient), stating that his suit "threatens to destroy the economic incentive to conduct important medical research." The U.S. Supreme Court agreed. The plaintiff's attorney argued that the ruling left patients "vulnerable to exploitation at the hands of the state." Do you think the plaintiff was treated fairly? Is there anything else you would like to know about this case that might help you make up your mind?

FURTHER READING

Alberts, B., D. Bray, J. Lewis, M. Raff, K. Roberts, and J. D. Watson. *Molecular Biology of the Cell,* 3rd ed. New York: Garland, 1994. The best comprehensive text; lucidly written and illustrated.
Becker, W. M., J. B. Reece, and M. F. Poenie. *The World of the Cell,* 3rd ed. Menlo Park, CA: Benjamin/Cummings, 1996. Readable and student-oriented.
Gorlich, D., and I. W. Mattaj. "Nucleocytoplasmic Transport." *Science,* March 15, 1996. How proteins and RNA get in and out of the nucleus.
Horwitz, A. F. "Integrins and Health." *Scientific American,* May 1997. These adhesive cell-surface molecules turn out to be crucial to body function.
Ingber, D. E. "The Architecture of Life." *Scientific American,* January 1998. The cytoskeleton and many other biological structures are assembled in accordance with the architectural principle of "tensegrity."
Lichtman, J.W. "Confocal Microscopy." *Scientific American,* August 1994. A powerful window to the cell.
Pelham, H. R. B. "Green Light for Golgi Traffic." *Nature,* September 4, 1997. How vesicles transport proteins from the ER to the Golgi, as revealed by time-lapse video.
Rothman, J. E., and L. Orci. "Budding Vesicles in Living Cells." *Scientific American,* March 1996. More about transport vesicles.
Stossel, T.P. "The Machinery of Cell Crawling." *Scientific American,* September 1994. Cells on the move.
Strange, C. J. "Biological Ties that Bind." *BioScience,* January 1997. The surprisingly varied and important roles of the extracellular matrix in animals.
Zambryski, P. "Plasmodesmata: Plant Channels for Molecules on the Move." *Science,* December 22, 1995. Macromolecular travel between plant cells.

WEB LINKS

Visit the special edition of *The Biology Place* for BIOLOGY, Fifth Edition, at **http://www.biology.com/campbell**. Go to Chapter 7 for online resources, including learning activities, practice exams, and links to the following web sites:

"The Virtual Cell"
Climb inside this virtual cell and explore its parts.

"The Dictionary of Cell Biology"
An online dictionary of cell biology with over 6000 terms defined.

"Cell and Molecular Biology Online"
A general resource for the biology community with an emphasis on information for cell and molecular biologists.

"CELLS alive"
This site has interesting free pictures and videos of various cell types and processes.

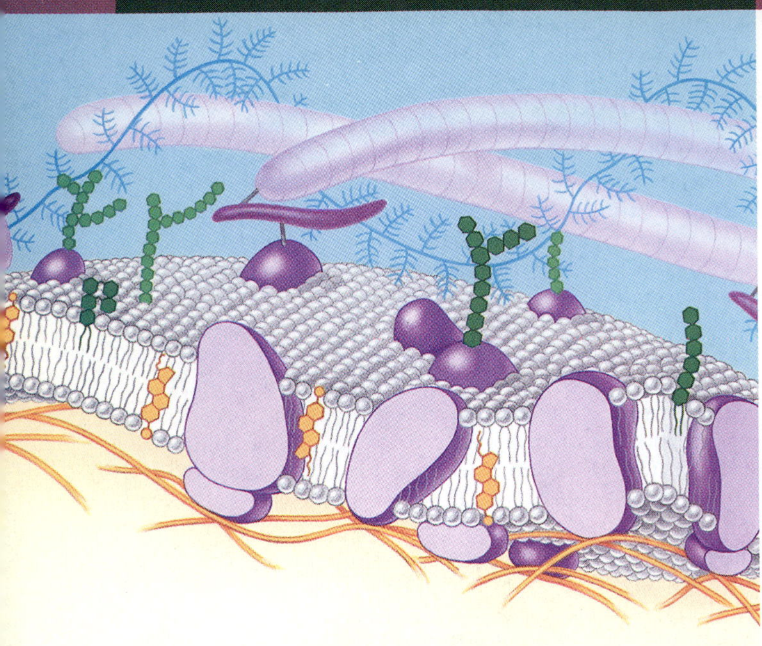

MEMBRANE STRUCTURE AND FUNCTION

The plasma membrane is the edge of life, the boundary that separates the living cell from its nonliving surroundings. A remarkable film only about 8 nm thick—it would take over 8000 to equal the thickness of this page—the plasma membrane controls traffic into and out of the cell it surrounds. Like all biological membranes, the plasma membrane has ***selective permeability****; that is, it allows some substances to cross it more easily than others. One of the earliest episodes in the evolution of life may have been the formation of a membrane that enclosed a solution of different composition from the surrounding solution while still permitting the uptake of nutrients and elimination of waste products. This ability of the cell to discriminate in its chemical exchanges with its environment is fundamental to life, and it is the plasma membrane that makes this selectivity possible.*

In this chapter you will learn how cellular membranes control the passage of substances. We will concentrate on the plasma membrane, the outermost membrane of the cell, represented in the drawing on this page. However, the same general principles of membrane traffic also apply to the many varieties of internal membranes that partition the eukaryotic cell. To understand how membranes work, we begin by examining their architecture.

MEMBRANE STRUCTURE

Lipids and proteins are the staple ingredients of membranes, although carbohydrates are also important. Currently, the accepted model for the arrangement of these molecules in membranes is the fluid mosaic model. We will trace the evolution of this model in some detail as an example of how scientists build on earlier observations and ideas and how they use models as hypotheses.

Membrane models have evolved to fit new data: *science as a process*

Scientists began building molecular models of the membrane decades before membranes were first seen with the electron microscope in the 1950s. In 1895, Charles Overton postulated that membranes are made of lipids, based on his observations that substances that dissolve in lipids enter cells much more rapidly than substances that are insoluble in lipids. Twenty years later, membranes isolated from red blood cells were chemically analyzed and found to be composed of lipids and proteins.

Phospholipids are the most abundant lipids in most membranes. The ability of phospholipids to form membranes is built into their molecular structure. A phospholipid is an **amphipathic** molecule, meaning it has both a hydrophilic region and a hydrophobic region (see FIGURE 5.12). Other types of membrane lipids are also amphipathic.

In 1917, Irving Langmuir made artificial membranes by adding phospholipids dissolved in benzene (an organic solvent) to water. After the benzene evaporated, the phospholipids remained as a film covering the surface of the water, with only the hydrophilic heads of the phospholipids immersed in the water (FIGURE 8.1a). In 1925, two Dutch scientists, E. Gorter and F. Grendel, reasoned that cell membranes must actually be phospholipid bilayers, two molecules thick. Such a bilayer could exist as a stable boundary between two aqueous compartments because the molecular arrangement shelters the hydrophobic tails of the phospholipids from water while exposing the hydrophilic heads to water (FIGURE 8.1b). Gorter and Grendel measured the phospholipid content of membranes isolated from red blood cells and found just enough of the lipid to cover the cells with two layers. (Ironically, Gorter and Grendel underestimated both the phospholipid content and the surface area of the cells, but the two errors canceled each other. Thus, what turned out to be a correct conclusion was based on flawed measurements.)

If we assume that a phospholipid bilayer is the main fabric of the membrane, where do we place the proteins? Although the heads of phospholipids are hydrophilic, the surface of an artificial membrane consisting of a phospholipid bilayer adheres less strongly to water than does the surface of an actual biological membrane. This difference could be accounted for if the membrane were coated on both sides with hydrophilic proteins. In 1935, Hugh Davson and James Danielli proposed a sandwich model: a phospholipid bilayer between two layers of globular protein (FIGURE 8.2a).

When researchers first used electron microscopes to study cells in the 1950s, the pictures seemed to support the Davson-Danielli model. In electron micrographs of cells stained with atoms of heavy metals, the plasma membrane is triple-layered, having two dark ("stained") bands separated by an unstained layer (see FIGURE 7.6a, p. 107). Most early electron

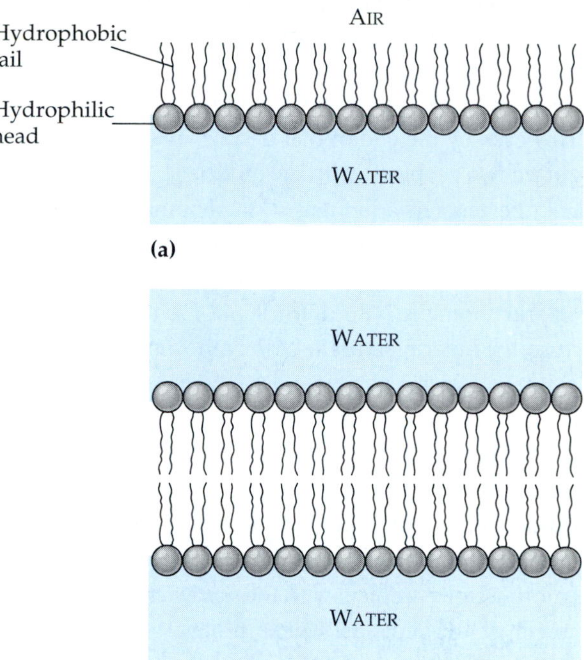

FIGURE 8.1 • Artificial membranes (cross sections). (a) Water can be coated with a single layer of phospholipid molecules. The hydrophilic heads of the phospholipids are immersed in water, and the hydrophobic tails are excluded from water. **(b)** A bilayer of phospholipids forms a stable boundary between two aqueous compartments. This arrangement exposes the hydrophilic parts of the molecules to water and shields the hydrophobic parts from water.

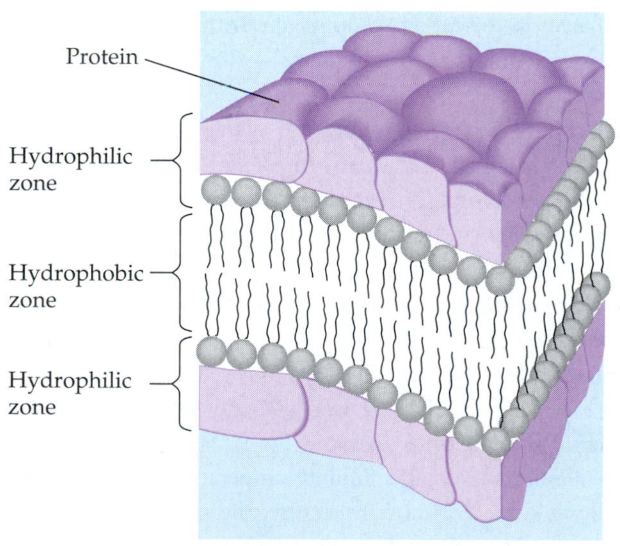

(a) Original Davson-Danielli model

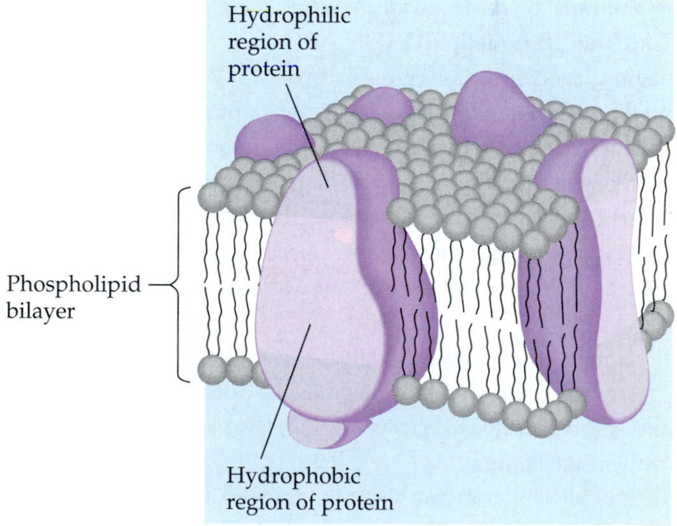

(b) Current fluid mosaic model

FIGURE 8.2 • Two generations of membrane models. (a) The Davson-Danielli model, proposed in 1935, sandwiched the phospholipid bilayer between two protein layers. With later modifications, this model was widely accepted until about 1970. **(b)** The fluid mosaic model disperses the proteins and immerses them in the phospholipid bilayer, which is in a fluid state. Shown here in simplified form, this is our present working model of the membrane.

microscopists assumed that the stain adhered to the proteins and hydrophilic heads of the phospholipids, leaving the hydrophobic core of the membrane unstained. In a somewhat circular manner, the electron micrographs and the membrane model became increasingly accepted as explanations for each other. By the 1960s, the Davson-Danielli sandwich had become widely accepted as the structure not only of the plasma membrane, but of all the internal membranes of the cell. However, by the end of that decade, many cell biologists recognized two problems with the model.

First, the generalization that all membranes of the cell are identical was challenged. Not all membranes look alike through the electron microscope. For example, whereas the plasma membrane is 7–8 nm thick and has the three-layered structure, the inner membrane of the mitochondrion is only 6 nm thick and in electron micrographs looks like a row of beads. Mitochondrial membranes also have a substantially greater percentage of proteins than plasma membranes, and there are differences in the specific kinds of phospholipids and other lipids. Membranes with different functions differ in chemical composition and structure.

A more serious problem with the sandwich model is in the placement of the proteins. Unlike proteins dissolved in the cytosol, membrane proteins are not very soluble in water. Membrane proteins have hydrophobic and hydrophilic regions; they are amphipathic, as are their phospholipid partners in membranes. If proteins were layered on the surface of the membrane, their hydrophobic parts would be in an aqueous environment.

In 1972, S. J. Singer and G. Nicolson advocated a revised membrane model that placed the proteins in a location compatible with their amphipathic character. They proposed that membrane proteins are dispersed and individually inserted into the phospholipid bilayer, with only their hydrophilic regions protruding far enough from the bilayer to be exposed to water. This molecular arrangement would maximize contact of hydrophilic regions of proteins and phospholipids with water while providing their hydrophobic parts with a nonaqueous environment. According to this model, the membrane is a mosaic of protein molecules bobbing in a fluid bilayer of phospholipids; hence the term **fluid mosaic model** (FIGURE 8.2b).

A method of preparing cells for electron microscopy called freeze-fracture has provided convincing evidence that proteins are embedded in the phospholipid bilayer of the membrane, rather than being spread upon the surface. Freeze-fracture can delaminate a membrane along the middle of the bilayer, splitting the membrane into outer and inner faces (see the Methods Box, p. 138). When the halves of the fractured membrane are viewed in the electron microscope, the interior of the bilayer appears cobblestoned, with protein particles interspersed in a smooth matrix. Proteins clearly penetrate into the hydrophobic interior of the membrane.

We have examined the evolution of our understanding of membrane structure as a case history of how science works. Models are proposed by scientists as hypotheses, ways of organizing and explaining existing information. Replacing one model of membrane structure with another does not imply that the original model was worthless. The acceptance or rejection of a model depends on how well it fits observations and explains experimental results. A good model also makes predictions that shape future research. Models inspire experiments, and few models survive these tests without modification. New findings may make a model obsolete; even then, it may not be totally scrapped, but revised to incorporate the new observations. Like its predecessor, which endured for 35 years, the fluid mosaic model is continually being refined and may one day undergo major revision.

A membrane is a fluid mosaic of lipids, proteins, and carbohydrates

What exactly does it mean to describe a membrane as a fluid mosaic? Let's begin with the word *fluid.*

The Fluid Quality of Membranes

Membranes are not static sheets of molecules locked rigidly in
8.1 place. A membrane is held together primarily by hydrophobic interactions, which are much weaker than covalent bonds (see Chapter 5). Most of the lipids and some of the proteins can drift about randomly in the plane of the membrane (FIGURE 8.3a). It is rare, however, for a molecule to flip-flop transversely across the membrane, switching from one phospholipid layer to the other; to do so, the hydrophilic part of the molecule would have to cross the hydrophobic core of the membrane.

Phospholipids move along the plane of the membrane rapidly, averaging about 2 μm—the length of a typical bacterial cell—per second. Proteins are much larger than lipids and move more slowly, but some membrane proteins do, in fact, drift (FIGURE 8.4). And some membrane proteins seem to move in a highly directed manner, perhaps driven along cytoskeletal fibers by motor proteins connected to the membrane proteins' cytoplasmic ends. However, many other membrane proteins seem to be held virtually immobile by their attachment to the cytoskeleton.

A membrane remains fluid as temperature decreases, until finally, at some critical temperature, the phospholipids settle into a closely packed arrangement and the membrane solidifies, much as bacon grease forms lard when it cools. The temperature at which a membrane solidifies depends on its lipid composition. The membrane remains fluid to a lower temperature if it is rich in phospholipids with unsaturated hydrocarbon tails (see FIGURES 5.11 and 5.12). Because of kinks where double

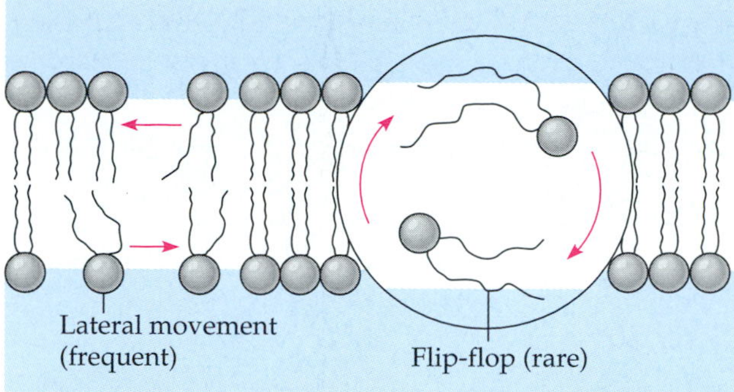

(a) Movement of phospholipids

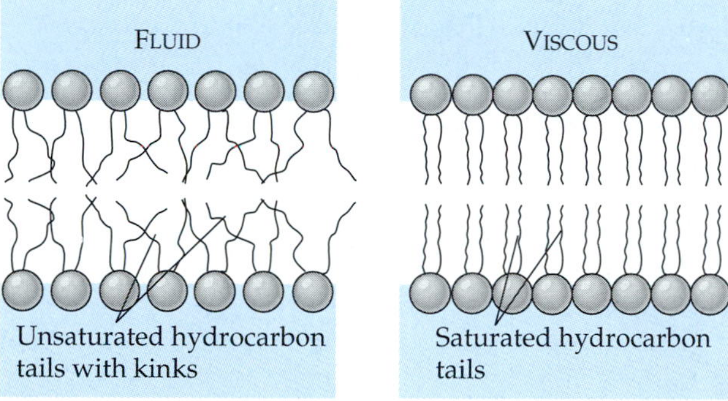

(b) Membrane fluidity

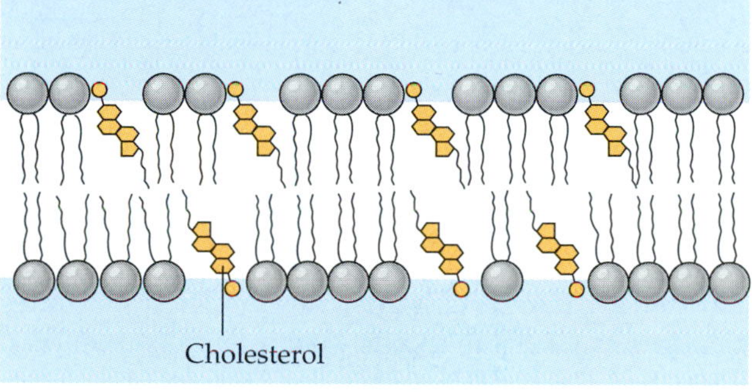

(c) Cholesterol within the membrane

FIGURE 8.3 ▪ The fluidity of membranes. (a) Lipids move laterally (that is, in two dimensions) in a membrane, but flip-flopping across the membrane (in the third dimension) is rare. **(b)** Unsaturated hydrocarbon tails of phospholipids have kinks that keep the molecules from packing together, enhancing membrane fluidity. **(c)** Cholesterol reduces membrane fluidity by reducing phospholipid movement at moderate temperatures but it also hinders solidification at low temperatures.

bonds are located, unsaturated hydrocarbons do not pack together as closely as saturated hydrocarbons (see FIGURE 8.3b).

The steroid cholesterol, which is wedged between phospholipid molecules in the plasma membranes of animals, helps stabilize the membrane (see FIGURE 8.3c). At relatively warm temperatures—at 37°C, the body temperature of humans, for example—cholesterol makes the membrane less fluid by restraining the movement of phospholipids. However, because cholesterol also hinders the close packing of phospholipids, it lowers the temperature required for the membrane to solidify.

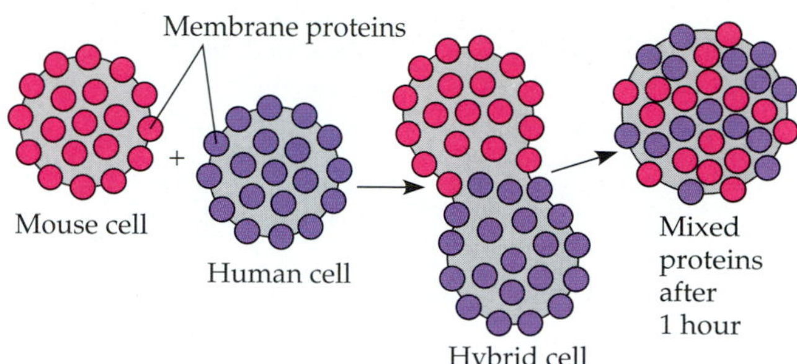

FIGURE 8.4 ▪ Evidence for the drifting of membrane proteins. When researchers fuse a human cell with a mouse cell, it takes less than an hour for the membrane proteins of the two species to completely intermingle in the membrane of the hybrid cell.

Membranes must be fluid to work properly; they are usually about as fluid as salad oil. When a membrane solidifies, its permeability changes, and enzymatic proteins in it may become inactive. A cell can alter the lipid composition of its membranes to some extent as an adjustment to changing temperature. For instance, in many plants that tolerate extreme cold, such as winter wheat, the percentage of unsaturated phospholipids increases in autumn, an adaptation that keeps the membranes from solidifying during winter.

Membranes as Mosaics of Structure and Function

Now we come to the word *mosaic.* A membrane is a collage of many different proteins embedded in the fluid matrix of the lipid bilayer (FIGURE 8.5, p. 134). The lipid bilayer is the main fabric of the membrane, but proteins determine most of the membrane's specific functions. The plasma membrane and the membranes of the various organelles each have unique collections of proteins. More than 50 kinds of proteins have been found to date in the plasma membrane of red blood cells, for example.

Notice in FIGURE 8.5 that there are two major populations of membrane proteins. **Integral proteins** are generally transmembrane proteins, with hydrophobic regions that completely span the hydrophobic interior of the membrane. The hydrophobic regions of an integral protein consist of one or more stretches of nonpolar amino acids (see FIGURE 5.15), usually coiled into α-helices (FIGURE 8.6, p. 134). The hydrophilic ends of the molecule are exposed to the aqueous solutions on either side of the membrane. **Peripheral proteins** are not embedded in the lipid bilayer at all; they are appendages loosely bound to the surface of the membrane, often to the exposed parts of integral proteins (see FIGURE 8.5).

On the cytoplasmic side of the plasma membrane, some membrane proteins are held in place by attachment to the cytoskeleton. And on the exterior side, certain membrane proteins are attached to fibers of the extracellular matrix (see Chapter 7). These attachments combine to give animal cells a stronger external framework than the plasma membrane itself could provide.

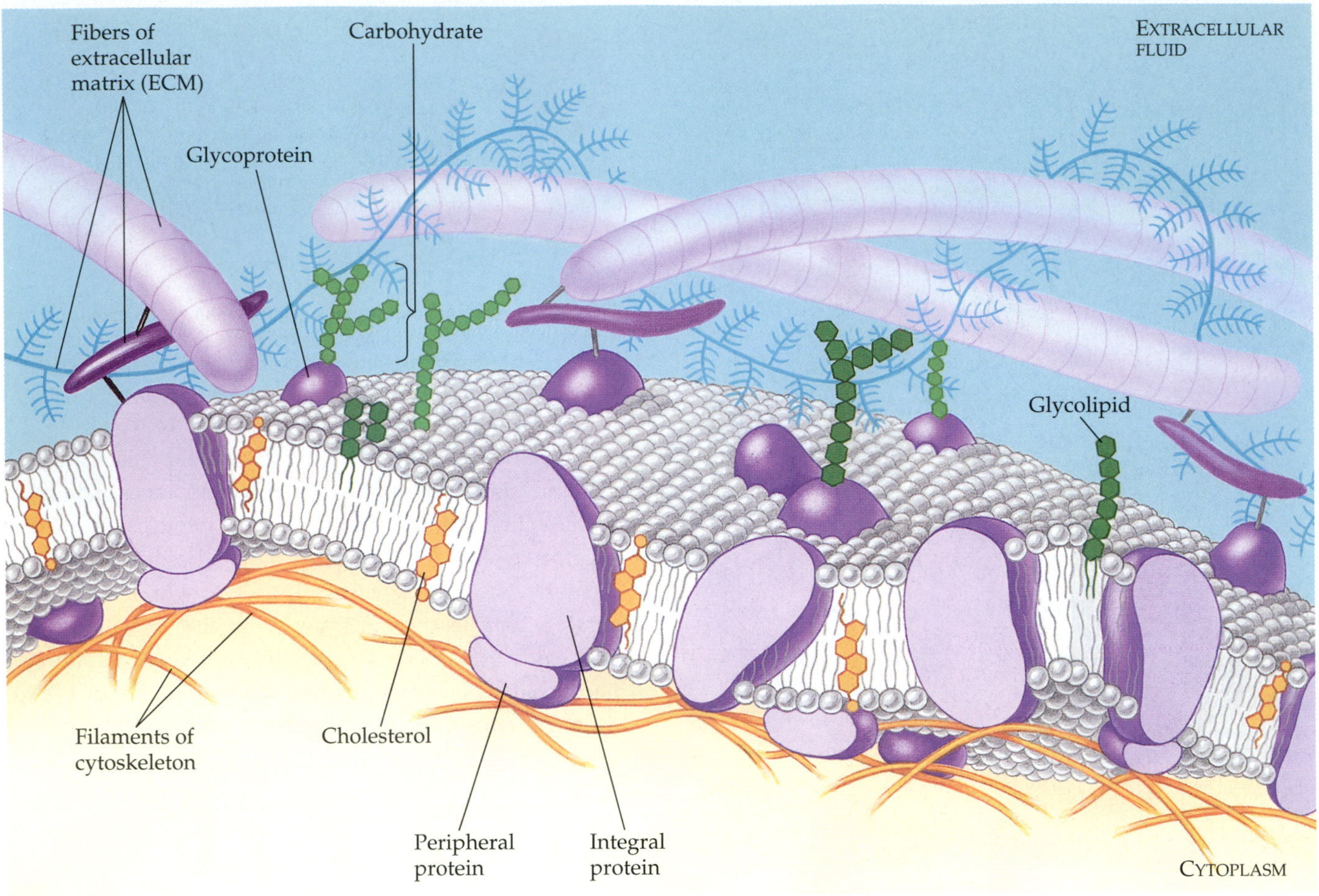

Fibers of extracellular matrix (ECM)
Carbohydrate
Extracellular fluid
Glycoprotein
Glycolipid
Filaments of cytoskeleton
Cholesterol
Peripheral protein
Integral protein
Cytoplasm

FIGURE 8.5 ▪ The detailed structure of an animal cell's plasma membrane, in cross section. See
8.1 Figure 7.29 for details of the ECM.

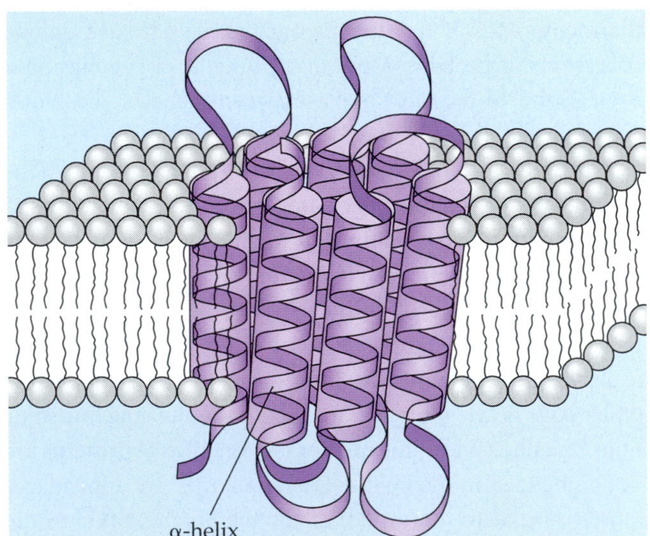

FIGURE 8.6 ▪ The structure of a transmembrane protein. This ribbon model highlights the α-helical secondary structure of the hydrophobic parts of the protein, which lie mostly within the hydrophobic core of the membrane. This particular protein, bacteriorhodopsin, has seven transmembrane helices (outlined with cylinders for emphasis). Joining the helices are hydrophilic polypeptide segments that together form the parts of the protein in contact with the aqueous solutions on either side of the membrane. Bacteriorhodopsin is a specialized transport protein found in certain bacteria.

Membranes have distinct inside and outside faces. The two lipid layers may differ in specific lipid composition, and each protein has directional orientation in the membrane. The plasma membrane also has carbohydrates, which are restricted to the exterior surface. This asymmetrical distribution of proteins, lipids, and carbohydrates is determined as the membrane is being built by the endoplasmic reticulum. Molecules that start out on the *inside* face of the ER end up on the *outside* face of the plasma membrane (FIGURE 8.7).

FIGURE 8.8 gives an overview of six major kinds of function exhibited by proteins of the plasma membrane. A single cell may have membrane proteins performing several of these functions, and a single protein may have multiple functions. Thus the membrane is indeed a functional mosaic, as well as a structural one.

Membrane Carbohydrates and Cell-Cell Recognition

Cell-cell recognition, a cell's ability to distinguish one type of neighboring cell from another, is crucial to the functioning of an organism. It is important, for example, in the sorting of cells into tissues and organs in an animal embryo. It is also the basis for the rejection of foreign cells (including those of

FIGURE 8.8 ▪ Some functions of membrane proteins. In many cases, a single protein performs some combination of these tasks.

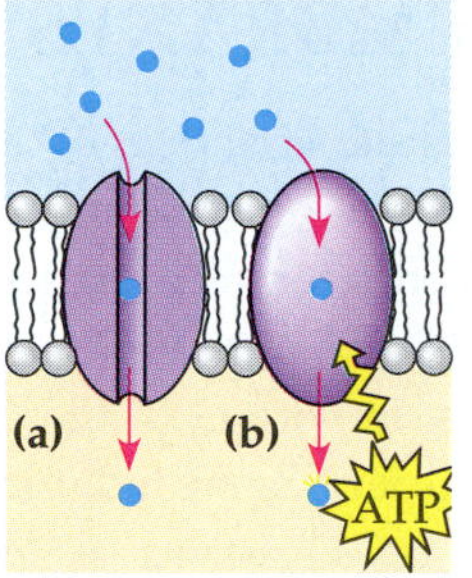

Transport **(a)** A protein that spans the membrane may provide a hydrophilic channel across the membrane that is selective for a particular solute. **(b)** Some transport proteins hydrolyze ATP as an energy source to actively pump substances across the membrane.

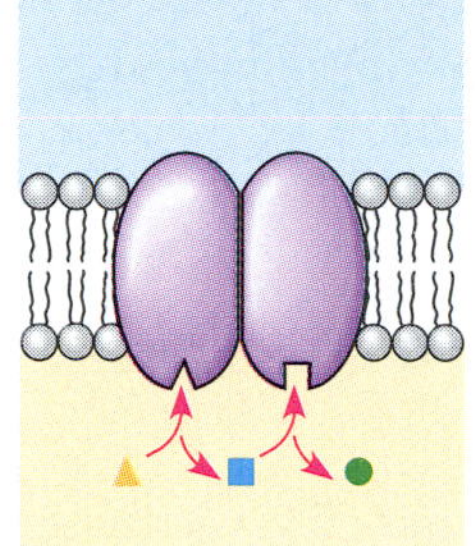

Enzymatic activity A protein built into the membrane may be an enzyme with its active site exposed to substances in the adjacent solution. In some cases, several enzymes in a membrane are ordered as a team that carries out sequential steps of a metabolic pathway.

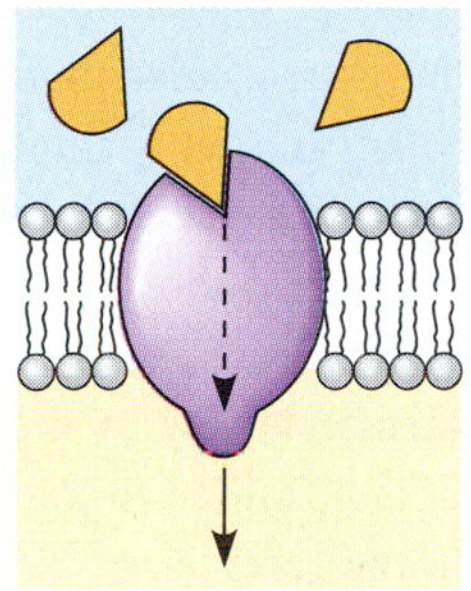

Signal transduction A membrane protein may have a binding site with a specific shape that fits the shape of a chemical messenger, such as a hormone. The external messenger (signal) may cause a conformational change in the protein that relays the message to the inside of the cell.

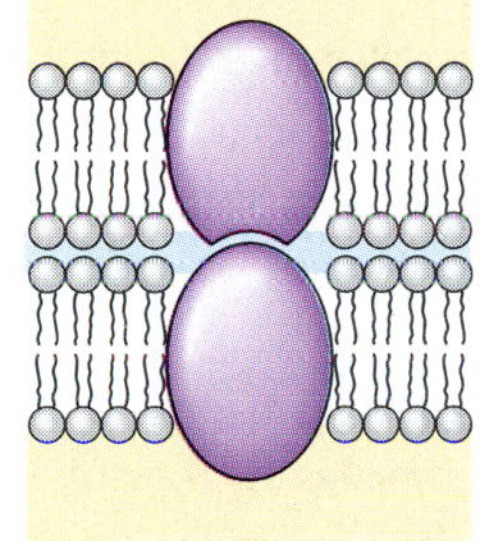

Intercellular joining Membrane proteins of adjacent cells may be hooked together in various kinds of junctions (see Figure 7.30).

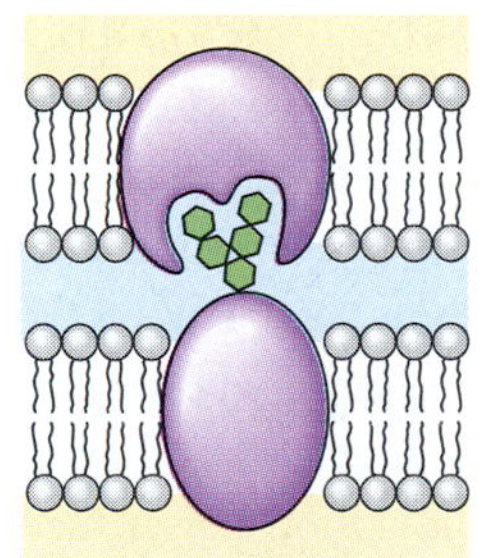

Cell-cell recognition Some glycoproteins (proteins with short chains of sugars) serve as identification tags that are specifically recognized by other cells.

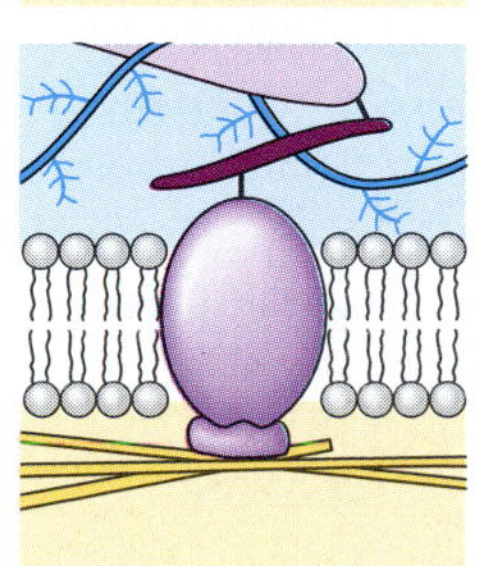

Attachment to the cytoskeleton and extracellular matrix (ECM) Microfilaments or other elements of the cytoskeleton may be bonded to membrane proteins, a function that helps maintain cell shape and fixes the location of certain membrane proteins. Proteins that adhere to the ECM can coordinate extracellular and intracellular changes.

transplanted organs) by the immune system, an important line of defense in vertebrate animals (see Chapter 43). The way cells recognize other cells is by keying on surface molecules, often carbohydrates, on the plasma membrane.

Membrane carbohydrates are usually branched oligosaccharides with fewer than 15 sugar units. (*Oligo* is Greek for "few"; an oligosaccharide is a short polysaccharide.) Some of these oligosaccharides are covalently bonded to lipids, forming molecules called glycolipids. (Recall that *glyco* refers to the presence of carbohydrate.) Most, however, are covalently bonded to proteins, which are thereby glycoproteins.

The oligosaccharides on the external side of the plasma membrane vary from species to species, among individuals of the same species, and even from one cell type to another in a single individual. The diversity of the molecules and their location on the cell's surface enable oligosaccharides to function as markers that distinguish one cell from another. For example, the four human blood groups designated A, B, AB, and O reflect variation in the oligosaccharides on the surfaces of red blood cells.

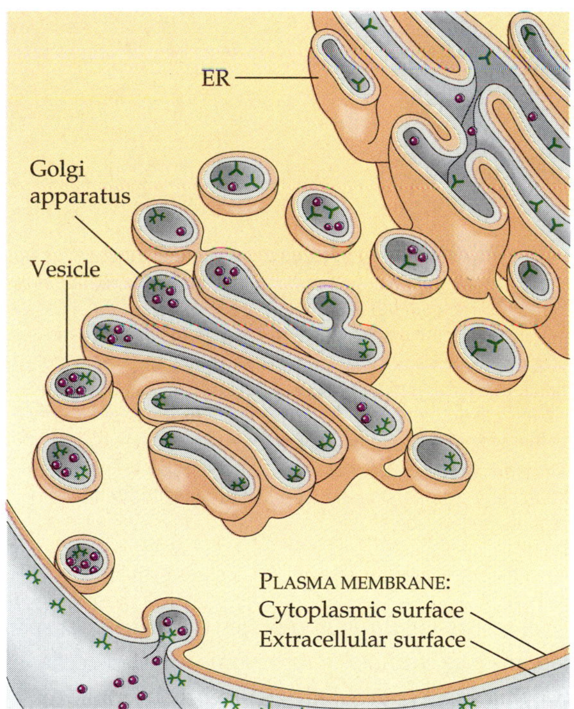

FIGURE 8.7 ▪ Sidedness of the plasma membrane. The membrane has distinct cytoplasmic and extracellular sides. This bifacial quality is determined when the membrane is first synthesized and modified by the ER and Golgi. The diagram color-codes the two sides of the membranes of the endomembrane system, to illustrate that the side facing the inside of the ER, Golgi, and vesicles is topologically equivalent to the extracellular surface of the plasma membrane. The other side always faces the cytosol, from the time the membrane is made by the ER to the time it is added to the plasma membrane by fusion of a vesicle. The small green "trees" represent membrane carbohydrates that are synthesized in the ER and modified in the Golgi. Vesicle fusion with the plasma membrane is also responsible for secretion of cell products (purple).

TRAFFIC ACROSS MEMBRANES

The biological membrane is an exquisite example of a supramolecular structure—many molecules ordered into a higher level of organization—with emergent properties beyond those of the individual molecules. The remainder of this chapter focuses on one of the most important of those properties: the ability to regulate transport across cellular boundaries, a function essential to the cell's existence as an open system. We will see once again that form fits function: The fluid mosaic model helps explain how membranes regulate the cell's molecular traffic.

A membrane's molecular organization results in selective permeability

A steady traffic of small molecules and ions moves across the plasma membrane in both directions. Consider the chemical exchanges between a muscle cell and the extracellular fluid that bathes it. Sugars, amino acids, and other nutrients enter the cell, and metabolic waste products leave. The cell takes in oxygen for cellular respiration and expels carbon dioxide. It also regulates its concentrations of inorganic ions, such as Na^+, K^+, Ca^{2+}, and Cl^-, by shuttling them one way or the other across the plasma membrane. Although traffic through the membrane is extensive, cell membranes are selectively permeable, and substances do not cross the barrier indiscriminately. The cell is able to take up many varieties of small molecules and ions and exclude others. Moreover, substances that move through the membrane do so at different rates.

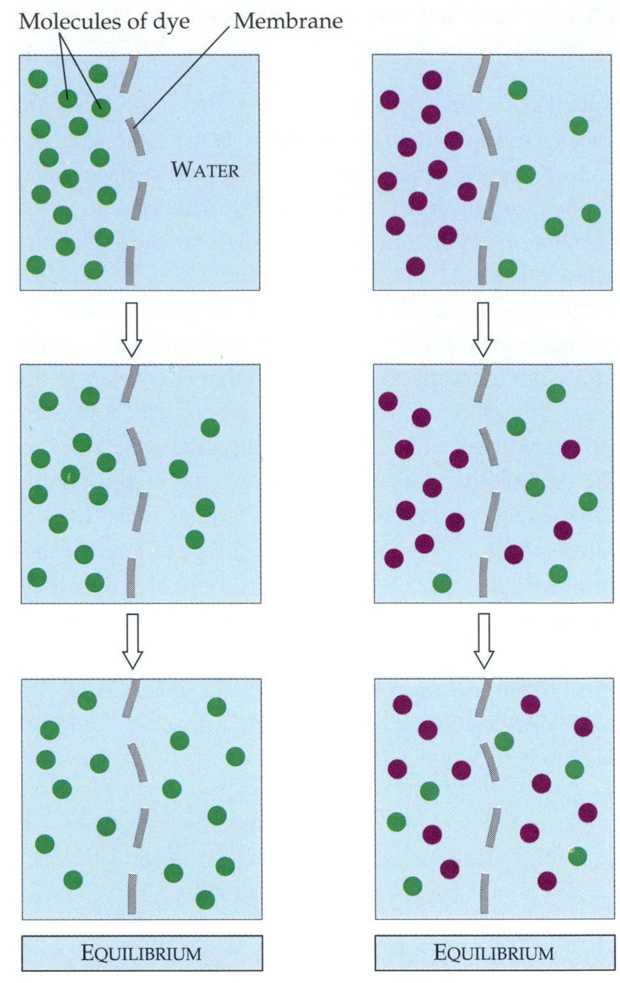

(a) Diffusion of one solute **(b)** Diffusion of two solutes

8.2 **FIGURE 8.9 ▪ The diffusion of solutes across membranes. (a)** A substance will diffuse from where it is more concentrated to where it is less concentrated. The membrane, viewed here in cross section, has pores large enough for molecules of dye to pass. Diffusion down the concentration gradient leads to a dynamic equilibrium; the solute molecules continue to cross the membrane, but at equal rates in both directions. **(b)** In this case, solutions of two different dyes are separated by a membrane that is permeable to both dyes. Each dye diffuses down its own concentration gradient. There will be a net diffusion of the green dye toward the left, even though the *total* solute concentration was initially greater on the left side.

Permeability of the Lipid Bilayer

The hydrophobic core of the membrane impedes the transport of ions and polar molecules, which are hydrophilic. Hydrophobic molecules, such as hydrocarbons, carbon dioxide, and oxygen, can dissolve in the membrane and cross it with ease. Very small molecules that are polar but uncharged can also pass through the membrane rapidly. Examples are water and ethanol, which are tiny enough to pass between the lipids of the membrane. The lipid bilayer is not very permeable to larger, uncharged polar molecules, such as glucose and other sugars. It is also relatively impermeable to all ions, even such small ones as H^+ and Na^+. A charged atom or molecule and its shell of water find the hydrophobic layer of the membrane difficult to penetrate. However, the lipid bilayer is only part of the story of a membrane's selective permeability. Proteins built into the membrane play a key role in regulating transport.

Transport Proteins

Cell membranes are permeable to specific ions and polar molecules. These hydrophilic substances avoid contact with the lipid bilayer by passing through **transport proteins** that span the membrane (see FIGURE 8.8, top). Some transport proteins function by having a hydrophilic channel that certain molecules use as a tunnel through the membrane. Other transport proteins bind to their passengers and physically move them across the membrane. In both cases, each transport protein is specific for the substances it translocates (moves), allowing only a certain substance or class of closely related substances to cross the membrane. For example, glucose carried in blood to the human liver enters liver cells rapidly through specific transport proteins in the plasma membrane. The protein is so selective that it even rejects fructose, a structural isomer of glucose.

Thus the selective permeability of a membrane depends on both the discriminating barrier of the lipid bilayer and the specific transport proteins built into the membrane. But what determines the *direction* of traffic across a membrane? At a given time, will a particular substance enter the cell or leave? And what mechanisms actually drive molecules across membranes? We will address these questions next as we explore two modes of membrane traffic: passive transport and active transport.

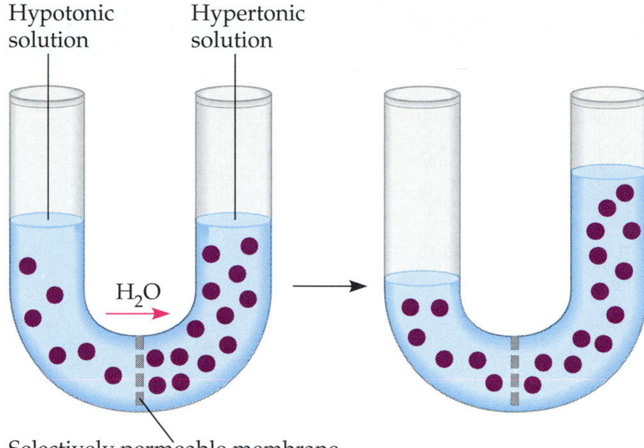

FIGURE 8.10 ▪ Osmosis. Two sugar solutions of different concentration are separated by a porous membrane that is permeable to the solvent (water) but not to the solute (sugar). Water diffuses from the hypotonic solution to the hypertonic solution. This passive transport of water, or osmosis, reduces the difference in sugar concentrations.

Passive transport is diffusion across a membrane

8.2 Molecules have intrinsic kinetic energy called thermal motion (heat). One result of thermal motion is **diffusion**, the tendency for molecules of any substance to spread out into the available space. Each molecule moves randomly, yet diffusion of a *population* of molecules may be directional. For example, imagine a membrane separating pure water from a solution of a dye in water. Assume that this membrane is permeable to the dye molecules (FIGURE 8.9a). Each dye molecule wanders randomly, but there will be a *net* movement of the dye molecules across the membrane to the side that began as pure water. The spreading of the dye across the membrane will continue until both solutions have equal concentrations of the dye. Once that point is reached, there will be a dynamic equilibrium, with as many dye molecules crossing the membrane in one direction as in the other each second.

We can now state a simple rule of diffusion: In the absence of other forces, a substance will diffuse from where it is more concentrated to where it is less concentrated. Put another way, any substance will diffuse down its **concentration gradient**. No work must be done to make this happen; diffusion is a spontaneous process because it decreases free energy (see FIGURE 6.4b). Recall that in any system there is a tendency for entropy, or disorder, to increase. Diffusion of a solute in water increases entropy by producing a more random mixture than exists when there are localized concentrations of the solute. It is important to note that each substance diffuses down its *own* concentration gradient, unaffected by the concentration differences of other substances (FIGURE 8.9b).

Much of the traffic across cell membranes occurs by diffusion. When a substance is more concentrated on one side of a membrane than on the other, there is a tendency for the substance to diffuse across the membrane down its concentration gradient (assuming that the membrane is permeable to that substance). One important example is the uptake of oxygen by a cell performing cellular respiration. Dissolved oxygen diffuses into the cell across the plasma membrane. As long as cellular respiration consumes the O_2 as it enters, diffusion into the cell will continue, because the concentration gradient favors movement in that direction.

The diffusion of a substance across a biological membrane is called **passive transport**, because the cell does not have to
8.3 expend energy to make it happen. The concentration gradient itself represents potential energy and drives diffusion. Remember, however, that membranes are selectively permeable and therefore affect the rates of diffusion of various molecules. One molecule that diffuses freely across most membranes is water, a fact that has important consequences for cells.

Osmosis is the passive transport of water

8.3 In comparing two solutions of unequal solute concentration, the solution with a higher concentration of solutes is said to be **hypertonic**. The solution with a lower solute concentration is **hypotonic**. (*Hyper* and *hypo* mean "more" and "less," respectively, referring here to solute concentration.) These are relative terms that are meaningful only in a comparative sense. For example, tap water is hypertonic to distilled water but hypotonic to seawater. In other words, tap water has a higher solute concentration than distilled water, but a lower concentration than seawater. Solutions of equal solute concentration are said to be **isotonic** (*iso* means "the same").

Picture a U-shaped vessel with a selectively permeable membrane separating two sugar solutions of different concentrations (FIGURE 8.10). Pores in this synthetic membrane are too small for sugar molecules to pass but large enough for water molecules to cross the membrane. In effect, the solution with higher solute concentration (hypertonic) has a lower water concentration.* Therefore the water will diffuse across

* Actually, at any instant, a fraction of the water molecules in a solution lose their freedom of independent movement by being bound to solute molecules in hydration shells. It is not really a difference in *total* water concentration that causes osmosis, but a difference in the concentration of *unbound* water molecules that are free to cross the membrane.

METHODS: Freeze-Fracture and Freeze-Etch

(a) A researcher freezes the specimen at the temperature of liquid nitrogen, then fractures the cells with a cold knife.

(b) The knife does not cut cleanly through the frozen cells; instead, it cracks the specimen, with the fracture plane following the path of least resistance. The fracture plane often follows the hydrophobic interior of a membrane, splitting the lipid bilayer down the middle into a P ("protoplasmic"—that is, cytoplasmic) face and an E (exterior) face. The membrane proteins are not split but go with one or the other of the phospholipid layers. The topography of the fractured surface may be enhanced by etching, the removal of water by sublimation (direct evaporation of frozen water to water vapor).

(c) A fine mist of platinum is sprayed from an angle onto the fractured surface of the cell. There will be "shadows" where elevated regions of the fractured cell block the platinum. Adding a film of carbon strengthens the platinum coat.

The original specimen is digested away with bleach, acids, and enzymes, leaving the platinum-carbon film as a replica of the fractured surface. It is this replica, not the membrane itself, that is examined through the electron microscope.

Electron micrographs have been inserted into this drawing of a delaminated membrane. Notice the protein particles (the "bumps").

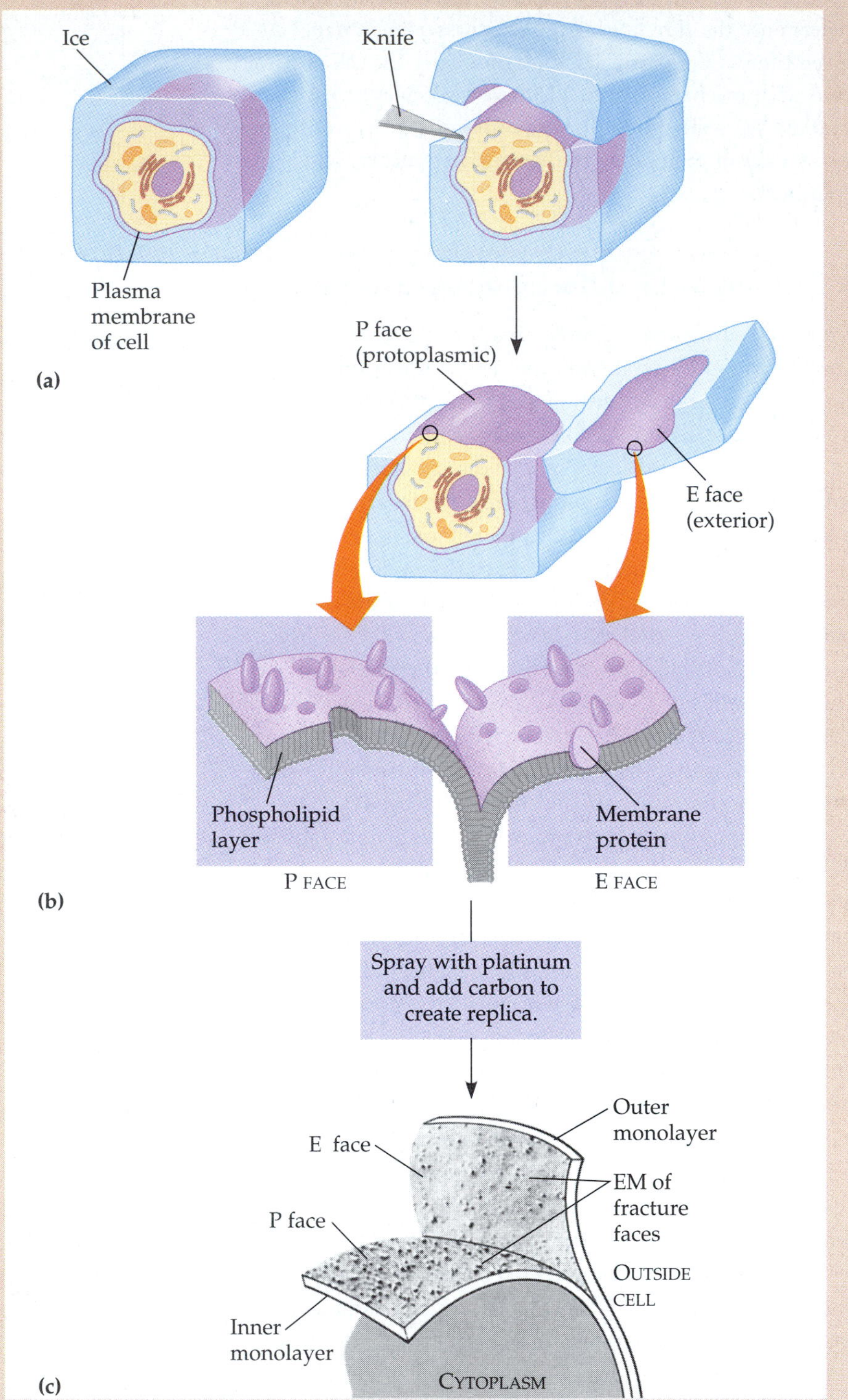

the membrane from the hypotonic solution to the hypertonic solution. This diffusion of water across a selectively permeable membrane is a special case of passive transport called **osmosis**.

The direction of osmosis is determined only by a difference in *total* solute concentration. Water moves from a hypotonic to a hypertonic solution even if the hypotonic solution has more *kinds* of solutes. Seawater, which has a great variety of solutes, will lose water to a very concentrated solution of a single sugar, because the total solute concentration of the seawater is lower. If two solutions are isotonic, water moves across a membrane separating the solutions at an equal rate in both directions; that is, there is no net osmosis between isotonic solutions. (In Units Six and Seven, we will look at osmosis in a more quantitative way.)

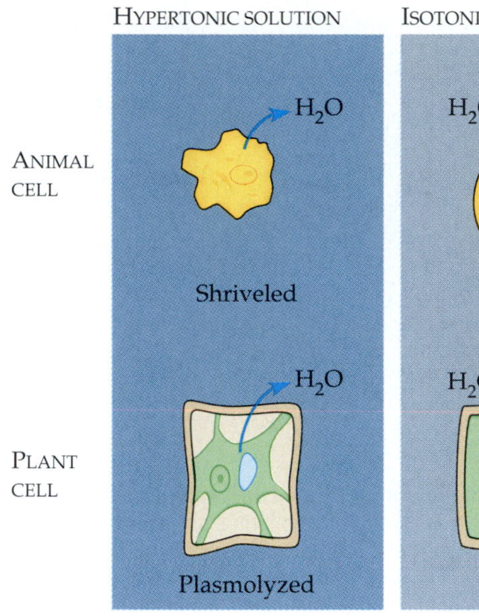
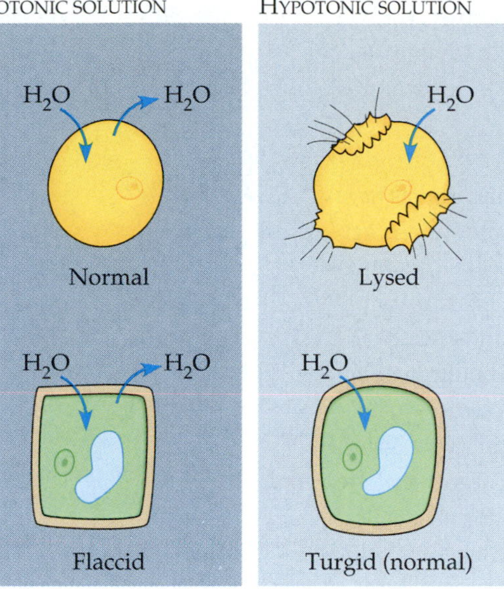

FIGURE 8.11 ▪ The water balance of living cells. How living cells react to changes in the solute concentrations of their environments depends on whether or not they have cell walls. Animal cells do not have cell walls; plant cells do. Unless it has special adaptations to offset the osmotic uptake or loss of water, an animal cell fares best in an isotonic environment. Plant cells are turgid (firm) and generally healthiest in a hypotonic environment, where the tendency for continued uptake of water is balanced by the elastic wall pushing back on the cell. (Arrows indicate net water movement when the cells are *first* placed in these solutions.)

Cell survival depends on balancing water uptake and loss

The movement of water across cell membranes and the balance of water between the cell and its environment are crucial to organisms. Let's now apply to living cells what we have learned about osmosis in artificial systems.

Water Balance of Cells Without Walls

If an animal cell is immersed in an environment that is isotonic to the cell, there will be no net movement of water across the plasma membrane. Water flows across the membrane, but at the same rate in both directions. In an isotonic environment, the volume of an animal cell is stable (FIGURE 8.11). Now let's transfer the cell to a solution that is hypertonic to the cell. The cell will lose water to its environment, shrivel, and probably die. This is one reason why an increase in the salinity (saltiness) of a lake can kill the animals there. However, taking up too much water can be just as hazardous to an animal cell as losing water. If we place the cell in a solution that is hypotonic to the cell, water will enter faster than it leaves, and the cell will swell and lyse (burst) like an overfilled water balloon.

A cell without rigid walls can tolerate neither excessive uptake nor excessive loss of water. This problem of water balance is automatically solved if such a cell lives in isotonic surroundings. Seawater is isotonic to many marine invertebrates. The cells of most terrestrial (land-dwelling) animals are bathed in an extracellular fluid that is isotonic to the cells. Animals and other organisms without rigid cell walls living in hypertonic or hypotonic environments must have special adaptations for **osmoregulation**, the control of water balance. For example, the protist *Paramecium* lives in pond water, which is hypotonic to the cell (FIGURE 8.12). Water continually tends to enter the cell, but *Paramecium* has a plasma membrane that is much less permeable to water than the membranes of most other cells. Also, *Paramecium* is equipped with a contractile vacuole, an organelle that functions as a bilge pump to force

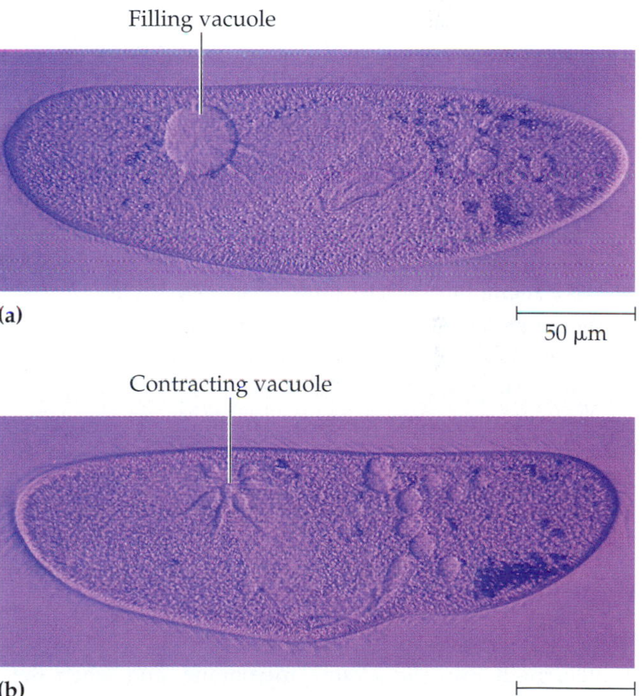

FIGURE 8.12 ▪ Evolutionary adaptations for osmoregulation in *Paramecium*. *Paramecium* is a freshwater protist. Pond water, which is hypotonic to the cell, surrounds a *Paramecium*, and water tends to enter the cell by osmosis. What keeps the cell, which lacks a cell wall, from bursting? Compared to the plasma membranes of most other organisms, the *Paramecium* membrane is less permeable to water. However, this adaptation only slows the uptake of water. The contractile vacuole offsets osmosis by bailing water out of the cell. **(a)** A contractile vacuole fills with fluid that enters from a system of canals radiating throughout the cytoplasm (LM). **(b)** When full, the vacuole and canals contract, expelling fluid from the cell (LM).

water out of the cell as fast as it enters by osmosis. We will examine other adaptations for osmoregulation in Chapter 44.

Water Balance of Cells with Walls

The cells of plants, prokaryotes, fungi, and some protists have walls. When such a cell is in a hypotonic solution—when bathed by rainwater, for example—the wall helps maintain the cell's water balance. Consider a plant cell. Like an animal cell, the plant cell swells as water enters by osmosis (see FIGURE 8.11). However, the elastic wall will expand only so much before it exerts a back pressure on the cell that opposes further water uptake. At this point, the cell is **turgid** (very firm), which is the healthy state for most plant cells. Plants that are not woody, such as most house plants, depend for mechanical support on cells kept turgid by a surrounding hypotonic solution. If a plant's cells and their surroundings are isotonic, there is no net tendency for water to enter, and the cells become **flaccid** (limp), causing the plant to wilt.

On the other hand, a wall is of no advantage if the cell is immersed in a hypertonic environment. In this case, a plant cell, like an animal cell, will lose water to its surroundings and shrink. As the plant cell shrivels, its plasma membrane pulls away from the wall. This phenomenon, called **plasmolysis**, is usually lethal. The walled cells of bacteria and fungi also plasmolyze in hypertonic environments.

Specific proteins facilitate the passive transport of selected solutes

8.4 Let's now turn our attention from the transport of water across a membrane to the traffic of specific solutes dissolved in the water. As mentioned earlier, many polar molecules and ions impeded by the lipid bilayer of the membrane diffuse with the help of transport proteins that span the membrane. This phenomenon is called **facilitated diffusion**.

A transport protein has many of the properties of an enzyme. Just as an enzyme is specific for its substrate, a transport protein is specialized for the solute it transports and may even have a specific binding site akin to the active site of an enzyme. Like enzymes, transport proteins can be saturated. There are only so many molecules of each type of transport protein built into the plasma membrane, and when these molecules are translocating passengers as fast as they can, transport is occurring at a maximum rate. Also like enzymes, transport proteins can be inhibited by molecules that resemble the normal "substrate." This occurs when the imposter competes with the normally transported solute by binding to the transport protein. Unlike enzymes, however, transport proteins do not usually catalyze chemical reactions. Their function is to catalyze a *physical* process—the faster transport of a molecule across a membrane that would otherwise be relatively impermeable to the substance.

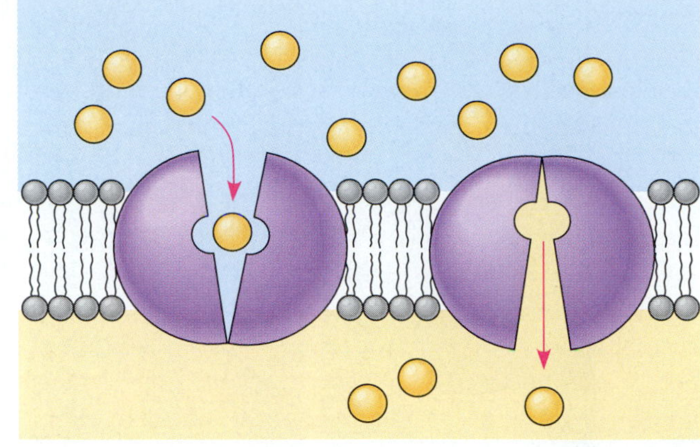

8.4 **FIGURE 8.13 ▪ One model for facilitated diffusion.** The transport protein (purple) alternates between two conformations, moving a solute across the membrane as the shape of the protein changes. The protein can transport the solute in either direction, with the net movement being down the concentration gradient of the solute.

Cell biologists are still trying to learn how various transport proteins facilitate diffusion. In many cases, the protein probably undergoes a subtle change in shape that translocates the solute-binding site from one side of the membrane to the other (FIGURE 8.13). The changes in shape could be triggered by the binding and release of the transported molecule. Other transport proteins simply provide selective corridors allowing a specific solute to cross the membrane (see FIGURE 8.8, top, part (a)). Some of these proteins function as **gated channels**; a stimulus causes them to open or close. The stimulus may be electrical or chemical; if chemical, it is a substance other than the one to be transported. For example, stimulation of a nerve cell by neurotransmitter molecules opens gated channels that allow sodium ions into the cell.

In certain inherited diseases, specific transport systems are either defective or missing altogether. An example is cystinuria, a human disease characterized by the absence of a protein that transports cystine and other amino acids across the membranes of kidney cells. Kidney cells normally reabsorb these amino acids from the urine and return them to the blood, but an individual afflicted with cystinuria develops painful stones from amino acids that accumulate and crystallize in the kidneys.

Active transport is the pumping of solutes against their gradients

8.5 Despite the help of a transport protein, facilitated diffusion is still considered passive transport because the solute is moving down its concentration gradient. Facilitated diffusion speeds the transport of a solute by providing an efficient passage through the membrane, but it does not alter the direction of transport. Some transport proteins, however, *can* move solutes against their concentration gradients, across the plasma membrane from the side where they are less concentrated to the side where they are more concentrated. This transport is "uphill" and therefore requires work. To pump a molecule across a membrane against its gradient, the cell must expend its own metabolic energy; therefore, this type of membrane traffic is called **active transport.**

Active transport is a major factor in the ability of a cell to maintain internal concentrations of small molecules that differ from concentrations in its environment. For example, compared to its surroundings, an animal cell has a much higher concentration of potassium ions and a much lower concentration of sodium ions. The plasma membrane helps maintain these steep gradients by pumping sodium out of the cell and potassium into the cell.

The work of active transport is performed by specific proteins embedded in membranes. As in other types of cellular work, ATP supplies the energy for most active transport. One way ATP can power active transport is by transferring its terminal phosphate group directly to the transport protein. This may induce the protein to change its conformation in a manner that translocates a solute bound to the protein across the membrane. One transport system that works this way is the **sodium-potassium pump,** which exchanges sodium (Na^+) for potassium (K^+) across the plasma membrane of animal cells (FIGURE 8.14). FIGURE 8.15 (p. 142) reviews the distinction between passive transport and active transport.

Some ion pumps generate voltage across membranes

All cells have voltages across their plasma membranes. Voltage is electrical potential energy—a separation of opposite charges. The cytoplasm of a cell is negative in charge compared to the extracellular fluid because of an unequal distribution of anions and cations on opposite sides of the

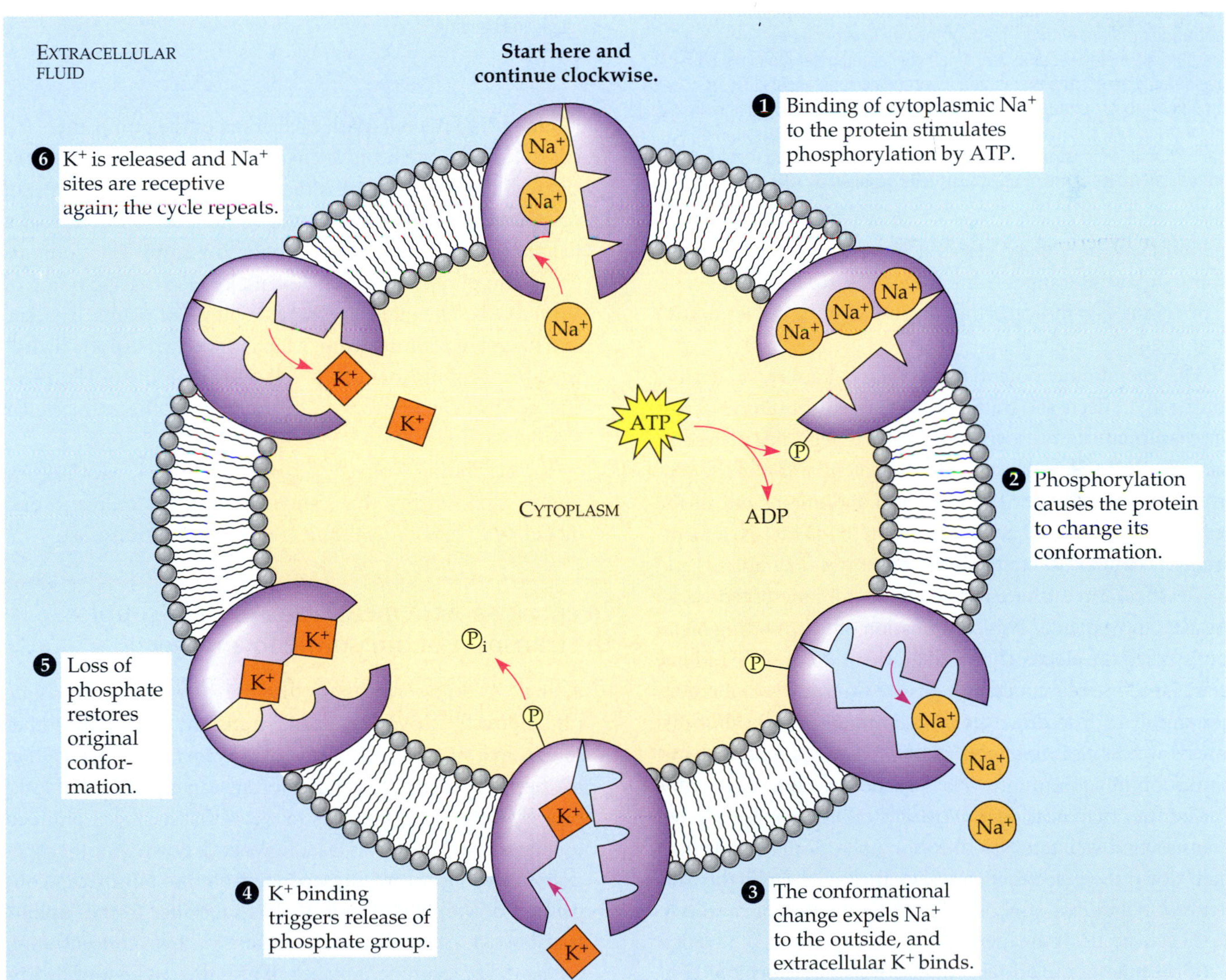

FIGURE 8.14 ▪ The sodium-potassium pump: a specific case of active transport. This transport system pumps ions against steep concentration gradients. The pump oscillates between two conformational states in a pumping cycle that translocates three Na^+ ions out of the cell for every two K^+ ions pumped into the cell. ATP powers the changes in conformation by phosphorylating the transport protein (that is, by transferring a phosphate group to the protein).

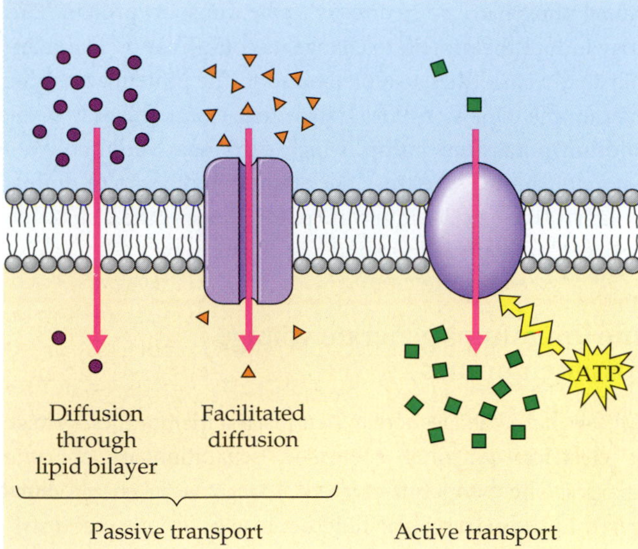

FIGURE 8.15 • Review: passive and active transport compared.
8.5 In passive transport, a substance diffuses spontaneously down its concentration gradient with no need for the cell to expend energy. Hydrophobic molecules and very small uncharged polar molecules diffuse directly across the membrane. Hydrophilic substances diffuse through transport proteins in a process called facilitated diffusion. In active transport, a transport protein moves substances across the membrane "uphill" against their concentration gradients. Active transport requires an expenditure of energy, usually supplied by ATP.

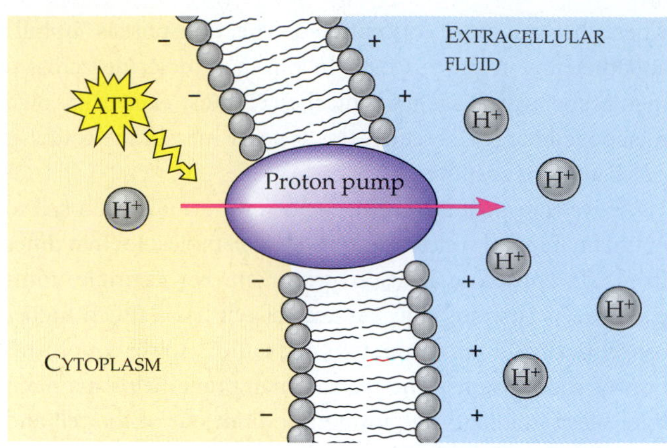

FIGURE 8.16 • An electrogenic pump. Proton pumps are examples of membrane proteins that store energy by generating voltage (charge separation) across membranes. Using ATP for power, a proton pump translocates positive charge in the form of hydrogen ions. The voltage and H^+ gradient represent a dual energy source that can be tapped by the cell to drive other processes, such as the uptake of sugar and other nutrients. Proton pumps are the main electrogenic pumps of plants, fungi, and bacteria.

membrane. The voltage across a membrane, called a **membrane potential**, ranges from about -50 to -200 millivolts. (The minus sign indicates that the inside of the cell is negative compared to the outside.)

The membrane potential acts like a battery, an energy source that affects the traffic of all charged substances across the membrane. Because the inside of the cell is negative compared to the outside, the membrane potential favors the passive transport of cations into the cell and anions out of the cell. Thus, *two* forces drive the diffusion of ions across a membrane: a chemical force (the ion's concentration gradient) and an electrical force (the effect of the membrane potential on the ion's movement). This combination of forces acting on an ion is called the **electrochemical gradient**. In the case of ions, we must refine our concept of passive transport: An ion does not simply diffuse down its *concentration* gradient, but diffuses down its *electrochemical* gradient. For example, the concentration of sodium ions (Na^+) inside a resting nerve cell is much lower than outside it. When the cell is stimulated, gated channels that facilitate Na^+ diffusion open. Sodium ions then "fall" down their electrochemical gradient, driven by the concentration gradient of Na^+ and by the attraction of cations to the negative side of the membrane.

Some membrane proteins that actively transport ions contribute to the membrane potential. An example is the sodium-potassium pump. Notice in FIGURE 8.14 that the pump does not translocate Na^+ and K^+ one for one, but actually pumps three sodium ions out of the cell for every two potassium ions it pumps into the cell. With each crank of the pump, there is a net transfer of one positive charge from the cytoplasm to the extracellular fluid, a process that stores energy in the form of voltage. A transport protein that generates voltage across a membrane is called an **electrogenic pump**. The sodium-potassium pump seems to be the major electrogenic pump of animal cells. The main electrogenic pump of plants, bacteria, and fungi is a **proton pump**, which actively transports hydrogen ions (protons) out of the cell. The pumping of H^+ transfers positive charge from the cytoplasm to the extracellular solution (FIGURE 8.16).

By generating voltage across membranes, electrogenic pumps store energy that can be tapped for cellular work, including a type of membrane traffic called cotransport.

In cotransport, a membrane protein couples the transport of one solute to another

A single ATP-powered pump that transports a specific solute can indirectly drive the active transport of several other solutes in a mechanism called **cotransport**. A substance that has been pumped across a membrane can do work as it leaks back by diffusion, analogous to water that has been pumped uphill and performs work as it flows back down. Another specialized transport protein, separate from the pump, can couple the "downhill" diffusion of this substance to the "uphill" transport of a second substance against its own concentration gradient. For example, a plant cell uses the gradient of hydrogen ions generated by its proton pumps to drive the active transport of amino acids, sugars, and several other nutrients into the cell. One specific transport protein couples the return of hydrogen ions to the transport of sucrose into the cell (FIG-

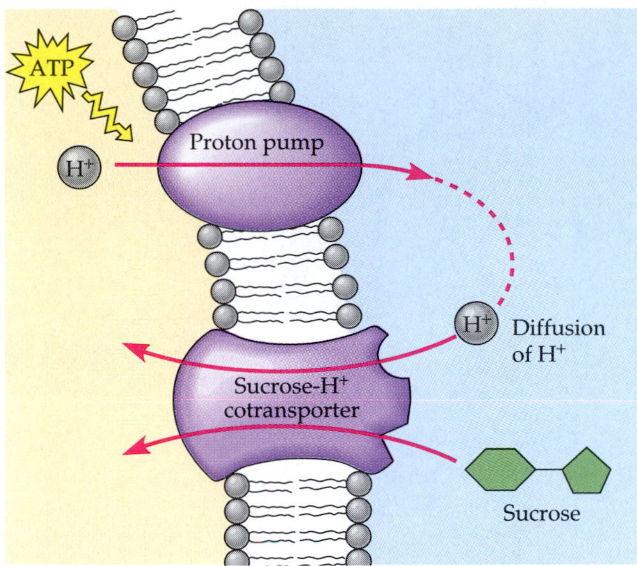

FIGURE 8.17 • Cotransport. An ATP-driven pump stores energy by concentrating a substance (H^+, in this case) on one side of the membrane. As the substance leaks back across the membrane through specific transport proteins, it escorts other substances into the cell. In this case, the proton pump of the membrane is indirectly driving sucrose accumulation by a plant cell, with the help of a protein that cotransports the two solutes.

URE 8.17). The protein can translocate sucrose into the cell against a concentration gradient, but only if the sucrose molecule travels in the company of a hydrogen ion. The sucrose rides on the coattails of the hydrogen ion, which uses the common transport protein as an avenue to diffuse down the concentration gradient maintained by the proton pump. Plants use the mechanism of sucrose-H^+ cotransport to load sucrose produced by photosynthesis into specialized cells in the veins of leaves. The sugar can then be distributed by the vascular tissue of the plant to nonphotosynthetic organs, such as roots.

Exocytosis and endocytosis transport large molecules

Water and small solutes enter and leave the cell by passing
8.6 through the lipid bilayer of the plasma membrane, or by being pumped or carried across the membrane by transport proteins. Large molecules, such as proteins and polysaccharides, generally cross the membrane by a different mechanism involving vesicles. As we described in Chapter 7, the cell secretes macromolecules by the fusion of vesicles with the plasma membrane; this is **exocytosis.** A transport vesicle budded from the Golgi apparatus is moved by the cytoskeleton to the plasma membrane. When the vesicle membrane and plasma membrane come into contact, the lipid molecules of the two bilayers rearrange themselves so that the two membranes fuse. The contents of the vesicle then spill to the outside of the cell (see FIGURE 8.7).

Many secretory cells use exocytosis to export their products. For example, certain cells in the pancreas manufacture the hormone insulin and secrete it into the blood by exocytosis. Another example is the neuron, or nerve cell, which uses exocytosis to release chemical signals that stimulate other neurons or muscle cells (see FIGURE 2.16). When plant cells are making walls, exocytosis delivers carbohydrates from Golgi vesicles to the outside of the cell.

In **endocytosis,** the cell takes in macromolecules and particulate matter by forming new vesicles from the plasma membrane. The steps are basically the reverse of exocytosis. A small area of the plasma membrane sinks inward to form a pocket. As the pocket deepens, it pinches in, forming a vesicle containing material that had been outside the cell.

There are three types of endocytosis: phagocytosis ("cellular eating"), pinocytosis ("cellular drinking"), and receptor-mediated endocytosis. In **phagocytosis,** a cell engulfs a particle by wrapping pseudopodia around it and packaging it within a membrane-enclosed sac large enough to be classified as a vacuole (FIGURE 8.18a, p. 144). The particle is digested after the vacuole fuses with a lysosome containing hydrolytic enzymes. In **pinocytosis,** the cell "gulps" droplets of extracellular fluid in tiny vesicles (FIGURE 8.18b). Because any and all solutes dissolved in the droplet are taken into the cell, pinocytosis is unspecific in the substances it transports. In contrast, **receptor-mediated endocytosis** is very specific (FIGURE 8.18c). Embedded in the membrane are proteins with specific receptor sites exposed to the extracellular fluid. The extracellular substances that bind to the receptors are called **ligands,** a general term for any molecule that binds specifically to a receptor site of another molecule (from the Latin *ligare,* "to bind"). The receptor proteins are usually clustered in regions of the membrane called coated pits, which are lined on their cytoplasmic side by a fuzzy layer of protein. These coat proteins probably help deepen the pit and form the vesicle.

Receptor-mediated endocytosis enables the cell to acquire bulk quantities of specific substances, even though those substances may not be very concentrated in the extracellular fluid. For example, human cells use the process to take in cholesterol for use in the synthesis of membranes and as a precursor for the synthesis of other steroids. Cholesterol travels in the blood in particles called low-density lipoproteins (LDLs), complexes of lipids and proteins. These particles bind to LDL receptors on membranes, and then enter the cell by endocytosis. In humans with familial hypercholesterolemia, an inherited disease characterized by a very high level of cholesterol in the blood, the LDL receptor proteins are defective, and the LDL particles cannot enter cells. Cholesterol accumulates in the blood, where it contributes to early atherosclerosis (the buildup of fat deposits on blood vessel linings).

Vesicles not only transport substances between the cell and its surroundings, they also provide a mechanism for rejuvenating or remodeling the plasma membrane. Endocytosis and

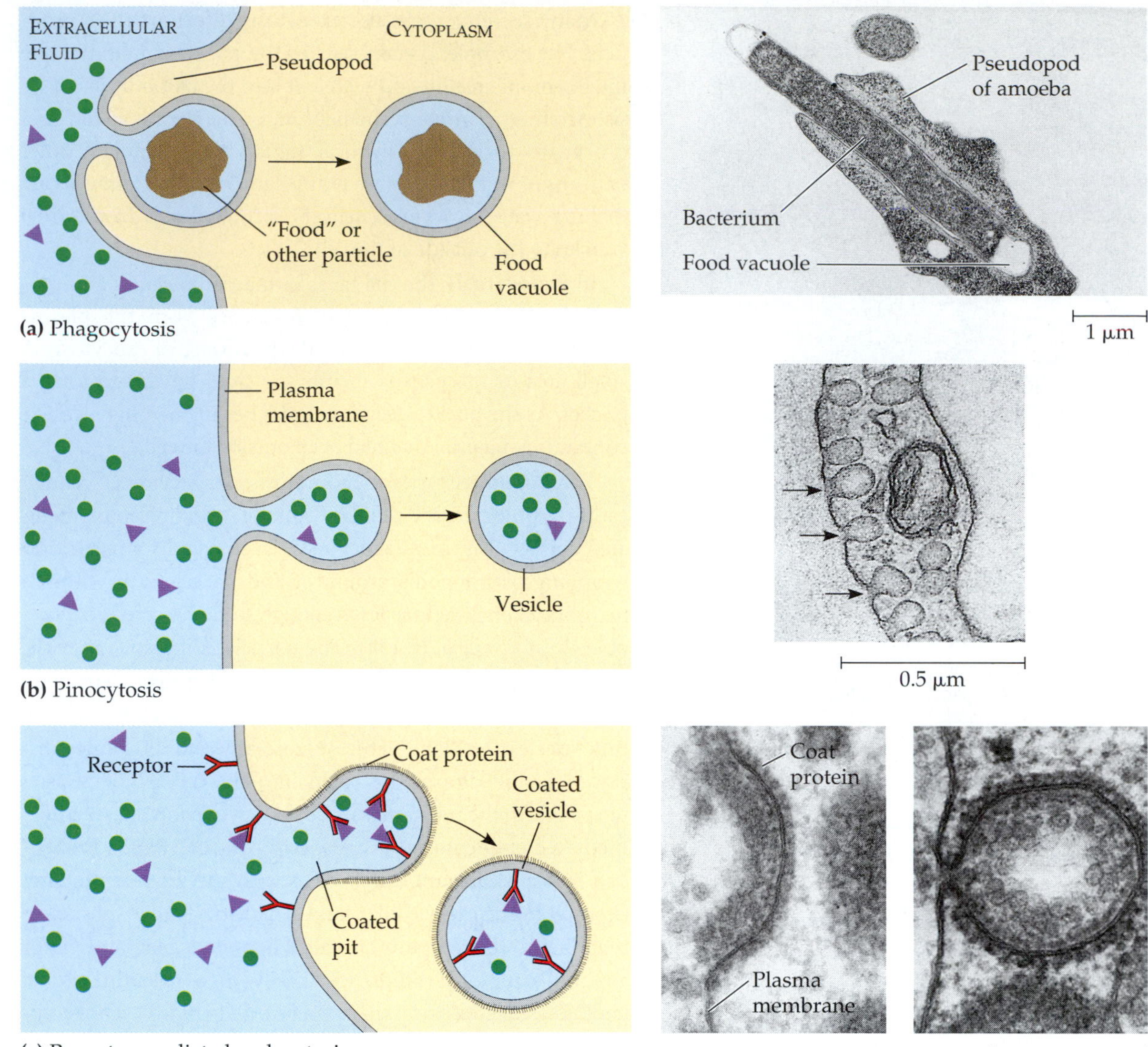

8.6 **FIGURE 8.18 ▪ The three types of endocytosis in animal cells.** **(a)** In phagocytosis, pseudopodia engulf a particle and package it in a vacuole. The micrograph shows an amoeba engulfing a bacterium (TEM). **(b)** In pinocytosis, droplets of extracellular fluid are incorporated into the cell in small vesicles. The micrograph shows pinocytotic vesicles forming (arrows) in a cell lining a small blood vessel (TEM). **(c)** In receptor-mediated endocytosis, coated pits form vesicles when specific molecules (ligands) bind to receptors on the cell surface. Notice that there is a greater relative number of bound molecules (purple) inside the vesicles, though other molecules (green) are also present. The micrographs show two progressive stages of receptor-mediated endocytosis (TEMs). After the ingested material is liberated from the vesicle for metabolism, the receptors are recycled to the plasma membrane.

exocytosis occur continually to some extent in most eukaryotic cells, yet the amount of plasma membrane in a nongrowing cell remains fairly constant over the long run. Apparently, the addition of membrane by one process offsets the loss of membrane by the other.

▪ ▪ ▪

Energy and cellular work have figured prominently in our study of membranes. We have seen, for example, that active transport is powered by ATP. In the next two chapters you will learn more about how cells acquire chemical energy to do the work of life. We'll call membranes back for an encore in these chapters, as they play a major role in how mitochondria and chloroplasts make energy available to cells.

CHAPTER REVIEW

REVIEW OF KEY CONCEPTS

(with page numbers and key figures)

MEMBRANE STRUCTURE

- **Membrane models have evolved to fit new data: *science as a process* (pp. 130–132, FIGURE 8.2)** The Davson-Danielli model, placing layers of proteins on either side of a phospholipid bilayer, has been replaced by the fluid mosaic model.

8.1 **A membrane is a fluid mosaic of lipids, proteins, and carbohydrates (pp. 132–135, FIGURE 8.5)** Integral proteins are embedded in the lipid bilayer; peripheral proteins are attached to the surface. The inside and outside membrane faces differ in composition. Carbohydrates linked to proteins and lipids in the plasma membrane are important for cell-cell recognition.

TRAFFIC ACROSS MEMBRANES

- **A membrane's molecular organization results in selective permeability (pp. 136–137)** A cell must exchange small molecules and ions with its surroundings, a process controlled by the plasma membrane. Hydrophobic substances such as CO_2 are soluble in lipid and pass through membranes rapidly. Small polar molecules such as H_2O also pass through the membrane. Larger polar molecules and ions require specific transport proteins to help them cross.

8.2 **Passive transport is diffusion across a membrane (p. 137, FIGURE 8.9)** Diffusion is the spontaneous movement of a substance down its concentration gradient.

8.3 **Osmosis is the passive transport of water (pp. 137–138, FIGURE 8.10)** Water flows across a membrane from the side where solute is less concentrated (hypotonic) to the side where solute is more concentrated (hypertonic). If the concentrations are equal (isotonic), no net osmosis occurs.

- **Cell survival depends on balancing water uptake and loss (pp. 139–140, FIGURE 8.11)** Cells lacking walls (as in animals and some protists) are isotonic with their environments or have adaptations for osmoregulation. Plants, prokaryotes, fungi, and some protists have elastic cell walls, so the cells don't burst in a hypotonic environment.

8.4 **Specific proteins facilitate the passive transport of selected solutes (p. 140, FIGURE 8.13)** In facilitated diffusion, a transport protein speeds movement of a solute across a membrane down its concentration gradient.

8.5 **Active transport is the pumping of solutes against their gradients (pp. 140–141, FIGURE 8.14)** Specific membrane proteins use energy, usually in the form of ATP, to do this work.

- **Some ion pumps generate voltage across membranes (pp. 141–142, FIGURE 8.16)** Ions can have both a concentration (chemical) gradient and an electric gradient (voltage). These forces combine in the electrochemical gradient, which determines the net direction of ionic diffusion. Electrogenic pumps, such as sodium-potassium pumps and proton pumps, are transport proteins that contribute to electrochemical gradients.

- **In cotransport, a membrane protein couples the transport of one solute to another (pp. 142–143, FIGURE 8.17)** One solute's "downhill" diffusion drives the other's "uphill" transport.

8.6 **Exocytosis and endocytosis transport large molecules (pp. 143–144, FIGURE 8.18)** In exocytosis, transport vesicles migrate to the plasma membrane, fuse with it, and release their contents. In endocytosis, large molecules enter cells within vesicles pinched inward from the plasma membrane. The three types of endocytosis are phagocytosis, pinocytosis, and receptor-mediated endocytosis.

SELF-QUIZ

1. In what way do the various membranes of a eukaryotic cell differ?
   a. Phospholipids are found only in certain membranes.
   b. Certain proteins are unique to each membrane.
   c. Only certain membranes of the cell are selectively permeable.
   d. Only certain membranes are constructed from amphipathic molecules.
   e. Some membranes have hydrophobic surfaces exposed to the cytosol, while others have hydrophilic surfaces facing the cytosol.

2. According to the fluid mosaic model of membrane structure, proteins of the membrane are mostly
   a. spread in a continuous layer over the inner and outer surfaces of the membrane
   b. confined to the hydrophobic core of the membrane
   c. embedded in a lipid bilayer
   d. randomly oriented in the membrane, with no fixed inside-outside polarity
   e. free to depart from the fluid membrane and dissolve in the surrounding solution

3. Which of the following factors would tend to increase membrane fluidity?
   a. a greater proportion of unsaturated phospholipids
   b. a lower temperature
   c. a relatively high protein content in the membrane
   d. a greater proportion of relatively large glycolipids compared to lipids having smaller molecular weights
   e. a high membrane potential

4. Which of the following processes includes all others in the list?
   a. osmosis
   b. diffusion of a solute across a membrane
   c. facilitated diffusion
   d. passive transport
   e. transport of an ion down its electrochemical gradient

5. Based on the model of sucrose uptake in FIGURE 8.17, which of the following experimental treatments would increase the rate of sucrose transport into the cell?
   a. decreasing extracellular sucrose concentration
   b. decreasing extracellular pH
   c. decreasing cytoplasmic pH
   d. adding an inhibitor that blocks the regeneration of ATP
   e. adding a substance that makes the membrane more permeable to hydrogen ions

Questions 6–10

An artificial cell consisting of an aqueous solution enclosed in a selectively permeable membrane has just been immersed in a beaker containing a different solution. The membrane is permeable to water and to the simple sugars glucose and fructose but completely impermeable to the disaccharide sucrose.

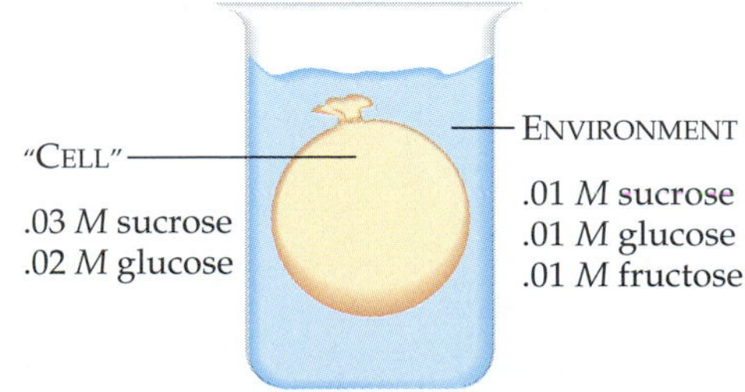

6. Which solute(s) will exhibit a net diffusion into the cell?
7. Which solute(s) will exhibit a net diffusion out of the cell?
8. Which solution—the *cell contents* or the *environment*—is hypertonic?
9. In which direction will there be a net osmotic movement of water?
10. After the cell is placed in the beaker, which of the following changes will occur?
    a. The artificial cell will become more flaccid.
    b. The artificial cell will become more turgid.
    c. The entropy of the system (cell plus surrounding solution) will decrease.
    d. The overall free energy stored in the system will increase.
    e. The membrane potential will decrease.

CHALLENGE QUESTIONS

1. The cells of plant seeds store oils in the form of droplets enclosed by membranes. Unlike the membranes you studied in this chapter, the oil droplet membrane probably consists of a single layer of phospholipids rather than a bilayer. Draw a model for a membrane around an oil droplet, and explain why this arrangement is more stable than a bilayer.
2. An experiment is designed to study the mechanism of sucrose uptake by plant cells. Cells are immersed in a sucrose solution, and the pH of the surrounding solution is monitored with a pH meter. The measurements show that sucrose uptake by the plant cells raises the pH of the surrounding solution. The magnitude of the pH change is proportional to the starting concentration of sucrose in the extracellular solution. A metabolic poison that blocks the ability of the cells to regenerate ATP also inhibits the pH changes in the surrounding solution. Propose a hypothesis accounting for these results. Suggest an additional experiment to test your hypothesis.
3. In an adaptation of the preceding experiment, the rates of sucrose uptake from solutions of different sucrose concentrations are compared.

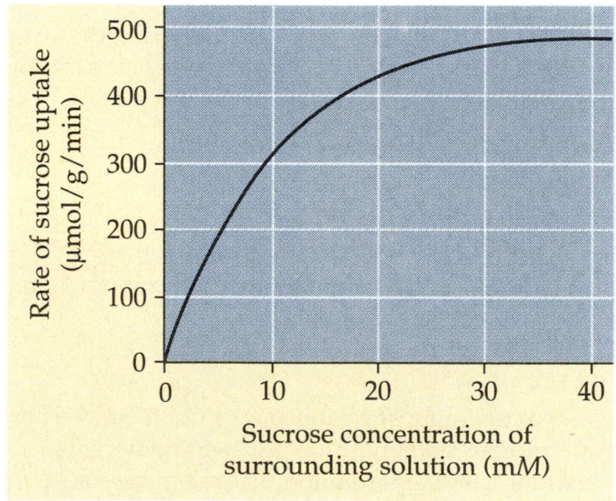

Explain the shape of the curve above in terms of what is happening at the membranes of the plant cells.

SCIENCE, TECHNOLOGY, AND SOCIETY

1. A U.S. government panel has recommended that all people in the United States over age 20 should have their blood cholesterol level measured, and that those with high cholesterol levels should be put on a special diet or drug therapy. The annual cost of this program could be \$10–\$50 billion. Research suggests that relatively few people benefit from drug treatment, mainly those suffering from abnormal conditions like familial hypercholesterolemia. Do you think it is worth the effort and expense of testing and medicating everyone with high cholesterol to help only 1% or 2% of the population? Why or why not?
2. Extensive irrigation in arid regions causes salts to accumulate in the soil. (The water contains low concentrations of salts, but when the water evaporates from the fields, the salts are left behind to concentrate in the soil.) Based on what you have learned about water balance in plant cells, explain why increasing soil salinity (saltiness) has an adverse effect on agriculture. Suggest some ways to minimize this damage. What costs are attached to your solutions?

FURTHER READING

Alberts, B., D. Bray, J. Lewis, M. Raff, K. Roberts, and J. D. Watson. *Molecular Biology of the Cell*, 3rd ed. New York: Garland, 1994. A description of the structure and functions of the plasma membrane is in Chapter 10.

De Camilli, P., S. D. Emr, P. S. McPherson, and P. Novick. "Phosphoinositides as Regulators in Membrane Traffic." *Science*, March 15, 1996. This issue also has other articles on macromolecular transport across membranes.

Jacobson, K., E. D. Sheets, and R. Simson. "Revisiting the Fluid Mosaic Model of Membranes." *Science*, June 9, 1995. A refinement of the current model.

Lang, F., and Waldegger, S. "Regulating Cell Volume." *American Scientist*, September–October 1997. The effects of osmosis on cell function.

Morré, D. J., and T. W. Keenan. "Membrane Flow Revisited." *BioScience*, September 1997. How molecules move through the endomembrane system.

Sharon, N., and H. Lis. "Carbohydrates in Cell Recognition." *Scientific American*, January 1993. How do membrane markers enable an organism to distinguish "self" cells from "nonself" cells?

WEB LINKS

Visit the special edition of *The Biology Place* for BIOLOGY, Fifth Edition, at **http://www.biology.com/campbell**. Go to Chapter 8 for online resources, including learning activities, practice exams, and links to the following web sites:

"Bioenergetics"
Membranes house proteins that carry out many important biochemical pathways. This well-written and nicely illustrated site describes the role of the inner mitochondrial membrane in cellular respiration, among other topics.

"Center of Membrane Sciences"
The Center of Membrane Sciences was established to foster multidisciplinary research on both biological and synthetic membranes.

"The UNESCO Centre for Membrane Science & Technology"
Based at the University of New South Wales, this site deals with research on biological membranes and with possible applications to the development of synthetic membranes for industrial purposes.

"Bioenergetics Links"
Many links to sites related to membrane-bound proteins that participate in bioenergetic pathways.

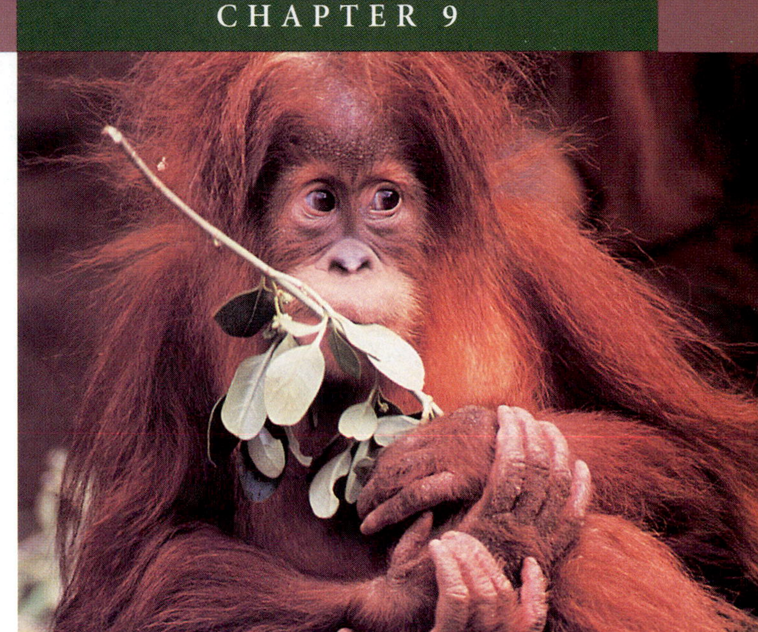

CHAPTER 9

CELLULAR RESPIRATION: HARVESTING CHEMICAL ENERGY

Principles of Energy Harvest

- Cellular respiration and fermentation are catabolic (energy-yielding) pathways
- Cells recycle the ATP they use for work
- Redox reactions release energy when electrons move closer to electronegative atoms
- Electrons "fall" from organic molecules to oxygen during cellular respiration
- The "fall" of electrons during respiration is stepwise, via NAD^+ and an electron transport chain

The Process of Cellular Respiration

- Respiration involves glycolysis, the Krebs cycle, and electron transport: *an overview*
- Glycolysis harvests chemical energy by oxidizing glucose to pyruvate: *a closer look*
- The Krebs cycle completes the energy-yielding oxidation of organic molecules: *a closer look*
- The inner mitochondrial membrane couples electron transport to ATP synthesis: *a closer look*
- Cellular respiration generates many ATP molecules for each sugar molecule it oxidizes: *a review*

Related Metabolic Processes

- Fermentation enables some cells to produce ATP without the help of oxygen
- Glycolysis and the Krebs cycle connect to many other metabolic pathways
- Feedback mechanisms control cellular respiration

*Living is work. A cell organizes small organic molecules into polymers, such as proteins and DNA. It pumps substances across membranes. Many cells move or change their shapes. They grow and reproduce. A cell must work just to maintain its complex structure, because order is intrinsically unstable. To perform their many tasks, cells require transfusions of energy from outside sources. Energy enters most ecosystems in the form of sunlight, the energy source for plants and other photosynthetic organisms (*FIGURE 9.1, *p. 148). Animals, such as the orangutan in the photograph to the right, obtain fuel by eating plants, or by eating other organisms that eat plants. In this chapter you will learn how cells harvest the chemical energy stored in organic molecules and use it to regenerate ATP, the molecule that drives most cellular work.*

PRINCIPLES OF ENERGY HARVEST

In harvesting chemical energy, cells usually employ metabolic pathways with many steps. Fortunately, we can organize this complexity with the help of just a handful of general principles. The first part of this chapter develops these principles, which will then enable you to make sense of cellular respiration and related pathways.

Cellular respiration and fermentation are catabolic (energy-yielding) pathways

Organic compounds store energy in their arrangements of atoms. With the help of enzymes, a cell systematically degrades complex organic molecules that are rich in potential energy to simpler waste products that have less energy. Some of the energy taken out of chemical storage can be used to do work; the rest is dissipated as heat. As you learned in Chapter 6, metabolic pathways that release stored energy by breaking down complex molecules are called catabolic pathways. One catabolic process, **fermentation**, is a partial degradation of sugars that occurs without the help of oxygen. However, the most prevalent and efficient catabolic pathway is **cellular respiration**, in which oxygen is consumed as a reactant along with the organic fuel. In eukaryotic cells, mitochondria house most of the metabolic equipment for cellular respiration.

Although very different in mechanism, respiration is in principle similar to the combustion of gasoline in an automobile engine after oxygen is mixed with the fuel (hydrocarbons). Food is the fuel for respiration, and the exhaust is carbon dioxide and water. The overall process can be summarized as follows:

$$\text{Organic compounds} + \text{Oxygen} \longrightarrow \text{Carbon dioxide} + \text{Water} + \text{Energy}$$

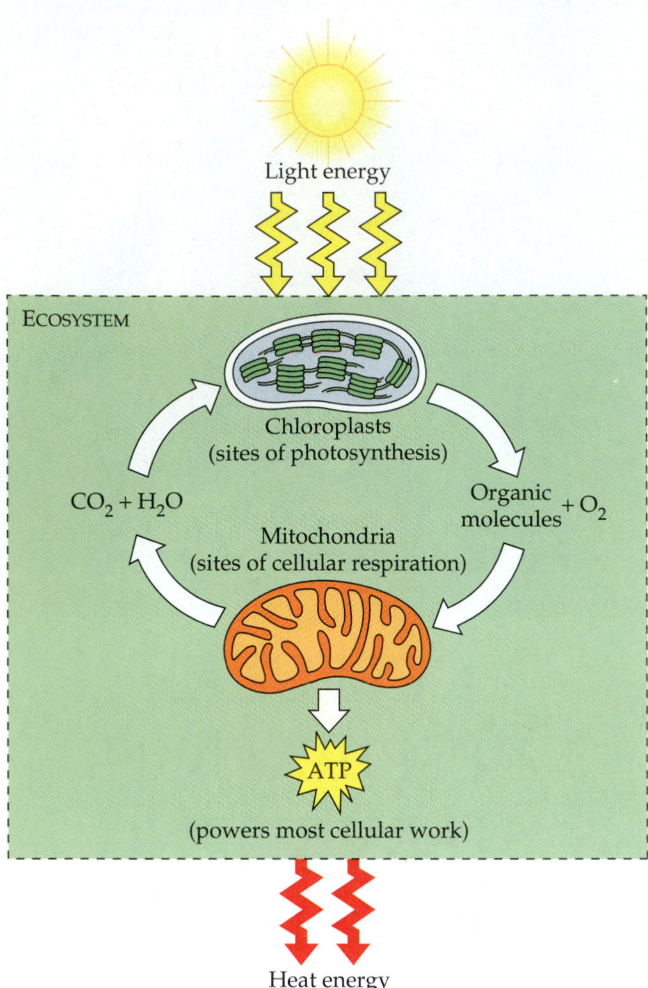

FIGURE 9.1 ▪ Energy flow and chemical recycling in ecosystems. The mitochondria of eukaryotes (including plants) use the organic products of photosynthesis as fuel for cellular respiration, which also consumes the oxygen produced by photosynthesis. Respiration harvests the energy stored in organic molecules to generate ATP, which powers most cellular work. The waste products of respiration, carbon dioxide and water, are the very substances that chloroplasts use as raw materials for photosynthesis. Thus, the chemical elements essential to life are recycled. But energy is not: It flows into an ecosystem as sunlight and leaves it as heat.

Although carbohydrates, fats, and proteins can all be processed and consumed as fuel, it is traditional to learn the steps of cellular respiration by tracking the degradation of the sugar glucose ($C_6H_{12}O_6$):

$$C_6H_{12}O_6 + 6\,O_2 \longrightarrow 6\,CO_2 + 6\,H_2O + \text{Energy (ATP + heat)}$$

This breakdown of glucose is exergonic, having a free-energy change of -686 kcal (-2870 kJ) per mole of glucose decomposed ($\Delta G = -686$ kcal/mol; recall that a negative ΔG indicates that the products of the chemical process store less energy than the reactants).

Catabolic pathways do not directly move flagella, pump solutes across membranes, polymerize monomers, or perform other cellular work. Catabolism is linked to work by a chemical drive shaft: ATP. The processes of cellular respiration and fermentation are complex and challenging to learn. Therefore, in studying this chapter it will help you to keep in mind your main objective: discovering how cells use the energy stored in food molecules to make ATP.

Cells recycle the ATP they use for work

The molecule known as ATP, short for adenosine triphosphate, is the central character in bioenergetics. Recall from Chapter 6 that the triphosphate tail of ATP is the chemical equivalent of a loaded spring; the close packing of the three negatively charged phosphate groups is an unstable, energy-storing arrangement (because like charges repel each other). The chemical "spring" tends to "relax" by losing the terminal phosphate (see FIGURE 6.6). The cell taps this energy source by using enzymes to transfer phosphate groups from ATP to other compounds, which are then said to be phosphorylated. Phosphorylation primes a molecule to undergo some kind of change that performs work, and the molecule loses its phosphate group in the process (FIGURE 9.2). The price of most cellular work, then, is the conversion of ATP to ADP and inorganic phosphate (abbreviated Ⓟ$_i$ in this book), products that store less energy than ATP. To keep working, the cell must regenerate its supply of ATP from ADP and inorganic phosphate. A working muscle cell, for example, recycles its ATP at a rate of about 10 million molecules per second. To understand how cellular respiration regenerates ATP, we need to examine the fundamental chemical processes known as oxidation and reduction.

Redox reactions release energy when electrons move closer to electronegative atoms

Just what happens when cellular respiration decomposes glucose and other organic fuels? And why does this metabolic pathway yield energy? The answers are based on the transfer of electrons during the chemical reactions. The relocation of electrons releases the energy stored in food molecules, and this energy is used to synthesize ATP.

In many chemical reactions, there is a transfer of one or more electrons (e^-) from one reactant to another. These electron transfers are called oxidation-reduction reactions, or **redox reactions** for short. In a redox reaction, the loss of electrons from one substance is called **oxidation**, and the addition of electrons to another substance is known as **reduction**.* Consider, for example, the reaction between sodium and chlorine to form table salt:

* This term defies intuition; *adding* electrons is called *reduction.* The term was derived from the electrical effects of adding electrons: Negatively charged electrons added to a cation *reduce* the amount of positive charge of the cation.

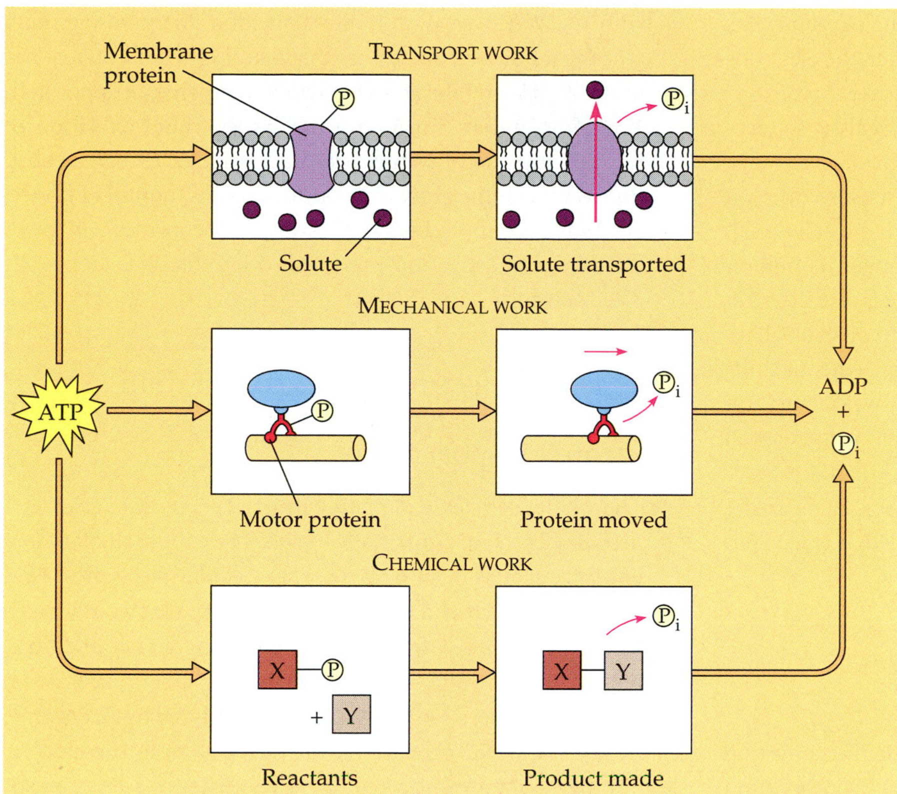

FIGURE 9.2 ▪ A review of how ATP drives cellular work. Phosphate-group transfer is the mechanism responsible for most types of cellular work. Enzymes shift a phosphate group (Ⓟ) from ATP to some other molecule, and this phosphorylated molecule undergoes a change that performs work. For example, ATP drives active transport by phosphorylating specialized proteins built into membranes; drives mechanical work by phosphorylating motor proteins, such as the ones that move organelles along cytoskeletal "tracks" in the cell; and drives chemical work by phosphorylating key reactants. The phosphorylated molecules lose the phosphate groups as work is performed, leaving ADP and inorganic phosphate ($Ⓟ_i$) as products. Cellular respiration replenishes the ATP supply by powering the phosphorylation of ADP.

Oxidation: Na → Na⁺; Reduction: Cl → Cl⁻

$$\mathrm{Na + Cl \longrightarrow Na^+ + Cl^-}$$

Or we could generalize a redox reaction this way:

Oxidation: Xe⁻ → X; Reduction: Y → Ye⁻

$$Xe^- + Y \longrightarrow X + Ye^-$$

In the generalized reaction, substance X, the electron donor, is called the **reducing agent**; it reduces Y. Substance Y, the electron acceptor, is the **oxidizing agent**; it oxidizes X.

Because an electron transfer requires both a donor and an acceptor, oxidation and reduction always go together.

Not all redox reactions involve the complete transfer of electrons from one substance to another; some change the degree of electron sharing in covalent bonds. The reaction between methane and oxygen to form carbon dioxide and water, shown in FIGURE 9.3, is an example. As explained in Chapter 2, the covalent electrons in methane are shared equally between the bonded atoms because carbon and hydrogen have about the same affinity for valence electrons; they are about equally electronegative. But when methane reacts with oxygen to form carbon dioxide, electrons are shifted away from the carbon atoms to their new covalent partner, oxygen, which is very electronegative. Methane has thus been oxidized. The two atoms of the oxygen molecule

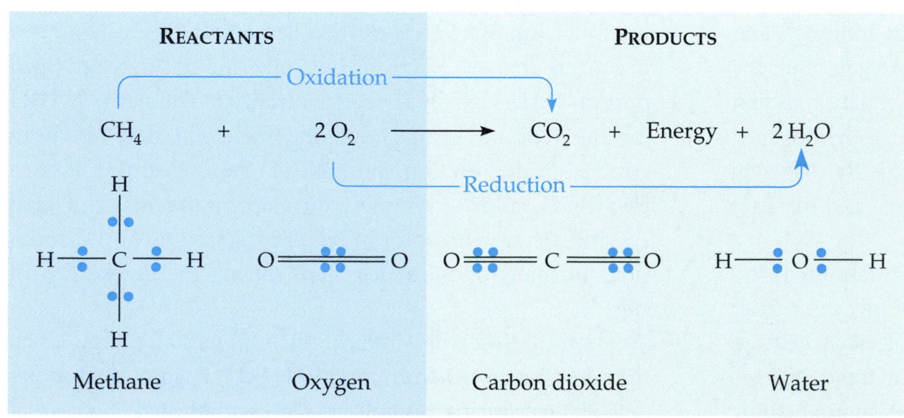

FIGURE 9.3 ▪ Methane combustion as an energy-yielding redox reaction. During the reaction, covalently shared electrons move away from carbon and hydrogen atoms and closer to oxygen, which is very electronegative. The reaction releases energy to the surroundings, because the electrons lose potential energy as they move closer to electronegative atoms.

also share their electrons equally. But when the oxygen reacts with the hydrogen from methane to form water, the electrons of the covalent bonds are drawn closer to the oxygen; the oxygen molecule has been reduced. Because oxygen is so electronegative, it is one of the most potent of all oxidizing agents.

Energy must be added to pull an electron away from an atom, just as energy must be added to push a large ball uphill. The more electronegative the atom (the stronger its pull on electrons), the more energy is required to keep the electron away from it, just as more energy is required to push a ball up a steeper hill. An electron *loses* potential energy when it shifts from a less electronegative atom (one with a weaker pull on electrons) *toward* a more electronegative one, just as a ball loses potential energy when it rolls downhill. A redox reaction that relocates electrons closer to oxygen, such as the burning of methane, releases chemical energy that can be put to work.

Electrons "fall" from organic molecules to oxygen during cellular respiration

The oxidation of methane by oxygen is the main combustion reaction that occurs at the burner of a gas stove. Combustion of gasoline in an automobile engine is also a redox reaction, and the energy released pushes the pistons. But the energy-yielding redox process of greatest interest here is respiration: the oxidation of glucose and other fuel molecules in food. Examine again the summary equation for cellular respiration, but this time think of it as a redox process:

$$\overbrace{C_6H_{12}O_6 + 6\,O_2 \longrightarrow 6\,CO_2 + 6\,H_2O}$$

Oxidation: $C_6H_{12}O_6 \rightarrow 6\,CO_2$; Reduction: $6\,O_2 \rightarrow 6\,H_2O$

As in the combustion of methane or gasoline, the fuel (sugar) is oxidized and oxygen is reduced, and the electrons lose potential energy along the way.

In general, organic molecules that have an abundance of hydrogen are excellent fuels because their bonds are a source of "hilltop" electrons with the potential to "fall" closer to oxygen. The summary equation for respiration indicates that hydrogen is transferred from glucose to oxygen. But the important point, not visible in the summary equation, is that the change in the covalent status of electrons as hydrogen is transferred to oxygen is what liberates energy. By oxidizing glucose, respiration takes energy out of storage and makes it available for ATP synthesis.

The main energy foods, carbohydrates and fats, are reservoirs of electrons associated with hydrogen. Only the barrier of activation energy holds back the flood of electrons to a lower energy state (see Chapter 6). Without this barrier, a food substance like glucose would combine spontaneously with O_2. When we supply the activation energy by igniting glucose, it burns in air, releasing 686 kcal (2870 kJ) of heat per mole of glucose (about 180 g). Body temperature is not high enough to initiate burning, which is the rapid oxidation of fuel accompanied by an enormous release of energy as heat. But swallow some glucose in the form of a spoonful of honey, and when the molecules reach your cells, enzymes will lower the barrier of activation energy, allowing the sugar to be oxidized slowly.

The "fall" of electrons during respiration is stepwise, via NAD^+ and an electron transport chain

The wholesale release of energy from a fuel is difficult to harness efficiently for constructive work: The explosion of a gasoline tank cannot drive a car very far. Cellular respiration does not oxidize glucose in a single explosive step that would transfer all the hydrogen from the fuel to the oxygen at one time. Rather, glucose and other organic fuels are broken down gradually in a series of steps, each one catalyzed by an enzyme. At key steps, hydrogen atoms are stripped from the glucose, but they are not transferred directly to oxygen. They are usually passed first to a coenzyme called **NAD^+** (nicotinamide adenine dinucleotide). Thus, NAD^+ functions as an oxidizing agent during respiration.

How does NAD^+ trap electrons from glucose and other fuel molecules? Enzymes called dehydrogenases remove a pair of hydrogen atoms from the substrate, a sugar or some other fuel. We can think of this as the removal of two electrons and two protons (the nuclei of hydrogen atoms). The enzyme delivers the *two* electrons along with *one* proton to its coenzyme, NAD^+ (FIGURE 9.4). The other proton is released as a hydrogen ion (H^+) into the surrounding solution:

$$H{-}\overset{|}{\underset{|}{C}}{-}OH + NAD^+ \xrightarrow{\text{Dehydrogenase}} \overset{|}{\underset{|}{C}}{=}O + NADH + H^+$$

Though the oxidized form, NAD^+, has a positive charge, the reduced form, NADH, is electrically neutral. By receiving two negatively charged electrons but only one positively charged proton, NAD^+ has its charge neutralized. The name NADH for the reduced form shows the hydrogen that has been received in the reaction. Since NAD^+ gains electrons, it is an electron acceptor (a synonym for oxidizing agent). The most versatile electron acceptor in cellular respiration, NAD^+ functions in many of the redox steps during the breakdown of sugar.

Electrons lose very little of their potential energy when they are transferred from food to NAD^+. Each NADH molecule formed during respiration represents stored energy that

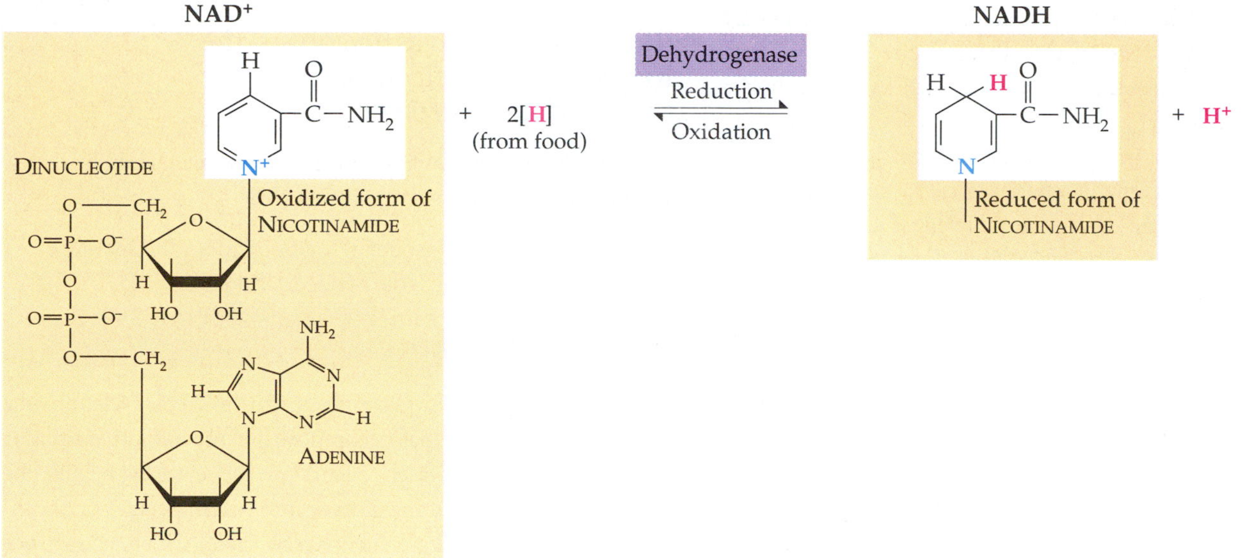

FIGURE 9.4 ▪ NAD^+ as an electron shuttle. The full name for NAD^+, nicotinamide adenine dinucleotide, describes its structure; the molecule consists of two nucleotides joined together. (Nicotinamide is a nitrogenous base, although not one that is present in DNA or RNA.) The enzymatic transfer of two electrons and one proton from some organic substrate to NAD^+ reduces the NAD^+ to NADH. Most of the electrons removed from food are transferred initially to NAD^+.

can be tapped to make ATP when the electrons complete their "fall" from NADH to oxygen.

How do electrons extracted from food and stored by NADH finally reach oxygen? It will help to compare this complex redox chemistry of cellular respiration to a much simpler reaction: the reaction between hydrogen and oxygen to form water (FIGURE 9.5a). Mix H_2 and O_2, provide a spark for activation energy, and the gases combine explosively. The explosion represents a release of energy as the electrons of hydrogen fall closer to the electronegative oxygen. Cellular respiration also brings hydrogen and oxygen together to form water, but there are two important differences. First, in cellular respiration, the hydrogen that reacts with oxygen is derived from organic molecules. Second, respiration uses an **electron transport chain** to break the fall of electrons to oxygen into several energy-releasing steps instead of one explosive reaction (FIGURE 9.5b). The transport chain consists of a number of molecules, mostly proteins, built into the inner membrane of a

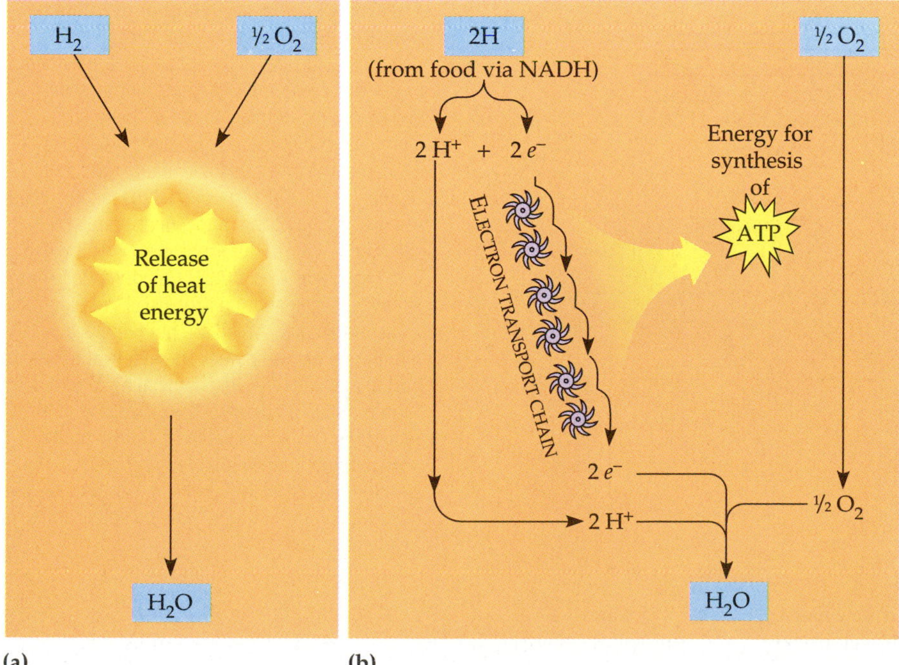

FIGURE 9.5 ▪ An introduction to electron transport chains. (a) The exergonic reaction of hydrogen with oxygen to form water releases a large amount of energy in the form of heat and light: an explosion. **(b)** In cellular respiration, an electron transport chain breaks the "fall" of electrons in this reaction into a series of smaller steps and stores some of the released energy in a form that can be used to make ATP (the rest of the energy is released as heat).

mitochondrion. Electrons removed from food are shuttled by NADH to the "top" end of the chain. At the "bottom" end, oxygen captures these electrons along with hydrogen nuclei (H^+), forming water.

Electron transfer from NADH to oxygen is an exergonic reaction with a free energy change of −53 kcal/mol (−222 kJ/mol). Instead of this energy being released and wasted in a single explosive step, electrons cascade down the chain from one carrier molecule to the next, losing a small amount of energy with each step until they finally reach oxygen, the terminal electron acceptor. What keeps the electrons moving is that each carrier is more electronegative than its "uphill" neighbor in the chain. At the bottom of the chain is oxygen, which has a very great affinity for electrons. Thus, electrons removed from food by NAD^+ fall down the electron transport chain to a far more stable location in the electronegative oxygen atom. Put another way, oxygen pulls electrons down the chain in an energy-yielding tumble analogous to gravity pulling objects downhill.

Thus, during cellular respiration, most electrons travel this "downhill" route: food ⟶ NADH ⟶ electron transport chain ⟶ oxygen. In the next major section of the chapter, you will learn more about how respiration uses the energy released from this exergonic electron fall to regenerate the cell's supply of ATP.

THE PROCESS OF CELLULAR RESPIRATION

Now that we have covered the basic redox mechanisms of respiration, let's look at the entire process.

Respiration involves glycolysis, the Krebs cycle, and electron transport: *an overview*

9.1 Respiration is a cumulative function of three metabolic stages, which are diagrammed in FIGURE 9.6:

1. Glycolysis (color-coded teal throughout the chapter).
2. The Krebs cycle (color-coded salmon).
3. The electron transport chain and oxidative phosphorylation (color-coded violet).

The first two stages, glycolysis and the Krebs cycle, are the catabolic pathways that decompose glucose and other organic fuels. **Glycolysis**, which occurs in the cytosol, begins the degradation by breaking glucose into two molecules of a compound called pyruvate. The **Krebs cycle**, which takes place within the mitochondrial matrix, completes the job by decomposing a derivative of pyruvate to carbon dioxide.

Thus, the carbon dioxide produced by respiration represents fragments of oxidized organic molecules. Some of the steps of glycolysis and the Krebs cycle are redox reactions in which dehydrogenase enzymes transfer electrons from substrates to NAD^+, forming NADH. In the third stage of respiration, the electron transport chain accepts electrons from the breakdown products of the first two stages (usually via NADH) and passes these electrons from one molecule to another. At the end of the chain, the electrons are combined with hydrogen ions and molecular oxygen to form water (see FIGURE 9.5b). The energy released at each step of the chain is stored in a form the mitochondrion can use to make ATP. This mode of ATP synthesis is called **oxidative phosphoryla-**

9.1
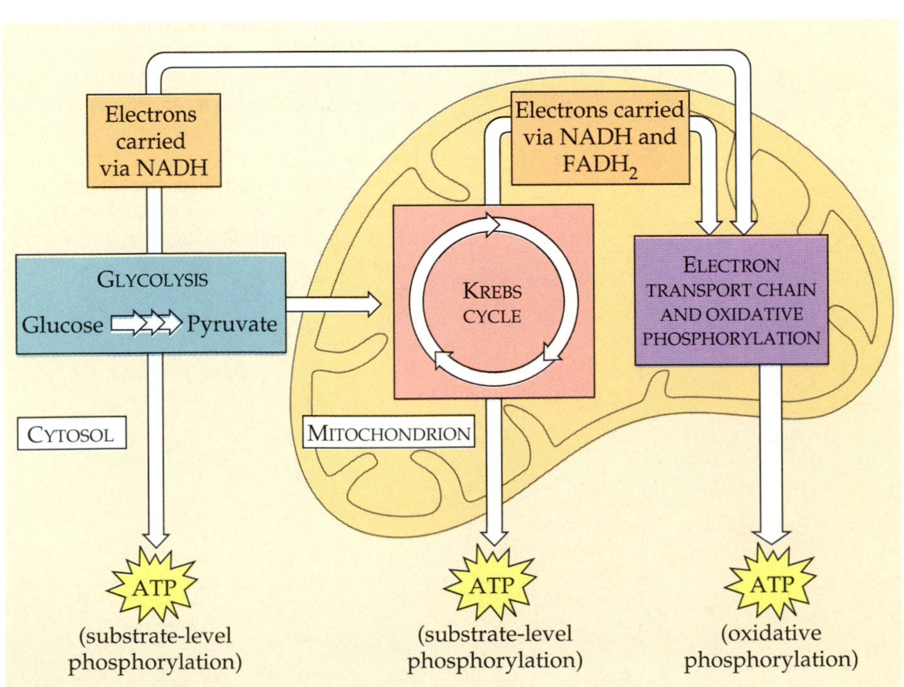

FIGURE 9.6 ▪ **An overview of cellular respiration.** In a eukaryotic cell, glycolysis occurs outside the mitochondria in the cytosol. The Krebs cycle and the electron transport chains are located inside the mitochondria. During glycolysis, each glucose molecule is broken down into two molecules of the compound pyruvate. The pyruvate crosses the double membrane of the mitochondrion to enter the matrix, where the Krebs cycle decomposes it to carbon dioxide. NADH transfers electrons from glycolysis and the Krebs cycle to electron transport chains, which are built into the membrane of the cristae. The electron transport chain converts the chemical energy to a form that can be used to drive oxidative phosphorylation, which accounts for most of the ATP generated by cellular respiration. A smaller amount of ATP is formed directly during glycolysis and the Krebs cycle by substrate-level phosphorylation.

tion because it is powered by the redox reactions that transfer electrons from food to oxygen.

The site of electron transport and oxidative phosphorylation is the inner membrane of the mitochondrion (see FIGURE 7.18). Oxidative phosphorylation accounts for almost 90% of the ATP generated by respiration. A smaller amount of ATP is formed directly in a few reactions of glycolysis and the Krebs cycle by a mechanism called **substrate-level phosphorylation** (FIGURE 9.7). This mode of ATP synthesis occurs when an enzyme transfers a phosphate group from a substrate to ADP. ("Substrate" here refers to an organic molecule generated during the sequential catabolism of glucose.)

Respiration cashes in the large denomination of energy banked in glucose for the small change of ATP, which is more practical for the cell to spend on its work. For each molecule of glucose degraded to carbon dioxide and water by respiration, the cell makes up to about 38 molecules of ATP.

This overview has introduced how glycolysis, the Krebs cycle, and electron transport fit into the overall process of cellular respiration. We are now ready to take a closer look at each of these three stages of respiration.

Glycolysis harvests chemical energy by oxidizing glucose to pyruvate: *a closer look*

9.2 The word *glycolysis* means "splitting of sugar," and that is exactly what happens during this pathway. Glucose, a six-carbon sugar, is split into two three-carbon sugars. These smaller sugars are then oxidized, and their remaining atoms rearranged to form two molecules of pyruvate. (Pyruvate is the ionized form of a three-carbon acid, pyruvic acid.)

The catabolic pathway of glycolysis consists of ten steps, each catalyzed by a specific enzyme. We can divide these ten steps into two phases. The energy-investment phase includes the first five steps, and the energy-payoff phase includes the next five steps. During the energy-investment phase, the cell actually spends ATP to phosphorylate the fuel molecules (FIGURE 9.8). This investment is repaid with dividends during the energy-payoff phase, when ATP is produced by substrate-level phosphorylation and NAD^+ is reduced to NADH by oxidation of the food. On the next two pages you will find a detailed diagram of glycolysis (FIGURE 9.9). Do not let the details obscure the main functions of glycolysis: Glycolysis is a source of ATP and NADH and also prepares organic molecules for further oxidation in the Krebs cycle.

As summarized in FIGURE 9.8, the net energy yield from glycolysis, per glucose molecule, is 2 ATP plus 2 NADH. Notice that all of the carbon originally present in glucose is accounted for in the two molecules of pyruvate; no CO_2 is

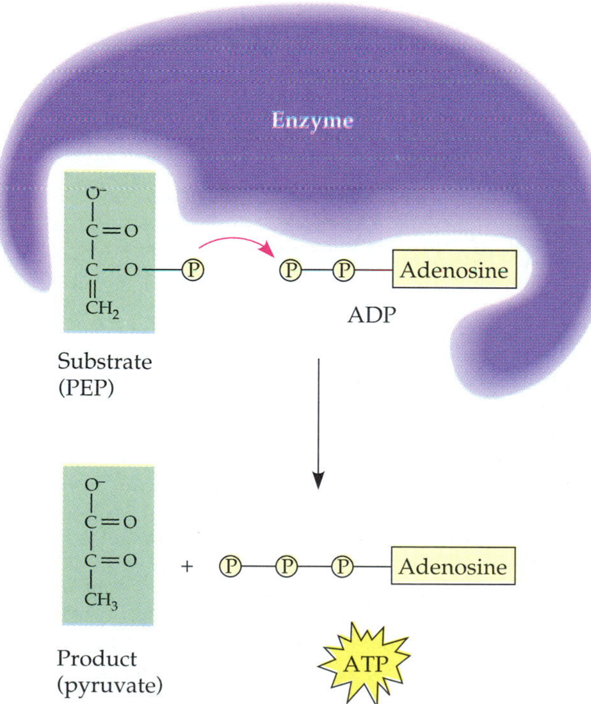

FIGURE 9.7 ▪ Substrate-level phosphorylation. Some ATP is made by direct enzymatic transfer of a phosphate group from a substrate to ADP. The phosphate donor in this case is phosphoenolpyruvate (PEP), which is formed from the breakdown of sugar during glycolysis.

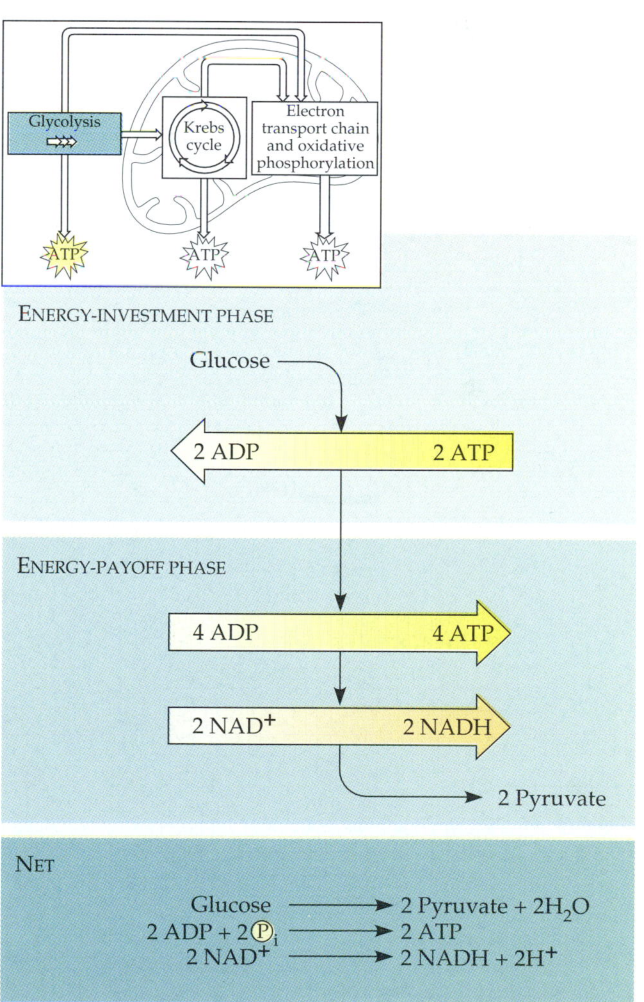

FIGURE 9.8 ▪ The energy input and output of glycolysis.

FIGURE 9.9 ▪ A closer look at glycolysis. 9.2 The orientation diagram at the right relates glycolysis to the whole process of respiration. Do not let the chemical detail in this diagram block your view of glycolysis as a source of ATP and NADH.

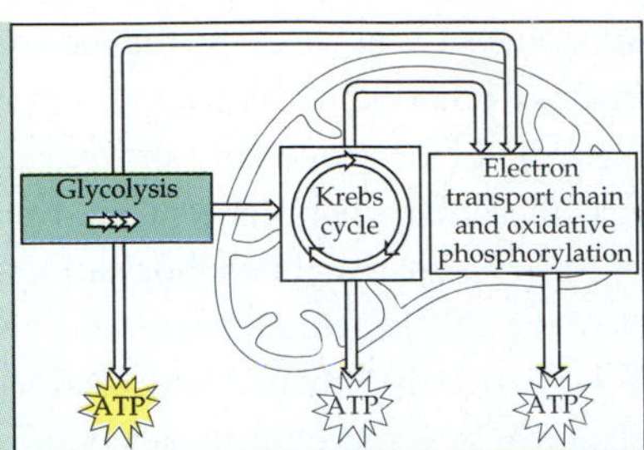

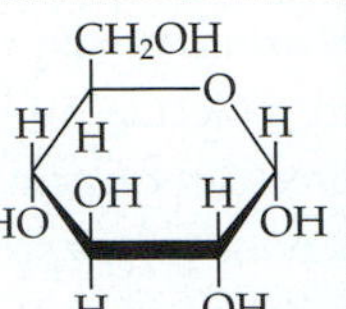

Glucose

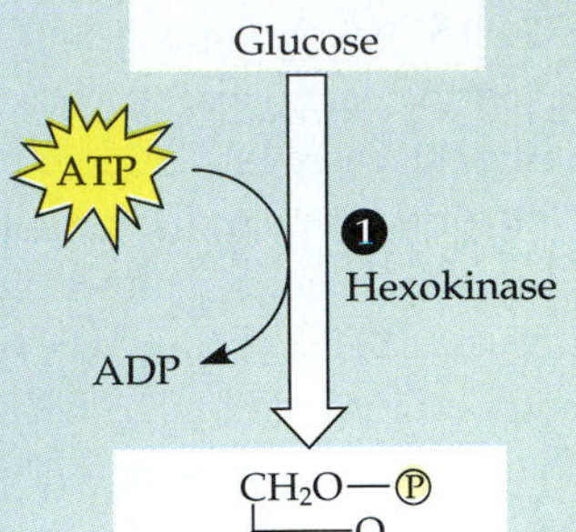

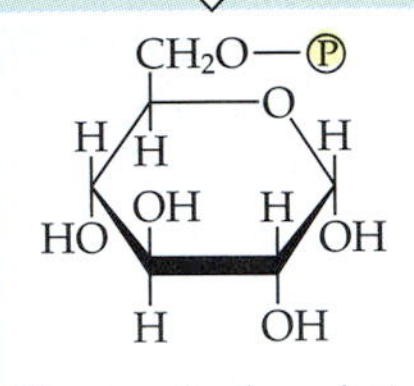

Glucose 6-phosphate

2 Phosphoglucoisomerase

CH₂O—Ⓟ
O
CH₂OH
H
HO
H
OH
HO
H

Fructose 6-phosphate

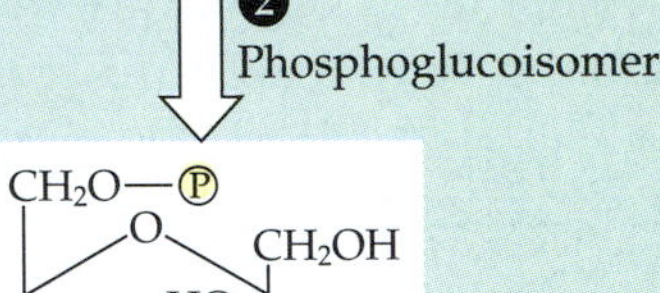

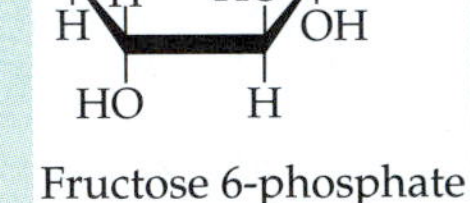

Fructose 1, 6-bisphosphate

4 Aldolase

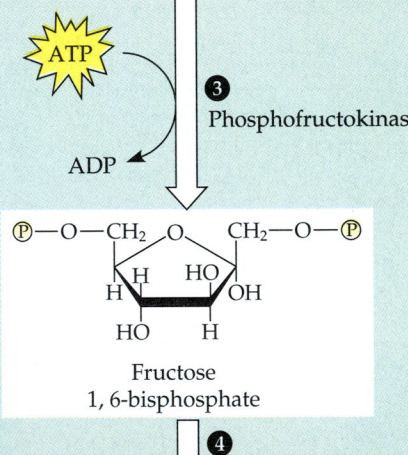

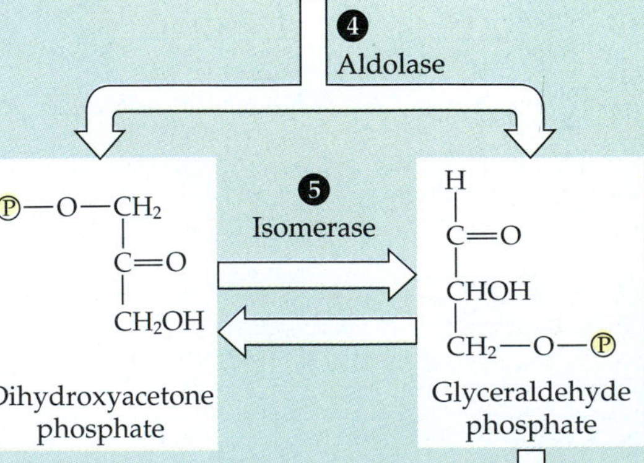

Dihydroxyacetone phosphate

Glyceraldehyde phosphate

ENERGY-INVESTMENT PHASE

1. Glucose enters the cell and is phosphorylated by the enzyme hexokinase, which transfers a phosphate group from ATP to the sugar. The electrical charge of the phosphate group traps the sugar in the cell because of the impermeability of the plasma membrane to ions. Phosphorylation of glucose also makes the molecule more chemically reactive. In this diagram, the transfer of a phosphate group or pair of electrons from one reactant to another is indicated by coupled arrows: ⟩⇓

2. Glucose 6-phosphate is rearranged to convert it to its isomer, fructose 6-phosphate.

3. In this step, still another molecule of ATP is invested in glycolysis. An enzyme transfers a phosphate group from ATP to the sugar. So far, the ATP ledger shows a debit of 2. With phosphate groups on its opposite ends, the sugar is now ready to be split in half.

4. This is the reaction from which glycolysis gets its name. An enzyme cleaves the sugar molecule into two different three-carbon sugars: glyceraldehyde phosphate and dihydroxyacetone phosphate. These two sugars are isomers of each other.

5. Another enzyme catalyzes the reversible conversion between the two three-carbon sugars, and if left alone in a test tube, the reaction reaches equilibrium. This does not happen in the cell, however, because the next enzyme in glycolysis uses only glyceraldehyde phosphate as its substrate and is unreceptive to dihydroxyacetone phosphate. This pulls the equilibrium between the two three-carbon sugars in the direction of glyceraldehyde phosphate, which is removed as fast as it forms. Thus, the net result of steps 4 and 5 is cleavage of a six-carbon sugar into two molecules of glyceraldehyde phosphate; each will progress through the remaining steps of glycolysis.

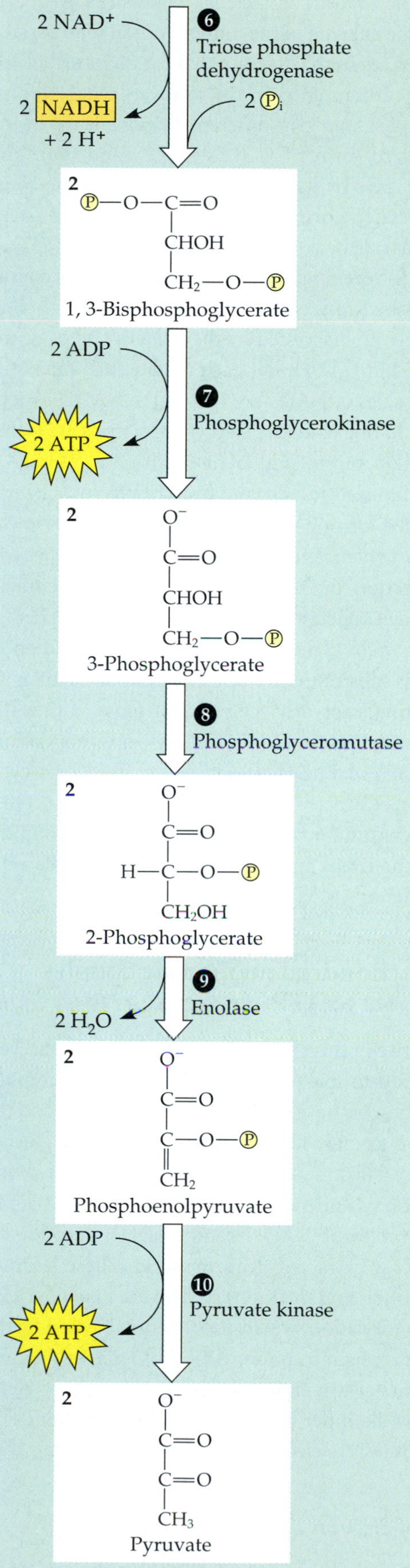

2 NAD^+

2 NADH + 2 H^+

2 Ⓟi

❻ Triose phosphate dehydrogenase

2 Ⓟ—O—C=O | CHOH | CH_2—O—Ⓟ

1, 3-Bisphosphoglycerate

ENERGY-PAYOFF PHASE

❻ An enzyme now catalyzes two sequential reactions while it holds glyceraldehyde phosphate in its active site. First, the sugar is oxidized by the transfer of electrons and H^+ to NAD^+, forming NADH. Here we see in metabolic context the type of redox reaction described earlier. This reaction is very exergonic, and the enzyme uses the energy released to attach a phosphate group to the oxidized substrate, making a product of very high potential energy. The source of the phosphate is inorganic phosphate, which is always present in the cytosol. Notice that the coefficient 2 precedes all molecules in the energy-payoff phase; these steps occur after glucose is split into two three-carbon sugars.

2 ADP

2 ATP

❼ Phosphoglycerokinase

2 O^- | C=O | CHOH | CH_2—O—Ⓟ

3-Phosphoglycerate

❼ Finally, glycolysis produces some ATP. The phosphate group added in the previous step is transferred to ADP in an exergonic reaction. For each glucose molecule that began glycolysis, step 7 produces two molecules of ATP, since every product after the sugar-splitting step (step 4) is doubled. Of course, two ATPs were invested to get sugar ready for splitting. The ATP ledger now stands at zero. By the end of step 7, glucose has been converted to two molecules of 3-phosphoglycerate. This compound is not a sugar. The carbonyl group that characterizes a sugar has been oxidized to a carboxyl group, the hallmark of an organic acid. The sugar was oxidized in step 6, and now the energy made available by that oxidation has been used to make ATP.

❽ Phosphoglyceromutase

2 O^- | C=O | H—C—O—Ⓟ | CH_2OH

2-Phosphoglycerate

❽ Next, an enzyme relocates the remaining phosphate group. This prepares the substrate for the next reaction.

❾ Enolase

2 H_2O

2 O^- | C=O | C—O—Ⓟ || CH_2

Phosphoenolpyruvate

❾ An enzyme forms a double bond in the substrate by extracting a water molecule to form phosphoenolpyruvate, or PEP. This results in the electrons of the substrate being rearranged in such a way that the remaining phosphate bond becomes very unstable, preparing the substrate for the next reaction.

2 ADP

2 ATP

❿ Pyruvate kinase

2 O^- | C=O | C=O | CH_3

Pyruvate

❿ The last reaction of glycolysis produces more ATP by transferring the phosphate group from PEP to ADP. Since this step occurs twice for each glucose molecule, the ATP ledger now shows a net gain of two ATPs. Steps 7 and 10 each produce two ATPs for a total credit of four, but a debt of two ATPs was incurred from steps 1 and 3. Glycolysis has repaid the ATP investment with 100% interest. Additional energy was stored by step 6 in NADH, which can be used to make ATP by oxidative phosphorylation if oxygen is present. In the meantime, glucose has been broken down and oxidized to two molecules of pyruvate, the end-product of the glycolytic pathway.

released during glycolysis. Also note that glycolysis occurs whether or not molecular oxygen (O_2) is present. However, if oxygen *is* present, the energy stored in NADH can be converted to ATP energy during oxidative phosphorylation. And in the presence of oxygen, the chemical energy left in pyruvate can be extracted by the Krebs cycle.

The Krebs cycle completes the energy-yielding oxidation of organic molecules: *a closer look*

9.3 Glycolysis releases less than a quarter of the chemical energy stored in glucose; most of the energy remains stocked in the two molecules of pyruvate. If molecular oxygen is present, the pyruvate enters the mitochondrion, where the enzymes of the Krebs cycle complete the oxidation of the organic fuel.

Upon entering the mitochondrion, pyruvate is first converted to a compound called **acetyl CoA** (FIGURE 9.10). This step, the junction between glycolysis and the Krebs cycle, is accomplished by a multienzyme complex that catalyzes three reactions: ① Pyruvate's carboxyl group, which has little chemical energy, is removed and given off as a molecule of CO_2. (This is the first step in respiration where CO_2 is released.) ② The remaining two-carbon fragment is oxidized to form a compound named acetate (the ionized form of acetic acid). An enzyme transfers the extracted electrons to NAD^+, storing energy in the form of NADH. ③ Finally, coenzyme A, a sulfur-containing compound derived from a B vitamin, is attached to the acetate by an unstable bond that makes the acetyl group (the attached acetate) very reactive. The product of this chemical grooming, acetyl CoA, is now ready to feed its acetate into the Krebs cycle for further oxidation.

The Krebs cycle is named after Hans Krebs, the German-British scientist who was largely responsible for elucidating the pathway in the 1930s. The cycle has eight steps, each catalyzed by a specific enzyme in the mitochondrial matrix (FIGURE 9.11). You can see in the diagram that for each turn of the Krebs cycle, two carbons enter in the relatively reduced form of acetate (step ①), and two different carbons leave in the completely oxidized form of CO_2 (steps ③ and ④). The acetate joins the cycle by its enzymatic addition to the compound oxaloacetate, forming citrate. Subsequent steps decompose the citrate back to oxaloacetate, giving off CO_2 as "exhaust." It is this regeneration of oxaloacetate that accounts for the "cycle" in the Krebs cycle.

Most of the energy harvested by the oxidative steps of the cycle is conserved in NADH. For each acetate that enters the cycle, three molecules of NAD^+ are reduced to NADH—steps ③, ④, and ⑧. In one oxidative step, ⑥, electrons are transferred not to NAD^+, but to a different electron acceptor, FAD (flavin adenine dinucleotide, derived from riboflavin, a B vitamin). The reduced form, $FADH_2$, donates its electrons to the electron transport chain, as does NADH. (However, as we will see in the next section, $FADH_2$ feeds its electrons to the transport chain at a lower energy level than does NADH.) There is also a step in the Krebs cycle, step ⑤, that forms an ATP molecule directly by substrate-level phosphorylation, similar to the ATP-generating steps of glycolysis. But most of the ATP output of respiration results from oxidative phosphorylation, when the NADH and $FADH_2$ produced by the Krebs cycle relay the electrons extracted from food to the electron transport chain. Use FIGURE 9.12, page 158, to review the inputs and outputs of the Krebs cycle before proceeding to the electron transport chain.

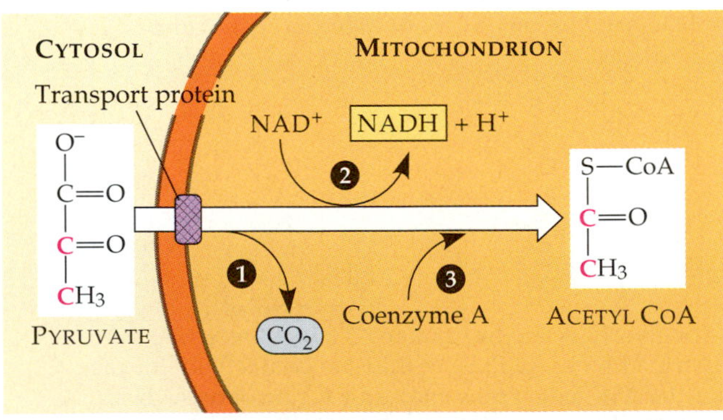

9.3 **FIGURE 9.10 ▪ Conversion of pyruvate to acetyl CoA, the junction between glycolysis and the Krebs cycle.** A protein built into the inner mitochondrial membrane translocates pyruvate from the cytosol into the mitochondrial matrix. Then ① the carboxyl group of pyruvate, already fully oxidized, is removed as a CO_2 molecule, which diffuses out of the cell. ② The remaining two-carbon fragment is oxidized while NAD^+ is reduced to NADH. ③ Finally, the two-carbon acetyl group is attached to coenzyme A (CoA). The coenzyme has a sulfur atom, which attaches to the acetyl fragment by an unstable bond. This activates the acetyl group for the first reaction of the Krebs cycle.

The inner mitochondrial membrane couples electron transport to ATP synthesis: *a closer look*

Our main objective in this chapter is to learn how cells harvest the energy of food to make ATP. But the metabolic components of respiration we have dissected so far, glycolysis and the Krebs cycle, produce only four molecules of ATP per glucose molecule, all by substrate-level phosphorylation: two net ATPs from glycolysis and two ATPs from the Krebs cycle. At this point, molecules of NADH (and $FADH_2$) account for most of the energy extracted from the food. These electron escorts link glycolysis and the Krebs cycle to the machinery for oxidative phosphorylation, which uses energy released by the electron transport chain to power ATP synthesis. In this section, you will first learn how the electron transport chain works, then how the inner membrane of the mitochondrion couples ATP synthesis to electron flow down the chain.

The Pathway of Electron Transport

You learned earlier that the electron transport chain is a collection of molecules embedded in the inner membrane of the

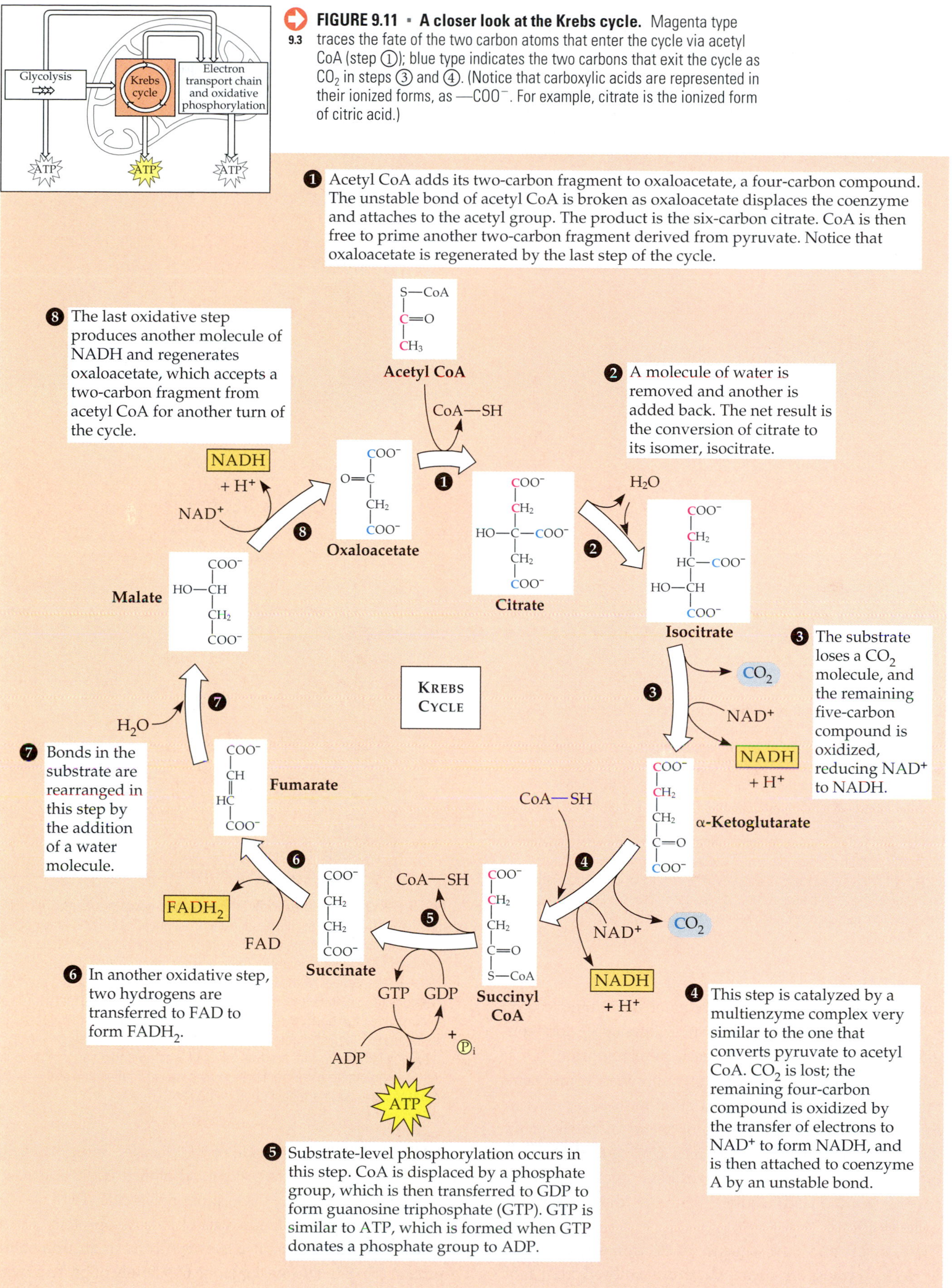

Glycolysis
Krebs cycle
Electron transport chain and oxidative phosphorylation
ATP
ATP
ATP
1 Acetyl CoA adds its two-carbon fragment to oxaloacetate, a four-carbon compound. The unstable bond of acetyl CoA is broken as oxaloacetate displaces the coenzyme and attaches to the acetyl group. The product is the six-carbon citrate. CoA is then free to prime another two-carbon fragment derived from pyruvate. Notice that oxaloacetate is regenerated by the last step of the cycle.
S—CoA
C=O
CH3
Acetyl CoA
CoA—SH
2 A molecule of water is removed and another is added back. The net result is the conversion of citrate to its isomer, isocitrate.
8 The last oxidative step produces another molecule of NADH and regenerates oxaloacetate, which accepts a two-carbon fragment from acetyl CoA for another turn of the cycle.
NADH
+ H+
NAD+
Oxaloacetate
Citrate
H2O
Isocitrate
Malate
KREBS CYCLE
CO2
NAD+
NADH
+ H+
3 The substrate loses a CO2 molecule, and the remaining five-carbon compound is oxidized, reducing NAD+ to NADH.
H2O
7 Bonds in the substrate are rearranged in this step by the addition of a water molecule.
Fumarate
CoA—SH
α-Ketoglutarate
CoA—SH
FADH2
FAD
Succinate
GTP
GDP
Succinyl CoA
NAD+
CO2
NADH
+ H+
ADP
+ Pi
ATP
6 In another oxidative step, two hydrogens are transferred to FAD to form FADH2.
4 This step is catalyzed by a multienzyme complex very similar to the one that converts pyruvate to acetyl CoA. CO2 is lost; the remaining four-carbon compound is oxidized by the transfer of electrons to NAD+ to form NADH, and is then attached to coenzyme A by an unstable bond.
5 Substrate-level phosphorylation occurs in this step. CoA is displaced by a phosphate group, which is then transferred to GDP to form guanosine triphosphate (GTP). GTP is similar to ATP, which is formed when GTP donates a phosphate group to ADP.

FIGURE 9.11 ▪ A closer look at the Krebs cycle. Magenta type traces the fate of the two carbon atoms that enter the cycle via acetyl CoA (step ①); blue type indicates the two carbons that exit the cycle as CO_2 in steps ③ and ④. (Notice that carboxylic acids are represented in their ionized forms, as —COO⁻. For example, citrate is the ionized form of citric acid.)

9.3

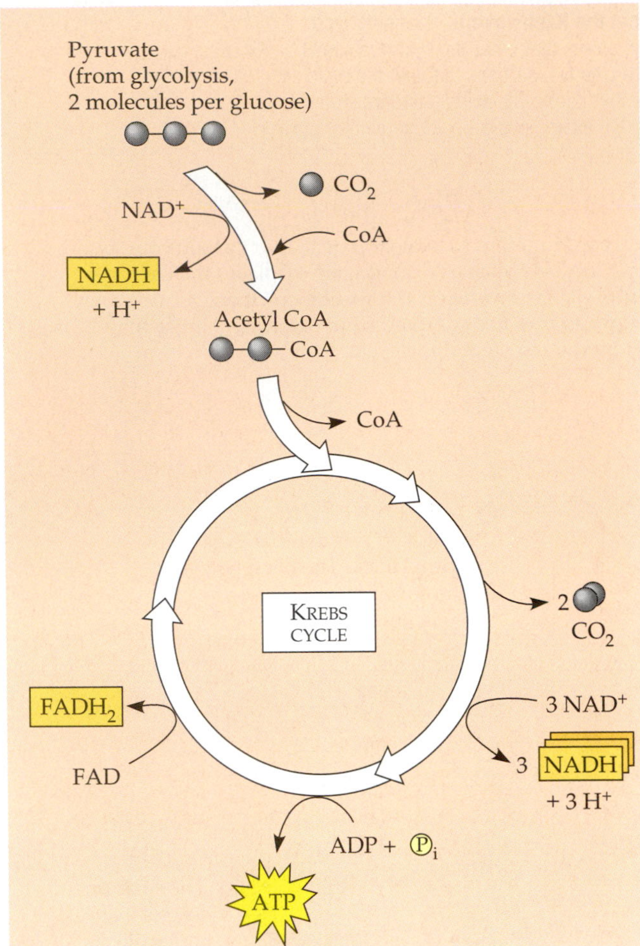

FIGURE 9.12 ▪ A summary of the Krebs cycle. The cycle functions as a metabolic "furnace" that oxidizes organic fuel derived from pyruvate, the product of glycolysis. This diagram summarizes the inputs and outputs as pyruvate is broken down to three molecules of CO_2. (The diagram includes the pre-Krebs cycle conversion of pyruvate to acetyl CoA.) The cycle generates 1 ATP per turn by substrate phosphorylation, but most of the chemical energy is transferred during the redox reactions to NAD^+ and FAD. The reduced coenzymes, NADH and $FADH_2$, shuttle their cargo of high-energy electrons to the electron transport chain, which uses the energy to synthesize ATP by oxidative phosphorylation. (To calculate the inputs and outputs on a "per-glucose" basis, multiply by 2, because each glucose molecule is split during glycolysis into two pyruvate molecules.)

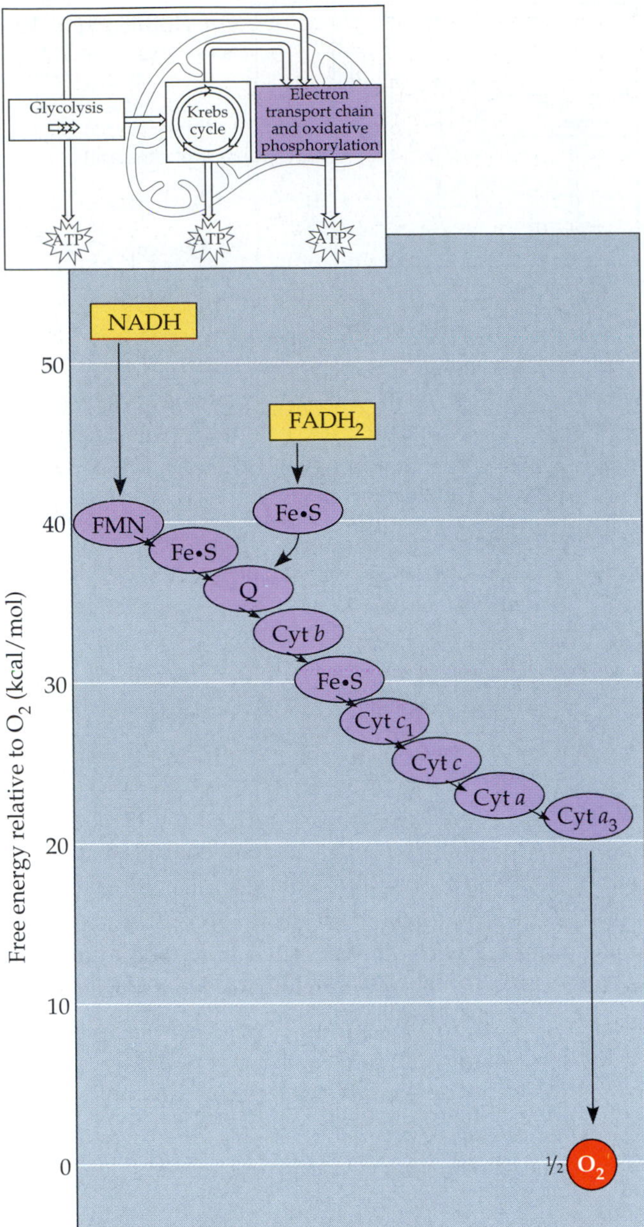

FIGURE 9.13 ▪ A closer look at the electron transport chain. Each member of the chain oscillates between a reduced state and an oxidized state. A component of the chain becomes reduced when it accepts electrons from its "uphill" neighbor (which has a lower affinity for the electrons). Each member of the chain returns to its oxidized form as it passes electrons to its "downhill" neighbor (which has a greater affinity for the electrons). At the bottom of the chain is oxygen, which is *very* electronegative. The overall energy drop for electrons traveling from NADH to oxygen is 53 kcal/mol, but this fall is broken up into a series of smaller steps by the electron transport chain. The various molecules of the electron transport chain are described in the text.

mitochondrion. The folding of the inner membrane to form cristae increases its surface area, providing space for thousands of copies of the chain in each mitochondrion. (Once again, we see that structure fits function.) Most components of the chain are proteins. Tightly bound to these proteins are prosthetic groups, nonprotein components essential for the catalytic functions of certain enzymes. During electron transport along the chain, these prosthetic groups alternate between reduced and oxidized states as they accept and donate electrons.

FIGURE 9.13 traces the sequence of electron transfers along the electron transport chain. Electrons removed from food during glycolysis and the Krebs cycle are transferred by NADH to the first molecule of the electron transport chain. This molecule is a flavoprotein, so named because it has a prosthetic group called flavin mononucleotide (FMN in FIGURE 9.13). In the next redox reaction, the flavoprotein returns to its oxidized form as it passes electrons to an iron-sulfur protein (Fe•S in FIGURE 9.13), one of a family of proteins with both iron and sulfur tightly bound. The iron-sulfur protein then passes the electrons to a compound called ubiquinone (Q in FIGURE 9.13). This electron carrier is a lipid, the only member of the electron transport chain that is not a protein.

Most of the remaining electron carriers between Q and oxygen are proteins called **cytochromes (cyt)**. Their prosthetic group, called a heme group, has four organic rings surrounding a single iron atom. It is similar to the iron-containing prosthetic group found in hemoglobin, the red protein of blood that transports oxygen. But the iron of cytochromes transfers electrons, not oxygen. The electron transport chain has several types of cytochromes, each a different protein with a heme group. The last cytochrome of the chain, cyt a_3, passes its electrons to oxygen, which also picks up a pair of hydrogen ions from the aqueous solution to form water. (An oxygen atom is represented in FIGURE 9.13 as ½ O_2 to emphasize that the electron transport chain reduces molecular oxygen, O_2, not individual oxygen atoms. For every two NADH molecules, one O_2 molecule is reduced to two molecules of water.)

Another source of electrons for the transport chain is $FADH_2$, the other reduced product of the Krebs cycle. Notice in FIGURE 9.13 that $FADH_2$ adds its electrons to the electron transport chain at a lower energy level than NADH does. Consequently, the electron transport chain provides about one-third less energy for ATP synthesis when the electron donor is $FADH_2$ rather than NADH.

The electron transport chain makes no ATP directly. Its function is to ease the fall of electrons from food to oxygen, breaking a large free-energy drop into a series of smaller steps that release energy in manageable amounts. How does the mitochondrion couple this electron transport and energy release to ATP synthesis? The answer is a mechanism called chemiosmosis.

Chemiosmosis: The Energy-Coupling Mechanism

Populating the inner membrane of the mitochondrion are many copies of a protein complex called an **ATP synthase**, the enzyme that actually makes ATP (FIGURE 9.14). It works like an ion pump running in reverse. Recall from Chapter 8 that ion pumps use ATP as an energy source to transport ions against their gradients. In the reverse of that process, an ATP synthase uses the energy of an existing ion gradient to power ATP synthesis. The ion gradient that drives oxidative phosphorylation is a proton (hydrogen ion) gradient; that is, the power source for the ATP synthase is a difference in the concentration of H^+ on opposite sides of the inner mitochondrial membrane. We can also think of this gradient as a difference in pH, since pH is a measure of H^+ concentration.

How does the mitochondrial membrane generate and
9.4 maintain an H^+ gradient? That is the function of the electron transport chain (FIGURE 9.15, p. 160). The chain is an energy converter that uses the exergonic flow of electrons to pump H^+ across the membrane, from the matrix into the intermembrane space. The H^+ leaks back across the membrane, diffusing down its gradient. But the ATP synthases are the only patches of the membrane that are freely permeable to H^+. The ions pass through a channel in an ATP synthase, and the complex of proteins functions as a mill that harnesses the exergonic flow of H^+ to drive the phosphorylation of ADP. Thus, an H^+ gradient couples the redox reactions of the electron transport chain to ATP synthesis. This coupling mechanism for oxidative phosphorylation is called **chemiosmosis** (Gr. *osmos*, "push"), a term that highlights the relationship between chemical reactions and transport across a membrane. We have previously used the word *osmosis* in discussing water transport, but here the word refers to the flowing of H^+ across a membrane.

If you have followed this complex story of chemiosmosis so far, you should have at least two questions. How does the electron transport chain pump hydrogen ions? And how does the ATP synthase use H^+ backflow to make ATP? On the first problem, researchers have found that certain members of the electron transport chain accept and release protons (H^+) along with electrons, whereas other carriers transport only electrons. Therefore, at certain steps along the chain, electron transfers cause H^+ to be taken up and released back into the surrounding solution. The electron carriers are spatially

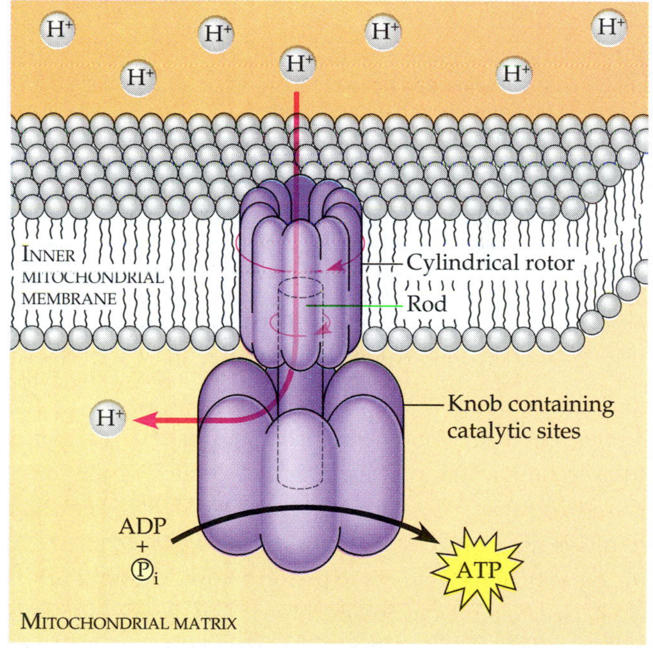

FIGURE 9.14 ▪ ATP synthase, a molecular machine. This protein complex, which uses the energy of an H^+ gradient to drive ATP synthesis, resides in mitochondrial and chloroplast membranes and in the plasma membranes of prokaryotes. ATP synthase has three main parts, each consisting of a number of protein subunits: a cylindrical component within the membrane, a protruding knob (which, in mitochondria, is in the matrix), and a rod (or "stalk") connecting the other two parts. The cylinder is a rotor that spins clockwise when H^+ flows through it down a gradient. The attached rod also spins, activating catalytic sites in the knob, the component that joins inorganic phosphate to ADP to make ATP.

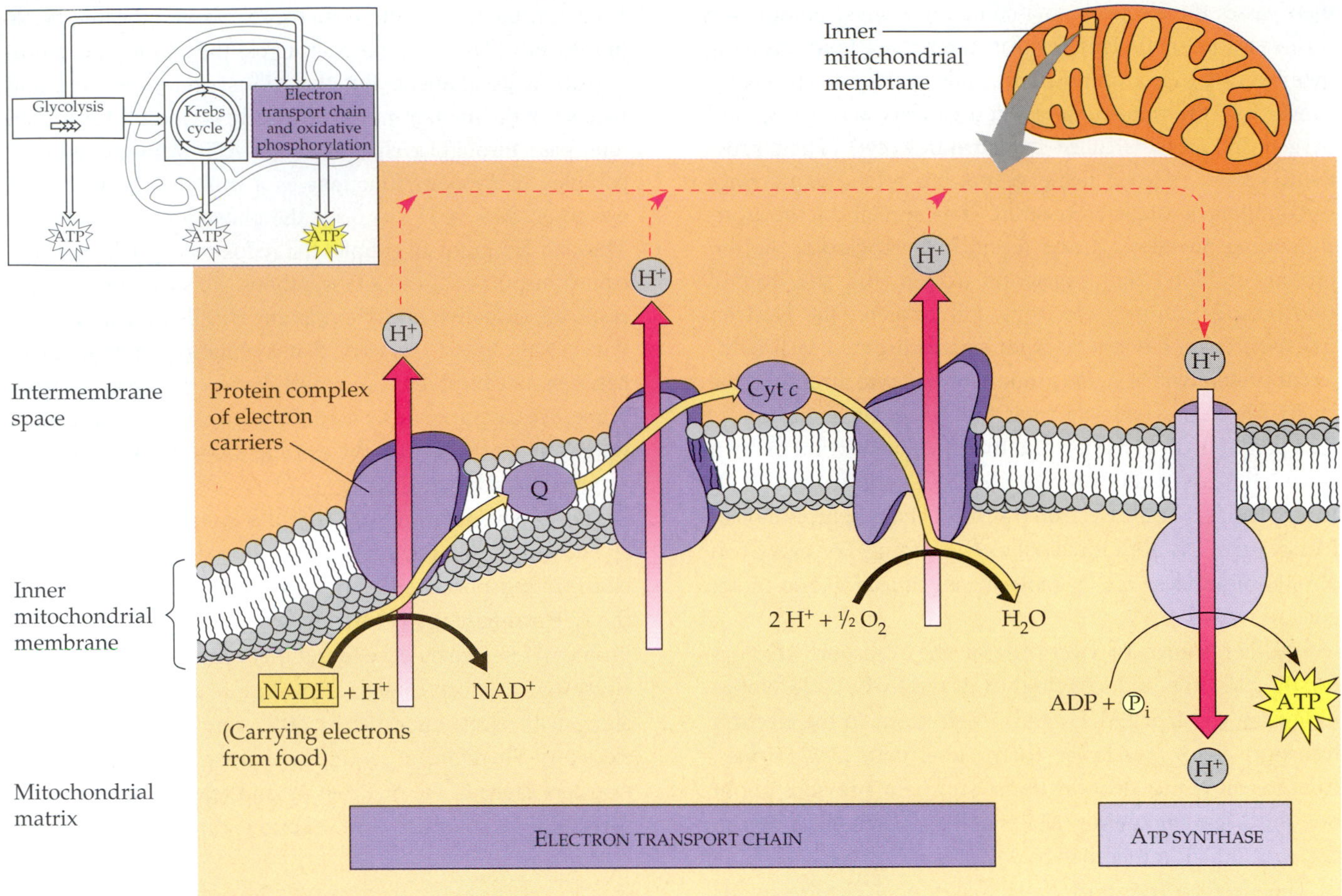

FIGURE 9.15 ▪ Chemiosmosis: How the mitochondrial membrane couples electron transport to oxidative phosphorylation. 9.4 NADH shuttles high-energy electrons extracted from food during glycolysis and the Krebs cycle to an electron transport chain, which is built into the inner mitochondrial membrane. The gold arrow in this diagram traces the transport of electrons, which pass to oxygen at the "downhill" end of the chain to form water. Most of the cytochromes and other electron carriers of the chain (see FIGURE 9.13) are collected into three complexes, each represented here by a purple blob embedded in the membrane. Two mobile carriers, ubiquinone (Q) and cytochrome *c,* move rapidly along the membrane, ferrying electrons between the three large complexes. As each complex of the chain accepts and then donates electrons, it pumps hydrogen ions (protons) from the mitochondrial matrix into the intermembrane space (magenta arrows trace H^+ transport). Thus, chemical energy harvested from food is transformed into a proton-motive force, a gradient of H^+ across the membrane. The hydrogen ions complete their circuit by flowing down their gradient through an H^+ channel in an ATP synthase, another protein complex built into the membrane. The ATP synthase harnesses the proton-motive force to phosphorylate ADP, forming ATP. (This is called oxidative phosphorylation because it is driven by the exergonic transfer of electrons from food to oxygen.) This mechanism for energy coupling—the use of an H^+ gradient (proton-motive force) to transfer energy from redox reactions to cellular work (ATP synthesis, in this case)—is called chemiosmosis.

arranged in the membrane in such a way that H^+ is accepted from the mitochondrial matrix and deposited in the intermembrane space (see FIGURE 9.15). The H^+ gradient that results is referred to as a **proton-motive force**, emphasizing the capacity of the gradient to perform work. The force drives H^+ back across the membrane through the specific H^+ channels provided by ATP synthase complexes.

Scientists are also learning how the flow of H^+ through ATP synthase powers ATP generation. This new understanding of ATP synthase function has emerged from studies of its molecular structure (see FIGURE 9.14). The ATP synthase complex consists of three main parts: a cylinder within the inner mitochondrial membrane, a knob protruding into the mitochondrial matrix, and an internal rod connecting the two. When hydrogen ions flow through the cylinder down their gradient, they cause the cylinder and the attached rod to rotate, much as a rushing stream turns a water wheel. The spinning rod brings about a conformational change in the knob, activating catalytic sites where ADP and inorganic phosphate combine to make ATP.

In general terms, *chemiosmosis is an energy-coupling mechanism that uses energy stored in the form of an H^+ gradient to drive cellular work.* In mitochondria, the energy for gradient formation comes from exergonic redox reactions, and ATP synthesis is the work performed. But chemiosmosis also occurs elsewhere, and in other variations. Chloroplasts use chemisosmosis to generate ATP during photosynthesis; in these organelles, light (rather than chemical energy) drives both electron flow down an electron transport chain and H^+-gradient formation. Prokaryotes, which lack both mitochondria and chloroplasts, generate H^+ gradients across their plasma membranes. They then tap the proton-motive force not only to make ATP but also to pump nutrients and waste products across the membrane, and to rotate their flagella.

Experiments with bacteria first led British biochemist Peter Mitchell to propose chemiosmosis as an energy-coupling mechanism in 1961. Nearly two decades later, after many scientists had confirmed the centrality of chemiosmosis in energy conversions within bacteria, mitochondria, and chloroplasts, Mitchell was awarded the Nobel Prize. Chemiosmosis has helped unify the study of bioenergetics.

Cellular respiration generates many ATP molecules for each sugar molecule it oxidizes: *a review*

Now that we have looked more closely at the key processes of cellular respiration, let's return to its overall function: harvesting the energy of food for ATP synthesis.

During respiration, most energy flows in this sequence: Glucose ⟶ NADH ⟶ electron transport chain ⟶ proton-motive force ⟶ ATP. We can do some bookkeeping to calculate the ATP profit when cellular respiration oxidizes a molecule of glucose to six molecules of carbon dioxide. The three main departments of this metabolic enterprise are glycolysis, the Krebs cycle, and the electron transport chain, which drives oxidative phosphorylation. FIGURE 9.16 gives a detailed accounting of the ATP yield per glucose molecule oxidized. The tally adds the few molecules of ATP produced directly by substrate-level phosphorylation during glycolysis and the Krebs cycle to the many more molecules of ATP generated by oxidative phosphorylation. Each NADH that transfers a pair of electrons from food to the electron transport chain contributes enough to the proton-motive force to generate a maximum of about three ATPs. (The average ATP yield per NADH is probably between two and three; we are rounding off to three here to simplify the bookkeeping.) The Krebs cycle also supplies electrons to the electron transport chain via $FADH_2$, but each molecule of this electron carrier is worth a maximum of only about two molecules of ATP.

In some eukaryotic cells, this lower ATP yield per electron pair also applies to the NADH produced by glycolysis in the cytosol. The mitochondrial inner membrane is impermeable to NADH, so NADH in the cytosol is segregated from the machinery of oxidative phosphorylation. The two electrons of NADH captured in glycolysis must be conveyed into the mitochondrion by one of several electron-shuttle systems. Depending on which shuttle is operating, the electrons are passed either to NAD^+ or to FAD. If the electrons are passed to FAD, only about 2 ATP can result from each cytosolic

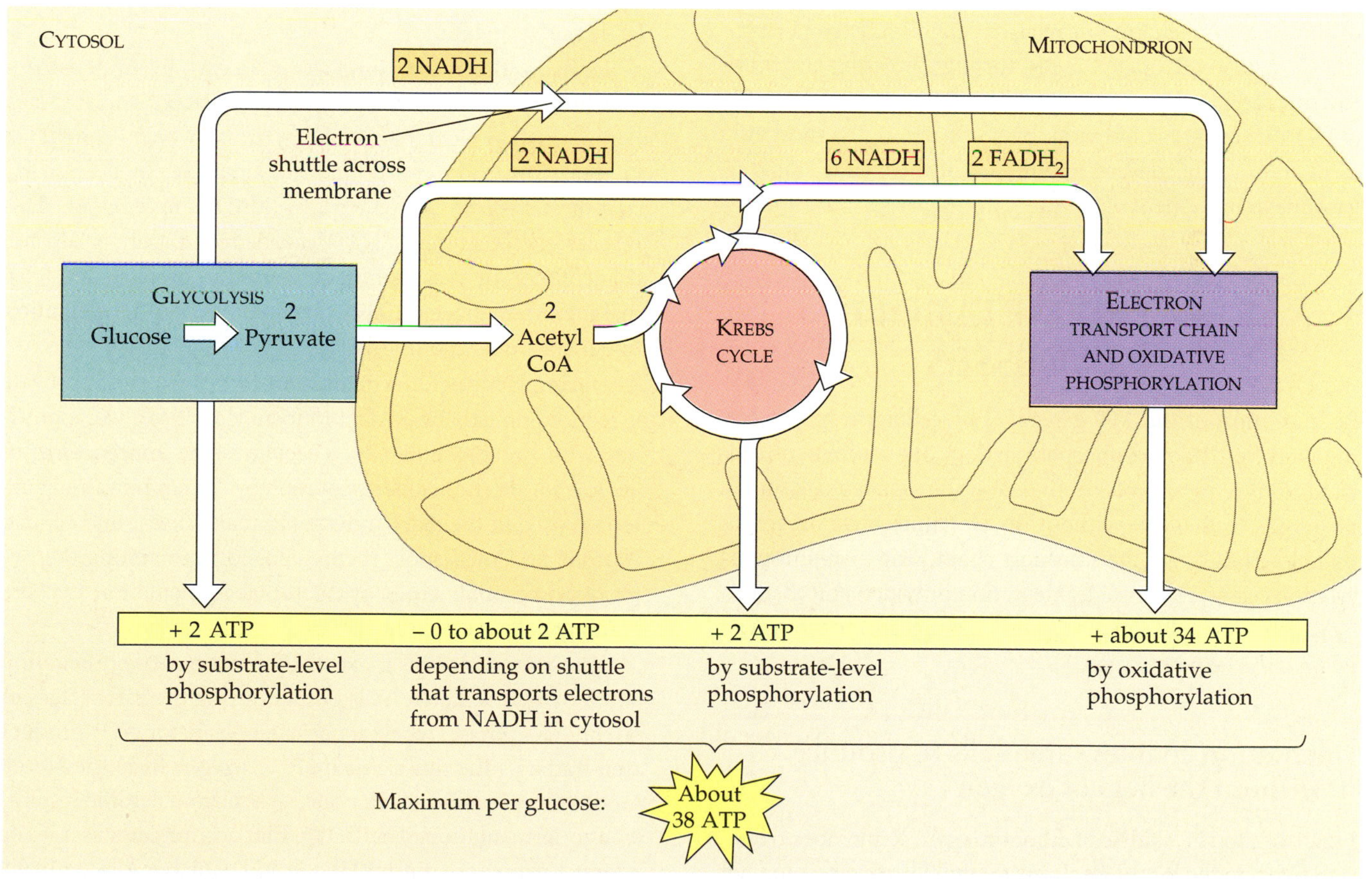

FIGURE 9.16 ▪ Review: Each molecule of glucose yields many ATP molecules during cellular respiration. The text explains why the yield of 38 ATP per glucose is only an estimate of the maximum output.

$NADH_2$. If passed to mitochondrial NAD^+, the yield is closer to 3 ATP.

Assuming that the more energy-efficient type of shuttle is active, we can add a maximum of 34 ATP produced by oxidative phosphorylation to the net of 4 ATP from substrate-level phosphorylation to arrive at a bottom line of 38 ATP. This is only an estimate of the maximum ATP yield from a glucose molecule and is undoubtedly somewhat high. One variable that reduces ATP yield is the use of the proton-motive force (generated by the redox reactions of respiration) to drive other kinds of work. For example, the proton-motive force powers the mitochondrion's uptake of pyruvate from the cytoplasm. Also, by rounding up the number of ATP molecules produced per NADH and $FADH_2$ to three and two, respectively, we have inflated the ATP yield of respiration by at least 10%.

We can now make a rough estimate of the efficiency of respiration—that is, the percentage of chemical energy stored in glucose that has been restocked in ATP. Recall that the complete oxidation of a mole of glucose releases 686 kcal of energy ($\Delta G = -686$ kcal/mol). Phosphorylation of ADP to form ATP stores at least 7.3 kcal per mole of ATP (p. 90 explains why this number is probably higher under cellular conditions). Therefore, the efficiency of respiration is 7.3 times 38 (maximum ATP yield per glucose) divided by 686, or about 40%. The rest of the stored energy is lost as heat. We use some of this heat to maintain our relatively high body temperature (37°C), and we dissipate the rest through sweating and other cooling mechanisms. Cellular respiration is remarkably efficient in its energy conversion. By comparison, the most efficient automobile converts about 25% of the energy stored in gasoline to movement of the car.

RELATED METABOLIC PROCESSES

Because most of the ATP generated by cellular respiration is the work of oxidative phosphorylation, our estimate of ATP yield from respiration is contingent upon an adequate supply of oxygen to the cell. Without the electronegative oxygen to pull electrons down the transport chain, oxidative phosphorylation ceases. However, fermentation provides a mechanism by which some cells can oxidize organic fuel and generate ATP *without* the help of oxygen.

Fermentation enables some cells to produce ATP without the help of oxygen

How can food be oxidized without oxygen? Remember, oxidation refers to the loss of electrons to *any* electron acceptor, not just to oxygen. Glycolysis oxidizes glucose to two molecules of pyruvate. The oxidizing agent of glycolysis is NAD^+, *not* oxygen. Overall, glycolysis is exergonic, and some of the energy made available is used to produce two ATPs (net) by substrate-level phosphorylation. If oxygen *is* present, then additional ATP is made by oxidative phosphorylation when NADH passes electrons removed from glucose to the electron transport chain. But glycolysis generates two ATPs whether oxygen is present or not—that is, whether conditions are **aerobic** or **anaerobic** (Gr. *aer,* "air," and *bios,* "life"; the prefix *an-* means "without").

Anaerobic catabolism of organic nutrients can occur by fermentation, as mentioned at the beginning of the chapter. Fermentation is an extension of glycolysis that can generate ATP solely by substrate-level phosphorylation—as long as there is a sufficient supply of NAD^+ to accept electrons during the oxidation step of glycolysis. Without some mechanism to recycle NAD^+ from NADH, glycolysis would soon deplete the cell's pool of NAD^+ and shut itself down for lack of an oxidizing agent. Under aerobic conditions, NAD^+ is recycled productively from NADH by the transfer of electrons to the electron transport chain. The anaerobic alternative is to transfer electrons from NADH to pyruvate, the end-product of glycolysis.

Fermentation consists of glycolysis plus reactions that regenerate NAD^+ by transferring electrons from NADH to pyruvate or derivatives of pyruvate. There are many types of fermentation, differing in the waste products formed from pyruvate. Two common types are alcohol fermentation and lactic acid fermentation.

In **alcohol fermentation** (FIGURE 9.17a), pyruvate is converted to ethanol (ethyl alcohol) in two steps. The first step releases carbon dioxide from the pyruvate, which is converted to the two-carbon compound acetaldehyde. In the second step, acetaldehyde is reduced by NADH to ethanol. This regenerates the supply of NAD^+ needed for glycolysis. Alcohol fermentation by yeast, a fungus, is used in brewing and wine making. Many bacteria also carry out alcohol fermentation under anaerobic conditions.

During **lactic acid fermentation** (FIGURE 9.17b), pyruvate is reduced directly by NADH to form lactate as a waste product, with no release of CO_2. (Lactate is the ionized form of lactic acid.) Lactic acid fermentation by certain fungi and bacteria is used in the dairy industry to make cheese and yogurt. Acetone and methanol (methyl alcohol) are among the by-products of other types of microbial fermentation that are commercially important.

Human muscle cells make ATP by lactic acid fermentation when oxygen is scarce. This occurs during the early stages of strenuous exercise, when sugar catabolism for ATP production outpaces the muscle's supply of oxygen from the blood. Under these conditions, the cells switch from aerobic respiration to fermentation. The lactate that accumulates as a waste product may cause muscle fatigue and pain, but it is gradually carried away by the blood to the liver. Lactate is converted back to pyruvate by liver cells.

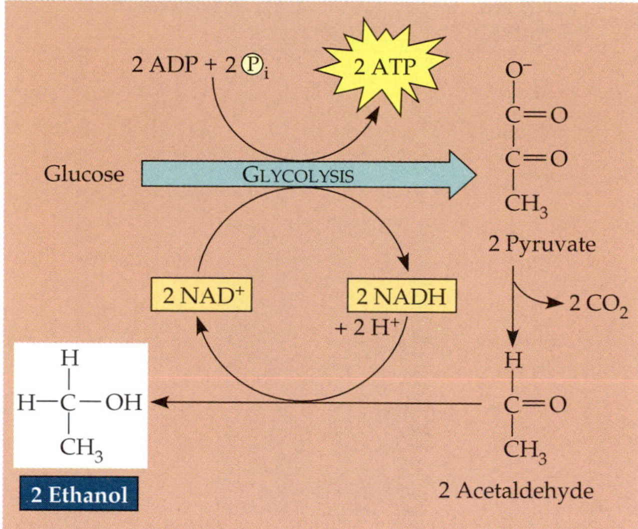

(a) Alcohol fermentation

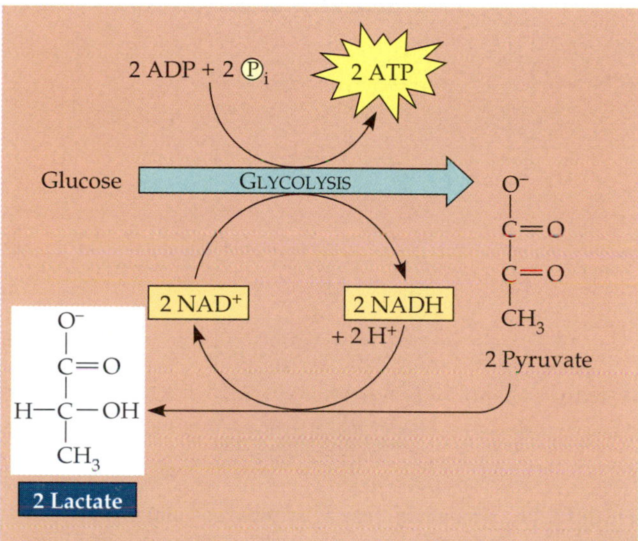

(b) Lactic acid fermentation

FIGURE 9.17 ▪ Fermentation. Pyruvate, the end-product of glycolysis, serves as an electron acceptor for oxidizing NADH back to NAD^+. The NAD^+ can then be reused to oxidize sugar during glycolysis, which yields two net molecules of ATP by substrate-level phosphorylation. Two of the common waste products formed from fermentation are **(a)** ethanol and **(b)** lactate, the ionized form of lactic acid.

Fermentation and Respiration Compared

Fermentation and cellular respiration are anaerobic and aerobic alternatives, respectively, for producing ATP by harvesting the chemical energy of food. Both pathways use glycolysis to oxidize glucose and other organic fuels to pyruvate, with a net production of 2 ATP by substrate phosphorylation. And in both fermentation and respiration, NAD^+ is the oxidizing agent that accepts electrons from food during glycolysis. A key difference is the contrasting mechanisms for oxidizing NADH back to NAD^+, which is required to sustain glycolysis. In fermentation, the final electron acceptor is an organic molecule such as pyruvate (lactic acid fermentation) or acetaldehyde (alcohol fermentation). In respiration, by contrast, the final acceptor for electrons from NADH is oxygen. This not only regenerates the NAD^+ required for glycolysis but pays an ATP bonus when the stepwise electron transport from NADH to oxygen drives oxidative phosphorylation. An even bigger ATP payoff comes from the oxidation of pyruvate in the Krebs cycle, which is unique to respiration. Without oxygen, the energy still stored in pyruvate is unavailable to the cell. Thus, cellular respiration harvests much more energy from each sugar molecule than fermentation can. In fact, respiration yields as much as 19 times more ATP per glucose molecule than does fermentation—38 ATP for respiration, compared to 2 ATP produced by substrate-level phosphorylation in fermentation.

Some organisms, including yeasts and many bacteria, can make enough ATP to survive by either fermentation or respiration. Such species are called **facultative anaerobes.** On the cellular level, our muscle cells behave as facultative anaerobes. In a facultative anaerobe, pyruvate is a fork in the metabolic road that leads to two alternative catabolic routes (FIGURE 9.18). Under aerobic conditions, pyruvate can be converted to acetyl CoA, and oxidation continues in the Krebs cycle. Under anaerobic conditions, pyruvate is diverted from the Krebs

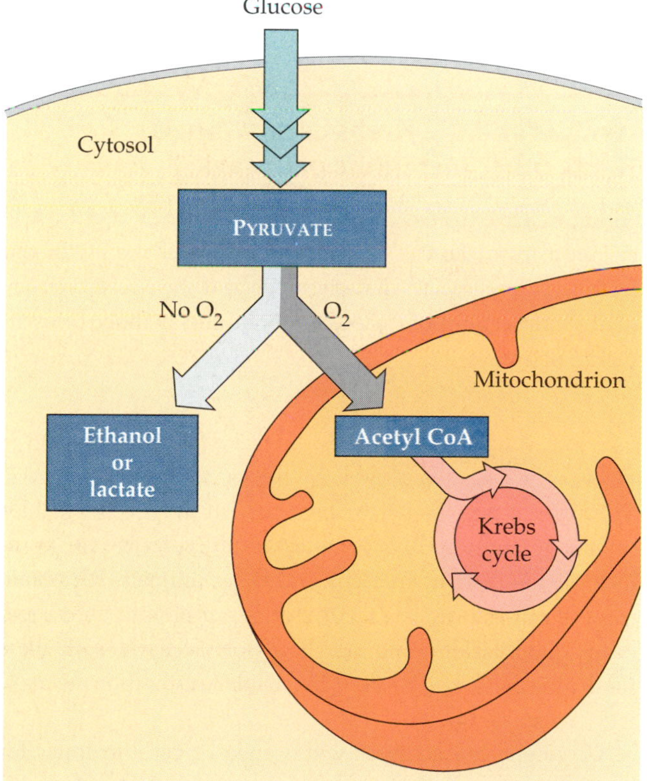

FIGURE 9.18 ▪ Pyruvate as a key juncture in catabolism. Glycolysis is common to fermentation and respiration. The end-product of glycolysis, pyruvate, represents a fork in the catabolic pathways of glucose oxidation. In a cell capable of both respiration and fermentation, pyruvate is committed to one of those two pathways, usually depending on whether or not oxygen is present.

cycle, serving instead as an electron acceptor to recycle NAD^+. To make the same amount of ATP, a facultative anaerobe would have to consume sugar at a much faster rate when fermenting than it would when respiring.

The Evolutionary Significance of Glycolysis

The role of glycolysis in both fermentation and respiration has an evolutionary basis. Ancient prokaryotes probably used glycolysis to make ATP long before oxygen was present in the Earth's atmosphere. The oldest known fossils of bacteria date back over 3.5 billion years, but appreciable quantities of oxygen probably did not begin to accumulate in the atmosphere until about 2.5 billion years ago. (According to fossil evidence, the cyanobacteria that produce O_2 as a by-product of photosynthesis had evolved by then.) Therefore, the first prokaryotes may have generated ATP exclusively from glycolysis, which does not require oxygen. In addition, glycolysis is the most widespread metabolic pathway, which suggests that it evolved very early in the history of life. The cytosolic location of glycolysis also implies great antiquity; the pathway does not require any of the membrane-enclosed organelles of the eukaryotic cell, which evolved nearly 2 billion years after the prokaryotic cell. Glycolysis is a metabolic heirloom from the earliest cells that continues to function in fermentation and as the first stage in the breakdown of organic molecules by respiration.

Glycolysis and the Krebs cycle connect to many other metabolic pathways

So far, we have treated the oxidative breakdown of glucose in isolation from the cell's overall metabolic economy. In this section, you will learn that glycolysis and the Krebs cycle are major intersections of various catabolic and anabolic (biosynthetic) pathways.

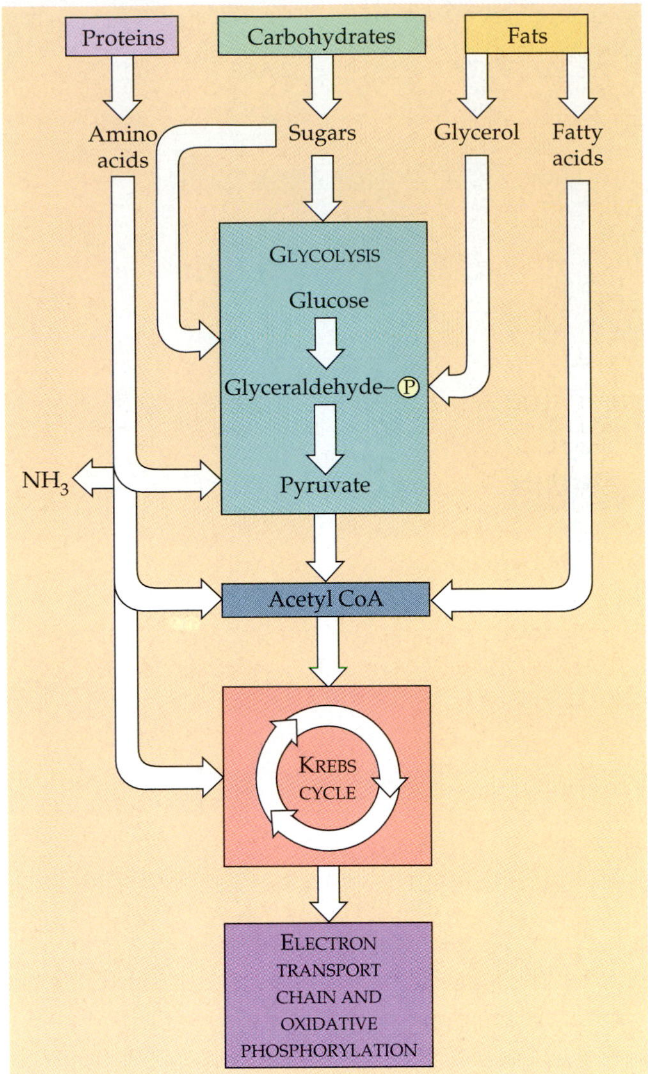

FIGURE 9.19 ▪ The catabolism of various food molecules. Carbohydrates, fats, and proteins can all be used as fuel for cellular respiration. Monomers of these food molecules enter glycolysis or the Krebs cycle at various points. Glycolysis and the Krebs cycle are catabolic funnels through which electrons from all kinds of food molecules flow on their exergonic fall to oxygen.

The Versatility of Catabolism

Throughout this chapter we have used glucose as the fuel for cellular respiration. But free glucose molecules are not common in the diets of humans and other animals. We obtain most of our calories in the form of fats, proteins, sucrose and other disaccharides, and starch, a polysaccharide. All these food molecules can be used by cellular respiration to make ATP (FIGURE 9.19).

Glycolysis can accept a wide range of carbohydrates for catabolism. In the digestive tract, starch is hydrolyzed to glucose, which can then be broken down in the cells by glycolysis and the Krebs cycle. Similarly, glycogen, the polysaccharide that humans and many other animals store in their liver and muscle cells, can be hydrolyzed to glucose between meals as fuel for respiration. The digestion of disaccharides, including sucrose, provides glucose and other monosaccharides as fuel for respiration.

Proteins can also be used for fuel, but first they must be digested to their constituent amino acids. Many of the amino acids, of course, are used by the organism to build new proteins. Amino acids present in excess are converted by enzymes to intermediates of glycolysis and the Krebs cycle. Before amino acids can feed into glycolysis or the Krebs cycle, their amino groups must be removed, a process called deamination. The nitrogenous refuse is excreted from the animal in the form of ammonia, urea, or other waste products.

Catabolism can also harvest energy stored in fats obtained either from food or from storage cells in the body. After fats are digested, the glycerol is converted to glyceraldehyde phosphate, an intermediate of glycolysis. Most of the energy of a fat is stored in the fatty acids. A metabolic sequence called **beta oxidation** breaks the fatty acids down to two-carbon fragments, which enter the Krebs cycle as acetyl CoA. Fats make excellent fuel. A gram of fat oxidized by respiration produces more than twice as much ATP as a gram of carbohydrate. Unfortunately, this also means that a dieter must be patient while using fat stored in the body, because so many calories are stockpiled in each gram of fat.

Biosynthesis (Anabolic Pathways)

Cells need substance as well as energy. Not all the organic molecules of food are destined to be oxidized as fuel to make ATP. In addition to calories, food must also provide the carbon skeletons that cells require to make their own molecules. Some organic monomers obtained from digestion can be used directly. For example, as previously mentioned, amino acids from the hydrolysis of proteins in food can be incorporated into the organism's own proteins. Often, however, the body needs specific molecules that are not present as such in food. Compounds formed as intermediates of glycolysis and the Krebs cycle can be diverted into anabolic pathways as precursors from which the cell can synthesize the molecules it requires. For example, humans can make about half of the 20 amino acids in proteins by modifying compounds siphoned away from the Krebs cycle. Also, glucose can be made from pyruvate, and fatty acids can be synthesized from acetyl CoA. Of course, these anabolic or biosynthetic pathways do not generate ATP, but instead consume it.

In addition, glycolysis and the Krebs cycle function as metabolic interchanges that enable our cells to convert some kinds of molecules to others as we need them. For instance, carbohydrates and proteins can be converted to fats through intermediates of glycolysis and the Krebs cycle. If we eat more food than we need, we store fat even if our diet is fat-free. Metabolism is remarkably versatile and adaptable.

Feedback mechanisms control cellular respiration

Basic principles of supply and demand regulate the metabolic economy. The cell does not waste energy making more of a particular substance than it needs. If there is a glut of a certain amino acid, for example, the anabolic pathway that synthesizes that amino acid from an intermediate of the Krebs cycle is switched off. The most common mechanism for this control is feedback inhibition: The end-product of the anabolic pathway inhibits the enzyme that catalyzes an early step of the pathway (see FIGURE 6.16). This prevents the needless diversion of key metabolic intermediates from uses that are more urgent.

The cell also controls its catabolism. If the cell is working hard and its ATP concentration begins to drop, respiration speeds up. When there is plenty of ATP to meet demand, respiration slows down, sparing valuable organic molecules for other functions. Again, control is based mainly on regulating the activity of enzymes at strategic points in the catabolic pathway. One important switch is phosphofructokinase, the enzyme that catalyzes step 3 of glycolysis (see FIGURE 9.9). That is the earliest step that commits substrate irreversibly to the glycolytic pathway. By controlling the rate of this step, the cell can speed up or slow down the entire catabolic process; phosphofructokinase is thus the pacemaker of respiration (FIGURE 9.20).

An allosteric enzyme with receptor sites for specific inhibitors and activators, phosphofructokinase is inhibited by ATP and stimulated by AMP, which the cell derives from ADP (see p. 95). As ATP accumulates, inhibition of the enzyme slows down glycolysis. The enzyme becomes active again as

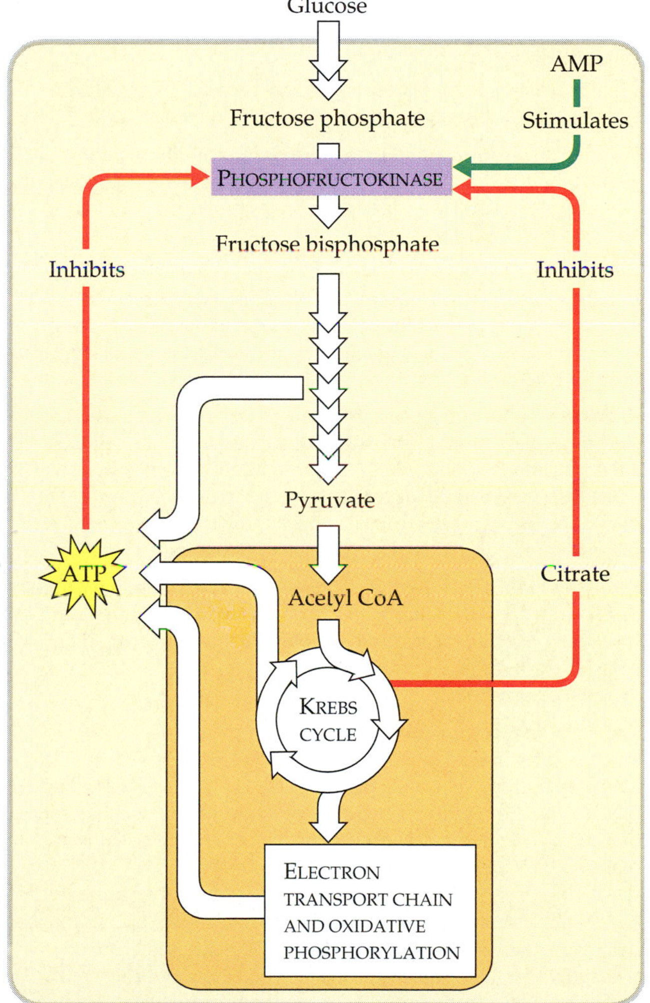

FIGURE 9.20 ▪ The control of cellular respiration. Allosteric enzymes at certain points in the respiratory pathway respond to inhibitors and activators that help set the pace of glycolysis and the Krebs cycle. Phosphofructokinase, the enzyme that catalyzes step 3 of glycolysis, is one such enzyme. It is stimulated by AMP that is derived from ADP, but it is inhibited by ATP and by citrate. This feedback regulation adjusts the rate of respiration as the cell's catabolic and anabolic demands change.

cellular work converts ATP to ADP (and AMP) faster than ATP is being regenerated. Phosphofructokinase is also sensitive to citrate, the first product of the Krebs cycle. If citrate accumulates in mitochondria, some of it passes into the cytosol and inhibits phosphofructokinase. This mechanism helps synchronize the rates of glycolysis and the Krebs cycle. As citrate accumulates, glycolysis slows down, and the supply of acetate to the Krebs cycle decreases. If citrate consumption increases, either because of a demand for more ATP or because anabolic pathways are draining off intermediates of the Krebs cycle, glycolysis accelerates and meets the demand. Metabolic balance is augmented by the control of other enzymes at other key locations in glycolysis and the Krebs cycle. Cells are thrifty, expedient, and responsive in their metabolism.

■ ■ ■

Examine FIGURE 9.1 again to put cellular respiration into the broader context of energy flow and chemical cycling in ecosystems. The energy that keeps us alive is *released,* but not *produced,* by cellular respiration. We are tapping energy that was stored in food by photosynthesis. In the next chapter you will learn how photosynthesis captures light and converts it to chemical energy.

CHAPTER REVIEW

REVIEW OF KEY CONCEPTS

(with page numbers and key figures)

PRINCIPLES OF ENERGY HARVEST

Chemical elements important to life are recycled by respiration and photosynthesis, but energy is not (p. 147, FIGURE 9.1).

- **Cellular respiration and fermentation are catabolic (energy-yielding) pathways (pp. 147–148)** The breakdown of glucose and other organic fuels to simpler products is exergonic, yielding energy for ATP synthesis.
- **Cells recycle the ATP they use for work (p. 148, FIGURE 9.2)** ATP transfers phosphate groups to various substrates, priming them to do work. To keep working, a cell must regenerate ATP. Starting with glucose or another organic fuel, and using O_2, cellular respiration yields H_2O, CO_2, and energy in the form of ATP and heat.
- **Redox reactions release energy when electrons move closer to electronegative atoms (pp. 148–150, FIGURE 9.3)** The cell taps the energy stored in food molecules through redox reactions, in which one substance partially or totally shifts electrons to another. The substance receiving electrons is reduced; the substance losing electrons is oxidized.
- **Electrons "fall" from organic molecules to oxygen during cellular respiration (p. 150)** Glucose ($C_6H_{12}O_6$) is oxidized to CO_2, and O_2 is reduced to H_2O. Electrons lose potential energy during their transfer from organic compounds to oxygen, and this energy drives ATP synthesis.
- **The "fall" of electrons during respiration is stepwise, via NAD^+ and an electron transport chain (pp. 150–152, FIGURE 9.5)** Electrons from food are usually passed to NAD^+, reducing it to NADH. NADH passes them to an electron transport chain, which conducts them to O_2 in energy-releasing steps. The energy released is used to make ATP by oxidative phosphorylation.

THE PROCESS OF CELLULAR RESPIRATION

9.1 **Respiration involves glycolysis, the Krebs cycle, and electron transport (pp. 152–153, FIGURE 9.6)** Glycolysis and the Krebs cycle supply electrons (via NADH) to the transport chain, which drives oxidative phosphorylation. Glycolysis occurs in the cytosol and the Krebs cycle in the mitochondrial matrix. The electron transport chain is built into the inner mitochondrial membrane.

9.2 **Glycolysis harvests chemical energy by oxidizing glucose to pyruvate (pp. 153–156, FIGURE 9.8)** Glycolysis nets 2 ATP, produced by substrate-level phosphorylation, and 2 NADH.

9.3 **The Krebs cycle completes the energy-yielding oxidation of organic molecules (p. 156, FIGURE 9.12)** The conversion of pyruvate to acetyl CoA links glycolysis to the Krebs cycle. The 2-carbon acetate of acetyl CoA joins the 4-carbon oxaloacetate to form the 6-carbon citrate, which is degraded back to oxaloacetate. The cycle releases CO_2, forms 1 ATP by substrate-level phosphorylation, and passes electrons to 3 NAD^+ and 1 FAD.

9.4 **The inner mitochondrial membrane couples electron transport to ATP synthesis (pp. 156–161, FIGURE 9.15)** Most of the ATP made in cellular respiration is produced by oxidative phosphorylation when NADH and $FADH_2$ donate electrons to the series of electron carriers in the electron transport chain. At the end of the chain, electrons are passed to O_2, reducing it to H_2O. Electron transport is coupled to ATP synthesis by chemiosmosis. At certain steps along the chain, electron transfer causes electron-carrying protein complexes to move H^+ from the matrix to the intermembrane space, storing energy as a proton-motive force (H^+ gradient). As H^+ diffuses back into the matrix through ATP synthase, its exergonic passage drives the endergonic phosphorylation of ADP.

- **Cellular respiration generates many ATP molecules for each sugar molecule it oxidizes (pp. 161–162, FIGURE 9.16)** The oxidation of glucose to CO_2 in respiration produces a maximum of about 38 ATP.

RELATED METABOLIC PROCESSES

- **Fermentation enables some cells to produce ATP without the help of oxygen (pp. 162–164, FIGURE 9.17)** Fermentation is anaerobic catabolism of organic nutrients. It yields ATP from glycolysis. The electrons from NADH made in glycolysis are passed to pyruvate, restoring the NAD^+ required to sustain glycolysis. Yeast and certain bacteria are facultative anaerobes, capable of making ATP by either aerobic respiration or fermentation. Of the two pathways, respiration is the more efficient in terms of ATP yield per glucose. Glycolysis occurs in nearly all organisms and probably evolved in ancient prokaryotes before there was O_2 in the atmosphere.
- **Glycolysis and the Krebs cycle connect to many other metabolic pathways (pp. 164–165, FIGURE 9.19)** These catabolic pathways combine to funnel electrons from all kinds of food molecules into cellular respiration. Carbon skeletons for anabolism (biosynthesis) come directly from digestion or from intermediates of glycolysis and the Krebs cycle.
- **Feedback mechanisms control cellular respiration (pp. 165–166, FIGURE 9.20)** Cellular respiration is controlled by allosteric enzymes at key points in glycolysis and the Krebs cycle. This helps the cell strike a moment-to-moment balance between catabolism and anabolism.

SELF-QUIZ

1. The *immediate* energy source that drives ATP synthesis during oxidative phosphorylation is
   a. the oxidation of glucose and other organic compounds
   b. the flow of electrons down the electron transport chain
   c. the affinity of oxygen for electrons
   d. a difference of H^+ concentration on opposite sides of the inner mitochondrial membrane
   e. the transfer of phosphate from Krebs cycle intermediates to ADP

2. What is the reducing agent in the following reaction?

$$\text{pyruvate} + \text{NADH} + H^+ \longrightarrow \text{lactate} + \text{NAD}^+$$

   a. oxygen
   b. NADH
   c. NAD^+
   d. lactate
   e. pyruvate

3. Which metabolic pathway is common to both fermentation and cellular respiration?
   a. Krebs cycle
   b. electron transport chain
   c. glycolysis
   d. synthesis of acetyl CoA from pyruvate
   e. reduction of pyruvate to lactate

4. In mitochondria, exergonic redox reactions
   a. are the source of energy driving prokaryotic ATP synthesis
   b. are directly coupled to substrate-level phosphorylation
   c. provide the energy to establish the proton gradient
   d. reduce carbon atoms to carbon dioxide
   e. are coupled via phosphorylated intermediates to endergonic processes

5. The final electron acceptor of the electron transport chain that functions in oxidative phosphorylation is
   a. oxygen
   b. water
   c. NAD^+
   d. pyruvate
   e. ADP

6. When electrons flow along the electron transport chains of mitochondria, which of the following changes occurs? (Explain your answer.)
   a. The pH of the matrix increases.
   b. ATP synthase pumps protons by active transport.
   c. The electrons gain free energy.
   d. The cytochromes of the chain phosphorylate ADP to form ATP.
   e. NAD^+ is oxidized.

7. In the presence of a metabolic poison that specifically and completely inhibits the function of mitochondrial ATP synthase, which of the following would you expect? (Explain your answer.)
   a. a decrease in the pH difference across the inner mitochondrial membrane
   b. an increase in the pH difference across the inner mitochondrial membrane
   c. increased synthesis of ATP
   d. oxygen consumption to cease
   e. proton pumping by the electron transport chain to cease

8. Cells do not catabolize carbon dioxide because
   a. its double bonds are too stable to be broken
   b. CO_2 has fewer bonding electrons than other organic compounds
   c. the carbon atom is already completely reduced
   d. most of the available electron energy was released by the time the CO_2 was formed
   e. the molecule has too few atoms

9. Which of the following is a true distinction between fermentation and cellular respiration?
   a. Only respiration oxidizes glucose.
   b. NADH is oxidized by the electron transport chain only in respiration.
   c. Fermentation, but not respiration, is an example of a catabolic pathway.
   d. Substrate-level phosphorylation is unique to fermentation.
   e. NAD^+ functions as an oxidizing agent only in respiration.

10. Most CO_2 from catabolism is released during
   a. glycolysis
   b. the Krebs cycle
   c. lactate fermentation
   d. electron transport
   e. oxidative phosphorylation

CHALLENGE QUESTION

In the 1940s, some physicians prescribed low doses of a drug called dinitrophenol (DNP) to help patients lose weight. This unsafe method was abandoned after a few patients died. DNP uncouples the chemiosmotic machinery by making the lipid bilayer of the inner mitochondrial membrane leaky to H^+. Explain how this causes weight loss.

SCIENCE, TECHNOLOGY, AND SOCIETY

Nearly all human societies use fermentation to produce alcoholic drinks such as beer and wine. The practice dates back to the earliest days of agriculture. How do you suppose this use of fermentation was first discovered? Why did wine prove to be a more useful beverage, especially to a preindustrial culture, than the grape juice from which it was made?

FURTHER READING

Becker, W. M., J. B. Reece, and M. F. Poenie. *The World of the Cell,* 3rd ed. Menlo Park, CA: Benjamin/Cummings, 1996. Provides particularly clear and comprehensive coverage of how cells harvest energy in Chapters 11 and 12.

Gennis, R. B. "Cytochrome *c* Oxidase: One Enzyme, Two Mechanisms?" *Science,* June 12, 1998. Describes recent discoveries about the third protein complex in the electron transport chain.

Kearney, J. T. "Training the Olympic Athlete." *Scientific American,* June 1996. Includes discussion of how athletes work to achieve an optimum balance between aerobic and anaerobic energy generation.

Mathews, C., and K. van Holde. *Biochemistry,* 2nd ed. Redwood City, CA: Benjamin/Cummings, 1996. Features effective diagrams of catabolism.

Service, R. F. "Chemistry Prize Taps the Energy of Life." *Science,* October 24, 1997. The awarding of the Nobel Prize for research on ATP synthase.

Weiss, R. "Blazing Blossoms." *Science News,* June 24, 1989. How skunk cabbage and other plants use their catabolism to generate heat.

WEB LINKS

Visit the special edition of *The Biology Place* for BIOLOGY, Fifth Edition, at http://www.biology.com/campbell. Go to Chapter 9 for online resources, including learning activities, practice exams, and links to the following web sites:

"DIY Glycolysis Home Page"
To complete "Do It Yourself Glycolysis," you must correctly select the reactions necessary to convert glucose to lactate.

"Metabolism of TCA Intermediates"
From the Argonne National Laboratory, this clickable image map takes you to metabolic pathways linked to the Krebs cycle.

"Biomolecular machines: ATP Synthase"
Animated information on the functioning of ATP synthase.

"The Biology Project—Biochemistry"
Interesting problem sets and tutorials from the Biology Project at the University of Arizona.

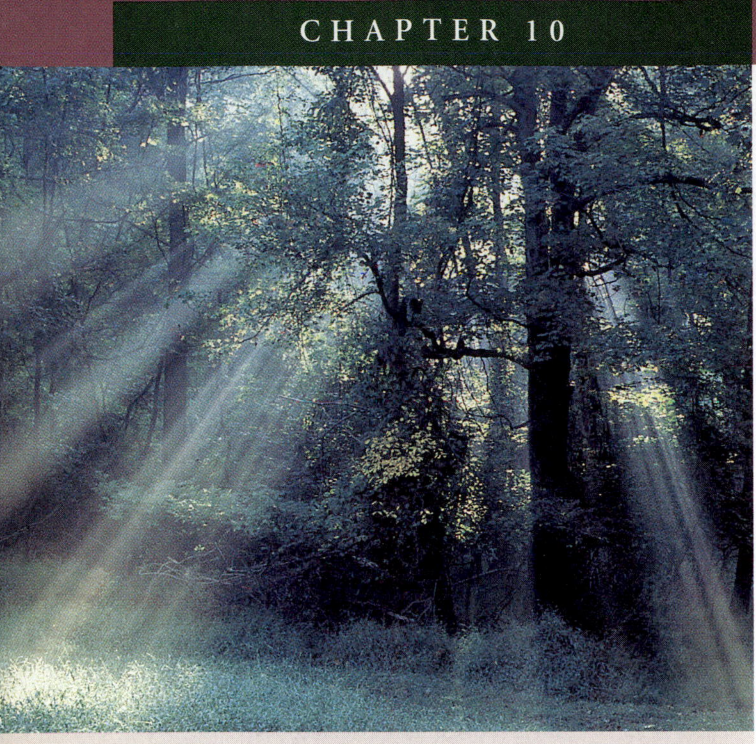

PHOTOSYNTHESIS

Life on Earth is solar-powered. The chloroplasts of plants capture light energy that has traveled 160 million kilometers from the sun and convert it to chemical energy stored in sugar and other organic molecules. The process is called **photosynthesis.** *In this chapter you will learn how photosynthesis works. We begin by placing photosynthesis in its ecological context.*

PHOTOSYNTHESIS IN NATURE

Plants and other autotrophs are the producers of the biosphere

Photosynthesis nourishes almost all of the living world directly or indirectly. An organism acquires the organic compounds it uses for energy and carbon skeletons by one of two major modes: autotrophic or heterotrophic nutrition. At first, the term *autotrophic* (Gr. *autos,* "self," and *trophos,* "feed") may seem to contradict the principle that organisms are open systems, taking in resources from their environment. **Autotrophs** are not totally self-sufficient, however; they are self-feeders only in the sense that they sustain themselves without eating or decomposing other organisms. They make their organic molecules from inorganic raw materials obtained from the environment. It is for this reason that biologists refer to autotrophs as the *producers* of the biosphere.

Plants are autotrophs; the only nutrients they require are carbon dioxide from the air, and water and minerals from the soil. Specifically, plants are *photo*autotrophs, organisms that use light as a source of energy to synthesize carbohydrates, lipids, proteins, and other organic substances. Photosynthesis also occurs in algae, including certain protists, and in some prokaryotes (FIGURE 10.1). In this chapter our emphasis will be on plants. Variations in photosynthesis that occur in algae and bacteria will be discussed in Unit Five. A much rarer form of self-feeding is unique to those bacteria that are *chemo*autotrophs. These organisms produce their organic compounds without the help of light, obtaining their energy by oxidizing inorganic substances, such as sulfur or ammonia. (We will postpone further discussion of this type of autotrophic nutrition until Chapter 27.)

Heterotrophs obtain their organic material by the second major mode of nutrition. Unable to make their own food, they live on compounds produced by other organisms; heterotrophs are the biosphere's *consumers.* The most obvious form of this "other-feeding" (*hetero* means "other, different") occurs when an animal eats plants or other animals. But heterotrophic nutrition may be more subtle. Some heterotrophs consume the remains of dead organisms, decomposing and feeding on organic litter such as carcasses, feces, and fallen

(a)

(b)

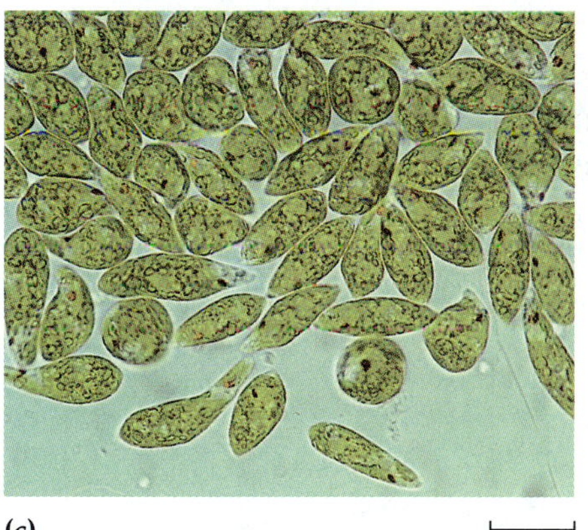

(c) 10 μm

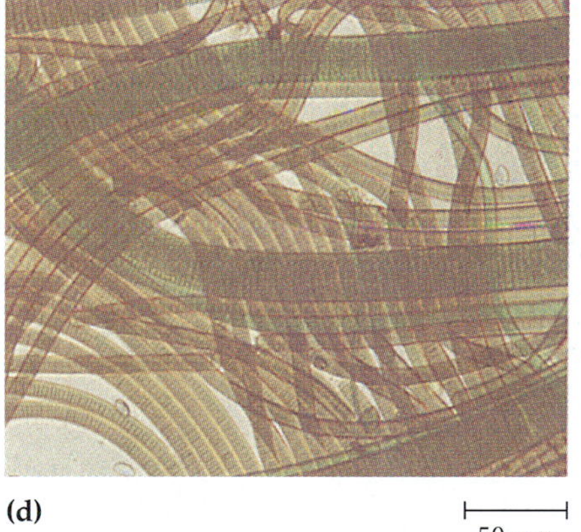

(d) 50 μm

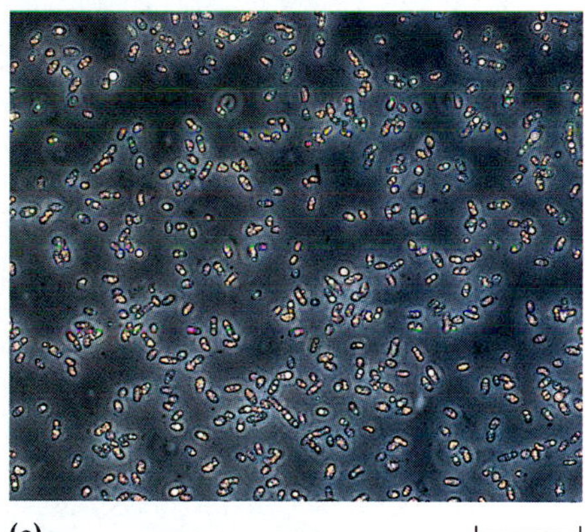

(e) 25 μm

FIGURE 10.1 ▪ Photoautotrophs: producers for most ecosystems. These organisms use light energy to drive the synthesis of organic molecules from carbon dioxide and (in most cases) water. They feed not only themselves, but the entire living world. **(a)** On land, plants are the predominant producers of food. Three major groups of land plants—mosses, ferns, and flowering plants—are represented in this scene. In oceans, ponds, lakes, and other aquatic environments, photosynthetic organisms include **(b)** multicellular algae, such as kelp; **(c)** some unicellular protists, such as *Euglena;* **(d)** the prokaryotes called cyanobacteria; and **(e)** other photosynthetic prokaryotes, such as these purple sulfur bacteria (c, d, e: LMs).

leaves; they are known as decomposers. Most fungi and many types of bacteria get their nourishment this way. Almost all heterotrophs, including humans, are completely dependent on photoautotrophs for food, and also for oxygen, a by-product of photosynthesis. Thus, we can trace the food we eat and the oxygen we breathe to the chloroplast.

Chloroplasts are the sites of photosynthesis in plants

All green parts of a plant, including green stems and unripened fruit, have chloroplasts, but the leaves are the major sites of photosynthesis in most plants (FIGURE 10.2). There are about half a million chloroplasts per square millimeter of leaf surface. The color of the leaf is from **chlorophyll**, the green pigment located within the chloroplasts. It is the light energy absorbed by chlorophyll that drives the synthesis of food molecules in the chloroplast. Chloroplasts are found mainly in the cells of the **mesophyll**, the tissue in the interior of the leaf. Carbon dioxide enters the leaf, and oxygen exits, by way of microscopic pores called **stomata** (singular, **stoma**; Gr. "mouth"). Water absorbed by the roots is delivered to the leaves in veins. Leaves also use veins to export sugar to roots and other nonphotosynthetic parts of the plant.

A typical mesophyll cell has about 30 to 40 chloroplasts, each a watermelon-shaped organelle measuring about 2–4 μm by 4–7 μm. An envelope of two membranes encloses the stroma, the dense fluid within the chloroplast. An elaborate system of interconnected thylakoid membranes segregates the stroma from another compartment, the thylakoid space (or lumen). In some places, thylakoid sacs are stacked in columns called grana. Chlorophyll resides in the thylakoid membranes. (Photosynthetic prokaryotes lack chloroplasts, but, as you will see in Chapter 27, they do have membranes that function in a manner similar to the thylakoid membranes of chloroplasts.) Now that we have looked at the sites of photosynthesis in plants, we are ready to see how these organelles convert the light energy absorbed by chlorophyll to chemical energy.

THE PATHWAYS OF PHOTOSYNTHESIS

Evidence that chloroplasts split water molecules enabled researchers to track atoms through photosynthesis: *science as a process*

Scientists have tried for centuries to piece together the process by which plants make food. Although some of the steps are

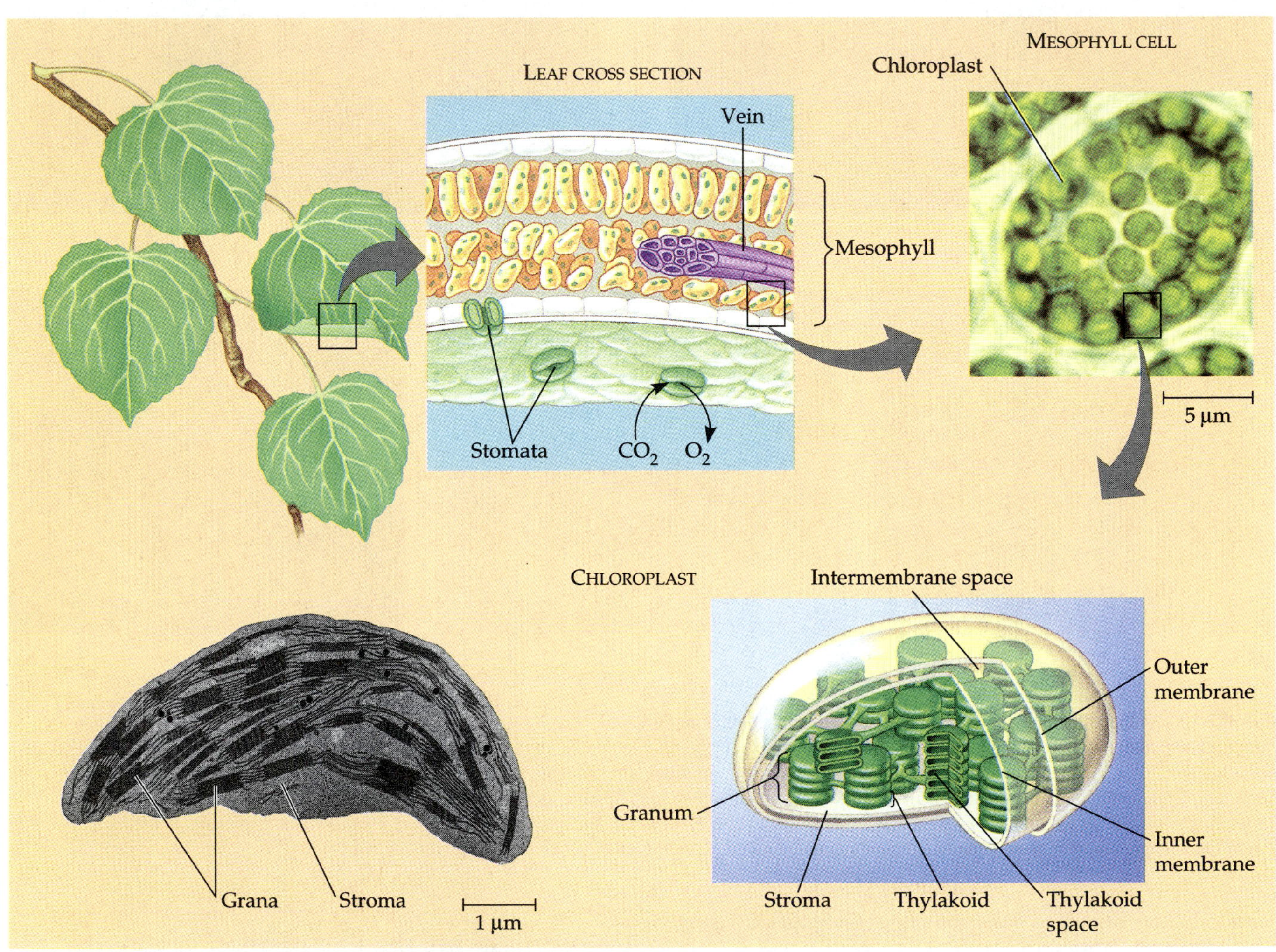

FIGURE 10.2 ▪ The site of photosynthesis in a plant. Leaves are the major organs of photosynthesis in plants. Gas exchange between the mesophyll and the atmosphere occurs through microscopic pores called stomata. Chloroplasts, found mainly in the mesophyll, are bounded by two membranes that enclose the stroma, the dense fluid content of the chloroplast. Membranes of the thylakoid system separate the stroma from the thylakoid space. Thylakoids are concentrated in stacks called grana. (Top right, LM; bottom left, TEM.)

still not completely understood, the overall photosynthetic equation has been known since the 1800s: In the presence of light, the green parts of plants produce organic material and oxygen from carbon dioxide and water. Using molecular formulas, we can summarize photosynthesis with this chemical equation:

$$6\,CO_2 + 12\,H_2O + \underset{\text{energy}}{\text{Light}} \longrightarrow C_6H_{12}O_6 + 6\,O_2 + 6\,H_2O$$

The carbohydrate $C_6H_{12}O_6$ is glucose.* Water appears on both sides of the equation because 12 molecules are consumed and 6 molecules are newly formed during photosynthesis. We can simplify the equation by indicating only the net consumption of water:

$$6\,CO_2 + 6\,H_2O + \text{Light energy} \longrightarrow C_6H_{12}O_6 + 6\,O_2$$

Writing the equation in this form, we can see that the overall chemical change during photosynthesis is the reverse of the one that occurs during cellular respiration. Both of these metabolic processes occur in plant cells. However, as you will soon learn, plants do not make food by simply reversing the steps of respiration.

Now let's divide the photosynthetic equation by 6 to put it in its simplest possible form:

$$CO_2 + H_2O \longrightarrow CH_2O + O_2$$

Here, CH_2O is not an actual sugar but represents the general formula for a carbohydrate. In other words, we are imagining the synthesis of a sugar molecule one carbon at a time. Six repetitions would produce a glucose molecule. Let's now use this simplified formula to see how researchers tracked the chemical elements (C, H, and O) from the reactants of photosynthesis to the products.

The Splitting of Water

One of the first clues to the mechanism of photosynthesis came from the discovery that the oxygen given off by plants is derived from water and not from carbon dioxide. The chloroplast splits water into hydrogen and oxygen. Before this discovery, the prevailing hypothesis was that photosynthesis split carbon dioxide and then added water to the carbon:

$$\text{Step 1: } CO_2 \longrightarrow C + O_2$$

$$\text{Step 2: } C + H_2O \longrightarrow CH_2O$$

This hypothesis predicted that the O_2 released during photosynthesis came from CO_2. This idea was challenged in the 1930s by C. B. van Niel of Stanford University. Van Niel was investigating photosynthesis in bacteria that make their carbohydrate from CO_2 but do not release O_2. Van Niel concluded that, at least in these bacteria, CO_2 is not split into carbon and oxygen. One group of bacteria used hydrogen sulfide (H_2S) rather than water for photosynthesis, forming yellow globules of sulfur as a waste product (these globules are visible in FIGURE 10.1e). Here is the chemical equation:

$$CO_2 + 2\,H_2S \longrightarrow CH_2O + H_2O + 2\,S$$

Van Niel reasoned that the bacteria split H_2S and used the hydrogen to make sugar. He then generalized that idea, proposing that all photosynthetic organisms require a hydrogen source, but that the source varies:

$$\text{General: } CO_2 + 2\,H_2X \longrightarrow CH_2O + H_2O + 2\,X$$

$$\text{Sulfur bacteria: } CO_2 + 2\,H_2S \longrightarrow CH_2O + H_2O + 2\,S$$

$$\text{Plants: } CO_2 + 2\,H_2O \longrightarrow CH_2O + H_2O + O_2$$

Thus, van Niel hypothesized that plants split water as a source of hydrogen, releasing oxygen as a by-product.

Nearly 20 years later, scientists confirmed van Niel's hypothesis by using oxygen-18 (^{18}O), a heavy isotope, as a tracer to follow the fate of oxygen atoms during photosynthesis. The O_2 that came from plants was labeled with ^{18}O *only* if water was the source of the tracer. If the ^{18}O was introduced to the plant in the form of CO_2, the label did not turn up in the released O_2. In the following summary of these experiments, red denotes labeled atoms of oxygen:

$$\text{Experiment 1: } CO_2 + 2\,H_2O \longrightarrow CH_2O + H_2O + O_2$$

$$\text{Experiment 2: } CO_2 + 2\,H_2O \longrightarrow CH_2O + H_2O + O_2$$

The most important result of the shuffling of atoms during photosynthesis is the extraction of hydrogen from water and its incorporation into sugar. The waste product of photosynthesis, O_2, restores the atmospheric oxygen consumed during cellular respiration. FIGURE 10.3 shows the fates of all atoms in photosynthesis.

Photosynthesis as a Redox Process

Let's briefly contrast photosynthesis with cellular respiration. During respiration, energy is released from sugar when electrons associated with hydrogen are transported by carriers to oxygen, forming water as a by-product. The electrons lose potential energy as electronegative oxygen pulls them down the electron transport chain, and the mitochondrion uses the energy to synthesize ATP. Photosynthesis, also a redox process, reverses the direction of electron flow. Water is split, and

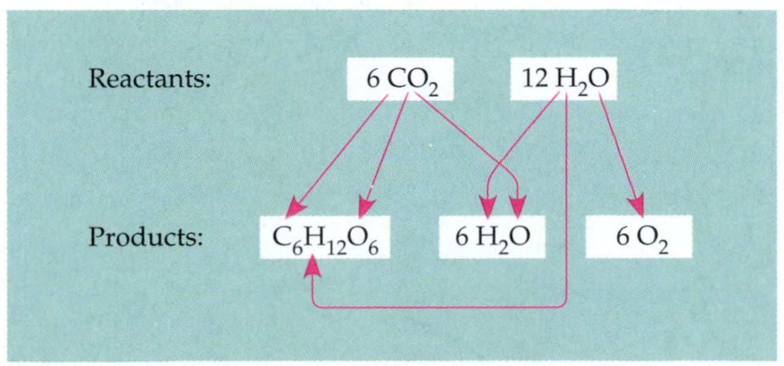

FIGURE 10.3 ▪ **Tracking atoms through photosynthesis.**

* The direct product of photosynthesis is actually a three-carbon sugar. Glucose is used here only to simplify the relationship between photosynthesis and respiration.

electrons are transferred along with hydrogen ions from the water to carbon dioxide, reducing it to sugar. The electrons increase in potential energy as they move from water to sugar. The required energy boost is provided by light.

The light reactions and the Calvin cycle cooperate in converting light energy to the chemical energy of food: *an overview*

10.1 The equation for photosynthesis is a deceptively simple summary of a very complex process. Actually, photosynthesis is not a single process, but two, each with multiple steps. These two stages of photosynthesis are known as the **light reactions** (the *photo* part of photosynthesis) and the **Calvin cycle** (the *synthesis* part) (FIGURE 10.4).

The light reactions are the steps of photosynthesis that convert solar energy to chemical energy. Light absorbed by chlorophyll drives a transfer of electrons and hydrogen from water to an acceptor called **$NADP^+$** (nicotinamide adenine dinucleotide phosphate), which temporarily stores the energized electrons. Water is split in the process, and thus it is the light reactions of photosynthesis that give off O_2 as a by-product. The electron acceptor of the light reactions, $NADP^+$, is first cousin to NAD^+, which functions as an electron carrier in cellular respiration; the two molecules differ only by the presence of an extra phosphate group in the $NADP^+$ molecule. The light reactions use solar power to reduce $NADP^+$ to NADPH by adding a pair of electrons along with a hydrogen nucleus, or H^+. The light reactions also generate ATP by powering the addition of a phosphate group to ADP, a process called **photophosphorylation**. Thus, light energy is initially converted to chemical energy in the form of two compounds: NADPH, a source of energized electrons ("reducing power"), and ATP, the versatile energy currency of cells. Notice that the light reactions produce no sugar; that happens in the second stage of photosynthesis, the Calvin cycle.

The Calvin cycle is named for Melvin Calvin, who along with his colleagues began to elucidate its steps in the late 1940s. The cycle begins by incorporating CO_2 from the air into organic molecules already present in the chloroplast. This initial incorporation of carbon into organic compounds is known as **carbon fixation**. The Calvin cycle then reduces the fixed carbon to carbohydrate by the addition of electrons. The reducing power is provided by NADPH, which acquired energized electrons in the light reactions. To convert CO_2 to carbohydrate, the Calvin cycle also requires chemical energy in the form of ATP, which is also generated by the light reactions. Thus, it is the Calvin cycle that makes sugar, but it can do so only with the help of the NADPH and ATP produced by the light reactions. The metabolic steps of the Calvin cycle are sometimes referred to as the dark reactions, or light-independent reactions, because none of the steps requires light *directly*. Nevertheless, the Calvin cycle in most plants occurs during daylight, for only then can the light reactions regenerate the NADPH and ATP spent in the reduction of CO_2 to sugar. In essence, the chloroplast uses light energy to make sugar by coordinating the two stages of photosynthesis.

As FIGURE 10.4 indicates, the thylakoids of the chloroplast are the sites of the light reactions, while the Calvin cycle occurs in the stroma. As molecules of $NADP^+$ and ADP bump into the thylakoid membrane, they pick up electrons and phosphate, respectively, and then transfer their high-energy cargo to the Calvin cycle. The two stages of photosynthesis are

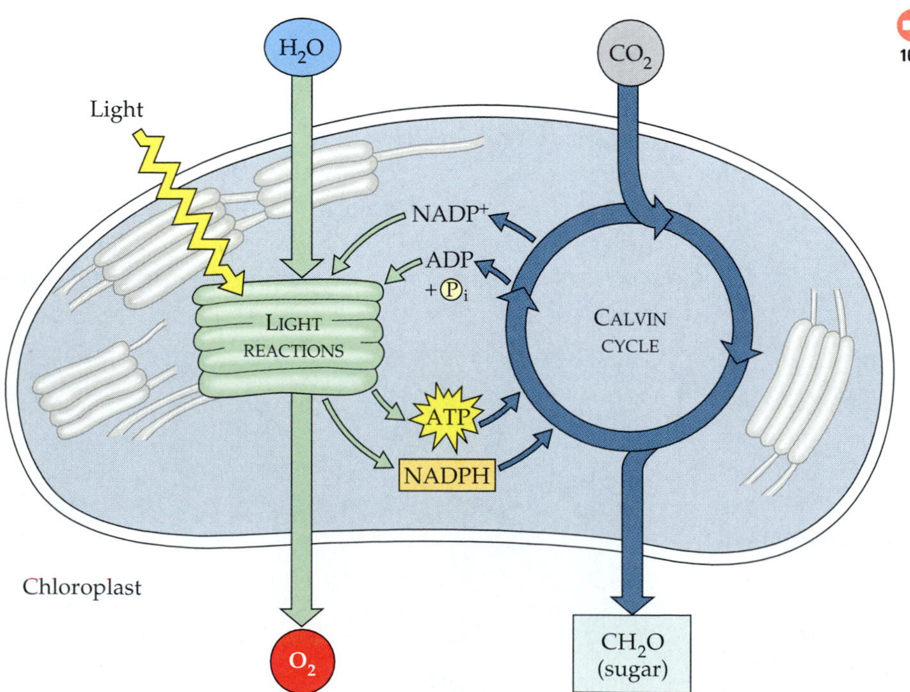

10.1 **FIGURE 10.4 • An overview of photosynthesis: cooperation of the light reactions and the Calvin cycle.** The light reactions use solar energy to make ATP and NADPH, which function as chemical energy and reducing power, respectively, in the Calvin cycle. In contrast to ATP generated by cellular respiration, ATP produced in the light reactions of photosynthesis is usually dedicated to a single kind of cellular work, driving the Calvin cycle. The Calvin cycle incorporates CO_2 into organic molecules, which are converted to sugar. (Recall from Chapter 5 that most simple sugars have formulas that are some multiple of CH_2O.) Thylakoid membranes, especially those of the grana, are the sites of the light reactions, whereas the Calvin cycle occurs in the stroma.

A smaller version of this diagram will reappear in several subsequent figures as a reminder of whether the events being described occur in the light reactions or in the Calvin cycle.

treated in this figure as metabolic modules that take in ingredients and crank out products. Our next step toward understanding photosynthesis is to look more closely at how the two stages work, beginning with the light reactions.

The light reactions convert solar energy to the chemical energy of ATP and NADPH: *a closer look*

Chloroplasts are chemical factories powered by the sun. Their thylakoids transform light energy into the chemical energy of ATP and NADPH. To understand this conversion better, it is first necessary to learn about some important properties of light.

The Nature of Sunlight

Light is a form of energy known as electromagnetic energy, also called radiation. Electromagnetic energy travels in rhythmic waves analogous to those created by dropping a pebble into a puddle of water. Electromagnetic waves, however, are disturbances of electrical and magnetic fields rather than disturbances of a material medium such as water.

The distance between the crests of electromagnetic waves is called the **wavelength**. Wavelengths range from less than a nanometer (for gamma rays) to more than a kilometer (for radio waves). This entire range of radiation is known as the **electromagnetic spectrum** (FIGURE 10.5). The segment most important to life is the narrow band that ranges from about 380 to 750 nm in wavelength. This radiation is known as **visible light**, because it is detected as various colors by the human eye.

The model of light as waves explains many of light's properties, but in certain respects light behaves as though it consists of discrete particles, called **photons**. Photons are not tangible objects, but they act like objects in that each of them has a fixed quantity of energy. The amount of energy is inversely related to the wavelength of the light; the shorter the wavelength, the greater the energy of each photon of that light. Thus, a photon of violet light packs nearly twice as much energy as a photon of red light.

Although the sun radiates the full spectrum of electromagnetic energy, the atmosphere acts like a selective window, allowing visible light to pass through while screening out a substantial fraction of other radiation. The part of the spectrum we can see—visible light—is also the radiation that drives photosynthesis. Blue and red, the two wavelengths most effectively absorbed by chlorophyll, are the colors most useful as energy for the light reactions.

Photosynthetic Pigments: The Light Receptors

As light meets matter, it may be reflected, transmitted, or absorbed. Substances that absorb visible light are called pigments. Different pigments absorb light of different wavelengths, and the wavelengths that are absorbed disappear. If a pigment is illuminated with white light, the color we see is the color most reflected or transmitted by the pigment. (If a pigment absorbs all wavelengths, it appears black.) We see green when we look at a leaf because chlorophyll absorbs red and blue light while transmitting and reflecting green light (FIGURE 10.6, p. 174). The ability of a pigment to absorb various wavelengths of light can be measured by placing a solution of the pigment in a **spectrophotometer** (see the Methods Box, p. 182). A graph plotting a pigment's light absorption versus wavelength is called an **absorption spectrum**.

FIGURE 10.7a, page 174, shows the absorption spectra of a type of chlorophyll called **chlorophyll *a*** and some other pigments in the chloroplast. The absorption spectrum for chlorophyll *a* provides clues to the relative effectiveness of different wavelengths for driving photosynthesis, since light can perform work in chloroplasts only if it is absorbed. As previously mentioned, blue and red light work best for photosynthesis, while green is the least effective color. An **action spectrum** (FIGURE 10.7b) profiles the relative performance of the different

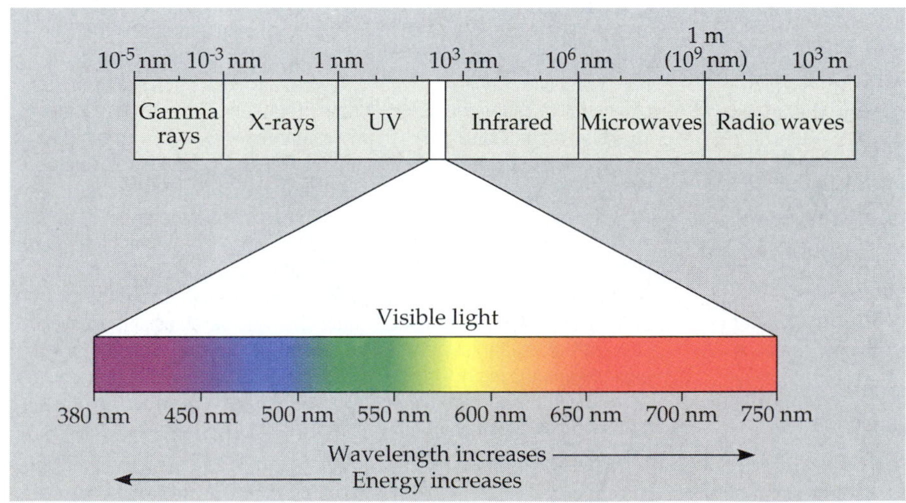

FIGURE 10.5 ▪ The electromagnetic spectrum. Visible light and other forms of electromagnetic energy radiate through space as waves of various lengths. We perceive different wavelengths of visible light as different colors. White light is a mixture of all wavelengths of visible light. A prism can sort white light into its component colors by bending light of different wavelengths varying degrees. Visible light drives photosynthesis.

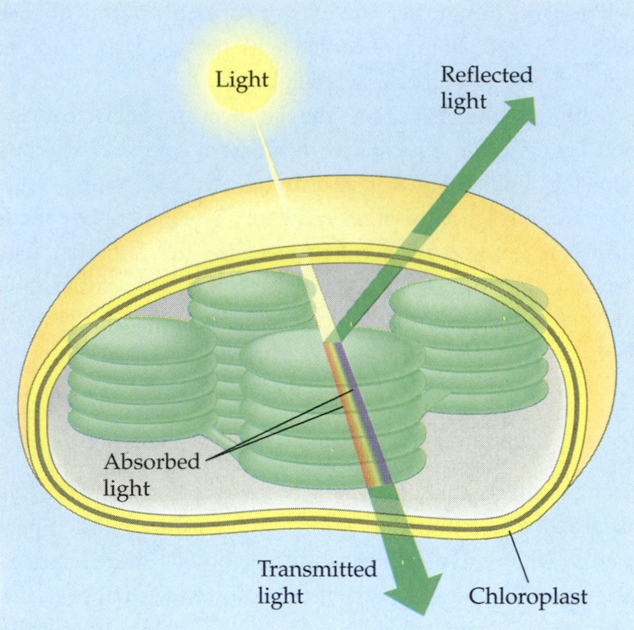

FIGURE 10.6 ▪ Interactions of light with matter in a chloroplast. The pigments of chloroplasts absorb blue and red light, the colors most effective in photosynthesis. The pigments reflect or transmit green light, which is why leaves appear green.

wavelengths more accurately than an absorption spectrum. An action spectrum is prepared by illuminating chloroplasts with different colors of light and then plotting wavelength against some measure of photosynthetic rate, such as carbon dioxide consumption or oxygen release (see also FIGURE 10.7c).

Notice by comparing FIGURE 10.7a to FIGURE 10.7b that the action spectrum for photosynthesis does not exactly match the absorption spectrum of chlorophyll *a*. The absorption spectrum underestimates the effectiveness of certain wavelengths in driving photosynthesis. This is partly because chlorophyll *a* is not the only photosynthetically important pigment in chloroplasts. Only chlorophyll *a* can participate directly in the light reactions, which convert solar energy to chemical energy. But other pigments in the thylakoid membrane can absorb light and transfer the energy to chlorophyll *a*, which then initiates the light reactions. One of these accessory pigments is another form of chlorophyll, **chlorophyll *b*.** Chlorophyll *b* is almost identical to chlorophyll *a* (FIGURE 10.8), but the slight structural difference between them is enough to give the two pigments slightly different absorption spectra and hence different colors. Chlorophyll *a* is blue-green, whereas chlorophyll *b* is yellow-green. If a photon of sunlight is absorbed by chlorophyll *b*, energy is conveyed to chlorophyll *a*, which then behaves just as though it had absorbed the photon. Other accessory pigments include **carotenoids**, hydrocarbons that are various shades of yellow and orange (see FIGURE 10.7a). Some carotenoids may broaden the spectrum of colors that can drive photosynthesis. However, other carotenoids seem to function primarily in *photoprotection:* Rather than transmit

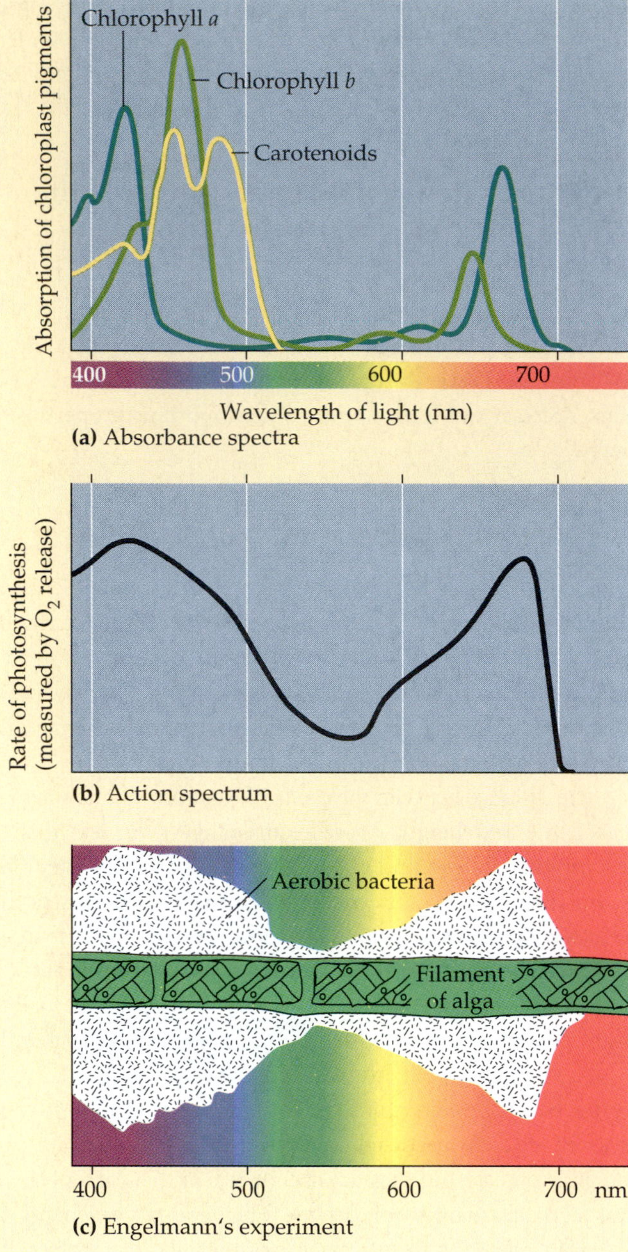

FIGURE 10.7 ▪ Absorption and action spectra for photosynthesis. **(a)** A comparison of the absorption spectra for chlorophyll *a* and accessory pigments extracted from chloroplasts. **(b)** An action spectrum, profiling the effectiveness of different wavelengths of light in driving photosynthesis. Compared to the peaks in the absorption spectrum for chlorophyll *a* (blue-green line in graph at top), the peaks in the action spectrum are broader, and the valley is narrower and not as deep. This is partly due to the absorption of light by accessory pigments, which broaden the spectrum of colors that can be used for photosynthesis. **(c)** An elegant experiment demonstrating the action spectrum for photosynthesis was first performed in 1883 by Thomas Engelmann, a German botanist. He illuminated a filamentous alga with light that had been passed through a prism, thus exposing different segments of the alga to different wavelengths of light. Engelmann used aerobic bacteria, which concentrate near an oxygen source, to determine which segments of the alga were releasing the most O_2. Bacteria congregated in greatest numbers around the parts of the alga illuminated with red or blue light. Notice the close match of the bacterial distribution to the action spectrum in part b.

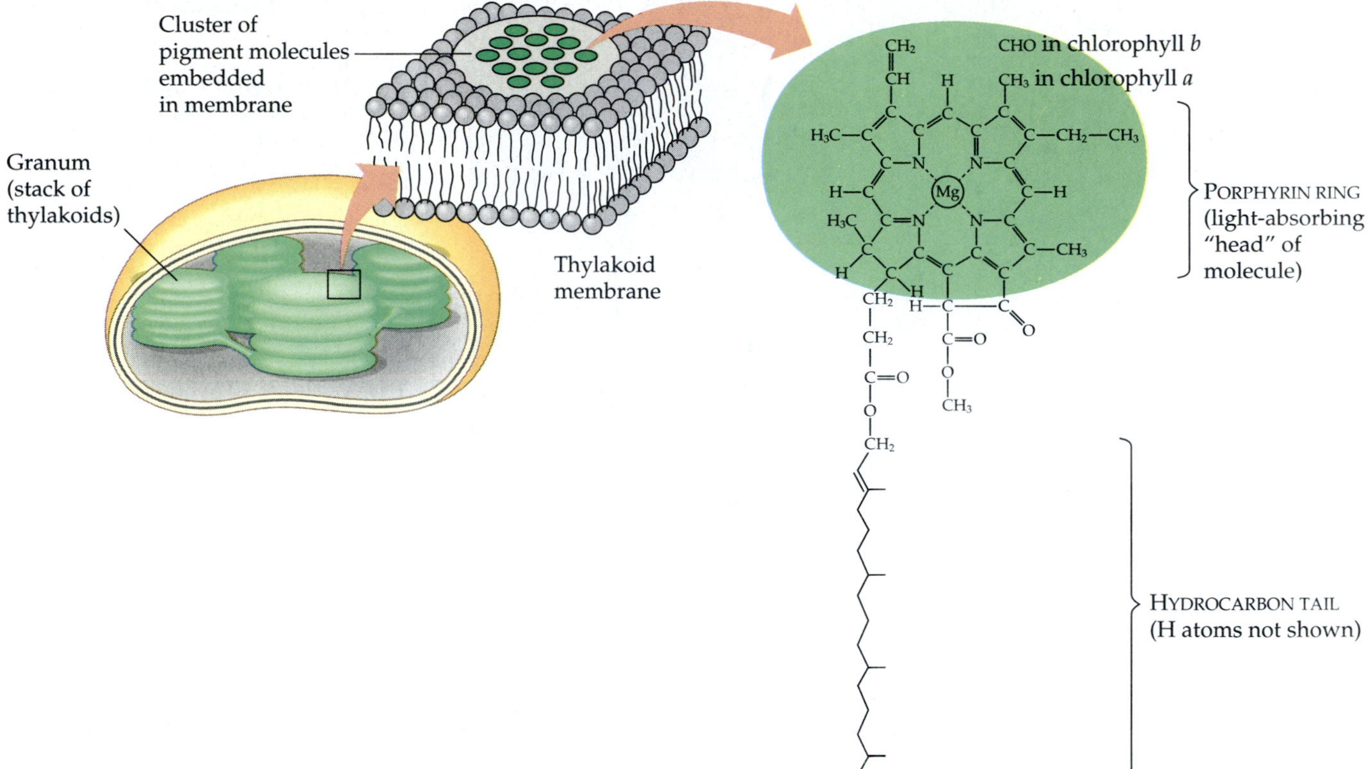

FIGURE 10.8 • Structure of chlorophyll. Chlorophyll *a*, the pigment that participates directly in the light reactions of photosynthesis, has a "head," called a porphyrin ring, with a magnesium atom at its center. Attached to the porphyrin is a hydrocarbon tail, which interacts with hydrophobic regions of proteins in the thylakoid membrane. Chlorophyll *b* differs from chlorophyll *a* only in one of the functional groups bonded to the porphyrin. This diagram simplifies by placing chlorophyll at the surface of the membrane; most of the molecule is actually immersed in the hydrophobic core of the membrane.

energy to chlorophyll, these compounds absorb and dissipate excessive light energy that would otherwise damage chlorophyll. (Interestingly, similar carotenoids may have a photoprotective role in the human eye.)

Photoexcitation of Chlorophyll

What exactly happens when chlorophyll and other pigments absorb photons? The colors corresponding to the absorbed wavelengths disappear from the spectrum of the transmitted and reflected light, but energy cannot disappear. When a molecule absorbs a photon, one of the molecule's electrons is elevated to an orbital where it has more potential energy. When the electron is in its normal orbital, the pigment molecule is said to be in its ground state. After absorption of a photon boosts an electron to an orbital of higher energy, the pigment molecule is said to be in an excited state. The only photons absorbed are those whose energy is exactly equal to the energy difference between the ground state and an excited state, and this energy difference varies from one kind of atom or molecule to another. Thus, a particular compound absorbs only photons corresponding to specific wavelengths, which is why each pigment has a unique absorption spectrum.

The energy of an absorbed photon is converted to the potential energy of an electron raised from the ground state to an excited state. But the electron cannot remain there long; the excited state, like all high-energy states, is unstable. Generally, when pigments absorb light, their excited electrons drop back down to the ground-state orbital in a billionth of a second, releasing their excess energy as heat. This conversion of light energy to heat is what makes the top of an automobile so hot on a sunny day. (White cars are coolest because their paint reflects all wavelengths of visible light, although it may absorb ultraviolet and other invisible radiation.) Some pigments, including chlorophyll, emit light as well as heat after absorbing photons. The electron jumps to a state of greater energy, and as it falls back to ground state, a photon is given off. This afterglow is called fluorescence. If a solution of chlorophyll isolated from chloroplasts is illuminated, it will fluoresce in the red part of the spectrum and also give off heat (FIGURE 10.9, p. 176).

Photosystems: Light-Harvesting Complexes of the Thylakoid Membrane

Chlorophyll excited by absorption of light energy produces very different results in an intact chloroplast than it does in

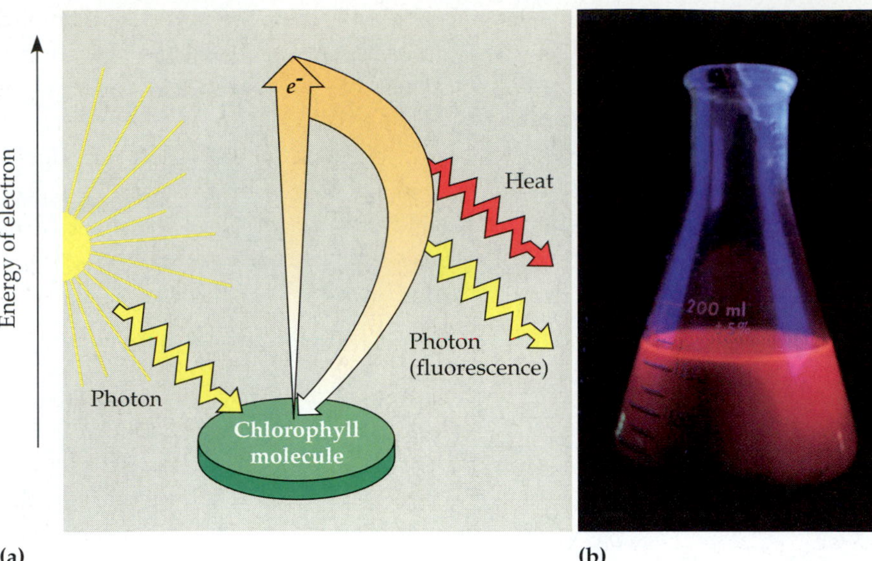

FIGURE 10.9 ▪ Photoexcitation of isolated chlorophyll. (a) Absorption of a photon causes a transition of the chlorophyll molecule from its ground state to its excited state. The photon boosts an electron to an orbital where it has more potential energy. If isolated chlorophyll is illuminated, its excited electron immediately drops back down to the ground-state orbital, giving off its excess energy as heat and fluorescence (light). **(b)** A chlorophyll solution excited with ultraviolet light will fluoresce, giving off a red-orange glow.

isolation. In its native environment of the thylakoid membrane, chlorophyll is organized along with proteins and other kinds of smaller organic molecules into **photosystems**.

A photosystem has a light-gathering "antenna complex" consisting of a cluster of a few hundred chlorophyll *a*, chlorophyll *b*, and carotenoid molecules (FIGURE 10.10). The number and variety of pigment molecules enable a photosystem to harvest light over a larger surface and a larger portion of the spectrum than any single pigment molecule could harvest. When any antenna molecule absorbs a photon, the energy is transmitted from pigment molecule to pigment molecule until it reaches a particular chlorophyll *a*. What is special about *this* chlorophyll *a* molecule is not its molecular structure, but its position. Only this chlorophyll molecule is located in the region of the photosystem called the **reaction center**, where the first light-driven chemical reaction of photosynthesis occurs.

Sharing the reaction center with the chlorophyll *a* molecule is a specialized molecule called the **primary electron acceptor**. In an oxidation-reduction reaction, the chlorophyll *a* molecule at the reaction center loses one of its electrons to the primary electron acceptor. This redox reaction occurs when light excites the electron to a higher energy level in chlorophyll and the electron acceptor traps the high-energy electron before it can return to the ground state in the chlorophyll molecule. Isolated chlorophyll fluoresces because there is no electron acceptor to prevent electrons of photoexcited chlorophyll from dropping right back to the ground state. In a chloroplast, the acceptor molecule functions like a dam that prevents this immediate plunge of high-energy electrons back to the ground state. Thus, each photosystem—reaction-center chlorophyll and electron acceptor surrounded by an antenna complex—functions in the chloroplast as a light-harvesting unit. The solar-powered transfer of electrons from chlorophyll to the primary electron acceptor is the first step of the light reactions.

The thylakoid membrane is populated by two types of photosystems that cooperate in the light reactions of photosynthesis. They are called **photosystem I** and **photosystem II**, in order of their discovery. Each has a characteristic reaction center—a particular kind of primary electron acceptor next to a chlorophyll *a* molecule associated with specific proteins. The reaction-center chlorophyll of photosystem I is known as P700 because this pigment is best at absorbing light having a wavelength of 700 nm (the far-red part of the spectrum). The chlorophyll at the reaction center of photosystem II is called

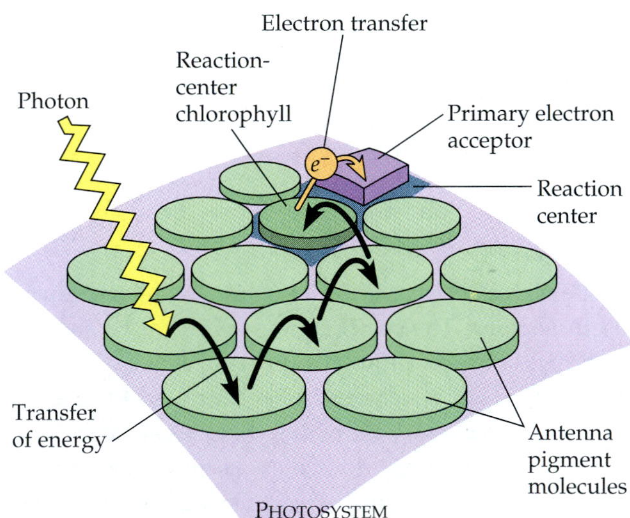

FIGURE 10.10 ▪ How a photosystem harvests light. Photosystems are the light-harvesting units of the thylakoid membrane. Each photosystem is a complex of proteins and other kinds of molecules, and includes an antenna consisting of a few hundred pigment molecules. When a photon strikes a pigment molecule, the energy is passed from molecule to molecule until it reaches the reaction center. At the reaction center, the energy drives an oxidation-reduction reaction. An excited electron from the reaction-center chlorophyll is captured by a specialized molecule called the primary electron acceptor.

P680 because its absorption spectrum has a peak at 680 nm (also in the red part of the spectrum). These two pigments, P700 and P680, are actually identical chlorophyll *a* molecules. However, their association with different proteins in the thylakoid membrane affects the electron distribution in the chlorophyll molecules and accounts for the slight differences in light-absorbing properties. Let's now see how the two photosystems work together in using light energy to generate ATP and NADPH, the two main products of the light reactions.

Noncyclic Electron Flow

Light drives the synthesis of NADPH and ATP by energizing the two photosystems embedded in the thylakoid membranes of chloroplasts. The key to this energy transformation is a flow of electrons through the photosystems and other molecular components built into the thylakoid membrane. During the light reactions of photosynthesis, there are two possible routes for electron flow: cyclic and noncyclic. **Noncyclic electron flow**, the predominant route, is shown in FIGURE 10.11. The numbers in the following description correspond to the numbered steps in the figure.

① When photosystem II absorbs light, an electron excited to a higher energy level in the reaction-center chlorophyll (P680) is captured by the primary electron acceptor. The oxidized chlorophyll is now a very strong oxidizing agent; its electron "hole" must be filled.

② An enzyme extracts electrons from water and supplies them to P680, replacing each electron that the chlorophyll molecule lost when it absorbed light energy. This reaction splits a water molecule into two hydrogen ions and an oxygen atom, which immediately combines with another oxygen atom to form O_2. This is the water-splitting step of photosynthesis that releases O_2.

③ Each photoexcited electron passes from the primary electron acceptor of photosystem II to photosystem I via an electron transport chain. This chain is very similar to the one that functions in cellular respiration. The chloroplast version consists of an electron carrier called plastoquinone (Pq), a complex of two cytochromes (closely related to the cytochromes of mitochondria), and a copper-containing protein called plastocyanin (Pc).

④ As electrons cascade down the chain, their exergonic "fall" to a lower energy level is harnessed by the thylakoid membrane to produce ATP. This ATP synthesis is called photophosphorylation because it is driven by light energy.

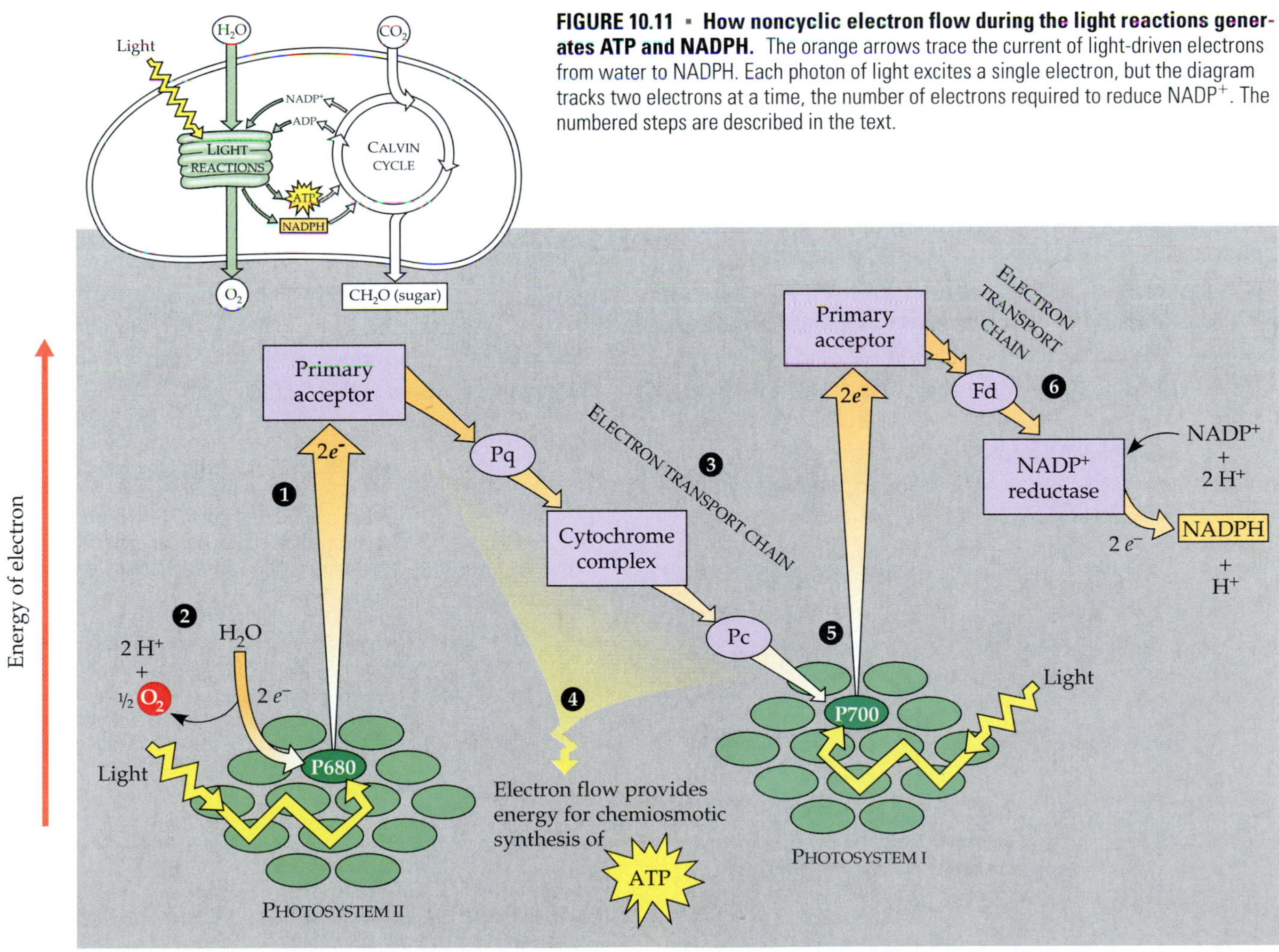

FIGURE 10.11 ▪ How noncyclic electron flow during the light reactions generates ATP and NADPH. The orange arrows trace the current of light-driven electrons from water to NADPH. Each photon of light excites a single electron, but the diagram tracks two electrons at a time, the number of electrons required to reduce $NADP^+$. The numbered steps are described in the text.

Specifically, ATP synthesis during noncyclic electron flow is called **noncyclic photophosphorylation**. (As we will discuss, the mechanism for photophosphorylation is chemiosmosis, the same process that operates in respiration.) This ATP generated by the light reactions will provide chemical energy for the synthesis of sugar during the Calvin cycle, the second major stage of photosynthesis.

⑤ When an electron reaches the "bottom" of the electron transport chain, it fills an electron "hole" in P700, the chlorophyll *a* molecule in the reaction center of photosystem I. This hole is created when light energy drives an electron from P700 to the primary acceptor of photosystem I.

⑥ The primary electron acceptor of photosystem I passes the photoexcited electrons to a second electron transport chain, which transmits them to ferredoxin (Fd), an iron-containing protein. An enzyme called $NADP^+$ reductase then transfers the electrons from Fd to $NADP^+$. This is the redox reaction that stores the high-energy electrons in NADPH, the molecule that will provide reducing power for the synthesis of sugar in the Calvin cycle.

The energy changes of electrons as they flow through the light reactions are analogous to the cartoon in FIGURE 10.12. As complicated as the scheme is, do not lose track of its functions: The light reactions use solar power to generate ATP and NADPH, which provide chemical energy and reducing power, respectively, to the sugar-making reactions of the Calvin cycle.

FIGURE 10.12 • A mechanical analogy for the light reactions.

Cyclic Electron Flow

Under certain conditions, photoexcited electrons take an alternative path called **cyclic electron flow**, which uses photosystem I but not photosystem II. You can see in FIGURE 10.13 that cyclic flow is a short circuit: The electrons cycle back from ferredoxin (Fd) to the cytochrome complex and from there continue on to the P700 chlorophyll. There is no production of NADPH and no release of oxygen. Cyclic flow does, however, generate ATP. This is called **cyclic photophosphorylation**, to distinguish it from noncyclic photophosphorylation.

What is the function of cyclic electron flow? Noncyclic electron flow produces ATP and NADPH in roughly equal quantities, but the Calvin cycle consumes more ATP than NADPH. Cyclic electron flow makes up the difference. The concentration of NADPH in the chloroplast may help regulate which pathway, cyclic versus noncyclic, electrons take through the light reactions. If the chloroplast runs low on ATP for the Calvin cycle, NADPH will begin to accumulate as the Calvin cycle slows down. The rise in NADPH may stimulate a temporary shift from noncyclic to cyclic electron flow until ATP supply catches up with demand.

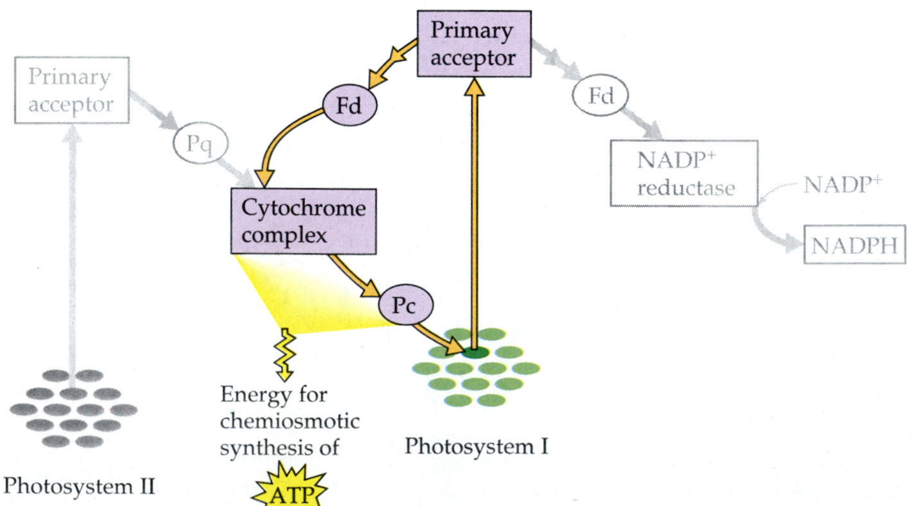

FIGURE 10.13 • Cyclic electron flow. Photoexcited electrons from photosystem I are occasionally shunted back from ferredoxin (Fd) to chlorophyll via the cytochrome complex and plastocyanin (Pc). This cyclic electron flow supplements the supply of ATP but produces no NADPH. (The two ferredoxin molecules shown in this diagram are actually one and the same—the final electron carrier in the electron transport chain of photosystem I. The "shadow" of noncyclic electron flow is included in the diagram for comparison with the cyclic route.)

Whether photophosphorylation is driven by noncyclic or cyclic electron flow, the actual mechanism for ATP synthesis is the same. This is a good time to review chemiosmosis, the basic process that uses membranes to couple redox reactions to ATP production.

A Comparison of Chemiosmosis in Chloroplasts and Mitochondria

Chloroplasts and mitochondria generate ATP by the same basic mechanism: chemiosmosis. An electron transport chain assembled in a membrane pumps protons across the membrane as electrons are passed through a series of carriers that are progressively more electronegative. Thus, electron transport chains transform redox energy to a proton-motive force, potential energy stored in the form of an H^+ gradient across a membrane. Built into the same membrane is an ATP synthase complex that couples the diffusion of hydrogen ions down their gradient to the phosphorylation of ADP. Some of the electron carriers, including the iron-containing proteins called cytochromes, are very similar in chloroplasts and mitochondria. The ATP synthase complexes of the two organelles are also very much alike. But there are noteworthy differences between oxidative phosphorylation in mitochondria and photophosphorylation in chloroplasts. In mitochondria, the high-energy electrons dropped down the transport chain are extracted by the oxidation of food molecules. Chloroplasts do not need food to make ATP; their photosystems capture light energy and use it to drive electrons to the top of the transport chain. In other words, mitochondria transfer chemical energy from food molecules to ATP, whereas chloroplasts transform light energy into chemical energy.

The spatial organization of chemiosmosis also differs in chloroplasts and mitochondria (FIGURE 10.14). The inner membrane of the mitochondrion pumps protons from the matrix out to the intermembrane space, which then serves as a reservoir of hydrogen ions that powers the ATP synthase. The thylakoid membrane of the chloroplast pumps protons from the stroma into the thylakoid space, which functions as the H^+ reservoir. The membrane makes ATP as the hydrogen ions diffuse from the thylakoid space back to the stroma through ATP synthase complexes, whose catalytic knobs are on the stroma side of the membrane. Thus, ATP forms in the stroma, where it is used to help drive sugar synthesis during the Calvin cycle.

The proton gradient, or pH gradient, across the thylakoid membrane is substantial. When chloroplasts are illuminated, the pH in the thylakoid space drops to about 5, and the pH in the stroma increases to about 8. This gradient of three pH units corresponds to a thousandfold difference in H^+ concentration. If in the laboratory the lights are turned off, the pH gradient is abolished, but it can quickly be restored by turning the lights back on. Such experiments add to the evidence described in Chapter 9 in support of the chemiosmotic model.

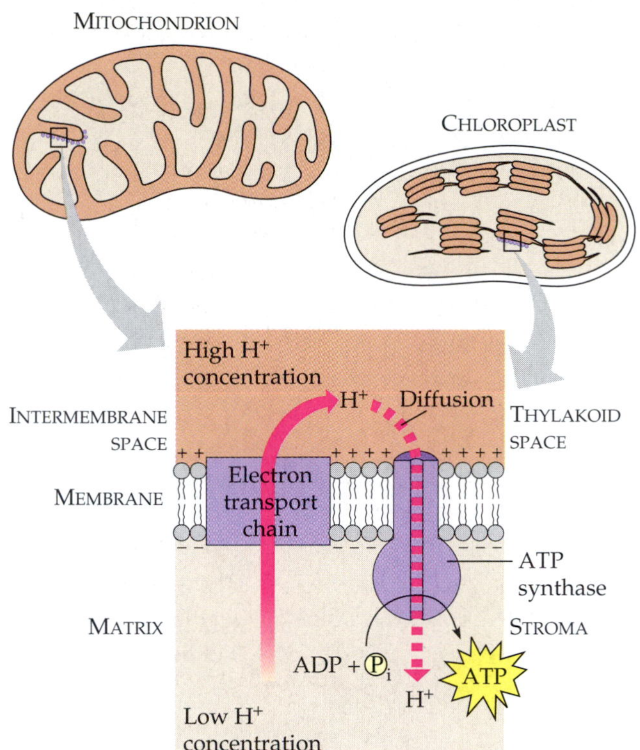

FIGURE 10.14 • The logistics of chemiosmosis in mitochondria and chloroplasts. The inner membrane of the mitochondrion pumps protons (H^+) from the matrix into the intermembrane space (darker brown). ATP is made on the matrix side of the membrane as hydrogen ions diffuse through ATP synthase complexes. In chloroplasts, the thylakoid membrane pumps protons from the stroma into the thylakoid space (lumen). As the hydrogen ions leak back across the membrane through the ATP synthase, phosphorylation of ADP occurs on the stroma side of the membrane.

Based on studies in several laboratories, FIGURE 10.15 on
10.2 page 180 shows a current model for the organization of the thylakoid membrane. Each of the molecules and molecular complexes in the figure are present in numerous copies in each thylakoid. Notice that NADPH, like ATP, is produced on the side of the membrane facing the stroma, where sugar is synthesized by the Calvin cycle.

Let's summarize the light reactions. Noncyclic electron flow pushes electrons from water, where they are at a low state of potential energy, to NADPH, where they are stored at a high state of potential energy. The light-driven electron current also generates ATP. Thus, the equipment of the thylakoid membrane converts light energy to the chemical energy stored in NADPH and ATP. (Oxygen is a by-product.) Let's now see how the Calvin cycle uses the products of the light reactions to synthesize sugar from CO_2.

The Calvin cycle uses ATP and NADPH to convert CO_2 to sugar: *a closer look*

The Calvin cycle is a metabolic pathway similar to the Krebs
10.3 cycle in that a starting material is regenerated after molecules enter and leave the cycle. Carbon enters the Calvin cycle in the

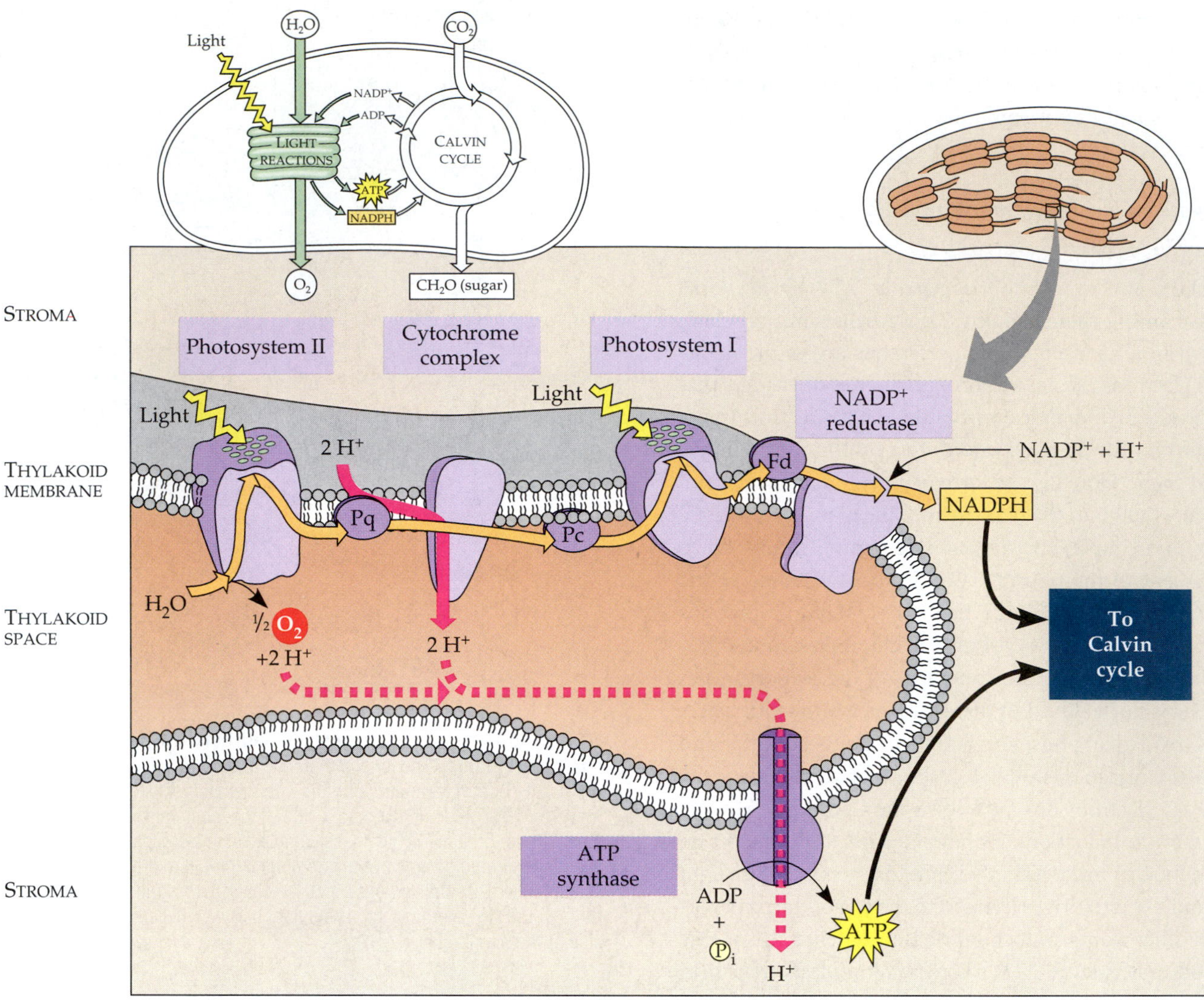

FIGURE 10.15 ▪ A tentative model for the organization of the thylakoid membrane. The orange arrows track electron flow. As electrons pass from carrier to carrier during redox reactions, hydrogen ions removed from the stroma are deposited in the thylakoid space, storing energy as a proton-motive force (H^+ gradient). At least three steps in the light reactions contribute to the proton gradient: Water is split by photosystem II on the side of the membrane facing the thylakoid space; as plastoquinone (Pq), a mobile carrier, transfers electrons to the cytochrome complex, protons are translocated across the membrane; and a hydrogen ion in the stroma is taken up by $NADP^+$ when it is reduced to NADPH. The diffusion of H^+ from the thylakoid space to the stroma (along the H^+ concentration gradient) powers the ATP synthase. These light-driven reactions store chemical energy in NADPH and ATP, which shuttle the energy to the sugar-producing Calvin cycle.

10.2

form of CO_2 and leaves in the form of sugar. The cycle spends ATP as an energy source and consumes NADPH as reducing power for adding high-energy electrons to make the sugar (FIGURE 10.16).

The carbohydrate produced directly from the Calvin cycle is actually not glucose, but a three-carbon sugar named **glyceraldehyde 3-phosphate (G3P)**. For the net synthesis of one molecule of this sugar, the cycle must take place three times, fixing three molecules of CO_2. (Recall that carbon fixation refers to the initial incorporation of CO_2 into organic material.) As we trace the steps of the cycle, keep in mind that we are following three molecules of CO_2 through the reactions.

FIGURE 10.16 divides the Calvin cycle into three phases:

Phase 1: Carbon fixation. The Calvin cycle incorporates each CO_2 molecule by attaching it to a five-carbon sugar named ribulose bisphosphate (abbreviated RuBP). The enzyme that catalyzes this first step is RuBP carboxylase, or **rubisco.** (It is the most abundant protein in chloroplasts, and probably the most abundant protein on Earth.) The product of the reaction is a six-carbon intermediate so unstable that it immediately splits in half to form two molecules of 3-phosphoglycerate (for each CO_2).

Phase 2: Reduction. Each molecule of 3-phosphoglycerate receives an additional phosphate group. An enzyme transfers the phosphate group from ATP, forming 1,3-bisphosphoglycerate as a product. Next, a pair of electrons donated from NADPH reduces 1,3-bisphosphoglycerate to G3P. Specifically, the electrons from NADPH reduce the carboyxl group of 3-phosphoglycerate to the carbonyl group of G3P, which stores more potential energy. G3P is a

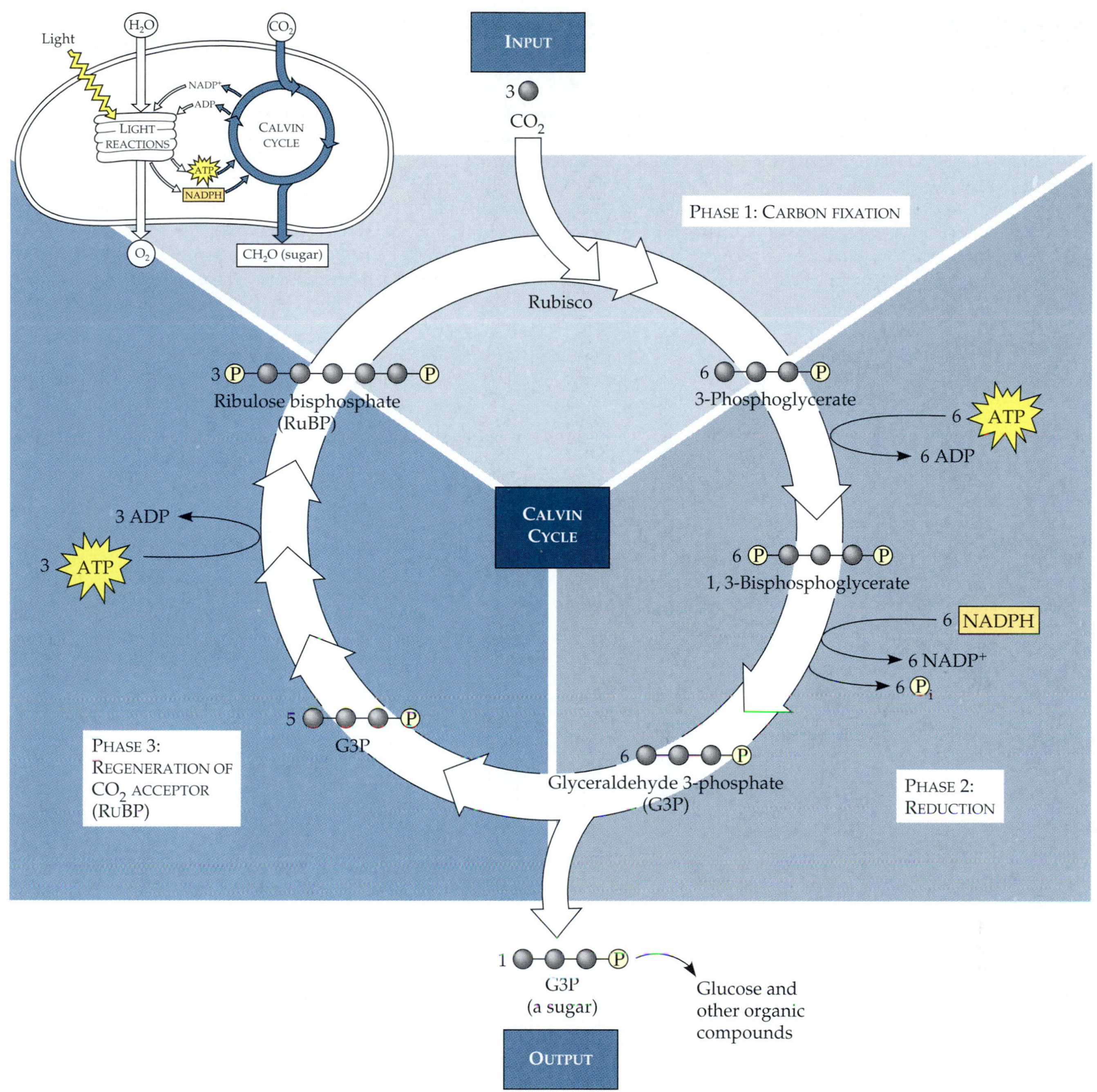

10.3 **FIGURE 10.16 ▪ The Calvin cycle.** This diagram tracks carbon atoms (gray balls) through the cycle. The three phases of the cycle correspond to the phases discussed in the text. For every three molecules of CO_2 that enter the cycle, the net output is one molecule of glyceraldehyde 3-phosphate (G3P), a three-carbon sugar. For each G3P synthesized, the cycle spends nine molecules of ATP and six molecules of NADPH. The light reactions sustain the Calvin cycle by regenerating ATP and NADPH.

sugar—the same three-carbon sugar formed in glycolysis by the splitting of glucose. Notice in FIGURE 10.16 that for every *three* molecules of CO_2, there are *six* molecules of G3P. But only one molecule of this three-carbon sugar can be counted as a net gain of carbohydrate. The cycle began with 15 carbons' worth of carbohydrate in the form of three molecules of the five-carbon sugar RuBP. Now there are 18 carbons' worth of carbohydrate in the form of six molecules of G3P. One molecule exits the cycle to be used by the plant cell, but the other five molecules must be recycled to regenerate the three molecules of RuBP.

Phase 3: Regeneration of CO_2 acceptor (RuBP). In a complex series of reactions, the carbon skeletons of five molecules of G3P are rearranged by the last steps of the Calvin cycle into three molecules of RuBP. To accomplish this, the cycle spends three more molecules of ATP. The RuBP is now prepared to receive CO_2 again, and the cycle continues.

For the net synthesis of one G3P molecule, the Calvin cycle consumes a total of nine molecules of ATP and six molecules

METHODS: Determining an Absorption Spectrum

Spectrophotometers are among the most widely used research instruments in biology. A spectrophotometer measures the relative amounts of light of different wavelengths absorbed and transmitted by a pigment solution. Inside the spectrophotometer, white light is separated into colors (wavelengths) by a prism. Then, one by one, the different colors of light are passed through the sample. The transmitted light strikes a photoelectric tube, which converts the light energy to electricity, and the electrical current is measured by a meter. Each time the wavelength of light is changed, the meter indicates the fraction of light transmitted through the sample or, conversely, the fraction of light absorbed. A graph that profiles absorption at different wavelengths is called an absorption spectrum. For example, the absorption spectrum for chlorophyll *a*, the form of chlorophyll most important in photosynthesis, has two peaks, corresponding to blue and red light. These are the colors chlorophyll *a* absorbs best (see FIGURE 10.7a). The absorption spectrum has a valley in the green region because the pigment transmits light of that color.

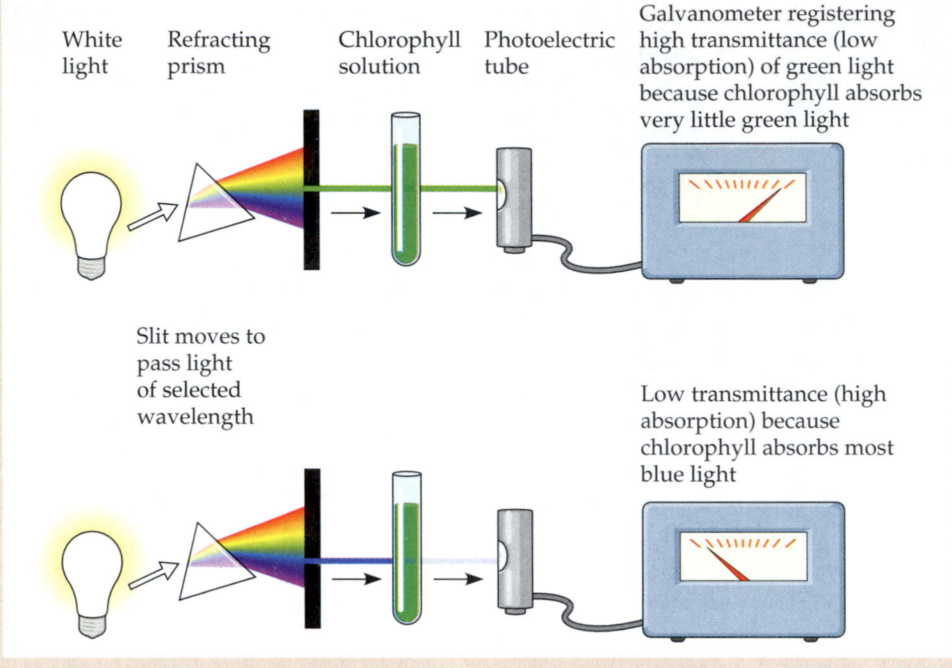

of NADPH. The light reactions regenerate the ATP and NADPH. The G3P spun off from the Calvin cycle becomes the starting material for metabolic pathways that synthesize other organic compounds, including glucose and other carbohydrates. Neither the light reactions nor the Calvin cycle alone can make sugar from CO_2. Photosynthesis is an emergent property of the intact chloroplast, which integrates the two stages of photosynthesis.

Alternative mechanisms of carbon fixation have evolved in hot, arid climates

Since plants first moved onto land about 425 million years ago, they have been adapting to the problems of terrestrial life, particularly the problem of dehydration. In Chapters 29 and 36 we will consider anatomical adaptations that help plants conserve water. Here we are concerned with metabolic adaptations. The solutions often involve trade-offs. An important example is the compromise between photosynthesis and the prevention of excessive water loss from the plant. The CO_2 required for photosynthesis enters a leaf via stomata, the pores through the leaf surface (see FIGURE 10.2). However, stomata are also the main avenues of transpiration, the evaporative loss of water from leaves. On a hot, dry day, most plants close their stomata, a response that conserves water. This response also reduces photosynthetic yield by limiting access to CO_2. With stomata even partially closed, CO_2 concentrations begin to decrease in the air spaces within the leaf, and the concentration of O_2 released from photosynthesis begins to increase. These conditions within the leaf favor a seemingly wasteful process called photorespiration.

Photorespiration: An Evolutionary Relic?

In most plants, initial fixation of carbon occurs via rubisco, the Calvin cycle enzyme that adds CO_2 to ribulose bisphosphate. Such plants are called **C_3 plants** because the first organic product of carbon fixation is a three-carbon compound, 3-phosphoglycerate (see FIGURE 10.16). Rice, wheat, and soybeans are among the C_3 plants that are important in agriculture. These plants produce less food when their stomata close on hot, dry days. The declining level of CO_2 in the leaf starves the Calvin cycle. Making matters worse, rubisco can accept O_2 in place of CO_2. As O_2 concentrations overtake CO_2 concentrations within the air spaces of the leaf, rubisco adds O_2 to the Calvin cycle instead of CO_2. The product splits, and one piece, a two-carbon compound, is exported from the chloroplast. Mitochondria and peroxisomes then break the two-carbon molecule down to CO_2. The process is called **photorespiration** because it occurs in the light *(photo)* and consumes O_2 *(respiration)*. However, unlike normal cellular respiration, photorespiration generates no ATP. And unlike photosynthesis, photorespiration produces no food. In fact, photorespiration *decreases* photosynthetic output by siphoning organic material from the Calvin cycle.

How can we explain the existence of a metabolic process that seems to be counterproductive to the plant? According to one hypothesis, photorespiration is evolutionary baggage—a metabolic relic from a much earlier time, when the atmosphere had less O_2 and more CO_2 than it does today. In the ancient atmosphere present when rubisco first evolved, the inability of the enzyme's active site to exclude O_2 would have made little difference. The hypothesis speculates that modern rubisco retains some of its ancestral affinity for O_2, which is now so concentrated in the atmosphere that a certain amount of photorespiration is inevitable.

It is not known whether photorespiration is beneficial to plants in any way. It *is* known that in many types of plants—including some of agricultural importance, such as soybeans—photorespiration drains away as much as 50% of the carbon fixed by the Calvin cycle. As heterotrophs that depend on carbon fixation in chloroplasts for our food, we naturally view photorespiration as wasteful. Indeed, if photorespiration could be reduced in certain plant species without otherwise affecting photosynthetic productivity, crop yields and food supplies would increase.

The environmental conditions that foster photorespiration are hot, dry, bright days—the conditions that cause stomata to close. In certain plant species, alternate modes of carbon fixation that minimize photorespiration—even in hot, arid climates—have evolved. The two most important of these photosynthetic adaptations are C_4 photosynthesis and CAM.

C_4 Plants

The **C_4 plants** are so named because they preface the Calvin cycle with an alternate mode of carbon fixation that forms a four-carbon compound as its first product. Several thousand species in at least 19 plant families use the C_4 pathway. Among the C_4 plants important to agriculture are sugarcane and corn, members of the grass family.

A unique leaf anatomy is correlated with the mechanism of C_4 photosynthesis (FIGURE 10.17a; compare to FIGURE 10.2). In C_4 plants, there are two distinct types of photosynthetic cells: bundle-sheath cells and mesophyll cells. **Bundle-sheath cells** are arranged into tightly packed sheaths around the veins of the leaf. Between the bundle sheath and the leaf surface are the more loosely arranged **mesophyll cells**. The Calvin cycle is confined to the chloroplasts of the bundle sheath. However, the cycle is preceded by incorporation of CO_2 into organic compounds in the mesophyll (FIGURE 10.17b). The first step is the addition of CO_2 to phosphoenolpyruvate (PEP) to form the four-carbon product oxaloacetate. The enzyme **PEP carboxylase** adds CO_2 to PEP. Compared to rubisco, PEP

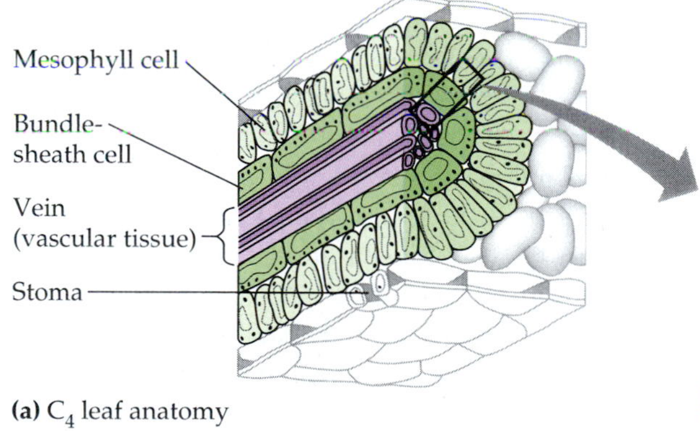

(a) C_4 leaf anatomy

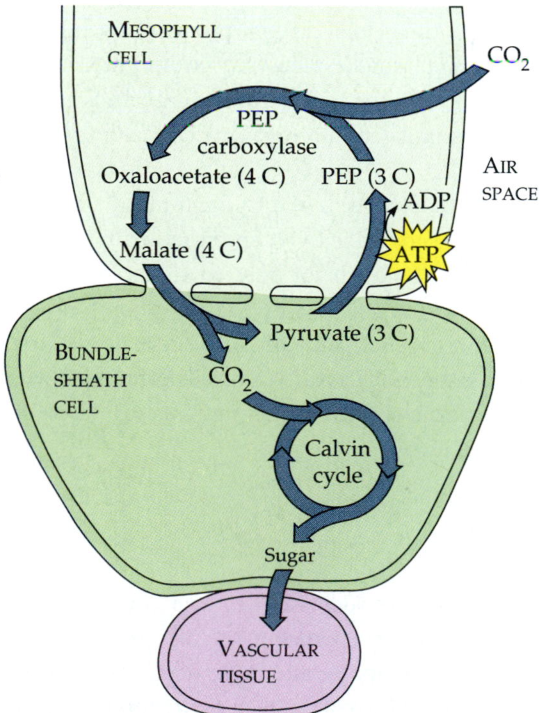

(b) The C_4 pathway

FIGURE 10.17 ▪ The C_4 anatomy and pathway. (a) Leaves of C_4 plants contain two types of photosynthetic cells: a cylinder of bundle-sheath cells surrounding the vein, and mesophyll cells located outside the bundle sheath. **(b)** Carbon dioxide is fixed in mesophyll cells by the enzyme PEP carboxylase. A four-carbon compound—malate, in this case—conveys the atoms of the CO_2 into a bundle-sheath cell, via plasmodesmata. There CO_2 is released and enters the Calvin cycle. This adaptation maintains a CO_2 concentration in the bundle sheath that favors photosynthesis over photorespiration.

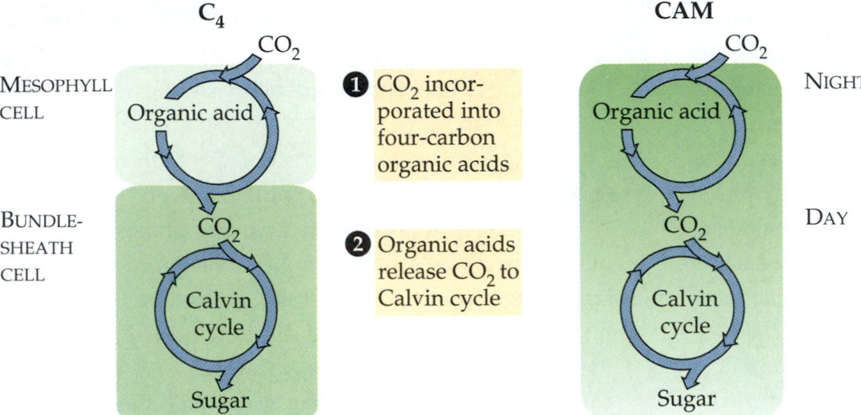

FIGURE 10.18 ▪ C_4 and CAM photosynthesis compared. Both adaptations are characterized by ① preliminary incorporation of CO_2 into organic acids, followed by ② transfer of the CO_2 to the Calvin cycle. In C_4 plants, such as sugarcane, these two steps are separated spatially; they are segregated into two cell types. In CAM plants, such as pineapple, the two steps are separated temporally; carbon fixation into organic acids occurs at night, and the Calvin cycle operates during the day. The C_4 and CAM pathways are two evolutionary solutions to the problem of maintaining photosynthesis with stomata partially or completely closed on hot, dry days.

carboxylase has a much higher affinity for CO_2. Therefore, PEP carboxylase can fix CO_2 efficiently when rubisco cannot—that is, when it is hot and dry and stomata are partially closed, causing CO_2 concentration in the leaf to fall and O_2 concentration to rise. After the C_4 plant fixes CO_2, the mesophyll cells export their four-carbon products to bundle-sheath cells through plasmodesmata (see FIGURE 7.28). Within the bundle-sheath cells, the four-carbon compounds release CO_2, which is reassimilated into organic material by rubisco and the Calvin cycle.

In effect, the mesophyll cells of a C_4 plant pump CO_2 into the bundle sheath, keeping the CO_2 concentration in the bundle-sheath cells high enough for rubisco to accept carbon dioxide rather than oxygen. In this way, C_4 photosynthesis minimizes photorespiration and enhances sugar production. This adaptation is especially advantageous in hot regions with intense sunlight, and it is in such environments that C_4 plants evolved and thrive today.

CAM Plants

A second photosynthetic adaptation to arid conditions has evolved in succulent (water-storing) plants (including ice plants), many cacti, pineapples, and representatives of several other plant families. These plants open their stomata during the night and close them during the day, just the reverse of how other plants behave. Closing stomata during the day helps desert plants conserve water, but it also prevents CO_2 from entering the leaves. During the night, when their stomata are open, these plants take up CO_2 and incorporate it into a variety of organic acids. This mode of carbon fixation is called **crassulacean acid metabolism**, or **CAM**, after the plant family Crassulaceae, the succulents in which the process was first discovered. The mesophyll cells of **CAM plants** store the organic acids they make during the night in their vacuoles until morning, when the stomata close. During the day, when the light reactions can supply ATP and NADPH for the Calvin cycle, CO_2 is released from the organic acids made the night before to become incorporated into sugar in the chloroplasts.

Notice in FIGURE 10.18 that the CAM pathway is similar to the C_4 pathway in that carbon dioxide is first incorporated into organic intermediates before it enters the Calvin cycle. The difference is that in C_4 plants, the initial steps of carbon fixation are separated structurally from the Calvin cycle, whereas in CAM plants the two steps occur at separate times. Keep in mind that CAM, C_4, and C_3 plants all eventually use the Calvin cycle to make sugar from carbon dioxide.

Photosynthesis is the biosphere's metabolic foundation: *a review*

In this chapter we have followed photosynthesis from photons to food (FIGURE 10.19). The light reactions capture solar

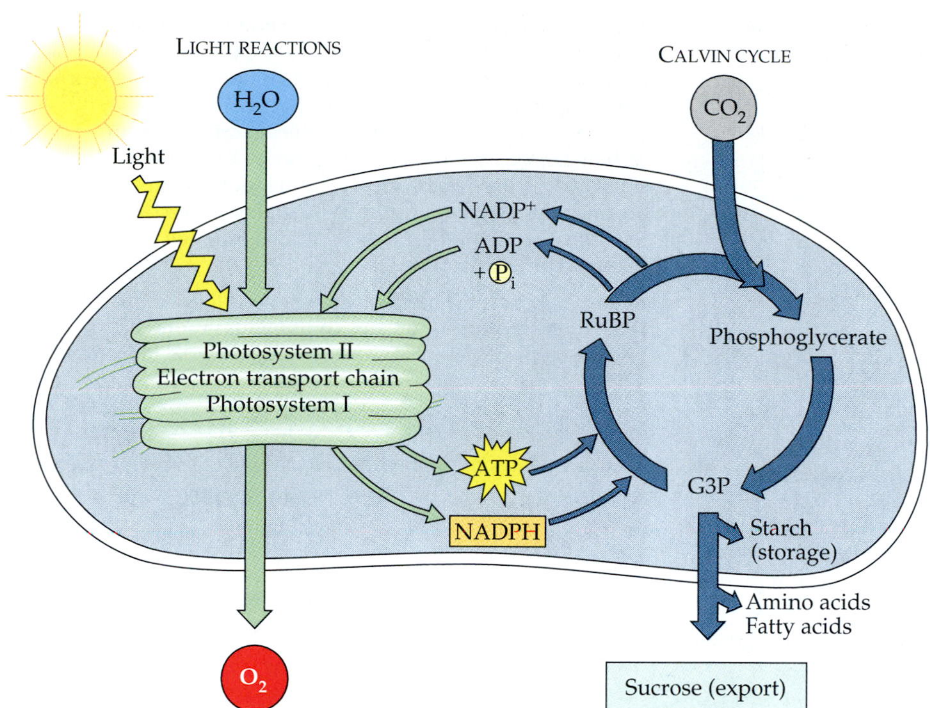

FIGURE 10.19 ▪ A review of photosynthesis. This diagram outlines the main reactants and products of photosynthesis as it occurs in the chloroplasts of plant cells. The light reactions convert light energy to the chemical energy of ATP and NADPH. The pigment and protein molecules that carry out the light reactions are found in the thylakoid membranes and include the molecules of two photosystems and electron transport chains. The light reactions split H_2O and release O_2 to Earth's atmosphere. The Calvin cycle, which takes place in the stroma of the chloroplast, uses ATP and NADPH to convert CO_2 to carbohydrate (three key compounds of the cycle are shown). The direct product of the Calvin cycle is the three-carbon sugar glyceraldehyde 3-phosphate (G3P). Enzymes in the chloroplast and cytosol convert this small sugar to a diversity of other organic compounds. The Calvin cycle returns ADP, inorganic phosphate, and $NADP^+$ to the light reactions. The entire ordered operation depends on the structural integrity of the chloroplast and its membranes.

energy and use it to make ATP and transfer electrons from water to $NADP^+$. The Calvin cycle uses the ATP and NADPH to produce sugar from carbon dioxide. The energy that entered the chloroplasts as sunlight becomes stored as chemical energy in organic compounds.

What are the fates of photosynthetic products? The sugar made in the chloroplasts supplies the entire plant with chemical energy and carbon skeletons to synthesize all the major organic molecules of cells. About 50% of the organic material made by photosynthesis is consumed as fuel for cellular respiration in the mitochondria of the plant cells. Sometimes there is a loss of photosynthetic products to photorespiration.

Technically, green cells are the only autotrophic parts of the plant. The rest of the plant depends on organic molecules exported from leaves via veins. In most plants, carbohydrate is transported out of the leaves in the form of sucrose, a disaccharide. After arriving at nonphotosynthetic cells, the sucrose provides raw material for cellular respiration and a multitude of anabolic pathways that synthesize proteins, lipids, and other products. A considerable amount of sugar in the form of glucose is linked together to make the polysaccharide cellulose, especially in plant cells that are still growing and maturing. Cellulose, the main ingredient of cell walls, is the most abundant organic molecule in the plant—and probably on the surface of the planet.

Most plants manage to make more organic material each day than they need to use as respiratory fuel and precursors for biosynthesis. They stockpile the extra sugar by synthesizing starch, storing some in the chloroplasts themselves and some in storage cells of roots, tubers, seeds, and fruits. In accounting for the consumption of the food molecules produced by photosynthesis, let's not forget that most plants lose leaves, roots, stems, fruits, and sometimes their entire bodies to heterotrophs, including humans.

On a global scale, the collective productivity of the minute chloroplasts is prodigious; it is estimated that photosynthesis makes about 160 billion metric tons of carbohydrate per year (a metric ton is 1000 kg, about 1.1 tons). That's organic matter equivalent to a stack of about 60 trillion copies of this textbook—17 stacks of books reaching from Earth to the sun! No other chemical process on the planet can match the output of photosynthesis. And no process is more important than photosynthesis to the welfare of life on Earth.

CHAPTER REVIEW

REVIEW OF KEY CONCEPTS

(with page numbers and key figures)

PHOTOSYNTHESIS IN NATURE

- **Plants and other autotrophs are the producers of the biosphere (pp. 168–169, FIGURE 10.1)** Autotrophs nourish themselves without ingesting organic molecules. Photoautotrophs use the energy of sunlight to synthesize organic molecules from CO_2 and H_2O. Heterotrophs ingest organic molecules from other organisms to get energy and carbon.
- **Chloroplasts are the sites of photosynthesis in plants (pp. 169–170, FIGURE 10.2)** In autotrophic eukaryotes, photosynthesis

occurs in chloroplasts, organelles containing thylakoid membranes that separate the thylakoid space from the chloroplast's stroma. Stacks of thylakoids form grana.

THE PATHWAYS OF PHOTOSYNTHESIS

- **Evidence that chloroplasts split water molecules enabled researchers to track atoms through photosynthesis (pp. 170–172, FIGURE 10.3)** Photosynthesis is summarized as:

$$6\,CO_2 + 12\,H_2O + \frac{\text{Light}}{\text{energy}} \longrightarrow C_6H_{12}O_6 + 6\,O_2 + 6\,H_2O$$

Experiments show that the chloroplast splits water into hydrogen and oxygen, incorporating the electrons of hydrogen into the bonds of sugar molecules. Photosynthesis is a redox process: H_2O is oxidized, CO_2 is reduced.

10.1 **The light reactions and the Calvin cycle cooperate in converting light energy to the chemical energy of food (pp. 172–173, FIGURE 10.4)** The light reactions in the grana produce ATP and split water, releasing O_2 and forming NADPH by transferring electrons from water to $NADP^+$. The Calvin cycle in the stroma forms sugar from CO_2, using ATP for energy and NADPH for reducing power.

10.2 **The light reactions convert solar energy to the chemical energy of ATP and NADPH (pp. 173–179, FIGURE 10.11)** Light is a form of electromagnetic energy, which travels in waves. The colors we see as visible light are a part of the electromagnetic spectrum. A pigment is a substance that absorbs visible light of specific wavelengths. The action spectrum of photosynthesis does not exactly match the absorption spectrum of chlorophyll *a*, the main photosynthetic pigment in plants, because accessory pigments (chlorophyll *b* and various carotenoids) absorb different wavelengths of light and pass the energy on to chlorophyll *a*.

A pigment goes from a ground state to an excited state when a photon boosts one of its electrons to a higher-energy orbital. The pigments of chloroplasts are built into the thylakoid membrane near molecules called primary electron acceptors, which trap the excited electrons before they return to the ground state. Pigment molecules are clustered in an antenna complex surrounding a chlorophyll *a* molecule at the reaction center. Photons absorbed anywhere in the antenna can pass their energy along to energize this chlorophyll *a*, which then passes an electron to a nearby primary electron acceptor. The antenna complex, the reaction-center chlorophyll, and the primary electron acceptor make up a photosystem, a light-harvesting unit built into the thylakoid membrane. There are two kinds of photosystems. Photosystem I contains P700 chlorophyll *a* molecules at the reaction center; photosystem II contains P680 molecules.

Noncyclic electron flow involves both photosystems and produces NADPH, ATP, and oxygen. Cyclic electron flow employs only photosystem I, producing ATP but no NADPH or O_2. ATP production during the light reactions is called photophosphorylation. The mechanism is chemiosmosis. The redox reactions of the electron transport chain that connects the two photosystems generate an H^+ gradient across the thylakoid membrane. An ATP synthase uses this proton-motive force to make ATP.

10.3 **The Calvin cycle uses ATP and NADPH to convert CO_2 to sugar (pp. 179–182, FIGURE 10.16)** The Calvin cycle is a metabolic pathway in the chloroplast stroma. An enzyme (rubisco) combines CO_2 with ribulose bisphosphate (RuBP), a five-carbon sugar. Then, using electrons from NADPH and energy from ATP, the cycle synthesizes the three-carbon sugar glyceraldehyde 3-phosphate. Most of the G3P is reused in the cycle to reconstitute RuBP, but some exits the cycle and is converted to glucose and other essential organic molecules.

- **Alternative mechanisms of carbon fixation have evolved in hot, arid climates (pp. 182–184, FIGURE 10.18)** On dry, hot days, plants close their stomata, conserving water. Oxygen from the light reactions builds up. When O_2 substitutes for CO_2 in the active site of rubisco, the product formed leaves the cycle and is oxidized to CO_2 and H_2O in the peroxisomes and mitochondria. This process, photorespiration, consumes organic fuel without producing ATP. C_4 plants avert photorespiration by incorporating CO_2 into four-carbon compounds in mesophyll cells. These compounds are exported to photosynthetic bundle-sheath cells, where they release carbon dioxide for use in the Calvin cycle. CAM plants open their stomata during the night, incorporating the CO_2 that enters into organic acids, which they store in mesophyll cells. During the day the stomata close, and the CO_2 is released from the organic acids for use in the Calvin cycle.

- **Photosynthesis is the biosphere's metabolic foundation (pp. 184–185, FIGURE 10.19)** The organic compounds produced by photosynthesis provide the energy and building material for ecosystems.

SELF-QUIZ

1. The light reactions of photosynthesis supply the Calvin cycle with
   a. light energy
   b. CO_2 and ATP
   c. H_2O and NADPH
   d. ATP and NADPH
   e. sugar and O_2

2. Which of the following sequences correctly represents the flow of electrons during photosynthesis?
   a. NADPH $\longrightarrow$ O_2 $\longrightarrow$ CO_2
   b. H_2O $\longrightarrow$ NADPH $\longrightarrow$ Calvin cycle
   c. NADPH $\longrightarrow$ chlorophyll $\longrightarrow$ Calvin cycle
   d. H_2O $\longrightarrow$ photosystem I $\longrightarrow$ photosystem II
   e. NADPH $\longrightarrow$ electron transport chain $\longrightarrow$ O_2

3. Which of the following conclusions does *not* follow from studying the absorption spectrum for chlorophyll *a* and the action spectrum for photosynthesis?
   a. Not all wavelengths are equally effective for photosynthesis.
   b. There must be accessory pigments that broaden the spectrum of light that contributes energy for photosynthesis.
   c. The red and blue areas of the spectrum are most effective in driving photosynthesis.
   d. Chlorophyll owes its color to the absorption of green light.
   e. Chlorophyll *a* has two absorption peaks.

4. Cooperation of the *two* photosystems of the chloroplast is required for
   a. ATP synthesis
   b. reduction of $NADP^+$
   c. cyclic photophosphorylation
   d. oxidation of the reaction center of photosystem I
   e. generation of a proton-motive force

5. In *mechanism*, photophosphorylation is most similar to
   a. substrate-level phosphorylation in glycolysis
   b. oxidative phosphorylation in cellular respiration
   c. the Calvin cycle
   d. carbon fixation
   e. reduction of $NADP^+$

6. In what respect are the photosynthetic adaptations of C_4 plants and CAM plants similar?
   a. In both cases, the stomata normally close during the day.
   b. Both types of plants make their sugar without the Calvin cycle.
   c. In both cases, an enzyme other than rubisco carries out the first step in carbon fixation.
   d. Both types of plants make most of their sugar in the dark.
   e. Neither C_4 plants nor CAM plants have grana in their chloroplasts.

7. Which of the following processes is most directly driven by light energy?

a. creation of a pH gradient by pumping protons across the thylakoid membrane
b. carbon fixation in the stroma
c. reduction of NADP molecules
d. removal of electrons from membrane-bound chlorophyll molecules
e. ATP synthesis

8. Which of the following statements is a correct distinction between cyclic and noncyclic photophosphorylation?
   a. Only noncyclic photophosphorylation produces ATP.
   b. In addition to ATP, noncyclic photophosphorylation also produces O_2 and NADPH.
   c. Only cyclic photophosphorylation utilizes light at 700 nm.
   d. Chemiosmosis is unique to noncyclic photophosphorylation.
   e. Only cyclic photophosphorylation can operate in the absence of photosystem II.

9. Which of the following statements is a correct distinction between autotrophs and heterotrophs?
   a. Only heterotrophs require chemical compounds from the environment.
   b. Cellular respiration is unique to heterotrophs.
   c. Only heterotrophs have mitochondria.
   d. Autotrophs, but not heterotrophs, can nourish themselves beginning with CO_2 and other nutrients that are entirely inorganic.
   e. Only heterotrophs require oxygen.

10. Which of the following processes could still occur in a chloroplast in the presence of an inhibitor that prevents H^+ from passing through ATP synthase complexes? (Explain your answer.)
   a. sugar synthesis
   b. generation of a proton-motive force
   c. photophosphorylation
   d. the Calvin cycle
   e. oxidation of NADPH

CHALLENGE QUESTION

The diagram below represents an experiment with isolated chloroplasts. The chloroplasts were first made acidic by soaking them in a solution at pH 4. After the thylakoid space reached pH 4, the chloroplasts were transferred to a basic solution at pH 8. The chloroplasts then made ATP, in the dark. Explain this result.

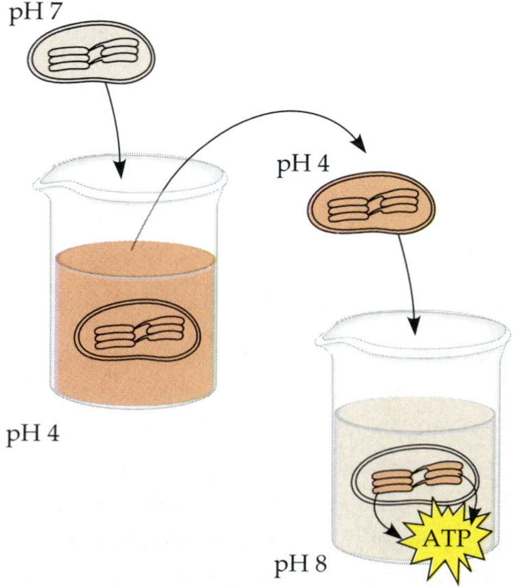

SCIENCE, TECHNOLOGY, AND SOCIETY

Tropical rain forests cover only about 3% of Earth's surface, but they are estimated to be responsible for more than 20% of global photosynthesis. It seems reasonable to expect that the lush growth of jungle foliage would produce large amounts of oxygen and reduce global warming by consuming carbon dioxide. But in fact, many experts now believe that rain forests make little or no *net* contribution to global oxygen production or reduction of global warming. Using your knowledge of photosynthesis and cellular respiration, explain what the basis of this hypothesis might be. What happens to the food produced by a rain forest tree when it is eaten by animals or the tree dies?

FURTHER READING

Bazzazz, F. A., and E. D. Fajer. "Plant Life in a CO_2-Rich World." *Scientific American*, January 1992. How will increasing atmospheric CO_2 and global warming affect the relative success of C_3 and C_4 plants?
Becker, W. M., J. B. Reece, and M. F. Poenie. *The World of the Cell*, 3rd ed. Menlo Park, CA: Benjamin/Cummings, 1996. Chapter 13 explains photosynthesis in some detail.
Caldwell, M. "The Amazing All-Natural Light Machine." *Discover*, December 1995. Discusses the elegant and efficient antenna complexes of a photosynthetic prokaryote.
Galston, A. W. "Photosynthesis as a Basis for Life Support on Earth and in Space." *Bioscience*, July/August 1992. Plants in space.
Govindjee and W. J. Coleman. "How Plants Make Oxygen." *Scientific American*, February 1990.
Hendry, G. "Oxygen, the Great Destroyer." *Natural History*, August 1992. Explores photorespiration and other problems associated with an oxygen-rich atmosphere.
Walker, D. *Energy, Plants, and Man*. Mill Valley, CA: University Science Books, 1992. Contains cartoons and stories that make bioenergetics fun.
Williams, N. "Mutant Alga Blurs Classic Picture of Photosynthesis." *Science*, July 19, 1996. Mutants with photosystem I disabled carry out photosynthesis with photosystem II alone.

WEB LINKS

Visit the special edition of *The Biology Place* for BIOLOGY, Fifth Edition, at **http://www.biology.com/campbell**. Go to Chapter 10 for online resources, including learning activities, practice exams, and links to the following web sites:

"Photosynthesis: An Interactive Study Guide"
Visit Campbell's Biology Place where you can use the activities of this tutorial to review photosynthesis. Your tutor is Dr. Graham Kent of Smith College.

"ASU Photosynthesis Center"
The main page of a comprehensive site on photosynthesis from the Arizona State University Photosynthesis Center (ASUPC). Includes links to global photosynthesis resources, including educational resources.

"Virtual Photosynthesis Experiments"
A chatty, interactive site at ASUPC that demonstrates some of the ways scientists study photosynthesis.

"Photosynthesis and Time"
An interactive "experiment" looking at the rate of the light reactions of photosynthesis.

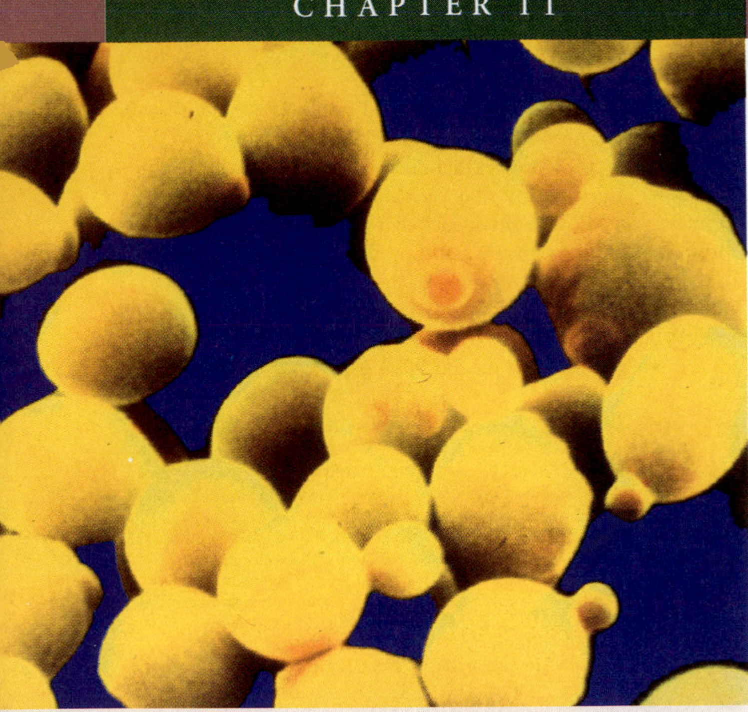

CHAPTER 11

CELL COMMUNICATION

An Overview of Cell Signaling
- Cell signaling evolved early in the history of life
- Communicating cells may be close together or far apart
- The three stages of cell signaling are reception, transduction, and response

Signal Reception and the Initiation of Transduction
- A signal molecule binds to a receptor protein, causing the protein to change shape
- Most signal receptors are plasma-membrane proteins

Signal-Transduction Pathways
- Pathways relay signals from receptors to cellular responses
- Protein phosphorylation, a common mode of regulation in cells, is a major mechanism of signal transduction
- Certain small molecules and ions are key components of signaling pathways (second messengers)

Cellular Responses to Signals
- In response to a signal, a cell may regulate activities in the cytoplasm or transcription in the nucleus
- Elaborate pathways amplify and specify the cell's response to signals

"Watch out! There's a car coming!"—We don't really need such a warning to remind us of the importance of communication in our lives as human beings. Perhaps less obvious is the critical role of communication in life at the cellular level. Cell-to-cell communication is absolutely essential for multicellular organisms. The billions of cells of a human or an oak tree must communicate in order to coordinate their activities in a way that enables the organism to develop from a fertilized egg and then survive and reproduce in turn. Communication among cells is also important for many unicellular organisms, such as the yeast pictured in the adjacent micrograph.

As you read in Chapter 1, regulation is a unifying theme in biology, and in recent years cell signaling has emerged as a universal mode of regulation in living things. In many lines of biological research, the same small set of cell-signaling mechanisms are showing up again and again. To the delight of scientists, studies of cell signaling are helping to answer some of the most important questions in biology and medicine—in areas ranging from embryological development to hormone action to the development of cancer and other kinds of disease.

The signals received by cells, whether originating from another cell or from some change in the organism's physical surroundings, take various forms. For instance, cells can sense and respond to electromagnetic signals, such as light, and to mechanical signals, such as touch. (In later units of this text, we discuss cellular detection of these and other types of physical signals.) However, cells most often communicate with each other using chemical signals. In this chapter we focus on the main mechanisms by which cells detect, process, and respond to chemical signals sent from other cells.

AN OVERVIEW OF CELL SIGNALING

What do cells talk about? What kinds of things does a "talking" cell say to a "listening" cell, and how does the latter cell respond to the message? Let's approach these questions by first looking at communication among microorganisms, for modern microbes are a window to the role of cell signaling in the evolution of life on Earth.

Cell signaling evolved early in the history of life

One topic of cell "conversation" is sex—at least for the yeast *Saccharomyces cerevisiae*, the fungus people have used for millennia for making bread, wine, and beer. Researchers have learned that cells of this yeast identify their mates by chemical signaling (FIGURE 11.1). There are two sexes, or mating types, called **a** and α. Cells of mating type **a** secrete a chemical signal called **a**-factor, which can bind to specific receptor proteins on

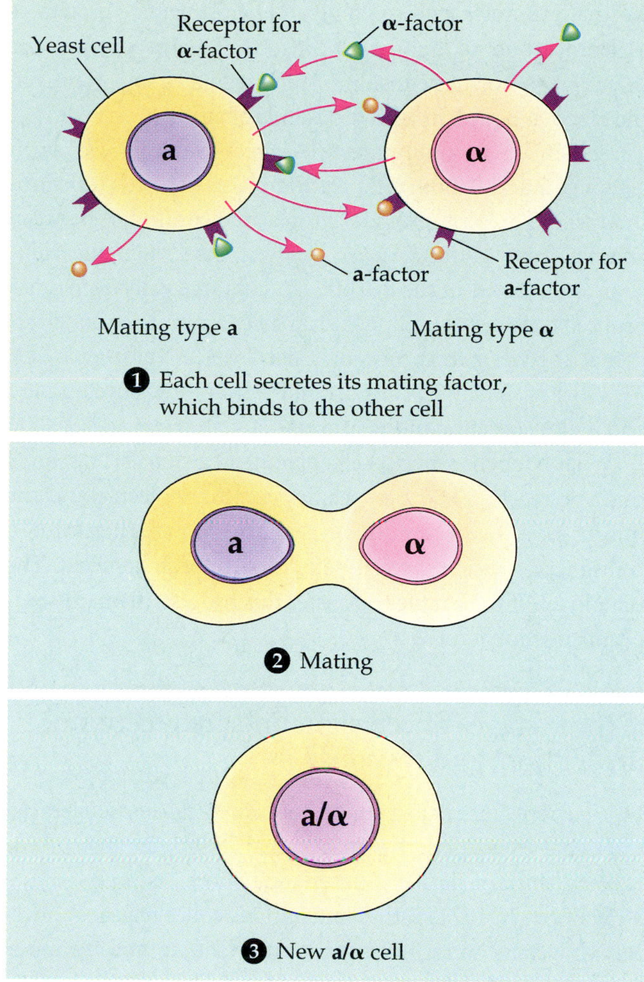

FIGURE 11.1 ▪ Communication between mating yeast cells. Cells of the yeast *Saccharomyces cerevisiae* use chemical signaling to identify cells of opposite mating type and initiate the mating process. ① Cells of mating type **a** release **a**-factor, which binds to receptors on nearby cells of mating type α. Meanwhile, α cells release α-factor, which binds to specific receptors on **a** cells. (Both factors are peptides about 12 amino acids in length.) ② Binding of the factors to receptors induces changes in the cells that lead to their fusion, or mating. ③ The resulting **a**/α cell combines in its nucleus all the genes from the **a** and α cells.

nearby α cells. At the same time, α cells secrete α-factor, which binds to receptors on **a** cells. Without actually entering the cells, the receptor-bound molecules of the two mating factors cause the cells to grow toward each other and bring about other cellular changes. The result is the fusion, or mating, of two cells of opposite type. The new **a**/α cell contains all the genes of both original cells, a combination of genetic resources that provides advantages to this cell's descendants.

How is the mating signal at the yeast cell surface transduced, or changed, into a form that brings about the cellular response of mating? The process by which a signal on a cell's surface is converted into a specific cellular response, a series of steps called a **signal-transduction pathway**, has been extensively studied in both yeast and animal cells. Amazingly, the molecular details of signal transduction in yeast and mammals are strikingly similar, even though the last common ancestor of these two groups of organisms lived over a billion years ago. These similarities—and others newly uncovered between signaling systems in bacteria and plants—suggest that early versions of the cell-signaling mechanisms used today evolved well before the first multicellular creatures appeared on Earth. Scientists think that signaling mechanisms evolved first in ancient prokaryotes and single-celled eukaryotes and were then adopted for new uses by their multicellular descendants. Meanwhile, cell signaling has remained important in the microbial world. FIGURE 11.2 shows an example in a sophisticated modern bacterium.

Communicating cells may be close together or far apart

Like microbes, cells in a multicellular organism usually communicate by releasing chemical messengers targeted for cells that may not be immediately adjacent. Some messengers travel only short distances: The transmitting cell secretes molecules of a **local regulator**, a substance that influences cells in

FIGURE 11.2 ▪ Communication among bacteria. Soil-dwelling bacteria called myxobacteria ("slime bacteria") use chemical signaling to share information about nutrient availability. When food is scarce, starving cells secrete a molecule that enters neighboring cells and stimulates them to aggregate. The cells form a structure that produces thick-walled spores capable of surviving until the environment improves. The bacteria shown here are *Myxococcus xanthus (SEMs)*.

the vicinity (FIGURE 11.3a). One class of local regulators in animals, growth factors, are compounds that stimulate nearby target cells to grow and multiply. Numerous cells can simultaneously receive and respond to the molecules of growth factor produced by a single cell in their vicinity. This type of local signaling in animals is called paracrine signaling.

Another, more specialized type of local signaling occurs in the animal nervous system. Here a nerve cell produces a chemical signal, a neurotransmitter, that diffuses to a single target cell that is almost touching the first cell. An electrical signal transmitted the length of the nerve cell triggers the secretion of neurotransmitter molecules into the synapse, the narrow space between the nerve cell and its target cell (often another nerve cell). Because specific nerve cells are so close together at synapses, a nerve signal can travel from your brain to your big toe, for example, without causing unwanted responses in other parts of your body.

Local signaling in plants is less well understood. Because of their cell walls, plants must use some mechanisms different from those operating locally in animals.

Both animals and plants use chemicals called **hormones** for signaling at greater distances. In hormonal signaling in animals, also known as endocrine signaling, specialized cells release hormone molecules into vessels of the circulatory system, by which they travel to target cells in other parts of the body (FIGURE 11.3b). In plants, hormones sometimes travel in vessels but more often reach their targets by moving through cells (see FIGURE 39.4) or by diffusion through the air as a gas. Hormones range widely in molecular size and type, as do local regulators. For instance, the plant hormone ethylene, a gas that promotes fruit ripening and helps regulate growth, is a hydrocarbon of only six atoms (C_2H_4). In contrast, the mammalian hormone insulin, which regulates sugar levels in the blood, is a protein with thousands of atoms.

Cells may also communicate by direct contact, as we saw in Chapters 7 and 8. Both animals and plants have cell junctions that, where present, provide cytoplasmic continuity between adjacent cells (FIGURE 11.4a). In these cases, signaling substances dissolved in the cytosol can pass freely between adjacent cells. Moreover, animal cells may communicate via direct contact between molecules on their surfaces (FIGURE 11.4b). This sort of signaling is important in embryonic development and in the operation of the immune system.

What happens when a cell encounters a signal? The signal must be recognized by a specific receptor molecule and the information it carries must be changed into another form—transduced—inside the cell before the cell can respond. The remainder of the chapter discusses this process, primarily as it occurs in animal cells.

The three stages of cell signaling are reception, transduction, and response

Our current understanding of how chemical messengers act via signal-transduction pathways had its origins in the pioneering work of Earl W. Sutherland, whose research led to a Nobel Prize in 1971. Sutherland and his colleagues at Vanderbilt University were investigating how the animal hormone epinephrine stimulates breakdown (depolymerization) of the storage polysaccharide glycogen within liver cells and skeletal

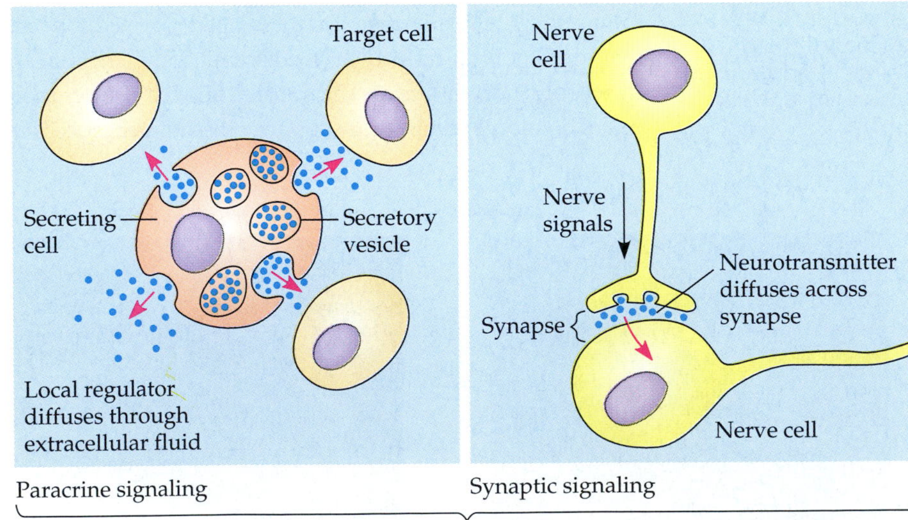

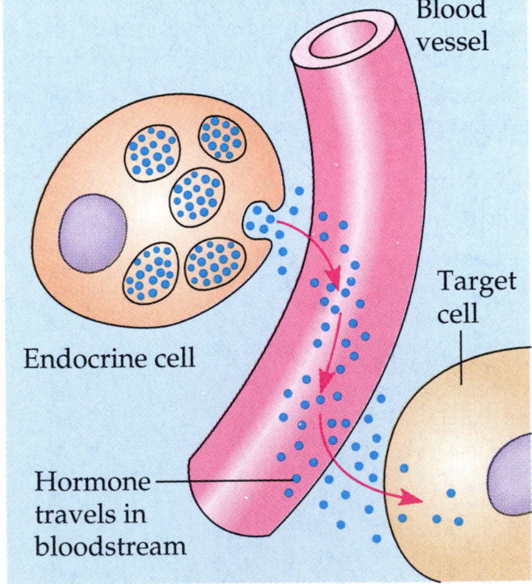

FIGURE 11.3 ▪ Local and distant cell communication in animals. (a) Animals have two main kinds of local chemical signaling. In paracrine signaling, a secreting cell acts on nearby target cells by discharging molecules of a local regulator into the extracellular fluid. In synaptic signaling, a nerve cell releases neurotransmitter molecules into a synapse, the narrow space between the transmitting cell and the target cell, here another nerve cell. **(b)** Hormones signal target cells at much greater distances. In animals, specialized endocrine cells secrete hormones into body fluids, often the blood. Hormones may reach virtually all body cells, but, as with local regulators, only specific target cells recognize and respond to a given chemical signal. (Plants also use hormones for signaling from one part of the plant to another.)

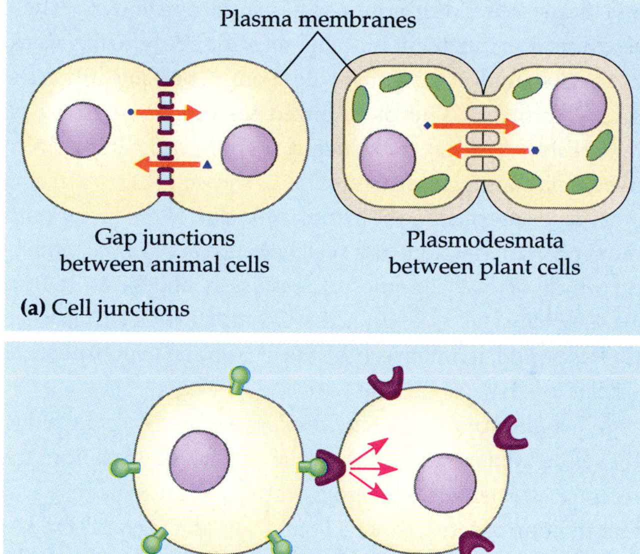

FIGURE 11.4 • Communication by direct contact between cells. **(a)** Both animals and plants have cell junctions that allow molecules to pass readily between adjacent cells without crossing plasma membranes. **(b)** Two cells in an animal may communicate by interaction between molecules protruding from their surfaces.

muscle cells. Glycogen depolymerization releases the sugar glucose-1-phosphate, which the cell converts to glucose-6-phosphate. The cell can then use this compound, an early intermediate in glycolysis, for energy production. Alternatively, the compound can be stripped of phosphate and released into the blood as glucose that can fuel cells throughout the body. Thus one effect of epinephrine, which is secreted from the adrenal gland during times of physical or mental stress, is the mobilization of fuel reserves.

Sutherland's research team discovered that epinephrine stimulates glycogen breakdown by somehow activating a cytosolic enzyme, glycogen phosphorylase. However, when epinephrine was added to a test-tube mixture containing the phosphorylase and its substrate, glycogen, no depolymerization occurred. Epinephrine could activate glycogen phosphorylase only when the hormone was added to a solution containing *intact* cells. This result told Sutherland two things. First, epinephrine does not interact directly with the enzyme responsible for glycogen breakdown; an intermediate step or series of steps must be occurring inside the cell. Second, the plasma membrane is somehow involved in transmitting the epinephrine signal.

11.1 Thus Sutherland's early work suggested that the process going on at the receiving end of a cellular conversation can be dissected into three stages: reception, transduction, and response (FIGURE 11.5):

1. Reception is the target cell's detection of a signal coming from outside the cell. A chemical signal is "detected" when it binds to a cellular protein, usually at the cell's surface.
2. The binding of the signal molecule changes the receptor protein in some way, thus initiating the process of transduction. The transduction stage converts the signal to a form that can bring about a specific cellular response. In Sutherland's system, the binding of epinephrine to the outside of a receptor protein in a liver cell's plasma membrane leads via a series of steps to activation of glycogen phosphorylase. Transduction sometimes occurs in a single step but more often requires a sequence of changes in a series of different molecules—a signal-transduction *pathway.* The molecules in the pathway are often called relay molecules.
3. In the third stage of cell signaling, the transduced signal finally triggers a specific cellular response. The response may be almost any imaginable cellular activity—such as catalysis by an enzyme (such as glycogen phosphorylase), rearrangement of the cytoskeleton, or activation of specific genes in the nucleus. The cell-signaling process helps ensure that crucial activities like these occur in the right cells, at the right time, and in proper coordination with the other cells of the organism. We'll now explore the mechanisms of cell signaling in more detail.

11.1

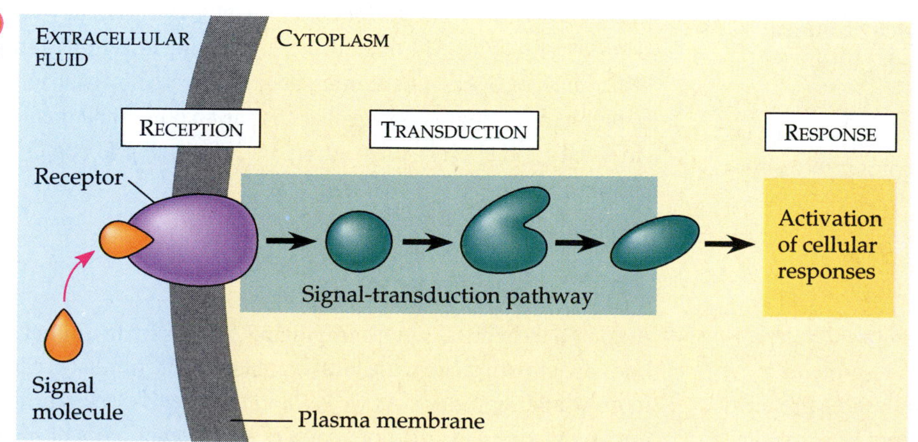

FIGURE 11.5 • Overview of cell signaling. From the perspective of the cell receiving the message, cell signaling can be divided into three stages: signal reception, signal transduction, and cellular response. When reception occurs at the plasma membrane, as shown here, the transduction stage is usually a pathway of several steps, with each molecule in the pathway bringing about a change in the next. The last molecule in the pathway triggers the cell's response.

SIGNAL RECEPTION AND THE INITIATION OF TRANSDUCTION

When we speak to someone, others nearby may hear our message, sometimes with unfortunate consequences. However, errors of this kind rarely occur between cells. The signals emitted by an α yeast cell are "heard" only by its prospective mates, **a** cells. Similarly, although epinephrine encounters many types of cells as it circulates in the blood, only certain target cells detect and react to the hormone. The signal receptor is the identity tag on the target cell.

A signal molecule binds to a receptor protein, causing the protein to change shape

A cell targeted by a particular chemical signal has molecules of a receptor protein that recognizes the signal molecule. The signal molecule is complementary in shape to a specific site on the receptor and attaches there, like a key in a lock—or like a substrate in the catalytic site of an enzyme. The signal molecule behaves as a **ligand**, the term for a small molecule that specifically binds to a larger one. Ligand binding generally causes a receptor protein to undergo a change in conformation—that is, to change shape. For many receptors, this shape change directly activates the receptor so that it can interact with another cellular molecule. For other kinds of receptors, as we'll see shortly, the immediate effect of ligand binding is more limited, mainly causing the aggregation of two or more receptor molecules.

Most signal receptors are plasma-membrane proteins

11.2 Most signal molecules are water-soluble and too large to pass freely through the plasma membrane. But, as Sutherland learned for epinephrine, they still influence cellular activity in major ways. Like yeast mating factors, most water-soluble signal molecules bind to specific sites on receptor proteins embedded in the cell's plasma membrane. Such a receptor transmits information from the extracellular environment to the inside of the cell by changing shape or aggregating when a specific ligand binds to it.

We'll see how membrane receptors work by looking at three major types: G-protein-linked receptors, tyrosine-kinase receptors, and ion-channel receptors.

G-Protein-Linked Receptors

A **G-protein-linked receptor** is a plasma-membrane receptor that works with the help of a protein called a G protein (FIGURE 11.6a). Many different signal molecules use G-protein-linked receptors, including yeast mating factors, epinephrine and many other hormones, and neurotransmitters. These receptors vary in their binding sites for recognizing signal molecules and for recognizing different G proteins inside the cell. Nevertheless, G-protein-linked receptor proteins are all remarkably similar in structure. They each have seven α-helices spanning the membrane, as shown in FIGURE 11.7.

Loosely attached to the cytoplasmic side of the membrane, the **G protein** functions as a switch that is on or off depending on which of two guanine nucleotides is attached, GDP or GTP. When GDP is bound, the G protein is inactive; when GTP is bound, it is active. (GTP, or guanosine triphosphate, is similar to ATP.)

When the appropriate chemical signal binds as a ligand to the extracellular side of a G-protein-linked receptor, the receptor is activated, changing conformation in such a way that it, in turn, can activate a G protein: The receptor binds a specific, inactive G protein and causes a GTP to displace the GDP (FIGURE 11.6b). The activated G protein then binds to another protein, usually an enzyme, and alters *its* activity. These changes are only temporary, however, for the G protein also functions as a GTPase enzyme and soon hydrolyzes its bound GTP to GDP (FIGURE 11.6c). Now inactive again, the G protein leaves the enzyme. The GTPase function of the G protein allows the pathway to shut down rapidly when the extracellular signal molecule is no longer present.

G-protein receptor systems are extremely widespread and diverse in their functions. In addition to the functions already mentioned, they are important in embryonic development, as indicated by genetic studies. For instance, mutant mouse embryos lacking a certain G protein do not develop normal blood vessels and die in utero. Furthermore, G proteins are involved in sensory reception; in humans, for example, both vision and smell depend on such proteins. Similarities in structure among G proteins and G-protein-linked receptors of modern organisms suggest that G proteins and G-protein-linked receptors evolved very early, possibly as sensory receptors of ancient microbes.

G-protein systems are involved in many human diseases, including bacterial infections. The bacteria that cause cholera, pertussis (whooping cough), and botulism, among others, make their victims ill by producing toxins that interfere with G-protein function. Although drugs for treating infections and other kinds of diseases have often been discovered by trial and error, pharmacologists now realize that up to 60% of all medicines used today exert their effects by influencing G-protein pathways.

Tyrosine-Kinase Receptors

Among the chemical signals impinging on cells in an animal's body are growth factors, the local regulators that stimulate cells to grow and reproduce. As we'll see in Chapter 12, cell reproduction involves a variety of activities by different parts of the

11.2

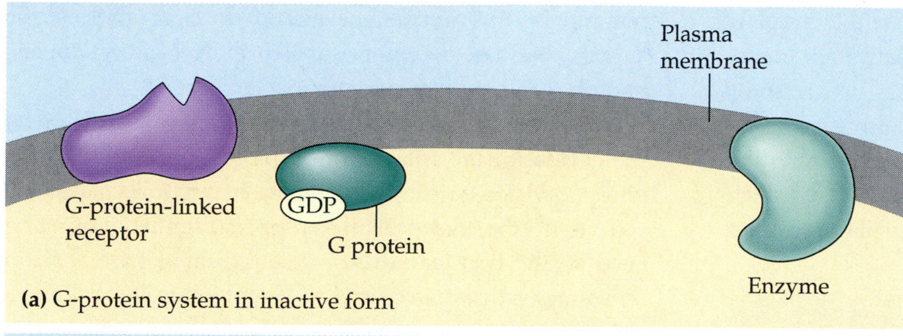

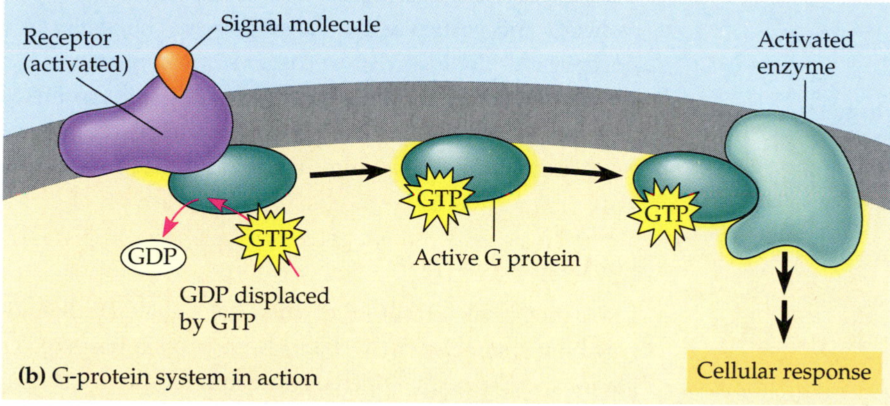

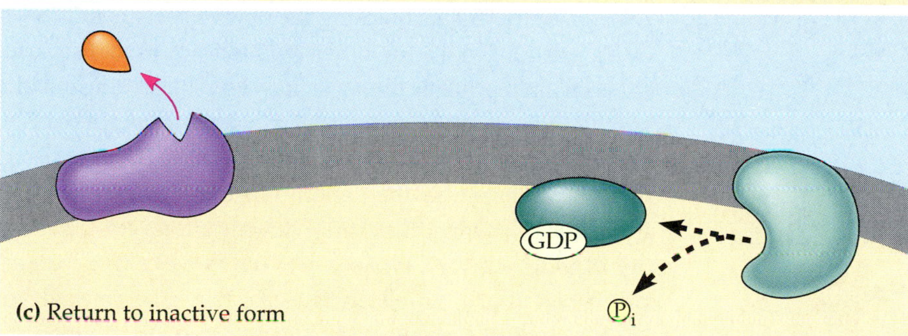

FIGURE 11.6 ▪ The functioning of a G-protein-linked receptor. (a) This type of receptor is a membrane protein that works in conjunction with a G protein and another protein, usually an enzyme. In the absence of the extracellular signal molecule specific for the receptor, all three proteins are in inactive form. The inactive G protein has a GDP molecule bound to it. **(b)** When the signal molecule binds to the receptor, the receptor changes shape in such a way that it binds and activates the G protein. A molecule of GTP replaces the GDP on the G protein. The active G protein (moving freely along the membrane) binds to and activates the enzyme, which triggers the next step in the pathway leading to the cell's responses. **(c)** The G protein then catalyzes the hydrolysis of its GTP and dissociates from the enzyme, becoming available for reuse. All three proteins remain attached to the plasma membrane.

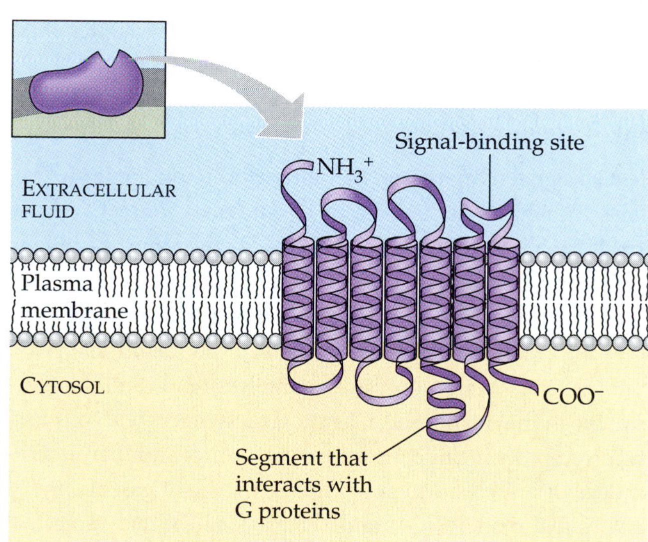

11.2 **FIGURE 11.7 ▪ The structure of a G-protein-linked receptor.** A large family of eukaryotic receptor proteins have this secondary structure: The single polypeptide, represented as a ribbon, has seven transmembrane α helices. Specific loops correspond to the sites where signal molecules and G-protein molecules bind. The α helices are depicted as cylinders for emphasis.

cell, including protein synthesis in the cytoplasm, chromosome duplication in the nucleus, and the rearrangement of elements of the cytoskeleton. Helping the cell regulate and coordinate these activities is a type of receptor specialized for triggering more than one signal-transduction pathway at once.

The receptor for a growth factor is often a tyrosine-kinase receptor, one of a major class of plasma-membrane receptors characterized by having enzymatic activity. Part of the receptor protein on the cytoplasmic side of the membrane functions as an enzyme, called a **tyrosine kinase**, that catalyzes the transfer of phosphate groups from ATP to the amino acid tyrosine on a substrate protein. Thus **tyrosine-kinase receptors** are membrane receptors that attach phosphates to protein tyrosines.

Tyrosine-kinase receptors often have the structure shown in rough schematic form in FIGURE 11.8, page 194. Before the signal molecule binds, the receptors exist as individual polypeptides. Notice that each has an extracellular signal-binding site, an intracellular tail containing a number of tyrosines, and a single α helix spanning the membrane. The binding of a signal molecule to such a receptor does not cause enough of a conformational change to activate the cytoplasmic side of the protein directly. Instead, the activation occurs in three steps: (1) The ligand binding causes two receptor polypeptides to aggregate, forming a dimer (a protein consisting of two polypeptides). (2) This aggregation activates the tyrosine-kinase parts of both polypeptides, each of which then (3) phosphorylates the tyrosines on the tail of the other polypeptide. In summary, the effect of the signal molecule on a tyrosine-kinase receptor is polypeptide aggregation and phosphorylation of the receptor.

The receptor protein is now recognized by specific relay proteins inside the cell. Each such protein binds to a specific phosphorylated tyrosine, undergoing a structural change that activates it (the relay protein may or may not be phosphorylated by the tyrosine kinase). One tyrosine-kinase receptor dimer may activate ten or more different intracellular proteins simultaneously, triggering as many different transduction pathways and particular cellular responses. The ability of a single ligand-binding event to trigger so many pathways is a key difference between these receptors and G-protein-linked receptors. Abnormal tyrosine-kinase receptors that aggregate even without ligand cause some kinds of cancer.

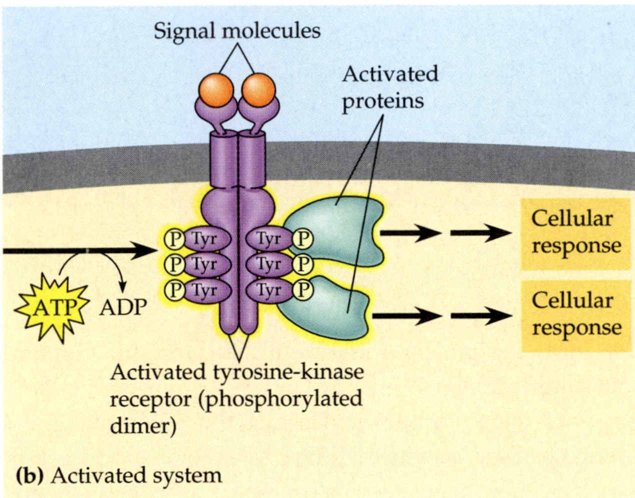

(a) Inactive tyrosine-kinase receptor system

(b) Activated system

11.2 **FIGURE 11.8 ▪ The structure and function of a tyrosine-kinase receptor. (a)** In the absence of specific signal molecules, tyrosine-kinase receptors exist as single polypeptides in the plasma membrane. The extracellular portion of the protein, with the signal-molecule binding site, is connected by a single transmembrane α helix to the protein's cytoplasmic portion. This part of the protein is responsible for the receptor's tyrosine-kinase activity and also has a series of tyrosine amino acids. **(b)** When signal molecules (such as a growth factor) attach to their binding sites, two polypeptides aggregate, forming a dimer. Using phosphate groups from ATP, the tyrosine-kinase region of each polypeptide phosphorylates the tyrosines on the other polypeptide. In other words, the dimer is both an enzyme and its own substrate. Now fully activated, the receptor protein can bind specific intracellular proteins, which attach to particular phosphorylated tyrosines and are themselves activated. Each can then initiate a signal-transduction pathway leading to a specific cellular response. Tyrosine-kinase receptors often activate several different signal-transduction pathways at once, helping regulate such complicated functions as cell reproduction (cell division). Inappropriate activation of these receptors can lead to uncontrolled cell growth—cancer (see Chapter 19).

Ion-Channel Receptors

Some membrane receptors of chemical signals are **ligand-gated ion channels**. These channels are protein pores in the plasma membrane that open or close in response to a chemical signal, allowing or blocking the flow of specific ions, such as Na^+ or Ca^{2+}. Like the other receptors we have discussed, these channel proteins bind a signal molecule as a ligand at a specific site on their extracellular side (FIGURE 11.9). The shape change produced in the channel protein immediately leads to a change in the concentration of a particular ion inside the cell. Often this change directly affects cell functioning in some way. At a synapse between nerve cells, for example, it may trigger an electrical signal that propagates down the length of the receiving cell. Ligand-gated ion channels are very important in the nervous system, as are gated ion channels that are controlled by electrical signals (see Chapter 48).

Intracellular Receptors

Not all signal receptors are membrane proteins. Some are proteins located in the cytoplasm or nucleus of target cells. To reach such a receptor, a chemical messenger must be able to pass through the target cell's plasma membrane. A number of important signaling molecules can do just that, either because they are small enough to pass between the membrane phospholipids or because they are themselves lipids and therefore soluble in the membrane. Chemical messengers with intracellular receptors include the steroid hormones and thyroid hormones of animals, which are lipids; the small gaseous molecule nitric oxide (NO); and certain small signal molecules used by bacteria.

The behavior of testosterone is representative of steroid hormones: Secreted by cells of the testis, the hormone travels through the blood and enters cells all over the body. In target cells—those with testosterone receptor molecules in their cytosol—the hormone binds to the receptor, activating it. The

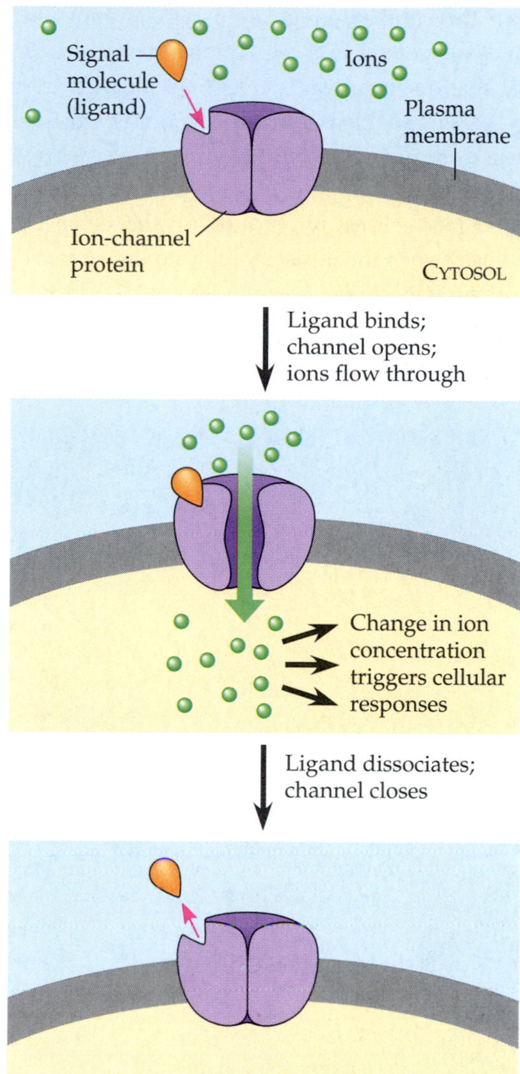

11.2 **FIGURE 11.9 ▪ A ligand-gated ion-channel receptor.** This signal receptor is a transmembrane protein in the plasma membrane that opens to allow the flow of a specific kind of ion across the membrane when a specific signal molecule binds to the extracellular side of the protein.

active form of the receptor protein can then bind to and turn on genes in the nucleus that control male sex characteristics. Thus, the receptor carries out the complete transduction of the signal. We will look more closely at steroids and other hormones with intracellular receptors in Chapter 45. In the next section we discuss the signal-transduction pathways often triggered by membrane receptors.

SIGNAL-TRANSDUCTION PATHWAYS

When signal receptors are plasma-membrane proteins, like most of those we have discussed, the transduction stage of cell signaling is usually a multistep pathway. One benefit of such pathways is signal amplification. If some of the molecules in a pathway transmit the signal to multiple molecules of the next component in the series, the result can be a large number of activated molecules at the end of the pathway. In other words, a very small number of extracellular signal molecules can produce a major cellular response. Moreover, multistep pathways provide more opportunities for coordination and regulation than simpler systems do, as we'll discuss later.

Pathways relay signals from receptors to cellular responses

The binding of a specific extracellular signal molecule to a receptor in the plasma membrane triggers the first step in the chain of molecular interactions—the signal-transduction pathway—that leads to a particular response within the cell. Like falling dominoes, the signal-activated receptor activates another protein, which activates another molecule, and so on, until the protein that produces the final cellular response is activated. The molecules that relay a signal from receptor to response, sometimes called relay molecules, are mostly proteins. The interaction of proteins is a major theme of cell signaling. Indeed, protein interaction is a unifying theme of all regulation at the cellular level.

Keep in mind that the original signal molecule is not physically passed along a signaling pathway; in most cases, it never even enters the cell. When we say that the signal is relayed along a pathway, we mean that certain *information* is passed on. At each step the signal is transduced into a different form, commonly a conformational change in a protein. Very often, the conformational change is brought about by phosphorylation.

Protein phosphorylation, a common mode of regulation in cells, is a major mechanism of signal transduction

11.3 Previous chapters introduced the concept of activating a protein by adding one or more phosphate groups to it (see FIGURE 9.2). In this chapter we have already seen how phosphorylation is involved in the activation of tyrosine-kinase receptors. In fact, the phosphorylation of proteins is a widespread cellular mechanism for regulating protein activity. The general name for an enzyme that transfers phosphate groups from ATP to a protein is **protein kinase**. Unlike receptor tyrosine kinases, most cytoplasmic protein kinases act not on themselves, but on other substrate proteins; also, they phosphorylate their substrates on either of two amino acids, serine or threonine. Such serine/threonine kinases are widely involved in signaling pathways in animals, fungi, and plants.

Many of the relay molecules in signal-transduction pathways are protein kinases, and they often act on each other. FIGURE 11.10, page 196, shows a hypothetical pathway containing three different protein kinases. This sequence is similar to many known pathways, including those triggered in yeast by mating factors and in animal cells by many growth factors.

The signal is transmitted by a cascade of protein phosphorylations, each bringing with it a conformational change. Each shape change results from the interaction of the charged phosphate groups with charged and polar amino acids (see FIGURE 5.15). The addition of phosphates often changes a protein from an inactive form to an active form (although in other cases phosphorylation *decreases* the activity of the protein).

The importance of protein kinases can hardly be overstated. Fully 1% of our own genes are thought to code for protein kinases. A single cell may have hundreds of different kinds, each with specificity for a different substrate protein. Together, they probably regulate a large proportion of the thousands of proteins in a cell. Among these are most of the proteins that, in turn, regulate cell reproduction. Abnormal activity of such a kinase frequently causes abnormal cell growth and contributes to the development of cancer.

For a cell to respond normally to an extracellular signal, it must have mechanisms for turning off the signal-transduction pathway when the initial signal is no longer present. The effects of protein kinases are rapidly reversed in the cell by **protein phosphatases**, enzymes that remove phosphate groups from proteins. At any given moment, the activity of a

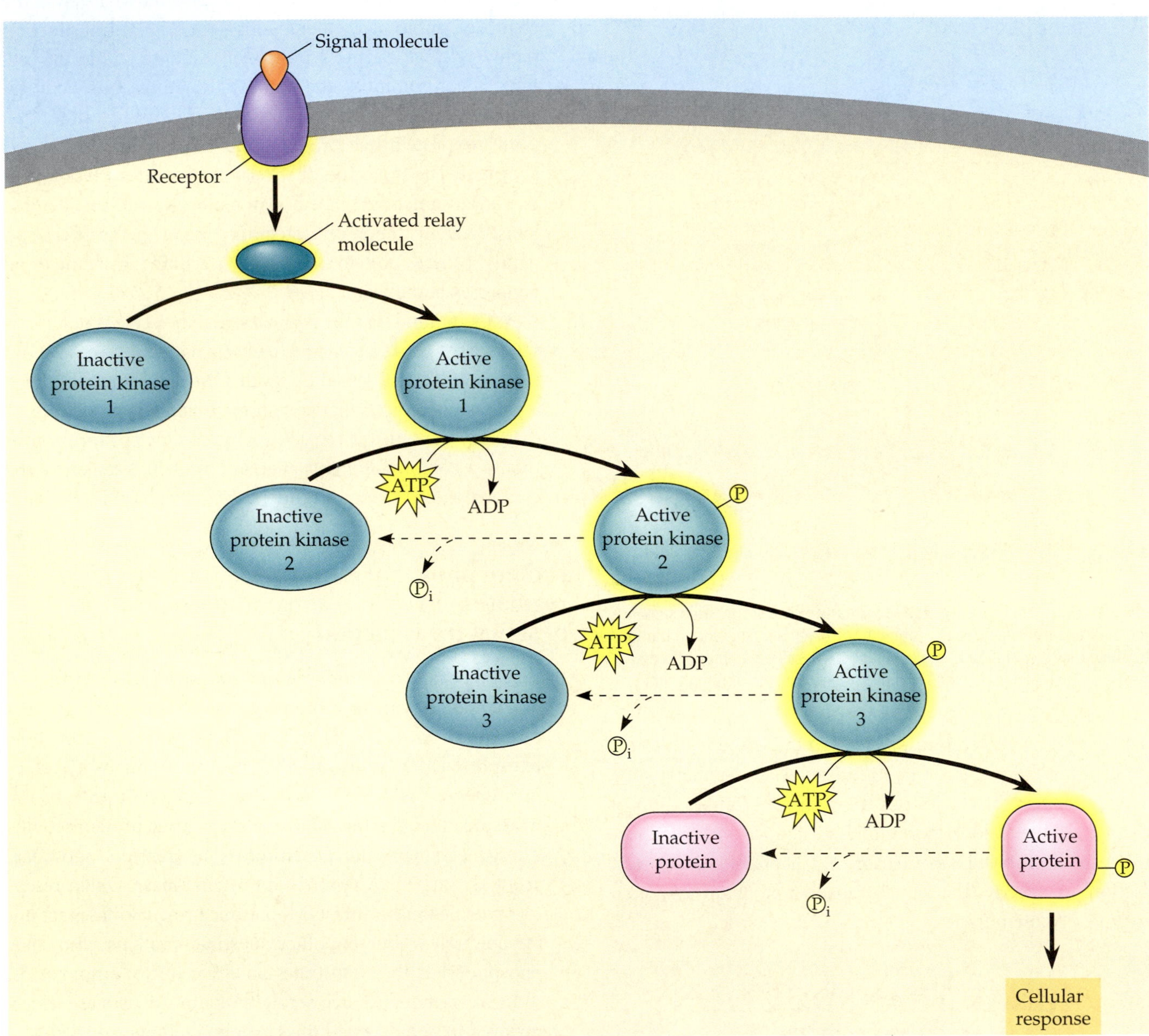

11.3 **FIGURE 11.10 ▪ A phosphorylation cascade.** This hypothetical signaling pathway begins when a signal molecule binds to a membrane receptor. The receptor then activates a relay molecule, which activates protein kinase 1. Active protein kinase 1 transfers a phosphate from ATP to an inactive molecule of protein kinase 2, thus activating this second kinase. In turn, active protein kinase 2 catalyzes the phosphorylation (and activation) of protein kinase 3. Finally, active protein kinase 3 phosphorylates a protein that brings about the cell's final response to the signal. The dashed arrows represent inactivation of the phosphorylated proteins, making them available for reuse; enzymes called phosphatases catalyze the removal of the phosphate groups. The active and inactive proteins are represented by different shapes to remind you that activation is usually associated with a change in molecular conformation.

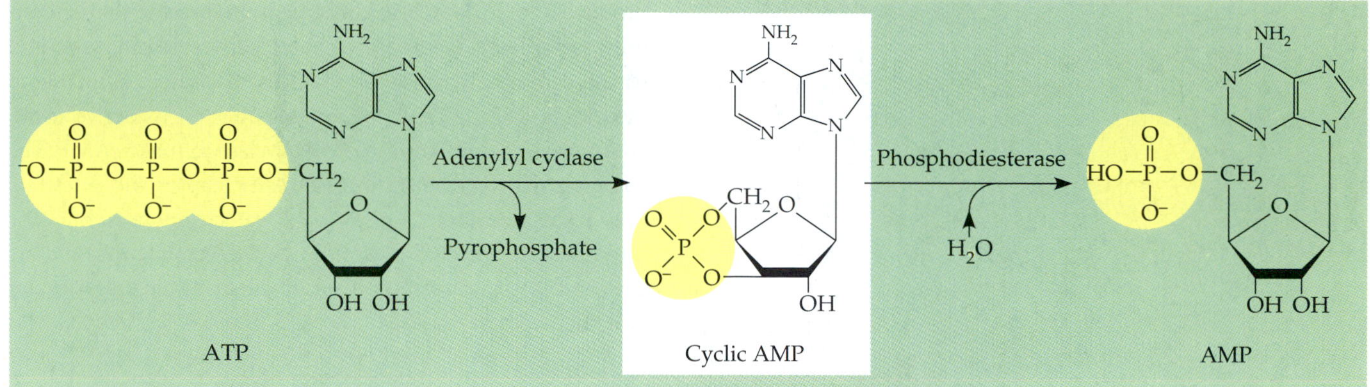

FIGURE 11.11 ▪ Cyclic AMP. Cyclic AMP (cAMP) is made from ATP by adenylyl cyclase, an enzyme embedded in the plasma membrane. The enzyme is activated as a consequence of binding of a signal molecule (such as the hormone epinephrine) to a membrane receptor. Cyclic AMP functions as a second messenger that relays the signal from the membrane to the metabolic machinery of the cytoplasm. Cyclic AMP is inactivated by phosphodiesterase, an enzyme that converts it to inactive AMP.

protein regulated by phosphorylation depends on the balance in the cell between active kinase and active phosphatase molecules. When the extracellular signal molecule is not present, active phosphatase molecules predominate, and the signaling pathway and cellular response shut down.

Certain small molecules and ions are key components of signaling pathways (second messengers)

Not all components of signal-transduction pathways are pro-
11.3 teins. Many signaling pathways also involve small, nonprotein, water-soluble molecules or ions, called **second messengers**. (The extracellular signal molecule that binds to the membrane receptor is a pathway's "first messenger.") Because second messengers are both small and water-soluble, they can readily spread throughout the cell by diffusion. For example, as we'll see shortly, it is a second messenger called cyclic AMP that carries the signal initiated by epinephrine from the plasma membrane of a liver or muscle cell into the cell's interior, where it brings about glycogen breakdown. Second messengers participate in pathways initiated by both G-protein-linked receptors and tyrosine-kinase receptors. The two most widely used second messengers are cyclic AMP and calcium ions, Ca^{2+}. A large variety of relay proteins are sensitive to the cytosolic concentration of one or the other of these second messengers.

Cyclic AMP

Once Earl Sutherland had established that epinephrine somehow causes glycogen breakdown without passing through the plasma membrane, the search began for the second messenger (he coined the term) that transmits the signal from the plasma membrane to the metabolic machinery in the cytoplasm.

Sutherland found that the binding of epinephrine to the plasma membrane of a liver cell elevates the cytoplasmic concentration of a compound called cyclic adenosine monophosphate, abbreviated **cyclic AMP** or **cAMP** (FIGURE 11.11). An enzyme built into the plasma membrane, **adenylyl cyclase**, converts ATP to cAMP in response to an extracellular signal—in this case, epinephrine. Adenylyl cyclase becomes active only after epinephrine binds to a specific receptor protein. Thus the first messenger, the hormone, causes a membrane enzyme to synthesize cAMP, which broadcasts the signal to the cytoplasm. The cAMP does not persist for long in the absence of the hormone, because another enzyme converts the cAMP to an inactive product, AMP. Another surge of epinephrine is needed to boost the cytosolic concentration of cAMP again.

Subsequent research revealed that epinephrine is only one of many hormones and other signal molecules that trigger pathways involving cAMP. It also brought to light the other components of cAMP pathways, including G proteins, G-protein-linked receptors, and protein kinases (FIGURE 11.12 on page 198). The relay molecule immediately after cAMP in a signaling pathway is usually *protein kinase A*, a serine/threonine kinase. Cyclic AMP activates this kinase. The active kinase then phosphorylates various other proteins, depending on the cell. (The complete pathway for epinephrine's stimulation of glycogen breakdown is shown in FIGURE 11.15.)

Further fine-tuning of cell metabolism is provided by other G-protein systems that *inhibit* adenylyl cyclase. In these systems, a different signal molecule activates a different receptor, which activates an inhibitory G protein.

Now that we know about the role of cAMP in G-protein-signaling pathways, we can explain in molecular detail how certain microbes cause disease. Consider cholera, a disease that is frequently epidemic in places where the water supply is contaminated with human feces. People acquire the cholera bacterium, *Vibrio cholerae,* by drinking contaminated water. The bacteria colonize the lining of the small intestine and produce a toxin, which in this case is an enzyme that chemically modifies a G protein involved in regulating salt and water secretion. Because the modified G protein is unable to hydrolyze GTP to GDP, it remains stuck in its active form, continuously stimulating adenylyl cyclase to make cAMP. The resulting high concentration of cAMP causes the intestinal cells to secrete large amounts of water and salts into the intestines. An infected person quickly develops profuse diarrhea and if left untreated can easily die from the loss of water and salts.

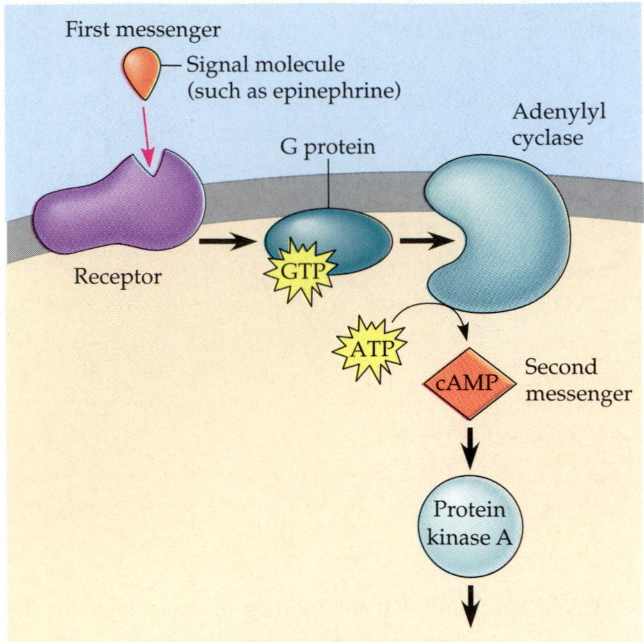

11.3 **FIGURE 11.12 ▪ cAMP as a second messenger.** Cyclic AMP is a component of many G-protein-signaling pathways. The signal molecule—the "first messenger"—activates a G-protein-linked receptor, which activates a specific G protein. In turn, the G protein activates adenylyl cyclase, which catalyzes the conversion of ATP to cAMP. The cAMP then activates another protein, most often protein kinase A. The role of cAMP was discovered in research on a hormonal signal molecule, epinephrine.

Calcium Ions and Inositol Trisphosphate

Many signal molecules in animals, including neurotransmitters, growth factors, and some hormones, induce responses in their target cells via signal-transduction pathways that increase the cytosolic concentration of calcium ions (Ca^{2+}). Calcium is even more widely used than cAMP as a second messenger. Increasing the cytosolic concentration of Ca^{2+} causes many responses in animal cells, including muscle cell contraction, secretion of certain substances, and cell division. In plant cells, calcium functions as a second messenger in signaling pathways plants have evolved for coping with environmental stresses, such as drought or cold. Cells use Ca^{2+} as a second messenger in both G-protein pathways and tyrosine-kinase receptor pathways.

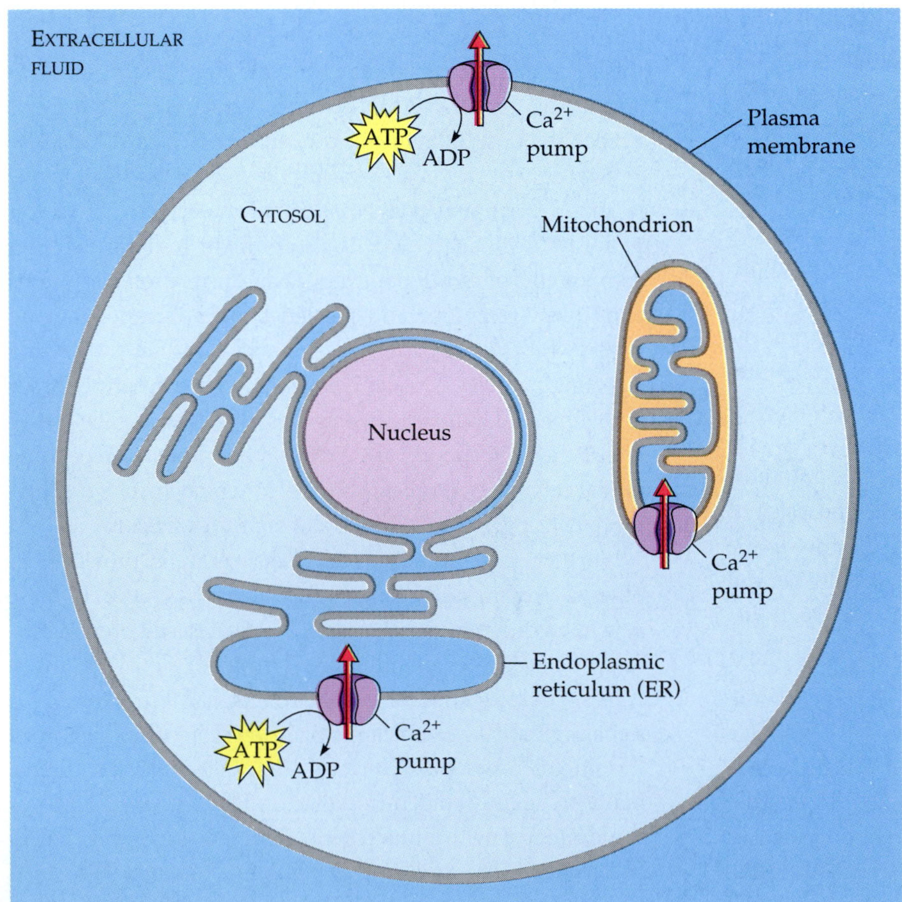

FIGURE 11.13 ▪ Calcium ion concentrations in an animal cell. Calcium ions (Ca^{2+}) are actively transported out of the cytosol by a variety of protein pumps. Pumps in the plasma membrane move Ca^{2+} into the extracellular fluid, and ones in the ER membrane move Ca^{2+} into the lumen of the ER. Consequently, the Ca^{2+} concentration in the cytosol is usually much lower (light blue) than in the extracellular fluid and ER (darker blue). Additional Ca^{2+} pumps in the mitochondrial inner membrane operate when the calcium level in the cytosol rises significantly. These pumps are driven by the proton-motive force generated across the membrane by mitochondrial electron transport chains (see Chapter 9).

Although cells always contain some Ca^{2+}, this ion can function as a second messenger because its concentration in the cytosol is normally much lower than the concentration outside the cell. In fact, the level of Ca^{2+} in the blood and extracellular fluid of an animal often exceeds that in the cytosol by more than 10,000 times. Calcium ions are actively transported out of the cell and are actively imported from the cytosol into the endoplasmic reticulum (and, under some conditions, into mitochondria and chloroplasts). As a result, the calcium concentration in the ER is usually much higher than in the cytosol (FIGURE 11.13). Because the cytosolic calcium level is low, a small change in absolute numbers of ions represents a relatively large percentage change in calcium concentration.

In response to a signal relayed by a signal-transduction pathway, the cytosolic calcium level may rise, usually by a mechanism that releases Ca^{2+} from the cell's ER. The pathways leading to calcium release involve still other second messengers, **diacylglycerol (DAG)** and **inositol trisphosphate (IP_3)**. These two messengers are produced by cleavage of a certain kind of phospholipid in the plasma membrane. FIGURE 11.14 shows how this occurs and how IP_3 stimulates the release of calcium from the ER. Because IP_3 acts before calcium in the pathway, calcium could be considered a *"third* messenger." However, scientists use the term *second messenger* for all small, nonprotein components of signal-transduction pathways.

In some cases, calcium ions activate a signal-transduction protein directly, but often they function by means of **calmodulin**, a Ca^{2+}-binding protein present at high levels in eukaryotic cells (in an animal cell, for example, calmodulin may

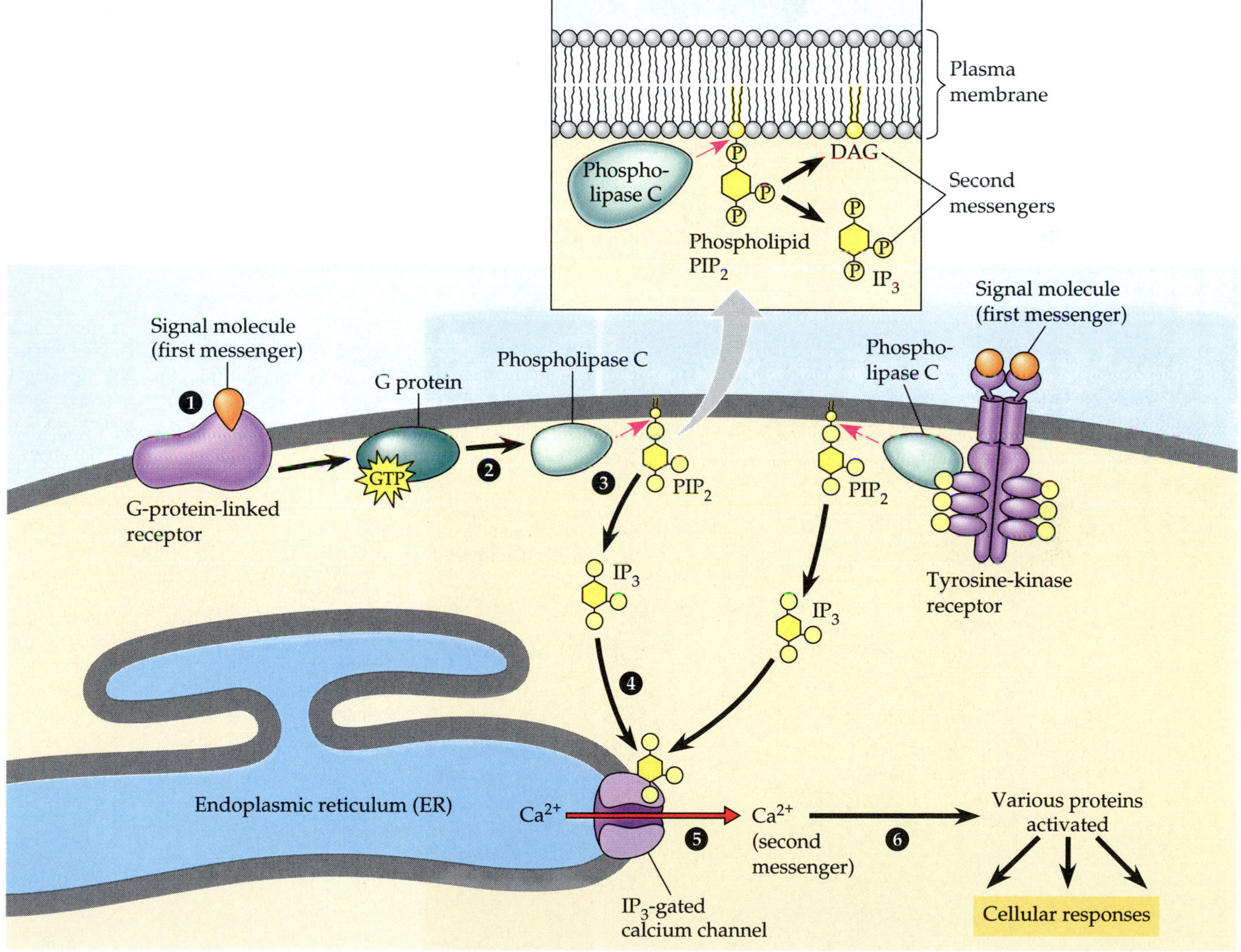

11.3 **FIGURE 11.14 • Calcium and inositol trisphosphate in signaling pathways.** Calcium ions (Ca^{2+}) and inositol trisphosphate (IP_3) function as second messengers in many signal-transduction pathways. The process is initiated by the binding of a signal molecule to either a G-protein-linked receptor (left) or a tyrosine-kinase receptor (right). The circled numbers trace the former pathway. ① A signal molecule binds to a receptor, leading to ② activation of an enzyme called phospholipase C. ③ This enzyme cleaves a plasma-membrane phospholipid called PIP_2 into DAG and IP_3 (inset). Both can function as second messengers. ④ IP_3, a small molecule, quickly diffuses through the cytosol and binds to a ligand-gated calcium channel in the ER membrane, causing it to open. ⑤ Calcium ions flow out of the ER (down their gradient), raising the Ca^{2+} level in the cytosol. ⑥ The calcium ions activate the next protein in one or more signaling pathways, often acting via calmodulin, a ubiquitous Ca^{2+}-binding protein. DAG functions as a second messenger in still other pathways.

represent as much as 1% of the total protein). Calmodulin mediates many calcium-regulated processes in cells. When calcium ions bind to it, calmodulin changes conformation and then binds to other proteins, activating or inactivating them. The proteins most often regulated by calmodulin are protein kinases and phosphatases—the most common relay proteins in signaling pathways.

CELLULAR RESPONSES TO SIGNALS

We now take a closer look at the cell's eventual response to an extracellular signal—what some researchers call the "output response." What is the nature of the final step in a signaling pathway?

In response to a signal, a cell may regulate activities in the cytoplasm or transcription in the nucleus

Ultimately, a signal-transduction pathway leads to the regula-
11.4 tion of one or more cellular activities. The regulated activities may occur in the cytoplasm, such as a rearrangement of the cytoskeleton, the opening or closing of an ion channel in the plasma membrane, or some aspect of cell metabolism. As we have discussed already, the response of liver cells to signaling by the hormone epinephrine helps regulate cellular energy metabolism. The final step in the signaling pathway activates the enzyme that catalyzes the breakdown of glycogen. FIGURE 11.15 shows the complete pathway leading to the release of glucose-1-phosphate from glycogen.

Many other signaling pathways ultimately regulate not the *activity* of an enzyme but the *synthesis* of enzymes or other proteins, usually by turning specific genes on or off. Recall that the genes in a cell's DNA function by being transcribed into an RNA version called messenger RNA (mRNA), which leaves the nucleus and is translated into a specific protein by ribosomes in the cytoplasm (see FIGURE 5.26). Special proteins called *transcription factors* control which genes are turned on—that is, transcribed into mRNA—in a particular cell at a particular time. The activity of a transcription factor may itself be regulated by a signaling pathway that extends into the cell nucleus (FIGURE 11.16). All the different kinds of signal receptors and relay molecules introduced in this chapter participate in gene-regulating pathways, as well as in pathways leading to other kinds of responses. The molecular messengers that produce gene-regulation responses include growth factors and certain plant and animal hormones. Malfunctioning of growth-factor pathways like the one in FIGURE 11.16 can cause cancer, as we will see in Chapter 19.

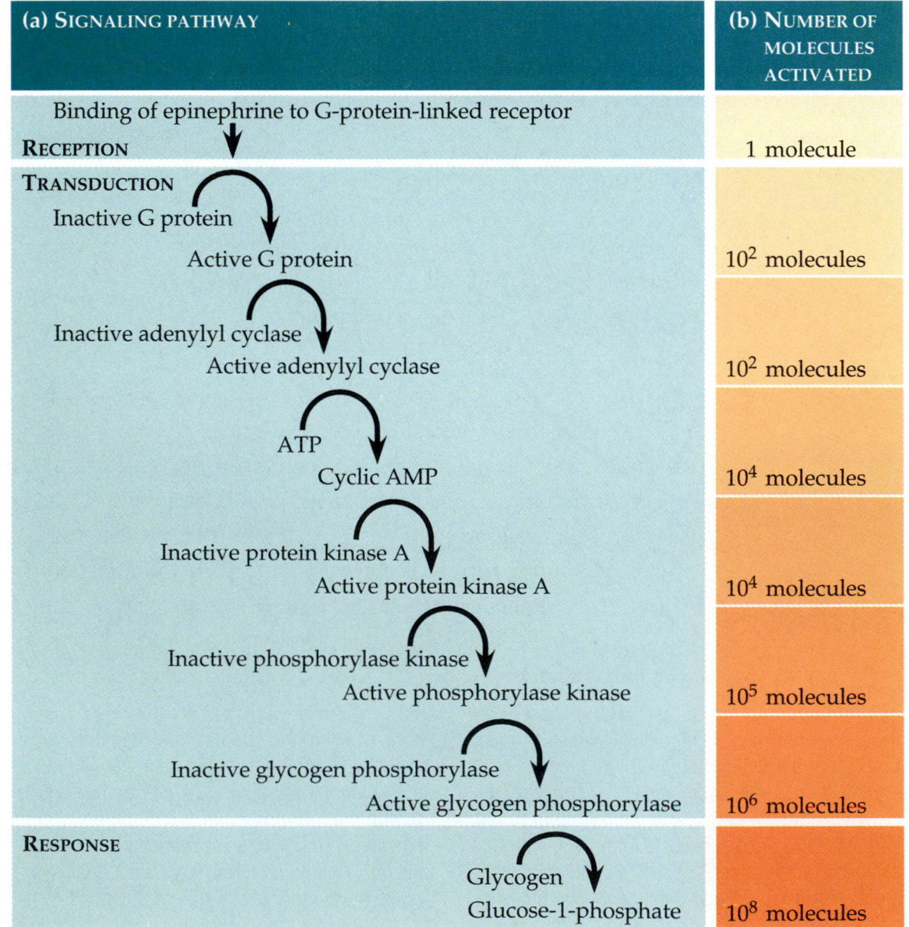

FIGURE 11.15 • Cytoplasmic response to a signal: The stimulation of glycogen breakdown by epinephrine. (a) In this signaling system, the hormone epinephrine acts through a G-protein-linked receptor to activate a succession of relay molecules, including cAMP and two protein kinases. The final protein to be activated is the cytosolic enzyme glycogen phosphorylase, which releases glucose-1-phosphate units from glycogen. **(b)** As discussed in the next section of the text, this pathway *amplifies* the hormonal signal, because the receptor protein can activate many molecules of G protein, and each enzyme molecule in the pathway can act on many molecules of its substrate, the next molecule in the cascade. The number of activated molecules given for each step is only approximate.

11.4

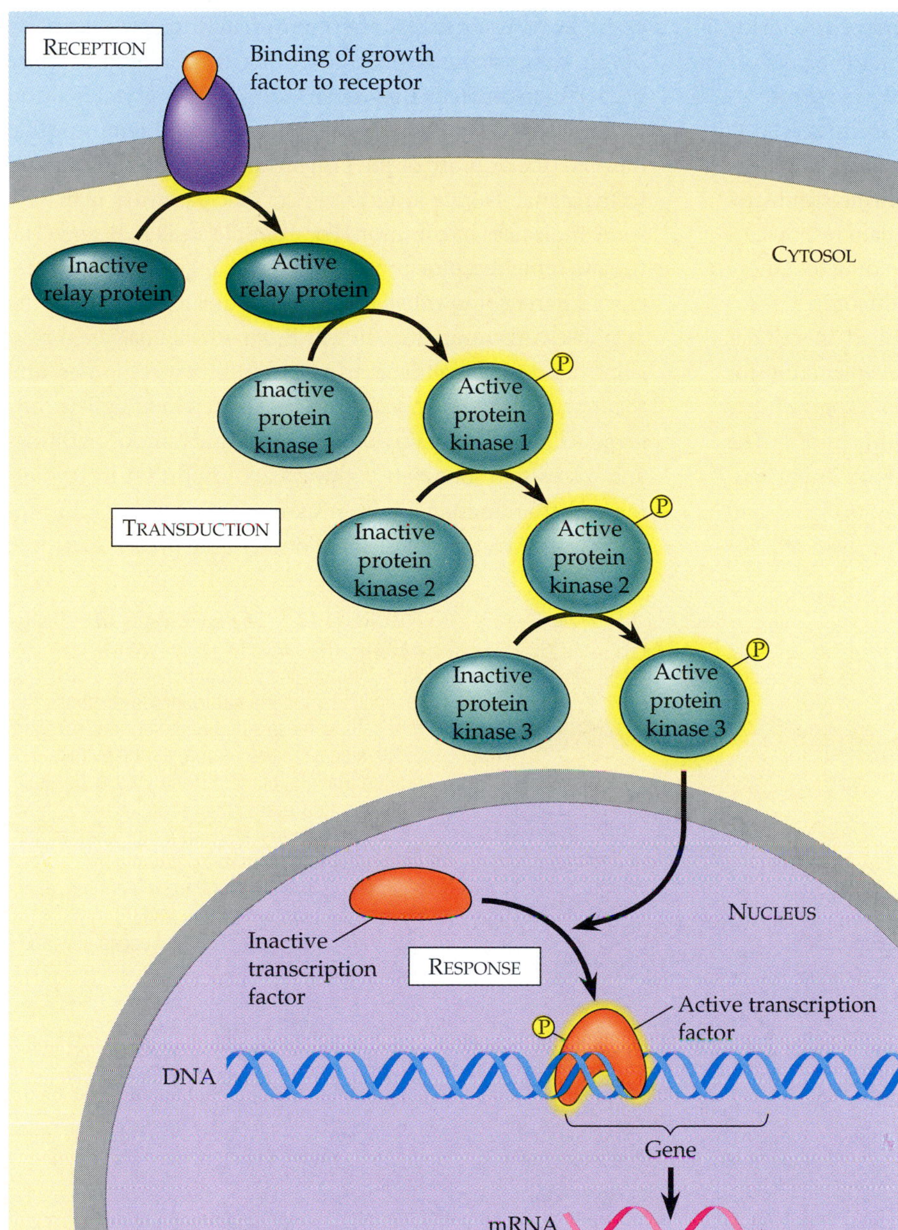

FIGURE 11.16 ▪ Nuclear response to a signal: The activation of a specific gene by a growth factor. This diagram is a simplified representation of a typical signaling pathway that leads to the regulation of gene activity in the cell nucleus. The initial signal molecule, a local regulator called a growth factor, triggers a phosphorylation cascade. The last kinase in the sequence enters the nucleus and there activates a gene-regulating protein, a transcription factor. This protein stimulates a specific gene to be transcribed into mRNA, which then directs the synthesis of a particular protein in the cytoplasm. Sometimes a transcription factor turns on several different genes.

Elaborate pathways amplify and specify the cell's response to signals

11.5 Why are there often so many steps between the original signaling event at the cell surface and the cell's response? As mentioned earlier, signaling pathways with a multiplicity of steps have two important benefits: They amplify the signal (and thus the response), and they contribute to the specificity of response.

Signal Amplification

Elaborate enzyme cascades amplify the cell's response to a signal. At each catalytic step in the cascade, the number of activated products is much greater than in the preceding step. For example, in the epinephrine-triggered pathway in FIGURE 11.15, each adenylyl cyclase molecule catalyzes the formation of many cAMP molecules, each molecule of protein kinase A phosphorylates many molecules of the next kinase in the pathway, and so on. The amplification effect depends on the fact that these proteins persist in active form long enough to process numerous molecules of substrate before they become inactive again. As a result of the signal's amplification, a small number of epinephrine molecules binding to receptors on the surface of a liver cell or skeletal muscle cell can lead to the release of hundreds of millions of glucose molecules from glycogen.

The Specificity of Cell Signaling

Consider two different cells in your body—a liver cell and a heart muscle cell, for example. Both are in contact with your bloodstream and are therefore constantly exposed to many different hormone molecules, as well as to local regulators secreted by nearby cells. Yet the liver cell responds to some signals but ignores others, and the same is true for the heart cell. And some kinds of signals trigger responses in both cells—but different responses. For instance, epinephrine stimulates the liver cell to break down glycogen, but the main response of the heart cell to epinephrine is contraction, leading to a more rapid heartbeat. How do we account for this difference?

The explanation for the specificity exhibited in cellular responses to signals is the same as the basic explanation for virtually all differences between cells: *Different kinds of cells have different collections of proteins* (FIGURE 11.17). The response of a particular cell to a signal depends on its particular collection of signal receptor proteins, relay proteins, and proteins needed to carry out the response. A liver cell, for example, is poised to respond appropriately to epinephrine by having the proteins shown in FIGURE 11.15, as well as those needed to manufacture glycogen.

Thus two cells that respond differently to the same signal differ in one or more of the proteins that handle and respond to the signal. Notice in FIGURE 11.17 that pathways may have some molecules in common. For example, cells A, B, and C all use the same receptor protein for the triangular signal molecule; differences in other proteins account for their differing responses. Also note that in cell B, a pathway triggered by a single kind of signal diverges to produce two responses, and that in cell C, two pathways triggered by separate signals converge to modulate a single response. Branching of pathways and "cross-talk" between pathways are important in regulating and coordinating a cell's responses to incoming information. Moreover, the use of some of the same proteins in more

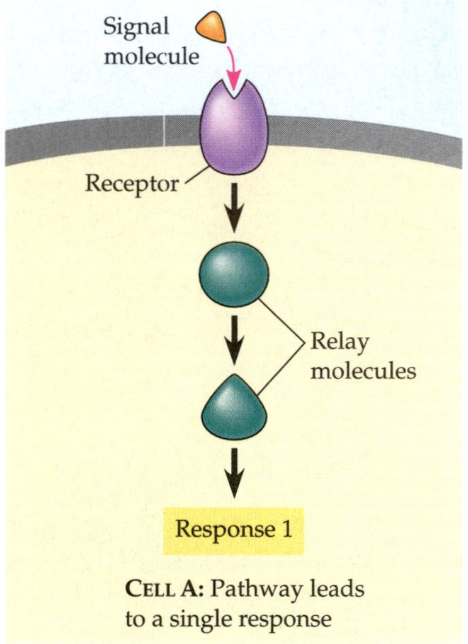

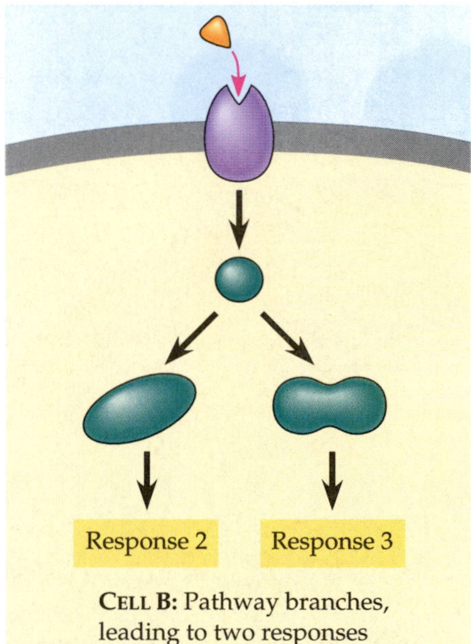

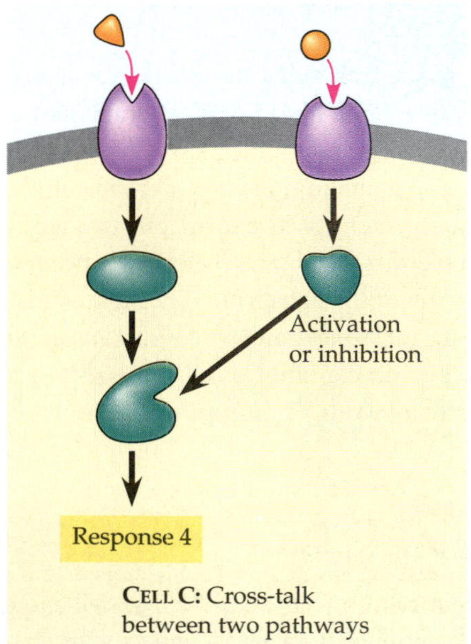

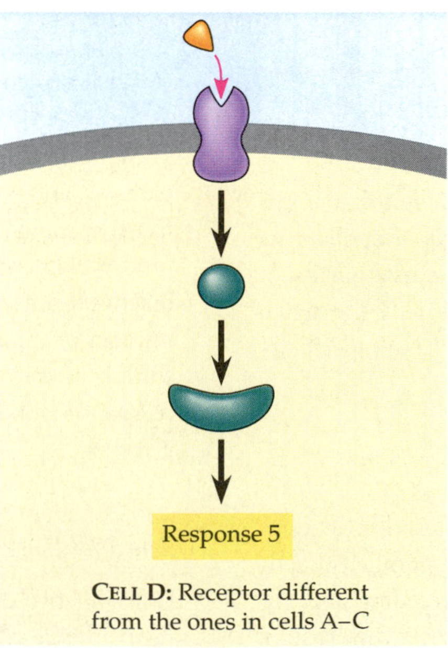

FIGURE 11.17 • The specificity of cell signaling. The particular proteins a cell possesses determine what signal molecules it responds to and the nature of the response. All four cells in these simplified diagrams respond to the signal molecule represented by the orange triangle, but in different ways, because each has a different set of proteins. Note, however, that the same kinds of molecules can participate in more than one pathway; for example, cells A, B, and C have identical receptors for the orange triangle. Cell B has a branched pathway, as is found especially often in pathways that use tyrosine-kinase receptors (which can activate multiple relay proteins) or second messengers (which can regulate numerous proteins). Cell C exhibits cross-talk between two pathways, enabling the cell to integrate information from two different signals. Cell D has a receptor for the orange triangle that differs from the ones in cells A, B, and C.

than one pathway allows the cell to economize on the number of different proteins it must make.

The importance of the relay proteins that serve as points of branching or intersection in signaling pathways is highlighted by the problems arising when these proteins are defective or missing. For instance, in an inherited disorder called Wiskott-Aldrich syndrome (WAS), the absence of a single relay protein leads to such diverse effects as abnormal bleeding, eczema, and a predisposition to infections and leukemia. These symptoms are thought to arise primarily from the absence of the protein in cells of the immune system. By studying normal cells, scientists found that the WAS protein is located just beneath the cell surface. The protein interacts both with microfilaments of the cytoskeleton and with several different components of signaling pathways that relay information from the cell surface, including pathways regulating immune-cell proliferation. This multifunctional relay protein is thus both a branch point and an important intersection point in a complex signal transduction network that controls immune-cell behavior. When the WAS protein is absent, the cytoskeleton is not properly organized and signaling pathways are disrupted, leading to the WAS symptoms.

To keep FIGURE 11.17 simple, we have not indicated the *inactivation* mechanisms that are an essential aspect of cell signaling. For a cell of a multicellular organism to remain alert and capable of responding to incoming signals, each molecular change in its signaling pathways must last only a short time. As we saw in the cholera example, if a signaling pathway component becomes locked into one state, whether active or inactive, dire consequences for the organism can result.

Thus a key to a cell's continuing receptiveness to regulation is the reversibility of the changes that signals produce. The binding of signal molecules to receptors is reversible, which means that the lower the concentration of signal molecules, the fewer will be bound at any given moment. When they leave the receptor, the receptor reverts to its inactive form. Then, by a variety of means, the relay molecules return to their inactive forms: The GTPase activity intrinsic to a G protein hydrolyzes its bound GTP; the enzyme phosphodiesterase converts cAMP to AMP; protein phosphatases inactivate phosphorylated kinases and other proteins; and so forth. As a result, the cell is soon ready to respond to a fresh signal.

■ ■ ■

This chapter has introduced you to a number of details of cellular signal processing. More important than the details, however, are the general mechanisms of cell communication—mechanisms involving ligand binding, conformational changes in proteins, interactions among proteins, cascades of interactions that transduce and amplify signals, and protein phosphorylations by kinases. As you continue through the text, you will encounter numerous examples of cell signaling. In the very next chapter, in fact, you'll learn about the enormously important role of cell signaling in the regulation of cell reproduction.

CHAPTER REVIEW

REVIEW OF KEY CONCEPTS

(with page numbers and key figures)

AN OVERVIEW OF CELL SIGNALING

- **Cell signaling evolved early in the history of life (pp. 188–189, FIGURES 11.1, 11.2)** Cell signaling in microorganisms has much in common with the process that goes on in multicellular organisms, indicating that the sending and receiving of signals arose early in the history of life. This chapter concentrates on chemical signaling between cells.
- **Communicating cells may be close together or far apart (pp. 189–190, FIGURES 11.3, 11.4)** Animal cells communicate with nearby cells by secreting local regulators or, if nerve cells, by secreting neurotransmitters at synapses. Both animal and plant cells use hormones for signaling over long distances. Cells can also communicate by direct contact—for example, in plant cells via plasmodesmata or in animal cells via gap junctions.

11.1 **The three stages of cell signaling are reception, transduction, and response (pp. 190–191, FIGURE 11.5)** The pioneering research of Earl Sutherland focused on the effect of the hormone epinephrine on liver and skeletal muscle cells. In this system, the signal molecule epinephrine binds to specific receptors on a cell's surface (reception), leading to a series of changes in the receptor and other molecules inside the cell (transduction, the change in signal form), and finally to the activation of an enzyme that breaks down glycogen to glucose-1-phosphate (response).

SIGNAL RECEPTION AND THE INITIATION OF TRANSDUCTION

- **A signal molecule binds to a receptor protein, causing the protein to change shape (p. 192).** The binding between signal molecule (ligand) and receptor is highly specific. A conformational change in a receptor is often the initial transduction of the signal.

11.2 **Most signal receptors are plasma-membrane proteins (pp. 192–195, FIGURES 11.6, 11.8, 11.9)** A G-protein-linked receptor is a membrane receptor that works with the help of a cytoplasmic G protein. Ligand binding activates the cytoplasmic side of the receptor, which then activates a specific G protein by causing it to exchange a bound GDP for a GTP. Until the G protein inactivates itself by hydrolyzing its GTP to GDP, it can activate yet another protein in a signal-transduction pathway. Epinephrine uses this sort of receptor.

Membrane receptors called tyrosine-kinase receptors react to the binding of signal molecules by forming dimers and then using an intrinsic enzyme to add phosphate groups to tyrosines on the cytoplasmic side of the receptor. A variety of relay proteins inside the cell can then be activated by binding to different phosphorylated tyrosines, allowing this receptor to trigger several different pathways simultaneously. Growth factors are a type of signal molecule that commonly uses tyrosine-kinase receptors. Specific signal molecules cause ligand-gated ion channels in a membrane to open or close, regulating the flow of specific ions across the membrane. Such channels are important at nerve cell synapses. Intracellular receptors are cytoplasmic or nuclear proteins. Signal molecules that can readily penetrate the plasma membrane, such as lipid hormones and the gas nitric oxide, use these receptors.

SIGNAL-TRANSDUCTION PATHWAYS

- **Pathways relay signals from receptors to cellular responses (p. 195)** At each step in a signal-transduction pathway, the signal is transduced into a different form, commonly a conformational change in a protein.

Protein phosphorylation, a common mode of regulation in cells,
11.3 **is a major mechanism of signal transduction (pp. 195–197, FIGURE 11.10)** Many signal-transduction pathways include phosphorylation cascades, in which a series of protein kinases successively add phosphate groups to the next one in line, activating it. Phosphatase enzymes soon remove the phosphates, keeping the pathway open for a new signal.

Certain small molecules and ions are key components of signaling
11.3 **pathways (second messengers) (pp. 197–200, FIGURES 11.11–11.14)** Second messengers, such as cyclic AMP (cAMP) and Ca^{2+}, diffuse readily through the cytosol and thus help broadcast signals throughout the cell quickly. Many G proteins, including the one operating in the epinephrine signal-transduction pathway, activate adenylyl cyclase, which makes cAMP from ATP. Although continually present in the fluids of organisms, Ca^{2+} can serve as a messenger because protein pumps usually keep it at low concentrations in the cytosol. Many G proteins and tyrosine-kinase receptors activate an enzyme that splits a plasma-membrane phospholipid into two second messengers, one of which is inositol trisphosphate (IP_3). IP_3 is the ligand for a gated calcium channel in the membrane of the ER, which stores Ca^{2+} at high concentrations. When IP_3 binds, Ca^{2+} flows into the cytosol, where it activates proteins of many signaling pathways, often via the protein calmodulin.

CELLULAR RESPONSES TO SIGNALS

In response to a signal, a cell may regulate activities in the cyto-
11.4 **plasm or transcription in the nucleus (p. 200, FIGURES 11.15, 11.16)** For example, signaling pathways regulate enzyme activity (such as the enzyme that breaks down glycogen) and cytoskeleton rearrangement in the cytoplasm. Other pathways regulate genes; they do this by activating transcription factors, proteins that turn on specific genes.

Elaborate pathways amplify and specify the cell's response to sig-
11.5 **nals (pp. 201–203, FIGURE 11.17)** Each catalytic protein in a signaling pathway amplifies the signal by activating multiple copies of the next component of the pathway; for long pathways, the total amplification may be a millionfold or more. The particular combination of proteins in a cell gives the cell great specificity in both the signals it detects and the responses it carries out. Pathway branching and cross-talk further help the cell coordinate the signals received and the responses produced.

SELF-QUIZ

1. Phosphorylation cascades involving a series of protein kinases are useful for cellular signal transduction because
   a. they are species specific
   b. they always lead to the same cellular response
   c. they amplify the original signal many fold
   d. they counter the harmful effects of phosphatases
   e. the number of molecules used is small and fixed

2. Binding of a signal molecule to which type of receptor leads to a change in membrane potential?
   a. tyrosine-kinase receptor
   b. G-protein-linked receptor
   c. phosphorylated tyrosine-kinase dimer
   d. ligand-gated ion channel
   e. intracellular receptor

3. The activation of tyrosine-kinase receptors is characterized by
   a. aggregation and phosphorylation
   b. IP_3 binding
   c. calmodulin formation
   d. GTP hydrolysis
   e. channel protein conformational change

4. Cell signaling is believed to have evolved early in the history of life because
   a. it is seen in "primitive" organisms such as bacteria
   b. yeast cells of different mating types signal one another
   c. signal-transduction molecules found in distantly related organisms are similar
   d. signaling can operate over large distances, a function required before the development of multicellular life
   e. signal molecules typically interact with the outer surface of the plasma membrane

5. Which observation suggested to Sutherland the involvement of a second messenger in epinephrine's effect on liver cells?
   a. Enzymatic activity was proportional to the amount of calcium added to a cell-free extract.
   b. Receptor studies indicated epinephrine was a ligand.
   c. Glycogen depolymerization was observed only when epinephrine was administered to intact cells.
   d. Glycogen depolymerization was observed when epinephrine and glycogen phosphorylase were combined.
   e. Epinephrine was known to have different effects on different types of cells.

6. Protein phosphorylation is commonly involved with all of the following *except*
   a. regulation of transcription by extracellular signal molecules
   b. enzyme activation
   c. activation of G-protein-linked receptors
   d. activation of tyrosine-kinase receptors
   e. activation of protein-kinase relay molecules

7. Amplification of a chemical signal occurs when
   a. a receptor in the plasma membrane activates several G-protein molecules while a signal molecule is bound to it
   b. a cAMP molecule activates one protein-kinase molecule before being converted to AMP
   c. phosphorylase and phosphatase activities are balanced
   d. numerous calcium ions flow through an open ligand-gated calcium channel
   e. both a and d occur

8. Lipid-soluble signal molecules, such as testosterone, cross the plasma membranes of all cells but affect only their target cells because
   a. only target cells retain the appropriate DNA segments
   b. intracellular receptors are present only in target cells
   c. most cells lack the *Y* chromosome required
   d. only target cells possess the cytosolic enzymes that transduce the testosterone
   e. only in target cells is testosterone able to initiate the phosphorylation cascade leading to activated transcription factor

9. Signal-transduction pathways benefit cells for all of the following reasons *except*
   a. they help cells respond to signal molecules that are too large or too polar to cross the plasma membrane
   b. they enable different cells to respond appropriately to the same signal
   c. they help cells use up phosphate generated by ATP breakdown
   d. they can amplify a signal
   e. variations in the signal-transduction pathways can enhance response specificity

10. Consider this pathway: epinephrine ⟶ G-protein-linked receptor ⟶ G protein ⟶ adenylyl cyclase ⟶ cAMP. Identify the "second messenger."
    a. cAMP
    b. G protein
    c. GTP
    d. adenylyl cyclase
    e. G-protein-linked receptor

CHALLENGE QUESTIONS

1. Cell biologists recently reported the discovery of orexin, a signaling molecule that appears to regulate appetite in humans and other mammals. Orexin concentrations were measurably higher in fasting individuals. Using your knowledge of membrane receptors and signal-transduction pathways, suggest ways in which the understanding of orexin function could lead to treatments for both anorexia and obesity.
2. Explain how the same chemical signal, such as a particular hormone, can have one effect on cell-type A, a different effect on cell-type B, and no effect on cell-type C.

SCIENCE, TECHNOLOGY, AND SOCIETY

A patient with severe epilepsy has recently received an injection of fetal pig cells into the brain. Physicians hope these cells will integrate with the patient's brain cells and eventually produce enough GABA, a signaling molecule, to prevent seizures. The recipient was enthusiastic about the transplant, but medical ethicists were less so, because the long-term effects of this treatment are unknown. In addition, undetected viruses present in the transplanted cells might infect the recipient. Should individuals with severely debilitating or terminal diseases be permitted to receive experimental treatments before they are known to be safe? Develop arguments both for and against such experimental procedures.

FURTHER READING

Alberts, B., et al. *Essential Cell Biology.* New York: Garland, 1998. See Chapter 15. Additional information is available in Chapter 15 of *Molecular Biology of the Cell,* 3rd ed (1994), the much larger book by the same authors.

Becker, W. M., J. B. Reece, and M. F. Poenie. *The World of the Cell,* 3rd ed. Menlo Park, CA: Benjamin/Cummings, 1996. Chapter 23 focuses on chemical signaling.

Bourne, H. R. "Pieces of the True Grail: A G Protein Finds Its Target." *Science,* December 12, 1997. Exactly how certain G proteins interact with adenylyl cyclase.

Cossart, P., P. Boquet, S. Normark, and R. Rappuoli. "Cellular Microbiology Emerging." *Science,* January 19, 1996. How pathogenic bacteria interfere with host-cell signaling pathways and other cell functions, and the use of these pathogens as tools for basic research.

Featherstone, C. "The Many Faces of WAS Protein." *Science,* January 3, 1997. More about the protein that is missing in people with Wiskott-Aldrich syndrome.

Harbron, P. "Conversation in a Cell." *Discover,* February 1996. Stuart Schreiber's research on cell signaling, which began with an investigation of the sex attractant of cockroaches.

Linder, E., and A. Gilman. "G Proteins." *Scientific American,* July 1992. Signal-transduction pathways in the actions of hormones.

Losick, R., and D. Kaiser. "Why and How Bacteria Communicate." *Scientific American,* February 1997. Bacteria "talk" with one another and with plants and animals by emitting and reacting to chemical signals.

Marx, J. "Medicine: A Signal Contribution to Cell Biology." *Science,* October 23, 1992. The Nobel Prize awarded for pioneering work on protein kinases and phosphatases. Also "Medicine: A Signal Award for Discovering G Proteins." *Science,* October 21, 1994. The Nobel Prize awarded for the discovery of G proteins.

WEB LINKS

Visit the special edition of *The Biology Place* for BIOLOGY, Fifth Edition, at **http://www.biology.com/campbell**. Go to Chapter 11 for online resources, including learning activities, practice exams, and links to the following web sites:

"Yeast (Budding, Fission and Candida)"
A compendium of resources about the yeasts used as model organisms in biological research.

"The G-Protein-Coupled Receptor Database"
Anything you ever wanted to know about G-protein-linked receptors can be found at this site.

"The Dictionary of Cell Biology"
An on-line dictionary of cell biology that defines over 6000 terms.

"The Virtual Cell"
Climb inside this virtual cell and explore its parts.

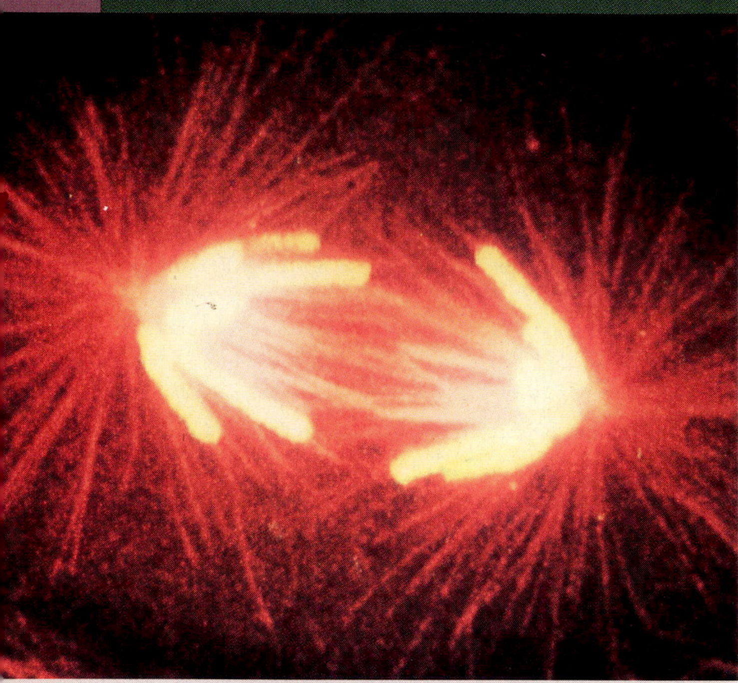

CHAPTER 12

THE CELL CYCLE

The Key Roles of Cell Division

- Cell division functions in reproduction, growth, and repair
- Cell division distributes identical sets of chromosomes to daughter cells

The Mitotic Cell Cycle

- The mitotic phase alternates with interphase in the cell cycle: *an overview*
- The mitotic spindle distributes chromosomes to daughter cells: *a closer look*
- Cytokinesis divides the cytoplasm: *a closer look*
- Mitosis in eukaryotes may have evolved from binary fission in bacteria

Regulation of the Cell Cycle

- A molecular control system drives the cell cycle
- Internal and external cues help regulate the cell cycle
- Cancer cells have escaped from cell-cycle controls

Photograph courtesy of J. M. Murray, University of Pennsylvania Medical School.

*The ability of organisms to reproduce their kind is the one characteristic that best distinguishes living things from nonliving matter. This unique capacity to procreate, like all biological functions, has a cellular basis. Rudolf Virchow, a German physician, put it this way in 1855: "Where a cell exists, there must have been a preexisting cell, just as the animal arises only from an animal and the plant only from a plant." He summarized with the Latin axiom, "*Omnis cellula e cellula,*" meaning "Every cell from a cell." The continuity of life is based on the reproduction of cells, or* ***cell division****. The micrograph on this page shows a cell dividing; only the chromosomes (yellow) and microtubules (red) are visible.*

In this chapter you will learn how cells reproduce to form genetically equivalent daughter cells. This process is an integral part of the* ***cell cycle****, the life of a cell from its origin in the division of a parent cell until its own division into two.*

THE KEY ROLES OF CELL DIVISION

Before describing the cell cycle or the mechanics of cell division, let's look at the roles that cellular reproduction plays in the lives of organisms.

Cell division functions in reproduction, growth, and repair

When a unicellular organism such as *Amoeba* divides to form duplicate offspring, the division of a cell reproduces an entire organism (FIGURE 12.1a). But cell division also enables multicellular organisms, including humans, to grow and develop from a single cell—the fertilized egg (FIGURE 12.1b). Even after the organism is fully grown, cell division continues to function in renewal and repair, replacing cells that die from normal wear and tear or accidents. For example, dividing cells in your bone marrow continuously supply new blood cells (FIGURE 12.1c).

The reproduction of an ensemble as complex as a cell cannot occur by a mere pinching in half; the cell is not like a soap bubble that simply enlarges and splits in two. Cell division involves the distribution of identical genetic material—DNA—to two daughter cells. What is most remarkable about cell division is the fidelity with which the DNA is passed along, without dilution, from one generation of cells to the next. A dividing cell duplicates its DNA, allocates the two copies to opposite ends of the cell, and only then splits into two daughter cells.

* Although the terms *daughter cells* and *sister chromatids* (a term you will encounter later in the chapter) are traditional and will be used throughout this book, the structures they refer to have no gender.

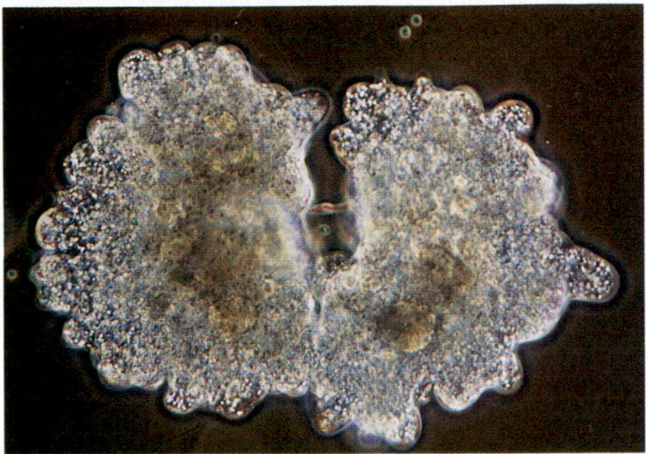

(a) Reproduction 100 µm

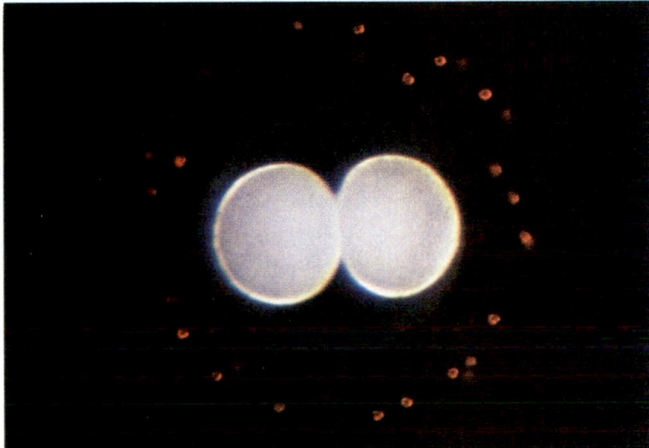

(b) Growth and development 100 µm

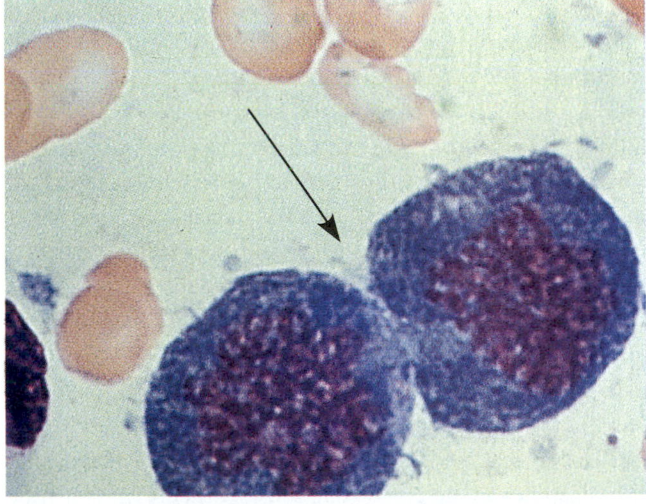

(c) Tissue renewal 10 µm

FIGURE 12.1 • The functions of cell division. (a) *Amoeba,* a single-celled eukaryote, divides to form two cells, each an individual organism. In this case, the function of cell division is reproduction. **(b)** For multicellular organisms, the division of embryonic cells is essential for growth and development. This darkfield micrograph shows a sand dollar embryo shortly after the fertilized egg divided to form two cells. **(c)** Even in a mature multicellular organism, cell division continues to function in the renewal and repair of tissues. For example, these dividing bone marrow cells (arrow) give rise to new blood cells. (All LMs.)

Cell division distributes identical sets of chromosomes to daughter cells

A cell's total hereditary endowment of DNA is called its **genome.** Although a prokaryotic genome is often a single long DNA molecule, eukaryotic genomes usually consist of several such molecules. The overall length of DNA in a eukaryotic cell is enormous. A typical human cell, for example, has about 3 meters of DNA—a length about 300,000 times greater than the cell's diameter. Yet before the cell can divide, all of this DNA must be copied and then apportioned to give each daughter cell a complete genome.

The replication and distribution of so much DNA is manageable because the DNA molecules are packaged into chromosomes (FIGURE 12.2). Every eukaryotic species has a

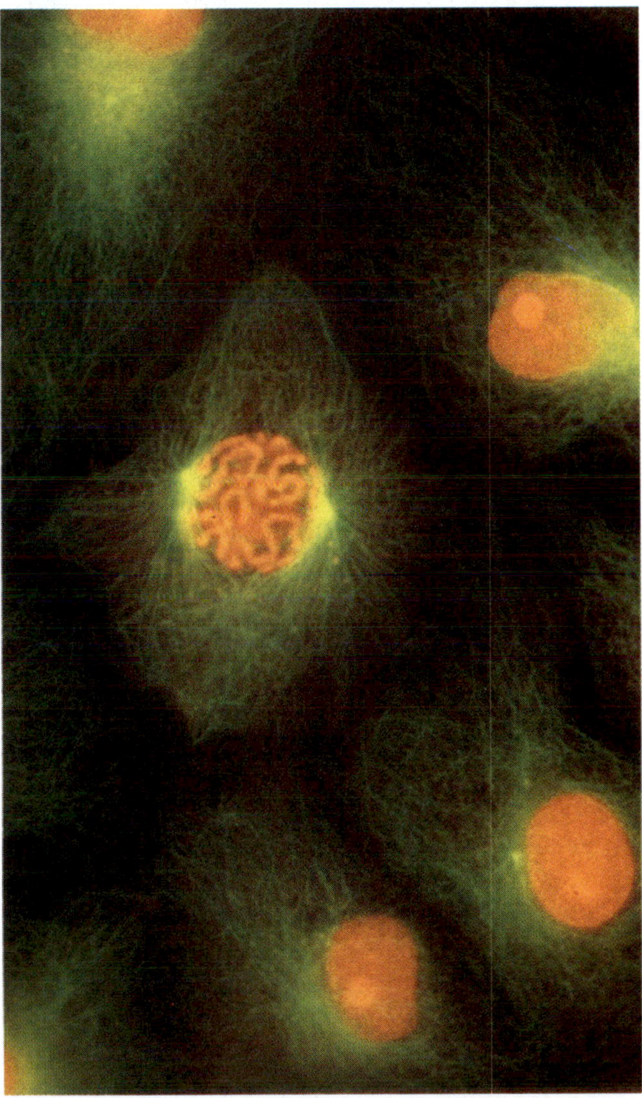

25 µm

FIGURE 12.2 • Eukaryotic chromosomes. A tangle of chromosomes (orange) is visible within the nucleus of the kangaroo rat epithelial cell in the center of this micrograph. The cell is preparing to divide (LM). Chromosomes are so named because they take up certain dyes used in microscopy (*chromo,* "colored," and *somes,* "bodies").

Photo courtesy of J. M. Murray, University of Pennsylvania Medical School.

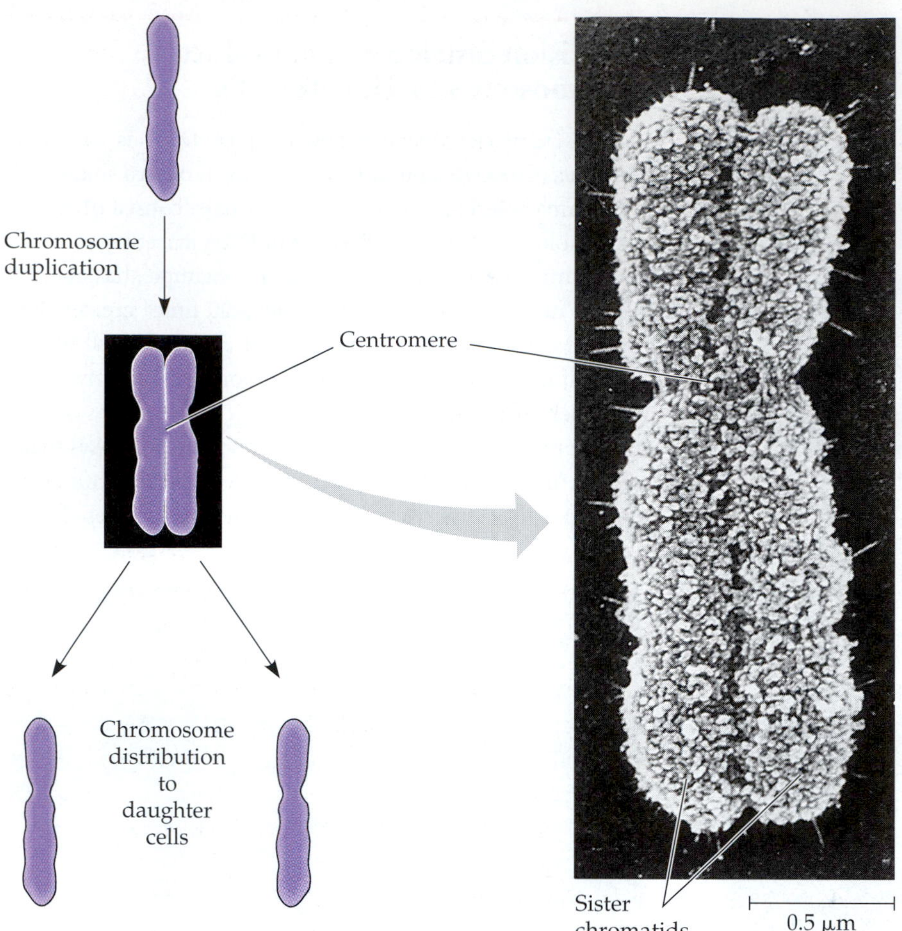

FIGURE 12.3 • Chromosome duplication and distribution during mitosis. A eukaryotic cell preparing to divide duplicates each of its multiple chromosomes, one of which is represented here. A duplicated chromosome consists of two sister chromatids, which narrow at their centromeres. The micrograph shows a human chromosome in this state (SEM). The chromosome looks hairy because each chromatid consists of a very long chromatin fiber folded and coiled in a compact arrangement. Each chromatin fiber, in turn, consists of one long DNA molecule associated with a variety of protein molecules. The DNA molecules of the sister chromatids are identical, products of a precise replication process. As mitosis continues, mechanical processes separate the sister chromatids and distribute them to two daughter cells.

characteristic number of chromosomes in each cell nucleus. For example, human **somatic cells** (all body cells except the reproductive cells) contain 46 chromosomes. Reproductive cells, or **gametes**—sperm cells and egg cells—have half as many chromosomes as somatic cells, or 23 chromosomes in humans.

Incorporated into each eukaryotic chromosome is one very long, linear DNA molecule representing thousands of genes, the units that specify an organism's inherited traits. The DNA is associated with various proteins that maintain the structure of the chromosome and help control the activity of the genes. The DNA-protein complex, called **chromatin**, is organized into a long, thin fiber. After a cell duplicates its genome in preparation for division, the chromatin condenses: It becomes densely coiled and folded, making the chromosomes so thick that we can see them with a light microscope.

Each duplicated chromosome consists of two **sister chromatids**. The two chromatids, containing identical copies of the chromosome's DNA molecule, are initially attached along their lengths. In its condensed form, the chromosome has a narrow "waist" at a specialized region called the **centromere** (FIGURE 12.3). Later in the cell division process, the sister chromatids of all the chromosomes are pulled apart and repackaged as complete chromosome sets in two new nuclei, one at each end of the cell. **Mitosis**, the division of the nucleus, is usually followed immediately by **cytokinesis**, the division of the cytoplasm. Where there was one cell, there are now two, each the genetic equivalent of the parent cell.

What happens to chromosome number as we follow the human life cycle through the generations? You inherited 46 chromosomes, 23 from each parent. They were combined in the nucleus of a single cell when a sperm cell from your father united with an egg cell from your mother to form a fertilized egg, or zygote. Mitosis and cytokinesis produced the trillions of somatic cells that now make up your body, and the same processes continue to generate new cells to replace dead and damaged ones. In contrast, your gametes—eggs or sperm cells—are produced by a variation of cell division called **meiosis**, which yields daughter cells that have half as many chromosomes as the parent cell. Meiosis occurs only in your gonads (ovaries or testes). In each generation of humans, meiosis reduces the chromosome number from 46 to 23. Fertilization joins gametes and doubles the chromosome number to 46 again, and mitosis conserves that number in every somatic cell of the new individual. In Chapter 13 we will examine the role of meiosis in reproduction and inheritance in more detail. In the remainder of this chapter we focus on mitosis and the rest of the mitotic cell cycle.

THE MITOTIC CELL CYCLE

The mitotic phase alternates with interphase in the cell cycle: *an overview*

Mitosis is just one part of the cell cycle (FIGURE 12.4). In fact,
12.1 the **mitotic (M) phase**, which includes both mitosis and cytokinesis, is usually the shortest part of the cell cycle. Successive mitotic cell divisions alternate with a much longer **interphase**, which often accounts for about 90% of the cycle. It is during interphase that the cell grows and copies its chromosomes in preparation for cell division. Interphase can be divided into subphases: **G_1 phase** ("first gap"), **S phase**, and **G_2 phase** ("second gap"). During all three subphases, the cell grows by producing proteins and cytoplasmic organelles. However, chromosomes are duplicated only during the S phase (S stands for synthesis of DNA). Thus a cell grows (G_1), continues to grow as it copies its chromosomes (S), grows more as it completes preparations for cell division (G_2), and divides (M). The daughter cells may then repeat the cycle.

Time-lapse films of living, dividing cells reveal the dynam-
12.2 ics of mitosis as a continuum of changes. For purposes of description, however, mitosis is conventionally broken down into five subphases: **prophase, prometaphase, metaphase, anaphase**, and **telophase**. FIGURE 12.5 on the following pages
12.3 describes these stages in an animal cell. Be sure to study this figure thoroughly before progressing to the next section, which examines mitosis more closely.

The mitotic spindle distributes chromosomes to daughter cells: *a closer look*

Many of the events of mitosis depend on the **mitotic spindle**, which begins to form in the cytoplasm during prophase. This structure consists of fibers made of microtubules and associated proteins. While the mitotic spindle assembles, the microtubules of the cytoskeleton partially disassemble, probably providing the material used to construct the spindle. The spindle microtubules elongate by incorporating more subunits of the protein tubulin (see TABLE 7.2, p. 120).

The assembly of spindle microtubules starts in the centrosome, also called the microtubule organizing center. In animal cells, a pair of centrioles is located at the center of the centrosome, but these structures are not essential for cell division. In fact, the centrosomes of most plants lack centrioles, and if a researcher destroys the centrioles of an animal cell with a laser microbeam, a spindle nevertheless forms during mitosis.

During interphase, the single centrosome replicates to form two centrosomes located just outside the nucleus (see FIGURE 12.5). The two centrosomes move farther apart during prophase and prometaphase, as spindle microtubules grow out from them. By the end of prometaphase, the two centrosomes, referred to as *spindle poles* in this context, are at opposite poles of the cell.

Each of the two joined chromatids of a chromosome has a **kinetochore**, a structure of proteins and specific sections of chromosomal DNA at the centromere. The chromosome's two kinetochores face in opposite directions. During prometaphase, some of the spindle microtubules attach to kinetochores. When one of a chromosome's kinetochores is

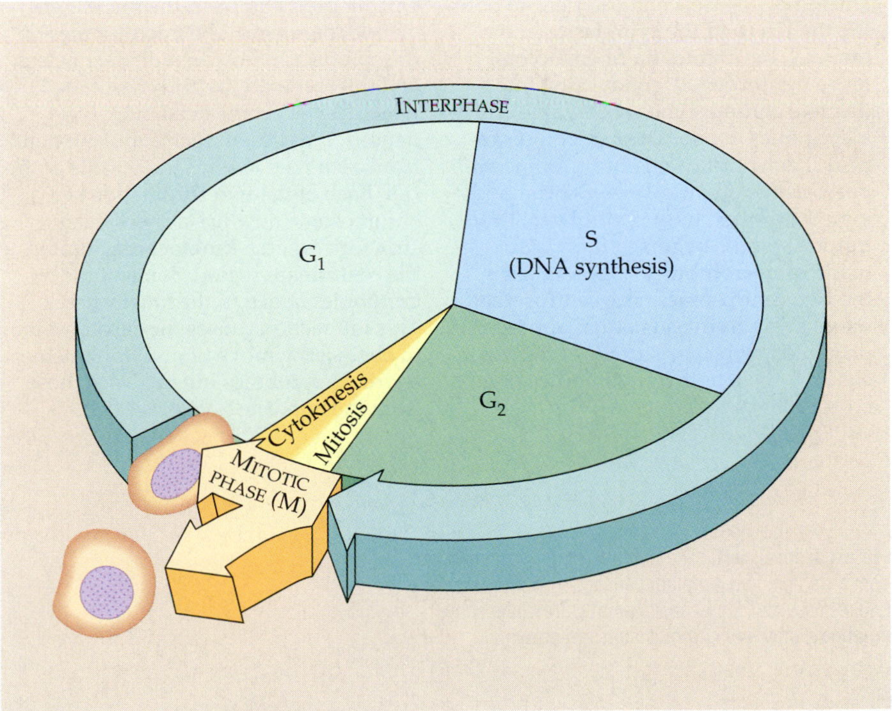

12.1 **FIGURE 12.4 ▪ The cell cycle.** In a dividing cell, the mitotic (M) phase alternates with interphase, a growth period. The first part of interphase, called G_1, is followed by the S phase, when the chromosomes replicate; the last part of interphase is called G_2. In the M phase, mitosis divides the nucleus and distributes its chromosomes to the daughter nuclei, and cytokinesis divides the cytoplasm, producing two daughter cells.

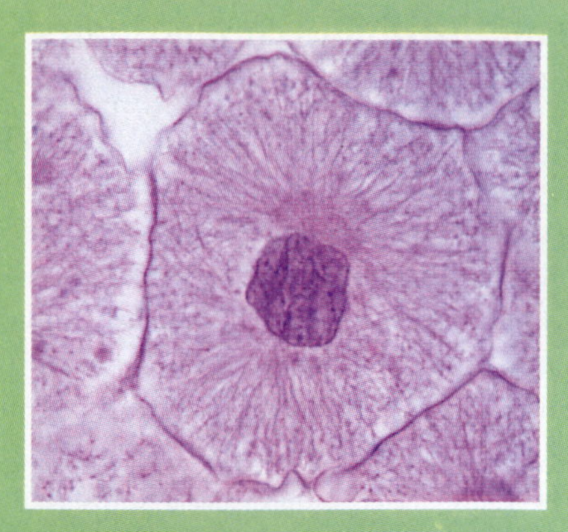

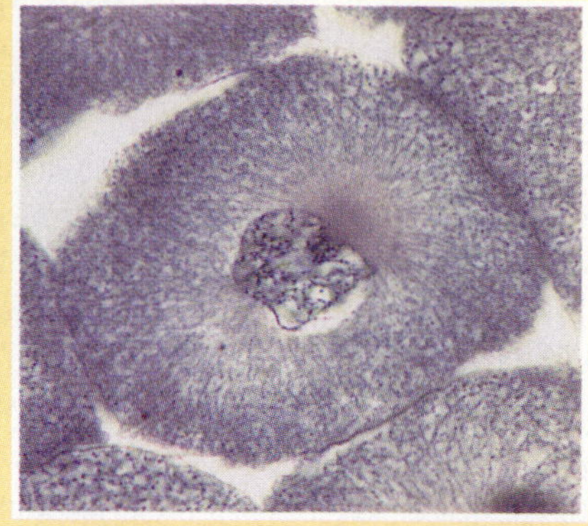

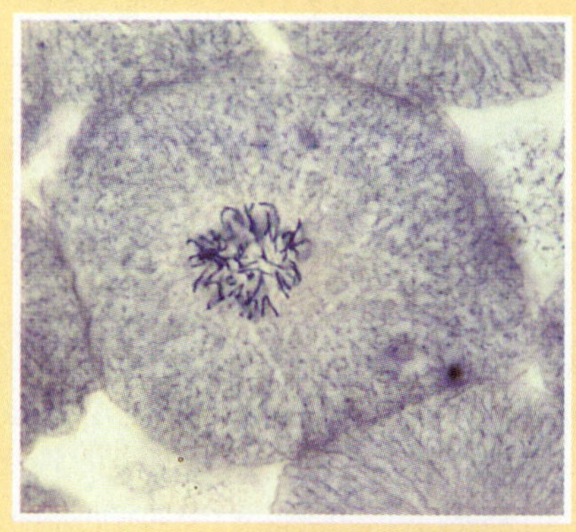

G₂ OF INTERPHASE | PROPHASE | PROMETAPHASE

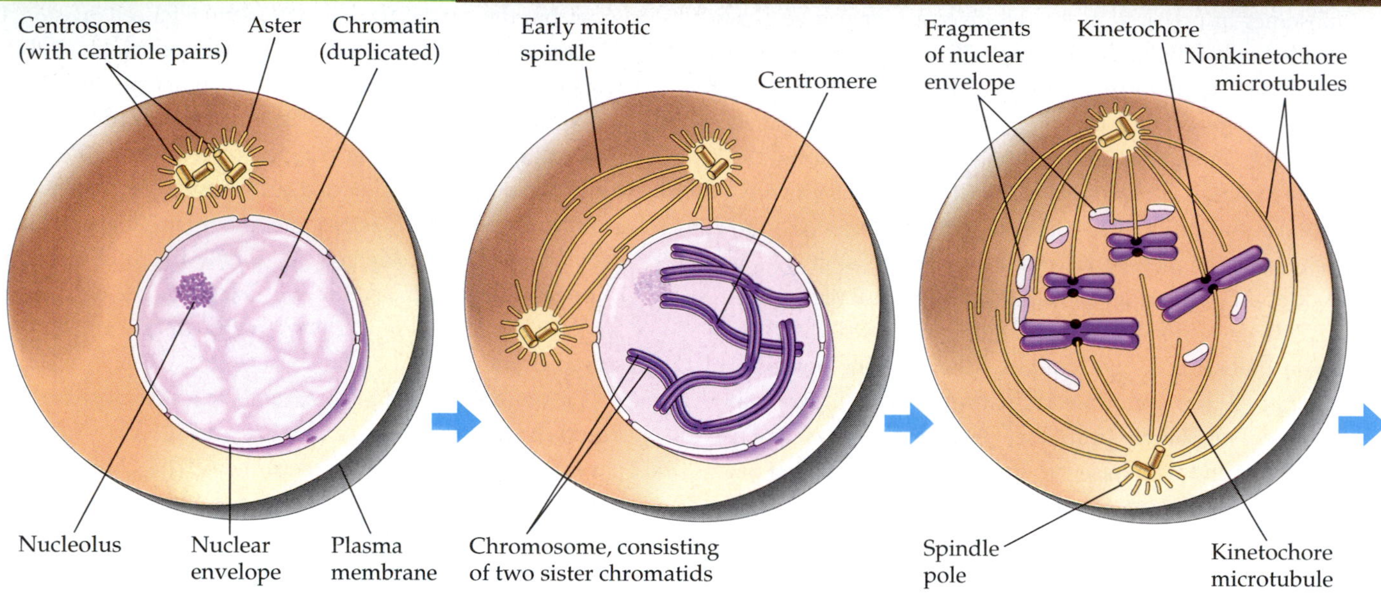

During late interphase, the nucleus is well defined and bounded by the nuclear envelope. It contains one or more nucleoli. Just outside the nucleus are two centrosomes, formed earlier by replication of a single centrosome. In animal cells, each centrosome features a pair of centrioles. Microtubules extend from the centrosomes in radial arrays called **asters** ("stars"). The chromosomes have already duplicated (during the S phase), but at this stage they cannot be distinguished individually because they are still in the form of loosely packed chromatin fibers.

During prophase, changes occur in both the nucleus and the cytoplasm. In the nucleus, the chromatin fibers become more tightly coiled, condensing into discrete chromosomes observable with a light microscope. The nucleoli disappear. Each duplicated chromosome appears as two identical sister chromatids joined together. In the cytoplasm, the mitotic spindle begins to form; it is made of microtubules radiating from the two centrosomes. The centrosomes move away from each other, apparently propelled along the surface of the nucleus by the lengthening bundles of microtubules between them.

During prometaphase, the nuclear envelope fragments. The microtubules of the spindle can now invade the nucleus and interact with the chromosomes, which have become even more condensed. Bundles of microtubules extend from each pole toward the middle of the cell. Each of the two chromatids of a chromosome now has a specialized structure called a **kinetochore,** located at the centromere region. Some of the microtubules attach to the kinetochores. This interaction causes the chromosomes to begin jerky movements. Nonkinetochore microtubules interact with those from the opposite pole of the cell.

12.2
12.3
FIGURE 12.5 ▪ The stages of mitotic cell division in an animal cell. The light micrographs, similar to slides you are likely to see in lab, show dividing cells from a fish embryo. (In plant cells, centrioles are lacking and cytokinesis occurs by a different mechanism; see FIGURES 12.8b and 12.9.) The schematic drawings show details not visible in the micrographs. For the sake of simplicity, only four chromosomes are drawn.

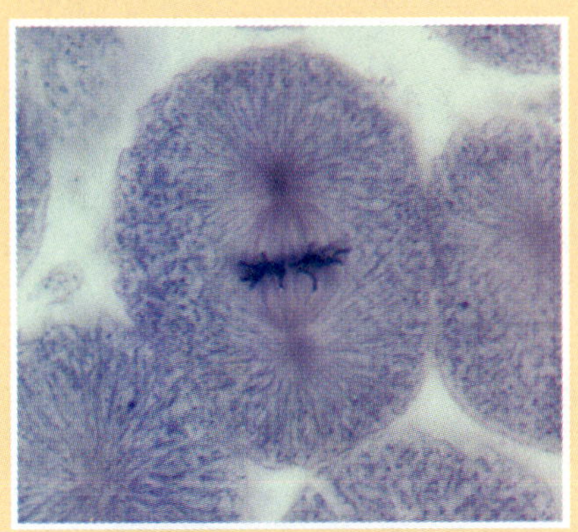

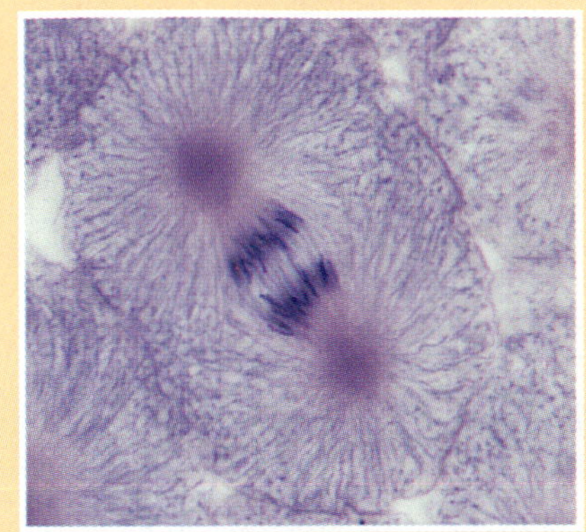

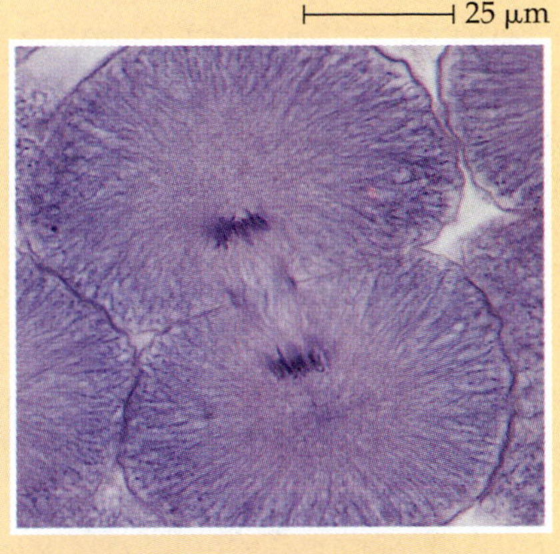

METAPHASE | **ANAPHASE** | **TELOPHASE AND CYTOKINESIS**

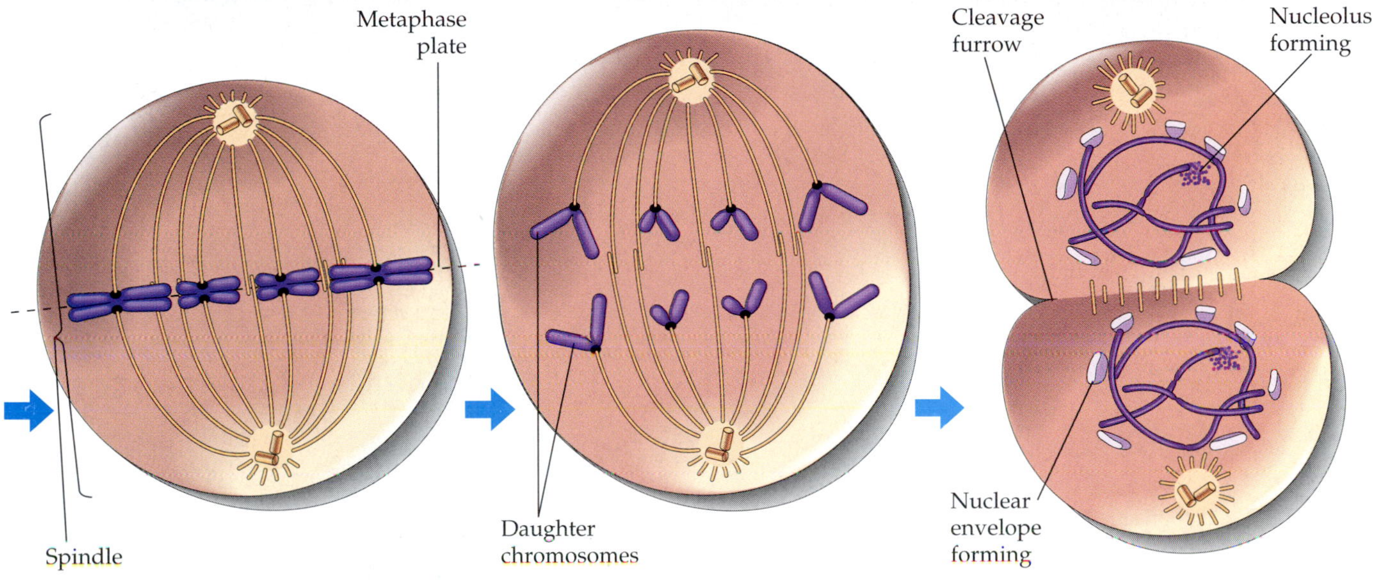

The centrosomes are now at opposite poles of the cell. The chromosomes convene on the **metaphase plate,** an imaginary plane that is equidistant between the spindle's two poles. The centromeres of all the chromosomes are aligned with one another, and the sister chromatids of each chromosome straddle the metaphase plate. For each chromosome, the kinetochores of the sister chromatids are attached to microtubules coming from opposite poles of the cell. The entire apparatus of microtubules is called the spindle because of its shape.

Anaphase begins when the paired centromeres of each chromosome separate, finally liberating the sister chromatids from each other. Each chromatid is now considered a full-fledged chromosome. The once-joined sisters begin moving toward opposite poles of the cell, as their kinetochore microtubules shorten. Because the kinetochore microtubules are attached at the centromere, the chromosomes move centromere first (their pace is about 1 μm/min). At the same time, the poles of the cell move farther apart, as the nonkinetochore microtubules lengthen. By the end of anaphase, the two poles of the cell have equivalent—and complete—collections of chromosomes.

At telophase, the nonkinetochore microtubules elongate the cell still more, and daughter nuclei form at the two poles of the cell. Nuclear envelopes arise from the fragments of the parent cell's nuclear envelope and other portions of the endomembrane system. In a further reversal of prophase and prometaphase events, the chromatin fiber of each chromosome becomes less tightly coiled. Mitosis, the equal division of one nucleus into two genetically identical nuclei, is now complete. Cytokinesis, the division of the cytoplasm, is usually well under way by this time, so the appearance of two separate daughter cells follows shortly after the end of mitosis. In animal cells, cytokinesis involves the formation of a cleavage furrow, which pinches the cell in two.

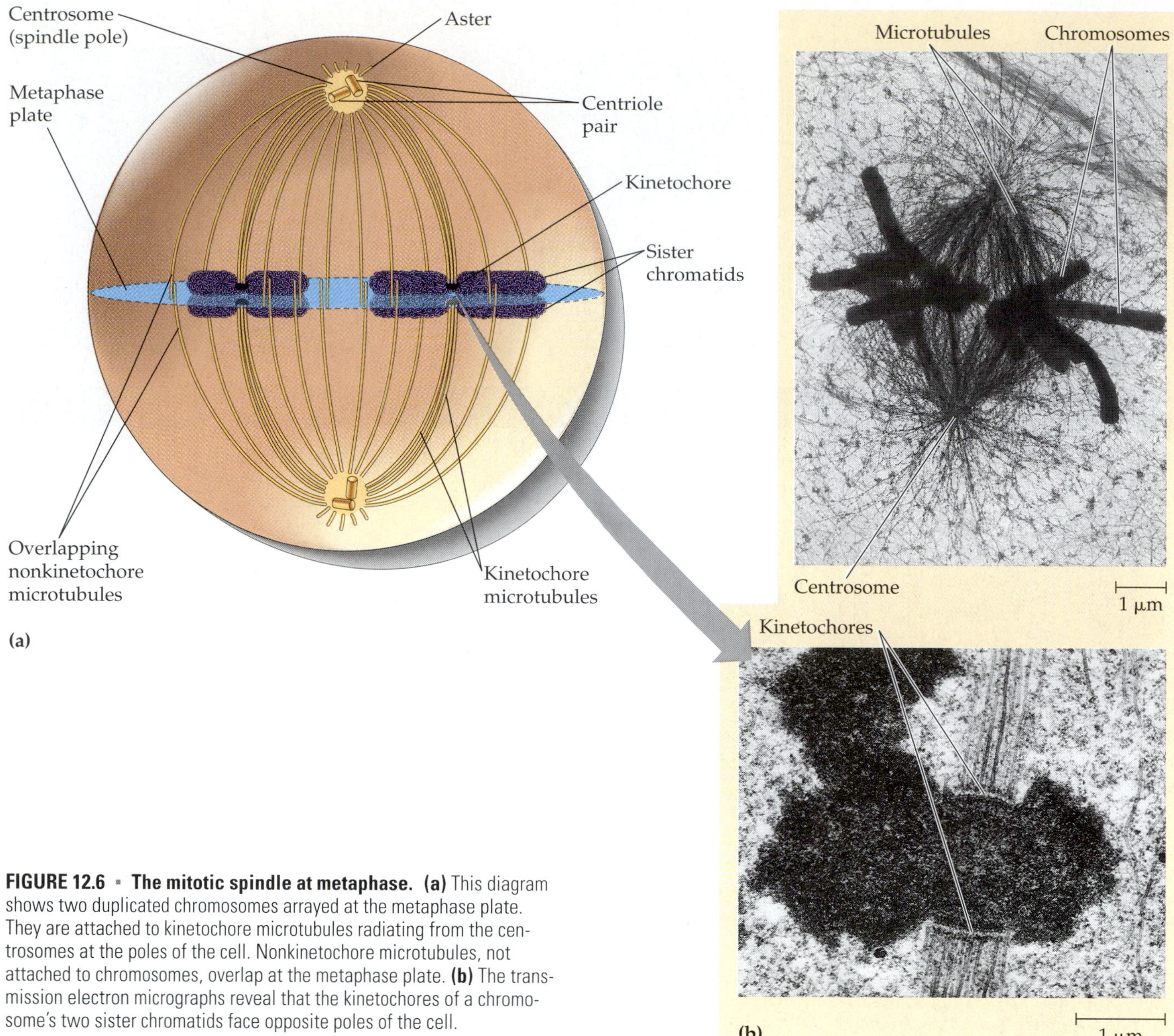

FIGURE 12.6 • The mitotic spindle at metaphase. (a) This diagram shows two duplicated chromosomes arrayed at the metaphase plate. They are attached to kinetochore microtubules radiating from the centrosomes at the poles of the cell. Nonkinetochore microtubules, not attached to chromosomes, overlap at the metaphase plate. **(b)** The transmission electron micrographs reveal that the kinetochores of a chromosome's two sister chromatids face opposite poles of the cell.

"captured" by microtubules, the chromosome begins to move toward the pole from which those microtubules come. However, this movement is checked as soon as microtubules from the opposite pole attach to the other kinetochore. What happens next is like a tug-of-war that ends in a draw. The chromosome moves first in one direction, then the other, back and forth, finally settling midway between the two poles of the cell (FIGURE 12.6). Meanwhile, microtubules that do not attach to kinetochores interact with nonkinetochore microtubules from the opposite pole of the cell. At metaphase, these microtubules overlap, and the centromeres of all the duplicated chromosomes are on a plane midway between the two poles. This plane is called the **metaphase plate** of the cell. The spindle is now complete.

Let's now see how the structure of the completed spindle correlates with its function during anaphase. Anaphase commences rather suddenly, with the separation of the sister chromatids of each chromosome. Now full-fledged chromosomes, they move toward opposite poles of the cell. How do the kinetochore microtubules function in this poleward movement of chromosomes? Experimental evidence indicates that kinetochore microtubules shorten during anaphase by depolymerizing at their kinetochore ends (FIGURE 12.7). A chromosome migrates poleward while its kinetochore somehow hangs onto the remaining tips of microtubules just ahead of the zone of depolymerization. The exact mechanism of this interaction between kinetochores and microtubules is not yet fully understood, but there is much evidence that the kinetochore is equipped with motor proteins that "walk" a chromosome along the shortening microtubules. (To review the relationship between motor proteins and the cytoskeleton, see Chapter 7.)

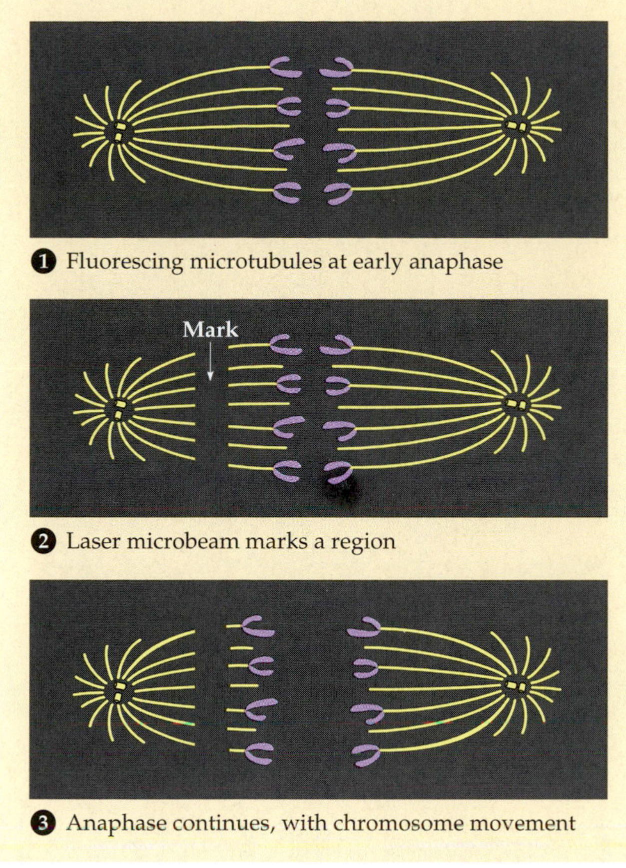

(a)

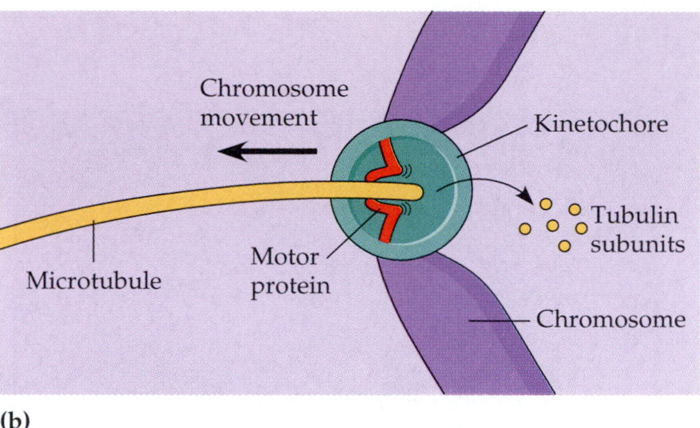

(b)

FIGURE 12.7 ▪ Testing a hypothesis for chromosome migration during anaphase. (a) In this experiment, ① the microtubules of a dividing cell were labeled with a fluorescent dye that glows in the microscope (yellow). ② During early anaphase, the researchers aimed a laser microbeam at the kinetochore microtubules about midway between one spindle pole and the chromosomes. The laser marked the microtubules by eliminating fluorescence at the targeted region, but the microtubules still functioned. The researchers could now monitor changes in the lengths of the microtubules on either side of the nonfluorescing mark. ③ As anaphase proceeded and the chromosomes moved toward the poles, the microtubule segments on the kinetochore side of the laser mark shortened while the portions of the microtubules on the centrosome side of the mark stayed the same length. This is one of the experiments supporting **(b)**, the hypothesis that chromosomes track along microtubules as the microtubules depolymerize at their kinetochore ends, releasing tubulin subunits.

What is the function of the nonkinetochore microtubules? In a dividing animal cell, these microtubules, which interact at the middle of the cell, are responsible for elongating the whole cell along the polar axis during anaphase (see FIGURE 12.5). The interdigitating microtubules move past each other away from the metaphase plate. The mechanism seems to be similar to the one that slides neighboring microtubules in a flagellum: Motor proteins attached to the nonkinetochore microtubules drive them past one another, using energy from ATP.

At the end of anaphase, duplicate sets of chromosomes have arrived at opposite poles of the elongated parent cell. Nuclei re-form during telophase. Cytokinesis generally begins during this last stage of mitosis.

Cytokinesis divides the cytoplasm: *a closer look*

In animal cells, cytokinesis occurs by a process known as cleavage. The first sign of cleavage is the appearance of a **cleavage furrow**, which begins as a shallow groove in the cell surface near the old metaphase plate (FIGURE 12.8a, p. 214). On the cytoplasmic side of the furrow is a contractile ring of actin microfilaments associated with molecules of the protein myosin. Actin and myosin are the same proteins that are responsible for muscle contraction, as well as many other kinds of cell movement. The contraction of the dividing cell's ring of microfilaments is like the pulling of drawstrings. The cleavage furrow deepens until the parent cell is pinched in two, producing two completely separated cells.

Cytokinesis in plant cells, which have walls, is markedly different. There is no cleavage furrow. Instead, during telophase, vesicles derived from the Golgi apparatus move along microtubules to the middle of the cell, where they coalesce, producing a **cell plate** (FIGURE 12.8b). Cell-wall materials carried in the vesicles collect in the cell plate as it grows. The cell plate enlarges until its surrounding membrane fuses with the plasma membrane along the perimeter of the cell. Two daughter cells result, each with its own plasma membrane. Meanwhile, a new cell wall has formed between them.

FIGURE 12.9 on page 215 is a series of micrographs of a dividing plant cell. Examining this figure will help you review mitosis and cytokinesis.

Mitosis in eukaryotes may have evolved from binary fission in bacteria

The complex cellular choreography of mitotic cell division solves the problem of correctly distributing copies of eukaryotic

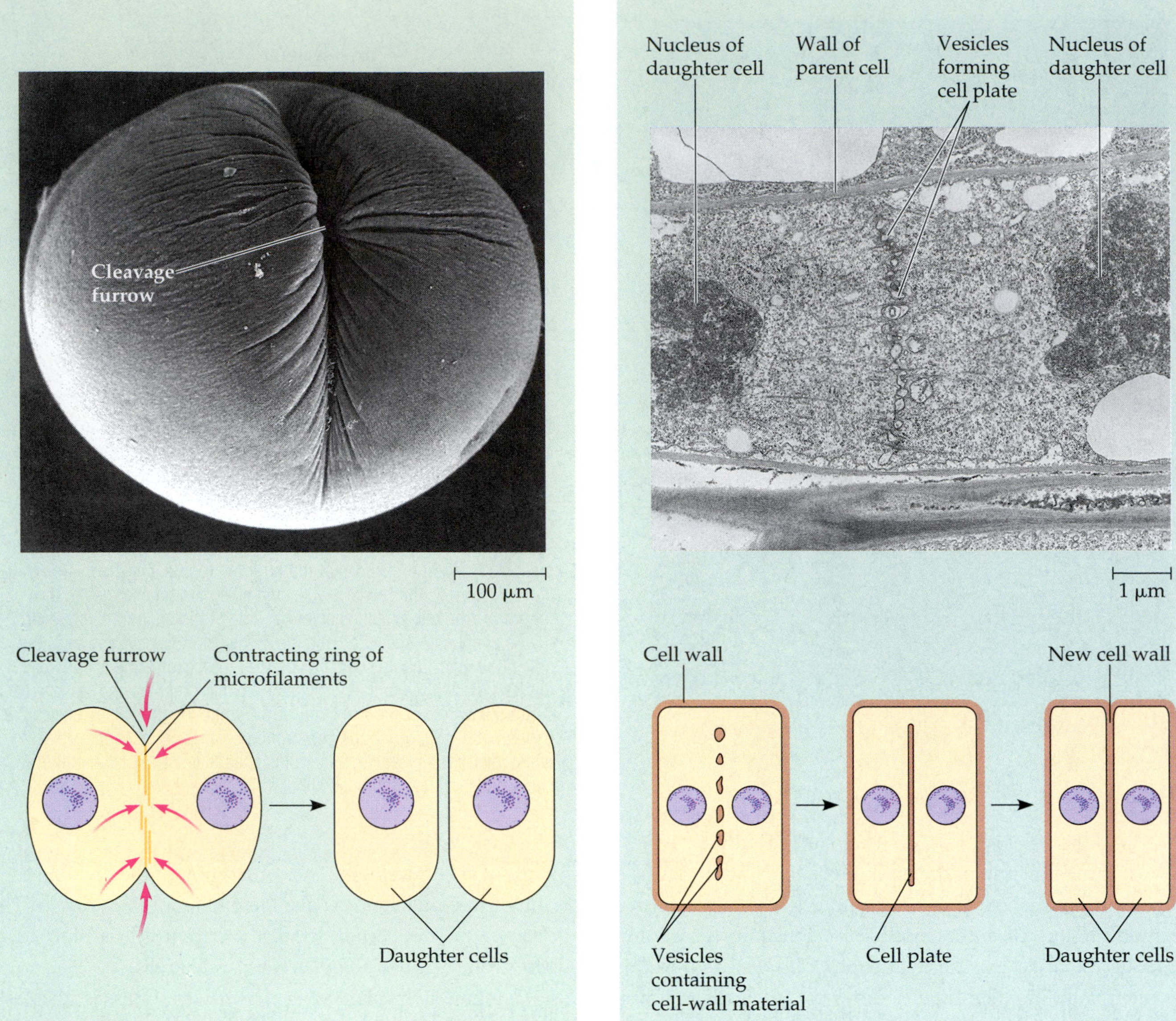

(a) Cleavage of an animal cell

(b) Cell plate formation in a plant cell

FIGURE 12.8 ▪ Cytokinesis in animal and plant cells. (a) The micrograph (SEM) shows the cleavage furrow of a dividing animal cell as it appears on the cell surface. Just inside the plasma membrane at the location of furrowing, microfilaments form a ring. These actin filaments interact with molecules of myosin, causing the ring to contract. The cleavage furrow deepens until the cell is pinched in two. **(b)** In this micrograph (TEM) of a soybean root cell at telophase, you can see the new nuclei at the two poles and vesicles from the Golgi apparatus that are coming together to form a cell plate at the equatorial plane of the cell. These vesicles, laden with cell-wall material, fuse to form a bigger and bigger cell plate, until the plate's membrane fuses at its edges with the plasma membrane, dividing the cell in two. New cell wall between the daughter cells arises from the contents of the cell plate.

genomes to daughter cells. How did mitosis evolve? Given that prokaryotes preceded eukaryotes on Earth by billions of years, we might hypothesize that mitosis had its origins in simpler bacterial mechanisms of cell reproduction.

Prokaryotes (bacteria) reproduce by a type of cell division called **binary fission**, meaning literally "division in half" (FIGURE 12.10). Most bacterial genes are carried on a single chromosome that consists of a circular DNA molecule and associated proteins. Although bacteria are smaller and simpler than eukaryotic cells, the problem of replicating their genomes in an orderly fashion and distributing the copies equally to two daughter cells is still formidable. The chromosome of the bacterium *Escherichia coli,* for example, when it is fully stretched out, is about 500 times longer than the length of the cell. Clearly, such a long chromosome must be highly folded within the cell.

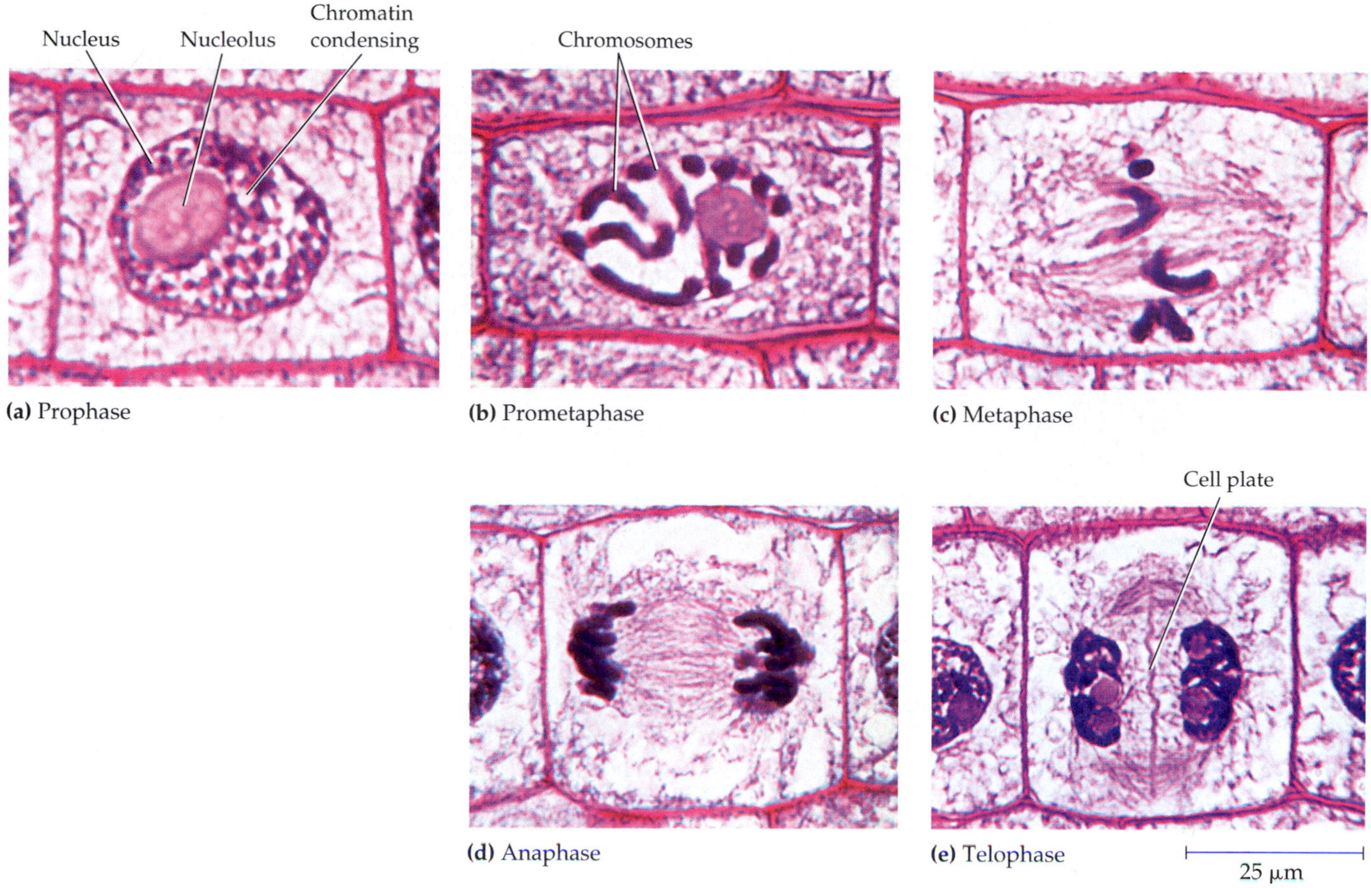

Nucleus
Nucleolus
Chromatin condensing

(a) Prophase

Chromosomes

(b) Prometaphase

(c) Metaphase

(d) Anaphase

Cell plate

(e) Telophase

25 μm

FIGURE 12.9 ▪ Mitosis in a plant cell. These light micrographs show the process of mitosis in cells of an onion root.
(a) Prophase. The chromatin is condensing. The nucleolus is still clearly present but will soon begin to disappear. Although not yet visible in the micrograph, the mitotic spindle is starting to form.
(b) Prometaphase. We now see discrete chromosomes; each consists of two identical sister chromatids attached all along their lengths. Later in prometaphase, the nuclear envelope will fragment, and spindle microtubules will attach to the kinetochores of the chromosomes.
(c) Metaphase. The spindle is complete, and the chromosomes, pulled equally by kinetochore microtubules coming from opposite poles of the cell, are lined up at the metaphase plate.
(d) Anaphase. The chromatids of each chromosome have separated, and the daughter chromosomes are moving to the poles of the cell as their kinetochore microtubules shorten.
(e) Telophase. Daughter nuclei are forming. Meanwhile, cytokinesis has started: The cell plate, which will divide the cytoplasm in two, is growing toward the perimeter of the parent cell.

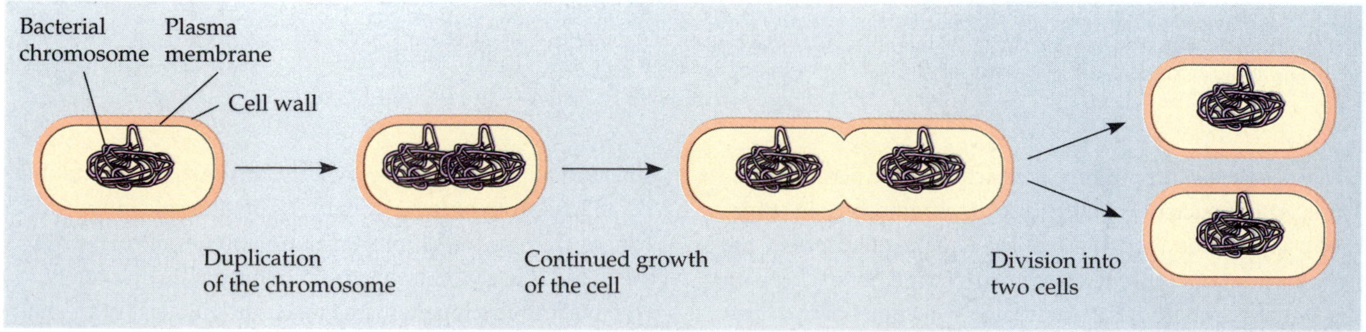

FIGURE 12.10 ▪ Bacterial cell division (binary fission). Allocation of identical genomes to daughter cells depends on the attachment of duplicated chromosomes to the plasma membrane of the parent cell. Continued growth of the cell gradually separates the chromosomes. Eventually the plasma membrane grows inward to divide the cell in two as a new cell wall is deposited between the daughter cells.

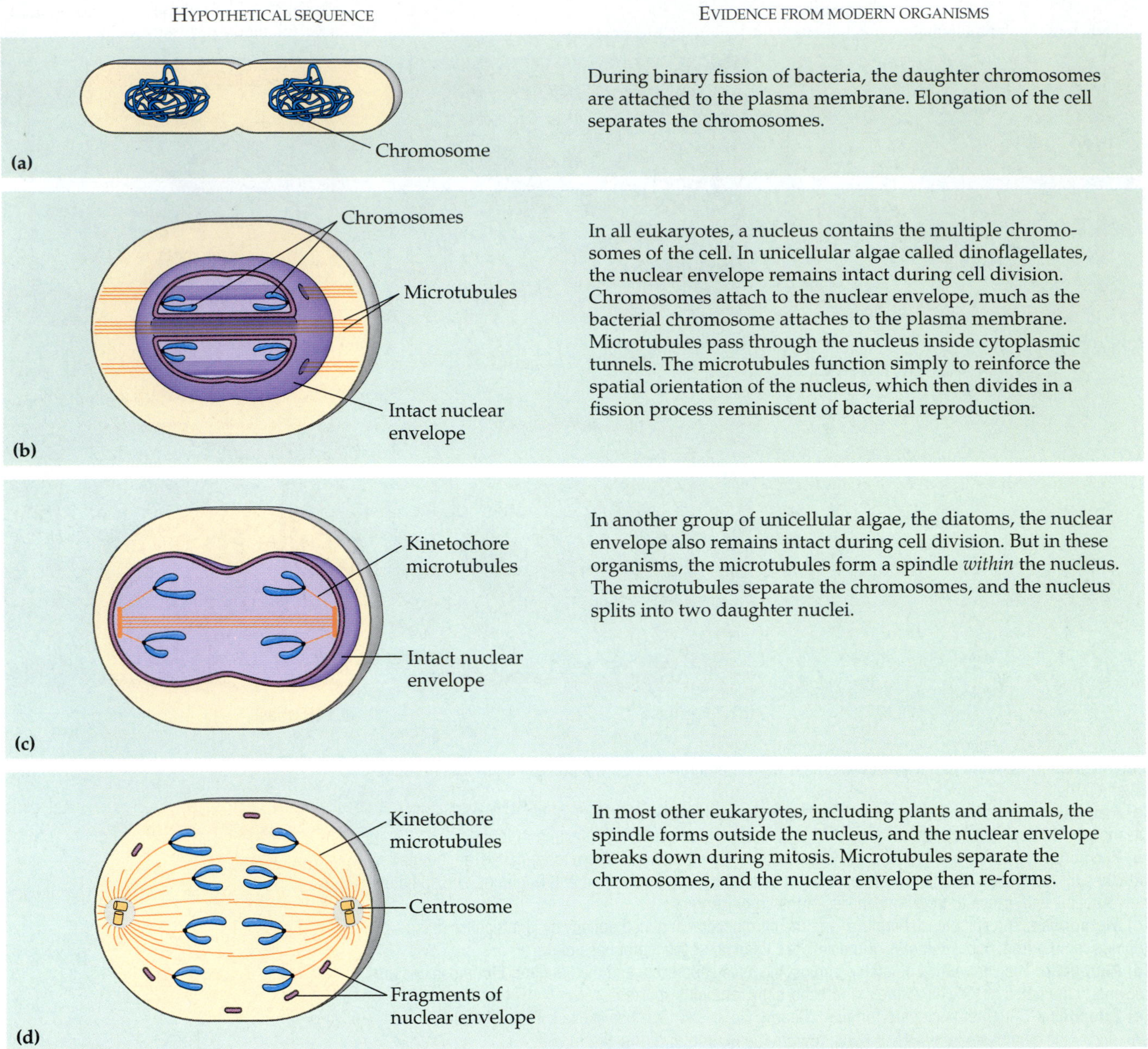

HYPOTHETICAL SEQUENCE

EVIDENCE FROM MODERN ORGANISMS

Chromosome

(a)

During binary fission of bacteria, the daughter chromosomes are attached to the plasma membrane. Elongation of the cell separates the chromosomes.

Chromosomes

Microtubules

Intact nuclear envelope

(b)

In all eukaryotes, a nucleus contains the multiple chromosomes of the cell. In unicellular algae called dinoflagellates, the nuclear envelope remains intact during cell division. Chromosomes attach to the nuclear envelope, much as the bacterial chromosome attaches to the plasma membrane. Microtubules pass through the nucleus inside cytoplasmic tunnels. The microtubules function simply to reinforce the spatial orientation of the nucleus, which then divides in a fission process reminiscent of bacterial reproduction.

Kinetochore microtubules

Intact nuclear envelope

(c)

In another group of unicellular algae, the diatoms, the nuclear envelope also remains intact during cell division. But in these organisms, the microtubules form a spindle *within* the nucleus. The microtubules separate the chromosomes, and the nucleus splits into two daughter nuclei.

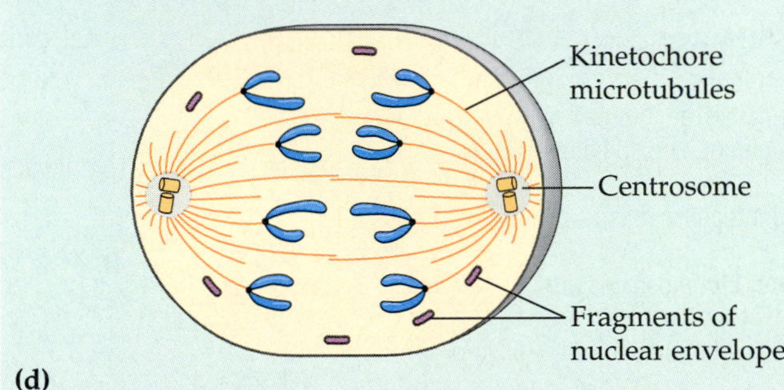

(d)

In most other eukaryotes, including plants and animals, the spindle forms outside the nucleus, and the nuclear envelope breaks down during mitosis. Microtubules separate the chromosomes, and the nuclear envelope then re-forms.

FIGURE 12.11 ▪ A hypothesis for the evolution of mitosis. Mitotic cell division is unique to eukaryotes. Prokaryotes (bacteria) have much smaller genomes and reproduce by the simpler process of binary fission **(a)**. As eukaryotes, with their larger genomes, evolved, the ancestral process of binary fission somehow gave rise to mitosis. Researchers interested in this evolutionary problem have observed what they believe are mechanisms of cell division intermediate between binary fission and mitosis. Two of these variations in cell division are illustrated in **(b)** and **(c)**, beneath bacterial fission and above mitosis as it occurs in plants and animals **(d)**. The schematic diagrams in this figure do not include cell walls.

The bacterial chromosome is attached to the plasma membrane. After a bacterial cell replicates its chromosome in preparation for fission, the two copies remain attached to the membrane at adjacent sites. Growth of the membrane between the two attachment sites separates the two copies of the chromosome. When the bacterium has reached about twice its initial size, its plasma membrane grows inward, dividing the parent cell into two daughter cells. Each cell inherits a complete genome.

FIGURE 12.11 traces a hypothesis for the stepwise evolution of mitosis from bacterial binary fission. Possible intermediate stages are represented by two unusual types of nuclear division found in certain modern unicellular algae. In both types, the nuclear envelope remains intact. In dinoflagellates, replicated chromosomes are attached to the nuclear envelope and separate as it elongates prior to cell division. In diatoms, a spindle within the nucleus separates the chromosomes.

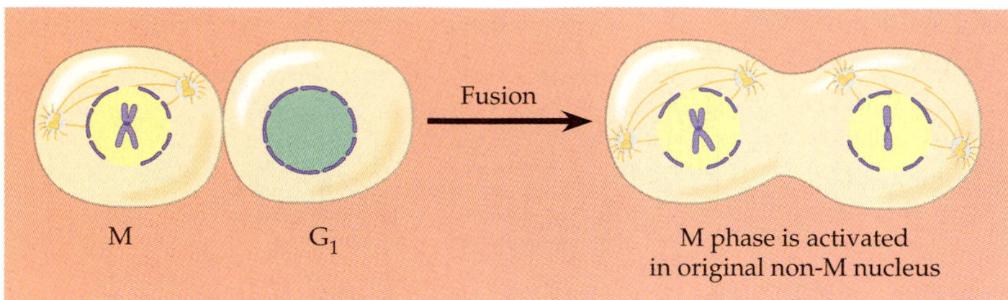

FIGURE 12.12 ▪ Evidence for cytoplasmic chemical signals in cell-cycle regulation. Cultured mammalian cells can be induced to fuse, forming a single cell with two nuclei. The results of fusing cells at two different phases of the cell cycle suggested that particular chemicals control the progression of phases. For example, when a cell in M phase was fused with one in any other phase, the nucleus from the latter cell immediately began mitosis. If the second cell was in G_1, the condensed chromosomes that appeared had single chromatids.

REGULATION OF THE CELL CYCLE

The timing and rate of cell division in different parts of a plant or animal are crucial to normal growth, development, and maintenance. The frequency of cell division varies with the type of cell. For example, human skin cells divide frequently throughout life, whereas liver cells maintain the ability to divide but keep it in reserve until an appropriate need arises—say, to repair a wound. Some of the most specialized cells, such as nerve cells and muscle cells, do not divide at all in a mature human. These cell-cycle differences result from regulation at the molecular level. The mechanisms of this regulation are of intense interest, not only for understanding the life cycles of normal cells but also for understanding how cancer cells manage to escape normal controls.

A molecular control system drives the cell cycle

What drives the cell cycle? One reasonable hypothesis might be that each event in the cycle triggers the next. According to this hypothesis, for example, the replication of chromosomes in S phase might cause cell growth during G_2, which might in turn directly trigger the onset of mitosis. However, this apparently logical hypothesis is not in fact correct.

In the early 1970s a variety of experiments suggested an alternative hypothesis: that the cell cycle is driven by specific chemical signals present in the cytoplasm. Some of the first strong evidence for this hypothesis came from experiments with mammalian cells grown in culture (see the Methods Box, p. 219). In these experiments, two cells in different phases of the cell cycle were fused to form a single cell with two nuclei. If one of the original cells was in S phase and the other was in G_1, the G_1 nucleus immediately entered S phase, as though stimulated by chemicals present in the cytoplasm of the first cell. Similarly, if a cell undergoing mitosis (M phase) was fused with another cell in any stage of its cell cycle, even G_1, the second nucleus immediately entered mitosis, with condensation of the chromatin and formation of a spindle (FIGURE 12.12).

These and other experiments demonstrated that the sequential events of the cell cycle are directed by a distinct **cell-cycle control system**, a cyclically operating set of molecules in the cell that both triggers and coordinates key events in the cell cycle. The cell-cycle control system has been compared to the control device of an automatic washing machine (FIGURE 12.13). Like the washer's device, the cell-cycle control system proceeds on its own, driven by a built-in clock. However, just as a washer's cycle is subject to both external adjustment (such as faucets controlling the water supply) and internal control (such as the sensor that detects when the tub is filled with water), the cell cycle is regulated at certain checkpoints by internal and external controls.

Cell-Cycle Checkpoints

A **checkpoint** in the cell cycle is a critical control point where stop and go-ahead signals can regulate the cycle. Animal cells generally have built-in stop signals that halt the cell cycle at checkpoints until overridden by go-ahead signals. Many signals registered at checkpoints come from cellular surveillance

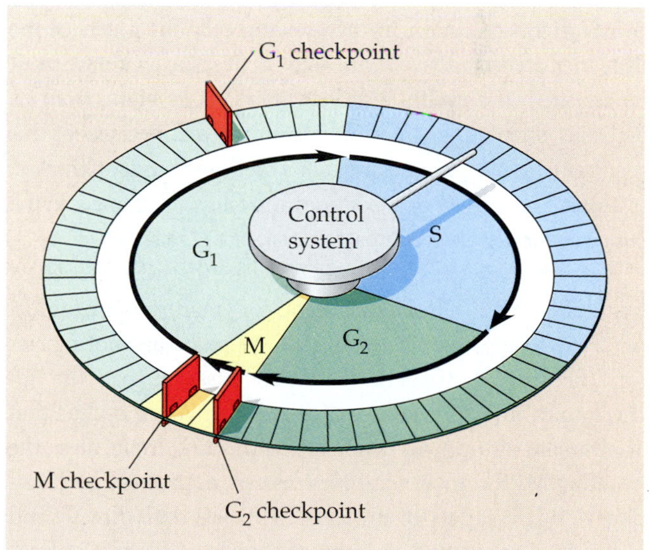

FIGURE 12.13 ▪ Mechanical analogy for the cell-cycle control system. In this diagram of the cell cycle, the flat blocks represent sequential events within each phase. Like the control device of an automatic washer, the cell-cycle control system proceeds on its own, driven by a built-in clock. However, the system is subject to regulation at various checkpoints, of which three are shown (red).

mechanisms; the signals report whether crucial cellular processes up to that point have been completed correctly, and thus whether or not the cell cycle should proceed. Checkpoints also register signals from outside the cell, as we will discuss later. Three major checkpoints are found in the G_1, G_2, and M phases.

For many cells, the G_1 checkpoint—dubbed the "restriction point" in mammalian cells—seems to be the most important. If a cell receives a go-ahead signal at the G_1 checkpoint, it will usually complete the cycle and divide. Alternatively, if it does not receive a go-ahead signal at that point, it will exit the cycle, switching into a nondividing state called the **G_0 phase**. Most cells of the human body are actually in the G_0 phase. As mentioned earlier, specialized nerve and muscle cells never divide. Other cells, such as liver cells, can be "called back" to the cell cycle by certain environmental cues, such as growth factors released during injury.

To understand how cell-cycle checkpoints work, we need to see what kinds of molecules make up the cell-cycle control system.

The Cell-Cycle Clock: Cyclins and Cyclin-Dependent Kinases

Rhythmic fluctuations in the abundance and activity of cell-cycle control molecules pace the sequential events of the cell cycle. These regulatory molecules are proteins of two main types. Some are protein kinases, enzymes that activate or inactivate other proteins by phosphorylating them (see Chapter 11). Particular protein kinases give the go-ahead signals at the G_1 and G_2 checkpoints.

The kinases that drive the cell cycle are actually present at a constant concentration in the growing cell, but much of the time they are in inactive form. To be active, such a kinase must be attached to a **cyclin**, a protein that gets its name from its cyclically fluctuating concentration in the cell. Because of this requirement, these kinases are called **cyclin-dependent kinases**, or **Cdks**. The activity of a Cdk rises and falls with changes in the concentration of its cyclin partner (FIGURE 12.14a).

Let's examine the cyclin-Cdk complex that was discovered first, called **MPF**. The initials stand for "maturation promoting factor," but we can think of it as "M-phase promoting factor," because it triggers the cell's passage past the G_2 checkpoint into M phase (FIGURE 12.14b). When cyclins that accumulate during G_2 associate with Cdk molecules, the resulting MPF complex initiates mitosis, apparently by phosphorylating a variety of proteins. MPF acts both directly and indirectly. For example, it causes the nuclear envelope to fragment by phosphorylating—*and* stimulating other kinases to phosphorylate—proteins of the nuclear lamina, the lining of the nuclear envelope (see FIGURE 7.9).

Later in M phase, MPF helps switch itself off by initiating a process that leads to the destruction of its cyclin by proteolytic (protein-hydrolyzing) enzymes. Proteolytic enzymes are also involved in driving the cycle past the M-phase checkpoint, which controls the onset of anaphase. Anaphase starts when proteins that hold sister chromatids together at metaphase are broken down. The noncyclin part of MPF, the Cdk, persists in the cell in inactive form until it associates with new cyclin molecules synthesized during interphase of the next round of the cycle.

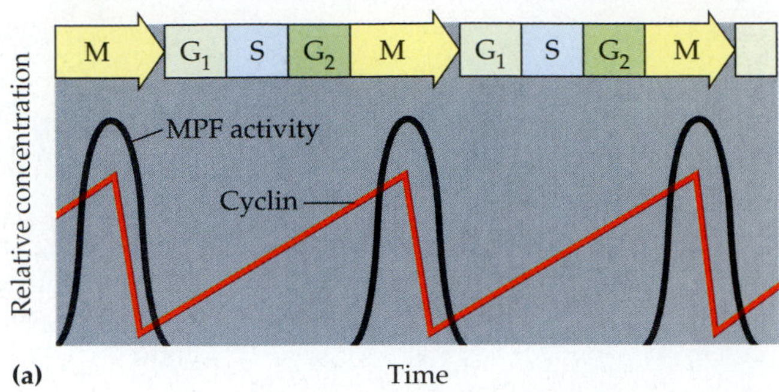

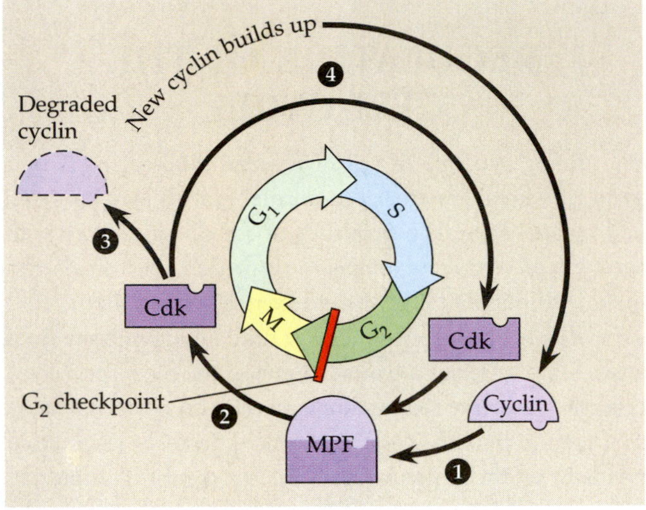

FIGURE 12.14 • The molecular basis of the cell-cycle control system: control at the G_2 checkpoint. The stepwise processes of the cell cycle are timed by rhythmic fluctuations in the activity of protein kinases, enzymes that switch target proteins on or off by phosphorylating them. These enzymes are called cyclin-dependent kinases (Cdks) because they are active only when bound to a cyclin, a protein whose concentration varies cyclically. Here we focus on a Cdk-cyclin complex called MPF, which acts at the G_2 checkpoint to trigger mitosis. **(a)** The graph shows how MPF activity fluctuates with the level of cyclin in the cell. The cyclin level rises throughout interphase (G_1, S, and G_2 phases), then falls abruptly during mitosis (M phase). The peaks of MPF activity and cyclin concentration correspond. The Cdk itself is present at a constant level (not shown). **(b)** ① By the G_2 checkpoint (red bar), enough cyclin is available to produce many molecules of MPF. ② MPF promotes mitosis by phosphorylating various proteins, including other enzymes. ③ One effect of MPF is the initiation of a sequence of events leading to the breakdown of its own cyclin. ④ The Cdk component of MPF is recycled. Its kinase activity will be restored by association with new cyclin that accumulates during interphase.

METHODS: Cell Culture

Many types of animal and plant cells can be removed from an organism and cultured in an artificial environment. This makes it possible to study the cell cycle and many other activities of cells under controlled conditions.

In the experiment illustrated here, we test the effect of a specific growth factor on the ability of human fibroblast cells to divide. Fibroblasts are the connective tissue cells that secrete collagen, the protein making up the extracellular fibers that strengthen our tissues and organs. Our source of fibroblasts is a small sample of tissue removed during a biopsy. Two steps are required to isolate these cells. First, small pieces of connective tissue are dissected from the tissue sample. Second, specific enzymes are added to digest the collagen fibers and other components of the extracellular matrix, resulting in a suspension of free fibroblast cells. The isolated fibroblasts are placed in a culture vessel, a flat flask that lies on its side. The cells adhere to the glass on the inside of the vessel, and they are bathed in a liquid culture medium of known composition. It is essential to sterilize the vessels and the culture medium before use, in order to kill microorganisms that would contaminate the culture. The culture medium is a complex mixture of glucose, amino acids, salts, and antibiotics (as a further precaution against bacterial growth). The cultures are incubated at an optimal temperature, 37°C for human fibroblast cultures.

In our experiment, some culture vessels contain only this basic medium while others contain the basic medium plus a protein called platelet-derived growth factor, or PDGF. In the intact animal, blood cells called platelets release this growth factor at the site of an injury, stimulating nearby fibroblasts to proliferate. This is an important step in the healing of wounds. Our experiment confirms that PDGF stimulates the cell division of cultured fibroblasts.

SOURCE: Art adapted from R. I. Freshney, *Culture of Animal Cells: A Manual of Basic Technique,* 2nd ed., fig. 1.2, p. 5. © 1989 John Wiley and Sons. Reprinted by permission of Wiley-Liss, a subsidiary of John Wiley and Sons, Inc.

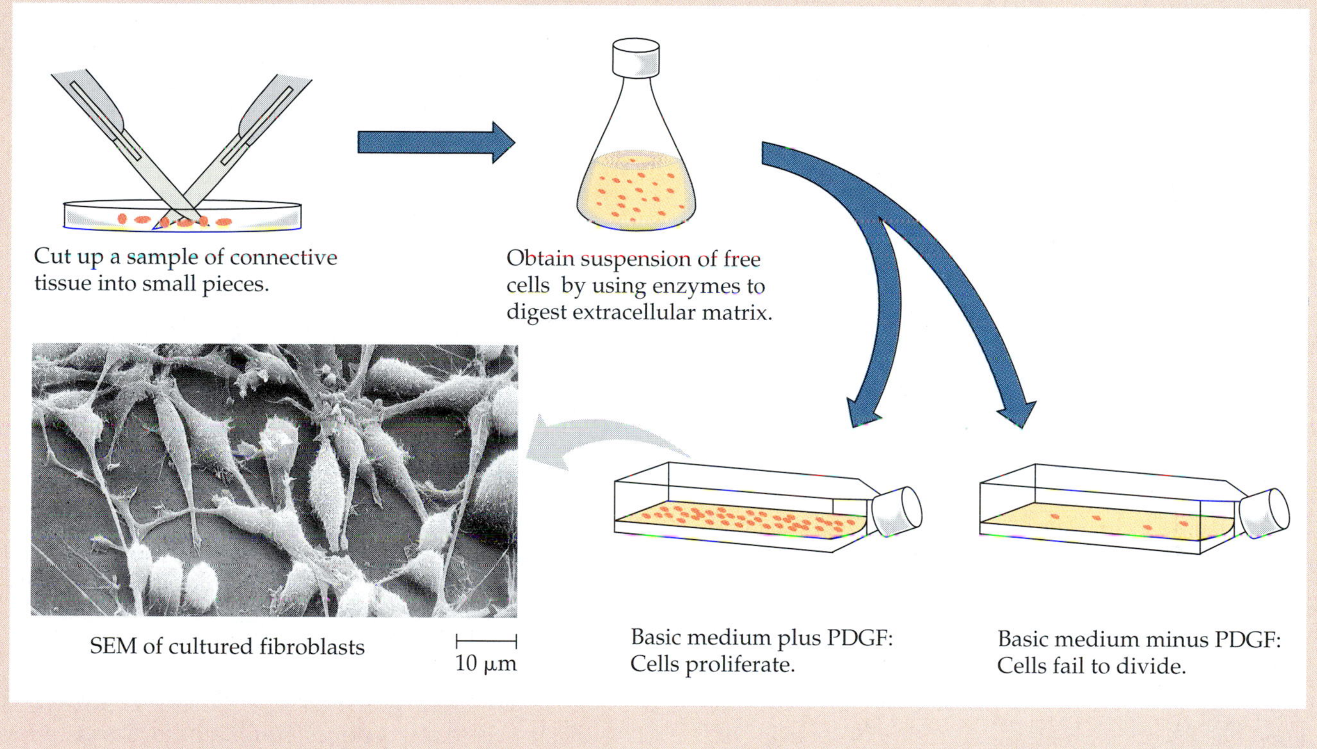

Cut up a sample of connective tissue into small pieces.
Obtain suspension of free cells by using enzymes to digest extracellular matrix.
SEM of cultured fibroblasts
10 μm
Basic medium plus PDGF: Cells proliferate.
Basic medium minus PDGF: Cells fail to divide.

What about the G_1 checkpoint? Recent research suggests the involvement of at least three Cdk proteins and several cyclins at this checkpoint. The fluctuating activities of different cyclin-Cdk complexes probably control all the stages of the cell cycle.

Internal and external cues help regulate the cell cycle

Research scientists are only in the early stages of working out the signaling pathways that link cyclin-dependent kinases to other molecules and events inside and outside the cell. For example, they know that active Cdks function by phosphorylating substrate proteins that affect particular steps in the cell cycle. But identifying the specific substrates of the various Cdks that become active at different phases of the cell cycle has proven difficult. In other words, scientists don't yet know what Cdks actually do in most cases. However, they have identified some steps of the signaling pathways that convey information to the cell-cycle machinery. We'll next discuss two examples of such signaling, one originating inside the cell and one outside.

Internal Signals: Messages from the Kinetochores

Anaphase, the separation of sister chromatids, does not begin until all the chromosomes are properly attached to the spindle at the metaphase plate. The gatekeeper is the M-phase checkpoint, and it ensures that daughter cells do not end up with missing or extra chromosomes. Researchers have learned that a signal that delays anaphase originates at the kinetochores that are not yet attached to spindle microtubules. Certain associated proteins trigger a signaling pathway that keeps an anaphase-promoting complex (APC) in an inactive state. Only when all the kinetochores are attached to the spindle does this "wait" signal cease. The APC becomes active, and its proteolytic enzymes break down cyclin and the proteins holding the sister chromatids together.

External Signals: Growth Factors

By growing animal cells in culture, researchers have been able to identify many external factors, both chemical and physical, that can influence cell division. For example, cells fail to divide if an essential nutrient is left out of the culture medium. Even if all other conditions are favorable, most types of mammalian cells divide in culture only if the growth medium includes specific growth factors. As mentioned in Chapter 11, a **growth factor** is a protein released by certain body cells that stimulates other cells to divide.

One example of a growth factor is *platelet-derived growth factor (PDGF),* which is made by blood cells called platelets. Important for wound healing, PDGF is required for the division of fibroblasts. These connective tissue cells have PDGF receptors on their plasma membranes. The binding of PDGF molecules to these receptors (which are tyrosine-kinase receptors; see Chapter 11) triggers a signal-transduction pathway that leads to stimulation of cell division. Presumably the pathway activates one or more components of the cell-cycle control system. PDGF stimulates fibroblast division not only in the artificial conditions of cell culture, but in an animal's body as well. When an injury occurs, platelets release PDGF in the vicinity. The resulting proliferation of fibroblasts helps heal the wound. Researchers have discovered a number of different growth factors. Each cell type probably responds specifically to a certain growth factor or combination of different growth factors.

The discovery of growth factors provided the key to understanding **density-dependent inhibition** of cell division, a phenomenon in which crowded cells stop dividing (FIGURE 12.15a). As first observed many years ago, cultured cells normally divide until they form a single layer of cells on the inner surface of the culture container, at which point the cells stop dividing. If some cells are removed, those bordering the open space begin dividing again and continue until the vacancy is filled. Apparently, when a cell population reaches a certain density, the amount of required growth factors and nutrients available to each cell becomes insufficient to allow continued cell growth.

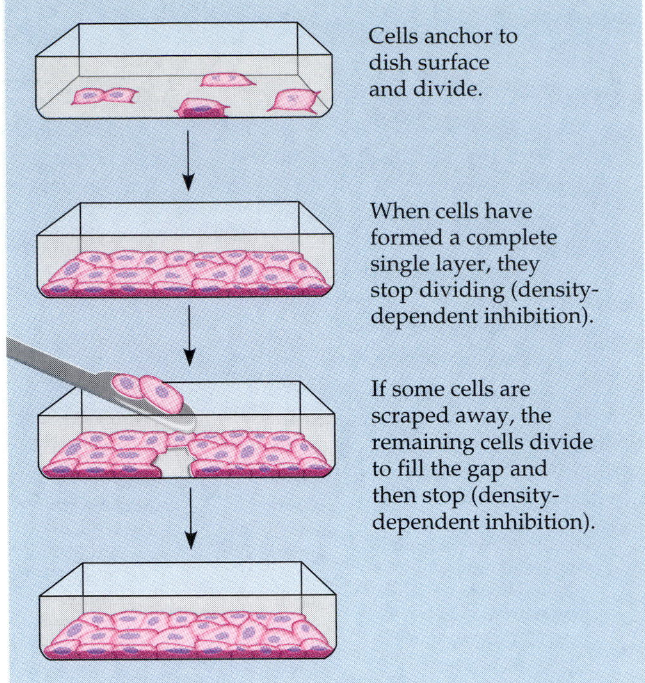

(a)

Cancer cells do not exhibit density-dependent inhibition.

(b)

FIGURE 12.15 ▪ Density-dependent inhibition of cell division. **(a)** When grown in cell culture, normal mammalian cells will multiply only until they form a single layer. The availability of nutrients, growth factors, and substratum for attachment limits the density of the cell population. If some cells are removed, those at the border of the resulting gap will divide until the gap is filled with a single layer. **(b)** In contrast, cancer cells are tolerant of crowding and usually continue to divide well beyond a single layer, forming a clump of overlapping cells. (Individual cells are shown disproportionately large in this figure.)

Most animal cells also exhibit **anchorage dependence.** To divide, they must be attached to a substratum, such as the inside of a culture jar or the extracellular matrix of a tissue. Experiments suggest that anchorage is signaled to the cell-cycle control system via pathways involving plasma membrane proteins and elements of the cytoskeleton linked to them. Density-dependent inhibition and anchorage dependence probably function in the body's tissues as well as in cell culture, checking the growth of cells at some optimal density and location. Cancer cells, which we discuss next, exhibit neither density-dependent inhibition nor anchorage dependence (FIGURE 12.15b).

Cancer cells have escaped from cell-cycle controls

Cancer cells do not respond normally to the body's control mechanisms. They divide excessively and invade other tissues. If unchecked, they can kill the organism.

By studying cells growing in culture, researchers have learned that cancer cells do not heed the normal signals that regulate the cell cycle. For example, as FIGURE 12.15b shows, cancer cells do not exhibit density-dependent inhibition when growing in culture; they do not stop dividing when growth factors are depleted. A logical hypothesis to explain this behavior is that cancer cells have no need for growth factors in their culture medium. They may make a required growth factor themselves or have an abnormality in the signaling pathway that conveys the growth factor's signal to the cell-cycle control system—or the cell-cycle control system itself may be abnormal. In fact, as you will learn in Chapter 19, these are some of the possible explanations.

There are other important differences between normal cells and cancer cells that reflect derangements of the cell cycle. If and when they stop dividing, cancer cells do so at random points in the cycle, rather than just at the normal checkpoints. Moreover, in culture, cancer cells can go on dividing indefinitely if they are given a continual supply of nutrients; they are said to be "immortal." A striking example is a cell line that has been reproducing in culture since 1951. Cells of this line are called HeLa cells because their original source was from a tumor removed from a woman named Henrietta Lacks. By contrast, nearly all normal mammalian cells growing in culture divide only about 20–50 times before they stop dividing, age, and die. (We'll see a possible reason for this phenomenon when we discuss chromosome replication in Chapter 16.)

The abnormal behavior of cancer cells can be catastrophic when it occurs in the body. The potential problem begins when a single cell in a tissue undergoes **transformation**, the process that converts a normal cell to a cancer cell. The body's immune system normally recognizes a transformed cell as an insurgent and destroys it. However, if the cell evades destruction, it may proliferate to form a **tumor**, a mass of abnormal cells within an otherwise normal tissue. If the abnormal cells remain at the original site, the lump is called a **benign tumor**. Most benign tumors do not cause serious problems and can be completely removed by surgery. In contrast, a **malignant tumor** becomes invasive enough to impair the functions of one or more organs (FIGURE 12.16). An individual with a malignant tumor is said to have cancer.

The cells of malignant tumors are abnormal in many ways besides their excessive proliferation. They may have unusual numbers of chromosomes. Their metabolism may be deranged, and they cease to function in any constructive way. Also, due to abnormal changes of the cells' surfaces, they lose their attachments to neighboring cells and the extracellular matrix, and can spread into nearby tissues. Cancer cells may also separate from the original tumor, enter the blood and lymph vessels of the circulatory system, and invade other parts of the body, where they proliferate to form more tumors. This spread of cancer cells beyond their original site is called **metastasis**. If a tumor metastasizes, treatments may include high-energy radiation and chemotherapy with toxic drugs that are especially harmful to actively dividing cells.

Researchers are beginning to understand how a normal cell is transformed into a cancer cell. Though the causes of cancer are diverse, cellular transformation always involves the alteration of genes that somehow influence the cell-cycle control system. Our knowledge of how changes in the genome lead to the various abnormalities of cancer cells remains rudimentary, however.

Perhaps the reason we have so many unanswered questions about cancer cells is that there is still so much to learn about how normal cells function. The cell, life's basic unit of structure and function, holds enough secrets to engage researchers well into the future.

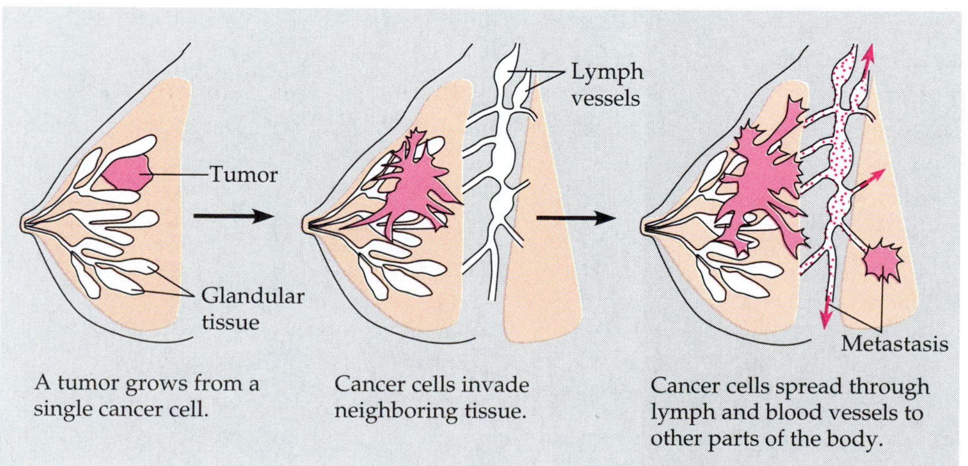
Tumor
Glandular tissue
Lymph vessels
Metastasis
A tumor grows from a single cancer cell.
Cancer cells invade neighboring tissue.
Cancer cells spread through lymph and blood vessels to other parts of the body.

FIGURE 12.16 ▪ The growth and metastasis of a malignant breast tumor. The cells of malignant (cancerous) tumors grow in an uncontrolled way and can spread to neighboring tissues and, via the circulatory system, to other parts of the body. The spread of cancer cells beyond their original site is called metastasis.

CHAPTER REVIEW

REVIEW OF KEY CONCEPTS

THE KEY ROLES OF CELL DIVISION

- **Cell division functions in reproduction, growth, and repair** (p. 206, FIGURE. 12.1) Unicellular organisms reproduce by cell division. Multicellular organisms depend on it for development from a fertilized egg, growth, and repair.
- **Cell division distributes identical sets of chromosomes to daughter cells** (pp. 206–208) Eukaryotic cell division consists of mitosis (division of the nucleus) and cytokinesis (division of the cytoplasm). DNA is partitioned among chromosomes, making it easier for the eukaryotic cell to replicate and distribute its huge amounts of DNA. Chromosomes consist of chromatin, a complex of DNA and protein that condenses during mitosis. When chromosomes replicate, they form identical sister chromatids. The chromatids separate during mitosis, becoming the chromosomes of the new daughter cells.

THE MITOTIC CELL CYCLE

- **The mitotic phase alternates with interphase in the cell cycle: *an***
12.1 ***overview*** (p. 209, FIGURES 12.4, 12.5) Mitosis and cytokinesis make
12.2 up the M (mitotic) phase of the cell cycle. Between divisions, cells
12.3 are in interphase: the G_1, S, and G_2 phases. The cell grows throughout interphase, but DNA is replicated only during the S (synthesis) phase. Mitosis is a continuous process, often described as occurring in five stages: prophase, prometaphase, metaphase, anaphase, and telophase.
- **The mitotic spindle distributes chromosomes to daughter cells: *a closer look*** (pp. 209–213, FIGURE 12.6) The mitotic spindle is an apparatus of microtubules that controls chromosome movement during mitosis. The spindle arises from the centrosomes, regions near the nucleus associated with centrioles in animal cells. Spindle microtubules attach to the kinetochores of chromatids and move the chromosomes to the metaphase plate. In anaphase, sister chromatids separate and move toward opposite poles of the cell. Using motor proteins, each kinetochore moves along shortening microtubules. Meanwhile, nonkinetochore microtubules slide, elongating the cell. In telophase, daughter nuclei form at opposite ends of the cell.
- **Cytokinesis divides the cytoplasm: *a closer look*** (p. 213, FIGURE 12.8) Mitosis is usually followed by cytokinesis, involving cleavage furrows in animals and cell plates in plants.
- **Mitosis in eukaryotes may have evolved from binary fission in bacteria** (pp. 213–216, FIGURES 12.10, 12.11)

REGULATION OF THE CELL CYCLE

- **A molecular control system drives the cell cycle** (pp. 217–219, FIGURES 12.13, 12.14, METHODS BOX) Cyclical changes in regulatory proteins work as a mitotic clock. The key molecules are cyclin-dependent kinases, complexes of cyclins (whose concentrations build during the cell cycle) and specific protein kinases.
- **Internal and external cues help regulate the cell cycle** (pp. 219–220, FIGURE 12.15) Cell culture has enabled researchers to study the molecular details of cell division. Both internal signals, such as those emanating from kinetochores not yet attached to the spindle, and external signals, such as growth factors, control the cell-cycle checkpoints via signal-transduction pathways. Growth factor depletion explains density-dependent inhibition.
- **Cancer cells have escaped from cell-cycle controls** (p. 221, FIGURE 12.16) Cancer cells elude normal regulation and divide out of control, forming tumors. Malignant tumors invade surrounding tissues and can metastasize, exporting cancer cells to other parts of the body.

SELF-QUIZ

1. During the cell cycle, increases in the enzymatic activity of protein kinases are due to
   a. kinase synthesis by ribosomes
   b. activation of inactive kinase by binding to cyclin
   c. conversion of inactive cyclin to the active kinase via phosphorylation
   d. cleavage of the inactive kinase molecules by cytoplasmic proteases
   e. a decline in external growth factors to a concentration below the inhibitory threshold

2. Through a microscope, you can see a cell plate beginning to develop across the middle of the cell and nuclei re-forming at opposite poles of the cell. This cell is most likely a (an)
   a. animal cell in the process of cytokinesis
   b. plant cell in the process of cytokinesis
   c. animal cell in the S phase of the cell cycle
   d. bacterial cell dividing
   e. plant cell in metaphase

3. Vinblastine is a standard chemotherapeutic drug used to treat cancer. Since it interferes with the assembly of microtubules, its effectiveness must be related to
   a. disruption of mitotic spindle formation
   b. inhibition of regulatory protein phosphorylation
   c. suppression of cyclin production
   d. myosin denaturation and inhibition of cleavage furrow formation
   e. inhibition of DNA synthesis

4. A particular cell has half as much DNA as some of the other cells in a mitotically active tissue. The cell in question is most likely in
   a. G_1
   b. G_2
   c. prophase
   d. metaphase
   e. anaphase

5. In the light micrograph below of dividing cells near the tip of an onion root, identify a cell in each of the following stages: interphase, prophase, metaphase, and anaphase. Describe the major events occurring at each stage.

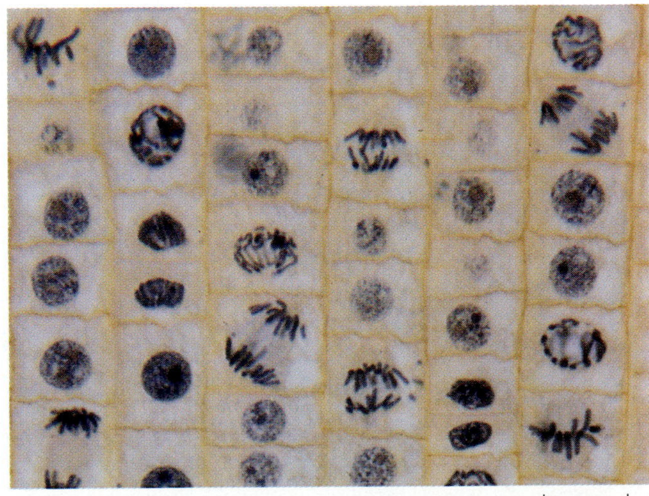

6. One difference between a cancer cell and a normal cell is that
   a. the cancer cell is unable to synthesize DNA
   b. the cell cycle of the cancer cell is arrested at the S phase
   c. cancer cells continue to divide even when they are tightly packed together
   d. cancer cells cannot function properly because they suffer from density-dependent inhibition
   e. cancer cells are always in the M phase of the cell cycle
7. The decline of MPF at the end of mitosis is caused by
   a. the destruction of the protein kinase (Cdk)
   b. decreased synthesis of cyclin
   c. the enzymatic destruction of cyclin
   d. synthesis of DNA
   e. an increase in the cell's volume-to-genome ratio
8. A red blood cell (RBC) has a 120-day lifespan. If an average adult has 5 L (5000 cm^3) of blood and each mm^3 contains 5 million RBCs, how many new cells must be produced each second to replace the entire RBC population?
   a. 30,000
   b. 2400
   c. 2,400,000
   d. 18,000
   e. 30,000,000
9. In *function,* the plant cell structure that is analogous to an animal cell's cleavage furrow is the
   a. chromosome
   b. cell plate
   c. nucleus
   d. centrosome
   e. spindle apparatus
10. In some organisms, mitosis occurs without cytokinesis occurring. This will result in
    a. cells with more than one nucleus
    b. cells that are unusually small
    c. cells lacking nuclei
    d. destruction of chromosomes
    e. cell cycles lacking an S phase

CHALLENGE QUESTION

When a population of cells is examined with a microscope, the percentage of the cells in the M phase is called the mitotic index. The greater the proportion of cells that are dividing, the higher the mitotic index. In a particular study, cells from a cell culture are spread on a slide, preserved and stained, and then inspected in the microscope. A hundred cells are examined: 9 cells are in prophase; 5 cells are in metaphase; 2 cells are in anaphase; 4 cells are in telophase; the remainder, 80 cells, are in interphase. Answer the following questions:

a. What is the mitotic index for this cell culture?

b. The average duration for the cell cycle in this culture is known to be 20 hours. What is the duration of interphase? Of metaphase?

c. Going back to the living culture of these cells, the average quantity of DNA per cell is measured. Of the cells in interphase, 50% contain 10 ng (1 nanogram = 10^{-9} g) of DNA per cell, 20% contain 20 ng DNA per cell, and the remaining 30% of the interphase cells have amounts of DNA between 10 and 20 ng. Based on these data, determine the duration of the G_1, S, and G_2 portions of the cell cycle.

SCIENCE, TECHNOLOGY, AND SOCIETY

Every year about a million Americans are diagnosed as having cancer. This means that about 75 million Americans now living will eventually have cancer, and one in five will die of the disease. There are many kinds of cancers and many causes of the disease. For example, smoking causes most lung cancers, and overexposure to ultraviolet rays in sunlight causes most skin cancers. There is evidence that a high-fat, low-fiber diet is a factor in breast, colon, and prostate cancers. And agents in the workplace such as asbestos and vinyl chloride are also implicated as causes of cancer. Hundreds of millions of dollars are spent each year in the search for effective treatments for cancer; far less money is spent on preventing cancer. Why might this be the case? What kinds of lifestyle changes could we make to help prevent cancer? What kinds of prevention programs could be initiated or strengthened to encourage these changes? What factors might impede such changes and programs?

FURTHER READING

Alberts, B., D. Bray, T. Lewis, M. Raff, K. Roberts, and J. D. Watson. *Molecular Biology of the Cell,* 3rd ed. New York: Garland, 1994. Chapter 13.

Baringa, M. "A New Twist to the Cell Cycle." *Science,* August 4, 1995. How the timed destruction of cyclins regulates cell division.

Becker, W. M., J. Reece, and M. F. Poenie. *The World of the Cell,* 3rd ed. Menlo Park, CA: Benjamin/Cummings, 1996. Chapter 15.

"Cancer." *Scientific American,* September 1996. The entire issue is devoted to the subject of cancer.

Elledge, S. J. "Mitotic Arrest: Mad2 Prevents Sleepy from Waking Up the APC." *Science,* February 13, 1998. How unattached kinetochores control the onset of anaphase. See also "Cell Division Gatekeepers Identified" by E. Pennisi in *Science,* January 23, 1998.

Glover, D. M., C. Gonzalez, and J. W. Raff. "The Centrosome." *Scientific American,* June 1993. Discusses the structure and function of the mitotic spindle.

McIntosh, J. R., and K. L. McDonald. "The Mitotic Spindle." *Scientific American,* October 1989.

Murray, A. W., and T. Hunt. *The Cell Cycle: An Introduction.* New York: W. H. Freeman, 1993. A thorough overview.

Murray, A. W., and M. W. Kirschner. "What Controls the Cell Cycle." *Scientific American,* March 1991.

Orr-Weaver, T. L., and R. A. Weinberg. "A checkpoint on the road to cancer." *Nature,* March 19, 1998. Mutations that disrupt a cell-cycle checkpoint have been identified in cancer cells.

Pluta, A. F., A. M. Mackay, A. M. Ainsztein, I. G. Goldberg, and W. C. Earnshaw. "The Centromere: Hub of Chromosomal Activities." *Science,* December 8, 1995. Discusses the functioning of centromeres and kinetochores.

WEB LINKS

Visit the special edition of *The Biology Place* for BIOLOGY, Fifth Edition, at **http://www.biology.com/campbell**. Go to Chapter 12 for online resources, including learning activities, practice exams, and links to the following web sites:

"Interactive Mitosis Tutorial"
A small but highly animated tutorial on mitosis. Requires a Shockwave plug-in.

"Online Onion Root Tips"
In this interactive experiment from the Biology Project at the University of Arizona, you score the number of onion root tip cells in various stages of the cell cycle and then calculate the time cells spend in each phase of the cycle.

"The Cell Cycle and Mitosis Tutorial"
Another well-conceived tutorial from the Biology Project at the University of Arizona. It explains the cycle with graphics and animation, and you can take a small interactive test at the end to see how well you understand the concepts.

"The Forsburg Lab Home Pages"
Visit the home pages of this research group at the Salk Institute in La Jolla, California, who study the cell cycle. This site has graphic explanations of their work and provides links to cell cycle resources worldwide.

UNIT THREE

GENETICS

Mary-Claire King is best known for her research on the genetics of breast cancer, a disease that strikes one out of every ten American women in her lifetime. In this work, and in her research on inherited deafness and inherited susceptibility to HIV infection (among other subjects), Dr. King has demonstrated the power of approaching biological questions from a variety of perspectives. Depending on the question at hand, she may focus on molecules, on cells, on human families, or on whole populations—using whatever techniques are most appropriate. In addition to basic research, King's laboratory group applies the concepts and methods of genetics to solving practical problems relating to human-rights abuses around the world. Jane Reece spoke with Dr. King on these and other topics at the University of Washington, where she is a professor in the Departments of Medicine and Genetics.

How did you get started in biology?

After majoring in math at Carleton College in Minnesota in the mid '60s, I went out to UC Berkeley to study statistics. I wanted to try to integrate math, which was something I liked and was reasonably good at, with some sort of work that mattered to people. In those days we were very much involved with the civil rights movement, and we were becoming increasingly involved in what was going on in Vietnam. It wasn't an era in which it made sense to ignore the world around you. My first quarter at Berkeley, I took Curt Stern's genetics course, just before he retired. I remember going to that class every day at 1 o'clock, and it was like Christmas every day. I couldn't imagine that people got paid to work on these fascinating kinds of problems! That course changed my life. Soon I transferred to Genetics.

What kind of research did you do as a graduate student?

I worked in Allan Wilson's lab on questions relating to human evolution. With little background in biology, I was initially a menace in the lab, but I didn't let that stop me. My dissertation asked, "How similar at the level of genes and proteins are people and chimpanzees?" We discovered, much to our amazement, that they are extraordinarily similar, that we differ in only about one percent of our DNA. Morphologically and behaviorally, humans and chimpanzees seem to be in different taxonomic families. But genetically we are closely related species. How does one resolve this paradox? One possibility is that similar genes in humans and chimps are turned on and off at different times during embryonic development. The idea that the evolutionary divergence of species might be at least partially explained by changes in the *timing* of gene activity is one that other researchers are exploring.

What led you to study breast cancer?

I finished my Ph. D. in '73 and then went to teach at the University of Chile in a Ford Foundation program. I was in Chile when the Allende government was overthrown. There had been wonderful opportunities for the integration of scientific and economic development of the country, all of which were destroyed in the coup. I came back to Berkeley at a loss, thinking, "What shall I do next?" I knew that I liked genetics very much, and I also knew that I needed to do work that was important to people in an everyday sort of way. The first thing that came up was the possibility of working at UC San Francisco setting up a modern genetics lab in their cancer epidemiology group. At that time a lot of money was available for bringing approaches from other fields into cancer research. I was hired to set up a cancer genetics lab and was captivated by the question of what causes cancer, in particular breast cancer.

INTERVIEW

MARY-CLAIRE KING

When scientists say that cancer is a genetic disease, what does that mean?

It means two things. Cancer is always genetic in the sense that cancer is always the consequence of changes in DNA. Cells that have cancer-causing mutations no longer divide and develop in the way that they should. Something goes wrong with cell division; it's no longer under normal control. The great majority of mutations that lead to cancer arise in the tissue where the cancer starts. The first occurrence is in the colon or in the breast, for example. These mutations are what we call somatic mutations; they are in the body but not in the germ line, the cells that give rise to eggs or sperm. But very rarely, in some families, there *are* germ line mutations in one or more of these same genes. These mutations are passed on from parent to child.

So breast cancer is not usually inherited?

The vast majority of breast-cancer cases seem to have nothing to do with inherited mutations. However, there are many accounts, going back to the ancient Greeks, of families in which breast cancer appears frequently, generation after generation. Of course, this doesn't prove that the disease is inherited. Lung cancer, for example, also clusters in families—but mainly because smoking clusters in families. So, does breast cancer cluster in families because of their exposure to some unidentified environmental agent, or because of an inherited genetic trait, or some combination of both? I thought it would be possible to sort that out by combining statistical studies of families with experimental genetics, and it was. But it took 20 years.

Did you first approach the problem, then, by collecting data on families in which breast cancer was common?

Yes. It occurred to me in 1974 that if mutations were involved in familial breast cancer, we might be able to identify some of the genes that are mutated and then figure out how these genes normally work. The results might give us insight into the more common, nonhereditary forms of the disease.

You eventually identified the gene *BRCA1*, which is mutated in many families with inherited breast cancer, and located the gene on chromosome 17. Are mutations in *BRCA1* also found in the tumors of women with *non*familial breast cancer?

If you test tumor cells from women with nonfamilial breast cancer, you find that cells from more than half of the tumors completely lack one copy of the *BRCA1* gene (of the two in each normal cell) and make reduced amounts of *BRCA1's* protein product. But the sorts of breast-cancer mutations inherited in families are not found in tumors from women with nonfamilial breast cancer. This paradox still has not been resolved.

How do you sort out hereditary influences on breast cancer from environmental influences?

We're now addressing that question in a project with Jewish families in New York City and Israel, families with known mutations in *BRCA1* or *BRCA2,* a gene found soon after *BRCA1.* (Among breast cancer patients of Jewish ancestry, about 10% have one of these mutations.) With the participants' permission, we test the DNA of breast cancer patients from these families for mutations in *BRCA1* and *BRCA2.* For each patient who carries one of these mutations, we trace the history of the mutation in her family back as far as possible. Then we ask, "Did all of the women in the family who inherited the mutation develop breast cancer? At what age did cancer appear?" If some women developed breast cancer at 70 and some at 30, were there any differences in their environmental exposures? If some women with the mutation lived to a very old age and never developed breast cancer at all, can we explain that?

What is your view of genetic testing?

A woman carrying an inherited mutation in *BRCA1* or *BRCA2* seems to have a very high risk of breast cancer (and ovarian cancer as well), a more than 80% risk of developing cancer during her lifetime. But does testing women for mutations in these genes make sense? To me there are three principles that are important to consider. One is that testing must be accompanied by patient education. Testing is useful only if it is presented in such a way that the person understands what the limitations of the test are and what the results mean. Genetic counselors specialize in conveying that information, and their participation is very helpful.

The second critical principle is one of social justice: The social context in which genetic testing takes place needs to respect absolutely the rights of the individual. A person's genetic background should have no bearing at all on his or her ability to obtain health insurance, for instance. It seems to me a very simple concept. We are all predisposed to something. Although researchers have sorted out only a relatively small number of disease-predisposing genes to date, the number is growing daily. If health insurance depended on all genes being "normal," there wouldn't be any insurable people! So, as a matter of both logic and justice, it makes sense to separate the availability of health insurance from a person's genetic makeup.

The third principle has to do with the pace of research in biology. For a woman today to know that she is predisposed to breast and ovarian cancer offers her a problem—but no solution except preventive surgical removal of her breasts and/or ovaries. I wish research moved so quickly that disease-predisposing genes could be identified and treatments developed within months of each other, but that is clearly not the case. The development of a useful intervention is enormously important but is also usually enormously difficult. So we need to work out a rational policy for genetic testing that recognizes the need for education and justice, and the realities of biological research.

Tell us about your lab's involvement in projects relating to human rights abuses.

Our first project, which continues now some 15 years later, was the identification of children who were kidnapped as infants during the Argentine military dictatorship between 1975 and 1983. The children's parents "disappeared." When the grandmothers of these children realized that kidnapping had been widely used as a political tool during that period, they organized and began a search for the missing children. Because the children would not know who they were, the grandmothers needed a geneticist. In 1984, I became that geneticist.

We originally used classical markers for the identifications—blood groups and HLA types (the protein types used to match tissues for organ transplants). But within a few years we began to use newer methods based on DNA technology, such as mitochondrial-DNA sequencing (determining the order of nucleotides in the small DNA molecules of mitochondria). Mitochondrial DNA is ideal for the purpose because it is inherited only through the mother and is highly variable from family to family.

The grandmothers know of about 220 children who were kidnapped, of whom 59 have been identified so far. We have a database of DNA sequences from families who lost children and grandchildren. Now, when young adults come to the grandmothers suspecting they may have been kidnapped, we can compare their mitochondrial DNA with the sequences in our database. We can often tell these young people who they are and put them back in touch with their biological families.

That work evolved in the early 1990s to the analysis of DNA samples from murder victims. Our first cases were some of the murdered mothers of the kidnapped children. And we've since extended our DNA-identification activities to victims in other countries. We've helped identify the remains of MIAs from the Vietnam War, for example, and now we're working as the Molecular Genetics Identification Laboratory for the UN War Crimes Tribunal. We're identifying victims of human rights abuses in Bosnia, Croatia, Rwanda, Somalia, Ethiopia, Chiapas, Guatemala, El Salvador, and Colombia. So we have many projects under way.

What advice would you offer an undergraduate biology student who is interested in a career in genetics?

The field is moving so quickly that the opportunities to answer fascinating, important questions are without limit. The work itself requires devotion, but the effort is easy when you are enchanted by the subject. Do not be discouraged if the first two or three or four things you try aren't a perfect match for your interests and abilities. Have the courage to go up to a professor and say, "The research going on in your lab intrigues me, and I would like to understand it better." It takes a lot of initiative to do that, but it is worth it. One really does need to move beyond courses and have real-life experience doing biology in a laboratory or field setting in order to get a sense of what it is about. In my own lab of 21 people, six are undergraduates.

Do the undergraduates get to make creative contributions to the research?

Oh yes. They come up with ideas, do experiments, publish papers, and go to scientific meetings. But no one does these things the first day! To do research, you have to master a set of skills, as with any other new activity. It's like learning sewing or woodworking. It takes a while to get to the point where you can be creative.

MEIOSIS AND SEXUAL LIFE CYCLES

Life's most exclusive distinction is the ability of organisms to reproduce their kind. Like begets like. Only oak trees produce oaks, and only elephants can make more elephants. Furthermore, offspring resemble their parents more than they do less closely related individuals of the same species. The transmission of traits from one generation to the next is called inheritance, or ***heredity*** *(Latin* heres, *"heir"). Along with inherited similarity, there is also* ***variation****: Offspring differ somewhat in appearance from parents and siblings. These observations have been exploited for the thousands of years that people have bred plants and animals. Curiosity about human similarities and differences is just as ancient. The photo on this page illustrates similarities and differences in the Bridges family (from left to right, Beau, Lloyd, and Jeff). The mechanisms of heredity and variation, however, eluded biologists until the development of genetics in the twentieth century.*

Genetics*, the scientific study of heredity and hereditary variation, is the subject of this unit. Here you will learn about genetics at the levels of organism, cell, and molecule. You will find out how geneticists are helping answer age-old questions about life, including the mystery of how multicellular animals and plants arise from a single cell, the fertilized egg. And you will learn that genetic methods and discoveries are catalyzing progress in every other biological field, including physiology, evolutionary biology, ecology, and even behavior. On the practical side, you will learn how modern genetics is revolutionizing medicine and agriculture. Finally, you will consider some social and ethical questions raised by our new ability to manipulate DNA, the genetic material. In this chapter we begin our study of genetics by examining how sexual reproduction passes chromosomes from parents to offspring.*

AN INTRODUCTION TO HEREDITY

Offspring acquire genes from parents by inheriting chromosomes

Family friends may tell you that you have your mother's freckles, even though *she* still *has* hers. Parents do not, in any literal sense, give their children freckles, eyes, hair, or any other traits. What, then, actually *is* inherited? Parents endow their offspring with coded information in the form of hereditary units called **genes**. The tens of thousands of genes we inherit from our mothers and fathers constitute our genome. Our genetic link to our parents accounts for family resemblance. Your genome may include the gene for freckles, which you inherited from your mother. Our genes program the specific traits that emerge as we develop from fertilized eggs into adults.

Genes are segments of DNA. You learned in Chapters 1 and 5 that DNA is a polymer of four different kinds of monomers called nucleotides. Inherited information is passed on in the form of each gene's specific sequence of nucleotides, much as printed information is communicated in the form of meaningful sequences of letters. Language is symbolic. The brain translates words and sentences into mental images and ideas; for example, the object you imagine when you read "apple" looks nothing like the word itself. Analogously, cells translate genetic "sentences" into freckles and other features with no resemblance to genes. Most genes program cells to synthesize specific enzymes and other proteins, and it is the cumulative action of these proteins that produces an organism's inherited traits. The programming of these traits in the form of DNA is one of the unifying themes of biology.

The transmission of hereditary traits has its molecular basis in the precise replication of DNA, which produces copies of genes that can be passed along from parents to offspring. In animals and plants, the cellular vehicles that transmit genes from one generation to the next are sperm and ova (unfertilized eggs). After a sperm cell unites with an ovum (a single egg), genes from both parents are present in the nucleus of the fertilized egg.

The DNA of a eukaryotic cell is subdivided into chromosomes within the nucleus. Every living species has a characteristic number of chromosomes. For example, humans have 46 chromosomes (except in their reproductive cells). Each chromosome consists of a single long DNA molecule elaborately coiled in association with various proteins. One chromosome includes hundreds or thousands of genes, each of which is a specific part of the DNA molecule. A gene's specific location along the length of a chromosome is called the gene's **locus** (plural, **loci**). Our genetic endowment consists of whatever genes happened to be part of the chromosomes we inherited from our parents.

Like begets like, more or less: a comparison of asexual and sexual reproduction

13.1 Strictly speaking, "Like begets like" applies only to organisms that reproduce asexually. In **asexual reproduction**, a single individual is the sole parent and passes copies of all its genes on to its offspring. For example, single-celled eukaryotic organisms can reproduce asexually by mitotic cell division, in which DNA is copied and allocated equally to two daughter cells. The genomes of the offspring are virtually exact copies of the parent's genome. Some multicellular organisms are also capable of reproducing asexually. *Hydra*, a relative of the jellyfish, can reproduce by budding (FIGURE 13.1). Because the cells

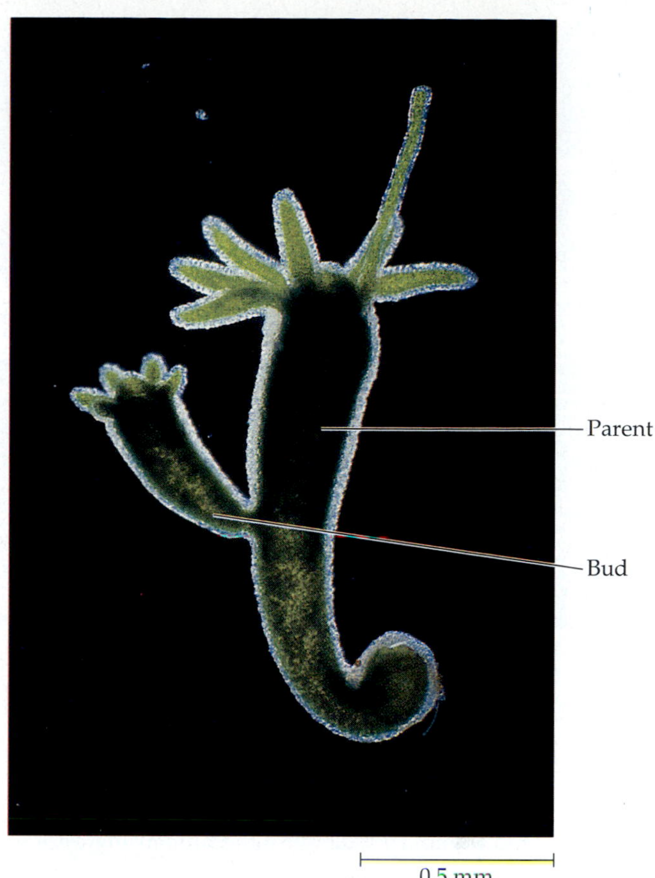

FIGURE 13.1 • The asexual reproduction of a hydra. This relatively simple multicellular animal reproduces by budding. The bud, a localized mass of mitotically dividing cells, develops into a small hydra, which detaches from the parent (LM).

of the bud were derived by mitosis in the parent, the "chip off the old block" is usually genetically identical to its parent. Occasional genetic differences are due to relatively rare changes in the DNA called mutations, which will be discussed in Chapter 17. An individual that reproduces asexually gives rise to a **clone**, a group of genetically identical individuals.

Compared to asexual reproduction, **sexual reproduction** usually results in greater variation; two parents give rise to offspring that have unique combinations of genes inherited from the two parents. In contrast to a clone, offspring of sexual reproduction vary genetically from their siblings and both parents (FIGURE 13.2, p. 228). What mechanisms generate this genetic variation? The key is the behavior of chromosomes during the sexual life cycle.

THE ROLE OF MEIOSIS IN SEXUAL LIFE CYCLES

A **life cycle** is the generation-to-generation sequence of stages in the reproductive history of an organism, from conception to production of its own offspring. In this section we track the behavior of chromosomes through sexual life cycles.

13.1 This symbol links topics in the text to interactive exercises in the CD-ROM that accompanies the book. The number indicates the appropriate activity in the CD.

Couple 1

Couple 2

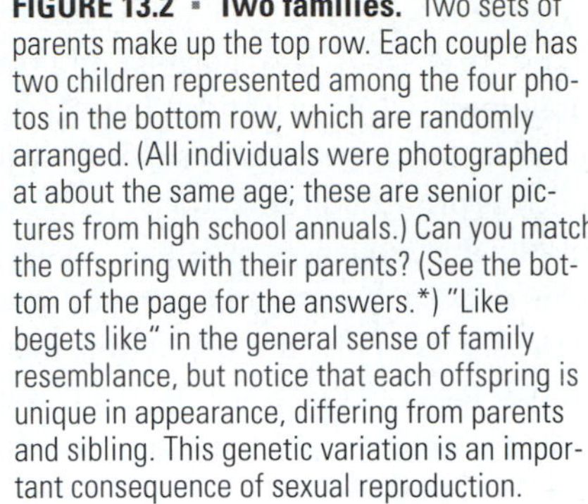

FIGURE 13.2 ▪ Two families. Two sets of parents make up the top row. Each couple has two children represented among the four photos in the bottom row, which are randomly arranged. (All individuals were photographed at about the same age; these are senior pictures from high school annuals.) Can you match the offspring with their parents? (See the bottom of the page for the answers.*) "Like begets like" in the general sense of family resemblance, but notice that each offspring is unique in appearance, differing from parents and sibling. This genetic variation is an important consequence of sexual reproduction.

Fertilization and meiosis alternate in sexual life cycles

We begin with a familiar example—the human life cycle—and use it to introduce some basic terms.

The Human Life Cycle

In humans, each **somatic cell**—any cell other than a sperm or ovum—has 46 chromosomes. With a light microscope, condensed (mitotic) chromosomes can be distinguished from one another by their appearance. The sizes of chromosomes and the positions of their centromeres differ. Each chromosome also has a distinctive pattern of bands when stained with certain dyes.

Careful examination of a micrograph of the 46 human chromosomes reveals that there are two of each type. This becomes clear when the chromosomes are arranged in pairs, starting with the longest chromosomes. The resulting display is called a **karyotype** (see the Methods Box on the facing page). The chromosomes that make up a pair—that have the same length, centromere position, and staining pattern—are called **homologous chromosomes**, or homologues. The two chromosomes of each pair carry genes controlling the same inherited characters. For example, if a gene for eye color is situated at a particular locus on a certain chromosome, then the homologue of that chromosome will also have a gene specifying eye color at the equivalent locus.

There is an important exception to the rule of homologous chromosomes for human somatic cells: the two distinct chromosomes referred to as *X* and *Y*. Human females have a homologous pair of *X* chromosomes *(XX)*, but males have one *X* and one *Y* chromosome *(XY)*. Because they determine an individual's sex, the *X* and *Y* chromosomes are called **sex chromosomes**. The other chromosomes are called **autosomes**.

The occurrence of homologous pairs of chromosomes in our karyotype is a consequence of our sexual origins. We inherit one chromosome of each pair from each parent. So the 46 chromosomes in our somatic cells are actually two sets of 23 chromosomes—a maternal set (from our mother) and a paternal set (from our father).

Sperm cells and ova are different from somatic cells in chromosome count. Each of these reproductive cells, or **gametes**, has a single set of the 22 autosomes plus a single sex chromosome, either *X* or *Y*. A cell with a single chromosome set is called a **haploid cell**. For humans, the haploid number, abbreviated n, is 23 ($n = 23$).

By means of sexual intercourse, a haploid sperm cell from the father reaches and fuses with a haploid ovum of the mother. This union of gametes is called **fertilization**, or **syngamy**. The resulting fertilized egg, or **zygote**, contains the two haploid sets of chromosomes bearing genes representing the maternal and paternal family lines. The zygote and all other cells having two sets of chromosomes are called **diploid cells**. For humans, the diploid number, abbreviated $2n$, is 46 ($2n = 46$).

As a human develops from a zygote to a sexually mature adult, the zygote's genes are passed on with precision to all somatic cells of the body by the process of mitosis. Thus somatic cells, like the zygote from which they are derived, are diploid.

The only cells of the human body *not* produced by mitosis are the gametes, which develop in the gonads (ovaries in

*Answer: The boy and girl in the center of the bottom row are children of Couple 1. The boy and girl on either end of the bottom row are children of Couple 2.

METHODS: Preparation of a Karyotype

Karyotypes, ordered displays of an individual's chromosomes, are useful in identifying certain abnormalities in the chromosomes. Medical technicians usually prepare karyotypes by using lymphocytes, a type of white blood cell.

After the cells are treated with a drug to stimulate mitosis, they are grown in culture for several days. Another drug is then added to arrest mitosis at metaphase, when the chromosomes, each consisting of two joined sister chromatids, are very condensed. This is the stage when chromosomes are easiest to identify in the microscope. The drawings here outline the further steps in the preparation of a karyotype from lymphocytes. The display in step 5 is the karyotype of a human male, with one *X* and one *Y* chromosome. (Human females have two *X* chromosomes.) The chromosomes have been stained to reveal band patterns, which help identify specific chromosomes and parts of chromosomes. Karyotyping can be used to screen for abnormal numbers of chromosomes or defective chromosomes associated with certain congenital disorders, such as Down syndrome. The causes and effects of chromosomal disorders are discussed in Chapter 15.

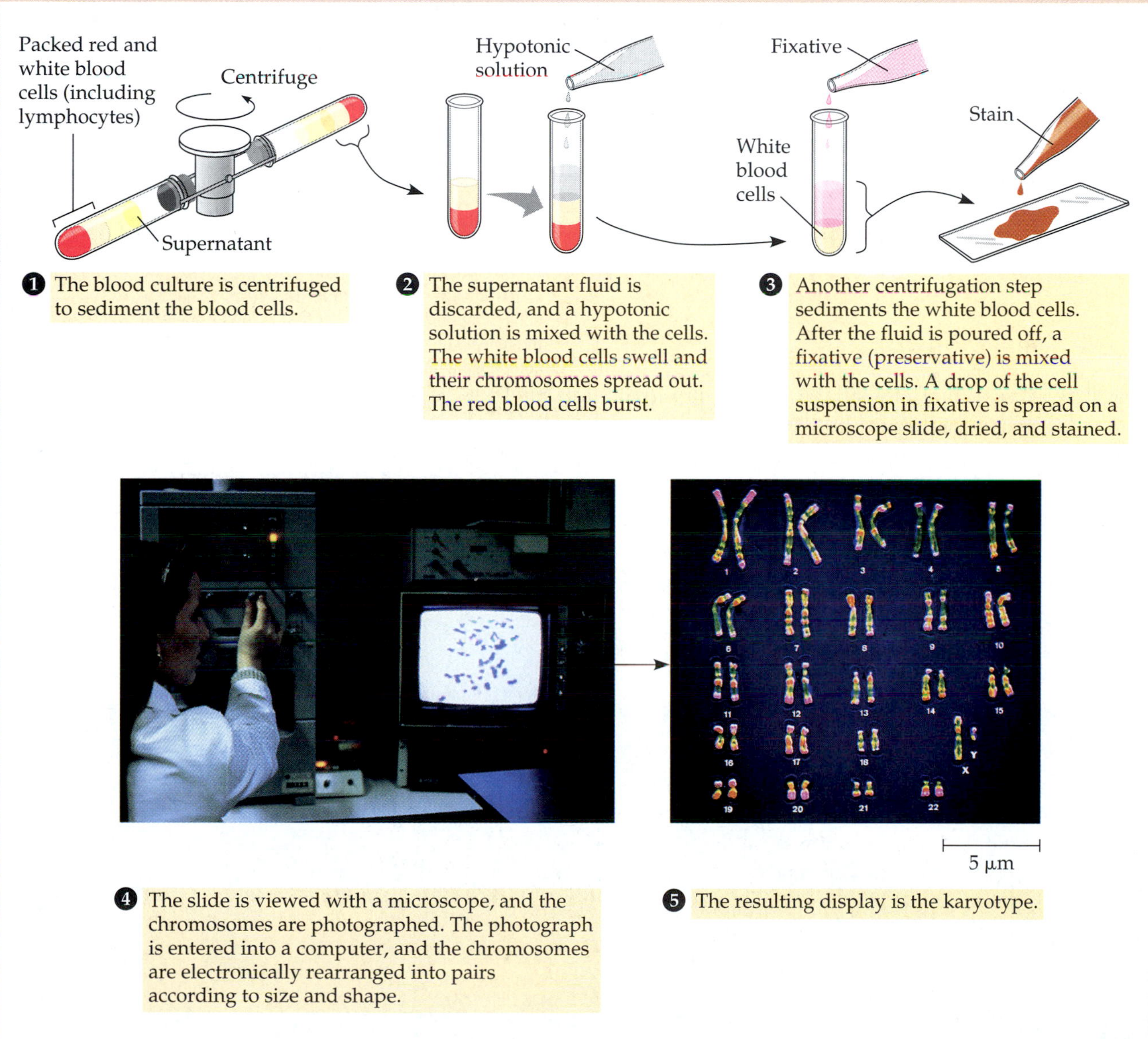

❶ The blood culture is centrifuged to sediment the blood cells.

❷ The supernatant fluid is discarded, and a hypotonic solution is mixed with the cells. The white blood cells swell and their chromosomes spread out. The red blood cells burst.

❸ Another centrifugation step sediments the white blood cells. After the fluid is poured off, a fixative (preservative) is mixed with the cells. A drop of the cell suspension in fixative is spread on a microscope slide, dried, and stained.

❹ The slide is viewed with a microscope, and the chromosomes are photographed. The photograph is entered into a computer, and the chromosomes are electronically rearranged into pairs according to size and shape.

❺ The resulting display is the karyotype.

females and testes in males). Imagine what would happen if human gametes *were* made by mitosis: They would be diploid like the somatic cells. At the next round of fertilization, when two gametes fused, the normal chromosome number of 46 would double to 92, and each subsequent generation would double the number of chromosomes yet again. But sexually reproducing organisms carry out a process that halves the chromosome number in the gametes, compensating for the doubling that occurs at fertilization. This process is a form of cell division called **meiosis**, and in animals it occurs only in the ovaries or testes. While mitosis conserves chromosome number, meiosis reduces the chromosome number by half. As a result, human sperm and ova have haploid sets of 23 different chromosomes. Fertilization restores the diploid condition, and the human life cycle is repeated, generation after generation (FIGURE 13.3).

In general outline, the human life cycle is typical of many animals. Indeed, the processes of fertilization and meiosis are the unique trademarks of sexual reproduction. Fertilization and meiosis alternate in sexual life cycles, offsetting each other's effects on the chromosome number and thus perpetuating a species' chromosome count.

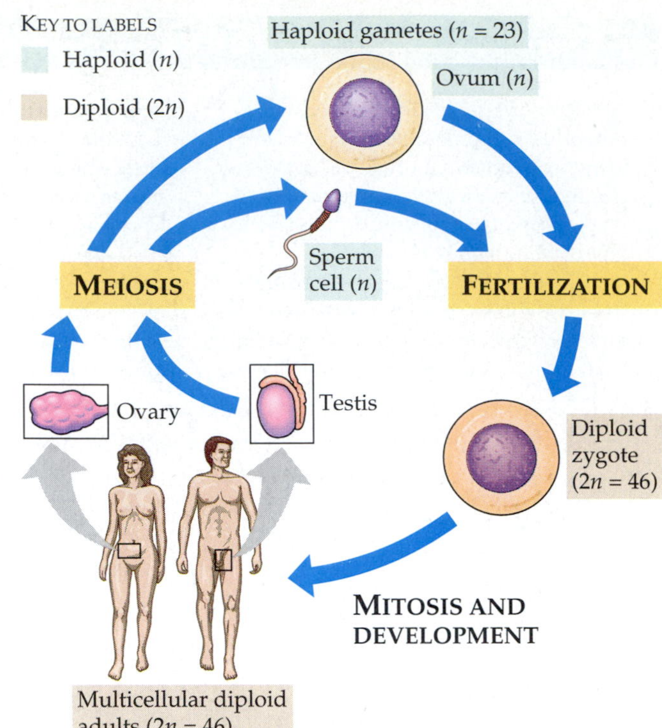

13.1 **FIGURE 13.3 • The human life cycle.** In each generation, the doubling of chromosome number that results from fertilization is offset by the halving of chromosome number that results from meiosis. For humans, the number of chromosomes in a haploid cell is 23 ($n = 23$); the number of chromosomes in the diploid zygote and all somatic cells arising from it is 46 ($2n = 46$).

Figure 13.3 introduces a color code that will be used for some other life cycles later in the book. A green background represents haploid stages of a life cycle, and a tan background represents diploid stages.

The Variety of Sexual Life Cycles

Although the alternation of meiosis and fertilization is common to all organisms that reproduce sexually, the timing of these two events in the life cycle varies, depending on the species. These variations can be grouped into three main types of life cycles (FIGURE 13.4). The human life cycle is an example of one type, characteristic of most animals. Gametes are the only haploid cells. Meiosis occurs during the production of gametes, which undergo no further cell division prior

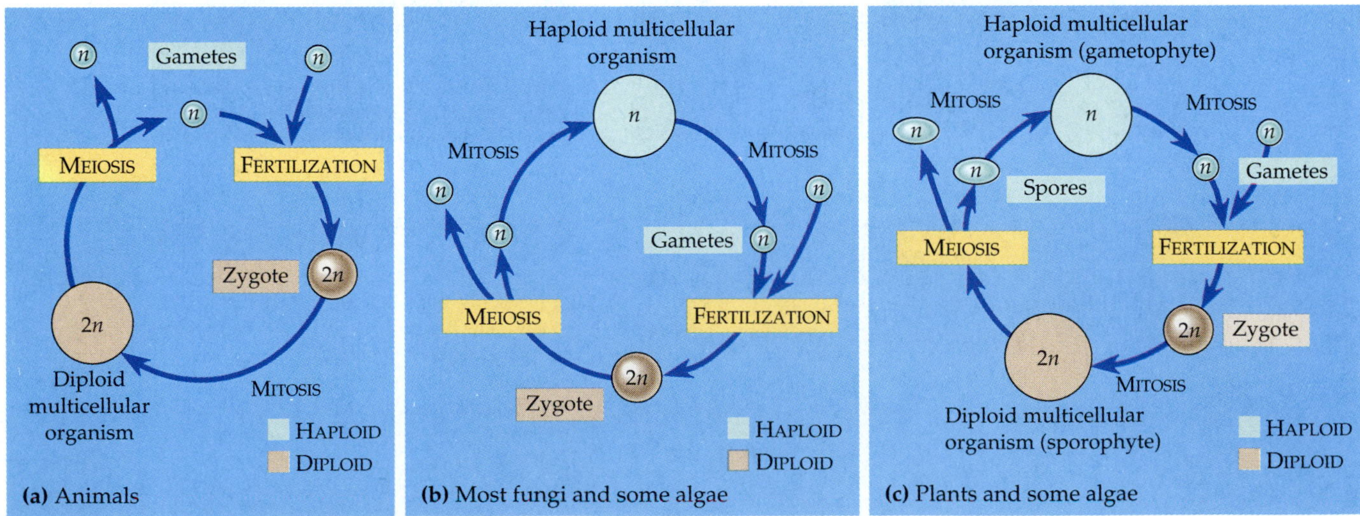

FIGURE 13.4 • Three sexual life cycles differing in the timing of meiosis and fertilization (syngamy). The common feature of all three cycles is the alternation of these two key events, which contribute to genetic variation among offspring.

to fertilization. The diploid zygote divides by mitosis, producing a multicellular organism that is diploid (FIGURE 13.4a).

A second type of life cycle occurs in most fungi and some protists, including some algae. After gametes fuse to form a diploid zygote, meiosis occurs before offspring develop. This meiosis produces not gametes but haploid cells that then divide by mitosis to give rise to a haploid multicellular adult organism. Subsequently, the haploid organism produces gametes by mitosis, rather than by meiosis. The only diploid stage is the zygote (FIGURE 13.4b). (Note that *either* haploid or diploid cells can divide by mitosis depending on the type of life cycle. Only diploid cells, however, can undergo meiosis.)

Plants and some species of algae exhibit a third type of life cycle called **alternation of generations**. This type of life cycle includes both diploid and haploid multicellular stages. The multicellular diploid stage is called the **sporophyte**. Meiosis in the sporophyte produces haploid cells called **spores**. Unlike a gamete, a spore gives rise to a multicellular individual without fusing with another cell. A spore divides mitotically to generate a multicellular haploid stage called the **gametophyte**. The haploid gametophyte makes gametes by mitosis. Fertilization results in a diploid zygote, which develops into the next sporophyte generation. In this type of life cycle, therefore, the sporophyte and gametophyte generations take turns reproducing each other (FIGURE 13.4c).

Though the three types of sexual life cycles differ in the timing of meiosis and fertilization, they share a fundamental result: Each cycle of chromosome halving and doubling contributes to genetic variation among offspring. A closer look at meiosis will reveal the sources of this variation.

Meiosis reduces chromosome number from diploid to haploid: *a closer look*

13.2 Many of the steps of meiosis closely resemble corresponding steps in mitosis. Meiosis, like mitosis, is preceded by the replication of chromosomes. However, this single replication is followed by two consecutive cell divisions, called **meiosis I** and **meiosis II** (FIGURE 13.5). These divisions result in four daughter cells (rather than the two daughter cells of mitosis), each with only half as many chromosomes as the parent cell. FIGURE 13.6 on page 232 describes in some detail the two divisions of meiosis for an animal cell whose diploid number is 4. Study FIGURE 13.6 thoroughly before going on to the next section.

Mitosis and Meiosis Compared

Now that you have followed chromosomes through meiosis in FIGURE 13.6, let's summarize the key differences between meiosis and mitosis. The chromosome number is reduced by half in meiosis but not in mitosis. The genetic consequences of this difference are important. Whereas mitosis produces daughter cells genetically identical to their parent cell and to

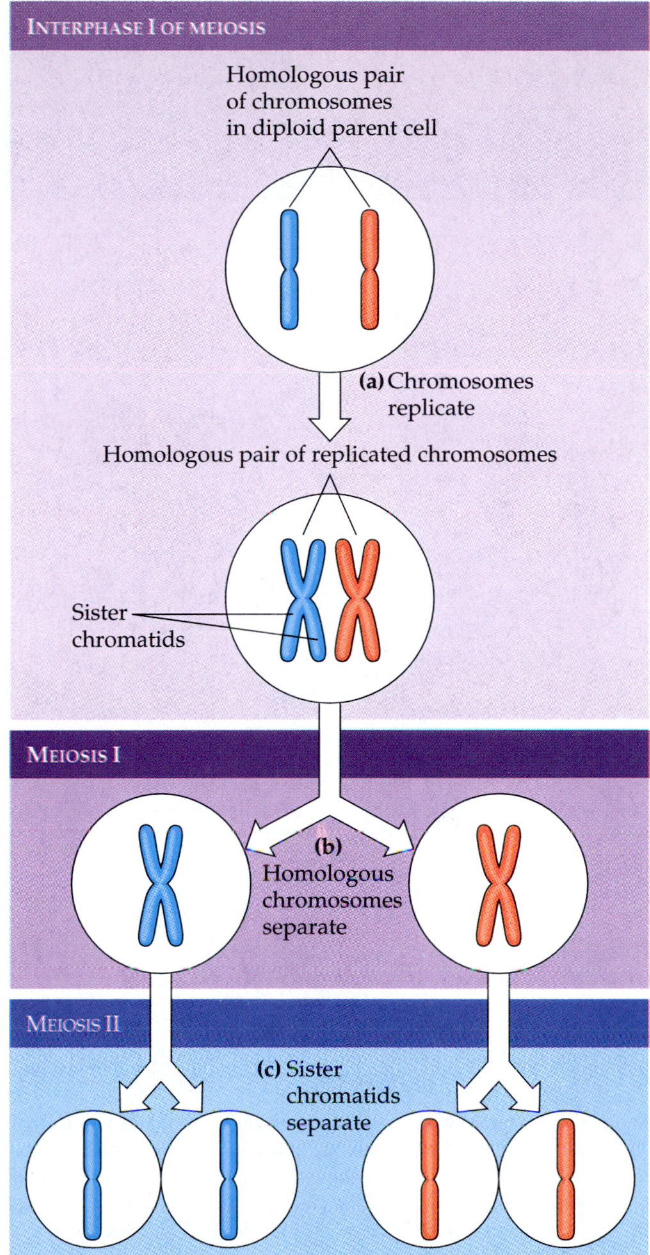

13.1 **FIGURE 13.5 • Overview of meiosis: how meiosis reduces chromosome number.** After the chromosomes replicate once, the diploid cell divides *twice*, yielding four haploid daughter cells. This simplified diagram tracks just one pair of homologous chromosomes. **(a)** First, each of the chromosomes replicates. **(b)** Then the first division (meiosis I) segregates the two chromosomes of the homologous pair, packaging them in separate (haploid) daughter cells. **(c)** The second division (meiosis II) separates the sister chromatids. Each daughter cell resulting from meiosis II is haploid, containing one single chromosome from the homologous pair.

Before proceeding to the more detailed description of meiosis in FIGURE 13.6, be sure you understand the difference between homologous chromosomes and sister chromatids. The two chromosomes of a homologous pair are individual chromosomes that were inherited from different parents. Homologues *appear* alike in the microscope, but they have different versions of genes at some of their corresponding loci (for example, a gene for freckles on one chromosome and a gene for the absence of freckles at the same locus of the homologue). Homologues in this and later figures are colored red and blue to remind you that they differ in this way. Before meiosis, *each* homologue replicates to form identical sister chromatids that remain together until meiosis II.

MEIOSIS I:
Separates homologous chromosomes

INTERPHASE I	PROPHASE I	METAPHASE I	ANAPHASE I

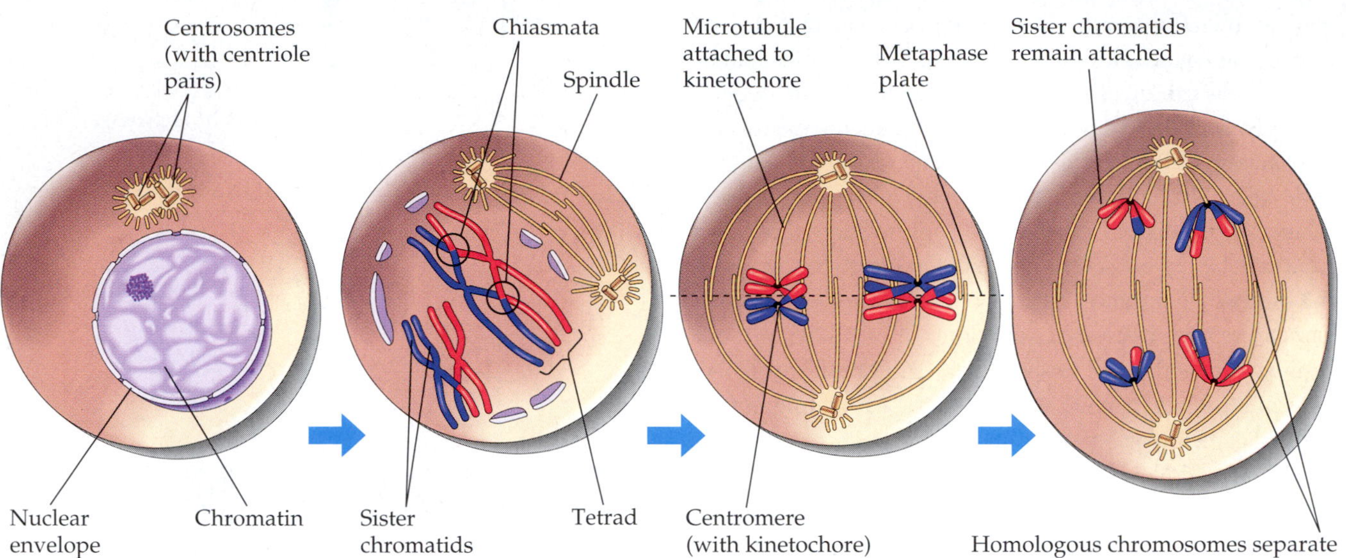

INTERPHASE I

Meiosis is preceded by an interphase, during which each of the chromosomes replicates. This process is similar to the chromosome replication preceding mitosis. For each chromosome, the result is two genetically identical sister chromatids, which remain attached at their centromeres. The centrosomes also replicate to form the two represented in this drawing.

PROPHASE I

Meiotic prophase I lasts longer and is more complex than prophase in mitosis. The chromosomes begin to condense. In a process called synapsis, homologous chromosomes, each made up of two sister chromatids, come together as pairs. Each chromosome pair is now visible in the microscope as a tetrad, a complex of four chromatids. At numerous places along their length, chromatids of homologous chromosomes are criss-crossed. These crossings, which help hold homologous chromosomes together, are called chiasmata (singular, chiasma). Notice that the chromosomes have traded segments. (You will learn the genetic significance of chiasmata later in the chapter.)

Meanwhile, other cellular components prepare for the division of the nucleus in a manner similar to that observed during mitosis. The centrosomes move away from each other, and spindle microtubules form between them. The nuclear envelope and nucleoli disperse. Finally, spindle microtubules capture the kinetochores that form on the chromosomes, and the chromosomes begin moving to the metaphase plate. Prophase I, which can last for days or even longer, typically occupies more than 90% of the time required for meiosis.

METAPHASE I

The chromosomes are now arranged on the metaphase plate, still in homologous pairs. Kinetochore microtubules from one pole of the cell are attached to one chromosome of each pair, while microtubules from the opposite pole are attached to the homologue.

ANAPHASE I

As in mitosis, the spindle apparatus moves the chromosomes toward the poles. However, sister chromatids remain attached at their centromeres and move as a single unit toward the same pole. The homologous chromosome moves toward the opposite pole. (This contrasts with the behavior of chromosomes during mitosis. In mitosis, chromosomes appear individually on the metaphase plate rather than in pairs, and the spindle separates sister chromatids of each chromosome.)

13.2 **FIGURE 13.6 ▪ The stages of meiotic cell division.** These diagrams show meiotic cell division for an animal cell with a diploid number of 4 ($2n = 4$). The behavior of the chromosomes is emphasized by using red and blue to differentiate the members of each homologous pair. For a discussion of spindle formation and other features common to mitosis and meiosis, see FIGURE 12.7.

MEIOSIS II: Separates sister chromatids

TELOPHASE I AND CYTOKINESIS	PROPHASE II	METAPHASE II	ANAPHASE II	TELOPHASE II AND CYTOKINESIS

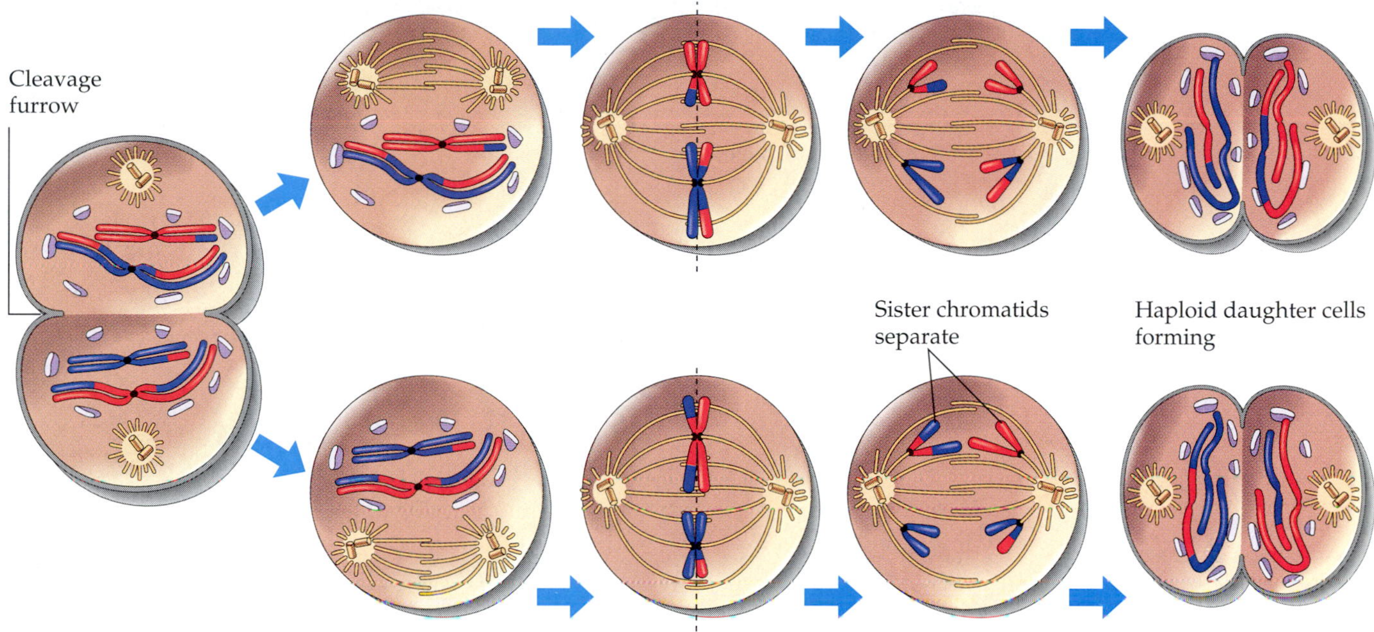

TELOPHASE I AND CYTOKINESIS

The spindle apparatus continues to separate the homologous chromosome pairs until the chromosomes reach the poles of the cell. Each pole now has a haploid chromosome set, but each chromosome still has two sister chromatids. Usually cytokinesis (division of the cytoplasm) occurs simultaneously with telophase I, forming two daughter cells. Cleavage furrows form in animal cells, and cell plates appear in plant cells. In some species, the chromosomes decondense, nuclear membranes and nucleoli re-form, and there is an interphase II before meiosis II. In other species, daughter cells of telophase I immediately begin preparation for the second meiotic division. In any case, there is no further replication of the genetic material prior to the second division of meiosis.

PROPHASE II

A spindle apparatus forms, and the chromosomes progress toward the metaphase II plate.

METAPHASE II

The chromosomes are positioned on the metaphase plate in mitosis-like fashion, with the kinetochores of sister chromatids of each chromosome pointing toward opposite poles.

ANAPHASE II

The centromeres of sister chromatids finally separate, and the sister chromatids of each pair, now individual chromosomes, move toward opposite poles of the cell.

TELOPHASE II AND CYTOKINESIS

Nuclei form at opposite poles of the cell, and cytokinesis occurs. At the completion of cytokinesis, there are four daughter cells, each with the haploid number of unreplicated chromosomes.

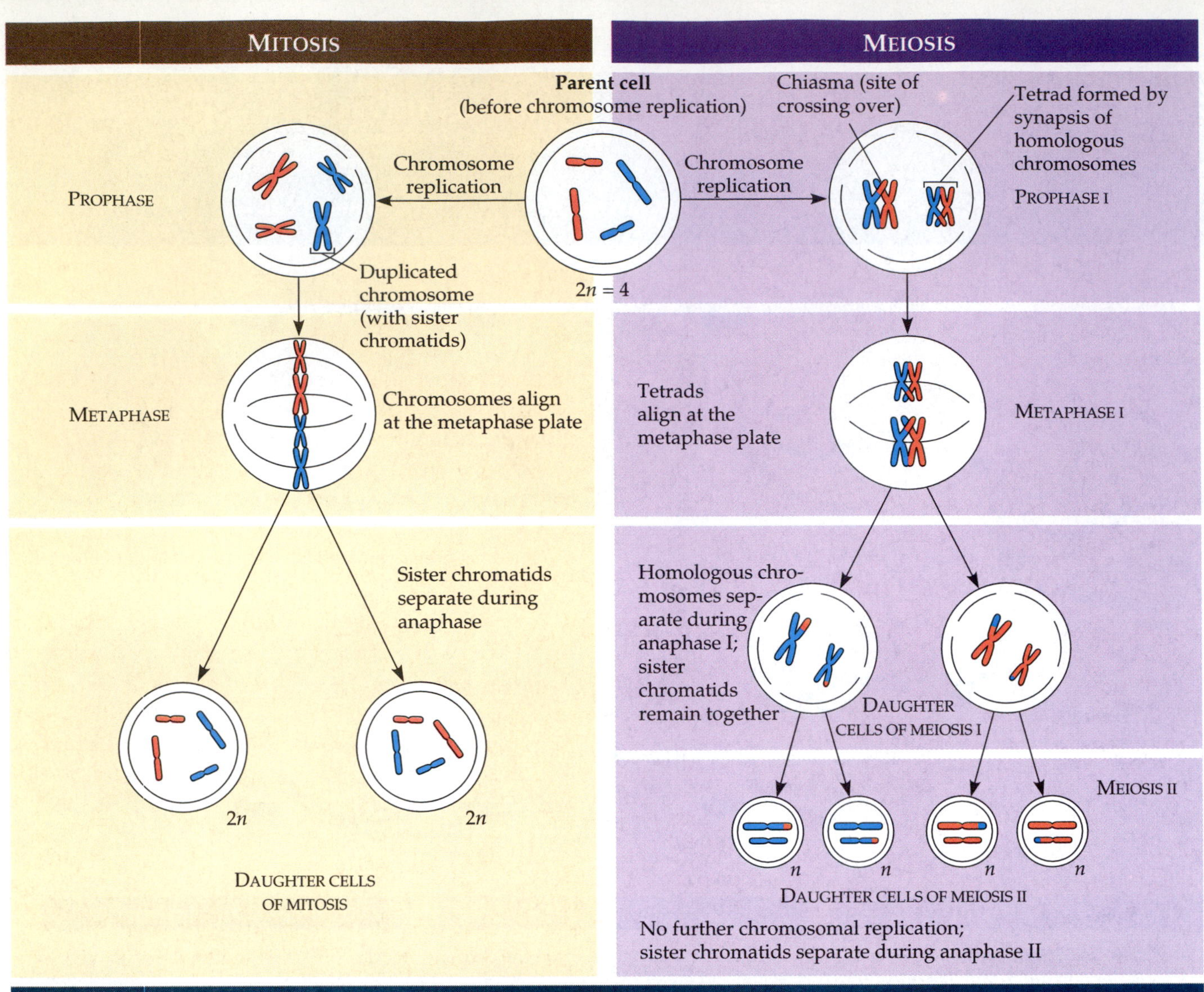

SUMMARY

Event	Mitosis	Meiosis
DNA replication	Occurs during interphase before nuclear division begins.	Occurs once, during the interphase before meiosis I begins.
Number of divisions	One, including prophase, metaphase, anaphase, and telophase.	Two, each including prophase, metaphase, anaphase, and telophase.
Synapsis of homologous chromosomes	Does not occur.	Synapsis is unique to meiosis: During prophase I, the homologous chromosomes join along their length, forming tetrads (groups of four chromatids); synapsis is associated with crossing over between nonsister chromatids.
Number of daughter cells and genetic composition	Two, each diploid ($2n$) and genetically identical to the parent cell.	Four, each haploid (n), containing half as many chromosomes as the parent cell; genetically nonidentical to the parent cell and to each other.
Role in the animal body	Enables multicellular adult to arise from zygote; produces cells for growth and tissue repair.	Produces gametes; reduces chromosome number by half and introduces genetic variability among the gametes.

FIGURE 13.7 ▪ A comparison of mitosis and meiosis.

each other, meiosis produces cells that differ genetically from their parent cell and from each other.

FIGURE 13.7 compares the key steps of mitosis and meiosis. Although meiosis involves two cell divisions, three events that are unique to meiosis all occur during the first division, meiosis I:

1. During prophase I of meiosis, the duplicated chromosomes pair with their homologues, a process called **synapsis**. The four closely associated chromatids of a homologous pair are visible in the light microscope as a **tetrad**. Also visible in the light microscope are X-shaped regions called **chiasmata** (singular, **chiasma**). They represent a crossing of *nonsister* chromatids, which are two chromatids belonging to separate but homologous chromosomes. Chiasmata are the physical manifestations of a genetic rearrangement called crossing over, discussed in the next section. Neither synapsis nor chiasma formation occurs during mitosis.
2. At metaphase I of meiosis, homologous pairs of chromosomes, rather than individual chromosomes, align on the metaphase plate.
3. At anaphase I of meiosis, sister chromatids do not separate, as they do in mitosis. Rather, the two sister chromatids of each chromosome remain attached and go to the same pole of the cell. *Meiosis I separates homologous pairs of chromosomes, not sister chromatids of individual chromosomes.*

The second meiotic division, meiosis II, separates sister chromatids and is virtually identical in mechanism to mitosis. However, since the chromosomes do not replicate between meiosis I and meiosis II, the final outcome of meiosis is a halving of the number of chromosomes per cell.

ORIGINS OF GENETIC VARIATION

How do we account for the genetic variation we observed in FIGURE 13.2? We are now ready to address this question.

Sexual life cycles produce genetic variation among offspring

In species that reproduce sexually, the behavior of chromo-
13.3 somes during meiosis and fertilization is responsible for most of the variation that arises each generation. Let's examine three mechanisms that contribute to the genetic variation arising from sexual reproduction: independent assortment of chromosomes, crossing over, and random fertilization.

Independent Assortment of Chromosomes

One way sexual reproduction generates genetic variation is shown in FIGURE 13.8, which features meiosis of a diploid cell with two homologous pairs of chromosomes. The red and blue colors distinguishing the maternal and paternal chromosomes of each homologous pair allow us to track individual chromosomes as meiosis proceeds and they are packaged in gametes. At metaphase I, the homologous pairs of chromosomes, each consisting of one maternal and one paternal chromosome, are situated on the metaphase plate. The orientations of the homologous pairs relative to the poles of the cell are random; there are two alternative possibilities for each pair. Thus there is a fifty-fifty chance that a particular daughter cell of meiosis I will get the maternal chromosome of a certain homologous pair, and a fifty-fifty chance that it will receive the paternal chromosome.

13.3

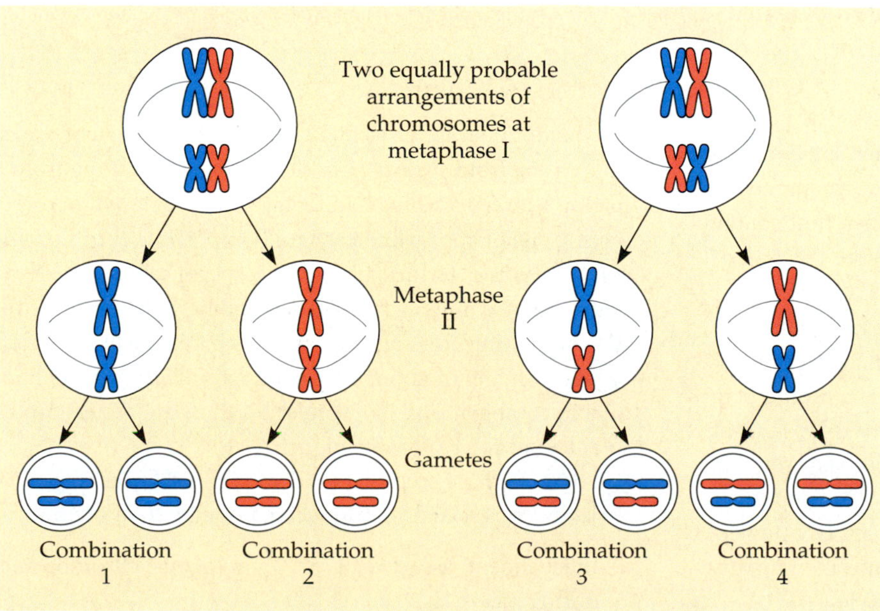

FIGURE 13.8 ▪ The results of alternative arrangements of two homologous chromosome pairs on the metaphase plate in meiosis I. In this figure we consider the consequences of meiosis in a hypothetical organism with a diploid chromosome number of 4 ($2n = 4$). The parental origins of the chromosomes are designated with different colors, blue for chromosomes inherited from one parent, red for chromosomes from the other parent. The positioning of each homologous pair of chromosomes at metaphase I is a matter of chance, like the flip of a coin. The arrangement of chromosomes at metaphase I determines which chromosomes will be packaged together in the haploid daughter cells.

Because each homologous pair of chromosomes is positioned independently of the other pairs at metaphase I—its orientation is as random as the flip of a coin—the first meiotic division results in independent assortment of maternal and paternal chromosomes into daughter cells. Each gamete represents one outcome of all possible combinations of maternal and paternal chromosomes. The number of combinations possible for gametes formed by meiosis starting with two homologous pairs of chromosomes ($2n = 4, n = 2$) is four, as shown in FIGURE 13.8. In the case of $n = 3$, eight combinations of chromosomes are possible for gametes. More generally, the number of combinations possible when chromosomes assort independently into gametes during meiosis is 2^n, where n is the haploid number of the organism.

In the case of humans, the haploid number (n) in the formula is 23. Thus the number of possible combinations of maternal and paternal chromosomes in the resulting gametes is 2^{23}, or about 8 million. The variety of gametes is analogous to the 8 million combinations of heads and tails possible for the simultaneous tossing of 23 coins. Thus each gamete that a human produces contains one of 8 million possible assortments of chromosomes inherited from that individual's mother and father.

Crossing Over

As a consequence of the independent assortment of chromosomes during meiosis, each of us produces a collection of gametes differing greatly in their combinations of the chromosomes we inherited from our two parents. But from what you have learned so far, it would seem that each *individual* chromosome in a gamete would be exclusively maternal or paternal in origin; that is, it would consist of DNA derived from our mother or father, but not from both. In fact, this is *not* the case. The process called **crossing over** produces individual chromosomes that combine genes inherited from our two parents.

Crossing over occurs during prophase of meiosis I. When homologous chromosomes first come together as pairs during prophase I, a protein apparatus called the synaptonemal complex joins the chromosomes tightly together, functioning something like a zipper. The pairing is precise, the homologues aligning with each other gene by gene. Crossing over takes place when homologous portions of two nonsister chromatids trade places. In the case of humans, an average of two or three such crossover events occur per chromosome pair. The locations of these genetic exchanges are visible in light micrographs as chiasmata, depicted in FIGURE 13.9. You will learn more about crossing over in Chapter 15. The important point for now is that crossing over, by combining DNA inherited from two parents into a single chromosome, is an important source of genetic variation in sexual life cycles.

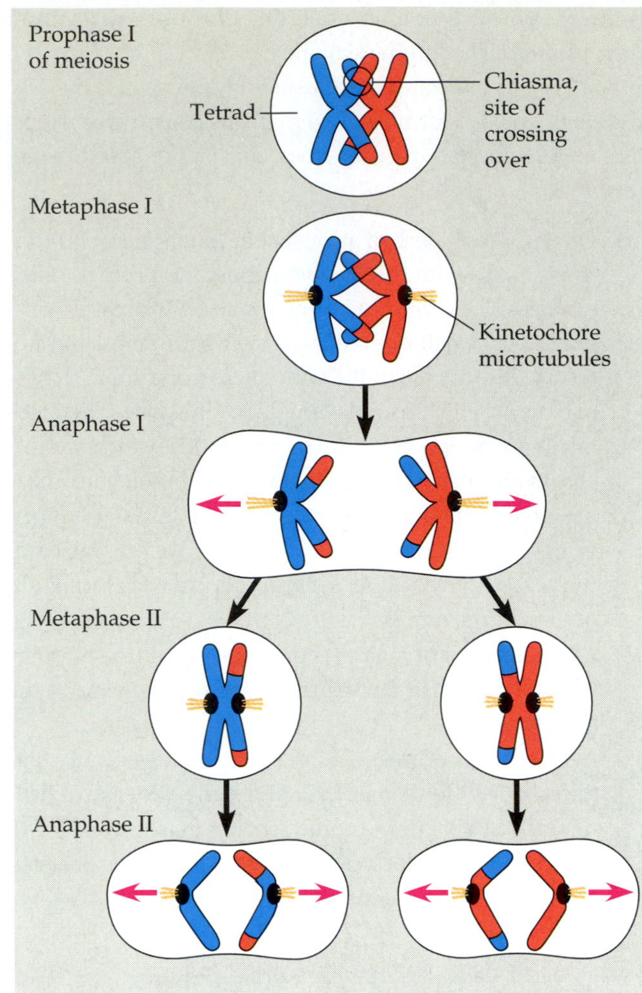

13.3 **FIGURE 13.9 ▪ The results of crossing over during meiosis.** During prophase of meiosis I, nonsister chromatids of homologous chromosomes exchange corresponding segments. Following these chromosomes through meiosis, we can see that crossing over gives rise to individual chromosomes that have some combination of DNA originally derived from two different parents.

Random Fertilization

The random nature of fertilization adds to the genetic variation arising from meiosis. Consider a zygote resulting from a mating between a woman and a man. A human ovum, representing one of approximately 8 million possible chromosome combinations, is fertilized by a single sperm cell, which represents one of 8 million *different* possibilities. Thus even without considering crossing over, any two parents will produce a zygote with any of about 64 trillion (8 million × 8 million) diploid combinations. No wonder brothers and sisters can be so different. You really *are* unique.

So far, we have seen that there are three sources of genetic variability in a sexually reproducing population of organisms:

- Independent assortment of homologous chromosomes during meiosis I.

- Crossing over between homologous chromosomes during prophase of meiosis I.
- Random fertilization of an ovum by a sperm.

All three mechanisms reshuffle the various genes carried by the individual members of a population. However, as you will learn in subsequent chapters, mutations are what ultimately create a population's diversity of genes.

Evolutionary adaptation depends on a population's genetic variation

Having considered how sexual reproduction contributes to genetic variation in a population, we can relate these concepts to evolution, biology's core theme. Darwin recognized the importance of genetic variation in the evolutionary mechanism he called natural selection. Recall from Chapter 1 that a population evolves through the differential reproductive success of its variant members. On average, those individuals best suited to the local environment leave the most offspring, transmitting their genes in the process. This natural selection results in adaptation, the accumulation of the genetic variations favored by the environment. As the environment changes or a population moves, the population may survive if, in each generation, at least some of its members can cope effectively with the new conditions. Different genetic variations may work better than those that prevailed in the old time or place. Sex and mutations are the two sources of this variation, and we have considered the sexual contribution in this chapter.

Although Darwin realized that heritable variation is what makes evolution possible, he could not explain why offspring resemble—but are not identical to—their parents. Ironically, Gregor Mendel, a contemporary of Darwin, published a theory of inheritance that helps explain genetic variation, but his discoveries had no impact on biologists until 1900, more than 15 years after Darwin (1809–1882) and Mendel (1822–1884) died. In the next chapter you will learn how Mendel discovered the basic rules governing the inheritance of specific traits.

CHAPTER REVIEW

REVIEW OF KEY CONCEPTS

(with page numbers and key figures)

AN INTRODUCTION TO HEREDITY

- **Offspring acquire genes from parents by inheriting chromosomes (pp. 226–227)** Genetics is the study of heredity and gene-based variation. Each gene in an organism's DNA has a specific locus on a certain chromosome.

13.1 **Like begets like, more or less: a comparison of asexual and sexual reproduction (p. 227, FIGURE 13.2)** In asexual reproduction, one parent produces genetically identical offspring by mitosis. Sexual reproduction combines genes from two different parents to form genetically diverse offspring.

THE ROLE OF MEIOSIS IN SEXUAL LIFE CYCLES

- **Fertilization and meiosis alternate in sexual life cycles (pp. 227–231, FIGURE 13.4)** Normal human somatic cells have 46 chromosomes, half from each parent. Each of the 22 maternal autosomes has a homologous paternal chromosome. The 23rd pair, the sex chromosomes, determines whether the person is female *(XX)* or male *(XY)*. Single, haploid *(n)* sets of chromosomes in ovum and sperm unite during fertilization to form a diploid (2*n*) single-celled zygote, which develops into a multicellular organism by mitosis. At sexual maturity, ovaries and testes (the gonads) produce haploid gametes by meiosis. Sexual life cycles differ in the timing of meiosis in relation to fertilization. Multicellular organisms may be diploid (as in animals), or haploid (as in most fungi), or may alternate between haploid and diploid generations (as in plants).

13.2 **Meiosis reduces chromosome number from diploid to haploid: *a closer look* (pp. 231–235, FIGURES 13.5–13.7)** The two cell divisions of meiosis, meiosis I and meiosis II, produce four haploid daughter cells. Meiosis is distinguished from mitosis by the events of meiosis I. In prophase I, replicated homologous chromosomes, each chromosome with two chromatids, undergo synapsis. Nonsister chromatids cross over, exchanging segments (the crossover sites appear as chiasmata). The paired chromosomes (tetrads) align on the metaphase plate, and at anaphase I the two chromosomes of each homologous pair (not the sister chromatids) move to separate poles. The cell divides, with half the chromosomes going to each daughter cell. Meiosis II separates the sister chromatids, yielding four haploid daughter cells.

ORIGINS OF GENETIC VARIATION

13.3 **Sexual life cycles produce genetic variation among offspring (pp. 235–237, FIGURES 13.8, 13.9)** The events of sexual reproduction that contribute to genetic variation in a population are independent assortment of chromosomes during meiosis I, crossing over between homologous chromosomes during meiosis I, and random fertilization of ova by sperm.

- **Evolutionary adaptation depends on a population's genetic variation (p. 237)** Genetic variation among a population's members is the raw material for evolution by natural selection. Sexual reproduction and mutations generate this variation.

SELF-QUIZ

1. A human cell containing 22 autosomes and a *Y* chromosome is
   a. a somatic cell of a male
   b. a zygote
   c. a somatic cell of a female
   d. a sperm cell
   e. an ovum

2. Homologous chromosomes move to opposite poles of a dividing cell during
   a. mitosis
   b. meiosis I
   c. meiosis II
   d. fertilization
   e. binary fission

3. Meiosis II is similar to mitosis in that
   a. homologous chromosomes synapse
   b. DNA replicates before the division
   c. the daughter cells are diploid
   d. sister chromatids separate during anaphase
   e. the chromosome number is reduced

4. The DNA content of a diploid cell in the G_1 phase of the cell cycle is measured (see Chapter 12). If this DNA content is x, then the DNA content of the same cell at metaphase of meiosis I would be
   a. $0.25x$
   b. $0.5x$
   c. x
   d. $2x$
   e. $4x$

5. If we continued to follow the cell lineage from question 4, then the DNA content at metaphase of meiosis II would be
   a. $0.25x$
   b. $0.5x$
   c. x
   d. $2x$
   e. $4x$

6. How many different combinations of maternal and paternal chromosomes can be packaged in gametes made by an organism with a diploid number of 8 ($2n = 8$)?
   a. 2
   b. 4
   c. 8
   d. 16
   e. 32

7. The immediate product of meiosis in a plant is a
   a. spore
   b. gamete
   c. zygote
   d. sporophyte
   e. gametophyte

8. Multicellular haploid organisms
   a. are typically called sporophytes
   b. produce new cells for growth by meiosis
   c. produce gametes by mitosis
   d. are found only in aquatic environments
   e. are the direct result of syngamy

9. Crossing over usually contributes to genetic variation by exchanging chromosomal segments between
   a. sister chromatids of a chromosome
   b. chromatids of nonhomologues
   c. nonsister chromatids of homologues
   d. nonhomologous loci of the genome
   e. autosomes and sex chromosomes

10. In comparing the typical life cycles of plants and animals, a stage found in plants but not in animals is a
    a. gamete
    b. zygote
    c. multicellular diploid
    d. multicellular haploid

CHALLENGE QUESTIONS

1. An error in meiosis (and occasionally in mitosis shortly after fertilization) can result in a condition called triploidy, in which each somatic cell in the offspring has *three* sets of chromosomes. Starting with a diploid cell with $2n = 4$, draw a diagram showing how an error in meiosis could produce a gamete that could lead to triploidy ($3n = 6$).

2. Many species can reproduce either asexually or sexually. It is often when the environment changes in some way that is unfavorable to an existing population that the organisms begin to reproduce sexually. Speculate about the evolutionary significance of this switch from asexual to sexual reproduction.

SCIENCE, TECHNOLOGY, AND SOCIETY

1. It is possible to grow seedlings of pine trees from short pieces of their needles. A few of the straightest, fastest-growing trees are selected for this treatment. By this method, thousands of genetically identical trees can be grown to create a forest that is a superior producer of lumber. What are the short-term and long-term advantages and disadvantages of this approach?

2. Studies of human characteristics indicate that most variation in the human population is due to the differences *within* racial groups. Only a small amount of variation is due to differences *between* what have traditionally been considered races. Based on these observations, write a paragraph or two evaluating whether the concept of "race" is valid.

FURTHER READING

Anderson, A. "The Evolution of Sexes." *Science,* July 17, 1992. Why do humans and many other organisms exist in only two sexes?

Becker, W. M., J. B. Reece, and M. F. Poenie. *The World of the Cell,* 3rd ed. Menlo Park, CA: Benjamin/Cummings, 1996. Chapter 16.

Cunningham, P. "The Genetics of Thoroughbred Horses." *Scientific American,* May 1991. Relates the history of how humans have exploited heritable variation in one species.

de Lange, T. "Ending Up with the Right Partner." *Nature,* April 23, 1998. Addresses the question of how homologous chromosomes find each other during meiosis.

Gould, S. J. "Ghosts of Bell Curves Past." *Natural History,* February 1995. A famous evolutionary biologist argues against a best-selling book about the genetic contribution to intelligence.

Griffiths, A. J. F., J. H. Miller, D. T. Suzuki, R. C. Lewontin, and W. M. Gelbart. *An Introduction to Genetic Analysis,* 6th ed. New York: W. H. Freeman, 1996. A good undergraduate genetics text.

Haber, J. E. "Searching for a Partner." *Science,* February 6, 1998. Contains a summary of recent discoveries about crossing over and synapsis in different organisms.

Ridley, M. "Is Sex Good For Anything?" *New Scientist,* December 4, 1993. Did disease play an important role in the evolution of sex?

WEB LINKS

Visit the special edition of *The Biology Place* for BIOLOGY, Fifth Edition, at http://www.biology.com/campbell. Go to Chapter 13 for online resources, including learning activities, practice exams, and links to the following web sites:

"Meiosis"
A meiosis tutorial site with pictures and animations. The section on meiotic errors is particularly good.

"The Kinsey Institute for Research in Sex, Gender, and Reproduction"
This interesting site is named after Dr. Alfred Kinsey, a pioneer in the scientific study of human sexuality, reproduction, and behavior.

"The Hawley Lab"
Visit the home pages of this research group at the University of California, Davis, which studies meiosis in the fruit fly *Drosophila melanogaster.*

CHAPTER 14

MENDEL AND THE GENE IDEA

Gregor Mendel's Discoveries

- Mendel brought an experimental and quantitative approach to genetics: *science as a process*
- By the law of segregation, the two alleles for a character are packaged into separate gametes
- By the law of independent assortment, each pair of alleles segregates into gametes independently
- Mendelian inheritance reflects rules of probability
- Mendel discovered the particulate behavior of genes: *a review*

Extending Mendelian Genetics

- The relationship between genotype and phenotype is rarely simple

Mendelian Inheritance in Humans

- Pedigree analysis reveals Mendelian patterns in human inheritance
- Many human disorders follow Mendelian patterns of inheritance
- Technology is providing new tools for genetic testing and counseling

Eyes of brown, blue, green, or gray; hair of black, brown, blond, or red—these are just a few examples of heritable variations that we may observe among individuals in a population. What genetic principles account for the transmission of such traits from parents to offspring?

One possible explanation of heredity is a "blending" hypothesis, the idea that genetic material contributed by the two parents mixes in a manner analogous to the way blue and yellow paints blend to make green. This hypothesis predicts that over many generations, a freely mating population will give rise to a uniform population of individuals. However, our everyday observations, and the results of breeding experiments with animals and plants, contradict such a prediction. The blending hypothesis also fails to explain other phenomena of inheritance, such as traits skipping a generation.

An alternative to the blending model is a "particulate" hypothesis of inheritance: the gene idea. According to this model, parents pass on discrete heritable units—genes—that retain their separate identities in offspring. An organism's collection of genes is more like a bucket of marbles than a pail of paint. Like marbles, genes can be sorted and passed along, generation after generation, in undiluted form.

Modern genetics had its genesis in an abbey garden, where a monk named Gregor Mendel documented a particulate mechanism of inheritance. In the painting on this page, Mendel works with his experimental organism, garden peas. In this chapter you will learn how Mendel developed his theory and how the Mendelian model applies to the inheritance of human variations.

GREGOR MENDEL'S DISCOVERIES

Mendel discovered the basic principles of heredity by breeding garden peas in carefully planned experiments. As we retrace his work in this and the following sections, we will be able to recognize the key elements of the scientific process that were introduced in Chapter 1.

Mendel brought an experimental and quantitative approach to genetics: *science as a process*

Mendel grew up on his parents' small farm in a region of Austria that is now part of the Czech Republic. At school in this agricultural area, Mendel and the other children received agricultural training along with basic education. Later, Mendel overcame financial hardship and illness to excel in high school and at the Olmutz Philosophical Institute.

In 1843 Mendel entered an Augustinian monastery. After three years of theological studies, he was assigned to a school as a temporary teacher but failed the teacher's examination.

An administrator sent Mendel to the University of Vienna, where he studied from 1851 to 1853. These were very important years for Mendel's development as a scientist. Two professors were especially influential. One was the physicist Doppler, who encouraged his students to learn science through experimentation and trained Mendel to use mathematics to help explain natural phenomena. The second was a botanist named Unger, who aroused Mendel's interest in the causes of variation in plants. These influences came together in Mendel's subsequent experiments with garden peas.

After attending the university, Mendel was assigned to teach at the Brünn Modern School, where several teachers shared his enthusiasm for scientific research. At the monastery where Mendel lived, he also found stimulating colleagues, many of them university professors and active researchers. Moreover, there had been a long tradition of interest in the breeding of plants, including peas, at the monastery. Thus, it was probably not extraordinary when, around 1857, Mendel began breeding garden peas in the abbey garden in order to study inheritance. What *was* extraordinary was Mendel's fresh approach to very old questions about heredity.

Mendel probably chose to work with peas because they are available in many varieties. For example, one variety has purple flowers, while a contrasting variety has white flowers. Geneticists use the term **character** for a heritable feature, such as flower color, that varies among individuals. Each variant for a character, such as purple or white color for flowers, is called a **trait**.

The use of peas also gave Mendel strict control over which plants mated with which. The sex organs of a pea plant are in its flowers, and each pea flower has both male and female organs—stamens and carpel, respectively. Normally, the plants self-fertilize: Pollen grains released from the stamens land on the carpel of the same flower, and sperm from the pollen fertilize ova in the carpel. To achieve cross-pollination (fertilization between different plants), Mendel removed the immature stamens of a plant before they produced pollen and then dusted pollen from another plant onto the emasculated flowers (FIGURE 14.1). Each resulting zygote then developed into a plant embryo encased in a seed (pea). Whether ensuring self-pollination or executing artificial cross-pollination, Mendel could always be sure of the parentage of new seeds.

Mendel chose to track only characters that varied in an "either-or" rather than a "more-or-less" manner. For example, his plants had either purple or white flowers; there was nothing intermediate between these two varieties. Had Mendel focused instead on characters that vary in a continuum among individuals—seed weight, for example—he would not have discovered the particulate nature of inheritance.

Mendel also made sure he started his experiments with varieties that were **true-breeding**, which means that when the plants self-pollinate, all their offspring are of the same variety. For example, a plant with purple flowers is true-breeding if its seeds produced by self-pollination all give rise to plants that also have purple flowers.

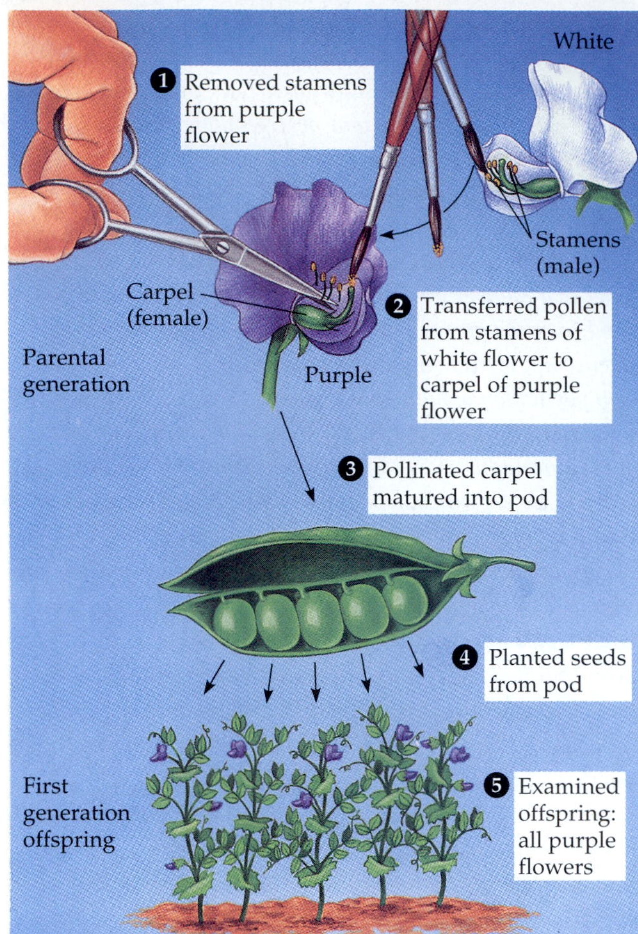

FIGURE 14.1 • A genetic cross. To mate (hybridize) two varieties of pea plants, Mendel used an artist's brush to transfer sperm-bearing pollen from one plant to the egg-bearing carpel of another. In this case, the character of interest is flower color, and the two varieties are purple-flowered and white-flowered. Seeds develop within the carpel, which develops into the fruit (pod). Seed germination produces the first generation hybrids, which all have purple flowers. The result is the same for the reciprocal cross, the transfer of pollen from purple flowers to white flowers.

In a typical breeding experiment, Mendel would cross-pollinate two contrasting, true-breeding pea varieties—purple-flowered plants and white-flowered plants, for example (see FIGURE 14.1). This mating, or crossing, of two varieties is called **hybridization**. Our example is specifically a **monohybrid cross**, the term for a cross that tracks the inheritance of a single character—flower color, in this case. The true-breeding parents are referred to as the **P generation** (for parental), and their hybrid offspring are the **F_1 generation** (for first filial, referring to offspring). Allowing these F_1 hybrids to self-pollinate produces an **F_2 generation** (second filial). Mendel usually followed traits for at least these three generations: the P, F_1, and F_2 generations. Had Mendel stopped his experiments with the F_1 generation, the basic patterns of inheritance would have eluded him. It was mainly Mendel's quantitative analysis of F_2 plants that

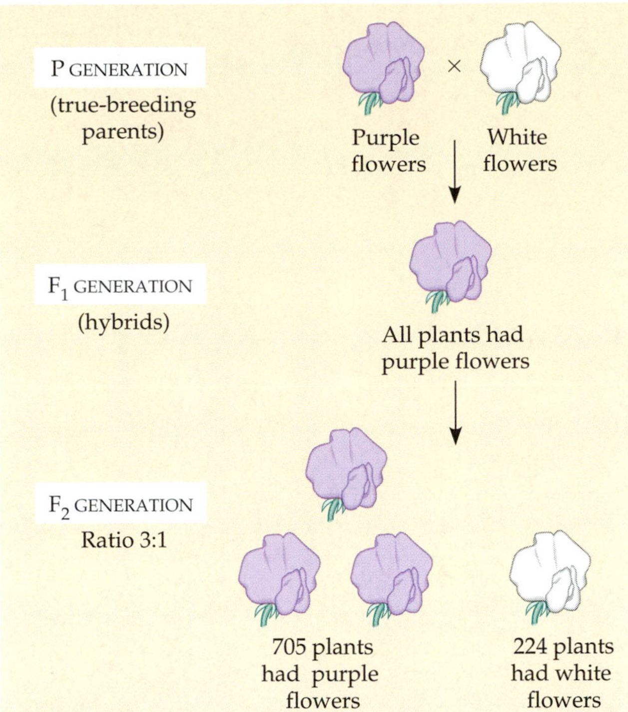

FIGURE 14.2 ▪ Mendel tracked heritable characters for three generations. When F_1 hybrids were allowed to self-pollinate, or when they were cross-pollinated with other F_1 hybrids, a 3:1 ratio of the two varieties occurred in the F_2 generation. An "×" sign symbolizes a genetic cross, or mating.

revealed the two fundamental principles of heredity that are now known as the law of segregation and the law of independent assortment.

By the law of segregation, the two alleles for a character are packaged into separate gametes

14.1 If the blending model of inheritance were correct, the F_1 hybrids from a cross between purple-flowered and white-flowered pea plants would have pale purple flowers, intermediate between the two varieties of the P generation. Notice in FIGURE 14.1 that the experiment produced a very different result: The F_1 offspring all had flowers just as purple as the purple-flowered parents. What happened to the white-flowered plants' genetic contribution to the hybrids? If it were lost, then the F_1 plants could produce only purple-flowered offspring in the F_2 generation. But when Mendel allowed the F_1 plants to self-pollinate and planted their seeds, the white-flower trait reappeared in the F_2 generation. Mendel used very large sample sizes and kept accurate records of his results: 705 of the F_2 plants had purple flowers, and 224 had white flowers. These data fit a ratio of about 3 purple to 1 white (FIGURE 14.2). Mendel reasoned that the heritable factor for white flowers did not disappear in the F_1 plants, but only the purple-flower factor was affecting flower color in these hybrids. In Mendel's terminology, purple flower is a dominant trait and

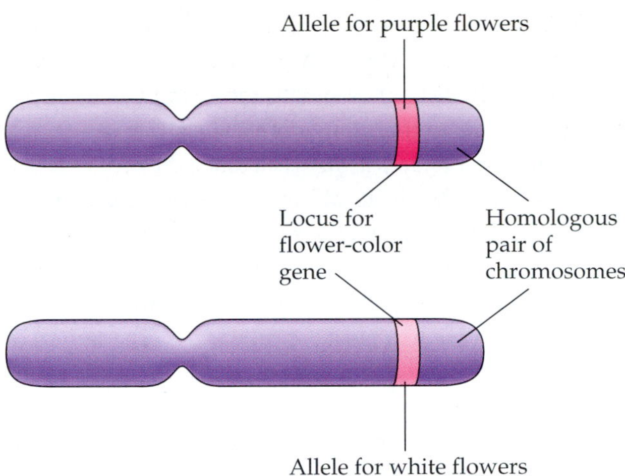

FIGURE 14.3 ▪ Alleles, contrasting versions of a gene. The gene for a particular inherited character, such as flower color in garden peas, resides at a specific locus (position) on a certain chromosome. Alleles are variants of that gene. In this case, the flower-color gene exists in two versions: the allele for purple flowers and the allele for white flowers. The homologous pair of chromosomes illustrated here represents an F_1 hybrid, which inherited the allele for purple flowers from one parent and the allele for white flowers from the other parent.

white flower is a recessive trait. The occurrence of white-flowered plants in the F_2 generation was evidence that the heritable factor causing that recessive trait had not been diluted in any way by coexisting with the purple-flower factor in the F_1 hybrids.

Mendel observed the same pattern of inheritance in six other characters, each represented by two contrasting varieties (TABLE 14.1, p. 243). For example, the parental pea seeds either had a smooth, round shape, or they were wrinkled. In a monohybrid cross for this character, all the F_1 hybrids produced round seeds; this is the dominant trait. In the F_2 generation, 75% of the seeds were round and 25% were wrinkled—the typical 3:1 ratio. How did Mendel explain this pattern, which he consistently observed in his monohybrid crosses? He developed a hypothesis that we can break down into four related ideas. (We will replace some of Mendel's original terms with modern words; for example, "gene" will be used in place of Mendel's "heritable factor.")

1. *Alternative versions of genes (different alleles) account for variations in inherited characters.* The gene for flower color, for example, exists in two versions, one for purple flowers and the other for white. These alternative versions of a gene are now called **alleles** (FIGURE 14.3). Today, we can relate this concept to chromosomes and DNA. As we mentioned in Chapter 13, each gene resides at a specific locus on a specific chromosome. The DNA at that locus, however, can vary somewhat in its sequence of nucleotides, and hence in its information content. The purple-flower allele and the white-flower allele are two DNA variations possible at the flower-color locus on one of a pea plant's chromosomes.

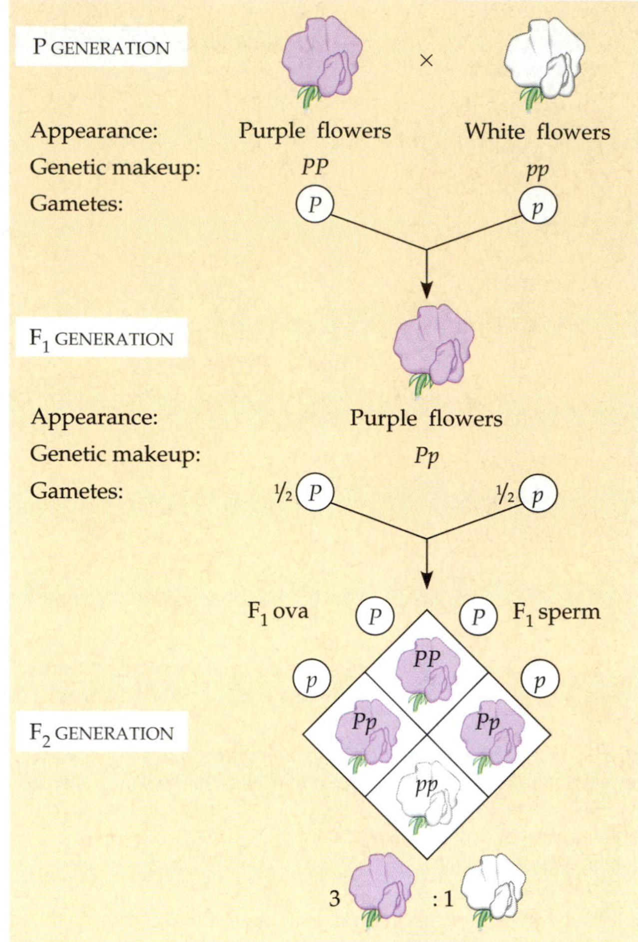

FIGURE 14.4 ▪ **Mendel's law of segregation.** A more detailed version of FIGURE 14.2, this diagram illustrates Mendel's model for monohybrid inheritance. The purple-flower allele *(P)* is dominant, and the white-flower allele *(p)* is recessive. Each plant has two alleles for the gene controlling flower color, one allele inherited from each parent. Each true-breeding plant of the parental generation has matching alleles, either *PP* (purple-flower parentals) or *pp* (white-flower parentals). Gametes, symbolized with circles, each contain only one allele for the flower color gene. Union of the parental gametes produces F_1 hybrids having nonmatching alleles, a *Pp* combination. Because the purple-flower allele is dominant, all these hybrids have purple flowers. When the hybrid plants produce gametes, the two alleles segregate, half the gametes receiving the *P* allele and the other half receiving the *p* allele. Random combination of these gametes results in the 3:1 ratio that Mendel observed in the F_2 generation. The box at the bottom of the figure is a Punnett square, a useful tool for showing all possible combinations of alleles in offspring. Each square represents an equally probable product of fertilization. For example, the box at the right corner of the Punnett square shows the genetic combination resulting from a (*p*) sperm fertilizing a (*P*) ovum.

14.1

2. *For each character, an organism inherits two alleles, one from each parent.* Mendel made this deduction without knowing about the role of chromosomes, but what you learned about chromosomes in Chapter 13 will help you understand Mendel's idea. Recall that a diploid organism has homologous pairs of chromosomes, one chromosome of each pair inherited from each parent. Thus, a genetic locus is actually represented twice in a diploid cell. These homologous loci may have identical alleles, as in the true-breeding plants of Mendel's P generation. Or, the two alleles may differ, as in the F_1 hybrids. In the flower-color example, the hybrids inherited a purple-flower allele from one parent and a white-flower allele from the other parent (see FIGURE 14.3). This brings us to the third part of Mendel's hypothesis.
3. *If the two alleles differ, then one, the* ***dominant allele****, is fully expressed in the organism's appearance; the other, the* ***recessive allele****, has no noticeable effect on the organism's appearance.* According to this part of the hypothesis, Mendel's F_1 plants had purple flowers because the allele for that variation is dominant and the allele for white flowers is recessive.
4. *The two alleles for each character segregate during gamete production.* Thus an ovum and a sperm each get only one of the two alleles that are present in the somatic cells of the organism. In terms of chromosomes, this segregation corresponds to meiotic reduction of chromosome count from the diploid to the haploid number. Note that if an organism has identical alleles for a particular character—that is, the organism is true-breeding for that character—then that allele exists in a single copy in all gametes. But if contrasting alleles are present, as in the F_1 hybrids, then 50% of the gametes receive the dominant allele, while 50% receive the recessive allele. It is this last part of the hypothesis, the separation of alleles into separate gametes, for which Mendel's **law of segregation** is named.

One test of Mendel's segregation hypothesis is whether or not it can account for the 3:1 ratio he observed in the F_2 generation of his numerous monohybrid crosses. The hypothesis predicts that the F_1 hybrids will produce two classes of gametes. When alleles segregate, half the gametes receive a purple-flower allele, while the other half get a white-flower allele. During self-pollination, the gametes of these two classes unite randomly. An ovum with a purple-flower allele has an equal chance of being fertilized by a sperm with a purple-flower allele or one with a white-flower allele. Since the same is true for an ovum with a white-flower allele, there are a total of four equally likely combinations of sperm and ovum. FIGURE 14.4 illustrates these combinations using a type of diagram called a Punnett square, a handy device for predicting the results of a genetic cross. Notice that a capital letter symbolizes a dominant allele and a lowercase letter represents a recessive allele. In our example, *P* is the purple-flower allele, and *p* is the white-flower allele.

What will be the physical appearance of these F_2 offspring? One-fourth of the plants have two alleles specifying purple flowers; clearly, these plants will have purple flowers. But one-half of the F_2 offspring have inherited one allele for purple flowers and one allele for white flowers; like the F_1 plants, these plants will also have purple flowers, the dominant trait.

Table 14.1 ■ The Results of Mendel's F_1 Crosses for Seven Characters in Pea Plants

CHARACTER	DOMINANT TRAIT	×	RECESSIVE TRAIT	F_2 GENERATION DOMINANT:RECESSIVE	RATIO
Flower color	Purple	×	White	705:224	3.15:1
Flower position	Axial	×	Terminal	651:207	3.14:1
Seed color	Yellow	×	Green	6022:2001	3.01:1
Seed shape	Round	×	Wrinkled	5474:1850	2.96:1
Pod shape	Inflated	×	Constricted	882:299	2.95:1
Pod color	Green	×	Yellow	428:152	2.82:1
Stem length	Tall	×	Dwarf	787:277	2.84:1

Finally, one-fourth of the F_2 plants have inherited two alleles specifying white flowers and will, in fact, express the recessive trait. Thus Mendel's model accounts for the 3:1 ratio that he observed in the F_2 generation.

Some Useful Genetic Vocabulary

An organism having a pair of identical alleles for a character is said to be **homozygous** for the gene controlling that character. A pea plant that is true-breeding for purple flowers *(PP)* is an example. Pea plants with white flowers are also homozygous, but for the recessive allele *(pp)*. If we cross dominant homozygotes with recessive homozygotes, as in the parental cross (P generation) of FIGURE 14.4, every offspring will have two different alleles—*Pp* in the case of the F_1 hybrids of our flower-color experiment. Organisms having two different alleles for a gene are said to be **heterozygous** for that gene. Unlike homozygotes, heterozygotes are not true-breeding, because they produce gametes having one *or* the other of the different alleles. We have seen that a *Pp* plant of the F_1 generation will produce both purple-flowered and white-flowered offspring when it self-pollinates.

Because of dominance and recessiveness, an organism's appearance does not always reveal its genetic composition. Therefore, we have to distinguish between an organism's appearance, called its **phenotype**, and its genetic makeup, its **genotype**. In the case of flower color in peas, *PP* and *Pp* plants have the same phenotype (purple) but different genotypes.

FIGURE 14.5 reviews these terms. Phenotype refers to physiological traits as well as traits relating directly to appearance. For example, there is a pea variety that lacks the normal trait of being able to self-pollinate.

The Testcross

Suppose we have a pea plant that has purple flowers. We cannot tell from its flower color if this plant is homozygous or heterozygous because the genotypes *PP* and *Pp* result in the same phenotype. But if we cross this pea plant with one having white flowers, the appearance of the offspring will reveal the genotype of the purple-flowered parent. The genotype of the plant with white flowers is known: Because this is the recessive trait, the plant must be homozygous. If all the offspring of the cross have purple flowers, then the other parent must be homozygous for the dominant allele; a *PP* × *pp* cross produces nothing but *Pp* offspring. But if both the purple and the white phenotypes appear among the offspring, the purple-flowered parent must be heterozygous. The offspring of a *Pp* × *pp* cross will have a 1:1 phenotypic ratio (FIGURE 14.6). This breeding of a recessive homozygote with an organism of dominant phenotype, but unknown genotype, is called a testcross. It was devised by Mendel and continues to be an important tool of geneticists.

By the law of independent assortment, each pair of alleles segregates into gametes independently

14.1 Gregor Mendel derived the law of segregation by performing monohybrid crosses, breeding experiments using parental varieties that differ in a single character, such as flower color. What would happen in a mating of parental varieties differing in *two* characters—a **dihybrid cross**? For instance, two of the seven characters Mendel studied were seed color and seed shape. Seeds may be either yellow or green. They also may be either round (smooth) or wrinkled. From monohybrid crosses, Mendel knew that the allele for yellow seeds is dominant *(Y)* and that the allele for green seeds is recessive *(y)*. For the seed-shape character, the allele for round is dominant *(R)*, and the allele for wrinkled is recessive *(r)*. Imagine breeding two pea varieties differing in *both* of these characters—a parental cross between a plant with yellow-round seeds *(YYRR)* and a plant with green-wrinkled seeds *(yyrr)*. Are these two characters, seed color and seed shape, transmitted from parents to offspring as a package? Put another way, will the *Y* and *R* alleles always stay together, generation after generation? Or are seed color and seed shape inherited independently of each other? FIGURE 14.7 illustrates how a dihybrid cross can determine which of these two hypotheses is correct.

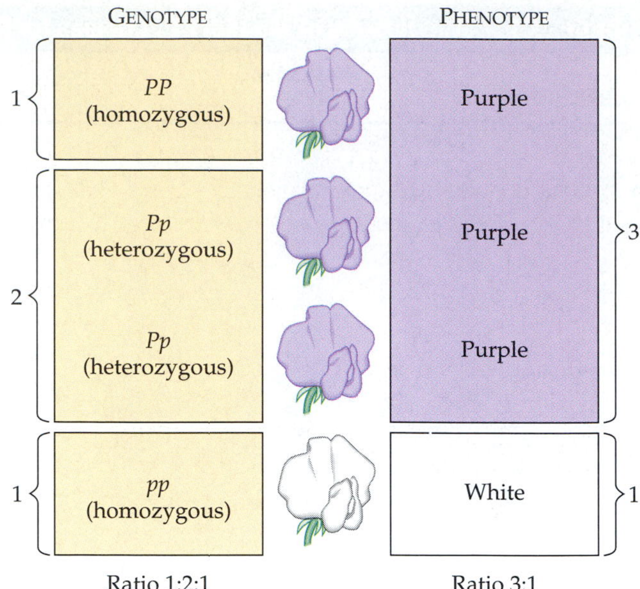

FIGURE 14.5 • Genotype versus phenotype. Grouping F_2 offspring from a monohybrid cross for flower color according to phenotype results in the typical 3:1 ratio. In terms of genotype, however, there are actually two categories of purple-flowered plants: *PP* (homozygous) and *Pp* (heterozygous).

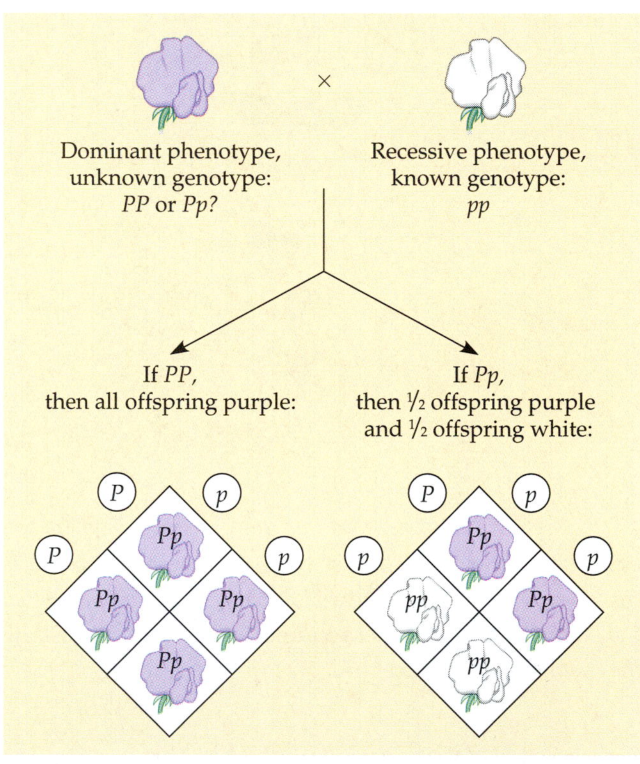

FIGURE 14.6 • A testcross. A testcross is designed to reveal the genotype of an organism that exhibits a dominant trait, such as purple flowers in pea plants. Such an organism could be either homozygous for the dominant allele or heterozygous. The most efficient way to determine the genotype is to cross the organism with an individual expressing the recessive trait. Since the genotype of the white-flowered parent must be homozygous, we can deduce the genotype of the purple-flowered parent by observing the phenotypes of the offspring.

In the F_1 generation of this dihybrid cross, the genotype is *YyRr*, and the plants exhibit both dominant phenotypes, yellow seeds with round shapes. The key step in the experiment is to see what happens when F_1 plants self-pollinate to produce F_2 offspring. If the hybrids must transmit their alleles in the same combinations in which they were inherited from the P generation, then there will only be two classes of gametes: *YR* and *yr*. This hypothesis predicts that the phenotypic ratio of the F_2 generation will be 3:1, just as in a monohybrid cross (FIGURE 14.7a).

The alternative hypothesis is that the two pairs of alleles segregate independently of each other. In other words, genes are packaged into gametes in all possible allelic combinations, as long as each gamete has one allele for each gene. In our example, four classes of gametes would be produced in equal quantities: *YR, Yr, yR,* and *yr.* If sperm of four classes are mixed with ova of four classes, there will be 16 (4×4) equally probable ways in which the alleles can combine in the F_2 generation, as shown in FIGURE 14.7b. These combinations make up four phenotypic categories with a ratio of 9:3:3:1 (nine yellow-round to three green-round to three yellow-wrinkled to one green-wrinkled). When Mendel did the experiment and "scored" (classified) the F_2 offspring, he obtained a ratio of 315:108:101:32, which is approximately 9:3:3:1.

The experimental results supported the hypothesis that each character is independently inherited; that in the dihybrids *(YyRr)*, the two alleles for seed color segregate independently of the two alleles for seed shape. Mendel tried his seven pea characters in various dihybrid combinations and always observed a 9:3:3:1 phenotypic ratio in the F_2 generation.

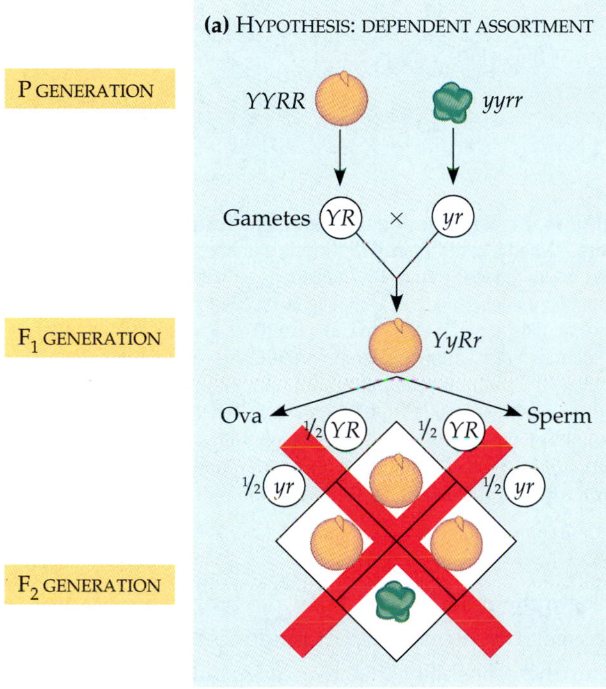

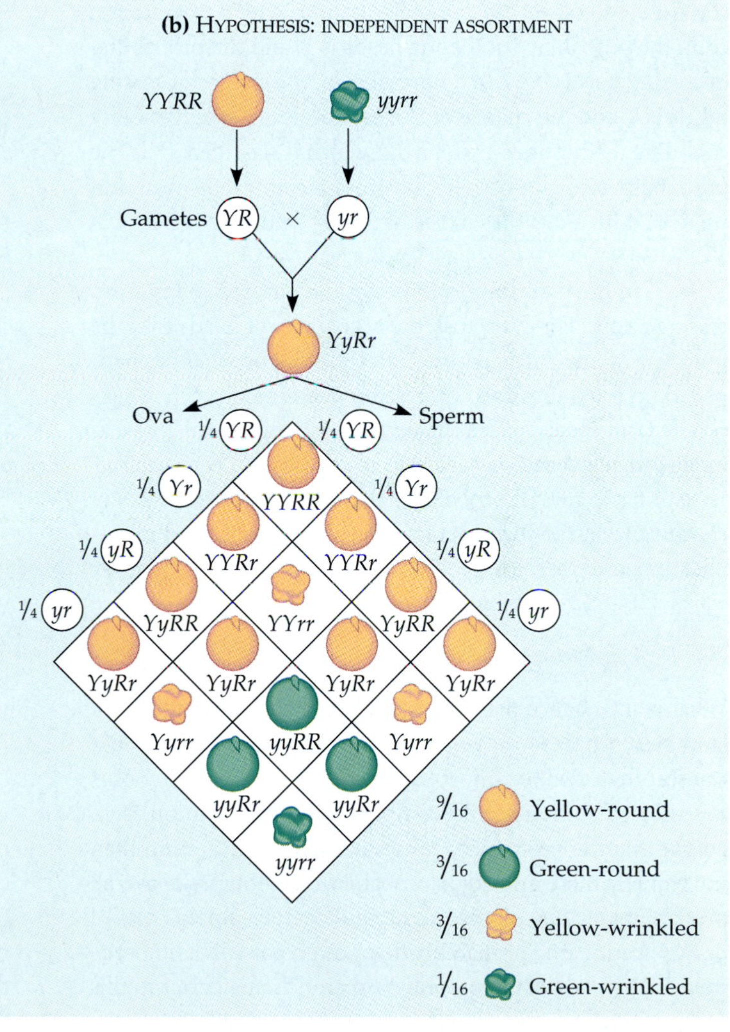

14.1 **FIGURE 14.7 • Testing two hypotheses for segregation in a dihybrid cross.** A cross between true-breeding parent plants that differ in two characters produces F_1 hybrids that are heterozygous for both characters. In this example, the two characters are seed color and seed shape. Yellow color *(Y)* and round shape *(R)* are dominant. **(a)** If the two characters segregate dependently (together), the F_1 hybrids can only produce the same two classes of gametes that they received from the parents, and the F_2 offspring will show a 3:1 phenotypic ratio. **(b)** If the two characters segregate independently, four classes of gametes will be produced by the F_1 generation, and there will be a 9:3:3:1 phenotypic ratio in the F_2 generation. Mendel's results supported this latter hypothesis, called independent assortment.

Notice in FIGURE 14.7b, however, that there remains a 3:1 phenotypic ratio for each character alone: three yellow to one green; three round to one wrinkled. As far as an individual character is concerned, the segregation behavior is the same as if this were a monohybrid cross. The independent segregation of each pair of alleles during gamete formation is now called Mendel's **law of independent assortment.**

Mendelian inheritance reflects rules of probability

Mendel's laws of segregation and independent assortment are specific applications of the same general rules of probability that apply to tossing coins or rolling dice. A basic understanding of these rules of chance is essential for genetic analysis.

The probability scale ranges from 0 to 1. An event that is certain to occur has a probability of 1, while an event that is certain *not* to occur has a probability of 0. With a two-headed coin, the probability of tossing heads is 1, and the probability of tossing tails is 0. With a normal coin, the chance of tossing heads is ½ and the chance of tossing tails is ½. The probability of rolling the number 3 with a die, which is six-sided, is ⅙. The probabilities of all possible outcomes for an event must add up to 1. With a die, the chance of rolling a number other than 3 is ⅚.

We can learn an important lesson about probability from tossing a coin. For every toss, the probability of heads is ½. The outcome of any particular toss is unaffected by what has happened on previous trials. We refer to phenomena such as successive coin tosses, or simultaneous tosses of several coins, as independent events.

Two basic laws of probability that can help us in games of chance and in solving genetic problems are the rule of multiplication and the rule of addition.

The Rule of Multiplication

What is the chance that two coins tossed simultaneously will land heads up? More generally, how do we determine the chance that two or more independent events will occur together in some specific combination? The solution is in computing the probability for each independent event, then multiplying these individual probabilities to obtain the overall probability of these events occurring together. By this rule of multiplication, the probability that both coins will land heads up is ½ × ½ = ¼. An F_1 monohybrid cross is analogous to this game of chance. With flower color as the heritable character, the genotype of a given F_1 plant is *Pp*. What is the probability that a particular F_2 plant will have white flowers? For this to happen, both the ovum and the sperm must carry the *p* allele, so we invoke the rule of multiplication. Segregation in the heterozygous plant is like flipping a coin. The probability that an ovum will have the *p* allele is ½. The chance that a sperm

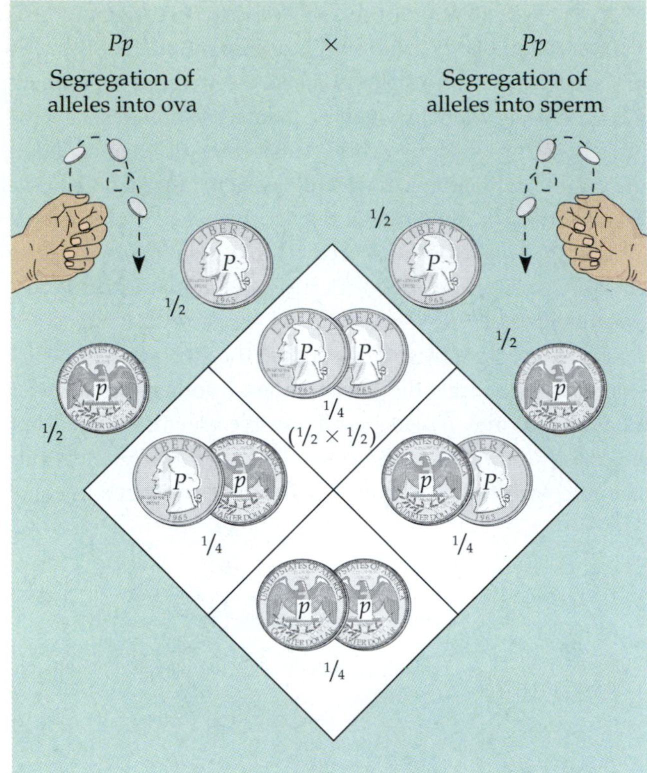

14.1 **FIGURE 14.8 • Segregation of alleles and fertilization as chance events.** When a heterozygote *(Pp)* forms gametes, segregation of alleles is like the toss of a coin. An ovum has a 50% chance of receiving the dominant allele and a 50% chance of receiving the recessive allele. The same odds apply to a sperm cell. Like two separately tossed coins, segregation during sperm and ovum formation occurs as two independent events. To determine the probability that an individual offspring will inherit the dominant allele from both parents, we multiply the probabilities of each required event: ½ × ½ = ¼. We can use the rule of multiplication to predict the probability for any genotype among offspring, as long as we know the genotypes of the parents.

will have the *p* allele is ½. Thus, the probability that two *p* alleles will come together at fertilization is ½ × ½ = ¼, equivalent to the probability that two independently tossed coins will land heads up (FIGURE 14.8).

We can also apply the rule of multiplication to dihybrid crosses, such as the one shown in FIGURE 14.7b. For a parent with the genotype *YyRr*, the probability that a gamete will carry the *Y* and *R* alleles is ¼. We can use the rule of multiplication to determine the probability of specific genotypes in the F_2 generation, without having to construct a 16-part Punnett square. For example, the probability of an F_2 plant having the genotype *YYRR* is 1/16 (¼ chance for a *YR* ovum × ¼ chance for a *YR* sperm). This corresponds to the top box in the Punnett square of FIGURE 14.7b.

The Rule of Addition

What is the probability that an F_2 plant from a monohybrid cross will be heterozygous? Notice in FIGURE 14.8 that there are two ways F_1 gametes can combine to produce a heterozygous result. The dominant allele can come from the ovum and the recessive allele from the sperm, or vice versa. By the rule of addition, the probability of an event that can occur in two or more different ways is the sum of the separate probabilities of those ways. Using this rule, we can calculate the probability of an F_2 heterozygote as $\frac{1}{4} + \frac{1}{4} = \frac{1}{2}$.

Using Rules of Probability to Solve Genetics Problems

We can combine the rules of multiplication and addition to solve complex problems in Mendelian genetics. For instance, imagine a cross of two pea varieties that differ in three characters—a *trihybrid* cross. Suppose we cross a trihybrid with purple flowers and yellow, round seeds (heterozygous for all three genes) with a plant with purple flowers and green, wrinkled seeds (heterozygous for flower color but homozygous recessive for the other two characters). Using Mendelian symbols, our cross is:

$$PpYyRr \times Ppyyrr$$

Let's use the rules of probability to calculate the fraction of offspring predicted to exhibit the recessive phenotypes for *at least two* of the three traits. We can start by listing all genotypes that fulfill this condition: *ppyyRr, ppYyrr, Ppyyrr, PPyyrr,* and *ppyyrr.* (Because the condition is *at least two* recessive traits, this last genotype, which produces all three recessive phenotypes, counts.) Next, use the rule of multiplication to calculate the individual probabilities for each of these genotypes from our $PpYyRr \times Ppyyrr$ cross. Finally, use the rule of addition to pool the probabilities for fulfilling the condition of at least two recessive traits:

ppyyRr	$\frac{1}{4}$ (probability of *pp*) $\times \frac{1}{2}$ (*yy*) $\times \frac{1}{2}$ (*Rr*)	$= \frac{1}{16}$
ppYyrr	$\frac{1}{4} \times \frac{1}{2} \times \frac{1}{2}$	$= \frac{1}{16}$
Ppyyrr	$\frac{1}{2} \times \frac{1}{2} \times \frac{1}{2}$	$= \frac{2}{16}$
PPyyrr	$\frac{1}{4} \times \frac{1}{2} \times \frac{1}{2}$	$= \frac{1}{16}$
ppyyrr	$\frac{1}{4} \times \frac{1}{2} \times \frac{1}{2}$	$= \frac{1}{16}$
Chance of *at least two* recessive traits		$= \frac{6}{16}$ or $\frac{3}{8}$

With practice, you'll be able to apply the rules of probability to solve such genetics problems faster than you could by filling in Punnett squares.

Mendel discovered the particulate behavior of genes: *a review*

If we plant a seed from the F_2 generation of FIGURE 14.4, we cannot predict with absolute certainty that the plant will yield white flowers, any more than we can predict with certainty that two tossed coins will both come up heads. What we *can* say is that there is exactly a $\frac{1}{4}$ chance that the plant will have white flowers. Stated another way, among a large sample of F_2 plants, one-fourth (25%) will have white flowers. Usually, the larger the sample size, the closer the results will conform to our predictions. The fact that Mendel counted so many offspring from his crosses suggests that he understood this statistical feature of inheritance and had a keen sense of the rules of chance.

Thus, Mendel's two laws, segregation and independent assortment, explain heritable variations in terms of alternative forms of genes (hereditary "particles") that are passed along, generation after generation, according to simple rules of probability. This particulate theory of inheritance, first discovered in garden peas, is equally valid for figs, flies, fish, birds, and human beings. Mendel's impact endures, not only on genetics, but on all of science, as a case study of the power of hypothetico-deductive thinking.

EXTENDING MENDELIAN GENETICS

The relationship between genotype and phenotype is rarely simple

In the twentieth century, geneticists have extended Mendelian principles not only to diverse organisms, but also to patterns of inheritance more complex than Mendel actually described. It was brilliant (or lucky) that Mendel chose pea plant characters that turned out to have a relatively simple genetic basis: Each character he studied is determined by one gene, for which there are only two alleles, one completely dominant to the other.* But these conditions are not met by all heritable characters, not even in garden peas. The relationship between genotype and phenotype is rarely so simple. This does not diminish the utility of Mendelian genetics, for the basic principles of segregation and independent assortment apply even to more complex patterns of inheritance. In this section we will extend Mendelian genetics to hereditary patterns that were not reported by Mendel.

Incomplete Dominance

The F_1 offspring of Mendel's classic pea crosses always looked like one of the two parental varieties because of the complete dominance of one allele over another. But for some genes, there is **incomplete dominance**, where the F_1 hybrids have an appearance somewhere in between the phenotypes of the two parental varieties. For instance, when red snapdragons are crossed with white snapdragons, all the F_1 hybrids have pink flowers. This third phenotype results from flowers of the heterozygotes having less red pigment than the red homozygotes

* There is one exception: Geneticists have found that Mendel's "flower position" character is actually determined by two genes.

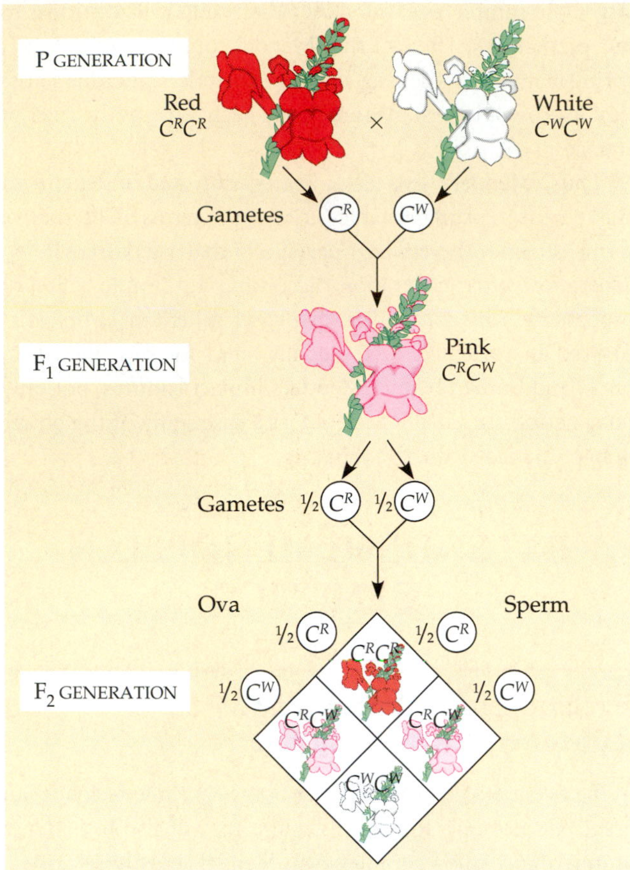

FIGURE 14.9 ▪ Incomplete dominance in snapdragon color. When red snapdragons are crossed with white ones, the F_1 hybrids have pink flowers. Segregation of alleles into the gametes of the F_1 plants results in an F_2 generation with a 1:2:1 ratio for both genotype and phenotype. C^R = allele for red flower color; C^W = allele for white flower color.

(FIGURE 14.9). We should not, however, regard incomplete dominance as evidence for the blending hypothesis, which would predict that the red or white traits could never be retrieved from the pink hybrids. In fact, breeding the F_1 hybrids produces F_2 offspring with a phenotypic ratio of 1 red to 2 pink to 1 white. (Because heterozygotes have a separate phenotype, the genotypic and phenotypic ratios for the F_2 generation are the same, 1:2:1.) The segregation of the red and white alleles in the gametes produced by the pink-flowered plants confirms that the alleles for flower color are heritable factors that maintain their identity in the hybrids; that is, inheritance is particulate.

What Is *a Dominant Allele?*

Now that you have learned about incomplete dominance, let's reexamine the meaning of dominance and recessiveness. What *is* a dominant allele? Or, more importantly, what is it *not?*

In **complete dominance**, the situation described by Mendel, the phenotypes of the heterozygote and dominant homozygote are indistinguishable. This represents one extreme of a spectrum in the dominance/recessiveness relationships of alleles. At the other extreme is **codominance**, in which *both* alleles are separately manifest in the phenotype. One example is the existence of three different human blood groups called the M, N, and MN blood groups. These groupings are based on two specific molecules located on the surfaces of red blood cells. People of group M have one of these two types of molecules, and people of group N have the other type. Group MN is characterized by the presence of *both* molecules on red blood cells. What is the genetic basis of these phenotypes? A single gene locus, at which two allelic variations are possible, determines these blood groups. M individuals are homozygous for one allele; N individuals are homozygous for the other allele. A heterozygous condition results in the blood of the MN group.

Notice that the MN phenotype is *not* intermediate between M and N phenotypes, but that both of these phenotypes are individually expressed by the presence of the two types of molecules on red blood cells. In contrast, incomplete dominance is characterized by an intermediate phenotype, as in the pink flowers of snapdragon hybrids. Thus the range of relationships between alleles includes complete dominance, codominance, and different degrees of incomplete dominance. These variations are reflected in the phenotypes of heterozygotes.

For any character, the dominance/recessiveness relationship we observe depends on the level at which we examine phenotype. For example, consider Tay-Sachs disease, an inherited disorder in humans. The brain cells of a baby with Tay-Sachs disease are unable to metabolize gangliosides, a type of lipid, because a crucial enzyme does not work properly. As the lipids accumulate in the brain, the brain cells gradually cease to function normally, leading to death. Only children who inherit two copies of the Tay-Sachs allele (homozygotes) have the disease. Thus on the *organismal* level of normal versus Tay-Sachs phenotype, the Tay-Sachs allele qualifies as a recessive. At the *biochemical* level, however, we observe an intermediate phenotype characteristic of incomplete dominance: The enzyme deficiency that causes Tay-Sachs disease can be detected in heterozygotes, who have an activity level of the lipid-metabolizing enzyme that is intermediate between individuals homozygous for the normal allele and individuals with Tay-Sachs disease. Heterozygotes lack symptoms of the disease, apparently because half the normal amount of functional enzyme is sufficient to prevent lipid accumulation in the brain. In fact, heterozygous individuals produce equal numbers of normal and dysfunctional enzyme molecules. Thus at the *molecular* level, the normal allele and the Tay-Sachs allele are codominant. As you can see, dominance/recessiveness relationships are rarely as straightforward as Mendel reported.

It is also important to understand that an allele is not termed *dominant* because it somehow subdues a recessive allele. Recall that alleles are simply variations in a gene's nucleotide sequence. When a dominant allele coexists with a

Phenotype (Blood Group)	Genotypes	Antibodies Present in Blood Serum	Reacts (Clumps) When Red Blood Cells from Groups Below Are Added to Serum from Groups at Left?			
			O	A	B	AB
O	ii	Anti-A Anti-B	No	Yes	Yes	Yes
A	$I^A I^A$ or $I^A i$	Anti-B	No	No	Yes	Yes
B	$I^B I^B$ or $I^B i$	Anti-A	No	Yes	No	Yes
AB	$I^A I^B$	—	No	No	No	No

FIGURE 14.10 ▪ Multiple alleles for the ABO blood groups. There are three alleles. Because each person carries two alleles, six genotypes are possible. Whenever the I^A or the I^B allele is present, the corresponding factor (A or B, respectively) is present on the surface of red blood cells. Both of these alleles, which are codominant, are dominant to the i allele, which does not code for any surface factor. A person produces antibodies against foreign blood factors, causing clumping of red blood cells if a transfusion is performed with incompatible blood.

recessive allele in a heterozygous genotype, they do not actually interact at all. It is in the pathway from genotype to phenotype that dominance and recessiveness come into play. We can use one of Mendel's characters—round versus wrinkled pea seed shape—as an example. The dominant allele codes for the synthesis of an enzyme that helps convert sugar to starch in the seed. The recessive allele codes for a defective form of this enzyme. Thus in a recessive homozygote, sugar accumulates in the seed because it is not converted to starch. As the seed develops, the high sugar concentration causes the osmotic uptake of water, and the seed swells. When the mature seed dries, it has wrinkles, analogous to the stretch marks of a person who has lost a lot of weight. In contrast, if a dominant allele is present, sugar is converted to starch, and the seeds do not wrinkle when they dry. One dominant allele results in enough of the enzyme to convert sugar to starch, and thus dominant homozygotes and heterozygotes have the same phenotype: round seeds. By exploring the mechanisms responsible for phenotype, we can demystify the concepts of dominance and recessiveness.

There is another important lesson about the meaning of the term *dominance.* Because an allele for a particular character is dominant does not necessarily mean that it is more common in a population than the recessive allele for that character. For example, about one baby out of 400 in the United States is born with extra fingers or toes, a condition known as polydactyly. The allele for polydactyly is *dominant* to the allele for five digits per appendage. In other words, 399 out of every 400 people are recessive homozygotes for this character; the recessive allele is far more prevalent than the dominant allele in the population. In Chapter 23 you will learn how the relative frequencies of alleles in a population are affected by natural selection.

Let's summarize three important points about dominance/recessiveness relationships:

1. They range from complete dominance, through various degrees of incomplete dominance, to codominance.
2. They reflect the mechanisms by which specific alleles are expressed in phenotype and do not involve the ability of one allele to subdue another at the level of the DNA.
3. They do not determine the relative abundance of alleles in a population.

Multiple Alleles

Most genes actually exist in populations in more than two allelic forms. The ABO blood groups in humans are one example of multiple alleles of a single gene. There are four possible phenotypes for this character: A person's blood group may be either A, B, AB, or O. These letters refer to two carbohydrates, the A substance and the B substance, which may be found on the surface of red blood cells. (These groups are based on blood-cell molecules different from those used for the MN classification discussed earlier.) A person's blood cells may have one substance or the other (type A or B), both (type AB), or neither (type O). Matching compatible blood groups is critical for blood transfusions. If the donor's blood has a factor (A or B) that is foreign to the recipient, specific proteins called antibodies produced by the recipient bind to the foreign molecules and cause the donated blood cells to agglutinate (clump together). This agglutination can kill the recipient.

The four blood groups result from various combinations of three different alleles of one gene, symbolized as I^A (for the A carbohydrate), I^B (for B), and i (giving rise to neither A nor B). Six genotypes are possible (FIGURE 14.10). Both the I^A and the I^B alleles are dominant to the i allele. Thus, I^AI^A and I^Ai individuals have type A blood, and I^BI^B and I^Bi individuals have type B. Recessive homozygotes, ii, have type O blood, because neither the A nor the B substance is produced. The I^A and I^B alleles are codominant; both are expressed in the phenotype of the I^AI^B heterozygote, who has type AB blood.

Pleiotropy

So far, we have treated Mendelian inheritance as though each gene affects one phenotypic character. Most genes, however, have multiple phenotypic effects. The ability of a gene to affect an organism in many ways is called **pleiotropy** (Gr. *pleion*, "more"). For example, alleles that are responsible for certain hereditary diseases in humans, such as sickle-cell disease, usually cause multiple symptoms (see FIGURE 14.15, p. 254). Considering the intricate molecular and cellular interactions responsible for an organism's development, it is not surprising that a gene can affect a number of an organism's characteristics.

Epistasis

In some cases, a gene at one locus alters the phenotypic expression of a gene at a second locus, a condition known as **epistasis** (Gr. for "standing upon"). An example will help clarify this concept. In mice and many other mammals, black coat color is dominant to brown. Let's designate *B* and *b* as the two alleles for this character. For a mouse to have brown fur, its genotype must be *bb*. But there is more to the story. A second gene locus determines whether or not pigment will be deposited in the hair. For this second gene, the dominant allele, symbolized by *C* (for color), results in the deposition of pigment. This allows either black or brown color, depending on the genotype at the first locus. But if the mouse is homozygous recessive for the second locus *(cc)*, then the coat is white (albino), regardless of the genotype at the black/brown locus.

What happens if we mate black mice that are heterozygous for both genes *(BbCc)*? Although the two genes affect the same phenotypic character (coat color), they follow the law of independent assortment (the two genes are inherited separately). Thus, our breeding experiment represents an F_1 dihybrid cross, like those that produced a 9:3:3:1 ratio in Mendel's experiments. In the case of coat color, however, the ratio of phenotypes among F_2 offspring is 9 black to 3 brown to 4 white. FIGURE 14.11 uses a Punnett square to account for this ratio in terms of epistasis. Other types of epistatic interactions produce different ratios.

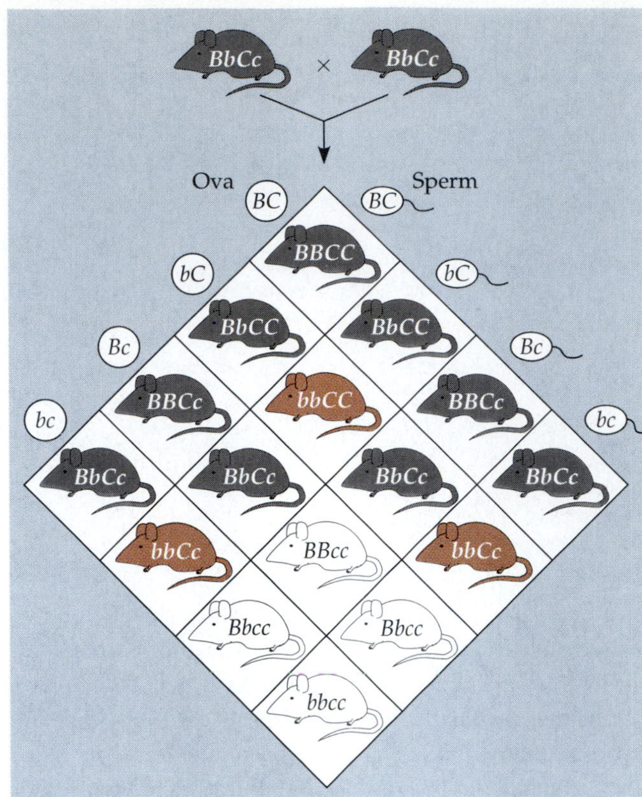

FIGURE 14.11 • An example of epistasis. This Punnett square illustrates the genotypes and phenotypes for offspring of matings between two black mice. The parents are heterozygous for two genes, which assort independently of each other. Thus our breeding experiment corresponds to a Mendelian F_1 cross between dihybrids. One gene determines whether the coat will be black (dominant, *B*) or brown (recessive, *b*). The second gene controls whether or not pigment of any color will be deposited in the hair, with the allele for the presence of color *(C)* dominant to the allele for the absence of color *(c)*. All offspring of the *cc* genotype are white (albino), regardless of the genotype for the black/brown genetic locus. This epistatic relationship of the color gene to the black/brown gene results in an F_2 phenotypic ratio of 9 black to 3 brown to 4 white.

Polygenic Inheritance

Mendel studied characters that could be classified on an either-or basis, such as purple versus white flower color. For many characters, however, such as human skin color and height, an either-or classification is impossible, because the characters vary in the population along a continuum (in gradations). These are called **quantitative characters.** Quantitative variation usually indicates **polygenic inheritance**, an additive effect of two or more genes on a single phenotypic character (the converse of pleiotropy, where a single gene affects several phenotypic characters).

There is evidence, for instance, that skin pigmentation in humans is controlled by at least three separately inherited genes (probably more, but we will simplify). Let's consider three genes, with a dark-skin allele for each gene *(A, B, C)* contributing one "unit" of darkness to the phenotype and being incompletely dominant to the other alleles *(a, b, c)*. An *AABBCC* person would be very dark, while an *aabbcc* individual would be very light. An *AaBbCc* person would have skin of an intermediate shade. Because the alleles have a cumulative effect, the genotypes *AaBbCc* and *AABbcc* would make the same genetic contribution (three units) to skin darkness. FIGURE 14.12 shows how this polygenic inheritance could result in a bell-shaped curve, called a normal distribution, for skin darkness among the members of a hypothetical population.

(You are probably familiar with the concept of a normal distribution for class curves of test scores.) Environmental factors, such as exposure to the sun, also affect the skin-color phenotype and help make the graph a smooth curve rather than a stairlike histogram.

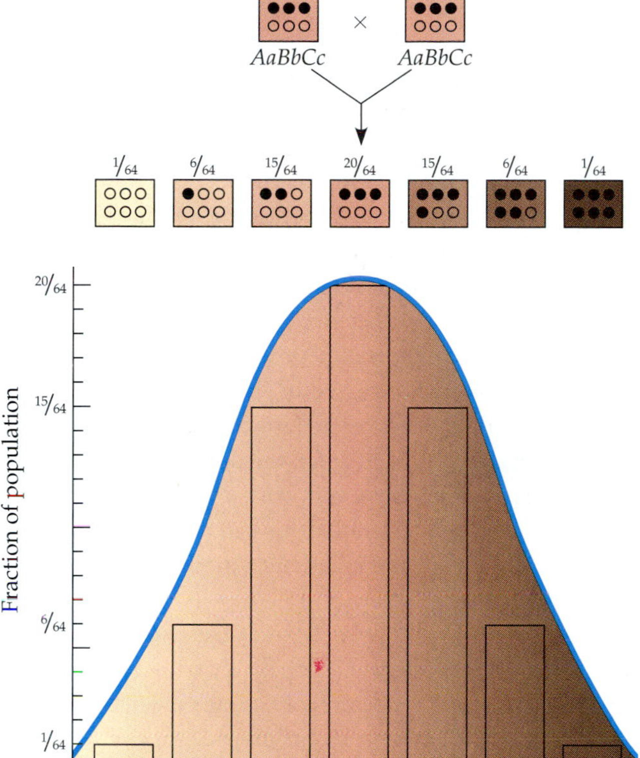

FIGURE 14.12 ▪ A simplified model for polygenic inheritance of skin color. According to this model, three separately inherited genes affect the darkness of skin. For each gene, an allele for dark skin is incompletely dominant to an allele for light skin. Thus, an individual who is heterozygous for all three genes *(AaBbCc)* has inherited three "units" of darkness, indicated in this figure by the three dots in the small square. An individual with the genotype *AABbcc* would also have a total of three units for dark skin. Imagine a large number of matings between individuals who are heterozygous for all three genes. Along the top of the graph are the variations that can occur among offspring. The *y*-axis represents the fractions of these variations among offspring of such trihybrid matings. The resulting histogram is smoothed into a bell-shaped curve by environmental factors, such as exposure to the sun, that affect skin color. This model is an oversimplification, but it should help clarify how polygenic inheritance can contribute to phenotypic characters that vary in gradations among the members of a population.

Nature Versus Nurture: The Environmental Impact on Phenotype

Phenotype depends on environment as well as on genes. A single tree, locked into its inherited genotype, has leaves that vary in size, shape, and greenness, depending on exposure to wind and sun. For humans, nutrition influences height, exercise alters build, sun-tanning darkens the skin, and experience improves performance on intelligence tests. Even identical twins, who are genetic equals, accumulate phenotypic differences as a result of their unique experiences.

Whether it is genes or the environment—nature or nurture—that most influences human characteristics is a very old and hotly contested debate that we will not attempt to settle here. We can say, however, that the product of a genotype is generally not a rigidly defined phenotype, but a range of phenotypic possibilities over which there may be variation due to environmental influence. This phenotypic range is called the **norm of reaction** for a genotype (FIGURE 14.13). There are cases where the norm of reaction has no breadth whatsoever; that is, a given genotype mandates a very specific phenotype. An example is the gene locus that determines a person's ABO blood group. In contrast, a person's blood count of red and white cells varies, depending on such factors as the altitude of one's home, the person's customary level of physical activity, and the presence of infectious agents.

Generally, norms of reaction are broadest for polygenic characters. Environment contributes to the quantitative nature of these characters, as we have seen in the continuous variation of skin color. Geneticists refer to such characters as **multifactorial**, meaning that many factors, both genetic and environmental, collectively influence phenotype.

Integrating a Mendelian View of Heredity and Variation

Over the past several pages we have broadened our view of Mendelian inheritance by exploring incomplete dominance and other variations in dominance/recessiveness relationships, multiple alleles, pleiotropy, polygenic inheritance, and the phenotypic impact of the environment. How can we integrate these refinements into a comprehensive theory of Mendelian genetics? The key is to make the transition from the reductionist

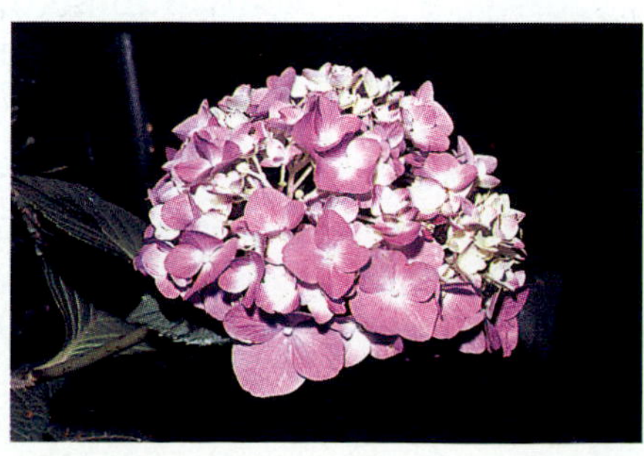

FIGURE 14.13 ▪ The effect of environment on phenotype. The outcome of a genotype lies within its norm of reaction, a phenotypic range that depends on the environment in which the genotype is expressed. For example, hydrangea flowers of the same genetic variety range in color from blue-violet to pink, depending on the acidity of the soil.

emphasis on single genes and phenotypic characters to the idea of the organism as a whole, one of the themes of this book. In fact, the term *phenotype* does double duty. We have been using the word in the context of specific characters, such as flower color and blood group. But phenotype is also used to describe the organism in its entirety—*all* aspects of its physical appearance, internal anatomy, physiology, and behavior. Similarly, the term *genotype* can also refer to an organism's entire genetic makeup, not just its alleles for a single genetic locus. In most cases, a gene's impact on phenotype is affected by other genes and by the environment. In this integrated view of heredity and variation, an organism's phenotype reflects its overall genotype and unique environmental history.

Considering all that can occur in the pathway from genotype to phenotype, it is indeed impressive that Mendel could simplify the complexities to reveal the fundamental principles governing the transmission of individual genes from parents to offspring. By extending the principles of segregation and independent assortment to help explain such hereditary patterns as epistasis and quantitative characters, we begin to see how broadly Mendelism applies. From Mendel's abbey garden came a theory of particulate inheritance that anchors modern genetics. In the last section of this chapter we will apply Mendelian genetics to human inheritance, especially the transmission of hereditary diseases.

MENDELIAN INHERITANCE IN HUMANS

Whereas peas are convenient subjects for genetic research, humans are not. The human generation span is about 20 years, and human parents produce relatively few offspring compared to peas and most other species. Furthermore, well-planned breeding experiments like the ones Mendel performed are impossible (or at least socially unacceptable) with humans. In spite of these difficulties, the study of human genetics continues to advance, powered by the incentive to understand our own inheritance. New techniques in molecular biology have led to many breakthrough discoveries, as we will see in Chapter 20, but basic Mendelism endures as the foundation of human genetics.

Pedigree analysis reveals Mendelian patterns in human inheritance

Unable to manipulate the mating patterns of people, geneticists must analyze the results of matings that have already occurred. As much information as possible is collected about a family's history for a particular trait, and this information is assembled into a family tree describing the interrelationships of parents and children across the generations—the family **pedigree**. As you read in the interview with geneticist Mary-Claire King, extensive pedigrees have been crucial to her research on breast cancer. A simpler example of a pedigree appears in FIGURE 14.14a, which traces the occurrence of widow's peak (a pointed contour of the hairline on the forehead). The trait is due to a dominant allele, which we symbolize as *W*.

We know that all individuals in this family who lack a widow's peak are homozygous recessive, and thus we can fill in their genotypes on the pedigree *(ww)*. We also know that both grandparents with widow's peaks are heterozygous *(Ww)*; if they were homozygous dominant *(WW)*, then all of their offspring would have had widow's peaks. The offspring in this second generation who *do* have widow's peaks must also be heterozygous, because they are the products of *Ww* × *ww* matings. The third generation in this pedigree consists of two sisters. The one who has a widow's peak could be either homozygous *(WW)* or heterozygous *(Ww)*, given what we know about the genotypes of her parents (both *Ww*).

FIGURE 14.14b is a pedigree of the same family, but this time we focus on a recessive trait, attached earlobes. We'll use *f* for the recessive allele and *F* for the dominant allele, which results in free earlobes. As you work your way through the pedigree, notice once again that you can apply what you have learned about Mendelian inheritance to fill in the genotypes for most individuals.

A pedigree not only helps us understand the past; it also helps us predict the future. Suppose that the couple represented in the second generation of FIGURE 14.14 decide to have one more child. What is the probability that the child will have a widow's peak? This is a Mendelian F_1 cross *(Ww* × *Ww)*, and thus the probability that a child will exhibit the dominant phenotype (widow's peak) is $\frac{3}{4}$. What is the probability that the child will have attached earlobes? Again, we can treat this as an F_1 monohybrid cross *(Ff* × *Ff)*, but this time we want to know the chance that the offspring would be homozygous recessive. The probability is $\frac{1}{4}$. What is the chance that the child will have a widow's peak *and* attached earlobes? If the two pairs of alleles assort independently in this dihybrid problem *(WwFf* × *WwFf)*, then we can use the rule of multiplication: $\frac{3}{4}$ (chance of widow's peak) × $\frac{1}{4}$ (chance of attached earlobes) = $\frac{3}{16}$ (chance of widow's peak and attached earlobes).

Many human disorders follow Mendelian patterns of inheritance

Pedigree analysis is a more serious matter when the alleles in question cause disabling or deadly hereditary diseases instead of innocuous human variations such as hairline or earlobe configuration. However, for disorders that are inherited as simple Mendelian traits, these same techniques of pedigree analysis apply.

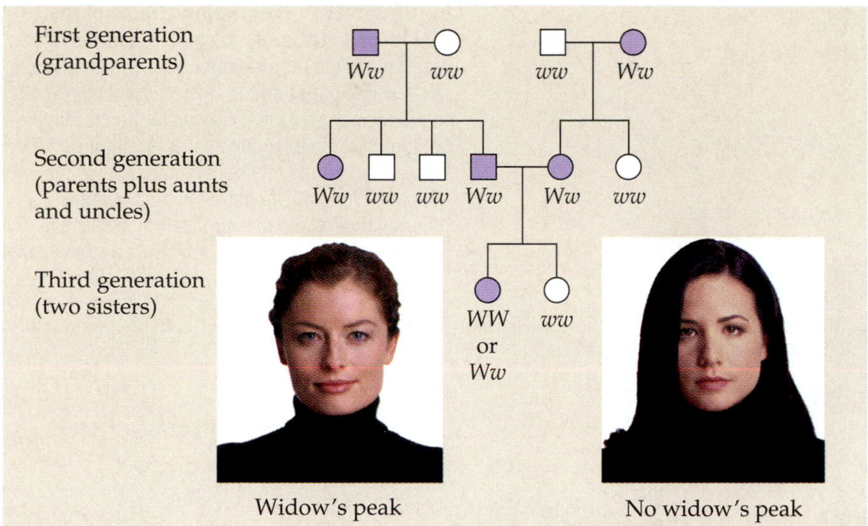

(a) Pedigree tracing a dominant trait (widow's peak)

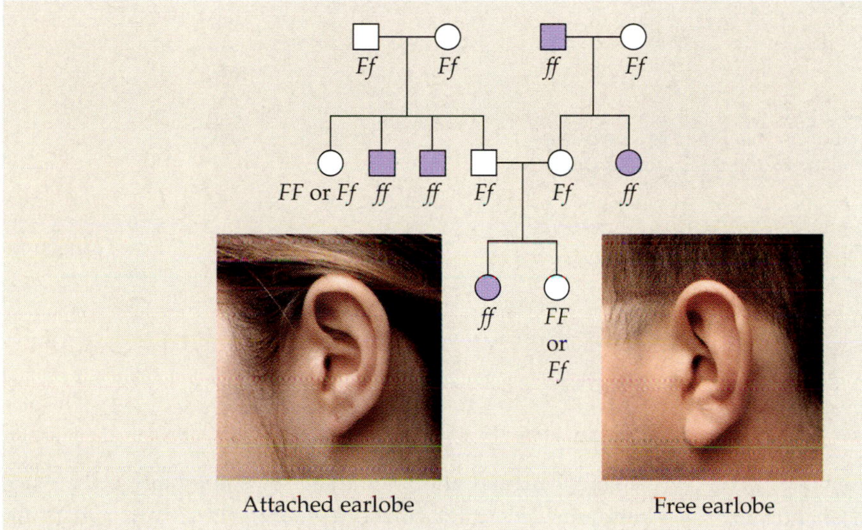

(b) Pedigree tracing a recessive trait (attached earlobes)

FIGURE 14.14 ▪ Pedigree analysis. In these family trees, squares symbolize males and circles represent females. A horizontal line connecting a male and female (□—○) indicates a mating, with offspring listed below in their order of birth, from left to right. Shaded symbols stand for individuals with the trait being traced. **(a) Dominant trait.** This pedigree traces a trait called widow's peak through three generations of a family. Notice in the third generation that the second-born daughter lacks a widow's peak, although both of her parents had the trait. Such a pattern of inheritance supports the hypothesis that the trait is due to a dominant allele. (If the trait were due to a *recessive* allele, and both parents had the recessive phenotype, then *all* of their offspring would also have the recessive phenotype.) **(b) Recessive trait.** This is the same family, but in this case we are tracing the inheritance of a recessive trait, attached earlobes. Notice that the first-born daughter in the third generation has attached lobes, although both of her parents lack that trait (they have free earlobes). Such a pattern is easily explained if the attached-lobe phenotype is due to a recessive allele. If it were due to a *dominant* allele, then at least one parent would also have had the trait. (As you may have already guessed, the ears shown in part b do not actually belong to the women shown in part a, but their earlobes will resemble those depicted here.)

Recessively Inherited Disorders

Thousands of genetic disorders are known to be inherited as simple recessive traits. These disorders range in severity from traits that are relatively harmless, such as albinism (lack of skin pigmentation), to life-threatening conditions, such as cystic fibrosis.

How can we account for the recessive behavior of the alleles causing these disorders? Recall that genes code for proteins of specific function. An allele that causes a genetic disorder codes either for a malfunctional protein or for no protein at all. In the case of disorders classified as recessive, heterozygotes are normal in phenotype because one copy of the "normal" allele produces a sufficient amount of the specific protein. Thus, a recessively inherited disorder shows up only in the homozygous individuals who inherit one recessive allele from each parent. We can symbolize the genotype of such people as *aa*, with individuals lacking the disorder being either *AA* or *Aa*. The heterozygotes *(Aa)*, who are phenotypically normal, are called **carriers** of the disorder because they may transmit the recessive allele to their offspring.

Most people who have recessive disorders are born to parents of normal phenotype who are both carriers. A mating between two carriers corresponds to a Mendelian F_1 cross *(Aa × Aa)*, with the zygote having a ¼ chance of inheriting a double dose of the recessive allele. A child of normal phenotype from such a cross has a ⅔ chance of being a carrier. (The genotypic ratio for the offspring is 1*AA*:2*Aa*:1*aa*. Thus two out of three offspring with *normal* phenotype—*AA* or *Aa*—are predicted to be heterozygous carriers.) Recessive homozygotes could also result from *Aa* × *aa* and *aa* × *aa* matings, but if the disorder is lethal before reproductive age or results in sterility, no *aa* individuals will reproduce. Even if recessive homozygotes are able to reproduce, such individuals will still account

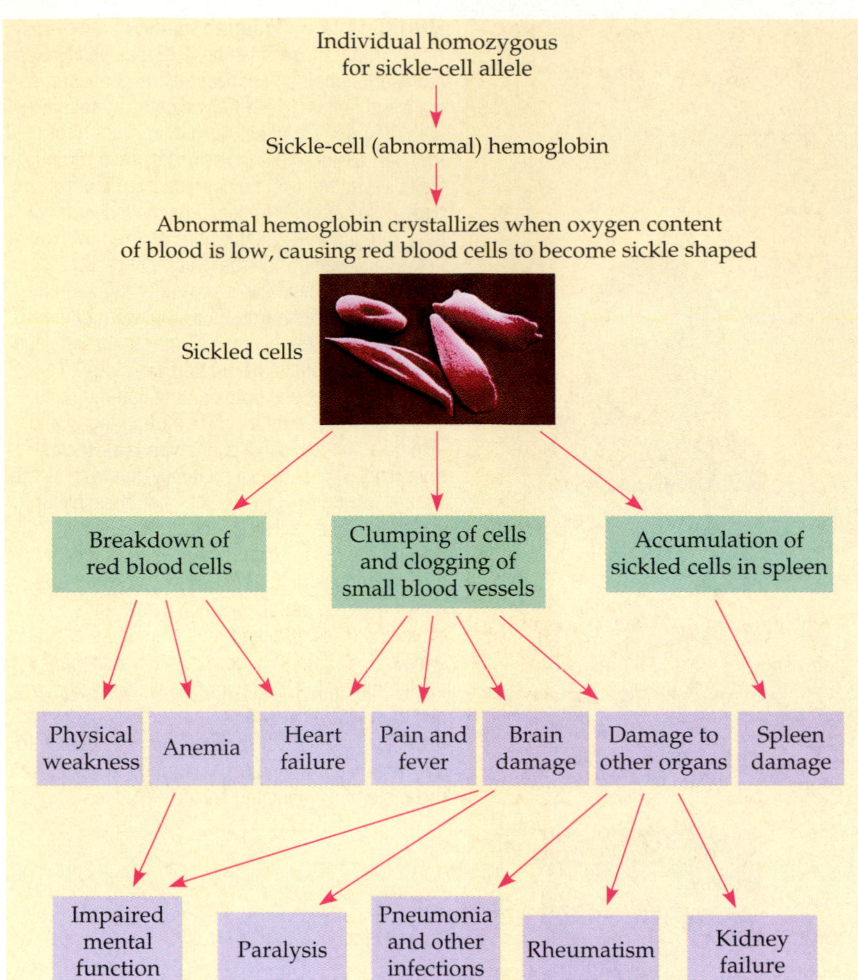

FIGURE 14.15 • Pleiotropic effects of the sickle-cell allele. At the molecular level, the recessive allele responsible for sickle-cell disease has a single direct effect: It causes red blood cells to produce an abnormal version of the protein hemoglobin. If sickle-cell alleles are inherited from both parents, then all the hemoglobin is of the abnormal variety. The abnormal hemoglobin deforms the red blood cells, starting a cascade of symptoms throughout the body. Thus at the level of the whole organism, the sickle-cell allele has multiple phenotypic effects.

for a much smaller percentage of the population than heterozygous carriers, for reasons we will examine in Chapter 23.

In general, a genetic disorder is not evenly distributed among all groups of humans. These disparities result from the different genetic histories of the world's peoples during less technological times, when populations were more geographically, hence genetically, isolated. We will now examine three examples of recessively inherited disorders.

The most common lethal genetic disease in the United States is **cystic fibrosis**, which strikes one out of every 2500 whites of European descent but is much rarer in other groups. One out of 25 whites (4%) is a carrier. The normal allele for this gene codes for a membrane protein that functions in chloride ion transport between certain cells and the extracellular fluid. These chloride channels are defective or absent in the plasma membranes of children who have inherited two of the recessive alleles that cause cystic fibrosis. The result is an abnormally high concentration of extracellular chloride, which causes the mucus that coats certain cells to become thicker and stickier than normal. The mucus builds up in the pancreas, lungs, digestive tract, and other organs, a condition that favors bacterial infections. Recent research indicates that the extracellular chloride also contributes to infection by disabling a natural antibiotic made by some body cells. When immune cells come to the rescue, their remains add to the mucus—creating a vicious cycle. Untreated, most children with cystic fibrosis die before their fifth birthday. Gentle pounding on the chest to clear mucus from clogged airways, daily doses of antibiotics to prevent infection, and other preventive treatments can prolong life. In the United States, more than half of the people with cystic fibrosis now survive into their late twenties or beyond.

Another lethal disorder inherited as a recessive allele is **Tay-Sachs disease**, described earlier in the chapter. Recall that the disease is caused by a dysfunctional enzyme that fails to break down brain lipids of a certain class. The symptoms of Tay-Sachs disease usually become manifest a few months after birth. The infant begins to suffer seizures, blindness, and degeneration of motor and mental performance. Inevitably, the child dies within a few years. There is a disproportionately high incidence of Tay-Sachs disease among Ashkenazic Jews, Jewish people whose ancestors lived in central Europe. In that population, the frequency of Tay-Sachs disease is one case of the disease out of 3600 births, about 100 times greater than the incidence among non-Jews or Mediterranean (Sephardic) Jews.

The most common inherited disease among blacks is **sickle-cell disease**, which affects one out of 400 African-Americans. Sickle-cell disease is caused by the substitution of a single amino acid in the hemoglobin protein of red blood cells (see FIGURE 5.19). When the oxygen content of an affected individual's blood is low (at high altitudes or under physical stress, for instance), the sickle-cell hemoglobin deforms the red cells to a sickle shape. Sickling of the cells, in turn, can lead to other symptoms. The multiple effects of a double dose of the sickle-cell allele exemplify pleiotropy (FIGURE 14.15). Doctors now use regular blood transfusions to ward off brain damage in children with sickle-cell disease, and new drugs can help prevent or treat other problems, but there is no cure.

Individuals who are heterozygous for the sickle-cell allele are said to have sickle-cell trait. These carriers are usually healthy, although a fraction of heterozygotes suffer some symptoms of sickle-cell disease when there is an extended reduction of blood oxygen. (The two alleles are codominant at the molecular level; both normal and abnormal hemoglobins are made.) About one out of ten African-Americans has sickle-cell trait, an unusually high frequency of heterozygotes for an allele with severe detrimental effects in homozygotes. The reason for the prevalence of this allele appears to be that while only individuals who are homozygous for the sickle-cell allele suffer from the disease, in certain environments heterozygotes have an advantage over people who carry no copies of the sickle-cell allele. A single copy of the sickle-cell allele increases resistance to malaria. Thus, in tropical Africa, where malaria is common, the sickle-cell allele is both boon and bane. The relatively high frequency of African-Americans with sickle-cell trait is a vestige of their African roots.

Although it is relatively unlikely that two carriers of the same rare harmful allele will meet and mate, the probability increases greatly if the man and woman are close relatives (for example, siblings or first cousins). These are called consanguineous ("same blood") matings, and they are indicated in pedigrees by double lines. Because people with recent common ancestors are more likely to carry the same recessive alleles than are unrelated people, it is more likely that a mating of close relatives will produce offspring homozygous for a harmful recessive trait. Such effects can be observed in many types of domesticated and zoo animals that have become inbred.

There is debate among geneticists about the extent to which human consanguinity increases the risk of inherited diseases. Many deleterious alleles have such severe effects that a homozygous embryo spontaneously aborts long before birth. Most societies and cultures have laws or taboos forbidding marriages between close relatives. These rules may have evolved out of empirical observation that in most populations, stillbirths and birth defects are more common when parents are closely related. But social and economic factors have also influenced the development of customs and laws against consanguineous marriages.

Dominantly Inherited Disorders

Although most harmful alleles are recessive, many human disorders are due to dominant alleles. One example is achondroplasia, a form of dwarfism with an incidence of one case among every 10,000 people. Heterozygous individuals have the dwarf phenotype. Therefore, all people who are not achondroplastic dwarfs—99.99% of the population—are homozygous for the recessive allele.

Lethal dominant alleles are much less common than lethal recessives. One reason for this difference is that the effects of lethal dominant alleles are not masked in heterozygotes. Many lethal dominant alleles are the result of new mutations (changes) in a gene of the sperm or egg that subsequently kill the developing offspring. An individual who does not survive to reproductive maturity will not pass on the new form of the gene. This is in contrast to lethal recessive mutations, which are perpetuated from generation to generation by the reproduction of heterozygous carriers who have normal phenotypes.

A lethal dominant allele can escape elimination if it is late-acting, causing death at a relatively advanced age. By the time the symptoms become evident, the individual may have already transmitted the lethal allele to his or her children. For example, **Huntington's disease**, a degenerative disease of the nervous system, is caused by a lethal dominant allele that has no obvious phenotypic effect until the individual is about 35 to 45 years old. Once the deterioration of the nervous system begins, it is irreversible and inevitably fatal. Any child born to a parent who has the allele for Huntington's disease has a 50% chance of inheriting the allele and the disorder. (The mating can be symbolized as $Aa \times aa$, with A being the dominant allele that causes Huntington's disease.) Until recently, it was impossible to tell before the onset of symptoms if a person at risk for Huntington's disease had actually inherited the allele, but that has changed. Analyzing DNA samples from a large family with a high incidence of the disorder, molecular geneticists have tracked the Huntington's allele to a locus near the tip of chromosome 4 (FIGURE 14.16). It is now possible to test for the presence of the allele in an individual's genome. (The methods that make such tests possible are discussed in Chapter 20.) For those with a family history of Huntington's disease, the availability of this test poses an agonizing dilemma: Under what circumstances is it beneficial for a presently healthy person to find out whether he or she has inherited a fatal and not yet curable disease?

Multifactorial Disorders

The hereditary diseases we have discussed so far are sometimes described as simple Mendelian disorders because they result

FIGURE 14.16 ▪ Large families provide excellent case studies of human genetics. Here, Nancy Wexler, of Columbia University and the Hereditary Disease Foundation, stands in front of a huge pedigree that traces Huntington's disease through several generations of one large family in Venezuela. Classical Mendelian analysis of this family, coupled with the techniques of molecular biology, enabled scientists to develop a test for the presence of the dominant allele that causes Huntington's disease—a test that can be used before symptoms appear. Dr. Wexler, who has studied the Venezuelan family, is herself at risk for developing Huntington's disease. Her mother died of the disorder, and there is a 50% chance that Dr. Wexler inherited the dominant allele that causes the disease.

from certain alleles at a single genetic locus. Many more people are susceptible to diseases that have a multifactorial basis—a genetic component plus a significant environmental influence. The long list of multifactorial diseases includes heart disease, diabetes, cancer, alcoholism, and certain mental illnesses, such as schizophrenia and manic-depressive disorder. In many cases the hereditary component is polygenic. For example, many genes affect our cardiovascular health, making some of us more prone than others to heart attacks and strokes. But our lifestyle intervenes tremendously between genotype and phenotype for cardiovascular health and other multifactorial characters. Exercise, a healthful diet, abstinence from smoking, and an ability to put stressful situations in perspective all reduce our risk of heart disease and some types of cancer.

At present, so little is understood about the genetic contributions to most multifactorial diseases that the best public-health strategy is to educate people about the importance of environmental factors and to promote healthful behavior.

Technology is providing new tools for genetic testing and counseling

A preventive approach to simple Mendelian disorders is sometimes possible, because in some cases the risk that a particular genetic disorder will occur can be assessed before a child is conceived or in the early stages of the pregnancy. Many hospitals have genetic counselors who can provide information to prospective parents concerned about a family history for a specific disease.

Let's consider the example of an imaginary couple, John and Carol, who are planning to have their first child and are seeking genetic counseling because of family histories of a lethal disease known to be recessively inherited. John and Carol each had a brother who died of the disorder, so they want to determine the risk of their having a child with the disease. From the information about their brothers, we know that both parents of John and both parents of Carol must have been carriers of the recessive allele. Thus, John and Carol are both products of *Aa* and *Aa* crosses, where *a* symbolizes the allele that causes this particular disease. We also know that John and Carol are not homozygous recessive *(aa)* because they do not have the disease. Therefore, their genotypes are either *AA* or *Aa*. Given a genotypic ratio of 1*AA*:2*Aa*:1*aa* for offspring of an $Aa \times Aa$ cross, John and Carol each have a $\frac{2}{3}$ chance of being carriers *(Aa)*. Using the rule of multiplication, we can determine that the overall probability of their firstborn having the disorder is $\frac{2}{3}$ (the chance that John is a carrier) multiplied by $\frac{2}{3}$ (the chance that Carol is a carrier) multiplied by $\frac{1}{4}$ (the chance of two carriers having a child with the disease), which equals $\frac{1}{9}$. Suppose that Carol and John decide to take the risk and have a child—after all, there is an $\frac{8}{9}$ chance that their baby will not have the disorder—but their child is born with the disease. We no longer have to guess about John's and Carol's genotypes. We now know that both John and Carol are, in fact, carriers. If the couple decides to have another child, they now know there is a $\frac{1}{4}$ chance that the second child will have the disease.

When we use Mendel's laws to predict possible outcomes of matings, it is important to keep in mind that chance has no memory: Each child represents an independent event in the sense that its genotype is unaffected by the genotypes of older siblings. Suppose that John and Carol have three more children, and *all three* have the hypothetical hereditary disease. This is an unfortunate family, for there is only one chance in 64 ($\frac{1}{4} \times \frac{1}{4} \times \frac{1}{4}$) that such an outcome will occur. But this run of misfortune will in no way affect the result if John and Carol decide to have still another child. There is still a $\frac{1}{4}$ chance that the additional child will have the disease and a $\frac{3}{4}$ chance that it will not. Mendel's laws, remember, are simply rules of probability applied to heredity.

Carrier Recognition

Because most children with recessive disorders are born to parents with normal phenotypes, the key to assessing the genetic risk for a particular disease is determining whether the prospective parents are heterozygous carriers of the recessive

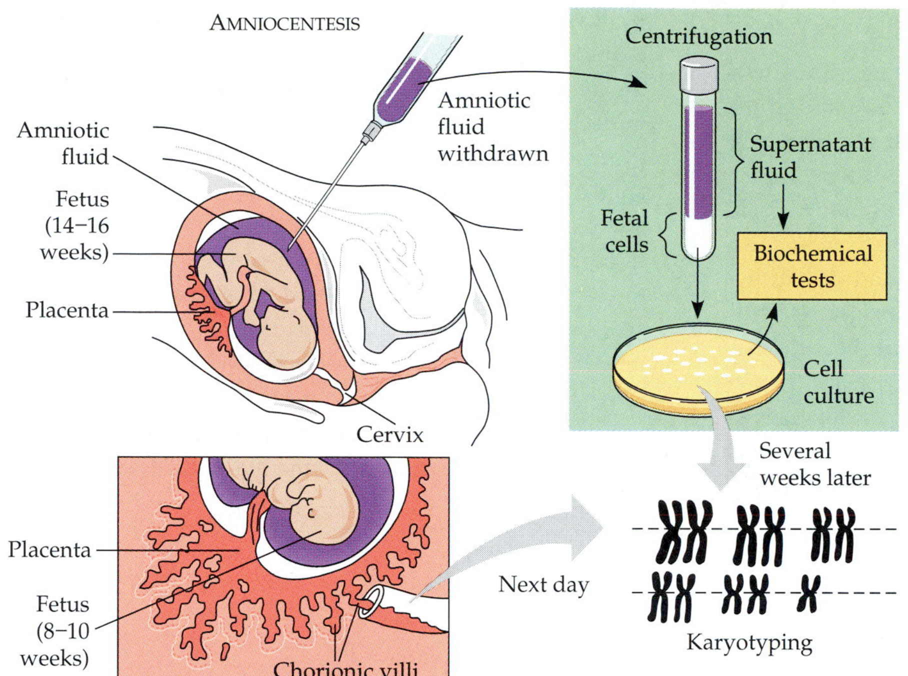

FIGURE 14.17 ▪ Fetal diagnosis. In amniocentesis, ultrasound is used to locate the fetus, and a small amount of amniotic fluid is extracted for testing. Physicians can diagnose some disorders from chemicals in the fluid itself, while other disorders may show up in tests performed on cells cultured from fetal cells present in the fluid. This analysis includes biochemical tests, to detect the presence of certain enzymes, and karyotyping, to determine whether the chromosomes of the fetal cells are normal in number and microscopic appearance. In chorionic villus sampling, a physician inserts a narrow tube through the cervix and uses suction to extract a tiny sample of fetal tissue from the placenta, the organ that transmits nutrients and fetal wastes between the fetus and the mother. The tissue sample is used for immediate karyotyping.

trait. For some heritable disorders, there are now tests that can distinguish individuals of normal phenotype who are dominant homozygotes from those who are heterozygotes, and the number of such tests increases each year. Examples are tests that can identify carriers of the alleles for Tay-Sachs disease, sickle-cell disease, and the most common form of cystic fibrosis. On one hand, these tests enable people with family histories of genetic disorders to make informed decisions about having children. On the other hand, these new methods for genetic screening could be abused. If confidentiality is breached, will carriers be stigmatized? Will they be denied health or life insurance, even though they are themselves healthy? Will misinformed employers equate "carrier" with disease? And will sufficient genetic counseling be available to help a large number of individuals understand their test results?

New biotechnology offers possibilities of reducing human suffering, but not before key ethical issues are resolved. The dilemmas posed by human genetics reinforce one of this book's themes: the immense social implications of biology.

Fetal Testing

Suppose a couple learns that they are both Tay-Sachs carriers, but they decide to have a child anyway. Tests performed in conjunction with a technique known as **amniocentesis** can determine, beginning at the fourteenth to sixteenth week of pregnancy, whether the developing fetus has Tay-Sachs disease (FIGURE 14.17). To perform this procedure, a physician inserts a needle into the uterus and extracts about 10 milliliters of amniotic fluid, the liquid that bathes the fetus. Some genetic disorders can be detected from the presence of certain chemicals in the amniotic fluid itself. Tests for other disorders, including Tay-Sachs disease, are performed on cells grown in the laboratory from the fetal cells that had been sloughed off into the amniotic fluid. These cultured cells can also be used for karyotyping to identify certain chromosomal defects (see the Methods Box in Chapter 13, p. 229).

In an alternative technique called **chorionic villus sampling (CVS)**, the physician suctions off a small amount of fetal tissue from the placenta. Because the cells of the chorionic villi are proliferating rapidly, enough cells are undergoing mitosis to allow karyotyping to be carried out immediately, giving results within 24 hours. This is an advantage over amniocentesis, in which the cells must be cultured for several weeks before karyotyping. Another advantage of CVS is that it can be performed as early as the eighth to tenth week of pregnancy.

Other techniques allow a physician to examine a fetus directly for major abnormalities. One such technique is ultrasound, which uses sound waves to produce an image of the fetus by a simple noninvasive procedure. This procedure has no known risk to either fetus or mother. With another technique, fetoscopy, a needle-thin tube containing a viewing scope and fiber optics (to transmit light) is inserted into the uterus. Fetoscopy enables the physician to examine the fetus for certain anatomical deformities.

In about 1% of the cases, amniocentesis or fetoscopy causes complications, such as maternal bleeding or fetal death. Thus, these techniques are usually reserved for cases in which the risk of a genetic disorder or other type of birth defect is relatively great. If the fetal tests reveal a serious disorder, the parents face the difficult choice of terminating the pregnancy or preparing to care for a child with a genetic disorder.

Newborn Screening

Some genetic disorders can be detected at birth by simple tests that are now routinely performed in most hospitals in the United States. One screening program is for the recessively inherited disorder called phenylketonuria (PKU), which occurs in about one out of every 10,000 births in the United States. Children with this disease cannot properly break down the amino acid phenylalanine. This compound and its byproduct, phenylpyruvate, can accumulate to toxic levels in the blood, causing mental retardation. However, if the deficiency is detected in the newborn, retardation can usually be prevented with a special diet, low in phenylalanine, that allows normal development. Thus, screening newborns for PKU and other treatable disorders is vitally important. Unfortunately, very few genetic disorders are treatable at the present time.

■ ■ ■

In this chapter you have learned about the Mendelian model of inheritance and its application to human genetics. We owe the "gene idea," the concept of heritable factors that are transmitted according to simple rules of chance, to the elegant experiments of Gregor Mendel. Mendel's quantitative approach was foreign to the biology of his era, and even the few biologists who read his papers apparently missed the importance of his discoveries. It wasn't until the beginning of the twentieth century that Mendelian genetics was rediscovered by biologists who studied the role of chromosomes in inheritance. In the next chapter you will learn how Mendel's laws have their physical basis in the behavior of chromosomes during sexual life cycles, and how the synthesis of Mendelism and a chromosome theory of inheritance catalyzed progress in genetics.

CHAPTER REVIEW

REVIEW OF KEY CONCEPTS

(with page numbers and key figures)

GREGOR MENDEL'S DISCOVERIES

- **Mendel brought an experimental and quantitative approach to genetics (pp. 239–241, FIGURE 14.1)** Gregor Mendel formulated a particulate theory of inheritance based on experiments with garden peas, carried out in the 1860s. He showed that parents pass on to their offspring discrete genes that retain their identity through the generations.
- 14.1 **By the law of segregation, the two alleles for a character are packaged into separate gametes (pp. 241–244, FIGURE 14.4)** Mendel arrived at this law by making hybrid offspring and letting them self-pollinate. The hybrids (F_1) exhibited the dominant trait. In the next generation (F_2), 75% of offspring had the dominant trait and 25% had the recessive trait, for a 3:1 ratio. Mendel's explanation was that genes have alternative forms (now called alleles) and that each organism inherits one allele for each gene from each parent. These separate (segregate) during gamete formation, so that a sperm or an egg carries only one allele. After fertilization, if the two alleles of a gene are different, one (the dominant allele) is expressed in the offspring and the other (the recessive allele) is masked. Homozygous individuals have identical alleles for a given character and are true-breeding. Heterozygous individuals have two different alleles for a given character.
- 14.1 **By the law of independent assortment, each pair of alleles segregates into gametes independently (pp. 244–246, FIGURE 14.7)** Mendel proposed this law based on dihybrid crosses between plants contrasting in two or more characters (e.g., flower color and seed shape). Alleles for each character segregate into gametes independently of alleles for other characters. The F_2 generation of a dihybrid cross has four possible phenotypes in a 9:3:3:1 ratio.
- **Mendelian inheritance reflects rules of probability (pp. 246–247, FIGURE 14.8)** The rule of multiplication states that the probability of a compound event is equal to the product of the separate probabilities of the independent single events. The rule of addition states that the probability of an event that can occur in two or more independent ways is the sum of the separate probabilities.
- **Mendel discovered the particulate behavior of genes: *a review* (p. 247)** Mendel's quantitative analysis of carefully planned experiments exemplifies the process of science.

EXTENDING MENDELIAN GENETICS

- **The relationship between genotype and phenotype is rarely simple (pp. 247–252, FIGURES 14.9–14.13)** In incomplete dominance, a heterozygous individual has a phenotype intermediate between those of the two types of homozygotes. In codominance, a heterozygote exhibits phenotypes for *both* its alleles. Many genes exist in *multiple* (more than two) alleles in a population. Pleiotropy is the ability of a single gene to affect multiple phenotypic traits. In epistasis, one gene affects the expression of another gene. Certain characters are quantitative; they vary continuously, indicating polygenic inheritance, an additive effect of two or more genes on a single phenotypic character. Quantitative characters also influenced by environment are called multifactorial.

MENDELIAN INHERITANCE IN HUMANS

- **Pedigree analysis reveals Mendelian patterns in human inheritance (p. 252, FIGURE 14.14)** Family pedigrees can be used to deduce the possible genotypes of individuals and make predictions about future offspring. Any predictions are usually statistical probabilities rather than certainties.
- **Many human disorders follow Mendelian patterns of inheritance (pp. 252–256)** Certain genetic disorders are inherited as simple recessive traits from phenotypically normal, heterozygous carriers. Some human disorders are due to dominant alleles. Medical researchers are beginning to sort out the genetic and environmental components of multifactorial disorders, such as heart disease and cancer.
- **Technology is providing new tools for genetic testing and counseling (pp. 256–258, FIGURE 14.17)** Using family histories, genetic counselors help couples determine the odds that their children will have genetic disorders. For certain diseases, tests that identify carriers define the odds more accurately. Once a child is conceived, amniocentesis and chorionic villus sampling can help determine whether a suspected genetic disorder is present.

GENETICS PROBLEMS

1. A rooster with gray feathers is mated with a hen of the same phenotype. Among their offspring, 15 chicks are gray, 6 are black, and 8 are white. What is the simplest explanation for the inheritance of these colors in chickens? What offspring would you predict from the mating of a gray rooster and a black hen?
2. In some plants, a true-breeding, red-flowered strain gives all pink flowers when crossed with a white-flowered strain: *RR* (red) × *rr* (white) ⟶ *Rr* (pink). If flower position (axial or terminal) is inherited as it is in peas (see TABLE 14.1), what will be the ratios of genotypes and phenotypes of the generation resulting from the following cross: axial-red (true-breeding) × terminal-white? What will be the ratios in the F_2 generation?
3. Flower position, stem length, and seed shape were three characters that Mendel studied. Each is controlled by an independently assorting gene and has dominant and recessive expression as follows:

Character	*Dominant*	*Recessive*
Flower position	Axial *(A)*	Terminal *(a)*
Stem length	Tall *(L)*	Dwarf *(l)*
Seed shape	Round *(R)*	Wrinkled *(r)*

If a plant that is heterozygous for all three characters were allowed to self-fertilize, what proportion of the offspring would be expected to be as follows? (*Note:* Use the rules of probability instead of a huge Punnett square.)

a. homozygous for the three dominant traits
b. homozygous for the three recessive traits
c. heterozygous
d. homozygous for axial and tall, heterozygous for seed shape

4. A black guinea pig crossed with an albino guinea pig produced 12 black offspring. When the albino was crossed with a second black one, 7 blacks and 5 albinos were obtained. What is the best explanation for this genetic situation? Write genotypes for the parents, gametes, and offspring.
5. In sesame plants, the one-pod condition *(P)* is dominant to the three-pod condition *(p)*, and normal leaf *(L)* is dominant to wrinkled leaf *(l)*. Pod type and leaf type are inherited independently. Determine the genotypes for the two parents for all possible matings producing the following offspring:
   a. 318 one-pod normal, 98 one-pod wrinkled
   b. 323 three-pod normal, 106 three-pod wrinkled
   c. 401 one-pod normal
   d. 150 one-pod normal, 147 one-pod wrinkled, 51 three-pod normal, 48 three-pod wrinkled
   e. 223 one-pod normal, 72 one-pod wrinkled, 76 three-pod normal, 27 three-pod wrinkled
6. A man with group A blood marries a woman with group B blood. Their child has group O blood. What are the genotypes of these individuals? What other genotypes, and in what frequencies, would you expect in offspring from this marriage?
7. Color pattern in a species of duck is determined by one gene with three alleles. Alleles *H* and *I* are codominant, and allele *i* is recessive to both. How many phenotypes are possible in a flock of ducks that contains all the possible combinations of these three alleles?
8. Phenylketonuria (PKU) is an inherited disease caused by a recessive allele. If a woman and her husband are both carriers, what is the probability of each of the following?
   a. all three of their children will be of normal phenotype
   b. one *or* more of the three children will have the disease
   c. all three children will have the disease
   d. *at least* one child will be phenotypically normal

   (*Note:* Remember that the probabilities of all possible outcomes always add up to 1.)
9. The genotype of F_1 individuals in a tetrahybrid cross is *AaBbCcDd*. Assuming independent assortment of these four genes, what are the probabilities that F_2 offspring would have the following genotypes?
   a. *aabbccdd*
   b. *AaBbCcDd*
   c. *AABBCCDD*
   d. *AaBBccDd*
   e. *AaBBCCdd*
10. In 1981, a stray black cat with unusual rounded, curled-back ears was adopted by a family in California. Hundreds of descendants of the cat have since been born, and cat fanciers hope to develop the "curl" cat into a show breed. Suppose you owned the first curl cat and wanted to develop a true-breeding variety. How would you determine whether the curl allele is dominant or recessive? How would you select for true-breeding cats? How would you know they are true-breeding?

11. What is the probability that each of the following pairs of parents will produce the indicated offspring? (Assume independent assortment of all gene pairs.)
    a. *AABBCC* × *aabbcc* ⟶ *AaBbCc*
    b. *AABbCc* × *AaBbCc* ⟶ *AAbbCC*
    c. *AaBbCc* × *AaBbCc* ⟶ *AaBbCc*
    d. *aaBbCC* × *AABbcc* ⟶ *AaBbCc*
12. Karen and Steve each have a sibling with sickle-cell disease. Neither Karen, Steve, nor any of their parents has the disease, and none of them has been tested to reveal sickle-cell trait. Based on this incomplete information, calculate the probability that if this couple has a child, the child will have sickle-cell disease.
13. Imagine that a newly discovered, recessively inherited disease is expressed only in individuals with type O blood, although the disease and blood group are independently inherited. A normal man with type A blood and a normal woman with type B blood have already had one child with the disease. The woman is now pregnant for a second time. What is the probability that the second child will also have the disease? Assume both parents are heterozygous for the "disease" gene.
14. In tigers, a recessive allele causes an absence of fur pigmentation (a "white tiger") and a cross-eyed condition. If two phenotypically normal tigers that are heterozygous at this locus are mated, what percentage of their offspring will be cross-eyed? What percentage will be white?

15. In corn plants, a dominant allele *I* inhibits kernel color, while the recessive allele *i* permits color when homozygous. At a different locus, the dominant gene *P* causes purple kernel color, while the homozygous recessive genotype *pp* causes red kernels. If plants heterozygous at both loci are crossed, what will be the phenotypic ratio of the F_1 generation?

16. The pedigree below traces the inheritance of alkaptonuria, a biochemical disorder. Affected individuals, indicated here by the filled-in circles and squares, are unable to break down a substance called alkapton, which colors the urine and stains body tissues. Does alkaptonuria appear to be caused by a dominant or recessive allele? Fill in the genotypes of the individuals whose genotypes you know. What genotypes are possible for each of the other individuals?

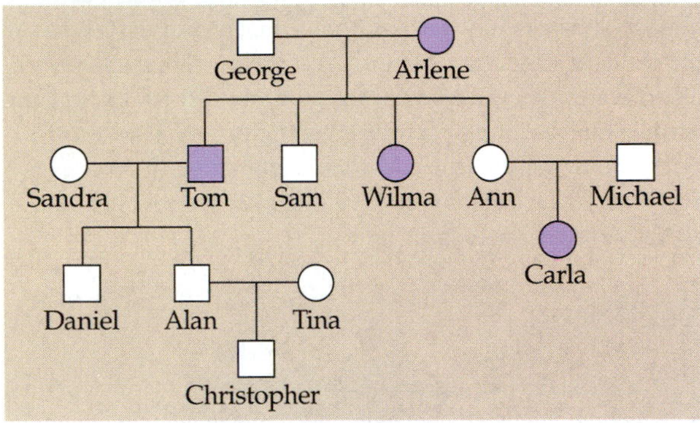

17. A man has six fingers on each hand and six toes on each foot. His wife and their daughter have the normal number of digits. Extra digits is a dominant trait. What fraction of this couple's children would be expected to have extra digits?

18. Imagine you are a genetic counselor, and a couple planning to start a family came to you for information. Charles was married once before, and he and his first wife had a child who has cystic fibrosis. The brother of his current wife Elaine died of cystic fibrosis. What is the probability that Charles and Elaine will have a baby with cystic fibrosis? (Neither Charles nor Elaine has cystic fibrosis.)

19. In mice, black color *(B)* is dominant to white *(b)*. At a different locus, a dominant allele *(A)* produces a band of yellow just below the tip of each hair in mice with black fur. This gives a frosted appearance known as agouti. Expression of the recessive allele *(a)* results in a solid coat color. If mice that are heterozygous at both loci are crossed, what will be the expected phenotypic ratio of their offspring?

20. The pedigree below traces the inheritance of a very rare biochemical disorder in humans. Affected individuals are indicated by filled-in circles and squares. Is the allele for this disorder dominant or recessive? What genotypes are possible for the individuals marked 1, 2, and 3?

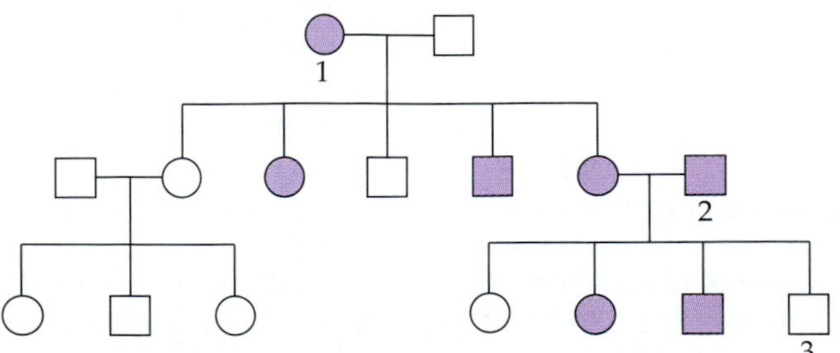

SCIENCE, TECHNOLOGY, AND SOCIETY

1. Imagine that one of your parents had Huntington's disease. What would be the probability that you, too, would someday manifest the disease? There is no cure for Huntington's. Would you want to be tested for the Huntington's allele? Why or why not?

2. In the future, gene therapy may be an option for the treatment and cure of some inherited disorders. What do you think are the most serious ethical issues that must be dealt with before human gene therapy is used on a large scale? Why do you think these issues are important?

FURTHER READING

Ahmad, W. et al. "Alopecia Universalis Associated with a Mutation in the Human *hairless* Gene." *Science*, January 30, 1998. A six-generation pedigree was used in this research.

Beardsley, T. "Fetal Checkup." *Scientific American*, January 1997. Fetal cells for testing can be obtained from the mother's blood, without the need for amniocentesis or CVS.

Horgan, J. "Eugenics Revisited." *Scientific American*, June 1993. Discusses controversies centered on the genetics of human behavior.

Kahn, P. "Gene Hunters Close in On Elusive Prey." *Science*, March 8, 1996. Combining family sample collection and DNA technology to study complex diseases caused by a combination of genes and environment.

Kolata, G. "Seeking Reasons for Disease Genes." *New York Times*, December 3, 1996. Genetic "defects" may offer evolutionary benefits.

Meistel, R. "Young Blood, New Life." *New Scientist*, November 16, 1996. Relates the search for drugs for treating sickle-cell disease.

Revkin, A. "Hunting Down Huntington's." *Discover*, December 1993. A team of genetic sleuths locates the gene for Huntington's disease.

Seachrist, L. "Testing Genes." *Science News*, December 6, 1995. Wrestling with the information from genetic tests.

Welsh, M. J., and A. E. Smith. "Cystic Fibrosis." *Scientific American*, December 1995. All about this disease, from molecules to treatments.

WEB LINKS

Visit the special edition of *The Biology Place* for BIOLOGY, Fifth Edition, at http://www.biology.com/campbell. Go to Chapter 14 for online resources, including learning activities, practice exams, and links to the following web sites:

"Hereditary Disease Foundation"
Current information about the genetic diseases mentioned in this chapter, and others.

"MendelWeb"
A comprehensive site about Gregor Mendel, constructed around his famous paper "Experiments in Plant Hybridization."

"The Virtual FlyLab"
This excellent interactive site allows you to learn the principles of genetic inheritance by "mating" virtual flies.

"Pea Soup—The Story of Mendel"
This is a well-written and simple account of Mendel's work and includes an interactive pea-breeding experiment.

"Experimenting with Mendel's Peas"
Visit Campbell's Biology Place and use this activity to investigate the basics of Mendelian inheritance.

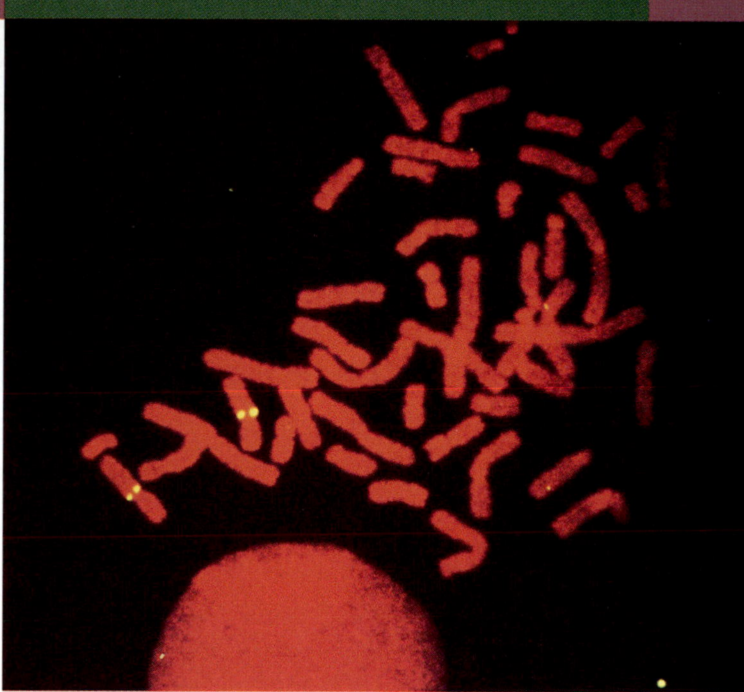

THE CHROMOSOMAL BASIS OF INHERITANCE

Relating Mendelism to Chromosomes

- Mendelian inheritance has its physical basis in the behavior of chromosomes during sexual life cycles
- Morgan traced a gene to a specific chromosome: *science as a process*
- Linked genes tend to be inherited together because they are located on the same chromosome
- Independent assortment of chromosomes and crossing over produce genetic recombinants
- Geneticists can use recombination data to map a chromosome's genetic loci

Sex Chromosomes

- The chromosomal basis of sex varies with the organism
- Sex-linked genes have unique patterns of inheritance

Errors and Exceptions in Chromosomal Inheritance

- Alterations of chromosome number or structure cause some genetic disorders
- The phenotypic effects of some genes depend on whether they were inherited from the mother or the father (imprinting)
- Extranuclear genes exhibit a non-Mendelian pattern of inheritance

It was not until the year 1900 that biology finally caught up with Gregor Mendel. At that time, three botanists, working independently on plant breeding experiments, reproduced Mendel's results. By searching the literature, the German Karl Correns, the Austrian Erich von Tschermak, and the Dutchman Hugo de Vries all found that Mendel had explained the same results 35 years before. During the intervening years, biology had grown more experimental and quantitative and thus more receptive to Mendelism. Nevertheless, many biologists remained incredulous about Mendel's laws of segregation and independent assortment until evidence had mounted that these principles of heredity had a physical basis in the behavior of chromosomes. Mendel's hereditary factors—genes—are located on chromosomes. For example, the yellow dots (a fluorescent dye) in the light micrograph on this page mark the locus of a specific gene on a homologous pair of human chromosomes (which have already replicated, hence the two dots per chromosome). This chapter integrates and extends what you have learned in the past two chapters by describing the chromosomal basis for the transmission of genes from parents to offspring, along with some important exceptions.

RELATING MENDELISM TO CHROMOSOMES

Mendelian inheritance has its physical basis in the behavior of chromosomes during sexual life cycles

Cytologists worked out the process of mitosis in 1875 and meiosis in the 1890s. Then, around 1900, cytology and genetics converged as biologists began to see parallels between the behavior of chromosomes and the behavior of Mendel's factors. For example, chromosomes and genes are both present in pairs in diploid cells, homologous chromosomes separate and alleles segregate during meiosis, and fertilization restores the paired condition for both chromosomes and genes. Around 1902, Walter S. Sutton, Theodor Boveri, and others independently noted these parallels, and a **chromosome theory of inheritance** began to take form. According to this theory, Mendelian genes have specific loci on chromosomes, and it is the chromosomes that undergo segregation and independent assortment (FIGURE 15.1, next page).

Morgan traced a gene to a specific chromosome: *science as a process*

Thomas Hunt Morgan, an embryologist at Columbia University, was the first to associate a specific gene with a specific chromosome, early in the twentieth century. Although Morgan was skeptical about both Mendelism and the chromosome theory,

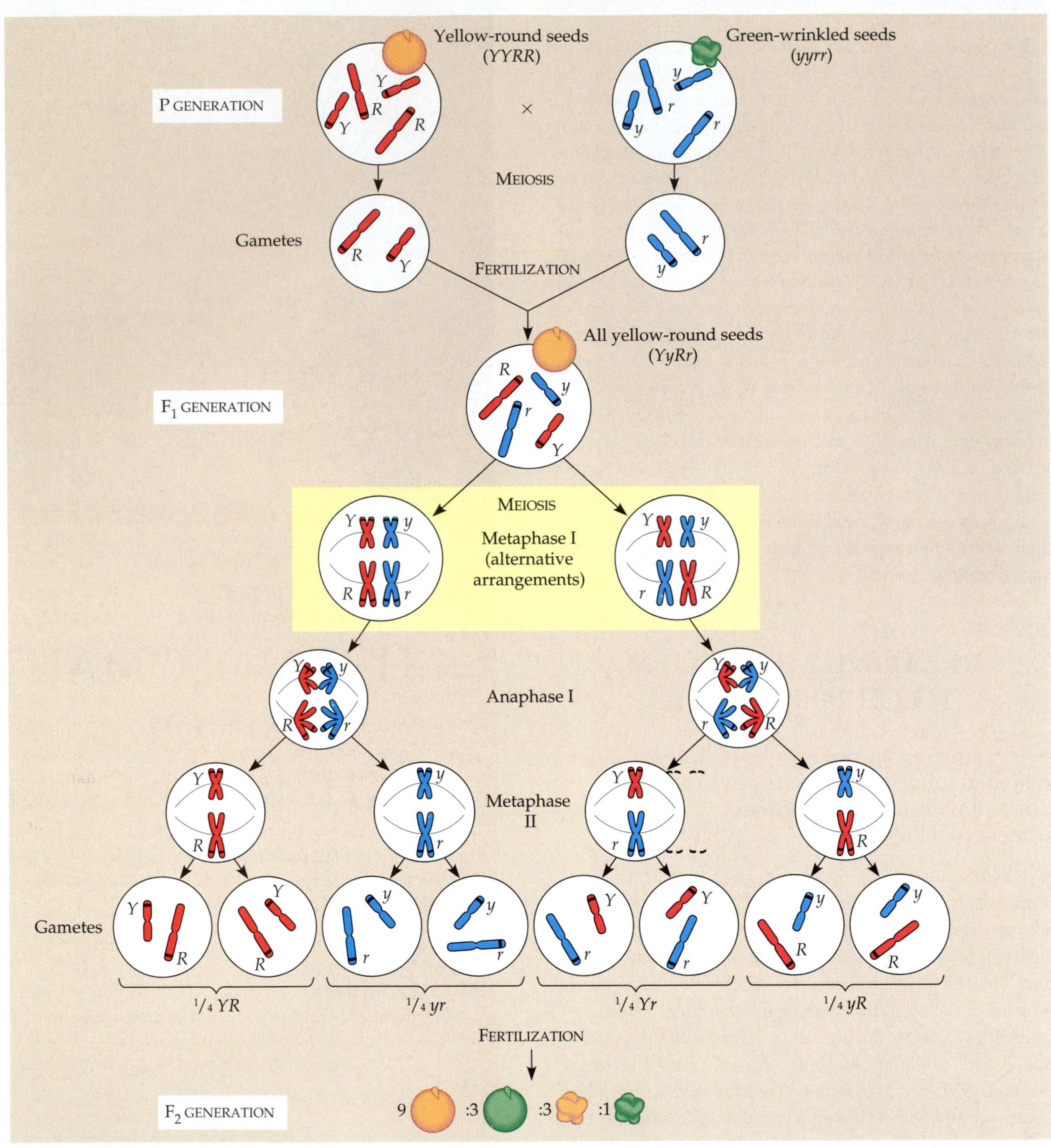

FIGURE 15.1 ▪ The chromosomal basis of Mendel's laws. Here we correlate the results of one of Mendel's dihybrid crosses with the behavior of chromosomes. The two characters we are tracking are the color and shape of pea seeds. The two genes are on different chromosomes, and black bands symbolize the loci in these drawings. (Peas actually have a total of seven chromosome pairs, but only two are illustrated here.) The arrangements of chromosomes at metaphase I of meiosis and their movement during anaphase I account for the segregation and independent assortment of the alleles for seed color and shape. The two alleles for each character segregate when homologous chromosomes move toward opposite poles of the cell and become packaged in different cells during the first meiotic division. Mendel's law of independent assortment has its physical basis in the random arrangement of chromosomes on the metaphase plate during meiosis I. For each homologous pair of chromosomes, the polar orientation of the maternal and paternal versions (color-coded red and blue, respectively) of the chromosome is unrelated to the orientation of the other chromosome pair. This results in the alternative metaphase I arrangements illustrated in the yellow box. Because the alternatives are equally likely, F_1 plants will produce equal numbers of eight kinds of gametes. If we continue to trace the results of these alternative alignments of chromosomes to the F_2 generation, we see the physical explanation for Mendel's 9:3:3:1 ratio of phenotypes.

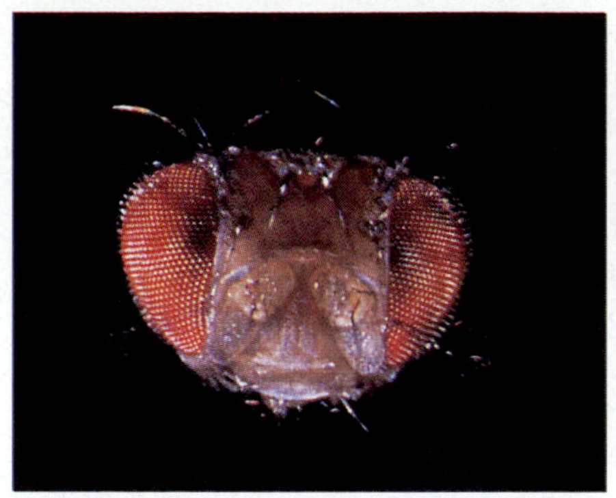

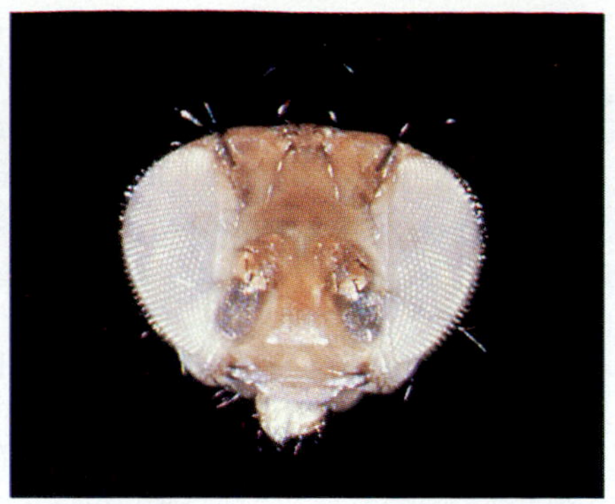

FIGURE 15.2 ▪ Morgan's first mutant. Wild-type *Drosophila* flies have red eyes (left). Among his flies, Morgan discovered a mutant male with white eyes (right). This variation made it possible for Morgan to trace a gene for eye color to a specific chromosome (LMs).

his early experiments provided convincing evidence that chromosomes are indeed the location of Mendel's heritable factors.

Morgan's Choice of Experimental Organism

Many times in the history of biology, important discoveries have come to those insightful enough or lucky enough to choose an experimental organism suitable for the research problem being tackled. Mendel chose the garden pea because a number of varieties were available. For his work, Morgan selected a species of fruit fly, *Drosophila melanogaster,* a common, generally innocuous insect that feeds on the fungi growing on fruit. Fruit flies are prolific breeders; a single mating will produce hundreds of offspring, and a new generation can be bred every two weeks. These characteristics make the fruit fly a convenient organism for genetic studies. Morgan's laboratory soon became known as "the fly room."

Another advantage of the fruit fly is that it has only four pairs of chromosomes, which are easily distinguishable with a light microscope. There are three pairs of autosomes and one pair of sex chromosomes. Female fruit flies have a homologous pair of *X* chromosomes, and males have one *X* chromosome and one *Y* chromosome.

While Mendel could readily obtain different pea varieties, there were no convenient suppliers of fruit fly varieties for Morgan to employ. Indeed, he was probably the first person to want different varieties of this common insect. After a year of breeding flies and looking for variant individuals, Morgan was rewarded with the discovery of a single male fly with white eyes instead of the usual red. The normal phenotype for a character (the phenotype most common in natural populations), such as red eyes in *Drosophila,* is called the **wild type** (FIGURE 15.2). Traits that are alternatives to the wild type, such as white eyes in *Drosophila,* are called **mutant phenotypes**, because they are due to alleles assumed to have originated as changes, or mutations, in the wild-type allele.

Discovery of Sex-Linkage

After Morgan discovered his white-eyed male fly, he mated it with a red-eyed female. All the F_1 offspring had red eyes, suggesting that the wild type was dominant. When Morgan bred the F_1 flies to each other, he observed the classical 3:1 phenotypic ratio among the F_2 offspring. However, there was a surprising result: The white-eye trait showed up only in males. All the F_2 females had red eyes, while half the males had red eyes and half had white eyes. Somehow, a fly's eye color was linked to its sex.

From this and other evidence, Morgan deduced that the gene affected in his white-eyed mutant is located exclusively on the *X* chromosome; there is no corresponding eye-color locus on the *Y* chromosome (FIGURE 15.3, p. 264). Thus, females *(XX)* carry two copies of the gene for this character, while males *(XY)* carry only one. Because the mutant allele is recessive, a female will have white eyes only if she receives that allele on both *X* chromosomes—an impossibility for the F_2 females in Morgan's experiment. For a male, on the other hand, a single copy of the mutant allele confers white eyes. Since a male has only one *X* chromosome, there can be no wild-type allele present to offset the recessive allele.

Genes located on a sex chromosome are called **sex-linked genes**. Morgan's evidence that a specific gene is carried on the *X* chromosome added credibility to the chromosome theory of inheritance. Recognizing the importance of this work, many bright students were attracted to Morgan's fly room, and his laboratory dominated genetics research for the next three decades. We will see the influence of Morgan and his colleagues as we consider some other important aspects of the chromosomal basis of inheritance.

Linked genes tend to be inherited together because they are located on the same chromosome

The number of genes in a cell is far greater than the number of chromosomes; in fact, each chromosome has hundreds or thousands of genes. Genes located on the same chromosome tend to be inherited together in genetic crosses because the chromosome is passed along as a unit. Such genes are said to be **linked genes**. (Note that the use of the word *linked* in this way is different from its use in the term *sex-linked.*) When

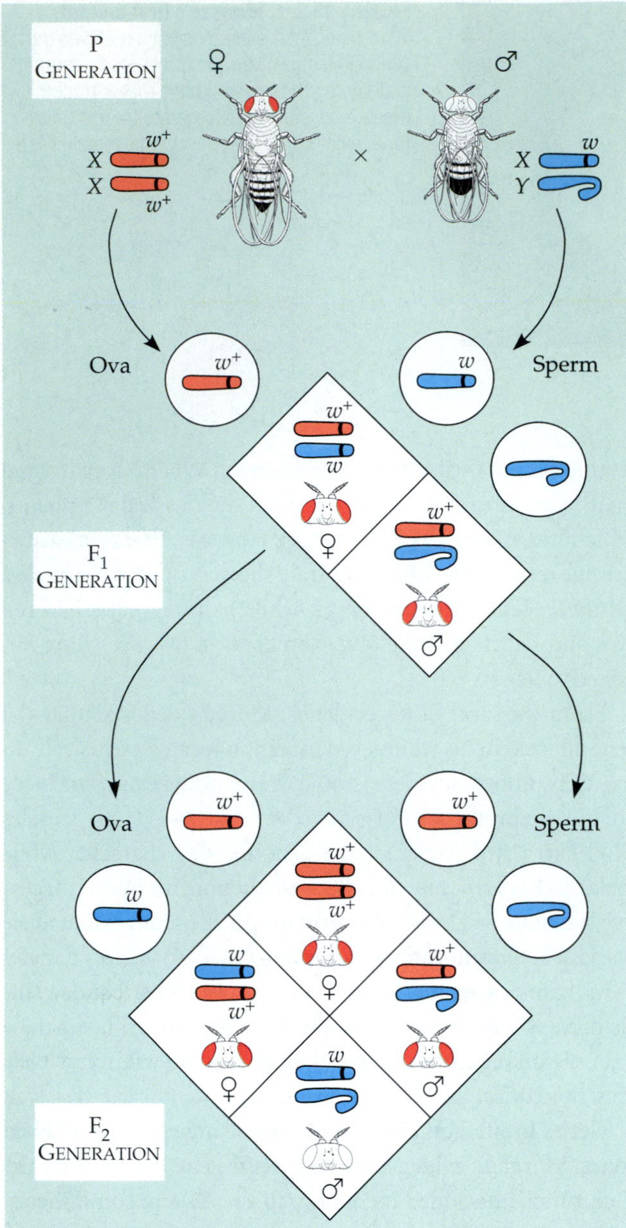

FIGURE 15.3 ▪ Sex-linked inheritance. When Morgan bred his white-eyed male to a wild-type female, all F_1 offspring had red eyes. The F_2 generation showed a typical Mendelian 3:1 ratio of traits, but the recessive trait—white eyes—was linked to sex. All females had red eyes, but half the males had white eyes. Morgan hypothesized that the gene responsible was located on the *X* chromosome (symbolized as a straight chromosome here) and that there was no corresponding locus on the *Y* chromosome (hooked, in this diagram). In this figure, the dominant allele (for red eyes) is symbolized by w^+, and the recessive allele (for white eyes) is symbolized by *w*. The symbols ♀ and ♂ stand for female and male, respectively.

geneticists follow linked genes in breeding experiments, the results deviate from those expected according to the Mendelian principle of independent assortment.

Let's examine another of Morgan's *Drosophila* experiments to see how linkage between genes affects the inheritance of two different characters. In this case, the two characters are body color and wing size. Wild-type flies have gray bodies and normal wings. Mutant phenotypes for these characters are black body and vestigial wings, which are much smaller than normal wings. The alleles for these traits are represented by the following symbols: b^+ = gray, b = black; vg^+ = normal wings, vg = vestigial wings.* (Neither gene is sex-linked; the loci are on an autosome.) Morgan crossed female dihybrids ($b^+\ b\ vg^+\ vg$) with males that had both mutant phenotypes, black bodies and vestigial wings ($b\ b\ vg\ vg$). Notice that this corresponds to a Mendelian testcross rather than to an F_1 cross between two dihybrids. According to Mendel's law of independent assortment, Morgan's *Drosophila* testcross would produce four phenotypic classes of offspring, approximately equal in number: 1 gray (color)-normal (wings) : 1 black-vestigial : 1 gray-vestigial : 1 black-normal. The actual results were very different (FIGURE 15.4). There were disproportionate numbers of wild-type (gray-normal) and double mutant (black-vestigial) flies among the offspring. Notice that these two phenotypes corresponded to the phenotypes of the two parents. Morgan reasoned that body color and wing shape are usually inherited together in a specific combination because the genes for these two characters are located on the same chromosome:

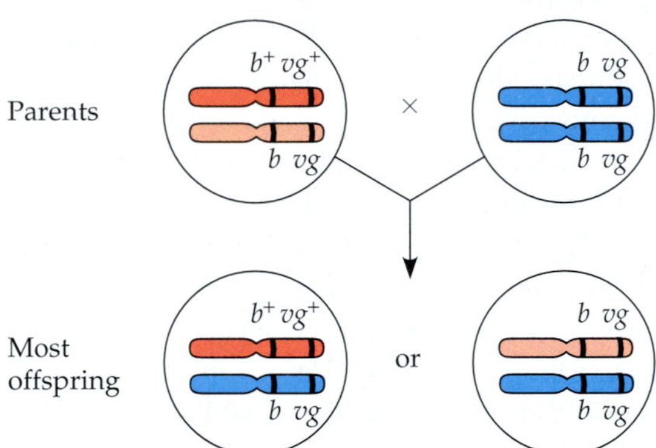

Although the other two phenotypes (gray-vestigial and black-normal) numbered fewer than expected based on independent assortment, these phenotypes *were* represented among the offspring of Morgan's cross. These new phenotypic variations resulted from crossing over, a source of genetic variation discussed in the next section.

* **A note on genetic symbols:** Morgan and his students invented a convention for symbolizing alleles that is now more widely used than Mendel's simple notation of uppercase and lowercase letters. For a given character, the gene takes its symbol from the first mutant (non-wild type) discovered. For example, the allele for white eyes in *Drosophilia* is symbolized by *w*. A superscript $^+$ identifies the allele for the wild-type trait—for red eyes, w^+. Gene nomenclature differs somewhat with the type of organism. For example, the symbols for human genes are usually written in all capital letters, as in *HD*, the Huntington's disease gene. Depending on the context, a symbol like *HD* or *w* can refer either specifically to a mutant allele or to the gene in general.

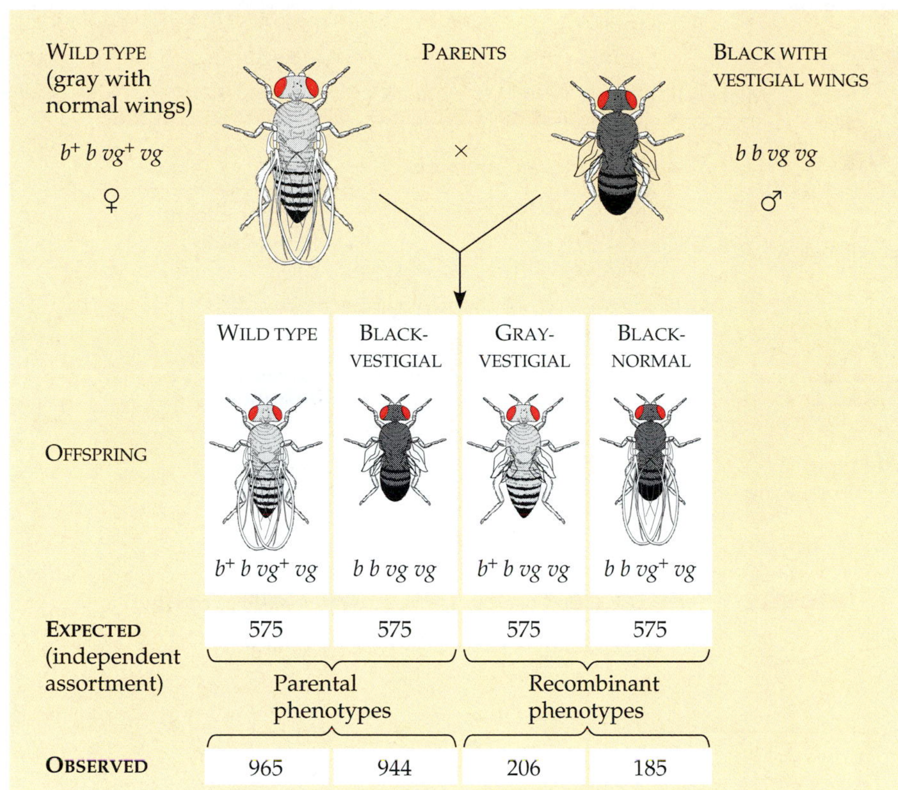

FIGURE 15.4 ▪ Evidence for linked genes in *Drosophila*. This is a testcross between flies differing in two characters, body color and wing size. The females are heterozygous for both genes, and their phenotypes are wild type (gray bodies and normal wings). The males are homozygous recessive and express the mutant phenotypes for both characters (black bodies and vestigial wings). Morgan "scored" (classified according to phenotype) 2300 offspring from such matings. The top row of numbers (Expected) represents the 1:1:1:1 ratio of phenotypes we would expect if the two genes assort independently of each other. The bottom row (Observed) shows the actual results. The parental phenotypes are disproportionately represented among offspring. Morgan concluded that genes for body color and wing size are usually transmitted together from parents to offspring because they are located on the same chromosome—that is, they are linked genes. Crossing over accounts for the recombinant phenotypes, offspring that have combinations of traits different from either parent.

Independent assortment of chromosomes and crossing over produce genetic recombinants

In Chapter 13, we saw that meiosis and random fertilization generate genetic variation among offspring of sexually reproducing organisms. The general term for the production of offspring with new combinations of traits inherited from two parents is **genetic recombination.** Here we will examine the chromosomal basis of recombination in more detail.

The Recombination of Unlinked Genes: Independent Assortment of Chromosomes

Mendel learned from his dihybrid crosses that some offspring have combinations of traits that do not match either parent. For example, draw a Punnett square for a testcross between a pea plant with yellow-round seeds that is heterozygous for both traits *(YyRr)* and a plant with green-wrinkled seeds (homozygous for both recessive traits, *yyrr*). The gene loci for these two characters are on separate chromosomes: Seed shape and seed color are unlinked, their alleles assorting independently. Notice that in your Punnett square, one-half of the offspring of the testcross inherit a phenotype that matches one of the parental phenotypes—either yellow-round seeds or wrinkled-green seeds. These offspring are called **parental types.** But other phenotypes are also represented: one-fourth of the offspring have green-round seeds, and one-fourth have yellow-wrinkled seeds. Because these offspring have different combinations of seed shape and color than either parent, they are called **recombinants.** When half of all offspring are recombinants, geneticists say that there is a 50% frequency of recombination.

A 50% frequency of recombination is observed for any two genes that are located on different chromosomes. The physical basis of recombination between unlinked genes is the random orientation of homologous chromosomes at metaphase I of meiosis, which leads to the independent assortment of alleles (see FIGURE 15.1).

The Recombination of Linked Genes: Crossing Over

Linked genes do not assort independently because they are
15.1 located on the same chromosomes and tend to move together through meiosis and fertilization. We would not expect linked genes to recombine into assortments of alleles not found in the parents. But, in fact, recombination between linked genes *does* occur. To see how, let's return to Morgan's fly room.

How can we explain the results of the *Drosophila* cross illustrated in FIGURE 15.4? The offspring of the testcross for body color and wing shape did not conform to the 1:1:1:1 phenotypic ratio we would expect if the genes for these two characters were on different chromosomes and assorted independently. But if the two genes were *completely* linked because their loci are on the same chromosome, then we should observe a 1:1 ratio, with only the parental phenotypes represented

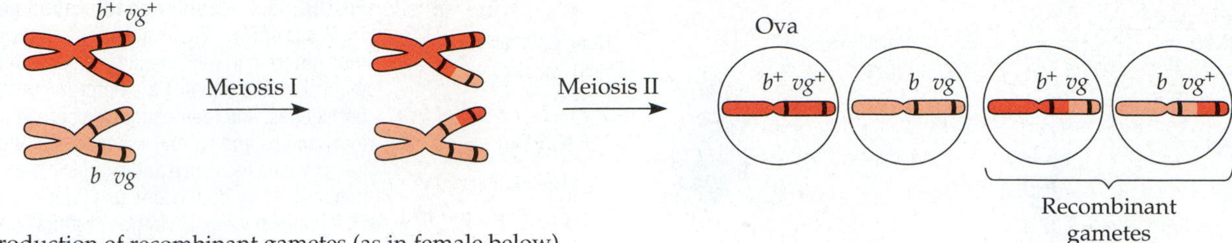

(a) Production of recombinant gametes (as in female below)

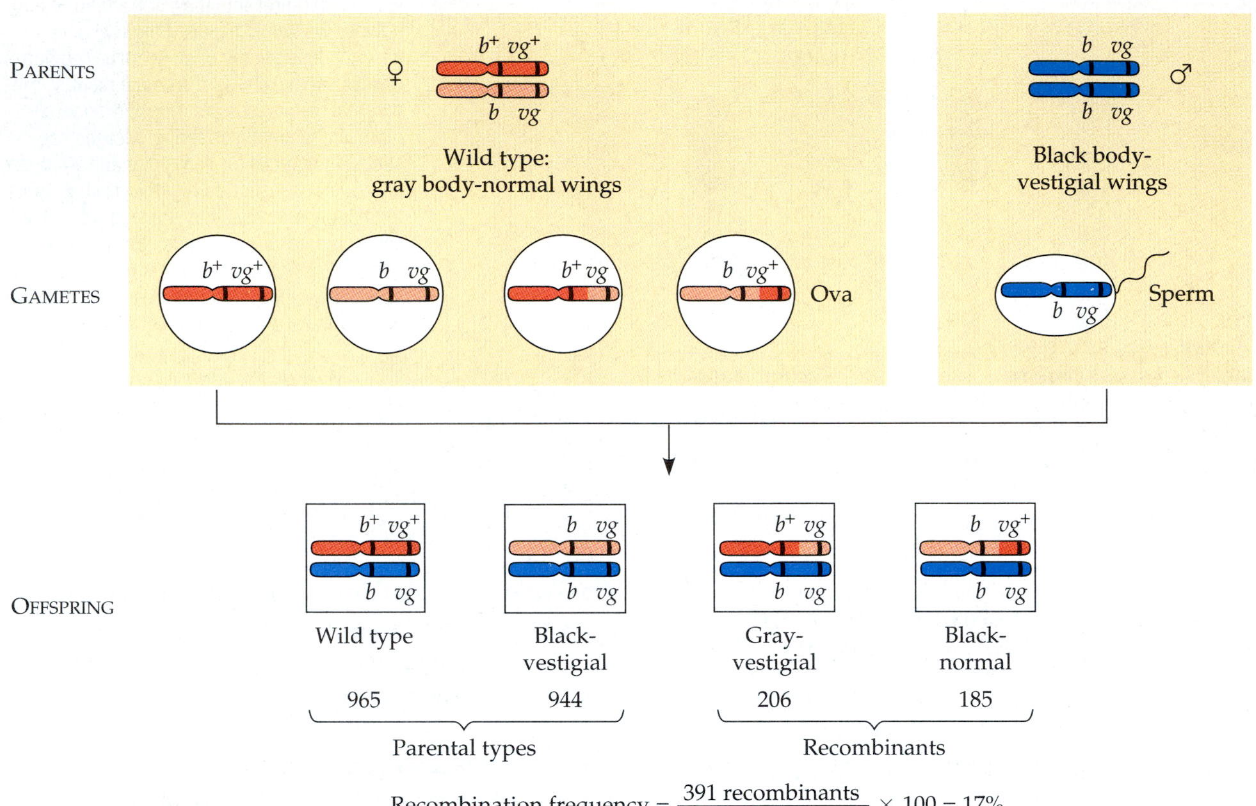

$$\text{Recombination frequency} = \frac{391 \text{ recombinants}}{2300 \text{ total offspring}} \times 100 = 17\%$$

(b) Production of recombinant offspring

15.1 **FIGURE 15.5 ▪ Recombination due to crossing over.** These diagrams recreate the testcross in FIGURE 15.4, but this time we track chromosomes as well as genes. The *b* and *vg* loci are linked; they are on the same chromosome. The maternal chromosomes are color-coded with two shades of red so that we can distinguish one homologue from the other. **(a) Production of recombinant gametes.** Crossing over occurs when homologous chromosomes pair during prophase of meiosis I. Nonsister chromatids break, and the fragments attach to the homologous chromosome. In this case, the crossover occurs between the *b* and *vg* loci. Two of the four kinds of ova resulting from meiosis have recombinant genotypes, b^+ *vg* and *b* vg^+. **(b) Production of recombinant offspring.** If we follow the recombinant chromosomes through fertilization of the ova by sperm, we see that they give rise to recombinant offspring, with genotypes and phenotypes different from either parent. To sum up, most of the offspring have parental phenotypes because the genes for body color and wing size are linked, but some have recombinant phenotypes because of crossing over. The recombination frequency is the percentage of recombinant flies in the total pool of offspring.

among offspring. The actual results satisfy neither of these expectations. Most of the offspring had parental phenotypes, suggesting linkage between the two genes, but about 17% of the flies were recombinants (FIGURE 15.5). Although there was linkage, it appeared incomplete. Morgan proposed that some mechanism that exchanges segments between homologous chromosomes must occasionally break the linkage between the two genes. Subsequent experiments have demonstrated that such an exchange—crossing over—accounts for the recombination of linked genes (see FIGURE 13.9). While homologous chromosomes are paired in synapsis during prophase of meiosis I, nonsister chromatids may break at corresponding points and switch fragments. A crossover between homologous chromosomes breaks linkages in the parental chromosomes to form recombinant chromosomes that may bring together alleles in new combinations. The subsequent events of meiosis distribute the recombinant chromosomes to gametes.

Geneticists can use recombination data to map a chromosome's genetic loci

The discovery of linked genes and recombination due to crossing over led one of Morgan's students, Alfred H. Sturtevant, to a method for constructing a **genetic map**, an ordered list of the genetic loci along a particular chromosome.

Sturtevant hypothesized that recombination frequencies calculated from experiments like the one in FIGURE 15.5 reflect the distances between genes on a chromosome. Assuming that the chance of crossing over is approximately equal at all points on a chromosome, he predicted that the farther apart two genes are, the higher the probability that a crossover will occur between them and therefore the higher the recombination frequency. His reasoning was simple: The greater the distance between two genes, the more points there are between them where crossing over can occur. With this idea in mind, Sturtevant began using recombination data from fruit fly crosses to assign to genes relative positions on chromosomes—that is, to *map* genes.

FIGURE 15.6 shows Sturtevant's map of three genes relative to each other, the body-color *(b)* and wing-size *(vg)* genes depicted in FIGURE 15.5 and a third gene on the same chromosome, called cinnabar *(cn)*. Cinnabar is one of many *Drosophila* genes affecting eye color. Cinnabar eyes, a mutant phenotype, are brighter red than the wild-type color. The recombination frequency between *cn* and *b* is 9%, and the recombination frequency between *cn* and *vg* is 9.5%. In other words, crossovers between *cn* and *b*, and between *cn* and *vg*, are only about half as frequent as crossovers between *b* and *vg* (17%). Only a map that locates *cn* about midway between *b* and *vg* is consistent with the data (as you can prove to yourself by drawing alternative maps). Sturtevant expressed the distances between genes in "map units," defining one map unit as equivalent to a 1% recombination frequency. Today, the word *centimorgan* is often used, in honor of Morgan.

Some genes on a chromosome are so far apart from each other that crossovers between them occur very often. The frequency of recombination measured between such genes can have a maximum value of 50%, a result indistinguishable from that for genes on different chromosomes. In fact, the seven characters that Mendel studied in his peas are not all on separate chromosomes, although the pea coincidentally has seven chromosome pairs. Seed color and flower color, for instance, are now known to be on chromosome 1. But they are so far apart on that chromosome that linkage is not observed in genetic crosses. Genes located far apart on a chromosome

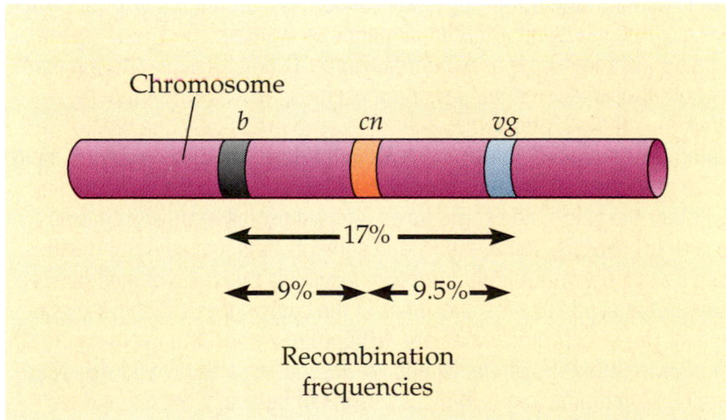

FIGURE 15.6 • Using recombination frequencies to construct a genetic map. A genetic map represents the linear sequence of genes along a chromosome. One method for constructing a genetic map uses recombination frequencies, which reflect how often crossovers occur in the region between two genes. This method is based on the assumption that the probability of a crossover between two genetic loci is proportional to the distance separating the loci. In this example, recombination frequencies have been used to map three *Drosophila* genes: *b, vg,* and *cn*. The data best fit a sequence with *cn* positioned about halfway between the other two genes. This type of genetic map is called a linkage map.

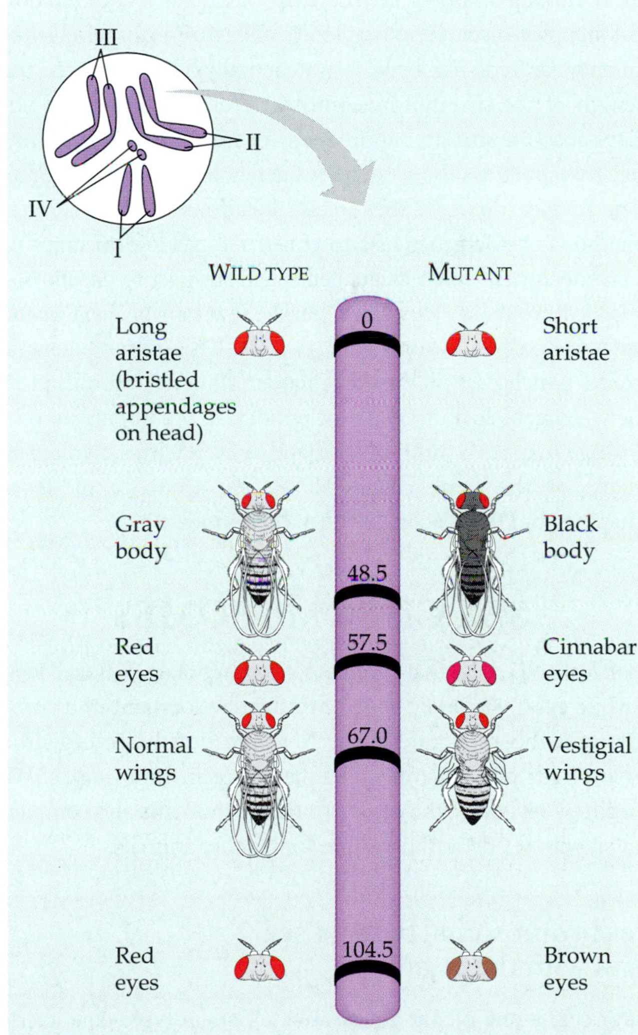

FIGURE 15.7 • A partial genetic map of a *Drosophila* chromosome. *Drosophila* has four pairs of chromosomes, and genetic cartographers have been able to map many loci along each chromosome. The simplified map shown here represents just a few of the genes that have been mapped on chromosome II. Included are the *b* (black body), *cn* (cinnabar eyes), and *vg* (vestigial wings) loci we mapped in FIGURE 15.6. (Notice that more than one gene can affect a given phenotypic characteristic, such as eye color.)

are mapped by adding the recombination frequencies from crosses involving each of the distant genes and an intermediate gene. Only for one pair of the genes Mendel studied, the genes for plant height and pod shape, do modern biologists observe linkage. Although Mendel observed segregation of alleles for each of these characters in monohybrid crosses, he did not report the results of dihybrid crosses for this particular combination of characters.

Using crossover data, Sturtevant and his colleagues were able to map the other identified *Drosophila* genes in linear arrays. They found that the genes clustered into four groups of linked genes. Because microscopists had found four pairs of chromosomes in *Drosophila* cells, this clustering of genes was additional evidence that genes are located on chromosomes. Each chromosome has a linear array of specific gene loci (FIGURE 15.7, p. 267).

A **linkage map**—a genetic map based on recombination frequencies—is not really a picture of a chromosome. The frequency of crossing over is not actually uniform over the length of the chromosome, and therefore map units do not have absolute size (in nanometers, for instance). Thus, a linkage map portrays the sequence of genes along a chromosome, but it does not give the precise locations of genes. Other methods enable geneticists to construct **cytological maps** of chromosomes, which locate genes with respect to chromosomal features, such as stained bands, that can be seen in the microscope (see FIGURE 15.13, p. 273). The ultimate genetic maps, which we'll discuss in Chapter 20, show the distances between gene loci in DNA nucleotides. In a comparison of a linkage map with this type of map (or even with a cytological map) of the same chromosome, the sequence of genes matches, but the spacing between them does not.

SEX CHROMOSOMES

You learned earlier that Morgan's discovery of a sex-linked trait (white eyes) was a key episode in the development of a chromosome theory of inheritance. In this section, we consider the role of sex chromosomes in inheritance in more detail. We begin by reviewing the genetics of sex in humans and comparing it with sex determination in some other animals.

The chromosomal basis of sex varies with the organism

Our sex is one of our more obvious phenotypic characters. Although the anatomical and physiological differences between women and men are numerous, the chromosomal basis of sex is rather simple. In humans and other mammals, as in fruit flies, there are two varieties of sex chromosomes, designated *X* and *Y*. A person who inherits two *X* chromosomes, one from each parent, usually develops as a female. A male usually develops from a zygote containing one *X* chromosome and one *Y* chromosome (FIGURE 15.8a, and see the karyotype on p. 229). When meiosis occurs in the testis, the *X* and *Y* chromosomes behave like homologous chromosomes, although they are only partially homologous and undergo very little crossing over with each other.

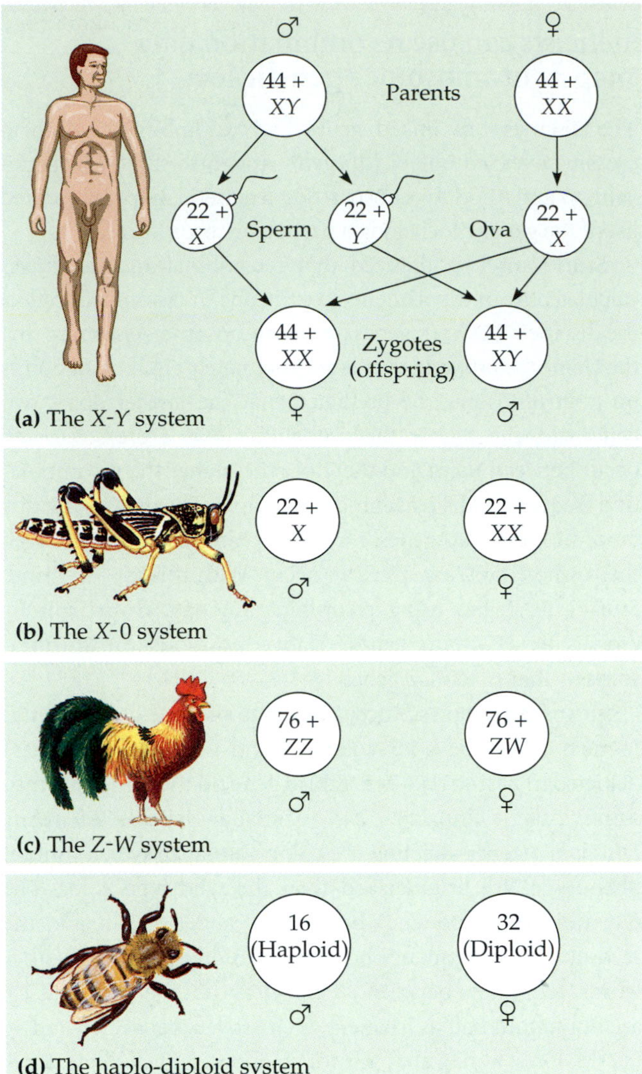

FIGURE 15.8 ▪ Some chromosomal systems of sex determination. **(a)** In humans and other mammals as well as fruit flies and some other insects, the sex of an offspring depends on whether the sperm cell carries an *X* chromosome or a *Y* chromosome. (Numerals indicate the number of autosomes, the nonsex chromosomes.) **(b)** In grasshoppers, crickets, roaches, and some other insects, there is only one type of sex chromosome, the *X*. Females are *XX;* males are *X*0 (the 0 is a zero; males have only one sex chromosome). Sex of the offspring is determined by whether the sperm cell contains an *X* chromosome or no sex chromosome. **(c)** In birds, some fishes, and some insects, including butterflies and moths, the variable that determines sex is the sex chromosome present in the ovum (not the sperm, as is the case in the *X-Y* and *X*-0 systems). The sex chromosomes are designated *Z* and *W* to avoid confusions with the *X-Y* system. Males are *ZZ* and females are *ZW.* **(d)** There are no sex chromosomes in most species of bees and ants. Females develop from fertilized ova and are thus diploid. Males develop from unfertilized eggs and are haploid; they are fatherless.

In both testes and ovaries, the two sex chromosomes segregate during meiosis, and each gamete receives one. Each ovum contains one X chromosome. In contrast, sperm fall into two categories: Half the sperm cells a male produces contain an X chromosome, and half contain a Y chromosome. We can trace the sex of each offspring to the moment of conception: If a sperm cell bearing an X chromosome happens to fertilize an ovum, the zygote is XX; if a sperm cell containing a Y chromosome fertilizes an ovum, the zygote is XY. Sex is a matter of chance—a fifty-fifty chance. FIGURE 15.8 compares this X-Y system of mammals to chromosomal systems of sex determination in other animal groups.

In humans, the anatomical signs of sex begin to emerge when the embryo is about two months old. Before then, the rudiments of the gonads are generic—they can develop into either ovaries or testes, depending on hormonal conditions within the embryo. Which of these two possibilities occurs depends on whether or not a Y chromosome is present. In 1990, a British research team identified a gene required for the development of testes. They named the gene *SRY*, for sex-determining region of Y. In the absence of *SRY*, the gonads develop into ovaries. The researchers emphasized that the presence (or absence) of *SRY* is just a trigger. The biochemical, physiological, and anatomical features of sex are complex, and many genes are involved in their development. It is likely that *SRY* codes for a protein that regulates many other genes. Recently, researchers have identified a number of additional genes on the Y chromosome that are required for normal testis functioning. In the absence of these genes, an XY individual is male but does not produce normal sperm.

Sex-linked genes have unique patterns of inheritance

In addition to their role in determining sex, the sex chromo-
15.2 somes, especially X chromosomes, have genes for many characters unrelated to sex. In humans, the term *sex-linked* usually refers to X-linked characters. These all follow the same pattern of inheritance that Morgan observed for the white-eye locus in *Drosophila*. Fathers pass X-linked alleles to all their daughters but to none of their sons (FIGURE 15.9). In contrast, mothers can pass sex-linked alleles to both sons and daughters.

If a sex-linked trait is due to a recessive allele, a female will express the phenotype only if she is a homozygote. Because males have only one locus, the terms *homozygous* and *heterozygous* lack meaning for describing their sex-linked genes (the term *hemizygous* is used in such cases). Any male receiving the recessive allele from his mother will express the trait. For this reason, far more males than females have disorders that are inherited as sex-linked recessives. However, even though the chance of a female inheriting a double dose of the mutant allele is much less than the probability of a male inheriting a single dose, there *are* females with sex-linked disorders. For instance, color blindness is a mild disorder

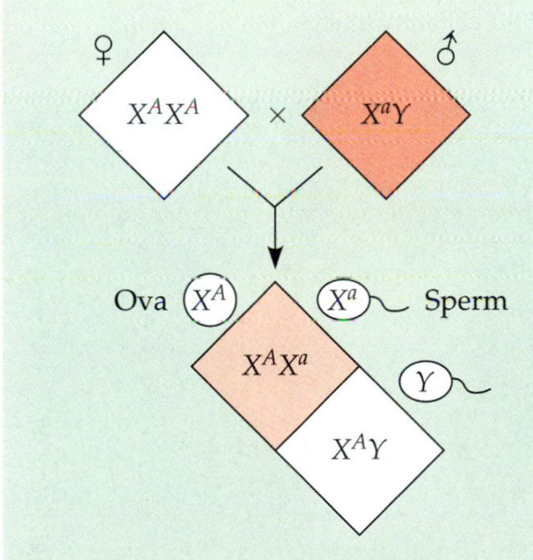

(a) A father with the trait will transmit the mutant allele to all daughters but to no sons. When the mother is a dominant homozygote, the daughters will have the normal phenotype but will be carriers of the mutation.

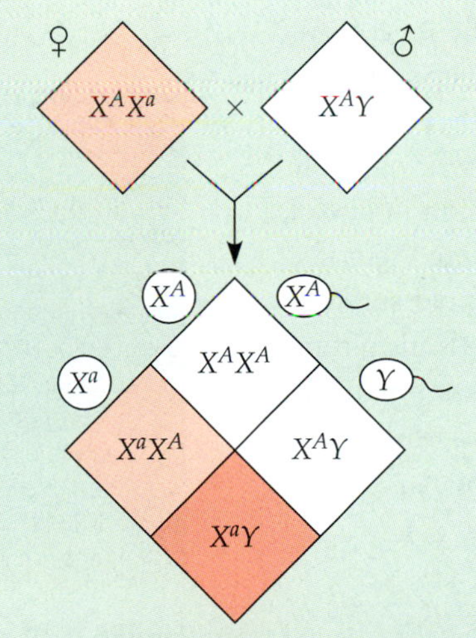

(b) A carrier who mates with a male of normal phenotype will pass the mutation to half her sons and half her daughters. The sons with the mutation will have the disorder. The daughters who have inherited the mutation in single dose will have the normal phenotype but will be carriers like their mother.

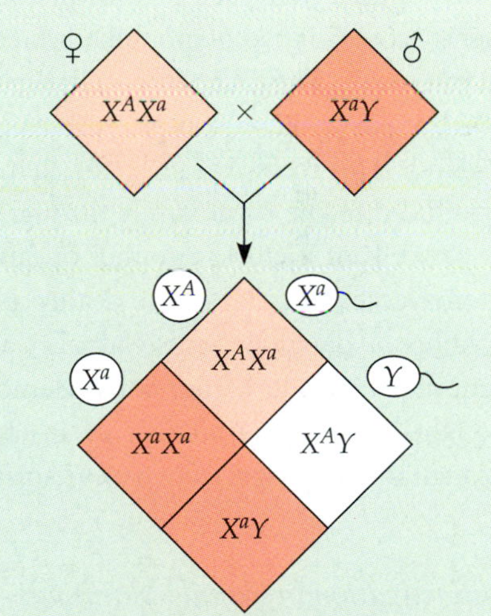

(c) If a carrier mates with a male who has the trait, there is a 50% chance that each child born to them will have the trait, regardless of sex. Daughters who do not have the trait will be carriers, whereas males without the trait will be completely free of the harmful recessive allele.

FIGURE 15.9 ▪ The transmission of sex-linked recessive traits. In this diagram, X and Y symbolize the sex chromosomes. The superscript A represents a dominant allele carried on the X chromosome, and the superscript a represents a recessive allele. Imagine that this recessive allele is a mutation that causes a sex-linked disease. White boxes indicate unaffected individuals, light-colored boxes indicate carriers, and dark-colored boxes indicate individuals with the sex-linked disorder.

inherited as a sex-linked trait. A color-blind daughter may be born to a color-blind father whose mate is a carrier (see FIGURE 15.9C). However, because the sex-linked allele for color blindness is relatively rare, the probability that such a man and woman will come together is low.

Sex-Linked Disorders in Humans

A number of human *X*-linked disorders are much more serious than color blindness. An example is **Duchenne muscular dystrophy**, which affects about one out of every 3500 males born in the United States. People with Duchenne muscular dystrophy rarely live past their early twenties. The disease is characterized by a progressive weakening of the muscles and loss of coordination. Researchers have traced the disorder to the absence of a key muscle protein called dystrophin and have tracked the gene for this protein to a specific locus on the *X* chromosome. These discoveries may eventually lead to treatments to prevent the disease from progressing.

Hemophilia is a sex-linked recessive trait defined by absence of a certain protein required for blood clotting. Hemophiliacs bleed excessively when injured; the most seriously affected individuals may bleed to death after relatively minor skin abrasions, bruises, or cuts.

Hemophilia has an interesting history. The ancient Hebrews must have had some understanding of its hereditary pattern because sons born to women having a family history of hemophilia were exempted from circumcision. A high frequency of sex-linked hemophilia has plagued the royal families of Europe. The first hemophiliac in the royal line seems to have been Leopold, a son of Queen Victoria (1819–1901) of England. The recessive allele for hemophilia was probably introduced to the royal family through a mutation in one of the sex cells of Victoria's mother or father, making Victoria a heterozygote, a carrier of the deadly allele. Leopold survived to father a daughter who was also a carrier, transmitting hemophilia to one of her sons. Hemophilia was eventually brought to the royal families of Prussia, Russia, and Spain through the marriages of two of Victoria's daughters.

X-*Inactivation in Female Mammals*

Although female mammals, including humans, inherit two *X* chromosomes, one *X* chromosome in each cell becomes almost completely inactivated during embryonic development. As a result, the cells of females and males have the same effective dose (one copy) of genes with loci on the *X* chromosome. The inactive *X* in each cell of a female condenses into a compact object, called a **Barr body**, which lies along the inside of the nuclear envelope. Most of the genes of the *X* chromosome that forms the Barr body are not expressed, although some remain active. (Barr-body chromosomes are reactivated in the ovary cells that give rise to ova.)

British geneticist Mary Lyon has demonstrated that the selection of which of the two *X*s will form the Barr body occurs randomly and independently in each of the embryonic cells present at the time of *X*-inactivation. As a consequence, females consist of a mosaic of two types of cells: those with the active *X* derived from the father and those with the active *X* derived from the mother. After an *X* chromosome is inactivated in a particular cell, all mitotic descendants of that cell have the same inactive *X*. Therefore, if the female is heterozygous for a sex-linked trait, approximately half her cells will express one allele, while the others will express the alternate allele. This mosaicism can be seen graphically in the coloration of a calico cat (FIGURE 15.10). In humans, an example is a recessive *X*-linked mutation that prevents the development of sweat glands. A woman who is heterozygous for this

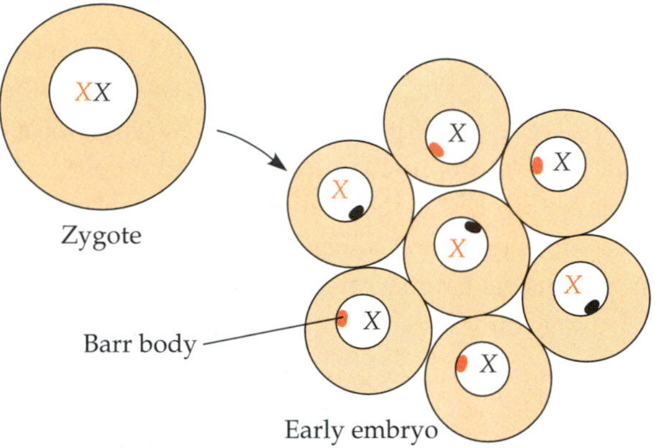

FIGURE 15.10 ▪ *X*-inactivation and the calico cat. On the *X* chromosome is a gene controlling fur color, with one allele causing black fur and another causing orange fur. (A separate gene is responsible for a patchlike pattern of colored and white fur.) A male *(XY)* can inherit one of these alleles, but not both. Thus, calicos, which have both black and orange patches, are almost always females. A female calico is heterozygous for the patch-color locus, inheriting an allele for black on one *X* chromosome and an allele for orange on the other *X*. The *X* chromosomes in the diagram are color coded according to the allele they carry. During the cat's early embryonic development, one or the other *X* chromosome is inactivated in each cell. The inactivated *X* condenses, forming a Barr body just inside the nuclear envelope. As the embryonic cells divide by mitosis, each gives rise to a clone of cells having a specific *X* chromosome active and the other inactive. The patchwork coat of the calico results from these dual cell populations.

trait has patches of normal skin and patches of skin lacking sweat glands.

Inactivation of an *X* chromosome involves attachment of methyl groups (—CH_3) to cytosine, one of the nitrogenous bases of DNA nucleotides. (The regulatory role of DNA methylation is discussed in more detail in Chapter 19.) But what determines which of the two *X* chromosomes is targeted for methylation? Researchers have discovered a gene that is active *only* on the Barr-body chromosome. The gene is called *XIST*, for *X*-inactive specific transcript. The gene's product, or "specific transcript," is an RNA molecule, and multiple copies of this molecule apparently attach to the *X* chromosome where they are made, almost covering it. Interaction of this RNA with the chromosome may initiate *X*-inactivation. This hypothesis leads to more questions. How does the *XIST* RNA initiate *X*-inactivation? And what determines which of the two *X* chromosomes in each of a female's cells will have an active *XIST* gene and become the Barr body? Our understanding of *X*-inactivation is still rudimentary.

ERRORS AND EXCEPTIONS IN CHROMOSOMAL INHERITANCE

Sex-linked traits are not the only notable deviation from the inheritance patterns observed by Mendel, and the gene mutations that create new alleles are not the only kind of changes to the genome that can affect phenotype. The last section of this chapter discusses major chromosomal errors and their consequences, as well as two types of normal inheritance that are exceptions to the standard chromosome theory.

Alterations of chromosome number or structure cause some genetic disorders

Physical and chemical disturbances, as well as errors during meiosis, can damage chromosomes in major ways or alter their number in a cell. Here, we survey these chromosomal alterations and see how this information applies to some important disorders in humans.

Alterations of Chromosome Number: Aneuploidy and Polyploidy

Ideally, the meiotic spindle distributes chromosomes to daughter cells without error. But there is an occasional mishap, called a **nondisjunction**, in which the members of a pair of homologous chromosomes do not move apart properly during meiosis I, or in which sister chromatids fail to separate during meiosis II. In these cases, one gamete receives two of the same type of chromosome and another gamete receives no copy (FIGURE 15.11). The other chromosomes are usually distributed normally. If either of the aberrant gametes unites with a normal one at fertilization, the offspring will have an abnormal chromosome number, known as **aneuploidy**. If the chromosome is present in triplicate in the fertilized egg (so that the cell has a total of $2n + 1$ chromosomes), the aneuploid cell is said to be **trisomic** for that chromosome. If a chromosome is missing (so that the cell has $2n - 1$ chromosomes), the

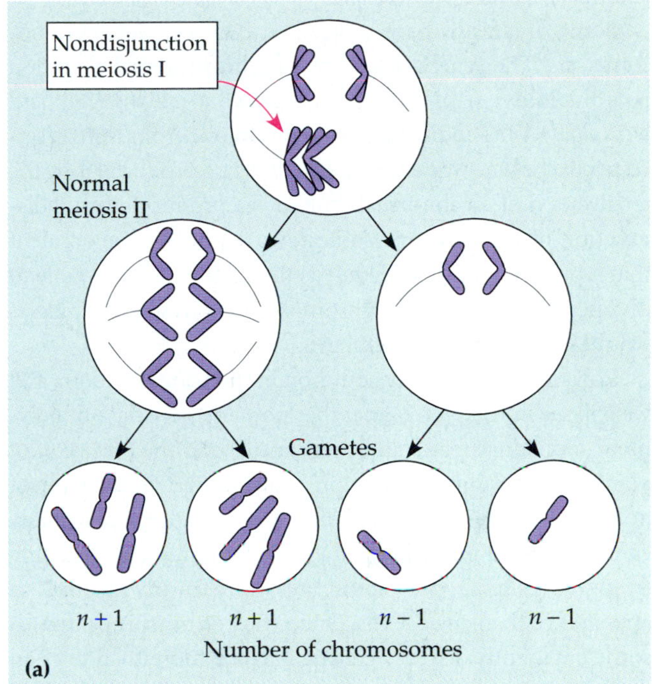

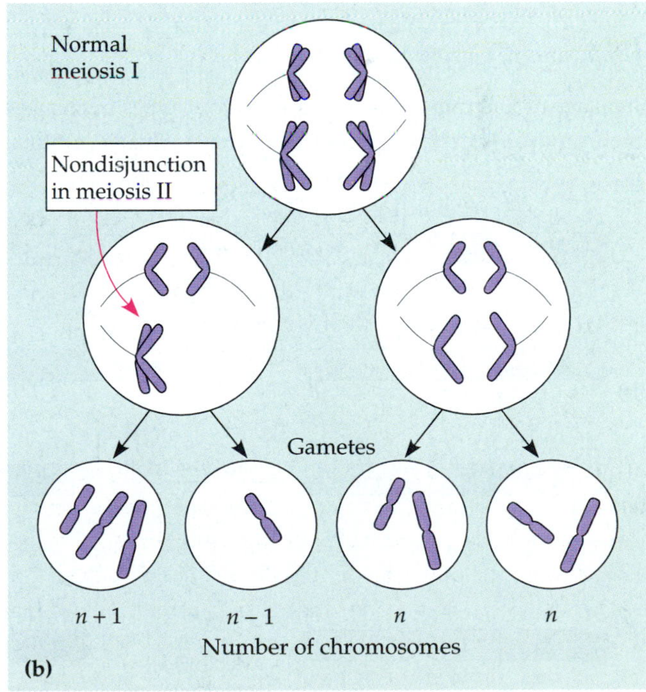

FIGURE 15.11 • Meiotic nondisjunction. (a) Homologues may fail to separate during anaphase of meiosis I, or **(b)** chromatids may fail to separate during anaphase of meiosis II. Either type of meiotic error will produce gametes with an abnormal chromosome number.

aneuploid cell is **monosomic** for that chromosome. Mitosis will subsequently transmit the anomaly to all embryonic cells. If the organism survives, it usually has a set of symptoms caused by the abnormal dose of genes located on the extra or missing chromosome. Nondisjunction can also occur during mitosis. If such an error takes place early in embryonic development, then the aneuploid condition is passed along by mitosis to a large number of cells and is likely to have a substantial effect on the organism.

Some organisms have more than two complete chromosome sets. The general term for this chromosomal alteration is **polyploidy**, with the specific terms *triploidy* ($3n$) and *tetraploidy* ($4n$) indicating three or four chromosomal sets respectively. One way a triploid cell may be produced is by the fertilization of an abnormal diploid egg produced by nondisjunction of all its chromosomes. An example of an accident that would result in tetraploidy is the failure of a $2n$ zygote to divide after replicating its chromosomes. Subsequent mitosis would then produce a $4n$ embryo.

Polyploidy is relatively common in the plant kingdom, and we will see in Chapter 24 that the spontaneous origin of polyploid individuals plays an important role in the evolution of plants. In the animal kingdom, the natural occurrence of polyploids seems to be extremely rare, although polyploidy can be induced experimentally in certain animals, including frogs. In general, polyploids are more nearly normal in appearance than aneuploids. One extra (or missing) chromosome apparently disrupts genetic balance more than does an entire extra set of chromosomes.

Alterations of Chromosome Structure

Breakage of a chromosome can lead to four types of changes in chromosome structure. A **deletion** occurs when a chromosomal fragment lacking a centromere is lost during cell division. The chromosome from which the fragment originated will then be missing certain genes. In some cases, however, the fragment may join to the homologous chromosome, producing a **duplication** there. It also may reattach to the original chromosome but in the reverse orientation, producing an **inversion**. A fourth possible result of chromosomal breakage is for the fragment to join a nonhomologous chromosome, a rearrangement called a **translocation**. FIGURE 15.12 illustrates these different types of structural alterations of chromosomes.

Deletions and duplications are especially likely to occur during meiosis. Homologous (nonsister) chromatids sometimes break and rejoin at "incorrect" places, so that one partner gives up more genes than it receives. The products of such a nonreciprocal crossover are one chromosome with a deletion and one chromosome with a duplication.

An organism that inherits a large homozygous deletion (or a single *X* chromosome with such a deletion, in a male) is usually missing a number of essential genes, a condition that is ordinarily lethal. Duplications and translocations also tend to have harmful effects. In reciprocal translocations, in which segments are exchanged between nonhomologous chromosomes, and in inversions, the balance of genes is not abnormal—all genes are present in their normal doses. Nevertheless, inversions and translocations can alter phenotype because a gene's expression can be influenced by its location among neighboring genes.

Human Disorders Due to Chromosomal Alterations

Alterations of chromosome number and structure are associated with a number of serious human disorders. When nondisjunction occurs in meiosis, the result is aneuploidy, an abnormal number of chromosomes in the gamete produced

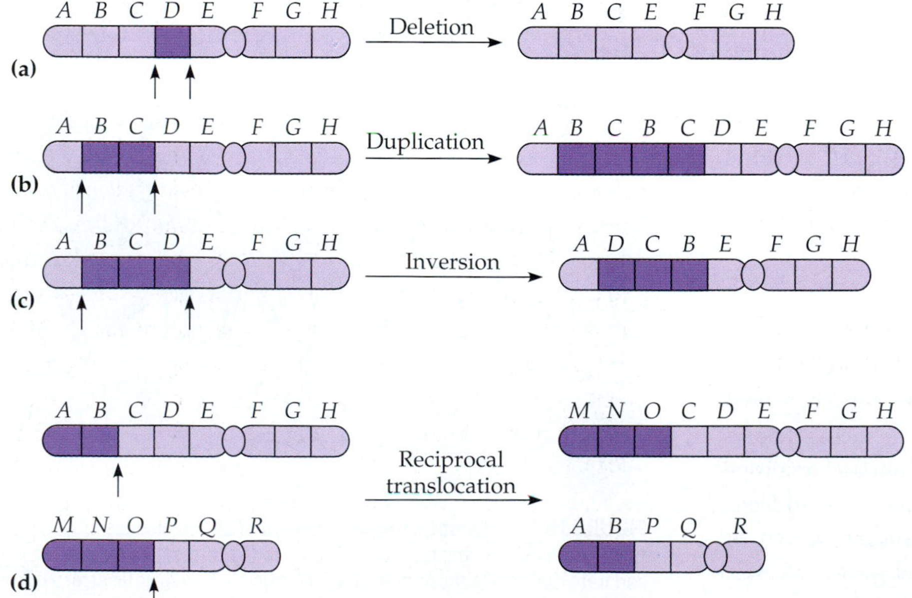

FIGURE 15.12 ▪ Alterations of chromosome structure. Vertical arrows indicate points of chromosome breakage. Dark purple highlights the genes affected by the chromosomal rearrangements. **(a)** A deletion removes a chromosomal segment. **(b)** A duplication repeats a segment. **(c)** An inversion reverses a segment within a chromosome. **(d)** A translocation moves a segment from one chromosome to another, nonhomologous one. The most common type of translocation is reciprocal, in which nonhomologous chromosomes exchange fragments. Nonreciprocal translocations, in which a chromosome transfers a fragment without receiving a fragment in return, also occur.

and, later, in the zygote. Although the frequency of aneuploid zygotes may be quite high in humans, most of these chromosomal alterations are so disastrous to development that the embryos are spontaneously (naturally) aborted long before birth. However, some types of aneuploidy appear to upset the genetic balance less than others, with the result that individuals with certain aneuploid conditions can survive to birth and beyond. These individuals have a set of symptoms—a syndrome—characteristic of the type of aneuploidy. Genetic disorders caused by aneuploidy can be diagnosed before birth by fetal testing (see Chapter 14).

One aneuploid condition, **Down syndrome**, affects approximately one out of every 700 children born in the United States. Down syndrome is usually the result of an extra chromosome 21, so that each body cell has a total of 47 chromosomes (FIGURE 15.13). In chromosomal terms, the cells are trisomic for chromosome 21. Although chromosome 21 is the smallest human chromosome, its trisomy severely alters the individual's phenotype. Down syndrome includes characteristic facial features, short stature, heart defects, susceptibility to respiratory infection, and mental retardation. Furthermore, individuals with Down syndrome are prone to developing leukemia and Alzheimer's disease. (It is probably not a coincidence that certain genes associated with the latter two diseases are located on chromosome 21.) Although people with Down syndrome, on average, have a lifespan much shorter than normal, some live to middle age or beyond. Most are sexually underdeveloped and sterile. Thus, most cases of Down syndrome result from nondisjunction during gamete production in one of the parents.

The frequency of Down syndrome correlates with the age of the mother. Down syndrome occurs in 0.04% of children born to women under age 30. The risk climbs to 1.25% for mothers in their early thirties and is even higher for older mothers. Because of this relatively high risk, pregnant women who are over 35 are candidates for fetal testing in order to check for trisomy 21 in the embryo. The correlation of Down syndrome with maternal age has not yet been explained. One hypothesis is that older women are more likely than younger women to carry Down babies to term rather than spontaneously aborting the trisomic embryos. No other chromosomal disorder is known to follow this pattern of increased incidence with maternal age.

Nondisjunction of sex chromosomes produces a variety of aneuploid conditions in humans. Most of these conditions appear to upset genetic balance less than aneuploid conditions involving autosomes. This may be because the *Y* chromosome carries relatively few genes and because extra copies of the *X* chromosome become inactivated as Barr bodies in the somatic cells.

An extra *X* chromosome in a male, producing *XXY*, occurs approximately once in every 2000 live births. People with this disorder, called Klinefelter syndrome, have male sex organs, but the testes are abnormally small and the man is sterile. The syndrome often includes breast enlargement and other feminine body characteristics. The affected individual is usually of normal intelligence. Males with an extra *Y* chromosome *(XYY)* are not characterized by any well-defined syndrome, although they tend to be somewhat taller than average. Females with trisomy *X (XXX)*, which occurs once in approximately 1000 live births, are healthy and cannot be distinguished from *XX* females except by karyotype. Monosomy *X*, called Turner syndrome, occurs about once in every 5000 births and is the only known viable monosomy in humans. Although these *X*0 individuals are phenotypically female, their sex organs do not mature at adolescence, and secondary sex characteristics fail to develop. Such individuals are sterile and of short stature. Most have normal intelligence.

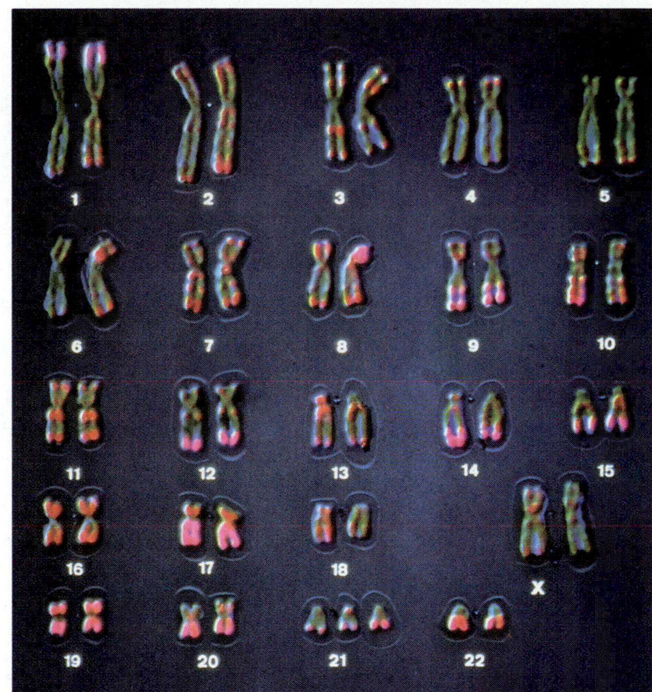

FIGURE 15.13 ▪ Down syndrome. The karyotype shows trisomy 21. The child exhibits the facial features characteristic of Down syndrome.

Even if chromosome *number* is normal, structural alterations of chromosomes can cause human disorders. Many deletions in human chromosomes (see FIGURE 15.12a), even in a heterozygous state, cause severe physical and mental problems. One such syndrome is known as *cri du chat* ("cry of the cat"). A child born with this specific deletion in chromosome 5 is mentally retarded, has a small head with unusual facial features, and has a cry that sounds like the mewing of a distressed cat. Such individuals usually die in infancy or early childhood.

Another type of chromosomal structural alteration associated with human disorders is chromosomal translocation, the attachment of a fragment from one chromosome to another, nonhomologous chromosome (see FIGURE 15.12d). Chromosomal translocations have been implicated in certain cancers. One example is chronic myelogenous leukemia (CML). Leukemia is a cancer affecting the cells that give rise to white blood cells, and in the cancerous cells of CML patients a reciprocal translocation has occurred. A portion of chromosome 22 has switched places with a small fragment from a tip of chromosome 9. (How such a switch might cause cancer will be discussed in Chapter 19.)

A small fraction of individuals with Down syndrome have a chromosomal translocation of a different sort. All the cells of such people have the normal number of chromosomes, 46. Close inspection of the karyotype, however, shows the presence of part or all of a third chromosome 21 attached to another chromosome by translocation.

The phenotypic effects of some genes depend on whether they were inherited from the mother or the father (imprinting)

Throughout our discussions of Mendelian genetics and the chromosomal basis of inheritance, we have assumed that a specific allele will have the same effect regardless of whether it was inherited from the mother or the father. This is probably a safe assumption most of the time. For example, when Mendel crossed purple-flowered peas with white-flowered peas, he observed the same results regardless of whether the purple-flowered parent supplied the ova or pollen. In recent years, however, geneticists have identified some traits, including some inherited disorders in humans, that depend on which parent passed along the alleles for those traits.

Consider the two disorders called Prader-Willi syndrome and Angelman syndrome. The symptoms are different. Prader-Willi syndrome is characterized by mental retardation, obesity, short stature, and unusually small hands and feet. People with Angelman syndrome exhibit spontaneous (uncontrollable) laughter, jerky movements, and other motor and mental symptoms. For both disorders, the genetic cause seems to be the same: deletion of a particular segment of chromosome 15. If a child inherits the abnormal chromosome from the father, the result is Prader-Willi syndrome. If the abnormal chromosome is inherited from the mother, the result is Angelman syndrome. It appears that the genes of the deleted region normally behave differently in offspring, depending on whether they belong to the maternal or the paternal chromosome.

A process called **genomic imprinting** can explain the Prader-Willi/Angelman enigma and some similar phenomena. In mammals, certain genes are imprinted in some way in each generation, with the imprinting status of a given gene depending on whether the gene resides in a female or in a male (FIGURE 15.14). In other words, the same alleles may have different effects on offspring depending on whether they arrive in the zygote via the ovum or via the sperm. In the new generation, both maternal and paternal imprints are apparently "erased" in gamete-producing cells, and all the chromosomes are re-imprinted according to the sex of the individual in which they now reside. There is much evidence that a parental imprint consists of methyl ($—CH_3$) groups that are added to nucleotides at specific loci on chromosomes and that genes with many methyl groups are inactivated (see Chapter 19). Thus, in most cases, the animal probably uses the allele that is *not* imprinted.

Researchers have so far identified about 20 mammalian genes subject to imprinting, and there may be a couple of hundred more. Most of the known imprinted genes are critical for embryonic development. In experiments with mice, embryos engineered to inherit both copies of certain chromosomes from the same parent inevitably die before birth, whether that parent was male or female. Normal development apparently requires that certain genes have exactly one active copy—not zero or two.

In humans, in addition to the Prader-Willi/Angelman case, genomic imprinting may help explain the inheritance pattern of other disorders, including one called **fragile *X* syndrome**. This disorder is named for the physical appearance of an abnormal *X* chromosome, the tip of which hangs on to the rest of the chromosome by a thin thread of DNA. Children with fragile *X* syndrome—about one in every 1500 males and about one in every 2500 females—are mentally retarded. Of all forms of mental retardation with a genetic basis, fragile *X* is the most common.

Inheritance of fragile *X* has a complex pattern, but the syndrome is more common when the abnormal chromosome is inherited from the mother rather than the father. This is consistent with the disorder being more common in males: If a male *(XY)* inherits a fragile *X* chromosome, it *has* to be from his mother. Fragile *X* is unusual in that maternal imprinting (methylation) of the abnormal allele, rather than "silencing" it, seems somehow to cause the syndrome. An unusual molecular characteristic of this abnormal allele will be described in Chapter 19.

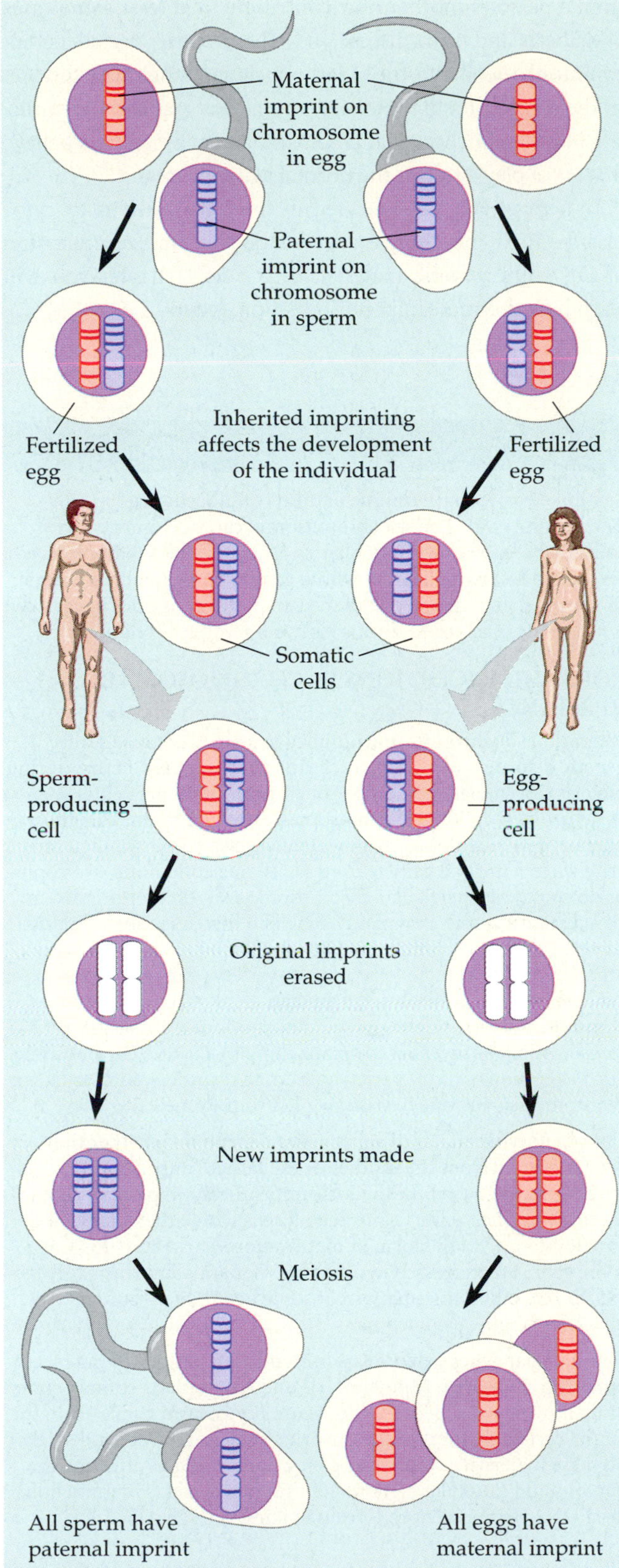

FIGURE 15.14 ▪ Genomic imprinting. The sperm and ova of mammals may convey chromosomes that are differently imprinted. The phenotypic effect of a particular allele in an offspring may therefore depend on whether the allele came from the mother or father. In each generation the old imprints are "erased" when sperm or ova are produced, and all the chromosomes are newly imprinted according to the sex of the individual.

Extranuclear genes exhibit a non-Mendelian pattern of inheritance

Although our focus in this chapter has been on the chromosomal basis of inheritance, we end with an important amendment: Not all of a eukaryotic cell's genes are located on nuclear chromosomes, or even in the nucleus. Extranuclear genes are found on small circles of DNA in mitochondria and in plants' plastids, including chloroplasts. Both mitochondria and plastids reproduce themselves and transmit their genes to daughter organelles. These cytoplasmic genes do not display Mendelian inheritance because they are not distributed to offspring according to the same rules that direct the distribution of nuclear chromosomes during meiosis.

Cytoplasmic genes were first observed in plants. In 1909, Karl Correns studied the inheritance of yellow or white patches on the leaves of an otherwise green plant. He found that the coloration of the offspring was determined only by the maternal parent (the source of seeds that germinate to give rise to the offspring) and not by the paternal parent (the pollen source). Subsequent research has shown that such coloration patterns are due to genes in the plastids that control pigmentation. In most plants, a zygote receives all its plastids from the cytoplasm of the ovum and none from pollen. Thus, as the zygote of these plants develops, its pattern of leaf coloration depends only on maternal plastid genes (FIGURE 15.15).

Maternal inheritance is also the rule for the mitochondrial genes in mammals, because the mitochondria passed on by the zygote all come from the cytoplasm of the ovum. In recent years, scientists have learned that mutations in mitochondrial DNA cause a number of rare human disorders. The products of most mitochondrial genes help make up the protein complexes

FIGURE 15.15 ▪ Cytoplasmic inheritance in tomato leaves. Variegated (striped or spotted) leaves result from genes located in the plastids rather than on the nuclear chromosomes of plant cells. Because only the ovum contributes plastids to a plant zygote, all plastid genes are inherited from the maternal parent.

of the electron transport chain and ATP synthase (see Chapter 9). Defects in one or more of these proteins therefore reduce the amount of ATP the cell can make. The parts of the body most susceptible to energy deprivation are the nervous system and the muscles, and most mitochondrial diseases affect these systems. For example, a person with the disease called mitochondrial myopathy suffers weakness, intolerance of exercise, and muscle deterioration.

In addition to the rare diseases clearly caused by defects in mitochondrial DNA, mitochondrial mutations inherited from a person's mother may contribute to at least some cases of diabetes and heart disease, as well as to other disorders that commonly debilitate the elderly, such as Alzheimer's disease. In the course of a lifetime, new mutations gradually accumulate in our mitochondrial DNA, and some researchers believe that these play a role in the normal aging process.

Wherever genes are located in the cell—nucleus or cytoplasm—their inheritance depends on the precise replication of DNA, the genetic material. In the next chapter you will learn how this molecular reproduction occurs.

CHAPTER REVIEW

REVIEW OF KEY CONCEPTS

(with page numbers and key figures)

RELATING MENDELISM TO CHROMOSOMES

- **Mendelian inheritance has its physical basis in the behavior of chromosomes during sexual life cycles** (p. 261, FIGURE 15.1, p. 262) In the early 1900s, geneticists showed that chromosomal movements in meiosis account for Mendel's laws.
- **Morgan traced a gene to a specific chromosome** (pp. 261–263, FIGURE 15.3, p. 264) His discovery that the *X* chromosome in *Drosophila* carries a gene for eye color supported the chromosome theory of inheritance.
- **Linked genes tend to be inherited together because they are located on the same chromosome** (pp. 263–265, FIGURE 15.4) Each chromosome has hundreds or thousands of genes. Linked genes do not assort independently.
- 15.1 **Independent assortment of chromosomes and crossing over produce genetic recombinants** (pp. 265–266, FIGURE 15.5) Recombinant offspring, which exhibit new combinations of traits inherited from two parents, result from events of meiosis and random fertilization. These events include crossing over and independent assortment of chromosomes during the first meiotic division. A recombination frequency under 50% indicates that the genes are linked but that crossing over has occurred. In this process, homologous chromosomes in synapsis during prophase of meiosis I break at corresponding points and switch fragments, creating new combinations of alleles that are then passed on to the gametes.
- **Geneticists can use recombination data to map a chromosome's genetic loci** (pp. 267–268, FIGURE 15.6) One way to map genes is to deduce their order and a rough indication of the relative distances between them from crossover data. The farther apart genes are on a chromosome, the more likely they are to be separated during crossing over. Cytological mapping is a technique that pinpoints the physical locus of a gene by associating a mutant phenotype with a chromosomal defect seen in the microscope.

SEX CHROMOSOMES

- **The chromosomal basis of sex varies with the organism** (pp. 268–269, FIGURE 15.8) Sex is an inherited phenotypic character usually determined by the presence or absence of special chromosomes; the exact mechanism varies among different species. Humans and other mammals, like fruit flies, have an *X-Y* system. An *XY* male gives either an *X* chromosome or a *Y* chromosome to the sperm, which combines with an ovum containing an *X* chromosome from an *XX* female. The offspring's sex is determined at conception by whether the sperm carries *X* or *Y*.
- **Sex-linked genes have unique patterns of inheritance** (pp. 269–271, FIGURE 15.9) The sex chromosomes carry certain genes for traits that are unrelated to maleness or femaleness. Hemophilia is a sex-linked recessive disorder whose gene is on the *X* chromosome. In mammalian females, one of the two *X* chromosomes in each cell is randomly inactivated during early embryonic development.

ERRORS AND EXCEPTIONS IN CHROMOSOMAL INHERITANCE

- **Alterations of chromosome number or structure cause some genetic disorders** (pp. 271–274, FIGURES 15.11, 15.12) Errors during meiosis can change the number of chromosomes per cell or the structure of individual chromosomes. Such alterations can affect phenotype. Aneuploidy, an abnormal chromosome number, can arise when a normal gamete unites with one containing two copies or no copies of a particular chromosome as a result of nondisjunction during meiosis. Polyploidy, in which there are more than two complete sets of chromosomes, can result from complete nondisjunction during gamete formation. A variety of rearrangements can result from chromosome breakage. A lost fragment leaves one chromosome with a deletion and may produce a duplication, translocation, or inversion by reattaching to another chromosome. Such alterations cause a variety of human disorders, such as Down syndrome, usually due to trisomy of chromosome 21.
- **The phenotypic effects of some genes depend on whether they were inherited from the mother or the father (imprinting)** (p. 274, FIGURE 15.14) Individuals imprint certain parts of chromosomes in their gamete-producing cells with either a male or a female "stamp," in the form of methylation. This affects the way some genes are expressed in offspring. Genomic imprinting helps explain the inheritance pattern of some hereditary disorders, including fragile *X* syndrome.
- **Extranuclear genes exhibit a non-Mendelian pattern of inheritance** (pp. 275–276) Mitochondria and chloroplasts contain some of their own genes. Because the zygote's cytoplasm comes from the ovum, certain features of the offspring's phenotype depend solely on these maternal cytoplasmic genes. Some diseases affecting the nervous and muscular systems are caused by defects in mitochondrial DNA that prevent cells from making enough ATP.

GENETICS PROBLEMS

1. A man with hemophilia (a recessive, sex-linked condition) has a daughter of normal phenotype. She marries a man who is normal for the trait. What is the probability that a daughter of this mating will be a hemophiliac? That a son will be a hemophiliac? If the couple has four sons, what is the probability that all four will be born with hemophilia?

2. Pseudohypertrophic muscular dystrophy is a disorder that causes gradual deterioration of the muscles. It is seen only in boys born to apparently normal parents and usually results in death in the early teens. Is this disorder caused by a dominant or a recessive allele? Is its inheritance sex-linked or autosomal? How do you know? Explain why this disorder is seen only in boys and never in girls.
3. Red-green color blindness is caused by a sex-linked recessive allele. A color-blind man marries a woman with normal vision whose father was color-blind. What is the probability that they will have a color-blind daughter? What is the probability that their first son will be color-blind? (*Note:* The two questions are worded a bit differently.)
4. A wild-type fruit fly (heterozygous for gray body color and normal wings) was mated with a black fly with vestigial wings. The offspring had the following phenotypic distribution: wild type, 778; black-vestigial, 785; black-normal, 158; gray-vestigial, 162. What is the recombination frequency between these genes for body color and wing type?
5. What pattern of inheritance would lead a geneticist to suspect that an inherited disorder of cell metabolism is due to a defective mitochondrial gene?
6. An aneuploid person is obviously female, but her cells have two Barr bodies. What is the probable complement of sex chromosomes in this individual?
7. Determine the sequence of genes along a chromosome based on the following recombination frequencies: *A–B*, 8 map units; *A–C*, 28 map units; *A–D*, 25 map units; *B–C*, 20 map units; *B–D*, 33 map units.
8. About 5% of individuals with Down syndrome are the result of chromosomal translocation in which one copy of chromosome 21 becomes attached to chromosome 14. How does this translocation lead to Down syndrome?
9. More common than completely polyploid animals are mosaic polyploids, animals that are diploid except for patches of polyploid cells. How might a mosaic tetraploid—an animal with some cells containing four sets of chromosomes—arise from an error in *mitosis*?
10. Assume genes *A* and *B* are linked and are 50 map units apart. An animal heterozygous at both loci is crossed with one that is homozygous recessive at both loci. What percentage of the offspring will show phenotypes resulting from crossovers? If you did not know genes *A* and *B* were linked, how would you interpret the results of this cross?
11. In *Drosophila*, the gene for white eyes and the gene that produces "hairy" wings have both been mapped to the same chromosome and have a crossover frequency of 1.5%. A geneticist noticed that in a particular stock of flies, these two genes assorted independently; that is, they behaved as though they were on different chromosomes. What explanation can you offer for this observation?
12. In another cross, a wild-type fruit fly (heterozygous for gray body color and red eyes) was mated with a black fruit fly with purple eyes. The offspring were as follows: wild type, 721; black-purple, 751; gray-purple, 49; black-red, 45. What is the recombination frequency between these genes for body color and eye color? Using information from problem 4, what fruit flies (genotypes and phenotypes) would you mate to determine the sequence of the body-color, wing-shape, and eye-color genes on the chromosome?
13. A space probe discovers a planet inhabited by creatures who reproduce with the same hereditary patterns as those in humans. Three phenotypic characters are height (T = tall, t = dwarf), head appendages (A = antennae, a = no antennae), and nose morphology (S = upturned snout, s = downturned snout). Since the creatures were not "intelligent," Earth scientists were able to do some controlled breeding experiments, using various heterozygotes in testcrosses. For a tall heterozygote with antennae, the offspring were: tall-antennae, 46; dwarf-antennae, 7; dwarf-no antennae, 42; tall-no antennae, 5. For a heterozygote with antennae and an upturned snout, the offspring were: antennae-upturned snout, 47; antennae-downturned snout, 2; no antennae-downturned snout, 48; no antennae-upturned snout, 3. Calculate the recombination frequencies for both experiments.
14. Using the information from problem 13, a further testcross was done using a heterozygote for height and nose morphology. The offspring were: tall-upturned nose, 40; dwarf-upturned nose, 9; dwarf-downturned nose, 42; tall-downturned nose, 9. Calculate the recombination frequency from these data; then use your answer from problem 13 to determine the correct sequence of the three linked genes.

SCIENCE, TECHNOLOGY, AND SOCIETY

Opinions differ about whether children with learning disorders should be tested by karyotyping for the presence of a fragile *X* chromosome. Some argue that it's always better to know the cause of the problem so that education specialized for that disorder can be prescribed. Others counter that attaching a specific biological cause to a learning disability stigmatizes a child and limits his or her opportunities. What is your evaluation of these arguments?

FURTHER READING

Chess, A. "Expansion of the Allelic Exclusion Principle?" *Science*, March 27, 1998. In addition to imprinted genes and genes on inactivated *X* chromosomes, there are other kinds of genes for which only one allele is expressed.
Mestel, R. "The Genetic Battle of the Sexes" *Natural History*, February 1998. Provides a possible evolutionary explanation for genomic imprinting.
Miller, K. R. "Whither the *Y*?" *Discover*, February 1995. How did sex chromosomes evolve?
Patterson, D. "The Causes of Down Syndrome." *Scientific American*, August 1987. Explains the identification and mapping of genes on chromosome 21.
Sapienza, C. "Parental Imprinting of Genes." *Scientific American*, October 1990.
Travis, J. "A Gene that Silences the *X* Chromosome." *Science News*, March 22, 1997. Discusses possible roles of the *XIST* gene.
Wallace, D. C. "Mitochondrial DNA in Aging and Disease." *Scientific American*, August 1997.

WEB LINKS

Visit the special edition of *The Biology Place* for BIOLOGY, Fifth Edition, at http://www.biology.com/campbell. Go to Chapter 15 for online resources, including learning activities, practice exams, and links to the following web sites:

"The Virtual FlyLab"
The Virtual FlyLab allows you to play the role of a research geneticist by carrying out genetic crosses using "virtual" fruit flies.

"Fraxa Research Foundation"
This organization supports research into, and education about, fragile X syndrome.

"National Down Syndrome Society (NDSS)"
The NDSS provides comprehensive information about Down syndrome.

"Laboratory of John W. Sedat"
Visit the home pages of this research group at the University of California, San Francisco, who have developed several new methods of studying chromosomes using 3-D microscopy.

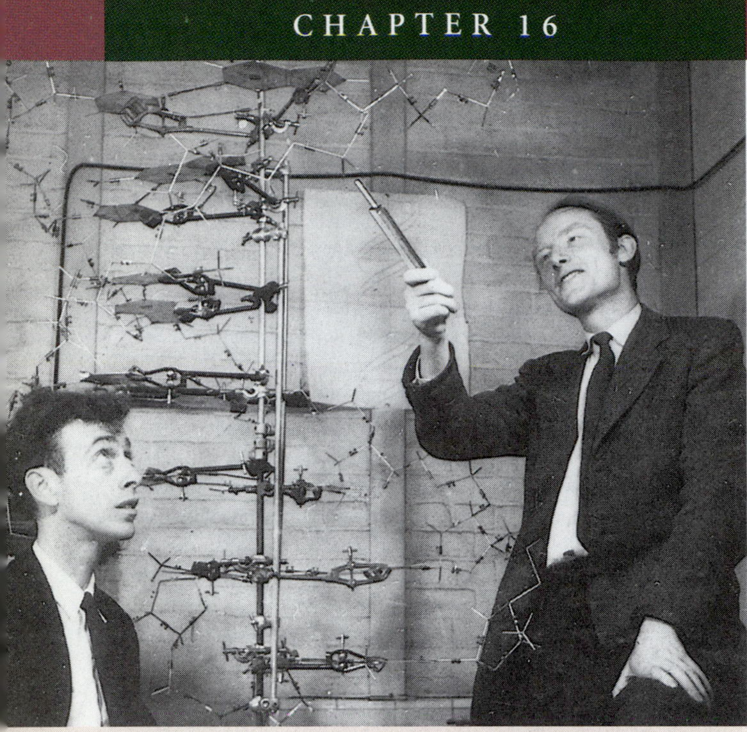

CHAPTER 16

THE MOLECULAR BASIS OF INHERITANCE

DNA as the Genetic Material

- The search for the genetic material led to DNA: *science as a process*
- Watson and Crick discovered the double helix by building models to conform to X-ray data: *science as a process*

DNA Replication and Repair

- During DNA replication, base pairing enables existing DNA strands to serve as templates for new complementary strands
- A large team of enzymes and other proteins carries out DNA replication
- Enzymes proofread DNA during its replication and repair damage in existing DNA
- The ends of DNA molecules pose a special problem

In 1953, James Watson and Francis Crick won a race to discover the three-dimensional molecular structure of deoxyribonucleic acid, or DNA. The adjacent photograph captures Watson and Crick admiring their DNA model. DNA is the most celebrated molecule of our time, for it is the substance of inheritance. Mendel's heritable factors and Morgan's genes on chromosomes are, in fact, composed of DNA. Chemically speaking, your genetic endowment is the DNA you inherited from your parents.

Of all nature's molecules, nucleic acids are unique in their ability to direct their own replication. Indeed, the resemblance of offspring to their parents has its basis in the precise replication of DNA and its transmission from one generation to the next. In other words, DNA is the substance behind the adage "Like begets like." Hereditary information is encoded in the chemical language of DNA and reproduced in all the cells of your body. It is this DNA program that directs the development of your biochemical, anatomical, physiological, and, to some extent, behavioral traits. In this chapter you will learn how biologists deduced that DNA is the genetic material, how Watson and Crick discovered its structure, and how cells replicate and repair their DNA—the molecular basis of inheritance.

DNA AS THE GENETIC MATERIAL

Today, even schoolchildren have heard of DNA, and scientists routinely manipulate DNA in the laboratory and use it to change the heritable characteristics of cells. Earlier in the twentieth century, however, the identification of the molecules of inheritance loomed as a major challenge to biologists.

The search for the genetic material led to DNA: *science as a process*

Once T. H. Morgan's group showed that genes are located on chromosomes, the two chemical components of chromosomes—DNA and protein—became the candidates for the genetic material. Until the 1940s, the case for proteins seemed stronger, especially since biochemists had identified them as a class of macromolecules with great heterogeneity and specificity of function, essential requirements for the hereditary material. Moreover, little was known about nucleic acids, whose physical and chemical properties seemed far too uniform to account for the multitude of specific inherited traits exhibited by every organism. This view gradually changed, as experiments with microorganisms yielded unexpected results. As with the work of Mendel and Morgan, a key factor in determining the identity of the genetic material was the choice of appropriate experimental organisms. Bacteria and the viruses that infect them are far simpler than pea plants, fruit flies, or humans, and the role of DNA in heredity was first worked out by studying such microbes. In this section we will trace the search for the genetic material in some detail, as a case study of the scientific process.

Evidence That DNA Can Transform Bacteria

We can trace the discovery of the genetic role of DNA back to 1928. Frederick Griffith, a British medical officer, was studying *Streptococcus pneumoniae,* a bacterium that causes pneumonia in mammals. Griffith had two strains (varieties) of the bacterium, a pathogenic (disease-causing) one and a variant that was harmless. He found that when he killed the pathogenic bacteria with heat and then mixed the cell remains with living bacteria of the harmless strain, some of the living cells were converted to the pathogenic form (FIGURE 16.1). Furthermore, this new trait of pathogenicity was inherited by all the descendants of the transformed bacteria. Clearly, some chemical component of the dead pathogenic cells caused this heritable change, although the identity of the substance was not known. Griffith called the phenomenon **transformation,** now defined as a change in genotype and phenotype due to the assimilation of external DNA by a cell. (This usage of *transformation* should not be confused with the conversion of a normal animal cell to a cancerous one, discussed in Chapter 12.)

Griffith's work set the stage for a 14-year search for the identity of the transforming substance by American bacteriologist Oswald Avery. He purified various chemicals from the heat-killed pathogenic bacteria, then tried to transform live nonpathogenic bacteria with each chemical. Only DNA worked. Finally, in 1944, Avery and his colleagues Maclyn McCarty and Colin MacLeod announced that the transforming agent was DNA. Their discovery was greeted with considerable skepticism, in part because of the lingering belief that proteins were better candidates for the genetic material. Moreover, many biologists were not convinced that the genes of bacteria would be similar in composition and function to those of more complex organisms. But the major reason for the continued doubt was that so little was known about DNA. No one could imagine how DNA could carry genetic information.

Evidence That Viral DNA Can Program Cells

Additional evidence for DNA as the genetic material came from studies of a virus that infects bacteria. Viruses are much simpler than cells. A virus is little more than DNA (or sometimes RNA) enclosed by a protective coat of protein. To reproduce, a virus must infect a cell and take over the cell's metabolic machinery.

Viruses that infect bacteria are widely used as research tools in molecular genetics. These viruses are called **bacteriophages** (meaning "bacteria-eaters"), or just **phages.** In 1952, Alfred Hershey and Martha Chase discovered that DNA is the genetic material of a phage known as T2. This is one of many phages that infect the bacterium *Escherichia coli (E. coli),* which normally lives in the intestines of mammals. At that time, biologists already knew that T2, like many other viruses, was composed almost entirely of DNA and protein. They also knew that the phage could quickly turn an *E. coli* cell into a T2-producing factory that released phages when the cell ruptured. Somehow, T2 could reprogram its host cell to produce viruses, but which viral component—protein or DNA—was responsible?

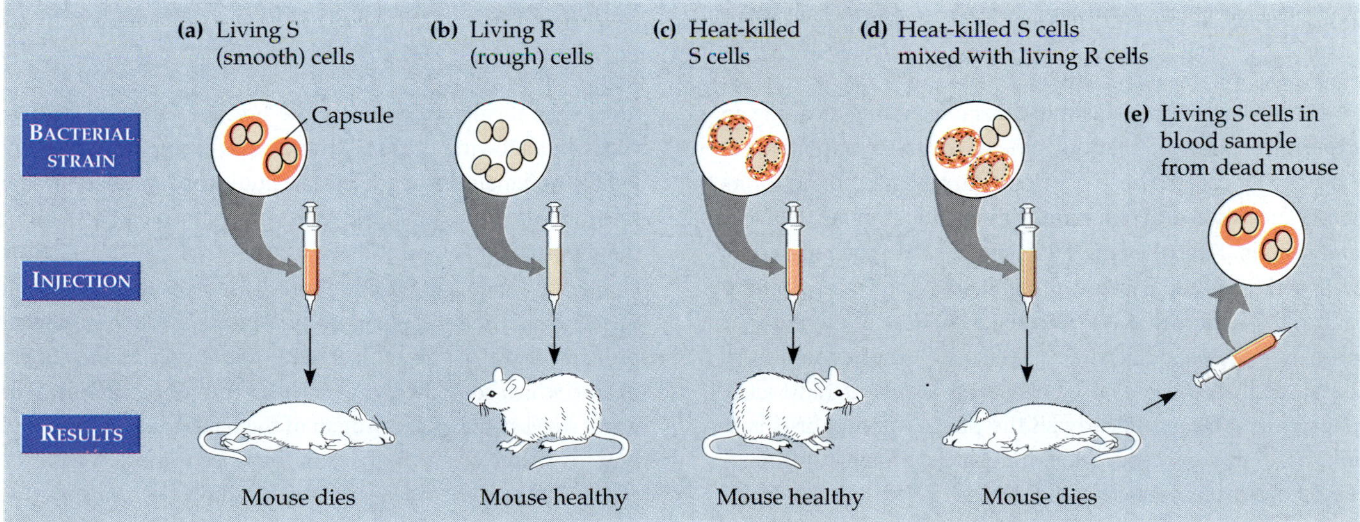

FIGURE 16.1 • Transformation of bacteria. Griffith discovered that **(a)** the S strain of the bacterium *Streptococcus pneumoniae,* which was protected from a mouse's defensive system by a capsule, was pathogenic; **(b)** the R strain, a mutant lacking the capsule, was nonpathogenic; **(c)** heat-killed S cells were harmless; but **(d)** a mixture of heat-killed S cells and live R cells caused pneumonia and death. **(e)** Live S bacteria could be retrieved from the dead mice that had been injected with the mixture. Griffith concluded that molecules from the dead S cells had genetically transformed some of the living R bacteria into S bacteria.

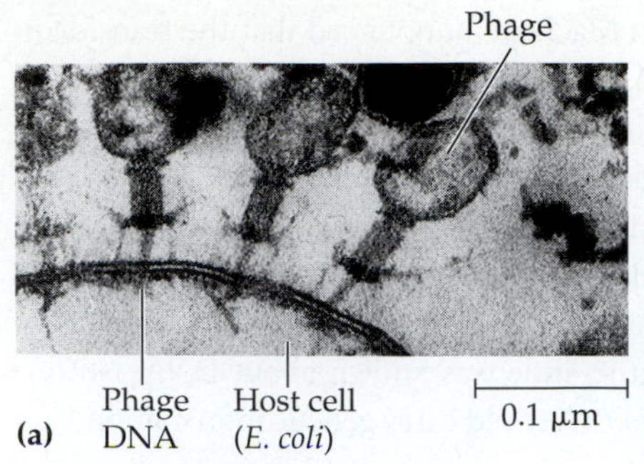
Phage
Phage DNA
Host cell (*E. coli*)
0.1 μm
(a)

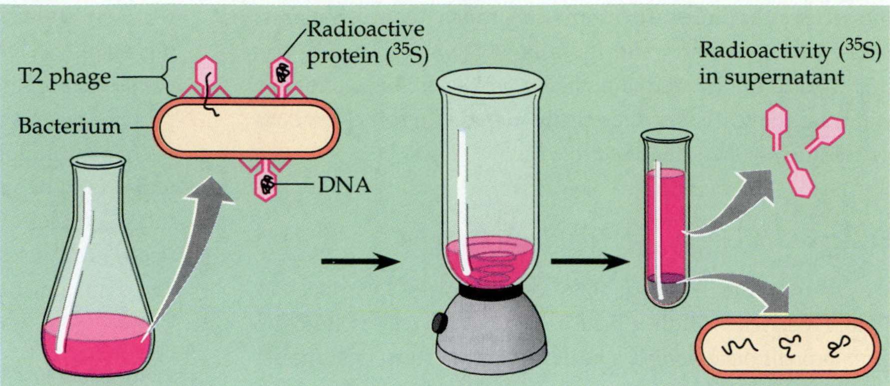
T2 phage
Bacterium
Radioactive protein (^{35}S)
DNA
Radioactivity (^{35}S) in supernatant
Mix radioactively labeled phage with bacteria. The phage infects the bacterial cells.
Agitate in a blender to separate phage outside the bacteria from the cells and their contents.
Centrifuge and measure the radioactivity in the pellet and supernatant.

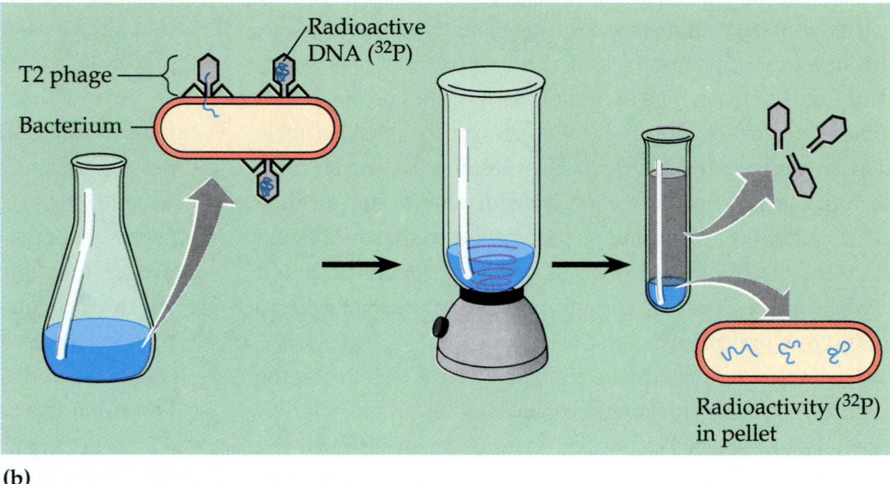
T2 phage
Bacterium
Radioactive DNA (^{32}P)
Radioactivity (^{32}P) in pellet
(b)

FIGURE 16.2 ▪ The Hershey-Chase experiment. (a) Phages are viruses that infect bacteria. T2 phages use their tail pieces to attach to the host cell and inject their genetic material (TEM). **(b)** In their famous 1952 experiment, Hershey and Chase demonstrated that it was DNA, not protein, that functioned as the T2 phage's genetic material. Viral proteins, labeled with radioactive sulfur, remained outside the host cell during infection. In contrast, viral DNA, labeled with radioactive phosphorus, entered the bacterial cell.

Hershey and Chase answered this question by devising an experiment showing that only one of the two components of T2 actually enters the *E. coli* cell during infection (FIGURE 16.2). They used different radioactive isotopes to tag the DNA and protein. First, they grew T2 with *E. coli* in the presence of radioactive sulfur. Because protein, but not DNA, contains sulfur, the radioactive atoms were incorporated only into the protein of the phage. Next, in a similar way, the DNA of a separate batch of phage was labeled with atoms of radioactive phosphorus; because nearly all the phage's phosphorus is in its DNA, this procedure left the phage protein unlabeled. Then the protein-labeled and DNA-labeled batches of T2 were each allowed to infect separate samples of nonradioactive *E. coli* cells. Shortly after the onset of infection, the cultures were whirled in a kitchen blender to shake loose any parts of the phages that remained outside the bacterial cells. The mixtures were then spun in a centrifuge, forcing the bacterial cells to form a pellet at the bottom of the centrifuge tubes, but allowing the viruses and parts of viruses, which are lighter, to remain suspended in the liquid, or supernatant. The scientists then measured the radioactivity in the pellet and in the supernatant.

Hershey and Chase found that when the bacteria had been infected with the T2 phage containing radioactively labeled proteins, most of the radioactivity was found in the supernatant, which contained viral particles (but not bacteria). This result suggested that the protein of the phage did not enter the host cells. But when the bacteria had been infected with T2 phage whose DNA was tagged with radioactive phosphorus, then the pellet of mainly bacterial material contained most of the radioactivity. Moreover, when these bacteria were returned to a culture medium, the infection ran its course, and the *E. coli* released phages that contained some radioactive phosphorus.

Hershey and Chase concluded that the DNA of the virus is injected into the host cell, while most of the proteins remain outside. The injected DNA molecules cause the cells to produce new viral DNA and proteins—indeed, additional intact viruses. Thus the Hershey-Chase experiment provided powerful evidence that nucleic acids, rather than proteins, are the hereditary material, at least for viruses.

Additional Evidence That DNA Is the Genetic Material of Cells

Additional circumstantial evidence pointed to DNA as the genetic material in eukaryotes. Prior to mitosis, a eukaryotic cell exactly doubles its DNA content, and during mitosis, this DNA is distributed exactly equally to the two daughter cells. Also, diploid sets of chromosomes have twice as much DNA as the haploid sets found in the gametes of the same organism.

Still more evidence came from the laboratory of biochemist Erwin Chargaff. It was already known that DNA is a polymer of nucleotides, each consisting of three components: a nitrogenous (nitrogen-containing) base, a pentose sugar called deoxyribose, and a phosphate group. (See FIGURE 16.3 for review.) The base can be adenine (A), thymine (T), guanine (G), or cytosine (C). Chargaff analyzed the base composition of DNA from a number of different organisms. In 1947, he reported that DNA composition varies from one species to another: In the DNA of any one species, the amounts of the four nitrogenous bases are not all equal but are present in a characteristic ratio. Such evidence of molecular diversity, which had been presumed absent from DNA, made DNA a more credible candidate for the genetic material.

Chargaff also found a peculiar regularity in the ratios of nucleotide bases. In the DNA of each species he studied, the number of adenines approximately equaled the number of thymines, and the number of guanines approximately equaled the number of cytosines. In human DNA, for example, the four bases are present in these percentages: A = 30.9% and T = 29.4%; G = 19.9% and C = 19.8%. The A = T and G = C equalities, later known as Chargaff's rules, remained unexplained until the discovery of the double helix.

Watson and Crick discovered the double helix by building models to conform to X-ray data: *science as a process*

16.1 Once most biologists were convinced that DNA was the genetic material, a race was under way to determine how the structure of DNA could account for its role in inheritance. By the beginning of the 1950s, the arrangement of covalent bonds in a nucleic acid polymer was well established (FIGURE 16.3), and the competition focused on discovering the three-dimensional structure of DNA. Among the scientists working on the problem were Linus Pauling, in California, and Maurice Wilkins and Rosalind Franklin, in London. First to the finish line, however, were two scientists who were relatively unknown at the time—the American James Watson and the Englishman Francis Crick.

The brief but celebrated partnership that solved the DNA puzzle began soon after the young Watson journeyed to Cambridge University, where Crick was studying protein structure with a technique called X-ray crystallography (see the Methods Box in Chapter 5, p. 80). While visiting the laboratory of Maurice Wilkins at King's College in London, Watson saw an X-ray photograph of DNA produced by Wilkins's colleague,

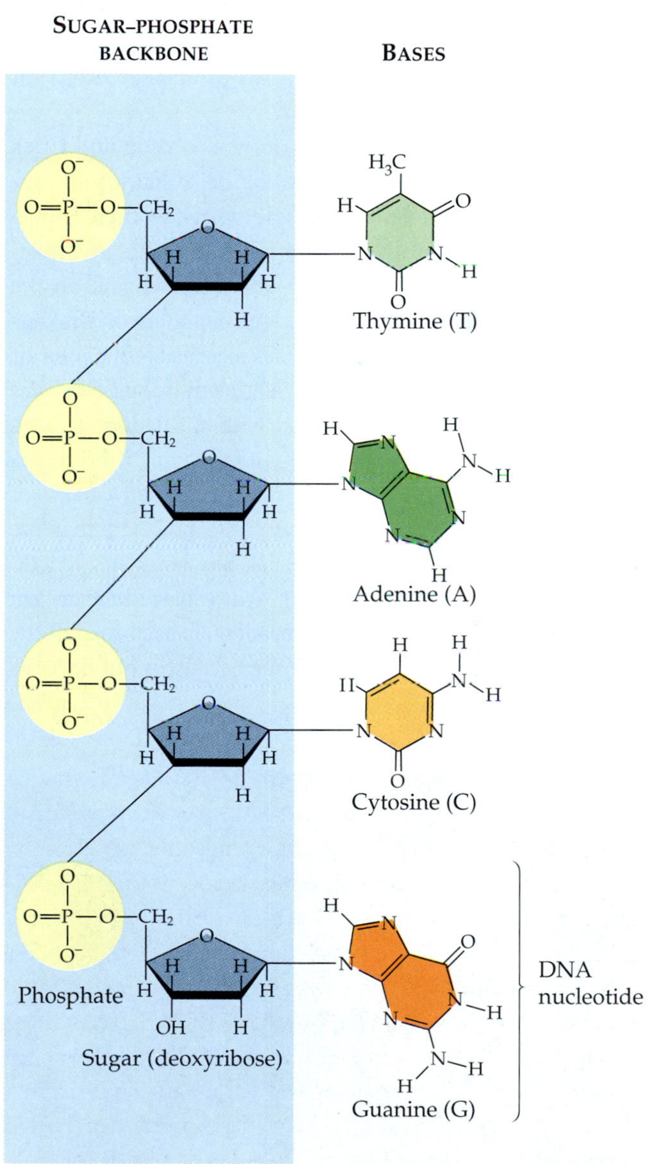

16.1 **FIGURE 16.3 • The structure of a DNA strand.** Each nucleotide unit of the polynucleotide chain consists of a nitrogenous base (T, A, C, or G), the sugar deoxyribose, and a phosphate group. The phosphate of one nucleotide is attached to the sugar of the next nucleotide in line. The result is a "backbone" of alternating phosphates and sugars, from which the bases project.

Rosalind Franklin (FIGURE 16.4a). Photographs produced by the X-ray crystallography method are not actually pictures of molecules. The spots and smudges in FIGURE 16.4b were produced by X-rays that were diffracted (deflected) as they passed through crystallized DNA. Crystallographers use mathematical equations to translate such patterns of spots into information about the three-dimensional shapes of molecules. Watson and Crick based their model of DNA on data they were able to extract from Franklin's X-ray diffraction photo. They interpreted the pattern of spots on the X-ray photograph to mean that DNA was helical in shape. Based on Watson's recollection of the photograph, he and Crick deduced that the helix had a uniform width of 2 nanometers (nm), with its nitrogenous bases stacked 0.34 nm apart. The width of the helix suggested that it was made up of two strands, contrary to a three-stranded model that Linus Pauling had recently proposed. The presence of two strands accounts for the now-familiar term **double helix**.

Using molecular models made of wire, Watson and Crick began building scale models of a double helix that would conform to the X-ray measurements and what was then known about the chemistry of DNA. After failing to make a satisfactory model that placed the sugar-phosphate chains on the inside of the molecule, Watson tried putting them on the outside and forcing the nitrogenous bases to swivel to the interior of the double helix (FIGURE 16.5). Imagine this double helix as a rope ladder having rigid rungs, with the ladder twisted into a spiral. The side ropes are the equivalent of the sugar-phosphate backbones, and the rungs represent pairs of nitrogenous bases. Franklin's X-ray data indicated that the helix makes one full turn every 3.4 nm along its length. Because the bases are stacked just 0.34 nm apart, there are ten layers of base pairs, or rungs on the ladder, in each turn of the helix. This arrangement was appealing because it put the relatively hydrophobic nitrogenous bases in the molecule's interior and thus away from the surrounding aqueous medium.

The nitrogenous bases of the double helix are paired in specific combinations: adenine (A) with thymine (T), and guanine (G) with cytosine (C). It was mainly by trial and error that Watson and Crick arrived at this key feature of DNA. At first, Watson imagined that the bases paired like-with-like—for example, A with A and C with C. But this model did not fit the X-ray data, which suggested that the double helix had a uniform diameter. Why is this requirement inconsistent with like-with-like pairing of bases? Adenine and guanine are purines, nitrogenous bases with two organic rings. In contrast, cytosine and thymine belong to the family of nitrogenous bases known as pyrimidines, which have a single ring. Thus, purines (A and G) are about twice as wide as pyrimidines (C and T). A purine-purine pair is too wide and a pyrimidine-pyrimidine pair too narrow to account for the 2-nm diameter of the double helix. The solution is to always pair a purine with a pyrimidine:

Purine + purine: too wide

Pyrimidine + pyrimidine: too narrow

Purine + pyrimidine: width consistent with X-ray data

Watson and Crick reasoned that there must be additional specificity of pairing dictated by the structure of the bases. Each base has chemical side groups that can form hydrogen bonds with its appropriate partner: Adenine can form two hydrogen bonds with thymine and only thymine; guanine

(a)

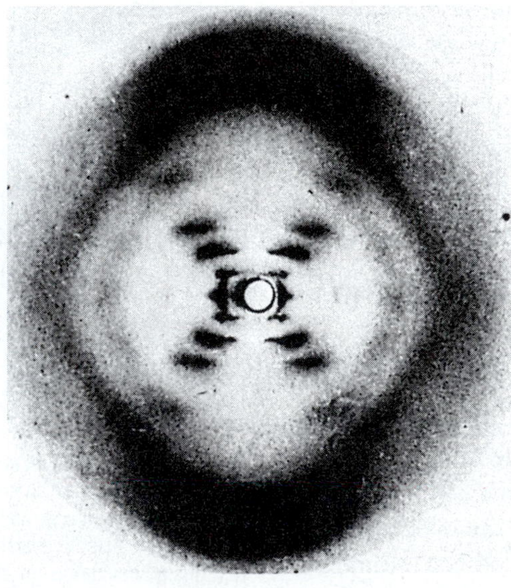
(b)

FIGURE 16.4 ▪ Rosalind Franklin and her X-ray diffraction photo of DNA. (a) Franklin was the X-ray crystallographer who took the photo that enabled Watson and Crick to deduce the double-helical structure of DNA. Franklin died of cancer when she was only 38. Her colleague, Maurice Wilkins, received the Nobel Prize in 1962 along with Watson and Crick. Because the Nobel Prize is not awarded posthumously, science historians can only speculate about whether the committee would have recognized Franklin's contribution to the discovery of the double helix. **(b)** Franklin's X-ray diffraction photograph of DNA.

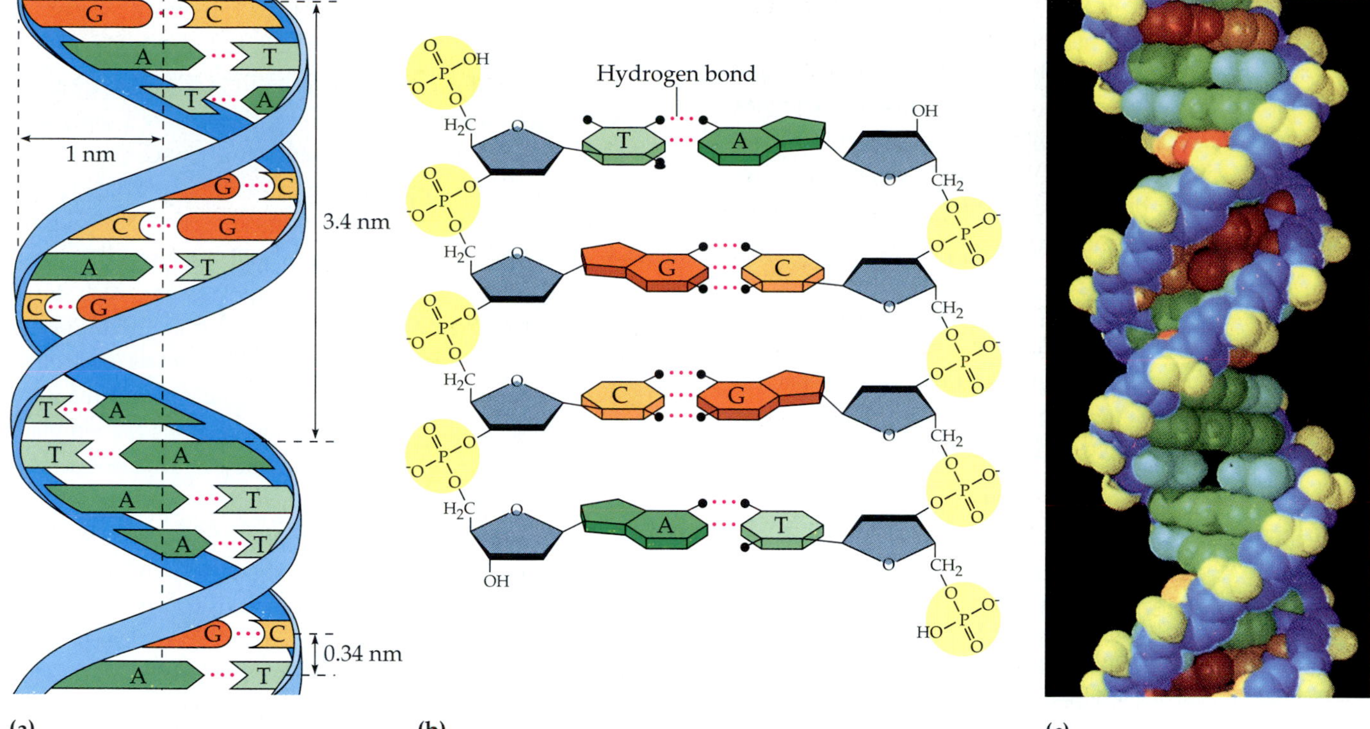

16.1 **FIGURE 16.5 ▪ The double helix. (a)** The "ribbons" in this diagram represent the sugar-phosphate backbones of the two DNA strands. The helix is "right-handed," curving up to the right. The two strands are held together by hydrogen bonds (dotted lines) between the nitrogenous bases, which are paired in the interior of the double helix. **(b)** Partial chemical structure, with the two strands untwisted. Notice that the strands are oriented in opposite directions. **(c)** The tight stacking of the base pairs is clear in this computer model. Van der Waals attractions between the stacked pairs play a major role in holding the molecule together.

forms three hydrogen bonds with cytosine and only cytosine. In shorthand, A pairs with T, and G pairs with C (FIGURE 16.6).

The Watson-Crick model explained Chargaff's rules. Wherever one strand of a DNA molecule has an A, the partner strand has a T. And a G in one strand is always paired with a C in the complementary strand. Therefore, in the DNA of any organism, the amount of adenine equals the amount of thymine, and the amount of guanine equals the amount of cytosine. Although the base-pairing rules dictate the combinations of nitrogenous bases that form the "rungs" of the double helix, they do not restrict the sequence of nucleotides *along* each DNA strand. Thus, the linear sequence of the four bases can be varied in countless ways, and each gene has a unique order, or base sequence.

In April 1953, Watson and Crick surprised the scientific world with a succinct, one-page paper in the British journal *Nature.** The paper reported their molecular model for DNA: the double helix, which has since become the symbol of molecular biology. The beauty of the model was that its structure suggested the basic mechanism of DNA replication.

* Watson, J. D., and F. H. C. Crick, "Molecular Structure of Nucleic Acids: A Structure for Deoxynucleic Acids." *Nature* 171 (1953), p. 738.

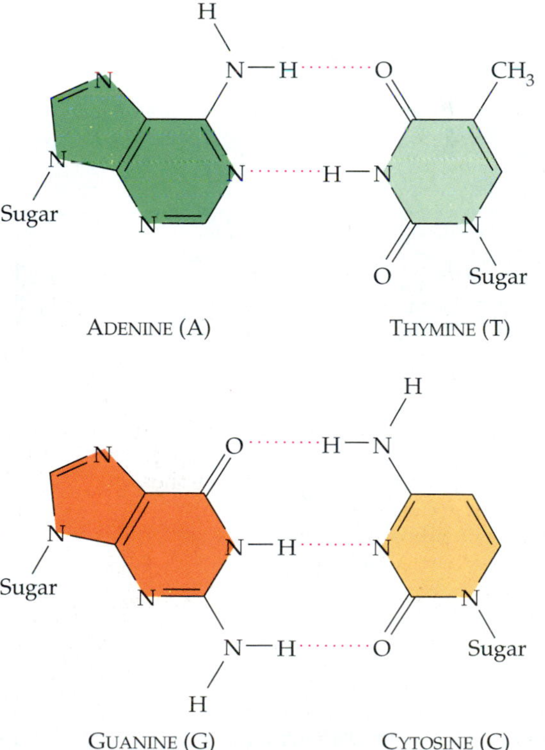

FIGURE 16.6 ▪ Base pairing in DNA. The pairs of nitrogenous bases in a DNA double helix are held together by hydrogen bonds as shown here.

DNA REPLICATION AND REPAIR

The relationship between structure and function, one of the themes of biology, is manifest in the double helix. The idea that there is specific pairing of nitrogenous bases in DNA was the flash of inspiration that led Watson and Crick to the correct double helix. At the same time, they saw the functional significance of the base-pairing rules. They ended their classic paper with this wry statement: "It has not escaped our notice that the specific pairing we have postulated immediately suggests a possible copying mechanism for the genetic material." In the next section you will learn about this basic mechanism of DNA replication. Some important details of the process will be presented in the section that follows.

During DNA replication, base pairing enables existing DNA strands to serve as templates for new complementary strands

16.2 In a second paper that followed their announcement of the double helix, Watson and Crick stated their hypothesis for how DNA replicates:

> Now our model for deoxyribonucleic acid is, in effect, a pair of templates, each of which is complementary to the other. We imagine that prior to duplication the hydrogen bonds are broken, and the two chains unwind and separate. Each chain then acts as a template for the formation onto itself of a new companion chain, so that eventually we shall have two pairs of chains, where we only had one before. Moreover, the sequence of the pairs of bases will have been duplicated exactly.*

* Crick, F. H. C., and J. D. Watson. "The Complementary Structure of Deoxyribonucleic Acid." *Proc. Roy. Soc.* (A) 223 (1954), p. 80.

FIGURE 16.7 illustrates Watson and Crick's basic idea. To make it easier to follow, the diagram shows only a short section of double helix, in untwisted form. Notice that if you cover one of the two DNA strands of FIGURE 16.7a with a piece of paper, you can still determine its linear sequence of bases by referring to the unmasked strand and applying the base-pairing rules. The two strands are complementary; each stores the information necessary to reconstruct the other. When a cell copies a DNA molecule, each strand serves as a template (mold) for ordering nucleotides into a new complementary strand. One at a time, nucleotides line up along the template strand according to the base-pairing rules. The nucleotides are linked to form the new strands. Where there was one double-stranded DNA molecule at the beginning of the process, there are now two, each an exact replica of the "parent" molecule. The copying mechanism is analogous to using a photographic negative to make a positive image, which can in turn be used to make another negative, and so on.

This model of gene replication remained untested for several years following publication of the DNA structure. The requisite experiments were simple in concept but difficult to perform. Watson and Crick's model predicts that when a double helix replicates, each of the two daughter molecules will have one old strand, derived from the parent molecule, and one newly made strand. This **semiconservative model** can be distinguished from a conservative model of replication, in which the parent molecule remains intact (is conserved) and the new molecule is formed entirely from scratch. In a third

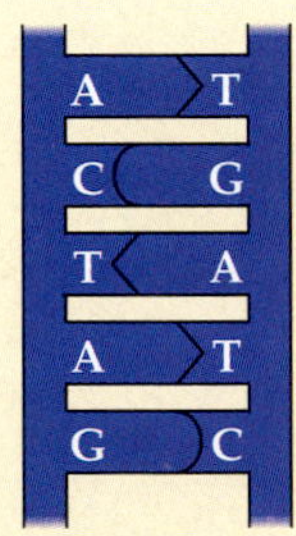

(a) Before replication, the parent molecule has two complementary strands of DNA. Each base is paired by hydrogen bonding with its specific partner, A with T and G with C.

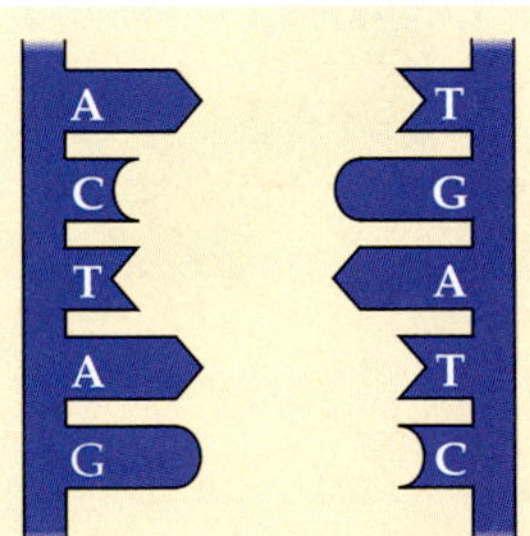

(b) The first step in replication is separation of the two DNA strands.

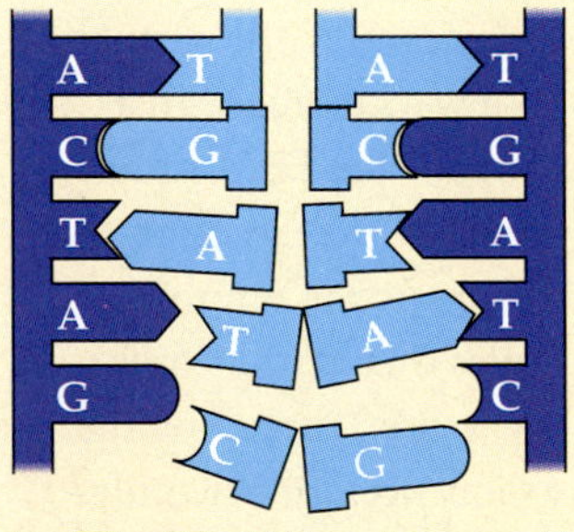

(c) Each "old" strand now serves as a template that determines the order of nucleotides along "new" complementary strands. Nucleotides plug into specific sites along the template surface according to the base-pairing rules.

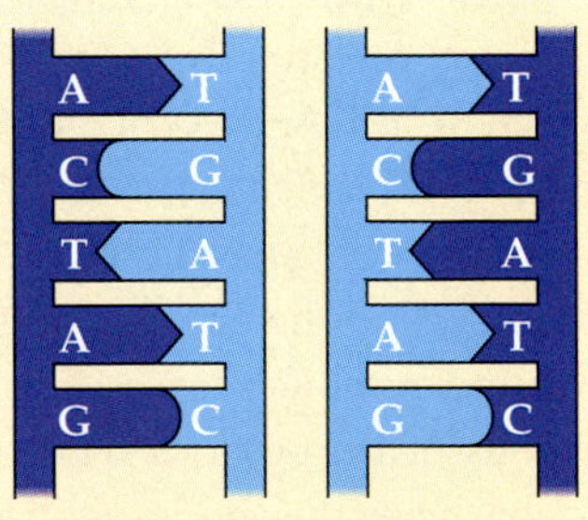

(d) The nucleotides are connected to form the sugar-phosphate backbones of the new strands. Each DNA molecule now consists of one "old" strand and one "new" strand. We have two DNA molecules identical to the one molecule with which we started.

16.2 **FIGURE 16.7 • A model for DNA replication: the basic concept.** In this simplification, a short segment of DNA has been untwisted to convert the double helix to a two-dimensional version of the molecule that resembles a ladder. The rails of the ladder are the sugar-phosphate backbones of the two DNA strands. The rungs are the pairs of nitrogenous bases. Simple shapes are used to symbolize the four kinds of bases. Dark blue represents DNA strands originally present in the parent cell. Newly synthesized DNA is represented by light blue.

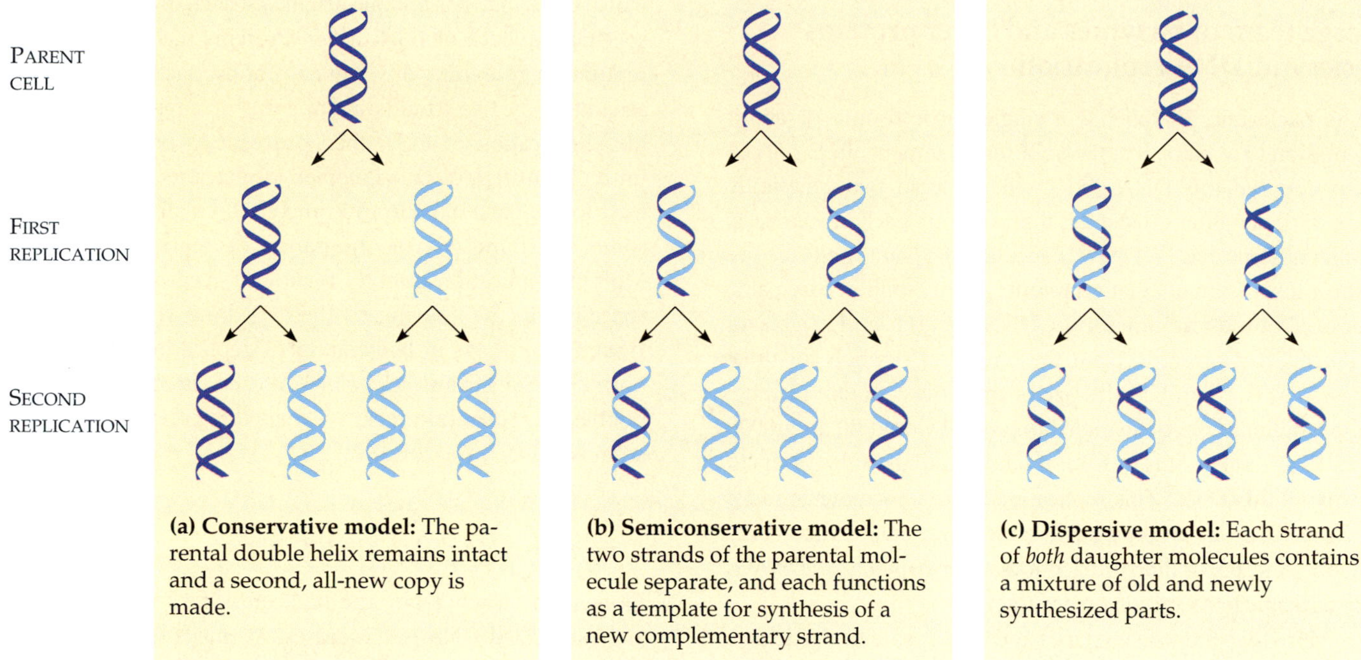

PARENT CELL
FIRST REPLICATION
SECOND REPLICATION
(a) Conservative model: The parental double helix remains intact and a second, all-new copy is made.
(b) Semiconservative model: The two strands of the parental molecule separate, and each functions as a template for synthesis of a new complementary strand.
(c) Dispersive model: Each strand of both daughter molecules contains a mixture of old and newly synthesized parts.

FIGURE 16.8 ▪ Three models of DNA replication. The short segments of double helix here symbolize the DNA within a cell. Beginning with a parent cell, we follow the DNA for two generations of cells—two rounds of DNA replication. Newly made DNA is lighter blue.

model, called the dispersive model, all four strands of DNA, after the double helix is replicated, have a mixture of old and new DNA (FIGURE 16.8). In the late 1950s, Matthew Meselson and Franklin Stahl devised experiments that tested these three alternative hypotheses for DNA replication. Their experiments supported the semiconservative model, as predicted by Watson and Crick (FIGURE 16.9).

The basic principle of DNA replication is elegantly simple. However, the actual process involves complex biochemical gymnastics, as we will now see.

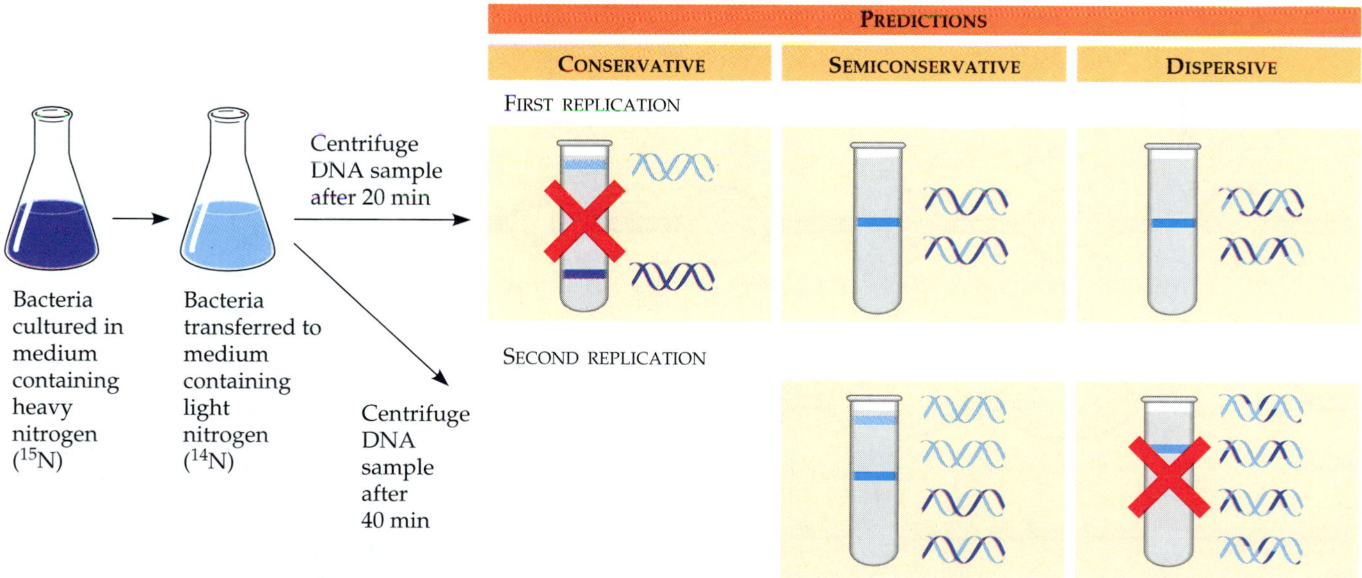

FIGURE 16.9 ▪ The Meselson-Stahl experiment tested three hypotheses of DNA replication. Meselson and Stahl cultured *E. coli* for several generations on a medium containing a heavy isotope of nitrogen, ^{15}N. The bacteria incorporated the heavy nitrogen into their nucleotides and then into their DNA. The scientists then transferred the bacteria to a medium containing ^{14}N, the lighter, more common isotope of nitrogen. Thus, any new DNA that the bacteria synthesized would be lighter than the "old" DNA made in the ^{15}N medium. Meselson and Stahl could distinguish DNA of different densities by centrifuging DNA extracted from the bacteria. The centrifuge tubes in this drawing represent the results predicted by the three hypotheses in FIGURE 16.8. The first replication in the ^{14}N medium produced a band of hybrid (^{15}N-^{14}N) DNA. This result eliminated the conservative hypothesis. A second replication produced both light and hybrid DNA, a result that eliminated the dispersive hypotheses and supported the semiconservative hypothesis.

A large team of enzymes and other proteins carries out DNA replication

16.3 The bacterium *E. coli* has a single chromosome of about 5 million base pairs. In a favorable environment, an *E. coli* cell can copy all this DNA and divide to form two genetically identical daughter cells in less than an hour. Each of *your* cells, in its nucleus, has 46 DNA molecules, one giant molecule per chromosome. In all, that represents about 6 billion base pairs, or over a thousand times more DNA than is found in a bacterial cell. If we were to print the one-letter symbols for these bases (A, G, C, and T) the size of the letters you are reading, the 6 billion bases of a single human cell would fill about 900 books as thick as this text. Yet it takes a cell just a few hours to copy all this DNA. This replication of an enormous amount of genetic information is achieved with very few errors—only about one per billion nucleotides. The copying of DNA is remarkable in its speed and accuracy.

More than a dozen enzymes and other proteins participate in DNA replication. Much more is known about how this "replication machine" works in bacteria than in eukaryotes. However, most of the process seems to be fundamentally similar for prokaryotes and eukaryotes. In this section we take a closer look at the basic steps.

Getting Started: Origins of Replication

The replication of a DNA molecule begins at special sites called **origins of replication**. The bacterial chromosome, which is circular, has a single origin, a stretch of DNA having a specific sequence of nucleotides. Proteins that initiate DNA replication recognize this sequence and attach to the DNA, separating the two strands and opening up a replication "bubble." Replication of DNA then proceeds in both directions, until the entire molecule is copied (see FIGURE 18.10). In contrast to the bacterial chromosome, each eukaryotic chromosome has hundreds or thousands of replication origins. Multiple replication bubbles form and eventually fuse, thus speeding up the copying of the very long DNA molecules (FIGURE 16.10). As in bacteria, DNA replication proceeds in both directions from each origin. At each end of a replication bubble is a **replication fork**, a Y-shaped region where the new strands of DNA are elongating.

Elongating a New DNA Strand

Elongation of new DNA at a replication fork is catalyzed by enzymes called **DNA polymerases.** As nucleotides align with complementary bases along a template strand of DNA, they are added by polymerase, one by one, to the growing end of the new DNA strand. The rate of elongation is about 500 nucleotides per second in bacteria and 50 per second in human cells.

What is the source of energy that drives the polymerization of nucleotides to form new DNA strands? The nucleotides that serve as substrates for DNA polymerase are actually nucleoside triphosphates, which are nucleotides with *three* phosphate groups (FIGURE 16.11). You have already encoun-

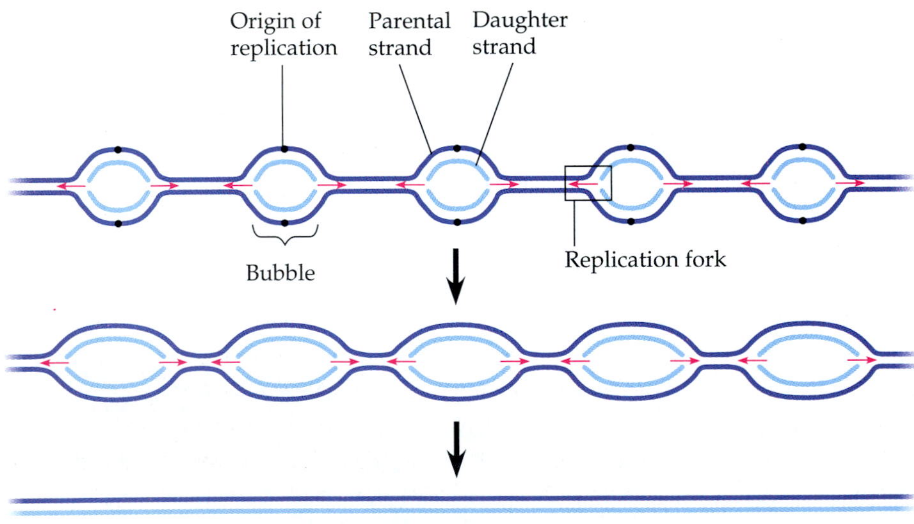

(b)

0.25 μm

FIGURE 16.10 ▪ Origins of replication. (a)
16.2 DNA replication begins at specific sites where the two parental strands of DNA separate to form replication "bubbles" (top). In eukaryotes, there are hundreds or thousands of origin sites along the giant DNA molecule of each chromosome. A replication bubble expands laterally, as DNA replication proceeds in both directions. Eventually, the replication bubbles fuse (center), and synthesis of the daughter strands of DNA is complete (bottom). **(b)** In this micrograph, three replication bubbles are visible along the DNA of cultured Chinese hamster cells. The arrows indicate the directions of DNA replication at the two ends of each bubble (TEM).

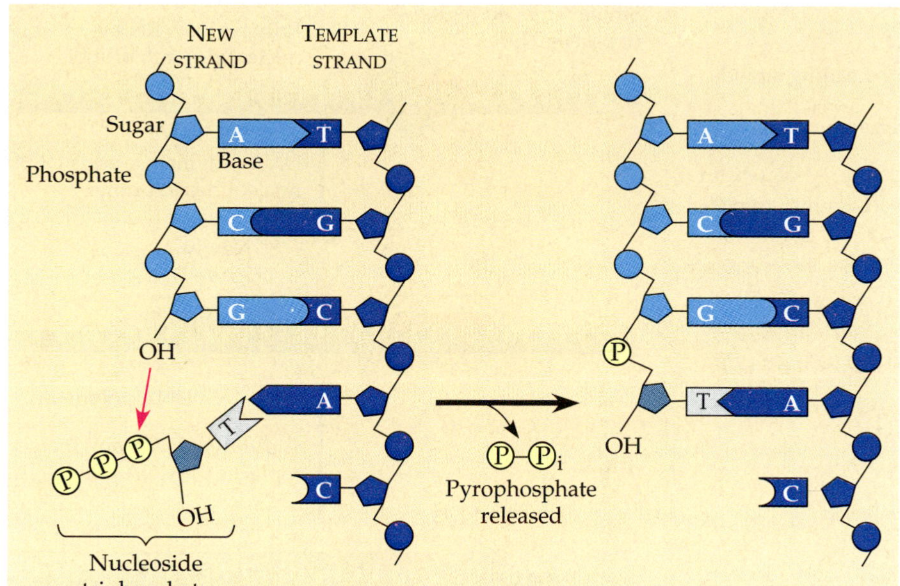

FIGURE 16.11 ▪ Incorporation of a nucleotide into a DNA strand. When a nucleoside triphosphate links to the sugar-phosphate backbone of a growing DNA strand, it loses two of its phosphates as a pyrophosphate molecule. The enzyme catalyzing the reaction is a DNA polymerase, and hydrolysis of the bonds between the phosphate groups provides the energy for the reaction.

tered such a molecule—ATP. The only difference between the ATP of energy metabolism and the nucleoside triphosphate that supplies adenine to DNA is the sugar component, which is deoxyribose in the building block of DNA, but ribose in ATP. (As you might guess, ATP is a substrate for *RNA* synthesis.) Like ATP, the triphosphate monomers used for DNA synthesis are chemically reactive, partly because their triphosphate tails have an unstable cluster of negative charge. As each monomer joins the growing end of a DNA strand, it loses two phosphate groups. Hydrolysis of the phosphate is the exergonic reaction that drives the polymerization of nucleotides to form DNA.

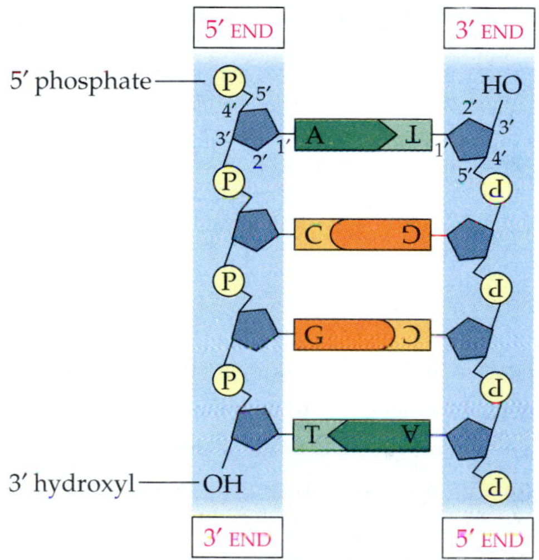

FIGURE 16.12 ▪ The two strands of DNA are antiparallel. The $5' \longrightarrow 3'$ direction of one strand runs counter to the $5' \longrightarrow 3'$ direction of the other strand. The carbon atoms of the deoxyribose sugars at the top are numbered for orientation.

The Problem of Antiparallel DNA Strands

There is more to the scenario of DNA synthesis at the replication fork. Until now, we have ignored an important feature of the double helix: The two DNA strands are antiparallel; that is, their sugar-phosphate backbones run in opposite directions. In FIGURE 16.12, the five carbons of one deoxyribose sugar of each DNA strand are numbered, from 1′ to 5′. (The prime sign is used to distinguish the carbon atoms of the sugar from the carbon and nitrogen atoms of the nitrogenous bases.) Notice in FIGURE 16.12 that a nucleotide's phosphate group is attached to the 5′ carbon of deoxyribose. Notice also that the phosphate group of one nucleotide is joined to the 3′ carbon of the adjacent nucleotide. The result is a DNA strand of distinct polarity. At one end, denoted the 3′ end, a hydroxyl group is attached to the 3′ carbon of the terminal deoxyribose. At the opposite end, the 5′ end, the sugar-phosphate backbone terminates with the phosphate group attached to the 5′ carbon of the last nucleotide. In the double helix, the two sugar-phosphate backbones are essentially upside-down (antiparallel) relative to each other.

How does the antiparallel structure of the double helix affect replication? DNA polymerases add nucleotides only to the free 3′ end of a growing DNA strand, never to the 5′ end. Thus, a new DNA strand can elongate only in the $5' \longrightarrow 3'$ direction. With this in mind, let's examine a replication fork (FIGURE 16.13, p. 288). Along one template strand, DNA polymerase can synthesize a continuous complementary strand by elongating the new DNA in the mandatory $5' \longrightarrow 3'$ direction. The polymerase simply nestles in the replication fork and moves along the template strand as the fork progresses. The DNA strand made by this mechanism is called the **leading strand.**

To elongate the other new strand of DNA, polymerase must work along the template *away from* the replication fork.

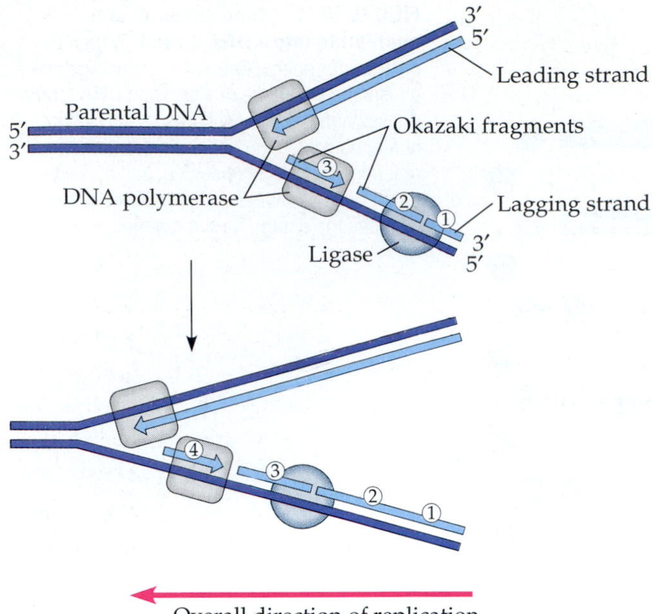

FIGURE 16.13 ▪ Synthesis of leading and lagging strands during DNA replication. A DNA polymerase elongates strands only in the 5′ ⟶ 3′ direction. One new strand, called the leading strand, can therefore elongate continuously in the 5′ ⟶ 3′ direction as the replication fork progresses. But the other new strand, the lagging strand, must grow in an overall 3′ ⟶ 5′ direction by the addition of short segments, Okazaki fragments, that individually grow 5′ ⟶ 3′. We have numbered the fragments in the order in which they were made. An enzyme called ligase connects the fragments.

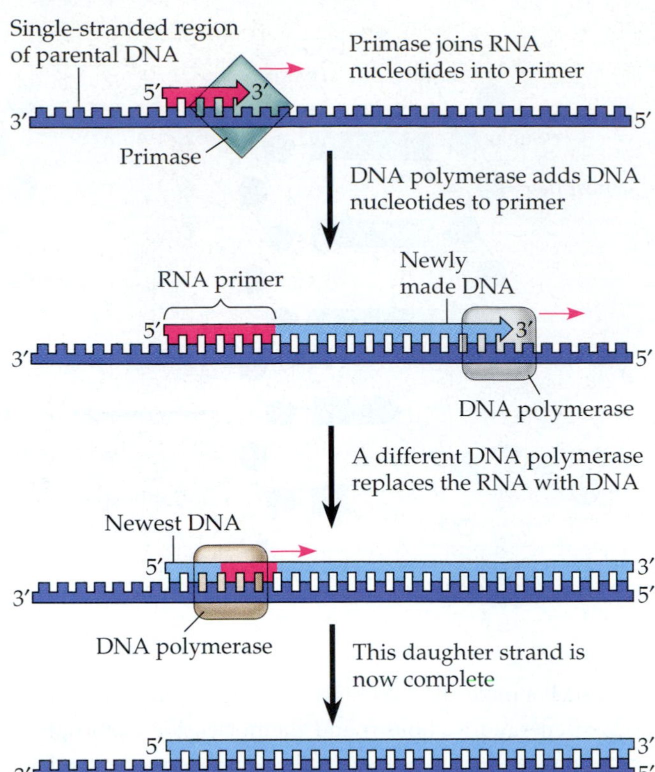

FIGURE 16.14 ▪ Priming DNA synthesis. DNA polymerase cannot initiate a polynucleotide strand; it can only add to the 3′ end of an already-started strand. The primer is a short segment of RNA synthesized by the enzyme primase. Each primer is eventually replaced by DNA.

The DNA synthesized in this direction is called the **lagging strand.** The process is analogous to a sewing method called back-stitching. As a replication bubble opens, a polymerase molecule can work its way away from a replication fork and synthesize a short segment of DNA. As the bubble grows, another short segment of the lagging strand can be made in a similar way. In contrast to the leading strand, which elongates continuously, the lagging strand is first synthesized as a series of segments. These pieces are called Okazaki fragments, after the Japanese scientist who discovered them. The fragments are about 100 to 200 nucleotides long in eukaryotes. Another enzyme, **DNA ligase**, joins the Okazaki fragments into a single DNA strand.

Priming DNA Synthesis

There is another important restriction for DNA polymerases. They can add a nucleotide only to an existing polynucleotide that is already paired with the complementary strand (see FIGURE 16.11). DNA polymerases cannot actually *initiate* synthesis of a polynucleotide; they can only add nucleotide to the end of an already existing chain. In the cell, the original preexisting chain, the **primer**, is not DNA, but a short stretch of RNA, the other class of nucleic acid. An enzyme called **primase** joins RNA nucleotides to make the primer, which is about 10 nucleotides long in eukaryotes (FIGURE 16.14). Another DNA polymerase later replaces the RNA nucleotides of the primers with DNA versions. Only one primer is required for a DNA polymerase to begin synthesizing the leading strand of new

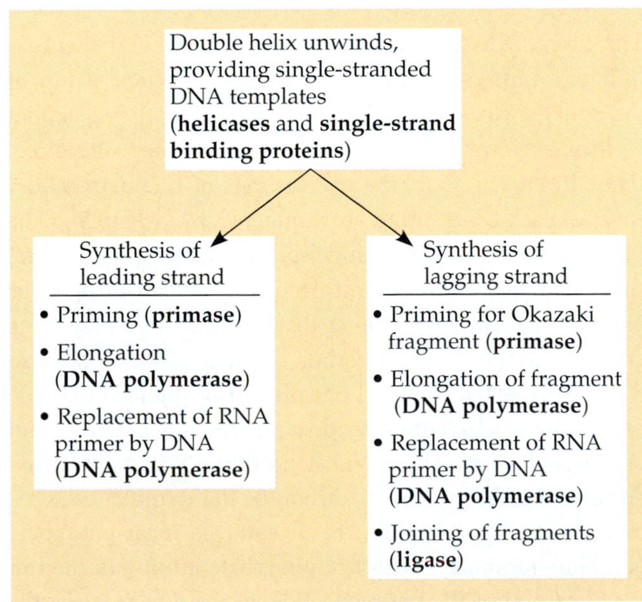

FIGURE 16.15 ▪ The main proteins of DNA replication and their functions.

DNA. For the lagging strand, each fragment must be primed; the primers are converted to DNA before ligase joins the fragments together.

Other Proteins Assisting DNA Replication

You have learned about three kinds of proteins that function in DNA synthesis: DNA polymerase, ligase, and primase. Other kinds of proteins also participate; two of them are helicase and single-strand binding protein. A **helicase** is an enzyme that untwists the double helix at the replication fork, separating the two old strands. Molecules of **single-strand binding protein** then line up along the unpaired DNA strands, holding them apart while they serve as templates for the synthesis of new complementary strands.

FIGURE 16.15 summarizes the functions of the main proteins that cooperate in DNA replication. FIGURE 16.16 is a visual summary of DNA replication.

Enzymes proofread DNA during its replication and repair damage in existing DNA

We cannot attribute the accuracy of DNA replication solely to the specificity of base pairing. Although errors in the completed DNA molecule amount to only one in 1 billion nucleotides, initial pairing errors between incoming nucleotides and those in the template strand are 100,000 times more common—an error rate of one in 10,000 base pairs. One DNA repair mechanism, **mismatch repair**, fixes mistakes that are made when DNA is copied. During DNA replication, DNA polymerase itself carries out mismatch repair. The polymerase proofreads each nucleotide against its template as soon as it is added to the strand. Upon finding an incorrectly paired nucleotide, the polymerase removes the nucleotide and then resumes synthesis. (This action resembles correcting a word-processing error by using the "delete" key and then entering the correction.) Proteins other than DNA polymerases also

16.3 **FIGURE 16.16 ▪ A summary of DNA replication.** The detailed diagram below shows one replication fork, but as indicated in the overview diagram at the right, DNA replication usually occurs simultaneously at two forks, one at either end of a replication bubble. Notice in the overview diagram that a leading strand is initiated by an RNA primer, just as each Okazaki fragment in a lagging strand is initiated. Viewing each daughter strand in its entirety, you can see that one half of it is made continuously, as a leading strand, and the other half, on the other side of the origin, is synthesized in fragments, as a lagging strand.

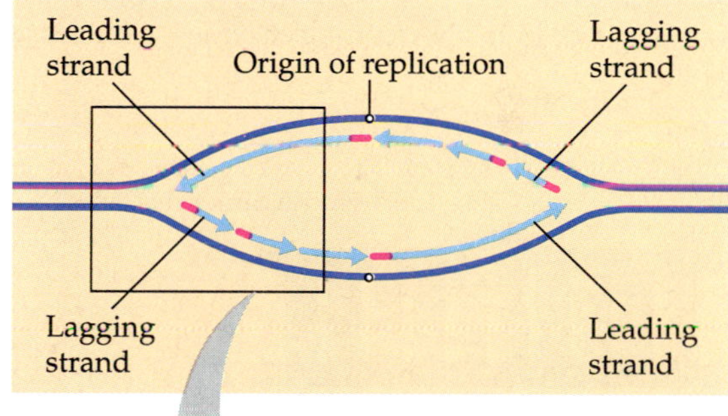
Leading strand
Origin of replication
Lagging strand
Lagging strand
Leading strand

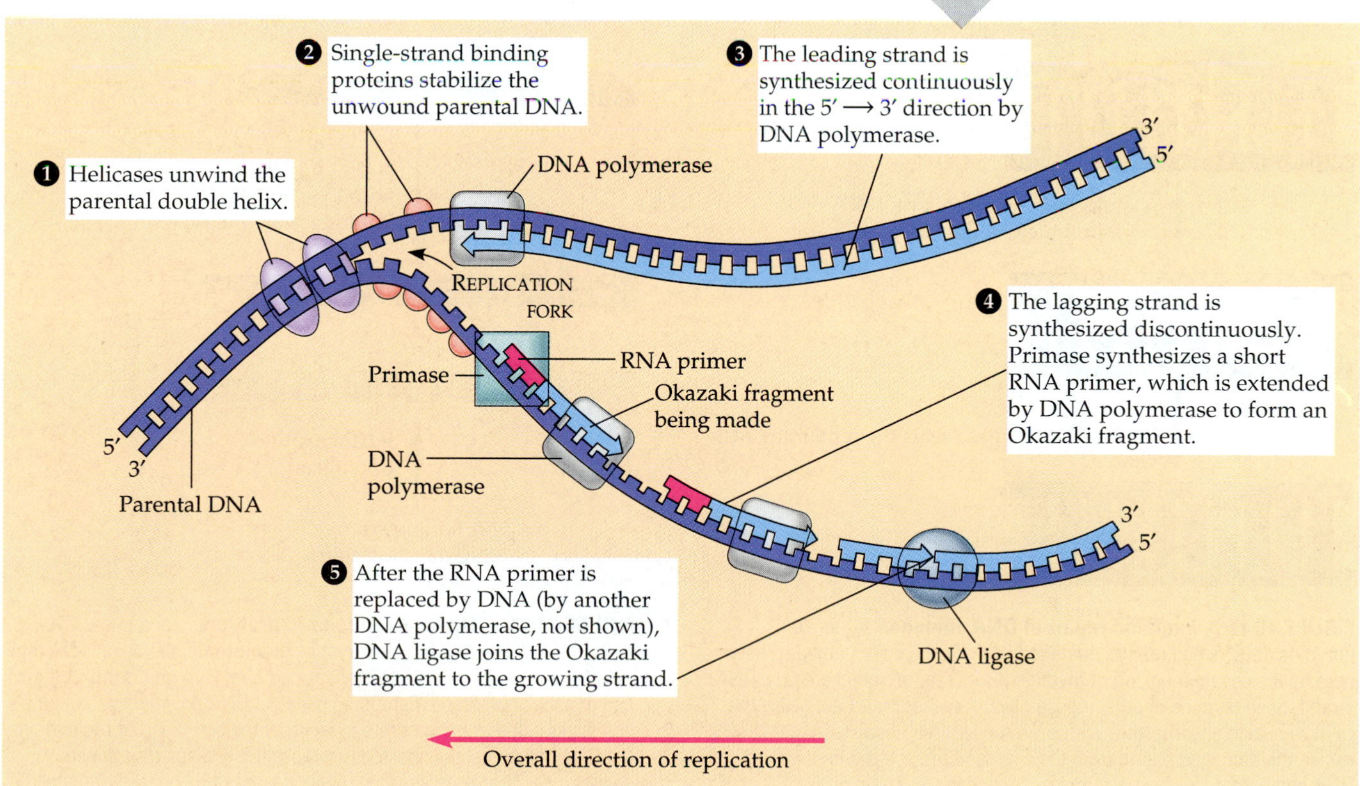
❶ Helicases unwind the parental double helix.
❷ Single-strand binding proteins stabilize the unwound parental DNA.
❸ The leading strand is synthesized continuously in the 5′ ⟶ 3′ direction by DNA polymerase.
DNA polymerase
3′
5′
Replication fork
❹ The lagging strand is synthesized discontinuously. Primase synthesizes a short RNA primer, which is extended by DNA polymerase to form an Okazaki fragment.
Primase
RNA primer
Okazaki fragment being made
5′
3′
Parental DNA
DNA polymerase
3′
5′
❺ After the RNA primer is replaced by DNA (by another DNA polymerase, not shown), DNA ligase joins the Okazaki fragment to the growing strand.
DNA ligase
Overall direction of replication

perform mismatch repair. Researchers spotlighted the importance of such proteins when they found that a hereditary defect in one of them is associated with a form of colon cancer. Apparently, this defect allows cancer-causing errors to accumulate in the DNA.

In addition to the repair of replication errors, maintenance of the genetic information encoded in DNA also requires repair of damage to existing DNA. DNA molecules are constantly subjected to potentially harmful physical and chemical agents. Reactive chemicals, radioactive emissions, X-rays, and ultraviolet light can change nucleotides in ways that can affect encoded genetic information, usually adversely. Fortunately, these changes, or mutations, are usually corrected. Each cell continuously monitors and repairs its genetic material. Because repair of damaged DNA is so important to the survival of an organism, it's no surprise that many different DNA-repair enzymes have evolved; almost 100 are known in *E. coli* alone.

Like mismatch repair, most mechanisms for repairing DNA damage take advantage of the base-paired structure of DNA. Usually, a segment of the strand containing the damage is cut out (excised) by a DNA-cutting enzyme—a **nuclease**—and the resulting gap is filled in with nucleotides properly paired with the nucleotides in the undamaged strand. The enzymes involved in filling the gap are a DNA polymerase and ligase. DNA repair of this type is called **excision repair** (FIGURE 16.17).

One function of the DNA repair enzymes in our skin cells is to repair genetic damage caused by the ultraviolet rays of sunlight. The importance of this function to healthy people is underscored by the disorder xeroderma pigmentosum, which is caused by an inherited defect in an excision-repair enzyme. Individuals with this disorder are hypersensitive to sunlight; mutations in their skin cells caused by ultraviolet light are left uncorrected and cause skin cancer.

The ends of DNA molecules pose a special problem

Most DNA-repair processes involve DNA polymerases, but these enzymes are helpless to fix a "defect" that results from their own limitations. For linear DNA, such as the DNA of eukaryotic chromosomes, the fact that *a DNA polymerase can only add nucleotides to the 3′ end of a preexisting polynucleotide* creates a serious problem. The usual replication machinery

Thymine dimer distorts DNA molecule
A nuclease enzyme cuts the damaged DNA strand at two points
Repair synthesis by a DNA polymerase fills the gap
DNA ligase seals the remaining nick

FIGURE 16.17 ▪ Excision repair of DNA damage. A team of enzymes detects and repairs damaged DNA. One type of damage, shown here, is the covalent linking of thymine bases that are adjacent on a DNA strand. Such thymine dimers, induced by ultraviolet radiation, cause the DNA to buckle and interfere with DNA replication. Repair enzymes can excise the damaged region from the DNA and replace it with a normal DNA segment.

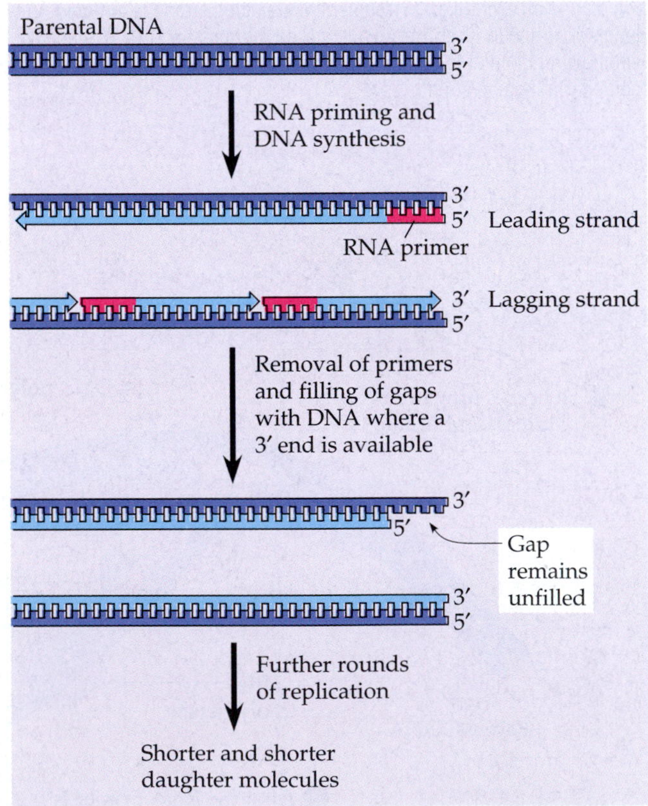
Parental DNA
3′
5′
RNA priming and DNA synthesis
3′
5′
Leading strand
RNA primer
3′
5′
Lagging strand
Removal of primers and filling of gaps with DNA where a 3′ end is available
3′
5′
Gap remains unfilled
3′
5′
Further rounds of replication
Shorter and shorter daughter molecules

FIGURE 16.18 ▪ The end-replication problem. For a linear DNA molecule, such as that of a eukaryotic chromosome, the usual DNA replication machinery is unable to replicate both ends. A gap is left at the 5′ end of each new strand (light blue) because DNA polymerase can only add nucleotides to a 3′ end. As a result, with each round of replication the DNA molecules get shorter, with potentially disastrous consequences. For simplicity we show only one end of a linear DNA molecule.

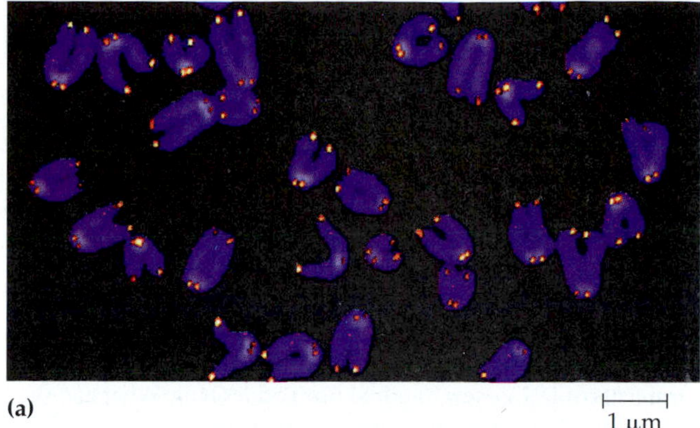

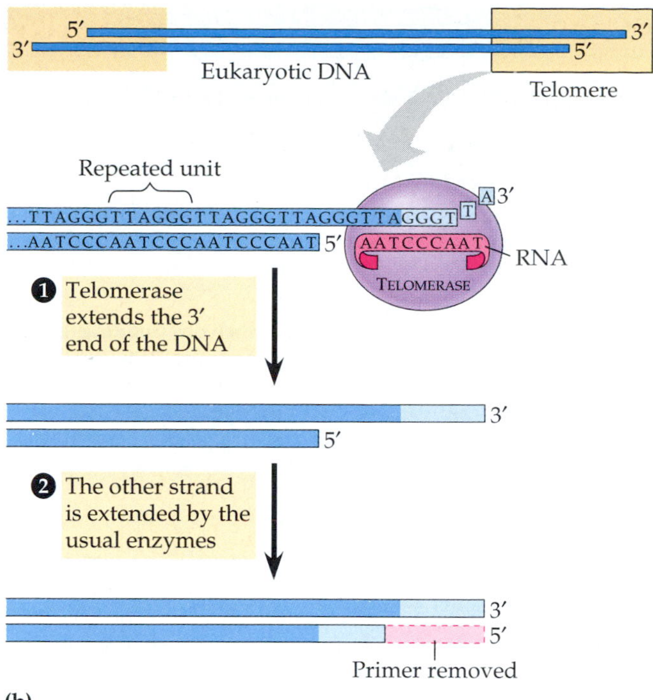

FIGURE 16.19 ▪ Telomeres and telomerase. Eukaryotes solve the end-replication problem by having expendable, noncoding sequences called telomeres at the ends of their DNA and the enzyme telomerase in some of their cells. **(a)** The bright orange stain marks the telomeres of these mouse chromosomes (LM). **(b)** Telomeric DNA consists of repeating six-nucleotide units. ① The enzyme telomerase has a short molecule of RNA with a sequence that serves as a template for extending the 3′ end of the telomere. ② The complementary strand of the telomere is extended by the usual combined actions of primase, DNA polymerase, and ligase. After the primer is removed, the result is a longer telomere with a 3′-end "overhang (LM)."

simply provides no way to complete the 5′ ends of daughter DNA strands; as a result, repeated replication produces shorter and shorter DNA molecules (FIGURE 16.18). If a cell divided enough times, essential genes would be deleted. Clearly, if this trend continued over generations, we would not be here today!

Prokaryotes avoid this problem by having circular DNA molecules (which have no ends), but what about eukaryotes? Eukaryotic chromosomal DNA molecules have special nucleotide sequences called **telomeres** at their ends. Telomeres do not contain genes; instead, the DNA consists of multiple repetitions of one short nucleotide sequence. The repeated unit in human telomeres, which is typical, is TTAGGG. The number of repetitions in a telomere varies between 100 and 1000 or so. Telomeric DNA protects the organism's genes from being eroded through successive rounds of DNA replication. In addition, telomeric DNA and special proteins associated with it somehow prevent the ends from activating the cell's systems for monitoring DNA damage. (A DNA molecule "seen" as broken may trigger signal-transduction pathways leading to cell-cycle arrest or cell death.)

There is another crucial piece to the eukaryotic solution to the problem of DNA ends: **telomerase**, a special enzyme that catalyzes the lengthening of telomeres. But how does telomerase synthesize DNA if the DNA template has been lost? Telomerase is unusual in having a molecule of RNA along with its protein. The RNA contains a nucleotide sequence that serves as the template for new telomere segments. FIGURE 16.19 shows how telomerase and DNA polymerase work together to lengthen telomeres.

Telomerase is *not* present in most cells of multicellular organisms like ourselves, and the DNA of dividing somatic cells does tend to be shorter in older individuals and in cultured cells that have divided many times. Thus, it is possible that telomeres are a limiting factor in the life span of certain tissues and even the organism as a whole. In any case, telomerase *is* present in germ line cells, those that give rise to gametes. The enzyme produces long telomeres in these cells and hence in the newborn.

Intriguingly, researchers have also found telomerase in somatic cells that are cancerous. Cells from large tumors often have unusually short telomeres, as one would expect for cells that have undergone many cell divisions. Progressive shortening would presumably lead eventually to self-destruction of the cancer, unless telomerase became available to stabilize telomere length. This is exactly what seems to happen in cancer cells and also in immortal strains of cultured cells (see Chapter 12). If telomerase is indeed an important factor in many cancers, it may provide a useful target for both cancer diagnosis and chemotherapy.

▪ ▪ ▪

DNA replication provides the copies of genes that parents pass to offspring via gametes. However, it is not enough that genes be copied and transmitted; they must also be expressed. How do genes manifest themselves in phenotypic characters such as eye color? In the next chapter we will examine the molecular basis of gene expression—how the cell translates genetic information encoded in DNA.

CHAPTER REVIEW

REVIEW OF KEY CONCEPTS

(with page numbers and key figures)

DNA AS THE GENETIC MATERIAL

- **The search for the genetic material led to DNA: *science as a process*** **(pp. 278–281, FIGURES 16.1, 16.2)** Experiments with bacteria and, later, with phages provided the first strong evidence that the genetic material is DNA.

- 16.1 **Watson and Crick discovered the double helix by building models to conform to X-ray data: *science as a process*** **(pp. 281–283, FIGURES 16.3, 16.5, 16.6)** Watson and Crick discovered that DNA is a double helix. Two antiparallel sugar-phosphate chains wind around the outside of the molecule; the nitrogenous bases project into the interior, where they hydrogen-bond in specific pairs, A with T and G with C.

DNA REPLICATION AND REPAIR

- 16.2 **During DNA replication, base pairing enables existing DNA strands to serve as templates for new complementary strands** **(pp. 284–285, FIGURE 16.7)** DNA replication is semiconservative: The parent molecule unwinds, and each strand then serves as a template for the synthesis of a new half-molecule according to base-pairing rules.

- 16.3 **A large team of enzymes and other proteins carries out DNA replication** **(pp. 286–289, FIGURES 16.15, 16.16)** Replication begins at origins of replication. Y-shaped replication forks form at opposite ends of a replication bubble, where the two DNA strands separate. DNA polymerases catalyze the synthesis of new DNA strands, working in the 5′ ⟶ 3′ direction. DNA synthesis at a replication fork yields a continuous leading strand and short, discontinuous segments of lagging strand. The fragments are then joined together by DNA ligase. DNA synthesis must start on the end of a primer, which is a short segment of RNA.

- **Enzymes proofread DNA during its replication and repair damage in existing DNA** **(pp. 289–290, FIGURE 16.17)** In mismatch repair, proteins proofread replicating DNA and correct errors in base pairing. In excision repair, repair enzymes fix DNA damaged by physical and chemical agents.

- **The ends of DNA molecules pose a special problem** **(pp. 290–291, FIGURE 16.19)** The ends of the linear DNA molecules of eukaryotic chromosomes, called telomeres, get shorter with each replication. The enzyme telomerase, present in certain cells, can re-extend the ends.

SELF-QUIZ

1. In his work with pneumonia-causing bacteria and mice, Griffith found that
   a. the protein coat from pathogenic cells was able to transform nonpathogenic cells
   b. heat-killed pathogenic cells caused pneumonia
   c. some chemical from pathogenic cells was transferred to non-pathogenic cells, making them pathogenic
   d. the polysaccharide coat of bacteria caused pneumonia
   e. bacteriophages injected DNA into bacteria

2. *E. coli* cells grown on ^{15}N medium are transferred to ^{14}N medium and allowed to grow for two generations (two rounds of DNA replication). DNA extracted from these cells is centrifuged. What density distribution of DNA would you expect in this experiment? Explain your answer.
   a. one high-density and one low-density band
   b. one intermediate-density band
   c. one high-density and one intermediate-density band
   d. one low-density and one intermediate-density band
   e. one low-density band

3. A biochemist isolated and purified various molecules needed for DNA replication. When she added some DNA, replication occurred, but the DNA molecules formed were defective. Each consisted of a normal DNA strand paired with numerous segments of DNA a few hundred nucleotides long. What had she probably left out of the mixture? Explain your answer.
   a. DNA polymerase
   b. ligase
   c. nucleotides
   d. Okazaki fragments
   e. primers

4. What is the basis for the difference in the synthesis of the leading and lagging strands of DNA molecules?
   a. The origins of replication occur only at the 5′ end.
   b. Helicases and single-strand binding proteins work at the 5′ end.
   c. DNA polymerase can join new nucleotides only to the 3′ end of a growing strand.
   d. DNA ligase works only in the 3′ ⟶ 5′ direction.
   e. Polymerase can only work on one strand at a time.

5. In analyzing the number of different bases in a DNA sample, which result would be consistent with the base-pairing rules? Explain your answer.
   a. $A = G$
   b. $A + G = C + T$
   c. $A + T = G + T$
   d. $A = C$
   e. $G = T$

6. The primer that initiates synthesis of a new DNA strand is
   a. RNA
   b. DNA
   c. an Okazaki fragment
   d. a structural protein
   e. a thymine dimer

7. A eukaryotic cell lacking telomerase would
   a. be unable to take up DNA from the surrounding solution
   b. be unable to identify and correct mismatched nucleotides in its daughter DNA strands
   c. experience a gradual reduction of chromosome length with each replication cycle
   d. have a greater potential to become cancerous
   e. incorporate one extraneous nucleotide for each Okazaki fragment added

8. The elongation of the *leading* strand during DNA synthesis
   a. progresses away from the replication fork
   b. occurs in the 3′ ⟶ 5′ direction
   c. produces Okazaki fragments
   d. depends on the action of DNA polymerase
   e. does not require a template strand

9. DNA is not completely stable; the spontaneous loss of amino groups from adenine, for example, results in hypoxanthine, an unnatural base, opposite thymine. What combination of molecules could the cell use to repair such damage?
   a. nuclease, DNA polymerase, DNA ligase
   b. telomerase, primase, DNA polymerase

c. telomerase, helicase, single-strand binding protein
d. DNA ligase, replication fork proteins, adenase
e. nuclease, telomerase, primase

10. Of the following, the most reasonable inference from the observation that defects in DNA repair enzymes contribute to some forms of cancer is that
a. cancer is generally inherited
b. uncorrected changes in DNA can cause cancer
c. cancer cannot occur when DNA repair enzymes work properly
d. mutations generally lead to cancer
e. cancer is caused by environmental factors that damage DNA repair enzymes

CHALLENGE QUESTIONS

1. Nerve cells do not replicate their DNA after reaching maturity. A cell biologist observed that there was x amount of DNA in a human nerve cell. In four other types of human cells, labeled A–D, she measured the following amounts of DNA: cell A, $2x$; cell B, $1.6x$; cell C, $0.5x$; cell D, x. Match the four cell types with the following choices: (a) sperm cell; (b) bone cell just beginning interphase; (c) skin cell in the S phase; (d) intestinal cell beginning mitosis. (See Chapter 12 to review the cell cycle and mitosis.)
2. Cells proofread and repair errors in the genetic information encoded in DNA. But some lasting mistakes, or mutations, still happen. The most error-prone stage in DNA replication occurs after parent DNA strands have separated but before new complementary strands are in place. Why might permanent errors be more likely then?
3. Write a short essay explaining the evidence that DNA is the genetic material.

SCIENCE, TECHNOLOGY, AND SOCIETY

1. Cooperation and competition are both common in science. What roles did these two social behaviors play in Watson and Crick's discovery of the double helix? How might competition between scientists accelerate progress in a scientific field? How might it slow progress?
2. DNA molecules with any desired base sequence can be synthesized and replicated in large quantities in the laboratory. It is also possible to determine the base sequence of a DNA molecule. For these reasons, it has been suggested that oil shipments could be tagged with DNA of known sequences. If an oil slick were found, the DNA could be sequenced and the offending ship tracked down. Suggest some other potential applications for "DNA labeling."

FURTHER READING

Greider, C. W., and E. H. Blackburn. "Telomeres, Telomerase and Cancer." *Scientific American*, February 1996.

Travis, J. "Crystal Clear: X-ray Snapshots Illuminate How Enzymes Stitch Together DNA." *Science News*, February 14, 1998.

Watson, J. D. *The Double Helix*. New York: Atheneum, 1968. The controversial bestseller by the codiscoverer of the double helix.

WEB LINKS

Visit the special edition of *The Biology Place* for BIOLOGY, Fifth Edition, at **http://www.biology.com/campbell**. Go to Chapter 16 for online resources, including learning activities, practice exams, and links to the following web sites:

"Primer on Molecular Genetics"
This introduction to molecular genetics contains numerous explanatory figures and is continuously being revised and updated by the Human Genome Management Information System at the Department of Energy.

"A Structure for Deoxyribose Nucleic Acid"
This site includes a full text and graphic reprint of Watson and Crick's groundbreaking paper on the structure of DNA published in *Nature* in 1953.

"James Dewey Watson"
Learn more about Francis Crick, James Watson, and Maurice Wilkins, who were awarded a Nobel Prize for their work in elucidating the structure of DNA.

"DNA Computing and Informatics at Surfaces"
This site gives a fascinating insight into possible future uses of DNA. Its information storage and retrieval powers may make it the best candidate for super computers of the future.

FROM GENE TO PROTEIN

The information content of DNA, the genetic material, is in the form of specific sequences of nucleotides along the DNA strands. But how is this information related to an organism's inherited traits? Put another way, what does a gene actually say? And how is its message translated by cells into a specific trait, such as brown hair or type A blood?

Consider, once again, Mendel's peas. One of the characters Mendel studied was stem length. Variation in a single gene accounts for the difference between the tall and dwarf varieties of pea plants in the drawing on this page. Mendel did not know the physiological basis of this phenotypic difference, but plant scientists have since worked out the explanation: Dwarf peas lack growth hormones called gibberellins, which stimulate the normal elongation of stems. A dwarf plant treated with gibberellins grows to normal height. Why do dwarf peas fail to make their own gibberellins? They are missing a key protein, an enzyme required for gibberellin synthesis. This example illustrates the main point of this chapter: The DNA inherited by an organism leads to specific traits by dictating the synthesis of proteins. Proteins are the links between genotype and phenotype. This chapter explores the flow of information from genes to proteins.

THE CONNECTION BETWEEN GENES AND PROTEINS

Before going into the details of how genes direct protein synthesis, let's step back and examine the fundamental relationship between genes and proteins and how this connection was discovered.

The study of metabolic defects provided evidence that genes specify proteins: *science as a process*

In 1909, British physician Archibald Garrod first suggested that genes dictate phenotypes through enzymes that catalyze specific chemical processes in the cell. Garrod postulated that the symptoms of an inherited disease reflect a person's inability to make a particular enzyme. He referred to such diseases as "inborn errors of metabolism." Garrod gave as one example the hereditary condition called alkaptonuria, in which the urine appears black because it contains the chemical alkapton, which darkens upon exposure to air. Garrod reasoned that normal individuals have an enzyme that breaks down alkapton, whereas alkaptonuric individuals have inherited an inability to make the enzyme that metabolizes alkapton.

How Genes Control Metabolism

Garrod's idea was ahead of its time, but research conducted several decades later supported his hypothesis that the function of a gene is to dictate the production of a specific enzyme. Biochemists accumulated much evidence that cells synthesize and degrade most organic molecules via metabolic pathways, in which each chemical reaction in a sequence is catalyzed by a specific enzyme. Such metabolic pathways lead, for instance, to the synthesis of the pigments that give *Drosophila* their eye color (see FIGURE 15.2). In the 1930s, George Beadle and Boris Ephrussi speculated that each of the various mutations affecting eye color in *Drosophila* blocks pigment synthesis at a specific step by preventing production of the enzyme that catalyzes that step. However, neither the chemical reactions nor the enzymes that catalyze them were known at the time.

A breakthrough in demonstrating the relationship between genes and enzymes came a few years later, after Beadle and Edward Tatum began to search for mutants of a bread mold, *Neurospora crassa.* They discovered mutants that differed from the wild-type mold in their nutritional needs (FIGURE 17.1). Wild-type *Neurospora* has modest food requirements. It can survive in the laboratory on agar (a moist support medium) mixed only with inorganic salts, sucrose, and the vitamin biotin. From this *minimal medium,* the mold uses its metabolic pathways to produce all the other molecules it needs. Beadle and Tatum identified mutants that could not survive on minimal medium, apparently because they were unable to synthesize certain essential molecules from the minimal ingredients. Such nutritional mutants are called **auxotrophs** (Gr. *auxely,* "to increase"; *trophikos,* "nourishment"). Most auxotrophs *can* survive on a *complete growth medium,* minimal medium supplemented with all 20 amino acids and a few other nutrients.

To pinpoint an auxotroph's metabolic defect, Beadle and Tatum took samples from the mutant growing on complete medium and distributed them to several different vials. Each vial contained minimal medium plus a single additional nutrient. The particular supplement that allowed growth indicated the metabolic defect. For example, if the only supplemented vial that supported growth of the mutant was the

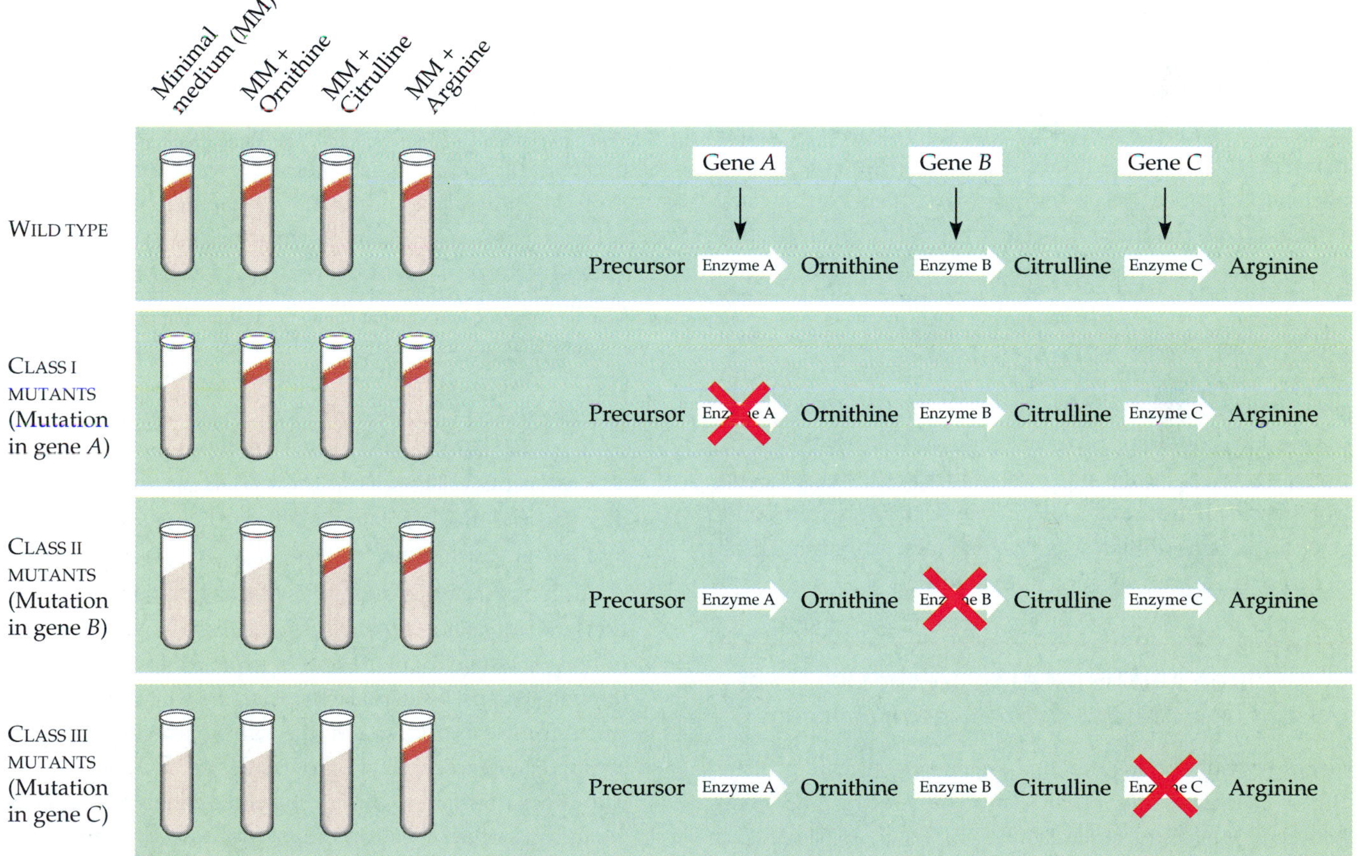

FIGURE 17.1 ▪ The one gene–one enzyme hypothesis. Beadle and Tatum based this idea on their study of nutritional mutants of the red bread mold *Neurospora crassa.* The wild-type strain requires only a minimal nutritional medium containing sucrose, essential minerals (inorganic salts), and one vitamin. The mold uses a multistep pathway to synthesize the amino acid arginine from a precursor. Beadle and Tatum identified three classes of mutants unable to synthesize arginine. Each mutant class had a metabolic block (X in this diagram) at a different step in the pathway. For example, class II mutants failed to grow on minimal medium or minimal medium supplemented with ornithine. Adding either citrulline or arginine to the nutritional medium enabled these mutants to grow. Beadle and Tatum deduced that class II mutants lacked the enzyme that converts ornithine to citrulline. Adding citrulline to the medium bypasses the metabolic block and allows the mold to survive. The other classes of mutants lacked different enzymes. Beadle and Tatum concluded that various mutations were abnormal variations of different genes, each gene dictating the production of one enzyme; hence, the one gene–one enzyme hypothesis.

one fortified with the amino acid arginine, the researchers could conclude that the mutant was defective in the pathway that normally synthesizes arginine.

With further experimentation, a defect could be described even more specifically. For instance, consider three of the steps in the synthesis of arginine: A precursor nutrient is converted to ornithine, which is converted to citrulline, which is converted to arginine (see FIGURE 17.1). Beadle and Tatum could distinguish among the various arginine auxotrophs they found. Some required arginine, others required either arginine or citrulline, and still others could grow when any of the three compounds—arginine, citrulline, or ornithine—was provided. These three classes of mutants, Beadle and Tatum reasoned, must be blocked at different steps in the pathway that synthesizes arginine. They concluded that each mutant lacked a different enzyme. Assuming that each mutant was defective in a single gene, they formulated the one gene–one enzyme hypothesis, which states that the function of a gene is to dictate the production of a specific enzyme.

One Gene–One Polypeptide

As researchers learned more about proteins, they made minor revisions in the one gene–one enzyme hypothesis. Not all proteins are enzymes. Keratin, the structural protein of animal hair, and the hormone insulin are two examples of nonenzyme proteins. Because proteins that are not enzymes are nevertheless gene products, molecular biologists began to think in terms of one gene–one protein. However, many proteins are constructed from two or more different polypeptide chains (see Chapter 5), and each polypeptide is specified by its own gene. For example, hemoglobin, the oxygen-transporting protein of vertebrate red blood cells, is built from two kinds of polypeptides, and thus *two* genes code for this protein. We can therefore restate Beadle and Tatum's idea as the **one gene–one polypeptide hypothesis**. Note, however, that it is common to refer to proteins, rather than polypeptides, as the gene products, a practice you will encounter in this book.

Transcription and translation are the two main processes linking gene to protein: *an overview*

Genes provide the instructions for making specific proteins. 17.1 But a gene does not build a protein directly. The bridge between DNA and protein synthesis is RNA. You learned in Chapter 5 that RNA is chemically similar to DNA, except that it contains ribose instead of deoxyribose as its sugar and has the nitrogenous base uracil rather than thymine (see FIGURE 5.27). Thus, each nucleotide along a DNA strand has deoxyribose as its sugar and A, G, C, or T as its base; each nucleotide along an RNA strand has ribose as its sugar and A, G, C, or U as its base. An RNA molecule almost always consists of a single strand.

It is customary to describe the flow of information from gene to protein in linguistic terms because both nucleic acids and proteins are polymers with specific sequences of monomers that convey information, much as specific sequences of letters communicate information in a written language. In DNA or RNA, the monomers are the four types of nucleotides, which differ in their nitrogenous bases. Genes are typically hundreds or thousands of nucleotides long, each gene having a specific sequence of bases. Each polypeptide of a protein also has monomers arranged in a particular linear order (the protein's primary structure; see Chapter 5), but its monomers are the 20 amino acids. Thus, nucleic acids and proteins contain information written in two different chemical languages. To get from DNA, written in one language, to protein, written in the other, requires two major stages, transcription and translation.

Transcription is the synthesis of RNA under the direction of DNA. Both nucleic acids use the same language, and the information is simply transcribed, or copied, from one molecule to the other. Just as a DNA strand provides a template for the synthesis of a new complementary strand during DNA replication, it provides a template for assembling a sequence of RNA nucleotides. The resulting RNA molecule is a faithful transcript of the gene's protein-building instructions. This type of RNA molecule is called **messenger RNA (mRNA)**, because it carries a genetic message from the DNA to the protein-synthesizing machinery of the cell (FIGURE 17.2a). (Transcription is the general term for the synthesis of *any* kind of RNA from DNA. Later in this chapter you will learn about other types of RNA produced by transcription.)

Translation is the actual synthesis of a polypeptide, which occurs under the direction of mRNA. During this stage there is a change in language: The cell must translate the base sequence of an mRNA molecule into the amino acid sequence of a polypeptide. The sites of translation are ribosomes, complex particles that facilitate the orderly linking of amino acids into polypeptide chains.

Although the basic mechanics of transcription and translation are similar for prokaryotes and eukaryotes, there is an important difference in the flow of genetic information within the cells. Because bacteria lack nuclei, their DNA is not segregated from ribosomes and the other protein-synthesizing equipment. Transcription and translation are coupled, with ribosomes attaching to the leading end of an mRNA molecule while transcription is still in progress (see FIGURE 17.20, p. 311). In a eukaryotic cell, by contrast, the nuclear envelope separates transcription from translation in space and time. Transcription occurs in the nucleus, and mRNA is dispatched to the cytoplasm, where translation occurs (FIGURE 17.2b). But before they can leave the nucleus, eukaryotic RNA transcripts are modified in various ways to produce the final, functional mRNA. Thus, in a two-step process, the transcription of a eukaryotic gene results in *pre-mRNA*, and **RNA processing**

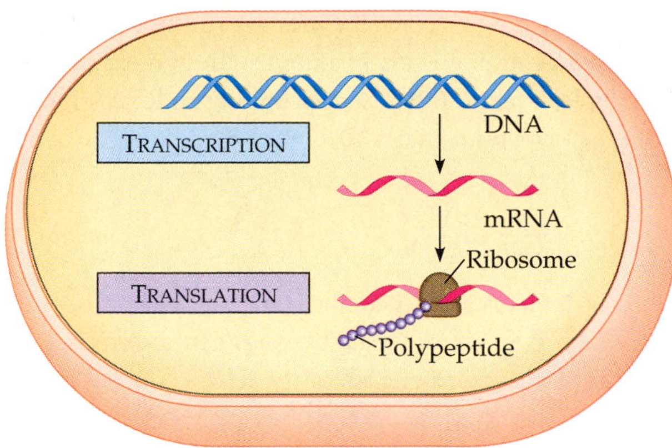

(a) Prokaryotic cell

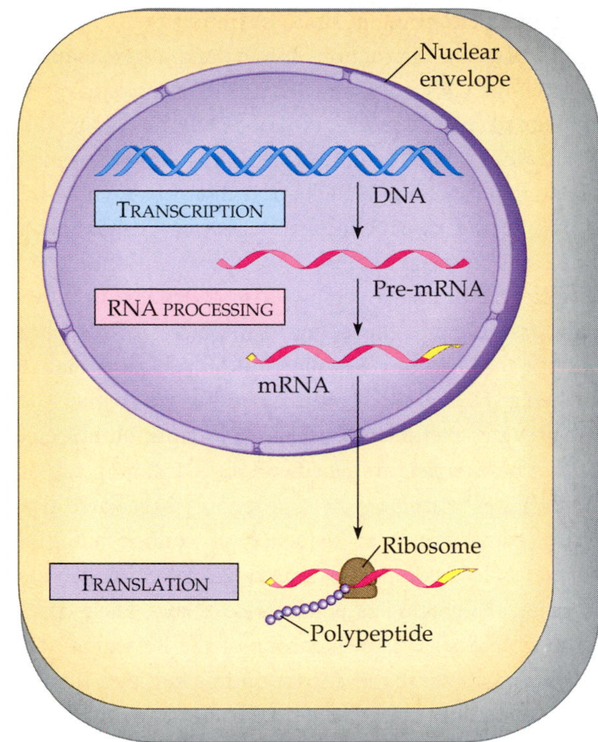

(b) Eukaryotic cell

17.1 **FIGURE 17.2 ▪ Overview: the roles of transcription and translation in the flow of genetic information.** In a cell's chain of command, inherited information flows from DNA to RNA to protein. The two main stages of information flow are transcription and translation. In transcription, a gene provides the instructions for synthesizing a messenger RNA (mRNA) molecule. In translation, the information encoded in mRNA determines the order of amino acids that are joined to form a specific polypeptide. Ribosomes are the sites of translation. **(a)** In a prokaryotic cell, which lacks a nucleus, mRNA produced by transcription is immediately translated without additional processing. **(b)** In a eukaryotic cell, transcription and translation occur in separate compartments, the nucleus and the cytoplasm. The mRNA moves from nucleus to cytoplasm via pores in the nuclear envelope. The original RNA transcript, called pre-mRNA, is processed in various ways by enzymes before leaving the nucleus as mRNA. A miniature version of this illustration (or, in some cases, part a) accompanies several figures later in the chapter as an orientation diagram to help you see where the content of a particular figure fits into the overall scheme.

yields the finished mRNA. A more general term for an initial RNA transcript is **primary transcript.**

Let's summarize the main point of our overview of protein synthesis: Genes program protein synthesis via genetic messages in the form of messenger RNA. Put another way, cells are governed by a molecular chain of command: DNA ⟶ RNA ⟶ protein. The next section discusses how the instructions for assembling amino acids into a specific order are encoded in nucleic acids.

In the genetic code, nucleotide triplets specify amino acids

When biologists began to suspect that the instructions for protein synthesis were encoded in DNA, they recognized a problem: There are only 4 nucleotides to specify 20 amino acids. Thus, the genetic code cannot be a language like Chinese, where each written symbol corresponds to a single word. If each nucleotide base were translated into an amino acid, only 4 of the 20 amino acids could be specified. Would a language of two-letter code words suffice? The base sequence AG, for example, could specify one amino acid, and GT could specify another. Since there are four bases, this would give us 16 (that is, 4^2) possible arrangements—still not enough to code for all 20 amino acids.

Triplets of nucleotide bases are the smallest units of uniform length that can code for all the amino acids. If each arrangement of three consecutive bases specifies an amino acid, there can be 64 ($=4^3$) possible code words—more than enough to specify all the amino acids. Experiments have verified that the flow of information from gene to protein is based on a **triplet code**: The genetic instructions for a polypeptide chain are written in the DNA as a series of three-nucleotide words. For example, the base triplet AGT at a particular position along a DNA strand says to place the amino acid serine at the corresponding position of the polypeptide to be produced.

A cell cannot directly translate a gene into amino acids. The intermediate step is transcription, during which the gene determines the sequence of base triplets along the length of an mRNA molecule. For each gene, only one of the two DNA strands is transcribed. This strand is called the **template strand**, because it provides the template for ordering the sequence of nucleotides in an RNA transcript. (The other strand provides the instructions for making a new template strand when the DNA replicates.) A given DNA strand can be the template strand in some regions of a DNA molecule, while in other regions along the double helix it is the complementary strand that functions as the template for RNA synthesis.

An mRNA molecule is complementary rather than identical to its DNA template because RNA bases are assembled on the template according to base-pairing rules (FIGURE 17.3). The pairs are similar to those that form during DNA replication, except that U, the RNA substitute for T, pairs with A. Thus, when a DNA strand is transcribed, the base triplet ACC in DNA provides a template for UGG in the mRNA molecule. The mRNA base triplets are called **codons**. For example, UGG is the codon for the amino acid tryptophan (abbreviated Trp).

During translation, the sequence of codons along an mRNA molecule is decoded, or translated, into a sequence of amino acids making up a polypeptide chain. Each codon along the mRNA specifies which one of the 20 amino acids will be incorporated at the corresponding position along a polypeptide. Because codons are base triplets, the number of nucleotides making up a genetic message must be three times the number of amino acids making up the protein product. For example, it takes 300 nucleotides along an RNA strand to code for a polypeptide that is 100 amino acids long.

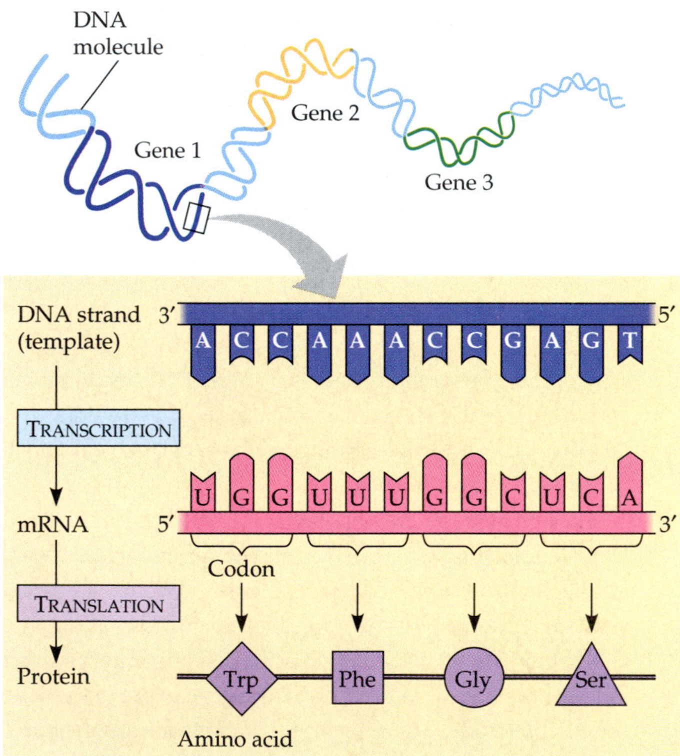

FIGURE 17.3 ▪ The triplet code. For each gene, one of the two strands of DNA functions as a template for transcription—the synthesis of an mRNA molecule of complementary sequence. The same base-pairing rules that apply to DNA synthesis also guide transcription, but the base uracil (U) takes the place of thymine (T) in RNA. During translation, the genetic message (mRNA) is read as a sequence of base triplets, analogous to three-letter code words. Each of these triplets, called a codon (bracketed in the figure), specifies the amino acid to be added at the corresponding position along a growing polypeptide chain. The gene, its mRNA transcript, and the polypeptide product are all much longer than the segments shown here.

Cracking the Genetic Code

Molecular biologists cracked the code of life in the early 1960s, when a series of elegant experiments disclosed the amino acid translations of each of the RNA codons. The first codon was deciphered in 1961 by Marshall Nirenberg of the National Institutes of Health. Nirenberg had synthesized an artificial mRNA by linking identical RNA nucleotides containing uracil as their base. No matter where this message started or stopped, it could contain only one codon in repetition: UUU. Nirenberg added this "poly U" to a test-tube mixture containing amino acids, ribosomes, and the other components required for protein synthesis. His artificial system translated the poly U into a polypeptide containing a single amino acid, phenylalanine (Phe), strung together as a long polyphenylalanine chain. Thus, Nirenberg determined that the mRNA codon UUU specifies the amino acid phenylalanine. Soon, the amino acids specified by the codons AAA, GGG, and CCC were also determined.

Although more elaborate techniques were required to decode mixed triplets such as AUA and CGA, all 64 codons were deciphered by the mid-1960s. As FIGURE 17.4 shows, 61 of the 64 triplets code for amino acids. Notice that the codon AUG has a dual function: It not only codes for the amino acid methionine (Met), but also functions as a "start" signal, or an initiation codon. Genetic messages begin with the mRNA codon AUG, which signals the protein-synthesizing machinery to begin translating the mRNA at that location. (Since AUG also stands for methionine, polypeptide chains begin with methionine when they are synthesized. However, an enzyme may subsequently remove this starter amino acid from a chain.) The remaining three codons do not designate amino acids. Instead, they are "stop" signals, or termination codons, marking the end of translation.

Notice in FIGURE 17.4 that there is redundancy in the genetic code, but no ambiguity. For example, although codons GAA and GAG both specify glutamic acid (redundancy), neither of them ever specifies any other amino acid (no ambiguity). The redundancy in the code is not altogether random. In many cases, codons that are synonyms for a particular amino acid differ only in the third base of the triplet. We will consider a possible explanation for this redundancy later in the chapter.

Our ability to extract the intended message from a written language depends on reading the symbols in the correct sequence and groupings—that is, in the correct **reading frame**. Consider this statement: "The red dog ate the cat." Read the words out of order, and you may get the unintended

First base (5′ end)	Second base: U	C	A	G	Third base (3′ end)
U	UUU Phe	UCU Ser	UAU Tyr	UGU Cys	U
	UUC	UCC	UAC	UGC	C
	UUA Leu	UCA	UAA Stop	UGA Stop	A
	UUG	UCG	UAG Stop	UGG Trp	G
C	CUU Leu	CCU Pro	CAU His	CGU Arg	U
	CUC	CCC	CAC	CGC	C
	CUA	CCA	CAA Gln	CGA	A
	CUG	CCG	CAG	CGG	G
A	AUU Ile	ACU Thr	AAU Asn	AGU Ser	U
	AUC	ACC	AAC	AGC	C
	AUA	ACA	AAA Lys	AGA Arg	A
	AUG Met or start	ACG	AAG	AGG	G
G	GUU Val	GCU Ala	GAU Asp	GGU Gly	U
	GUC	GCC	GAC	GGC	C
	GUA	GCA	GAA Glu	GGA	A
	GUG	GCG	GAG	GGG	G

FIGURE 17.4 ▪ The dictionary of the genetic code. The three bases of an mRNA codon are designated here as the first, second, and third bases, reading in the 5′⟶3′ direction along the mRNA. Practice using this dictionary by finding the codon UGG, the first codon in FIGURE 17.3. This is the only codon for the amino acid tryptophan, but most amino acids are specified by two or more codons. For example, both UUU and UUC stand for the amino acid phenylalanine (Phe). When either of these codons is read along an mRNA molecule, phenylalanine will be incorporated into the growing polypeptide chain. We can think of UUU and UUC as synonyms in the genetic code. Notice that the codon AUG not only stands for the amino acid methionine (Met) but also functions as a "start" signal for ribosomes, the cell organelles that actually assemble polypeptides, to begin translating the mRNA at that location. Three of the 64 codons function as "stop" signals. Any one of these termination codons marks the end of a genetic message.

message, "The red cat ate the dog." Group the letters incorrectly, and you may think that the first word of the message is "There." The reading frame is also important in the molecular language of cells. The short stretch of polypeptide shown in FIGURE 17.3, for instance, will only be made correctly if the mRNA nucleotides UGG UUU GGC UCA are read from left to right in groups of three. Although a genetic message is written with no spaces between the codons, the cell's protein-synthesizing machinery reads the message as a series of nonoverlapping three-letter words. The message is *not* read as a series of overlapping words—UGGUUU, and so on—which would convey a very different message.

Let's summarize what we have just covered. Genetic information is encoded as a sequence of nonoverlapping base triplets, or codons, each of which is translated into a specific amino acid during protein synthesis.

The genetic code must have evolved very early in the history of life

The genetic code is nearly universal, shared by organisms from the simplest bacteria to the most complex plants and animals. The RNA codon CCG, for instance, is translated as the amino acid proline in all organisms whose genetic code has been examined. In laboratory experiments, genes can be transcribed and translated after they are transplanted from one species to another (FIGURE 17.5). One important application is that bacteria can be programmed by the insertion of human genes to synthesize certain human proteins that have important medical uses. Such applications have produced many exciting developments in biotechnology, which you will learn about in Chapter 20.

There are some exceptions to the universality of the genetic code, translation systems where a few codons differ from the standard ones. The main examples are found in certain single-celled eukaryotes, such as *Paramecium*, an organism you may know from the lab. Other examples are found in certain

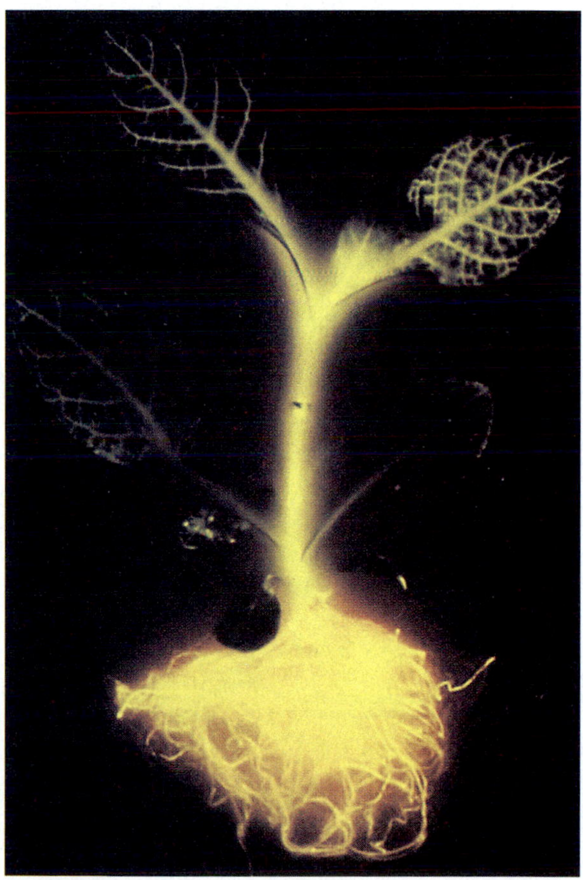

FIGURE 17.5 ▪ A tobacco plant expressing a firefly gene. Because diverse forms of life share a common genetic code, it is possible to program one species to produce proteins characteristic of another species by transplanting DNA. In this experiment, researchers were able to incorporate a gene from a firefly into the DNA of a tobacco plant. The gene codes for the firefly enzyme that catalyzes the chemical reaction that releases energy in the form of light.

mitochondria and chloroplasts, which transcribe and translate the genes carried by their small amount of DNA. However, the evolutionary significance of the code's *near* universality is clear. A language shared by all living things must have been operating very early in the history of life—early enough to be present in the organisms that were the common ancestors of all modern organisms. A shared genetic vocabulary is a reminder of the kinship that bonds all life on Earth.

Now that we have considered the linguistic logic and evolutionary significance of the genetic code, we are ready to reexamine transcription, translation, and related topics in more detail.

THE SYNTHESIS AND PROCESSING OF RNA

Transcription is the DNA-directed synthesis of RNA: *a closer look*

17.2 Messenger RNA, the carrier of information from DNA to the cell's protein-synthesizing machinery, is transcribed from the template strand of a gene. An enzyme called an **RNA polymerase** pries the two strands of DNA apart and hooks together the RNA nucleotides as they base-pair along the DNA tem-

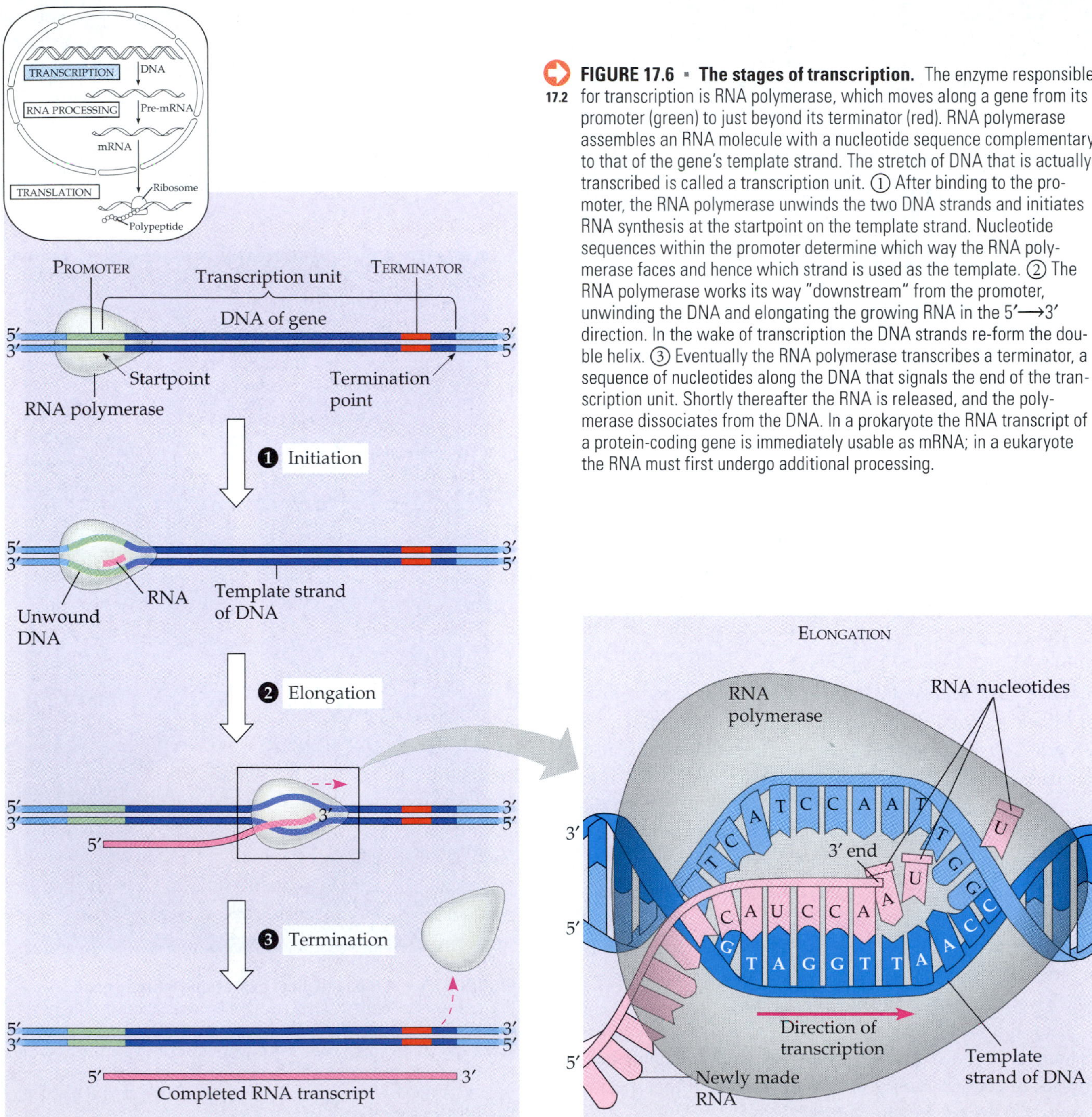

17.2 **FIGURE 17.6 ▪ The stages of transcription.** The enzyme responsible for transcription is RNA polymerase, which moves along a gene from its promoter (green) to just beyond its terminator (red). RNA polymerase assembles an RNA molecule with a nucleotide sequence complementary to that of the gene's template strand. The stretch of DNA that is actually transcribed is called a transcription unit. ① After binding to the promoter, the RNA polymerase unwinds the two DNA strands and initiates RNA synthesis at the startpoint on the template strand. Nucleotide sequences within the promoter determine which way the RNA polymerase faces and hence which strand is used as the template. ② The RNA polymerase works its way "downstream" from the promoter, unwinding the DNA and elongating the growing RNA in the 5′⟶3′ direction. In the wake of transcription the DNA strands re-form the double helix. ③ Eventually the RNA polymerase transcribes a terminator, a sequence of nucleotides along the DNA that signals the end of the transcription unit. Shortly thereafter the RNA is released, and the polymerase dissociates from the DNA. In a prokaryote the RNA transcript of a protein-coding gene is immediately usable as mRNA; in a eukaryote the RNA must first undergo additional processing.

plate. Like the DNA polymerases that function in DNA replication, RNA polymerases can add nucleotides only to the 3′ end of the growing polymer. Thus, an RNA molecule elongates in its 5′ ⟶ 3′ direction. Specific sequences of nucleotides along the DNA mark where transcription of a gene begins and ends. The stretch of DNA that is transcribed into an RNA molecule is called a **transcription unit** (FIGURE 17.6).

Bacteria have a single type of RNA polymerase that synthesizes not only mRNA but also other types of RNA that function in protein synthesis. In contrast, eukaryotes have three types of RNA polymerase in their nuclei, numbered I, II, and III. The one used for mRNA synthesis is RNA polymerase II. In the discussion of transcription that follows, we start with the features of mRNA synthesis common to both prokaryotes and eukaryotes and then describe some key differences.

The three stages of transcription, as shown in FIGURE 17.6, are initiation, elongation, and termination of the RNA chain.

RNA Polymerase Binding and Initiation of Transcription

A region of DNA where RNA polymerase attaches and initiates transcription is known as a **promoter**. A promoter includes the transcription startpoint (the nucleotide where RNA synthesis actually begins) and typically extends several dozen nucleotide pairs "upstream" from the startpoint. In addition to determining where transcription starts, the promoter determines which of the two strands of the DNA helix is used as the template.

Certain sections of a promoter are especially important for binding RNA polymerase. In prokaryotes, the RNA polymerase itself specifically recognizes and binds to the promoter. In eukaryotes, by contrast, a collection of proteins called **transcription factors** mediate the binding of RNA polymerase and the initiation of transcription. Only after certain transcription factors are bound to the promoter does the RNA polymerase bind to it. The completed assembly of transcription factors and RNA polymerase bound to the promoter is called a **transcription initiation complex**.

The interaction between eukaryotic RNA polymerase and transcription factors is an example of the special importance of protein-protein interactions in controlling eukaryotic transcription (as we will discuss further in Chapter 19). FIGURE 17.7 shows the role of transcription factors and a crucial promoter DNA sequence called a **TATA box** in forming the initiation complex. Once the polymerase is firmly attached to the promoter DNA, the two DNA strands unwind there, and the enzyme starts transcribing the template strand.

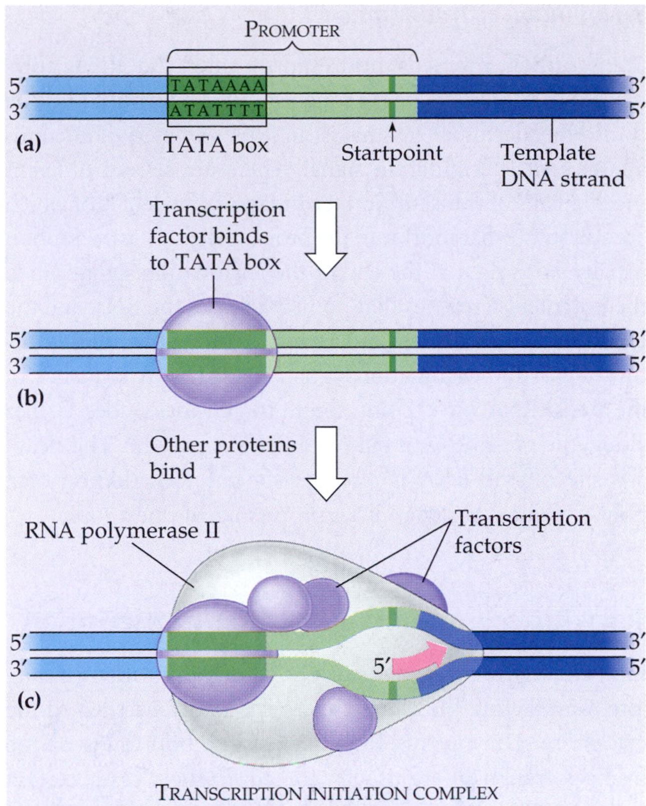

FIGURE 17.7 ▪ The initiation of transcription at a eukaryotic promoter. In eukaryotic cells, the enzyme that transcribes protein-coding genes into pre-mRNA is RNA polymerase II. This enzyme initiates RNA synthesis at promoters that commonly include a *TATA box,* a nucleotide sequence typically something like TATAAAA. (By convention, nucleotide sequences are given as they occur on the *non*-template strand.) **(a)** Within the promoter, the TATA box is located about 25 nucleotides upstream from the transcriptional startpoint. **(b)** RNA polymerase II cannot recognize the TATA box and other landmarks of the promoter on its own. Another protein, a transcription factor that recognizes the TATA box, binds to the DNA before the RNA polymerase can do so. **(c)** Additional transcription factors (purple) join the polymerase on the DNA. Protein-protein interactions are critical in the formation of the eukaryotic transcription initiation complex. The DNA double helix unwinds, and RNA synthesis begins at the startpoint on the template strand.

Elongation of the RNA Strand

As RNA polymerase moves along the DNA, it continues to untwist the double helix, exposing about 10–20 DNA bases at a time for pairing with RNA nucleotides (see FIGURE 17.6). The enzyme adds nucleotides to the 3′ end of the growing RNA molecule as it continues along the double helix. In the wake of this advancing wave of RNA synthesis, the DNA double helix re-forms and the new RNA molecule peels away from its DNA template. Transcription progresses at a rate of about 60 nucleotides per second in eukaryotes.

A single gene can be transcribed simultaneously by several molecules of RNA polymerase following each other like trucks in a convoy. The growing strands of RNA trail off from each polymerase, with the length of each new strand reflecting how far along the template the enzyme has traveled from the startpoint (see FIGURE 17.20, p. 311). The congregation of many polymerase molecules simultaneously transcribing a single gene increases the number of mRNA molecules and helps a cell make a protein in large amounts.

Termination of Transcription

Transcription proceeds until shortly after the RNA polymerase transcribes a DNA sequence called a **terminator**. The transcribed terminator—that is, an *RNA* sequence—functions as the actual termination signal. There are several different mechanisms of transcription termination, the details of which are still somewhat murky. In the prokaryotic cell, transcription usually stops right at the end of the termination signal; when the polymerase reaches that point it releases the RNA and the DNA. By contrast, in the eukaryotic cell the polymerase continues past the termination signal, an AAUAAA sequence in the pre-mRNA. At a point about 10–35 nucleotides farther along, the pre-mRNA is cut free from the enzyme. The cleavage site on the RNA is also the site for the addition of a poly(A) tail—one step of RNA processing, our next topic.

Eukaryotic cells modify RNA after transcription

Enzymes in the eukaryotic nucleus modify pre-mRNA in var-
17.3 ious ways before the genetic messages are dispatched to the cytoplasm. During this RNA processing, both ends of the primary transcript are usually altered. In most cases, certain interior sections of the molecule are then cut out, and the remaining parts spliced together.

Alteration of mRNA Ends

Each end of a pre-mRNA molecule is modified in a particular way. The 5′ end, the end made first during transcription, is immediately capped off with a modified form of a guanine (G) nucleotide. This **5′ cap** has at least two important functions. First, it helps protect the mRNA from degradation by hydrolytic enzymes. Second, after the mRNA reaches the cytoplasm, the 5′ cap functions as part of an "attach here" sign for ribosomes. The other end of an mRNA molecule, the 3′ end, is also modified before the message exits the nucleus. To the 3′ end an enzyme adds a **poly(A) tail** consisting of 30 to 200 adenine nucleotides. Like the 5′ cap, the poly(A) tail inhibits degradation of the RNA and helps the ribosome attach to it. The poly(A) tail also seems to facilitate the export of mRNA from the nucleus. FIGURE 17.8 shows a eukaryotic mRNA molecule with cap and tail; it also shows the nontranslated *leader* and *trailer* segments of RNA to which they are attached.

Split Genes and RNA Splicing

The most remarkable stage of RNA processing in the eukaryotic nucleus is the removal of a large portion of the RNA molecule that is initially synthesized—a cut-and-paste job called **RNA splicing** (FIGURE 17.9). The average length of a transcription unit along a eukaryotic DNA molecule is about 8000 nucleotides, so the primary RNA transcript is also that long. But it takes only about 1200 nucleotides to code for an average-sized protein of 400 amino acids. (Remember, each amino acid is encoded by a *triplet* of nucleotides.) This means that most eukaryotic genes and their RNA transcripts have long noncoding stretches of nucleotides, regions that are not translated. Even more surprising is that most of these noncoding sequences are interspersed between coding segments of the gene, and thus between coding segments of the pre-mRNA. In other words, the sequence of DNA nucleotides that codes for a eukaryotic polypeptide is not continuous. The noncoding segments of nucleic acid that lie between coding regions are called intervening sequences, or **introns** for short. The other regions are called **exons**, because they are eventually expressed—that is, translated into amino acid sequences. (The exceptions are the leader and trailer portions of the exons at the two ends of the RNA; see FIGURE 17.9.) Richard Roberts and Phillip Sharp, who independently found evidence of "split genes" in 1977, shared a Nobel Prize in 1993 for this discovery.

RNA polymerase transcribes both introns and exons from the DNA, creating an oversized RNA molecule. But this pre-mRNA never leaves the nucleus; the mRNA molecule that enters the cytoplasm is an abridged version of the primary transcript. The introns are excised from the molecule and the exons joined together to form an mRNA molecule with a continuous coding sequence.

17.3 **FIGURE 17.8 • RNA processing: addition of the 5′ cap and poly(A) tail.** Enzymes modify the two ends of a eukaryotic pre-mRNA molecule. A cap consisting of a modified guanosine triphosphate is added to the 5′ end of the RNA as soon as it is made. A poly(A) tail consisting of up to 200 adenine nucleotides is attached to the 3′ end, the end created by cleavage downstream of the AAUAAA termination signal. The modified ends help protect the RNA from degradation, and the poly(A) tail may facilitate the export of mRNA from the nucleus. When the mRNA reaches the cytoplasm, the modified ends, in conjunction with certain cytoplasmic proteins, signal a ribosome to attach to the mRNA. The leader and trailer segments of RNA are not translated.

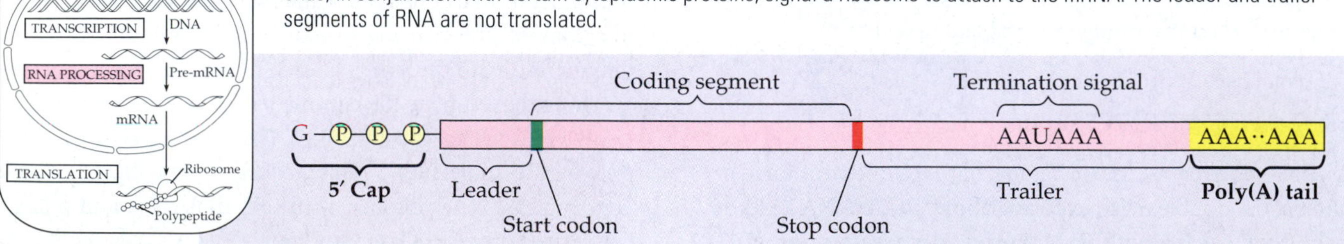

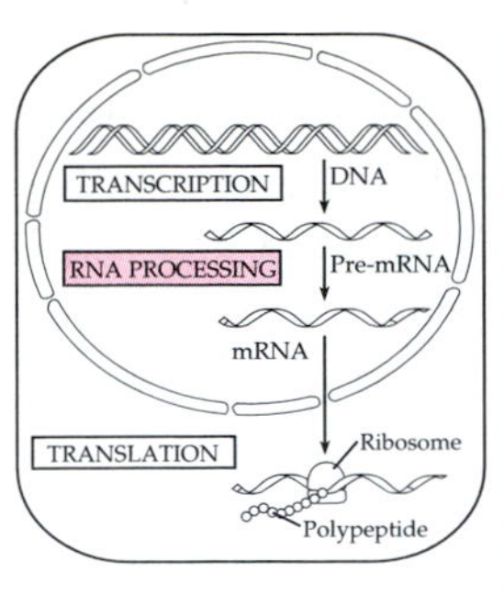

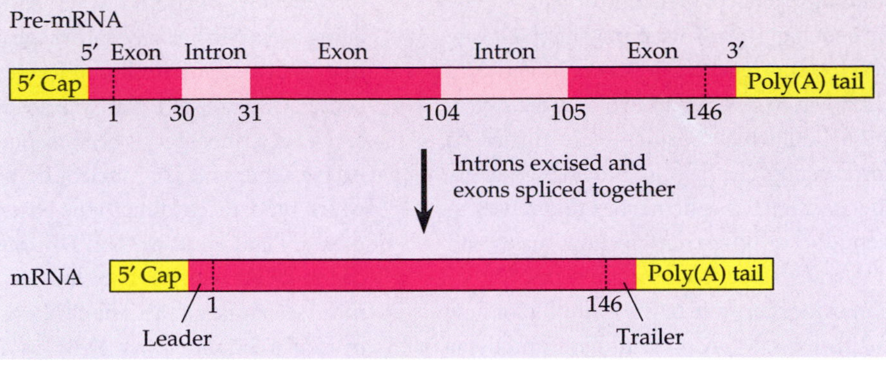

17.3 **FIGURE 17.9 ▪ RNA processing: RNA splicing.** The gene depicted here codes for β-globin, one of the polypeptides of hemoglobin. β-globin is 146 amino acids long. (The numbers under the RNA refer to codons.) Its gene has three segments containing coding regions, called exons, that are separated by noncoding regions, called introns. The entire gene is transcribed to form a molecule of pre-mRNA. However, before the RNA leaves the nucleus as mRNA, the introns are excised and the exons are spliced together. Notice that the exons at the ends of the mRNA include noncoding sections (leader and trailer), to which are attached the cap and tail.

How is mRNA splicing carried out? Researchers have learned that the signals for RNA splicing are short nucleotide sequences at the ends of introns. Particles called *small nuclear ribonucleoproteins,* or *snRNPs* (pronounced "snurps"), recognize these splice sites. As the name implies, snRNPs are located in the cell nucleus and are composed of RNA and protein molecules. The RNA in a snRNP particle is called a *small nuclear RNA (snRNA);* each molecule is about 150 nucleotides long. Several different snRNPs join with additional proteins to form an even larger assembly called a **spliceosome**, which is almost as big as a ribosome. The spliceosome interacts with the splice sites at the ends of an intron. It cuts at specific points to release the intron, then immediately joins together the two exons that flanked the intron (FIGURE 17.10). There is strong evidence that snRNA plays a role in the catalytic process, as well as in spliceosome assembly and splice-site recognition. The idea of a catalytic role for snRNA arose from the discovery of **ribozymes**, RNA molecules that function as enzymes.

Ribozymes

Like pre-mRNA, other kinds of primary transcripts may also be spliced, but by diverse mechanisms that do not involve spliceosomes. However, as with mRNA splicing, RNA is often involved in catalyzing the reactions. In a few cases the splicing occurs completely without proteins or even extra RNA molecules: The intron RNA itself catalyzes the process! For example, in the protozoan *Tetrahymena,* self-splicing occurs in the production of an RNA component of the organism's ribosomes (ribosomal RNA). Like enzymes, ribozymes function as catalysts. The discovery of ribozymes rendered the statement "All biological catalysts are proteins" obsolete.

The Functional and Evolutionary Importance of Introns

What are the biological functions of introns and RNA splicing? One idea is that introns play regulatory roles in the cell. At

17.3 **FIGURE 17.10 ▪ The roles of snRNPs and spliceosomes in mRNA splicing.** After a eukaryotic gene containing exons and introns is transcribed, the RNA transcript combines with small nuclear ribonucleoproteins (snRNPs) and other proteins to form a molecular complex called a spliceosome. Within the spliceosome the RNA of certain snRNPs base-pairs with nucleotides at the ends of the intron, the RNA transcript is cut to release the intron, and the exons are spliced together. The spliceosome then comes apart, releasing mRNA, which now contains only exons. The diagram shows only a portion of the RNA transcript; additional introns and exons lie downstream from the ones pictured here.

least some introns contain sequences that control gene activity in some way, and the splicing process itself may help regulate the passage of mRNA from nucleus to cytoplasm. Moreover, a number of genes are known to give rise to two or more different proteins, depending on which segments are treated as exons during RNA processing. The fruit fly provides an interesting example of alternative RNA splicing: Sex differences in this animal are largely due to differences in how males and females splice the RNA transcribed from certain genes.

Introns also play an important role in the evolution of new and potentially useful proteins. Many proteins have a modular architecture consisting of discrete structural and functional components called **domains**. One domain of an enzymatic protein, for instance, might include the active site, while another might attach the protein to a cellular membrane. In many cases the exons of a "split gene" code for the different domains of a protein. It is relatively easy for genetic recombination to modify the function of such a protein by changing just one of its domains without altering the other domains. Because the coding regions (exons) for the protein are separated by stretches of noncoding (intron) DNA, the frequency of recombination within a split gene can be higher than for a gene lacking introns. Introns increase the opportunity for crossing over between two alleles of a gene, raising the probability that a crossover will switch one version of an exon for another version found on the homologous chromosome. We can also imagine the occasional mixing and matching of exons between completely different (nonallelic) genes. Exon shuffling of either sort could lead to new proteins with novel combinations of functions.

THE SYNTHESIS OF PROTEIN

We will now examine more closely how genetic information flows from mRNA to protein—the process of translation. As we did for transcription, we'll concentrate on the basic steps of translation that occur in both prokaryotes and eukaryotes while pointing out key differences.

Translation is the RNA-directed synthesis of a polypeptide: *a closer look*

In the process of translation a cell interprets a genetic message
17.4 and builds a protein accordingly. The message is a series of codons along an mRNA molecule, and the interpreter is **transfer RNA (tRNA)**. The function of tRNA is to transfer amino acids from the cytoplasm's amino acid pool to a ribosome. A cell keeps its cytoplasm stocked with all 20 amino acids, either by synthesizing them from other compounds or by taking them up from the surrounding solution. The ribosome adds each amino acid brought to it by tRNA to the growing end of a polypeptide chain (FIGURE 17.11).

Molecules of tRNA are not all identical. The key to translating a genetic message into a specific amino acid sequence is that each type of tRNA molecule links a particular mRNA codon with a particular amino acid. As a tRNA molecule arrives at a ribosome, it bears a specific amino acid at one end. At the other end is a nucleotide triplet called an **anticodon**, which binds, according to the base-pairing rules, to a complementary codon on mRNA. For example, consider the mRNA codon UUU, which is translated as the amino acid phenylalanine (see FIGURE 17.4). The tRNA that plugs into this codon by hydrogen bonding has AAA as its anticodon and carries phenylalanine at its other end. As an mRNA molecule slides through a ribosome, phenylalanine will be added to the polypeptide chain whenever the codon UUU is presented for translation. Codon by codon, the genetic message is translated, as tRNAs deposit amino acids in the order prescribed, and the ribosome joins the amino acids into a chain. The tRNA molecule is like a flashcard with a nucleic acid word

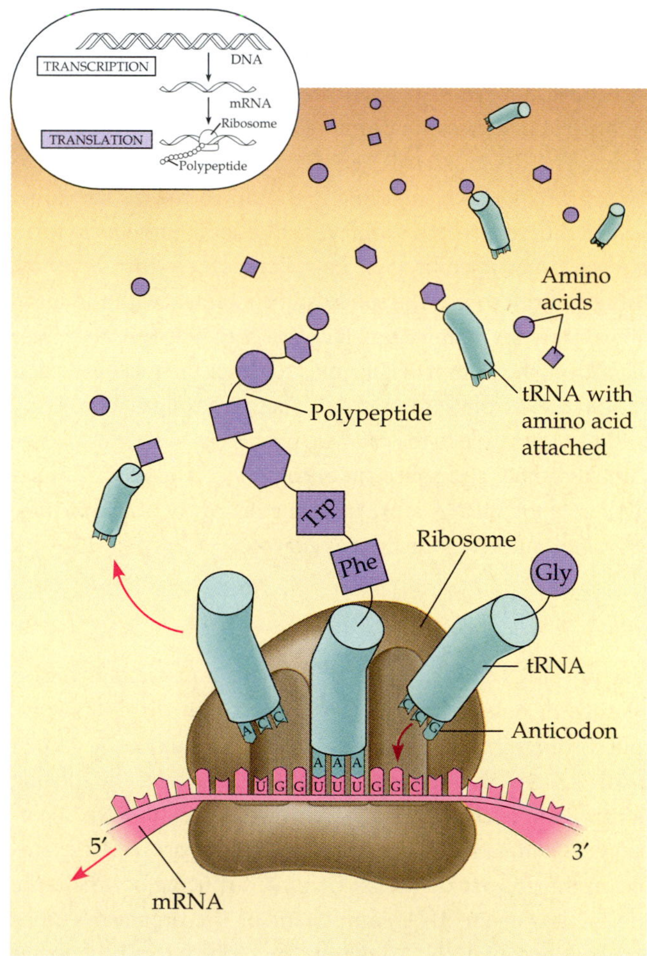

17.4 **FIGURE 17.11 ▪ Translation: the basic concept.** As a molecule of mRNA slides through a ribosome, codons are translated into amino acids, one by one. The interpreters are tRNA molecules, each type with a specific anticodon at one end and a certain amino acid at the other end. A tRNA adds its amino acid cargo to a growing polypeptide chain when the anticodon bonds to a complementary codon on the mRNA. The figures that follow show some of the details of translation in the prokaryotic cell.

(anticodon) on one side and a protein word (amino acid) on the other.

Translation is simple in principle but complex in its biochemistry and mechanics, especially in the eukaryotic cell. In dissecting translation we'll concentrate on the slightly less complicated version of the process that occurs in prokaryotes. Let's first look at some of the major players in this cellular drama, then see how they act together to make a polypeptide.

The Structure and Function of Transfer RNA

Like mRNA and other types of cellular RNA, transfer RNA molecules are transcribed from DNA templates. In a eukaryotic cell, tRNA, like mRNA, is made in the nucleus and must travel from the nucleus to the cytoplasm, where translation occurs. In both prokaryotic and eukaryotic cells, each tRNA molecule is used repeatedly, picking up its designated amino acid in the cytosol, depositing this cargo at the ribosome, and leaving the ribosome to pick up another load.

As illustrated in FIGURE 17.12, a tRNA molecule consists of a single RNA strand that is only about 80 nucleotides long (compared to hundreds of nucleotides for most mRNA molecules). This RNA strand folds back upon itself to form a molecule with a three-dimensional structure reinforced by interactions between different parts of the nucleotide chain. Nucleotide bases in certain regions of the tRNA strand form hydrogen bonds with complementary bases of other regions. Flattened into one plane to reveal this hydrogen bonding, a tRNA molecule looks like a cloverleaf. The tRNA actually twists and folds into a compact three-dimensional structure that is roughly L-shaped. The loop protruding from one end of the L includes the anticodon, the specialized base triplet that binds to a specific mRNA codon. From the other end of the L-shaped tRNA molecule protrudes its 3′ end, which is the attachment site for an amino acid. Thus, the structure of a tRNA molecule fits its function

If one tRNA variety existed for each of the mRNA codons that specifies an amino acid, there would be 61 tRNAs (see FIGURE 17.4). The actual number is smaller: about 45. This number is sufficient because some tRNAs have anticodons that can recognize two or more different codons. Such versatility is possible because the rules for base pairing between the third base of a codon and the corresponding base of a tRNA anticodon are not as strict as those for DNA and mRNA codons. For example, the base U of a tRNA anticodon can pair with either A or G in the third position of an mRNA codon. This relaxation of the base-pairing rules is called **wobble.** The most versatile tRNAs are those with inosine (I), a modified base, in the wobble position of the anticodon. Inosine is formed by enzymatic alteration of adenine after tRNA is synthesized. When anticodons associate with codons, the base I can hydrogen-bond with any one of three bases: U, C, or A. Thus, the tRNA molecule that has CCI as its anticodon can bind to the codons GGU, GGC, and GGA, all of which code for the amino acid glycine. Wobble explains why the synonymous codons for a given amino acid can differ in their third base, but usually not in their other bases.

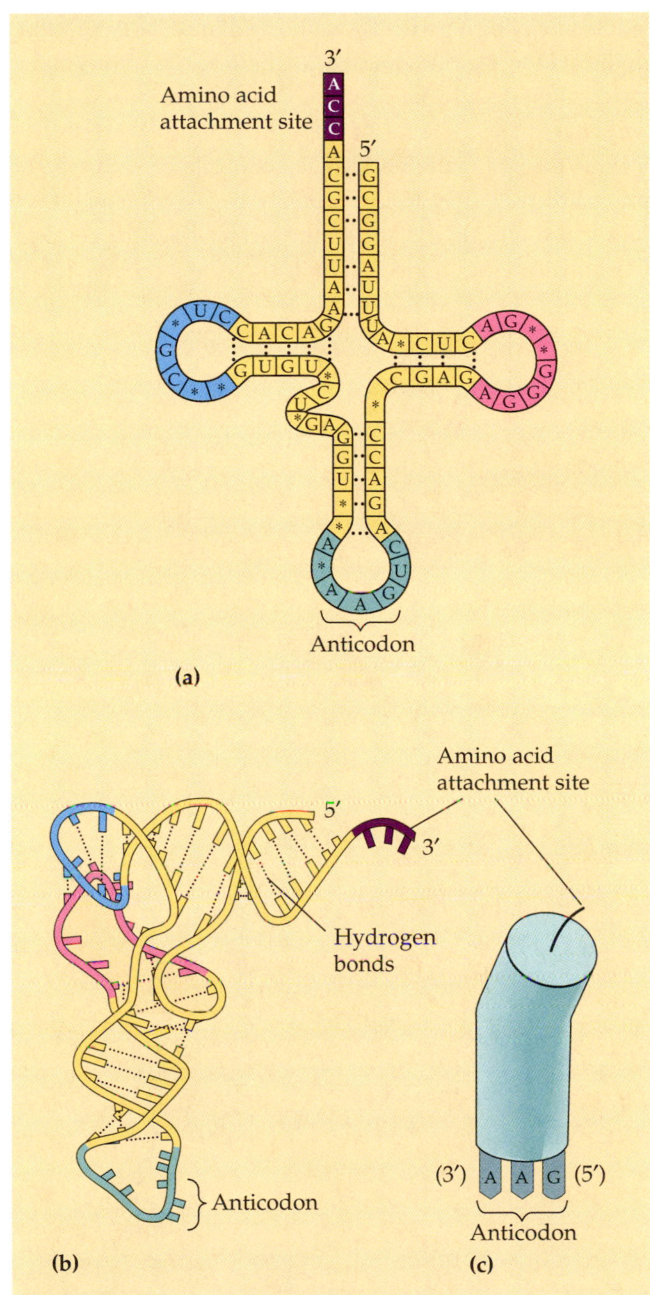

FIGURE 17.12 ▪ The structure of transfer RNA. (a) The two-dimensional structure of a tRNA molecule specific for the amino acid phenylalanine. Notice the four base-paired regions and three loops characteristic of all tRNAs. At the 3′ end of the molecule is the amino acid attachment site, which has the same base sequence for all tRNAs; within the middle loop is the anticodon triplet, which is unique to each tRNA type. (The asterisks mark bases that have been chemically modified, a characteristic of tRNA.) **(b)** The three-dimensional, L-shaped structure of a tRNA. **(c)** This simplified shape is used for tRNA in the figures that follow. Anticodons are conventionally written 3′⟶5′ in order to align properly with codons written 5′⟶3′ (see FIGURE 17.11). For base-pairing, RNA strands must be antiparallel, like DNA (see FIGURE 16.12). For example, anticodon 3′-AAG-5′ pairs with mRNA codon 5′-UUC-3′.

Aminoacyl-tRNA Synthetases

Codon-anticodon bonding is actually the second of two recognition steps required for the accurate translation of a genetic message. It must be preceded by a correct match between tRNA and an amino acid. A tRNA that binds to an mRNA codon specifying a particular amino acid must carry *only* that amino acid to the ribosome. Each amino acid is joined to the correct tRNA by a specific enzyme called an **aminoacyl-tRNA synthetase.** There are 20 of these enzymes in the cell, one enzyme for each amino acid. The active site of each type of aminoacyl-tRNA synthetase fits only a specific combination of amino acid and tRNA. The synthetase catalyzes the covalent attachment of the amino acid to its tRNA in a process driven by the hydrolysis of ATP (FIGURE 17.13). The resulting aminoacyl tRNA is released from the enzyme and delivers its amino acid to a growing polypeptide chain on a ribosome.

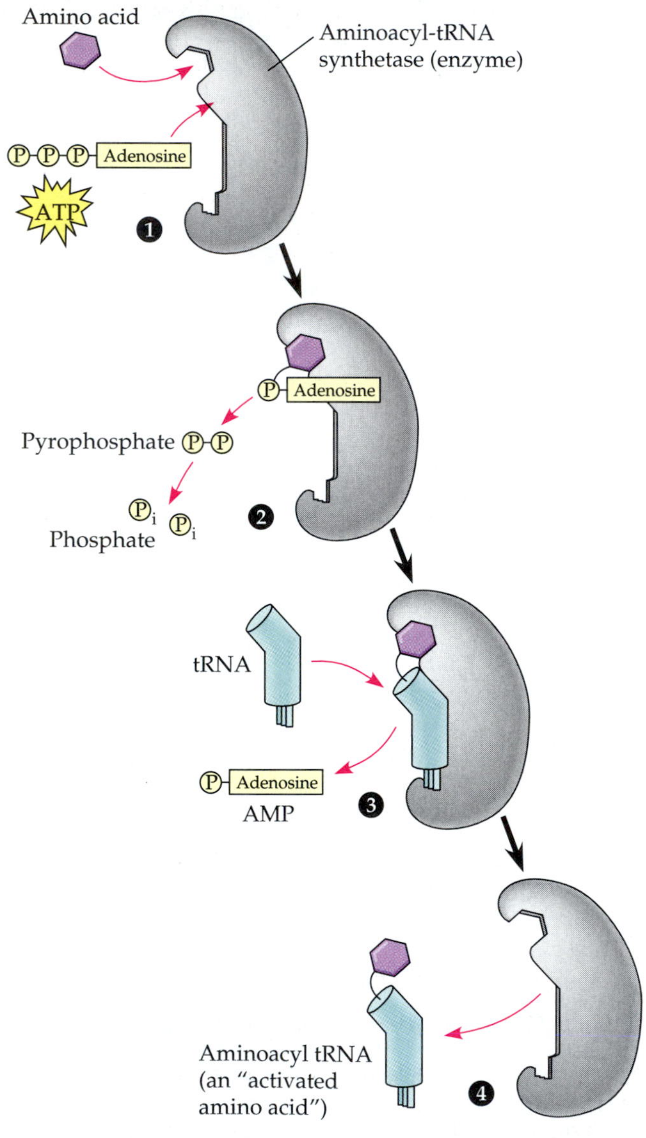

17.4 **FIGURE 17.13 • An aminoacyl-tRNA synthetase joins a specific amino acid to a tRNA.** Linkage of the tRNA and amino acid is an endergonic process that occurs at the expense of ATP. ① The active site of the enzyme binds the amino acid and an ATP molecule. ② The ATP loses two phosphate groups and joins to the amino acid as AMP (adenosine monophosphate). ③ The appropriate tRNA covalently bonds to the amino acid, displacing the AMP from the enzyme's active site. ④ The enzyme releases the aminoacyl tRNA, also called an "activated amino acid."

Ribosomes

Ribosomes facilitate the specific coupling of tRNA anticodons with mRNA codons during protein synthesis. A ribosome, which can be seen with the electron microscope, is made up of two subunits, termed the large and small subunits (FIGURE 17.14a). The ribosomal subunits are constructed of proteins and RNA molecules called **ribosomal RNA (rRNA).** In eukaryotes, the subunits are made in the nucleolus. Ribosomal RNA genes on the chromosomal DNA are transcribed, and the RNA is processed and assembled with proteins imported from the cytoplasm. The resulting ribosomal subunits are then exported via nuclear pores to the cytoplasm. In both prokaryotes and eukaryotes, large and small subunits join to form a functional ribosome only when they attach to an mRNA molecule. About 60% of the weight of each ribosome is rRNA. Because most cells contain thousands of ribosomes, rRNA is the most abundant type of RNA.

Although the ribosomes of prokaryotes and eukaryotes are very similar in structure and function, those of eukaryotes are slightly larger and differ somewhat from prokaryotic ribosomes in their molecular composition. The differences are medically significant. Certain drugs can paralyze prokaryotic ribosomes without inhibiting the ability of eukaryotic ribosomes to make proteins. These drugs, including tetracycline and streptomycin, are used as antibiotics to combat bacterial infection.

The structure of a ribosome reflects its function of bringing mRNA together with amino acid–bearing tRNAs. In addition to a binding site for mRNA, each ribosome has three binding sites for tRNA (FIGURE 17.14b). The **P site** (peptidyl-tRNA site) holds the tRNA carrying the growing polypeptide chain, while the **A site** (aminoacyl-tRNA site) holds the tRNA carrying the next amino acid to be added to the chain. Discharged tRNAs leave the ribosome from the more recently discovered **E site** (exit site). Acting like a vise, the ribosome holds the tRNA and mRNA close together and positions the new amino acid for addition to the carboxyl end of the growing polypeptide (FIGURE 17.14c).

Building a Polypeptide

We can divide translation, the synthesis of a polypeptide chain, into three stages (analogous to those of transcription): initiation, elongation, and termination. All three stages require protein "factors" that aid mRNA, tRNA, and ribosomes in the translation process. For chain initiation and elongation, energy is also required. It is provided by GTP (guanosine triphosphate), a molecule closely related to ATP.

Initiation. The initiation stage of translation brings together mRNA, a tRNA bearing the first amino acid of the polypeptide, and the two subunits of a ribosome. First, a small ribosomal subunit binds to both mRNA and a special initiator tRNA (FIGURE 17.15). The small ribosomal subunit attaches to the leader segment at the 5′ (upstream) end of the mRNA. In bacteria, rRNA of the subunit base-pairs with a specific sequence of nucleotides within the mRNA leader; in eukaryotes the 5′ cap first tells the small subunit to attach to the 5′ end of the mRNA. Downstream on the mRNA is the initiation codon, AUG, which signals the start of translation. The initiator tRNA, which carries the amino acid methionine, attaches to the initiation codon.

The union of mRNA, initiator tRNA, and a small ribosomal subunit is followed by the attachment of a large ribosomal subunit, completing a translation initiation complex. Proteins called *initiation factors* are required to bring all these components together. The cell also spends energy in the form of a GTP molecule to form the initiation complex. At the completion of the initiation process, the initiator tRNA sits in the P site of the ribosome, and the vacant A site is ready for the next aminoacyl tRNA. The synthesis of a polypeptide is initiated at its amino end.

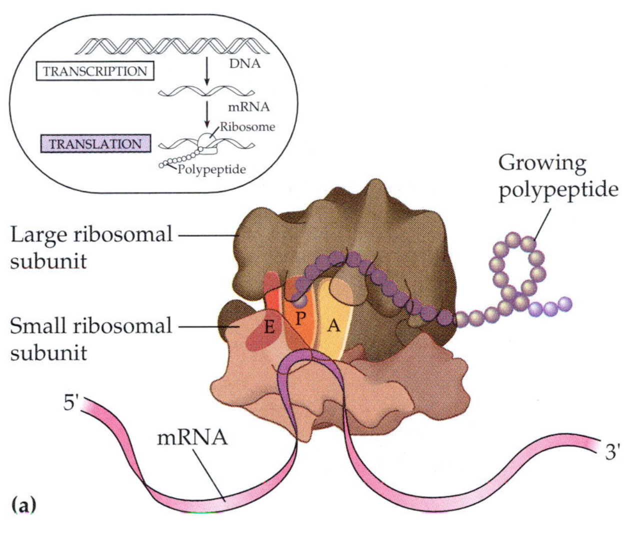

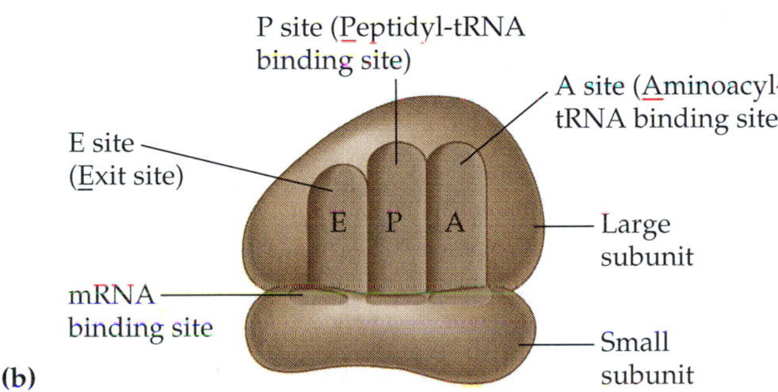

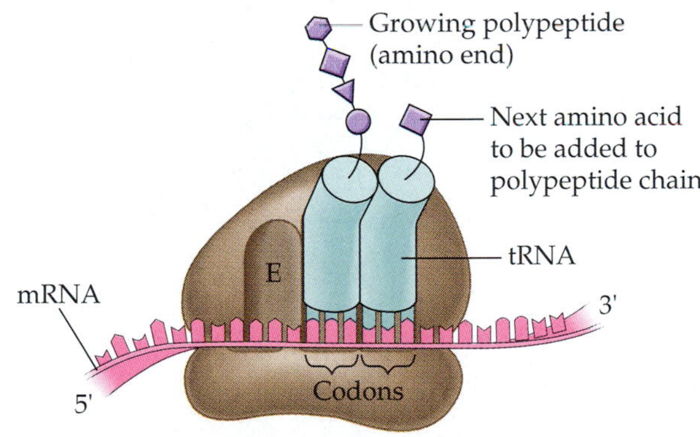

FIGURE 17.14 ▪ The anatomy of a ribosome. **(a)** A functional ribosome consists of two subunits, each an aggregate of ribosomal RNA and many proteins. This is a model of a bacterial ribosome. The eukaryotic ribosome is roughly similar, but larger, with more proteins and rRNA molecules. **(b)** A ribosome has an mRNA-binding site and three tRNA-binding sites, known as the P, A, and E sites. This is a simplified version of the ribosomal shape that will appear in the next several figures. **(c)** A tRNA fits into a binding site when its anticodon base-pairs with an mRNA codon. The P site holds the tRNA attached to the growing polypeptide. The A site holds the tRNA carrying the next amino acid to be added to the polypeptide chain. Discharged tRNA leaves via the E site.

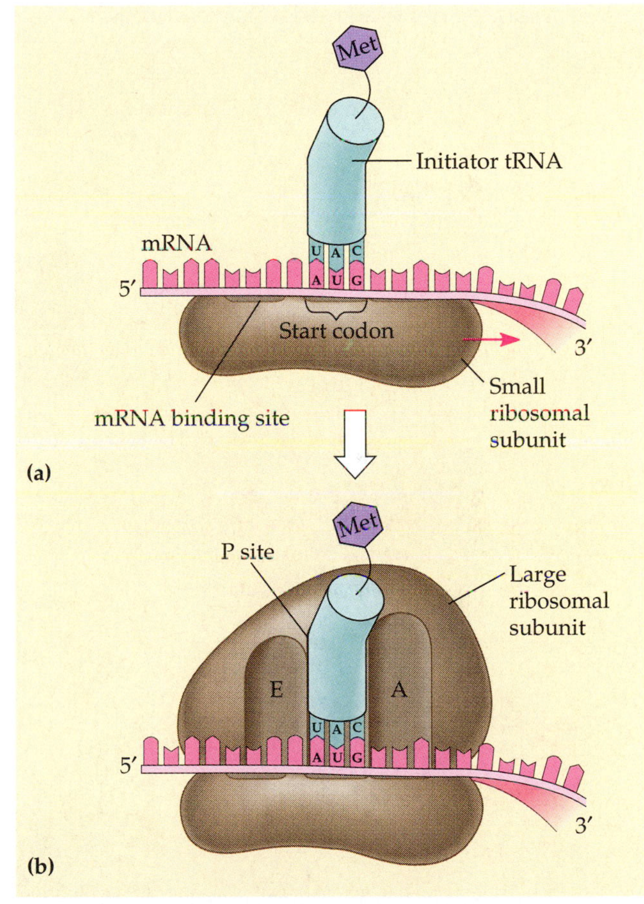

17.4 **FIGURE 17.15 ▪ Initiation of translation.** **(a)** A small ribosomal subunit binds to a molecule of mRNA. In a prokaryotic cell, the mRNA-binding site on this subunit recognizes a specific nucleotide sequence on the mRNA just upstream of the start codon. An initiator tRNA, with the anticodon UAC, base-pairs with the start codon, AUG. This tRNA carries the amino acid methionine (Met). **(b)** The arrival of a large ribosomal subunit completes the initiation complex. The initiator tRNA is in the P site. The A site is available to the tRNA bearing the next amino acid. Proteins called initiation factors (not shown) are required to bring all the translation components together. GTP provides the energy for the initiation process.

Elongation. In the elongation stage of translation, amino acids are added one by one to the first amino acid. Each addition involves the participation of several proteins called *elongation factors* and occurs in a three-step cycle (FIGURE 17.16):

① *Codon recognition.* The mRNA codon in the A site of the ribosome forms hydrogen bonds with the anticodon of an incoming molecule of tRNA carrying its appropriate amino acid. An elongation factor ushers the tRNA into the A site. This step also requires the hydrolysis of a GTP.

② *Peptide bond formation.* An rRNA molecule of the large ribosomal subunit, functioning as a ribozyme, catalyzes the formation of a peptide bond that joins the polypeptide extending from the P site to the newly arrived amino acid in the A site. In this step, the polypeptide separates from the tRNA to which it was attached, and the amino acid at its carboxyl end bonds to the amino acid carried by the tRNA in the A site.

③ *Translocation.* The tRNA in the A site, now attached to the growing polypeptide, is translocated to the P site. As the tRNA moves, its anticodon remains hydrogen-bonded to the mRNA codon; the mRNA moves along with it and brings the next codon to be translated into the A site. Meanwhile, the tRNA that was in the P site moves to the E (exit) site and from there leaves the ribosome. The translocation step requires energy, which is provided by hydrolysis of a GTP. The mRNA moves through the ribosome in one direction only, 5′ end first; this is equivalent to the ribosome moving 5′ ⟶ 3′on the mRNA. The important point is that the ribosome and the mRNA move relative to each other, unidirectionally, codon by codon.

17.4 **FIGURE 17.16 ▪ The elongation cycle of translation.** Not shown in this diagram are the proteins called elongation factors and GTP, whose hydrolysis drives the process.

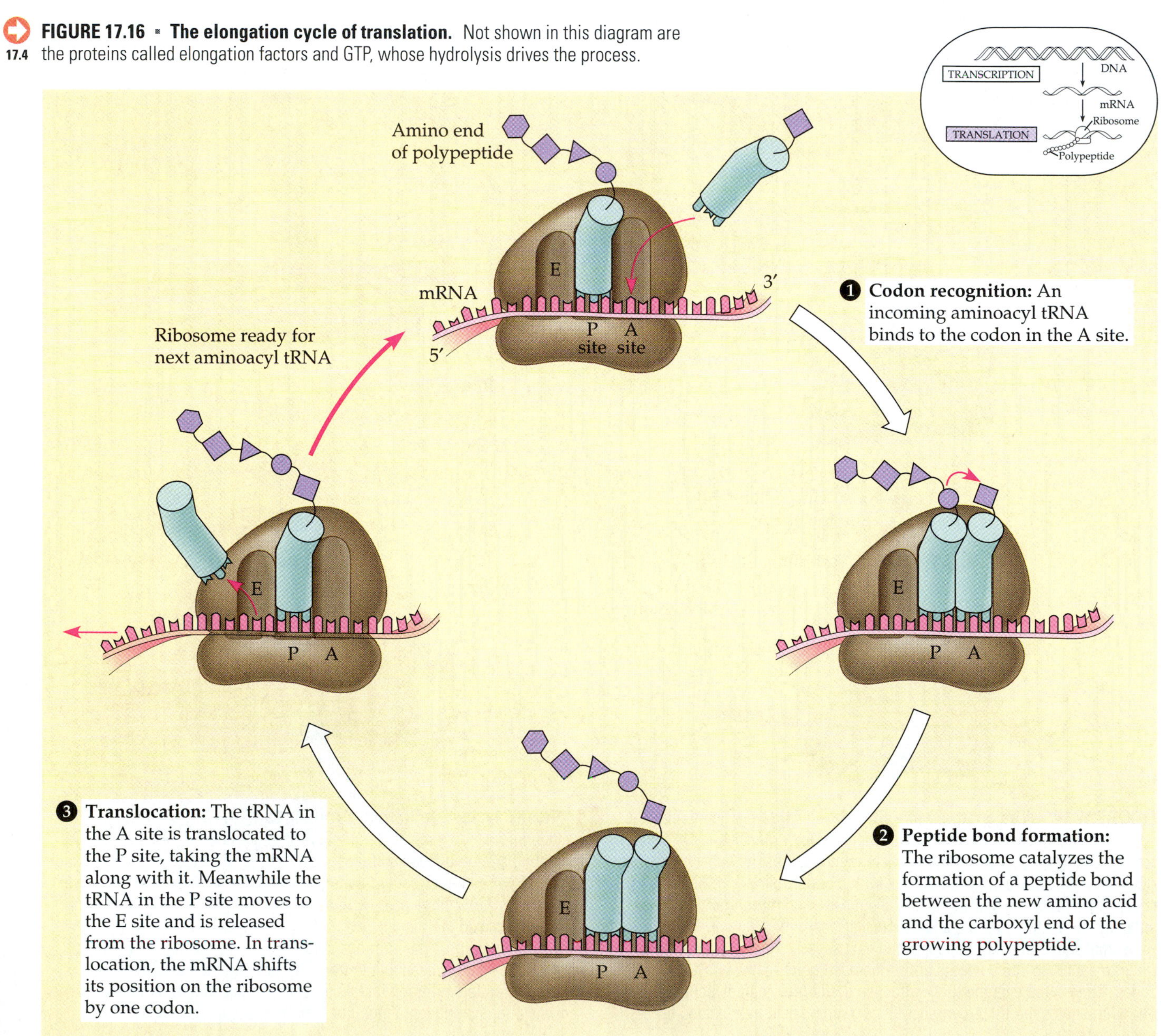
TRANSCRIPTION
DNA
mRNA
Ribosome
TRANSLATION
Polypeptide
Amino end of polypeptide
mRNA
3′
5′
E
P site
A site
Ribosome ready for next aminoacyl tRNA
❶ Codon recognition: An incoming aminoacyl tRNA binds to the codon in the A site.
E
P
A
❷ Peptide bond formation: The ribosome catalyzes the formation of a peptide bond between the new amino acid and the carboxyl end of the growing polypeptide.
E
P
A
❸ Translocation: The tRNA in the A site is translocated to the P site, taking the mRNA along with it. Meanwhile the tRNA in the P site moves to the E site and is released from the ribosome. In translocation, the mRNA shifts its position on the ribosome by one codon.
E
P
A

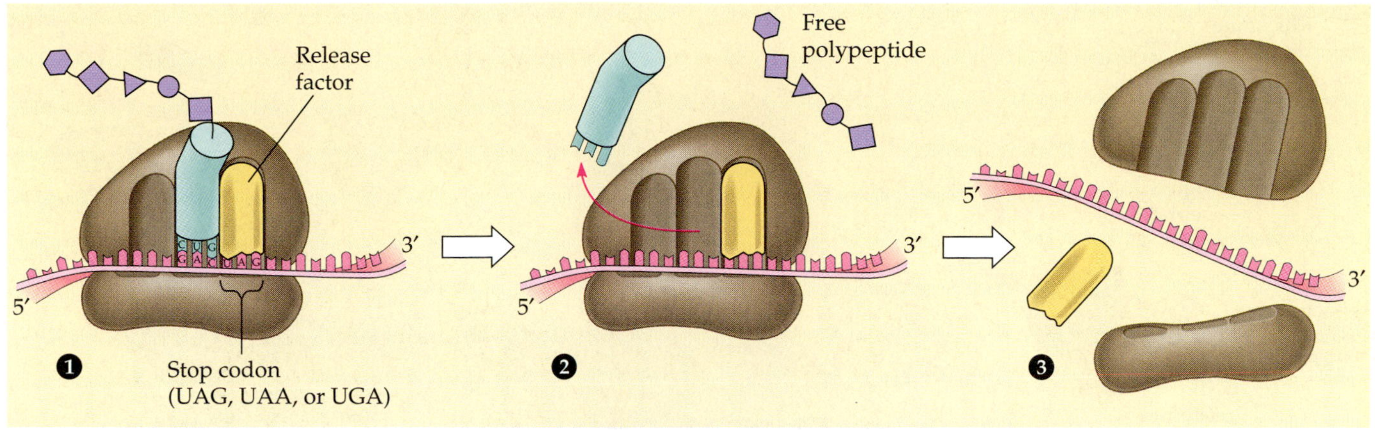

FIGURE 17.17 ▪ Termination of translation. ① When a ribosome reaches a termination codon on a strand of mRNA, the A site of the ribosome accepts a protein called a release factor instead of tRNA. ② The release factor hydrolyzes the bond between the tRNA in the P site and the last amino acid of the polypeptide chain. The polypeptide is thus freed from the ribosome. ③ The two ribosomal subunits and the other components of the assembly dissociate.

17.4

The elongation cycle takes less than a tenth of a second and is repeated as each amino acid is added to the chain until the polypeptide is completed.

Termination. The final stage of translation is termination (FIGURE 17.17). Elongation continues until a stop codon reaches the A site of the ribosome. These special base triplets—UAA, UAG, and UGA—do not code for amino acids but instead act as signals to stop translation. A protein called a *release factor* binds directly to the stop codon in the A site. The release factor causes the addition of a water molecule instead of an amino acid to the polypeptide chain. This reaction hydrolyzes the completed polypeptide from the tRNA that is in the P site, freeing the polypeptide from the ribosome. The remainder of the translation assembly then comes apart.

Polyribosomes

A single ribosome can make an average-sized polypeptide in less than a minute. Typically, however, a single mRNA is used to make many copies of a polypeptide simultaneously, because several ribosomes work on translating the message at the same time. Once a ribosome moves past the initiation codon, a second ribosome can attach to the mRNA, and thus several ribosomes may trail along the same mRNA. Such strings of ribosomes, called **polyribosomes**, can be seen with the electron microscope (FIGURE 17.18). Polyribosomes are found in both prokaryotic and eukaryotic cells.

From Polypeptide to Functional Protein

During and after its synthesis a polypeptide chain begins to coil and fold spontaneously, forming a functional protein of specific conformation: a three-dimensional molecule with secondary and tertiary structure. A gene determines primary structure, and primary structure in turn determines conformation. In many cases, chaperone proteins help the polypeptide fold correctly (see Chapter 5).

Additional steps—*posttranslational modifications*—may be required before the protein can begin doing its particular job

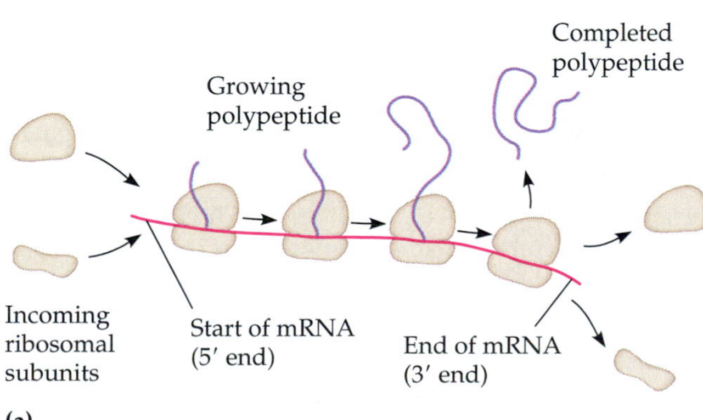

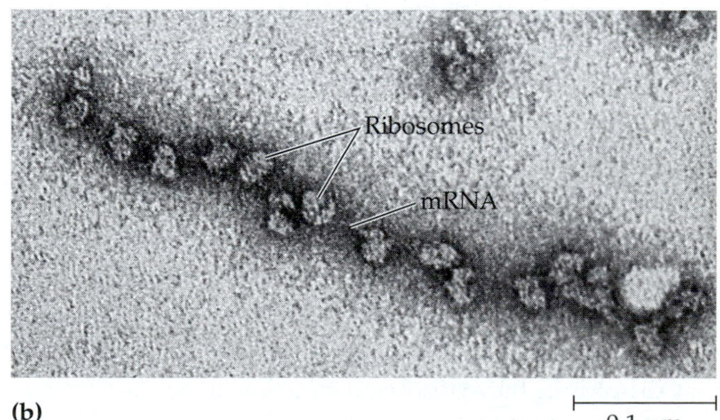

FIGURE 17.18 ▪ Polyribosomes. (a) An mRNA molecule is generally translated simultaneously by several ribosomes in clusters called polyribosomes. **(b)** This micrograph shows a large polyribosome in a prokaryotic cell (TEM).

17.4

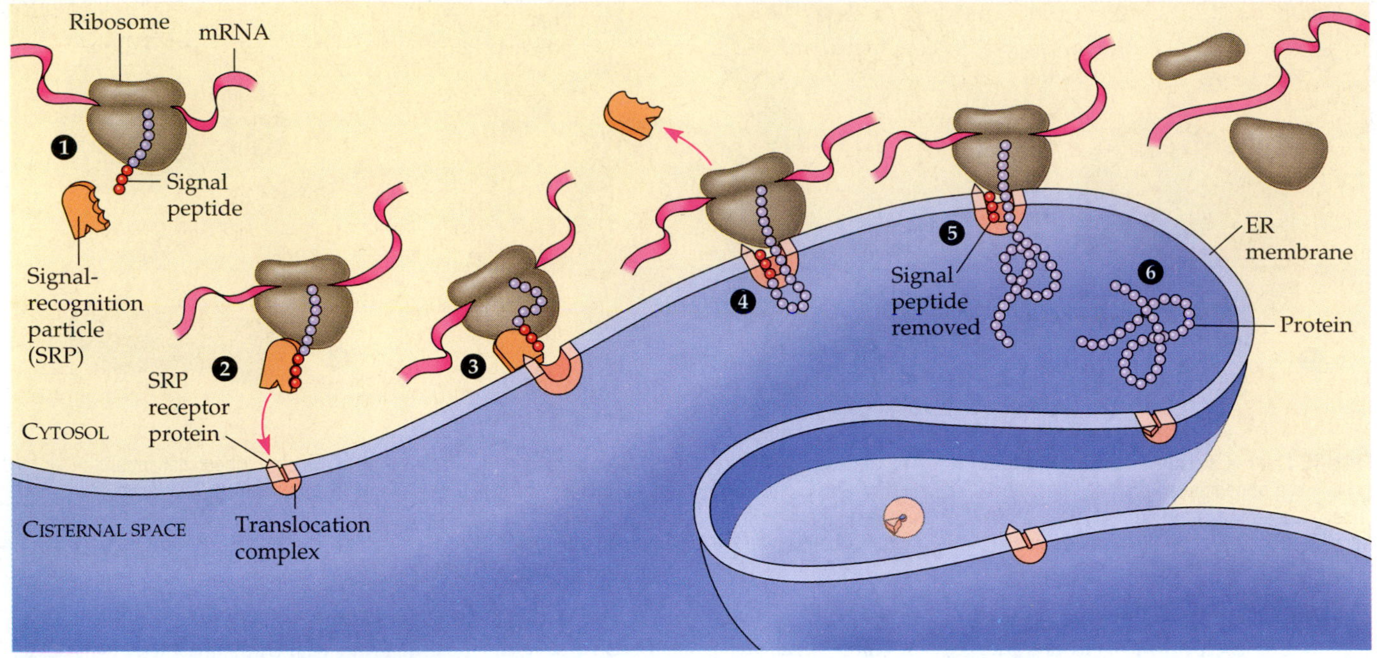

FIGURE 17.19 • The signal mechanism for targeting proteins to the ER. A polypeptide destined for the endomembrane system or for secretion from the cell begins with a signal peptide, a stretch of amino acids that targets it for the ER. This figure shows the synthesis of a secretory protein and its simultaneous import into the ER. ① Polypeptide synthesis begins on a free ribosome in the cytosol. ② A signal-recognition particle (SRP) binds to the signal peptide. ③ The SRP then binds to a receptor protein in the ER membrane. This receptor is part of a protein complex, here called a translocation complex, that also includes a membrane pore and a signal-cleaving enzyme. ④ The SRP is released, and the growing polypeptide translocates across the membrane. The signal peptide stays attached to the membrane. ⑤ The signal-cleaving enzyme cuts off the peptide. ⑥ The rest of the completed polypeptide leaves the ribosome and folds into its final conformation. The protein's secretion is accomplished as described in Chapter 7.

in the cell. Certain amino acids may be chemically modified by the attachment of sugars, lipids, phosphate groups, or other additions. Enzymes may remove one or more amino acids from the leading (amino) end of the polypeptide chain. In some cases, a single polypeptide chain may be enzymatically cleaved into two or more pieces. For example, the protein insulin is first synthesized as a single polypeptide chain but becomes active only after an enzyme excises a central part of the chain, leaving a protein made up of two polypeptide chains connected by disulfide bridges. In other cases, two or more polypeptides that are synthesized separately may join to become the subunits of a protein that has quaternary structure. (To review the levels of protein structure, see Chapter 5.)

Signal peptides target some eukaryotic polypeptides to specific destinations in the cell

In electron micrographs of eukaryotic cells active in protein synthesis, two populations of ribosomes (and polyribosomes) are evident: free and bound (see FIGURE 7.10). Free ribosomes are suspended in the cytosol and mostly synthesize proteins that dissolve in the cytosol and function there. In contrast, bound ribosomes are attached to the cytosolic side of the endoplasmic reticulum (ER). They make proteins of the endomembrane system (the nuclear envelope, ER, Golgi apparatus, lysosomes, vacuoles, and plasma membrane) and proteins that are secreted from the cell. Insulin is an example of a secretory protein. The ribosomes themselves are identical and can switch their status from free to bound.

What determines whether a ribosome will be free in the cytosol or bound to rough ER at any particular time? The synthesis of all proteins begins in the cytosol, when a free ribosome starts to translate an mRNA molecule. There the process continues to completion—*unless* the growing polypeptide itself cues the ribosome to attach to the ER. The polypeptides of proteins destined for the endomembrane system or for secretion are marked by a **signal peptide**, which targets the protein to the ER (FIGURE 17.19). The signal peptide, a sequence of about 20 amino acids at or near the leading (amino) end of the polypeptide, is recognized as it emerges from the ribosome by a protein-RNA complex called a **signal-recognition particle (SRP)**. This particle functions as an adaptor that brings the ribosome to a receptor protein built into the ER membrane. This receptor is part of a multiprotein complex. Polypeptide synthesis continues there, and the growing polypeptide snakes across the membrane into the cisternal space via a protein pore. The signal peptide is usually removed by an enzyme. The rest of the completed polypeptide, if it is to be a secretory protein, is released into solution within the cisternal space (as in FIGURE 17.19). Alternatively, if the polypeptide is to be a membrane protein, it remains partially embedded in the ER membrane.

Other kinds of signal peptides are used to target polypeptides to mitochondria, chloroplasts, the interior of the nucleus, and other organelles that are not part of the endomembrane system. The critical difference in these cases is that translation is completed in the cytosol before the polypeptide is imported into the organelle. The mechanisms of translocation also vary, but in all cases studied to date, the "postal" codes that address proteins to cellular locations are signal peptides of some sort.

RNA plays multiple roles in the cell: *a review*

As we have seen, the cellular machinery of protein synthesis (and ER targeting) is dominated by RNA of various kinds. In addition to mRNA, these include tRNA, rRNA, and, in eukaryotes, snRNA and SRP RNA (TABLE 17.1). The diverse functions of these molecules range from structural to informational to catalytic. The ability of RNA to perform so many different functions is based on two related characteristics of this kind of molecule: RNA can hydrogen-bond to other nucleic acid molecules (DNA or RNA), and it can assume a specific three-dimensional shape by forming hydrogen bonds between bases in different parts of its polynucleotide chain (you saw an example of this intramolecular bonding in tRNA, FIGURE 17.12). DNA may be the genetic material of all living cells today, but RNA is much more versatile. You will learn in Chapter 18 that many viruses even use RNA rather than DNA as their genetic material.

Comparing protein synthesis in prokaryotes and eukaryotes: *a review*

Although bacteria and eukaryotes carry out transcription and translation in very similar ways, we have noted certain differences in cellular machinery and in details of the processes. Prokaryotic and eukaryotic RNA polymerases are different, and those of eukaryotes depend on transcription factors. Transcription is terminated differently in the two kinds of cells. Also prokaryotic and eukaryotic ribosomes are slightly different. The most important differences, however, arise from the eukaryotic cell's compartmental organization. Like a one-room workshop, a prokaryotic cell assures a streamlined operation. In the absence of a nucleus, it can simultaneously transcribe and translate the same gene (FIGURE 17.20), and the newly made protein can quickly diffuse to its site of function. In contrast, the eukaryotic cell's nuclear envelope segregates transcription from translation and provides a compartment for extensive RNA processing. This processing stage provides an additional way to regulate and coordinate the eukaryotic cell's elaborate activities (see Chapter 19). Finally, as we have seen, eukaryotic cells have complicated mechanisms for targeting proteins to the appropriate cellular compartment (organelle).

Table 17.1 ■ Types of RNA in a Eukaryotic Cell

TYPE OF RNA	FUNCTION
Messenger RNA (mRNA)	Carries information specifying amino acid sequences of proteins from DNA to ribosomes.
Transfer RNA (tRNA)	Serves as adaptor molecule in protein synthesis; translates mRNA codons into amino acids.
Ribosomal RNA (rRNA)	Plays structural and catalytic (ribozyme) roles in ribosomes.
Primary transcript	Serves as a precursor to mRNA, rRNA, or tRNA; this RNA molecule is processed by cleavage or splicing. In eukaryotes, pre-mRNA commonly contains introns, noncoding segments that are spliced out as the primary transcript is processed. Some intron RNA acts as a ribozyme, catalyzing its own splicing.
Small nuclear RNA (snRNA)	Plays structural and catalytic roles in spliceosomes, the complexes of protein and RNA that splice pre-mRNA in the eukaryotic nucleus.
SRP RNA	Acts as a component of the signal-recognition particle (SRP), the protein-RNA complex that recognizes the signal peptides of polypeptides targeted to the ER.

Where did eukaryotes—and prokaryotes, for that matter—get the genes that encode the huge diversity of proteins they synthesize? For the past few billion years the ultimate source of new genes has been the mutation of preexisting genes, the topic of the next section.

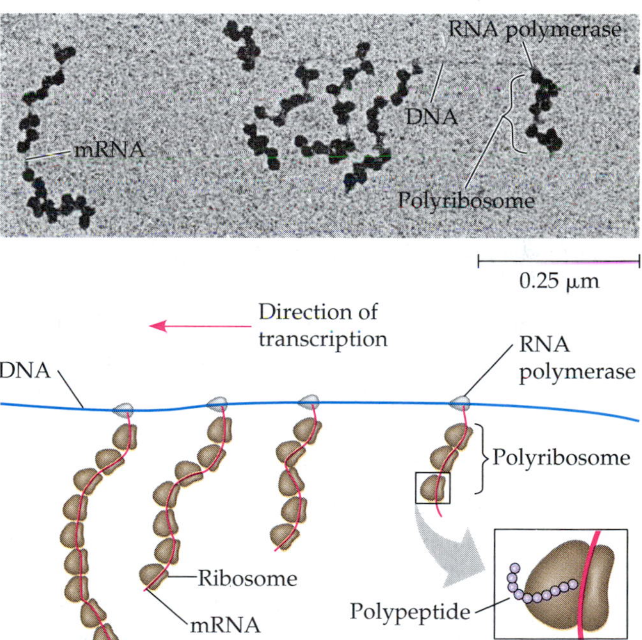

FIGURE 17.20 • Coupled transcription and translation in bacteria. In prokaryotic cells, transcription and translation are not separated by a nuclear envelope. The translation of mRNA can begin as soon as the leading (5′) end of the mRNA molecule peels away from the DNA template. The micrograph shows a strand of *E. coli* DNA being transcribed by RNA polymerase molecules. Attached to each RNA polymerase molecule is a growing strand of mRNA, which is already being translated by ribosomes. The newly synthesized polypeptides are not visible in the micrograph (TEM).

Point mutations can affect protein structure and function

Mutations are changes in the genetic material of a cell (or virus). In Chapter 15 we considered large-scale mutations, chromosomal rearrangements that affect long segments of DNA. Now that you have learned about the genetic code and its translation, we can examine **point mutations**, which are chemical changes in just one or a few base pairs in a single gene.

If a point mutation occurs in a gamete, or in a cell that gives rise to gametes, it may be transmitted to offspring and to a succession of future generations. If the mutation has an adverse effect on the phenotype of a human or other animal, the mutant condition is referred to as a genetic disorder, or hereditary disease. For example, we can trace the genetic basis of sickle-cell disease to a mutation of a single base pair in the gene that codes for one of the polypeptides of hemoglobin. The change of a single nucleotide in the DNA's template strand leads to the production of an abnormal protein (FIGURE 17.21). Let's see how different types of point mutations translate into altered proteins.

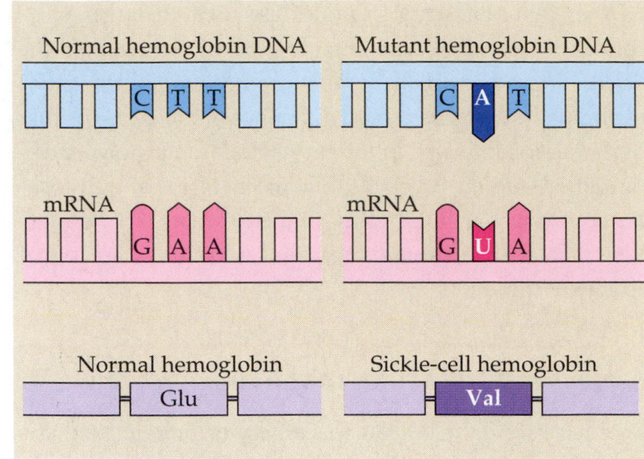

FIGURE 17.21 • The molecular basis of sickle-cell disease. The allele that causes sickle-cell disease differs from the normal allele by a change in a single base pair—a point mutation. The top row of this figure shows the template strand of the gene for one of the polypeptides making up the protein hemoglobin. Where the normal allele has the base thymine, the sickle-cell allele has adenine. This alters one of the codons in the mRNA transcribed from the gene, with the result that the amino acid valine appears in the polypeptide in place of the glutamic acid found in the normal polypeptide. In individuals who are homozygous for the mutant allele, the sickling of red blood cells caused by the altered hemoglobin produces the multiple symptoms associated with sickle-cell disease (see FIGURE 14.16).

Types of Point Mutations

Point mutations within a gene can be divided into two general categories: base-pair substitutions and base-pair insertions or deletions. While reading about how these mutations affect proteins, refer to the appropriate parts of FIGURE 17.22.

Substitutions. A **base-pair substitution** is the replacement of one nucleotide and its partner in the complementary DNA strand with another pair of nucleotides. Some substitutions are called *silent mutations* because, due to the redundancy of the genetic code, they have no effect on the encoded protein. In other words, a change in a base pair may transform one codon into another that is translated into the same amino acid. For example, if CCG mutated to CCA, the mRNA codon that used to be GGC would become GGU, and a glycine would still be inserted at the proper location in the protein (see FIGURE 17.4). Other changes of a single nucleotide pair may switch an amino acid but have little effect on the protein. The new amino acid may have properties similar to those of the amino acid it replaces, or it may be in a region of the protein where the exact sequence of amino acids is not essential to the protein's function.

However, the base-pair substitutions of greatest interest are those that cause a readily detectable change in a protein. The alteration of a single amino acid in a crucial area of a protein—in the active site of an enzyme, for example—will significantly alter protein activity. Occasionally, such a mutation leads to an improved protein or one with novel capabilities that enhance the success of the mutant organism and its descendants. But much more often such mutations are detrimental, creating a useless or less active protein that impairs cellular function.

Substitution mutations are usually **missense mutations**; that is, the altered codon still codes for an amino acid and thus makes sense, although not necessarily the *right* sense. But if a point mutation changes a codon for an amino acid into a stop codon, translation will be terminated prematurely, and the resulting polypeptide will be shorter than the polypeptide encoded by the normal gene. Alterations that change an amino acid codon to a stop signal are called **nonsense mutations**, and nearly all nonsense mutations lead to nonfunctional proteins.

Insertions and Deletions. **Insertions** and **deletions** are additions or losses of one or more nucleotide pairs in a gene. These mutations have a disastrous effect on the resulting protein more often than substitutions do. Because mRNA is read as a series of nucleotide triplets during translation, the insertion or deletion of nucleotides may alter the reading frame (triplet grouping) of the genetic message. Such a mutation, called a **frameshift mutation**, will occur whenever the number of nucleotides inserted or deleted is not a multiple of 3. All the nucleotides that are downstream of the deletion or insertion will be improperly grouped into codons, and the result will be extensive missense ending sooner or later in nonsense—premature termination. Unless the frameshift is very near the end of the gene, it will produce a protein that is almost certain to be nonfunctional.

Wild type

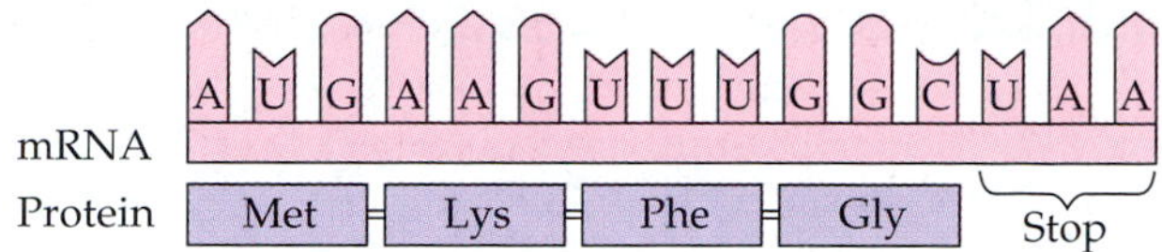

Base-pair insertion or deletion

Frameshift causing extensive missense

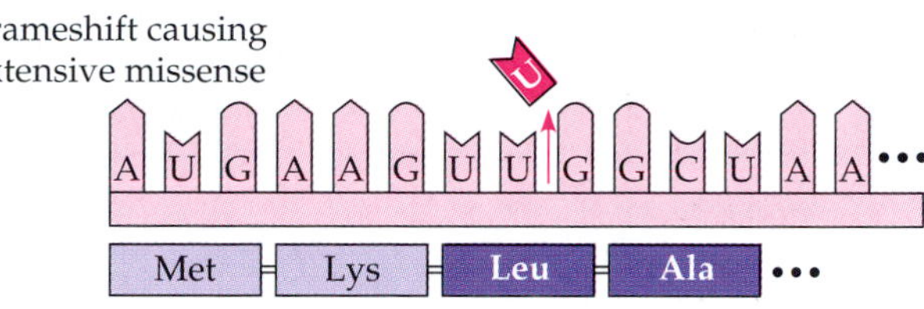

Frameshift causing immediate nonsense

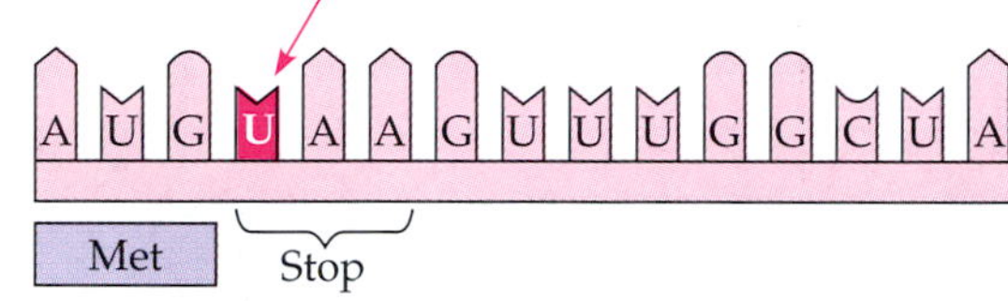

Insertion or deletion of 3 nucleotides: no extensive frameshift

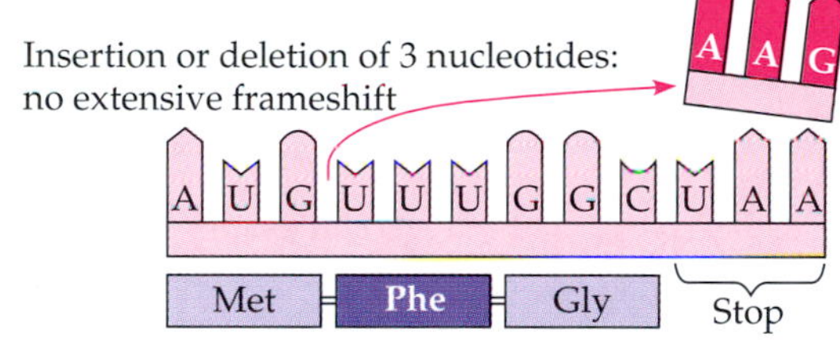

Base-pair substitution

No effect on amino acid sequence

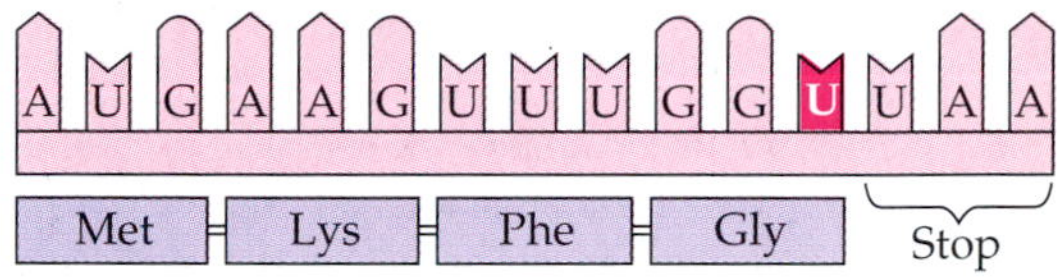

Missense

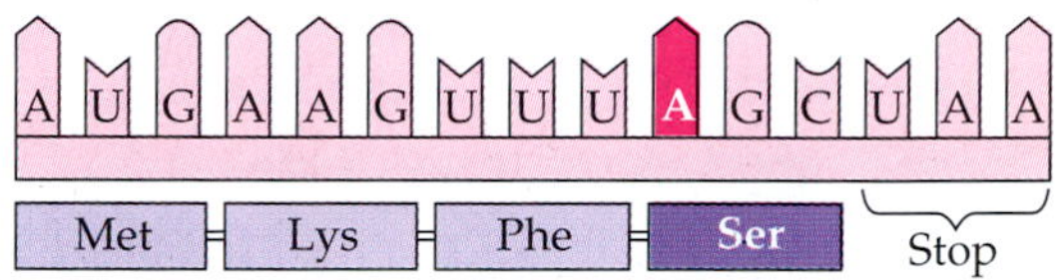

Nonsense

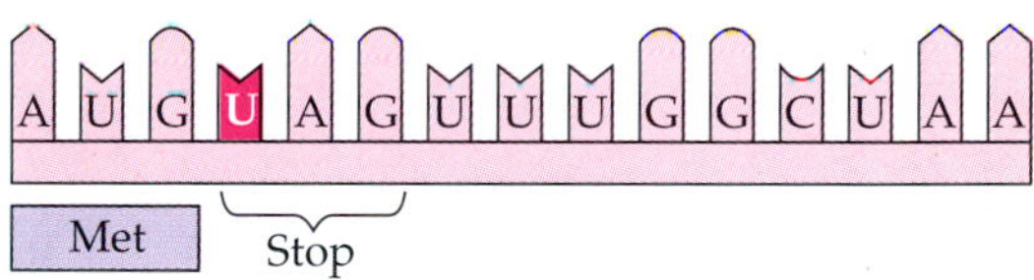

FIGURE 17.22 ▪ Categories and consequences of point mutations. Mutations are changes in DNA, but they are represented here as they are reflected in mRNA and its protein product.

Mutagens

The production of mutations can occur in a number of ways. Errors during DNA replication, repair, or recombination can lead to base-pair substitutions, insertions, or deletions, as well as to mutations affecting longer stretches of DNA. Mutations resulting from such errors are called *spontaneous mutations.*

A number of physical and chemical agents, called **mutagens**, interact with DNA to cause mutations. In the 1920s, Hermann Muller discovered that if he subjected fruit flies to X-rays, genetic changes increased in frequency. Using this method, Muller was able to obtain mutants of *Drosophila* that he could use in his genetic studies. But he also recognized an alarming implication of his discovery: Because they are mutagens, X-rays and other forms of high-energy radiation pose hazards to the genetic material of people as well as laboratory organisms. Mutagenic radiation, a physical mutagen, includes ultraviolet (UV) light, which can produce disruptive thymine dimers in DNA (see FIGURE 16.17).

Chemical mutagens fall into several categories. Base analogues are chemicals that are similar to normal DNA bases but that pair incorrectly during DNA replication. Some other chemical mutagens interfere with correct DNA replication by inserting themselves into the DNA and distorting the double helix. Still other mutagens cause chemical changes in bases that change their pairing properties.

Researchers have developed various methods to test the mutagenic activity of different chemicals. One of the simplest and most popular methods is the **Ames test**, named for its developer, Bruce Ames (see the Methods Box on the next page). The Ames test uses easily grown bacteria as test organisms. A suspected mutagenic chemical is added to a culture of bacteria that already carry a point mutation rendering them unable to synthesize the amino acid histidine. The only bacteria that will grow to form colonies on a medium lacking histidine will be those that have undergone a back-mutation, a second mutation that restores the ability to make histidine. Thus, the number of colonies that grow on the histidine-lacking medium is a measure of the strength of the mutagen.

A major application of the Ames test is the preliminary screening of chemicals to identify those that may cause cancer. This approach makes sense because most carcinogens

METHODS: The Ames Test

The Ames test, named for its developer, microbiologist Bruce Ames, measures the mutagenic strength of various chemicals. The suspected mutagen is mixed with a culture of bacteria. It is also necessary to add a rat liver extract, which contains enzymes that convert certain chemicals from nonmutagenic to mutagenic forms. (These enzymes, also present in our own livers, normally function to metabolize toxic substances, but unfortunately the chemical modification sometimes makes a bad situation worse by increasing the mutagenicity of the substances.) The bacteria, *Salmonella,* carry a mutation preventing them from synthesizing the amino acid histidine. These bacteria fail to grow when plated on a growth medium lacking histidine. Some of the bacteria, however, undergo a back-mutation that restores the ability to make histidine. Such revertant bacteria give rise to colonies on the histidine-free medium. A mutagen will increase the frequency of these back-mutations, thereby bringing about the appearance of more colonies on the histidine-free medium. The mutagenic activity of a chemical can be quantified by comparing the number of colonies that grow after chemical treatment to control samples not treated with the suspected mutagen. One application of the Ames test is the screening of chemicals to identify those that can cause cancer. Most carcinogens (cancer-causing chemicals) are mutagenic and, conversely, most mutagens are carcinogenic.

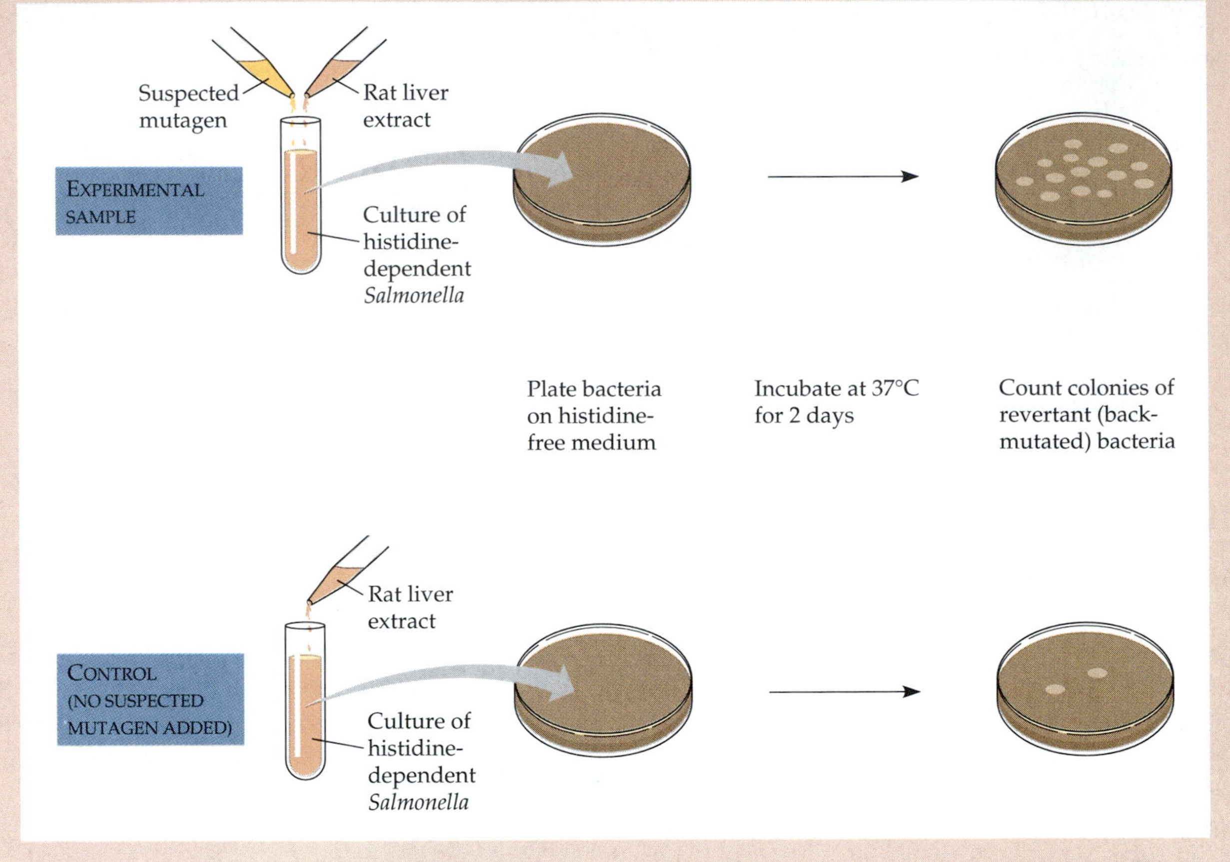

(cancer-causing chemicals) are mutagenic and, conversely, most mutagens are carcinogenic.

■ ■ ■

What is a gene?: *revisiting the question*

Our definition of a gene has evolved over the past few chapters, as it has through the history of genetics. We began with the Mendelian concept of a gene as a discrete unit of inheritance that affects a phenotypic character (Chapter 14). We saw that Morgan and his colleagues assigned such genes to specific loci on chromosomes and that geneticists sometimes use the term *locus* as a synonym for gene (Chapter 15). We went on to view a gene as a region of specific nucleotide sequence along the length of a DNA molecule (Chapter 16). Finally, in this chapter, we have moved toward a functional definition of a gene as a DNA sequence coding for a specific polypeptide chain. All these definitions are useful, depending on the context in which genes are being studied. (FIGURE 17.23 summarizes the path from gene to polypeptide in a eukaryotic cell.)

Even the one gene–one polypeptide definition must be refined and applied selectively. Most eukaryotic genes contain noncoding segments (introns), so large portions of these

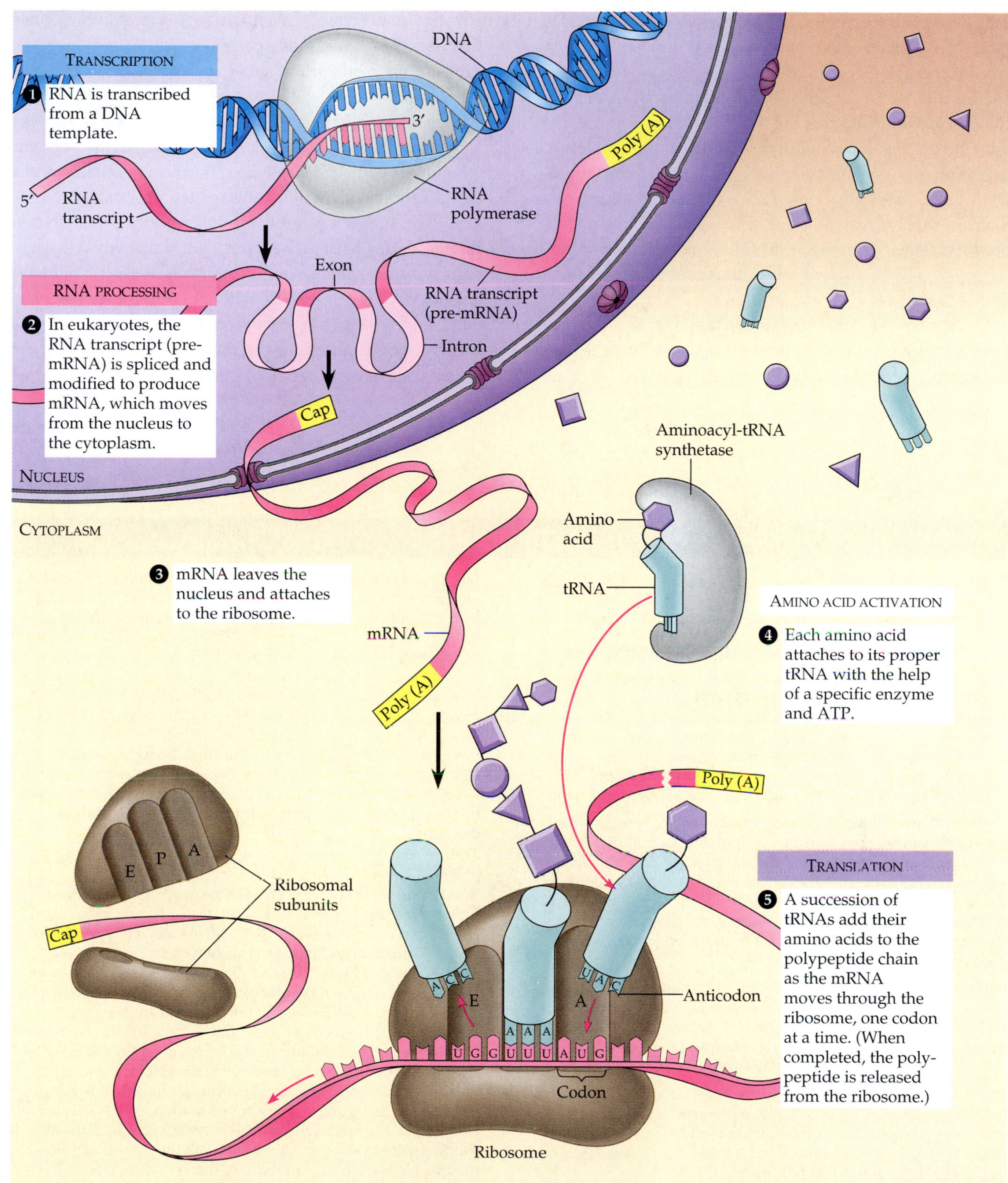
TRANSCRIPTION
1 RNA is transcribed from a DNA template.
DNA
3′
5′
RNA transcript
RNA polymerase
Poly (A)
RNA PROCESSING
2 In eukaryotes, the RNA transcript (pre-mRNA) is spliced and modified to produce mRNA, which moves from the nucleus to the cytoplasm.
Exon
RNA transcript (pre-mRNA)
Intron
Cap
NUCLEUS
CYTOPLASM
3 mRNA leaves the nucleus and attaches to the ribosome.
mRNA
Poly (A)
Aminoacyl-tRNA synthetase
Amino acid
tRNA
AMINO ACID ACTIVATION
4 Each amino acid attaches to its proper tRNA with the help of a specific enzyme and ATP.
Poly (A)
E
P
A
Ribosomal subunits
Cap
TRANSLATION
5 A succession of tRNAs add their amino acids to the polypeptide chain as the mRNA moves through the ribosome, one codon at a time. (When completed, the polypeptide is released from the ribosome.)
A C C
U A C
A A A
U G G U U U A U G
E
A
Anticodon
Codon
Ribosome

17.2 17.4 **FIGURE 17.23 • A summary of transcription and translation in a eukaryotic cell.** Keep in mind that each gene in the DNA can be transcribed repeatedly into many RNA molecules, and that each mRNA can be translated repeatedly to yield numerous protein molecules. (Also, remember that the final products of some genes are RNA molecules, including tRNA and rRNA.) In general, the steps of transcription and translation are similar in prokaryotic and eukaryotic cells. The major difference is the occurrence of mRNA processing in the eukaryotic nucleus. Other significant differences are found in the initiation stages of both transcription and translation and in the termination of transcription.

genes have no corresponding segments in polypeptides. Molecular biologists also often include promoters and certain other regulatory regions of DNA within the boundaries of a gene. These DNA sequences are not transcribed, but they can be considered part of the functional gene because they must be present for transcription to occur. Our molecular definition of a gene must also be broad enough to include the DNA that is transcribed into rRNA, tRNA, and other RNAs that are not translated. These genes have no polypeptide products. Thus, we arrive at the following definition: *A gene is a region of DNA whose final product is either a polypeptide or an RNA molecule.*

For most genes, however, it is still useful to retain the one gene–one polypeptide idea. In this chapter you have learned in molecular terms how a typical gene is expressed—by transcription into RNA and then translation into a polypeptide that forms a protein of specific structure and function. Proteins, in turn, bring about an organism's observable phenotype.

Genes are subject to regulation. The control of gene expression enables a bacterium, for example, to vary the amounts of particular enzymes as the metabolic needs of the cell change. In multicellular eukaryotes, the control of gene expression makes it possible for cells with the same DNA to diverge during their development into different cell types, such as muscle and nerve cells in animals. We will explore the regulation of gene expression in eukaryotes in Chapters 19 and 21. In the next chapter we begin our discussion of gene regulation by focusing on the simpler molecular biology of bacteria and viruses.

CHAPTER REVIEW

REVIEW OF KEY CONCEPTS

(with page numbers and key figures)

THE CONNECTION BETWEEN GENES AND PROTEINS

- **The study of metabolic defects provided evidence that genes specify proteins: *science as a process* (pp. 294–296, FIGURE 17.1)** DNA controls metabolism by commanding cells to make specific enzymes and other proteins. Beadle and Tatum's experiments on mutant strains of *Neurospora* gave rise to the one gene–one enzyme hypothesis, later modified to one gene–one polypeptide. A gene determines the amino acid sequence of a polypeptide chain.

- **Transcription and translation are the two main processes linking**
17.1 **gene to protein: *an overview* (pp. 296–297; FIGURE 17.2)** Both nucleic acids and proteins are informational polymers with linear sequences of monomers—nucleotides and amino acids, respectively. Transcription is the nucleotide-to-nucleotide transfer of information from DNA to RNA, while translation is the informational transfer from nucleotide sequence in RNA to amino acid sequence in a polypeptide.

- **In the genetic code, nucleotide triplets specify amino acids (pp. 297–299, FIGURES 17.3, 17.4)** The three-nucleotide units in DNA are transcribed into mRNA nucleotide triplets called codons. Of the 64 codons, 61 code for amino acids, with many synonyms. A few codons are start and stop signals for the genetic message.

- **The genetic code must have evolved very early in the history of life (pp. 299–300)** The near universality of the genetic code suggests that it was present in ancestors common to all kingdoms of life.

THE SYNTHESIS AND PROCESSING OF RNA

- **Transcription is the DNA-directed synthesis of RNA: *a closer look***
17.2 **(pp. 300–302, FIGURES 17.6, 17.7)** RNA synthesis on a DNA template is catalyzed by RNA polymerase. It follows the same base-pairing rules as DNA replication, except that in RNA, uracil substitutes for thymine. Promoters, specific nucleotide sequences at the start of a gene, signal the initiation of RNA synthesis. Transcription factors (proteins) help eukaryotic RNA polymerase recognize promoter sequences. Transcription continues until a particular RNA sequence signals termination.

- **Eukaryotic cells modify RNA after transcription (pp. 302–304,**
17.3 **FIGURES 17.8–17.10)** Eukaryotic mRNA molecules are processed before leaving the nucleus by modification of their ends and by RNA splicing. The 5′ end receives a modified nucleotide cap, and the 3′ end a poly(A) tail. These probably protect the molecule from degradation and enhance translation. Most eukaryotic genes have introns, noncoding regions interspersed among the coding regions, exons. In RNA splicing, introns are removed and exons joined. RNA splicing is catalyzed by small nuclear ribonucleoproteins (snRNPs), operating within larger assemblies called spliceosomes. In some cases, RNA alone catalyzes splicing. Catalytic RNA molecules are called ribozymes. The shuffling of exons by recombination may contribute to the evolution of protein diversity.

THE SYNTHESIS OF PROTEIN

- **Translation is the RNA-directed synthesis of a polypeptide:**
17.4 ***a closer look* (pp. 304–310, FIGURE 17.15–17.17)** After picking up specific amino acids, transfer RNA (tRNA) molecules line up by means of their anticodon triplets at complementary codons on mRNA. The attachment of a specific amino acid to its particular tRNA is an ATP-driven process catalyzed by an aminoacyl-tRNA synthetase enzyme. Ribosomes coordinate the three stages of translation: initiation, elongation, and termination. Each ribosome is composed of two subunits made of protein and ribosomal RNA (rRNA). Ribosomes have a binding site for mRNA; P and A sites that hold adjacent tRNAs as amino acids are linked in the growing polypeptide chain; and an E site for release of tRNA. A number of ribosomes can work on a single mRNA molecule simultaneously, forming a polyribosome. After translation the protein may be modified in ways that affect its three-dimensional shape.

- **Signal peptides target some eukaryotic polypeptides to specific destinations in the cell (pp. 310–311, FIGURE 17.19)** Proteins that will remain in the cytosol are made on free ribosomes. Proteins destined for membranes or for export from the cell are synthesized on ribosomes bound to the endoplasmic reticulum. A signal-recognition particle binds to a signal sequence on the leading end of the growing polypeptide, enabling the ribosome to bind to the ER. Other signal sequences target proteins for mitochondria or chloroplasts.

- **RNA plays multiple roles in the cell: *a review*** (p. 311, TABLE 17.1) More versatile than DNA, RNA performs structural, informational and catalytic roles.
- **Comparing protein synthesis in prokaryotes and eukaryotes: *a review*** (p. 311) In a bacterial cell, which lacks a nuclear envelope, translation of an mRNA can begin while transcription is still in progress. In a eukaryotic cell, the nuclear envelope separates transcription from translation; extensive RNA processing occurs in the nucleus.
- **Point mutations can affect protein structure and function** (pp. 312–314, FIGURE 17.22) Point mutations are changes in one or a few sequential base pairs. Base-pair substitutions can cause missense or nonsense mutations, which are often detrimental to protein function. Base-pair insertions or deletions may produce frameshift mutations that disrupt the codon messages downstream of the mutation. Spontaneous mutations can occur during DNA replication or repair. Various chemical and physical mutagens can also alter genes.
- **What is a gene?: *revisiting the question*** (pp. 314–316, FIGURE 17.23) A gene is usually a region of DNA encoding a polypeptide, but some genes have RNA molecules as their final products.

SELF-QUIZ

1. Base-pair substitutions involving the third base of a codon are unlikely to result in an error in the polypeptide. This is because
   a. base-pair substitutions are corrected before transcription begins
   b. base-pair substitutions are restricted to introns, and these regions are later deleted from the mRNA
   c. most tRNAs bind tightly to a codon with only the first two bases of the anticodon
   d. a signal-recognition particle corrects coding errors before the mRNA reaches the ribosome
   e. transcribed errors attract snRNPs, which then stimulate splicing and correction

2. In eukaryotic cells, transcription cannot begin until
   a. the two DNA strands have completely separated and exposed the promoter
   b. the appropriate transcription factors have bound to the promoter
   c. the 5′ caps are removed from the mRNA
   d. the DNA introns are removed from the template
   e. DNA nucleases have isolated the transcription unit from the noncoding DNA

3. Which of the following is *not* true of a codon?
   a. It consists of three nucleotides.
   b. It may code for the same amino acid as another codon does.
   c. It never codes for more than one amino acid.
   d. It extends from one end of a tRNA molecule.
   e. It is the basic unit of the genetic code.

4. Beadle and Tatum discovered several classes of *Neurospora* mutants that were able to grow on minimal medium with arginine added. Class I mutants were also able to grow on medium supplemented with either ornithine or citrulline, whereas class II mutants could grow on citrulline medium but not on ornithine medium. The metabolic pathway of arginine synthesis is as follows:

   Precursor ⟶ Ornithine ⟶ Citrulline ⟶ Arginine
   (A) (B) (C)

   From the behavior of their mutants, Beadle and Tatum could conclude that
   a. one gene codes for the entire metabolic pathway
   b. the genetic code of DNA is a triplet code
   c. class I mutants have their mutations later in the nucleotide chain than do class II mutants and thus have more functional enzymes
   d. class I mutants have a nonfunctional enzyme at step A, and class II mutants have a nonfunctional enzyme at step B
   e. class I mutants have a nonfunctional enzyme at step B, and class II mutants have a nonfunctional enzyme at step C

5. The anticodon of a particular tRNA molecule is
   a. complementary to the corresponding mRNA codon
   b. complementary to the corresponding triplet in rRNA
   c. the part of tRNA that bonds to a specific amino acid
   d. changeable, depending on the amino acid that attaches to the tRNA
   e. catalytic, making the tRNA a ribozyme

6. Which of the following is *not* true of RNA processing?
   a. Exons are excised and hydrolyzed before mRNA moves out of the nucleus.
   b. The existence of exons and introns may facilitate crossing over between regions of a gene that code for polypeptide domains.
   c. Ribozymes function in RNA splicing.
   d. RNA splicing may be catalyzed by spliceosomes.
   e. A primary transcript is often much longer than the final RNA molecule that leaves the nucleus.

7. Which of the following terms includes all others in the list?
   a. ribozyme
   b. enzyme
   c. catalyst
   d. snRNP
   e. aminoacyl-tRNA synthetase

8. Using the genetic code in FIGURE 17.4, identify a possible 5′⟶3′ sequence of nucleotides in the *DNA template strand* for an mRNA coding for the polypeptide sequence Phe-Pro-Lys.
   a. UUU-GGG-AAA
   b. GAA-CCC-CTT
   c. AAA-ACC-TTT
   d. CTT-CGG-GAA
   e. AAA-CCC-UUU

9. Which of the following mutations would be *most* likely to have a harmful effect on an organism? Explain your answer.
   a. a base-pair substitution
   b. a deletion of three bases near the middle of the gene
   c. a single base deletion near the middle of an intron
   d. a single base deletion close to the end of the coding sequence
   e. a single base insertion near the start of the coding sequence

10. Which component is *not directly* involved in the process known as translation?
    a. mRNA
    b. DNA
    c. tRNA
    d. ribosomes
    e. GTP

CHALLENGE QUESTIONS

1. The base sequence of the template strand of an unusual gene, which codes for an extremely short polypeptide, is CTACGCTAGGCGATTCAT. Which is the 5′ end of this DNA sequence? How do you know? What would be the base sequence of the mRNA transcribed

from this gene? Using the genetic code chart (FIGURE 17.4), give the amino acid sequence of the polypeptide translated from this mRNA.

2. A biologist inserted a gene from a human liver cell into the chromosome of a bacterium. The bacterium then transcribed this gene into mRNA and translated the mRNA into protein. The protein produced was useless; it contained many more amino acids than the protein made by the eukaryotic cell, and the amino acids were in a different sequence. Explain why.

3. Choose one catalyst, either an enzyme or a ribozyme, and write a paragraph explaining how it functions in the pathway from gene to protein.

SCIENCE, TECHNOLOGY, AND SOCIETY

1. Our civilization generates many potentially mutagenic chemicals (pesticides, for example) and modifies the environment in ways that increase exposure to other mutagens, notably UV radiation. What role should government play in identifying mutagens and regulating their release to the environment?

2. As part of the Human Genome Project (discussed in Chapter 20), researchers are determining the nucleotide sequences of human genes and identifying the proteins coded by the genes. Labs of the U.S. National Institutes of Health (NIH), for example, have worked out thousands of sequences, and similar analysis is being carried out by many private companies. Knowing the nucleotide sequence of a gene and identifying its product can be useful; this information might be used to treat genetic defects or produce life-saving medicines. U.S. law allows the first person or research group to isolate a pure protein or a gene to patent it, whether or not a practical use for the discovery has been demonstrated. The NIH and biotechnology companies have applied for patents on their discoveries. What are the purposes of a patent? How might the discoverer of a gene benefit from a patent? How might the public benefit? What kinds of negative impacts might result from patenting genes? Do you think individuals and companies should be able to patent genes and gene products? Why or why not? Under what conditions should such patenting be permitted?

FURTHER READING

Becker, W. M., J. B. Reece, and M. F. Poenie. *The World of the Cell,* 3rd ed. Menlo Park, CA: Benjamin/Cummings, 1996. Chapters 17 and 18 present the topics of this chapter in more detail.

Hartl, F. U. "Molecular Chaperones in Cellular Protein Folding." *Nature,* June 13, 1996. Describes the critically important roles of chaperones in protein folding and related activities, including protein translocation into organelles.

Hentze, M. W. "elF4G: A Multipurpose Ribosome Adapter?" *Science,* January 24, 1997. Discusses the roles of 5′ cap, poly(A) tail, leader sequences, and protein factors in the binding of the eukaryotic small ribosomal subunit to mRNA.

Landick, R., and J. W. Roberts. "The Shrewd Grasp of RNA Polymerase." *Science,* July 12, 1996. Provides the details of the transcription elongation stage in bacteria.

Pennisi, E. "How the Nucleus Gets It Together." *Science,* June 6, 1997. Eukaryotic mRNA synthesis and processing seem to occur at particular locations within the nucleus.

Radetsky, P. "Genetic Heretic." *Discover,* November 1990. The story of how the discovery of ribozymes led to a Nobel Prize.

Stoltzfus, A., et al. "Testing the Exon Theory of Genes: The Evidence from Protein Structure." *Science,* July 8, 1994. Did introns first arise in prokaryotes or eukaryotes? This article reports evidence for a eukaryotic origin.

Tijan, R. "Molecular Machines That Control Genes." *Scientific American,* February 1995. Transcription is a team effort.

WEB LINKS

Visit the special edition of *The Biology Place* for BIOLOGY, Fifth Edition, at **http://www.biology.com/campbell**. Go to Chapter 17 for online resources, including learning activities, practice exams, and links to the following web sites:

"From Gene to Function"

The first topic in Genetech's Access Excellence Graphics Gallery, this is a sequence of illustrated pages describing how cells use the information encoded in genes to make proteins.

"The Central Dogma Directory"

This is a chapter from the MIT HyperTextbook that deals with DNA replication, transcription, and translation.

"Nucleic Acids Problem Sets"

Use this problem set to test your understanding of nucleic acids.

"From Gene to Protein"

Visit Campbell's Biology Place for access to these interactive exercises on transcription and translation by Peter Russell of Reed College.

CHAPTER 18

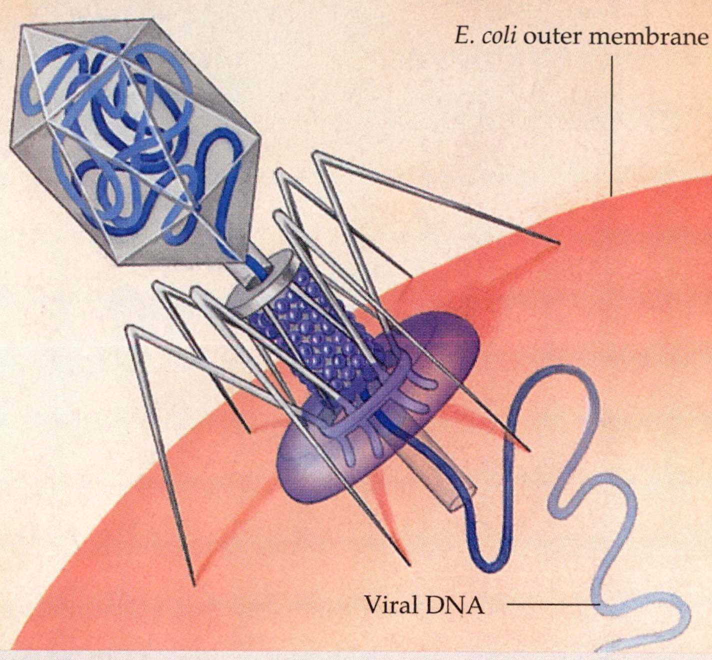

MICROBIAL MODELS: THE GENETICS OF VIRUSES AND BACTERIA

The drawing that opens this chapter dramatizes a remarkable event: the genetic takeover of a cell by a virus. In this case, the cell is the bacterium E. coli, *and the virus, looking something like a miniature lunar landing craft, is the bacteriophage T4. The phage is infecting the cell by injecting its DNA. Earlier you met a closely related virus, T2, which helped prove that DNA is the genetic material (see* FIGURE 16.2*). Molecular biology was born in the laboratories of microbiologists studying such viruses and bacteria. Microbiologists provided most of the evidence that genes are made of DNA, and they worked out the major steps of DNA replication, transcription, and translation, the three main processes in the flow of genetic information. Viruses and bacteria are the simplest biological systems—microbial models where scientists find life's fundamental molecular mechanisms in their most basic, accessible forms.*

The value of viruses and bacteria as model systems in biological research is just one reason to learn about these microbes. While microbial models have helped biologists understand the molecular genetics of more complex organisms, viruses and bacteria also have unique features that make their genetics interesting in its own right. These specialized mechanisms have important applications for understanding how viruses and bacteria cause disease. In addition, techniques enabling scientists to manipulate genes and transfer them from one organism to another have emerged from the study of microbes. These techniques are having an important impact on both basic research and biotechnology (see Chapter 20).

*In this chapter we explore the genetics of viruses and bacteria. Recall that bacteria are prokaryotic organisms, with cells much smaller and more simply organized than those of eukaryotes, such as plants and animals. Viruses are smaller and simpler still, lacking the structures and most of the metabolic machinery found in cells (*FIGURE 18.1, p. 320*). In fact, most viruses are little more than aggregates of nucleic acid and protein—genes packaged in protein coats. It is with these simplest of all genetic systems that we begin.*

THE GENETICS OF VIRUSES

Researchers discovered viruses by studying a plant disease: *science as a process*

Microbiologists were able to detect viruses indirectly long before they were actually able to see them. The story of how viruses were discovered begins in 1883 with Adolf Mayer, a German scientist who was seeking the cause of tobacco mosaic disease. This disease stunts the growth of tobacco plants and gives their leaves a mottled, or mosaic, coloration (see FIGURE 18.8a, p. 329). Mayer discovered that the disease was contagious when he found he could transmit it from

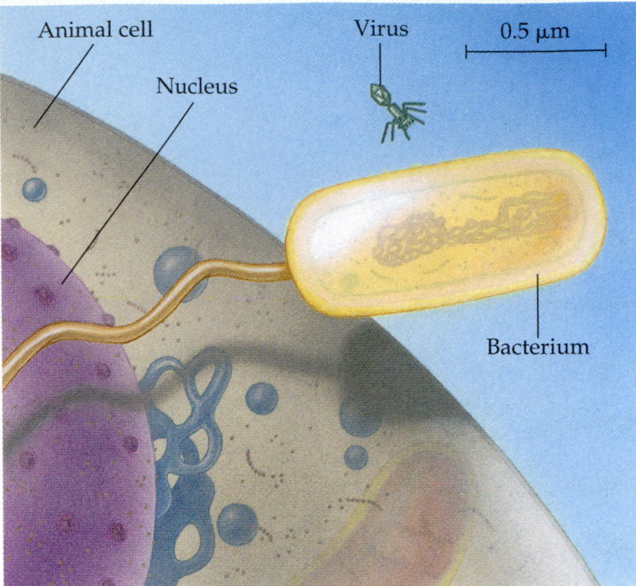

FIGURE 18.1 ▪ **Comparing the sizes of a virus, a bacterium, and a eukaryotic cell.** By studying viruses and bacteria, the simplest biological systems, scientists caught their first glimpses of the elegant molecular mechanisms of heredity.

plant to plant by spraying sap extracted from diseased leaves onto healthy plants. He searched for a microbe in the infectious sap but found none. Mayer concluded that the disease was caused by unusually small bacteria that could not be seen with the microscope. This hypothesis was tested a decade later by Dimitri Ivanowsky, a Russian who passed sap from infected tobacco leaves through a filter designed to remove bacteria. After filtering, the sap still produced mosaic disease.

Ivanowsky clung to the hypothesis that bacteria caused tobacco mosaic disease. Perhaps, he reasoned, the pathogenic bacteria were so small they could pass through the filter. Or perhaps the bacteria made a filterable toxin that caused the disease. This latter possibility was ruled out in 1897 when the Dutch botanist Martinus Beijerinck discovered that the infectious agent in the filtered sap could reproduce. Beijerinck sprayed plants with the filtered sap, and after these plants developed mosaic disease he used their sap to infect more plants, continuing this process through a series of infections. The pathogen must have been reproducing, for its ability to cause disease was undiluted after several transfers from plant to plant.

In fact, the pathogen could reproduce only within the host it infected. Unlike bacteria, the mysterious agent of mosaic disease could not be cultivated on nutrient media in test tubes or petri dishes. Also, the pathogen was not inactivated by alcohol, which is generally lethal to bacteria. Beijerinck imagined a reproducing particle much smaller and simpler than bacteria. His suspicions were confirmed in 1935, when the American scientist Wendell Stanley crystallized the infectious particle, now known as tobacco mosaic virus (TMV). Subsequently, TMV and many other viruses were actually seen with the help of the electron microscope.

A virus is a genome enclosed in a protective coat

The tiniest viruses are only 20 nm in diameter—smaller than a ribosome. Millions could easily fit on a pinhead. Even the largest viruses can barely be resolved with the light microscope. Stanley's discovery that some viruses could be crystallized was exciting and puzzling news. Not even the simplest of cells can aggregate into regular crystals. But if viruses are not cells, then what are they? They are infectious particles consisting of nucleic acid enclosed in a protein coat and, in some cases, a membranous envelope. Let's examine the structure of viruses more closely; then we'll look at how viruses replicate.

Viral Genomes

We usually think of genes as being made of double-stranded DNA—the conventional double helix—but viruses often defy this convention. Their genomes (sets of genes) may consist of double-stranded DNA, single-stranded DNA, double-stranded RNA, or single-stranded RNA, depending on the specific type of virus. A virus is called a DNA virus or an RNA virus, according to the kind of nucleic acid that makes up its genome. In either case, the genome is usually organized as a single linear or circular molecule of nucleic acid. The smallest viruses have only four genes, while the largest have several hundred.

Capsids and Envelopes

The protein shell that encloses the viral genome is called a **capsid**. Depending on the type of virus, the capsid may be rod-shaped (more precisely, helical), polyhedral, or more complex in shape. Capsids are built from a large number of protein subunits called capsomeres, but the number of different *kinds* of proteins is usually small. Tobacco mosaic virus, for example, has a rigid, rod-shaped capsid made from over a thousand molecules of a single type of protein (FIGURE 18.2a). Adenoviruses, which infect the respiratory tracts of animals, have 252 identical protein molecules arranged into a polyhedral capsid with 20 triangular facets—an icosahedron (FIGURE 18.2b).

Some viruses have accessory structures that help them infect their hosts. Influenza viruses, and many other viruses found in animals, have **viral envelopes**, membranes cloaking their capsids (FIGURE 18.2c). These envelopes are derived from membrane of the host cell, but in addition to host-cell phospholipids and proteins, they also contain proteins and glycoproteins (proteins with carbohydrate covalently attached) of viral origin. In addition to virus-coded envelope proteins and the capsid proteins, some viruses also carry a few viral enzyme molecules within their capsids.

The most complex capsids are found among viruses that infect bacteria. As you learned in Chapter 16, bacterial viruses are called **bacteriophages**, or simply **phages**. The first phages

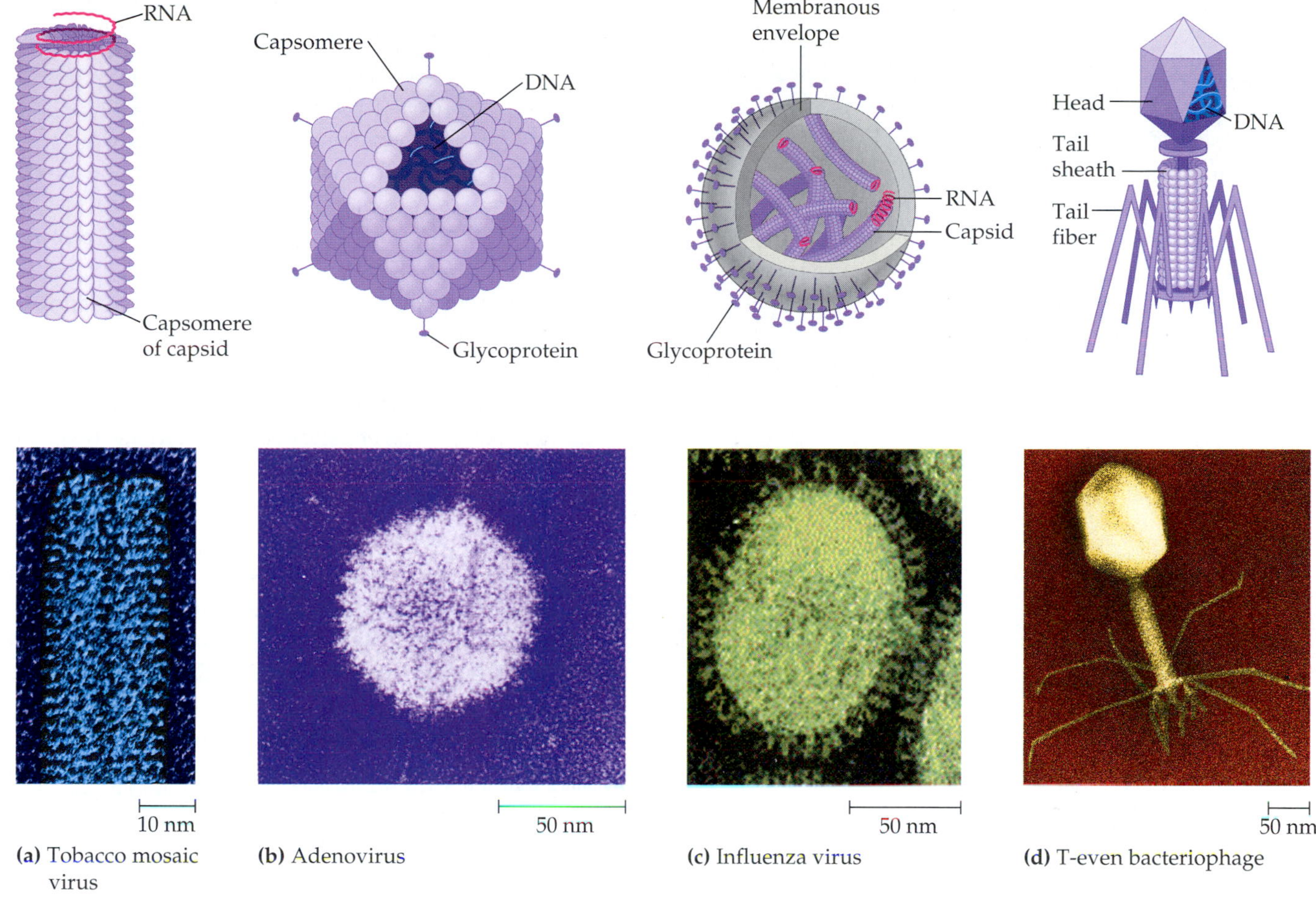

(a) Tobacco mosaic virus **(b)** Adenovirus **(c)** Influenza virus **(d)** T-even bacteriophage

FIGURE 18.2 ▪ Viral structure. Viruses are made up of nucleic acid (DNA or RNA) enclosed in a protein coat (the capsid) and sometimes further wrapped in a membranous envelope. The individual protein subunits making up the capsid are called capsomeres. Although viruses are diverse in size and shape, there are common structural motifs, most of which appear in the four examples shown here. **(a)** Tobacco mosaic virus has a helical capsid with the overall shape of a rigid rod. **(b)** Adenoviruses have a polyhedral capsid with a glycoprotein spike at each vertex. **(c)** Influenza virus has an outer envelope studded with glycoprotein spikes. Its genome consists of eight RNA molecules, each wrapped in a helical capsid. **(d)** Phages are viruses that infect bacteria. A T-even phage, such as T4, has a complex capsid consisting of a polyhedral head and a tail apparatus. DNA is stored in the head, and the tail piece functions in the injection of this DNA into a bacterium (see the drawing on p. 319). (All TEMs.)

studied included seven that infect the bacterium *Escherichia coli*. These seven phages were named type 1 (T1), type 2 (T2), and so forth, in the order of their discovery. By coincidence, the three T-even phages—T2, T4, and T6—turned out to be very similar in structure. Their capsids have elongated icosahedral heads that enclose their DNA. Attached to the head is a protein tail piece with tail fibers that the phage uses to attach to a bacterium (FIGURE 18.2d).

Viruses can reproduce only within a host cell: *an overview*

Viruses are obligate intracellular parasites; they can reproduce only within a host cell. An isolated virus is unable to reproduce—or do anything else, for that matter, except infect an appropriate host cell. Viruses lack the enzymes for metabolism and have no ribosomes or other equipment for making their own proteins. Thus, isolated viruses are merely packaged sets of genes in transit from one host cell to another.

Each type of virus can infect and parasitize only a limited range of host cells, called its **host range**. This host specificity depends on the evolution of recognition systems by the virus. Viruses identify their host cells by a "lock-and-key" fit between proteins on the outside of the virus and specific receptor molecules on the surface of the cell. (Presumably the receptors first evolved because they carried out functions of benefit to the organism.) Some viruses have host ranges broad enough to include several species. Swine flu virus, for example, can infect both hogs and humans, and the rabies virus can infect a number of mammalian species, including racoons, skunks, dogs, and humans. In other cases, viruses have host ranges so narrow that they infect only a single species. For instance, there are several phages that can parasitize only *E. coli*.

Viruses of eukaryotes are usually tissue-specific. Human cold viruses infect only the cells lining the upper respiratory tract, ignoring other tissues. And the AIDS virus binds to a specific receptor on certain types of white blood cells.

A viral infection begins when the genome of a virus makes
18.1 its way into a host cell (FIGURE 18.3). The mechanism by which this nucleic acid enters the cell varies, depending on the type of virus. For example, the T-even phages use their elaborate tail apparatus to inject DNA into a bacterium (see the drawing on p. 319). Once inside, the viral genome can commandeer its host, reprogramming the cell to copy the viral nucleic acid and manufacture viral proteins. Most DNA viruses use the DNA polymerases of the host cell to synthesize new genomes along the templates provided by the viral DNA.

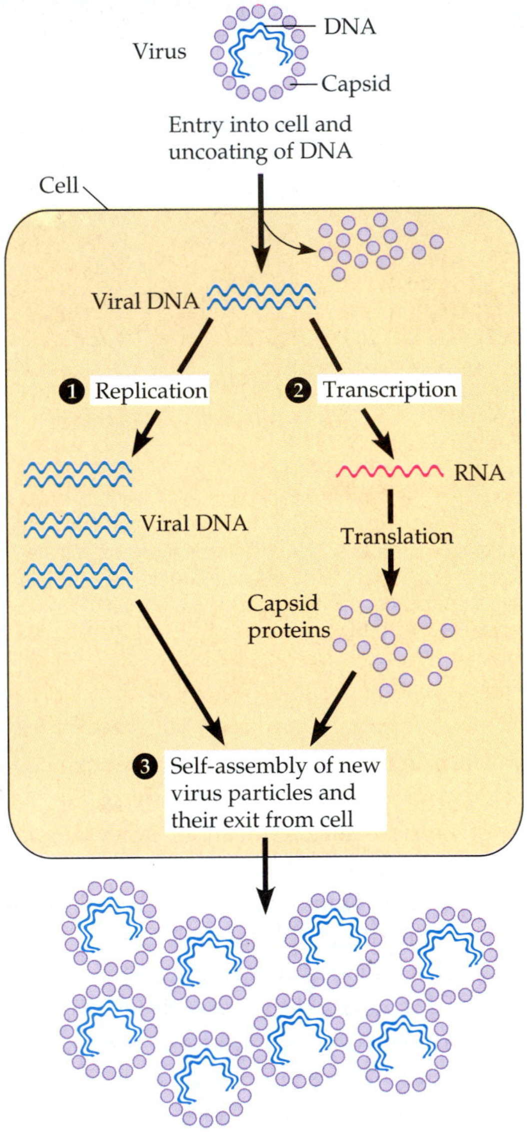

FIGURE 18.3 • A simplified viral reproductive cycle. A virus is an
18.1 obligate intracellular parasite that uses the equipment of its host cell to reproduce. In this simplest of all possible viral cycles, the parasite is a DNA virus with a capsid consisting of a single type of protein. ① After entering the cell, the viral DNA uses host nucleotides and enzymes to replicate iself. ② It uses other host materials and machinery to produce its capsid proteins. ③ Viral DNA and capsid proteins then assemble into new virus particles, which leave the cell.

In contrast, to replicate their genomes, RNA viruses must use special virus-encoded polymerases, ones that can use RNA as a template. (Cells generally have no native enzymes for carrying out such a process.) We will describe the replication of DNA and RNA viruses in more detail later in the chapter.

Regardless of the type of viral genome, the parasite diverts its host's resources for viral production. The host provides the nucleotides for nucleic acid synthesis. It also provides enzymes, ribosomes, tRNAs, amino acids, ATP, and other components needed for making the viral proteins dictated by mRNA transcribed from viral genes.

After the viral nucleic acid molecules and capsomeres are produced, their assembly into new viruses is often a spontaneous process, a process of self-assembly. In fact, the RNA and capsomeres of TMV can be separated in the laboratory and then reassembled to form complete viruses simply by mixing the components together again.

The simplest type of viral reproductive cycle is completed when hundreds or thousands of viruses emerge from the infected host cell. The cell is often destroyed in the process. In fact, some of the symptoms of human viral infections, such as colds and influenza, result from cellular damage and death and from the body's responses to this destruction. The viral progeny that exit a cell have the potential to infect additional cells, spreading the viral infection.

There are many variations on the simplified viral reproductive cycle we have traced in this overview. We will see several examples as we take a closer look at some bacterial viruses (phages), animal viruses, and plant viruses.

Phages reproduce using lytic or lysogenic cycles

The phages are the best understood of all viruses, although some of them are also among the most complex. Research on phages led to the discovery that some double-stranded DNA viruses can reproduce by two alternative mechanisms: the lytic cycle and the lysogenic cycle.

The Lytic Cycle

A viral reproductive cycle that culminates in death of the host
18.2 cell is known as a **lytic cycle**. The term refers to the last stage of infection, during which the bacterium lyses (breaks open) and releases the phages that were produced within the cell. Each of these phages can then infect a healthy cell, and a few successive lytic cycles can destroy an entire bacterial colony in just hours. A virus that reproduces only by a lytic cycle is a **virulent virus**. FIGURE 18.4 uses the virulent phage T4 to illustrate the steps of a lytic cycle. The figure and legend describe the process, which you should study before proceeding.

After reading about the lytic cycle, you may wonder why phages haven't exterminated all bacteria. Actually, bacteria are not defenseless. Natural selection favors bacterial mutants

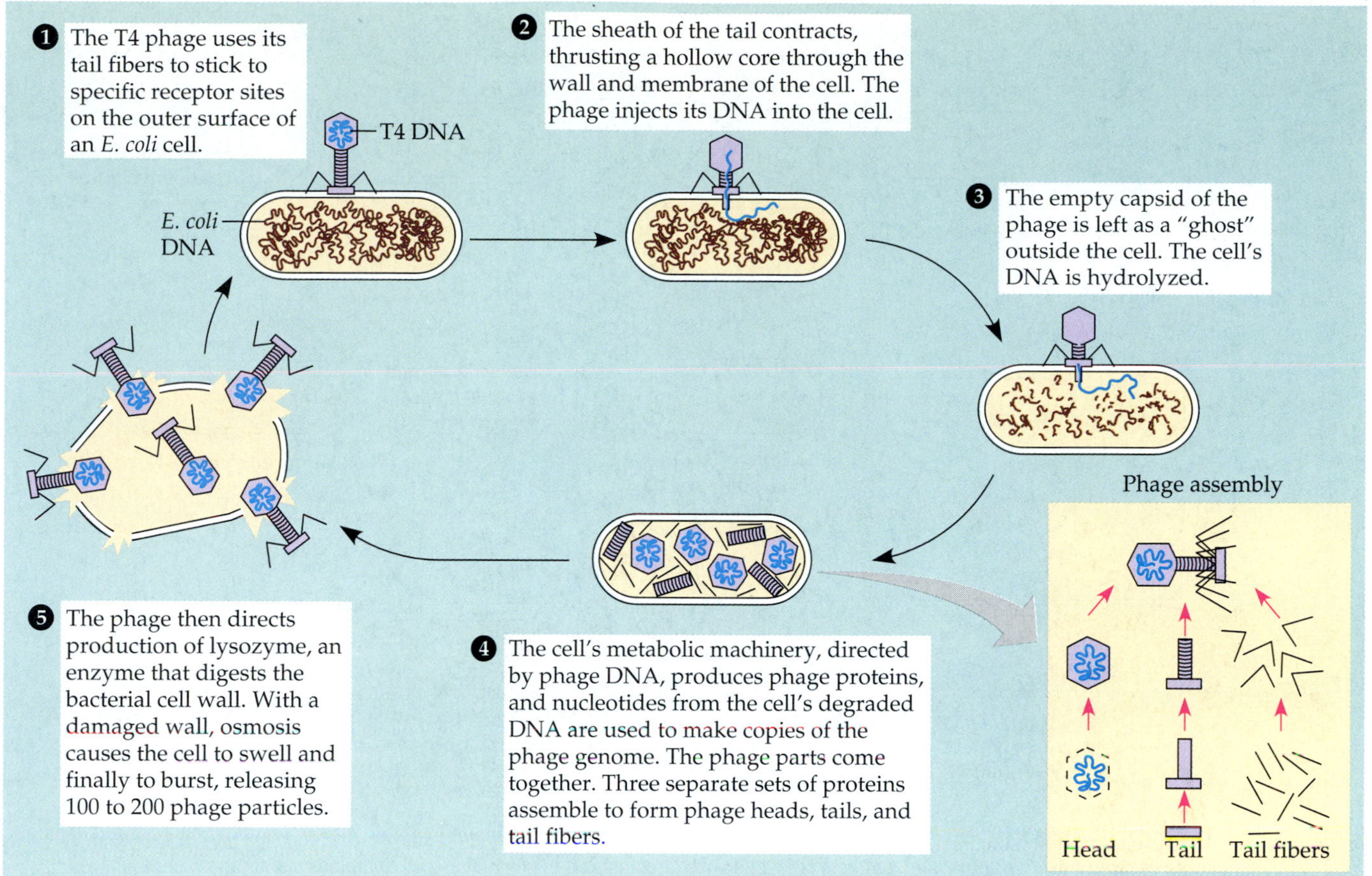

1 The T4 phage uses its tail fibers to stick to specific receptor sites on the outer surface of an E. coli cell.
T4 DNA
E. coli DNA
2 The sheath of the tail contracts, thrusting a hollow core through the wall and membrane of the cell. The phage injects its DNA into the cell.
3 The empty capsid of the phage is left as a "ghost" outside the cell. The cell's DNA is hydrolyzed.
Phage assembly
4 The cell's metabolic machinery, directed by phage DNA, produces phage proteins, and nucleotides from the cell's degraded DNA are used to make copies of the phage genome. The phage parts come together. Three separate sets of proteins assemble to form phage heads, tails, and tail fibers.
5 The phage then directs production of lysozyme, an enzyme that digests the bacterial cell wall. With a damaged wall, osmosis causes the cell to swell and finally to burst, releasing 100 to 200 phage particles.
Head
Tail
Tail fibers

18.2 **FIGURE 18.4 ▪ The lytic cycle of phage T4.** Phage T4 has about 100 genes, which are transcribed and translated using the host cell's machinery. One of the first phage genes translated after infection codes for an enzyme that chops up the host cell's DNA. The phage DNA is protected because it contains a modified form of cytosine that is not recognized by the enzyme. Nucleotides salvaged from the cell's DNA are used in making copies of the phage genome. The entire lytic cycle, from the phage's first contact with the cell surface to lysis, takes only 20–30 minutes at 37°C.

with receptor sites that are no longer recognized by a particular type of phage. And when phage DNA successfully enters a bacterium, various degradative enzymes may break it down. Enzymes called *restriction nucleases,* for example, recognize and cut up DNA that is foreign to the cell, including certain phage DNA. The bacterial cell's own DNA is chemically modified in a way that prevents attack by restriction enzymes. But just as natural selection favors bacteria with effective restriction enzymes, natural selection favors phage mutants that are resistant to these enzymes. Thus, the parasite-host relationship is in constant evolutionary flux.

There is still another important reason bacteria have been spared from extinction as a result of phage activity. Many phages can check their own destructive tendencies and, instead of lysing their host cells, coexist with them in what is called the lysogenic cycle.

The Lysogenic Cycle

18.3 In contrast to the lytic cycle, which kills the host cell, the **lysogenic cycle** replicates the viral genome without destroying the host. Viruses that are capable of using both modes of reproducing within a bacterium are called **temperate viruses.** To compare the lytic and lysogenic cycles, we will examine a temperate phage called lambda, abbreviated with the Greek letter λ. Phage λ resembles T4, but its tail has only a single, short tail fiber (not shown in FIGURE 18.5, p. 324).

Infection of an *E. coli* cell by λ begins when the phage binds to the surface of the cell and injects its DNA (FIGURE 18.5). Within the host, the λ DNA molecule forms a circle. What happens next depends on the reproductive mode: lytic cycle or lysogenic cycle. During a lytic cycle, the viral genes immediately turn the host cell into a λ-producing factory, and the cell soon lyses and releases its viral products. The viral genome behaves differently during a lysogenic cycle. The λ DNA molecule is incorporated by genetic recombination (crossing over) into a specific site on the host cell's chromosome. It is then known as a **prophage.** One prophage gene codes for a protein that represses most of the other prophage genes. Thus, the phage genome is mostly silent within the bacterium. How, then, does the phage reproduce? Every time the *E. coli* cell prepares to divide, it replicates the phage DNA

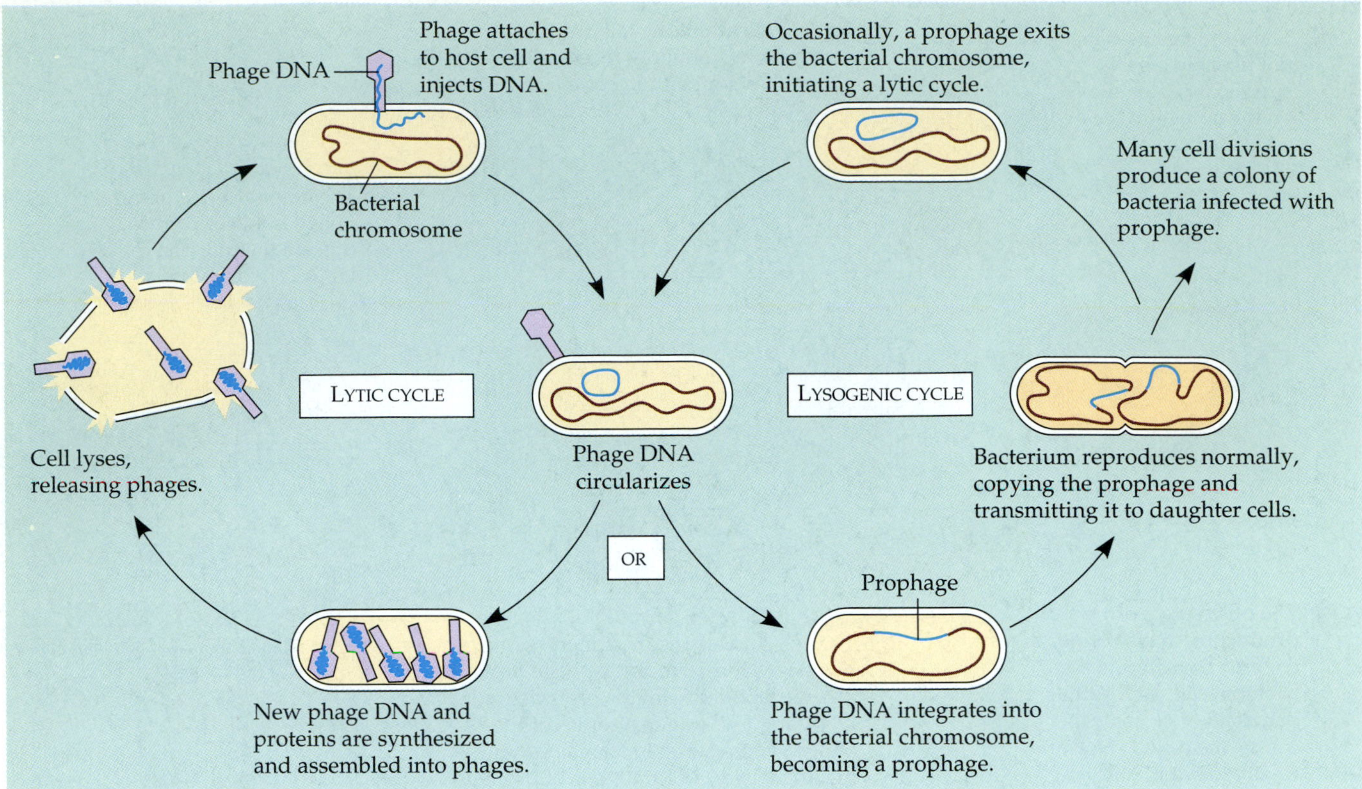

18.3 **FIGURE 18.5 ▪ The lysogenic and lytic reproductive cycles of phage λ, a temperate virus.** After entering the bacterial cell, the λ DNA can either integrate into the bacterial chromosome (lysogenic cycle) or immediately initiate the production of a large number of progeny phages (lytic cycle). In most cases the lytic pathway is followed, but once a lysogenic cycle is started, the prophage may be carried in the host cell's chromosome for many generations.

along with its own and passes the copies on to daughter cells. A single infected cell can soon give rise to a large population of bacteria carrying the virus in prophage form. This mechanism enables viruses to propagate without killing the host cells upon which they depend.

The term *lysogenic* implies that prophages can, at some point, give rise to active phages that lyse their host cells. This occurs when the λ genome exits the bacterial chromosome. At this time, the λ genome commands the host cell to manufacture complete phages and then self-destruct, releasing the infectious phage particles. It is usually an environmental trigger, such as radiation or the presence of certain chemicals, that switches the virus from the lysogenic to the lytic mode.

In addition to the gene for the repressor protein, a few other prophage genes may also be expressed during lysogenic cycles, and the expression of these genes may alter the phenotype of the host bacteria. This phenomenon can have important medical significance. For example, the bacteria that cause the human diseases diphtheria, botulism, and scarlet fever would be harmless to humans if it were not for certain prophage genes that induce the host bacteria to make toxins.

Animal viruses are diverse in their modes of infection and replication

Everyone has suffered from viral infections, whether chicken pox, influenza, or the common cold. TABLE 18.1 lists some important classes of animal viruses. Like all viruses, those that cause illness in humans and other animals can reproduce (replicate) only inside host cells.

Reproductive Cycles of Animal Viruses

Many variations on the basic scheme of viral infection and reproduction are represented among the animal viruses. One key variable is the type of nucleic acid that serves as a virus's genetic material (the basis for classification in TABLE 18.1). Another is the presence or absence of a membranous envelope. Rather than consider all the mechanisms of viral infection and reproduction, we will focus on the roles of viral envelopes and on the functioning of RNA as the genetic material of many viruses.

Viral Envelopes. Some animal viruses are equipped with an outer membrane, or viral envelope, outside the capsid.

Table 18.1 ▪ Classes of Animal Viruses, Grouped by Type of Nucleic Acid

CLASS*	EXAMPLES/DISEASES
I. dsDNA**	
Papovavirus	Papilloma (human warts, cervical cancer); polyoma (tumors in certain animals)
Adenovirus	Respiratory diseases; some cause tumors in certain animals
Herpesvirus	Herpes simplex I (cold sores); herpes simplex II (genital sores); varicella zoster (chicken pox, shingles); Epstein-Barr virus (mononucleosis, Burkitt's lymphoma)
Poxvirus	Smallpox; vaccinia; cowpox
II. ssDNA (parvovirus)	
	Roseola; most parvoviruses depend on coinfection with adenoviruses for growth
III. dsRNA (reovirus)	
	Diarrhea; mild respiratory diseases
IV. ssRNA that can serve as mRNA	
Picornavirus	Poliovirus; rhinovirus (common cold); enteric (intestinal) viruses
Togavirus	Rubella virus; yellow fever virus; encephalitis viruses
V. ssRNA that is a template for mRNA	
Rhabdovirus	Rabies
Paramyxovirus	Measles; mumps
Orthomyxovirus	Influenza viruses
VI. ssRNA that is a template for DNA synthesis (retrovirus)	
	RNA tumor viruses (e.g., leukemia viruses); HIV (AIDS virus)

*The subclasses within each class differ mainly in capsid structure and in the presence or absence of a membranous envelope.
**ds = double-stranded; ss = single-stranded.

The viral envelope helps the parasite enter the host cell (FIGURE 18.6). This membrane is generally a lipid bilayer, like a cellular membrane, with glycoproteins protruding from the outer surface. The glycoprotein spikes bind to specific receptor molecules on the surface of a host cell. The viral envelope then fuses with the host's plasma membrane, transporting the capsid and viral genome into the cell. After cellular enzymes remove the capsid, the viral genome can replicate and direct the synthesis of viral proteins, including glycoproteins for new viral envelopes. The endoplasmic reticulum of the host cell makes these membrane proteins, which in most cases are transported to the plasma membrane. There they are clustered in patches that serve as exit points for the viral

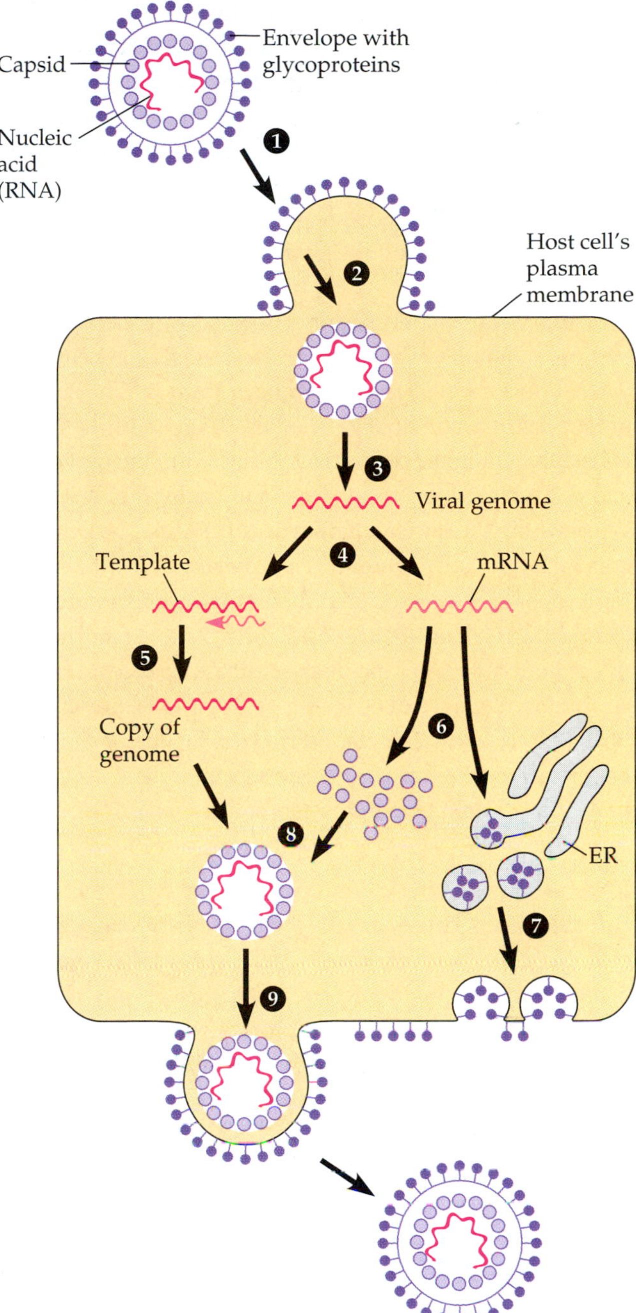

FIGURE 18.6 ▪ The reproductive cycle of an enveloped virus. The virus shown here has a genome of single-stranded RNA (of class V in TABLE 18.1.) ① Glycoproteins projecting from the viral envelope recognize and bind to specific receptor molecules (not shown) on the surface of the host cell. ② The viral envelope fuses with the cell's plasma membrane, and the capsid and viral genome enter the cell. ③ Cellular enzymes remove the capsid. ④ The viral genome functions as a template for making complementary RNA strands (lighter color), which have two functions: ⑤ they serve as templates for making new copies of genome RNA, and ⑥ they serve as mRNA. The mRNA is translated into both capsid proteins and glycoproteins for the viral envelope. The cell's endoplasmic reticulum (ER) synthesizes the glycoproteins. ⑦ Vesicles transport the glycoproteins to the cell's plasma membrane. ⑧ A capsid assembles around each viral genome molecule. ⑨ The virus buds from the cell. Its envelope, studded with viral glycoproteins, is derived from the cell's plasma membrane.

progeny. The new viruses bud from the cell surface at these points, in a process much like exocytosis, wrapping themselves in membrane as they go. In other words, the viral envelope is derived from the host cell's plasma membrane, although some of the molecules of this membrane are specified by viral genes. The enveloped viruses are now free to spread the infection to other cells. This reproductive cycle does not necessarily kill the host cell, in contrast to the lytic cycles of phages.

Some viruses have envelopes that are not derived from plasma membrane. Herpesviruses, for example, have envelopes derived from the nuclear membrane of the host. The genomes of herpesviruses are double-stranded DNA, and these viruses reproduce within the cell nucleus, using a combination of viral and cellular enzymes to replicate and transcribe their DNA. While within the nucleus, herpesvirus DNA may become integrated into the cell's genome as a **provirus**, similar to a bacterial prophage. Once acquired, herpes infections (which may cause cold sores or genital sores, for instance) tend to recur throughout a person's life. Between these episodes, the virus apparently remains latent within the host cells' nuclei. From time to time, physical or emotional stress may cause the herpes proviruses to leave the host's genome and initiate active virus production, resulting in the blisters of active infections.

RNA as Viral Genetic Material. Although some phages and most plant viruses are RNA viruses, the broadest variety of RNA genomes is found among the viruses that infect animals. As TABLE 18.1 indicates, RNA viruses are classified according to the strandedness of their RNA and how it functions in a host cell. Notice that there are three types of single-stranded RNA genomes (classes IV–VI). The genome of class IV viruses can directly serve as mRNA and thus can be translated into viral protein immediately after infection. FIGURE 18.6 shows a virus of class V, in which the RNA genome serves as a *template* for mRNA synthesis. The RNA genome is transcribed into a strand of complementary RNA, which serves both as mRNA and as template for the synthesis of additional copies of genome RNA. Like all viruses that require RNA ⟶ RNA synthesis to make mRNA, this one uses a viral enzyme that is packaged with the genome inside the capsid.

The RNA viruses with the most complicated reproductive cycles are the **retroviruses** (class VI). *Retro,* meaning "backward," refers to the reverse direction in which genetic information flows for these viruses. Retroviruses are equipped with an enzyme called **reverse transcriptase**, which transcribes DNA from an RNA template, providing an RNA ⟶ DNA information flow. The newly made DNA then integrates as a provirus into a chromosome within the nucleus of the animal cell. The host's RNA polymerase transcribes the viral DNA into RNA molecules, which can function both as mRNA for the synthesis of viral proteins and as genomes for new virus particles released from the cell. A retrovirus of particular importance is **HIV (human immunodeficiency virus)**, the virus that causes **AIDS (acquired immunodeficiency syndrome)**. FIGURE 18.7 traces the reproductive cycle of HIV as
18.4 an example of a retrovirus. We will postpone a detailed discussion of AIDS until Chapter 43.

Important Viral Diseases in Animals

The link between a viral infection and the symptoms it produces is often obscure. Some viruses damage or kill cells by causing the release of hydrolytic enzymes from lysosomes. Some viruses cause the infected cells to produce toxins that lead to disease symptoms, and some have toxic components themselves, such as envelope proteins. How much damage a virus causes depends partly on the ability of the infected tissue to regenerate by cell division. We usually recover completely from colds because the epithelium of the respiratory tract, which the viruses infect, can efficiently repair itself. In contrast, the poliovirus attacks nerve cells, which do not divide and cannot be replaced. Polio's damage to such cells, unfortunately, is permanent. Many of the temporary symptoms associated with viral infections, such as fever, aches, and inflammation, actually result from the body's own efforts at defending itself against the infection.

As we will see in Chapter 43, the immune system is a complex and critical part of the body's natural defense mechanisms. The immune system is also the basis for the major medical weapon for preventing viral infections—vaccines. **Vaccines** are harmless variants or derivatives of pathogenic microbes that stimulate the immune system to mount defenses against the actual pathogen.

The term *vaccine* is derived from *vacca,* the Latin word for cow; the first vaccine, against smallpox, consisted of cowpox virus. In the late 1700s, Edward Jenner, an English physician, learned from his patients in farm country that milkmaids who had contracted cowpox (a milder disease that usually infects cows) were resistant to subsequent smallpox infections. In his famous experiment in 1796, Jenner scratched a farmboy with a needle bearing fluid from a sore of a milkmaid who had cowpox. When the boy was later exposed to smallpox, he resisted the disease.

The cowpox and smallpox viruses are so similar that the immune system cannot distinguish them. Vaccination with the cowpox virus sensitizes the immune system to react vigorously in defending the body, if it is ever exposed to actual smallpox virus. This strategy has eradicated smallpox, which was once a devastating scourge in many parts of the world. Effective vaccines have also been developed against many other viral diseases, including polio, rubella, measles, and mumps.

Although vaccines can prevent certain viral illnesses, medical technology can do little, at present, to cure most viral infections once they occur. The antibiotics that help us recover from bacterial infections are powerless against viruses.

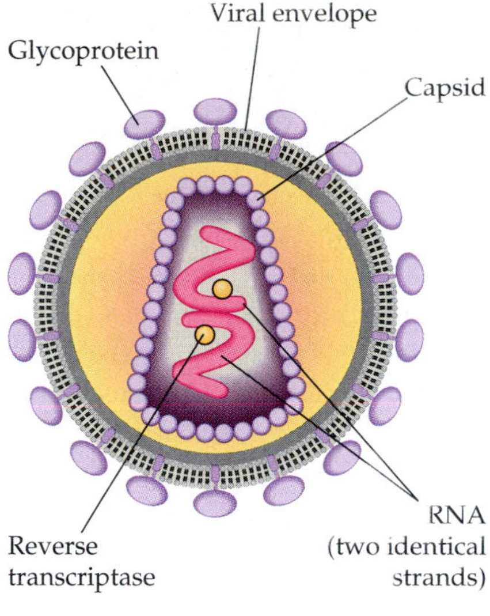

(a)

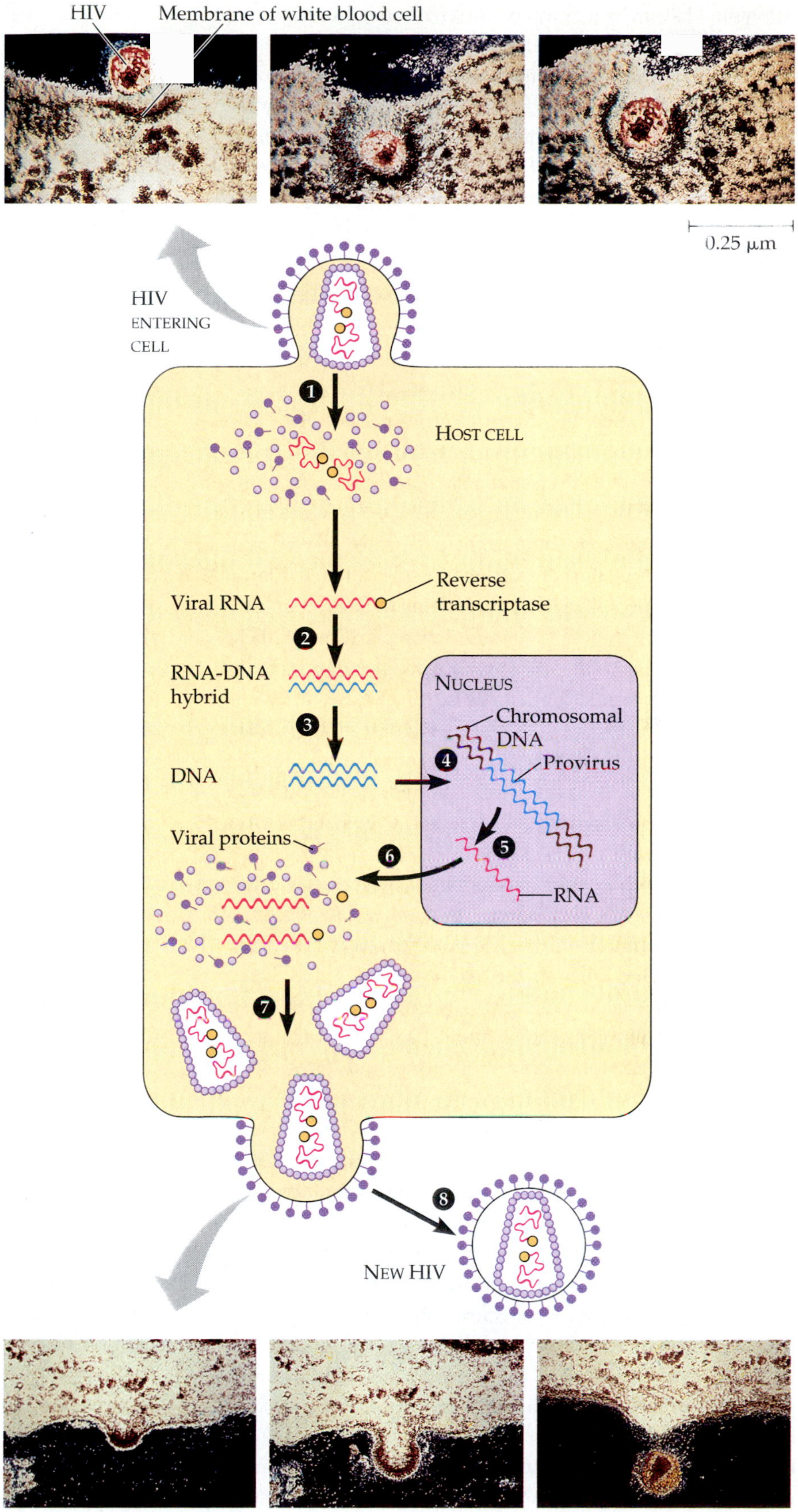

HIV
Membrane of white blood cell
0.25 µm
HIV
ENTERING
CELL
HOST CELL
Viral RNA
Reverse
transcriptase
RNA-DNA
hybrid
NUCLEUS
Chromosomal
DNA
Provirus
DNA
Viral proteins
RNA
NEW HIV

(b)

18.4 **FIGURE 18.7 ▪ HIV, a retrovirus. (a) The structure of HIV, the virus that causes AIDS.** The glycoproteins of the envelope enable the virus to bind to specific receptors on the surface of certain white blood cells. The capsid contains two identical single-stranded RNA molecules and two molecules of the enzyme reverse transcriptase. **(b) The reproductive cycle of HIV.** ① The genome enters a host cell when the virus fuses with the plasma membrane and the proteins of the capsid are enzymatically removed. ② Reverse transcriptase catalyzes the synthesis of a DNA strand complementary to the RNA template provided by the viral genome and then ③ the synthesis of a complementary DNA strand. ④ The resulting double-stranded DNA is incorporated as a provirus into the host cell's genome. ⑤ Proviral genes are transcribed into mRNA molecules that are ⑥ translated in the cytoplasm into HIV proteins. RNA transcribed from the provirus also serves as genomes for the next generation of viruses. ⑦ The assembly of capsids around genomes is followed by ⑧ budding of the new viruses from the host cell. Transmission electron micrographs (artificially colored) depict HIV entering (top) and leaving (bottom) a human white blood cell.

Antibiotics kill bacteria by inhibiting enzymes or processes specific to the pathogens, but viruses have few or no enzymes of their own. However, a few drugs have been found that do combat viruses, mostly by interfering with viral nucleic acid synthesis. One such drug is AZT, which inhibits HIV reproduction by interfering with the action of reverse transcriptase. Another is acyclovir, which inhibits herpesvirus DNA synthesis.

Emerging Viruses

HIV, the AIDS virus, seemed to make a sudden appearance in the early 1980s. In 1993, dozens of people in the southwestern United States died from hantavirus infection, which the news media first described as a "new" disease. The deadly Ebola virus menaces the peoples of central Africa periodically. Each year, new strains of influenza virus cause millions to miss work or classes, and deaths from flu are not uncommon. From where or what do these and other "emerging viruses" arise?

Three processes contribute to the emergence of viral diseases. First, the mutation of existing viruses is a major source of new viral diseases. The RNA viruses tend to have an unusually high rate of mutation because the replication of their nucleic acid does not involve the proofreading steps of DNA replication. Some mutations enable existing viruses to evolve into new genetic varieties that can cause disease in individuals who had developed immunity to the ancestral virus. Flu epidemics are caused by viruses that are genetically different enough from earlier years' viruses that people have little immunity to them.

Another source of new viral diseases is the spread of existing viruses from one host species to another. For example, hantavirus is common in rodents, especially deer mice. The population of deer mice in the southwestern United States exploded in 1993 after unusually wet weather increased the rodents' food supply. Humans acquired hantavirus when they inhaled dust containing traces of urine and feces from infected mice.

Finally, the dissemination of a viral disease from a small, isolated population can lead to widespread epidemics. AIDS, for example, went unnamed and virtually unnoticed for decades before it began to spread around the world. In this case, technological and social factors, including affordable international travel, blood transfusion technology, sexual promiscuity, and the abuse of intravenous drugs, allowed a previously rare human disease to become a global scourge.

Thus, emerging viruses are generally not new, but are existing viruses that expand their host territory by evolving, by spreading to new host species, or by disseminating to a larger proportion of the host species. Environmental change, natural or caused by humans, can increase the viral traffic responsible for emerging diseases. For example, new roads through remote areas can allow viruses to spread between previously isolated human populations. Another problem is the destruction of forests to expand cropland, an environmental disturbance that brings humans into contact with other animals that may host viruses capable of infecting humans.

Viruses and Cancer

Since 1911, when Peyton Rous discovered a virus that causes cancer in chickens, scientists have recognized that some viruses can cause cancer in animals. We know that these tumor viruses include members of the retrovirus, papovavirus, adenovirus, and herpesvirus groups (see TABLE 18.1).

There is strong evidence that viruses cause certain types of human cancer. The virus responsible for hepatitis B also seems to cause liver cancer in individuals with chronic hepatitis. And the Epstein-Barr virus, the herpesvirus that causes infectious mononucleosis, has been linked to several types of cancer prevalent in parts of Africa, notably Burkitt's lymphoma. Papilloma viruses (of the papovavirus group) have been associated with cancer of the cervix. Among the retroviruses, one called HTLV-1 causes a type of adult leukemia. Most tumor viruses transform cells through the integration of viral nucleic acid into host-cell DNA.

Scientists have identified a number of viral genes directly involved in triggering cancerous characteristics in cells. Many of these genes, called *oncogenes,* are not unique to tumor viruses or tumor cells; versions of these genes are also found in normal cells, where they are called *proto-oncogenes.* All the proto-oncogenes identified thus far code for proteins that affect the cell cycle, such as growth factors and proteins involved in growth factor action (for example, growth-factor receptors). In some cases the tumor virus lacks oncogenes and transforms the cell simply by turning on or increasing the expression of one or more of the cell's proto-oncogenes. Whatever the mechanism by which a particular virus causes cancer, there is evidence that more than one change must occur in a cell's genome to transform the cell into a fully cancerous state. It is likely that most tumor viruses cause cancer only in combination with other, mutagenic events, such as exposure to mutagenic chemicals or mistakes in DNA replication or repair. (Cancer is covered more thoroughly in Chapter 19.)

Plant viruses are serious agricultural pests

Plant viruses can stunt plant growth and diminish crop yields (FIGURE 18.8a). Most plant viruses discovered thus far are RNA viruses. Many of them, including the tobacco mosaic virus, have rod-shaped capsids with protein arranged in a spiral (see FIGURE 18.2a).

There are two major routes by which a plant viral disease can spread. By the first route, called *horizontal transmission,* a plant is infected from an external source of the virus. Since the invading virus must get past the plant's outer protective layer of cells (the epidermis), the plant becomes more susceptible to viral infections if it has been damaged by wind, chilling,

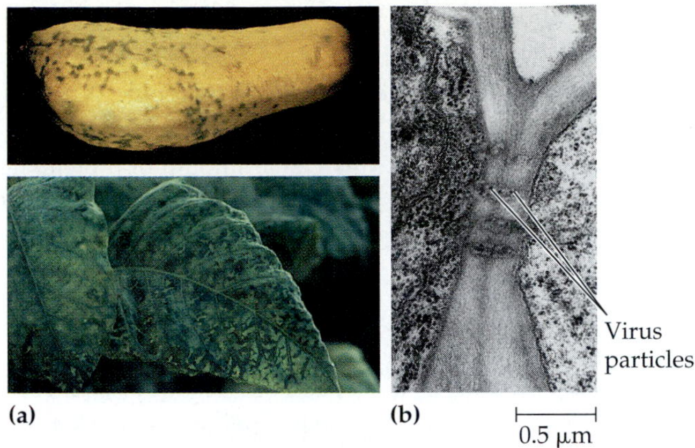

FIGURE 18.8 ▪ Viral infection of plants. (a) Mosaic viruses caused the mottling of this summer squash (top) and tobacco leaf (bottom). **(b)** Viruses that infect a plant, such as these rice yellow mottle viruses, can spread throughout the plant body via the plasmodesmata that connect cells (TEM).

injury, or insects. Insects are a double threat, because they often also act as carriers of viruses, transmitting disease from plant to plant. Farmers and gardeners may transmit plant viruses inadvertently on pruning shears and other tools. The other route of viral infection is *vertical transmission,* in which a plant inherits a viral infection from a parent. Vertical transmission can occur in asexual propagation (for example, by taking cuttings) or in sexual reproduction via infected seeds.

Once a virus enters a plant cell and begins reproducing, virus particles can spread throughout the plant by passing through plasmodesmata, the cytoplasmic connections that penetrate the walls between adjacent plant cells (FIGURE 18.8b). Agricultural scientists have not yet devised cures for most viral diseases of plants. Therefore, their efforts have focused largely on reducing the incidence and transmission of such diseases and on breeding genetic varieties of crop plants that are relatively resistant to certain viruses.

Viroids and prions are infectious agents even simpler than viruses

As small and simple as viruses are, they dwarf another class of pathogens, **viroids**. These are tiny molecules of naked circular RNA that infect plants. Only several hundred nucleotides long, viroids do not encode proteins but can replicate in host plant cells, apparently using cellular enzymes. Somehow, these RNA molecules can disrupt the metabolism of a plant cell and stunt the growth of the whole plant. One viroid disease has killed over 10 million coconut palms in the Philippines. Viroids seem to cause errors in the regulatory systems that control plant growth, and the symptoms that are typically associated with viroid diseases are abnormal development and stunted growth.

An important lesson from viroids is that a *molecule* can be an infectious agent that spreads a disease. But viroids are nucleic acid, whose ability to replicate is well known. More difficult to explain is the evidence for infectious *proteins,* called **prions**. Prions appear to cause a number of degenerative brain diseases, including scrapie in sheep, the "mad cow disease" that has plagued the British beef industry in recent years, and Creutzfeldt-Jakob disease in humans. How can a protein, which cannot replicate itself, be a transmissible pathogen? According to one hypothesis, a prion is a misfolded form of a protein normally present in brain cells. When the prion gets into a cell containing the normal form of the protein, the prion somehow converts the normal protein to the prion version (FIGURE 18.9). In this way, prions might repeatedly trigger chain reactions that increase their numbers. American scientist Stanley Prusiner has championed the prion hypothesis and in 1997 was awarded a Nobel Prize for his research in this area. However, full acceptance of prions as pathogens still awaits their production in a system guaranteed to be free of viral nucleic acid.

Viruses may have evolved from other mobile genetic elements

Viruses are in the semantic fog between life and nonlife. Do we think of them as nature's most complex molecules or as the simplest forms of life? Either way, we must bend our usual definitions. An isolated virus is biologically inert, unable to replicate its genes or regenerate its own supply of ATP. Yet it has a genetic program written in the universal language of life. Although viruses are obligate intracellular parasites that cannot reproduce independently, it is hard to deny their evolutionary connection to the living world.

How did viruses originate? Because they depend on cells for their own propagation, it is reasonable to assume that viruses

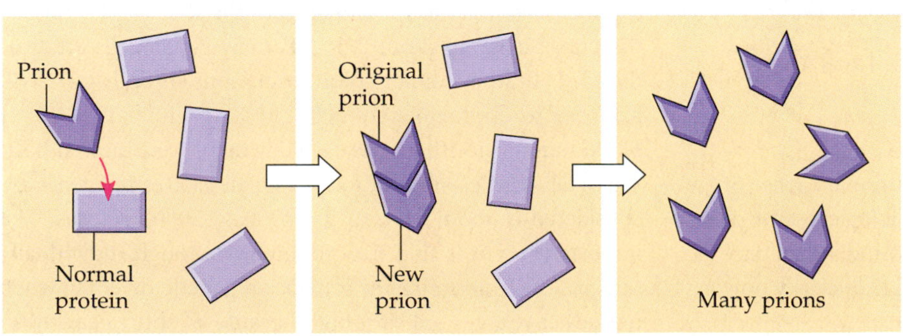

FIGURE 18.9 ▪ A hypothesis to explain how prions may propagate. Prions seem to be misfolded versions of normal brain proteins. When a prion contacts its normal "twin," it may induce the normal protein to assume the abnormal shape. The resulting chain reaction may continue until prions accumulate to dangerous levels, causing cellular malfunction and eventually degeneration of the brain.

are not the descendants of precellular prototypes of life, but evolved *after* the first cells appeared. Most molecular biologists favor the hypothesis that viruses originated from fragments of cellular nucleic acids that could move from one cell to another. Consistent with this idea is the observation that a viral genome usually has more in common with the genome of its host than with the genomes of viruses infecting other hosts. Indeed, some viral genes are essentially identical to genes of the host, as in the case of oncogenes, for example. Perhaps the earliest viruses were naked bits of nucleic acid, similar to plant viroids, that made it from one cell to another via injured cell surfaces. The evolution of genes coding for capsid proteins may have facilitated the infection of undamaged cells.

The most likely candidates for potential sources of viral genomes are two kinds of cellular genetic elements, called plasmids and transposons. Plasmids are small, circular DNA molecules that are separate from chromosomes. They are found in bacteria and yeasts, which are unicellular eukaryotes. Plasmids, like most viruses, can replicate independently of the rest of the cell's genome and are occasionally transferred between cells. Transposons are DNA segments that can move from one location to another within a cell's genome. Thus, plasmids, transposons, and viruses all share an important feature: They are mobile genetic elements. (We will discuss plasmids and transposons in more detail later in the chapter.)

It is the evolutionary relationship between viruses and the genomes of their host cells that makes viruses such useful model systems in molecular biology. By studying how the replication of viruses is controlled, researchers are learning more about the mechanisms that regulate DNA replication and gene expression (transcription and translation) in cells. Bacteria are equally valuable as microbial models in genetics research, but for different reasons. Unlike viruses, bacteria are true cells. But as prokaryotic cells, bacteria provide researchers with the opportunity to investigate molecular genetics in the simplest organisms. In fact, *E. coli,* the bacterium sometimes called "the laboratory rat of molecular biology," is the most completely understood of all organisms at the molecular level. Let's move from the topic of viruses and learn more about the genetics of bacteria.

THE GENETICS OF BACTERIA

The short generation span of bacteria facilitates their evolutionary adaptation to changing environments

Bacteria are very adaptable, in both the evolutionary sense of adaptation via natural selection and the physiological sense of adjustment to changes in the environment by individual bacteria. These sections on bacterial genetics will help clarify how these microbes can be so malleable.

The major component of the bacterial genome is one double-stranded, circular DNA molecule. Although we will refer to this structure as the bacterial chromosome, it is very different from eukaryotic chromosomes, which have linear DNA molecules associated with a large amount of protein. In the case of the well-studied intestinal bacterium *E. coli,* the chromosomal DNA consists of about 4.6 million nucleotide pairs, representing about 4300 genes. This is 100 times more DNA than is found in a typical virus, but only about one-thousandth as much DNA as in an average eukaryotic cell. Still, this is a lot of DNA to be packaged in such a small container. Stretched out, the DNA of an *E. coli* cell would measure about a millimeter in length, 500 times longer than the cell. Within a bacterium, however, the chromosome is so tightly packed that it does not even fill the whole cell but forms a structure something like a long loop of yarn tangled into a ball. This dense region of DNA, called the **nucleoid**, is not bounded by membrane like the nucleus of a eukaryotic cell. In addition to the chromosome, many bacteria also have plasmids, much smaller circles of DNA. Each plasmid has only a small number of genes, from just a few to several dozen. You will learn about the structure and function of plasmids in the next section.

Bacterial cells divide by binary fission, which is preceded by replication of the bacterial chromosome (see FIGURE 12.10). From a single origin of replication, the copying of DNA progresses in both directions around the circular chromosome (FIGURE 18.10).

Bacteria can proliferate very rapidly in a favorable environment, whether in a natural habitat or in a laboratory culture. For example, *E. coli* growing under optimal conditions can divide every 20 minutes. A laboratory culture started with a single cell can produce a colony of 10^7–10^8 bacteria overnight (12 hours). Reproductive rates in the organism's natural habitat, the large intestines (colons) of mammals, are just as impressive. For example, in the human colon, *E. coli* reproduces rapidly enough to replace the 2×10^{10} bacteria lost each day in feces.

Since fission is an asexual process—the production of offspring from a single parent—most of the bacteria in a colony are genetically identical to the parent cell. As a result of mutation, however, some of the offspring *do* differ slightly in genetic makeup. For a given *E. coli* gene, the probability of a spontaneous mutation averages only about 1×10^{-7} per cell division. But since about 2×10^{10} new *E. coli* cells are produced daily in a human colon, mutations in each gene will give rise to approximately 2000 [$= (2 \times 10^{10})(1 \times 10^{-7})$] *E. coli* mutants for that gene each day in just one human host. The total number of mutations when all 4300 *E. coli* genes are considered is about 9×10^6 ($= 4300 \times 2000$) per day. The important point is that new mutations, though individually rare, can have a significant impact on genetic diversity when reproductive rates are very high because of short generation

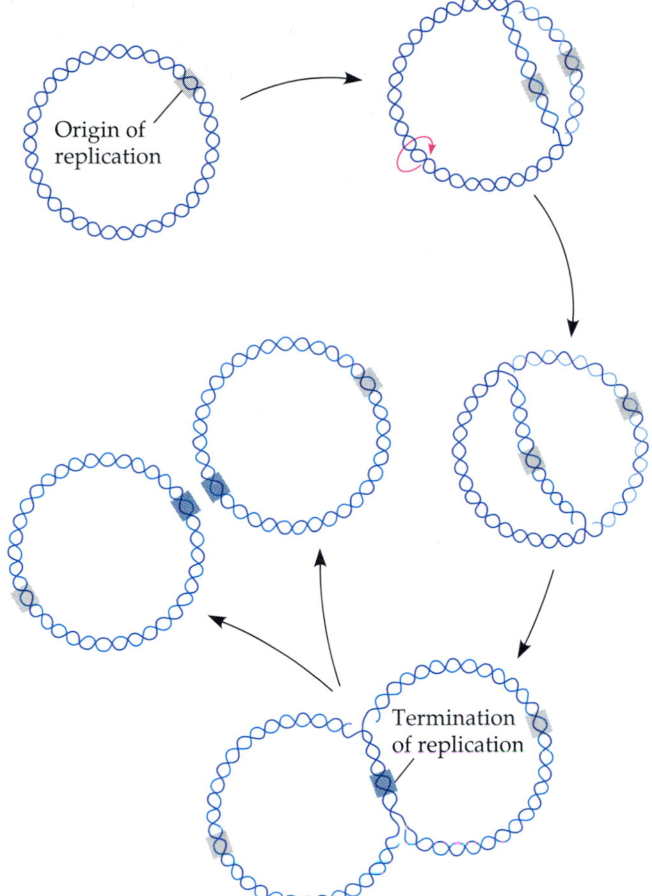

FIGURE 18.10 ▪ Replication of the bacterial chromosome. From one origin, DNA replication progresses in both directions around the circular chromosome until the entire chromosome has been reproduced. Enzymes that cut, twirl (magenta arrow), and reseal the double helix prevent the DNA from knotting.

spans. This diversity, in turn, affects the evolution of bacterial populations: Individual bacteria that are genetically well equipped for the local environment clone themselves more prolifically than do less fit individuals.

In contrast, new mutations make a relatively small contribution to genetic variation among individuals of a population of slowly reproducing organisms, such as humans. Most of the heritable variation we observe in a human population is due not to the creation of novel alleles by *new* mutations, but to the sexual recombination of existing alleles (see Chapter 15). Even in bacteria, where new mutations *are* a major source of individual variation, genetic recombination adds more diversity to a population, as we will now see.

Genetic recombination produces new bacterial strains

Natural selection depends on heritable variation among the individuals of a population (see Chapter 1). In addition to mutations, genetic recombination generates diversity within bacterial populations. We will define recombination here simply as the combining of DNA from two individuals into the genome of a single individual.

How can we detect genetic recombination in bacteria? Consider two mutant *E. coli* strains (genetic varieties), each unable to synthesize one of its required amino acids. Wild-type *E. coli* can grow on a minimal medium containing only glucose as a source of organic carbon. The mutant strains cannot grow on this culture medium of minimal nutrients because one of them cannot synthesize tryptophan and the other cannot synthesize arginine (FIGURE 18.11).

Let's suppose we mix bacteria from the two strains together in a liquid medium and allow them to incubate for an hour or so. We then spread a small sample of this culture on solid minimal medium in a petri dish and incubate the dish overnight. The next morning we observe numerous colonies of bacteria on the minimal medium. Each of these colonies must have started with a cell capable of making *both* tryptophan *and* arginine, but their number far exceeds what can be accounted for by mutation. Most of the cells that can synthesize both amino acids must have acquired one or more genes from the other strain. Genetic recombination has occurred.

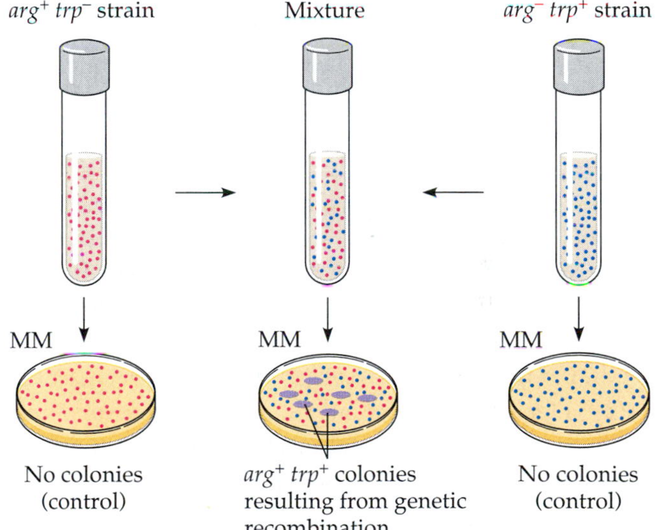

FIGURE 18.11 ▪ Detecting genetic recombination in bacteria. Each of the two mutant strains of *E. coli* in this experiment is unable to synthesize a particular amino acid, either arginine or tryptophan. The arg^+ trp^- bacteria can make arginine but not tryptophan, and the arg^- trp^+ bacteria can make tryptophan but not arginine. These mutant bacteria cannot grow on minimal medium, which is simply a solution of glucose and salts. They can grow only on a medium supplemented with the amino acid they are unable to make. The middle tube of the diagram contains a mixture of bacteria of the two mutant strains. Samples containing equal numbers of bacteria from the three tubes are plated onto agar (a gelatinous substance) containing minimal medium (MM). The original mutant cells fail to produce colonies. However, some of the cells in the sample from the mixed culture are able to grow on minimal medium. Those cells are recombinant bacteria, arg^+ trp^+ cells produced by gene transfer between the two types of mutant cells. (You will learn about the mechanism in FIGURE 18.14c and d.)

Bacteria are different from eukaryotes in the ways DNA from two individuals can come together in one cell. In eukaryotes, the sexual processes of meiosis and fertilization combine DNA from two individuals in a single zygote (see Chapter 13). But sex, as defined in eukaryotes, is absent in prokaryotes: Meiosis and fertilization do not occur. Instead, other processes bring together bacterial DNA from different individuals. These processes are transformation, transduction, and conjugation.

Transformation

In the context of bacterial genetics, **transformation** is the alteration of a bacterial cell's genotype by the uptake of naked, foreign DNA from the surrounding environment. For example, we saw in Chapter 16 that harmless *Streptococcus pneumoniae* bacteria could be transformed to pneumonia-causing cells by the uptake of naked DNA from a medium containing dead cells of the pathogenic strain. This transformation occurs when a live nonpathogenic cell takes up a piece of DNA that happens to include the allele for pathogenicity (the gene for a cell coat that protects the bacterium from a host's immune system). The foreign allele is then incorporated into the bacterial chromosome, replacing the native allele (for the "coatless" condition, in this case). The process is genetic recombination—an exchange of DNA segments by crossing over. The transformed cell now has a chromosome containing DNA derived from two different cells.

For many years after transformation was discovered in laboratory cultures, most biologists believed the process to be too rare and haphazard to play an important role in natural bacterial populations. But researchers have since learned that many bacterial species possess on their surfaces proteins that are specialized for the uptake of naked DNA from the surrounding solution. These proteins specifically recognize and transport only DNA from closely related species of bacteria. Not all bacteria have such membrane proteins. For instance, *E. coli* does not seem to have any specialized mechanism for the uptake of foreign DNA. However, placing *E. coli* in a culture medium containing a relatively high concentration of calcium ions will artificially stimulate the cells to take up small pieces of DNA. In biotechnology, this technique is applied to introduce foreign genes into *E. coli*—genes coding for valuable proteins, such as human insulin and growth hormone.

Transduction

In the DNA-transfer process known as **transduction**, phages (the viruses that infect bacteria) carry bacterial genes from one host cell to another. There are two forms of transduction: generalized transduction and specialized transduction (FIGURE 18.12). Both result from aberrations in phage reproductive cycles.

First let's consider *generalized* transduction, shown in FIGURE 18.12a. Recall that near the end of a phage's lytic cycle, viral nucleic acid molecules are packaged within capsids, and the completed phages are released when the host cell lyses.

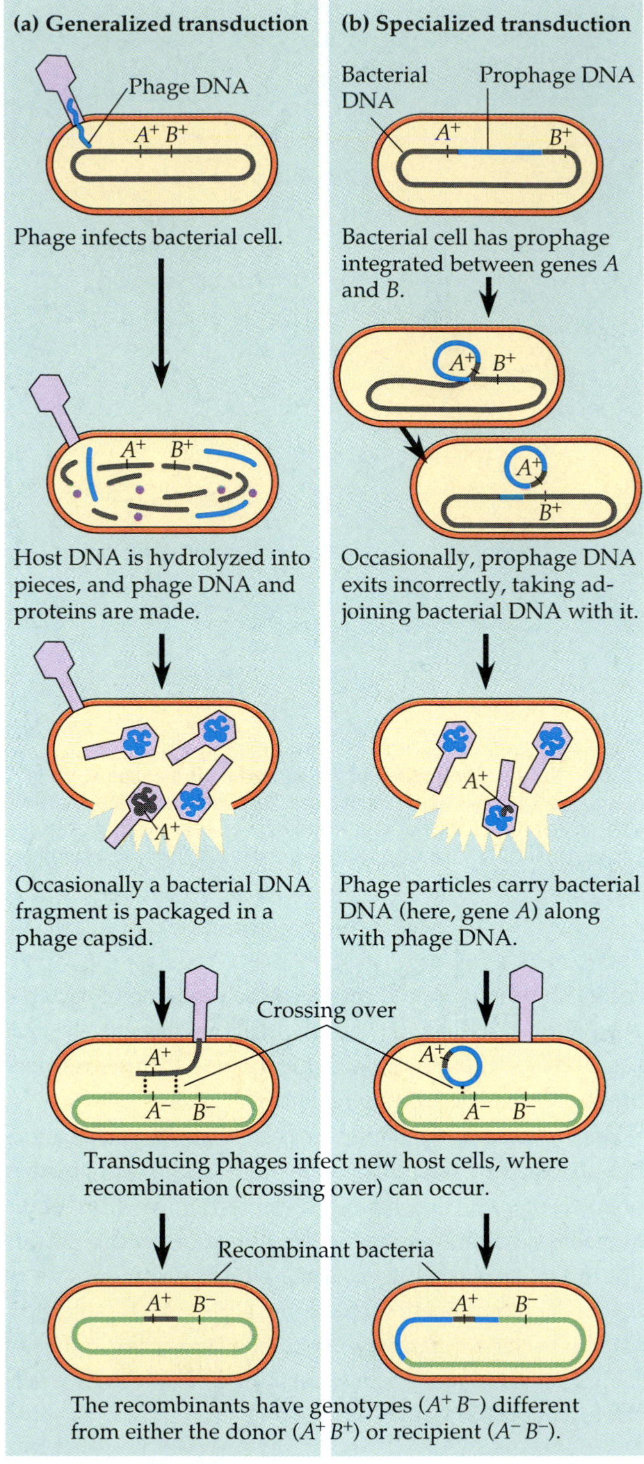

(a) Generalized transduction
(b) Specialized transduction
Phage DNA
A^+ B^+
Phage infects bacterial cell.
Bacterial DNA
Prophage DNA
Bacterial cell has prophage integrated between genes *A* and *B*.
Host DNA is hydrolyzed into pieces, and phage DNA and proteins are made.
Occasionally, prophage DNA exits incorrectly, taking adjoining bacterial DNA with it.
Occasionally a bacterial DNA fragment is packaged in a phage capsid.
Phage particles carry bacterial DNA (here, gene *A*) along with phage DNA.
Crossing over
A^- B^-
Transducing phages infect new host cells, where recombination (crossing over) can occur.
Recombinant bacteria
A^+ B^-
The recombinants have genotypes ($A^+ B^-$) different from either the donor ($A^+ B^+$) or recipient ($A^- B^-$).

FIGURE 18.12 ▪ Transduction. Phages occasionally carry bacterial genes from one cell to another. **(a)** In generalized transduction, random pieces of the host chromosome are packaged within a phage capsid. **(b)** In specialized transduction, a prophage exits the chromosome in such a way that it carries adjacent bacterial genes along with it. In both types of transduction, the transferred DNA may recombine with the genome of the new host cell.

Occasionally a small piece of the host cell's degraded DNA is accidentally packaged within a phage capsid in place of the phage genome. Such a virus is defective because it lacks its own genetic material. However, after its release from the lysed host, the phage can attach to another bacterium and inject the piece of bacterial DNA acquired from the first cell. Some of this DNA can subsequently replace the homologous region of the second cell's chromosome. The cell's chromosome now has a combination of DNA derived from two cells; genetic recombination has occurred. This type of transduction is called generalized transduction because bacterial genes are transferred at random.

Let's contrast this process with *specialized* transduction, shown in FIGURE 18.12b. This form of transduction requires infection by a temperate phage. Recall that in the lysogenic cycle, the genome of a temperate phage integrates as a prophage into the host bacterium's chromosome, usually at a specific site. Later, when the phage genome is excised from the chromosome, it sometimes takes with it a small region of the bacterial DNA that was adjacent to the prophage. When such a virus carrying bacterial DNA infects another host cell, the bacterial genes are injected along with the phage's genome. Specialized transduction transfers only certain genes, those near the prophage site on the chromosome.

Conjugation and Plasmids

Conjugation is the direct transfer of genetic material between two bacterial cells that are temporarily joined. This process, the bacterial version of sex, has been studied most extensively in *E. coli.* The DNA transfer is one-way: one cell donating DNA, and its "mate" receiving the genes. The DNA donor, referred to as the "male," uses appendages called sex pili to attach to the DNA recipient, the "female" (FIGURE 18.13). Then a temporary cytoplasmic bridge forms between the two cells, providing an avenue for DNA transfer. In most cases, "maleness," the ability to form sex pili and donate DNA during conjugation, results from the presence of a special piece of DNA called an **F factor** (F for fertility). An F factor can exist either as a segment of DNA within the bacterial chromosome or as a plasmid. Before discussing the role of the F factor in conjugation, we should examine plasmids more generally.

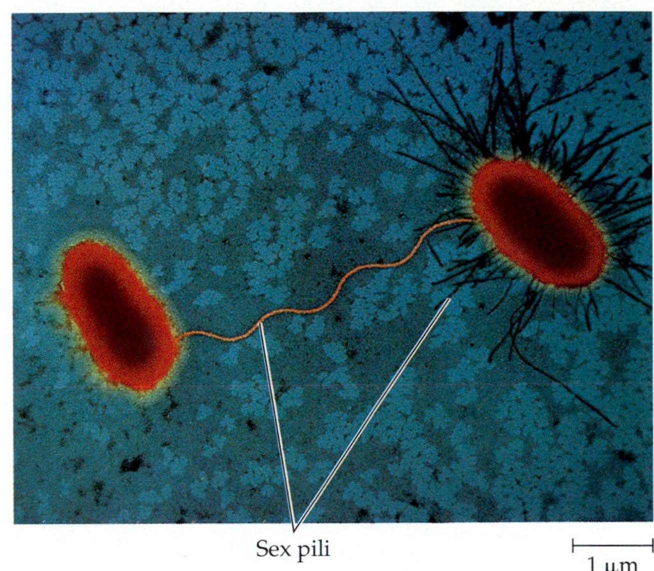

FIGURE 18.13 ▪ Bacterial mating. The *E. coli* "male" (right) extends sex pili, one of which is attached to a "female" cell. Later, the mates will briefly join by a cytoplasmic bridge through which the "male" transfers DNA to the "female." This mechanism of DNA transfer is called conjugation (colorized TEM).

General Characteristics of Plasmids. A **plasmid** is a small, circular, self-replicating DNA molecule separate from the bacterial chromosome. Certain plasmids, such as F plasmids, can undergo reversible incorporation into the cell's chromosome. A genetic element that can replicate either as a plasmid or as part of the bacterial chromosome is called an **episome.** In addition to some plasmids, temperate viruses, such as phage λ, also qualify as episomes. Recall that the genomes of these phages replicate separately in the cytoplasm during a lytic cycle, and as an integral part of the host's chromosome during a lysogenic cycle. The hypothesis that some viruses evolved from plasmids was introduced earlier. Of course, there are important differences between plasmids and viruses. Plasmids, unlike viruses, lack protein coats and do not normally exist outside the cell. And plasmids are generally beneficial to the bacterial cell, while viruses are parasites that usually harm their hosts.

A plasmid has only a small number of genes, and these genes are not required for the survival and reproduction of the bacterium under normal conditions. However, the genes of plasmids can confer advantages on bacteria living in stressful environments. For example, the F plasmid facilitates genetic recombination, which may be advantageous in a changing environment that no longer favors existing strains in a bacterial population.

The F Plasmid and Conjugation. The F factor and its plasmid form, the **F plasmid**, consist of about 25 genes, most required for the production of sex pili. Geneticists use the symbol F^+ to denote a cell that contains the F plasmid (a "male" cell). The F^+ condition is heritable: The F plasmid replicates in synchrony with the chromosomal DNA, and division of an F^+ cell usually gives rise to two offspring that are both F^+. Cells lacking the F factor in either form are designated F^-, and they function as DNA recipients ("females") during conjugation. The F^+ condition is "contagious" in the sense that an F^+ cell converts an F^- cell to F^+ when the two cells conjugate. The F plasmid replicates within the "male" cell, and a copy is transferred to the "female" through the conjugation tube joining the cells (FIGURE 18.14a, p. 334). In such an $F^+ \times F^-$ mating, only an F plasmid is transferred.

Under what circumstances are genes of the bacterial chromosome transferred during conjugation? This can occur when the donor cell's F factor is integrated into the chromosome (FIGURE 18.14b). A cell with the F factor built into its chromosome is called an Hfr cell (for high frequency of recombination). An Hfr cell continues to function as a male during conjugation, replicating the F factor DNA and transferring the copy to its F^- partner. But now, the F factor takes a copy of some of the chromosomal DNA along with it (FIGURE 18.14c). Random movements of the bacteria usually disrupt conjugation before an entire copy of the Hfr chromosome can be passed to the F^- cell. Temporarily, the recipient cell is a partial diploid, containing its own chromosome plus DNA copied from part of the donor's chromosome. Then recombination can occur: If part of the newly acquired DNA aligns with the homologous region of the F^- chromosome, segments of DNA can be exchanged (FIGURE 18.14d). Binary fission of this cell gives rise to a colony of recombinant bac-

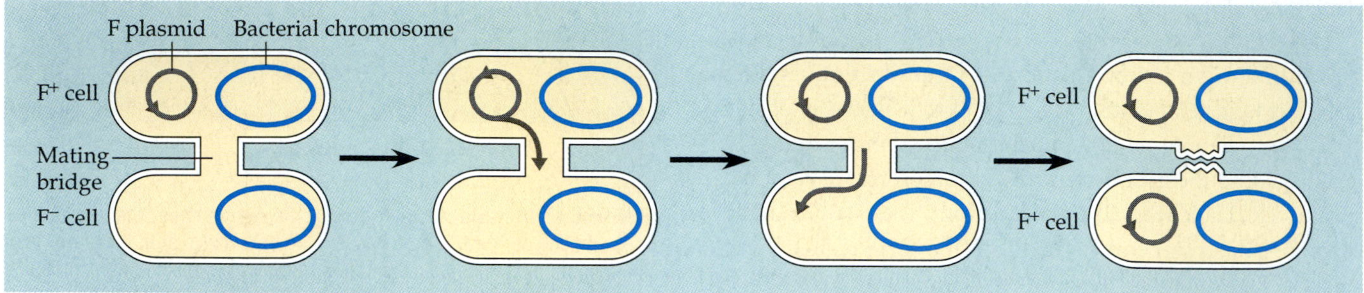

(a) Conjugation between an F⁺ (male) and an F⁻ (female) bacterium. Cells that carry an F plasmid are called F⁺ cells. They are "male" in that they can transfer an F plasmid to a "female" F⁻ cell. In this way, an F⁻ cell can become F⁺. The F plasmid replicates as it is transferred, so that the donor cell remains F⁺. The arrowhead marks the point where replication and transfer begin.

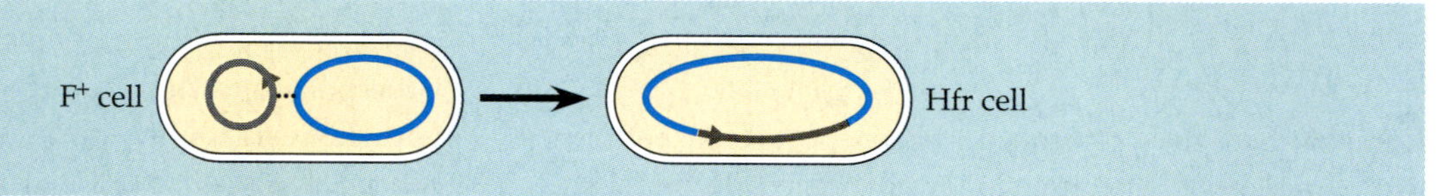

(b) Conversion of an F⁺ male into an Hfr male by integration of the F plasmid (an episome) into the chromosome. This process is similar to phage DNA joining the host chromosome as a prophage: Crossing over occurs between the two DNA circles at a specific site on each.

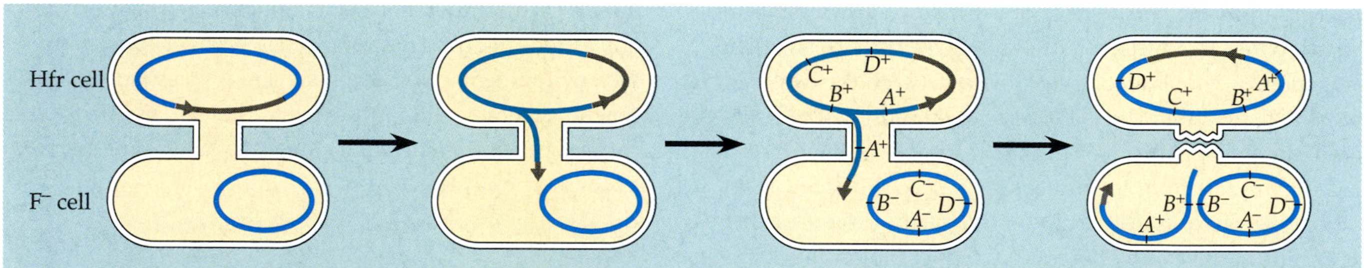

(c) Conjugation between an Hfr and an F⁻ bacterium. Replication and transfer of the "male's" chromosome begins at a fixed point (arrowhead) within the F episome. The location and orientation of the F factor in the chromosome determine the sequence in which genes are transferred during conjugation. In this *E. coli* strain, the transfer sequence for four genes is *A-B-C-D*. The conjugation bridge usually breaks before the entire chromosome and the tail end of the F episome are transferred.

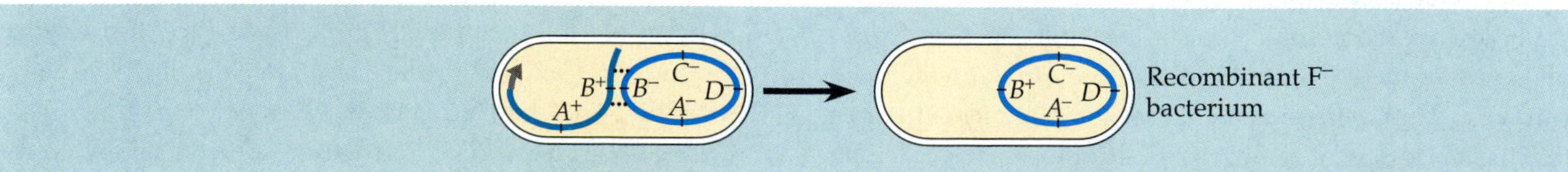

(d) Recombination between the Hfr chromosome fragment and the F⁻ chromosome. Crossing over can occur between genes on the fragment of bacterial chromosome transferred from the Hfr cell and the same (homologous) genes on the recipient (F⁻) cell's chromosome. A recombinant F⁻ cell will result. Pieces of DNA ending up outside the bacterial chromosome will eventually be degraded by the cell's enzymes or lost in cell division.

FIGURE 18.14 ▪ Conjugation and recombination in *E. coli*.

teria with genes derived from two different cells. (This is what happened in the experiment described in FIGURE 18.11, where one of the bacterial strains was actually an Hfr and the other an F$^-$.)

R Plasmids and Antibiotic Resistance. In the 1950s, Japanese physicians began to notice that some hospital patients suffering from bacterial dysentery, which produces severe diarrhea, did not respond to antibiotics that had generally been effective in treating this type of infection. Apparently, resistance to these antibiotics had evolved in certain strains of *Shigella,* the pathogen. Eventually, researchers began to identify the specific genes that confer antibiotic resistance in *Shigella* and other pathogenic bacteria. Some of these genes, for example, code for enzymes that specifically destroy certain antibiotics, such as tetracycline or ampicillin. The genes conferring resistance, it turned out, are carried by plasmids, now known as **R plasmids** (R for resistance).

Exposure of a bacterial population to a specific antibiotic—whether in a laboratory culture or within a host organism—will kill antibiotic-sensitive bacteria but not those that happen to have R plasmids that counter that antibiotic. The theory of natural selection predicts that, under these circumstances, an increasing number of bacteria will inherit genes for antibiotic resistance, and that is exactly what happens. The medical consequences are also predictable: Resistant strains of pathogens are becoming more common, making the treatment of certain bacterial infections more difficult. The problem is compounded by the fact that R plasmids, like F plasmids, can transfer from one bacterial cell to another by conjugation. Making the problem still worse, some R plasmids carry as many as ten genes for resistance to that many antibiotics. How do so many antibiotic-resistant genes become part of a single plasmid? The answer involves another type of mobile genetic element, called a transposon, which we investigate next.

Transposons

A **transposon,** also called a transposable genetic element, is a piece of DNA that can move from one location to another in a cell's genome. Unlike an episome or prophage, transposons never exist independently. Instead, transposon movement (transposition) occurs as a type of recombination between the transposon and another DNA site—a target site—that comes in contact with the transposon. In a bacterial cell, a transposon may move within the chromosome, from a plasmid to the chromosome (or vice versa), or from one plasmid to another. For example, transposons are responsible for bringing multiple genes for antibiotic resistance into a single R plasmid by moving the genes to that location from different plasmids.

Transposons are sometimes called "jumping genes," but the phrase is slightly misleading. Some transposons *do* jump from one genomic location to another, in what is called cut-and-paste transposition. However, in another type of transposition, called replicative transposition, the transposon replicates at its original site, and a *copy* inserts elsewhere; that is, the transposon is added at some new site without being lost from the old site.

Although transposons vary in their selectivity for target sites, most can move to many alternative locations in the DNA. This ability to scatter certain genes throughout the genome makes transposition fundamentally different from other mechanisms of genetic shuffling. The genetic recombination that occurs in bacterial transformation, generalized transduction, and conjugation (and during meiosis in eukaryotes as well) depends on base-pairing between homologous regions of DNA, regions of identical or very similar base sequence. In contrast, the insertion of a transposon in a new site does not depend on complementary base sequences. A transposon can move genes to a site where genes of that sort have never before existed.

Insertion Sequences. The simplest bacterial transposons are **insertion sequences.** They consist of only the DNA necessary for the act of transposition. The one gene found in an insertion sequence codes for a transposase, an enzyme that catalyzes transposition. The transposase gene is bracketed by a pair of DNA sequences called *inverted repeats,* noncoding sequences about 20–40 nucleotides long (FIGURE 18.15). Inverted repeats are upside-down, backward versions of each other. In FIGURE 18.15, note that each base sequence is repeated, in reverse, along the opposite DNA strand of the inverted repeat at the other end of the transposon. Transposase recognizes these inverted repeats as the boundaries of the transposon. Molecules of the enzyme bind to the inverted repeats and to a target site elsewhere in the genome and catalyze the DNA cutting and resealing required for transposition. Other enzymes also participate in transposition. For example, DNA polymerase helps create identical regions of

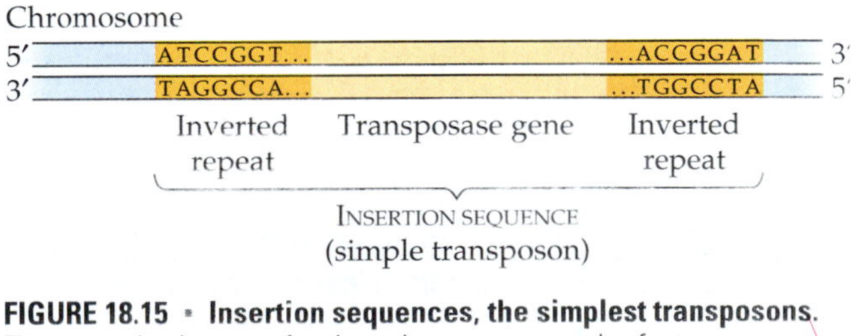

FIGURE 18.15 • Insertion sequences, the simplest transposons. The one and only gene of an insertion sequence codes for transposase, an enzyme that catalyzes movement of the transposon from one location to another in the genome. On either end of the transposase gene are inverted repeats, sequences about 20–40 nucleotide pairs in length that are backward, upside-down versions of each other. In transposition, transposase molecules bind to the inverted repeats and catalyze the cutting and resealing of DNA required for insertion of the transposon at a target site. (This diagram and the ones that follow are not to scale.)

DNA, called *direct repeats,* that flank a transposon in its new site (FIGURE 18.16).

Insertion sequences cause mutations when they happen to land within the coding sequence of a gene or within a DNA region that regulates gene expression. But notice that this mechanism of mutation is intrinsic to the cell, in contrast to mutagenesis by extrinsic factors such as environmental radiation and chemicals. Amazingly, insertion sequences account for about 1.5% of the *E. coli* genome. However, mutation of a given gene by transposition occurs only rarely—about once in every ten million generations. This is about the same as the spontaneous mutation rate due to other factors.

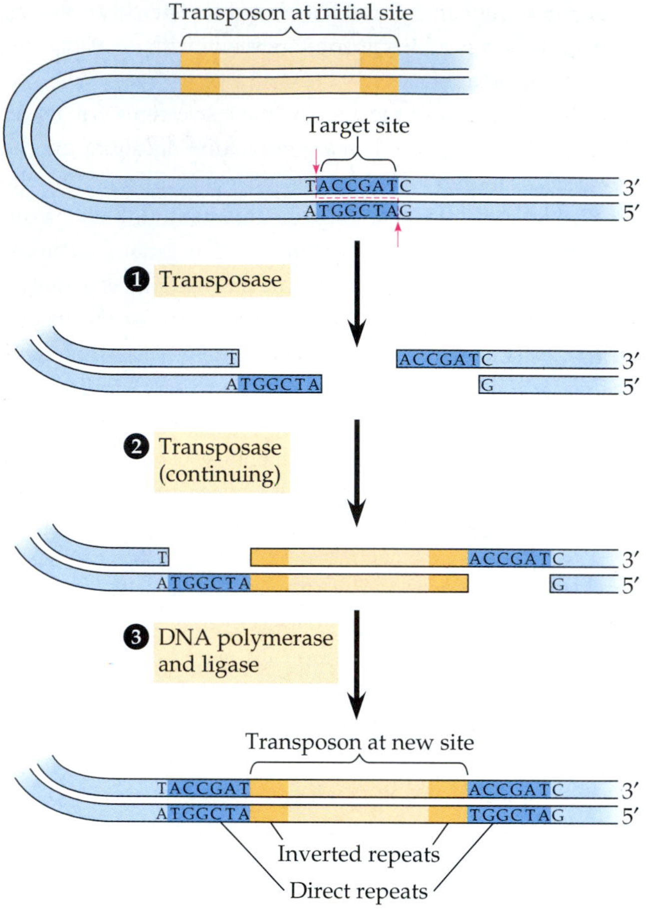

FIGURE 18.16 ▪ Insertion of a transposon and creation of direct repeats. ① The transposase enzyme makes staggered cuts (red arrows) in the two DNA strands at a target site, leaving short segments of unpaired DNA as shown. Meanwhile, the transposon is cut out or copied at its initial site. ② The transposon is joined to the single-stranded ends at the target site. Presumably, the transposase holds all the components together during this process. ③ The gaps in the DNA strands are filled in by DNA polymerase and sealed by ligase. This results in *direct repeats,* identical segments of DNA on either side of the transposon. The direct repeats flank the transposon's inverted repeats (not to scale).

Composite Transposons. Transposons longer and more complex than insertion sequences also move about in the bacterial genome. In addition to the DNA required for transposition, these *composite transposons* (also called complex transposons) include extra genes that go along for the ride, such as genes for antibiotic resistance. The extra genes are sandwiched between two insertion sequences (FIGURE 18.17). It is as though two insertion sequences happened to land relatively close together in the genome and now travel together, along with all the DNA between them, as a single transposon. As is characteristic of transposons, there is an inverted repeat and a direct repeat at each end. (The direct repeat is not considered part of the transposon, however.)

In contrast to insertion sequences, which are not known to benefit bacteria in any specific way, composite transposons may help bacteria adapt to new environments. We have already mentioned the example of transposition packaging several genes for resistance to different antibiotics into a single plasmid. In an antibiotic-rich environment, natural selection favors bacterial clones that have built up these composite R plasmids through a series of transpositions.

Transposable genetic elements are not unique to bacteria but are important components of eukaryotic genomes as well. In fact, the first evidence for such wandering DNA segments came from the American geneticist Barbara McClintock's breeding experiments with Indian corn in the 1940s and 1950s. McClintock identified changes in the color of corn kernels that made sense only if she postulated the existence of mobile genetic elements capable of moving from other locations in the genome to the genes for kernel color. She called these mobile elements "controlling elements" because they

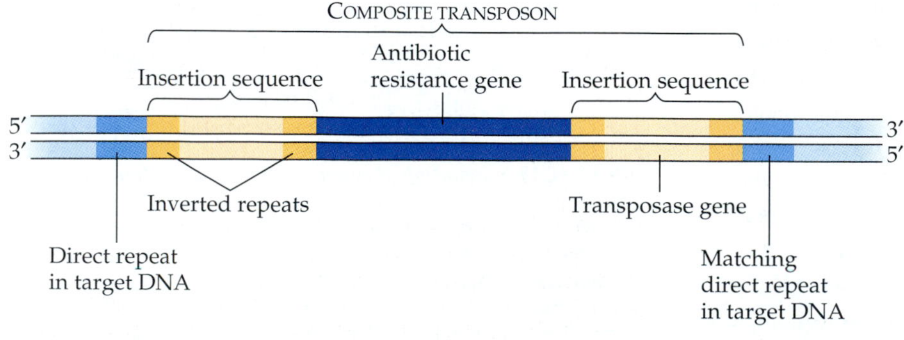

FIGURE 18.17 ▪ Anatomy of a composite transposon. A composite transposon consists of one or more genes located between twin insertion sequences. Here a gene for resistance to a specific antibiotic is carried along as part of the transposon when this DNA inserts at some new site in the genome. This sort of transposition can add a gene for antibiotic resistance to a plasmid already carrying genes for resistance to other antibiotics. The transmission of such composite plasmids to other bacterial cells by cell division and/or conjugation can then spread resistance to a variety of antibiotics throughout a bacterial population.

seemed to insert next to the genes responsible for kernel color, either activating or inactivating those genes. McClintock's discovery received little attention until transposons were discovered in bacteria many years later, and microbial geneticists learned more about the molecular basis of transposition. In 1983, more than 30 years after she discovered transposable genetic elements, Barbara McClintock was awarded a Nobel Prize, at age 81. McClintock continued her experiments at Cold Spring Harbor Laboratory in New York until her death in 1992.

You will learn more about transposons in eukaryotes in Chapter 19. We end this chapter by examining how bacterial genes are regulated in different environments.

The control of gene expression enables individual bacteria to adjust their metabolism to environmental change

Mutations and the various types of genetic transfer we have been studying generate the genetic variation that makes natural selection possible. And natural selection, acting over many generations, can increase the proportion of individuals in a bacterial population that are adapted to some new environmental condition. But how can an individual bacterium, locked into the genome it has inherited, cope with environmental fluctuation?

Think, for instance, of an *E. coli* cell living in the erratic environment of a human colon, dependent for its nutrients on the whimsical eating habits of its host. If the bacterium is deprived of the amino acid tryptophan, which it needs in order to survive, it responds by activating a metabolic pathway to make its own tryptophan from another compound. Later, if the human host eats a tryptophan-rich meal, the bacterial cell stops producing tryptophan for itself, thus saving the cell from squandering its resources to produce a substance that is available from the surrounding solution in prefabricated form. This is just one example of how bacteria tune their metabolism to changing environments.

Metabolic control occurs on two levels (FIGURE 18.18). First, cells can vary the numbers of specific enzyme molecules made; that is, they can regulate the expression of a gene. Second, cells can adjust the activity of enzymes already present. The latter mode of control, which is more immediate, depends on the sensitivity of many enzymes to chemical cues that increase or decrease their catalytic activity (see Chapter 6). For example, activity of the first enzyme of the tryptophan-synthesis pathway is inhibited by the pathway's end-product. Thus, if tryptophan accumulates in a cell, it shuts down its own synthesis. Such feedback inhibition, typical of anabolic (biosynthetic) pathways, allows a cell to adapt to short-term fluctuations in levels of a substance it needs.

If, in our example, the environment continues to provide all the tryptophan the cell needs, the regulation of gene expression also comes into play: The cell stops making enzymes of the tryptophan pathway. This control of enzyme production occurs at the level of transcription, the synthesis of messenger RNA coding for these enzymes. More generally, many genes of the bacterial genome are switched on or off by changes in the metabolic status of the cell. The basic mechanism for this control of gene expression in bacteria, described as the operon model, was discovered in 1961 by François Jacob and Jacques Monod at the Pasteur Institute in Paris. Let's see what an operon is and how it works, using the control of tryptophan synthesis as our first example.

Operons: The Basic Concept

E. coli synthesizes tryptophan from a precursor molecule in a series of steps, each reaction catalyzed by a specific enzyme (see FIGURE 18.18). The five genes coding for the polypeptide chains that make up these enzymes are clustered together on the chromosome. A single promoter serves all five genes, which constitute a transcription unit. (Recall from Chapter 17 that a promoter is a site where RNA polymerase can bind to DNA and begin transcribing genes.) Thus, transcription gives rise to one long mRNA molecule representing all five genes for the tryptophan pathway. The cell can translate this transcript into separate polypeptides because the mRNA is

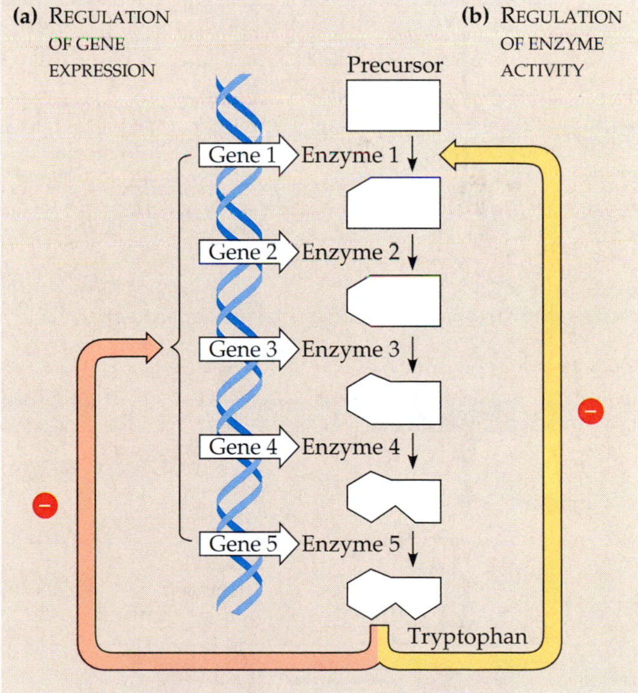

FIGURE 18.18 ▪ Regulation of a metabolic pathway. Cells can adjust the rates of specific metabolic pathways by regulating gene expression (the synthesis of new enzyme molecules) or by regulating the catalytic activity of existing enzymes. In the pathway for tryptophan synthesis, an abundance of tryptophan can both **(a)** repress expression of the genes for all the enzymes needed for the pathway, and **(b)** inhibit the activity of the first enzyme in the pathway (feedback inhibition).

punctuated with start and stop codons signaling where the coding sequence for each polypeptide begins and ends.

A key advantage of grouping genes of related function into one transcription unit is that a single "on-off switch" can control the whole cluster of functionally related genes. When an *E. coli* cell must make tryptophan for itself because the nutrient medium lacks this amino acid, all the enzymes for the metabolic pathway are synthesized at one time. The switch is a segment of DNA called an **operator.** Its location and name both suit its function: Positioned within the promoter or between the promoter and the enzyme-coding genes, the operator controls the access of RNA polymerase to the genes. All together, the operator, the promoter, and the genes they control—the entire stretch of DNA required for enzyme production for the tryptophan pathway—is called an **operon** (FIGURE 18.19). Here we are dissecting one of many operons that have been discovered in *E. coli:* the *trp* operon (*trp* for tryptophan).

If the operator is the control point for transcription, what determines whether the operator is in the on or off mode? By itself, the operator is on; RNA polymerase can bind to the promoter and transcribe the genes of the operon. The operon can be switched off by a protein called the **repressor.** The repressor binds to the operator and blocks attachment of RNA polymerase to the promoter, preventing transcription of the genes. Repressor proteins are specific; that is, they recognize and bind only to the operator of a certain operon. The repressor that switches off the *trp* operon has no effect on other operons in the *E. coli* genome.

The repressor is the product of a gene called a **regulatory gene.** The regulatory gene encoding the *trp* repressor, *trpR*, is located some distance away from the operon it controls and has its own promoter. Transcription of *trpR* produces an mRNA molecule that is translated into the repressor protein, which can then reach the operator of the *trp* operon by diffusion. Regulatory genes are transcribed continuously, although at a low rate, and a few *trp* repressor molecules are always present in the cell. Why, then, is the *trp* operon not switched off permanently? First, the binding of repressors to operators is reversible. An operator vacillates between the on and off

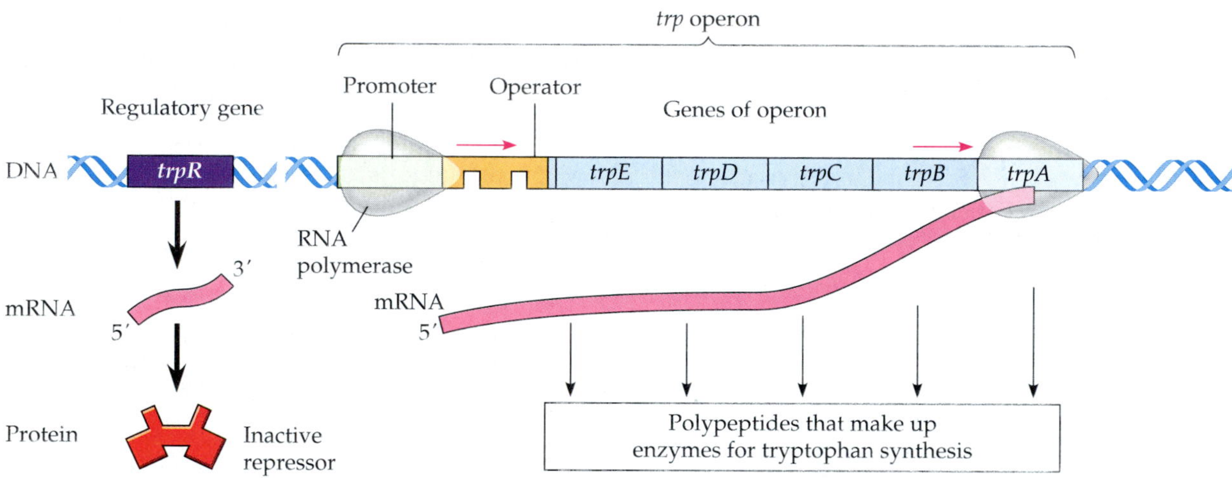

(a) Tryptophan absent, repressor inactive, operon on

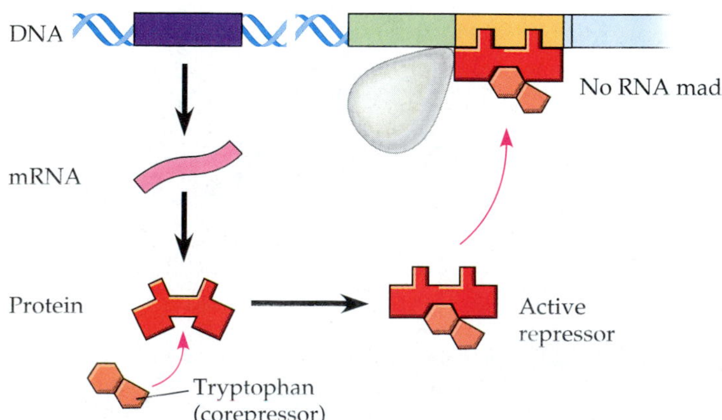

(b) Tryptophan present, repressor active, operon off

FIGURE 18.19 • The *trp* operon: regulated synthesis of repressible enzymes. (a) Tryptophan is an amino acid produced by an anabolic pathway catalyzed by repressible enzymes. Accumulation of tryptophan, the end-product of the pathway, represses synthesis of the enzymes. The mechanism for this regulation in an *E. coli* cell is shown here. Five genes encoding the polypeptides that make up the enzymes of the pathway are grouped into an operon, along with a promoter and an operator. (The *trp* operator is actually located within the *trp* promoter.) When the operon is in the "on" mode, RNA polymerase molecules attach to the DNA at the promoter region and transcribe the operon's genes. A regulatory gene, located outside the operon, codes for a repressor protein (the gene has its own promoter, not shown). The repressor can switch the *trp* operon off by binding to the operator and blocking access of RNA polymerase to the promoter. However, the repressor protein is synthesized in an inactive form and remains inactive in the absence of tryptophan. With no repressor bound to the operator, the operon is on, producing mRNA for the enzymes that synthesize tryptophan. **(b)** As tryptophan accumulates in the cell, it inhibits its own production by activating the repressor protein. The tryptophan binds to an allosteric site on the protein, causing its conformation to change. The repressor can now bind to the operator and switch the operon off.

modes, with the relative duration of each state depending on the number of active repressor molecules around. Secondly, the *trp* repressor, like most regulatory proteins, is an allosteric protein, with two alternative shapes, active and inactive (see FIGURES 6.15 and 11.5). The *trp* repressor is synthesized in an inactive form with little affinity for the *trp* operator. Only if tryptophan binds to the repressor at an allosteric site does the repressor protein change to the active form that can attach to the operator.

Tryptophan functions in this system as a **corepressor**, a small molecule that cooperates with a repressor protein to switch an operon off. As tryptophan accumulates, more tryptophan molecules associate with *trp* repressor molecules, which can then bind to the *trp* operator and shut down tryptophan production. If the cell's tryptophan level drops, transcription of the operon's genes resumes. This is one example of how gene expression responds rapidly to changes in the cell's internal and external environment.

Repressible Versus Inducible Operons: Two Types of Negative Gene Regulation

The *trp* operon is said to be a *repressible operon* because its transcription is *inhibited* when a specific small molecule (tryptophan) binds allosterically to a regulatory protein. In contrast, an *inducible operon* is *stimulated* when a specific small molecule interacts with a regulatory protein. Let's investigate an example (FIGURE 18.20).

18.5 The disaccharide lactose (milk sugar) is available to *E. coli* if the host human drinks milk. The bacteria can absorb the lactose and break it down for energy or use it as a source of organic carbon for synthesizing other compounds. Lactose metabolism begins with hydrolysis of the disaccharide into its two component monosaccharides, glucose and galactose. The enzyme that catalyzes this reaction is called β-galactosidase. Only a few molecules of this enzyme are present in an *E. coli* cell that has been growing in the absence of lactose—in the intestines of a person who does not drink milk, for example. But if lactose is added to the bacterium's nutrient medium, it takes only about 15 minutes for the number of β-galactosidase molecules in the cell to increase a thousandfold.

The gene for β-galactosidase is part of an operon, the *lac* operon (*lac* for lactose metabolism), that includes two other genes coding for proteins that function in lactose metabolism (see FIGURE 18.20). This entire transcription unit is under the command of a single operator and promoter. The regulatory gene, *lacI*, located outside the operon, codes for an allosteric repressor protein that can switch off the *lac* operon by binding

18.5

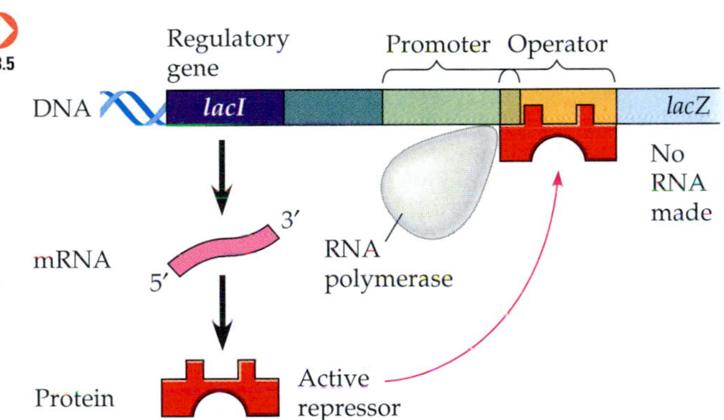

(a) Lactose absent, repressor active, operon off

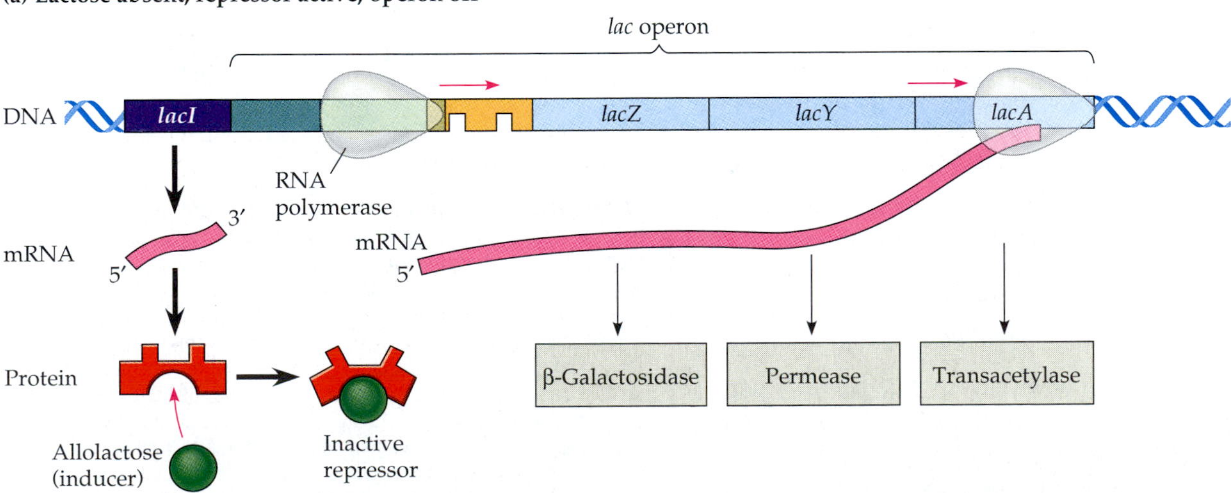

(b) Lactose present, repressor inactive, operon on

FIGURE 18.20 ▪ The *lac* operon: regulated synthesis of inducible enzymes. *E. coli* uses three enzymes to take up and metabolize lactose. The genes for these three enzymes are clustered in the *lac* operon. One gene, *lacZ*, codes for β-galactosidase, which hydrolyzes lactose to glucose and galactose. The second gene, *lacY*, codes for a permease, the membrane protein that transports lactose into the cell. The third gene, *lacA*, codes for an enzyme called transacetylase, whose function in lactose metabolism is still unclear. The gene for the *lac* repressor, *lacI*, happens to be adjacent to the *lac* operon, an unusual situation. (The function of the DNA between *lacI* and the promoter is revealed in FIGURE 18.21.) **(a)** The *lac* repressor is innately active, and in the absence of lactose it switches off the operon by binding to the operator. **(b)** Allolactose, an isomer formed from lactose, derepresses the operon by inactivating the repressor. In this way the enzymes for lactose metabolism are induced.

to the operator. So far, this sounds just like regulation of the *trp* operon, but there is one important difference. Recall that the *trp* repressor was innately inactive and required tryptophan as a corepressor in order to bind to the operator. The *lac* repressor, in contrast, is active all by itself, binding to the operator and switching the *lac* operon off. In this case, a specific small molecule, called an **inducer**, *inactivates* the repressor. For the *lac* operon, the inducer is allolactose, an isomer of lactose formed in small amounts from lactose that enters the cell. In the absence of lactose (and hence allolactose), the *lac* repressor is in its active configuration, and the genes of the *lac* operon are silent. If lactose is added to the cell's nutrient medium, allolactose binds to the *lac* repressor and alters its conformation, nullifying the repressor's ability to attach to the operator. Now, on demand, the *lac* operon produces mRNA for the enzymes of the lactose pathway. In the context of gene regulation, these enzymes are referred to as inducible enzymes, because their synthesis is induced by a chemical signal (allolactose, in this case). Analogously, the enzymes for tryptophan synthesis are said to be repressible.

Let's contrast repressible enzymes and inducible enzymes in terms of the metabolic economy of the *E. coli* cell. Repressible enzymes generally function in anabolic pathways, which synthesize essential end-products from raw materials (precursors). By suspending production of an end-product when it is already present in sufficient quantity, the cell can allocate its organic precursors and energy for other uses. In contrast, inducible enzymes usually function in catabolic pathways, which break a nutrient down to simpler molecules. By producing the appropriate enzymes only when the nutrient is available, the cell avoids making proteins that have nothing to do. Why bother, for example, to make the enzymes that break down milk sugar when no milk is present?

In comparing repressible and inducible enzymes, there is one more important point: Both systems are examples of the *negative* control of genes, because the operons are switched *off* by the active form of the repressor protein. It may be easier to see this in the case of the *trp* operon, but it is true for the *lac* operon as well. Allolactose induces enzyme synthesis not by acting directly on the genome, but by freeing the *lac* operon from the negative effect of the repressor. Technically, allolactose is more of a *derepressor* than an inducer of genes. Gene regulation is termed positive only when an activator molecule interacts directly with the genome to switch transcription on. Let's look at an example, again involving the *lac* operon.

An Example of Positive Gene Regulation

For the enzymes that break down lactose to be synthesized in appreciable quantity, it is not enough that lactose be present in the bacterial cell. The other requirement is that the simple sugar glucose be in short supply. Given a choice of substrates for glycolysis and other catabolic pathways, *E. coli* preferentially uses glucose, the sugar most reliably present in its environment.

How does the *E. coli* cell sense the glucose concentration, and how is this information relayed to the genome? Again, the mechanism depends on the interaction of an allosteric regulatory protein with a small organic molecule. The small molecule is **cyclic AMP (cAMP)**, which accumulates when glucose is absent (see FIGURE 11.11 for the structure of cAMP). The regulatory protein is **cAMP receptor protein (CRP)**, and it is an *activator* of transcription. When cAMP binds to the allosteric site on CRP, the protein assumes its active shape, and can bind to a specific site next to the *lac* promoter (FIGURE 18.21). The attachment of CRP to the DNA makes it easier for RNA polymerase to bind to the adjacent promoter and start transcription of the operon. Because CRP is a regulatory protein that directly stimulates gene expression, this mechanism qualifies as positive regulation.

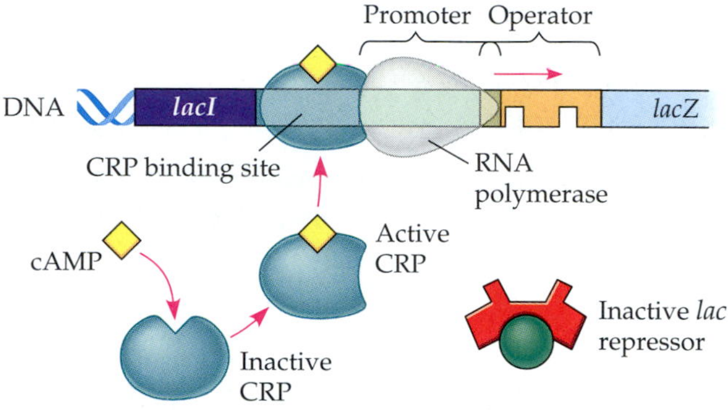

(a) Lactose present, glucose absent (cAMP level high): abundant *lac* mRNA synthesized

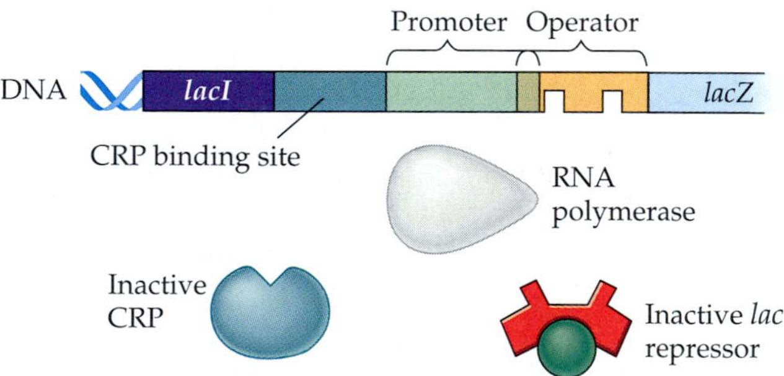

(b) Lactose present, glucose present (cAMP level low): little *lac* mRNA synthesized

FIGURE 18.21 • Positive control: cAMP receptor protein. RNA polymerase has a low affinity for the promoter of the *lac* operon unless helped by a regulatory protein called the cAMP receptor protein (CRP), which binds to a site on the DNA next to the promoter. The CRP molecule can attach to the DNA only when associated with cyclic AMP (cAMP), whose concentration in the cell rises when the glucose concentration falls. **(a)** If glucose is scarce, cAMP activates CRP, and the *lac* operon produces abundant mRNA for the lactose pathway. **(b)** But when glucose is present, cAMP is scarce, and CRP is unable to stimulate transcription. Thus, even if lactose is available, the cell preferentially catabolizes glucose, using enzymes that are always present. This regulatory system ensures that *E. coli* will gear up for consumption of lactose and other secondary catabolites only when glucose is unavailable.

If the amount of glucose in the cell increases, the cAMP concentration falls, and CRP disengages from the *lac* operon. Thus, the *lac* operon is under dual control: negative control by the *lac* repressor (described earlier) and positive control by CRP. The state of the *lac* repressor (with or without allolactose) determines whether or not transcription of the *lac* operon's genes can occur; the state of CRP (with or without cAMP) controls the rate of transcription if the operon is repressor-free. It is as though the operon has both an on-off switch and a volume control.

Although we have used the *lac* operon as an example, CRP, unlike repressor proteins, works on several different operons that encode enzymes used in catabolic pathways. When glucose is present and CRP is inactive, there is a general slowdown in the synthesis of enzymes required for the catabolism of compounds other than glucose. The cell's ability to catabolize other compounds, such as lactose, provides backup systems that enable a cell deprived of glucose to survive. The specific compounds present at the moment determine which operons are switched on. These elaborate contingency mechanisms suit an organism that cannot control what its host eats. Bacteria are remarkable in their ability to adapt—over the longer term by evolutionary changes in their genetic makeup, and over the shorter term by the control of gene expression in individual cells. Of course, the various control mechanisms are also evolutionary products that exist because they have been favored by natural selection.

■ ■ ■

Molecular genetics was founded on the study of viruses and bacteria, the microbial models that have been the subjects of this chapter. Eukaryotic organisms and their genomes are much more complex, and only in the past decade or so have researchers begun to learn how the control of gene expression helps bring about this complexity. In the next chapter we begin to explore this topic.

CHAPTER REVIEW

REVIEW OF KEY CONCEPTS

(with page numbers and key figures)

THE GENETICS OF VIRUSES

- **Researchers discovered viruses by studying a plant disease (pp. 319–320)** In the late 1800s, they found an infectious agent much smaller than bacteria.
- **A virus is a genome enclosed in a protective coat (pp. 320–321, FIGURE 18.2)** A virus is a small nucleic acid genome enclosed in a protein capsid and sometimes a membranous envelope. The genome may be single- or double-stranded DNA, or single- or double-stranded RNA.
- 18.1 **Viruses can reproduce only within a host cell (pp. 321–322, FIGURE 18.3)** They use enzymes, ribosomes, and small molecules of host cells to synthesize progeny viruses. Each type of virus has a characteristic host range, determined by specific receptors on host cells.
- 18.2 18.3 **Phages reproduce using lytic or lysogenic cycles (pp. 322–324, FIGURE 18.5)** In the lytic cycle, injection of a phage genome into a bacterium programs destruction of host DNA, production of new phages, and digestion of the host's cell wall, releasing the progeny phages. In a lysogenic cycle, a temperate phage inserts its genome into the bacterial chromosome as a prophage, which is passed on to host daughter cells until it is stimulated to leave the chromosome and initiate a lytic cycle.
- 18.4 **Animal viruses are diverse in their modes of infection and replication (pp. 324–328, FIGURES 18.6, 18.7)** They often have an envelope acquired from host cell membrane, which allows them to enter and exit the host cell. Retroviruses (such as HIV) are RNA viruses that use the enzyme reverse transcriptase to synthesize DNA from their RNA template. The DNA can then integrate into the host genome as a provirus. Vaccines against specific viruses stimulate the immune system to defend the host against an infection. Emerging viruses that cause new outbreaks of disease are usually existing viruses that manage to expand their host territory. Tumor viruses insert viral DNA into host cell DNA, triggering cancerous changes through their own or host cell oncogenes.
- **Plant viruses are serious agricultural pests (pp. 328–329)** Most are single-stranded RNA viruses. They enter plant cells through damaged cell walls or are inherited from a parent.
- **Viroids and prions are infectious agents even simpler than viruses (p. 329)** Plant diseases can be caused by viroids, naked RNA molecules that disrupt plant growth. Prions are infectious proteins that cause brain diseases in mammals.
- **Viruses may have evolved from other mobile genetic elements (pp. 329–330)** Evidence points to their origin as packaged fragments of cellular nucleic acid.

THE GENETICS OF BACTERIA

- **The short generation span of bacteria facilitates their evolutionary adaptation to changing environments (pp. 330–331)** The bacterial chromosome is a circular DNA molecule with few associated proteins. Plasmids are smaller rings of DNA with accessory genes. Because bacteria proliferate rapidly and have a short generation span, new mutations can affect a population's genetic variation quickly.
- **Genetic recombination produces new bacterial strains (pp. 331–337, FIGURES 18.11, 18.12, 18.14, 18.17)** The mechanisms of gene transfer between bacteria are transformation, transduction, and conjugation. In transformation, naked DNA enters the cell from the surroundings. In transduction, bacterial DNA is carried from one cell to another by phages. In conjugation, an F-factor–containing "male" cell transfers DNA to an F^- cell. F^+ cells transfer only the F plasmid. The F factor of an Hfr cell, which is integrated into the bacterial chromosome, brings some chromosomal DNA along with it when it transfers to a F^- cell. R plasmids confer resistance to various antibiotics. Their transfer between bacterial cells poses medical problems. Transposons, DNA segments that can insert at multiple sites in a cell's DNA, also contribute to genetic shuffling in bacteria (and eukaryotes). Insertion sequences, the simplest bacterial transposons, can affect gene function as they move about. They consist of inverted repeats of DNA flanking a gene for transposase. Composite transposons have additional genes, such as ones for antibiotic resistance.
- 18.5 **The control of gene expression enables individual bacteria to adjust their metabolism to environmental change (pp. 337–341,**

FIGURES 18.18–18.21) Cells control metabolism by regulating enzyme activity or by regulating enzyme synthesis through activating or inactivating genes. In bacteria, coordinately regulated genes are often clustered into operons, with one promoter serving several adjacent genes. An operator site on the DNA switches the operon on or off. In a repressible operon, binding of a specific repressor protein to the operator shuts off transcription by blocking the attachment of RNA polymerase. The repressor is activated by binding a small corepressor molecule, usually the end-product of an anabolic pathway. In an inducible operon, an innately active repressor is inactivated by binding an inducer, thereby turning on the genes of the operon only when necessary. Inducible enzymes usually function in catabolic pathways. Operons can also be subject to positive control via a stimulatory activator protein. For example, the active form of cAMP receptor protein (CRP) stimulates transcription by binding to a site next to the promoter and enhancing its ability to bind RNA polymerase.

SELF-QUIZ

1. Scientists have discovered how to put together a bacteriophage with the protein coat of phage T2 and the DNA of phage T4. If this composite phage were allowed to infect a bacterium, the phages produced in the host cell would have
   a. the protein of T2 and the DNA of T4
   b. the protein of T4 and the DNA of T2
   c. a mixture of the DNA and proteins of both phages
   d. the protein and DNA of T2
   e. the protein and DNA of T4

2. Horizontal transmission of a plant viral disease could be caused by
   a. the movement of viral particles through plasmodesmata
   b. the inheritance of an infection from a parent plant
   c. the spread of an infection by vegetative (asexual) propagation
   d. insects as vectors carrying viral particles between plants
   e. the transmission of proviruses via cell division

3. RNA viruses require their own supply of certain enzymes because
   a. the viruses are rapidly destroyed by host cell defenses
   b. host cells do not have RNA ⟶ RNA or RNA ⟶ DNA enzymes
   c. the enzymes translate viral mRNA into proteins
   d. the viruses use these enzymes to penetrate host cell membranes
   e. these enzymes cannot be made in host cells

4. Researchers recently identified a strain of *Staphylococcus* resistant to vancomycin. If the gene for vancomycin resistance were acquired by conjugation, it would be reasonable to expect
   a. the presence of an R plasmid within the bacterium
   b. to identify genes for capsid proteins belonging to an infective phage
   c. the resistance gene to be bracketed by inverted repeats
   d. that the bacterium could grow on a tryptophan-deficient medium
   e. that the bacterium would become F^+

5. Transposition differs from other mechanisms of genetic shuffling because it
   a. occurs only in bacteria
   b. moves genes between homologous regions of the DNA
   c. plays little or no role in evolution
   d. occurs only in eukaryotes
   e. scatters genes to new loci in the genome

6. A particular operon produces enzymes that manufacture an important amino acid. If this process works as it does in other operons,
   a. the amino acid inactivates the repressor
   b. the enzymes produced are called inducible enzymes
   c. the repressor binds to the operator in the absence of the amino acid
   d. the amino acid acts as a corepressor
   e. the amino acid "turns on" enzyme synthesis

7. A mutation that renders the regulatory gene of a repressible operon nonfunctional would result in
   a. continuous transcription of the operon's genes
   b. continuous synthesis of an inducer
   c. accumulation of large quantities of a substrate for the catabolic pathway controlled by the operon
   d. irreversible binding of the repressor to the promoter
   e. excessive synthesis of a catabolic activator protein

8. Which of the following information transfers is catalyzed by reverse transcriptase?
   a. RNA ⟶ RNA
   b. DNA ⟶ RNA
   c. RNA ⟶ DNA
   d. protein ⟶ DNA
   e. RNA ⟶ protein

9. Which of the following characteristics or processes is common to *both* bacteria *and* viruses?
   a. nucleic acid as the genetic material
   b. binary fission
   c. mitosis
   d. ribosomes in the cytoplasm
   e. conjugation

10. Which of the following processes would never contribute to genetic variation within a bacterial population?
    a. transduction
    b. transformation
    c. conjugation
    d. mutation
    e. meiosis

CHALLENGE QUESTIONS

1. Mutations can alter the function of an operon; in fact, it was the effects of various mutations that enabled Jacob and Monod to figure out how the *lac* operon works. Predict how the following mutations would affect *lac* operon function in the presence and absence of allolactose.
   a. Mutation of regulatory gene; repressor will not bind to lactose.
   b. Mutation of operator; repressor will not bind to operator.
   c. Mutation of regulatory gene; repressor will not bind to operator.
   d. Mutation of promoter; RNA polymerase will not attach to promoter.

2. When bacteria infect an animal, the number of bacteria in the body increases gradually. A graph of the growth of the bacterial population is a smoothly increasing curve, as shown in graph a at the top of the next page. A viral infection shows a different pattern. For a while, there is no evidence of infection, and then there is a sudden rise in the number of viruses. The number stays constant for a time; then there is another sudden increase. The curve of viral population growth looks like a series of steps, as shown in graph b. Explain the difference in the growth curves.

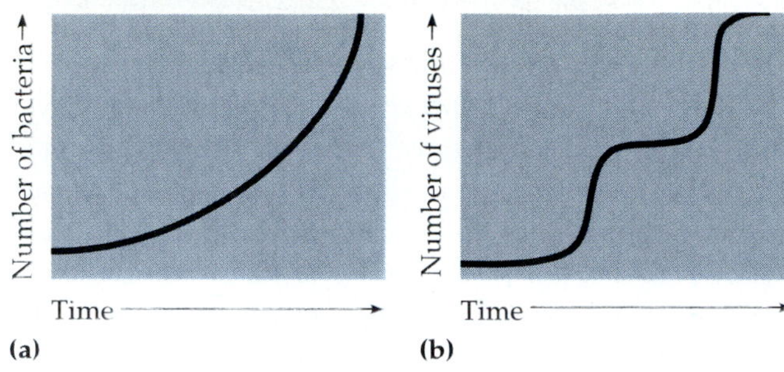

3. The influenza outbreak of 1918 is considered by most epidemiologists to have been the worst pandemic in human history. Over 21 million people died from the disease worldwide, with more than 700,000 deaths reported in the U.S. alone. Recently, researchers realized that genetic material from the virus might be obtainable. Using tissue samples collected in 1918 from deceased soldiers and some of the techniques of molecular biology, they were able to isolate and amplify several viral genes. Why the virus was so virulent is not yet entirely clear, but many critics have suggested that this information is best left undiscovered. How would you defend the action of the researchers? What benefits can be expected from this type of research?

SCIENCE, TECHNOLOGY, AND SOCIETY

1. Explain how the excessive or inappropriate use of antibiotics poses a health hazard for a human population.

2. AIDS is not a new disease, just new to the Western world. HIV probably originated in central Africa and may have been infecting humans for decades. In what ways might modern technology and social changes contribute to the emergence of viruses that make new epidemics like AIDS possible?

FURTHER READING

Beardsley, T. "Better Than a Cure." *Scientific American,* January 1995. Discusses the importance of immunizing children.

Diamond, J. "The Arrow of Disease." *Discover,* October 1992. Describes how Columbus and other explorers brought European pathogenic viruses and bacteria to the New World.

Larson, E. "The Flu Hunters." *Time,* February 22, 1998. Explains how medical detectives tracked the mysterious and deadly 1997 flu.

Le Guenno, B. "Emerging Viruses." *Scientific American,* October 1995. Describes newly discovered viral diseases.

Levy, S. B. "The Challenge of Antibiotic Resistance." *Scientific American,* March 1998. Explores how certain bacterial infections now defy all antibiotics.

Miller, R. V. "Bacterial Gene Swapping in Nature." *Scientific American,* January 1998. Gene swapping occurs more often than once assumed.

Vogel, G. "Prusiner Recognized for Once-Heretical Prion Theory." *Science,* October 10, 1997. A balanced discussion of the theory of and the controversies surrounding prions. The same issue has an article by Prusiner on the possibility of human illness due to the prions responsible for "mad cow disease."

WEB LINKS

Visit the special edition of *The Biology Place* for BIOLOGY, Fifth Edition, at **http://www.biology.com/campbell**. Go to Chapter 18 for online resources, including learning activities, practice exams, and links to the following web sites:

"The Virtual Virus"
Created as part of the ThinkQuest Project, this site is an easy-to-use introduction to the study of viruses.

"Virus Ultrastructure"
This site introduces you to the architecture of viruses and how scientists obtain images of them.

"All About Hantavirus"
The Centers for Disease Control provide updated information on the deadly disease caused by hantavirus.

"Plant Viruses Online"
This site contains comprehensive information on a large number of plant viruses.

CHAPTER 19

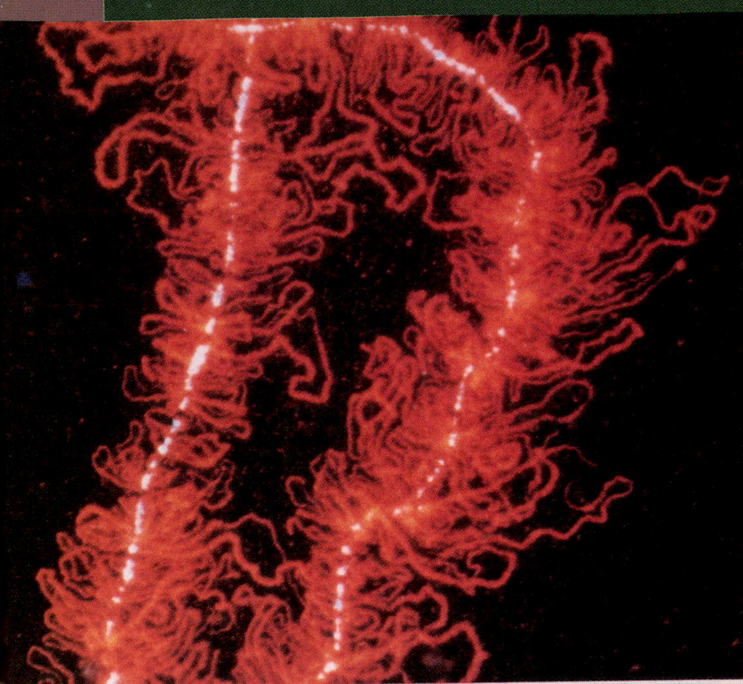

THE ORGANIZATION AND CONTROL OF EUKARYOTIC GENOMES

Eukaryotic cells face the same challenges as prokaryotic cells in expressing their genes, but with two main differences: the vastly greater size of the typical eukaryotic genome and the importance of cell specialization in multicellular eukaryotes. These two complications present a formidable information-processing task for the eukaryotic cell.

Consider the genome of a human cell. It has an estimated 50,000 to 100,000 genes—about 20 times more than a typical bacterium. Moreover, the genomes of humans and other eukaryotes also include an enormous amount of DNA that does not program the synthesis of RNA or protein. The entire mass of DNA must be precisely replicated with each turn of the cell cycle. Managing such a large amount of DNA requires that it be elaborately organized. In both prokaryotes and eukaryotes, DNA is associated with proteins, but in the eukaryotic cell the DNA-protein complex, called chromatin, is ordered into higher structural levels. The light micrograph that introduces this chapter gives a sense of the complex organization of chromatin in a developing salamander ovum. Part of the chromatin (stained white) is packed into the main axis of each of the chromosome's two chromatids, while other parts (red stained loops) are being actively transcribed. As in prokaryotes, transcription is the stage at which eukaryotic gene expression is most often regulated. In this chapter you will learn more about the structure of eukaryotic chromosomes, the organization of nucleotide sequences in eukaryotic genomes, and the mechanisms by which eukaryotic genes are controlled. You will also learn about DNA changes that disrupt gene regulation and lead to cancer.

THE STRUCTURE OF CHROMATIN

Let's start by looking at how eukaryotic cells package their chromosomal DNA into chromatin.

Chromatin structure is based on successive levels of DNA packing

Even bacterial DNA has a degree of packing. Typically containing only several million nucleotide pairs of DNA, the bacterial genome was once thought to be a naked circle of DNA with only trivial amounts of associated protein and without any particular folding pattern. Now we know that the DNA of the bacterial chromosome is associated with specific proteins and that it is coiled and looped in a complex but orderly manner. In fact, the loops are about the same size as those in the salamander chromosomes seen above.

Nevertheless, it remains true that eukaryotic chromatin is considerably more complex than bacterial chromatin (FIGURE 19.1). Eukaryotic DNA is precisely combined with a large amount of protein, and the resulting chromatin undergoes striking changes in the course of the cell cycle. During interphase, chromatin fibers are usually highly extended within the nucleus. When interphase cells are stained, the chromatin appears as a diffuse, colored mass. As you learned in Chapter 12, however, when a cell prepares for mitosis, its chromatin coils and folds up ("condenses") to form a characteristic number of short, thick chromosomes that are visible with a light microscope.

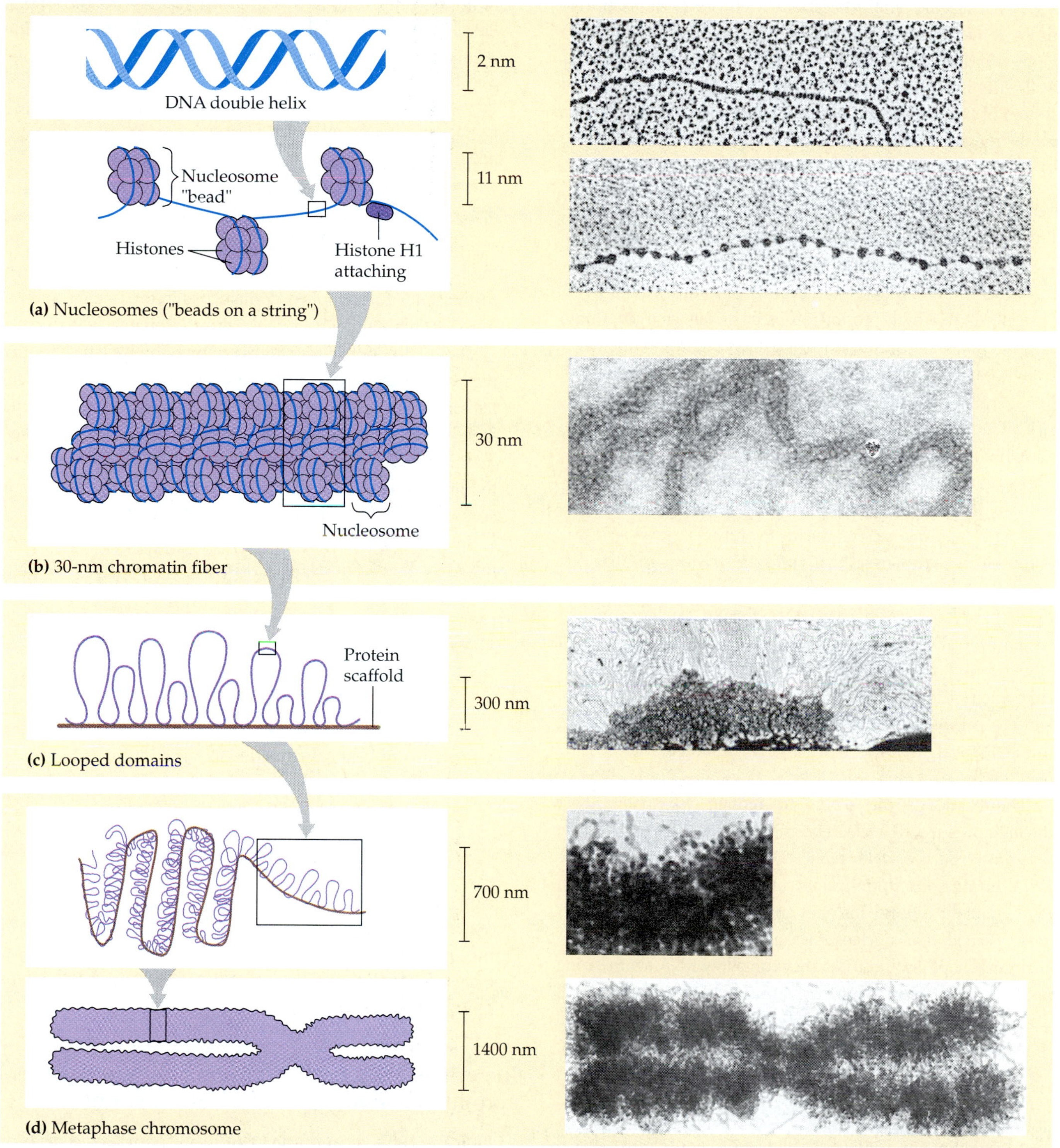

FIGURE 19.1 • Levels of chromatin packing. This series of diagrams and transmission electron micrographs depicts a current model for the progressive stages of DNA coiling and folding that culminate in a highly condensed metaphase chromosome. **(a)** DNA and histone molecules form "beads on a string," consisting of nucleosomes in an extended configuration. Each nucleosome has two molecules each of four types of histone, around which the DNA wraps. The fifth histone, called H1, can bind to DNA adjacent to a "bead." **(b)** With the help of histone H1, the string of nucleosomes coils to form a chromatin fiber that is 30 nm in diameter. **(c)** At the next level are looped domains of the 30-nm fiber. The loops are attached to a scaffold of nonhistone proteins. **(d)** The chromatin folds further, resulting in the maximally compacted chromosome seen at metaphase. (Recall that each metaphase chromosome consists of two chromatids.)

Eukaryotic chromosomes contain an enormous amount of DNA relative to their condensed length. Each chromosome contains a single linear DNA double helix that, in humans, typically has about 2×10^8 nucleotide pairs. If extended, such a DNA molecule would be about 6 cm long, thousands of times longer than the diameter of a cell nucleus. All this DNA—and the DNA of the other 45 human chromosomes, as well—fits into the nucleus through an elaborate, multilevel system of DNA packing. FIGURE 19.1 (preceding page) shows a model for this packing.

Nucleosomes, or "Beads on a String"

Proteins called **histones** are responsible for the first level of DNA packing in eukaryotic chromatin. In fact, the mass of histone in chromatin is approximately equal to the mass of DNA. Histones have a high proportion of positively charged amino acids (lysine and arginine), and they bind tightly to the negatively charged DNA. (Recall that the phosphate groups of DNA give it a negative charge all along its length; see FIGURE 16.5.) The DNA-histone complex is chromatin in its most fundamental form. There are five types of histone. Histones are very similar from one eukaryote to another, and similar proteins are found even in bacteria. Apparently, the histone genes have been highly conserved during evolution.

In electron micrographs, unfolded chromatin has the appearance of beads on a string, as shown in FIGURE 19.1a. Each "bead" and its adjacent DNA form a **nucleosome**, the basic unit of DNA packing. The nucleosome bead consists of DNA wound around a protein core composed of two molecules each of four different types of histone: H2A, H2B, H3, and H4. A molecule of the fifth histone, called H1, attaches to the DNA near the bead when the chromatin undergoes the next level of packing.

The beaded string seems to remain essentially intact throughout the cell cycle. The histones leave the DNA only transiently during DNA replication, and they stay with the DNA during transcription. How can DNA be transcribed when it is wrapped around histones in nucleosomes? Researchers have learned that nucleosomes are dynamic structures. By changing shape and position, they can allow RNA-synthesizing polymerases to move along the DNA. Later in this chapter we'll discuss some recent discoveries about the roles of nucleosomes in the regulation of gene expression.

Higher Levels of DNA Packing

The beaded string undergoes higher-order packing. This is clearest when the extended interphase chromatin condenses to produce the thickened, compact chromosomes we see during mitosis. In the laboratory, mitotic chromosomes can be isolated and unraveled to reveal several orders of chromatin coiling. FIGURES 19.1b–d illustrate the various structures in order of increasing compactness. With the aid of histone H1, the beaded string coils or folds to form a fiber roughly 30 nm in thickness, known as the *30-nm chromatin fiber* (FIGURE 19.1b). The 30-nm fiber, in turn, forms loops called *looped domains*, which are attached to a chromosome scaffold made of nonhistone proteins. In a mitotic chromosome the looped domains themselves coil and fold, further compacting all the chromatin to produce the characteristic metaphase chromosome you see in the micrograph at the bottom of FIGURE 19.1. The packing steps are apparently highly specific and precise, for particular genes always end up located at the same places in metaphase chromosomes.

Though interphase chromatin is generally much less condensed than the chromatin of mitotic chromosomes, it shows several of these same levels of higher-order packing. Much of the beaded string is compacted into a 30-nm fiber, and then further folded into looped domains (as in the micrograph on p. 345). Although an interphase chromosome lacks a discrete "scaffold," its looped domains seem to be attached to the nuclear lamina, on the inside of the nuclear envelope, and perhaps also to matrix fibers (see pp. 109–110). These attachments may help organize regions of active transcription. The chromatin of each chromosome occupies a restricted area within the interphase nucleus, and the chromatin fibers of different chromosomes do not become entangled with each other.

Even during interphase, portions of certain chromosomes in some cells exist in the highly condensed state represented in FIGURE 19.1d. This type of interphase chromatin, which is visible with a light microscope, is called **heterochromatin**, to distinguish it from the less compacted **euchromatin** ("true chromatin"). What is the function of this selective condensation in interphase cells? The formation of heterochromatin may be a sort of coarse adjustment in the control of gene expression, for heterochromatin DNA is not transcribed.

GENOME ORGANIZATION AT THE DNA LEVEL

Now we are ready to examine how genes and other DNA sequences are organized within the genome. Because most of this unit has focused on genes, you may be surprised to learn that genes make up only a tiny portion of the genomes of most multicellular eukaryotes.

Repetitive DNA and other noncoding sequences account for much of a eukaryotic genome

In prokaryotes, most of the DNA in a genome codes for protein (or tRNA and rRNA), with the small amount of noncoding DNA consisting mainly of regulatory sequences, such as promoters. The coding sequence of nucleotides along a prokaryotic gene proceeds from start to finish without interruption. In eukaryotic genomes, by contrast, most of the DNA—about 97% in humans—does *not* encode protein or

RNA. What does this DNA consist of? Some of it is known to be regulatory sequences, but much of it consists of sequences whose functions, if any, are not yet understood. This DNA includes introns, the stretches of noncoding DNA that often interrupt the coding sequences of eukaryotic genes (see Chapter 17). Even more of the noncoding DNA consists of **repetitive DNA**, nucleotide sequences that are present in many copies in a genome, usually not within genes.

Tandemly Repetitive DNA

TABLE 19.1 lists the main categories of repetitive DNA, which occur in different amounts in different species. In mammals, about 10–15% of the genome is *tandemly repetitive DNA*, short sequences repeated in series, as in the following example (showing one strand only):

...GTTACGTTACGTTACGTTACGTTACGTTAC...

The number of repetitions of the GTTAC unit at a site in the genome could be as high as several hundred thousand. Repeated units are up to 10 base pairs long.

The nucleotide composition of tandemly repetitive DNA is often different enough from the rest of the cell's DNA to give the repetitive DNA an intrinsically different density, so that researchers can isolate it by differential ultracentrifugation (as in the Meselson-Stahl experiment; see FIGURE 16.9). If the genomic DNA is cut into pieces before centrifugation, segments of different density migrate to different positions in the centrifuge tube. Repetitive DNA isolated in this way was originally called **satellite DNA**, because it appeared as a "satellite" band in the centrifuge tube, separate from the rest of the DNA. Now the term is often used for all tandemly repetitive DNA.

As the table shows, satellite DNA is classified into three types, depending on the total length of the DNA at each site.

Table 19.1 ■ Types of Repetitive DNA

TANDEMLY REPETITIVE DNA (SATELLITE DNA)	
Proportion of mammalian DNA:	10–15%
Length of each repeated unit:	1–10 base pairs
Total length of repetitive DNA per site, in base pairs:	
Regular satellite DNA	100,000–10 million
Minisatellite DNA	100–100,000
Microsatellite DNA	10–100
Repeated units at a site are usually identical	
INTERSPERSED REPETITIVE DNA	
Proportion of mammalian DNA:	25–40%
Length of each repeated unit:	100–10,000 base pairs
Number of repetitions per genome:	10–1 million
"Copies" are very similar but not identical	

Regular satellite DNA occurs in stretches over 100,000 base pairs long, while *minisatellite* and *microsatellite* subcategories have been created for sequences that appear in shorter stretches. Microsatellite DNA, with short units repeated only 10–100 times, has turned out to be extremely useful for DNA fingerprinting, as we'll discuss in Chapter 20.

A number of genetic disorders, including two you read about in Chapter 15, are caused by abnormally long stretches of tandemly repeated nucleotide triplets—a sort of microsatellite—within the affected gene. The first such "triplet repeat" sequence to be identified was the one responsible for fragile *X* syndrome, a major cause of mental retardation. In the normal allele for the fragile *X* gene, within the 5′ untranslated leader region of the first exon, the triplet CGG is repeated about 30 times; in a fragile *X* allele the same triplet is repeated hundreds or even thousands of times, creating the fragile site. The lengthening of the repetitive DNA occurs in steps over a number of generations. Huntington's disease is also a triplet-repeat disorder, with a stretch of repeated CAG triplets that lengthens in a similar fashion. In this case, the triplet repeat is actually translated, and the resulting protein has a long string of glutamines, the amino acid encoded in DNA by CAG. The dozen triplet-repeat disorders identified to date all affect the nervous system, and in all cases the number of repeats seems to correlate with the severity of the disease and the age of onset. Scientists do not yet know exactly how the extra repeats cause disorders.

Much of a genome's regular satellite DNA is located at chromosomal telomeres and centromeres, suggesting that this DNA plays a structural role for chromosomes. The DNA at centromeres is essential for the separation of chromatids in cell division (see Chapter 12), and, along with other tandemly repetitive sequences, it may help organize the chromatin within the interphase nucleus. We discussed the DNA of telomeres, the tips of chromosomes, in Chapter 16. Besides protecting genes from being lost as the DNA shortens with each round of replication, telomeric DNA protects the chromosome by binding proteins that stop the ends from "fraying" and from sticking to other chromosomes.

The importance of telomeres and centromeres has been highlighted by the creation of *artificial chromosomes* in the laboratory. The minimal requirements for a functional chromosome are simple: a centromere, two telomeres, and an origin of DNA replication. If an artificial chromosome has these features—no matter what the nature of its other DNA—a cell will be able to replicate it and distribute the copies properly to daughter cells.

Interspersed Repetitive DNA

In addition to tandemly repetitive DNA, eukaryotic genomes have huge amounts of *interspersed repetitive DNA* (see TABLE 19.1). The repeated units of this type of DNA are not next to each other; instead, they are scattered about the genome. A single unit is usually hundreds or even thousands of base pairs long, and the dispersed "copies" are similar but usually not identical to each other. Interspersed repetitive DNA makes up 25–40% of most mammalian genomes. In humans and other primates, a large portion of this DNA (at least 5% of the genome) consists of a family of similar sequences called ***Alu* elements**, for which each unit is about 300 nucleotide pairs long. Like the Huntington's disease triplet repeats, *Alu* elements are an exception to the notion that repetitive DNA is noncoding. Many *Alu* elements are transcribed into RNA molecules; their cellular function, if any, is unknown.

Although we know little about the functions of interspersed repetitive DNA, we do know how it comes to be both abundant and variable in location. Most such sequences, at least in mammals, seem to be transposons, the mobile genetic elements introduced in Chapter 18. We'll have more to say about eukaryotic transposons later in this chapter.

Gene families have evolved by duplication of ancestral genes

As in prokaryotes, most eukaryotic genes are present as unique sequences, with only one copy per haploid set of chromosomes. However, some genes are present in more than one copy, and others closely resemble each other in nucleotide sequence. A collection of identical or very similar genes is called a **multigene family**. It is likely that the members of each family evolved from a single ancestral gene.

Multigene families can be thought of as repetitive DNA with very long (gene-length) repeating units. The members of a multigene family may be clustered or dispersed in the genome. Some multigene families consist of identical genes, usually clustered tandemly. Although the genes for histone proteins are a notable exception, multigene families of identical genes usually consist of genes for RNA products. An example is the family of identical genes for the three largest ribosomal RNA (rRNA) molecules (FIGURE 19.2). These rRNA molecules are encoded in a single transcription unit that is repeated tandemly hundreds to thousands of times in the genome of a multicellular eukaryote. The many copies enable cells to make the millions of ribosomes needed for active protein synthesis. The primary transcript is cleaved to yield the three rRNA molecules. These are then combined with proteins and one other kind of rRNA to form ribosomal subunits.

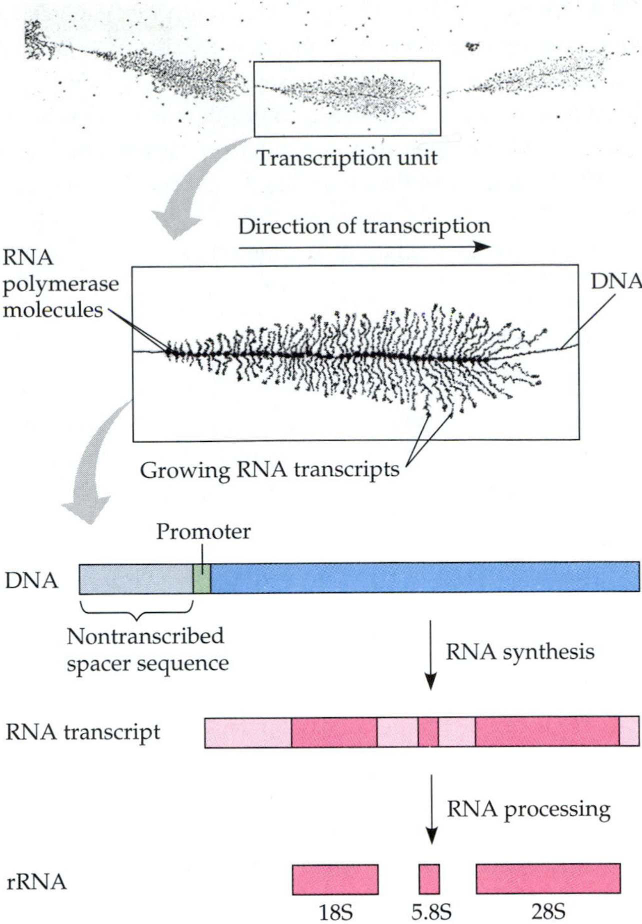

FIGURE 19.2 • Part of a family of identical genes for ribosomal RNA. The transmission electron micrograph at the top shows three of the hundreds of copies of rRNA genes in a salamander genome. Each "feather" corresponds to a single transcription unit being transcribed by about 100 molecules of RNA polymerase (the dark dots along the DNA), moving left to right. The growing RNA transcripts extend out from the DNA. The transcription units are separated from one another by a spacer sequence of nontranscribed DNA. The RNA transcripts are processed by cleavage to yield three kinds of rRNA molecules: 18*S*, 5.8*S*, and 28*S*. (The *S* designations refer to their sedimentation rates in the ultracentrifuge, which depend on molecular size.)

The classic examples of multigene families of *nonidentical* genes are the two related families of genes that encode globins, the α and β polypeptide subunits of hemoglobin. One family, located on chromosome 16 in humans, encodes various versions of α-globin; the other, on chromosome 11, encodes versions of β-globin (FIGURE 19.3). Similarities in the sequences of the various globin genes indicate that the α-like globins and β-like globins all evolved from one common ancestral globin. The different versions of each globin subunit are expressed at different times in development, allowing hemoglobin to function effectively in the changing environment of the developing animal. In humans, for example, the embryonic and fetal forms of hemoglobin have a higher affinity for oxygen than the adult forms, ensuring the efficient transfer of oxygen from mother to developing fetus.

How do families of genes arise from a single gene? The most likely explanation is that they arise by repeated gene duplication, which can result from errors during DNA replication and recombination. The differences among the genes in families of nonidentical genes probably arise from mutations that accumulate in the gene copies over many genera-

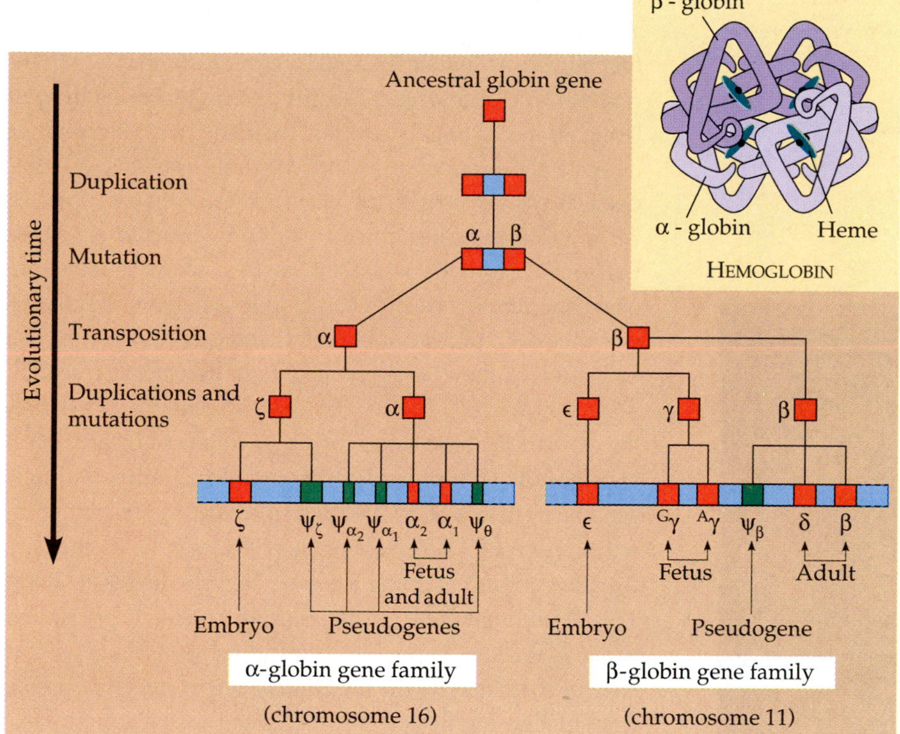

FIGURE 19.3 ▪ The evolution of human α-globin and β-globin. Each of the two gene families consists of a group of similar, but not identical, genes clustered together on a chromosome. The various genes (red) are represented by different Greek letters. In each gene family the genes are arranged in order of their expression during development; individual genes are turned on or off in response to the organism's changing environment as it develops from an embryo into a fetus, and then into an adult. At all times during development, functional hemoglobin consists of a total of four polypeptide subunits, two from the α family and two from the β family (see insert). Separating the functional genes within each family cluster are long stretches of noncoding DNA, which include pseudogenes (green), nonfunctional nucleotide sequences very similar to the functional genes. Presumably, the different genes and pseudogenes in each family arose from gene duplication of an original α or β gene followed by mutation. In fact, the original α and β genes themselves undoubtedly arose in much the same way from a common ancestral globin gene. Transposition put the α-globin and β-globin families on different chromosomes, probably early in their evolution.

tions. The existence of DNA segments called pseudogenes is evidence for this process of gene duplication and mutation. **Pseudogenes** have sequences very similar to real (functional) genes but lack the regulatory sequences (for example, promoters) necessary for gene expression. The globin gene families include several pseudogenes within the stretches of noncoding DNA between the functional genes.

The location of the α and β globin families on different human chromosomes, as well as the dispersion of the genes of certain other gene families, probably arose by transposition, one of the topics in the next section.

Gene amplification, loss, or rearrangement can alter a cell's genome

We are used to the idea that, except for rare mutations, the nucleotide sequence of an organism's DNA is constant during its lifetime. However, there are a few important exceptions in which the DNA of somatic cells is altered in a systematic way. Since these changes do not affect gametes, they are not passed on to offspring, but they can and do have major effects on gene expression within particular cells and tissues.

Gene Amplification and Selective Gene Loss

Sometimes the number of copies of a gene or gene family temporarily increases in the cells of some tissues during a particular stage of development. For instance, consider the genes for ribosomal RNA in amphibians. As in most eukaryotes, multiple copies of these genes are built into the genome of every cell. A developing ovum, however, synthesizes a million or more additional copies of the rRNA genes, which exist in nucleoli as tiny DNA circles separate from the chromosomes. This selective replication of certain genes, or **gene amplification**, is a potent way of increasing expression of the rRNA genes, enabling the developing egg cell to make enormous numbers of ribosomes. These ribosomes make possible a burst of protein synthesis once the egg is fertilized. The extra copies of the rRNA genes cannot replicate and are broken down during early embryonic development.

Gene amplification has also been observed in cancer cells exposed to high concentrations of chemotherapeutic drugs. While such drugs may kill a great many cells in a tumor, invariably some cells are resistant. These cells contain amplified segments of DNA carrying genes conferring drug resistance. In the laboratory, increasing the concentration of such drugs leads to increasing resistance in the cell population by selecting for cells that have amplified these genes.

In certain insects, genes are selectively *lost* in certain tissues (although not in the cells that give rise to gametes, of course). In fact, whole chromosomes or parts of chromosomes may be eliminated from certain cells early in insect development.

Rearrangements in the Genome

More common than gene amplification or loss is the shuffling of substantial stretches of DNA. Here we are talking not about the genetic recombination that goes on in meiosis but about rearrangements that change the loci of genes in somatic cells of an organism. Such rearrangements may have powerful effects on gene expression.

Transposons and Retrotransposons. All organisms seem to have transposons, stretches of DNA that can move from one location to another within the genome. Transposons were discussed in detail in Chapter 18, and transposition was mentioned as a source of scattered gene families earlier in this chapter. Recall that if a transposon "jumps" into the middle of a coding sequence of another gene, it prevents the normal functioning of the interrupted gene (FIGURE 19.4). If the transposon inserts within a sequence that is involved in regulating transcription, the transposition may increase or decrease the production of one or more proteins. In some cases, the transposon itself carries a gene that is activated when it is inserted just downstream from an active promoter. Barbara McClintock was the first to discover transposons when she found evidence for mobile genetic elements that affect the color of developing corn kernels.

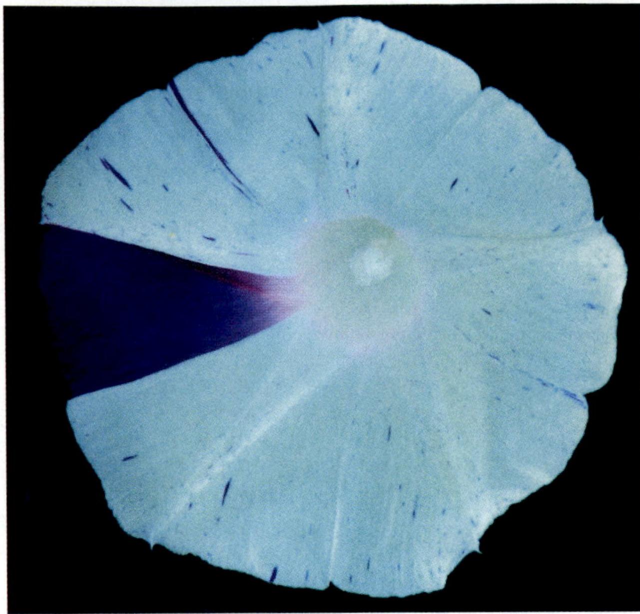

FIGURE 19.4 ▪ The effect of a transposon on flower color. Without the movement of a transposon, this morning glory flower would be entirely purple. The white portion of the flower (light blue-green in this photo) resulted from the movement of a transposon in the genome of one cell to a locus that determines purple flower color, destroying the function of the flower-color gene. The white portion of the flower consists of the cells descended from the single cell in which the transposition occurred.

Courtesy of Evelyne Cudel-Epperson, MSU.

Recent studies have revealed that transposons actually make up over 50% of the corn (maize) genome and 10% of the human genome. Most of these transposons are **retrotransposons**, transposable elements that move within a genome by means of an RNA intermediate, a transcript of the retrotransposon DNA (FIGURE 19.5). To insert at another site, the RNA retrotransposon must be converted back to DNA. This is accomplished by the enzyme reverse transcriptase, which is encoded in the retrotransposon itself, along with an enzyme that catalyzes the insertion at a new site. Thus, reverse transcriptase can be present in cells not infected with retroviruses. (In fact, retroviruses may have evolved from escaped and packaged retrotransposons.) The *Alu* elements, mentioned earlier, are retrotransposons that do not code for reverse transcriptase but can move using enzymes encoded elsewhere in the genome.

Immunoglobulin Genes. In vertebrates, at least one set of genes normally undergoes permanent rearrangements of DNA segments. These rearrangements occur during cellular *differentiation*, the specialization in cellular structure and function that is part of embryonic development. Several kinds of genes become rearranged as cells of the immune system differentiate. Let's look at the genes encoding antibodies, or

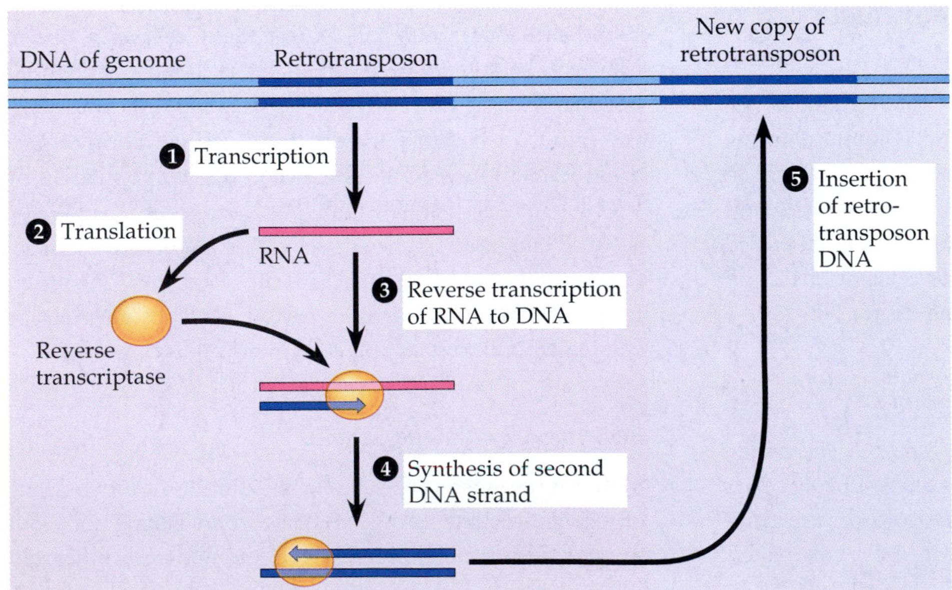

FIGURE 19.5 ▪ Retrotransposon movement. The process ① starts with transcription of the retrotransposon DNA by RNA polymerase. ② Translation of part of the RNA yields reverse transcriptase, which ③ catalyzes the synthesis of a DNA strand on the retrotransposon RNA template and ④ the replacement of the RNA strand with DNA. ⑤ The resulting double-stranded DNA version of the retrotransposon is then inserted into the genome at some other location. By this sort of replicative transposition, a retrotransposon can populate the genome of a multicellular eukaryote in huge numbers. For example, at least 10% of the human genome consists of retrotransposons and related elements. Notice that the retrotransposition mechanism is essentially identical to part of the retrovirus reproductive cycle (see FIGURE 18.7).

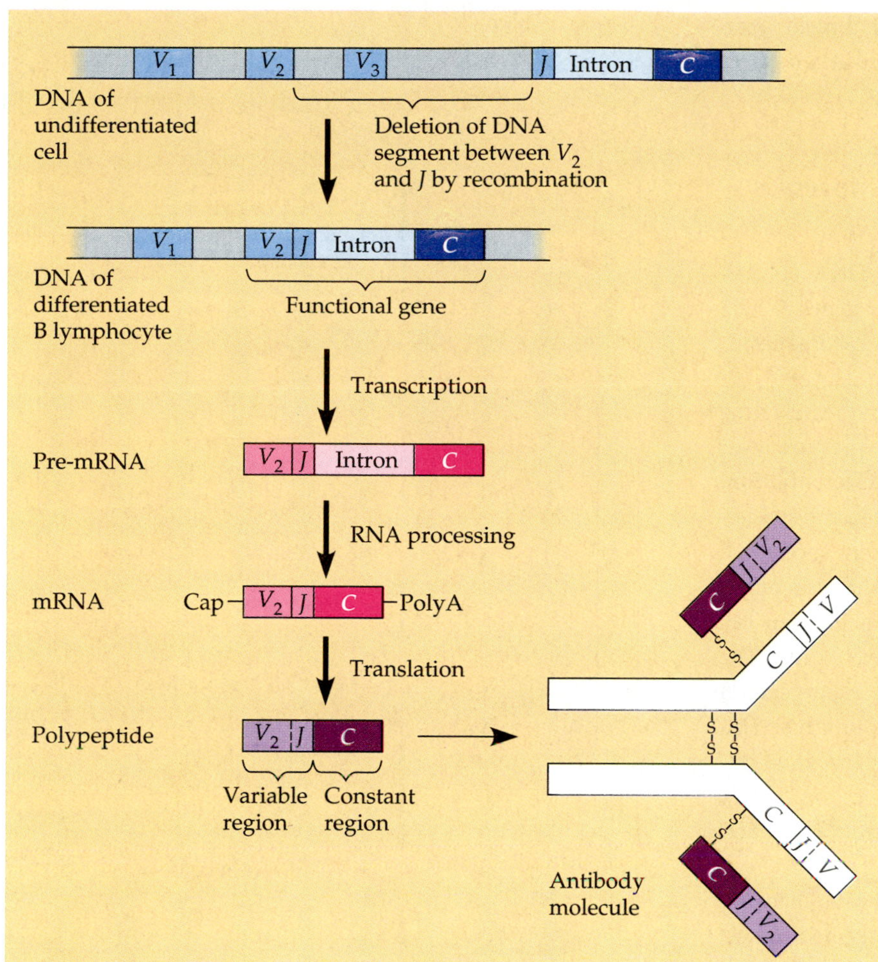

FIGURE 19.6 • DNA rearrangement in the maturation of an immunoglobulin (antibody) gene. The DNA of an immunoglobulin gene in an undifferentiated cell carries coding segments for hundreds of different antibody variable *(V)* regions, for several different junction *(J)* regions, and for one or more different constant *(C)* regions. This simplified diagram shows only three *V* segments, one *J* segment, and one *C* segment. During differentiation of a B lymphocyte, a long stretch of DNA, from the end of one of the *V* segments to the beginning of a *J* segment, is deleted. This deletion brings a *V* segment (in this case V_2) adjacent to a *J* segment and produces a gene that can be transcribed. The RNA transcript is processed in the usual way to remove introns, and the resulting mRNA is translated into one of the polypeptides for an immunoglobulin molecule. The amino acids coded by the *J* segment are considered part of the variable region of the polypeptide. The joining of *V*, *J*, and *C* regions of DNA in random combinations (many more than would be possible from the version in this diagram) arms the immune system with diverse antibody-producing lymphocytes, each able to recognize a different foreign invader (see Chapter 43). Different combinations of the two types of polypeptides that make up each antibody molecule are another source of diversity in these proteins.

immunoglobulins, proteins that specifically recognize and help combat viruses, bacteria, and other invaders of the body.

Immunoglobulins are made by cells of the immune system called B lymphocytes, a type of white blood cell. B lymphocytes are highly specialized, with each differentiated cell and its mitotic descendants producing one specific type of antibody that attacks a specific invader. As an unspecialized cell differentiates into a B lymphocyte, segments of antibody genes are pieced together from several DNA regions that are physically separated in the genome of an embryonic cell (FIGURE 19.6). One segment is taken at random from each type of region. As a result, the mature immune system, with millions of subpopulations of B lymphocytes, can make millions of different kinds of antibody molecules.

The basic immunoglobulin molecule is represented at the bottom right of FIGURE 19.6. It consists of four polypeptide chains held together by disulfide bridges. Each chain has two major parts: a constant region *(C)*, which is the same for all antibodies of a particular class, and a variable region *(V)*, which gives a particular antibody its unique function—the ability to recognize and bind to a specific foreign molecule. In the genome of an embryonic cell, the DNA region coding for the constant part of each type of antibody polypeptide is separated by a long stretch of DNA from a location containing hundreds of variable region–coding segments. As a B lymphocyte differentiates, a variable segment of the DNA is connected to a constant segment by the deletion of intervening DNA. This occurs by a type of genetic recombination within the DNA (remember that this is not RNA splicing!). The joined segments form the continuous sequence of nucleotides that functions as the gene for one of the immunoglobulin polypeptides. Thus, much antibody variation arises from different combinations of variable and constant regions in immunoglobulin polypeptides, as well as from different combinations of polypeptides that form the complete antibody molecules.

THE CONTROL OF GENE EXPRESSION

The rearrangement of antibody genes and other processes that alter DNA sequences in somatic cells are bona fide mechanisms for regulating gene expression. They are limited, however, to certain genes in certain cells. We now examine the more general mechanisms of gene control that are used to regulate most eukaryotic genes. As we see next, a major gene regulation challenge in cells of multicellular eukaryotes is to keep genes turned off.

Each cell of a multicellular eukaryote expresses only a small fraction of its genes

Interspersed within their masses of noncoding DNA, eukaryotic genomes may contain tens of thousands of genes. Which of these genes are to be expressed? Like unicellular organisms, the cells of multicellular organisms must continually turn certain genes on and off in response to signals from their external and internal environments. In addition, gene expression must be controlled on a long-term basis for cellular **differentiation**, the divergence in form and function as cells become specialized during an organism's development. Highly specialized cells, such as those of muscle or nervous tissue, express only a tiny fraction of their genes. In fact, a typical human cell expresses only 3–5% of its genes at any given time. The enzymes that transcribe DNA must locate the right genes at the right time, which is like finding a needle in a haystack. When gene expression goes awry, serious imbalances and diseases, including cancer, can arise. Thus, the question of how eukaryotic genes are regulated is paramount for medical research as well as for basic biology.

Only 30 years ago, an understanding of the mechanisms that control gene expression in eukaryotes seemed almost hopelessly out of reach. Since then, new research methods have empowered molecular biologists to begin to solve these once impenetrable mysteries. Equipped with DNA technology for isolating individual genes and sequencing DNA (discussed in Chapter 20), biologists are now uncovering many of the details of eukaryotic gene regulation.

Despite the extra "nuts and bolts" it requires, the control of gene activity in eukaryotes involves some of the same basic principles as prokaryotic gene regulation. In all organisms the expression of specific genes is most commonly regulated at the level of transcription, by DNA-binding proteins that also interact with other proteins and often with external signals. For that reason, the term *gene expression* is often equated with gene activity—that is, transcription—for both prokaryotes and eukaryotes. However, the greater complexity of eukaryotic cell structure and function offer opportunities for controlling gene expression at additional stages.

The control of gene expression can occur at any step in the pathway from gene to functional protein: *an overview*

19.1 FIGURE 19.7 is an overview of the entire process of gene expression in a eukaryotic cell. The figure highlights key stages in the expression of a protein-coding gene, from chromatin changes that unpack the DNA, through transcription, RNA

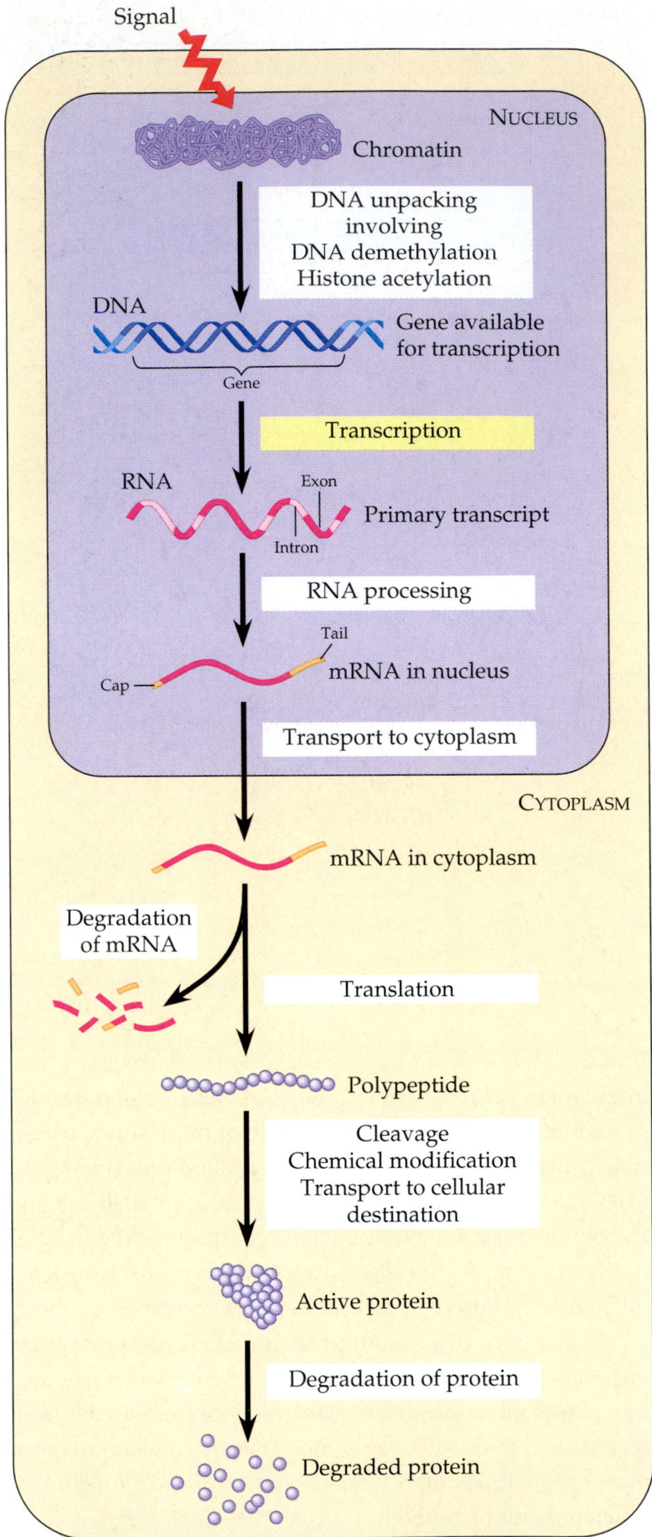

19.1 **FIGURE 19.7 • Opportunities for the control of gene expression in eukaryotic cells.** The nuclear envelope separating transcription from translation in eukaryotic cells offers opportunity for posttranscriptional control in the form of RNA processing that is absent in bacteria. In addition, eukaryotes have a greater variety of control mechanisms operating before transcription and after translation. In this diagram, the processes that offer opportunities for regulation are highlighted by white boxes. As in prokaryotes, transcription initiation is the most important control point.

processing, and translation, to various alterations of the protein product. The expression of a given gene will not necessarily involve every stage shown; for example, not every polypeptide is cleaved. The main lesson is that each stage is a potential control point where gene expression can be turned on or off, speeded up, or slowed down. The figure omits details and does not convey the web of control that connects *different* genes and their products. In the following three sections we'll examine some of the important control points more closely.

Chromatin modifications affect the availability of genes for transcription

The organization of chromatin discussed earlier in the chapter serves a dual purpose. Certainly one function is to pack the DNA into a compact form that fits inside the nucleus of a cell. The other main function is regulatory: The physical state of DNA in or near a gene is important in helping control whether the gene is available for transcription. Thus, the genes of heterochromatin, which is highly condensed, are usually not expressed. Also, a gene's location relative to nucleosomes and to the sites where the DNA attaches to the chromosome scaffold or nuclear lamina can affect whether it is transcribed. A flurry of recent research indicates that chemical modifications of chromatin play key roles in both chromatin structure and the regulation of transcription. Especially important are DNA methylation and histone acetylation. Both are catalyzed by specific enzymes.

DNA Methylation

DNA methylation is the attachment of methyl groups ($—CH_3$) to DNA bases after DNA is synthesized. The DNA of most plants and animals has methylated bases, usually cytosine. About 5% of the cytosine bases in methylated eukaryotic DNA have methyl groups. Inactive DNA, such as that of inactivated mammalian *X* chromosomes, is generally highly methylated compared to DNA that is actively transcribed, although there are exceptions. Comparison of the same genes in different types of tissues shows that the genes are usually more heavily methylated in cells where they are not expressed. In addition, demethylating certain inactive genes (removing their extra methyl groups) turns them on.

At least in some species, DNA methylation seems to be essential for the long-term inactivation of genes that occurs during cellular differentiation in the embryo. In organisms as different as mice and *Arabidopsis* (a plant), deficient DNA methylation—resulting from lack of a methylating enzyme, for example—causes abnormalities in embryonic development. Once methylated, genes usually stay that way through successive cell divisions. Methylation enzymes act at DNA sites where one strand is already methylated and thus correctly methylate the daughter strand after each round of DNA replication. In this way, methylation patterns are passed on, and cells forming specialized tissues keep a chemical record of what occurred during embryonic development. A methylation pattern maintained in this way also accounts for **genomic imprinting** in mammals, where methylation permanently turns off either the maternal or paternal allele of certain genes at the start of development (see Chapter 15).

Histone Acetylation

There is mounting evidence for a direct role for histone acetylation and deacetylation in the regulation of gene transcription. **Histone acetylation** is the attachment of acetyl groups ($—COCH_3$) to certain amino acids of histone proteins; deacetylation is the removal of acetyl groups. When a nucleosome's histones are acetylated, they change shape so that they grip the DNA less tightly. As a result, transcription proteins have easier access to genes in the acetylated region. Researchers have recently shown that some enzymes that acetylate or deacetylate histones are closely associated with or even components of the transcription factors that bind to promoters (see FIGURE 17.7). Thus, histone acetylation and the initiation of gene transcription seem to be coupled structurally as well as functionally. In summary, key enzymes that modify chromatin structure are apparently integral parts of the cell's machinery for regulating transcription.

Transcription initiation is controlled by proteins that interact with DNA and with each other

If chromatin-modifying enzymes provide a coarse adjustment
19.1 of gene expression by making a region of DNA either more or less available for transcription, then fine-tuning begins with the interaction of transcription factors with DNA sequences that control specific genes. Indeed, once a gene is "unpacked," the initiation of transcription is the most important and universally used control point in gene expression. Before looking at control mechanisms, let's review the structure of a eukaryotic gene and its transcript.

Organization of a Typical Eukaryotic Gene

A eukaryotic gene and the DNA elements (segments) that control it are typically organized as shown in FIGURE 19.8, p. 354, which reviews and extends what you learned about eukaryotic genes in Chapter 17. One striking difference between this gene and a prokaryotic gene is the presence of introns, noncoding sequences that are interspersed between coding sequences, the exons. Recall from Chapter 17 that a cluster of proteins called a *transcription initiation complex* assembles on the promoter sequence at the "upstream" end of the gene, and one of the proteins, the RNA polymerase, proceeds to transcribe the gene. The introns are removed from the primary transcript during RNA processing, so that they do not appear in the mature mRNA. The eukaryotic processing of pre-mRNA usually also includes the addition of a modified guanosine triphosphate cap to its 5′ end and a poly(A) tail to its 3′ end. Another distinctive feature of the eukaryotic gene illustrated in FIGURE 19.8 is the relatively large number of control elements associated with it. **Control elements** are simply segments of noncoding DNA that help regulate transcription of a gene by binding proteins—*transcription factors.* (Strictly speaking, control elements include promoters.)

The Roles of Transcription Factors

As you learned in Chapter 17, eukaryotic RNA polymerase alone cannot initiate transcription of a gene; it is dependent on transcription factors. The transcription factors mentioned in Chapter 17 are essential for the transcription of *all* protein-coding genes (see FIGURE 17.7). Recall that only one of these transcription factors independently recognizes a DNA sequence, the TATA box within the promoter; the other transcription factors primarily recognize proteins, including each other and RNA polymerase. Protein-protein interactions are crucial to the initiation of eukaryotic transcription. Only when the complete initiation complex has assembled can the polymerase begin to move along the DNA template strand, producing a complementary strand of RNA.

However, the interaction of transcription factors and RNA polymerase with the promoter as summarized so far usually initiates transcription inefficiently, producing only a low rate of initiation and few RNA transcripts. The keys to high levels of eukaryotic transcription are the control elements. These DNA sequences greatly improve the efficiency of promoters by binding additional transcription factors.

As you can see in FIGURE 19.8, some of the control elements on the DNA are close to the promoter; in fact, some biologists include these *proximal control elements* in what they call the promoter. The more distant *distal control elements,* called **enhancers,** may be thousands of nucleotides away from the promoter and may even be downstream of the gene. The associations between transcription factors and enhancers play an important role in the control of gene expression in eukaryotes. How do segments of DNA so far away from the promoter influence transcription? Apparently, bending of the DNA enables the transcription factors bound to enhancers to contact proteins of the transcription initiation complex at the promoter (FIGURE 19.9). A transcription factor that binds to an enhancer and stimulates transcription of a gene is termed an **activator.** Activators help position the initiation complex on the promoter.

Do eukaryotic cells also have transcription factors that function as *repressors,* as bacteria do? Yes, there is evidence for eukaryotic repressors. Some of them bind selectively to DNA

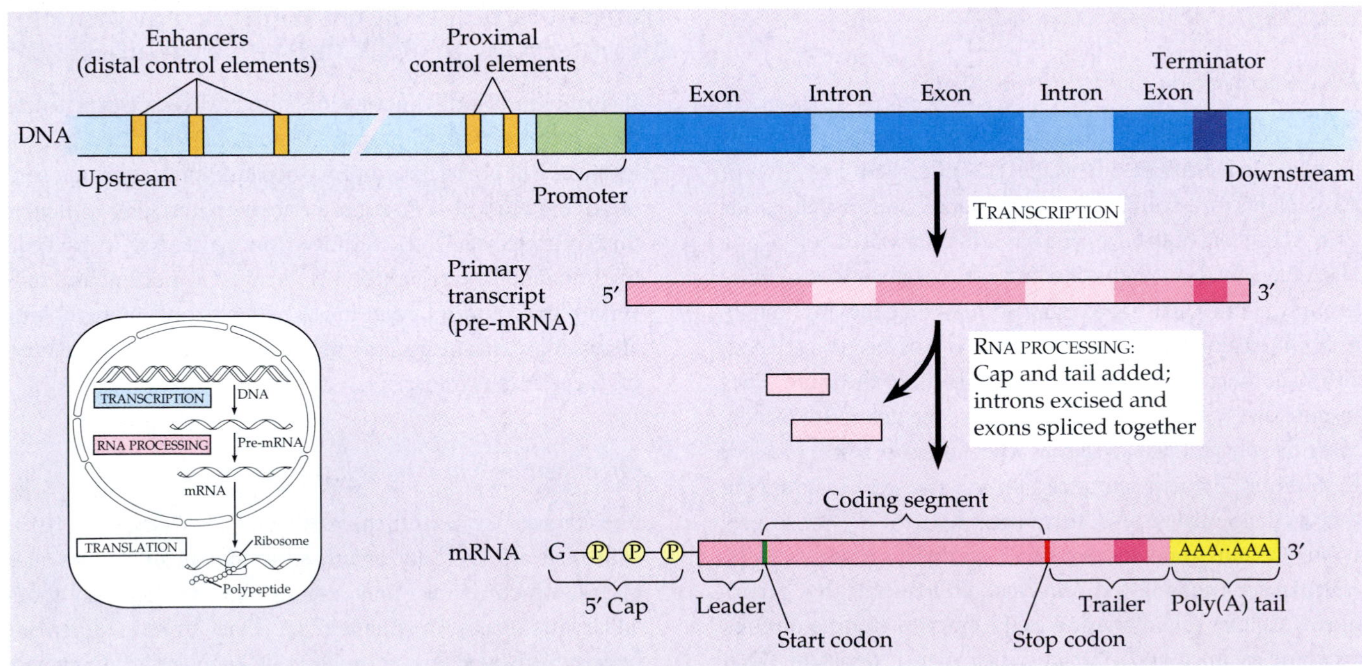

FIGURE 19.8 • A eukaryotic gene and its
19.1 **transcript.** Each functional eukaryotic gene has a promoter, a DNA sequence where RNA polymerase binds (with the help of transcription factors) and starts transcription, proceeding "downstream." The 3′ end of the transcript is determined by an RNA sequence encoded by a terminator located near the end of the last exon. RNA-processing enzymes immediately add a 5′ cap and then a poly(A) tail to the primary transcript, and a spliceosome excises the introns. The resulting mRNA, ready for export to the cytoplasm, includes leader and trailer regions that will not be translated. Involved in regulating the initiation of transcription are a number of control elements (orange), DNA sequences located near (*proximal* to) or far from (*distal* to) the promoter. Distal control elements called enhancers can lie either upstream or downstream of the gene.

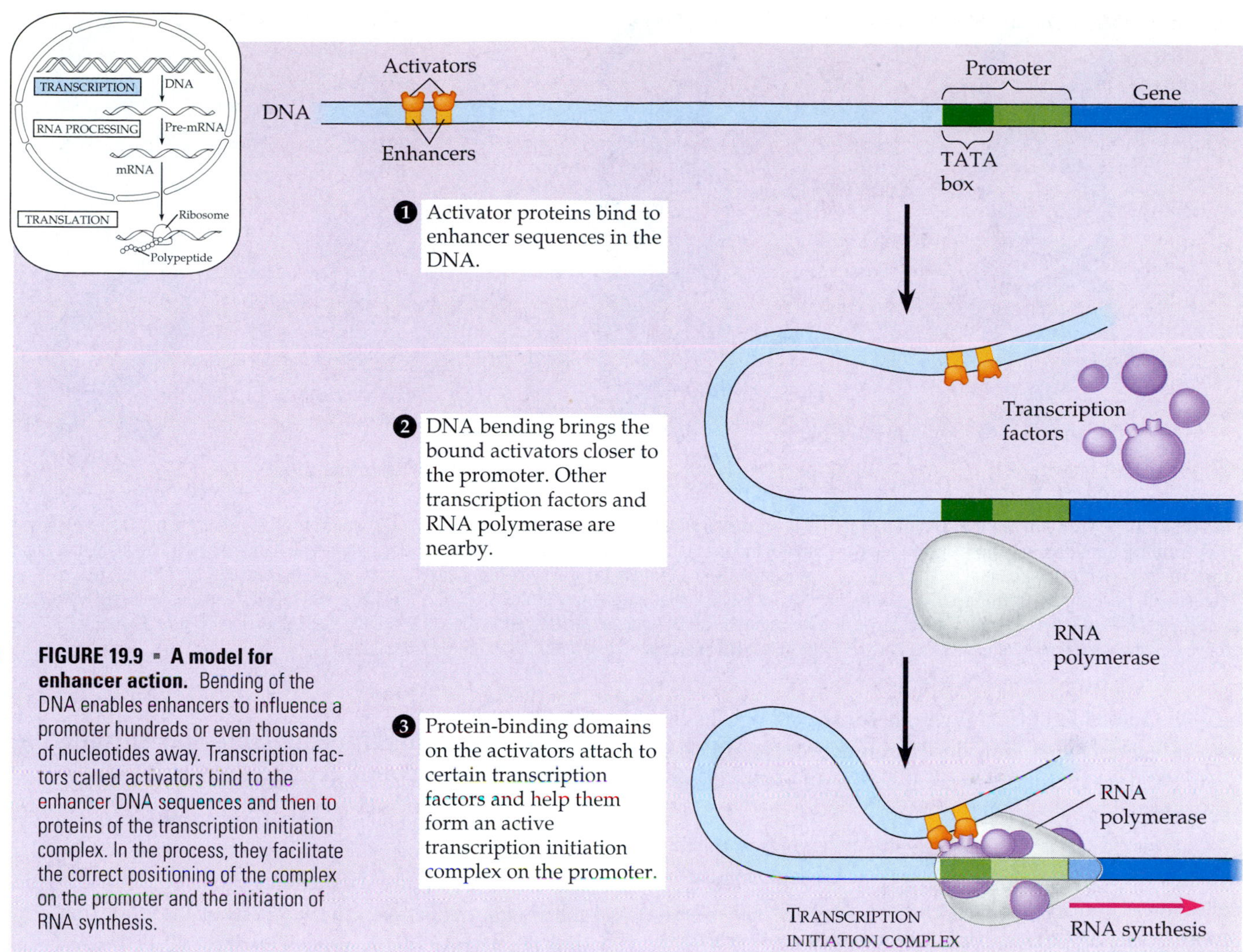
TRANSCRIPTION
DNA
RNA PROCESSING
Pre-mRNA
mRNA
TRANSLATION
Ribosome
Polypeptide
Activators
DNA
Enhancers
Promoter
Gene
TATA box
1 Activator proteins bind to enhancer sequences in the DNA.
Transcription factors
2 DNA bending brings the bound activators closer to the promoter. Other transcription factors and RNA polymerase are nearby.
RNA polymerase
3 Protein-binding domains on the activators attach to certain transcription factors and help them form an active transcription initiation complex on the promoter.
RNA polymerase
TRANSCRIPTION INITIATION COMPLEX
RNA synthesis

19.1 **FIGURE 19.9 ▪ A model for enhancer action.** Bending of the DNA enables enhancers to influence a promoter hundreds or even thousands of nucleotides away. Transcription factors called activators bind to the enhancer DNA sequences and then to proteins of the transcription initiation complex. In the process, they facilitate the correct positioning of the complex on the promoter and the initiation of RNA synthesis.

control elements called *silencers*, which may be analogous to enhancers. Activators, though, are probably more important than such repressors, because the main regulatory mode in eukaryotic cells seems to be transcriptional activation of otherwise silent genes. The mechanisms responsible for eukaryotic gene silencing—that is, repression—may operate mostly at the level of chromatin modification (for example, DNA methylation).

In any case, the direct control of transcription largely depends on regulatory proteins that bind selectively to DNA and to other proteins. Hundreds of transcription factors have been discovered in eukaryotes. As numerous as these proteins are, however, they follow a few basic structural principles. A transcription factor generally has a **DNA-binding domain**, a part of its three-dimensional structure that binds to DNA. There are a few types of these domains (FIGURE 19.10, p. 356). In addition, each transcription factor has a protein-binding domain that recognizes another transcription factor.

Considering the challenge of regulating the tens of thousands of genes in a typical animal or plant cell, the number of completely different nucleotide sequences found in DNA control elements is surprisingly small. Members of a dozen or so sequences about 4–10 base pairs long appear again and again in the control elements for different genes. For many genes the particular *combination* of control elements associated with the gene may be more important than the presence of a control element unique to the gene.

Coordinately Controlled Genes

How does the eukaryotic cell deal with genes of related function that need to be turned on or off at the same time? In Chapter 18 you learned that in prokaryotes, such coordinately controlled genes are often clustered into an operon; they lie adjacent to each other in the DNA molecule and share a promoter and other control elements located at the upstream end of the cluster. The genes of the operon are transcribed sequentially into a single mRNA molecule and are translated together. Such operons have not been found in eukaryotic cells. Genes coding for the enzymes of a metabolic pathway, for example, are often scattered over different chromosomes in a eukaryotic genome. Even when genes for related functions are located

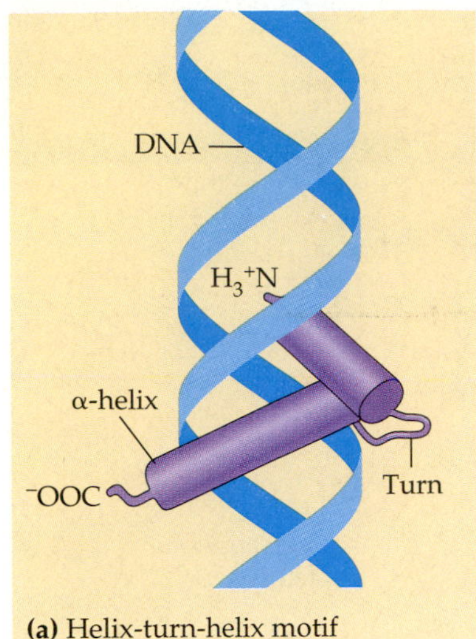

(a) Helix-turn-helix motif

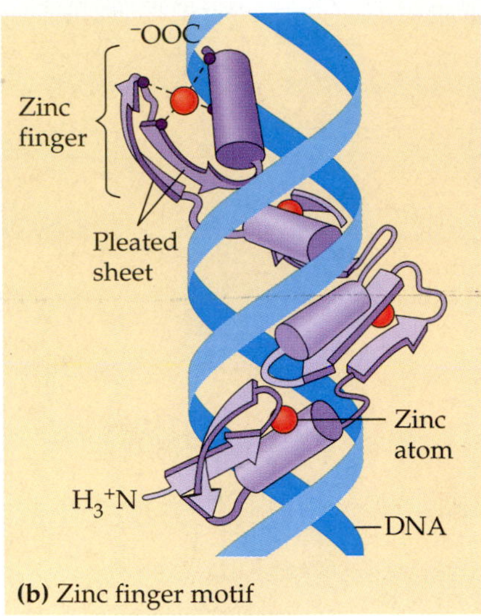

(b) Zinc finger motif

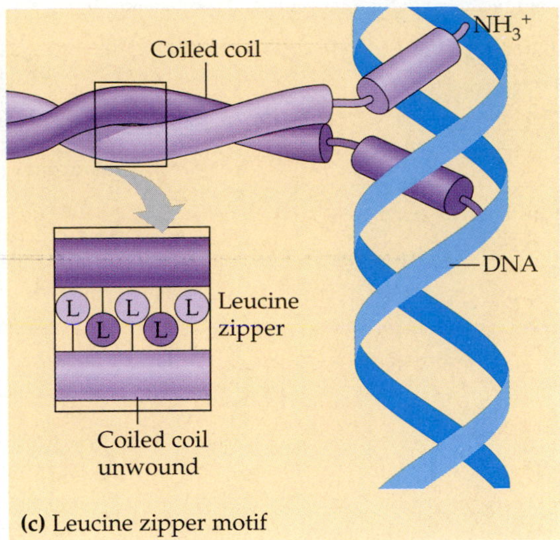

(c) Leucine zipper motif

FIGURE 19.10 ▪ Three of the major types of DNA-binding domains in transcription factors. The parts of these structural motifs that interact with DNA are mostly α helices (shown as cylinders) that fit into the larger ("major") groove of the double helix. Variations in amino acid sequence within these helices are responsible for the recognition of specific DNA sequences by the protein. **(a)** The helix-turn-helix motif is found in many regulatory proteins, including the prokaryotic *lac* and *trp* repressor proteins. **(b)** In the zinc finger motif, each "finger" consists of an α helix and a β sheet (shown as ribbons) held together with the help of a zinc atom. **(c)** In the leucine zipper motif, two α helices with regularly spaced leucines wrap around each other, joining two polypeptides together.

near one another on the same chromosome, each gene has its own promoter and is individually transcribed. Nevertheless, scattered collections of eukaryotic genes are often coordinately expressed.

Coordinate gene expression in eukaryotes probably depends on the association of a specific control element or collection of control elements with every gene of a dispersed group. Copies of the transcription factors that recognize these control elements bind to them, promoting simultaneous transcription of the genes. One example of such coordinate control in eukaryotes is the activation of a variety of genes by a steroid hormone. Steroid hormones—sex hormones, for example—have multiple effects on the body. As we'll discuss further in Chapter 45, a steroid hormone functions as a chemical signal by entering cells and binding to a specific receptor protein in the cytoplasm or nucleus. When activated by steroid binding, a steroid receptor functions as a transcription activator. Every gene to be turned on by that hormone has a control element recognized by that activator.

Most other kinds of signal molecules, such as nonsteroid hormones or the extracellular signals exchanged by cells in a developing embryo, bind to receptors on the receiving cell's surface and never actually enter the cell. But they can control gene expression indirectly by triggering signal-transduction pathways that lead to the activation of particular transcription factors (see Chapter 11). The principle of coordinate regulation remains the same: Genes with the same control elements are activated by the same chemical signals. Systems for coordinating gene regulation probably evolved by the duplication and distribution of control elements within the genome.

Posttranscriptional mechanisms play supporting roles in the control of gene expression

19.1 Transcription alone does not constitute gene expression. The expression of a protein-coding gene is ultimately measured in terms of the amount of functional protein a cell makes, and much happens between the synthesis of the RNA transcript and the activity of the protein in the cell. In theory, gene expression may be blocked or stimulated at any posttranscriptional step (see FIGURE 19.7). By using regulatory mechanisms that operate after transcription, a cell can rapidly fine-tune gene expression in response to environmental changes, without altering its transcription patterns.

RNA processing in the nucleus and the export of mature RNA to the cytoplasm provide several opportunities for controlling gene expression that are not available in bacteria. One example of regulation at the RNA-processing level is **alternative splicing**, in which different mRNA molecules are produced from the same primary transcript depending on which RNA segments are treated as exons and which as introns. Regulatory proteins specific to a cell type control intron-exon choices by binding to regulatory sequences within the primary transcript.

After RNA processing, other stages of gene expression that the cell may regulate are mRNA degradation, translation initiation, and protein processing and degradation.

Regulation of mRNA Degradation

The lifespan of mRNA molecules in the cytoplasm is an important factor in determining the pattern of protein synthesis in a cell. Prokaryotic mRNA molecules typically have very short lives; they are degraded by enzymes after only a few minutes. This is one reason bacteria can vary their patterns of protein synthesis so quickly in response to environmental changes. In contrast, mRNA lifetimes in multicellular eukaryotes are typically hours and can be days or even weeks. A striking example of long-lived mRNA is found in developing red blood cells, which are factories for the production of the protein hemoglobin. The mRNAs for the hemoglobin polypeptides (α-globin and β-globin) are unusually stable and are translated repeatedly in these cells.

Recent research on yeast suggests that a common pathway of mRNA breakdown begins with the enzymatic shortening of the poly(A) tail. This helps trigger the action of enzymes that remove the 5′ cap (the two ends of the mRNA may be briefly held together by the proteins involved). The removal of the cap, a critical step, is also regulated by particular nucleotide sequences in the mRNA. Once the cap is removed, nuclease enzymes rapidly chew up the mRNA from its 5′ end.

Nucleotide sequences that affect mRNA stability are often found in the untranslated trailer region at the 3′ end of the molecule (see FIGURE 19.8). In one experiment, researchers transferred such a sequence from the short-lived mRNA for a growth factor to the 3′ end of a normally stable globin mRNA. The globin mRNA was quickly degraded.

Control of Translation

Most translational control mechanisms block the initiation stage of polypeptide synthesis, when ribosomal subunits and the initiator tRNA attach to an mRNA (see FIGURE 17.15). Translation of specific mRNAs can be blocked by regulatory proteins that bind to specific sequences or structures within the leader region at the 5′ end of the mRNA, preventing the attachment of ribosomes.

Protein factors required to initiate translation in eukaryotes offer targets for simultaneously controlling the translation of *all* the mRNA in a cell. One example of such "global" control involves hemoglobin. A functional hemoglobin molecule has four heme groups, one attached to each polypeptide (see FIGURE 19.3). If hemes are in short supply in a developing red blood cell, a regulatory protein inactivates an essential translation initiation factor by phosphorylating it. This inhibits all translation, but mainly affects translation of hemoglobin mRNA, which constitutes most of the mRNA in the cell.

Global control of translation is important in embryonic development. For example, the egg cells of many organisms synthesize and store large numbers of mRNA molecules that are not translated until just after fertilization. At that point, translation is triggered by the sudden activation of translation initiation factors. The response is a burst of synthesis of particular proteins. Some plants and algae store mRNAs during periods of darkness; light then triggers the reactivation of the translational apparatus.

Protein Processing and Degradation

The final opportunities for controlling gene expression occur after translation. Often, eukaryotic polypeptides must be processed to yield functional protein molecules. The post-translational cleavage of the hormone insulin is an example (see p. 310). In addition, many proteins require chemical modifications to function. For instance, animal cell surface proteins acquire sugars, and regulatory proteins are commonly activated or inactivated by the reversible addition of phosphate groups. Polypeptides must often be transported to targeted destinations in the cell in order to function. Regulation might occur at any of the steps involved in modifying or transporting a protein.

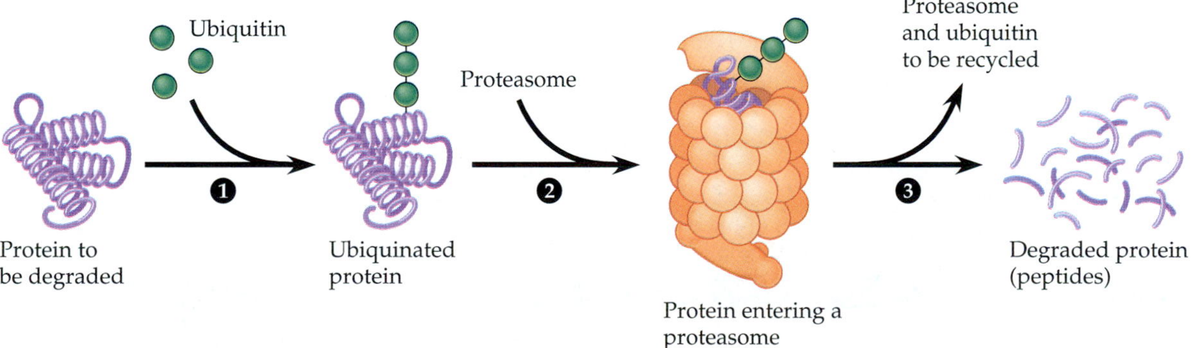

FIGURE 19.11 ▪ Degradation of a protein by
19.1 **a proteasome.** A proteasome is an enormous protein complex, with a shape suggesting a trash can, that chops up unneeded proteins in the cell. In most cases the proteins attacked by a proteasome have been tagged with short chains of ubiquitin, a small protein. ① Enzymes in the cytosol attach ubiquitin molecules to a protein. (It is not known how the protein is selected.) ② A proteasome recognizes the ubiquitinated protein, unfolds it, and sequesters it in its central cavity. ③ Enzymatic components of the proteasome cut the protein into small peptides, which can be further degraded by cytosolic enzymes. Steps 1 and 3 require ATP. Eukaryotic proteasomes are as massive as ribosomal subunits and are distributed throughout the cell. Their barrel-like shape somewhat resembles that of chaperone proteins, which *nurture* protein structure rather than destroy it (see Chapter 5).

Abnormal targeting of a protein can have serious consequences, as, for example, in cystic fibrosis (described in Chapter 14). This disease results from mutations in the gene for a protein that functions as a chloride-ion channel. The defective protein never reaches its final destination in the cell, the plasma membrane, and is rapidly degraded.

In addition to degrading defective and damaged proteins, the cell limits the lifetimes of normal proteins by selective degradation. Many proteins, such as the cyclins involved in regulating the cell cycle, must be relatively short-lived if the cell is to function appropriately (see Chapter 12). To mark a particular protein for destruction, the cell commonly attaches molecules of a small protein called ubiquitin to the protein. Giant protein complexes called **proteasomes** then recognize the ubiquitin and degrade the tagged protein (FIGURE 19.11, p. 357). The importance of proteasomes is underscored by the finding that mutations making cell-cycle proteins impervious to proteasome degradation can lead to cancer.

THE MOLECULAR BIOLOGY OF CANCER

In Chapter 12 we considered cancer as a set of diseases in which cells escape from the control mechanisms normally limiting their growth. Now that we have discussed the molecular basis of gene expression and its regulation, we are ready to look at cancer more closely.

Cancer results from genetic changes that affect the cell cycle

Certain genes normally regulate cell growth and division—the cell cycle—and mutations that alter those genes in somatic cells can lead to cancer. The agent of such change can be random spontaneous mutation. However, it is likely that many cancer-causing mutations result from environmental influences such as chemical carcinogens, physical mutagens such as X-rays, or certain viruses. In fact, a breakthrough in understanding cancer came from the study of tumors induced by viruses. This research led to the discovery of cancer-causing genes called **oncogenes** in certain retroviruses (*onco-* comes from the Greek for "tumor"). Subsequently, close counterparts of these oncogenes were found in the genomes of humans and other animals. The normal cellular genes, called **proto-oncogenes**, code for proteins that stimulate normal cell growth and division. (To review the cell cycle, see Chapter 12.)

How might a proto-oncogene—a gene that has an essential function in normal cells—become an oncogene, a cancer-causing gene? In general, an oncogene arises from a genetic change that leads to an increase in either the amount of the proto-oncogene's protein product or the intrinsic activity of each protein molecule. The genetic changes that convert proto-oncogenes to oncogenes fall into three main categories: movement of DNA within the genome, amplification of a proto-oncogene, and point mutation in a proto-oncogene (FIGURE 19.12). Malignant cells are frequently found to contain chromosomes that have broken and rejoined incorrectly, translocating fragments from one chromosome to another (see FIGURE 15.12). A proto-oncogene ending up in the joint region may now lie adjacent to an especially active promoter (or other control element) that increases transcription of the gene, making it an oncogene. An increase in gene expression can also arise when a proto-oncogene comes under the control of a more active promoter by transposition of either the gene or the promoter within a chromosome. The second main type of genetic change, amplification, increases the number of copies of the gene in the cell. The third possibility is a point mutation that changes the gene's protein product to one that is more active or more resistant to degradation than the normal protein. All these mechanisms can lead to abnor-

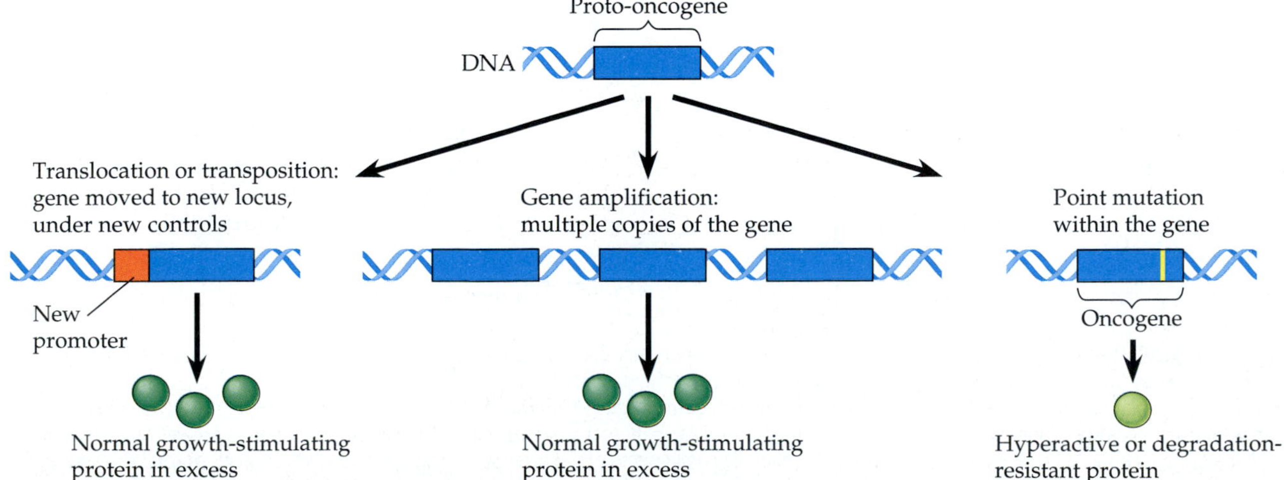

FIGURE 19.12 • Genetic changes that can turn proto-oncogenes into oncogenes.

mal stimulation of the cell cycle and put the cell on the path to malignancy.

In addition to mutations affecting growth-stimulating proteins, changes in genes whose normal products *inhibit* cell division also contribute to cancer. Such genes are called **tumor-suppressor genes** because the proteins they encode normally help prevent uncontrolled cell growth. Any mutation that decreases the normal activity of a tumor-suppressor protein may contribute to the onset of cancer, in effect stimulating growth through the absence of suppression. The protein products of tumor-suppressor genes have various functions. For instance, some tumor-suppressor proteins normally repair damaged DNA, a function that prevents the cell from accumulating cancer-causing mutations. Other tumor-suppressor proteins control the adhesion of cells to each other or to an extracellular matrix; proper cell anchorage is crucial in normal tissues—and often absent in cancers. Still other tumor-suppressor proteins are components of cell-signaling pathways that inhibit the cell cycle.

We'll encounter examples of some important tumor-suppressor proteins and oncogene proteins as we examine their roles in cell-signaling pathways.

Oncogene proteins and faulty tumor-suppressor proteins interfere with normal signaling pathways

Let's take a closer look at what the protein products of cancer genes do—or fail to do—in the cell. We will focus on the products of two key genes, the *ras* proto-oncogene and the *p53* tumor-suppressor gene. Mutations in these genes are very common in human cancers: *ras* is mutated in about 30% of human cancers; for *p53* the frequency is close to 50%.

Both the Ras protein and the p53 protein are components of signal-transduction pathways that convey external signals to the DNA in the cell's nucleus. While learning about cell signaling in Chapter 11 you saw a generalized version of such a pathway in FIGURE 11.16. Now, in FIGURE 19.13a on p. 360 you can see that Ras, the product of the ***ras* gene**, is a G protein that relays a growth signal from a growth-factor receptor on the plasma membrane to a cascade of protein kinases. The cellular response at the end of the pathway is the synthesis of a protein that stimulates the cell cycle. Normally, such a pathway will not operate unless triggered by the appropriate growth factor. However, an oncogene protein that is a hyperactive version of a protein in the pathway can increase cell division even in the absence of growth factor. Many *ras* oncogenes have a point mutation that leads to a hyperactive version of the Ras protein, one that issues signals on its own. In fact, hyperactive versions or excess amounts of any of the pathway's components can have the same outcome: excessive cell division.

FIGURE 19.13b shows a comparable growth-*inhibiting* pathway, in which a growth-inhibiting signal leads to the synthesis of a protein that suppresses the cell cycle. Thus, the genes for the components of the pathway act as tumor-suppressor genes (as do genes whose products interfere with the growth-stimulating pathway). The tumor-suppressor protein encoded by the wild-type *p53* gene is a transcription factor that promotes the synthesis of growth-inhibiting proteins. That is why a mutation knocking out the *p53* gene can lead to excessive cell growth and cancer.

The modestly titled ***p53* gene**, named for the 53,000-dalton molecular weight of its protein product, is often called the "guardian angel of the genome." Damage to the cell's DNA (by high-energy radiation or toxic chemicals, for example) acts as a signal that leads to the expression of the *p53* gene. Once made, the p53 protein functions as a transcription factor for several genes. Often it activates a gene called *p21*, whose product halts the cell cycle by binding to cyclin-dependent kinases, allowing time for the cell to repair the DNA; p53 can also turn on genes directly involved in DNA repair. When DNA damage is irreparable, p53 activates "suicide" genes, whose protein products cause cell death by a process called *apoptosis* (see Chapter 21). Thus, in at least three ways, p53 prevents a cell from passing on mutations due to DNA damage. If mutations do accumulate and the cell survives through many divisions—as is more likely if the *p53* tumor-suppressor gene is defective or missing—cancer may ensue.

Multiple mutations underlie the development of cancer

More than one somatic mutation is generally needed to produce all the changes characteristic of a full-fledged cancer cell. This may help explain why the incidence of cancer increases greatly with age. If cancer results from an accumulation of mutations and if mutations occur throughout life, then the longer we live, the more likely we are to develop cancer.

The model of a multistep path to cancer is well supported by studies of one of the best understood types of human cancer, colorectal cancer. About 135,000 new cases of colorectal cancer are diagnosed each year in the United States. Like most cancers, colorectal cancer develops gradually (FIGURE 19.14, p. 361). The first sign is often a polyp, which is a small, benign growth in the colon lining. The cells of the polyp look normal, although they divide unusually frequently. The tumor grows and may eventually become malignant. The development of a malignant tumor is paralleled by a gradual accumulation of mutations that activate oncogenes and knock out tumor-suppressor genes. A *ras* oncogene and a mutated *p53* tumor-suppressor gene are often involved.

About a half dozen changes must occur at the DNA level for a cell to become fully cancerous. These usually include the appearance of at least one active oncogene and the mutation or loss of several tumor-suppressor genes. Furthermore, since mutant tumor-suppressor alleles are usually recessive,

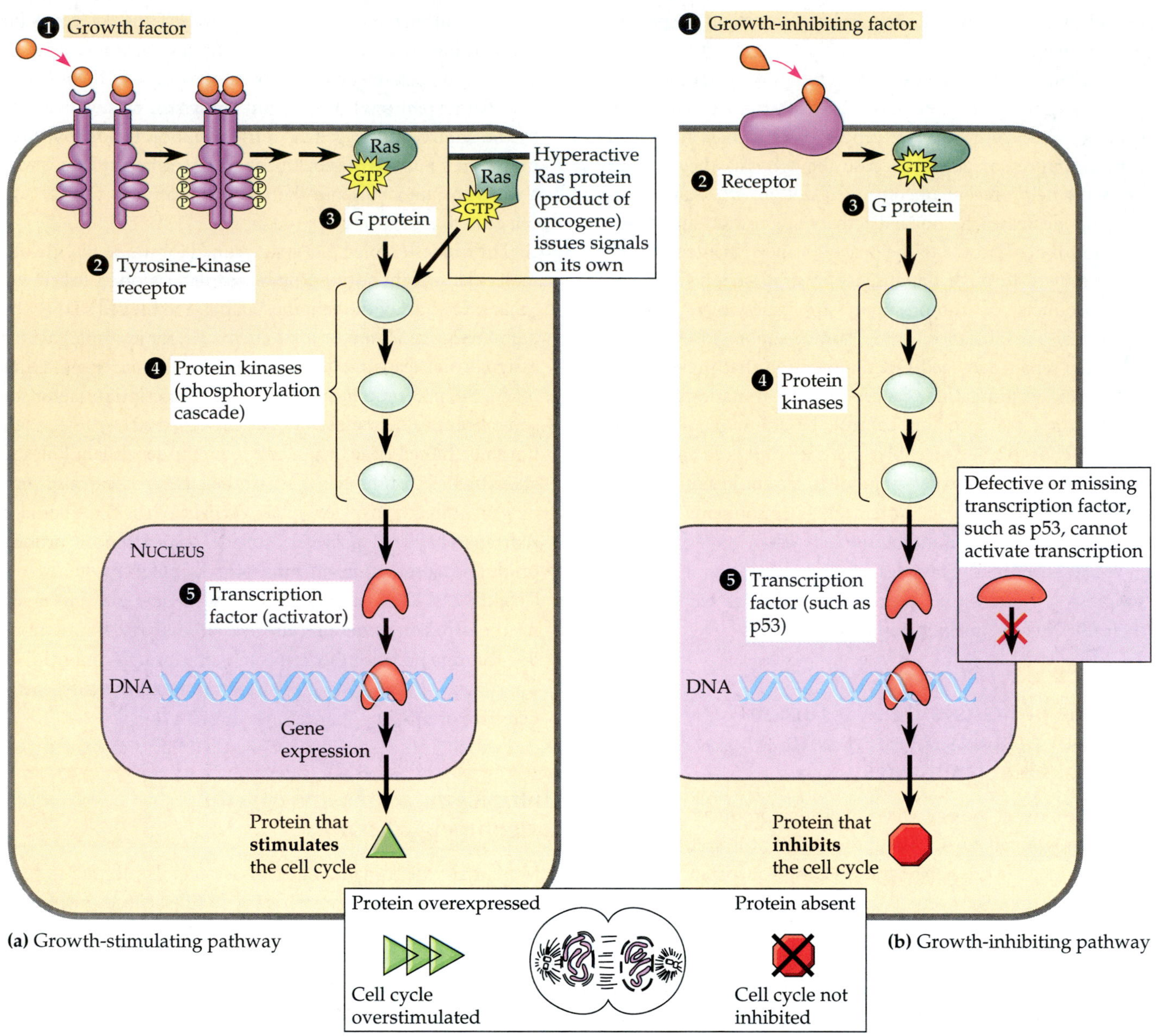

FIGURE 19.13 ▪ Signaling pathways that regulate cell growth. Both stimulatory and inhibitory pathways regulate the cell cycle, commonly by influencing transcription. Cancer can result from aberrations in such pathways. **(a)** This typical stimulatory pathway is triggered by ① a growth factor that binds to ② a tyrosine-kinase receptor (see FIGURE 11.8). The signal is relayed to ③ a G protein called Ras. (Like all G proteins, Ras is active when GTP is bound to it.) Ras passes the signal to ④ a series of protein kinases. The last kinase activates ⑤ a transcription factor that turns on genes for one or more proteins that stimulate the cell cycle. If a mutation makes Ras—or any other component of the pathway—abnormally active, excessive cell division and cancer may result. **(b)** This inhibitory pathway leads to the *suppression* of cell division. In this case, mutations causing *deficiencies* in any pathway component can lead to cancer. The p53 protein is an example of a transcription factor that participates in an inhibitory pathway. The signal that puts p53 into action is usually DNA damage (normally ensuring that damaged DNA does not replicate). **(c)** Whether the cell cycle is overstimulated or not inhibited when it normally would be, the result is the same: increased cell division.

mutations must knock out *both* alleles in a cell's genome to block tumor suppression. (Most oncogenes, on the other hand, behave as dominant alleles.) Finally, in many malignant tumors the gene for telomerase is activated. This enzyme prevents the erosion of the ends of the chromosomes, thus removing a natural limit on the number of times the cells can divide (see Chapter 16).

Viruses seem to play a role in about 15% of human cancer cases worldwide. For example, retroviruses are involved in some types of leukemia, hepatitis viruses can cause liver cancer, and certain wart viruses promote the development of cancer of the cervix. Viruses contribute to cancer development by integrating their genetic material into the DNA of infected cells. By this process, a retrovirus may donate an oncogene to the cell. Alternatively, viral DNA may insert into a cellular genome in a way that disrupts a tumor-suppressor gene or converts a proto-oncogene to an oncogene.

The fact that multiple genetic changes are required to produce a cancer cell helps explain the predispositions to cancer that run in some families. An individual inheriting an oncogene or a mutant allele of a tumor-suppressor gene will be one step closer to accumulating the necessary mutations for cancer to develop.

Geneticists are now devoting much effort to finding inherited cancer alleles, so that predisposition to certain cancers can be detected early in life. About 15% of colorectal cancers, for example, involve inherited mutations. Some of these affect DNA-repair genes. Many others affect a tumor-suppressor gene called *APC* (see FIGURE 19.14). This gene seems to have multiple functions in the cell, including regulation of cell migration and adhesion. In 1997, researchers discovered a previously unrecognized mutation in *APC* carried by 6% of Ashkenazic Jews, making it the most common cancer-predisposing mutation yet found in a defined ethnic group.

Inherited alleles involved in breast cancer are also being scrutinized. Understanding the genetic basis of breast cancer is particularly important because it is the second most common type of cancer in the United States, striking over 180,000 women (and some men) each year. In 5–10% of breast cancer cases, there is evidence of a strong inherited predisposition. In 1994 and 1995, researchers identified two genes involved in these breast cancers, *BRCA1* and *BRCA2* (*BRCA* stands for BReast CAncer). Mutations in either gene increase the risk of developing breast cancer, and of ovarian cancer as well, as Mary-Claire King mentions in the interview opening this unit. Both genes are considered tumor-suppressor genes because their wild-type alleles protect against breast cancer and because their mutant alleles are recessive. However, determining what the normal products of *BRCA1* and *BRCA2* actually do in the cell has proven very difficult. One hypothesis is that one or both proteins are involved in DNA repair. Whatever the answer, the study of these and other genes associated with inherited cancer may lead to new methods for early diagnosis and treatment of all cancers. The DNA technologies you will learn about in the next chapter figure prominently in this research.

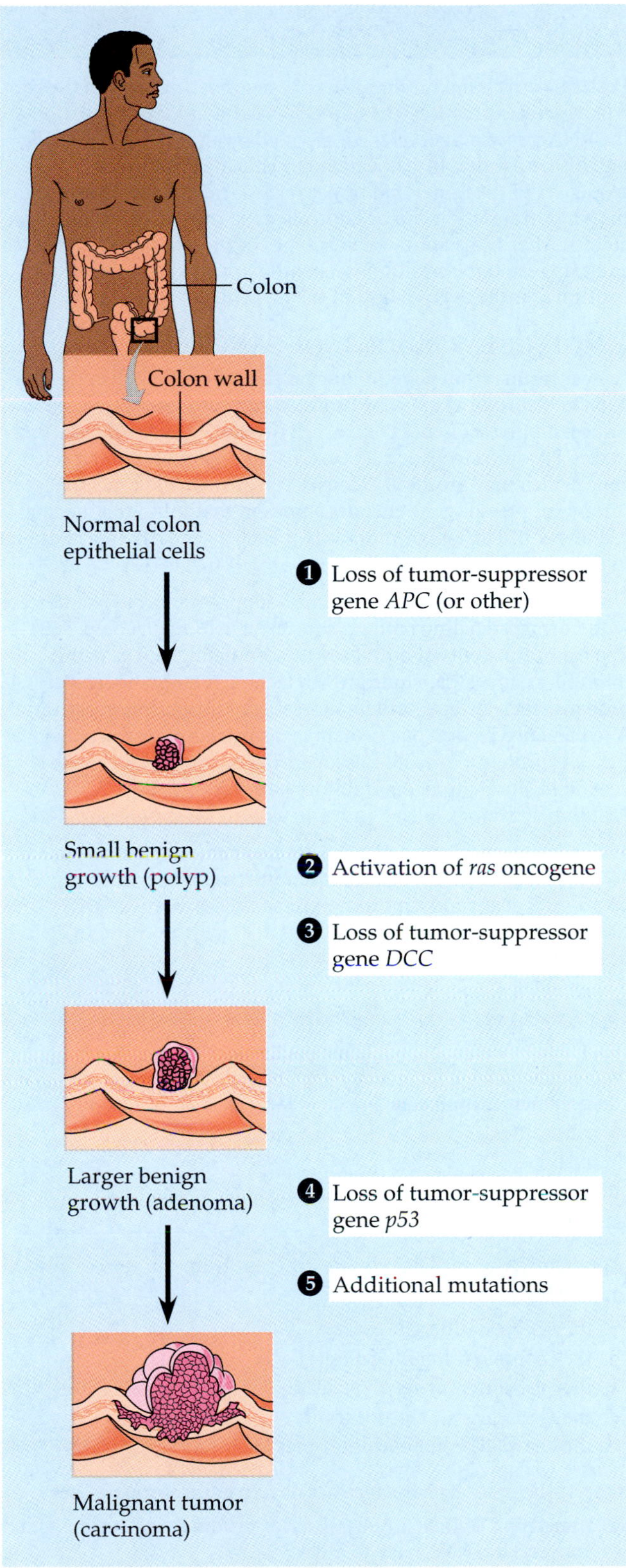

FIGURE 19.14 ▪ A multi-step model for the development of colorectal cancer. Affecting the colon and/or rectum, this type of cancer is one of the best understood. Changes in a tumor parallel a series of genetic changes, including mutations affecting several tumor-suppressor genes (such as *p53*) and the *ras* proto-oncogene. Mutations of tumor-suppressor genes often entail loss (deletion) of the gene. *APC* stands for "adenomatous polyposis coli"; mutant *APC* alleles are carried in many families predisposed to colorectal cancer. *DCC* stands for "deleted in colorectal cancer." Other mutation sequences can also lead to cancer.

CHAPTER REVIEW

REVIEW OF KEY CONCEPTS

(with page numbers and key figures)

THE STRUCTURE OF CHROMATIN

- **Chromatin structure is based on successive levels of DNA packing (pp. 344–346, FIGURE 19.1)** Eukaryotic chromatin is composed mostly of DNA and histone proteins that bind to the DNA to form nucleosomes, the most basic units of DNA packing. Additional folding leads ultimately to highly compacted heterochromatin, the form of chromatin in a metaphase chromosome. In interphase cells, most chromatin is in a highly extended form, euchromatin.

GENOME ORGANIZATION AT THE DNA LEVEL

- **Repetitive DNA and other noncoding sequences account for much of a eukaryotic genome (pp. 346–348; TABLE 19.1)** Sequences of up to 10 nucleotide pairs tandemly repeated thousands of times are especially prominent in the DNA of centromeres and telomeres, where they probably play structural roles in the chromosome. The units of interspersed repetitive DNA are typically longer, and most may be transposons or retrotransposons.
- **Gene families have evolved by duplication of ancestral genes (pp. 348–349; FIGURE 19.3)** The ribosomal RNA genes, a tandemly arranged family of identical genes, enable the cell to make millions of ribosomes quickly. The nonidentical genes in the two globin gene families encode polypeptides used at different developmental stages of the animal.
- **Gene amplification, loss, or rearrangement can alter a cell's genome (pp. 349–351; FIGURES 19.5, 19.6)** The selective replication of certain genes, such as rRNA genes, occurs during development in some species, further bolstering the cells' ability to make rRNA in large amounts. The cells of some species selectively lose entire chromosomes or parts of chromosomes in certain cells. Rearrangements of DNA in a somatic cell by transposons or retrotransposons can alter the cell's genome in ways that affect gene expression. In vertebrates, a rearrangement and selective deletion of DNA segments in differentiating B lymphocytes accounts for antibody diversity.

THE CONTROL OF GENE EXPRESSION

- **Each cell of a multicellular eukaryote expresses only a small fraction of its genes (p. 352)** In particular, selective control of genes is required for cellular differentiation.
- 19.1 **The control of gene expression can occur at any step in the pathway from gene to functional protein: *an overview* (pp. 352–353, FIGURE 19.7)** Key stages include chromatin changes, initiation of transcription, RNA processing, translation, modification of the protein, and degradation of mRNA and protein.
- **Chromatin modifications affect the availability of genes for transcription (p. 353)** DNA methylation seems to diminish transcription of that DNA. Histone acetylation seems to loosen nucleosome structure and thereby enhance transcription.
- 19.1 **Transcription initiation is controlled by proteins that interact with DNA and with each other (pp. 353–356, FIGURES 19.8, 19.9)** Multiple DNA control elements that bind specific transcription factors regulate the formation of a transcription initiation complex at a eukaryotic promoter. But protein-protein interactions are as important as protein-DNA binding in regulating eukaryotic genes. Bending of DNA enables transcription factors bound to control elements (enhancers) distant from the promoter to contact proteins at the promoter. Unlike the genes of a prokaryotic operon, coordinately controlled eukaryotic genes each have a promoter and control elements; the same regulatory sequences are common to all the genes of a group, enabling recognition by the same transcription factors.
- 19.1 **Posttranscriptional mechanisms play supporting roles in the control of gene expression (pp. 356–358, FIGURE 19.11)** Regulation at the RNA-processing level is exemplified by alternative RNA splicing. Also, each mRNA has a characteristic lifespan, which sequences in the leader and trailer regions help determine. The initiation of translation can be controlled via regulation of initiation factors. After translation, various types of protein processing (such as cleavage and the addition of chemical groups) are subject to control, as is the degradation of the protein.

THE MOLECULAR BIOLOGY OF CANCER

- **Cancer results from genetic changes that affect the cell cycle (pp. 358–359, FIGURE 19.12)** The products of proto-oncogenes and tumor-suppressor genes control cell division. A DNA change that makes a proto-oncogene excessively active converts it to an oncogene, which may promote excessive cell growth and cancer. A tumor-suppressor gene encodes a protein that inhibits abnormal cell division. The loss or mutation of such a gene has effects similar to the activation of an oncogene.
- **Oncogene proteins and faulty tumor-suppressor proteins interfere with normal signaling pathways (p. 359, FIGURE 19.13 on p. 360)** Typical components of both growth-stimulating and growth-inhibiting pathways include growth factors, membrane receptors, G proteins (such as Ras), protein kinases, and transcription activators. A hyperactive version of a protein in a stimulatory pathway, such as Ras (a G protein), functions as an oncogene protein. A defective version of a protein in an inhibitory pathway, such as p53 (a transcription activator), fails to function as a tumor suppressor.
- **Multiple mutations underlie the development of cancer (pp. 359–361, FIGURE 19.14)** Tumor cells accumulate changes affecting proto-oncogenes and tumor-suppressor genes. Some of these mutations can be inherited, resulting in a predisposition to developing certain types of cancer.

SELF-QUIZ

1. In a nucleosome, the DNA is wrapped around
   a. polymerase molecules
   b. ribosomes
   c. histones
   d. the nucleolus
   e. satellite DNA

2. Apparently, our muscle cells are different from our nerve cells mainly because
   a. they express different genes
   b. they contain different genes
   c. they use different genetic codes
   d. they have unique ribosomes
   e. they have different chromosomes

3. One of the unique characteristics of retrotransposons is that
   a. translation of their RNA transcript produces an enzyme that converts the RNA back to DNA
   b. they are found only in animal cells
   c. once removed from the DNA, the gene segments for an antibody variable region are rejoined to the constant region
   d. they contribute a significant portion of the genetic variability seen within a population of gametes
   e. their amplification is dependent upon a concurrent retrovirus infection

4. The functioning of enhancers is an example of
   a. transcriptional control of gene expression

b. a posttranscriptional mechanism for editing mRNA
c. the stimulation of translation by initiation factors
d. posttranslational control that activates certain proteins
e. a eukaryotic equivalent of prokaryotic promoter functioning

5. Multigene families are
   a. groups of enhancers that control transcription
   b. usually clustered at the telomeres
   c. equivalent to the operons of prokaryotes
   d. collections of genes whose expression is controlled by the same regulatory proteins
   e. identical or similar genes that have evolved by gene duplication

6. Which of the following statements about the DNA in one of your brain cells is true?
   a. Some DNA sequences are present in multiple copies.
   b. Most of the DNA codes for protein.
   c. The majority of genes are likely to be transcribed.
   d. Each gene lies immediately adjacent to an enhancer that helps control transcription.
   e. Many genes are grouped into operonlike clusters.

7. Rearrangement of DNA segments is known to occur in genes coding for
   a. ribosomal RNA
   b. most proteins in eukaryotes
   c. hemoglobin
   d. histone proteins
   e. antibodies

8. Which of the following is an example of a possible step in the posttranscriptional control of gene expression?
   a. the addition of methyl groups to cytosine bases of DNA
   b. the binding of transcription factors to a promoter
   c. the removal of introns and splicing together of exons
   d. gene amplification during a stage in development
   e. the folding of DNA to form heterochromatin

9. The amount of protein made from a given mRNA molecule depends partly on
   a. the degree of DNA methylation
   b. the rate at which the mRNA is degraded
   c. the presence of certain transcription factors
   d. the number of introns present in the mRNA
   e. the types of ribosomes present in the cytoplasm

10. All our cells contain proto-oncogenes, which can change into oncogenes that cause cancer. Which of the following is the best explanation for the presence of these potential time bombs in our cells?
   a. Proto-oncogenes first arose from viral infections.
   b. Proto-oncogenes normally help regulate cell division.
   c. Proto-oncogenes are genetic "junk."
   d. Proto-oncogenes are mutant versions of normal genes.
   e. Cells produce proto-oncogenes as a by-product of the aging process.

CHALLENGE QUESTION

Researchers have long puzzled over the observation that some prostate cancer cells thrive despite treatments that eliminate testosterone and other androgens usually necessary for prostate cell survival. It has recently been suggested that estrogen, often considered a female hormone, may be activating genes that are normally controlled by an androgen receptor. Describe an experiment, or experiments, to test this hypothesis. (The hormones mentioned are all steroids; to review their mode of action, see pp. 194–195.)

SCIENCE, TECHNOLOGY, AND SOCIETY

A chemical called dioxin, or TCDD, is produced as a by-product of some chemical manufacturing processes. Trace amounts of this substance were present in Agent Orange, a defoliant sprayed on vegetation during the Vietnam War. There has been a continuing controversy over its effects on soldiers exposed to it during the war. Animal tests have suggested that dioxin can be lethal and can cause birth defects, cancer, liver and thymus damage, and immune system suppression. But its effects on humans are unclear, and even animal tests are equivocal; a hamster is not affected by a dose that can kill a guinea pig. Researchers have discovered that dioxin acts somewhat like a steroid hormone. It enters a cell and binds to a receptor protein, which in turn attaches to the cell's DNA. How might this mechanism help explain the variety of dioxin's effects on different body systems and in different animals? How might you determine whether a type of illness is related to dioxin exposure or whether a particular individual became ill as a result of exposure to dioxin? Which would be more difficult to demonstrate? Why?

FURTHER READING

Baylin, S. B. "Tying It All Together: Epigenetics, Genetics, Cell Cycle, and Cancer." *Science*, September 26, 1997. In this article, "epigenetics" refers to DNA methylation, whose role in cell regulation is emphasized.

"Cancer." *Scientific American*, September 1996. An entire issue devoted to the subject of cancer.

Featherstone, C. "Researchers Get Their First Good Look at the Nucleosome." *Science*, September 19, 1997.

Graves, B. J. "Inner Workings of a Transcription Factor Partnership." *Science*, February 13, 1998. Describes how transcription factors cooperate to recognize their sites of action on the DNA.

Gura, T. "New Kind of Cancer Mutation Found." *Science*, August 29, 1997. A novel kind of mutation has been found in the *APC* gene, which is associated with one type of inherited colorectal cancer.

Tjian, R. "Molecular Machines That Control Genes." *Scientific American*, February 1995. Discusses how elaborate complexes of transcription factors regulate the initiation of transcription in eukaryotic cells.

Travis, J. "Guardians of the Genome?" *Science News*, June 21, 1997. The normal forms of *BRCA1* and *BRCA2* may help safeguard the cell's DNA.

WEB LINKS

Visit the special edition of *The Biology Place* for BIOLOGY, Fifth Edition, at **http://www.biology.com/campbell**. Go to Chapter 19 for online resources, including learning activities, practice exams, and links to the following web sites:

"The Making of the Nucleosome"
This site contains 3-D renderings of nucleosomes and links to other chromatin resources. Because of the graphic nature of this site, loading can sometimes be slow, but it is well worth any wait.

"A Few Words about DNA and Chromatin"
By Dr. R. D. Stewart of the Pacific Northwest National Laboratory, this site is an excellent introduction to chromatin structure.

"Chromatin Structure & Function Page"
This site contains lists of sites relating to chromatin structure, nucleosomes, and chromatin-associated proteins.

"CancerNet"
The National Cancer Institute's CancerNet contains a wide range of accurate cancer information that is based on the latest research and is reviewed regularly by oncology experts.

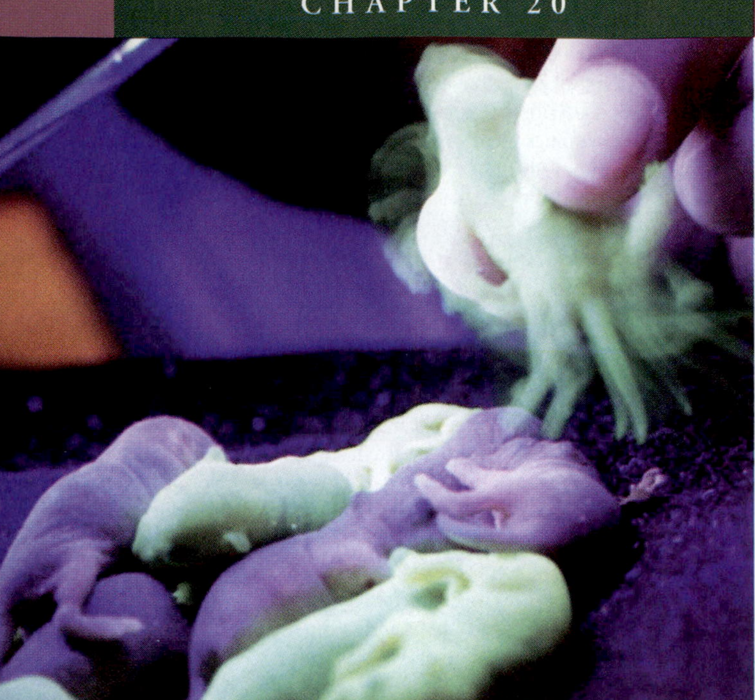

DNA TECHNOLOGY

Biologists in Japan have developed the world's first light-emitting mammals. The researchers injected DNA from a light-emitting marine organism into mouse zygotes to create genetically altered mice that glow green under UV rays (see photo at left; normal mice appear purple). The DNA of the green mice is ***recombinant DNA****, DNA in which genes from two different sources—in this case, different species—are linked. The Japanese scientists hope to use the process to monitor new cancer drugs by devising a way of limiting the green glow to cancer cells.*

This research is only one example of how molecular biologists are using ***genetic engineering****, the direct manipulation of genes for practical purposes. In addition to research, the applications of genetic engineering include the manufacture of hundreds of useful products. Using the biochemical and mechanical tools of DNA technology, scientists can make recombinant DNA in vitro ("in glass"—that is, in a test tube or flask). They can then introduce the DNA into cultured cells that replicate the DNA and may express its genes, yielding a desired product. The bacterium* E. coli *is often used as a "host" organism for recombinant DNA because it is easy to grow and its biochemistry is well understood.*

DNA technology has launched a revolution in biotechnology. In broad terms, ***biotechnology*** *is the manipulation of organisms or their components to perform practical tasks or provide useful products. Practices that go back centuries, such as the use of microbes to make wine and cheese and the selective breeding of livestock and crops, are examples of biotechnology. These practices have always relied on natural genetic processes, such as mutation and genetic recombination. But biotechnology based on the manipulation of DNA in vitro is different from earlier practices in that it enables scientists to modify specific genes and move them between organisms as distinct as bacteria, plants, and animals.*

DNA technology is now applied in areas ranging from agriculture to criminal law, but its most important achievements have been in basic research. DNA technology has stimulated important discoveries in virtually all fields of biology by giving researchers new tools for tackling age-old questions. And it has put within reach detailed knowledge of the human genome, knowledge largely inaccessible only a few decades ago.

In this chapter we examine the main techniques of DNA manipulation and analysis, survey the practical applications of DNA technology, and, finally, consider some of the social and ethical issues that arise as gene technology becomes more powerful.

DNA CLONING

The molecular biologist studying a particular gene faces a challenge. Naturally occurring DNA molecules are very long, and a single molecule usually carries many genes. Moreover, for multicellular eukaryotes, genes occupy only a small proportion of the chromosomal DNA, the rest being noncoding, repetitive nucleotide sequences. A particular human gene, for example, might constitute only 1/100,000 of the chromosomal DNA molecule where it resides. As a further complication, DNA molecules are structurally and chemically very homogeneous; the distinctions between a gene and the surrounding DNA are subtle, consisting only of differences in nucleotide sequence. To work directly with specific genes, scientists needed to develop methods for preparing well-defined, gene-sized pieces of DNA in multiple identical copies. In other words, they needed techniques for **gene cloning**.

DNA technology makes it possible to clone genes for basic research and commercial applications: *an overview*

Scientists have developed a number of ways to clone pieces of DNA in the laboratory, but almost all share certain general features. In FIGURE 20.1, an overview of gene cloning and its applications, we consider an approach that uses bacteria and

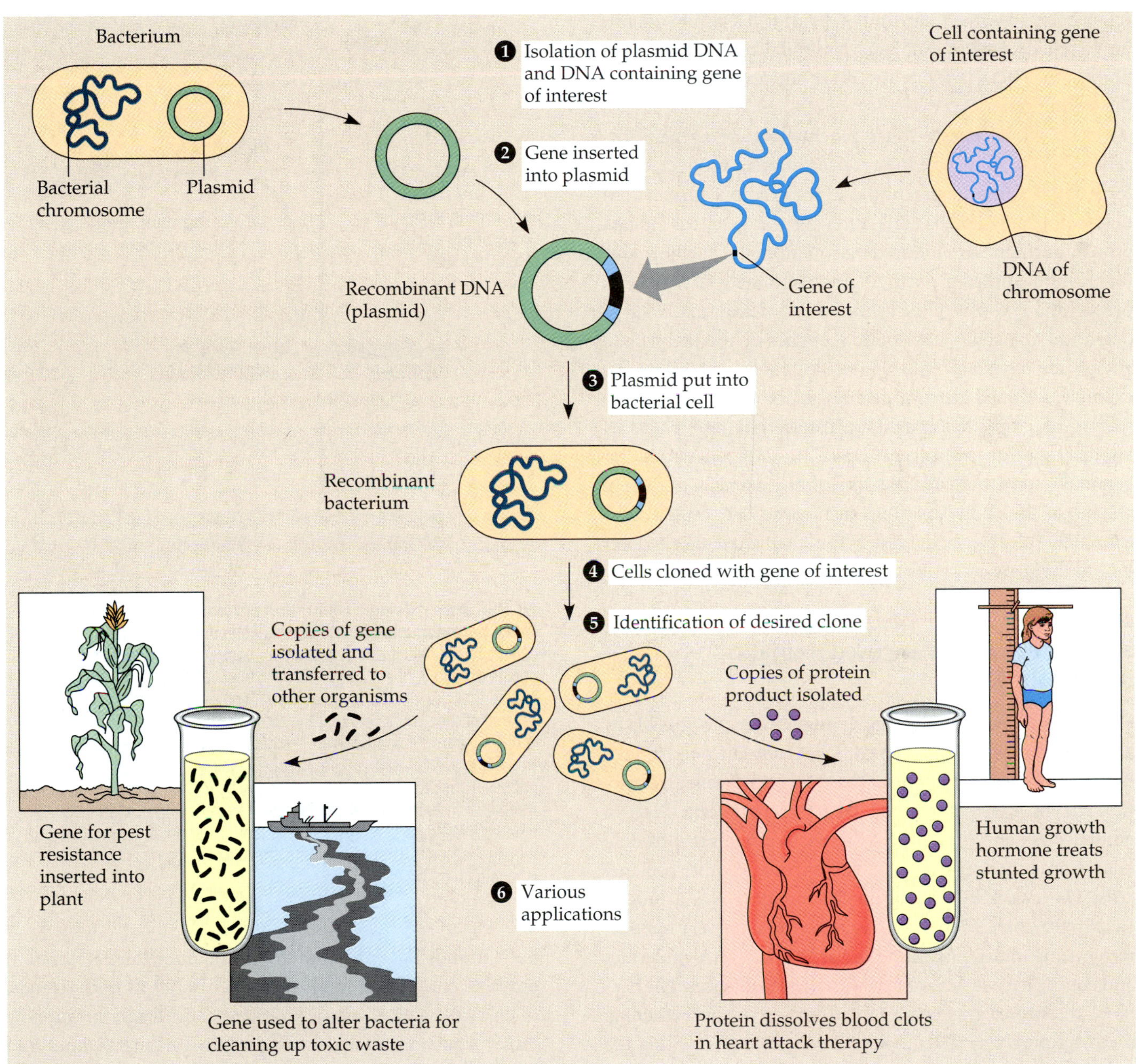

FIGURE 20.1 ▪ An overview of how bacterial plasmids are used to clone genes for biotechnology. ① A scientist isolates plasmid DNA from bacteria and DNA containing a gene of interest from cells of another kind. The cells might be animal cells, as shown here, and the gene of interest might, for instance, be one that encodes a hormone. ② A piece of DNA containing the gene is inserted into one of the plasmids, producing recombinant DNA, and ③ the recombinant plasmid is put back into a bacterial cell. ④ This bacterial cell is then grown in culture, forming a clone of cells. The foreign DNA spliced into the plasmid does not impair the plasmid's ability to replicate within a bacterial cell, and the gene of interest is replicated with the rest of the plasmid as the host cell multiplies. Focusing on the gene itself, we can say that it, too, has been "cloned." ⑤ A critical step in gene cloning is the identification of the bacterial clone carrying the gene of interest. ⑥ The drawings show some current applications of gene cloning in bacteria. In the examples on the left, a gene originating in one organism is used to give another organism a new metabolic capability. In the cases on the right, useful protein products are harvested in large quantities from bacterial cultures.

their plasmids. Recall from Chapter 18 that plasmids are small, circular DNA molecules that replicate within bacterial cells. For cloning genes or other pieces of DNA, plasmids are first isolated from bacterial cells. FIGURE 20.1 follows one plasmid as a foreign gene—from a eukaryotic cell, in this example—is inserted into it. The plasmid is now a recombinant DNA molecule combining DNA from two sources. The plasmid is returned to a bacterial cell, which then reproduces to form a cell clone. The foreign gene carried by the plasmid is "cloned" at the same time, for the dividing bacterium continues to replicate the recombinant plasmid. Under suitable conditions, the bacterial clone will make the protein encoded by the foreign gene.

The potential uses of cloned genes fall into two general categories. The goal may be to produce a protein product, either for study or for some practical use. For example, pharmaceutical companies use bacteria carrying the gene for human growth hormone to produce large quantities of the hormone for treating stunted growth. Alternatively, the goal may be to prepare many copies of the gene itself. A scientist may want to determine the gene's nucleotide sequence or use the gene to endow an organism with a new metabolic capability. For example, a cloned gene for pest resistance originating in one species of crop plant might be transferred into plants of another species. Most genes exist in only one copy per genome—something on the order of one part in a million of DNA—so the ability to clone such rare DNA fragments is extremely valuable. In the rest of this chapter we take a closer look at the steps in FIGURE 20.1 and at related methods.

Restriction enzymes are used to make recombinant DNA

Gene cloning and genetic engineering were made possible by the discovery of enzymes that cut DNA molecules at a limited number of specific locations. These enzymes, called **restriction enzymes**, were discovered in bacteria in the late 1960s. In nature, these enzymes protect the bacteria against intruding DNA from other organisms, such as viruses or other bacterial cells. They work by cutting up the foreign DNA, a process called *restriction*. Most restriction enzymes are very specific, recognizing short nucleotide sequences in DNA molecules and cutting at specific points within these sequences. The bacterial cell protects its own DNA from restriction by adding methyl groups (—CH_3) to adenines or cytosines within the sequences recognized by the restriction enzyme. Hundreds of different restriction enzymes have been identified and isolated, and many are available commercially.

The top of FIGURE 20.2 is a diagram of a DNA molecule containing a recognition sequence, or **restriction site**, for a particular restriction enzyme. As shown in this example, most restriction sites (darker green) are symmetrical: The same sequence of four to eight nucleotides (here, six) is found on

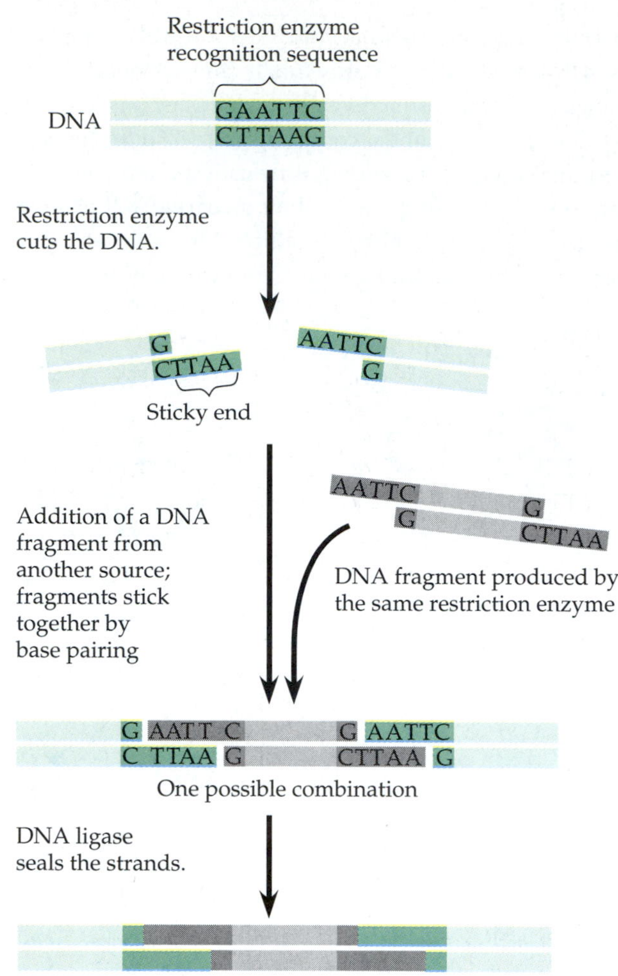

FIGURE 20.2 • Using a restriction enzyme and DNA ligase to make recombinant DNA. The restriction enzyme in this example (called *Eco*RI) recognizes a specific six-base-pair sequence and makes staggered cuts in the sugar-phosphate backbone within this sequence. Notice that the recognition sequence along one DNA strand is the exact reverse of the sequence along the complementary strand. Because the sequences are symmetrical and the cuts staggered, the enzyme produces fragments of DNA with single-stranded "sticky" ends. Complementary ends will stick to each other by hydrogen bonding, transiently rejoining fragments in their original combinations or in new, recombinant arrangements. DNA ligase can then catalyze the formation of covalent bonds joining their ends. If the fragments are from different sources, the result is recombinant DNA.

both strands but running in opposite directions. Restriction enzymes cut covalent phosphodiester bonds of both strands, often in a staggered way, as indicated in the diagram. Since the target sequence usually occurs (by chance) many times in a long DNA molecule, an enzyme will make many cuts. Copies of a DNA molecule always yield the same set of **restriction fragments** when exposed to that enzyme. In other words, a restriction enzyme cuts a DNA molecule in a reproducible way. (Later you will learn how the different fragments can be separated.)

Notice in FIGURE 20.2 that the restriction fragments are double-stranded DNA fragments with at least one single-

20.1 **FIGURE 20.3 ▪ Cloning a human gene in a bacterial plasmid: *a closer look.*** In this figure, double-stranded DNA is represented as a single band. (Refer to FIGURE 20.2 to see steps 2a–2c in more detail.) The circled numbers correspond to numbered steps in the text. The diagram focuses on a fragment of human DNA carrying a gene of interest (black), but steps 2a–2c actually produce an enormous variety of recombinant plasmids, each of which shows up as a separate colony in step 4. Identifying the desired colony, step 5, is often the greatest challenge in this "shotgun" procedure.

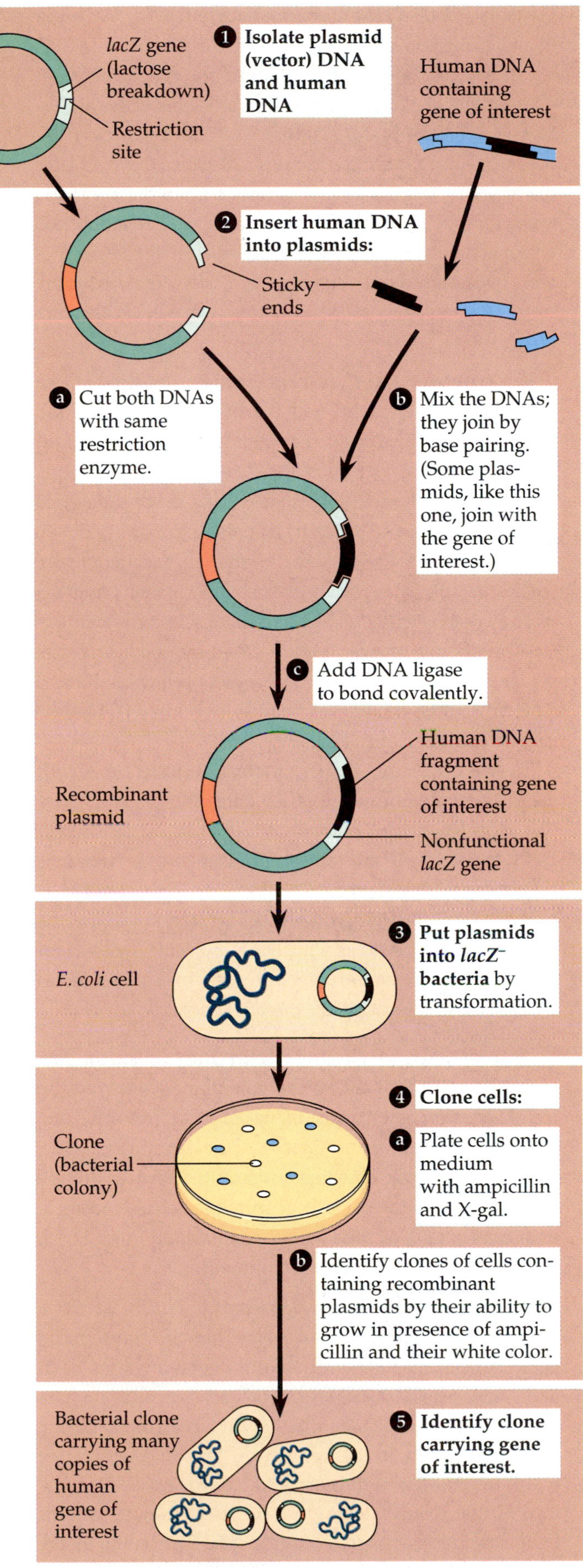

Bacterial plasmid
lacZ gene (lactose breakdown)
1 Isolate plasmid (vector) DNA and human DNA
Human DNA containing gene of interest
amp^R (ampicillin resistance)
Restriction site
2 Insert human DNA into plasmids:
Sticky ends
a Cut both DNAs with same restriction enzyme.
b Mix the DNAs; they join by base pairing. (Some plasmids, like this one, join with the gene of interest.)
c Add DNA ligase to bond covalently.
Recombinant plasmid
Human DNA fragment containing gene of interest
Nonfunctional *lacZ* gene
E. coli cell
3 Put plasmids into *lacZ*$^-$ bacteria by transformation.
4 Clone cells:
a Plate cells onto medium with ampicillin and X-gal.
Clone (bacterial colony)
b Identify clones of cells containing recombinant plasmids by their ability to grow in presence of ampicillin and their white color.
Bacterial clone carrying many copies of human gene of interest
5 Identify clone carrying gene of interest.

stranded end, called a **sticky end**. These short extensions will form hydrogen-bonded base pairs with complementary single-stranded stretches on other DNA molecules cut with the same enzyme. The unions formed in this way are only temporary, because only a few hydrogen bonds hold the fragments together. The DNA fusions can be made permanent, however, by the enzyme **DNA ligase**, which seals the strands by catalyzing the formation of phosphodiester bonds. (Recall from Chapter 16 that DNA ligase is a key enzyme in DNA replication and repair.) We now have recombinant DNA, DNA that has been spliced together from two different sources.

Genes can be cloned in recombinant DNA vectors: *a closer look*

With a restriction enzyme and DNA ligase in our toolkit, we can make the recombinant plasmid depicted in step 2 of FIGURE 20.1. The original plasmid is called a **cloning vector**, defined as a DNA molecule that can carry foreign DNA into a cell and replicate there. Bacterial plasmids are widely used as cloning vectors. Recombinant plasmids produced by splicing restriction fragments from foreign DNA into plasmids isolated from bacteria can be returned relatively easily to bacteria. Then, as a bacterium carrying a recombinant plasmid reproduces, the plasmid replicates within it. The resulting cell clone, which appears as a colony on solid nutrient medium, contains multiple copies—a "clone"—of the foreign DNA.

Bacteria are the most commonly used host cells for gene cloning, primarily because of the ease with which DNA can be isolated from and reintroduced into such cells. Bacterial cultures also grow quickly, rapidly replicating any foreign genes they carry.

Procedure for Cloning a Eukaryotic Gene in a Bacterial Plasmid

Suppose we want to clone a particular eukaryotic gene using the method sketched out in steps 1–5 of FIGURE 20.1. The more detailed diagram in FIGURE 20.3 shows how we might proceed. The step numbers in the text on the next page correspond to the numbers in the figure.

20.1

① **Isolation of vector and gene-source DNA.** We begin by preparing two kinds of DNA: the bacterial plasmid we'll use as a vector and DNA containing the gene of interest. This latter DNA comes from human tissue cells that we have grown in laboratory culture. The plasmid comes from the bacterium *E. coli* and carries two genes that will later prove useful: amp^R, conferring resistance to the antibiotic ampicillin on its *E. coli* host cell, and *lacZ*, encoding the enzyme β-galactosidase, which hydrolyzes the sugar lactose. The plasmid has a single recognition sequence for the restriction enzyme used, and the sequence lies within the *lacZ* gene.

② **Insertion of DNA into the vector.** In step 2a, we digest both the plasmid and the human DNA with the same restriction enzyme. The enzyme cuts the plasmid DNA at its single restriction site, disrupting the *lacZ* gene. It also cuts the human DNA, generating many thousands of fragments; one of these fragments carries the gene we want. In making the cuts, the restriction enzyme creates sticky ends on both the human DNA fragments and the plasmid. For simplicity, the figure here shows the step-by-step processing of one human DNA fragment and one plasmid, but actually millions of copies of the plasmid and a heterogeneous mixture of millions of human DNA fragments are treated simultaneously.

In step 2b we mix the fragments of human DNA with the clipped plasmids. The sticky ends of a plasmid base-pair with the complementary sticky ends of the human DNA fragment we are following. (Many other combinations form as well, such as two plasmids together, or a plasmid with several DNA fragments.) In step 2c we use the enzyme DNA ligase to join the DNA molecules by covalent bonds. The result is a mixture of recombinant DNA molecules, some of which are like the one shown.

③ **Introduction of cloning vector into cells.** In this step, bacterial cells take up the recombinant plasmids by transformation (the uptake of naked DNA from the surrounding solution; see Chapter 18). The bacteria are $lacZ^-$; a mutation in gene *lacZ* makes them unable to hydrolyze lactose. Some of the bacteria acquire the desired recombinant plasmid DNA; many other cells take up other DNA, both recombinant and nonrecombinant.

④ **Cloning of cells (and foreign genes).** Finally, we arrive at the actual cloning step. We plate out the transformed bacteria on solid nutrient medium containing ampicillin and a sugar called X-gal (step 4a). Each reproducing bacterium forms a cell clone that appears as a colony on the nutrient medium. In the process, any human genes carried by recombinant plasmids are also cloned. We take advantage of the plasmid's own genes to select colonies of cells that carry recombinant plasmids. The ampicillin in the medium ensures that only plasmid-containing cells grow, for only they have the amp^R gene conferring ampicillin resistance. The X-gal in the medium makes it easy to identify colonies of bacteria whose plasmids carry foreign DNA. X-gal is hydrolyzed by β-galactosidase to yield a blue product, so bacterial colonies containing plasmids with intact β-galactosidase genes will be blue. But if a plasmid has foreign DNA inserted into its *lacZ* gene, then a colony of cells containing it will be white, because the cells cannot produce β-galactosidase. In summary, bacteria with recombinant plasmids carrying foreign DNA will form white colonies on medium containing ampicillin and X-gal.

Our procedure to this point will have cloned many different human-DNA fragments, not just the one that interests us. The next step, screening colonies for the desired human gene, is the most difficult.

⑤ **Identification of cell clones carrying the gene of interest.** How can we distinguish the colony containing our gene of interest from the many thousands of colonies carrying other pieces of human DNA? We can look either for the gene itself or for its protein product. All the methods for detecting the DNA of a gene directly depend on base pairing between the gene and a complementary sequence on another nucleic acid molecule, a process called **nucleic acid hybridization**. The complementary molecule, a short single-stranded nucleic acid that can be either RNA or DNA, is called a **nucleic acid probe**. If we know at least part of the nucleotide sequence of our gene, we can synthesize a probe complementary to it.

We trace the probe, which will hydrogen-bond specifically to the desired gene, by labeling it with a radioactive isotope or a fluorescent tag. FIGURE 20.4 shows how a number of bacterial clones, growing as colonies on a plate of solid medium (agar), can be simultaneously screened for the presence of DNA complementary to a DNA probe. An essential step is the **denaturation** of the cells' DNA—that is, the separation of its two strands. As with protein denaturation, DNA denaturation is routinely accomplished with chemicals or heat. The radioactive probe tags the correct clone—finds the needle in the haystack—by hydrogen bonding to its single-stranded DNA complement.

Once we've identified a cell clone carrying the desired gene, we can grow the cells in liquid culture in a large tank and then easily isolate large amounts of the gene. Also, we can use the cloned gene itself as a probe to identify similar or identical genes in DNA from other sources.

If the cells containing a desired gene translate the gene into protein, then it is sometimes possible to identify them by screening all clones for the protein. Detection of the protein can be based on either its activity (as with an enzyme) or its structure, using antibodies that combine specifically with it. However, for bacterial cells to make a eukaryotic protein, we need to take special measures, as discussed next.

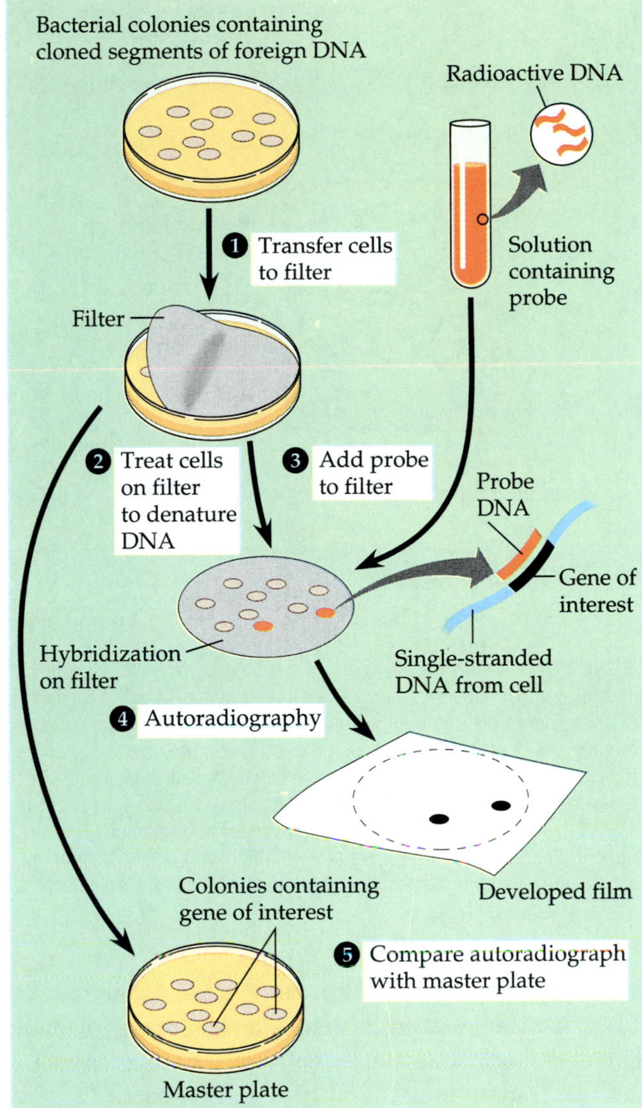

FIGURE 20.4 • Using a nucleic acid probe to identify a cloned gene. This technique depends on the fact that complementary nucleotide sequences will base-pair, stuck together by hydrogen bonds. Here the cloned gene of interest is carried on a bacterial plasmid, and the probe is a short length of radioactive single-stranded DNA complementary to part of the gene. ① Bacterial colonies on agar are pressed against special filter paper, transferring cells to the filter. ② The filter is treated to break open the cells and denature their DNA; the resulting single-stranded DNA molecules stick to the filter. ③ A solution of probe molecules is incubated with the filter. The probe DNA hybridizes (base-pairs) with any complementary DNA on the filter; excess DNA is rinsed off. ④ The filter is laid on photographic film, allowing any radioactive areas to expose the film (autoradiography). ⑤ The developed film, an autoradiograph, is compared with the master culture plate to determine which colonies carry the desired gene.

Cloning and Expressing Eukaryotic Genes: Problems and Solutions

Getting a cloned eukaryotic gene to function in a prokaryotic setting can be difficult because certain details of gene expression are different in the two kinds of cells (see Chapters 18 and 19). To overcome differences in promoters and other DNA control sequences, scientists usually employ an **expression vector**, a cloning vector that contains the requisite prokaryotic promoter just upstream of a restriction site where the eukaryotic gene can be inserted. The bacterial host cell will then recognize the promoter and proceed to express the foreign gene that has been linked to it. Expression vectors allow the synthesis of many eukaryotic proteins in bacterial cells.

Another problem with cloning and expressing eukaryotic DNA in bacteria is the presence of long noncoding regions (introns) in many eukaryotic genes. Introns can make a eukaryotic gene very long and unwieldy, and they prevent correct expression of the gene by bacterial cells, which do not have RNA-splicing machinery. Fortunately, scientists can make artificial eukaryotic genes that lack introns (FIGURE 20.5). The starting material for this procedure is fully processed mRNA, from which cells have already removed the introns. We can extract mRNA from the eukaryotic cells that made it and then use the enzyme reverse transcriptase

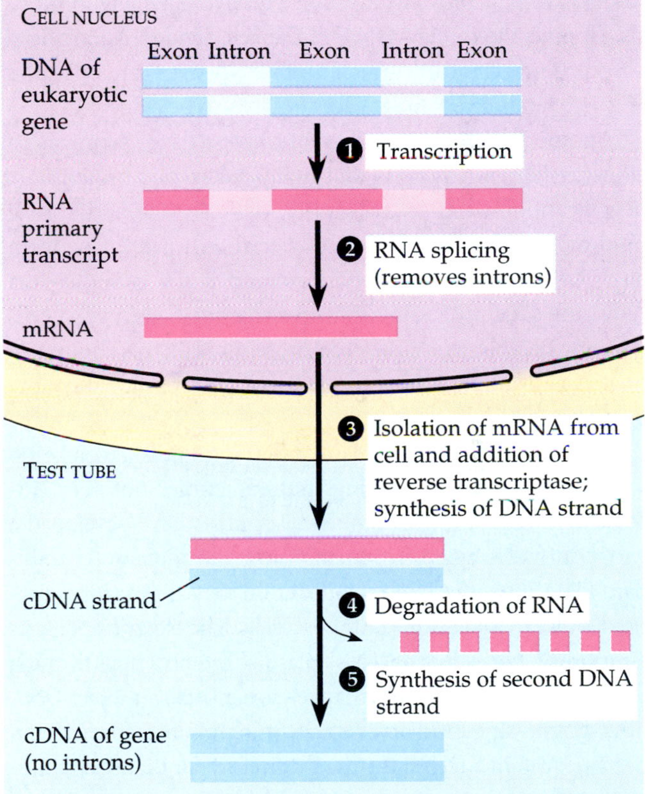

FIGURE 20.5 • Making complementary DNA (cDNA) for a eukaryotic gene. Complementary DNA is DNA made in vitro using mRNA as a template and the enzyme reverse transcriptase. The synthesis of an mRNA molecule in a cell nucleus is shown in steps ① and ②. In step ③, the molecular biologist isolates the mRNA molecules from the cell and adds reverse transcriptase, which makes a DNA strand using the RNA as a template. The biologist provides ④ another enzyme to degrade the RNA and ⑤ DNA polymerase to synthesize a second DNA strand. The result is cDNA, which carries the coding sequence of the gene but no introns. Molecules of cDNA can be cloned by inserting them into vectors. Lacking introns, cDNA for eukaryotic genes can be correctly expressed in bacterial cells. However, to be transcribed, the cDNA has to be joined to a bacterial promoter.

(obtained from retroviruses) to make DNA transcripts of this RNA. Each DNA molecule produced carries the coding sequence for a gene but no introns. This DNA is called **complementary DNA**, or **cDNA**, and can be attached to vector DNA for replication inside a cell. Bacteria can express a eukaryotic cDNA gene if the vector provides a bacterial promoter and any other control elements necessary for the gene's transcription and translation.

Molecular biologists can avoid eukaryotic-prokaryotic incompatibility by using eukaryotic cells rather than bacteria as hosts for cloning and/or expressing eukaryotic genes of interest. Yeast cells, single-celled fungi, offer two advantages: They are as easy to grow as bacteria, and they have plasmids, a rarity among eukaryotes. Scientists have even constructed recombinant plasmids that combine yeast and bacterial DNA and can replicate in either type of cell. Scientists have also constructed vectors called **artificial chromosomes** that combine the essentials of a eukaryotic chromosome—an origin for DNA replication, a centromere, and two telomeres—with foreign DNA. These chromosomes behave normally in mitosis, cloning the foreign DNA as the cell divides. An artificial chromosome can carry much more DNA than can a plasmid vector, enabling very long pieces of DNA to be cloned.

Another reason to use eukaryotic host cells for expressing a cloned eukaryotic gene is that many eukaryotic proteins are heavily modified after translation, often by the addition of lipid or carbohydrate groups. Bacterial cells cannot perform any of these processing functions, and if the gene product requiring such processing is from a mammal, even yeast cells will not be able to modify the protein correctly. The use of host cells from an animal or plant cell culture may therefore be necessary.

Like bacteria, many kinds of eukaryotic cells can take up DNA from their surroundings but sometimes not very efficiently. To deal with this problem, scientists have developed a variety of more aggressive methods for introducing recombinant DNA into cells. In **electroporation** they apply a brief electrical pulse to a solution containing cells. The electricity creates temporary holes in the cells' plasma membranes, through which DNA can enter. Alternatively, scientists can inject DNA directly into single eukaryotic cells using microscopically thin needles. And in a technique used primarily for plant cells, they can attach DNA to microscopic particles of metal and fire the particles into cells with a gun (see FIGURE 38.15). Once inside the cell, DNA has a chance to be incorporated into the cell's DNA by natural genetic recombination.

Cloned genes are stored in DNA libraries

Because the gene-cloning procedure in FIGURE 20.3 starts with a mixture of fragments from the entire genome of an organism, it is called a "shotgun" approach—no single gene is targeted for cloning. Thousands of different recombinant plasmids are actually produced in step 2 of FIGURE 20.3, and a clone of each ends up in a (white) colony in step 4. The complete set of thousands of recombinant-plasmid clones, each carrying copies of a particular segment from the initial genome, is referred to as a **genomic library** (FIGURE 20.6, left). A researcher can save such a library and use it as a source of other genes of interest or for genome mapping (as we'll discuss later).

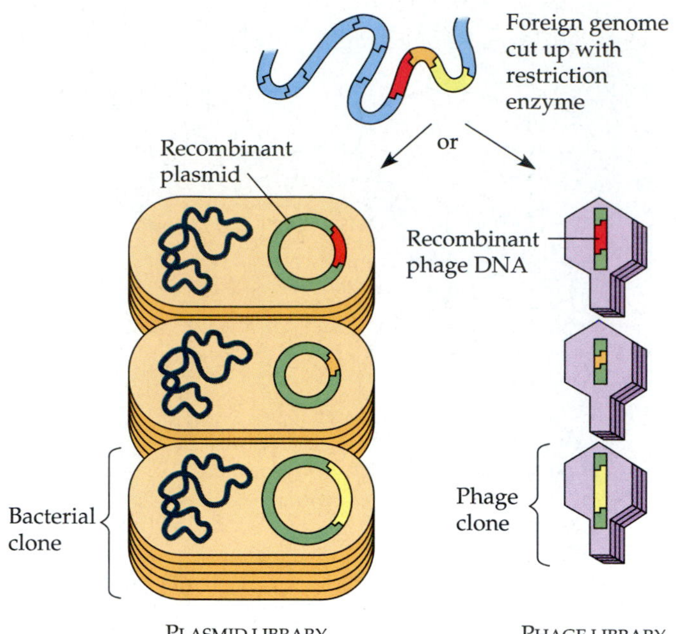

FIGURE 20.6 • Genomic libraries. A genomic library is a collection of a large number of bacterial or phage clones, each containing copies of a DNA segment from a foreign genome. In a complete genomic library, the foreign DNA segments represented cover the entire genome of an organism. This diagram shows parts of two genomic libraries. On the left are three of the thousands of "books" in a plasmid library. Each "book" is a bacterial clone containing one particular variety of foreign genome fragment—red, orange, or yellow here—in its recombinant plasmid. On the right, the same three foreign genome fragments are shown in three "books" of a phage library.

In addition to plasmids, certain bacteriophages are also common cloning vectors for making genomic libraries. Fragments of foreign DNA can be spliced into a phage genome, as into a plasmid, by using a restriction enzyme and ligase. The recombinant phage DNA is then packaged into capsids in vitro and introduced into a bacterial cell through the normal infection process. Once inside the cell, the phage DNA replicates and produces new phage particles, each carrying the foreign DNA. A genomic library made using phage is stored as collections of phage clones (FIGURE 20.6, right). Whatever the cloning vector, restriction enzymes do not respect gene boundaries in cutting up genomic DNA, so some genes in a genomic library may be divided up among two or more clones.

Researchers can make a more limited kind of gene library by using complementary DNA (cDNA). When they isolate mRNA from cells (step 3 in FIGURE 20.5), they actually obtain a mixture of all the mRNA molecules in the cell, transcribed from a number of different genes. Therefore, the cDNA that is

METHODS: The Polymerase Chain Reaction (PCR)

The polymerase chain reaction (PCR) is a method for making many copies of a specific segment of DNA. It is much faster than gene cloning with plasmid or phage DNA and is performed completely in vitro. The starting material for PCR is a solution of double-stranded DNA containing the nucleotide sequence that is "targeted" for copying. The scientist adds a heat-resistant type of DNA polymerase (which catalyzes the reaction), a supply of all four nucleotides (for assembly into new DNA), and primers. The primers are required for the DNA polymerase to initiate DNA synthesis (see Chapter 16). The primers used in PCR are short, synthetic molecules of single-stranded DNA; they are complementary to the ends of the targeted DNA and thus determine the particular segment of DNA to be amplified.

The figure outlines the PCR procedure. ① The DNA is briefly heated to separate its strands and then ② cooled to allow the primers to bind by hydrogen bonding to the ends of the target sequence, one primer on each strand. Then ③ the DNA polymerase adds nucleotides to the 3′ ends of the primers, using the longer DNA strands as templates. Within about 5 minutes the targeted DNA sequence—even one hundreds of base pairs long—has been doubled. The solution is then heated again, starting another cycle of strand separation, primer binding, and DNA synthesis. The cycle runs again and again, until the targeted sequence has been duplicated enough times. By about 20 cycles, virtually all the product DNA molecules will consist of the exact targeted sequence (like two of the molecules at the bottom of the figure).

There are practical limits to the number of copies that can be made, often imposed by the accumulation of rare errors in the DNA copies (in vitro versions of mutations). To prepare larger amounts of a DNA segment, PCR can be followed by cloning of the DNA in cells.

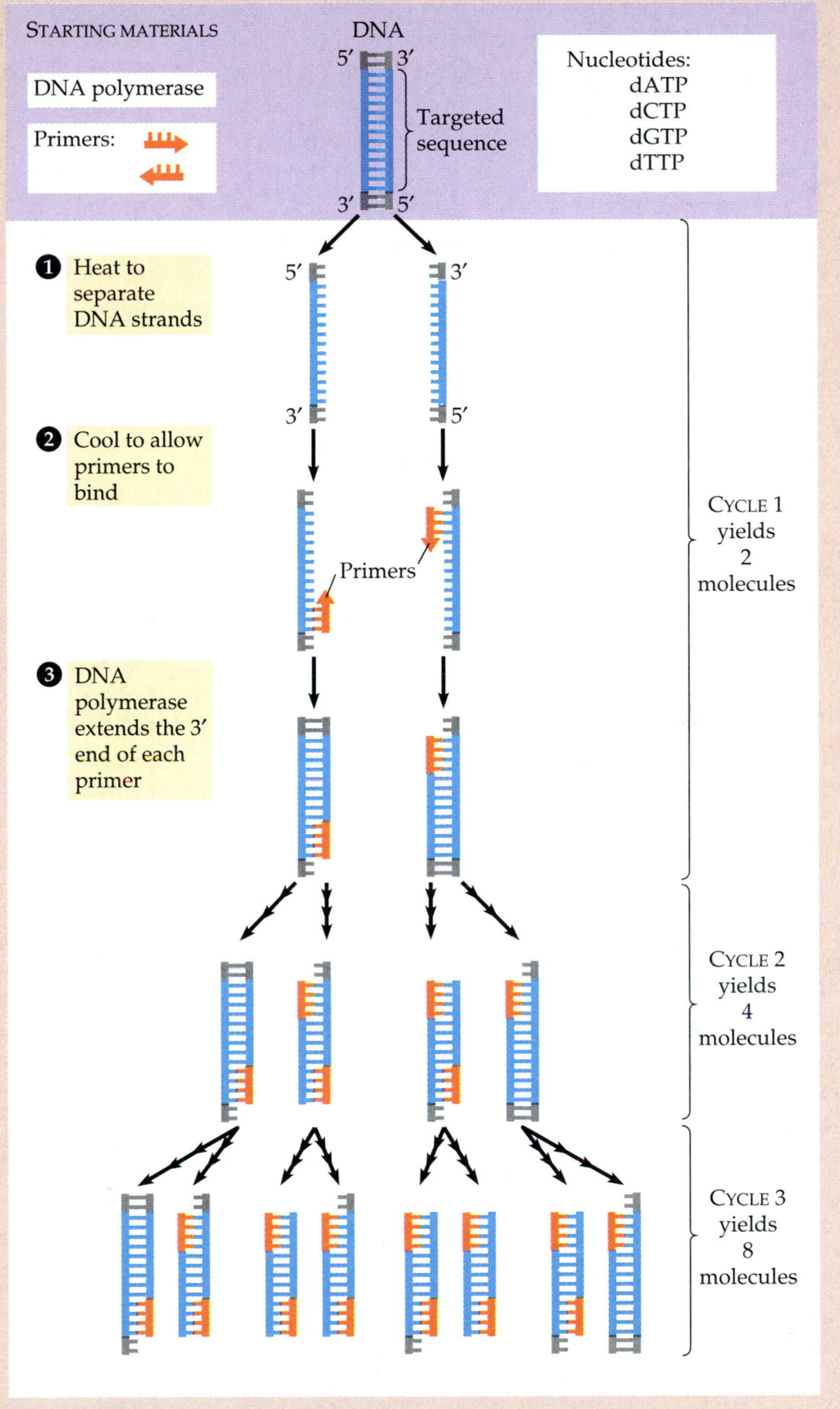

made is a library containing a collection of genes. Such a **cDNA library** represents only part of a cell's genome—only the genes that were transcribed in the starting cells. This is an advantage if a researcher wants to study the genes responsible for specialized functions of a particular kind of cell, such as a brain or liver cell. Also, by making cDNA from cells of the same type at different times in the life of an organism, one can trace changes in patterns of gene expression.

The polymerase chain reaction (PCR) clones DNA entirely in vitro

DNA cloning in cells remains the best method for preparing large quantities of a particular gene or other DNA sequence. However, when the source of DNA is scanty or impure, a method called PCR is quicker and more selective. **PCR**, the

polymerase chain reaction, is a technique by which any piece of DNA can be quickly amplified (copied many times) without using cells. The DNA is incubated in a test tube with a special kind of DNA polymerase, a supply of nucleotides, and short pieces of synthetic single-stranded DNA that serve as primers for DNA synthesis. (Recall from Chapter 16 that DNA polymerases need primers because they can add nucleotides only to a preexisting nucleotide chain.) With automation, PCR can make billions of copies of a targeted segment of DNA in a few hours, significantly faster than the days it takes to clone a piece of DNA by making a recombinant plasmid and letting it replicate within bacteria. Amplifying a tiny DNA sample a billionfold does not produce much DNA, but it can be enough for some purposes, such as DNA fingerprinting in a murder trial.

The Methods Box on p. 371 describes the PCR procedure. A three-step cycle brings about a chain reaction that produces an exponentially growing population of DNA molecules. The key to easy PCR automation is an unusual DNA polymerase that was first isolated from bacteria living in hot springs. Unlike most proteins, this enzyme can withstand the heat needed to separate the DNA strands at the start of each cycle.

Just as impressive as the speed of PCR is its specificity. The primers determine the DNA sequence that is amplified. PCR can be used, for example, to amplify a specific gene prior to further cloning in cells. The PCR makes the gene by far the most abundant DNA fragment, thus simplifying the later task of finding a clone carrying that gene. In fact, PCR is so specific and powerful that its starting material does not even have to be purified DNA. Only minute amounts of DNA need be present in the starting material, and this DNA can be in a partially degraded state.

Devised in 1985, PCR has already had a major impact on biological research and biotechnology. PCR has been used to amplify DNA from a wide variety of sources: fragments of ancient DNA from a 40,000-year-old frozen woolly mammoth; DNA from tiny amounts of blood, tissue, or semen found at the scenes of violent crimes; DNA from single embryonic cells for rapid prenatal diagnosis of genetic disorders; and DNA of viral genes from cells infected with such difficult-to-detect viruses as HIV. We'll return to applications of PCR later in the chapter.

ANALYSIS OF CLONED DNA

Once we have techniques for preparing homogeneous samples of DNA, each containing large numbers of identical segments of a genome, we can begin to ask some far-ranging questions. Suppose we have cloned a DNA segment carrying a human gene of interest. Does the gene differ in different people, and are there certain alleles associated with a hereditary disorder? Where in the body and when is the gene expressed? What is the location of the gene within the genome? We can also ask evolutionary questions, such as how does the gene differ from species to species?

To answer these and other questions fully, we will eventually need to know the complete nucleotide sequence of the gene and its counterparts in other individuals and species. But we can start by applying more indirect methods that can rapidly provide useful comparative information. Most of these methods, and DNA sequencing as well, make use of a technique called **gel electrophoresis.** This technique, described in the Methods Box on p. 374, separates macromolecules—nucleic acids or proteins—on the basis of size, electrical charge, and other physical properties. For linear molecules of DNA, separation depends mainly on size. Gel electrophoresis sorts a mixture of DNA molecules into bands, each consisting of DNA molecules of the same length.

Restriction fragment analysis detects DNA differences that affect restriction sites

Restriction fragment analysis indirectly detects certain differences in the nucleotide sequences of DNA molecules. In this method we use gel electrophoresis to sort by size the DNA fragments that result from treating a long DNA molecule with a restriction enzyme. Just viewing the stained pattern of bands on the gel can provide scientifically useful information. When the mixture of restriction fragments from a particular DNA molecule undergoes electrophoresis, it yields a band pattern characteristic of the starting molecule and the restriction enzyme used. In fact, the relatively small DNA molecules of viruses and plasmids can be identified simply by their restriction-fragment patterns. (Larger DNA molecules, such as those of chromosomes, yield too many fragments to appear as distinct bands.) Because DNA can be recovered undamaged from gels, the procedure also provides a way to prepare pure samples of individual fragments.

Let's use restriction fragment analysis to compare two different DNA molecules representing, for example, different alleles of a gene. We start by digesting each DNA sample with the same restriction enzyme. Because the two alleles must differ slightly in DNA sequence, they may differ in one or more restriction sites. Thus, the collections of restriction fragments from the two alleles may produce different band patterns in gel electrophoresis. FIGURE 20.7 shows how restriction-fragment analysis by electrophoresis can distinguish between two alleles of a gene that differ in sequence by only one base pair, if a restriction site is affected.

The starting materials in FIGURE 20.7 were samples of cloned and purified genes (alleles). However, by combining gel electrophoresis with nucleic acid hybridization, we can make the same comparison without preliminary gene cloning. We can start with DNA of the whole genome. Electrophoresis will yield too many bands to distinguish individually, but we can

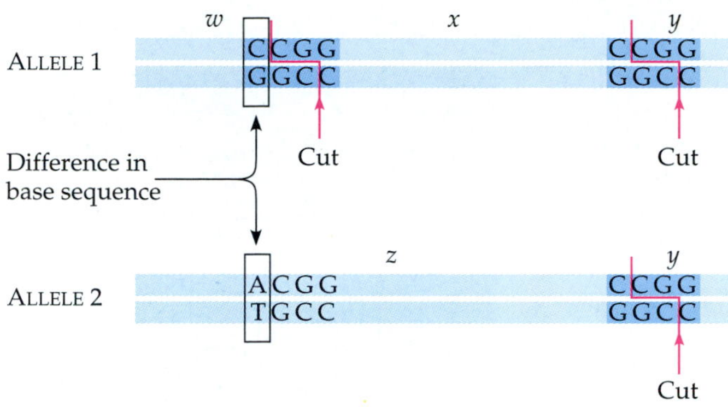

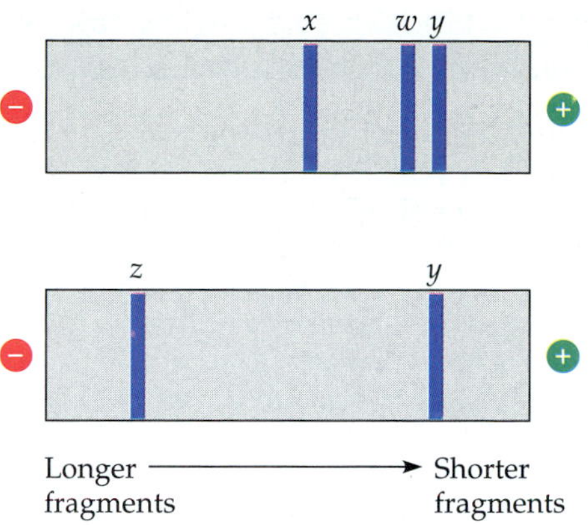

FIGURE 20.7 ▪ Using restriction fragment patterns to distinguish DNA from different alleles. (a) Two homologous segments of DNA that carry different alleles of a gene are depicted; only the relevant bases are shown. A single base-pair difference results in allele 2 having one less recognition sequence (restriction site) for a particular restriction enzyme. This enzyme cuts the DNA from allele 1 into three pieces (*w, x,* and *y*) but cuts the DNA from allele 2 into only two pieces (*z* and *y*). **(b)** Electrophoresis separates the restriction fragments formed from each allele. A clear difference between the two alleles is revealed by their band patterns on the gel. Allele 1 has three bands, corresponding to fragments *w, x,* and *y;* allele 2 has two bands, corresponding to *z* and *y.*

use nucleic acid hybridization to label discrete bands that derive from our gene of interest.

The principle is the same as for the nucleic acid hybridization in FIGURE 20.4: A radioactive single-stranded nucleic acid probe selectively hydrogen-bonds with a targeted, complementary DNA sequence, which is then detected by autoradiography. Now, however, we use a probe in combination with an additional technique called Southern hybridization or **Southern blotting**. The Methods Box on p. 375 outlines the entire procedure and demonstrates how it can be used to compare DNA samples from three individuals. This method goes beyond FIGURE 20.4 by revealing not only whether a particular sequence is present in a sample of DNA but also the restriction fragments that contain the sequence.

Southern blotting proved invaluable when biologists turned their attention to *noncoding* DNA, which comprises most of the DNA of animal and plant genomes. Are there differences in noncoding DNA sequences analogous to the differences in alleles of genes? Using noncoding DNA in procedures like that of FIGURE 20.7, researchers were excited to discover many differences in band patterns. Differences in DNA sequence on homologous chromosomes that result in different restriction fragment patterns turn out to be scattered abundantly throughout genomes, including the human genome. Such differences have been named **restriction fragment length polymorphisms** (**RFLPs**, pronounced "RIF-lips"). This type of difference is conceptually the same as a difference in coding sequence; it too can serve as a genetic marker for a particular location (locus) in the genome. A given RFLP marker frequently occurs in numerous variants in a population (the word *polymorphisms* comes from the Greek for "many forms").

RFLPs are detected and analyzed by Southern blotting. The example shown in the Methods Box on p. 375 could as easily represent the detection of a RFLP in noncoding DNA as a difference in the coding sequences of two alleles. Because of the sensitivity of DNA hybridization, the entire genome can be used as the DNA starting material. (Samples of human DNA are typically obtained from white blood cells.) The result of the procedure shown in the Methods Box indicates that individuals I and II carry the same version of the marker (RFLP or gene), but individual III carries a different version.

Because RFLP markers are inherited in a Mendelian fashion, they can serve as genetic markers for making linkage maps. The geneticist uses the same reasoning you learned about in Chapter 15: The frequency with which two RFLP markers—or a RFLP marker and a certain allele for a gene—are inherited together is a measure of the closeness of the two loci on a chromosome. The discovery of RFLPs greatly increased the number of markers available for mapping the human genome. No longer are human geneticists limited to genetic variations that lead to obvious phenotypic differences (such as genetic diseases) or even to differences in protein products. In the next section we discuss current efforts and achievements in genome mapping.

Entire genomes can be mapped at the DNA level

Using the tools and techniques discussed so far in this chapter—restriction enzymes, DNA cloning, gel electrophoresis, labeled probes, and so forth—along with some additional techniques, researchers have made tremendous advances in the mapping of eukaryotic genes. One approach is to focus on genes that have been cloned. If you want to identify the

METHODS: Gel Electrophoresis of Macromolecules

Gel electrophoresis separates macromolecules on the basis of their rate of movement through a gel under the influence of an electric field. As the diagram indicates, mixtures of nucleic acids or proteins are placed in wells near one end of a thin slab of a polymeric gel. The gel is supported by glass plates and bathed in an aqueous solution. Electrodes are attached to both ends, and voltage is applied. Each macromolecule then migrates toward the electrode of opposite charge at a rate determined mostly by the molecule's charge and size. Usually several different samples, each a mixture of molecules, are run simultaneously in multiple "lanes" on a slab gel, as shown here.

For nucleic acids, the rate of migration—how far a molecule travels while the current is on—is inversely proportional to molecular size. Nucleic acids carry negative charges (on phosphate groups) proportional to their lengths, but the thicket of polymer fibers in the gel impedes longer fragments more than it does shorter ones. The gel in the photograph has been treated after electrophoresis with a DNA-binding dye that fluoresces pink in ultraviolet light. In each lane you can see a number of pink bands, which correspond to DNA molecules of different sizes. The larger molecules move more slowly through the gel and are located toward the right in the photograph. The bands contain DNA restriction fragments. Six samples were run, each a mixture of fragments from a DNA preparation digested with a restriction enzyme.

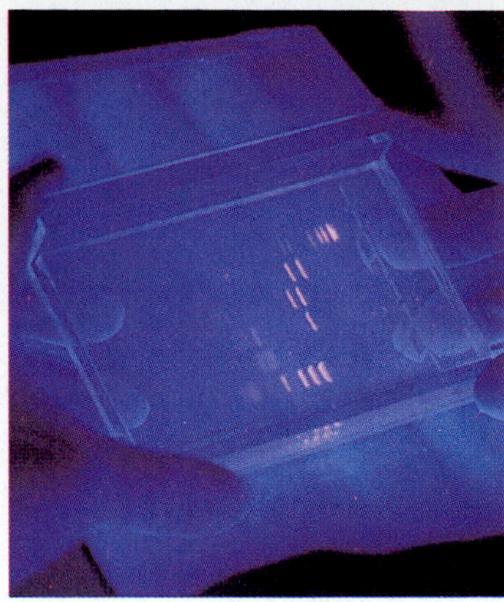

(b)

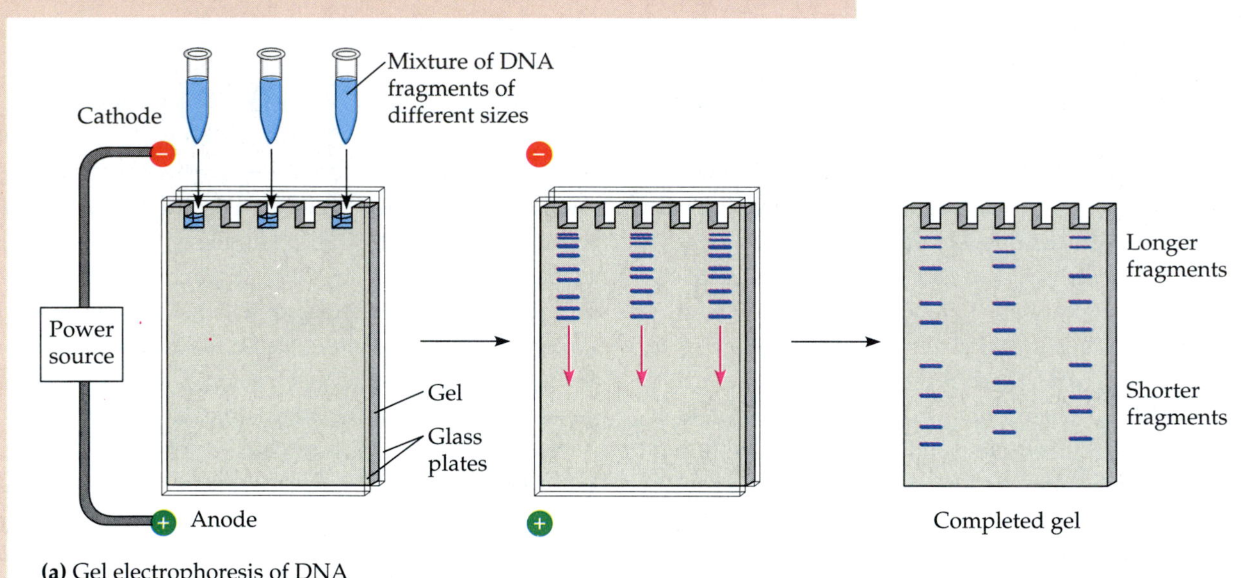

(a) Gel electrophoresis of DNA

chromosome that carries a known gene, nucleic acid hybridization is the key.

Locating Genes by In Situ Hybridization

A nucleic acid probe can be used to find the general location of a gene on a eukaryotic chromosome. In the technique called **in situ hybridization**, a radioactive probe is allowed to base-pair with complementary sequences in the denatured DNA of intact chromosomes on a microscope slide (*in situ* means "in place"). Autoradiography and chromosome staining are then used to reveal where the probe has attached. Alternatively, the probe can be labeled with a fluorescent dye (see the micrograph on p. 261).

Although in situ hybridization is sensitive enough to find a "needle in a haystack," it is not absolutely specific. The same probe can be used to recognize homologous DNA sequences that differ in a number of nucleotides. Suppose, for example, that you have just identified and cloned a gene that plays an important regulatory role during mitosis in yeast cells. Do other organisms—for example, frogs or humans—also have this gene? You can address this question by using the yeast gene as a probe in an in situ hybridization experiment with frog and human cells.

METHODS: Restriction Fragment Analysis by Southern Blotting

The procedure below, combining five laboratory techniques, enables researchers to detect and analyze particular DNA sequences. The basis for detecting specific sequences is nucleic acid hybridization. The results can show not only whether certain sequences exist in different samples but also the number of such sequences within a genome and the size of the restriction fragments that contain them. In this way it is possible to compare the DNA of different individuals or even species.

Because of the selective power of nucleic acid hybridization, the starting material for analysis can be the entire genome of an organism. This huge length of DNA will produce so many restriction fragments that, if all were made visible (with a dye, for example), they would appear as a smear in gel electrophoresis, rather than as discrete bands. Instead, only the DNA bands of interest are made visible by using a labeled probe. The probe consists of multiple copies of a radioactively (or fluorescently) labeled piece of single-stranded DNA that will hybridize (base-pair) with the DNA of interest. Before hybridization, the DNA being tested is transferred by capillary action (blotting) from the gel to a solid support, a sheet of nitrocellulose or nylon paper. The entire hybridization procedure is known as *Southern blotting,* after E. M. Southern, who developed it in 1975. The sensitivity of the hybridization can be adjusted either to detect only those sequences that are perfectly complementary to the probe or to detect similar (homologous) sequences as well.

In a similar procedure, dubbed *Northern blotting,* mRNA (whole molecules, rather than fragments) is subjected to hybridization analysis. Typically the goal is to determine whether a particular gene has been transcribed and how much of that mRNA is present. Northern blotting has become a mainstay of research on the control of gene expression.

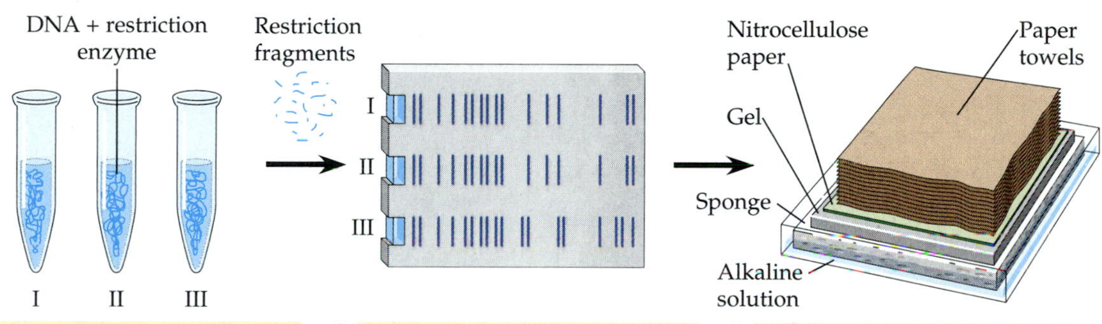

❶ **Restriction fragment prepartion.** DNA samples to be tested (in this case identified as samples I, II, and III) are prepared from the appropriate sources. A restriction enzyme is added to the three samples of DNA to produce restriction fragments.

❷ **Electrophoresis.** The mixtures of restriction fragments from each sample are separated by electrophoresis. Each sample forms a characteristic pattern of bands. (There would be many more bands than shown here, and they would be invisible unless stained.)

❸ **Blotting.** Capillary action pulls an alkaline solution upward through the gel and through a sheet of nitrocellulose paper laid on top of it, transferring the DNA to the paper and denaturing it in the process. The single strands of DNA stick to the paper, positioned in bands exactly as on the gel.

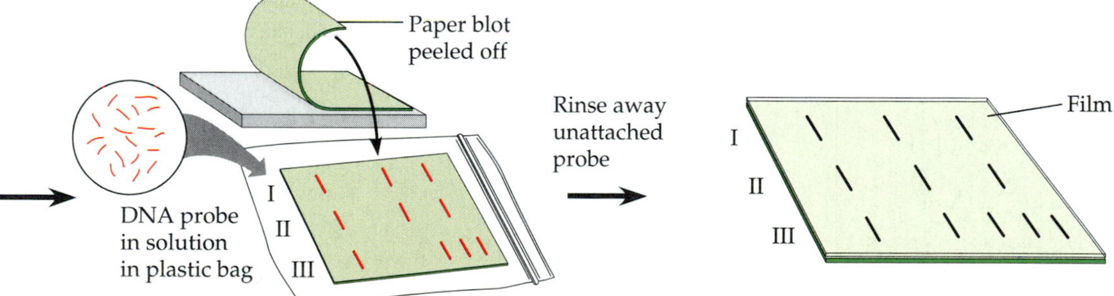

❹ **Hybridization with radioactive probe.** The paper blot is exposed to a solution containing radioactively labeled probe. The probe is single-stranded DNA complementary to the DNA sequence of interest, and it attaches by base pairing to restriction fragments of complementary sequence.

❺ **Autoradiography.** A sheet of photographic film is laid over the paper. The radioactivity in the bound probe exposes the film to form an image corresponding to specific DNA bands—the bands containing DNA that base pairs with the probe. The band patterns for samples I and II are identical, but III is different.

The Mapping of Entire Genomes

The information obtained from studying individual genes is interesting and valuable, but in recent years geneticists have also undertaken a much broader approach, with the goal of systematically mapping entire genomes. They aim to determine the precise locations of all of an organism's genes—and noncoding DNA segments, as well—along the DNA molecules of its genome. The most ambitious research project made possible by DNA technology is the **Human Genome Project**, officially begun in 1990. This is an international effort to map the entire human genome, ultimately by determining the complete nucleotide sequence of the DNA of each human chromosome (the 22 autosomes and the *X* and *Y* sex chromosomes). The project is proceeding through three stages that focus on the DNA more and more closely. These stages are genetic (or linkage) mapping, physical mapping, and, finally, DNA sequencing. We'll describe these stages shortly.

In addition to mapping human DNA, the Human Genome Project includes mapping of the genomes of other species important in biological research, including *E. coli* and other prokaryotes, *Saccharomyces cerevisiae* (yeast), *Caenorhabditis elegans* (nematode), *Drosophila melanogaster* (fruit fly), and *Mus musculus* (mouse). These genomes are of great interest in their own right, and the sequences completed so far have already brought us important biological insights, as we'll discuss later. In addition, work on these genomes is useful for developing the strategies, methods, and new technologies necessary for deciphering the human genome, which is much larger. The technological power for this daunting task will come in large part from advances in automation and from utilization of electronic technology, including computer software.

Genetic (Linkage) Mapping. In mapping a large genome, the first stage is to construct a linkage map of several thousand genetic markers spaced evenly throughout the chromosomes. As you learned in Chapter 15, the order of the markers on such a map and the relative distances between them are based on recombination frequencies. The markers can be genes (as in Chapter 15) or any other identifiable sequences in the DNA, such as RFLPs or the short repetitive sequences called microsatellites (see Chapter 19). Relying primarily on microsatellites, which are abundant in the human genome and highly variable from person to person, researchers have completed a human genetic map with over 5000 markers. Such a map enables researchers to locate other markers, including genes, by testing for genetic linkage to the known markers.

Physical Mapping: Ordering DNA Fragments. In a physical map, the distances between markers are expressed in some physical measure, usually as number of nucleotides along the DNA. For whole-genome mapping, a physical map is made by cutting the DNA of each chromosome into a number of identifiable restriction fragments, and then determin-

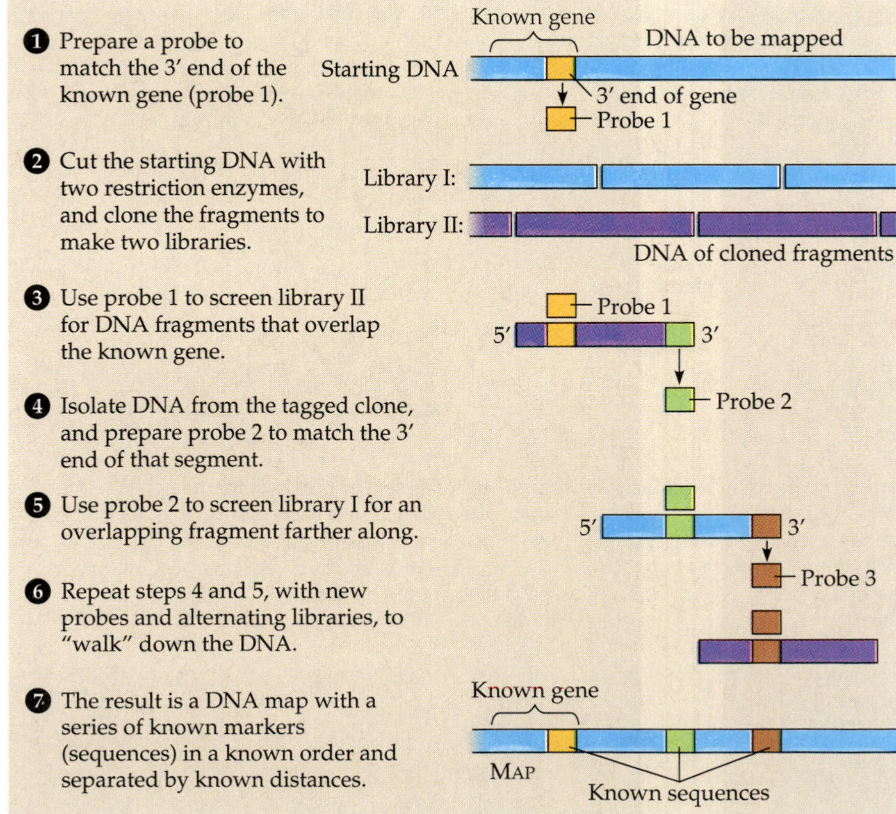

1 Prepare a probe to match the 3′ end of the known gene (probe 1).
Known gene
DNA to be mapped
Starting DNA
3′ end of gene
Probe 1
2 Cut the starting DNA with two restriction enzymes, and clone the fragments to make two libraries.
Library I:
Library II:
DNA of cloned fragments
3 Use probe 1 to screen library II for DNA fragments that overlap the known gene.
Probe 1
5′
3′
4 Isolate DNA from the tagged clone, and prepare probe 2 to match the 3′ end of that segment.
Probe 2
5 Use probe 2 to screen library I for an overlapping fragment farther along.
5′
3′
Probe 3
6 Repeat steps 4 and 5, with new probes and alternating libraries, to "walk" down the DNA.
7 The result is a DNA map with a series of known markers (sequences) in a known order and separated by known distances.
Known gene
MAP
Known sequences

FIGURE 20.8 ▪ Chromosome walking. The researcher starts with a known gene or other sequence—one that has already been cloned, mapped, and sequenced—and "walks" along the chromosomal DNA from that locus, producing a map of overlapping restriction fragments. To prepare the fragments the scientist cuts two samples of chromosomal DNA with different restriction enzymes and then clones both sets of fragments to form two libraries. This and other methods of ordering DNA fragments involve determining the nucleotide sequences of short, unique segments within each fragment. The resulting physical map is thus a linear array of fragments, each identifiable by two or more sequenced regions. All the DNA in this diagram is shown as single strands.

ing the original order of the fragments in the chromosomal DNA. The key is to make fragments that overlap and then use probes or automated nucleotide sequencing to find the overlapping ends. A method called **chromosome walking**, which uses probes, has been very useful (FIGURE 20.8).

Supplies of the DNA fragments used for physical mapping are prepared by cloning. In working with large genomes, researchers carry out several rounds of DNA cutting, cloning, and mapping. The first cloning vector is often a yeast artificial chromosome (YAC), which can carry inserted fragments a million base pairs long. The scientists determine the order of these long fragments (by chromosome walking, for instance) and then cut each fragment into smaller pieces, which are cloned and ordered in turn. The final sets of fragments, cloned in plasmids or phage, are only about 1500 base pairs long, short enough to be sequenced easily.

DNA Sequencing. The complete nucleotide sequence of a genome is the ultimate map. To repeat a key point, it is the technology for cloning fragments of DNA that has made possible the sequencing of the very long DNA molecules of genomes. Starting with a pure preparation of many copies of a DNA fragment, the nucleotide sequence of the fragment can be determined by a sequencing machine. The usual procedure, described in the Methods Box on p. 378, combines DNA labeling, DNA synthesis involving special chain-terminating nucleotides, and high-resolution gel electrophoresis. But even with automation, the sequencing of all 3 billion nucleotide pairs in a haploid set of human chromosomes remains a formidable challenge. In fact, a major thrust of the Human Genome Project is the development of technology for faster sequencing. Faster sequencing will be necessary not only to complete the human sequence by the targeted date of 2005 but also to allow easy sequencing of additional DNA for comparison with this first sequence—DNA from different human populations or individuals, for example.

As of mid-1998, only 3% of the human genome has been sequenced. However, about fifteen genomes of other species have been completed. These include *E. coli* and a number of other bacteria (including several of medical importance), several archaea (prokaryotes now regarded as a domain separate from the true bacteria), and yeast, a eukaryote. On one level these sequences are simply dry lists of nucleotide bases—millions of A's, T's, C's, and G's in mind-numbing succession. But on another level, reached with the help of computer analysis, these sequences are already the basis of some exciting discoveries, which we discuss next.

Genome Analysis

The beauty of DNA technology is that it enables geneticists to study genes directly, without having to infer genotype from phenotype as in classical genetics. But the newer approach poses the opposite problem, determining the phenotype from the genotype. Starting with a long DNA sequence, how does one identify genes and determine their function?

Analyzing DNA Sequences. DNA sequences are being collected in computer data banks that are shared among researchers. The scientists use software to scan the sequences for the telltale signs of protein-coding genes: the nucleotide triplets that, transcribed into RNA, serve as start and stop codons. (They also look for known gene-control sequences, such as promoters, which give clues about how genes are regulated.) In this way, they assemble a list of nucleotide sequences for putative genes. They then compare these sequences with those of known genes from the same or other organisms. In some cases a newly determined gene sequence will match, at least partially, a gene whose function is well known. For example, part of a new gene may match a known gene that encodes a protein kinase, suggesting that the new gene does, too. (Many similarities involve *parts* of proteins, perhaps reflecting the shuffling of exons during evolution; see p. 304.) In other cases the new gene sequence will be similar to a sequence encountered before—perhaps in the genome of another organism—but of unknown function. In still other cases the sequence may be entirely unlike anything ever seen before.

One of the humbling aspects of the sequences compiled to date is how many of the putative genes are entirely new to science—about 55% of the genes found in the archaeon *Methanococcus*, 38% of the genes of *E. coli* (the most well-studied of research organisms!), and at least 35% of yeast genes, for instance. Even so, the analysis of genome sequences is already yielding important and useful information. DNA sequences confirm the evolutionary connections between even distantly related organisms and the relevance of research on simpler organisms to our understanding of human biology. Similarities between genes of disparate organisms can be surprising, to the point that one researcher has said that he now views fruit flies as "little people with wings." Yeast, for example, has a number of genes close enough to the human versions that they can substitute for them in a human cell. In fact, researchers can sometimes work out what a human disease gene does by studying its normal counterpart in yeast. Bacterial sequences reveal unsuspected metabolic pathways that may have industrial or medical uses. Moreover, comparisons of the completed genome sequences of bacteria, archaea, and eukarya strongly support the theory that these are the three fundamental domains of life (see Chapters 1 and 27).

Studying Gene Expression. We study genomes not only to discover individual genes and their functions but to learn how genes act together to produce and maintain a functioning organism. While many scientists are laboring to sequence and analyze genomes, others are devising new ways to study patterns of gene expression. Without waiting for the genomes

METHODS: Sequencing of DNA by the Sanger Method

Developed by British scientist Frederick Sanger, this method for determining the nucleotide sequence of DNA molecules involves synthesizing in vitro DNA strands complementary to one of the strands of the DNA being sequenced. The method is based on the random incorporation of a modified nucleotide, a dideoxyribonucleotide, which terminates a growing DNA chain because it lacks both a 2′—OH and the 3′—OH needed for attachment of the next nucleotide (see p. 287). DNA strands are synthesized that reflect all possible positions of the dideoxy nucleotides, and thus ultimately the entire DNA sequence. DNA molecules that are sequenced are cloned restriction fragments, typically several hundred nucleotides long.

Most of the work of DNA sequencing is now automated. As an alternative to labeling the fragments with radioactive primers, the newest sequencing machines use fluorescent dyes to tag the dideoxyribonucleotides, one color for each of the four types of nucleotide. This approach allows the four reactions to be performed in a single tube; the dideoxy ends of the reaction-product DNA strands can be distinguished by the color of their fluorescence.

① A preparation of one of the strands of the DNA fragment is divided into four portions, and each portion is incubated with all the ingredients needed for the synthesis of complementary strands: a labeled primer (made to base-pair with the known 3′ end of the template), DNA polymerase, and the four deoxyribonucleoside triphosphates. In addition, each reaction mixture contains a different *one* of the four nucleotides in the modified, dideoxy (dd) form.

② Synthesis of the new strands starts at the primer and continues until a dideoxyribonucleotide is incorporated, which prevents further synthesis. A dideoxynucleotide is inserted every so often, at random, instead of its normal equivalent. Eventually, a set of radioactive strands of various lengths is generated. This is shown here only for the reaction mixture with ddGTP.

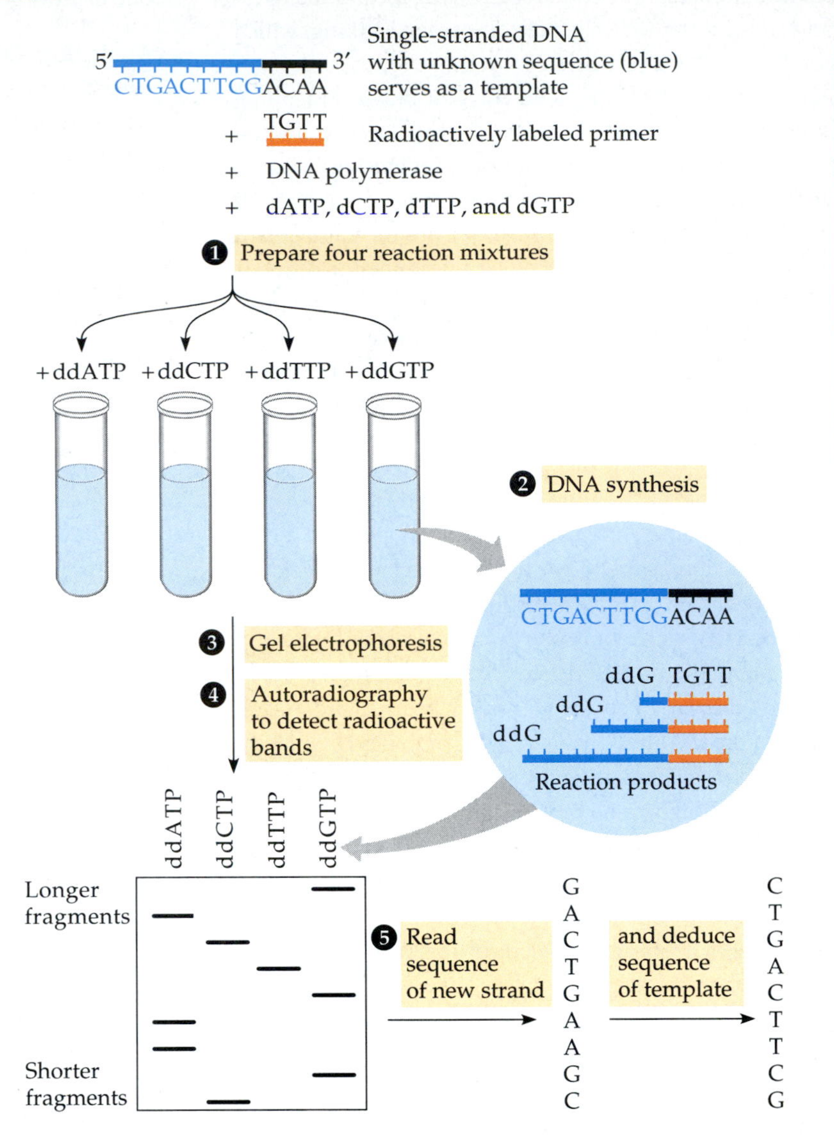

③ The new DNA strands in each reaction mixture are separated by electrophoresis on a polyacrylamide gel, which can separate strands differing by as little as one nucleotide in length. ④ An autoradiograph is made. ⑤ The sequence of the newly synthesized strands can be read directly from the bands on the autoradiograph, and from that the sequence of the original template strand is deduced. In this example, because the longest fragment terminates with ddG, G must be the last base in the new DNA strand. The second longest fragment terminates with ddA, meaning that A is the second to last base, and so on.

of multicellular eukaryotes to be completely sequenced, these researchers are focusing in on the genes that are transcribed in different situations. Their basic strategy is to isolate the mRNA made in particular cells, use these molecules as templates for making a cDNA library by reverse transcription, and then compare this cDNA with other collections of DNA by hybridization. Researchers can see what genes are active (transcribed) at different stages of development, in different tissues, or in tissues in different states of health.

DNA technology makes such gene-expression studies possible; automation allows them to be easily performed on a large scale. Scientists can now detect and measure the expression of thousands of genes at one time. One method, described in FIGURE 20.9, uses **DNA microarray assays.** Tiny amounts of a large number of single-stranded DNA fragments representing different genes are fixed to a glass microscope slide in a tightly spaced array. Ideally, these fragments represent all the genes of an organism, as is already possible for organisms whose genomes have been completely sequenced. The fragments are tested for hybridization with various samples of cDNA molecules, which have been labeled with fluorescent dyes. The first test of this method compared the genes expressed in the roots and leaves of *Arabidopsis.* The researchers went on to sequence the genes that were more highly expressed in one tissue than in the other. The results were as predicted; genes encoding known photosynthetic enzymes, for instance, were turned on in the leaves but not in the roots. But more important than confirming predictions is the method's potential for uncovering new genes and gene interactions and providing clues to gene function. For example, DNA microarray assays are being used to compare cancerous with noncancerous tissues. Studying the differences in gene expression may lead researchers to new diagnostic techniques and biochemically targeted treatments.

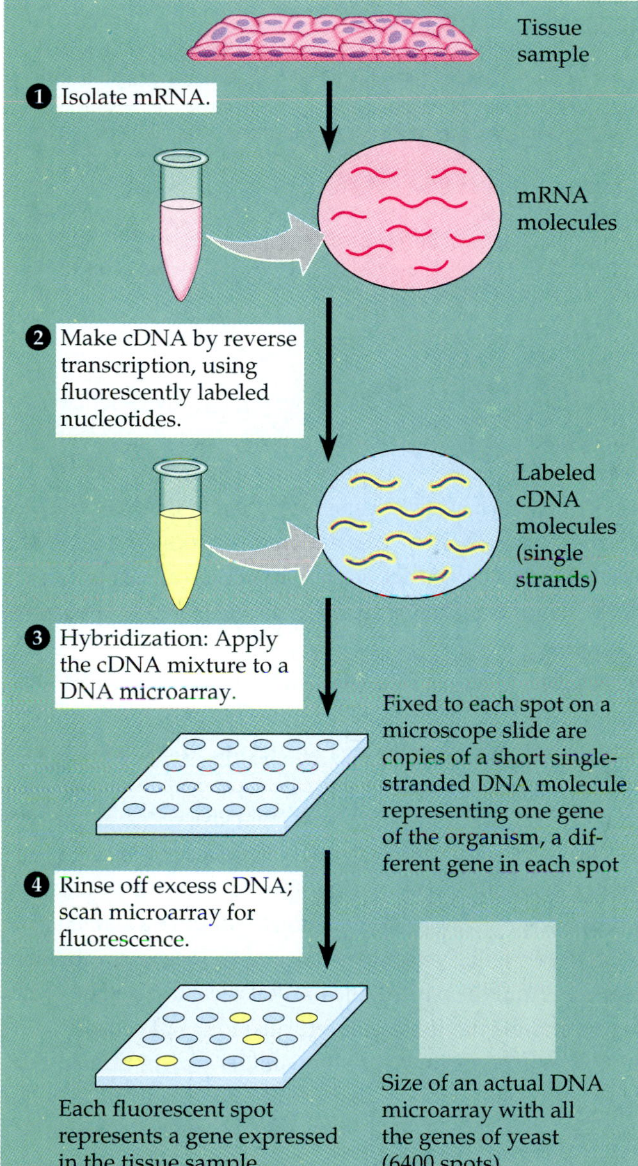

FIGURE 20.9 ▪ DNA microarray assay for gene expression. With this method a researcher can simultaneously test all the genes expressed in a particular tissue for hybridization with an array of short DNA sequences representing thousands of genes, even all the genes of the organism. ① Messenger RNA is isolated from the tissue sample and ② used as template for making cDNA by reverse transcription; the cDNA thus represents the genes that were active in the tissue. ③ The mixture of cDNA molecules is applied to the entire microarray. ④ Because the cDNA is fluorescently labeled, the spots where any of it hybridizes fluoresce with an intensity indicating the relative amount of the mRNA that was in the tissue. Molecules of cDNA representing more than one tissue (or the same tissue under different conditions) can be tested together by using a different colored label for each.

Determining Gene Function. Perhaps the most interesting genes discovered in genome sequencing and gene-expression studies are those whose functions are completely mysterious. How do researchers discover the functions of these genes? One powerful approach is **in vitro mutagenesis,** a technique that can be used to introduce specific changes into the sequence of a cloned gene. Such mutations can alter or destroy the function of the protein product. Then, when the mutated gene is returned to the host cell, it may be possible to determine the function of the missing normal protein by examining the cell physiology of the mutant. Researchers can even put such a mutated gene into cells from the early embryo of a multicelled organism (such as a mouse), enabling them to study the role of the gene in the development and functioning of the whole organism.

Genomics, the study of genomes and genes based on DNA sequencing, is yielding new insights into fundamental questions about genome organization, the control of gene expression, growth and development, and evolution. This information and the technology used to obtain it can also benefit our daily lives. We devote the rest of the chapter to a discussion of some of these applications and related issues.

PRACTICAL APPLICATIONS OF DNA TECHNOLOGY

Hardly a day goes by that DNA technology is not in the news. Commonly the topic of the story is a new and promising application of the technology, most often an application to a medical problem.

DNA technology is reshaping medicine and the pharmaceutical industry

Modern biotechnology is making enormous contributions to medicine, both in the diagnosis of diseases and in the development of pharmaceutical products. One obvious benefit of DNA technology and of the Human Genome Project is the identification of genes whose mutation is responsible for genetic diseases, for these discoveries should lead to ways of diagnosing, treating, and preventing those conditions.

Equally important are the potential benefits of DNA technology for dealing with other kinds of diseases, from arthritis to AIDS. Susceptibility to many "nongenetic" diseases is influenced by a person's genes. Furthermore, diseases of all sorts involve changes in gene expression within the affected cells and often within the patient's immune system. By using DNA microarray assays or other techniques to compare gene expression in healthy and diseased tissues, researchers hope to find many of the genes that are turned on or off in particular diseases. These genes and their products are potential targets for prevention or therapy.

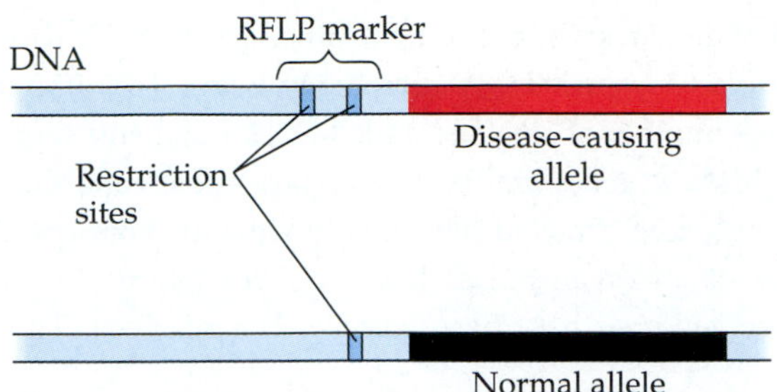

FIGURE 20.10 ▪ RFLP markers close to a gene. Even if a disease-causing allele has not been cloned and its precise locus is unknown, its presence can sometimes be detected with high (though not perfect) accuracy by testing for the presence of RFLP markers that are very close to the gene in question. This diagram depicts homologous segments of DNA from a family in which some members have a genetic disease. In this family, different versions of a RFLP marker are associated with the different alleles, allowing the test to be applied. If a family member has inherited the version of the RFLP marker with two restriction sites (rather than one), there is a high probability that the individual has also inherited the disease-causing allele.

Diagnosis of Diseases

A new chapter in the diagnosis of infectious diseases has been opened by DNA technology, in particular the use of PCR and labeled nucleic acid probes to track down certain pathogens. For example, because the sequence of HIV DNA is known, PCR can be used to amplify, and thus detect, HIV DNA in blood or tissue samples. This is often the best way to detect an otherwise elusive infection.

Medical scientists can now diagnose hundreds of human genetic disorders using DNA technology. Increasingly, they can identify individuals with genetic diseases before the onset of symptoms, even before birth. It is also possible to identify symptomless carriers of potentially harmful recessive alleles. Genes have already been cloned for many human diseases, including hemophilia, phenylketonuria (PKU), cystic fibrosis, and Duchenne muscular dystrophy (see Chapters 14 and 15).

Hybridization analysis makes it possible to detect abnormal allelic forms of genes present in DNA samples. Even in cases where the gene has not yet been cloned, the presence of an abnormal allele can be diagnosed with reasonable accuracy if a closely linked RFLP marker has been found (FIGURE 20.10). Alleles for Huntington's disease and a number of other genetic diseases were previously detected in this indirect way. Once a gene is mapped more precisely, it can be cloned for study and for use as a probe for finding identical or similar DNA, as is now the case for Huntington's disease, cystic fibrosis, and many other diseases.

Human Gene Therapy

Improved techniques for gene manipulation combined with a deeper understanding of gene function in the body may some day enable medical scientists to correct genetic disorders in individuals. Attempts at human gene therapy have not yet produced any proven benefits to patients, contrary to some claims in the popular media. However, for any genetic disorder traceable to a single defective allele, it should theoretically be possible to replace or supplement the defective allele with a functional, normal allele using recombinant DNA techniques. The new allele could be inserted into the somatic cells of the tissue affected by the disorder in a child or an adult, or perhaps even into germ line (gamete-producing) cells or embryonic cells.

For gene therapy of somatic cells to be permanent, the cells that receive the normal allele must be ones that multiply throughout the patient's life, so that the transplanted allele will be replicated and continually expressed. Bone marrow cells, which include the "stem cells" that give rise to all the cells of the blood and immune system, are prime candidates. FIGURE 20.11 outlines one possible procedure for a situation in which bone marrow cells are failing to produce a vital enzyme because of a single defective gene. Some bone marrow cells are removed from the patient, the normal allele inserted by a viral vector, and the modified cells returned to the patient.

Of the gene therapy trials now under way in humans, the most promising are ones that involve bone marrow cells but are not necessarily aimed at correcting genetic defects. For example, a number of researchers are seeking to enhance the abilities of immune cells to fight off cancer; others aim to engineer immune cells that are resistant to HIV. Most experiments to date have been preliminary, designed to test the safety and feasibility of a procedure rather than to attempt a cure.

Many technical questions are posed by gene therapy. For example, how can the proper genetic control mechanisms be made to operate on the transferred gene so that cells make appropriate amounts of the gene product at the right time and in the right place? How can we be sure that the insertion of the therapeutic gene does not harm some other necessary cell function? Information about DNA control elements from

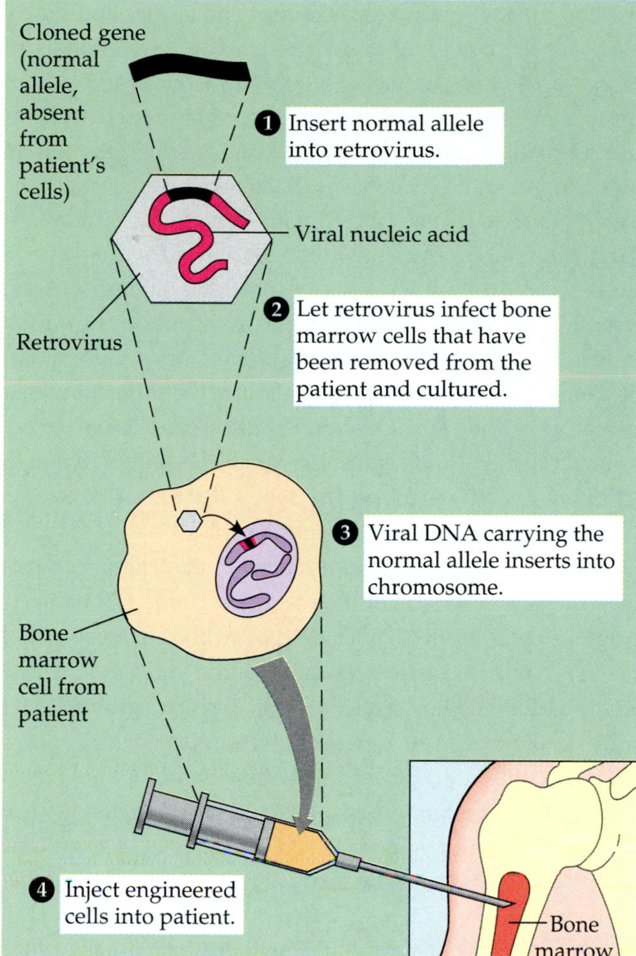
Cloned gene (normal allele, absent from patient's cells)
1 Insert normal allele into retrovirus.
Viral nucleic acid
Retrovirus
2 Let retrovirus infect bone marrow cells that have been removed from the patient and cultured.
3 Viral DNA carrying the normal allele inserts into chromosome.
Bone marrow cell from patient
4 Inject engineered cells into patient.
Bone marrow

FIGURE 20.11 ▪ One type of gene therapy. In this procedure, a disarmed retrovirus is used as a vector to introduce a normal allele of a gene into the cells of a patient who lacks it. The method takes advantage of the fact that a retrovirus inserts a copy of its nucleic acid (a DNA transcript of its RNA genome; see Chapter 18) into the chromosomal DNA of its host cell. If its nucleic acid includes a foreign gene and that gene is expressed, the cell—and its mitotic descendants—will possess the product of the gene and will potentially be cured. Cells that continue to reproduce throughout life, such as bone marrow cells, are ideal candidates for gene therapy.

genome sequencing and results of large-scale gene-expression studies that teach us about gene interactions may help answer such questions.

Gene therapy raises some difficult ethical and social questions. The most difficult ethical question is whether we should try to treat human germ line cells in the hope of correcting the defect in future generations. In mice at least, transferring foreign genes into the germ line (egg cells) is now a routine procedure in experiments. For example, scientists have created mice with sickle cell anemia, animals that will be immensely valuable in the study of this disease (see pp. 254–255). The researchers transplanted the human hemoglobin genes, including a double dose of the sickle-cell allele, into mice whose own hemoglobin genes had been deleted. Thus, despite the challenges, it is clear that the technical problems relating to similar genetic engineering in humans will eventually be solved. We will then have to face the question of whether it is advisable, under any circumstances, to alter the genomes of human germ lines or embryos.

Some critics suggest that tampering with human genes in any way, even in somatic cells and even to treat individuals who have life-threatening diseases, is wrong. They argue that it will inevitably lead to the practice of eugenics, a deliberate effort to control the genetic makeup of human populations. Other observers see no fundamental difference between genetic engineering of somatic cells and other conventional medical interventions to save lives. They compare the transplantion of genes to the transplantation of organs.

Pharmaceutical Products

DNA technology has been used to create many useful pharmaceutical products, mostly proteins. By transferring the gene for a desired protein product into a bacterium, yeast, or other kind of cell that is easy to grow in culture, one can produce large quantities of proteins that are present naturally in only minute amounts.

Using DNA technology to put highly active promoters (and other gene control elements) into vector DNA, scientists create expression vectors that enable the host cell to make large amounts of the product of a gene inserted into the vector. In addition, host cells can be engineered to secrete a protein as it is made, thereby simplifying the task of purifying it by traditional biochemical methods.

One of the first practical applications of gene splicing was the production of mammalian hormones and other mammalian regulatory proteins in bacteria. Human insulin and human growth hormone were among the earliest examples. The insulin produced in this way has greatly benefited the two million diabetics in the United States who depend on insulin treatment to control their disease; previously they had to rely on insulin from pigs and cattle, which is not identical to human insulin. Human growth hormone (HGH) has been a boon to children born with hypopituitarism, a form of dwarfism caused by inadequate amounts of HGH, and may prove to have other uses, such as the healing of injuries.

Another important pharmaceutical product produced by genetic engineering is tissue plasminogen activator (TPA). This protein helps dissolve blood clots and reduces the risk of subsequent heart attacks if administered very shortly after an initial attack. However, TPA illustrates a problem with many genetically engineered products: Because the development costs were high and the market is relatively limited, it is expensive.

The most recent developments in pharmaceutical products involve truly novel ways to fight certain diseases that do not respond to traditional drug treatments. One approach uses **antisense nucleic acid**, single-stranded molecules of DNA or

RNA that have been constructed explicitly to base-pair with mRNA molecules and block their translation. Interfering with crucial mRNAs involved in viral replication or the transformation of cells into a cancerous state could prevent the spread of the diseases. Another approach is the use of genetically engineered proteins that either block or mimic surface receptors on cell membranes. One such experimental drug mimics a receptor protein that HIV binds to in entering white blood cells. The HIV binds to the drug molecules instead and fails to enter the blood cells.

For many viral diseases for which there are no effective drug treatments, prevention by vaccination is virtually the only way to fight the disease. A **vaccine** is a harmless variant or derivative of a pathogen that stimulates the immune system to fight the pathogen. Traditional vaccines for viral diseases are of two types: particles of a virulent virus that have been inactivated by chemical or physical means, and active virus particles of an attenuated (nonpathogenic) viral strain. In both cases the virus particles are similar enough to the active pathogen to trigger an immune response (see Chapter 43).

Recombinant DNA techniques can generate large amounts of a specific protein molecule normally found on the surface of a pathogen. If the protein, referred to as a subunit, is one that triggers an immune response against the intact pathogen, then it can be used as a vaccine. Alternatively, genetic engineering methods can be used to modify the genome of the pathogen in order to attenuate it. A vaccine consisting of an attenuated microbe is often more effective than a subunit vaccine, because it usually triggers a greater response by the immune system. Pathogens attenuated by gene-splicing techniques may be safer than the natural mutants traditionally used.

DNA technology offers forensic, environmental, and agricultural applications

Forensic Uses of DNA Technology

In violent crimes, blood or small amounts of other tissue may be left at the scene or on the clothes or other possessions of the victim or assailant. If rape is involved, small amounts of semen may be recovered from the victim's body. If enough tissue or semen is available, forensic laboratories can determine the blood type or tissue type by using antibodies to test for specific cell-surface proteins. However, such tests require fairly fresh tissue in relatively large amounts. Also, because there are many people in the population with the same blood or tissue type, this approach can only exclude a suspect; it cannot provide strong evidence of guilt.

DNA testing, on the other hand, can identify the guilty individual with a much higher degree of certainty, because the DNA sequence of every person is unique (except for identical twins). RFLP analysis by Southern blotting is a powerful method for the forensic detection of similarities and differences in DNA samples and requires only tiny amounts of blood or other tissue (about 1000 cells). For example, in a murder case this method can be used to compare DNA samples from the suspect, the victim, and a small amount of blood found at the crime scene. Radioactive probes mark the electrophoresis bands that contain certain RFLP markers. Usually the forensic scientist tests for about five markers; in other words, only a few selected portions of the DNA are tested. However, even such a small set of markers from an individual can provide a **DNA fingerprint**, or specific pattern of bands, that is of forensic use, because the probability that two people (who are not identical twins) would have the exact same set of RFLP markers is very small. The autoradiograph in FIGURE 20.12 resembles the type of evidence presented (with explanation) to juries in murder trials.

Today, instead of RFLPs, variations in the lengths of satellite DNA are increasingly used as markers for DNA fingerprinting. Recall from Chapter 19 that satellite DNA consists of tandemly repeated base sequences within the genome. The most useful satellite sequences for forensic purposes are microsatellites, which are roughly 10–100 base pairs long, have repeating units of only a few base pairs, and are highly variable from person to person. For example, one individual may have the unit ACA repeated 65 times at one genome locus, 118 times at a second locus, and so on, whereas another individual is likely to have different numbers of repeats at these loci. Such polymorphic genetic loci are usually called **simple tandem repeats (STRs)**. Restriction fragments containing STRs vary in size among individuals because of differences in STR lengths, rather than because of different numbers of restriction sites within that region of the genome, as in RFLP analysis. The greater the number of markers exam-

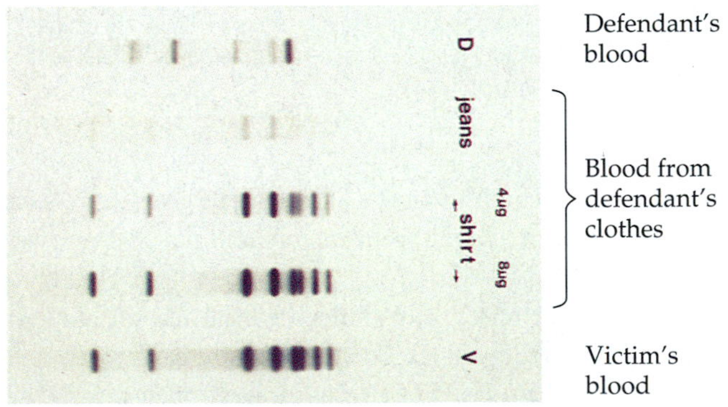

FIGURE 20.12 ▪ DNA fingerprints from a murder case. As revealed by RFLP analysis, DNA from bloodstains on the defendant's clothes matches the DNA fingerprint of the victim but differs from the DNA fingerprint of the defendant. This is evidence that the blood on the defendant's clothes came from the victim, not the defendant. Ten different RFLP markers appear at different positions on this one "autorad" (autoradiograph). It is now more common to make separate autorads for each marker tested. The DNA bands resulting from electrophoresis are exposed to various probes in succession, with the previous probe washed off before the next one is applied.

Cellmark Diagnostics, Germantown, MD.

ined in a DNA sample, the more likely it is that the DNA fingerprint is unique to one individual. PCR is often used to selectively amplify particular STRs or other markers before electrophoresis. Because of its selective power, PCR is especially valuable when the DNA is in poor condition or available only in minute quantities. A tissue sample as small as 20 cells can be sufficient for PCR.

Just how reliable is DNA fingerprinting? The DNA fingerprint of an individual would be truly unique if it were feasible to perform restriction fragment analysis on the person's entire genome. In practice, as already mentioned, forensic DNA tests focus on only about five tiny regions of the genome. However, the DNA regions chosen are ones known to be highly variable from one person to another. In most forensic cases, the probability of two people having identical DNA fingerprints is between one chance in 100,000 and one in a billion. The exact figure depends on the number of markers compared and on the frequency of those markers in the population. Information on how common various markers are in different ethnic groups is key because these marker frequencies may be very different from frequencies in the population as a whole. Such data now enable forensic scientists to make extremely accurate statistical calculations. Thus, despite problems that can still arise from insufficient statistical data, human error, or flawed evidence, DNA fingerprints are now accepted as compelling evidence by legal experts and scientists alike. Many argue that DNA evidence is more reliable than eyewitnesses in placing a suspect at the scene of a crime. The O. J. Simpson murder trial of 1995 made "DNA fingerprint" a household term, and this type of evidence will have an increasing forensic impact.

Environmental Uses of DNA Technology

Increasingly, genetic engineering is being applied to environmental work. The ability of microorganisms to transform chemicals is remarkable, and scientists are now engineering these metabolic capabilities into organisms that will help cope with some environmental problems. For example, many bacteria can extract heavy metals, such as copper, lead, and nickel, from their environments and incorporate the metals into compounds such as copper sulfate or lead sulfate, which are readily recoverable. Genetically engineered microbes may become important in both mining minerals (especially as ore reserves are depleted) and cleaning up highly toxic mining wastes.

The metabolic diversity of microbes is also employed in dealing with wastes from other sources. Sewage treatment plants rely on the ability of microbes to degrade many organic compounds into nontoxic form. However, an increasing number of potentially harmful compounds being released into the environment are not readily degraded by naturally occurring microbes; chlorinated hydrocarbons are a prime example. Biotechnologists are trying to engineer microbes to degrade these compounds. These microbes could be used in waste-water treatment plants or by manufacturers before the compounds are ever released into the environment.

A related research area is the identification and engineering of microbes capable of detoxifying specific toxic wastes found in spills and waste dumps. For example, bacterial strains have been developed that can degrade some of the compounds released during oil spills. The ability to move the genes responsible for these transformations into different organisms allows the development of strains that can survive the harsh conditions of these environmental disasters and still help detoxify the wastes.

Agricultural Uses of DNA Technology

Scientists are working to learn more about the genomes of agriculturally important plants and animals, and are already using genetic engineering to improve agricultural productivity.

Animal Husbandry. For over a decade, farm animals have been treated with products made by recombinant DNA methods. These products include new or redesigned vaccines, antibodies, and growth hormones. For example, some milk cows are being injected with bovine growth hormone (BGH), made by *E. coli*, in order to raise milk production (it usually increases by about 10%). BGH also improves weight gain in beef cattle. BGH has passed all safety tests so far, and it is now being used extensively in dairy herds. (However, some countries presently refuse to import such milk.) Another protein made by engineered *E. coli* that is useful for agriculture is the enzyme cellulase, which hydrolyzes cellulose and makes it possible for virtually all parts of a plant to be used for animal feed.

A number of **transgenic organisms**, organisms that contain genes from another species, have been developed for potential agricultural use. Transgenic animals, including beef and dairy cattle, hogs, sheep, and several species of commercially raised fishes, are being produced by injecting foreign DNA into the nuclei of egg cells or early embryos. For instance, rainbow trout and salmon that are given a foreign growth hormone gene can reach in one year a size that usually requires two or three years of growth. (Transgenic organisms are also used in basic research, as mentioned earlier.)

Genetic Engineering in Plants. In one striking way, plants have thus far proved easier to engineer than most animals. For many plant species, an adult plant can be regenerated from a single cell grown in culture (see FIGURE 38.14). Thus, genetic manipulations can be performed on a single cell, which can then be used to regenerate a new organism with new traits. Commercially important plants that readily grow from single somatic cells include cabbage, citrus fruits, carrots, alfalfa, tomatoes, potatoes, and tobacco.

The usual DNA vector for moving genes into plants is a plasmid of the bacterium *Agrobacterium tumefaciens*. In

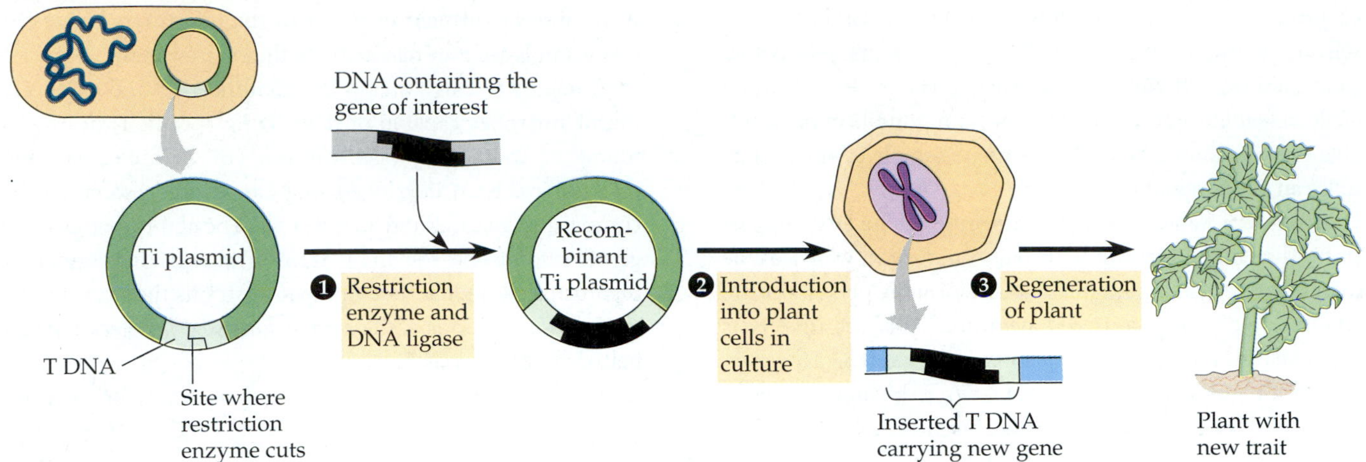

FIGURE 20.13 ▪ Using the Ti plasmid as a vector for genetic engineering in plants. ① The Ti plasmid is isolated from the bacterium *Agrobacterium tumefaciens,* and a fragment of foreign DNA is inserted into its T region by standard recombinant DNA techniques. ② When the recombinant plasmid is introduced into cultured plant cells, the T DNA integrates into the plant's chromosomal DNA. ③ As the plant cell divides, each of its descendants receives a copy of the T DNA and any foreign genes it carries. If an entire plant is regenerated, all its cells will carry—and may express—the new genes.

nature, *A. tumefaciens* infects plants and causes tumors called crown galls. The tumors are induced by the plasmid, called the **Ti plasmid** (Ti for tumor-inducing). The Ti plasmid integrates a segment of its DNA, known as T DNA, into the chromosomal DNA of its host plant cells. For vector purposes, researchers work with a version of the plasmid that does not cause disease.

Foreign genes can be inserted into the Ti plasmid using recombinant DNA techniques. The recombinant plasmid is either put back into *Agrobacterium,* which can then be used to infect plant cells growing in culture, or introduced directly into plant cells, where it inserts itself into the plant's chromosomes. Then, taking advantage of the capacity of those cells to regenerate whole plants, it is possible to produce plants that contain and express the foreign gene and pass it on to their offspring (FIGURE 20.13).

A major drawback to using the Ti plasmid as a vector is that only dicots (plants with two seed leaves) are susceptible to infection by *Agrobacterium.* Monocots, including agriculturally important grasses such as corn and wheat, cannot be infected by *Agrobacterium.* Fortunately, scientists can use newer techniques, such as electroporation and DNA guns (see FIGURE 38.15), to get DNA into cells of these plants.

Genetic engineering is quickly replacing traditional plant breeding programs, especially in cases where useful traits are determined by one gene or only a few genes. Of the genetically engineered varieties of plants now in field trials, over 40% have received genes for herbicide resistance. For example, several companies have developed strains of cotton carrying a bacterial gene that makes the plants resistant to the herbicides used by many farmers to control weeds. This gene should make it easier to grow crops while still ensuring that weeds are destroyed. In addition, a number of crop plants are being engineered to resist infectious pathogens and pest insects. Growing insect-resistant plants will reduce the need to apply chemical insecticides to crops.

The first genetically engineered fruits approved by the FDA for human consumption were tomatoes engineered with antisense genes that retard spoilage (see FIGURE 39.9). After cloning the tomato gene that codes for the enzyme mainly responsible for ripening, researchers prepared a gene whose template strand had a base sequence complementary to the normal gene—that is, an antisense version of the gene. When spliced into the DNA of a tomato plant, the antisense gene is transcribed into RNA that is complementary to the ripening gene's mRNA. The antisense RNA binds to the normal mRNA, blocking the synthesis of the enzyme. The engineered tomatoes produce only about 1% of the normal amount of enzyme and are less likely to spoil before reaching the market.

Many crop plants will soon be made more productive by the genetically engineered enlargement of their agriculturally valuable parts—roots, leaves, flowers, or stems. Researchers are also making strides toward improving the food value of plants, as in engineering grains to produce storage proteins containing a mix of amino acids more suitable for the human diet.

The Nitrogen-Fixation Challenge. Perhaps the most exciting potential use of DNA technology in agriculture involves nitrogen fixation, a bacterial process that benefits plants and their consumers. Nitrogen fixation is the conversion of atmospheric nitrogen gas (N_2), which cannot be utilized by plants, into nitrogen-containing compounds that plants can take up from the soil and use for making essential organic molecules, such as amino acids and nucleotides (see Chapter 37). The bacteria that fix nitrogen live either in the soil or within the roots of certain plants, such as beans and alfalfa. Since the availability of appropriate nitrogen compounds is often the limiting factor in plant growth and crop yield, modern agriculture makes extensive use of chemically synthesized nitrogen fertilizers to supplement bacterial nitrogen fixation. DNA technology offers ways to reduce the use of expensive and polluting fertilizers by increasing the nitrogen-fixing capability of bacteria. In the future it may also be possible to design nitrogen-fixing bacteria that can live in the tissues of nitrogen-demanding plants such as corn and wheat. The ultimate challenge is to engineer crop plants that fix nitrogen themselves.

DNA technology raises important safety and ethical questions

As soon as its potential power became apparent, DNA technology raised questions about possibly dangerous consequences. The earliest concerns were that genetic manipulations of microorganisms could create hazardous new pathogens, which might escape from the laboratory. In response to these concerns, scientists developed a self-monitoring approach—an honor system whereby researchers would adhere to a set of voluntary and self-imposed guidelines to ensure safety. This early approach soon led to formal regulatory programs administered by federal agencies. Today, governments and regulatory agencies throughout the world are grappling with how to promote the potential industrial, medical, and agricultural revolution engendered by biotechnology while ensuring that new products are safe. In the United States, the FDA, the National Institutes of Health Recombinant DNA Advisory Committee, the Department of Agriculture (USDA), and the Environmental Protection Agency (EPA) share responsibility for setting policies and regulating new developments in genetic engineering.

In the past decade, hundreds of genetically designed products and new strains of organisms have been developed. Many have been exhaustively tested, and these, at least, seem to pose little or no threat to humans or the environment. The tests also confirm that genetic engineering holds enormous potential for improving human health and increasing agricultural productivity. Nevertheless, safety remains an important question. The benefits of gene therapy, for example, must be weighed against the need to assure that gene vectors are safe. In the case of environmental problems—oil spills, for example, and chemical wastes that threaten our soil, water, and air—genetically engineered organisms may be part of the solution, but their own impact on the environment must be considered before they are widely used.

With new medical products, the main cause for concern is the potential for harmful side effects, both short-term and long-term. Hundreds of new genetically engineered diagnostic products, vaccines, and drugs, including some designed to treat AIDS and certain forms of cancer, currently await federal approval. Before the FDA considers a new medical product for general marketing, the substance must pass exhaustive tests in laboratory animals and humans.

There is heated debate about genetically engineered agricultural products because of the potential dangers of introducing new organisms into the environment. Some scientists argue that producing transgenic organisms by gene splicing is only an extension of traditional crossbreeding, or hybridization—the procedure that has given us the tangelo (a tangerine-grapefruit hybrid) and beefalo (a cow-buffalo hybrid). Since no hybrid crops or animals were ever tested for safety before they were marketed, it is argued that genetically engineered organisms need not be treated differently. In general, the FDA has held that if the result of genetic engineering is not significantly different from a product already on the market, testing is not required.

On the other side of the argument are scientists who believe that creating transgenic organisms by splicing genes from bacteria or animals into plants, and vice versa, is radically different from hybridizing closely related species of plants or animals. They fear that foods produced by gene splicing will contain new proteins that are toxic or allergenic in some people.

Another consideration is that genetically engineered crop plants may pass their new genes to close relatives in nearby wild areas. Lawn and crop grasses, for example, commonly exchange genes with wild relatives via pollen transfer, and there is little doubt that if some species received new genes, they would pass them to wild plants. If domestic plants carrying genes for resistance to herbicides, natural diseases, and/or insect pests pollinate wild ones, the offspring could become "superweeds" that are exceedingly difficult to control. Researchers are looking for ways to prevent the escape of engineered plant genes. Among the possibilities are crop isolation and the engineering of plants so that they cannot hybridize.

As with all new technologies, developments in DNA technology have ethical overtones. Obtaining a complete map of the human genome, for example, will open the door for significant advances in gene therapy. It will also raise significant ethical questions. Who should have the right to examine someone else's genes? How should that information be used? Should a person's genome be a factor in suitability for a job or eligibility for insurance? Ethical considerations, as well as concerns about potential environmental and health hazards, will likely slow the application of the products of the new biotechnology. There is always a danger that too much regulation will stifle basic research and its potential benefits. However, the power of genetic engineering—our ability to profoundly and rapidly alter species that have been evolving for millennia—demands that we proceed with humility and caution.

REVIEW OF KEY CONCEPTS

(with page numbers and key figures)

DNA CLONING

- **DNA technology makes it possible to clone genes for basic research and commercial applications: *an overview* (pp. 364–366, FIGURE 20.1)** DNA technology is a powerful set of techniques that enables biologists to manipulate and analyze DNA. It can help make useful new products and organisms.

- **Restriction enzymes are used to make recombinant DNA (pp. 366–367, FIGURE 20.2)** A variety of bacterial restriction enzymes recognize short, specific nucleotide sequences in DNA and cut the sequences at specific points on both strands to yield a set of double-stranded DNA fragments with single-stranded sticky ends. The sticky ends readily form base pairs with complementary single-stranded segments on other DNA molecules. The enzyme DNA ligase can seal the strands to produce recombinant DNA molecules.

- **Genes can be cloned in recombinant DNA vectors: *a closer look***
20.1 **(pp. 367–370, FIGURES 20.3–20.5)** Plasmids can serve as vectors (carriers) to introduce foreign genes into host bacteria. Recombinant DNA is made by inserting restriction fragments from DNA containing a gene of interest into the vector DNA, which has been cut open by the same enzyme. Gene cloning results when the foreign genes replicate inside the host bacterial cell as part of the recombinant vector. Certain eukaryotic cells can also serve as host cells for gene cloning. Cell clones carrying the gene of interest can be identified with a radioactively labeled nucleic acid probe, which has a sequence complementary to the gene.

- **Cloned genes are stored in DNA libraries (pp. 370–371, FIGURE 20.6)** When the starting material for DNA (gene) cloning is an entire genome, the collection of recombinant-vector clones produced is called a genomic library. Alternatively, a cDNA (complementary DNA) library can be made by cloning DNA made in vitro by reverse transcription of all the mRNA produced by a particular kind of cell. Libraries of cDNA are especially useful for working with eukaryotic genes (whose introns are not present in the cDNA versions) and for studying gene expression.

- **The polymerase chain reaction (PCR) clones DNA entirely in vitro (pp. 371–372, Methods Box)** For quickly making many copies of a particular segment of DNA, this easily automated method uses primers that bracket the desired sequence and a special heat-resistant DNA polymerase.

ANALYSIS OF CLONED DNA

- **Restriction fragment analysis detects DNA differences that affect restriction sites (pp. 372–373, FIGURE 20.7, 2 Methods Boxes pp. 374 and 375)** Gel electrophoresis makes it possible to separate and isolate DNA restriction fragments of different lengths. Restriction fragment length polymorphisms (RFLPs) are differences in DNA sequence on homologous chromosomes that result in different patterns of restriction fragment lengths. These patterns are visualized as bands on gel electrophoresis. Specific fragments can be identified by Southern blotting, using labeled probes that hybridize to the DNA stuck to a "blot" of the gel. RFLPs are prevalent genetic markers, being present throughout the masses of eukaryotic noncoding DNA. RFLP analysis has many applications, including genetic mapping and diagnosis of genetic disorders.

- **Entire genomes can be mapped at the DNA level (pp. 373–379, FIGURES 20.8, 20.9, Methods Box p. 378)** An international research effort, the Human Genome Project involves linkage mapping, physical mapping, and sequencing of the entire human genome and the genomes of other organisms. The DNA sequencing depends on fast sequencing machines and the analysis of the data by computer software. The results will help researchers answer questions about molecular evolution, probe details of gene organization and control, and produce and catalog proteins of interest. Learning the functions of the genes discovered is a major task.

PRACTICAL APPLICATIONS OF DNA TECHNOLOGY

- **DNA technology is reshaping medicine and the pharmaceutical industry (pp. 379–382; FIGURES 20.10, 20.11)** Medical applications of DNA technology include diagnostic tests for genetic and other diseases; safer, more effective vaccines; the large-scale production of many new, and some previously scarce, pharmaceutical products; and the ultimate prospect of curing and preventing certain genetic disorders.

- **DNA technology offers forensic, environmental, and agricultural applications (pp. 382–385; FIGURES 20.12, 20.13)** DNA "fingerprints" obtained from RFLP or STR analysis of tissue found at the scenes of violent crimes provide evidence in trials; such fingerprints are also useful in paternity disputes. Genetic engineering can modify the metabolism of microorganisms such that they can be used to extract minerals from the environment or degrade waste materials. In agriculture, transgenic plants and animals (those containing genes from other species) are being designed to improve food productivity and quality.

- **DNA technology raises important safety and ethical questions (p.385)** Several U.S. government agencies are responsible for setting policies and regulating recombinant DNA technology. The potential benefits of genetic engineering must be carefully weighed against the potential hazards of creating products or developing procedures that are harmful to humans or the environment.

SELF-QUIZ

1. Which of the following tools of recombinant DNA technology is *incorrectly* paired with its use?
   a. restriction enzyme—production of RFLPs
   b. DNA ligase—enzyme that cuts DNA, creating the sticky ends of restriction fragments
   c. DNA polymerase—used in a polymerase chain reaction to amplify sections of DNA
   d. reverse transcriptase—production of cDNA from mRNA
   e. electrophoresis—DNA sequencing

2. Which of the following would *not* be true of cDNA produced using human brain tissue as the starting material?
   a. It could be amplified by the polymerase chain reaction.
   b. It could be used to create a complete genomic library.
   c. It is produced from mRNA using reverse transcriptase.
   d. It could be used as a probe to locate a gene of interest.
   e. It lacks the introns of the human genes and thus can probably be introduced into phage vectors.

3. Plants are more readily manipulated by genetic engineering than are animals because
   a. plant genes do not contain introns
   b. more vectors are available for transferring recombinant DNA into plant cells
   c. a somatic plant cell can often give rise to a complete plant
   d. genes can be inserted into plant cells by microinjection
   e. plant cells have larger nuclei

4. A paleontologist has recovered a bit of tissue from the 400-year-old preserved skin of an extinct dodo (bird). The researcher would like to compare DNA from the sample with DNA from

living birds. Which of the following would be most useful for increasing the amount of dodo DNA available for testing?

a. RFLP analysis
b. polymerase chain reaction (PCR)
c. electroporation
d. gel electrophoresis
e. Southern hybridization

5. Expression of a cloned eukaryotic gene in a prokaryotic cell involves many difficulties. The use of mRNA and reverse transcriptase is part of a strategy to solve the problem of

a. posttranscriptional processing
b. electroporation
c. posttranslational processing
d. nucleic acid hybridization
e. restriction fragment ligation

6. DNA technology has many medical applications. Which of the following is *not yet* done routinely?

a. production of hormones for treating diabetes and dwarfism
b. production of viral subunits of viruses for vaccines
c. introduction of genetically engineered genes into human gametes
d. prenatal identification of genetic disease genes
e. genetic testing for carriers of harmful alleles

7. RFLP analysis is being used as evidence to link suspects with blood and tissues found at crime scenes. DNA fingerprints look something like bar codes. The pattern of bars in a DNA fingerprint shows

a. the order of bases in a particular gene
b. the individual's genotype
c. the order of genes along particular chromosomes
d. the presence of dominant or recessive alleles for particular traits
e. the presence of certain DNA restriction fragments

8. Which of the following sequences in double-stranded DNA is most likely to be recognized as a cutting site for a restriction enzyme that creates fragments with sticky ends?

a. A A G G
   T T C C
b. A G T C
   T C A G
c. G G C C
   C C G G
d. A C C A
   T G G T
e. A A A A
   T T T T

9. In recombinant DNA methods, the term *vector* can refer to

a. the enzyme that cuts DNA into restriction fragments
b. the sticky end of a DNA fragment
c. a RFLP marker
d. a plasmid used to transfer DNA into a living cell
e. a DNA probe used to identify a particular gene

10. The template used to make cDNA is

a. DNA
b. mRNA
c. a plasmid
d. a DNA probe
e. a restriction fragment

CHALLENGE QUESTION

A neurophysiologist hopes to study a gene that codes for a neurotransmitter protein in human brain cells. She knows the amino acid sequence of the protein. Briefly explain how she might (a) identify only the genes that are expressed in a specific type of brain cell, (b) identify the gene that codes for the neurotransmitter, (c) produce multiple copies of the gene for study, and (d) produce a quantity of the neurotransmitter for evaluation as a potential medication.

SCIENCE, TECHNOLOGY, AND SOCIETY

Do you think there is a potential in our society for genetic discrimination based on testing for "harmful" genes? What policies can you suggest that would prevent abuses of genetic testing?

FURTHER READING

Beardsley, T. "Genes in the Not So Public Domain." *Scientific American*, April 1995. Who owns the base sequences in DNA databases?

Pennisi, E. "Genome Data Shakes Tree of Life." *Science*, May 1, 1998. New genome sequences reveal unexpected evolutionary connections.

Rennie, J. "Grading the Gene Tests." *Scientific American*, June 1994. Discusses genetic testing and its ethical dilemmas.

Rothenberg, K., et al. "Genetic Information and the Workplace: Legislative Approaches and Policy Challenges." *Science*, March 21, 1997.

Science, October 24, 1997. "Genome Issue." Every October, *Science* publishes an issue devoted to the Human Genome Project and related topics.

Scientific American, June 1997. "Making Gene Therapy Work." A special section consisting of five articles on current progress toward gene therapy.

Travis, J. "Inner Strength." *Science News*, March 14, 1998. Explores the potential use of gene therapy to build cells that thwart HIV replication.

Velander, W. H., H. Lubon, and W. N. Drohan. "Transgenic Livestock as Drug Factories." *Scientific American*, January 1997. Describes how dairy animals can be made to produce therapeutic human proteins in their milk.

Watzman, H. "DNA to Unravel Secrets of the Dead Sea Scrolls." *New Scientist*, January 14, 1995. Discusses DNA technology as a tool in historical research.

WEB LINKS

Visit the special edition of *The Biology Place* for BIOLOGY, Fifth Edition, at http://www.biology.com/campbell. Go to Chapter 20 for online resources, including learning activities, practice exams, and links to the following web sites:

"Biotechnology Information Center"
This searchable database provides access to a variety of information services and publications covering many aspects of agricultural biotechnology.

"FoodFuture"
An interesting site that makes modern DNA technology easily understandable.

"BioTech Life Science Dictionary"
This biotechnology dictionary currently contains definitions of more than 7300 biotechnological terms.

"Information Systems for Biotechnology (ISB)"
Funded by the USDA, ISB provides a balanced view of the potential benefits and dangers of DNA technology.

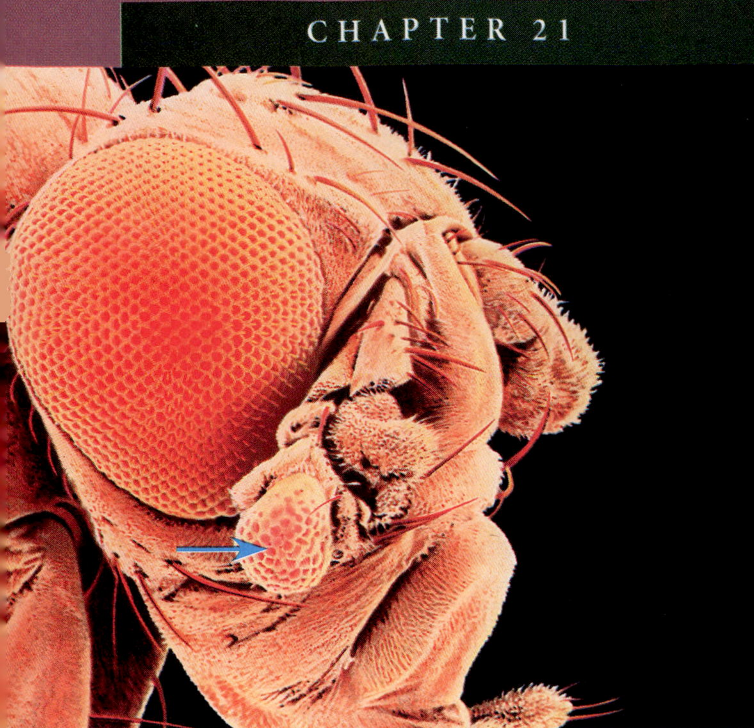

CHAPTER 21

THE GENETIC BASIS OF DEVELOPMENT

From Single Cell to Multicellular Organism

- Embryonic development involves cell division, cell differentiation, and morphogenesis
- Researchers study development in model organisms to identify general principles: *science as a process*

Differential Gene Expression

- Different types of cells in an organism have the same DNA
- Different cell types make different proteins, usually as a result of transcriptional regulation
- Transcriptional regulation is directed by maternal molecules in the cytoplasm and signals from other cells

Genetic and Cellular Mechanisms of Pattern Formation

- Genetic analysis of *Drosophila* reveals how genes control development: *an overview*
- Gradients of maternal molecules in the early embryo control axis formation
- A cascade of gene activations sets up the segmentation pattern in *Drosophila: a closer look*
- Homeotic genes direct the identity of body parts
- Homeobox genes have been highly conserved in evolution
- Neighboring cells instruct other cells to form particular structures: cell signaling and induction in the nematode
- Plant development depends on cell signaling and transcriptional regulation: *science as a process*

This chapter applies much of what you've learned about molecules, cells, and genes to one of biology's most important questions—how a complex multicellular organism develops from a single cell. The application of genetic analysis and DNA technology to the study of development has brought about a revolution in the field. In much the same way that researchers have used mutations to deduce pathways of cellular metabolism, they now use mutations to dissect developmental pathways. In one striking example, Swiss researchers demonstrated in 1995 that a particular gene functions as a master switch that triggers the development of an eye in Drosophila. *The scanning electron micrograph on this page shows the head of an abnormal fly with small extra eyes on its antennae (the blue arrow points to one of these eyes). Expression of the master gene for eye development in an abnormal location in the fly caused the extra eyes. The same gene also triggers eye development in mice and other mammals. In fact, developmental biologists are discovering remarkable similarities in the mechanisms that shape diverse organisms.*

The scientific study of development got under way about a century ago, at roughly the same time as genetics. But for decades the two disciplines proceeded along mostly separate paths. We have seen how genetics advanced from Mendel's laws to an understanding of the molecular basis of inheritance. Meanwhile, developmental research focused on embryology, the study of the stages of development leading from fertilized egg to fully formed organism. Only recently have the concepts and tools of molecular genetics reached the point where a real synthesis has been possible. The synthesis is a challenge, for it means relating the linear information in genes to a process of development that takes place in four dimensions, three of space and one of time.

This chapter introduces some of the basic genetic and cellular mechanisms that control development. It focuses on principles that apply to both animals and plants, with emphasis on two invertebrate animals: Drosophila melanogaster *and* Caenorhabditis elegans *(a nematode, or roundworm). In later chapters you will learn much more about the development of plants (Chapter 38) and vertebrate animals (Chapter 47).*

A capstone to the genetics unit, this chapter also serves as a bridge to the rest of the book, for understanding development is crucial to understanding the evolution, diversity, structure, function, and ecology of organisms.

FROM SINGLE CELL TO MULTICELLULAR ORGANISM

In the development of most multicellular organisms, a single-celled zygote (fertilized egg) gives rise to cells of many different types, each type with a different structure and corresponding function. For example, an animal will have muscle

(a) (b)

FIGURE 21.1 ▪ From fertilized egg to animal: what a difference a week makes. It took just one week for cell division, differentiation, and morphogenesis to transform this fertilized frog egg (left) into a hatching tadpole (right).

cells that enable it to move and nerve cells that transmit signals to the muscle cells; a plant's cells will include mesophyll cells that carry out photosynthesis and stomatal cells that regulate the passage of gases into and out of leaves (see FIGURE 10.2). As you know, cells are only one level in the hierarchy of biological order within a multicellular organism (see FIGURES 1.2 and 2.1). Cells of similar types are organized into tissues, tissues into organs, organs into organ systems, and organ systems into the whole organism. Thus, the process of embryonic development must not only give rise to cells of different types but to higher-level structures arranged in a particular way in three dimensions.

Embryonic development involves cell division, cell differentiation, and morphogenesis

An organism arises from a fertilized egg cell as the result of three interrelated processes: cell division, cell differentiation, and morphogenesis (FIGURE 21.1). Through a succession of mitotic divisions, the zygote gives rise to a large number of cells. Cell division alone, however, would produce a great ball of identical cells, which is nothing like an animal or plant. During embryonic development, cells not only increase in number, they also undergo **differentiation**, becoming specialized in structure and function. Moreover, the different kinds of cells aren't just mixed up randomly but are organized into tissues and organs. The physical processes that actually give shape to the organism and its various parts are called **morphogenesis**, which means "creation of form."

The processes of cell division, differentiation, and morphogenesis overlap in time. Early events of morphogenesis lay out the basic body plan very early in embryonic development, establishing, for example, which end of an animal embryo will be the head or which end of a plant embryo will become the roots. Cell division and differentiation play important roles in morphogenesis in all organisms, as does the death of certain cells. However, the overall schemes of morphogenesis in animals and plants are very different, and the mechanisms differ in two major ways (FIGURE 21.2, p. 390). In animals, but not in plants, *movements* of cells and tissues are necessary to convert the cell mass of the early embryo into the characteristic three-dimensional form of the organism. We will discuss morphogenetic movements in animal development in Chapter 47. The second major difference is that in plants, growth in overall size and morphogenesis are not limited to embryonic and juvenile periods but occur throughout the life of the plant. **Apical meristems**, which are perpetually embryonic regions in the tips of shoots and roots, are responsible for the plant's continual growth and formation of new organs such as leaves and roots.

The importance of precise regulation of morphogenesis is evident in human disorders that result from morphogenesis gone awry. For example, cleft palate, in which the upper wall of the mouth cavity fails to close completely, is a defect of morphogenesis.

As humans we may naturally be most interested in developmental processes in our own species. However, many aspects of development are much easier to study in other kinds of organisms.

Researchers study development in model organisms to identify general principles: *science as a process*

Much of the early research on animal development focused on animals that lay their eggs in water, in particular on amphibians such as frogs. Frogs have large eggs (2–3 mm in diameter) that are easy to observe and manipulate, and fertilization and development occur outside the mother's body. By studying these animals and others, biologists were able to work out a description of animal development at the macroscopic and microscopic levels, making a number of important discoveries in the process (see Chapter 47). Research on a variety of plants led to a basic understanding of plant development (see Chapter 38). When the primary research goal is to understand broad biological principles—of animal or plant development, in this case—the organism chosen for study is called a **model organism**. Researchers select model organisms that lend themselves to the study of a particular question and that are representative of a larger group. Frogs, for example, are useful model organisms for elucidating the role of cell movement in morphogenesis because frog development is easy to observe and fairly typical of vertebrate animals.

However, because of its complex genetics, the frog is not an ideal model organism for the more recent research efforts aimed at uncovering the connections between genes and

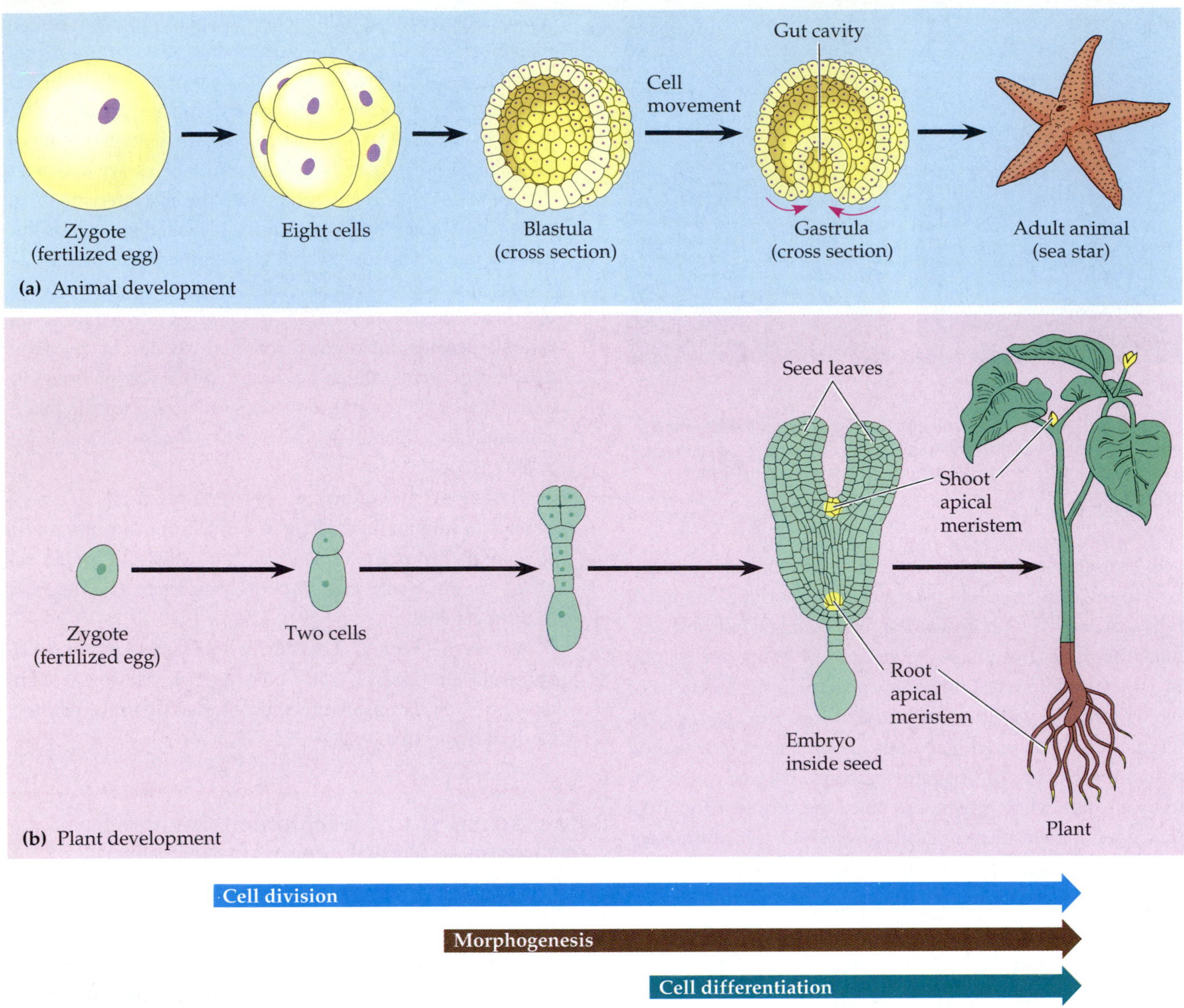

FIGURE 21.2 ▪ Some key stages of development in animals and plants. Cell division, morphogenesis, and cell differentiation are involved in both animal development and plant development. **(a)** Most animals go through some variation of the blastula and gastrula stages shown here in simplified form. The blastula is a sphere of cells surrounding a fluid-filled cavity. In forming the gastrula, a region of the blastula folds inward, forming a rudimentary gut cavity. The movement of cells and tissues plays an important role in animal morphogenesis. Biochemical events that will lead to cellular differentiation actually start before the gastrula forms. Once the animal is mature, differentiation occurs in only a limited way—for the replacement of damaged or lost cells. **(b)** In plants with seeds, a complete embryo develops within the seed. Morphogenesis, which does not involve cell or tissue movement, occurs throughout the plant's lifetime. Apical meristems (yellow) continuously arise and develop into the various plant organs as the plant grows to an indeterminate size.

development. Instead, many developmental biologists have turned to organisms that are more convenient for genetic analysis. For developmental genetics, the criteria for choosing a model organism include, in addition to readily observable embryos, short generation times, relatively small genomes, and, ideally, preexisting knowledge about the organism and its genes. Several organisms have emerged as favorites, including *Drosophila*, the nematode *C. elegans*, the mouse, the zebrafish, and the plant *Arabidopsis* (FIGURE 21.3).

First chosen as a model organism by the pioneering geneticist T. H. Morgan and intensively studied by generations of geneticists after him, the fruit fly *Drosophila melanogaster* is small and easily grown in the laboratory. As you read in Chapter 15, *Drosophila* has a generation time of only two weeks and produces many offspring. Embryos develop outside the mother's body, an asset for developmental studies. And researchers can draw on a vast amount of information and experience relating to this animal's genes and other aspects of its biology. One disadvantage of the fruit fly as a model organism for developmental research is that its early development is at least superficially quite different from the process depicted in FIGURE 21.2: The first rounds of mitosis occur without

(a) *Drosophila melanogaster* (fruit fly)

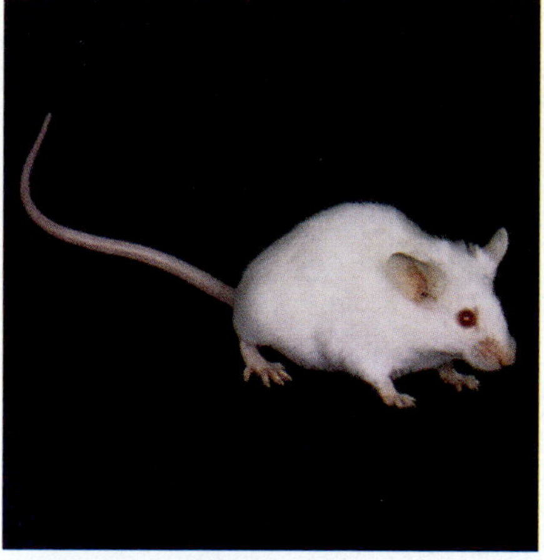
(c) *Mus musculus* (mouse)

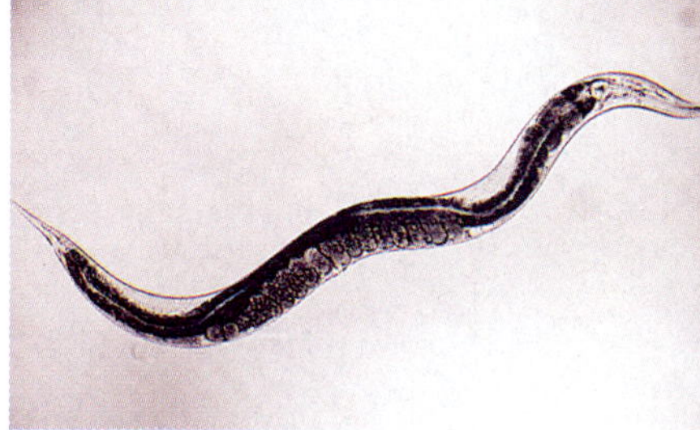
(b) *Caenorhabditis elegans* (nematode)

(d) *Danio rerio* (zebrafish)

(e) *Arabidopsis thaliana* (common wall cress)

FIGURE 21.3 ▪ Model organisms. Each of these organisms offers particular advantages for research on the genetics of development.

cytokinesis, leading to an early blastula containing a large number of nuclei within a single mass of cytoplasm. Nevertheless, research on *Drosophila* development has yielded deep insights into basic principles of animal development, as we will see.

The nematode *Caenorhabditis elegans* normally lives in soil but is easily grown in the laboratory in petri dishes. It is only about a millimeter long, has a simple, transparent body with only a few types of cells, and grows from zygote to mature adult in only three and a half days. One of its advantages for genetic studies is that most individuals are hermaphrodites, which produce both eggs and sperm. Hermaphrodites are convenient for genetic studies because recessive mutations are easy to detect. An individual with a recessive mutation in one copy of a gene will be a heterozygote (genotype *Aa*, for example, if *A* is the wild-type allele and *a* is the mutant allele) with the wild-type phenotype. But following self-fertilization (*Aa* × *Aa*), one-fourth of the offspring will be homozygous for the mutant allele *(aa)* and have a mutant phenotype.

For developmental biologists, a more important feature of *C. elegans* is that every adult (hermaphrodite) has exactly 959 somatic cells (some of which are fused together), and the cells always arise from the zygote in exactly the same way. Using a microscope to follow all the cell divisions starting immediately after a zygote forms, biologists have been able to reconstruct the entire ancestry of every cell in the adult body, the organism's complete **cell lineage**. A cell-lineage diagram like the one in FIGURE 21.4 (p. 392) is a type of *fate map*, a representation of the fates of various parts of a developing embryo. (You'll see more-traditional kinds of fate maps in Chapter 47.)

Among vertebrate animals, two in particular lend themselves to the genetic analysis of development, the mouse and the zebrafish. The mouse *Mus musculus* has a long history as a mammalian model, and much is known about its biology, including its genes. Moreover, researchers are now adept at manipulating mouse genes to make transgenic mice and mice in which particular genes are "knocked out" by mutation. But mice are complex animals with large genomes, and their embryos develop in the mother's uterus, hidden from view.

Many of the disadvantages of the mouse are absent in a newer vertebrate model, the zebrafish *Danio rerio*. These small fish (2–4 cm long) are easy to breed in the laboratory in large numbers, and transparent embryos develop outside the mother's body. Although the generation time is relatively long (2–4 months), the early stages of development proceed quickly: By 24 hours after fertilization, most of the tissues and early versions of the organs have formed, and by 2 days a tiny fish hatches out of the egg case. However, zebrafish embryos

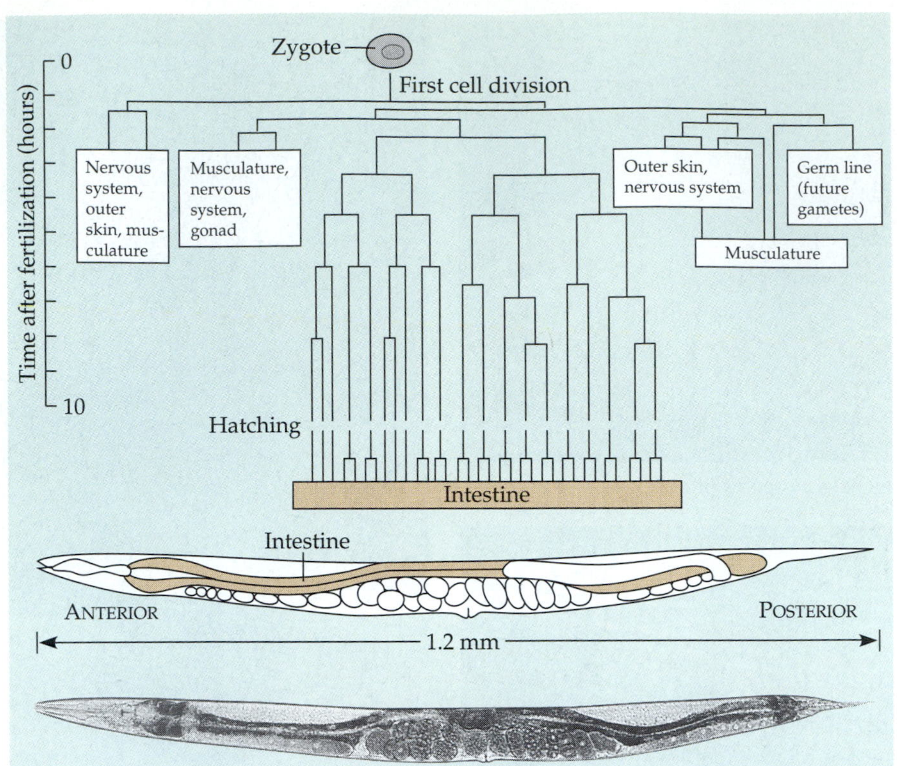

FIGURE 21.4 • Cell lineage in *C. elegans*. The nematode *Caenorhabditis elegans* is transparent at all stages of its development, making it possible for researchers to trace the lineage of every cell, from the zygote to the adult worm (LM). The diagram shows a detailed lineage only for the intestine, which is derived exclusively from one of the first four cells formed from the zygote. The intestinal cell lineage does not happen to include any cell death, an important aspect of the lineages for some other parts of the animal.

are somewhat difficult to manipulate experimentally, and a detailed genetic map does not yet exist for this organism.

For studying the molecular genetics of plant development, researchers are focusing on a small weed called *Arabidopsis thaliana* (the common wall cress, a member of the mustard family). One of these plants can grow in a test tube and produce thousands of progeny after 8–10 weeks; as in Mendel's pea plants, each flower makes both ova and sperm (in pollen). For gene-manipulation research, scientists can grow *Arabidopsis* cells in culture and can get these cells to take up foreign DNA (genetic transformation). Another advantage of *Arabidopsis* is its unusually small genome; at 70 million nucleotide pairs, it is even smaller than those of *C. elegans* and *Drosophila.*

Later in this chapter you will learn about important discoveries that researchers have made using some of these model organisms.

DIFFERENTIAL GENE EXPRESSION

We have stated on several previous occasions that differences between cells in a multicellular organism come almost entirely from differences in gene *expression,* not from differences in the cells' genomes. (There are a few exceptions, such as antibody-producing cells; see Chapter 19.) Furthermore, we have mentioned that these differences arise during development, as regulatory mechanisms turn specific genes on and off. Let's now look at some of the evidence for this assertion.

Different types of cells in an organism have the same DNA

Much evidence supports the conclusion that nearly all cells of an organism have *genomic equivalence*—that is, that they all have the same genes. What happens to these genes as a cell begins to differentiate? We can shed some light on this question by asking whether genes are irreversibly inactivated during differentiation. For example, does an epidermal cell in your finger contain a viable gene specifying eye color, or has the eye-color gene been destroyed or permanently inactivated there?

Totipotency in Plants

One experimental approach to the question of genomic equivalence is to try to generate a whole organism from differentiated cells of a single type. In many plant species, whole new individuals *can* develop from differentiated somatic cells. This was first demonstrated during the 1950s by F. C. Steward and his students at Cornell University, working with carrot plants (FIGURE 21.5). They found that cells removed from the root (the carrot) and placed in culture medium could grow into normal adult plants. In other words, they were able to clone carrot plants from differentiated cells. (Plant cloning is now used extensively in agriculture.) The fact that a mature plant cell can dedifferentiate and then give rise to all the different kinds of specialized cells of a new plant shows that differentiation does not necessarily involve irreversible changes

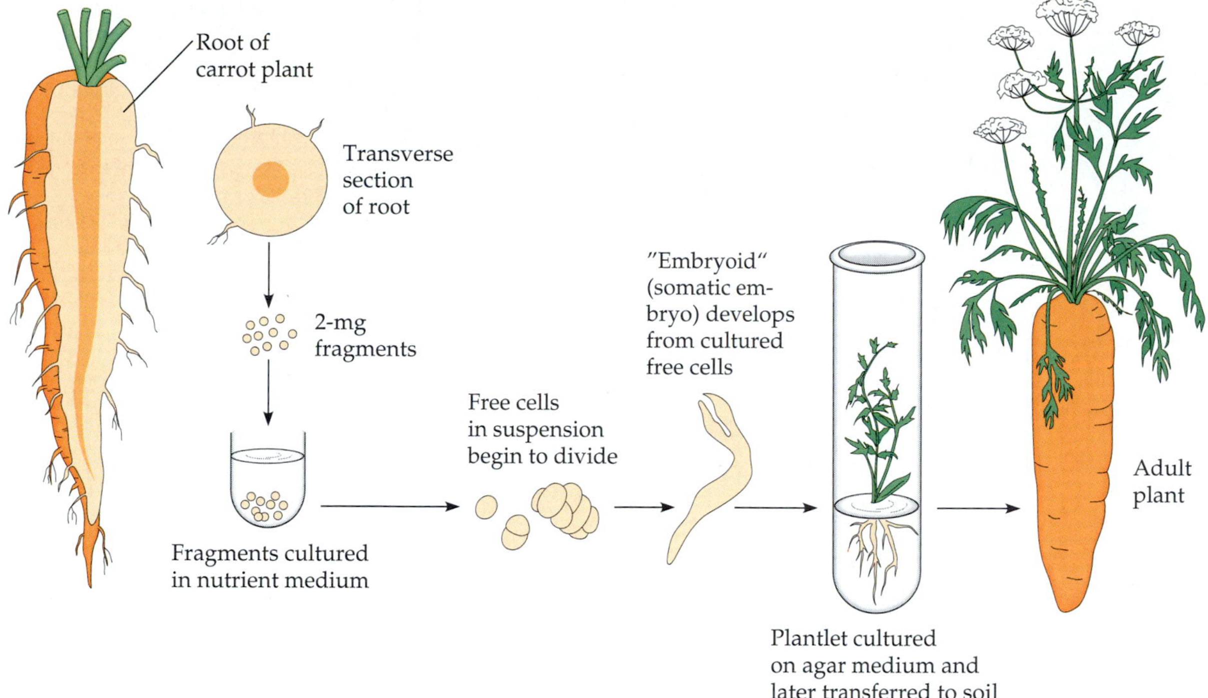

FIGURE 21.5 • Test-tube cloning of carrots. In classic experiments conducted during the 1950s, F. C. Steward and his students at Cornell University demonstrated that whole plants could be regenerated from somatic (nonreproductive) cells dissected from a carrot. The new plants that result are genetic duplicates of the parent plant.

in the DNA. In plants, at least, cells can remain **totipotent**; that is, they can retain the zygote's potential to form all parts of the mature organism.

Nuclear Transplantation in Animals

Differentiated cells from animals will often fail to divide in culture, much less develop into a new organism. Therefore, animal researchers have approached the genomic-equivalence question by replacing the nucleus of an unfertilized egg cell or zygote with the nucleus of a differentiated cell. Will a normal animal develop? The pioneering experiments in nuclear transplantation were carried out by American embryologists Robert Briggs and Thomas King during the 1950s and were later extended by British embryologist John Gurdon. These investigators removed or destroyed the nuclei of frog egg cells, then transplanted nuclei from embryonic and tadpole cells of the same species into the enucleated eggs (FIGURE 21.6). The ability of the transplanted nuclei to support normal development turned out to be inversely related to the age of the donor embryos. If the nuclei came from the relatively undifferentiated cells of an early embryo, most of the recipient eggs developed into tadpoles. But with nuclei from the differentiated intestinal cells of a tadpole, fewer than 2% of the eggs developed into normal tadpoles, and most of the embryos failed to make it through even the earliest stages of embryonic development.

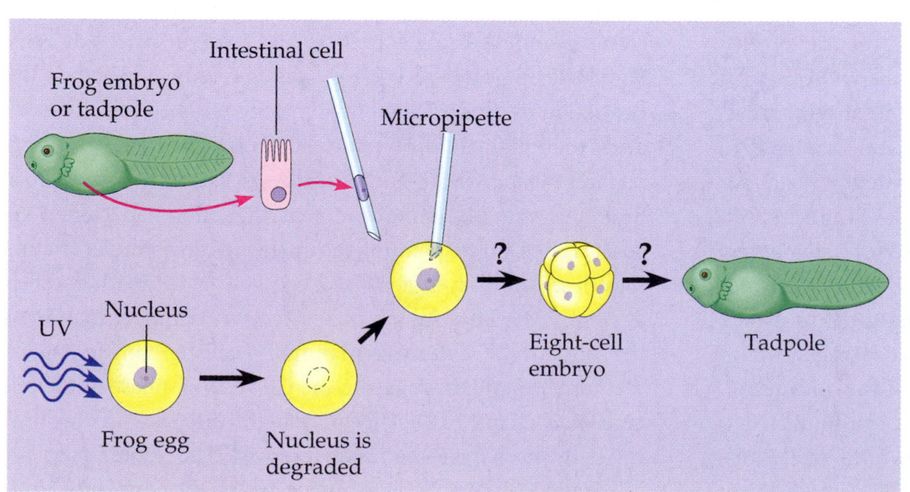

FIGURE 21.6 • Nuclear transplantation. After the frog egg nucleus is destroyed by ultraviolet (UV) radiation, a nucleus from a more advanced developmental stage is inserted into the egg to test whether nuclei change irreversibly as cells begin to differentiate. The earlier the developmental stage from which the nucleus comes, the more likely it will support development. Nuclei from very early embryonic stages frequently prove to be totipotent, whereas nuclei from late developmental stages (such as a tadpole) rarely are.

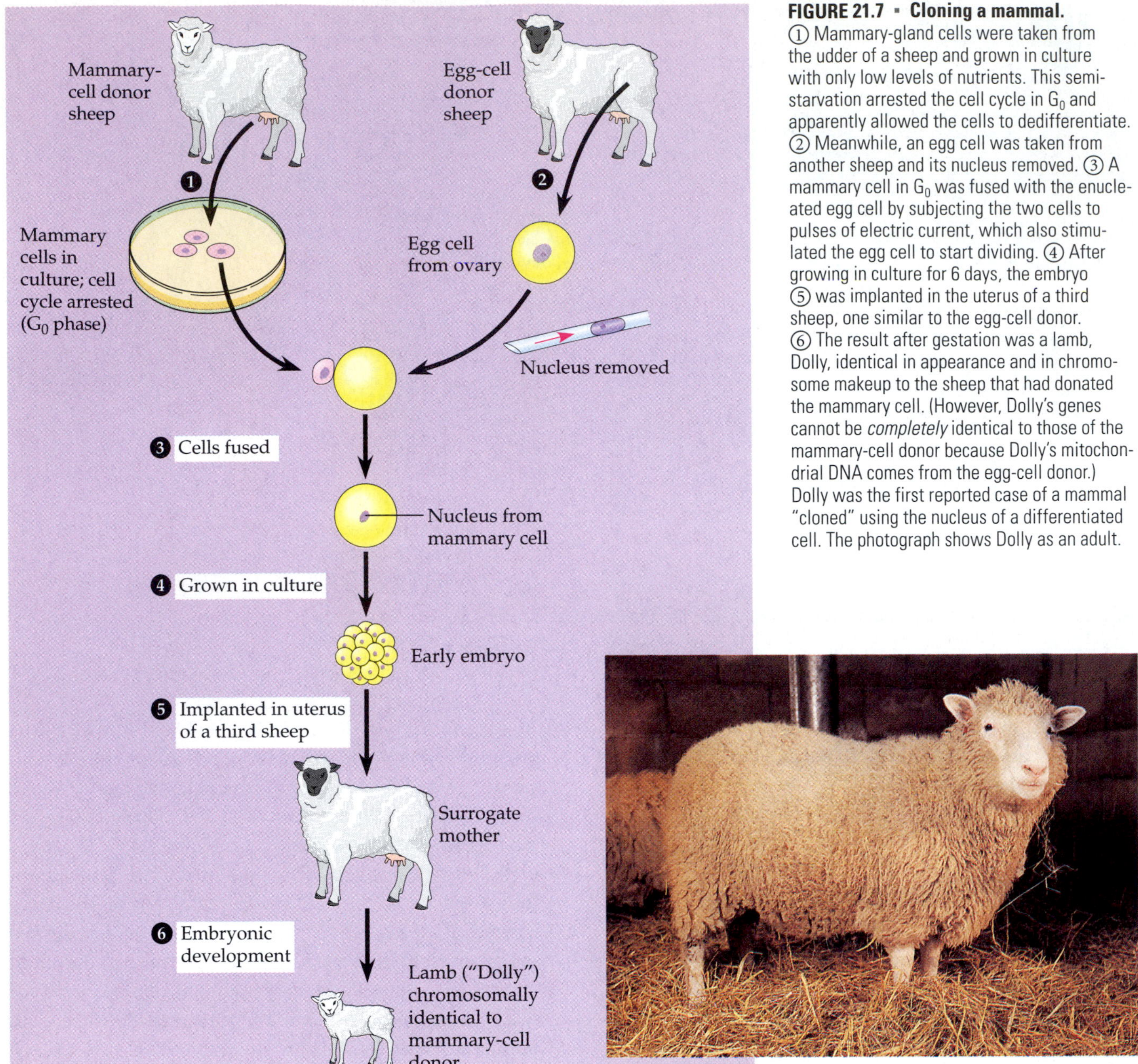

FIGURE 21.7 ▪ Cloning a mammal. ① Mammary-gland cells were taken from the udder of a sheep and grown in culture with only low levels of nutrients. This semi-starvation arrested the cell cycle in G_0 and apparently allowed the cells to dedifferentiate. ② Meanwhile, an egg cell was taken from another sheep and its nucleus removed. ③ A mammary cell in G_0 was fused with the enucleated egg cell by subjecting the two cells to pulses of electric current, which also stimulated the egg cell to start dividing. ④ After growing in culture for 6 days, the embryo ⑤ was implanted in the uterus of a third sheep, one similar to the egg-cell donor. ⑥ The result after gestation was a lamb, Dolly, identical in appearance and in chromosome makeup to the sheep that had donated the mammary cell. (However, Dolly's genes cannot be *completely* identical to those of the mammary-cell donor because Dolly's mitochondrial DNA comes from the egg-cell donor.) Dolly was the first reported case of a mammal "cloned" using the nucleus of a differentiated cell. The photograph shows Dolly as an adult.

Developmental biologists agree on several conclusions about these results. First, nuclei *do* change in some way as cells differentiate. Although the base sequence of the DNA usually does not change, chromatin structure alters in specific ways (for example, the DNA may be methylated; see Chapter 19). In frogs and most other animals, nuclear "potency" tends to be restricted more and more as embryonic development and cell differentiation progress. However, biologists also agree that these chromatin changes are sometimes reversible and that the nuclei of most differentiated animal cells probably have all the genes required for making the entire organism. In other words, biologists believe that the cells of the body differ in structure and function not because they contain different genes, but because they express different portions of a common genome.

Researchers working with mammals have succeeded in cloning animals using nuclei or cells from a variety of early embryos, but until recently it was not known whether the restriction of genomic potential in differentiated cells from an adult mammal could be reversed. However, in 1997, Scottish researcher Ian Wilmut and his colleagues captured newspaper headlines with the announcement that they had cloned an adult sheep by transplanting the nucleus from an udder (mammary) cell into an unfertilized egg cell from another sheep (FIGURE 21.7). They achieved the necessary dedifferentiation of the nucleus by culturing mammary cells in nutrient-poor medium, forcing the cells into the G_0 "resting" phase of the cell cycle (see Chapter 12). They then fused these cells with sheep egg cells whose nuclei had been removed. The resulting diploid cells divided to form early embryos, which they implanted into surrogate mothers. One of several hundred of these embryos, they reported, successfully completed normal development. DNA analyses have since shown that the chromosomal DNA of this sheep, "Dolly," is indeed identical to that of the nucleus donor. And in July 1998, researchers in Hawaii reported cloning over 50 mice using nuclei from ovary cells.

Different cell types make different proteins, usually as a result of transcriptional regulation

As the tissues and organs of an embryo take shape, differentiation of their cells becomes apparent; the cells become obviously different in structure and function. Cellular differentiation is actually the outcome of a cell's developmental history extending back to the first mitotic divisions of the zygote. However, the earliest changes that set a cell on a path to specialization are subtle ones, showing up only at the molecular level. Before biologists knew much about the molecular changes occurring in embryos, they coined the term **determination** to refer to the process that leads up to the observable differentiation of a cell. At the point in the process when the cell is irreversibly committed to its final fate, it is said to be "determined." Today we understand determination in terms of molecular changes. The outcome of determination, differentiation, is heralded by the expression of genes that encode *tissue-specific proteins,* which are found only in a certain type of cell and give the cell its characteristic structure and function. The first evidence of differentiation is the appearance of the mRNA for these proteins. Eventually, differentiation is observable with a microscope as changes in cellular structure. In most cases the pattern of gene expression in a differentiated cell—what proteins the cell makes—is controlled at the level of transcription.

Differentiated cells are specialists at making tissue-specific proteins. These are the proteins that allow them to carry out their specialized roles in the organism. Developing lens cells in vertebrates, for example, synthesize large quantities of crystallins, proteins that aggregate to form transparent fibers that give the lens the ability to transmit and focus light. Because no other vertebrate cell type makes crystallins, these proteins are signposts of lens-cell differentiation. Lens cells devote 80% of their capacity for protein synthesis to making this one type of protein.

The differentiation of skeletal muscle cells is another instructive example (FIGURE 21.8, p. 396). The "cells" of skeletal muscle, which we use for walking and lifting and other voluntary movements, are long fibers containing many nuclei within a single plasma membrane. They contain very high concentrations of proteins specific to muscle tissue, such as muscle-specific versions of the contractile proteins myosin and actin and membrane receptor proteins that detect signals from nerve cells. Muscle cells develop from embryonic precursor cells that have the potential to develop into a number of alternative cell types, including cartilage cells or fat cells, but particular conditions commit them to becoming muscle cells. Although the committed cells appear unchanged under the microscope, determination has occurred, and they are now *myoblasts.* Eventually myoblasts start to churn out large amounts of muscle-specific proteins and fuse to form mature, elongated, multinucleated skeletal muscle cells.

What actually happens in muscle-cell determination? Researchers have been able to answer this question by growing myoblasts in culture (see p. 219) and applying some of the techniques you learned about in Chapter 20. To test the hypothesis that certain muscle-specific regulatory genes are active in myoblasts, researchers isolated mRNA from cultured myoblasts and used reverse transcriptase to prepare a library of cDNA genes (see FIGURE 20.6). These genes were thus intron-lacking versions of the genes expressed in myoblasts. In cloning the genes they positioned them next to a viral promoter that would turn on transcription in any kind of cell. The researchers then inserted the cloned genes into embryonic precursor cells and looked for differentiation into myoblasts and muscle cells. In this way they identified several crucial muscle-determination genes, "master regulatory genes" that are transcribed (and translated) only when cells become committed to becoming skeletal muscle. Thus, in the muscle-cell case, the molecular basis of determination is the expression of one or more of these regulatory genes. We will focus on the master regulatory gene called *myoD.*

Researchers learned that the protein product of *myoD,* called MyoD, is a transcription factor. It is a regulatory protein that binds to specific control elements in the DNA and stimulates the transcription of genes encoding still other muscle-specific transcription factors (see FIGURE 19.9). Presumably all these target genes have enhancers recognized by MyoD and are thus coordinately controlled. Finally, the secondary transcription factors activate the muscle-protein genes.

The MyoD protein is powerful. Researchers have been able to use it to change some kinds of fully differentiated nonmuscle cells, such as fat cells and liver cells, into muscle cells. Why doesn't it work on *all* kinds of cells? One likely explanation is that activation of the muscle-specific genes is not solely dependent on MyoD but requires a particular *combination* of regulatory proteins, some of which are lacking in cells that don't respond to MyoD. It is probable that the determination and differentiation of other kinds of tissues play out in a similar fashion.

Transcriptional regulation is directed by maternal molecules in the cytoplasm and signals from other cells

Explaining the role of *myoD* in muscle-cell differentiation is a long way from explaining the development of an organism. The

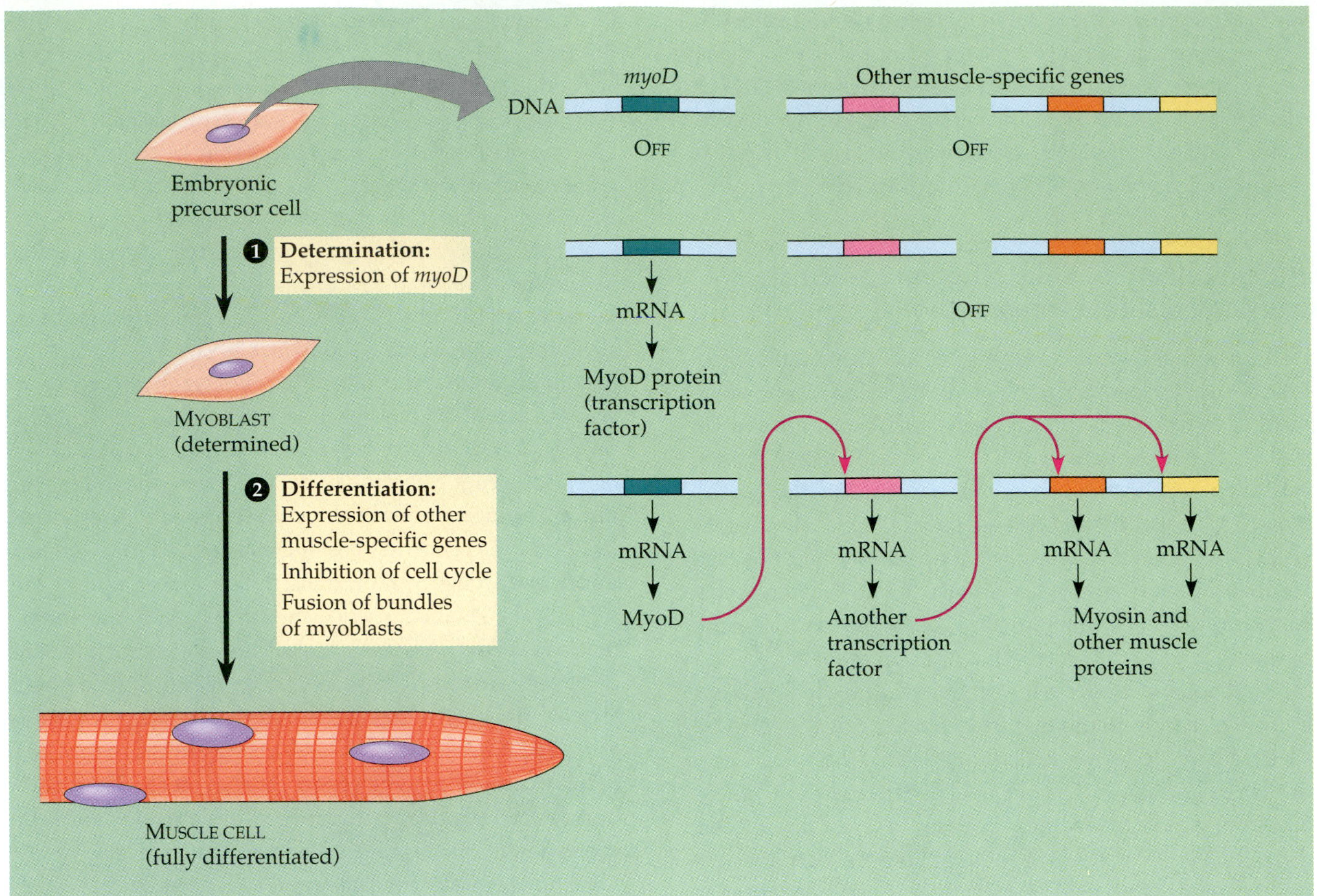

FIGURE 21.8 ▪ Determination and differentiation of muscle cells. This figure depicts a simplified version of how skeletal muscle cells arise from ordinary-looking embryonic cells that resemble fibroblasts (see p. 219). ① When this kind of embryonic precursor cell receives certain signals from other cells, a master control gene called *myoD* is activated, and the cell makes the MyoD protein. Although the appearance of the cell in the microscope does not change, determination has occurred; the activation of *myoD* commits the cell, now called a myoblast, to becoming a skeletal muscle cell. ② The MyoD protein is a transcription factor that activates genes encoding other muscle-specific transcription factors. MyoD also turns on genes such as *p21* that block the cell cycle and thus stop cell division (see Chapter 19). The various muscle-specific transcription factors activate the genes for muscle proteins such as myosin and actin. Meanwhile, the myoblasts fuse to become multinucleated mature muscle cells, also called muscle fibers.

myoD story immediately raises the question of what triggers the expression of *that* gene, and then a series of similar questions leading back to the zygote. What generates the *first* differences that arise among the cells in an early embryo? And what controls morphogenesis and the differentiation of all the different cell types as development proceeds? As we saw in the muscle-cell case, this question comes down to which genes are transcribed in the cells of a developing organism. Two sources of information "tell" a cell what genes to express at any given time.

One important source of developmental information that operates early in development is the cytoplasm of the unfertilized egg cell, which contains both RNA and protein molecules encoded by the mother's DNA. The cytoplasm of an egg cell and even its cytosolic fluid are not homogeneous. Messenger RNA, proteins, other substances, and organelles are distributed unevenly in the unfertilized egg, and this heterogeneity has a profound impact on the development of the future embryo in many species. After fertilization, the cell nuclei resulting from mitotic division of the zygote are exposed to different cytoplasmic environments. The maternal substances in the egg that influence the course of early development, called **cytoplasmic determinants**, regulate the expression of genes that affect the developmental fate of cells (FIGURE 21.9).

The other important source of developmental information, which becomes increasingly important as the number of embryonic cells increases, is the environment surrounding a cell. Most important are the signals impinging on an embryonic cell from other embryonic cells in its vicinity. The synthesis of the molecules conveying these signals is controlled by the embryo's own genome. The signal molecules cause changes in nearby target cells, a process called **induction**. Thus, interactions among embryonic cells eventually induce differentiation of the many specialized cell types making up a new organism. Induction may be accomplished by the diffu-

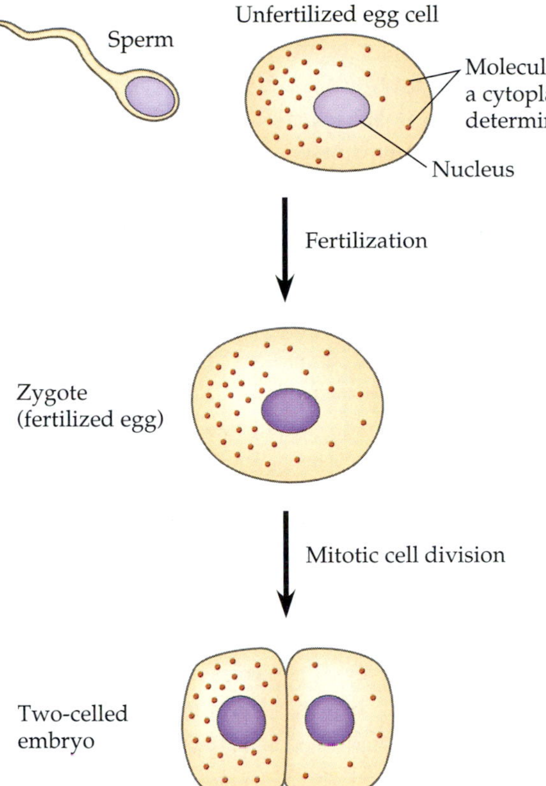

FIGURE 21.9 ▪ Nuclei in the early embryo are exposed to different concentrations of cytoplasmic determinants. The unfertilized egg cell has molecules in its cytoplasm, encoded by the mother's genes, that will influence development. Many of these cytoplasmic determinants, like the one shown here, are unevenly distributed in the egg. After fertilization and mitotic division, the cell nuclei of the embryo have different environments with respect to cytoplasmic determinants and, as a result, will express different genes.

sion of chemical signals or, if the cells are actually in contact, by cell-surface interactions.

You'll learn more about cytoplasmic determinants and induction as we take a closer look at some important genetic and cellular mechanisms of development in three model organisms: *Drosophila, C. elegans,* and *Arabidopsis.*

GENETIC AND CELLULAR MECHANISMS OF PATTERN FORMATION

How do cytoplasmic determinants, inductive signals, and their effects on embryonic cells contribute to morphogenesis, the shaping of the organism and its parts? We'll explore this question in the context of **pattern formation**, the development of a *spatial organization* in which the tissues and organs of an organism are all in their characteristic places. In the life of a plant, pattern formation occurs continually, in the apical meristems (see FIGURE 21.2). In animals, pattern formation is mostly limited to embryos and juveniles.

Pattern formation in animals begins in the early embryo, when the animal's basic body plan—its overall three-dimensional arrangement—is established. Just as the outline of a building is laid out before construction actually begins, the major axes of an animal are established very early. Before specialized tissues or organs appear, the relative positions of the animal's head and tail (for example) are established. The molecular cues that control pattern formation, collectively called **positional information**, tell a cell its location relative to the body axes and to neighboring cells and determine how the cell and its progeny will respond to future molecular signals.

Genetic analysis of *Drosophila* reveals how genes control development: *an overview*

Pattern formation has been most extensively studied in *Drosophila melanogaster*, where genetic approaches have had spectacular success. These studies have established that genes control development and have led to an understanding of the key roles specific molecules play in defining position and directing differentiation. Combining anatomical, genetic, and biochemical approaches to the study of *Drosophila* development, researchers have discovered developmental principles common to many other species, including humans.

The Life Cycle of Drosophila

Fruit flies and other arthropods have a modular construction, an ordered series of segments. These segments make up the body's three major parts: the head, the thorax (midbody, from which the wings and legs extend), and the abdomen (see FIGURE 21.10, bottom, p. 398). Like other bilaterally symmetrical animals, *Drosophila* has an anterior-posterior (head-tail) axis and a dorsal-ventral (back-belly) axis. In *Drosophila*, cytoplasmic determinants that are present in the unfertilized egg provide positional information for the placement of the two axes even before fertilization. After fertilization, positional information operating on a finer and finer scale establishes a specific number of correctly oriented segments and finally triggers the formation of each segment's characteristic structures.

The developmental stages of *Drosophila* are illustrated in FIGURE 21.10. The egg cell develops in the mother's ovary, surrounded by ovarian cells called nurse cells and follicle cells. These supply the egg cell with nutrients and other substances needed for development and make the egg shell. ① Following fertilization and laying of the egg, mitosis begins. The early mitotic divisions have two notable features. First, the amount of cytoplasm does not change; the first ten divisions, which occur very quickly, consist of S and M phases only. Second, cytokinesis does not occur; the early *Drosophila* embryo is one big multinucleated cell (in contrast to vertebrate embryos; see Chapter 47). ② At the tenth nuclear division, the nuclei migrate to the periphery of the embryo, and ③ at division 13, plasma membranes finally partition the 6000 or so nuclei into separate cells. Although not yet apparent under the microscope, the basic body plan—including body axes and segment boundaries—has already been determined by this time. A central yolk nourishes the embryo, and the egg shell continues to protect it.

④ Subsequent events in the embryo create clearly visible segments, which at first look very much alike. ⑤ Then, some cells move to new positions, organs form, and a wormlike larva (juvenile form) hatches out of the shell. *Drosophila* goes through three larval stages, during which the larva eats, grows, and molts (sheds its tough outer layer), ⑥ eventually forming a pupa enclosed in a case. ⑦ Metamorphosis, the change from larva to adult fly, occurs inside the pupal case, and the fly emerges. In the adult, each segment is anatomically distinct, with characteristic appendages. For example, the first thoracic segment bears a pair of legs, the second thoracic segment has a pair of legs plus a pair of wings, and the third thoracic segment bears a pair of legs plus a pair of balancing organs called halteres.

Genetic Analysis of Early Development in Drosophila: *science as a process*

During the first half of the twentieth century, biologists made detailed anatomical observations of embryonic development in a number of species. Working with larger animals such as frogs, they also performed experiments in which they manipulated embryonic tissues (see Chapter 47). This research laid a groundwork for understanding the mechanisms of development. However, the "classical" embryologists were unable to identify the specific molecules that guide development or determine how patterns were established. In the 1940s, one visionary biologist, Edward B. Lewis of Caltech, showed that a genetic approach—the study of mutants—could be used to investigate *Drosophila* development. Lewis studied bizarre mutants, flies with developmental defects that led to extra wings or legs in the wrong places (see FIGURE 21.13). He located the mutations on the fly's genetic map, thus connecting the developmental abnormalities to specific genes. This research supplied the first concrete evidence that genes some-

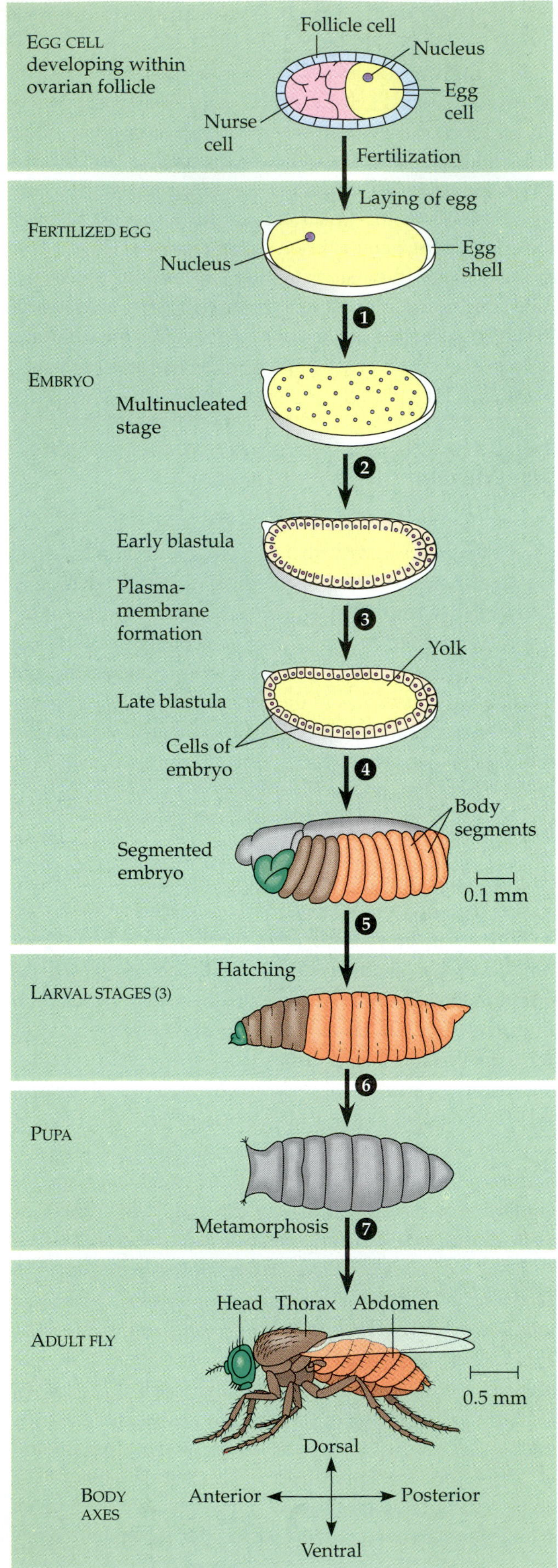

FIGURE 21.10 • Key developmental events in the life cycle of *Drosophila*. In the top drawing, the yellow egg cell is surrounded by other cells, which form a structure called the follicle within one of the mother's ovaries. The egg cell grows as it matures and eventually fills the egg shell that is secreted by the follicle cells; the nurse cells shrink and disappear. The egg is fertilized within the mother and then laid. The embryo develops within the protective egg shell, as described in the text. The cell layer forming the equivalent of the blastula in *Drosophila* is called the blastoderm.

how direct the developmental processes studied by embryologists. The genes Lewis discovered control development in the late embryo; we'll return to them shortly.

Meanwhile, the mystery of pattern formation during early development remained. But in the late 1970s, two researchers in Germany, Christiane Nüsslein-Volhard and Eric Wieschaus, undertook an ambitious quest that finally pushed the understanding of early pattern formation to the molecular level. These scientists set out to identify *all* the genes that affect segment formation in *Drosophila.* The project was daunting for several reasons. The first was the sheer number of *Drosophila* genes, about 12,000. The genes affecting segmentation might be just a few needles in a haystack or might be so numerous and varied that the scientists would be unable to make sense of them. Second, mutations affecting a process as fundamental as segmentation would surely be **embryonic lethals**, mutations with phenotypes leading to death at the embryo or larval stage. Because such mutant organisms never reproduce, they cannot be bred for study. Third, because cytoplasmic determinants in the egg were known to play a role in axis formation, the researchers would have to study maternal genes as well as those of the embryo.

As described in part (a) of the Methods Box on p. 400, Nüsslein-Volhard and Wieschaus dealt with the problem of embryonic lethality by focusing their search on recessive mutations, which could be propagated in heterozygous flies. Their basic strategy was to expose flies to a mutagenic chemical to create mutations in the flies' gametes and then look among the flies' descendants for dead embryos or larvae with abnormal segmentation. By doing the appropriate crosses, they would be able to identify living heterozygotes carrying embryonic lethal mutations. To find as many segmentation genes as possible, they planned a *saturation screen:* They would make enough mutants to "saturate" the fly genome with mutations. The researchers hoped that the segmentation abnormalities visible in the dead embryos would suggest how the affected genes normally functioned.

After a year of laboriously performing thousands of crosses and examining thousands of dead embryos, Nüsslein-Volhard and Wieschaus succeeded in identifying about 1200 genes essential for embryonic development, of which about 120 were essential for pattern formation leading to normal segmentation. Over several years they were able to group the genes by general function, to map them, and to clone many of them. The result was a detailed molecular understanding of the early steps in pattern formation in *Drosophila.* When their results were combined with Lewis's earlier work, a coherent picture of *Drosophila* development emerged. In recognition of their discoveries, Nüsslein-Volhard, Wieschaus, and Lewis were awarded a Nobel Prize in 1995. Before we discuss how the segmentation genes function, we need to back up and look at cytoplasmic determinants, for these control the expression of the segmentation genes.

Gradients of maternal molecules in the early embryo control axis formation

21.1 As previously mentioned, cytoplasmic determinants are the substances that initially establish the axes of the *Drosophila* body. Already present in the unfertilized egg, they are encoded by genes of the mother called **maternal effect genes.** Because they control the orientation (polarity) of the egg and consequently the fly, they are also called **egg-polarity genes.** One group of these genes sets up the anterior-posterior axis of the embryo, while a second group establishes the dorsal-ventral axis. Part (b) of the Methods Box (p. 401) shows how mutations in egg-polarity genes have been identified; like mutations in segmentation genes, these are generally embryonic lethals.

How do products of maternal effect genes determine the body axes of the offspring? Let's focus on one important egg-polarity gene, *bicoid,* and see how it works (FIGURE 21.11).

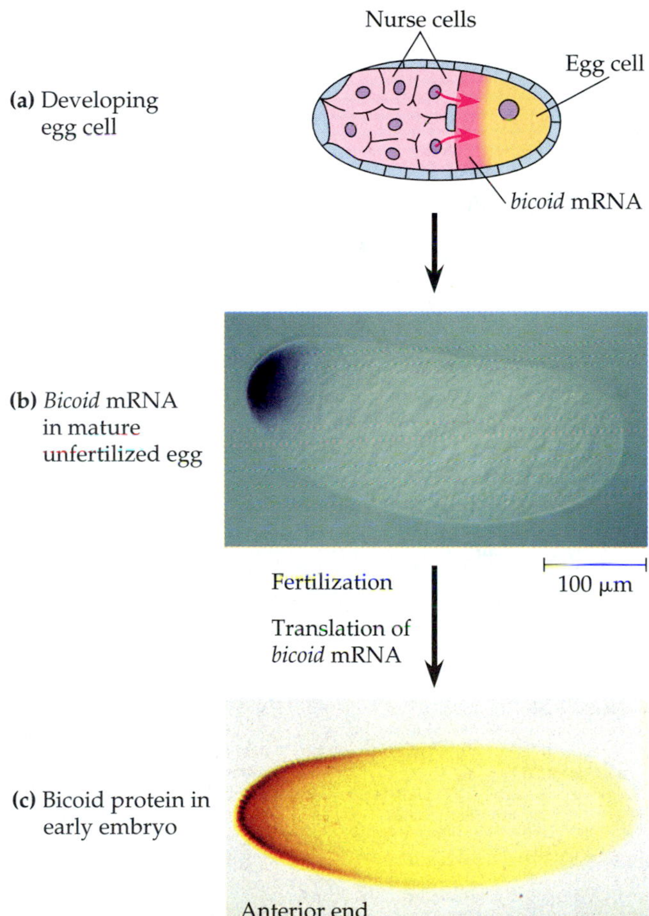

21.1 **FIGURE 21.11 • *Bicoid* mRNA and protein in *Drosophila.* (a)** Transcribed from the maternal *bicoid* gene in nurse cells, *bicoid* mRNA reaches the egg cell via cytoplasmic bridges. It becomes anchored to the cytoskeleton at the anterior end of the egg as the egg grows and the nurse cells disappear. **(b)** Labeled *bicoid* DNA (purple) was used as a probe to locate *bicoid* mRNA in this unfertilized egg. The hybridization pattern indicates that the mRNA is concentrated at the anterior end of the egg (LM). **(c)** After fertilization, the mRNA is translated. In this early embryo, the gradient of color reveals a gradient of bicoid protein, with its highest concentration at the anterior end. Bicoid protein is only one of several morphogens involved in axis specification (LM).

METHODS: Using Mutation to Study the Genetic Basis of *Drosophila* Development

Taking a genetic approach to the analysis of fruit fly development, Christiane Nüsslein-Volhard and Eric Wieschaus carried out "saturation screening" for mutations in the genes that direct the formation of the early embryo. They exposed flies to a chemical mutagen and then performed many thousands of genetic crosses to detect recessive mutations causing abnormal segmentation and death in embryos or larvae descended from these flies. (Such mutations are called embryonic lethals.) To hit as many different genes as possible, they made enough mutants to "saturate" the fly genome with mutations. The abnormalities visible in the dead embryos provided clues to how the mutated genes normally functioned and thus to the normal process of development. Here we describe simplified versions of their procedures.

(a) Finding segmentation mutants (mutations in the embryo's DNA). ① Cross a male, mutation-carrying fly with a wild-type female. All the F_1 progeny will be phenotypically normal, but some may be heterozygous carriers of *s*, an embryonic lethal mutation (genotype *sS*). ② Cross the F_1 flies in all possible combinations. Most crosses will produce only phenotypically normal progeny, but 1/4 of the progeny of some of the crosses will die as embryos. ③ Examine the dead embryos—which are homozygous for a recessive embryonic lethal mutation *(ss)*—for segmentation defects. Both parents of these flies must be heterozygous carriers of the mutation.

The photos compare a wild-type larva with one homozygous for an embryonic lethal mutation in a segmentation gene called *knirps*. The mutant larva lacks abdominal segments A2–A7. These are ventral views.

(b) Finding maternal effect mutants (mutations in the mother's DNA). Maternal effect genes, also called egg-polarity genes, are genes in the mother's genome that control the polarity of the egg and the embryo that develops from it. An embryo whose mother is homozygous for a recessive maternal-effect mutation—genotype mat^- mat^-—will be defective and die, whatever its own genotype. The mat^- mat^- female cannot make one of the cytoplasmic determinants needed for normal development of the eggs she produces. An embryo arising from one of her eggs will die even if the allele it inherits from its father is the dominant wild-type allele (mat^+).

In saturation screening, the phenotypic effect of a recessive maternal effect mutation does not show up in any of the flies of the F_1 or F_2 generations, which all develop normally because their mothers are mat^- mat^+ or mat^+ mat^+. However, some of the F_3 generation will die as embryos or larvae. Thus, to isolate *mat* mutations, researchers

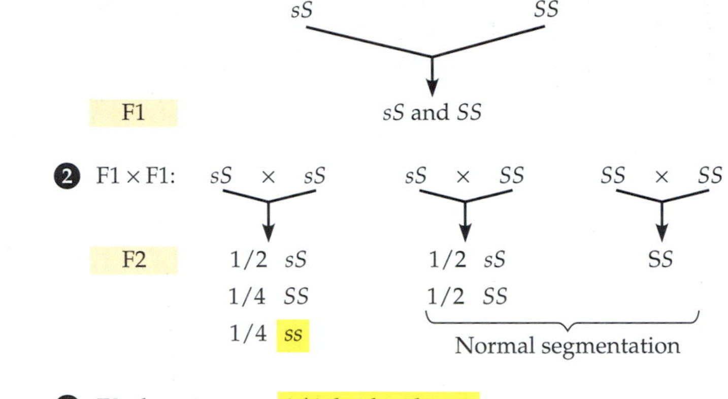

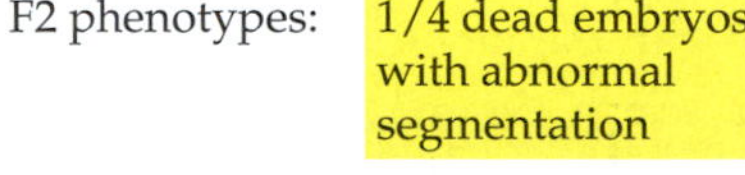

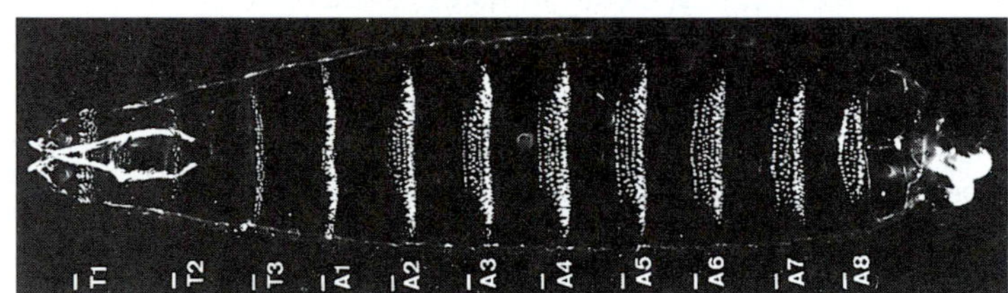

Wild-type larva

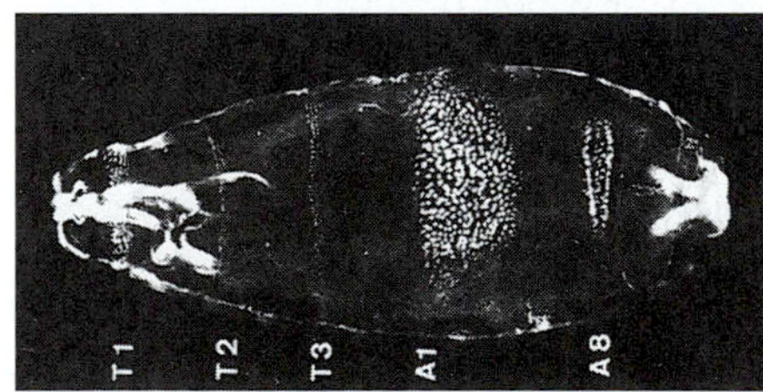

Mutant larva (*knirps*), an example of a segmentation mutant

(a) Finding segmentation mutants (mutations in the embryo's DNA)

Bicoid means "two-tailed," and an embryo whose mother was defective in this gene lacks the front half of its body. Instead, the embryo has duplicate posterior structures at both ends (see the photos in the Methods Box, part b). This phenotype suggested the hypothesis that the product of the mother's *bicoid* gene is essential for setting up the anterior end of the fly and that the gene's product is concentrated at the future anterior end.

This hypothesis is a more specific version of the *gradient hypothesis* first proposed by embryologists a century ago. According to this idea, gradients of substances called **morphogens** establish an embryo's axes and other features of its must perform more crosses than required for finding mutations in the embryo's own genome.

Mutations in egg-polarity genes disrupt the formation of the embryo's body axes, leading to abnormal segmentation and other defects in the embryonic offspring. These defects are similar to the ones observed in segmentation mutants (part a).

The photos compare a wild-type larva with a larva mutated in an important egg-polarity gene called *bicoid.* All offspring of females homozygous for recessive *bicoid* mutations die early in development. The larva of a *bicoid* ("two-tailed") mutant lacks anterior structures. It is missing a head, all thoracic segments, and abdominal segments A1–A5. Instead, it has duplicate posterior structures at both ends (arrows). The photos show side views.

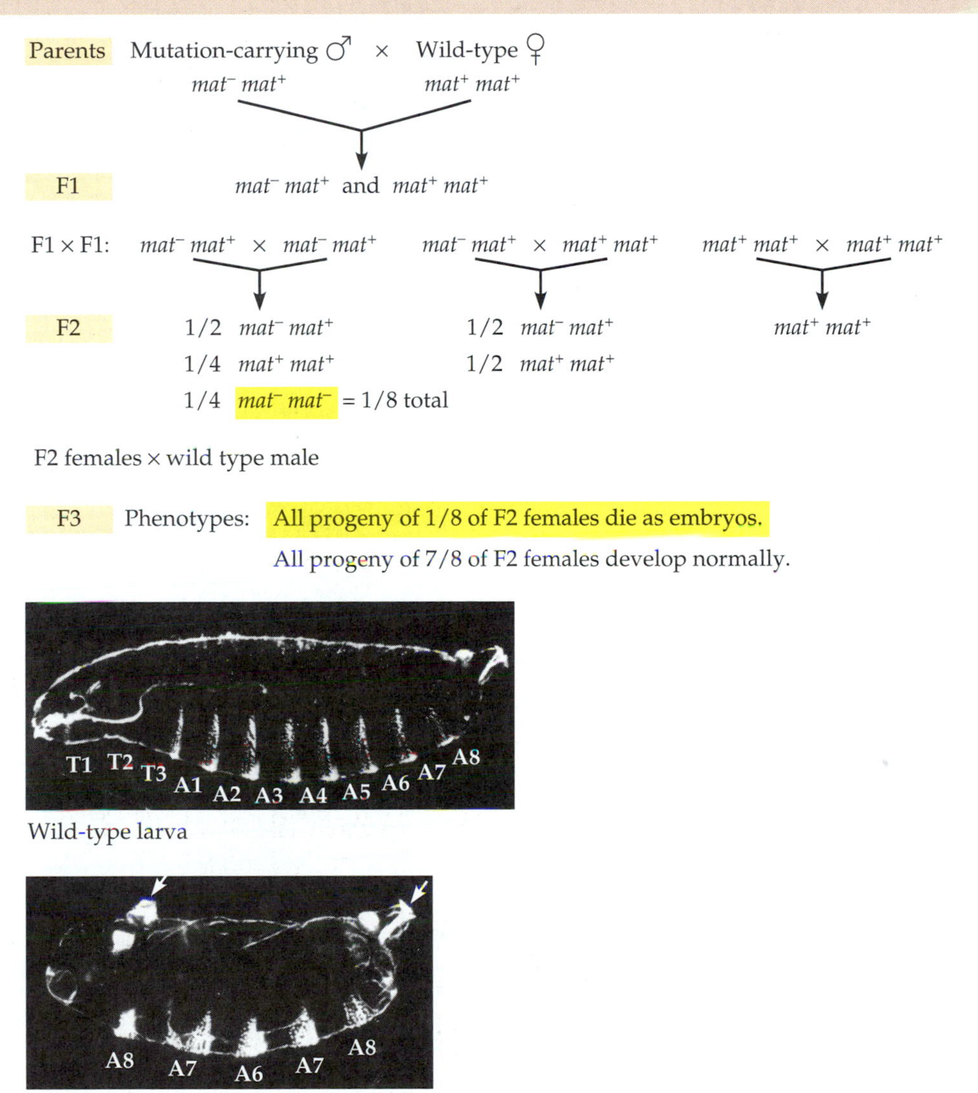

(b) Finding maternal effect (egg-polarity) mutants (mutations in the mother's DNA)

form. Providing evidence that the function of the *bicoid* product depends on such a gradient, fruit-fly embryologists had shown that a late embryo would develop the two-tailed phenotype if they removed cytoplasm from its anterior end very early in development. In other experiments they had transplanted anterior cytoplasm to other regions of embryos and found that anterior structures developed at the sites of injection. But they had been unable to find out what the anterior-determining morphogen was.

DNA technology and other modern biochemical methods enabled researchers to test the hypothesis that the *bicoid* product is in fact a morphogen that determines the anterior end of the fly. The researchers cloned the *bicoid* gene and used it as a DNA probe to learn the location of *bicoid* mRNA in the eggs produced by wild-type female flies (in situ hybridization; see Chapter 20). As predicted by the hypothesis, the *bicoid* mRNA is concentrated at the extreme anterior end of the egg cell (see FIGURE 21.11b). After the egg is fertilized, the mRNA is translated into protein, which diffuses from the anterior end toward the posterior, resulting in a gradient of protein within the early embryo (see FIGURE 21.11c). These results were consistent with but did not directly support the hypothesis that bicoid protein was responsible for specifying the fly's anterior end.

The scientists could now perform a more precise version of the earlier cytoplasmic transplantation experiment. They injected pure *bicoid* mRNA into various regions of early embryos. The protein that resulted from its translation caused anterior structures to form at the injection sites.

The bicoid research is important for several reasons. First, it led to the identification of a specific protein required for some of the earliest steps in pattern formation. Second, it increased our understanding of the mother's role in the development of an embryo. (As one developmental biologist puts it, "Mom tells Junior which way is up.") Finally, the principle that a gradient of molecules can determine polarity and position has proven to be a key developmental concept, as early embryologists had thought. In *Drosophila,* gradients of specific proteins determine the posterior end as well as the anterior and also are responsible for establishing the dorsal-ventral axis. Saturation screening has led to the identification of most of the genes and proteins involved.

A cascade of gene activations sets up the segmentation pattern in *Drosophila:* *a closer look*

The bicoid protein and other morphogens that are products of egg-polarity genes are transcription factors, proteins that regulate the activity (transcription) of some of the embryo's own genes. Gradients of these morphogens bring about regional differences in the expression of **segmentation genes**, the genes of the embryo that direct the actual formation of segments after the embryo's major axes are defined. These are the genes mutated in the saturation screening described in part (a) of the Methods Box on p. 400.

In a cascade of gene activations, sequential activation of three sets of segmentation genes provides the positional information for increasingly fine details of the animal's modular body plan. First, products of the **gap genes** map out the basic subdivisions along the anterior-posterior axis of the embryo (FIGURE 21.12a). Mutations in these genes cause "gaps" in the animal's segmentation. For example, one gap mutation results in an embryo lacking six abdominal segments (see the photos

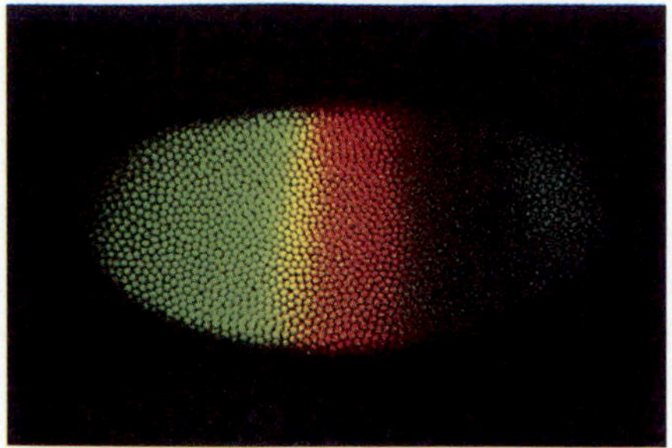

(a) Products of two gap genes

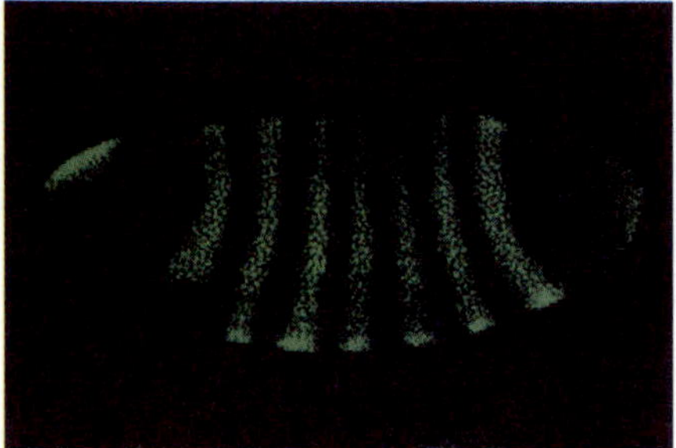

(b) Product of a pair-rule gene

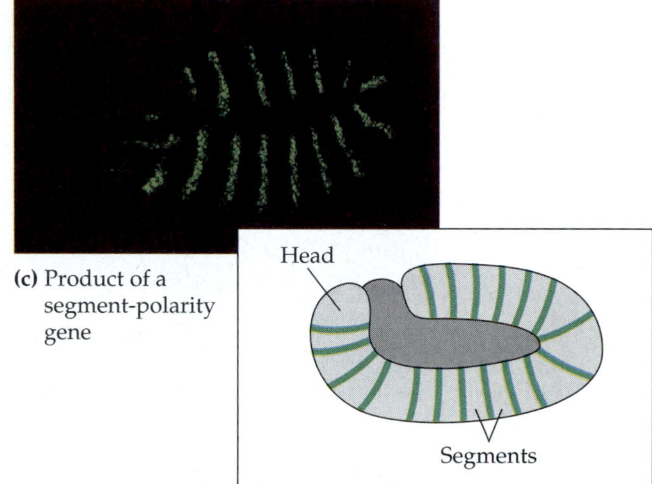

(c) Product of a segment-polarity gene

FIGURE 21.12 ▪ Segmentation genes in *Drosophila.* Soon after fertilization, once the mother's egg-polarity genes have set up the body axes, bicoid protein and other maternal gene products initiate a cascade of activity by segmentation genes in the embryo's nuclei. These micrographs of developing fly embryos illustrate successive bands of regulatory proteins encoded by segmentation genes. These are DNA-binding transcription factors that direct the division of the body into the segments characteristic of flies and other arthropods. The colors result from fluorescent dyes on antibodies bound to the segmentation-gene proteins. **(a)** Gap genes, the first set of segmentation genes turned on, produce these broad bands of gene-regulating proteins, prescribing a coarse subdivision of the embryo. The green and red colors represent products of two different gap genes; the yellow band is the region of overlap. **(b)** Next, localized products of gap genes turn on a second set of segmentation genes, the pair-rule genes. The protein product of a pair-rule gene produced these green bands. The pair-rule genes prescribe further subdivision of the embryo. **(c)** The final directions for subdividing the embryo into segments are in place when the pair-rule proteins bring about localized expression of various segment-polarity genes, the third set of segmentation genes. The product of a segment-polarity gene accounts for the green bands visible in this micrograph. Each of the compartments between these protein bands represents a body segment of the embryo, which at this stage is folded over on itself.

in the Methods Box, part a). **Pair-rule genes** are the next segmentation genes to act. They define the modular pattern in terms of pairs of segments (FIGURE 21.12b). Mutations in pair-rule genes result in embryos having half the normal segment number because every other segment (odd or even, depending on the mutation) fails to develop. The third group of segmentation genes to act are the **segment-polarity genes**, which set the anterior-posterior axis of each segment (FIGURE 21.12c). Embryos with mutations in segment-polarity genes have the normal number of segments, but a part of each segment is replaced by a mirror-image repetition of some other part of the segment.

The products of many of the segmentation genes, like those of egg-polarity genes, are transcription factors that directly activate the next set of genes in the hierarchical scheme of pattern formation. Other segmentation genes operate more indirectly, supporting the functioning of the transcription factors in various ways. For example, some are components of cell-signaling pathways, including signal molecules used in cell-cell communication and the membrane receptors that recognize them (see Chapter 11). Cell-signaling molecules are critically important once plasma membranes have divided the embryo into separate cellular compartments.

Working together, the products of egg-polarity genes regulate the regional expression of gap genes, which control the localized expression of pair-rule genes, which in turn activate specific segment-polarity genes in different parts of each segment. The boundaries and axes of the segments are now set. In the hierarchy of gene activations responsible for pattern formation, the next genes to be expressed determine the specific anatomy of each segment along the embryo.

Homeotic genes direct the identity of body parts

In a normal fly, structures such as antennae, legs, and wings develop on the appropriate segments. The anatomical identity of the segments is set by master regulatory genes called **homeotic genes**. These are the genes discovered by Edward Lewis. Once the segmentation genes have staked out the fly's segments, homeotic genes specify the types of appendages and other structures that each segment will form. Mutations in homeotic genes produce flies with such strange traits as an extra set of wings or legs growing from the head in place of antennae (FIGURE 21.13). Thus, as Lewis found, homeotic mutations replace structures characteristic of one part of the animal with structures normally found at some other location.

Like many of the egg-polarity and segmentation genes preceding them in the developmental cascade, the homeotic genes encode transcription factors. These regulatory proteins control the expression of the genes responsible for specific anatomical structures. For example, a homeotic protein made in the cells of a particular thoracic segment may selectively activate genes that bring about leg development. In contrast, a homeotic protein active in a certain head segment specifies "antennae go here." A mutant version of this protein may label the segment as "thoracic" instead of "head," causing legs to develop in place of antennae. Scientists are now busy identifying the genes activated by the homeotic proteins—the genes specifying the proteins that actually build the fly structures. The following flowchart summarizes the cascade of gene activity in the *Drosophila* embryo:

Hierarchy of Gene Activity in *Drosophila* Development

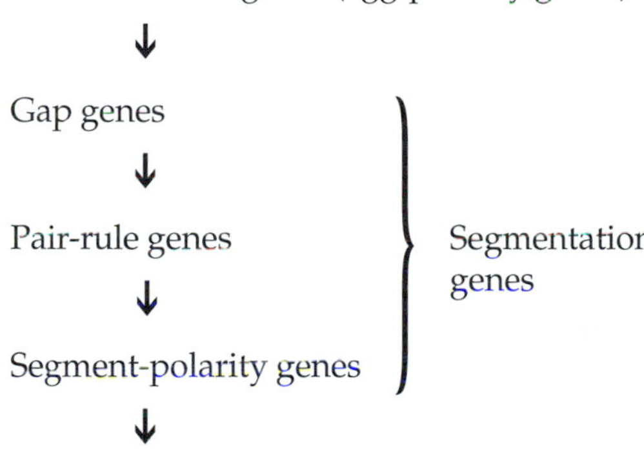

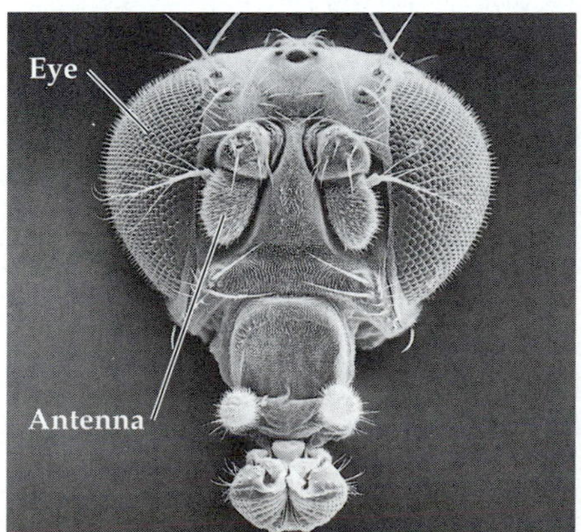

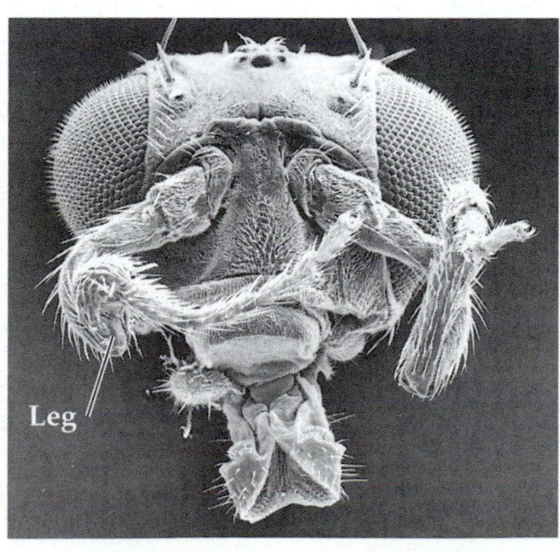

FIGURE 21.13 ▪ Homeotic mutations and abnormal pattern formation in *Drosophila*. Homeotic mutations cause a misplacement of structures in an animal. These micrographs contrast the heads of two fruit flies (SEMs). Where small antennae are located in the normal fruit fly (left photo), one homeotic mutant has legs (right photo).

Amazingly, many of the molecules and mechanisms revealed by research on fly pattern formation have turned out to have close counterparts throughout the animal kingdom. And nowhere are the similarities more striking than in the homeotic genes and their products.

Homeobox genes have been highly conserved in evolution

The homeotic genes of *Drosophila* all include a 180-nucleotide sequence called the **homeobox.** (For this reason they are often called *Hox genes.*) An identical or very similar sequence has been discovered in genes of many other animals, including other insects, nematodes, mollusks, fish, frogs, birds, and mammals, including humans. Furthermore, related sequences have been found in regulatory genes of much more distantly related eukaryotes, such as yeast, and even in prokaryotes. From these similarities we can deduce that the homeobox DNA sequence evolved very early in the history of life, and that it is sufficiently valuable to organisms to have been conserved in animals virtually unchanged for hundreds of millions of years. In fact, the animal genes homologous to the homeotic genes of fruit flies have even kept their chromosomal arrangement (FIGURE 21.14).

Homeobox-containing genes are not all homeotic genes; that is, they don't all directly control the identity of body parts. However, most are associated with development, suggesting their ancient and fundamental importance in that process. For example, in *Drosophila,* homeoboxes are present not only in the homeotic genes but also in the egg-polarity gene *bicoid,* in several of the segmentation genes, and in the master regulatory gene for eye development (see the photograph on p. 388).

What is the role of the homeobox sequence in a protein? When a homeobox-containing gene is translated, the homeobox nucleotide sequence specifies a 60-amino-acid *homeodomain.* This polypeptide segment is the part of the protein that binds to DNA when the protein functions as a transcription factor. (The homeodomain forms three alpha helices, one of which fits neatly into the major groove of the DNA helix.) However, the shape of the homeodomain allows it to bind to any DNA segment; by itself it cannot select a specific sequence. It is other, more variable domains of such a protein that determine which genes the protein regulates. These latter domains interact with other transcription factors, which help the protein recognize specific enhancers or promoters in the DNA. Presumably, homeodomain-containing proteins regulate development by coordinating the transcription of batteries of developmental genes, switching them on or off. In *Drosophila* embryos, different combinations of homeobox genes are active in different parts of the embryo. This selective expression of regulatory genes, varying over time and space, is central to pattern formation.

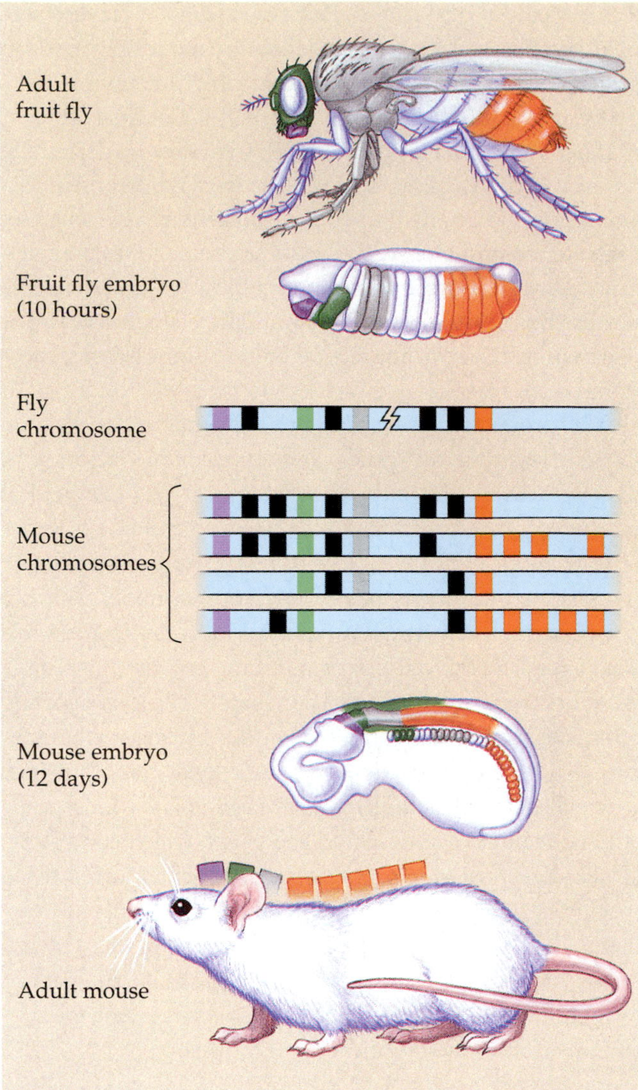

FIGURE 21.14 • Homologous genes that affect pattern formation in a fruit fly and a mouse. Homeotic, homeobox-containing genes that control the form of the anterior and posterior structures of the body occur in the same linear sequence on chromosomes in *Drosophila* and mice. Each small box represents a homeobox-containing gene. In fruit flies, all of these genes are found on one chromosome. The mouse and other mammals have the same or similar sets of genes on four chromosomes. Genes represented by purple and green boxes code for proteins that regulate pattern formation of anterior body parts; those represented by gray and orange code for proteins regulating the formation of middle and posterior body parts. The color code indicates the parts of the body of a fruit fly embryo and mouse embryo in which these genes are expressed and the adult body regions that result. Colored boxes represent genes with homeoboxes that are essentially identical in flies and mice. The black boxes represent genes with homeoboxes that are similar but not identical between the two animals.

Neighboring cells instruct other cells to form particular structures: *cell signaling and induction in the nematode*

The development of a multicellular organism requires close communication among cells. Even when the *Drosophila* embryo is still one multinucleated "cell," signaling between cells has already played key roles. For example, an exchange of signals between the unfertilized egg cell and neighboring follicle cells guided the establishment of the egg's anterior end; it was a follicle-cell signal that triggered the localization of *bicoid* mRNA there. Once the embryo is truly multicellular, signaling among the embryo's own cells becomes increasingly important. In the process called induction, cells signal nearby cells to change in some specific way, often by expressing particular genes. As we've seen, the ultimate basis for the differences that arise among the cells is transcriptional regulation—the turning on and off of specific genes. It is induction, signaling from one group of cells to an adjacent group, that brings about differentiation.

Induction in Vulval Development

The nematode *C. elegans* has proven to be a very useful model organism for investigating the roles of cell signaling and induction in development. Particularly revealing has been research on the development of the nematode *vulva,* the tiny opening through which the worm lays its eggs. Researchers have combined genetic, biochemical, and embryological approaches to learn how this organ forms.

The pathway from fertilized egg to an adult nematode capable of reproduction involves four larval stages (the larvae look much like smaller versions of the adult). Already present on the ventral surface of the second-stage larva are six cells from which the vulva will arise (FIGURE 21.15a). A single cell of the embryonic gonad, the *anchor cell,* initiates a cascade of signals that establishes the fates of the vulval precursor cells. If an experimenter destroys the anchor cell with a laser beam, the vulva fails to form and the precursor cells simply become part of the worm's epidermis.

Researchers have figured out the molecular details of vulval induction by isolating mutants in which vulval development is abnormal. Such mutants can grow to adulthood because a normal egg-laying apparatus is not essential for viability. The phenotypes of the mutants ranged from multiple vulvas to none at all. (In the latter mutants, offspring develop internally within self-fertilizing hermaphrodites, eventually eating their way out of the parent's body!) From studying the mutants, researchers have identified a number of genes involved in vulval development and have worked out where and how their products function (FIGURE 21.15b).

The anchor cell secretes an inducer, a signal protein that binds to a receptor protein in the plasma membrane of vulval precursor cells. Initially, all the precursor cells are equivalent:

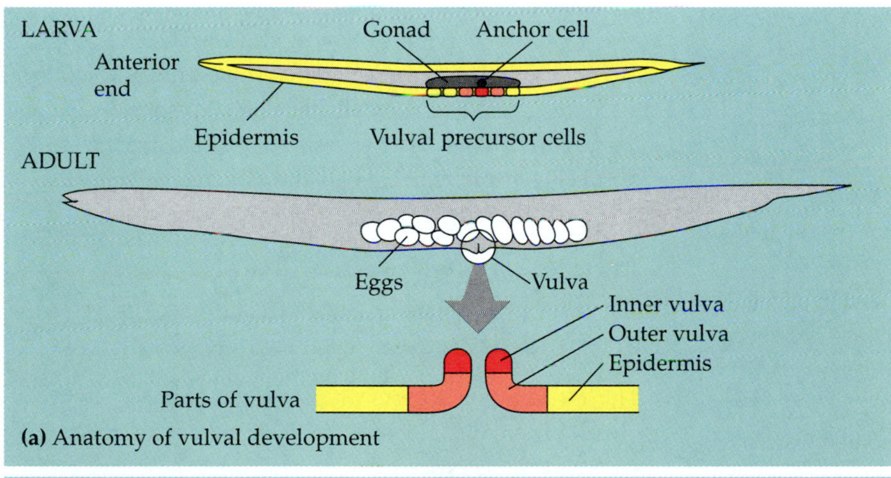

(a) Anatomy of vulval development

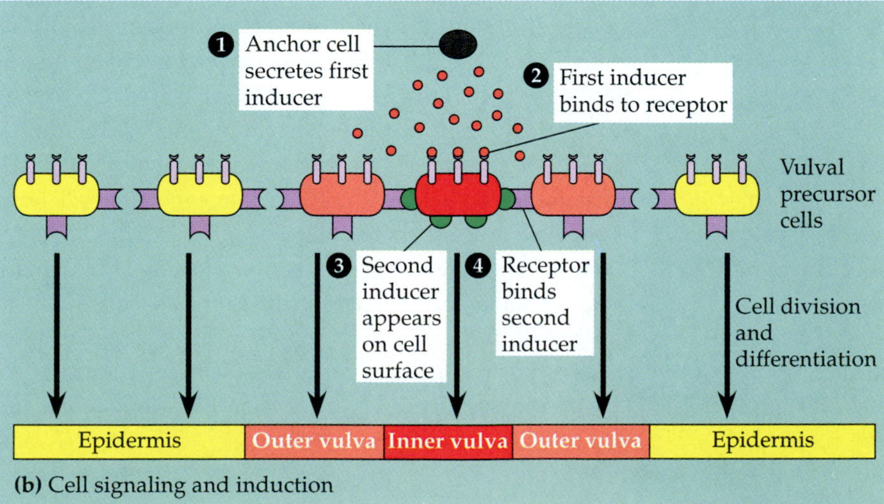

(b) Cell signaling and induction

FIGURE 21.15 ▪ Cell signaling and induction in the development of the nematode vulva. The vulva is the opening through which the nematode lays its eggs. **(a)** The vulval precursor cells of the larva have three possible fates: They can develop into the inner part of the vulva, the outer part, or adjacent epidermis. **(b)** The actual fates of these cells are determined by a series of inductions initiated by ① a chemical signal from a nearby cell called the anchor cell. Orange circles represent this first inducer. ② The binding of these molecules to receptors on the nearest vulval precursor cell triggers a signal-transduction pathway leading to ③ the synthesis of a second inducer, a cell-surface protein (green). ④ This inducer, in turn, binds to specific receptors (purple) on the two adjacent vulval precursor cells.

All make a basic set of proteins necessary for vulval formation, including receptors for the anchor cell's signal molecules. However, these cells do not all respond to the inducer. The cell that normally becomes the inner part of the vulva receives high levels of inducer since it is closest to the anchor cell. The high levels of inducer probably produce two effects: (1) division and differentiation of the cell to form the inner part of the vulva, and (2) activation of a gene for a cell-surface signaling protein, which acts as a second inducer. Receptors on the two adjacent vulval precursor cells bind the second inducer, which stimulates these cells to divide and develop into the outer vulva. The three remaining vulval precursor cells are too far away to receive either signal; they give rise to epidermal cells.

The inducer released by the anchor cell and the signal-transduction pathway it triggers in the closest vulval precursor cell are similar to ones you have encountered in previous chapters. The signal is a growth-factor-like protein, and it is transduced within its target cell via a tyrosine-kinase receptor, a Ras protein, and a cascade of protein kinases—a common pathway leading to transcriptional regulation in many organisms (see FIGURE 19.13a). (In fact, the nematode research provided the first in vivo evidence linking the Ras pathway with a growth factor.)

In summary, vulval development in the nematode illustrates a number of important concepts that apply elsewhere in the development of *C. elegans* and many other animals:

- In the developing embryo, sequential inductions drive the formation of organs.
- The effect of an inducer can depend on its concentration (just as we saw with cytoplasmic determinants in *Drosophila*).
- Inducers produce their effects via signal-transduction pathways similar to those operating in adult cells.
- The induced cell's response is often the activation (or inactivation) of genes—transcriptional regulation—which, in turn, establishes the pattern of gene activity characteristic of a particular kind of differentiated cell.
- Genetics is a powerful approach for elucidating the mechanisms of development.

Programmed Cell Death (Apoptosis)

Lineage analysis of *C. elegans* has highlighted another outcome of cell signaling that is crucial in animal development—programmed cell death, or **apoptosis**. The timely suicide of cells occurs exactly 131 times in the course of *C. elegans*'s normal development. At precisely the same points in each new worm's cell lineage, signals trigger the activation of a cascade of "suicide" proteins in the cells destined to die. The cells shrink, their nuclei condense, and neighboring cells quickly engulf and digest them, leaving no trace (FIGURE 21.16a).

Genetic screening of *C. elegans* has revealed two key suicide genes, *ced-3* and *ced-4* (*ced* stands for cell death), which encode proteins essential for apoptosis. These and most other proteins involved in apoptosis are continually present in cells, but in inactive form; thus protein *activity* must be regulated in this case, not transcription or translation. A gene called *ced-9* is the master control gene for apoptosis (FIGURE 21.16b). At the end of the suicide-protein pathway, enzymes are activated that cut up the proteins and DNA of the cell, killing it. Scientists are still working out the details of these suicide pathways and of the signaling pathways that regulate them. But it seems likely that a cell makes its life-or-death "decision" by integrating the signals it receives, both "death" signals and "life" signals such as growth factors (see Chapter 11).

A built-in cell-suicide mechanism is essential to development in all animals, and similarities between apoptosis genes in nematodes and mammals indicate that the basic mechanism evolved early in animal evolution. The timely activation of apoptosis proteins in some cells functions in normal development and growth in both embryos and adults. In vertebrates, programmed cell death is essential for normal development of the nervous system, for normal operation of the immune system, and for the normal morphogenesis of human hands and feet. In the last case, the failure of normal cell death can result in webbed fingers and toes. Researchers are also investigating the possibility that certain degenerative diseases of the nervous system result from the inappropriate activation of suicide genes, and that some cancers result from a failure of cell suicide. Cells that have suffered irreparable damage, including DNA damage that could lead to cancer, normally generate *internal* signals that trigger apoptosis.

Plant development depends on cell signaling and transcriptional regulation: *science as a process*

The last common ancestor of plants and animals was probably a single-celled microbe living hundreds of millions of years ago, so the processes of development must have evolved independently in the two lineages of organisms. Plants evolved with rigid cell walls that make the movement of cells and tissue layers virtually impossible, ruling out a morphogenetic mechanism that is important in animals. Plant morphogenesis relies more heavily on differing planes of cell division and on selective cell enlargement. (You will learn about these processes in Chapter 38.)

But despite the differences between plants and animals, there are some basic similarities in their molecular, cellular, and genetic mechanisms of development—legacies of their shared cellular origins. In particular, plant development, like that of animals, depends on cell signaling (induction) and transcriptional regulation.

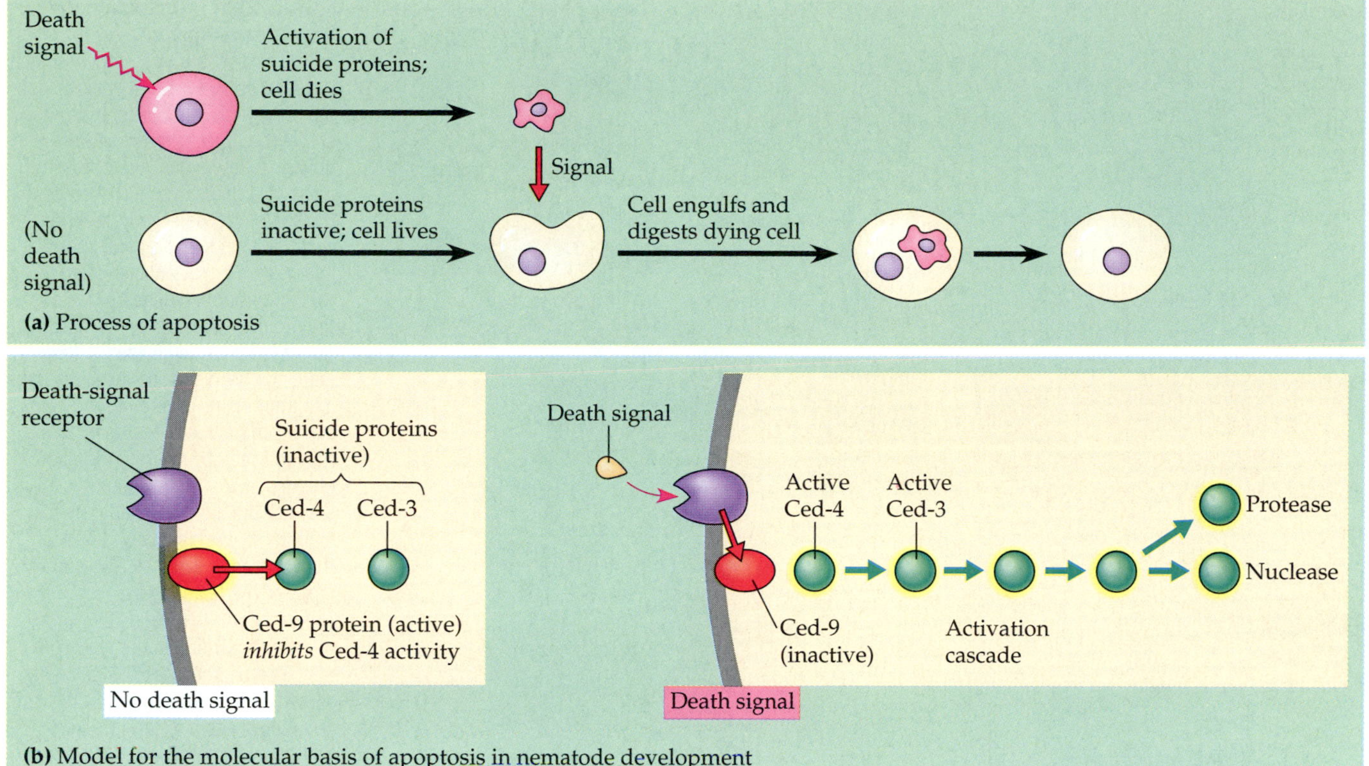

FIGURE 21.16 ▪ Apoptosis (programmed cell death). Many cells must die during normal development in all animals. **(a)** When a cell fated to die receives the appropriate signal, a cascade of protein activations occurs, leading to the chopping up of cellular proteins and DNA, and cell death. Signals from the dying cell instruct neighboring cells to engulf and digest the remains, keeping the embryo free of harmful enzymes and metabolites. **(b)** In *C. elegans,* apoptosis is controlled by the protein products of three main genes. A cell remains alive as long as Ced-9, the product of the master-control gene *ced-9,* remains active. This protein directly inhibits the activity of Ced-4, the protein encoded by the suicide gene *ced-4,* preventing it from activating Ced-3, the product of the suicide gene *ced-3.* When a cell receives a "death signal," thereby inactivating Ced-9, the suicide proteins are active. The active Ced-3 is a protease whose protein cutting activates the next protein in an activation cascade. The final result is the activation of protein-cutting (protease) and DNA-cutting (nuclease) enzymes that kill the cell.

We do not yet understand the molecular basis of plant development in detail. But thanks to DNA technology, clues from animal research, and model organisms such as *Arabidopsis,* plant research is now progressing rapidly. The embryonic development of most plants occurs inside the seed and is relatively inaccessible to study (a mature seed already contains a fully formed embryo). However, important aspects of plant development are observable throughout the plant's life in its meristems, and in particular in the apical meristems at the tips of shoots. There cell division, morphogenesis, and differentiation give rise to new organs such as leaves or the petals of flowers. We'll discuss two examples of research on floral meristems, the apical meristems that produce flowers.

Cell Signaling in Flower Development

Environmental signals, such as day length and temperature, trigger signal-transduction pathways that convert ordinary shoot meristems to floral meristems. Researchers have combined a genetic approach with tissue transplantation to study induction in the development of tomato flowers. As shown in FIGURE 21.17a (p. 408), a floral meristem is a "bump" consisting of three layers of cells (L1–L3). All three layers participate in the formation of a flower, a reproductive structure with four types of organs: carpels (containing egg cells), petals, stamens (containing sperm-bearing pollen), and sepals (leaf-like structures outside the petals).

Researchers discovered that tomato plants homozygous for a mutant allele called *fasciated (f)* produce flowers with an abnormally large number of organs. Taking advantage of the totipotency of plant cells, they performed an experiment in which they grafted stems from this mutant onto wild-type plants and then grew new plants from the shoots that emerged at the graft sites. The new plants were **chimeras**, organisms with a mixture of genetically different cells. Some of the chimeras produced floral meristems in which the three cell layers did not all come from the same "parent" (FIGURE 21.17b). The researchers could identify the parental sources of the meristem layers by monitoring other genetic markers, such as an unrelated mutation that caused yellow leaves. They found that the number of organs per flower depends on the genes of the L3 (innermost) cell layer, which somehow

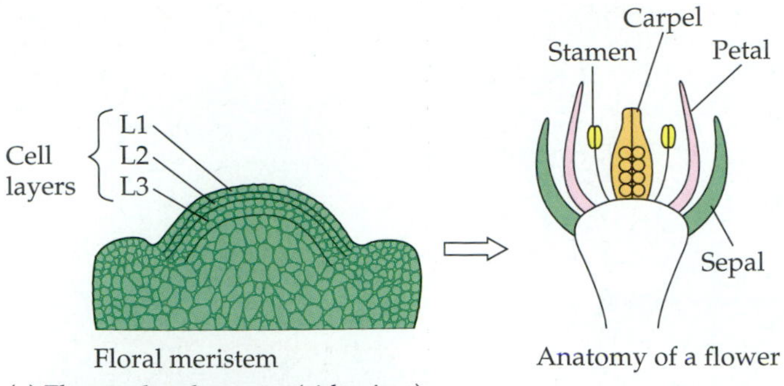

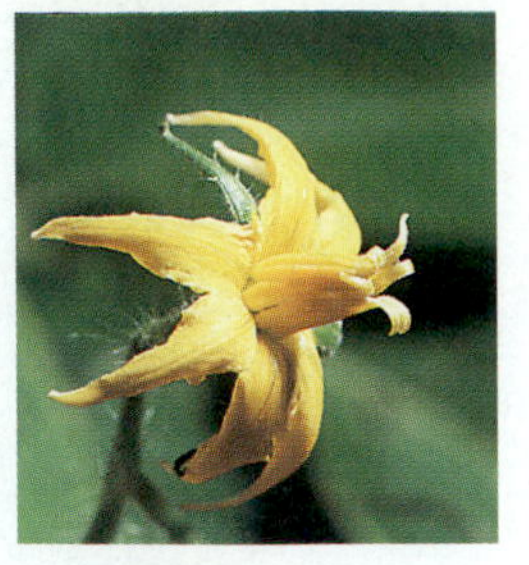

(a) Flower development (side view)

Plant	Genotypes of floral meristem layers			Flower phenotype
	L1	L2	L3	
Wild-type "parent"	++	++	++	Wild type
Fasciated "parent"	*ff*	*ff*	*ff*	Extra organs (=fasciated)
Chimera 1	++	*ff*	*ff*	Extra organs
Chimera 2	++	++	*ff*	Extra organs
Chimera 3	++	*ff*	++	Wild type
Chimera 4	*ff*	++	++	Wild type

(b) Evidence of induction in the floral meristem

FIGURE 21.17 ▪ Induction in flower development. (a) A flower develops from a meristem with three layers of cells, L1–L3. A specific pattern of cell division, differentiation, and enlargement produces the flower. A flower consists of four types of organs arranged in concentric circles: carpels (containing egg cells), stamens (containing sperm-bearing pollen), petals, and sepals. **(b)** Experiments with chimeras of the tomato plant (*Lycopersicum esculentum*) have provided evidence that induction between cell layers of the meristem helps determine the number of organs that develop in a flower. Made by grafting together plants of two different genotypes, some of the chimeras had floral meristems with layers originating from different "parents." The phenotypes of the resulting flowers showed that the L1 and L2 layers develop according to the genotype of the L3 layer. Thus, the L3 cells must somehow signal the L1 and L2 layers and induce them to form flowers with a particular number of organs.

induces the overlaying layers L2 and L1 to form that number of organs. The mechanism of cell-cell signaling leading to this induction is not yet known.

Organ-Identity Genes in Plants

In contrast to genes controlling organ *number* in flowers are genes controlling organ *identity*. An **organ-identity gene** determines the type of structure that will grow from a meristem—for instance, whether a particular outgrowth from a floral meristem becomes a petal or a stamen. Most of what we know about plant organ-identity genes comes from research on flower development in *Arabidopsis*. Organ-identity genes are analogous to homeotic genes in animals and are sometimes referred to as plant homeotic genes. Just as a mutation in a fruit fly homeotic gene can cause legs to grow in place of antennae, a mutation in a plant organ-identity gene can cause carpels to grow in place of sepals.

By collecting and studying mutants with abnormal flowers, researchers have been able to identify and clone a number of floral organ-identity genes, and they are beginning to determine how they act. To follow their work we need a top view of the floral meristem (FIGURE 21.18a). Viewed from above, the meristem can be divided into four concentric circles, or whorls, each of which develops into a circle of identical organs. The organ-identity genes fall into three classes (*A*, *B*, and *C*), each of which affects two adjacent whorls. A simple model, diagrammed in FIGURE 21.18a, explains how the three kinds of genes can direct the formation of the four types of organs.

This model (hypothesis) predicts that each organ-identity gene will be transcribed in two specific whorls of the floral meristem, and once the genes were cloned, the predictions could be tested. The technique employed was in situ hybridization (see Chapter 20). Using nucleic acid from cloned genes as probes, researchers showed that the mRNA resulting from the transcription of each class of organ-identity gene is present in the appropriate whorls of the developing floral meristem. For example, nucleic acid from a *C* gene hybridized appreciably only to cells in whorls 3 and 4 because only those cells were making RNA complementary to it (FIGURE 21.18b).

The model also can account for the phenotypes of mutants lacking *A*, *B*, or *C* gene activity with one addition: Where *A* gene activity is present, it inhibits *C* and vice versa; if either *A* or *C* is missing, the other, not inhibited, takes its place. FIGURE 21.18c shows the floral patterns of mutants lacking each of the three classes of organ-identity genes and explains how the model accounts for the floral phenotypes.

Presumably the organ-identity genes are acting as master regulatory genes, each controlling the activity of a battery of other genes that more directly bring about an organ's structure and function. Indeed, like the homeotic genes of animals, the organ-identity genes of plants probably encode transcription factors that regulate other genes by binding to specific

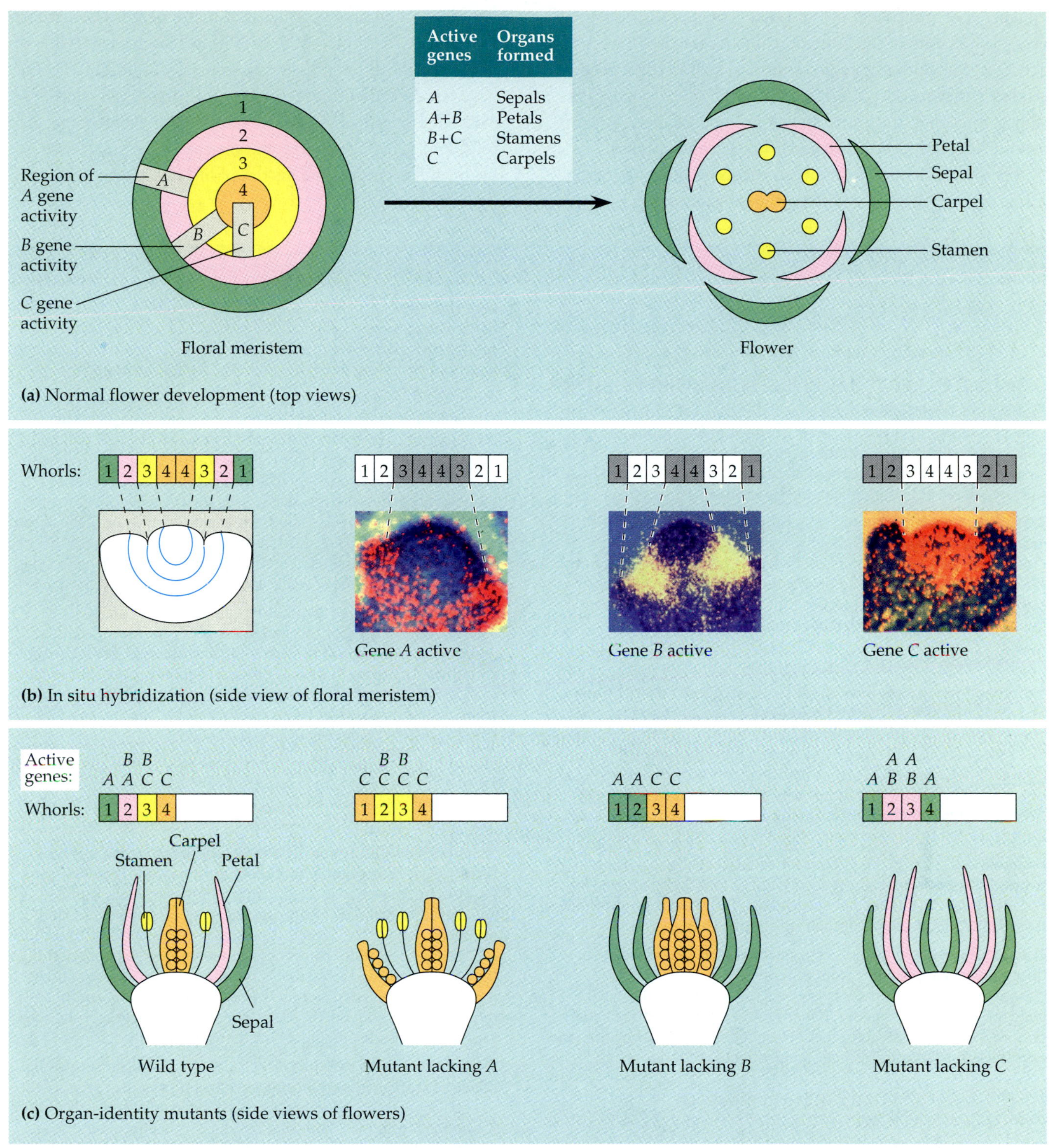

FIGURE 21.18 • Organ-identity genes and pattern formation in flower development. This is a model for how genes specify the floral organs in *Arabidopsis*, with supporting evidence. **(a)** The top view of the floral meristem indicates the four whorls of tissue that give rise to the four whorls of organs in the flower. (The whorls extend down through meristem cell layers L1–L3; see FIGURE 21.17.) There are three classes of organ-identity genes, each normally active in two adjacent whorls. *A* genes alone direct the formation of sepals, *A* and *B* genes that of petals, *B* and *C* genes that of stamens, and *C* genes alone that of carpels. **(b)** Evidence that *A, B,* and *C* genes are active in particular whorls has come from in situ hybridization experiments, in which nucleic acid from the cloned genes, labeled with dyes, was used to tag the location of complementary mRNA in developing flowers. In these micrographs you can see that an *A* gene is expressed in the outside whorls (red), a *B* gene in the middle whorls (yellow), and a *C* gene in the central whorls (orange). The products of these genes seem to be transcription factors, DNA-binding proteins that probably regulate the transcription of genes responsible for actually building the organs. **(c)** Combining the model in part (a) with the rule that if *A*-gene or *C*-gene activity is missing, the other spreads through all four whorls, we can explain the phenotypes of mutants lacking a functional *A, B,* or *C* gene. For example, a plant with a mutation in an *A* gene substitutes carpels for sepals and stamens for petals.

enhancers or promoters in the DNA. The plant genes, however, do not contain the homeobox DNA sequence that characterizes homeotic genes in animals. Instead they have a sequence that encodes a different DNA-binding domain. Possibly as ancient as the homeobox, this sequence is also present in some transcription-factor genes in yeast and animals.

The more scientists learn about the molecules and mechanisms of development in different organisms, the more striking are the underlying similarities that emerge. The similarities reflect the unity—the common ancestry—of life on Earth. But the differences are also crucially important, for they have created the huge diversity of organisms that have evolved. The remainder of the book goes beyond molecules, cells, and genes to explore life on the organismal level. We begin with evolution, the history of life, in the next unit.

CHAPTER REVIEW

REVIEW OF KEY CONCEPTS

(with page numbers and key figures)

FROM SINGLE CELL TO MULTICELLULAR ORGANISM

- **Embryonic development involves cell division, cell differentiation, and morphogenesis (p. 389)** In addition to mitosis, embryonic cells undergo differentiation, becoming specialized in structure and function. Morphogenesis encompasses the processes that give shape to the organism and its various parts.
- **Researchers study development in model organisms to identify general principles: *science as a process* (pp. 389–392; FIGURE 21.3)** Organisms used to study the genetic basis of development include the fruit fly *Drosophila melanogaster,* the nematode *Caenorhabditis elegans,* the mouse *Mus musculus,* the zebrafish *Danio rerio,* and the common wall cress *Arabidopsis thaliana.*

DIFFERENTIAL GENE EXPRESSION

- **Different types of cells in an organism have the same DNA (pp. 392–395; FIGURES 21.5, 21.6)** Cells differ in structure and function not because they contain different genes, but because they express different portions of a common genome; they have genomic equivalence. Differentiated cells from mature plants are often totipotent: capable of generating a complete new plant. The nucleus from a differentiated animal cell can sometimes give rise to a new animal if transplanted to an enucleated egg cell.
- **Different cell types make different proteins, usually as a result of transcriptional regulation (p. 395)** Differentiation is heralded by the appearance of tissue-specific proteins. These proteins allow differentiated cells to carry out their specialized roles.
- **Transcriptional regulation is directed by maternal molecules in the cytoplasm and signals from other cells (pp. 395–397)** Cytoplasmic determinants in the cytoplasm of the unfertilized egg regulate the expression of genes that affect the developmental fate of cells. In the process called induction, signal molecules from embryonic cells cause transcriptional changes in nearby target cells.

GENETIC AND CELLULAR MECHANISMS OF PATTERN FORMATION

Pattern formation, the development of a spatial organization of tissues and organs, occurs continually in plants but is mostly limited to embryos and juveniles in animals. Positional information, the molecular cues that control pattern formation, tell a cell its location relative to the body's axes and to other cells.

- **Genetic analysis of *Drosophila* reveals how genes control development: *an overview* (pp. 397–399, FIGURE 21.10)** After fertilization, positional information on an increasingly fine scale specifies the segments in *Drosophila* and finally triggers the formation of each segment's characteristic structures. By analyzing *Drosophila* mutants, Christiane Nüsslein-Volhard and Eric Wieschaus identified and mapped about 120 genes essential for normal segmentation.
- **Gradients of maternal molecules in the early embryo control axis**
21.1 **formation (pp. 399–402, FIGURE 21.11)** Studies of one important maternal effect gene in *Drosophila, bicoid,* established that its product is a morphogen, a substance whose concentration gradient influences an embryo's axes and other features. In effect, the products of maternal genes tell the embryo "which way is up."
- **A cascade of gene activations sets up the segmentation pattern in *Drosophila: a closer look* (pp. 402–403, FIGURE 21.12)** Gradients of morphogens encoded by maternal effect genes produce regional differences in the expression of segmentation genes, the products of which direct the actual formation of segments. A cascade sequentially activates three sets of segmentation genes. First, products of gap genes map out subdivision along the anterior-posterior axis of the embryo. Next, pair-rule genes define the pattern in terms of pairs of segments. Finally, segment-polarity genes set the anterior-posterior axis of each segment.
- **Homeotic genes direct the identity of body parts (pp. 403–404)** The anatomical identity of segments is set by master regulatory genes called homeotic genes, which specify the type of appendages and other structures that form on each segment. Transcription factors encoded by the homeotic genes are regulatory proteins that control the expression of genes responsible for specific anatomical structures.
- **Homeobox genes have been highly conserved in evolution (p. 404, FIGURE 21.14)** A nucleotide sequence called the homeobox is part of homeotic genes. Similar sequences are found in the genes of diverse animals and of more distantly related eukaryotes.
- **Neighboring cells instruct other cells to form particular structures: *cell signaling and induction in the nematode* (pp. 404–406, FIGURES 21.15, 21.16)** In the process of induction, cells signal nearby cells to change in some specific way, often by expressing particular genes. In *C. elegans,* an inducer produced by the anchor cell, a single cell of the embryonic gonad, initiates a chain of inductions that results in the formation of the vulva. Thus, sequential inductions in the embryo can drive organ formation. In apoptosis, or programmed cell death, precisely timed signals trigger the activation of a cascade of "suicide" proteins in the cells destined to die.
- **Plant development depends on cell signaling and transcriptional regulation: *science as a process* (pp. 406–410, FIGURES 21.17, 21.18)** Cell-cell signaling (induction) seems to be involved in determining the *number* of floral organs that develop from a floral meristem. Organ-identity genes determine the *type* of structure (stamen, carpal, sepal, or petal) that grows from each whorl of a floral meristem. The organ-identity genes apparently act as master regulatory genes, each controlling the activity of other genes that more directly bring about an organ's structure and function.

SELF-QUIZ

1. The establishment of the dorsal-ventral axis in a developing fruitfly embryo is a crucial aspect of
   a. pattern formation
   b. transcriptional regulation
   c. apoptosis
   d. cell division
   e. induction

2. The criteria for a good model organism for studying development would probably include all of the following *except*
   a. observable embryonic development
   b. short generation time
   c. a relatively small genome
   d. preexisting knowledge of the organism's life history
   e. abundant local populations for specimen collection

3. Totipotency is demonstrated when
   a. mutations in homeotic genes result in the development of misplaced appendages
   b. a cell isolated from a plant leaf grows into a normal adult plant
   c. an embryonic cell divides and differentiates
   d. the replacement of the nucleus of an unfertilized egg with that of an intestinal cell converts the egg to an intestinal cell
   e. segment-specific organs develop along the anterior-posterior axis of a *Drosophila* embryo

4. Cell differentiation always involves
   a. the production of tissue-specific proteins, such as muscle actin
   b. the formation of a gastrula
   c. the transcription of the *myoD* gene
   d. the selective loss of certain genes from the genome
   e. the cell's sensitivity to environmental cues such as light or heat

5. The development of *Drosophila* is somewhat unusual in that
   a. the early mitotic divisions proceed without cytokinesis
   b. metamorphosis occurs during the larval stage rather than the pupal stage as with other insects
   c. homeotic genes are mutated
   d. cell migration within the embryo does not occur
   e. the initial cell divisions have lengthy G_1 phases

6. In *Drosophila*, which genes initiate a cascade of gene activation that includes all other genes in the list?
   a. homeotic genes
   b. gap genes
   c. pair-rule genes
   d. egg-polarity genes
   e. segment-polarity genes

7. Select the correct statement concerning the *bicoid* gene and its role in fruit-fly development.
   a. Defects in the gene result in embryos with heads at both ends.
   b. Maternal *bicoid* RNA is deposited within the unfertilized egg and translated only after fertilization.
   c. The bicoid protein is made by the mother.
   d. The *bicoid* gene was the first segmentation gene identified.
   e. The production of bicoid protein and its diffusion to adjacent cells demonstrates induction.

8. Homeotic genes
   a. encode transcription factors that control the expression of genes responsible for specific anatomical structures
   b. are found only in *Drosophila* and other arthropods
   c. specify the anterior-posterior axis for each fruit-fly segment
   d. create the basic subdivisions of the anterior-posterior axis of the fly embryo
   e. are responsible for the programmed cell death occurring during morphogenesis

9. The embryonic development of *C. elegans* illustrates all of the following developmental concepts *except*
   a. the effect of an inducer can depend on its concentration gradient
   b. the response of an induced cell involves the establishment of a unique pattern of gene activity
   c. the signal-transduction pathways activated by inducers are unique to embryonic cells
   d. sequential inductions direct the formation of complex structures in the developing embryo
   e. inducers promote their effects via the activation or inactivation of genes that code for transcriptional regulators

10. Although quite different in structure, plants and animals share some basic similarities in their development, such as
    a. the importance of cell and tissue movements
    b. the importance of selective cell enlargement
    c. the activation of signal-transduction pathways by environmental cues
    d. the retention of meristematic tissues in the adult
    e. master regulatory genes that encode DNA-binding proteins

CHALLENGE QUESTION

The cascades of induction events leading to organ formation are vulnerable to error. Using vulval development in *C. elegans*, construct a flowchart of induction events beginning with the anchor cell. Indicate each point where an error in transcription, post-transcriptional processing, translation, or post-translational events could disrupt the cascade.

SCIENCE, TECHNOLOGY, AND SOCIETY

Spina bifida, a spinal cord disorder, is a common and serious birth defect resulting from an error in morphogenesis. The United States Public Health Service recommends that all women of childbearing age in the U.S. who are capable of becoming pregnant consume 0.4 mg of folic acid per day to guard against such defects. It has been suggested that the most expeditious way to increase folic acid consumption is through fortification of food staples such as cereal grains. What are your thoughts about this suggestion? Is it reasonable for the entire population to receive an additive that primarily benefits certain individuals? Is there a difference between additives that improve the safety of the food supply for everyone and those that affect only a subset of the population?

FURTHER READING

Gilbert, S. F. *Developmental Biology*, 5th ed. Sunderland, MA: Sinauer Associates, 1997. A widely used upper-division text.

Nüsslein-Volhard, C. "Gradients That Organize Embryo Development." *Scientific American*, August 1996. More about the bicoid protein and other fruit-fly morphogens.

Wolpert, L. *Principles of Development*. London: Oxford University Press, 1998. A new, well-written and attractively illustrated upper-division text oriented around basic concepts.

WEB LINKS

Visit the special edition of *The Biology Place* for BIOLOGY, Fifth Edition, at http://www.biology.com/campbell. Go to Chapter 21 for online resources, including learning activities, practice exams, and links to the following web sites:

"The Society for Developmental Biology"
This site provides information on many aspects of developmental biology. The "Interactive Fly" and the "Developmental Biology Cinema" are well worth a visit.

"The Virtual Embryo"
Development in a number of model organisms, including *C. elegans, Xenopus,* and *Danio rerio*, is described here.

"THE FISH NET"
This site provides access to a wealth of zebrafish information.

"Bill Wasserman's Developmental Biology Page"
Dr. Wasserman's site links to numerous web resources. The movie and picture page is particularly good.

UNIT FOUR

MECHANISMS OF EVOLUTION

Oxford professor Richard Dawkins has helped explain Darwin and natural selection to the public through a series of award-winning books, beginning with The Selfish Gene *in 1976. He followed that international best-seller with* The Blind Watchmaker, The Extended Phenotype, River Out of Eden, *and* Climbing Mount Improbable. *While making the beauty and power of natural selection accessible to nonscientists, Dawkins's books have also stirred controversy among evolutionary biologists. Particularly provocative has been Dawkins's argument that genes, not whole organisms, are the units of natural selection. In this view of Darwinism, organisms are "survival machines" for the "replicators," selfish molecules known as genes. Many biologists disagree, countering that a strictly gene-centered concept of natural selection is too simplistic. Richard Dawkins is among the very few scientists who can stimulate thoughtful debate and research in a scientific field at the same time as he engages and challenges nonscientists.*

One of your books, *The Blind Watchmaker,* argues the case for the cumulative power of natural selection in the adaptation of organisms. Tell us about the metaphorical title of that book.

The "watchmaker" comes from William Paley, the eighteenth- and early nineteenth-century theologian who was one of the most famous exponents of the argument of design. Paley famously said that if you are wandering along and stumble upon a watch and you pick it up and open it, you realize that the internal mechanism—the way in which it's all meshed together—is detailed perfection. Add this to the fact that the watch mechanism has a purpose—namely, telling the time—then this compels you to conclude that the watch had to have a designer. Paley then went on throughout his book giving example after example of detailed structure of living organisms—eyes, heart, bowels, joints, and everything about animals—showing how beautifully designed they apparently are, how well they work, how intricately the parts mesh together, just like the cog wheels of a watch. And if the watch had to have a watchmaker, then of course these biological structures also had to have a designer.

My reason for beginning *The Blind Watchmaker* was Paley. He really saw the magnitude of the problem of adaptation when most people just didn't see how elegant, how beautiful, apparent design in life is. Paley saw that, and Darwin saw that. And Darwin was introduced to it at least partly by Paley. All undergraduates at Cambridge had to read William Paley. He at least put the question right. So the only thing Paley got wrong, which is quite a big thing, was the answer to the question. And nobody got the right answer until Charles Darwin in the nineteenth century.

Is it that answer that puts the "blind" in your book title?

The "blind" watchmaker is natural selection. Natural selection is totally blind to the future. It's always just picking the best of available alternatives in the present. It can never say, "Well, if we tolerate a bit of unpleasantness for the next 2 million years, then that will set us up wonderfully to improve in 2 million years' time." Natural selection can't do that. It's always optimizing in the short term, though it's blind to the future.

INTERVIEW

RICHARD DAWKINS

In *Climbing Mount Improbable,* you built on the theme of how natural selection can lead to elegant structure by optimizing for the short term. Tell us how that book title relates to the problem of adaptation.

"Mount Improbable" is a metaphor for adaptation occurring gradually, in increments. The metaphor is that of a mountain which has an absolutely sheer cliff face. If we relate this cliff to adaptation, to the most complicated piece of biological machinery you can think of, which for many people is an eye, then you say to yourself that it's impossible to leap from the bottom of this mountain to the top, which indeed it is. Leaping from the bottom of the cliff to the top would correspond to having the sheer luck to get that eye coming into place in one fell swoop. Many people wrongly think that Darwinism is a theory of chance, that it means that eyes and other complex organs come about by sheer luck. So no wonder these people don't believe natural selection. Of course an eye couldn't possibly come about like that. But on the other side of the mountain, you've got a slow, gradual slope, and it is very easy to get to the top of the mountain if you go around the other side and just walk up the slope. Relating this to adaptation, you have gradual, incremental improvement. You begin with hardly any eye at all and then each step of the way up the mountain is a gradual improvement. It may not be much, but it's enough to be better than your predecessors, who didn't have even that improvement. *Climbing Mount Improbable* emphasizes that evolution of complex adaptations has got to be gradual.

In another of your books, *The Selfish Gene,* you argue that genes are the units upon which natural selection acts and that organisms are "survival machines" for genes. To what extent are humans exceptions to this mechanistic view of life?

Humans are fundamentally not exceptional because we came from the same evolutionary source as every other species. It is natural selection of selfish genes that has given us our bodies and our brains. However, the brains that natural selection gave us are exceptionally big brains, so big that they have done a rather unusual thing. Using language and culture, humans have formed societies in which there is something like Darwinian evolution going on, though it is not really Darwinian. We live in a highly domestic environment, largely governed by technology, largely divorced from the environment in which our genes were originally naturally selected. So what is different about us is that it is no longer possible to look at a human the way one might look at a wildebeest or a kangaroo and ask, "Why is that? What's that kangaroo doing that increases its gene survival?" If you see a wild animal doing something in the wild, then it's sensible to ask the question, "What is it about that behavior, or what is it about that morphological structure, which improves its survival, or more particularly the survival of its genes?"

And you can't do that for humans?

No, you can't look at humans playing the violin, or trying to run a company, or writing a book or writing a symphony, and ask, "In what way does writing this symphony benefit survival and replication of that human's genes?" because it doesn't. You have to be more sophisticated and ask, "In what way does the behavior of a brain which was originally built by natural selection for surviving in Africa in the Pleistocene and Pliocene translate into the behavior of this brain, now that it finds itself in this very different, artificial environment?" For example, we still enjoy sex, even though today we often use contraceptives, which separate the enjoyment of sex from the reproductive function. Before contraceptives, enjoying sex translated into Darwinian reproductive success. Back then, as Steven Pinker has argued, if there had been a tree with berries that were the equivalents of today's contraceptive pills, we would probably have treated those trees with the same terror as we would a poisonous snake or poisonous spider. Now that's obviously not the case. All the questions that you might wish to ask about humans have to be rewritten in a more sophisticated way.

This example of modern humans choosing to limit their reproduction seems like the most counter-Darwinian of all human behaviors. Is it possible that there will be long-term, global limiting of population growth by choice rather than by famine and disease?

If we were relying purely on Darwinian mechanisms, there would be no hope. There is nothing in Darwinism that explains limiting reproduction for the good of future populations. But where natural selection has no foresight, human brains do have foresight. The human brain is an adaptive mechanism which, having been put together for quite different reasons, now finds itself capable of gazing into the distant future and working out the long-term consequences of our actions. So it is possible for us now to do what has never been possible before. We can look into the long-term future; we can see it would be a very bad idea to do all sorts of things, such as over-reproducing, overfishing, depleting our natural resources. But being able to do that by no means gets us out of the woods. It's one thing for individuals to sit down and say, "Well, I can see that it's a bad thing to over-reproduce and overfish," but it's quite another matter to persuade humanity at large to see that. I don't mean that humanity at large is stupid. Anybody can see that it would benefit everybody if we all limited our greed, but as long as others are not limiting their greed, people quite understandably say, "Why on Earth should I? Why should I be the first to limit my own appetite, limit my own greed, limit my own desire for selfish advancement?" But having recognized this behavior is at least a first step. We do have institutions like government, we do pay our taxes, we have police forces. We pass laws to limit selfish overexploitation. Most of us accept paying taxes and obeying laws if we are assured that everybody else is paying their taxes and obeying the same laws. So in many countries it does seem to work. We're kind of teetering on the edge when you look at things like overexploiting whales or fish on an international level. There are international conferences and conventions that are aimed at avoiding overexploitation. To a limited extent they seem to work. I mean they are constantly on the brink of not working, with people constantly walking out and refusing to abide by the rules. But there is something like a social machinery coming into place for limiting greed. And I see hope in that because this tendency to restrain individual advantage for the sake of group advantage is something no other species has ever even come close to.

Earlier, you mentioned that in human culture, there may be something like Darwinian evolution. Tell us more about that.

There's no doubt, of course, that cultural evolution happened, and it has some analogy to evolution, perhaps even Darwinian evolution. In *The Selfish Gene* I constantly emphasized the importance of genes as the central units of natural selection. Genes are the only things that pass down through generations. But what's special about genes is that they are replicated, they function as what I called "replicators" in *The Selfish Gene.* So, it's replicators that matter, not specifically genes. Anything that is self-replicating anywhere in the universe is fair game for natural selection. There probably is life on other planets, and if there is, then I absolutely bet my shirt that it's based upon natural selection. That would require a replicator. It doesn't have to be DNA, but it would have the fundamental property of self-replication. In making this point in *The Selfish Gene* that replicators are the units of selection, I also gave the example of what I called the "meme," the unit of cultural evolution. You can look at human culture and ask yourself, "Is there something which is passing from brain to brain and perhaps from generation to generation by nongenetic means?" I think there is. For example, when we were at school, I think we all had the experience of some craze that spreads through a school like a measles epidemic—a new kind of toy or a new style of wearing a hat. It literally does spread from person to person and then it may die away, or it may perhaps leap to another school. Well, that's a trivial example, but it's enough to show that there is some replicator which is analogous to a virus but does not consist of DNA.

The phenotypes of most memes are behavioral, such as religious traditions. So there we have memes that pass longitudinally down the generations. The school craze is a meme that passes horizontally across one generation. And we live in an environment that is saturated by both kinds of memes. You then have to ask, "Do some memes survive better than others because they have what it takes to survive?" If they do, that is all that you need to establish that there is a Darwinian–like component to cultural evolution.

What if you could travel in time and visit with Darwin? What would you ask him?

First, I'd be so overawed that I wouldn't know quite what to say. But I think maybe if I were really forced to ask him one question it would be, "Why did you wait so long after you had this brilliantly simple yet powerful idea? Didn't it seem to you so fantastically simple yet so fantastically powerful that if you didn't write it down quickly, somebody else would?" I'm genuinely baffled about that because it's as though Darwin thought he had all the time in the world, and pretty nearly he did. I mean Wallace did get there, but still Darwin had about 20 years before that. What I find remarkable is that Aristotle didn't get it and Plato didn't get it, nor did Pythagorus, Archimedes, Newton, even though you don't need any technical know-how to get the idea. Natural selection is a bewilderingly simple idea. And yet what it explains is the whole of life, the diversity of life, the complexity of life, the apparent design of life. It all flows from this one remarkably simple idea.

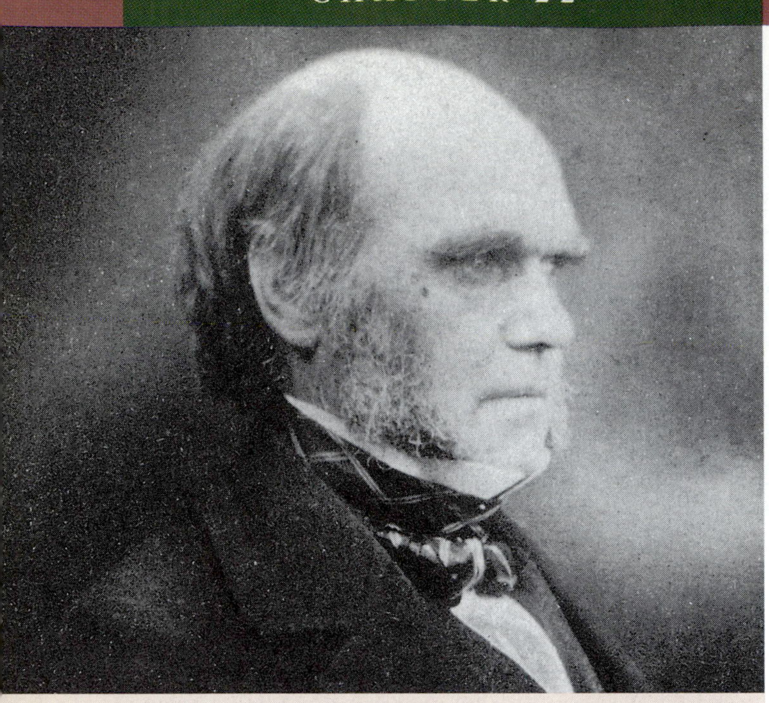

CHAPTER 22

DESCENT WITH MODIFICATION: A DARWINIAN VIEW OF LIFE

Historical Context for Evolutionary Theory

- Western culture resisted evolutionary views of life
- Theories of geological gradualism helped clear the path for evolutionary biologists
- Lamarck placed fossils in an evolutionary context

The Darwinian Revolution

- Field research helped Darwin frame his view of life: *science as a process*
- *The Origin of Species* developed two main points: the occurrence of evolution and natural selection as its mechanism

Evidence of Evolution

- Evidence of evolution pervades biology
- What is theoretical about the Darwinian view of life?

Biology came of age on November 24, 1859, the day Charles Darwin published On the Origin of Species by Means of Natural Selection. *Darwin's book, which presented a convincing case for evolution, connected what had previously seemed a bewildering array of unrelated facts into a cohesive view of life. In biology,* **evolution** *refers to the processes that have transformed life on Earth from its earliest forms to the vast diversity that characterizes it today. Darwin addressed the sweeping issues of biology: the great diversity of organisms, their origins and relationships, their similarities and differences, their geographical distribution, and their adaptations to the surrounding environment. Thus, evolution is the most pervasive principle in biology, and a thematic thread woven throughout this book. This unit focuses specifically on the mechanisms by which life evolves.*

Darwin made two points in The Origin of Species. *First, he argued from evidence that species were not created in their present forms but had evolved from ancestral species. Second, he proposed a mechanism for evolution, which he termed* **natural selection**. *According to Darwin's concept of natural selection, a population of organisms can change over time as a result of individuals with certain heritable traits leaving more offspring than other individuals. Darwin's first point—that evolution occurs—can stand on its own, whether or not natural selection is the cause.*

A recurrent theme runs throughout the chapters of this unit: Evolutionary change is based mainly on the interactions between populations of organisms and their environments. These interactions result in **evolutionary adaptations**, *inherited characteristics that enhance an organism's ability to survive and reproduce in a particular environment. This first chapter defines the Darwinian view of life and traces its historical development.*

HISTORICAL CONTEXT FOR EVOLUTIONARY THEORY

To put the Darwinian view in perspective, we must compare it to earlier ideas about Earth and its life. The impact of an intellectual revolution such as Darwinism depends on timing as well as logic. Let's explore the historical context of Darwin's life and ideas.

Western culture resisted evolutionary views of life

The Origin of Species was truly radical for its time; not only did it challenge prevailing scientific views, it also shook the deepest roots of Western culture. Darwin's view of life contrasted sharply with the conventional paradigm of an Earth only a few thousand years old, populated by unchanging forms of life that had been individually made during the single week in which the Creator formed the entire universe. Darwin's book challenged a worldview that had been taught for centuries.

The Scale of Nature and Natural Theology

A number of classical Greek philosophers believed in the gradual evolution of life. But the philosophers who influenced Western culture most, Plato (427–347 B.C.) and his student Aristotle (384–322 B.C.), held opinions that opposed any concept of evolution. Plato believed in two worlds: a real world that is ideal and eternal, and an illusory world of imperfection that we perceive through our senses. Evolution would be counterproductive in a world where ideal organisms were already perfectly adapted to their environments.

Aristotle believed that all living forms could be arranged on a scale, or ladder, of increasing complexity, later called the *scala naturae* ("scale of nature"). Each form of life had its allotted rung on this ladder, and every rung was taken. In this view of life, which prevailed for over 2000 years, species are permanent, are perfect, and do not evolve.

In Judeo-Christian culture, the Old Testament account of creation fortified the idea that species were individually designed and permanent. In the 1700s, biology in Europe and America was dominated by **natural theology**, a philosophy dedicated to discovering the Creator's plan by studying nature. Natural theologians saw the adaptations of organisms as evidence that the Creator had designed each and every species for a particular purpose. A major objective of natural theology was to classify species in order to reveal the steps of the scale of life that God had created.

Carolus Linnaeus (1707–1778), a Swedish physician and botanist, sought to discover order in the diversity of life "for the greater glory of God." Linnaeus was the founder of **taxonomy**, the branch of biology concerned with naming and classifying the diverse forms of life. He developed the two-part, or binomial, system of naming organisms according to genus and species that is still used today. In addition, Linnaeus adopted a system for grouping similar species into a hierarchy of increasingly general categories. For example, similar species are grouped in the same genus, similar genera (plural of genus) are grouped in the same family, and so on (see FIGURE 1.10, p. 10). To Linnaeus, clustering similar species together implied no evolutionary kinship, but a century later his taxonomic system would become a focal point in Darwin's arguments for evolution (FIGURE 22.1).

Cuvier, Fossils, and Catastrophism

The study of fossils also helped lay the groundwork for Darwin's ideas. **Fossils** are relics or impressions of organisms from the past, mineralized in rock. Most fossils are found in **sedimentary rocks** formed from the sand and mud that settle to the bottom of seas, lakes, and marshes. New layers of sediment cover older ones and compress them into superimposed layers of rock called strata. Later, erosion may scrape or carve through upper (younger) strata and reveal more ancient strata that had been buried. Fossils within the layers show that

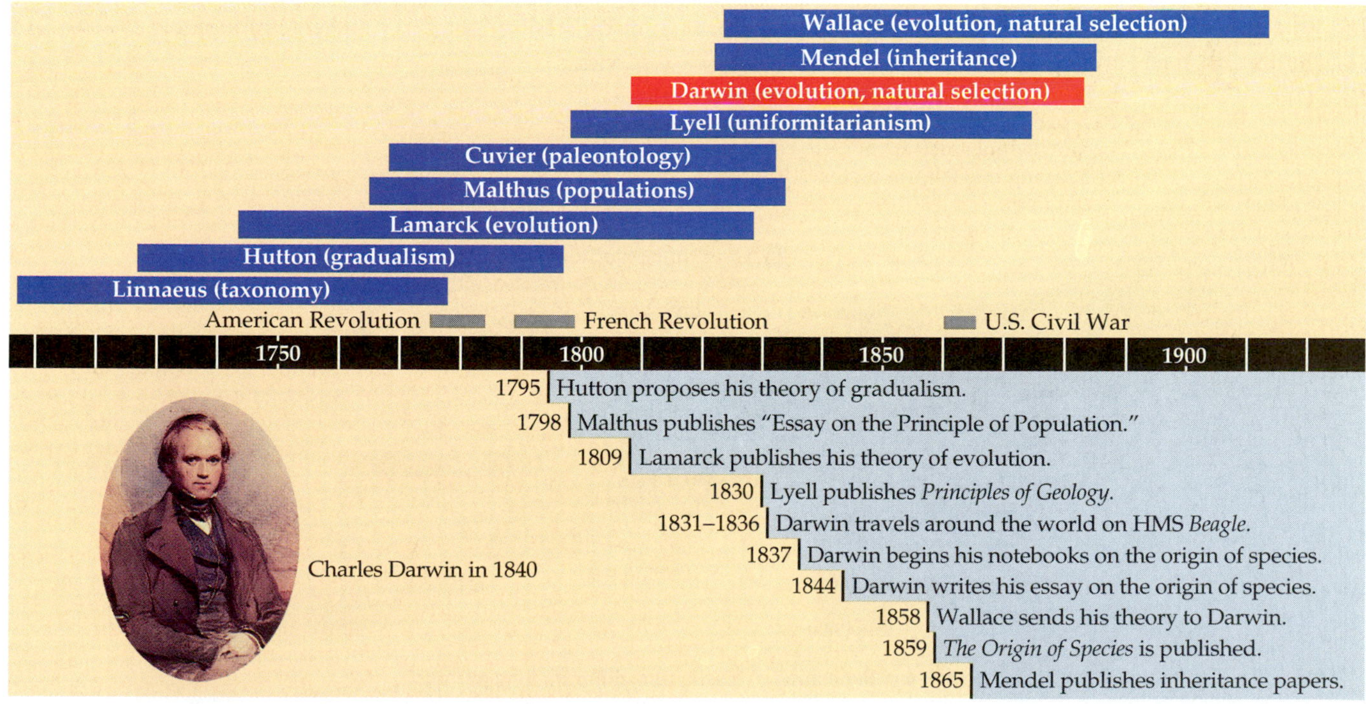

FIGURE 22.1 ▪ The historical context of Darwin's life and ideas.

a succession of organisms has populated Earth throughout time (FIGURE 22.2).

Paleontology, the study of fossils, was largely developed by French anatomist Georges Cuvier (1769–1832). Realizing that the history of life is recorded in strata containing fossils, he documented the succession of fossil species in the Paris Basin. He noted that each stratum is characterized by a unique group of fossil species, and that the deeper (older) the stratum, the more dissimilar the flora (plant life) and fauna (animal life) are from modern life. Cuvier even recognized that extinction had been a common occurrence in the history of life. From stratum to stratum, new species appear and others disappear. Yet Cuvier was a staunch opponent to the evolutionists of his day. Instead, he advocated **catastrophism**, speculating that each boundary between strata corresponded in time to a catastrophe, such as a flood or drought, that had destroyed many of the species living there at that time. He proposed that these periodic catastrophes were usually confined to local geographical regions, and that the ravaged region was repopulated by species immigrating from other areas.

Theories of geological gradualism helped clear the path for evolutionary biologists

Competing with Cuvier's theory of catastrophism was a very different idea of how geological processes had shaped the Earth's crust. In 1795, Scottish geologist James Hutton (1726–1797) proposed that it was possible to explain the various land forms by looking at mechanisms currently operating in the world. For example, he suggested that canyons were formed by rivers cutting down through rocks, and that sedimentary rocks with marine fossils were built of particles that had eroded from the land and been carried by rivers to the sea. Hutton explained Earth's geological features by the theory of **gradualism**, which holds that profound change is the cumulative product of slow but continuous processes.

The leading geologist of Darwin's era, a Scot named Charles Lyell (1797–1875), incorporated Hutton's gradualism into a theory known as **uniformitarianism**. The term refers to Lyell's idea that geological processes have not changed throughout Earth's history. Thus, for example, the forces that build mountains and erode mountains and the rates at which these forces operate are the same today as in the past. Darwin was strongly influenced by two conclusions that followed directly from the observations of Hutton and Lyell. First, if geological change results from slow, continuous actions rather than sudden events, then Earth must be very old, certainly much older than the 6000 years assigned by many theologians on the basis of biblical inference. Second, very slow and subtle processes persisting over a long period of time can cause substantial change. Darwin was not the first to apply the principle of gradualism to biological evolution, however.

Lamarck placed fossils in an evolutionary context

Toward the end of the eighteenth century, several naturalists suggested that life had evolved along with the evolution of Earth. But only one of Darwin's predecessors developed a comprehensive model that attempted to explain how life evolves: Jean Baptiste Lamarck (1744–1829).

Lamarck published his theory of evolution in 1809, the year Darwin was born. Lamarck was in charge of the invertebrate collection at the Natural History Museum in Paris. By comparing current species to fossil forms, Lamarck could see what appeared to be several lines of descent, each a chronological series of older to younger fossils leading to a modern species.

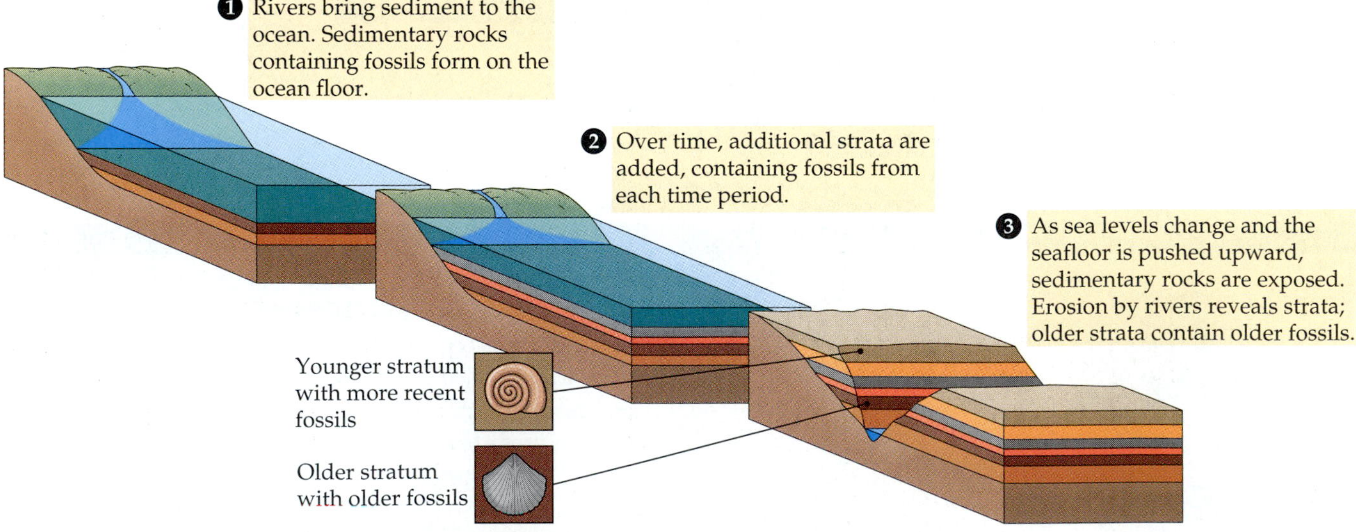
1 Rivers bring sediment to the ocean. Sedimentary rocks containing fossils form on the ocean floor.
2 Over time, additional strata are added, containing fossils from each time period.
3 As sea levels change and the seafloor is pushed upward, sedimentary rocks are exposed. Erosion by rivers reveals strata; older strata contain older fossils.
Younger stratum with more recent fossils
Older stratum with older fossils

FIGURE 22.2 ▪ Formation of sedimentary rock and deposition of fossils from different time periods. Each stratum, or layer, represents a particular period in Earth's history and is characterized by a collection of fossils of organisms that lived at that time.

Where Aristotle saw one ladder of life, Lamarck saw many, and he thought that species could move up the ladders toward greater complexity. On the lowest rungs were the microscopic organisms, which Lamarck believed were continually generated spontaneously from nonliving material. At the top of the evolutionary ladders were the most complex plants and animals. Evolution was driven by an innate tendency toward greater and greater complexity, which Lamarck seemed to equate with perfection. As organisms attained perfection, they became better and better adapted to their environments. Thus Lamarck believed that evolution responded to organisms' *sentiments interieurs,* or "felt needs."

Lamarck is remembered most for the mechanism he proposed to explain how specific adaptations evolve. It incorporates two ideas that were popular during Lamarck's era. The first was use and disuse, the idea that those parts of the body used extensively to cope with the environment become larger and stronger while those that are not used deteriorate. Among the examples Lamarck cited were a blacksmith developing a bigger bicep in the arm that wields the hammer and a giraffe stretching its neck to reach leaves on high branches. The second idea Lamarck adopted was called the inheritance of acquired characteristics. In this concept of heredity, the modifications an organism acquires during its lifetime can be passed along to its offspring. The long neck of the giraffe, Lamarck reasoned, evolved gradually as the cumulative product of a great many generations of ancestors stretching ever higher.

There is, however, no evidence that acquired characteristics can be inherited. Blacksmiths may increase strength and stamina by a lifetime of pounding with a heavy hammer, but these acquired traits do not change genes transmitted by gametes to offspring. Even though the Lamarckian theory of evolution is ridiculed by some today because of its erroneous assumption that acquired characteristics are inherited, in Lamarck's time that concept of inheritance was generally accepted (and, indeed, Darwin could offer no acceptable alternative). To most of Lamarck's contemporaries, however, the mechanism of evolution was an irrelevant issue because they firmly believed that species were fixed and that no theory of evolution could be taken seriously. Lamarck was vilified, especially by Cuvier, who would have no part of evolution. In retrospect, Lamarck deserves much credit for his theory, which was visionary in many respects: in its claim that evolution is the best explanation for both the fossil record and the current diversity of life, in its recognition of the great age of Earth, and especially in its emphasis on *adaptation to the environment* as a primary product of evolution.

THE DARWINIAN REVOLUTION

We have set the scene for the Darwinian revolution. Natural theology, with its view of an ordered world where each living form fit its environment perfectly because it had been specially created, still dominated the intellectual climate as the nineteenth century dawned. A few clouds of doubt about the permanence of species were beginning to gather, but no one could have forecast the thundering storm just over the horizon.

Charles Darwin (1809–1882) was born in Shrewsbury in western England. Even as a boy he had a consuming interest in nature. When he was not reading nature books, he was fishing, hunting, and collecting insects. Darwin's father, an eminent physician, could see no future for a naturalist and sent Charles to the University of Edinburgh to study medicine. Only 16 years old at the time, Charles found medical school boring and distasteful, although he managed decent grades. He left Edinburgh without a degree and shortly thereafter enrolled at Christ College at Cambridge University, with the intent of becoming a clergyman. At that time in Great Britain, most naturalists and other scientists belonged to the clergy, and nearly all saw the world in the context of natural theology. Darwin became the protégé of the Reverend John Henslow, professor of botany at Cambridge. Soon after Darwin received his B.A. degree in 1831, Professor Henslow recommended the young graduate to Captain Robert FitzRoy, who was preparing the survey ship *Beagle* for a voyage around the world.

Field research helped Darwin frame his view of life: *science as a process*

The Voyage of the Beagle

Darwin was 22 years old when he sailed from Great Britain
22.1 aboard HMS *Beagle* in December 1831. The primary mission of the voyage was to chart poorly known stretches of the South American coastline. While the ship's crew surveyed the coast, Darwin spent most of his time on shore, observing and collecting thousands of specimens of the exotic and diverse faunas and floras of South America. As the ship worked its way around the continent, Darwin observed the various adaptations of plants and animals that inhabited such diverse environments as the Brazilian jungles, the expansive grasslands of the Argentine pampas, the desolate lands of Tierra del Fuego near Antarctica, and the towering heights of the Andes Mountains.

Darwin wrote extensively about the fauna and flora of the different regions of South America. He noted that plants and animals on the continent had definite South American characteristics, very distinct from those of Europe. That in itself may not have been surprising. But Darwin also noted that the plants and animals in temperate regions of South America were more closely related to species living in tropical regions of that continent than to species in temperate regions of

22.1 This symbol links topics in the text to interactive exercises in the CD-ROM that accompanies the book. The number indicates the appropriate activity in the CD.

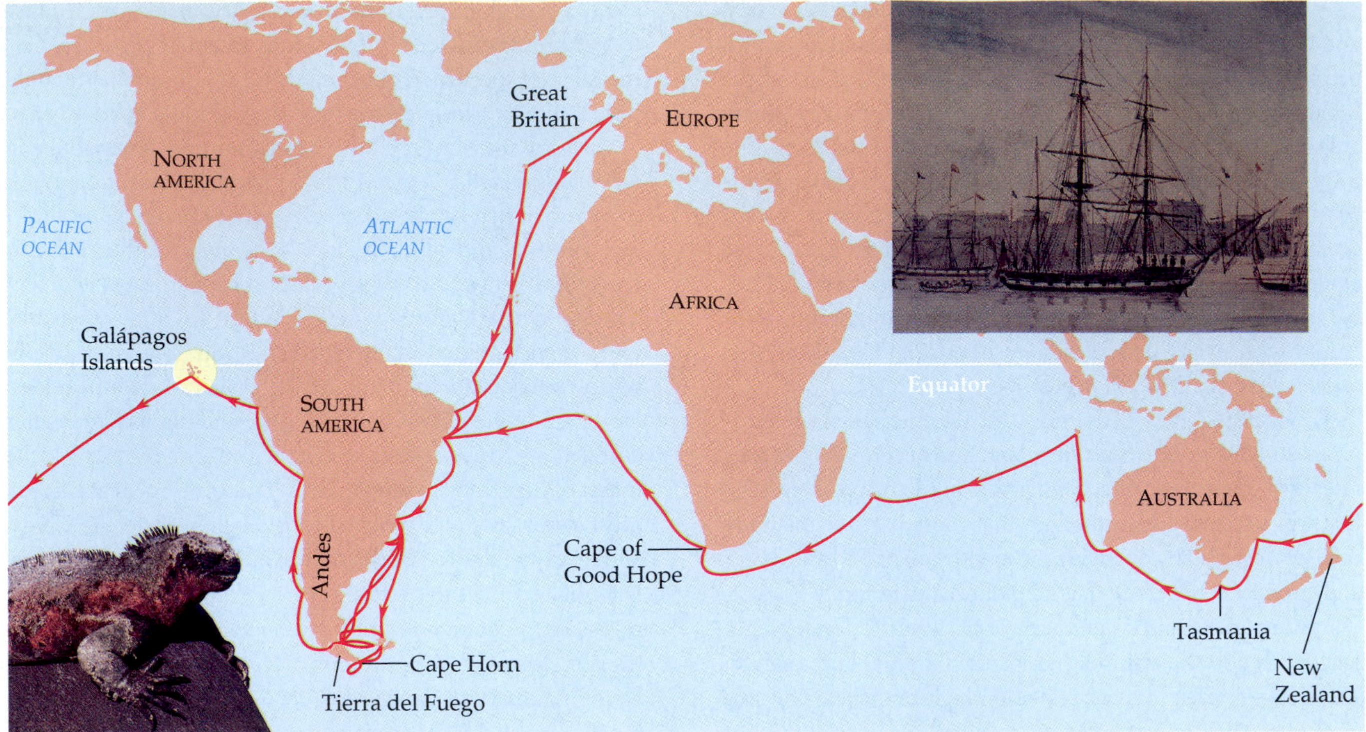

FIGURE 22.3 ▪ **The Voyage of HMS *Beagle.*** On its round-the-world journey, HMS *Beagle* spent time at the Galápagos Islands, located about 900 km off the west coast of Ecuador. Many plant and animal species inhabiting the islands are found nowhere else in the world. This marine iguana, for example, is among the species that evolved from mainland ancestors that colonized the volcanic islands.

22.1

Europe. Furthermore, the South American fossils that Darwin found, though clearly different from modern species, were distinctly South American in their resemblance to the living plants and animals of that continent.

The geographical distribution of species perplexed Darwin. Especially puzzling to him was the fauna of the Galápagos, islands of relatively recent volcanic origin that lie on the equator about 900 km west of the South American coast (FIGURE 22.3). Most of the animal species on the Galápagos live nowhere else in the world, although they resemble species living on the South American mainland. It was as though the islands had been colonized by plants and animals that strayed from the South American mainland and then diversified on the different islands. Among the birds Darwin collected on the Galápagos were 13 types of finches that, although quite similar, seemed to be different species. Some were unique to individual islands, while other species were distributed on two or more islands that were close together.

By the time the *Beagle* sailed from the Galápagos, Darwin had read Lyell's *Principles of Geology.* Lyell's ideas, together with his experiences on the Galápagos, had Darwin doubting the church's position that the Earth was static and had been created only a few thousand years ago. By acknowledging that Earth was very old and constantly changing, Darwin had taken an important step toward recognizing that life on Earth had also evolved.

Darwin Focuses on Adaptation

Soon after returning to Great Britain in 1836, Darwin started reassessing all that he had observed during the voyage of the *Beagle.* He began to perceive the origin of new species and adaptation to the environment as closely related processes. It occurred to him that a new species could arise from an ancestral form by the gradual accumulation of adaptations to a different environment. For example, if a geographical barrier—such as a channel separating oceanic islands—isolates two populations of a single species, the populations would diverge more and more in appearance as each adapted to local environmental conditions. This hypothesis for the origin of species predicted that over many generations, two populations could become dissimilar enough to be designated separate species. From studies made years after Darwin's voyage, biologists have concluded that this is what happened to the Galápagos finches. Among the differences between the finches are their beaks, which are adapted to the specific foods available on their home islands (FIGURE 22.4). Darwin anticipated that explaining how such adaptations arise was essential to understanding evolution.

By the early 1840s, Darwin had worked out the major features of his theory of natural selection as the mechanism of evolution. However, he had not yet published his ideas. He was in poor health, and he rarely left home. Despite his reclusiveness, Darwin was not isolated from the scientific commu-

(a) Seed eater

(b) Insect eater

(c) Insect eater

FIGURE 22.4 ▪ Galápagos finches. The Galápagos Islands have a total of 13 species of closely related finches, some found on only a single island. The most striking difference among species is in their beaks, which are adapted for specific diets. **(a)** The large ground finch (*Geospiza magnirostris*) has a large beak adapted for cracking seeds. **(b)** The small tree finch (*Camarhynchus parvulus*) uses its beak to grasp insects. **(c)** The woodpecker finch (*Camarhynchus pallidus*) uses a cactus spine or small twig as a tool to probe for termites and other wood-boring insects.

nity. Already famous as a naturalist because of the letters and specimens he sent to Great Britain during the voyage of the *Beagle,* Darwin had frequent correspondence and visits from Lyell, Henslow, and other scientists.

In 1844, Darwin wrote a long essay on the origin of species and natural selection. Realizing the significance of this work, he asked his wife to publish the essay if he died before writing a more thorough dissertation on evolution. Evolutionary thinking was emerging in many areas by this time, but Darwin was reluctant to introduce his theory publicly. Apparently, he anticipated the uproar it would cause. While he procrastinated, he continued to compile evidence in support of his theory. Lyell, not yet convinced of evolution himself, nevertheless advised Darwin to publish on the subject before someone else came to the same conclusions and published first.

In June 1858, Lyell's prediction came true. Darwin received a letter from Alfred Wallace (1823–1913), a young British naturalist working in the East Indies. The letter was accompanied by a manuscript in which Wallace developed a theory of natural selection essentially identical to Darwin's. Wallace asked Darwin to evaluate the paper and forward it to Lyell if it merited publication. Darwin complied, writing to Lyell: "Your words have come true with a vengeance. . . . I never saw a more striking coincidence . . . so all my originality, whatever it may amount to, will be smashed." Lyell and a colleague presented Wallace's paper, along with extracts from Darwin's unpublished 1844 essay, to the Linnaean Society of London on July 1, 1858. Darwin quickly finished *The Origin of Species* and published it the next year. Although Wallace wrote up his ideas for publication first, Darwin developed and supported the theory of natural selection so much more extensively than Wallace that he is known as its main author. Darwin's notebooks also prove that he formulated his theory of natural selection 15 years before reading Wallace's manuscript.

Within a decade, Darwin's book and its proponents had convinced the majority of biologists that biological diversity was the product of evolution. Darwin succeeded where previous evolutionists had failed, partly because science was beginning to shift away from natural theology, but mainly because he convinced his readers with immaculate logic and an avalanche of evidence in support of evolution.

The Origin of Species developed two main points: the occurrence of evolution and natural selection as its mechanism

Darwinism has a dual meaning. One facet is recognition of evolution as the explanation for life's unity and diversity. The second facet is the Darwinian concept of natural selection as the cause of adaptive evolution. This section highlights these two main points of Darwin's book.

Descent with Modification

In the first edition of *The Origin of Species,* Darwin did not use the word *evolution* until the last paragraph, referring instead to **descent with modification**, a phrase that condensed his view of life. Darwin perceived unity in life, with all organisms related through descent from some unknown prototype that lived in the remote past. As the descendants of that inaugural organism spilled into various habitats over millions of years, they accumulated diverse modifications, or adaptations, that fit them to specific ways of life.

In the Darwinian view, the history of life is like a tree, with multiple branching and rebranching from a common trunk all the way to the tips of the youngest twigs, symbolic of the diversity of living organisms. At each fork of the evolutionary tree is an ancestor common to all lines of evolution branching from that fork. Closely related species, such as lions and tigers, share many characteristics because their lineage of common descent extends to the smallest branches of the tree of life. Most branches of evolution, even some major ones, are dead ends; about 99% of all species that have ever lived are extinct.

Ironically, Linnaeus, who apparently believed that species are fixed, provided Darwin with a connection to evolution by recognizing that the great diversity of organisms could be ordered into "groups subordinate to groups" (Darwin's phrase). FIGURE 1.10 introduced the major taxonomic categories; let's review them here: kingdom > phylum > class > order > family > genus > species.

To Darwin, the natural hierarchy of the Linnaean scheme reflected the branching genealogy of the tree of life, with organisms at the different taxonomic levels related through descent from common ancestors. If we acknowledge that lions and tigers are more closely related than are lions and horses, then we have recognized that evolution has left signs in the form of different degrees of kinship among modern species. Because taxonomy is a human invention, it cannot, by itself, confirm common descent. But taken along with the many other types of evidence, the evolutionary implication of taxonomy is unmistakable. Genetic analysis, for example, reveals that such species as lions and tigers, thought to be closely related on the basis of anatomical features and other criteria, are indeed close relatives with a very similar hereditary background.

Natural Selection and Adaptation

Despite the title of his book, Darwin actually devoted little space to the origin of species, concentrating instead on how populations of individual species become better adapted to their local environments through natural selection (FIGURE 22.5).

(a) Flower mantid

(b) Green leaf mantid

FIGURE 22.5 ▪ Evolutionary adaptation shaped by natural selection: camouflage. Related species of mantids have diverse shapes and colors that evolved in different environments. **(a)** A flower mantid in Malaya. **(b)** A Central American mantid that resembles a green leaf.

Evolutionary biologist Ernst Mayr has dissected the logic of Darwin's theory of natural selection into three inferences based on five observations:*

OBSERVATION #1: All species have such great potential fertility that their population size would increase exponentially if all individuals that are born reproduced successfully.

OBSERVATION #2: Populations tend to remain stable in size, except for seasonal fluctuations.

OBSERVATION #3: Environmental resources are limited.

<u>INFERENCE #1:</u> Production of more individuals than the environment can support leads to a struggle for existence among individuals of a population, with only a fraction of offspring surviving each generation.

OBSERVATION #4: Individuals of a population vary extensively in their characteristics; no two individuals are exactly alike.

OBSERVATION #5: Much of this variation is heritable.

<u>INFERENCE #2:</u> Survival in the struggle for existence is not random, but depends in part on the hereditary constitution of the surviving individuals. Those individuals whose inherited characteristics best fit them to their environment are likely to leave more offspring than less-fit individuals.

<u>INFERENCE #3:</u> This unequal ability of individuals to survive and reproduce will lead to a gradual change in a population, with favorable characteristics accumulating over the generations.

Summarizing Darwin's main ideas:

Natural selection is differential success in reproduction (unequal ability of individuals to survive and reproduce).

Natural selection occurs through an interaction between the environment and the variability inherent among the individual organisms making up a population.

* Adapted from Mayr, E. *The Growth of Biological Thought: Diversity, Evolution and Inheritance.* Cambridge, MA: Harvard University Press, 1982.

The product of natural selection is the adaptation of populations of organisms to their environment.

Let's elaborate on the important connections Darwin perceived among natural selection, the struggle for existence, and the capacity of organisms to "overreproduce." Darwin already had recognized the struggle for existence when he read an influential essay on human population written by the Reverend Thomas Malthus in 1798 (see FIGURE 22.1). Malthus contended that much of human suffering—disease, famine, homelessness, and war—was the inescapable consequence of the potential for the human population to increase faster than food supplies and other resources. The capacity to overproduce seems to be characteristic of all species. Of the many eggs laid, young born, and seeds spread, only a tiny fraction complete their development and leave offspring of their own. The rest are eaten, frozen, starved, diseased, unmated, or unable to reproduce for some other reason.

In each generation, environmental factors screen heritable variations, favoring some over others. Differential reproduction results in the favored traits being disproportionately represented in the next generation. But can selection actually cause substantial change in a population? Darwin found evidence that it could in **artificial selection**, the breeding of domesticated plants and animals. Humans have modified other species over many generations by selecting individuals with the desired traits as breeding stock. The plants and animals we grow for food often bear little resemblance to their wild ancestors (FIGURE 22.6). The power of selective breeding is especially apparent in our pets, which have been bred more for fancy than for utility.

If so much change can be achieved by artificial selection in a relatively short period of time, Darwin reasoned, then natural selection should be capable of considerable modification of species over hundreds or thousands of generations. Even if the advantages of some heritable traits over others are slight, the advantageous variations will accumulate in the population after many generations of natural selection eliminating less favorable variations.

Darwin incorporated gradualism, a concept so important in Lyell's geology, into evolutionary theory. He envisioned life as evolving by a gradual accumulation of minute changes, and he postulated that natural selection operating in varying contexts over vast spans of time could account for the entire diversity of life.

We can now summarize the two main features of the Darwinian view of life: The diverse forms of life have arisen by descent with modification from ancestral species, and the mechanism of modification has been natural selection working continuously over enormous tracts of time.

Some Subtleties of Natural Selection. At this point, we need to emphasize several subtleties of natural selection. One is the importance of populations in evolution. For now, we will define a population as a group of interbreeding individuals belonging to a particular species and sharing a common geographic area. A population is the smallest unit that can evolve. Natural selection involves interactions between individual organisms and their environment, but individuals do not evolve. Evolution can be measured only as changes in relative proportions of variations in a population over a succession of generations.

Another key point about natural selection is that it can amplify or diminish only heritable variations. As we have seen, an organism may become modified through its own

FIGURE 22.6 • Artificial selection. These vegetables all have a common ancestor in one species of wild mustard. By selecting different parts of the plant to accentuate, breeders have obtained these divergent results.

experiences during its lifetime, and such acquired characteristics may even adapt the organism to its environment, but there is no evidence that characteristics acquired during a lifetime can be inherited. We must distinguish between adaptations an organism acquires by its own actions, and inherited adaptations that evolve in a population over many generations as a result of natural selection.

It must also be emphasized that the specifics of natural selection are situational; environmental factors vary from place to place and from time to time. An adaptation in one situation may be useless or even detrimental in different circumstances. Some examples will reinforce this situational quality of natural selection.

Examples of Natural Selection in Action. In one investigation of natural selection in action, scientists are testing Darwin's hypothesis that the beaks of Galápagos finches are evolutionary adaptations to different food sources. For over 20 years, Peter and Rosemary Grant of Princeton University have been studying the population of medium ground finches (*Geospiza fortis*) on Daphne Major, a tiny islet of the Galápagos. These birds use their strong beaks to crush seeds. They feed preferentially on small seeds, which are produced in abundance by certain plant species during wet years. In dry years all seeds are in short supply, and the finches must resort to eating both small seeds and larger seeds, which are more difficult to crush. The Grants discovered that the average beak depth (the dimension from top to bottom of the beak) in this finch population changes over the years (FIGURE 22.7). During droughts, average beak depth increases, only to decrease again during wet periods. This is a heritable trait. The Grants attribute the change to the relative availability of small seeds from year to year. Birds with stronger beaks may have an advantage during dry periods, when survival and reproduction depend on cracking large seeds. In contrast, a smaller beak is apparently more efficient equipment for feeding on the small seeds that are abundant during wet periods.

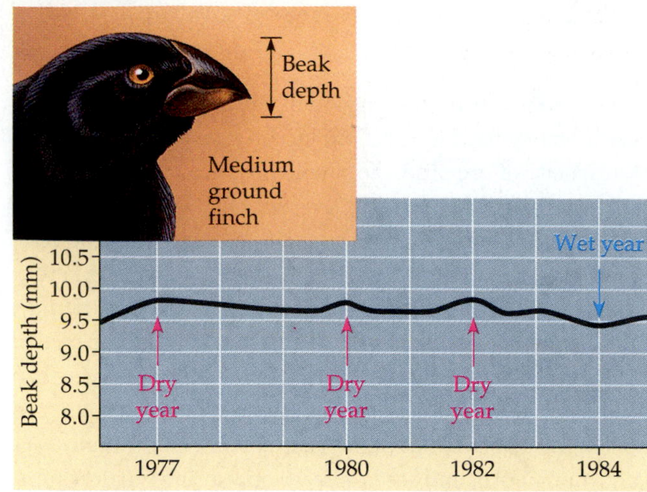

FIGURE 22.7 ▪ Natural selection in action: beak evolution in one of Darwin's finches. The medium ground finch (*Geospiza fortis*), one of the birds Darwin found on the Galápagos, uses its strong beak to crush seeds. Given a choice of small seeds or large seeds, the birds eat mostly small ones, which are easier to crush. During wet years, small seeds are produced in such abundance that ground finches consume relatively few large seeds. However, during dry years all seeds are in short supply, and the birds eat proportionately more large seeds. This change in diet is correlated with a change in the average depth (top-to-bottom dimension) of the birds' beaks. This trait is inherited rather than acquired (by exercising the beak on large seeds, for example). The most likely explanation is that those birds that happen to have stronger beaks have a feeding advantage and so have greater reproductive success during droughts. Thus, they pass the genes for thicker beaks on to their offspring.

The Grants' studies of beak evolution reinforce the point that natural selection is situational: What works best in one environmental context may be less suitable in some other situation. It is also important to understand that beak evolution on Daphne Major does not result from the inheritance of acquired characteristics. The environment did not *create* beaks specialized for larger or smaller seeds, depending on annual rainfall. The environment only acted on the inherited variations in the population, favoring the survival and reproductive success of some individuals over others. Natural selection edits populations. The proportion of thicker-beaked finches increased during dry periods because, on average, those individuals with thicker beaks transmitted their genes to more offspring than did thinner-beaked birds.

Researchers have published more than 100 other accounts of natural selection in the wild. Two examples we examined earlier are change in the body size of guppies exposed to different predators (see Chapter 1) and the evolution of antibiotic resistance in bacteria (see Chapter 18). Hundreds of other studies have documented natural selection in laboratory populations of such organisms as *Drosophila*.

Thus, scientists have shown natural selection to be a pervasive mechanism of change in populations; the process has been confirmed repeatedly by careful scientific studies, in which predictions based on hypotheses are tested by observation and experimentation (see Chapter 1). Ironically, Darwin himself thought that natural selection always operated too slowly to be observed. He was also unable to satisfactorily account for the genetics of variation. We now understand that variations arise by the chance mechanisms of mutation and genetic recombination (see Chapter 15). In Chapter 23 we will address the relationship between variation in populations and natural selection.

EVIDENCE OF EVOLUTION

Biological evolution leaves observable signs, evidence of its effects on life in the past and present. In this section we briefly survey some of these signs of evolution (which we will discuss more thoroughly in Chapter 24). Darwin documented evolution mainly with evidence from the geographical distribution of species and from the fossil record, but we will not limit ourselves to these two categories of evidence.

Evidence of evolution pervades biology

Scientific theories are subject to constant evaluation and updating. Indeed, scientists would have discarded the concept of evolution if it were inconsistent with observations. However, as biology progresses, new discoveries, including the revelations of molecular biology, continue to validate the Darwinian view of life.

Biogeography

The geographical distribution of species—**biogeography**—first suggested evolution to Darwin. Islands have many species of plants and animals that are indigenous (native, found nowhere else) yet closely related to species of the nearest mainland or neighboring island. Some revealing questions arise. Why are two islands with similar environments in different parts of the world populated not by closely related species, but by species taxonomically affiliated with the plants and animals of the nearest mainland, where the environment is often quite different? Why are the tropical animals of South America more closely related to species of South American deserts than to species of the African tropics? Why is Australia home to a great diversity of pouched mammals (marsupials) but relatively few placental mammals (eutherians), those in which embryonic development is completed in the uterus? Apparently, it is not because Australia is inhospitable to placental mammals; in recent years, humans have introduced rabbits to Australia, and the rabbit population exploded. The prevailing hypothesis is that the unique Australian fauna evolved on that island continent in isolation from places where the ancestors of placental mammals lived.

Although such biogeographical patterns are incongruous if one imagines that species were individually placed in suitable environments, they make sense in the historical context of evolution. In the evolutionary view, we find modern species where they are because they evolved from ancestors that inhabited those regions. Consider armadillos, the armored mammals that live only in the Americas. The evolutionary view of biogeography predicts that modern armadillos are modified descendants of earlier species that occupied these continents, and the fossil record confirms that such ancestors existed. This example brings us to the more general importance of the fossil record as a chronicle of evolution.

The Fossil Record

The succession of fossil forms is compatible with what is known from other types of evidence about the major branches of descent in the tree of life. For instance, evidence from biochemistry, molecular biology, and cell biology places prokaryotes as the ancestors of all life and predicts that bacteria should precede all eukaryotic life in the fossil record. Indeed, the oldest known fossils are prokaryotes. Another example is the chronological appearance of the different classes of vertebrate animals in the fossil record. Fossil fishes predate all other vertebrates, with amphibians next, followed by reptiles, then mammals and birds. This sequence is consistent with the history of vertebrate descent as revealed by many other types of evidence. In contrast, the idea that all species were individually created at about the same time predicts that all vertebrate classes would make their first appearance in the fossil record in rocks of the same age, a prediction at odds with what paleontologists actually observe.

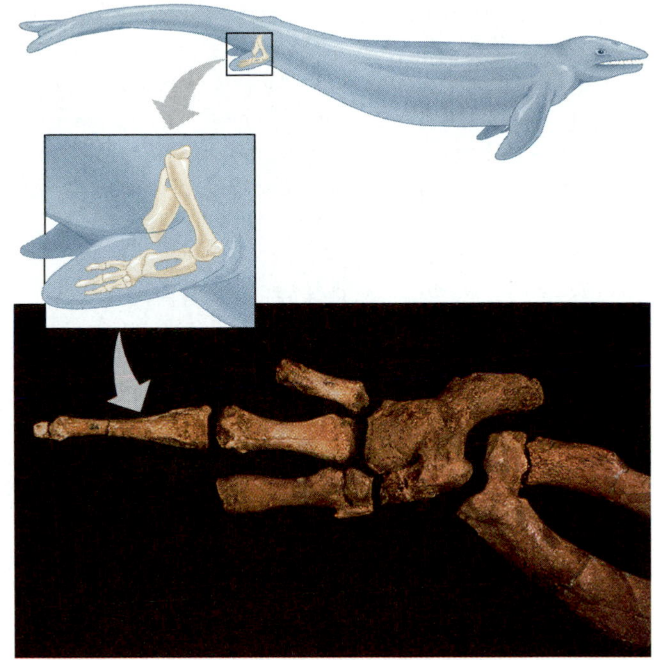

FIGURE 22.8 ▪ Transitional fossils linking past and present. Whales evolved from terrestrial ancestors, an evolutionary transition that left many signs, including fossil evidence. Paleontologists digging in Egypt and Pakistan have identified extinct whales that had hind limbs. Shown here are the fossilized leg bones of *Basilosaurus,* one of those ancient whales. These whales were already aquatic animals that no longer used their legs to support their weight and walk. The leg bones of an even older fossilized whale named *Ambulocetus* are heftier. *Ambulocetus* may have been amphibious, living on land and in water.

The Darwinian view of life also predicts that evolutionary transitions should leave signs in the fossil record. Paleontologists have discovered many transitional forms that link even older fossils to modern species. For example, a series of fossils documents the changes in skull shape and size that occurred as mammals evolved from reptiles. Every year, paleontologists turn up other important links between modern forms and their ancestors. In the past few years, for instance, researchers have found fossilized whales that link these aquatic mammals to their terrestrial predecessors (FIGURE 22.8).

Comparative Anatomy

Descent with modification is evident in anatomical similarities between species grouped in the same taxonomic category. For example, many of the same skeletal elements make up the

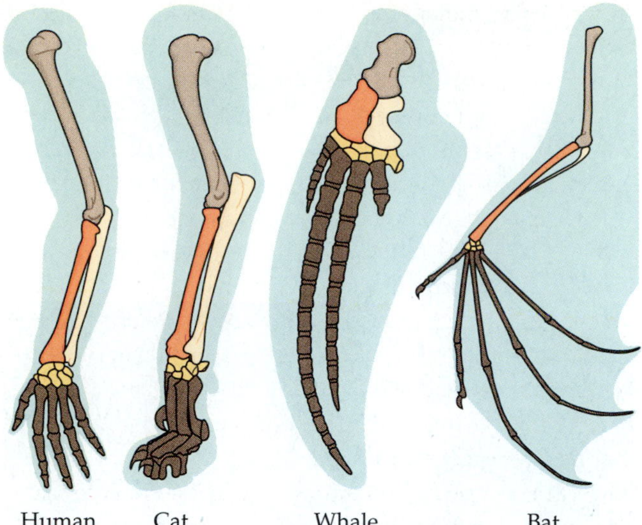

FIGURE 22.9 ▪ Homologous structures: anatomical signs of evolution. The forelimbs of all mammals are constructed from the same skeletal elements, an architectural relationship we would expect if a common ancestral forelimb became modified for many different functions.

forelimbs of humans, cats, whales, bats, and all other mammals, although these appendages have very different functions (FIGURE 22.9). Surely, the best way to construct the infrastructure of a bat's wing is not also the best way to build a whale's flipper. Such anatomical peculiarities make no sense if the structures are uniquely engineered and unrelated. A more likely explanation is that the basic similarity of these forelimbs is the consequence of the descent of all mammals from a common ancestor. The forelegs, wings, flippers, and arms of different mammals are variations on a common structural theme. In taking on different functions in each species, the basic structures were modified.

Similarity in characteristics resulting from common ancestry is known as **homology**, and such anatomical signs of evolution are called **homologous structures.** Comparative anatomy is consistent with all other evidence in testifying that evolution is a remodeling process in which ancestral structures that functioned in one capacity become modified as they take on new functions.

Some of the most interesting homologous structures are **vestigial organs**, structures of marginal, if any, importance to the organism. Vestigial organs are historical remnants of structures that had important functions in ancestors. For instance, the whales of today lack hindlimbs but have vestiges of the pelvis and leg bones of their four-footed terrestrial ancestors (see FIGURE 22.8). On a superficial level, vestigial organs may seem to support Lamarck's use and disuse concept, but as we have discussed, the effects of an individual's use of a body structure are not inherited by that individual's offspring. On the contrary, vestigial organs are evidence of evolution by natural selection. Because it would be wasteful to continue providing blood, nutrients, and space to an organ that no longer has a major function, natural selection would tend to favor individuals with reduced versions of those organs, and thereby tend to phase out obsolete structures. Ultimately, structural changes—such as the adaptation of the tail as the main propulsive structure and the reduction of hindlimbs in whales—involve changes in patterns of gene expression during embryonic development. Because embryonic processes affect the functioning of the adult organism, they are themselves subject to natural selection. Thus, vestigial organs represent changes in an organism's embryonic development wrought by natural selection.

Comparative Embryology

Closely related organisms go through similar stages in their embryonic development. For example, all vertebrate embryos go through a stage in which they have gill pouches on the sides of their throats (FIGURE 22.10). Indeed, at this stage of development, similarities between fishes, frogs, snakes, birds, humans, and all other vertebrates are much more apparent than differences. As development progresses, the various vertebrates diverge more and more, taking on the distinctive characteristics of their classes. In fish, for example, the gill pouches develop into gills; in terrestrial vertebrates, these embryonic structures become modified for other functions, such as the Eustachian tubes that connect the middle ear with

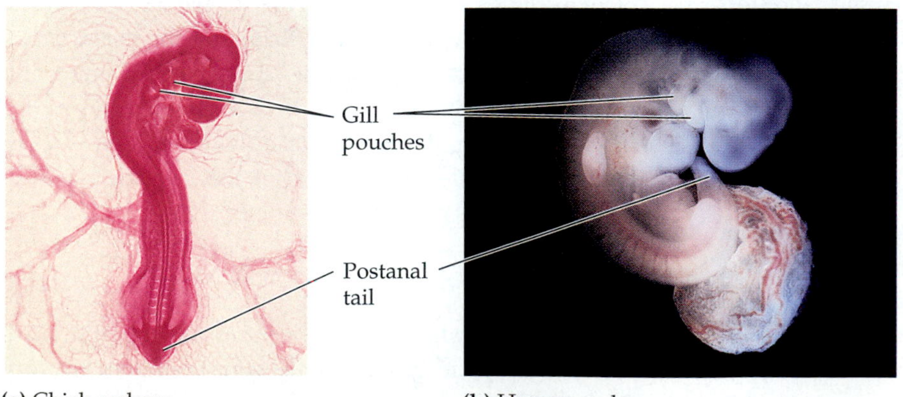

FIGURE 22.10 ▪ Evolutionary signs from comparative embryology. At this early stage of development, the kinship of vertebrates is unmistakable. Notice, for example, the gill pouches in both **(a)** the bird embryo and **(b)** the human embryo. Comparative embryology helps biologists identify anatomical homology that is less apparent in adults because the structures are extensively modified in different ways during the organisms' subsequent development.

the throat in humans. Comparative embryology often establishes homology among structures, such as the gill pouches, that become so altered in later development that their common origin is not apparent by comparing their fully developed forms.

Inspired by the Darwinian principle of descent with modification, many embryologists in the late nineteenth century proposed the extreme view that "ontogeny recapitulates phylogeny." This notion holds that the development of an individual organism, **ontogeny**, is a replay of the evolutionary history of the species, **phylogeny**. The theory of recapitulation is an overstatement. Although vertebrates share many features of embryonic development, it is not as though a mammal first goes through a "fish stage," then an "amphibian stage," and so on. Ontogeny can provide clues to phylogeny, but it is important to remember that all stages of development may become modified over the course of evolution.

Molecular Biology

As we discussed in Chapter 5, evolutionary relationships among species are reflected in their DNA and proteins—in their genes and gene products. If two species have libraries of genes and proteins with sequences of monomers that match closely, the sequences must have been copied from a common ancestor. If two long paragraphs are identical except for the substitution of a letter here and there, we would surely attribute them both to a single source.

Molecular biology supports Darwin's boldest speculation—that *all* forms of life are related to some extent through branching descent from the earliest organisms. Even taxonomically remote organisms, such as humans and bacteria, have some proteins in common. An example is cytochrome *c*, a protein involved in cellular respiration in all aerobic species (see Chapter 9). Mutations have substituted amino acids at some places in the protein during the long course of evolution, but the cytochrome *c* molecules of all species are clearly akin in structure and function. Similarly, a comparison of the number of amino acid differences in hemoglobin in several vertebrates corroborates evidence from paleontology and comparative anatomy about the evolutionary relationships among these species (FIGURE 22.11).

A common genetic code is overwhelming evidence that all life is related. Evidently, the language of the genetic code has been passed along through all branches of the tree of life ever since the code's inception in an early life form. Molecular biology has thus added the latest chapter to the evidence that evolution is the basis for the unity and diversity of life.

What is theoretical about the Darwinian view of life?

Some people dismiss Darwinism as "just a theory." This tactic for nullifying the evolutionary view of life has two flaws. First, it fails to separate Darwin's two claims: that modern species evolved from ancestral forms, and that natural selection is the

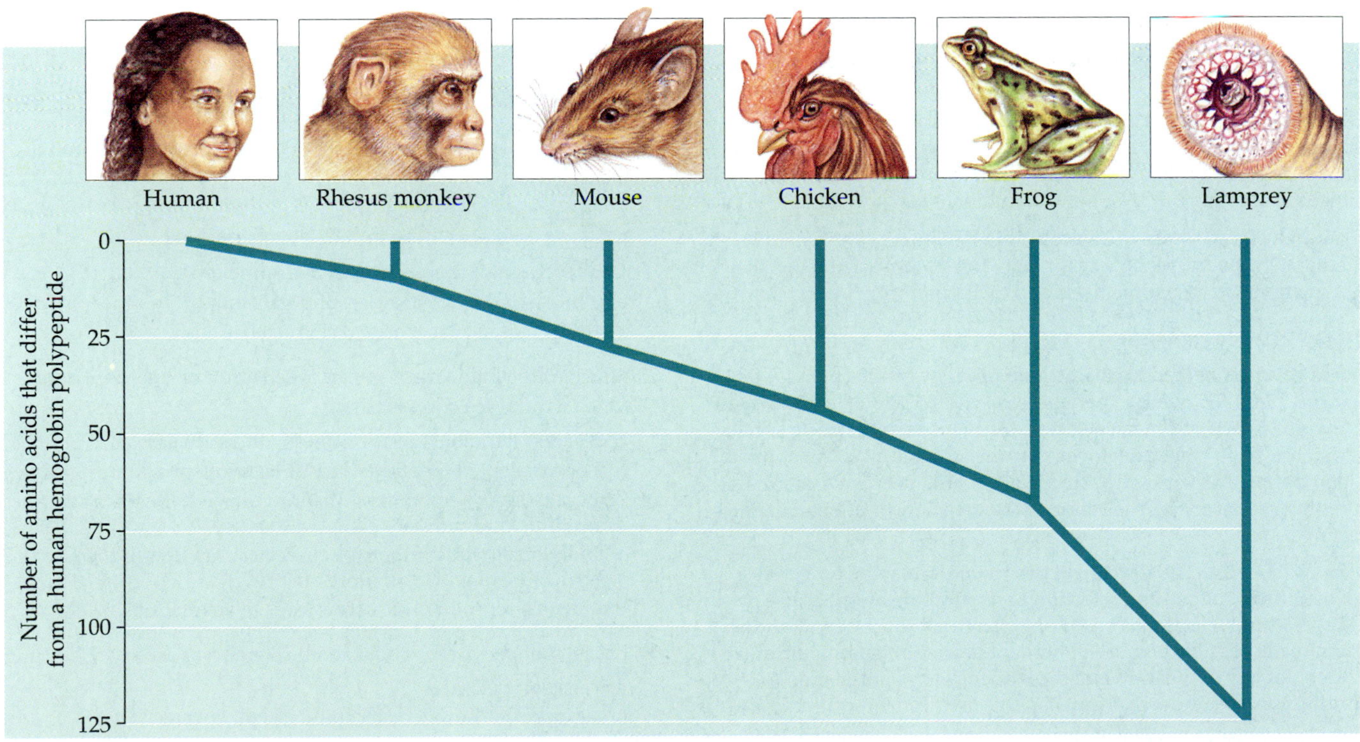

FIGURE 22.11 ▪ Molecular data and the evolutionary relationships of vertebrates. Branch points in this vertebrate tree are defined by the number of amino acids that differ from a polypeptide of human hemoglobin, a protein that functions in oxygen transport. The molecular data are compatible with vertebrate relationships deduced from the fossil record and comparative anatomy.

main mechanism for this evolution. The conclusion that life has evolved is based on historical evidence—the signs of evolution discussed in the previous section.

What, then, is theoretical about evolution? Theories are our attempts to explain facts and integrate them with overarching concepts. To biologists, "Darwin's theory of evolution" is natural selection—the mechanism Darwin proposed to explain the historical facts of evolution documented by fossils, biogeography, and other types of evidence.

So the "just a theory" argument concerns Darwin's second claim, his theory of natural selection. This brings us to the second flaw in the "just a theory" case. The term *theory* has a very different meaning in science compared to our general use of the word. The colloquial use of "theory" comes close to what scientists mean by a "hypothesis." In science, a theory is more comprehensive than a hypothesis. A theory, such as Newton's theory of gravitation or Darwin's theory of natural selection, accounts for many facts and attempts to explain a great variety of phenomena. Such a unifying theory does not become widely accepted in science unless its predictions stand up to thorough and continual testing by experiments and observations. Even then, good scientists do not allow theories to become dogma. For example, many evolutionary biologists now question whether natural selection alone accounts for the evolutionary history observed in the fossil record.

The study of evolution is more robust and livelier than ever, and we will evaluate some of the current debates in the next three chapters. But these questions about *how* life evolves in no way imply that most biologists consider evolution itself to be "just a theory." Debates about evolutionary theory are like arguments over competing theories about gravity; we know that objects keep right on falling while we debate the cause.

By attributing the diversity of life to natural causes rather than to supernatural creation, Darwin gave biology a sound, scientific basis. Nevertheless, the diverse products of evolution are elegant and inspiring. As Darwin said in the closing paragraph of *The Origin of Species*, "There is grandeur in this view of life."

CHAPTER REVIEW

REVIEW OF KEY CONCEPTS

(with page numbers and key figures)

HISTORICAL CONTEXT FOR EVOLUTIONARY THEORY

- **Western culture resisted evolutionary views of life (pp. 414–416, FIGURE 22.1)** Darwin's theory that natural selection is responsible for evolutionary change was a radical departure from the dominant religious and philosophical climate of the nineteenth-century western worldview.
- **Theories of geological gradualism helped clear the path for evolutionary biologists (p. 416)** Geologists Hutton and Lyell perceived that changes in Earth's surface can result from slow, continuous actions.
- **Lamarck placed fossils in an evolutionary context (pp. 416–417)** Lamarck helped pave the way for Darwin by emphasizing the interactions between organisms and their environment.

THE DARWINIAN REVOLUTION

- **Field research helped Darwin frame his view of life (pp. 417–419,**
22.1 **FIGURE 22.3)** Darwin's experiences on the voyage of HMS *Beagle* provided much of the background for his theory that new species originate from ancestral forms by the gradual accumulation of adaptations. Darwin and Alfred Wallace independently arrived at the theory that natural selection is the mechanism of evolutionary change.
- ***The Origin of Species* developed two main points: the occurrence of evolution and natural selection as its mechanism (pp. 419–422)** Natural selection is based on differential success in reproduction, made possible because of heritable variation among the individuals of any population and the tendency for a population to produce many more offspring than the environment can support.

EVIDENCE OF EVOLUTION

- **Evidence of evolution pervades biology (pp. 422–425, FIGURES 22.8–22.11)** Evolution is validated by evidence from the fossil record and from many fields of biology, including biogeography, comparative anatomy, comparative embryology, and molecular biology.
- **What is theoretical about the Darwinian view of life? (pp. 425–426)** Darwin's theory of natural selection unified biology and placed it in the realm of science.

SELF-QUIZ

1. The ideas of Hutton and Lyell that Darwin incorporated into his theory concerned
   a. the age of Earth and gradual change
   b. extinctions evident in the fossil record
   c. adaptation of species to the environment
   d. a hierarchical classification of organisms
   e. the inheritance of acquired characteristics

2. Which of the following is *not* an observation or inference upon which natural selection is based?
   a. There is heritable variation among individuals.
   b. Poorly adapted individuals never leave offspring.
   c. Because only a fraction of offspring survive, there is a struggle for limited resources.
   d. Individuals whose inherited characteristics best fit them to the environment will leave more offspring.
   e. Unequal reproductive success leads to adaptations.

3. The gill pouches of mammal and bird embryos are
   a. vestigial structures
   b. support for "ontogeny recapitulates phylogeny"
   c. homologous structures
   d. used by the embryos to breathe
   e. evidence for the degeneration of unused body parts

4. Which of the following observations provided the most support for the concept of descent with modification?
   a. Species diversity declined as distance from the equator increased.
   b. Fewer species were found living on islands than on the nearest continents.
   c. Birds could be found on islands more distant from the mainland than their maximum nonstop flight distance.
   d. South American temperate plants were more similar to the tropical plants of South America than to the temperate plants of Europe.
   e. Finches on the Galápagos fed on seeds whereas finches on the South American mainland were insectivorous.
5. Darwin's theory, as presented in *The Origin of Species*, mainly concerned
   a. how new species arise
   b. the origin of life
   c. how adaptations evolve
   d. how extinctions occur
   e. the genetics of evolution
6. In science, the term *theory* generally applies to an idea that
   a. is a speculation lacking supportive observations or experiments
   b. attempts to explain many related phenomena
   c. is synonymous with what biologists mean by a hypothesis
   d. is so widely accepted that it is considered a fact
   e. cannot be tested
7. Who conceived a theory of evolution identical to Darwin's?
   a. Cuvier
   b. Wallace
   c. Hutton
   d. Linnaeus
   e. Malthus
8. The smallest biological unit that can evolve over time is
   a. a cell
   b. an individual organism
   c. a population
   d. a species
   e. an ecosystem
9. Which of the following ideas is common to both Darwin's and Lamarck's theories of evolution?
   a. Adaptation results from differential reproductive success.
   b. Evolution drives organisms to greater and greater complexity.
   c. Evolutionary adaptation results from interactions between organisms and their environments.
   d. Adaptation results from the use and disuse of anatomical structures.
   e. The fossil record supports the view that species are fixed.
10. Which of the following pairs of structures is least likely to represent homology?
   a. the wings of a bat and the forelimbs of a human
   b. the cytochrome *c* protein of a bacterium and cytochrome *c* of a cat
   c. the mitochondria of a plant and those of an animal
   d. the bark of a tree and the protective covering of a lobster
   e. the brain of a frog and the brain of a dog

CHALLENGE QUESTION

An orange grower discovered that most of his trees were infested with destructive mites. He sprayed the trees with insecticide, which killed 99% of the mites. Five weeks later, most of the trees were infested again, so he sprayed again, using the same quantity of the same insecticide. This time, only about half the mites were killed. Explain why the spray did not work as well the second time.

SCIENCE, TECHNOLOGY, AND SOCIETY

Is the concept of natural selection relevant in a political or economic context? In other words, if a particular nation or corporation achieves success or dominance, does this mean that it is fitter than its competitors and that unregulated dominance is justified? Why or why not?

FURTHER READING

Browne, J. *Charles Darwin Voyaging*. Princeton, NJ: Princeton University Press, 1995. A comprehensive and engaging biography.

Darwin, C. *On the Origin of Species by Means of Natural Selection, or The Preservation of Favored Races in the Struggle for Life*. New York: New American Library, 1963. A modern printing of the historical book that revolutionized biology.

Futuyma, D. J. *Evolutionary Biology*, 3rd ed. Sunderland, MA: Sinauer Associates. 1998. An authoritative upper-level text.

Gould, S. J. "Creating the Creators." *Discover*, October 1996. Musings on natural selection as a refitting process.

Mayr, E. *This is Biology: The Science of the Living World*. Cambridge, MA: Harvard University Press, 1997. The nature of life science discussed by one of the twentieth century's foremost evolutionary biologists.

Milner, R. "Charles Darwin and Associates, Ghostbusters." *Scientific American*, October 1996. Interesting glimpses of Darwin and Wallace and their era.

Moore, R. "The Persuasive Mr. Darwin." *Bioscience*, February 1997. A critical analysis of the way Darwin presented his case in *The Origin of Species*.

Weiner, J. "Evolution Made Visible." *Science*, January 6, 1995. Observing evolution in action.

Wise, D. U. "Creationism's Geological Time Scale." *American Scientist*, March-April 1998. A critical look at the creationists' model of Earth history and a call for some evidence that could withstand unbiased scientific scrutiny.

Zimmer, C. "How the Snake Lost Its Legs." *Discover*, July 1997. A new hypothesis of snake evolution.

WEB LINKS

Visit the special edition of *The Biology Place* for BIOLOGY, Fifth Edition, at **http://www.biology.com/campbell**. Go to Chapter 22 for online resources, including learning activities, practice exams, and links to the following web sites:

"The C. Warren Irvin, Jr., Collection of Charles Darwin and Darwiniana"
This unique and interesting collection from the University of South Carolina was formed to reflect Darwin's writings and interests carefully and to place them in context with his peers and predecessors.

"Charles Darwin's Texts Online"
You can read online the complete texts of Darwin's *The Descent of Man, Voyage of the Beagle*, and *The Origin of Species*.

"Virtual Galápagos"
Take a virtual tour of the Galápagos islands. This richly illustrated site contains many high quality pictures of the islands and island wildlife.

"Galápagos Conservation Trust"
The Galápagos Conservation trust is affiliated with the Charles Darwin Foundation that operates the Charles Darwin Research Station on the Galápagos island of Santa Cruz.

"Voyage of the *Beagle*"
This investigative project about the voyage of HMS *Beagle* is available at Campbell's Biology Place.

CHAPTER 23

THE EVOLUTION OF POPULATIONS

Population Genetics

- The modern evolutionary synthesis integrated Darwinian selection and Mendelian inheritance
- The genetic structure of a population is defined by its allele and genotype frequencies
- The Hardy-Weinberg theorem describes a nonevolving population

Causes of Microevolution

- Microevolution is a generation-to-generation change in a population's allele or genotype frequencies
- The five causes of microevolution are genetic drift, gene flow, mutation, nonrandom mating, and natural selection

Genetic Variation, the Substrate for Natural Selection

- Genetic variation occurs within and between populations
- Mutation and sexual recombination generate genetic variation
- Diploidy and balanced polymorphism preserve variation

Natural Selection as the Mechanism of Adaptive Evolution

- Evolutionary fitness is the relative contribution an individual makes to the gene pool of the next generation
- The effect of selection on a varying characteristic can be stabilizing, directional, or diversifying
- Sexual selection may lead to pronounced secondary differences between the sexes
- Natural selection cannot fashion perfect organisms

One obstacle to understanding evolution is the common misconception that individual organisms evolve, in the Darwinian sense, during their lifetimes. In fact, natural selection does *act on individuals; their characteristics affect their chances of survival and their reproductive success. But the evolutionary impact of this natural selection is only apparent in tracking how a* population *of organisms changes over time. Consider, for example, the African swallowtail butterflies* (Papilio dardanus) *in the photograph on this page. These female specimens were all collected from one population; their different patterns of coloration represent genetic variations within that population. If birds prey preferentially on* Papilio *butterflies having a particular coloration, then the proportion of individuals with that coloration probably will decline from one generation to the next because such butterflies will produce fewer offspring. Thus it is the population, not its individuals, that evolves; some characteristics become more common within the overall population while other characteristics decline.* FIGURE 23.1 *illustrates another example of this principle. Evolution on the smallest scale, or microevolution, can be defined as a change in the genetic makeup of a population. We will refine this definition as we examine in more detail how natural selection and other mechanisms cause populations to evolve. We begin by tracing how biologists finally began to understand Darwin's theory of natural selection during the first half of the twentieth century.*

POPULATION GENETICS

The Origin of Species convinced most biologists that species are products of evolution, but Darwin was not nearly so successful in gaining acceptance for his idea that natural selection is the mechanism of evolution. Natural selection requires hereditary processes that Darwin could not explain. His theory was based on what seems like a paradox of inheritance: Like begets like—but not exactly. What was missing in Darwin's explanations was an understanding of inheritance that could explain how chance variations arise in a population while also accounting for the precise transmission of these variations from parents to offspring. Although Gregor Mendel and Charles Darwin were contemporaries, Mendel's discoveries were unappreciated at the time, and apparently no one noticed that he had elucidated the very principles of inheritance that could have resolved Darwin's paradox and given credibility to natural selection.

The modern evolutionary synthesis integrated Darwinian selection and Mendelian inheritance

Ironically, when Mendel's research article was rediscovered and reassessed at the beginning of the twentieth century, many geneticists believed that the laws of inheritance were at

FIGURE 23.1 ▪ Individuals are selected, but populations evolve. The bent grass (*Agrostis tenuis*) in the foreground is growing on the tailings of an abandoned mine in Wales. These plants tolerate concentrations of heavy metals that are toxic to other plants of the same species growing just meters away, in the pasture on the other side of the fence. Each year, many seeds land on the mine tailings, but most are unable to grow successfully there. The only plants that germinate, grow, and reproduce are those that inherited genes enabling them to tolerate metallic soil. Thus, this adaptation does not evolve by individual plants becoming more metal-tolerant during their lifetimes. We can only see the evolution of this population by observing the proportions of metal-tolerant plants in successive generations. Natural selection works by favoring the survival and reproductive success of certain individuals over others among the varying members of a population. But the impact of this individual selection is a generation-to-generation change in the prevalence of certain traits within the whole population.

odds with Darwin's theory of natural selection. As the raw material for natural selection, Darwin emphasized quantitative characters, those characteristics in a population that vary along a continuum, such as the fur length of mammals or the speed with which an animal can flee from a predator. We know today that quantitative characters are influenced by multiple genetic loci. (To review polygenic inheritance and quantitative characters, see Chapter 14.) But Mendel, and later the geneticists of the early twentieth century, recognized only discrete "either-or" traits, such as purple or white flowers in pea plants, as heritable. Thus, there seemed to be no genetic basis for natural selection to work on the more subtle variations within a population that were central to Darwin's theory.

An important turning point for evolutionary theory was the birth of **population genetics**, which emphasizes the extensive genetic variation within populations and recognizes the importance of quantitative characters. With progress in population genetics in the 1930s, Mendelism and Darwinism were reconciled, and the genetic basis of variation and natural selection was worked out.

A comprehensive theory of evolution that became known as the **modern synthesis** was forged in the early 1940s. It is called a synthesis because it integrated discoveries and ideas from many different fields, including paleontology, taxonomy, biogeography, and, of course, population genetics. Among the architects of the modern synthesis were geneticist Theodosius Dobzhansky, biogeographer and taxonomist Ernst Mayr, paleontologist George Gaylord Simpson, and botanist G. Ledyard Stebbins. The modern synthesis emphasizes the importance of populations as the units of evolution, the central role of natural selection as the most important mechanism of evolution, and the idea of gradualism to explain how large changes can evolve as an accumulation of small changes occurring over long periods of time. No scientific paradigm is likely to endure without modification for half a century. Many evolutionary biologists are now challenging some of the assumptions of the modern synthesis. Still, twentieth-century biology has been profoundly affected by the modern synthesis, which has shaped most of our ideas about how populations evolve.

The genetic structure of a population is defined by its allele and genotype frequencies

A **population** is a localized group of individuals belonging to the same species. For now, we will define a **species** as a group of populations whose individuals have the potential to interbreed and produce fertile offspring in nature (this definition will be examined more critically in Chapter 24). Each species has a geographical range within which individuals are unevenly distributed, but are usually concentrated in several localized populations. A population may be isolated from others of the same species, exchanging genetic material only rarely. Such isolation is particularly common for populations confined to widely separated islands, unconnected lakes, or mountain ranges separated by lowlands. However, populations are not always isolated, nor do they necessarily have sharp boundaries. One dense population center may blur into another in an intermediate region where members of the species occur but are less numerous. Although these populations are not isolated, individuals are still concentrated in centers and are more likely to breed with members of the same population than with members of other populations. Therefore, individuals near a population center are, on average, more closely related to one another than to members of other populations (FIGURE 23.2, p. 430).

The total aggregate of genes in a population at any one time is called the population's **gene pool**. It consists of all alleles at all gene loci in all individuals of the population. For a diploid species, each locus is represented twice in the genome of an individual, who may be either homozygous or heterozygous for those homologous loci. (Recall that homozygous individuals have two identical alleles for a given character, whereas heterozygous individuals have two different alleles for that character.) If all members of a population are homozygous for the same allele, that allele is said to be *fixed* in the gene pool. Usually, however, there are two or more alleles for a gene, each having a relative frequency (proportion) in the gene pool.

FIGURE 23.2 • Population distribution. Populations are localized groups of individuals belonging to the same species. Here, two dense populations of Douglas fir (*Pseudotsuga menziesii*) are separated by a river bottom, where firs are uncommon. The two populations are not totally isolated; interbreeding occurs when wind blows pollen between the populations. Nevertheless, trees are more likely to breed with members of the same population than with trees on the other side of the river.

An example will make the concept of allele frequency in a gene pool less abstract. Imagine a wildflower population with two varieties contrasting in flower color. An allele for pink flowers, which we will symbolize by *A*, is completely dominant to an allele for white flowers, symbolized by *a*. For our simplified situation, these are the only two alleles for this locus in the population. Suppose an imaginary population has 500 plants, and 20 of these plants have white flowers because they are homozygous for the recessive allele; their genotype is *aa*. The other 480 plants have pink flowers; 320 are homozygous (*AA*) and 160 are heterozygous (*Aa*). Because these are diploid organisms, there are a total of 1000 copies of genes for flower color in the population of 500 individuals. The dominant allele (*A*) accounts for 800 of these genes ($320 \times 2 = 640$ for *AA* plants, plus $160 \times 1 = 160$ for *Aa* individuals). Thus, the frequency of the *A* allele in the gene pool of this population is $800/1000 = 0.8 = 80\%$. And because there are only two allelic forms of the gene, the *a* allele must have a frequency of 0.2, or 20%.

Related to these allele frequencies are the frequencies of genotypes. In our imaginary wildflower population, these frequencies are: $AA = 0.64$ (320 out of 500 plants), $Aa = 0.32$ (160/500), and $aa = 0.04$ (20/500). Population geneticists use the term **genetic structure** to refer to a population's frequencies of alleles and genotypes.

The Hardy-Weinberg theorem describes a nonevolving population

Before we consider the mechanisms that cause a population to evolve, it will be helpful to examine, for comparison, the genetic structure of a nonevolving population. Such a gene pool is described by the **Hardy-Weinberg theorem**, named for the two scientists who derived the principle independently in 1908. It states that the frequencies of alleles and genotypes in a population's gene pool remain constant over the generations unless acted upon by agents other than sexual recombination. Put another way, the sexual shuffling of alleles due to meiosis and random fertilization has no effect on the overall genetic structure of a population.

To apply the Hardy-Weinberg theorem, let's return to our imaginary wildflower population of 500 plants (FIGURE 23.3a). Recall that 80% (0.8) of the flower-color loci in the gene pool have the *A* allele and 20% (0.2) have the *a* allele. How will genetic recombination during sexual reproduction affect the frequencies of the two alleles in the next generation of our wildflower population? We will assume that the union of sperm and ova in the population is completely random; that is, all male-female mating combinations are equally likely. The situation is analogous to mixing all gametes in a sack and then drawing them randomly, two at a time, to determine the genotype for each zygote (fertilized egg). Each gamete has one allele for flower color, and the allele frequencies of the gametes will be the same as the allele frequencies in the parent population. Every time a gamete is drawn from the pool at random, the chance that the gamete will bear an *A* allele is 0.8, and the chance that the gamete will have an *a* allele is 0.2.

Using the rule of multiplication (see Chapter 14), we can calculate the frequencies of the three possible genotypes in the next generation of the population (FIGURE 23.3b). The probability of picking two *A* alleles from the pool of gametes is 0.64 (0.8×0.8). Thus, about 64% of the plants in the next generation will have the genotype *AA*. The frequency of *aa* individuals will be about 4%, or 0.04 (0.2×0.2). And 32%, or 0.32, of the plants will be heterozygous—that is, *Aa* or *aA*, depending on whether it is the sperm or ovum that supplies the dominant allele (frequency of $Aa = 0.8 \times 0.2 = 0.16$; frequency of $aA = 0.2 \times 0.8 = 0.16$; frequency of $Aa + aA = 0.32$).

Notice in FIGURE 23.3 that the sexual processes of meiosis and random fertilization have maintained the same allele and genotype frequencies that existed in the previous generation of the wildflower population. For the flower-color locus, the population's genetic structure is in a state of equilibrium—referred to as **Hardy-Weinberg equilibrium**. (In this example, the wildflower population was at equilibrium initially. If we had started with the population not yet at equilibrium, only a single generation would be required for equilibrium to be attained. You will have a chance to prove this in the Challenge Question at the end of the chapter.)

We can use our imaginary wildflower population to describe the Hardy-Weinberg theorem in more general terms. We will restrict our analysis to the simplest case of only two alleles, one dominant over the other. However, the Hardy-Weinberg theorem also applies to situations in which there are

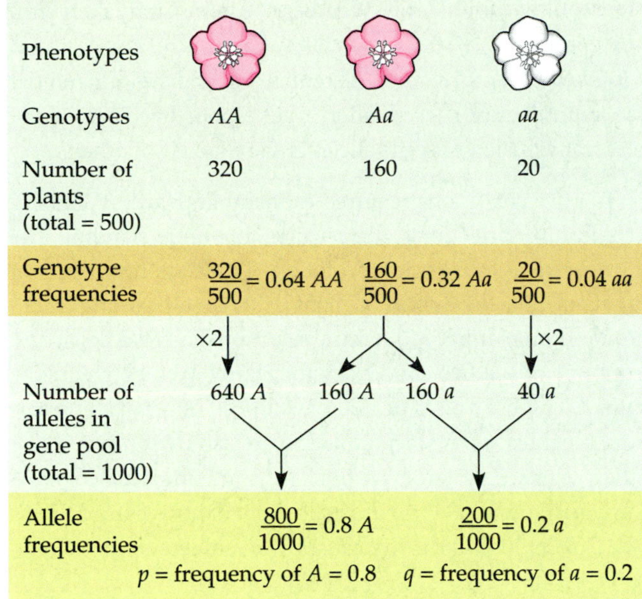

(a) Genetic structure of parent population

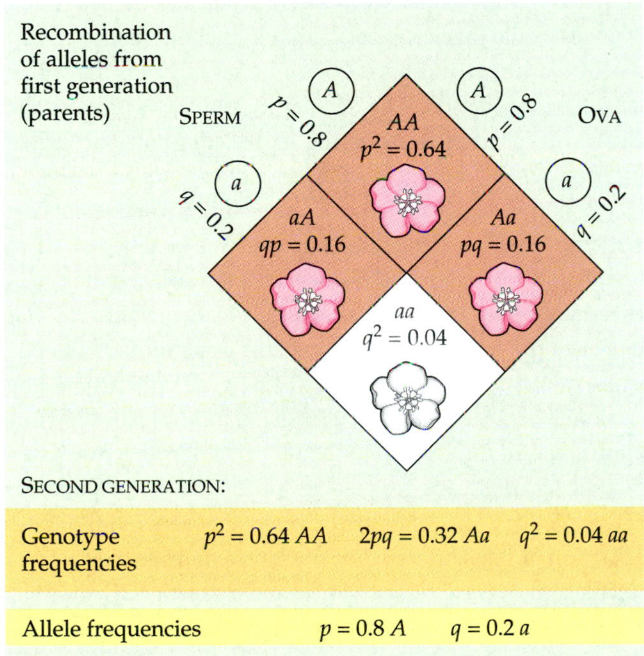

(b) Genetic structure of second generation

FIGURE 23.3 • The Hardy-Weinberg theorem. The genetic structure of a nonevolving population remains constant over the generations; sexual recombination alone will not alter the relative frequencies of alleles or genotypes. In this model (a hypothetical flower population), note that the frequencies of alleles and genotypes remain the same between **(a)** parents and **(b)** their offspring.

three or more alleles for a particular locus and no clear-cut dominance.

For a gene locus where only two alleles occur in a population, population geneticists use the letter p to represent the frequency of one allele and the letter q to represent the frequency of the other allele. In our imaginary wildflower population, $p = 0.8$ and $q = 0.2$ (see FIGURE 23.3). Note that $p + q = 1$; the combined frequencies of all possible alleles must account for 100% of the genes for that locus in the population. If there are only two alleles and we know the frequency of one, the frequency of the other can be calculated:

$$\text{If} \quad p + q = 1 \quad \text{then} \quad p = 1 - q \quad \text{and} \quad q = 1 - p$$

When gametes combine their alleles to form zygotes, the probability of generating an *AA* genotype is p^2 (an application of the rule of multiplication). In our wildflower population, $p = 0.8$, and $p^2 = 0.64$, the probability of an *A* sperm fertilizing an *A* ovum to produce an *AA* zygote. The frequency of individuals homozygous for the other allele *(aa)* is q^2, or $0.2 \times 0.2 = 0.04$ for the wildflower population. There are two ways in which an *Aa* genotype can arise, depending on which parent contributes the dominant allele. Therefore, the frequency of heterozygous individuals in the population is $2pq$ ($2 \times 0.8 \times 0.2 = 0.32$, in our example). If we have included all possible genotypes, the genotype frequencies add up to 1:

$$\underset{\text{Frequency of } AA \text{ genotype}}{p^2} + \underset{\text{Frequency of } Aa \text{ plus } aA \text{ genotype}}{2pq} + \underset{\text{Frequency of } aa \text{ genotype}}{q^2} = 1$$

For our wildflowers, this is $0.64 + 0.32 + 0.04 = 1$.

Population geneticists sometimes refer to this general expression as the **Hardy-Weinberg equation.** The equation enables us to calculate frequencies of alleles in a gene pool if we know frequencies of genotypes, and vice versa. One application is to calculate the percentage of the human population that carries the allele for a particular inherited disease. For instance, one out of approximately 10,000 babies in the United States is born with phenylketonuria (PKU), a metabolic disorder that, left untreated, results in mental retardation and other problems. (Newborn babies are now routinely tested for PKU, and symptoms can be prevented by following a strict diet.) The disease is caused by a recessive allele, and thus the frequency of individuals in the U.S. population born with PKU corresponds to q^2 in the Hardy-Weinberg equation (q^2 = frequency of the homozygous recessive genotype). Given one PKU occurrence per 10,000 births, $q^2 = 0.0001$. Therefore, the frequency of the recessive allele for PKU in the population is

$$q = \sqrt{0.0001} = 0.01$$

And the frequency of the dominant allele is

$$p = 1 - q = 1 - 0.01 = 0.99$$

The frequency of carriers, heterozygous people who do not have PKU but may pass the PKU allele on to offspring, is

$$2pq = 2 \times 0.99 \times 0.01 = 0.0198 \text{ (approximately 2\%)}$$

Thus, about 2% of the U.S. population carries the PKU allele.

CAUSES OF MICROEVOLUTION

Microevolution is a generation-to-generation change in a population's allele or genotype frequencies

If the Hardy-Weinberg theorem describes a gene pool in equilibrium—that is, a *nonevolving* population—then how is it relevant to our study of evolution? The concept of Hardy-Weinberg equilibrium tells us what to expect if a population is *not* evolving. The equilibrium values for allele and genotype frequencies we calculate from the Hardy-Weinberg equation provide a baseline for tracking the genetic structure of a population over a succession of generations. If the frequencies of alleles or genotypes deviate from values expected from Hardy-Weinberg equilibrium, then the population is evolving. We can now refine our definition of evolution at the population level: *Evolution is a generation-to-generation change in a population's frequencies of alleles or genotypes—a change in a population's genetic structure.* Because such change in a gene pool is evolution on the smallest scale, it is referred to more specifically as **microevolution**.

Microevolution is occurring even if the frequencies of alleles are changing for only a single genetic locus. If we track allele and genotype frequencies in a population over a succession of generations, some loci may be at equilibrium while allele frequencies at other loci are changing. Such a population is evolving. For example, our imaginary wildflower population would be evolving if the frequencies of the pink-flower and white-flower alleles were changing from generation to generation, even if Hardy-Weinberg equilibrium were maintained for all other genetic loci.

The five causes of microevolution are genetic drift, gene flow, mutation, nonrandom mating, and natural selection

If we think of microevolution as a departure from Hardy-Weinberg equilibrium, then it is important to understand the five conditions that are required for Hardy-Weinberg equilibrium to be maintained in a population:

1. *Very large population size.* In a small population, genetic drift, which is chance fluctuation in the gene pool, can alter the frequencies of alleles.
2. *Isolation from other populations.* Gene flow, the transfer of alleles between populations due to the movement of individuals or gametes, can change gene pools.
3. *No net mutations.* By changing one allele into another, mutations alter the gene pool.
4. *Random mating.* If individuals select mates having certain heritable traits, then the random mixing of gametes required for Hardy-Weinberg equilibrium does not occur.
5. *No natural selection.* Differential survival and differential reproductive success alter a gene pool by favoring the transmission of some alleles at the expense of others.

The five conditions required to maintain Hardy-Weinberg
23.1 equilibrium provide a framework for understanding the processes that cause microevolution. There are five potential agents of microevolution: genetic drift, gene flow, mutation, nonrandom mating, and natural selection. Each is a departure from one of the five conditions for Hardy-Weinberg equilibrium. Of all the causes of microevolution, only natural selection generally adapts a population to its environment. The other agents of microevolution are sometimes called non-Darwinian because of their usually nonadaptive nature.* Let's take a closer look at the five causes of microevolution.

Genetic Drift

Flip a coin a thousand times, and a result of 700 heads and 300 tails would make you very suspicious about that coin. Flip a coin ten times, and an outcome of seven heads and three tails is within reason. The smaller a sample, the greater the chance of deviations from the expected result—an equal number of heads and tails, in the case of coin tosses. This disproportion of results in a small sample is known as sampling error, and it is an important factor in the genetics of small populations of organisms. If a new generation draws its alleles at random, then the larger the sample size, the better it will represent the gene pool of the previous generation. If a population of organisms is small, its existing gene pool may not be accurately represented in the next generation because of sampling error. In the small population of wildflowers in FIGURE 23.4, for example, the frequencies of the alleles for pink *(A)* and white *(a)* flowers fluctuate over several generations. Only a fraction of the plants in the population manage to leave offspring, and over successive generations, genetic variation has been reduced. This is a case of microevolution caused by **genetic drift**, changes in the gene pool of a small population due to chance. Only luck could result in random drift improving the population's adaptiveness to its environment.

Ideally, a population must be infinitely large for genetic drift to be ruled out completely as an agent of evolution. Although that is impossible, many populations are so large that drift may be negligible. However, some populations are small enough for significant genetic drift to occur. Two situa-

* "Nonadaptive" does not mean "maladaptive." Microevolution due to causes other than natural selection may affect populations in positive, negative, or neutral ways. In contrast, the effects of natural selection are almost always positive because selection favors the disproportionate propagation of favorable traits.

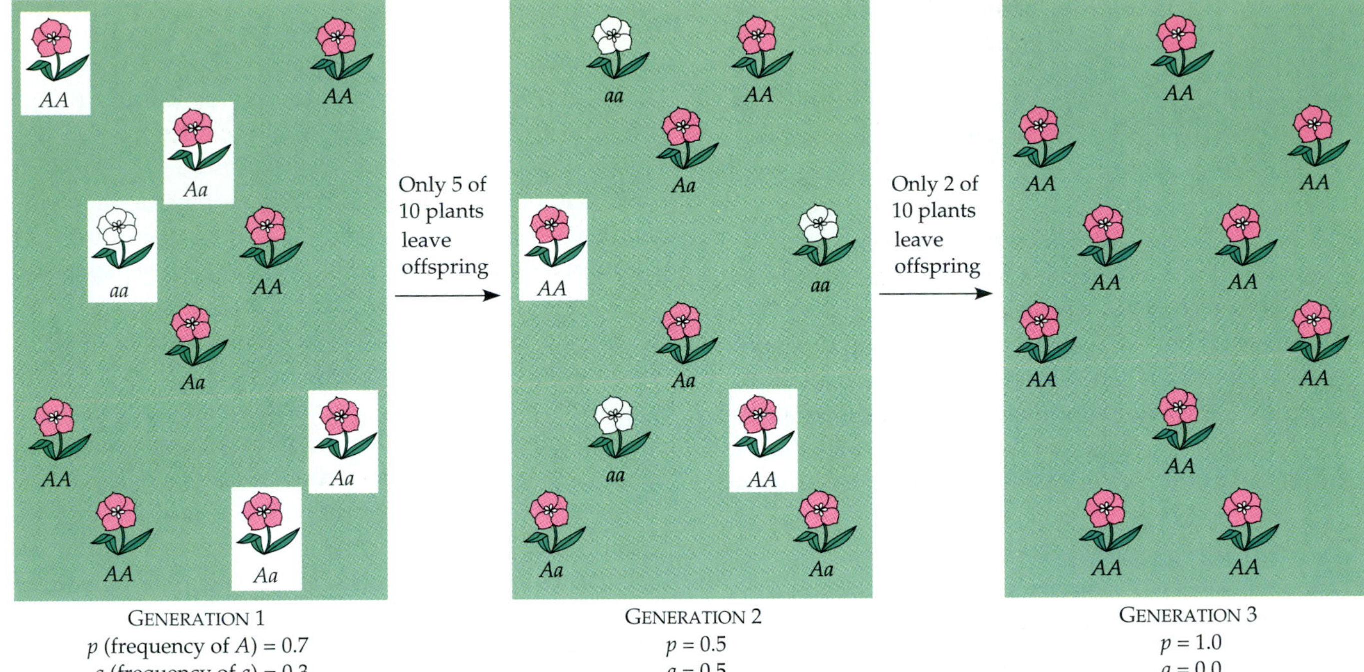

23.1 **FIGURE 23.4 • Genetic drift.** This small wildflower population has a stable size of about ten plants. For generation 1, only the five boxed plants produce seeds that germinate and give rise to offspring that survive to flowering. Only two of the plants of generation 2 manage to leave fertile offspring. Such random change in the frequencies of alleles and genotypes due to sampling error in a small population is called genetic drift. Eventually, genetic drift usually reduces genetic variability by fixing alleles, as is the case for the *A* allele in generation 3 of this imaginary population.

tions that can lead to genetic drift are population bottlenecks and the founding of new colonies by a small number of individuals.

The Bottleneck Effect. Disasters such as earthquakes, floods, and fires may reduce the size of a population drastically, killing victims rather unselectively. The result is that the genetic makeup of the small surviving population is unlikely to be representative of the makeup of the original population—a situation known as the **bottleneck effect**. By chance, certain alleles will be overrepresented among survivors, other alleles will be underrepresented, and some alleles may be eliminated completely (FIGURE 23.5). Genetic drift may continue to affect the population for many generations, until the population is again large enough for sampling errors to be insignificant.

Bottlenecking and the genetic drift that follows usually reduce the overall genetic variability in a population because alleles for at least some loci are likely to be lost from the gene pool. One example concerns the population of northern elephant seals, which passed through a bottleneck in the 1890s when hunters reduced it to about 20 individuals. Since then, the animal has become a protected species, and the population has grown to over 30,000. Researchers have examined 24 gene loci in many individuals of the northern elephant seal population, and no genetic variation has been found; a single allele has been fixed at each of the 24 loci, probably due in large part to genetic drift. By comparison, genetic variation abounds in populations of the southern elephant seal, which have not been bottlenecked. Bottlenecking may also explain why the South African population of cheetahs displays less genetic variation than do inbred strains of laboratory mice. The cheetah population was probably severely reduced during the last

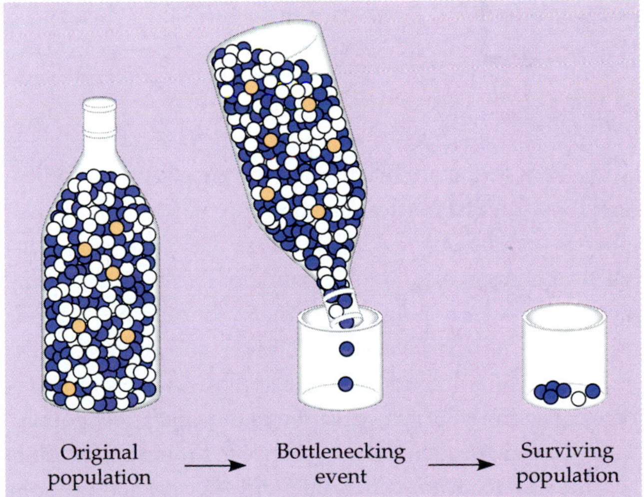

FIGURE 23.5 • The bottleneck effect. A gene pool can begin to drift by chance after the population is drastically reduced by a disaster that kills unselectively. In this analogy, a population of blue, orange, and white marbles in a bottle is reduced in the bottleneck. Notice that the composition of the surviving population emerging from the bottleneck is not representative of the makeup of the larger, original population. By chance, blue marbles are overrepresented in the new population, and orange marbles are absent.

ice age about 10,000 years ago, and a second time when the animals were hunted to near extinction in the early 1900s.

The Founder Effect. Genetic drift is also likely whenever a few individuals colonize an isolated island, lake, or some other new habitat. The smaller the sample size, the less the genetic makeup of the colonists will represent the gene pool of the larger population they left. The most extreme case would be the founding of a new population by one pregnant animal or a single plant seed. If the colony is successful, random drift will continue to affect the frequency of alleles in the gene pool until the population is large enough for sampling errors from generation to generation to be minimal. Genetic drift in a new colony is known as the **founder effect**. The effect undoubtedly contributed to the evolutionary divergence of Darwin's finches after strays from the South American mainland reached the remote Galápagos Islands.

The founder effect probably accounts for the relatively high frequency of certain inherited disorders among human populations established by a small number of colonists. In 1814, fifteen people founded a British colony on Tristan da Cunha, a group of small islands in the Atlantic Ocean midway between Africa and South America. Apparently, one of the colonists carried a recessive allele for retinitis pigmentosa, a progressive form of blindness that afflicts homozygous individuals. Of the 240 descendants who still lived on the island in the late 1960s, four had retinitis pigmentosa, and at least nine others were known to be carriers, based on pedigree analysis. The frequency of this allele is much higher on Tristan da Cunha than in the populations from which the founders came. In addition to inherited diseases, the founder effect also alters the frequencies of many alleles that affect more subtle characteristics.

Mutations

A **mutation** is a change in an organism's DNA. A new mutation that is transmitted in gametes can immediately change the gene pool of a population by substituting one allele for another. For example, a mutation that causes a white-flowered plant *(aa)* in our hypothetical wildflower population to produce gametes bearing the dominant allele for pink flowers *(A)* would decrease the frequency of the *a* allele in the population and increase the frequency of the *A* allele.

For any one gene locus, however, mutation alone does not have much quantitative effect on a large population in a single generation. This is because a mutation at any given gene locus is a very rare event; although mutation rates vary, depending on the species and the gene locus, rates of one mutation per locus per 10^5 to 10^6 gametes are typical. If an allele has a frequency of 0.50 in the gene pool and mutates to another allele at a rate of 0.00001 mutations per generation, it would take 2000 generations to reduce the frequency of the original allele from 0.50 to 0.49. If some new allele produced by mutation increases its frequency by a significant amount in a population, it is not because mutation is generating the allele in abundance, but because individuals carrying the mutant allele are producing a disproportionate number of offspring as a result of natural selection or genetic drift.

Although mutations at a particular gene locus are rare, the cumulative impact of mutations at *all* loci can be significant. This is because each individual has thousands of genes, and many populations have thousands or millions of individuals. Certainly over the long term, mutation is, in itself, very important to evolution because it is the original source of the genetic variation that serves as raw material for natural selection.

Gene Flow

Hardy-Weinberg equilibrium requires the gene pool to be a closed system, but most populations are not completely isolated. A population may gain or lose alleles by **gene flow**, genetic exchange due to the migration of fertile individuals or gametes between populations. Perhaps, for example, a population near our hypothetical wildflower population consists entirely of white-flowered individuals *(aa)*. A wind storm may blow pollen from the *aa* population to our wildflower population, and allele frequencies may change in the next generation.

Gene flow tends to reduce differences between populations that have accumulated because of natural selection or genetic drift. If it is extensive enough, gene flow can eventually amalgamate neighboring populations into a single population with a common genetic structure. As humans began to move about the world more freely, gene flow undoubtedly became an important agent of microevolutionary change in populations that were previously quite isolated.

Nonrandom Mating

For Hardy-Weinberg equilibrium to hold, an individual of any genotype must choose its mates at random from the population. In actuality, individuals usually mate more often with close neighbors than with more distant members of the population, especially in species that do not disperse far. Other individuals in the same "neighborhood" within a larger population tend to be closely related. This promotes **inbreeding**, mating between closely related partners. The most extreme case of inbreeding is self-fertilization ("selfing"), particularly common in plants.

Inbreeding causes the relative frequencies of genotypes to deviate from what is expected from Hardy-Weinberg equilibrium. For example, in our imaginary wildflower population, self-pollination would tend to increase the frequencies of homozygous genotypes at the expense of heterozygotes. If *AA* individuals and *aa* individuals "self," then their offspring must also be homozygous. If *Aa* plants "self," however, only half

FIGURE 23.6 ▪ Assortative mating in toads. Males and females of this species (*Bufo bufo*), found in North America, tend to pair off according to size during mating.

their offspring will be heterozygous. With each generation, the proportion of heterozygotes decreases and the proportions of dominant and recessive homozygotes increase. Even in less extreme cases of inbreeding without selfing, the proportion of heterozygotes decreases, though more slowly. One visible effect of this change in genotype frequencies is a greater proportion of individuals expressing recessive phenotypes; the frequency of white-flowered individuals in our wildflower population would be greater than the Hardy-Weinberg equation predicts. Regardless of the impact of inbreeding on the ratio of genotypes and phenotypes in the population, the frequencies of the two alleles remain the same. It is just that a smaller proportion of recessive alleles are "masked" in heterozygous individuals.

Another type of nonrandom mating is **assortative mating**, in which individuals select partners that are like themselves in certain phenotypic characters. For example, some toads (*Bufo*) most commonly mate with individuals of similar size (FIGURE 23.6). To some extent, humans also use size as a criterion for assortative mating; for example, tall women commonly (but not always) pair with tall men.

Notice again that nonrandom mating—inbreeding or assortative mating—increases the number of gene loci in the population that are homozygous, but nonrandom mating does not in itself alter the overall frequencies of alleles in a population's gene pool. Remember, however, that a population's genetic structure is defined by its frequencies of alleles *and genotypes.* Any change in a population's inbreeding or assortative mating behaviors will shift the frequencies of different genotypes. Thus, nonrandom mating can cause a population to evolve.

Natural Selection

Hardy-Weinberg equilibrium requires that all individuals in a population be equal in their ability to survive and produce viable, fertile offspring. This condition is probably never completely met. Populations consist of varied individuals, and on average, some variants leave more offspring than others. This differential success in reproduction is **natural selection**. Selection results in alleles being passed along to the next generation in numbers disproportionate to their relative frequencies in the present generation. For example, in our imaginary wildflower population, plants with pink flowers (*AA* or *Aa* genotypes) may for some reason produce more offspring on average than plants having white flowers *(aa);* perhaps white flowers are more visible to herbivorous insects that eat the flowers. This would disturb Hardy-Weinberg equilibrium; the frequency of the *A* allele would increase, and the frequency of the *a* allele would decline in the gene pool.

Of all the agents of microevolution that change a gene pool, only selection is likely to adapt a population to its environment. Natural selection accumulates and maintains favorable genotypes in a population. If the environment should change, selection responds by favoring genotypes adapted to the new conditions. But the degree of adaptation can be extended only within the realm of the genetic variability present in the population. Before we examine the process of adaptation by natural selection more closely, let's look at the genetic basis of the variation that makes it possible for populations to evolve.

GENETIC VARIATION, THE SUBSTRATE FOR NATURAL SELECTION

Heritable variation is at the heart of Darwin's theory of evolution, for variation provides the raw material—the substrate—on which natural selection works.

Genetic variation occurs within and between populations

You have no trouble recognizing your friends in a crowd. Each person has a unique genome, reflected in individual variations of appearance and temperament. Individual variation occurs in populations of all species of sexually reproducing organisms. We are very conscious of human diversity; we are less sensitive to individuality in populations of other animals and of plants, and the diversity may escape our notice because the variations are subtle. But these slight differences between individuals in a population are the variations Darwin wrote most about as the raw material for natural selection.

Not all the variation we observe in a population is heritable. Phenotype is the cumulative product of an inherited genotype and a multitude of environmental influences. For example, bodybuilders alter their phenotypes dramatically. It is important to remember that only the genetic component of variation can have evolutionary consequences as a result of natural selection, because it is the only component that transcends generations.

Variation Within Populations

Both quantitative and discrete characters contribute to variation *within* a population. Most heritable variation consists of *quantitative characters* that vary along a continuum within a population. For example, plant height may vary continuously in our hypothetical wildflower population, from very short individuals to very tall individuals and everything in between. Quantitative variation usually indicates polygenic inheritance, an additive effect of two or more genes on a single phenotypic character. *Discrete characters,* such as pink versus white flowers, can be classified on an either-or basis, usually because they are determined by a single gene locus with different alleles that produce distinct phenotypes.

Polymorphism. When two or more forms of a discrete character are represented in a population, the contrasting forms are called *morphs*—as in the pink-flowered and white-flowered morphs of our wildflower population. A population is said to be *polymorphic* for a character if two or more distinct morphs are each represented in high enough frequencies to be readily noticeable. (Obviously, this definition is arbitrary, but a population is not termed polymorphic if it consists almost exclusively of a single morph, with other morphs extremely rare.) FIGURE 23.7 illustrates a striking example of **polymorphism** (the existence of polymorphic characters) in garter snakes. Polymorphism is extensive in human populations, both in physical characters, such as the presence or absence of freckles, and in biochemical characters, such as ABO blood groups (for which there are four morphs: type A, type B, type AB, and type O). Polymorphism applies only to discrete characters, not to characters such as human height that vary continuously in a population.

FIGURE 23.7 ▪ Polymorphism. Some populations consist of two or more discrete varieties of individuals. These four garter snakes (*Thamnophis ordinoides*), which differ markedly in their patterns of coloration, were captured in the same Oregon field. Edmund Brodie of the University of Chicago has discovered that the behavior of each morph is keyed to its coloration. Spotted snakes generally blend into their background better than striped snakes, but stripes make it more difficult to judge the speed of the snake in motion. When approached, spotted garter snakes usually freeze, whereas snakes of the striped morph move away rapidly.

Measuring Genetic Variation. Population biologists use several quantitative definitions of genetic variation. Two common measures are the percentage of gene loci represented by two or more alleles in a population (loci at which a single allele is not fixed), and the average percentage of loci that are heterozygous in the individuals of a population.

By these two measures of genetic diversity, the reservoir of heritable variation in a population is much more extensive than Darwin realized. Although much of the variation is invisible, it is manifest in molecular differences that can be detected by biochemical methods. For example, some evolutionary biologists use gel electrophoresis to study variations in the protein products of specific gene loci among individuals in a population. Such variations represent different alleles at a locus. Scores of loci have been studied in many different species. In populations of the fruit fly *Drosophila,* for example, the gene pool typically has two or more alleles for about 30% of the loci examined, and each fly is heterozygous at about 12% of its loci; that amounts to 700 to 1200 heterozygous loci per fly. The extent of genetic variation in human populations, as revealed by electrophoresis, is similar.

Variation Between Populations

Most species exhibit **geographical variation**, differences in genetic structure *between* populations. Because at least some environmental factors are likely to be different from one place to another, natural selection can contribute to geographical variation. For example, one population of our now-familiar wildflower species may have a higher frequency of recessive alleles at the flower-color locus than other populations, perhaps because of a local prevalence of pollinators that key on white flowers (recessive homozygotes). Genetic drift can also cause chance variations among different populations. On a more local scale, geographical variation can also occur within a population, either because the environment has patchlike diversity or because the population is differentiated into subpopulations resulting from the limited dispersal of individuals.

One particular type of geographical variation, called a **cline**, is a graded change in some trait along a geographic axis. In some cases, a cline may represent a graded region of overlap where individuals of neighboring populations are interbreeding. In other cases, a gradation in some environmental variable may produce a cline. For example, the average body size of many North American species of birds and mammals

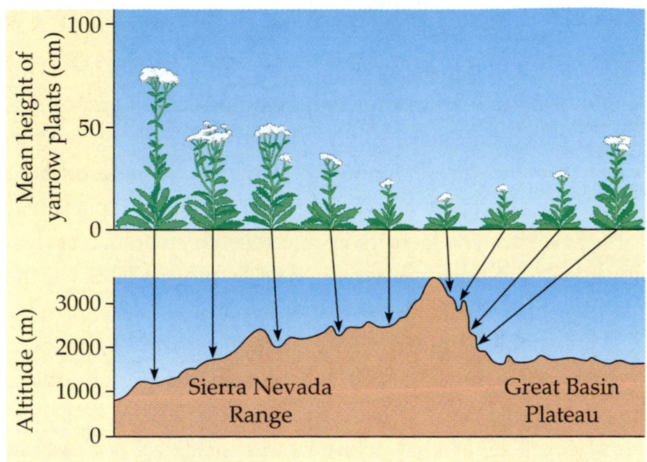

FIGURE 23.8 ▪ Clinal variation in a plant. Average size of yarrow plants (*Achillea*) growing on the slopes of the Sierra Nevada mountains gradually decreases with increasing elevation. Although the environment affects growth rates directly to some extent, some of the variation has a genetic basis. Researchers collected seeds at different elevations and grew plants in a common garden; the average sizes of the plants were correlated with the altitude at which the seeds were collected.

increases gradually with increasing latitude. Presumably, the reduced ratio of surface area to volume that accompanies larger size is an adaptation that helps animals conserve body heat in cold environments. Experimental studies of some clines confirm the role of genetic variation in the spatial differences of phenotype (FIGURE 23.8).

Mutation and sexual recombination generate genetic variation

Two random processes, mutation and sexual recombination (see Chapter 15), create variation in the genetic composition of a population.

Mutation

New alleles originate only by mutation, or change in the nucleotide sequence of DNA. A mutation affecting any gene locus is an accident that is rare and random. Most mutations occur in somatic cells and are lost when the individual dies. Only mutations that occur in cell lines that produce gametes can be passed along to offspring.

A mutation is like a shot in the dark: Chance determines where it will strike and how it will alter a gene. Most point mutations, those affecting a single base in DNA, are probably relatively harmless. Much of the DNA in the eukaryotic genome does not code for protein products, and it is uncertain how a change of a single nucleotide base in this silent DNA will affect the well-being of the organism. Even mutations of structural genes, which do code for proteins, may occur with little or no effect on the organism, partly because of redundancy in the genetic code. Of course, a single point mutation can have a significant impact on phenotype, as in sickle-cell disease, for example.

A mutation that alters a protein enough to affect its function is more often harmful than beneficial. Organisms are the refined products of thousands of generations of past selection, and a random change is not likely to improve the genome any more than blindly firing a gunshot through the hood of a car is likely to improve engine performance. On rare occasions, however, a mutant allele may actually fit its bearer to the environment better and enhance the reproductive success of the individual. This is not especially likely in a stable environment but becomes more probable when the environment is changing and mutations that were once selected against are now favorable under the new conditions. For example, some mutations that happen to endow houseflies with resistance to DDT also reduce growth rate. Before DDT was introduced, such mutations were harmful. But once DDT was part of the environment, the mutant alleles were favored, and natural selection increased their frequency in fly populations.

Because chromosomal mutations usually affect many gene loci, they are almost certain to disrupt the development of the organism. But even rearrangements of chromosomes may in rare instances bring benefits. For example, the translocation of a chromosomal piece from one chromosome to another could link alleles that affect the organism in some positive way when they are inherited together as a package.

Duplications of chromosome segments, like other chromosomal mutations, are nearly always harmful. But if the repeated segment does not disrupt genetic balance severely, it can persist over the generations and provide an expanded genome with superfluous loci that may eventually take on new functions by mutation. New genes may also arise from existing DNA sequences by the shuffling of exons within the genome, either within a single locus or between loci (see Chapter 19).

For bacteria and other microorganisms that have very short generation spans, mutation can have a noticeable effect on a population's variation in a relatively short time. Bacteria reproduce asexually by dividing as often as once every 20 minutes; a single cell can potentially give rise to a billion descendants in just 10 hours. A new mutation that happens to be beneficial can increase its frequency in a bacterial population very rapidly. Imagine, for example, exposing a bacterial population to an antibiotic. If a single individual in the population happens to harbor a mutation that renders it resistant to the poison, in just a few hours there may be millions of resistant bacteria, while bacteria sensitive to the antibiotic may have been almost completely eliminated. Bacterial populations can evolve, one mutation at a time, by the explosive asexual expansion of clones favored by the local environment. On a generation-to-generation time scale, animals and plants depend mainly on sexual recombination for the genetic variation that makes adaptation possible. But even most bacteria increase genetic variation by occasionally exchanging and recombining genes through processes that resemble sex (see Chapter 18).

Sexual Recombination

Members of a sexually reproducing population owe nearly all their genetic differences to the unique recombinations of existing alleles each individual receives from the gene pool. (Of course, this allele variation has its ultimate basis in past mutations.)

Sex shuffles alleles and deals them at random to determine individual genotypes. During meiosis, homologous chromosomes, one inherited from each parent, trade some of their genes by crossing over, and then the homologous chromosomes and the alleles they carry segregate randomly into separate gametes. Gametes from one individual vary extensively in their genetic makeup, and each zygote made by a mating pair has a unique assortment of alleles resulting from the random union of a sperm and an ovum. A population, of course, contains a vast number of possible mating combinations, each bringing together the gametes of individuals that are likely to have different genetic backgrounds. Sexual reproduction recombines old alleles into fresh assortments every generation.

Diploidy and balanced polymorphism preserve variation

What prevents natural selection from extinguishing a population's variation by culling unfavorable genotypes? The tendency for natural selection to reduce variation is countered by several mechanisms that preserve or restore variation.

Diploidy

The diploid nature of most eukaryotes hides a considerable amount of genetic variation from selection in the form of recessive alleles in heterozygotes. Recessive alleles that are less favorable than their dominant counterparts, or even harmful in the present environment, can persist in a population through their propagation by heterozygous individuals. This latent variation is exposed to selection only when both parents carry the same recessive allele and combine two copies in one zygote. This happens only rarely if the frequency of the recessive allele is very low. For example, if the frequency of the recessive allele is 0.01 and the frequency of the dominant allele is 0.99, then 99% of the copies of that recessive allele are protected from selection in heterozygotes, and only 1% of the recessive alleles are present in homozygotes. The rarer the recessive allele, the greater the degree of protection. Heterozygote protection maintains a huge pool of alleles that may not be suitable for present conditions but that could bring new benefits when the environment changes.

Balanced Polymorphism

Selection itself may preserve variation at some gene loci. This ability of natural selection to maintain diversity in a population is called **balanced polymorphism**. One of the mechanisms for this preservation of variation is **heterozygote advantage.** If individuals who are heterozygous at a particular locus have greater survivorship and reproductive success than any type of homozygote, then two or more alleles will be maintained at that locus by natural selection.

An example of heterozygote advantage involves the locus in humans for one chain of hemoglobin, the protein of red blood cells that transports oxygen. As we have seen, a specific recessive allele at that locus causes sickle-cell disease in homozygous individuals. Heterozygotes, however, are resistant to malaria, an important advantage in tropical regions where that disease is a major cause of death. The environment in these regions favors the heterozygotes over homozygous dominant individuals, who are susceptible to malaria, and homozygous recessive individuals, who are harmed by sickle-cell disease. The frequency of the sickle-cell allele in Africa is generally highest in areas where the malaria parasite is most common. In some tribes, the recessive allele accounts for 20% of the hemoglobin loci in the gene pool, a very high frequency for an allele that is disastrous in homozygotes. But at this frequency ($q = 0.2$), 32% of the population consists of heterozygotes resistant to malaria ($2pq$), and only 4% of the population suffers from sickle-cell disease (q^2).

Another example of heterozygote advantage is found in the crossbreeding of crop plants. When corn, for instance, is highly inbred, the number of homozygous gene loci increases,

FIGURE 23.9 ▪ A balanced polymorphism in a finch population. Two distinctly different beak sizes occur in a single population of black-bellied seedcrackers, a species of finch that lives in Cameroon, West Africa. There are no individuals with bills of intermediate size. Small-billed individuals (left) feed mainly on soft seeds, whereas large-billed birds specialize in cracking hard seeds. Each morph shows a higher feeding efficiency on each respective seed. Thomas Smith of San Francisco State University, who has been studying this population in the field, hypothesizes that natural selection maintains the polymorphism by selecting against intermediate-sized bills, which crack both classes of seeds relatively inefficiently.

and the corn may gradually become stunted in growth and increasingly susceptible to a variety of diseases. Crossbreeding between two different inbred varieties often produces hybrids that are much more vigorous than either parent stock. This **hybrid vigor** is probably due to two factors: the segregation of harmful recessives that were homozygous in the inbred varieties, and the heterozygote advantage at many loci in the hybrids.

A patchy environment, where natural selection favors different phenotypes in different subregions within a population's geographical boundaries, can also result in balanced polymorphism. For example, protective coloration suited to different backgrounds may help explain the morphs of garter snakes in FIGURE 23.7. FIGURE 23.9 illustrates another example of a polymorphism that may be due to environmental patchiness—in this case, specialization for slightly different foods.

Still another cause of balanced polymorphism is **frequency-dependent selection**, in which the reproductive success of any one morph declines if that phenotypic form becomes too common in the population. A particularly intricate example is a balanced polymorphism that has been observed in populations of *Papilio dardanus*, the African swallowtail butterfly in the photograph on p. 428. All males have similar coloration, but females occur in several different morphs, each resembling another butterfly species that is noxious to predators. *Papilio* females are not noxious, but birds avoid them because they look so much like the distasteful butterflies. This mimicry would be less effective if all *Papilio* females resembled the same noxious species, because birds would be slow to associate a particular pattern of coloration with bad taste if they encountered good-tasting mimics as often as the noxious models. FIGURE 23.10 illustrates another example of frequency-dependent selection.

Neutral Variation

Some of the genetic variations observed in populations are probably trivial in their impact on reproductive success. The diversity of human fingerprints is an example of what is called **neutral variation**, which seems to confer no selective advantage for some individuals over others. Much of the protein variation detectable by electrophoresis may represent chemical "fingerprints" that are neutral in their adaptive qualities. The relative frequencies of neutral variations will not be affected by natural selection; some neutral alleles will increase in the gene pool, and others will decrease by the chance effects of genetic drift.

There is no consensus among evolutionary biologists on how much genetic variation is neutral, or even if any variation can be considered truly neutral. Variations appearing to be neutral may, in fact, influence survival and reproductive success in ways that are difficult to measure. It is possible to show that a particular allele is detrimental, but it is impossible to demonstrate that an allele brings no benefits at all to an organism. Furthermore, a variation may be neutral in one environment but not in another. We can never know the

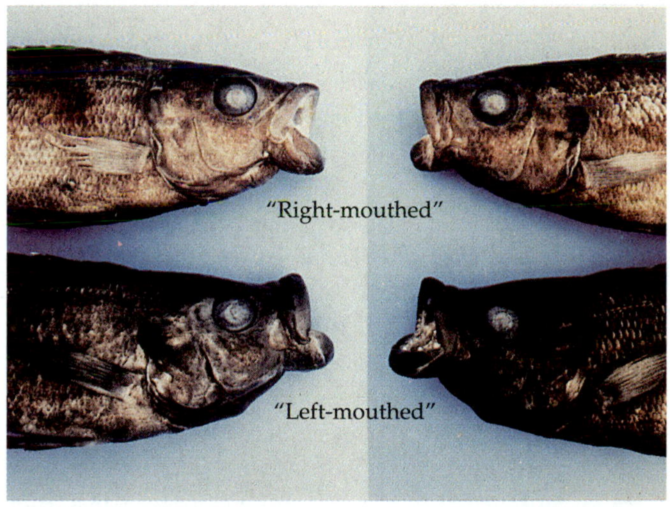

(a)

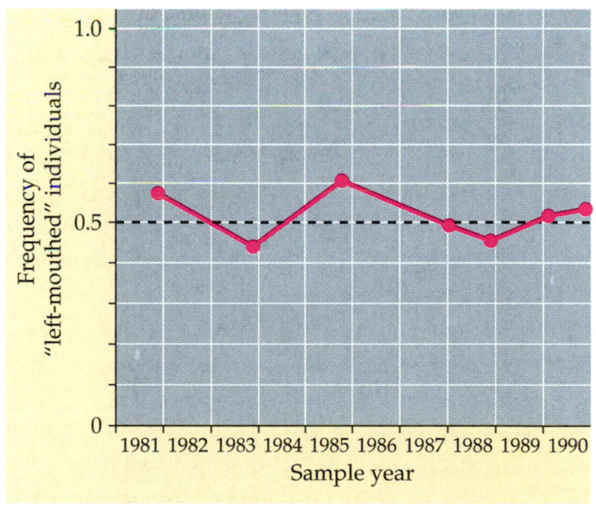

(b)

FIGURE 23.10 • Frequency-dependent selection. Michio Hori of Wakayama Medical College in Japan has correlated a balanced polymorphism in mouth anatomy with the feeding habits of *Perissodus microlepis*, a small cichlid fish that lives in Africa's Lake Tanganyika. These fish feed on the scales of other fishes by approaching their prey from behind and using their mouths to snatch scales from the flanks of the prey. **(a)** *P. microlepis* has an asymmetrical mouth oriented either toward the right (top) or left (bottom). The two morphs attack opposite sides of their prey. Simple Mendelian inheritance determines this character. **(b)** The two morphs, "right-mouthed" and "left-mouthed," occur in approximately equal numbers among individuals in a population of *P. microlepis*, fluctuating only slightly in relative frequency from year to year, as we can conclude from this graph of "left-mouthed" frequency over a 10-year period. Hori offers this hypothesis: Prey fish guard more effectively against attack from either the left side or the right side, depending on which *P. microlepis* morph is more common at a particular time. Thus, the polymorphism is balanced by the less common morph having a feeding advantage that enhances survival and reproductive success.

degree to which genetic variation is neutral. But we can be certain that even if only a fraction of the extensive variation in a gene pool significantly affects the organisms, that is still an enormous reservoir of raw material for natural selection and the adaptive evolution it causes.

NATURAL SELECTION AS THE MECHANISM OF ADAPTIVE EVOLUTION

Adaptive evolution is a blend of chance and sorting—chance in the origin of new genetic variations by mutation and sexual recombination, and sorting in the workings of selection as it favors the propagation of some chance variations over others. From the range of variations available to it, natural selection increases the frequencies of certain genotypes and fits organisms to their environments.

Evolutionary fitness is the relative contribution an individual makes to the gene pool of the next generation

The phrases *struggle for existence* and *survival of the fittest* are loaded with misleading connotations. There are, of course, animal species in which individuals, usually the males, lock horns or otherwise do combat to determine mating privilege. But reproductive success is generally more subtle and passive. A barnacle may produce more eggs than its neighbors because it is more efficient at collecting food from the water. In a population of moths, certain variants may average more offspring than others because their body and wing colors hide them from predators better. Plants in a wildflower population may differ in reproductive success because some are better able to attract pollinators, owing to slight variations in flower color, shape, or fragrance. ***Darwinian fitness,*** *the measure that is critical to selection, is the relative contribution an individual makes to the gene pool of the next generation.*

In a more quantitative approach to natural selection, population geneticists define **relative fitness** as the contribution of a genotype to the next generation compared to the contributions of alternative genotypes for the same locus. For example, consider our wildflower population, in which *AA* and *Aa* plants have pink flowers and *aa* plants have white flowers. Let's assume that, on average, individuals with pink flowers produce more offspring than those with white flowers. The relative fitness of the most reproductively successful variants is set at 1 as a basis for comparison; so in this case, the relative fitness of an *AA* or *Aa* plant is 1. If plants with white flowers average only 80% as many offspring, their relative fitness is 0.8.

Survival alone does not guarantee reproductive success. Relative fitness is zero for a sterile plant or animal, even if it is robust and outlives other members of the population. But, of course, survival is a prerequisite for reproducing, and longevity increases fitness if it results in certain individuals leaving more descendants than other individuals leave. Then again, an individual that matures quickly and becomes fertile at an early age may have a greater reproductive potential than individuals that live longer but mature late. Thus, many factors that affect both survival and fertility determine an individual's evolutionary fitness.

An organism exposes its phenotype—its physical traits, metabolism, physiology, and behavior—not its genotype, to the environment. Acting on phenotypes, selection indirectly adapts a population to its environment by increasing or maintaining favorable genotypes in the gene pool.

The entity subjected to natural selection is the whole organism, which is an integrated composite of its many phenotypic features, not a collage of individual parts. Thus, the relative fitness of an allele depends on the entire genetic context in which it works. For example, alleles that enhance the growth of the trunk and limbs of a tree may be useless or even detrimental in the absence of alleles at other loci that enhance the growth of roots required to support the tree. On the other hand, alleles that contribute nothing to an organism's success, or may even be slightly maladaptive, may be perpetuated because they are present in individuals whose overall fitness is high. The whole baseball team wins the league pennant, even the player with the worst batting average and the most errors.

The effect of selection on a varying characteristic can be stabilizing, directional, or diversifying

Natural selection can affect the frequency of a heritable trait in a population in three different ways, depending on which phenotypes in a varying population are favored. These three modes of selection are called stabilizing selection, directional selection, and diversifying selection. They can be depicted with graphs that show how the frequencies of different phenotypes change with time (FIGURE 23.11). This is most meaningful for quantitative traits that depend on many gene loci.

Stabilizing selection acts against extreme phenotypes and favors the more common intermediate variants. This mode of selection reduces variation and maintains the status quo for a particular phenotypic character. For example, stabilizing selection keeps the majority of human birth weights in the 3–4 kg range. For babies much smaller or larger than this, infant mortality is greater.

Directional selection is most common during periods of environmental change or when members of a population migrate to some new habitat with different environmental conditions. Directional selection shifts the frequency curve for variations in some phenotypic character in one direction or the other by favoring what are initially relatively rare individuals that deviate from the average for that character. For instance, fossil evidence indicates that the average size of black

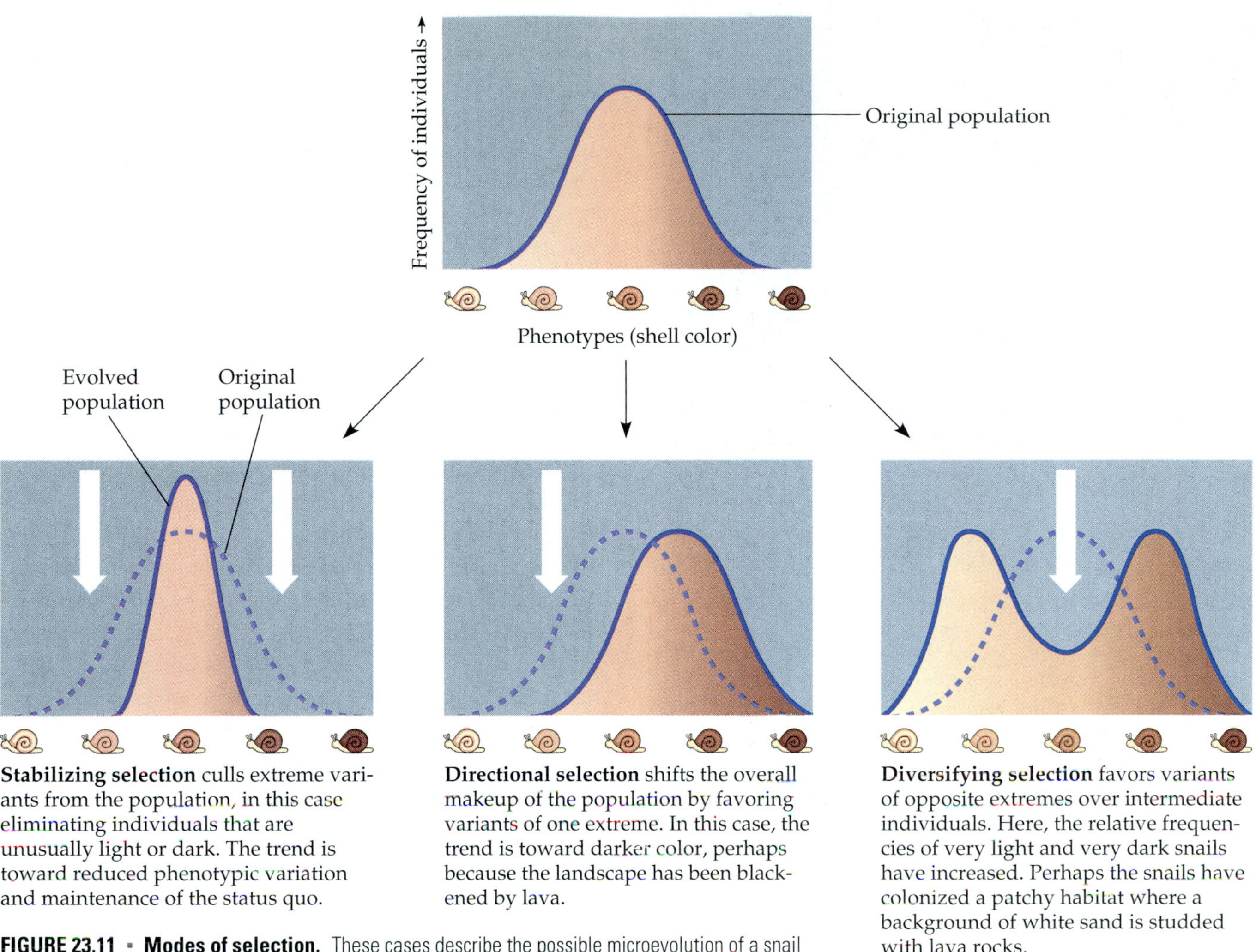

Stabilizing selection culls extreme variants from the population, in this case eliminating individuals that are unusually light or dark. The trend is toward reduced phenotypic variation and maintenance of the status quo.

Directional selection shifts the overall makeup of the population by favoring variants of one extreme. In this case, the trend is toward darker color, perhaps because the landscape has been blackened by lava.

Diversifying selection favors variants of opposite extremes over intermediate individuals. Here, the relative frequencies of very light and very dark snails have increased. Perhaps the snails have colonized a patchy habitat where a background of white sand is studded with lava rocks.

FIGURE 23.11 ▪ Modes of selection. These cases describe the possible microevolution of a snail population in which there is quantitative variation in shell coloration from light to dark. The graphs show how the frequencies of individuals of varying darkness change over time. The large arrows symbolize natural selection working against certain phenotypes.

bears in Europe increased with each glacial period of the ice ages, only to decrease again during the warmer interglacial periods.

Diversifying selection occurs when environmental conditions are varied in a way that favors individuals on both extremes of a phenotypic range over intermediate phenotypes. Diversifying selection can result in balanced polymorphism, as in the finch population with the two bill sizes shown in FIGURE 23.9.

Although we refer to these three selection trends as "modes of selection," the basic mechanism of natural selection is the same for each case. Selection favors certain heritable traits via differential reproductive success.

Sexual selection may lead to pronounced secondary differences between the sexes

Males and females of many animal species exhibit marked differences in addition to the differences in the reproductive organs that define the sexes. This distinction between the secondary sex characteristics of males and females is known as **sexual dimorphism**. It is often expressed as a difference in size, with the male usually larger, but it also involves such features as colorful plumage in male birds, manes on male lions, antlers on male deer, and other adornments. In most cases of sexual dimorphism, at least among vertebrates, the male is the showier sex. In some cases, the males with the most impressive masculine features may be the most attractive to females. There are also species in which the secondary sex structures may be used in direct competition with other males; this is particularly common in species where a single male garners a harem of females. These males may succeed because they defeat smaller, weaker, or less fierce males in combat; more often, however, they are effective in ritualized displays that discourage would-be competitors (see Chapter 51).

Darwin was intrigued by **sexual selection**, which he saw as a separate selection process leading to sexual dimorphism. Many secondary sex features do not seem to be adaptive in the

FIGURE 23.12 • Sexual selection and the evolution of male appearance in jungle fowl. Marlene Zuk of the University of California, Riverside, has been studying mate selection in Asian jungle fowl (*Gallus gallus*), the ancestors of domesticated chickens. Hens generally choose roosters like the one in this photograph—those with bright eyes and large red combs and wattles (the flap of skin hanging from the throat). These traits are correlated with good health and resistance to pathogens. Thus, according to Dr. Zuk's hypothesis, a hen keys on rooster features that advertise vitality. Over the generations, such choices by females shape the appearance of males and may play an active role in favoring genotypes that contribute to healthy offspring.

general sense; showy plumage probably does not help male birds cope with their environment and may even attract predators. If such accoutrements give the individual an edge in gaining a mate, however, they will be favored for the most Darwinian of reasons—because they enhance reproductive success. Thus, in many cases, the ultimate evolutionary outcome is a compromise between the two selection forces. In some species, the line between sexual selection and ordinary natural selection blurs because the secondary sex feature does double duty as an adaptation to the environment. For example, a stag may use his antlers to defend himself against a predator.

Many researchers who study sexual selection and dimorphism have recently shifted their attention to the roles of females in the evolution of these characters. Every time a female chooses a mate based on particular phenotypic traits, she perpetuates the alleles that caused her to make that choice and allows a male with a particular phenotype to perpetuate his alleles (FIGURE 23.12).

Natural selection cannot fashion perfect organisms

There are at least four reasons why natural selection cannot produce perfection.

1. *Organisms are locked into historical constraints.* As we saw in Chapter 22, each species has a legacy of descent with modification from a long line of ancestral forms. Evolution does not scrap ancestral anatomy and build each new complex structure from scratch, but co-opts existing structures and adapts them to new situations. For example, the excruciating back problems some humans endure result in part because the skeleton and musculature modified from the anatomy of four-legged ancestors are not fully compatible with upright posture.
2. *Adaptations are often compromises.* Each organism must do many different things. A seal spends part of its time on rocks; it could probably walk better if it had legs instead of flippers, but it would not swim nearly as well. We humans owe much of our versatility and athleticism to our prehensile hands and flexible limbs, which also make us prone to sprains, torn ligaments, and dislocations; structural reinforcement has been compromised for agility.
3. *Not all evolution is adaptive.* Chance probably affects the genetic structure of populations to a greater extent than was once believed. For instance, when a storm blows insects hundreds of miles over an ocean to an island, the wind does not necessarily transport the specimens that are best suited to the new environment. Thus, not all alleles fixed by genetic drift in the gene pool of the small founding population are better suited to the environment than alleles that are lost. Similarly, the bottleneck effect can cause nonadaptive or even maladaptive evolution.
4. *Selection can only edit variations that exist.* Natural selection favors only the fittest variations from the phenotypes that are available, which may not be the ideal traits. New alleles do not arise on demand.

With all these constraints, we cannot expect evolution to craft perfect organisms. Natural selection operates on a "better than" basis. We can see evidence for evolution in the subtle imperfections of the organisms it produces.

■ ■ ■

Natural selection is usually thought of as an agent of change, but it can also act to maintain the status quo. Stabilizing selection probably prevails most of the time, resisting change that may be maladaptive. Evolutionary spurts occur when a population is stressed by a change in the environment, migration to a new place, or a change in the genome. When challenged with a new set of problems, a population either adjusts through natural selection or becomes extinct. The fossil record indicates that extinction is the more common outcome. Those populations that do survive crises often change enough to become new species, as we will see in Chapter 24.

CHAPTER REVIEW

REVIEW OF KEY CONCEPTS

(with page numbers and key figures)

POPULATION GENETICS

- **The modern evolutionary synthesis integrated Darwinian selection and Mendelian inheritance (pp. 428–429)** The development of population genetics, with its emphases on quantitative inheritance and variation, brought Darwinian theory and Mendelian principles of inheritance together. The modern synthesis focuses on populations as units of evolution.
- **The genetic structure of a population is defined by its allele and genotype frequencies (pp. 429–430)** A population, a localized group of organisms belonging to the same species, is united by its gene pool, the aggregate of all alleles in the population.
- **The Hardy-Weinberg theorem describes a nonevolving population (pp. 430–431, FIGURE 23.3)** According to the Hardy-Weinberg theorem, the frequencies of alleles in a population will remain constant if sexual reproduction is the only process that affects the gene pool. If p and q represent the relative frequencies of the dominant and recessive alleles of a two-allele locus, respectively, then $p^2 + 2pq + q^2 = 1$, where p^2 and q^2 are the frequencies of the homozygous genotypes, and $2pq$ is the frequency of the heterozygous genotype.

CAUSES OF MICROEVOLUTION

- **Microevolution is a generation-to-generation change in a population's allele or genotype frequencies (p. 432)** For Hardy-Weinberg equilibrium to apply, the population must be very large, be totally isolated, have no net mutations, show random mating, and have equal reproductive success for all individuals. Microevolution can occur when one or more of the conditions required for Hardy-Weinberg equilibrium are not met.
- **The five causes of microevolution are genetic drift, gene flow,**
23.1 **mutation, nonrandom mating, and natural selection (pp. 432–435, FIGURES 23.4–23.6)** Natural selection—differential success in reproduction—is the only agent of microevolution that tends to be adaptive.

GENETIC VARIATION, THE SUBSTRATE FOR NATURAL SELECTION

- **Genetic variation occurs within and between populations (pp. 435–437)** Genetic variation includes individual variation in discrete and quantitative characters within a population, as well as geographical variation between populations.
- **Mutation and sexual recombination generate genetic variation (pp. 437–438)** Most mutations have no effect or are harmful, but some are adaptive. Sexual recombination produces most of the genetic variation that makes adaptation possible in populations of sexually reproducing organisms.
- **Diploidy and balanced polymorphism preserve variation (pp. 438–440)** Diploidy maintains a reservoir of latent variation in heterozygotes. Balanced polymorphism may maintain variation at some gene loci as a result of heterozygote advantage or frequency-dependent selection. Some genetic variation may be unaffected by natural selection.

NATURAL SELECTION AS THE MECHANISM OF ADAPTIVE EVOLUTION

- **Evolutionary fitness is the relative contribution an individual makes to the gene pool of the next generation (p. 440)** Darwinian fitness is measured only by reproductive success. One genotype has a greater relative fitness than another if it leaves more descendants. Selection favors certain genotypes in a population by acting on the phenotype of individual organisms. The whole organism is the object of selection.
- **The effect of selection on a varying characteristic can be stabilizing, directional, or diversifying (pp. 440–441, FIGURE 23.11)** Natural selection can act against extreme phenotypes (stabilizing selection), favor relatively rare individuals on one end of the phenotypic range (directional selection), or favor individuals at both extremes of the range over intermediate phenotypes (diversifying selection).
- **Sexual selection may lead to pronounced secondary differences between the sexes (pp. 441–442)** Sexual selection leads to the evolution of secondary sex characteristics, which can give individuals an advantage in mating.
- **Natural selection cannot fashion perfect organisms (p. 442)** The reasons are: Structures result from modified ancestral anatomy, adaptations are often compromises, the gene pool can be affected by genetic drift, and natural selection can act only on available variation.

SELF-QUIZ

1. A gene pool consists of
   a. all the alleles exposed to natural selection
   b. the total of all alleles present in a population
   c. the entire genome of a reproducing individual
   d. the frequencies of the alleles for a gene locus within a population
   e. all the gametes in a population

2. In a population with two alleles for a particular locus, *B* and *b*, the allele frequency of *B* is 0.7. What would be the frequency of heterozygotes if the population is in Hardy-Weinberg equilibrium?
   a. 0.7
   b. 0.49
   c. 0.21
   d. 0.42
   e. 0.09

3. In a population that is in Hardy-Weinberg equilibrium, 16% of the individuals show the recessive trait. What is the frequency of the dominant allele in the population?
   a. 0.84
   b. 0.36
   c. 0.6
   d. 0.4
   e. 0.48

4. The average length of jackrabbit ears decreases the farther north the rabbits live. This variation is an example of
   a. a cline
   b. discrete variation
   c. polymorphism
   d. genetic drift
   e. diversifying selection

5. Which of the following is an example of a polymorphism in humans?
   a. variation in height
   b. variation in intelligence
   c. the presence or absence of a widow's peak (see FIGURE 14.14)
   d. variation in the number of fingers
   e. variation in fingerprints

6. Selection acts *directly* on
   a. phenotype
   b. genotype
   c. the entire genome
   d. each allele
   e. the entire gene pool

7. As a mechanism of microevolution, natural selection can be most closely equated with
   a. assortative mating
   b. genetic drift
   c. differential reproductive success
   d. mutation
   e. gene flow

8. Most of the variation we see in coat coloration and pattern in a population of wild mustangs in any generation is probably due to
   a. new mutations that occurred in the preceding generation
   b. sexual recombination of alleles
   c. genetic drift due to the small size of the population
   d. geographical variation within the population
   e. environmental effects

9. In terms of the algebraic symbols used to represent the Hardy-Weinberg theorem (p and q), the most likely effect of assortative mating on the frequencies of alleles and genotypes for a gene locus would be
   a. a decrease in p^2 compared to q^2
   b. a trend toward zero for q^2
   c. convergence of p^2 and q^2 toward equal values
   d. a change in p and q, the relative frequencies of the two alleles in the gene pool
   e. a decrease in $2pq$ below the value expected by the Hardy-Weinberg theorem

10. A founder event favors microevolution in the founding population mainly because
   a. mutations are more common in a new environment
   b. a small founding population is subject to extensive sampling error in the composition of its gene pool
   c. the new environment is likely to be patchy, favoring diversifying selection
   d. gene flow increases
   e. members of a small population tend to migrate

CHALLENGE QUESTION

Let's return to the wildflowers we used to illustrate the Hardy-Weinberg theorem. The frequency of *A*, the dominant allele for pink flowers, is 0.8, and the frequency of *a*, the recessive allele for white flowers, is 0.2. In one starting population, the frequencies of genotypes do not conform to Hardy-Weinberg equilibrium: 60% of the plants are *AA* and 40% of the plants are *Aa*. (At this point, the population has no plants with white flowers.) Assuming that all conditions for the Hardy-Weinberg theorem are met, prove that genotypes will reach equilibrium in the next generation.

SCIENCE, TECHNOLOGY, AND SOCIETY

Some scientists have suggested that the increased appreciation of the importance of females in sexual selection is one consequence of increasing numbers of women making contributions to animal behavior and evolutionary biology. Do you think gender makes a difference in the kinds of questions scientists ask and the information they uncover? Why or why not?

FURTHER READING

Brookes, M. "Day of the Mutators." *New Scientist,* February 14, 1998. Discusses evidence supporting the hypothesis that prokaryotes have some genes with unusually high mutation rates.

Dawkins, R. *The Blind Watchmaker.* New York: Norton, 1986. A master of metaphor explains how complexity can arise in the absence of design.

Gibbons, A. "The Mystery of Humanity's Missing Mutations." *Science,* January 6, 1995. Evidence for a bottleneck in our history.

Gillis, A. M. "Getting a Picture of Human Diversity." *BioScience,* January 1994. The application of population genetics to anthropology.

Levy, S. B. "The Challenge of Antibiotic Resistance." *Scientific American,* March 1998. A thoughtful analysis of how resistant prokaryotes arise and the health hazards they pose.

Morell, V. "Sex Frees Viruses From Genetic Ratchet." *Science,* November 28, 1997. Interesting experiments testing the hypothesis that sexual reproduction in itself enhances fitness.

Vergano, D. "Butterfly Wings It With a Few Genes." *Science News,* November 23, 1996. Some genes are expressed differently depending on the stage of development.

Williams, N. "Streetcar Carries Evolution Modelers Around Roadblocks." *Science,* March 8, 1996. Application of game theory to evolutionary models.

WEB LINKS

Visit the special edition of *The Biology Place* for BIOLOGY, Fifth Edition, at **http://www.biology.com/campbell**. Go to Chapter 23 for online resources, including learning activities, practice exams, and links to the following web sites:

"The Tree of Life"
The Tree of Life is a web-based "phylogenetic navigator" that contains information about the phylogenetic relationships and characteristics of organisms, and illustrates the diversity and unity of living organisms.

"Paleontology Without Walls"
Three exhibits at this site allow you to study the macroevolution of life on Earth.

"The Hardy–Weinberg Equilibrium Simulator"
Using this simulator you can observe the effects on Hardy–Weinberg equilibrium of different genotype survival rates, allele percentages, population sizes, and the number of generations.

"The WWW Virtual Library: Evolution"
This site contains many links to global resources on evolution.

CHAPTER 24

THE ORIGIN OF SPECIES

What Is a Species?

- The biological species concept emphasizes reproductive isolation
- Prezygotic and postzygotic barriers isolate the gene pools of biological species
- The biological species concept does not work in all situations
- Other species concepts emphasize features and processes that identify and unite species members

Modes of Speciation

- Geographical isolation can lead to the origin of species: allopatric speciation
- A new species can originate in the geographical midst of the parent species: sympatric speciation
- Genetic change in populations can account for speciation
- The punctuated equilibrium model has stimulated research on the tempo of speciation

The Origin of Evolutionary Novelty

- Most evolutionary novelties are modified versions of older structures
- Genes that control development play a major role in evolutionary novelty
- An evolutionary trend does not mean that evolution is goal oriented

When Darwin saw that the geologically young Galápagos Islands had already become populated with many plants and animals known nowhere else in the world, he realized that he was visiting a place of genesis. The islands are named for the giant tortoises, such as the Geochelone elephantopus *shown here, that are among the unique inhabitants. (Galápago is the Spanish word for tortoise.) After visiting the Galápagos, Darwin wrote in his diary: "Both in space and time, we seem to be brought somewhat near to that great fact—that mystery of mysteries—the first appearance of new beings on this Earth." The beginning of new forms of life—the origin of species—is at the focal point of evolutionary theory, for it is in new species that biological diversity arises. It is not enough to explain how adaptations evolve in populations, a topic covered in Chapter 23. Evolutionary theory must also explain* ***macroevolution****, the origin of new taxonomic groups (new species, new genera, new families, even new kingdoms).* ***Speciation*** *(the origin of new species) is the keystone process because any genus, family, or higher taxon originates with a new species that is novel enough to be the inaugural member of the higher taxon.*

*The fossil record chronicles two processes involved in speciation: anagenesis and cladogenesis (*FIGURE 24.1*).* ***Anagenesis*** *(Gr.* ana*, "up," and* genesis*, "origin"), also known as* ***phyletic evolution****, is the accumulation of changes associated with the transformation of one species into another.* ***Cladogenesis*** *(Gr.* clados*, "branch"), also called* ***branching evolution****, is the*

FIGURE 24.1 ▪ Two patterns of speciation. (a) Anagenesis (phyletic evolution) is the accumulation of heritable changes in a population associated with the process of speciation. **(b)** Cladogenesis is branching evolution, in which a new species arises from a population that buds from a parent species. Cladogenesis is the basis for biological diversity.

budding of one or more new species from a parent species that continues to exist. Only cladogenesis can promote biological diversity by increasing the number of species.

Our objectives in this chapter are to evaluate definitions of species and mechanisms of speciation, and to examine the possible origins of some novel features that define higher taxonomic groups. Our first step is to appraise the assumption that species actually exist in nature as discrete biological units distinct from all others.

WHAT IS A SPECIES?

Species is a Latin word meaning "kind" or "appearance." Indeed, we learn to distinguish between the kinds of plants or animals—between dogs and cats, for instance—from differences in their appearance. Linnaeus, the founder of modern taxonomy, described individual species in terms of their physical form; the study of form or structure, called morphology, is still the method most often used to characterize species. Modern taxonomists also consider differences in body functions, biochemistry, behavior, and genetic makeup. However, dividing organisms into different species based on comparative data is only part of an extensive effort to better understand the nature of species and the factors that maintain their distinctiveness in nature.

The biological species concept emphasizes reproductive isolation

A classical and widely accepted species definition known as the biological species concept was first enunciated by evolutionary biologist Ernst Mayr in 1942. Mayr's definition addresses the question, What factors divide biological diversity into separate forms that we identify as species?

The **biological species concept** defines a species as a population or group of populations whose members have the potential to interbreed with one another in nature to produce viable, fertile offspring, but who cannot produce viable, fertile offspring with members of other species (FIGURE 24.2). In other words, a biological species is the largest unit of population in which genetic exchange is possible and that is genetically isolated from other such populations. Members of a biological species are united by being reproductively compatible, at least potentially. A businesswoman in Manhattan has little probability of producing offspring with a dairyman in Mongolia, but if the two should get together, they could have viable babies that develop into fertile adults. All humans belong to the same biological species. In contrast, humans and chimpanzees remain distinct biological species even where they share territory, because the two species do not interbreed.

One important exception to the rule of reproductive compatibility stems from biological species being defined in *natural* environments. In the laboratory or in zoos, fertile hybrids can sometimes be produced between species that do not interbreed in nature.

(a) Similarity between different species

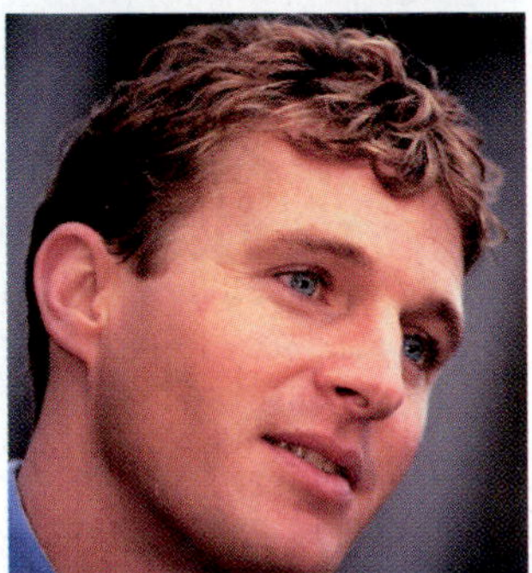

(b) Diversity within a species

FIGURE 24.2 ▪ The biological species concept is based on interfertility rather than physical similarity. (a) The eastern meadowlark (*Sturnella magna,* left) and the western meadowlark (*Sturnella neglecta*) have very similar body shapes and colorations, but they represent different species. Their songs are distinct, a behavioral difference that helps prevent interbreeding between the two species. **(b)** In contrast, all humans, as seemingly diverse as we are, belong to a single species (*Homo sapiens*), defined by our capacity to interbreed.

In summary, the biological species concept hinges on reproductive isolation, with each species isolated by factors (barriers) that prevent interbreeding, thereby blocking genetic mixing with other species.

Prezygotic and postzygotic barriers isolate the gene pools of biological species

Any factor that impedes two species from producing viable, fertile hybrids contributes to reproductive isolation. No single barrier may be completely impenetrable to genetic exchange, but many species are genetically sequestered by more than one type of barrier. Here, we are considering only biological barriers to reproduction, which are intrinsic to the organisms. Of course, if two species are geographically segregated, they cannot possibly interbreed, but a geographical barrier is not considered equivalent to reproductive isolation because it is not intrinsic to the organisms themselves. Reproductive isolation prevents populations belonging to different species from interbreeding, even if their ranges overlap.

Clearly, a fly will not mate with a frog or a fern, but what prevents biological species that are very similar—that is, closely related—from interbreeding? The various reproductive barriers that isolate the gene pools of species can be categorized as prezygotic or postzygotic, depending on whether they function before or after the formation of zygotes, or fertilized eggs.

Prezygotic Barriers

Prezygotic barriers impede mating between species or hinder the fertilization of ova if members of different species attempt to mate.

Habitat Isolation. Two species that live in different habitats within the same area may encounter each other rarely, if at all, even though they are not technically geographically isolated. For example, two species of garter snakes in the genus *Thamnophis* occur in the same areas, but one lives mainly in water and the other is primarily terrestrial. Habitat isolation also affects parasites, which are generally confined to certain plant or animal host species. Two species of parasites living on different hosts will not have a chance to mate.

Behavioral Isolation. Special signals that attract mates, as well as elaborate behavior unique to a species, are probably the most important reproductive barriers among closely related animals. Male fireflies of various species signal to females of their kind by blinking their lights in particular patterns. The females respond only to signals characteristic of their own species, flashing back and attracting the males.

The eastern and western meadowlarks shown in FIGURE 24.2a are almost identical in shape, coloration, and habitat, and their ranges overlap in the central United States. Yet they remain two separate biological species, partly because of the differences in their songs, which enable them to recognize individuals of their own kind. Still another form of behavioral isolation is a courtship ritual specific to a species (FIGURE 24.3).

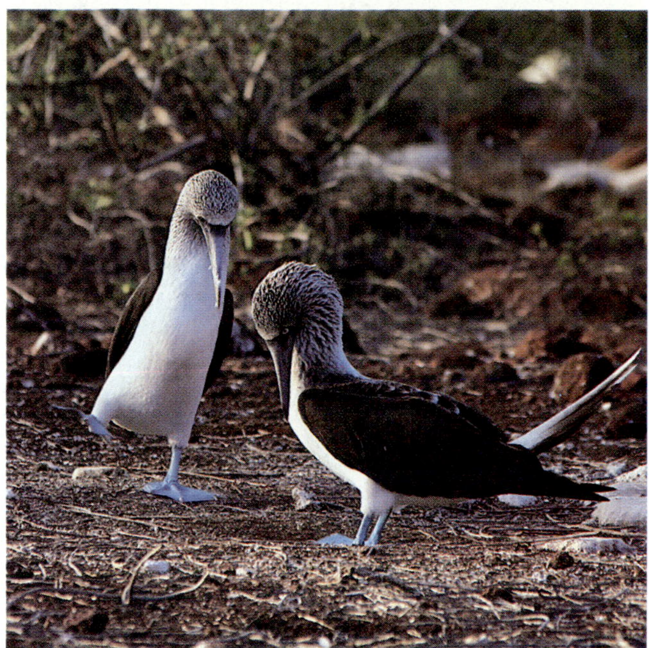

FIGURE 24.3 • Courtship ritual as a behavioral barrier between species. These blue-footed boobies, inhabitants of the Galápagos, will mate only after a specific ritual of courtship displays. Part of the "script" calls for the male to high-step, a behavior that advertises the bright blue feet characteristic of the species.

Temporal Isolation. Two species that breed during different times of the day, different seasons, or different years cannot mix their gametes. The geographical ranges of the western spotted skunk (*Spilogale gracilis*) and the eastern spotted skunk (*Spilogale putorius*) overlap, but these two very similar species do not interbreed because *S. gracilis* mates in late summer and *S. putorius* mates in late winter. Three species of the orchid genus *Dendrobium* living in the same rain forest do not hybridize because they flower on different days. Pollination of each species is limited to a single day because the flowers open in the morning and wither that evening.

Mechanical Isolation. Closely related species may attempt to mate, but fail to consummate the act because they are anatomically incompatible. For example, mechanical barriers contribute to reproductive isolation of flowering plants that are pollinated by insects or other animals. Floral anatomy is often adapted to certain pollinators that transfer pollen only among plants of the same species.

Gametic Isolation. Even if the gametes of different species meet, they rarely fuse to form a zygote. For animals whose eggs are fertilized within the female reproductive tract (internal fertilization), the sperm of one species may not be able to survive in the environment of the female reproductive tract of another species. Many aquatic animals release their gametes into the surrounding water, where the eggs are fertilized (external fertilization). Even when two closely related species release their gametes at the same time in the same place, cross-specific fertilization is uncommon. Gamete recognition may be based on the presence of specific molecules on the coats around the egg, which adhere only to complementary molecules on sperm cells of the same species. A similar mechanism of molecular recognition enables a flower to discriminate between pollen of the same species and pollen of different species.

Postzygotic Barriers

If a sperm cell from one species does fertilize an ovum of another species, then **postzygotic barriers** prevent the hybrid zygote from developing into a viable, fertile adult.

Reduced Hybrid Viability. When prezygotic barriers are crossed and hybrid zygotes are formed, genetic incompatibility between the two species may abort development of the hybrid at some embryonic stage. Of the numerous species of frogs belonging to the genus *Rana,* some live in the same regions and habitats, where they may occasionally hybridize. But the hybrids generally do not complete development, and those that do are frail.

Reduced Hybrid Fertility. Even if two species mate and produce hybrid offspring that are vigorous, reproductive isolation is intact if the hybrids are completely or largely sterile. Since the infertile hybrid cannot backbreed with either parental species, genes cannot flow freely between the species. One cause of this barrier is a failure of meiosis to produce normal gametes in the hybrid if chromosomes of the two parent species differ in number or structure. A familiar case of a sterile hybrid is the mule, a robust cross between a horse and a donkey; horses and donkeys remain distinct species because, except very rarely, mules cannot backbreed with either species (FIGURE 24.4).

Hybrid Breakdown. In some cases when species cross-mate, the first-generation hybrids are viable and fertile, but when these hybrids mate with one another or with either parent species, offspring of the next generation are feeble or sterile. For example, different cotton species can produce fertile hybrids, but breakdown occurs in the next generation when offspring of the hybrids die as seeds or grow into weak and defective plants.

FIGURE 24.5 summarizes the reproductive barriers between closely related species.

FIGURE 24.4 ▪ Hybrid sterility, a postzygotic barrier. A horse (left) can mate with a donkey (center) to produce a viable hybrid offspring in the form of a mule (right). This does not blur the distinction between horses and donkeys as separate species because the mule is sterile, and thus genes cannot flow between the two parental species.

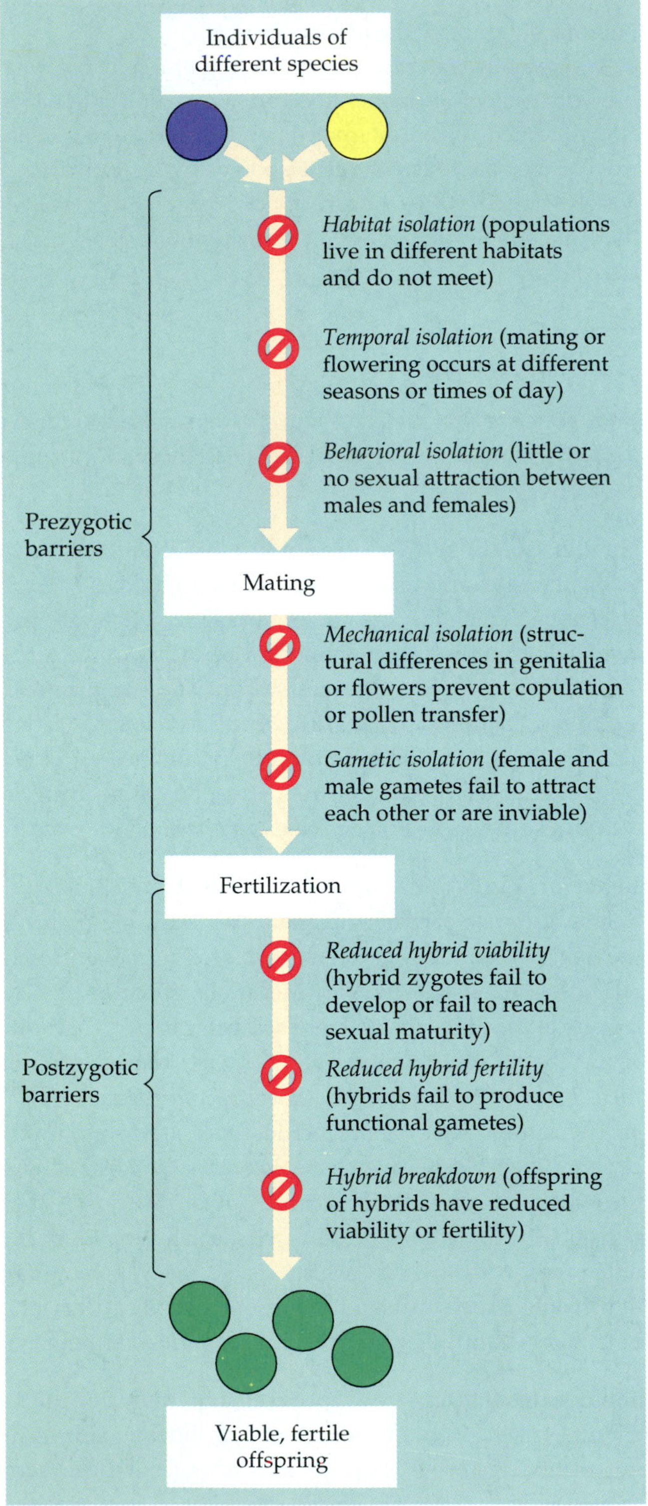

FIGURE 24.5 ▪ A summary of reproductive barriers between closely related species.

The biological species concept does not work in all situations

Having examined the biological species concept and the reproductive barriers upon which it depends, let's now examine the concept's limitations. In doing so, we will begin to see why a number of evolutionary biologists question whether the biological species concept, with its emphasis on reproductive isolation, is too rigid to be widely useful.

Clearly, the biological species concept does not work in all situations. For instance, it is inadequate for grouping extinct forms of life, the fossils of which must be classified according to morphology. The criterion of interbreeding is meaningless for organisms that are completely asexual in their reproduction, as are all prokaryotes, some protists, some fungi, and even some plants (such as the commercial banana) and some animals (including certain lizards and other vertebrates). Many bacteria do transfer genes on a limited scale by conjugation and other processes, but there is nothing akin to the equal contribution of genetic material from two parents that occurs in sexual reproduction. Different lineages of descent give rise to clones, which, genetically speaking, represent single individuals. Asexual organisms can be assigned to species only by the grouping of clones that have the same structural and biochemical characteristics.

Even for populations that are sexual, contemporaneous, and geographically contiguous, in some cases the biological species concept may be too rigid to apply. For example, even though coyotes interbreed with domestic dogs and wolves and produce fertile hybrid offspring, all three remain distinct.

Four subspecies of the deer mouse (*Peromyscus maniculatus*) illustrate a more complex situation (FIGURE 24.6). A subspecies is a population or group of populations that lives in one area and has minor differences from populations of the same species found elsewhere in the species' geographical range. The deer mouse subspecies overlap at certain locations, and some interbreeding occurs in these zones of cohabitation; we would therefore consider the populations to belong to the same species according to the biological species concept. The exceptions are the subspecies *Peromyscus maniculatus artemisiae* and *P. m. nebrascensis.* Though their ranges overlap, these two subspecies do not interbreed. Nevertheless, their gene pools are not completely isolated from each other, because each subspecies interbreeds with other neighboring subspecies, and genes could conceivably flow between the *artemisiae* and the *nebrascensis* populations. If the corridors for gene flow were eliminated by extinction of the other two subspecies, then *artemisiae* and *nebrascensis* could be named separate species.

Population biologists are discovering more and more cases where the distinction between subspecies with limited genetic exchange and full biological species with segregated gene pools blurs. It is as though we are catching populations at different stages in their evolutionary descent from common ancestors. This is to be expected if new species usually arise by the gradual divergence of populations.

With an understanding that defining species as discrete units in nature can pose serious problems, let's examine some alternative species concepts that evolutionary biologists are evaluating.

Other species concepts emphasize features and processes that identify and unite species members

There are many cases where the biological species concept clearly applies—that is, reproductive isolation is complete. However, there are numerous examples where reproductive isolation is not complete, and even more situations where we do not yet know enough to say. In fact, most of the species recognized by taxonomists have been designated as

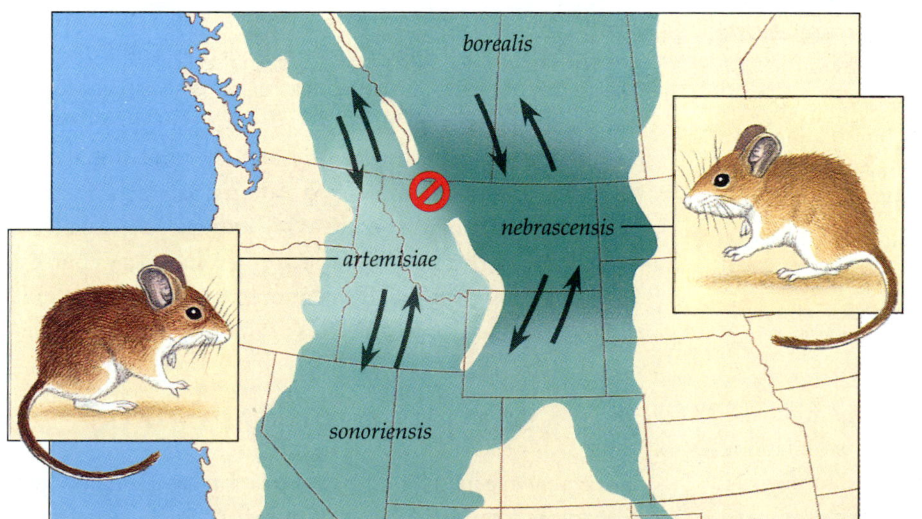

FIGURE 24.6 ▪ Speciation in progress? These four subspecies of the deer mouse (*Peromyscus maniculatus*), found in and around the Rocky Mountains, display geographical variation. Interbreeding between subspecies (indicated by arrows) occurs where their ranges overlap, except for the subspecies *artemisiae* and *nebrascensis.* These two subspecies will not interbreed, yet gene flow between them is possible via the other neighboring subspecies. Such cases may represent speciation in progress; for example, *artemisiae* and *nebrascensis* could legitimately be called different species if, in the future, the populations connecting them become extinct.

separate species based on measurable physical features, not reproductive isolation. This approach, called the **morphological species concept**, is practical to apply in the field, even to fossils.

Several other species concepts have been proposed in recent years. One idea, called the **recognition species concept**, emphasizes mating adaptations that are fixed in a population. A species is defined by a unique set of characteristics that maximize successful mating—the molecular, morphological, and behavioral characteristics that enable an individual to recognize a mate. In contrast to the biological species concept, the recognition concept tends to focus on characteristics that are in fact subject to natural selection. However, its focus on mating limits its application to sexually reproducing species.

Another proposal for defining species, the **cohesion species concept**, focuses on the mechanisms that maintain species as discrete phenotypic entities. Depending on the species, such mechanisms may include reproductive barriers, stabilizing selection, and linkages among sets of genes that make zygotes develop into adult organisms with species-specific characteristics. Asexual reproduction is an effective cohesion mechanism, and in contrast to the biological species concept and the recognition concept, the cohesion concept is applicable to organisms that reproduce without sex. The cohesion concept also acknowledges that interbreeding between some species produces fertile hybrids, and that hybrids sometimes mate successfully with one of their parent species. For example, corn (*Zea mays*) has some alleles that can be traced to a closely related wild grass called teosinte (*Zea mexicana*). Transplanted to corn by hybrids that managed to cross with corn plants, the teosinte alleles increase the reservoir of genetic variation that can be exploited by breeders trying to produce new corn varieties by artificial selection. But occasional hybridization does not erase the boundary between corn and teosinte. The cohesion species concept emphasizes the adaptations that keep parent species intact despite some gene flow between them. We could apply the biological species concept to the corn/teosinte example by making an exception to the rule of reproductive isolation; we could argue that as long as reproductive barriers hold gene transfer to a trickle, the isolation of the two gene pools is not seriously breached, and the two species remain distinct. Posing a larger problem for the isolation rule are situations where hybrid populations dominate part of two species' geographic range (which we will discuss later). By contrast, the cohesion species concept is applicable to all cases involving hybridization.

Yet another definition, the **ecological species concept**, defines species on the basis of where they live and what they do instead of what they look like. An ecological species is defined by a unique role it plays—or its specific position and function—in its environment. For example, two populations of animals that look identical would qualify as two ecological species if each is found only in a specific kind of environment (for example, a freshwater pond with a unique set of chemical, physical, and biological conditions).

Table 24.1 ■ Six Concepts of Species Compared

Biological species concept	Emphasizes reproductive isolation, the potential of members of a species to interbreed with each other but not with members of other species.
Morphological species concept	Emphasizes measurable anatomical differences between species. Most species recognized by taxonomists have been designated as separate species based on morphological criteria.
Recognition species concept	Emphasizes mating adaptations that become fixed in a population as individuals "recognize" certain characteristics of suitable mates.
Cohesion species concept	Emphasizes cohesion of phenotype as the basis of species integrity, with each species defined by its integrated complex of genes and set of adaptations.
Ecological species concept	Emphasizes species' roles (niches), their positions and functions in the environment.
Evolutionary species concept	Emphasizes evolutionary lineages and ecological roles.

Also emphasizing the relationship between organisms and their environment, the **evolutionary species concept** defines a species as a sequence of ancestral and descendant populations that are evolving independently of other such groups. Each evolutionary species has its own separate and unique role in the environment; any particular role involves a specific set of forces of natural selection (called selective pressures). Thus, the populations making up a species are subject to, and united by, a unique set of selective pressures.

TABLE 24.1 reviews the species concepts we have discussed. It is unlikely that any single definition of a species can be extended to cover all cases. The classical biological species concept requires nearly complete isolation of the gene pool, whereas other species concepts generally recognize a greater degree of genetic exchange among species. Each concept may have utility in particular situations, depending on the kinds of questions we are asking about species.

However we attempt to formally define species, we are talking about populations of organisms that are genetically discrete units in nature. Crucial to the origin of such units are processes that promote genetic isolation—that initiate and perpetuate the isolation of a population's gene pool from the gene pool of a parent species. With its gene pool isolated, a splinter population can follow its own evolutionary course as selection, genetic drift, and mutations produce changes in its allele frequencies. What processes promote genetic isolation?

MODES OF SPECIATION

There are two general modes of speciation based on how gene flow among populations is interrupted. In a speciation mode termed **allopatric speciation** (Gr. *allos,* "other," and L. *patria,* "homeland"), a geographical barrier that physically isolates populations initially blocks gene flow. Populations segregated by a geographical barrier are known as allopatric populations. In the second speciation mode, called **sympatric speciation** (Gr. *sym,* "together"), intrinsic factors, such as chromosomal changes (in plants) and nonrandom mating (in animals), alter gene flow. Sympatric populations become genetically isolated even though their ranges overlap.

Geographical isolation can lead to the origin of species: allopatric speciation

Geographical Barriers

Geological processes can fragment a population into two or more isolated populations. A mountain range may emerge and gradually split a population of organisms that can inhabit only lowlands; a creeping glacier may gradually divide a population; a land bridge, such as the Isthmus of Panama, may form and separate the marine life on either side; or a large lake may subside until there are several smaller lakes with their populations now isolated. If individuals colonize a new, geographically isolated area, the colonizing population may become isolated from the parent population.

Just how formidable a geographical barrier must be to keep allopatric populations apart depends on the organisms' ability to disperse—on the mobility of animals or the dispersibility of spores, pollen, and seeds of plants. The Grand Canyon is easily crossed by hawks and many other birds, but it is an impassable barrier to populations of small rodents confined to either the north or south rim of the canyon. Indeed, the same bird species populate both rims of the canyon, but each rim has several unique species of rodents (FIGURE 24.7).

Let's consider an example of how geographical isolation can lead to allopatric speciation. About 50,000 years ago, during an ice age, what is now the Death Valley region of California and Nevada had a very rainy climate and a system of interconnected lakes and rivers. A drying trend began about 10,000 years ago, and by 4000 years ago the region had become a desert. Today, all that is left of the network of lakes and rivers is isolated springs scattered in the desert, mostly in deep clefts between rocky walls. The springs vary extensively in water temperature and salinity. Living in many of the springs are tiny fishes, called pupfishes, of the genus *Cyprinodon.* Each inhabited spring, often no more than a few meters in diameter, is home to its own species of pupfish adapted to that pool and found nowhere else in the world. The various pupfishes probably descended from a single ancestral species whose range was broken up when the region became arid, cloistering several small populations that subsequently diverged from one another.

Conditions Favoring Allopatric Speciation

Whenever populations become allopatric, it is possible for speciation to occur as the isolated gene pools accumulate genetic differences by microevolution. But an isolated population that is small is more likely than a large population to change substantially enough to become a new species.

The geographical isolation of a small population usually occurs at the fringe of the parent population's range. The splinter population, or *peripheral isolate,* is a good candidate for speciation for three reasons:

1. The gene pool of the peripheral isolate probably differs from that of the parent population from the outset. Living near the border of the range, the peripheral isolate represents the extremes of any genotypic clines that existed in the original population. And if the peripheral isolate is small, there will be a founder effect resulting in a gene pool that is not representative of the gene pool of the parent population.
2. Until the peripheral isolate becomes a large population, genetic drift will continue to change its gene pool at random. New mutations or combinations of existing alleles that are neutral in adaptive value may become fixed in

FIGURE 24.7 ▪ Allopatric speciation of squirrels in the Grand Canyon. One example of allopatric speciation is the two species of antelope squirrels that inhabit opposite rims of the Grand Canyon. On the south rim is Harris's antelope squirrel (*Ammospermophilus harrisi*). A few miles away on the north rim is the closely related white-tailed antelope squirrel (*Ammospermophilus leucurus*), which is slightly smaller and has a shorter tail that is white underneath. Birds and other organisms that can disperse easily across the canyon have not diverged into different species on opposite rims.

the population by chance alone, causing genotypic and phenotypic divergence from the parent population.

3. Evolution caused by natural selection may take a different direction in the peripheral isolate than in the parent population. Because the peripheral isolate inhabits a frontier, where the environment is somewhat different, the peripheral isolate will probably encounter selection factors that are different from, and generally more severe than, those operating on the parent population.

These factors may cause peripheral isolates to follow an evolutionary course that diverges from that of the parent population, as long as the gene pools remain isolated. This does not mean that all peripheral isolates persist long enough or change enough to become new species. Life on the frontier is usually harsh, and most pioneer populations probably become extinct. As evolutionary biologist Stephen Jay Gould puts it: "Status as a peripheral isolate merely gives a lottery ticket to a small population. A population can't win (speciate) without a ticket, but there are very few winning tickets."

Adaptive Radiation on Island Chains

Islands are living laboratories for the study of speciation. Flurries of allopatric speciation have occurred on island chains where organisms that have strayed or become passively dispersed from their parent populations have founded new populations that evolved in isolation (FIGURE 24.8). The many indigenous species of the Galápagos descended from stragglers that floated, flew, or were blown over the sea from the South American mainland. For example, consider the Galápagos finches. A single dispersal event may have seeded one island with a small population of the ancestral finch, and this peripheral isolate formed a new species. Later, a few individuals of this island species may have reached neighboring islands, where geographical isolation permitted additional speciation episodes. After diverging on one of these other islands, a new species could recolonize the island from which its founding population emigrated and coexist there with its parent species, or form still another species. Multiple invasions of islands by peripheral isolates of species from neighboring islands would eventually lead to the coexistence of several species on each island. The islands are far enough apart to permit populations to evolve in isolation, but close enough together for occasional dispersion events to occur.

The evolution of many diversely adapted species from a common ancestor is called **adaptive radiation** (FIGURE 24.9). Adaptive radiation of the Galápagos finches is evident in the many types of bills specialized for different foods (see FIGURE 22.4).

The Hawaiian Archipelago is one of the world's great showcases of evolution. The volcanic islands are about 3500 km from the nearest continent. They become progressively younger to the southeast, terminating with the youngest and largest island, Hawaii, which is less than a million years old and still has active volcanoes. Each island was born naked and was gradually populated by species derived from strays that rode the ocean currents and winds either from distant islands and continents or from older islands of the archipelago itself. The physical diversity of each island, including a range of altitudes and extensive differences in rainfall, provides many environmental opportunities for evolutionary divergence by natural selection. Multiple invasions and allopatric speciations have ignited an explosion of adaptive radiation; most of the thousands of species of plants and animals that now inhabit the islands are found nowhere else in the world. In contrast, there are no indigenous species on the Florida Keys. Apparently, those islands are so close to the mainland that founding populations are not sufficiently sequestered for their gene pools to become isolated from the steady stream of immigrants from the parent populations on the mainland.

FIGURE 24.8 ▪ Long-distance dispersal. Plant seeds (the black dots) cling to this sea bird, which is capable of long flights. This is one mechanism that can disperse terrestrial organisms to isolated islands.

A new species can originate in the geographical midst of the parent species: sympatric speciation

24.1 In sympatric speciation, new species arise within the range of parent populations; genetic isolation evolves in a variety of ways without geographical isolation. For instance, in a single generation a new species (defined by the biological species concept) can be generated if a genetic change results in a reproductive barrier between the mutants and the parent population. Many plant species have their origins in accidents during cell division that result in extra sets of chromosomes, a mutant condition called **polyploidy**. An **autopolyploid** (Gr. *autos*, "self") is an individual that has more than two chromosome sets, all derived from a single species. For example, a failure of meiosis during gamete production can double chromosome number from the diploid count ($2n$) to a tetraploid number ($4n$) (FIGURE 24.10a, p. 454). The tetraploid can then fertilize itself (self-pollinate) or mate with other tetraploids. However, the mutants cannot interbreed successfully with diploid plants of the original population; the offspring, which would be triploid ($3n$), would be sterile because unpaired chromosomes result in abnormal meiosis. In just one generation, a postzygotic barrier has caused reproductive isolation and has interrupted gene flow between a tiny population of tetraploids (maybe only a single plant initially) and the parent diploid population that surrounds it.

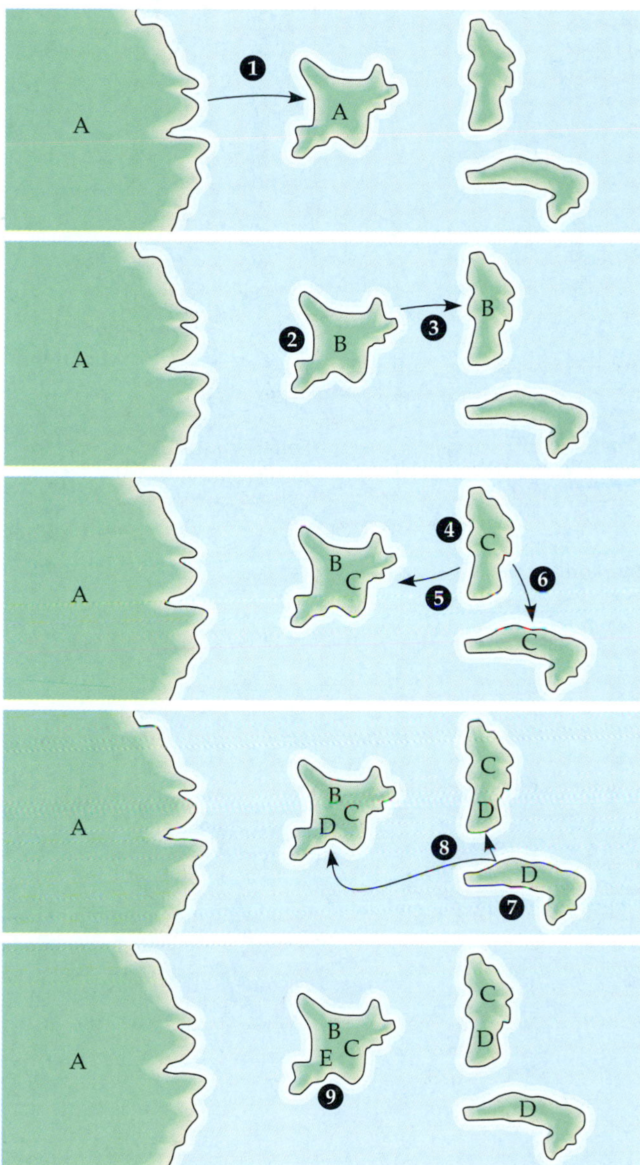

FIGURE 24.9 ▪ A model for adaptive radiation on island chains. ① One island in this cluster of three is seeded by a small colony founded by individuals of species A, blown over from a mainland population. ② Its gene pool now isolated from the parent species, the island population evolves into species B as it adapts to its new environment. ③ Storms or other agents of dispersion spread species B to a second island, ④ where the isolated colony evolves into species C. ⑤ Later, individuals from species C recolonize the first island and cohabit with species B, but reproductive barriers keep the species distinct. ⑥ A colony of species C may also populate a third island, ⑦ where it adapts and forms species D. ⑧ Species D is dispersed to the two islands of its ancestors, ⑨ and evolves into a new species, E, on one of those islands. The story could go on and on, with a series of allopatric speciation episodes made possible by the combination of isolation and occasional dispersal.

Sympatric speciation by autopolyploidy was first discovered in the early 1900s by geneticist Hugo de Vries while he was studying the genetics of the evening primrose, *Oenothera lamarckiana,* a diploid species with 14 chromosomes. One day, de Vries noticed an unusual variant that had appeared among his plants, and microscopic inspection revealed that it was a tetraploid with 28 chromosomes. He found that the plant was unable to breed with the diploid primrose, and he named the new species *Oenothera gigas.*

Another type of polyploid species, much more common than autopolyploids, is called an **allopolyploid**, referring to the contribution of two *different* species to a polyploid hybrid (FIGURE 24.10b). The origin of an allopolyploid begins when two different species interbreed and combine their chromosomes. Interspecific hybrids are usually sterile because the haploid set of chromosomes from one species cannot pair during meiosis with the haploid set from the other species. Though infertile, a hybrid may actually be more vigorous than its parents and propagate itself asexually (which many plants can do). At least two mechanisms, illustrated in FIGURE 24.10b, can transform the sterile hybrids into fertile polyploids.

Some allopolyploids are especially vigorous, perhaps because they combine the best qualities of their two parent species. Speciation of polyploids, especially allopolyploids, accounts for 25% to 50% of plant species. Some of these new polyploid species have originated and spread within historic times. An example is a new species of salt-marsh grass that originated along the coast of southern England in the 1870s. It is an allopolyploid derived from a European species (*Spartina maritima*) and an American species (*Spartina alternaflora*). Seeds of the American species, stowaways in the ballast of ships, were accidentally introduced to England in the early nineteenth century. The invaders hybridized with the local species, and eventually a third species (*Spartina anglica*), morphologically distinct and reproductively isolated from its two parent species, evolved as an allopolyploid. Chromosome numbers are consistent with this mechanism of speciation: For *S. maritima,* $2n = 60$; for *S. alternaflora,* $2n = 62$; and for the new species, *S. anglica,* $2n = 122$. Since its inception, the new marsh grass has spread around the coast of Great Britain, clogging estuaries and becoming something of a pest.

Many of the plants we grow for food are polyploids. Oats, cotton, potatoes, tobacco, and wheat are among the polyploid species important to agriculture. The wheat used for bread, *Triticum aestivum,* is an allopolyploid that probably originated about 8000 years ago as a spontaneous hybrid of a

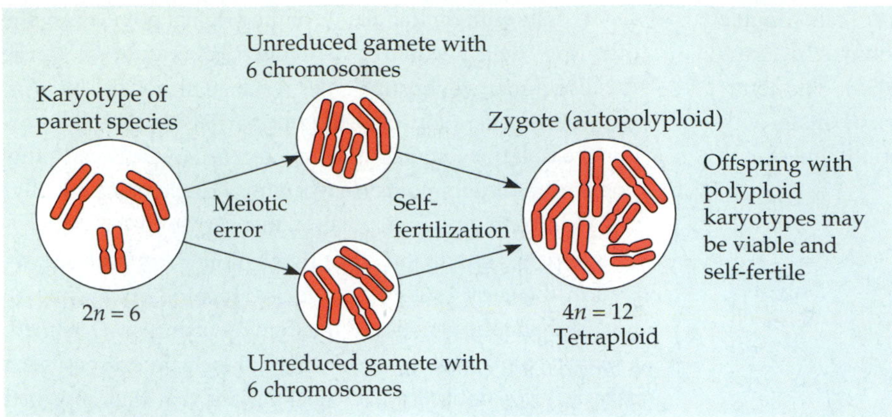

(a) Autopolyploidy

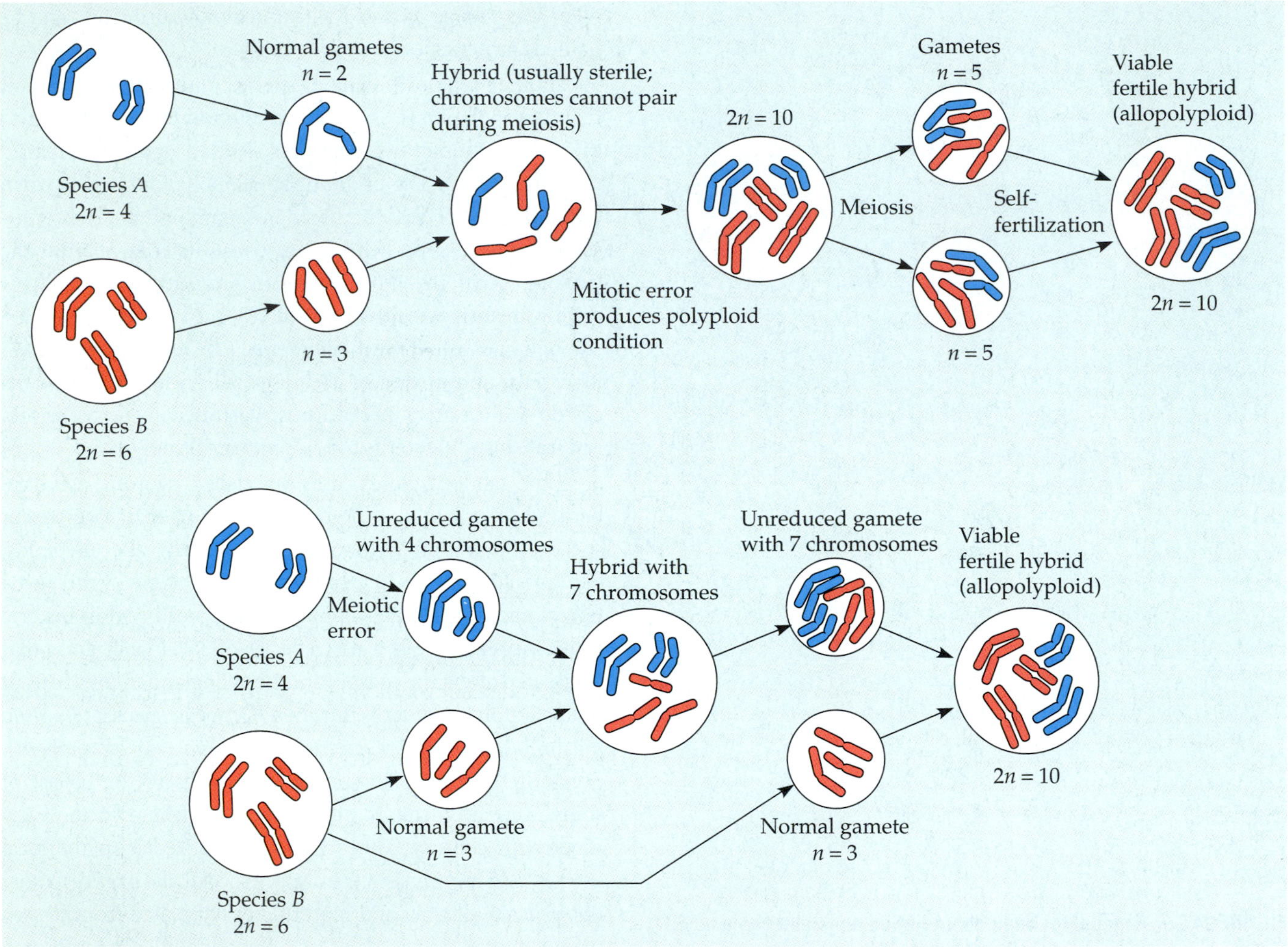

Normal gametes
$n = 2$
Hybrid (usually sterile;
chromosomes cannot pair
during meiosis)
$2n = 10$
Gametes
$n = 5$
Viable
fertile hybrid
(allopolyploid)
Species *A*
$2n = 4$
Meiosis
Self-
fertilization
Mitotic error
produces polyploid
condition
$2n = 10$
$n = 3$
$n = 5$
Species *B*
$2n = 6$
Unreduced gamete
with 4 chromosomes
Unreduced gamete
with 7 chromosomes
Hybrid with
7 chromosomes
Viable
fertile hybrid
(allopolyploid)
Meiotic
error
Species *A*
$2n = 4$
$2n = 10$
Normal gamete
$n = 3$
Normal gamete
$n = 3$
Species *B*
$2n = 6$

(b) Allopolyploidy

24.1 **FIGURE 24.10 • Sympatric speciation by polyploidy in plants. (a) Autopolyploidy.** An error during meiosis (or mitosis) in the germ (reproductive) cell line results in gametes in which the number of chromosomes is not reduced from the number of chromosomes in the parent cell. Self-fertilization produces a tetraploid zygote. **(b) Allopolyploidy.** A hybrid between two species is normally sterile because its chromosomes are not homologous and cannot pair during meiosis. However, the hybrids may be able to reproduce asexually. Two different mechanisms can produce allopolyploid species from such hybrids. (Top) At some instant in the history of the hybrid clone, a mitotic error affecting the reproductive tissue of an individual may double chromosome number. The hybrid will then be able to make gametes because each chromosome has a homologue with which to synapse during meiosis. Union of gametes from this hybrid may give rise to a new species of interbreeding plants, reproductively isolated from both parent species. (Bottom) This diagram illustrates a more complex, but probably more common, origin of allopolyploid species. Both mechanisms produce new species having a chromosome number equal to the sum of the chromosomes in the two parent species. These various kinds of accidents during cell division make it possible for a new species to arise without geographical isolation from parent species.

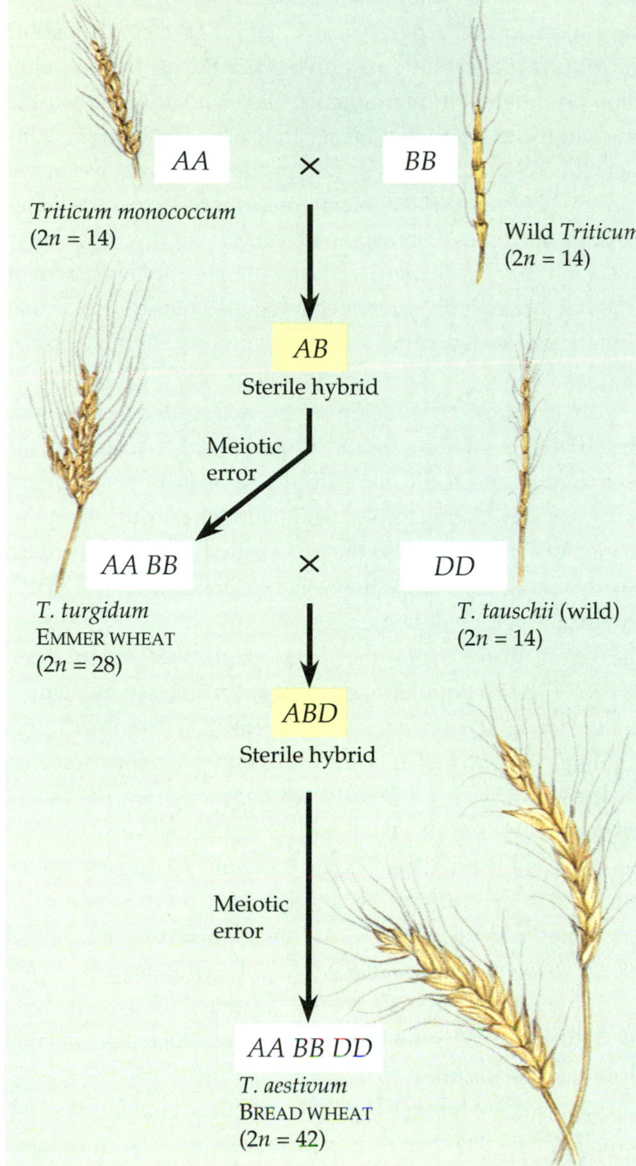

FIGURE 24.11 • The evolution of wheat (*Triticum*). Bread wheat is the evolutionary product of two hybridization/meiotic error episodes producing allopolyploids. The first sympatric speciation produced emmer wheat, derived from hybridization between wild *Triticum* and a domesticated species (*T. monococcum*) that has been cultivated for food in the Middle East for at least 11,000 years. A second hybridization, between emmer wheat and a wild species (*T. tauschii*), gave rise to bread wheat about 8000 years ago. The letters *A, B,* and *D* represent chromosomes that can be traced to a particular species. Thus, bread wheat has chromosomes derived from three different ancestral species.

cultivated wheat and a wild grass (FIGURE 24.11). Plant geneticists now hybridize plants and use chemicals that induce meiotic and mitotic errors to help create new polyploids with special qualities. For example, artificial hybrids combine the high yield of wheat with the ability of rye to resist disease.

Sympatric speciation may also occur in animal evolution, though the mechanisms are different from the chromosome doublings of plants. Animals may become reproductively isolated within the geographical range of a parent population if genetic factors cause them to become fixed on resources not used by the parent population. Consider, for example, the wasps that pollinate figs. Each fig species is pollinated by a particular species of wasp, which mates and lays its eggs in the figs. A genetic change that caused wasps to select a different fig species would segregate mating individuals of this new phenotype from the parent population. This would set the stage for further evolutionary divergence. A balanced polymorphism coupled with assortative mating could also result in sympatric speciation (see Chapter 23). For example, if birds in a population that is dimorphic for bill size began to mate selectively with birds of the same morph, then eventual speciation is possible. Lake Victoria in Africa, which is less than a million years old, is home to almost 200 species of closely related fishes belonging to the cichlid family. Subdivision of populations into groups specialized for exploiting different food sources and other resources in the lake is one process that probably contributed to the explosive radiation of these fishes.

Genetic change in populations can account for speciation

Classifying speciation as allopatric or sympatric accents biogeographical factors and genetic mechanisms that can differentiate populations into separate species. Now we examine the changes required for the origin of species in the context of the population genetics presented in Chapter 23.

Adaptive Divergence

When two populations adapt to disparate environments, they accumulate differences in their gene pools—differences in the frequencies of alleles and genotypes. In the course of this gradual adaptive divergence of two gene pools, reproductive barriers between the two populations may evolve coincidentally, differentiating the populations into two species.

A key idea in evolution by divergence is that reproductive barriers can arise without being favored directly by natural selection—there is no drive toward speciation for its own sake. Reproductive isolation is usually a secondary result of the divergence of two populations as they adapt to separate environments. For example, postzygotic barriers may result from genes with multiple phenotypic effects (see pleiotropy, Chapter 14). In one such case, laboratory hybrids between two very similar species of *Drosophila, D. melanogaster* and *D. simulans,* inherit only one active set of two sets of genes needed for the synthesis of ribosomal RNA. The side effect—a very low viability for the hybrids—is a postzygotic barrier between the two species.

Prezygotic barriers can also evolve as side effects of the gradual genetic divergence of two populations. For instance, if a change in gene frequency enables one population of an insect species to adapt to a different host plant than that of other populations, then a habitat barrier to interbreeding with the other populations is a side effect.

FIGURE 24.12 ▪ The role of sexual selection in the radiation of Hawaiian *Drosophila* species. Unique courtship rituals, usually dependent on specific morphological features of the males, contribute to prezygotic barriers between closely related species. In *D. heteroneura,* shown here as an example, males have "hammerheads," with eyes set far apart. This trademark probably functions in species recognition as the male faces the female during a courtship dance. The "pictures" on the wings, species-unique patterns of dots, may also contribute to mate recognition.

There *are* cases where reproductive isolation evolves more directly by sexual selection in isolated populations. Such selection may have helped separate *Drosophila* on the Hawaiian Islands into hundreds of species. For example, the wide head of the *D. heteroneura* male enhances reproductive success with females of the same species, but makes successful courtship with females of other species very unlikely (FIGURE 24.12). However, even when sexual selection results in reproductive barriers, such barriers evolve as adaptations that enhance reproductive success within a single population, not as safeguards against interbreeding with other populations.

As we have discussed, a major criticism of the biological species concept is its emphasis on reproductive isolation. We now see that reproductive barriers can be side effects or spin-offs of speciation by adaptive divergence. By contrast, the characteristics emphasized by the recognition species concept—those that maximize successful mating with members of the same population—are actually subject to natural selection. For some evolutionary biologists, this is a reason to favor the recognition concept.

Hybrid Zones and the Cohesion Species Concept

What happens when two closely related populations that have been allopatric for some time come back into contact? Several outcomes are possible. The two populations may interbreed freely and produce fertile hybrid offspring, in which case their gene pools will fuse into one; speciation has not occurred during the period of geographical isolation. A second possible outcome is that evolutionary divergence during the period of allopatry resulted in reproductive barriers that keep the gene pools of the two populations separate even if they are in contact; the speciation process has produced a new biological species. Between these two extremes, there are several other possible outcomes. For instance, evolutionary divergence may not result in reproductive isolation but in the establishment of a set of behavioral and morphological characteristics that define a species according to the recognition concept. Another possible outcome is the establishment of a hybrid zone.

A **hybrid zone** is a region where two related populations that diverged after becoming geographically isolated make secondary contact and interbreed where their geographical ranges overlap. For example, two phenotypically distinct populations of woodpeckers, the red-shafted flicker in western North America and the yellow-shafted flicker in central North America, interbreed along a hybrid zone stretching through the Great Plains from the Texas panhandle to southern Alaska. The two populations probably became separated during the ice ages, renewing contact at least a few centuries ago. Although the two flicker populations have been interbreeding in that hybrid zone for at least two hundred years, the movement of alleles between the populations does not penetrate far beyond the hybrid zone. The hybrid zone is relatively stable, rather than expanding. Put another way, the frequencies of genotypes and phenotypes that distinguish the red-shafted flicker from the yellow-shafted flicker grade sharply in the hybrid zone as steep clines (see Chapter 23). But the two flicker populations remain distinct deep in their ranges, away from the hybrid zone.

Evolutionary biologists debate whether populations that remain distinct in spite of hybrid zones should be considered subspecies or full, separate species. Some researchers who favor species status for such populations argue that stable hybrid zones pose a problem for the biological species concept. On the other hand, species with stable hybrid zones fit the cohesion species concept; if two species can maintain their taxonomic identity even though they hybridize, there must be cohesive forces other than reproductive isolation that hold a species together and prevent its complete merger with a closely related species. Stabilizing selection would restrict phenotypic variation to a range narrow enough to define the species as separate from other species.

How Much Genetic Change Is Required for Speciation?

It is not possible to generalize about the "genetic distance" between closely related species. Some cases involving complete reproductive isolation may result from the cumulative divergence of populations at many gene loci. In other cases, changes at only a few loci produce separate species. For example, two species of Hawaiian *Drosophila, D. silvestris* and

D. heteroneura, differ in alleles at a gene locus determining head shape, a character important in mate recognition by these flies (see FIGURE 24.12). However, the phenotypic effect of different alleles at this locus is amplified by at least ten other gene loci that interact in an epistatic system. (For a review of the genetic interactions known as epistasis, see Chapter 14.) Thus, it took more than one mutation to differentiate these two species of *Drosophila*. However, it is also clear from such examples that massive genetic change involving hundreds of loci is not mandatory for speciation. This conclusion is relevant to a debate among evolutionary biologists about the tempo of speciation, an issue we will now examine.

The punctuated equilibrium model has stimulated research on the tempo of speciation

Traditional evolutionary trees that diagram the descent of species from ancestral forms sprout branches that diverge gradually, each new species evolving continuously over long spans of time (FIGURE 24.13a). Such trees are based on the idea that big changes occur by the accumulation of many small ones. This concept extends the processes of microevolution to the divergence of species. However, paleontologists rarely find gradual transitions of fossil forms. Instead, they often observe species appearing as new forms rather suddenly (in geological terms) in a layer of rocks, persisting essentially unchanged for their tenure on Earth, and then disappearing from the fossil record as suddenly as they appeared. To Darwin, the origin of species was an extension of adaptation by natural selection, with isolated populations from common ancestral stock evolving differences gradually as they adapted to their local environments. But Darwin himself was bewildered by the dearth of connecting fossils and wrote: "Although each species must have passed through numerous transitional stages, it is probable that the periods during which each underwent modification, though many and long as measured by years, have been short in comparison with the periods during which each remained in an unchanged condition."

Within the past 50 years, some evolutionary biologists have addressed the nongradual appearance of species in the fossil record by developing a model now known as **punctuated equilibrium**. According to this model, species diverge in spurts of relatively rapid change, instead of slowly and gradually (FIGURE 24.13b). In other words, species undergo most of their morphological modification as they first bud from parent species, and then change little, even as they give rise to additional species. The term *punctuated equilibrium* is derived from the idea of long periods of stasis (equilibrium) punctuated by episodes of speciation.

Is it likely that most species evolve abruptly and then remain essentially unchanged for most of their existence? One mechanism of sudden speciation is a change in a genome, such as by polyploidy. Also, mutation of the genes that regulate embryonic development may be associated with changes that can generate new species. Proponents of punctuated equilibrium point out that allopatric speciation can also be quite rapid; genetic drift and natural selection can cause significant change in just a few hundred to a few thousand generations in the gene pool of a population cloistered in a challenging new environment.

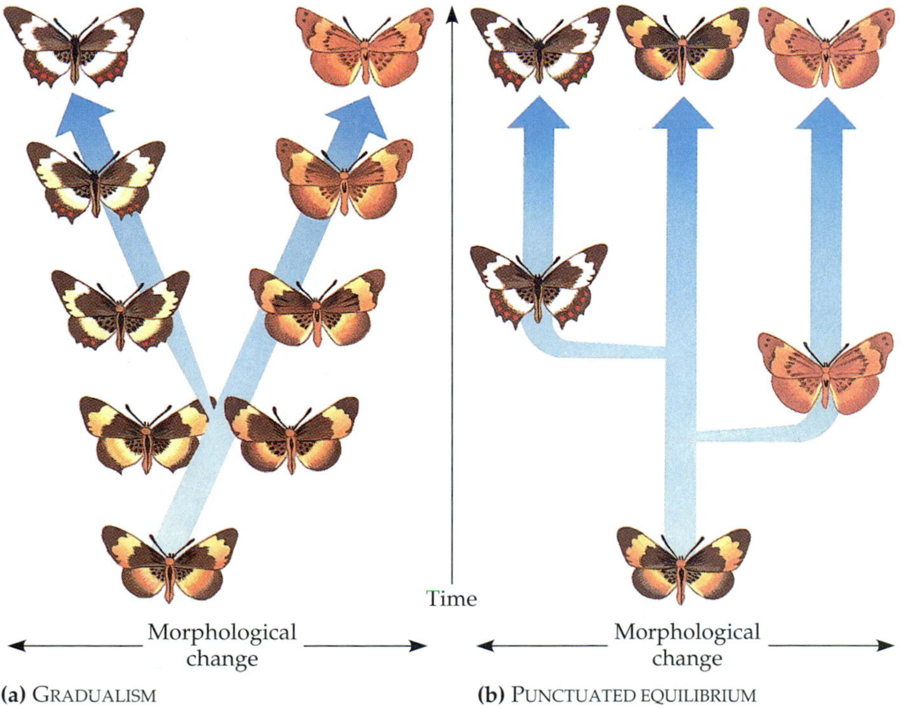

FIGURE 24.13 ▪ Two models for the tempo of speciation. A new species buds from its parent species as a small isolated population, indicated in this figure by the narrow bases of the branching arrows. **(a)** In gradualism, species descended from a common ancestor gradually diverge more and more in morphology as they acquire unique adaptations. **(b)** According to the theory of punctuated equilibrium, a new species changes most as it buds from a parent species, and then changes little for the rest of its existence.

How can speciation in a few thousand generations, which may require several thousand years, be called an abrupt episode? The fossil record indicates that successful species last for a few million years, on average. Suppose that a particular species survives for 5 million years, but most of its morphological changes occurred during the first 50,000 years of its existence. In this case, the evolution of the species-defining characteristics was compressed into just 1% of the lifetime of the species. On the time scale that can generally be determined in fossil strata, the species will appear suddenly in rocks of a certain age and then linger with little or no change before becoming extinct. During its formative millennia, the species may have accumulated its modifications gradually, but relative to the overall history of the species, its inception was abrupt. This scenario of an evolutionary spurt preceding a much longer period of morphological stasis would explain why paleontologists find relatively few smooth transitions in the fossil record of species.

Once it is acknowledged that "sudden" may be many thousands of years on the vast scale of geological time, the debate over the rate of speciation is muted somewhat. The degree to which a species changes after its origin is another issue. If the species is adapted to an environment that stays the same, then natural selection would counter changes in the gene pool. In this view, stabilizing selection tends to hold a population in a long period of stasis.

Some gradualists retort that stasis is an illusion. Many species may continue to change after they come into existence, but in ways that cannot be detected from fossils. By necessity, paleontologists base their theories of descent almost entirely on external anatomy and skeletons. Changes in internal anatomy and functions may go unnoticed, as would modifications in behavior.

Controversy over the punctuated equilibrium model has stimulated research and catalyzed a new interest in paleontology. Researchers generally acknowledge a need for more studies of fossils where specific lineages are well preserved. Only then will we be able to assess the relative importance of gradual and punctuated tempos in the origin of new species and higher taxonomic groups.

THE ORIGIN OF EVOLUTIONARY NOVELTY

What processes cause the large-scale evolutionary changes that we can trace through the fossil record? How, for instance, do the novel features that define taxonomic groups above the species level, such as the flight adaptations of birds, arise? What accounts for evolutionary trends that appear from the fossil record to be progressive, such as the increase in brain size during human evolution? The next three concepts focus on processes relevant to such questions about macroevolution.

Most evolutionary novelties are modified versions of older structures

The fossil record indicates that birds evolved from a lineage of earthbound dinosaurs. How could flying vertebrates evolve from flightless ancestors? In a more general sense, how do major evolutionary novelties evolve? One mechanism is the gradual refinement of existing structures for new functions.

Most biological structures have an evolutionary plasticity that makes alternative functions possible. The term **exaptation** refers to a structure that evolved in one context and became co-opted for another function. This concept does not imply that a structure somehow evolves in anticipation of future use. Natural selection cannot predict the future and can only improve a structure in the context of its current utility. The lightweight, honeycombed bones of birds (see FIGURE 1.6) are homologous to the bones of the earthbound ancestors of birds. However, honeycombed bones could not have evolved in the ancestors as an adaptation for upcoming flights. If light bones predated flight, as is clearly indicated by the fossil record, then they must have had some function on the ground. The probable ancestors of birds were agile, bipedal dinosaurs that also would have benefited from a light frame. It is possible that winglike forelimbs and feathers, which increased the surface area of these forelimbs, were also co-opted for flight after functioning in some other capacity, such as social displays—in courtship, for example. The first flights may have been only extended hops in pursuit of prey or escape from a predator. Once flight itself became an advantage, natural selection would have remodeled feathers and wings to better fit their additional function.

Exaptation offers one explanation for how novel features can arise gradually through a series of intermediate stages, each of which has some function in the organism's current context. Harvard zoologist Karel Liem puts it this way: "Evolution is like modifying a machine while it's running." The concept that evolutionary novelties can evolve by the remodeling of old structures for new functions is in the Darwinian tradition of large changes being an accumulation of many small changes crafted by natural selection.

Genes that control development play a major role in evolutionary novelty

The evolution of complex structures, such as wings and feathers, from their antecedents required so much remodeling that they may have involved a large number of gene loci. In other cases, relatively few changes in the genome can cause major structural modifications. How can slight genetic divergence become magnified into major differences between organisms? Scientists working at the interface of developmental biology and evolutionary biology are finding some answers to this question.

Genes that program development of an organism control the rate, timing, and spatial pattern of changes in an organism's form as it is transfigured from a zygote into an adult. For example, **allometric growth** (Gr. *allos*, "other," and *metron*, "measure"), a difference in the relative rates of growth of various parts of the body, helps shape an organism. FIGURE 24.14a tracks how allometric growth alters human body proportions during development. Change these relative rates of growth even slightly, and you change the adult form substantially. For example, different allometric patterns contribute to the contrasting shapes of human and chimpanzee skulls (FIGURE 24.14b). Allometry is one mechanism by which a subtle alteration of development becomes compounded in its effects on the adult.

In addition to affecting growth rates, genetic changes can also alter the timing of developmental events—the sequence in which different body parts start and stop developing. In some species, changes in developmental timing result in **paedomorphosis** (Gr. *paedos*, "child," and *morphosis*, "shaping"), in which a sexually mature adult retains features that were juvenile structures in its evolutionary ancestors. For example, most salamander species have a larval stage that undergoes metamorphosis to become an adult. But some species grow to adult size and become sexually mature while retaining gills and certain other larval features (FIGURE 24.15). Such an evolutionary alteration of developmental timing can produce animals that appear very different from their ancestors, even though the overall genetic change may be small.

Changes in developmental chronology have also been important in human evolution. Humans and chimpanzees are very closely related through descent from a common ancestor. Of the numerous anatomical differences between these two modern primates that can be attributed to different allometric properties and variations in developmental timing, the most significant contrast is brain size. The human brain is proportionately larger than the chimpanzee brain because growth of the organ is switched off much later in human development.

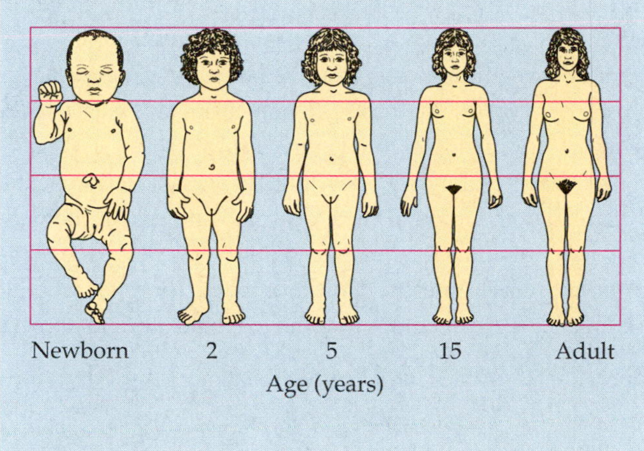

(a) Differential growth rates in a human

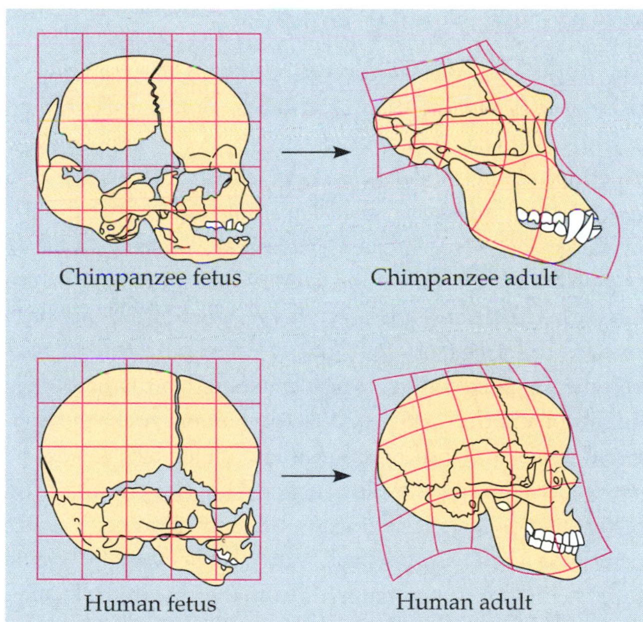

(b) Comparison of chimpanzee and human skull growth

FIGURE 24.14 ▪ Allometric growth. Different growth rates for some parts of the body compared to others determine body proportions. **(a)** During growth of a human, the arms and legs grow faster than the head and trunk, as can be seen in this conceptualization of different-aged individuals all rescaled to the same height. **(b)** The fetal skulls of humans and chimpanzees are similar in shape. Allometric growth of the bones transforms the rounded skull of a newborn chimpanzee into the sloping skull characteristic of adult apes. The same allometric pattern occurs in humans, but to a lesser degree; the adult human skull has departed less than the chimpanzee skull from the fetal shape common to primates.

FIGURE 24.15 ▪ Paedomorphosis. Some species retain as adults features that were juvenile in ancestors. This salamander is an axolotl, which grows to full size, becomes sexually mature, and reproduces while retaining certain larval (tadpole) characteristics, including gills. This figure and FIGURE 24.14 illustrate examples of heterochrony, evolutionary change in the timing or rate of development.

Compared to the brain of chimpanzees, our brain continues to grow for several more years, which can be interpreted as the prolonging of a juvenile process. We owe our culture to this evolutionary novelty, coupled with an extended childhood during which parents and teachers can influence what the young, growing brain stores.

All these examples of temporal changes in development that create evolutionary novelties fit into the category of **heterochrony**, a general term for evolutionary changes in the timing or rate of development.

At least equally important in evolution is **homeosis**, alteration in what biologists are fond of calling the *bauplan* of an organism, its basic body design, or spatial arrangement of body parts. Relatively small sets of genes function as developmental master switches. For instance, homeotic genes initiate developmental events that determine such basic features as where a pair of wings and a pair of legs will develop on a bird, and how a plant's flower parts are arranged. In many cases, experimentally induced mutations in homeotic genes create drastic changes in the bauplan. Similar changes in genes regulating developmental events may have played important roles in evolutionary history. For example, some 520 million years ago, a duplication of a cluster of homeotic genes called the *Hox* complex may have been a seminal event in the origin of vertebrates (animals with backbones) from invertebrates. Vertebrates have multiple sets of these homeotic genes, whereas most invertebrates seem to have only a single cluster of *Hox* genes (FIGURE 24.16; see also Figure 21.14).

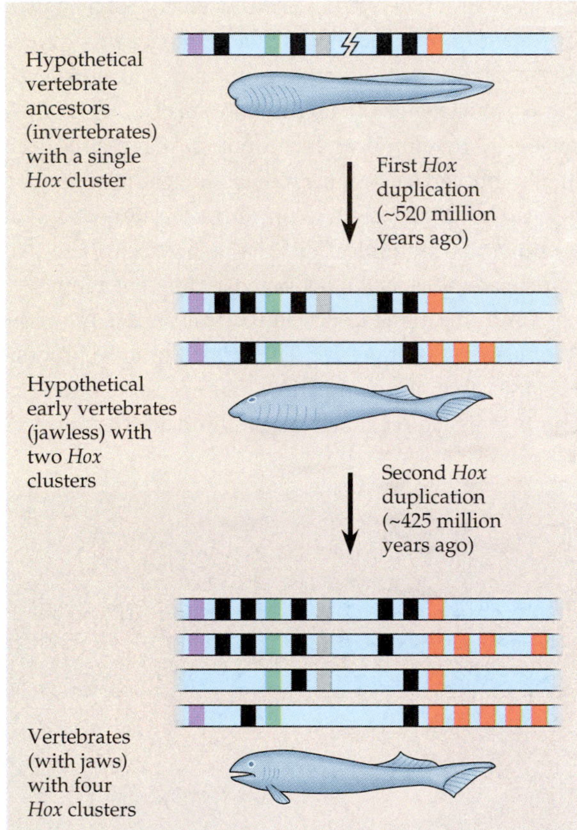

FIGURE 24.16 • Possible macroevolutionary changes associated with mutations in genes regulating development. Most invertebrates have a single cluster of homeotic genes (the *Hox* complex), shown here as colored boxes on a chromosome. The *Hox* genes direct development of major body parts. Researchers postulate that a mutation (duplication) of the single *Hox* complex occurred about 520 million years ago and may have provided genetic material associated with the origin of the first vertebrates. In an early vertebrate, the duplicate set of genes could have taken on entirely new roles, such as directing the development of a backbone, the hallmark of the vertebrates. A second duplication of the *Hox* complex, yielding the four clusters found in most vertebrates, may have occurred later and may have allowed the development of the first jaws in the vertebrate lineage. The vertebrate *Hox* complex contains many of the same genes, which occur in virtually the same linear order on chromosomes, and they direct the sequential development of the same body regions of the animal as the single invertebrate cluster; thus, the vertebrate *Hox* complex appears to be homologous to the single cluster in invertebrates.

In creating evolutionary novelties, changes in developmental dynamics, both temporal (heterochrony) and spatial (homeosis), have undoubtedly played important roles in macroevolution. A lively research effort promises to tell us more about the link between mutations in genes that regulate development and evolutionary history.

An evolutionary trend does not mean that evolution is goal oriented

The fossil record seems to reveal trends in the evolution of many species and lineages. For instance, some lineages exhibit a trend toward larger or smaller body size. A case in point is the evolution of the modern horse, which is a descendant of a much smaller ancestor named *Hyracotherium*. About the size of a large dog, *Hyracotherium* browsed in woodlands some 40 million years ago. In comparison to this ancestor, modern horses (genus *Equus*) are larger, the number of toes has been reduced from four on each foot to one, and the teeth have become modified for grazing on grasses rather than browsing on shrubs and trees. Do these macroevolutionary changes represent trends, and if so, how do we account for such trends?

Extracting a single evolutionary progression from a fossil record that is likely to be incomplete is misleading; it is like describing a bush as growing toward a single point by tracing the system of branches that leads from the base of the bush to one particular twig. For instance, by selecting certain species from the available fossils, it is possible to arrange a succession of animals intermediate between *Hyracotherium* and modern horses that shows trends toward increased size, reduced number of toes, and modification of teeth for grazing (yellow line in FIGURE 24.17). We might interpret this series of fossils as an unbranched lineage leading directly from *Hyracotherium* to modern horses through a continuum of intermediate stages. If we include all fossil horses known today, however, the illusion of coherent, progressive evolution leading directly to modern horses vanishes. The genus *Equus* is the only surviv-

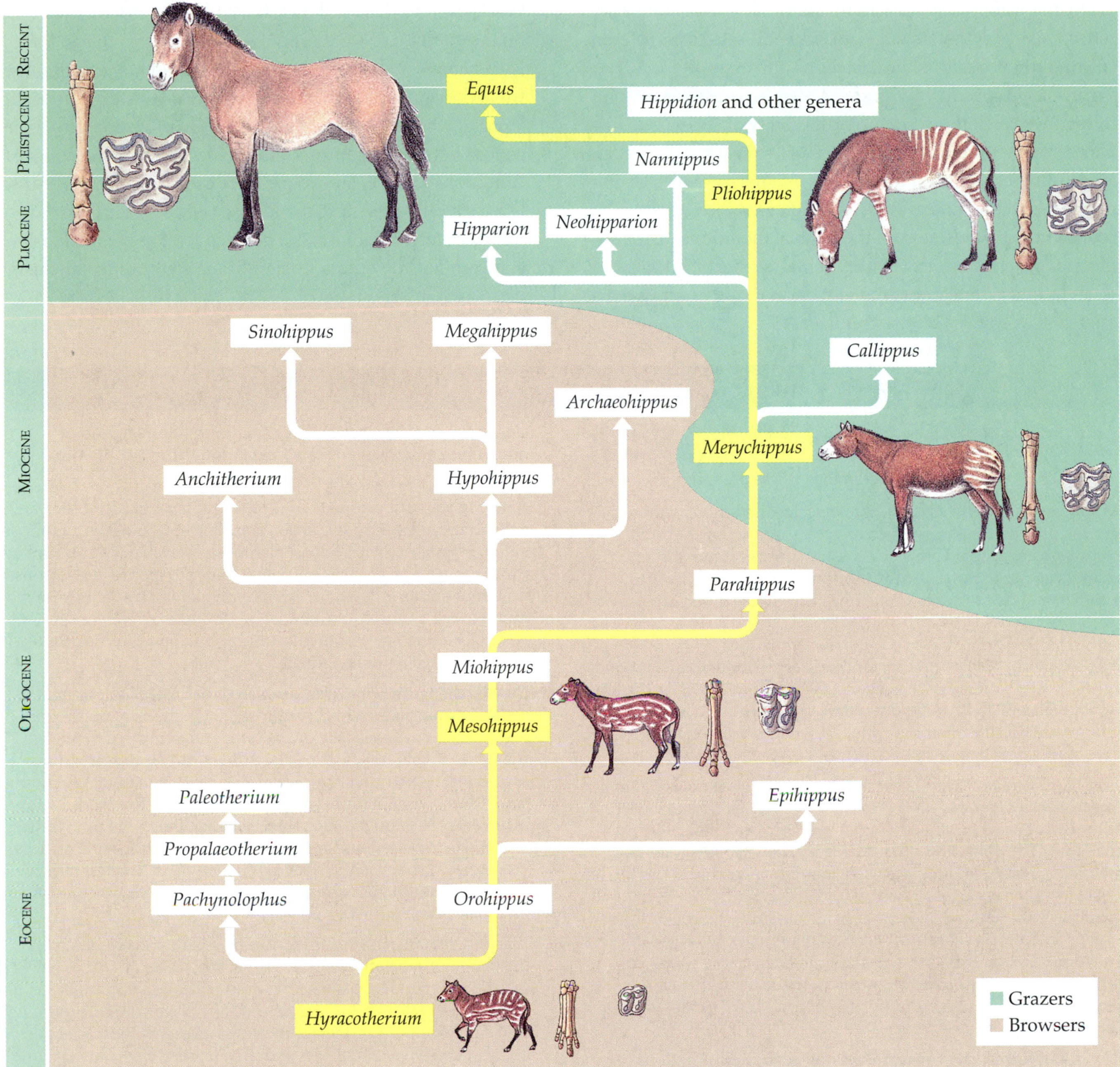

FIGURE 24.17 ▪ The branched evolution of horses. If we use a yellow highlighter to trace one sequence of fossil horses that are intermediate in form between the modern horse and its Eocene ancestor *Hyracotherium,* we create the illusion of a progressive trend toward larger size, reduced number of toes, and teeth modified for grazing. In fact, the modern horse is the only surviving twig of an evolutionary bush with many divergent trends.

ing twig of an evolutionary tree that is so branched it is more like a bush. *Equus* descended through a series of speciation episodes that included several adaptive radiations, not all of which led to large, one-toed, grazing horses. For instance, notice in FIGURE 24.17 that only those lineages derived from *Parahippus* include grazers; other lineages derived from *Miohippus* produced browsers.

Cladogenesis (branching evolution) can produce a macroevolutionary trend even if some new species counter the trend. In a view of macroevolution enunciated by Steven Stanley of Johns Hopkins University, species are analogous to individuals: Speciation is their birth, extinction is their death, and new species are their offspring. According to this model, an evolutionary trend is produced by **species selection,** which is analogous to the production of a trend within a population by natural selection. The species that endure the longest and generate the greatest number of new species determine the direction of major evolutionary trends. This concept suggests that differential speciation plays a role in macroevolution similar to the role of differential reproduction in microevolution.

To the extent that speciation rates and species longevity reflect success, the analogy to natural selection is even stronger. But qualities unrelated to the overall success of organisms in specific environments may be equally important in species selection. For example, the ability of a species to disperse to new locations may contribute to its giving rise to a large number of "daughter species." The species selection model has many critics who argue that evolutionary trends more commonly result from the gradual modification of populations in response to environmental change. The value of such debates is that they stimulate research, and many paleontologists and other evolutionary biologists are now focusing on the question of what produces evolutionary trends.

Whatever its cause, the appearance of an evolutionary trend does not imply that there is some intrinsic drive toward a preordained state of being. Evolution is a response to interactions between organisms and their current environments. If conditions change, an evolutionary trend may cease or even reverse itself. In the next chapter we will continue our study of macroevolution with a more detailed look at the fossil record and some of the effects of significant environmental change on species and speciation.

CHAPTER REVIEW

REVIEW OF KEY CONCEPTS

(with page numbers and key figures)

Macroevolution is the origin of new species and other taxonomic groups. Two patterns of species change chronicled by the fossil record are anagenesis (phyletic evolution), the accumulation of changes associated with the transformation of one species into another, and cladogenesis, branching evolution (FIGURE 24.1).

WHAT IS A SPECIES?

- **The biological species concept emphasizes reproductive isolation (p. 446)** The biological species concept defines a species as a group of populations whose individuals have the potential to interbreed and produce fertile offspring with each other but not with members of other species.
- **Prezygotic and postzygotic barriers isolate the gene pools of biological species (pp. 446–448, FIGURE 24.5)** Prezygotic barriers prevent mating or fertilization between species. Species that occupy the same geographic area often live in separate habitats (habitat isolation); possess unique, exclusive mating signals and courtship behaviors (behavioral isolation); breed at different times (temporal isolation); and/or have anatomically incompatible reproductive organs (mechanical isolation) or incompatible sex cells (gametic isolation). Even if two different species manage to mate, postzygotic barriers usually prevent the interspecific hybrids from developing into adults, breeding with either parent species, or producing viable, fertile offspring.
- **The biological species concept does not work in all situations (p. 449)** For instance, it is not applicable to fossils or to organisms that reproduce only asexually.
- **Other species concepts emphasize features and processes that identify and unite species members (pp. 449–450, TABLE 24.1)** Alternative concepts include the morphological species concept, which defines species by phenotypic characteristics; the cohesion concept, recognizing species as populations with discrete clusters of genetic traits; the recognition species concept, emphasizing mating adaptations that are fixed in a population; the ecological species concept, which stresses the role of species in the environment; and the evolutionary species concept, which recognizes species as independently evolving units performing unique ecological roles.

MODES OF SPECIATION

- **Geographical isolation can lead to the origin of species: allopatric speciation (pp. 451–452, FIGURES 24.7, 24.9)** Allopatric speciation occurs when a splinter population diverges from its parent population after becoming geographically isolated. Small splinter populations are better candidates for allopatric speciation than large ones because genetic drift and natural selection can change a small gene pool faster. Adaptive radiation is the evolution of numerous species with diverse adaptations from a common ancestor.
- **A new species can originate in the geographical midst of the parent species: sympatric speciation (pp. 452–455, FIGURES 24.10, 24.11)** Sympatric speciation in plants often involves doubling of the chromosome number. Autopolyploids are species derived this way from one ancestral species. Allopolyploids are species with multiple sets of chromosomes derived from two different species. Assortative mating in a polymorphic population can contribute to sympatric speciation in animals.

24.1

- **Genetic change in populations can account for speciation (pp. 455–457)** Reproductive barriers may arise as side effects of speciation as two populations genetically diverge while adapting to different environments. Sexual selection may lead directly to reproductive barriers. According to the recognition species concept, natural selection would amplify adaptations that enhance reproductive success with members of the same species. Consistent with the cohesion species concept, two species may interbreed freely in a hybrid zone without losing their distinctiveness away from the zone. Speciation may be associated with change at just a few gene loci or with the cumulative divergence at many loci.
- **The punctuated equilibrium model has stimulated research on the tempo of speciation (pp. 457–458, FIGURE 24.13)** One view of speciation is that it usually occurs gradually by an accumulation of microevolutionary changes in gene pools. In contrast, the punctuated equilibrium model views species as changing most when they bud from an ancestral species and then undergoing relatively little change for the rest of their existence.

THE ORIGIN OF EVOLUTIONARY NOVELTY

Speciation is the basis of all macroevolutionary changes. The origin of some new species involves changes so novel that the new species is an inaugural member of a higher taxon.

- **Most evolutionary novelties are modified versions of older structures (p. 458)** Most biological structures that evolve in one context can become co-opted for another function, and many features have arisen by the evolutionary remodeling of old structures.
- **Genes that control development play a major role in evolutionary novelty (pp. 458–460, FIGURES 24.14, 24.16)** Many macroevolutionary changes may have been associated with mutations in genes that regulate development.
- **An evolutionary trend does not mean that evolution is goal oriented (pp. 460–462, FIGURE 24.17)** According to the species selection hypothesis, macroevolutionary trends result when species with certain characteristics endure longer and speciate more often than those with other characteristics.

SELF-QUIZ

1. Most of biological diversity has probably arisen by
   a. anagenesis
   b. cladogenesis
   c. phyletic evolution
   d. hybridization
   e. sympatric speciation

2. The largest unit in which gene flow is possible is a
   a. population
   b. species
   c. genus
   d. subspecies
   e. phylum

3. Bird guides once listed the myrtle warbler and Audubon's warbler as distinct species, but applying the biological species concept, recent books show them as eastern and western forms of a single species, the yellow-rumped warbler. Experts must have found that the two kinds of warblers
   a. live in the same areas
   b. successfully interbreed in nature
   c. look enough alike to be considered one species
   d. are reproductively isolated from each other
   e. are allopatric

4. Among allopatric species of *Anopheles* mosquito, some live in brackish water, some in running fresh water, and others in stagnant water. What type of reproductive barrier is most obviously separating these different species?
   a. habitat isolation
   b. temporal isolation
   c. behavioral isolation
   d. gametic isolation
   e. postzygotic barriers

5. In *Drosophila* a mutation to a particular gene locus can result in a four-winged fly. If this locus were conserved, its addition to the gene pool would be described as
   a. heterochrony
   b. homeosis
   c. allometric growth
   d. paedomorphosis
   e. exaptation

6. According to advocates of the punctuated equilibrium model,
   a. natural selection is unimportant as a mechanism of evolution
   b. given enough time, most existing species will branch gradually into new species
   c. a new species accumulates most of its unique features as it comes into existence and changes little for the rest of its duration as a species
   d. most evolution is anagenic
   e. speciation is usually due to a single mutation

7. The biological species concept cannot be applied to two putative species that are
   a. sympatric
   b. nearly indistinguishable in morphology
   c. reproductively isolated
   d. capable of forming viable hybrids
   e. exclusively asexual

8. Where the ranges of blue-winged and golden-winged warblers overlap, hybridization is common, with hybrids able both to mate with one another and the parental types. If the cohesion species concept is applicable to these warblers it could be expected that
   a. the zone of hybrid distribution would be geographically stable
   b. DNA analysis would indicate a progressive reduction in genotypic variation
   c. hybrid fertility would be reduced in comparison to parental types
   d. hybrids would disappear if allopatric isolation were reintroduced
   e. distinguishing phenotypic frequencies would be represented as a gradual cline across the area of range overlap

9. Plant species A has a diploid number of 12. Plant species B has a diploid number of 16. A new species, C, arises as an allopolyploid from hybridization of A and B. The diploid number of C would probably be
   a. 12
   b. 14
   c. 16
   d. 28
   e. 56

10. The speciation episode described in question 9 is most likely a case of
    a. allopatric speciation
    b. sympatric speciation
    c. speciation based on sexual selection
    d. adaptive radiation
    e. anagenic speciation

CHALLENGE QUESTION

Write a paragraph explaining why remote islands have a proportionately greater number of indigenous species than do islands close to the mainland.

SCIENCE, TECHNOLOGY, AND SOCIETY

What is the biological basis for assigning all human populations to a single species? Explain why it is unlikely that a second human species could arise by cladogenesis in the future.

FURTHER READING

Goldschmidt, T. *Darwin's Dreampond, Drama in Lake Victoria.* Cambridge, MA: MIT Press, 1996. A researcher's popular account of field work on speciation.

Kerr, R. A. "Did Darwin Get It All Right?" *Science,* March 10, 1995. Studies of invertebrate fossils are adding support for punctuated equilibrium.

Kerr, R. A. "Does Evolutionary History Take Million-Year Breaks?" *Science,* October 24, 1997. Examines the hypothesis that whole communities of organisms may undergo stasis.

Kluger, T. "Go Fish." *Discover,* March 1992. Describes rapid speciation in Lake Victoria.

Knowlton, N. "A Tale of Two Seas." *Natural History,* June 1994. Discusses allopatric speciation resulting from the Panamanian land bridge separating the Caribbean from the Pacific Ocean.

Majerus, M., A. William, and G. Hurst. *Evolution, The Four Billion Year War.* Essex, UK: Longman, 1996.

Mesterton-Gibbons, M. and E. S. Adams. "Animal Contests as Evolutionary Games." *American Scientist,* July–August 1998. Examines the evolutionary basis for conflicts among animals.

Price, P. *Biological Evolution.* Philadelphia: Saunders, 1996. A strong, accessible text.

WEB LINKS

Visit the special edition of *The Biology Place* for BIOLOGY, Fifth Edition, at http://www.biology.com/campbell. Go to Chapter 24 for online resources, including learning activities, practice exams, and links to the following web sites:

"On the Origin of Species by Means of Natural Selection, or the Preservation of Favored Races in the Struggle for Life"
The on-line version of Darwin's historic book of the same name.

"Charles Darwin Research Station, Galápagos"
Research scientists from around the world visit the Charles Darwin Research Station to perform research on topics such as evolutionary biology, geology, ecotourism, climatology, and population genetics.

"Principia Cybernetica: Evolutionary Theory"
This site examines the philosophy of evolutionary systems. Of particular interest are the meticulous definitions of much of the terminology associated with evolution and evolutionary theory.

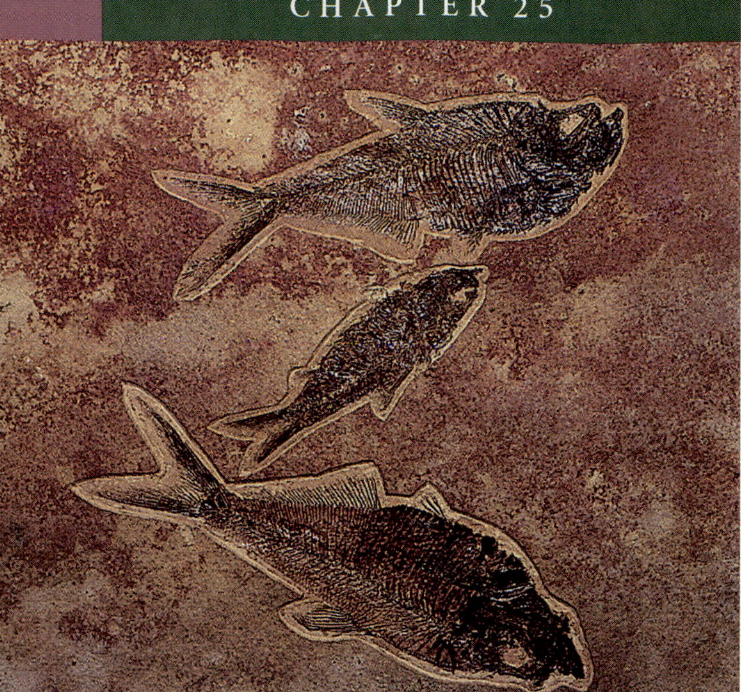

CHAPTER 25

TRACING PHYLOGENY

The Fossil Record and Geological Time

- Sedimentary rocks are the richest source of fossils
- Paleontologists use a variety of methods to date fossils
- The fossil record is a substantial, albeit incomplete, chronicle of evolutionary history
- Phylogeny has a biogeographical basis in continental drift
- The history of life is punctuated by mass extinctions followed by adaptive radiations of the survivors

Phylogeny and Systematics

- Taxonomy employs a hierarchical system of classification
- The branching pattern of a phylogenetic tree represents the taxonomic hierarchy
- Determining monophyletic taxa is key to classifying organisms according to their evolutionary history
- Molecular biology provides powerful tools for systematics
- The search for fossilized DNA continues despite recent setbacks: *science as a process*

The Science of Phylogenetic Systematics

- Phenetics increased the objectivity of systematic analysis
- Cladistic analysis uses novel homologies to define branch points on phylogenetic trees
- Phylogenetic systematics relies on both morphology and molecules

In broad terms, macroevolution is the story of the major events in the history of life as revealed by the fossil record. Evolution on this grand scale encompasses the origin of novel designs, such as the jaws of vertebrates (seen in the adjacent photograph of fossilized fishes) and the upright posture of humans; evolutionary trends, such as increasing brain size in mammals; the explosive diversification of certain groups of organisms following some evolutionary breakthrough, such as the adaptive radiation of flowering plants; and mass extinctions, which cleared the way for new episodes of adaptive radiation, such as the radiation of mammals in the past 65 million years.

This chapter describes how biologists trace ***phylogeny*** *(Gr.* phylon, *"tribe," and* genesis, *"origin"), the evolutionary history of a species or group of related species. Reconstructing phylogeny is part of the scope of* ***systematics****, the study of biological diversity in an evolutionary context. As we study systematics and phylogeny, especially in the context of the fossil record, the theme that evolution is a consequence of interactions between organisms and their environments will be extended to environmental and biological changes of global proportions.*

THE FOSSIL RECORD AND GEOLOGICAL TIME

Fossils, the preserved remnants or impressions left by organisms that lived in the past, are the historical documents of biology. The **fossil record** is the ordered array in which fossils appear within layers, or strata, of sedimentary rocks that mark the passing of geological time. Paleontologists collect and interpret fossils (FIGURE 25.1). We begin this section by studying how fossils form and how paleontologists determine their age. We then turn to the contributions and limitations of the fossil record in the study of phylogeny.

Sedimentary rocks are the richest sources of fossils

Sedimentary rocks form from layers of minerals that settle out of water. Sand and silt that are weathered and eroded from the land are carried by rivers to seas and swamps, where the particles settle to the bottom. Deposits pile up and compress the older sediments below into rock—sand into sandstone and mud into shale. When aquatic life-forms and terrestrial organisms swept into the seas and swamps die, they settle along with the sediments. A tiny fraction of them is then preserved as fossils. At any particular location, sedimentation is not continuous but occurs in intervals when the sea level changes or lakes and swamps dry up and refill. Even when a region is submerged, the rate of sedimentation and the types of sedimentary particles vary over time. As a result of these different periods of sedimentation, the rock forms in strata (see FIGURE 22.2).

(a)

(b)

(c)

(d)

(e)

(f)

(g)

FIGURE 25.1 ▪ A gallery of fossils. (a) Sedimentary rocks are the richest hunting grounds for paleontologists. Numerous invertebrate shells and other fossils about 15 million years old are visible in the sediments exposed on the face of this cliff. **(b)** The hard parts of organisms, such as this skull of *Australopithecus africanus,* an ancestor of humans that lived about 2.5 million years ago, are the most common fossils. The fossils may become even harder if minerals replace their organic matter. **(c)** These petrified (stone) trees in Arizona are about 190 million years old. **(d)** Some sedimentary fossils, such as this 40-million-year-old leaf, retain organic material. **(e)** Buried organisms, such as these invertebrates called brachiopods, which lived about 375 million years ago, decay and leave molds that may be filled by minerals dissolved in water. The casts that form when the minerals harden are replicas of the organisms. **(f)** Trace fossils are footprints, burrows, and other remnants of an ancient organism's behavior. The boy is standing in a 150-million-year-old dinosaur track in Colorado. **(g)** This 30-million-year-old scorpion, embedded in amber (hardened resin from a tree), is almost perfectly preserved.

The organic substances of a dead organism buried in sediments usually decay rapidly. However, hard parts that are rich in minerals, such as the shells of many invertebrates and protists and the bones and teeth of vertebrates, may remain as fossils (FIGURE 25.1a and b, p. 465). Paleontologists have unearthed nearly complete skeletons of dinosaurs and other forms, but more often the finds consist of parts of skulls, bone fragments, or teeth. Many of these relics are hardened even more and preserved by chemical changes; under the right conditions, minerals dissolved in groundwater seep into the tissues of a dead organism and replace its organic material. The plant or animal turns to stone (FIGURE 25.1c).

Rarer than mineralized fossils are those that retain organic material. They are sometimes discovered as thin films pressed between layers of sandstone or shale. For example, paleontologists have discovered plant leaves millions of years old that are still green with chlorophyll and well enough preserved for their organic composition to be analyzed and the ultrastructure of their cells to be explored with the electron microscope (FIGURE 25.1d). The most common fossilized plant material is pollen, which has a hard organic case that resists degradation.

The fossils that paleontologists find in many of their digs are not the actual remnants of organisms at all, but rocks that form as replicas of the organisms. These fossils result when a dead organism captured in sediments decays and leaves an empty mold that becomes filled with minerals dissolved in water. The minerals may subsequently crystallize, forming a cast in the shape of the organism (FIGURE 25.1e).

Trace fossils form in footprints, animal burrows, or other impressions left in sediments by the activities of animals. These rocks are in essence fossilized behavior; they tell paleontologists something about how the animals that left the trace fossils lived. For example, dinosaur tracks provide clues about the animal's locomotion—its gait (pattern of leg movements), stride length, and speed (FIGURE 25.1f).

If an organism happens to die in a place where bacteria and fungi cannot decompose the corpse, the entire body, including soft parts, may be preserved as a fossil. For example, the scorpion in FIGURE 25.1g got stuck in a drop of resin from a tree about 30 million years ago. The resin eventually hardened into amber, entombing the animal. There are other mechanisms that preserve whole organisms. Explorers have discovered mammoths, bison, and even prehistoric humans frozen in ice and preserved in acid bogs, where conditions retard decomposition. Such rare discoveries make the news, but biologists rely mainly on more common sedimentary fossils to reconstruct the history of life.

Paleontologists use a variety of methods to date fossils

Fossils are reliable historical data only if we can determine their ages. Next we examine the methods for determining where fossils fit on the geological time scale.

Relative Dating

The trapping of dead organisms in sediments freezes fossils in time. Thus, the fossils in each stratum of sedimentary rock are a local sample of the organisms that existed at the time that sediment was deposited. Because younger sediments are superimposed upon older ones, this book of sedimentary pages tells the relative ages of fossils.

The strata at one location can often be correlated with strata at another location by the presence of similar fossils, known as *index fossils.* The best index fossils for correlating strata that are far apart are the shells of sea animals that were widespread. At any one location where a roadcut or canyon wall reveals layered rocks, there are likely to be gaps in the sequence. That area may have been above sea level during different periods, and thus no sedimentation occurred. Some of the sedimentary layers that were deposited when the area *was* submerged may have been scraped away by subsequent periods of erosion.

By studying many different sites, geologists have established a **geological time scale** with a consistent sequence of historical periods (TABLE 25.1). These periods are grouped into four eras: the Precambrian, Paleozoic, Mesozoic, and Cenozoic eras. Each era represents a distinct age in the history of Earth and its life; the boundaries are marked in the fossil record by explosive radiations of many new forms of life. Mass extinctions also mark many of the boundaries between periods and eras. For example, the beginning of the Cambrian period is delineated by a great diversity of fossilized animals that are absent in rocks of the late Precambrian. And most of the animals that lived during the late Precambrian became extinct at the end of that era. The periods within each era are further subdivided into finer intervals called epochs (only the epochs of the current era, the Cenozoic, appear in TABLE 25.1). The timeline to the left of TABLE 25.1 shows that the geological eras were not equal in duration. The periods also varied in length. For example, the Jurassic period lasted almost twice as long as the Triassic period. Scientists have not divided geological time in an arbitrary manner, but have located boundaries in the record of the rocks that correspond to times of great change.

The record of the rocks is a serial that chronicles the *relative* ages of fossils; it tells us the order in which groups of species present in a sequence of strata evolved. However, the series of sedimentary rocks does not tell the *absolute* ages of the embedded fossils. The difference is analogous to peeling the layers of wallpaper from the walls of a very old house that has been inhabited by many owners. You could determine the sequence in which the papers had been applied, but not the year that each layer was added.

Absolute Dating

"Absolute" dating does not mean errorless dating, but only that age is given in years instead of relative terms such as

RELATIVE TIME SPAN OF ERAS

- Cenozoic
- Mesozoic
- Paleozoic
- Pre-Cambrian

Table 25.1 ■ The Geological Time Scale

ERA	PERIOD	EPOCH	AGE (MILLIONS OF YEARS AGO)	SOME IMPORTANT EVENTS IN THE HISTORY OF LIFE
Cenozoic	Quaternary	Recent	0.01	Historic time
		Pleistocene	1.8	Ice ages; humans appear
	Tertiary	Pliocene	5	Apelike ancestors of humans appear
		Miocene	23	Continued radiation of mammals and angiosperms
		Oligocene	35	Origins of many primate groups, including apes
		Eocene	57	Angiosperm dominance increases; origins of most modern mammalian orders
		Paleocene	65	Major radiation of mammals, birds, and pollinating insects
Mesozoic	Cretaceous		145	**Flowering plants (angiosperms) appear; many groups of organisms, including most dinosaur lineages, become extinct at end of period (Cretaceous extinctions)**
	Jurassic		208	Gymnosperms continue as dominant plants; dinosaurs dominant
	Triassic		245	Cone-bearing plants (gymnosperms) dominate landscape; radiation of dinosaurs, early mammals, and birds
Paleozoic	Permian		290	**Extinction of many marine and terrestrial organisms (Permian extinctions); radiation of reptiles; origins of mammal-like reptiles and most modern orders of insects**
	Carboniferous		363	Extensive forests of vascular plants; first seed plants; origin of reptiles; amphibians dominant
	Devonian		409	Diversification of bony fishes; first amphibians and insects
	Silurian		439	Diversity of jawless fishes; first jawed fishes; colonization of land by vascular plants and arthropods
	Ordovician		510	Origin of plants; marine algae abundant
	Cambrian		570	**Origin of most modern animal phyla (Cambrian explosion)**
Precambrian			610	Diverse soft-bodied invertebrate animals; first vertebrates; diverse algae
			700	Oldest animal fossils
			1700	Oldest eukaryotic fossils
			2500	Oxygen begins accumulating in atmosphere
			3500	Oldest fossils known (prokaryotes)
			4600	Approximate time of origin of Earth

before and *after, early* and *late.* **Radiometric dating** is the method most often used to determine the ages of rocks and fossils on a scale of absolute time. Fossils contain isotopes of elements that accumulated in the organisms when they were alive. Because each radioactive isotope has a fixed rate of decay, it can be used to date a specimen (see Chapter 2). An isotope's **half-life**, the number of years it takes for 50% of the original sample to decay, is unaffected by temperature, pressure, and other environmental variables. For example, carbon-14 has a half-life of 5600 years, a reliable rate of decay that can be used to date relatively young fossils (FIGURE 25.2, p. 468). Paleontologists use radioactive isotopes with longer half-lives to date older fossils. For example, uranium-238, which has a half-life of 4.5 billion years, has been used as a radiometric clock to date fossil-containing rocks of the Cambrian period (see TABLE 25.1).

Methods other than radiometric dating can be used on some fossils. Amino acids exist in two isomers with either left-handed or right-handed symmetry, designated the L and D forms, respectively. Organisms synthesize only L-amino acids,

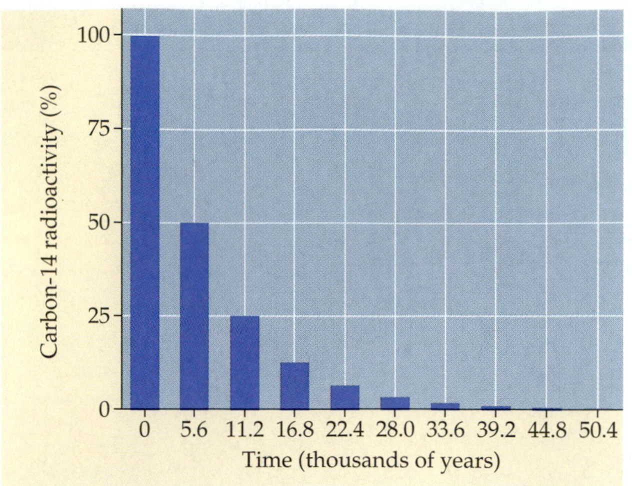

(a)

FIGURE 25.2 ▪ Radiometric dating. Radioactive isotopes can be detected and their amounts measured by the radiation they emit as they decompose to more stable atoms (see Chapter 2). Paleontologists use the clocklike decay of radioactive isotopes to date fossils and rocks. **(a)** The exponential decline in radioactivity of carbon-14. This isotope has a half-life of 5600 years, meaning that half the carbon-14 in a specimen will be gone in 5600 years, half the remainder will be gone in another 5600 years, and so on. **(b)** The use of carbon-14 dating to determine the age of a fossilized clam shell. Because the half-life of carbon-14 is relatively short, this isotope is only reliable for dating fossils less than about 50,000 years old. To date older fossils, paleontologists use radioactive isotopes with longer half-lives. Radiometric dating has an error factor of less than 10%.

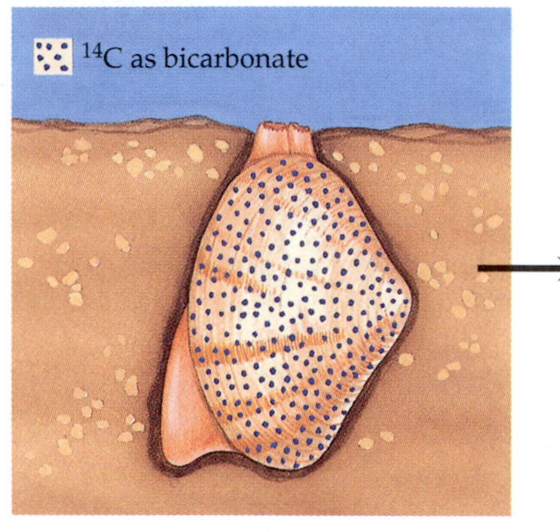

❶ While an organism, in this case a clam, is alive, it assimilates the different isotopes of each element in proportions determined by their relative abundances in the environment. Carbon-14 is taken up in trace quantities, along with much larger quantities of the more common carbon-12.

❷ After the clam dies, it is covered with sediment, and its shell eventually becomes consolidated into a layer of rock as the sediment is compressed. From the time the clam dies and ceases to assimilate carbon, the amount of carbon-14 relative to carbon-12 in the fossil declines due to radioactive decay.

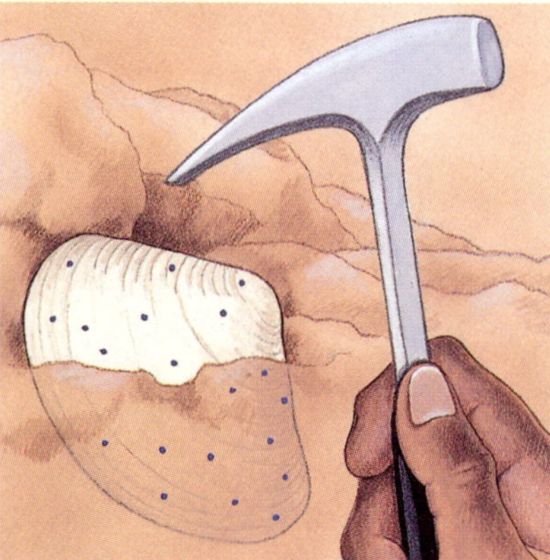

❸ After the clam fossil is found, its age can be determined by measuring the ratio of the two isotopes to learn how many half-life reductions have occurred since it died. For example, if the ratio of carbon-14 to carbon-12 in this fossil clam was found to be one-fourth that of a living organism, this fossil would be about 11,200 years old.

(b)

which are incorporated into proteins. After an organism dies, however, its population of L-amino acids is slowly converted, resulting in a mixture of L- and D-amino acids. In a fossil, the ratio of L- and D-amino acids can be measured. Knowing the rate at which this chemical conversion, called racemization, takes place, we can determine how long the organism has been dead. For instance, archeologists have used this method to date ostrich eggshells found along with fossils of early humans. The humans probably ate the eggs and used the shells for water bowls. Racemization, unlike radioactive decay, is temperature sensitive, meaning that past changes in climate made the racemization clock run faster or slower. But for fossils found in locations where climate apparently has not changed significantly since the fossils formed, the two dating methods agree closely on the age of the fossils.

The fossil record is a substantial, albeit incomplete, chronicle of evolutionary history

The discovery of a fossil is the culmination of a sequence of improbable coincidences. First, the organism had to die in the right place at the right time for burial conditions to favor fossilization. Then the rock layer containing the fossil had to escape geological processes that destroy or severely distort rocks, such as erosion, pressure from superimposed strata, or the melting of rocks that occurs at some locations. If the fossil *was* preserved, there is only a slight chance that a river carving a canyon or some other process will expose the rock containing the fossil. The chance that someone will find the fossil is even more remote, although discovery is more probable for people who are purposefully looking for fossils. No wonder the fossil record is incomplete: A substantial fraction of species that have lived probably left no fossils, most fossils that formed have been destroyed, and only a fraction of the existing fossils has been discovered.

The fossil record, far from being a complete sample of organisms of the past, is slanted in favor of species that existed for a long time, were abundant and widespread, and had shells or hard skeletons. Paleontologists, like all historians, must reconstruct the past from incomplete records. Even with its limitations, however, the fossil record is a remarkably detailed document of phylogeny over the vast scale of geological time. The order of sedimentary strata records the sequence of biological change, and dating methods tell us approximately how long ago the changes occurred. Also recorded in rocks is the chronology of environmental changes associated with evolutionary changes in organisms. We now turn to some of the major changes in Earth's environments that have caused significant transformations of life.

Phylogeny has a biogeographical basis in continental drift

Evolution has dimension in space as well as in time. Indeed, it was biogeography, even more than fossils, that first nudged Darwin and Wallace toward an evolutionary view of life. The history of Earth helps explain the current geographical distribution of species. For example, the emergence of volcanic islands such as the Galápagos opens new environments for founders that reach the outposts, and adaptive radiation fills many of the available niches with new species. On a global scale, the drifting of continents is the major geographical factor correlated with the spatial distribution of life and with such evolutionary episodes as mass extinctions and explosive increases in biological diversity.

The continents are not fixed, but drift about Earth's surface like passengers on great plates of crust floating on the hot, underlying mantle (FIGURE 25.3a). Unless two land masses are embedded in the same plate, their positions relative to each

(a)

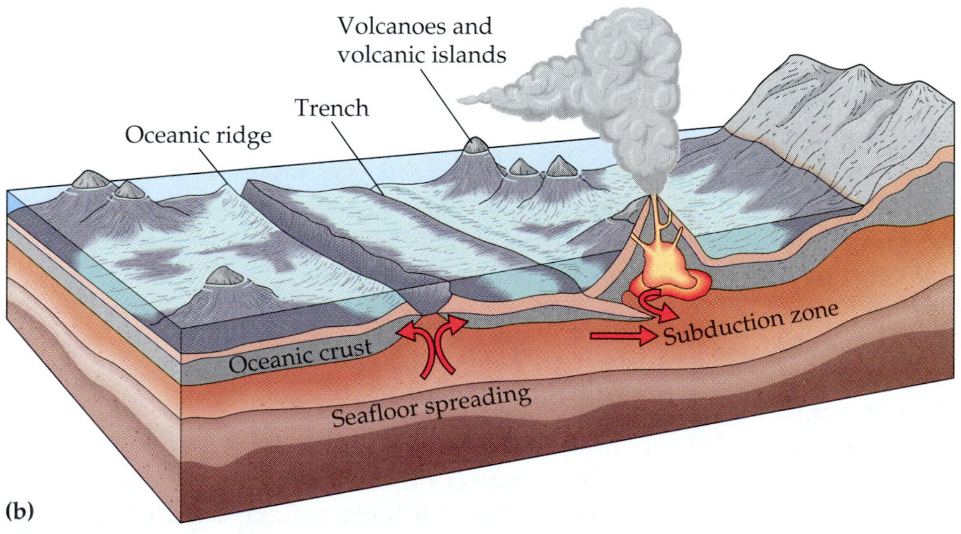

(b)

FIGURE 25.3 ▪ Earth's crustal plates and plate tectonics (geological processes resulting from plate movements). (a) The modern continents are passengers on crustal plates that are swept across Earth's surface by convection currents of the hot mantle below. This map identifies only the major plates. Red arrowheads indicate zones of violent tectonic events; many of these are subduction zones where the edge of one plate is being pushed over the edge of an adjacent plate. **(b)** At some plate boundaries, such as oceanic ridges (indicated by paired, opposing arrows in part a), the plates separate, and molten rock wells up in the gap. The rock solidifies and adds crust symmetrically to both plates, a phenomenon called seafloor spreading. At subduction zones, where plates move toward each other, the denser plate dives below the less dense one, creating a trench. The Marianas Trench in the South Pacific is a subduction zone over 11,000 m deep. The abrasion at subduction zones causes earthquakes and volcanic eruptions. When continents riding on different plates collide, they pile up and build mountains.

other change. For example, North America and Europe are presently drifting apart at a rate of about 2 cm per year. Many important geological processes, including mountain building, volcanism, and earthquakes, occur at plate boundaries (FIGURE 25.3b). California's infamous San Andreas Fault is part of a border where two plates slide past each other.

Plate movements rearrange geography incessantly, but two chapters in the continuing saga of continental drift had an especially strong influence on life. About 250 million years ago, near the end of the Paleozoic era, plate movements brought all the land masses together into a supercontinent that has been named **Pangaea**, meaning "all land" (FIGURE 25.4). Imagine some of the possible effects on life. Species that had been evolving in isolation came together and competed. When the land masses coalesced, the total amount of shoreline was reduced, and there is evidence that the ocean basins increased in depth, which lowered sea level and drained the shallow coastal seas. Then, as now, most marine species inhabited shallow waters, and the formation of Pangaea destroyed a considerable amount of that habitat. It was probably a long, traumatic period for terrestrial life as well. The continental interior, which has a drier and more erratic climate than coastal regions, increased in area substantially when the land came together. Changing ocean currents also would have affected land life as well as sea life. The formation of Pangaea surely had a tremendous environmental impact that reshaped biological diversity by causing extinctions and providing new opportunities for taxonomic groups that survived the crisis.

The second dramatic chapter in the history of continental drift was written about 180 million years ago, during the Mesozoic era. Pangaea began to break up, causing geographical isolation of colossal proportions. As the continents drifted apart, each became a separate evolutionary arena, and the faunas and floras of the different biogeographical realms diverged.

The pattern of continental separations is the solution to many biogeographical puzzles. For example, paleontologists have discovered matching fossils of Triassic reptiles in Ghana (West Africa) and Brazil. These two parts of the world, now separated by 3000 km of ocean, were contiguous during the early Mesozoic era. Continental drift also explains much about the current distribution of organisms, such as why the Australian fauna and flora contrast so sharply with the rest of the world. The great diversity of marsupials (pouched mammals), which fill ecological roles in Australia analogous to those filled by placental mammals on other continents, is just one example of Australia's unique collection of species. Marsupials probably evolved first in what is now North America and reached Australia via South America and Antarctica while the continents were still joined. The subsequent breakup of the southern continents set Australia "afloat" like a great ark of marsupials. On Australia, marsupials evolved and diversified, while placental mammals evolved and diversified on other continents. Australia has been isolated for 50 million years. Bats, rats, mice, and humans (and their domesticated animals) are the only placental mammals that have managed to populate the island continent.

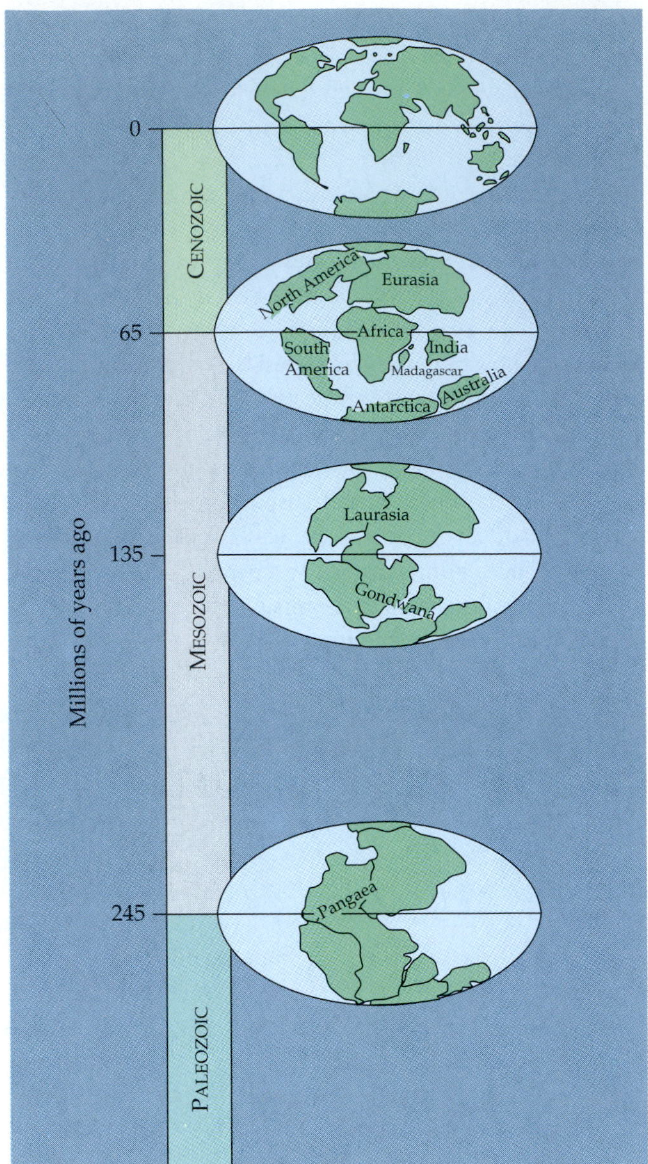

FIGURE 25.4 ▪ The history of continental drift. About 200 to 250 million years ago, all of Earth's land masses were locked together in a supercontinent named Pangaea. About 180 million years ago, Pangaea began to split into northern (Laurasia) and southern (Gondwana) land masses, which later separated into the modern continents. India collided with Eurasia just 10 million years ago, forming the Himalayas, the tallest and youngest of Earth's mountain ranges. The continents continue to drift.

The history of life is punctuated by mass extinctions followed by adaptive radiations of the survivors

The evolutionary byways from ancient to modern life have not been smooth. The fossil record reveals an episodic history, with long, relatively quiescent periods punctuated by briefer intervals when the turnover in species composition was much more extensive. The episodes include explosive adaptive radiations of major taxonomic groups as well as mass extinctions.

Examples of Major Adaptive Radiations

Taxonomic groups often diversify extensively early in their history. In many cases, major adaptive radiations seem to have occurred after the evolution of some novel characteristic that opened a new **adaptive zone**, a set of new living conditions and resources that presents many previously unexploited opportunities. For example, the evolution of wings provided many new possibilities for insects, such as rapid access to treetops, islands, and other new feeding and breeding areas. Adaptive radiation in this new adaptive zone produced hundreds of thousands of variations on the basic insect body plan.

One particularly impressive adaptive radiation known as the Cambrian explosion marks the boundary between the Precambrian era and the Paleozoic era (see TABLE 25.1). At this time, the diversity of sea animals increased dramatically. The oldest animal fossils are found in late Precambrian rocks about 700 million years old. Fossilized imprints discovered at several sites around the world indicate that these early animals were shell-less invertebrates with body plans that seem quite different from those of their Paleozoic successors. The fate of these lineages is unknown. A popular hypothesis is that the Precambrian animal lineages became extinct (although some apparently lived on into the Cambrian), and that all the animal phyla that exist today evolved during a span of less than 10 million years during the Cambrian. An alternate hypothesis is that at least some modern animal phyla arose during the Precambrian.

One key evolutionary novelty behind the remarkable diversification during the Cambrian may have been the origin of hard body parts—shells, skeletons, and claws—in a few phyla. These innovations would have made many new complex body designs possible and altered many predator-prey relations. More generally, the evolution of genes controlling development may have increased the range of morphological complexity and diversity that was possible. We will examine the Cambrian explosion in more detail in Chapter 32.

Empty adaptive zones probably exist even today. An empty adaptive zone can be exploited only if the appropriate evolutionary novelties arise. For example, flying insects existed at least 100 million years before the flying reptiles and birds that ate the insects evolved. Conversely, an evolutionary novelty cannot enable organisms to take advantage of adaptive zones that do not exist or are already occupied. Mammals, with the many unique features characteristic of their class, existed at least 75 million years before their first major adaptive radiation. The rise in mammalian diversity during the early Cenozoic era may have been associated with the opening of adaptive zones following the extinction of most of the dinosaur lineages. New adaptive radiations have often followed mass extinctions that swept away old tenants of adaptive zones. Most likely these new adaptive radiations did not simply fill old adaptive zones emptied by mass extinctions. Rather, the environmental changes that caused mass extinctions created new adaptive zones that were filled by new species.

Examples of Mass Extinctions

A species may become extinct because its habitat has been destroyed or because the environment has changed in a direction unfavorable to the species. If ocean temperatures fall by even a few degrees, many species that are otherwise beautifully adapted will perish. Even if physical factors in the environment are stable, biological factors may change; the environment in which a species lives includes the other organisms that live there, and evolutionary change in one species is likely to have some impact on other species in the community. For example, the evolution by some Cambrian animals of hard body parts, such as jaws and shells, may have made some organisms lacking hard parts more vulnerable to predation and thereby more prone to extinction.

Extinction is in fact inevitable in a changing world, and there have been crises in the history of life when global environmental changes have been so rapid and disruptive that a majority of species were swept away. Mass extinctions are known primarily from the decimation of hard-bodied animals of shallow seas, the organisms for which the fossil record is most complete. Of the dozen or so mass extinctions chronicled in the fossil record, two have received the most attention: the Permian extinctions, which occurred about 250 million years ago, and the Cretaceous extinctions of about 65 million years ago (FIGURE 25.5, p. 472).

The Permian extinctions, which define the boundary between the Paleozoic and Mesozoic eras, claimed about 90% of the species of marine animals. Terrestrial life also crashed; for example, 8 out of 27 orders of Permian insects did not survive into the Triassic, the next geological period. This mass extinction occurred in less than 5 million years—possibly much less—an instant in the context of geological time.

Several factors may have combined to cause radical environmental change during the late Permian. That was about the time the continents merged to form Pangaea, which disturbed many marine and terrestrial habitats and altered climate. Massive volcanic eruptions also occurred in what is now Siberia, constituting the most extreme episode of volcanism of the past half-billion years. Besides spewing lava and ash into the atmosphere, the Siberian volcanoes may have produced enough carbon dioxide to warm the global climate. Reduced temperature differences between the equator and the

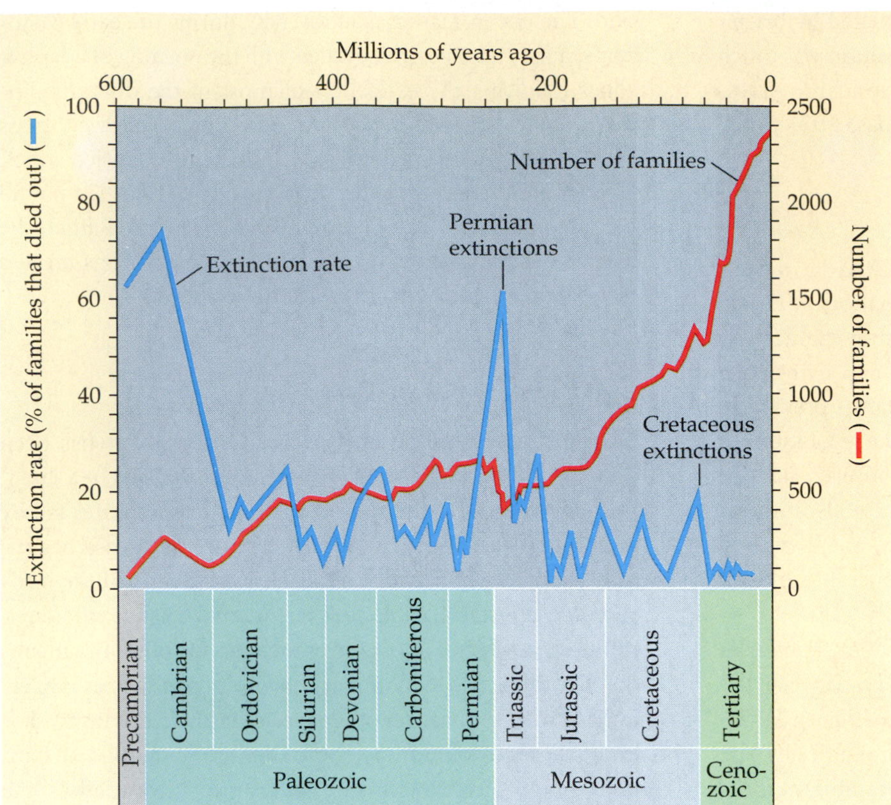

FIGURE 25.5 ▪ Diversity of life and periods of mass extinction. This graph, based on the fossil record of terrestrial and marine organisms, reveals a general increase in the diversity of organisms over time (red line and right vertical axis). Mass extinctions on land and in the oceans interrupted the buildup of diversity during many periods of geological time (blue line and left vertical axis). In this figure, extinction events are estimated as percentages of taxonomic families that died out relative to the number extant in each period of geological time. Of the two mass extinction periods labeled in the figure, the Permian extinctions claimed more than 90% of species on land and in the seas, whereas the extinctions at the end of the Cretaceous probably wiped out more than half of all species, including nearly all dinosaur lineages. Other methods of estimating extinctions indicate similar timing for mass extinctions, but the magnitudes of extinctions are more difficult to determine. Accuracy tends to be lowest in the oldest time periods and when diversity is very low. For instance, the extinction rate peak in the early Cambrian may be largely a function of the low diversity at the time. Paleontologists question whether mass extinctions actually occurred during the early Cambrian.

poles may have slowed the mixing of ocean water, which in turn may have reduced the amounts of oxygen available to marine organisms. An oxygen deficit in the oceans may have played a large role in the Permian extinctions.

Another emphatic punctuation, the Cretaceous extinctions of 65 million years ago, delineates the boundary between the Mesozoic and Cenozoic eras. That debacle doomed more than half the marine species and exterminated many families of terrestrial plants and animals, including nearly all of the dinosaur lineages. The climate became cooler at that time, and shallow seas receded from continental lowlands. Large volcanic eruptions in what is now India may have contributed to the cooling by releasing material into the atmosphere that blocked sunlight. However, many scientists now favor the so-called impact hypothesis, which postulates that the main cause of the Cretaceous extinctions was a collision of an asteroid or large comet with Earth. Separating Mesozoic from Cenozoic sediments is a thin layer of clay enriched in iridium, an element very rare on Earth, but common in meteorites and other extraterrestrial debris that occasionally fall to Earth. Walter and Luis Alvarez and their colleagues at the University of California, Berkeley, studied the clay and suggested it is fallout from a huge cloud of debris that billowed into the atmosphere when an asteroid hit Earth. The Alvarezes proposed that the great cloud would have blocked sunlight and severely disturbed climate for several months.

The impact hypothesis really has two parts: that such a collision occurred and that the event caused the Cretaceous extinctions. Much evidence in addition to the iridium layer supports the first part of the hypothesis—that a large comet or small asteroid crashed into Earth 65 million years ago. Earth is pocked with enough craters to tell us that many large objects have fallen on the planet in the past. Recent research has focused on the Chicxulub crater, a 65-million-year-old scar located beneath sediments on the Yucatán coast of Mexico. About 180 km in diameter, the Chicxulub crater is about the right size to have been caused by an asteroid with a diameter of about 10 km.

Critical evaluation of the impact hypothesis now focuses on the second claim—that the collision caused the Cretaceous extinctions. Advocates of the impact hypothesis argue that the impact was large enough to darken the Earth for years, not months, and that the resulting reduction of photosynthesis would have lasted long enough for food chains to collapse. Some of the minerals in the dust cloud would also have caused severe acid precipitation (see Chapter 3). The crater's horseshoelike shape and evidence from sediments deposited during the late Cretaceous indicate to some scientists that the Chicxulub asteroid struck Earth at a low angle from the southeast (FIGURE 25.6). Such an impact would have sent white-hot debris to the northwest, setting off a firestorm that within minutes would have killed most land plants and animals on the North American continent. This scenario could explain fossil evidence that extinction rates of species in North America were more severe and occurred in a briefer time than elsewhere. Less severe global effects would have developed more slowly after the initial catastrophe, which would explain why extinction rates during the late Cretaceous do not seem to have been uniform around the globe.

Although the debate over the impact hypothesis has muted somewhat, researchers maintain a healthy skepticism about the link between the apparent Chicxulub impact event and the period of mass extinctions at the end of the Cretaceous. The original Alvarez hypothesis triggered a strong research effort, fueled in part by its opponents, who contend that changes in climate due to continental drift, increased volcanism, and other processes on Earth, rather than extraterrestrial causes, could have caused mass extinctions 65 million years ago. As some researchers point out, hypotheses about the Cretaceous extinctions may not be mutually exclusive. It is possible that an asteroid impact was the sudden final blow in an environmental assault on late Cretaceous life that included more gradual processes.

Whatever the causes, mass extinctions affect biological diversity profoundly. But there is a creative side to the destruction. The species that manage to survive these crises, whether by their adaptive qualities or by sheer luck, become the stock for new radiations that fill many of the adaptive zones vacated or newly created by the extinctions. The world might be a very different place today if several dinosaur lineages had escaped the Cretaceous extinctions. Or imagine if none of the primates that lived in the Cretaceous period had survived.

PHYLOGENY AND SYSTEMATICS

So far in this chapter we have examined evolutionary history (phylogeny) in light of the fossil record and the geological time scale. Paleontology is closely related to the science of systematics, which studies biological diversity, past and present. One of the main goals of systematics is to make biological classification reflect phylogeny. Biologists traditionally represent the genealogies of organisms as **phylogenetic trees**, diagrams that trace evolutionary relationships as best they can be determined (FIGURE 25.7, p. 474).

Systematists use evidence derived from the fossil record and extant organisms to reconstruct phylogeny. Because the genetic makeup and phenotypic appearance of organisms living today reflect past episodes of macroevolution, systematists

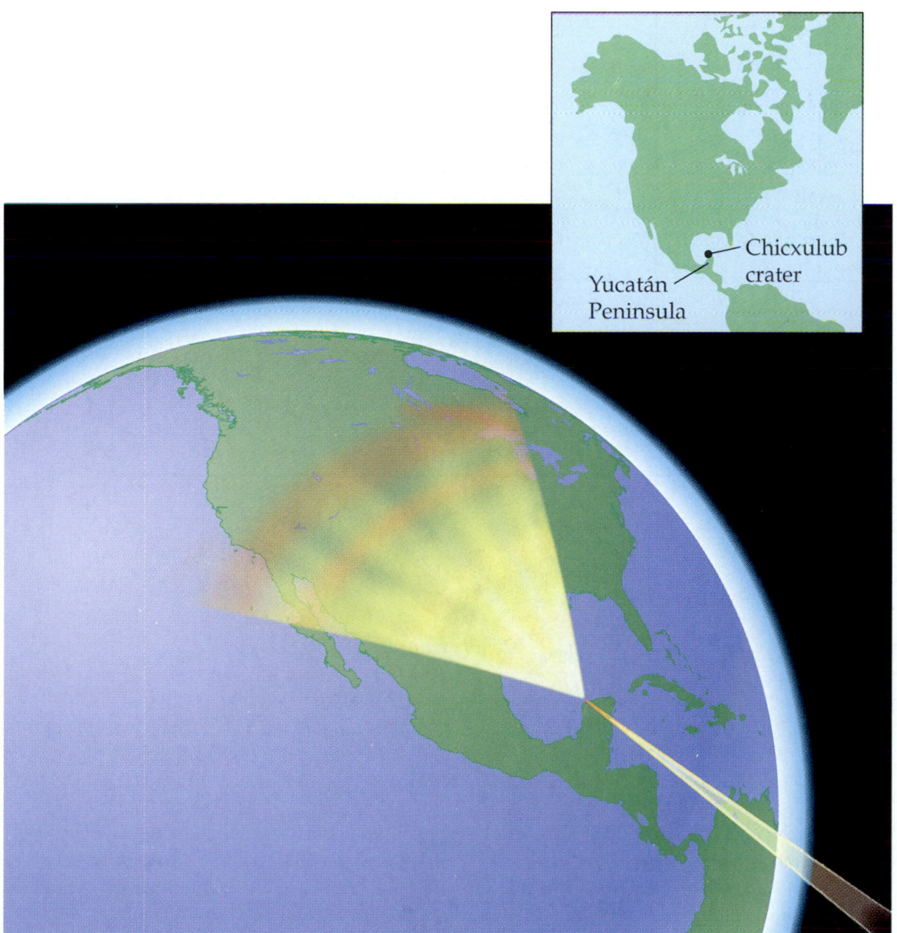

FIGURE 25.6 ▪ Possible correlation between mass extinctions and the nature of an asteroid impact at the end of the Cretaceous period. The 65-million-year-old Chicxulub impact crater, located in the Caribbean Sea near the Yucatán Peninsula of Mexico, is shaped like a horseshoe. The crater's shape and the pattern of debris found in sedimentary rocks indicate that an asteroid or comet struck the site at a low angle from the southeast. This artist's interpretation shows the impact and its immediate effect—a cloud of hot vapor and debris that would have killed most of the plants and animals in North America within minutes. This proposed impact scenario may have been only part of a series of events that triggered mass extinctions at the end of the Cretaceous period, but it generally fits the higher extinction rates of land animals and plants in North America than the rates elsewhere around the globe. Fossil evidence also indicates that in North America the survival rate of freshwater organisms, which would have been somewhat protected from a searing blast, was higher than that of terrestrial species on the continent.

obtain phylogenetic information by comparing modern species. Fossils are especially useful in determining the relative ages of taxa, but they often provide very limited information about the body features of extinct organisms. We will look in more detail at some of the methods used to reconstruct phylogeny after we examine biological classification.

Taxonomy employs a hierarchical system of classification

Systematists use taxonomy, the identification and classification of species, in their attempts to arrange organisms in categories that reflect phylogeny. The taxonomy developed by

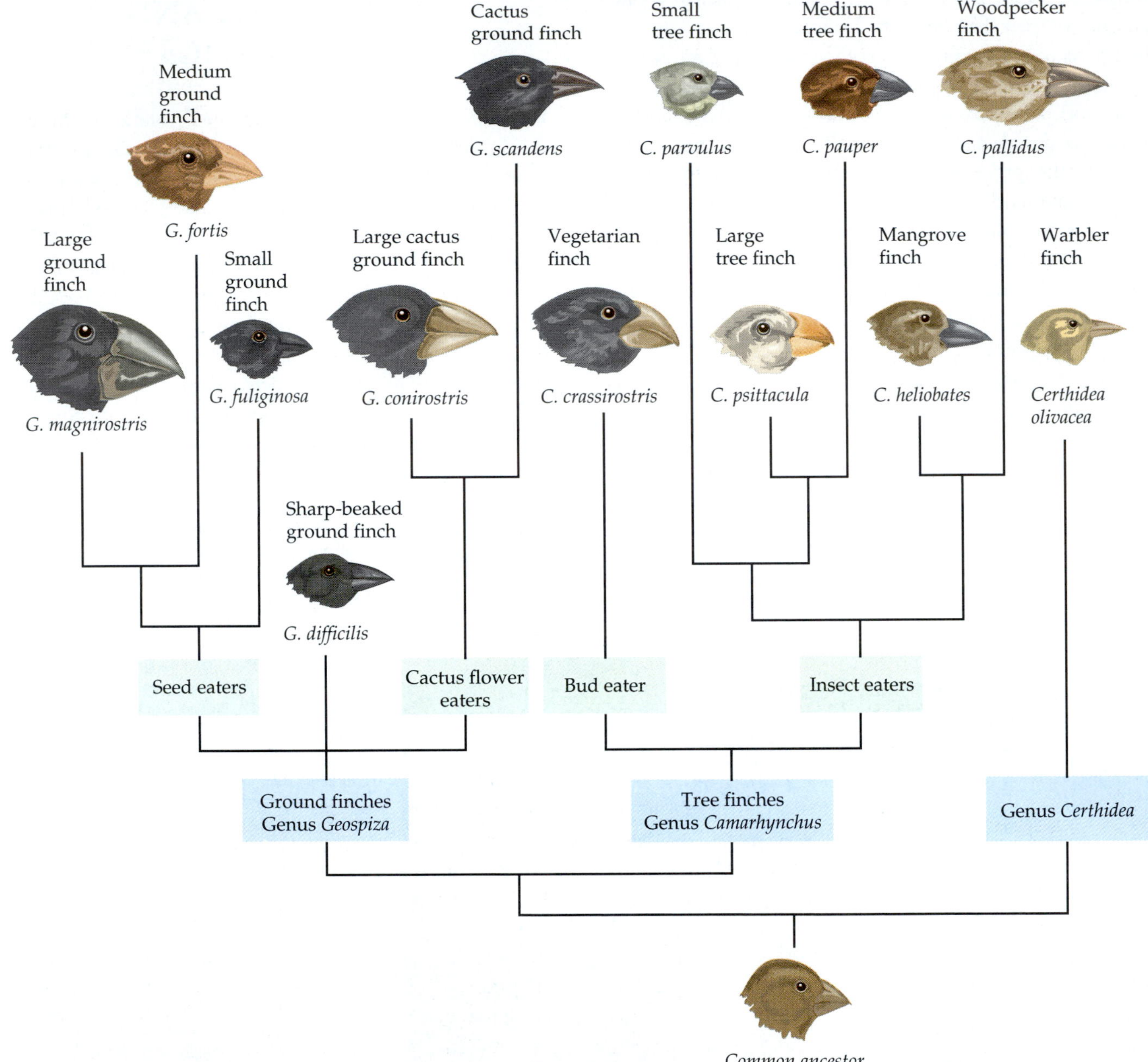

FIGURE 25.7 • A phylogenetic tree for the Galápagos finches. Phylogenetic trees are genealogies of probable evolutionary relationships among taxonomic groups. This tree of the 13 species of finches inhabiting the Galápagos Islands is based on comparative studies of body structures, especially beak shape and size, and on extensive field studies of reproductive isolation and feeding behavior (see Chapter 22). As indicated at the bottom of the tree, all 13 species were probably derived from a single ancestral population that colonized the Galápagos from the South American mainland several million years ago. Each branch point on a phylogenetic tree has meaning. For example, the lowest branch on the finch tree indicates that the warbler finch lineage (far right) diverged early in the history of the Galápagos finches. The next branch point upward shows a more recent divergence of two large groups. One of these, the ground finches (classified in the genus *Geospiza*), consists of six ground-feeding species. The other lineage (genus *Camarhynchus*) gave rise to six species that feed in trees. Among the tree finches, five species are insect eaters, more closely related to one another than to the vegetarian finch, which eats mainly tree buds. In fact, some scientists put the vegetarian finch in a separate genus, *Platyspiza*. Among the insect eaters, the mangrove finch's closest relative is the woodpecker finch. Both of these species have the unusual habit of probing insects out of tree bark using a small twig or cactus spine (see FIGURE 22.4).

Linnaeus in the eighteenth century had two main features. First, it assigned to each species a two-part Latin name, or **binomial**. The first word of the name is the **genus** (plural, **genera**) to which the species belongs. For example, all the ground finches of the Galápagos Islands (see FIGURE 25.7) are members of the genus *Geospiza* (a Latinized name derived from the Greek words *ge*, "land," and *spiza*, "finch"). The second part of the binomial, the **specific epithet**, refers to one species within the genus. For example, the scientific name for the large ground finch of the Galápagos is *Geospiza magnirostris* (*magnirostris* is Latin for "large beak"). Any given genus may include several related species, each with its own binomial. For instance, the large cactus ground finch of the Galápagos, *Geospiza conirostris* (Latin for "conelike beak"), belongs to the same genus as the large ground finch.

In contrast to scientific names (binomials), common names—such as finch, cat, bear, and lilac—often work well in informal communication, but they can be ambiguous because there are many species of each of these kinds of organisms. When biologists publish their research, they refer to the organisms they have studied with scientific names to avoid ambiguity. Many of the scientific names in use today date back to Linnaeus, who assigned binomials to over 11,000 species of plants and animals.

The second major contribution Linnaeus made to taxonomy was adopting a filing system for grouping species into a hierarchy of increasingly general categories. The first step in grouping species is built into binomial nomenclature. Species that are very similar, such as the Galápagos ground finches, are placed in the same genus. Grouping species is natural for us, at least in concept. We lump together several trees we know as oaks and distinguish them from several other species of trees we call maples. Indeed, oaks and maples belong to separate genera. The Linnaean system formalizes this grouping of species into genera and extends the scheme to progressively broader categories of classification, some of which have been added since the time of Linnaeus. Taxonomists place related genera in the same **family**, group families into **orders**, orders into **classes**, classes into **phyla** (singular, **phylum**), and phyla into **kingdoms** (see FIGURE 1.10). Many taxonomists, and this book, also group kingdoms into a higher taxonomic category, the **domain** (see FIGURE 1.11). TABLE 25.2 illustrates the various taxonomic levels for the large ground finch *(Geospiza magnirostris)* and the common buttercup *(Ranunculus acris)*. The genus *Geospiza* is lumped with six species of the genus *Camarhynchus* (see FIGURE 25.7) and numerous other genera of finches, including goldfinches and some grosbeaks, in the family Fringillidae, the finch family. This family belongs to the order Passeriformes (perching birds), which also includes the Hirundinidae (swallows), Sturnidae (starlings), and several other related families of birds. The order Passeriformes is grouped with many other orders in the class Aves, the birds. The class Aves is one of several belonging to the phylum Chordata in the kingdom Animalia, and the kingdom Animalia is nested in the domain Eukarya. Each taxonomic level is more comprehensive than the previous one. All finches are birds—that is, all members of the finch family, the Fringillidae, also belong to the order Passeriformes and class Aves, but not all birds are finches. The named taxonomic unit at any level is called a **taxon** (plural, **taxa**). For example, *Pinus* is a taxon at the genus level, the generic name for the various species of pine trees. Mammalia, a taxon at the class level, includes all the many orders of mammals. Only the genus name and specific epithet are italicized, and all taxa at the genus level and beyond are capitalized.

Table 25.2 ■ Classification of the Large Ground Finch and Common Buttercup

LEVEL	LARGE GROUND FINCH	COMMON BUTTERCUP
Species	*Geospiza magnirostris*	*Ranunculus acris*
Genus	*Geospiza*	*Ranunculus*
Family	Fringillidae (finches)	Ranunculaceae (crowfoot family)
Order	Passeriformes (perching birds)	Ranunculales
Class	Aves (birds)	Dicotyledones (dicots)
Subphylum	Vertebrata (vertebrates)	—
Phylum (animals) or division (plants)	Chordata (chordates)	Anthophyta (flowering plants)
Kingdom	Animalia	Plantae
Domain	Eukarya	Eukarya

Classifying a species by phylum, class, and so on is analogous to sorting mail, first by ZIP codes and then by streets and house numbers. Appendix Three gives the taxonomic classification of the major groups of organisms discussed in this book from the level of domain to the level of class.

As a component of systematics, taxonomy has two main objectives. The first is to sort out closely related organisms and assign them to species, describing the diagnostic characteristics that distinguish one species from another. The second major objective of taxonomy is classification, the arranging of species into the broader taxonomic categories, from genera to domains. In some cases, there are intermediate categories, such as subfamilies (a category between families and genera). For instance, the Galápagos finches are grouped together in their own subfamily, Geospizinae.

The branching pattern of a phylogenetic tree represents the taxonomic hierarchy

Systematics has a goal beyond simple organization: to have classification reflect the evolutionary affinities of species. The taxonomic hierarchy is set up to fit evolutionary history. For example, within the domain Eukarya, for the common buttercup (see TABLE 25.2), the most inclusive taxon (Kingdom Plantae) is the oldest in the hierarchy; the inaugural member of the plant kingdom (the earliest species of plant) arose about 450 million years ago (see TABLE 25.1). The next most inclusive taxon, Division Anthophyta, is the next oldest; its first member species arose some 100 million years ago, during the Cretaceous period. If we were to construct a phylogenetic tree of plants, Kingdom Plantae would be the main trunk at the bottom; Division Anthophyta would be one of several lineages derived from the Plantae; Class Dicotyledones would be one of several lineages derived from the Anthophyta; and so on up to the species *Ranunculus acris*, which is the youngest taxon in the hierarchy and so would occupy a branch at the top of the tree. The phylogenetic tree in FIGURE 25.8 further illustrates the connection between classification and phylogeny.

Classification schemes and phylogenetic trees are hypotheses of past history based on available data. Like all hypotheses, they make predictions that can be tested by further study. Phylogenetic trees and classification are refined whenever new information does not fit the existing hypotheses.

Determining monophyletic taxa is key to classifying organisms according to their evolutionary history

The ideal in systematics is to create a classification that reflects the evolutionary history of organisms. Doing so requires that species be grouped into more inclusive taxa that are monophyletic ("a single tribe"). A taxon is **monophyletic** if a single ancestor gave rise to all species in that taxon and to no species in any other taxon (FIGURE 25.9a).

By contrast, other kinds of taxa do not accurately reflect evolutionary history. For instance, a taxon is **polyphyletic** if its members are derived from two or more ancestral forms not common to all members. A taxon is **paraphyletic** if it excludes species that share a common ancestor that gave rise to the species included in the taxon (FIGURE 25.9b and c). As we will see, achieving the goal of a classification scheme made up of only monophyletic taxa is often easier said than done.

Sorting Homology from Analogy

To a large extent, systematics is a comparative science. A systematist classifies species into more inclusive, preferably monophyletic, taxa based on the extent of similarities in morphology and other characteristics. Recall from Chapter 22 that likeness attributed to shared ancestry is called **homology**. The forelimbs of mammals are homologous; that is, the similarity in the intricate skeleton that supports the limbs has a genealogical basis (see FIGURE 22.9).

There is a wild card in this game of making evolutionary connections by evaluating similarity: Not all likeness is inherited from a common ancestor. Species from different evolutionary branches may come to resemble one another if they have similar ecological roles and natural selection has shaped analogous adaptations. This is called **convergent evolution**, and similarity due to convergence is termed **analogy**, not

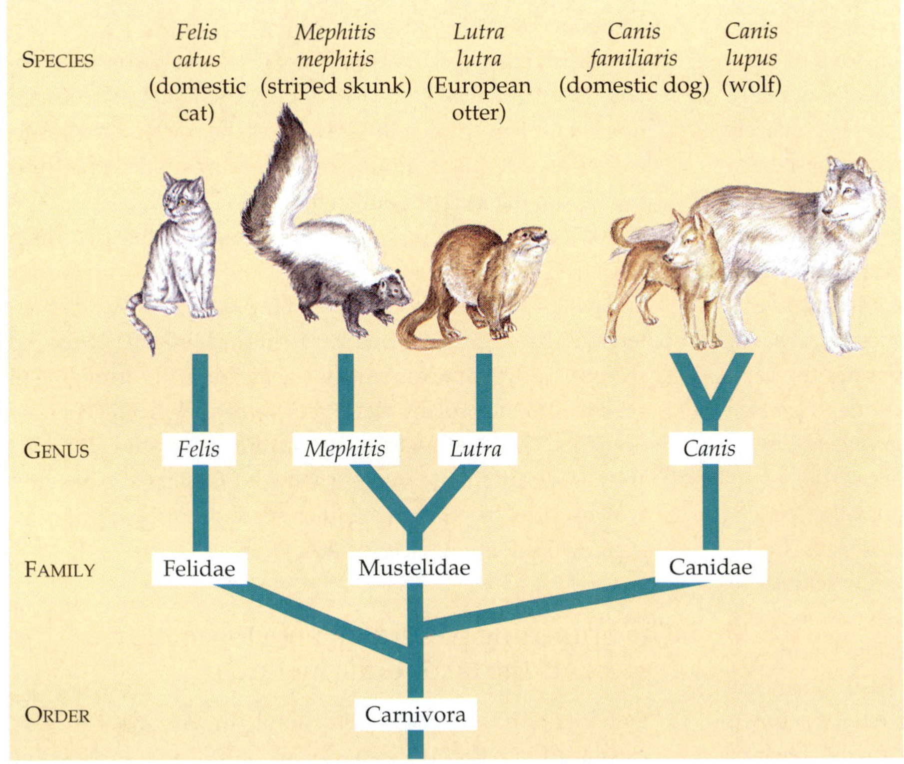

FIGURE 25.8 ▪ The connection between classification and phylogeny. The branching nature of evolutionary trees reflects the hierarchical arrangement of taxa. This tree suggests possible genealogical affinities among some of the taxa subordinate to the order Carnivora, itself a branch of the class Mammalia. The position of the tree's branches also indicates the relative age of the evolutionary divergences; thus species, the most recently derived taxa, are on the topmost branches of this tree. Whenever possible, systematists use the fossil record and comparative anatomy to help construct phylogenetic trees, but they also apply other methods, such as comparisons of the species' DNA and proteins.

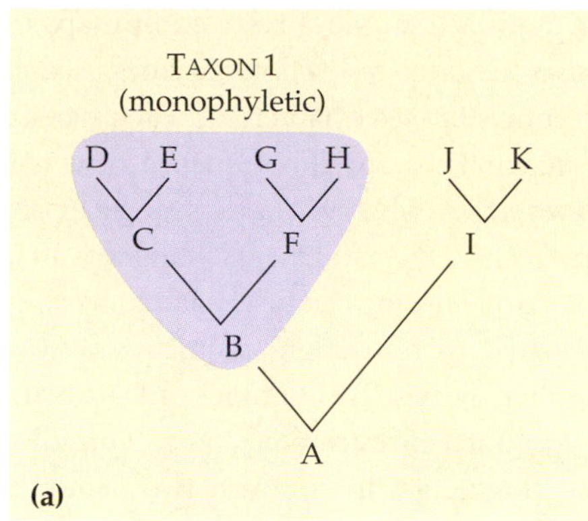

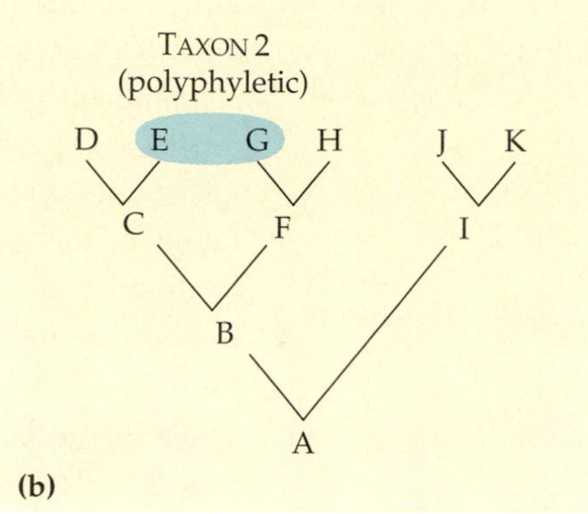

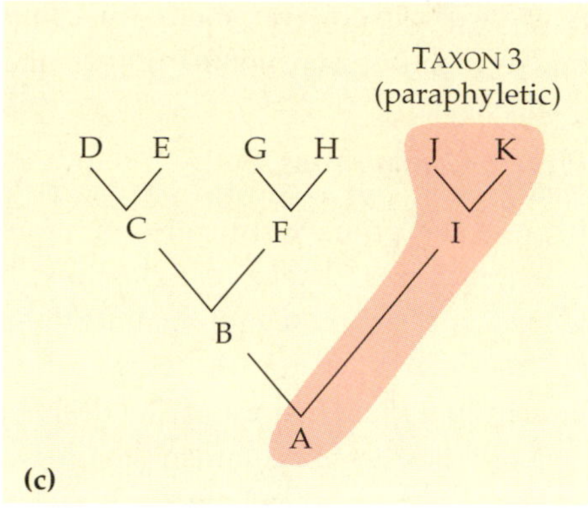

FIGURE 25.9 ▪ Monophyletic versus polyphyletic and paraphyletic taxa. **(a)** Taxon 1, with seven species (B–H), qualifies as a monophyletic grouping, the taxonomic ideal. The taxon includes all descendant species along with their immediate common ancestor (species B). **(b)** Taxon 2, a subgrouping within taxon 1, is polyphyletic—species E and G are derived from two different immediate ancestors (species C and species F). **(c)** Taxon 3 is paraphyletic; it includes species A without incorporating all other descendants of that ancestor.

homology (FIGURE 25.10). The wings of insects and those of birds, for example, are analogous flight equipment that evolved independently and are built from entirely different structures. Convergent evolution has also produced the analogous resemblances between certain Australian marsupials and placental look-alikes that have evolved independently on other continents—the Tasmanian devil and the coyote, for example.

To develop phylogenetic trees and classify organisms according to evolutionary history, we must use only homologous similarities. As a general rule, the greater the number of homologous parts between two species, the more closely the species are related, and this should be reflected in their classification. This guideline is simpler in principle than it is in practice. Adaptation can obscure homologies, and convergence can create misleading analogies. As we saw in Chapter 22, comparing the embryonic development of the features in question can often expose homology that is not apparent in the mature structures.

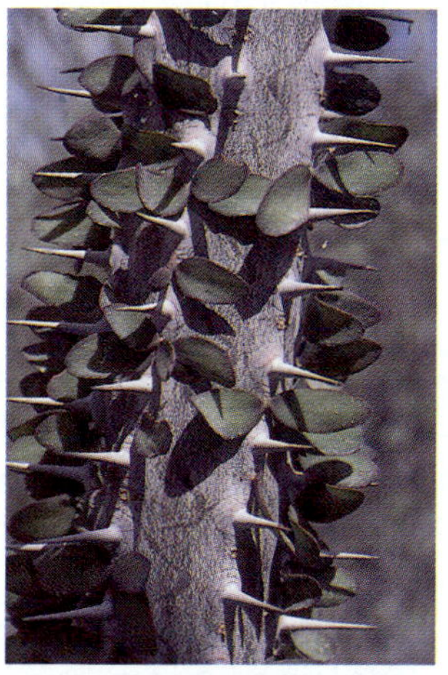

FIGURE 25.10 ▪ Convergent evolution and analogous structures. The ocotillo of southwestern North America (left) looks remarkably similar to the allauidia (right) found in Madagascar. The plants are not closely related and owe their resemblance to analogous adaptations that evolved independently in response to similar environmental pressures.

There is another clue to identifying homology and sorting it from analogy: The more complex two similar structures are, the less likely it is they evolved independently. Consider the skulls of a human and a chimpanzee, for example. The skulls are not single bones, but a fusion of many, and the two skulls match almost perfectly, bone for bone. It is highly improbable that such complex structures matching in so many details could have separate origins. Most likely, the genes required to build these skulls were inherited from a common ancestor.

Molecular biology provides powerful tools for systematics

When classifying organisms, it is useful to compare macro-
25.1 molecules as well as anatomical features. Sequences of nucleotides in DNA are inherited, and they program corresponding sequences of amino acids in proteins. Species diverge as changes occur in nucleotide bases, with each species acquiring its own set of genetic mutations. Thus, we would predict that species that are phylogenetically closely related have more similar nucleotide sequences in their nucleic acids and a greater number of similar amino acid sequences in their proteins than do distantly related species.

The comparison of nucleic acids and proteins has become a powerful addition to the other comparative methods systematists use to measure evolutionary relationships between species. Molecular tools are objective and quantitative. They can be used to compare species that are morphologically indistinguishable and to assess relationships between groups of organisms that are so phylogenetically distant that they share very few, if any, morphological similarities. Today, both the amino acid sequences for many proteins and the nucleotide sequences for a rapidly increasing number of genomes are in electronic databases that are available via the Internet for comparative studies. The use of molecular comparisons in developing and testing hypotheses about classification and phylogeny is one of the fastest growing areas in biology today.

As we will see, however, molecular comparisons, like structural comparisons, are not free of potential pitfalls.

Protein Comparisons

Because the primary structures of proteins are genetically determined (see Chapter 5), a close match in the amino acid sequences of two proteins from different species indicates that the genes for those proteins evolved from a common gene present in a shared ancestor. The degree of similarity is evidence of the extent of common genealogy.

Some of the first applications of molecular biology in systematics and phylogenetic analyses were comparisons of amino acid sequences. For example, the differences in amino acid sequence of the blood protein hemoglobin and cytochrome *c*, an ancient protein common to all aerobic organisms, were shown to be greater the more phylogenetically distant the species (see FIGURE 22.11). The amino acid sequence of human cytochrome *c*, for instance, is identical to that of our close relative the chimpanzee but differs from that of a dog (a distant relative) by 13 amino acids, of a rattlesnake (a more distant relative) by 20 amino acids, and of a much more distant relative, the tuna, by 31 amino acids. These differences are consistent with evidence from comparative anatomy and the fossil record.

Though still widely used, amino acid sequence data generally are not the preferred source of molecular information for phylogenetic studies. One drawback is that amino acid sequences provide information about only those genes that code for proteins; this is often only a small fraction of the total amount of molecular information that is potentially available for comparing species. In humans, for example, protein-coding genes make up only about 2% of the genome.

DNA and RNA Comparisons

Comparing the genes or genomes of two species is the most direct measure of inheritance from shared ancestors. Comparisons can be made by three methods: DNA-DNA hybridization, restriction maps, and DNA or RNA sequence analysis.

Whole genomes can be compared by **DNA-DNA hybridization**, which measures the extent of hydrogen bonding between single-stranded DNA obtained from two sources. How tightly the DNA of one species can bind to the DNA of the other depends on the degree of similarity, as base pairing between complementary sequences holds the two strands together. DNA-DNA hybridization has helped settle some old debates. For example, mammalogists once disagreed about whether pandas are true bears or members of the raccoon family; the evidence from DNA-DNA hybridization puts the giant panda with the bears but places the lesser panda in the raccoon family (FIGURE 25.11).

DNA-DNA hybridization can estimate the overall similarity of two genomes, but it does not give precise information about the matchup in specific nucleotide sequences of the DNA. An alternative approach employs **restriction maps** of DNA. This method uses the same restriction enzymes used in recombinant DNA technology (see Chapter 20). Each type of restriction enzyme recognizes a specific sequence of a few nucleotides and cleaves DNA wherever such sequences are found in the genome. The DNA fragments obtained after treatment with a restriction enzyme can be separated by electrophoresis and compared to restriction fragments derived from the DNA of another species. Two samples of DNA with similar maps for the locations of restriction sites will produce similar collections of fragments. In contrast, two genomes that have diverged extensively since their last common ancestor will have a very different distribution of restriction sites, and the DNA will not match closely in the sizes of restriction fragments.

Because so many fragments are obtained from the genome of the nucleus, restriction mapping is more practical for comparing smaller segments of DNA, usually a few thousand nucleotides long. Several laboratories have used restriction maps to compare mitochondrial DNA (mtDNA), which is relatively small. There is the added benefit that mtDNA changes by mutation about ten times faster than does the nuclear genome, which makes it possible to sort out phylogenetic relationships between very closely related species, or even between different populations of the same species. In one application of mtDNA comparison, a research team measured the relationships between different groups of Native Americans. Their results corroborate evidence from the study of languages that the Pima of Arizona, the Maya of Mexico, and the Yanomami of Venezuela are closely related, probably descending from the first of three waves of immigrants to cross the Bering Land Bridge from Asia to the Americas during the glaciation of the late Pleistocene epoch.

The most precise and potentially powerful method for comparing DNA (nuclear and mitochondrial) from two species, and the one most widely used today, is **DNA sequence analysis**—that is, comparing the actual nucleotide sequences of DNA segments. The application of polymerase chain reaction (PCR; see Chapter 20) to clone traces of DNA, coupled with automated sequencing, has made the collection of DNA sequence data relatively simple and fast. Systematists now use nucleotide sequence data from nuclear DNA, mtDNA, or both to deduce phylogenies, construct trees, and classify organisms. DNA sequencers have become commonplace in research laboratories, and Internet databases are expanding rapidly.

Related to DNA sequence analysis is the sequencing of ribosomal RNA (rRNA). Because DNA coding for rRNA changes slowly relative to most other DNA, differences in rRNA sequences can be used to trace some of the earliest branching in the tree of life. We will see in Chapter 27 that comparisons of ribosomal RNA sequences have been especially useful in sorting out the phylogenetic relationships among bacteria.

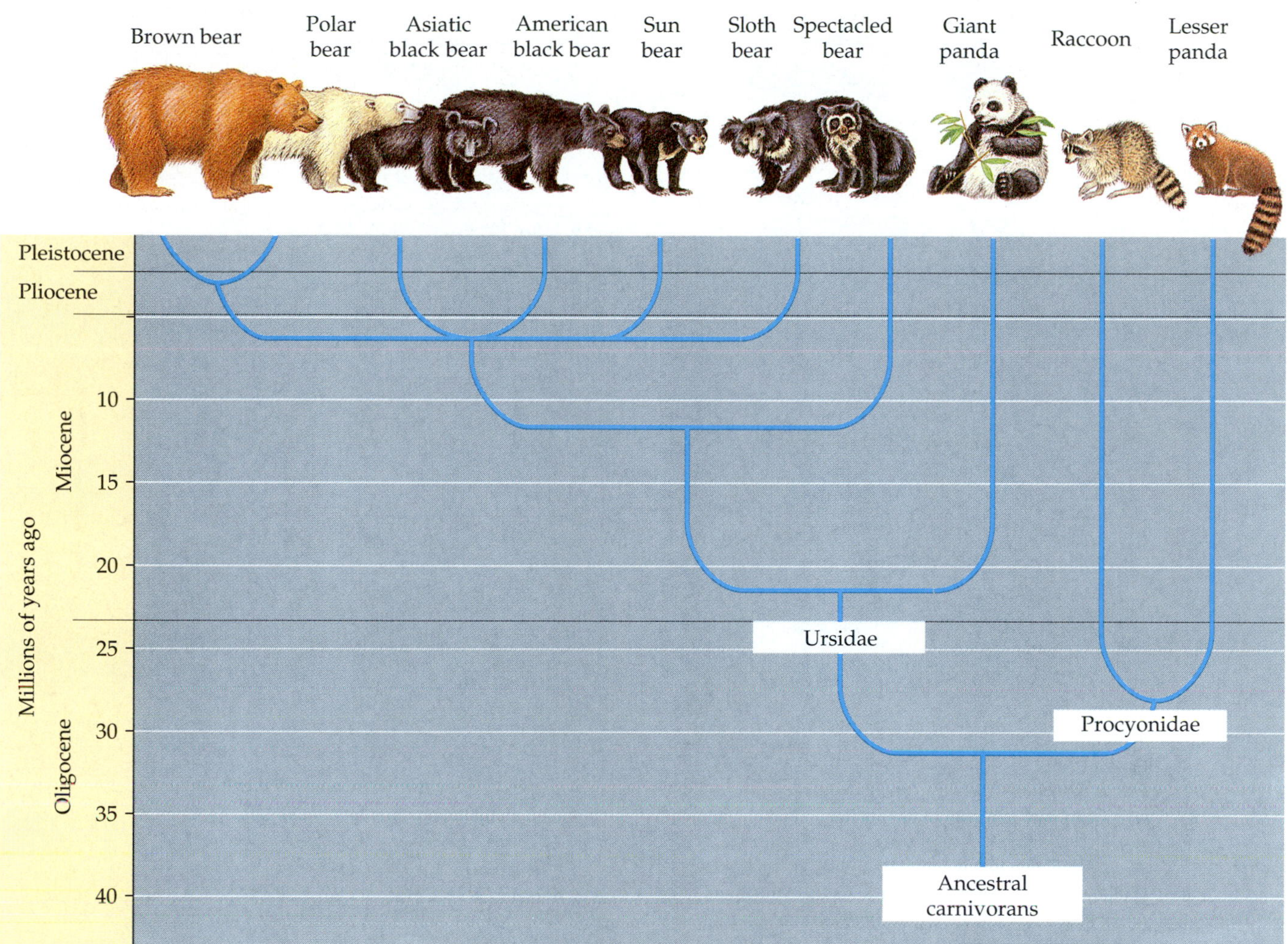

FIGURE 25.11 ▪ A phylogenetic tree based on systematics. Two methods, comparison of blood proteins and comparison of genomes by DNA-DNA hybridization, were used as taxonomic criteria to construct this tree of species of bears and raccoons. These methods helped determine that the giant panda should be in the Ursidae (bear) family, but the lesser panda should be classified with raccoons in the Procyonidae family.

Identifying and Comparing Homologous DNA Sequences

Comparing nucleotide sequences between corresponding DNA segments from different species has the potential of telling us how much divergence there has been in the evolution of two genes derived from the same ancestral gene. To measure differences between two species, it is necessary to identify homologous nucleotide sequences (those found in both species and having a common ancestry).

Identifying homologous nucleotide sequences can be as difficult in molecular systematics as it is in morphological studies. A researcher first selects a portion of a genome that is appropriate to compare (often mtDNA segments for comparing species that have recently diverged, and rRNA genes for distantly related species). After sequencing, the genomic segments are aligned side by side (FIGURE 25.12, p. 480). Common ancestry is clear when two sequences of the same gene from two species that have diverged very recently are the same or differ in only a few nucleotide bases. Determination of homology is more difficult when sequences have many unmatched bases.

Because mutations tend to accumulate as species diverge, the number of differences in nucleotide bases between homologous sequences is a measure of evolutionary distance. The accuracy of the measure, however, depends on proper alignment of the sequences. Sophisticated computer programs are used to arrive at the most likely fit based on what is known about the frequencies and types of mutations that occur in the DNA segments under study. Simply counting the number of differences between sequences does not provide accurate comparison because different kinds of mutations occur at different rates and often are not independent of one another. Natural selection may also affect some mutations more than

1. Ancestral homologous DNA segments are identical in species A and B

A CCATCAGAGTCC
B CCATCAGAGTCC

2. Mutations occur; species A and B diverge

Deletion
A CCATCAGAGTCC
B CCATCAGAGTCC
GTA Insertion

3. Homologous segments do not align because of mutations

A CCATCAAGTCC
B CCATGTACAGAGTCC

4. Homologous segments align after addition of gaps in sequence A

A CCAT CA AGTCC
B CCATGTACAGAGTCC

FIGURE 25.12 ▪ **Aligning segments of DNA.** This simplified example indicates how mutations create problems in aligning homologous sequences and how realignments can be made. ① Two homologous DNA segments (each from a different species) originally have exactly the same order of bases. ② As the two species diverge, the homologous sequences are altered, one by a point mutation that deletes a single nucleotide from the sequence, the other by the insertion of three adjacent nucleotides. ③ As a result, the nucleotide sequences of the two segments do not match, and the differences mask the fact that the sequences are homologous. ④ After a computer program identifies the similarities between the segments, it reestablishes a match by inserting gaps in one sequence that compensate for the mutations. Attaining the correct alignment between sequences is essential to accurate studies of species relationships. Computer programs have been developed to determine the best possible fit of two sequences assumed to be homologous. The best fit between two sequences is attained when the bases match with the fewest gaps added by the computer. Alignment programs are highly effective, but the placement of gaps is based on the laws of probability and can be more or less subjective, depending on the frequency and types of mutations in each species.

others. In comparing species, homologous sequences are generally evaluated according to the number of differences they exhibit per unit length of nucleotide segment. Factored into the evaluation is a ranking of the differences based on current knowledge of how genomes change over time.

Once a match is achieved between homologous nucleotide sequences, their differences can be evaluated, and phylogenetic hypotheses can be developed in the form of trees (FIGURE 25.13). Each hypothesis is tested repeatedly using all the appropriate sequence data that is available. Computer programs designed to build phylogenetic trees from sequence data are sophisticated but not foolproof. Whenever possible, hypotheses developed from sequence data are compared with those based on morphological comparisons, including any information from the fossil record.

1. In a table, arrange genetic distances among taxa A–D.

	A	B	C	D
A		6	8	12
B	6		10	14
C	8	10		16
D	12	14	16	

2. Look for the two taxa that are the most closely related (shortest genetic distance). Taxon A and taxon B are closest (6 is the shortest genetic distance). Begin the tree with branches connecting A and B. Each branch should be 6 ÷ 2 = 3 units long.

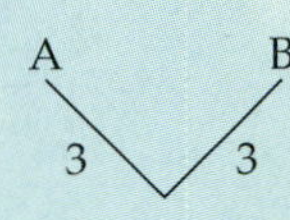

3. Estimate the distance between unit AB and the other taxa.

(a) Calculate the average distance from AB to C:

Distance from A to C = 8
Distance from B to C = 10
Average distance from AB to C = 18 ÷ 2 = 9

(b) Calculate the average distance from AB to D:

A to D = 12
B to D = 14
26 ÷ 2 = 13

4. Since 9 is less than 13, C appears to be more closely related to AB than D does. Thus, C should be next on the tree. Each branch should be 9 ÷ 2 = 4.5 units long (so that the distance from C to A is 9, and the distance from C to B is 9).

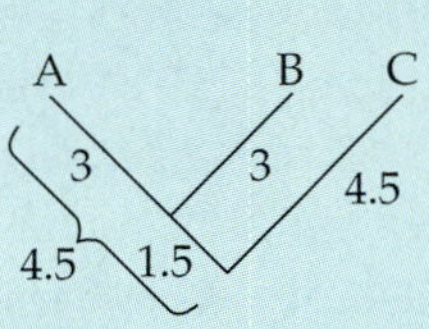

5. Estimate the distance between unit ABC and D.

A to D = 12
B to D = 14
C to D = 16
42 ÷ 3 = 14

6. Add D to tree; each branch should be 14 ÷ 2 = 7 units long.

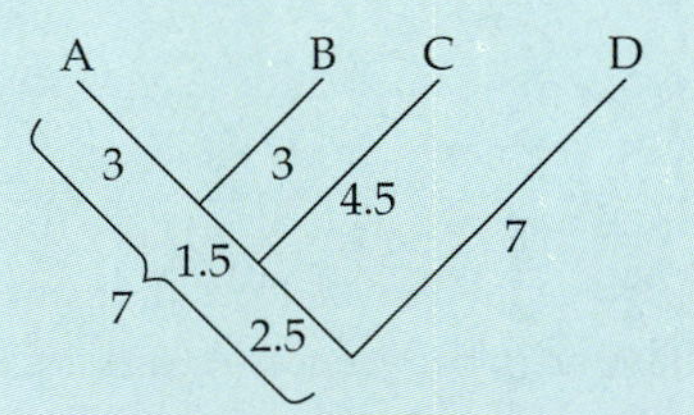

25.1 **FIGURE 25.13** ▪ **Developing a phylogenetic hypothesis based on DNA sequences.** The numbers in the table represent genetic distances (determined from DNA sequence comparisons) among four taxa (A, B, C, and D). The more similar their DNA, the less the genetic distance. This figure shows one way to construct a phylogenetic tree to illustrate possible evolutionary relationships among taxa. Typical of phylogenetic hypotheses, the tree developed here seems to show some inconsistencies with the data in the table. For instance, the sequence data indicate that A and B are both closer to D than is C. But if we switch A and C or B and C on the tree, other inconsistencies appear. Phylogenetic trees usually point to the need for further analysis.

Molecular Clocks

Proteins evolve at different rates, and so do nucleic acids. However, for some proteins (such as cytochrome *c*) and for some mitochondrial genomes, the rate of evolution seems to be quite constant over time. If homologous proteins and DNA sequences are compared for taxa that are known to have diverged from common ancestors during certain periods in the past, the number of amino acid and nucleotide substitutions is proportional to the time that has elapsed since the lineages branched. The homologous proteins of bats and dolphins are much more alike than are those of sharks and tuna, which is consistent with the fossil evidence that sharks and tuna have been on separate evolutionary lines much longer than bats and dolphins. In this case, molecular divergence has kept better track of the time than have changes in body form.

Molecular clocks are calibrated in actual time by graphing the number of amino acid or nucleotide differences against the times for a series of evolutionary branch points known from the fossil record. The graphs can then be used to estimate the time of divergence for species and higher taxa, such as kingdoms and phyla.

The validity of molecular clocks rests on the assumption that mutation rates for the genes (and hence proteins) of interest are relatively constant. Especially for distantly related groups, differences in generation time and metabolic rates are two factors that may affect mutation rates and make molecular clocks less reliable. Within a closely related group of species, the consistent rate of DNA divergence implies that there is a significant background of neutral mutations that gradually changes the whole genome more than the specific genetic changes associated with adaptation. Evolutionary biologists disagree about the extent of neutral variations (see Chapter 23). Many biologists doubt that neutral evolution is prevalent, so they also question the credibility of molecular clocks as tools for *absolute* dating of the origins of taxa. There is less skepticism about the value of molecular clocks for determining the *relative* sequence of branch points in phylogeny.

The search for fossilized DNA continues despite recent setbacks: *science as a process*

Like all areas of science, molecular biology and its applications advance by an ebb and flow of ideas. Hypotheses are proposed, predictions made, and test results evaluated. Too often we highlight advances as though they involve only a forward march, and thus the importance of disproving hypotheses is lost. Some recent studies of fossilized DNA illustrate this key aspect of science.

Using PCR, a mainstay of molecular biology, systematists have attempted to extend nucleotide sequence analysis to DNA traces recovered from some types of fossils that retain organic material (see FIGURE 25.1d). The first apparent success came in 1990, when researchers reported using PCR to amplify DNA extracted from magnolia leaves found in 17-million-year-old sediments in Idaho. Since then, research teams have reported finding DNA fragments from many other sources, including a 40,000-year-old frozen mammoth, a 40-million-year-old insect fossilized in amber, bone marrow in a 65-million-year-old fossil of *Tyrannosaurus rex* in Montana, a 30,000-year-old fossilized arm bone of an extinct member of the human family tree (a Neanderthal from Germany), and a frozen 5000-year-old Stone Age man discovered by hikers in the Tyrolean Alps (FIGURE 25.14).

Each of these reports stimulated more analyses to confirm or disprove the DNA findings. Skeptics, including some of the same scientists who first reported the ancient DNA, searched for ways to ensure that the DNA traces were not contamination from bacteria or some other organism. Scientists realize it is unlikely that DNA would remain intact except when organisms fossilize in extremely dry or cold places, where organic material tends to be preserved. Otherwise, like other components of cells, nucleic acids start decomposing within hours of death and are completely gone in a few thousand years.

It now appears that most of the DNA first reported to be ancient, including that from dinosaurs, is contamination from bacteria or fungi. The conversion of amino acids from the biologically active L forms to the inactive D forms (see the discussion of racemization on pages 467–468) occurs at rates

FIGURE 25.14 ▪ Tyrolean Ice Man, a source of prehistoric human DNA. Hikers discovered the frozen 5000-year-old Stone Age man in a melting glacier in the Alps in 1991. Some skepticism that the man was a fraudulently transplanted South American mummy persisted until 1994, when researchers analyzed mitochondrial DNA recovered from the fossil. The DNA closely matches that of northern Europeans, not Native Americans. Continuing analysis of the Stone Age man's DNA may provide more clues about his place in human evolution.

similar to the breakdown rate of DNA. Racemization tests run on samples from fossils with DNA traces show such high percentages of D forms that it is very unlikely that any ancient DNA could remain in them. Low racemization percentages associated with the Neanderthal arm bone, the Tyrolean "ice man," and some but not all fossils in amber indicate that these are authentic cases of old DNA.

Does isolating DNA from fossils mean that we can recreate extinct organisms—dinosaurs, for example? Pieces of DNA mined from any fossil represent only tiny fractions of an organism's whole genome. Even if we had the whole genome of a dinosaur or some other ancient creature, we do not yet understand the developmental steps that translate a library of genetic instructions into an organism.

THE SCIENCE OF PHYLOGENETIC SYSTEMATICS

Charles Darwin defined the goals of modern systematics in *The Origin of Species:* "Our classifications will come to be, as far as they can be so made, genealogies." Spurred by Darwin's vision, systematists soon assembled enough information from the fossil record and detailed studies in comparative anatomy and embryology to construct a classification system and an associated tree of life rooted in phylogeny. The main trunks of the phylogenetic tree of life, as well as the relationships among its finer branches, emerged from a classical (post-Darwinian) period in systematics based largely on morphology. By the 1950s, however, efforts in systematics and phylogeny had slowed considerably. To some biologists, the analytical tools then available seemed intellectually stagnant, making systematics too subjective to be part of rigorous science. Concerning the bottom of life's tree, the scant fossil record offered little hope that more could be learned about life's origins. Was there really much more that could be learned about the history of life?

The science of phylogenetic systematics, with the goal of making classification more objective and consistent with evolutionary history, entered a vigorous new era in the 1960s. Just as molecular methods became readily available, computational technology helped usher in two new analytical approaches, phenetics and cladistics.

Phenetics increased the objectivity of systematic analysis

Phenetics (Gr. *phainein,* "to appear"; the term *phenotype* is derived from this same root) makes no phylogenetic assumptions and bases taxonomic affinities entirely on measurable similarities and differences. Phenetics compares as many anatomical characteristics (known as characters) as possible and makes no attempt to sort homology from analogy. Pheneticists contend that if enough phenotypic characters are examined, the contribution of analogy to overall similarity will be swamped by the degree of homology. Critics of phenetics argue that overall phenotypic similarity is not a reliable index of phylogenetic proximity. Although a strictly phenetic approach has few proponents today, the methods used in phenetics, especially the emphasis on multiple quantitative comparisons analyzed statistically with the help of computers, has had an important impact on systematics. Objective "number crunching" is especially useful for analyzing DNA sequence data and other molecular comparisons between species.

Cladistic analysis uses novel homologies to define branch points on phylogenetic trees

Phylogenetic trees may have two significant structural features. One feature is the location of branch points along the tree, symbolizing the relative time of origin of different taxa. The second is the extent of divergence between branches, representing how different two taxa have become since branching from a common ancestor.

Cladistic analysis has become synonymous with phylogenetic systematics. A **clade** (Gr. *clados,* "branch") is an evolutionary branch. Cladistic analysis classifies organisms according to the order in time that branches arose along a dichotomous phylogenetic tree. Each branch point in a tree is defined by novel homologies unique to the various species on that branch. Because it views the extent of divergence among organisms as uninformative in assessing evolutionary relationships, cladistic analysis considers *only* homologies in developing hypotheses about classification and phylogeny.

Outgroup Comparison

In a simplified example, suppose our goal is to divide four hypothetical taxa (*A, B, C,* and *D*) among evolutionary branches (FIGURE 25.15). Each taxon has a mixture of primitive characters that already existed in the common ancestor, along with characters that have evolved more recently. The sharing of primitive characters tells us nothing about the pattern of evolutionary branching from a common ancestor. So if we wish to develop a hypothesis about the branching sequence among these four taxa, it helps to know which characters are primitive for all four.

How do we start a phylogenetic tree—that is, define its roots—and is there an objective way to distinguish primitive characters from those that have evolved more recently? Cladistic analysis uses a concept called outgroup comparison to recognize primitive characters for all members of a group of interest and to establish a starting point for a phylogenetic tree. An **outgroup** (taxon O in FIGURE 25.15) is a species or group of species that is relatively closely related to the group of species actually being studied, but clearly not as closely related as any study-group members are to each other. All members of the study group are compared as a group to the outgroup. Characters common to both the outgroup and the study group are likely to have been present in a common ancestor and are thus *shared primitive characters.*

Characters	Outgroup *O*	Taxa *A*	*B*	*C*	*D*
1		1	1	1	1
2			2	2	2
3				3	3
4				4	4
5					5

(a)

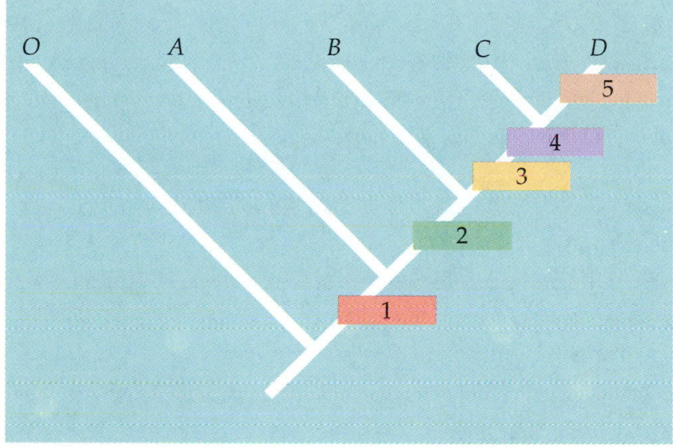

(b)

FIGURE 25.15 ▪ Constructing a phylogenetic tree using cladistic analysis. This idealized example illustrates how phylogenetic hypotheses are generated. **(a)** The set of data is arranged to compare taxa A–D using five characters. An outgroup (taxon O) possesses the primitive condition for all five characters (no color). What are the most likely phylogenetic relationships among the four taxa A–D? **(b)** A tree constructed from the data illustrates the use of the historical pattern of character change to postulate relationships among the taxa. The outgroup, with five primitive character states, defines the root of the phylogenetic tree. The branches on the tree are determined by comparing the number of derived characters in each taxon. Each node on the tree represents an ancestor common to all species above that node. The colored rectangles with the numbers 1 to 5 on the tree represent the change in an ancestor from the primitive to the derived character state. The derived condition represented by each rectangle distinguishes the group of taxa above from the taxa below. For instance, the change from the primitive to derived state of character 1 distinguishes the clade ABCD from the outgroup. The tree is constructed using the parsimony principle: The simplest tree that is consistent with the comparative data is most likely the correct one. For instance, parsimony in this example leads to the hypothesis that B, C, and D are the most closely related taxa because they all share characters 1 and 2; parsimony also leads to the hypothesis that taxa C and D are closer to each other than either is to taxon B, because C and D share characters 3 and 4. Typical cladistic analyses involve much more complex data sets and are usually handled by computer programs that also construct parsimonious trees.

Use of Synapomorphies (Shared Derived Characters) and Parsimony

Using the outgroup to help identify primitive characters of the study group, cladistic analysis looks for **synapomorphies**—*shared derived characters*—to construct a phylogenetic tree. These characters are homologies that evolved in an ancestor that is common to all species on *one* branch of a fork in the tree but not common to the other branch. The data set in FIGURE 25.15 is arranged to compare taxa A, B, C, and D and the outgroup O using five characters. (In an actual cladistic analysis, a researcher would compile a much larger set of comparative data.) Helping to define the root of the tree, outgroup taxon O possesses the primitive condition for all five characters. (The first node of the tree represents the ancestor common to the clade containing both the outgroup and the study group.) The branches on the tree are determined by comparing the number of derived characters in each taxon.

Another key component of cladistic analysis is the concept of **parsimony**, the quest for the simplest (and thus probably the most likely) explanation for observed phenomena. Useful in many areas of biology, parsimony in systematics means that phylogenetic trees using the fewest changes to illustrate evolutionary relationships have the greatest likelihood of being correct. FIGURE 25.15 illustrates the comparative use of primitive and shared derived characters and parsimony in constructing phylogenetic trees. FIGURE 25.16 on page 484 illustrates an actual example of cladistic analysis of several genera of dinosaurs.

Acceptance of Only Monophyletic Taxa

By focusing on phylogenetic branching, cladistic analysis accepts only monophyletic taxa. As a result, cladistic systematics produces some taxonomic surprises. For example, the branch point between birds and crocodiles is more recent than the branch point between crocodiles and the other reptiles; that is, birds and crocodiles have synapomorphies not present in snakes and lizards. Indeed, the fossil record supports the hypothesis that both birds and crocodiles evolved during the Mesozoic era from a common ancestor different from the ancestor of snakes and lizards. In the *strictly* cladistic view, there are no such taxa as Class Aves (birds) and Class Reptilia, as we conventionally know them, because classifying snakes, lizards, and crocodiles in the same taxonomic group and excluding birds from that group is at odds with evolutionary history. Such a taxon is paraphyletic, and only monophyletic taxa reflect phylogeny (see FIGURE 25.9). Apparently, birds seem so different because of the extensive morphological remodeling associated with flight that has occurred since birds branched from their ancestors. While recognizing the inconsistencies with cladistic analysis, many biologists still use the classes Aves and Reptilia as convenient groupings for these familiar vertebrates.

	Outgroup *Allosaurus*	*Stegosaurus*	*Parasaurolophus*	*Pachycephalosaurus*	*Triceratops*
Hole in hip socket					
Posterior process of pubis	absent				
Unequal enamel layer on tooth	absent	absent			
Shelf at base of skull	absent	absent	absent		

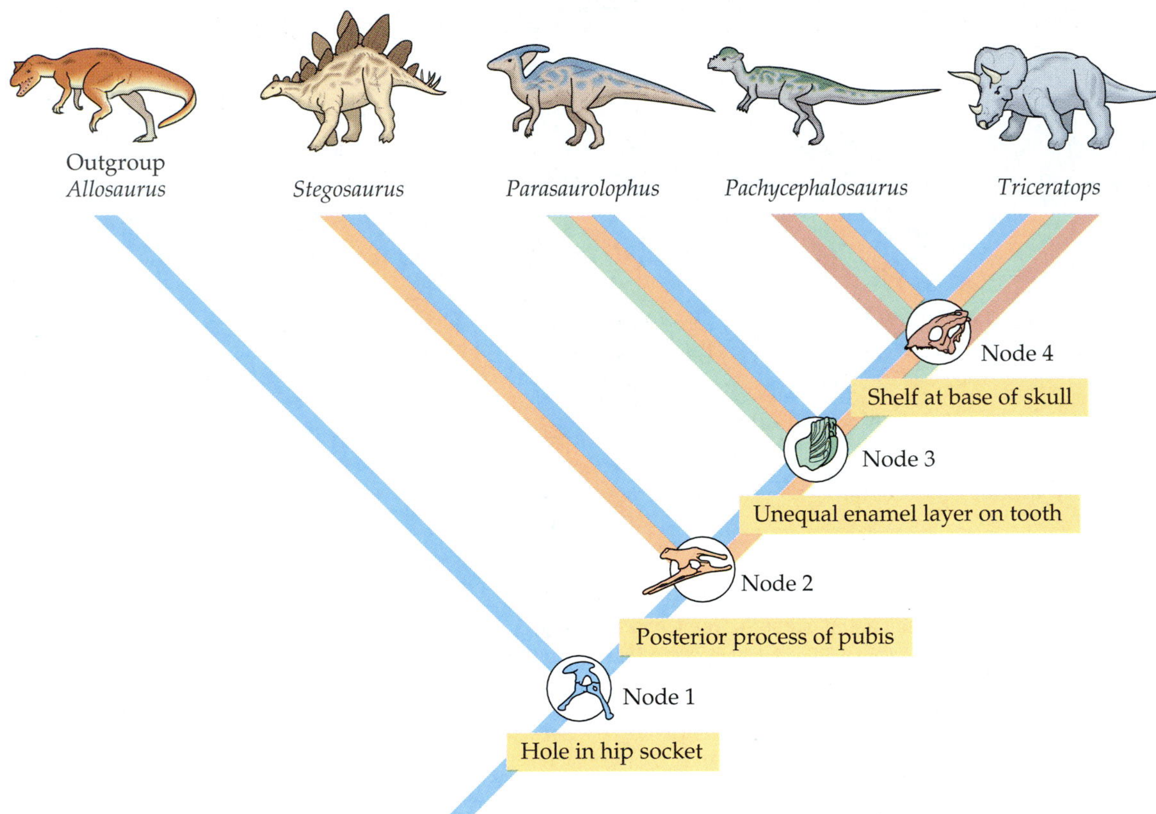

FIGURE 25.16 ▪ Building a phylogenetic tree of dinosaurs. The top part of this illustration tabulates the presence or absence of morphological characters in a group of dinosaurs. The outgroup in this analysis, *Allosaurus,* is a member of the so-called lizard-hipped (saurischian) dinosaurs (their pelvises resembled those of lizards). Saurischians were fundamentally different from the other four genera illustrated here, which are all bird-hipped (ornithischian) dinosaurs. The tree is the simplest (most parsimonious) one that is consistent with the comparative data in the table. Each node, representing an ancestor common to all species above that node, is defined by a character present in all dinosaurs above that branch point and in their common ancestor. The character at node 1, a hole in the hip socket, is an adaptation associated with the erect posture of dinosaurs. It is present in all dinosaurs (notice that the blue color representing this character runs throughout the tree). Because no other four-legged vertebrate has this type of hip socket, it is said to be a derived character of dinosaurs, and it distinguishes the dinosaurs *as a group.* However, if we are comparing groups of dinosaurs, the hole in the hip socket is a primitive character that cannot help us distinguish one dinosaur from another. Node 2 represents the common ancestor of all four dinosaurs whose phylogeny is the focus of this example. The defining character of node 2, a backward-pointing extension of the pubis (a pelvic bone), is a shared derived character that distinguishes all dinosaurs above node 2 from all those below it. This character is primitive when we are comparing the dinosaurs above node 2. These same principles apply for the characters defining the nodes on any phylogenetic tree developed by cladistic analysis. This particular tree does not show any features that distinguish *Pachycephalosaurus* from *Triceratops.* However, a number of characters (such as a neck shield and three horns on the head of *Triceratops*) distinguish these two genera. Notice also that *Pachycephalosaurus* and *Triceratops* share more derived characters than any other pair of genera in this example. This suggests a relatively close evolutionary relationship. Once systematists have constructed a tree such as this, they can use it to group species into a hierarchy of monophyletic taxa.

Phylogenetic systematics relies on both morphology and molecules

Modern systematics owes much of its vitality to the power of cladistic analysis to formulate hypotheses about the history of life. Biologists have been accumulating phenotypic (mainly morphological) information about organisms, living and extinct, for centuries, and cladistics was largely developed to objectively compare phenotypic characteristics. The amount of morphological information available for phylogenetic comparisons far outweighs the available molecular data, and most of our current phylogenetic hypotheses and classification schemes are derived from morphological data.

Systematics gained considerable strength and popularity when evolutionary biologists began using cladistic methods to generate phylogenetic hypotheses from DNA, RNA, and protein sequence data. Molecular studies have barely begun, but they have already made significant contributions to our understanding of evolutionary history. Molecular databases are expanding rapidly but currently exist for only a small fraction of species. Though hardly a complete measure of any organism's evolutionary history, a genome contains much more information than do anatomical features, which represent a relatively small number of expressed genes. As we will see in the next unit, researchers are pursuing questions about the earliest branchings of life, using slow-changing molecular sequences such as rRNAs. In the phylogenetic short term, anatomical structures take too long to change to be useful in assessing life's most recent branchings, but useful comparisons can be made among rapidly changing sequences, such as mtDNAs. As we discussed, molecular sequence comparisons can also assess the relative and absolute timing of phylogenetic events.

Initially, molecular methods were hailed as a panacea for phylogenetic studies, but as evolutionary biologist Wayne Maddison puts it, "As large as the genome is, history is even larger, and for many of our attempts to understand it, we will be left with too few and too ambiguous data." Indeed, with greater use of sequence data came the understanding that molecular comparisons present their own set of challenges (for example, determining homologies and choosing appropriate sequences). Most researchers today appreciate that both morphology and molecular biology are essential sources of data for phylogenetic study. The strongest support for any phylogenetic hypothesis is agreement between results obtained from repeated tests comparing any available molecular data with all other types of evidence.

Two decades ago, phylogenetic trees appeared only occasionally in a few research journals in evolutionary biology. Today, **phylogenetic biology**, the application of cladistic analysis in the study of evolutionary history and its relations to all aspects of life, pervades virtually all subfields of life science. To mention but a few examples, cladistic analysis is used in animal physiology to study the evolution of body temperature regulation and brain structures; plant developmental biologists apply cladistics to the evolution of genes regulating flower development; behavioral biologists use cladistics and phylogenetic trees to plot the evolutionary history of animal social systems; and conservation biologists apply these methods to sort out genetic differences among populations of endangered species.

■ ■ ■

In this unit, we have seen how the Darwinian theme of descent with modification has shaped the entire science of biology. Phylogenetic biology is just one example of the Darwinian legacy. We have also seen that evolutionary theory itself has evolved, and we have examined some of the progress evolutionary biologists have made since Darwin's time toward understanding evolutionary processes. Vigorous debate is vital to that progress and is a healthy sign that evolutionary biology is a robust science.

CHAPTER REVIEW

REVIEW OF KEY CONCEPTS

(with page numbers and key figures)

THE FOSSIL RECORD AND GEOLOGICAL TIME

- **Sedimentary rocks are the richest source of fossils (pp. 464–466, FIGURE 25.1)** The fossil record provides the historical archives biologists use to study the history of life.
- **Paleontologists use a variety of methods to date fossils (pp. 466–468, TABLE 25.1, FIGURE 25.2)** Sedimentary strata reveal the relative ages of fossils in successive geological periods. The absolute ages of fossils in years can be determined by radiometric dating and other methods. The geological eras and periods correspond to major transitions in the composition of fossil species. The chronology of geological periods and eras makes up the geological time scale.
- **The fossil record is a substantial, albeit incomplete, chronicle of evolutionary history (pp. 468–469)** The fossil record favors species that existed for a long time, were abundant and widespread, and had shells or hard skeletons.
- **Phylogeny has a biogeographical basis in continental drift (pp. 469–470, FIGURE 25.4)** Continental drift has had a significant impact on the history of life by causing major geographical rearrangements affecting biogeography and evolution. The formation of the supercontinent Pangaea during the late Paleozoic era and its subsequent breakup during the early Mesozoic era explain many biogeographical puzzles.
- **The history of life is punctuated by mass extinctions followed by adaptive radiations of the survivors (pp. 470–473, FIGURES 25.5, 25.6)** Evolutionary history has involved long, relatively stable periods interrupted by intervals of extensive species turnover—mass extinctions followed by grand episodes of adaptive radiation.

PHYLOGENY AND SYSTEMATICS

- **Taxonomy employs a hierarchical system of classification (pp. 473–475, TABLE 25.2)** Systematics, the study of biological diversity in an evolutionary context, includes taxonomy, the identification and classification of species.
- **The branching pattern of a phylogenetic tree represents the taxonomic hierarchy (pp. 475–476, FIGURE 25.8)** Phylogenetic trees represent hypotheses developed by systematists whose goal is to have classification reflect evolutionary history.
- **Determining monophyletic taxa is key to classifying organisms according to their evolutionary history (pp. 476–477, FIGURE 25.9)** A taxon is monophyletic if a single ancestor gave rise to all species in that taxon and to no species placed in any other taxon. Phylogenetic hypotheses are based on homologies (similarities due to common ancestry).
- **Molecular biology provides powerful tools for systematics (pp.
25.1 477–481, FIGURES 25.11–25.13)** Evolutionary relationships can be revealed by comparing amino acid sequences of proteins and nucleotide sequences of DNA and RNA.
- **The search for fossilized DNA continues despite recent setbacks (pp. 481–482)** It turns out that most DNA thought to be ancient is contamination from bacteria or fungi.

THE SCIENCE OF PHYLOGENETIC SYSTEMATICS

- **Phenetics increased the objectivity of systematic analysis (p. 482)** Phenetics classifies organisms solely on the basis of similar characteristics.
- **Cladistic analysis uses novel homologies to define branch points on phylogenetic trees (pp. 482–483, FIGURES 25.15, 25.16)** Cladistic analysis highlights phylogeny by basing classification on the timing of branch points as defined by synapomorphies, or shared derived characters.
- **Phylogenetic systematics relies on both morphology and molecules (p. 485)** The strongest support for any phylogenetic hypothesis is agreement between molecular data and other evidence.

SELF-QUIZ

1. A paleontologist estimates that when a particular rock formed, it contained 12 mg of the radioactive isotope potassium-40. The rock now contains 3 mg of potassium-40. The half-life of potassium-40 is 1.3 billion years. About how old is the rock?
   a. 0.4 billion years
   b. 0.3 billion years
   c. 1.3 billion years
   d. 2.6 billion years
   e. 5.2 billion years

2. If humans and pandas belong to the same class, then they must also belong to the same
   a. order
   b. phylum
   c. family
   d. genus
   e. species

3. In the case of comparing birds to other vertebrates, having four appendages is (explain your answer)
   a. a shared primitive character
   b. a synapomorphy
   c. a character useful for distinguishing the birds from other vertebrates
   d. an example of analogy rather than homology
   e. a character useful for sorting the avian (bird) class into orders

4. The evidence linking the Cretaceous extinctions and the Chicxulub impact event is
   a. charcoalized fossils found in Central America
   b. extinction rates that are globally uniform
   c. the disproportionately high number of North American species to become extinct
   d. the large number of marine species that became extinct
   e. the geological evidence indicating the extinction period to be within a single year

5. The greatest adaptive radiation of the animal kingdom occurred during the
   a. early Precambrian era
   b. late Precambrian era
   c. early Paleozoic era
   d. early Mesozoic era
   e. early Cenozoic era

6. The DNA from two species is compared by the method of restriction mapping. Extensive similarity between the species in the collection of DNA fragments from treatment with a restriction enzyme indicates that
   a. the genes being compared have the same functions
   b. most sites recognized by the restriction enzyme have equivalent locations in the DNA samples from the two species
   c. the two species normally possess the same restriction enzyme
   d. the DNA fragments that match in size between the two species have identical base sequences
   e. the genomes of the same species are about the same size in their total amount of DNA

7. Extensive adaptive radiations have usually followed in the wake of mass extinctions mainly because
   a. many adaptive zones are available
   b. conditions of the physical environment are usually at their most favorable after some crisis has passed
   c. the survivors have superior adaptations that enable them to move into many environments when conditions improve after the extinction episode
   d. genomes tend to expand when adaptive zones are vacated
   e. given a stable environment, biological diversity tends to increase

8. Which of the following would be most useful for constructing a phylogenetic tree emphasizing evolutionary branchings among several fish species? (explain your answer)
   a. several analogous characteristics shared by all the fishes
   b. a single homologous characteristic shared by all the fishes
   c. the total degree of morphological similarity among various fish species
   d. several characteristics thought to have evolved after different fish diverged from one another
   e. a single characteristic that is different in all the fishes

9. Microsatellites are short repetitive sequences of nuclear DNA. Microsatellite analysis of ten DNA loci from red wolves, gray wolves, and coyotes identified different microsatellites in gray wolves and coyotes. In red wolves, however, some microsatellites matched those of gray wolves, while others matched coyote microsatellites. This finding suggests that
   a. DNA sequence analysis is unreliable for extant species
   b. the red wolf is a coyote–gray wolf hybrid rather than a true species
   c. incorrect restriction enzymes were used to cleave the DNA
   d. the red wolf is the common ancestor of both the gray wolf and the coyote
   e. the red wolf is the outgroup for this group of species

10. Swifts and swallows are aerial foragers. Although the birds are similar in appearance, biologists place them in different orders; thus their similarities
   a. must represent homologies
   b. must have resulted from convergence
   c. support the hypothesis of a polyphyletic origin for birds
   d. are synapomorphic
   e. represent a clade

CHALLENGE QUESTIONS

1. In the "DNA clock," some nucleotide changes cause amino acid substitutions in the encoded protein (nonsynonymous changes), and others do not (synonymous changes). In a comparison of rodent and human genes, rodents were found to accumulate synonymous changes 2.0 times faster than humans, and nonsynonymous substitutions 1.3 times as fast. What factors could explain this difference? How do such data complicate the use of molecular clocks in absolute dating?

2. Imagine that Pangaea were re-formed today. What might be some of the environmental effects, and how might these changes influence life on Earth?

SCIENCE, TECHNOLOGY, AND SOCIETY

Experts estimate that human activities cause the extinction of hundreds of species every year. The natural rate of extinction is thought to be a few species per year. As we continue to alter the global environment, especially by cutting down tropical rain forests, the resulting extinction will probably rival that at the end of the Cretaceous period. Most biologists are alarmed at this prospect. What are some reasons for their concern? Consider that life has endured numerous mass extinctions and has always bounced back. How is the present mass extinction different from previous extinctions? Why? What might be some of the consequences for the surviving species?

FURTHER READING

Benton, M., and D. Harper. *Basic Paleontology.* Essex, England: Addison Wesley Longman, 1997. An introductory text.

Briggs, D. E. G., D. H. Erwin, and F. J. Collier. *The Fossils of the Burgess Shale.* Washington, DC: Smithsonian Institution Press, 1994. A detailed manual illustrating fossils of the Cambrian radiation.

Eldridge, N. *Fossils, The Evolution and Extinction of Species.* Princeton, NJ: Princeton University Press, 1991. Spectacular photos and concise, lucid text illuminate high points of the fossil record.

Gee, H. *Before the Backbone: Views on the Origin of Vertebrates.* New York: Chapman and Hall, 1996. Lucid analyses of hypotheses about vertebrate beginnings.

Gibbons, A. "Calibrating the Mitochondrial Clock." *Science,* January 2, 1998. A brief, critical review of molecular dating techniques.

Hecht, J. "The Geological Timescale." *New Scientist,* May 20, 1995. How to date fossils.

Lewin, R. *Patterns in Evolution, The New Molecular View.* New York: Scientific American Library, 1997. Emphasis on the influx of molecular biology in the study of evolution.

Majerus, M., W. Amos, and G. Hurst. *Evolution, The Four Billion Year War.* Essex, England: Addison Wesley Longman, 1996. An accessible, up-to-date text; includes a lucid chapter on phylogenetic analysis.

Monastersky, R. "Life's Closest Call, What Caused the Spectacular Extinctions at the End of the Permian Period?" *Science News,* February 1, 1997.

Officer, C., and J. Page. *The Great Dinosaur Extinction Controversy.* Reading, MA: Addison-Wesley, 1996. A scathing critique of the impact hypothesis and its general acceptance by the scientific and popular media.

Pääbo, S. "Ancient DNA." *Scientific American,* November 1993. Analyzing DNA from fossils.

Vogel, G. "Phylogenetic Analysis: Getting Its Day in Court." *Science,* March 14, 1997. Forensic use of phylogenetic analysis.

WEB LINKS

Visit the special edition of *The Biology Place* for BIOLOGY, Fifth Edition, at http://www.biology.com/campbell. Go to Chapter 25 for online resources, including learning activities, practice exams, and links to the following web sites:

"Paleontology at the U.S. Geological Survey (USGS)"
This is the home page of the group of the USGS that focuses on fossils and the fossil record. Of particular interest is their extensive glossary.

"Fossils, Rocks, and Time" and "Geologic Time"
These online publications from the United States Geological Service explore geologic time, sedimentary rocks, and fossils.

"Learning from the Fossil Record"
This site is a nice primer for understanding fossils and paleontology.

"The Tree of Life"
The Tree of Life is a web-based "phylogenetic navigator" that contains information about the phylogenetic relationships and characteristics of organisms and illustrates the diversity and unity of living organisms.

UNIT FIVE

THE EVOLUTIONARY HISTORY OF BIOLOGICAL DIVERSITY

Elisabeth Vrba always knew she would be a scientist. Her undergraduate degree at the University of Cape Town was in mathematical statistics and zoology. Currently professor of paleontology and biology at Yale University, Dr. Vrba describes her career and research in the spirit of two grand principles: Earth and life evolved together, and nothing in biology makes sense except in the light of evolution. Her controversial turnover-pulse hypothesis has enlivened debate and research on the effect of past climatic changes on macroevolution, especially seminal events in human evolution. Larry Mitchell interviewed Dr. Vrba in her office at the Peabody Museum of Natural History at Yale.

This unit on biological diversity follows our chapters on evolutionary principles. Could you help us make the connection between evolution and diversity?

I find that the best evidence for evolution is the diversity of living organisms, the distribution of characteristics among species, and the hierarchical patterns of diversity. For instance, all of life has DNA and RNA, and all eukaryotes have nuclei and organelles. You could think of these common features as creations of a higher power or, scientifically, as the result of life having evolved on a single tree. This is a very simple hypothesis, but the evidence supporting it is overwhelming.

Tell us some of the highlights of your career since you began graduate studies.

In 1966, my husband and I were living in Pretoria, and I was looking around for a Ph.D. topic. I remember one day my old friend Bob Brain said to me, "Well look, there's this pile of rocks full of fossils from the Transvaal caves, and they still have to be cleaned. This is hard work, and nobody wants to do it." Not having much money, I asked, "Can you pay me something?" and he said, "No. Not in the beginning." I still had some money, so I said I would do it for 6 months to see if I could manage. And then I got really keen on paleontology! You see, I was working with materials from some of the most famous fossil sites in South Africa, especially in human evolution. Some of the fossil specimens you see here on my desk are plastic casts of actual fossils that came from two caves in which I did my own excavating. You will recognize some of the names: a robust-bodied australopithecine, sometimes called *Paranthropus*, and one that many workers think is closer to the lineage of the genus *Homo*; that is *Australopithecus africanus*.

About how old are these fossils?

This specimen of *Australopithecus africanus* would have lived about 2.5 million years ago, and *Paranthropus*, which was found with fossils of *Homo*, would be about 1.8 million years old. At that point, there were two prominent lineages, and we can see them side by side in the same deposits in South Africa and also in East Africa.

INTERVIEW

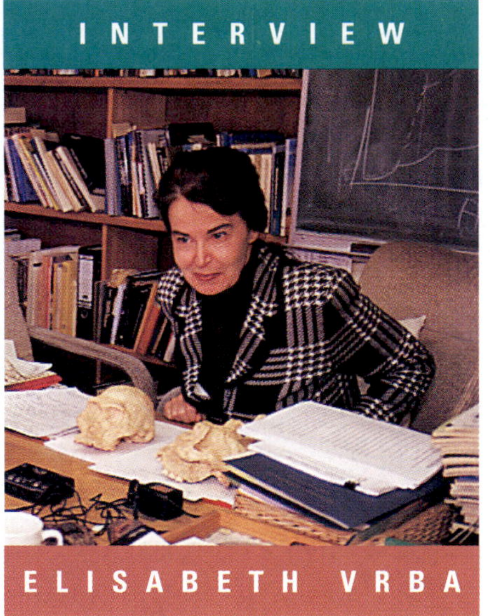

ELISABETH VRBA

What exactly does the term *lineage* mean in an evolutionary sense? Is it something different (more specific) than a line of descendants from a particular ancestor?

A lineage, as many of us now use the term, is synonymous with a clade, or a monophyletic group—a single ancestral species and all of its descendants. For instance, by the lineage of the genus *Homo*, we mean that there is a common ancestor that belongs to *Homo*, and a number of descendants from that. There may have been a lot of branching after that first ancestral *Homo*—for instance, into *Homo habilis, Homo erectus, Homo neanderthalensis*, and *Homo sapiens*—and that entire clade, or monophyletic group, from a single ancestor would be called the lineage of the genus *Homo*. The term *lineage* can also be used for a single species (of sexually reproducing organisms). For instance, a single lineage would be from the time *Homo sapiens* started, through and including the present.

A major focus of your research has been to use the fossil record to analyze the effects of past tectonic and climatic events on species diversity. Tell us more about this.

We are very interested in macroevolutionary processes, speciation, and evolutionary change in relation to paleoclimates and paleoecology. In other words, we look for evidence in the fossil record of a relationship between morphology, phylogeny, and patterns of associations of fossil species and major changes in the physical environment, such as climate. Why did some lineages change hardly at all, while others show massive long-term trends? I am especially interested in the environmental context of assemblages of fossil mammals, and in environmental influences on human evolution.

Please explain your turnover-pulse hypothesis.

Quite simply, my idea of turnover-pulse is that major physical changes on earth, such as climatic shifts and tectonism, underlie periods of major change in the biota. In other words, environmental pulses correlate with abrupt increases in speciations and extinctions, specific times in Earth's history when many species vanished and others appeared. With my special interest in human evolution, I have looked in the fossil record of phylogenetic patterns in other species that were contemporary with our early ancestors. Working in Africa, I found that antelopes make up a large portion—60 to 80 percent—of the mammalian fossils in the same strata where we find our early ancestors, the australopithecines. Antelopes are good indicators of paleoclimates. For instance, some evidence in the fossil record indicates that about 2.5 to 3 million years ago many species of antelopes that inhabited moist woodlands in southern Africa disappeared and were replaced by new species that lived in drier, more open areas we call savannas. Fossil evidence also points to a radiation of hominid species on African savannas at about the same time. I would predict that as we learn more about macroevolution, we will find that recurrent pulses of varying intensity are an important part of the history of life.

It seems that the turnover-pulse idea closely resembles the concept of punctuated equilibrium.

Punctuated equilibrium was articulated by Eldredge and Gould in a 1972 paper (Eldredge, N. & Gould, S.J. 1972. Punctuated equilibria: an alternative to phyletic gradualism. Pp. 82–115, in *Models in Paleobiology*, T. Schopf, ed., W.H. Freeman, San Francisco). Punctuated equilibrium and turnover-pulse resemble each other only in some respects and differ substantially in others. Resemblances include that both argue that speciation most often occurs in small allopatric populations ("allopatric speciation," that is, speciation in geographically separated populations). Both predict a pattern of morphological change in a given lineage through time that is episodic and mostly associated with speciation (that is, with lineage splitting).

Could you summarize some of the current debate about turnover-pulses?

As any scientist knows, controversy is essential, and the turnover-pulse idea has generated a great deal. The current debate is mainly about the timing of events and in some cases about the methods used to determine whether pulses actually occurred. There will always be new information challenging what has gone before, but the important thing is to keep the research moving forward. We have much, much more to learn about the evolutionary events that gave rise to all the diversity of life than we know at present.

Often scientists seem to work on one small branch of a research area. But your work involves a great range of subjects.

It's sad to see specialists in different fields, each with fantastic information and ideas, limiting themselves because they never speak to each other. When you cross that boundary and say, "Tell me what you've got" and they say, "Well, look at this," you suddenly realize, Oh, that is so relevant to what I'm doing!

From the beginning, I made up my mind that if there's ever a subject that I'm interested in I'm going to do my level best to find out about it, even if I'm not, so-called, trained. After all, what is training? You've got to start somewhere. I'm very interested in how Earth and life evolved together, and so I've had to learn about geology and climates as well as paleontology.

Great discoveries have always been made at the intersections of disciplines. Reading Malthus, Charles Darwin was looking at demography; he was also looking at the works of an economist, Adam Smith—quite different—and at the works of the Belgian statistician Adolphe Quetelet. And then, out came the idea—natural selection. There are many examples of people working at intersections, simply to get a different perspective. It's like getting in a boat and traveling to the other side of a river and looking to see what it's like from there. I keep telling my students they've got to be open and try to learn about different things. I want them to have more than one kind of skill. I firmly believe in interdisciplinary work, and that's where all the excitement lies.

Speaking of Charles Darwin, ecologists are fond of claiming him as one of their own, so to speak. What do you think about that?

Yes, that's absolutely correct. Chief among Darwin's many contributions is natural selection, which he saw resulting from an interaction between organisms, their particular features, and the environment. This is the very core of ecology. Also, Darwin was not just an arm-chair philosopher; he went out and looked at nature, at the interactions, and he had a very broad view of the living world and the environment.

Do you have any advice or counsel for students beginning a career in life science?

I think the first and most important thing is that you have got to love what you do. And what I mean by that is don't be pushed by your parents or by what your peers think. Don't ask, "Is there money in it? Is it going to land me a job?" First ask, "Can I love this? Is my heart in it?" Don't get sidelined, don't get worried about difficulties. Just work hard. There are always difficulties and hurdles, not least of which is one's own inability to concentrate, to focus, to commit oneself. That's perhaps the biggest hurdle you ever have to overcome. You've got to personally make a commitment: I love this and I want to do it, and it's worth it. I'm going to put my energy there.

CHAPTER 26

EARLY EARTH AND THE ORIGIN OF LIFE

Introduction to the History of Life

- Life on Earth originated between 3.5 and 4.0 billion years ago
- Major episodes in the history of life: *a preview*

Prebiotic Chemical Evolution and the Origin of Life

- The first cells may have originated by chemical evolution on a young Earth: *an overview*
- Abiotic synthesis of organic monomers is a testable hypothesis: *science as a process*
- Laboratory simulations of early Earth conditions have produced organic polymers
- Protobionts can form by self-assembly
- RNA was probably the first genetic material
- The origin of hereditary information made Darwinian evolution possible
- Debate about the origin of life abounds

The Major Lineages of Life

- Arranging the diversity of life into the highest taxa is a work in progress

Life is a continuum extending from the earliest organisms through various lineages to the great variety of forms alive today. In Unit Five we will survey the diversity of life today and trace the evolution of this diversity over 3.8 billion years of history.

One recurrent theme in these chapters is the association between biological and geological history. Geological events that alter environments change the course of biological evolution. The formation and subsequent breakup of the supercontinent Pangaea, for instance, affected the diversity of life tremendously (see Chapter 25). Conversely, life has changed the planet it inhabits. For example, the evolution of photosynthetic organisms that release oxygen into the air completely altered Earth's atmosphere. (This early photosynthetic life included cyanobacteria similar to those in the dense mats that resemble stepping stones in the painting of early Earth at the left.) Much more recently, the emergence of Homo sapiens *has changed the land, water, and air on a scale and at a rate unprecedented for a single species. The histories of Earth and its life are inseparable.*

These chapters also emphasize key junctures in evolution that have punctuated the history of biological diversity. Geological history and biological history have been episodic, marked by what were in essence revolutions that opened many new ways of life. Historical study of any sort is an inexact discipline that depends on the preservation, reliability, and interpretation of past records. The fossil record of past life is generally less and less complete the farther into the past we delve. Fortunately, each organism alive today carries traces of its evolutionary history in its molecules, metabolism, and anatomy. As we saw in Unit Four, such traces are clues to the past that augment the fossil record. Still, the evolutionary episodes of greatest antiquity are generally the most obscure.

This chapter is the most speculative of the unit, for its main subject is the origin of life on a young Earth, and no fossil record of that seminal episode exists. We begin with the oldest known fossils, in order to certify the antiquity of life on Earth. Then we preview some of the seminal milestones in the history of life and examine hypotheses about how natural processes on the youthful planet could have created life. The last section of the chapter introduces the various kingdoms of life, as a prelude to the survey of biological diversity in Chapters 27 through 34.

INTRODUCTION TO THE HISTORY OF LIFE

Life began remarkably early in Earth's history, and those first organisms were ancestral to the kaleidoscope of biological diversity we see today. The organisms most familiar to us are macroscopic and multicellular—mainly plants and animals. However, for the first three-quarters of evolutionary history, Earth's only organisms were microscopic and unicellular.

(a)

(b)

(c)

FIGURE 26.1 ▪ Bacterial mats and stromatolites. (a) Kenneth Nealson and Lynn Margulis, who study the history of life, are collecting bacterial mats in a Baja California lagoon. The mats are sedimentary structures produced by colonies of bacteria and cyanobacteria that live, uncropped by predators, in environments inhospitable to most other life. **(b)** The bands seen in this section of a mat are layers of sediment that adhere to the sticky prokaryotes, which produce the succession of layers by migrating upward. **(c)** Fossilized mats known as stromatolites resemble the layered structures formed by modern-day bacterial colonies (shown in b). This stromatolite is a Western Australian specimen that is about 3.5 billion years old.

Life on Earth originated between 3.5 and 4.0 billion years ago

Earth formed about 4.5 billion years ago, and life probably began only a few hundred million years later. Scientists have found isotopes of carbon that seem to represent the metabolic activity of organisms in 3.8-billion-year-old rocks in Greenland.

One would guess from the relatively simple structure of the prokaryotic cell (compared with the eukaryotic cell) that the earliest organisms were prokaryotes, and the fossil record now supports that presumption. Evidence of ancient prokaryotic life has been found in rocks called stromatolites (Gr. *stroma,* "bed," and *lithos,* "rock"). **Stromatolites** are banded domes of sedimentary rock that are strikingly similar to layered mats formed today in salt marshes and some warm-ocean lagoons by colonies of bacteria and cyanobacteria. The layers are sediments that stick to the jellylike coats of the motile microbes, which continually migrate out of one layer of sediment and form a new one above, producing the banded pattern (FIGURE 26.1). Although some stromatolites may have formed from mineral sedimentation without the presence of life, fossils resembling spherical and filamentous prokaryotes have been found in 3.5-billion-year-old stromatolites in southern Africa and Western Australia (FIGURE 26.2). These

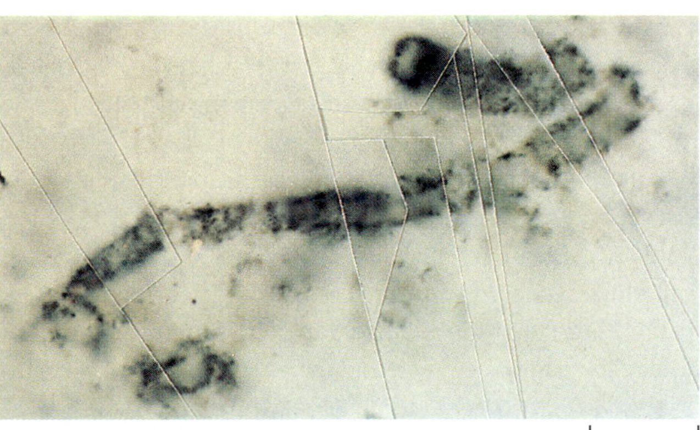

FIGURE 26.2 ▪ An early prokaryote. Fossils in rocks older than about 500 million years escaped notice until about 35 years ago, partly because they are microscopic. This fossilized filamentous prokaryote, about 3.5 billion years old, was found in a stromatolite collected in Western Australia (LM).

are currently the oldest known fossils of living organisms. However, the Western Australia fossils appear to be those of photosynthetic organisms, perhaps oxygen producers. If so, it is likely that life had been evolving long before these organisms lived, possibly as early as 4.0 billion years ago.

Major episodes in the history of life: *a preview*

Before we focus on the question of how life on Earth began, let's preview some major episodes of biological history in the context of geological time (FIGURE 26.3). Prokaryotes originated within a few hundred million years after Earth's crust cooled and solidified, beginning a succession of microorganisms that were the only forms of life on Earth for about 2 billion years. RNA sequence data show that two distinct groups of prokaryotes—Bacteria and Archaea—diverged early in the history of life. According to the fossil record, the split occurred about 3 billion years ago; according to some molecular data, the time was closer to 2 billion years ago.

About 2.5 billion years ago, the production of oxygen (O_2) by early photosynthetic prokaryotes created an aerobic atmosphere, setting the stage for the evolution of aerobic life. As prokaryotic evolution continued, some organisms were able to tolerate the corrosive effects of atmospheric oxygen and became the first to use oxygen, and some species even used oxygen to metabolize organic molecules. In terms of metabolic diversity, more variations exist among prokaryotes than among all eukaryotes combined.

The oldest fossils that are definitely eukaryotic are about 1.7 billion years old, but eukaryotic cells probably evolved several hundred million years earlier. Strong evidence supports the hypothesis that eukaryotic cells evolved from a symbiotic community of prokaryotes; in Chapter 28 we will trace the evolutionary steps that created membranous organelles (mitochondria and chloroplasts) from small prokaryotic cells living within larger prokaryotic cells.

The name *protists* refers to a large, disparate group of unicellular eukaryotes and some closely related multicellular organisms. By using nucleic acid sequencing and phylogenetic analyses, researchers are beginning to unravel the complex history of protist lineages. We will take a closer look at protistan evolution in Chapter 28.

Plants, fungi, and animals arose from distinct groups of unicellular eukaryotes during the Precambrian. Plants evolved from a lineage of green algae; fungi and animals arose from different groups of heterotrophic unicellular organisms. Molecular evidence indicates that fungi are more closely related to animals than to plants.

So far, the oldest animal fossils, mainly soft-bodied invertebrates, date from about 700 million years ago in the late Precambrian. The Precambrian fauna was less diverse than that of the Cambrian, but the basic body plans of most modern animal phyla probably arose in the late Precambrian.

For nearly 90% of its existence (about 3.5 billion years), life on Earth was confined to aquatic environments, and the colonization of land was one of the pivotal milestones in the history of life. Plants, in the company of fungi, led the way about 475 million years ago. The roots of most plants are associated with fungi in a mutually beneficial symbiotic relationship. Fossils indicate that early land plants were closely associated with fungi, and it is likely that the move onto land depended on this association. Plants transformed the landscape, creating new environmental opportunities for all life, especially herbivorous animals and their predators. We devote Chapters 29 and 30 in this unit to plant diversity. Chapter 31 explores fungi, and Chapters 32–34 examine the major lineages of animals.

PREBIOTIC CHEMICAL EVOLUTION AND THE ORIGIN OF LIFE

The question of how life began is more specifically about the genesis of prokaryotes. Sometime between about 4.0 billion years ago, when Earth's crust began to solidify, and 3.5 billion years ago, when the planet was inhabited by bacteria advanced enough to build stromatolites, the first organisms came into being. What was their origin?

The first cells may have originated by chemical evolution on a young Earth: *an overview*

Most biologists subscribe to the hypothesis that life on Earth developed from nonliving materials that became ordered into molecular aggregates that were eventually capable of self-replication and metabolism. As far as we know, life cannot arise by spontaneous generation from inanimate material today, but conditions were very different when Earth was only a billion years old. The atmosphere was different (there was little atmospheric O_2, for instance), and lightning, volcanic activity, meteorite bombardment, and ultraviolet radiation were all more intense than what we experience today. In that ancient environment, the origin of life was evidently possible, and it is likely that at least the early stages of biological inception were inevitable. However, debate abounds about what occurred during these early stages.

According to one hypothetical scenario, the first organisms were products of a chemical evolution in four stages: (1) the abiotic (nonliving) synthesis and accumulation of small organic molecules, or monomers, such as amino acids and nucleotides; (2) the joining of these monomers into polymers, including proteins and nucleic acids; (3) the aggregation of

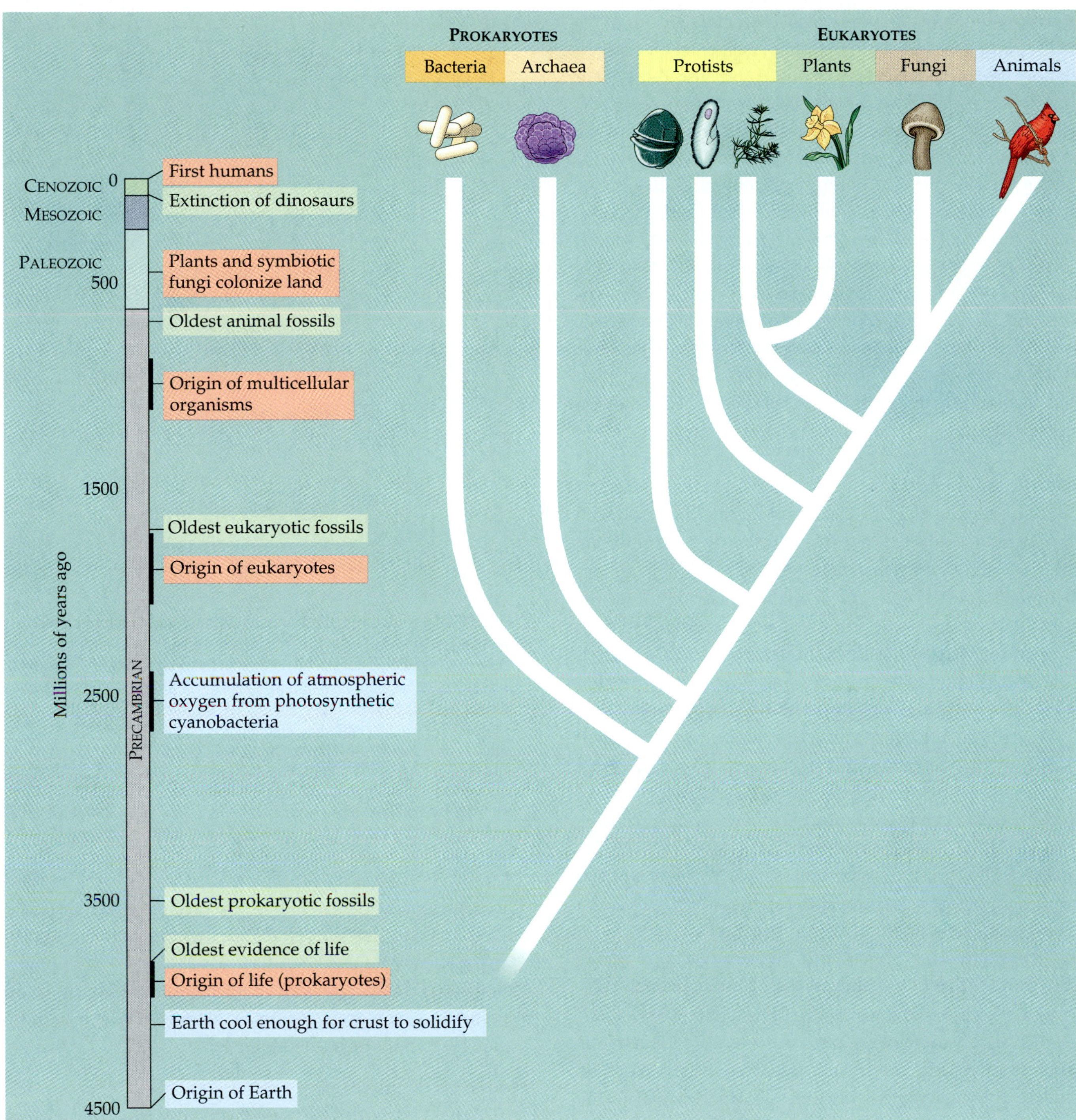

FIGURE 26.3. ▪ Some major episodes in the history of life. The timing of events is based on fossil evidence and molecular analyses. Colors indicate the nature of the milestones on the geological time scale. Blue represents basic geological and environmental events—one of which, the switch from an anaerobic to an aerobic atmosphere, was caused by organisms (photosynthetic prokaryotes). Green indicates major milestones dated by fossils; light red represents estimates of when major lineages arose.

abiotically produced molecules into droplets, called protobionts, that had chemical characteristics different from their surroundings; and (4) the origin of heredity (which may well have been under way even before the "droplet" stage). It is possible to test the plausibility of these stages of chemical evolution in laboratory experiments.

Abiotic synthesis of organic monomers is a testable hypothesis: *science as a process*

In the 1920s, A. I. Oparin of Russia and J. B. S. Haldane of Great Britain independently postulated that conditions on the primitive Earth favored chemical reactions that synthesized organic compounds from inorganic precursors present in the early atmosphere and seas. This cannot happen in the modern world, Oparin and Haldane reasoned, because the present atmosphere is rich in oxygen produced by photosynthetic life. The oxidizing atmosphere of today is not conducive to the spontaneous synthesis of complex molecules because the oxygen attacks chemical bonds, extracting electrons. Before oxygen-producing photosynthesis, Earth had a much less oxidizing atmosphere, derived mainly from volcanic vapors. Such a reducing (electron-adding) atmosphere would have enhanced the joining of simple molecules to form more complex ones. Even with a reducing atmosphere, making organic molecules would require considerable energy, which was probably provided by lightning and the intense UV radiation that penetrated the primitive atmosphere. The modern atmosphere has a layer of ozone produced from oxygen, and this ozone shield screens out most UV radiation. There is also evidence that young suns emit more UV radiation than older suns. Oparin and Haldane envisioned an ancient world with the chemical conditions and energy resources needed for the abiotic synthesis of organic molecules.

In 1953, Stanley Miller and Harold Urey tested the Oparin-Haldane hypothesis by creating, in the laboratory, conditions comparable to those of early Earth. Their apparatus produced a variety of amino acids and other organic compounds found in living organisms today (FIGURE 26.4; also see FIGURE 4.1).

The atmosphere in the Miller-Urey model was made up of H_2O, H_2, CH_4 (methane), and NH_3 (ammonia), the gases that researchers in the 1950s believed prevailed in the ancient world. This atmosphere was probably more strongly reducing than the actual atmosphere of early Earth. Modern volcanoes emit CO, CO_2, N_2, and water vapor, and it is likely that these gases were abundant in the ancient atmosphere. Hydrogen (H) was probably not a major component, and traces of O_2 may even have been present, formed from reactions among other gases as they baked under powerful UV radiation.

Many laboratories have repeated the Miller-Urey experiment using a variety of recipes for the atmosphere. Abiotic synthesis of organic compounds occurred in these modified models, although yields were generally smaller than in the original experiment. Laboratory analogs of primeval Earth have produced all 20 amino acids commonly found in organisms, several sugars, lipids, the purine and pyrimidine bases present in the nucleotides of DNA and RNA, and even ATP (if phosphate is added to the flask).

The Miller-Urey experiments still stimulate debate and research on the abiotic synthesis of organic compounds. Today, research focuses on where chemicals needed for organic syntheses came from and where the reactions most likely occurred. Many scientists doubt that the early atmosphere played a significant role in early chemical reactions. Indeed, submerged volcanoes and deep-sea vents—gaps in Earth's crust where hot water and minerals gush into deep oceans—may have provided the essential resources. Evidence is also building that life could have begun in a much simpler chemical environment than formerly thought. For instance, the first cells may have used inorganic sulfur and iron compounds as energy sources to make their own ATP instead of taking it up from their surroundings.

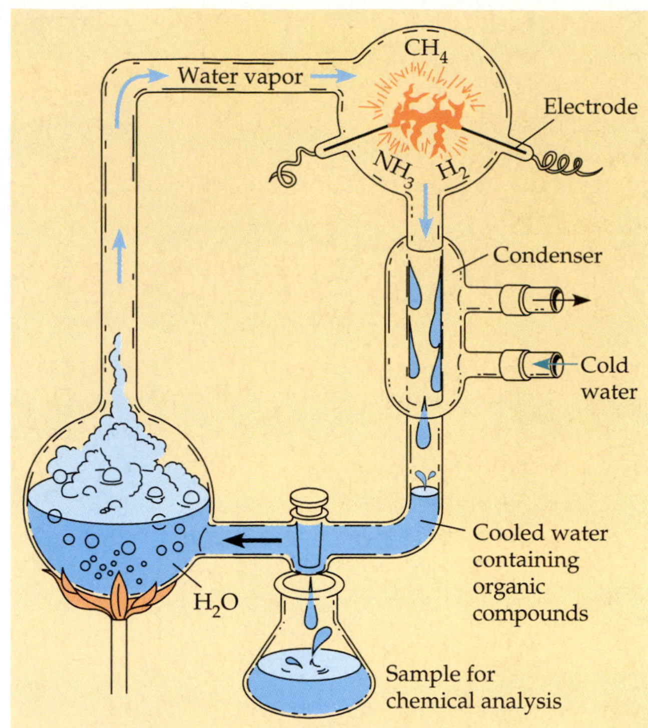

FIGURE 26.4 ▪ Abiotic synthesis of organic molecules in a model system. Stanley Miller and Harold Urey used an apparatus similar to this one to simulate chemical dynamics on the primitive Earth. A warmed flask of water simulated the primeval sea. The "atmosphere" consisted of H_2O, H_2, CH_4, and NH_3. Sparks were discharged in the synthetic atmosphere to mimic lightning. A condenser cooled the atmosphere, raining water and any dissolved compounds back to the miniature sea. As material circulated through the apparatus, the solution in the flask changed from clear to murky brown. After one week, Miller and Urey analyzed the contents of the solution and found a variety of organic compounds, including some of the amino acids that make up the proteins of organisms.

Laboratory simulations of early Earth conditions have produced organic polymers

Before there was life, its chemical building blocks may have been accumulating as a natural stage in the chemical evolution of the planet. In this setting, the abiotic synthesis of more complex organic molecules by the joining together of smaller ones also may have been inevitable. Organic polymers such as proteins are chains of similar building blocks, or monomers. In the living cell, specific enzymes catalyze the reactions. Abiotic synthesis of polymers would have had to occur without the help of these efficient enzymes, and the dilute concentrations of the monomers dissolved in an excess of water would not favor spontaneous reactions. Polymerization *does* occur in laboratory experiments when dilute solutions of organic monomers are dripped onto hot sand, clay, or rock. This process vaporizes water and concentrates the monomers on the substratum. Using this method, Sidney Fox of the University of Miami has made what he calls proteinoids, which are polypeptides produced by abiotic means. Perhaps waves or rain splashed dilute solutions of organic monomers onto fresh lava or other hot rocks on the early Earth and then rinsed proteinoids and other polymers back into the water.

Clay, even cool clay, may have been especially important as a substratum for the polymerization reactions prerequisite to life. Clay concentrates amino acids and other organic monomers from dilute solutions because the monomers bind to charged sites on the clay particles. At some of the binding sites, metal atoms, such as iron and zinc, function as catalysts facilitating the reactions that link monomers. Clay, having many of these binding sites, could have functioned as a lattice that brought monomers close together and then assisted in joining them into polymers. As an alternative to clay, iron pyrite (fool's gold), which consists of iron and sulfur, has been proposed as the substratum of organic synthesis. This hypothesis is advocated by Günter Wachtershäuser of Germany, who points out that certain properties of pyrite could have catalyzed abiotic synthesis of organic polymers. Pyrite provides a charged surface, and the formation of this mineral from iron and sulfur yields electrons that could support bonding between organic molecules to form more complex products.

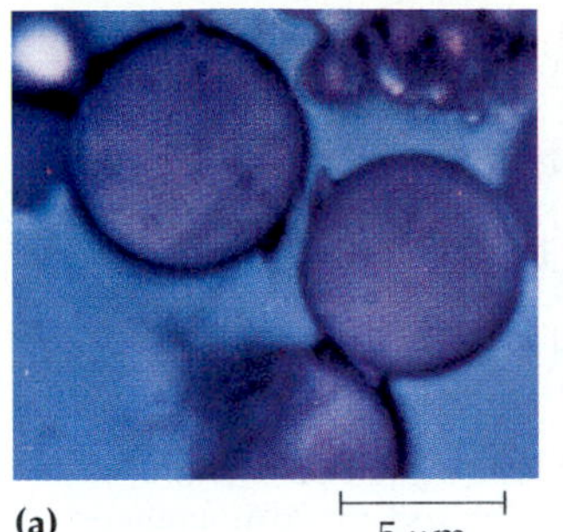

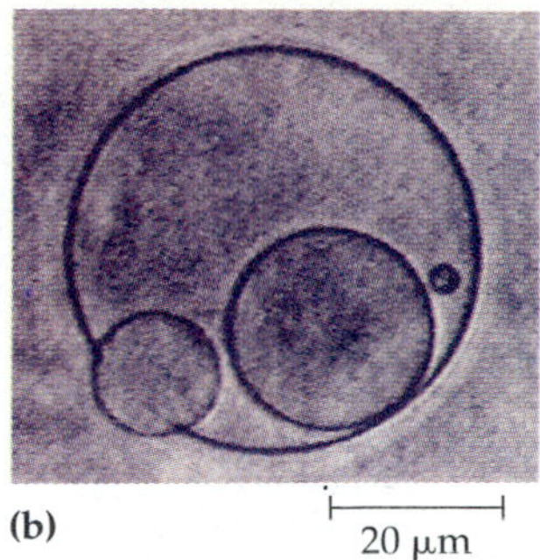

FIGURE 26.5 ▪ Laboratory versions of protobionts, aggregates of organic molecules with some biological properties. (a) These microspheres (LM) are made by cooling solutions of proteinoids, polypeptides created abiotically from amino acids polymerized on hot surfaces. Microspheres grow by absorbing free proteinoids until they reach an unstable size, when they split to form daughter microspheres. Of course, this division lacks the precision of cellular reproduction. **(b)** In an aqueous environment, certain kinds of lipids self-assemble to form liposomes. As shown in this micrograph (LM), one type of liposome made by scientists at Emory University gives rise to smaller liposomes.

Protobionts can form by self-assembly

The properties of life emerge from an interaction of molecules organized into higher levels of order. Living cells may have been preceded by **protobionts**, aggregates of abiotically produced molecules. Protobionts are not capable of precise reproduction, but they maintain an internal chemical environment different from their surroundings and exhibit some of the properties associated with life, including metabolism and excitability.

One type of protobiont (named a coacervate by Oparin) is a stable droplet that tends to self-assemble when a suspension of macromolecules (polypeptides, nucleic acids, and polysaccharides) is shaken. Each coacervate is an aggregate of macromolecules, which are mostly hydrophobic, surrounded and stabilized by a shell of water molecules. If enzymes are included among the ingredients, they are incorporated into the coacervates. The coacervates are then able to absorb substrates from their surroundings and release the products of the reactions catalyzed by the enzymes.

Laboratory experiments demonstrate that protobionts could have formed spontaneously from abiotically produced organic compounds. When mixed with cool water, proteinoids self-assemble into tiny droplets called microspheres (FIGURE 26.5a). Coated by a protein membrane that is selectively permeable, the microspheres undergo osmotic swelling or shrinking when placed in solutions of different salt concentrations. Some microspheres also store energy in the form of a membrane potential, a voltage across the surface. The protobionts can discharge the voltage in nervelike fashion; such excitability is characteristic of all life (which is not to say that microspheres are alive, only that they display *some* of the properties of life). Droplets of another type, called liposomes, form spontaneously when the organic ingredients include certain lipids. These lipids organize into a molecular bilayer at the surface of the droplet, much like the lipid bilayer of cell membranes. The liposomes behave dynamically, sometimes growing by engulfing smaller liposomes and then splitting, other times "giving birth" to smaller liposomes (FIGURE 26.5b).

Unlike some laboratory models, protobionts that formed in the ancient seas would not have possessed refined enzymes, which are made in cells according to inherited instructions. Some molecules produced abiotically, however, do have weak catalytic capacities, and there could well have been protobionts that had a rudimentary metabolism that allowed them to modify substances they took in across their membranes.

RNA was probably the first genetic material

Imagine a tidepool, pond, or moist clay on the primeval Earth with a suspension of protobionts varying in chemical composition, permeability, and catalytic capabilities. Those droplets most stable and best able to accumulate organic molecules from the environment would grow and split, distributing their chemical components to the "baby" droplets. Other droplets would fall apart or fail to grow and divide. In this way, the environment may have selected in favor of some molecular aggregates and against others. But competition among the

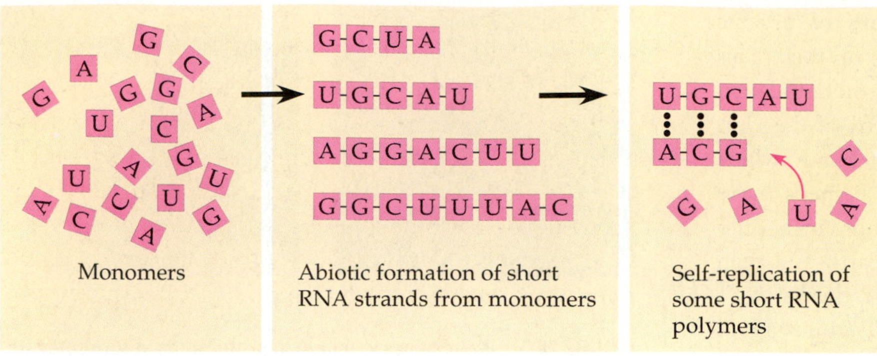

FIGURE 26.6 ▪ Abiotic replication of RNA. According to this hypothesis, the first genes were RNA molecules that polymerized abiotically and replicated themselves autocatalytically while bound to clay surfaces. The letters A, G, C, and U symbolize the four RNA nucleotide bases, which pair specifically in A-U and G-C combinations (see Chapter 17).

various protobionts could not lead to long-range improvement because there was no way to perpetuate success. As prolific droplets grew, split, grew, and split again, their unique catalysts and other functional molecules would become increasingly diluted. The chemical aggregates that were the forerunners of cells could not build on the past and evolve until the development of some mechanism for replicating their characteristics—some mechanism of heredity. Genetic information would make it possible for molecular aggregates to pass along not just samples of key molecules but also instructions for making more of those molecules.

A cell stores its genetic information as DNA, transcribes the information into RNA, and then translates the messages into specific enzymes and other proteins (discussed in Chapters 16 and 17). Instructions are transmitted by the replication of DNA when a cell divides. The DNA ⟶ RNA ⟶ protein mechanism of cellular control employs intricate machinery that could not have evolved all at once, but instead must have emerged bit by bit as improvements to much simpler processes. Even before DNA, some primitive mechanism may have existed for aligning amino acids along strands of RNA that could replicate themselves. According to this hypothesis, the first genes were not DNA molecules but short strands of RNA that began self-replicating in the prebiotic world.

Several scientists have tested the hypothesis of RNA self-replication. Short polymers of ribonucleotides have been produced abiotically in laboratory experiments. If RNA is added to a solution containing monomers for making more RNA, sequences about five to ten nucleotides long are copied from the template according to the base-pairing rules (FIGURE 26.6). If zinc is added as a catalyst, sequences up to 40 nucleotides long are copied with less than 1% error.

In the 1980s, Thomas Cech and his co-workers at the University of Colorado, Boulder, revolutionized thinking about the evolution of life when they discovered that RNA molecules are important catalysts in modern cells. This finding disproved the long-held view that only proteins (enzymes) serve as biological catalysts. Cech and others found that modern cells use RNA catalysts, called **ribozymes**, to do such things as remove introns from RNA (see Chapter 17). Ribozymes also help catalyze the synthesis of new RNA, notably rRNA, tRNA, and mRNA. Thus, RNA is autocatalytic, and in the prebiotic world, long before there were enzymes (proteins) or DNA, RNA molecules may have been fully capable of self-replication (see FIGURE 26.6). Many biologists now envision an "RNA world," an early period in the evolution of life when RNA molecules served as both rudimentary genes and organic catalysts.

Natural selection on the molecular level has been observed operating on RNA populations in the laboratory. Unlike double-stranded DNA, which takes the form of a uniform helix, single-stranded RNA molecules assume a variety of specific three-dimensional shapes mandated by their nucleotide sequences. Each sequence folds into a unique conformation reinforced by hydrogen bonding between regions of the strand having complementary sequences of bases. The molecule thus has both a genotype (its nucleotide sequence) and a phenotype (its conformation, which interacts with surrounding molecules in specific ways). In a particular environment, RNA molecules of certain base sequences are more stable and replicate faster with fewer errors than other sequences. Beginning with a diversity of RNA molecules that must compete for monomers to replicate, the sequence best suited to the temperature, salt concentration, and other features of the surrounding solution and having the greatest autocatalytic activity will prevail. Its descendants will be not a single RNA species but a family of closely related sequences (because of copying errors). Selection screens mutations in the original sequence, and occasionally a copying error results in a molecule that folds into a shape that is even more stable or more adept at self-replication than the ancestral sequence. Similar selection events may have occurred in the prebiotic RNA world.

The rudiments of RNA-directed protein synthesis may have been the weak binding of specific amino acids to bases along RNA molecules, which functioned as simple templates holding a few amino acids together long enough for them to be linked. (Indeed, this is one function of rRNA in modern ribosomes.) If RNA happened to synthesize a short polypeptide that in turn behaved as an enzyme helping the RNA molecule to replicate, then the early chemical dynamics included molecular cooperation as well as competition. These first steps toward the replication and translation of genetic information may have been taken by molecular evolution even before RNA and polypeptides became packaged within membranes (FIGURE 26.7).

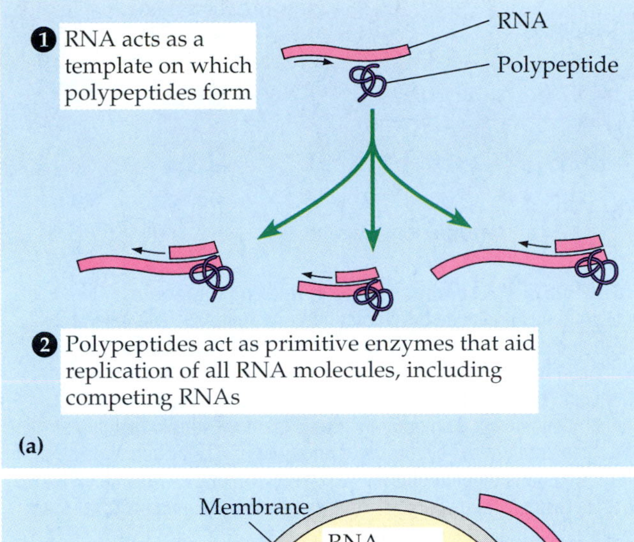
RNA
Polypeptide
1 RNA acts as a template on which polypeptides form
2 Polypeptides act as primitive enzymes that aid replication of all RNA molecules, including competing RNAs
(a)

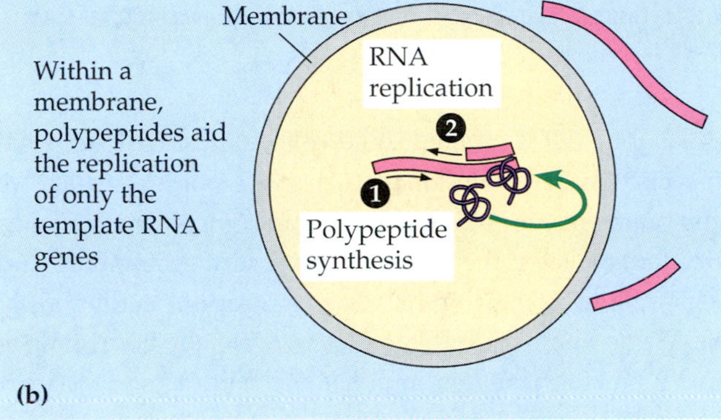
Membrane
Within a membrane, polypeptides aid the replication of only the template RNA genes
RNA replication
2
1
Polypeptide synthesis
(b)

FIGURE 26.7 ▪ Hypotheses for the beginnings of molecular cooperation. (a) An RNA strand acts as a template both for self-replication and for ordering and bonding amino acids, forming polypeptides. In this model, an RNA molecule enhances its own replication by directing the synthesis of a polypeptide that in turn functions as an enzyme helping the RNA molecule to replicate. In such cooperation may have been the rudiments for the translation of genetic information into protein structure. If these molecules were not segregated within a protobiont, however, the benefits of this protein would have been shared by competing RNA molecules. **(b)** This model shows an RNA/polypeptide cooperative enclosed in a membrane. In this setting, RNA genes would have benefited exclusively from their protein products.

The packaging of primitive RNA genes and their polypeptide products within a membrane would have been a significant milestone in the early history of life. Once this occurred, protobionts could have evolved as units, and molecular cooperation could be refined because components that interacted in ways favorable to the success of the protobiont as a whole were concentrated together in a microscopic volume rather than being spread throughout the surroundings. Suppose, for example, that an RNA molecule ordered amino acids into a primitive enzyme that extracted energy from inorganic sulfur compounds taken up from the surroundings. This energy could be used for other reactions within the protobiont, including replication of RNA. Natural selection could favor such a gene only if its product were kept close by, rather than being shared with competing RNA sequences in its environment.

The origin of hereditary information made Darwinian evolution possible

In the preceding scenario we have built a hypothetical antecedent of the cell by incorporating genetic information into an aggregate of molecules that selectively accumulates monomers from its surroundings and uses enzymes programmed by genes to make polymers and carry out other chemical reactions. The protobiont grows and splits, distributing copies of its genes to offspring. Even if only one such protobiont arose initially by the abiotic processes that have been described, its descendants would vary because of mutations, errors in the copying of RNA. Evolution in the true Darwinian sense—differential reproductive success of varied individuals—presumably accumulated many refinements to primitive metabolism and inheritance. One trend apparently led to DNA becoming the hereditary material. Initially, RNA could have provided the template on which DNA nucleotides were assembled. But DNA is a much more stable repository for genetic information than RNA, and once DNA appeared, RNA molecules would have begun to take on their modern roles as intermediates in the translation of genetic programs. The "RNA world" gave way to a "DNA world."

Debate about the origin of life abounds

Laboratory simulations cannot prove that the kind of chemical evolution that has been described here actually created life on the primitive Earth, but only that some of the key steps *could* have happened. The origin of life remains a matter of scientific speculation, and there are alternative views of how several key processes occurred.

Some researchers question whether abiotic synthesis of organic monomers on Earth was necessary as a first step in the origin of life. It is possible that at least some organic compounds reached the early Earth from space. This idea, called *panspermia,* holds that hundreds of thousands of meteorites and comets hitting the early Earth brought with them organic molecules formed by abiotic reactions in outer space. Extraterrestrial organic compounds, including amino acids, have been found in modern meteorites, and it seems likely that these bodies could have seeded the early Earth with organic compounds. Biochemists recently demonstrated that organic molecules extracted from a meteorite produce small vesicles when mixed with water. Both panspermia and chemical evolution could have contributed to the pool of organic molecules that formed the earliest life, but many scientists who study the origin of life believe that extraterrestrial sources made only a minor contribution.

Some biologists interested in the origin of life challenge the idea of an "RNA world." They argue that even short RNA strands may be too complicated to be the first self-replicating molecules. In 1991, Julius Rebek, Jr., and his colleagues at the Massachusetts Institute of Technology synthesized a simple organic molecule that acts as a template for producing copies of itself (FIGURE 26.8). This breakthrough strengthened an alternative hypothesis that nucleic acid genes were preceded by simpler heredity systems.

Where life began is another issue. Until recently, most researchers favored shallow water or moist sediments as the most likely sites for life's origin. Some scientists now question this view, arguing that Earth's surface was very inhospitable during the period when life probably began. Asteroids and comets, debris left over from the formation of the solar system, pounded Earth and the other young planets. Some scientists speculate that incipient life could not survive this cosmic assault—unless, that is, life began on the less exposed sea floor. The discovery of deep-sea vents in the late 1970s raised the possibility that earlier vents supplied the energy and chemical precursors for the origin of protobionts. Molecular phylogenetic analyses indicate that the ancestors of modern prokaryotes thrived in very hot conditions and may have lived on inorganic sulfur compounds that are common in deep-sea vent environments. Laboratory studies by Günter Wachtershäuser and his colleagues also point to the environments of deep-sea vents and volcanoes as abiotic sources of some of the organic compounds, such as acetyl coenzyme A, that cells use in energy metabolism. Sulfides of iron and nickel, common in volcanic and deep-sea vent areas, catalyze the formation of acetic acid and precursors of acetyl coenzyme from CO and H_2S. Wachtershäuser considers it likely that life arose in an environment where these reactions were common and then adopted them as the earliest form of metabolism.

On a broader scale, the hypothesis that life is not restricted to Earth is becoming more accessible to scientific testing. Photographs taken by the Galileo spacecraft of the ice-covered surface of Europa, one of Jupiter's moons, have led to hypotheses that liquid water lies beneath the surface and may support prokaryotic life. An ongoing debate centers on whether structures resembling fossil prokaryotes in a Martian meteorite found in Antarctica are geological artifacts or evidence of early life on Mars. The surface of Mars is a cold, dry, lifeless desert, but billions of years ago it was probably relatively warm, and may have held liquid water and had a CO_2-rich atmosphere. It is possible that prebiotic chemistry, similar to that on early Earth, also occurred on Mars before the planet became senescent. Scientists anticipate that exploration in the next decade will provide clues about microbial life on Mars. Did life evolve there and then die out, or was prebiotic chemistry snuffed out by planetary changes before any life forms developed? Many scientists also see Mars as an ideal place to test hypotheses about Earth's prebiotic chemistry.

Debate about the origin of terrestrial and extraterrestrial life abounds, and we have sampled only a few of the issues. Whatever way prebiotic chemicals accumulated, polymerized, and eventually reproduced on Earth, the leap from an aggregate of molecules that reproduces to even the simplest prokaryotic cell is immense, and change must have occurred in many smaller evolutionary steps. The point at which we stop calling membrane-enclosed compartments that metabolize and replicate their genetic programs protobionts and begin calling them living cells is as fuzzy as our definitions of life. We do know that prokaryotes were already flourishing at least 3.5 billion years ago, and that all the lineages of life arose from those ancient prokaryotes.

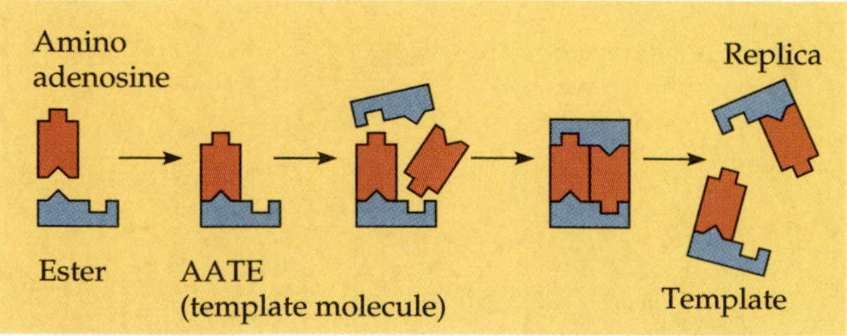

FIGURE 26.8 ▪ A synthetic self-replicating molecule. Were RNA or DNA genes preceded by simpler heredity systems? The laboratory creation of a relatively simple organic molecule that makes copies of itself is influencing ideas on the origin of life. The molecule is amino-adenosine triacid ester (AATE), which consists of two components, amino adenosine and an ester. From a pool of these two building blocks, AATE can catalyze the synthesis of additional AATE by acting as a template. No one is arguing that the first genetic material resembled AATE, but this research does demonstrate that molecules much smaller than nucleic acids can replicate.

THE MAJOR LINEAGES OF LIFE

In Chapter 25 we looked at systematics as the study of biological diversity in an evolutionary context. Now that we have gone back in time to the very origin of life on Earth, systematics is once again relevant as we attempt to reconstruct evolutionary relationships among the immense diversity of forms that arose from those early organisms.

Arranging the diversity of life into the highest taxa is a work in progress

26.1 Systematists have traditionally considered the kingdom to be the highest—the most inclusive—taxonomic category. Many of us grew up with the notion that there are only two kingdoms of life—plants and animals—because we live in a macroscopic, terrestrial realm where we rarely see organisms that do not fit neatly into a plant-animal dichotomy. The two-kingdom scheme also had a long tradition in formal taxonomy; Linnaeus divided all known forms of life between the plant and animal kingdoms.

Even with the discovery of the diverse microbial world, the two-kingdom system persisted. Bacteria were placed in the

26.1 This symbol links topics in the text to interactive exercises in the CD-ROM that accompanies the book. The number indicates the appropriate activity in the CD.

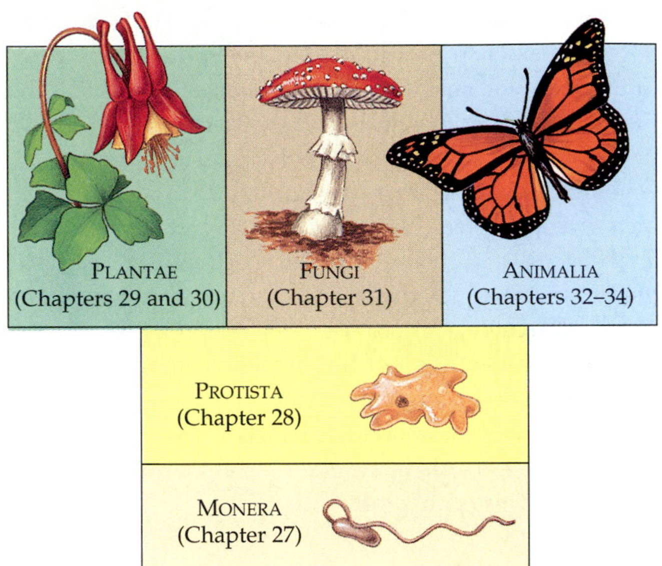

FIGURE 26.9 ▪ The traditional five-kingdom system.

plant kingdom, their rigid cell walls cited as justification. Eukaryotic unicellular organisms with chloroplasts were also considered plants. Fungi, too, fell under the plant banner, partly because they are sedentary, even though no fungi are photosynthetic and they have little in common structurally with green plants. In the two-kingdom system, unicellular creatures that move and ingest food—protozoa—were called animals. Microbes such as *Euglena* that move but are photosynthetic were claimed by both botanists and zoologists, and showed up in the taxonomies of plant and animal kingdoms. Schemes with additional kingdoms were proposed, but none became popular with the majority of biologists until Robert H. Whittaker of Cornell University argued effectively for a five-kingdom system in 1969. Whittaker designated these five kingdoms as Monera, Protista, Plantae, Fungi, and Animalia (FIGURE 26.9 and 26.10a).

The five-kingdom system recognizes the two fundamentally different types of cell, prokaryotic and eukaryotic, and sets the prokaryotes (commonly called bacteria) apart from all eukaryotes by placing them in their own kingdom, Monera. Lumping all prokaryotes together in Kingdom Monera, the five-kingdom system differs from alternative classification schemes that have been proposed more recently (FIGURE 26.10b and c).

A common feature of the five-kingdom system and other classification schemes is the recognition of three kingdoms of multicellular eukaryotes, the Plantae, Fungi, and Animalia. Plants, fungi, and animals generally differ in structure and life cycles and in their modes of nutrition, the criterion Whittaker used to define these kingdoms. Plants are autotrophic in nutritional mode, making their food by photosynthesis. Fungi are heterotrophic organisms that are absorptive in nutritional mode. Most fungi are decomposers that live embedded in their food source, secreting digestive enzymes and absorbing the small organic molecules that are the products of digestion. Most animals live by ingesting food and digesting it within specialized cavities.

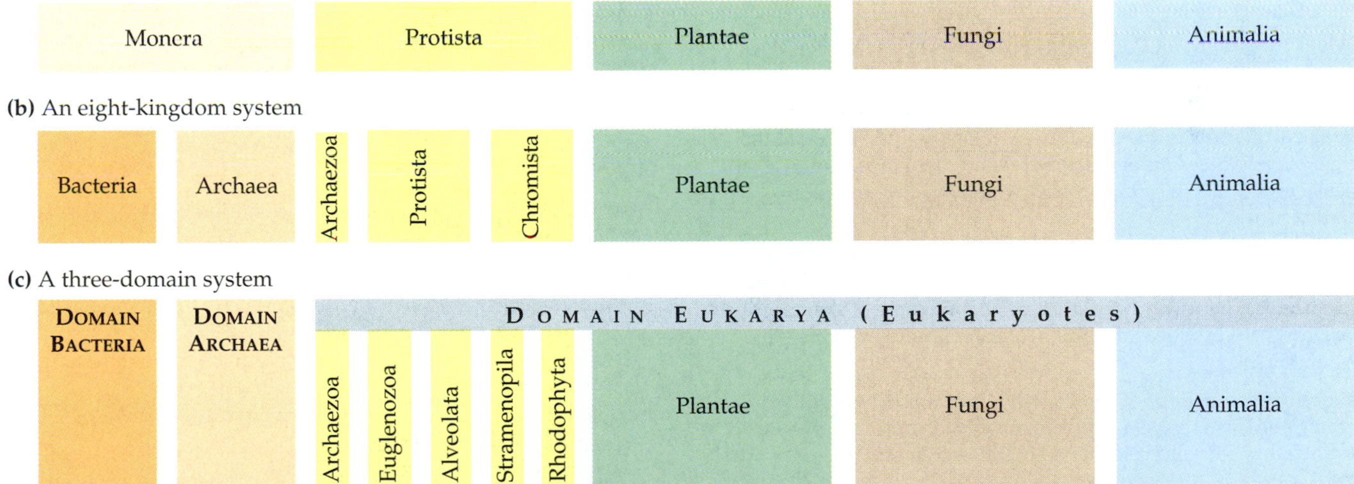

26.1 **FIGURE 26.10 ▪ The five-kingdom system compared with two alternative classification schemes. (a)** The five-kingdom system is color-coded here to match FIGURE 26.9. This color coding is continued throughout to make comparisons among the schemes easier. **(b)** This eight-kingdom system is one example of a "kingdom-splitting" solution to the taxonomic problems posed by certain distinct lineages of microorganisms. It divides the prokaryotes into two kingdoms, a modification based on molecular evidence for an early evolutionary divergence between bacteria (eubacteria; most bacteria) and archaea (archaebacteria), a distinct lineage of prokaryotes. In addition to two separate prokaryotic kingdoms, this system also splits the protists into three kingdoms. **(c)** The three-domain scheme assigns more significance to the ancient evolutionary split between bacteria and archaea by using a superkingdom taxon called the domain. Systematists are in the process of sorting out kingdoms of prokaryotes within the domains Bacteria and Archaea. The domain Eukarya consists of all kingdoms of eukaryotic organisms. This system emphasizes the biological diversity among protists even more than the eight-kingdom system in part b. We will discuss the rationale for the domain system and the status of protist systematics in Chapters 27 and 28. What is important to understand here is that new information has reopened issues concerning biological diversity at the highest taxonomic levels.

We are left with the kingdom Protista. In the five-kingdom system, Protista became a grab bag containing all eukaryotes that did not fit the definitions of plants, fungi, or animals. Most protists are unicellular forms, but Whittaker's Kingdom Protista also includes relatively simple multicellular organisms that are believed to be direct descendants of unicellular protists.

Like any classification scheme, the five-kingdom system is not a natural fact, but a human construct. It is one attempt to order the diversity of life into a scheme that is useful and, hopefully, phylogenetically reasonable. During the past two decades, systematists using comparisons of nucleic acids and proteins to probe the relationships between different groups of organisms have been pointing out problems with the traditional five-kingdom scheme. The alternative classification systems outlined in FIGURE 26.10 have emerged from these challenges to the five-kingdom system. A widely accepted three-domain system, which we follow in this book, recognizes three fundamentally different lineages, classified at the superkingdom level as the domains Bacteria, Archaea, and Eukarya (see FIGURE 26.10C). We will consider some of the arguments for these alternative taxonomic constructs in the next several chapters. Classifying life is a work in progress, an evolving view of biodiversity that reflects our increased understanding of the characteristics and evolutionary histories of different organisms.

From this discussion of biological kingdoms, we turn to prokaryotes, the first forms of life and the *only* ones for at least 2 billion years. Prokaryotes remain tremendously important on modern-day Earth. In the next chapter we will address the diversity and history of prokaryotic life.

CHAPTER REVIEW

REVIEW OF KEY CONCEPTS

(with page numbers and key figures)

INTRODUCTION TO THE HISTORY OF LIFE

- **Life on Earth originated between 3.5 and 4.0 billion years ago (pp. 490–492, FIGURE 26.2)** Earth formed 4.5 billion years ago. The oldest available evidence of life is carbon isotopes of cellular origin found in 3.8-billion-year-old rocks. The oldest fossils are 3.5 billion years old.
- **Major episodes in the history of life: *a preview* (p. 493, FIGURE 26.3)** Throughout the approximately 4-billion-year tenure of life on Earth, geological and biological history have been closely intertwined.

PREBIOTIC CHEMICAL EVOLUTION AND THE ORIGIN OF LIFE

- **The first cells may have originated by chemical evolution on a young Earth: *an overview* (pp. 492–493)** One hypothesis about the origin of life is based on the chemical evolution of protobionts, abiotically produced molecular droplets with distinctive chemical characteristics.
- **Abiotic synthesis of organic monomers is a testable hypothesis: *science as a process* (pp. 493–494; FIGURE 26.4)** Laboratory experiments performed under conditions simulating those of the primitive Earth have produced diverse organic molecules from inorganic precursors.
- **Laboratory simulations of early Earth conditions have produced organic polymers (pp. 494–495)** Small organic molecules polymerize when they are concentrated on hot sand, rock, or clay.
- **Protobionts can form by self-assembly (p. 495, FIGURE 26.5)** Organic molecules synthesized in the laboratory have spontaneously assembled into a variety of droplets—coacervates, microspheres, and liposomes—with some of the properties associated with life.
- **RNA was probably the first genetic material (pp. 495–497, FIGURES 26.6, 26.7)** The first genes may have been abiotically produced RNA, whose base sequences served as templates for both alignment of amino acids in polypeptide synthesis and alignment of complementary nucleotide bases in a primitive form of self-replication. Once genetic information became incorporated inside membrane-enclosed compartments, protobionts would have acquired heritability and the ability to evolve as units.
- **The origin of hereditary information made Darwinian evolution possible (p. 497)** The transition from an RNA world to a DNA world, a prerequisite for true Darwinian evolution, may have begun with RNA serving as the template on which DNA nucleotides were assembled.
- **Debate about the origin of life abounds (pp. 497–498)** Researchers continue to debate how and where life originated. Three alternative views explore abiotic synthesis of organic molecules on Earth versus import via meteorites, RNA genes versus simpler self-replicating molecules, and origins of life in shallow water versus deep-sea vents.

THE MAJOR LINEAGES OF LIFE

- **Arranging the diversity of life into the highest taxa is a work in
26.1 progress (pp. 498–500, FIGURES 26.9, 26.10)** The traditional five-kingdom system classifies organisms as Monera (prokaryotes), Protista (relatively simple eukaryotes), Plantae, Fungi, and Animalia. New information is leading to taxonomic alternatives to the five-kingdom system.

SELF-QUIZ

1. The *main* explanation for the lack of a continuing abiotic origin of life on Earth today is that
   a. there is not sufficient lightning to provide an energy source
   b. our oxidizing atmosphere is not conducive to the spontaneous formation of complex molecules
   c. much less visible light is reaching Earth to serve as an energy source
   d. there are no molten surfaces on which weak solutions of organic molecules would polymerize
   e. all habitable places are already filled
2. All of the following support the hypothesis that RNA functioned as the first genetic material of early protobionts *except*
   a. short RNA sequences can self-assemble when combined with nucleotide monomers
   b. catalytic activity has been demonstrated for RNA in modern cells
   c. variations in base sequences produce molecules with variable stabilities in different environments
   d. modern cells use an RNA template when synthesizing proteins
   e. in modern cells, RNA provides the template on which DNA nucleotides are assembled

3. The sedimentary assemblages called stromatolites
   a. contain prokaryotes in their lowest levels and eukaryotes above, supporting the hypothesis of the earlier appearance of prokaryotes
   b. currently produced in restricted habitats are strikingly similar to the fossil assemblages called bacterial mats
   c. concentrated many of the essential abiotic components needed for protobiont assembly
   d. contain fossils resembling spherical and filamentous prokaryotes
   e. contain quantities of oxidized metals, thereby lending support for the hypothesis that an oxygen-rich atmosphere developed early in Earth's history

4. Clays and other fine mineral particles may have contributed all of the following to abiotic origins of organic polymers *except*
   a. a charged surface holding reactants close together
   b. catalysis
   c. enzymatic action on reactants
   d. a supply of electrons
   e. the ability to concentrate reactants from the surrounding solution

5. The formation of microspheres, liposomes, and coacervates all require
   a. RNA
   b. a membrane potential
   c. self-assembly
   d. phospholipids
   e. primitive genes

6. Competition among various protobionts would have led to evolutionary improvement only when
   a. they were able to catalyze chemical reactions
   b. some kind of heredity mechanism developed
   c. they were able to grow and reproduce
   d. the protobionts acquired selectively permeable membranes
   e. DNA first appeared

7. Which of the following represents a probable order of evolution of prelife components on Earth?
   a. protobionts before mitosis
   b. an oxidizing atmosphere followed by a reducing atmosphere
   c. amino acids and sugars after the corresponding polymers
   d. DNA before RNA
   e. eukaryotic cell structure before photosynthesis

8. One current debate raises the issue that, rather than beginning in shallow pools, life could have begun
   a. as plate tectonics changed the surface of our planet
   b. near deep-sea vents
   c. from viruses
   d. in northern Africa
   e. when chunks that broke off from the moon bombarded Earth

9. Which of the following steps has *not* yet been accomplished by scientists studying the origin of life?
   a. abiotic synthesis of small RNA polymers
   b. abiotic synthesis of polypeptides
   c. formation of molecular aggregates with selectively permeable membranes
   d. formation of protobionts that use DNA to direct the polymerization of amino acids
   e. abiotic synthesis of organic monomers

10. Current debates about the number and boundaries of the kingdoms of life center *mainly* on which groups of organisms?
   a. plants and animals
   b. plants and fungi
   c. prokaryotes and relatively simple eukaryotes
   d. fungi and animals
   e. bacteria and the most complex eukaryotes

CHALLENGE QUESTION

Describe the minimum structural, metabolic, and genetic equipment of a postprotobiont that you would consider to be a true primitive cell.

SCIENCE, TECHNOLOGY, AND SOCIETY

There is no reason to believe that the processes that led to the origin of life are unique to Earth. Many scientists think that life may be present on many other planets throughout the universe. During the next decade, several nations (including the United States) plan to explore the Martian environment with orbiting laboratories and landers. One of the main goals is to test the hypothesis that chemical conditions on ancient Mars were similar to those on ancient Earth and may have led to prokaryotic life. In what ways do you think our perspective would be changed by convincing evidence of extraterrestrial life?

FURTHER READING

Aldhous, P. "New Ingredient for the Primeval Soup." *New Scientist,* February 25, 1995. One of the abiotically produced organic molecules may have been a versatile catalyst.
Balter, M. "Looking for Clues to the Mystery of Life on Earth." *Science,* August 16, 1996. Summary of an international conference on the origin of life.
Cohen, P. "Let There Be Life." *New Scientist,* July 6, 1996. Contains interviews with several leaders in research on the origin of life.
Cowen, R. "C'est la Vie, Searching for Life in the Solar System." *Science News,* November 1, 1997. Describes methods and possible extraterrestrial sources.
Gibson, E. K. Jr., D. S. McKay, K. Thomas-Keprta, and C. S. Romanek. "The Case for Relic Life on Mars." *Scientific American,* December 1997. Proponents of the hypothesis that Mars once supported microbial life present their case.
Kerr, R. A. "Putative Martian Microbes Called Microscopy Artifacts." *Science,* December 5, 1997. Presents arguments refuting the idea that Mars once supported life.
Landweber, L. F., P. J. Simon, and T. A. Wagner. "Ribozyme Engineering and Early Evolution." *BioScience,* February 1998. Explores modern evidence for an RNA world.
Monastersky, R. "When Earth Tipped, Life Went Wild." *Science News,* July 26, 1997. Explores the hypothesis that reorientation of Earth's surface triggered rapid diversification of lineages in the Cambrian.
Rebek, J. "Synthetic Self-Replicating Molecules." *Scientific American,* July 1994. Discusses the possibility of molecular reproduction before nucleic acids.

WEB LINKS

Visit the special edition of *The Biology Place* for BIOLOGY, Fifth Edition, at http://www.biology.com/campbell. Go to Chapter 26 for online resources, including learning activities, practice exams, and links to the following web sites:

"The Talk.Origins Archive"
Talk.origins <news:talk.origins> is a usenet newsgroup devoted to the discussion and debate of biological and physical origins. Discussions at this interesting site mainly center around the creation/evolution controversy.

"From Primordial Soup to the Prebiotic Beach"
An interview with Dr. Stanley Miller who pioneered experimental studies on the origin of life.

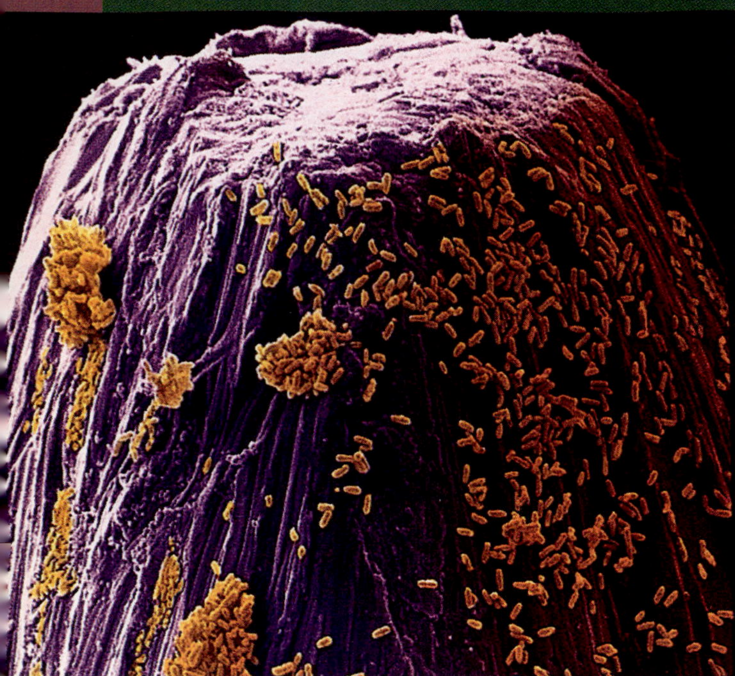

CHAPTER 27

PROKARYOTES AND THE ORIGINS OF METABOLIC DIVERSITY

The history of prokaryotic life is a success story spanning more than 3.5 billion years. Prokaryotes were the earliest organisms, and they lived and evolved all alone on Earth for 2 billion years. They have continued to adapt and flourish on an evolving Earth, and in turn they have helped to change the Earth. In this chapter you will become more familiar with prokaryotes by studying their structure and function, their origins and evolution, their diversity, and their ecological significance.

THE WORLD OF PROKARYOTES

They're (almost) everywhere! *an overview of prokaryotic life*

In terms of metabolic impact and numbers, prokaryotes still dominate the biosphere, outnumbering all eukaryotes combined. More prokaryotes inhabit a handful of dirt or the human mouth or skin than the total number of people who have ever lived. Prokaryotes are not only the most numerous organisms by far but also the most pervasive. Wherever we find life of any kind, prokaryotes are among the organisms present. Prokaryotes also thrive in habitats too hot, too cold, too salty, too acidic, or too alkaline for any eukaryote. Incomparably bountiful and omnipresent, prokaryotes have endured and expanded their numbers and kinds through billions of years of descent from the first cells that were the beginnings of all life on Earth.

Most prokaryotes are relatively small. The colorized scanning electron micrograph on this page demonstrates their general size (orange structures) in relation to a pinpoint (purple). Although individual prokaryotes are microscopic, their collective impact on Earth and all of life is gigantic. We rarely notice these ubiquitous microbes because they are usually invisible to the unaided eye. Illness caused by infectious prokaryotes occasionally reminds us that these tiny organisms exist, but prokaryotic life is no rogues' gallery. Only a minority of prokaryotes cause disease in humans or any other organisms, and the great majority are essential to all life on Earth. For example, certain prokaryotes decompose matter from dead organisms and return vital chemical elements to the environment in the form of inorganic compounds required by plants, which in turn are consumed by animals. If for some reason all prokaryotes were suddenly to perish, the chemical cycles that sustain life would halt, and all other forms of life would also be doomed. In contrast, prokaryotic life would undoubtedly persist in the absence of eukaryotes, as it did for about 2 billion years in the early history of life on Earth.

Prokaryotes often live in close associations among themselves and with eukaryotes in what are called symbiotic relationships. In the most historically important case of such symbiosis, mitochondria and chloroplasts evolved from prokaryotes that became residents within larger host cells. Thus, animals, plants, fungi, and protists probably evolved from symbiotic associations of ancestral cells. (We will examine this theory of eukaryotic origins further in Chapter 28.)

Modern prokaryotes are diverse in structure and physiology. About 5000 species of prokaryotes are known, and estimates of actual prokaryotic diversity range from about 400,000 to 4 million species. As Harvard University biologist E. O. Wilson puts it, a true sense of biodiversity requires a "downward adjustment of scale."

Bacteria and archaea are the two main branches of prokaryote evolution

In the traditional five-kingdom system of classification, prokaryotes make up the kingdom Monera, and the four eukaryotic kingdoms are Protista, Plantae, Fungi, and Animalia (see FIGURE 26.9). This scheme emphasizes the structural differences between the cells of prokaryotes and eukaryotes. In the past decade, however, systematists have determined that a single kingdom incorporating all prokaryotes is not consistent with evolutionary history. By comparing ribosomal RNA and the completely sequenced genomes of several extant (living today) species, researchers have identified two major branches of prokaryote evolution. The names for these two groups are *bacteria* (formerly called eubacteria) and *archaea* (formerly archaebacteria). The term *archaea* refers to the antiquity of the group's origin from the earliest cells (Gr. *archaio,* "ancient"). Most species of archaea inhabit extreme environments, such as hot springs and salt ponds. Few, if any, other modern organisms can survive in these environments, which may resemble habitats on the early Earth. Most prokaryotes, however, are bacteria. They differ from archaea in many key structural, biochemical, and physiological characteristics, differences that will be highlighted later in this chapter.

When researchers, led by Carl Woese of the University of Illinois, first recognized the distinction between bacteria and archaea, they proposed a six-kingdom system: two prokaryotic kingdoms along with the four eukaryotic kingdoms. Because bacteria and archaea diverged so early in the history of life and are so fundamentally different, Woese and many other systematists now favor organizing the diversity of life into three **domains**, a taxonomic level above kingdom (FIGURE 27.1). Prokaryotes account for two of the domains: Domain Bacteria and Domain Archaea. (Even though the term *bacteria* is still commonly used in biology to refer to all prokaryotes, in this unit the term refers specifically to members of Domain Bacteria.)

Taxonomic issues aside, bacteria and archaea are both structurally organized at the prokaryotic level, which is the rationale for combining them in this chapter. The distinction between bacteria and archaea will become more apparent after we examine some of the structural, genetic, and metabolic adaptations that contribute to the pervasiveness of prokaryotes on Earth.

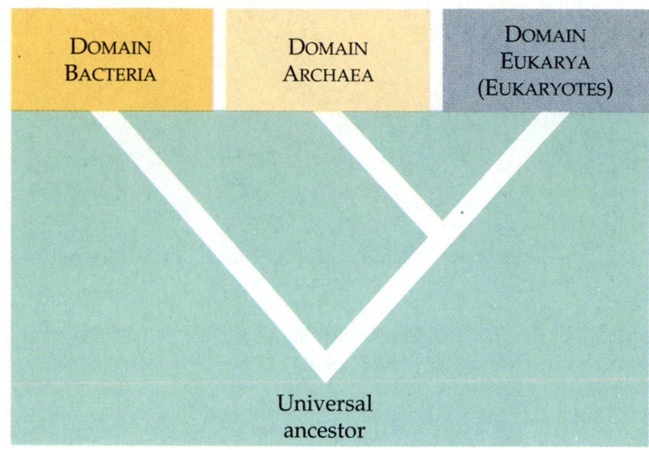

FIGURE 27.1 ▪ The three major lineages of life. Molecular sequencing studies strongly support the hypothesis that organisms have been evolving in three independent lineages for over 1.5 billion years. Thus, many systematists favor organizing the diversity of life into three domains, groupings above kingdoms. Two distinct groups of prokaryotes are the domains Bacteria and Archaea. All eukaryotes are placed in the third domain, the Eukarya (or Eucarya). This phylogenetic tree incorporates the hypothesis that Domains Eukarya and Archaea share a common ancestor that lived more recently than the ancestor common to archaea and bacteria. Molecular studies support this hypothesis that the archaea are more closely related to eukaryotes than they are to bacteria.

STRUCTURE, FUNCTION, AND REPRODUCTION OF PROKARYOTES

Most prokaryotes are unicellular. However, some species tend to aggregate transiently in groups of two or more cells. Others have the form of true colonies, which are permanent aggregates of identical cells. And some species of prokaryotes even exhibit a simple multicellular organization in which there is a division of labor between two or more specialized types of cells.

There is a diversity of cell shapes among prokaryotes, the three most common being spheres (cocci), rods (bacilli), and helices (including the bacteria known as spirilla and spirochetes). An important step in identifying prokaryotes is determining their shapes by microscopic examination (FIGURE 27.2, p. 504).

Most prokaryotes have diameters in the range of 1–5 μm, compared to 10–100 μm for the majority of eukaryotic cells. There are, however, notable exceptions; the largest prokaryote discovered so far is a rod-shaped species that measures about a half-millimeter in length, dwarfing most eukaryotic cells (FIGURE 27.3, p. 504).

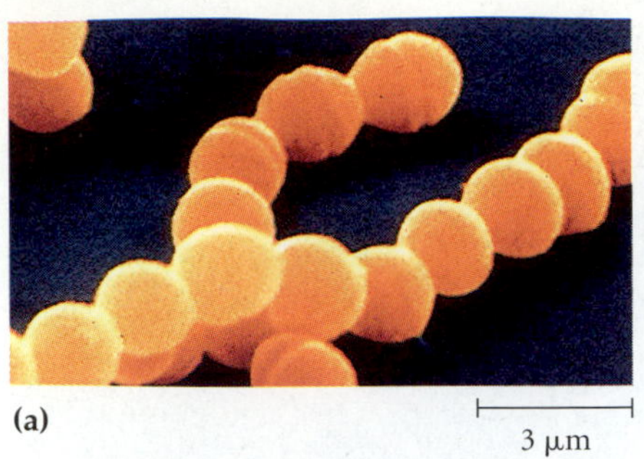

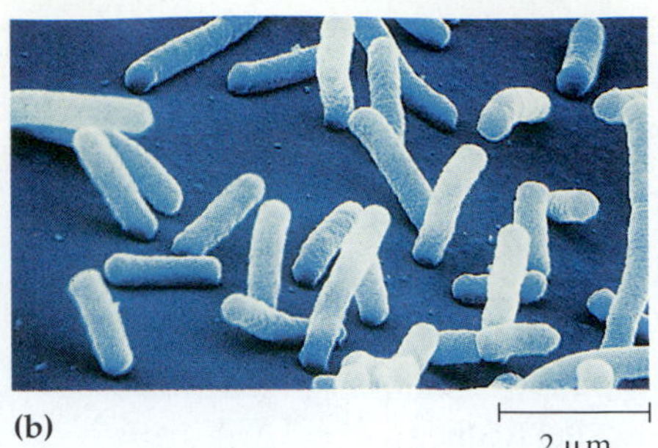

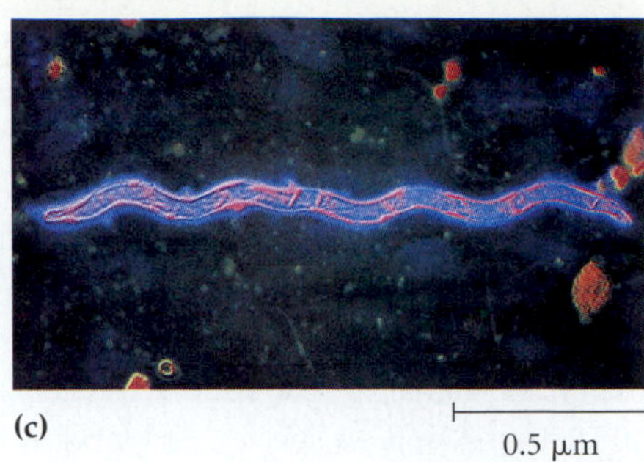

FIGURE 27.2 ▪ The most common shapes of prokaryotes. **(a)** Cocci (singular, coccus), or spherical prokaryotes, occur singly, in pairs (diplococci), in chains of many cells (streptococci, shown here), and in clusters resembling bunches of grapes (staphylococci). **(b)** Rod-shaped prokaryotes, or bacilli (singular, bacillus), are most commonly solitary, but in some forms the rods are arranged in chains. **(c)** Helical prokaryotes include the spirilla and the corkscrew-shaped spirochetes. (All SEMs.)

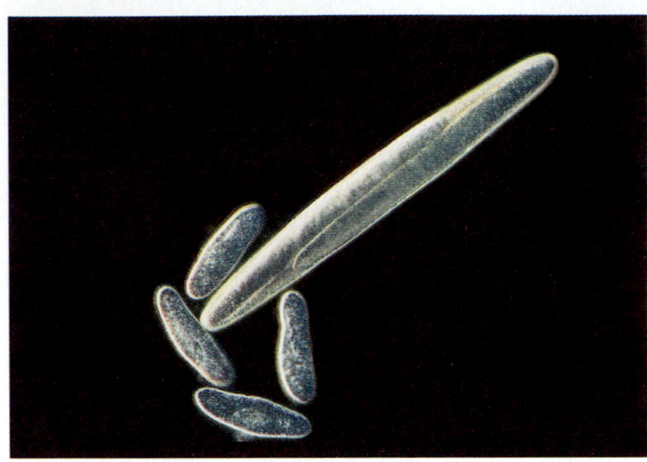

FIGURE 27.3 ▪ The largest known prokaryote. In this light micrograph, *Epulopiscium fishelsoni,* about a half-millimeter long, dwarfs four eukaryotes (the protist *Paramecium*). The prokaryotic giant lives as a symbiont in the gut of surgeonfish.

Nearly all prokaryotes have cell walls external to their plasma membranes

Present in nearly all prokaryotes, the cell wall maintains the shape of the cell, affords physical protection, and prevents the cell from bursting in a hypotonic environment (see Chapter 8). Like other walled cells, however, prokaryotes plasmolyze and may die in a hypertonic medium, which is why heavily salted meat can be kept so long without being spoiled by prokaryotes. The cell walls of prokaryotes differ in molecular composition and construction from those of plants, fungi, and protists.

The presence of a cell wall is one reason prokaryotes were grouped with plants in the old two-kingdom system. Instead of cellulose, the staple of plant walls, most bacterial walls contain a unique material called **peptidoglycan**, which consists of polymers of modified sugars cross-linked by short polypeptides that vary from species to species (the walls of archaea lack peptidoglycan). The effect is a single molecular network enclosing and protecting the entire cell. External to this fabric are other substances that also differ from species to species.

One of the most valuable tools in microbial taxonomy is the **Gram stain**, which can be used to separate many members of the domain Bacteria into two groups based on differences in their cell walls. **Gram-positive** bacteria have simpler walls, with a relatively large amount of peptidoglycan. The walls of **gram-negative** bacteria have less peptidoglycan and are structurally more complex. An outer membrane on the gram-negative cell wall contains lipopolysaccharides, carbohydrates bonded to lipids (see the Methods Box).

Among pathogenic, or disease-causing, bacteria, gram-negative species are generally more threatening than gram-positive species. The lipopolysaccharides on the walls of gram-negative bacteria are often toxic, and the outer membrane helps protect the pathogens against the defenses of their hosts. Furthermore, gram-negative bacteria are commonly more resistant than gram-positive species to antibiotics because the outer membrane impedes entry of the drugs.

Many antibiotics, including penicillins, inhibit the synthesis of cross-links in peptidoglycan and prevent the formation of a functional wall, particularly in gram-positive species. These drugs are like selective bullets that cripple many species

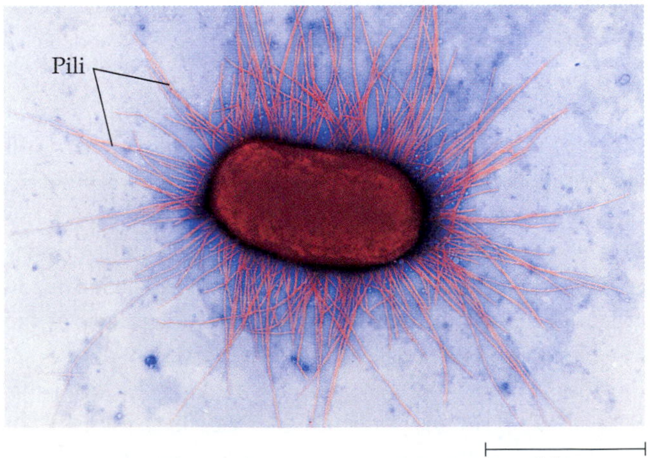

FIGURE 27.4 ▪ Pili. These appendages enable some prokaryotes to attach to surfaces or to other prokaryotes (TEM). Some pili function during conjugation, holding partners together while DNA is transferred (see FIGURE 18.13).

METHODS: The Gram Stain

This method, named for Hans Christian Gram, a Danish physician who developed the technique in the late 1800s, distinguishes between two different kinds of bacterial cell walls. Bacteria are stained with a violet dye and iodine, rinsed in alcohol, and then stained again with a red dye. The structure of the cell wall determines the staining response. Gram-positive bacteria (top) have cell walls with a large amount of peptidoglycan that traps the violet dye (LM). Gram-negative bacteria (bottom) have less peptidoglycan, which is located in a periplasmic gel between the plasma membrane and an outer membrane. The violet dye used in the Gram stain is easily rinsed from gram-negative bacteria, but the cells retain the red dye (LM). (The colors in the diagrams do not represent the stains.)

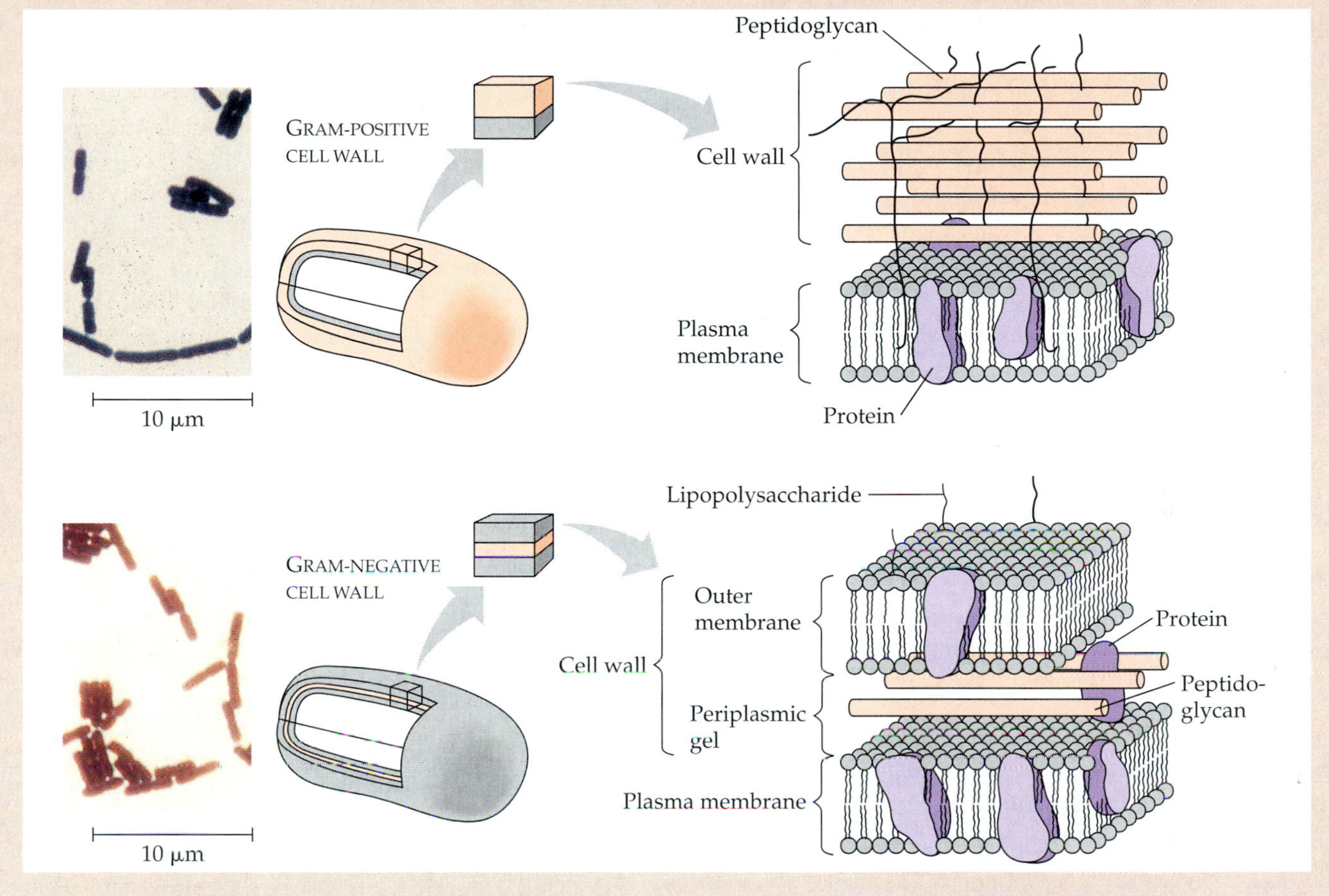

of infectious bacteria without adversely affecting humans and other eukaryotes, which do not make peptidoglycan.

Many prokaryotes secrete sticky substances that form still another protective layer called a **capsule** outside the cell wall. Capsules enable the organisms to adhere to their substrate and provide additional protection, including increased resistance of pathogenic prokaryotes to host defenses. Gelatinous capsules glue together the cells of many prokaryotes that live as colonies.

Another way some prokaryotes adhere to one another or to some substratum is by means of surface appendages called **pili** (singular, **pilus**; FIGURE 27.4). For example, *Neisseria gonorrhoeae,* the pathogen that causes gonorrhea, uses pili to fasten itself to mucous membranes of its host. Some pili are specialized for holding prokaryotes together long enough for the cells to transfer DNA during conjugation.

Many prokaryotes are motile

About half of all prokaryotes are capable of directional movement. Relative to our experience, many of them move through solutions of water, ions, and other solutes that are as thick as molasses. Other environments, such as the mucous linings of lungs, are even more viscous. Despite these conditions, many motile prokaryotes can move about 50 μm/sec, or about 100 times their body length per second.

Flagellar action is the most common mechanism of movement among prokaryotes. Flagella may be scattered over the entire cell surface or concentrated at one or both ends of the cell. The flagella of prokaryotes and eukaryotes differ in both structure and function. (To review eukaryotic flagella, see Chapter 7.) Prokaryotic flagella are one-tenth the width of those of eukaryotes and are not covered by an extension of the plasma membrane (FIGURE 27.5).

A second motility mechanism characterizes a group of helical-shaped bacteria called spirochetes. Two or several helical filaments just under the outer sheath of the cell wall are much like prokaryotic flagella in structure. Each has a basal motor attached at one or the other end of the cell. When the filaments rotate, the flexible cell moves like a corkscrew. This mechanism is particularly effective in moving spirochetes through the highly viscous environments in which they sometimes live. In a third mechanism of motility, some prokaryotes secrete slimy chemicals and move by a gliding motion that may result from the presence of flagellar motors that lack flagellar filaments.

In a relatively uniform environment, flagellated prokaryotes may wander randomly. In a heterogeneous environment, however, many prokaryotes are capable of **taxis**, movement toward or away from a stimulus (Gr. *taxis*, "to arrange"). With chemotaxis, for example, prokaryotes respond to chemical stimuli, perhaps moving toward food or oxygen (a positive chemotaxis) or away from some toxic substance (a negative chemotaxis). Several kinds of receptor molecules that detect specific substances are located on the surfaces of chemotactic prokaryotes. Motile prokaryotes that are photosynthetic generally display a positive phototaxis, a behavior that keeps them in the light. Some prokaryotes even contain a row of tiny magnetic particles that allow the cells to orient in Earth's magnetic field. These particles may help cells distinguish up from down and cause the prokaryotes to migrate toward the nutrient-rich sediments at the bottoms of ponds and shallow seas.

The cellular and genomic organization of prokaryotes is fundamentally different from that of eukaryotes

Recall that prokaryotes are named for their lack of true nuclei enclosed by membranes (see FIGURE 7.4). The cells of prokaryotes lack the extensive compartmentalization by internal membranes characteristic of eukaryotes. However, various prokaryotes do have a variety of specialized membranes that perform many of their metabolic functions. These membranes are usually infolded regions of the plasma membrane (FIGURE 27.6).

Compared to eukaryotes, prokaryotes also have smaller, simpler genomes. On average, prokaryotes have only about one-thousandth as much DNA as a eukaryotic cell. In most prokaryotes, the DNA is concentrated as a snarl of fibers in a **nucleoid region** that stains less dense than the surrounding cytoplasm in electron micrographs. The mass of fibers is actually the prokaryotic chromosome, one double-stranded DNA molecule in the form of a ring. The DNA has relatively little protein associated with it. The term *genophore* is sometimes used for the prokaryotic chromosome to distinguish it from eukaryotic chromosomes, which have a very different structure. The genome of eukaryotes consists of linear DNA molecules packaged along with proteins into a number of chromosomes characteristic of the species.

In addition to its one major chromosome, the prokaryotic cell may also have much smaller rings of DNA called plasmids, most consisting of only a few genes. In most environments, prokaryotes can survive without their plasmids because all essential functions are programmed by the chromosome. However, plasmids endow the cell with genes for resistance to antibiotics, for the metabolism of unusual nutrients not present in the normal environment, and for other special contingencies. Plasmids replicate independently of the main chromosome, and many can be readily transferred between partners when prokaryotes conjugate (see FIGURE 18.14).

Although the general processes for DNA replication and the translation of genetic messages into proteins are alike for eukaryotes and prokaryotes, some of the details differ. For example, the prokaryotic ribosome is slightly smaller than the eukaryotic version and differs in its protein and RNA content.

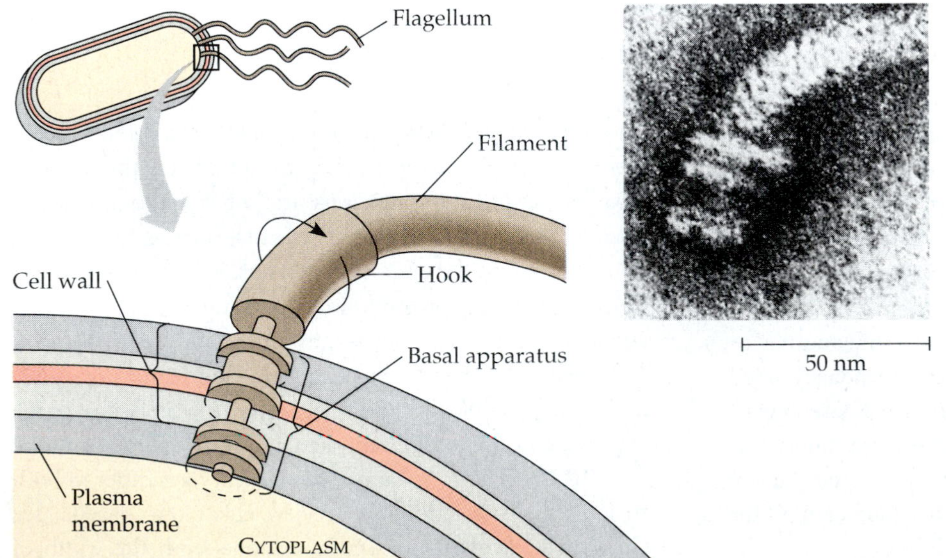

FIGURE 27.5 • Form and function of prokaryotic flagella. Entirely different from the flagellum of eukaryotes, the prokaryotic flagellum is a naked protein structure that lacks microtubules. Chains of a globular protein, flagellin, are wound in a tight spiral to form a semirigid, helical filament. The filament is attached to another protein that forms a curved hook, which is inserted into a basal apparatus. Composed of about 35 different proteins, the basal apparatus includes a system of rings in the layers of the cell wall. (The structures shown in this figure are characteristic of gram-negative bacteria.) The basal apparatus—the flagellar motor—rotates the filament, which propels the cell by pushing against the outside medium. The motor is powered by the diffusion of protons (H^+) into the cell after they have been pumped outward across the plasma membrane at the expense of ATP.

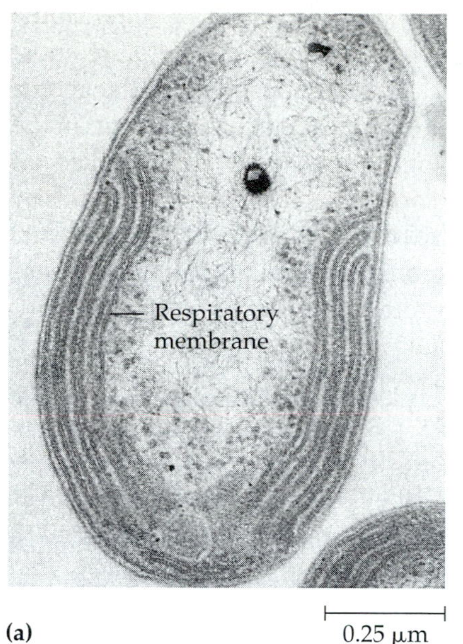

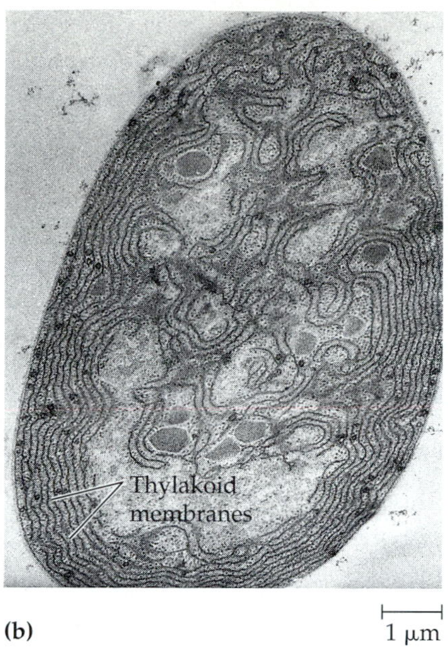

FIGURE 27.6 ▪ Specialized membranes of prokaryotes. (a) These infoldings of the plasma membrane, reminiscent of the cristae of mitochondria, function in cellular respiration of some aerobic prokaryotes (TEM). **(b)** Members of Domain Bacteria called cyanobacteria have thylakoid membranes, much like those in chloroplasts, that function in photosynthesis (TEM).

The disparity is great enough that selective antibiotics, including tetracycline and chloramphenicol, bind to the ribosomes and block protein synthesis in many prokaryotes but not in eukaryotes.

Populations of prokaryotes grow and adapt rapidly

Prokaryotes reproduce only asexually by the mode of cell division called **binary fission**, synthesizing DNA almost continuously. (Binary fission is described in Chapter 12.) A single prokaryotic cell in a favorable environment will give rise by repeated divisions to a colony of offspring (FIGURE 27.7). Note that neither mitosis nor meiosis occurs among prokaryotes; this is another fundamental difference between prokaryotes and eukaryotes.

Without meiosis, prokaryotes lack a sexual cycle, which includes syngamy (the union of haploid nuclei; see Chapter 13), an important source of genetic variation in eukaryotes. However, as you learned in Chapter 18, prokaryotes have three mechanisms of genetic recombination: **transformation**, in which genes are taken up from the surrounding environment, allowing for considerable genetic transfer between prokaryotes; **conjugation**, in which genes are transferred directly from one prokaryote to another; and **transduction**, in which genes are transferred between prokaryotes by viruses. These processes, however, involve the unilateral passage of a variable amount of DNA—nothing like the meiotic sex of eukaryotes, in which two parents each contribute homologous genomes to a zygote. Mutation is the major source of genetic variation in prokaryotes. Because generation times are measured in minutes or hours, a favorable mutation can be rapidly propagated to a large number of offspring.

A short generation span enables prokaryotic populations to adapt very rapidly to environmental change, as natural selection screens new mutations and novel genomes resulting from recombination. This adaptive evolution is important to

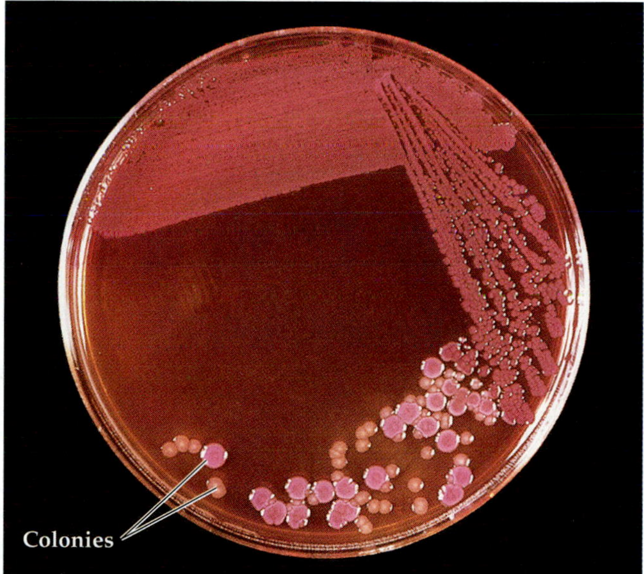

FIGURE 27.7 ▪ Prokaryote colonies in culture. Laboratory cultures of prokaryotes are grown in liquid or solid media of known composition. The media, in containers such as petri dishes or test tubes, are sterilized to ensure that no unwanted microbes will grow. Then a sample of prokaryotes, sometimes just a single cell, is introduced, and the containers are incubated at an appropriate temperature. When the prokaryotes are grown on solid media, colonies are usually large enough to be visible to the unaided eye after a day or two. The size, shape, texture, and color of a colony provide clues to the identity of the prokaryotes, as do the nutrients and physical conditions required for their growth. Colonies of several species of bacteria are shown here. Microscopic examination of cells taken from a colony is another step in identification.

the continuing success of prokaryotes, as it was when prokaryotic life began to diversify billions of years ago.

The word *growth* as applied to prokaryotes refers more to the multiplication of cells and an increase in population size than to the enlargement of individual cells. The conditions for optimal growth—temperature, pH, salt concentrations, nutrient sources, and so on—vary according to species. Refrigeration retards food spoilage because most microorganisms grow only very slowly at such low temperatures.

In an environment without limiting resources, the growth of prokaryotes is effectively geometric: One cell divides to form 2, which divide again to produce a total of 4 cells, then 8, 16, and so on, the number of cells in a colony doubling with each generation. Many prokaryotes have generation times in the range of 1 to 3 hours, but some species can double every 20 minutes in an optimal environment. If the latter growth rate were sustained, a single cell would give rise to a colony weighing 1 million kg in just 24 hours. However, prokaryotic growth both in the laboratory and in nature is usually checked at some point, when the cells exhaust some nutrient or the colony poisons itself with an accumulation of metabolic wastes.

The ability of some prokaryotes to withstand harsh conditions is impressive. Some bacteria form resistant cells called **endospores** (see the micrograph of *Bacillus* in TABLE 27.3, pp. 514–515). The original cell replicates its chromosome, and one copy becomes surrounded by a durable wall. The outer cell disintegrates, but the endospore it contained survives all sorts of trauma, including lack of nutrients and water, extreme heat or cold, and most poisons. Boiling water is not hot enough to kill most endospores in a relatively short length of time. Home canners and the food-canning industry, therefore, must take extra precautions to kill endospores of dangerous bacteria. To sterilize media, glassware, and utensils in the laboratory, microbiologists use an appliance called an autoclave, a pressure cooker that kills even endospores by heating to temperatures 120°C. In less hostile environments, endospores may remain dormant for centuries. If placed in a hospitable environment, they will hydrate and revive to the vegetative (colony-producing) state.

In most natural environments, prokaryotes must compete for space and nutrients. A general feature of many microorganisms (including certain species of prokaryotes, protists, and fungi) is the release of **antibiotics**, chemicals that inhibit the growth of other microorganisms. Humans have discovered some of these compounds and use them to combat pathogenic bacteria.

NUTRITIONAL AND METABOLIC DIVERSITY

One of the most profound results of the adaptive radiation of prokaryotes is a great diversity of modes of nutrition and metabolism. Every type of nutrition observed in eukaryotes is represented among prokaryotes, plus some nutritional modes are unique to prokaryotes. Metabolic diversity is greater among prokaryotes than among all eukaryotes combined.

Prokaryotes can be grouped into four categories according to how they obtain energy and carbon

Nutrition refers here to how an organism obtains two
27.1 resources for synthesizing organic compounds: energy and a source of carbon. Species that use light energy are termed *phototrophs. Chemotrophs* obtain their energy from chemicals taken from the environment. If an organism needs only the inorganic compound CO_2 as a carbon source, it is called an *autotroph. Heterotrophs* require at least one organic nutrient—glucose, for instance—as a source of carbon for making other organic compounds. We can combine the phototroph-versus-chemotroph (energy source) and autotroph-versus-heterotroph (carbon source) criteria to group prokaryotes according to four major modes of nutrition:

1. **Photoautotrophs** are photosynthetic organisms that harness light energy to drive the synthesis of organic compounds from carbon dioxide. The specialized metabolic machinery of these organisms includes internal membranes with light-harvesting pigment systems (see Chapter 10 and FIGURE 27.6b). Among the diverse groups of photosynthetic prokaryotes are the cyanobacteria. All photosynthetic eukaryotes—plants and certain protists—also fit into this nutritional category.
2. **Chemoautotrophs** need only CO_2 as a carbon source, but instead of using light for energy, these prokaryotes obtain energy by oxidizing inorganic substances. Chemical energy is extracted from hydrogen sulfide (H_2S), ammonia (NH_3), ferrous ions (Fe^{2+}), or some other chemical, depending on the species. This mode of nutrition is unique to certain prokaryotes. For instance, archaea of the genus *Sulfolobus* oxidize sulfur.
3. **Photoheterotrophs** can use light to generate ATP but must obtain their carbon in organic form. This mode of nutrition is restricted to certain prokaryotes.
4. **Chemoheterotrophs** must consume organic molecules for both energy and carbon. This nutritional mode is found widely among prokaryotes, protists, fungi, animals, and even some plants.

TABLE 27.1 reviews the four major modes of nutrition.

Nutritional Diversity Among Chemoheterotrophs

The majority of prokaryotes are chemoheterotrophs. This category includes **saprobes**, decomposers that absorb their nutrients from dead organic matter, and **parasites**, which absorb their nutrients from the body fluids of living hosts.

27.1

Table 27.1 ■ Major Nutritional Modes

MODE OF NUTRITION	ENERGY SOURCE	CARBON SOURCE	TYPES OF ORGANISMS
Autotroph			
Photoautotroph	Light	CO_2	Photosynthetic prokaryotes, including cyanobacteria; plants; certain protists
Chemoautotroph	Inorganic chemicals	CO_2	Certain prokaryotes (e.g., *Sulfolobus*)
Heterotroph			
Photoheterotroph	Light	Organic compounds	Certain prokaryotes
Chemoheterotroph	Organic compounds	Organic compounds	Most prokaryotes and protists; fungi; animals; some plants

The specific organic nutrients needed for growth vary extensively among chemoheterotrophic prokaryotes. Some species are very exacting in their requirements; for example, bacteria of the genus *Lactobacillus* will grow well only in a medium containing all 20 amino acids, several vitamins, and other organic compounds. Among species less particular in their nutritional needs, *E. coli* can grow on a medium containing glucose as the only organic ingredient, and the organism's metabolism is so versatile that many other compounds can substitute for glucose as the sole organic nutrient.

There is such a diversity of chemoheterotrophs that almost any organic molecule can serve as food for at least some species. For example, some bacteria are capable of metabolizing petroleum; they are used to clean up oil spills. Those few classes of synthetic organic compounds (including some kinds of plastics) that cannot be broken down by any chemoheterotrophs are said to be nonbiodegradable.

Nitrogen Metabolism

Nitrogen metabolism is another facet of nutritional diversity among prokaryotes. Nitrogen is an essential component of proteins and nucleic acids. While animals, plants, and other eukaryotes are limited in the forms of nitrogen they can use, diverse prokaryotes are able to metabolize most nitrogenous compounds.

Key steps in the cycling of nitrogen through ecosystems are performed only by prokaryotes. (For an overview of the nitrogen cycle, see FIGURE 54.11.) Some chemoautotrophic bacteria, such as *Nitrosomonas,* convert NH_3 to NO_2^+. Other bacteria, such as a few species of *Pseudomonas,* "denitrify" NO_2^+ or NO_3^+ to atmospheric N_2 gas. And diverse species of prokaryotes, including some cyanobacteria, are able to use atmospheric nitrogen directly as a source of nitrogen. In this process, called **nitrogen fixation**, prokaryotes convert atmospheric N_2 to NH_3 (ammonia). Nitrogen fixation, unique to certain prokaryotes, is the only biological mechanism that makes atmospheric nitrogen available to organisms for incorporation into organic compounds. In terms of nutrition, nitrogen-fixing cyanobacteria are the most self-sufficient of all organisms. These photoautotrophs require only light energy, CO_2, N_2, water, and some minerals in order to grow.

Metabolic Relationships to Oxygen

Another metabolic variation among prokaryotes is in the effect that oxygen has on growth (see Chapter 9). **Obligate aerobes** use O_2 for cellular respiration and cannot grow without it. **Facultative anaerobes** will use O_2 if it is present but can also grow by fermentation in an anaerobic environment. **Obligate anaerobes** are poisoned by O_2. Some obligate anaerobes live exclusively by fermentation; other species extract chemical energy by **anaerobic respiration**, in which inorganic molecules other than O_2 accept electrons at the "downhill" end of electron transport chains.

Now that we have surveyed variation in nutrition and metabolism among prokaryotes, let's trace the evolutionary roots of this metabolic diversity.

The evolution of prokaryotic metabolism was both cause and effect of changing environments on Earth

All forms of nutrition and nearly all metabolic pathways evolved among prokaryotes before eukaryotes arose. As early prokaryotes evolved, they were met with constantly changing physical and biological environments. In response to these changes, new metabolic capabilities evolved that, in turn, changed the environment faced by the next community of prokaryotes. All the major metabolic capabilities seen among modern prokaryotes probably evolved in the first billion years of life. Reasonable hypotheses about the early history of prokaryotes and the origins of metabolic diversity are now possible. The hypothetical scenarios described here are based on inferences from molecular systematics, from comparisons of energy metabolism among extant prokaryotes, and from geological evidence about conditions on the early Earth.

The Origins of Metabolism

The presence of a particular metabolic process in all or nearly all modern organisms may imply that the process developed in a common ancestor and is thus ancient. For example, the universal role of ATP as an energy currency in all extant organisms implies that prokaryotes adopted its use very early.

Likewise, glycolysis, a metabolic pathway that breaks organic molecules down to simpler waste products and uses the energy to generate ATP by substrate phosphorylation, is common to nearly all modern organisms. Glycolysis does not require O_2, and fermentation, in which electrons extracted from nutrients during glycolysis are transferred to organic recipients, may have become a way of life on the anaerobic Earth. The chemiosmotic mechanism of ATP synthesis is also common to most organisms, implying an early origin (see Chapter 9).

The first prokaryotes, which originated between 3.5 and 4.0 billion years ago, undoubtedly had few enzymes and were very simple metabolically. Living in an environment with virtually no molecular oxygen, they would have been anaerobes.

A traditional hypothesis proposes that the earliest cells were chemoheterotrophs that absorbed free organic compounds, including ATP, generated in the primordial seas by abiotic synthesis, similar to the Miller-Urey simulations (see Chapter 26; FIGURE 27.8a). As early chemoheterotrophs began to deplete the supply of free ATP, natural selection would have favored cells with enzymes that could regenerate ATP from ADP using energy extracted from other available organic nutrients. The result may have been the step-by-step evolution of glycolysis and the generation of ATP by substrate phosphorylation. Chemiosmotic synthesis of ATP may have evolved somewhat later.

Many biologists now consider it unlikely that environmental conditions on the primordial Earth generated enough ATP or other organic molecules to support chemoheterotrophs. The hypothesis most widely favored today is that the earliest prokaryotes were chemoautotrophs that obtained energy from inorganic chemicals and made their own energy-currency molecules instead of absorbing ATP. Hydrogen sulfide (H_2S) and compounds of iron (Fe^{2+}) were abundant on the early Earth, and primitive cells may have obtained energy from reactions involving such compounds. Some modern archaea that thrive in hot sulfur springs can carry out the reaction

$$FeS + H_2S \longrightarrow FeS_2 + H_2 + \text{free energy}$$

and exergonic reactions such as this may have been an early source of energy. Membrane proteins in an early prokaryote may have used some of the resulting free energy to split the product H_2 into protons and electrons and establish a proton gradient across its plasma membrane. In a primitive form of chemiosmosis, the gradient may have driven the synthesis of ATP (FIGURE 27.8b). Natural selection would have favored cells with membrane proteins capable of manipulating hydrogen, leading to the evolution of electron transport chains in the plasma membrane (see Chapter 9).

Although the earliest cells were probably chemoautotrophic, they may also have been opportunistic, obtaining some nutrients by absorbing organic compounds when they were available. The ability to use light energy may also have evolved relatively early. Having the enzyme systems to use a variety of energy sources would have been highly advantageous.

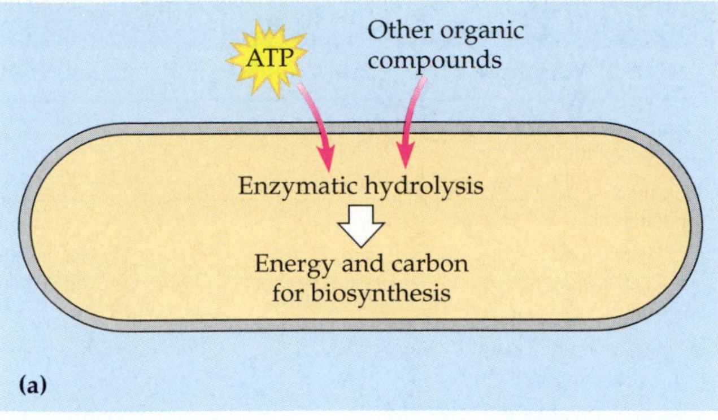

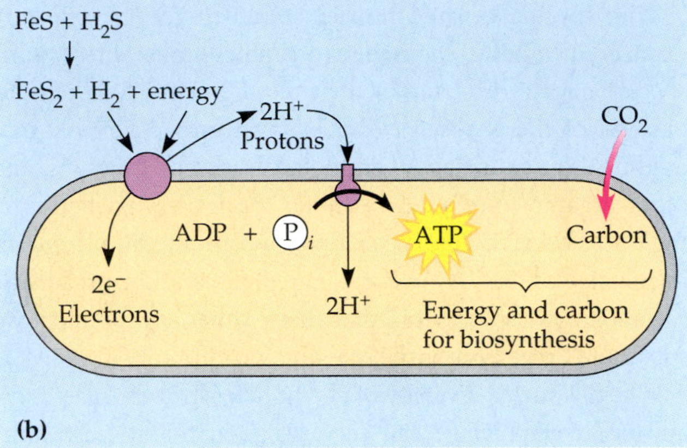

FIGURE 27.8 • Two hypotheses for the origin of energy metabolism. Both scenarios postulate that the use of ATP as energy currency evolved very early. **(a)** In one scenario, an early chemoheterotroph lived on ATP and other organic compounds absorbed from its surroundings. This scenario assumes that the first cells lived in a soup of abiotically produced organic molecules. Enzymatic breakdown of the organic compounds provided energy and carbon. **(b)** A different, more plausible scheme proposes that the earliest mode of energy metabolism was chemoautotrophic, with ancestral cells obtaining carbon from CO_2 and synthesizing their own ATP by a simple chemiosmotic mechanism (see Chapter 9). In the model shown here, enzymes from a chemoautotroph promote chemical reactions among simple inorganic compounds such as FeS and H_2S, which were abundant on the early Earth. Hydrogenase (purple ball), an enzyme in the organism's membrane, couples the release of energy from the inorganic compounds to the splitting of H_2 and to the maintenance of a proton (H^+) gradient across the plasma membrane. In turn, the potential energy of the proton gradient drives the enzymatic synthesis of ATP.

The Origin of Photosynthesis

In some early prokaryotes, light-absorbing pigments may have absorbed excess light energy (particularly ultraviolet) that was harmful to cells growing at the surface of their aquatic environments. Later, these energized pigments may have been coupled with membrane proteins involved in ATP synthesis. In a group of modern archaea called extreme halophiles, a pigment that captures light energy, known as **bacteriorhodopsin**, is built into the plasma membrane. (This

(a) Cyanobacterial bloom

(b) *Anabaena*

FIGURE 27.9 ▪ A bloom of cyanobacteria. **(a)** The larger lake in this photograph, located in Cape Cod, Massachusetts, is undergoing a population explosion—often called a "bloom"—of cyanobacteria. The lake's blue-green color results from the presence of trillions of cyanobacterial cells. **(b)** A micrograph of *Anabaena*, the predominant cyanobacterium in the lake (LM). Cyanobacteria are the only prokaryotes that split water and release O_2 in their light-harvesting reactions. Earth's relationship with life changed when O_2 produced by early cyanobacteria began to accumulate in the atmosphere over 2.5 billion years ago.

molecule is structurally related to visual pigments in the retina of the eye.) Bacteriorhodopsin absorbs light and uses the energy to pump hydrogen ions (H^+) out of the cell. The H^+ gradient then drives the synthesis of ATP. This is the simplest known mechanism of photophosphorylation. Researchers are studying the halophiles as model systems for solar energy conversion.

In some early prokaryotes, pigments and photosystems evolved that use light to drive electrons from hydrogen sulfide (H_2S) to $NADP^+$. These organisms could have generated the reducing power needed to fix CO_2 photosynthetically (see Chapter 10). The membrane proteins (electron transport chains) that powered ATP synthesis may have been co-opted to provide the reducing power. The modern organisms with nutrition most like the early photosynthetic prokaryotes are believed to be the purple sulfur bacteria (Proteobacteria; see TABLE 27.3, p. 514) and green sulfur bacteria. These prokaryotes owe their color to bacteriochlorophyll, which functions instead of chlorophyll *a* as their main photosynthetic pigment. Purple and green sulfur bacteria split H_2S instead of H_2O as a source of electrons; thus, they produce no O_2.

Cyanobacteria, the Oxygen Revolution, and the Origin of Cellular Respiration

Some of the photosynthetic bacteria eventually had the metabolic machinery to use plentiful H_2O instead of H_2S or other compounds as a source of electrons and hydrogen for reducing CO_2. These were the first **cyanobacteria**, formerly known as blue-green algae. Capable of making organic compounds from water and CO_2, they flourished and changed the world by releasing O_2 as a by-product of their photosynthesis (FIGURE 27.9).

Cyanobacteria evolved between 2.5 and 3.4 billion years ago, living along with other prokaryotes in colonies that probably built the stromatolites that have been found all over the world (see FIGURE 26.1). Some marine sediments that are 2.5 billion years old are banded iron formations, red layers rich in iron oxide and valuable as a source of iron ore today. The sediments probably formed during a period when abundant cyanobacteria released oxygen that reacted with dissolved iron ions in the oceans to produce iron oxide, which precipitated out. This reaction would have prevented any accumulation of free O_2 for perhaps a few hundred million years, until precipitation exhausted the dissolved iron. Only then would the seas become saturated with O_2, resulting in the release of oxygen gas into the atmosphere. Beginning about 2 billion years ago, terrestrial rocks rich in iron were rusted red by oxidation with atmospheric O_2.

The gradual change to a more oxidizing atmosphere created a crisis for Precambrian prokaryotes, because oxygen attacks the bonds of organic molecules. The corrosive atmosphere probably caused the extinction of many prokaryotes unable to cope. Other species survived in habitats that remained anaerobic, where we find their descendants living today as obligate anaerobes. The evolution of antioxidant mechanisms enabled other prokaryotes to tolerate the rising oxygen levels. Among photosynthetic prokaryotes, some species went a step beyond mere oxygen tolerance to actually using the oxidizing power of O_2 to pull electrons from organic molecules down existing transport chains. Thus, aerobic respiration may have originated as modifications of electron transport chains co-opted from photosynthesis. Among photoheterotrophs, the purple nonsulfur bacteria still use an electron transport system that is a hybrid of photosynthetic and respiratory equipment. Several other bacterial lineages gave up photosynthesis and reverted to chemoheterotrophic nutrition, their electron transport chains adapted to function exclusively in aerobic respiration.

PHYLOGENY OF PROKARYOTES

Now that we have surveyed the structural and metabolic adaptations of prokaryotes, we have the evolutionary context we need for a closer look at the two prokaryote domains, Bacteria and Archaea.

Molecular systematics is leading to a phylogenetic classification of prokaryotes

27.2 Researchers' first hint of the early bacteria/archaea split was the correlation of each of the two prokaryote domains with unique **signature sequences**, taxon-specific base sequences at comparable locations in ribosomal RNA or other nucleic acids. TABLE 27.2 highlights other ways in which bacteria differ from archaea. The third domain, Eukarya, is included in the table to reinforce the point that archaea have at least as much in common with eukaryotes as they do with bacteria. However, the archaea also have many unique traits, as should be expected of a taxon that has followed a separate evolutionary path for so long.

Table 27.2 ▪ A Comparison of the Three Domains of Life

	DOMAIN		
CHARACTERISTIC	**Bacteria**	**Archaea**	**Eukarya**
Nuclear envelope	Absent	Absent	Present
Membrane-enclosed organelles	Absent	Absent	Present
Peptidoglycan in cell wall	Present	Absent	Absent
Membrane lipids	Unbranched hydrocarbons	Some branched hydrocarbons	Unbranched hydrocarbons
RNA polymerase	One kind	Several kinds	Several kinds
Initiator amino acid for start of protein synthesis	Formyl-methionine	Methionine	Methionine
Introns (noncoding parts of genes)	Absent	Present in some genes	Present
Response to the antibiotics streptomycin and chloramphenicol	Growth inhibited	Growth not inhibited	Growth not inhibited

Domain Archaea

You learned earlier in this chapter that most archaea inhabit the more extreme environments of Earth. Biologists who study the prokaryotic life of such habitats have identified three main groups of archaea: the methanogens, the extreme halophiles, and the extreme thermophiles.

The **methanogens** are named for their unique form of energy metabolism, in which H_2 is used to reduce CO_2 to methane (CH_4). Methanogens, among the strictest of anaerobes, are poisoned by oxygen. They live in swamps and marshes where other microbes have consumed all the oxygen; the methane that bubbles out at these sites is known as marsh gas. Methanogens are also important decomposers used in sewage treatment. Some farmers have experimented with the use of these microbes to convert garbage and dung to methane, a valuable fuel. Other species of methanogens inhabit the anaerobic environment within the guts of animals, playing an important role in the nutrition of cattle, termites, and other herbivores that subsist mainly on a diet of cellulose.

The **extreme halophiles** (Gr. *halo,* "salt," and *philos,* "lover") live in such saline places as the Great Salt Lake and

FIGURE 27.10 ▪ Extreme halophiles. These archaea live in extremely saline waters. The colors of these seawater evaporating ponds at the edge of San Francisco Bay result from a dense growth of extreme halophiles that thrive in the ponds when the water reaches a salinity of 15% to 20%. (Before evaporation, the salinity of seawater is about 3%.) The ponds are used for commercial salt production; the halophilic archaea are harmless.

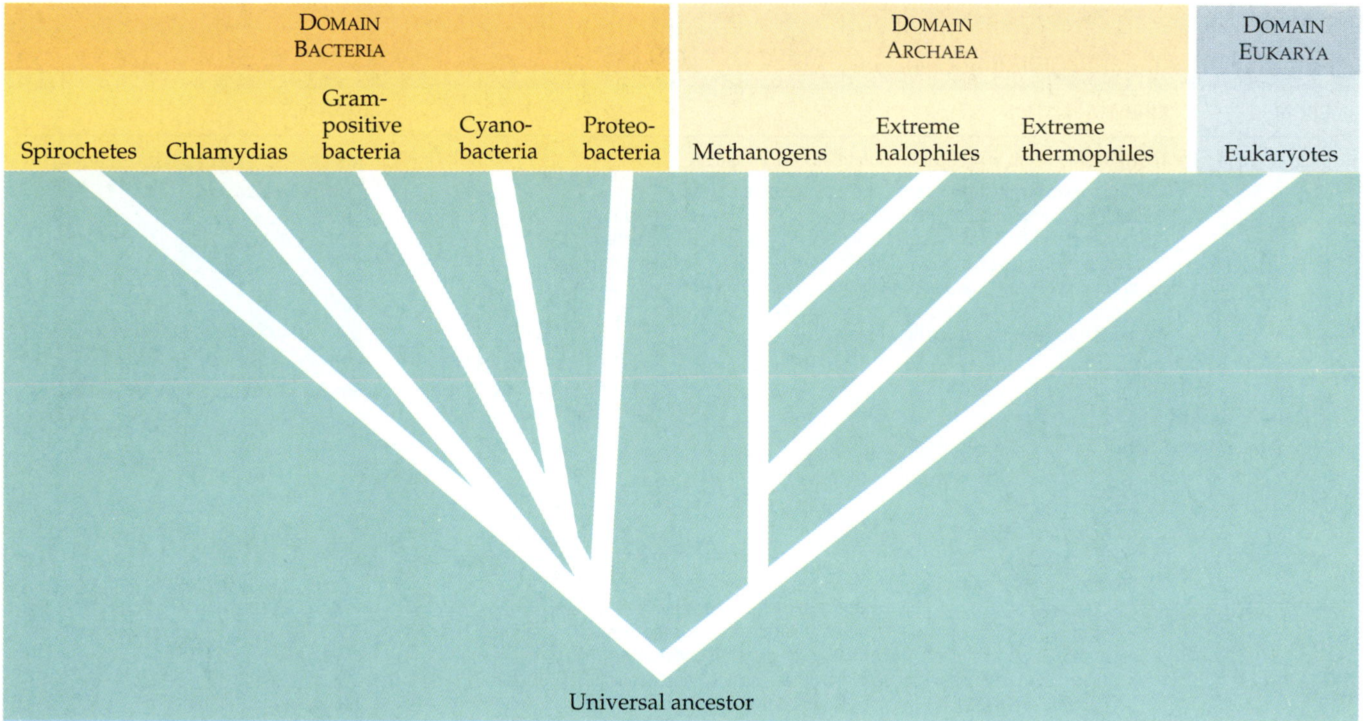

27.2 **FIGURE 27.11 ▪ Evolutionary relationships of the prokaryotes.** This tentative phylogeny, like all such trees, is a hypothesis about evolutionary history. This particular hypothesis is based mainly on molecular systematics, particularly comparisons of rRNA signature sequences. (Of the approximately 12 groups of Domain Bacteria, only those featured in TABLE 27.3 are included in this simplified tree.) The groups shown here might be considered kingdoms within Domains Bacteria and Archaea. However, most prokaryote systematists prefer to wait for more data before assigning the groups to concrete taxonomic categories.

the Dead Sea. Some species merely tolerate salinity, whereas others actually require an environment ten times saltier than seawater to grow (FIGURE 27.10). Colonies of halophiles form a purple-red scum that owes its color to bacteriorhodopsin (see p. 510).

As their name implies, the **extreme thermophiles** thrive in hot environments. The optimal conditions for these archaea are temperatures of 60°C to 80°C. *Sulfolobus* inhabits hot sulfur springs in Yellowstone National Park, obtaining its energy by oxidizing sulfur. Another sulfur-metabolizing thermophile lives in the 105°C water near deep-sea hydrothermal vents. Comparisons of key proteins have convinced James Lake of the University of California, Los Angeles, that the extreme thermophiles are the prokaryotes most closely related to eukaryotes. He highlights this evolutionary significance by calling the extreme thermophiles *eocytes,* meaning "dawn cells."

Domain Bacteria

Bacteria account for most prokaryotes, with every major mode of nutrition and metabolism represented among the thousands of known species. The bacteria diversified so long ago that evolutionary ties between the various taxonomic groups were, until recently, hazy. Molecular systematics offers the most powerful tool for tracing prokaryote evolution, and researchers can now propose taxonomic subdivisions of Domain Bacteria that are phylogenetically reasonable. Systematists recognize about a dozen bacterial groups, five of which are featured in TABLE 27.3 (pp. 514–515). FIGURE 27.11, an expanded version of the tree in FIGURE 27.1, shows the evolutionary relationships of these bacteria to the archaea and eukaryotes.

ECOLOGICAL IMPACT OF PROKARYOTES

One of the themes of this unit is changing life on a changing planet—the interactions of geological history and evolving biological forms. Organisms as pervasive, abundant, and diverse as the prokaryotes have had a tremendous impact on the Earth and all its inhabitants.

Prokaryotes are indispensable links in the recycling of chemical elements in ecosystems

Not too long ago, in geological terms, the atoms of the organic molecules in our bodies were parts of the inorganic compounds of soil, air, and water, as they will be again. Ongoing life depends on the recycling of chemical elements between the biological and physical components of ecosystems. Prokaryotes play essential roles in these chemical cycles (to be discussed in detail in Chapter 54). If it were not for such

27.2

Table 27.3 ▪ Five Major Phylogenetic Groups of Bacteria (based mainly on comparisons of signature sequences in ribosomal RNA)

GROUP	CHARACTERISTICS	EXAMPLE
Proteobacteria	The most diverse group of bacteria, with three main subgroups: 1. Purple bacteria: photoautotrophs or photoheterotrophs with bacteriochlorophylls built into inpocketings of the plasma membrane; extract electrons from molecules other than H_2O, such as H_2S, and thus release no oxygen (the yellow globules in the cells at the right consist of sulfur produced as a waste product of H_2S-splitting); most species are obligate anaerobes; found in the sediments of ponds, lakes, and mudflats; many species are flagellated. (LM)	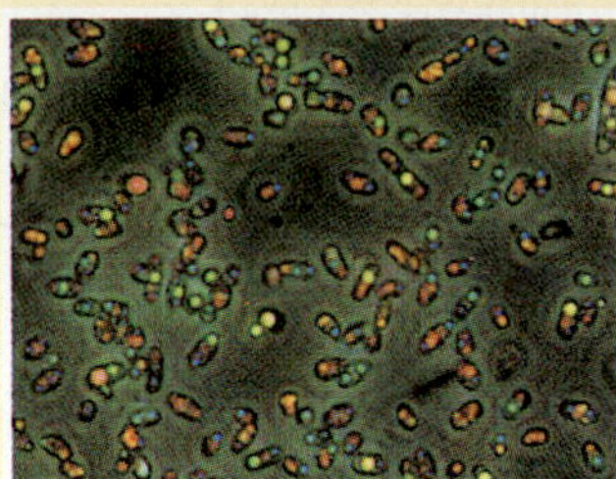 *Chromatium* 1 μm
	2. Chemoautotrophic proteobacteria: free-living and symbiotic species; many play key roles in the chemical cycles of ecosystems, including nitrogen fixation (conversion of atmospheric N_2 to nitrogenous minerals that plants can use); for example, the genus *Rhizobium* lives symbiotically in root nodules of peas and other legumes, contributing to the nutrition of those plants (you can see these bacteria within the vesicles of a root nodule cell in the photograph, at the arrows). (TEM)	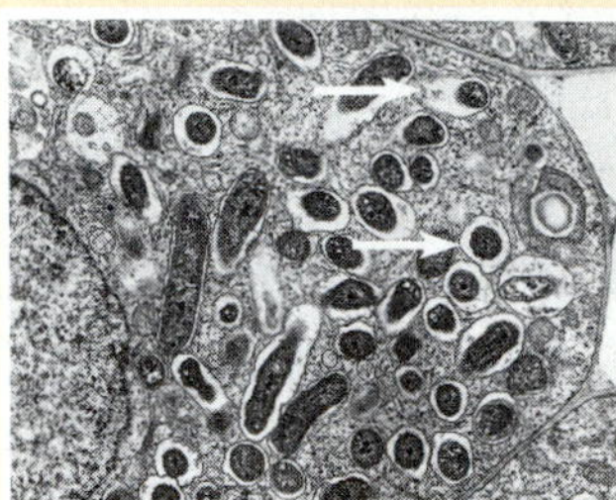 *Rhizobium* 2.5 μm
	3. Chemoheterotrophic proteobacteria: includes the enteric bacteria, which inhabit the intestinal tracts of animals; most enterics are rod-shaped facultative anaerobes; many, such as *E. coli*, are usually harmless; others are generally pathogenic, including *Salmonella*, one of the micro-organisms that causes food poisoning. (SEM)	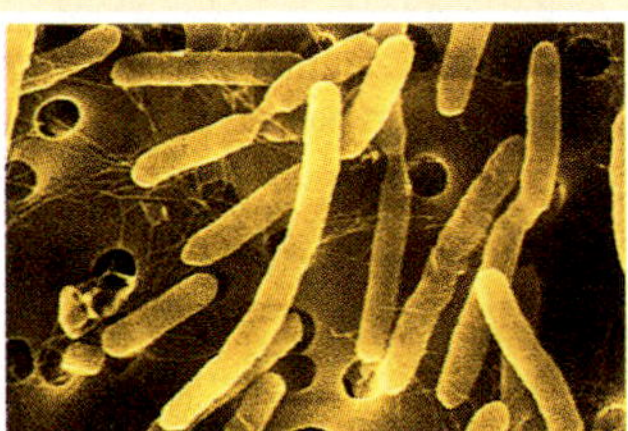 *Salmonella* 2.5 μm
Gram-positive bacteria	Most *are* gram-positive, but the name of the group is misleading because some species are actually gram-negative and are grouped in this taxon because molecular systematics indicates a close relationship to the gram-positive bacteria. Some are photosynthetic, but most species are chemoheterotrophs; many, including *Clostridium* and *Bacillus*, form endospores (boxed area in the photograph) that are resistant to harsh conditions. (TEM)	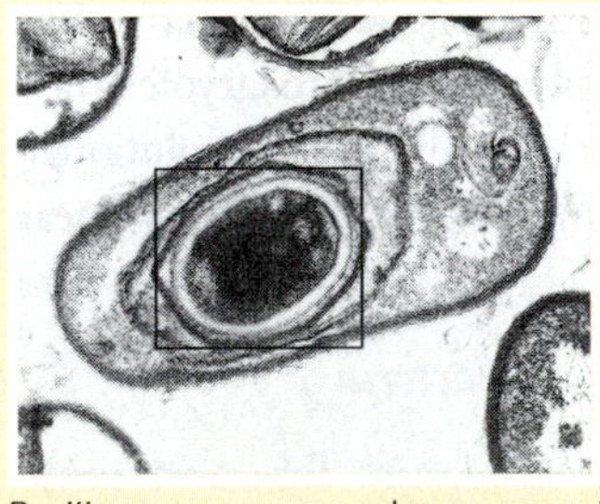 *Bacillus* 1 μm

Table 27.3 ■ (continued)

GROUP	CHARACTERISTICS	EXAMPLE
Gram-positive bacteria, continued	Among those gram-positive bacteria that do not form spores are the mycoplasmas, the smallest of all known cells, with diameters of only 0.10–0.25 μm; the only bacteria that lack cell walls, they are not actually gram-positive but are genetically closely related to *Clostridium* species, which are gram-positive; mycoplasmas are common in soil, and some are pathogenic in animals (e.g., *Mycoplasma pneumoniae* causes "walking pneumonia" in humans). (SEM)	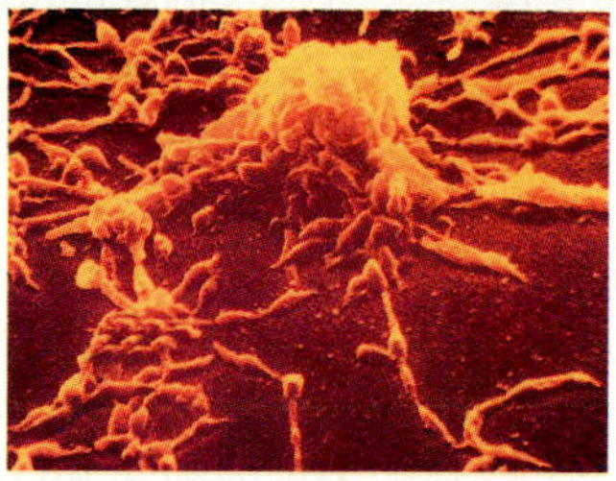 *Mycoplasma* 1 μm
	The gram-positive group also includes the actinomycetes, soil bacteria that form branching colonies resembling fungi; many actinomycetes, including *Streptomyces,* are important commercial sources of antibiotics. (LM)	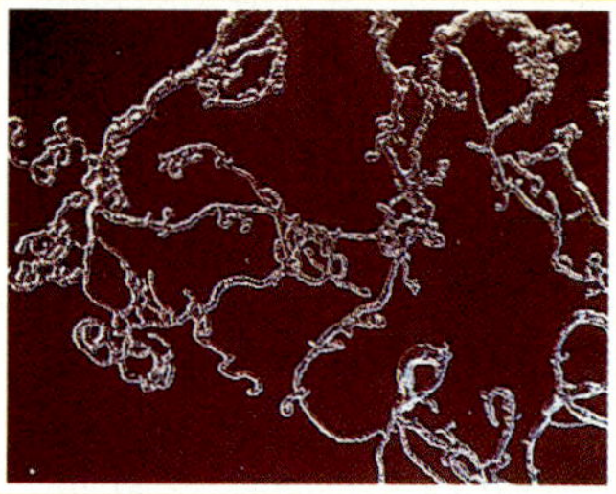 *Streptomyces* 2.5 μm
Cyanobacteria	Photoautotrophs with plantlike photosynthesis; have chlorophyll *a* and use two photosystems to split water, yielding O_2 as a by-product; most inhabit fresh water, but there are also marine species and symbionts that live along with fungi as lichens; some aquatic species fix nitrogen; cell walls are often thick and gelatinous; flagella absent, motile forms glide; among cyanobacteria are single-celled forms, colonial species, and truly multicellular organisms with a division of labor between specialized cells (boxed area in the photograph highlights a heterocyst, a cell specialized for nitrogen fixation). (LM)	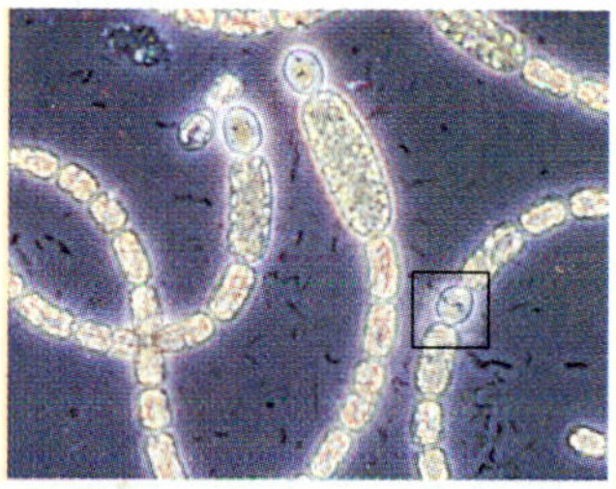 *Anabaena* 50 μm
Spirochetes	Helical chemoheterotrophs, sometimes very long (up to 0.25 mm, but too thin to be resolved without a microscope); rotation of internal, flagellalike filaments produce corkscrewlike movements; includes both free-living species and pathogens such as *Treponema pallidum* (the cause of syphilis) and *Borrelia burgdorferi* (the cause of Lyme disease). (TEM)	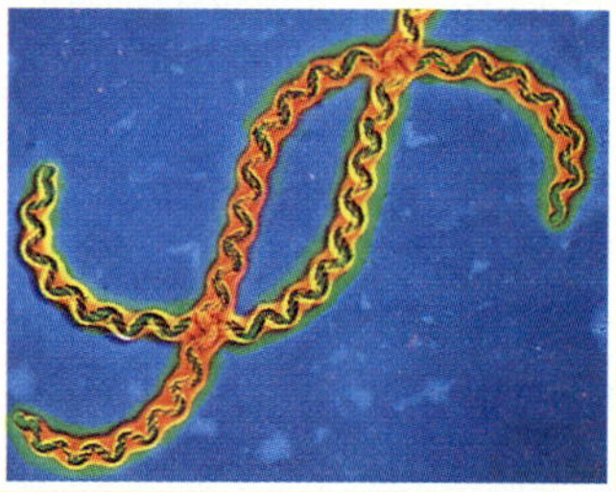 *Leptospira* 0.5 μm
Chlamydias	Obligate intracellular parasites of animals (see arrow); obtain all their ATP from host cells; gram-negative cell walls, but unusual among bacteria in lacking peptidoglycan; *Chlamydia trachomatis* is the most common cause of blindness in the world and also causes the most common form of sexually transmitted disease (nongonococcal urethritis) in the United States. (TEM)	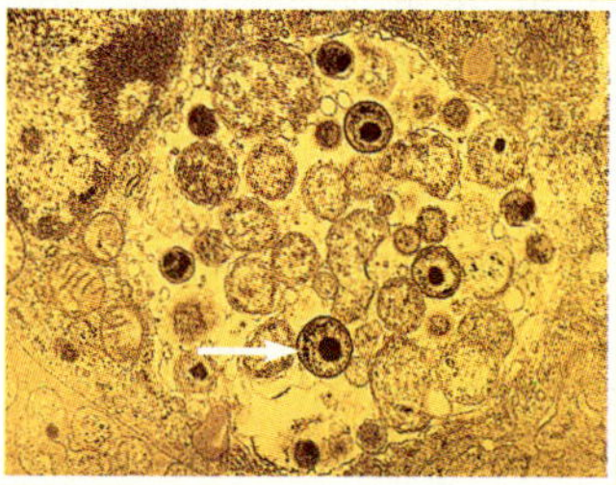 *Chlamydia* 2.5 μm

decomposers, carbon, nitrogen, and other elements essential to life would become locked in the organic molecules of corpses and waste products.

Prokaryotes also mediate the return of elements from the nonliving components of the environment (air, inorganic soil, and water) to the pool of organic compounds. Autotrophic prokaryotes fix CO_2, supporting food chains through which organic nutrients pass from the prokaryotes to prokaryote-eaters, then on to secondary consumers. Because of their many unique metabolic capabilities, prokaryotes are the only organisms able to metabolize inorganic molecules containing elements such as iron, sulfur, nitrogen, and hydrogen. Cyanobacteria not only synthesize food and restore oxygen to the atmosphere but also fix nitrogen, stocking the soil and water with nitrogenous compounds that other organisms can use to make proteins. And when plants and the animals that eat them die, soil prokaryotes return the nitrogen to the atmosphere. All life on Earth depends on prokaryotes and their unparalleled metabolic diversity.

Many prokaryotes are symbiotic

Prokaryotes rarely function singly in the environment. More often they interact in groups, which often include other species of prokaryotes or eukaryotes with complementary metabolisms.

Symbiosis (Gr. for "living together") is the term used to describe ecological relationships between organisms of different species that are in direct contact. The organisms involved are known as **symbionts**. If one of the symbionts is much larger than the other, the larger is also termed the **host**. There are three categories of symbiotic relationships: mutualism, commensalism, and parasitism. In **mutualism**, both symbionts benefit. In **commensalism**, one symbiont receives benefits while neither harming nor helping the other in any significant way. In **parasitism**, one symbiont, called a **parasite** in this case, benefits at the expense of the host.

Symbiosis is common among prokaryotes and probably has been for billions of years. Symbiosis likely played a major role in prokaryote evolution and also in the origin of the early eukaryotes (as you will see in Chapter 28). Prokaryotes are represented in all three categories of symbiosis with modern eukaryotes. For example, plants of the legume family (peas, beans, alfalfa, and others) have lumps on their roots called nodules, which are home to mutualistic prokaryotes that fix nitrogen used by the host (see *Rhizobium* in TABLE 27.3, p. 514). The plant reciprocates by providing the prokaryotes with a steady supply of sugar and other organic nutrients. Bacteria inhabiting the inner and outer surfaces of the human body consist mostly of commensal species, but some species are mutual symbionts. For instance, fermenting bacteria living in the vagina produce acids that maintain a pH between 4.0 and 4.5, suppressing the growth of yeast and other potentially harmful microorganisms. Among parasitic bacteria are those classified as pathogens because they cause disease in their hosts.

Prokaryotes and Disease

Exposure to pathogenic prokaryotes is a certainty. Most of us are well most of the time because our defenses check the growth of pathogens to which we are exposed. Occasionally the balance shifts in favor of a pathogen, and we become ill. To be pathogenic, a parasite must invade the host, resist internal defenses well enough to begin growing, then harm the host in some way. Pathogenic prokaryotes cause about one half of all human diseases.

Some pathogens are **opportunistic**, meaning they are normal residents of a host but can cause illness when the host's defenses are weakened by such factors as poor nutrition or a recent bout with the flu. For example, *Streptococcus pneumoniae* (a gram-positive bacterium) lives in the throats of most healthy people, but this opportunist can multiply and cause pneumonia when the host's defenses are down.

Louis Pasteur, Joseph Lister, and other scientists began linking disease to pathogenic microbes in the late 1800s. The first to actually connect certain diseases to specific bacteria was Robert Koch, a German physician who identified the bacteria responsible for anthrax and tuberculosis. His methods established four criteria, now called **Koch's postulates**, that are still the guidelines for medical microbiology. To establish that a specific pathogen is the cause of a disease, the researcher must (1) find the same pathogen in each diseased individual investigated, (2) isolate the pathogen from a diseased subject and grow the microbe in a pure culture, (3) induce the disease in experimental animals by transferring the pathogen from the culture, and (4) isolate the same pathogen from the experimental animals after the disease develops. The postulates work for most pathogens, but exceptions must be made for some cases. For example, no one has yet cultured the spirochete *Treponema pallidum* on artificial media, but circumstantial evidence leaves no doubt that this organism causes syphilis.

How do pathogenic prokaryotes actually produce symptoms of disease? Some species, such as the gram-positive actinomycete that causes tuberculosis, disrupt the health of the host by invading tissues. More commonly, pathogens cause illness by producing poisons called exotoxins and endotoxins.

Exotoxins are proteins secreted by prokaryotes. Exotoxins can produce disease symptoms even without the prokaryotes being present. For example, when the gram-positive bacterium *Clostridium botulinum* grows anaerobically in improperly canned foods, one of the by-products of its fermentation is an exotoxin that causes the potentially fatal disease botulism. Some exotoxins are among the most potent poisons known; one gram of botulism toxin would be sufficient to kill

FIGURE 27.12 ▪ Treatment of an oil spill. In this photograph, workers are spraying fertilizers on an oil-soaked beach in Alaska. The fertilizers stimulate growth of indigenous bacteria that can initiate the breakdown of the oil—in some cases, speeding the natural process of breakdown some five-fold. This technique is the fastest and least expensive way yet devised to clean up oil spills on beaches.

a million people. Another exotoxin-producing species is the enteric proteobacterium *Vibrio cholerae,* which causes cholera, a dangerous disease characterized by severe diarrhea. Resulting from the consumption of water contaminated with human feces, cholera is often epidemic during wars and famines. Even *E. coli* can be an exotoxin-releasing culprit. Some cases of traveler's diarrhea result from toxins released by strains of *E. coli* obtained from another person via contaminated food or water.

In contrast to exotoxins, **endotoxins** are components of the outer membranes of certain gram-negative bacteria. Examples of endotoxin-producing bacteria include nearly all members of the genus *Salmonella,* which are not normally present in healthy animals. *Salmonella typhi* causes typhoid fever, and several other species of *Salmonella,* some of which are commonly found in poultry, cause food poisoning (see TABLE 27.3, p. 514).

Since the discovery in the nineteenth century that "germs" cause disease, improved sanitation has significantly reduced infant mortality and extended life expectancy in developed countries. Antibiotics have also saved many lives and reduced disease incidence. More than half of our antibiotics (including streptomycin, neomycin, erythromycin, aureomycin, and tetracycline) come from soil bacteria of the genus *Streptomyces* (an actinomycete; see TABLE 27.3, p. 515, in the gram-positive group). In nature, these compounds prevent encroachment by competing microbes.

Bacterial diseases are a continuing threat. Their general decline over the past century is probably due more to public-health policies than to "wonder drugs." Today, the rapid evolution of antibiotic-resistant strains of pathogenic bacteria is a serious health threat aggravated by imprudent and excessive antibiotic use. Although declared illegal by the United Nations, the production and stockpiling of deadly bacterial disease agents for use as biological weapons remains a threat to world peace.

Humans use prokaryotes in research and technology

Humans have learned many ways of exploiting the diverse metabolic capabilities of prokaryotes, both for scientific research and for practical purposes. Much of what we know about metabolism and molecular biology has been learned in laboratories using prokaryotes as relatively simple model systems. In fact, *E. coli,* the prokaryotic "white rat" of so many research labs, is the best understood of all organisms. Methanogens are important decomposers utilized for sewage treatment. We are just beginning to explore the great potential prokaryotes have for helping us solve some of our environmental problems. Soil bacteria called pseudomonads (proteobacteria) decompose pesticides, petroleum products, and other synthetic compounds (FIGURE 27.12). The chemical industry grows immense cultures of bacteria that produce acetone, butanol, and several other products. Pharmaceutical companies culture bacteria that make vitamins and antibiotics. Bacteria containing magnetic particles may soon become a commercial source for manufacturing magnetic tapes and other recording devices. The food industry uses bacteria to convert milk to yogurt and various kinds of cheese. And recombinant DNA techniques have opened a new era in the commercial use of prokaryotes (see Chapter 20).

▪ ▪ ▪

In this chapter we have surveyed the prokaryotes and traced their history. On an ancient Earth inhabited only by prokaryotes, all the diverse forms of nutrition and metabolism evolved. Most subsequent evolutionary breakthroughs were structural rather than metabolic. The most significant development was the origin of eukaryotic cells from prokaryotic ancestors, a juncture in the history of life we will explore in the next chapter.

CHAPTER REVIEW

REVIEW OF KEY CONCEPTS

(with page numbers and key figures)

THE WORLD OF PROKARYOTES

- **They're (almost) everywhere!** ***an overview of prokaryotic life*** **(pp. 502–503)** Prokaryotes were the first organisms, and they persist today as the most numerous and pervasive of all living things.
- **Bacteria and archaea are the two main branches of prokaryote evolution (p. 503, FIGURE 27.1)** Classification of prokaryotes in two domains, Bacteria and Archaea, and all eukaryotes in a third domain, Eukarya, signifies that organisms probably have been evolving in three independent lineages for over 1.5 billion years.

STRUCTURE, FUNCTION, AND REPRODUCTION OF PROKARYOTES

- Prokaryotes are generally single-celled organisms, although some occur as aggregates, colonies, or simple multicellular forms. Most prokaryotes are spherical (cocci), rod-shaped (bacilli), or helical in shape (FIGURE 27.2).
- **Nearly all prokaryotes have cell walls external to their plasma membranes (pp. 504–505, Methods Box)** Gram-positive and gram-negative bacteria differ in the structure of their walls and other surface layers. Many species have capsules and pili outside the cell wall, which help the cells adhere to one another. Some pili are specialized for conjugation.
- **Many prokaryotes are motile (pp. 505–506, FIGURE 27.5)** Motile bacteria propel themselves by flagella, use flagella-like filaments positioned inside the cell wall (spirochetes), or glide on slime secretions.
- **The cellular and genomic organization of prokaryotes is fundamentally different from that of eukaryotes (pp. 506–507, FIGURE 27.6)** The cells of prokaryotes are not compartmentalized by internal membranes, but infoldings of the plasma membrane in some species provide internal membrane surfaces. The genome of prokaryotes is a single circular DNA molecule unbounded by a membrane. Smaller separate rings of DNA called plasmids code for special metabolic pathways and resistance to antibiotics in some species.
- **Populations of prokaryotes grow and adapt rapidly (pp. 507–508, FIGURE 27.7)** Prokaryotes grow in number and colony size by binary fission. Genetic variation occurs through mutation and through gene transfer by transformation, conjugation, or viral transduction.

NUTRITIONAL AND METABOLIC DIVERSITY

All major types of nutrition and metabolism evolved among prokaryotes, which are the most metabolically diverse organisms on Earth.

27.1 **Prokaryotes can be grouped into four categories according to how they obtain energy and carbon (pp. 508–509, TABLE 27.1)** Photoautotrophs use light energy, and chemoautotrophs use inorganic substances to synthesize their organic compounds from carbon dioxide. Photoheterotrophs use light energy and require organic molecules. Most prokaryotes are chemoheterotrophs, which require organic molecules as a source of both energy and carbon. Obligate aerobes require O_2, obligate anaerobes are poisoned by it, and facultative anaerobes can survive with or without O_2.

- **The evolution of prokaryotic metabolism was both cause and effect of changing environments on Earth (pp. 509–512, FIGURES 27.8, 27.9)** The first prokaryotes were likely chemoautotrophs that obtained energy from reactions involving inorganic chemicals. Early photosynthetic prokaryotes used pigments and light-powered photosystems to fix carbon dioxide. The first cyanobacteria made organic compounds from water and carbon dioxide, releasing O_2 as a by-product. This drastically changed Earth's ancient atmosphere and affected subsequent biological evolution.

PHYLOGENY OF PROKARYOTES

27.2 **Molecular systematics is leading to a phylogenetic classification of prokaryotes (pp. 512–513, TABLES 27.2, 27.3, FIGURE 27.11)** Prokaryotes of Domain Archaea live in extreme environments reminiscent of conditions on the primordial Earth. Most prokaryotes are members of Domain Bacteria.

ECOLOGICAL IMPACT OF PROKARYOTES

- **Prokaryotes are indispensable links in the recycling of chemical elements in ecosystems (pp. 513–516)**
- **Many prokaryotes are symbiotic (pp. 516–517)** Many prokaryotes live with other species in symbiotic relationships: mutualism, commensalism, or parasitism. Some parasitic species are pathogenic, causing disease in the host by invading its tissues or poisoning it with endotoxins or exotoxins.
- **Humans use prokaryotes in research and technology (p. 517, FIGURE 27.12)**

SELF-QUIZ

1. Home canners pressure-cook vegetables as a precaution primarily against
   a. mycoplasmas
   b. endospore-forming bacteria
   c. enteric bacteria
   d. pseudomonads
   e. actinomycetes

2. Photoautotrophs use
   a. light as an energy source and can use water or hydrogen sulfide as a source of electrons for producing organic compounds
   b. light as an energy source and oxygen as an electron source
   c. inorganic substances for energy and CO_2 as a carbon source
   d. light to generate ATP but need organic molecules for a carbon source
   e. light as an energy source and CO_2 to reduce organic nutrients

3. Which of the following statements about the domains of prokaryotes is *not* true?
   a. The lipid composition of the plasma membrane found in archaea is different from that of bacteria.
   b. The archaea and bacteria probably diverged very early in evolutionary history.
   c. Both archaea and bacteria have cell walls, but those of archaea lack peptidoglycan.
   d. Of the two groups, bacteria are more closely related to Domain Eukarya.
   e. Bacteria include the cyanobacteria.

4. A prokaryotic genome is different from a eukaryotic genome in that
   a. it has only one-half as much DNA as does a typical eukaryotic genome
   b. it consists of a single-stranded DNA molecule
   c. it has less protein associated with its DNA and is not enclosed in a nuclear envelope
   d. it is made of ribosomes that are smaller and chemically distinct
   e. it consists of RNA rather than DNA

5. Chemoautotrophs probably predate chemoheterotrophs because
   a. the use of ATP is seen in all extant organisms
   b. glycolysis is observed in nearly all modern organisms
   c. the probable concentrations of organic molecules were initially too low to support chemoheterotrophs
   d. the oxygen needed to oxidize organic molecules was initially unavailable
   e. the earliest organisms utilized a type of enzyme-independent metabolism
6. Banded iron formations in marine sediments indicate that
   a. early cyanobacteria were probably producing oxygen during that time period
   b. the early atmosphere was very reducing
   c. aerobic bacteria were the dominant life form in the seas at that time
   d. the pH of the early seas was quite low because of early prokaryotic excretion of organic acids from fermentation
   e. mats of bacterial colonies were forming stromatolites
7. Which of the following statements about demonstrating the pathogenicity of a particular bacterial species is *not* true?
   a. The bacteria must be capable of inducing the disease when transferred to an experimental host.
   b. The bacteria isolated from a diseased host must be grown in pure culture.
   c. The same bacteria must be present in each diseased host investigated.
   d. The bacteria isolated from the experimental host must be capable of reinducing the disease when returned to the original host.
   e. The artificially infected experimental host must produce the bacteria after the disease develops.
8. Penicillins function as antibiotics mainly by inhibiting the ability of some bacteria to
   a. form spores
   b. replicate DNA
   c. synthesize normal cell walls
   d. produce functional ribosomes
   e. synthesize ATP
9. Plantlike photosynthesis that releases O_2 occurs in the
   a. cyanobacteria
   b. chlamydias
   c. archaea
   d. actinomycetes
   e. chemoautotrophic bacteria
10. According to the phylogenetic tree in FIGURE 27.11, cyanobacteria share the most recent ancestor with the
   a. eukaryotes
   b. proteobacteria
   c. gram-positive bacteria
   d. methanogens
   e. chlamydias

CHALLENGE QUESTIONS

1. Some prokaryotes are capable of reproducing once every half-hour. Suppose you placed a single bacterium in a suitable culture medium and it (and its descendants) proceeded to reproduce at this rate. How many cells would be present after 2 hrs? After a total of 12 hrs? What factors determine how long this kind of increase can continue?
2. Health officials worldwide are concerned by a resurgence of diseases caused by prokaryotes that are resistant to standard antibiotics. For instance, antibiotic-resistant bacteria are now causing an epidemic of tuberculosis (TB), a lung disease spread by airborne droplets. Drugs can relieve TB symptoms in a few weeks, but it takes much longer to halt the infection, and patients are likely to discontinue treatment while bacteria are still present. Why can prokaryotes quickly reinfect a patient if they are not wiped out? How might this result in the evolution of drug-resistant pathogens?

SCIENCE, TECHNOLOGY, AND SOCIETY

Many local newspapers publish a weekly list of restaurants that have been cited by inspectors for poor sanitation. Locate such a report and highlight the cases that are likely associated with potential food contamination by pathogenic prokaryotes.

FURTHER READING

Fackelman, K. "Tuberculosis Outbreak, An Ancient Killer Strikes a New Population." *Science News,* January 31, 1998. Reports on the recent introduction by gold miners of deadly tuberculosis to Amazonia.

Levy, S. B. "The Challenge of Antibiotic Resistance." *Scientific American,* March 1998. Details the threat, its cause, and possible solutions.

Madigan, M. T., and B. L. Marrs. "Extremophiles." *Scientific American,* April 1997. An introduction to prokaryotes that thrive in harsh environments and the enzymes that enable them to do so.

Miller, R. V. "Bacterial Gene Swapping in Nature." *Scientific American,* January 1998. Describes the conditions that promote or prevent exchange of genes between species.

Monastersky, R. "Deep Dwellers." *Science News,* March 29, 1997. Examines prokaryotes that live deep within Earth's crust.

The Race Against Lethal Microbes. Chevy Chase, MD: The Howard Hughes Medical Institute. 1996. A multi-authored, well-illustrated survey of infectious diseases that threaten modern human societies.

Tortora, G. J., B. R. Funke, and C. L. Case. *Microbiology: An Introduction,* 6th ed. Redwood City, CA: Benjamin/Cummings, 1998. A general text.

Travis, J. "Spying Diseases From the Sky." *Science News,* August 2, 1997. Describes the use of satellite data to identify landscapes where the risk of diseases is high.

WEB LINKS

Visit the special edition of *The Biology Place* for BIOLOGY, Fifth Edition, at http://www.biology.com/campbell. Go to Chapter 27 for online resources, including learning activities, practice exams, and links to the following web sites:

"American Society for Microbiology (ASM)"
This is the home page of the ASM. With 40,000 members around the world, it is the oldest and largest single life science organization. The site includes the Biofilm Imaging project and the National Collection for Microbiology Teaching and Learning.

"Antoni van Leeuwenhoek (1632–1723)"
Online biography of the inventor of the microscope and the discoverer of bacteria.

"TIGR Microbial Database"
This site contains a listing of microbial genomes that have been published or are in the process of being sequenced. Visit The Biology Place to try out the learning activity called "Investigating the Bacterial Genome" by Tom Terry, University of Connecticut, that makes extensive use of the TIGR database.

"Introduction to the Archaea: Life's Extremists"
An introduction to Domain Archaea with links to Archaea resources around the world.

CHAPTER 28

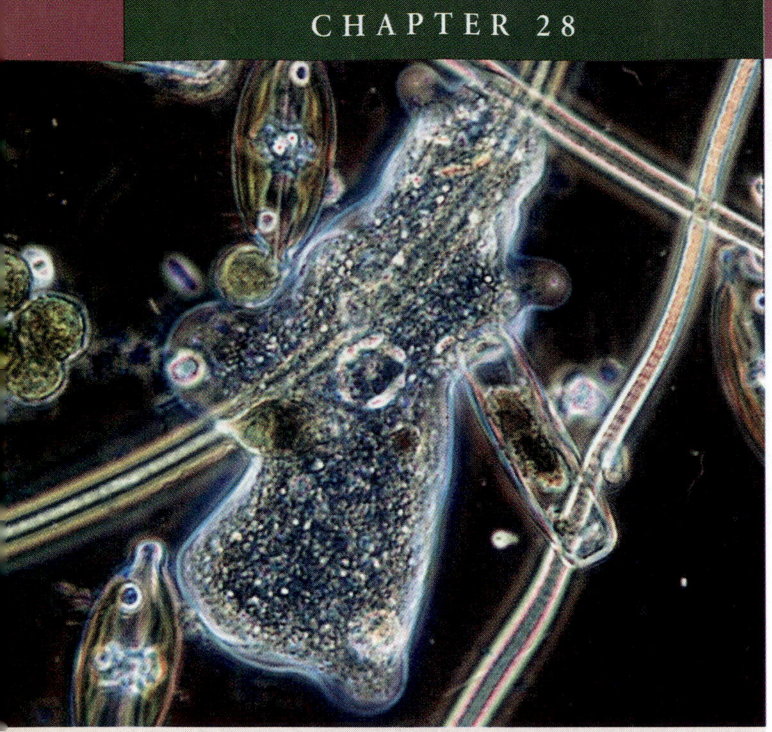

THE ORIGINS OF EUKARYOTIC DIVERSITY

No more pleasant sight has met my eye than this of so many thousands of living creatures in one small drop of water," wrote Anton van Leeuwenhoek after his discovery of the microbial world more than three centuries ago. It is a world every biology student rediscovers by peering through a microscope into a droplet of pond water filled with diverse creatures we call protists (see the photograph on this page).

Protists are eukaryotic, and thus even the simplest protists are much more complex than the prokaryotes. The first eukaryotes to evolve from prokaryotic ancestors were probably unicellular and would therefore be called protists. The very word implies great antiquity (Gr. protos, *"first"). The primal eukaryotes were not only the predecessors of the great variety of modern protists but were also ancestral to all other eukaryotes—plants, fungi, and animals. Two of the most significant chapters in the history of life—the origin of the eukaryotic cell and the subsequent emergence of multicellular eukaryotes—unfolded during the evolution of protists.*

This chapter traces the origins of eukaryotic cells; examines the diversity, evolution, and ecology of protists; and considers the evolutionary origins of multicellular organization.

INTRODUCTION TO THE PROTISTS

At least a billion years before the origin of plants, fungi, and animals, there were protists, the earliest eukaryotic descendants of prokaryotes. The oldest putative fossils of protists are Precambrian objects called **acritarchs** (Gr., "of uncertain origin") that are 2.1 billion years old. Some are the right size and structure to be the ruptured coats of cysts (protective capsules) similar to those made by certain protists today. The first billion years of eukaryote evolution apparently produced several divergent lineages.

Protists are the most diverse of all eukaryotes

All protists are eukaryotes, but protists are so diverse that few other general characteristics can be cited without exceptions. In fact, protists vary in structure and function more than any other group of organisms. Most of the approximately 60,000 known species of extant protists are unicellular, but there are some colonial and multicellular species. Because most protists are unicellular, they are justifiably considered to be the simplest eukaryotic organisms. But at the *cellular* level, many protists are exceedingly complex—the most elaborate of all cells. We should expect this of organisms that must carry out within the boundaries of a single cell all the basic functions performed by the collective of specialized cells that makes up the bodies of plants and animals. Each unicellular protist is not at all analogous to a single cell from a human or other multicellular organism, but is itself an organism as complete as any whole plant or animal.

Protists are metabolically diverse, and as a group they are the most nutritionally diverse of all eukaryotes. Most protists are aerobic in their metabolism, using mitochondria for cellular respiration (a few lack mitochondria and either live in anaerobic environments or contain mutualistic respiring bacteria). Some protists are photoautotrophs with chloroplasts, some are heterotrophs that absorb organic molecules or ingest larger food particles, and still others, called mixotrophs, combine photosynthesis *and* heterotrophic nutrition (an example is *Euglena,* shown in FIGURE 28.1). It is useful in an ecological context to divide this nutritional diversity into three categories: ingestive (animal-like) protists, or **protozoa** (singular, **protozoan**); absorptive (funguslike) protists (these have no other general name); and photosynthetic (plantlike) protists, or **algae** (singular, **alga**). Though commonly used, the terms *protozoa* and *algae* have no basis in phylogeny and no significance in taxonomy. (The term *alga* refers to relatively simple aquatic photoautotrophs, including some organisms sometimes classified in the plant kingdom.) Although all algae have chlorophyll *a,* the same "primary" pigment found in plants, they differ considerably in their accessory pigments—pigments that trap wavelengths of light to which chlorophyll *a* is not as sensitive. These pigments include other forms of chlorophyll (greenish), carotenes (yellow-orange), xanthophylls (brownish), and phycobilins (red and blue varieties). The mixture of pigments in the chloroplasts lends characteristic colors to algae. Many of the scientific and common names of algae are based on these colors (for example, the Chlorophyta are green algae).

Most protists are motile, having flagella or cilia at some time in their life cycles. It is important to understand that prokaryotic and eukaryotic versions of flagella are not homologous structures. Prokaryotic flagella are attached to the cell surface (see FIGURE 27.5). In contrast, eukaryotic flagella and cilia are extensions of the cytoplasm, with bundles of microtubules covered by the plasma membrane (see FIGURE 7.24). Eukaryotic cilia and flagella have the same basic ultrastructure (the 9 + 2 arrangement of microtubules), but cilia are shorter and more numerous. They move a cell with their rhythmic power strokes, analogous to the oars of a boat (see FIGURE 7.23b).

Reproduction and life cycles are highly varied among protists. Mitosis occurs in most protists, but there are many variations in the process unknown in other eukaryotes. Some protists are exclusively asexual; others can also reproduce sexually or at least employ the sexual processes of meiosis and **syngamy** (the union of two gametes), thereby shuffling genes between two individuals that then go on to reproduce asexually. In Chapter 13 you learned about three basic types of sexual life cycles that differ in the timing of meiosis and syngamy (see FIGURE 13.4). All three types are represented among protists, along with some variations that do not quite fit any of the three basic life cycle patterns. At some point in the life cycle of many protists, resistant cells called **cysts** that can survive harsh conditions are formed. Microfossils resembling the ruptured cysts of some algae are among the acritarchs found in Precambrian rocks.

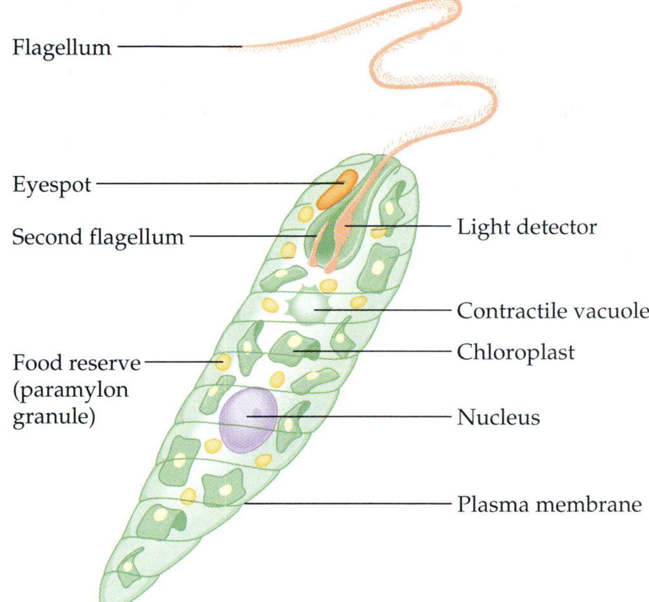

FIGURE 28.1 ▪ *Euglena:* an example of eukaryotic complexity. This protist is one of the most common inhabitants of murky pond water. The drawing illustrates the greater structural complexity of eukaryotic microorganisms compared to prokaryotes. *Euglena* uses its long "tinseled" flagellum to pull itself through the water. The eyespot, a pigmented organelle, functions as a light shield. Depending on the position of the organism, the eyespot allows light from only a certain direction to strike a light detector, the swelling near the base of the long flagellum. The result is movement of the organism toward light of appropriate intensity, an important adaptation that enhances photosynthesis. *Euglena* lacks a cell wall but has strong, flexible plates made of protein beneath its plasma membrane. The contractile vacuole functions as a "bilge pump," expelling excess water that enters the cell by osmosis from the hypotonic environment. Surplus food made in the cell's chloroplasts is stored in granules as a polysaccharide called paramylon. *Euglena* is mixotrophic: if placed in the dark, it can live as a heterotroph by absorbing organic nutrients from the environment. Some related species lack chloroplasts and ingest food by phagocytosis.

Protists are found almost anywhere there is water. They are common inhabitants of damp soil, leaf litter, and other terrestrial habitats that are sufficiently moist. In oceans, ponds, and lakes, many protists are bottom-dwellers that attach themselves to rocks and other anchorages or creep through the sand and silt. Protists are also important constituents of **plankton** (Gr. *planktos,* "wandering"), the communities of organisms, mostly microscopic, that drift passively or swim weakly near the water surface. As a large group of autotrophs, the eukaryotic algae are extremely important ecologically. *Phytoplankton* (planktonic algae, along with the prokaryotic cyanobacteria) are the bases of most marine and freshwater food webs.

Accounting for at least half the photosynthetic production of organic material globally, they support an enormous abundance and diversity of heterotrophic protists and animals.

In addition to free-living protists are the many symbionts that inhabit the body fluids, tissues, or cells of hosts. These symbiotic relationships span the continuum from mutualism to parasitism. Some parasitic protists are important pathogens of animals, including many that cause potentially fatal diseases in humans. As we see next, symbiosis has probably always been an important feature of protistan life.

Symbiosis was involved in the genesis of eukaryotes from prokaryotes

The many differences between prokaryotic and eukaryotic cells represent a distinction much greater than that between the cells of plants and animals. During the genesis of eukaryotes, the cellular structures and processes unique to eukaryotic cells arose: a membrane-enclosed nucleus, mitochondria, chloroplasts, the endomembrane system, the cytoskeleton, 9 + 2 flagella, multiple chromosomes consisting of linear DNA molecules compactly arranged with proteins, and life cycles that include mitosis, meiosis, and sex. Among the most fundamental questions in biology is how the complex eukaryotic cell evolved from much simpler prokaryotic cells.

The small size and relatively simple construction of a prokaryote have many advantages but also impose limits on the number of different metabolic activities that can be handled at one time. The relatively small size of the prokaryotic genome limits the number of genes coding for the enzymes that control these activities. This is not to say that prokaryotes are less successful than eukaryotes. Prokaryotes have been evolving and adapting since the dawn of life, and they are the most widespread organisms even today. In at least some prokaryotic groups, natural selection favored increasing complexity—higher levels of organization with emergent properties. One trend was the evolution of multicellular prokaryotes, such as some filamentous cyanobacteria, where different cell types are specialized for different functions. A second trend was the evolution of complex communities of prokaryotes, where each species benefited from the metabolic specialties of other species. A third trend was the compartmentalization of different functions within single cells, an evolutionary solution that produced the first eukaryotes.

How did compartmental organization of the eukaryotic cell evolve from the simpler prokaryotic condition? In one process, the endomembrane system of eukaryotic cells—the nuclear envelope, endoplasmic reticulum, Golgi apparatus, and related structures—may have evolved from specialized infoldings of the prokaryotic plasma membrane (FIGURE 28.2a). Another process, called endosymbiosis, probably led to mitochondria, chloroplasts, and perhaps some of the other features of eukaryotic cells.

An idea originated by the early twentieth-century Russian biologist C. Mereschkovsky and developed extensively by Lynn Margulis of the University of Massachusetts, the hypothesis of **serial endosymbiosis** proposes that mitochondria and chloroplasts were formerly small prokaryotes living within larger cells. (The term *endosymbiont* is used for the cell that lives within another cell, termed the host cell.) The proposed ancestors of mitochondria were aerobic heterotrophic prokaryotes that became endosymbionts. The proposed ancestors of chloroplasts in early eukaryotes were photosynthetic prokaryotes, probably cyanobacteria, that became endosymbionts. Perhaps the prokaryotic ancestors of mitochondria and chloroplasts first gained entry to the host cell as undigested prey or internal parasites. As indicated in FIGURE 28.2b, the host cell may have already had an endomembrane system. By whatever means the relationships began, it is not hard to imagine the symbiosis eventually becoming mutually beneficial. A heterotrophic host could derive nourishment from photosynthetic endosymbionts. And in a world that was becoming increasingly aerobic, a cell that was itself an anaerobe would have benefited from aerobic endosymbionts that turned the oxygen to advantage. In the process of becoming more interdependent, the host and endosymbionts would have become a single organism, its parts inseparable. Almost all eukaryotes, whether heterotrophic or autotrophic, have mitochondria or genetic remnants of these organelles. Only photosynthetic eukaryotes, however, have chloroplasts. Thus the hypothesis of serial endosymbiosis (a sequence of endosymbiotic events) supposes that mitochondria evolved before chloroplasts.

As an evolutionary mechanism contributing to the origin of eukaryotes, serial endosymbiosis is very different from the mechanisms of evolution you learned about in Unit Four of this book. In those chapters we studied how a lineage of organisms changes over time (anagenesis), and how lineages sometimes split into two or more divergent lineages of organisms (cladogenesis). In the case of endosymbiosis, a *merger* of evolutionary lineages gave rise to a new form of life.

The evidence supporting an endosymbiotic origin of chloroplasts and mitochondria includes the existence of endosymbiotic relationships in the modern world. Another line of evidence is the similarity between bacteria and the chloroplasts and mitochondria of eukaryotes. Chloroplasts and mitochondria are the appropriate size to be descendants of bacteria. The inner membranes of chloroplasts and mitochondria, perhaps derived from the membranes of endosymbiotic prokaryotes, have several enzymes and transport systems that resemble those found on the plasma membranes of modern prokaryotes. Mitochondria and chloroplasts replicate by a splitting process reminiscent of binary fission in bacteria. Chloroplasts and mitochondria contain a genome consisting of circular DNA molecules not associated with histones or other proteins, as in most prokaryotes. The organelles contain the transfer RNAs, ribosomes, and other

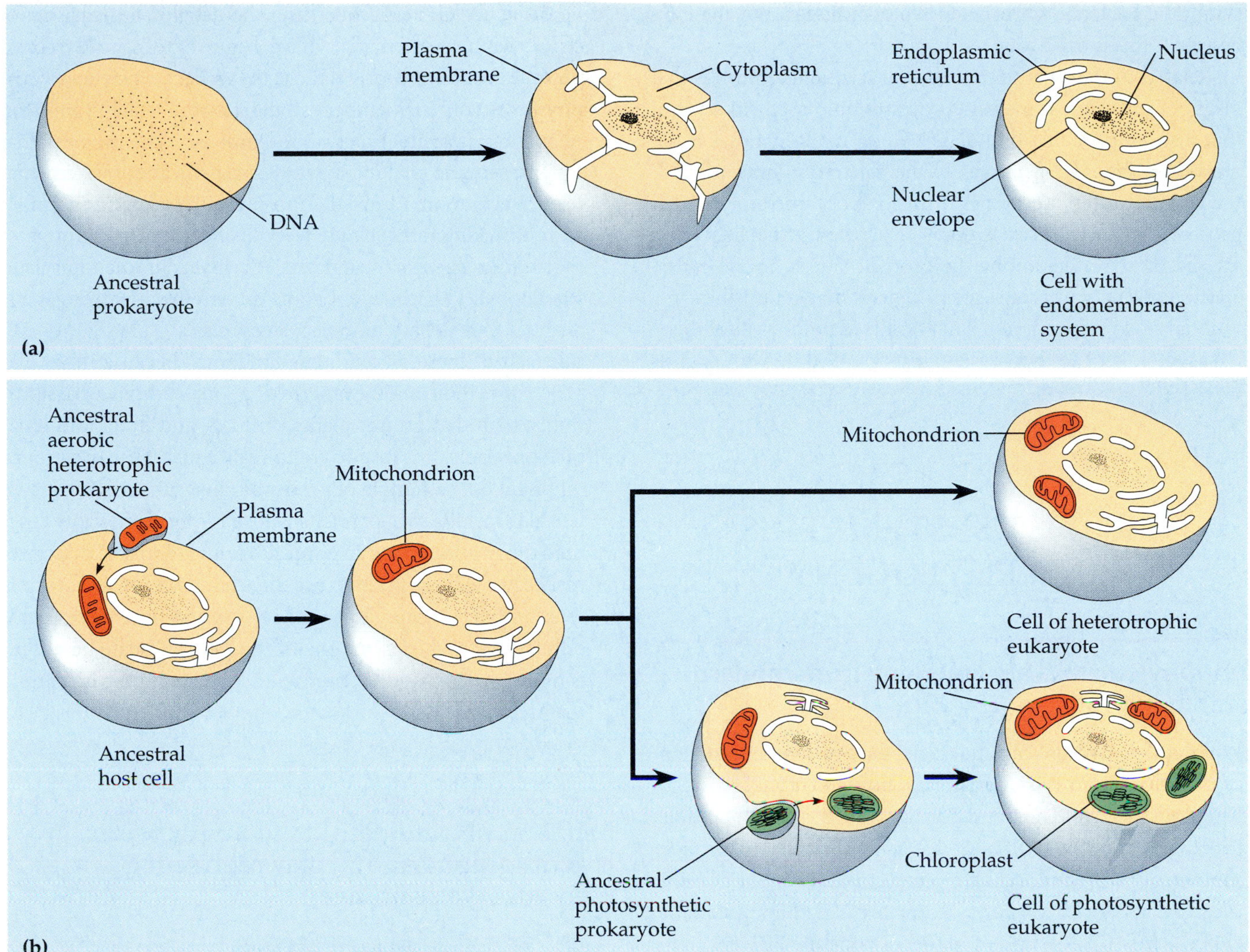

FIGURE 28.2 ▪ A model of the origin of eukaryotes. Two main processes may have contributed to eukaryotic origins. **(a)** The endomembrane system of eukaryotes, including the nuclear envelope, may have evolved by infolding of the plasma membrane of an ancestral prokaryote. **(b)** Strong circumstantial evidence supports the theory of serial endosymbiosis—that mitochondria and chloroplasts evolved from prokaryotes that lived symbiotically within some ancestral eukaryotes. Mitochondria originated well before chloroplasts, and many lineages of eukaryotes never evolved chloroplasts. Molecular evidence indicates that chloroplasts evolved from cyanobacteria in the eukaryotic lineage that gave rise to green algae.

equipment needed to transcribe and translate their DNA into proteins. In terms of size, biochemical characteristics, and sensitivity to certain antibiotics, the ribosomes of chloroplasts are more similar to prokaryotic ribosomes than they are to the ribosomes outside the chloroplast in the cytosol of the eukaryotic cell. Mitochondrial ribosomes vary extensively from one group of eukaryotes to another, but they are generally more similar to prokaryotic ribosomes than to their counterparts in the eukaryotic cytoplasm.

Molecular systematics also points to bacterial origins for chloroplasts and mitochondria. Ribosomal RNA of chloroplasts, which is transcribed from genes within the organelles, is more similar in base sequence to the RNA of certain photosynthetic bacteria than it is to the ribosomal RNA in the eukaryotic cytosol, which is transcribed from nuclear DNA. In most eukaryotes, the majority of mitochondrial proteins are encoded by nuclear DNA. However, base-sequence comparisons support a bacterial origin for the ribosomal RNA of mitochondria. Recently, B. Franz Lang and Gertraud Burger of the University of Montreal found that the mitochondrial genome of the protist *Reclinomonas americana* closely matches that of bacteria in both structure and function.

A comprehensive theory for the origin of the eukaryotic cell must also account for the evolution of 9 + 2 flagella and cilia, which are *analogous,* not *homologous,* to the flagella of prokaryotes. (See Chapter 22 to review the distinction between analogous and homologous similarity.) Some researchers have speculated that eukaryotic flagella and cilia evolved from symbiotic bacteria (spirochetes; see TABLE 27.3), but the evidence for this is not strong.

Related to the evolution of the eukaryotic flagellum is the origin of mitosis and meiosis, processes unique to eukaryotes that also employ microtubules. Mitosis made it possible to reproduce the large genomes of the eukaryotic nucleus, and the closely related mechanics of meiosis became an essential process in eukaryotic sex. Among eukaryotes, sexual life cycles are the most varied among the protists. Variety in life cycles and the diversity of protist life in general may reflect the evolutionary "experimentation" that occurred among the earliest eukaryotes. Let's see how systematists are dealing with protist diversity.

PROTIST SYSTEMATICS AND PHYLOGENY

Monophyletic taxa are emerging from modern research in protist systematics

When Robert Whittaker popularized the five-kingdom system of classification in 1969, he assigned unicellular eukaryotes to Kingdom Protista. The trend during the 1970s and 1980s was to expand the boundaries of Kingdom Protista to include some groups of multicellular organisms, such as seaweeds, classified in earlier versions of the five-kingdom system as either plants (in the case of seaweeds) or fungi. These taxonomic transfers were based mainly on comparisons of cell structure and details of life cycles. In its expanded form, Kingdom Protista also encompassed phyla of funguslike organisms, such as the forms known as slime molds and water molds, which may have their closest relatives among the unicellular eukaryotes called amoebas. (Slime molds are funguslike only in the sense that a whale is fishlike; the resemblance is due to convergent evolution of morphological adaptations, not to common ancestry.) The tendency was to treat Kingdom Protista as the taxonomic home for all eukaryotes that did not fit comfortably into the definitions of plants, fungi, or animals.

Most systematists currently working on the puzzle of eukaryotic relationships consider Kingdom Protista and the five-kingdom system obsolete. Kingdom Protista is polyphyletic. (Recall from FIGURE 25.9 that polyphyletic taxa include members derived from two or more ancestral forms not common to all members and thus do not reflect phylogeny.) Among several alternate classifications, a popular eight-kingdom system recognizes three protist kingdoms (Archaezoa, Protista, and Chromista) in place of the single kingdom Protista (FIGURE 28.3b). However, the less inclusive version of Kingdom Protista in the eight-kingdom system is still polyphyletic.

28.1
Using nucleic acid sequencing and detailed comparisons of cell structure, systematists have begun sorting out monophyletic groups of protists, those that reflect evolutionary history because their members all derive from a single ancestor. At present, it seems practical to emphasize the progress that systematists are making toward a classification system that reflects evolution, rather than to prescribe a specific number of protist kingdoms. Thus, we will focus on five groups of protists (Archaezoa, Euglenozoa, Alveolata, Stramenopila, and Rhodophyta) that are emerging from current phylogenetic analyses and debate among systematists (FIGURE 28.3c). We call each of these a "candidate kingdom" because molecular systematics indicates they merit kingdom status in a classification system that groups plants, animals, and fungi into separate kingdoms. It is important to realize that these five groups represent only a sample of protist biodiversity, and our survey will also include some protists whose phylogeny is more enigmatic than that of the five candidate kingdoms. (For convenience, we will continue to use the term *protist* in an informal sense to refer to unicellular eukaryotes and their close multicellular relatives.) All but one of the five candidate kingdoms is monophyletic, and we begin our survey with the exception, the Archaezoa.

Members of candidate kingdom Archaezoa lack mitochondria and may represent early eukaryotic lineages

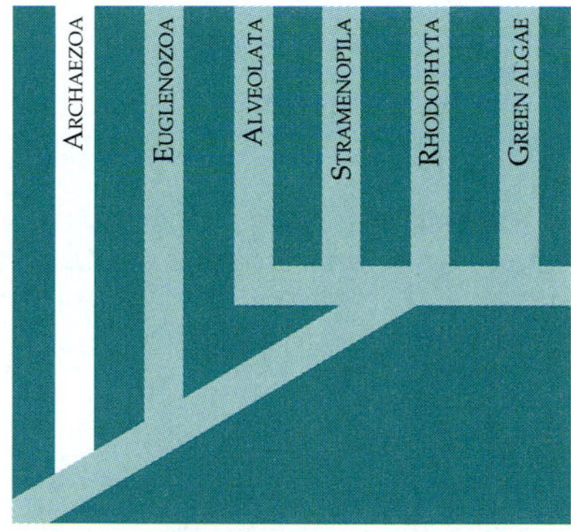

Several protists lack mitochondria, an observation leading to the hypothesis that the lineages of these organisms diverged before the endosymbiotic event that gave rise to mitochondria (see FIGURE 28.2b). Both the eight-kingdom system and the candidate kingdom plan unite protists that lack mitochondria in Kingdom Archaezoa. The name *archaezoa* denotes an ancient ancestry (Gr. *arkhaios,* "ancient"), and this kingdom was proposed in the 1980s to unite protists whose lineages may predate the endosymbiotic origin of mitochondria. Of course, the extant archaezoans are not ancient organisms; they are as modern as humans. But like all organisms, they have evolved in the context of historical constraints, adapting the equipment that was in place in their ancestors.

One subgroup of archaezoans called the diplomonads have flagella, two separate nuclei, no mitochondria, no plastids, and a simple cytoskeleton (compared to other eukaryotes). One of the diplomonads, a parasite called *Giardia lamblia,*

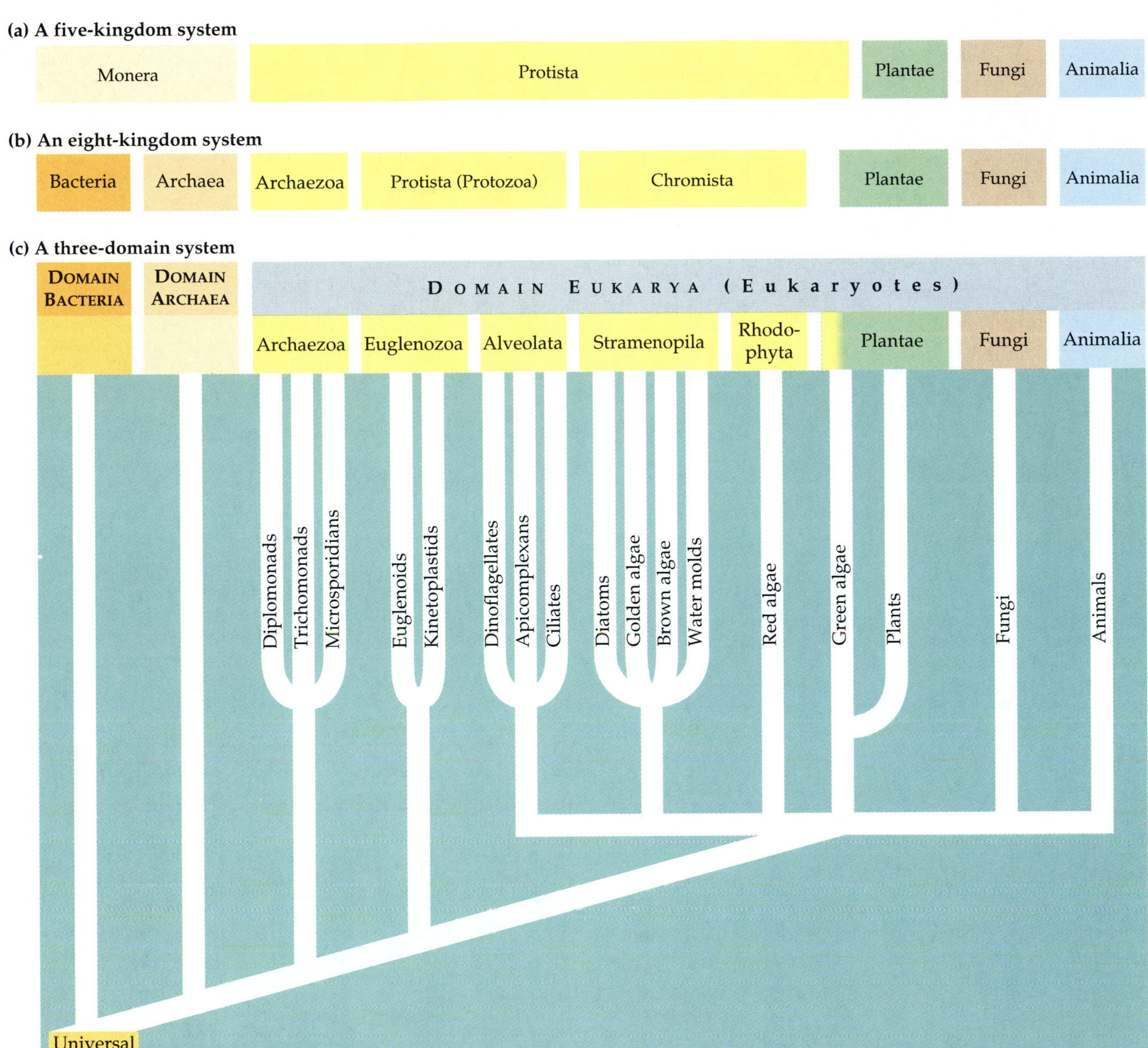

FIGURE 28.3 ▪ Systematics and phylogeny of eukaryotes. This diagram compares how three systems of classification represent biological diversity, with an emphasis on the unicellular eukaryotes and their close multicellular relatives (groups commonly called protists; yellow color). **(a)** The five-kingdom system classifies all such organisms in a single kingdom, the Protista. **(b)** An eight-kingdom system recognizes three kingdoms of protists (Archaezoa, Protista, and Chromista). **(c)** Coupled with the three-domain system, the scheme followed in this chapter recognizes five candidate kingdoms: Archaezoa, Euglenozoa, Alveolata, Stramenopila, and Rhodophyta. These groups represent a sample of protist diversity; many other lineages are not shown on the tree. Of the five groups, the Archaezoa (protists that lack mitochondria) is the most controversial; evolutionary relationships among the diplomonads (e.g., *Giardia*), trichomonads, and microsporidians have not yet been resolved. Some archaezoans may have never had mitochondria, and others may have lost these organelles in the course of evolution. Nucleic acid sequence data and cellular structure indicate that the Euglenozoa, Alveolata, Stramenopila, and Rhodophyta are monophyletic taxa. The Euglenozoa includes photosynthetic organisms (e.g., *Euglena*) and a closely related group of parasites called the kinetoplastids (e.g., *Trypanosoma*). Candidate kingdom Alveolata includes the dinoflagellates (one of many groups of unicellular eukaryotes commonly called flagellates because they have flagella), the apicomplexans (all parasitic), and the ciliates. In the eight-kingdom system, the Euglenozoa and Alveolata are grouped together with several other protists in a polyphyletic group, the kingdom Protista, or Protozoa. Candidate kingdom Stramenopila includes a large monophyletic assemblage of photosynthetic eukaryotes (diatoms, golden algae, and brown algae) and several groups of heterotrophs, including the water molds (formerly allied with fungi). In the eight-kingdom system, the stramenopiles are called the Chromista, a name that implies pigmentation—that the organisms bear chloroplasts. Because the group has many nonpigmented heterotrophs (and the name Chromista also has other historical baggage), the more cumbersome name Stramenopila is gaining popularity. Distinct from the stramenopiles, the red algae (Rhodophyta) and green algae (Chlorophyta) probably derived their chloroplasts from cyanobacterial symbionts. (Stramenopile chloroplasts were probably derived from endosymbionts that were eukaryotic cells.) In our survey of protists in this chapter, miniature versions of this tree will appear as orientation diagrams to help you keep sight of evolutionary relationships.

28.1

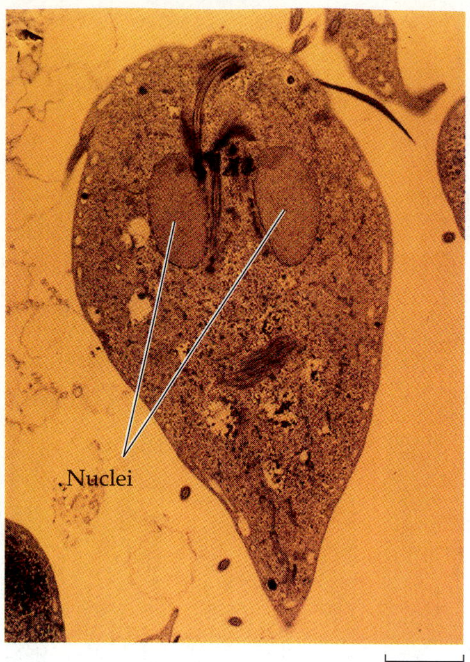

FIGURE 28.4 ▪ ***Giardia*: a modern representative of an early branch of eukaryotes.** Molecular systematics and ultrastructural evidence place *Giardia,* a parasitic protist, as a descendant of a lineage that originated very early during the prokaryote-eukaryote transition more than 2 billion years ago (colorized SEM).

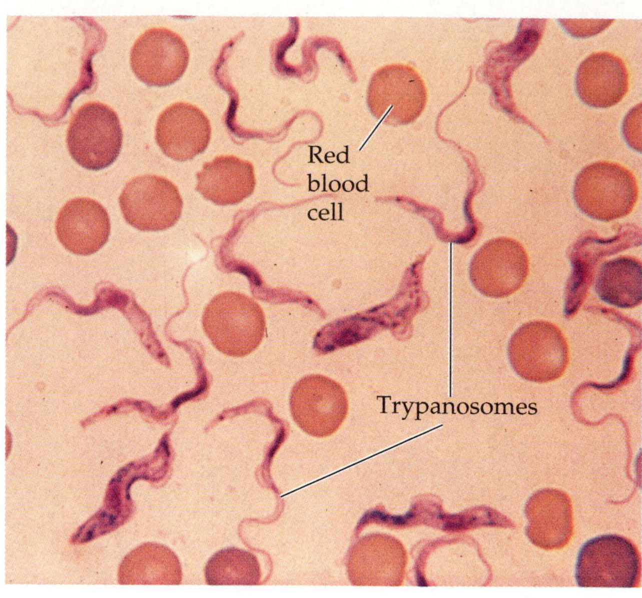

FIGURE 28.5 ▪ **Euglenozoans.** *Trypanosoma,* seen here in human blood, is the flagellate that causes African sleeping sickness. The molecular composition of these pathogens' coats changes frequently, preventing immunity from developing in hosts. Trypanosomes are members of a eukaryotic (protist) group called the kinetoplastids, symbiotic flagellates closely related to the free-living autotroph *Euglena* (LM).

infects the human intestine, causing abdominal cramps and severe diarrhea (FIGURE 28.4). *Giardia* is transmitted mainly in water contaminated with human feces.

Ribosomal RNA studies strongly support the hypothesis that *Giardia* and other diplomonads are living relicts of an early lineage of eukaryotes—modern representatives of evolutionary "experiments" that occurred near the base of the eukaryotic tree. Whether their ancestry precedes the origin of mitochondria, however, remains questionable. Recent evidence indicates that *Giardia* and several other protists that lack mitochondria have genes that once coded mitochondrial proteins. Thus, the ancestors of *Giardia* and certain other archaezoans may have possessed mitochondria that were lost in the organisms' long evolutionary history.

Though many biologists find the kingdom Archaezoa useful, it seems destined to be replaced. The evolutionary relationships of diplomonads and two other groups of parasitic archaezoans, called trichomonads and microsporidians, have not yet been resolved. Like *Giardia,* these organisms lack mitochondria, but this may or may not indicate a close phylogenetic relationship among them. However we classify these protists, as molecular systematist Mitchell Sogin of the Marine Biological Laboratory at Woods Hole, Massachusetts, puts it, "*Giardia* and other representatives of early eukaryote branches can tell us a great deal about what was present in early common ancestors."

Candidate kingdom Euglenozoa includes both autotrophic and heterotrophic flagellates

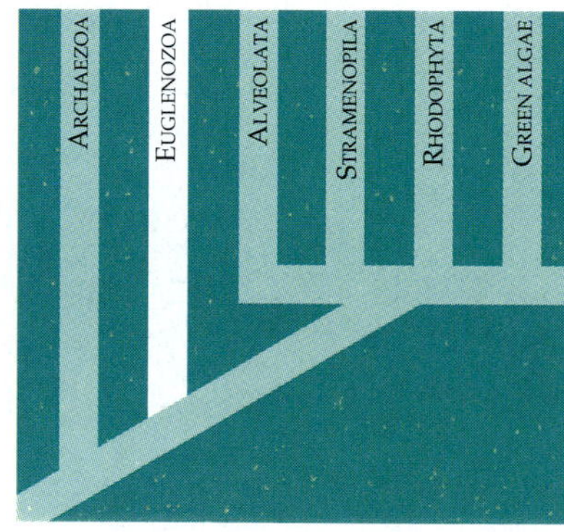

Protists with flagella are commonly called **flagellates**, a term that is not used in formal taxonomy. Molecular systematics indicates that two groups of flagellates, called the euglenoids and the kinetoplastids, make up the monophyletic candidate kingdom **Euglenozoa.** The **euglenoids** (*Euglena* and its close relatives) are characterized by an anterior pocket, or chamber, from which one or two flagella emerge. Paramylum, a glucose polymer that functions as a storage molecule, is also characteristic of euglenoids. *Euglena* is chiefly autotrophic, but many euglenozoans are mixotrophic or heterotrophic, absorbing organic molecules from their surroundings or engulfing prey by phagocytosis.

The **kinetoplastids** have a single large mitochondrion associated with a unique organelle, the kinetoplast, that houses extranuclear DNA. The kinetoplastids are symbiotic, and some are pathogenic to their hosts. For example, species of *Trypanosoma* cause African sleeping sickness, a human disease that is spread by the bite of the tsetse fly (FIGURE 28.5).

Subsurface cavities (alveoli) are diagnostic of candidate kingdom Alveolata

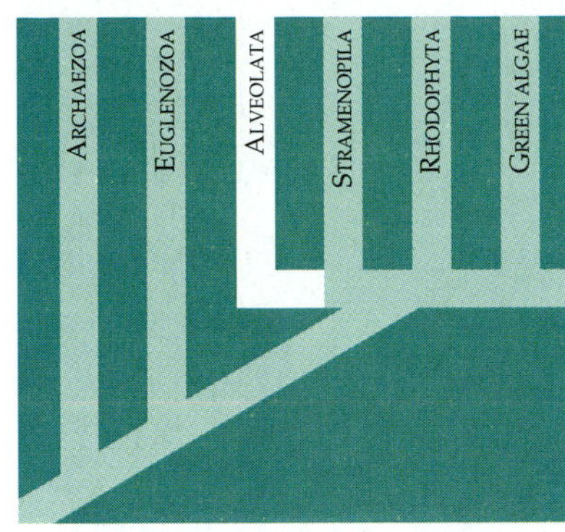

Another monophyletic candidate kingdom that is emerging from molecular systematics, the **Alveolata**, draws together a group of photosynthetic flagellates (the dinoflagellates), a group of parasites (apicomplexans), and a distinctive group of eukaryotes that move by means of cilia (the ciliates). Alveolates have small membrane-bounded cavities (alveoli) under their cell surfaces. The function of alveoli is unknown; they may help stabilize the cell surface and regulate the cell's water and ion content.

Dinoflagellates

Dinoflagellates are abundant components of the vast aquatic pastures of phytoplankton that float near the water surface and provide the foundation of most marine and many freshwater food chains (FIGURE 28.6).

Dinoflagellate blooms—episodes of explosive population growth—cause red tides in coastal waters. These blooms are brownish-red or pinkish-orange because of the predominant pigments (xanthophylls) in the chloroplasts of dinoflagellates. Toxins produced by some red-tide organisms have produced massive invertebrate and fish kills and can be deadly to humans as well. One especially dangerous species called *Pfiesteria piscicida* is actually carnivorous. During blooms, its toxin stuns fish, and the dinoflagellates feed on their prey's body fluids. In the past decade the frequency of *Pfiesteria* blooms and fish kills have increased in coastal waters of the U.S. mid-Atlantic states, possibly a result of pollution of the water with fertilizers (nitrates and phosphates).

Of the several thousand known dinoflagellate species, most are unicellular, but some are colonial forms. Each dinoflagellate species has a characteristic shape reinforced by internal plates of cellulose. The beating of two flagella in perpendicular grooves in this "armor" produces a spinning movement for which these organisms are named (Gr. *dinos*, "whirling"). The structure of the dinoflagellate nucleus and its division during asexual reproduction are unusual (see FIGURE 12.11b).

Some dinoflagellates live as mutualistic symbionts of animals called cnidarians that build coral reefs; the photosynthetic output of these dinoflagellates is the main food source for reef communities. Other dinoflagellates lack chloroplasts and live as parasites within marine animals. Some heterotrophic dinoflagellates, such as *Pfiesteria*, can become temporarily autotrophic by extracting chloroplasts from photosynthetic protists.

Apicomplexans

All **apicomplexans** (formerly called sporozoans) are parasites of animals. Some cause serious human diseases. The parasites disseminate as tiny infectious cells called **sporozoites**. As seen with the electron microscope, one end (the apex) of the sporozoite cell contains a complex of organelles specialized for penetrating host cells and tissues; thus the name Apicomplexa. Most apicomplexans have intricate life cycles with both sexual and asexual stages, and these cycles often require two or more different host species for completion. An example is

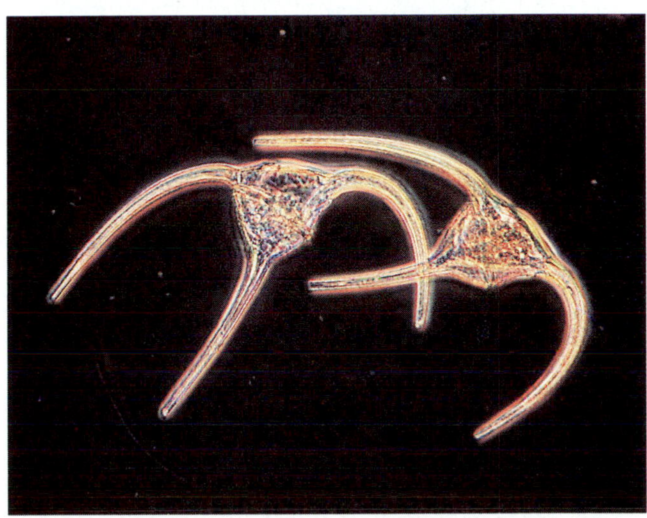

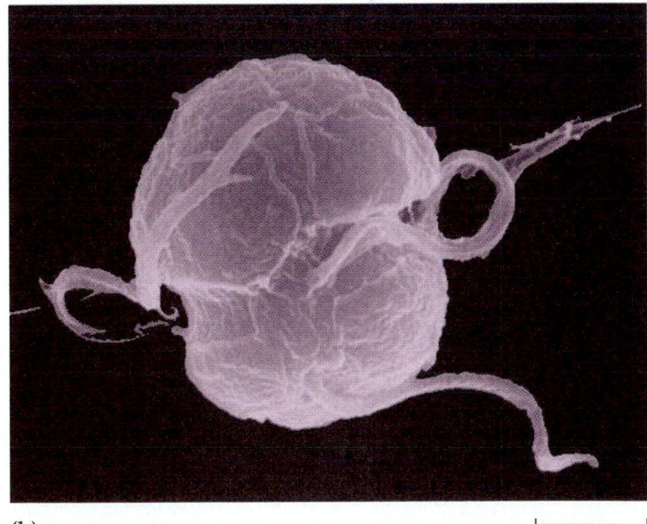

FIGURE 28.6 ▪ Dinoflagellates. These unicellular eukaryotes are characterized by a pair of flagella in perpendicular grooves. Beating of the flagella causes the cell to spin as it swims. Each species has a distinctively shaped internal wall. **(a)** *Ceratium* is common in freshwater lakes (LM). **(b)** *Pfiesteria piscicida* produces toxins that are lethal to fish and dangerous to humans (SEM). So-called red tides result from population explosions of *Pfiesteria* and several other marine dinoflagellates.

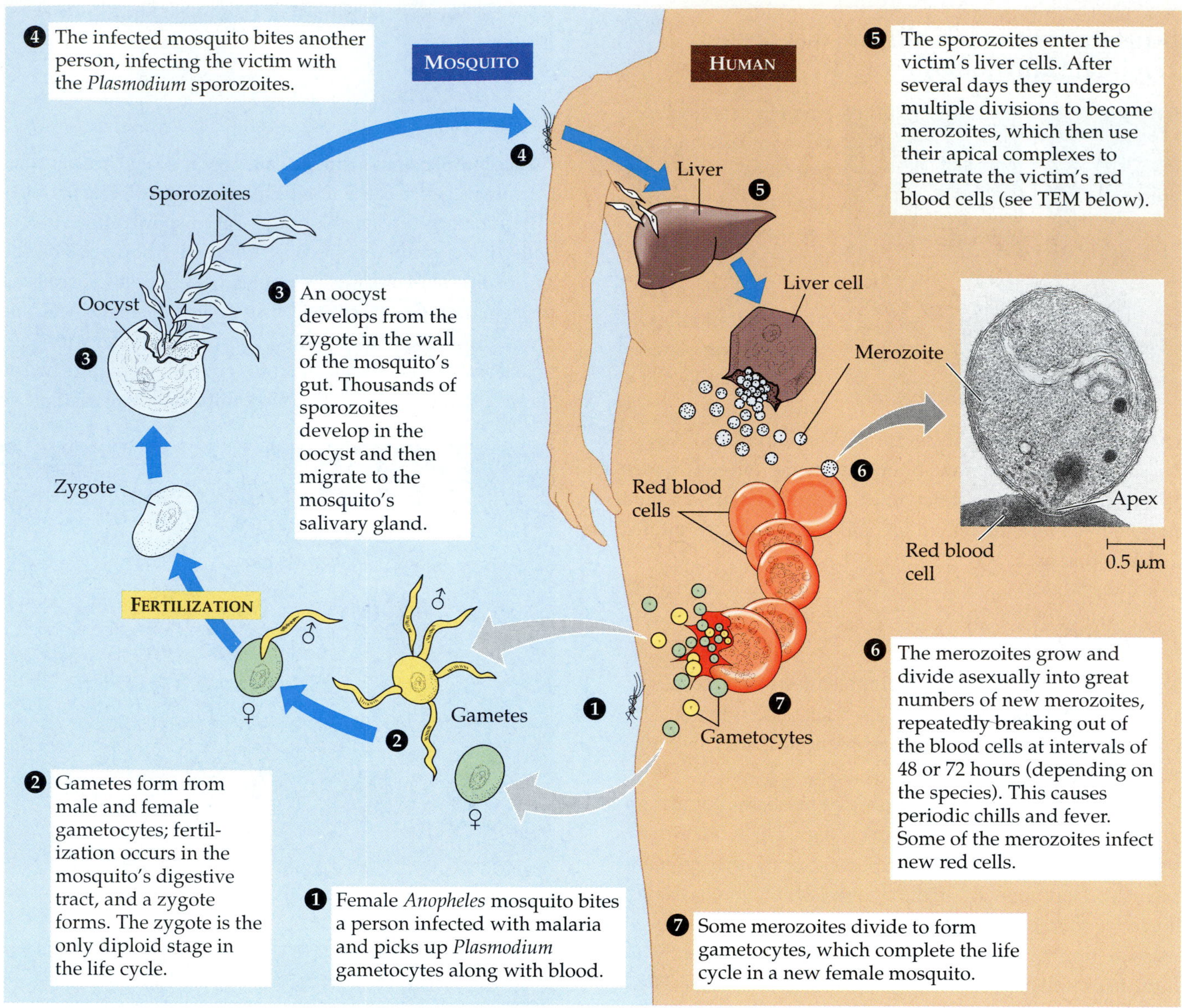

MOSQUITO
HUMAN
4 The infected mosquito bites another person, infecting the victim with the Plasmodium sporozoites.
5 The sporozoites enter the victim's liver cells. After several days they undergo multiple divisions to become merozoites, which then use their apical complexes to penetrate the victim's red blood cells (see TEM below).
Sporozoites
Liver
Liver cell
Oocyst
3 An oocyst develops from the zygote in the wall of the mosquito's gut. Thousands of sporozoites develop in the oocyst and then migrate to the mosquito's salivary gland.
Merozoite
Zygote
Red blood cells
Apex
Red blood cell
0.5 µm
FERTILIZATION
Gametes
Gametocytes
6 The merozoites grow and divide asexually into great numbers of new merozoites, repeatedly breaking out of the blood cells at intervals of 48 or 72 hours (depending on the species). This causes periodic chills and fever. Some of the merozoites infect new red cells.
2 Gametes form from male and female gametocytes; fertilization occurs in the mosquito's digestive tract, and a zygote forms. The zygote is the only diploid stage in the life cycle.
1 Female Anopheles mosquito bites a person infected with malaria and picks up Plasmodium gametocytes along with blood.
7 Some merozoites divide to form gametocytes, which complete the life cycle in a new female mosquito.

FIGURE 28.7 ▪ The life history of *Plasmodium*, the apicomplexan that causes malaria. (Colors are not true to life.)

Plasmodium, the parasite that causes malaria (FIGURE 28.7). The incidence of malaria was greatly diminished in the 1960s by the use of insecticides that reduced populations of *Anopheles* mosquitoes, which spread the disease, and by drugs that killed the parasites in humans. However, the multiplication of resistant varieties of both the mosquitoes and *Plasmodium* species have caused a resurgence of the disease. Each year about 300 million people are infected in the tropics, and up to 2 million die from the disease.

Considerable research on possible malarial vaccines has been carried out, with little success. *Plasmodium* is an extremely evasive parasite. It spends most of its time inside human liver and blood cells, thus hiding from the host's immune system. The problem is compounded by changes in the surface proteins of *Plasmodium*—the parasite continually changes the "face" that it shows to the infected person's immune system. Recent identification of a gene that seems to confer resistance to chloroquine, an important antimalarial drug, may lead to ways of blocking drug resistance in *Plasmodium*. Another recent discovery has exciting evolutionary and medical implications. A team of molecular biologists led by Sabine Kohler at the University of Pennsylvania found that *Plasmodium* and several other apicomplexans contain a plastid that an apicomplexan ancestor probably acquired from a green alga by endosymbiosis. Once researchers discover the critical functions the plastids perform for their host cells, it may be possible to develop new antimalarial drugs that target those functions.

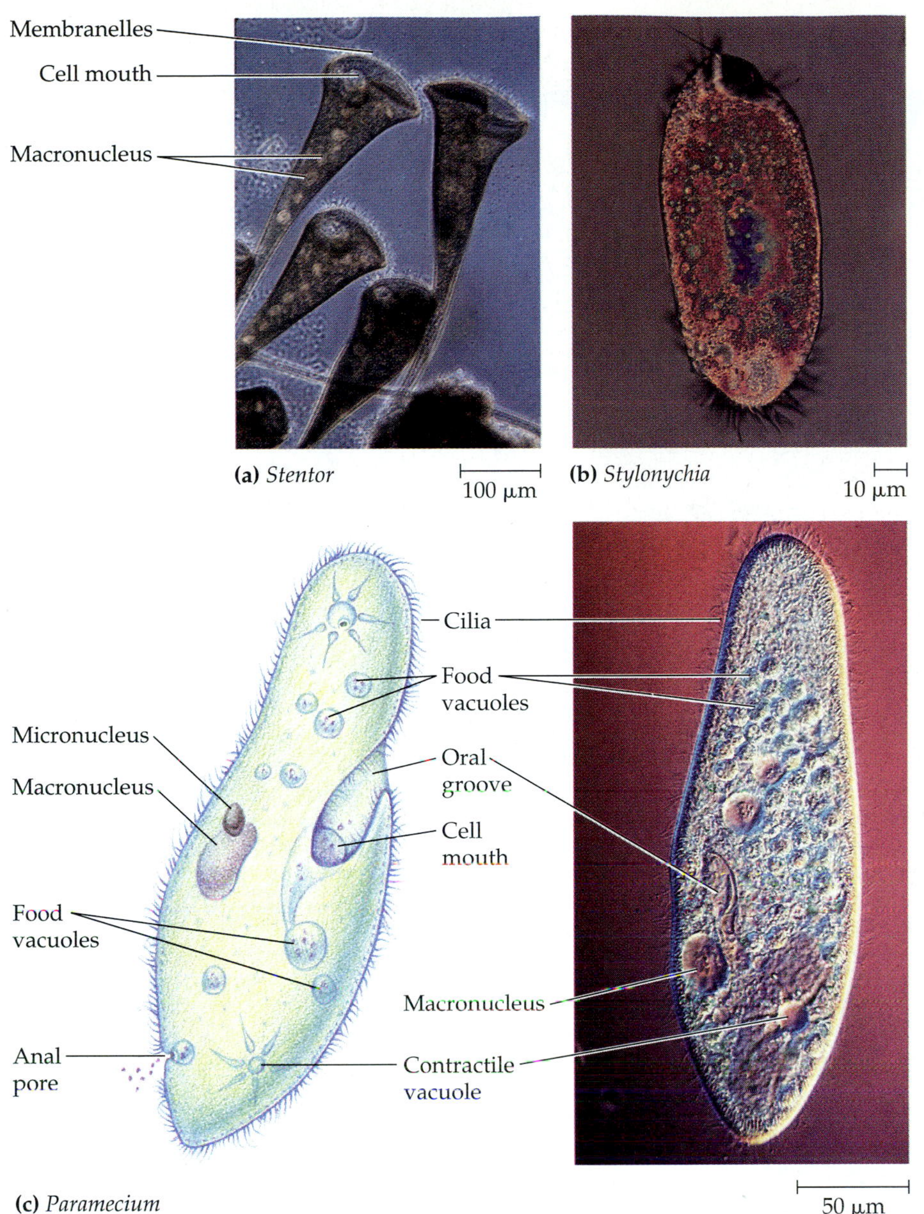

FIGURE 28.8 ▪ Ciliates (ciliophorans). **(a)** The beautiful freshwater ciliate *Stentor* moves by individual cilia along the cell sides and by rows of finlike membranelles that spiral around the broad anterior end of the cell. *Stentor* often attaches its narrow (posterior) end to debris and, by beating the anterior membranelles, causes a whirlpool-like current that moves food into the cell mouth. The macronucleus can be seen as light beaded strands running the length of these cells. **(b)** *Stylonychia* belongs to a group of ciliates that often have no individual cilia at all. Cilia on the cell-mouth side of the cell are united into leglike cirri. These ciliates scurry around on sediments, feeding among bits of organic debris. **(c)** *Paramecium* is covered by thousands of individual cilia. *Paramecium* feeds mainly on bacteria. Rows of cilia along a funnel-shaped oral groove move food into the cell mouth, where the food is engulfed by phagocytosis. The food vacuoles combine with lysosomes and, as the food is digested, the vacuoles follow a looping path, first to the anterior end and finally to the posterior end. The undigested contents are released when the vacuoles fuse with a specialized region of the plasma membrane that functions as an anal pore. *Paramecium,* like other freshwater protists, constantly takes in water by osmosis from the hypotonic environment. Bladderlike contractile vacuoles accumulate the excess water from radial canals and periodically expel it through the plasma membrane by contractions of the surrounding cytoplasm (see FIGURE 8.12). (All LMs.)

Ciliates (Ciliophorans)

The diverse protists usually called ciliates are characterized by their use of cilia to move and feed. Most **ciliates,** or ciliophorans, live as solitary cells in fresh water. In contrast to most flagella, cilia are relatively short. They are associated with a submembrane system of microtubules that may coordinate the movement of the thousands of cilia.

The organization of the cytoskeleton and other components in the cell's cortex (outer layer of cytoplasm) also provides the information that orders cilia into a specific pattern as the cell grows. When researchers use microsurgery to remove a piece of cortex and then reinsert the piece with its polarity relative to the ends of the cell reversed, this also reverses the pattern of cilia that develops in that location, even after many generations of cell division. In this developmental mechanism, called *cytotaxis,* it is cytoplasmic organization, not genes (at least not directly), that provides heritable information.

Some ciliates are completely covered by rows of cilia, whereas others have their cilia clustered into fewer rows or tufts. The specific arrangements adapt the ciliates for their diverse lifestyles. Some species, for instance, scurry about on leglike structures constructed from many cilia bonded together. Other forms, such as *Stentor,* have rows of tightly packed cilia that function collectively as locomotor membranelles. Ciliates are among the most complex of all cells (FIGURE 28.8).

A unique feature of ciliate genetics is the presence of two types of nuclei, a large macronucleus and usually several tiny micronuclei. The macronucleus has 50 or more copies of the genome. The genes are not distributed in typical chromosomes but are instead packaged into a much larger number of small units, each with hundreds of copies of just a few genes. The macronucleus controls the everyday functions of the cell by synthesizing RNA and is also necessary for asexual

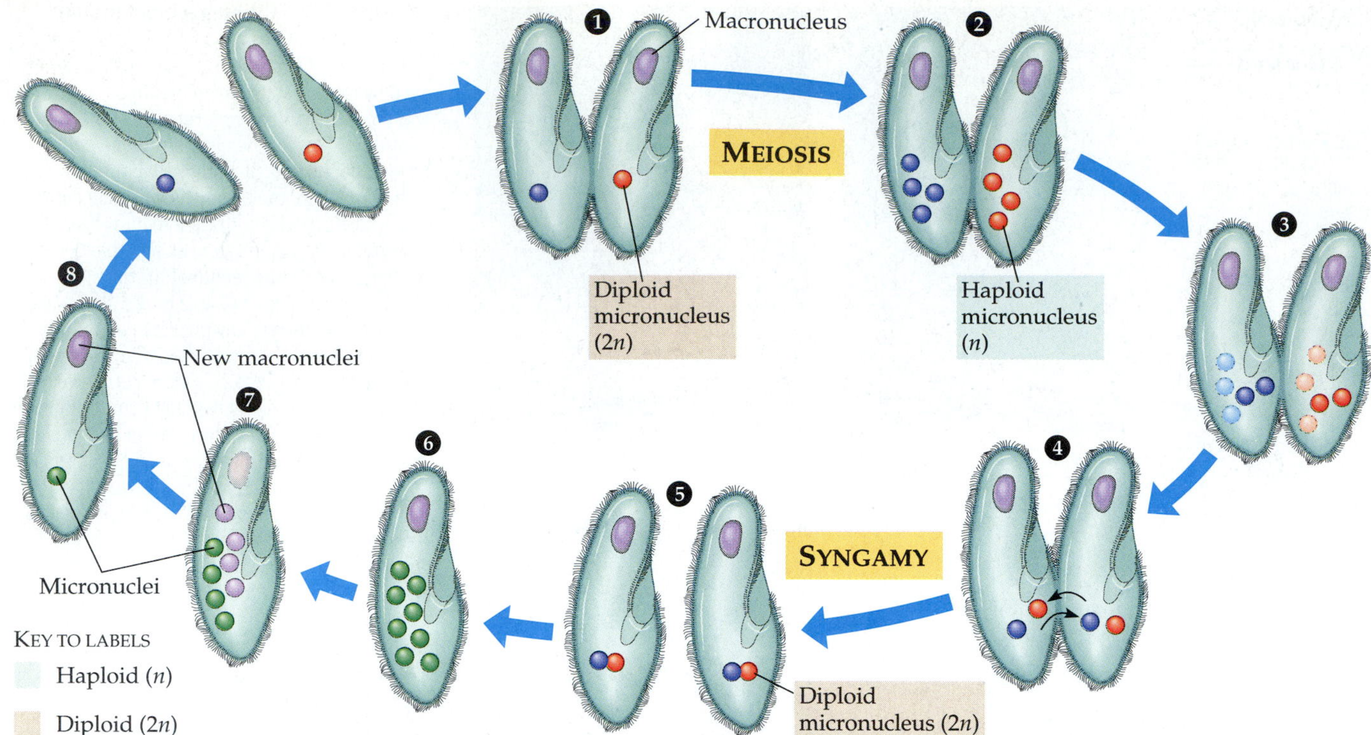

FIGURE 28.9 ▪ Conjugation and genetic recombination in *Paramecium caudatum.* ① Two individuals of compatible mating strains align side by side and partially fuse. The two cells differ in genetic makeup. ② The micronucleus in each individual undergoes meiosis to produce four haploid micronuclei. ③ One of these divides by mitosis, and the other three disintegrate. ④ Mates then swap one micronucleus. ⑤ Syngamy occurs when the micronucleus a cell acquired from its partner fuses with its remaining micronucleus, forming a fresh diploid nucleus with a mixture of chromosomes derived from the two individuals. The partners separate. ⑥ From now on, only one partner is shown. From the single new micronucleus in each individual, a sequence of three mitotic divisions produces eight identical micronuclei (colored green). ⑦ Subsequently, in each cell the original macronucleus disintegrates. Four micronuclei develop into new macronuclei by repeated replication of the DNA without nuclear division. The other four remain as micronuclei. ⑧ After two cell divisions (without nuclear divisions), the new macronuclei and micronuclei are parceled out into four new individuals for each original conjugating cell. (Only one individual is shown here.) Note that all eight individuals ultimately resulting from one conjugation are identical genetically. However, they represent a genetic makeup different from either of the two original conjugating cells.

reproduction. Ciliates generally reproduce by binary fission, during which the macronucleus elongates and splits, rather than undergoing mitotic division. Some species of *Paramecium* have from one to as many as 80 micronuclei, which do not function in growth, maintenance, and asexual reproduction of the cell but are required for sexual processes that generate genetic variation. The sexual shuffling of genes occurs during the process known as **conjugation** (FIGURE 28.9). Notice in the diagram that in ciliates, sexual mechanisms of meiosis and syngamy are processes separate from reproduction.

A diverse assemblage of unicellular eukaryotes move by means of pseudopodia

The three groups we discuss in this section represent some of the immense diversity of unicellular eukaryotes that move and often feed by means of cellular extensions called **pseudopodia**. Most of these organisms are heterotrophs that actively seek and consume bacteria, other protists, and **detritus** (dead organic matter). There are also symbiotic species, including some parasites that cause human diseases. We discuss these three groups together only for convenience. Little is known about their phylogeny, although it is clear that they represent several distinct eukaryotic lineages. Because of taxonomic uncertainties, we have not included these protists on the phylogenetic tree in FIGURE 28.3. What is important to take from this brief survey is not so much the formal taxonomy as the diversity of locomotor and feeding mechanisms (TABLE 28.1).

Rhizopods (Amoebas)

Rhizopods, more often called **amoebas**, are all unicellular and use pseudopodia to move and to feed (FIGURE 28.10). You may have observed their mode of motility in *Amoeba proteus* in the laboratory. Pseudopodia may bulge from virtually anywhere on the cell surface. When an amoeba moves, it extends a pseudopodium and anchors its tip, and then more cytoplasm streams into the pseudopodium (*rhizopoda* means "rootlike feet"). The cytoskeleton, consisting of microtubules and microfilaments, functions in amoeboid movement (see FIGURE 7.27b). Pseudopodial activity may appear chaotic, but in fact amoebas show directed movement as they creep toward a

Table 28.1 ▪ A Partial List of Protists That Move by Pseudopodia and the Slime Molds

GENERAL GROUP	BRIEF DESCRIPTION
Protists That Move by Pseudopodia	
Rhizopods (p. 530)	Naked and shelled (testate) amoebas, with broad pseudopodia for motility and feeding; a polyphyletic group; e.g., *Amoeba*
Actinopods (p. 531)	Occupy planktonic habitats; usually spherically symmetrical; feed with axopodia (slender, radiating pseudopodia supported by internal microtubules); radiolarians possess fused siliceous skeletons. The skeleton of heliozoans is not fused
Foraminiferans (p. 532)	"Forams": feed and move with slender, interconnected pseudopodia exuding from spirally arranged calcareous compartments
The Slime Molds	
Myxomycotes (p. 532)	Plasmodial slime molds; terrestrial heterotrophic decomposers with a multinucleate amoeboid feeding stage (plasmodium)
Acrasiomycotes (p. 532)	Cellular slime molds; terrestrial heterotrophic decomposers with amoeboid unicellular and multicellular stages

food source. Some amoebas live inside a protein shell that they secrete. Pseudopodia extend outward through an opening in the shell, which in some species is coated with fine sand grains.

Meiosis and sex are not known to occur in amoebas. The organisms reproduce asexually by various mechanisms of cell division. Spindle fibers form, but the typical stages of mitosis are not apparent in most amoebas. In many genera, for instance, the nuclear envelope persists during cell division.

Amoebas inhabit both freshwater and marine environments and are also abundant in soils. The majority of amoebas are free-living, but some are important parasites, including *Entamoeba histolytica*, which causes amoebic dysentery in humans. These organisms spread via contaminated drinking water, food, or eating utensils.

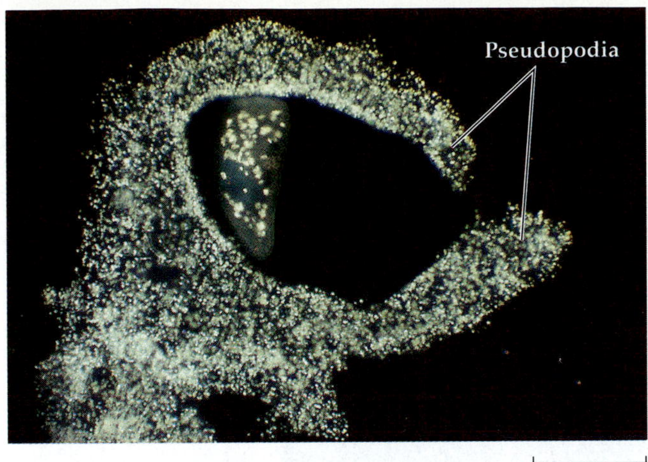

FIGURE 28.10 ▪ Rhizopods. Amoebas use pseudopodia to move and feed. This light micrograph shows the pseudopodia of an amoeba engulfing a ciliate. The food, engulfed by the process of phagocytosis (see Chapter 8), will be packaged in a food vacuole, where digestion will occur.

Actinopods (Heliozoans and Radiolarians)

Actinopod means "ray foot," a reference to the slender pseudopodia called axopodia that radiate from these exceptionally beautiful protists (FIGURE 28.11). The axopodia of most actinopods are part of an ornate skeleton that is mainly internal. Each axopodium is reinforced by a bundle of microtubules covered by a thin layer of cytoplasm. Most actinopods are planktonic, and their projections place an extensive area of cellular surface in contact with the surrounding water, helping the organisms float and feed. Smaller protists and other

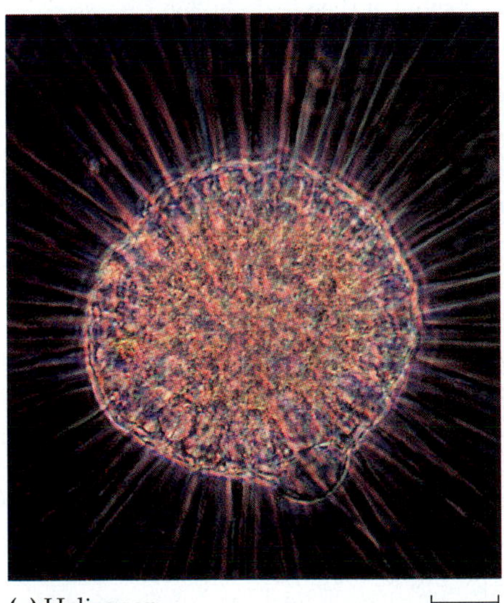

(a) Heliozoan

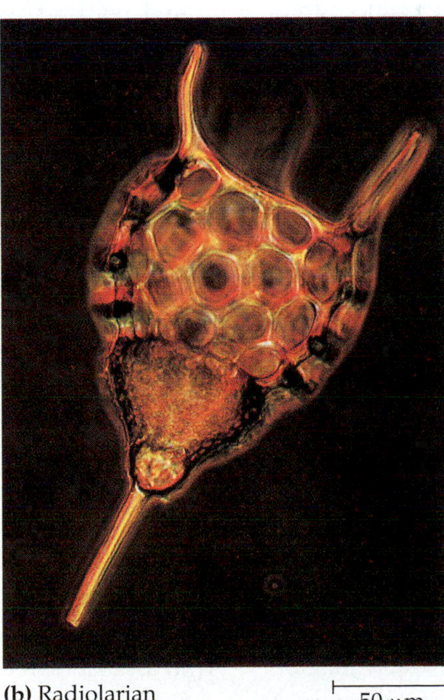

(b) Radiolarian

FIGURE 28.11 ▪ Actinopods. (a) Heliozoans are mainly freshwater protists with stiff axopodia used for feeding. **(b)** Radiolarians are mostly marine forms with glassy shells that differ in shape for each species. (Both LMs.)

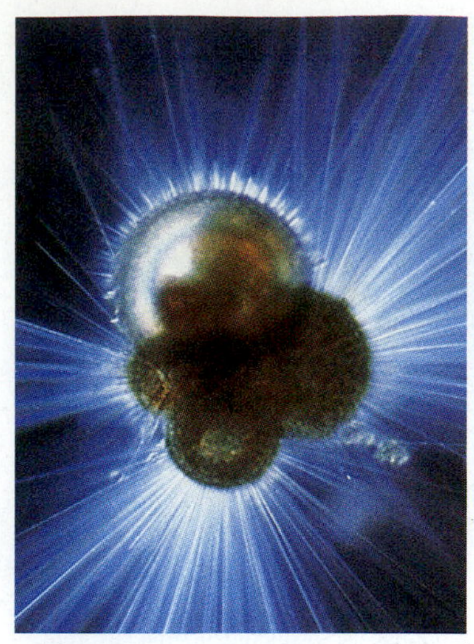

10 μm

FIGURE 28.12 ▪ Foraminiferan. This living foram has a snail-like shell. The largest forams grow to diameters of several centimeters. The calcium carbonate shells of these protists have left an excellent fossil record (LM).

microorganisms stick to the axopodia and are phagocytized by the thin layer of cytoplasm. Cytoplasmic streaming then carries the engulfed prey into the main part of the cell.

Most **heliozoans** ("sun animals") live in fresh water. Heliozoan skeletons consist of siliceous (glassy) or chitinous unfused plates. The term **radiolarian** (not used in formal taxonomy) refers to several groups of mostly marine actinopods with skeletons fused into one delicate piece, most commonly made of silica. After actinopods die, their skeletons settle to the seafloor, where they have accumulated as an ooze that is hundreds of meters thick in some locations.

Foraminiferans (Forams)

Foraminiferans, or **forams**, are almost all marine. Most species live in the sand or attach themselves to rocks and algae, but some are also abundant in plankton. Foraminiferans are named for their porous shells (L. *foramen,* "little hole," and *ferre,* "to bear"). The shells are generally multichambered and consist of organic material hardened with calcium carbonate. Strands of cytoplasm (pseudopodia) extend through the pores, functioning in swimming, shell formation, and feeding (FIGURE 28.12). Many forams also derive nourishment from the photosynthesis of symbiotic algae that live within the shells.

Ninety percent of all identified species of forams are fossils. Along with the calcareous remains of other protists, the fossilized shells of forams are components of marine sediments, including sedimentary rocks that are now land formations. Foram fossils are excellent markers for correlating the ages of sedimentary rocks in different parts of the world.

Slime molds have structural adaptations and life cycles that enhance their ecological role as decomposers

Two groups of protists called slime molds resemble fungi in appearance and lifestyle, but the similarities are the result of convergence. In their cellular organization, reproduction, and life cycles, slime molds depart from the true fungi and probably have their closest relatives among the amoeboid protists. The superficial similarity of the slime molds to true fungi is partly due to convergent evolution of filamentous body structure, a morphological adaptation that increases exposure to the environment and enhances the ecological role of these organisms as decomposers. Slime molds have complex life cycles that are adaptations that contribute to survival in changing habitats and facilitate dispersal to new food sources.

Plasmodial Slime Molds (Myxomycota)

The **plasmodial slime molds** are more attractive than their name implies. Many species are brightly pigmented, usually yellow or orange, but all are heterotrophic. The feeding stage of the life cycle is an amoeboid mass called a **plasmodium**, which may grow to a diameter of several centimeters (FIGURE 28.13). Large as it is, the plasmodium is not multicellular; it is a single mass of cytoplasm undivided by membranes and containing many nuclei. In most species, the nuclei of the plasmodium are diploid and divisions are synchronous, with each of thousands of nuclei going through each phase of mitosis at the same time. Because of this characteristic, plasmodial slime molds have been used to study the molecular details of mitosis. Within the fine channels of the plasmodium, cytoplasm streams first one way, then the other, in pulsing flows that are beautiful to watch through a microscope. The cytoplasmic streaming apparently helps distribute nutrients and oxygen. The plasmodium engulfs food particles by phagocytosis as it grows by extending pseudopodia through moist soil, leaf mulch, or rotting logs. If the habitat of a slime mold begins to dry up or there is no food left, the plasmodium ceases growth and differentiates into a stage of the life cycle that functions in sexual reproduction.

Cellular Slime Molds (Acrasiomycota)

Cellular slime molds pose a semantic question about what it means to be an individual organism. Although the feeding stage of the life cycle consists of solitary cells that function individually, when food is depleted the cells form an aggregate that functions as a unit (FIGURE 28.14, p. 534). Though the mass of cells resembles a plasmodial slime mold, the important distinction is that the cells of a cellular slime mold maintain their identity and remain separated by their membranes.

Cellular slime molds also differ from plasmodial slime molds in being haploid organisms (only the zygote is diploid), whereas the diploid condition predominates in the life cycles of most plasmodial slime molds (compare FIGURES 28.13 and 28.14). Cellular slime molds have fruiting bodies that function in asexual reproduction. Also, most cellular slime molds have no flagellated stages.

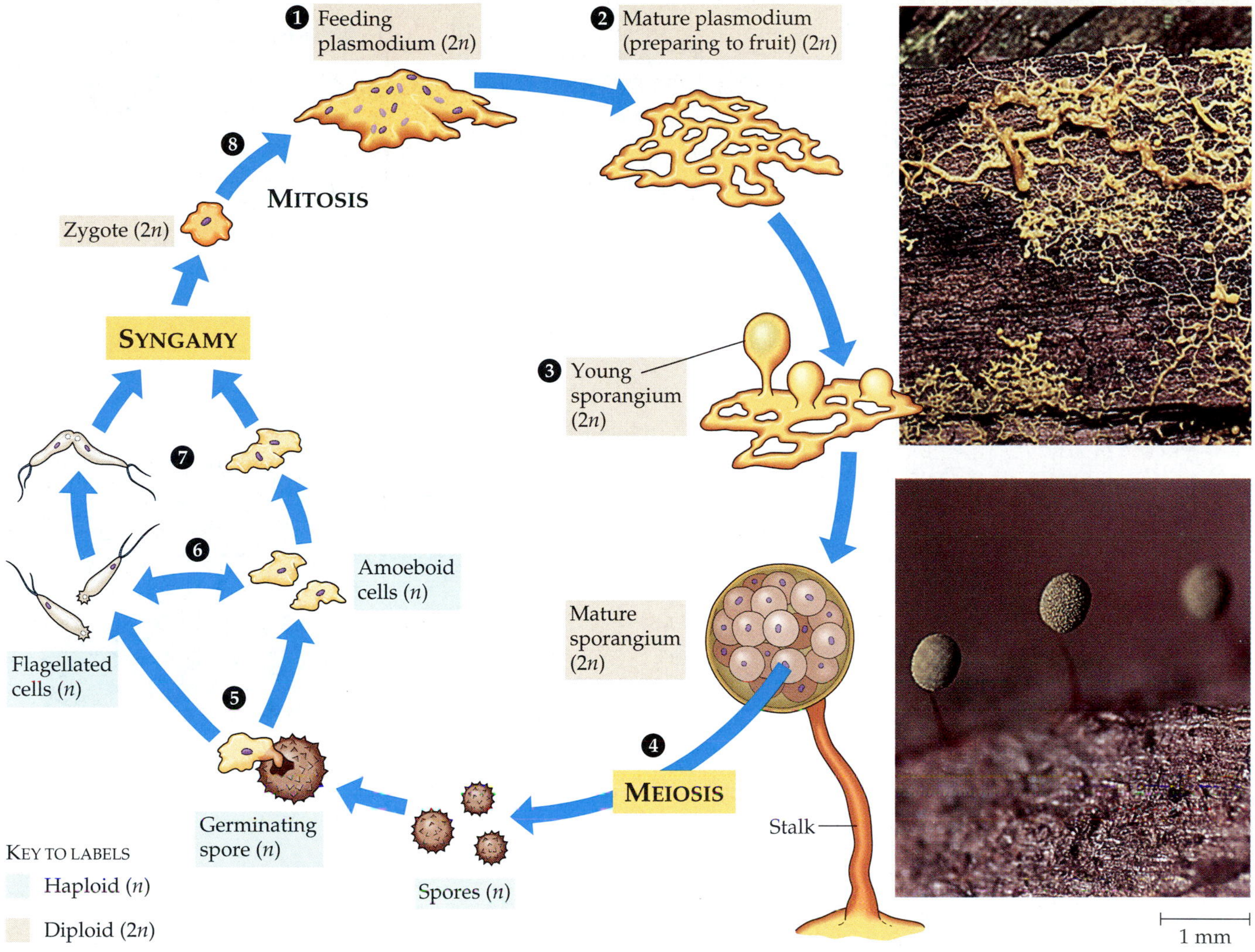

1 Feeding plasmodium (2*n*)
2 Mature plasmodium (preparing to fruit) (2*n*)
3 Young sporangium (2*n*)
Mature sporangium (2*n*)
Stalk
4 MEIOSIS
Spores (*n*)
Germinating spore (*n*)
5
Flagellated cells (*n*)
6
Amoeboid cells (*n*)
7
SYNGAMY
Zygote (2*n*)
8
MITOSIS
KEY TO LABELS
Haploid (*n*)
Diploid (2*n*)
1 mm

28.2 **FIGURE 28.13 ▪ The life cycle of a plasmodial slime mold.** (1) The feeding stage is a multinucleate plasmodium that lives on organic refuse. (2) The plasmodium often takes a weblike form, an adaptation that increases the surface area that contacts food, water, and oxygen (upper photo). (3) The plasmodium erects stalked fruiting bodies called sporangia when conditions become harsh (lower photo, LM). (4) Within the bulbous tips of the sporangia, meiosis produces haploid spores. (5) The resistant spores germinate to become active haploid cells when conditions are again favorable. (6) These cells are either amoeboid or flagellated; the two forms readily revert from one to the other. (7) These cells unite in like pairs (flagellated with flagellated and amoeboid with amoeboid) to form diploid zygotes. (8) Repeated division of the nucleus of the zygote by mitosis, without cytoplasmic division, forms a feeding plasmodium and completes the life cycle.

Diatoms, golden algae, brown algae, and water molds are members of candidate kingdom Stramenopila

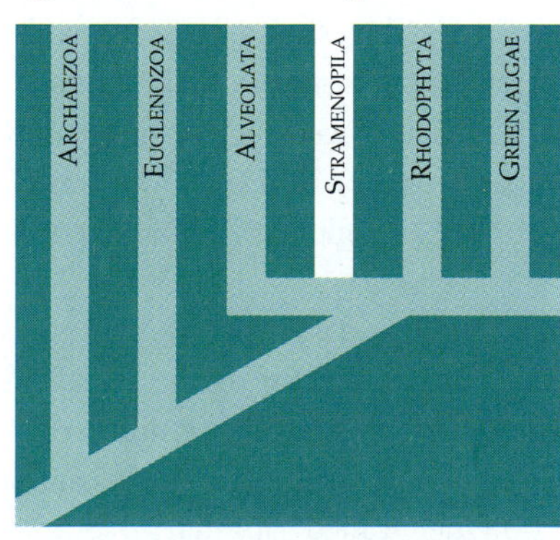

A diverse candidate kingdom, the **Stramenopila**, includes several groups of photosynthetic autotrophs (algae) and numerous heterotrophs. The term *stramenopila* refers to numerous fine, hairlike projections on the flagella that are characteristic of these organisms (L. *stramen*, "straw"or "flagellum," and *pilos*, "hair"). Molecular systematics indicates that the stramenopiles make up a monophyletic taxon.

In the eight-kingdom classification system, stramenopiles are classified in Kingdom Chromista (see FIGURE 28.3). However, the term *chromista* has some historical baggage and can be misleading because it implies that pigmented plastids (chromoplasts) characterize the kingdom. In fact, many stramenopiles, such as the water molds, are heterotrophic and lack plastids.

Photosynthetic stramenopiles have unusual chloroplasts, with two additional membranes outside the usual chloroplast envelope, a small amount of cytoplasm, and a vestigial nucleus. Increasing evidence suggests that these chloroplasts did not evolve directly from cyanobacteria but descended

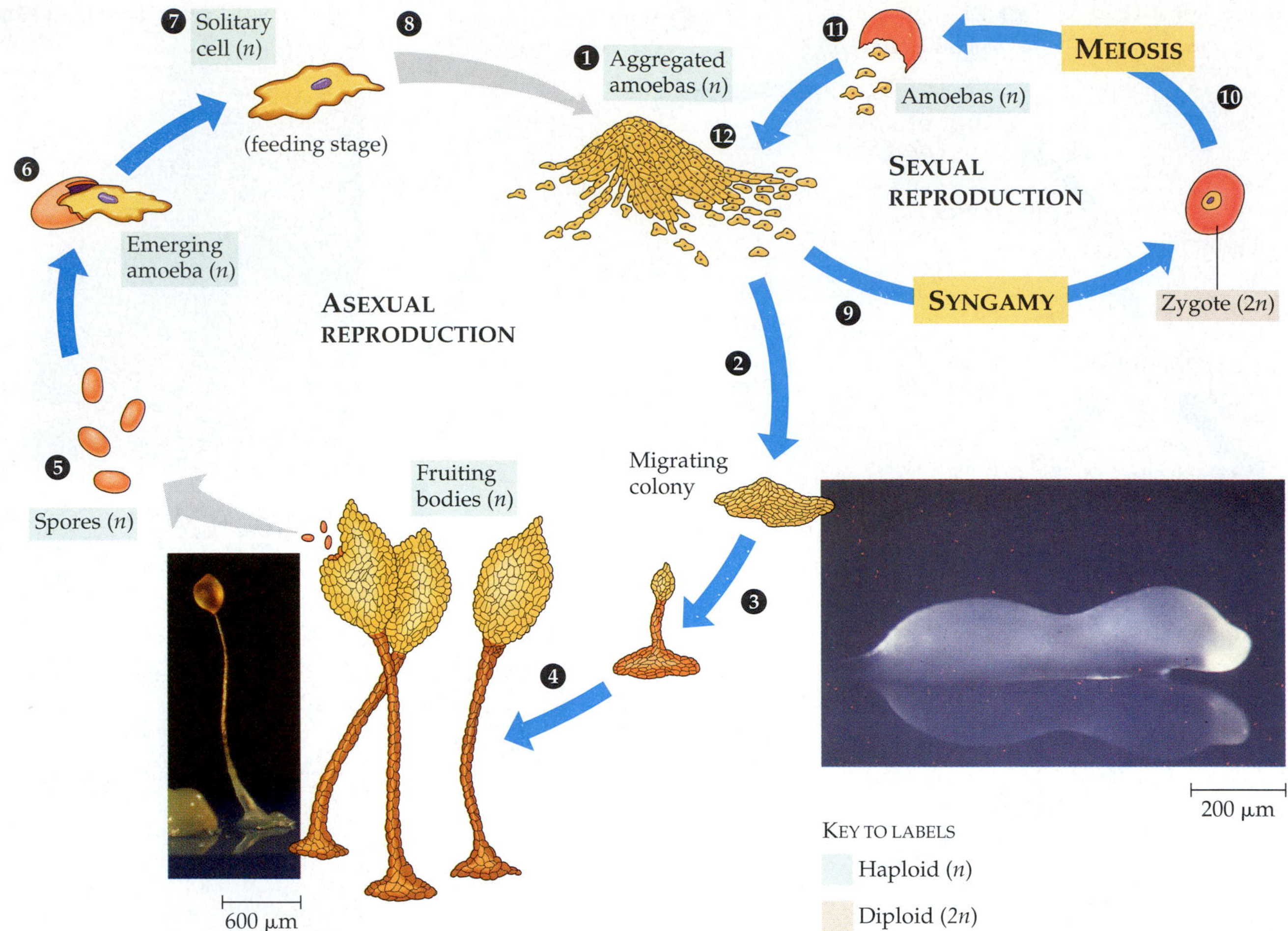

FIGURE 28.14 ▪ The life cycle of a cellular slime mold (*Dictyostelium*). ① The asexual portion of the life cycle includes an aggregate stage composed of a mass of amoeboid cells. The aggregate functions as a sluglike unit (right photo) that may ② migrate for a while before ③ settling down and developing stalked asexual fruiting bodies. As fruiting bodies form ④ some cells dry up and form a supportive stalk, while others continue to crawl up over the dried cells, amass, then turn into spores. Thus, a cluster of somewhat resistant spores forms at the tip of each fruiting body. ⑤ After the spores are released and exposed to a favorable environment, ⑥ amoeboid cells emerge from their protective coats and begin feeding. ⑦ The feeding stage of the life cycle consists of solitary cells that engulf bacteria while creeping through damp compost. ⑧ When food is depleted, the amoeboid cells migrate toward an aggregation center, where hundreds of the cells congregate in response to a chemical attractant they secrete. In the sexual phase of *Dictyostelium,* a pair of haploid amoebas fuse ⑨ to form a zygote, which is the only diploid stage in the life cycle. ⑩ A zygote becomes a giant cell by consuming surrounding haploid amoebas. The giant cell (not shown) then becomes surrounded by a resistant wall. The encysted giant cell undergoes meiosis, followed by several mitotic divisions. ⑪ Completing the sexual portion of the life cycle, new haploid amoebas are released when the cyst ruptures. These amoebae feed on bacteria and ⑫ may eventually form an aggregate as do the amoeboid cells resulting from asexual reproduction.

from endosymbionts that were eukaryotic cells, probably red algae. (The chloroplasts of red algae did begin as endosymbiotic cyanobacteria.) Thus, stramenopile chloroplasts may be derived from endosymbionts containing endosymbionts!

Diatoms (Bacillariophyta)

Yellow or brown in color, **diatoms** have unique glasslike walls consisting of hydrated silica embedded in an organic matrix (FIGURE 28.15). Each wall is in two parts that overlap like a shoe box and lid. Many diatoms are capable of a gliding movement apparently caused by physical interactions in the cytoplasm between the contractile protein actin and polysaccharide filaments; the interactions produce locomotor movements of the cell's outer membrane along a longitudinal slit in the cell wall.

Most of the year, diatoms reproduce asexually by mitotic cell divisions, with each daughter cell receiving half of the cell wall of its parent and regenerating a new second half. Cysts are formed by some species as resistant stages. Sexual stages are not common and involve the formation of eggs and sperm; sperm cells are amoeboid or have a single flagellum, depending on the species.

Both freshwater and marine plankton are rich in diatoms; a bucket of water scooped from the surface of the sea may have millions of these microscopic algae. In common with golden algae and brown algae, diatoms store food reserves in the form of a glucose polymer called laminarin. Some

(a)

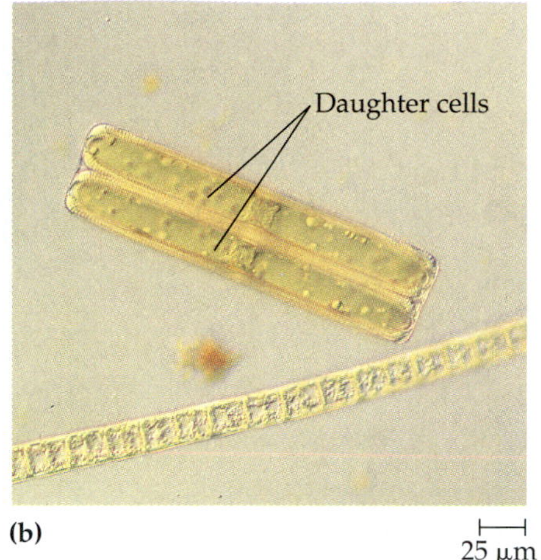

(b)

FIGURE 28.15 ▪ Diatoms. (a) The glasslike shells consist of two halves that fit like the bottom and lid of a shoe box. Tiny pores in the ornate shells allow for the exchange of gases and other substances between the cell and its environment. The shape of the shell and its pattern of pores are used for classifying diatoms. This species is *Navicula monilifera* (SEM). **(b)** In this side view of a species of *Pinnularia,* the cell has just divided by mitosis. Each daughter cell keeps half of the parent cell's wall and builds a new complementary half (LM).

diatoms also store food in the form of oil. Buoyancy maintained by cellular regulation of ions counteracts the relatively heavy weight of the walls and keeps diatoms near the water surface during daylight hours.

Massive accumulations of fossilized diatom walls are major constituents of the sediments known as diatomaceous earth, which is mined for its quality as a filtering medium and for many other uses.

Golden Algae (Chrysophyta)

Golden algae (Gr. *chrysos,* "golden") are named for their color, which results from yellow and brown carotene and xanthophyll accessory pigments. Their cells are typically biflagellated, with both flagella attached near one end of the cell. Many golden algae live among freshwater and marine plankton. Some species are mixotrophic, absorbing dissolved organic compounds or extending pseudopodia to ingest food particles and bacteria. Most golden algae are unicellular, but some, such as the freshwater genus *Dinobryon,* are colonial (FIGURE 28.16). If cell density reaches a certain high level, many species form resistant cysts that can remain viable for decades.

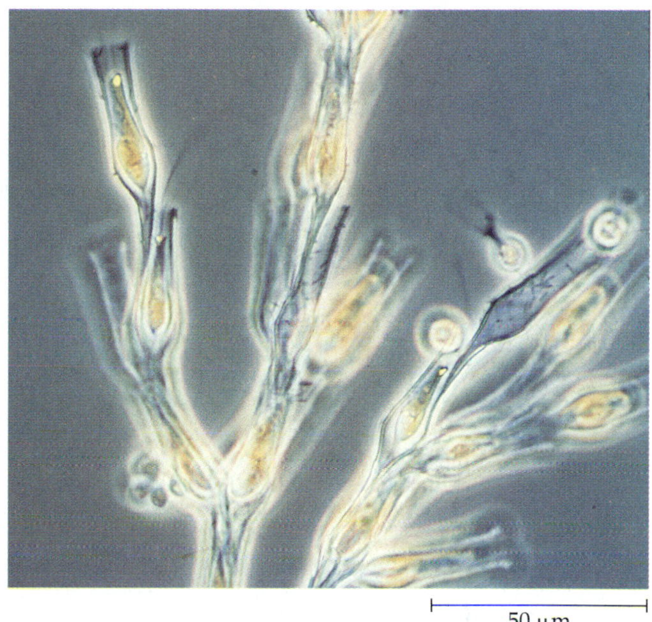

FIGURE 28.16 ▪ A golden alga. *Dinobryon* is a freshwater member of this algal group. This species is colonial, but the genus also contains unicellular species. Most golden algae are unicellular (LM).

Water Molds and Their Relatives (Oomycota)

Oomycotes are the **water molds**, **white rusts**, and **downy mildews**, all examples of heterotrophic stramenopiles that lack chloroplasts. Some of these organisms are unicellular; others consist of coenocytic hyphae (fine, branching filaments). Hyphae in this group are analogous to the branching filaments (also called hyphae) of true fungi. However, water molds and their relatives typically have cell walls made of cellulose, while the walls of true fungi are made of another polysaccharide, chitin. The diploid condition, which is reduced in true fungi, prevails in the life cycles of most members of Oomycota (FIGURE 28.17). Biflagellated cells occur in the life cycles of oomycotes, while almost all true fungi lack flagella.

Oomycota means "egg fungi," a reference to the mode of sexual reproduction in water molds. A relatively large egg cell is fertilized by a smaller "sperm nucleus," forming a resistant zygote.

Most water molds are decomposers that grow as cottony masses on dead algae and animals, mainly in fresh water. They are important decomposers in aquatic ecosystems. There are also parasitic water molds, such as those that grow on the skin and gills of fish in ponds or aquariums, but they usually attack only injured tissue.

White rusts and downy mildews generally live on land as parasites of plants. They are dispersed primarily by wind-blown spores, but they also form flagellated zoospores at some point during their life cycles. Among the most devastating plant pathogens are a downy mildew that threatened the French vineyards in the 1870s and a species of oomycote that causes late potato blight, which contributed to the Irish famine in the nineteenth century.

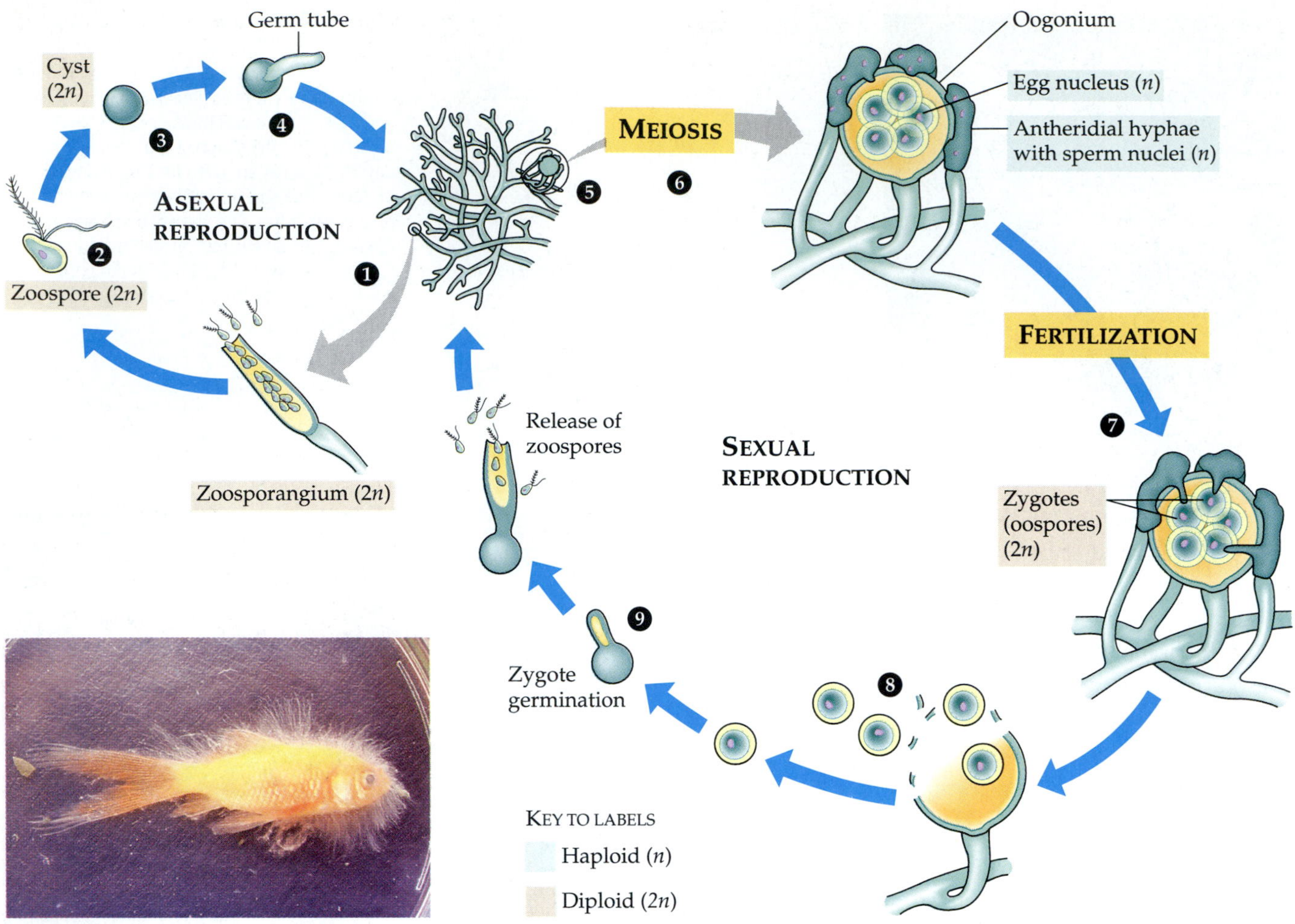

FIGURE 28.17 ▪ The life cycle of a water mold. Water molds commonly help decompose dead insects, fish, and other animals. ① The ends of the coenocytic hyphae form tubular zoosporangia, ② each of which produces about 30 biflagellated zoospores asexually (the hyphal mass on the goldfish in the inset is in the early zoosporangial stage). ③ Encysted zoospores land on a substrate and ④ germinate, growing into the tufted body of coenocytic hyphae. ⑤ Several days later, the organism begins to form sexual structures. ⑥ Meiosis produces eggs within structures called oogonia (singular, oogonium). On separate branches of the same or different individuals, meiosis produces several haploid "sperm nuclei" contained within compartments called antheridial hyphae. ⑦ These hyphae grow like hooks around the oogonium and deposit their nuclei through fertilization tubes that lead to the eggs. The resulting zygotes (oospores) may develop resistant walls but are also protected within the walls of the old oogonia. ⑧ After a period of dormancy, during which the oogonium wall usually disintegrates, ⑨ the zygotes germinate and form short hyphae tipped by zoosporangia, and the cycle is completed.

Brown Algae (Phaeophyta)

The largest and most complex algae are phaeophytes, or **brown algae** (Gr. *phaios,* "dusky," "brown"). All are multicellular, and most are marine. Brown algae are especially common along temperate coasts, where the water is cool. They owe their characteristic brown or olive color to accessory pigments in the chloroplasts. The chloroplast structure and pigment composition of brown algae are homologous to the photosynthetic equipment of golden algae and diatoms.

Many of the eukaryotes commonly called seaweeds are brown algae. *Seaweeds* are large marine algae, and the largest members of this important ecological group are brown algae. Two other groups, the red algae and green algae, also include seaweeds. We will examine these algal groups after we look at some of the adaptations, human uses, and life cycles of seaweeds.

Structural and biochemical adaptations help seaweeds survive and reproduce at the ocean's margins

Seaweeds, along with many animals and other heterotrophs these algae support, inhabit the intertidal and subtidal zones of coastal waters. The intertidal zone presents unique challenges to life. At times it is violently active, churned by waves and wind. Two times each day, at low tide, intertidal seaweeds are exposed to the drying atmosphere and rays of the sun. Twice each day, at high tide, the same seaweeds are submerged. Seaweeds have unique structural and biochemical adaptations that enable them to survive and thrive in this rough-and-tumble environment.

Seaweeds have the most complex multicellular anatomy of all algae. Some even have differentiated tissues and organs that resemble those we find in plants. The similarities evolved independently in the algal and plant lineages, and are thus analogous, not homologous. The term **thallus** (plural, **thalli**; Gr. *thallos,* "sprout") is used for a seaweed body that is plant-like but lacks true roots, stems, and leaves. A typical seaweed thallus consists of a rootlike **holdfast**, which anchors the alga, and a stemlike **stipe**, which supports leaflike **blades** (FIGURE 28.18). The blades provide most of the surface for photosynthesis. Some brown algae are equipped with floats, which keep the blades near the water surface. Beyond the intertidal zone in deeper waters live the giant seaweeds known as kelps (FIGURE 28.19). The stipes of these brown algae may be as long as 60 m.

In addition to these structural adaptations of thalli, some seaweeds are also endowed with biochemical adaptations to intertidal and subtidal conditions. For example, their cell walls are composed of cellulose and gel-forming polysaccharides, accounting for the slimy and rubbery feel of many seaweeds. These substances help cushion the thalli against the agitation of the waves. Some red algae in the lower intertidal and subtidal zones incorporate large amounts of calcium carbonate in the cell walls, making them virtually unpalatable to marine invertebrate grazers.

Seaweeds are also an important source of food and other commodities for humans. Coastal people, particularly in Asia, harvest seaweeds for food. For example, in Japan and Korea, the brown alga *Laminaria* is used in soups (Japanese "kombu"), and the red alga *Porphyra* is used to wrap sushi (Japanese "nori"). Marine algae are rich in iodine and other essential minerals, but much of their organic material consists of unusual polysaccharides that humans cannot digest, which prevents seaweeds from becoming staple foods. They are used mostly for their rich tastes and unusual textures. The gel-forming substances in their cell walls (algin in brown algae, agar and carageenan in red algae) are extracted in commercial operations. These substances are widely used in the manufacture of thickeners for such processed foods as puddings and salad dressing, and as lubricants in oil drilling. Agar is also used as the gel-forming base for microbiological culture media.

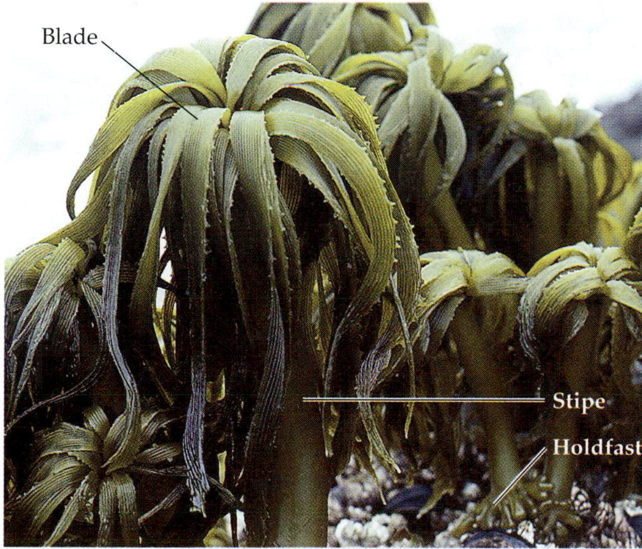

FIGURE 28.18 ▪ Seaweeds: adapted to life at the ocean's margins. The sea palm, *Postelsia,* lives on rocks in the crashing surf along the northwest coast of the United States and Canada. Its thallus is well adapted to maintaining a firm foothold in this extreme environment. *Postelsia* is a brown alga (Phylum Phaeophyta).

Some algae have life cycles with alternating multicellular haploid and diploid generations

A variety of life cycles has evolved among the multicellular brown, red, and green algae. The most complex cycles include an **alternation of generations**, the alternation of multicellular haploid forms and multicellular diploid forms. (Notice that haploid and diploid conditions alternate in *all* sexual life cycles—human gametes, for example, are a haploid stage—but the term *alternation of generations* is reserved for life cycles that include haploid and diploid stages that are both multicellular organisms.) As we will see in Chapter 29, alternation of generations is also a key feature in the life cycles of all plants.

We can examine the brown alga *Laminaria* as an example of a complex life cycle with an alternation of generations. The diploid individual is called the **sporophyte** because it produces

FIGURE 28.19 ▪ A kelp forest. The great kelp beds of temperate coastal waters provide habitat and food for a variety of organisms, including many fish caught by humans. The kelps, brown algae (Phaeophyta), are prodigiously productive. This brown alga, *Macrocystis,* common along the U.S. Pacific coast, grows to a length of more than 60 m in a single season, the fastest linear growth of any organism. Kelp is a renewable resource reaped by special boats that cut and collect the tops of the algae.

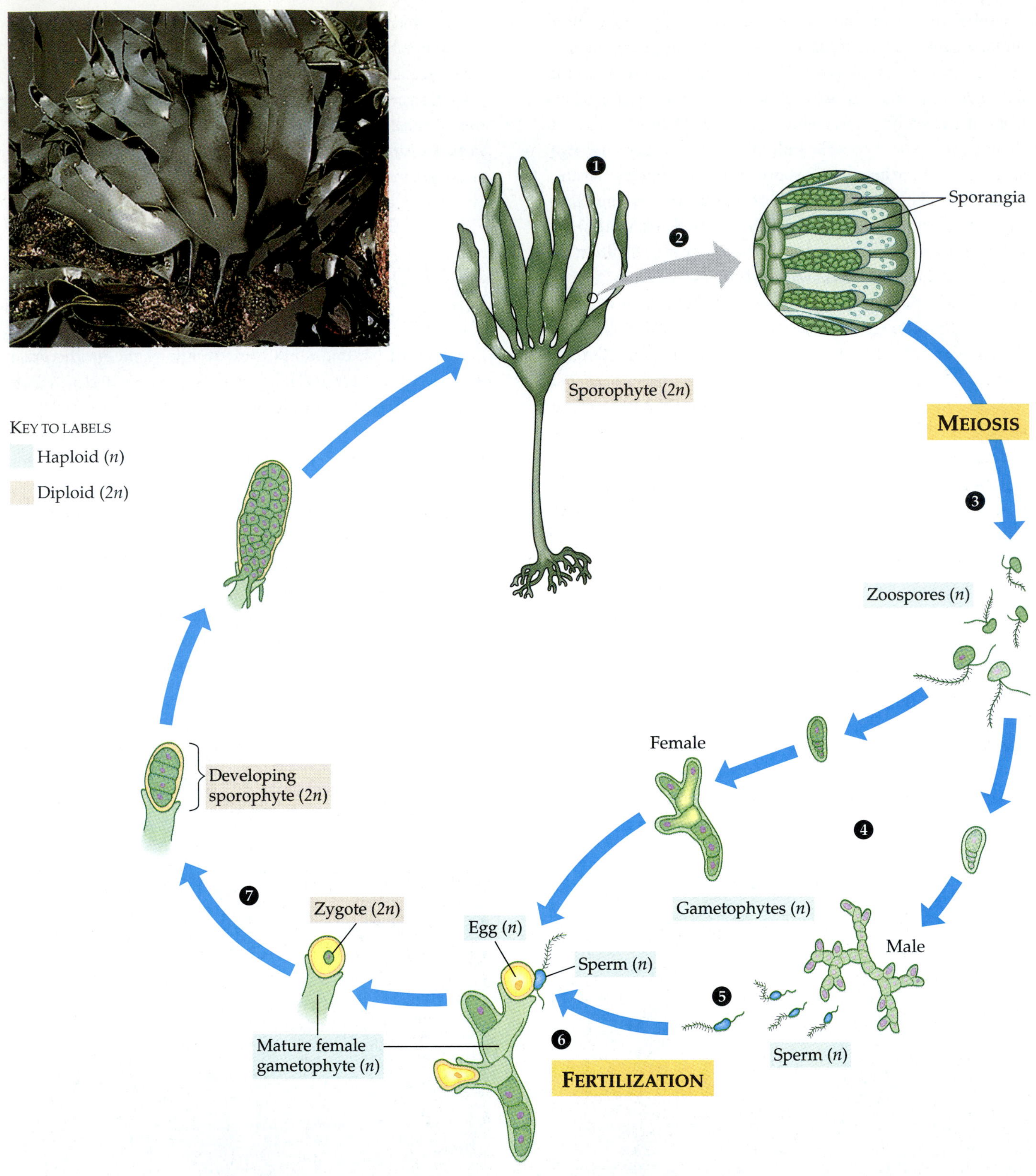

FIGURE 28.20 ▪ The life cycle of *Laminaria:* an example of heteromorphic alternation of generations. ① The sporophytes of this seaweed are usually found in water just below the line of lowest tides, attached to rocks by branching holdfasts. ② In early spring, at the end of the main growing season, cells on the surface of the blade develop into sporangia, which ③ produce zoospores by meiosis. ④ The zoospores are all structurally alike, but about half of them are capable of developing into a male gametophyte and half into a female gametophyte. The gametophytes look nothing like the sporophytes, being short, branched filaments that grow on the surface of subtidal rocks, often entangled in one another. ⑤ Male gametophytes release sperm and female gametophytes produce eggs, which remain attached to the gametophyte. Eggs secrete a chemical signal that attracts sperm of the same species, thereby increasing the probability of gametic union in the ocean. ⑥ Sperm fertilize the eggs, and ⑦ the zygotes grow into new sporophytes, starting life attached to the remains of the old female gametophyte. The *Laminaria* life cycle is an example of heteromorphic alternation of generations, in which sporophyte and gametophyte forms are noticeably different in appearance.

reproductive cells called spores (zoospores). The haploid individual is called the **gametophyte**, named for its production of gametes. Notice in FIGURE 28.20 that these two generations alternate—that is, they take turns producing one another. Spores released from the sporophyte develop into gametophytes, which in turn produce gametes. The union of two gametes (fertilization, or syngamy) results in a diploid zygote, which gives rise to a new sporophyte. In the case of *Laminaria,* the two generations are **heteromorphic**, meaning that the sporophytes and gametophytes are structurally different. Other algal life cycles have an alternation of **isomorphic** generations, meaning that the sporophytes and gametophytes look alike, although they differ in chromosome number. An example is the life cycle of *Ulva,* a green alga (see FIGURE 28.25).

Red algae (candidate kingdom Rhodophyta) lack flagella

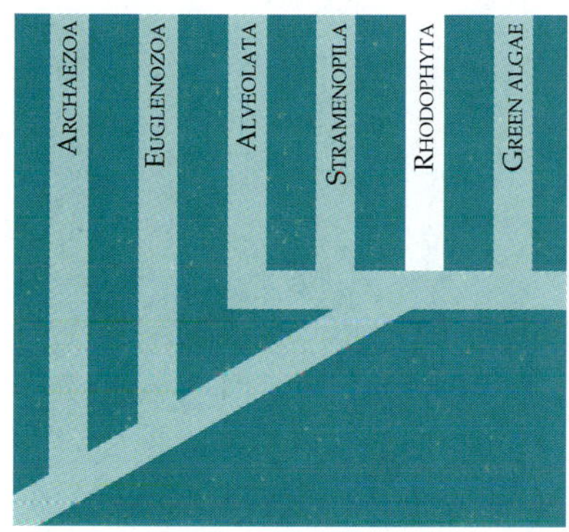

Unlike other eukaryotic algae, **red algae** (Rhodophyta; Gr. *rhodos,* "red") have no flagellated stages in their life cycle. Does this mean that their lineage arose before the origin of eukaryotic flagella? Nucleic acid sequencing indicates that red algae are not that ancient and that flagella were lost during their evolution. The red algae lineage seems to be about as old as that of the Stramenopila, and it is appropriate to recognize the Rhodophyta, a monophyletic taxon, as a candidate kingdom (see FIGURE 28.3).

Red algae are commonly reddish because of an accessory pigment called phycoerythrin. It belongs to a family of pigments known as phycobilins, found in red algae, cyanobacteria, and other organisms whose plastids were probably derived from red algae or cyanobacteria.

Despite their name, not all rhodophytes are red. Species adapted to different water depths differ in their proportions of accessory pigments. Rhodophytes may be almost black in deep water, bright red at more moderate depths, and greenish in very shallow water because less phycoerythrin masks the green of chlorophyll. Some species lack pigmentation altogether and function heterotrophically as parasites on other red algae.

Red algae are most abundant in the warm coastal waters of tropical oceans, but there are also some freshwater and soil species. The phycobilins and other accessory pigments allow some species to absorb filtered wavelengths (blues and greens) in deep water. A species of red alga has recently been discovered living near the Bahamas at a depth of more than 260 m.

Most red algae are multicellular, and the largest share the designation "seaweeds" with the brown algae, although none of the reds are as big as the giant browns (kelps). The thalli of many red algae are filamentous, often branched and interwoven in delicate lacy patterns (FIGURE 28.21). The base of the thallus is usually differentiated as a simple holdfast.

Life cycles are especially diverse among the red algae. Lacking flagella, the gametes rely on water currents to get together. Alternation of generations is common in red algae.

(a)

(b)

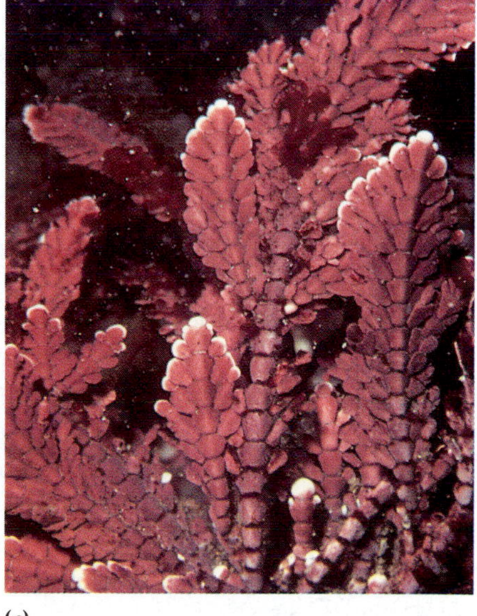
(c)

FIGURE 28.21 ▪ Red algae. (a) Dulse (*Palmaria*), an edible species with a "leafy" form. **(b)** *Polysiphonia,* a filamentous red alga. **(c)** A coralline alga, one of several species whose cell walls are hardened by calcium carbonate. Some coralline algae are members of the biological communities called coral reefs.

Green algae and plants probably had a common photoautotrophic ancestor

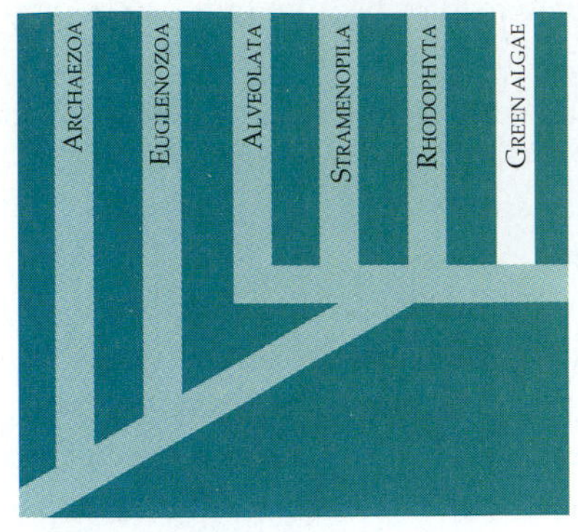

Green algae (Chlorophyta) are named for their grass-green chloroplasts (Gr. *chloros,* "green"), which are much like those of organisms we traditionally call plants in ultrastructure and pigment composition. Molecular systematics and cellular morphology leave little doubt that green algae and plants are closely related; in fact, some systematists now advocate inclusion of green algae in the plant kingdom (see FIGURE 28.3). Strong evidence supports the hypothesis that green algae and plants were derived from a common ancestor different from ancestors of the stramenopiles and the red algae. This common ancestor was probably a photosynthetic autotroph that arose by an endosymbiotic association between a flagellated, heterotrophic eukaryote and a cyanobacterium.

More than 7000 species of green algae have been identified. Most live in fresh water, but there are also many marine species. Various species of unicellular green algae live as plankton or inhabit damp soil or snow; some species live symbiotically within other eukaryotes, contributing portions of their photosynthetic products to the food supply of the hosts (see FIGURE 33.5d). Chlorophytes are among the algae that live symbiotically with fungi in the mutualistic collectives known as **lichens** (see FIGURE 31.14).

The simplest chlorophytes are biflagellated unicells such as *Chlamydomonas,* which resemble the gametes of more complex green algae. In addition to unicellular chlorophytes, there are colonial species, as well as many filamentous forms that contribute to the stringy masses known as pond scum. There are even some truly multicellular chlorophytes, with bodies large and complex enough that marine species qualify as seaweeds along with the large brown algae and red algae.

Three separate evolutionary trends have probably produced the diverse forms of colonial and multicellular chlorophytes from unicellular, flagellated ancestors. Larger size and greater complexity have evolved by: (1) the formation of colonies of individual cells, as seen in species of *Volvox* (FIGURE 28.22a); (2) the repeated division of nuclei with no cytoplasmic division, as seen in multinucleate filaments of *Caulerpa* (FIGURE 28.22b); and (3) the formation of true multicellular forms, as in *Ulva* (FIGURE 28.22c).

Most green algae have complex life histories, with both sexual and asexual reproductive stages. Nearly all reproduce sexually by way of biflagellated gametes having cup-shaped

(a)

(b)

(c)

FIGURE 28.22 ▪ Colonial and multicellular green algae. (a) Species of *Volvox* are colonial chlorophytes that inhabit fresh water. The colony is a hollow ball, with its wall composed of hundreds or thousands of biflagellated cells embedded in a gelatinous matrix. The cells are usually connected by strands of cytoplasm; if isolated, these cells cannot reproduce. The large colonies seen here will eventually release the small "daughter" colonies within them (LM). **(b)** Species of *Caulerpa* are found in the marine intertidal zone. The branched filaments of these green algae lack cross-walls and thus are multinucleate. In effect, the thallus is one huge "supercell." **(c)** *Ulva,* or sea lettuce, is an edible seaweed with a thallus differentiated into leaflike blades and a rootlike holdfast that anchors the alga against turbulent waves and tides. The thallus is truly multicellular, consisting of specialized cells organized into tissues.

chloroplasts. The exceptions are the conjugating algae, such as *Spirogyra* (FIGURE 28.23), which produce amoeboid gametes.

The life cycle of *Chlamydomonas* includes asexual and sexual stages and is believed to be primitive, meaning that it evolved early in the chlorophyte lineage (FIGURE 28.24). Except for the zygote, which is diploid, all life cycle stages are haploid. The gametes of *Chlamydomonas* are structurally indistinguishable, and their fusion is known as **isogamy**, which literally means a "marriage of equals."

Extrapolating from the basic features of the *Chlamydomonas* life cycle, we can identify many refinements that have evolved among chlorophytes. Some green algae, for instance, produce gametes that differ structurally from vegetative cells, and in some species the male and female gametes differ in size

FIGURE 28.23 ▪ *Spirogyra*, a conjugating alga. The cells of *Spirogyra* are arranged in long, unbranched filaments. Each cell contains one or more large, spiral-shaped chloroplasts. The filaments grow in length and reproduce asexually by fragmentation as the cells divide by mitosis. During sexual reproduction, as seen here, cells of adjacent mating filaments join by conjugation tubes (LM).

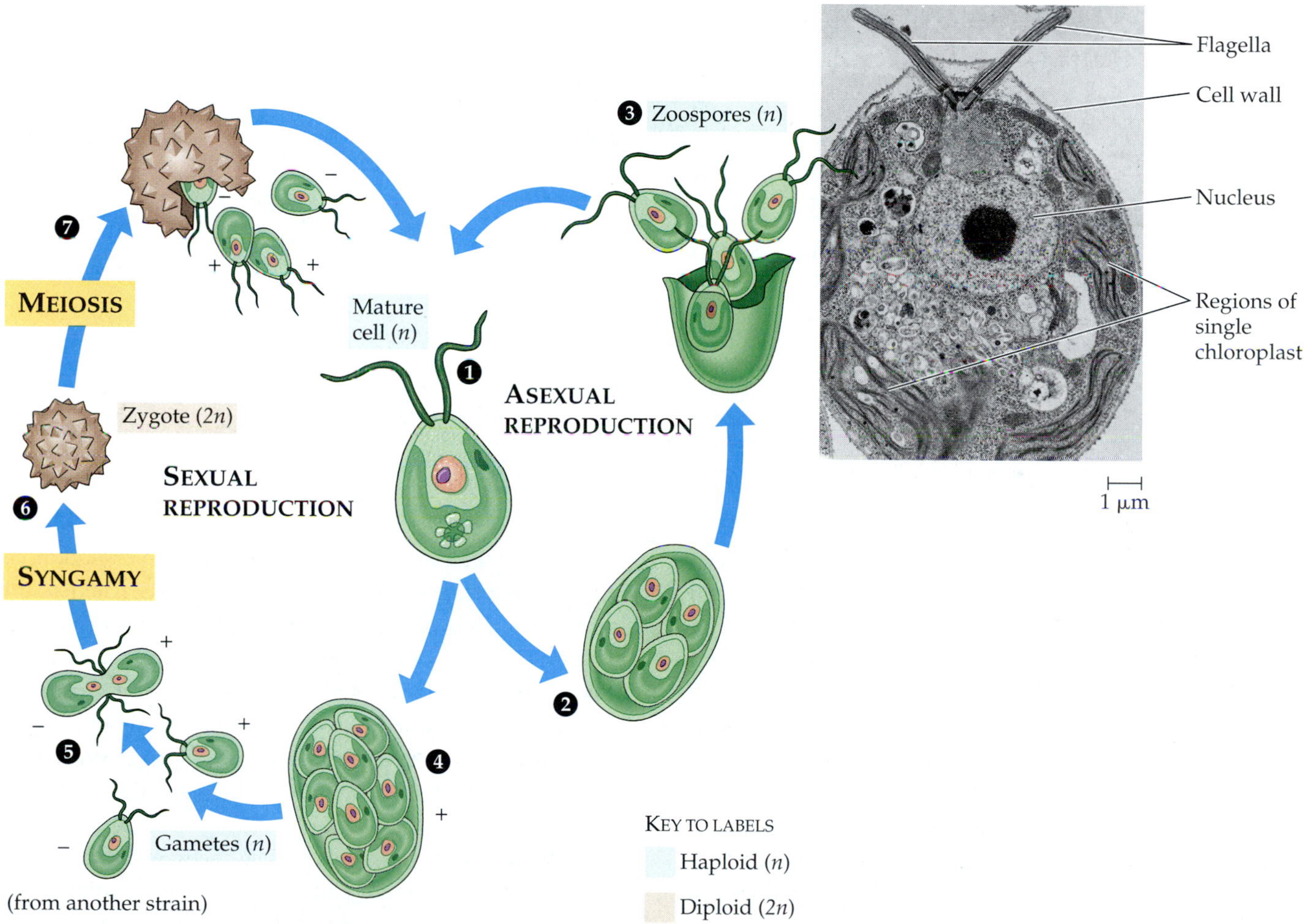

FIGURE 28.24 ▪ The life cycle of *Chlamydomonas.* This unicellular green alga exhibits sexual as well as asexual reproduction. The gametes are structurally identical, a condition known as isogamy. ① A mature cell of *Chlamydomonas* is haploid. ② When a mature cell reproduces asexually, it resorbs its flagella and then divides twice by mitosis, forming four cells (more in some species). ③ These daughter cells develop flagella and cell walls and then emerge as swimming zoospores from the wall of the parent cell that had enclosed them. The zoospores grow into mature haploid cells, completing the asexual life cycle. The inset shows a mature cell before reproduction; each *Chlamydomonas* cell contains only one chloroplast (TEM). Sexual reproduction is triggered by a shortage of nutrients, drying of the pond, or some other stress. ④ Within the wall of the parent cell, mitosis produces many haploid gametes. ⑤ After their release, gametes from opposite mating types (designated + and −) pair off and cling together. Fusion (syngamy) of the isogametes occurs slowly, forming a diploid zygote ⑥, which secretes a durable coat that protects the cell against harsh conditions. ⑦ When the zygote breaks dormancy, meiosis produces four haploid individuals (two of each mating type) that emerge from the coat and grow into mature cells.

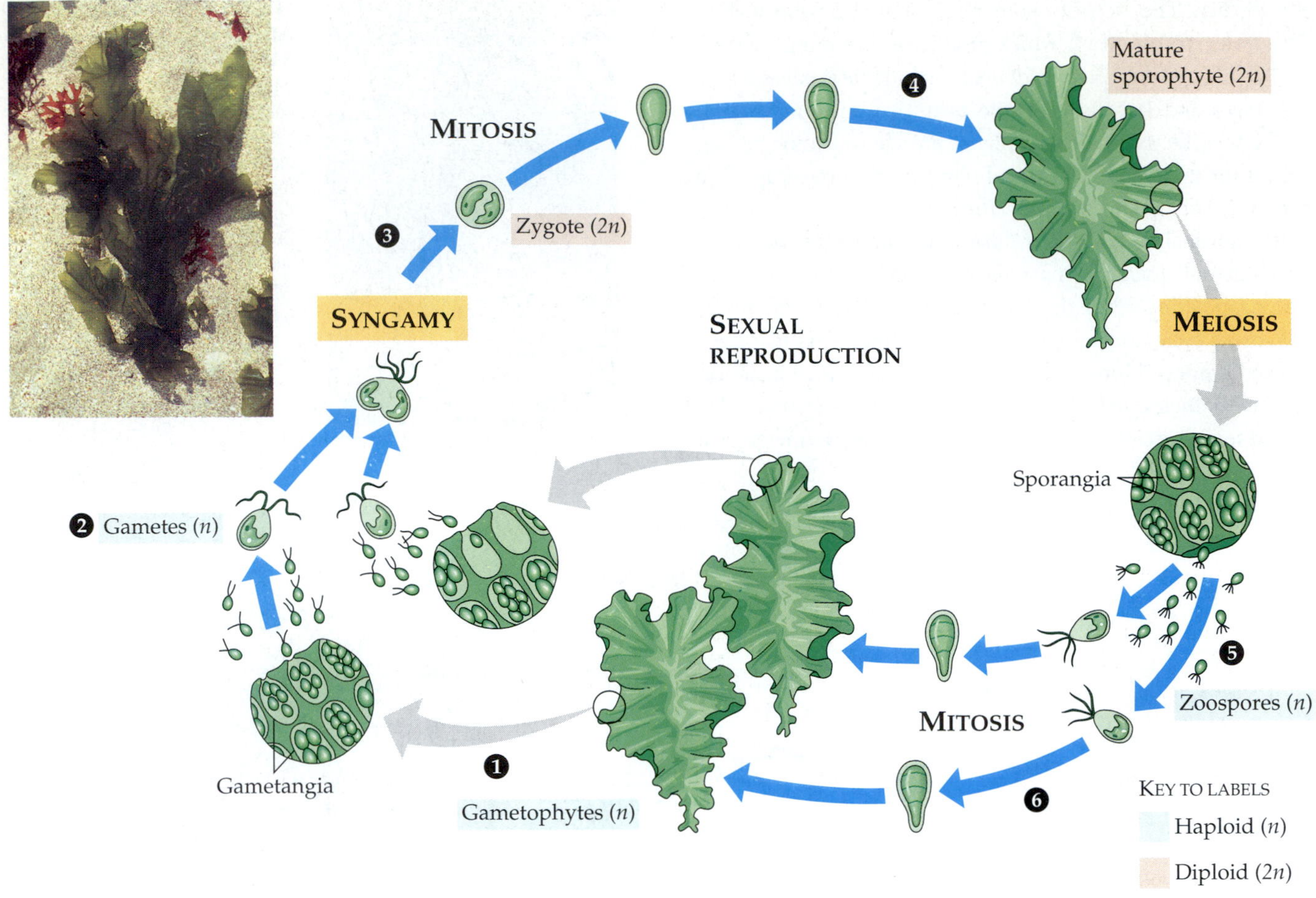

FIGURE 28.25 ▪ The life cycle of *Ulva*: an example of isomorphic alternation of generations. The haploid, sexual generation (gametophyte) and the diploid, asexual generation (sporophyte), are identical in appearance (isomorphic). ① Gametophytes produce ② gametes, which form ③ zygotes by syngamy. ④ Zygotes develop into sporophytes, which develop sporangia. ⑤ The sporangia produce reproductive cells called zoospores. ⑥ These cells develop directly into gametophytes. Compare the life cycle of *Ulva* to *Laminaria*, which illustrates heteromorphic alternation of generations (see FIGURE 28.20).

or shape, and their fusion is called **anisogamy**. Many species exhibit a type of anisogamy called **oogamy**, a flagellated sperm fertilizing a nonmotile egg. Another refinement is an alternation of generations, with alternate haploid and diploid individuals producing each other in turn. An example is the life cycle of *Ulva*, in which the sporophyte (diploid thallus) and gametophyte (haploid thallus) are structurally identical, or isomorphic (FIGURE 28.25).

This completes our survey of diversity and phylogenetic relationships among the eukaryotes traditionally classified as protists. TABLE 28.2 summarizes the candidate kingdoms we have studied. We complete this chapter with a brief discussion of the origins of multicellularity.

Multicellularity originated independently many times

Put simply, more variations are possible for complex structures than for simpler ones. Thus, the origin of eukaryotes ignited an explosion of biological diversification, and unicellular eukaryotes are much more diverse in structure than the simpler prokaryotes. The evolution of multicellular bodies broke through another threshold in structural organization and became the stock for new waves of adaptive radiations. Multicellularity evolved several times among early eukaryotes, with the diverse multicellular algae as examples of the results. The evolution of multicellularity also gave rise to plants, fungi, and animals.

A widely held view is that links between multicellular organisms and their unicellular ancestors were colonies, or loose aggregates of interconnected cells (FIGURE 28.26). Multicellular algae, plants, fungi, and animals probably descended from several lineages of colonial eukaryotes that formed by amalgamations of individual cells. The evolution of multicellularity from colonial aggregates involved increasing cellular specialization and division of labor. Initially, in colonial aggregates ancestral to multicellular algae, plants, and animals, all cells may have been motile with flagella. As cells in the colony tended to become increasingly interdependent, some of the

Table 28.2 ▪ Summary of the Candidate Kingdoms of Protists

CANDIDATE KINGDOM (WITH SUBGROUPS)	BRIEF DESCRIPTION
Archaezoa (p. 524)	Mostly parasitic; lack mitochondria
Diplomonads (e.g., *Giardia*) Trichomonads Microsporidians	
Euglenozoa (p. 526)	A monophyletic group of flagellates
Euglenoids (e.g., *Euglena*)	Autotrophic; flagella arise from an anterior pocket
Kinetoplastids (e.g., *Trypanosoma*)	Parasites with a single mitochondrion and associated extranuclear DNA
Alveolata (p. 527)	Cells with subsurface cavities (alveoli)
Dinoflagellates (e.g., *Pfiesteria*)	Flagella arise from grooves formed by internal cellulose plates
Apicomplexans (e.g., *Plasmodium*)	Parasites whose infectious stages have an apical complex of organelles that penetrate the host cell
Ciliates (e.g., *Paramecium*)	Have two types of nuclei; use cilia for motility and feeding
Stramenopila (p. 533)	Unicellular and multicellular; cells often have flagella of unequal length; long flagella usually have rows of fine, hairlike projections; chloroplasts probably derived from endosymbiotic eukaryotic cells
Diatoms (bacillariophytes)	Freshwater and marine unicellular photoautotrophs with two-part siliceous cell walls
Golden algae (chrysophytes)	Mostly freshwater unicells; some colonial; cell walls pectic/siliceous; most are photoautotrophic, some are mixotrophic
Brown algae (phaeophytes)	Mostly marine, multicellular; cell walls of cellulose and other polysaccharides; photoautotrophic
Water molds, white rusts, downy mildews (oomycotes)	Unicellular or coenocytic heterotrophs (decomposers and parasites); freshwater and terrestrial; cell walls made of cellulose
Rhodophyta (p. 539)	Red algae; most are multicellular photoautotrophs; cell walls of cellulose and other polysaccharides; no flagella; some heterotrophic parasites
Chlorophyta* (p. 540)	Green algae; mostly freshwater, some marine and terrestrial; unicellular, colonial, and multicellular photoautotrophs; cell walls of cellulose; free-living and symbiotic

*Candidate member of Kingdom Plantae.

cells may have lost their flagella and become more proficient in performing functions other than locomotion.

Another early form of division of labor probably involved the distinction of sex cells (gametes) from somatic (nonreproductive) cells. We see this type of specialization and intercellular cooperation today in several colonial organisms, such as the green alga *Volvox* (see FIGURE 28.22a). Gametes are specialized for reproduction, and they depend on somatic cells while developing. One of the events involved in the formation of gametes may have been the evolution of the enzyme telomerase, which catalyzes the lengthening of special nucleotide sequences (telomeres) on DNA (Chapter 19). Present at the ends of DNA molecules in eukaryotes but not on the circular DNA of prokaryotes, telomeres protect the organism's genes from erosion during DNA replication. Active in the cells that give rise to sperm and eggs but not in most somatic cells, telomerase endows the gametes (and thus the somatic cells of the next generation) with an adequate supply of telomeres.

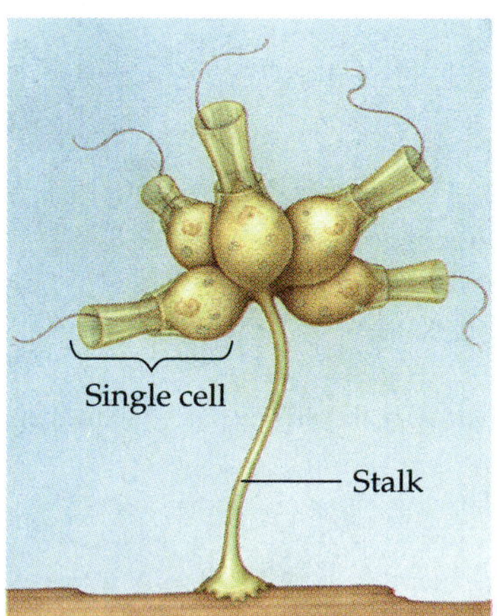

FIGURE 28.26 ▪ The colonial connection between unicellular and multicellular life. The organism represented here is a choanoflagellate, a colonial protist that many zoologists believe is related to the ancestor of animals. More generally, colonial organisms of diverse kinds may have been intermediates in the many origins of multicellular life.

Evolution of the extensive division of labor required to perform all the nonreproductive functions in multicellular organisms involved many additional steps in somatic cell specialization. In seaweeds, for example, there is extensive division of labor among the different tissues that make up the thalli. Multicellular forms of life more complex than these algae did not appear until about 700 million years ago, during the twilight of the Precambrian era. A variety of animal fossils has been found in late Precambrian strata, and many new forms evolved after the Paleozoic era dawned with the Cambrian period, about 570 million years ago. Seaweeds and other complex algae were also abundant in Cambrian oceans and lakes. The land, however, was barren. About 440 million years ago, certain green algae living along the edges of lakes gave rise to primitive plants. Strong evidence supports the hypothesis that land plants evolved from a particular group of green algae. In the next chapter we will trace the long evolutionary trek of plants onto land.

CHAPTER REVIEW

REVIEW OF KEY CONCEPTS

(with page numbers and key figures)

INTRODUCTION TO THE PROTISTS

Diverse unicellular eukaryotes (protists) predate plants, animals, and fungi by over a billion years.

- **Protists are the most diverse of all eukaryotes (pp. 520–522, FIGURE 28.1)** All protists are eukaryotes, and the diversity of protists represents different "experiments" in the evolution of eukaryotic organization. Most protists are unicellular, but colonial and simple multicellular forms also exist. Protists are found wherever there is water, living as plankton, submerged bottom-dwellers, or inhabitants of moist soil or the body fluids of other organisms. Of all eukaryotes, protists are the most nutritionally diverse; photoautotrophs, heterotrophs, and mixotrophs are all represented. Protists exhibit the most diverse spectrum of structure and life cycles of all known organisms.
- **Symbiosis was involved in the genesis of eukaryotes from prokaryotes (pp. 522–524, FIGURE 28.2)** The endomembrane system of eukaryotes may have evolved from specialized infoldings of the plasma membrane of ancestral prokaryotes. Chloroplasts and mitochondria are descendants of cyanobacteria and aerobic, heterotrophic prokaryotes, respectively, that took up residence within evolving eukaryote cells.

PROTIST SYSTEMATICS AND PHYLOGENY

28.1 **Monophyletic taxa are emerging from modern research in protist systematics (p. 524, FIGURE 28.3)** General agreement among biologists that classifying all eukaryotes traditionally called protists in a single kingdom does not reflect phylogeny has prompted alternative proposals, such as an eight-kingdom system. Monophyletic taxa are beginning to emerge from current research and debate. Candidate kingdoms Euglenozoa, Alveolata, Stramenopila, and Rhodophyta are monophyletic.

- **Members of candidate kingdom Archaezoa lack mitochondria and may represent early eukaryotic lineages (pp. 524–526, FIGURE 28.4, TABLE 28.2)** *Giardia* and other archaezoans, which lack mitochondria, seem to represent the most ancient eukaryotic lineages. Organisms in these lineages may never have had mitochondria or may have lost these organelles in the course of evolution.
- **Candidate kingdom Euglenozoa includes both autotrophic and heterotrophic flagellates (p. 526, FIGURES 28.1, 28.5, TABLE 28.2)** Euglenozoans include the euglenoids (e.g., *Euglena*), characterized by an anterior pocket housing the base of flagella, and the kinetoplastids, parasites with a single large mitochondrion and an associated DNA-containing organelle.
- **Subsurface cavities (alveoli) are diagnostic of candidate kingdom Alveolata (pp. 527–530, FIGURES 28.6–28.8, TABLE 28.2)** Alveolates include dinoflagellates that move in a spinning motion by the beating of flagella; apicomplexans, parasites with complex life cycles characterized by both sexual and asexual stages that often require two or more host species; and ciliates, which use cilia to move and feed, and are among the most complex of cells.
- **A diverse assemblage of unicellular eukaryotes move by means of pseudopodia (pp. 530–532, FIGURES 28.10–28.12, TABLE 28.1)** Rhizopods (unicellular amoebas and their relatives) move by cellular extensions called pseudopodia. Actinopods (heliozoans and radiolarians) have slender, raylike pseudopodia called axopodia that help them float and feed. The mostly marine forams have porous shells through which strands of cytoplasm extend.

28.2 **Slime molds have structural adaptations and life cycles that enhance their ecological role as decomposers (p. 532, FIGURES 28.13, 28.14, TABLE 28.1)** Plasmodial slime molds feed by means of an amoeboid plasmodium capable of differentiating into sexually reproducing sporangia when moisture or food is scarce. Cellular slime molds are haploid organisms with life cycles that include a multicellular amoeboid stage that erects asexual fruiting bodies.

- **Diatoms, golden algae, brown algae, and water molds are members of candidate kingdom Stramenopila (pp. 533–536, FIGURES 28.15–28.17, TABLE 28.2)** A diverse group of photoautotrophs and heterotrophs, the stramenopiles are united by molecular systematics, flagella bearing hairlike projections, and chloroplasts that were probably derived from endosymbiotic eukaryotic cells. Diatoms (bacillariophytes) are primarily unicellular organisms with unique glasslike walls of silica. Golden algae (chrysophytes) are biflagellated freshwater and marine plankton. Oomycotes, (water molds, white rusts, and downy mildews) are heterotrophs with cell walls of cellulose and biflagellated stages in their life cycles. Brown algae (phaeophytes), which are multicellular and primarily marine, include kelps. Most species show some type of alternation of generations.
- **Structural and biochemical adaptations help seaweeds survive and reproduce at the ocean's margins (pp. 536–537, FIGURES 28.18, 28.19)** Seaweeds include thallus-forming marine species among the brown, red, and green algae. They are well adapted to life along the turbulent margins of the oceans.
- **Some algae have life cycles with alternating multicellular haploid and diploid generations (pp. 537–539, FIGURE 28.20)**
- **Red algae (candidate kingdom Rhodophyta) lack flagella (p. 539, FIGURE 28.21, TABLE 28.2)** Mostly multicellular and mostly marine, red algae possess the red accessory pigment phycoerythrin.
- **Green algae and plants probably had a common photoautotrophic ancestor (pp. 540–542, FIGURES 28.22–28.25, TABLE 28.2)** The common ancestor of green algae probably arose from an endosymbiotic association between a flagellated, heterotrophic eukaryote and a cyanobacterium.
- **Multicellularity originated independently many times (pp. 542–543, FIGURE 28.26)**

SELF-QUIZ

1. Which of the following terms is *not* applicable to the life cycle of *Laminaria*?
   a. isogamy
   b. heteromorphy
   c. oogamy
   d. gametophyte
   e. multicellular
2. Which of the following organisms are *incorrectly* paired with their description?
   a. rhizopods—naked and shelled amoebas
   b. actinopods—planktonic with slender, raylike axopodia
   c. forams—flagellated algae, free-living or symbiotic
   d. apicomplexans—parasites with complex life cycles
   e. ciliates—complex, unicellular organisms with macronucleus and micronuclei
3. Which of the following organisms are *incorrectly* paired with their description?
   a. dinoflagellates—marine plankton; whirling, spinning movement; characteristic wall
   b. chrysophytes—golden algae; xanthophylls predominant; flagellated, freshwater plankton

c. bacillariophytes—diatoms; two-piece shells of silica
d. phaeophytes—multicellular brown algae, seaweeds
e. rhodophytes—cause red tides

4. Which of the following is *not* characteristic of the Rhodophyta?
   a. flagellated gametes
   b. syngamy
   c. meiosis
   d. heteromorphic generations
   e. plastids containing phycobilin

5. Biologists suspect that endosymbiosis gave rise to mitochondria before chloroplasts because
   a. the products of photosynthesis could not be metabolized without mitochondrial enzymes
   b. almost all eukaryotes have mitochondria, while only autotrophic eukaryotes have chloroplasts
   c. mitochondrial DNA is less similar to prokaryotic DNA than is the DNA from chloroplasts
   d. without mitochrondrial CO_2 production, photosynthesis could not occur
   e. chloroplasts utilize indigenous ribosomes, while mitochondrial proteins are synthesized in the cytosol

6. Which of the following characteristics provides support for combining the dinoflagellates, apicomplexans, and ciliates in the monophyletic kingdom Alveolata?
   a. Their flagella or cilia are organized with the 9 + 2 microtubular ultrastructure.
   b. All are pathogenic.
   c. All are found exclusively in freshwater or marine habitats.
   d. All possess either mitochondria or carry mitochondrial genes within their genome.
   e. The three groups have small membrane-bound alveoli under their cell surfaces.

7. Which of the following is an *incorrect* statement about the possible endosymbiotic origins of chloroplasts and mitochondria?
   a. They are the appropriate size to be descendants of bacteria.
   b. They contain their own genome and produce all their own proteins.
   c. They contain circular DNA molecules not associated with histones.
   d. Their membranes have enzymes and transport systems that resemble those found in the plasma membranes of prokaryotes.
   e. Their ribosomes are more similar to those of bacteria than to those of eukaryotes.

8. The organism that contributed to the Irish potato famine is
   a. an actinopod
   b. a ciliate
   c. an oomycote
   d. a plasmodial slime mold
   e. a cellular slime mold

9. Many multicellular chlorophytes are isomorphic; what distinguishes the sporophyte from the gametophyte?
   a. Mitosis can be observed only within the sporophyte.
   b. Both mitosis and meiosis can be observed within the gametophyte.
   c. The chromosome number differs.
   d. The sporophyte produces zoospores that fuse during syngamy to produce the gametophyte.
   e. The sporangia and gametangia are morphologically distinct.

10. Which of the following groups probably represents the earliest branch in the evolution of eukaryotes?
    a. archaea
    b. archaezoa
    c. fungi
    d. amoebas
    e. diatoms

CHALLENGE QUESTION

Write a short essay that contrasts the ways the five-kingdom and three-domain systems classify the eukaryotes included in the candidate kingdoms discussed in this chapter.

SCIENCE, TECHNOLOGY, AND SOCIETY

Different genes are expressed at different stages of the life cycle of the malaria-causing apicomplexan *Plasmodium*, causing different proteins to appear on the outer coat of the infecting cells. The continual sloughing and replacing of the protein coat prevents destruction of the parasites by host antibodies. This ability to evade the immune system is one reason that it is difficult to develop a malaria vaccine. An additional problem is that fewer scientists are engaged in malaria research, and less money is spent on it, than on diseases (such as cystic fibrosis) that affect far fewer people than does malaria. What are the possible reasons for this imbalance in research effort?

FURTHER READING

Fenchel, T., and B. J. Finlay. "The Evolution of Life Without Oxygen." *American Scientist*, January/February 1994. Reveals clues to the origins of eukaryotes.

Jacobs, W. P. "*Caulerpa*." *Scientific American*, December 1994. This tropical alga may be the largest single-celled organism.

Knoll, A. "The Early Evolution of Eukaryotes: A Geological Perspective." *Science*, May 1, 1992. What can Precambrian rocks tell us about the origin of eukaryotic cells?

Macilwain, C. "Scientists Close in on 'Cell From Hell' Lurking in Chesapeake Bay." *Nature*, September 25, 1997. A brief, informative article about the deadly dinoflagellate *Pfiesteria*.

Monastersky, R. "The Rise of Life on Earth." *National Geographic*, March 1998. A superbly illustrated overview of prokaryote and eukaryote evolution.

Palmer, J. D. "Organelle Genomes: Going, Going, Gone?" *Science*, February 7, 1997. Details the interesting relationship between mitochondria and hydrogenosomes (energy-generating organelles in anaerobic protists).

Radmer, R. J. "Algal Diversity and Commercial Algal Products." *BioScience*, April 1996.

Vogel, G. "Parasites Shed Light on Cellular Evolution." *Science*, March 7, 1997. Presents evidence that apicomplexans carry a plastid probably derived from a green alga.

WEB LINKS

Visit the special edition of *The Biology Place* for BIOLOGY, Fifth Edition, at http://www.biology.com/campbell. Go to Chapter 28 for online resources, including learning activities, practice exams, and links to the following web sites:

"Introduction to the Eukaryota"
From this starting point at the UCMP you can navigate to information on various groups of protists.

"The Protist Image Database (PID)"
The intent of the PID is to provide up-to-date on-line information on the morphology, taxonomy, and phylogenetic relationships of protists.

"Welcome to the Cambrian"
This web site graphically illustrates early eukaryotic diversity.

"The Protist Information Server"
This Japanese information server contains more than 6700 images of protists.

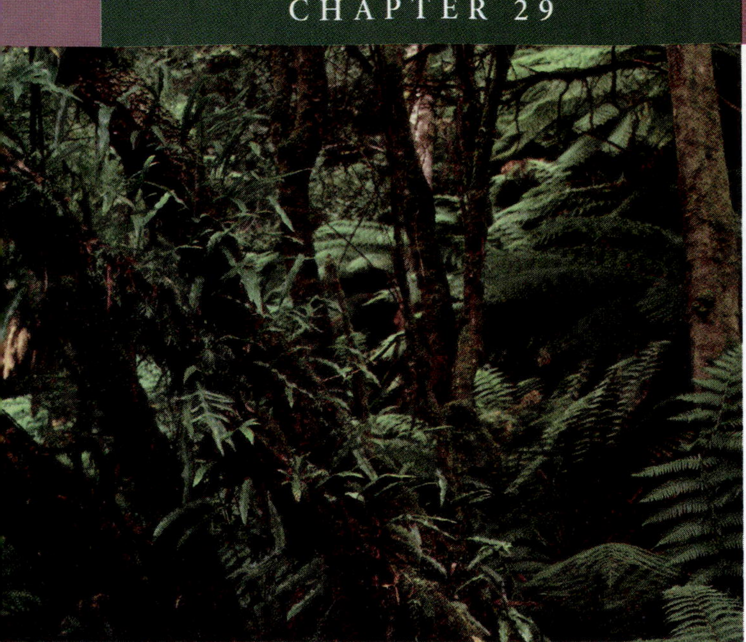

PLANT DIVERSITY I: THE COLONIZATION OF LAND

When you view a lush landscape, such as the Tasmanian forest scene in the photo on this page, it is difficult to picture the land barren, totally uninhabited by macroscopic life. But that is how we must imagine Earth for almost the first 90% of the time life has existed. Life was cradled in the seas and ponds, and there it evolved in confinement for 3 billion years. Paleobiologists have recently discovered fossils of cyanobacteria that may have coated moist soil about 1.2 billion years ago, but the long evolutionary pilgrimage of more complex organisms onto land did not begin until about 475 million years ago. The terrestrial communities founded by plants transformed the biosphere. Consider, for example, that humans would not exist had it not been for the chain of evolutionary events that began when certain descendants of green algae first colonized land.

The evolutionary history of the plant kingdom is a story of adaptation to changing terrestrial conditions. That is the historical context in which this chapter and the next survey the current diversity of plants and trace their origins.

AN OVERVIEW OF PLANT EVOLUTION

Structural, chemical, and reproductive adaptations enabled plants to colonize land

All plants, as defined in this book, are multicellular eukaryotes that are photosynthetic autotrophs. However, not all organisms with these characteristics are plants; such characteristics also apply to some algae, including the giant brown seaweeds classified as protists (see Chapter 28). Plant cells have walls made mostly of cellulose, and plants store their surplus carbohydrate in the form of starch. However, these characteristics too are shared with some algae. Plants share even more characteristics with their closest algal relatives, the green algae. For example, the chloroplasts of both green algae and plants contain chlorophyll *b* as an accessory photosynthetic pigment. (All photosynthetic eukaryotes, remember, use chlorophyll *a* as the pigment directly involved in conversion of light energy to chemical energy; see Chapter 10.)

So, how *do* we distinguish plants from multicellular algae? First, plants as we are defining them are nearly all terrestrial organisms, although some plants, such as water lilies, have returned secondarily to water during their evolution. Living on land poses very different problems from living in water, and it is a set of structural, chemical, and reproductive adaptations for terrestrial living that distinguishes plants from algae.

In terrestrial habitats, the resources a photosynthetic organism needs are found in two very different places: Light and carbon dioxide are mainly available above ground, while water and mineral nutrients are found mainly in the soil. Thus

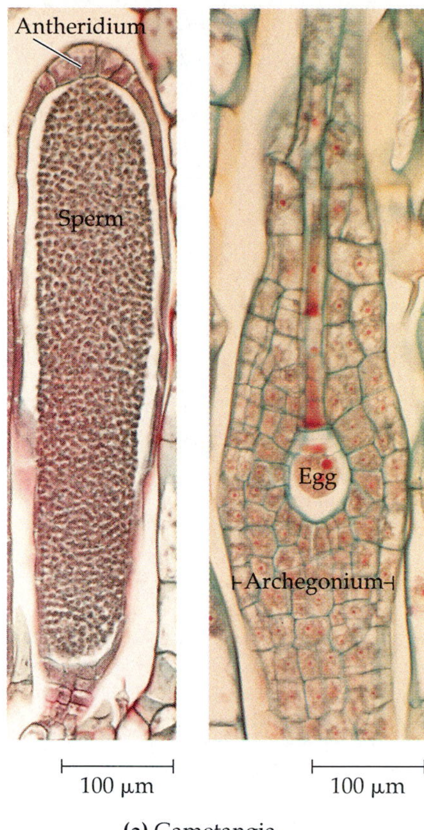

(a) Gametangia

(b) The embryophyte condition

FIGURE 29.1 ▪ Plant adaptations for reproducing on land. Specialized structures prevent the desiccation of developing gametes and embryos in terrestrial environments. **(a) Gametangia of a moss.** In early land plants, gametes developed within gametangia, moist chambers on the parent plants. This adaptation is retained by modern seedless plants, including mosses and ferns. On the left is a male gametangium, or antheridium, in which sperm develop. The gametangium on the right is an archegonium, in which an ovum (unfertilized egg) develops. Flagellated sperm released from antheridia fertilize eggs within the archegonia, where the zygotes develop into embryos within a protected environment. **(b) The embryophyte condition.** The protection and nourishing of embryos by maternal tissues of the parent plant is a terrestrial adaptation common to the entire plant kingdom. This is an embryo of shepherd's purse, a flowering plant (LMs).

the complex bodies of plants show varying degrees of structural specialization into subterranean and aerial organs—roots and leaf-bearing shoots, respectively. In most plants, exchange of carbon dioxide and oxygen between the atmosphere and the photosynthetic interior of leaves occurs via **stomata**, microscopic pores through the surfaces of leaves (see FIGURE 10.2, p. 170).

Terrestrial adaptations of plant structure are complemented by chemical adaptations. For example, aerial parts of most plants, such as leaves, have a waxy coating called a **cuticle**, which helps prevent excessive water loss, a major problem on land.

The waxes of the cuticle are examples of **secondary products**, so named because they are not produced by the primary, mainstream metabolic pathways common to all plants. Amino acids synthesized by plants are examples of *primary* products, as is cellulose, the product of a metabolic pathway required for the growth and development of all plants. Secondary products are synthesized by side branches of the main metabolic pathways and give rise to compounds that might be essential to the survival of the plant but are not central to its most basic metabolic needs, such as photosynthesis, amino acid synthesis, and cellulose production. Most secondary products are unique to a particular plant species or group of species, such as a plant family.

Many secondary products help protect the plant against excessive damage by herbivores. For example, plants of the milkweed family produce cardiac glycosides, poisons that apparently taste bad to vertebrates and impair heart function of herbivorous vertebrates if consumed. Other secondary products, such as the waxes of cuticles, are much more widespread in the plant kingdom. Another example of secondary products as terrestrial adaptations is **lignin**, the substance that hardens the cell walls of "woody" tissues in many plants.

A secondary product particularly important in the evolutionary move of plants onto land was **sporopollenin**, a polymer that is resistant to almost all kinds of environmental damage. In fact, the fossil record of plants is due mainly to the durability of sporopollenin, lignin, and the materials of cuticles. Sporopollenin did not originate in plants, for it is also found in the walls of resistant zygotes of some algae. But this toughest of all substances was co-opted by plants in a way that enhances reproduction on land, where resistance to harsh environments is a crucial challenge. Plants have sporopollenin in the walls of their spores and, in plants that have pollen, in the textured coats of pollen grains (hence the name, sporopollenin). Here we see an interplay among structural, chemical, and reproductive adaptations for terrestrial living. We next examine the reproductive adaptation that perhaps most defines the plant kingdom.

Plants as Embryophytes

The move onto land paralleled a new mode of reproduction (FIGURE 29.1). In contrast to the environment in which algae reproduce, gametes now had to be dispersed in a nonaquatic environment, and embryos, like mature body structures, had to be protected against desiccation.

Early plants produced their gametes within **gametangia**, organs having protective jackets of sterile (nonreproductive) cells that prevent the delicate gametes from drying out during their development (FIGURE 29.1a). The egg is fertilized within the female gametangium, where the zygote develops into an embryo that is retained and nourished for some time within the jacket of protective cells. In contrast, developing algae are not retained as embryos within a parent. This difference is so fundamental that plants are sometimes referred to as **embryophytes**, a term that emphasizes a key adaptation that contributed to success on land (FIGURE 29.1b).

Alternation of Generations: A Review

The life cycles of all plants feature an alternation of generations. One generation is the **gametophyte**, a multicellular individual with haploid cells; the other generation is the **sporophyte**, a multicellular individual with diploid cells. Gametophytes produce haploid gametes (hence the name), which fuse to form zygotes. The zygotes develop into diploid sporophytes. Meiosis in the sporophytes, in turn, produces haploid spores (hence the name *sporophyte*), which divide by mitosis to give rise to the next generation of gametophytes. This alternation of generations continues, haploid and diploid versions of the plant taking turns producing one another (FIGURE 29.2; also see Chapters 13 and 28 to review the concept of alternation of generations).

Alternation of generations occurs only in plants and certain groups of algae. It is worth reinforcing a distinction here. *All* sexually reproducing organisms alternate haploid and diploid conditions. In humans, for instance, meiosis forms haploid gametes, which restore the diploid state by fusing to form zygotes. In this example, the haploid stage consists of single cells, the gametes. What distinguishes alternation of generations as a special case of the haploid-diploid cycle is that both the haploid and diploid stages include multicellular individuals.

In the life cycles of all extant plants, the sporophyte and gametophyte individuals are heteromorphic—that is, they are different in morphology, or form. In the mosses and their closest relatives, the gametophyte, or haploid generation, is the larger, more elaborate plant—the stage we usually notice (a mat of moss, in this case). But in all other plant groups, including the ferns, conifers (pines, for example), and flowering plants, the diploid generation, or sporophyte, is the dominant, more noticeable stage (a rose bush, for example).

Throughout our study of the evolution of plant diversity, we will compare the life cycles of different plant groups. It is not important for you to memorize the details of these life cycles. What *is* important is to view these diverse life cycles as variations on a common ancestral cycle of alternation of generations. The plant kingdom is monophyletic, meaning that it

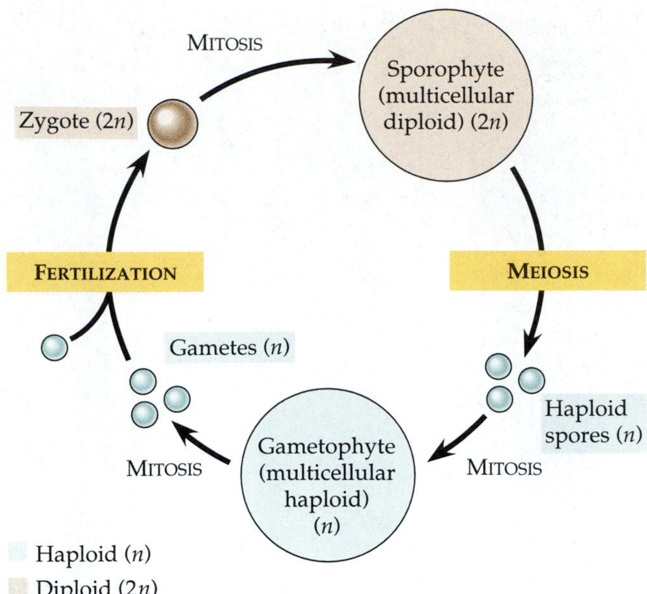

FIGURE 29.2 ▪ Alternation of generations: a generalized scheme. The life cycles of all plants include a gametophyte (haploid generation) and a sporophyte (diploid generation). The two generations alternate, each producing the other. The two plant forms are named for the type of reproductive cells they produce: Gametophytes form gametes by mitosis; sporophytes produce spores by meiosis. A spore is defined as a reproductive cell that develops directly into an organism without fusing with another cell. Gametes, in contrast, *cannot* develop directly into organisms but unite, sperm and egg, to form a zygote. The zygote, in turn, gives rise to an organism, the new sporophyte.

is derived from a common ancestor (see Chapter 25). Thus we can interpret the differences in the life cycles we will observe as special reproductive adaptations of the various plant groups as they diversified from the first plants. Descent with modification is the theme. In the study of plants, as in the study of all life, an evolutionary perspective connects what would otherwise be an overwhelming collection of seemingly unrelated facts. And in the case of the plant kingdom, the evolutionary history is about adaptation to living on land.

The history of terrestrial adaptation is the key to modern plant diversity

Some Highlights of Plant Phylogeny

The fossil record chronicles four major periods of plant evo-
29.1 lution, which are also evident in the diversity of modern plants (FIGURE 29.3). Each period was an adaptive radiation that followed the evolution of structures that opened new opportunities on the land.

The first period of evolution was associated with the origin of plants from their aquatic ancestors, a group of green algae known as charophytes, during the Ordovician period of the Paleozoic era, about 475 million years ago. The first terrestrial adaptations included spores toughened by sporopollenin and jacketed gametangia that protected gametes and embryos.

These adaptations made it possible for the plants known as bryophytes, including the mosses, to diversify from the first plants. **Vascular tissue**, consisting of cells joined into tubes that transport water and nutrients throughout the plant body, also evolved relatively early in plant history. Most bryophytes lack vascular tissue, and hence they are sometimes categorized as "nonvascular" plants. This distinction is not entirely accurate, however, for water-conducting tubes *are* present in some bryophytes.

The second major period of plant evolution was the diversification of vascular plants during the early Devonian period, about 400 million years ago. The earliest vascular plants lacked seeds, a condition still found in ferns and a few other groups of seedless vascular plants.

The third major period of plant evolution began with the origin of the seed, a structure that advanced the colonization of land by further protecting plant embryos from desiccation and other hazards. A seed consists of an embryo packaged along with a store of food within a protective covering. The first vascular plants with seeds arose about 360 million years ago, near the end of the Devonian period. Their seeds were not enclosed in any specialized chambers. Early seed plants gave rise to many types of **gymnosperms** (Gr. *gymnos*, "naked," and *sperma*, "seed"), including the conifers, which are the pines and other plants with cones. Gymnosperms coexisted with ferns and other seedless plants in great forests that dominated the landscape for more than 200 million years.

The fourth major episode in the evolutionary history of plants was the emergence of flowering plants during the early Cretaceous period in the Mesozoic era, about 130 million years ago. The flower is a complex reproductive structure that bears seeds within protective chambers called ovaries. This contrasts with the bearing of naked seeds by gymnosperms. The great majority of modern-day plants are flowering plants, or **angiosperms** (Gr. *angion*, "container," referring to the ovary, and *sperma*, "seed").

Classification of Plants

Plant biologists use the term **division** for the major plant groups within the plant kingdom. This taxonomic category corresponds to phylum, the highest unit of classification within the animal kingdom. Divisions, like phyla, are further subdivided into classes, orders, families, and genera.

29.1

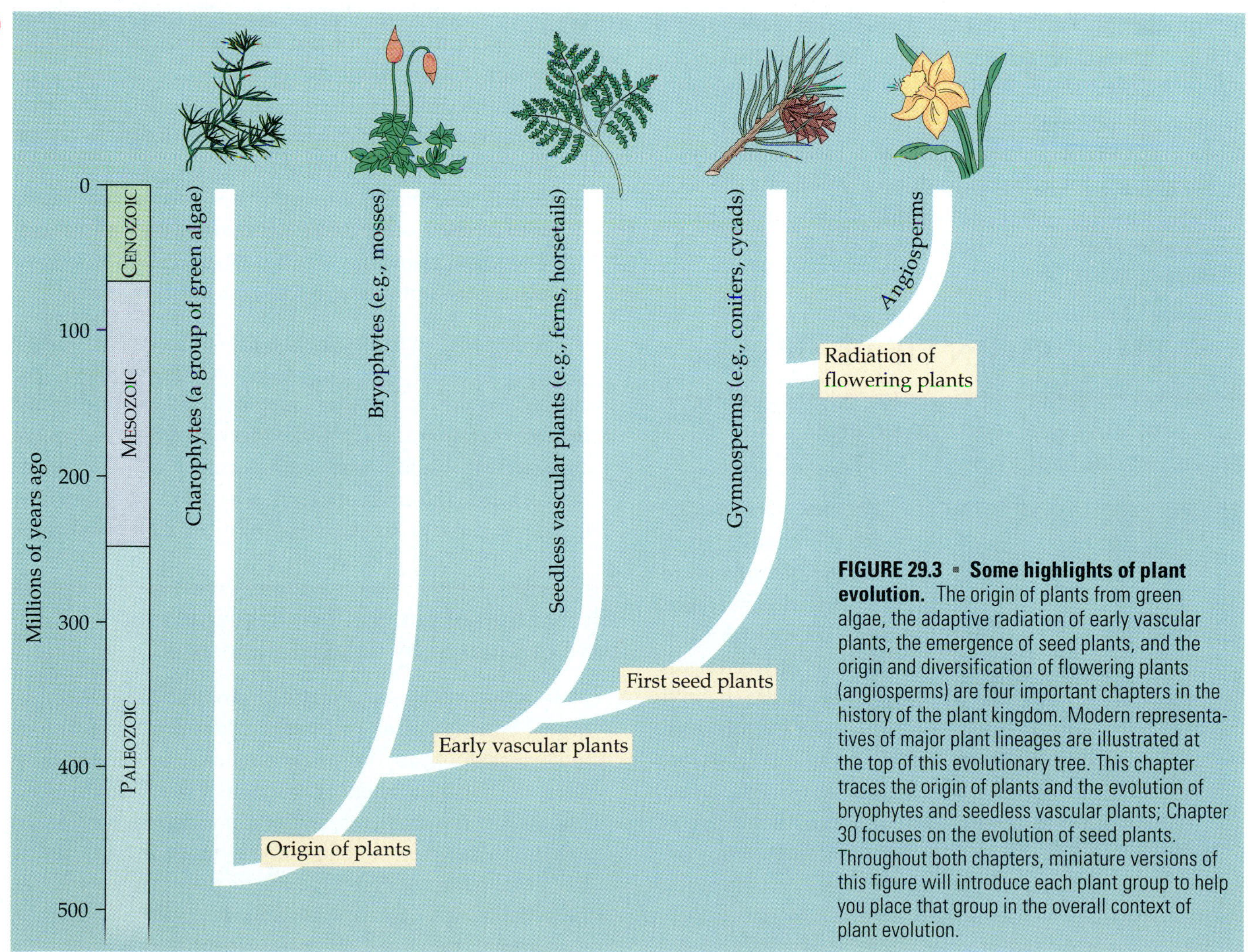

FIGURE 29.3 ▪ Some highlights of plant evolution. The origin of plants from green algae, the adaptive radiation of early vascular plants, the emergence of seed plants, and the origin and diversification of flowering plants (angiosperms) are four important chapters in the history of the plant kingdom. Modern representatives of major plant lineages are illustrated at the top of this evolutionary tree. This chapter traces the origin of plants and the evolution of bryophytes and seedless vascular plants; Chapter 30 focuses on the evolution of seed plants. Throughout both chapters, miniature versions of this figure will introduce each plant group to help you place that group in the overall context of plant evolution.

Table 29.1 ■ A Classification of Plants

	COMMON NAME	APPROXIMATE NUMBER OF EXTANT SPECIES
Nonvascular Plants (Bryophytes)*		
Division Bryophyta	Mosses	12,000
Division Hepatophyta	Liverworts	6,500
Division Anthocerophyta	Hornworts	100
Vascular Plants		
*Seedless Vascular Plants***		
Division Lycophyta	Lycophytes	1,000
Division Sphenophyta	Horsetails	15
Division Pterophyta	Ferns	12,000
Seed Plants		
Gymnosperms		
Division Coniferophyta	Conifers	550
Division Cycadophyta	Cycads	100
Division Ginkgophyta	Ginkgo	1
Division Gnetophyta	Gnetae	70
Angiosperms		
Division Anthophyta	Flowering plants	250,000

*Use of the term *nonvascular* for bryophytes must be qualified: Water-conducting tissue is present in some species.

**Seedless vascular plants called psilophytes, or whiskferns, were placed in a separate division until recently. Most botanists now include these plants in the fern division (see Challenge Question 1, p. 560).

The classification scheme used in this book recognizes eleven divisions within the kingdom Plantae. In addition to listing these divisions, TABLE 29.1 associates them with the stages of evolution discussed in the preceding section. Although these broader groupings do not have the status of formal taxonomic categories, they help fit the current diversity of plants into the historical context of refinement of terrestrial adaptations.

THE ORIGIN OF PLANTS

Plants probably evolved from green algae called charophytes

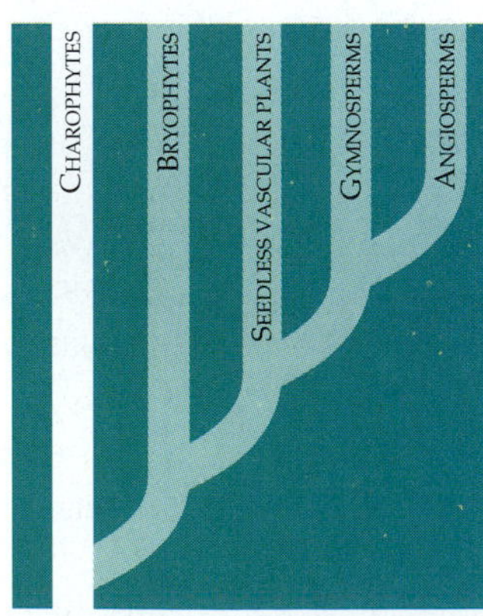

For decades, systematists have recognized that green algae are the photosynthetic protists most closely related to plants. Because there is a great diversity of green algae, recent research has focused on which group of these aquatic organisms represents the plant kingdom's closest algal relatives. Abundant evidence now points to the green algae called **charophytes** (FIGURE 29.4). By comparing cell ultrastructure, biochemistry, and hereditary information (DNA and its RNA and protein products), researchers have turned up homologies between charophytes and plants, including:

1. *Homologous chloroplasts.* Of all photosynthetic protists, only the green algae match plants in having chlorophyll *b* and beta-carotene as accessory pigments. The chloroplasts of green algae are also similar to plant chloroplasts in having their thylakoid membranes stacked as grana. When molecular systematists compared the chloroplast DNA of various green algae with that of plants, the closest match was between charophytes and plants.
2. *Biochemical similarity.* Cellulose is a structural component in the cell walls of most green algae, a characteristic shared with plants. Among green algae, charophytes are the most plantlike in wall composition, with cellulose making up 20% to 26% of the wall material in both charophytes and plants. Charophytes are also the only algae with peroxisomes matching the enzyme composition of plant peroxisomes.
3. *Similarity in the mechanisms of mitosis and cytokinesis.* During cell division in charophytes and plants, the nuclear envelope completely disperses during late prophase, and the mitotic spindle persists until cytokinesis begins. In some charophytes, as in plants, cytokinesis involves the cooperation of microtubules, actin microfilaments, and vesicles in the formation of a cell plate.
4. *Similarity in sperm ultrastructure.* In the details of sperm ultrastructure, charophytes are more similar to certain plants than to other green algae.
5. *Genetic relationship.* Molecular systematists have examined certain nuclear genes and ribosomal RNA in charophytes and plants, and the data agree with other evidence placing charophytes as the closest relatives of plants.

It is important to understand that *modern* charophytes, such as those in FIGURE 29.4, are *not* the ancestors of plants. The available evidence, however, supports the hypothesis that modern charophytes and plants both evolved from a common ancestor that would probably be classified as a charophyte. Researchers may learn more about how plants work by studying the physiology of their simpler relatives, the charophytes.

Alternation of generations in plants may have originated by delayed meiosis

How did alternation of generations evolve in the ancestors of plants? Alternation of generations does not occur among modern charophytes, but we can find clues in some of those algae, including members of the genus *Coleochaete* (see FIGURE 29.4b). The thallus (body) of *Coleochaete* is haploid. Its mode of sexual reproduction is very unusual compared to that of other algae. Most algae release their gametes into the surrounding water, where fertilization takes place. In contrast, the parental thallus of *Coleochaete* retains the eggs, and after

(a)

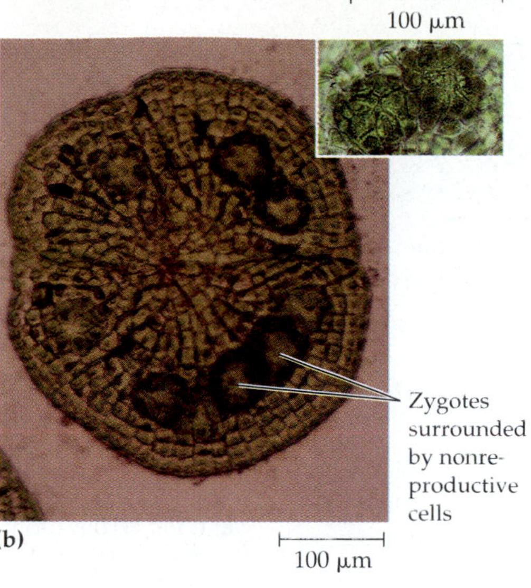

(b)

FIGURE 29.4 ▪ Charophytes, closest algal relatives of the plant kingdom. (a) *Chara braunii* is a pond organism popular with researchers because of its giant cells (the cylinder between consecutive whorls of "branches" is a single cell). For example, physiologists studying the role of electricity in membrane transport impale these giant cells with electrodes to measure voltage changes. **(b)** *Coleochaete orbicularis* is a charophyte that retains zygotes on the parent after fertilization. Nonreproductive cells grow around and over the zygotes, perhaps transporting nutrients that enable the zygotes to enlarge before they undergo meiosis. In the inset, you can see these nonreproductive cells covering two zygotes (LM).

fertilization occurs, the zygotes remain attached to the parent. Nonreproductive cells of the thallus grow around each zygote, which then enlarges, perhaps nourished by the surrounding cells. The grown zygote then undergoes meiosis, releasing haploid swimming spores that develop into new individuals.

Notice that the only diploid stage in the life cycle of *Coleochaete* is the zygote; alternation of *multicellular* diploid and haploid generations does not occur. But imagine a plant ancestor in which meiosis was delayed until after the zygote first divided by mitosis to give rise to a mass of diploid cells still attached to the haploid parent. Such a life cycle would fit the definition of alternation of generations. In this case, the rudimentary sporophyte (the clump of diploid cells) would be dependent on the gametophyte (the haploid parent). If specialized cells of the gametophyte formed protective layers around the tiny sporophyte, such a hypothetical ancestor would also qualify as a primitive embryophyte (FIGURE 29.5).

What would be the advantage of delaying meiosis and forming a mass of diploid cells? If the zygote undergoes meiosis directly, each fertilization event results in only a few haploid spores. But *mitotic* division of the zygote to form a sporophyte amplifies the sexual product, with meiosis by the many diploid cells producing a large number of haploid spores. This may have been an important adaptation for maximizing the output of sexual reproduction in environments where a shortage of water decreased the probability that swimming sperm would fertilize eggs.

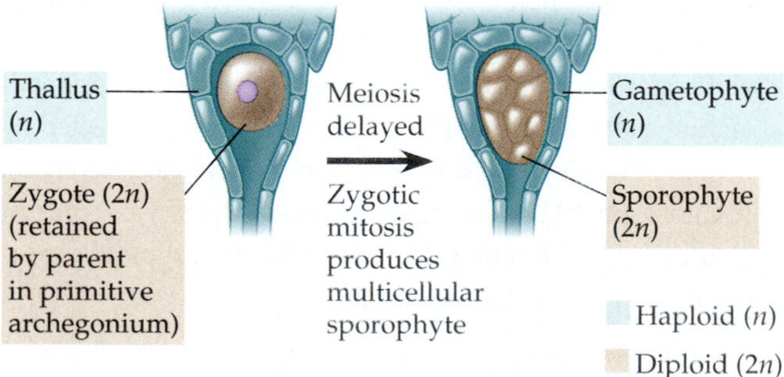

FIGURE 29.5 ▪ A hypothetical mechanism for the origin of alternation of generations in the ancestor of plants. According to the hypothesis illustrated by these cutaway views, alternation of generations evolved in the ancestor of plants by the postponement of meiosis until after mitotic division of the zygote produced a mass of diploid cells, the sporophyte. Subsequent meiosis by multiple diploid cells would have increased the number of haploid offspring (gametophytes) possible from each sexual union of sperm and ovum.

Adaptations to shallow water preadapted plants for living on land

Many species of modern charophytes are found in shallow water around the edges of ponds and lakes. Some of the ancient charophytes that lived about the time land was first colonized may have inhabited shallow-water habitats subject to occasional drying. Natural selection would have favored individual algae that could survive through periods when they were not submerged. The protection of developing gametes and embryos within jacketed organs (gametangia) on the parent is one example of adaptations to living in shallow water that would also prove useful on land. The resistance that sporopollenin adds to spores is another example.

At least one lineage of organisms that evolved from Ordovician algae accumulated adaptations enabling the organisms to live permanently above the water line. The evolutionary novelties of these first plants opened an adaptive zone that had never before been occupied. The new frontier was spacious, the bright sunlight was unfiltered by water and algae, the atmosphere had an abundance of carbon dioxide (the carbon source for photosynthesis), the soil was rich in minerals, and, at least at first, there were no herbivores on land.

As we survey the diversity of modern plants, remember that the past is the key to the present. Continue to think about the problems and opportunities facing organisms that began living on land.

BRYOPHYTES

The embryophyte adaptation evolved in bryophytes

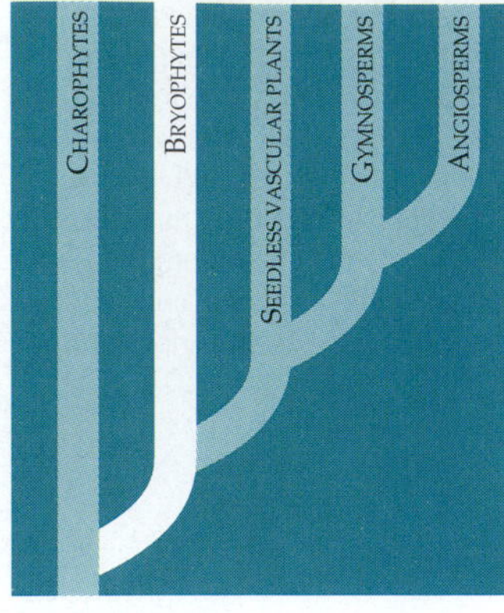

Until recently, the nonvascular plants—mosses, liverworts, and hornworts—were grouped together in a single division, Bryophyta (Gr. *bryon*, "moss"). This book adopts the current view that mosses, liverworts, and hornworts deserve separate divisions because they are probably not closely related. In this taxonomy, only mosses belong to the formal division Bryophyta. However, most plant biologists (and this book) still use the common name *bryophyte* to refer to liverworts and hornworts as well as mosses. This informal usage is appropriate because these three plant groups share many characteristics.

Bryophytes display a key adaptation that first made the move onto land possible: the embryophyte condition. Their gametes develop within gametangia (see FIGURE 29.1a). The male gametangium, known as an **antheridium**, produces flagellated sperm. In each female gametangium, or **archegonium**, one egg (ovum) is produced. The egg is fertilized within the archegonium, and the zygote develops into an embryo within the protective jacket of the female organ. The retention of the zygote and the sporophyte that develops from the zygote is a refined version of the process illustrated in FIGURE 29.5.

FIGURE 29.6 ▪ **A peat-moss bog.** Lacking rigid supporting tissue, bryophytes are low-profile plants most common in damp habitats. The matlike plants of this landscape are gametophytes, the dominant generation in the life cycles of mosses and other bryophytes. This particular bryophyte belongs to the genus *Sphagnum*, or peat moss.

Even with their protected embryos, bryophytes are not totally liberated from their ancestral aquatic habitat. First of all, these plants need water to reproduce; their sperm, like those of most green algae, are flagellated and must swim from the antheridium to the archegonium to fertilize the egg. For many bryophyte species, a film of rainwater or dew is sufficient for fertilization to occur, and thus some bryophyte species can even live in deserts. In addition, most bryophytes have no vascular tissue to carry water from the soil to the aerial parts of the plant (the exceptions, as mentioned earlier, are certain bryophytes with elongated water-conducting cells). As water moves over the surface of most bryophytes, they must imbibe it like sponges and distribute it throughout the plant by the relatively slow processes of diffusion, capillary action, and cytoplasmic streaming. This mode of hydration helps explain why damp, shady places are the most common habitats of bryophytes.

Bryophytes lack the lignin-fortified tissue required to support tall plants on land. Although they may sprawl horizontally as mats over a large surface, bryophytes always have a low profile (FIGURE 29.6). Most are only 1–2 cm in height, and even the largest are usually less than 20 cm tall.

The gametophyte is the dominant generation in the life cycles of bryophytes

29.2 In the life cycle of a bryophyte, such as moss, we see a specific example of an alternation of haploid and diploid generations (FIGURE 29.7). The haploid gametophyte, remember, is the dominant generation in mosses and other bryophytes. The sporophyte is generally smaller and shorter lived, and it depends on the gametophyte for water and nutrients. The diploid sporophyte produces haploid spores via meiosis in a structure called a **sporangium**. The tiny spores, protected by sporopollenin, disperse and give rise to new gametophytes. The gametophyte-dominant life cycles of bryophytes contrast with the life cycles of vascular plants, where the diploid sporophyte is the dominant generation.

The three divisions of bryophytes are mosses, liverworts, and hornworts

Mosses (Division Bryophyta)

The most familiar bryophytes are **mosses**. A mat of moss actually consists of many plants growing in a tight pack, helping to hold one another up. The mat has a spongy quality that enables it to absorb and retain water. Each plant of the mat grips the substratum with elongate cells or cellular filaments called rhizoids. Most photosynthesis occurs in the upper part of the plant, which has many small stemlike and leaflike appendages. The "stems," "leaves," and "roots" (rhizoids) of a moss, however, are not homologous to these structures in vascular plants.

Although mosses are short in stature, their collective impact on Earth is huge. For example, peat mosses, or *Sphagnum,* carpet at least 3% of Earth's terrestrial surface, with greatest density in high northern latitudes (see FIGURE 29.6).

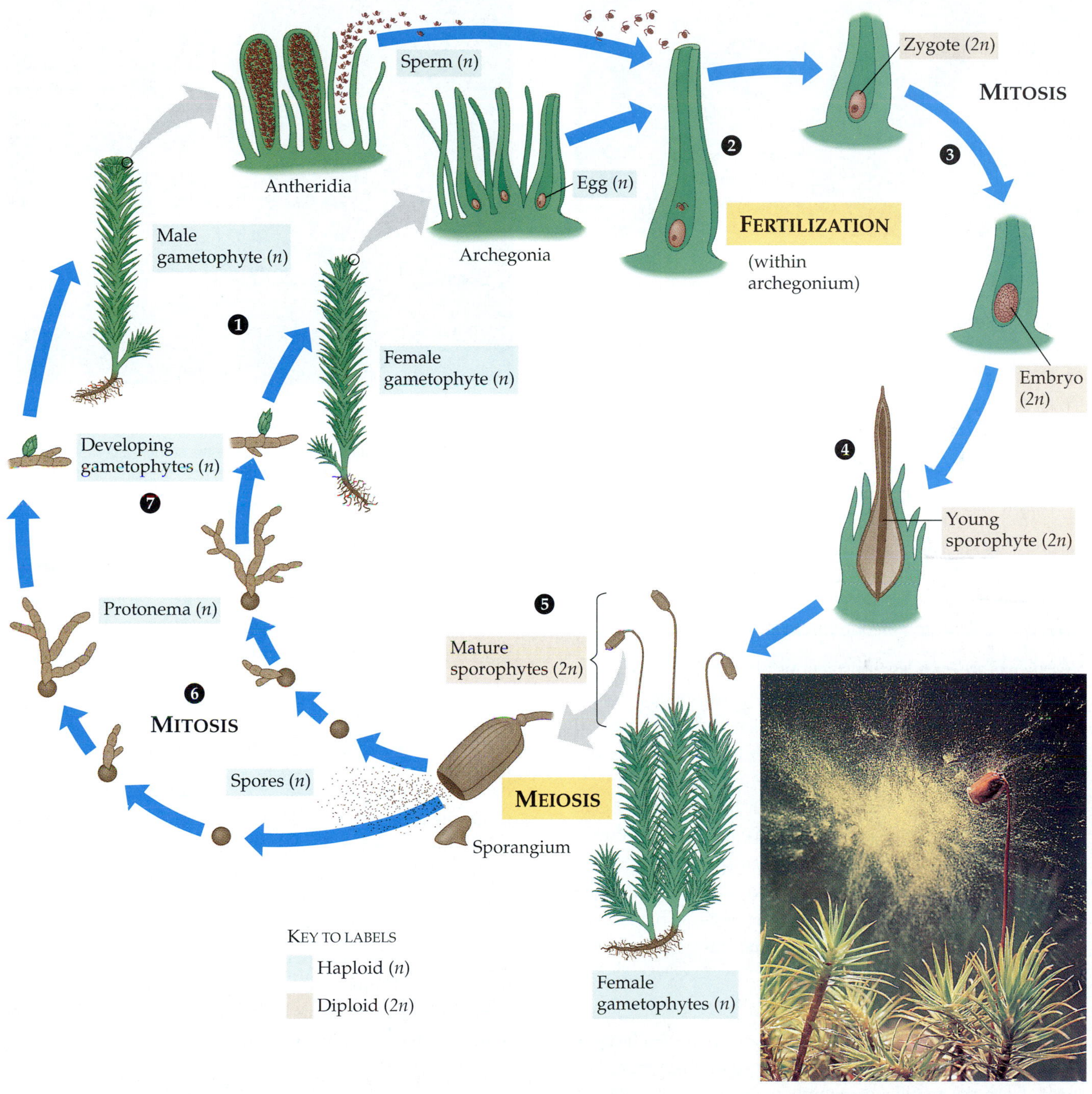

FIGURE 29.7 ▪ The life cycle of a moss. The gametophyte is the prevalent generation in the bryophyte life cycle, a characteristic that contrasts with other plants. ① Most species of moss have separate male and female gametophytes, which have antheridia and archegonia, respectively. ② After a sperm swims through a film of moisture to an archegonium and fertilizes the egg, ③ the diploid zygote divides by mitosis and develops into an embryonic sporophyte within the archegonium. ④ During the next stage of its development, the sporophyte grows a long stalk that emerges from the archegonium, but the base of the sporophyte remains attached to the female gametophyte. ⑤ At the tip of the stalk is a sporangium, a capsule in which meiosis occurs and haploid spores develop. When the lid of the sporangium pops off, the spores scatter. ⑥ Spores germinate by mitotic division, forming small, green, threadlike protonemata (sing.: protonema) resembling green algae. ⑦ The haploid protonemata continue to grow and differentiate, eventually forming sexually mature gametophytes, completing the life cycle.

29.2

FIGURE 29.8 ▪ Liverworts. The gemmae cups function in asexual reproduction. The tiny plantlets (gemmae) within the cups are dispersed by the impact of raindrops.

FIGURE 29.9 ▪ Hornworts. You can see the hornlike sporophytes, for which these plants are named, growing from the parental gametophytes.

The accumulation of "peat," the thick mat of living and dead plants in wetlands, ties up an enormous amount of organic carbon because peat's abundance of resistant materials is not easily degraded by microbes. As carbon reservoirs, peat bogs play an important role in stabilizing Earth's atmospheric carbon dioxide concentrations, and hence climate, through the CO_2-related greenhouse effect (see Chapter 54).

Liverworts (Division Hepatophyta)

Liverworts are even less conspicuous plants than mosses. The bodies of some are divided into lobes, giving an appearance that must have reminded someone of the lobed liver of an animal (*wort* means "herb"). Tropical forests are home to the greatest diversity of liverwort species.

The life cycle of a liverwort is much like that of a moss. Within the sporangia of some liverworts are coil-shaped cells that spring out of the capsule when it opens, helping to disperse the spores. Liverworts can also reproduce asexually from little bundles of cells called gemmae, which are bounced out of cups on the surface of the gametophyte by raindrops (FIGURE 29.8).

Hornworts (Division Anthocerophyta)

Hornworts resemble liverworts but are distinguished by their sporophytes, which are elongated capsules that grow like horns from the matlike gametophyte (FIGURE 29.9). Recent evidence based on nucleic acid sequences suggests that hornworts, of all bryophytes, are the most closely related to vascular plants.

The three divisions of bryophytes—mosses, liverworts, and hornworts—continue a long success story on land, enduring and adapting for over 450 million years. Even today, *Sphagnum,* the peat mosses, may be the most abundant plants on Earth. And for at least the first 50 million years that terrestrial communities existed, bryophytes were probably the only plants. Then the landscape began to change again, with the vegetation taking on a taller profile as vascular plants evolved.

THE ORIGIN OF VASCULAR PLANTS

As we have seen, the bryophyte adaptations to land include gametangia, embryos, and sporopollenin-walled spores. Stomata also evolved in bryophytes, and some bryophytes, including the sporophytes of hornworts, have cuticles similar in composition to those of vascular plants. (It is still unknown whether the cuticles and stomata of vascular plants evolved independently or are derived from similar structures in bryophytes.) Plants had a monophyletic origin from green algae, with vascular plants adding new terrestrial adaptations to those that evolved first in bryophytes.

Additional terrestrial adaptations evolved as vascular plants descended from bryophyte-like ancestors

The bodies of most vascular plants are differentiated into subterranean root systems, which absorb water and minerals, and aerial shoot systems of stems and leaves, where photosynthesis occurs. Vascular tissue, consisting of tubular chains of cells, transports materials between the distant organs of the plant.

The two conducting tissues of the vascular system are **xylem** and **phloem**. Tube-shaped cells in the xylem carry water and minerals up from roots. These water-conducting cells are actually dead, with only their walls remaining to provide a system of microscopic water pipes. Phloem is a living tissue with food-conducting cells arranged into tubes that distribute sugars, amino acids, and other organic nutrients throughout the plant.

Another important terrestrial adaptation of vascular plants is lignin, a hard material embedded in the cellulose matrix of the walls of cells that function in mechanical support. In contrast to aquatic habitats, where even large organisms such as seaweeds are buoyed by the surrounding water, terrestrial environments provide no significant external support for organisms. Imagine what would happen to you if your skeleton were to disappear suddenly. A tree would also collapse if it were not for its version of a "skeleton," its framework of lignified cells. Some of those cells, called fibers, are specialized for support; in addition, xylem cells have lignified walls, and thus xylem does double duty as a vascular/support tissue. Turgor pressure (see Chapter 8) helps support small plants, but it is the skeleton of lignified walls that holds up a tree or any other large vascular plant.

The branched sporophytes of vascular plants amplified the production of spores and made more complex bodies possible

Encased in the sedimentary strata of the late Silurian and early Devonian periods are fossils of a variety of early vascular plants. Many of these fossilized plants are beautifully preserved, right down to the microscopic organization of their tissues. The oldest is *Cooksonia*, which has been discovered in Silurian rocks in both Europe and North America, which were joined during the Silurian period. Some of the branches of *Cooksonia* terminated in bulbous sporangia, the structures on sporophytes that produce spores (FIGURE 29.10).

Compared to bryophytes, fossils of *Cooksonia* and related plants reveal two major changes. First, the sporophyte is the dominant stage in early vascular plants (indicated by the presence of sporangia), while the gametophyte, remember, is the dominant stage in the life cycles of bryophytes. A second major change is also evident in FIGURE 29.10. The sporophyte of *Cooksonia* was branched, in contrast to the unbranched sporophytes of bryophytes (see FIGURE 29.7). The branching increased the number of sporangia, and hence spores, an individual plant could produce. And perhaps the branching provided structural raw material for the evolution of more complex plant bodies. For example, the leaves of most vascular plants probably evolved by the formation of webbing between many branches growing close together.

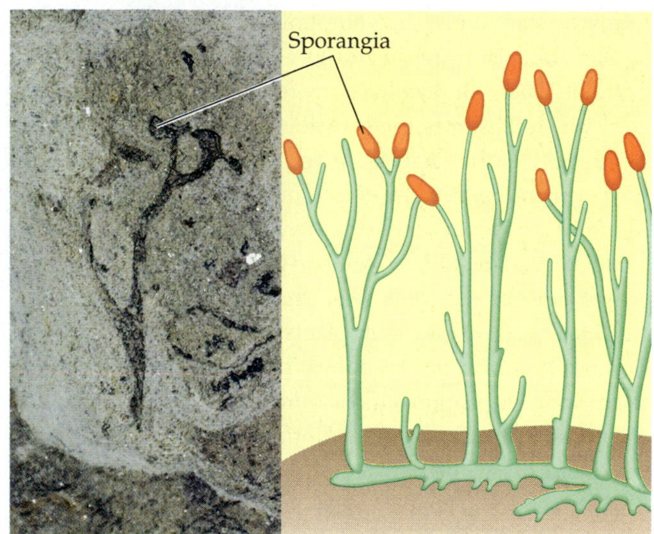

FIGURE 29.10 ▪ *Cooksonia*, a vascular plant of the Silurian. The dichotomous branching and terminal sporangia characteristic of *Cooksonia* is evident in the photograph of the fossil on the left. True roots and leaves were absent. The plant was anchored by a rhizome, a horizontal stem. *Cooksonia* grew in dense stands around marshes. The largest species was about 50 cm tall.

Early vascular plants did not form seeds, and a diversity of seedless vascular plants still inhabits Earth today.

SEEDLESS VASCULAR PLANTS

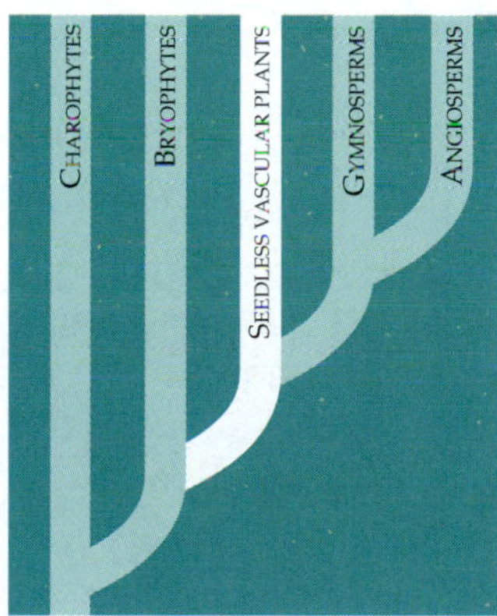

Seedless vascular plants dominated the forest landscapes of the Carboniferous period, which began about 360 million years ago. Among the descendants of those organisms are three divisions of extant seedless vascular plants: the lycophytes, the horsetails, and the ferns (see TABLE 29.1). We can examine the life cycle of a fern to reinforce an important contrast between vascular plants and bryophytes.

A sporophyte-dominant life cycle evolved in seedless vascular plants

From *Cooksonia* and the other early vascular plants to all of the vascular plants that live today, the sporophyte (diploid) generation is the larger and more complex plant in the alternation of generations. For example, the leafy fern plants familiar to us are the sporophytes. You would have to get down on your hands and knees and explore with careful hands and sharp eyes to find fern gametophytes, tiny plants growing on or just below the soil surface. Until you have a chance to do that, you can study the sporophyte-dominant life cycle of seedless vascular plants in FIGURE 29.11, which uses a fern as an example. Then, for review, compare this life cycle with FIGURE 29.7, which illustrates the gametophyte-dominated life cycles of bryophytes. In Chapter 30 you will see that the gametophyte generation became even more reduced during the evolution of seed plants, and it is there that we will consider this trend as an adaptation to living on land.

29.3 We can also use ferns to illustrate a key variation among the life cycles of vascular plants: the distinction between homosporous and heterosporous plants. The sporophyte of a **homosporous** plant produces a single type of spore. The fern in FIGURE 29.11 is an example. Notice that each spore develops into a bisexual gametophyte having both female and male sex organs, the gametangia called archegonia and antheridia, respectively. In contrast, the sporophyte of a **heterosporous** plant produces two kinds of spores: **Megaspores** develop into female gametophytes bearing archegonia; **microspores** develop into male gametophytes with antheridia. Among ferns, those that returned to aquatic habitats during their evolution—the water ferns—are the only heterosporous species. However, we will see in Chapter 30 that the heterosporous condition was very important in the evolution of seeds.

The diagrams below will help you compare the homosporous and heterosporous conditions:

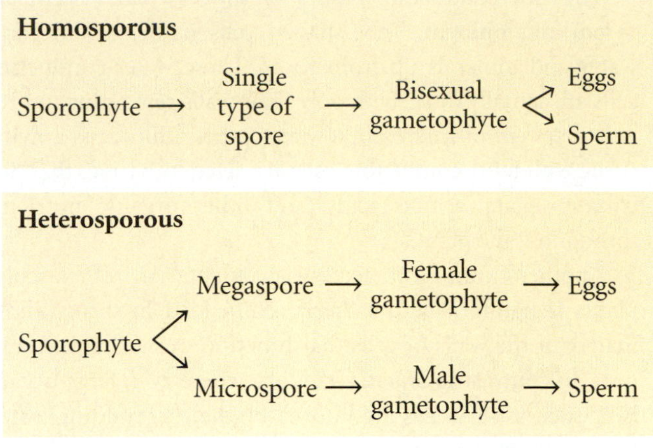

The life cycle in FIGURE 29.11 illustrates one more point: The sperm cells of ferns and all other seedless vascular plants (and even some seed plants) are flagellated and must swim through a film of water to reach eggs, a characteristic shared with bryophytes. With their swimming sperm and fragile gametophytes, seedless vascular plants are most common in relatively damp habitats.

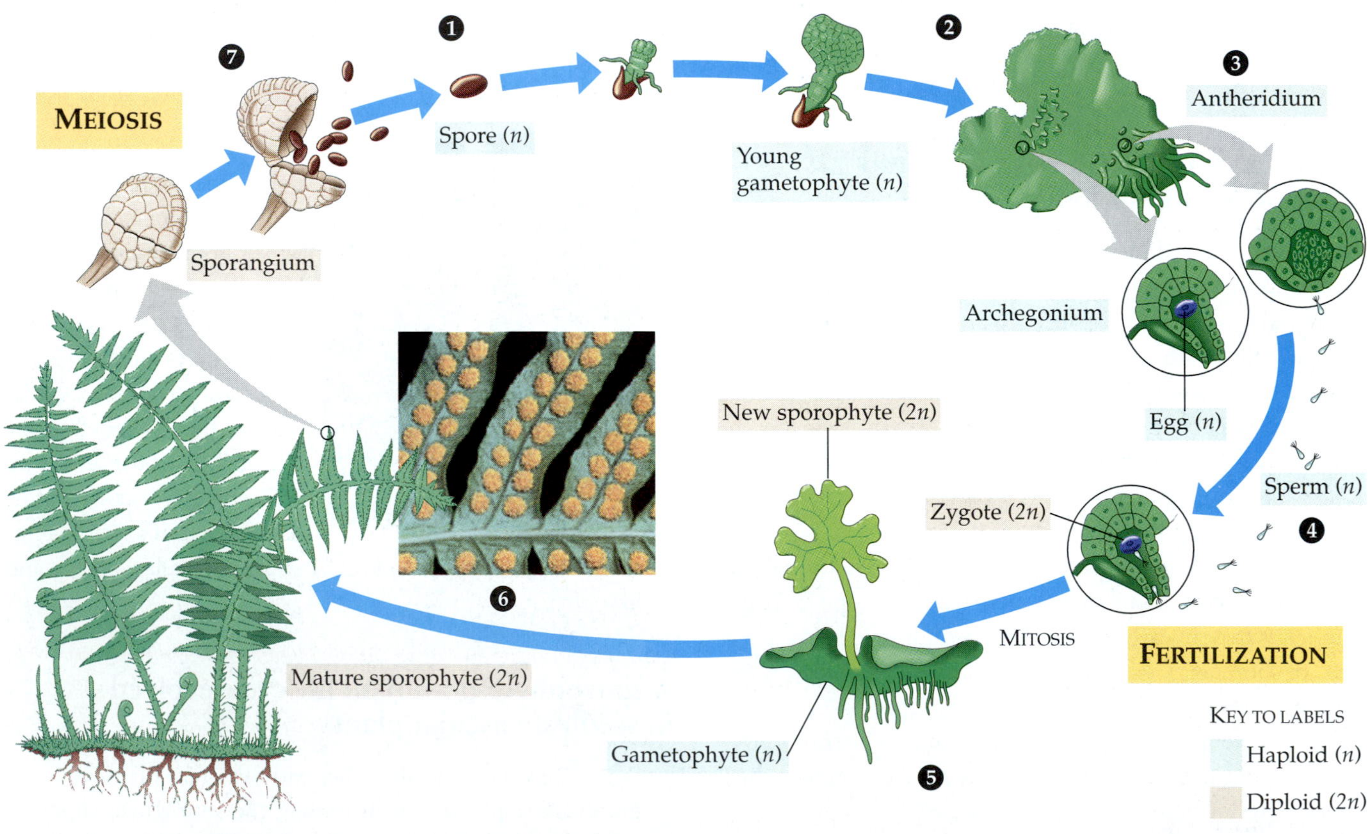

29.3 **FIGURE 29.11 • The life cycle of a fern.** ① Most ferns are homosporous, meaning that they produce a single type of spore. After a fern spore settles in a favorable place, ② it develops into a small, heart-shaped gametophyte that sustains itself by photosynthesis. ③ Each gametophyte has both male and female sex organs, but the archegonia and antheridia usually mature at different times, ensuring cross-fertilization between gametophytes. ④ Fern sperm, like those of all seedless vascular plants, use flagella to swim through moisture from antheridia to eggs in the archegonia and then fertilize the eggs. ⑤ A fertilized egg develops into a new sporophyte, and the young plant grows out from an archegonium of its parent, the gametophyte. ⑥ The spots on the underside of reproductive leaves (sporophylls) are called sori (see photograph). Each sorus is a cluster of sporangia. ⑦ Sporangia release spores, which give rise to gametophytes.

The three divisions of seedless vascular plants are lycophytes, horsetails, and ferns

Lycophytes (Division Lycophyta)

The extant **lycophytes**, of the division Lycophyta, are relicts of a far more eminent past. Lycophytes first evolved during the Devonian period and became a major part of the landscape during the Carboniferous period. By that time, the division Lycophyta split into two evolutionary lines. One group evolved into woody trees that had diameters as large as 2 m and heights of more than 40 m. A second line of lycophytes remained small and herbaceous (nonwoody). The giant lycophytes thrived in the Carboniferous swamps for millions of years but became extinct when the swamps began to dry up at the end of that geological period. The small lycophytes survived, and they are represented today by about a thousand species. Common names for these plants are club mosses or ground pines, though they are neither mosses nor pines.

Many species of lycophytes are tropical plants that grow on trees as **epiphytes**—plants that use another organism as a substratum but are not parasites. Other lycophyte species grow close to the ground on forest floors in temperate regions, including the northeastern United States.

The lycophyte in FIGURE 29.12 is the sporophyte, the diploid generation. The sporangia are borne on **sporophylls**, leaves specialized for reproduction. After their discharge, the spores develop into inconspicuous gametophytes that may live underground for ten years or longer. These tiny haploid plants are nonphotosynthetic and are nurtured by symbiotic fungi. In homosporous species, each gametophyte develops archegonia with eggs and antheridia that make flagellated sperm. After a swimming sperm fertilizes an egg, the diploid zygote gives rise to a new sporophyte. Heterosporous lycophytes that form separate male and female gametophytes also exist.

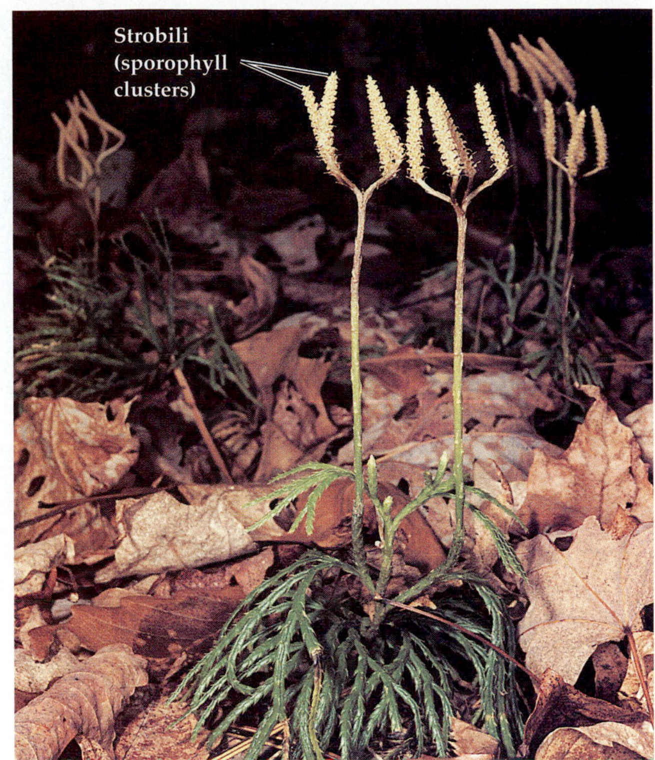

FIGURE 29.12 ▪ A lycophyte. Club mosses are common inhabitants of forest floors in the northeastern United States. The small plant has a horizontal rhizome that gives rise to roots and vertical branches and has true leaves containing strands of vascular tissue. The sporangia of lycophytes are borne by specialized leaves called sporophylls. In some species, such as the one shown here, the sporophylls are clustered at the tips of branches into club-shaped structures called strobili (hence the common name club mosses).

Horsetails (Division Sphenophyta)

Sphenophyta, whose members are commonly called **horsetails**, is another ancient lineage of seedless plants dating back to the Devonian radiation of early vascular plants. The group reached its zenith during the Carboniferous period, when many species grew as tall as 15 m. All that survives of this division of plants are about 15 species of a single, widely distributed genus, *Equisetum*, that is most common in the Northern Hemisphere, generally in damp locations such as stream banks (FIGURE 29.13).

The conspicuous horsetail plant is the sporophyte generation. Meiosis occurs in the sporangia, and haploid spores are released. Horsetails are homosporous. The bisexual gametophytes that develop from the spores are only a few millimeters long, but they are photosynthetic and free-living (not dependent on the sporophyte for food).

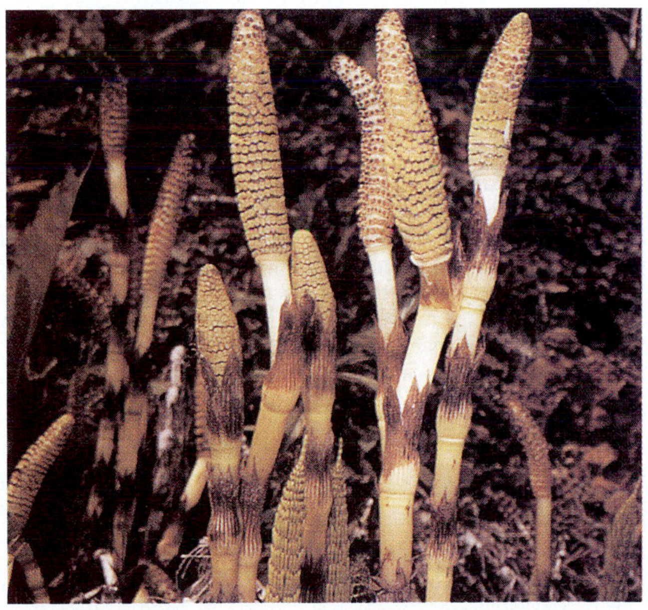

FIGURE 29.13 ▪ *Equisetum* (horsetail). *Equisetum* has an underground rhizome from which vertical stems arise. The straight, hollow stems are jointed, and whorls of small leaves or branches emerge at the joints. At the tips of some stems of *Equisetum* are conelike structures bearing sporangia. The epidermis, the outer layer of cells, is embedded with silica, which gives the plants a gritty texture. Before scouring pads, people used the abrasive stems of horsetails to scrub pots and pans, which is why these plants are also known as scouring rushes.

FIGURE 29.14 ▪ Artist's conception of a Carboniferous forest based on fossil evidence. Most of the large trees with straight trunks are lycophytes. On the left, the tree with numerous feathery branches is a horsetail. Tree ferns, though not featured in this painting, were also abundant in the "coal forests" of the Carboniferous. Animals, including giant dragonflies like the one near the horsetail, also thrived. Seed plants, early gymnosperms, evolved during the Carboniferous and began to dominate landscapes when the climate turned drier and cooler at the end of the period.

Ferns (Division Pterophyta)

From their Devonian beginning, **ferns** (division Pterophyta) radiated into the many species that stood alongside tree lycophytes and horsetails in the great forests of the Carboniferous period. Of all seedless vascular plants, ferns are by far the most extensively represented in modern floras. More than 12,000 species of ferns live today. They are most diverse in the tropics, but a variety of species is also found in temperate forests.

The leaves of ferns are generally much larger than those of lycophytes and probably evolved in a different way. The small leaves of lycophytes (see FIGURE 29.12) probably evolved as emergences from the stem that contained a single strand of vascular tissue. Leaves with this origin are called microphylls. Each leaf of a fern, termed a megaphyll, has a branched system of veins. Megaphylls probably evolved by the formation of webbing between many separate branches growing close together. (This hypothesis was mentioned earlier as an example of how the branched shoots of early vascular plants made more complex plants possible.)

Most ferns have leaves, commonly called fronds, that are compound, meaning each leaf is divided into several leaflets (see FIGURE 29.11). The frond grows as its coiled tip, the fiddlehead, unfurls. The leaves may sprout directly from a prostrate stem, as they do in brackens and sword ferns. Large tropical tree ferns, by contrast, have upright stems many meters tall.

Return to FIGURE 29.11 to review the life cycle of a fern.
29.3 Notice that some of the leaves are specialized sporophylls with sporangia on their undersides. The sporangia of many ferns are arranged in clusters called sori and are equipped with springlike devices that catapult spores several meters. Once airborne, spores can be blown by the wind far from their origin. Spores, protected by sporopollenin, are the means of dispersal for all seedless plants.

Seedless vascular plants formed vast "coal forests" during the Carboniferous period

The three divisions of plants we have just surveyed represent the extant lineages of seedless vascular plants that formed forests during the Carboniferous period about 290 to 360 million years ago (FIGURE 29.14). Seedless vascular plants of the Carboniferous forests left not only living relicts but also fossilized fuel in the form of coal.

Coal formed during several geological periods, but the most extensive beds of coal are found in strata deposited during the Carboniferous, a time when most of the continents were flooded by shallow swamps. Europe and North America, near the equator at that time, were covered by tropical swamp forests. Dead plants did not completely decay in the stagnant waters, and great depths of organic rubble called peat accumulated by a process similar to that occurring today in peat-moss bogs. The swamps were later covered by the sea, and marine sediments piled on top of the peat. Heat and pressure gradually converted the peat to coal, a "fossil fuel." Coal powered the Industrial Revolution, and a resurgence in its use is inevitable as we continue to deplete nonrenewable oil and gas reserves.

Growing along with the seedless plants in the Carboniferous swamps were primitive seed plants. Though seed plants were not the dominant plants at that time, they rose to prominence after the swamps began to dry up at the end of the Carboniferous period. The next chapter traces the origin and diversification of seed plants, continuing our theme of terrestrial adaptation.

CHAPTER REVIEW

REVIEW OF KEY CONCEPTS

(with page numbers and key figures)

AN OVERVIEW OF PLANT EVOLUTION

- **Structural, chemical, and reproductive adaptations enabled plants to colonize land (pp. 546–548, FIGURES 29.1, 29.2)** Gametangia, the embryophyte condition, and resistant polymers such as sporopollenin are among the terrestrial adaptations of plants. All plants have life cycles with an alternation of gametophyte and sporophyte generations.
- 29.1 **The history of terrestrial adaptation is the key to modern plant diversity (pp. 548–550, FIGURE 29.3)** Four key periods in the evolution of plants were the origin of embryophytes from green algae, the evolution of vascular plants, the origin of the seed, and the diversification of flowering plants.

THE ORIGIN OF PLANTS

- **Plants probably evolved from green algae called charophytes (p. 550, FIGURE 29.4)** Homology at the cellular and molecular levels points to charophytes as the algae most closely related to plants.
- **Alternation of generations in plants may have originated by delayed meiosis (pp. 550–551, FIGURE 29.5)** The sporophyte may have its origins in mitotic division of a zygote before the meiotic production of spores, which would have amplified the output of offspring from each fertilization event.
- **Adaptations to shallow water preadapted plants for living on land (p. 551)** Periodic drops in water level would have selected for characteristics, such as the embryophyte condition, that contributed to eventual success on land.

BRYOPHYTES

- **The embryophyte adaptation evolved in bryophytes (p. 552)** The embryo develops within the archegonium, the maternal gametangium.
- 29.2 **The gametophyte is the dominant generation in the life cycles of bryophytes (p. 552, FIGURE 29.7)** A mat of moss consists of haploid gametophytes, with the diploid sporophytes growing out of archegonia, dependent on the gametophyte for nourishment. Flagellated sperm require a film of water to reach eggs.
- **The three divisions of bryophytes are mosses, liverworts, and hornworts (pp. 552–554, FIGURES 29.6, 29.8, 29.9)** Mosses (Bryophyta) are the most diverse and widespread bryophytes. Liverworts (Hepatophyta) are named for their shape and reproduce asexually via gemmae. Hornworts (Anthocerophyta) may be the bryophytes most closely related to vascular plants.

THE ORIGIN OF VASCULAR PLANTS

- **Additional terrestrial adaptations evolved as vascular plants descended from bryophyte-like ancestors (pp. 554–555)** One vascular tissue, xylem, mainly transports water and minerals; a second vascular tissue, phloem, mainly transports organic nutrients. Another adaptation, lignin, hardens cell walls, providing vascular plants with mechanical support.
- **The branched sporophytes of vascular plants amplified the production of spores and made more complex bodies possible (p. 555, FIGURE 29.10)** *Cooksonia* is an example of a Silurian plant that had dichotomous branching, with sporangia at the tips of some branches.

SEEDLESS VASCULAR PLANTS

- 29.3 **A sporophyte-dominant life cycle evolved in seedless vascular plants (pp. 555–556, FIGURE 29.11)** One variation on this life cycle is the contrast between homosporous and heterosporous plants. The ancestral condition of flagellated sperm is retained by all seedless vascular plants.
- **The three divisions of seedless vascular plants are lycophytes, horsetails, and ferns (pp. 557–558, FIGURES 29.11–29.13)** Lycophytes (Lycophyta) includes the club mosses, with their clusters of sporophylls at the tips of some shoots. Horsetails (Sphenophyta), like lycophytes, date back to the Devonian, and both groups were far more diverse during the Carboniferous than they are today. Ferns (Pterophyta) are by far the most diverse seedless vascular plants in the modern flora.
- **Seedless vascular plants formed vast "coal forests" during the Carboniferous period (pp. 558–559, FIGURE 29.14)** Coal formed from peat, the partially decomposed bodies of swamp plants. Seed plants also evolved during this period but did not become the prevalent plants until the end of the Carboniferous.

SELF-QUIZ

1. Which of the following characteristics of plants is *absent* in their closest relatives, the charophytes?
   a. chlorophyll *b*
   b. cellulose in cell walls
   c. alternation of multicellular generations
   d. sexual reproduction
   e. formation of a cell plate during cytokinesis

2. All bryophytes (mosses, liverworts, and hornworts) share certain characteristics. These are
   a. reproductive cells in gametangia; embryos
   b. branched sporophytes
   c. vascular tissues, true leaves, and a waxy cuticle
   d. seeds
   e. lignified walls

3. Which of the following is *not* common to all divisions of vascular plants?
   a. the development of seeds
   b. alternation of generations
   c. dominance of the diploid generation
   d. xylem and phloem
   e. the addition of lignin to cell walls

4. A heterosporous plant is one that
   a. produces a gametophyte that bears both sex organs

b. produces microspores and megaspores in separate sporangia, giving rise to separate male and female gametophytes
c. is a seedless vascular plant
d. produces two kinds of spores, one asexually by mitosis and the other sexually by meiosis
e. reproduces only sexually

5. During the Carboniferous period, the dominant plants, which later formed the great coal beds, were mainly
   a. giant lycophytes, horsetails, and ferns
   b. conifers
   c. angiosperms
   d. charophytes
   e. early seed plants

6. A land plant that produces flagellated sperm and has a diploid-dominant generation is most likely a
   a. fern
   b. moss
   c. liverwort
   d. charophyte
   e. hornwort

Questions 7–10. For each of the following structures or life cycle stages, indicate whether the cells are haploid or diploid in chromosome number.

7. The thallus of a charophyte
8. The nonreproductive cells that line the gametangia of a moss
9. The cells that make up the stalk of a moss sporophyte
10. The spores produced by the sporophyte of a fern

CHALLENGE QUESTIONS

1. The plant in this photo is *Psilotum,* a whisk fern. The dichotomous (repeated "Y") branching and relatively simple form are reminiscent of *Cooksonia,* the fossil plant in FIGURE 29.10. In fact, until recently, most botanists considered *Psilotum* to be a primitive seedless vascular plant and placed it in its own division, Psilophyta. Other botanists postulated that the plant is actually a fern in which a relatively simple body structure evolved secondarily. Most of the evidence now supports this second hypothesis, and so it is reasonable to include *Psilotum* in the fern division. Based on what you have learned about the methods of systematics and taxonomy, how would you test this hypothesis that *Psilotum* belongs with the ferns?
2. Contrast a fern to a green alga in terms of adaptations for life on land versus life in the water.

SCIENCE, TECHNOLOGY, AND SOCIETY

1. In 1986, an explosion at a nuclear reactor in Chernobyl, Ukraine, released clouds of radioactive material. Since then, researchers have been using mosses and other bryophytes as "living radioactivity meters" to monitor biologically harmful radiation in the regions most likely affected by the Chernobyl fallout. What characteristics of bryophytes make them especially sensitive to radioactivity and other harmful agents in the environment?
2. Plants have coexisted with parasitic organisms such as certain bacteria and fungi for millions of years. Defensive adaptations include many plant chemicals that inhibit the growth of these pathogens. How might these chemicals be useful to humans?

FURTHER READING

Graham, L. E. *Origin of Land Plants.* New York: Wiley, 1993. What can plant biologists learn by studying the algal relatives of plants?

Kenrick, P., and P. R. Crane. "The Origin and Early Evolution of Plants on Land." *Nature,* September 4, 1997. New discoveries and new methods are challenging some old hypotheses.

Moore, R., W. D. Clark, and D. S. Vodopich. *Botany,* 2d ed. Dubuque, IA: W. C. Brown, 1998. Contains beautifully illustrated chapters on bryophytes and seedless vascular plants.

Raven, P. H., R. F. Evert, and S. E. Eichhorn. *Biology of Plants,* 6th ed. New York: Worth Publishers, 1998. A widely used botany text that is especially strong on plant diversity.

WEB LINKS

Visit the special edition of *The Biology Place* for BIOLOGY, Fifth Edition, at http://www.biology.com/campbell. Go to Chapter 29 for online resources, including learning activities, practice exams, and links to the following web sites:

"The Internet Directory for Botany"
A comprehensive listing of plant-related sites around the world.

"Flora of North America"
The Flora of North America project is gathering and making available, in a variety of media, scientifically authoritative, up-to-date information on the approximately 20,000 species of vascular plants and bryophytes of North America, north of Mexico.

"Bryophytes"
This site is a resource devoted to bryology, the branch of plant science concerned with the study of mosses, liverworts, and hornworts, and contains links to other bryophyte resources.

"Fern Resource Hub"
Provided by the San Diego Fern Society, this site encourages access to sources for studying, collecting, buying, and growing ferns.

CHAPTER 30

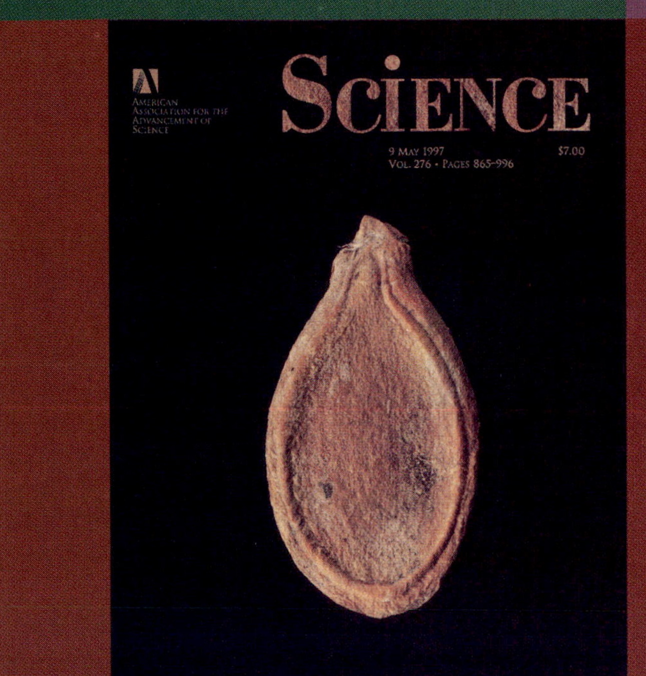

PLANT DIVERSITY II: THE EVOLUTION OF SEED PLANTS

Overview of Reproductive Adaptations of Seed Plants

- The gametophytes of seed plants became even more reduced than the gametophytes of seedless vascular plants
- In seed plants, the seed replaced the spore as the main means of dispersing offspring
- Pollen became the vehicles for sperm cells in seed plants

Gymnosperms

- The Mesozoic era was the age of gymnosperms
- The four divisions of extant gymnosperms are the cycads, the ginkgo, the gnetophytes, and the conifers
- The life cycle of a pine demonstrates the key reproductive adaptations of seed plants

Angiosperms (Flowering Plants)

- Terrestrial adaptation continued with the refinement of vascular tissue in angiosperms
- The flower is the defining reproductive adaptation of angiosperms
- Fruits help disperse the seeds of angiosperms
- The life cycle of an angiosperm is a highly refined version of the alternation of generations common to all plants
- The radiation of angiosperms marks the transition from the Mesozoic era to the Cenozoic era
- Angiosperms and animals have shaped one another's evolution
- Agriculture is based almost entirely on angiosperms

The Global Impact of Plants

- Plants transformed the atmosphere and the climate
- Plant diversity is a nonrenewable resource

This chapter continues the saga of how plants adapted to land and transformed Earth. The beautifully preserved squash seed in the photo, discovered in 1997 in a cave in Oaxaca, Mexico, symbolizes two important landmarks in the evolution of plants: (1) the evolution of seed plants, which led to the gymnosperms and angiosperms, the plants that dominate most modern landscapes, and (2) the emergence of the importance of seed plants to animals, specifically to humans in this case.

Why was this discovery front-page news for many newspapers and scientific journals? The cave had been occupied by humans some 8000–10,000 years ago, and the seeds and pieces of squash fruit found there differ enough from wild varieties of the same species to suggest that humans had already begun cultivating the plants, perhaps more for gourds than for food. If this interpretation is correct, then agriculture started at about the same time—approximately 10,000 years ago—in Asia, Europe, and the Americas. (Before this discovery, scientists had been able to trace American agriculture back only about 5000 years.) The invention of agriculture, which depends almost entirely on the cultivation and harvest of seed plants, was the single most important cultural change in the history of humanity, for it made possible the transition from hunter-gatherer societies to permanent settlements.

More generally, the seeds and other adaptations of gymnosperms and angiosperms enhanced the ability of plants to survive and reproduce in diverse terrestrial environments, and these plants became the main producers supporting the food webs of most ecosystems on land. Our study of gymnosperms and angiosperms begins with an overview of some of the key terrestrial adaptations that seed plants added to those already in place in the bryophytes and seedless vascular plants you learned about in Chapter 29.

OVERVIEW OF REPRODUCTIVE ADAPTATIONS OF SEED PLANTS

Modifications in life cycles catalyzed the success of seed plants as terrestrial organisms. Here we examine the three most important reproductive adaptations: reduction of the gametophyte, the advent of the seed, and the evolution of pollen.

The gametophytes of seed plants became even more reduced than the gametophytes of seedless vascular plants

The gametophytes of ferns and other seedless vascular plants develop in the soil as an independent generation, but the minute gametophytes of seed plants are protected from desiccation by being retained within the moist reproductive tissue of the sporophyte generation. Notice that this evolutionary trend reverses the gametophyte-sporophyte relationship as it occurs in bryophytes (FIGURE 30.1, p. 562).

(a) Sporophyte dependent on gametophyte (e.g., bryophytes)

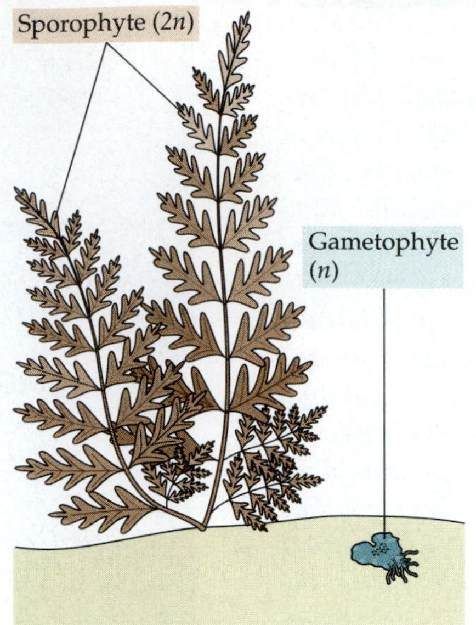

(b) Large sporophyte and small, independent gametophyte (e.g., ferns)

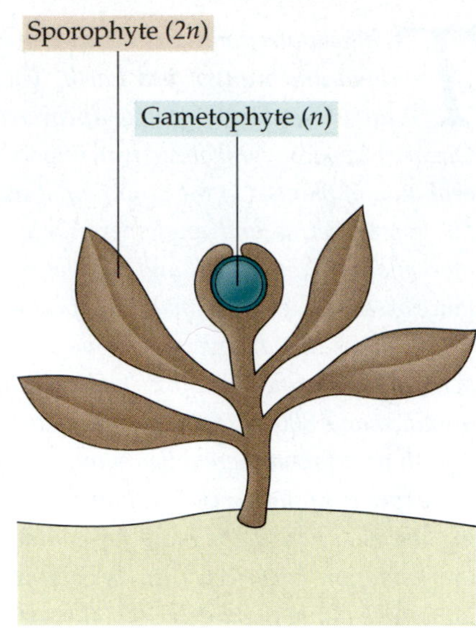

(c) Reduced gametophyte dependent on sporophyte (seed plants)

FIGURE 30.1 • Overview: three variations on alternation of generations in plants. **(a)** In the gametophyte-dominant life cycle of mosses and other bryophytes, the dependent sporophyte is nourished by the gametophyte as it grows out of the archegonium. **(b)** The sporophyte is the dominant generation in the life cycles of all vascular plants. The gametophytes of ferns, though small, are photosynthetic and free-living (not dependent on the sporophyte for their nutrition). **(c)** The gametophytes of seed plants are surrounded by tissues of the sporophyte, from which the gametophyte derives its nutrition.

Some plant biologists speculate that the shift toward diploidy in land plants was related to the harmful impact of the sun's ionizing radiation, which causes mutations. This damaging radiation is more intense on land than in aquatic habitats, where organisms are somewhat protected by the light-filtering properties of water. Of the two generations of land plants—gametophyte and sporophyte—the diploid form (sporophyte) may cope better with mutagenic radiation. A diploid organism homozygous for a particular essential allele has a "spare tire" in the sense that one copy of the allele may be sufficient for survival if the other is damaged. According to this hypothesis, the increasing prevalence of sporophytes during the evolution of vascular plants can be interpreted as another adaptation to terrestrial conditions.

Why has the gametophyte generation not been completely eliminated from the plant life cycle? One speculation is that small, relatively simple gametophytes provide a mechanism for "screening" alleles, including new mutations, with a minimum investment of the parent plant's resources. Since the gametophyte is haploid, no alleles can "hide" from the environment. Only those gametophytes that are genetically competent to at least perform basic metabolism and cell division will survive and produce gametes that combine to start new sporophytes.

Another reason the gametophyte has not gone completely "out of style" in seed plants is that all sporophyte embryos are dependent, at least to some extent, on tissues of the maternal gametophyte. You learned, for example, that in bryophytes the embryonic sporophyte is nourished by a gametophyte as it grows from the archegonium. You'll soon learn that even in seed plants, the gametophyte continues to play a role in nourishing the sporophyte embryo, at least during its early development.

In seed plants, the seed replaced the spore as the main means of dispersing offspring

In bryophytes and seedless vascular plants such as ferns, spores produced by sporophytes are the resistant stage in the life cycle that can withstand harsh environments. (A spore, recall from Chapter 29, is a resistant cell that can develop into a new organism.) For example, the spores of a moss might be able to survive even if the local environment becomes too cold, too hot, or too dry for the moss plants themselves to live. And because of their tiny size, the spores might also be dispersed in a dormant state to a new area, where they will germinate to give rise to new moss gametophytes if and when the environment is favorable enough for them to break dormancy. This mechanism spread plants over the Earth for almost the first 100 million years of life on land.

The seed represents a different solution to resisting harsh environments and dispersing offspring. In contrast to a spore, which is single-celled, a seed is a resistant structure that is multicellular and much more complex. A **seed** consists of a sporophyte embryo packaged along with a food supply within a protective coat. There are evolutionary and developmental relationships between spores and seeds. Recall from the previous section that the reduced gametophytes of seed plants develop within tissues of the parental sporophyte (see FIGURE 30.1c). This occurs because the parent sporophyte does not

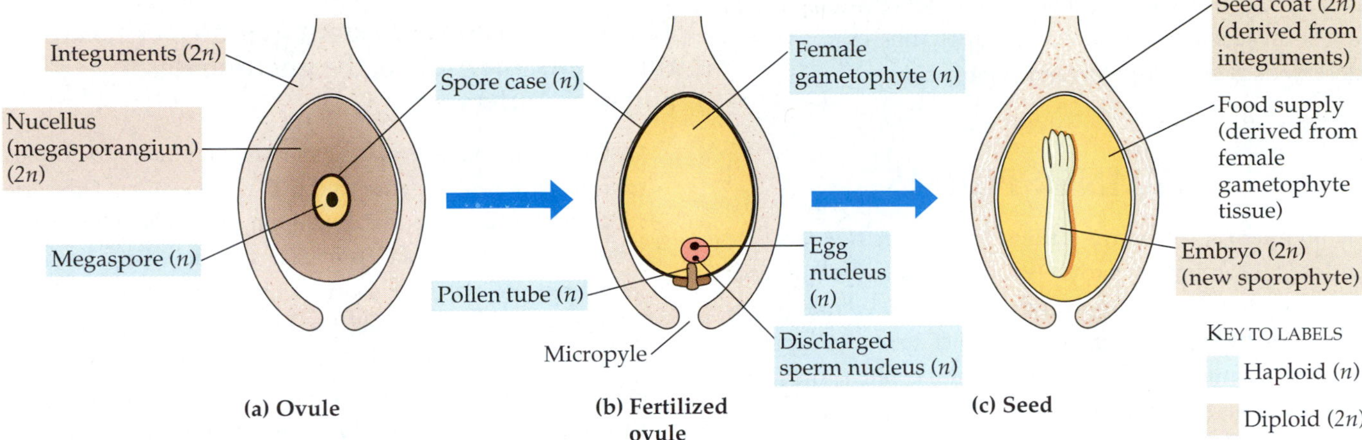

FIGURE 30.2 ▪ From ovule to seed. (a) In this sectional view through a generalized ovule of a sporophyte, the nucellus, a fleshy megasporangium, is surrounded by a protective layer of tissue called the integuments. **(b)** A megaspore develops into a multicellular female gametophyte. The micropyle, the only opening through the integuments, allows entry of a pollen grain, which develops a pollen tube that discharges sperm. **(c)** Fertilization initiates the transformation of the ovule into a seed, which consists of a sporophyte embryo, a food supply, and a protective seed coat derived from the integuments. Note that the evolution of the ovule represents a variation on the ancestral spore-forming characteristic that evolved in the first bryophytes. The main differences are that the spores are retained within the sporangia instead of being released, and that the gametophytes develop within the spore case while still protected and nourished by the parent sporophyte.

release its spores, but instead retains them within its sporangia. The spores are not only retained within the sporangia instead of being released, but the gametophyte develops within the wall of the spore from which it is derived.

All seed plants are heterosporous, meaning they have two different types of sporangia that produce two types of spores: Megasporangia produce megaspores, which give rise to female (egg-containing) gametophytes; microsporangia produce microspores, which give rise to male (sperm-containing) gametophytes. Recall from Chapter 29 that some seedless vascular plants, including water ferns, are also heterosporous. The gametophytes of those plants also develop within the spore cases, as they do in seed plants. What distinguishes seed plants is that the spores, and hence the gametophytes, are retained on the parent sporophyte.

The evolution of the seed is associated with the megasporangium. In seed plants, the megasporangium is not a chamber, but is instead a solid, fleshy structure called the **nucellus**. Another contrast with seedless plants is that additional layers of sporophyte tissues, called **integuments**, envelop the megasporangium of seed plants. Thus the megaspores formed within the megasporangium are very well protected. The whole structure—integuments, megasporangium (nucellus), and megaspores—is called an **ovule** (FIGURE 30.2a). Inside the ovule, the female gametophyte develops within the wall of the megaspore, nourished by the nucellus. The female gametophyte contains an egg cell, and if the egg cell is fertilized by a sperm cell (FIGURE 30.2b), the zygote develops into a sporophyte embryo. The whole ovule develops into a seed (FIGURE 30.2c).

A seed's protective seed coat is derived from the integuments of the ovule. Once released from the parent plant, the resistant seed can remain dormant for days, months, or even years. Under favorable conditions, the seed can then germinate, its sporophyte embryo emerging through the seed coat as a seedling. Some seeds drop close to their parents; others are carried far by wind or animals. Thus it is the seed, not the spore, that is the resistant and dispersible stage in the life cycle of seed plants.

Pollen became the vehicles for sperm cells in seed plants

We have seen the relationship of the megasporangium to the ovule and the seed, but what goes on in a microsporangium of a seed plant? Microspores develop into pollen grains, which mature to become the male gametophytes of seed plants. The pollen grains, protected by tough sporopollenin-containing coats (see Chapter 29), can be carried away by wind or animals after their release from the microsporangium. If a pollen grain, or male gametophyte, lands in the vicinity of an ovule, it will elongate a tube that discharges one or more sperm into the female gametophyte within the ovule (see FIGURE 30.2b). In some gymnosperms, the sperm cells retain the ancestral flagellated condition. But in the most common gymnosperms (the conifers) and in all angiosperms (flowering plants), the sperm cells lack flagella.

This mechanism for transfer of sperm contrasts sharply with what we observed in seedless plants. Recall that in bryophytes and seedless vascular plants such as ferns, flagellated sperm released from antheridia must swim through a film of water to reach egg cells in archegonia. In seed plants, the use of resistant, airborne pollen to bring gametes together is a terrestrial adaptation that led to even greater success and diversity of plants on land.

The roles of seeds and pollen as reproductive adaptations will seem less abstract now as we apply this overview to a closer look at gymnosperms and angiosperms.

GYMNOSPERMS

Gymnosperms (the term means "naked seeds") lack the enclosed chambers (ovaries) in which angiosperm seeds develop. Of the two groups of seed plants, gymnosperms appear in the fossil record much earlier than angiosperms. The most familiar gymnosperms are the conifers, the cone-bearing plants such as pines.

The Mesozoic era was the age of gymnosperms

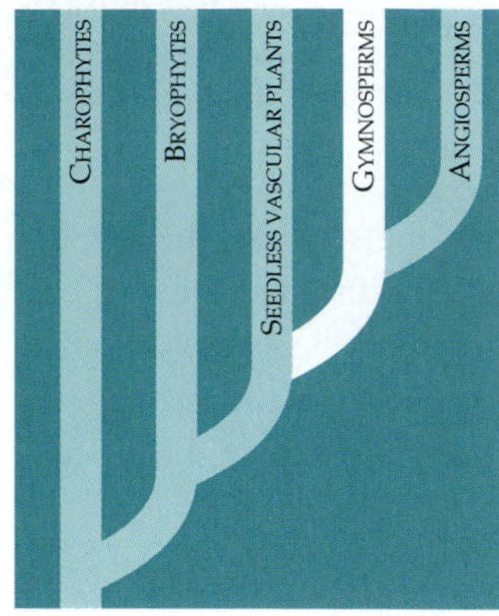

The gymnosperms probably descended from progymnosperms, a group of Devonian plants. Progymnosperms were originally seedless plants, but by the end of the Devonian period, seeds had evolved. Adaptive radiation during the Carboniferous and early Permian produced the various divisions of gymnosperms.

In the history of life, the Permian period was one of great crises. Formation of the supercontinent Pangaea (see Chapter 25) may have been one reason that continental interiors became warmer and drier as the Permian progressed. The flora and fauna of Earth changed dramatically, as many groups of organisms disappeared and others emerged as their successors. The changeover was most pronounced in the seas, but terrestrial life was affected as well. In the animal kingdom, amphibians decreased in diversity and were replaced by reptiles, which were especially well adapted to the arid conditions. Similarly, the lycophytes, horsetails, and ferns that dominated the Carboniferous swamps were largely replaced by gymnosperms, which were more suited to the drier climate. The world and its life had changed so markedly that geologists use the end of the Permian period, about 245 million years ago, as the boundary between the Paleozoic and Mesozoic eras. (This boundary was originally defined by the changeover in marine fossils.) The Mesozoic is sometimes referred to as the age of dinosaurs, but those giant reptiles were supported by a vegetation consisting mostly of conifers and great palmlike cycads, two divisions of gymnosperms. When the climate changed again at the end of the Mesozoic, becoming cooler, the dinosaurs became extinct. Many of the gymnosperms, particularly conifers, persisted, however, and are still an important part of Earth's flora.

The four divisions of extant gymnosperms are the cycads, the ginkgo, the gnetophytes, and the conifers

Of the eleven divisions in the plant kingdom (see TABLE 29.1), four are grouped as gymnosperms (FIGURE 30.3). Three are relatively small divisions: Cycadophyta, Ginkgophyta, and Gnetophyta. The cycads (division Cycadophyta) resemble palms but are not true palms, which are flowering plants. Being gymnosperms, cycads bear naked seeds on sporophylls, leaves specialized for reproduction. The ginkgo is the only extant species of division Ginkgophyta. It has fanlike leaves that turn gold and are deciduous in autumn, an unusual trait for a gymnosperm. Division Gnetophyta consists of three genera that are probably not closely related. One, *Weltwitschia,* is shown in FIGURE 30.3c. Plants of the second genus, *Gnetum,* grow in the tropics as trees or vines. *Ephedra* (Mormon tea), the third genus of Gnetophyta, is a shrub in the American deserts.

By far the largest of the four gymnosperm divisions is Coniferophyta, the conifers. The term **conifer** (L. *conus,* "cone," and *ferre,* "to carry") comes from the reproductive structure of these plants, the cone, which is a cluster of scale-like sporophylls. Pines, firs, spruces, larches, yews, junipers, cedars, cypresses, and redwoods all belong to this division of gymnosperms. Most are large trees. Although there are only about 550 species, conifers dominate vast forested regions of the Northern Hemisphere, where the growing season is relatively short because of latitude or altitude.

Nearly all conifers are evergreens, meaning they retain leaves throughout the year. Even during winter, a limited amount of photosynthesis occurs on sunny days. And when spring comes, conifers already have fully developed leaves that can take advantage of the sunnier days.

The needle-shaped leaves of pines and firs are adapted to dry conditions. A thick cuticle covers the leaf, and the stomata are located in pits, further reducing water loss.

We get most of our lumber and paper pulp from the wood of conifers. What we call wood is actually an accumulation of lignified xylem tissue, which gives the tree structural support.

Coniferous trees are among the tallest, largest, and oldest organisms on Earth. Redwoods, found only in a narrow coastal strip of northern California, grow to heights of more than 110 m; only certain eucalyptus trees in Australia are taller. The largest (most massive) organisms alive are the giant sequoias, relatives of redwoods that grow in the Sierra Nevada mountains of California (see FIGURE 30.3d). One, known as the General Sherman tree, has a trunk with a circumference of 26 m and weighs more than the combined weight of a dozen space shuttles. Bristlecone pines, another species of California conifer, are among the oldest organisms alive. One bristlecone, named Methuselah, is more than 4600 years old; it was a young tree when humans invented writing.

(a)

(b)

(c)

(d)

FIGURE 30.3 ▪ The four divisions of extant gymnosperms. (a) Cycadophyta. Cycads, such as this *Cycas revoluta,* resemble palms and in fact are sometimes called sago palms. **(b) Ginkgophyta.** The ginkgo, also known as the maidenhair tree, is a popular ornamental tree in cities because it can survive air pollution and other environmental insults. **(c) Gnetophyta.** *Welwitschia,* shown here, lives only in the deserts of southwestern Africa. Its straplike leaves are the largest known leaves. **(d) Coniferophyta.** Pines, firs, and redwoods are among the cone-bearing plants, or conifers. This giant sequoia (*Sequoiadendron giganteum*), the General Grant tree in California's Kings Canyon National Park, is over 80 m tall.

The life cycle of a pine demonstrates the key reproductive adaptations of seed plants

30.1 You learned earlier in the chapter that the evolution of seed plants added three key terrestrial adaptations in reproduction: the increasing dominance of the sporophyte generation; the advent of the seed as a resistant, dispersible stage in the life cycle; and the evolution of pollen as an agent that brings gametes together. Examining the life cycle of a pine, one of the most familiar of conifers and gymnosperms, will reinforce your understanding of these adaptations.

The pine tree is a sporophyte. Its sporangia are located on scalelike sporophylls that are packed densely in the structures called cones. The gametophyte generation develops from haploid spores that are retained within the sporangia. Conifers, like all seed plants, are heterosporous; male and female gametophytes develop from different types of spores produced by separate cones. Each tree usually has both types of cones. Small pollen cones produce microspores that develop into the male gametophytes, or pollen grains. Larger ovulate cones usually develop on separate branches of the tree and make megaspores that develop into female gametophytes (FIGURE 30.4, p. 566). From the time young cones appear on the tree, it takes nearly three years to produce the male and female gametophytes, bring them together via pollination, and form mature seeds from the fertilized ovules. The scales of the ovulate cone then separate, and the winged seeds travel on the wind. A seed that lands in a habitable place germinates, its embryo emerging as a pine seedling.

ANGIOSPERMS (FLOWERING PLANTS)

Today, angiosperms, or flowering plants, are by far the most diverse and geographically widespread of all plants. About 250,000 species are now known, compared with about 720 gymnosperm species. All angiosperms are placed in a single division, Anthophyta (Gr. *antho,* "flower"). The division is split into two classes: Monocotyledones (monocots) and Dicotyledones (dicots), which differ in several ways that are described in Chapter 35. Examples of monocots are lilies, orchids, yuccas, palms, and grasses, including lawn grasses,

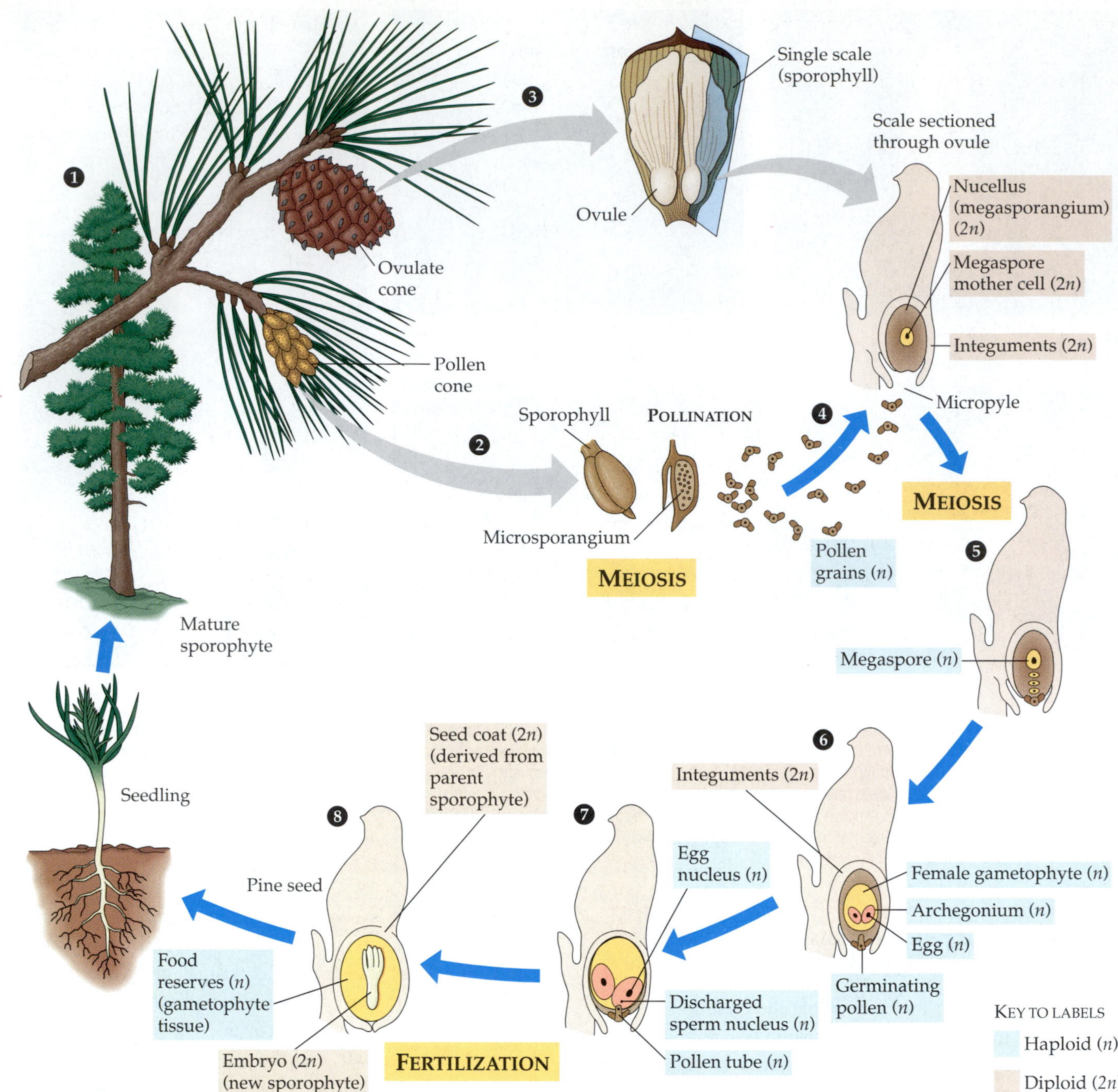

30.1 **FIGURE 30.4 ▪ The life cycle of a pine.** ① Trees (sporophytes) of most species bear both pollen cones and ovulate cones. ② A pollen cone contains hundreds of microsporangia held in small reproductive leaves, or sporophylls. Cells in the microsporangia undergo meiosis, giving rise to haploid microspores that develop into pollen grains (immature male gametophytes). ③ An ovulate cone consists of many scales, each a sporophyll with two ovules. Each ovule includes a megasporangium, called the nucellus, enclosed in protective integuments with a single opening, the micropyle. ④ During pollination, windblown pollen falls on the ovulate cone and is drawn into the ovule through the micropyle. The pollen grain germinates in the ovule, forming a pollen tube that begins to digest its way through the nucellus. Fertilization usually occurs more than a year after pollination. During that year, ⑤ a megaspore mother cell in the nucellus undergoes meiosis to produce four haploid cells. One of these cells survives as a megaspore, which grows and divides repeatedly, giving rise to the immature female gametophyte. Notice that the gametophyte develops within the wall of the spore. ⑥ Two or three archegonia, each with an egg, then develop within the gametophyte. ⑦ By the time eggs are ready to be fertilized, two sperm cells have developed in the male gametophyte (pollen grain) and the pollen tube has grown through the nucellus to the female gametophyte. Fertilization occurs when one of the sperm nuclei, injected into an egg cell by the pollen tube, unites with the egg nucleus. All the eggs in an ovule may be fertilized, but usually only one zygote develops into an embryo. ⑧ The pine embryo, or the new sporophyte, has a rudimentary root and several embryonic leaves. A food supply, consisting of the female gametophyte, surrounds and nourishes the embryo. The ovule has developed into a pine seed, which consists of an embryo (new sporophyte), its food supply (derived from gametophyte tissue), and a surrounding seed coat derived from the integuments of the parent tree (parent sporophyte). Notice that three plant generations—one gametophyte and two sporophyte generations—are represented in a gymnosperm seed.

sugar cane, and grain crops (corn, wheat, rice, and others). Among the many dicot families are roses, peas, buttercups, sunflowers, oaks, and maples.

Terrestrial adaptation continued with the refinement of vascular tissue in angiosperms

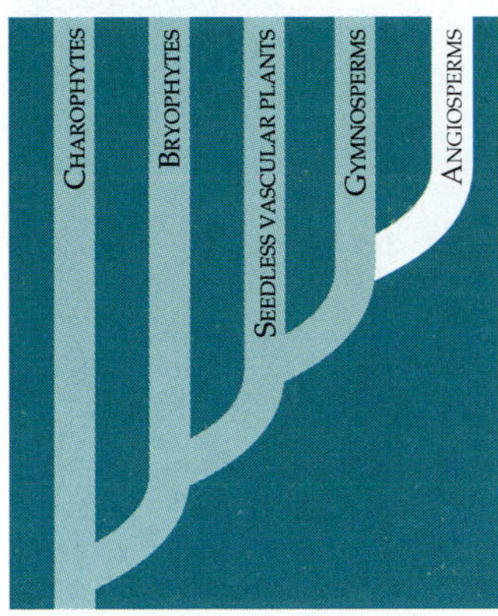

Xylem became more specialized for water transport during angiosperm evolution. The cells that conduct water in conifers are **tracheids**, believed to be a relatively early type of xylem cell (FIGURE 30.5). The tracheid is an elongated, tapered cell that functions in both mechanical support and movement of water up the plant. In most angiosperms, shorter, wider cells called **vessel elements** evolved from tracheids. Vessel elements are arranged end to end, forming continuous tubes that are more specialized than tracheids for transporting water but less specialized for support. The xylem of angiosperms is reinforced by a second cell type, the **fiber**, which also evolved from the tracheid. With their thick lignified walls, the xylem fibers are specialized for support. Fiber cells evolved in conifers, but vessel elements did not.

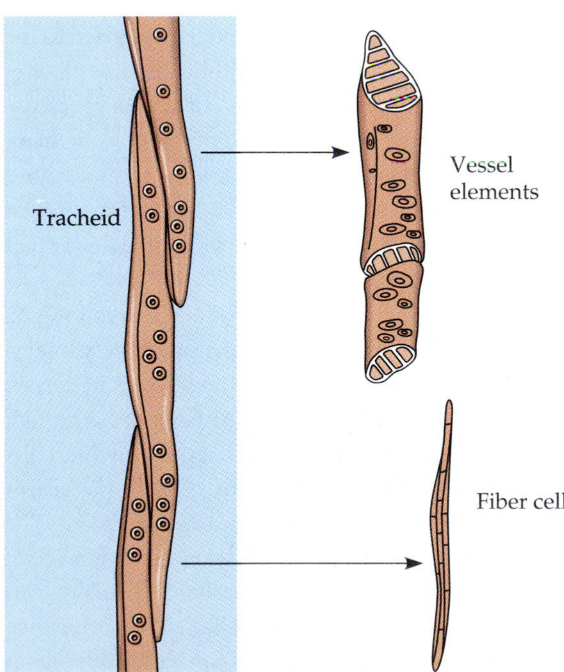

FIGURE 30.5 ▪ The evolution of xylem cells. In angiosperm evolution, tracheids gave rise to vessel elements specialized for conducting water and to fiber cells specialized for support.

Refinements in vascular tissue and other structural advances surely contributed to the success of angiosperms, but the greatest factor in the rise of angiosperms was probably the evolution of the flower, a remarkable apparatus that enhances the efficiency of reproduction by attracting and rewarding pollen-carrying animals.

The flower is the defining reproductive adaptation of angiosperms

The **flower** is the reproductive structure of an angiosperm. In most angiosperms, insects and other animals transfer pollen from one flower to female sex organs on another flower, which makes pollination less random than the wind-dependent pollination of gymnosperms. (Some flowering plants are wind-pollinated, but we do not know whether this condition is primitive or whether it evolved secondarily from ancestors that were pollinated by animals.)

A flower is a compressed shoot with four whorls of modified leaves: sepals, petals, stamens, and carpels (FIGURE 30.6). Starting at the bottom of the flower are the **sepals**, which are usually green. They enclose the flower before it opens (think of a rosebud). Above the sepals are the **petals**, brightly colored in most flowers. They aid in attracting insects and other pollinators. Flowers that are wind-pollinated, such as those of many grasses, are generally drab in color. The sepals and petals are sterile floral parts, meaning that they are not directly involved in reproduction. Within the ring of petals are the reproductive organs, **stamens** and **carpels**, the "male" and

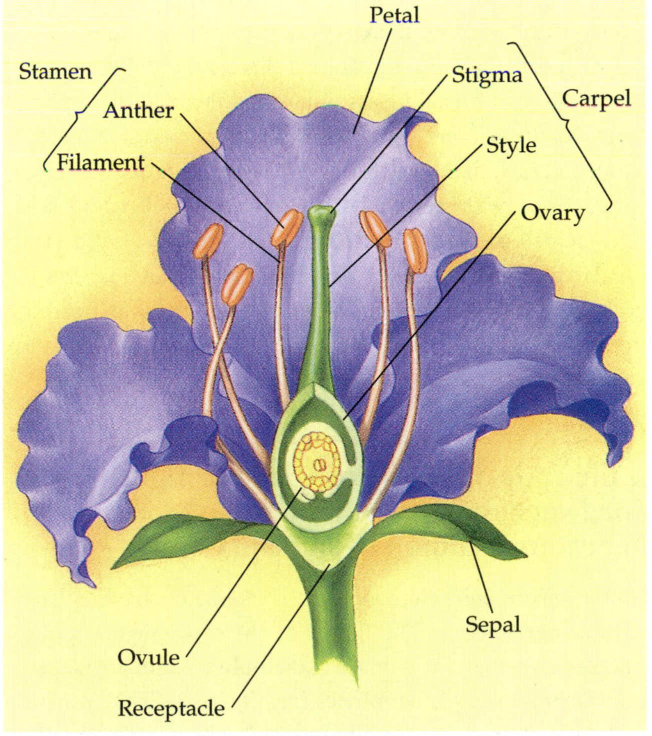

FIGURE 30.6 ▪ The structure of a flower.

"female" flower parts, respectively. A stamen consists of a stalk called the **filament** and a terminal sac, the **anther**, where pollen is produced. At the tip of the carpel is a sticky **stigma** that receives pollen. A **style** leads to the **ovary** at the base of the carpel. Protected within the ovary are the ovules, which develop into seeds after fertilization.

Recall that the enclosure of seeds within the ovary is one of the features that distinguishes angiosperms from gymnosperms. The carpel probably evolved from a seed-bearing leaf that became rolled into a tube. Some angiosperms, such as garden peas, have flowers with single carpels. Others, such as magnolias, have several separate carpels. Still other species, such as lilies, have two or more fused carpels, usually forming an ovary with multiple ovule-containing chambers. As an alternative to the term *carpel*, some botanists prefer *pistil*, which refers to either a single carpel or the more complex organs consisting of fused carpels.

FIGURE 30.7 ▪ Fruit as an adaptation for seed dispersal. After pollination and fertilization, the ovules of flowers develop into seeds and the ovaries develop into fruits, such as these mountain ash berries. Edible fruits are ingested by animals, most commonly mammals or birds, such as this cedar waxwing. Seed coats usually prevent digestion of the seeds, which the animal may deposit in its feces some distance from the parent plant.

Fruits help disperse the seeds of angiosperms

A **fruit** is a mature ovary. As seeds develop after fertilization, the wall of the ovary thickens. A pea pod is an example of a fruit, with seeds (mature ovules, the peas) encased in the ripened ovary (the pod). Fruits protect dormant seeds and aid in their dispersal (FIGURE 30.7).

Various modifications in fruits help disperse seeds. Some flowering plants, such as dandelions and maples, have seeds within fruits that are shaped something like kites or propellers, which enhances dispersal by wind. However, most angiosperms use animals to carry seeds. Some of these plants have fruits modified as burrs that cling to animal fur (or to the clothes of humans). Other angiosperms produce edible fruits. When it eats the fruit, the animal digests the fleshy part, but the tough seeds usually pass unharmed through the digestive tract. Mammals and birds may deposit seeds, along with a supply of fertilizer, miles from where the fruit was eaten. Interactions with animals that transport pollen and seeds have helped angiosperms become the most successful plants on Earth. However, as we will now see, the angiosperm life cycle was not a new evolutionary "invention," but was built upon adaptive themes we have tracked throughout our study of plant diversity.

The life cycle of an angiosperm is a highly refined version of the alternation of generations common to all plants

Angiosperms are heterosporous, a characteristic they share
30.2 with all seed plants. The flower of the sporophyte produces microspores that form male gametophytes and megaspores that form female gametophytes (FIGURE 30.8). The immature male gametophytes are **pollen grains**, which develop within the anthers of stamens. Each pollen grain has two haploid cells. **Ovules**, which develop in the ovary, contain the female gametophyte, which is called the **embryo sac**. It consists of only a few cells. In most angiosperms, the megaspore divides three times to form eight haploid nuclei in seven cells (a large central cell has two haploid nuclei). One of the cells is the egg. (The development of pollen and the embryo sac is described in more detail in Chapter 38.) Notice that the evolutionary trend toward reduction of the gametophyte generation in vascular plants continued with the angiosperms.

After its release from the anther, the pollen is carried to the sticky stigma at the tip of a carpel. Although some flowers self-pollinate, most have mechanisms that ensure **cross-pollination**, the transfer of pollen from flowers of one plant to flowers of another plant of the same species. In some cases, stamens and carpels of a single flower may mature at different times, or the organs may be so arranged within the flower that self-pollination is unlikely.

The pollen grain germinates after it adheres to the stigma of a carpel. The pollen grain, now a mature male gametophyte, extends a tube that grows down the style of the carpel. After it reaches the ovary, the pollen tube penetrates through the micropyle, a pore in the integuments of the ovule, and discharges two sperm cells into the embryo sac. One sperm nucleus unites with the egg, forming a diploid zygote. The other sperm nucleus fuses with two nuclei in the center cell of the embryo sac. This central cell now has a triploid ($3n$) nucleus. The pollen of conifers, remember, also releases two sperm nuclei, but one disintegrates. In contrast, both sperm nuclei of angiosperm pollen fertilize cells in the embryo sac. This phenomenon, known as **double fertilization**, is characteristic of angiosperms. (Double fertilization also occurs in

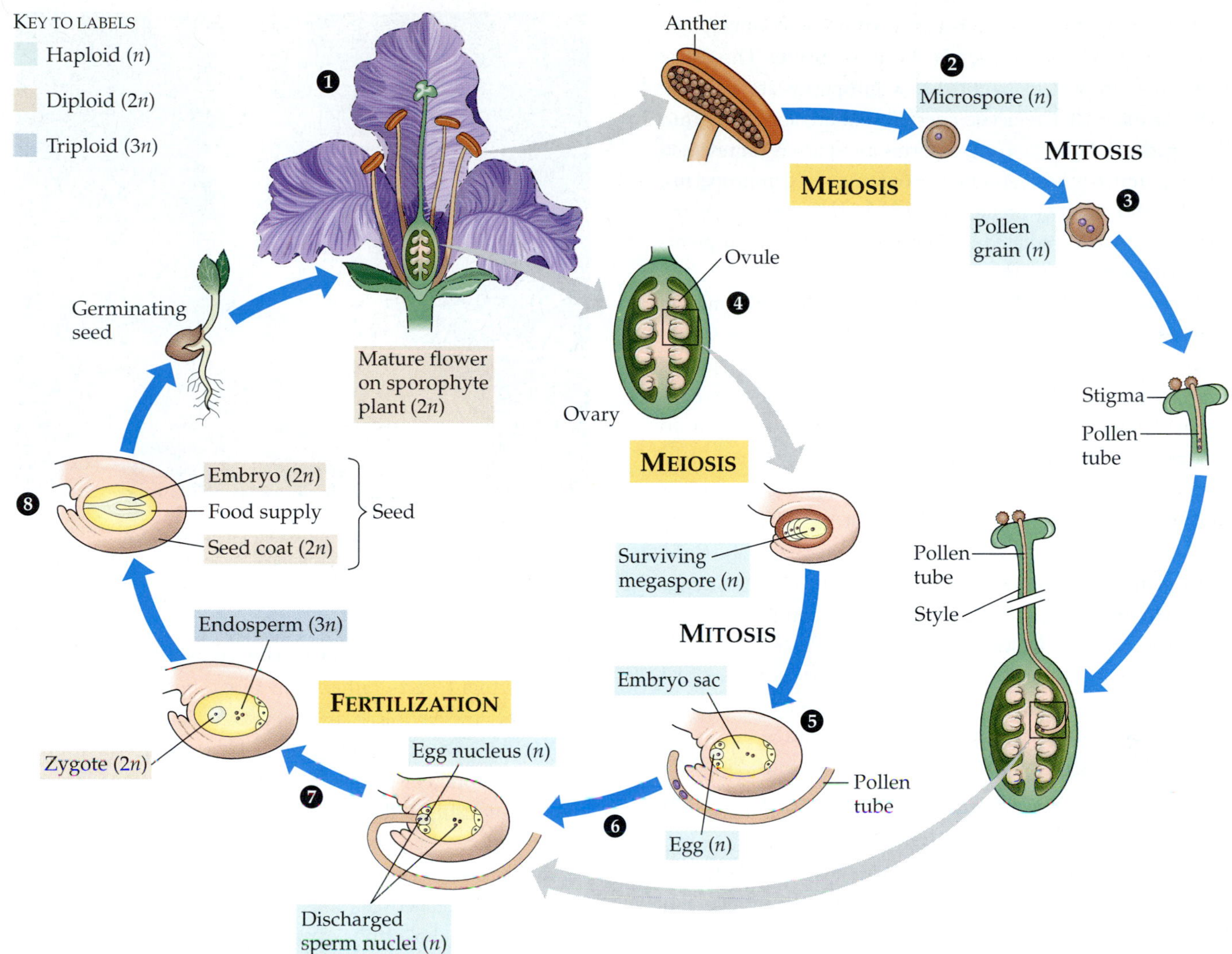

FIGURE 30.8 • The life cycle of an angiosperm. ① The anthers of the flower produce ② microspores that form ③ male gametophytes (pollen). ④ The ovules produce megaspores that form ⑤ female gametophytes (embryo sacs) within ovules. ⑥ Pollination brings the gametophytes together in the ovary. ⑦ Fertilization occurs, and ⑧ zygotes develop into sporophyte embryos that are packaged along with food into seeds. When a seed germinates, the embryo grows and develops into a sporophyte.

30.2

Ephedra, a member of division Gnetophyta, the gymnosperm division most closely related to angiosperms.)

After double fertilization, the ovule matures into a seed. The zygote develops into a sporophyte embryo with a rudimentary root and either one or two seed leaves, the **cotyledons** (monocots have one seed leaf and dicots have two, hence their names). The triploid nucleus in the center of the embryo sac divides repeatedly, giving rise to a triploid tissue called **endosperm**, rich in starch and other food reserves. Monocot seeds such as corn store most of their food in the endosperm. Beans and many other dicots transfer most of the nutrients from the endosperm to the developing cotyledons.

What is the function of double fertilization? According to one hypothesis, double fertilization synchronizes the development of food storage in the seed with development of the embryo. If a particular flower is not pollinated or sperm cells are not discharged into the embryo sacs, fertilization does not occur and neither embryo nor endosperm forms. Perhaps the requirement for double fertilization prevents flowering plants from squandering nutrients on infertile ovules.

The seed is a mature ovule, consisting of the embryo, endosperm, and a seed coat derived from the integuments (outer layers of the ovule). An ovary develops into a fruit as its ovules develop into seeds. In a suitable environment, a seed germinates. The coat ruptures and the embryo emerges as a seedling, using the food stored in the endosperm and cotyledons.

The radiation of angiosperms marks the transition from the Mesozoic era to the Cenozoic era

Earth's landscapes changed dramatically with the origin and radiation of flowering plants. The ancestry of angiosperms is still uncertain, but recent cladistic analysis of homologous features points to the gymnosperms of division Gnetophyta as the closest living relatives of flowering plants. The oldest fossils that are widely accepted as angiosperms are found in rocks of the early Cretaceous period that are about 130 million years old. Fossilized angiosperms are sparsely represented among a much greater abundance of ferns and gymnosperms. By the end of the Cretaceous, 65 million years ago, the angiosperms had radiated and become the dominant plants on Earth, as they are today.

Just as the end of the Permian period almost 200 million years earlier featured mass extinctions, the end of the Cretaceous was a crisis period when many old groups of organisms were replaced by new ones. Cooler climates may have contributed to the changeover. Again, the frequency of extinctions was greatest in the seas, but significant changes in terrestrial fauna and flora also occurred. The dinosaurs disappeared, as did many of the cycads and conifers that had thrived during the Mesozoic era. They were replaced by mammals and flowering plants. The change in fossils during the late Cretaceous is so extreme that geologists use the end of that period as the boundary between the Mesozoic and Cenozoic eras.

Angiosperms and animals have shaped one another's evolution

Ever since they followed plants onto the land, animals have influenced the evolution of terrestrial plants, and vice versa. The fact that animals must eat affects the natural selection of both animals and plants. For instance, with animals crawling and foraging for food on the forest floor, natural selection must have favored plants that kept their spores and gametophytes up in the treetops, rather than dropping these crucial structures to within reach of hungry animals on the ground. This, in turn, may have been a selection factor in the evolution of flying insects. On the other hand, as plants with flowers and fruits evolved, some herbivores became beneficial to the plants by carrying the pollen and seeds of plants they used as food. Certain animals became specialists at these tasks, feeding on specific plants. Natural selection reinforced these interactions, for they improved the reproductive success of both partners: The plant got pollinated and the animal got fed. The mutual evolutionary influence between two species is termed **coevolution**. (This definition is refined in Chapter 53.)

Coevolution of angiosperms and their pollinators is partly responsible for the diversity of flowers. Some flowers are pollinated by a specific animal, such as a particular type of bee, beetle, bird, or bat (FIGURE 30.9). These exclusive relationships ensure that the plant's pollen will not be wasted by being carried to the flower of a different species. At the same time, the pollinator has a monopoly on a food source.

In most cases, relationships between plants and their pollinators are less specific than in the extreme coevolution between one plant species and one animal species. For example, the flowers of a particular plant species may be adapted for attracting insects rather than birds, but many different insect species may serve as pollinators. Conversely, a single animal species—a honeybee species, for example—may pollinate many different plant species. But even in these less specific relationships, flower color, fragrance, and structure usually reflect specialization for a particular group of pollinators, such as various species of bees or hummingbirds.

FIGURE 30.9 • A relationship between an angiosperm and its pollinator. This Scotch broom has a tripping mechanism that dusts pollen onto the back of a visiting bee.

Relationships between angiosperms and animals are also evident in the edible fruits of angiosperms. Fruits that are not yet ripe are usually green, hard, and distasteful (at least to humans). This helps the plant retain its fruit until the seeds are mature and ready for dispersal. As it ripens, the fruit becomes softer and its sugar content increases. Many fruits also become fragrant and brightly colored, advertising their ripeness to animals. One of the most common colors for ripe fruit is red, which vertebrates can probably distinguish better than insects can. Thus, most of the fruit is saved for birds and mammals, animals large enough to disperse the seeds (see FIGURE 30.7). Again, we see that one of the keys to angiosperm success has been interaction with animals.

Agriculture is based almost entirely on angiosperms

Flowering plants provide nearly all our food. All of our fruit and vegetable crops are angiosperms. Corn, rice, wheat, and the other grains are grass fruits. The endosperm of the grain seeds is the main food source for most of the people of the world and their domesticated animals. We also grow angiosperms for fiber, medications, perfumes, and decoration.

Like other animals, early humans probably collected wild seeds and fruits. Agriculture was gradually invented as humans began sowing seeds and cultivating plants to have a more dependable food source. As they domesticated certain plants, such as the squash represented by the seed in the photo that introduced this chapter, humans began to intervene in plant evolution by selective breeding designed to improve the quantity and quality of the foods the crops produced. Agriculture is a unique kind of evolutionary relationship between plants and animals. And the dependence of animals on plants is just one facet of the plant kingdom's significance for life on Earth.

THE GLOBAL IMPACT OF PLANTS

As a capstone to our two-chapter survey of plant diversity, this last section highlights a few other examples of the crucial ecological importance of plants.

Plants transformed the atmosphere and the climate

Clearly, the evolution of plants had a global impact by providing a nutritional foundation for terrestrial ecosystems. What is less obvious is that plants changed the physical environment on a global scale. One of the most important effects was a large decrease in the amount of carbon dioxide in the atmosphere, which in turn made Earth cooler.

About the time plants colonized land, atmospheric CO_2 concentrations dropped, and some researchers, including Robert Berner of Yale University, propose that the correlation was no coincidence (FIGURE 30.10). The decline in CO_2 resumed about the time vascular plants evolved and spread to uplands that were probably less habitable by bryophytes.

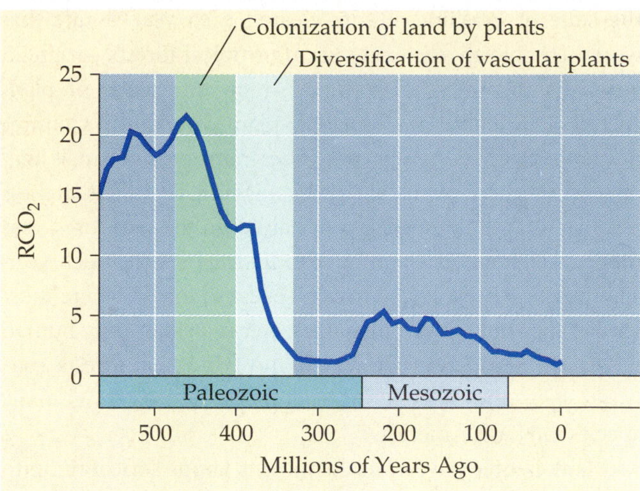

FIGURE 30.10 ▪ A history of atmospheric CO_2 concentration. The curve represents averages of estimates of past concentrations based on several different methods. RCO_2 is the ratio of the mass of atmospheric CO_2 at different times in the past relative to the amount today. Atmospheric CO_2 was about 20 times greater during the early Paleozoic than it is today but began to decline about the time plants became established on land. Plants could have contributed to this global change by locking up carbon in resistant organic polymers and by making minerals that react with CO_2 available through the weathering actions of roots on soil. By reducing the greenhouse effect, the drop in atmospheric CO_2 probably caused global cooling during the Paleozoic era.

Carbon dioxide in the atmosphere contributes to warming of the Earth's surface because of what is called the greenhouse effect. The glass of a greenhouse traps heat because it is transparent to sunlight, which warms the interior of the greenhouse but impedes the escape of heat back to the outside. Similarly, CO_2 allows solar radiation to penetrate the atmosphere at a faster rate than it allows heat from the warmed surface of the Earth to radiate back into space. You'll learn more about the greenhouse effect in Chapter 54. For now, what is important to understand is that by lowering atmospheric CO_2 concentration, plants likely contributed to a significant cooling of the Earth that occurred during the late Paleozoic era. And that cooling may have made more terrestrial locations habitable by plants and animals.

Related to this hypothesis about the role of plants in the global changes of the late Paleozoic are ideas about how plants lowered the amount of CO_2 in the atmosphere. Your first guess might be that plants had this effect through their use of CO_2 as a carbon source for photosynthesis. But plants also return some of that CO_2 through their own respiration, and the heterotrophs they support add more CO_2. Much of the assimilated carbon, however, becomes locked up for relatively long periods in resistant polymers such as sporopollenin, lignin, and waxes that remain in the soil long after the plants that produced them have died. By forming these resistant polymers in great abundance, plants would have lowered the amount of CO_2 in the atmosphere. However, Berner postulates that plants had their biggest effect on the atmosphere by their action on soil, especially as vascular plants, with their true roots, spread to rockier uplands. Roots break up rocks and secrete acids that free some of the minerals from soil particles. Carbon dioxide can react with some of these minerals, especially after they run off into the oceans, and such reactions would lower the amount of atmospheric CO_2.

All these hypotheses require much more testing, but there is little doubt that a change so significant as the founding of terrestrial ecosystems by plants had a huge impact on the planet.

Plant diversity is a nonrenewable resource

The exploding human population and its demand for space and natural resources are extinguishing plant species at an unprecedented rate. The problem is especially critical in the tropics, where more than half the human population lives and population growth is fastest. Tropical rain forests are being destroyed at a frightening pace. The most common cause of this destruction is slash-and-burn clearing of the forest for agricultural use. Fifty million acres, an area about the size of the state of Washington, are cleared each year, a rate that would completely eliminate Earth's tropical forests within 25 years. As the forest disappears, so do thousands of plant species. Extinction is irrevocable; plant diversity is a nonrenewable resource. Insects and other rain forest animals that depend on these plants are also vanishing. In all, researchers estimate that the destruction of habitat in the rain forest and other ecosystems is claiming hundreds of species each year. The toll is greatest in the tropics, because that is where most species live; but environmental assault is a generically human tendency. Europeans eliminated most of their forests centuries ago, and habitat destruction is endangering many species in North America.

Many people have ethical concerns about contributing to the extinction of living forms. But there are also practical reasons to be concerned about the loss of plant diversity. We depend on plants for thousands of products, including food, building materials, and medicines. TABLE 30.1 lists only a few examples of how we use the unique secondary products of plants (see p. 547 to review secondary products). So far, we have explored the potential uses of only a tiny fraction of the 300,000 known plant species. For example, almost all our food is based on the cultivation of only about two dozen species. More than 120 prescription drugs are extracted from plants. However, researchers have investigated fewer than 5000 plant species as potential sources of medicine. Pharmaceutical companies were led to most of these species by local people who use the plants in preparing their traditional medicines.

Table 30.1 ▪ A Sampling of Plant Secondary Products and Their Uses by Humans*

COMPOUND	EXAMPLE OF SOURCE	EXAMPLE OF USE
Atropine	Belladonna plant	Pupil dilator in eye exams
Digitalin	Foxglove	Heart medication
Menthol	Eucalyptus tree	Ingredient in cough medicines
Morphine	Opium poppy	Pain reliever
Quinine	Quinine tree	Malaria preventive
Rubber	Rubber tree	Tires and many other products
Taxol	Pacific yew	Ovarian cancer drug
Tubocurarine	Curare tree	Muscle relaxant during surgery
Vinblastine	Periwinkle	Leukemia drug

*Adapted from Randy Moore et al., *Botany*, 2nd ed. Dubuque, IA: Brown, 1998. Table 2.2, p. 37. ©1998 The McGraw-Hill Companies. Reprinted by permission of The McGraw-Hill Companies.

The tropical rain forest may be a medicine chest of healing plants that could be extinct before we even know they exist. This is only one reason to value what is left of plant diversity and to search for ways to slow the loss. The solutions we propose must be economically realistic. If the goal is only profit for the short term, then we will continue to slash and burn until the forests are gone. If, however, we begin to see rain forests and other ecosystems as living treasures that can regenerate only slowly, we may learn to harvest their products at sustainable rates. What else can we do to preserve plant diversity? Few questions are as important.

CHAPTER REVIEW

REVIEW OF KEY CONCEPTS

(with page numbers and key figures)

OVERVIEW OF REPRODUCTIVE ADAPTATIONS OF SEED PLANTS

- **The gametophytes of seed plants became even more reduced than the gametophytes of seedless vascular plants (pp. 561–562, FIGURE 30.1)** The gametophytes of seed plants develop within the cases of spores retained within tissues of the parent sporophyte.
- **In seed plants, the seed replaced the spore as the main means of dispersing offspring (pp. 562–563, FIGURE 30.2)** A seed, which is derived from a fertilized ovule, consists of a sporophyte embryo packaged along with a food supply within a seed coat.
- **Pollen became the vehicles for sperm cells in seed plants (p. 563)** A pollen grain, which is an immature male gametophyte, can be dispersed through the air by wind or transported by animals.

GYMNOSPERMS

- **The Mesozoic era was the age of gymnosperms (p. 564)** Gymnosperms bear their seeds "naked" on the surfaces of sporophylls.
- **The four divisions of extant gymnosperms are the cycads, the ginkgo, the gnetophytes, and the conifers (p. 564, FIGURE 30.3)** The cone-bearing conifers, including pines, firs, and spruces, are by far the most diverse gymnosperms today.

30.1 **The life cycle of a pine demonstrates the key reproductive adaptations of seed plants (pp. 564–565, FIGURE 30.4)** Dominance of the sporophyte generation, the development of seeds from fertilized ovules, and the role of pollen in transferring sperm between plants are key features of the life cycle of a conifer.

ANGIOSPERMS (FLOWERING PLANTS)

- **Terrestrial adaptation continued with the refinement of vascular tissue in angiosperms (p. 567, FIGURE 30.5)** In angiosperms, both vessel elements and fibers evolved from tracheids, a type of xylem.
- **The flower is the defining reproductive adaptation of angiosperms (pp. 567–568, FIGURE 30.6)** Sepals, petals, stamens (which produce pollen), and carpels (which produce ovules) are the whorls of modified leaves that make up flowers.
- **Fruits help disperse the seeds of angiosperms (p. 568, FIGURE 30.7)** Ovaries ripen into fruits, which are often carried by wind or animals to new locations.

30.2 **The life cycle of an angiosperm is a highly refined version of the alternation of generations common to all plants (pp. 568–569, FIGURE 30.8)** Double fertilization occurs when a pollen tube discharges two sperm into the embryo sac, the female gametophyte within an ovule. One sperm fertilizes the egg, while the other combines with two nuclei in the center cell of the embryo sac to initiate development of food-storing endosperm.

- **The radiation of angiosperms marks the transition from the Mesozoic era to the Cenozoic era (pp. 569–570)** The origin of angiosperms is still unresolved.
- **Angiosperms and animals have shaped one another's evolution (p. 570, FIGURE 30.9)** Pollination of flowers by animals and transport of seeds by animals are two important relationships in terrestrial ecosystems.
- **Agriculture is based almost entirely on angiosperms (pp. 570–571)** Human cultures depend on the cultivation and harvest of angiosperms, especially the fruits of grains.

THE GLOBAL IMPACT OF PLANTS

- **Plants transformed the atmosphere and the climate (p. 571, FIGURE 30.10)** By lowering the concentration of atmospheric CO_2, plants probably contributed to the cooling of Earth during the Paleozoic era.
- **Plant diversity is a nonrenewable resource (pp. 571–572)** Destruction of tropical forests is an especially urgent problem because they contain the greatest diversity of plants on Earth.

SELF-QUIZ

1. The male gametophyte of an angiosperm is the
   a. anther
   b. embryo sac
   c. microspore
   d. germinated pollen grain
   e. ovule

2. A fruit is most commonly
   a. a mature ovary
   b. a thickened style
   c. an enlarged ovule
   d. a modified root
   e. a mature female gametophyte

3. Important terrestrial adaptations that evolved *exclusively* in seed plants include all of the following *except*
   a. pollination
   b. transport of water through vascular tissue
   c. retention of the gametophyte plant within the sporophyte
   d. dispersal of new plants by seeds
   e. retention of spores by the parent sporophyte

4. Plant diversity is greatest in
   a. tropical forests
   b. deserts
   c. salt marshes
   d. the temperate forests of Europe
   e. farmlands

5. Microspore is to ______________ as ______________ is to embryo sac.
   a. anther; pollen
   b. pollen; megaspore
   c. pollen; gametophyte
   d. pollen; ovule
   e. gymnosperm; angiosperm

6. Gymnosperms and angiosperms have the following in common *except*
   a. seeds
   b. pollen
   c. vascular tissue
   d. ovaries
   e. ovules

Questions 7–10

In the following cladogram, match the derived characters (see Chapter 25) with the correct branch points:

a. double fertilization
b. embryos
c. seeds
d. vascular tissue

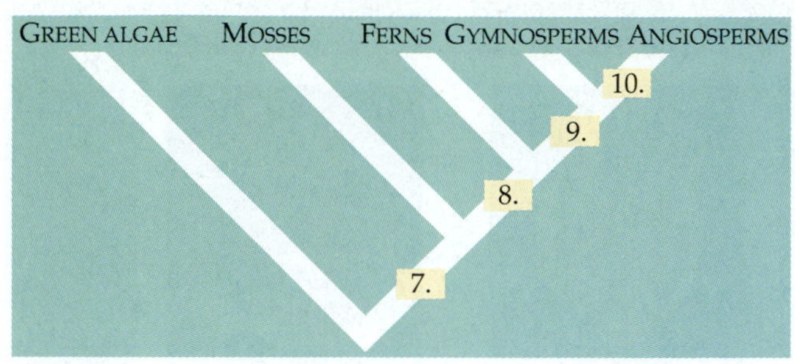

CHALLENGE QUESTION

The history of life has been punctuated by several mass extinctions. The impact of a meteorite may have wiped out the dinosaurs and many forms of marine life at the end of the Cretaceous period. Fossils indicate that plants were much less severely affected by this and other mass extinctions. What adaptations may have enabled plants to withstand these disasters better than animals?

SCIENCE, TECHNOLOGY, AND SOCIETY

Why are tropical rain forests being destroyed at such an alarming rate? What kinds of social, technological, and economic factors are responsible? Most forests in developed Northern Hemisphere countries have already been cut. Do the developed nations have a right to ask the developing nations in the Southern Hemisphere to slow or stop the destruction of their forests? Defend your answer. What kinds of benefits, incentives, or programs might slow the assault on rain forests?

FURTHER READING

Berner, R.A. "The Rise of Plants and Their Effect on Weathering and Atmospheric CO_2." *Science*, April 25, 1997. How plants changed Earth.

Cox, P. A., and M. J. Balick. "The Ethnobotanical Approach to Drug Discovery." *Scientific American*, June 1994. What can we learn about medicinal plants from other cultures?

Diamond, J. "How to Tame a Wild Plant." *Discover*, September 1994. Describes how humans domesticated food plants.

Doyle, R. "Plants at Risk in the U.S." *Scientific American*, August 1997. A state-by-state tally of loss of plant diversity.

Holden, C. "Ancient Trees Down Under." *Science*, January 20, 1995. Discovery of a "living fossil" near Sydney, Australia, a conifer thought to be extinct for 50 million years.

Niklas, K. *The Evolutionary Biology of Plants*. Chicago: The University of Chicago Press, 1997. A synthesis of what researchers have learned about the history of plants.

WEB LINKS

Visit the special edition of *The Biology Place* for BIOLOGY, Fifth Edition, at http://www.biology.com/campbell. Go to Chapter 30 for online resources, including learning activities, practice exams, and links to the following web sites:

"National Plant Data Center"
Information on many North American plants.

"Botanic Gardens Conservation International"
Much of the world's plant diversity is under threat, and the 1600 botanical gardens around the world are now working together to conserve plant diversity. You can visit botanical gardens on all five continents by following this link.

"Centre for Plant Biodiversity Research and Australian National Herbarium"
Australia's separation from the other continents gave rise to a diverse and unique flora. Of particular interest at this site are the many color images of plants native to Australia.

"Medicinal and Poisonous Plant Databases"
Links to databases of poisonous and medicinal plants.

CHAPTER 31

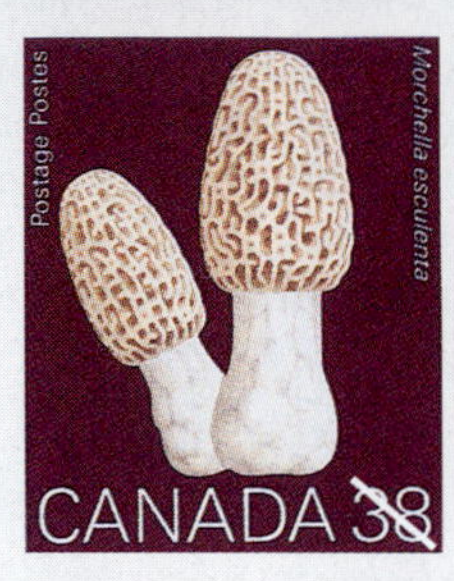

FUNGI

Introduction to the Fungi

- Absorptive nutrition enables fungi to live as decomposers and symbionts
- Extensive surface area and rapid growth adapt fungi for absorptive nutrition
- Fungi reproduce by releasing spores that are produced either sexually or asexually

Diversity of Fungi

- Division Chytridiomycota: Chytrids may provide clues about fungal origins
- Division Zygomycota: Zygote fungi form resistant dikaryotic structures during sexual reproduction
- Division Ascomycota: Sac fungi produce sexual spores in saclike asci
- Division Basidiomycota: Club fungi have long-lived dikaryotic mycelia and a transient diploid stage
- Molds, yeasts, lichens, and mycorrhizae represent unique lifestyles that evolved independently in three fungal divisions

Ecological Impacts of Fungi

- Ecosystems depend on fungi as decomposers and symbionts
- Some fungi are pathogens
- Many animals, including humans, eat fungi

Phylogenetic Relationships of Fungi

- Fungi and animals probably evolved from a common protistan ancestor

The words fungus and mold may evoke some unpleasant images. Fungi rot timbers, attack plants, spoil food, and afflict humans with athlete's foot and worse maladies. However, ecosystems would collapse without fungi to decompose dead organisms, fallen leaves, feces, and other organic materials, thus recycling vital chemical elements back to the environment in forms other organisms can assimilate. Virtually all plants depend on symbiotic fungi that help their roots absorb minerals and water from the soil. In addition to these ecological roles, fungi have been used by humans in various ways for centuries. We eat some fungi (mushrooms, for instance), culture fungi to produce antibiotics and other drugs, add them to dough to make bread rise, and use them to ferment beer and wine. Many fungi are also quite beautiful, as Canada recognized in issuing the stamps shown on this page. Whatever our initial subjective perceptions may be, fungi are fascinating as objects of study. They are a form of life so distinctive that they have been accorded their own taxonomic kingdom.

In this chapter we will characterize the members of the kingdom Fungi, survey their diversity, discuss their ecological and commercial impact, and consider hypotheses about their evolutionary origin. As we did with the plant kingdom, we will look in some detail at life cycles, mainly for what they tell us about the phylogeny and evolutionary adaptations of fungi.

INTRODUCTION TO THE FUNGI

Fungi are eukaryotes, and most are multicellular. Although fungi were once grouped with plants, they are unique organisms that generally differ from other eukaryotes in nutritional mode, structural organization, and growth and reproduction. In fact, molecular studies indicate that fungi and animals probably arose from a common ancestor.

Absorptive nutrition enables fungi to live as decomposers and symbionts

Fungi are heterotrophs that acquire their nutrients by **absorption**. In this mode of nutrition, small organic molecules are absorbed from the surrounding medium. A fungus digests food outside its body by secreting powerful hydrolytic enzymes into the food. The enzymes decompose complex molecules to the simpler compounds that the fungus can absorb and use.

The absorptive mode of nutrition specializes fungi as decomposers (saprobes), parasites, or mutualistic symbionts. *Saprobic fungi* absorb nutrients from nonliving organic material, such as fallen logs, animal corpses, or the wastes of live organisms. In the process of this saprobic nutrition, fungi decompose the organic material. *Parasitic fungi* absorb nutrients from the cells of living hosts. Some of these fungi, such as certain species infecting the lungs of humans, are pathogenic. *Mutualistic fungi* also absorb nutrients from a host organism,

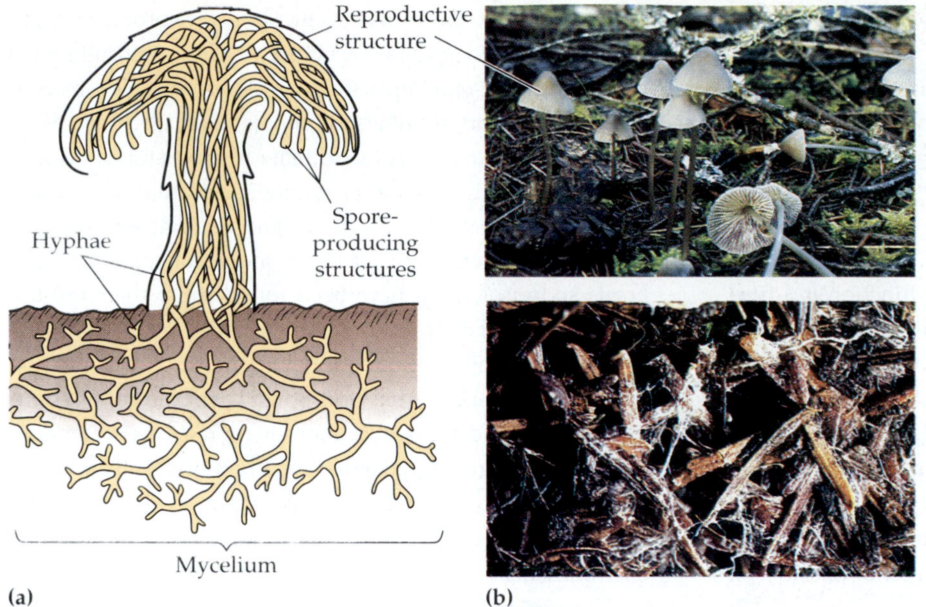

FIGURE 31.1 ▪ Fungal mycelia. (a) The drawing illustrates the relationship of the thin hyphae that make up a mycelium to the visible structure we recognize as a mushroom. The mushroom functions in reproduction by producing tiny cells called spores. A fungal mycelium begins with the germination of a fungal spore in a suitable habitat. (Other kinds of fungi have different types of reproductive structures.) **(b)** The bottom photo shows the vegetative part of a mycelium (white, cottony threads) decomposing brown conifer needles. The top photo shows the mushrooms (*Mycena*) that grow up from this mycelium each fall.

but they reciprocate with functions beneficial to their partners in some way, such as aiding a plant in the uptake of minerals from the soil.

Fungi inhabit diverse environments and are associated symbiotically with many organisms. Although most common in terrestrial habitats, some fungi inhabit aquatic environments, where they are associated with both marine and freshwater organisms and their remains. Lichens, symbiotic associations of fungi and algae, are widespread and are found in some of the most inhospitable habitats on Earth: cold, dry deserts on Antarctica and alpine and arctic tundras. Other symbiotic fungi live inside the healthy tissues of plants, and still other species form cellulose-consuming mutualisms with insects, including ants and termites.

Extensive surface area and rapid growth adapt fungi for absorptive nutrition

The vegetative (nutritionally active) bodies of most fungi are usually hidden, being diffusely organized around and within the tissues of their food sources. Except for yeasts, which are unicellular, the bodies of fungi are constructed of units called **hyphae** (singular, **hypha**). Hyphae are minute threads composed of tubular walls surrounding plasma membranes and cytoplasm. The cytoplasm contains the usual eukaryotic organelles. The hyphae form an interwoven mat called a **mycelium** (plural, **mycelia**), the "feeding" network of a fungus (FIGURE 31.1). Fungal mycelia can be huge, although they usually escape our notice—because they are subterranean, for instance. The mycelium of a giant individual of the fungus *Armillaria ostoyae* in Washington State spreads through 6.5 km^2 of land located in three counties. This fungus is probably thousands of years old and hundreds of tons in weight, qualifying it among Earth's oldest and largest organisms.

Most fungi are multicellular with hyphae divided into cells by cross-walls, or **septa** (singular, **septum**). The septa generally have pores large enough to allow ribosomes, mitochondria, and even nuclei to flow from cell to cell (FIGURE 31.2a). The cell walls of fungi differ from the cellulose walls of plants. Most fungi build their cell walls mainly of **chitin**, a strong but

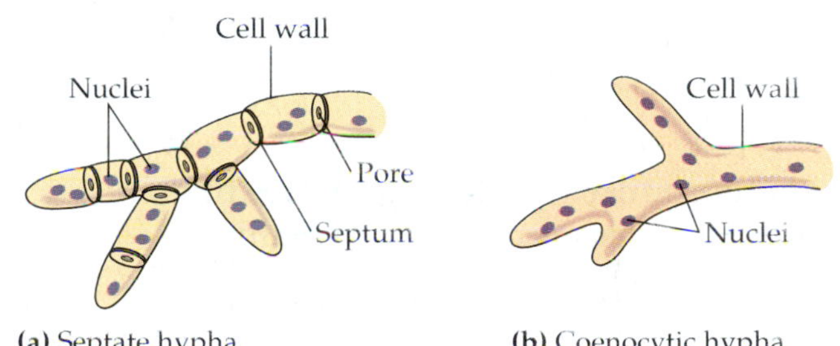

(a) Septate hypha

(b) Coenocytic hypha

Fungal hypha
Plant cell wall
Plant cell
Plant plasma membrane
Haustorium

(c) Haustoria

FIGURE 31.2 ▪ Characteristics of fungal hyphae. (a) Septate hypha. **(b)** Coenocytic hypha. **(c)** Specialized hyphae called haustoria parasitize plant cells from *outside*, separated from the host cell's cytoplasm by the plasma membrane of the plant cell (blue). (The effect is like sinking your fingers into an underinflated balloon.)

flexible nitrogen-containing polysaccharide similar to the chitin found in the external skeletons of insects and other arthropods. Some fungi are aseptate; that is, their hyphae are not divided into cells by cross-walls. Known as **coenocytic** fungi, they consist of a continuous cytoplasmic mass with hundreds or thousands of nuclei (FIGURE 31.2b). The coenocytic condition results from the repeated division of nuclei without cytoplasmic division.

The correlation between structure and function is a fundamental theme of biology. The filamentous structure of the mycelium provides an extensive surface area that suits the absorptive nutrition of fungi: 10 cm^3 of rich organic soil may contain as much as 1 km of hyphae. If these hyphae were 10 μm in diameter, about 314 cm^2 of fungal surface area would interface with that 10 cm^3 of soil. Parasitic fungi usually have some of their hyphae modified as **haustoria**, nutrient-absorbing hyphal tips that penetrate the tissues of the host (FIGURE 31.2c).

A fungal mycelium grows rapidly, adding as much as a kilometer of hyphae each day as it branches within a food source. Such fast growth is possible because proteins and other materials synthesized by the entire mycelium are channeled by cytoplasmic streaming to the tips of the extending hyphae. The fungus concentrates its energy and resources on adding hyphal length and thus overall absorptive surface area, rather than girth. Fungi are nonmotile organisms; they cannot run, swim, or fly in search of food or mates. But the mycelium makes up for the lack of mobility by swiftly extending the tips of its hyphae into new territory.

Fungi reproduce by releasing spores that are produced either sexually or asexually

Fungal spores come in a variety of shapes and sizes, and may
31.1 be produced sexually or asexually. Usually they are unicellular, but multicellular spores also occur. Spores are produced in, or from, specialized hyphal structures. When environmental conditions favor rapid growth, fungi clone themselves by producing enormous numbers of spores asexually. Carried by wind or water, the spores germinate if they land in a moist place where there is an appropriate substratum, or surface. Spores thus function in dispersal and account for the wide geographical distribution of many species of fungi. The airborne spores of fungi have been found more than 160 km (100 mi) above Earth.

For many fungi, sex is a contingency mode of reproduction that occurs when there has been some change in the environment. Compared to asexual reproduction, sexual reproduction results in greater genetic diversity among offspring. As raw material for natural selection, this individual variation among offspring may contribute to adaptation in changing environments.

The nuclei of fungal hyphae and spores of most species are haploid, except for transient diploid stages that form during sexual life cycles. However, some mycelia may become genetically heterogeneous through the fusion of two hyphae that have genetically different nuclei. In some cases, the different nuclei stay in separate parts of the same mycelium, which is then a mosaic in terms of genotype and phenotype. In other cases, the different nuclei mingle and may even exchange chromosomes and genes in a process similar to crossing over.

A special case of genetic heterogeneity occurs during the sexual cycle of many fungi. Syngamy, the sexual union of cells from two individuals, occurs in two stages that are separated in time. These two stages of syngamy are called **plasmogamy** (the fusion of cytoplasm) and **karyogamy** (the fusion of nuclei). After plasmogamy, the nuclei from each parent pair up but do not fuse, forming a **dikaryon** ("two nuclei"). The pairs of nuclei may coexist and divide in tandem in a dikaryotic cell or mycelium for months or years. This condition has

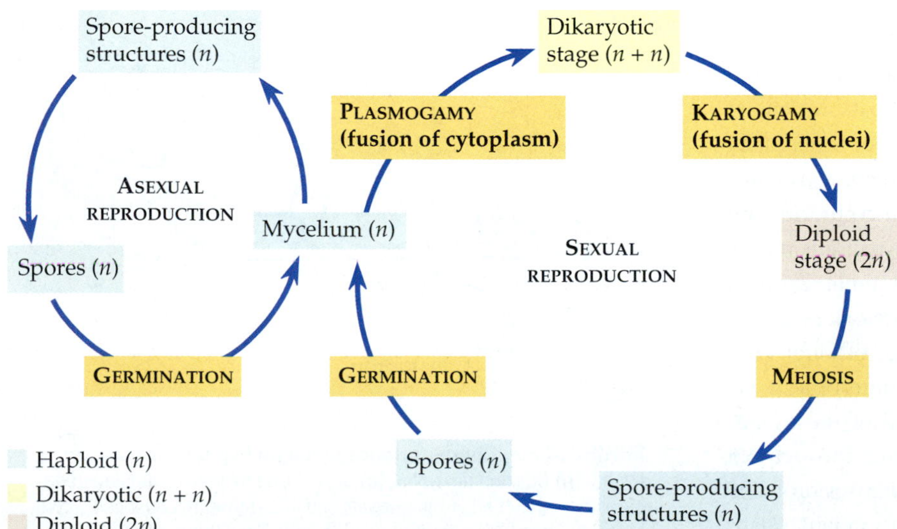

FIGURE 31.3 ▪ Generalized life cycle of
31.1 **fungi.** Haploid spores are produced asexually and sexually. Produced by specialized hyphae of a mycelium, large numbers of asexual spores are disseminated widely by wind or water and germinate on moist surfaces. Sexual reproduction in many fungal species includes two distinct stages of syngamy (sexual union of cells). The first stage, called plasmogamy, is the fusion of cytoplasm (hyphae) of two neighboring mycelia. Plasmogamy results in a dikaryotic stage denoted $n + n$ because the haploid nuclei from each parent pair up but do not fuse. During the second stage of syngamy, called karyogamy, the haploid nuclei fuse, producing the diploid stage. Meiosis then leads to the formation of genetically varied haploid spores.

some of the advantages of diploidy; one haploid genome may be able to compensate for harmful mutations in the other nucleus, and vice versa. Finally, in the second stage of syngamy, the nuclei fuse (karyogamy), forming a diploid cell that undergoes immediate meiosis. As we will see in the next section, the life cycles of many fungi include three distinct phases: haploid, dikaryotic, and diploid (FIGURE 31.3).

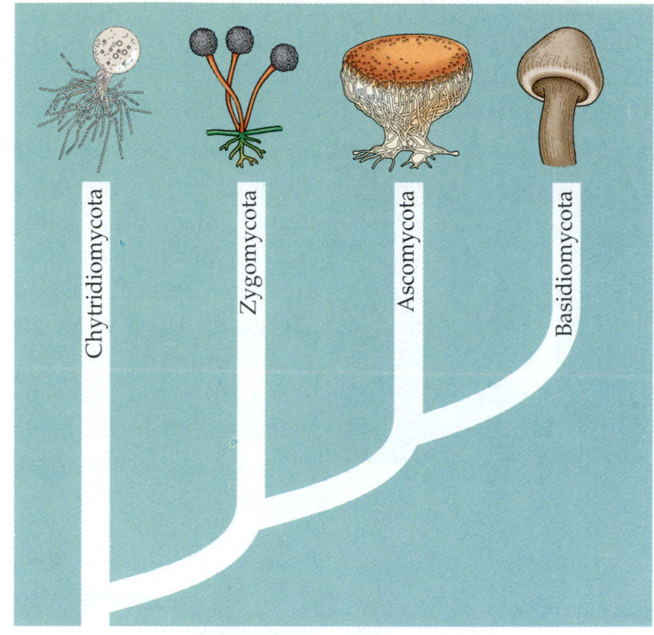

FIGURE 31.4 ▪ Phylogeny of fungi. This phylogenetic tree, supported by molecular evidence, indicates probable evolutionary relationships among the four divisions of Kingdom Fungi. Representing the oldest lineage of fungi, members of Division Chytridiomycota (chytrids) are mainly aquatic and have flagellated cells. Fungi in the other three divisions lack flagellated stages.

DIVERSITY OF FUNGI

More than 100,000 species of fungi are known, and mycologists (biologists who study fungi) estimate that there are about 1.5 million species worldwide. The taxonomic scheme used in this chapter classifies fungi into four divisions (FIGURE 31.4). Use of the botanical term *division* instead of *phylum* is a vestige of the taxonomic scheme that grouped fungi with plants.

Division Chytridiomycota: Chytrids may provide clues about fungal origins

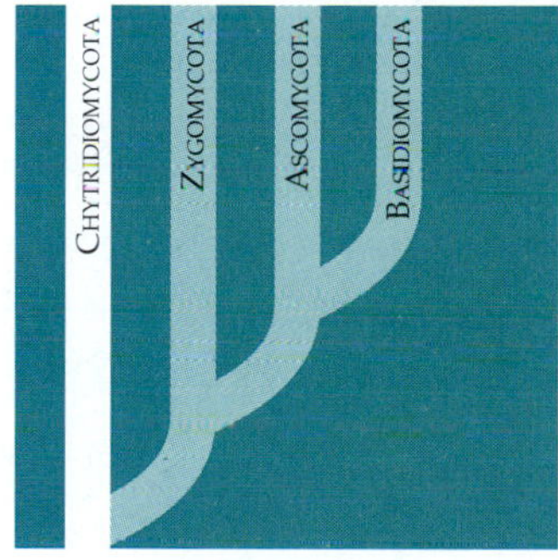

Systematists are making rapid progress in sorting out the phylogenetic relationships between fungi and other eukaryotes. A link between fungi and protists may be a group of organisms called **chytrids** (FIGURE 31.5). The chytrids are mainly aquatic. Some are saprobes; others parasitize protists, plants, and aquatic invertebrates.

Until recently, some systematists emphasized the absence of flagellated cells as a membership requirement for Kingdom Fungi. By that criterion, chytrids were excluded and placed instead in Kingdom Protista (of the five-kingdom system) because they form uniflagellated spores called zoospores. However, in the past decade, molecular systematists comparing the sequences of proteins and nucleic acids in chytrids and fungi uncovered strong support for combining the chytrids with the fungi as a monophyletic branch of the eukaryotic tree. Other key fungal characteristics of chytrids are an absorptive mode of nutrition and cell walls made of chitin. Most chytrids form coenocytic hyphae, although some are unicellular. Chytrids also have some key enzymes and metabolic pathways that are common among fungi but are not found in the so-called funguslike protists (slime molds and water molds; see Chapter 28). The weight of evidence leads many biologists to classify the chytrids in the division Chytridiomycota within Kingdom Fungi, and this is the taxonomic view we follow.

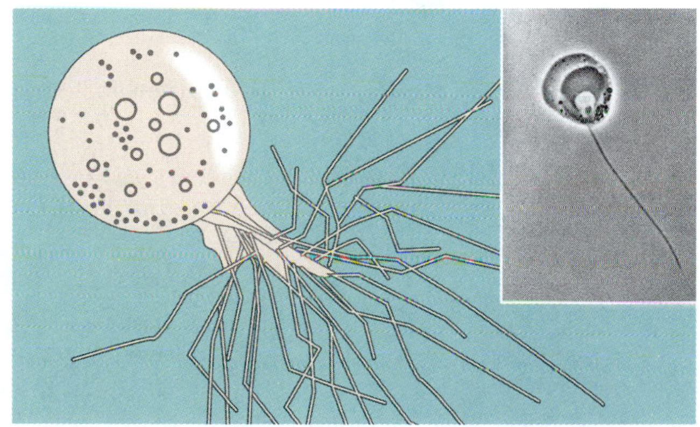

FIGURE 31.5 ▪ Chytridiomycota (chytrids). The branched hyphae of *Chytridium* expose a large surface to the surrounding medium, from which the organism absorbs nutrients. Chytrids are the only fungi with a flagellated stage, the zoospore shown in the inset (TEM).

Molecular evidence also supports the hypothesis that chytrids are the most primitive fungi, meaning that they belong to the lineage that diverged earliest in the phylogeny of fungi. A reasonable extension of this hypothesis is that fungi evolved from protists that had flagella, a feature retained in the fungal kingdom only by the chytrids.

Division Zygomycota: Zygote fungi form resistant dikaryotic structures during sexual reproduction

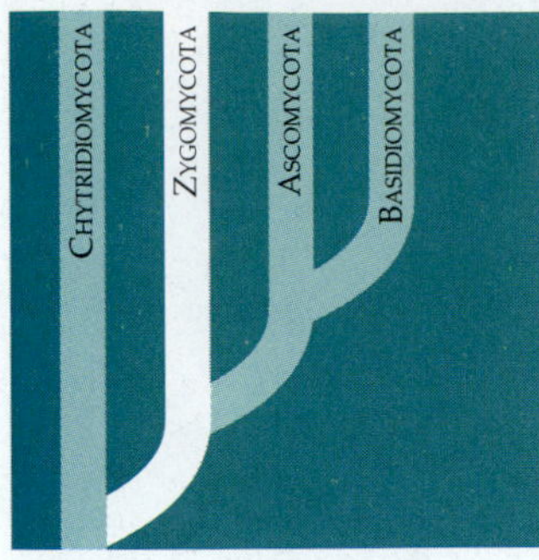

Mycologists have described about 600 zygomycetes, or **zygote fungi**. (The suffix *-mycete,* which occurs many times in this chapter, means "fungus.") These fungi are mostly terrestrial and live in soil or on decaying plant and animal material. One group of major importance forms **mycorrhizae**, mutualistic associations with the roots of plants (see FIGURE 31.16). Zygomycete hyphae are coenocytic, with septa found only where reproductive cells are formed. The name of this division comes from zygosporangia, resistant structures formed during sexual reproduction (FIGURE 31.6, upper right).

A common zygomycete is black bread mold, *Rhizopus stolonifer,* still an occasional household pest despite the addition of preservatives to most processed foods. Horizontal hyphae spread out over the food, penetrate it, and absorb nutrients. In the asexual phase, bulbous black sporangia develop at the tips of upright hyphae (see FIGURE 31.6, lower left). Within each sporangium, hundreds of haploid spores develop and are dispersed through the air. Spores that happen

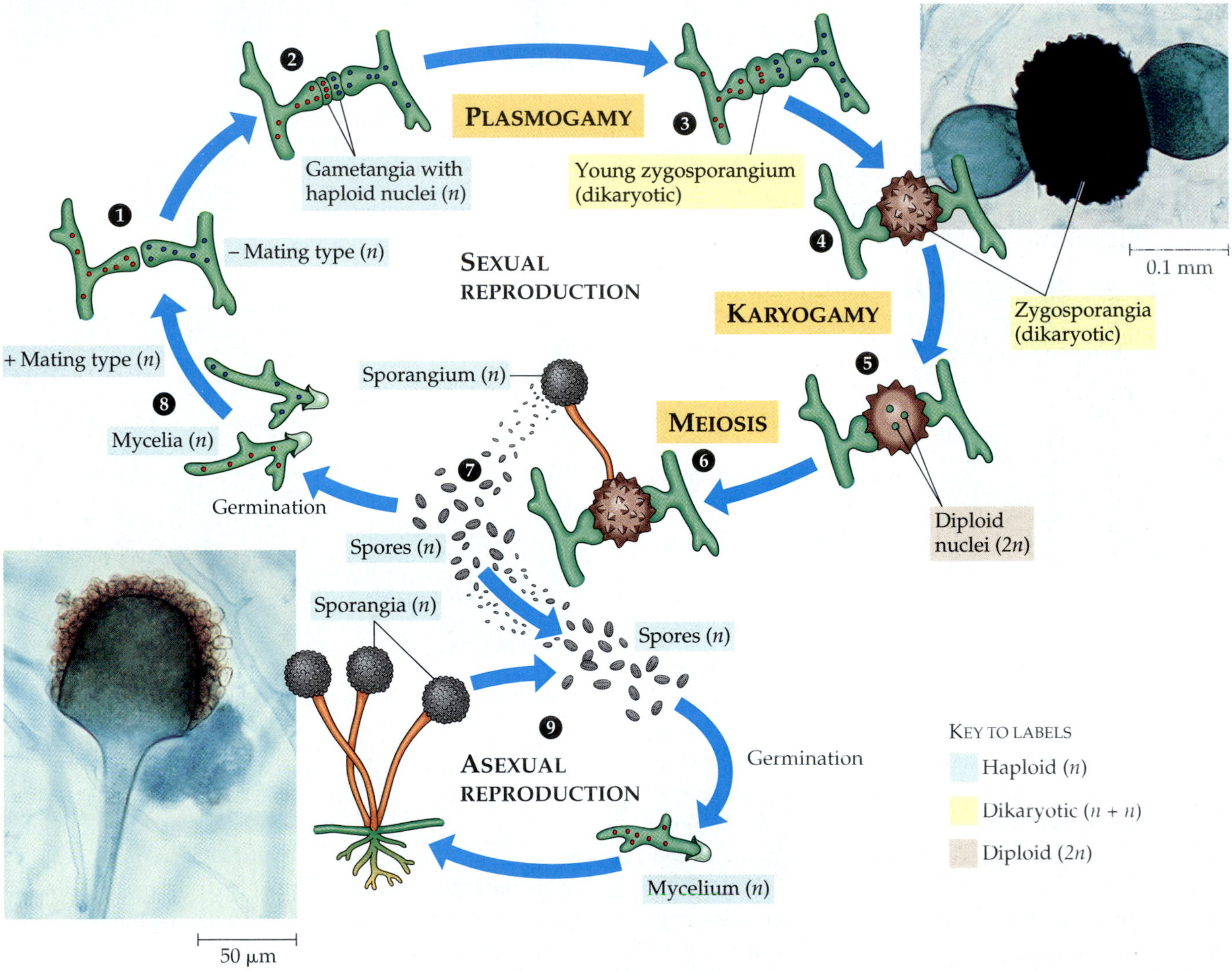

31.1 **FIGURE 31.6 • Zygomycota (zygote fungi): the life cycle of the zygomycete *Rhizopus* (black bread mold).** ① Neighboring mycelia of opposite mating types (designated + and −) ② form hyphal extensions called gametangia, each walled off around several haploid nuclei by a septum. ③ The gametangia undergo plasmogamy (fusion of cytoplasm) and the haploid nuclei pair off, forming a dikaryotic zygosporangium. ④ This cell develops a rough, thick-walled coating (upper right LM) that can resist dry conditions and other harsh environments for months. ⑤ When conditions are favorable again, karyogamy occurs, paired nuclei fuse, followed rapidly by meiosis. ⑥ The zygosporangium then breaks dormancy, germinating into a short sporangium that ⑦ disperses genetically diverse, haploid spores. ⑧ These spores germinate and grow into new mycelia. ⑨ Mycelia of *Rhizopus* can also reproduce asexually by forming sporangia (lower left LM) that produce genetically identical haploid spores.

to land on moist food germinate, growing into new mycelia. If environmental conditions deteriorate—for instance, if all the food is used up—and mycelia of opposite mating types (with genetically different nuclei) are present, this species of *Rhizopus* reproduces sexually (see FIGURE 31.6, upper part). The zygosporangia formed are resistant to freezing and drying and are metabolically inactive. When conditions improve, the zygosporangia release genetically diverse haploid spores that colonize the new substrate.

Some zygomycetes can actually aim their spores. One is *Pilobolus*, a fungus that decomposes animal dung. *Pilobolus* bends its sporangium-bearing hyphae toward light, a direction where grass is likely to be growing. The whole sporangium is then shot out the end of the hypha, sometimes landing as far away as 2 m. This adaptation disperses the spores away from the mass of dung and onto surrounding grass, which will be eaten by a herbivore such as a cow. The asexual life cycle is completed when the animal scatters the spores in feces.

Division Ascomycota: Sac fungi produce sexual spores in saclike asci

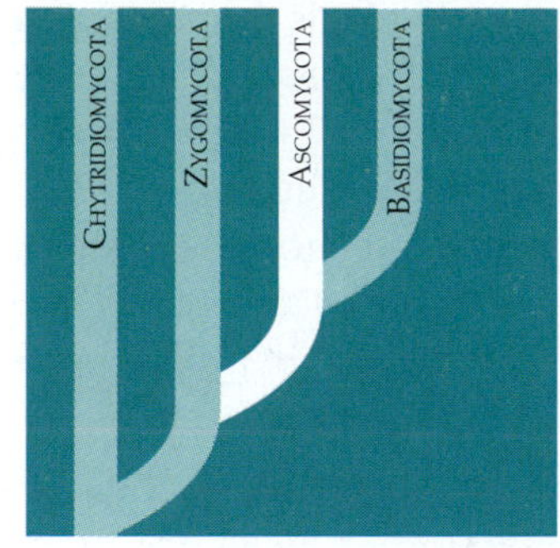

Over 60,000 species of ascomycetes, or **sac fungi**, have been described from a wide variety of marine, freshwater, and terrestrial habitats. They range in size and complexity from unicellular yeasts to minute leaf-spot fungi to elaborate cup fungi and morels (FIGURE 31.7). Ascomycetes include some of the most devastating plant pathogens, which will be discussed later in the chapter. However, many are important saprobes, particularly of plant material. About half the ascomycete species live with algae in the symbiotic associations called lichens. Some ascomycetes, including the morels, form mycorrhizae with plants. Others

(a)

(b)

(c)

FIGURE 31.7 ▪ Ascomycota (sac fungi). Ascomycetes range from yeasts to such gastronomical delicacies as truffles and morels. **(a)** Most ascomycetes form ascocarps in which sexual stages develop. In the carbon fungus, *Hypoxylon multiforme*, the ascocarps appear in clusters on the wood the mycelium is decomposing. **(b)** Truffles are ascocarps that fruit underground and emit strong odors, which attract animals that eat the fungi and disperse the ascospores. *Tuber melanosporum* is highly prized for its flavor by gourmet cooks, who pay over $600 a pound for this truffle. **(c)** This ascocarp of *Morchella esculenta*, the succulent morel, is edible. It is often found under trees in orchards. Many, if not all, morels are mycorrhizal, forming mutualistic associations with plant roots.

live in leaves on the surface of mesophyll cells, where the fungi apparently help protect these plant tissues from insects by releasing toxic compounds.

The defining feature of the Ascomycota is the production of sexual spores in saclike **asci** (singular, **ascus**). Unlike the zygote fungi, most sac fungi bear their sexual stages in macroscopic fruiting bodies, or **ascocarps** (see FIGURE 31.7). Ascomycetes reproduce asexually by producing enormous numbers of asexual spores, which are often dispersed by wind. The asexual spores are produced at the ends of hyphae, often in long chains or clusters. These spores are not formed inside sporangia, as in the Zygomycota. Such naked spores are called **conidia**, from the Greek for "dust."

Compared to zygomycetes, ascomycetes are characterized by a more extensive dikaryotic stage, which is associated with the formation of ascocarps (FIGURE 31.8). Plasmogamy gives rise to the dikaryotic hyphae, and the cells at the tips of these hyphae become the asci. Within the asci, karyogamy combines the two parental genomes, and then meiosis forms genetically varied ascospores. In many asci, eight ascospores are lined up in a row in the order in which they formed from a single zygote. This arrangement provides geneticists with a unique opportunity to study genetic recombination. Genetic differences between mycelia grown from ascospores taken from one ascus reflect crossing over and independent assortment of chromosomes during meiosis.

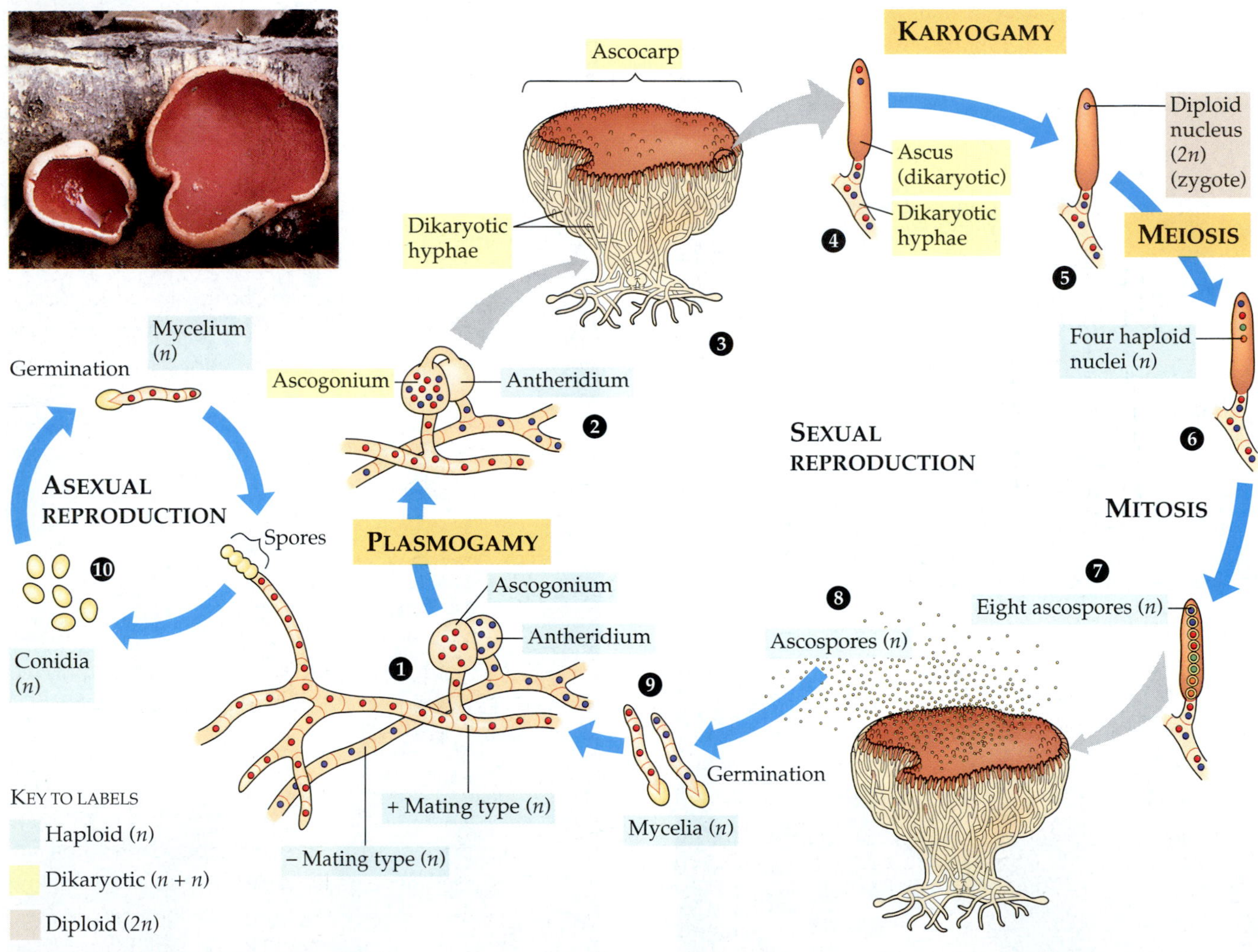

31.1 **FIGURE 31.8 ▪ The life cycle of an ascomycete.** ① Haploid mycelia of opposite mating types become intertwined and form an ascogonium and an antheridium. ② A cytoplasmic bridge forms, allowing plasmogamy (cytoplasmic fusion) to occur. The ascogonium acts as a "female," receiving haploid nuclei from the "male" antheridium. The ascogonium then has a pool of nuclei from both parents, but karyogamy (fusion of nuclei) does not occur at this time. ③ The ascogonium gives rise to dikaryotic hyphae that are incorporated into an ascocarp, the cup of a cup fungus. The photograph features ascocarps of the scarlet cup (*Sarcoscypha coccinea*) growing on a fallen branch in early spring. ④ The tips of the ascocarps' dikaryotic hyphae are partitioned into asci (singular, ascus). ⑤ Karyogamy occurs within these asci, and the diploid nucleus divides by meiosis, ⑥ yielding four haploid nuclei. ⑦ Each of these haploid nuclei divides once by mitosis, and the ascus now contains eight nuclei. Cell walls develop around these nuclei to form ascospores. ⑧ When mature, all ascospores in an ascus are dispersed at once out the end of the ascus. A collapsing ascus jars neighboring asci and causes them to release their spores. The chain reaction releases a visible cloud of spores with an audible hiss. ⑨ Germinating ascospores give rise to new haploid mycelia. ⑩ Ascomycetes can also reproduce asexually by producing airborne spores called conidia.

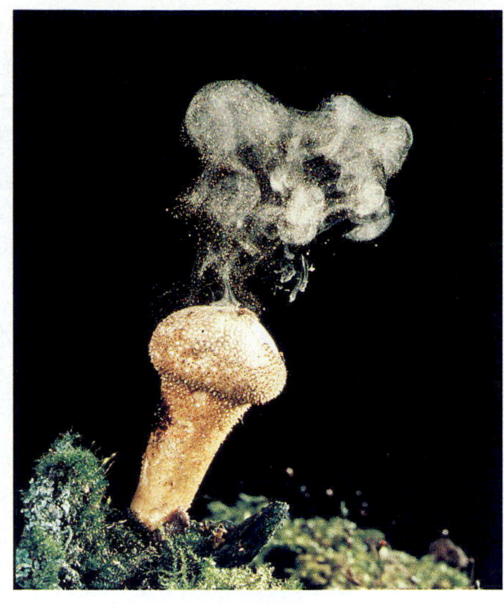

FIGURE 31.9 ▪ Basidiomycota (club fungi). These photographs showcase diverse basidiocarps, the fruiting bodies that produce sexual spores. **(a)** A miniature waxy cap (*Hygrophorus*) is mycorrhizal with oaks. **(b)** A shelf fungus growing on the trunk of a tree. Shelf fungi are important decomposers of wood. **(c)** The common puffball (*Lycoperdon gemmatum*) is a saprobe on buried plant litter. Some puffballs can be over 1 m in diameter and produce trillions of spores, which are released when the basidiocarp breaks apart from animal activity, wind, or rain.

Division Basidiomycota: Club fungi have long-lived dikaryotic mycelia and a transient diploid stage

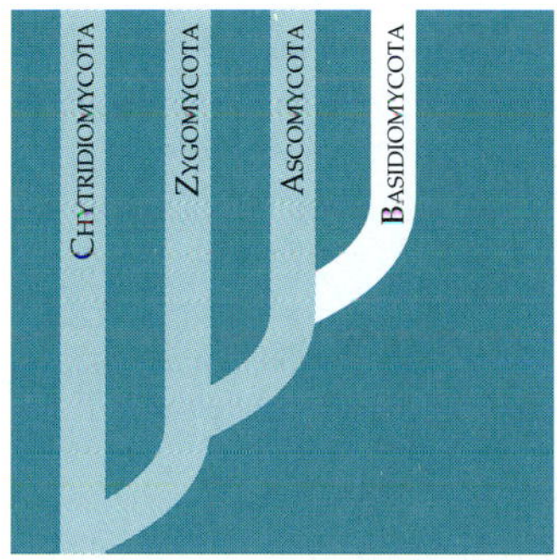

Approximately 25,000 fungi, including mushrooms, shelf fungi, puffballs, and rusts, are classified in the division Basidiomycota (FIGURE 31.9). The name derives from the **basidium** (L., "little pedestal"), a transient diploid stage in the organism's life cycle. The clublike shape of the basidium also gives rise to the common name **club fungus**.

Basidiomycetes are important decomposers of wood and other plant material. The division also includes mycorrhiza-forming mutualists and plant parasites. Of all fungi, the saprobic basidiomycetes are best at decomposing the complex polymer lignin, an abundant component of wood. Many shelf fungi (FIGURE 31.9b) parasitize the wood of weak or damaged trees and decompose the wood after the trees die. Two groups of basidiomycetes, the rusts and smuts, include particularly destructive plant parasites.

The life cycle of a club fungus usually includes a long-lived dikaryotic mycelium (FIGURE 31.10, p. 582). Periodically, in response to environmental stimuli, this mycelium reproduces sexually by producing elaborate fruiting bodies called **basidiocarps.** The numerous basidia of a basidiocarp are the sources of sexual spores. Asexual reproduction in basidiomycetes is much less common than in ascomycetes.

A mushroom is an example of a basidiocarp. The cap of the mushroom supports and protects a large surface area of basidia on gills; each common, store-bought mushroom has a gill surface area of about 200 cm^2. Such a mushroom may release a billion basidiospores, which drop beneath the cap and are blown away. Basidiomycetes called stinkhorns produce basidiospores in a slimy, stinky mass on upper surfaces of the basidiocarp, which resembles rotting meat in odor and appearance. Flies and other carrion-eating insects attracted to stinkhorns disperse the sticky spores.

Molds, yeasts, lichens, and mycorrhizae represent unique lifestyles that evolved independently in three fungal divisions

Certain ways of living that involve both morphological and ecological specialization have evolved independently among the zygote fungi, sac fungi, and club fungi. These lifestyles allow the fungi to take advantage of unusual habitats. Humans have learned to exploit the abilities of these fungi for a variety of commercial purposes. This section discusses four fungal forms with unique ways of life: molds, yeasts, lichens, and mycorrhizae.

Molds

Mention of fungi may bring the ubiquitous molds to mind. A **mold** is a rapidly growing, asexually reproducing fungus. The mycelia of these fungi grow as saprobes or parasites on a great

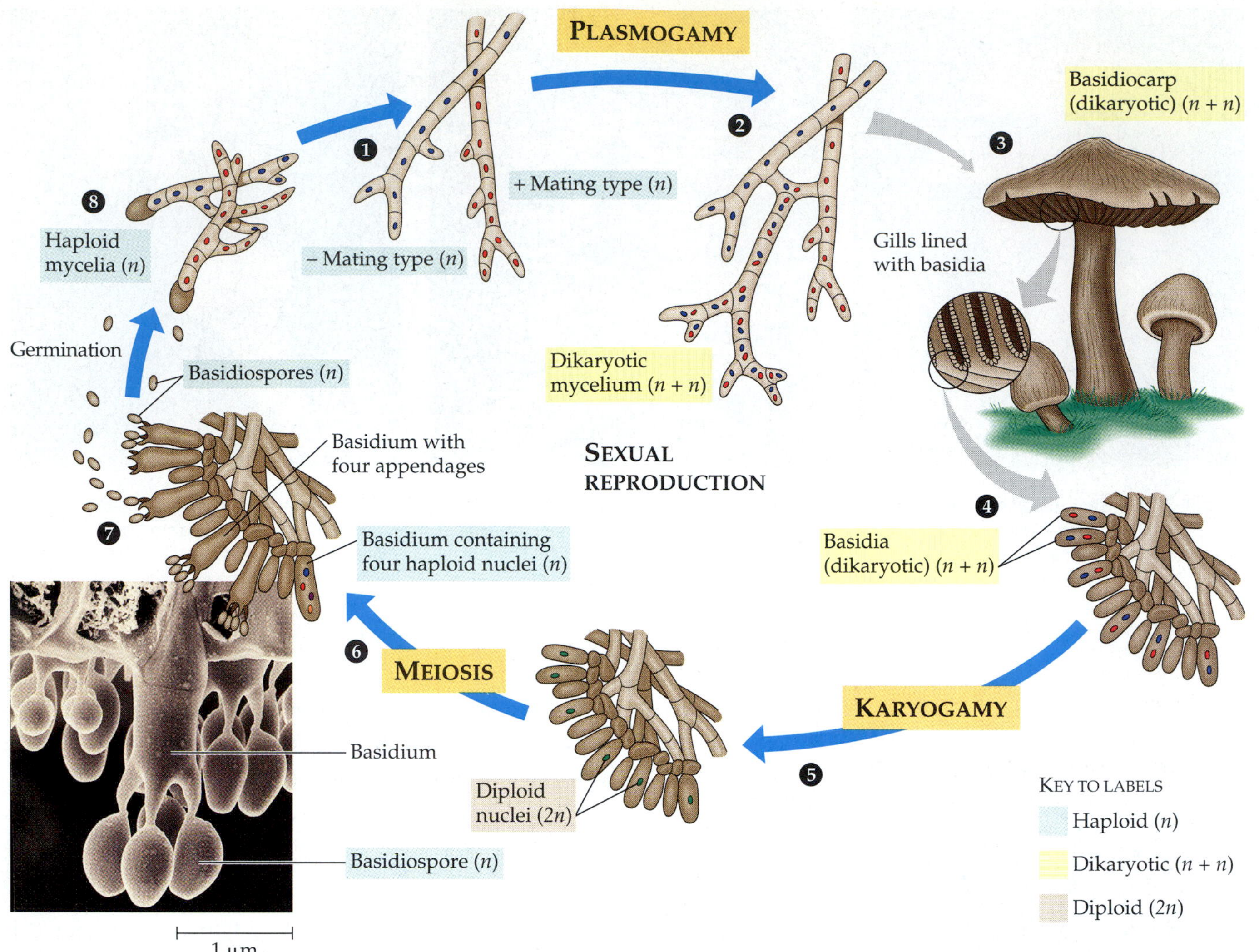

FIGURE 31.10 ▪ The life cycle of a mush-
31.1 **room-forming basidiomycete.** ① Two haploid mycelia of opposite mating type undergo plasmogamy, ② creating a dikaryotic mycelium that grows faster than, and ultimately crowds out, the parent haploid mycelia. The mycelia of some basidiomycetes, including the genus illustrated here (*Cortinarius*), form mycorrhizae with trees. ③ Environmental cues such as rain, temperature changes, and (for mycorrhizal species) seasonal changes in the plant host induce the dikaryotic mycelium to form compact masses that develop into basidiocarps (mushrooms, in this case). Cytoplasm streaming in from the mycelium and from the attached mycorrhizae swells the hyphae of mushrooms, causing them to "pop up" overnight. The dikaryotic mycelia of basidiomycetes are long-lived, generally producing a new crop of basidiocarps each year. ④ The surfaces of the basidiocarp's gills are lined with terminal dikaryotic cells called basidia. ⑤ Karyogamy produces diploid nuclei, which rapidly undergo meiosis, ⑥ each yielding four haploid nuclei. Each basidium grows four appendages, and one haploid nucleus enters each appendage and develops into a basidiospore (SEM). ⑦ When mature, the basidiospores are propelled slightly (by electrostatic forces) into the spaces between the gills. After the spores drop below the cap, they are dispersed by the wind. ⑧ The haploid basidiospores germinate in a suitable environment and grow into short-lived haploid mycelia.

variety of substrates. You are already familiar with one example, bread mold (*Rhizopus;* see FIGURE 31.6). Molds may go through a series of different reproductive stages. Early in life, a mold produces asexual spores. The term *mold* applies only to these asexual stages (FIGURE 31.11). Later, the same fungus *may* reproduce sexually, producing zygosporangia, ascocarps, or basidiocarps.

There are also molds that cannot be classified as zygomycetes, ascomycetes, or basidiomycetes, because they have no known sexual stages. Those molds are collectively called deuteromycetes, or **imperfect fungi** (from the botanical use of the term *perfect* to refer to the sexual stages of life cycles). Imperfect fungi reproduce asexually by producing spores. Among the more unusual imperfect fungi are some predatory fungi in soil that trap, kill, and consume protists and small animals, especially roundworms, or nematodes (FIGURE 31.12). This provides additional nitrogen-containing compounds, which are in short supply in the wood these fungi decompose.

Humans have found many commercial uses for molds. Some are sources of antibiotics; pharmaceutical companies grow these molds in large liquid cultures, then extract the antibiotics. Penicillin is produced by some species of *Penicil-*

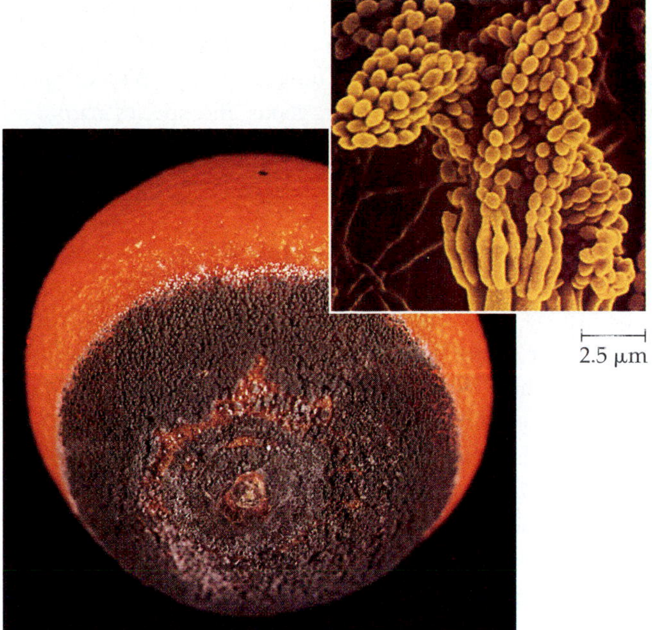

FIGURE 31.11 ▪ **A moldy orange.** "Blue mold" is usually caused by saprobic species of *Penicillium,* an ascomycete. This mold reproduces asexually by producing chains of conidia on hyphae called conidiophores (right, SEM).

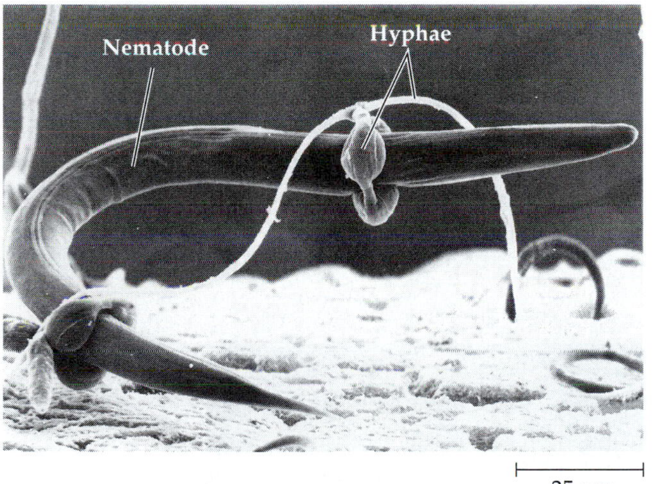

FIGURE 31.12 ▪ ***Arthrobotrys,* a predatory deuteromycete (imperfect fungus).** Portions of the hyphae are modified as hoops that constrict around nematodes (roundworms) in less than a second, when a worm rubs the inside of the hoop. The fungus then penetrates its prey with hyphae and digests its inner tissues (SEM). Many nematode-trapping fungi are basidiomycetes.

lium, which are ascomycetes. Other *Penicillium* species are important fermenters on the surfaces of blue cheese, Brie, and Camembert and produce the unusual colors and flavors we associate with each type of cheese. Roquefort is goat milk cheese that has been incubated in certain caves in the Roquefort region of France, where wild *Penicillium roquefortii* is allowed to "infect" the cheese naturally.

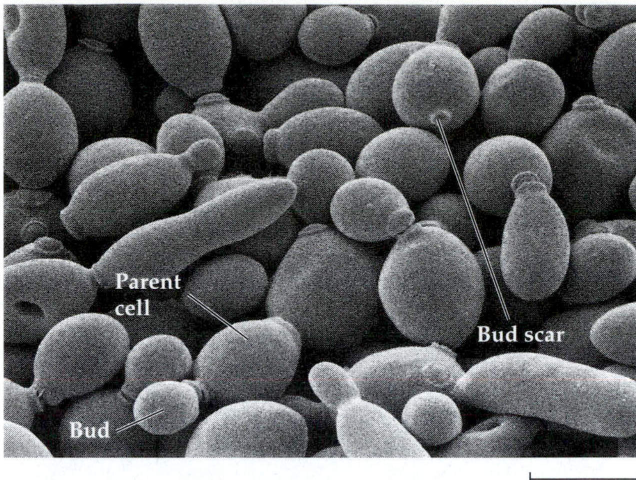

FIGURE 31.13 ▪ **Budding yeast.** This micrograph shows *Saccharomyces cerevisiae* in various stages of budding (SEM).

Yeasts

Yeasts are unicellular fungi that inhabit liquid or moist habitats, including plant sap and animal tissues. Yeasts reproduce asexually, by simple cell division or by the pinching of small "bud cells" off a parent cell (FIGURE 31.13). Some yeasts reproduce sexually, by forming asci or basidia, and are classified as Ascomycota or Basidiomycota. Others are placed in the imperfect fungi because no sexual stages are known. Some fungi can grow as either single cells (yeasts) or as a filamentous mycelium, depending on the availability of nutrients.

Humans have used yeasts to raise bread and ferment alcoholic beverages for thousands of years. Only relatively recently have the yeasts involved been separated into pure culture for more controlled human use. The yeast *Saccharomyces cerevisiae,* an ascomycete, is the most important of all domesticated fungi (see FIGURE 31.13). The tiny yeast cells, available as many strains of baker's yeast and brewer's yeast, are very active metabolically. The cells release small bubbles of CO_2 that leaven dough. Cultured anaerobically in breweries and wineries, *Saccharomyces* ferments sugars to alcohol. Researchers use *Saccharomyces* to study the molecular genetics of eukaryotes because these microbes are easy to culture and manipulate (see Chapter 19).

Some yeasts cause problems for humans. A pink yeast, *Rhodotorula,* grows on shower curtains and other moist surfaces in our homes. Another yeast is *Candida,* one of the normal inhabitants of moist human epithelial tissue, such as the vaginal lining. Certain circumstances can cause *Candida* to become pathogenic by growing too rapidly and releasing harmful substances. This can occur, for example, with an environmental change such as a pH change, or when the immune system of the human host is compromised—by AIDS, for instance.

FIGURE 31.14 ▪ Lichens. Lichens representing three growth forms inhabit the bark of this maple branch. *Parmelia* (lower left) is foliose (having a flattened, leaflike appearance). *Ramalina* (upper right) is fruticose (shrublike). Elsewhere, several crustose (paint smearlike) species in the genera *Lecanora* and *Bacidia* are forming disk-shaped ascocarps. The minute orange lichen is *Xanthoria,* a foliose form.

Lichens

At a distance, lichens are often mistaken for mosses or other simple plants growing on rocks, rotting logs, trees, and roofs (FIGURE 31.14). In fact, lichens are *not* mosses or any other kind of plant, nor are they even individual organisms. A **lichen** is a symbiotic association of millions of photosynthetic microorganisms held in a mesh of fungal hyphae. The fungal component is most commonly an ascomycete, but several basidiomycete lichens are known. The photosynthetic partners are usually unicellular or filamentous green algae or cyanobacteria. The merger of fungus and alga is so complete that lichens are actually given genus and species names, as though they were single organisms. The scientific name of the lichen is the name of the fungus. Over 25,000 species have been described.

The fungus usually gives the lichen its overall shape and structure, and tissues formed by hyphae account for most of the lichen's mass. The algal component usually occurs in an inner layer below the lichen surface (FIGURE 31.15). In most cases that have been examined, each partner provides things the other could not obtain on its own. The alga provides the fungus with food. Cyanobacteria in lichens fix nitrogen (see Chapter 27) and provide organic nitrogen. The fungus provides the alga with a suitable physical environment for growth. Lichens absorb most of the minerals they need either from dust in the air or from rain. The physical arrangement of hyphae retains water and minerals, allows for gas exchange, and protects the algae. Fungal pigments help shade the algae from intense sunlight. Some fungal compounds are toxic and prevent lichens from being eaten by consumers. The fungi also secrete acids, which aid in the uptake of minerals.

The fungi of many lichens reproduce sexually by forming ascocarps or basidiocarps. Lichen algae reproduce independently of the fungus by asexual cell division. As might be expected of "dual organisms," asexual reproduction as symbiotic units also occurs commonly, either by fragmentation of the parental lichen or by the formation of specialized structures called **soredia** (see FIGURE 31.15). Soredia are small clusters of hyphae with embedded algae.

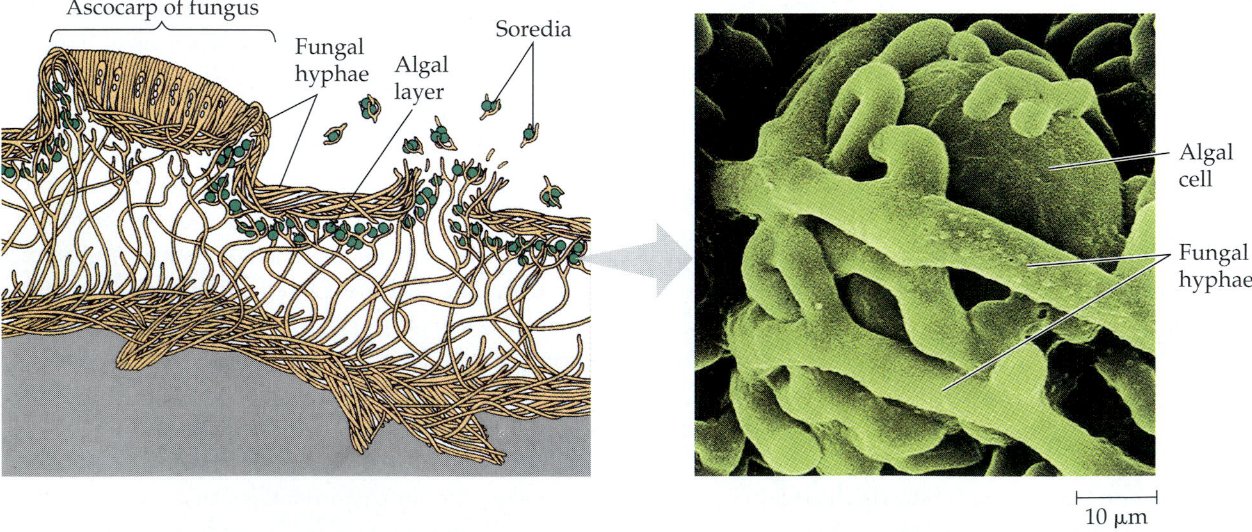

FIGURE 31.15 ▪ Anatomy of a lichen. The upper and lower surfaces are protective layers of tightly packed fungal hyphae. Just beneath the upper surface are the algae, enmeshed in a net of hyphae (SEM, right). The middle of a lichen generally consists of loosely woven hyphae of the fungus. Reproductive structures generally form on the upper surface. A sexual ascocarp of the fungus and several asexual soredia that disperse both fungal and algal components are shown here.

The nature of the lichen symbiosis is complex and is probably best described as mutual exploitation instead of mutual benefit. Lichens are able to live in environments where neither fungi nor algae could live alone. While the fungal components do not grow alone in the wild, some lichen algae also occur as free-living organisms. The fungi and algae of some lichens have been experimentally separated and cultured. Such cultures look like free-living molds and algae; the cultured fungi do not produce lichen compounds, and the cultured lichen algae do not "leak" carbohydrate from their cells as they do in lichens. In some lichens, the fungus invades the algal cells with haustoria and may kill some of them, though not as fast as the alga replenishes its numbers by reproduction.

Lichens are important pioneers on newly cleared rock and soil surfaces, such as burned forests and volcanic flows. Physical penetration of the outer crystals of rocks and chemical attack of rock by lichen acids help break down the rock and establish soil-trapping lichens. This process makes it possible for a succession of plants to grow. Nitrogen-fixing lichens also add organic nitrogen to some ecosystems.

Some lichens tolerate severe cold. In the arctic tundra, herds of caribou and reindeer graze on carpets of reindeer lichen at times of the year when other foods are unavailable. Lichens can also survive desiccation. When it is foggy or rainy, lichens may absorb more than ten times their weight in water. Photosynthesis begins almost immediately when the water content reaches 65% to 75%. In dry air, lichens rapidly dehydrate, and photosynthesis stops. Thus, in arid climates lichens grow very slowly, often in spurts of less than a millimeter per year. Some lichens are thousands of years old, rivaling the oldest plants as the elder organisms on Earth.

As tough as lichens are, many do not stand up very well to air pollution. Their passive mode of mineral uptake from rain and moist air makes them particularly sensitive to sulfur dioxide and other aerial poisons. The death of sensitive lichens and an increase in hardier species in an area can be an early warning that air quality is deteriorating.

Mycorrhizae

Mycorrhizae are mutualistic associations of plant roots and fungi. The word *mycorrhizae* means "fungus roots," referring to the structures formed by both root cells and hyphae from the associated fungus (FIGURE 31.16). The anatomy of this symbiosis varies, depending on the type of fungus. The extensions of the fungal mycelium from the hyphae forming the mycorrhizae greatly increase the absorptive surface of the plant roots. The partners exchange minerals accumulated from the soil by the fungus for organic nutrients synthesized by the plant.

Mycorrhizae are enormously important in natural ecosystems and agriculture. Over 95% of all vascular plants have mycorrhizae. The fungi involved are permanent associates with their hosts and periodically form fruiting bodies (structures for sexual reproduction). Basidiomycota, Ascomycota, and Zygomycota all have members that form mycorrhizae. In fact, half of all species of mushroom-forming basidiomycetes live as mycorrhizae with oak, birch, and pine trees. The mushrooms that sprout around the bases of these trees are the surface evidence of the underground symbiotic relationship between plants and fungi. (The structure and physiology of mycorrhizae will be described in more detail in Chapter 36.)

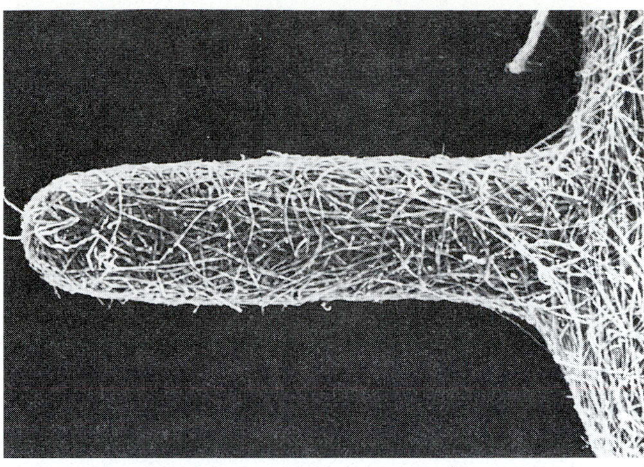

FIGURE 31.16 ▪ Mycorrhizae. Symbiotic associations between fungi and roots, mycorrhizae enhance the absorption of minerals. This scanning electron micrograph of fungal hyphae on a small root of a eucalyptus reveals the intimate association between the symbiotic partners of mycorrhizae. Some mycorrhizal fungi surround the living cells of the root, while others actually penetrate the root cells.

ECOLOGICAL IMPACTS OF FUNGI

The fossil record indicates that from their inception, terrestrial communities have been dependent on fungi as decomposers and symbionts. Much of the diversity of fungi we observe today may have had its phylogenetic origin in adaptive radiation when life began to colonize land. The oldest undisputed fossils of fungi are about 440 million years old, about the time plants began to colonize land. Fossils of the first vascular plants dating back to the late Silurian period have petrified mycorrhizae. Plants probably moved onto land in the company of mycorrhizae.

Ecosystems depend on fungi as decomposers and symbionts

Wood-rotting fungi apparently became a dominant group of organisms about 250 million years ago during the mass extinctions that wiped out extensive forests of vascular plants at the end of the Permian. Breakdown of the wood, which made nutrients available to living organisms, probably was a major factor in the recolonization of land by survivors after the Permian.

Fungi and bacteria are the principal decomposers that keep ecosystems stocked with the inorganic nutrients essential for plant growth. Without decomposers, carbon, nitrogen, and other elements would accumulate in organic matter. Plants and the animals they feed would starve because elements taken from the soil would not be returned (see Chapter 54).

Fungi are well adapted as decomposers of plant material. Their invasive hyphae enter the tissues and cells of dead organic matter and hydrolyze polymers, including the cellulose and lignin of plant cell walls. A succession of fungi, in concert with bacteria and, in some environments, invertebrate animals, are responsible for the complete breakdown of plant litter. The air is so loaded with fungal spores that as soon as a leaf falls or an insect dies, it is covered with spores and is soon infiltrated by saprobic hyphae.

We may applaud fungi that decompose forest litter or dung, but it is a different story when molds attack our fruit or our shower curtains. A wood-digesting saprobe does not distinguish between a fallen oak limb and the oak planks of a boat. During the Revolutionary War, the British lost more ships to fungal rot than to enemy attack. Soldiers stationed in the tropics during World War II watched their tents, clothing, boots, and binoculars be destroyed by molds. Some fungi can even decompose certain plastics.

Between 10% and 50% of the world's fruit harvest is lost each year to fungal attack. Ethylene, a plant hormone that causes fruit to ripen, also stimulates fungal spores on the fruit surface to germinate. This timing mechanism allows fungi to invade when fruit is most vulnerable and nutritious.

Some fungi are pathogens

Among the diseases that fungi cause in humans are athlete's foot, vaginal yeast infections, and lung infections that may be fatal.

Plants are particularly susceptible to fungal diseases. For example, the ascomycete that causes Dutch elm disease has drastically changed the landscape of the northeastern United States. Accidentally introduced to the United States on logs that were sent from Europe to help pay World War I debts, the fungus is carried from tree to tree by bark beetles. Another ascomycete has almost eliminated the native American chestnut.

Some of the fungi that attack food crops are toxic to humans. For example, some species of the mold *Aspergillus* contaminate improperly stored grain by secreting compounds called aflatoxins, which are carcinogenic. In another example, one type of ascomycete forms purple structures called ergots on rye. If diseased rye is inadvertently milled into flour and consumed, poisons from the ergots cause gangrene, nervous spasms, burning sensations, hallucinations, and temporary insanity. One epidemic in about 944 A.D. killed more than 40,000 people. One of the hallucinogens that has been isolated from ergots is lysergic acid, the raw material from which LSD is made. On the other hand, toxins extracted from fungi often have medicinal uses when administered in weak doses. For example, an ergot compound is helpful in treating high blood pressure and stopping maternal bleeding after childbirth.

Many animals, including humans, eat fungi

People in many countries consume a variety of cultivated and wild mushrooms. There are no simple rules to help the novice distinguish poisonous from edible mushrooms. Only experts in mushroom identification can safely collect wild mushrooms for eating.

The fungi most prized by gourmets are truffles, the underground fruiting bodies (ascocarps) of mycelia that are mycorrhizal on the roots of trees (FIGURE 31.17). Their complex flavor is variously described as nutty, musky, cheesy, or all three. The fruiting bodies (ascocarps) release strong odors that attract mammal and insect consumers that excavate the truffles and disperse their spores. In some cases, the odors mimic sex attractants of certain mammals. Truffle hunters traditionally used pigs to locate their prizes, although dogs are now more commonly used, because they have the nose for the scent without the fondness for the flavor.

FIGURE 31.17 • Gourmet fungi. Truffles, subterranean fruiting bodies of ascomycetes, are highly prized for their flavor by gourmet cooks.

PHYLOGENETIC RELATIONSHIPS OF FUNGI

Fungi and animals probably evolved from a common protistan ancestor

Molecular evidence supports the widely held view that the four fungal divisions are monophyletic (see FIGURE 31.4). The occurrence of flagella in the chytrids, representing the oldest lineage of fungi, indicates that fungal ancestors were aquatic flagellated organisms. Flagellated cells were probably lost in the chytrid lineage that led to the Zygomycota, as ancestral fungi became increasingly adapted for life on land. Many of the differences among the divisions Zygomycota, Ascomycota, and Basidiomycota center on contrasting solutions to the problem of reproducing and dispersing on land. These fungi may have diverged during the transition from aquatic to terrestrial habitats.

Animals probably evolved from aquatic flagellated organisms too, and there is compelling evidence that animals and fungi diverged from a common protistan ancestor. Comparisons of several proteins and ribosomal RNA indicate that fungi are more closely related to animals than to plants. We discuss animal phylogeny in more detail in Chapter 32.

CHAPTER REVIEW

REVIEW OF KEY CONCEPTS

(with page numbers and key figures)

INTRODUCTION TO THE FUNGI

Fungi are structurally and nutritionally distinctive eukaryotes. Most are multicellular.

- **Absorptive nutrition enables fungi to live as decomposers and symbionts (pp. 574–575)** All fungi are heterotrophs (decomposers and symbionts), acquiring their nutrients by absorption.
- **Extensive surface area and rapid growth adapt fungi for absorptive nutrition (pp. 575–576, FIGURES 31.1, 31.2)** The vegetative bodies of fungi consist of mycelia, netlike collections of branched hyphae adapted for absorption. Parasitic fungi penetrate their hosts with specialized hyphae called haustoria. Most fungi have cell walls made of chitin. Although aseptate (coenocytic) forms occur, most fungi have their hyphae partitioned into cells by septa, with pores allowing cell-to-cell continuity.
- **Fungi reproduce by releasing spores that are produced either sex-**
31.1 **ually or asexually (pp. 576–577, FIGURE 31.3)** Fungi produce reproductive structures (spores) by sexual and asexual means. The sexual cycle involves cell fusion (plasmogamy) and nuclear fusion (karyogamy), with an intervening dikaryotic stage (with two haploid nuclei). The diploid phase is short-lived and rapidly undergoes meiosis to produce haploid spores.

DIVERSITY OF FUNGI

- **Division Chytridiomycota: Chytrids may provide clues about fungal origins (p. 577, FIGURE 31.5)** Chytrids are fungi that retain a flagellated condition; they may represent a link between the fungal kingdom and the protists.
- **Division Zygomycota: Zygote fungi form resistant dikaryotic**
31.1 **structures during sexual reproduction (pp. 578–579, FIGURE 31.6)** Zygote fungi, such as the black bread mold, are named for their sexually produced zygosporangia, which are dikaryotic structures capable of persisting through unfavorable conditions.
- **Division Ascomycota: Sac fungi produce sexual spores in saclike**
31.1 **asci (pp. 579–580, FIGURES 31.7, 31.8)** Sexual reproduction in the sac fungi involves the formation of spores in sacs, or asci, at the ends of dikaryotic hyphae, usually in ascocarps.
- **Division Basidiomycota: Club fungi have long-lived dikaryotic**
31.1 **mycelia and a transient diploid stage (p. 581, FIGURES 31.9, 31.10)** The club fungi include mushrooms, shelf fungi, puffballs, and rusts. Mycelia of club fungi can last years as dikaryons. Sexual reproduction involves the formation of spores on club-shaped basidia at the ends of dikaryotic hyphae in fruiting bodies, such as mushrooms.
- **Molds, yeasts, lichens, and mycorrhizae represent unique lifestyles that evolved independently in three fungal divisions (pp. 581–585, FIGURES 31.11, 31.13–31.16)** Molds are rapidly growing, asexually reproducing fungi that are important in the commercial production of antibiotics, such as penicillin. Yeasts are unicellular fungi adapted to life in liquids such as plant saps. Lichens are such highly integrated symbiotic associations of algae and fungi that they are classified as single organisms. Mycorrhizae are mutualistic associations of fungi with the roots of vascular plants.

ECOLOGICAL IMPACTS OF FUNGI

- **Ecosystems depend on fungi as decomposers and symbionts (pp. 585–586)** Without fungi and bacteria as decomposers, biological communities would be deprived of the essential recycling of chemical elements between the biological and nonbiological world. Fungi are particularly important decomposers of wood. Fungi also decompose food and other useful objects.
- **Some fungi are pathogens (p. 586)** Some fungi cause disease, harming humans with a variety of ills. Plants are especially vulnerable to fungal infections.
- **Many animals, including humans, eat fungi (p. 586)**

PHYLOGENETIC RELATIONSHIPS OF FUNGI

- **Fungi and animals probably evolved from a common protistan ancestor (p. 587)** Molecular evidence supports the hypothesis that fungi and animals diverged from a common ancestor, probably a flagellated aquatic protist.

SELF-QUIZ

1. *All* fungi share which one of the following characteristics?
   a. symbiotic
   b. heterotrophic
   c. flagellated
   d. pathogenic
   e. saprobic

2. Which feature seen in chytrids supports the hypothesis that they represent the most primitive fungi? Explain your answer.
   a. the absence of chitin within the cell wall
   b. coenocytic hyphae
   c. a flagellated zoospore
   d. formation of resistant zygosporangia
   e. all representatives are parasitic

3. Which of the following cells or structures are associated with *asexual* reproduction in fungi?
   a. ascospores
   b. basidiospores
   c. conidia
   d. zygosporangia
   e. ascogonia
4. Which of the following statements about the Zygomycota is true?
   a. Meiosis immediately precedes karyogamy.
   b. Haploid hyphae are present.
   c. Diploid mycelia are present.
   d. Hyphae are exclusively septate.
   e. Sexual reproduction is never observed.
5. The adaptive advantage associated with the filamentous nature of the mycelium is primarily related to
   a. the ability to form haustoria and parasitize other organisms
   b. avoiding sexual reproduction until environmental change occurs
   c. the potential to inhabit almost all terrestrial habitats
   d. the increased probability of contact between different mating types
   e. an extensive surface area well suited for absorptive nutrition
6. Sporangia on erect hyphae that produce asexual spores are characteristic of
   a. Ascomycota
   b. Basidiomycota
   c. Ascocarps
   d. Zygomycota
   e. lichens
7. Which of the following statements is true about Basidiomycota?
   a. Hyphae fuse to give rise to a dikaryotic mycelium.
   b. Spores line up in a sac after they are formed by meiosis.
   c. The vast majority of spores are formed asexually.
   d. Basidiomycetes are the most important commercial sources of antibiotics.
   e. Basidiomycetes have no known sexual stages.
8. Mycorrhizae are
   a. asexual reproductive structures formed by lichens
   b. thin hyphae that grow directly into host tissues
   c. the mycelium that forms fairy rings
   d. compact, dikaryotic hyphae that form a basidiocarp
   e. mutualistic associations between plant roots and fungi
9. The photosynthetic symbiont of a lichen is most commonly a(n)
   a. moss
   b. green alga
   c. red alga
   d. ascomycete
   e. small vascular plant
10. The closest relatives of fungi are probably
   a. animals
   b. vascular plants
   c. mosses
   d. brown algae
   e. slime molds

CHALLENGE QUESTIONS

1. In what way might the use of a wide-spectrum fungicide that kills all fungal species have a harmful effect on vegetation, such as a forest?
2. Zygomycetes fuse dikaryotic nuclei to form zygotes, only to restore the haploid state again by meiosis before the growth of new mycelia. What does this formation of a transient diploid stage accomplish?
3. Many fungi produce antibiotics, such as penicillin, that are valuable in medicine. Of what possible value are antibiotics to the fungi that produce them? Similarly, fungi often produce compounds with unpleasant tastes and odors as they digest their food. What might be the value of these chemicals to the fungi? To other organisms that might eat the decomposing food? How might the production of antibiotics and odors have evolved?

SCIENCE, TECHNOLOGY, AND SOCIETY

1. Many fungi and lichens in the United States are threatened with extinction due to the destruction of habitats. Should they come under legal protection by their inclusion on lists of endangered species? Defend your answer.
2. American chestnut trees once made up more than 25% of the hardwood forests of the eastern United States. These trees were wiped out by a fungus accidentally introduced on imported Asian chestnuts, which are not affected. More recently, a fungus has killed large numbers of dogwood trees from New York to Georgia; some experts suspect the parasite was accidentally introduced from elsewhere. Why are plants particularly vulnerable to fungi imported from other regions? What kinds of human activities might contribute to the spread of plant diseases? Do you think the introduction of plant pathogens like chestnut blight are more or less likely to occur in the future? Why?

FURTHER READING

Bradley, D. "How Ripening Fruit Invite Fungal Attack." *New Scientist,* August 27, 1994.

Cilaghi, C., and P. L. Nimis. "Lichens, Air Pollution, and Lung Cancer." *Nature,* May 29, 1997. Explores the possible correlation between low lichen diversity and poor air quality in ecosystems and high cancer incidence.

Jaenike, J. "Behind-the-Scenes Role of Parasites." *Natural History,* June 1994. Mushrooms are key junctures in some food webs.

Kendrick, B. *The Fifth Kingdom,* 2nd ed. Newburyport, MA: Focus Information Group, 1992. A lively introduction to the fungi.

Lewis, R. "A New Place for Fungi?" *BioScience,* June 1994. Makes the case for an animal-fungus link in evolution.

Newhouse, J. R. "Chestnut Blight." *Scientific American,* July 1990. A new parasite may help control the blight.

Radetsky, P. "The Yeast Within." *Discover,* March 1994. Presents surprising new discoveries about how yeasts develop.

Sternberg, S. "The Emerging Fungal Threat," *Science,* December 9, 1994. Pathogenic fungi are especially deadly to AIDS patients and others with compromised immune systems.

Walsh, R. "Seeking the *Truffe,*" *Natural History,* January 1996. Describes the cultivation and harvest of gourmet fungi.

Wood, W. "Deadly Fungus Is Kind to Hearts," *New Scientist,* February 11, 1995. Discusses the pharmaceuticals made from fungi.

WEB LINKS

Visit the special edition of *The Biology Place* for BIOLOGY, Fifth Edition, at http://www.biology.com/campbell. Go to Chapter 31 for online resources, including learning activities, practice exams, and links to the following web sites:

"Fungi Online"
This site, devoted to the taxonomy, nomenclature, and protection of fungi, contains links to mycological resources around the world.

"International Culture Collection of Arbuscular and Vesicular-Arbuscular Mycorrhizal Fungi"
Information on the taxonomy and culture of the fungi involved in mycorrhizal associations can be found at this site, as well as links to global mycorrhizal resources.

"Treasures from the kingdom of fungi"
This is an interesting site well worth the visit for stunning photographs of fungi.

INTRODUCTION TO ANIMAL EVOLUTION

Animal life began in Precambrian seas with the evolution of multicellular forms that lived by eating other organisms. This new way of life allowed the exploitation of previously untapped resources and led to an evolutionary radiation of diverse forms. Early animals populated the seas, fresh water, and eventually the land. The dazzling diversity of animal life on Earth today, some of it illustrated in the photograph of a coral reef on this page, stems from over a half-billion years of evolution from Precambrian ancestors.

This chapter begins with the general characteristics of animals. We then discuss possible relationships among the animal phyla and examine hypotheses about the origin and early radiation of animals. This overview provides an orientation for our closer look at animal phyla in the next two chapters.

WHAT IS AN ANIMAL?

Constructing a good definition of an animal is not as easy as it might first appear. There are exceptions to nearly every criterion for distinguishing an animal from other life forms. However, when taken together, the following characteristics of animals will serve our purposes.

1. Animals are multicellular, heterotrophic eukaryotes. In contrast to the autotrophic nutrition of plants and algae, animals must take into their bodies preformed organic molecules; they cannot construct them from inorganic chemicals. Most animals do this by **ingestion**—eating other organisms or organic material that is decomposing.
2. Animal cells lack the cell walls that provide strong support in the bodies of plants and fungi. The multicellular bodies of animals are held together by structural proteins, the most abundant being collagen (see Chapter 7 and FIGURE 40.2). In addition to collagen, which is found mainly in extracellular matrices, animal tissues have unique types of intercellular junctions—tight junctions, desmosomes, and gap junctions—that are composed of other structural proteins (see FIGURE 7.30).
3. Also unique among animals are two types of tissues responsible for impulse conduction and movement: nervous tissue and muscle tissue.
4. A few key features of life history also distinguish animals. Most animals reproduce sexually, with the diploid stage usually dominating the life cycle. In most species, a small flagellated sperm fertilizes a larger, nonmotile egg to form a diploid zygote. The zygote then undergoes **cleavage**, a succession of mitotic cell divisions. During the development of most animals, cleavage leads to the formation of a multicellular stage called a **blastula**, which in many animals takes the form of a hollow ball. Following the blastula stage is the process of **gastrulation**, during which layers of embryonic tissues that will develop into

adult body parts are produced. The resulting developmental stage is called a **gastrula** (FIGURE 32.1). Some animals develop directly through transient stages of maturation into adults, but the life cycles of many animals include larval stages. The **larva** is a sexually immature form. It is morphologically distinct from the adult stage, usually eats different food, and may even have a different habitat than the adult, as in the case of a frog tadpole. Animal larvae eventually undergo **metamorphosis**, a resurgence of development that transforms the animal into an adult.

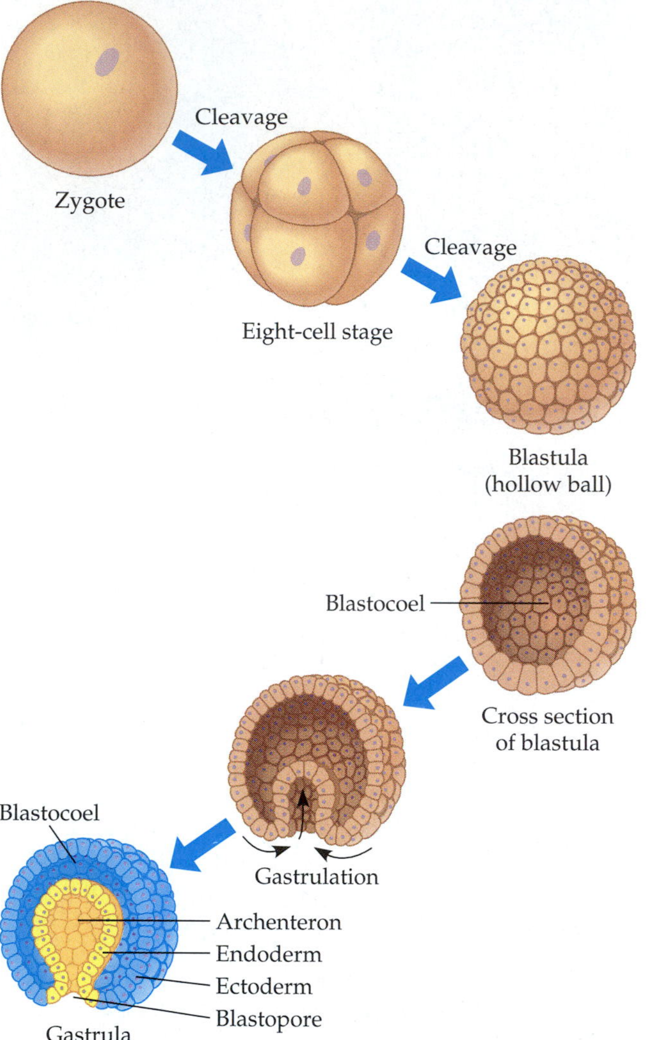

AN OVERVIEW OF ANIMAL PHYLOGENY AND DIVERSITY

Although lively debate continues, most systematists now agree that the animal kingdom is monophyletic; that is, if we could trace all animal lineages back to their origin, they would converge on a common ancestor. That ancestor was probably a colonial flagellated protist that lived over 700 million years ago in the Precambrian era. This protist was probably related to choanoflagellates, a group that arose about a billion years ago (see FIGURE 28.26). FIGURE 32.2 diagrams one hypothesis for how such an ancestor may have evolved into simple animals with specialized cells arranged in two or more layers.

With its branches radiating from a multicellular ancestor, the evolutionary tree in FIGURE 32.3 depicts one set of hypotheses about animal phylogeny. The circled numbers on the tree highlight four key evolutionary branch points, which we discuss in the following sections.

FIGURE 32.1 ▪ Early embryonic development. The zygote of an animal undergoes a succession of mitotic cell divisions called cleavage. (Only one cleavage stage—the eight-cell embryo—is shown here.) In most animals, cleavage results in the formation of a multicellular stage called the blastula. As shown in the cross section, the blastula of many animals is a hollow ball of cells. The cavity of the blastula, which is filled with fluid, is called the blastocoel. Most animals also undergo gastrulation, a rearrangement of the embryo that produces layers of embryonic tissues; shown here are the ectoderm (outer layer) and an inner layer, the endoderm. In the mechanism of gastrulation illustrated here, endoderm at one end of the embryo folds inward, expands, and eventually fills the blastocoel. The blind pouch formed by the infolding endoderm, called the archenteron, opens to the outside via the blastopore. The endoderm of the archenteron develops into the tissue lining the animal's digestive tract.

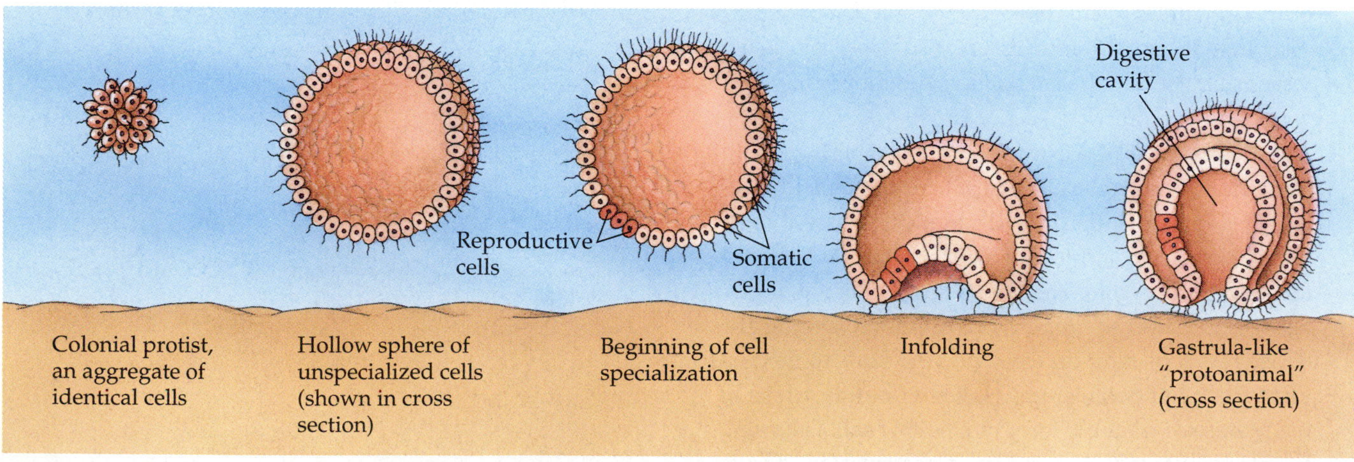

FIGURE 32.2 ▪ One hypothesis for the origin of animals from a flagellated protist.

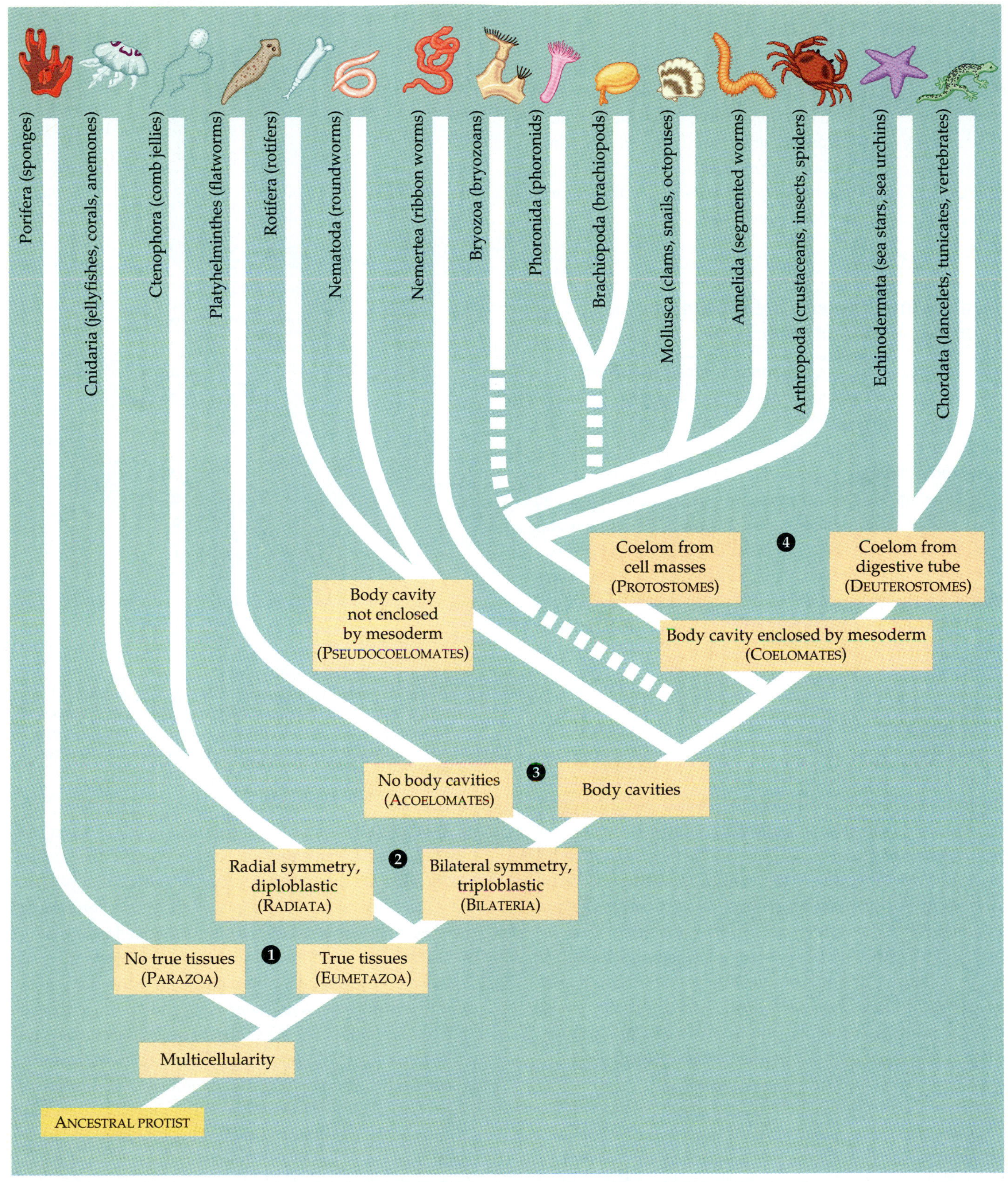

32.1 **FIGURE 32.3 ▪ A phylogenetic tree of animals.** The circled numbers key four main branch points to the headings and discussion in the text. In common with all phylogenetic trees, this one represents hypotheses based on current evidence. The branches with broken lines indicate relationships that are particularly uncertain. In Chapter 33 you will see a miniature version of this tree next to the headings to help you keep sight of evolutionary relationships among the animal phyla.

① Parazoans lack true tissues

Among the extant phyla, sponges (Phylum Porifera) represent an early branch (① in FIGURE 32.3) of the animal kingdom. Sponges have unique development and a structural simplicity that separates them from all other animal phyla. They lack true tissues and are called the **parazoa** (meaning "beside the animals"). Tissues are a basic feature of nearly all other animal phyla, collectively called the **eumetazoa**.

② Radiata and bilateria are the major branches of eumetazoans

The eumetazoa are divided into two major branches, partly on the basis of body symmetry (② in FIGURE 32.3). Members of Phylum Cnidaria (hydras; jellies, also called "jellyfishes"; and their relatives) and Phylum Ctenophora (comb jellies) have **radial symmetry** (FIGURE 32.4a) and are collectively called the **radiata**. A radial animal has a top and bottom, or an oral (mouth) and an aboral side, but no head end and rear end and no left and right. The other major branch of eumetazoan evolution led to animals with **bilateral** (two-sided) **symmetry** (FIGURE 32.4b). A bilateral animal has not only a **dorsal** (top) side and a **ventral** (bottom) side, but also an **anterior** (head) end and a **posterior** (tail) end and a left and right side. Animals of this evolutionary branch are collectively called the **bilateria**.

Associated with bilateral symmetry is **cephalization**, an evolutionary trend toward the concentration of sensory equipment on the anterior end, the end of a traveling animal that is usually first to encounter food, danger, and other stimuli. In most bilateral animals, cephalization also includes the development of a central nervous system concentrated in the head and extending toward the tail as a longitudinal nerve cord. A head end is an adaptation for movement, such as crawling, burrowing, or swimming. The symmetry of an animal generally fits its lifestyle. Many radial animals are sessile forms (attached to a substratum) or plankton (drifting or weakly swimming aquatic forms). Their symmetry equips them to meet the environment equally well from all sides. Most animals that move actively from place to place are bilateral. These two fundamentally different kinds of symmetry probably arose very early in the history of animal life.

Body symmetry alone is not a foolproof criterion for assigning an animal phylum to a particular evolutionary line. The radial symmetry of some animals has apparently evolved secondarily from a bilateral condition as an adaptation to a more sedentary lifestyle. For example, sea urchins (Phylum Echinodermata) are radially symmetrical, but their developmental genetics and embryology show clearly that they arose from a bilaterally symmetrical ancestor and belong with the bilateria.

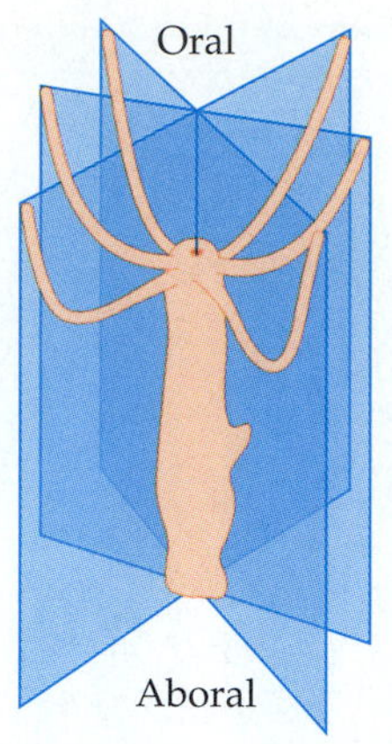

(a) Radial symmetry

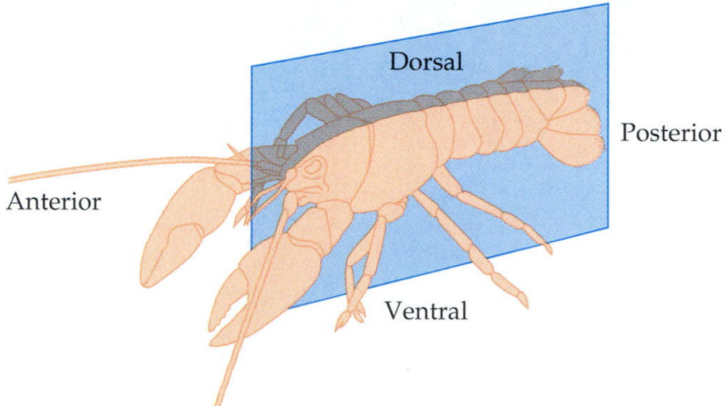

(b) Bilateral symmetry

FIGURE 32.4 ▪ Body symmetry. (a) The parts of a radial animal, such as a hydra, are arranged like spokes of a wheel that radiate from the center. Any imaginary slice through the central axis would divide the animal into mirror images. **(b)** A bilateral animal has a left and right side, and only one imaginary cut would divide the animal into mirror-image halves.

Another difference in body plan helps define the radiata-bilateria split: In all animals except sponges, the embryo becomes layered through the process of gastrulation (see FIGURE 32.1). As development progresses, these concentric layers, called **germ layers**, form the various tissues and organs of the body. **Ectoderm**, covering the surface of the embryo, gives rise to the outer covering of the animal and, in some phyla, to the central nervous system. **Endoderm**, the innermost germ layer, lines the developing digestive tube, or **archenteron**, and gives rise to the lining of the digestive tract and organs derived from it, such as the liver and lungs of vertebrates. All eumetazoans except cnidarians and ctenophores (the radiata) have a third germ layer, the **mesoderm**, between the ectoderm and endoderm. Mesoderm forms the muscles and most other organs between the digestive tube and the outer covering of the animal. Cnidarians and ctenophores have only two germ layers (ectoderm and endoderm) or have a third layer that is not homologous with the mesoderm of bilateral animals. As a group, the radiata are said to be **diploblastic** (having two germ layers). All other eumetazoa, the bilateria, are **triploblastic** (have three germ layers).

③ Evolution of body cavities led to more complex animals

Triploblastic animals with solid bodies—that is, without a cavity between the digestive tract and outer body wall—are collectively called **acoelomates** (Gr. *a*, "without," and *koilos*, "a hollow"). This group includes Phylum Platyhelminthes, the flatworms (FIGURE 32.5a). In contrast to the acoelomates, most phyla of bilateral, triploblastic animals have tube-within-a-tube body plans, with a **body cavity**, a fluid-lined space separating the digestive tract from the outer body wall (③ in FIGURE 32.3). Most animals with a body cavity also have a complete digestive tract (one with a mouth and an anus) and some form of circulatory system, a network of spaces or vessels through which blood or other body fluid circulates. In contrast, the acoelomates either lack a digestive tract (for example, tapeworms) or have a gastrovascular cavity (said to be an incomplete digestive tract because it lacks an anus). Also present in cnidarians and ctenophores, the gastrovascular cavity functions in both digestion and circulation (see FIGURES 33.3 and 33.9).

Among animals with a body cavity, there are differences in how the cavity develops. If the cavity is not completely lined by tissue derived from mesoderm, it is termed a **pseudocoelom**. Animals with this body plan, such as rotifers (Phylum Rotifera) and roundworms (Phylum Nematoda), are called **pseudocoelomates** (FIGURE 32.5b). **Coelomates** are animals with a true **coelom**, a fluid-filled body cavity completely lined by tissue derived from mesoderm. The inner and outer layers of tissue that surround the cavity connect dorsally and ventrally to form mesenteries, which suspend the internal organs (FIGURE 32.5c).

A body cavity has many functions. Its fluid cushions the suspended organs, helping to prevent internal injury. The cavity also enables the internal organs to grow and move independently of the outer body wall. If it were not for your coelom, every beat of your heart or ripple of your intestine could deform your body surface, and exercise would distort the shapes of the internal organs. In soft-bodied coelomates such as earthworms, the noncompressible fluid of the body cavity is under pressure and functions as a hydrostatic skeleton against which muscles can work. Though they may have originated as adaptations for burrowing by soft-bodied animals, coeloms evolved independently at least twice, as depicted on the tree in FIGURE 32.3, in the protostomes and in the deuterostomes.

④ Coelomates branched into protostomes and deuterostomes

The coelomate phyla are divided into two distinct evolutionary lines (④ in FIGURE 32.3). Mollusks, annelids, arthropods, and several other phyla represent one of these lines and are collectively called **protostomes**. Echinoderms, chordates, and some other phyla, collectively called **deuterostomes**, represent the other line. Protostomes and deuterostomes are distinguished by several fundamental differences in their development.

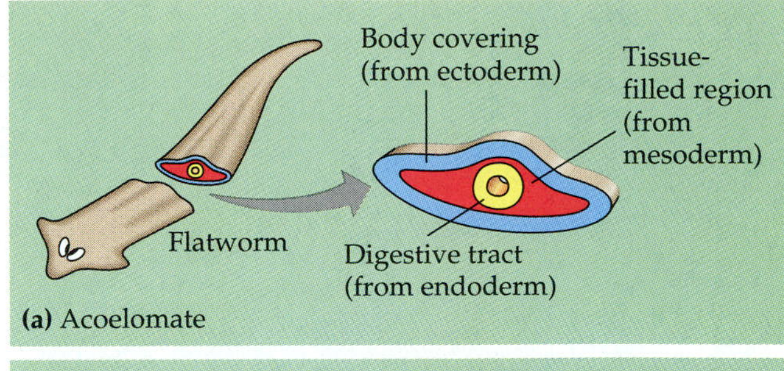

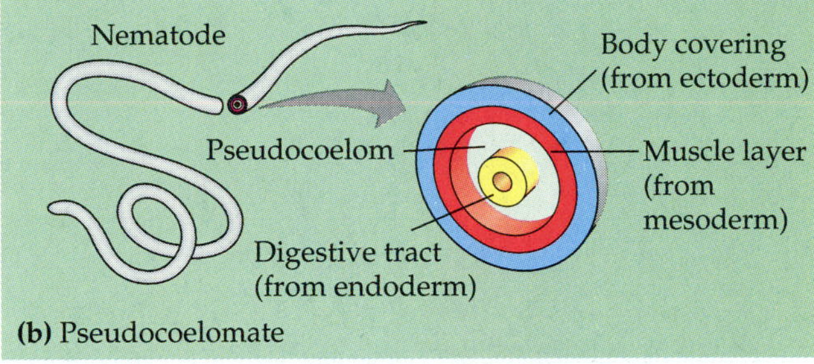

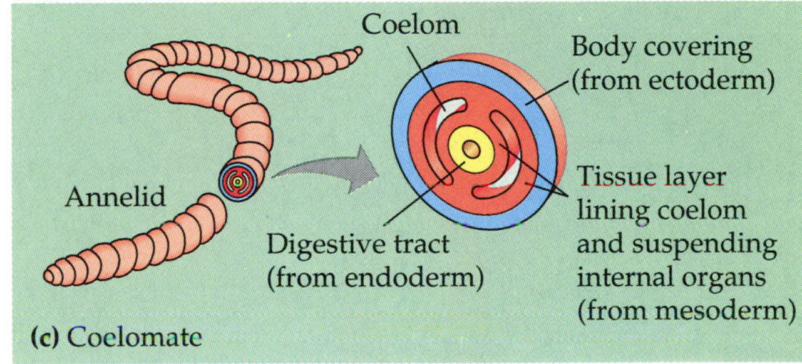

FIGURE 32.5 ▪ Body plans of the bilateria. The various organ systems of the animal develop from the three germ layers that form in the embryo. Colors traditionally used to represent the germ layers and their derivatives are blue for ectoderm, red for mesoderm, and yellow for endoderm. **(a)** Acoelomates lack a body cavity between the digestive tract and outer body wall. **(b)** Pseudocoelomates have a body cavity only partially lined by mesodermally derived tissue. **(c)** Coelomates have a true coelom, a body cavity completely lined by mesodermally derived tissue.

Cleavage

Differences between animals of the two coelomate lineages are evident as early as the cleavage divisions that transform the zygote into a ball of cells. Many protostomes undergo **spiral cleavage**, in which planes of cell division are diagonal to the vertical axis of the embryo. As seen in the eight-cell stage resulting from spiral cleavage, small cells lie in the grooves between larger, underlying cells (FIGURE 32.6a, p. 594). Furthermore, the so-called **determinate cleavage** of some protostomes rigidly casts the developmental fate of each embryonic cell very early. A cell isolated at the four-cell stage from a protostome, such as a snail, forms an inviable embryo that lacks parts.

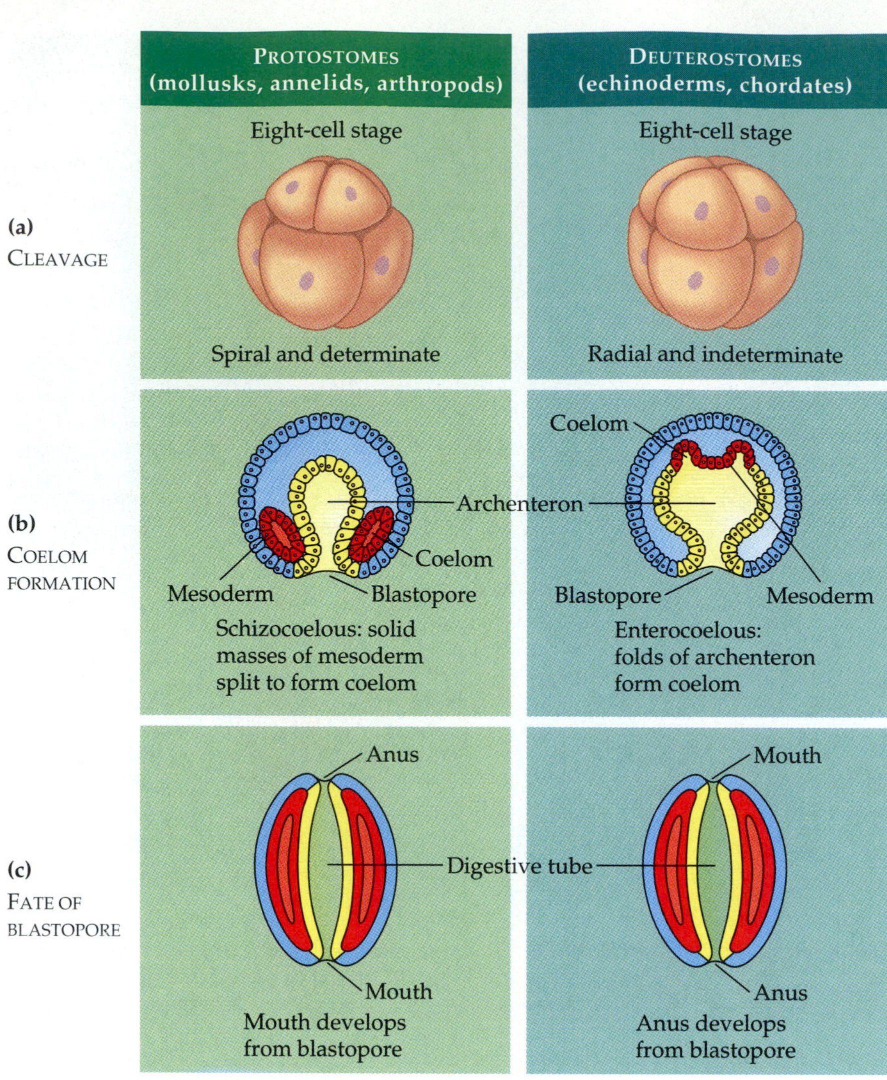

FIGURE 32.6 ▪ A comparison of early development in protostomes and deuterostomes. **(a)** Protostomes have spiral, determinate cleavage; deuterostomes have radial, indeterminate cleavage. **(b)** Coelom formation begins in the gastrula stage. In the schizocoelous development of protostomes, the coelom forms from splits in the mesoderm. In the enterocoelous development of deuterostomes, the coelom forms from mesodermal outpocketings of the archenteron (blue = ectoderm, yellow = endoderm, red = mesoderm). **(c)** The blastopore forms the mouth in protostomes; the mouth forms from a secondary opening in deuterostomes.

In contrast to the protostome pattern, the zygotes of many deuterostomes undergoes **radial cleavage**. Here, the cleavage planes are either parallel or perpendicular to the vertical axis of the egg; as seen in the eight-cell stage, the cells are aligned, one directly above the other. Most deuterostomes are further characterized by **indeterminate cleavage**, meaning that each cell produced by early cleavage divisions retains the capacity to develop into a complete embryo. If the cells of a sea star embryo, for example, are separated at the four-cell stage, each will go on to form a normal larva. It is the indeterminate cleavage of the human zygote that makes identical twins possible.

Coelom Formation

Another difference between protostomes and deuterostomes is apparent later in development. In gastrulation, the developing digestive tube of an embryo initially forms as a blind pouch, the archenteron, which has a single opening to the outside known as the **blastopore**. As the archenteron forms in a protostome, initially solid masses of mesoderm split to form the coelomic cavities; this is called **schizocoelous** development (Gr. *schizo*, "split"). Development of the body cavities of deuterostomes is termed **enterocoelous**: The mesoderm buds from the wall of the archenteron and hollows to become the coelomic cavities (FIGURE 32.6b).

Blastopore Fate

A third fundamental difference between protostomes and deuterostomes is in the fate of the blastopore. After the archenteron develops, a second opening forms at the opposite end of the gastrula. Ultimately, the blastopore and this second opening become the two ends of the digestive tube (the mouth and the anus). The mouth of many protostomes develops from the first opening, the blastopore, and it is for this characteristic that the protostome line is named (Gr. *protos*, "first," and stoma, "mouth"). By contrast, the mouth of a deuterostome (Gr. *deuteros*, "second") is derived from the secondary opening, and the blastopore usually forms the anus, not the mouth (FIGURE 32.6c).

■ ■ ■

Table 32.1 ■ Organization of Animal Phyla According to Major Features of Body Plans

CATEGORY	MAJOR FEATURES OF BODY PLAN	PHYLA
Kingdom Animalia (Metazoa)		
① Parazoa	Multicellular, without true tissues	Porifera (sponges)
① Eumetazoa	True tissues	
② Radiata	Radial symmetry; diploblastic (two germ layers: ectoderm, endoderm)	Cnidaria (hydras, jellies, sea anemones, corals) Ctenophora (comb jellies)
② Bilateria	Bilateral symmetry; triploblastic (three germ layers: ectoderm, endoderm, mesoderm)	
③ Acoelomates	Solid body, without a body cavity	Platyhelminthes (flatworms)
③ Pseudocoelomates	Pseudocoelom (body cavity between digestive tract and body wall incompletely lined by mesoderm)	Rotifera (rotifers) Nematoda (roundworms)
③ Coelomates	Body cavity (coelom) completely lined by mesoderm	
④ Protostomes	Spiral, determinate cleavage; mouth develops from blastopore; schizocoelous body cavity (formed by splitting of mesodermal mass of tissue)	Nemertea (proboscis worms) (phylogenetic position uncertain) Lophophorates: Bryozoa, Brachiopoda, Phoronida (phylogenetic position uncertain) Mollusca (clams, snails, squids) Annelida (segmented worms) Arthropoda (crustaceans, insects, spiders)
④ Deuterostomes	Radial, indeterminate cleavage; anus develops from blastopore; enterocoelous body cavity (formed by folding of mesodermal archenteron wall)	Echinodermata (sea stars, sea urchins) Chordata (lancelets, tunicates, vertebrates)

We have now mapped out the main animal lineages; TABLE 32.1 summarizes the main features of the animal body plans and phyla, and you may find it useful, along with FIGURE 32.3, as you review this chapter. We are now ready to explore the origins of animal diversity.

THE ORIGINS OF ANIMAL DIVERSITY

Most animal phyla originated in a relatively brief span of geological time

Based largely on anatomical and embryological criteria, animals are grouped into about 35 phyla, the exact number depending on the views of different systematists. Animals in each phylum exhibit a distinctive combination of body features, a unique body plan, markedly different from that of other phyla. For instance, the basic features of the body plan of Phylum Arthropoda (for example, crabs, spiders, insects) include jointed legs, an external skeleton, and segmentation (repetition of body parts).

The fossil record and molecular studies indicate that the diversification that produced the many animal phyla occurred rapidly on the vast scale of geological time. This relatively brief evolutionary episode probably lasted about 40 million years (about 565 to 525 million years ago) during the late Precambrian and early Cambrian (which began about 545 million years ago).

Paleontologists have named the last period of the Precambrian era the **Ediacaran period**, for the Ediacara Hills of Australia, where fossils of Precambrian animals were first discovered. Similar animals of similar vintage have since been found on other continents. Most of the Ediacaran fossils appear to represent cnidarians (animals similar to hydras), but soft-bodied mollusks (similar to a modern group called the chitons) were also present, and numerous fossilized burrows and tracks indicate the activities of several groups of worms. Did these ancient animals belong to modern phyla or to lineages that died out by the early Cambrian? Paleontologists are still debating the phylogenetic connections between the animals on opposite sides of the Cambrian boundary.

Nearly all the major animal body plans appear in Cambrian rocks dating from 545 to 525 million years ago. During this relatively short span, a burst of animal origins called the **Cambrian explosion** left a rich fossil assemblage that includes the first animals with hard, mineralized skeletons. The Burgess Shale in British Columbia, Canada, is the most famous fossil bed documenting the diversity of Cambrian animals. Two other fossil sites, one in Greenland and the other in the Yunnan region of China, predate the Burgess Shale by more than 10 million years. Burgess Shale fossils are rather bizarre-looking in the context of the marine animals we know today (FIGURE 32.7, p. 596). Some of these Cambrian forms probably represent extinct "experiments" in animal diversity. However, some researchers believe that most of the Cambrian fossils, as strange as they may appear to us, are simply ancient variations within the taxonomic boundaries of phyla still represented in the modern fauna. Indeed, the number of

FIGURE 32.7 ▪ A sample of some of the animals that evolved during the Cambrian explosion. This drawing is based on fossils collected from the Burgess Shale in British Columbia, Canada. Taxonomic debate about the Cambrian fauna centers on whether it includes a large number of extinct phyla or consists mainly of variations on the anatomical themes of modern phyla.

exclusively Cambrian phyla seems to be dropping as the fossils are studied more closely and are classified in extant phyla.

On the scale of geological time, animals diversified so rapidly that it is difficult from the fossil record to sort out the sequence of branching in animal phylogeny. Thus, when reconstructing the evolutionary history of animal phyla, systematists depend largely on clues from the comparative anatomy, embryology, developmental genetics, and molecular systematics of extant species.

Developmental genetics may clarify our understanding of the Cambrian diversification

Much of the Cambrian explosion involved bilateral, triploblastic lineages. What might have caused those rapid diversifications? One hypothesis emphasizes the emergence during the Cambrian of predator-prey relationships, which triggered diverse evolutionary adaptations, including various kinds of shells and different modes of locomotion. Some researchers argue that a major environmental change set the stage for the Cambrian explosion. Perhaps atmospheric oxygen finally reached a high enough concentration during the Cambrian to support the more active metabolism required for feeding and other activities by mobile animals. Actually, a combination of environmental forces may have triggered the Cambrian explosion, but what about the organisms themselves?

A traditional view is that the starting point of bilateral lineages—the common ancestor of all the triploblastic phyla—was a simple bilateral animal with little specialization of the three germ layers. Other features of body plans are thought to have evolved later in separate lineages. Segmentation, for instance, is a basic feature of the body plan of arthropods and vertebrates, and according to the traditional hypothesis, segmentation evolved independently in these two distinct lineages.

A contrasting view is emerging from molecular systematics and developmental genetics. Genes that direct the development of body plans, such as the *Hox* complex (see FIGURE 24.16), are common to diverse phyla and may have been present in the common ancestor of bilateral animals. If so, it is likely that the ancestor was structurally complex, with body features associated with expression of these regulatory genes. In fact, some researchers speculate that the ancestral bilateral animal had a well-developed head with light sensors, a nervous system with a longitudinal nerve cord, and a segmented body with legs. Today, variation in the expression of genes regulating development is associated with some of the body-plan differences that distinguish the phyla. Perhaps this developmental plasticity was also associated with the relatively rapid origin of diverse bilateral animals during the Cambrian. Some researchers who favor this hypothesis also speculate that the phyla became locked into developmental patterns that constrained evolution enough that no additional phyla evolved after the Cambrian explosion. Of course, this does not imply that animal evolution came to a halt; variations in developmental patterns continue to allow subtle changes in body structures and functions, leading to speciation and the origin of taxa below the phylum level.

Continuing research will help test these hypotheses. But even as the Cambrian explosion becomes less mysterious, it will seem no less wonderful. In the last half-billion years, animal evolution has mainly generated new variations on old "designs." In this chapter we traced the evolution of these body plans as a way to make phylogenetic sense of the vast diversity of living animals. In Chapters 33 and 34 we take a closer look at the extant animal phyla and their evolutionary history.

CHAPTER REVIEW

REVIEW OF KEY CONCEPTS

(with page numbers and key figures)

WHAT IS AN ANIMAL?

Animals are multicellular, heterotrophic eukaryotes that ingest their food. Animal cells lack walls. The main structural protein is collagen. Nervous tissue and muscle tissue are unique to animals. Animal reproduction is primarily sexual, and the life cycle is dominated by the diploid stage, gametes generally being the only haploid cells. Fertilization initiates cleavage (mitotic division) of the zygote and the formation of a blastula (a hollow ball of cells in many species). Gastrulation follows the blastula, resulting in formation of embryonic tissue layers (FIGURE 32.1). Many animals undergo metamorphosis, a second stage of development that transforms a sexually immature larva into a morphologically distinct sexual adult.

AN OVERVIEW OF ANIMAL PHYLOGENY AND DIVERSITY

Animals probably evolved from colonial flagellates (FIGURE 32.2).

32.1 Four key evolutionary branch points account for much of animal diversity (FIGURE 32.3).

- **Parazoans lack true tissues (p. 592)** The first major branching in the tree of animals produced the parazoans, represented by the sponges (Phylum Porifera). Parazoans lack true tissues, whereas all other animals (the eumetazoa) have tissues.
- **Radiata and bilateria are the major branches of eumetazoans (p. 592, FIGURE 32.4)** In the second major branch point in animal phylogeny, eumetazoa diverged early into two major branches, the radiata and the bilateria. Members of the branch radiata are jellies, sea anemones, and their relatives, sedentary and planktonic forms with radial symmetry. Bilaterians are characterized by bilateral symmetry and cephalization. Another important difference is that radial animals are diploblastic (two-layered: ectoderm and endoderm), while bilaterian animals are triploblastic (three-layered: ectoderm, mesoderm, and endoderm).
- **Evolution of body cavities led to more complex animals (p. 593, FIGURE 32.5)** The third major branch point on the animal tree splits bilaterians into animals lacking a body cavity (acoelomates) and animals having a body cavity. In pseudocoelomates, the cavity is incompletely lined by embryonic mesoderm. Coelomates have a true coelom, a body cavity completely lined by mesoderm.
- **Coelomates branched into protostomes and deuterostomes (pp. 593–595, FIGURE 32.6)** Coelomate phyla are divided into two main groups: protostomes, including the annelids, mollusks, and arthropods; and deuterostomes, including echinoderms and chordates.

THE ORIGINS OF ANIMAL DIVERSITY

- **Most animal phyla originated in a relatively brief span of geological time (pp. 595–596, FIGURE 32.7)** Animals are grouped into about 35 phyla, each distinguished by a particular body plan. The oldest known fauna consisted of soft-bodied animals that lived during the Ediacaran period of the Precambrian. During the ensuing Cambrian period, a much more diverse fauna evolved, which included many species with hard shells and skeletons. All the basic body plans observed in modern animals arose during the Cambrian radiation.
- **Developmental genetics may clarify our understanding of the Cambrian diversification (p. 596)** Developmental genetics indicates that the common ancestor of bilateral animals was complex, with a head, light detectors, a nerve cord, and segmentation. Evolutionary changes in molecular mechanisms that regulate gene activity during development may underlie much of the diversity of the bilateria.

SELF-QUIZ

1. The distinction between the parazoans and eumetazoans is based mainly on the absence versus the presence of
   a. body cavities
   b. a complete digestive tract
   c. true tissues
   d. a circulatory system
   e. mesoderm

2. As a group, acoelomates are characterized by
   a. gastrovascular cavities
   b. the absence of mesoderm
   c. deuterostome development
   d. a coelom that is not completely lined with mesoderm
   e. a solid body without a cavity surrounding internal organs

3. Which of the following is *not* descriptive of deuterostomes?
   a. radial cleavage
   b. includes humans
   c. formation of the coelom from outpocketings of archenteron
   d. development of the blastopore into the mouth
   e. echinoderms and chordates

4. The radiata and bilateria of the eumetazoa both exhibit
   a. cephalization
   b. bilateral symmetry of larval forms
   c. dominance of the diploid stage in the life cycle
   d. a complete digestive tract with separate mouth and anus
   e. three germ layers in embryonic development

5. Bilateral symmetry in the animal kingdom is best correlated with
   a. an ability to sense equally in all directions
   b. the presence of a skeleton
   c. motility and active predation and escape
   d. development of a true coelom
   e. adaptation to terrestrial environments

6. A direct consequence of indeterminate cleavage is
   a. formation of the archenteron
   b. the ability of cells isolated from the early embryo to develop into viable individuals
   c. the arrangement of cleavage planes perpendicular to the egg's vertical axis
   d. the unpredictable formation of either a schizocoelous or enterocoelous body cavity
   e. a mouth that forms in association with the blastopore

7. Many biologists suspect that the rapid diversification of bilateral phyla during the Cambrian
   a. was triggered by declining concentrations of atmospheric carbon dioxide
   b. was associated with variation in patterns of embryonic development
   c. followed the development of gastrovascular cavities
   d. was possible once cleavage evolved
   e. was the result of increased solar radiation and an accelerated mutation rate

8. Among the characteristics unique to animals is
   a. gastrulation
   b. multicellularity
   c. sexual reproduction
   d. flagellated sperm
   e. heterotrophy

9. Which of the following combinations of phylum and description is *incorrect?*
   a. Echinodermata—branch bilateria, coelom from archenteron
   b. Nematoda—roundworms, pseudocoelomate
   c. Cnidaria—radial symmetry, gastrovascular cavity
   d. Platyhelminthes—flatworms, gastrovascular cavity (in most species), acoelomate
   e. Porifera—gastrovascular cavity, mouth from blastopore

10. Which of the following subdivisions of the animal kingdom encompasses all the others in the list?
   a. protostomes
   b. bilateria
   c. pseudocoelomates
   d. coelomates
   e. deuterostomes

CHALLENGE QUESTIONS

1. Write a paragraph defining "animal."
2. Compare and contrast the body plans of a planarian (a flatworm) and an earthworm.
3. Lynn Margulis of the University of Massachusetts has suggested that observing an explosion of animal diversity in Cambrian strata is like viewing Earth from a satellite over a long period of time and noticing the emergence of cities only after they are large enough to be evident at that distance. What do you think Dr. Margulis was saying about the Cambrian "explosion"?

SCIENCE, TECHNOLOGY, AND SOCIETY

1. The study of animal phylogeny is sometimes viewed as "science for science's sake," and some organizations that fund scientific research tend to favor projects that have more obvious applications to human needs. On the other hand, general interest in the history of life seems to remain high, with articles on new discoveries frequently appearing in popular journals. Suppose you have the opportunity to join a research team studying various aspects of the Cambrian explosion. Write a paragraph that could convince nonbiologists that this research is worth funding.
2. Give examples from at least three phyla of animals that humans use for food.

FURTHER READING

Bengston, S. "Animal Embryos in Deep Time." *Nature,* February 5, 1998. A description of unusually well-preserved fossil embryos of late Precambrian multicellular animals.

Briggs, D. "Giant Predators from the Cambrian of China." *Science,* May 27, 1994. Jawed invertebrates 2 meters long were part of the Cambrian seascape.

Brusca, R. G., and G. J. Brusca. *Invertebrates.* Sunderland, MA: Sinauer Associates, 1990. An evolutionary approach to the animal kingdom.

Fedonkin, M. A., and B. M. Waggoner. "The Late Precambrian Fossil *Kimberella* Is a Mollusc-like Bilaterian Organism." *Nature,* August 28, 1997. The process of science is illustrated by the reevaluation of fossils formerly described as the remains of cnidarians.

Levinton, J. "The Big Bang of Animal Evolution." *Scientific American,* November 1992. What evolutionary mechanisms made the Cambrian explosion possible?

Monastersky, R. "The Rise of Life on Earth: Life Grows Up." *National Geographic,* April 1998. An exploration of late Precambrian/early Cambrian fossils in Australia.

Morris, S. C. *Crucible of Creation: The Burgess Shale and the Rise of Animals.* New York: Oxford University Press, 1998. A lively chronicle of scientific discovery and Cambrian fossils.

Pennisi, E., and W. Roush. "Developing a New View of Evolution." *Science,* July 4, 1997. Describes the infusion of new discoveries in developmental genetics into concepts of animal evolution.

Wright, K. "When Life Was Odd." *Discover,* March 1997. An illustrated essay on Precambrian animals.

WEB LINKS

Visit the special edition of *The Biology Place* for BIOLOGY, Fifth Edition, at **http://www.biology.com/campbell**. Go to Chapter 32 for online resources, including learning activities, practice exams, and links to the following web sites:

"Animal Diversity Web"
This site, sponsored in part by the University of Michigan Museum of Zoology, contains a wealth of information about a huge number of animal species.

"Diversity of Life Web Index"
This site is an excellent primer for understanding animal diversity.

"Wired for Conservation"
This is the homepage of the Nature Conservancy, a non-government organization dedicated to maintaining biological diversity by providing protected habitats for wildlife.

"United States Fish and Wildlife Service"
The mission of the USFWS is to protect fish and wildlife and their habitats.

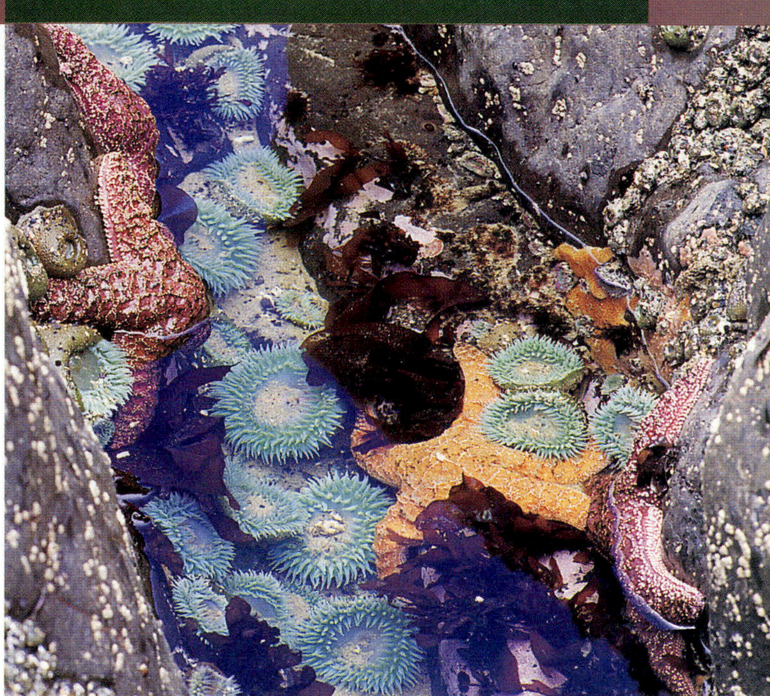

CHAPTER 33

INVERTEBRATES

The Parazoa

- Phylum Porifera: Sponges are sessile with porous bodies and choanocytes

The Radiata

- Phylum Cnidaria: Cnidarians have radial symmetry, a gastrovascular cavity, and cnidocytes
- Phylum Ctenophora: Comb jellies possess rows of ciliary plates and adhesive colloblasts

The Acoelomates

- Phylum Platyhelminthes: Flatworms are dorsoventrally flattened acoelomates

The Pseudocoelomates

- Phylum Rotifera: Rotifers have jaws and a crown of cilia
- Phylum Nematoda: Roundworms are unsegmented and cylindrical with tapered ends

The Coelomates: Protostomes

- Phylum Nemertea: The phylogenetic position of proboscis worms is uncertain
- The lophophorate phyla: Bryozoans, phoronids, and brachiopods have ciliated tentacles around their mouths
- Phylum Mollusca: Mollusks have a muscular foot, a visceral mass, and a mantle
- Phylum Annelida: Annelids are segmented worms
- Phylum Arthropoda: Arthropods have regional segmentation, jointed appendages, and an exoskeleton

The Coelomates: Deuterostomes

- Phylum Echinodermata: Echinoderms have a water vascular system and secondary radial symmetry
- Phylum Chordata: The chordates include two invertebrate subphyla and all vertebrates

More than a million extant species of animals are known, and at least as many more will probably be identified by future generations of biologists. Animals are grouped into about 35 phyla, the exact number depending on the views of different systematists. Animals inhabit nearly all environments on Earth, but most phyla consist mainly of aquatic species. The seas, where the first animals probably arose, are still home to the greatest number of animal phyla. The freshwater fauna is extensive, but not nearly as rich in diversity as the marine fauna.

Terrestrial habitats pose special problems for animals, as they do for plants (see Chapter 29), and few animal phyla have made successful evolutionary treks onto land. Earthworms (Phylum Annelida) and land snails (Phylum Mollusca) are generally confined to moist soil and vegetation. Only the vertebrates and arthropods, including insects and spiders, are represented by a great diversity of animal species adapted to various terrestrial environments.

Living as we do on land, our sense of animal diversity is biased in favor of vertebrates, the animals with backbones, which are well represented in terrestrial environments. But vertebrates make up one subphylum within Phylum Chordata, less than 5% of all animal species. If we were to sample the animals inhabiting a tidepool (as in the photograph on this page), a coral reef, or the rocks on a stream bottom, we would find ourselves in the realm of ***invertebrates****, the animals without backbones. The diversity of invertebrates is the main subject of this chapter. Our survey of phyla will follow the branching pattern of the phylogenetic tree discussed in Chapter 32 (see* FIGURE 32.3 *and* TABLE 32.1*).*

THE PARAZOA

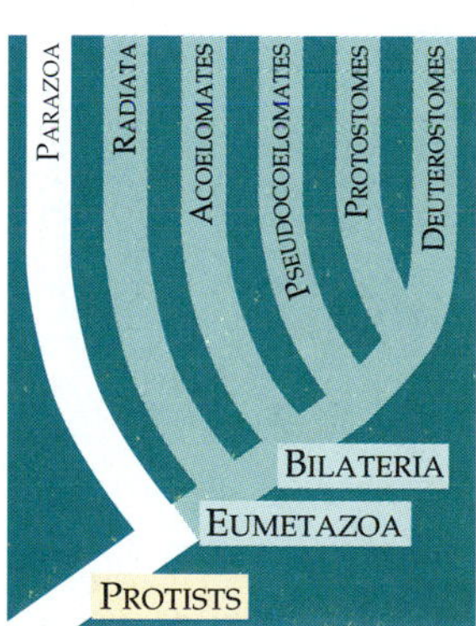

Of all the animals we will discuss, sponges (on branch parazoa of the phylogenetic tree) represent the lineage closest to the multicellular organisms (protists) that gave rise to the animal kingdom. The cell layers of sponges are loose federations of cells, not really tissues because the cells are relatively unspecialized.

Phylum Porifera: Sponges are sessile with porous bodies and choanocytes

33.1 Sponges are sessile animals that appear so sedate to the human eye that the ancient Greeks believed them to be plants. Sponges have no nerves or muscles, but the individual cells can sense and react to changes in the environment.

FIGURE 33.1 ▪ A sponge. Sessile animals without specialized organs and tissues, sponges filter food from water pumped through their porous bodies. The diverse species of sponges vary in shape, color, and structural complexity. Some species are brightly pigmented by symbiotic algae. This is an azure vase sponge, *Callyspongia plicifera.*

Sponges range in height from about 1 cm to 2 m. Of the 9000 or so species of sponges, only about 100 live in fresh water; the rest are marine. The body of a simple sponge resembles a sac perforated with holes (*Porifera* means "pore bearer"; FIGURES 33.1 and 33.2). Water is drawn through the pores into a central cavity, the **spongocoel**, then flows out of the sponge through a larger opening called the **osculum.** More complex sponges have folded body walls, and many contain branched water canals and several oscula. Under certain conditions, the cells around the pores and osculum contract, closing the openings.

Nearly all sponges are suspension-feeders (also known as filter-feeders), which are animals that collect food particles from water passed through some type of food-trapping equipment. Sponges trap food from the water circulated through the porous body. Lining the inside of the spongocoel or internal water chambers are flagellated **choanocytes,** or collar cells (for the membranous collar around the base of the flagellum). The flagella generate a water current, the collars trap food particles and the choanocytes phagocytose them. The similarity between choanocytes and the cells of choanoflagellates (the colonial protists in FIGURE 28.26) supports the hypothesis that sponges shared an ancestor with the choanoflagellates.

The body of a sponge consists of two layers of cells separated by a gelatinous region called the **mesohyl.** Wandering through the mesohyl are cells called **amoebocytes,** named for their use of pseudopodia. Amoebocytes have many functions. They take up food from the water and from choanocytes, digest it, and carry nutrients to other cells. Amoebocytes also form tough skeletal fibers within the mesohyl. In some groups of sponges, these fibers are sharp spicules made from calcium carbonate or silica; other sponges produce more flexible fibers composed of a collagen called spongin.

Most sponges are **hermaphrodites** (Gr. *Hermes,* the god, and *Aphrodite,* the goddess), meaning that each individual functions as both male and female in sexual reproduction by producing sperm *and* eggs. Gametes arise from choanocytes or amoebocytes. Eggs reside in the mesohyl, but sperm cells

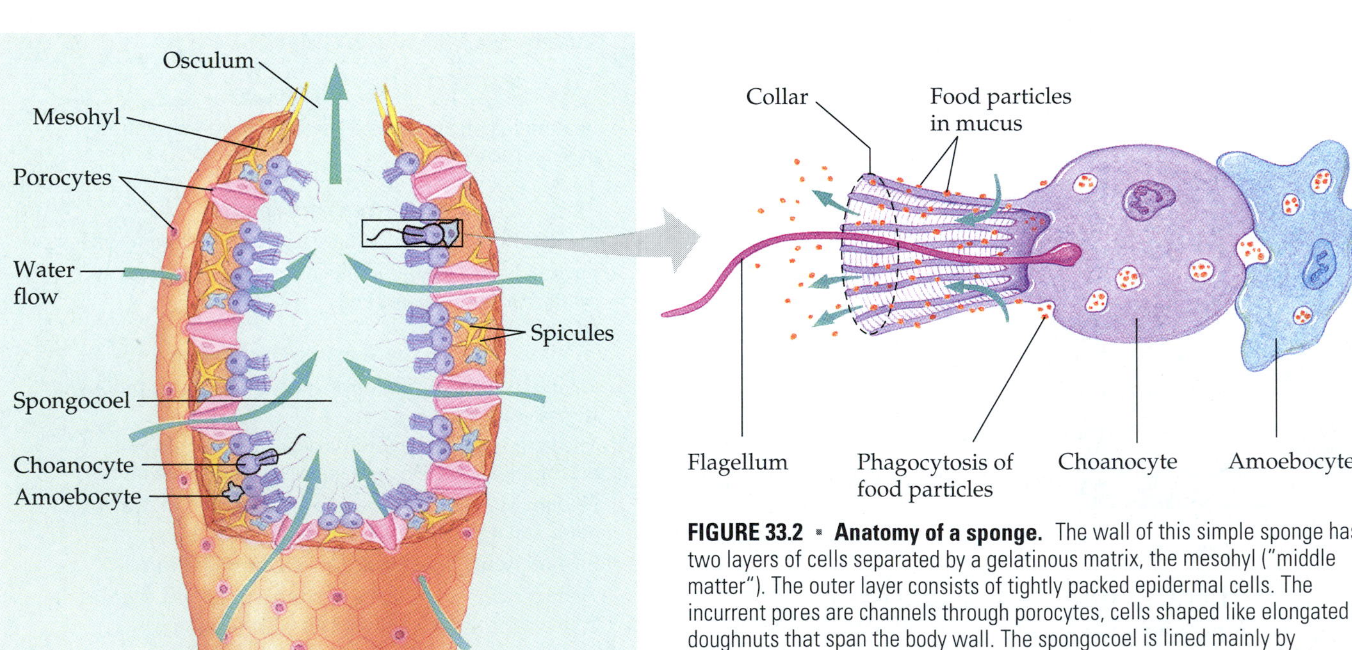

FIGURE 33.2 ▪ Anatomy of a sponge. The wall of this simple sponge has two layers of cells separated by a gelatinous matrix, the mesohyl ("middle matter"). The outer layer consists of tightly packed epidermal cells. The incurrent pores are channels through porocytes, cells shaped like elongated doughnuts that span the body wall. The spongocoel is lined mainly by choanocytes, each with a flagellum ringed by a collar of fingerlike projections coated with mucus. Beating flagella sweep large volumes of water (arrows) into the body through the incurrent pores. Food particles are trapped in mucus on the collar, phagocytosed, and digested within choanocytes and adjacent amoebocytes. Mobile amoebocytes transport nutrients to other cells of the body and also produce materials for skeletal fibers (spicules).

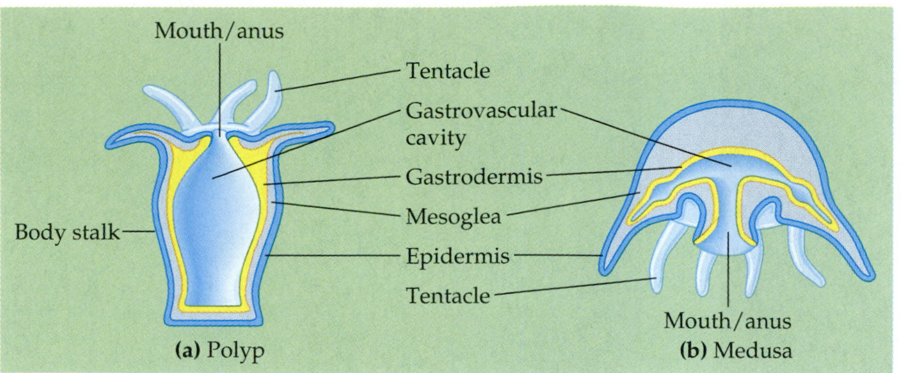

FIGURE 33.3 • Polyp and medusa forms of cnidarians. The body wall of **(a)** a polyp or **(b)** a medusa has two layers of cells, an outer layer of epidermis (from ectoderm) specialized for protection and an inner layer of gastrodermis (from endoderm) for digestion. Digestion begins in the gastrovascular cavity and is completed within the gastrodermal cells in food vacuoles. Flagella on the gastrodermal cells keep the contents of the gastrovascular cavity agitated and help distribute nutrients. Sandwiched between the epidermis and gastrodermis is a gelatinous layer of mesoglea. In many medusas, the mesoglea is thick and jellylike—thus the name jellies.

are carried out of the sponge by the water current. Cross-fertilization results from some of the sperm being drawn into neighboring individuals. Fertilization occurs in the mesohyl, where the zygotes develop into flagellated, swimming larvae that disperse from the parent. Upon settling on a suitable substratum, a larva develops into the sessile adult with internal choanocytes. Sponges are capable of extensive regeneration, the replacement of lost parts. They use regeneration not only for repair but also to reproduce asexually from fragments broken off a parent sponge.

THE RADIATA

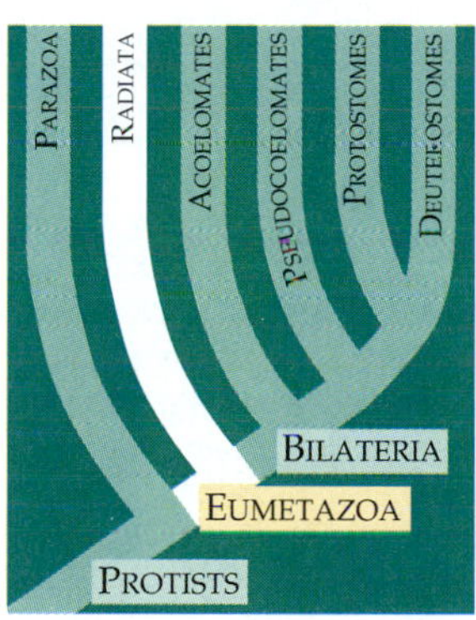

Representing another lineage that branched very early in animal (eumetazoan) history, radiate animals are diploblastic (have only ectoderm and endoderm) and exhibit radial symmetry. The two phyla of branch radiata are Cnidaria and Ctenophora.

Phylum Cnidaria: Cnidarians have radial symmetry, a gastrovascular cavity, and cnidocytes

33.1 Lacking mesoderm, the cnidarians (hydras, jellies, sea anemones, and coral animals) have a relatively simple body construction. Nonetheless, they are a diverse group with over 10,000 living species, most of which are marine.

The basic body plan of a cnidarian is a sac with a central digestive compartment, the **gastrovascular cavity**. A single opening to this cavity functions as both mouth and anus. This basic body plan has two variations: the sessile polyp and the floating medusa (FIGURE 33.3). **Polyps** are cylindrical forms that adhere to the substratum by the aboral end of the body and extend their tentacles, waiting for prey. Examples of the polyp form are hydras and sea anemones. A **medusa** is a flattened, mouth-down version of the polyp. It moves freely in the water by a combination of passive drifting and contractions of its bell-shaped body. The animals we generally call jellies are medusas. The tentacles of a jelly dangle from the oral surface, which points downward. Some cnidarians exist only as polyps, others only as medusas, and still others pass sequentially through both a medusa stage and a polyp stage in their life cycle.

Cnidarians are carnivores that use tentacles arranged in a ring around the mouth to capture prey and push the food into the gastrovascular cavity, where digestion begins. The undigested remains are egested through the mouth/anus. The tentacles are armed with batteries of **cnidocytes**, unique cells that function in defense and the capture of prey (FIGURE 33.4). Cnidocytes contain **cnidae**, organelles (capsules) capable of everting, giving Phylum Cnidaria its name (Gr. *cnide*, "nettle"). Cnidae called **nematocysts** are stinging capsules.

Muscles and nerves occur in their simplest forms in cnidarians. Cells of the epidermis (outer layer) and gastrodermis

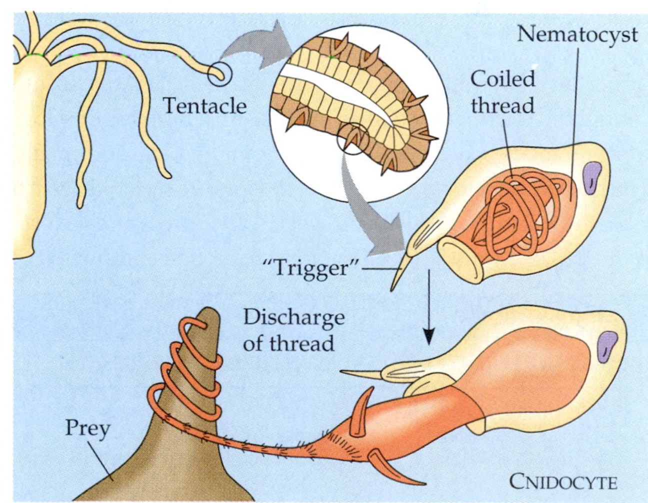

FIGURE 33.4 • A cnidocyte of a hydra. This type of cnidocyte contains a stinging capsule, the nematocyst, which itself contains an inverted thread. When a "trigger" is stimulated by touch or by certain chemicals, the thread shoots out from the nematocyst, puncturing and injecting poison into prey. Other kinds of cnidocytes have longer threads that stick to prey or entangle small animals that bump into the tentacles.

FIGURE 33.5 ▪ Representatives of the cnidarian classes. **(a)** Polyps of a colonial species belonging to Class Hydrozoa. **(b)** Jellies of Class Scyphozoa. The medusa is the conspicuous stage of the scyphozoan life cycle. The largest scyphozoan species have tentacles over 100 m long dangling from umbrellas up to 2 m in diameter. **(c)** Sea anemones and other members of Class Anthozoa exist only as polyps. **(d)** A colony of coral polyps (Class Anthozoa). Many corals harbor symbiotic algae that contribute to the food supply of the polyps. Coral reefs, which provide habitats for an enormous variety of invertebrates and fishes, are restricted to warm, shallow seas. This is a star coral.

(inner layer) have bundles of microfilaments arranged into contractile fibers (see Chapter 7). True muscle tissue develops from mesoderm and does not appear in diploblastic animals. The gastrovascular cavity acts as a hydrostatic skeleton against which the contractile cells can work. When the animal closes its mouth, the volume of the cavity is fixed, and contraction of selected cells causes the animal to change shape. Movements are coordinated by a nerve net. Cnidarians have no brain, and the noncentralized nerve net is associated with simple sensory receptors that are distributed radially around the body. Thus, the animal can detect and respond to stimuli equally from all directions.

Table 33.1 ▪ Classes of Phylum Cnidaria

CLASS AND EXAMPLES	MAIN CHARACTERISTICS
Hydrozoa (Portuguese man-of-war, hydras, *Obelia*, some corals) (see FIGURES 33.5a and 33.6)	Most marine, a few freshwater; both polyp and medusa stages in most species; polyp stage often colonial
Scyphozoa (jellies, sea wasp, sea nettle) (see FIGURE 33.5b)	All marine; polyp stage reduced; free-swimming; medusas up to 2 m in diameter
Anthozoa (sea anemones, most corals, sea fans) (see FIGURE 33.5c and d)	All marine; medusa stage completely absent; sessile, many colonial

The phylum Cnidaria is divided into three major classes: Hydrozoa, Scyphozoa, and Anthozoa (FIGURE 33.5, TABLE 33.1).

Class Hydrozoa

Most hydrozoans alternate polyp and medusa forms, as in the life cycle of *Obelia* (FIGURE 33.6). The polyp stage, a colony of interconnected polyps in the case of *Obelia,* is more conspicuous than the medusas. Hydras, among the few cnidarians found in fresh water, are unusual members of Class Hydrozoa in that they exist only in the polyp form. When environmental conditions are favorable, a hydra reproduces asexually by

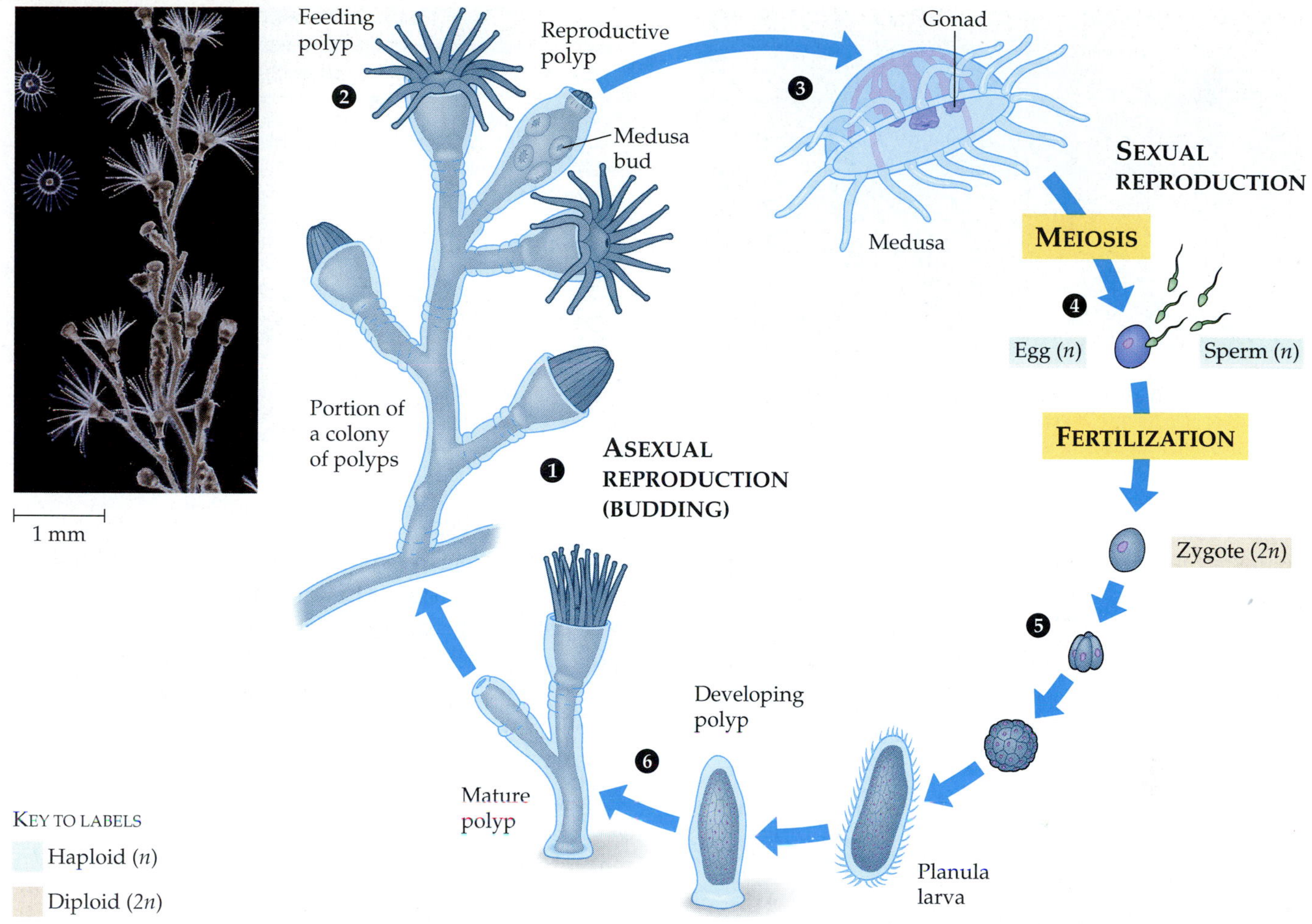

FIGURE 33.6 ▪ The life cycle of the hydrozoan *Obelia*. ① A colony of interconnected polyps results from asexual reproduction by budding (inset, LM). ② Some of the colony's polyps, equipped with tentacles, are specialized for feeding. ③ Other polyps, specialized for reproduction, lack tentacles and produce tiny medusas by asexual budding. ④ The medusas swim off, grow, and reproduce sexually. ⑤ The zygote develops into a solid ciliated larva called a planula. ⑥ The planula eventually settles and develops into a new polyp. The polyp stage is asexual, the medusa stage is sexual, and these two stages alternate, one producing the other. But do not confuse this with the alternation of generations that occurs in the plant kingdom. Both polyp and medusa are diploid organisms. (Typical of animals, only the gametes of *Obelia* are haploid.) By contrast, one of the plant generations is haploid, the other is diploid.

budding, the formation of outgrowths that pinch off from the parent to live independently (see FIGURE 13.1). When environmental conditions deteriorate, hydras reproduce sexually, forming resistant zygotes that remain dormant until conditions improve.

Class Scyphozoa

The medusa generally prevails in the life cycle of Class Scyphozoa. The medusas of most species live among the plankton as jellies. Most coastal scyphozoans go through a small polyp stage during their life cycle, but jellies that live in the open ocean generally lack the sessile polyp.

Class Anthozoa

Sea anemones and corals belong to Class Anthozoa ("flower animals"). They occur only as polyps. Coral animals live as solitary or colonial forms and secrete hard external skeletons of calcium carbonate. Each polyp generation builds on the skeletal remains of earlier generations to construct "rocks" with shapes characteristic of the species. It is these skeletons that we call coral.

Phylum Ctenophora: Comb jellies possess rows of ciliary plates and adhesive colloblasts

Comb jellies, or ctenophores, superficially resemble cnidarian medusas. However, the relationship between ctenophores and cnidarians is uncertain. There are only about 100 species of comb jellies, all of which are marine. Ctenophores range in diameter from about 1 to 10 cm. Most are spherical or ovoid, but there are elongate and ribbonlike forms up to 1 m long. *Ctenophora* means "comb-bearer," and these animals are named for their eight rows of comblike plates composed of fused cilia. They are the largest animals to use cilia for locomotion. An aboral sensory organ functions in orientation, and nerves running from the sensory organ to the combs of cilia

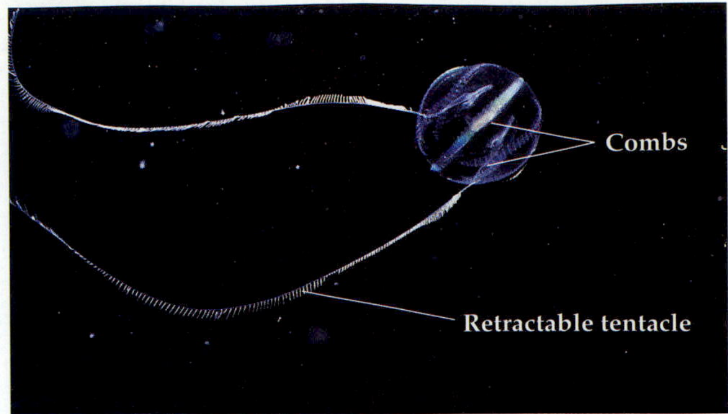

FIGURE 33.7 ▪ A ctenophore, or comb jelly. This planktonic marine animal is named for its eight combs of cilia, used for locomotion. The retractable tentacles capture food. Ctenophores and cnidarians are radiate, diploblastic animals, but many zoologists believe these groups have little else in common.

Table 33.2 ▪ Classes of Phylum Platyhelminthes

CLASS AND EXAMPLES	MAIN CHARACTERISTICS
Turbellaria (mostly free-living flatworms; e.g., *Dugesia*) (see FIGURES 33.8 and 33.9)	Most marine, some freshwater, a few terrestrial; predators and scavengers; body surface ciliated
Monogenea (monogeneans)	Marine and freshwater parasites; most infect external surfaces of fishes; life history simple; a ciliated larva starts an infection on a host
Trematoda (trematodes, also called flukes) (see FIGURE 33.10)	Parasites, almost always of vertebrates; two suckers attach to host; most life histories include intermediate hosts
Cestoidea (tapeworms) (see FIGURE 33.11)	Parasites of vertebrates; scolex attaches to host; proglottids produce eggs and break off after fertilization; no head or digestive system; life history with one or more intermediate hosts

coordinate movement. Most comb jellies have a pair of long, retractable tentacles (FIGURE 33.7). The tentacles bear adhesive structures called **colloblasts** (also called lasso cells). When prey (mostly small plankton) contact a tentacle, colloblasts burst open. A sticky thread released by each colloblast captures the food, which is then wiped off the tentacle into the mouth.

THE ACOELOMATES

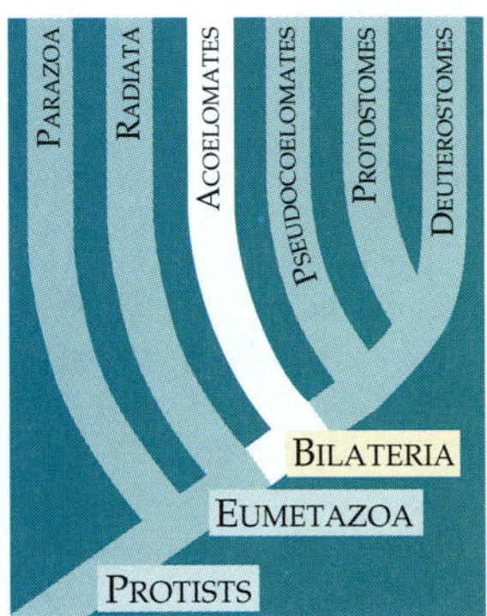

Representing an early branch of bilaterally symmetrical animals, the acoelomates lack a body cavity, a space between the body wall and the digestive tract. Acoelomates represent several evolutionary developments compared to the radiate animals. In common with all bilateral animals, they are triploblastic (have ectoderm, endoderm, and mesoderm). As bilateral animals, they exhibit forward movement and some degree of cephalization.

Phylum Platyhelminthes: Flatworms are dorsoventrally flattened acoelomates

33.1 There are about 20,000 species of flatworms living in marine, freshwater, and damp terrestrial habitats. In addition to many free-living forms, flatworms include many parasitic species, such as the flukes and the tapeworms. Flatworms are so named because their bodies are thin between the dorsal and ventral surfaces (flattened dorsoventrally; *platyhelminth* means "flat worm"). They range in size from nearly microscopic free-living species to certain tapeworms over 20 m long.

The third embryonic layer, mesoderm, contributes to the development of more complex organs and organ systems, and true muscle tissue. Thus the flatworms are structurally more complex than cnidarians or ctenophores. However, in common with the radiate animals, flatworms have a gastrovascular cavity with only one opening. (Tapeworms lack a digestive tract altogether and absorb nutrients across their body surface.)

Flatworms are divided into four classes: Turbellaria (mostly free-living flatworms), Monogenea (monogeneans), Trematoda (trematodes, or flukes), and Cestoidea (tapeworms) (TABLE 33.2). Parasitic flatworms (mainly monogeneans, trematodes, and tapeworms) are notorious for the diseases that some of them cause, and many flatworms play important roles in the structure and function of ecosystems, as we will see in Unit Eight.

Class Turbellaria

Turbellarians are nearly all free-living (nonparasitic) and mostly marine (FIGURE 33.8). Members of the genus *Dugesia*, commonly known as planarians, abound in unpolluted ponds and streams. **Planarians** are carnivores that prey on smaller animals or feed on dead animals (FIGURE 33.9).

Planarians and other flatworms lack organs specialized for gas exchange and circulation. The flat shape of the body

FIGURE 33.8 ▪ A flatworm. Class Turbellaria consists mainly of free-living marine flatworms, such as this colorful species.

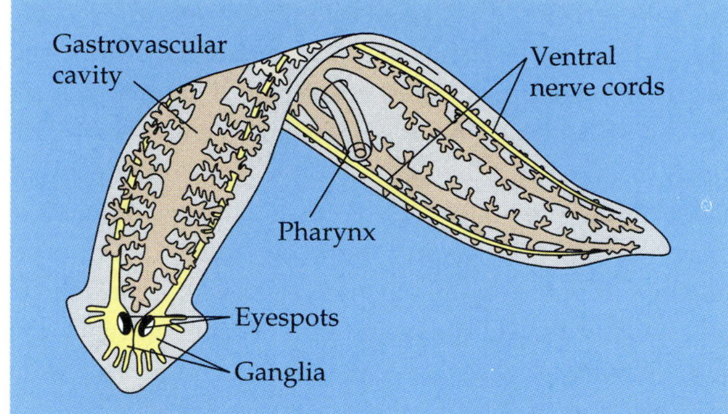

FIGURE 33.9 ▪ Anatomy of a planarian. The mouth is at the tip of a muscular pharynx that extends from the middle of the ventral side of the animal. Digestive juices are spilled onto prey, and the pharynx sucks small pieces of food into the gastrovascular cavity, where digestion continues. Digestion is completed within the cells lining the gastrovascular cavity, which has three branches, each with fine sub-branches that provide an extensive surface area. Undigested wastes are egested through the mouth. Located at the anterior end of the worm, near the main sources of sensory input, is a pair of ganglia, dense clusters of nerve cells. From the ganglia, a pair of ventral nerve cords runs the length of the body.

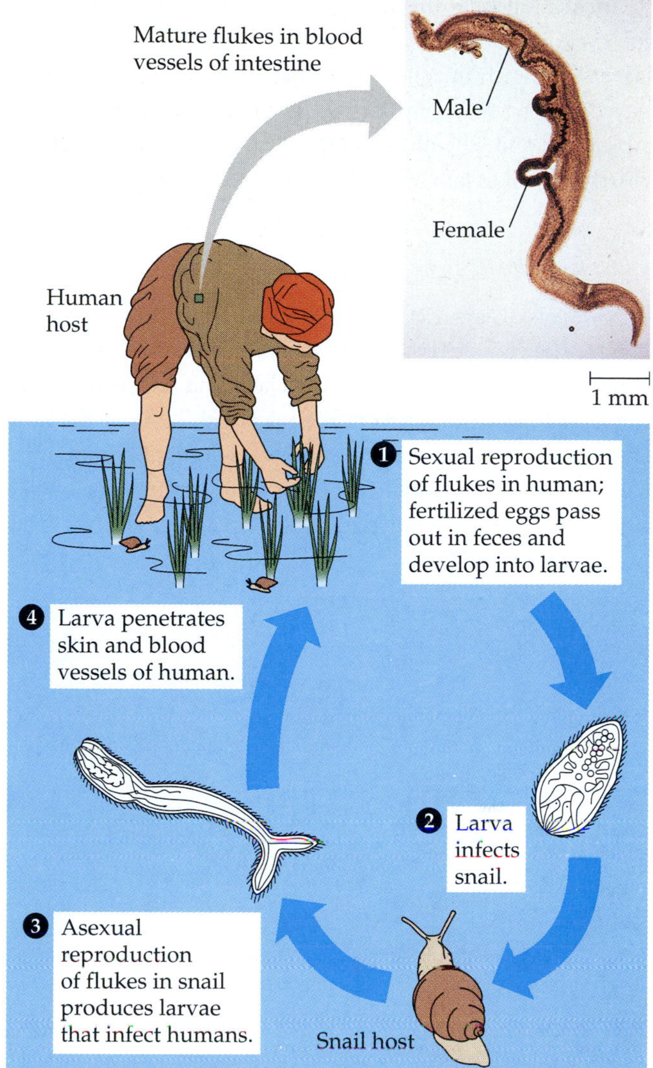

FIGURE 33.10 ▪ The life history of a blood fluke (*Schistosoma mansoni*). ① Blood flukes reproduce sexually in the human host. A female fluke fits into a groove running the length of the larger male's body (inset, LM). The fertilized eggs exit the host in feces, and ② develop in water into ciliated larvae. These larvae can enter snails, the intermediate hosts in the life cycle of blood flukes. ③ Asexual reproduction within a snail results in another type of motile larva, which escapes from the snail host. ④ People working in irrigated fields contaminated with human feces may be exposed to these larvae, which can penetrate the skin of the human host.

places all cells close to the surrounding water, and fine branching of the gastrovascular cavity distributes food throughout the animal. Nitrogenous waste in the form of ammonia diffuses directly from the cells into the surrounding water. Flatworms also have a relatively simple excretory apparatus that functions mainly to maintain osmotic balance between the animal and its surroundings. This system consists of ciliated cells called flame cells that waft fluid through branched ducts opening to the outside (see FIGURE 44.15). The evolution of osmoregulatory structures was a major factor in allowing some turbellarians to invade freshwater and even moist terrestrial environments.

Planarians move by using cilia on the ventral epidermis, gliding along a film of mucus they secrete. Some turbellarians also use their muscles to swim through water with an undulating motion.

A planarian has a head (is cephalized) with a pair of eyespots that detect light and lateral flaps that function mainly for smell. The planarian nervous system is more complex and centralized than the nerve nets of cnidarians. Planarians can learn to modify their responses to stimuli.

Planarians can reproduce asexually through regeneration. The parent constricts in the middle, and each half regenerates the missing end. Sexual reproduction also occurs. Although planarians are hermaphrodites, copulating mates cross-fertilize.

Classes Monogenea and Trematoda

The monogeneans and the trematodes (sometimes called flukes) live as parasites in or on other animals. Many have suckers for attaching to internal organs or to the outer surfaces of the host, and a tough covering helps protect the parasites. Reproductive organs nearly fill the interior of these worms.

As a group, trematodes parasitize a wide range of hosts, and most species have complex life cycles with an alternation of sexual and asexual stages. Many trematodes require an intermediate host in which larvae develop before infecting the final host (usually a vertebrate), where the adult worm lives. For example, trematodes that parasitize humans spend parts of their life histories in snails (FIGURE 33.10). The 200 million people around the world who are infected with blood flukes (*Schistosoma*) suffer body pains, anemia, and dysentery.

Most monogeneans are external parasites of fishes. Their life cycle is relatively simple, with a ciliated, free-swimming larva starting an infection on a host. Although monogeneans have been traditionally aligned with the trematodes, some structural and chemical evidence suggests they are more closely related to tapeworms.

Class Cestoidea

Tapeworms (Class Cestoidea) are also parasitic. The adults live mostly in vertebrates, including humans. The tapeworm head, or scolex, is armed with suckers and often menacing hooks that lock the worm to the intestinal lining of the host (FIGURE 33.11). Posterior to the scolex is a long ribbon of units called proglottids, which are little more than sacs of sex organs. Lacking a digestive tract, the tapeworm absorbs food predigested by the host.

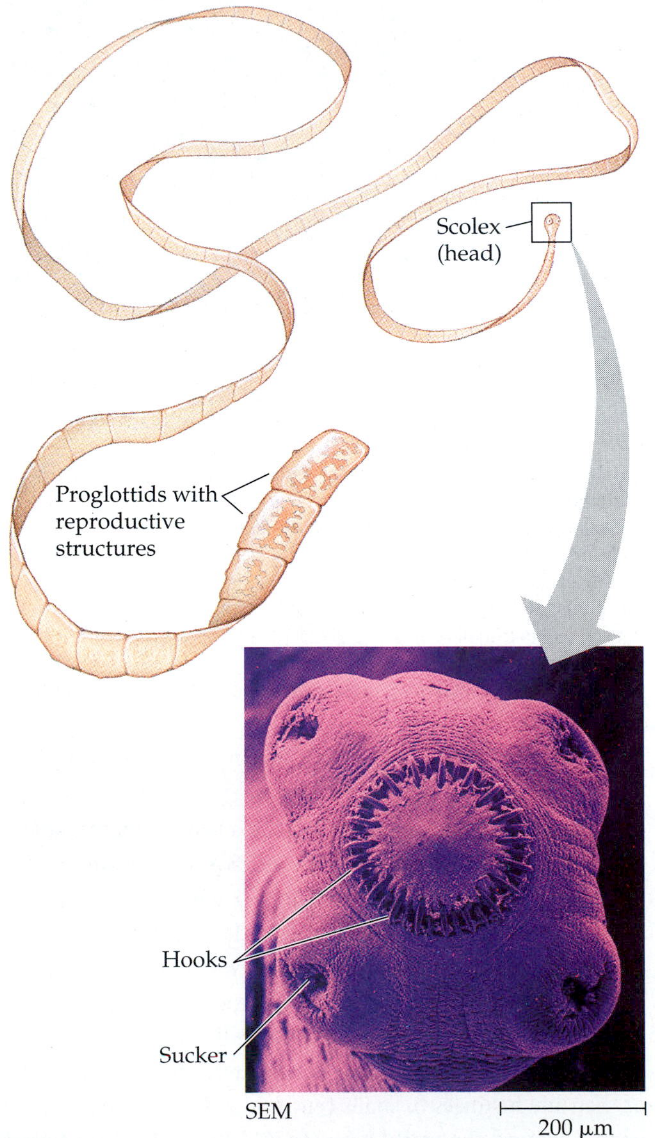

FIGURE 33.11 ▪ Anatomy of a tapeworm.

Mature proglottids, loaded with thousands of eggs, are released from the posterior end of a mature tapeworm and leave the host's body with feces. In one type of life cycle, human feces contaminate the food or water of intermediate hosts, such as pigs or cattle, and the tapeworm eggs develop into larvae that encyst in muscles of these animals. Humans acquire the larvae by eating undercooked meat contaminated with cysts, and the worms develop into mature adults within the human. Large tapeworms, which may be 20 m or more in length, can cause intestinal blockage and can rob enough nutrients from the human host to cause nutritional deficiencies.

THE PSEUDOCOELOMATES

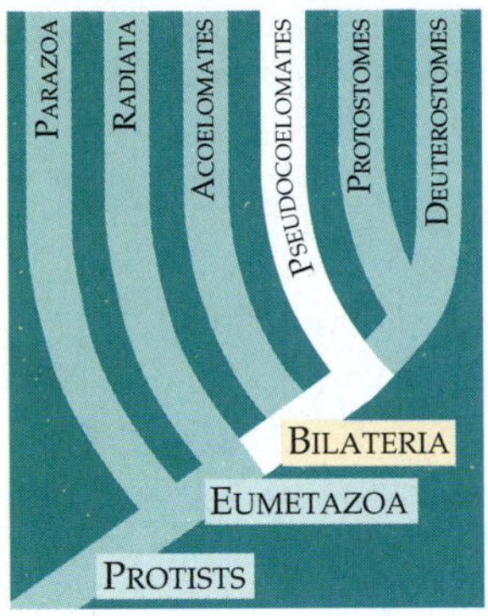

A pseudocoelomate body plan (see FIGURE 32.5b) has evolved in several phyla of small animals. Their evolutionary relationships with other groups and among themselves are still unclear. In fact, the pseudocoelomate condition probably arose independently several times. We discuss here only two of these phyla: Rotifera and Nematoda.

Phylum Rotifera: Rotifers have jaws and a crown of cilia

Rotifers (about 1800 species) are tiny animals that mainly inhabit fresh water, although some live in the sea or in damp soil. Ranging in size from only about 0.05 to 2.0 mm, smaller than many protists, rotifers are nevertheless truly multicellular and have specialized organ systems, including a **complete digestive tract** (a digestive tube with a separate mouth and anus) (FIGURE 33.12). Internal organs lie within the pseudocoelom. Fluid in the pseudocoelom serves as a hydrostatic skeleton and as a medium for the internal transport of nutrients and wastes in these tiny animals. Movement of a rotifer's body distributes the fluid within the pseudocoelom, and thus this body cavity and its fluid function as a circulatory system.

The word *rotifer,* derived from Latin, means "wheel-bearer," a reference to the crown of cilia that draws a vortex of water into the mouth. Posterior to the mouth, a region of the digestive tract called the pharynx bears jaws (trophi) that grind food, mostly microorganisms suspended in the water.

Rotifer reproduction is unusual. Some species consist only of females that produce more females from unfertilized eggs, a type of reproduction called **parthenogenesis**. Other species produce two types of eggs that develop by parthenogenesis, one type forming females and the other type developing into degenerate males that cannot even feed themselves. The males survive long enough to produce sperm that fertilize eggs,

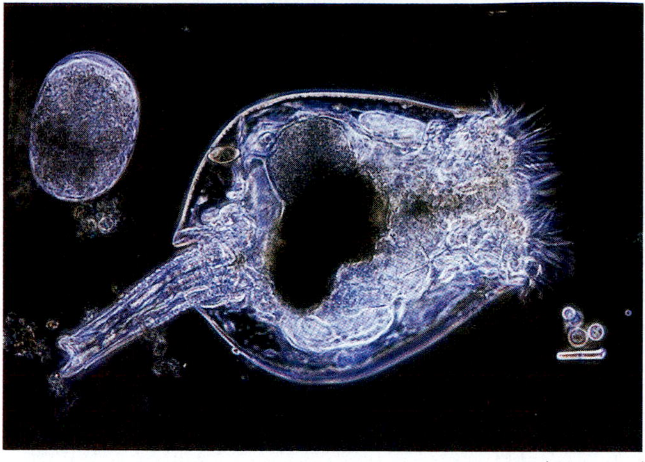

FIGURE 33.12 ▪ **A rotifer.** These pseudocoelomates, smaller than many protists, are much more anatomically complex than flatworms (LM).

forming resistant zygotes that can survive when a pond dries up. When conditions are favorable again, the zygotes break dormancy and develop into a new female generation that then reproduces by parthenogenesis until conditions become unfavorable again.

Phylum Nematoda: Roundworms are unsegmented and cylindrical with tapered ends

Among the most widespread of all animals, roundworms
33.1 (nematodes) are found in most aquatic habitats, in wet soil, in the moist tissues of plants, and in the body fluids and tissues of animals. About 90,000 species are known, and perhaps ten times that number actually exist. Roundworms range from less than 1 mm to more than 1 m in length. Covered by a tough, transparent cuticle, their cylindrical, unsegmented body tapers to a fine tip posteriorly and to a blunt tip at the head end (FIGURE 33.13a). Nematodes have a complete digestive tract. They lack a circulatory system, but nutrients are transported throughout the body via fluid in the pseudocoelom. The muscles of nematodes are all longitudinal, and their contraction produces a thrashing motion.

Nematode reproduction is usually sexual. The sexes are separate in most species, females generally being larger than males. Fertilization is internal, and a female may deposit 100,000 or more fertilized eggs per day. The zygotes of most species are resistant cells capable of surviving harsh conditions.

Great numbers of nematodes live in moist soil and in decomposing organic matter on the bottoms of lakes and oceans. These extremely numerous free-living worms play an important role in decomposition and nutrient cycling, but little is known about most species. One species of soil nematode, *Caenorhabitis elegans,* is widely cultured and has become a model research organism in developmental biology (see Chapter 21).

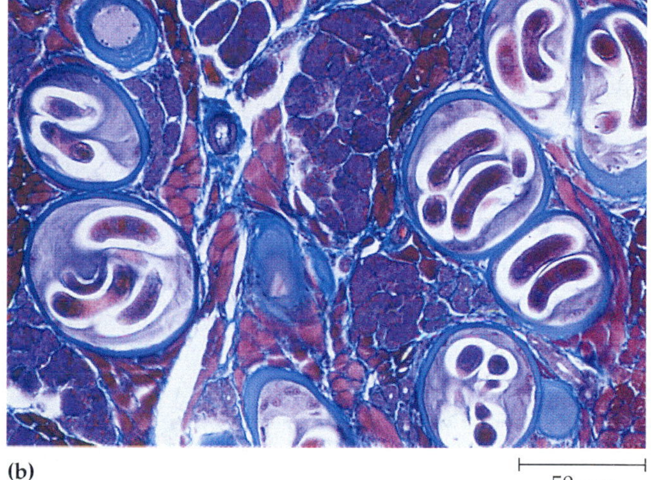

FIGURE 33.13 ▪ **Nematodes. (a)** A free-living nematode. Some of the internal organs of this freshwater worm are visible through the transparent skin. **(b)** Juveniles of *Trichinella spiralis* encysted in human muscle tissue. This parasitic nematode causes trichinosis, characterized by severe nausea and sometimes death when large numbers of the juveniles penetrate heart muscle. (Both LMs.)

Phylum Nematoda also includes many important agricultural pests that attack the roots of plants. Other species of roundworms parasitize animals. Humans host at least 50 nematode species, including various pinworms and hookworms. One notorious nematode is *Trichinella spiralis,* the worm that causes trichinosis (FIGURE 33.13b). Humans acquire this nematode by eating undercooked infected pork or other meat with juvenile worms encysted in the muscle tissue. Within the human intestine, the juveniles develop into sexually mature adults. Females burrow into the intestinal muscles and produce more juveniles, which bore through the body or travel in lymph vessels to encyst in other organs, including skeletal muscles.

THE COELOMATES: PROTOSTOMES

The protostome lineage of coelomate animals gave rise to several phyla, including the mollusks, annelids, and arthropods. We begin our discussion of the protostomes with a look at four phyla whose phylogenetic positions are still widely debated. Before reading further, you might want to review the key protostome characteristics in FIGURE 32.6 and TABLE 32.1.

Phylum Nemertea: The phylogenetic position of proboscis worms is uncertain

Members of the phylum Nemertea are called proboscis worms or ribbon worms (FIGURE 33.14). Their phylogenetic position is presently being debated, although molecular systematics supports anatomical evidence that they are related to the protostome lineage. A proboscis worm's body is structurally acoelomate, like that of a flatworm, but it contains a small fluid-filled sac that some biologists think is homologous to the body cavity (coelom) of protostomes. The sac and fluid hydraulically operate an extensible proboscis by which the worm captures prey.

Proboscis worms range in length from less than 1 mm to more than 30 m. Nearly all of the 900 or so members of this phylum are marine, but a few species inhabit fresh water or damp soil. Some are active swimmers, and others burrow in the sand.

Proboscis worms and flatworms have similar excretory, sensory, and nervous systems. But, in addition to the unique proboscis apparatus, two anatomical features not found in flatworms have evolved in the phylum Nemertea: a complete digestive tract (with mouth and anus) and a **closed circulatory system**—the blood is contained in vessels and is therefore distinct from fluid in the body cavity. Proboscis worms have no heart, but the blood is propelled by muscles squeezing the vessels.

FIGURE 33.14 ▪ A proboscis worm, Phylum Nemertea.

The lophophorate phyla: Bryozoans, phoronids, and brachiopods have ciliated tentacles around their mouths

In subdividing coelomate animals into protostomes and deuterostomes, three phyla—Bryozoa, Phoronida, and Brachiopoda—have been the subject of a good deal of controversy. These phyla are collectively called the **lophophorate animals** after the most distinctive structure they share, the lophophore (FIGURE 33.15). The **lophophore** is a horseshoe-shaped or circular fold of the body wall bearing ciliated tentacles that surround the mouth. The anus lies outside the whorl of tentacles. The cilia draw water toward the mouth between the tentacles, which help trap food particles for these suspension-feeders. The common occurrence of this complex apparatus in the lophophorate animals suggests that the three phyla are related. Other similarities, such as a U-shaped digestive tract and the absence of a distinct head, are adaptations to a sessile existence that also may have evolved convergently.

Molecular systematics places the lophophorate phyla closer to the protostomes than to the deuterostomes, as we do here. In their embryonic development, however, the lophophorates generally resemble deuterostomes. Because of these inconsistencies, the evolutionary position of these animals remains uncertain (hence the dashed lines of the phylogenetic tree of FIGURE 32.3).

Bryozoans are colonial animals that superficially resemble mosses. (*Bryozoa* means "moss animals.") In most species, the colony is encased in a hard exoskeleton with pores through which the lophophores of the animals extend (FIGURE 33.15a). Of the 5000 species of bryozoans, most live in the sea, where they are among the most widespread and numerous sessile animals. Several species are important reef builders.

Phoronids are tube-dwelling marine worms ranging from 1 mm to 50 cm in length. Some live buried in the sand within tubes made of chitin, extending their lophophore from the opening of the tube and withdrawing it into the tube when threatened. There are only about 15 species of phoronid worms in two genera.

Brachiopods, or lamp shells, superficially resemble clams and other bivalve mollusks, but the two halves of the brachiopod shell are dorsal and ventral to the animal rather than lateral, as in clams (FIGURE 33.15b). A brachiopod lives attached to its substratum by a stalk, opening its shell slightly to allow water to flow between the shells and the lophophore. All brachiopods are marine. The living brachiopods are remnants of

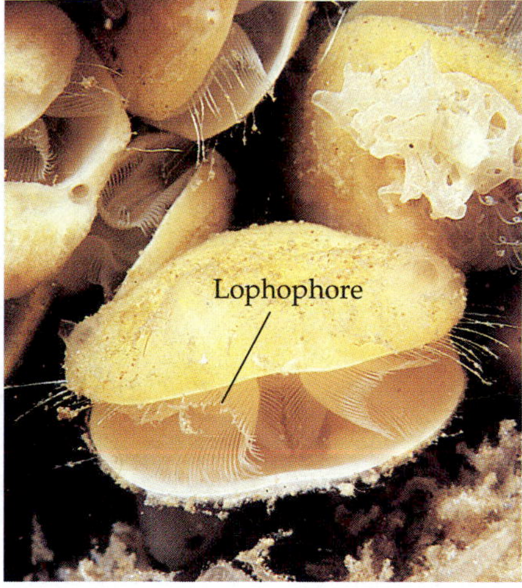

(a) (b)

FIGURE 33.15 ▪ The lophophorate animals. The most distinctive characteristic of these phyla is the lophophore, an organ that functions in suspension-feeding. **(a)** Bryozoans, such as this common sea mat (*Membranipora membranacea*), are colonial lophophorates, many with hard exoskeletons. **(b)** Brachiopods are lophophorates with a hinged shell. The two parts of the shell are dorsal and ventral, in contrast to the lateral shells of bivalve mollusks.

a much richer past; only about 330 extant species are known, but there are 30,000 species of Paleozoic and Mesozoic fossils. A tie to the past is *Lingula,* a living brachiopod genus that has changed little in 400 million years.

Phylum Mollusca: Mollusks have a muscular foot, a visceral mass, and a mantle

33.1 Snails and slugs, oysters and clams, and octopuses and squids are all mollusks. In all, the phylum Mollusca has more than 150,000 known species. Most mollusks are marine, though some inhabit fresh water, and there are snails and slugs that live on land. Mollusks are soft-bodied animals (L. *molluscus,* "soft"), but most are protected by a hard shell made of calcium carbonate. Slugs, squids, and octopuses have reduced shells, most of which are internal, or they have lost their shells completely during their evolution.

Despite their apparent differences, all mollusks have a similar body plan (FIGURE 33.16). The body has three main parts: a muscular **foot**, usually used for movement; a **visceral mass** containing most of the internal organs; and a **mantle**, a fold of tissue that drapes over the visceral mass and secretes a shell (if one is present). In many mollusks, the mantle extends beyond the visceral mass, producing a water-filled chamber, the **mantle cavity**, which houses the gills, anus, and excretory pores. Many mollusks feed by using a straplike rasping organ called a **radula** to scrape up food. Most mollusks have separate sexes, with gonads (ovaries or testes) located in the visceral mass. Many snails, however, are hermaphrodites.

Zoologists debate the relationship of mollusks to other coelomate protostomes. The life cycle of many marine mollusks includes a ciliated larva called the **trochophore**, also characteristic of marine annelids (segmented worms) and some other protostomes. But mollusks lack the one trait that most defines an annelid heritage—true segmentation. The mollusk *Neopilina* (Class Monoplacophora) has some of its internal organs repeated, but this condition may have evolved secondarily from an ancestral mollusk with unrepeated organs. As we discussed in Chapter 32, molecular and genetic evidence indicates that segmentation evolved in the ancestor

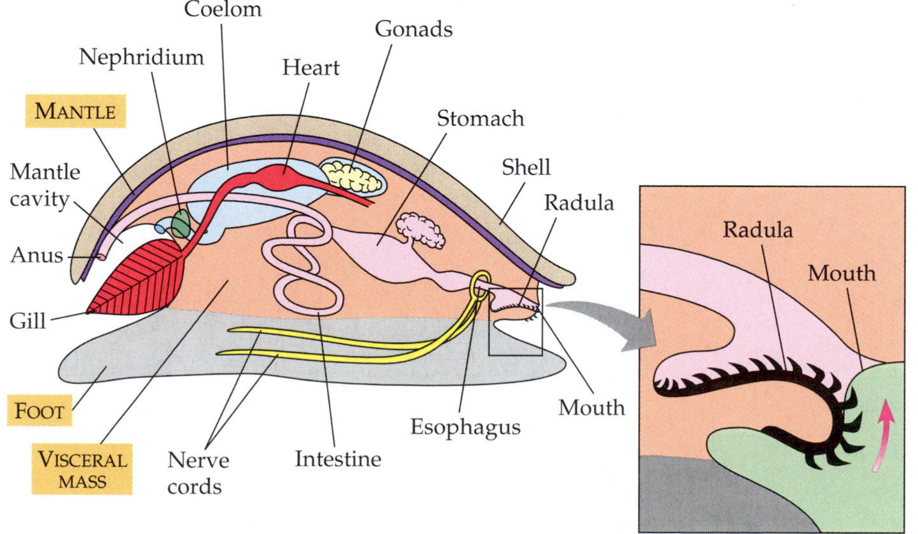

FIGURE 33.16 ▪ The basic body plan of mollusks. Three hallmarks of the phylum are the mantle, visceral mass, and foot. A mantle cavity houses gills in many species. The long digestive tract is coiled in the visceral mass. Most mollusks have an open circulatory system with a dorsal heart that pumps circulatory fluid (hemolymph) through arteries into sinuses (body spaces) bathing the organs. Excretory organs called nephridia remove metabolic wastes from the hemolymph. The nervous system consists of a nerve ring around the esophagus, from which nerve cords extend. The enlargement shows the mouth region with the radula, a rasplike feeding organ present in many mollusks. The radula is a belt of backward-curved teeth that extends from the mouth and slides back and forth, scraping and scooping like a backhoe.

of all bilateral animals. If so, this feature was lost during the early evolution of mollusks. Some zoologists doubt that mollusks evolved from a segmented ancestor and favor a traditional hypothesis that segmentation arose not in a bilateral ancestor but in an ancestral annelidlike protostome. According to this view, segmentation evolved independently in protostomes and deuterostomes.

The basic body plan of mollusks has evolved in various ways in the different classes of the phylum. Of the eight classes, we examine four here: Polyplacophora (chitons), Gastropoda (snails and slugs), Bivalvia (clams, oysters, and other bivalves), and Cephalopoda (squids, octopuses, and nautiluses) (TABLE 33.3).

Class Polyplacophora

Chitons are marine animals with oval shapes and shells divided into eight dorsal plates (the body itself, however, is unsegmented). You can find chitons clinging to rocks along the shore during low tide (FIGURE 33.17). Try to dislodge a chiton by hand, and you will be surprised at how well its foot, acting as a suction cup, grips the rock. Using this muscular foot, a chiton can creep slowly over the rock surface. Chitons are grazers that use their radulas to cut and ingest algae.

FIGURE 33.17 ▪ A chiton. Clinging tenaciously to rocks in the intertidal zone, this chiton (Class Polyplacophora) displays the eight-plate shell characteristic of this class of mollusks.

Class Gastropoda

The largest of the molluscan classes, Gastropoda has more than 40,000 living species. Most gastropods are marine, but there are also many freshwater species. Garden snails and slugs have adapted to land.

The most distinctive characteristic of the class Gastropoda is a process known as **torsion**. During embryonic development, an asymmetrical muscle forms, and one side of the visceral mass grows faster than the other. Contraction of the muscle and uneven growth causes the visceral mass to rotate up to 180 degrees, so that the anus and mantle cavity are placed above the head in the adult (FIGURE 33.18). Some zoologists speculate that the advantage of torsion is to place the visceral mass and heavy shell more centrally over the snail's body.

Most gastropods are protected by single, spiraled shells into which the animals can retreat when threatened (FIGURE 33.19). The shell is often conical, but abalones and limpets have somewhat flattened shells. Many gastropods have distinct heads with eyes at the tips of tentacles. Gastropods inch along literally at a snail's pace by a rippling motion of the elongated foot. Most gastropods use their radula to graze on algae or plant material. Several groups, however, are predators, and the radula is modified for boring holes in the shells of other mollusks or for tearing apart tough animal tissue. In one group, the cone snails, the teeth of the radula form separate poison darts, which penetrate prey, including fishes.

Gastropods are among the few invertebrate groups to have successfully populated the land. Terrestrial snails lack the gills typical of most aquatic gastropods, and instead the lining of the mantle cavity functions as a lung, exchanging respiratory gases with the air.

Table 33.3 ▪ Major Classes of Phylum Mollusca

CLASS AND EXAMPLES	MAIN CHARACTERISTICS
Polyplacophora (chitons) (see FIGURE 33.17)	Marine; shell with eight plates; foot used for locomotion; head reduced
Gastropoda (snails, slugs) (see FIGURES 33.18 and 33.19)	Marine, freshwater, or terrestrial; asymmetric body, usually with a coiled shell; shell reduced or absent in some; foot for locomotion; radula present
Bivalvia (clams, mussels, scallops, oysters) (see FIGURES 33.20 and 33.21)	Marine and freshwater; flattened shell with two valves; head reduced; paired gills; most are filter-feeders; mantle forms siphons
Cephalopoda (squids, octopuses, chambered nautiluses) (see FIGURE 33.22)	Marine; head surrounded by grasping tentacles, usually with suckers; shell external, internal, or absent; mouth with or without radula; locomotion by jet propulsion using siphon made from mantle

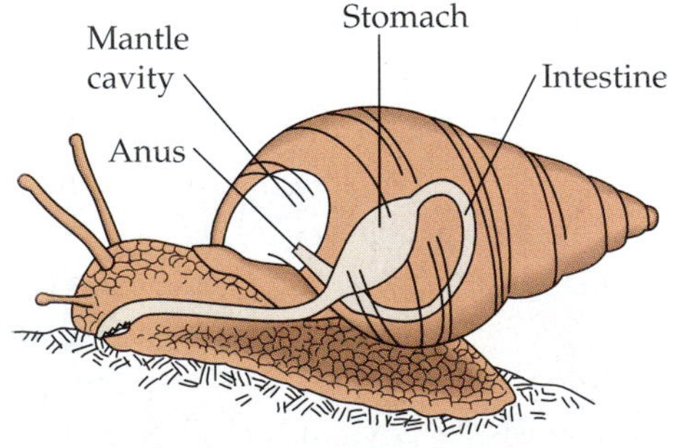

FIGURE 33.18 ▪ The results of torsion in a gastropod. Because of torsion (twisting of the visceral mass) during embryonic development, the digestive tract is coiled and the anus is near the mouth at the head end of the animal. After torsion, some of the organs that were bilateral are reduced in size or lost on one side of the body. Torsion should not be confused with the formation of a coiled shell, which is an independent process.

(a)

(b)

FIGURE 33.19 ▪ Gastropods. (a) Shell collectors find delight in the variety of Gastropoda, one of the most diverse animal classes. **(b)** Nudibranchs, or sea slugs, have lost the shell during their evolution.

Class Bivalvia

The mollusks of Class Bivalvia include many species of clams, oysters, mussels, and scallops. Bivalves have shells divided into two halves (FIGURE 33.20). The two parts of the shell are hinged at the mid-dorsal line, and powerful adductor muscles draw the two halves tightly together to protect the soft-bodied animal. When the shell is open, the bivalve may extend its hatchet-shaped foot for digging or anchoring.

The mantle cavity of a bivalve contains gills that are used for feeding as well as gas exchange (FIGURE 33.21). Most bivalves are suspension-feeders. They trap fine food particles in mucus that coats the gills, and cilia then convey the particles to the mouth. Water flows into the mantle cavity through an incurrent siphon, passes over the gills, and then exits the mantle cavity through an excurrent siphon. Bivalves have no distinct head, and the radula has been lost.

Being suspension-feeders, most bivalves lead rather sedentary lives. Sessile mussels secrete strong threads that tether

FIGURE 33.20 ▪ A bivalve. This scallop has many eyes peering out between the two halves of the hinged shell.

them to rocks, docks, boats, and the shells of other animals. Clams can pull themselves into the sand or mud, using the muscular foot for an anchor. In addition to digging, scallops can also skitter along the seafloor by flapping their shells, rather like the mechanical false teeth sold in novelty shops.

Class Cephalopoda

Cephalopods are built for speed, an adaptation that fits their carnivorous diet. Squids and octopuses use beaklike jaws to bite their prey; they then inject poison to immobilize the victim. The mouth is at the center of several long tentacles. A mantle covers the visceral mass, but the shell is reduced and internal (squids) or missing altogether (many octopuses)

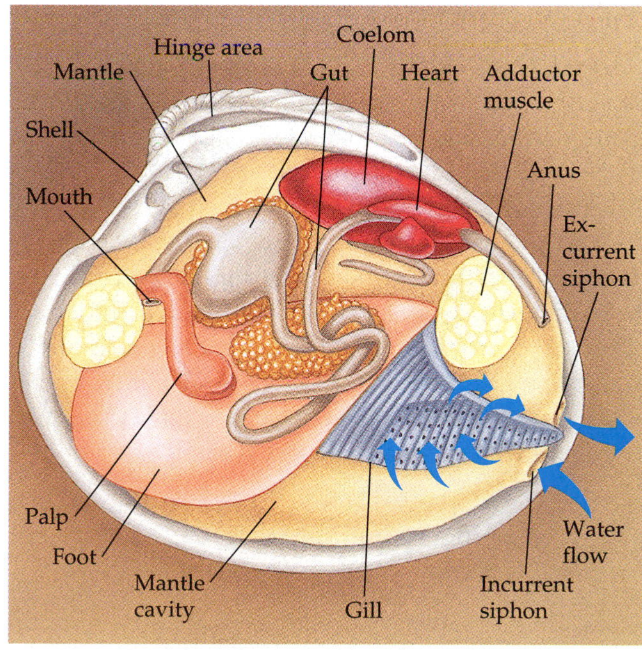

FIGURE 33.21 ▪ Anatomy of a clam. The left half of the bivalve shell has been removed. Food particles suspended in water that enters through the incurrent siphon are collected by the gills and passed via cilia and elongated flaps called palps to the mouth.

(a)

(b)

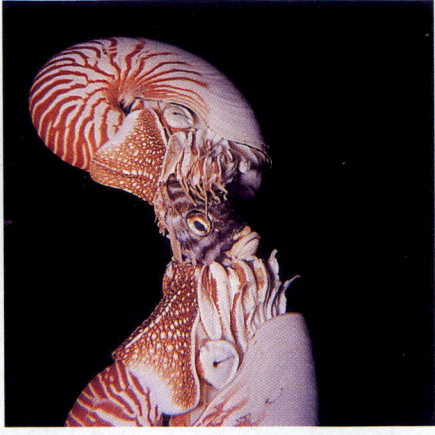

(c)

FIGURE 33.22 ▪ Cephalopods. (a) Squids, such as this California market squid (*Loligo opalescens*), are speedy carnivores with beaklike jaws and well-developed eyes. **(b)** Octopuses are believed to be among the most intelligent invertebrates. **(c)** Chambered nautiluses are the only extant cephalopods with external shells.

(FIGURES 33.22a and b). One small group of shelled cephalopods, the chambered nautiluses, survives today (FIGURE 33.22c).

A squid darts about, usually backward, by drawing water into its mantle cavity and then firing a jet stream of water through the excurrent siphon that points anteriorly. The animal steers by pointing the siphon in different directions. The foot of a cephalopod has become modified into this muscular siphon and parts of the tentacles and head. (*Cephalopod* means "head foot.") Most species of squid are less than 75 cm long, but there are also giant squids, the largest of all invertebrates. The biggest specimen on record was 17 m long (including the tentacles) and weighed about 2 tons.

Rather than swimming as squids do in the open seas, most octopuses live on the seafloor, where they creep and scurry about in search of crabs and other food.

Cephalopods are the only mollusks with a closed circulatory system. They also have a well-developed nervous system with a complex brain. The ability to learn and behave in a complex manner is probably more critical to fast-moving predators than to sedentary animals such as clams. Squids and octopuses also have well-developed sense organs.

The ancestors of octopuses and squids were probably shelled mollusks that took up a predaceous lifestyle, the loss of the shell occurring in later evolution. Shelled cephalopods called **ammonites**, many of them very large, were the dominant invertebrate predators of the seas for hundreds of millions of years until their disappearance during the mass extinctions at the end of the Cretaceous period.

Phylum Annelida: Annelids are segmented worms

Annelida means "little rings," and a segmented body resem-
33.1 bling a series of fused rings is a hallmark of the annelid worms. There are about 15,000 annelid species, ranging in length from less than 1 mm to the 3-m length of a giant Australian earthworm. Annelids live in the sea, most freshwater habitats, and damp soil. We can describe the anatomy of annelids in terms of a well-known member of the phylum, the earthworm (FIGURE 33.23).

The coelom of the earthworm is partitioned by septa, but the digestive tract, longitudinal blood vessels, and nerve cords penetrate the septa and run the length of the animal (the major vessels have segmental branches). The digestive system has several specialized regions: the pharynx, the esophagus, the crop, the gizzard, and the intestine. The closed circulatory system consists of a network of vessels containing blood with oxygen-carrying hemoglobin. Dorsal and ventral vessels are connected by segmental pairs of vessels. The dorsal vessel and five pairs of vessels that circle the esophagus of an earthworm are muscular and pump blood through the circulatory system. Tiny blood vessels are abundant in the earthworm's skin, which functions as its respiratory organ.

In each segment of the worm is a pair of excretory tubes called **metanephridia** with ciliated funnels, called nephrostomes, that remove wastes from the blood and coelomic fluid. The metanephridia lead to exterior pores, through which the metabolic wastes are discharged.

A brainlike pair of cerebral ganglia lies above and in front of the pharynx. A ring of nerves around the pharynx connects to a subpharyngeal ganglion, from which a fused pair of nerve cords runs posteriorly. All along these ventral nerve cords are segmental ganglia, also fused.

Earthworms are hermaphrodites, but they cross-fertilize. Two earthworms mate by aligning themselves in such a way that they exchange sperm, and then they separate. The received sperm cells are stored temporarily while a special organ, the clitellum, secretes a mucous cocoon. The cocoon slides along the worm, picking up the eggs and then the stored sperm. The cocoon then slips off the worm's head and resides in the soil while the embryos develop. Some earthworms can also reproduce asexually by fragmentation followed by regeneration.

Some aquatic annelids swim in pursuit of food, but most are bottom-dwellers that burrow in the sand and silt; earthworms, of course, are burrowers.

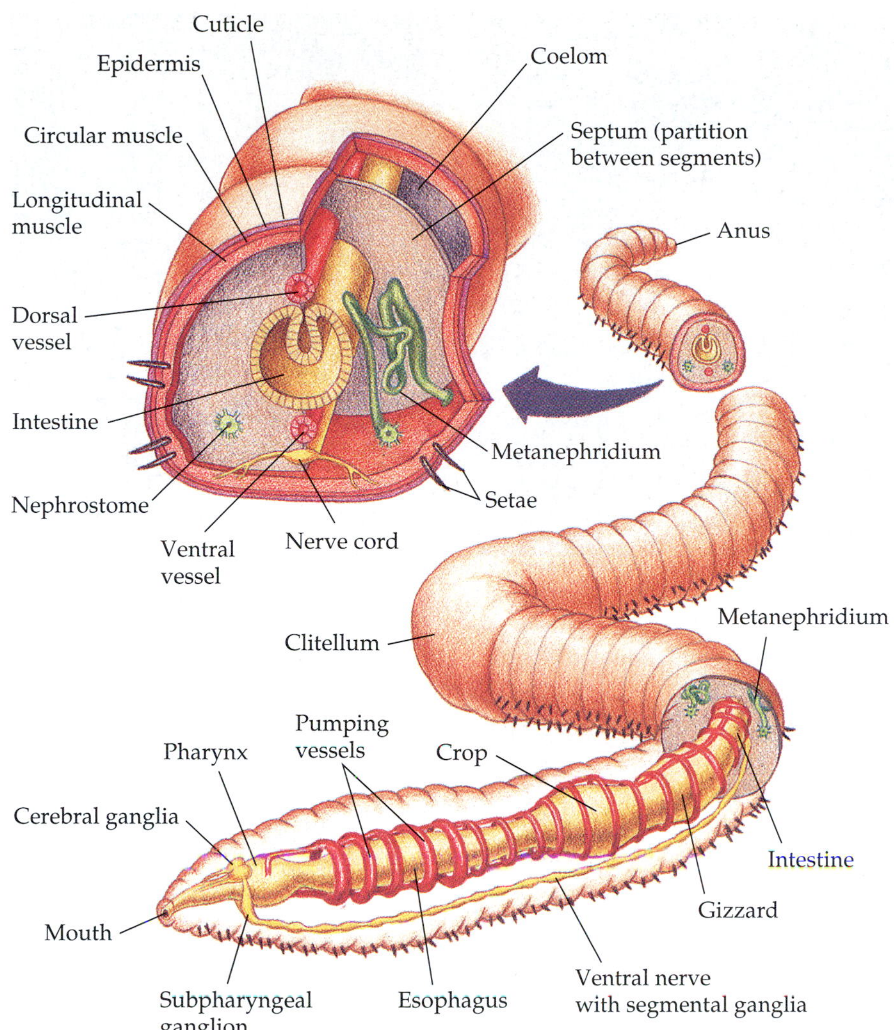

FIGURE 33.23 ▪ Anatomy of an earthworm. Annelids are segmented both externally and internally. Many of the internal structures are repeated, segment by segment. Externally, each segment has four pairs of setae, bristles that provide traction for burrowing. Earthworms and many other annelids creep along or burrow by coordinating two sets of muscles, one longitudinal and the other circular (see FIGURE 49.23). These muscles work against the noncompressible coelomic fluid, a hydrostatic skeleton. The muscles can alter the shape of each segment individually because the coelom is divided into separate compartments. When the circular muscles of a segment contract, that segment becomes thinner and elongates. Contraction of the longitudinal muscles causes the segment to shorten and thicken. The worm probes forward as alternating contractions of circular and longitudinal muscles progress along the segments like waves. The coelom and segmentation may have first functioned as adaptations for this type of movement.

The phylum Annelida is divided into three classes: Oligochaeta (earthworms and their relatives), Polychaeta (polychaetes), and Hirudinea (leeches) (TABLE 33.4).

Table 33.4 ▪ Classes of Phylum Annelida

CLASS AND EXAMPLES	MAIN CHARACTERISTICS
Oligochaeta (terrestrial and freshwater segmented worms; e.g., earthworms) (see FIGURE 33.23)	Reduced head; no parapodia, but setae present
Polychaeta (mostly marine segmented worms) (see FIGURES 33.24a and b)	Well-developed head; each segment usually has parapodia with setae; tube-dwelling and free-living
Hirudinea (leeches) (see FIGURE 33.24c)	Body usually flattened, with reduced coelom and segmentation; setae absent; suckers at anterior and posterior ends; parasites, predators, and scavengers

Class Oligochaeta

This class of segmented worms includes the earthworms and a variety of aquatic species. An earthworm eats its way through the soil, extracting nutrients as the soil passes through the digestive tube. Undigested material, mixed with mucus secreted into the digestive tract, is egested as castings through the anus. Farmers value earthworms because the animals till the earth, and the castings improve the texture of the soil. Darwin estimated that 1 acre of British farmland had about 50,000 earthworms that produced 18 tons of castings per year.

Class Polychaeta

Each segment of a polychaete ("many setae," for the bristles on each segment) has a pair of paddlelike or ridgelike structures called parapodia ("almost feet") that function in locomotion. Each parapodium has several setae made of the polysaccharide chitin. In many polychaetes the parapodia are richly supplied with blood vessels and function as gills (FIGURE 33.24a, p. 614).

Most polychaetes are marine. A few drift and swim among the plankton, many crawl on or burrow in the seafloor, and many others live in tubes, which the worms make by mixing mucus with bits of sand and broken shells. The tube-dwellers include the brightly colored fanworms, which trap

(a)

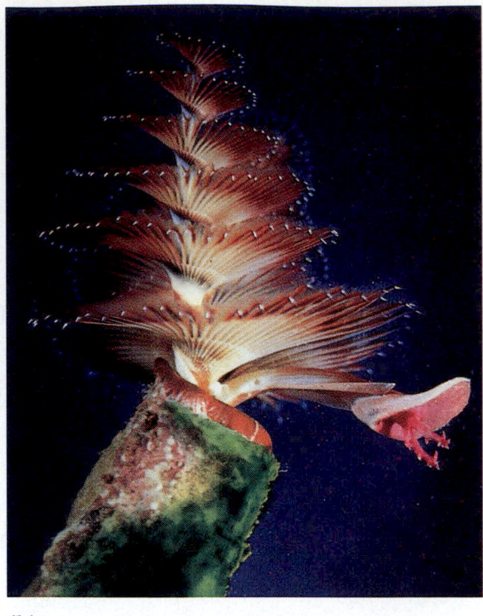

(b)

(c)

FIGURE 33.24 ▪ Annelids, the segmented worms. (a) Most annelids of the class Polychaeta are marine worms. Each segment has a pair of lateral flaps that function in movement and as gills for the exchange of respiratory gases with the surrounding water. **(b)** Fanworms (Class Polychaeta) are tube-dwellers that use their feathery headdresses for gas exchange and to extract suspended food particles from the seawater. This species is known as a Christmas-tree worm. **(c)** Leeches (Class Hirudinea) are free-living carnivores or parasites that suck blood from other animals. This species inhabits the Malaysian rain forest.

microscopic food particles in feathery tentacles that extend from the opening of the tube (FIGURE 33.24b).

Class Hirudinea

The majority of leeches inhabit fresh water, but there are also land leeches that move through moist vegetation. Many leeches feed on other small invertebrates, but some are bloodsucking parasites that feed by attaching temporarily to other animals, including humans (FIGURE 33.24c). Leeches range in length from about 1 to 30 cm. Some parasitic species use bladelike jaws to slit the skin of the host, whereas others secrete enzymes that digest a hole through the skin. The host is usually oblivious to this attack because the leech secretes an anesthetic. After making the incision, the leech secretes another chemical, hirudin, which keeps the blood of the host from coagulating. The parasite then sucks as much blood as it can hold, often more than ten times its own weight. After this gorging, a leech can last for months without another meal. Until this century, leeches were frequently used by physicians for bloodletting. Leeches are still used for treating bruised tissues and for stimulating the circulation of blood to fingers or toes that have been sewn back to hands or feet after accidents.

▪ ▪ ▪

Before leaving annelids, let's highlight two evolutionary adaptations that are well developed in this phylum: the coelom and segmentation. The evolutionary significance of the coelom cannot be overemphasized. In addition to providing a hydrostatic skeleton and allowing new and diverse methods of locomotion, it also provides body space for storage and for complex organ development. The coelom also serves as a cushion that protects internal structures, and it allows a functional separation of the action of body wall muscles from those of internal organs, such as the muscles of the digestive tract.

Segmentation allows a high degree of specialization of body regions. This regional specialization—groups of segments modified for different functions—is seen to some degree in annelids but is a true hallmark of the body plan of arthropods.

Phylum Arthropoda: Arthropods have regional segmentation, jointed appendages, and an exoskeleton

33.1 It is estimated that the arthropod population of the world, including crustaceans, spiders, and insects, numbers about 1 billion billion (10^{18}) individuals. Nearly 1 million arthropod species have been described, mostly insects. In fact, two out of every three organisms known are arthropods, and the phylum is represented in nearly all habitats of the biosphere. On the criteria of species diversity, distribution, and sheer numbers, Arthropoda must be regarded as the most successful of all animal phyla.

General Characteristics of Arthropods

The diversity and success of arthropods are largely related to their segmentation, hard exoskeleton, and jointed appendages. (*Arthropoda* means "jointed feet.") Groups of segments

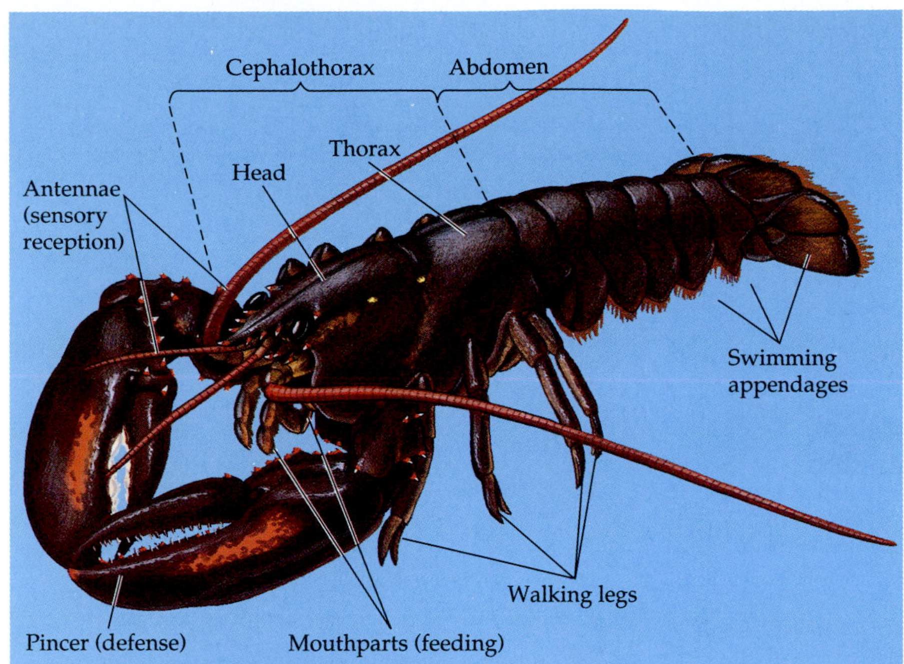

FIGURE 33.25 • External anatomy of an arthropod. Many of the distinctive features of arthropods are apparent in this dorsal view of a lobster. The whole body, including appendages, is covered by an exoskeleton. Distinct regions of the body are the head and thorax (together forming the cephalothorax) and the abdomen. The body is segmented, but this characteristic is obvious only in the abdomen. All body regions bear jointed appendages (including pincers, mouthparts, walking legs, and swimming appendages). Two pairs of sensory antennae are also jointed. The head bears a pair of eyes, each situated on a movable stalk.

and their appendages have become specialized for a great variety of functions. This evolutionary flexibility resulted not only in great diversification but also in an efficient body plan by the division of labor among regions. For example, the appendages are variously modified for walking, feeding, sensory reception, copulation, and defense. FIGURE 33.25 illustrates the diverse appendages and other arthropod characteristics of a lobster.

The body of an arthropod is completely covered by the **cuticle**, an **exoskeleton** (external skeleton) constructed from layers of protein and chitin. The cuticle can be a thick, hard armor over some parts of the body and paper-thin and flexible in other locations, such as the joints. The exoskeleton protects the animal and provides points of attachment for the muscles that move the appendages. The skeleton of arthropods is both strong and relatively impermeable to water. As we will see, both of these qualities were largely responsible for the move onto land by various arthropod groups. The rigid exoskeleton also posed some evolutionary problems. For example, in order to grow, an arthropod must occasionally shed its old exoskeleton and secrete a larger one. This process, called **molting**, is energetically expensive and leaves the animal temporarily vulnerable to predators and other dangers.

Arthropods tune in to their environment with well-developed sensory organs, including eyes, olfactory receptors for smell, and antennae for touch and smell. Cephalization is extensive, with most sensory organs concentrated at the anterior end of the animal.

Arthropods have **open circulatory systems** in which fluid called hemolymph is propelled by a heart through short arteries and then into spaces called sinuses surrounding the tissues and organs. (The term *blood* is best reserved for fluid in a closed circulatory system.) Hemolymph reenters the arthropod heart through pores that are usually equipped with valves. The body sinuses are collectively called the hemocoel, which is not part of the coelom. In most arthropods, the coelom that forms in the embryo becomes much reduced as development progresses, and the hemocoel becomes the main body cavity in adults. Although this condition resembles the open circulatory system of mollusks, the two systems probably arose independently.

A variety of organs specialized for gas exchange have evolved in arthropods. These organs must allow the diffusion of respiratory gases in spite of the exoskeleton. Most aquatic species have gills with thin feathery extensions that place an extensive surface area in contact with the surrounding water. Terrestrial arthropods generally have internal surfaces specialized for gas exchange. Most insects, for instance, have tracheal systems, branched air ducts leading into the interior from pores in the cuticle.

Arthropod Phylogeny and Classification

Arthropods and annelids are segmented, and many biologists think that arthropods and annelids shared a common ancestor. Cambrian fossils of jointed-legged animals that resemble segmented worms add to the evidence of an evolutionary link between Annelida and Arthropoda. One hypothesis is that early arthropods resembled animals called onychophorans, sometimes called walking worms.

Unlike arthropods, walking worms have unjointed appendages, and most zoologists assign the walking worms to

FIGURE 33.26 ▪ *Peripatus* (an onychophoran, or walking worm). *Peripatus* is distinctly segmented and has excretory organs, musculature, and certain other features that are annelidlike. *Peripatus* resembles arthropods in its respiratory and circulatory systems, its cuticle made of chitin, and its jaws modified from appendages. Most animal systematists assign *Peripatus* and a few related genera to a separate phylum, Onychophora. However, recent analysis by molecular systematists argues in favor of placing these animals within the Arthropoda.

Table 33.5 ▪ Major Classes of Phylum Arthropoda

CLASS AND EXAMPLES	MAIN CHARACTERISTICS
Arachnida (spiders, scorpions, ticks, mites) (see FIGURES 33.29 and 33.30)	Body having one or two main parts; six pairs of appendages (chelicerae, pedipalps, and four pairs of walking legs); mostly terrestrial
Diplopoda (millipedes) (see FIGURE 33.31a)	Body with distinct head bearing antennae and chewing mouthparts, segmented body with two pairs of walking legs per segment; terrestrial; herbivorous
Chilopoda (centipedes) (see FIGURE 33.31b)	Body with distinct head bearing large antennae and three pairs of mouthparts; appendages of first body segment modified as poison claws; trunk segments bear one pair of walking legs each; terrestrial; carnivorous
Insecta (insects) (see FIGURES 33.32–33.34 and TABLE 33.6)	Body divided into head, thorax, and abdomen; antennae present; mouthparts modified for chewing, sucking, or lapping; usually with two pairs of wings and three pairs of legs; mostly terrestrial
Crustacea (crabs, lobsters, crayfish, shrimp) (see FIGURE 33.35)	Body of two or three parts; antennae present; chewing mouthparts; three or more pairs of legs; mostly marine

a separate phylum, Onychophora. In counterpoint, however, evidence from molecular systematics indicates that onychophorans are true arthropods, not "missing links" (FIGURE 33.26). Some molecular evidence also indicates that annelids and arthropods are not so closely related as formerly thought, and that their segmentation is a primitive character derived from an ancestor common to all bilaterians.

Debate continues about the origin of arthropods, and about the subsequent diversification of Phylum Arthropoda. One widely held view based largely on anatomical and fossil evidence is that arthropods diverged into the following four subgroups: **trilobites** (all extinct), **chelicerates** (eurypterids, horseshoe crabs, scorpions, spiders, ticks), **uniramians** (centipedes, millipedes, insects), and **crustaceans** (crabs, lobsters, shrimps, barnacles, and many others) (TABLE 33.5).

Crustaceans are primarily aquatic and are believed to have evolved in the ocean. Insects, millipedes, and centipedes and most extant chelicerates diversified on land.

Chelicerates (Gr. *cheilos,* "lips," and *cheir,* "arm") are named for clawlike feeding appendages called **chelicerae.** In contrast, uniramians and crustaceans have jawlike **mandibles.** They are also distinguished from chelicerates in having one or two pairs of sensory **antennae** and usually a pair of **compound eyes** (multifaceted eyes with many separate focusing elements). Uniramians have one pair of antennae and uniramous (unbranched) appendages; crustaceans have two pairs of antennae and typically biramous (branched) appendages. Chelicerates lack antennae, and most have simple eyes (eyes with a single lens).

Expanding evidence from developmental genetics and nucleic acid sequencing challenges the uniramian grouping and supports a somewhat different view of arthropod classification and phylogeny. Insects and crustaceans may be more closely related than traditionally thought; in fact, they may constitute a monophyletic group distinct from the millipedes and centipedes. We discuss millipedes, centipedes, insects, and crustaceans separately in the sections that follow.

The move onto land by arthropods was made possible in part by the exoskeleton. When the exoskeleton first evolved in the seas, its main functions were probably protection and anchorage for muscles, but it eventually helped certain arthropods live on land by solving the problems of water loss and structural support. The arthropod cuticle is relatively impermeable to water, helping prevent desiccation. The firm exoskeleton also solved the problem of support when arthropods left the buoyancy of water. Chelicerates, insects, millipedes, and centipedes diversified on land during the late Silurian and early Devonian periods, following the colonization by plants. The oldest fossil evidence of terrestrial animals is tracks of extinct chelicerates (apparently eurypterids that spent at least some of their time on land) about 450 million years old. Fossilized arachnids almost as old have also been found.

Trilobites

Among the early arthropods were the **trilobites** (FIGURE 33.27). They were common denizens of the shallow seas throughout the Paleozoic era but disappeared with the great Permian extinctions that ended that era, about 250 million years ago. Trilobites had pronounced segmentation, but their appendages showed little variation from segment to segment.

FIGURE 33.27 ▪ A fossil arthropod. Trilobites were prevalent arthropods throughout the Paleozoic era. About 4000 trilobite species have been described from fossils.

FIGURE 33.28 ▪ The horseshoe crab (*Limulus polyphemus*). This "living fossil," which has changed little in hundreds of millions of years, has survived from a rich diversity of chelicerates that once filled the seas. The horseshoe crab is common on the Atlantic and Gulf coasts of the United States.

As arthropods continued to evolve, the segments tended to fuse and become fewer in number, and the appendages became specialized for a variety of functions. (Compare the trilobite in FIGURE 33.27 with the lobster in FIGURE 33.25.)

Spiders and Other Chelicerates

The trilobites were outlasted by the **eurypterids**, or water scorpions. These mainly marine and freshwater predators, up to 3 m long, were chelicerates. A chelicerate has an anterior cephalothorax and a posterior abdomen. The appendages are more specialized than those of trilobites, and the most anterior appendages are modified as chelicerae (either pincers or fangs). Most of the marine chelicerates, including all of the eurypterids, are extinct; only four marine species survive today, one being the horseshoe crab (FIGURE 33.28).

The bulk of modern chelicerates are found on land in the form of **Class Arachnida**, which includes scorpions, spiders, ticks, and mites (FIGURE 33.29). Ticks and many mites are

(a)

(b)

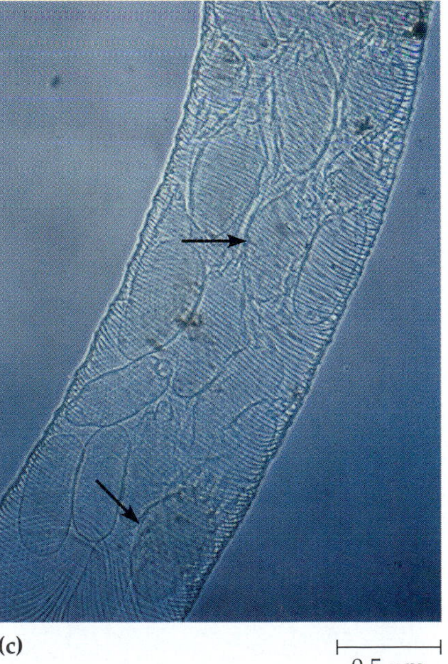

(c)

FIGURE 33.29 ▪ Arachnids. (a) Scorpions, which hunt by night, were among the first terrestrial carnivores, preying on other arthropods that fed on the early land plants. The pedipalps of scorpions are pincers specialized for defense and the capture of food. The tip of the tail bears a poisonous stinger. **(b)** This magnified house-dust mite is a ubiquitous scavenger in human dwellings (colorized SEM). Unlike some mites that carry disease-causing bacteria, dust mites are harmless except to people who are allergic to them. **(c)** Many arthropods parasitize other arthropods. In this light micrograph, you can see parasitic mites inhabiting the tracheae (air tubes) of a honeybee (an insect).

among a large group of parasitic arthropods. Nearly all ticks are blood-sucking parasites on the body surfaces of reptiles, birds, or mammals. Parasitic mites live on or in a wide variety of vertebrates and invertebrates, including other arthropods (FIGURE 33.29c).

Arachnids have a cephalothorax with six pairs of appendages: the chelicerae, a pair of appendages called pedipalps that usually function in sensing or feeding, and four pairs of walking legs (FIGURE 33.30). Spiders use their fanglike chelicerae, equipped with poison glands, to attack prey. As the chelicerae masticate (chew) the prey, the spider spills digestive juices onto the torn tissues. The food softens, and the spider sucks up the liquid meal.

In most spiders, gas exchange is carried out by **book lungs**, stacked plates contained in an internal chamber (see FIGURE 33.30b). The extensive surface area of these respiratory organs is a structural adaptation that enhances the exchange of O_2 and CO_2 between the hemolymph and air.

A unique adaptation of many spiders is the ability to catch flying insects by stringing webs of silk, a protein produced as a liquid by special abdominal glands. The silk is spun by organs called spinnerets into fibers that solidify. Each spider engineers a style of web characteristic of its species and constructs the web perfectly on the first try. This complex behavior is apparently inherited. Besides building their webs from silk, various spiders use these fibers in other ways: as droplines for rapid escape, as cloth that covers eggs, and even as "gift wrapping" for food that certain male spiders offer females during courtship.

Millipedes and Centipedes

Millipedes **(Class Diplopoda)** are wormlike, with a large number of walking legs (two pairs per segment), though fewer than the thousand their name implies (FIGURE 33.31a). They eat decaying leaves and other plant matter. Millipedes may have been among the earliest animals on land, living on mosses and primitive vascular plants.

Centipedes **(Class Chilopoda)** are terrestrial carnivores. The head has a pair of antennae and three pairs of appendages modified as mouthparts, including the jawlike mandibles. Each segment of the trunk region has one pair of walking legs (FIGURE 33.31b). Centipedes have poison claws on the anteriormost trunk segment that paralyze prey and aid in defense.

Insects

In species diversity, insects **(Class Insecta)** outnumber all other forms of life combined. They live in almost every terrestrial habitat and in fresh water, and flying insects fill the air. Insects are rare, though not absent, in the seas, where crustaceans are the dominant arthropods. Class Insecta is divided into about 26 orders, some of which are described in TABLE 33.6 (pp. 620–621). **Entomology**, the study of insects, is a vast field with many subspecialties, including physiology, ecology, and taxonomy.

The oldest insect fossils date back to the Devonian period, which began about 400 million years ago. However, when flight evolved during the Carboniferous and Permian periods, it spurred an explosion in insect variety. A fossil record of diverse

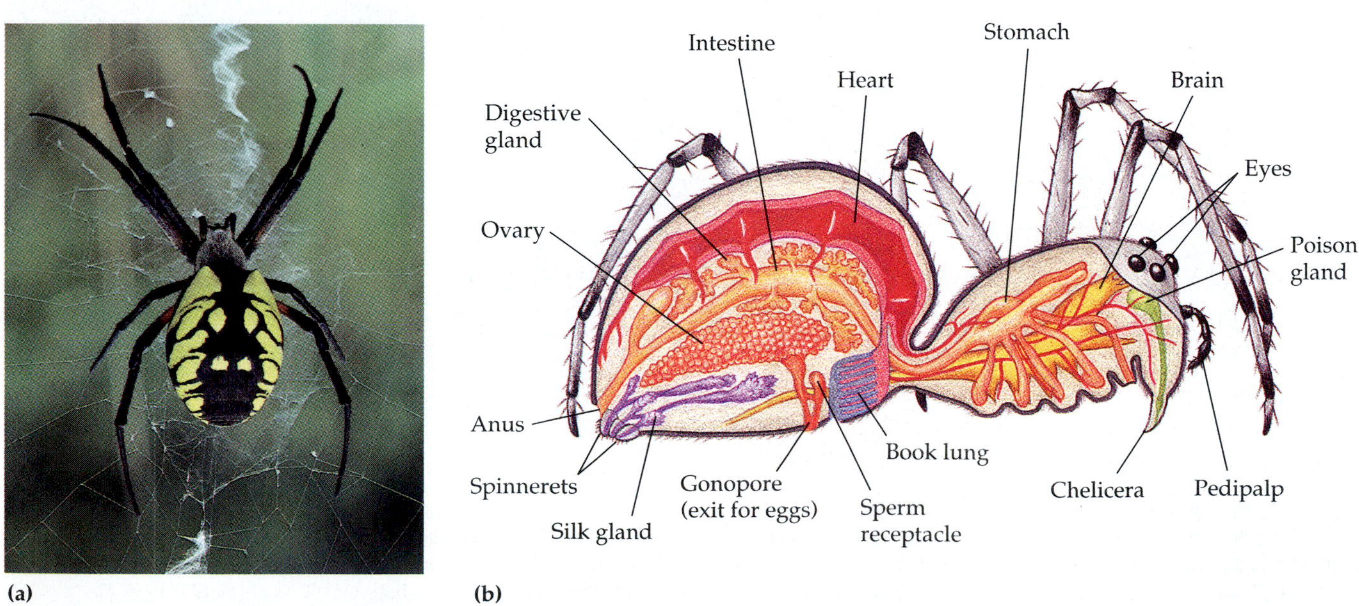

FIGURE 33.30 ▪ Spiders (Class Arachnida). (a) Spiders are generally most active during the daytime, when they hunt for prey or trap insects in webs. **(b)** Anatomy of a spider.

(a)

(b)

FIGURE 33.31 ▪ Class Diplopoda (millipedes) and Class Chilopoda (centipedes). (a) Millipedes feed on decaying plant matter. **(b)** The house centipede (*Scutigera coleoptrata*), a fast-moving carnivore, feeds on insects, including cockroaches, and other small invertebrates.

insect mouthparts indicates that specialized feeding on gymnosperms and other Carboniferous plants also contributed to the adaptive radiation of insects. A widely held hypothesis is that the greatest diversification of insects paralleled the evolutionary radiation of flowering plants during the Cretaceous and early Tertiary periods about 60 to 65 million years ago. This view is challenged by new research suggesting that insects diversified extensively before the angiosperm radiation. Thus, during the coevolution of flowering plants and the herbivorous insects that pollinated them, insect diversity may have been more a cause of angiosperm radiation than an effect.

Flight is obviously one key to the great success of insects. An animal that can fly can escape many predators, find food and mates, and disperse to new habitats much faster than an animal that must crawl about on the ground. Many insects have one or two pairs of wings that emerge from the dorsal side of the thorax (FIGURE 33.32). Because the wings are extensions of the cuticle and not true appendages, insects can fly without sacrificing any walking legs. (By contrast, the flying vertebrates—birds and bats—have one of their two pairs of walking legs modified for wings and are generally quite clumsy on the ground.)

Insect wings may have first evolved as extensions of the cuticle that helped the insect body absorb heat, only later becoming organs for flight. Other views suggest that wings allowed the animals to glide from vegetation to the ground, or even served as gills in aquatic insects. Still another hypothesis is that insect wings functioned for swimming before they functioned for flight.

Dragonflies, with two similar pairs of wings, were among the first insects to fly (see FIGURE 33.32). Several insect orders that evolved later than dragonflies have modified flight equipment. The wings of bees and wasps, for instance, are hooked together and move as a single pair. Butterfly wings operate in a similar fashion because the anterior pair overlaps the posterior wings. In beetles, the posterior wings function in flight, while the anterior ones are modified as covers that protect the flight wings when the beetle is on the ground or when the beetle is burrowing.

FIGURE 33.32 ▪ Insect flight. Insect wings, such as this dragonfly's, are not modified appendages but extensions of the cuticle. Some insects beat their wings at speeds of several hundred cycles per second by using muscles to warp the shape of the entire cuticle covering the thorax. As the wings flap, they change angles, producing lift on both the up and down strokes.

Table 33.6 ■ Some Major Orders of Insects

ORDER	APPROXIMATE NUMBER OF SPECIES	MAIN CHARACTERISTICS	EXAMPLES
Anoplura	2,400	Wingless ectoparasites; sucking mouthparts; small with flattened body, reduced eyes; legs with clawlike tarsi for clinging to skin; incomplete metamorphosis; very host-specific.	 Sucking lice Human body louse
Coleoptera	500,000	Two pairs of wings (one pair thick and leathery, one pair membranous), armored exoskeleton; biting and chewing mouthparts; complete metamorphosis.	 Beetles Japanese beetle
Dermaptera	1,000	Two pairs of wings (one pair leathery, one pair membranous) or wingless; biting mouthparts; large posterior pincers; incomplete metamorphosis.	 Earwigs Earwig
Diptera	120,000	One pair of wings and halteres (balancing organs); sucking, piercing or lapping mouthparts; complete metamorphosis	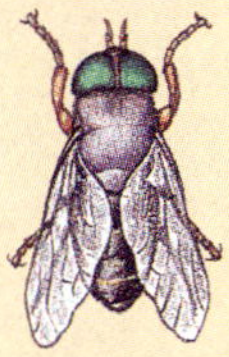 Flies, mosquitoes Horsefly
Hemiptera	55,000	Two pairs of wings (one pair partly leathery, one pair membranous); piercing or sucking mouthparts; incomplete metamorphosis.	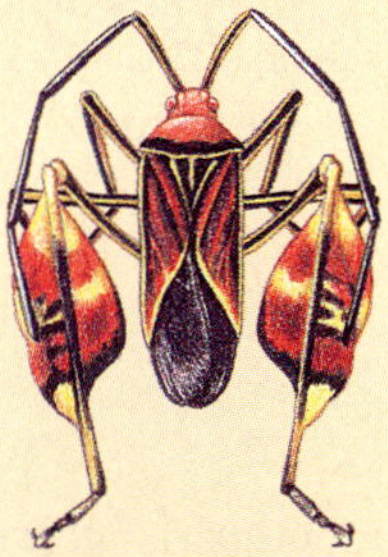 True bugs; assassin bug, bedbug, chinch bug Leaf-footed bug
Hymenoptera	100,000	Two pairs of membranous wings; head mobile; chewing or sucking mouthparts; posterior stinging organ on females; complete metamorphosis; many species social.	 Ants, bees, wasps Cicada-killer wasp

Table 33.6 ■ Some Major Orders of Insects

ORDER	APPROXIMATE NUMBER OF SPECIES	MAIN CHARACTERISTICS	EXAMPLES	
Isoptera	2,000	Two pairs of membranous wings (some stages wingless); chewing mouthparts; highly social; incomplete metamorphosis.	Termites	 Termite
Lepidoptera	140,000	Two pairs of wings covered with tiny scales; long coiled tongue for sucking; complete metamorphosis.	Butterflies, moths	 Swallowtail butterfly
Odonata	5,000	Two pairs of membranous wings; biting mouthparts; incomplete metamorphosis.	Damselflies, dragonflies	 Dragonfly
Orthoptera	30,000	Two pairs of wings (one pair leathery, one pair membranous); biting and chewing mouthparts; incomplete metamorphosis.	Crickets, roaches, grasshoppers, mantids	 Katydid
Siphonaptera	2,000	Wingless, laterally compressed; adults are bloodsuckers on birds and mammals; piercing and sucking mouthparts; jumping legs; complete metamorphosis.	Fleas	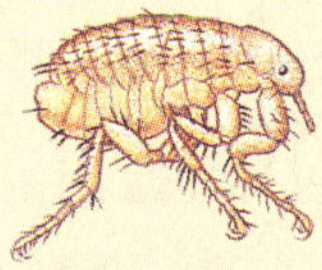 Flea
Trichoptera	7,000	Two pairs of hairy wings; chewing or lapping mouthparts; complete metamorphosis; aquatic larvae build silken nets or cases (of sand, gravel, and wood) bound together by silk.	Caddisflies	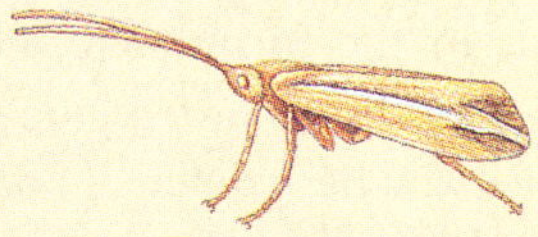 Caddisfly

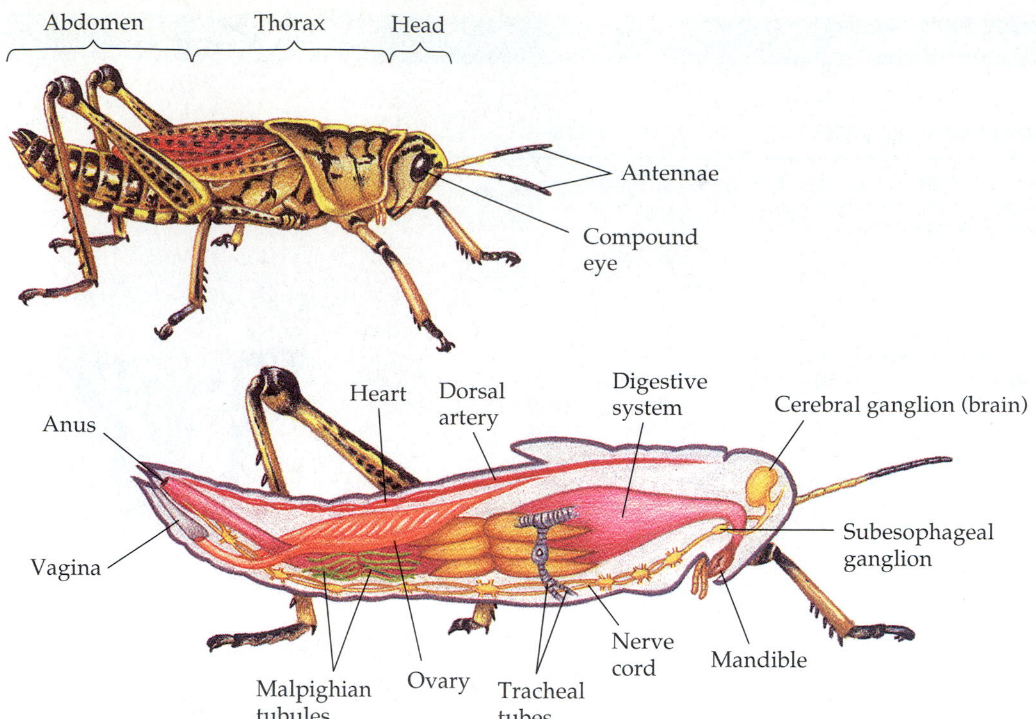

FIGURE 33.33 ▪ Anatomy of a grasshopper, an insect. The insect body has three regions: a head, a thorax, and an abdomen (top drawing). Segmentation is apparent along the thorax and abdomen, but the head segments are fused. On the insect head are one pair of antennae and a pair of compound eyes. Several pairs of appendages modified for chewing (as in grasshoppers) or for lapping, piercing, or sucking (in certain other insects) form the mouthparts. The thorax of the insect bears three pairs of walking legs. The bottom drawing identifies major internal organs.

The internal anatomy of an insect includes several complex organ systems (FIGURE 33.33). The complete digestive system is regionally specialized, with discrete organs functioning in the breakdown of food and the absorption of nutrients. Like other arthropods, an insect has an open circulatory system, with a heart pumping hemolymph through the sinuses of the hemocoel. Metabolic wastes are removed from the hemolymph by unique excretory organs called **Malpighian tubules**, which are outpocketings of the digestive tract. Gas exchange in insects is accomplished by a **tracheal system** of branched, chitin-lined tubes that infiltrate the body and carry oxygen directly to cells. The tracheal system opens to the outside of the body through spiracles, pores that can open or close to regulate air flow and limit water loss.

The insect nervous system consists of a pair of ventral nerve cords with several segmental ganglia. The two cords meet in the head, where the ganglia of several anterior segments are fused into a cerebral ganglion (brain) close to the antennae, eyes, and other sense organs concentrated on the head.

Many insects undergo metamorphosis in their development. In the **incomplete metamorphosis** of grasshoppers and some other orders, the young resemble adults but are smaller and have different body proportions. The animal goes through a series of molts, each time looking more like an adult, until it reaches full size. Insects with **complete metamorphosis** have larval stages specialized for eating and growing that are known by such names as maggot, grub, or caterpillar. The larval stage looks entirely different from the adult stage, which is specialized for dispersal and reproduction. Metamorphosis from the larval stage to the adult occurs during a pupal stage (FIGURE 33.34).

Reproduction in insects is usually sexual, with separate male and female individuals. Adults come together and recognize each other as members of the same species by advertising with bright colors (butterflies), sound (crickets), or odors (moths). Fertilization is generally internal. In most species, sperm cells are deposited directly into the female's vagina at the time of copulation, though in some species the male deposits a sperm packet outside the female, and the female picks it up. An internal structure in the female called the spermatheca stores the sperm, usually enough to fertilize more than one batch of eggs. Many insects mate only once in a lifetime. After mating, a female lays her eggs on an appropriate food source where the next generation can begin eating as soon as it hatches.

Animals so numerous, diverse, and widespread as insects are bound to affect the lives of all other terrestrial organisms, including humans. On one hand, we depend on bees, flies, and many other insects to pollinate our crops and orchards. On the other hand, insects are carriers for many diseases, including malaria (spread by mosquitos that carry *Plasmodium*; see FIGURE 28.7) and African sleeping sickness (spread by tsetse flies that carry *Trypanosoma*; see FIGURE 28.5). Furthermore, insects compete with humans for food. In parts of Africa, for instance, insects claim about 75% of the crops. Trying to minimize their losses, farmers in the United States spend billions of dollars each year on pesticides, spraying crops with massive doses of some of the deadliest poisons ever invented. Try as they may, not even humans have challenged the preeminence of insects and their arthropod kin. Or, as Cornell University's Thomas Eisner puts it: "Bugs are not going to inherit the Earth. They own it now. So we might as well make peace with the landlord."

(a) Larva (caterpillar)

(b) Pupa

(c) Pupa

(d) Emerging adult

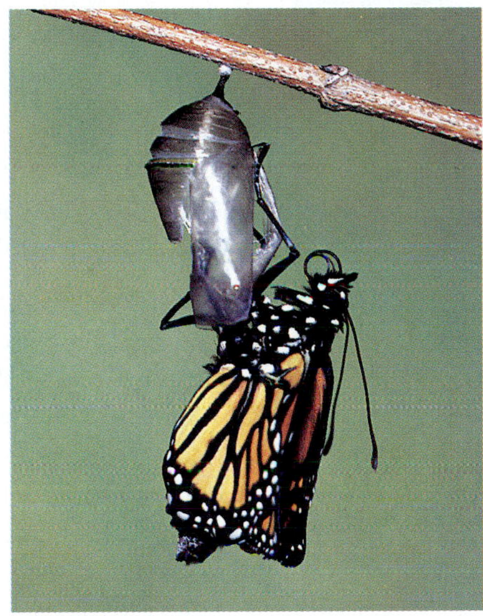

(e) Adult

FIGURE 33.34 ▪ Metamorphosis of a butterfly. (a) The larva (caterpillar) spends its time eating and growing, molting as it grows. **(b)** After several molts, the larva encases itself in a cocoon and becomes a pupa. **(c)** Within the pupa, the larval tissues are broken down, and the adult is built by the division and differentiation of cells that were quiescent in the larva. **(d)** Finally, the adult emerges from the cocoon. **(e)** Fluid is pumped into veins of the wings and then withdrawn, leaving the hardened veins as struts supporting the wings. Now the insect flies off and reproduces, deriving much of its nourishment from the calories stored by the feeding larva.

Crustaceans

While arachnids and insects thrived on land, crustaceans, for the most part, remained in marine and freshwater environments, where they are now represented by about 40,000 species. Crabs, lobsters, crayfish, and shrimp are the most familiar crustaceans (FIGURE 33.35, p. 624).

The multiple appendages of crustaceans are extensively specialized. Lobsters and crayfish, for instance, have a toolkit of 19 pairs of appendages (see FIGURE 33.25). Crustaceans are the only arthropods with two pairs of antennae. Three or more pairs of appendages are modified as mouthparts, including the hard mandibles. Walking legs are present on the thorax, and, unlike insects, crustaceans have appendages on the abdomen. A lost appendage can be regenerated.

Small crustaceans exchange gases across thin areas of the cuticle, but larger species have gills. The circulatory system is open, with a heart pumping hemolymph through arteries into sinuses that bathe the organs. Crustaceans excrete nitrogenous wastes by diffusion through thin areas of the cuticle, but a pair of glands regulates the salt balance of the hemolymph.

Sexes are separate in most crustaceans. In the case of lobsters and crayfish, the male uses a specialized pair of appendages to transfer sperm to the reproductive pore of the female during copulation. Most aquatic crustaceans go through one or more swimming larval stages.

One of the largest groups of crustaceans (about 10,000 species), the **isopods** are mostly small marine species. Some are numerous at the bottom of deep oceans. Isopods also include the land-dwelling pill bugs, or wood lice, common on the undersides of moist logs and leaves.

Another group of small crustaceans, the **copepods**, are among the most numerous of all animals. They are important members of marine and freshwater plankton communities,

(a)

(b)

(c)

FIGURE 33.35 ▪ Crustaceans. (a) This is a Sally lightfoot crab from the Galápagos Islands. **(b)** Planktonic crustaceans known as krill are consumed by the ton by whales and other large suspension-feeders. **(c)** Barnacles are sessile crustaceans with a shell (exoskeleton) hardened by calcium carbonate. The jointed appendages projecting from the shell capture small plankton and organic particles suspended in the water.

eating microscopic algae, protists, and bacteria, and being eaten by many fishes.

Lobsters, crayfish, crabs, and shrimp are all relatively large crustaceans called **decapods** (FIGURE 33.35a). The exoskeleton, or cuticle, is hardened by calcium carbonate; the portion that covers the dorsal side of the cephalothorax forms a shield called the carapace. Most decapods are marine. Crayfish, however, live in fresh water, and some tropical crabs live on land.

The larvae of many larger-bodied crustaceans are also planktonic. A marine group known as krill are shrimplike planktonic organisms that grow to about 3 cm long (FIGURE 33.35b). A major food source for many species of whales, krill are now being harvested in great numbers by humans for food and agricultural fertilizer.

Barnacles are sessile crustaceans with parts of their cuticles hardened into shells by calcium carbonate. They feed by using their appendages to strain food from the water (FIGURE 33.35c).

THE COELOMATES: DEUTEROSTOMES

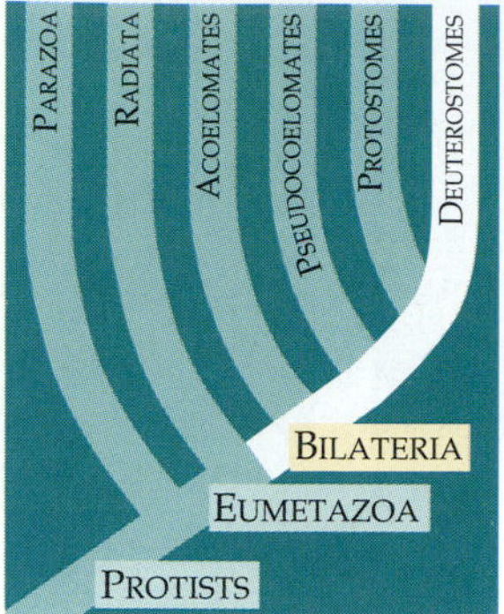

At first glance, sea stars and other echinoderms may seem to have little in common with the phylum Chordata, which includes the vertebrates: fishes, amphibians, reptiles, birds, and mammals. These animals, however, share features characteristic of deuterostomes: radial cleavage, development of the coelom from the archenteron, and formation of the mouth at the end of the embryo opposite the blastopore (see FIGURE 32.6).

Phylum Echinodermata: Echinoderms have a water vascular system and secondary radial symmetry

Sea stars and most other **echinoderms** (Gr. *echin,* "spiny," and
33.1 *derma,* "skin") are sessile or slow-moving animals with radial symmetry as adults. The internal and external parts of the animal radiate from the center, often as five spokes. A thin skin covers an endoskeleton of hard calcareous plates. Most echinoderms are prickly from skeletal bumps and spines that have various functions. Unique to echinoderms is the **water vascular system**, a network of hydraulic canals branching into extensions called **tube feet** that function in locomotion, feeding, and gas exchange.

Sexual reproduction of echinoderms usually involves separate male and female individuals that release their gametes into the seawater. The radial adults develop by metamorphosis from bilateral larvae. The early embryology of echinoderms clearly aligns them with the deuterostomes.

The 7000 or so echinoderms, all marine, are divided into six classes (FIGURE 33.36): Asteroidea (sea stars), Ophiuroidea (brittle stars), Echinoidea (sea urchins and sand dollars), Crinoidea (sea lilies and feather stars), Holothuroidea (sea cucumbers), and Concentricycloidea (sea daisies). The sea daisies, discovered only recently, live on waterlogged wood in the deep sea.

Class Asteroidea

Sea stars have five arms (sometimes more) radiating from a central disk (FIGURE 33.36a). The undersurfaces of the arms bear tube feet, each of which can act like a suction disk. By a

(a) Sea star

(b) Sea star feeding

(c) Brittle star

(d) Sea urchin

(e) Sea lily

(f) Sea cucumber

FIGURE 33.36 • Echinoderms. Exclusively marine, the mostly sedentary or slow-moving echinoderms are characterized by radial symmetry, bony endoskeletons, and hydraulic tube feet that function in movement, feeding, and, in some species, gas exchange. **(a)** A sea star (Class Asteroidea) on coral. **(b)** A sea star eating a clam. **(c)** A brittle star (Class Ophiuroidea). **(d)** A sea urchin (Class Echinoidea). **(e)** A sea lily (Class Crinoidea). **(f)** A sea cucumber (Class Holothuroidea).

complex set of hydraulic and muscle actions, the suction can be created or released (FIGURE 33.37). The sea star coordinates its tube feet to adhere firmly to rocks or to creep along slowly as the tube feet extend, grip, contract, release, extend, and grip again. Sea stars also use their tube feet to grasp prey, such as clams and oysters (FIGURE 33.36b). The arms of the sea star embrace the closed bivalve, hanging on tightly by the tube feet. The sea star then turns its stomach inside-out, everting it through its mouth and into the narrow opening between the shells of the bivalve. The digestive system of the sea star secretes juices that begin digesting the soft body of the mollusk within its own shell.

Sea stars and some other echinoderms are capable of regeneration. Sea stars can regrow lost arms, and members of one genus can even regrow an entire body from a single arm.

Class Ophiuroidea

Brittle stars have distinct central disks, and the arms are long and highly mobile (FIGURE 33.36c). Their tube feet lack suckers, and they move by serpentine lashing of the arms. Some species are suspension-feeders; others are predators or scavengers.

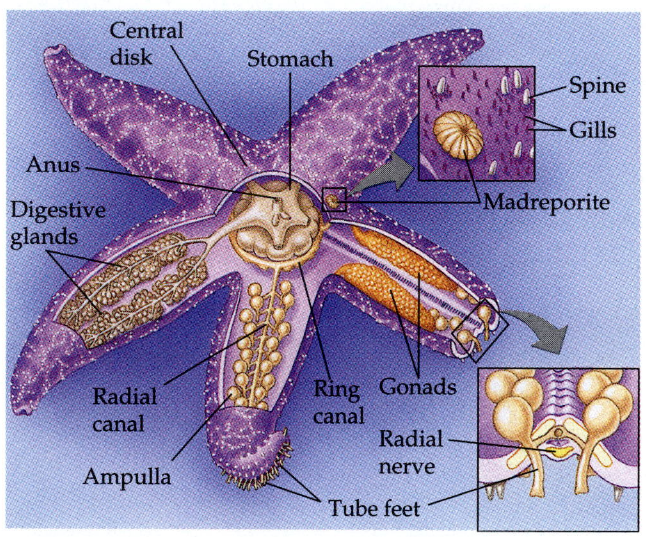

FIGURE 33.37 ▪ Anatomy of a sea star. The surface of a sea star is covered by spines that help defend against predators and by small gills for gas exchange. Internal organs are suspended by mesenteries in a well-developed coelom. A short digestive tract runs from the mouth on the bottom of the central disk to the anus on the top of the disk. Digestive glands secrete digestive juices and aid in the absorption and storage of nutrients. The central disk has a nerve ring and nerve cords radiating from the ring into the arms. The water vascular system consists of a ring canal in the central disk and five radial canals, each running the length of an arm in a groove. The system connects to the outside by way of the sievelike madreporite. Branching from each radial canal are hundreds of hollow, muscular tube feet filled with fluid continuous with the rest of the water vascular system. Each tube foot consists of a bulblike ampulla and suckered podium (foot portion). The podium expands and extends to contact the substratum when the ampulla squeezes water into it. The podium shortens and bends when muscles in its wall contract, forcing water back into the ampulla.

Class Echinoidea

Sea urchins and sand dollars have no arms, but they do have five rows of tube feet that function in slow movement (FIGURE 33.36d). Sea urchins also have muscles that pivot their long spines, which aids in moving. The mouth of an urchin is ringed by complex jawlike structures adapted for eating seaweeds and other food. Sea urchins are roughly spherical in shape, whereas sand dollars are flattened and disk-shaped.

Class Crinoidea

Sea lilies live attached to the substratum by stalks; feather stars crawl about by using their long, flexible arms. As a group, crinoids use their arms in suspension-feeding (FIGURE 33.36e). The arms circle the mouth, which is directed upward, away from the substratum. Crinoidea is an ancient class whose evolution has been very conservative; fossilized sea lilies some 500 million years old could pass for modern members of the class.

Class Holothuroidea

On casual inspection, sea cucumbers do not look much like other echinoderms (FIGURE 33.36f). They lack spines, and the hard endoskeleton is much reduced. Sea cucumbers are elongated in the oral-aboral axis, giving them the shape for which they are named and further disguising their relationship to sea stars and sea urchins. Closer examination, however, reveals five rows of tube feet, part of the water vascular system found only in echinoderms. Some of the tube feet around the mouth are developed as feeding tentacles.

Phylum Chordata: The chordates include two invertebrate subphyla and all vertebrates

33.1 This phylum, our own, consists of two subphyla of invertebrate animals plus the subphylum Vertebrata, the animals with backbones. Grouping the chordates with echinoderms as deuterostomes on the basis of similarities in early embryonic development is not meant to imply that one phylum evolved from the other. Chordates and echinoderms have existed as distinct phyla for at least half a billion years; if the developmental similarities stem from shared ancestry, then the evolutionary paths of the two phyla must have diverged very early. We will trace the phylogeny of chordates in Chapter 34, focusing on the history of vertebrates.

▪ ▪ ▪

TABLE 33.7 summarizes the animal phyla we have discussed in this chapter.

33.1

Table 33.7 ■ Animal Phyla

CATEGORY	PHYLA	DESCRIPTION OF PHYLA
Kingdom Animalia (Metazoa) Parazoa	Porifera (sponges)	Choanocytes (collar cells—unique flagellated cells that ingest bacteria and tiny food particles); cells tend to be totipotent (retain zygote's potential to form the whole animal)
Eumetazoa Radiata	Cnidaria (hydras, jellies, sea anemones, corals)	Unique stinging structures (cnidae), each housed in a specialized cell (cnidocyte); gastrovascular cavity (incomplete digestive tract with a mouth but no anus)
	Ctenophora (comb jellies)	Colloblasts (adhesive structures) for prey capture; eight rows of comblike ciliary plates; gastrovascular cavity
Bilateria Acoelomates	Platyhelminthes (flatworms)	Dorsoventrally flattened, unsegmented worms; gastrovascular cavity or no digestive tract
Pseudocoelomates	Rotifera (rotifers)	Jaws in pharynx structures (trophi); head with a ciliated crown (corona); no circulatory system
	Nematoda (roundworms)	Cylindrical, unsegmented worms with tapered ends; body wall has only longitudinal muscles; no circulatory system
Coelomates Protostomes	Nemertea (proboscis worms) (phylogenetic position uncertain)	Unique anterior proboscis surrounded by fluid-filled cavity (rhynchocoel); complete digestive tract (mouth and anus); circulatory system with closed vessels
	Lophophorates: Bryozoa, Brachiopoda, Phoronida (phylogenetic position uncertain)	Lophophore (feeding structure bearing ciliated tentacles)
	Mollusca (clams, snails, squids)	Three main body parts (muscular foot, visceral mass, mantle); coelom reduced; main body cavity is a hemocoel
	Annelida (segmented worms)	Body wall and internal organs (except digestive tract) segmented
	Arthropoda (crustaceans, insects, spiders)	Segmented body, jointed appendages, exoskeleton from ectoderm
Deuterostomes	Echinodermata (sea stars, sea urchins)	Secondary radial symmetry (larvae bilateral; adults radial); unique water vascular system; endoskeleton from mesoderm
	Chordata (lancelets, tunicates, vertebrates)	Notochord; dorsal hollow nerve cord; pharyngeal slits; muscular postanal tail

CHAPTER REVIEW

REVIEW OF KEY CONCEPTS

(with page numbers and key figures)

THE PARAZOA

- **Phylum Porifera: Sponges are sessile with porous bodies and**
33.1 **choanocytes (pp. 599–601, FIGURE 33.1)** Sponges lack tissues and organs. They filter-feed by drawing water through pores; choanocytes (flagellated collar cells) ingest bacteria and particulate food suspended in the water.

THE RADIATA

- **Phylum Cnidaria: Cnidarians have radial symmetry, a gastrovas-**
33.1 **cular cavity, and cnidocytes (pp. 601–603, FIGURE 33.5)** Cnidarians are mainly marine carnivores possessing tentacles armed with cnidocytes (cells containing evertible capsules) that aid in defense and the capture of prey. Two body forms are sessile polyps and floating medusas. The digestive tract (gastrovascular cavity) is incomplete (has a single opening, the mouth, that also functions as an anus). Class Hydrozoa usually alternates polyp and medusa forms, although the polyp is more conspicuous. In Class Scyphozoa, jellies (medusas) are the prevalent forms of the life cycle. Class Anthozoa contains the sea anemones and corals, which occur only as polyps.

- **Phylum Ctenophora: Comb jellies possess rows of ciliary plates and adhesive colloblasts (pp. 603–604, FIGURE 33.7)**

THE ACOELOMATES

- **Phylum Platyhelminthes: Flatworms are dorsoventrally flattened**
33.1 **acoelomates (pp. 604–606, FIGURE 33.8)** Most flatworms are ribbonlike animals with a gastrovascular cavity. Class Turbellaria is made up of mostly free-living, primarily marine species. Members of the classes Trematoda and Monogenea live as parasites in or on animals. Class Cestoidea (tapeworms), all parasites, lack a digestive tract.

THE PSEUDOCOELOMATES

- **Phylum Rotifera: Rotifers have jaws and a crown of cilia (pp. 606–607, FIGURE 33.12)** Found mainly in freshwater, rotifers have a complete digestive tract and a pseudocoelom. Many species are parthenogenetic.

- **Phylum Nematoda: Roundworms are unsegmented and cylindrical**
33.1 **with tapered ends (p. 607, FIGURE 33.13)** Among the most widespread and numerous animals, nematodes inhabit most aquatic habitats. Some species are important parasites of animals and plants.

THE COELOMATES: PROTOSTOMES

- **Phylum Nemertea: The phylogenetic position of proboscis worms is uncertain (p. 608, FIGURE 33.14)** The proboscis worms have a unique retractable tube (proboscis) used for defense and prey capture. A fluid-filled cavity surrounds the proboscis. Molecular and anatomical evidence indicates a relationship with protostome coelomates.

- **The lophophorate phyla: Bryozoans, phoronids, and brachiopods have ciliated tentacles around their mouths (pp. 608–609, FIGURE 33.15)** The phyla Bryozoa, Phoronida, and Brachiopoda are grouped together on the basis of their lophophores, horseshoe-shaped, suspension-feeding organs bearing ciliated tentacles. Their protostome-deuterostome affinity is uncertain, although molecular systematics places the phyla closer to protostomes.

- **Phylum Mollusca: Mollusks have a muscular foot, a visceral mass,**
33.1 **and a mantle (pp. 609–612, FIGURE 33.16)** Class Polyplacophora is composed of the chitons, oval-shaped marine animals encased in an armor of dorsal plates. Most members of Class Gastropoda, the snails and their relatives, possess a single, spiraled shell; sea slugs lack a shell; embryonic torsion of the body is a distinctive characteristic. Class Bivalvia (clams and their relatives) have hinged shells divided into two halves. Most bivalves are sedentary suspension-feeders that use gills for both gas exchange and feeding. Class Cephalopoda includes squids and octopuses, carnivores with beaklike jaws surrounded by tentacles of the modified foot. Consistent with their active lifestyles, the shell is either reduced or absent in most genera.

- **Phylum Annelida: Annelids are segmented worms (pp. 612–614,**
33.1 **FIGURE 33.24)** Body segmentation characterizes the annelid worms. Their progressive, wavelike locomotion results from alternating contractions of circular and longitudinal muscles against a fluid-filled, compartmentalized coelom. Class Oligochaeta includes earthworms and various aquatic species. Members of Class Polychaeta possess paddlelike parapodia that function as gills and aid in locomotion. Class Hirudinea consists of the leeches.

- **Phylum Arthropoda: Arthropods have regional segmentation,**
33.1 **jointed appendages, and an exoskeleton (pp. 614–624, FIGURE 33.25)** Phylum Arthropoda has more known species than all other phyla combined. Chelicerates, arthropods with pincer or fanglike feeding appendages, include Class Arachnida (spiders, ticks, scorpions, and mites). A traditional classification scheme groups the insects (Class Insecta), centipedes (Class Chilopoda), and millipedes (Class Diplopoda) as uniramians—all with one pair of antennae and unbranched (uniramous) appendages. Crustaceans (lobsters, crayfish, crabs, shrimps, barnacles) are primarily aquatic arthropods with two pairs of antennae and branched appendages. In terms of species diversity, insects outnumber all other forms of life combined. New data indicate that crustaceans and insects are more closely related than either is to the millipedes and centipedes. Insect success is partly due to the evolution of flight. In many insect species, young go through complete or incomplete metamorphosis.

THE COELOMATES: DEUTEROSTOMES

- **Phylum Echinodermata: Echinoderms have a water vascular sys-**
33.1 **tem and secondary radial symmetry (pp. 624–626, FIGURE 33.36)** Sea stars and their relatives make up six classes of the marine Phylum Echinodermata, radially symmetrical animals with a unique water vascular system ending in tube feet used for locomotion and feeding. A thin, bumpy, or spiny skin covers a calcareous endoskeleton.

- **Phylum Chordata: The chordates include two invertebrate sub-**
33.1 **phyla and all vertebrates (p. 626)**

SELF-QUIZ

1. What feature do nematodes, polychaetes, and mollusks have in common?
   a. All are pseudocoelomates.
   b. Each can reproduce by alternation of generations.
   c. All are protostomes.
   d. Each has a complete digestive system.
   e. Parasitic representatives are common.

2. Complete metamorphosis is advantageous for all of the following reasons *except*
   a. it permits a larval stage specialized for feeding and growth
   b. it permits an adult stage specialized for dispersal and reproduction
   c. pupae can suspend development and become dormant during periods of environmental stress

d. adults can be more mobile, improving the probability of locating a mate
e. feeding adults can disperse to new food resources, thus avoiding competition with larvae

3. Although adults of many species display radial symmetry, echinoderms are included with the bilateria because
   a. the embryo undergoes radial cleavage
   b. both endoderm and ectoderm are present
   c. gametes are produced by meiosis
   d. calcium carbonate is incorporated into the skeletal elements
   e. the digestive system is incomplete

4. Water movement through a sponge would follow what path?
   a. porocyte ⟶ spongocoel ⟶ osculum
   b. blastopore ⟶ gastrovascular cavity ⟶ protostome
   c. choanocyte ⟶ mesohyl ⟶ spongocoel
   d. porocyte ⟶ choanocyte ⟶ mesohyl
   e. colloblast ⟶ coelom ⟶ porocyte

5. Although a diverse group, all cnidaria are characterized by
   a. a gastrovascular cavity
   b. an alteration between a medusa and a polyp stage
   c. some degree of cephalization
   d. muscle tissue of mesodermal origin
   e. the complete absence of asexual reproduction

6. A land snail, a clam, and an octopus all share
   a. a mantle
   b. a radula
   c. gills
   d. embryonic torsion
   e. distinct cephalization

7. Which of the following is *not* a characteristic of most members of the phylum Annelida?
   a. hydrostatic skeleton
   b. segmentation
   c. metanephridia
   d. pseudocoelom
   e. closed circulatory system

8. Which of the following is *not* true of the chelicerates?
   a. They have antennae.
   b. Their body is divided into a cephalothorax and an abdomen.
   c. The horseshoe crab is one surviving marine member.
   d. They include ticks, scorpions, and spiders.
   e. Their anterior appendages are modified as pincers or fangs.

9. Which of the following combinations of phylum and description is *incorrect?*
   a. Echinodermata—bilateral and radial symmetry, coelom from archenteron
   b. Nematoda—roundworms, pseudocoelomate
   c. Cnidaria—radial symmetry, polyp and medusa body forms
   d. Platyhelminthes—flatworms, gastrovascular cavity, acoelomate
   e. Porifera—gastrovascular cavity, mouth from blastopore

10. Which of the following features distinguishes a member of Phylum Brachiopoda from a member of the molluscan class Bivalvia?
    a. bivalve shell
    b. bilateral symmetry
    c. lophophore
    d. complete digestive tract
    e. segmentation

CHALLENGE QUESTIONS

1. A marine biologist has dredged up a previously unknown animal species from the seafloor. Describe some of the important characteristics the biologist should look for to determine the animal phylum to which the creature should be assigned.
2. Compare and contrast spiders and insects.

SCIENCE, TECHNOLOGY, AND SOCIETY

Coral reefs harbor a greater diversity of animals than any other environment in the sea. Australia's Great Barrier Reef has been protected as a marine reserve and is a mecca for scientists and nature enthusiasts. In other parts of the world, such as Indonesia and the Philippines, coral reefs are in danger. Many reefs have been depleted of fish, and runoff from the shore has covered coral heads with sediment. Nearly all the changes in the reefs can be traced back to human activities. What kinds of activities do you think might be contributing to the decline of the reefs? What are some reasons to be concerned about this decline? What might be done on a local and global scale to halt the decline?

FURTHER READING

Brusca, R. G., and G. J. Brusca. *Invertebrates.* Sunderland, MA: Sinauer Associates, 1990. Presents an evolutionary approach to the animal kingdom.

Craig, C. L. "Webs of Deceit." *Natural History,* March 1995. Some spiderwebs actually *attract* prey.

Konop, D. "New Species of Ocean Bottom Animal Discovered Living on Frozen Natural Gas." *Science Journal* (Pennsylvania State University), Fall 1997. Describes a newly discovered group of annelid worms that lives in and on deep-sea methane ice mounds.

Marden, J. H. "Flying Lessons from a Flightless Insect." *Natural History,* February 1995. Did insects use their wings to swim before they could fly?

Nielsen, C. "Sequences Lead to Tree of Worms." *Nature,* March 5, 1998. Explains how nematode phylogeny was elucidated by ribosomal RNA sequencing studies.

Osorio, D., J. P. Bacon, and P. M. Whitington. "The Evolution of Arthropod Nervous Systems." *American Scientist,* May–June 1997. Discusses the remarkable similarity in the basic plan of an organ system in diverse groups of arthropods.

Pain, S. "Foul Medicine." *New Scientist,* February 7, 1998. Discusses prospects of new anticancer drugs from bryozoans that grow on ship hulls.

Svitil, K. A. "Bringing Tube Worms Back Alive." *Discover,* May 1998. Describes new cultivation techniques for deep-sea invertebrates.

Xerces Society. *Wings, Essays on Invertebrate Conservation.* Fall 1996. Six brief essays explore the ecological roles of aquatic insects.

WEB LINKS

Visit the special edition of *The Biology Place* for BIOLOGY, Fifth Edition, at http://www.biology.com/campbell. Go to Chapter 33 for online resources, including learning activities, practice exams, and links to the following web sites:

"Porifera Web Page"
The goals of the Porifera page are to promote sponge research, to enhance communication among sponge biologists, and to form a species, specimen, and bibliographic database on sponges.

"USDA Nematology Lab Homepage"
Nematodes are important agriculturally both as plant parasites and as beneficial organisms in insect control.

"BIODIDAC–Zoology"
This resource, primarily for French-speaking biology teachers, contains an impressive collection of pictures and diagrams of many invertebrates.

"Introduction to the Cnidaria"
Visit the section on Cnidaria at the University of California Museum of Paleontology.

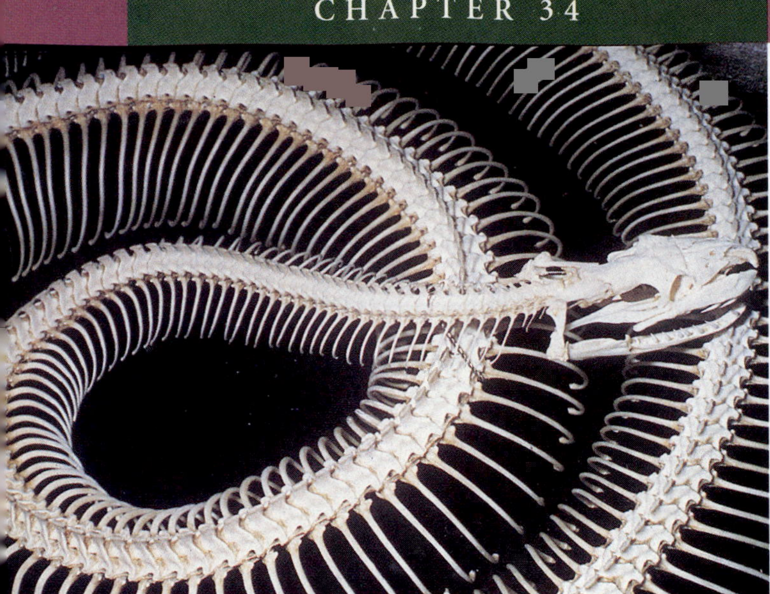

VERTEBRATE EVOLUTION AND DIVERSITY

Most of us are curious about our genealogies. On the personal level, we wonder about our family ancestry. As biology students, we are interested in retracing human ancestry within the broader scope of the evolutionary history of the entire animal kingdom. The questions we must ask are: What were our ancestors like? How are we related to other animals? What are our closest relatives? In this chapter, we trace the evolution of the vertebrates, the group that includes humans and their closest relatives. Mammals, birds, reptiles, amphibians, and the various classes of fishes are all classified as ***vertebrates****. They share many features unique to the vertebrates, including the cranium and backbone, a series of vertebrae for which the group is named. You can see these vertebrate hallmarks in the photograph of a snake skeleton that opens this chapter. Our first step in tracking the vertebrate genealogy is to determine where vertebrates fit in the animal kingdom.*

INVERTEBRATE CHORDATES AND THE ORIGIN OF VERTEBRATES

Based on certain similarities in early embryonic development, chordates are grouped as deuterostomes along with the echinoderms (see Chapter 32). The vertebrates make up one subphylum within the phylum Chordata. **Chordates** also include two subphyla of invertebrates, the urochordates and the cephalochordates.

Four anatomical features characterize Phylum Chordata

34.1 Although chordates vary widely in appearance, they are distinguished as a phylum by the presence of four anatomical structures that appear at some point during the animal's lifetime, often only during embryonic development. These four chordate characteristics are: a notochord; a dorsal, hollow nerve cord; pharyngeal slits; and a muscular, postanal tail (FIGURE 34.1).

1. Notochord

Chordates are named for a skeletal structure, the notochord, present in all chordate embryos. The **notochord** is a longitudinal, flexible rod located between the digestive tube and the nerve cord. Composed of large, fluid-filled cells encased in fairly stiff, fibrous tissue, it provides skeletal support throughout most of the length of the animal. The notochord persists in the adult of some invertebrate chordates and primitive vertebrates. However, in most vertebrates a more complex, jointed skeleton develops, and the adult retains only remnants of the embryonic notochord—as gelatinous material of the disks between the vertebrae of humans, for example.

2. *Dorsal, Hollow Nerve Cord*

The nerve cord of a chordate embryo develops from a plate of ectoderm that rolls into a tube located dorsal to the notochord. The result is a dorsal, hollow nerve cord unique to chordates. Other animal phyla have solid nerve cords, usually ventrally located. The nerve cord of a chordate embryo develops into the central nervous system: the brain and spinal cord.

3. *Pharyngeal Slits*

The digestive tube of chordates extends from the mouth to the anus. The region just posterior to the mouth is the pharynx, which opens to the outside of the animal through several pairs of slits. These pharyngeal slits allow water that enters the mouth to exit without continuing through the entire digestive tract. The pharyngeal slits function as suspension-feeding devices in many invertebrate chordates. The slits and structures supporting them have become modified for gas exchange (in aquatic vertebrates), jaw support, hearing, and other functions during vertebrate evolution.

4. *Muscular, Postanal Tail*

Most chordates have a tail extending posterior to the anus. By contrast, most nonchordates have a digestive tract that extends nearly the whole length of the body. The chordate tail contains skeletal elements and muscles and provides much of the propulsive force in many aquatic species.

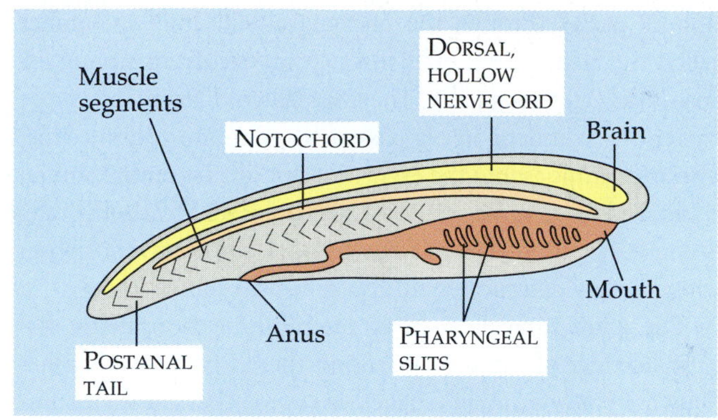

FIGURE 34.1 ▪ Chordate characteristics. All chordates possess the four trademarks of the phylum: a notochord; a dorsal, hollow nerve cord; pharyngeal slits; and a muscular, postanal tail.

Invertebrate chordates provide clues to the origin of vertebrates

Members of the two subphyla of invertebrate chordates, Urochordata and Cephalochordata, illustrate the chordate body plan in its most "stripped-down" versions—without the additional features that evolved in vertebrates. The study of urochordates and cephalochordates also provides clues to the origin of vertebrates.

Subphylum Urochordata

Urochordates are commonly called **tunicates.** Most tunicates are sessile marine animals that adhere to rocks, docks, and boats (FIGURE 34.2A). Others are planktonic. Some species are colonial. Seawater enters the animal through an incurrent

(a)

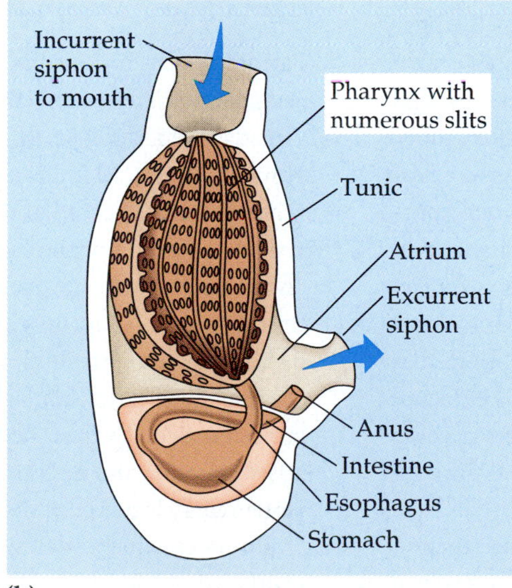

(b)

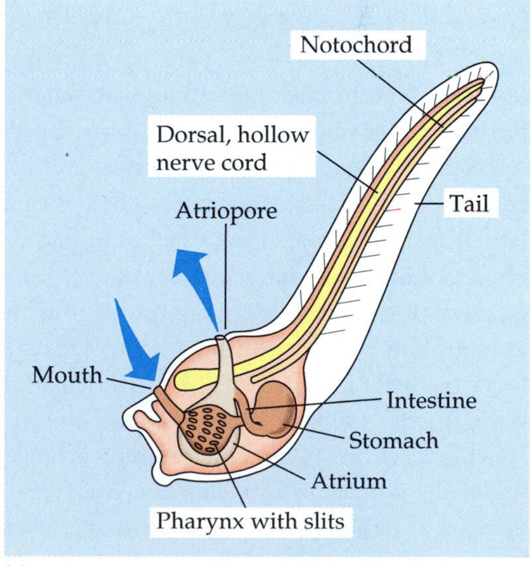

(c)

FIGURE 34.2 ▪ Subphylum Urochordata: a tunicate. (a) An adult tunicate, or sea squirt, is a sessile animal commonly arranged in a U shape (approximately life size). **(b)** In the adult, prominent pharyngeal slits function in suspension-feeding, but the other chordate characteristics are not obvious. **(c)** In the tunicate larva, which is a free-swimming suspension-feeder, the chordate characteristics are evident. Some urochordates lack the sessile adult stage, existing as free-swimming, larva-like forms throughout life.

siphon, passes through the pharyngeal slits into a chamber called the atrium, and exits through an excurrent siphon, or atriopore (FIGURE 34.2b). The food filtered from this water current by a mucus net is passed by cilia into the intestine. The anus empties into the excurrent siphon. The entire animal is cloaked in a tunic made of a celluloselike carbohydrate. Because they shoot a jet of water through the excurrent siphon when molested, tunicates are also called sea squirts.

The adult tunicate scarcely resembles a chordate. It displays no trace of a notochord, nor is there a nerve cord or tail. Only the pharyngeal slits suggest a link to other chordates. But all four chordate trademarks are manifest in the larval form of some groups of tunicates (FIGURE 34.2c). The larva swims until it attaches by its head to a surface and undergoes metamorphosis, during which most of its chordate characteristics disappear. In at least some tunicates, if a single gene (called *Manx* for the tailless cat breed) is turned off during development, a normally tailed tunicate larva will be tailless. Expression of *Manx* is also essential for notochord and dorsal nerve cord development. Once again, we are led to the hypothesis that a relatively small number of genes regulating development may have been associated with the evolution of some basic features of an animal body plan—in this case, the suite of characteristics that defines all chordates.

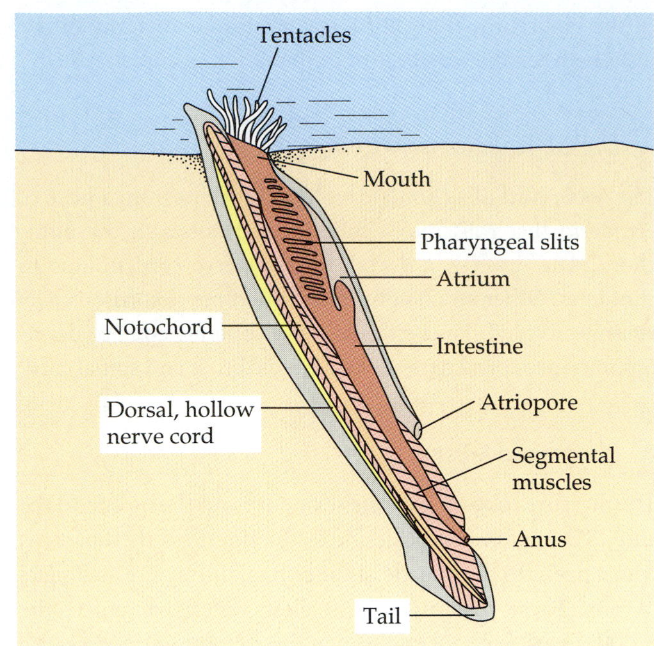

FIGURE 34.3 ▪ Subphylum Cephalochordata: the lancelet *Branchiostoma.* This small invertebrate displays all four chordate characteristics. The pharyngeal slits function in suspension-feeding. Water passes into the pharynx and through slits into the atrium, a chamber that vents to the outside via the atriopore. Food particles trapped by mucus are swept by cilia into the digestive tract. The muscle segments produce the sinusoidal swimming of these animals.

Subphylum Cephalochordata

Known as **lancelets** because of their bladelike shape, **cephalochordates** closely resemble the idealized chordate in FIGURE 34.1. The notochord; dorsal, hollow nerve cord; numerous gill slits; and postanal tail all persist into the adult stage (FIGURE 34.3). Small marine animals only a few centimeters long, lancelets wriggle backward into the sand, leaving only their anterior end exposed. A mucous net secreted across the pharyngeal slits traps tiny food particles from seawater drawn into the mouth by ciliary pumping. The water exits through the slits, and the trapped food passes down the digestive tube.

A lancelet frequently leaves its burrow and swims to a new location. Though feeble swimmers, these invertebrate chordates display, in simple form, the swimming mechanism of fishes. Coordinated contraction of muscles serially arranged like rows of chevrons («««) along the sides of the notochord flexes the notochord, producing side-to-side undulations that thrust the body forward. This serial musculature is evidence of the lancelet's segmentation. The muscle segments develop from blocks of mesoderm called **somites** arranged along each side of the notochord of a chordate embryo. Chordates are segmented animals (FIGURE 34.4). As we discussed in Chapter 33, a popular hypothesis that chordate segmentation evolved independently of segmentation in annelids and arthropods is currently being challenged. Molecular evidence indicates that segmentation may have evolved in the earliest bilateral animals, well before the lineage that led to annelids and arthropods diverged from the lineage that led to chordates. It seems likely that the set of regulatory genes that directs the development of segmentation in chordates, annelids, and arthropods was present in the ancestor of all bilateral animals.

The Relationship Between Invertebrate Chordates and Vertebrates

Paleontologists have found fossilized invertebrates resembling cephalochordates in the Burgess Shale of British Columbia, Canada (see FIGURE 32.7). Those fossils are about 545 million years old, about 50 million years older than the oldest known vertebrates. The record of the rocks is too incomplete for us to retrace the origin of the earliest vertebrates from invertebrate ancestors, but we can propose reasonable hypotheses about this evolution based on comparative anatomy and embryology. Many biologists think that the ancestor of vertebrates was a suspension-feeder similar to a cephalochordate, with all four basic chordate features. Recent research by molecular systematists supports the hypothesis that cephalochordates are the vertebrates' closest relatives.

Both cephalochordates and vertebrates may have evolved from a common sessile ancestor by **paedogenesis**, the precocious development of sexual maturity in a larva. Notice that a cephalochordate appears more akin to a urochordate larva than to an adult urochordate (compare FIGURES 34.2 and 34.3). Changes in genes controlling development can alter the

(a)

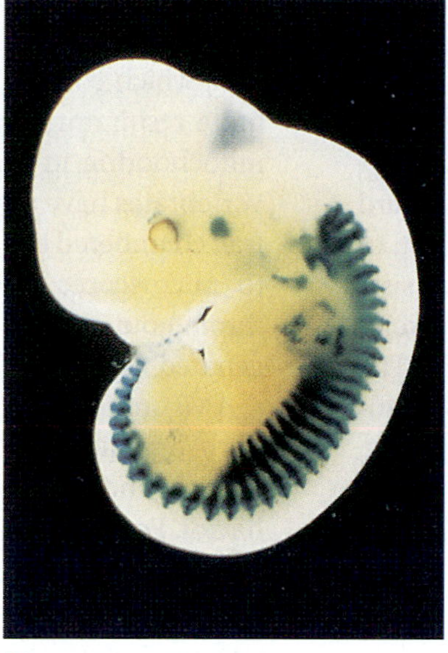
(b)

(c)

FIGURE 34.4 ▪ Chordate segmentation. Segmentation evolved in the phyla Annelida, Arthropoda, and Chordata. Here are three examples of the chordate version. **(a)** The segmental muscles of cephalochordates (lancelets) (shown here) and vertebrates develop from segmental blocks of mesoderm called somites. **(b)** Somites are also visible in this mouse embryo. **(c)** Much of the segmental anatomy so apparent in vertebrate embryos is disguised in the adult. Evidence of segmentation persists in the serial arrangement of vertebrae and in patterns of certain muscle groups, such as the abdominal muscles.

timing of developmental events, such as the maturation of gonads. Perhaps such a change occurred in the ancestor of cephalochordates and vertebrates, causing the gonads to mature in swimming larvae before the onset of metamorphosis to the sessile adult form. If reproducing larvae were very successful, natural selection may have reinforced paedogenesis and eliminated metamorphosis.

Although cephalochordates and vertebrates probably evolved from a common chordate ancestor, they diverged about a half-billion years ago and differ in many important ways.

INTRODUCTION TO THE VERTEBRATES

Vertebrates retain the primitive chordate characteristics but have additional specializations, shared derived features (synapomorphies; see Chapter 25) that distinguish the subphylum from invertebrate chordates. Many of the features that distinguish vertebrates are associated with larger size and an active lifestyle.

Neural crest, pronounced cephalization, a vertebral column, and a closed circulatory system characterize Subphylum Vertebrata

The dorsal, hollow nerve cord found in all chordates develops when the edges of an ectodermal plate on the surface of the embryo roll together to form the neural tube. In vertebrates, a group of embryonic cells called the **neural crest** forms near the dorsal margins of the closing neural tube (FIGURE 34.5). A distinctive feature of vertebrates, the neural crest contributes to the formation of certain skeletal elements, such as some of the bones and cartilage of the cranium (braincase), and many other structures that distinguish vertebrates from other chordates.

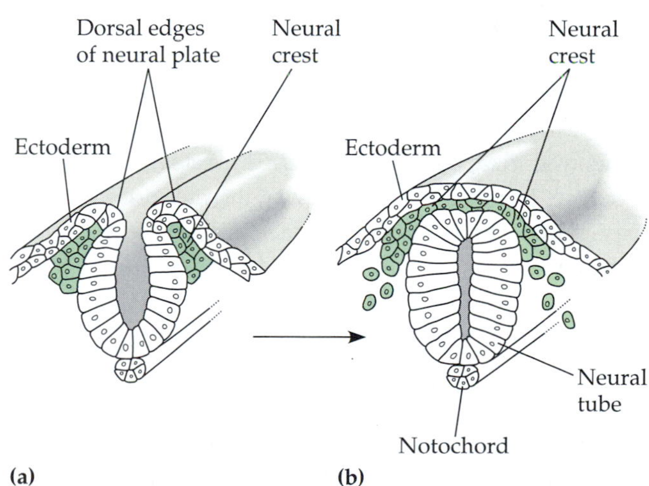

FIGURE 34.5 ▪ The neural crest, embryonic source of many vertebrate synapomorphies. (a) The neural crest consists of bilateral bands of cells near the margins of the embryonic folds that meet to form the dorsal, hollow nerve cord (neural tube). **(b)** Cells from the neural crest migrate to distant sites in the embryo, where they give rise to some of the anatomical structures unique to vertebrates, including some of the bones and cartilage of the skull.

The vertebrate cranium and brain (which is the enlarged anterior end of the dorsal, hollow nerve cord), along with the eyes, ears, and nose, are evidence of an important evolutionary feature of vertebrates—a high degree of cephalization, the concentration of sensory and neural equipment in the head (see Chapter 32).

The cranium and vertebral column, surrounding and protecting the nerve cord, are parts of the vertebrate axial skeleton, the main support structure for the axis, or central trunk, of the body. The axial skeleton helps make large body size and strong, fast movement possible. The axial skeleton of most vertebrates also includes ribs, which anchor muscles and protect internal organs. Vertebrates also have an appendicular skeleton, supporting their two pairs of appendages (fins, legs, or arms).

The vertebrate endoskeleton is made of bone or more flexible cartilage, or most commonly some combination of these two materials. Although the skeleton consists mostly of a nonliving matrix, living cells within the skeleton secrete and maintain the materials of the matrix. The endoskeleton of a vertebrate can grow with the animal, unlike the nonliving exoskeleton of arthropods, which must be periodically shed and rebuilt as the animal grows.

When vertebrates move in pursuit of food or when escaping predators, they regenerate their ATP supply mainly by cellular respiration, which consumes oxygen. Adaptations of the vertebrate respiratory and circulatory systems support busy mitochondria in muscle cells and other active tissues. Vertebrates have a closed circulatory system, with a ventral, chambered heart that pumps blood through arteries to microscopic vessels called capillaries that branch throughout every tissue in the body. The blood is oxygenated as it passes through capillaries in gills or lungs.

A vigorous lifestyle also requires a relatively large supply of organic fuel. Vertebrate adaptations for feeding, digestion, and nutrient absorption help support active behavior. For example, muscles in the walls of the digestive tract propel food from organ to organ along the tract. These are all examples of vertebrate form and function having their historical basis in the transition from a relatively sedentary lifestyle to a more active one.

Overview of vertebrate diversity

34.1 The taxonomic scheme we have adopted recognizes two extant superclasses of the subphylum Vertebrata (TABLE 34.1, p. 636). Members of Superclass Agnatha, the hagfishes and lampreys, lack jaws. The other superclass, the Gnathostomata, includes six classes of jawed vertebrates: Class Chondrichthyes (cartilaginous fishes, the sharks and rays); Class Osteichthyes (bony fishes); Amphibia (frogs and salamanders), Reptilia (reptiles), Aves (birds), and Mammalia (mammals). Amphibia, Reptilia, Aves, and Mammalia are collectively called **tetrapods** (Gr. *tetra,* "four," and *pod,* "foot") because most animals in these classes have two pairs of limbs that support them on land. Reptiles, birds, and mammals have additional terrestrial adaptations that distinguish them from amphibians. One of these is the **amniotic egg**, a shelled, water-retaining egg. The amniotic egg functions as a "self-contained pond" that enables these vertebrates to complete their life cycles on land. Although most mammals don't lay eggs, they retain other key features of the amniotic condition. In recognition of this important evolutionary breakthrough, reptiles, birds, and mammals are collectively called **amniotes**.

In the following pages, we will survey the diversity of organisms in each vertebrate class, highlight some of the major evolutionary trends among the vertebrates, and examine some of the taxonomic debate about the vertebrate classes. The survey also includes a brief description of placoderms, an extinct class of early jawed fishes. As a prelude to learning about the various groups of vertebrates, examine the evolutionary tree in FIGURE 34.6. Throughout the chapter you will find smaller versions of this tree as orientation diagrams, with particular branches highlighted to help you keep track of where each class fits in the vertebrate genealogy.

SUPERCLASS AGNATHA: JAWLESS VERTEBRATES

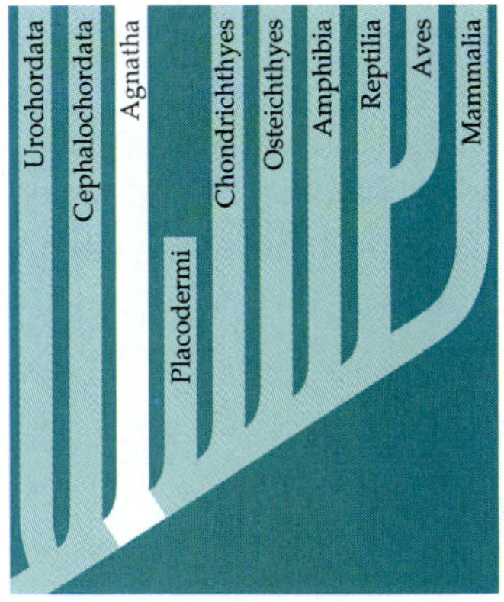

Vertebrates probably originated in the late Precambrian or early Cambrian. The oldest vertebrate fossils are jawless creatures, members of **Superclass Agnatha** ("without jaws"). Traces of these early vertebrates are found in Cambrian strata, but most date back to the Ordovician and Silurian periods, about 400 to 500 million years ago.

Superclass Agnatha includes extinct fishlike animals called **ostracoderms** ("shell skinned") that were encased in an armor of bony plates. These and other early agnathans were generally small, less than 50 cm in length. Most lacked paired fins and apparently were bottom-dwellers that wiggled along streambeds or the seafloor, but there were also some more active species with paired fins. Their mouths were circular or slitlike openings that lacked jaws. Most agnathans were probably mud-suckers or suspension-feeders that took in sediments or suspended organic debris through their mouths and then passed it through the gill slits, where food was trapped. Thus, the pharyngeal apparatus retained the primitive feeding function, although gills in agnathans were probably also the major

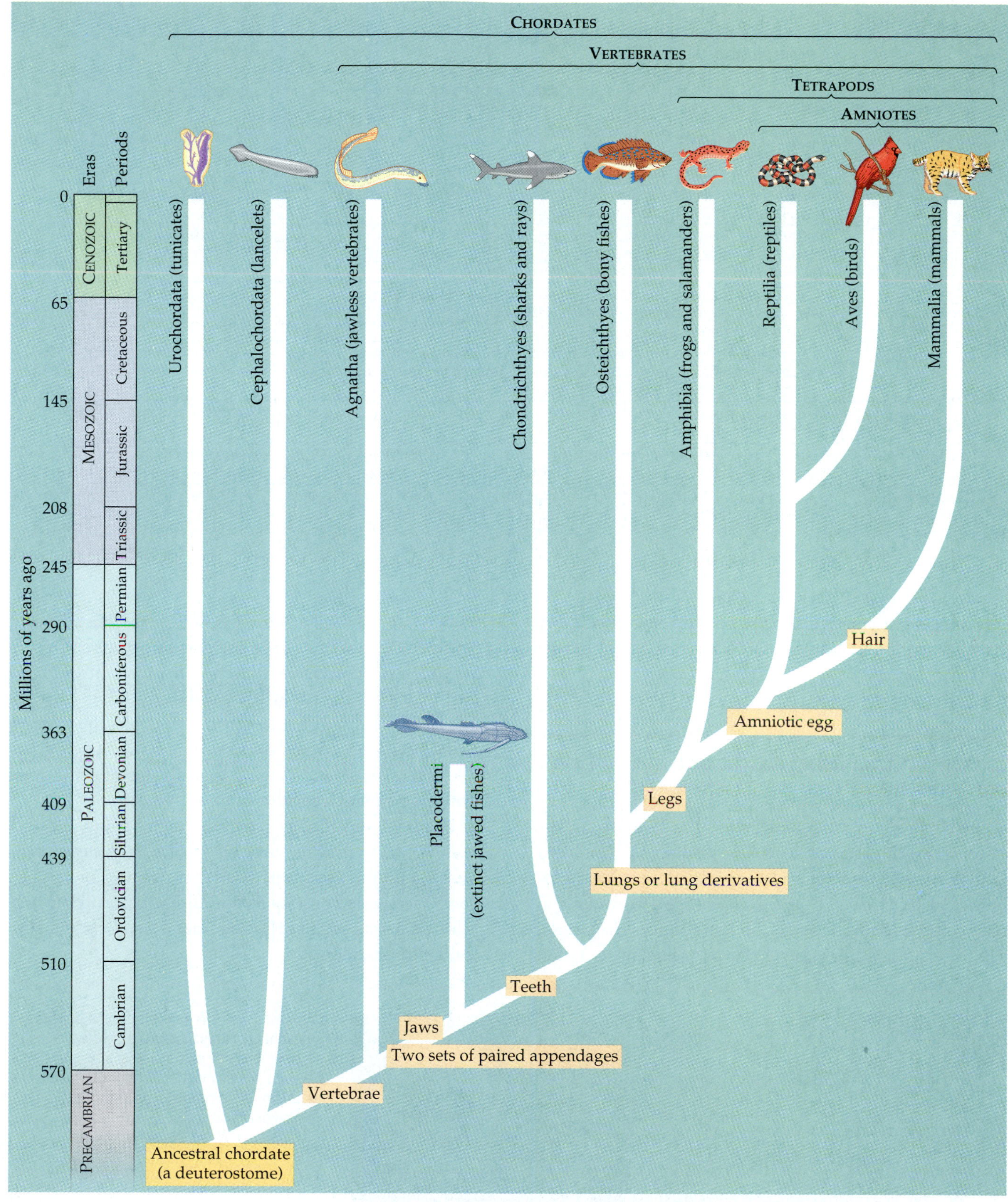

34.1 **FIGURE 34.6 ▪ One hypothesis for the evolutionary relationships among the chordates.** The characteristics along the diagonal lines are derived features (synapomorphies) that are shared by the vertebrate classes beyond that point.

 34.1

Table 34.1 ▪ The Extant Groups of Vertebrates

	MAIN CHARACTERISTICS	EXAMPLES
Superclass Agnatha	Jawless vertebrates: cartilaginous skeleton; rasping tongue; notochord persists throughout life; marine and freshwater; living species lack paired appendages.	Lampreys, hagfishes
Class Myxini	Marine scavengers; mouth surrounded by short tentacles; no larval stage.	Hagfishes
Class Cephalaspidomorphi	Marine and freshwater; mouth surrounded by adhesive sucker; larvae (ammocoetes) are suspension-feeders; adults parasitic or nonfeeding.	Lampreys
Superclass Gnathostomata	Vertebrates with hinged jaws; notochord largely or completely replaced by vertebrae in the adult of most species; paired appendages.	All extant vertebrates except lampreys and hagfishes
Class Chondrichthyes	Cartilaginous fishes: cartilaginous skeleton; jaws; respiration through gills; internal fertilization, may lay eggs or give birth to young; acute senses, including lateral line system.	Sharks, skates, rays, chimaeras
Class Osteichthyes	Bony fishes: bony skeletons and jaws; most species have external fertilization and lay large numbers of eggs; respiration mainly through gills; many have a swim bladder; marine and freshwater.	Bass, trout, perch, tuna
Class Amphibia	Appendages adapted for moving on land (tetrapod condition); aquatic larval stage metamorphosing into terrestrial adult (many species); may lay eggs or give birth to young; respiration through lungs and/or skin.	Salamanders, newts, frogs, caecilians
Class Reptilia	Terrestrial tetrapods with scaly skin: respiration via lungs; lay amniotic shelled eggs or give birth to young.	Snakes, lizards, turtles, crocodiles
Class Aves	Tetrapods with feathers: forelimbs modified as wings; respiration through lungs; endothermic; internal fertilization; shelled amniotic eggs; acute vision.	Owls, sparrows, penguins, eagles
Class Mammalia	Tetrapods with young nourished from mammary glands of females; hair; diaphragm that ventilates lungs; endothermic; amniotic sac; most give birth to young.	Monotremes (such as platypuses); marsupials (such as kangaroos); eutherians (such as rodents)

sites of gas exchange. The ostracoderms and most other agnathan groups declined and finally disappeared during the Devonian period.

Lampreys and hagfishes are the only extant agnathans

About 60 species of jawless vertebrates are alive today in Class Myxini (hagfishes) and Class Cephalaspidomorphi (lampreys) (FIGURE 34.7). In common with many extinct agnathans, lampreys and hagfishes lack paired appendages. In contrast to many ostracoderms, however, they have no external armor. The eel-shaped sea lamprey feeds by clamping its round mouth onto the flank of a live fish, using a rasping tongue to penetrate the skin of its prey, and ingesting the prey's blood. Sea lampreys live as larvae for years in freshwater streams and then migrate to the sea or lakes as they mature into adults. The larva is a suspension-feeder that resembles the lancelets (cephalochordates). Some species of lampreys feed only as larvae. Following several years in streams, they attain sexual maturity, reproduce, and die within a few days.

Hagfishes superficially resemble lampreys, but they are mainly scavengers rather than blood-suckers or suspension-feeders, and their mouthparts are not adapted for rasping. Some species feed on sick or dead fish, whereas other hagfishes eat marine worms. Hagfishes lack a larval stage and live entirely in salt water.

FIGURE 34.7 ▪ A sea lamprey, a jawless vertebrate (Class Agnatha). Acting as both a predator and a parasite, a lamprey uses this rasping mouth (inset) to bore a hole in the side of a fish, living on the blood and other tissues of its host.

SUPERCLASS GNATHOSTOMATA I: THE FISHES

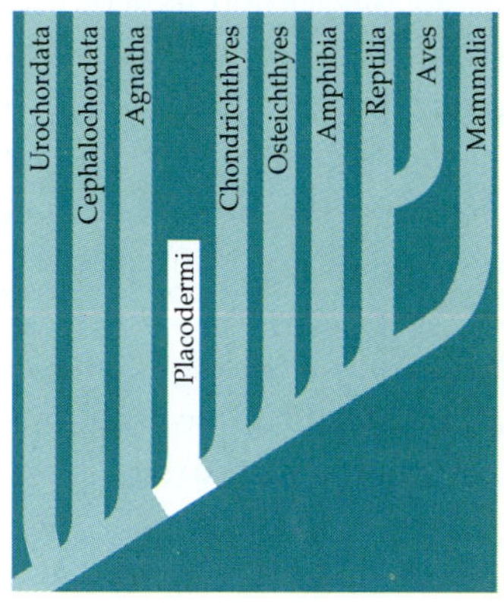

During the late Silurian and early Devonian periods, vertebrates with jaws, members of **Superclass Gnathostomata** ("jaw mouth") largely replaced the agnathans. The extant classes of fishes (Chondrichthyes and Osteichthyes) first appeared at this time, along with a group called the **placoderms** ("plate skinned") that has no living descendants. Jawed vertebrates also have two pairs of paired appendages, whereas agnathans either lacked paired appendages or had only a single pair. Judging from studies on extant vertebrates, differential expression of some of the *Hox* genes may have determined whether one or two pairs of paired appendages developed.

The largest placoderms were more than 10 m long, but most were less than 1 m long. With two pairs of paired fins and hinged jaws, many species were active predators, capable of chasing prey and biting off chunks of flesh. Thus, these two modifications of the early vertebrate body plan allowed the diversification of both lifestyles and nutrient sources.

Vertebrate jaws evolved from skeletal supports of the pharyngeal slits

The origin of jaws was a major adaptive event in early vertebrate phylogeny (see FIGURE 34.6). Vertebrate jaws are hinged and work up and down (dorsoventrally). Hinged jaws also evolved in arthropods, but these had a different origin than vertebrate jaws. Arthropod jaws are modified appendages that work from side to side.

Vertebrate jaws evolved by modification of the skeletal rods that had previously supported the anterior pharyngeal (gill) slits (FIGURE 34.8). The remaining gill slits, no longer required for suspension-feeding, remained as the major sites of respiratory gas exchange with the external environment. The origin of vertebrate jaws from these skeletal parts illustrates a general feature of evolutionary change: New adaptations usually evolve by the modification of existing structures. As a mechanism of adaptation, evolution is limited by the raw material with which it must work; evolution is generally more of a remodeling process than a creative one.

The Devonian period (about 360 to 400 million years ago) has been called the age of fishes, when the placoderms and another group of jawed fishes called acanthodians (placed in a separate class) radiated and many new forms evolved in both fresh and salt water. Placoderms and acanthodians dwindled and disappeared almost completely by the beginning of the Carboniferous period, about 360 million years ago. Ancestors of the placoderms and acanthodians may also have given rise to sharks (Class Chondrichthyes). Another group of large, jawed predators—the bony fishes (Class Osteichthyes)—diverged from the common ancestor some 425 to 450 million years ago. Sharks and bony fishes have teeth in their jaws, and the teeth are replaced regularly. In contrast, the placoderms and acanthodians had plates of external armor forming the edges of the jaws. Sharks and bony fishes are the reigning vertebrates over the watery two-thirds of the surface of Earth.

A cartilaginous endoskeleton reinforced by calcified granules is diagnostic of Class Chondrichthyes

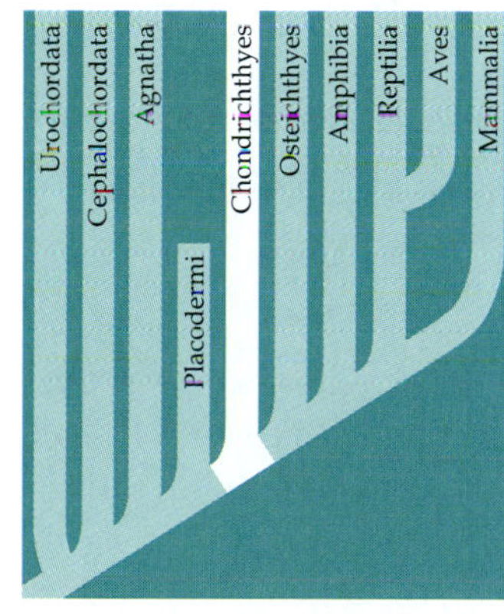

The vertebrates of **Class Chondrichthyes**, sharks and their relatives, are called cartilaginous fishes because they have relatively flexible endoskeletons made of cartilage rather than bone. However, in most species, parts of the skeleton are strengthened by calcified granules. There are about 750 extant species in this class. Jaws and paired fins are well developed in the cartilaginous fishes. The largest and most diverse subclass consists of the sharks and rays. A second subclass is composed of a few dozen species of unusual fishes called chimaeras, or ratfishes.

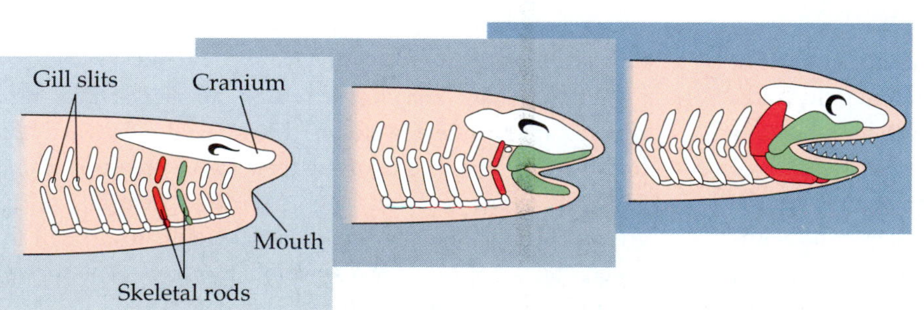

FIGURE 34.8 • The evolution of vertebrate jaws. The skeleton of the jaws and their supports evolved from two pairs of skeletal rods located between gill slits that were near the mouth. Pairs of rods anterior to those that formed the jaws were either lost or incorporated into the jaws.

(a)

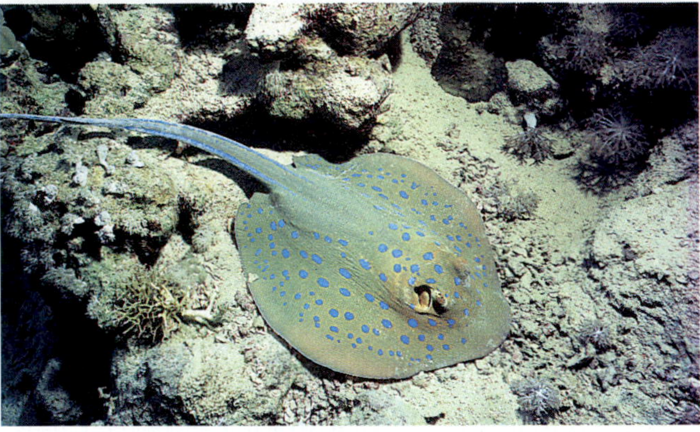

(b)

FIGURE 34.9 ▪ Cartilaginous fishes (Class Chondrichthyes). **(a)** Fast swimmers with acute senses and powerful jaws, sharks, such as this blacktip reef shark, are well adapted to their predatory way of life. Sharks have paired pectoral and pelvic fins. **(b)** Most rays, such as the blue-spotted stingray, are flattened bottom-dwellers that crush mollusks and crustaceans for food. Some other species cruise in open water, scooping food into their gaping mouths.

The cartilaginous skeleton of these fishes is a derived characteristic, not a primitive one; that is, the ancestors of Chondrichthyes had bony skeletons, and the cartilaginous skeleton characteristic of the class evolved secondarily. During the development of most vertebrates, the skeleton is first cartilaginous and then becomes bony (ossified) as hard calcium phosphate matrix replaces the leathery matrix of cartilage. Apparently, some modification in the developmental process of cartilaginous fishes prevents the replacement process.

Most sharks have streamlined bodies (FIGURE 34.9a). They are swift swimmers, but they do not maneuver very well. Powerful swimming muscles, especially in the caudal (tail) fin, propel them forward. The dorsal fins function mainly as stabilizers, and the paired pectoral (fore) and pelvic (hind) fins provide lift in the water. Although a shark gains additional buoyancy by storing a large amount of oil in its huge liver, the animal is still denser than water, and it sinks if it stops swimming. Continual swimming also ensures that water will flow into the mouth and out through the gills, where gas exchange occurs. However, some sharks and many skates and rays spend a good deal of time resting on the seafloor. When doing so, these fishes use muscles of the jaws and pharynx to pump water over the gills.

The largest sharks and rays are suspension-feeders that feed on plankton. Most sharks, however, are carnivores that swallow their prey whole or use their powerful jaws and sharp teeth to tear flesh from animals too large to swallow in one piece. Shark teeth probably evolved from the jagged scales that cover the abrasive skin. The digestive tract of many sharks is proportionately shorter than the digestive tube of many other vertebrates. Within the shark intestine is a **spiral valve**, a corkscrew-shaped ridge that increases surface area and prolongs the passage of food along the short digestive tract.

Acute senses are adaptations that go along with the active, carnivorous lifestyle of sharks. Sharks have sharp vision but cannot distinguish colors. The nostrils of sharks, like those of most fishes, open into dead-end cups. They function only for olfaction (smelling), not for breathing. Along with eyes and nostrils, the shark head also has a pair of regions in the skin that can detect electrical fields generated by the muscle contractions of nearby animals. Running the length of each flank of the shark is the **lateral line system**, a row of microscopic organs sensitive to changes in the surrounding water pressure. A primitive feature and present in most species of aquatic vertebrates, the lateral line system enables the animal to detect minor vibrations. Sharks and other fishes have no eardrums, structures in terrestrial vertebrates that convey sound waves traveling through air into the ear toward the auditory organs. Sound reaches the shark through water, and the animal's entire body conveys the sound to the hearing organs of the inner ear.

Shark eggs are fertilized internally. The male has a pair of claspers on its pelvic fins that transfer sperm into the reproductive tract of the female. Some species of sharks are **oviparous**; they lay eggs that hatch outside the mother's body. These sharks release their eggs after encasing them in protective coats. Other species are **ovoviviparous**; they retain the fertilized eggs in the oviduct. Nourished by the egg yolk, the embryos develop into young that are born after hatching within the uterus. A few species are **viviparous**; the young develop within the uterus, nourished prior to birth by nutrients received from the mother's blood through a placenta. The reproductive tract of the shark empties along with the excretory system and digestive tract into the **cloaca**, a common chamber that expels through a single vent.

Although rays are closely related to sharks, they have adopted a very different lifestyle. Most rays are flattened bottom-dwellers that feed by using their jaws to crush mollusks and crustaceans (FIGURE 34.9b). A ray's pectoral fins are greatly enlarged and used to propel the animal through the water. The tail of many rays is whiplike and, in some species, bears venomous barbs that function in defense.

A bony endoskeleton, operculum, and swim bladder are hallmarks of Class Osteichthyes

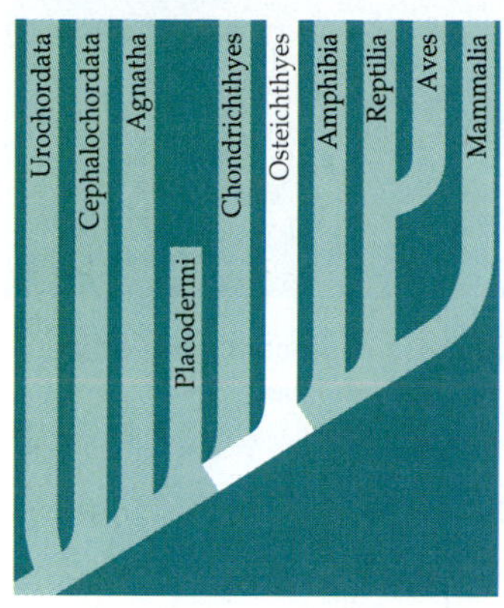

Of all vertebrate classes, bony fishes **(Class Osteichthyes)** are the most numerous, both in individuals and in species (about 30,000). Ranging in size from about 1 cm to more than 6 m long, bony fishes are abundant in the seas and in nearly every freshwater habitat (FIGURE 34.10).

Nearly all bony fishes have an endoskeleton with a hard matrix of calcium phosphate. The skin is often covered by flattened bony scales that differ in structure from the toothlike scales of sharks. Glands in the skin of a bony fish secrete a mucus that gives the animal its characteristic sliminess, an adaptation that reduces drag during swimming. In common with sharks, bony fishes have a lateral line system clearly evident as a row of tiny pits in the skin on either side of the body (FIGURE 34.11).

Bony fishes breathe by drawing water over four or five pairs of gills located in chambers covered by a protective flap called the **operculum**. Water is drawn into the mouth, through the pharynx, and out between the gills by movement of the operculum and contraction of muscles surrounding the gill chambers. This process enables a bony fish to breathe while stationary.

Another adaptation of most bony fishes not found in sharks is the **swim bladder**, an air sac that helps control the buoyancy of the fish. The transfer of gases between the swim bladder and the blood varies the inflation of the bladder and adjusts the density of the fish. Thus, many bony fishes, in contrast to most sharks, can conserve energy by remaining almost motionless.

Bony fishes are generally maneuverable swimmers, their flexible fins better for steering and propulsion than the stiffer

(a)

(b)

FIGURE 34.10 ▪ Bony fishes (Class Osteichthyes). These two species belong to the largest subclass of bony fishes, the ray-finned fishes. **(a)** A yellow trumpetfish. **(b)** A yellow perch.

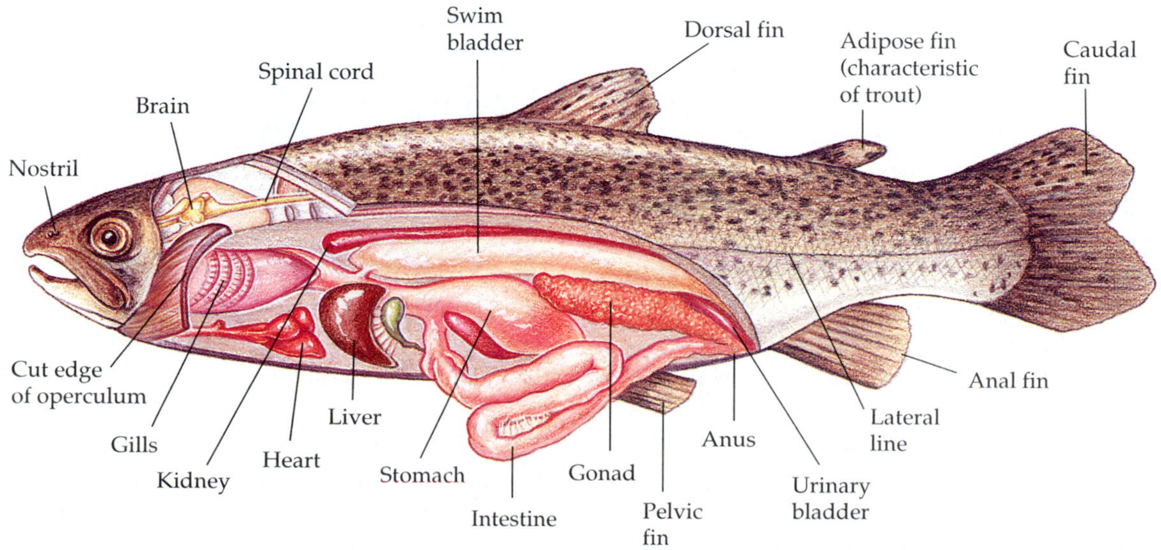

FIGURE 34.11 ▪ Anatomy of a trout, a representative bony fish.

fins of sharks. The fastest bony fishes, which can swim in short bursts of up to 80 km/hr, have the same basic body shape as a shark. In fact, this body shape, termed fusiform (tapering on both ends), is common to all fast fishes and aquatic mammals such as seals and whales. Water is about a thousand times denser than air, and thus the slightest bump that causes drag is even more impeding to a fish than to a bird. Regardless of their different origins, we should expect speedy fishes and marine mammals to have similar streamlined shapes, because the laws of hydrodynamics are universal. This is an example of convergent evolution.

Details about the reproduction of bony fishes vary extensively. Most species are oviparous, reproducing by external fertilization after the female sheds large numbers of small eggs. However, internal fertilization and birthing characterize other species.

Both cartilaginous and bony fishes diversified extensively during the Devonian and Carboniferous periods, but whereas sharks arose in the sea, bony fishes probably originated in fresh water. The swim bladder was modified from simple lungs that had augmented the gills in gas exchange, perhaps in stagnant swamps with low oxygen content. The two major groups (subclasses) of bony fishes that exist today had diverged by the end of the Devonian.

Nearly all the families of fishes familiar to us are **ray-finned fishes (Subclass Actinopterygii**; Gr. *aktin,* "ray," and *pteryg,* "wing" or "fin"). The various species of bass, trout, perch, tuna, and herring are examples. The fins, supported mainly by long flexible rays, are modified for maneuvering, defense, and other functions. Ray-finned fishes spread from fresh water to the seas during their long history. (Adaptations that solve the osmotic problems of this move to salt water will be discussed in Chapter 44.) Numerous species of ray-finned fishes returned to fresh water at some point in their evolution. Some of these, including salmon and sea-run trout, replay their evolutionary round-trip from fresh water to seawater back to fresh water during their life cycle.

In contrast to the ray-fins, most members of the other extant subclass of bony fishes, the **lobe-finned fishes** and **lungfishes (Subclass Sarcopterygii**; Gr. *sarkodes,* "fleshy"), remained in fresh water. Their lungs continued to aid the gills in breathing air. Two major groups of lobe-finned fishes, called coelocanths and rhipidistians, are characterized by muscular pectoral and pelvic fins supported by extensions of the bony skeleton. Many lobefins were large, apparently bottom-dwellers that may have used their paired, muscular fins as aids to "walking" on the substrate under water. Some may also have been able to waddle occasionally on land.

Lobe-finned fishes are represented today by only one known species, a coelocanth named *Latimeria chalumnae* (FIGURE 34.12). Although most Devonian lobe-fins were probably freshwater animals with lungs, *Latimeria* belongs to a lineage that entered the seas at some point in its evolution.

FIGURE 34.12 ▪ *Latimeria chalumnae* (a coelocanth), the only extant lobe-finned fish. Coelocanths live in the deep sea off Madagascar.

Three genera of lungfishes live today in the Southern Hemisphere. They generally inhabit stagnant ponds and swamps, surfacing to gulp air into lungs connected to the pharynx of the digestive tract. When ponds shrink during the dry season, lungfishes can burrow in the mud and aestivate (wait in a state of torpor).

Although most lineages of lungfishes and lobe-fins died out, these groups were dominant predators in shallow freshwater habitats during the Devonian. They were also of great importance in vertebrate evolution; the ancestor of amphibians was probably one of these Devonian bony fishes (FIGURE 34.13).

SUPERCLASS GNATHOSTOMATA II: THE TETRAPODS

Amphibians are the oldest class of tetrapods

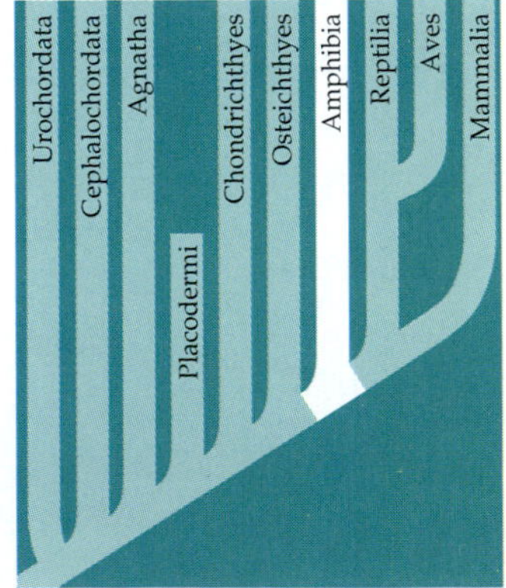

The first vertebrates on land were members of **Class Amphibia.** Today the class is represented by a total of about 4000 species of frogs, salamanders, and caecilians (limbless creatures that burrow in tropical forests and freshwater lakes).

Early Amphibians

As is the case today, parts of the Devonian world were subject to cycles of drought, followed by heavy rainfall and then drought again. In such areas, some lobe-finned fishes with lungs may have prevailed. The skeletal structure of their fins suggests that these paired appendages could have assisted in movement on land (FIGURE 34.14a). Some of the fossil lobe-fins, including a rhipidistian named *Eusthenopteron,* exhibited many other anatomical similarities to the earliest amphibians. A long-standing hypothesis is that Devonian lobe-finned

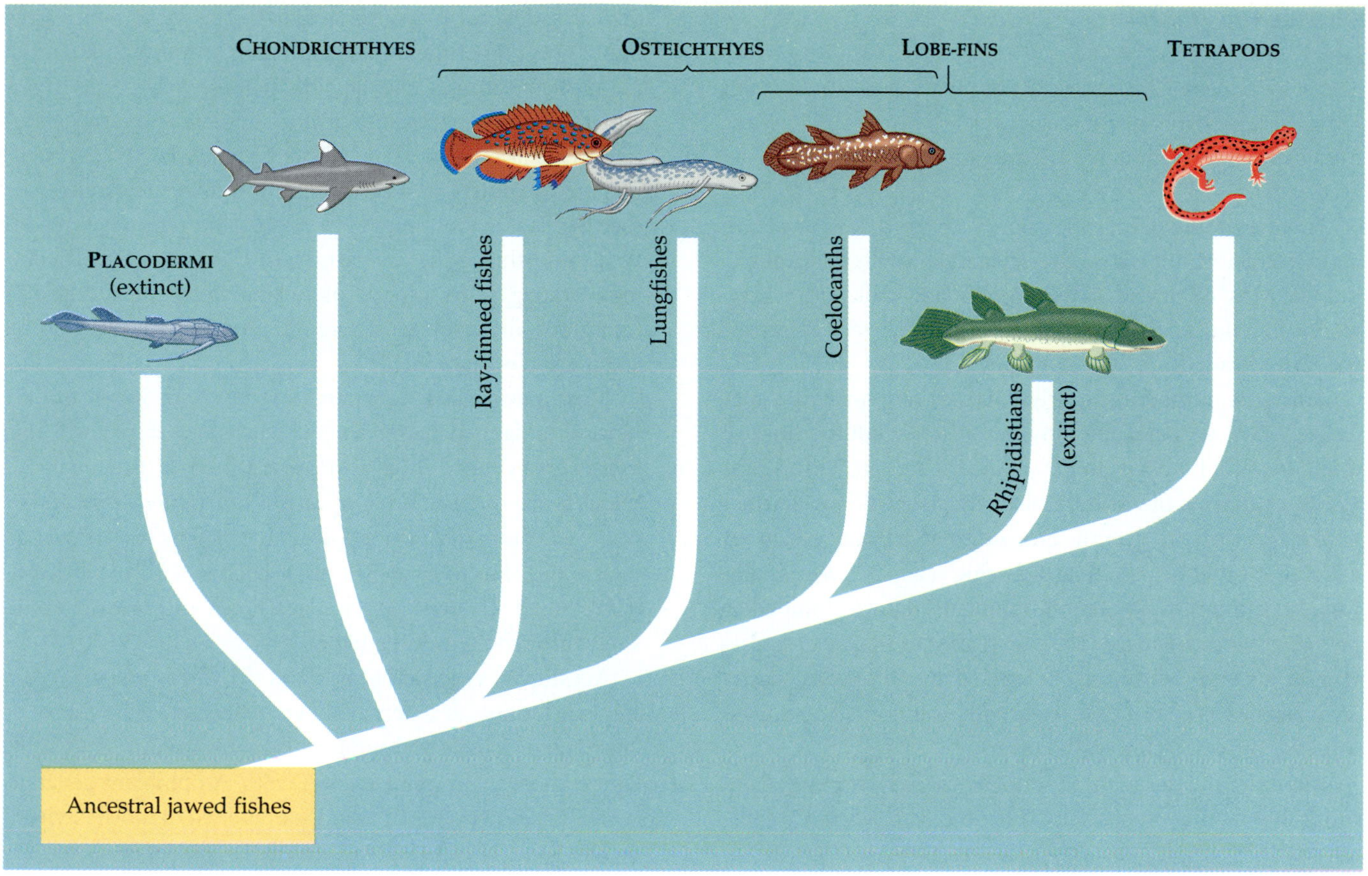

CHONDRICHTHYES

OSTEICHTHYES

LOBE-FINS

TETRAPODS

PLACODERMI (extinct)

Ray-finned fishes

Lungfishes

Coelocanths

Rhipidistians (extinct)

Ancestral jawed fishes

FIGURE 34.13 ▪ The Devonian radiation of fishes.

fishes were closely related to the ancestor of tetrapods. However, some molecular evidence suggests that lungfishes are phylogenetically closer to amphibians than are the lobe-fins.

The oldest amphibian fossils date back to late Devonian times, about 365 million years ago (FIGURE 34.14b). The first amphibians may have been primarily aquatic, occasionally venturing on land to escape carnivorous fishes or to exploit a cornucopia of food (insects and other invertebrates) that preceded them onto land. Adaptive radiation of the earliest tetrapods gave rise to a diversity of new forms. Many Carboniferous amphibians superficially resembled reptiles. Some reached 4 m in length. Because amphibians were the only vertebrates on land in late Devonian and early Carboniferous times, the age of amphibians is an appropriate name for the Carboniferous period. Amphibians began to decline during the late Carboniferous. As the Mesozoic era dawned with the Triassic period, about 245 million years ago, most survivors of the amphibian lineage resembled modern species.

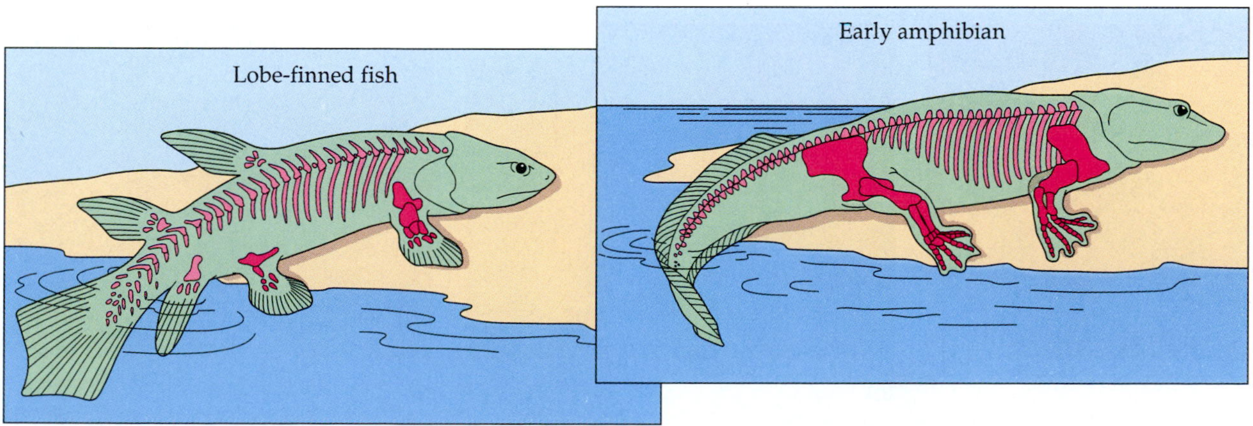

FIGURE 34.14 ▪ The origin of tetrapods. (a) The ancestor of tetrapods may have resembled a lobe-finned fish. As indicated in this drawing, muscular fins with extensions of the skeleton would have provided some support for the animal on land. **(b)** This reconstruction, based on Devonian fossils, depicts an early amphibian.

Modern Amphibians

There are three extant orders of amphibians (FIGURE 34.15): Urodela ("tailed ones"—salamanders); Anura ("tail-less ones"—frogs, including toads); and Apoda ("legless ones"—caecilians).

There are only about 400 species of **urodeles**. Some are entirely aquatic, but others live on land as adults or throughout life. Most salamanders that live on land walk with a side-to-side bending of the body that may resemble the swagger of the early tetrapods.

Anurans, numbering nearly 3500 species, are more specialized than urodeles for moving on land. Adult frogs use their powerful hind legs to hop along the terrain. A frog nabs insects by flicking out its long sticky tongue, which is attached to the front of the mouth. Frogs display a great variety of adaptations that help them avoid being eaten by larger predators. In common with other amphibians, many exhibit color patterns that camouflage. The skin glands of frogs secrete distasteful, or even poisonous, mucus. Many poisonous species have bright coloration that apparently warns predators, who associate the coloration with danger.

Apodans, the caecilians (about 150 species), are legless and nearly blind, and superficially resemble earthworms. Caecilians inhabit tropical areas where most species burrow in moist forest soil; a few South American apodans live in freshwater ponds and streams.

Amphibian means "two lives," a reference to the metamorphosis of many frogs (FIGURE 34.16). The tadpole, the larval stage of a frog, is usually an aquatic herbivore with gills, a lateral line system resembling that of fishes, and a long finned tail. The tadpole lacks legs and swims by undulating like its fishlike ancestors. During the metamorphosis that leads to the "second life," legs develop, and the gills and lateral line system disappear. The young tetrapod with air-breathing lungs, a pair of external eardrums, and a digestive system adapted to a carnivorous diet crawls onto shore and begins life as a terrestrial hunter. In spite of the name amphibian, however, many frogs do not go through the aquatic tadpole stage, and many amphibians do not live a dualistic—aquatic and terrestrial—life. There are some strictly aquatic and strictly terrestrial frogs, salamanders, and caecilians. Moreover, salamander and caecilian larvae look much like adults, and typically both the larvae and the adults are carnivorous. Paedogenesis is common among some groups of salamanders; the mudpuppy (*Necturus*), for instance, retains gills and other larval features when sexually mature.

Most amphibians maintain close ties with water and are most abundant in damp habitats such as swamps and rain forests. Even those frogs that are adapted to drier habitats spend much of their time in burrows or under moist leaves, where the humidity is high. Most amphibians rely heavily on their moist skin to carry out gas exchange with the environment. Some terrestrial species lack lungs and breathe exclusively through their skin and oral cavity.

Amphibian eggs lack a shell and dehydrate quickly in dry air. Fertilization is external in most species, with the male grasping the female and spilling his sperm over the eggs as the female sheds them (see FIGURE 34.16a). Amphibians generally lay their eggs in ponds or swamps or at least in moist environments. Some species lay vast numbers of eggs in temporary pools, and mortality is high. In contrast, some species display various types of parental care and lay relatively few eggs. Depending on the species, either males or females may incubate eggs on their back, in the mouth, or even in the stomach. Certain tropical tree frogs stir their egg masses into moist foamy nests that resist drying. There are also some ovoviviparous and viviparous species that retain the eggs in the female reproductive tract, where embryos can develop without drying out.

Many amphibians exhibit complex and diverse social behavior, especially during their breeding seasons. Frogs are

(a)

(b)

(c)

FIGURE 34.15 ▪ Amphibian orders.
(a) Urodeles (salamanders) retain their tails as adults. Some are entirely aquatic, but others live on land. This is a spotted salamander. **(b)** Anurans, such as this poison arrow frog, lack tails as adults. Poison arrow frogs inhabit tropical forests; their skin glands secrete deadly nerve toxins used by Central and South American natives to coat arrow tips. **(c)** Apodans, also called caecilians, are legless, mainly burrowing amphibians.

(a)

(b)

(c)

FIGURE 34.16 ▪ The "dual life" of a frog (*Rana temporaria*). **(a)** The male grasps the female, stimulating her to release eggs. The eggs are laid and fertilized in water. They have a jelly coat but lack shells and would desiccate in air. **(b)** The tadpole is an aquatic herbivore with a fishlike tail and internal gills. **(c)** During metamorphosis, the gills and tail are resorbed, and walking legs develop.

usually quiet creatures, but many species fill the air with their mating calls during the breeding season. Males may vocalize to defend breeding territory or to attract females. In some terrestrial species, migrations to specific breeding sites may involve vocal communication, celestial navigation, or chemical signaling.

Evolution of the amniotic egg expanded the success of vertebrates on land

The evolution of reptiles from an amphibian ancestor involved many adaptations for terrestrial living. The amniotic egg, a reproductive adaptation that enabled terrestrial vertebrates to complete their life cycles on land and sever their last ties with their aquatic origins, was a particularly important breakthrough. (Seeds played a similar role in the evolution of land plants, as you learned in Chapter 29.) In contrast to the shell-less eggs of amphibians, amniotic eggs (of birds, most reptiles, and some mammals) have a shell that retains water and can therefore be laid in a dry place. Most mammals have dispensed with the shell; instead, the embryo implants in the wall of the uterus and obtains its nutrients from the mother.

Reptiles, birds, and mammals all have specialized membranes within the amniote egg (FIGURE 34.17). Called **extraembryonic membranes** because they are not part of the body of the developing animal, these structures function in gas exchange, waste storage, and the transfer of stored nutrients to the embryo. They develop from tissue layers that grow out from the embryo. The amniotic egg is named for one of these

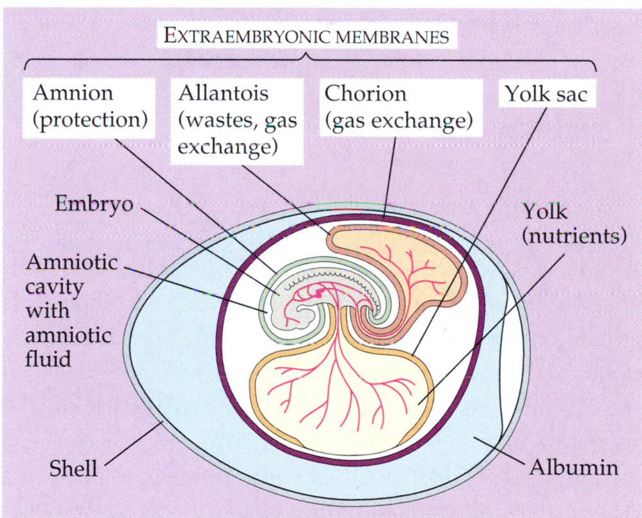

FIGURE 34.17 ▪ The amniotic egg. The embryos of reptiles, birds, and mammals form four extraembryonic membranes: the amnion, yolk sac, allantois, and chorion. The amnion protects the embryo in a fluid-filled cavity that prevents dehydration and cushions mechanical shocks. The yolk sac expands over the yolk, a stockpile of nutrients stored in the egg. Blood vessels in the yolk sac membrane transport nutrients from the yolk into the embryo. Other nutrients are stored in the albumin (the "egg white"). The allantois functions as a disposal sac for certain metabolic wastes produced by the embryo. The membrane of the allantois also functions with the chorion as a respiratory organ that exchanges gases between the embryo and the surrounding air. O_2 and CO_2 diffuse freely across the egg's shell. However, the shell is waterproof, an adaptation that prevents the embryo and the fluid compartments around it from drying out on land. This drawing represents the extraembryonic membranes in the shelled egg of a reptile or bird. Mammals have the four extraembryonic membranes, but in most species the embryos develop without shells in the reproductive tract of the mother.

membranes, the amnion, which encloses a compartment of amniotic fluid that bathes the embryo and acts as a hydraulic shock absorber. Together, reptiles, birds, and mammals make up a monophyletic group, the amniotes.

A reptilian heritage is evident in all amniotes

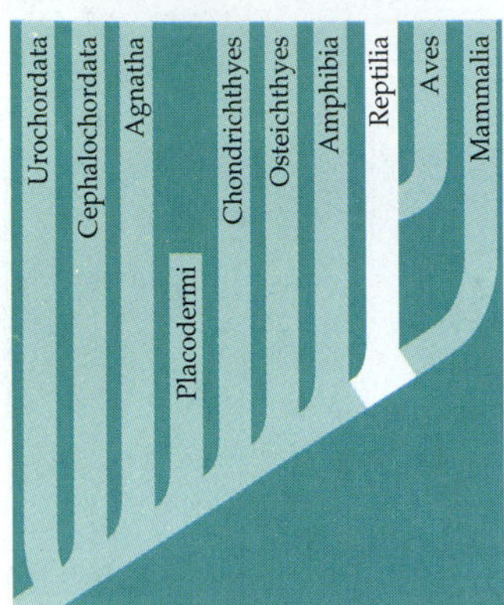

Class Reptilia, a diverse group with many extinct lineages, is represented today by about 7000 species, mainly lizards, snakes, turtles, and crocodilians. This grouping is traditional and based on the apparent similarity of all these tetrapods. Cladistic analysis indicates, however, that grouping all these vertebrates in a class that does not include birds is inconsistent with phylogeny. Birds appear to be more closely related to crocodiles than are turtles. In fact, Class Reptilia cannot be defined except by the absence of features that distinguish birds (flight feathers) and mammals (hair and mammary glands). Despite these taxonomic problems, studying reptiles (in the traditional sense) provides a context for understanding all amniotes. Hence, we follow tradition in this case and recognize the familiar classes Reptilia, Aves, and Mammalia.

Reptilian Characteristics

Reptiles have several adaptations for terrestrial living not generally found in amphibians. Scales containing the protein keratin waterproof the skin of a reptile, helping prevent dehydration in dry air. Keratinized skin is the vertebrate analogue of the chitinized cuticle of insects and the waxy cuticle of land plants. Because they cannot breathe through their dry skin, most reptiles obtain all their oxygen with lungs. Many turtles also use the moist surfaces of their cloaca for gas exchange.

Most reptiles lay shelled amniotic eggs on land (FIGURE 34.18). Fertilization in reptiles must occur internally, before the shell is secreted as the egg passes through the reproductive tract of the female. Some species of snakes and lizards are viviparous, their extraembryonic membranes forming a placenta that enables the embryo to obtain nutrients from its mother.

Reptiles are sometimes labeled "cold-blooded" animals because they do not use their metabolism extensively to control body temperature. But reptiles do regulate body temperature by using behavioral adaptations. For example, many lizards regulate their internal temperature by basking in the sun when the air is cool and seeking shade when the air is too warm. Because they absorb external heat rather than generating much of their own, reptiles are said to be **ectotherms**, a term more appropriate than cold-blooded. (The control of body temperature will be discussed in more detail in Chapter 44.) By heating directly with solar energy rather than through the metabolic breakdown of food, a reptile can survive on less than 10% of the calories required by a mammal of equivalent size.

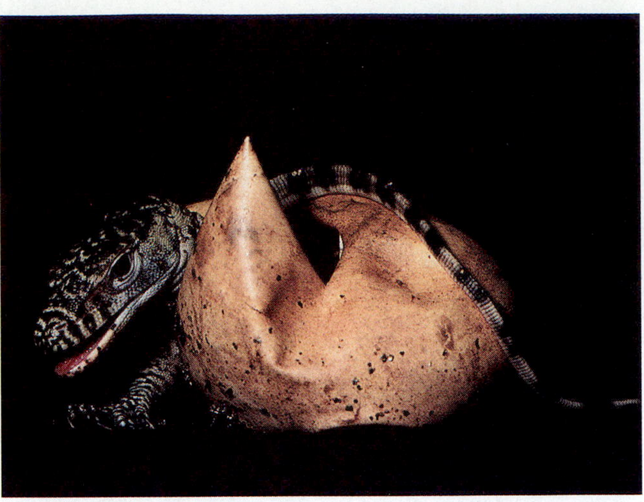

FIGURE 34.18 ▪ **A hatching reptile.** This Komodo dragon is breaking out of a parchmentlike shell, a common type of shell among reptiles. The shells of some reptiles are harder because of an abundance of calcium carbonate.

The Age of Reptiles

Reptiles were far more widespread, numerous, and diverse during the Mesozoic era than they are today. Let's look briefly at this period, known as the age of reptiles.

The Origin and Early Evolutionary Radiation of Reptiles. The oldest reptilian fossils are found in rocks from the late Carboniferous period; they are about 300 million years old. Their ancestor was among the Devonian amphibians. In two great waves of adaptive radiation, reptiles became the dominant terrestrial vertebrates in a dynasty that lasted more than 200 million years.

The first major reptilian radiation occurred by the dawn of the Permian, the last period of the Paleozoic era, giving rise to two main evolutionary branches (FIGURE 34.19):

1. The **synapsids**. This branch included a diversity of mammal-like reptiles called therapsids, including the probable ancestor of mammals.
2. The **sauropsids**. This branch gave rise to all modern amniotes *except* the mammals. Relatively early in their history, the sauropsids split into two subbranches:
   a. The **anapsids**. Turtles are the only survivors of this reptilian group.
   b. The **diapsids**. Lizards, snakes, and crocodilians are the extant diapsids classified as reptiles. The dinosaurs

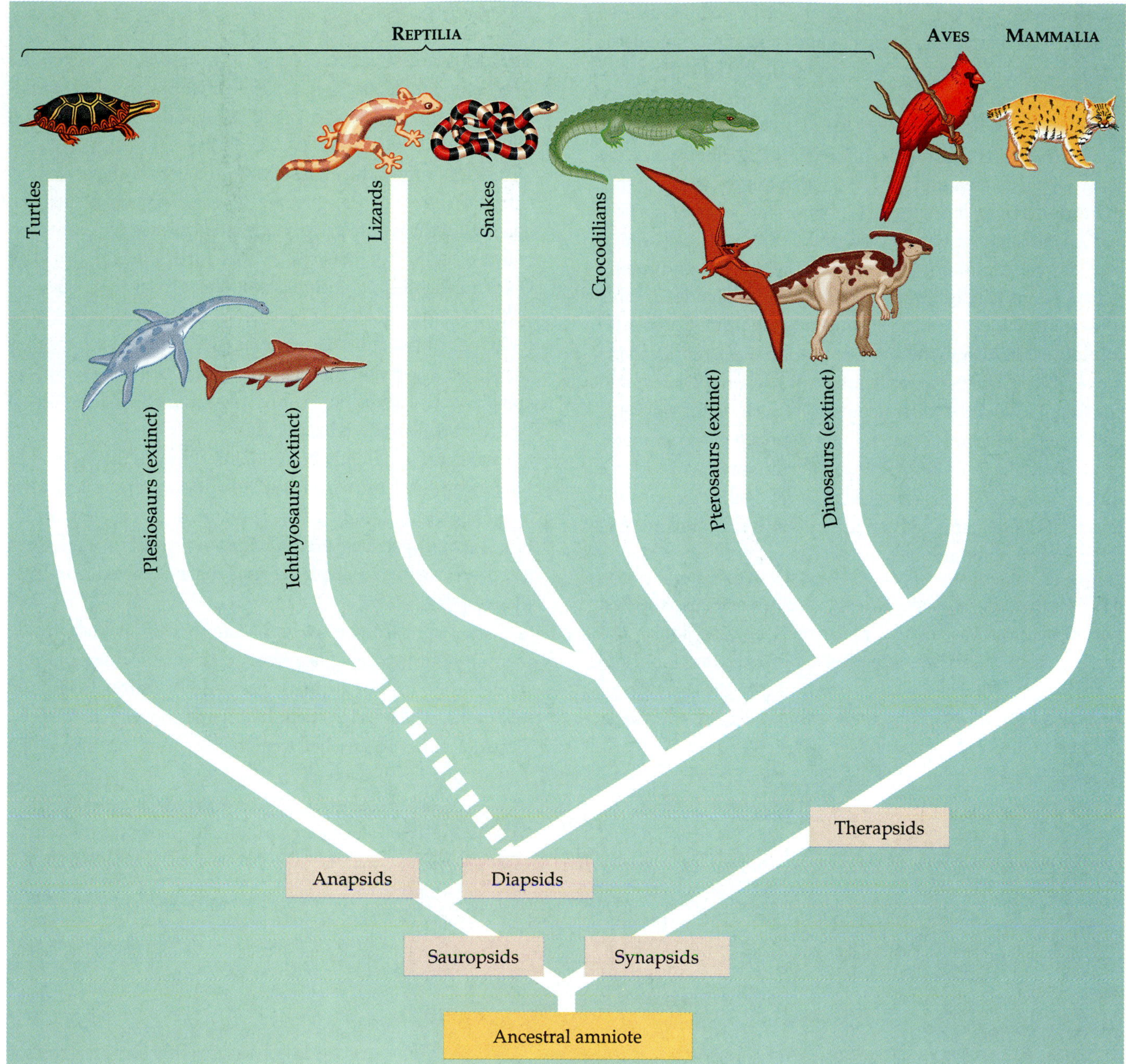

FIGURE 34.19 ▪ Phylogeny of the amniotes. The radiation of amniotes during the early Mesozoic era gave rise to a diversity of terrestrial forms, including the dinosaurs. Various reptiles also returned to aquatic habitats (plesiosaurs and ichthyosaurs) and took to the air (pterosaurs). The birds seem to be an extant group of flying "reptiles," although this book follows the taxonomic tradition of placing birds in their own class (Aves). Another class of amniotes, the mammals, evolved from the synapsids, a reptilian branch that diverged from the other reptiles very early. Notice that as a taxonomic cluster, the amniotes are monophyletic; modern reptiles, birds, and mammals all descended from a common ancestor. In Chapter 25 you learned that a monophyletic taxon includes a common ancestor *and all of its descendants*. Thus, as depicted here, Class Reptilia is *not* a monophyletic taxon because it excludes birds. Zoologists recognize the cladistic inconsistency, but most still find it convenient to classify reptiles, birds, and mammals as three classes of amniotic vertebrates.

and some other groups of extinct reptiles were also diapsids. Cladistic analysis provides strong evidence that birds are the closest living relatives of the extinct dinosaurs.

Dinosaurs and Pterosaurs. The second great reptilian radiation was under way by late Triassic times (a little more than 200 million years ago) and was marked mainly by the origin and diversification of two groups of reptiles: the dinosaurs, which lived on land, and the pterosaurs, or flying reptiles (see FIGURE 34.19). These groups were the dominant vertebrates on Earth for millions of years. Pterosaurs had wings formed from a membrane of skin stretched from the body wall, along the forelimb, to the tip of an elongated finger. Stiff fibers provided support for the skin of the wing. Dinosaurs, an extremely diverse group varying in body shape, size, and habitat, included the largest animals ever to inhabit land. Fossils of gigantic dinosaurs that were 45 m long have recently been unearthed in New Mexico and Utah.

Contradicting the long-standing view that dinosaurs were slow, sluggish creatures, there is increasing evidence that many dinosaurs were agile, fast-moving, and, in some species, social. Paleontologists have also discovered signs of parental care among dinosaurs (FIGURE 34.20). There is continuing debate about whether dinosaurs were **endothermic**, capable of keeping the body warm through metabolism. Some anatomical evidence supports this hypothesis, but most experts are skeptical. The Mesozoic climate was relatively warm and consistent, and behavioral adaptations such as basking may have been sufficient for maintaining a suitable body temperature, especially for the land-dwelling dinosaurs. Also, large dinosaurs had low surface-to-volume ratios that reduced the effects of daily fluctuations in air temperature on the internal temperature of the animal.

The Cretaceous Crisis. During the Cretaceous, the last period of the Mesozoic era, the climate became cooler and more variable. This was a period of mass extinctions, and except for a few dinosaurs that survived into the early Cenozoic, all of these reptiles were gone by the end of the Cretaceous, about 65 million years ago (see Chapter 25).

Modern Reptiles

The three largest and most diverse extant orders of reptiles (totaling about 6500 species) are **Chelonia** (turtles), **Squamata** (lizards and snakes), and **Crocodilia** (alligators and crocodiles).

Turtles evolved during the Mesozoic era and have scarcely changed since (FIGURE 34.21a). The usually hard shell, an adaptation that protects against predators, has certainly contributed to this long success. Those turtles that returned to water during their evolution crawl ashore to lay their eggs.

Lizards are the most numerous and diverse reptiles alive today (FIGURE 34.21b). Most are relatively small; perhaps they were able to survive the Cretaceous "crunch" by nesting in crevices and decreasing their activity during cold periods, a practice many modern lizards use.

Snakes are apparently descendants of lizards that adopted a burrowing lifestyle (FIGURE 34.21c). Today, most snakes live above the ground, but they have retained the limbless condition. Vestigial pelvic and limb bones in primitive snakes such as boas, however, are evidence that snakes evolved from reptiles with legs.

Snakes are carnivorous, and a number of adaptations aid them in hunting prey. They have acute chemical sensors, and though they lack eardrums, snakes are sensitive to ground vibrations, which helps them detect the movements of prey. Heat-detecting organs between the eyes and nostrils of pit

FIGURE 34.20 ▪ Dinosaur social behavior and parental care. This artist's re-creation of duck-billed dinosaurs is based on 80-million-year-old fossils discovered in Montana. Dinosaurs of this species were relatively large, about 7 m in length. They used their bills to browse on plants. Fossilized footprints indicate that duck-billed dinosaurs traveled in large social groups, or herds. Fossilized nests contain both eggs and offspring up to a few months old, evidence that these dinosaurs took care of their young. In fact, the genus name for this dinosaur, *Maiasaura*, means "good mother lizard."

(a)

(b)

(c)

(d)

FIGURE 34.21 ▪ Extant reptiles. (a) Turtles have changed little since their origin early in the Mesozoic era. This species is an eastern box turtle. **(b)** Lizards, such as this Australian frillneck, are the most numerous and diverse of the extant reptiles. A frillneck only spreads its frill when threatened, a response that probably discourages many predators. **(c)** Snakes may have evolved from lizards that adapted to a burrowing lifestyle. This is a bull snake laying eggs. **(d)** Crocodiles and alligators are the reptiles most closely related to dinosaurs. An alligator is shown here.

vipers, including rattlesnakes, are sensitive to minute temperature changes, enabling these night hunters to locate warm animals. Poisonous snakes inject their toxin through a pair of sharp hollow or grooved teeth. The flicking tongue is not poisonous but helps fan odors toward olfactory organs on the roof of the mouth. Loosely articulated jaws enable most snakes to swallow prey larger than the diameter of the snake itself.

Crocodiles and alligators (crocodilians) are among the largest living reptiles (some turtles are heavier) (FIGURE 34.21d). They spend most of their time in water, breathing air through their upturned nostrils. Crocodilians are confined to the warm regions of Africa, China, Indonesia, India, Australia, South America, and the southeastern United States, where alligators have made a strong comeback after spending years on the endangered species list. Among the modern animals traditionally classified as reptiles, crocodilians are the most closely related to the dinosaurs. However, the modern animals that seem to share the most recent ancestor with the dinosaurs are the birds.

Birds began as flying reptiles

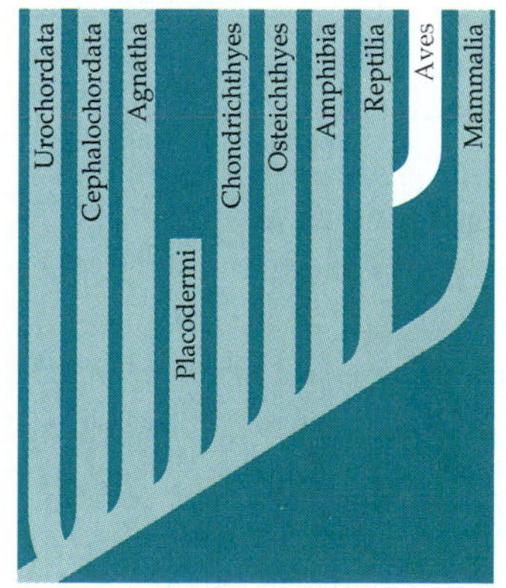

Class Aves (birds) evolved during the great reptilian radiation of the Mesozoic era (see FIGURE 34.19). Amniotic eggs and scales on the legs are just two of the reptilian features we see in birds. But modern birds look quite different from modern reptiles because of their feathers and other distinctive flight equipment.

Characteristics of Birds

Almost every part of a typical bird's anatomy is modified in some way that enhances flight. The bones have an internal

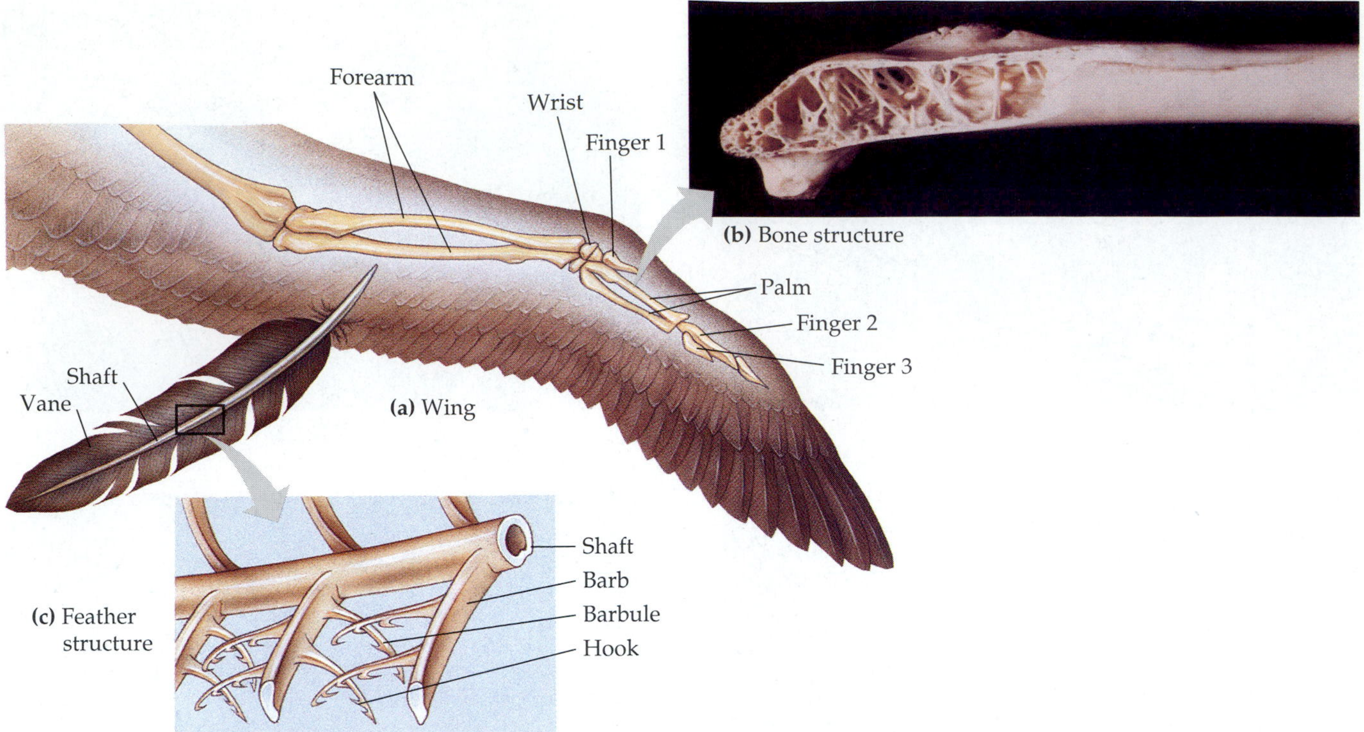

FIGURE 34.22 ▪ Form fits function: the avian wing and feather. (a) A wing is supported by a remodeled version of the tetrapod forelimb. **(b)** The honeycombed anatomy of the bones (photo) is a flight adaptation that reduces weight. **(c)** A feather consists of a central hollow shaft, from which radiate the vanes. The vanes are made up of barbs, in turn bearing even smaller branches called barbules. Birds have contour feathers and downy feathers. Contour feathers are the stiff ones that contribute to the aerodynamic shapes of the wings and body. Their barbules have hooks that cling to barbules on neighboring barbs. When a bird preens, it runs the length of a feather through its bill, engaging the hooks and uniting the barbs into a precisely shaped vane. Downy feathers lack hooks, and the free-form arrangement of barbs produces a fluffiness that provides excellent insulation because of the trapped air.

structure that is honeycombed, making them strong but light (FIGURE 34.22). The skeleton of a frigate bird, for instance, has a wingspan of more than 2 m but weighs only about 113 g (4 oz). Another adaptation reducing the weight of birds is the absence of some organs. Females, for instance, have only one ovary. Also, modern birds are toothless, an adaptation that trims the weight of the head. Food is not chewed in the mouth but ground in the gizzard, a digestive organ near the stomach. (Crocodiles also have gizzards, as did some dinosaurs.) The bird's beak, made of keratin, has proven to be very adaptable during avian evolution, taking on a great variety of shapes suitable for different diets.

Flying requires a great expenditure of energy from an active metabolism. Birds are endothermic; they use their own metabolic heat to maintain a warm, constant body temperature. Feathers and in some species a layer of fat provide insulation that enables birds to retain their metabolically generated heat. An efficient respiratory system and a circulatory system with a four-chambered heart keep tissues well supplied with oxygen and nutrients, supporting a high rate of metabolism. The efficient lungs have tiny tubes leading to and from elastic air sacs that help dissipate heat and reduce the density of the body.

For safe flight, senses, especially vision, must be acute. Birds have excellent eyes, perhaps the best of all vertebrates. The visual areas of the brain are well developed, as are the motor areas; flight also requires excellent coordination.

With brains proportionately larger than those of reptiles and amphibians, birds generally display very complex behavior. Avian behavior is particularly intricate during breeding season, when birds engage in elaborate rituals of courtship. Since eggs are shelled when laid, fertilization must be internal. Copulation involves contact between the mates' vents, the openings to their cloacas. After eggs are laid, the avian embryo must be kept warm through brooding by the mother, father, or both, depending on the species.

A bird's most obvious adaptation for flight is its wings. Bird wings are airfoils that illustrate the same principles of aerodynamics as the wings of an airplane (FIGURE 34.23). Providing power for flight, birds flap their wings by contractions

FIGURE 34.23 ▪ A bald eagle in flight. Bird wings are airfoils, structures whose shape creates lift by altering air currents. The leading edge of an airfoil is thicker than the trailing edge, its upper surface is somewhat convex, and its under surface is flattened or concave. This shape makes the air passing over the wing travel farther than the air passing under the wing. As a result, air molecules are spaced farther apart above the wing than below it. The air pressure underneath is therefore greater, and this greater pressure lifts the wing. Birds can reach great speeds and cover enormous distances. Swifts, which can fly 170 km/hr are the fastest birds. The bird that travels farthest in its annual migration is the arctic tern, which flies round-trip between the North Pole and South Pole each year.

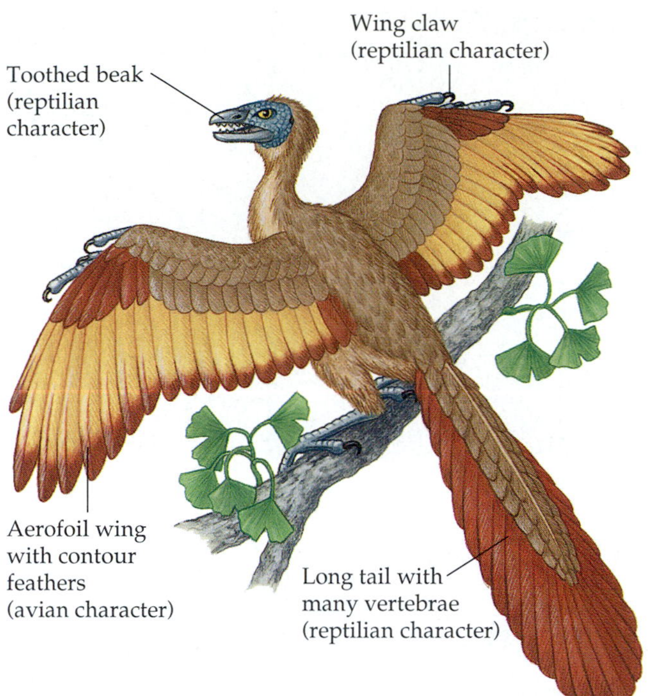

FIGURE 34.24 ▪ *Archaeopteryx*, a Jurassic bird-reptile.

of large pectoral (breast) muscles anchored to a keel on the sternum (breastbone). Some birds, such as eagles and hawks, have wings adapted for soaring on air currents and flap their wings only occasionally; other birds, including hummingbirds, must flap continuously to stay aloft. In either case, it is the shape and arrangement of the feathers that form the wing into an airfoil.

In being both extremely light and strong, feathers are among the most remarkable of vertebrate adaptations. Feathers are made of keratin, the same protein that forms our hair and fingernails and the scales of reptiles. Feathers may have functioned first as insulation during the evolution of endothermy, only later being co-opted as flight equipment. Besides providing support and wing shape, feathers can be manipulated to control air movements around the wing.

The evolution of flight required radical alteration in body form, but flight provides many benefits. It enhances hunting and scavenging; many birds exploit flying insects, an abundant, highly nutritious food resource. Flight also provides ready escape from earthbound predators and enables some birds to migrate great distances to utilize different food resources and seasonal breeding areas.

The Origin of Birds

The presence of feathers may be enough to classify an animal as a bird, although some recently discovered dinosaur fossils appear to have short, downlike feathers. Certainly, if we want to trace the ancestry of birds, we must search for the oldest fossils with feathers that could have functioned in flight. Fossils of an ancient bird named *Archaeopteryx lithographica* have been found in Bavarian limestone in Germany, dating back some 150 million years, into the Jurassic period. Unlike modern birds, but similar to reptiles, *Archaeopteryx* had clawed forelimbs, teeth, and a long tail containing vertebrae (FIGURE 34.24). Indeed, if it were not for the preservation of its feathers, *Archaeopteryx* would be regarded as a member of a diverse group of small, carnivorous, bipedal dinosaurs called theropods. In common with birds, many dinosaurs, including some theropods, built nests and cared for their young.

Archaeopteryx is not considered the ancestor of modern birds, and paleontologists place it on a side branch of the avian lineage. Nonetheless, *Archaeopteryx* probably was derived from ancestral forms that also gave rise to modern birds. Its skeletal anatomy indicates that it was a weak flyer, perhaps mainly a tree-dwelling glider. A combination of gliding downward and jumping into the air from the ground may have been the earliest mode of flying in the bird lineage.

Extensive cladistic analyses have convinced many researchers that birds arose from theropod dinosaurs. Heated debate continues, however, with some biologists arguing that the bird lineage diverged from a group of early Mesozoic reptiles, called thecodonts, that also gave rise to the dinosaurs.

Modern Birds

There are about 8600 extant species of birds classified in about 28 orders. Flight is typical of birds, but there are several

(a)

(b)

(c)

(d)

FIGURE 34.25 ▪ Birds. **(a)** Flightless birds, a pair of emus in Australia. **(b)** In common with many species, the harlequin duck, which inhabits mountain streams and coastal areas of the Pacific Northwest, exhibits pronounced color differences between the sexes. **(c)** Most birds, as exhibited here by a pair of laysan albatrosses, have specific courtship and mating behavior. **(d)** The barn swallow is a member of the order Passeriformes. Passeriformes are called perching birds because the toes of their feet can lock around a tree branch, enabling the bird to rest in place for long periods of time.

flightless species, including the ostrich, kiwi, and emu (FIGURE 34.25a). Flightless birds are collectively called **ratites** (L., "flat-bottomed") because their breastbone lacks a keel and the large breast muscles that attach to it in flying birds.

In contrast to ratites, other birds are called **carinates** because they have a carina, or sternal keel, supporting their large breast muscles. These muscles provide flight power in flying birds. The demands of flight have rendered the general body form of carinate birds similar to one another, yet experienced bird watchers can distinguish many species by their body profile. Carinate birds also exhibit great variety in feather colors, beak and foot shape, behavior, and flying style (FIGURE 34.25b–d). Among the most unusual carinate birds are the penguins, which do not fly, but use their powerful breast muscles in swimming. Nearly 60% of the living bird species belong in one carinate order called the **passeriformes**, or perching birds. These are the familiar jays, swallows, sparrows, warblers, and many others (FIGURE 34.25d).

Mammals diversified extensively in the wake of the Cretaceous extinctions

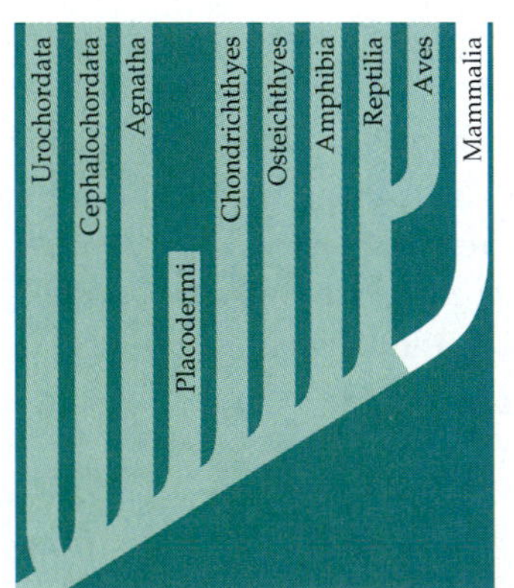

The extinction of the dinosaurs and the fragmentation of continents that occurred at the close of the Mesozoic era opened many adaptive zones, and mammals underwent an extensive adaptive radiation. There are about 4500 species of mammals on Earth today. As mammals ourselves, we naturally have a special interest in this class of vertebrates. Let's examine some of the features we share with all other mammals.

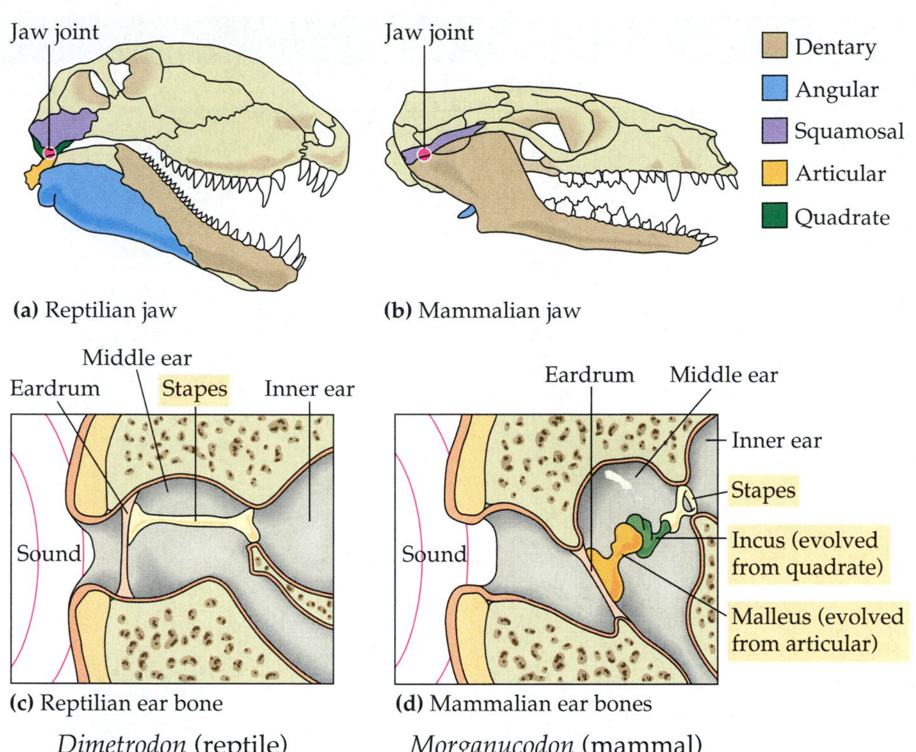

FIGURE 34.26 ▪ The evolution of the mammalian jaw and ear bones. The reptile *Dimetrodon* (left) was an early synapsid, the lineage that eventually gave rise to the mammals. *Morganucodon* (right) was one of the first mammals. **(a)** Notice that the lower jaw of the reptile is composed of several fused bones. In the reptile, two small jawbones, the quadrate and articular, form part of the jaw joint. **(b)** The mammalian lower jaw is reduced to a single bone, the dentary, and the location of the jaw joint has shifted. The single-bone lower jaw of mammals is stronger than the multibone jaw of reptiles. **(c)–(d)** During the evolutionary remodeling of the mammalian skull, the quadrate and articular jawbones became incorporated into the middle ear as two of the three bones that transmit sound from the eardrum to the inner ear. These ear bones enhance hearing by amplifying sound. The steps in this evolutionary remodeling are evident in a succession of fossils of mammal-like reptiles.

Mammalian Characteristics

Vertebrates of **Class Mammalia** have hair, a characteristic as diagnostic as the flight feathers of birds. Hair, like feathers, is made of keratin, although zoologists are uncertain about its evolutionary origin. Most mammals have an active metabolism and are endothermic. Efficient respiratory and circulatory systems (including a four-chambered heart) support a high metabolic rate. A sheet of muscle called the diaphragm helps ventilate the lungs. Hair and a layer of fat under the skin help the body retain metabolic heat.

Mammary glands that produce milk are as distinctively mammalian as hair. All mammalian mothers nourish their babies with milk, a balanced diet rich in fats, sugars, proteins, minerals, and vitamins.

Most mammals are born rather than hatched. Fertilization is internal, and the embryo develops inside the uterus of the female's reproductive tract. In eutherian (placental) mammals and marsupials, the lining of the mother's uterus and extraembryonic membranes arising from the embryo collectively form a **placenta**, where nutrients diffuse into the embryo's blood. Eutherians are commonly called placental mammals because their placentas are more complex and provide a more intimate and long-lasting association between the mother and her developing young.

Mammals generally have larger brains than other vertebrates of equivalent size, and many species are capable learners. The relatively long duration of parental care extends the time for offspring to learn important survival skills by observing their parents.

Differentiation of teeth is another important mammalian trait. Whereas the teeth of reptiles are generally conical and uniform in size, the teeth of mammals come in a variety of sizes and shapes adapted for chewing many kinds of foods. Our own dentition, for example, includes teeth modified for shearing (incisors and canine teeth) and for crushing and grinding (premolars and molars). Another part of the feeding apparatus, the jaw, was also remodeled during the evolution of mammals from reptiles, and two of the jawbones were incorporated into the mammalian inner ear (FIGURE 34.26).

The Evolution of Mammals

Mammals evolved from reptilian stock even earlier than birds. The oldest fossils believed to be mammalian date back 220 million years, into the Triassic period. The ancestor of mammals was among the **therapsids**, part of the synapsid branch of reptilian phylogeny (see FIGURE 34.19). The therapsids disappeared during the reign of the dinosaurs, but their mammalian offshoots coexisted with the dinosaurs throughout the Mesozoic era. Most Mesozoic mammals were very small—about the size of shrews—and most probably ate insects. A variety of evidence, such as the size of the eye sockets, suggests that these tiny mammals were nocturnal.

As the Cenozoic era dawned in the wake of the Cretaceous extinctions, mammals were in the midst of a great adaptive radiation. That diversity is represented today in the three major groups: monotremes (egg-laying mammals), marsupials (mammals with pouches), and eutherian (placental) mammals (TABLE 34.2, pp. 652–653).

Table 34.2 ■ Major Orders of Mammals

	ORDER	MAIN CHARACTERISTICS	EXAMPLES	
MONOTREMES	Monotremata	Lay eggs; have no nipples; suck milk from fur of mother	Platypuses, echidnas	 Echidna
MARSUPIAL MAMMALS	Marsupialia	Embryonic development completed in marsupial pouch	Kangaroos, opossums, koalas	 Koala
EUTHERIAN MAMMALS	Artiodactyla	Possess hooves with an even number of toes on each foot; herbivorous.	Sheep, pigs, cattle, deer, giraffes	 Bighorn sheep
	Carnivora	Carnivorous; possess sharp, pointed canine teeth and molars for shearing.	Dogs, wolves, bears, cats, weasels, otters, seals, walruses	 Coyote
	Cetacea	Marine forms with fish-shaped bodies, paddlelike forelimbs and no hind limbs; thick layer of insulating blubber.	Whales, dolphins, porpoises	 Pacific white-sided porpoise
	Chiroptera	Adapted for flying; possess a broad skinfold that extends from elongated fingers to body and legs.	Bats	 Frog-eating bat
	Edentata	Have reduced or no teeth.	Sloths, anteaters, armadillos	 Tamandua

Table 34.2 ■ Major Orders of Mammals (Continued)

ORDER	MAIN CHARACTERISTICS	EXAMPLES	
Insectivora	Insect-eating mammals	Moles, shrews, hedgehogs	 Star-nosed mole
Lagomorpha	Possess chisel-like incisors, hind legs longer than forelegs and adapted for running and jumping.	Rabbits, hares, pikas	 Jackrabbit
Perissodactyla	Possess hooves with an odd number of toes on each foot; herbivorous.	Horses, zebras, tapirs, rhinoceroses	 Indian rhinoceros
Primates	Opposable thumb; forward-facing eyes; well-developed cerebral cortex; omnivorous.	Lemurs, monkeys, apes, humans	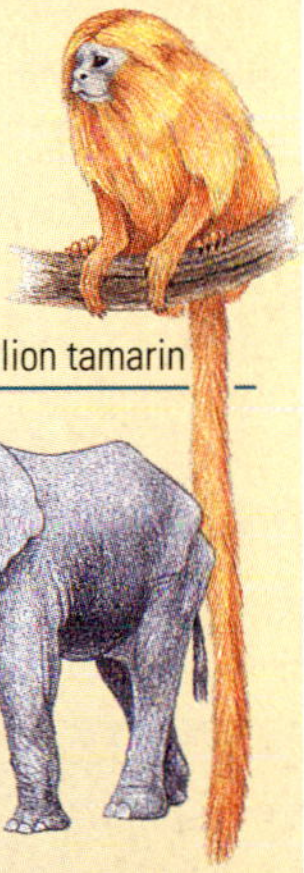 Golden lion tamarin
Proboscidea	Have a long, muscular trunk; thick, loose skin; upper incisors elongated as tusks.	Elephants	 African elephant
Rodentia	Possess chisel-like, continuously growing incisor teeth.	Squirrels, beavers, rats, porcupines, mice	 Red squirrel
Sirenia	Aquatic herbivores; possess finlike forelimbs and no hind limbs.	Sea cows (manatees)	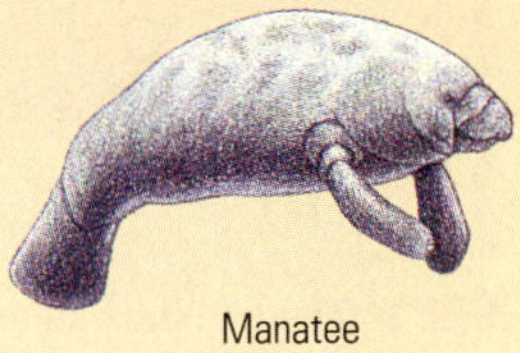 Manatee

Monotremes

Monotremes—the platypuses and the echidnas (spiny anteaters)—are the only living mammals that lay eggs (FIGURE 34.27a). The egg, which is reptilian in structure and development, contains enough yolk to nourish the developing embryo. Monotremes have hair and produce milk, two of the most important trademarks of Mammalia. On the belly of a monotreme mother are specialized glands that secrete milk. After hatching, the baby sucks the milk from the fur of the mother, who has no nipples. The mixture of ancestral reptilian characters and derived characters of mammals suggests that monotremes descended from a very early branch in the mammalian genealogy. Today, monotremes are found only in Australia and New Guinea.

Marsupials

Opossums, kangaroos, bandicoots, and koalas are examples of marsupial mammals. A **marsupial** is born very early in its development and completes its embryonic development while nursing (FIGURE 34.27b). In most species, the nursing young are held within a maternal pouch called a marsupium. A red kangaroo, for instance, is about the size of a honeybee at its birth, just 33 days after fertilization. Its hind legs are merely buds, but the forelimbs are strong enough for the offspring to crawl from the exit of the reproductive tract to the mother's pouch, a journey lasting a few minutes.

In Australia, marsupials have radiated and filled niches occupied by eutherian (placental) mammals in other parts of the world. Convergent evolution has resulted in a diversity of marsupials that resemble their eutherian counterparts occupying similar ecological roles (FIGURE 34.28).

The opossums of North and South America are the most diverse of only three families of extant marsupials outside the Australian region, though South America had a diverse marsupial fauna throughout the Tertiary period. The distribution of modern and fossil marsupials begins to make sense in the context of plate tectonics and continental drift. According to recent fossil evidence, marsupials probably originated in what is now North America, spreading southward when the land masses were still joined. After the breakup of Pangaea, South America and Australia became island continents, and their marsupials diversified in isolation from the placental mammals that began an adaptive radiation on the northern continents. Australia has not been in contact with another continent since early in the Cenozoic era, about 65 million years ago. The South American fauna, meanwhile, has not remained cloistered. Placental mammals reached South America throughout the Cenozoic. The most important migrations occurred about 12 million years ago and then again about 3 million years ago, when North and South America joined at the Panamanian isthmus. Extensive two-way traffic of animals took place over the land bridge. The biogeography of mammals is another example of the interplay between biological and geological evolution.

(a) A monotreme

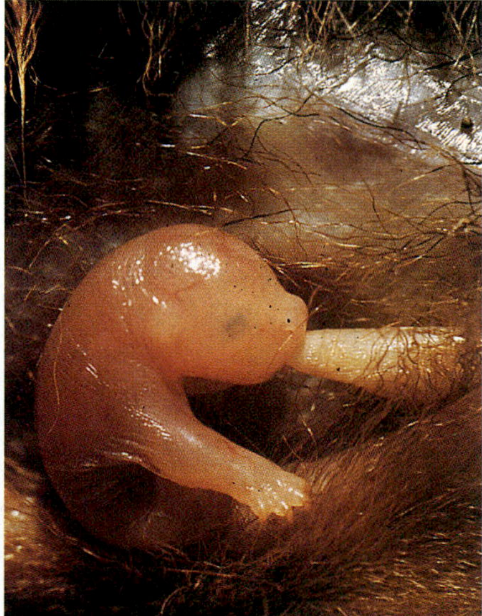

(b) A young marsupial in its mother's pouch

(c) The long-nosed bandicoot, a marsupial

FIGURE 34.27 • Australian monotremes and marsupials. (a) Monotremes, such as this short-beaked echidna, are the only mammals that lay eggs (inset). Monotremes have hair and milk, but no nipples. **(b)** The young of marsupials, such as this Australian brushtail possum, are born very early in their development. They finish their growth while nursing from a nipple (in their mother's pouch in most species). **(c)** Bandicoots have a pouch that opens to the rear, whereas in other marsupials, such as the kangaroos, it opens to the front. Most bandicoots are diggers and burrowers that eat mainly insects but also some small vertebrates and plant material. Their rear-opening pouch helps protect the young from dirt as the mother digs. Bandicoots have a placenta like that of eutherian mammals. Young bandicoots remain in the uterus for only about two weeks and then nurse for nearly two months.

FIGURE 34.28 ▪ Comparing marsupial and eutherian (placental) mammals. Adaptive radiation in Australia has fit marsupials to many of the ecological roles filled by eutherian mammals on other continents. Convergent evolution has produced a number of remarkable look-alikes, but marsupials and eutherians evolved on separate mammalian lineages. Among their differences, a developing marsupial spends most of its time outside the uterus attached to a nipple, whereas a eutherian completes its embryonic development in the uterus nourished by the placenta. (Drawings are not to scale.)

Eutherian (Placental) Mammals

Compared to marsupials, **eutherian mammals** have a longer period of pregnancy. Young eutherians complete their embryonic development within the uterus, joined to the mother by the placenta.

Adaptive radiation during late Cretaceous and early Tertiary periods (about 70 to 45 million years ago) produced the orders of eutherian mammals we recognize today (see TABLE 34.2, pp. 652–653). The phylogenetic relationship between marsupials and eutherian mammals is somewhat obscure, but scientists believe they are more closely related to each other than either is to the monotremes. Fossil evidence indicates that eutherians and marsupials may have diverged from a common ancestor about 80 to 100 million years ago.

The evolutionary relationships among the many orders of eutherian mammals are, for the most part, also unsettled. Most mammalogists now favor a tentative genealogy that recognizes at least four main evolutionary lines of eutherian (placental) mammals.

One branch of eutherian mammals consists of the order Insectivora (shrews), which resemble early mammals, and the order Chiroptera (bats). The bats, whose forelimbs are modified as wings, probably evolved from insectivores that fed on flying insects. In addition to the insect-eaters, some bat species feed on small vertebrates or fruit; three species of vampire bats found in Central and South America bite mammals and lap up their blood. Most bats are nocturnal.

The second branch started with a lineage of medium-sized herbivores that underwent a spectacular adaptive radiation during the Tertiary period, eventually giving rise to such modern orders as Lagomorpha (rabbits and their relatives); Perissodactyla (odd-toed ungulates, including horses and rhinoceroses; ungulates walk on the tips of toes); Artiodactyla (even-toed ungulates, including deer and swine); Sirenia (sea cows); Proboscidea (elephants); and Cetacea (porpoises and whales).

A third branch in the evolution of eutherian mammals produced the order Carnivora, which includes cats, dogs, raccoons, skunks, and the pinnipeds (seals, sea lions, and walruses). Carnivorans probably first appeared during the early Cenozoic era. Seals and their relatives apparently evolved from middle Cenozoic carnivorans that became adapted for swimming.

A fourth and very extensive adaptive radiation of eutherian mammals produced the primate-rodent complex. Order Rodentia includes rats, mice, squirrels, and beavers. With about 1770 species, this is by far the largest order of mammals. The name *Rodentia*, from the Latin word for "gnawing," refers to one of the distinctive features of the group: both the upper jaw and the lower jaw have a pair of large front teeth (incisors) that resist heavy wear by growing continually. Order Primates includes monkeys, apes, and humans, which we discuss next.

PRIMATES AND THE PHYLOGENY OF *Homo sapiens*

Having tracked the vertebrate genealogy to the mammalian orders, we now begin to trace our own specific ancestry. Order Primates includes *Homo sapiens* and its closest kin. We are primates. To learn what that means, we must trace our ancestry back to the trees, where some of our most treasured traits had their beginnings as adaptations to an arboreal (tree-dwelling) existence.

Primate evolution provides a context for understanding human origins

Dental structure suggests that the first primates descended from insectivores in the Cretaceous period. These early primates were probably small arboreal (tree-dwelling) mammals. Thus, during the Mesozoic era, our order was defined by characteristics that had been shaped, through natural selection, by the demands of living in the trees. For example, primates have limber shoulder joints, which make it possible to brachiate (swing from one hold to another). The dexterous hands of primates can hang on to branches and manipulate food. Claws have been replaced by nails in many species, and the fingers are very sensitive. The eyes of primates are close together on the front of the face. The overlapping fields of vision of the two eyes enhance depth perception, an obvious advantage when brachiating. Excellent eye-hand coordination is also important for arboreal maneuvering. Parental care is essential for young animals in the trees. Mammals devote more energy to caring for their young than most other vertebrates, and primates are among the most attentive parents of all mammals. Most primates have single births and nurture their offspring for a long time. Though humans do not live in trees, we retain in modified form many of the traits that evolved there.

Modern Primates

The two suborders of the Primates are the Prosimii and Anthropoidea. The **prosimians** ("premonkeys") probably resemble early arboreal primates (FIGURE 34.29). The lemurs of Madagascar and the lorises, pottos, and tarsiers that live in tropical Africa and southern Asia are examples of prosimians. The **anthropoids** include monkeys, apes, and humans.

The evolution of anthropoids from prosimians is a topic of active debate. Paleontologists divide prosimian fossils into two groups: one ancestral to tarsiers, the other ancestral to lemurs, lorises, and pottos (FIGURE 34.30). This split occurred at least 50 million years ago. Until recently, the choice was

FIGURE 34.29 ▪ **A crowned lemur, a prosimian.**

between these two branches as the lineage that also gave rise to anthropoids. However, some recently discovered fossils in Asia and Africa may be more similar to anthropoids than are either of the other two groups of prosimian fossils. The age of these fossils is still in dispute, but their discoverers date them back at least 50 million years. These discoveries raise the possibility that the prosimians had already split into at least three lineages by 50 million years ago, with the anthropoid ancestry already differentiated from the two branches leading to modern prosimians. Thus, the current debate about anthropoid origins can be phrased this way: Are anthropoids more closely related to tarsiers, to lemurs and lorises, or to some third group of prosimians lacking extant representatives? More fossils will help answer this question.

Fossils of monkeylike primates indicate that anthropoids were already established in Africa and Asia by 40 million years ago. Africa and South America had already drifted apart, and it is not clear whether ancestors of New World monkeys reached South America by rafting on logs or other debris from Africa or by migration southward from North America. What is certain is that New World monkeys and Old World monkeys have been evolving along separate pathways for

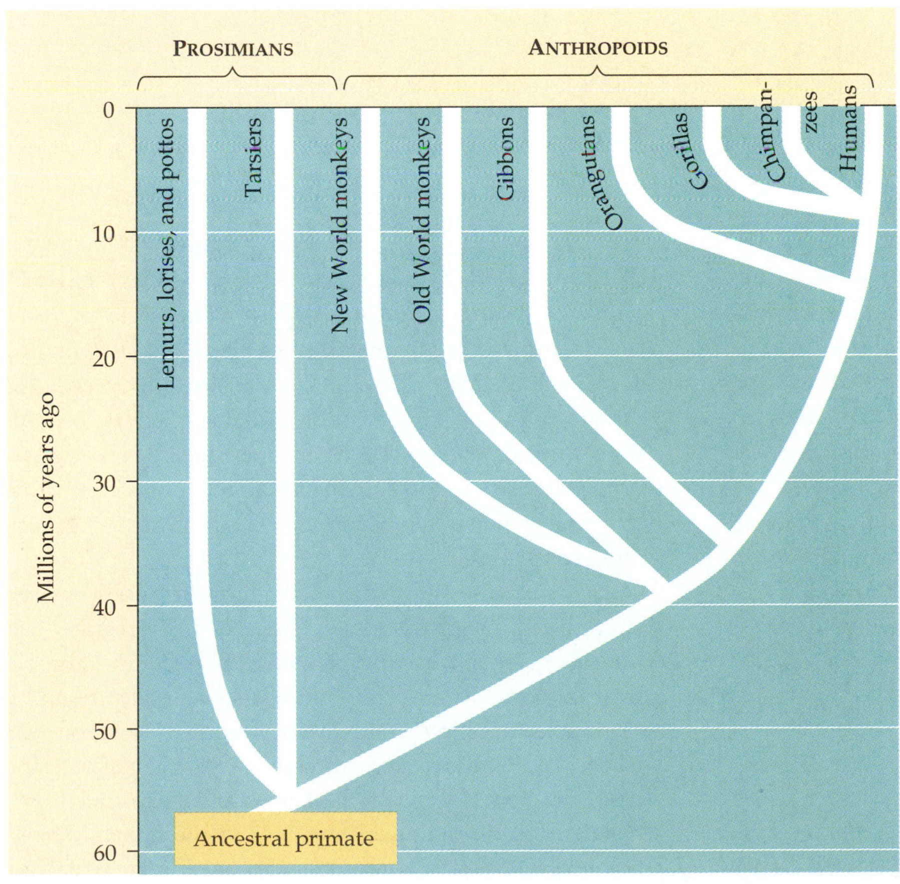

FIGURE 34.30 ▪ **A phylogenetic tree of primates.** The fossil record indicates that the prosimian and anthropoid lineages had diverged by 50 million years ago. Among the anthropoids, New World monkeys, Old World monkeys, and the apes (represented today by several species of gibbons, the orangutan, the gorilla, and chimpanzees) have been evolving on separate lineages for well over 30 million years. Molecular evidence shows that chimpanzees are more closely related to humans than are any other apes. The time of divergence of the hominid lineage (human branch of the evolutionary tree) is controversial; fossil evidence indicates that it occurred between 4 and 8 million years ago.

(a) New World monkey (squirrel monkey)

(b) Old World monkeys (pig-tailed macaques)

FIGURE 34.31 • New World monkeys and Old World monkeys compared. (a) New World monkeys, such as spider monkeys, squirrel monkeys, and capuchins, have prehensile tails and nostrils that open to the sides. This is a squirrel monkey. **(b)** Old World monkeys lack prehensile tails, and their nostrils open downward. The tough seat pad is unique to the Old World group, which includes macaques, mandrills, baboons, and rhesus monkeys. These are pig-tailed macaques.

many millions of years (FIGURE 34.31). All New World monkeys are arboreal, whereas Old World monkeys include ground-dwelling as well as arboreal species. Most monkeys of both groups are diurnal (active during the day) and usually live in bands held together by social behavior.

In addition to monkeys, the anthropoid suborder also includes the four genera of apes, shown in FIGURE 34.32: *Hylobates* (gibbons), *Pongo* (orangutans), *Gorilla* (gorillas), and *Pan* (chimpanzees). Modern apes are confined exclusively to tropical regions of the Old World. With the exception of gibbons, modern apes are larger than monkeys, with relatively long arms and short legs and no tails. Although all the apes are capable of brachiation, only gibbons and orangutans are primarily arboreal. Social organization varies among the genera of apes; gorillas and chimpanzees are highly social. Apes have larger brains proportionate to body size than monkeys, and their behavior is consequently more adaptable.

Humanity is one very young twig on the vertebrate tree

In the continuum of life spanning over 3.5 billion years, humans and apes have shared ancestry for all but the last few million years. **Paleoanthropology**, the study of human origins and evolution, focuses on this tiny fraction of geological time during which humans and chimpanzees diverged from a common ancestor.

Some Common Misconceptions

Paleoanthropology has a checkered history. Until about 20 years ago, researchers often gave new names to fossil forms that were undoubtedly the same species as fossils found by competing scientists. Elaborate theories have often been based on a few teeth or a fragment of jawbone. Many misconceptions about human evolution generated during the early part of the twentieth century still persist in the minds of much of the general population, long after these myths have been replaced by fossil discoveries.

Let's first dispose of the myth that our ancestors were chimpanzees or any other modern apes. Chimpanzees and humans represent two divergent branches of the anthropoid tree that evolved from a common, less-specialized ancestor.

Another misconception envisions human evolution as a ladder with a series of steps leading directly from an ancestral anthropoid to *Homo sapiens*. This is often illustrated as a parade of fossil hominids (members of the human family) becoming progressively more modern as they march across the page. If human evolution is a parade, then it is a disorderly one, with many splinter groups having traveled down dead

(a) Gibbon

(b) Orangutan

(c) Gorilla with infant

(d) Chimpanzee

FIGURE 34.32 ▪ Apes. **(a)** Gibbons have long arms and are among the most acrobatic of all primates. These Asian primates are also the only monogamous apes. **(b)** The orangutan is a shy and solitary ape that lives in the rain forests of Sumatra and Borneo. Orangutans spend most of their time in trees, but they do venture onto the forest floor occasionally. **(c)** The gorilla is the largest ape, with some males almost 2 m tall and weighing about 200 kg. These herbivores are confined to Africa, where they usually live in small groups of about 10 to 20 individuals. **(d)** The chimpanzee lives in tropical Africa. Chimpanzees feed and sleep in trees but also spend a great deal of time on the ground. Chimpanzees are intelligent, communicative, and social. Some molecular evidence indicates that chimpanzees are more closely related to humans than they are to other apes.

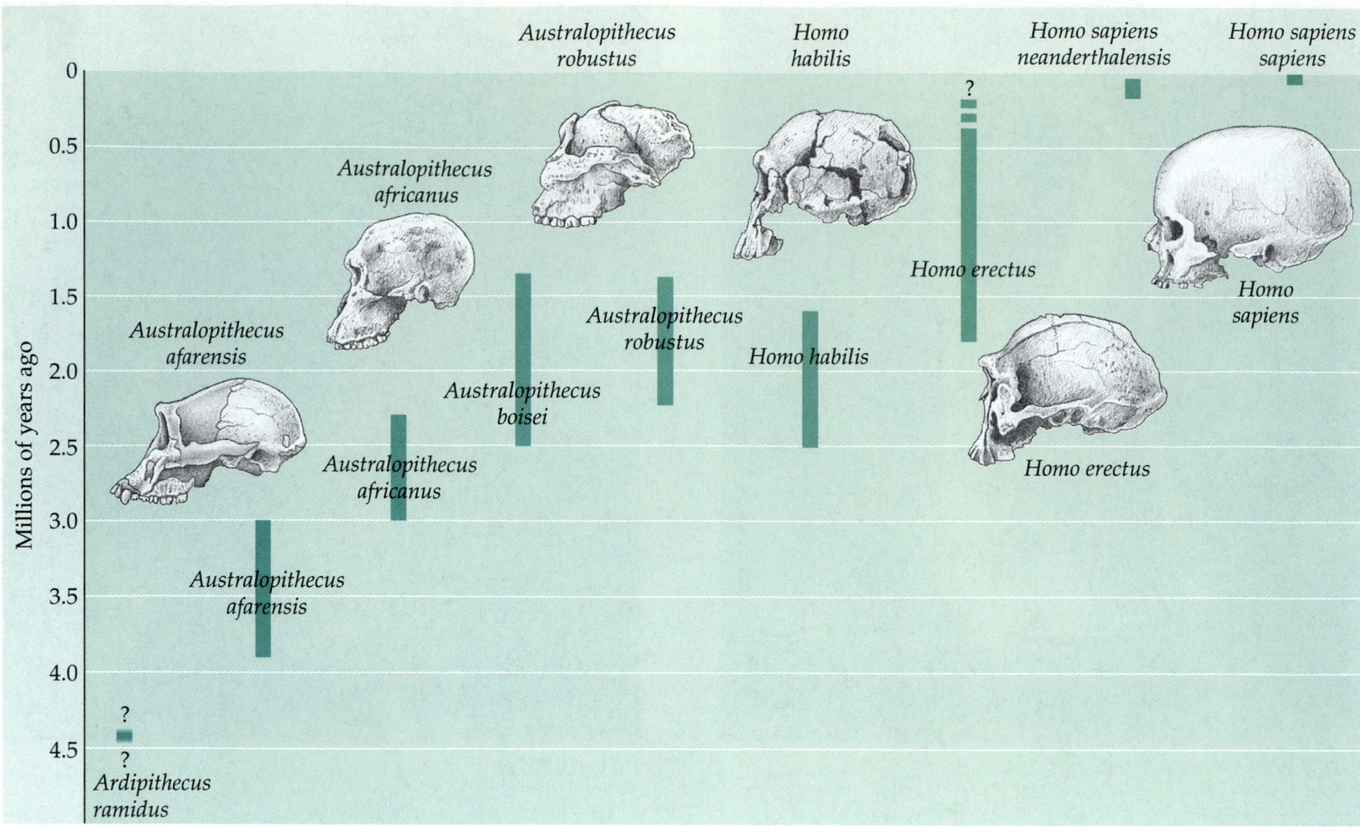

FIGURE 34.33 ▪ A timeline of some hominid species. Notice that there have been times in the history of human evolution when two or more hominids coexisted.

ends. At times in hominid history, several different human species coexisted (FIGURE 34.33). Human phylogeny is more like a multibranched bush than a ladder, our species being the tip of the only twig that still lives. If the theory of punctuated equilibrium applies to humans, then most change occurred as new hominid species came into existence, not by phyletic (anagenic) change within an unbranched hominid lineage.

One more myth we must bury is the notion that various human characteristics, such as upright posture and an enlarged brain, evolved in unison. A popular image is of early humans as half-stooped, half-witted cave-dwellers. Different features evolved at different rates—a phenomenon known as **mosaic evolution**—with erect posture, or bipedalism, leading the way. Our pedigree includes ancestors who walked upright but had a brain much less developed than ours.

After dismissing some of the folklore on human evolution, however, we must admit that many questions about our ancestry remain.

Early Anthropoids

The oldest known fossils of apes have been assigned to the genus *Aegyptopithecus*, the "dawn ape." This anthropoid was a cat-sized tree-dweller that lived about 35 million years ago. During the Miocene epoch, which began about 23 million years ago, descendants of the first apes diversified and spread into Eurasia. About 20 million years ago, the Indian plate collided with Asia and thrust up the Himalayan range. The climate became drier and the forests of what is now Africa and Asia contracted, isolating these two regions of anthropoid evolution. Among the African anthropoids was the common ancestor of humans and chimpanzees. Based on the fossil record and comparisons of DNA between humans and chimpanzees, most anthropologists now agree that humans and apes diverged from this common ancestor only about 6 to 8 million years ago.

The First Humans

In 1924, British anthropologist Raymond Dart announced that a fossilized skull discovered in a South African quarry was the remains of an early human. He named his "ape-man" *Australopithecus africanus* ("southern ape of Africa"). With the discovery of more fossils, it became clear that *Australopithecus* was in fact a hominid that walked fully erect (was bipedal) and had humanlike hands and teeth. However, the brain of *Australo-*

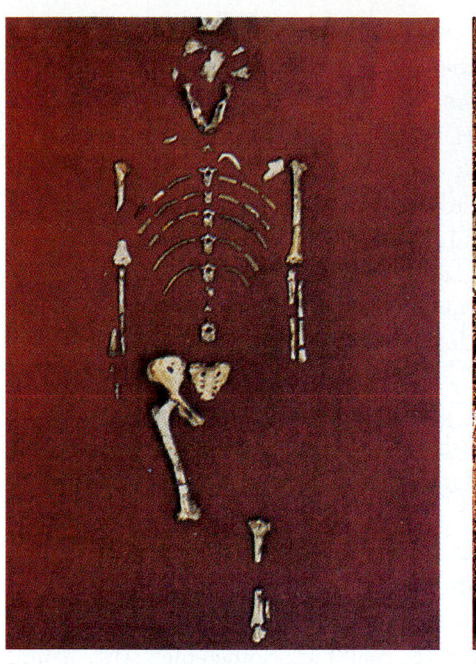
(a)

(b)

(c)

FIGURE 34.34 ▪ The antiquity of upright posture. (a) Lucy, a 3.18-million-year-old skeleton, represents the hominid species *Australopithecus afarensis.* Fragments of the pelvis and skull put *A. afarensis* on two feet. **(b)** The Laetoli footprints, over 3.5 million years old, confirm that upright posture evolved quite early in hominid history. **(c)** This skull, discovered in 1993, is among the *A. afarensis* fossils that now span almost a million years, from 3.9 to 3.0 million years ago.

pithecus was only about one-third the size of a modern human's brain. Various species of *Australopithecus* existed for over 3 million years, probably beginning about four and a half million years ago.

In 1974, in the Afar region of Ethiopia, paleoanthropologists discovered an *Australopithecus* skeleton that was 40% complete. "Lucy," as the fossil was named, was petite—only about 1 m tall, with a head about the size of a softball. The skeleton is 3.2 million years old. Lucy and similar fossils have been considered sufficiently different from *Australopithecus africanus* to be named a separate species, *Australopithecus afarensis* (for the Afar region). The fragments of the pelvis and skull bones indicate that *A. afarensis* walked on two legs (FIGURE 34.34). Fossils discovered in the early 1990s extend the longevity of *A. afarensis* as a species to a span of at least one million years.

Since 1994, paleoanthropologists working at other sites in East Africa have unearthed hominid fossils that predate *A. afarensis.* One, called *Australopithecus anamensis,* dates from about 4 million years ago; another very different hominid, named *Ardipithecus ramidus,* probably lived about 4.4 million years ago. With the discovery of these oldest known hominids, the fossil record of humanity is creeping closer to the ape-human split estimated to have occurred 5 to 8 million years ago.

The phylogenetic relationships among early hominids is stirring much debate. One issue is whether *Ardipithecus ramidus* is the ancestor of *Australopithecus* or an evolutionary side-branch that faded into extinction. Also at issue is the origin of bipedalism. The oldest hominid fossils challenge a longstanding hypothesis that bipedalism evolved when hominids began living on the African savanna (grasslands dotted with trees). *Australopithecus anamensis,* which was completely bipedal, apparently lived in lakeside forests. *Ardipithecus,* whose posture is still being debated, inhabited dense forests. Recent evidence also suggests that some populations of *Australopithecus afarensis* lived in wooded areas.

Another debate centers on the evolutionary fate of *Australopithecus afarensis.* For the million years that it is represented in the fossil record, *Australopithecus afarensis* changed little. Then, beginning about three million years ago, an adaptive radiation began that produced several new hominid species, including *Australopithecus africanus* and several heavier-boned *Australopithecus* species. This period of speciation also produced *Homo habilis,* the first member of our genus, which debuts in the fossil record about 2.5 million years ago. Was *Australopithecus africanus* ancestral to any or all of these diverse hominids? Or was Lucy a member of a species that branched early from an unknown ancestor that also gave rise to *Homo?*

Whatever the outcome of debates about the phylogeny of early hominids, the fossil record makes one fact of our history clear: Hominids walked upright for about two million years with no substantial enlargement of the brain. Perhaps bipedalism freed the hands for gathering food or caring for babies; the making of sophisticated tools came much later. Evolutionary biologist Stephen Jay Gould puts it this way: "Mankind stood up first and got smart later."

Homo habilis

Enlargement of the human brain is first evident in fossils dating back to the latter part of australopithecine times, about 2.5 million years ago. Simple handmade stone tools are sometimes found with the larger-brained fossils, named *Homo habilis* ("handy man"). After walking upright for at least two million years, hominids were finally beginning to use their brains and hands to fashion tools.

The hominid radiation that included *Homo habilis* is the focus of yet another important debate in paleoanthropology. One hypothesis is that the hominid radiation occurred very rapidly (from about 2.8 to 2.5 million years ago) and was part of a speciation pulse caused by a continent-wide drying trend in Africa. A related hypothesis invokes global climate change and expands the time frame to the period from 3 million to 2 million years ago. Yale University paleontologist Elisabeth Vrba is a central figure in this debate (see the interview at the beginning of this unit).

Homo habilis coexisted for a million years with the smaller-brained *Australopithecus,* including *Australopithecus robustus,* named for its stocky features and heavy skull. Early *Homo* and australopithecines may not have competed directly, although all may have scavenged for food, hunted animals, and gathered fruits and vegetables. According to one hypothesis of human origins, the late australopithecines and *Homo habilis* were distinct lines of hominids, neither evolving from the other. If this view is correct, then *Australopithecus africanus* was an evolutionary dead end, while *H. habilis* may have been on the path to modern humans, leading first to *Homo erectus,* which later gave rise to *Homo sapiens.*

Homo erectus *and Descendants*

What species first extended humanity's range from its birthplace in Africa to other continents? Some recent fossil evidence from sites in Europe and Asia indicates that *Homo habilis* was the first hominid to migrate out of Africa, perhaps as early as 1.8 million years ago. According to most accounts, however, a younger species, *Homo erectus* ("upright man"), was the first hominid to do so. The fossils known as Java Man and Beijing Man are examples of this species. *H. erectus* lived from about 1.8 million years ago until about 250,000 years ago. Fossils covering that entire range are found in Africa, where *H. erectus* continued to live contemporaneously with *H. erectus* populations on other continents. We need not picture this migration as a mad dash for new territory, or even a casual stroll. If *H. erectus* simply expanded its range from Africa by about a mile per year, it would take only about 15,000 years to reach Java and other parts of Asia and Europe. The gradual spread may have been associated with a change in diet to include a larger proportion of meat. In general, animals that hunt require more geographical territory than animals that feed exclusively on vegetation.

Homo erectus was taller than *H. habilis* and had a larger brain capacity. During the 1.5 million years the species existed, the *H. erectus* brain increased to as large as 1200 cc, a brain capacity that overlaps the normal range for modern humans.

The intelligence that evolved during the African origins of *H. habilis* enabled these humans to survive in the colder climates of the north, once migration began. *Homo erectus* resided in huts or caves, built fires, clothed themselves in animal skins, and designed stone tools that were more refined than the tools of *Homo habilis.* In anatomical and physiological adaptations, *H. erectus* was poorly equipped for life outside the tropics but made up for the deficiencies with intelligence and social cooperation.

Some African, Asian, European, and Australasian (Indonesia, New Guinea, and Australia) populations of *H. erectus* gave rise to regionally diverse descendants that had even larger brains. Among these descendants of *H. erectus* were the Neanderthals, who lived in Europe, the Middle East, and parts of Asia from about 130,000 years ago to about 35,000 years ago. (They are named Neanderthals because their fossils were first found in the Neander Valley of Germany.) Compared with us, Neanderthals had slightly heavier brow ridges and less pronounced chins, but their brains, on average, were slightly larger than ours. Neanderthals were skilled toolmakers, and they participated in burials and other rituals that required abstract thought. Much current research on Neanderthal skulls addresses an intriguing question: Did Neanderthals have the anatomical equipment necessary for speech?

The oldest known post–*H. erectus* fossils, dating back over 300,000 years, are found in Africa. Many paleoanthropologists group these African fossils along with Neanderthals and various Asian and Australasian fossils as the earliest forms of our species, *Homo sapiens.* These regionally diverse descendants of *H. erectus* are sometimes referred to as "archaic *Homo sapiens.*"

The Emergence of Homo sapiens: *Out of Africa . . . But When?*

What was the fate of Neanderthals and their contemporaries who populated different parts of the world? In the view of some anthropologists, these archaic *Homo sapiens* gave rise to fully modern humans. According to this model, called the *multiregional model*, modern humans evolved in parallel in different parts of the world (FIGURE 34.35a). If this view is correct, then the geographic diversity of humans originated relatively early, when *Homo erectus* spread from Africa into the other continents between 1 and 2 million years ago. This model accounts for the great genetic similarity of all modern people by pointing out that occasional interbreeding among neighboring populations has always provided corridors for gene flow throughout the entire geographical range of humanity.

A number of paleoanthropologists, including Christopher Stringer of University College, London, interpret the fossil record differently and have proposed an alternative to the multiregional model. The debate has focused partly on the relationship between Neanderthals and the modern humans of Europe and the Middle East. Perhaps the most famous fossils of modern *Homo sapiens*—skulls and other bones that look essentially like those of today's humans—are the remains of Cro-Magnons. Named for the French cave in which these fossils were first discovered, Cro-Magnons date back about 35,000 years. But the oldest fully modern fossils of *Homo sapiens*, about 100,000 years old, are found in Africa; similar fossils almost as ancient have also been discovered in caves in Israel. The Israeli fossils were found not far from other caves containing Neanderthal-like fossils ranging in vintage from about 120,000 to 60,000 years old. These two human types apparently coexisted in this region for at least 40,000 years, from the time modern forms appeared 100,000 years ago. And, according to Stringer and his supporters, the persistence of distinct Neanderthal and modern forms during their time of coexistence means that these two human types did not interbreed. If this interpretation of the Israeli fossils is correct, then the Neanderthals of that region could not have been the ancestors of the modern humans that also lived there. In fact, based on the Israeli bones and other fossil data, Stringer postulates that Neanderthals and the other so-called archaic *Homo sapiens* outside of Africa were evolutionary dead ends.

Thus, Stringer is among the anthropologists who contend that modern humans evolved from *Homo erectus* first in Africa, then migrated to other continents, replacing Neanderthals and the other regional descendants of *H. erectus*. This model, called the "out of Africa" or *monogenesis model*, contrasts sharply with the multiregional model. According to the monogenesis model, modern humanity did not emerge in many different parts of the world, but only in Africa, from which modern humans dispersed relatively recently (FIGURE 34.35b). If this view is correct, then the geographical diversification of modern humans occurred within just the past

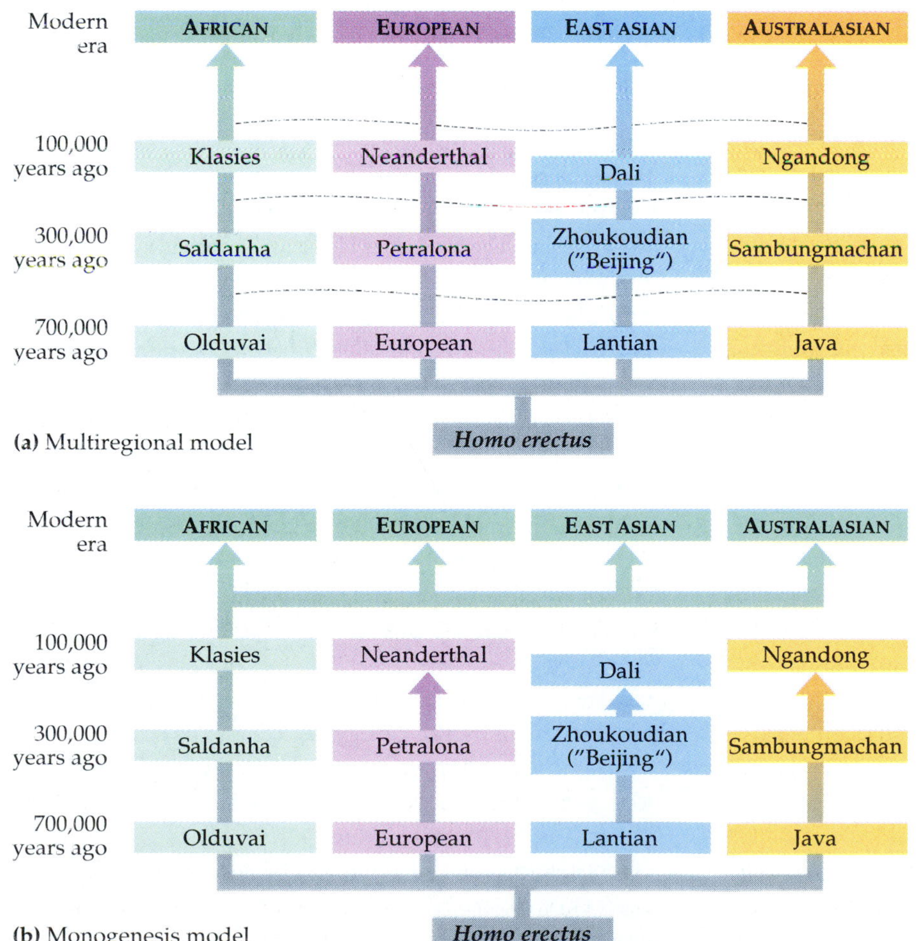

FIGURE 34.35 • Two models for the origin of modern humans. (a) According to the multiregional model, modern humans evolved in many parts of the world from regional descendants of *Homo erectus*, who dispersed from Africa between 1 and 2 million years ago. The boxed names indicate various fossils associated with each region. The dashed lines symbolize interbreeding and gene flow between different populations. **(b)** According to the monogenesis model, only the African descendants of *Homo erectus* gave rise to modern humans. All other regional descendants of *H. erectus*, including Neanderthals, became extinct without contributing to the gene pool of modern humanity. Advocates of the monogenesis model argue that modern humans began spreading from Africa just 100,000 years ago, giving rise to all the diverse populations of modern humans.

100,000 years (compared to at least 1 million years for the multiregional model). To Stringer, the strongest argument for an exclusively African genesis for modern humans is that only African fossils chronicle the complete transition from archaic *Homo sapiens* to modern humans.

In the late 1980s, Rebecca Cann and other geneticists in the laboratory of the late Allan Wilson at the University of California, Berkeley, made news with research results that seemed to support the monogenesis model of human origins. Instead of digging in the field for bones, Wilson's group probed in the DNA of living humans for clues about our ancestry. Specifically, these scientists compared the mitochondrial DNA (mtDNA) of a multiethnic sample of more than 100 people representing four continents. The greater the difference between the mtDNA of two people, the longer ago that mtDNA diverged from a common source. Using computers to analyze their data, the geneticists traced the source of all human mtDNA back to Africa. The surprise was the calculation that the divergence of mtDNA from this common source began just 200,000 years ago, much too late to represent the dispersal of *Homo erectus*. Advocates of the monogenesis model cheered the genetic evidence for a second, much later dispersal of Africans in the form of modern humans.

In 1992, several researchers challenged the Berkeley group's interpretation of the mtDNA data. At issue are the methods used to construct phylogenetic trees from the mtDNA comparisons and the reliability of changes in mtDNA as a molecular clock for timing branch points in these trees (see Chapter 25). The criticism encouraged proponents of the multiregional model, who continue to argue that multiregional evolution of modern humans fits the fossil evidence better than does African monogenesis. These scientists interpret certain fossils in various parts of the world as links between the regional versions of archaic *Homo sapiens* and the current people native to those continents.

The debate about the origin of modern humans continues, largely fueled by molecular studies. Strengthening the case for a spread of modern humans out of Africa, one research group has found greater genetic diversity within African populations south of the Sahara than in other parts of the world. The result is predicted by the hypothesis that modern humans originated in southern Africa and therefore have the longest history of genetic diversification in that region. If populations in other regions began by later migration, the founder effect and genetic drift (see Chapter 23) could account for the smaller genetic diversity.

In a more recent study, molecular systematists obtained a sample of mtDNA from an unusually well-preserved Neanderthal fossil in Europe. After a painstaking effort to eliminate contamination of the mtDNA with DNA from other sources, they used polymerase chain reaction (PCR) to amplify the DNA traces and produced enough for comparison to the DNA of living humans. Their analysis indicated that the Neanderthal individual was genetically very different from modern Europeans. This result supports the monogenesis model, but generalizations cannot be made from one specimen. However, studies of DNA from the *Y* chromosome of modern-day humans also provide support for the out of Africa scenario. In contrast, other research points to an archaic human population in Asia as the source of genes coding for some blood proteins. Perhaps, as some researchers have proposed, an intermediate model for the origin of modern humans is more realistic than the monogenesis and multiregional models: Modern *H. sapiens* may be the result of a migration out of Africa as well as some genetic contribution from non-African archaic groups.

Cultural Evolution: A New Force in the History of Life

An erect stance was the most radical anatomical change in our evolution; it required major remodeling of the foot, pelvis, and vertebral column. Enlargement of the brain was a secondary alteration made possible by prolonging the growth period of the skull and its contents. The brains of nonprimate mammalian fetuses grow rapidly, but the growth usually slows down and stops not long after birth. The primate brain continues to grow after birth, and the period of growth is longer for a human than for any other primate. The extended period of human development also lengthens the time parents care for their offspring, which contributes to the child's ability to benefit from the experiences of earlier generations. This is the basis of culture—the transmission of accumulated knowledge over the generations. The major means of this transmission is language, written and spoken.

Cultural evolution is continuous, but there have been three major stages. The first stage began with nomads who hunted and gathered food on the African grasslands 2 million years ago. They made tools, organized communal activities, and divided labor. The second stage came with the development of agriculture in Africa, Eurasia, and the Americas about 10,000 to 15,000 years ago. Along with agriculture came permanent settlements and the first cities. The third major stage in our cultural evolution was the Industrial Revolution, which began in the eighteenth century. Since then, new technology has escalated exponentially; a single generation spanned the flight of the Wright brothers and Neil Armstrong's walk on the moon. Through all this cultural evolution, from simple hunter-gatherers to high-tech societies, we have not changed biologically in any significant way. We are probably no more intelligent than our forebears in Africa and in Eurasian caves. The same toolmaker who chipped away at stones now fashions microchips. The know-how to build skyscrapers, computers, and spaceships is stored not in our genes but in the cumulative product of hundreds of generations of human experience, passed along by parents, teachers, books, and, most recently, by electronic means.

Evolution of the human brain may have been anatomically simpler than acquiring an upright stance, but the consequences of cerebral expansion have been enormous. Cultural evolution made *Homo sapiens* a new force in the history of life—a species that could defy its physical limitations and shortcut biological evolution. We do not have to wait to adapt to an environment through natural selection; we simply change the environment to meet our needs. We are the most numerous and widespread of all large animals, and everywhere we go, we bring environmental change. There is nothing new about environmental change. As we have seen throughout this unit, the history of life is the story of biological evolution on a changing planet. But it is unlikely that change has ever been as rapid as in the age of humans. Cultural evolution outpaces biological evolution by orders of magnitude. We are changing the world faster than many species can adapt; the rate of extinctions in the twentieth century is 50 times greater than the average for the past 100,000 years.

This rapid rate of extinction is mainly a result of habitat destruction, which is a function of human cultural changes and overpopulation. Feeding, clothing, and housing the approximately 6 billion people who now exist imposes an enormous strain on Earth's capacity to sustain life. If all these people suddenly assumed the high standard of living enjoyed by many people in developed nations, it is likely that Earth's support systems would be overwhelmed. Already, for example, current rates of fossil-fuel consumption, mainly by developed nations, are so great that waste carbon dioxide may be causing the temperature of the atmosphere to increase enough to alter world climates. Today, not just individual species, but entire ecosystems, the global atmosphere, and the oceans, are seriously threatened. Tropical rain forests, which play a vital role in moderating global weather, are being cut down at a startling rate. Scientists have hardly begun to study these ecosystems, and many species in them may become extinct before they are even discovered.

Of the many crises in the history of life, the impact of one species, *Homo sapiens,* is the latest and potentially the most devastating. We take a closer look at our environmental impacts in Chapters 54 and 55.

CHAPTER REVIEW

REVIEW OF KEY CONCEPTS

(with page numbers and key figures)

INVERTEBRATE CHORDATES AND THE ORIGIN OF VERTEBRATES

34.1 **Four anatomical features characterize Phylum Chordata (pp. 630–631, FIGURE 34.1)** Chordates are a diverse phylum of deuterostomes, classified together by virtue of shared structures: a notochord; a dorsal, hollow nerve cord; pharyngeal slits; and a postanal tail.

- **Invertebrate chordates provide clues to the origin of vertebrates (pp. 631–633, FIGURES 34.2, 34.3)** Subphylum Urochordata includes the marine, suspension-feeding tunicates, or sea squirts. Members of Subphylum Cephalochordata (the lancelets) are exemplary chordates. Cephalochordates and vertebrates probably evolved from a common ancestor by the elaboration of larval forms.

INTRODUCTION TO THE VERTEBRATES

- **Neural crest, pronounced cephalization, a vertebral column, and a closed circulatory system characterize Subphylum Vertebrata (pp. 633–634, FIGURE 34.5)** A mass of embryonic cells called the neural crest gives rise to many of the uniquely vertebrate characteristics. Vertebrates have a well-developed head with a braincase (skull). Their segmentally arranged vertebrae enclose the dorsal hollow nerve cord (spinal cord). Vertebrate organ systems, such as their closed circulatory system, support an active metabolism.

34.1 **Overview of vertebrate diversity (pp. 634–636, FIGURE 34.6, TABLE 34.1)** Except for a few species that lack jaws, all extant vertebrates have hinged jaws. Jawed vertebrates living today are classified in two extant classes of fishes and four classes of tetrapods.

SUPERCLASS AGNATHA: JAWLESS VERTEBRATES

- **Lampreys and hagfishes are the only extant agnathans (p. 636, FIGURE 34.7)** Lampreys and hagfishes living today represent the oldest vertebrates.

SUPERCLASS GNATHOSTOMATA I: THE FISHES

Vertebrates with jaws largely replaced the agnathans during the Silurian and Devonian periods. Placoderms, now extinct, were a large group of Devonian fishes with hinged jaws.

- **Vertebrate jaws evolved from skeletal supports of the pharyngeal slits (p. 637, FIGURE 34.8)** The evolution of jaws from the skeletal supports of gills was a pivotal milestone in vertebrate evolution. Vertebrate jaws are hinged dorsoventrally and work up and down. In contrast to the placoderms, two extant classes of fishes (Chondrichthyes and Osteichthyes) have teeth in their jaws.
- **A cartilaginous endoskeleton reinforced by calcified granules is diagnostic of Class Chondrichthyes (pp. 637–638, FIGURE 34.9)** Sharks, rays, and chimaeras (ratfishes) make up the class Chondrichthyes.
- **A bony endoskeleton, operculum, and swim bladder are hallmarks of Class Osteichthyes (pp. 639–640, FIGURE 34.13)** The bony fishes, the most species-rich vertebrate class, have skeletons reinforced by calcium phosphate. In contrast to the cartilaginous fishes, bony fishes have opercula (bony gill covers) and can adjust their density and thus control their buoyancy by means of a swim bladder. Ray-finned fishes, the most numerous group today, have fins supported by flexible rays and modified for various functions. Lungfishes and most lobe-finned fishes have lungs that aid their gills in breathing. Devonian lobe-finned fishes gave rise to tetrapods.

SUPERCLASS GNATHOSTOMATA II: THE TETRAPODS

- **Amphibians are the oldest class of tetrapods (pp. 640–643, FIGURE 34.15)** The first terrestrial vertebrates, amphibians radiated during early Carboniferous times. Attesting to their aquatic heritage, most modern amphibians have moist skin that complements the lungs in gas exchange. Most species lay their unshelled eggs in wet environments. Most frogs and their relatives undergo metamorphosis of an aquatic larval stage into a terrestrial adult. Extant amphibians are the urodeles (salamanders), anurans (frogs), and apodans (legless, burrowing amphibians).

- **Evolution of the amniotic egg expanded the success of vertebrates on land (pp. 643–644, FIGURE 34.17)** The amniotic egg has extraembryonic membranes and fluids that protect and hydrate the embryo. This adaptation, characteristic of reptiles, birds, and mammals, and permitting development in a dry environment, made it possible for vertebrates to reproduce on land.
- **A reptilian heritage is evident in all amniotes (pp. 644–647, FIGURE 34.19)** Reptiles, a diverse group represented today by lizards, snakes, turtles, and crocodilians, have numerous terrestrial adaptations, including lungs, waterproof scales, and a shelled amniotic egg.
- **Birds began as flying reptiles (pp. 647–650, FIGURES 34.23–34.25)** Flight feathers are diagnostic of birds, whereas their amniotic eggs and scaled legs give testimony to their reptilian heritage. Birds probably descended from a group of small, carnivorous dinosaurs. Almost every part of avian anatomy enhances flight.
- **Mammals diversified extensively in the wake of the Cretaceous extinctions (pp. 650–656, FIGURES 34.27, 34.28; TABLE 34.2)** Hair and mammary glands are the two diagnostic characteristics of mammals. Monotremes are egg-laying mammals, today represented only by the platypuses and echidnas found in Australia and New Guinea. Marsupials include opossums, kangaroos, and koalas, animals whose young complete their embryonic development inside a maternal pouch, the marsupium. Marsupials isolated in Australia show convergent evolution with eutherian (placental) mammals in other parts of the world. The most widespread and diverse modern mammals are the eutherians (placentals), a group whose young complete their embryonic development attached to a placenta inside the mother's uterus.

PRIMATES AND THE PHYLOGENY OF *Homo sapiens*

- **Primate evolution provides a context for understanding human origins (pp. 656–658, FIGURE 34.30)** The first primates were probably small arboreal animals, which probably descended from insectivores. Two subgroups of modern primates are the prosimians—lemurs and their relatives—and the anthropoids. Anthropoids diverged early into New World and Old World monkeys. Modern apes—gibbons, orangutans, gorillas, and chimpanzees—are confined to the Old World.
- **Humanity is one very young twig on the vertebrate tree (pp. 658–665, FIGURE 34.33)** The oldest humanlike fossil yet discovered is called *Ardipithecus ramidus*, a forest-dweller dating from about 4.4 million years ago in Africa. Upright posture evolved before enlargement of the brain in the hominids. The first species of *Homo*, *H. habilis*, dating from about 2.5 million years ago, coexisted with smaller-brained hominids of the genus *Australopithecus*. *Homo erectus*, apparently a descendant of *H. habilis*, may have been the first hominid to venture out of the tropics and into colder climates. *Homo erectus* gave rise to regionally diverse versions of archaic *Homo sapiens*, including Neanderthals. According to the multiregional model, modern humans evolved in several locations from Neanderthals and other archaic *Homo sapiens*. In contrast, the monogenesis model views all but the African archaic *Homo sapiens* as evolutionary dead ends. According to this model, a relatively recent dispersal (100,000 years ago) of modern Africans gave rise to today's human diversity. Some researchers favor an intermediate model—that modern humans were derived from a recent migration out of Africa and obtained some genes from non-African archaic groups. Human cultural evolution began in Africa with wandering hunter-gatherers, progressed to the development of agriculture, and eventually to the Industrial Revolution, which today continues as accelerating technological change. The exploding human population now threatens Earth's ecosystems.

SELF-QUIZ

1. Vertebrates and tunicates may seem as different as two animal groups can be, yet they share
   a. the same type of body symmetry as adults
   b. a high degree of cephalization
   c. the formation of structures from the neural crest
   d. an endoskeleton that includes a cranium
   e. the presence of a notochord; a dorsal, hollow nerve cord; and pharyngeal slits

2. Which of the following groups is entirely extinct?
   a. cephalochordates
   b. agnathans
   c. placoderms
   d. lobe-finned fishes
   e. ratite birds

3. In addition to skeletal differences, cartilaginous fishes can be distinguished from bony fishes
   a. by the presence in bony fishes of the spiral valve
   b. by the presence in bony fishes of eardrums
   c. by the presence in cartilaginous fishes of unpaired fins
   d. by the absence of a swim bladder in cartilaginous fishes
   e. because cartilaginous fishes retain the lateral line system, which bony fishes have lost

4. Mammals and extant birds share all of the following characteristics *except*
   a. endothermy
   b. descent from reptiles
   c. a four-chambered heart
   d. teeth specialized for diverse diets
   e. the ability of some species to fly

5. If you were to observe a monkey in a zoo, which characteristic would indicate a New World origin for that monkey species?
   a. distinct "seat pads"
   b. eyes close together on the front of the skull
   c. use of the tail to hang from a tree limb
   d. occasional bipedal walking
   e. downward orientation of the nostrils

6. Only an animal species with a diaphragm can be expected to have
   a. hair
   b. feathers
   c. scales
   d. lungs
   e. moist skin

7. Unlike eutherian (placental) mammals, both monotremes and marsupials
   a. lack nipples
   b. have some embryonic development outside the mother's uterus
   c. lay eggs
   d. are found in Australia and Africa
   e. include only insectivores and herbivores

8. Which of the following is *not* thought to be ancestral to humans?
   a. a reptile
   b. a bony fish
   c. a primate
   d. an amphibian
   e. a bird

9. As humans diverged from other primates, which of the following most likely appeared first?
   a. the development of culture
   b. language
   c. an erect stance
   d. toolmaking
   e. an enlarged brain

10. The multiregional and monogenesis hypotheses for the origin of modern humans agree that
    a. *Homo erectus* had an African origin
    b. modern *Homo sapiens* originated only in Africa
    c. Neanderthals are the ancestors of modern humans in Europe
    d. Australopithecines migrated out of Africa
    e. North America had the first population of modern humans

CHALLENGE QUESTIONS

1. Compare and contrast bats and birds, two flying vertebrates, in as many ways as you can.
2. Throughout chordate evolution, many existing structures have been co-opted for new functions. Describe several such cases of exaptation.
3. Write a short essay comparing and contrasting vertebrates with invertebrate chordates.
4. Evidence supports the hypothesis that alligators, dinosaurs, mammals, and birds are more closely related to each other than any of these groups is to turtles. If we seek to have taxonomic groups that reflect evolutionary relationships—all members of a group having a common ancestor and the group including all the descendants of that common ancestor—how should we revise the classification of vertebrates?
5. Using the hypothesis of tetrapod phylogeny mentioned in challenge question 4, construct a phylogenetic tree of the two classes of fishes, the amphibians, reptiles, birds, and mammals. Your tree should include only monophyletic groupings.

SCIENCE, TECHNOLOGY, AND SOCIETY

1. There are 90 million acres in the U.S. National Wildlife Refuge System, representing most of the habitats within the United States and protecting thousands of animal species. Many refuges permit logging, grazing, farming, mineral exploration, and recreational activities. Each year 400,000 animals are legally taken from refuges by hunters and trappers. Pressure is mounting to permit oil and gas exploration in the Arctic National Wildlife Refuge of Alaska. What level of human activity do you think should be permitted on wildlife refuges? Should they be open to "consumptive" uses, such as logging, farming, and hunting? Or should refuges be sanctuaries where animals are touched as little as possible by human activities? Give reasons for your answers.
2. A lot of money and attention have been directed toward saving a few endangered animals, such as the black-footed ferret and the California condor. (Millions of dollars have been spent in efforts to maintain and breed these two species—more than has been spent on all endangered invertebrates, reptiles, and plants combined.) Condors and ferrets have recently been returned to the wild, but it is questionable whether they can withstand the human encroachments that endangered them in the first place. Should we work so hard to save these animals? Why or why not? Would it be better to use our resources to save more species that have a better chance of survival? What are some biological arguments for focusing preservation efforts on whole ecosystems instead of individual endangered species?
3. Long vilified as heartless killing machines, sharks have until recently been overlooked while other fish species were given protection from overfishing. Many species are now in decline, and large coastal species, such as the great white shark, appear to be most seriously affected. California has passed legislation that prohibits taking great whites in state waters. The National Marine Fisheries Service has proceeded more cautiously, imposing commercial quotas and recreational restrictions. Was California overreacting with complete great white protection, or has the NMFS acted too slowly? What information would you need to be able to support either decision?

FURTHER READING

Agnew, N. and M. Demas. "Preserving the Laetoli Footprints." *Scientific American*, September 1998. New analyses and efforts to conserve these early human imprints; vividly illustrated.

Blaustein, A. R., and D. B. Wake. "The Puzzle of Declining Amphibian Populations." *Scientific American*, April 1995. Is it habitat destruction or damage to the ozone shield in the atmosphere that's causing loss of amphibians?

Carroll, R. L. *Patterns and Processes of Vertebrate Evolution*. New York: Cambridge University Press. 1997. An authoritative yet accessible text.

Currie, P. J., and K. Padian (eds). *Encyclopedia of Dinosaurs*. San Diego, CA: Academic Press, 1997. The most comprehensive single volume on dinosaurs.

Gore, R. "The Dawn of Humans, Expanding Worlds." *National Geographic*, May 1997. A pictorial essay on early human migrations.

Jeffery, W. R. "Evolution of Ascidian Development." *BioScience*, July/August 1997. Genetic changes that determine whether urochordates develop a tail may have been important in chordate evolution.

Kahn, P., and A. Gibbons. "DNA From an Extinct Human." *Science*, July 11, 1997. A news brief about the analysis of Neanderthal mtDNA.

Leakey, M., and A. Walker. "Early Hominid Fossils From Africa." *Scientific American*, June 1997. An illustrated essay on the oldest hominid fossils.

Padian, K., and L. M. Chiappe. "The Origin of Birds and Their Flight." *Scientific American*, February 1998. A summary of the evidence supporting the dinosaurian origin of birds.

Unwin, D. M. "Feathers, Filaments, and Theropod Dinosaurs." *Nature*, January 8, 1998. A discussion of the origin of feathers in dinosaurs.

WEB LINKS

Visit the special edition of *The Biology Place* for BIOLOGY, Fifth Edition, at http://www.biology.com/campbell. Go to Chapter 34 for online resources, including learning activities, practice exams, and links to the following web sites:

"Fossil Vertebrates in the Burke Museum"
An interesting collection of vertebrate fossils, including many photographs, from the Burke Museum of Natural History, Seattle, Washington.

"Fossils—A Tour Through the Evolution of Vertebrates"
Let the American Museum of Natural History take you on an expedition through the evolution of vertebrates. Requires a frames-capable browser.

"The Fish Out of Time....."
This web site, run by the Coelacanth Rescue Mission, is a comprehensive archive of information on the coelacanth, the only extant member of the lobe-finned fishes.

"The Amphibian Embryology Tutorial"
This site studies early vertebrate development in some detail and has many pictures, diagrams, and movies. Shockwave and Quicktime plug-ins are required to get the most out of this site.

UNIT SIX

PLANT FORM AND FUNCTION

This is a golden age for plant biology, and Gloria Coruzzi is one of the leaders of a new wave of researchers using modern methods of molecular biology to answer questions about how plants work. Dr. Coruzzi, a lifelong New Yorker, is a Carroll and Milton Petrie Professor of Biology at New York University. Her research applies genetic analysis and DNA manipulation to dissect the regulation of plant metabolism, especially the synthesis of amino acids, a process of great importance to agriculture and human nutrition. Professor Coruzzi is also strongly committed to promoting modern plant biology in the undergraduate curriculum; she teaches in NYU's introductory biology course and provides opportunities for undergraduate students to gain research experience in her lab. Gloria Coruzzi's dual achievements as a scientist and an educator confirm that excellent research and inspiring teaching are complementary activities of university faculty.

Tell us how your interest in science and plant biology developed.

I was interested in science way back, and as a high school student I did research at Rockefeller University on a project using rats as a model to study obesity. I remember one day chasing a fat rat around the lab and making an almost subconscious decision not to work on animals. By the time I did my Ph.D. at NYU Medical School, I was interested in molecular and cell biology and worked on mitochondrial DNA in yeast. It was about the time DNA sequencing was invented, and we sequenced the mitochondrial DNA and discovered that yeast mitochondria use a different genetic code. I began working on plant molecular biology as a postdoctoral researcher at Rockefeller University. At the time, scientists were saying that plant genes couldn't be cloned, and then we cloned some of the first plant genes and started to demonstrate how they are regulated by light. At the time I was what is called a "gene jock." I could isolate DNA and sequence it. Plants were just organisms to grind up to get out their DNA. But I've come to appreciate plant biology, and rather than just studying a gene, my lab now studies how a gene works in the context of the whole plant. The model organism we work on is *Arabidopsis*, which makes it possible for me to work on plants in New York City and which has brought plants back into many biology departments.

***Arabidopsis* is a tiny weed in the mustard family. How did it become such an important model organism in plant research?**

It was a problem for plant biology, especially at the biochemical level, that researchers traditionally worked on such a diversity of species. For example, somebody purifies an enzyme from one plant, and then another lab purifies the enzyme from a different species, and because everyone is working on different species, there may be slight differences and you never come to a synthesis of what's happening in a particular biochemical pathway within any single organism. So, there are benefits in focusing research on a single species. *Arabidopsis* has become to plant research what *E. coli* is to microbiology or *Drosophila* is to animal research. As a model organism, one advantage of *Arabidopsis* is that it has a relatively short generation span of four to six weeks, which makes genetic research easier. The plant also has a relatively small genome, and researchers have already sequenced about 20% to 30% of the *Arabidopsis* genome. This is a treasure trove of information for biologists. Genes for enzymes that we could never purify are popping out, and some of these genes have homology to genes in *E. coli* or humans. It's really a renaissance time in biology to be able to make these kinds of connections between organisms. *Arabidopsis* has brought plants back into biology departments that had almost eliminated plants during the early days of molecular biology. Here at NYU, biologists work on diverse organisms, but mostly at the cellular and molecular levels, and so we can all talk to each other, whether we're working on plants, or worms, or bacteria, or flies, or humans. Molecular biology is bringing biology departments back together. There are still plant biologists that say, "*Arabidopsis* is just a weed with no commercial value," but I think we're showing that genetic analysis is revealing certain processes that are common to all plants. *Arabidopsis* has also been a good model to show that plants can be genetically engineered. One of the reasons I went into plant biology is that I believed transgenic plants would have an important impact on society.

INTERVIEW

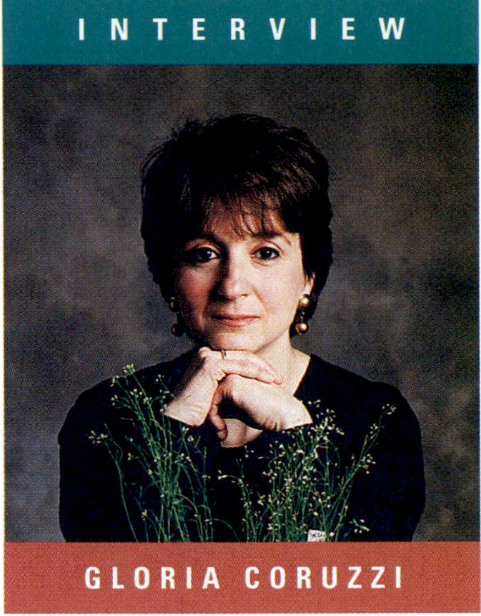

GLORIA CORUZZI

Producing transgenic plants involves moving genes from one organism into another. Give us an example from your current research.

We're studying how plants take inorganic nitrogen and convert it to organic form, such as amino acids. Gardeners know that nitrogen is an important element that limits plant growth, but I thought it would be possible to engineer plants to assimilate more nitrogen into organic compounds. When we started, there was really nothing known about the genes involved in amino acid biosynthesis in plants. One possibility was that plants just make amino acids continuously, with no regulation. But we discovered that light regulates synthesis in either a positive or a negative way, depending on the amino acid, by regulating transcription of genes for the metabolic pathways that assimilate nitrogen into organic form. We've been studying different mutants to dissect the molecular components of this sensing mechanism. Also, the amino acids themselves turn off transcription of the genes for their synthesis when the plant has made enough of those amino acids. This is very reminiscent of regulation in bacteria and yeast. We've screened for *Arabidopsis* mutants that can't sense excess amino acids. We have also altered this gene regulation by making transgenic plants in which nitrogen-assimilatory genes are continuously expressed. In some cases, we're successful in getting plants to actually assimilate more inorganic nitrogen into organic form because, I think, we've impaired the feedback mechanism that normally shuts off further assimilation once the plants have made enough. I believe we are able to improve on Mother Nature, because Mother Nature's *modus operandi* is for the plant just to live and reproduce, not to make a bigger plant with seeds that are more nutritious for humans. Transgenic plants may enable farmers to grow crops with higher yields of organic nitrogen and other nutrients.

You mentioned earlier that you first tasted research as a high school student. Tell us more about that.

I was a student at Hunter College High School, a public magnet school in New York City that at the time was an all-girls school. It wasn't a science magnet school, but was based more on literature and history, which is what women were supposed to be interested in. But there was a subgroup of us that said, "Hey, we want to do science," and we kind of broke the mold. We did internships in our senior year, and since the school was very close to Rockefeller University, I found myself there in the lab studying obesity. Although I realized that animal research was not my thing, working in a lab really changed my life and convinced me that I wanted to do research. The biology I had read about started to jell, and the experience changed the way I thought about science. I continued research as an undergraduate at Fordham, working on a botanical project and learning electron microscopy at the New York Botanical Gardens, and that experience just totally affirmed my interest in science. I also worked one summer at the Public Health Research Institute, and that was my first foray into yeast genetics. I was hooked.

Is that why you now include high school students and undergraduates in your own lab?

Yes. The high school students are very smart, but they also have to be good with their hands because working in a lab is a combination of being smart and dexterous. Some of my best high school researchers are musicians, which requires dexterity. We also have very smart undergrads. Some are thinking about graduate school in science, and others want to go to medical school, but even the pre-med students benefit from the experience in a plant biology lab because the kind of research transcends the organism. Mapping a gene is the same in plants and animals, and so studying a mutant gene in *Arabidopsis* helps students understand how disease genes are studied in humans. The students learn genetics and molecular biology, and they have a window into the future of biotechnology and medicine. Whenever high school students or undergraduates are significantly involved in a project, I encourage their co-authorship on research articles. It gives them a sense of empowerment, and it's just the fair thing to do.

Can we include some of your undergraduate students in the interview? I think other students will be interested in their experience.

Yes, I'd like you to meet two of my undergrads, Dimitrios Bliagos and Cliff Ransom. (Note: Dimitrios and Cliff are the students with Dr. Coruzzi in the photo on p. 668.)

Dimitrios, tell me a little about yourself, about your work in Dr. Coruzzi's lab, and about your plans for the future.

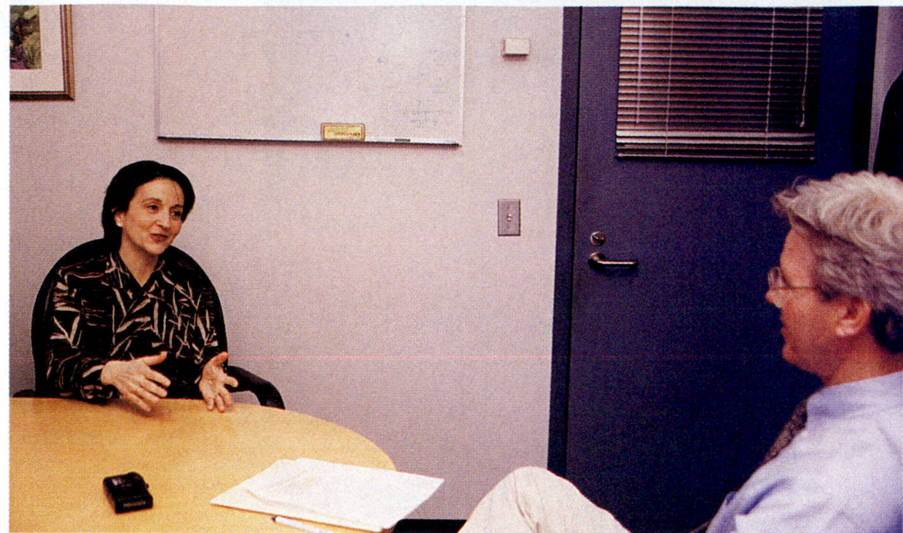

I was born and raised in Brooklyn, and now I'm a pre-med in my junior year here at NYU. I took the MCAT last week. I think I did all right, and I'll apply to medical school in the summer.

Why do you want to become a physician?

It started in high school. I liked biology, and first I wanted to be a biomedical engineer. But I didn't really like physics, and so engineering wasn't a good choice. During my senior year in high school I kind of zoned in on the biomedical more because my AP bio teacher stressed medical applications of physiology and other topics. Then I volunteered at St. Vincent's Hospital, and that was the final push.

What led you to undergraduate research?

I started looking for labs in my sophomore year because I really wanted to go beyond what I was learning in class. The theory I learned in class was very important and we went into different lab techniques in detail, but it didn't really reflect reality. In class, you do a procedure, and you get a certain result, but I see now that it's not so easy to select a gene clone or to get this working or get that working. My first project here was to map the genes for two of the enzymes involved in synthesis of the amino acid glutamine in *Arabidopsis.* I didn't realize that science is so involved in terms of the amount of work you have to do get the results, and I didn't understand that at the molecular level, working with plants isn't much different than working with animals and that what we're doing here with plant genes relates to what is going on in medical research. Now I think, "Wow, I could use these techniques to map a gene for a human disease."

Cliff, are you working on the same project?

I'm also working on glutamine synthesis, but a different aspect. I'm working with transgenic plants and selecting for mutants that can't sense glutamine and continue to make it because there is no negative feedback.

Tell us about your background and your plans.

I'm from Baltimore, and I came to NYU because it's very strong in molecular biology. I'm a senior now. I'll take a year off from school after I graduate, and then I'll probably go to grad school. I just finished my GREs. I also work at the New York Botanical Garden and really like that, so maybe I'll combine molecular biology and plant systematics.

How does Dr. Coruzzi pick the undergraduate students who work in her lab?

We have to initiate it by asking if there are any available positions, and then Dr. Coruzzi and a postdoc interview us to be sure that we're interested in the lab's work and that we work hard. Then I guess she asks for recommendations from our professors.

Most of the students who read this textbook are in their first or second year of college. Now that you're upper-division students in biology, what advice can you share?

Dimitrios: I just have to say to have a blast with biology, because if you don't try to get into what you're being taught and what you're reading, you are really not going to do very well. It's very important to get involved in learning and give it your all.

Cliff: I think in an introductory course, it's easy to get overwhelmed because you cover so many different fields of biology. But I think you can use that experience to choose one or two of the areas that interest you the most, and go with them after the first year, because it's impossible to take on the whole realm of biology. You need to kind of set a focus and let that guide your path.

Thank you, and best of success in your future studies.

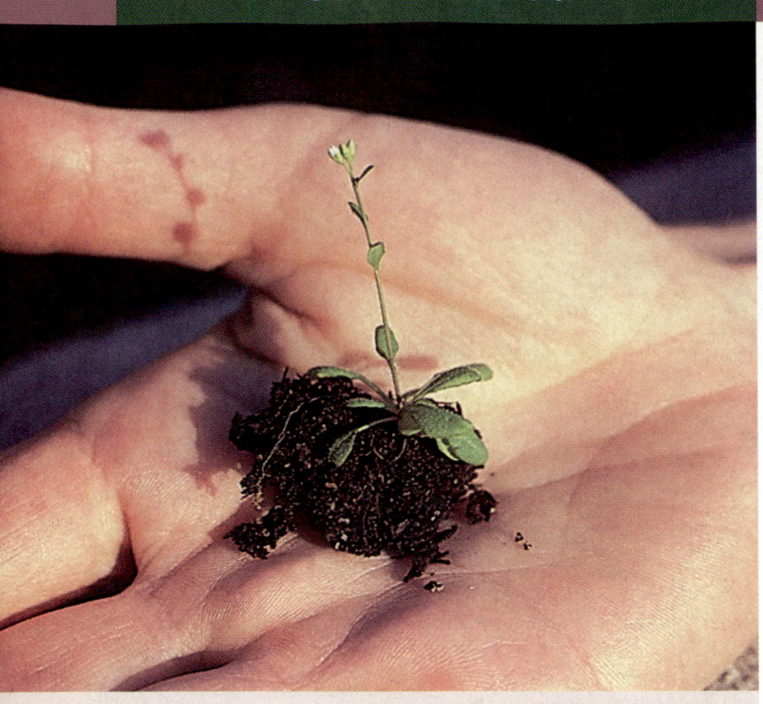

PLANT STRUCTURE AND GROWTH

Introduction to Modern Plant Biology

- Molecular biology is revolutionizing the study of plants
- Plant biology reflects the major themes in the study of life

The Angiosperm Body

- A plant's root and shoot systems are evolutionary adaptations to living on land
- Structural adaptations of protoplasts and walls equip plant cells for their specialized functions
- The cells of a plant are organized into dermal, vascular, and ground tissue systems

Plant Growth

- Meristems generate cells for new organs throughout the lifetime of a plant: *an overview of plant growth*
- Primary growth: Apical meristems extend roots and shoots by giving rise to the primary plant body
- Secondary growth: Lateral meristems add girth by producing secondary vascular tissue and periderm

Plants are the pillars of most terrestrial ecosystems. Their photosynthesis supports their own growth and maintenance, and it feeds, directly or indirectly, an ecosystem's various consumers, including animals. Although most humans now live far from natural ecosystems and farms, our dependence on plants is evident in our lumber, fabrics, paper, medicines, and—most importantly—our food. The study of plants arose from early humans' abilities to distinguish edible plants from poisonous ones and to make things from wood and other plant products. Modern plant biology still has its pragmatic side—in research aimed at improving crop productivity, for example—but the pure fun of discovery is what motivates most plant scientists. Plant biology, perhaps the oldest branch of science, is driven by a combination of curiosity and need—curiosity about how plants work and a need to apply this knowledge judiciously to feed, clothe, and house a burgeoning human population.

INTRODUCTION TO MODERN PLANT BIOLOGY

Molecular biology is revolutionizing the study of plants

Plant biology is in the midst of a renaissance. New laboratory and field methods coupled with clever choices of experimental organisms have catalyzed a research explosion. For example, many scientists interested in the genetic control of plant development are focusing their research on *Arabidopsis thaliana*, a little weed that belongs to the mustard family. *Arabidopsis*, shown in the photo that opens this chapter, is small enough to be grown in test tubes, allowing researchers to cultivate thousands of the plants in a few square meters of lab space. *Arabidopsis* also has a short generation span of about six weeks, making it an excellent model for genetic studies. Plant biologists are also attracted to *Arabidopsis* by its tiny genome; the amount of DNA per cell ranks among the least of all known plants. Efforts to associate specific genetic functions with certain regions of DNA are simplified, and it is likely that researchers will map the entire *Arabidopsis* genome within the next few years. Already, plant biologists have pinpointed some of the genes that control the development of flowers and have learned the functions of these genes (FIGURE 35.1).

The application of molecular biology to the study of flower development is just one example of what has become a major thrust of modern plant biology: relating processes that occur at the molecular and cellular levels to what we observe at the level of the whole plant. Some of the questions are very old. How does a plant produce a flower? How does a tree cause water to flow upward more than a hundred meters? How do plants protect themselves against disease? What makes roots

FIGURE 35.1 ▪ *Arabidopsis*, a little weed that is helping plant biologists answer big questions about the development of plant form. (a) Plant biologists have made this tiny member of the mustard family one of the organisms of choice for the experimental study of how genes control plant development. Some of this research has centered on flower development. *Arabidopsis* normally has four whorls of flower parts: sepals (Se), petals (Pe), stamens (St), and carpels (Ca). **(b)** Plant biologists can map the activity of genes that specify which organs develop in different zones of a floral bud (see Chapter 21). In this light micrograph, the orange dots mark the locations of a molecular probe that binds to RNA produced by a specific gene that is expressed in those locations. In this case, the RNA is produced by a gene with activity required for formation of stamens and carpels in this particular region of a floral bud. **(c)** Researchers have identified several mutations that cause abnormal flowers to develop. This flower, for example, has an extra set of petals in place of stamens and an internal flower where normal plants have carpels.

grow downward and shoots grow upward? By studying such processes at the molecular and cellular levels, plant biologists are beginning to understand how plants work.

Plant biology reflects the major themes in the study of life

The relationships of molecular and cellular mechanisms to the biology of a plant as a whole, integrated organism reflects one of the overarching themes introduced in Chapter 1: A living system has a hierarchy of structural levels, with emergent properties arising from the ordered arrangement and interactions of component parts. A floral bud, such as the one in FIGURE 35.1b, doesn't become a flower through isolated activities of its thousands of cells, but through a complex set of signals and responses within the bud, between the bud and other parts of the plant, and between the plant and its surrounding environment. This theme of emergent properties helps us keep our reductionist investigation of molecules and cells in perspective. The other biological themes of this book will also guide our study of plants. The structure and function of plants are shaped by interactions with the environment on two time scales. In their long evolutionary journey from water onto land, plants became adapted by natural selection to the specific problems posed by terrestrial environments (see Chapters 29 and 30). The evolution of tissues, such as the woody tissue of trees, that can support plants against the pull of gravity is just one example. Over the short term, individual plants exhibit structural and physiological responses to environmental stimuli. As an example of how individual plants, far more than individual animals, respond *structurally* to their environment, look at how wind has affected the growth of branches of the trees illustrated in FIGURE 35.2, p. 672. Plants, like animals, also have many evolutionary adaptations in the form of *physiological* responses to short-term change. For example, if a plant's leaves have a water deficiency, they close their stomata, which are the pores of the leaf surface. This emergency response helps the plant conserve water by reducing transpiration, the evaporative escape of water from leaves. As we analyze the evolutionary adaptations of plants and the responses of individual plants to their environment, our main tool will be a related theme: the correlation between structure and function. We will learn, for example, how the opening and closing of stomata are consequences of the structure of the guard cells that border these pores.

In studying the structure and function of plants, comparisons with the animal kingdom are inevitable and sometimes useful. Plants and animals confront many of the same problems, which they may solve in different ways. For example, an elephant and a redwood tree support their enormous weight

FIGURE 35.2 ▪ The effect of the environment on plant form. The "flagging" of these firs growing on a windy ridge on Mt. Hood, Oregon, resulted partly from the mechanical disturbance of the prevailing winds inhibiting limb growth on the windward sides of the trees. This growth response reduces the number of limbs that are broken during strong winds. Although environment affects the growth and development of all organisms, environmental impact on plant form is particularly impressive. In contrast, animal form is much less adjustable; wind, for example, does not alter the number, size, or placement of human limbs.

on land by means of very different kinds of "skeletons": bones in the elephant and woody cell walls in the tree. In some cases, plant and animal solutions to a common problem seem similar. For example, large organisms require internal transport systems to carry water and other substances between body parts, and networks of tubes have evolved in both plants and animals. But on closer examination, we will see that any anatomical similarities between plants and animals are superficial. These similar adaptations are, in the lexicon of biology, analogous, not homologous (see Chapter 25). The multicellular organization of plants evolved independently from that of animals, from unicellular ancestors with an entirely different mode of nutrition. The autotrophic/heterotrophic distinction has placed plants and animals on separate evolutionary paths.

In Table 35.1 on p. 674, you can review some overarching principles that distinguish plants and their role in the biosphere. This list of principles of plant biology was developed by the American Society of Plant Physiologists, a large organization of researchers and educators.

Setting the stage for our study of plant biology, this chapter introduces the general body plan of flowering plants, beginning with external structure. We will see that the architecture of plants is dynamic, continuously shaped by the plants' genetically directed growth patterns and their responses to the environment.

THE ANGIOSPERM BODY

A plant's root and shoot systems are evolutionary adaptations to living on land

In this section, we look at the morphology (body form) and anatomy (internal structure) of flowering plants, or angiosperms. (The structure of algae, mosses, ferns, and gymnosperms was covered in Unit Five, along with the evolutionary relationships of these plant groups to the angiosperms.)

With about 275,000 known species, angiosperms are by far the most diverse and widespread group of plants. Taxonomists split the angiosperms into two classes: **monocots**, named for their single cotyledon (seed leaf), and **dicots**, which have two cotyledons. Monocots and dicots have several other structural differences as well (FIGURE 35.3).

The basic morphology of plants reflects their evolutionary history as terrestrial organisms. A terrrestial plant must inhabit two very different environments, soil and air, at the same time and must draw resources from both. Soil provides water and minerals, but air is the main source of CO_2, and light does not penetrate far into the soil. The evolutionary solution to this separation of resources was differentiation of the plant body into two main systems: a subterranean **root system** and an aerial **shoot system** consisting of stems, leaves, and flowers (FIGURE 35.4). Neither system can live without the other. Lacking chloroplasts and living in the dark, roots would starve without sugar and other organic nutrients imported from the photosynthetic tissues of the shoot system. Conversely, the shoot system depends on water and minerals absorbed from the soil by roots. Vascular tissues, continuous throughout the plant, transport materials between roots and shoots. The two types of vascular tissue are **xylem**, which conveys water and dissolved minerals upward from roots into the shoots, and **phloem**, which transports food made in mature leaves to the roots and to parts of the shoot system, such as developing leaves and fruits. As we take closer looks at the morphology of roots and shoots, try to view these systems from the evolutionary perspective of adaptation to living on land.

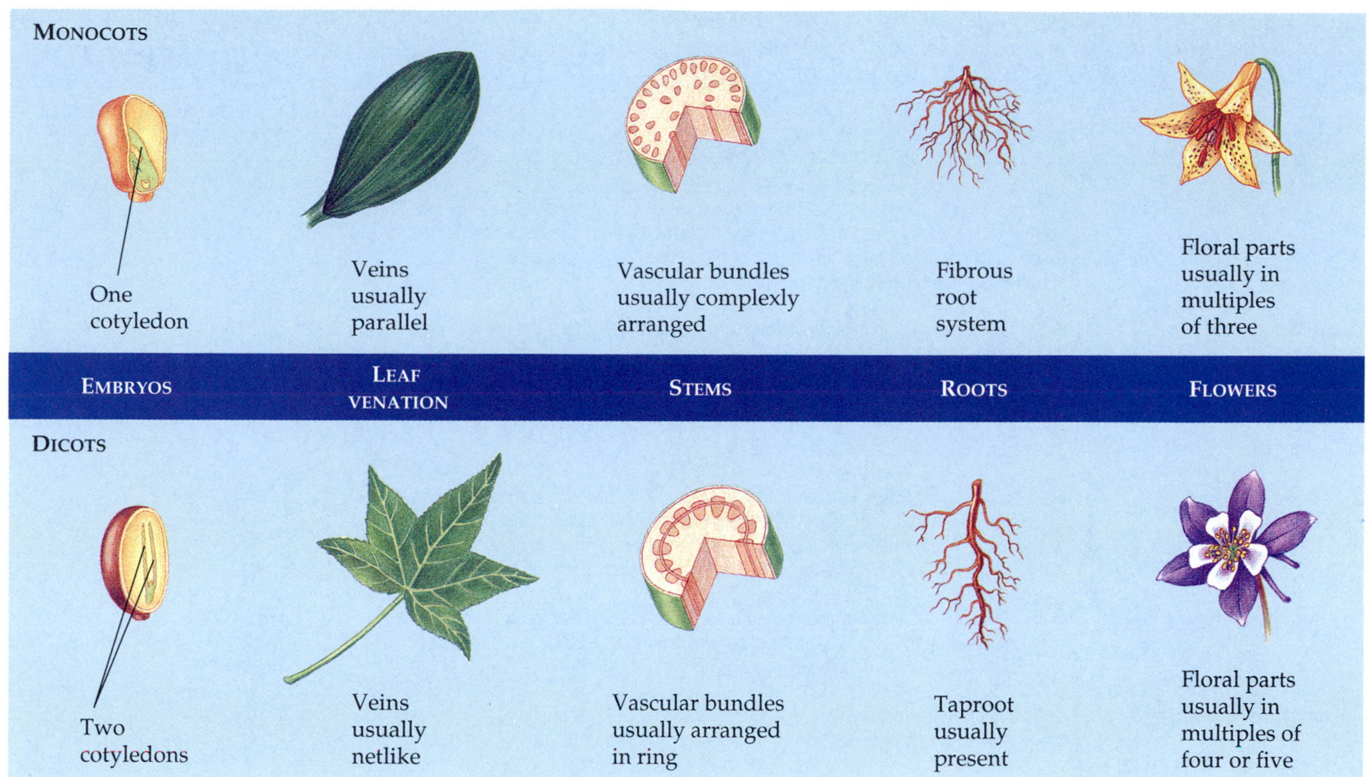

FIGURE 35.3 ▪ A comparison of monocots and dicots. These classes of angiosperms are named for the number of cotyledons, or seed leaves, present on the embryo of the plant. Monocots include orchids, bamboos, palms, lilies, and yuccas, as well as the grasses, such as wheat, corn, and rice. A few examples of dicots are roses, beans, sunflowers, maples, and oaks.

FIGURE 35.4 ▪ Morphology of a flowering plant: an overview. The plant body is divided into a root system and a shoot system, connected by vascular tissue that is continuous throughout the plant. The root system of this dicot consists of a taproot and several lateral roots. Shoots consist of stems, leaves, and flowers. The blade, the expanded portion of a leaf, is attached to a stem by a petiole. Nodes, the regions of a stem where leaves attach, are separated by internodes. At a shoot's tip is the terminal bud, the main growing point of the shoot. Axillary buds are located in the upper angles of leaves. Most of these axillary buds are dormant, but they have the potential to develop into vegetative (leaf-bearing) branches or flowers.

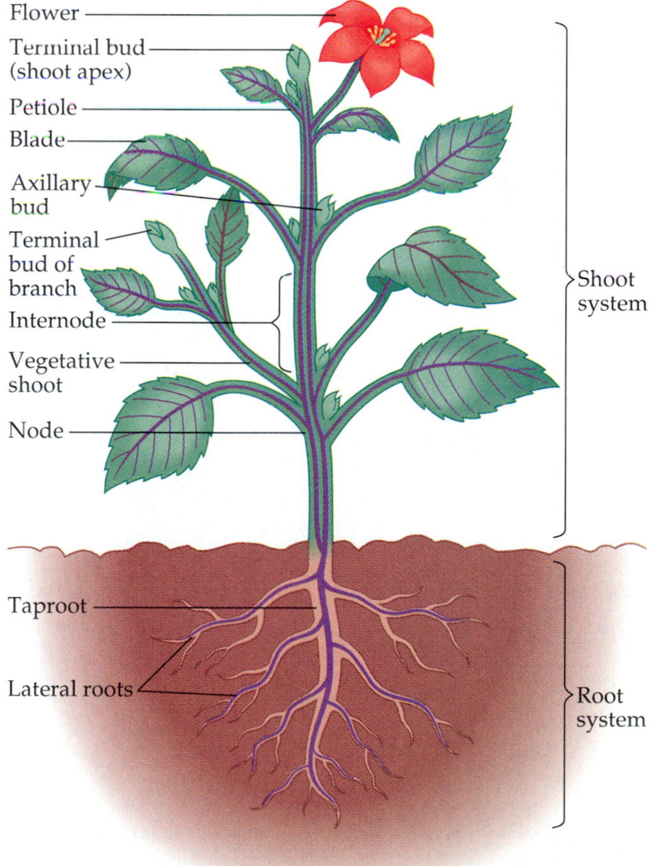

Table 35.1 ■ Some Principles of Plant Biology (Adapted from ASPP's *Principles of Plant Biology: Concepts for Science Education*, © 1998, by the American Society of Plant Physiologists)

CHARACTERISTICS	EXPLANATION	CAMPBELL, REECE, MITCHELL, *BIOLOGY* CHAPTER
Photosynthesis	Plants perform many biological processes and metabolic pathways that are also found in microbes and animals. However, unlike animals and most microbes, plants have the ability to use energy from sunlight to convert inorganic compounds into organic compounds required for growth. This process, photosynthesis, is the source of most of the biosphere's chemical energy and organic matter.	Ch. 10, 37
Chemical cycling	Plants require certain inorganic elements for growth and play an essential role in cycling these nutrients within the biosphere.	Ch. 37, 54
Evolutionary impact	Land plants evolved from aquatic algae and had an enormous impact on the history of life. Plant evolution established terrestrial ecosystems and altered the atmosphere by increasing oxygen and ozone concentrations.	Ch. 29, 30
Reproductive versatility	Most plants can reproduce both sexually and asexually. In flowering plants and gymnosperms, sexual reproduction produces seeds, an evolutionary adaptation that enhances reproduction in terrestrial environments. A seed consists of a resistant coat enclosing a plant embryo and a food supply that is consumed during germination of the seed.	Ch. 30, 38
Use of chemical energy	Plants, like animals and many microbes, are aerobic organisms that extract energy from organic molecules through the process of cellular respiration and use the energy for growth and reproduction. Plants have chloroplasts *and* mitochondria.	Ch. 9, 37
Cell walls	Cell walls provide structural support for plants and are also used as sources of fiber and building materials by humans, insects, birds, and many other organisms.	Ch. 7, 35
Diversity	The hundreds of thousands of plant species range in size from microscopic plants to gigantic trees.	Ch. 29, 30
Human dependence on plants	In addition to depending on plants for food and building materials, humans also extract medicines and many other useful chemicals from plants.	Ch. 29, 30, 35
Defense systems	Plants, like animals, are subject to injury and death due to infectious diseases and predation. These threats are countered by unique defense systems that have evolved in plants.	Ch. 39
Importance of water	Water is the most abundant substance in plant cells, and its uptake accounts for most of the size increase of growing cells. Water is the solvent in which most cells' chemical reactions occur, and it also serves as the medium for long-distance transport of mineral nutrients and organic compounds throughout the plant.	Ch. 3, 36, 37
Regulation	Plant growth and development are controlled by an interplay of internal regulators, such as hormones, and environmental cues, such as light, gravity, touch, and stressful conditions.	Ch. 11, 39
Plants and the diversity of biological communities	Plants have adapted to a wide variety of environments and dominate most terrestrial landscapes, providing diverse habitats for birds, insects, and other wildlife.	Ch. 29, 30, 50, 53, 55

The Root System

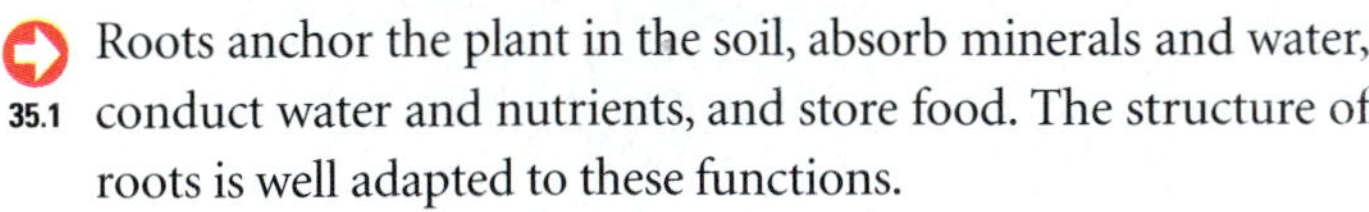

35.1 Roots anchor the plant in the soil, absorb minerals and water, conduct water and nutrients, and store food. The structure of roots is well adapted to these functions.

Many dicots have a **taproot** system, consisting of one large, vertical root (the taproot) that produces many smaller lateral roots (see FIGURES 35.3 and 35.4). Penetrating deep into the soil, the taproot is a firm anchor, as you know if you have ever tried to pull up a dandelion. The taproots of some plants adapted to arid environments are able to "tap" water sources located far below ground. Some taproots, such as carrots, turnips, and sugar beets, are modified roots that store exceptionally large amounts of food. The plant consumes these food reserves when it flowers and produces fruit. For this reason, root crops are harvested before the plants flower.

Monocots, including grasses, generally have **fibrous root** systems consisting of a mat of threadlike roots that spread out

35.1 This symbol links topics in the text to interactive exercises in the CD-ROM that accompanies the book. The number indicates the appropriate activity in the CD.

FIGURE 35.5 ▪ Root hairs of a radish seedling. Growing by the thousands just behind the tip of each root, the hairs cling tightly to soil particles and increase the surface area for the absorption of water and minerals by the roots.

below the soil surface. (Large monocots, including palms and bamboo, have much thicker roots—ropelike rather than threadlike.) The fibrous root system gives the plant extensive exposure to soil water and minerals and anchors it tenaciously to the ground (see FIGURE 35.3). Because their root systems are concentrated in the upper few centimeters of the soil, grasses hold the top layer in place and make excellent ground cover for preventing erosion.

Although the entire root system helps anchor a plant, most absorption of water and minerals in both monocots and dicots occurs near the root tips, where vast numbers of tiny **root hairs** increase the surface area of the root tremendously (FIGURE 35.5). Root hairs are extensions of individual epidermal cells on the root surface, not to be confused with branch roots, which are multicellular organs. Mycorrhizae, symbiotic associations between roots and fungi, account for even more water and mineral absorption than do root hairs in many plants (see FIGURE 36.7). The roots of many plants also have bumps called root nodules, which contain symbiotic bacteria that convert atmospheric nitrogen (N_2) to nitrogenous compounds the plant can incorporate into proteins and other organic molecules. You will learn more about the symbiotic relationships between plant roots and fungi and bacteria in Chapter 37.

In addition to roots that extend from the base of the shoot, some plants have roots arising aboveground from stems or even from leaves. Such roots are said to be **adventitious** (L. *adventicius*, "not belonging to"), a term that describes any plant part that grows in an unusual location. The adventitious roots of some plants, including corn, function as props that help support tall stems.

The Shoot System

The shoot system consists of vegetative shoots, which bear leaves, and floral shoots, which terminate in flowers. We will postpone discussion of the structure and function of flowers until Chapter 38 and focus here on vegetative shoots. A vegetative shoot consists of a stem and the attached leaves; it may be the plant's main shoot or a side shoot, called a vegetative branch (see FIGURE 35.4).

Stems. A stem is an alternating system of **nodes**, the points at which leaves are attached, and **internodes**, the stem segments between nodes (see FIGURE 35.4). In the angle formed by each leaf and the stem is an **axillary bud**, which has the potential to form a branch shoot. Most axillary buds of a young shoot are dormant. Thus, growth of a young shoot is usually concentrated at its apex (tip), where there is a **terminal bud** with developing leaves and a compact series of nodes and internodes. The presence of the terminal bud is partly responsible for inhibiting the growth of axillary buds, a phenomenon called **apical dominance.** By concentrating resources on growing taller, apical dominance is an evolutionary adaptation that increases the plant's exposure to light, especially in a location with dense vegetation. However, branching is also important for increasing the exposure of the shoot system to the environment, and under certain conditions, axillary buds begin growing. After breaking dormancy, an axillary bud gives rise to a vegetative branch complete with its own terminal bud, leaves, and axillary buds. In some cases, the growth of axillary buds can be stimulated by removing the terminal bud. This is the rationale for pruning trees and shrubs and "pinching back" houseplants to make them bushy.

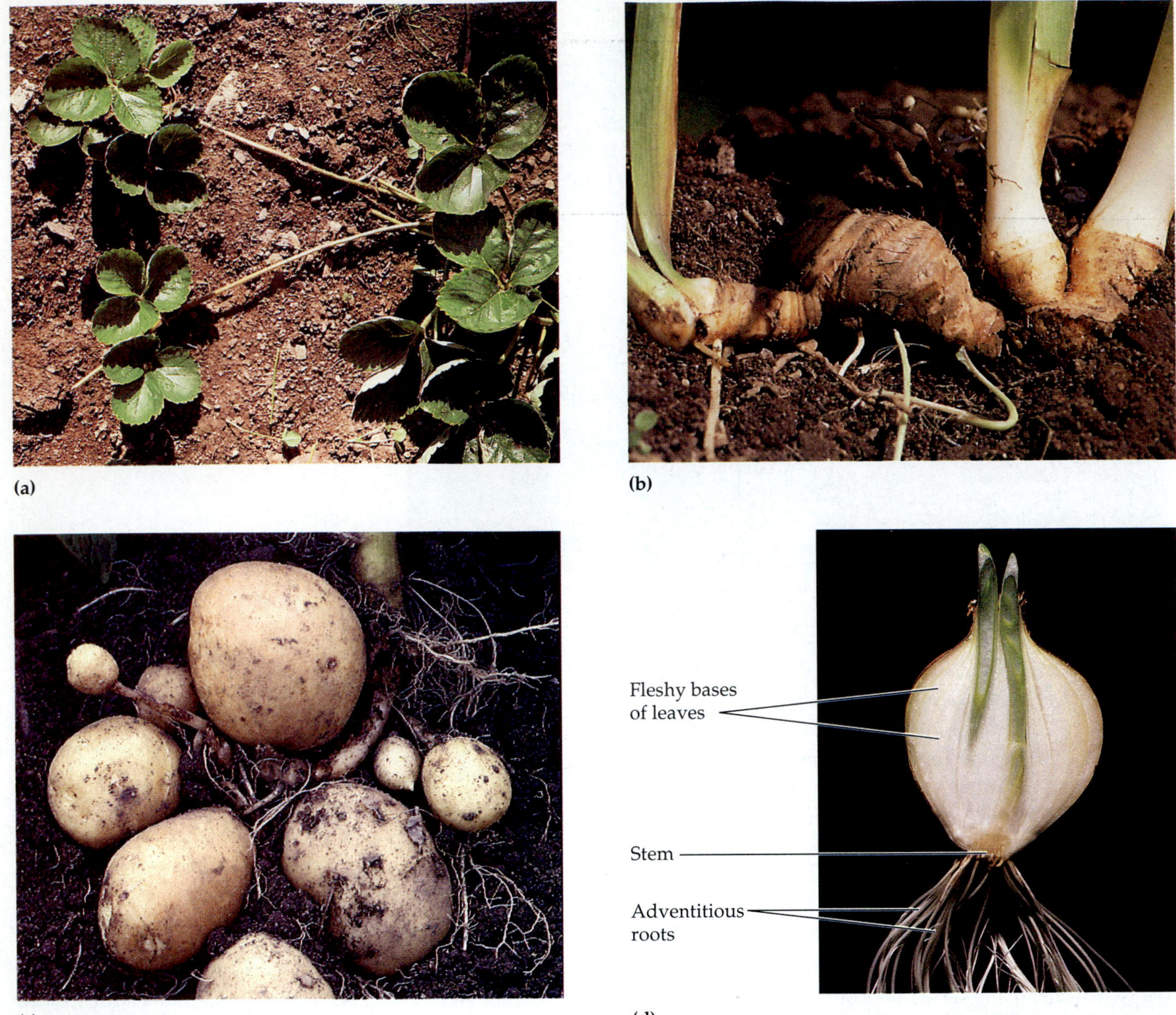

FIGURE 35.6 ▪ Modified stems. (a) Stolons, shown here on a strawberry plant, grow on the surface of the ground. These "runners" enable a plant to colonize a large area and to reproduce asexually if the single parent plant fragments into many smaller offspring. **(b)** Rhizomes, like the edible base of this ginger plant, are horizontal stems that grow underground. **(c)** Tubers, such as these white potatoes, are swollen ends of rhizomes specialized for storing food. The "eyes" arranged in a spiral pattern around a potato are clusters of axillary buds that mark the nodes. **(d)** Bulbs are vertical, underground shoots consisting mostly of the swollen bases of leaves that store food. You can see the many layers of modified leaves attached to the short stem by slicing an onion bulb lengthwise.

Modified stems with diverse functions have evolved in many plants. These modified stems, which include stolons, rhizomes, tubers, and bulbs, are often mistaken for roots (FIGURE 35.6).

Leaves. Leaves are the main photosynthetic organs of most plants, although green stems also perform photosynthesis. Leaves vary extensively in form, but they generally consist of a flattened **blade** and a stalk, the **petiole**, which joins the leaf to a node of the stem (see FIGURE 35.4). Grasses and many other monocots lack petioles; instead, the base of the leaf forms a sheath that envelopes the stem. Some monocots, including palm trees, do have petioles.

The leaves of monocots and dicots differ in the arrangement of their major veins (see FIGURE 35.3). Most monocots have parallel major veins that run the length of the leaf blade. In contrast, dicot leaves generally have a multibranched network of major veins. Since leaf morphology varies extensively among plant species, plant taxonomists use characteristics such as leaf shape, spatial arrangement of leaves on a stem, and the pattern of a leaf's veins to help identify and classify plants. FIGURE 35.7 illustrates one variation, simple versus compound leaves. Although most leaves are specialized for photosynthesis, some plants have leaves that have become adapted by evolution for other functions (FIGURE 35.8).

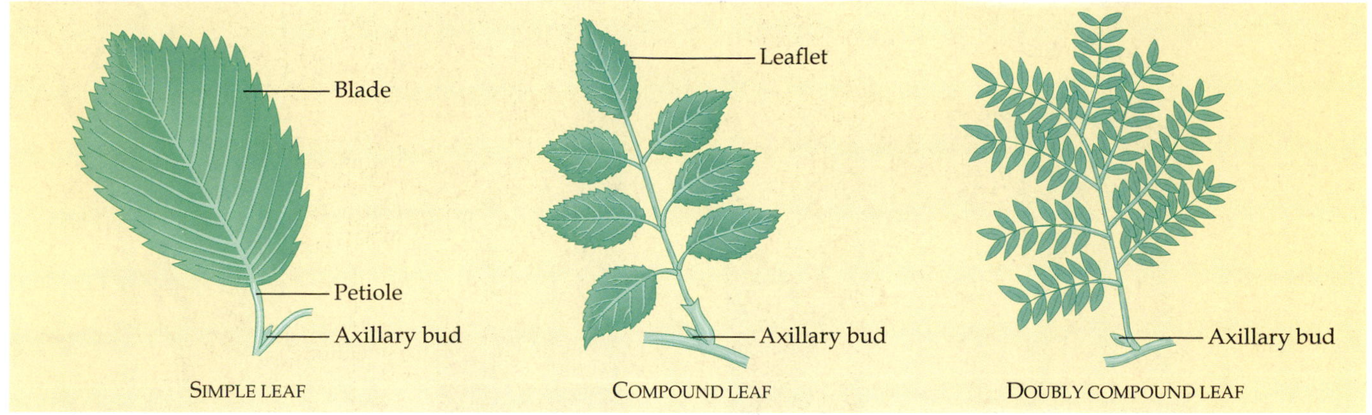

FIGURE 35.7 ▪ Simple versus compound leaves. A simple leaf has a single, undivided blade. An axillary bud is located where the petiole joins the stem. The blade of a compound leaf is divided into several leaflets, which are themselves divided in a doubly compound leaf. You can distinguish a compound leaf from a stem with several closely spaced simple leaves by examining the locations of axillary buds. There is only one axillary bud per leaf, whether the leaf is simple or compound. Thus, a compound leaf has a bud at its base where its petiole attaches to the stem, but not at the bases of the individual leaflets. Most very large leaves are compound or doubly compound. This structural adaptation enables large leaves to withstand strong wind with less tearing and also confines some pathogens that invade the leaf to a single leaflet, rather than allowing the pathogens to spread to the entire leaf.

(a)

(c)

(b)

(d)

FIGURE 35.8 ▪ Modified leaves. (a) The tendrils used by this pea plant to cling to supports are modified leaflets. **(b)** The spines of cacti, such as this prickly pear, are actually leaves, and photosynthesis is carried out mainly by the fleshy green stems. **(c)** Most succulents, such as ice plant, have leaves modified for storing water. **(d)** In many plants, brightly colored leaves help attract pollinators to the flower. The red "petals" of the poinsettia are actually leaves that surround a group of flowers.

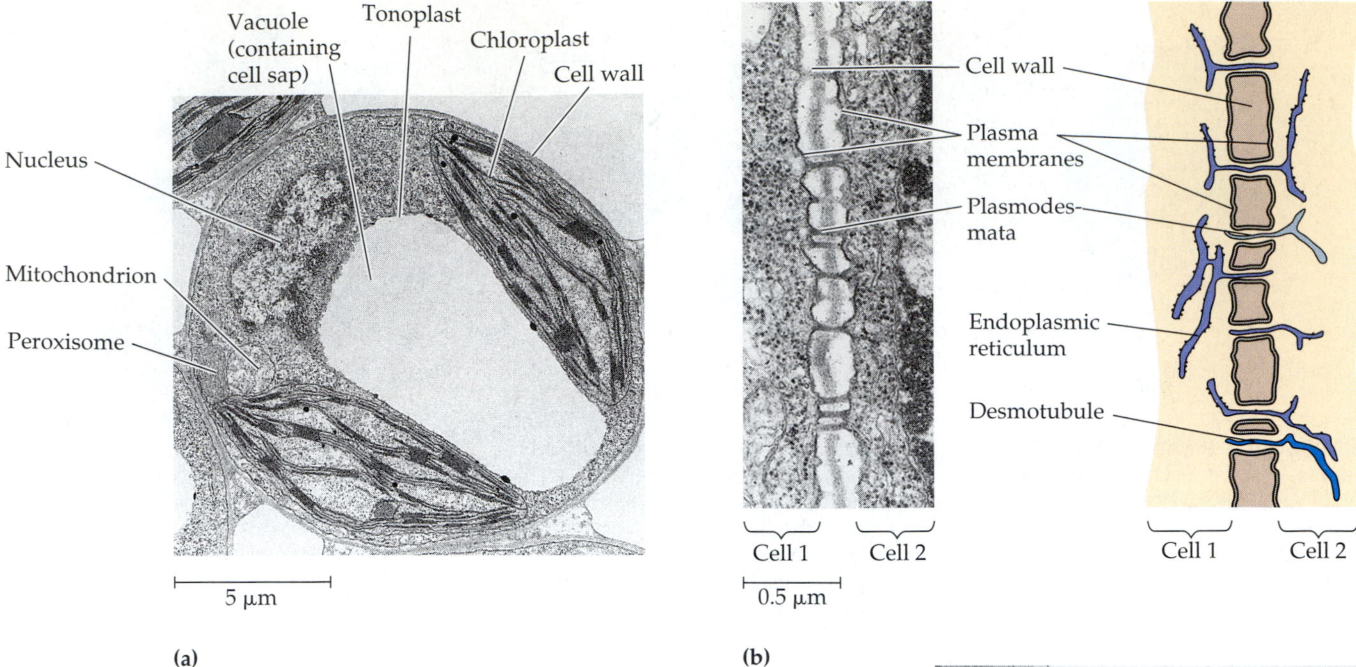

FIGURE 35.9 ▪ Review of plant cell structure. (a) In addition to general features of eukaryotic cells, you can see three structures of plant cells *not* found in animal cells: chloroplasts, the sites of photosynthesis; a central vacuole containing a fluid called cell sap and bounded by the tonoplast, a specialized membrane that regulates the traffic of molecules between the cell sap and the cytosol; and a cell wall external to the plasma membrane. The part of the cell interior to the cell wall—that is, the plasma membrane, the cytoplasm, and nucleus—is called the protoplast. Thus, a plant cell consists of a protoplast enclosed by a cell wall (TEM). **(b)** The protoplasts of neighboring cells are generally connected by plasmodesmata, cytoplasmic channels that pass through pores in the walls. The endoplasmic reticulum is also continuous through the plasmodesmata in the form of structures called desmotubules (left, TEM). **(c)** An adhesive layer called the middle lamella cements together the cell walls of adjacent cells. All plant cells have a primary cell wall secreted as the cell grows and develops. In addition, many specialized plant cells produce a secondary wall. Notice that the secondary wall is closer to the protoplast than the primary wall, because it forms later, usually after the cell has stopped growing. Most cell walls are relatively porous at the molecular level, with water and many small solutes moving freely through the spaces between cellulose fibrils.

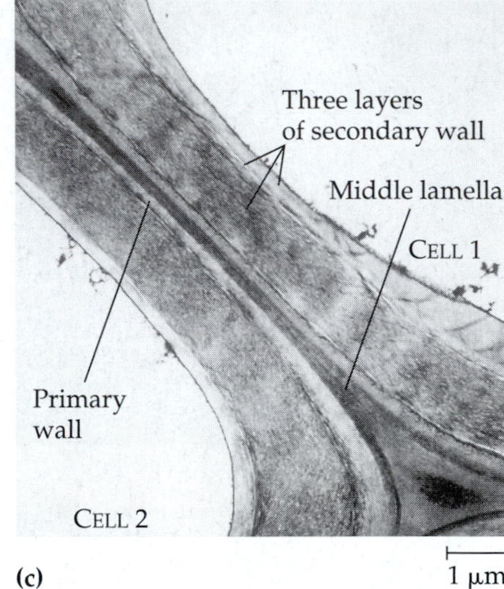

So far we have examined the structural organization of the whole plant as we see it with the unaided eye. We can now dissect the plant and explore its microscopic organization.

Structural adaptations of protoplasts and walls equip plant cells for their specialized functions

What distinguishes a multicellular organism from a colony of cells is a division of labor among cells differing in structure and function. As you consider each major type of plant cell, notice the structural adaptations that make specific functions possible. In some cases, we will find distinguishing characteristics within the **protoplast**, the contents of the cell exclusive of the cell wall. For example, only the protoplasts of photosynthetic cells contain chloroplasts. But also notice that modifications of cell walls are important in how the specialized cells of a plant function. FIGURE 35.9 will help you review the general structure of plant cells before you proceed to the following survey of specific cell types.

Parenchyma Cells

Because they are the least specialized of all plant cells, **parenchyma cells** are often depicted as "typical" plant cells (FIGURE 35.10a). Mature parenchyma cells have primary walls that are relatively thin and flexible. Most parenchyma cells lack secondary walls. The protoplast generally has a large central vacuole.

(a) Parenchyma cells are relatively unspecialized, with thin, flexible primary walls. These cells carry on most of the plant's metabolic functions. These parenchyma cells are from the root of a buttercup, and the purple granules are starch-storing organelles.

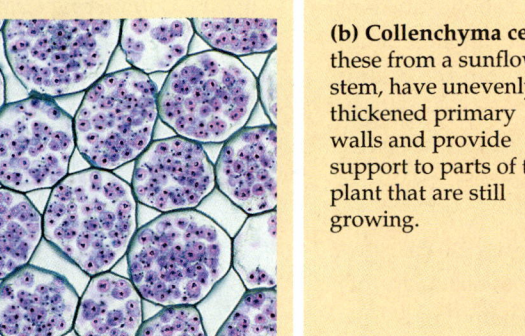

Parenchyma 50 μm

(b) Collenchyma cells, these from a sunflower stem, have unevenly thickened primary walls and provide support to parts of the plant that are still growing.

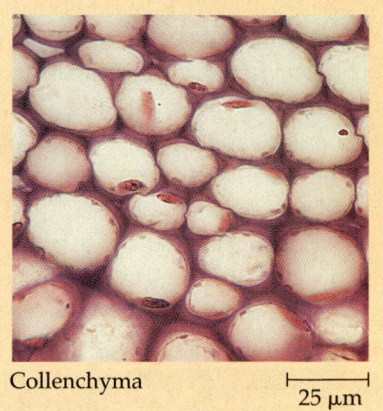

Collenchyma 25 μm

(c) Sclerenchyma cells, specialized for support, have secondary walls hardened with lignin and may be dead (lacking protoplasts) at functional maturity. The **fiber cells** in the left micrograph are elongated sclerenchyma cells, as viewed in cross section. **Sclereids** (right) are irregularly shaped sclerenchyma cells with very thick, lignified secondary walls. These sclereids, or stone cells, are from a pear fruit.

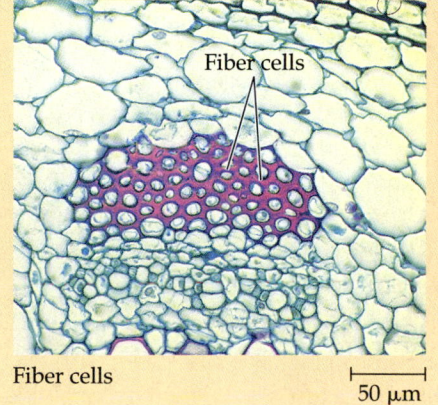

Fiber cells

Fiber cells 50 μm

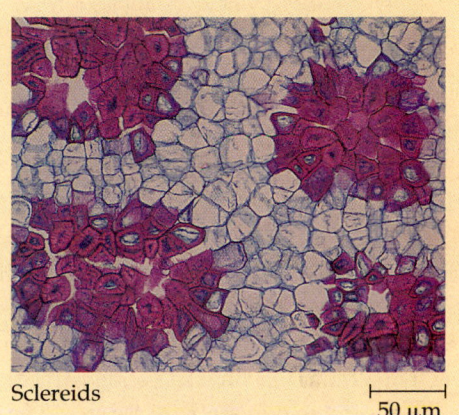

Sclereids 50 μm

(d) The **water-conducting cells** of xylem include tapered **tracheids** (left) and **vessel elements** arranged end to end, forming vessels (right). Both cell types have secondary walls and are dead at functional maturity. In gymnosperms, tracheids have the dual functions of water transport and structural support. In most angiosperms, both vessel elements and tracheids conduct water, and support is provided mainly by fiber cells.

Vessel elements

Tracheids

Pits

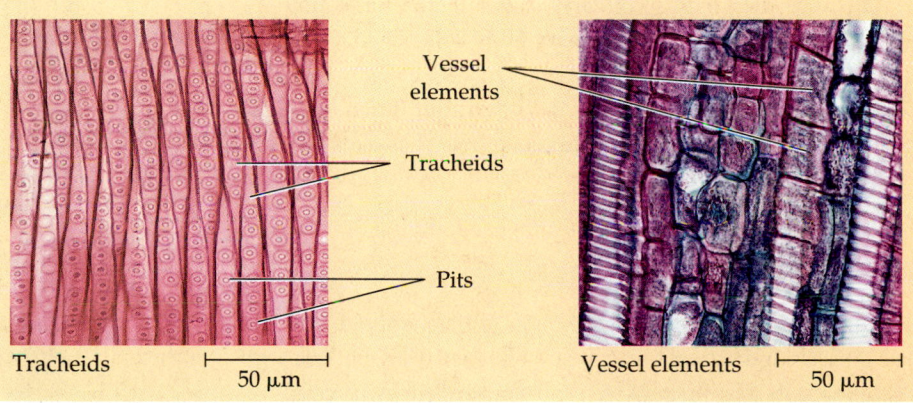

Tracheids 50 μm

Vessel elements 50 μm

(e) The **food-conducting cells** of phloem are **sieve-tube members,** which are arranged end to end with porous walls (sieve plates) between them. (In the transverse section on the right, two sieve-tube members, including one of the labeled cells, are sectioned through sieve plates.) The cells are living at functional maturity, but lack nuclei. Alongside each sieve-tube member is a nucleated **companion cell**.

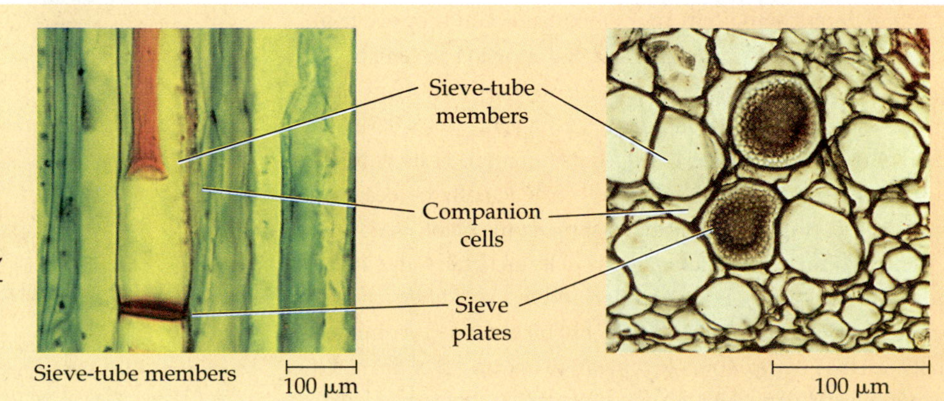

FIGURE 35.10 ▪ Survey of plant cells.

Parenchyma cells perform most of the metabolic functions of the plant, synthesizing and storing various organic products. For example, photosynthesis occurs within the chloroplasts of parenchyma cells in the leaf. Some parenchyma cells in stems and roots have colorless plastids that store starch. The fleshy tissue of most fruit is composed mostly of parenchyma cells.

Developing plant cells of all types usually have the generalized structure of parenchyma cells before specializing further in structure and function. Those cells that retain the less specialized condition to become mature parenchyma cells do not generally undergo cell division, but most of them retain the ability to divide and differentiate into other types of plant cells under special conditions—during the repair and replacement of organs after injury to the plant, for instance. It is even possible in the laboratory to regenerate an entire plant from a single parenchyma cell.

Collenchyma Cells

Compared to parenchyma cells, collenchyma cells have thicker primary walls, though the walls are unevenly thickened (FIGURE 35.10b). Grouped in strands or cylinders, **collenchyma cells** help support young parts of the plant. Young stems, for instance, often have a cylinder of collenchyma just below their surface (the "strings" of a celery stalk, for example). Because they lack secondary walls and the hardening agent lignin is absent in their primary walls, collenchyma cells provide support without restraining growth. Unlike sclerenchyma cells, which we discuss next, mature, functioning collenchyma cells are living and elongate with the stems and leaves they support.

Sclerenchyma Cells

Also functioning as supporting elements in the plant but with their thick secondary walls usually strengthened by lignin, **sclerenchyma cells** are much more rigid than collenchyma cells. Mature sclerenchyma cells cannot elongate, and they occur in regions of the plant that have stopped growing in length. So specialized are sclerenchyma cells for support that many lack protoplasts at functional maturity, the stage in a cell's development when it is fully specialized for its function. Thus, at functional maturity a sclerenchyma cell may actually be dead, its rigid wall serving as scaffolding to support the plant.

The two forms of sclerenchyma cells are **fibers** and **sclereids** (FIGURE 35.10c). Long, slender, and tapered, fibers usually occur in bundles. Some plant fibers are used commercially, such as hemp fibers for making rope and flax fibers for weaving into linen. Sclereids are shorter than fibers and irregular in shape. Nutshells and seed coats owe their hardness to sclereids, and sclereids scattered among the soft parenchyma tissue give the pear fruit its gritty texture.

Water-Conducting Cells of Xylem: Tracheids and Vessel Elements

The water-conducting elements of xylem are elongated cells of two types: **tracheids** and **vessel elements** (FIGURE 35.10d). Both types of cells are dead at functional maturity, but they produce secondary walls before the protoplast dies. In parts of the plant that are still elongating, the secondary walls are deposited unevenly in spiral or ring patterns that enable them to stretch like springs as the cell grows. Like the wire that rein-

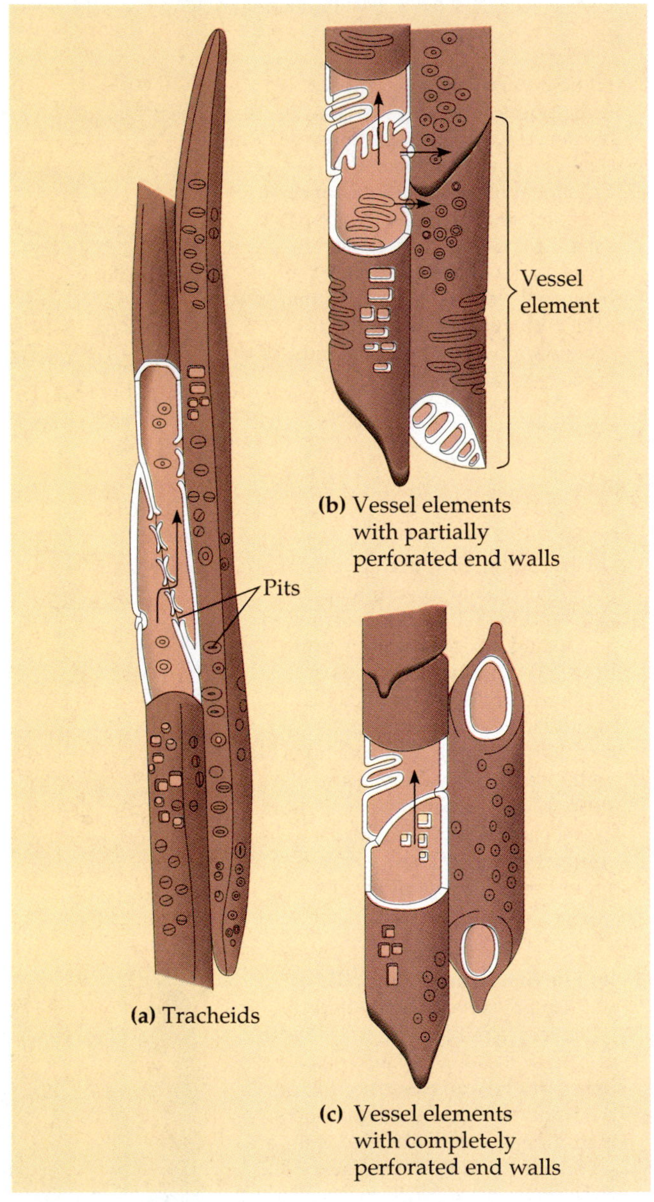

FIGURE 35.11 ▪ Water-conducting cells of xylem. Arrows indicate the flow of water. **(a)** Tracheids are spindle-shaped cells with pits through which water flows from cell to cell. **(b)** Vessel elements are individual cells linked together end to end, forming long tubes, or xylem vessels. Water streams from element to element through perforated end walls. Water can also migrate laterally between neighboring vessels through pits. **(c)** Resistance to water flow in some xylem vessels is further lowered by the complete perforation of walls between the vessel elements.

forces the wall of a garden hose, these wall thickenings strengthen the water-conducting cells of the plant. Tracheids and vessel elements that form in parts of the plant that are no longer elongating usually have secondary walls interrupted only by **pits**, thinner regions where only primary walls are present. A tracheid or vessel element completes its differentiation when its protoplast disintegrates, leaving behind a nonliving conduit through which water can flow (FIGURE 35.11).

Tracheids are long, thin cells with tapered ends. Water moves from cell to cell mainly through pits, where the water does not have to cross thick secondary walls. Because their secondary walls are hardened with lignin, tracheids function in support as well as water transport.

Vessel elements are generally wider, shorter, thinner-walled, and less tapered than tracheids. Vessel elements are aligned end to end, forming long micropipes, the **xylem vessels**. The end walls of vessel elements are perforated, enabling water to flow freely through xylem vessels.

Food-Conducting Cells of Phloem: Sieve-Tube Members

Sucrose, other organic compounds, and some mineral ions are transported within the phloem of a plant through tubes formed by chains of cells called **sieve-tube members** (see FIGURE 35.10e). In contrast to the water-conducting cells of xylem, sieve-tube members are alive at functional maturity, although their protoplasts lack such organelles as the nucleus, ribosomes, and a distinct vacuole. In angiosperms, the end walls between sieve-tube members, called **sieve plates**, have pores that presumably facilitate the flow of fluid from cell to cell along the sieve tube.

Alongside each sieve-tube member is at least one **companion cell**, which is connected to the sieve-tube member by numerous plasmodesmata. The nucleus and ribosomes of the companion cell may serve not only that cell but also the adjacent sieve-tube member, which has no nucleus or ribosomes of its own. In some plants, companion cells also help load sugar produced in the leaf into the sieve-tube members; the phloem then transports the sugar to other parts of the plant.

The cells of a plant are organized into dermal, vascular, and ground tissue systems

Each young organ of a plant—a leaf, stem, or root—has three tissue systems: the dermal, vascular, and ground tissue systems. Each tissue system is continuous throughout the plant body, although the specific characteristics of the tissues and their spatial relationships to one another vary in different organs of the plant (FIGURE 35.12). Here we survey the three tissue systems as they occur in a young, nonwoody plant.

The **dermal tissue system**, or **epidermis**, is generally a single layer of tightly packed cells that covers and protects all young parts of the plant—the "skin" of the plant. In addition

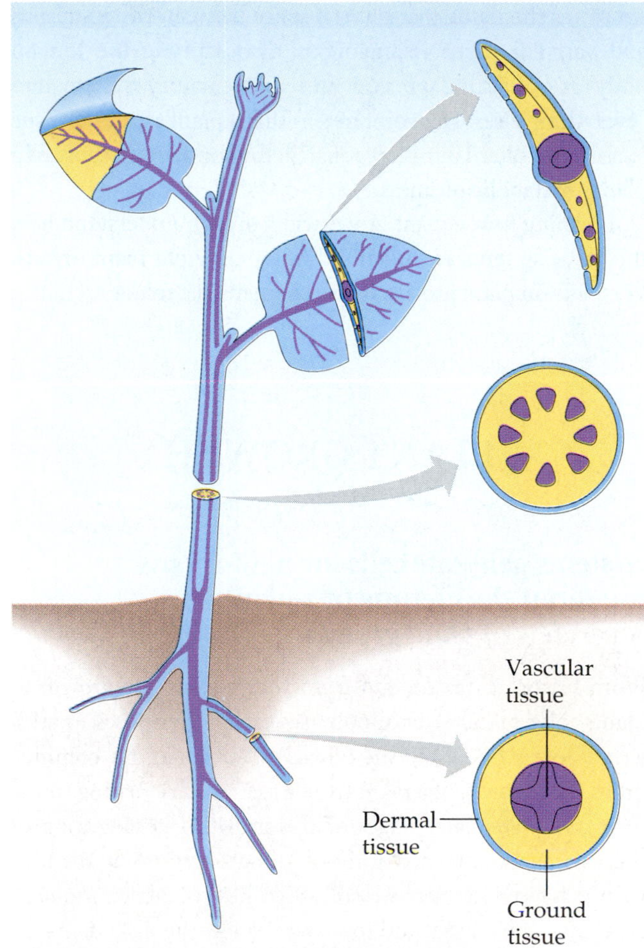

FIGURE 35.12 ▪ The three tissue systems. The dermal tissue system, or epidermis, is a single layer of cells that covers the entire body of a young plant. The vascular tissue system is also continuous throughout the plant, but it is arranged differently in each organ. The ground tissue system, responsible for most of the plant's metabolic functions, is located between the dermal tissue and the vascular tissue in each organ.

to the general function of protection, the epidermis has more specialized characteristics consistent with the function of the particular organ it covers. For example, the root hairs so important in the absorption of water and minerals are extensions of epidermal cells near the tips of roots. The epidermis of leaves and most stems secretes a waxy coating called the **cuticle** that helps the aerial parts of the plant retain water, an important adaptation to living on land.

The continuum of xylem and phloem throughout the plant forms the vascular tissue system, which functions in transport and support. The specific organization of vascular tissue in stems and roots is discussed in the next section.

The ground tissue system makes up the bulk of a young plant, occupying the space between the dermal and vascular tissue systems. Ground tissue is predominantly parenchyma, but collenchyma and sclerenchyma are also commonly present. Among the diverse functions of ground tissue are photosynthesis, storage, and support. Describing ground tissue simply as the tissue that fills the space between the epidermis and vascular tissue is analogous to describing the human body as consisting of skin and a circulatory system and "everything else." The point here is that a plant's ground tissue is not just "filler," but is responsible for conducting most of a plant's metabolic business.

Learning how a plant grows will help you understand how the tissue systems are organized in the different plant organs. We focus on plant growth in the chapter's last major section.

PLANT GROWTH

Meristems generate cells for new organs throughout the lifetime of a plant: *an overview of plant growth*

From season to season and from year to year, the growth of plants alters our surroundings—yards, campuses, parks, vacant lots, woods, and the other landscapes in our communities. The growth of a plant from a seed is a fascinating transformation. The early stages of this growth—germination of the seed and emergence of the seedling—are among the topics of Chapter 38. Here we will study how plants continue to grow after their shoot and root systems are established.

Most plants continue to grow as long as they live, a condition known as indeterminate growth. Most animals, in contrast, are characterized by determinate growth; that is, they cease growing after reaching a certain size. While whole plants usually show indeterminate growth, certain plant organs, such as leaves and flowers, exhibit determinate growth.

Indeterminate growth does not imply immortality. Although they continue to grow throughout their lives, plants, of course, do die. Plants known as **annuals** complete their life cycle—from germination through flowering and seed production to death—in a single year or less. Many wildflowers are annuals, as are the most important food crops, including the cereal grains and legumes. A plant is called a biennial if its life generally spans two years. In many cases, plants with this biennial life cycle are those that live through an intervening cold period (winter) between vegetative growth (first spring/summer) and flowering (second spring/summer). Beets and carrots are biennials, but we rarely leave them in the ground long enough to see them flower. Plants that live many years, including trees, shrubs, and some grasses, are known as **perennials**. Some of the buffalo grass of the North American plains is believed to have been growing for 10,000 years from seeds that sprouted at the close of the last ice age. When a perennial finally dies, it is not generally from old age, but from an infection or some environmental trauma, such as fire or severe drought.

For as long as it survives, a plant is capable of indeterminate growth because it has perpetually embryonic tissues called **meristems** in its regions of growth. Meristematic cells divide to generate additional cells. Some of the products of this division remain in the meristematic region to produce still more cells, while others become specialized and are incorporated into the tissues and organs of the growing plant. Cells that remain as wellsprings of new cells in the meristem are called initials. The new cells that are displaced from the meristem, called derivatives, continue to divide for some time, until the cells they produce begin to specialize within developing tissues.

The pattern of plant growth depends on the locations of the meristems (FIGURE 35.13). **Apical meristems**, located at the tips of roots and in the buds of shoots, supply cells for the plant to grow in length. This elongation, called **primary growth**, enables roots to ramify throughout the soil and shoots to increase their exposure to light and carbon dioxide. In herbaceous (nonwoody) plants, only primary growth occurs. In woody plants, however, there is also **secondary growth**, a progressive thickening of the roots and shoots formed earlier by primary growth. Secondary growth is the product of **lateral meristems**, cylinders of dividing cells

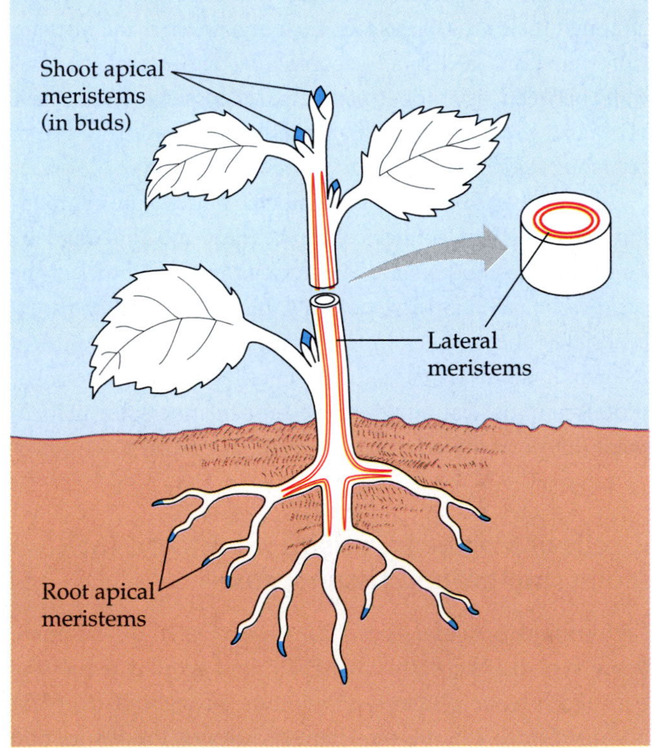

FIGURE 35.13 • Locations of major meristems: an overview of plant growth. Meristems are self-renewing populations of cells that divide and provide cells for plant growth. Apical meristems (color-coded blue in this diagram) are located near the tips of roots and shoots and are responsible for primary growth, or growth in length. Woody plants also have lateral meristems (color-coded red here) that function in secondary growth, which adds girth to roots and shoots.

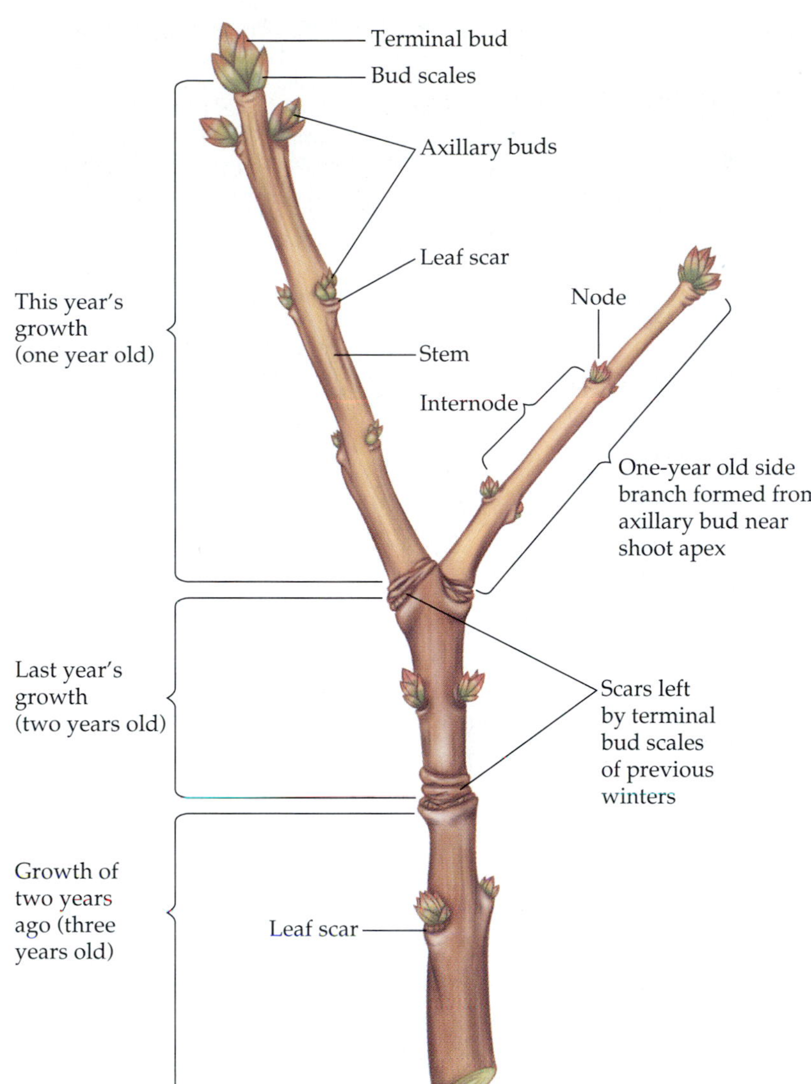

FIGURE 35.14 ▪ **Morphology of a winter twig.** At the tip of this lilac twig is the dormant terminal bud, enclosed by scales that protect its apical meristem. In spring, the bud will shed its scales and begin a new spurt of primary growth, producing a series of nodes and internodes. Along each growth segment, nodes are marked by scars left when leaves fell during autumn. Above each leaf scar is either an axillary bud or a branch twig formed by an axillary bud. Farther down the twig are whorls of scars left by the scales that enclosed the terminal bud during the previous winter. Each spring and summer, as primary growth extends the shoot, secondary growth thickens the parts of the shoot that formed in previous years.

extending along the length of roots and shoots. These lateral meristems replace the epidermis with a secondary dermal tissue, such as bark, that is thicker and tougher, and they also add layers of vascular tissue. Wood is the secondary xylem that accumulates over the years.

In woody plants, primary and secondary growth occur at the same time but in different locations. Primary growth is restricted to the youngest parts of the plant—the tips of roots and shoots, where the apical meristems are located. The lateral meristems develop in slightly older regions of the roots and shoots, some distance away from the tips. There, secondary growth adds girth to the organs. The oldest region of a root or shoot—the base of a tree branch, for example—has the greatest accumulation of secondary tissues formed by the lateral meristems. Each growing season, primary growth produces young extensions of roots and shoots, while secondary growth thickens and strengthens the older parts of the plant. Examine a winter twig of a deciduous tree, and you can see the relationship between primary and secondary growth (FIGURE 35.14).

Primary growth: Apical meristems extend roots and shoots by giving rise to the primary plant body

Primary growth produces what is called the **primary plant**
35.1 **body**, which consists of the three tissue systems: dermal, vascular, and ground tissues (see FIGURE 35.12). A herbaceous plant and the youngest parts of a woody plant represent the primary plant body. Although apical meristems are responsible for the extension of both roots and shoots, there are important differences in the primary growth of these two kinds of organs.

Primary Growth of Roots

Primary growth pushes roots through the soil. The root tip is covered by a thimblelike **root cap**, which physically protects the delicate meristem as the root elongates through the abrasive soil. The cap also secretes a polysaccharide slime that

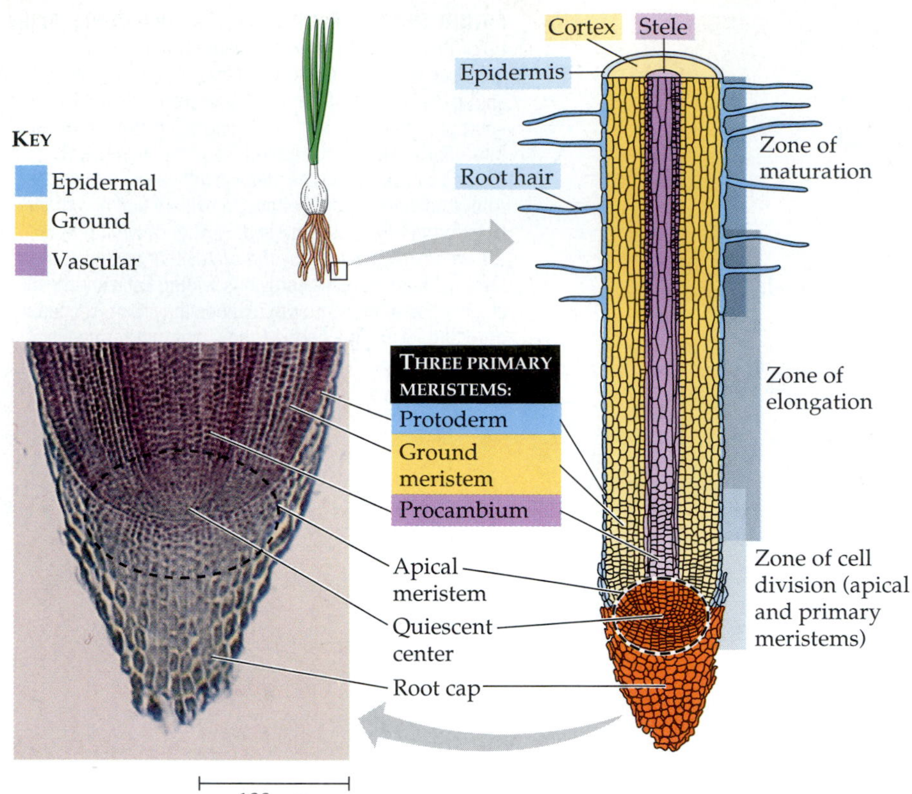

FIGURE 35.15 ▪ Primary growth of a root. The diagram and light micrograph take us into the tip of an onion root. Mitosis is concentrated in the zone of cell division, where the apical meristem and its products, the three primary meristems, are located. The apical meristem also maintains the root cap by generating new cells that replace those that are sloughed off. If the apical meristem is damaged, its quiescent center is activated and restores the meristem by means of cell division. Most lengthening of the root is concentrated in the zone of elongation. Cells become functionally mature in the zone of maturation. The zones of the root grade into one another without sharp boundaries.

lubricates the soil around the growing root tip. Growth in length is concentrated near the root's tip, where three zones of cells at successive stages of primary growth are located. From the root tip upward, they are the zone of cell division, the zone of elongation, and the zone of maturation. These regions grade together, with no sharp boundaries (FIGURE 35.15).

The **zone of cell division** includes the apical meristem and its derivatives, called primary meristems. The apical meristem, at the heart of the zone of cell division, produces the cells of the primary meristems and also replaces cells of the root cap that are sloughed off. Near the center of the apical meristem is the **quiescent center**, a population of cells that divide much more slowly than the other meristematic cells. Cells of the quiescent center are relatively resistant to damage from radiation and toxic chemicals, and they may function as reserves that can be recruited to restore the meristem if it is somehow damaged. In experiments where part of the apical meristem is removed, cells of the quiescent center become more mitotically active and produce a new meristem. Just above the apical meristem, the products of its cell division form three concentric cylinders of cells that continue to divide for some time. These are the primary meristems—the **protoderm**, **procambium**, and **ground meristem**—which will produce the three primary tissue systems of the root: dermal, vascular, and ground tissues.

The zone of cell division blends into the **zone of elongation**. Here the cells elongate to more than ten times their original length. Although the meristem provides the new cells for growth, the elongation of cells is mainly responsible for pushing the root tip, including the meristem, ahead. The meristem sustains growth by continuously adding cells to the youngest end of the zone of elongation.

Even before they finish elongating, the cells of the root begin to specialize in structure and function where the zone of elongation grades into the **zone of maturation**. In this region of the root, the three tissue systems produced by primary growth complete their differentiation.

Primary Tissues of Roots. The three primary meristems give rise to the three primary tissues of roots, shown in FIGURE 35.16. The protoderm, the outermost primary meristem, gives rise to the epidermis, a single layer of cells covering the root. Water and minerals that enter the plant from the soil must enter through the epidermis. The root hairs enhance this process by greatly increasing the surface area of epidermal cells.

The procambium gives rise to the **stele**, which is the vascular bundle where both xylem and phloem develop. In roots, the stele is a single central cylinder. In most dicots, the xylem cells radiate from the center of the stele in two or more spokes, with phloem developing in the wedges between the spokes. The stele of a monocot generally has a central core of parenchyma cells, often called the **pith**, which is ringed by vascular tissue with an alternating pattern of xylem and phloem.

Between the protoderm and procambium is the ground meristem, which gives rise to the ground tissue system. The

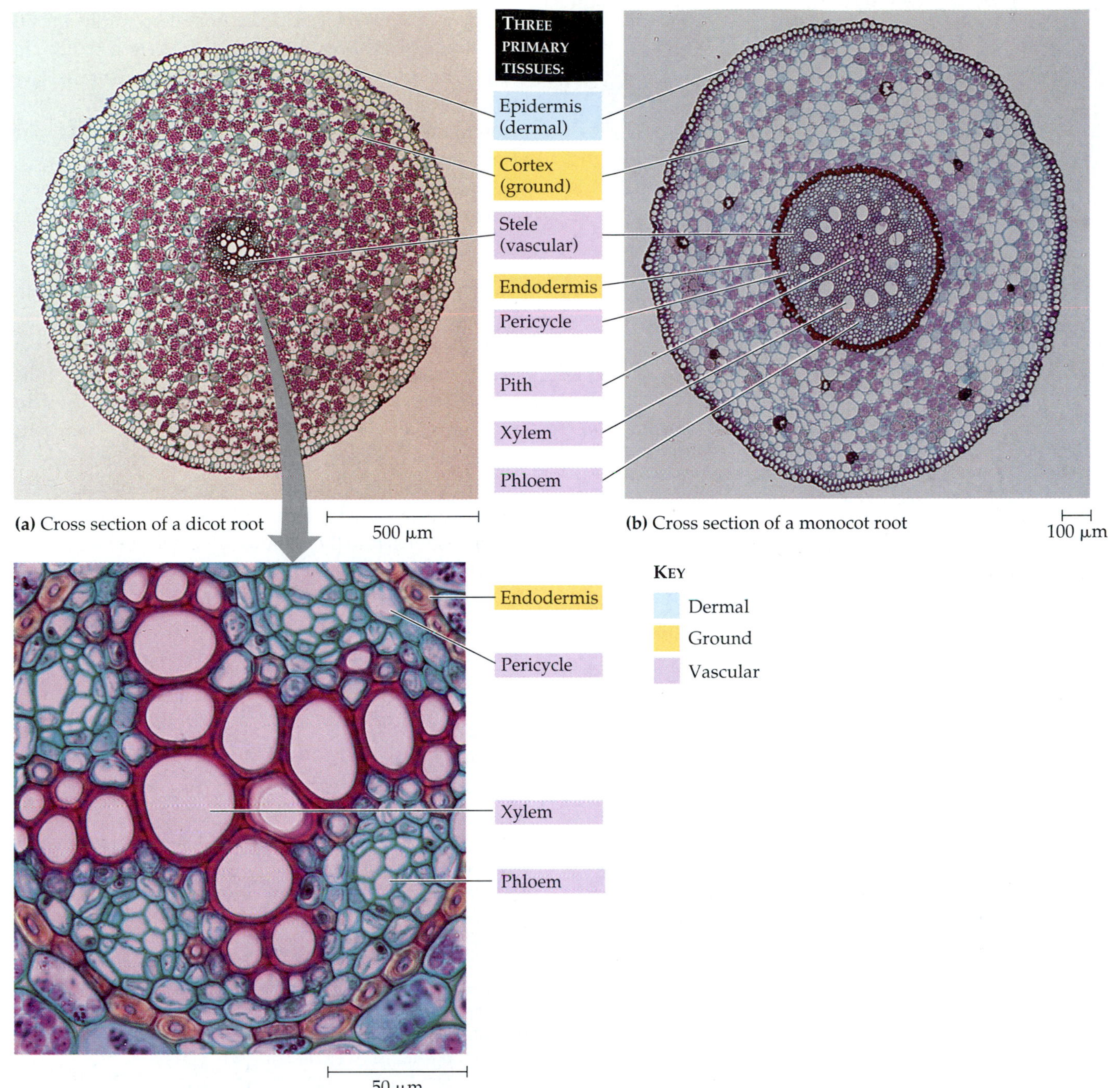

35.1 **FIGURE 35.16 ▪ Organization of primary tissues in young roots.** Parts **(a)** and **(b)** show, in transverse (cross) section, the three primary tissue systems in the roots of a dicot (*Ranunculus*, a buttercup) and a monocot (*Zea*, corn). The main difference between the dicot and monocot here is the organization of tissues within the stele, the aggregate of vascular tissue, which takes the form of a cylinder in roots. The enlargement of the dicot stele shows that the xylem vessels radiate like spokes from the center. Wedges of phloem are located between these xylem spokes. Xylem and phloem also alternate within the stele of the monocot root, but there the vascular tissues surround a core of parenchyma cells called the pith. In both dicots and monocots, the stele is encircled by the endodermis, the innermost region of cells of the cortex. Just inside the endodermis is the pericycle, a layer of cells with the potential to divide and give rise to lateral roots. (All LMs.)

ground tissue, which consists mostly of parenchyma cells, fills the **cortex**, the region of the root between the stele and epidermis. Ground tissue cells of the root store food, and their plasma membranes are active in the uptake of minerals that enter the root with the soil solution. The innermost layer of the cortex is the **endodermis**, a cylinder one cell thick that forms the boundary between the cortex and the stele. You will learn in Chapter 36 how the endodermis functions as a selective barrier that regulates the passage of substances from the soil solution into the vascular tissue of the stele.

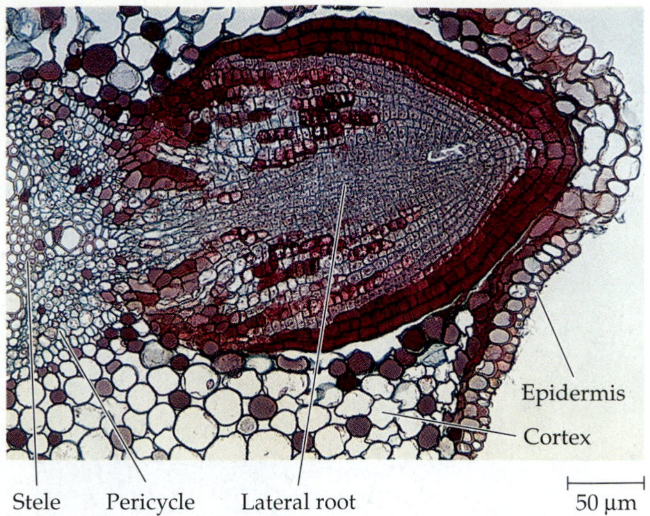

FIGURE 35.17 ▪ The formation of lateral roots. In this transverse section of a willow root, lateral roots emerge from the pericycle, the outermost layer of the stele (LM).

An established root may sprout **lateral roots**, which arise from the outermost layer of the stele, the **pericycle** (FIGURE 35.17). Just inside the endodermis, the pericycle is a layer of cells that may become meristematic and begin dividing again. Originating as a clump of cells formed by mitosis in the pericycle, a lateral root elongates and pushes through the cortex until it emerges from the main root. The stele of the lateral root retains its connection with the stele of the primary root, making the vascular tissue continuous throughout the root system.

Primary Growth of Shoots

The apical meristem of a shoot is a dome-shaped mass of dividing cells at the tip of the terminal bud (FIGURE 35.18). As in the root, the apical meristem of the shoot tip gives rise to the primary meristems—protoderm, procambium, and ground meristem—which will differentiate into the three tissue systems. Leaves arise as leaf primordia on the flanks of the apical meristem. Axillary buds develop from islands of meristematic cells left by the apical meristem at the bases of the leaf primordia.

Within a bud, nodes, with their leaf primordia, are crowded close together, because internodes are very short. Most of the actual elongation of the shoot occurs by the growth of slightly older internodes below the shoot apex. This growth is due to both cell division and cell elongation within the internode. In some plants, including grasses, internodes continue to elongate all along the length of the shoot over a prolonged period. This is possible because these plants have meristematic regions, called intercalary meristems, at the base of each internode.

Axillary buds have the potential to form branches of the shoot system at some later time (see FIGURE 35.4). Thus, there is an important difference in how roots and shoots form lateral organs. Lateral roots originate from deep within a main root as outgrowths from the pericycle (see FIGURE 35.17). In contrast, branches of the shoot system originate from axillary buds, located at the surface of a main shoot. Only by extending from the pericycle can a lateral root be connected to the plant's vascular system. The vascular tissue of a stem, however, is near the surface, and branches can develop with connections to the vascular tissue without having to originate from deep within the main shoot.

Primary Tissues of Stems. Vascular tissue runs the length of a stem in several strands called **vascular bundles** (FIGURE 35.19). This arrangement contrasts with the root, where the vascular tissue forms a vascular cylinder in the center of the root (see FIGURE 35.16). At the transition zone where the stem grades into the root, the vascular bundles of the stem converge as the root's vascular cylinder. (The term *stele* refers to the entire collection of vascular tissue in both roots and stems. In roots, the stele is in the form of the single vascular cylinder; in stems, *stele* refers to the entire set of vascular bundles.)

Each vascular bundle of the stem is surrounded by ground tissue. In most dicots the vascular bundles are arranged in a ring, with pith to the inside of the ring and cortex external to the ring. Both pith and cortex are part of the ground tissue system. The vascular bundles have their xylem facing the pith

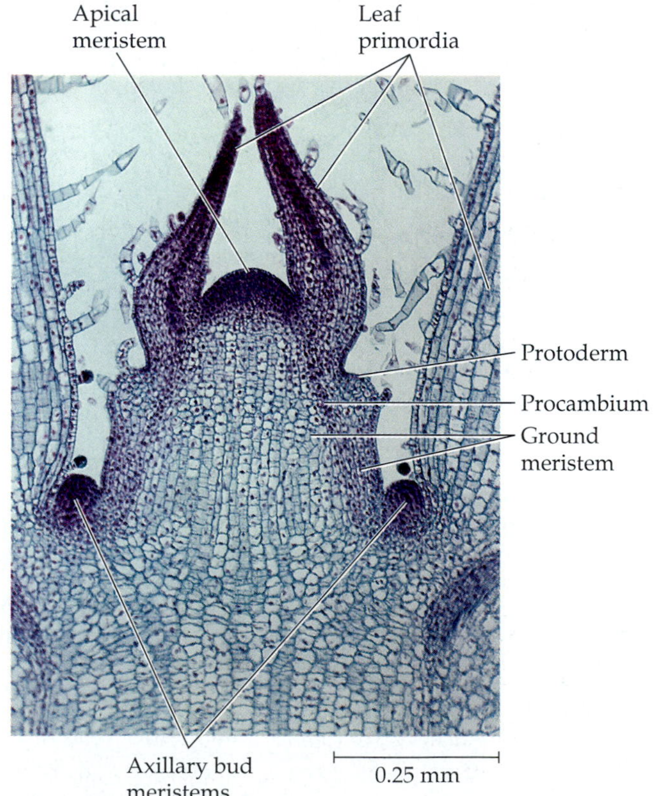

FIGURE 35.18 ▪ The terminal bud and primary growth of a shoot. Leaf primordia arise from the flanks of the apical dome. The apical meristem gives rise to protoderm, procambium, and ground meristem, which in turn develop into the three tissue systems. This is a longitudinal section of the shoot tip of *Coleus* (LM).

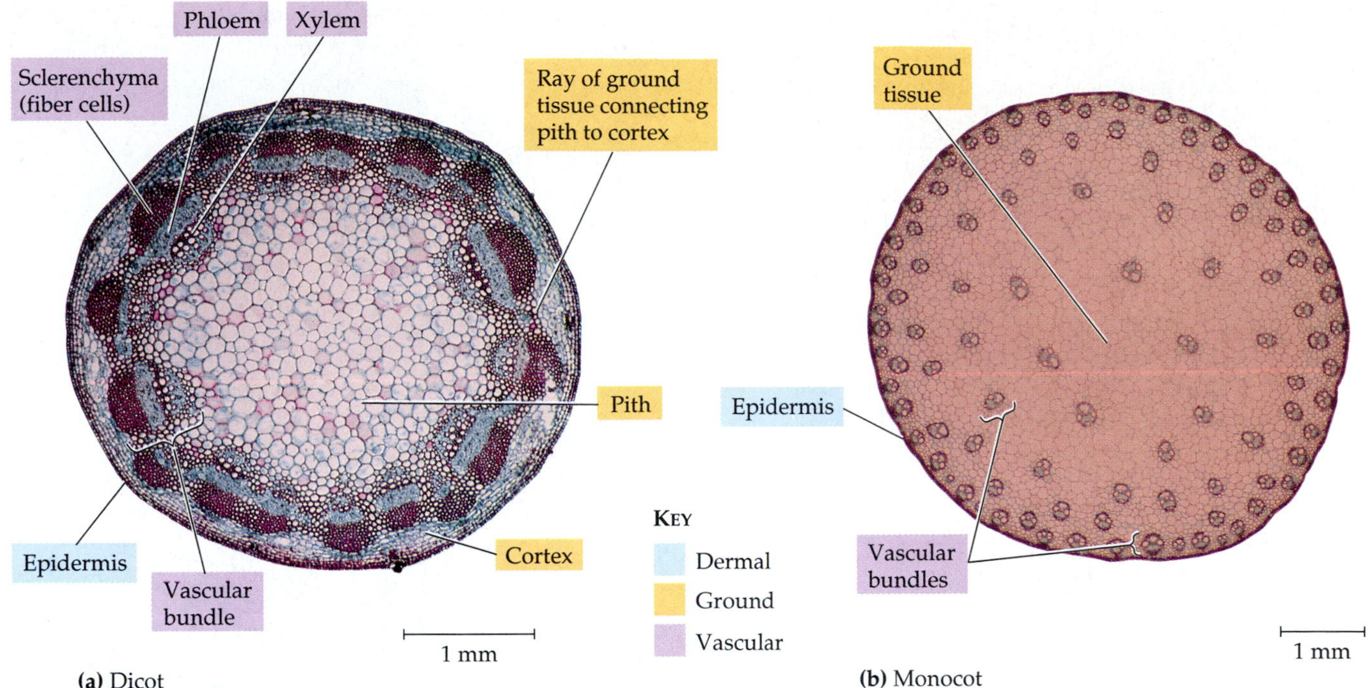

FIGURE 35.19 ▪ Organization of primary tissues in young stems. (a) A dicot stem (sunflower) with vascular bundles arranged in a ring. The ground tissue system consists of an outer cortex and an inner pith surrounded by vascular bundles. **(b)** A monocot stem (corn) with vascular bundles located throughout the ground tissue. (Both LMs of transverse sections.)

35.1

and their phloem facing the cortex side. The pith and cortex are connected by thin rays of ground tissue between the vascular bundles. In the stems of most monocots the vascular bundles are scattered throughout the ground tissue rather than being arranged in a ring. In both monocots and dicots the ground tissue of the stem is mostly parenchyma, but many stems are strengthened by collenchyma located just beneath the epidermis. Sclerenchyma, in the form of fiber cells within vascular bundles, also helps support stems.

The protoderm of the terminal bud gives rise to the epidermis, which covers stems and leaves as part of the continuous dermal tissue system.

Tissue Organization of Leaves. The leaf is cloaked by its epidermis, with cells tightly interlocked like pieces of a puzzle (FIGURE 35.20, p. 688). This epidermis, like our own skin, is a first line of defense against physical damage and pathogenic organisms. Also, the waxy cuticle of the epidermis is a barrier to the loss of water from the plant. The epidermal barrier is interrupted only by the **stomata**, tiny pores flanked by specialized epidermal cells called **guard cells**. Each stoma is actually a gap between a pair of guard cells. The stomata allow gas exchange between the surrounding air and the photosynthetic cells inside the leaf. Stomata are also the major avenues for the loss of water from the plant by evaporation, a process called **transpiration**.

The ground tissue of a leaf is sandwiched between the upper and lower epidermis in the region called **mesophyll** (Gr. *mesos*, "middle," and *phyll*, "leaf"). It consists mainly of parenchyma cells equipped with chloroplasts and specialized for photosynthesis. The leaves of many dicots have two distinct regions of mesophyll. On the upper half of the leaf are one or more layers of palisade parenchyma, made up of cells that are columnar in shape. Below the palisade region is the spongy parenchyma, which gets its name from the labyrinth of air spaces through which carbon dioxide and oxygen circulate around the irregularly shaped cells and up to the palisade region. The air spaces are particularly large in the vicinity of stomata, where gas exchange with the outside air occurs. In most plants, stomata are more numerous on the bottom surface of a leaf than on top. This adaptation minimizes water loss, which occurs more rapidly through stomata on the sunny upper side of a leaf. Once again, it helps to view the functional structure of plants in the evolutionary context of adaptation to land.

The vascular tissue of a leaf is continuous with the xylem and phloem of the stem. Leaf traces, which are branches from vascular bundles in the stem, pass through petioles and into leaves. Within a leaf, veins subdivide repeatedly and branch throughout the mesophyll. This brings xylem and phloem into close contact with the photosynthetic tissue, which obtains water and minerals from the xylem and loads its sugars and other organic products into the phloem for shipment to other parts of the plant. The vascular infrastructure also functions as a skeleton that reinforces the shape of the leaf.

Modular Shoot Construction and Phase Changes During Development. Serial development of nodes and internodes

35.1 **FIGURE 35.20 ▪ Leaf anatomy. (a)** This cutaway drawing of a leaf illustrates the organization of the three tissue systems: dermal tissue (epidermis), vascular tissue, and ground tissue (the mesophyll region consisting of palisade parenchyma and spongy parenchyma). **(b)** This surface view of a *Tradescantia* leaf shows the cells of the epidermis and stomata, with their guard cells (LM). The lower leaf surface generally has more stomata than the upper surface, an adaptation that helps reduce water loss by transpiration. **(c)** Palisade and spongy regions of mesophyll are present within the leaf of a lilac, a dicot (LM).

Guard cells
Epidermal cell
Stoma
(b)
50 μm
Cuticle
Collenchyma
Upper epidermis
Palisade parenchyma
Spongy parenchyma
Lower epidermis
Guard cells
Xylem
Phloem
Cuticle
Vein
Stoma
(a)
Vein
Air spaces
Guard cells
100 μm
(c)

within the shoot apex, followed by elongation of the internodes, produces a shoot having a modular construction—a series of segments, each consisting of a stem, one or more leaves, and an axillary bud associated with each leaf (FIGURE 35.21). The development of this modular morphology should not be confused with the development of the segmented anatomy of certain animals such as earthworms. In animal development, the rudiments of all organs form in the embryo. Plants, in contrast, add organs at their tips for as long as they live. Unlike the segments of an earthworm, which are all the same age, the modules of a plant vary in age in proportion to their distance from an apical meristem.

From what you have learned so far about the shoot apex and primary growth, it would seem as if the meristem lays down a series of identical modules for as long as the shoot lives. In fact, the apical meristem can change from one developmental phase to another during its history. One of these **phase changes** is a gradual transition in vegetative (leaf-producing) growth from a juvenile state to a mature state. Usually the most obvious sign of this phase change is a change in the morphology of the leaves produced. The leaves of juvenile versus mature shoot modules differ in shape and other features (FIGURE 35.22). Once the meristem has laid down juvenile nodes and internodes, they retain that status even as the shoot continues to elongate and the meristem eventually changes to the mature phase. If axillary buds give rise to branches, those shoots reflect the developmental phase of the main shoot modules from which they arise, even if the main shoot apex subsequently makes the phase change and begins laying down modules of the mature phase. Ironically, this means that a branch with mature leaves may actually be *younger* than a branch with juvenile leaves.

The juvenile-to-mature phase transition is another case where it is misleading to compare plant and animal development. In an animal this transition occurs at the level of the entire organism. In plants, phase changes during the history of apical meristems can result in juvenile and mature regions coexisting along the axis of each shoot.

In some cases a shoot apex undergoes a second phase transition from a mature vegetative state to a reproductive (flower-producing) state. Unlike vegetative growth, which is self-renewing, the production of a flower by an apical meristem

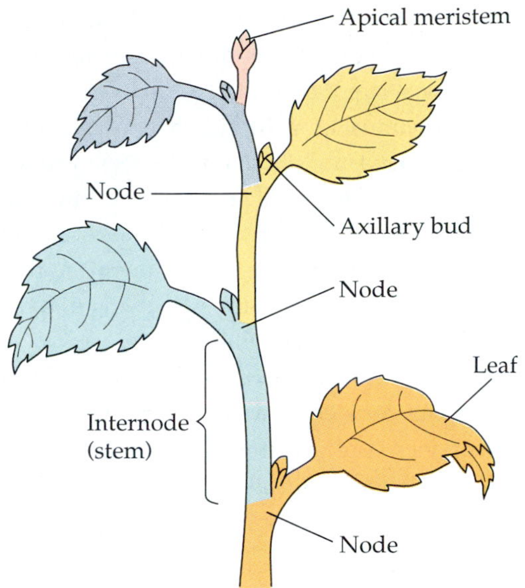

FIGURE 35.21 ▪ Modular construction of a shoot. Primary growth lays down a series of segments (different colors in this drawing), each consisting of a node with one or more leaves, an axillary bud in the axil of each leaf, and an internode. Miniature modules develop within the apical meristem and then grow, pushing the apex onward, where it forms the next module, and so on. This serial addition of segments at the growing end of the shoot contrasts with the development of segmentation in certain animals, in which all the segments form at about the same time in the embryo. (Imagine if our arms and legs were different ages, like the appendages of a plant!)

(a)

(b)

FIGURE 35.22 ▪ Phase change in the shoot system of *Eucalyptus*. A silver-dollar eucalyptus (*Eucalyptus polyanthemos*) has both **(a)** juvenile leaves (the round "silver dollars") and **(b)** mature leaves (the lance-shaped leaves). This dual foliage reflects a phase change in the development of the apical meristem of each shoot. In its juvenile vegetative phase, a meristem lays down modules on which round leaves develop. As the meristem changes gradually to the mature vegetative phase, the leaves become more and more lance-shaped. Once a module forms, its developmental phase—juvenile or mature—is fixed; that is, round leaves do not mature into lanceolate leaves. The developmental status of the apical meristem also sets the phase of the axillary buds it forms.

terminates primary growth of that shoot tip; the apical meristem is consumed in the production of the flower's organs.

In Chapter 38 we will study flower development in more detail, and in Chapter 39 we will examine how this phase change from vegetative growth of a shoot to the reproductive growth of flowering is controlled.

Secondary growth: Lateral meristems add girth by producing secondary vascular tissue and periderm

35.2 Most vascular plants undergo secondary growth, increasing in girth as well as length. The **secondary plant body** consists of the tissues produced during this secondary growth in diameter. Two lateral meristems function in secondary growth: the **vascular cambium**, which produces secondary xylem (wood) and phloem, and the **cork cambium**, which produces a tough, thick covering for stems and roots that replaces the epidermis. Secondary growth occurs in all gymnosperms. Among angiosperms, secondary growth takes place in most dicot species but is rare in monocots.

Secondary Growth of Stems

Vascular Cambium and the Production of Secondary Vascular Tissue. The vascular cambium is a cylinder of meristematic cells that forms secondary vascular tissue, and the accumulation of this secondary vascular tissue over the years accounts for most of the increase in the diameter of a woody plant (see FIGURE 35.13). The vascular cambium produces secondary xylem to its interior and secondary phloem to its exterior (FIGURE 35.23). A tree grows in girth as the

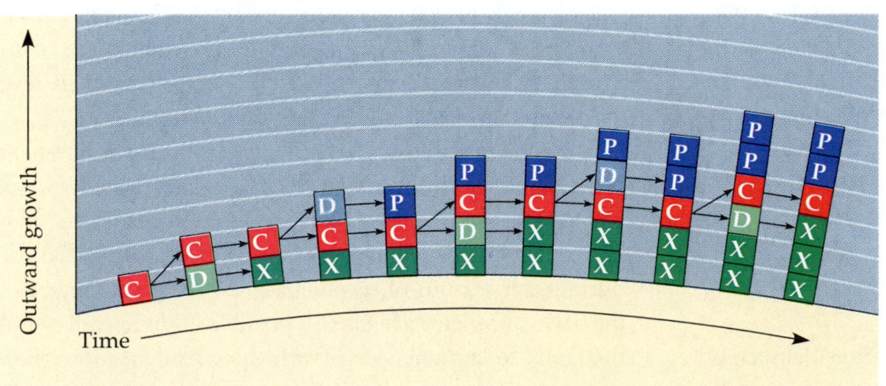

FIGURE 35.23 ▪ Production of secondary xylem and phloem by the vascular cambium. This diagram traces the radial file of cells that develops from the meristematic activity of a single cell of the vascular cambium as viewed in transverse section. The cambium cell (C) gives rise to secondary xylem (X) on the inside and secondary phloem (P) on the outside. Each time a cambium cell divides, one daughter cell retains its status as an initial (see p. 682), and the other, the derivative (D), differentiates into a xylem or phloem cell. As layers of xylem are added, vascular cambium itself increases in diameter.

diameter of the cylinder of vascular cambium increases over time, laying down successive layers of secondary tissues, each with a larger diameter than the last.

It is important to understand that primary and secondary growth occur simultaneously but in different regions of a stem. While the apical meristem is elongating the stem by giving rise to the primary plant tissues, including primary xylem and phloem in the form of vascular bundles, secondary growth commences farther down the shoot. By adding secondary vascular tissue, secondary growth transforms the structure of the older regions of a stem.

After the apical meristem extends a shoot, how does the primary plant body of that young shoot make the transition from primary growth to secondary growth? The vascular cambium forms from parenchyma cells that regain the capacity to divide; that is, the cells become meristematic. This meristem forms in a layer between the primary xylem and primary phloem of each vascular bundle and in the rays of ground tissue between the bundles. The meristematic bands within the vascular bundles and rays unite to form the vascular cambium as a continuous cylinder of dividing cells surrounding the primary xylem and pith of the stem (FIGURE 35.24).

Viewed in cross section, the cylinder of vascular cambium appears as a ring. If we trace around the ring, there are alternating regions of cambium cells called ray initials and fusiform initials. The **ray initials** are cambium cells that produce radial files of parenchyma cells known as xylem rays and phloem rays. These rays separate wedge-shaped sections of secondary vascular tissue. Xylem and phloem rays, consisting mainly of parenchyma, function as living avenues for the radial transport of water and nutrients within a woody stem and in the storage of starch and other reserves. The cambium cells within the vascular bundles are the **fusiform initials**, a name that refers to the shape of these cells, which have tapered (fusiform) ends and are elongated along the axis of the stem. Fusiform initials produce new vascular tissue, forming secondary xylem to the inside of the vascular cambium and secondary phloem to the outside (see FIGURE 35.23).

As secondary growth continues over the years, layer upon layer of secondary xylem accumulates, producing what we call wood. Wood consists mainly of tracheids, vessel elements (in angiosperms), and fibers. These cells, dead at functional maturity, have thick, lignified walls that give wood its hardness and strength. In temperate regions of the world, secondary growth in perennial plants is interrupted each year when the vascular cambium becomes dormant during winter. When secondary growth resumes in the spring, the first tracheids and vessel elements to develop usually have relatively large diameters and thin walls compared to the secondary xylem produced later in the summer. Thus, it is usually possible to distinguish spring wood from summer wood (see FIGURE 35.24).

The structure of the spring wood maximizes delivery of water to new, expanding leaves during the start of the growth season. Though the thick-walled cells of summer wood do not transport as much water as those of spring wood, they add more physical support to the tree than do the thinner-walled cells. The annual growth rings that are evident in cross sections of most tree trunks in temperate regions result from the yearly activity of the vascular cambium: cambium dormancy, spring wood production, and summer wood production. The boundary between one year's growth and the next is usually quite conspicuous, sometimes allowing us to estimate the age of a tree by counting its annual rings.

The secondary phloem, external to the vascular cambium, does not accumulate as extensively over the years as the secondary xylem does. As a tree grows in girth, the older (outermost) secondary phloem, and all tissues external to it, develop into bark, which eventually splits and sloughs off the tree trunk.

Cork Cambium and the Production of Periderm. During secondary growth, the epidermis produced by primary growth splits, dries, and falls off the stem. It is replaced by new protective tissues produced by the cork cambium, a cylinder of meristematic tissue that first forms in the outer cortex of the stem (see FIGURE 35.24C). Cork cambium produces cork cells, which accumulate exterior to the cork cambium. As the cork cells mature, they deposit a waxy material called suberin in their walls and then die. The cork tissue then functions as a barrier that helps protect the stem from physical damage and pathogens. And because cork is waxy, it impedes water loss from the stems. Together, the layers of cork plus the cork cambium make up the **periderm**. This is the protective coat of the secondary plant body that replaces the epidermis of the primary body. The term **bark**, more inclusive than periderm, refers to all tissues external to the vascular cambium. Thus, in an outward direction, bark consists of phloem, cork cambium, and cork. Put another way, bark is phloem plus periderm (FIGURE 35.25, p. 692).

Unlike the vascular cambium, which grows in diameter, the original cork cambium is a cylinder of fixed size. After a few weeks of cork production, the cork cambium loses its meristematic activity, and its remaining cells differentiate into cork. Expansion of the stem splits the original periderm. How is it renewed to keep pace with continued secondary growth? New cork cambium forms deeper and deeper in the cortex. Eventually, no cortex is left, and the cork cambium then develops from parenchyma cells in the secondary phloem.

Only the youngest secondary phloem, which is internal to the cork cambium, functions in sugar transport. The older secondary phloem, outside the cork cambium, dies and helps protect the stem until it is sloughed off as part of the bark during later seasons of secondary growth. Spongy regions in the bark called **lenticels** make it possible for living cells within the trunk to exchange gases with the outside air for cellular respiration.

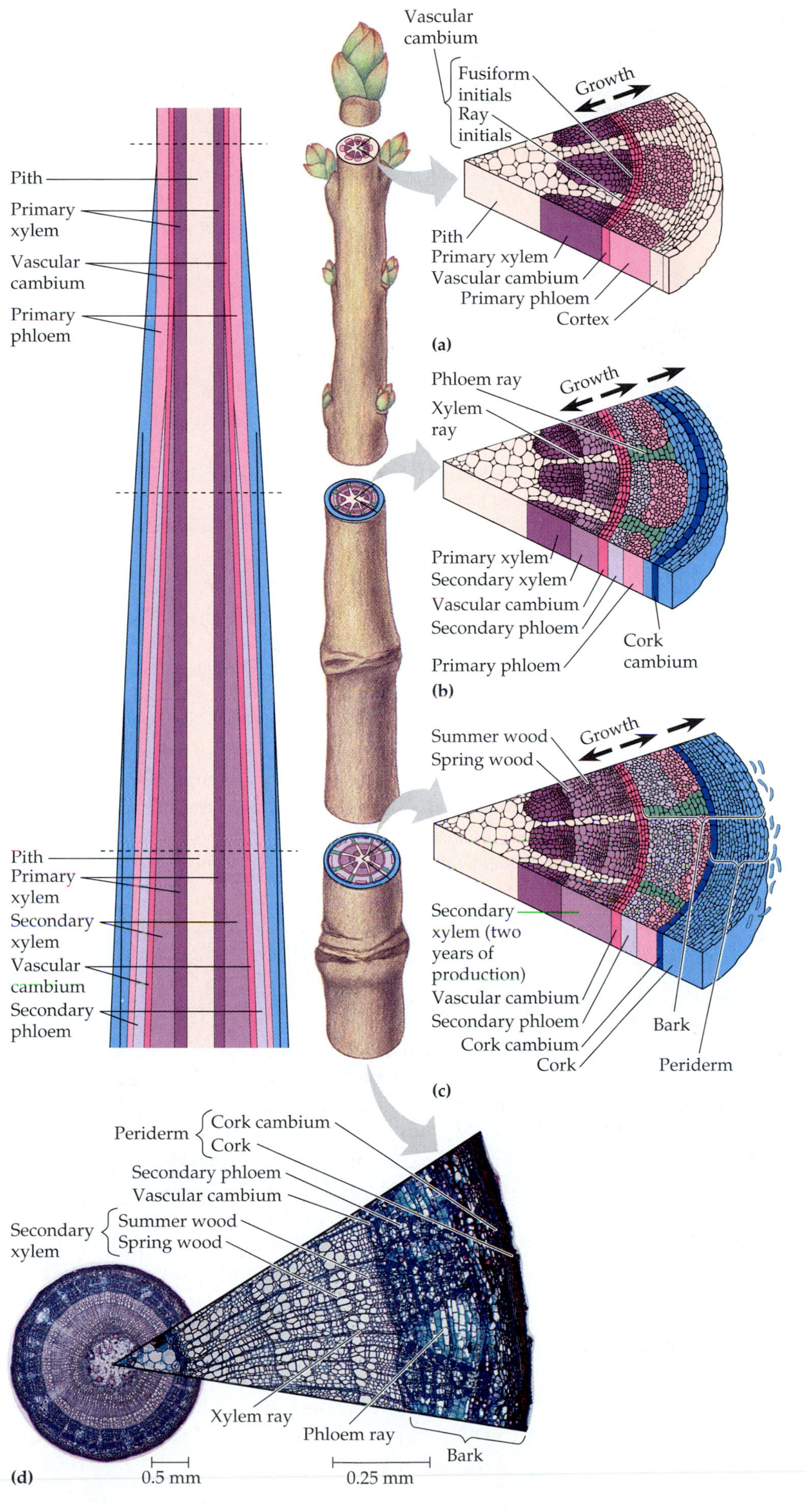

FIGURE 35.24 ▪ Secondary growth of a stem. You can track the progress of secondary growth by examining the sections through sequentially older parts of the stem. (You would observe the same changes if you could follow the youngest region, near the apex, for the next three years.) **(a)** In the youngest part of the stem, you can see the primary plant body, as formed by the apical meristem during primary growth. The vascular cambium is beginning to develop. **(b)** As primary growth continues to elongate the stem, the portion of the stem formed earlier the same year has already started its secondary growth. This portion of the stem increases in girth as fusiform initials of the vascular cambium form secondary xylem to the inside and secondary phloem to the outside. The ray initials of the cambium give rise to the parenchyma tissue of the xylem and phloem rays. As the diameter of the vascular cambium increases and secondary growth continues, the secondary phloem and other tissues external to the cambium cannot keep pace with the expansion because the cells no longer divide. A second lateral meristem, the cork cambium, develops from parenchyma cells in the stem's cortex. The cork cambium produces cork cells, which replace the epidermis as the protective covering at the stem's surface. **(c)** In year two of secondary growth, the vascular cambium adds to the secondary xylem and phloem, and the cork cambium produces cork. As the diameter of the stem continues to increase, the outermost tissues exterior to the cork cambium rupture and slough off from the stem. The cork cambium re-forms in progressively deeper layers of the cortex. When none of the original cortex is left, the cork cambium develops from parenchyma cells in the secondary phloem. The epidermis of the primary plant body has been replaced by the protective tissue of the secondary plant body, which is called periderm. Periderm consists of the cork cambium and the cork to its exterior. What we call bark consists of all tissues exterior to the vascular cambium: secondary phloem plus periderm. **(d)** This light micrograph shows a three-year-old section of the stem. Notice the annual growth rings and the contrast between spring wood and summer wood.

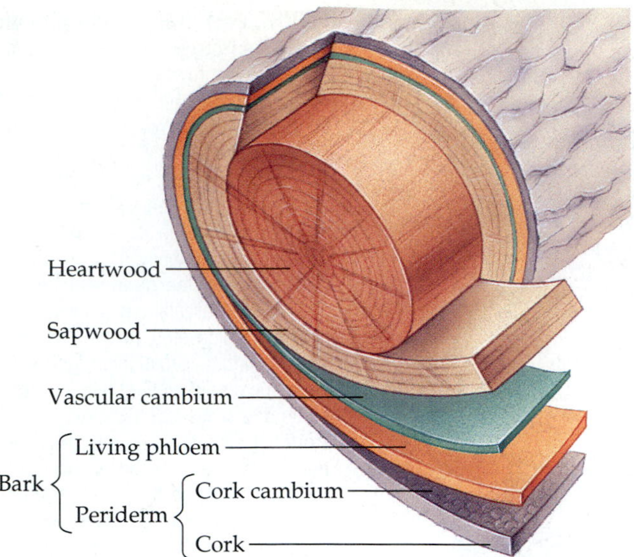

35.2 **FIGURE 35.25 • Anatomy of a tree trunk.** Beginning at the center of the tree and tracing outward, we can distinguish several zones. Heartwood and sapwood both consist of secondary xylem. Heartwood is older and no longer functions in water transport; the lignified walls of its dead cells form a central column that supports the tree. This wood owes its rich color to resins and other compounds that clog the cell cavities and help protect the core of the tree from fungi and insects. Sapwood is so named because its secondary xylem cells still function in the upward transport of water and minerals (xylem sap). Since each new layer of secondary xylem has a larger circumference, secondary growth enables the xylem to transport more sap each year, providing water and minerals to an increasing number of leaves.

The result of many years of secondary stem growth can be seen by examining an old tree trunk in cross section (FIGURE 35.25). FIGURE 35.26 will help you review the relationships among the primary and secondary tissues of a woody plant.

Secondary Growth of Roots

The two lateral meristems, vascular cambium and cork cambium, also develop and produce secondary growth in roots. The vascular cambium forms within the stele and produces secondary xylem to its inside and secondary phloem to its outside. As the stele grows in diameter, the cortex and epidermis are split and shed. A cork cambium forms from the pericycle of the stele and produces the periderm, which becomes the secondary dermal tissue. Unlike the primary epidermis of a younger root, periderm is impermeable to water. Therefore, it is only the youngest roots, those representing the primary plant body, that absorb water and minerals from the soil. Older roots, with secondary growth, function mainly to anchor the plant and to transport water and solutes between the younger roots and the shoot system.

Over the years the root becomes woodier, and annual rings are usually evident in the secondary xylem. The tissues external to the vascular cambium form a thick, tough bark. After extensive secondary growth, old stems and old roots are quite similar.

In dissecting the plant to examine its parts, as we have done in this chapter, we must remember that the whole plant functions as an integrated organism. In the following chapters you will learn more about how materials are transported within the plant (Chapter 36), how plants obtain nutrients (Chapter 37), how plants reproduce and develop (Chapter 38), and how the various functions of the plant are coordinated (Chapter 39). Your understanding of the working plant will be enhanced by remembering that structure fits function, and that plant anatomy and physiology reflect evolutionary adaptation to the problems of living on land.

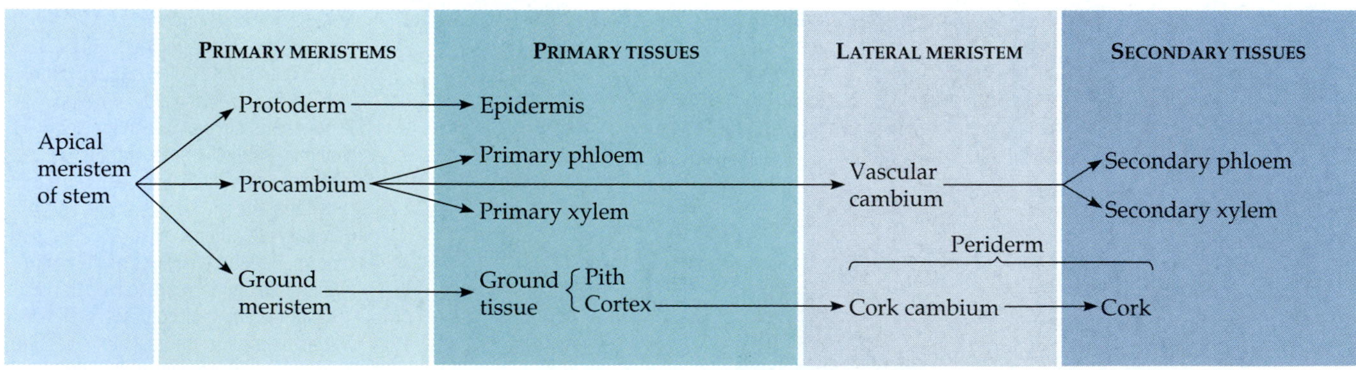

FIGURE 35.26 • A summary of primary and secondary growth in a woody stem.

CHAPTER REVIEW

REVIEW OF KEY CONCEPTS

(with page numbers and key figures)

INTRODUCTION TO MODERN PLANT BIOLOGY

- **Molecular biology is revolutionizing the study of plants (pp. 670–671, FIGURE 35.1)** New techniques and new model systems, including *Arabidopsis,* are catalyzing explosive progress in our understanding of plants.
- **Plant biology reflects the major themes in the study of life (pp. 671–672, FIGURE 35.2)** The correlation between structure and function and the evolutionary context of adaptation to the environment illuminate our study of plants.

THE ANGIOSPERM BODY

- **A plant's root and shoot systems are evolutionary adaptations to living on land (pp. 672–678, FIGURES 35.3 and 35.4)** Roots anchor the plant, absorb and conduct water and minerals, and store food. The shoot system consists of stems, leaves, and flowers. Leaves are attached by their petioles to the nodes of the stem, with internodes of the stem separating the nodes. Axillary buds, located in the axils of petioles and stems, have the potential to extend as vegetative or floral branches. Vascular tissues integrate the parts of the plant. Water and minerals move up from roots in the xylem. Sugar is exported from leaves or storage organs in the phloem. The two classes of angiosperms, monocots and dicots, differ in anatomical details.
- **Structural adaptations of protoplasts and walls equip plant cells for their specialized functions (pp. 678–681, FIGURE 35.10)** Parenchyma cells, relatively unspecialized cells that retain the ability to divide, perform most of the plant's metabolic functions of synthesis and storage. Collenchyma cells, which have unevenly thickened walls, support young, growing parts of the plant. Sclerenchyma cells, fibers and sclereids, have thick, lignified walls that help support mature, nongrowing parts of the plant. Tracheids and vessel elements, the water-conducting cells of xylem, have thick walls and are dead at functional maturity. Sieve-tube members are the sugar-transporting cells of phloem. Though alive at functional maturity, sieve-tube members depend on the services of neighboring companion cells.
- **The cells of a plant are organized into dermal, vascular, and ground tissue systems (pp. 681–682, FIGURE 35.12)** Dermal tissue (epidermis), vascular tissue (xylem and phloem), and ground tissue (mostly parenchyma cells) are continuous throughout the plant, although in the various plant organs they differ in arrangement and in some specialized functions.

PLANT GROWTH

- **Meristems generate cells for new organs throughout the lifetime of a plant: *an overview of plant growth* (pp. 682–683, FIGURE 35.13)** Apical meristems elongate shoots and roots through primary growth. Lateral meristems add girth to woody plants through secondary growth.
- 35.1 **Primary growth: Apical meristems extend roots and shoots and give rise to the primary plant body (pp. 683–689, FIGURES 35.15–35.20)** Apical meristems produce cells that continue to divide as meristematic cells of the protoderm, procambium, and ground meristem. These primary meristems give rise to the dermal, vascular, and ground tissues of the primary plant body. In roots, the apical meristem is located near the tip, where it regenerates the root cap as well as producing the primary meristems. The apical meristem of a shoot is located in the terminal bud, where it gives rise, module by module, to a repetition of internodes and leaf-bearing nodes.
- 35.2 **Secondary growth: Lateral meristems add girth by producing secondary vascular tissue and periderm (pp. 689–692, FIGURES 35.23–35.26)** The vascular cambium develops from parenchyma cells into a meristematic cylinder that produces secondary xylem and secondary phloem. The cork cambium gives rise to the secondary plant body's protective covering, or periderm, which consists of the cork cambium plus the layers of cork cells it produces. Bark is periderm plus secondary phloem—all the tissues external to the vascular cambium.

SELF-QUIZ

1. Which structure is *incorrectly* paired with its tissue system?
   a. root hair—dermal tissue
   b. palisade parenchyma—ground tissue
   c. guard cell—dermal tissue
   d. companion cell—ground tissue
   e. tracheid—vascular tissue

2. A vessel element would likely lose its protoplast in which zone of growth in a root?
   a. zone of cell division
   b. zone of elongation
   c. zone of maturation
   d. root cap
   e. quiescent center

3. Sieve-tube elements of the primary plant body originate from the
   a. protoderm
   b. procambium
   c. ground meristem
   d. vascular cambium
   e. cork cambium

4. Wood consists of
   a. bark
   b. periderm
   c. secondary xylem
   d. secondary phloem
   e. cork

5. Which of the following is *not* part of an older tree's bark?
   a. cork
   b. cork cambium
   c. lenticels
   d. secondary xylem
   e. secondary phloem

6. Each module of a primary plant body's shoot system consists of a
   a. stem, leaf, and axillary bud
   b. stem, axillary bud, and apical bud
   c. leaf, flower, and stem
   d. stem, flower, and axillary bud
   e. node, internode, and apical bud

7. Which of the following cell types or structures is *incorrectly* paired with its meristematic origin?
   a. epidermis—protoderm
   b. stele—procambium
   c. secondary xylem—vascular cambium
   d. secondary phloem—cork cambium
   e. three primary meristems—apical meristem

8. A tree will eventually die if it is girdled, meaning that a ringlike cut has been made all the way around the trunk to a depth just below the bark. The cause of death is mainly
   a. destruction of the procambium
   b. destruction of axillary buds
   c. the killing of bark cells
   d. destruction of the cork cambium and vascular cambium
   e. destruction of the plant's ability to continue primary growth

9. Which of the following cell types is *least* likely to have a secondary wall?
   a. sclerenchyma cell
   b. parenchyma cell
   c. fiber cell
   d. tracheid
   e. scleried

10. ______________ is to primary xylem as vascular cambium is to ______________.
    a. primary phloem; secondary xylem
    b. tracheid; vessel element
    c. procambium; secondary xylem
    d. apical meristem; lateral meristem
    e. stele; primary phloem

CHALLENGE QUESTIONS

1. Starting with a derivative cell produced by division of an initial in the apical meristem of a shoot, write an essay tracing the cellular lineage that would account for descendants of that cell eventually being sloughed from the plant as secondary phloem in bark.
2. If you were to live for the next several decades in a treehouse built on the large, lower branches of a tree, would you gain much altitude as the tree grew? Explain your answer.
3. Starting at the surface of a tree trunk and working to the center, describe the structure and function of the tissue layers.
4. Describe some important differences in the growth and development of plants and animals.
5. Choose three specialized types of plant cells and describe how their structures are adapted for their specific functions.

SCIENCE, TECHNOLOGY, AND SOCIETY

1. On your next trip to the grocery store, take a notepad and list the types of produce (fruits and vegetables) in one column and the parts of plants they represent in a parallel column.
2. Make a list of the plants and plant products you use in a typical day. How do you use these various plant products? Do you think the number of plants and plant products used in everyday life has increased or decreased in the last century? Do you think the number is likely to increase or decrease in the future? Why?

FURTHER READING

Dale, J. "How Do Leaves Grow?" *BioScience,* June 1992. Using new techniques of molecular and cell biology, researchers are answering long-standing questions about plant structure and growth.

Galston, A. W. *Life Processes of Plants.* New York: W. H. Freeman, 1994. Plant structure and physiology, with an emphasis on interactions with light.

Gillis, A. M. "Using a Mousy, Little Flower to Understand Flamboyant Ones." *BioScience,* May 1995. Explains the value of *Arabidopsis* as a model for developmental genetics.

Meyorowitz, E. M. "The Genetics of Flower Development." *Scientific American,* November 1994. Discusses the value to research of mutations.

Moore, R., W. D. Clark, and D. S. Vodopich. *Botany,* 2e. Dubuque, IA: W. C. Brown, 1998. An excellent introduction to plant biology.

Niklas, K. J. "How to Build a Tree." *Natural History,* February 1996. Explains how the structure of a tree makes sense in an engineering context.

Pearce, F. "Lure of the Rings." *New Scientist,* December 14, 1996. Reveals how the annual growth rings of trees might help us understand the history of global climate change.

Poethig, R. S. "Phase Change and the Regulation of Shoot Morphogenesis in Plants." *Science,* November 16, 1990. Describes the molecular and cellular basis of shoot development.

WEB LINKS

Visit the special edition of *The Biology Place* for BIOLOGY, Fifth Edition, at **http://www.biology.com/campbell**. Go to Chapter 35 for online resources, including learning activities, practice exams, and links to the following web sites:

"Glossary of Roots of Botanical Names"
Explains how to find meaning in the foreign-sounding terms describing plants and their parts.

"Virtual Cell"
The cell is from a plant, and you will be able to climb inside and explore its parts.

"Missouri Botanical Garden"
Visit one of the world's greatest showcases of plants.

"Internet Directory for Botany: Images"
Provides links to a worldwide collection of photographs and art depicting plant structure.

TRANSPORT IN PLANTS

The algal ancestors of plants were completely immersed in water and dissolved minerals, and none of their cells was far from these ingredients. The evolutionary journey onto land involved the differentiation of the plant body into roots, which absorb water and minerals from soil, and shoots, which are exposed to light and atmospheric CO_2. This body plan enabled plants to survive in an environment where chemical resources are divided between two media, soil and air. But the morphological solution to a dual environment posed a new problem: the need to transport materials between roots and shoots, sometimes over long distances. For example, the leaves of the eucalyptus trees in the photograph that opens this chapter are more than 100 m from the roots. These remote organs are bridged by vascular tissues that transport sap throughout the plant body. Water and minerals absorbed by roots are transported upward in the xylem to shoots. Sugar produced by photosynthesis is exported from mature leaves to other organs via the phloem. The whole plant depends on this commerce between organs to integrate the activities of its specialized parts. The mechanisms responsible for this internal transport are the subjects of this chapter.

AN OVERVIEW OF TRANSPORT MECHANISMS IN PLANTS

Transport in plants occurs on three levels: (1) the uptake and release of water and solutes by individual cells, such as the absorption of water and minerals from the soil by cells of a root; (2) short-distance transport of substances from cell to cell at the level of tissues and organs, such as the loading of sugar from photosynthetic cells of a mature leaf into the sieve tubes of phloem; and (3) long-distance transport of sap within xylem and phloem at the level of the whole plant. FIGURE 36.1, p. 696, provides an overview of these transport functions in plants.

Transport at the cellular level depends on the selective permeability of membranes

The transport of solutes and water across biological membranes was covered in detail in Chapter 8. Here we reexamine a few of these transport processes in the specific context of plant cells.

The selective permeability of a plant cell's plasma membrane controls the movement of solutes between the cell and the extracellular solution. Recall from Chapter 8 that solutes tend to diffuse down their gradients, and when this occurs across a membrane, the process is termed passive transport (*passive* because it happens without the direct expenditure of metabolic energy by the cell). Most solutes, however, cross a membrane very slowly unless they can pass through **transport proteins** embedded in the membrane. Some of these

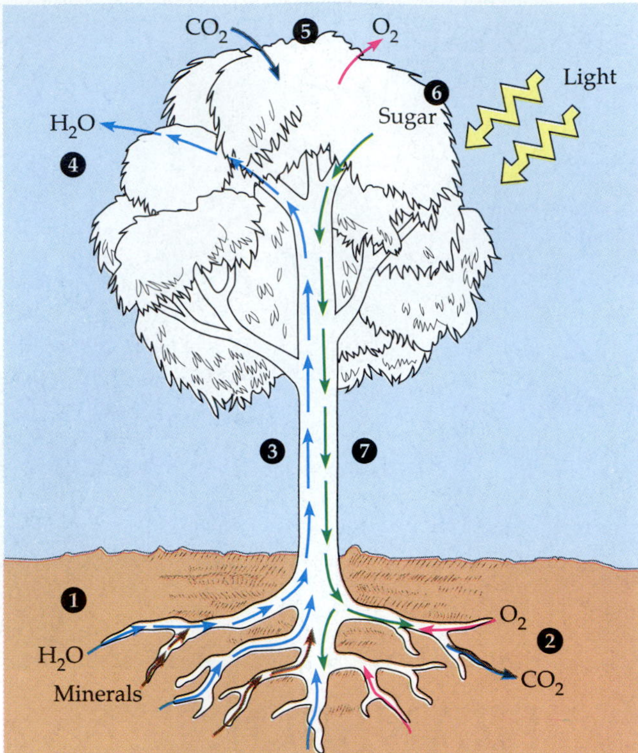

FIGURE 36.1 ▪ An overview of transport in plants. ① Roots absorb water and dissolved minerals from the soil. ② Roots also exchange gases with the air spaces of soil, taking in O_2 and discharging CO_2. This gas exchange supports the cellular respiration of root cells. ③ Water and minerals are transported upward as xylem sap within xylem, from the roots into the shoot system. ④ Transpiration, the loss of water vapor from leaves (mostly through stomata), creates a force within leaves that pulls xylem sap upward. ⑤ Leaves also exchange carbon dioxide and oxygen through stomata, taking in the CO_2 that provides carbon for photosynthesis and expelling O_2. ⑥ Sugar is produced by photosynthesis in the leaves and ⑦ is transported within phloem in a solution called phloem sap to roots and other parts of the plant.

facilitate diffusion by binding selectively to a solute on one side of the membrane and releasing the solute on the opposite side. Transfer of the solute across the membrane involves a conformational (shape) change by the transport protein. Other transport proteins function as **selective channels**, which are simply selective passageways across the membrane. For example, the membranes of most plant cells have potassium channels that allow potassium ions (K^+) to pass, but not similar ions, such as sodium (Na^+). Some channels are gated; that is, certain environmental stimuli can cause the channels to open or close. For instance, we will see later in this chapter how the regulation of K^+ gates in the membranes of guard cells functions in the opening and closing of stomata.

Recall also from Chapter 8 that active transport is the pumping of solutes across membranes against their electrochemical gradients, the combined effects of the solute's concentration gradient and the voltage (charge difference) across the membrane. It is termed *active* because the cell must expend metabolic energy, usually in the form of ATP, to transport a solute "uphill"—that is, counter to the direction in which the solute diffuses. Transport proteins that facilitate diffusion, such as selective channels, cannot perform active transport. The active transporters are a special class of membrane proteins, each responsible for pumping specific solutes.

Proton pumps play a central role in transport across plant membranes

One important active transporter in plant cells is the **proton pump**, which hydrolyzes ATP and uses the released energy to pump hydrogen ions (H^+) out of the cell. This results in a proton gradient, with the H^+ concentration higher outside the cell than inside the cell. The gradient is a form of stored energy, because the hydrogen ions tend to diffuse "downhill," back into the cell. And because the proton pump moves positive charge, in the form of H^+, out of the cell, the pump also generates a membrane potential. Membrane potential is a voltage, a separation of opposite charges across a membrane. Proton pumping makes the inside of a plant cell negative in charge relative to the outside. This voltage is called a membrane *potential* because the charge separation is a form of potential (stored) energy that can be harnessed to perform cellular work.

Plant cells use energy stored in the proton gradient and the membrane potential to drive the transport of many different solutes (FIGURE 36.2). Consider, for example, one mechanism root cells use to absorb potassium from the soil solution. Because potassium ions are positively charged and the inside of the cell is negatively charged compared to the outside, the membrane potential helps drive K^+ into the cell. Because K^+ is diffusing down its electrochemical gradient, accumulation of the ion by this mechanism represents passive transport. But it is the active transport of H^+ that maintains the membrane potential and makes it possible for the cell to accumulate K^+. In other cases, energy stored by H^+ pumping can actually be used to drive the transport of solutes *against* their electrochemical gradients. For example, many negatively charged minerals, such as nitrate (NO_3^-), enter root cells through carriers that also allow H^+ to reenter the cells. This mechanism is called **cotransport**. A transport protein couples the downhill passage of one solute (H^+) to the uphill passage of another (NO_3^-, in this case). This "coattail" effect is also responsible for the uptake of the sugar sucrose by plant cells. A membrane protein cotransports sucrose with the H^+ that is moving down its gradient through the protein.

The role of proton pumps in the transport processes of plant cells is a specific application of the general mechanism called **chemiosmosis** (see Chapter 9). The key feature of chemiosmosis is a transmembrane proton gradient, which links energy-releasing processes to energy-consuming processes in cells. For example, you learned in Chapters 9 and 10 that mitochondria and chloroplasts use proton gradients gen-

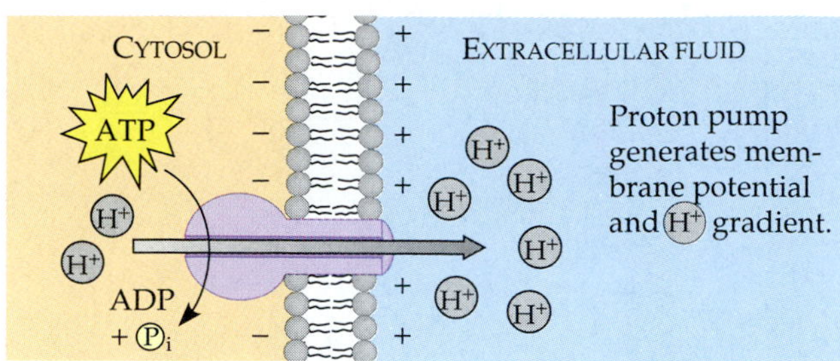

(a) Proton pump

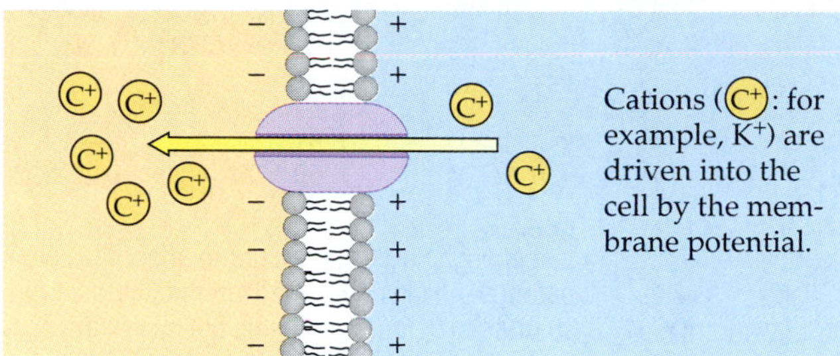

(b) Cation uptake

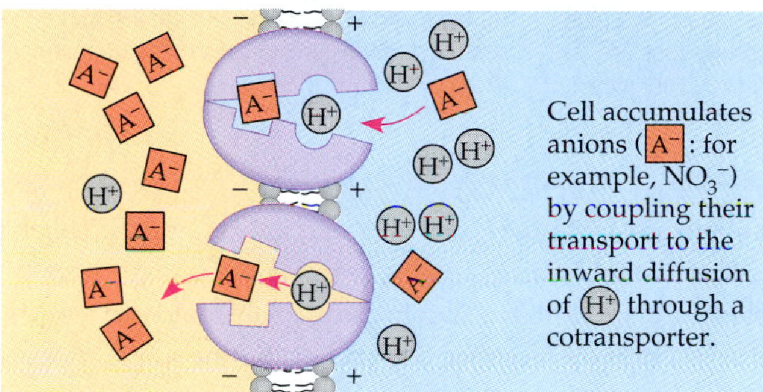

(c) Anion uptake

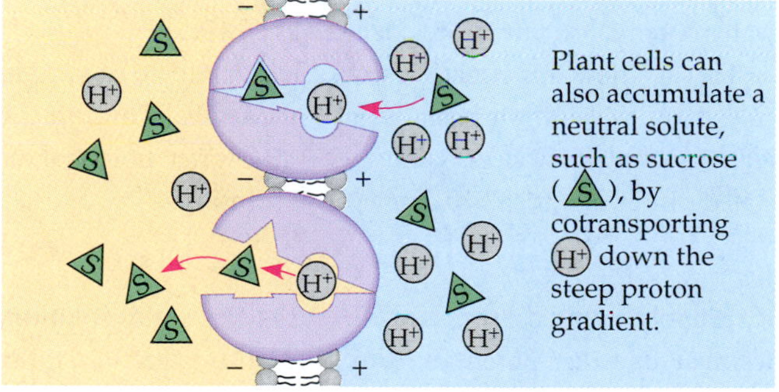

(d) Transport of a neutral solute

FIGURE 36.2 ▪ A chemiosmotic model of solute transport in plant cells.

erated by electron transport chains (which release energy) to drive ATP synthesis (which consumes energy). The ATP synthases that couple H^+ diffusion to ATP synthesis during cellular respiration and photosynthesis function somewhat like the proton pumps embedded in the plasma membranes of plant cells. But compared to ATP synthases, proton pumps normally run in reverse, using ATP energy to pump H^+ against its gradient. In both cases, proton gradients are the metabolic gears that enable one process to drive another. Chemiosmosis is a unifying principle of cellular energetics.

Differences in water potential drive water transport in plant cells

The survival of plant cells depends on their ability to balance water uptake and loss. The net uptake or loss of water by a cell occurs by **osmosis**, the passive transport of water across a membrane (see Chapter 8). How can we predict the direction of osmosis when a cell is surrounded by a particular solution? In the case of an animal cell, it is enough to know whether the extracellular solution is hypotonic (has a lower solute concentration) or hypertonic (has a higher solute concentration) relative to the cell; water will move by osmosis in the hypotonic ⟶ hypertonic direction. But in the case of a plant cell, the presence of a cell wall adds a second factor affecting osmosis: physical pressure. The combined effects of these two factors—solute concentration and pressure—are incorporated into a single measurement called **water potential**, abbreviated by the Greek letter psi (ψ). The most important thing for you to learn about water potential is that water will move across a membrane from the solution with the higher water potential to the solution with the lower water potential. For example, if a plant cell is immersed in a solution having a higher water potential than the cell, osmotic uptake of the water will cause the cell to swell. By moving, water can perform work (expanding a cell, for instance). The *potential* in water potential refers to this potential energy, the capacity to perform work when water moves from a region of higher ψ to a region of lower ψ. It is a special case of the general tendency of systems to change spontaneously to a state of lowest free energy (see Chapter 6).

Plant biologists measure ψ in units of pressure called **megapascals** (abbreviated MPa). An MPa is equal to about 10 atmospheres of pressure. (An atmosphere is the pressure exerted at sea level by an imaginary column of air—about 1 kg of pressure per cm^2.) A couple of nonbiological examples will give you some idea of the magnitude of a megapascal: A car tire is usually inflated to a pressure of about 0.2 MPa; the water pressure in home plumbing is about 0.25 MPa.

How Solutes and Pressure Affect Water Potential

Let's see how solute concentration and pressure affect water potential. For purposes of comparison, the water potential of pure water in a container open to the atmosphere is defined as zero megapascals ($\psi = 0$ MPa). The addition of solutes lowers the water potential. Since ψ is standardized as 0 MPa for pure water, any solution at atmospheric pressure has a negative water potential due to the presence of solutes. For instance, a 0.1 molar *(M)* solution of any solute has a water potential of −0.23 MPa. If this solution is separated from pure water by a selectively permeable membrane, water will move by osmosis

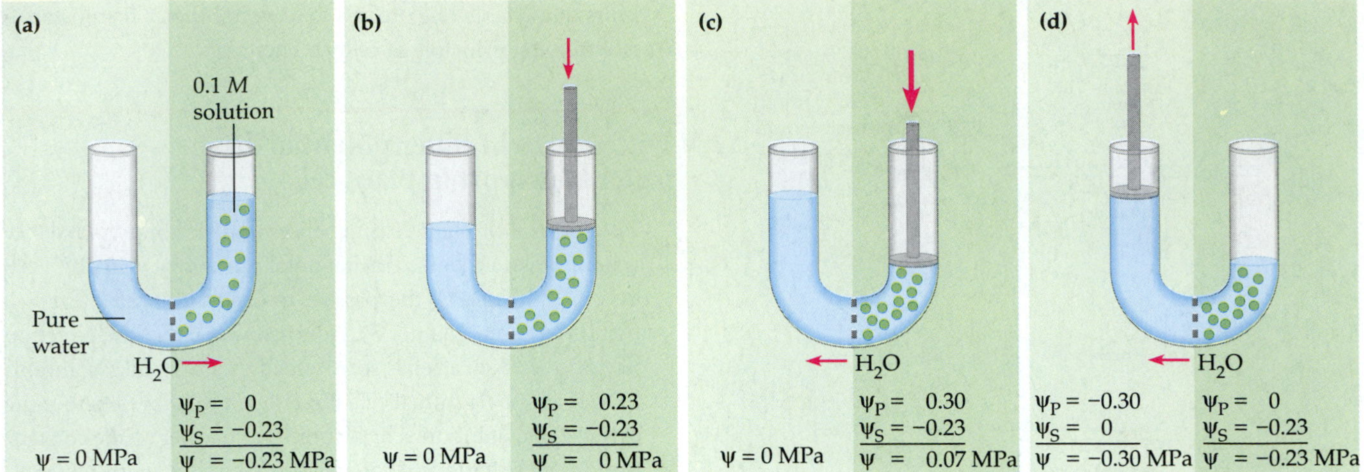

FIGURE 36.3 ▪ Water potential and water movement: a mechanical model. Water moves across a selectively permeable membrane from where water potential is higher to where it is lower. The water potential (ψ) of pure water at atmospheric pressure is 0 MPa. The addition of solutes reduces water potential (to a negative value); application of physical pressure increases water potential. If we know the values of pressure potential (ψ_P) and solute potential (ψ_S), we can calculate water potential: $\psi = \psi_P + \psi_S$. **(a)** In this U-shaped apparatus, a selectively permeable membrane separates pure water from a 0.1 molar *(M)* solution containing a particular solute that cannot pass freely across the membrane. Water will move by osmosis into the solution, increasing its volume. (The values for ψ and ψ_S in the left and right arms of the U-tube are given for *initial* conditions, *before* any net movement of water.) **(b)** If we use a piston to apply just enough physical pressure to offset the solute potential and increase the water potential of the solution to 0, there will be no net movement of water across the membrane. **(c)** If physical pressure exceeds the opposing effect of solute potential, we can force water from the solution into the reservoir of pure water. **(d)** A negative pressure, or tension, reduces water potential.

into the solution, from the region of higher ψ (0 MPa) to the region of lower ψ (-0.23 MPa). But we also have to factor in the influence of physical pressure on ψ.

In contrast to the inverse relationship of ψ to solute concentration, water potential is directly proportional to pressure; increasing pressure raises ψ. Remember that ψ measures the relative tendency for water to leave one location in favor of another. Physical pressure—pressing the plunger of a syringe filled with water, for example—causes water to escape via any available exits. If a solution is separated from pure water by a selectively permeable membrane, external pressure on the solution can counter its tendency to take up water due to the presence of solutes. In fact, even greater pressure will force water across the membrane from the solution to the compartment containing pure water. It is also possible to create a negative pressure, or **tension**, on water or solutions. For example, if you pull up on the plunger of a syringe, the negative pressure within the syringe draws a solution through the needle.

Quantitative Analysis of Water Potential

The combined effects of pressure and solute concentration on water potential are incorporated into the following equation:

$$\psi = \psi_P + \psi_S$$

where ψ_P is the pressure potential (physical pressure on a solution) and ψ_S is the solute potential, which is proportional to the solute concentration of a solution. (ψ_S is also called osmotic potential.) Pressure on a solution (ψ_P) can be either a positive number or a negative number (tension, a negative pressure). In contrast, a solution's solute potential (ψ_S) is always a negative number, and the greater the solute concentration, the more negative the value of ψ_S. This makes sense if you think of ψ_S as the effect solutes have on a solution's overall water potential (ψ), which is always to lower ψ below the water potential of pure water, defined as 0 MPa.

Let's see how this equation works, with the help of FIGURE 36.3. A 0.1 *M* solution has a ψ_S of -0.23 MPa. Thus, in the absence of a physical pressure ($\psi_P = 0$), water potential, as stated earlier, is -0.23 MPa for a 0.1 *M* solution:

$$\psi = \psi_P + \psi_S = 0 + (-0.23) = -0.23 \text{ MPa}$$

If we apply a physical pressure of $+0.23$ MPa to this solution, we raise its water potential from a negative value to 0 MPa ($\psi = 0.23 - 0.23$). If this pressurized solution is separated from water by a selectively permeable membrane, there will be no net water flow between the two compartments. If we increase ψ_P to $+0.3$ MPa, then the solution has a water potential of $+0.07$ MPa ($\psi = 0.3 - 0.23$), and this solution will lose water to a compartment containing pure water. In dissecting water potential to see these opposing effects of pressure and solutes, it is important to remember the key point: Water will move across a membrane in the direction of lower (more negative) water potential.

Let's now apply what we have learned about water potential to the uptake and loss of water by plant cells. First, imagine a flaccid cell (that is, $\psi_P = 0$) bathed in a solution of

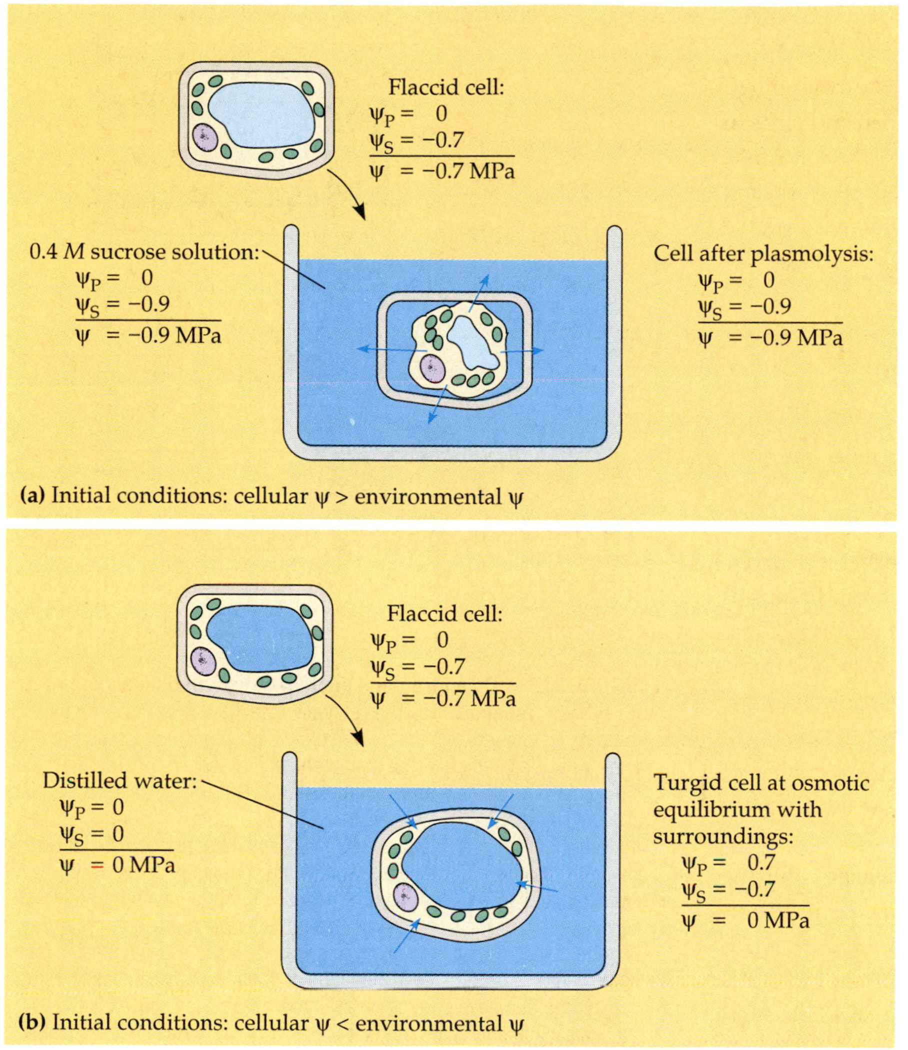

FIGURE 36.4 ▪ Water relations of plant cells. In these experiments, identical cells, initially flaccid, are placed in two different environments, one a relatively concentrated sugar solution and the other distilled water. (Flaccid cells are in contact with their walls but lack turgor pressure.) The blue arrows indicate the direction of water movement that precedes a dynamic equilibrium, at which no further net movement of water occurs. **(a)** In the concentrated sucrose solution, the cell initially has a greater water potential than its surroundings. The cell loses water and plasmolyzes. After plasmolysis is complete, the water potentials of the cell and its surroundings are the same. Plasmolysis kills most plant cells. **(b)** In distilled water, the cell initially has a lower water potential than its surroundings. There is a net uptake of water by osmosis, causing the cell to become turgid. When this tendency for water to enter is offset by the back pressure of the elastic wall, water potentials are equal for the cell and its surroundings. (The volume change of the cell is exaggerated in this diagram. The osmotic uptake of a relatively small amount of water does not actually increase the volume of the cell much. This explains why the solute potential, ψ_S, does not change by a significant amount when a cell becomes more turgid.) Healthy plant cells are generally turgid, and their turgor (firmness) contributes support to nonwoody parts of a plant. You can see the consequences of turgor loss in a wilted house plant, which has flaccid cells.

higher solute concentration than the cell itself (FIGURE 36.4a). Since the external solution has the lower (more negative) water potential, water will leave the cell by osmosis, and the cell will **plasmolyze**, or shrink and pull away from its wall. Now let's place the same flaccid cell in pure water ($\psi = 0$; FIGURE 36.4b). The cell has a lower water potential because of the presence of solutes, and water enters the cell by osmosis. The cell begins to swell and push against the wall, producing a **turgor pressure.** The partially elastic wall pushes back against the turgid cell. When this wall pressure is great enough to offset the tendency for water to enter because of the solutes in the cell, then ψ_P and ψ_S are equal in magnitude and thus $\psi = 0$. This matches the water potential of the extracellular environment—in this example, 0 MPa. A dynamic equilibrium has been reached, and there is no further net movement of water, although an equal exchange of water across the membrane continues.

In contrast to a flaccid cell, a walled cell that has a greater solute concentration than its surroundings will be **turgid,** referring to the turgor pressure that keeps it firm. Healthy plant cells are turgid most of the time.

Aquaporins

We have been examining water potential as the force that moves water across the membranes of plant cells, but how do the water molecules actually cross the membranes? Because water molecules are so small, they move relatively freely across the lipid bilayer of membranes, even though the middle zone of that bilayer is hydrophobic (see Chapter 8). Until recently, most biologists accepted the hypothesis that leakage of water across the lipid bilayer was enough to account for water fluxes across membranes. That hypothesis was challenged in the early 1990s, when careful measurements indicated that water transport across biological membranes is too specific and too rapid to be explained entirely by diffusion through the lipid bilayer. This observation suggested the possibility of selective channels for water, and such channels have since been discovered in the membranes of both plant and animal cells.

The channels are water-specific transport proteins named **aquaporins.** Aquaporins cannot actively transport water; instead, they facilitate water diffusion (osmosis), increasing the transport rate over the rate due only to flow across the lipid bilayer. The existence of aquaporins raises the possibility that a cell can regulate the rate of water uptake or loss when it has a water potential different from that of its surroundings. Aquaporins may form *gated* channels that open and close in response to variables such as turgor pressure of the cell. Many laboratories are now investigating how aquaporins work, and this research is likely to increase our understanding of how the cells of plants and other organisms regulate water balance.

Vacuolated plant cells have three major compartments

Recall from Chapter 35 that a plant cell includes the wall. Thus, the plasma membrane regulates the traffic of molecules between two major compartments, the wall and the cytosol of the protoplast (FIGURE 36.5a). Most mature plant cells have a third major compartment, the vacuole, which can occupy as much as 90% of the protoplast's volume. The membrane that bounds the vacuole, the **tonoplast**, regulates molecular traffic between the cytosol and the vacuolar contents, called cell sap. For example, the tonoplast has proton pumps that expel H^+ from the cytosol into the vacuole. This augments the ability of the proton pumps of the plasma membrane to maintain a low cytosolic concentration of H^+.

In most plant tissues, two of the three cellular compartments are continuous from cell to cell. Plasmodesmata connect the cytosolic compartments of neighboring cells, forming a continuous pathway for transport of certain molecules between cells. This cytoplasmic continuum is called the **symplast** (see FIGURE 36.5a). The walls of adjacent plant cells are also in contact, forming a second compartment at the tissue level. This compartment, the continuum of cell walls, is called the **apoplast**. The third cellular compartment, the vacuole, is not shared with neighboring cells.

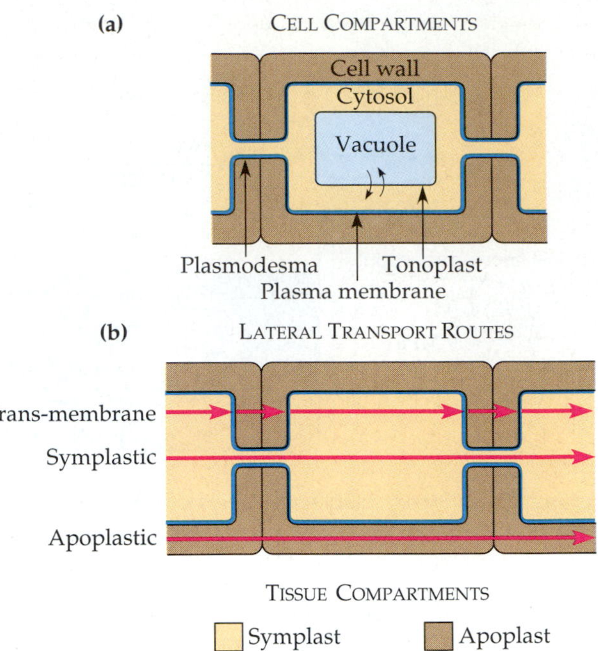

FIGURE 36.5 ▪ Compartments of plant cells and tissues and routes for lateral transport. (a) The cell wall, cytosol, and vacuole are the three compartments of most mature plant cells. Specific transport proteins imbedded in the plasma membrane and tonoplast regulate traffic of molecules between the three compartments. **(b)** At the tissue level, there are two compartments, the symplast and the apoplast. The symplast is the continuum of cytosol based on plasmodesmata, protoplast-connecting channels through walls. The apoplast is the continuum of cell walls and extracellular spaces. This anatomy provides three routes for lateral transport in a plant tissue or organ. In a transmembrane route, solutes and water move across an organ by the repeated crossing of the plasma membranes and walls of the cells along the pathway. In the symplastic route, substances that have entered one cell move across an organ via the cytosolic continuum. The complex structure of plasmodesmata probably regulates transport via the symplast, even permitting the passage of certain proteins and other large molecules between cells. In the apoplastic route, water and solutes travel across a tissue or organ via the cell walls and extracellular spaces. In this diagram, substances seem confined to one of the three routes; in fact, substances may transfer from one route to another during their commute across an organ.

The symplast and apoplast both function in transport within tissues and organs

How do water and solutes move from one location to another within plant tissues and organs? For example, what mechanisms transport water and minerals absorbed by a root from the outer cells to the inner cells of the root? Such short-distance transport is sometimes called lateral transport because its usual direction is along the radial axis of plant organs, rather than up and down along the length of the plant.

Three routes are available for lateral transport. By the first route, substances move out of one cell, across the cell wall, and into the neighboring cell, which may then pass the substances along to the next cell in the pathway by the same mechanism (FIGURE 36.5b). This transmembrane route requires repeated crossings of plasma membranes, as the solutes exit one cell and enter the next.

The second route, via the symplast, the continuum of cytosol within a plant tissue, requires only one crossing of a plasma membrane (FIGURE 36.5b). After entering one cell, solutes and water can then move from cell to cell via plasmodesmata.

The third route for lateral transport within a plant tissue or organ is along the apoplast, the extracellular pathway consisting of cell walls and extracellular spaces (FIGURE 36.5b). Before ever entering a cell, water and solutes can move from one location to another within a root or other organ along the byways provided by the continuum of cell walls.

Bulk flow functions in long-distance transport

Diffusion is much too slow to function in long-distance transport within a plant—the transport of water and minerals from roots to leaves, for example. Water and solutes move through xylem vessels and sieve tubes by **bulk flow**, the movement of a fluid driven by pressure. In phloem, for example, hydrostatic pressure is generated at one end of a sieve tube, and this forces sap to the opposite end of the tube. In xylem, it is actually tension, a negative pressure, that drives long-distance transport. Transpiration, the evaporation of water from a leaf, reduces pressure in the leaf xylem. This creates a tension that pulls xylem sap upward from the roots.

Now that we have an overview of the basic mechanisms of transport at the cellular, tissue, and whole-plant levels, we are ready to take a closer look at how these mechanisms work together in the overall transport functions that enable a plant to survive on land. For example, bulk flow due to a pressure difference is the mechanism of long-distance transport of phloem sap, but it is active transport of sugar at the cellular level that maintains this pressure difference. The four transport functions we will examine in more detail are the absorption of water and minerals by roots, the ascent of xylem sap, the control of transpiration, and the transport of organic nutrients within phloem.

ABSORPTION OF WATER AND MINERALS BY ROOTS

Water and mineral salts from soil enter the plant through the epidermis of roots, cross the root cortex, pass into the stele, and then flow up xylem vessels to the shoot system. In this section we focus on the soil ⟶ epidermis ⟶ root cortex ⟶ xylem segments of this transport pathway. Use FIGURE 36.6 to reinforce the key concepts.

Root hairs, mycorrhizae, and a large surface area of cortical cells enhance water and mineral absorption

Adaptations that increase surface area are associated with the root system's function of absorbing water and minerals from the soil. Much of this absorption occurs near root tips, where the epidermis is permeable to water and where root hairs are located. Root hairs, extensions of epidermal cells, account for much of the surface area of roots. Soil particles, which are

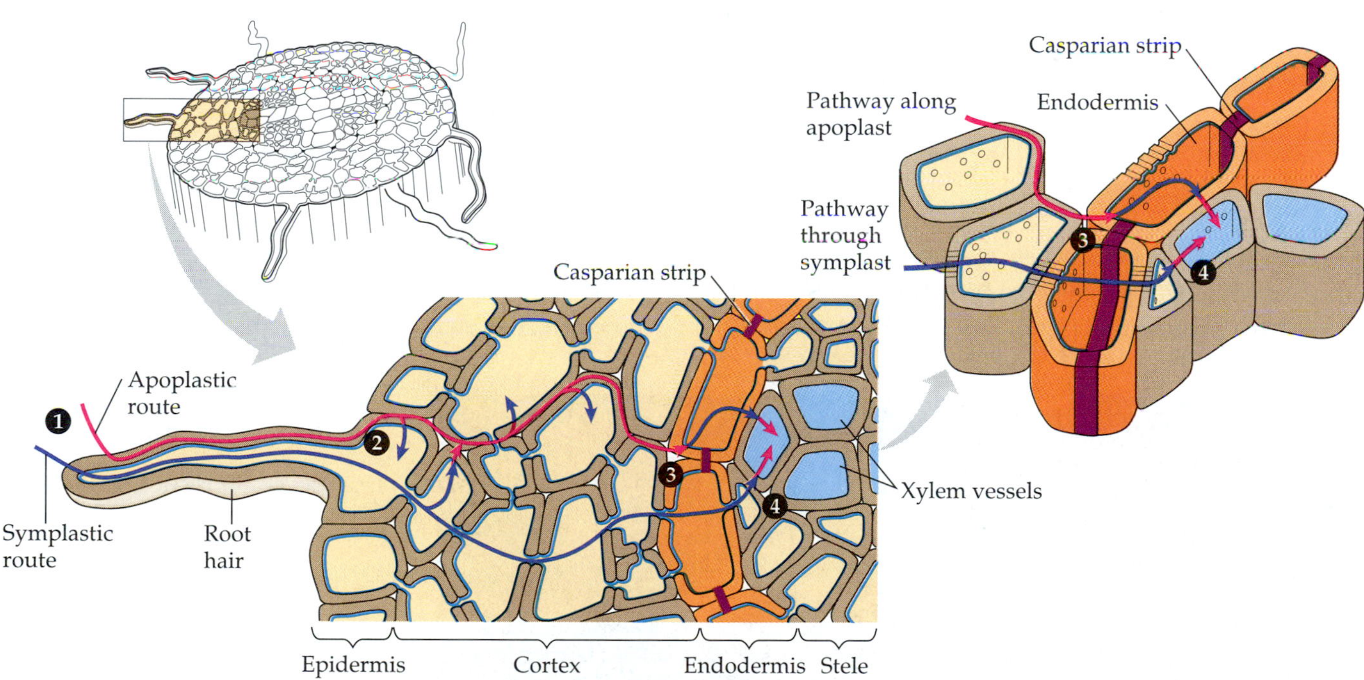

FIGURE 36.6 ▪ Lateral transport of minerals and water in roots. Minerals are absorbed with the soil solution by the root surface, especially by root hairs and mycorrhizae (symbiotic associations of roots and fungi, not shown in this diagram). The water and minerals then move across the root cortex to the vascular cylinder by a combination of the apoplastic and symplastic routes (see FIGURE 36.5). ① The uptake of soil solution by the hydrophilic walls of the epidermis provides access to the apoplast, and water and minerals can soak into the cortex along this matrix of cell walls. Minerals and water that cross the plasma membranes of root hairs enter the symplast. ② As soil solution moves along the apoplast, some water and minerals are transported into cells of the epidermis and cortex and then move inward via the symplast. ③ Water and minerals that move all the way to the endodermis along cell walls cannot continue into the stele via the apoplastic route. Within the wall of each endodermal cell is the Casparian strip, a belt of waxy material (purple band) that blocks the passage of water and dissolved minerals. Only minerals that are already in the symplast or enter that pathway by crossing the plasma membrane of an endodermal cell can detour around the Casparian strip and pass into the stele. Thus, the transport of minerals into the stele is selective; only those minerals that are admitted into cells by membranes gain access to the vascular tissue. ④ Endodermal cells and parenchyma cells within the stele discharge water and minerals into their walls, which, as part of the apoplast, are continuous with the xylem vessels. Water and minerals absorbed from soil are now ready for upward transport into the shoot system.

usually coated with water and dissolved minerals, adhere tightly to the hairs. The soil solution flows into the hydrophilic walls of epidermal cells and passes freely along the apoplast into the root cortex. This exposes all the parenchyma cells of the cortex to soil solution, providing a much greater surface area of membrane than the surface area of the epidermis alone.

As the soil solution moves along cell walls, some of the water and solutes are taken up by cells of the epidermis and cortex, switching pathways from the apoplast to the symplast. It is this crossing of plasma membranes that makes mineral absorption selective. The soil solution is usually very dilute, and roots can accumulate essential minerals to concentrations that are hundreds of times higher than the concentrations of these minerals in soil. For example, selective transport proteins of the plasma membrane and tonoplast enable root cells to extract K^+, an essential mineral nutrient. In contrast, the cells exclude most Na^+, which may be much more concentrated than K^+ in the soil solution.

In their essential business of absorbing water and minerals from the soil, most plants are not on their own, but have partners in the form of symbiotic fungi. "Infected" roots form **mycorrhizae**, the term for the symbiotic structures consisting of the plant's roots intermingled with the hyphae (filaments) of the fungi (FIGURE 36.7). The hyphae absorb water and selected minerals, transferring much of these resources to the host plant. Chapter 37 highlights the role of mycorrhizae in the mineral nutrition of plants, and Chapter 31 featured the fungal partners in these mutually symbiotic relationships. What is important to understand here is that the mycelium (network of hyphae) of the fungus endows mycorrhizae with an enormous surface area for absorption of water and minerals. As much as 3 meters of hyphae can extend from each centimeter along the length of a root, reaching a far greater volume of soil than the root alone could penetrate. Mycorrhizae enable even older regions of roots, far from the abundance of root hairs near the root tips, to supply water and minerals to the plant.

The endodermis functions as a selective sentry between the root cortex and vascular tissue

Water and minerals that pass from the soil into the root cortex cannot be transported to the rest of the plant until they enter the xylem of the vascular cylinder (stele). The **endodermis**, the innermost layer of cells in the root cortex, surrounds the stele and functions as a last checkpoint for the selective passage of minerals from the cortex into the vascular tissue (see FIGURE 36.6). Minerals already in the symplast when they reach the endodermis continue through the plasmodesmata of the endodermal cells and pass into the stele. These minerals were already screened by the selective membrane they had to cross to enter the symplast in the cortex. Those minerals that reach the endodermis via the apoplast encounter a dead end that blocks their passage into the stele. In the wall of each endodermal cell is the **Casparian strip**, a belt made of suberin, a waxy material that is impervious to water and dissolved minerals. Thus, water and minerals cannot cross the endodermis and enter vascular tissue via the apoplast. The only way past this barrier is for the water and minerals to cross the plasma membrane of an endodermal cell and enter the stele via the symplast. The endodermis, with its Casparian strip, ensures that no minerals can reach the vascular tissue of the root without crossing membranes. Recall that transport across membranes is selective. If minerals do not enter cells in the cortex, they must enter endodermal cells or be excluded

FIGURE 36.7 ▪ Mycorrhizae, symbiotic associations of fungi and roots. The white mycelium of the fungus ensheathes these roots of a red pine tree. The fungal hyphae provide an extensive surface area for the absorption of water and minerals.

from the vascular tissue. The structure of the endodermis and its strategic location in the root fit its function as sentry of the cortex-stele border, a function that contributes to the ability of roots to preferentially transport certain minerals from the soil into the xylem.

The last segment in the soil ⟶ xylem pathway is the passage of water and minerals into the tracheids and vessel elements of the xylem. These water-conducting cells lack protoplasts, and thus the lumen of the cells, as well as their walls, are part of the apoplast. But you have just learned that water and minerals enter the stele from the cytoplasm of endodermal cells. Thus, the entry of water and minerals into the xylem requires their transfer from the symplast to the apoplast. Endodermal cells and parenchyma cells within the stele discharge minerals into their walls. Both diffusion and active transport are probably involved in this transfer of solutes from symplast to apoplast, and the water and minerals are now free to enter the tracheids and xylem vessels. The water and mineral nutrients we have tracked from the soil to root xylem can now be transported upward as xylem sap to the shoot system.

TRANSPORT OF XYLEM SAP

Xylem sap flows upward in vessels at rates of 15 m per hour or faster. Veins branch throughout each leaf, placing xylem vessels close to every cell. Leaves depend on this efficient delivery system for their supply of water. Plants lose an astonishing amount of water by **transpiration**, the loss of water vapor from leaves and other aerial parts of the plant. An average-sized maple tree, for instance, loses more than 200 L of water per hour during the summer. Unless the transpired water is replaced by water transported up from the roots in xylem, leaves wilt and eventually die. The flow of xylem sap upward in the plant also brings mineral nutrients to the shoot system.

The ascent of xylem sap depends mainly on transpiration and the physical properties of water

36.1 Xylem sap rises against gravity, without the help of any mechanical pump, to reach heights of more than 100 m in the tallest trees. Is the sap *pushed* upward from the roots, or is it *pulled* upward by the leaves? Let's evaluate the relative contributions of these two possible mechanisms.

Pushing Xylem Sap: Root Pressure

At night, when transpiration is very low or zero, the root cells are still expending energy to pump mineral ions into the xylem. The endodermis surrounding the stele of the root helps prevent the leakage of these ions back out of the stele. The accumulation of minerals in the stele lowers water potential there. Water flows in from the root cortex, generating a positive pressure that forces fluid up the xylem. This upward push of xylem sap is called **root pressure**.

Root pressure causes **guttation**, the exudation of water droplets that can be seen in the morning on tips of grass blades or the leaf margins of some small, herbaceous (nonwoody) dicots. During the night, when the rate of transpiration is low, the roots of some plants keep accumulating minerals, and root pressure pushes xylem sap into the shoot system. More water enters leaves than is transpired, and the excess is forced as guttation through specialized structures called hydathodes, which function as escape valves.

In most plants, root pressure is not the major mechanism driving the ascent of xylem sap. At most, root pressure can force water upward only a few meters, and many plants, including some of the tallest trees, generate no root pressure at all. Even in most small plants that display guttation, root pressure cannot keep pace with transpiration after sunrise. For the most part, xylem sap is not pushed from below by root pressure but pulled upward by the leaves themselves.

Pulling Xylem Sap: The Transpiration-Cohesion-Tension Mechanism

If we want to move material upward, we can either push it from below or pull it from above. It is less obvious that something as seemingly fluid as water could be pulled up a pipe. Nevertheless, that is what happens in the xylem vessels of plants. As we investigate this mechanism of transport, we will see that transpiration provides the pull, and the cohesion of water due to hydrogen bonding transmits the upward pull along the entire length of the xylem to the roots.

Transpirational Pull. Stomata, the microscopic pores on the surface of a leaf, lead to a maze of internal air spaces that expose the mesophyll cells to the carbon dioxide they need for photosynthesis. The air in these spaces is saturated with water vapor because it is in contact with the moist walls of the cells. On most days the air outside the leaf is drier; that is, it has a lower water concentration than the air inside the leaf. Therefore, gaseous water, diffusing down its concentration gradient, exits the leaf via the stomata. It is this loss of water vapor from the leaf that we call transpiration.

How is transpiration translated into a pulling force for the movement of water in a plant? The mechanism depends on the generation of a negative pressure (tension) in the leaf due to the unique physical properties of water. Evaporation from the thin film of water that coats the mesophyll cells replaces the water vapor that is lost from the leaf's air spaces by transpiration. As water evaporates, the remaining film of liquid water retreats into the pores of the cell walls, attracted by adhesion to

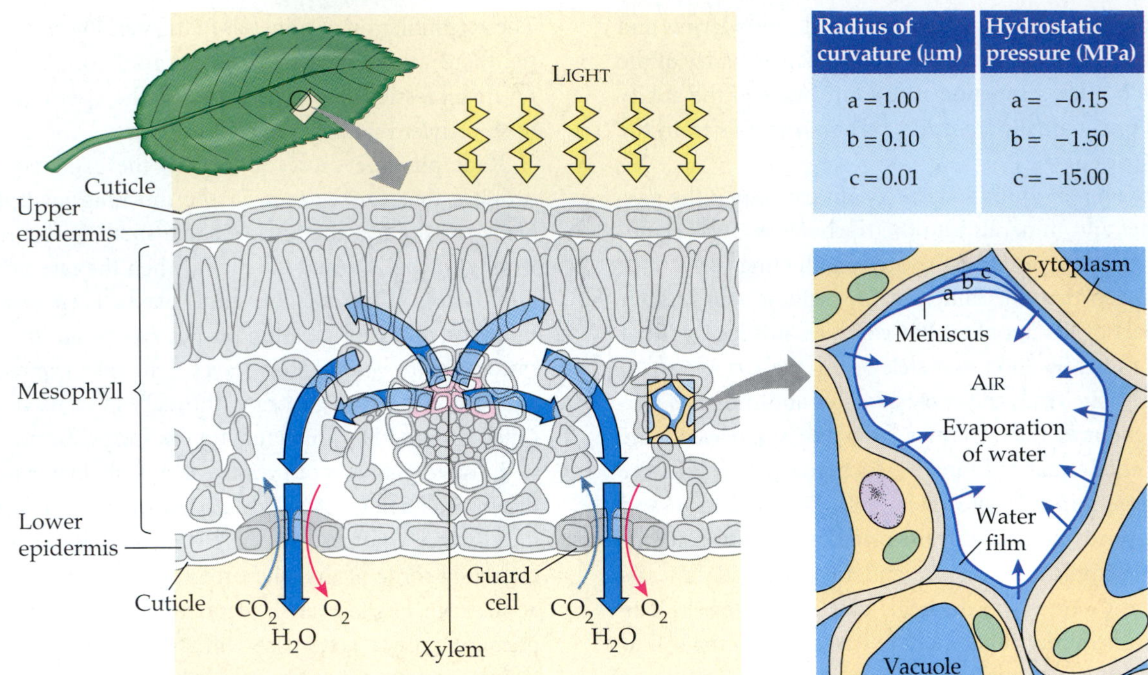

FIGURE 36.8 ▪ The generation of transpirational pull in a leaf. Water vapor diffuses from the moist air spaces of the leaf to the drier air outside via stomata. (Stomata are also avenues of the exchange of CO_2 and O_2 between the photosynthetic tissue and the atmosphere.) Evaporation from the water film coating the mesophyll cells maintains the high humidity of the air spaces. This loss of water causes the water film to form menisci (plural), which become more and more concave as the rate of transpiration (evaporative water loss from the leaf) increases. A meniscus has a tension that is inversely proportional to the radius of the curved water surface. Thus, as the water film recedes and its menisci become more concave, the tension of the water film increases. Tension is a negative pressure—a force that pulls water from locations where hydrostatic pressure is greater. The negative pressure (tension) of water lining the air spaces of the leaf is the physical basis of transpirational pull, which draws water out of xylem and through the mesophyll tissue to the surfaces near stomata. The cohesion of water due to hydrogen bonding enables transpiration to pull water up the narrow xylem vessels and tracheids without these columns of water breaking apart. In fact, transpirational pull, with the help of water's cohesion, is transmitted all the way from leaves to roots. The bulk flow of water to the top of a tree is solar-powered, because it is the absorption of sunlight by leaves that causes the evaporation responsible for transpirational pull.

the hydrophilic walls (FIGURE 36.8). At the same time, cohesive forces in the water resist an increase in the surface area of the film (a surface-tension effect; see Chapter 3). The combination of the two forces acting on the water—adhesion to the wall and surface tension—causes the surface of the water film to form a meniscus, or concave shape. In a sense, the water is being "pulled on" by the adhesive and cohesive forces. Thus, the water film at the surface of leaf cells has a negative pressure, a pressure less than atmospheric pressure. And the more concave the meniscus, the more negative the pressure of the water film. This negative pressure, or tension, is the pulling force that draws water out of the leaf xylem, through the mesophyll, and toward the cells and surface film bordering the air spaces near stomata. Transit across the mesophyll occurs via both symplastic and apoplastic routes. This water flow fits with what you learned earlier about water potential. In the equation for water potential, a tension (negative pressure) *lowers* the potential. And since water moves from where its potential is higher to where it is lower, mesophyll cells will lose water to the surface film lining air spaces, which in turn loses water by transpiration. The water lost via the stomata is replaced by water that is pulled out of the leaf xylem. FIGURE 36.9 illustrates a method for measuring the negative water potential of leaves, the "pull" in transpirational pull.

Cohesion and Adhesion of Water. The transpirational pull on xylem sap is transmitted all the way from the leaves to the root tips and even into the soil solution (FIGURE 36.10). The cohesion of water due to hydrogen bonding makes it possible to pull a column of sap from above without the water separating. Water molecules exiting the xylem in the leaf tug on adjacent water molecules, and this pull is relayed, molecule by molecule, down the entire column of water in the xylem. Also helping to fight gravity is the strong adhesion of water molecules (again by hydrogen bonds) to the hydrophilic walls of the xylem cells. The very small diameter of tracheids and vessel elements contributes to the importance of adhesion in overcoming the downward force of gravity.

The upward pull on the cohesive sap creates tension within the xylem. Pressure will cause an elastic pipe to swell, but tension will pull the walls of the pipe inward. (You can actually measure a decrease in the diameter of a tree trunk on a warm day, when transpirational pull puts the xylem under tension.) The rings of secondary walls prevent xylem vessels from col-

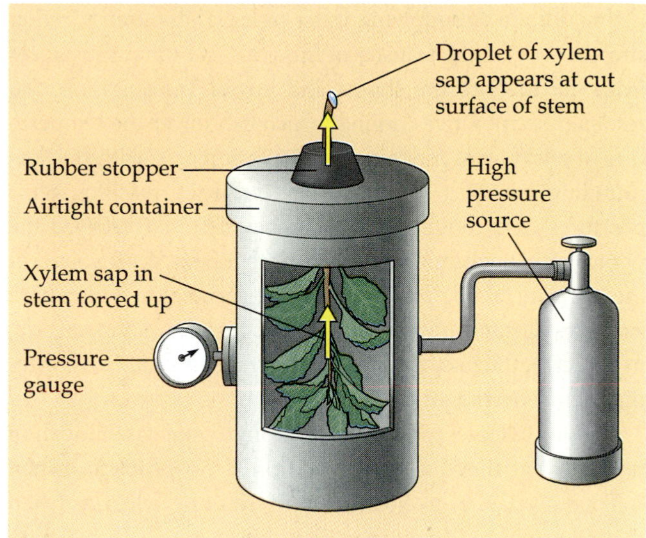

FIGURE 36.9 ▪ Measuring the water potential of leaves. When a branch is cut, the very negative potential in the leaves due to tension draws water in xylem away from the cut surface. The branch is placed in a pressure chamber (pressure "bomb"), and external pressure is increased until it is just great enough to counteract the negative water potential in leaves and push water back to the cut surface. The amount of pressure that must be applied is equal to the average water potential of the leaves.

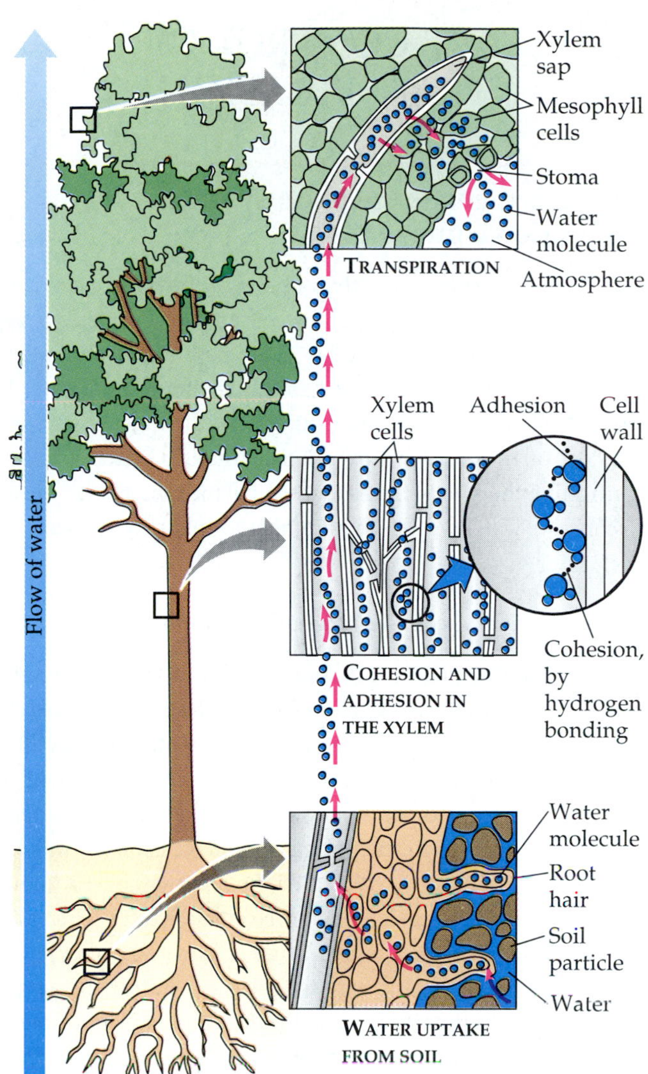

36.1 **FIGURE 36.10 ▪ The roles of cohesion and adhesion in the ascent of xylem sap.**

lapsing, much as wire rings maintain the shape of a vacuum hose. Transpirational pull puts the xylem under tension all the way down to the root tips, even in the tallest trees. This tension lowers water potential in the root xylem to such an extent that water flows passively from the soil, across the root cortex, and into the stele.

Transpirational pull can extend down to the roots only through an unbroken chain of water molecules. Cavitation, the formation of a water vapor pocket in a xylem vessel, such as when xylem sap freezes in winter, breaks the chain. Small plants can use root pressure to refill xylem vessels in spring, but in trees, root pressure cannot push water to the top, so a vessel with a water vapor pocket can never function as a water pipe again. However, the transpiration stream can detour around the water vapor pocket through pits between adjacent xylem vessels, and secondary growth adds a layer of new xylem vessels each year. In some angiosperm trees, including oaks and elms, the youngest, outermost growth ring of xylem transports most of the water. The older xylem functions to support the tree (see Chapter 35).

Xylem sap ascends by solar-powered bulk flow: *a review*

36.1 The transpiration-cohesion-tension mechanism that transports xylem sap against gravity is an excellent example of how physical principles apply to biological problems. Long-distance transport of water from roots to leaves occurs by bulk flow, the movement of fluid driven by a pressure difference at opposite ends of a conduit. In a plant, the conduits are xylem vessels or chains of tracheids. The pressure difference is generated at the leaf end by transpirational pull, which lowers pressure (increases tension) at the "upstream" end of the xylem. On a smaller scale, gradients of water potential drive the osmotic movement of water from cell to cell within root and leaf tissue. Differences in solute concentration and pressure both contribute to this microscopic transport. In contrast, bulk flow, the mechanism for long-distance transport up xylem vessels, depends only on pressure. Another contrast with osmosis, which moves only water, is that bulk flow moves the whole solution, water plus minerals and any other solutes dissolved in the water. The plant expends none of its own metabolic energy to lift water up to the leaves by bulk flow. The absorption of sunlight drives transpiration by causing water to evaporate from the moist walls of mesophyll cells and by maintaining a high humidity in the air spaces within a leaf. Thus, the ascent of xylem sap is ultimately solar-powered.

THE CONTROL OF TRANSPIRATION

Guard cells mediate the photosynthesis-transpiration compromise

A leaf may transpire more than its weight in water each day.
36.1 Leaves are kept from wilting by a transpiration stream in xylem vessels that flows as fast as 75 cm per minute, about the speed of the tip of a second hand sweeping around a wall clock. A plant's tremendous requirement for water is part of the cost of making food by photosynthesis. Guard cells, by controlling the size of stomata, help balance the plant's need to conserve water with its requirement for photosynthesis.

The Photosynthesis-Transpiration Compromise

To make food, a plant must spread its leaves to the sun and obtain CO_2 from the air. Carbon dioxide diffuses into the leaf, and oxygen produced as a by-product of photosynthesis diffuses out of the leaf through the stomata (see FIGURE 36.8). The stomata lead to the honeycomb of air spaces through which CO_2 diffuses to the photosynthetic cells of the mesophyll. The internal surface area of the leaf may be 10 to 30 times greater than the external surface area we see when we look at the leaf. This structural feature of leaves amplifies photosynthesis by increasing exposure to CO_2, but at the same time it increases the surface area for the evaporation of water, which exits the plant freely through open stomata. About 90% of the water a plant loses escapes through stomata, though these pores account for only 1% to 2% of the external leaf surface. The waxy cuticle limits water loss through the remaining surface of the leaf. Recall from Chapter 35 that the stomata of most plants are more concentrated on the lower surface of leaves, reducing transpiration because the bottom surface receives less sunlight than the top surface.

One way to evaluate how efficiently a plant uses water is to determine its **transpiration-to-photosynthesis ratio**, the amount of water lost per gram of CO_2 assimilated into organic material by photosynthesis. For many plant species this ratio is about 600:1, meaning the plant transpires 600 g of water for each gram of CO_2 that becomes incorporated into carbohydrate. However, corn and other plants that assimilate atmospheric CO_2 by the C_4 pathway have transpiration-to-photosynthesis ratios of 300:1 or less. With the same concentration of CO_2 within the air spaces of the leaf, C_4 plants can assimilate that CO_2 at a greater rate than C_3 plants can (see Chapter 10). Since water loss is the trade-off for enabling CO_2 to diffuse into the leaf, the photosynthetic return for each gram of water sacrificed is greater for plants that can assimilate CO_2 at greater rates when stomata are partially closed.

In addition to supplying water to leaves, the transpiration stream assists in the transfer of minerals and other substances from the roots to the shoots and leaves. Transpiration also results in evaporative cooling, which can lower the temperature of a leaf by as much as 10–15°C compared with the surrounding air. This prevents the leaf from reaching temperatures that could inhibit the enzymes that catalyze the photosynthetic reactions, as well as other enzymes involved in the leaf's metabolic processes. Desert succulents, which have low rates of transpiration, can tolerate high leaf temperatures; in this case, the loss of water due to transpiration is a greater threat than overheating.

As long as leaves can pull water from the soil fast enough to replace what they lose, transpiration is no problem. When transpiration exceeds the delivery of water by xylem, as when the soil begins to dry out, the leaves begin to wilt as their cells lose turgor pressure. The potential rate of transpiration will be greatest on a day that is sunny, warm, dry, and windy, because these are the environmental factors that increase the evaporation of water. Plants are not helpless against the elements, however, for they are capable of adjusting to their environment. In the photosynthesis-transpiration compromise, mechanisms that regulate the size of the stomatal openings strike a balance.

How Stomata Open and Close

Each stoma is flanked by a pair of guard cells, which are kidney-shaped in dicots and dumbbell-shaped in monocots. The guard cells are suspended by their epidermal neighbors over an air chamber, which leads to the leaf's maze of air spaces.

Guard cells control the diameter of the stoma by changing shape, thereby widening or narrowing the gap between the two cells (FIGURE 36.11a). When guard cells take in water by osmosis, they become more turgid and swell. In most dicots the cell walls of guard cells are not uniformly thick, and the cellulose microfibrils are oriented in a direction that causes the guard cells to buckle outward when they are turgid. This increases the size of the gap between the cells. When the cells lose water and become flaccid, they sag together and close the space between them. This basic mechanism also applies to the stomata of monocots.

The changes in turgor pressure that open and close stomata result primarily from the reversible uptake and loss of potassium ions (K^+) by the guard cells. Stomata open when guard cells actively accumulate K^+ from neighboring epidermal cells (FIGURE 36.11b). This uptake of solute causes the water potential to become more negative within the guard cells, and the cells become more turgid as water enters by osmosis. Most of the K^+ and water are stored in the vacuole, and thus the tonoplast also plays a role. The cell's increase in positive charge due to the influx of K^+ is offset by the uptake of chloride (Cl^-), by

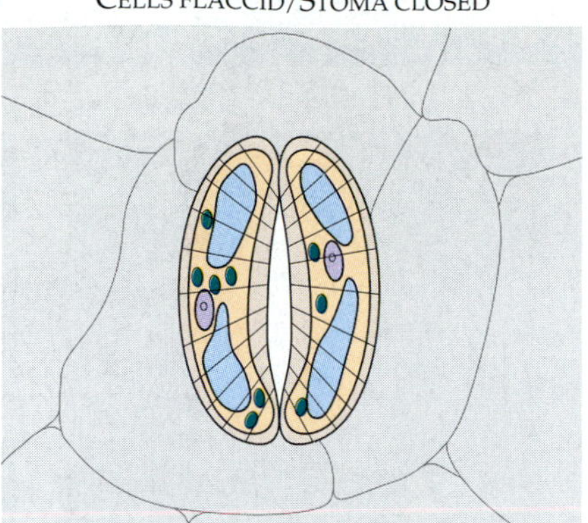

(a) Changes in guard cell shape and stomatal opening and closing (surface view)

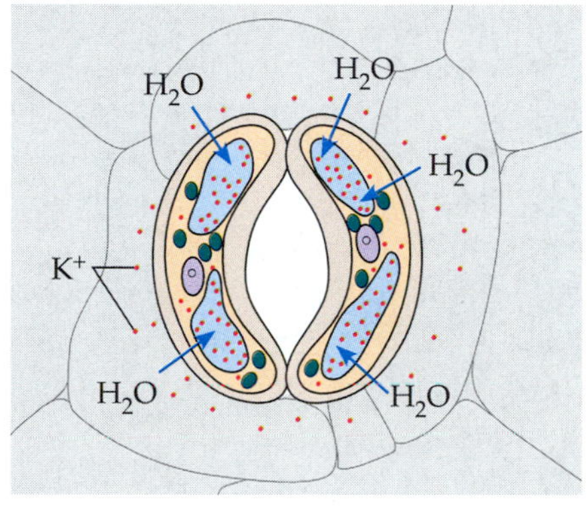

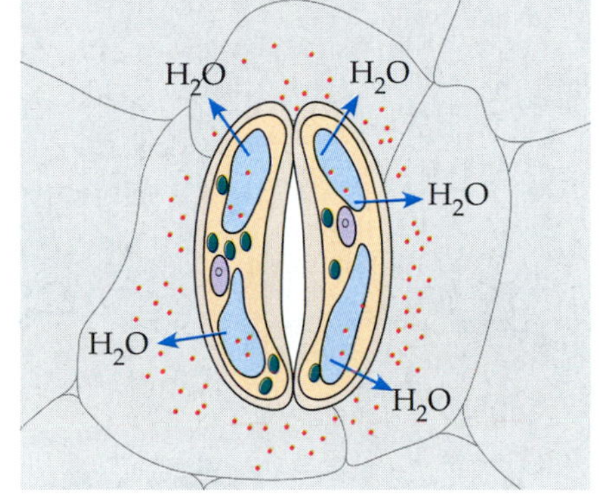

(b) Role of potassium in stomatal opening and closing

FIGURE 36.11 ▪ The control of stomatal opening and closing.
(a) Guard cells of a dicot are illustrated in their turgid (stoma open) and flaccid (stoma closed) states. The pair of guard cells buckle outward when turgid. Cellulose microfibrils in the walls resist stretching and compression in the direction parallel to the microfibrils. Thus, the radial orientation of the microfibrils causes the cells to increase in length more than width when turgor increases. The two guard cells are attached at their tips, so the increase in length causes buckling. **(b)** The transport of K^+ (potassium ions) across the plasma membrane and tonoplast causes the turgor changes of guard cells. Stomata open when guard cells accumulate potassium (red dots), which lowers the cells' water potential and causes them to take up water by osmosis. The cells become turgid. An exodus of K^+ from guard cells causes the stomatal closure.

the pumping out of the cell of hydrogen ions released from organic acids, and by the negative charges of these organic acids after they lose their hydrogen ions. Stomatal closing results from an exodus of K^+ from guard cells, which leads to an osmotic loss of water. Regulation of aquaporins may also be involved in the swelling and shrinking of guard cells by varying the permeability of the membranes to water.

The K^+ fluxes across the guard cell membrane are probably passive, being coupled to the generation of membrane potentials by proton pumps. Stomatal opening correlates with active transport of H^+ out of the guard cell. The resulting voltage (membrane potential) drives K^+ into the cell through specific membrane channels (see FIGURE 36.2). Plant physiologists are using a method called patch clamping to study regulation of the guard cell's proton pumps and K^+ channels (see the Methods Box, p. 708).

In general, stomata are open during the day and closed at night. This prevents the plant from needlessly losing water when it is too dark for photosynthesis. At least three cues contribute to stomatal opening at dawn. First, light itself stimulates guard cells to accumulate potassium and become turgid. This response is triggered by the illumination of a blue-light receptor in a guard cell, perhaps built into the plasma membrane. Activation of these blue-light receptors stimulates the activity of ATP-powered proton pumps in the plasma membrane of the guard cells, which, in turn, promotes the uptake of K^+ (see the Methods Box, page 708). Light may also stimulate stomatal opening by driving photosynthesis in guard cell chloroplasts, making ATP available for the active transport of hydrogen ions. (Guard cells are the only epidermal cells equipped with chloroplasts.) A second factor causing stomata to open is depletion of CO_2 within air spaces of the leaf, which occurs when photosynthesis begins in the mesophyll. A plant can be tricked into opening its stomata at night by placing it in a chamber devoid of CO_2. A third cue in stomatal opening is an internal clock located in the guard cells. Even if you keep a plant in a dark closet, stomata will continue their daily rhythm of opening and closing. All eukaryotic organisms have internal clocks that somehow keep track of time and regulate cyclical processes. Cycles that have intervals of approximately 24 hours are called **circadian rhythms**. You will learn more about circadian rhythms and the biological clocks that control them in Chapter 39.

Environmental stresses of various kinds can cause stomata to close during the daytime. When the plant is suffering a water deficiency, guard cells may lose turgor. In addition, a hormone called abscisic acid, which is produced in the mesophyll cells in response to water deficiency, signals guard cells to close stomata. This response reduces further wilting, but it also slows down photosynthesis; this is one reason droughts

METHODS: Patch-clamp Recording

Patch clamping makes it possible to isolate a very tiny "patch" of membrane and study ion movement through a single type of channel—a K^+ or H^+ channel, for instance.

To apply the patch-clamp method to plant cells, it is first necessary to remove the cell walls. This is usually accomplished by using enzymes to digest the walls, resulting in naked (wall-less) cells, or protoplasts. FIGURE a is a light micrograph of guard cell protoplasts from tobacco *(Nicotiana)* leaves.

The tool for making a patch is a micropipette with a tip only about 1 μm in diameter. A larger pipette is used to hold the cell in place (FIGURE b). Slight suction draws a membrane "blister" into the opening of the micropipette, and the rim of the micropipette seals to the membrane. This partitions a small patch from the rest of the plasma membrane. The micropipette, hooked up to appropriate equipment, now functions as an electrode to record ion fluxes across the tiny patch of membrane.

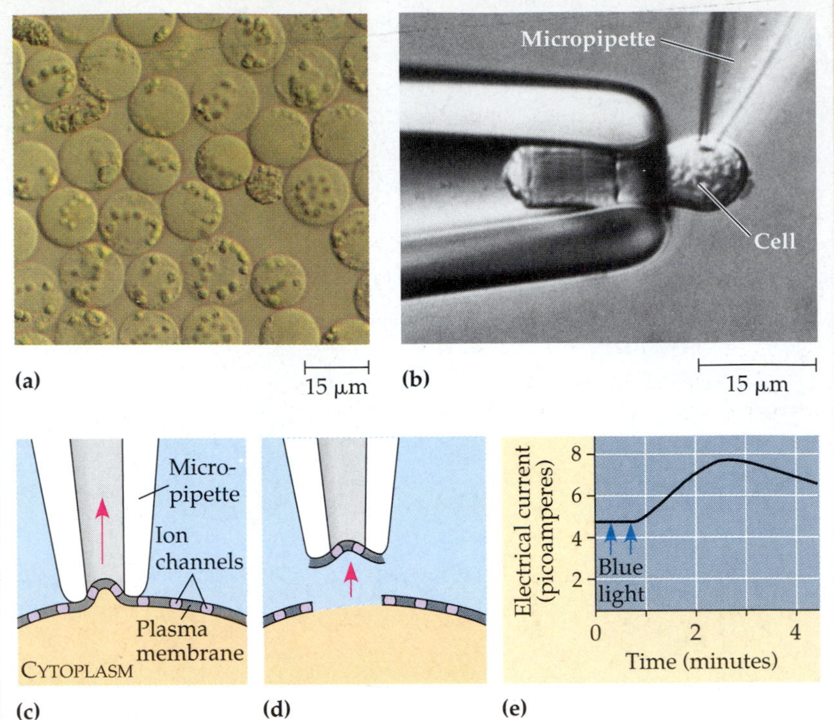

With the whole cell still attached to the micropipette by the tight seal around the membrane patch, it is possible to record ion transport between the cytosol and the artificial solution that fills the micropipette, as shown in FIGURE c. An alternative is to pull the micropipette away from the cell, separating the membrane patch from the rest of the cell, as shown in FIGURE d. This enables the researcher to control the composition of solutions on *both* sides of the membrane patch. Because the patch is so small, and because the ions in the bathing solutions are known, it is possible to record the transport of a single kind of ion through selective channels or pumps in the membrane. FIGURE e shows such a recording for H^+ transport across a membrane patch of a guard cell. Passage of H^+ across the membrane represents an electrical current that can be measured. The graph indicates current due to H^+ flux during a recording that lasts several minutes. This particular experiment measures the effect of blue light, which increases H^+ current across the membrane.

Along with other evidence, patch-clamp studies support the hypothesis that blue light and other stimuli regulate guard cells by affecting the proton (H^+) pumps of the membrane.

FIGURE b from T. D. Lamb, H. R. Matthews, and V. Torre, *Journal of Physiology* 372(1986):315–349. Reprinted by permission.

reduce crop yields. High temperatures also induce stomatal closure, probably by stimulating cellular respiration and increasing CO_2 concentration within the air spaces of the leaf. High temperature and excessive transpiration may combine to cause stomata to close briefly during midday. Thus, guard cells arbitrate the photosynthesis-transpiration compromise on a moment-to-moment basis by integrating a variety of internal and external stimuli.

Xerophytes have adaptations that reduce transpiration

Plants adapted to arid climates, called xerophytes, have various leaf modifications that reduce the rate of transpiration. Many xerophytes have small, thick leaves, an adaptation that limits water loss by reducing surface area relative to leaf volume. A thick cuticle gives some of these leaves a leathery consistency. The stomata are concentrated on the lower leaf surface, and they are often located in depressions that shelter the pores from the dry wind (FIGURE 36.12). During the driest months, some desert plants shed their leaves. Others, such as cacti, subsist on water the plant stores in its fleshy stems during the rainy season. (These modified stems are the photosynthetic organs of cacti; the spines are modified leaves.)

One of the most elegant adaptations to arid habitats is found in ice plants and other succulent plants of the family Crassulaceae and in representatives of many other plant families. These plants assimilate CO_2 by an alternative photosynthetic pathway known as CAM, for crassulacean acid metabolism (see Chapter 10). Mesophyll cells in a CAM plant have enzymes that can incorporate CO_2 into organic acids during the night. During the daytime, the organic acids are

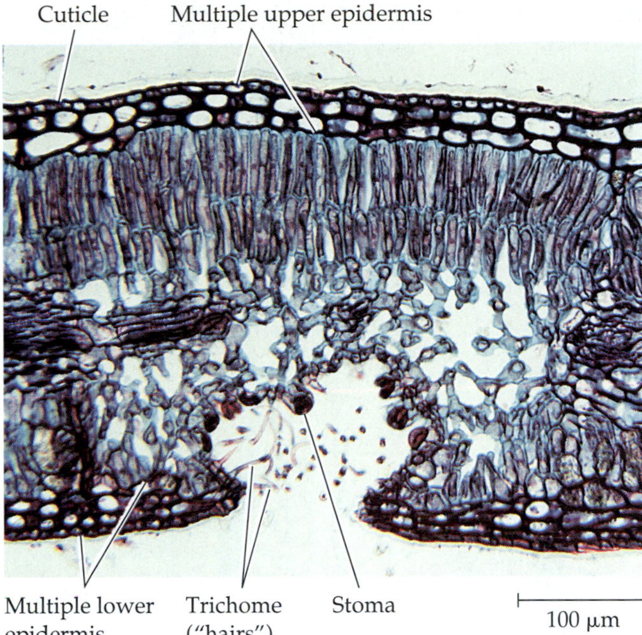

FIGURE 36.12 ▪ Structural adaptations of a xerophyte leaf. The thick cuticle and multiple layers of epidermal cells reduce water loss across the general surface of this oleander leaf. Stomata are recessed in "crypts," an adaptation that reduces the rate of transpiration by protecting the stomata from hot, dry wind. The trichomes ("hairs") also help minimize transpiration by breaking up the flow of air, allowing the chamber of the crypt to have a higher humidity than that of the surrounding atmosphere.

broken down to release CO_2 in the same cells, and sugars are synthesized by the conventional (C_3) photosynthetic pathway. Since the leaf takes in its CO_2 at night, the stomata can close during the day, when transpiration is most severe.

TRANSLOCATION OF PHLOEM SAP

Xylem sap generally flows in the wrong direction to function in exporting sugar from leaves to other parts of the plant. A second vascular tissue, the phloem, transports the organic products of photosynthesis throughout the plant. This transport of food in the plant is called **translocation**. In angiosperms, the specialized cells of phloem that function in translocation are the sieve-tube members, which are arranged end to end to form long sieve tubes (see FIGURE 35.10e). Between the members are sieve plates, porous cross walls that allow the flow of sap along the sieve tube.

Phloem sap is an aqueous solution that differs markedly in composition from xylem sap. By far the prevalent solute in phloem sap is sugar, primarily the disaccharide sucrose. The sucrose concentration may be as high as 30% by weight, giving the sap a syrupy thickness. Phloem sap may also contain minerals, amino acids, and hormones in transit from one part of the plant to another.

Phloem translocates its sap from sugar sources to sugar sinks

In contrast to the unidirectional transport of xylem sap from
36.2 roots to leaves, the direction that phloem sap travels is variable. The one generalization that holds is that sieve tubes carry food from a sugar source to a sugar sink. A **sugar source** is a plant organ in which sugar is being produced by either photosynthesis or the breakdown of starch. Mature leaves are the primary sugar sources. A **sugar sink** is an organ that consumes or stores sugar. Growing roots, shoot tips, stems, and fruits are sugar sinks supplied by phloem. A storage organ, such as a tuber or a bulb, may be either a source or a sink depending on the season. When the storage organ is stockpiling carbohydrates during the summer, it is a sugar sink. After breaking dormancy in the early spring, however, the storage organ becomes a source as its starch is broken down to sugar, which is carried away in the phloem to the growing buds of the shoot system.

Other solutes may be transported to sinks along with sugar. For example, minerals that reach leaves in xylem may later be transferred in the phloem to developing fruit.

A sugar sink usually receives its sugar from the sources nearest to it. The upper leaves on a branch may send sugar to the growing shoot tip, whereas the lower leaves export sugar to roots. A growing fruit requires so much food that it may monopolize the sugar sources all around it. One sieve tube in a vascular bundle may carry phloem sap in one direction while sap in a different tube in the same bundle flows in the opposite direction. For each sieve tube, the direction of transport depends only on the locations of the source and sink connected by that tube, and the direction may change with the season or developmental stage of the plant.

Phloem Loading and Unloading

Sugar from the mesophyll cells of a leaf and other sources must be loaded into sieve-tube members before it can be exported to sugar sinks. In some species, including certain plants of the mint and squash families, sucrose moves all the way from mesophyll cells to sieve-tube members via the symplast, passing from cell to cell through plasmodesmata. In other species, sucrose reaches sieve-tube members by a combination of symplastic and apoplastic pathways (FIGURE 36.13a, p. 710). For example, in corn leaves, sucrose diffuses through the symplast from mesophyll cells into small veins. Much of the sugar then moves out of the cells into the apoplast (walls) in the vicinity of sieve-tube members and companion cells. This sucrose is accumulated from the apoplast (walls) by the sieve-tube members and their companion cells, which pass the sugar into the sieve-tube members through plasmodesmata linking the cells. In some plants, companion cells have numerous ingrowths of

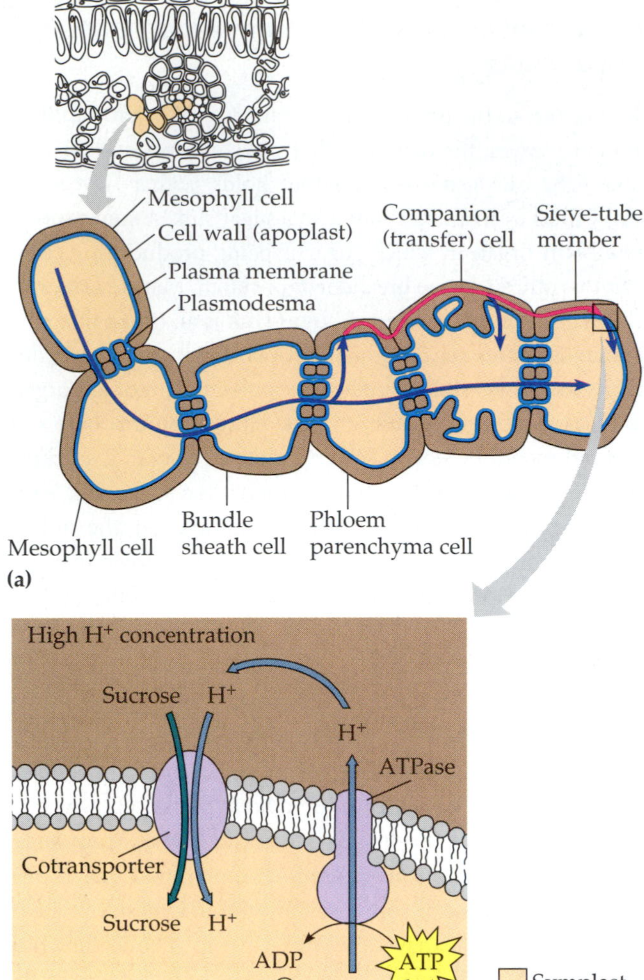

FIGURE 36.13 ▪ Loading of sucrose into phloem. (a) Sucrose manufactured in mesophyll cells can travel via the symplast (blue arrows) to sieve-tube members. In some species, sucrose exits (magenta arrow) the symplast near sieve tubes and is actively accumulated from the apoplast by sieve-tube members and their companion cells. Some companion cells are modified as transfer cells, which have wall ingrowths that increase the surface area for the transport of solutes. **(b)** A chemiosmotic mechanism is responsible for the active transport of sucrose into companion cells and sieve-tube members. Proton pumps (ATPase) generate an H^+ gradient, which drives sucrose accumulation with the help of a cotransport protein that couples sucrose transport to the diffusion of H^+ back into the cell.

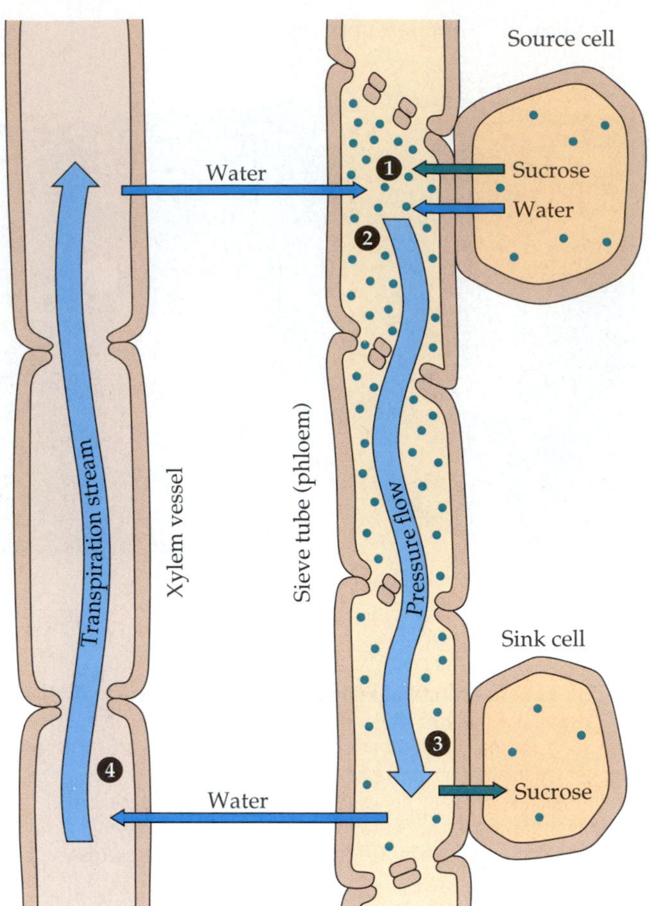

36.2 **FIGURE 36.14 ▪ Pressure flow in a sieve tube.** ① Loading of sugar into the sieve tube at the source reduces the water potential inside sieve-tube members. This causes the tube to take up water from surrounding tissues by osmosis. ② This absorption of water generates a hydrostatic pressure that forces the sap to flow along the tube. ③ The pressure gradient in the tube is reinforced by the unloading of sugar and the consequent loss of water from the tube at the sink. ④ Xylem recycles water from sink to source.

their walls, an adaptation that increases the cells' surface area and enhances the transfer of solutes between apoplast and symplast. Such modified cells are called **transfer cells**.

In corn and many other plants, sieve-tube members accumulate sucrose to concentrations two to three times higher than concentrations in mesophyll, and thus phloem loading requires active transport. Proton pumps do the work that enables the cells to accumulate sucrose (FIGURE 36.13b).

Downstream, at the sink end of a sieve tube, phloem unloads its sucrose. This involves active transport, again through cotransport with H^+ down a proton gradient.

Pressure flow is the mechanism of translocation in angiosperms

36.2 Phloem sap flows from source to sink at rates as great as 1 m per hour, which is much too fast to be accounted for by either diffusion or cytoplasmic streaming. Phloem sap moves by bulk flow, which is driven by pressure (thus the synonym, pressure flow). Phloem loading results in a high solute concentration at the source end of a sieve tube, which lowers the water potential and causes water to flow into the tube (FIGURE 36.14). Hydrostatic pressure develops within the sieve tube, and the pressure is greatest at the source end of the tube. At the sink end, the pressure is relieved by the loss of water, owing to water potential being lowered outside the sieve tube

(a)

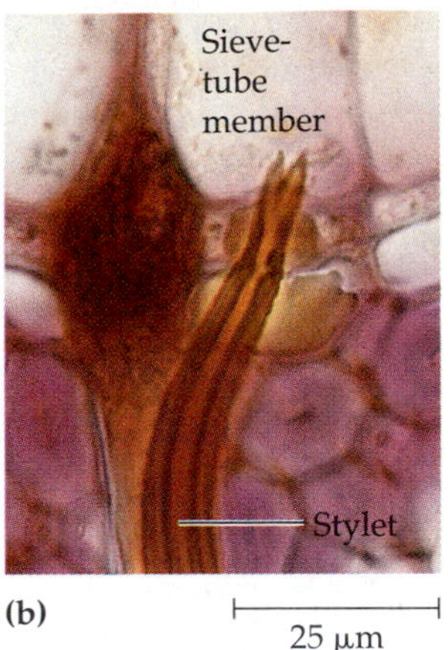

(b)

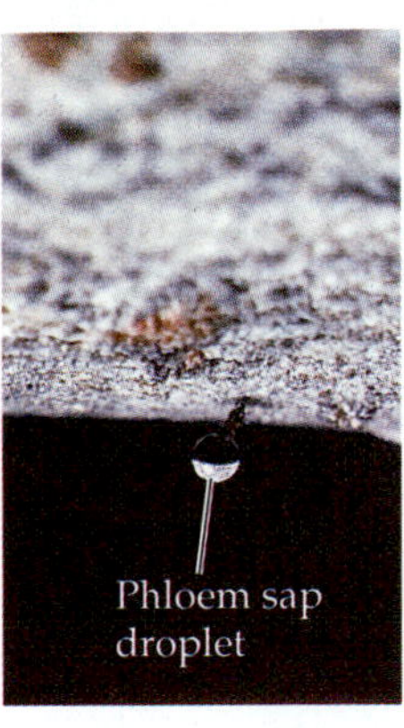

(c)

FIGURE 36.15 ▪ Tapping phloem sap with the help of an aphid. (a) The "honeydew" droplet exuded from the anus of this aphid consists of phloem sap minus some nutrients absorbed by the insect. **(b)** The aphid inserts a modified mouthpart called a stylet into the plant and probes until the tip of this hypodermic-like organ penetrates a single sieve-tube member (LM). The pressure within the sieve tube force-feeds the aphid, swelling it to several times its original size. **(c)** While the aphid is feeding, it can be anesthetized and severed from its stylet, which then serves the researcher as a miniature tap that exudes phloem sap for hours. The closer the stylet is to a sugar source, the faster the sap will flow out and the greater its sugar concentration. These results are predicted by the pressure-flow hypothesis.

by the exodus of sucrose. The building of pressure at one end of the tube (source) and reduction of that pressure at the opposite end (sink) cause water to flow from source to sink, carrying the sugar along. Water is recycled back from sink to source by xylem vessels.

The pressure-flow model explains why phloem sap always flows from a sugar source to a sugar sink, regardless of their locations in the plant. Researchers have devised several kinds of experiments to test the model. FIGURE 36.15 illustrates an innovative experiment that takes advantage of natural phloem probes: aphids that feed on phloem sap. The case for pressure flow as the mechanism of translocation in angiosperms is convincing. It is not yet known, however, if this model applies to gymnosperms.

In our study of how sugar moves in plants, we have seen examples of plant transport on all three levels: at the cellular level of transport across plasma membranes (sucrose accumulation by active transport in phloem cells); at the short-distance level of lateral transport within organs (sucrose migration from mesophyll to phloem via the symplast and apoplast); and at the long-distance level of transport between organs (bulk flow within sieve tubes).

▪ ▪ ▪

Plant physiologists still have much to learn about the mechanisms of transport in the elaborate vascular systems of xylem and phloem. William Harvey, the great seventeenth-century physiologist, speculated that plants and animals have similar circulatory systems. The idea was abandoned after careful dissection failed to turn up a heart in plants. We are only now beginning to understand how a plant keeps sap flowing through its veins without the help of moving parts.

CHAPTER REVIEW

REVIEW OF KEY CONCEPTS

(with page numbers and key figures)

AN OVERVIEW OF TRANSPORT MECHANISMS IN PLANTS

- **Transport at the cellular level depends on the selective permeability of membranes (pp. 695–696)** Specific transport proteins enable plant cells to maintain an internal environment different from their surroundings.
- **Proton pumps play a central role in transport across plant membranes (pp. 696–697, FIGURE 36.2)** The membrane potential and H^+ gradient generated by proton pumps are harnessed to drive the transport of a variety of solutes.
- **Differences in water potential drive water transport in plant cells (pp. 697–700, FIGURE 36.4)** Solutes decrease water potential, while pressure increases water potential. Water flows by osmosis from a compartment with a higher water potential to one with a lower potential. Aquaporins, water-specific channels in membranes, might help regulate the rate of osmosis. A turgid cell matches the water potential of the surroundings when the cell's wall exerts a pressure that counters the cell's tendency to take up water because of its solute potential.
- **Vacuolated plant cells have three major compartments (p. 700, FIGURE 36.5)** The plasma membrane regulates transport between the cytosol and the wall solution, while the tonoplast regulates transport between the cytosol and the vacuole.
- **The symplast and apoplast both function in transport within tissues and organs (p. 700, FIGURE 36.5)** The symplast is the continuum of cytosol linked by plasmodesmata. The apoplast is the continuum of cell walls.
- **Bulk flow functions in long-distance transport (pp. 700–701)** Transport of xylem sap and phloem sap is due to pressure differences at opposite ends of conduits, the xylem vessels and sieve tubes.

ABSORPTION OF WATER AND MINERALS BY ROOTS

- **Root hairs, mycorrhizae, and a large surface area of cortical cells enhance water and mineral absorption (pp. 701–702)** Root hairs are the most important avenues of absorption near root tips, but mycorrhizae, symbiotic associations of fungi and roots, are responsible for most absorption by the whole root system. Once soil solution enters the root, the extensive surface area of cortical cell membranes enhances uptake of water and selected minerals into cells.

- **The endodermis functions as a selective sentry between the root cortex and vascular tissue (pp. 702–703, FIGURE 36.6)** Water can cross the cortex via the symplast or apoplast, but minerals that reach the endodermis via the apoplast must finally cross the selective membranes of endodermal cells. The waxy Casparian strip of the endodermal wall blocks apoplastic transfer of minerals from the cortex to the stele.

TRANSPORT OF XYLEM SAP

36.1 **The ascent of xylem sap depends mainly on transpiration and the physical properties of water (pp. 703–705, FIGURE 36.8)** Loss of water vapor (transpiration) lowers water potential in the leaf by producing a negative pressure (tension). This low water potential draws water from the xylem. Cohesion and adhesion of the water transmits the pulling force all the way down to the roots.

36.1 **Xylem sap ascends by solar-powered bulk flow: *a review* (p. 705, FIGURE 36.10)** The movement of xylem sap against gravity is maintained by transpiration.

THE CONTROL OF TRANSPIRATION

36.1 **Guard cells mediate the photosynthesis-transpiration compromise (pp. 706–708, FIGURE 36.11)** Stomata support photosynthesis by allowing CO_2 and O_2 exchange between the leaf and atmosphere, but these pores are also the main avenues for transpirational loss of water from the plant. Turgor changes in guard cells, which depend on K^+ and water transport into and out of the cells, regulate the size of the stomatal openings.

- **Xerophytes have adaptations that reduce transpiration (pp. 708–709, FIGURE 36.12)** Protection of stomata within leaf indentations and other structural adaptations enable certain plants to survive in arid environments.

TRANSLOCATION OF PHLOEM SAP

36.2 **Phloem translocates its sap from sugar sources to sugar sinks (pp. 709–710, FIGURE 36.13)** Mature leaves are the main sources, though storage organs such as bulbs can be sources during certain seasons. Developing roots and shoot tips are examples of sugar sinks. Phloem loading and unloading depend on active transport of sucrose. The sucrose is cotransported along with H^+, which is diffusing down a gradient generated by proton pumps.

36.2 **Pressure flow is the mechanism of translocation in angiosperms (pp. 710–711, FIGURE 36.14)** Loading of sugar at the source end of a sieve tube and unloading at the sink end maintain a pressure difference that keeps the sap flowing through the tube.

SELF-QUIZ

1. Which of the following would *not* contribute to water uptake by a plant cell?
   a. an increase in the water potential (ψ) of the surrounding solution
   b. a decrease in pressure on the cell exerted by the wall
   c. the uptake of solutes by the cell
   d. a decrease in ψ of the cytoplasm
   e. an increase in tension on the surrounding solution

2. Stomata open when guard cells
   a. sense an increase in CO_2 in the air spaces of the leaf
   b. flop open because of a decrease in turgor pressure
   c. become more turgid because of an influx of K^+, followed by the osmotic entry of water
   d. close aquaporins, preventing uptake of water
   e. accumulate water by active transport

3. Which of the following is *not* part of the transpiration-cohesion-tension mechanism for the ascent of xylem sap?
   a. loss of water from the mesophyll cells, which initiates a pull of water molecules from neighboring cells and eventually from the xylem
   b. the transfer of transpirational pull from one water molecule to the next due to the cohesion caused by hydrogen bonds
   c. the hydrophilic walls of the narrow tracheids and xylem vessels that help maintain the column of water against the force of gravity
   d. active pumping of water into the xylem of roots
   e. the reduction of water potential in the surface film of mesophyll cells due to increased surface tension

4. Which of the following cell types lacks a plasma membrane?
   a. endodermal cell
   b. sieve tube member
   c. tracheid
   d. guard cell
   e. companion cell

5. The movement of sap from a sugar source to a sugar sink
   a. occurs through the apoplast of sieve-tube members
   b. may translocate sugars from the breakdown of stored starch in a root up to developing shoots
   c. is similar to the flow of xylem sap in depending on tension, or negative pressure
   d. depends on the active pumping of water into sieve tubes at the source end
   e. results mainly from diffusion

6. The productivity of a crop declines when leaves begin to wilt mainly because (explain your answer)
   a. the chlorophyll of wilting leaves decomposes
   b. flaccid mesophyll cells are incapable of photosynthesis
   c. stomata close, preventing CO_2 from entering the leaf
   d. photolysis, the water-splitting step of photosynthesis, cannot occur when there is a water deficiency
   e. an accumulation of CO_2 in the leaf inhibits the enzymes required for photosynthesis

7. Imagine cutting a live twig from a tree and examining the cut surface of the twig with a magnifying glass. You locate the vascular tissue and observe a growing droplet of fluid exuding from the cut surface. This fluid is probably (explain your answer)
   a. phloem sap
   b. xylem sap
   c. guttation fluid
   d. fluid of the transpiration stream
   e. cell sap from the broken vacuoles of cells

8. Which structure or compartment is not part of the plant's apoplast?
   a. the lumen of a xylem vessel
   b. the lumen of a sieve tube
   c. the cell wall of a mesophyll cell
   d. the cell walls of transfer cells
   e. the cell walls of root hairs

9. Which of the following is *not* an adaptation that enhances the uptake of water and minerals by roots?
   a. mycorrhizae, the symbiotic associations of roots and fungi
   b. root hairs, which increase surface area near root tips
   c. selective uptake of minerals by xylem vessels
   d. selective uptake of minerals by cortical cells
   e. plasmodesmata, which facilitate symplastic transport from the cortex into the stele
10. A plant cell with a solute potential of −0.65 MPa maintains a constant volume when bathed in a solution that has a solute potential of −0.3 MPa and is in an open container. What do we know about the cell?
    a. The cell has a pressure potential of +0.65 MPa.
    b. The cell has a water potential of −0.65 MPa.
    c. The cell has a pressure potential of +0.35 MPa.
    d. The cell has a pressure potential of +0.3 MPa.
    e. The cell has a water potential of 0 MPa.

CHALLENGE QUESTIONS

1. Compare and contrast the transport of xylem sap and phloem sap. Your discussion should include the roles of water potential and bulk flow, the pathways by which solutes reach the vascular tissue, the processes that generate the forces driving the transport, and the factors that determine the direction of transport.
2. Wind ventilates the leaf surface, increases transpiration, and replenishes the supply of carbon dioxide for photosynthesis. Stomata close when the carbon dioxide concentration inside the leaf increases. How do you think stomatal response to carbon dioxide is related to wind speed, and what might be the adaptive significance of this relationship?
3. Describe the environmental conditions that would minimize the transpiration-to-photosynthesis ratio for a C_3 plant.
4. A tip for making cut flowers last longer without wilting is to cut off the ends of the stems under water and then transfer the flowers to a vase full of water while drops of water are still present on the cut ends of the stems. Explain why this works.

SCIENCE, TECHNOLOGY, AND SOCIETY

1. In a reverse osmosis unit, pressure is applied to a solution, forcing water through a selectively permeable membrane but leaving solute molecules behind. Why is this called reverse osmosis? Relate reverse osmosis to the water potential concept discussed in this chapter. What might be some possible commercial applications of reverse osmosis?
2. Many scientists believe that increased carbon dioxide in the atmosphere may result in warming of the global climate in coming decades. This could have a serious impact on agriculture. Imagine that you are in charge of a project to use selective breeding and recombinant DNA techniques to "engineer" crop plants that can grow under hotter, drier conditions. What traits would you try to build into the "ideal" corn or wheat plant?

FURTHER READING

Canny, M. J. "Transporting Water in Plants." *American Scientist,* March–April 1998. Discusses how plants with embolisms, air pockets in their xylem vessels, still manage to transport water.

Heinrich, B. "Nutcracker Sweets." *Natural History,* February 1991. Describes how red squirrels obtain maple syrup from the sap flowing within trees.

Moore, R., W. D. Clark, and D. S. Vodopich. *Botany,* 2nd ed. New York: McGraw Hill, 1998. Chapter 21 surveys transport mechanisms in plants.

Neher, E., and B. Sakmann. "The Patch Clamp Technique." *Scientific American,* March 1992. The Nobel Prize–winning developers of this method explain how it can be used to study membrane transport.

Salisbury, F. B., and C. W. Ross. *Plant Physiology,* 4th ed. Belmont, CA: Wadsworth, 1992. A rigorous text with excellent chapters on transport.

Szabo, M. "Plant Sugars Move in Mysterious Ways." *New Scientist,* December 4, 1993. Discusses some agricultural applications of understanding source-sink transport in plants.

WEB LINKS

Visit the special edition of *The Biology Place* for BIOLOGY, Fifth Edition, at http://www.biology.com/campbell. Go to Chapter 36 for online resources, including learning activities, practice exams, and links to the following web sites:

"The Botanical Society of America"
The Botanical Society of America "exists to promote research and teaching in all fields of plant biology, to facilitate cooperation among plant biologists and other scientists worldwide, and to disseminate knowledge of plants, algae, and fungi to all groups of society for ultimate application to solving practical problems of humanity."

"Internet Directory for Botany"
This is a comprehensive list of links to botanical resources around the world.

"Vignettes from the History of Plant Morphology"
A rare find, this site looks at the life and times of three preeminent plant morphologists from the first half of the twentieth century.

"The Harmful Algae Page"
Modern-day relatives of the prehistoric ancestors of land plants lack a sophisticated transport system, relying on simple direct exchange of materials with the environment. At this site you can learn about the potential deadly consequences of this exchange for both human and wildlife communities.

PLANT NUTRITION

Every organism is an open system connected to its environment by a continuous exchange of energy and materials. In the energy flow and chemical cycling that keep an ecosystem alive, plants and other photosynthetic autotrophs perform the key step of transforming inorganic compounds into organic ones. Autotrophic does not mean autonomous, however. Plants need sunlight as the energy source for photosynthesis. But to synthesize organic matter, plants also require raw materials in the form of inorganic substances: carbon dioxide, water, and a variety of minerals present as inorganic ions in the soil. With its ramifying root system and shoot system (see the photograph of a bean seedling on this page), a plant is extensively networked with its environment—the soil and air, which are the reservoirs of the plant's inorganic nutrients. In this chapter you will learn more about the nutritional requirements of plants and examine some of the structural and physiological adaptations for plant nutrition that have evolved.

NUTRITIONAL REQUIREMENTS OF PLANTS

The chemical composition of plants provides clues to nutritional requirements

Watch a large plant grow from a tiny seed, and you cannot help wondering where all the mass comes from. Aristotle thought soil provided the substance for plant growth because plants seemed to spring from the ground. Leaves, he believed, functioned only to shade the developing fruit. In the seventeenth century a Belgian physician named Jean-Baptiste van Helmont performed an experiment to test the hypothesis that plants grew by absorbing soil. He planted a willow seedling in a pot that contained 90.9 kg of soil. After five years the willow had grown into a tree weighing 76.8 kg, but only 0.06 kg of soil had disappeared from the pot. Van Helmont concluded that the willow had grown mainly from the water he had added regularly. A century later, Stephen Hales, an English physiologist, postulated that plants were nourished mostly by air.

As it turns out, none of the early ideas about plant nutrition is entirely incorrect. Plants *do* extract minerals from the soil. **Mineral nutrients** are essential chemical elements absorbed from the soil in the form of inorganic ions. For example, plants need nitrogen, which they acquire from the soil mainly in the form of nitrate ions (NO_3^-). However, as we can conclude from van Helmont's data, mineral nutrients from the soil make only a small contribution to the overall mass of the plant. About 80% to 85% of a herbaceous (nonwoody) plant is water, and plants grow mainly by accumulating water in the central vacuoles of their cells. Furthermore, water can truly be considered a nutrient because it supplies most of the hydrogen atoms and some of the oxygen atoms that are incorporated into organic compounds by photosynthesis. However, only a small fraction of the water that enters a plant contributes atoms to organic molecules. Generally, more than 90% of the water absorbed by plants is lost by transpiration, and most of the water that is retained by the plant actually functions as a solvent, makes cell elongation possible, and serves to maintain the form of soft tissue by keeping cells turgid. By weight, the bulk of the organic material of a plant is derived not from water, but from the CO_2 that is assimilated from the atmosphere (FIGURE 37.1).

We can measure water content by comparing the weight of plant material before and after it is dried. We can then analyze the chemical composition of the dry residue. Organic substances account for about 95% of the dry weight, with inorganic substances making up the remaining 5%. Most of the organic material is carbohydrate, including the cellulose of cell walls. Thus, carbon, oxygen, and hydrogen, the ingredients of carbohydrates, are the most abundant elements in the dry weight of a plant. Because some organic molecules contain nitrogen, sulfur, or phosphorus, these elements are also relatively abundant in plants.

More than 50 chemical elements have been identified among the inorganic substances present in plants, but it is unlikely that all these elements are essential. Roots are able to absorb minerals somewhat selectively, enabling the plant to accumulate essential elements that may be present in the soil in very minute quantities. To a certain extent, however, the minerals in a plant reflect the composition of the soil in which the plant is growing. Plants growing on mine tailings, for instance, may contain gold or silver. Studying the chemical composition of plants provides clues about their nutritional requirements, but we must distinguish elements that are essential from those that are merely present in the plant.

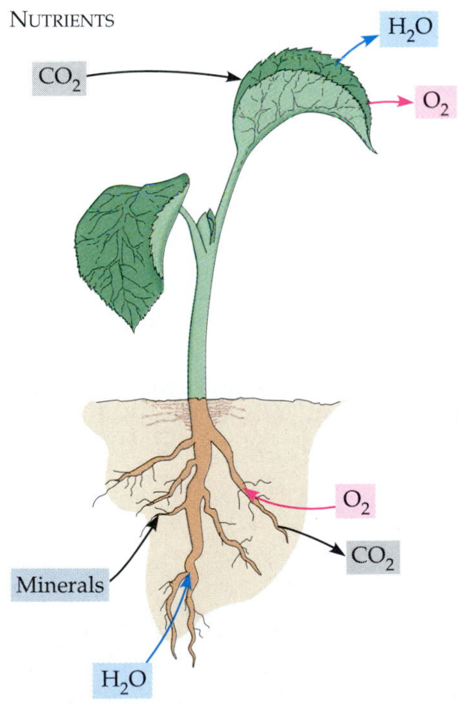

FIGURE 37.1 ▪ The uptake of nutrients by a plant: an overview. Roots absorb water and minerals from the soil, with mycorrhizae and root hairs greatly increasing surface area for absorption. Carbon dioxide, the source of carbon for photosynthesis, diffuses into leaves from the surrounding air through stomata. (Plants also need O_2 for cellular respiration, although the plant is a net producer of O_2.) From these inorganic nutrients the plant can produce all of its own organic material.

Plants require nine macronutrients and at least eight micronutrients

A particular chemical element is considered to be an **essential nutrient** if it is required for a plant to grow from a seed and complete the life cycle, producing another generation of seeds. Researchers can use a method known as hydroponic culture to determine which of the mineral elements are actually essential nutrients (FIGURE 37.2). Such studies have helped identify 17 elements that are essential nutrients in all plants and a few other elements that are essential to certain groups of plants (TABLE 37.1, p. 716). Most research has involved crop plants; little is known about the specific nutritional needs of uncultivated plants, even some of the most commercially important conifers.

FIGURE 37.2 ▪ Using hydroponic culture to identify essential nutrients. In the technique called hydroponic culture, a researcher bathes the roots of plants in solutions of various minerals dissolved in known concentrations. Aerating the water provides the roots with oxygen for cellular respiration. A particular mineral, such as potassium, can be omitted from the culture medium to test whether it is essential to the plants. If the element deleted from the mineral solution is an essential nutrient, then the incomplete medium will cause plants to become abnormal in appearance compared with controls grown on a complete mineral medium. The most common symptoms of a mineral deficiency are stunted growth and discolored leaves.

Table 37.1 ▪ Essential Nutrients in Plants

ELEMENT	FORM AVAILABLE TO PLANTS	MAJOR FUNCTIONS
Macronutrients		
Carbon	CO_2	Major component of plant's organic compounds
Oxygen	CO_2	Major component of plant's organic compounds
Hydrogen	H_2O	Major component of plant's organic compounds
Nitrogen	NO_3^-, NH_4^+	Component of nucleic acids, proteins, hormones, and coenzymes
Sulfur	SO_4^{2-}	Component of proteins, coenzymes
Phosphorus	$H_2PO_4^-$, HPO_4^{2-}	Component of nucleic acids, phospholipids, ATP, several coenzymes
Potassium	K^+	Cofactor that functions in protein synthesis; major solute functioning in water balance; operation of stomata
Calcium	Ca^{2+}	Important in formation and stability of cell walls and in maintenance of membrane structure and permeability; activates some enzymes; regulates many responses of cells to stimuli
Magnesium	Mg^{2+}	Component of chlorophyll; activates many enzymes
Micronutrients		
Chlorine	Cl^-	Required for water-splitting step of photosynthesis; functions in water balance
Iron	Fe^{3+}, Fe^{2+}	Component of cytochromes; activates some enzymes
Boron	$H_2BO_3^-$	Cofactor in chlorophyll synthesis; may be involved in carbohydrate transport and nucleic acid synthesis
Manganese	Mn^{2+}	Active in formation of amino acids; activates some enzymes; required for water-splitting step of photosynthesis
Zinc	Zn^{2+}	Active in formation of chlorophyll; activates some enzymes
Copper	Cu^+, Cu^{2+}	Component of many redox and lignin-biosynthetic enzymes
Molybdenum	MoO_4^{2-}	Essential for nitrogen fixation; cofactor that functions in nitrate reduction
Nickel	Ni^{2+}	Cofactor for an enzyme functioning in nitrogen metabolism

Elements required by plants in relatively large amounts are called **macronutrients**. There are nine macronutrients in all, including the six major ingredients of organic compounds: carbon, oxygen, hydrogen, nitrogen, sulfur, and phosphorus. The other three macronutrients are potassium, calcium, and magnesium. (TABLE 37.1 lists some of their functions.)

Elements that plants need in very small amounts are called **micronutrients**. The eight micronutrients are iron, chlorine, copper, manganese, zinc, molybdenum, boron, and nickel. These elements function in plants mainly as cofactors of enzymatic reactions (see Chapter 6). Iron, for example, is a metallic component of cytochromes, the proteins that function in the electron transport chains of chloroplasts and mitochondria. It is because micronutrients generally play catalytic roles that plants need only minute quantities of these elements. The requirement for molybdenum, for example, is so modest that there is only one atom of this rare element for every 16 million atoms of hydrogen in dried plant material. Yet a deficiency of molybdenum or any other micronutrient can weaken or kill a plant.

The symptoms of a mineral deficiency depend on the function and mobility of the element

The symptoms of a mineral deficiency depend partly on the
37.1 function of that nutrient in the plant. For example, a deficiency of magnesium, an ingredient of chlorophyll, causes yellowing of the leaves, or chlorosis (FIGURE 37.3). In some cases the relationship between a mineral deficiency and its symptoms is less direct. For instance, iron deficiency can cause chlorosis even though chlorophyll contains no iron, because this metal is required as a cofactor in one of the steps of chlorophyll synthesis.

Mineral deficiency symptoms depend not only on the role of the nutrient in the plant but also on its mobility within the

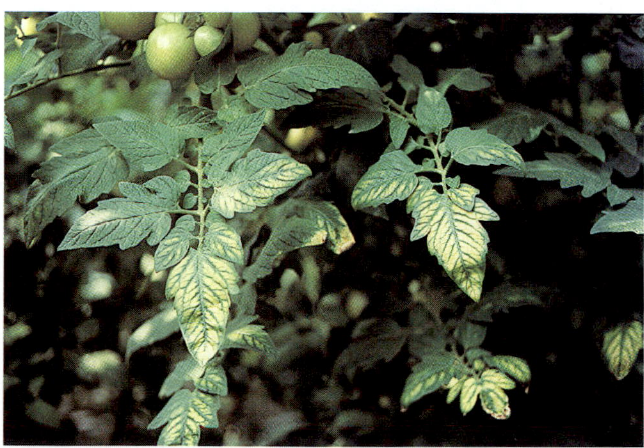

FIGURE 37.3 ▪ Magnesium deficiency in a tomato plant. Yellow-
37.1 ing of the leaves (chlorosis) is the result of an inability to synthesize chlorophyll, which contains magnesium.

plant. If a nutrient moves about freely from one part of the plant to another, symptoms of a deficiency will show up first in older organs. This is because young, growing tissues have more "drawing power" than old tissues for nutrients that are in short supply. (The mechanism for this preferential routing is the source-to-sink translocation in phloem that you learned about in Chapter 36.) A plant starved for magnesium, for example, will show signs of chlorosis first in its older leaves. Magnesium, which is relatively mobile in the plant, is shunted preferentially to young leaves. In contrast, a deficiency of a nutrient that is relatively immobile within a plant will affect young parts of the plant first. Older tissues may have adequate amounts of the mineral, which they are able to retain during periods of short supply. A deficiency of iron, which does not move freely in the plant, will cause yellowing of young leaves before any effect on older leaves is visible.

The symptoms of a mineral deficiency are often distinctive enough for a plant physiologist or farmer to diagnose its cause. One way to confirm the diagnosis of a specific deficiency is to analyze the mineral content of the plant and soil. Deficiencies of nitrogen, potassium, and phosphorus are the most common problems. Shortages of micronutrients are less common and tend to be geographically localized because of differences in soil composition. The amount of a micronutrient needed to correct a deficiency is usually quite small. For example, a zinc deficiency in fruit trees can usually be cured by hammering a few zinc nails into each tree trunk. Moderation is important because overdoses of some micronutrients can be toxic to plants.

One way to ensure optimal mineral nutrition is to grow plants hydroponically on nutrient solutions that can be precisely regulated (FIGURE 37.4). Hydroponics is currently practiced commercially, but only on a limited scale because the requirements for equipment and labor make hydroponic farming relatively expensive compared with growing crops in soil.

FIGURE 37.4 ▪ Hydroponic farming. In this apparatus, a nutrient solution flows over the roots of lettuce growing on a slat. Perhaps astronauts living in a space station will one day grow their vegetables hydroponically, but because of the expense it is unlikely that this type of farming will soon relieve hunger here on Earth.

THE ROLE OF SOIL IN PLANT NUTRITION

Soil characteristics are key environmental factors in terrestrial ecosystems

The texture and chemical composition of soil are major factors determining what kinds of plants can grow well in a particular location, be it a natural ecosystem or an agricultural region. (Climate, of course, is another important factor.) Plants that grow naturally in a certain type of soil are adapted to its mineral content and texture and are able to absorb water and extract essential nutrients from that soil. In interacting with the soil that supports their growth, plants, in turn, affect the soil, as we will soon see. The soil-plant interface is a critical component of the nutrient cycles that sustain terrestrial ecosystems.

Texture and Composition of Soils

Soil has its origin in the weathering of solid rock. Water that seeps into crevices and freezes in winter fractures the rock, and acids dissolved in the water also help to break down the rock. Once organisms are able to invade the rock, they accelerate the breakdown. Lichens, fungi, bacteria, mosses, and the roots of vascular plants all secrete acids, and the expansion of roots growing in fissures cracks rocks and pebbles. The eventual result of all this activity is **topsoil**, a mixture of particles derived from rock, living organisms, and humus, a residue of partially decayed organic material. The topsoil and other distinct soil layers, or **horizons**, are often visible in vertical profile where there is a roadcut (FIGURE 37.5, p. 718).

The texture of topsoil depends on the size of its particles, which are classified in a range from coarse sand to microscopic clay particles. The most fertile soils are usually **loams**, made up of roughly equal amounts of sand, silt (particles of intermediate size), and clay. Loamy soils have enough fine particles to provide a large surface area for retaining minerals and water, which adhere to the particles. But loams also have enough coarse particles to provide air spaces containing oxygen that can be used by roots for cellular respiration. If soil does not drain adequately, roots suffocate because the air spaces are replaced by water; the roots may also be attacked by molds favored by the soaked soil. These are common hazards

FIGURE 37.5 ▪ Soil horizons. This roadcut exposes a vertical profile of three soil layers, or horizons. The A horizon is the topsoil, a mixture of broken-down rock of various textures, living organisms, and decaying organic matter. The B horizon contains much less organic matter than the A horizon and is less weathered. The C horizon, composed mainly of partially broken-down rock, serves as the "parent" material for the upper layers of soil.

for houseplants that are overwatered in pots that do not drain. Some plants, however, are adapted to waterlogged soil. For example, mangroves and many other plants that inhabit swamps and marshes have some of their roots modified as hollow tubes that grow upward and function as snorkels, bringing down oxygen from the air.

Topsoil is home to an astonishing number and variety of organisms. A teaspoon of soil has about five billion bacteria that cohabit with various fungi, algae and other protists, insects, earthworms, nematodes, and the roots of plants. The activities of all these organisms affect the physical and chemical properties of the soil. Earthworms, for instance, aerate the soil by their burrowing and add mucus that holds fine soil particles together. The metabolism of bacteria alters the mineral composition of the soil. Plant roots extract water and minerals but also affect soil pH and reinforce the soil against erosion.

Humus is the decomposing organic material formed by the action of bacteria and fungi on dead organisms, feces, fallen leaves, and other organic refuse. Humus prevents clay from packing together and builds a crumbly soil that retains water but is still porous enough for the adequate aeration of roots. Humus is also a reservoir of mineral nutrients that are returned gradually to the soil as microorganisms decompose the organic matter.

The Availability of Soil Water and Minerals

37.1 After a heavy rainfall, water drains away from the larger spaces of the soil, but smaller spaces retain water because of its attraction for the soil particles, which have electrically charged surfaces. Some of this water adheres so tightly to the hydrophilic soil particles that it cannot be extracted by plants. The film of water bound less tightly to the particles is the water generally available to plants (FIGURE 37.6a). It is not pure water, but a soil solution containing dissolved minerals. Roots absorb this soil solution.

Many minerals in soil—especially those that are positively charged, such as potassium (K^+), calcium (Ca^{2+}), and magnesium (Mg^{2+})—adhere by electrical attraction to the negatively charged surfaces of clay particles. The presence of clay in a soil helps prevent the leaching (draining away) of mineral nutrients during heavy rain or irrigation because the finely divided particles provide so much surface area for binding minerals. On the other hand, clay particles must release their bound minerals to the soil solution in order for roots to absorb the nutrients. Minerals that are negatively charged, such as nitrate (NO_3^-), phosphate ($H_2PO_4^-$), and sulfate (SO_4^{2-}), are usually not bound tightly to soil particles and thus tend to leach away more quickly. Positively charged minerals are made available to the plant when hydrogen ions in the soil displace the mineral ions from the clay particles. This process, called **cation exchange**, is stimulated by the roots themselves, which secrete H^+ and compounds that form acids in the soil solution (FIGURE 37.6b).

Soil conservation is one step toward sustainable agriculture

It may take centuries for a soil to become fertile through the breakdown of rock and the accumulation of organic material, but human mismanagement can destroy that fertility within a few years. Good soil conservation, on the other hand, can preserve the soil's fertility, resulting in sustained agricultural productivity.

To understand soil conservation, we must begin with the premise that agriculture is unnatural. In forests, grasslands, and other natural ecosystems, mineral nutrients are usually recycled by the decomposition of dead organic material in the soil. In contrast, when we harvest a crop, essential elements are diverted from the chemical cycles going on in that location. In general, agriculture depletes the mineral content of the soil. To grow a ton of wheat grain, the soil gives up 18.2 kg of nitrogen, 3.6 kg of phosphorus, and 4.1 kg of potassium. Each year the fertility of the soil diminishes, unless fertilizers are applied to replace the lost minerals. Many crops also use far more water than the natural vegetation that once grew on that land, forcing farmers to irrigate. Prudent fertilization, thoughtful irrigation, and the prevention of erosion are three of the most important goals of soil conservation.

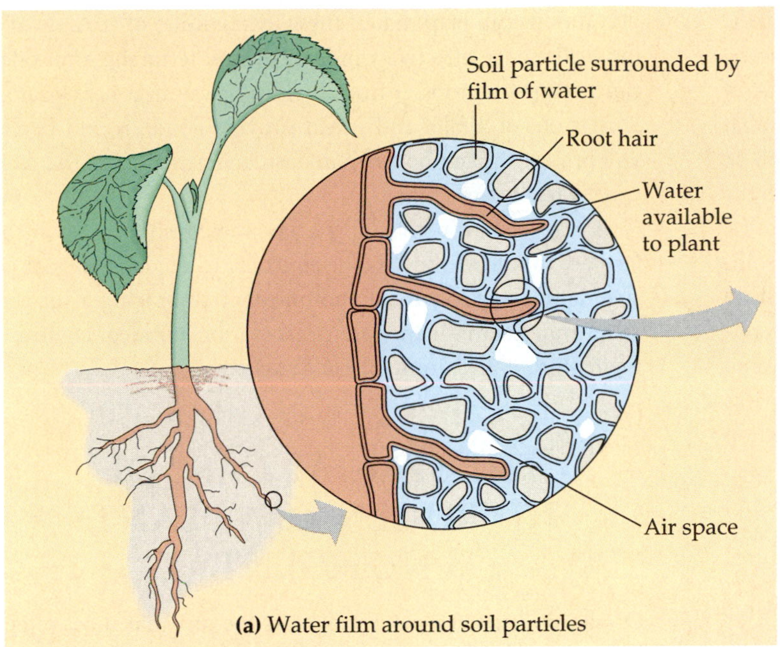

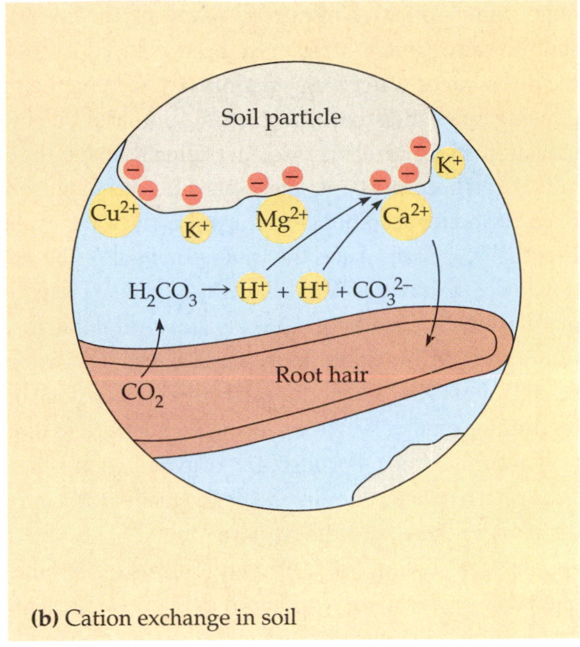

37.1 **FIGURE 37.6 ▪ The availability of soil water and minerals. (a)** A plant cannot extract all the water in the soil because some of it is tightly held by hydrophilic soil particles. Water bound less tightly to soil particles can be imbibed by the root. **(b)** Hydrogen ions in the soil solution help make certain nutrients available to plants by displacing positively charged minerals (cations) that were bound tightly to the surface of fine soil particles. In addition to secreting H^+, plants contribute to the pool of H^+ in the soil in the following way: Cellular respiration in roots releases CO_2 to the soil solution, where the CO_2 reacts with water to form carbonic acid (H_2CO_3). Dissociation of this acid adds hydrogen ions to the soil

Fertilizers

Prehistoric farmers may have started fertilizing their fields after noticing that grass grew faster and greener where animals had defecated. The Romans used manure to fertilize their crops, and Native Americans buried fish along with seeds when they planted corn. In developed nations today, most farmers use commercially produced fertilizers containing minerals that are either mined or prepared by industrial processes. These fertilizers are usually enriched in nitrogen, phosphorus, and potassium, the three mineral elements that are most commonly deficient in farm and garden soils. Commercial fertilizers, such as those you can buy in a garden shop, are labeled with a three-number code that indicates the mineral content. A fertilizer marked "10-12-8," for instance, is 10% nitrogen (as ammonium or nitrate), 12% phosphorus (as phosphoric acid), and 8% potassium (as the mineral potash).

Manure, fishmeal, and compost are referred to as "organic" fertilizers because they are of biological origin and contain organic material that is in the process of decomposing. However, before the elements in compost can be of any use to plants, the organic material must be decomposed to the inorganic nutrients that roots can absorb. In the end, the minerals a plant extracts from the soil are in the same form whether they came from organic fertilizer or from a chemical factory. Compost releases minerals gradually, however, whereas the minerals in commercial fertilizers are available immediately but may not be retained by the soil for long. Excess minerals not taken up by the plants are wasted because they may be rapidly leached from the soil by rainwater or irrigation. To make matters worse, this mineral runoff may enter the groundwater and eventually pollute streams and lakes. Agricultural researchers are attempting to develop ways to reduce the use of fertilizers while maintaining crop yields.

To fertilize judiciously, a farmer must pay close attention to the pH of the soil. Soil pH not only affects cation exchange but also influences the chemical form of all minerals. Even though an essential element may be abundant in the soil, plants may be starving for that element because it is bound too tightly to clay or is in a chemical form the plant cannot absorb. Managing the pH of soil is tricky; a change in hydrogen ion concentration may make one mineral more available to the plant while causing another mineral to become less available. At pH 8, for instance, the plant can absorb calcium, but iron is almost completely unavailable. The pH of the soil should be matched to the specific mineral needs of the crop. If the soil is too alkaline, sulfate can be added to lower the pH. Soil that is too acidic can be adjusted by liming (adding calcium carbonate or calcium hydroxide).

Irrigation

Even more than mineral deficiencies, the availability of water most often limits the growth of plants. Irrigation can transform a desert into a garden, but farming in arid regions is a huge drain on water resources. Many of the rivers in the southwestern United States have been reduced to trickles by the diversion of water for irrigation. (Quenching the thirst of growing cities adds to the problem.) Another problem is that irrigation in an arid region can gradually make the soil so salty that it becomes completely infertile. Salts dissolved in the irrigation water accumulate in the soil as the water evaporates. Eventually, the salt makes the water potential of the soil solution lower than that of root cells, which then *lose* water instead of absorbing it (see Chapter 36). As the world population continues to grow, more and more acres of arid land will have to be cultivated. New methods of irrigation may reduce the risks of running out of water or losing farmland to salinization (salt accumulation). For instance, drip irrigation is now used as an alternative to flooding fields for many of the crops and orchards in Israel and the western United States. In another approach to solving some of the problems of dryland farming, plant breeders are working to develop varieties of plants that require less water.

Erosion

Topsoil from thousands of acres of farmland is lost to water and wind erosion each year in the United States alone. Certain precautions can help reduce these losses. Rows of trees dividing fields make effective windbreaks, and terracing a hillside can prevent the topsoil from washing away in a heavy rain. Such crops as alfalfa and wheat provide good ground cover and protect the soil better than corn and other crops that are usually planted in rows.

If managed properly, soil is a renewable resource on which farmers can grow food for generations to come. The goal is **sustainable agriculture**, a commitment embracing a variety of farming methods that are conservation-minded, environmentally safe, and profitable (FIGURE 37.7).

FIGURE 37.7 ▪ Sustainable agriculture and "green manure." The "green manure" being mulched into the soil of this Washington state farm is sweet clover. Every third year, clover is planted and the crop is plowed under. This improves the physical structure and nutrient content of the soil for growing wheat and other crops during the other two years of the crop-rotation cycle. Substituting green manure for chemical additives, using beneficial insects that prey on crop pests as an alternative to insecticides, and choosing crops that are suited to the rainfall and other climatic conditions of the locale are examples of actions aimed at developing sustainable agriculture.

THE SPECIAL CASE OF NITROGEN AS A PLANT NUTRIENT

Of all mineral elements, nitrogen is the one that most often limits the growth of plants and the yields of crops. Plants require nitrogen as an ingredient of proteins, nucleic acids, and other important organic molecules.

The metabolism of soil bacteria makes nitrogen available to plants

It is ironic that plants sometimes suffer nitrogen deficiencies, for the atmosphere is nearly 80% nitrogen. This atmospheric nitrogen, however, is gaseous N_2, and plants cannot use nitrogen in that form. For plants to absorb nitrogen, it must first be converted to ammonium (NH_4^+) or nitrate (NO_3^-). In contrast to other minerals, the NH_4^+ and NO_3^- in soil are not derived from the breakdown of parent rock. Over the short term, the main source of nitrogenous minerals is the decomposition of humus by microbes, including ammonifying bacteria (FIGURE 37.8). In this way, nitrogen present in organic compounds, such as proteins, is repackaged in inorganic compounds that can be recycled when they are absorbed as minerals by roots. However, nitrogen is lost from this local cycle when soil microbes called denitrifying bacteria convert NO_3^- to N_2, which diffuses from the soil to the atmosphere. Still other bacteria, called **nitrogen-fixing bacteria**, restock nitrogenous minerals in the soil by converting N_2 to NH_3 (ammonia), a metabolic process called **nitrogen fixation**. The complex cycling of nitrogen in ecosystems is traced in detail in Chapter 54. Here we focus on nitrogen fixation and the other steps that lead directly to nitrogen assimilation by plants.

All life on Earth depends on nitrogen fixation, an exclusive function of certain prokaryotes. Soil is populated by several species of free-living bacteria that are among the nitrogen-fixing prokaryotes. (Other species of nitrogen-fixing bacteria live in plant roots in symbiotic relationships you will learn more about in the next section.) The conversion of atmospheric nitrogen (N_2) to ammonia (NH_3) is a complicated, multistep process, but we can simplify nitrogen fixation by just indicating the reactants and products:

$$N_2 + 8\,e^- + 8\,H^+ + 16\,ATP \longrightarrow 2\,NH_3 + H_2 + 16\,ADP + 16\,Ⓟ_i$$

One enzyme complex, called **nitrogenase**, catalyzes the entire reaction sequence, which reduces N_2 to NH_3 by adding electrons along with hydrogen ions. Notice that nitrogen fixation is very expensive in terms of metabolic energy, costing the bacteria 8 ATP molecules for each ammonia molecule synthesized. Nitrogen-fixing bacteria are most abundant in soils rich in organic material, which provides fuel for cellular respiration.

In the soil solution, ammonia picks up another hydrogen ion to form ammonium (NH_4^+), which plants can absorb. However, plants acquire their nitrogen mainly in the form of nitrate (NO_3^-), which is produced in the soil by nitrifying bacteria that oxidize ammonium (see FIGURE 37.8). After nitrate is absorbed by roots, a plant enzyme can reduce the nitrate back to ammonium, which other enzymes can then incorporate into amino acids and other organic compounds. Most plant species export nitrogen from roots to shoots, via the xylem, in the form of nitrate or organic compounds that have been synthesized in the roots.

Improving the protein yield of crops is a major goal of agricultural research

The ability of plants to incorporate fixed nitrogen into proteins and other organic substances has a major impact on human welfare; the most common form of malnutrition in humans is protein deficiency. Either by choice or by economic necessity, the majority of people in the world have a predominantly vegetarian diet and thus depend mainly on plants for protein, particularly in developing countries. Unfortunately, many plants have a low protein content, and the proteins that are present may be deficient in one or more of the amino acids that humans need from their diet. Improving the quality and quantity of proteins in crops is a major goal of agricultural research.

Plant breeding has resulted in new varieties of corn, wheat, and rice that are enriched in protein. However, many of these "super" varieties have an extraordinary demand for nitrogen, which is usually supplied in the form of commercial fertilizer. The industrial production of ammonia and nitrate from atmospheric nitrogen is, like biological nitrogen fixation, very expensive in energy costs. A chemical factory making fertilizer consumes large quantities of fossil fuels. Generally, the countries that most need high-protein crops are the ones least able to afford to pay the fuel bill. The use of new catalysts based on the mechanism by which nitrogenase fixes nitrogen may make commercial fertilizer production less costly in the future. Biochemists determined the structure of *Rhizobium* nitrogenase in 1992, providing a model for chemical engineers to design catalysts by imitating nature. Another strategy that could potentially increase protein yields of crops is to improve the productivity of symbiotic nitrogen fixation, a process we examine in the next section.

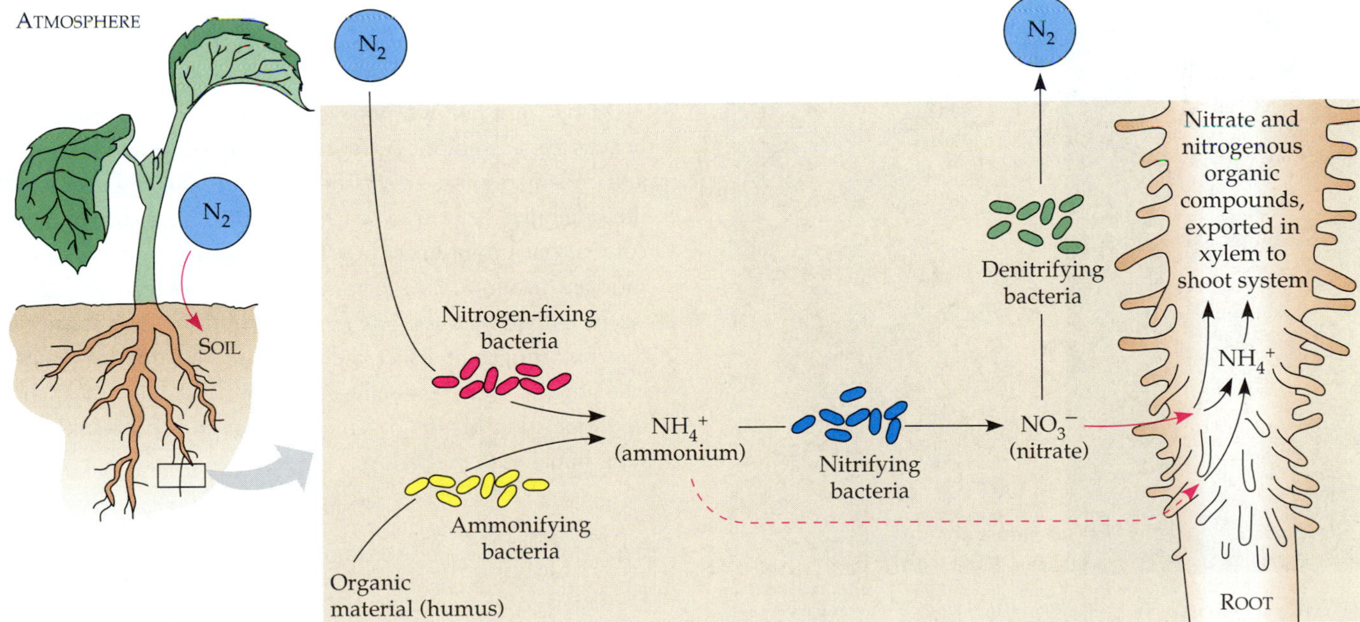

FIGURE 37.8 • The role of soil bacteria in the nitrogen nutrition of plants. Ammonium is made available to plants by two types of soil bacteria: those that fix atmospheric N_2 (nitrogen-fixing bacteria) and those that decompose organic material (ammonifying bacteria). Although plants absorb some ammonium from the soil, they absorb mainly nitrate, which is produced from ammonium by nitrifying bacteria. Plants reduce nitrate back to ammonium before incorporating the nitrogen into organic compounds. Xylem transports nitrogen from roots to shoots in the form of nitrate, amino acids, and various other organic compounds, depending on the species.

NUTRITIONAL ADAPTATIONS: SYMBIOSIS OF PLANTS AND SOIL MICROBES

The roots of plants belong to subterranean communities that include a diversity of other organisms. Among those organisms are certain species of bacteria and fungi that have coevolved with specific plants, forming symbiotic relationships with roots that enhance the nutrition of both partners. The two most important examples are symbiotic nitrogen fixation between roots and bacteria, and the formation of mycorrhizae, symbiotic associations of roots with fungi.

Symbiotic nitrogen fixation results from intricate interactions between roots and bacteria

Many plant families include species that form symbiotic relationships with nitrogen-fixing bacteria that give roots a built-in source of fixed nitrogen for assimilation into organic compounds. Most of the research on symbiotic nitrogen fixation has focused on agriculturally important members of the legume family, including peas, beans, soybeans, peanuts, alfalfa, and clover. Legumes' roots have swellings called **nodules** composed of plant cells that contain nitrogen-fixing bacteria of the genus *Rhizobium* ("root living"). Inside the nodule, the *Rhizobium* assume a form called **bacteroids**, which are contained within vesicles formed by the root cell (FIGURE 37.9). Each legume is associated with a particular species of *Rhizobium*. FIGURE 37.10 describes the steps in the development of root nodules.

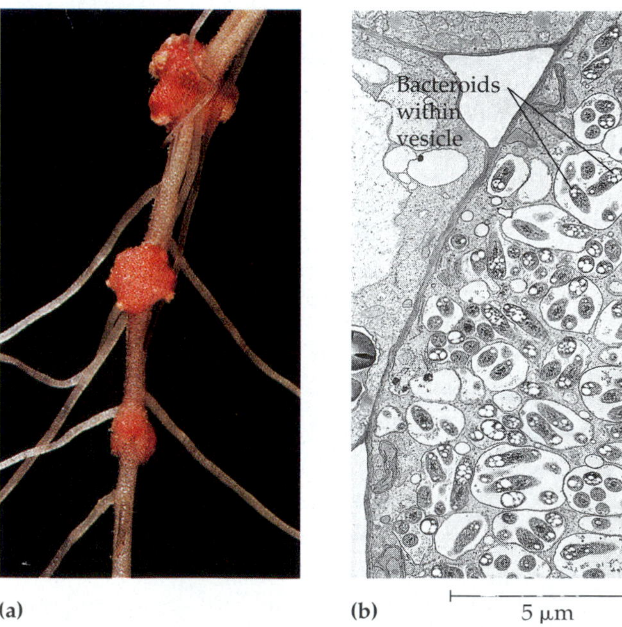

FIGURE 37.9 • Root nodules on a legume. (a) The red bumps on this soybean root are nodules containing symbiotic bacteria. The bacteria fix nitrogen and obtain photosynthetic products supplied by the plant. **(b)** In this transmission electron micrograph, a cell from a root nodule of soybean is filled with bacteroids. The adjoining cell remains uninfected.

The symbiotic relationship between a legume and its nitrogen-fixing bacteria is mutualistic, with both partners benefiting. The bacteria supply the legume with fixed nitrogen, and the plant provides the bacteria with carbohydrates and other organic compounds. Most of the ammonium produced by symbiotic nitrogen fixation is used by the nodules to make amino acids, which are then transported to the shoot and leaves via the xylem.

The exquisite coevolution of partners is evident in their cooperative synthesis of a molecule named leghemoglobin, with the plant and the bacteria each making part of the molecule. Leghemoglobin is an iron-containing protein that, like the hemoglobin of human red blood cells, binds reversibly to oxygen (*leg-* is for legume). The redness of the soybean nodules in FIGURE 37.9 is due to leghemoglobin. The root nodule's leghemoglobin acts as an oxygen "buffer," regulating the supply of oxygen for the intense respiration required by the bacteria to produce ATP for nitrogen fixation.

Symbiotic Nitrogen Fixation and Agriculture

Now that you have learned about symbiotic nitrogen fixation, you can understand the agricultural practice of crop rotation. One year a nonlegume such as corn is planted, and the following year alfalfa or some other legume is planted to restore the concentration of fixed nitrogen in the soil. Instead of being harvested, the legume crop is often plowed under so that it will decompose as "green manure" (see FIGURE 37.7). To ensure that the legume encounters its specific *Rhizobium*, the seeds are soaked in a culture of the bacteria or dusted with bacterial spores before sowing.

Many plant families besides legumes include species that benefit from symbiotic nitrogen fixation. For example, alders and certain tropical grasses host nitrogen-fixing bacteria of the actinomycete group (see Chapter 27). Rice, a crop of great commercial importance, benefits indirectly from symbiotic nitrogen fixation. Rice farmers culture a water fern called *Azolla* in their paddies. The fern has symbiotic cyanobacteria that fix nitrogen and increase the fertility of the rice paddy. The growing rice eventually shades and kills the *Azolla*, and decomposition of this organic material adds more nitrogenous minerals to the paddy.

The Molecular Biology of Root Nodule Formation in Legumes

How does a legume species recognize a certain species of *Rhizobium* among the many bacterial species inhabiting a root's soil environment? And how does an encounter with that specific *Rhizobium* lead to the development of a nodule? These two questions have led researchers to a chemical dialogue

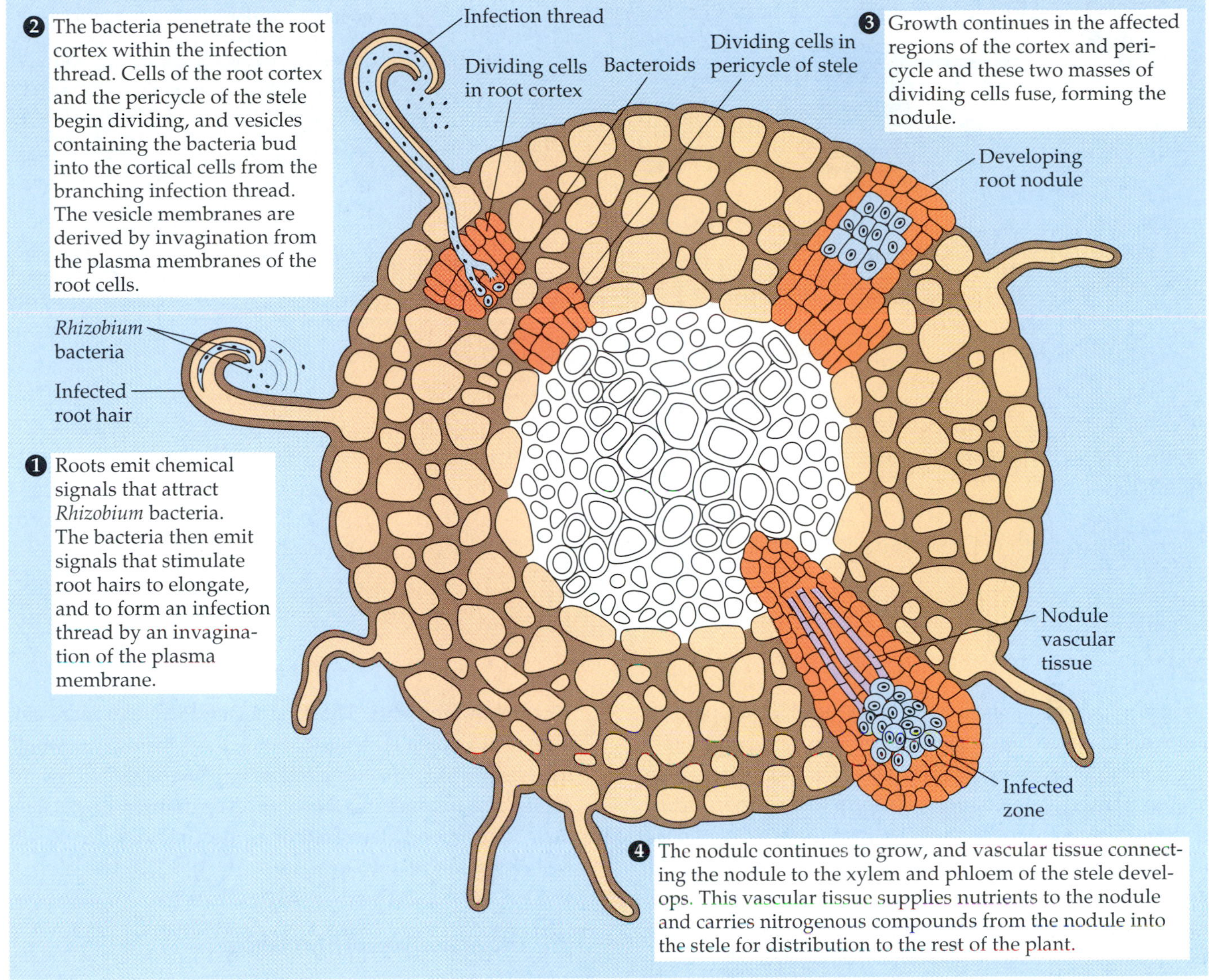

Infection thread
2 The bacteria penetrate the root cortex within the infection thread. Cells of the root cortex and the pericycle of the stele begin dividing, and vesicles containing the bacteria bud into the cortical cells from the branching infection thread. The vesicle membranes are derived by invagination from the plasma membranes of the root cells.
Dividing cells in root cortex
Bacteroids
Dividing cells in pericycle of stele
3 Growth continues in the affected regions of the cortex and pericycle and these two masses of dividing cells fuse, forming the nodule.
Developing root nodule
Rhizobium bacteria
Infected root hair
1 Roots emit chemical signals that attract Rhizobium bacteria. The bacteria then emit signals that stimulate root hairs to elongate, and to form an infection thread by an invagination of the plasma membrane.
Nodule vascular tissue
Infected zone
4 The nodule continues to grow, and vascular tissue connecting the nodule to the xylem and phloem of the stele develops. This vascular tissue supplies nutrients to the nodule and carries nitrogenous compounds from the nodule into the stele for distribution to the rest of the plant.

FIGURE 37.10 ▪ Development of a soybean root nodule. The coordinated activities of the legume and the *Rhizobium* bacteria depend on a chemical dialogue between the symbiotic partners.

between the bacteria and the root, with each partner responding to the chemical signals from the other by expressing certain genes whose products contribute to nodule formation (FIGURE 37.11, p. 724). The plant initiates the communication when its roots secrete molecules called flavenoids, which enter *Rhizobium* cells living in the vicinity of the roots. The specificity of this signal arises from variations in flavenoid structure, with a particular legume species secreting a type of flavenoid that only a certain *Rhizobium* species will detect and absorb. The plant's signal triggers the production of an answering molecule by the bacterium. Specifically, the plant's signal molecule activates a gene-regulating protein, which switches on a cluster of bacterial genes called *nod,* for "nodulation" genes. The products of these genes are enzymes that catalyze production of species-specific molecules called Nod factors. Secreted by the bacterial cells, the Nod factors return the "hello" and signal the root to form the infection thread the *Rhizobium* will enter, and to begin forming a new organ, the root nodule (see FIGURE 37.10). The plant's responses require activation of genes called early nodulin genes, probably by gene-regulating signal transduction pathways, such as those you learned about in Chapter 11.

Researchers have analyzed the molecular structure of the Nod factors for clues to these bacterial molecules' ability to influence genes within the cells of another species, the plant. At first they made the puzzling discovery that Nod factors are very similar to chitins, the main substances in the cell walls of fungi and the exoskeletons of arthropods (see Chapter 5). But now there is evidence that plants themselves produce chitinlike substances that probably function as growth regulators. This discovery leads to the hypothesis that Nod factors mimic certain plant growth regulators in stimulating the roots to grow new organs—nodules, in this case. Researchers have also learned that the plant genes that must be expressed for nodules to form

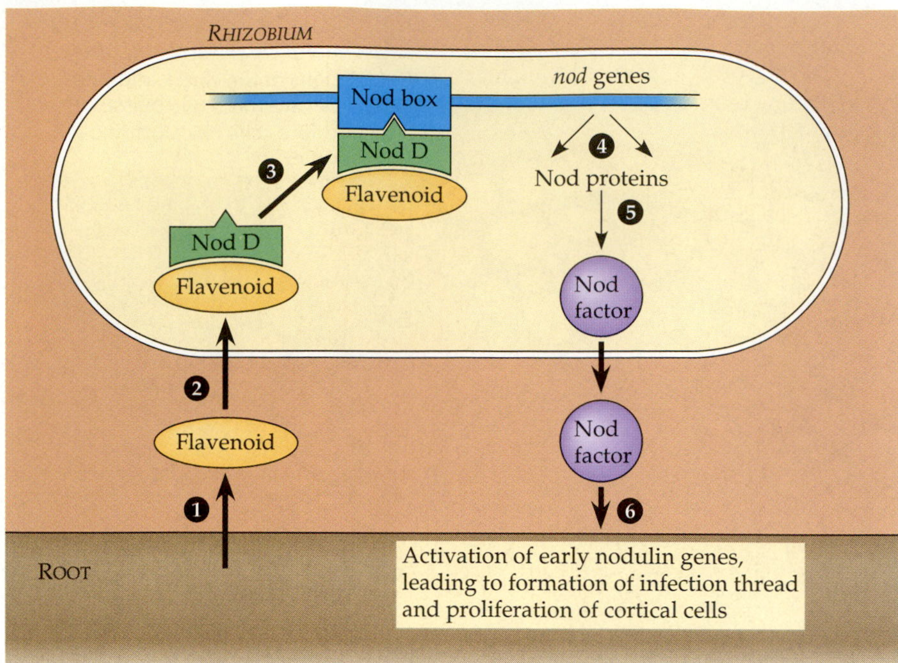

FIGURE 37.11 ▪ Molecular biology of root nodule formation. ① The root secretes a signal in the form of a specific flavenoid that is detected only by the plant's *Rhizobium* partner. ② The flavenoid activates a gene-regulator named Nod D. ③ Nod D, a transcription factor, then binds to a DNA region called the Nod box, activating transcription of the *nod* genes. ④ The products of the *nod* genes are enzymes for a metabolic pathway that ⑤ produces a chitinlike substance called a Nod factor. ⑥ The Nod factor functions as a specific signal between the *Rhizobium* and the root, triggering development of the infection thread and nodule.

are genes that also function in many other developmental processes in plants. This suggests that researchers might someday learn to induce *Rhizobium* uptake and nodule formation in crop plants that do not normally form such nitrogen-fixing symbiotic relationships. A more likely outcome of continued research, however, is that scientists will learn to manipulate the molecular biology of the root-*Rhizobium* relationships of legumes and other natural nodule formers to improve the efficiency of nitrogen fixation and protein production by crops.

Mycorrhizae are symbiotic associations of roots and fungi that enhance plant nutrition

Mycorrhizae ("fungus roots") are modified roots consisting of symbiotic associations of fungi and roots (see Chapters 31 and 36). The symbiosis is mutualistic. The fungus benefits from a hospitable environment and a steady supply of sugar donated by the host plant. In return, the fungus increases the surface area for water uptake and selectively absorbs phosphate and other minerals from the soil and supplies them to the plant. The fungi of mycorrhizae also secrete growth factors that stimulate roots to grow and branch. And the fungi produce antibiotics that may help protect the plant from pathogenic bacteria and pathogenic fungi in the soil.

Mycorrhizae are not oddities; they are formed by almost all plant species. In fact, this plant-fungus symbiosis might have been one of the evolutionary adaptations that made it possible for plants to colonize land in the first place; fossilized roots from some of the earliest plants include mycorrhizae. When terrestrial ecosystems were young, the soil was probably not very rich in nutrients. The fungi of mycorrhizae, more efficient at absorbing minerals than the roots themselves, would have helped nourish the pioneering plants. Even today, the plants that first become established on nutrient-poor soils, such as abandoned farmland or eroded hillsides, are usually well endowed with mycorrhizae.

The Two Main Types of Mycorrhizae

The modified roots formed from symbiosis of fungi and plants take two major forms: ectomycorrhizae and endomycorrhizae.

In **ectomycorrhizae**, the mycelium (mass of branching hyphae—see Chapter 31) forms a dense sheath, or mantle, over the surface of the root (FIGURE 37.12a). Hyphae extend from the mantle into the soil, greatly increasing the surface area for water and mineral absorption. Fungal hyphae also grow into the cortex of the root. These hyphae do not penetrate the root cells but form a network in the extracellular spaces that facilitates nutrient exchange between the fungus and plant. Compared to "uninfected" roots, ectomycorrhizae are generally thicker, shorter, and more branched. Ectomycorrhizae do not form root hairs, which would be superfluous given the extensive surface area of the fungal mycelium. Ectomycorrhizae are especially common in woody plants, including trees of the pine, spruce, oak, walnut, birch, willow, and eucalyptus families.

In contrast to ectomycorrhizae, endomycorrhizae do not have a dense mantle ensheathing the root (FIGURE 37.12b). It takes a microscope to see the fine fungal hyphae that extend

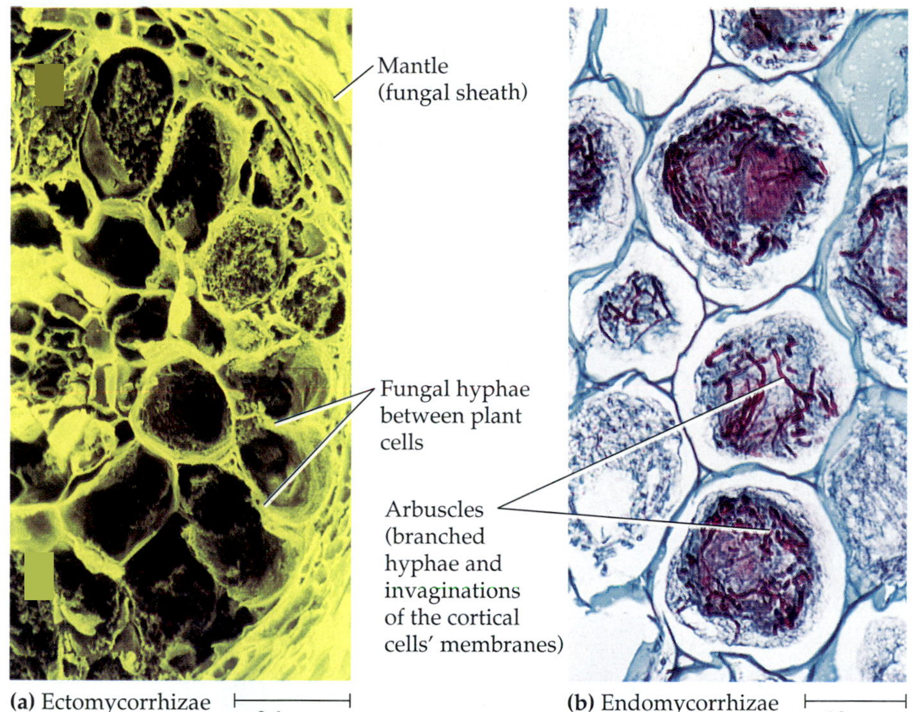

FIGURE 37.12 ▪ Mycorrhizae. (a) Ectomycorrhizae. The mantle of the fungal mycelium ensheathes this aspen root. Fungal hyphae extend from the mantle into the soil, absorbing water and minerals, especially phosphate. Hyphae also extend into the extracellular spaces of the root cortex, providing extensive surface area for nutrient exchange between the fungus and its host plant. The fungi of ectomycorrhizae are usually basidiomycetes, visible aboveground by the mushrooms they produce in the vicinity of the host tree (see Chapter 31). (Colorized SEM). **(b)** Endomycorrhizae. No mantle forms around the root, although microscopic hyphae of the fungus do extend into the soil. Within the root cortex the fungus makes extensive contact with the plant through branching of hyphae accommodated by invagination of the root cells' plasma membranes. In this micrograph of sectioned cortical cells, the arbuscles (knots of hyphae), which appear to reside in the cytoplasm, are actually outside the extensively invaginated plasma membrane. The arbuscles provide an enormous surface area for nutrient swapping between the symbionts. The fungal partners in endomycorrhizae are usually zygomycetes (see Chapter 31). (LM.)

from the root into the soil. Hyphae also extend inward (hence the term *endomycorrhizae*) by digesting small patches of the root cell walls. The hypha does not actually pierce the plasma membrane and enter the cytoplasm of the host cell, but instead grows into a tube formed by invagination of the root cell's membrane. The action is analogous to you poking a finger gently into a balloon; your finger is like the fungal hypha and the skin of the balloon is like the plant cell's membrane. Once they have penetrated in this way, some of the fungal hyphae become highly branched to form dense knotlike structures called arbuscles. These arbuscles are important sites of nutrient transfer between the fungus and the plant. To the unaided eye, endomycorrhizae look like "normal" roots complete with root hairs, but a microscope reveals a symbiotic relationship of enormous importance to plant nutrition. Endomycorrhizae, much more common than ectomycorrhizae, are found in over 90% of plant species, including important crop plants such as corn, wheat, and legumes.

Agricultural Importance of Mycorrhizae

Roots can be transformed into mycorrhizae only if they are exposed to the appropriate species of fungus. In most natural ecosystems these fungi are present in soil, and seedlings develop mycorrhizae. But if seeds are collected in one environment and planted in foreign soil, the plants may show signs of malnutrition resulting from the absence of the plants' mycorrhizal partners. Researchers observe similar results in experiments in which soil fungi are poisoned. Farmers and foresters are already applying the lessons of such research. For example, inoculating pine seeds with spores of mycorrhizal fungi promotes the formation of mycorrhizae by the seedlings. Pine seedlings so infected grow more vigorously than trees without the fungal association.

Mycorrhizae and root nodules may have an evolutionary relationship

There is growing evidence that the molecular biology of root nodule formation is closely related to mechanisms that evolved first in mycorrhizae. In 1997, researchers reported that the nodulin genes activated in the plant during the early stages of root nodule formation are the very same genes that are activated during the early development of endomycorrhizae. In fact, mutations in these early nodulin genes block development of both root nodules and mycorrhizae in legumes that form both structures. In addition, the signal-transduction pathways that relay messages from the microorganisms to the gene-regulating equipment involved in the development of both root nodules and mycorrhizae may share at least some components. For example, experimental application of plant hormones called cytokinins to root cells of legumes activates expression of the early nodulin genes even in the absence of the bacterial or fungal symbionts. (You will learn more about cytokinins and other plant hormones in Chapter 39.) "Infection" by either the bacteria or fungi causes the concentration of cytokinins in root tissue to increase naturally. These experiments suggest that the hormone is one of the links between the "I'm here" announcement of the microbes and the changes in gene expression that lead to structural modification of the roots. Even the chemical cues from the microbes may be similar. Recall that the Nod factors secreted by *Rhizobium* bacteria are related to chitins, the same compounds that make up the cell walls of fungi. A reasonable hypothesis is that root cells have a closely related family of specific receptors that detect their particular bacterial and fungal symbionts.

Mycorrhizae, as you have learned, evolved very early, probably over 400 million years ago in the earliest vascular plants. In contrast, the root nodules of legumes originated only 65 to 135 million years ago, during the early evolution of angiosperms. The recent experiments revealing common molecular mechanisms in roots' two major symbiotic relationships suggest that root nodule development is at least partly adapted from a signaling pathway that was already in place in mycorrhizae. This is one more example of how evolution can co-opt existing equipment for new functions.

NUTRITIONAL ADAPTATIONS: PARASITISM AND PREDATION BY PLANTS

Symbiotic nitrogen fixation and mycorrhizae underscore the relationship between plants and their environment, which includes the other organisms that interact with plants. We conclude this chapter by exploring predation and parasitism as two other types of plant adaptations that enhance nutrition through interactions with other organisms.

Parasitic plants extract nutrients from other plants

The mistletoe we find tacked above doorways during the holiday season lives in nature as a parasite on oaks and other trees. Mistletoe is photosynthetic, but it supplements its nutrition by using projections called haustoria to siphon xylem sap from the vascular tissue of the host tree. Some parasitic plants, such as dodder (FIGURE 37.13a), do not perform photosynthesis at all, drawing all their nutrients from other plants by tapping into the hosts' vascular tissue. In another version of parasitism, Indian pipe obtains its nutrients from trees indirectly via fungal hyphae of the host trees' mycorrhizae (FIGURE 37.13b).

Plants called epiphytes (Gr. *epi,* "upon," and *phyton,* "plant") are sometimes mistaken for parasites. An epiphyte is a plant that nourishes itself but grows on the surface of another plant, usually on the branches or trunks of trees. An epiphyte is anchored to its living substratum, but it absorbs water and minerals mostly from rain that falls on its leaves. Examples of epiphytes are staghorn ferns, some mosses, Spanish moss (actually an angiosperm), and many species of bromeliads and orchids.

(a)

(b)

FIGURE 37.13 ▪ Parasitic plants. (a) Dodder (the orange "strings") growing on a California pickleweed. The micrograph, a transverse section of a host stem supporting dodder, shows a haustorium (modified root) of the parasite tapping the host plant's vascular tissue for water and nutrients (LM). **(b)** Indian pipe is nutritionally bridged to its host tree by fungal hyphae extending from the host's mycorrhizae.

Carnivorous plants supplement their mineral nutrition by digesting animals

Living in acid bogs and other habitats where soil conditions are poor (especially in nitrogen) are plants that fortify themselves by occasionally feeding on animals. These carnivorous plants make their own carbohydrates by photosynthesis, but they obtain some of their nitrogen and minerals by killing and digesting insects. Various kinds of insect traps have evolved by the modification of leaves (FIGURE 37.14). The traps are usually equipped with glands that secrete digestive juices. Fortunately for the animal kingdom, this ironic turnabout is a relatively rare exception to the standard ecosystem dynamics of animals eating plants!

(a)

FIGURE 37.14 • Carnivorous plants. (a) The Venus flytrap is a modified leaf with two lobes that close together rapidly enough to capture an insect. Prey that enters the trap touches sensory hairs, initiating an electrical impulse that triggers closure of the trap. Movement of the trap is essentially a very rapid growth response in which cells in the outer region of each lobe accumulate water and enlarge. This changes the shape of the lobes, bringing their margins together. Glands in the trap then secrete digestive enzymes, and nutrients are later absorbed by the modified leaf. In spite of its name, the flytrap catches more ants and grasshoppers than it does flies. **(b)** Pitcher plants use a pitfall to capture insects. Insects slip into a long water-filled funnel. After the insect drowns, it is digested by enzymes secreted into the water.

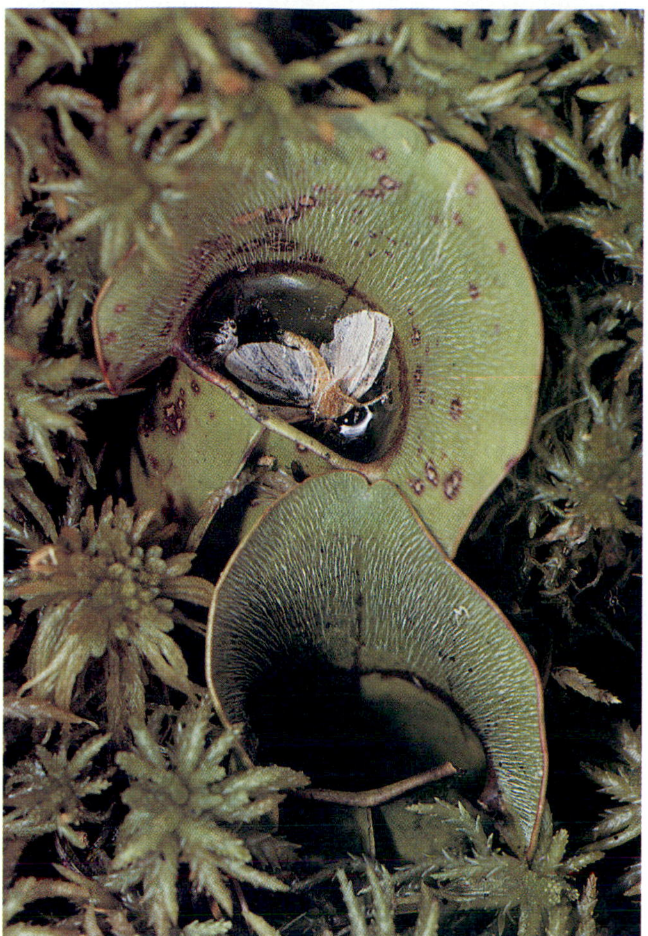

(b)

CHAPTER REVIEW

REVIEW OF KEY CONCEPTS

(with page numbers and key figures)

NUTRITIONAL REQUIREMENTS OF PLANTS

- **The chemical composition of plants provides clues to nutritional requirements (pp. 714–715, FIGURE 37.1)** Plants derive most of their organic mass from the CO_2 of air, but they also depend on soil nutrients in the form of water and minerals.
- **Plants require nine macronutrients and at least eight micronutrients (pp. 715–716, TABLE 37.1)** Macronutrients include C, H, O, N, and other major ingredients of organic compounds. Many micronutrients have catalytic functions as cofactors of enzymes.
- 37.1 **The symptoms of a mineral deficiency depend on the function and mobility of the element (pp. 716–717)** Deficiency of a mobile nutrient usually affects older organs more than younger ones; the reverse is true for nutrients that are less mobile within a plant.

THE ROLE OF SOIL IN PLANT NUTRITION

- 37.1 **Soil characteristics are key environmental factors in terrestrial ecosystems (pp. 717–718, FIGURES 37.5, 37.6)** Various sizes of particles derived from the breakdown of rock are found in soil, along with organic material (humus) in various stages of decomposition. Acids derived from roots contribute to plants' uptake of minerals when H^+ displaces mineral cations from clay particles.
- **Soil conservation is one step toward sustainable agriculture (pp. 718–720)** In contrast to natural ecosystems, agriculture depletes the mineral content of soil, taxes water reserves, and encourages erosion. The goal of soil conservation strategies is to minimize this damage.

THE SPECIAL CASE OF NITROGEN AS A PLANT NUTRIENT

- **The metabolism of soil bacteria makes nitrogen available to plants (pp. 720–721, FIGURE 37.8)** Nitrogen-fixing bacteria convert atmospheric N_2 to nitrogenous minerals that plants can absorb as a nitrogen source for organic synthesis.
- **Improving the protein yield of crops is a major goal of agricultural research (p. 721)** Such research addresses the most widespread form of human malnutrition: protein deficiency.

NUTRITIONAL ADAPTATIONS: SYMBIOSIS OF PLANTS AND SOIL MICROBES

- **Symbiotic nitrogen fixation results from intricate interactions between roots and bacteria (pp. 723–724, FIGURES 37.10, 37.11)** The development of nitrogen-fixing root nodules depends on chemical cross-talk between *Rhizobium* bacteria and root cells of their specific plant hosts. The bacteria of a nodule obtain sugar from the plant and supply the plant with fixed nitrogen.

- **Mycorrhizae are symbiotic associations of roots and fungi that enhance plant nutrition (pp. 724–725, FIGURE 37.12)** The fungal hyphae of both ectomycorrhizae and endomycorrhizae absorb water and minerals, which they supply to their plant hosts.
- **Mycorrhizae and root nodules may have an evolutionary relationship (pp. 725–726)** There is evidence that root nodule development depends on molecular mechanisms of signaling and root-cell responses that evolved first in mycorrhizae.

NUTRITIONAL ADAPTATIONS: PARASITISM AND PREDATION BY PLANTS

- **Parasitic plants extract nutrients from other plants (p. 726, FIGURE 37.13)** They do so either directly by tapping into the host's vascular tissue or indirectly via mycorrhizae.
- **Carnivorous plants supplement their mineral nutrition by digesting animals (p. 726, FIGURE 37.14)** This predation is most common in ecosystems with nutrient-poor soil.

SELF-QUIZ

1. Most of the mass of organic material of a plant comes from
   a. water
   b. carbon dioxide
   c. soil minerals
   d. atmospheric oxygen
   e. nitrogen

2. Micronutrients are needed in very small amounts because
   a. most of them are mobile in the plant
   b. most function as cofactors of enzymes
   c. most are supplied in large enough quantities in seeds
   d. they play only a minor role in the health of the plant
   e. only the growing regions of the plants require them

3. Two groups of tomatoes were grown under laboratory conditions, one with humus added to the soil and one a control without the humus. The leaves of the plants grown without humus were yellowish (less green) than those of the plants growing in humus-enriched soil. The best explanation for this difference is that
   a. the healthy plants used the food in the decomposing leaves of the humus for energy to make chlorophyll
   b. the humus made the soil more loosely packed, so the plants' roots would grow with less resistance
   c. the humus contained minerals such as magnesium and iron needed for the synthesis of chlorophyll
   d. the heat released by the decomposing leaves of the humus caused more rapid growth and chlorophyll synthesis
   e. the plants absorbed chlorophyll from the humus

4. We would expect the greatest difference in size and general appearance between two groups of plants of the same species, one group with mycorrhizae and one without, in an environment
   a. where nitrogen-fixing bacteria are abundant
   b. that has soil with poor drainage
   c. that has hot summers and cold winters
   d. in which the soil is relatively deficient in mineral nutrients
   e. that is near a body of water such as a pond or a river

5. A mineral deficiency is likely to affect older leaves more than younger leaves if
   a. the mineral is a micronutrient
   b. the mineral is very mobile within the plant
   c. the mineral is required for chlorophyll synthesis
   d. the deficiency persists for a long time
   e. the older leaves are in direct sunlight

6. Carnivorous adaptations of plants mainly compensate for soil that has a relatively low content of
   a. potassium
   b. nitrogen
   c. calcium
   d. water
   e. phosphate

7. Based on our retrospective view, the most reasonable conclusion to draw from van Helmont's famous experiment on the growth of a willow tree is that
   a. the tree increased in mass mainly by producing its own matter
   b. the increase in the mass of the tree could not be accounted for by the consumption of soil
   c. most of the increase in the mass of the tree was due to the uptake of O_2
   d. soil simply provides physical support for the tree without providing nutrients
   e. trees do not require water to grow

8. It is valid to consider water a plant nutrient because
   a. plants die without a water source
   b. cell elongation depends mainly on the osmotic absorption of water by cells
   c. hydrogen atoms from water molecules are incorporated into organic molecules
   d. transpiration depends on a continuous supply of water to leaves
   e. most of a plant's mass of organic compounds is derived from water

9. The specific relationship between a legume and its symbiotic *Rhizobium* species probably depends on
   a. each legume having a specific set of early nodulin genes
   b. each *Rhizobium* species having a form of nitrogenase that only works in the appropriate legume host
   c. each legume being found where the soil has only the *Rhizobium* specific to that legume
   d. specific recognition between the chemical signals and signal receptors of the *Rhizobium* and legume species
   e. destruction of all incompatible *Rhizobium* species by enzymes secreted from the legume's roots

10. Mycorrhizae enhance plant nutrition mainly by
    a. absorbing water and minerals through the fungal hyphae
    b. providing sugar to the root cells, which have no chloroplasts of their own
    c. converting atmospheric nitrogen to ammonia
    d. enabling the roots to parasitize neighboring plants
    e. stimulating the development of root hairs

CHALLENGE QUESTIONS

1. Compare and contrast legume root nodules and mycorrhizae as two symbiotic relationships of roots that enhance plant nutrition.
2. Explain why an acre of corn actually yields more total protein than an acre of soybeans. (*Hint:* Recall that nitrogen fixation requires large amounts of metabolic energy in the form of ATP.)
3. Researchers at the University of Sydney artificially induced root nodules on the roots of a wheat seedling and treated the roots with a chemical that softens cell walls. This enabled *Azospirillum*, a genus of nitrogen-fixing bacteria that normally live free in the soil, to infect the root cortex of seedlings. The nodules fix nitrogen. Design an experiment that uses radioactive N_2 to determine whether the nitrogen fixed in these nodules is incorporated into leaf proteins.

4. Choose five chemical elements and explain: (a) how each is assimilated from the environment by a plant; (b) how the element is used by the plant.

SCIENCE, TECHNOLOGY, AND SOCIETY

1. About 10% of U.S. cropland is irrigated. Agriculture is by far the biggest user of water in arid western states, including Colorado, Arizona, and California. The populations of these states are growing, and there is an ongoing conflict between cities and farm regions over water. To ensure water supplies for urban growth, cities are purchasing water rights from farmers. This is often the least expensive way for a city to obtain more water, and it is possible for some farmers to make more money selling water than growing crops. Discuss the possible consequences of this trend. Is this the best way to allocate water for all concerned? Why or why not?
2. Based on what you have learned in this chapter, identify some of the problems the world faces for future food production and outline some possible solutions. What economic and environmental costs are associated with each of these solutions?

FURTHER READING

Aldhous, P. "Ecologists Draft Plan to Dig in the Dirt." *Science*, September 9, 1994. Most soil organisms have not yet been identified.

Holmes, B. "Can Sustainable Farming Win the Battle of the Bottom Line?" *Science*, June 25, 1993. The relationship between environmental goals and economic realities.

Mestel, R. "Let's Make Nodules." *New Scientist*, January 11, 1997. Describes how molecular biologists are decoding the chemical communication between legumes and their *Rhizobium* partners.

"Reseeding the Green Revolution." *Science*, August 22, 1997. A special report on the relationship of agricultural research to human welfare.

Stone, J. "Little Bog Man." *Discover*, February 1990. A profile of a high school student who is one of the world's experts on carnivorous plants.

Strauss, E. "Communism in Trees Goes Underground." *Science News*, August 9, 1997. Discusses how mycorrhizal fungi shared between trees, even of different species, transfer nutrients between the plants.

WEB LINKS

Visit the special edition of *The Biology Place* for BIOLOGY, Fifth Edition, at **http://www.biology.com/campbell**. Go to Chapter 37 for online resources, including learning activities, practice exams, and links to the following web sites:

"Something to Grow On"
This site from the Cornell Cooperative Extension and Department of Floriculture and Ornamental Horticulture is an excellent introduction to plant nutrition.

"Soil Science Society of America"
The web site of the Soil Science Society of America contains news items, soil science links, and an extensive glossary of soil science terms.

"Soils and Substrates"
From the World Wide Web virtual library, "soils and substrates" contains a wealth of information about soils worldwide.

"The International Fertilizer Development Center (IFDC)"
The IFDC is a public, international, nonprofit organization dedicated to increasing agricultural productivity and food production in the tropics and subtropics through the appropriate use of plant nutrients in sustainable crop production systems. The site contains many links to plant nutrition resources worldwide.

PLANT REPRODUCTION AND DEVELOPMENT

It has been said that an oak is an acorn's way of making more acorns. Indeed, in a Darwinian view of life, the fitness of an organism is measured only by its ability to replace itself with healthy, fertile offspring. Consider the century plant (Agave) *in the photograph on this page. It lives for decades without flowering, and then one spring it grows a floral stalk as tall as a telephone pole. That season the plant produces seeds and then withers and dies, its food reserves, minerals, and water spent in the formation of its massive bloom. Although not all flowering plants are as completely consumed as the century plant in leaving offspring, most of their other functions can be interpreted, in the broadest Darwinian sense, as mechanisms contributing to propagation.*

This chapter focuses on the reproduction and development of flowering plants. (The life cycles of other plant and algal groups are covered in Chapters 28–30.) After comparing the sexual and asexual modes of reproduction, we will examine some of the cellular mechanisms responsible for plant development. The complex interactions of hormones and environmental cues that control these events are discussed in Chapter 39.

SEXUAL REPRODUCTION

Sporophyte and gametophyte generations alternate in the life cycles of plants: *a review*

38.1 In Chapter 30 you learned about the life cycle of a flowering plant from an evolutionary perspective. The life cycles of angiosperms and other plants are characterized by an **alternation of generations**, in which haploid (n) and diploid ($2n$) generations take turns producing each other (see FIGURE 29.2). The diploid plant, called the **sporophyte**, produces haploid spores by meiosis. Spores divide by mitosis, giving rise to multicellular male and female **gametophytes**, the haploid generation. Mitosis in the gametophytes produces gametes—sperm and eggs. Fertilization results in diploid zygotes, which divide by mitosis and form new sporophytes. FIGURE 38.1 follows the main stages of the angiosperm life cycle. In angiosperms, the sporophyte is the dominant generation in the sense that it is the conspicuous plant we see. Gametophytes became reduced during evolution to tiny structures totally contained within and dependent upon their sporophyte parents.

Male and female gametophytes develop within a flower's anthers and ovaries, respectively

Flowers develop from compressed shoots with four whorls of modified leaves separated by very short internodes. These four floral organs, in sequence from the outside to the inside of the flower, are the **sepals**, **petals**, **stamens**, and **carpels**.

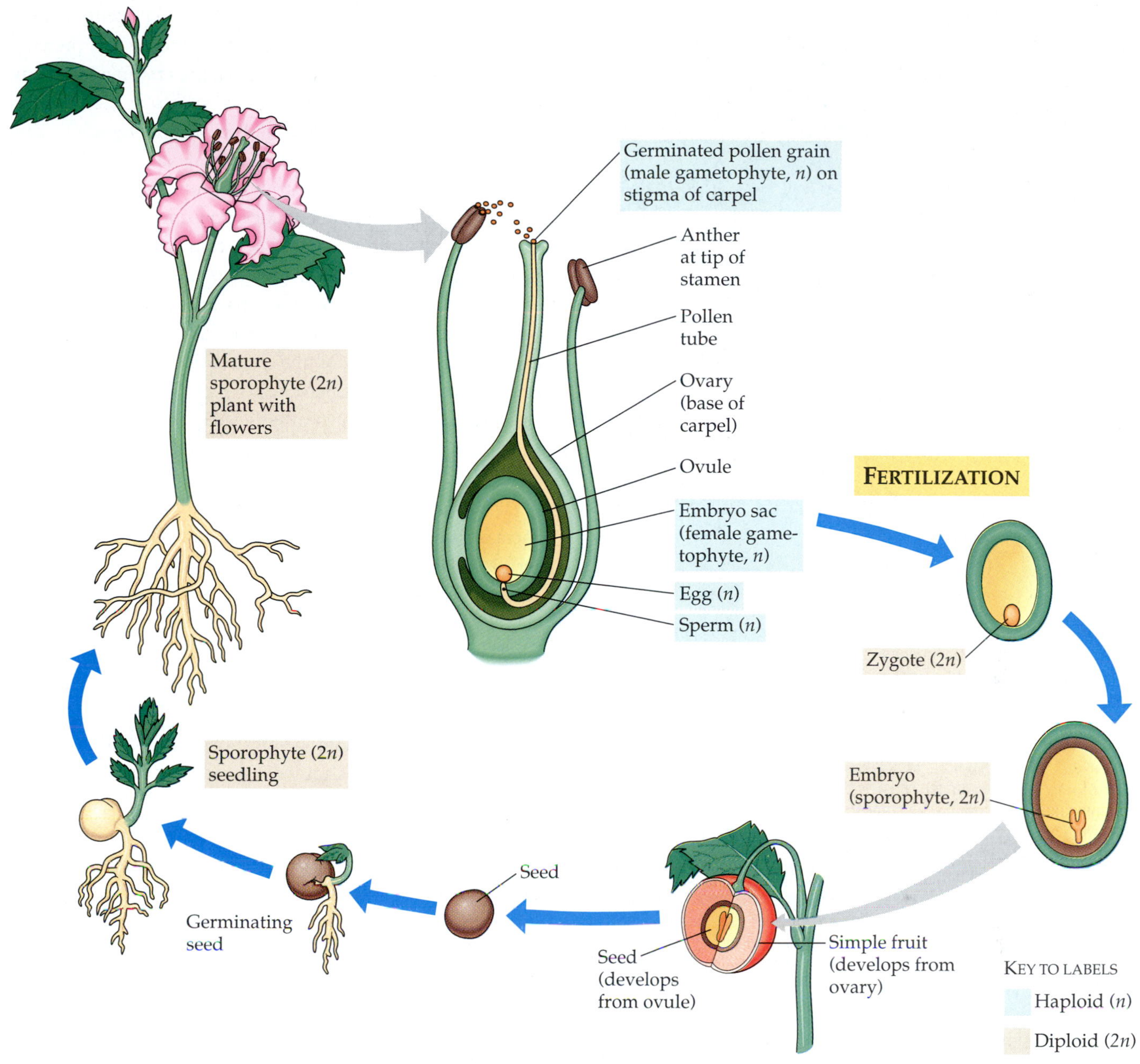

FIGURE 38.1 ▪ Overview of angiosperm life cycle. Within the ovary of a flower, the egg of an ovule is fertilized by a sperm cell released from a pollen tube. The egg is part of the embryo sac, the female gametophyte, and the sperm-bearing pollen grain is the male gametophyte. After fertilization, the ovule matures into a seed containing the embryo, and the ovary develops into a fruit, which aids in dispersal of the seed. In a suitable habitat the seed germinates, its embryo developing into a seedling.

38.1

Their structure and function are reviewed in FIGURE 38.2, p. 732. The stamens and carpels of flowers contain the sporangia, the chambers where male and female gametophytes, respectively, develop. The male gametophytes are sperm-containing pollen grains, which form within the chambers of anthers at the tips of stamens. The female gametophytes are egg-containing structures called embryo sacs. Embryo sacs develop inside structures called **ovules**, which are enclosed by the ovaries (the bases of carpels). Thus, stamens and carpels are the reproductive organs of flowers, while sepals and petals are nonreproductive organs.

Numerous floral variations evolved during the 130 million years of angiosperm history (see Chapter 30). In certain flowers, one or more of the four basic floral organs—sepals, petals, stamens, and carpels—has been eliminated. Plant biologists distinguish between **complete flowers**, those having all four organs, and **incomplete flowers**, those lacking one or more of the four floral parts. For example, most grasses have incomplete flowers lacking petals.

A flower equipped with both stamens and carpels is termed a **perfect flower**, even if it is incomplete because it lacks sepals or petals. **Imperfect flowers** are incomplete

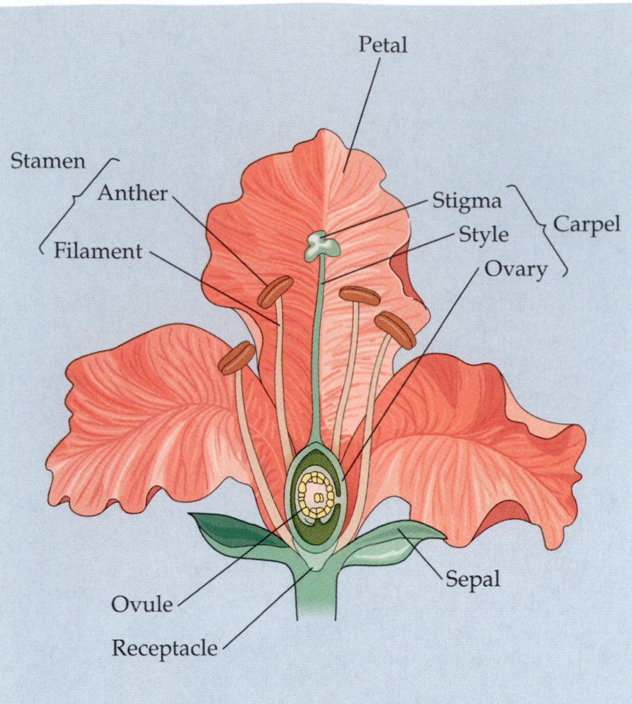

FIGURE 38.2 ▪ A review of an idealized flower. Sepals, petals, stamens, and carpels are arranged in four whorls, all attached to the receptacle at the end of a modified stem. Usually green, the sepals of most flowers have retained a more leaflike appearance than the other floral parts. Sepals enclose and protect a floral bud before it opens. Petals, generally more brightly colored than sepals, advertise the flower to insects and other pollinators. Stamens and carpels are the reproductive parts of flowers. Each stamen consists of a stalk called the filament and a terminal structure called the anther. Within the anther are chambers where the pollen grains (male gametophytes) develop. A carpel has a slender neck, the style, leading to an ovary located at the base of the carpel. Developing within the ovary are one or more ovules, in which egg-containing embryo sacs (female gametophytes) develop. This flower has a single carpel, but many plants have multiple carpels that are fused, forming an ovary with two or more chambers where ovules develop. At the tip of the carpel is a sticky stigma, which serves as a landing platform for pollen brought from other flowers by wind or animals.

flowers missing either stamens or carpels. These unisexual flowers are called staminate or carpellate, depending on which set of reproductive organs is present. If staminate and carpellate flowers are located on the same individual plant, then that plant species is said to be **monoecious** (Gr., "one house"). Corn is an example. The "ears" are derived from clusters of carpellate flowers. The tassels of a corn plant consist of staminate flowers. In contrast, a **dioecious** ("two houses") species has staminate flowers and carpellate flowers on separate plants, analogous to the presence of testes and ovaries on separate male and female animals. Date palms are dioecious. Because dates develop only on the carpellate (female) palms, commercial date growers plant mostly carpellate individuals. A few males (staminate plants) provide pollen enough for hundreds of females.

In addition to these differences based on the presence of floral organs, flowers have many variations in size, shape, and color (FIGURE 38.3). Much of this diversity represents adaptations of flowers to different pollinators. Indeed, the presence of animals in the environment was a key factor in angiosperm evolution (see Chapter 30).

Pollination occurs when pollen grains released from anthers and carried by wind or animals land on the sticky stigmas at the tips of carpels (though not necessarily on the same flower or plant). Pollen tubes grow down the carpels and discharge sperm into embryo sacs, resulting in the fertilization of eggs. Each zygote gives rise to an embryo, and as the embryo grows, the ovule develops into a seed. The entire ovary, meanwhile, develops into a fruit containing one or more seeds, depending on the species. Fruits, carried by wind or by animals, help disperse seeds some distance from their source plants. If deposited in sufficiently moist soil, seeds germinate; that is, their embryos start growing into seedlings, a new generation of flowering sporophytes.

The following sections describe in more detail the development of pollen and ovules, then trace how pollination leads to fertilization and the development of seeds and fruits. Keep in mind, however, that there are many variations in details in these processes, depending on species.

Development of Male Gametophyte (Pollen)

Within the sporangia (pollen sacs) of an anther, diploid cells called microsporocytes undergo meiosis, each forming four haploid **microspores** (FIGURE 38.4a, p. 734). Each microspore eventually divides once by mitosis and produces two cells, a generative cell and a tube cell. The two-celled structure is encased in a thick, resistant wall that becomes sculptured into an elaborate pattern unique to the particular plant species. Together, the two cells and their wall constitute a pollen grain, or immature male gametophyte.

Development of Female Gametophyte (Embryo Sac)

Ovules, each containing a sporangium, form within the chambers of the ovary. One cell in the sporangium of each ovule, the megasporocyte, grows and then goes through meiosis, producing four haploid **megaspores** (FIGURE 38.4b). The details of the next steps vary extensively, depending on the species. In many angiosperms, only one of the megaspores survives. This megaspore continues to grow, and its nucleus divides by mitosis three times, resulting in one large cell with eight haploid nuclei. Membranes then partition this mass into a multicellular structure called the **embryo sac**, which is the female gametophyte. At one end of the embryo sac are three cells: the egg cell, or female gamete, and two cells called synergids that flank the egg cell. At the opposite end are three antipodal cells. The other two nuclei, called polar nuclei, are not partitioned into separate cells but share the cytoplasm of the large central cell of the embryo sac. The ovule now

(a)

(b)

(c)

(d)

(e)

(f)

FIGURE 38.3 ▪ A few examples of floral diversity. **(a)** This lily flower is complete, meaning that sepals, petals, stamens, and carpels are all present. In the case of the lily, the sepals and petals look the same. **(b)** Lupines are examples of plants with inflorescences, clusters of flowers. **(c)** This sunflower represents a plant family characterized by composite flowers. What appears to be a single flower is actually a collection of hundreds. The central disk consists of tiny complete flowers. What appear to be petals ringing the central disk are actually imperfect flowers called ray flowers. **(d)** The diverse shapes, colors, and odors of flowers also reflect adaptations to different modes of pollination. For example, *Hibiscus* is pollinated by hummingbirds, which are attracted to the red color. (The yellow anthers look like they are attached to the carpel, but stamen filaments are actually fused as a tube that ensheathes the style of the carpel.) As the hummingbird probes deep in the flower for nectar, its feathers become dusted with pollen, while pollen carried from other *Hibiscus* flowers is transferred to the sticky stigma. **(e)** Corn is a monoecious species with inflorescences of staminate (male) and carpellate (female) flowers on the same individual plant. The staminate inflorescences are the tassels at the tip of a plant. An "ear" of corn is a collection of kernels (one-seeded fruits) that develops from an inflorescence of fertilized carpellate flowers. **(f)** *Sagittaria* is dioecious, its staminate (left) and carpellate (right) flowers on separate plants.

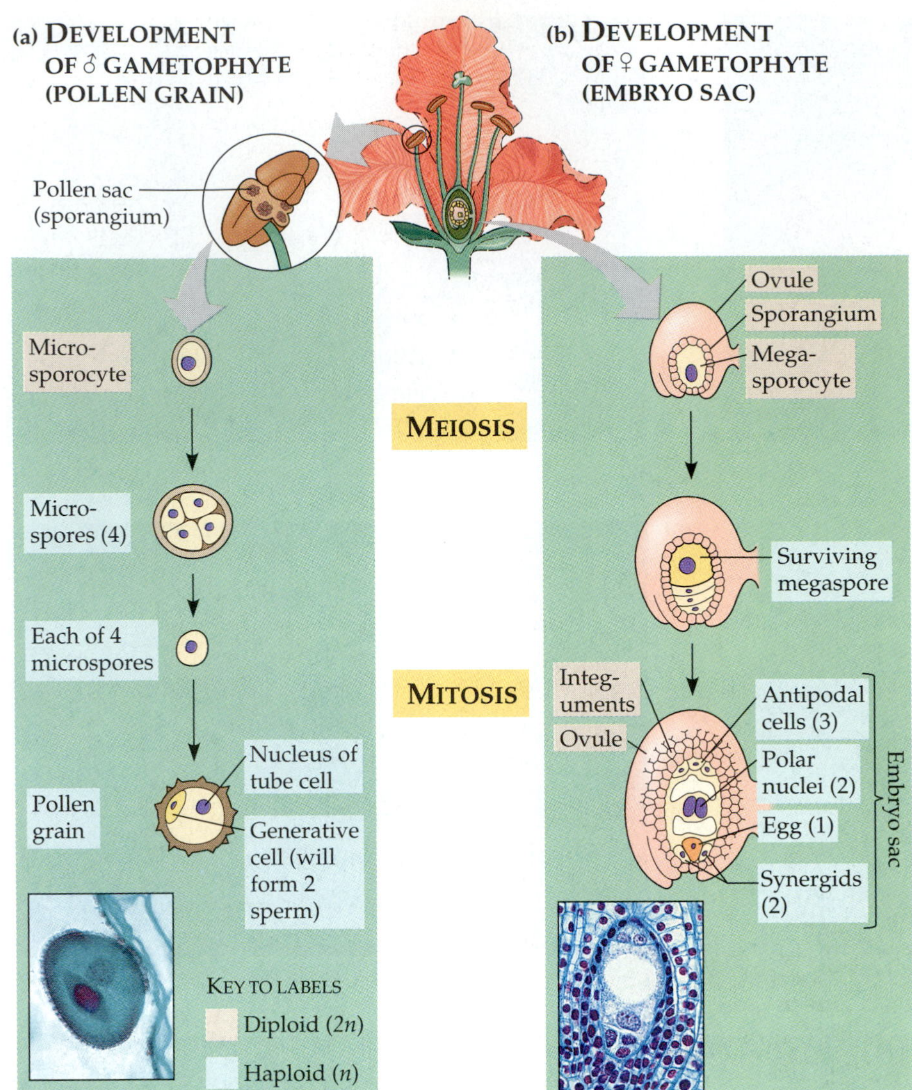

FIGURE 38.4 ▪ The development of angiosperm gametophytes (pollen and embryo sacs). (a) Pollen grains develop within the sporangia (pollen sacs) of anthers at the tips of stamens. Within each sac are numerous microsporocytes, the diploid cells that give rise to pollen. Meiosis forms four haploid microspores from each microsporocyte. Each microspore then undergoes a mitotic division, giving rise to a pollen grain, an immature male gametophyte consisting of a generative cell and a tube cell. The pollen grain has a thick, tough wall. This pollen becomes a *mature* male gametophyte when the generative cell divides to form two sperm. This usually occurs after a pollen grain lands on the stigma of a carpel and the pollen tube begins to grow (see FIGURE 38.1). **(b)** The embryo sac (female gametophyte) develops within an ovule, itself enclosed by the ovary at the base of a carpel. Within the ovule's sporangium is a large, diploid cell called the megasporocyte. This is the cell that gives rise to the embryo sac, but the details of that development and the structure of the resulting embryo sac differ extensively among plant species. In the case illustrated here, the megasporocyte divides by meiosis and gives rise to four haploid cells, but only one of these survives as the megaspore. (This contrasts with pollen formation, in which all four products of meiosis go on to form gametophytes.) Three mitotic divisions of the megaspore form the embryo sac, a multicellular female gametophyte. At one end of the embryo sac are the egg and two synergid cells. Three antipodal cells are located at the opposite end. The large central cell has two nuclei called polar nuclei. The ovule now consists of the embryo sac along with the surrounding integuments (protective tissues).

consists of the embryo sac (female gametophyte) and the integuments, protective layers of sporophyte tissue around the embryo sac.

Pollination brings male and female gametophytes together

For the egg to be fertilized, the male and female gametophytes must meet and unite their gametes. The first step is **pollination**, the placing of pollen onto the stigma of a carpel. Some plants, including grasses and many trees, use wind as a pollinating agent. They compensate for the randomness of this dispersal by releasing enormous quantities of tiny pollen grains. At certain times of the year the air is loaded with pollen, as anyone plagued with pollen allergies can attest. Many angiosperms, however, do not rely on the aimless wind to carry pollen but interact with animals that transfer pollen directly between flowers.

Some flowers self-pollinate, but the majority of angiosperms have mechanisms that make it difficult or impossible for a flower to pollinate itself. The various barriers that prevent self-pollination contribute to genetic variety by ensuring that sperm and eggs come from different parents. Dioecious plants, of course, cannot self-pollinate because they are unisexual, being either staminate or carpellate. In some plants with perfect flowers, the stamens and carpels mature at different times. Many flowers that are pollinated by animals are structurally arranged in such a way that it is unlikely the pollinator could transfer pollen from the anthers to the stigma of the same flower. Other flowers are **self-incompatible**; if a pollen grain from an anther happens to land on a stigma of a flower on the same plant, a biochemical block prevents the pollen from completing its development and fertilizing an egg.

Researchers are unraveling the molecular mechanisms of self-incompatibility

Self-incompatibility is the ability of flowers of some species to reject their own pollen and the pollen of closely related individuals. This plant response is analogous to the immune

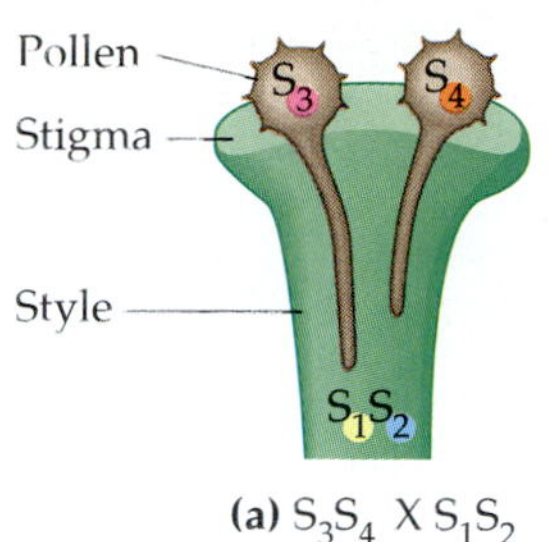

(a) S_3S_4 X S_1S_2

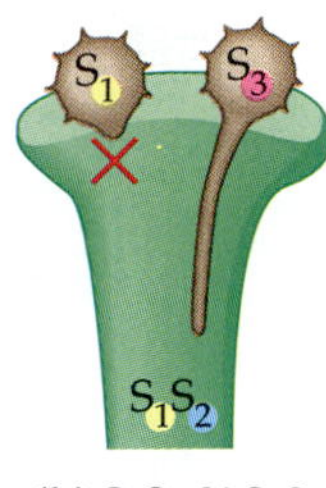

(b) S_1S_3 X S_1S_2

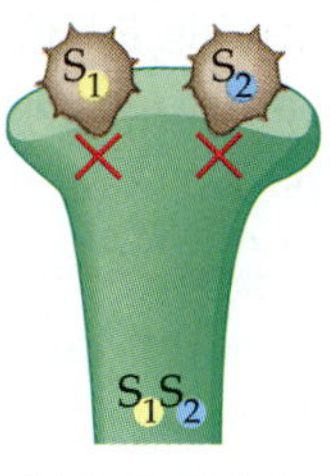

(c) S_1S_2 X S_1S_2

FIGURE 38.5 ▪ Genetic basis of self-incompatibility. In the gene pool of a plant population, there can be dozens of alleles of the S-gene. If a pollen grain has an allele that matches an allele of the stigma upon which it lands, then a pollen tube fails to grow. **(a)** In this cross, the pollen comes from a plant with an S_3S_4 genotype. (Segregation during meiosis results in half of the haploid pollen having the S_3 allele, the other half having the S_4 allele.) The pollen alleles do not match alleles of the stigma, and the pollen "germinates" (grows pollen tubes that can deliver sperm to ovules in the ovary at the base of the carpel). **(b)** In this cross, half of the pollen grains have an S_1 allele that matches an allele in the stigma, and those pollen grains fail to germinate. **(c)** In a cross between plants having the same S genotypes, none of the pollen can germinate.

response of animals, in the sense that both are based on the ability of organisms to distinguish the cells of "self" from those of "nonself." The key difference is that the animal immune system rejects nonself, as when the system mounts a defense against a pathogen or attempts to reject a transplanted organ. Self-incompatibility in plants, by contrast, is a rejection of self.

Recognition of "self" pollen is based on what are called S-genes, for self-incompatibility. In a particular plant population, as many as 50 different alleles can occur at the S-locus. If a pollen grain and the carpel's stigma upon which it lands have matching alleles at the S-locus, then the pollen grain fails to initiate or complete formation of a pollen tube, and thus no fertilization occurs. The pollen grain is haploid, and it will be recognized as "self" if its one S-allele matches either of the two S-alleles of the stigma, which is diploid (FIGURE 38.5).

Although self-incompatibility genes are all referred to as S-loci, such genes have actually evolved independently in various plant families. As a consequence, self-recognition blocks pollen tube growth by different molecular mechanisms. In some cases the block occurs in the pollen grain itself; this is called gametophytic self-incompatibility, because the pollen grain is a gametophyte. For example, in some members of the tobacco, rose, and bean (legume) families, self-recognition leads to enzymatic destruction of RNA within a rudimentary pollen tube. The RNA-hydrolyzing enzymes, or RNases, are present in the style of the carpel, but apparently they can enter a pollen tube and attack its RNA only if the pollen is of a "self" type. In other cases the block is a response by the cells of the carpel's stigma; this is called sporophytic self-incompatibility, because the carpel is part of the sporophyte. In members of the mustard family, for example, self-recognition activates a signal-transduction pathway in epidermal cells of the stigma that prevents germination of the pollen grain (FIGURE 38.6).

Basic research on mechanisms of self-incompatibility may lead to agricultural applications. Plant breeders sometimes hybridize between different varieties of a crop plant to combine the best traits of the varieties and counter the loss of vigor that can result from excessive inbreeding (see Chapter 14). Many agriculturally important plants are self-compatible. To maximize the number of hybrid seeds, breeders currently must prevent self-fertilization by laboriously removing the

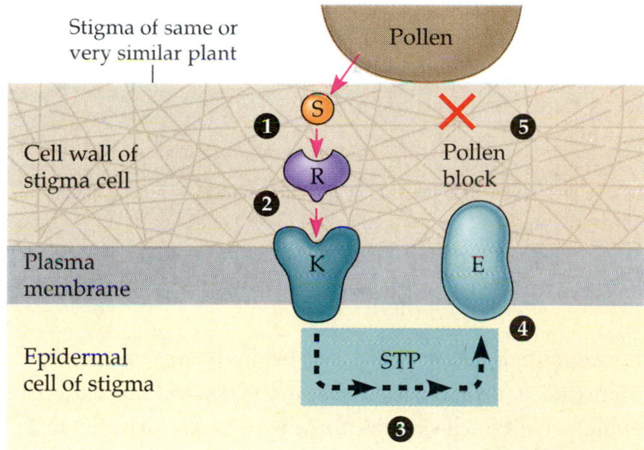

FIGURE 38.6 ▪ A possible mechanism of sporophytic self-incompatibility. In plants of the mustard family, there are at least two protein products of the S-locus. (The genes coding for these proteins are so tightly linked that they are inherited as though they were a single gene.) One of the protein products, labeled R in this diagram, is a receptor protein located in the extracellular matrix (wall) of the stigma epidermal cell. Another product of the S-locus is a protein kinase (K) embedded in the plasma membrane of the stigma cell. (A protein kinase, recall from Chapter 11, is an enzyme that activates other proteins by phosphorylating them.) How do these proteins function in blocking germination of "self" pollen? According to this model, ① pollen produces a chemical signal (labeled S here) that is specifically detected by the receptor protein of the same or closely related plant. ② This causes the receptor to bind to the membrane's kinase. ③ The kinase triggers a signal-transduction pathway (STP), which ④ activates one or more effector proteins (E). ⑤ The effectors block formation of a tube by the pollen grain. Recent research suggests that one of the effectors is an aquaporin, a membrane protein that functions in water transport (see Chapter 36). Uptake of additional water by the stigma cell might block pollen germination by preventing the stigma from hydrating the relatively dry pollen, a step required for growth of a pollen tube. This model of self-incompatibility will certainly evolve as researchers learn more about how certain plants reject their own pollen.

anthers from the parent plants that provide the seeds. Eventually, it may be possible instead to impose self-incompatibility on crop species that are normally self-compatible.

Double fertilization gives rise to the zygote and endosperm

A pollen grain produces a tube that extends down between the cells of the style toward the ovary (FIGURE 38.7). The generative cell divides by mitosis and forms two sperm, the male gametes. The pollen grain, now with a tube containing two sperm, is the mature male gametophyte. Directed by a chemical attractant, possibly calcium, the tip of the pollen tube enters the ovary, probes through the micropyle (a gap in the integuments), and discharges its two sperm within the embryo sac. One sperm fertilizes the egg to form the zygote. The other combines with the two polar nuclei to form a triploid ($3n$) nucleus in the center of the large central cell of the embryo sac. This large cell will give rise to the **endosperm**, a food-storing tissue. The union of two sperm cells with different cells of the embryo sac is termed **double fertilization**. Double fertilization ensures that the endosperm will develop only in ovules where the egg has been fertilized, thereby preventing angiosperms from squandering nutrients. After double fertilization, the ovule develops into a seed, and the ovary develops into a fruit enclosing the seed (or seeds, depending on the species).

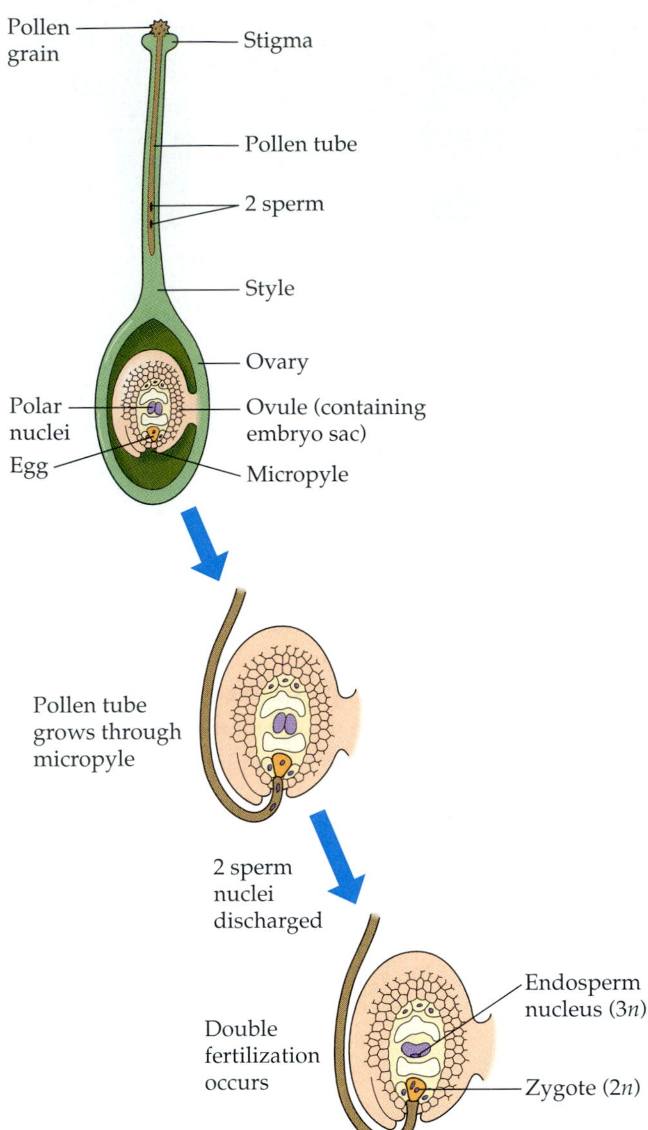

FIGURE 38.7 ▪ Growth of the pollen tube and double fertilization. After a pollen grain is carried by wind or an animal to the stigma, a long pollen tube begins growing down the style toward the ovary. The tube discharges two sperm into the embryo sac of an ovule. One sperm fertilizes the egg, forming the zygote. The other combines with the two polar nuclei of the embryo sac's large central cell and forms a triploid cell that will develop into a nutritive tissue called endosperm.

The ovule develops into a seed containing an embryo and a supply of nutrients

Endosperm Development

38.2 Endosperm development usually begins before embryo development. After double fertilization, the triploid nucleus of the ovule's central cell divides, forming a multinucleate "supercell" having a milky consistency. This mass, the endosperm, becomes multicellular and more solid when cytokinesis forms membranes and walls between the nuclei.

The endosperm is rich in nutrients, which it provides to the developing embryo. In most monocots, endosperm also stocks nutrients that can be used by the seedling after germination. In many dicots, the food reserves of the endosperm are exported to the cotyledons (seed leaves) before the seed completes its development, and consequently the mature seed lacks endosperm.

Embryo Development

The first mitotic division of the zygote is transverse, splitting the fertilized egg into a basal cell and a terminal cell (FIGURE 38.8). The terminal cell eventually gives rise to most of the embryo. The basal cell continues to divide transversely, producing a thread of cells called the suspensor, which will anchor the embryo to the ovule integuments and transfer nutrients to it from the parent plant and, in some plants, from the endosperm. Meanwhile, the terminal cell divides several times and forms a spherical proembryo attached to the suspensor. The cotyledons, or seed leaves, begin to form as bumps on the proembryo. A dicot, with its two cotyledons, is heart-shaped at this stage. Only one cotyledon develops in monocots.

Soon after the rudimentary cotyledons appear, the embryo elongates. Cradled between the cotyledons is the apical meristem of the embryonic shoot. At the opposite end of the embryo's axis, where the suspensor attaches, is the apex of the embryonic root, also with a meristem. (In some species, the basal cell gives rise to part of the root meristem.) After the seed germinates, the apical meristems at the tips of shoot and root will sustain primary growth as long as the plant lives. The three primary meristems—protoderm, ground meristem, and procambium—are also present in the embryo. Thus, development of the embryo establishes two features of plant form: the root-shoot axis, with meristems at opposite ends; and a radial pattern of protoderm, ground meristem, and procambium, set to give rise to the three tissue systems (dermal, ground, and vascular tissues).

As the embryo develops, the seed stockpiles proteins, oils, and starch and holds these nutrients in storage until the seed germinates.

38.2 **FIGURE 38.8 • The development of a dicot plant embryo.** By the time the ovule becomes a mature seed, the zygote has given rise to an embryonic plant with rudimentary organs.

Structure of the Mature Seed

During the last stages of its maturation, the seed dehydrates until its water content is only about 5% to 15% of its weight. The embryo stops growing and developing until the seed germinates. It is surrounded by its enlarged cotyledons, by endosperm, or by both. The embryo and its food supply are enclosed by a **seed coat** formed from the integuments of the ovule, the progenitor of the seed.

We can take a closer look at one type of dicot seed by splitting open the seed of a common bean (FIGURE 38.9a). At this stage, the embryo is an elongate structure, the embryonic axis,

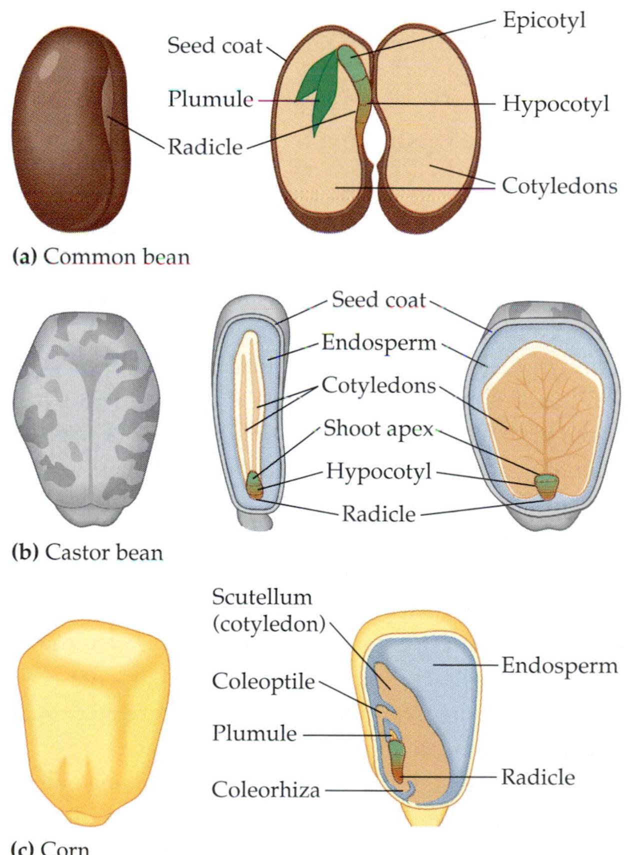

38.2 **FIGURE 38.9 • Seed structure. (a)** The fleshy cotyledons of the common garden bean, a dicot, store food that was absorbed from the endosperm when the seed developed. **(b)** The castor bean has membranous cotyledons that will absorb food from the endosperm when the seed germinates. **(c)** Corn, a monocot, has only one cotyledon (the scutellum of corn and other grasses). The rudimentary shoot is sheathed in a structure called the coleoptile.

attached to fleshy cotyledons. Below the point at which the cotyledons are attached, the embryonic axis is called the **hypocotyl** (Gr. *hypo*, "under"). The hypocotyl terminates in the **radicle**, or embryonic root. The portion of the embryonic axis above the cotyledons is the **epicotyl** (Gr. *epi*, "on" or "over"). At its tip is the plumule, consisting of the shoot tip with a pair of miniature leaves.

The cotyledons of the common bean are fleshy before the seed germinates because they absorbed food from the endosperm when the seed developed. However, the seeds of some dicots, such as castor beans, retain their food supply in the endosperm and have cotyledons that are very thin (FIGURE 38.9b). The cotyledons will absorb nutrients from the endosperm and transfer them to the embryo when the seed germinates.

The seed of a monocot has a single cotyledon (FIGURE 38.9c). Members of the grass family, including corn and wheat, have a specialized type of cotyledon called a **scutellum** (L. *scutella*, "small shield," a reference to the scutellum's shape). The scutellum is very thin, with a large surface area pressed against the endosperm, from which the scutellum absorbs nutrients during germination. The embryo of a grass seed is enclosed by a sheath consisting of a **coleorhiza**, which covers the root, and a **coleoptile**, which cloaks the embryonic shoot.

The ovary develops into a fruit adapted for seed dispersal

While the seeds develop from ovules, the ovary of the flower
38.2 develops into a **fruit**, which protects the enclosed seeds and aids in their dispersal by wind or animals. In some angiosperms, other floral parts also contribute to what we call a fruit in grocery store vernacular. The fleshy part of an apple, for instance, is derived mainly from the fusion of flower parts located at the base of the flower; only the core of the apple develops from the ovary.

The fruit begins to develop after pollination triggers hormonal changes that cause the ovary to grow (FIGURE 38.10). The wall of the ovary becomes the **pericarp**, the thickened wall of the fruit. As the ovary grows, the other parts of the flower wither away in many plants. This transformation of the flower parallels the development of the seeds. If a flower has not been pollinated, fruit usually does not develop, and the entire flower withers and falls away.

Fruits are classified into several types, depending on their developmental origin (TABLE 38.1). A fruit derived from a single ovary is called a **simple fruit**. A simple fruit may be fleshy, such as a cherry, or dry, such as a soybean pod. An **aggregate fruit**, such as a blackberry, results from a single flower that has several carpels. A **multiple fruit**, such as a pineapple, develops from an inflorescence, a group of flowers tightly clustered together. When the walls of the many ovaries start to thicken, they fuse together and become incorporated into one fruit.

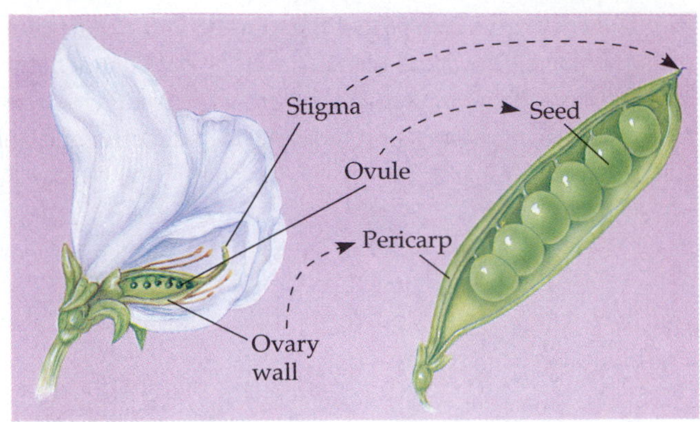

FIGURE 38.10 ▪ Relationship between a pea flower and a fruit (pea pod). At the same time that the ovules are transformed into seeds, other flower parts are transformed into a fruit we call a pod.

The fruit usually ripens about the time the seeds it contains are completing their development. For a dry fruit such as a soybean pod, ripening is little more than senescence (aging) of the fruit tissues, which allows the fruit to open and release

Table 38.1 ▪ Classification of Fleshy Fruits

TYPE OF FRUIT	FLORAL ORIGIN	EXAMPLE
Simple	Single ovary of one flower	Cherry
Aggregate	Many ovaries of one flower	Blackberry (unripe)
Multiple	Many ovaries of many flowers	Pineapple

the seeds. The ripening of a fleshy fruit is more elaborate, its steps guided by the complex interactions of hormones. In this case, ripening results in an edible fruit that serves as an enticement to the animals that help spread the seeds. The "pulp" of the fruit becomes softer as a result of enzymes digesting components of the cell walls. There is usually a color change from green to some other color such as red, orange, or yellow. The fruit becomes sweeter as organic acids or starch molecules are converted to sugar, which may reach a concentration of as much as 20% in a ripe fruit.

By selectively breeding plants, humans have capitalized on the production of edible fruits. The apples, oranges, and other fruits in grocery stores are exaggerated versions of much smaller natural varieties of fleshy fruits. However, the staple foods for humans are the dry, wind-dispersed fruits of grasses, which are harvested while still on the parent plant. The cereal grains of wheat, rice, corn, and other grasses are easily mistaken for seeds, but each is actually a fruit with a dry pericarp that adheres tightly to the seed coats of the single seed within.

Evolutionary adaptations of seed germination contribute to seedling survival

To many people the germination of a seed symbolizes the beginning of life, but in fact the seed already contains a miniature plant, complete with an embryonic root and shoot. At germination the plant does not begin life but rather resumes the growth and development that was temporarily suspended when the seed matured and its embryo became quiescent. Some seeds germinate as soon as they are in a suitable environment. Other seeds are dormant and will not germinate, even if sown in a favorable place, until a specific environmental cue causes them to break dormancy.

Seed Dormancy

"Dormant" means "sleeping" or "resting." Biologists use the term for a life cycle stage, such as a dormant seed, that has a very low metabolic rate and is not growing and developing.

Seed dormancy increases the chances that germination will occur at a time and place most advantageous to the seedling. Breaking dormancy generally requires certain environmental conditions. Seeds of desert plants, for instance, germinate only after a substantial rainfall. If they were to germinate after a modest drizzle, the soil might soon be too dry to support the seedlings. Where natural fires are common, many seeds require intense heat to break dormancy; seedlings are therefore most abundant after fire has cleared away competing vegetation. Where winters are harsh, seeds may require extended exposure to cold; seeds sown during summer or fall do not germinate until the following spring. This assures a long growth season before the *next* winter. Very small seeds, such as those of some lettuce varieties, require light for germination and will break dormancy only if they are buried shallow enough for the seedlings to poke through the soil surface. Some seeds have coats that must be weakened by chemical attack as they pass through an animal's digestive tract and thus are likely to be carried some distance before germinating.

The length of time a dormant seed remains viable and capable of germinating varies from a few days to decades or even longer, depending on the species and environmental conditions. Most seeds are durable enough to last a year or two until conditions are favorable for germinating. Thus, the soil has a pool of ungerminated seeds that may have accumulated for several years. This is one reason vegetation can come back so rapidly after a fire, drought, flood, or some other environmental disruption.

From Seed to Seedling

Germination of seeds depends on **imbibition**, the absorption
38.2 of water due to the low water potential of the dry seed. Imbibing water causes the seed to expand and rupture its coat and also triggers metabolic changes in the embryo that cause it to resume growth (FIGURE 38.11). Enzymes begin digesting the storage materials of the endosperm or cotyledons, and the

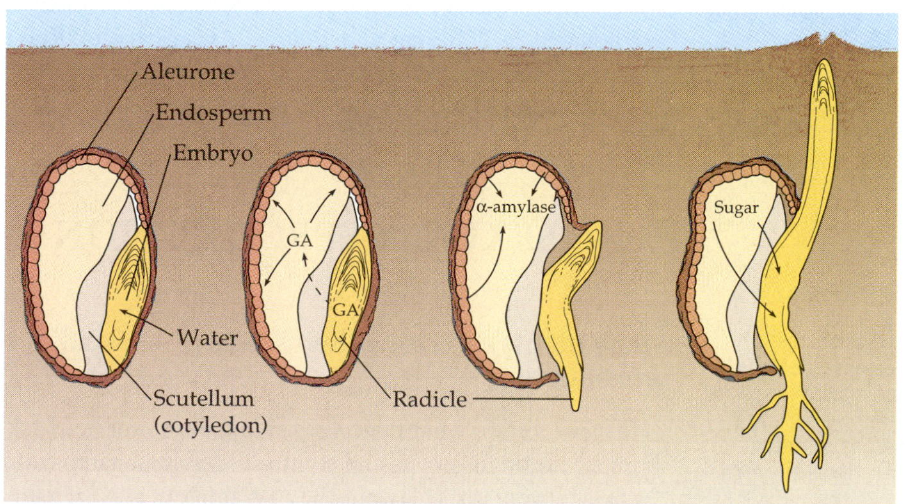

FIGURE 38.11 ▪ Mobilization of nutrients during the germination of a barley seed. After the seed imbibes water, the embryo releases hormones called gibberellins (GA) as signals to the aleurone, the thin outer layer of the endosperm. The aleurone responds by synthesizing and secreting digestive enzymes that hydrolyze stored foods in the endosperm, producing small, soluble molecules. One example is α-amylase, an enzyme that hydrolyzes starch. (A similar enzyme in our saliva helps us digest bread and other foods made from the starchy endosperm of ungerminated seeds.) Sugars and other nutrients absorbed from the endosperm by the scutellum (cotyledon) are consumed during growth of the embryo into a seedling.

nutrients are transferred to the growing regions of the embryo. This mobilization of food reserves has been studied most extensively in the seeds of barley and other grasses.

The first organ to emerge from the germinating seed is the radicle, the embryonic root. Next, the shoot tip must break through the soil surface. In garden beans and many other dicots, a hook forms in the hypocotyl, and growth pushes the hook aboveground (FIGURE 38.12a). Stimulated by light, the hypocotyl straightens, raising the cotyledons and epicotyl. Thus, the delicate shoot apex and bulky cotyledons are pulled aboveground, rather than being pushed tip-first through the abrasive soil. The epicotyl now spreads its first foliage leaves, which expand, become green, and begin making food by photosynthesis. The cotyledons shrivel and fall away from the seedling, their food reserves having been consumed by the germinating embryo.

Light seems to be the main cue that tells the seedling it has broken ground. We can trick a bean seedling into behaving as though it is still buried by germinating the seed in darkness. The unilluminated seedling extends an exaggerated hypocotyl with a hook at its tip, and the foliage leaves fail to turn green. After it exhausts its food reserves, the spindly seedling stops growing and dies.

Peas, although in the same family as beans, have a different style of germinating (FIGURE 38.12b). A hook forms in the epicotyl rather than the hypocotyl, and the shoot tip is lifted gently out of the soil by elongation of the epicotyl and straightening of the hook. Pea cotyledons, unlike those of beans, remain behind underground.

Corn and other grasses, which are monocots, use yet a different method for breaking ground when they germinate (FIGURE 38.12c). The coleoptile, the sheath enclosing and protecting the embryonic shoot, pushes upward through the soil and into the air. The shoot tip then grows straight up through the tunnel provided by the tubular coleoptile.

Germination of a plant seed, like the birth or hatching of an animal, is a critical stage in the life cycle. The tough seed gives rise to a fragile seedling that will be exposed to predators, parasites, wind, and other hazards. In the wild, only a small fraction of seedlings endure long enough to become parents themselves. Production of enormous numbers of seeds and fruits compensates for the odds against individual survival and gives natural selection ample material to screen for the most successful genetic combinations. However, this is a very expensive means of reproduction in terms of the resources consumed in flowering and fruiting. Asexual reproduction, generally simpler and less hazardous for offspring than sexual reproduction, is an alternative means of plant propagation.

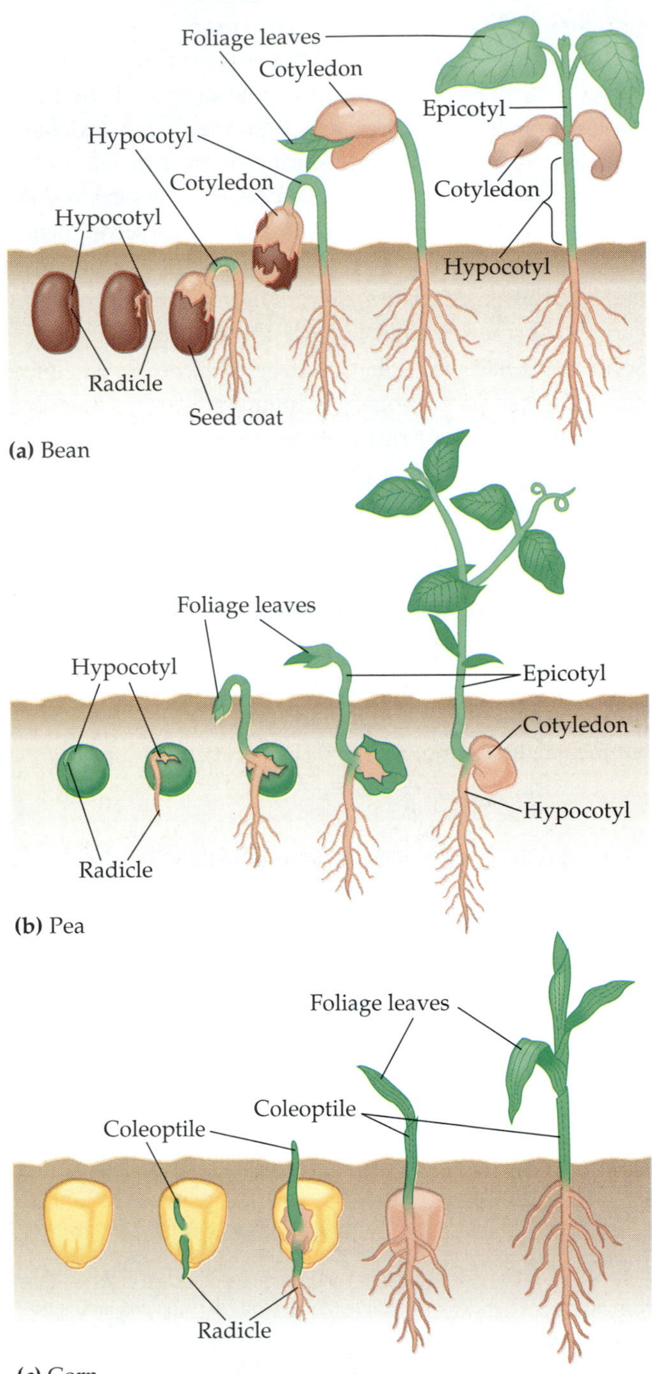

FIGURE 38.12 ▪ Seed germination. The radicle, the root of the embryo, emerges from the seed first. Then the shoot breaks the soil surface by one of the following mechanisms. **(a)** In beans, straightening of a hook in the hypocotyl pulls the shoot and cotyledons from the soil. **(b)** In peas, the hook is above the cotyledons on the epicotyl, and the cotyledons remain underground. **(c)** In corn and other grasses, the shoot grows straight up through the tube of the coleoptile.

38.2

ASEXUAL REPRODUCTION

Many plants can clone themselves by asexual reproduction

Imagine some of your fingers separating from your body, taking up life on their own, and eventually developing into entire copies of yourself. This would be an example of asexual reproduction, offspring derived from a single parent without genetic recombination. The result would be a clone, a population of asexually produced, genetically identical organisms. Some animals *can* reproduce asexually (though not humans, of course). And many plant species clone themselves by asexual reproduction, also called **vegetative reproduction**.

Vegetative reproduction is an extension of the capacity of plants for indeterminate growth. Plants, remember, have meristematic tissues of dividing, undifferentiated cells that can sustain or renew growth indefinitely. In addition, parenchyma cells throughout the plant can divide and differentiate into the various types of specialized cells, enabling plants to regenerate lost parts. Detached fragments of some plants can develop into whole offspring; a severed stem, for instance, may develop adventitious roots and become a whole plant. **Fragmentation**, the separation of a parent plant into parts that re-form whole plants, is one of the most common modes of vegetative reproduction (FIGURE 38.13a). A variation of this process occurs in some species of dicots, in which the root system of a single parent gives rise to many adventitious shoots that become separate shoot systems. The result is a clone formed by asexual reproduction from one parent (FIGURE 38.13b). Such asexual propagation has produced the oldest of all known plant clones, a ring of creosote bushes in the Mojave Desert of California, believed to be at least 12,000 years old.

An entirely different mechanism of asexual reproduction has evolved in dandelions and some other plants, which produce seeds without their flowers being fertilized. This asexual production of seeds is called **apomixis**. A diploid cell in the ovule gives rise to the embryo, and the ovules mature into seeds, which in the dandelion are dispersed by windblown fruit. Thus, though these plants clone themselves by an asexual process, they also have the advantage of seed dispersal, an adaptation usually associated with sexual reproduction of plants.

(a)

(b)

FIGURE 38.13 • Natural mechanisms of vegetative reproduction. **(a)** *Kalanchoe* is known as the maternity plant because of the numerous plantlets it produces along its leaf margins. The asexually produced plantlets fragment from their parent and become independent plants. **(b)** Some aspen groves, such as those shown here, are actually clones of thousands of trees descended by asexual reproduction from the root system of one parent. Notice that genetic differences among the clones result in different timing for the development of fall color and the loss of leaves.

Vegetative propagation of plants is common in agriculture

With the objective of improving crops, orchards, and ornamental plants, humans have devised various methods for propagating plants by vegetative reproduction. Most of these methods are based on the ability of plants to form adventitious roots or shoots.

Clones from Cuttings

Most houseplants, woody ornamentals, and orchard trees are asexually reproduced from plant fragments called cuttings. In some cases, shoot or stem cuttings are used. At the cut end of the shoot a mass of dividing, undifferentiated cells called a **callus** forms, and then adventitious roots develop from the callus. If the shoot fragment includes a node, then adventitious roots form without a callus stage. Some plants, including African violets, can be propagated from single leaves rather than stems. For still other plants, cuttings are taken from specialized storage stems. For example, a potato can be cut up into several pieces, each with a vegetative bud, or "eye," that regenerates a whole plant.

In a modification of vegetative reproduction from cuttings, a twig or bud from one plant can be grafted onto a plant of a closely related species or a different variety of the same species. Grafting makes it possible to combine the best qualities of different species or varieties into a single plant. The graft is usually done when the plant is young. The plant that provides the root system is called the **stock**; the twig grafted onto the stock is referred to as the **scion**. For example, scions from French varieties of vines that produce superior wine grapes are grafted onto root stock of American varieties, which are more resistant to certain diseases. The quality of the fruit, determined by the genes of the scion, is not diminished by the genetic makeup of the stock. In some cases of grafting, however, the stock can alter the characteristics of the shoot system that develops from the scion. For example, dwarf fruit trees are made by grafting normal twigs onto dwarf stock varieties that retard the vegetative growth of the shoot system. Since seeds are produced by the part of the plant derived from the scion, they would give rise to plants of the scion species if planted.

Test-Tube Cloning and Related Techniques

Agritechnologists have adopted test-tube methods to create and clone novel plant varieties. It is possible to grow whole plants by culturing small explants (pieces of tissue cut from the parent), or even single parenchyma cells, on an artificial medium containing nutrients and hormones (FIGURE 38.14). The cultured cells divide and form an undifferentiated callus. When the hormonal balance is manipulated in the culture medium, the callus can sprout shoots and roots with fully differentiated cells. The test-tube plantlets can then be transferred to soil, where they continue their growth. A single plant can be cloned into thousands of copies by subdividing calluses as they grow. This method is used for propagating orchids and also for cloning pine trees that deposit wood at unusually fast rates.

Plant tissue culture also facilitates genetic engineering in plants. Most techniques for the introduction of foreign genes into plants require the use of small pieces of plant tissue or single plant cells as the starting material. Test-tube culture makes it possible to regenerate genetically altered plants from a single plant cell into which the foreign DNA has been incorporated. For example, researchers have used recombinant DNA technology to transfer a gene for bean protein into cultured cells from a sunflower plant. The experiment improved the protein quality of sunflower seeds harvested from the transgenic plants. Firing DNA-coated pellets from a gun is one method researchers use to insert foreign DNA into plant cells (FIGURE 38.15). (The potential agricultural impact of genetic engineering is discussed in more detail in Chapter 20.)

A technique known as **protoplast fusion** is being coupled with tissue culture methods to actually invent new plant varieties that can be cloned. Protoplasts are plant cells that have had their cell walls removed. Before they are cultured, the protoplasts can be screened for mutations that may improve the agricultural value of the plant. It is also possible in some cases to fuse two protoplasts from different plant species that would otherwise be sexually incompatible, and then culture the hybrid protoplasts. Each of the many protoplasts can regenerate a wall and eventually form a hybrid plantlet. One success of this method has been a hybrid between a potato and a wild relative called black nightshade. The nightshade is resistant to an herbicide that is commonly used to kill weeds. The hybrids are also resistant, and this makes it possible to "weed" a potato field with the herbicide without killing the potato plants.

Benefits and Risks of Monoculture

By conscious effort, genetic variability in many crops has been virtually eliminated. For grains and other crops grown from seed, plant breeders have selected for varieties that self-pollinate. Whenever practical, vegetative reproduction is used to clone exceptional plants. Genetic uniformity ensures that all plants in a field grow at the same rate, fruit on all the trees in an orchard ripens in unison, and yields at harvest time are dependable. There is no doubt that **monoculture**, the cultivation of large areas of land with a single plant variety, has helped farmers feed human populations. But modern farms are very fragile ecosystems: Where there is little genetic variability, there is also little adaptability.

A clone is a kind of superorganism; in terms of natural selection, it is one genetic individual. What is good for one

(a)

(b)

FIGURE 38.14 ▪ Test-tube cloning of carrots. (a) Just a few parenchyma cells from a carrot gave rise to this callus, a mass of undifferentiated cells. **(b)** The callus differentiates into an entire plant, with leaves, stems, and roots. (see FIGURE 21.5)

(a)

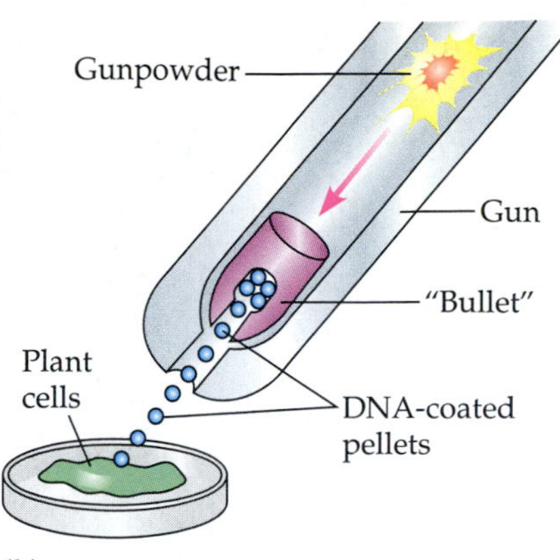

(b)

FIGURE 38.15 ▪ A DNA gun. (a) This researcher is preparing to use a modified .22-caliber gun to shoot foreign DNA into cultured plant cells. **(b)** The gun fires a plastic bullet loaded with tiny metallic pellets coated with DNA. (A different type of DNA gun uses a burst of gas rather than an explosion to propel pellets.) When a plate stops the bullet shell at the end of the gun, the pellets continue toward the cellular targets. (This drawing represents the pellets as much larger than they actually are, compared to the size of the gun and petri dish.) The projectiles penetrate cell walls and membranes, introducing foreign DNA to the nuclei of some cells. A cell that integrates this DNA into its genome can be cultured to produce a plantlet, and it is then possible to clone the transgenic plant.

plant is good for all, and what is bad for one threatens the entire clone. For example, monoculture was a major factor in the nineteenth-century Irish famine, which was caused by potato blight (a water mold). Spanish explorers brought potatoes to Europe from South America in the late sixteenth century. Europeans used vegetative reproduction to grow the "new" food, which became the staple of the Irish diet. When the blight hit almost two centuries later, the potato plants, which were genetically quite uniform, were all susceptible to the parasite. Plant scientists fear that some plant disease could again devastate thousands of acres of an important monoculture. Plant breeders have responded by maintaining "gene banks," where they store seeds of many plant varieties that can be used to breed new hybrids.

Sexual and asexual reproduction are complementary in the life histories of many plants: *a review*

The dilemma of monoculture provides us with insight about sexual and asexual reproduction in the wild. Many plants are capable of both modes of reproduction, and each offers advantages in certain situations. Sex generates variation in a population, an asset in an environment where evolving pathogens and other variables affect survival and reproductive success. An additional benefit of sexual reproduction in plants is the seed, which can disperse to new locations and can also wait to grow until hostile environmental conditions have improved.

An advantage of asexual reproduction is that a plant well suited to a particular environment can clone many copies of itself rapidly. Moreover, the offspring of vegetative reproduction, usually mature fragments of the parent plant, are not as frail as the seedlings produced by sexual reproduction. A sprawling clone of prairie grass may cover an area so thoroughly that seedlings of the same or other species have little chance of competing. But in the soil is a pool of seeds, waiting in the wings for some cue to germinate. After a fire, drought, or some other disturbance clears patches of the turf, seedlings can finally get a foothold when conditions improve. The seedlings are unequal in their traits, for their genotypes are products of the sexual recombination of genes. A new competition ensues, in which certain plants excel and reproduce themselves asexually. Both modes of reproduction, sexual and asexual, have had featured roles in the evolutionary adaptation of plant populations to their environments.

CELLULAR MECHANISMS OF PLANT DEVELOPMENT

Reproduction and development are closely related topics. Whether a plant arises from a sexually produced zygote or by vegetative reproduction, its transformation into a whole new individual depends on mechanisms that shape organs such as leaves and roots and generate specific patterns of specialized cells and tissues within those organs. **Development** is the sum of all of the changes that progressively elaborate an organism's body. These last sections of the chapter present what researchers are learning about the cellular mechanisms responsible for plant development. This builds on the general concepts of development you learned in Chapter 21.

Growth, morphogenesis, and differentiation produce the plant body: *an overview of plant development*

A plant zygote is a single cell that bears no resemblance to the organism it becomes. Three overlapping developmental processes transform the fertilized egg into a plant: growth, morphogenesis, and cellular differentiation.

Growth, an irreversible increase in size, results from cell division and cell enlargement. By a series of mitotic divisions, the zygote gives rise to the multicellular embryo within a seed.

After germination, mitosis resumes, concentrated mostly in the apical meristems near the tips of roots and shoots. But it is the enlargement of these newly made cells that accounts for most of the actual increase in the size of a plant.

If development were simply a matter of growth, then the zygote would give rise to an expanding ball of cells. In reality, growth is accompanied by **morphogenesis**, the development of form. The embryo encased in a seed has cotyledons and a rudimentary root and shoot, products of morphogenetic mechanisms that begin operating with the first division of the zygote. After the seed germinates, morphogenesis continues to shape the root and shoot systems of the growing plant (FIGURE 38.16). For example, morphogenesis at the shoot tip establishes the shapes of leaves and other morphological features.

A fundamental difference between plants and animals is reflected in their morphogenesis. Most animals *move* through their environments; plants, in contrast, *grow* through their environments. The indeterminate growth of a plant, as you have learned, is a function of regions at its shoot and root tips that remain embryonic for the life of the plant. These regions are centers not only of continuing growth, but of continuing morphogenesis as well.

Shape alone, of course, does not enable a plant organ to perform its various functions. Each organ—a leaf, for example—has a diversity of cell types specialized for certain functions and fixed in certain locations. For example, the guard cells that border stomata differ markedly in structure and function from the surrounding cells of the epidermis. Another example is the development of xylem and phloem from the vascular cambium during secondary growth. Recall from Chapter 35 that the cambium gives rise to xylem on the inside and phloem on the outside. This acquisition of a cell's specific structural and functional features is called **cellular differentiation** (see Chapter 21).

Although we have dissected plant development into growth, morphogenesis, and differentiation, it is important to realize that these processes occur in concert as the plant develops. Their integration will become apparent as we take a closer look at the cellular mechanisms of plant development.

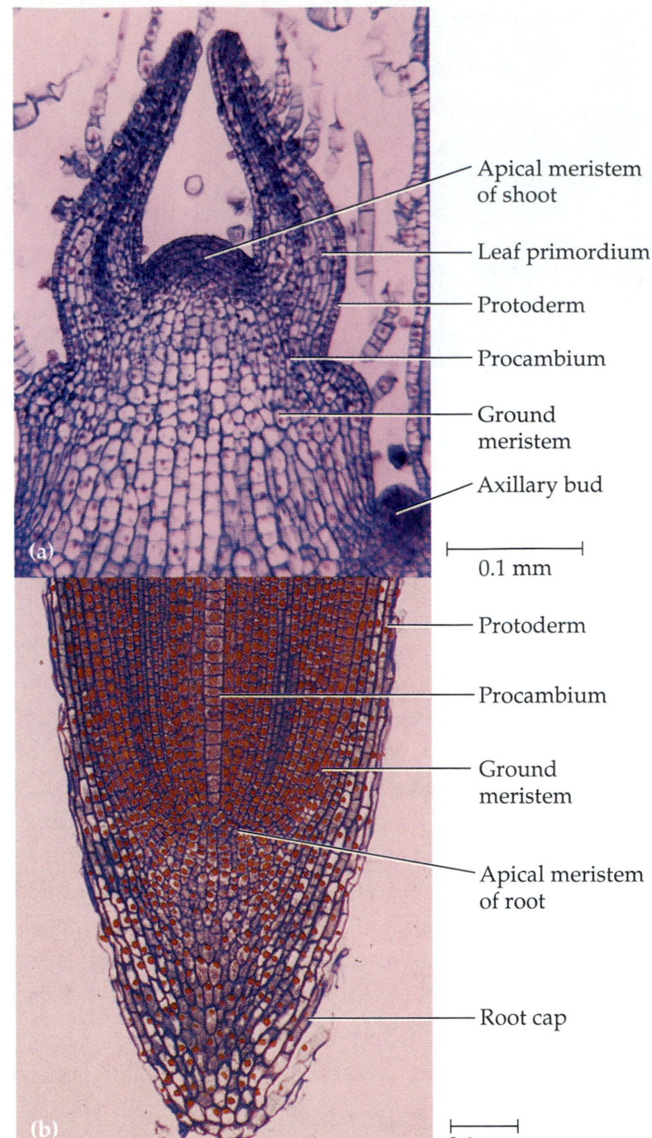

FIGURE 38.16 • Lifelong morphogenesis in plants. In contrast to animals, which cease growing and change shape little after reaching maturity, plants exhibit indeterminate growth and persistent morphogenesis for as long as they live. For elaboration of the primary plant body, the centers of ongoing morphogenesis are at a plant's tips, in the regions of the apical meristems of **(a)** shoots and **(b)** roots. Morphogenesis is especially evident at the shoot tip, where the meristem gives rise to a succession of modules (see Chapter 35), each consisting of a leaf-bearing node, internode, and axillary bud. (Both LMs.)

The cytoskeleton guides the geometry of cell division and expansion

Reexamine FIGURE 38.16, and you can see evidence of a basic principle of plant morphology: The shape of a plant organ depends mostly on the spatial orientations of cell divisions and cell expansions. In the case of the root tip, for example, notice that the young cells issue from the apical meristem in files. This is because most of the cell divisions in this region are oriented in a transverse plane, perpendicular to the long axis of the root. When the new cells begin growing, they mainly elongate; that is, most of their expansion parallels the long axis of the root.

Here we can see another important difference between plant and animal development. Animal morphogenesis also involves oriented cell division and growth, but the migration of cells plays a major role as well. In contrast, plant cells cannot move about individually within a developing organ, because they are immobilized by cell walls that are cemented to neighboring cells. This means that oriented cell division and expansion are the chief mechanisms of plant morphogenesis. Plant biologists have learned that the cytoskeleton controls these processes.

Orienting the Plane of Cell Division

The plane in which a cell will divide is determined during late interphase (G_2 of the cell cycle; see Chapter 12). The first sign of this spatial orientation is a rearrangement of the cytoskeleton. Microtubules in the cortex (outer cytoplasm) of the cell become concentrated into a ring called the **preprophase band** (FIGURE 38.17). The band disappears before metaphase, but it has already set the future plane of cell division. The "imprint" consists of an ordered array of actin microfilaments that remain after the microtubules of the preprophase band disperse. These microfilaments hold the nucleus in a fixed orientation until the spindle forms, and later they direct movement of the vesicles that produce the cell plate. (To review the role of the cell plate in the division of plant cells, see Chapter 12.) When the cell finally divides, the walls separating the daughter cells form in the plane defined earlier by the preprophase band. Researchers are now investigating what controls the placement of the preprophase band.

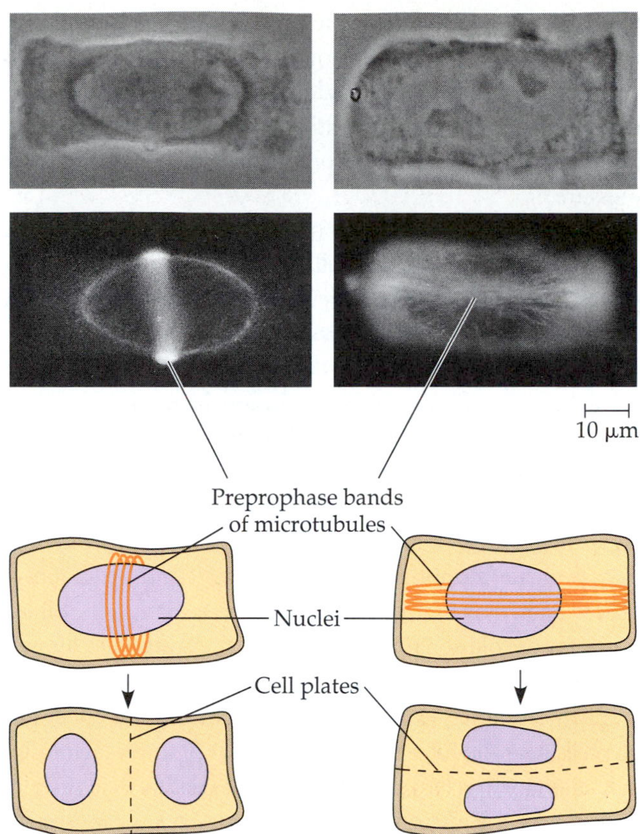

FIGURE 38.17 • The preprophase band and the plane of cell division. The band is a ring of microtubules that forms just inside the plasma membrane during late interphase. Its location predicts the future plane of cell division. Although the left and right cells are similar in shape, they will divide in different planes. Each cell is represented by two light micrographs, one unstained (top) and the other stained (bottom) with a fluorescent dye that binds specifically to microtubules. The microtubules form a "halo" (preprophase band) around the nucleus in the cortex of the cytoplasm.

Orienting the Direction of Cell Expansion

The shape of a plant organ is the outcome of the oriented growth of its cells, and we will see how the cytoskeleton controls this differential enlargement. But first we must take a closer look at *how* plant cells grow.

In a region of active growth, such as the zone of elongation in roots, cells can expand to a volume as much as 50 times their original size. This expansion occurs when the cell wall yields to the turgor pressure of the cell (FIGURE 38.18). Acid secreted by the cell activates enzymes in the cell wall that break hydrogen bonds between the cellulose microfibrils (see Chapter 5). This weakening of cross-links between

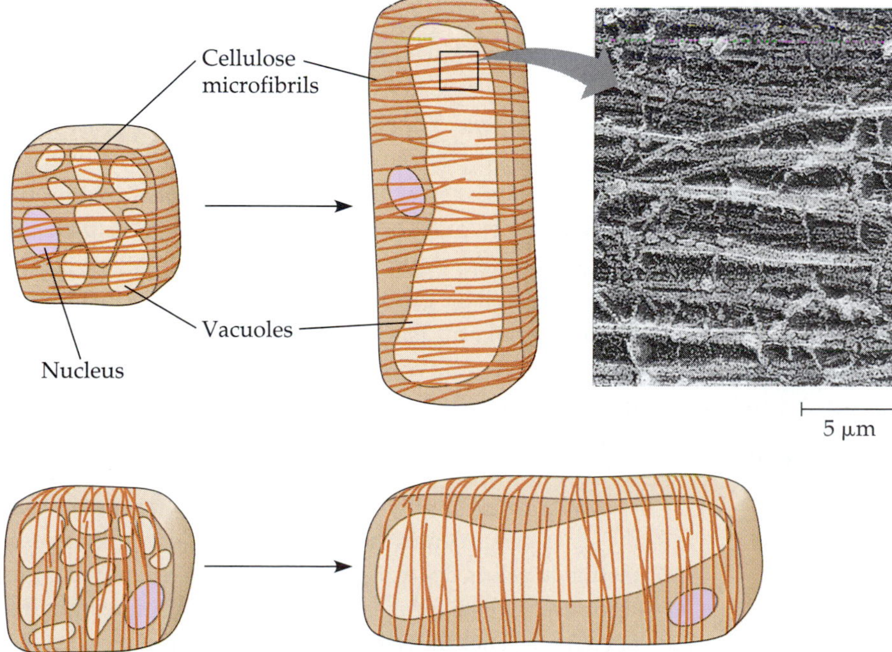

FIGURE 38.18 • The orientation of plant cell expansion. Growing plant cells expand mainly through water uptake. In a growing cell, enzymes weaken cross-links in the cell wall, allowing it to expand as water flows in by osmosis. The orientation of cell growth is mainly in the plane perpendicular to the orientation of cellulose microfibrils in the wall. The microfibrils are embedded in a matrix of other (noncellulose) polysaccharides, some of which form the cross-links visible in the micrograph (TEM). Loosening of the wall occurs when hydrogen ions secreted by the cell activate cell-wall enzymes that break the cross-links between polymers in the wall. This reduces restraint on the turgid cell, which can then take up more water and expand. Small vacuoles, which accumulate most of this water, coalesce and form the cell's central vacuole.

microfibrils enables the wall to stretch. The cell has a lower water potential than the surrounding solution, and with its wall loosened it can take up additional water by osmosis and expand. Growth continues until the wall again becomes knit tightly enough by hydrogen bonding to offset the cell's turgor pressure.

Once again, it is useful to highlight a difference between plants and animals. Animal cells grow mainly by synthesizing protein-rich cytoplasm, a metabolically expensive process. Growing plant cells also produce additional organic material in their cytoplasm, but the uptake of water typically accounts for about 90% of a plant cell's expansion. Most of this water is packaged in the large vacuole, which forms by the coalescence of numerous smaller vacuoles as a cell grows. A plant can grow rapidly and economically because a small amount of cytoplasm can go a long way. Bamboo shoots, for instance, may elongate more than 2 m per week. Rapid extension of shoots and roots increases the exposure to light and soil, an important evolutionary adaptation to the immobile lifestyle of plants.

Plant cells rarely expand equally in all directions. For example, cells near the root tip may elongate to 20 times their original length, with relatively little increase in width. The orientation of cellulose microfibrils in the innermost layers of the cell wall causes this differential growth (see FIGURE 38.18). The microfibrils cannot stretch much, so the cell expands mainly in the direction perpendicular to the "grain" of the microfibrils.

Microfibrils are synthesized by a complex of enzymes built into the plasma membrane (FIGURE 38.19). The pattern of microfibrils in the wall mirrors the orientation of microtubules located just across the plasma membrane in the cortex of the cell. According to one hypothesis, the microtubules confine the flow of the cellulose-producing enzymes to a specific direction along the membrane. This specifies the alignment of microfibrils in the wall, which in turn determines the direction of cell expansion. As with the role of the cytoskeleton in the plane of cell division, an important question remains: What regulates the orientation of microtubules in the cell's cortex? Many researchers are searching for answers.

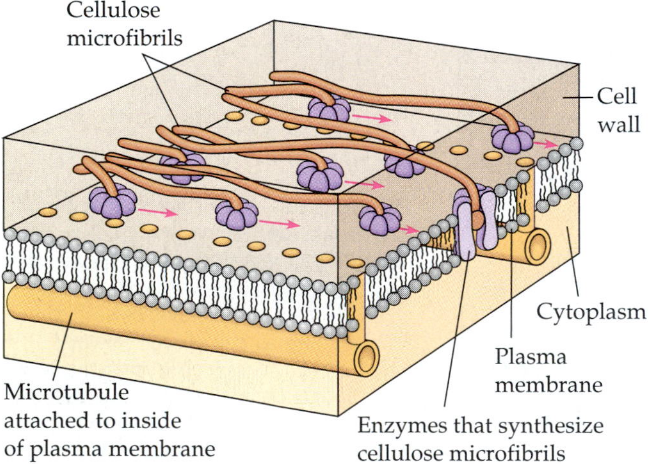

FIGURE 38.19 ▪ A hypothetical mechanism for how microtubules orient cellulose microfibrils. Cellulose microfibrils are synthesized at the cell surface by complexes of enzymes that can move in the plane of the plasma membrane. According to one hypothesis, microtubules form "banks" that confine the movement of the enzymes to channels of specified direction. Each enzyme complex advances along one of these channels as the microfibril it extends becomes locked in place by cross-linking to other microfibrils.

Cellular differentiation depends on the control of gene expression

It is remarkable that cells as diverse as guard cells, sieve-tube members (phloem), and xylem vessel elements all descend from a common cell, the zygote. Such differentiation continues throughout a plant's life, as meristems sustain indeterminate growth (FIGURE 38.20). Cellular differentiation, the progressive development of a cell's specialized structure and function, is one of the most engaging research problems in modern biology.

Differentiation reflects the synthesis of different proteins in different types of cells. For example, a developing xylem cell makes the enzymes that produce lignin, which hardens the cell wall. In contrast, guard cells do not make enzymes for lignin synthesis and have flexible walls. This difference in enzyme production contributes to a structural difference correlated with the contrasting functions of these two cell types. Xylem vessels function in support (as well as transport), and their hard cell walls fit this function. Guard cells, on the other hand, control the size of stomata by changing their shape, a function that requires flexible walls. In its final stage of differentiation, a xylem cell unleashes hydrolytic enzymes that destroy the protoplast, leaving only the wall. No such metabolic episode occurs during the differentiation of a guard cell.

What makes differentiation so fascinating is that the cells of a developing organism synthesize different proteins and diverge in structure and function even though they share a common genome. The cloning of whole plants from somatic cells supports the conclusion that the genome of a differentiated cell remains intact (see FIGURE 21.5). If a mature cell removed from a root or leaf can "dedifferentiate" in tissue culture and give rise to the diverse cell types of a plant, then it must possess all the genes necessary to make these many kinds of cells. This means that cellular differentiation depends, to a large extent, on the control of gene expression—the regulation of transcription and translation leading to specific proteins. Cells with the same genomes follow different developmental pathways because they selectively express certain genes at specific times during their differentiation. A guard cell has the genes that program the self-destruction of a xylem protoplast, but it does not express those genes. A xylem vessel element *does* express them, but only at a specific time in its differentiation, after the cell has elongated and produced its sec-

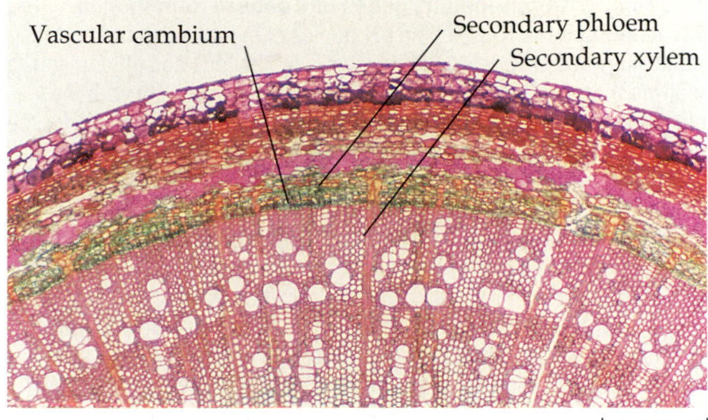

FIGURE 38.20 ▪ Example of cellular differentiation. The vascular cambium is a lateral meristem that provides cells for secondary growth, which increases the diameter of woody stems and roots (see Chapter 35). In this cross section of a stem, you can see the secondary phloem and secondary xylem on opposite sides of the cambium, which gives rise to these vascular tissues. Cellular differentiation is the process by which cells of common developmental origin diverge in structure and function.

ondary wall. Researchers are beginning to unravel the molecular mechanisms that switch specific genes on and off at critical times in a cell's development (see Chapters 19 and 21).

Pattern formation determines the location and tissue organization of plant organs

Cellular differentiation has a spatial component: Each plant organ has a characteristic pattern of tissues and cell types within those tissues. On a larger scale, spatial organization is also apparent in the placement of a plant's organs—in the arrangement of a flower's parts, for example. The development of specific structures in specific locations is called **pattern formation**. It is a key element in the overall development of an organism's form.

Positional Information

Pattern formation depends on **positional information**, signals of some kind that indicate each cell's location within an embryonic structure, such as a shoot tip. By affecting localized rates and planes of cell division and expansion, these signals apparently cause organs such as leaf primordia to emerge as "bumps" at certain places on the plant. Within a developing organ, each cell continues to detect positional information and respond by differentiating into a particular type of cell.

Developmental biologists are accumulating evidence that gradients of specific molecules, generally proteins, provide positional information. Perhaps, for example, a substance diffusing from a shoot's apical meristem "informs" the cells below of their distance from the shoot tip. We can imagine that the cells gauge their radial positions within the developing organ by detecting a second chemical signal that emanates from the outermost cells. The gradients of these two substances would be sufficient for each cell to "get a fix" on its position relative to the long axis and radial axis of the rudimentary organ. This idea of diffusible chemical signals is just one of several alternative hypotheses that developmental biologists are testing to learn how an embryonic cell detects its location.

Clonal Analysis of the Shoot Apex

In the process of shaping a rudimentary organ, patterns of cell division and cell expansion also affect the differentiation of cells by placing them in specific locations relative to other cells. Thus, positional information underlies all the processes of development: growth, morphogenesis, and differentiation. One approach to studying the relationships among these processes is clonal analysis, in which the cell lineages (clones) derived from each cell in an apical meristem are mapped as organs develop. Researchers can do this by using radiation or chemicals to induce somatic mutations that alter chromosome number or otherwise tag a cell in some way that distinguishes it from its neighbors in the shoot tip. The lineage of cells derived by mitosis from the mutant meristematic cell will also be "marked."

One of the important questions that clonal analysis can address is: How early is the developmental fate of a cell determined by its position in an embryonic structure? To some extent, the developmental fates of cells in the shoot apex are predictable. For example, almost all the cells derived from division of the outermost meristematic cells end up as part of the dermal tissue of leaves and stems. But it is not possible to pinpoint precisely which cells of the meristem will give rise to specific organs and tissues. Apparently random changes in rates and planes of cell division can reorganize the meristem. For example, the outermost cells *usually* divide in a plane perpendicular to the surface of the shoot tip, resulting in the addition of cells to the surface layer. But occasionally a cell at the surface divides in a plane parallel to this meristematic layer, placing one daughter cell beneath the surface among cells derived from different lineages. Thus, the cells of the meristem are not dedicated early to forming specific organs and tissues. Put another way, a cell's developmental fate is not determined by its membership in a lineage derived from a particular meristematic cell. Rather, it is the cell's *final* position in an emerging organ that determines what kind of cell it will become, presumably as a result of positional information.

The Genetic Basis of Pattern Formation in Flower Development

A particularly striking passage in plant development is the transition of a vegetative shoot tip into a floral meristem. A

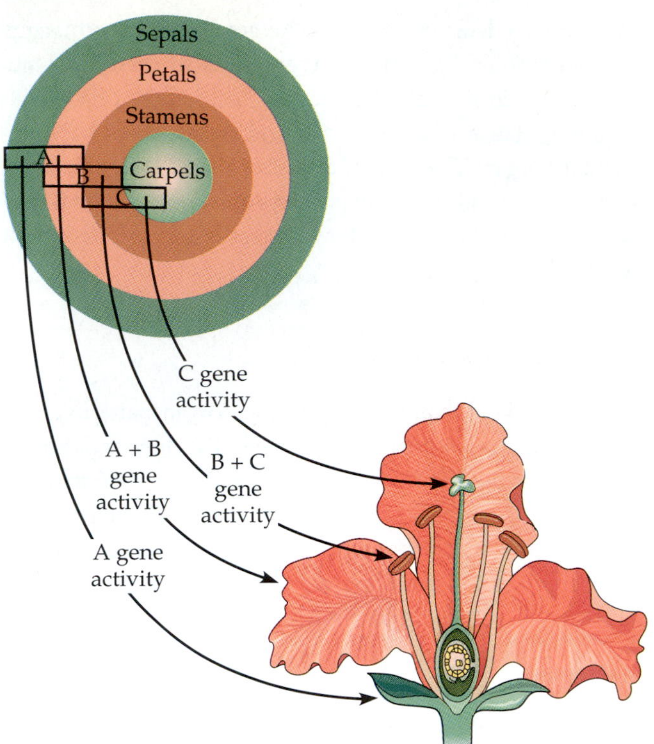

FIGURE 38.21 ▪ Organ-identity genes and pattern formation in flower development. Based on their studies of mutations that affect floral morphology in *Arabidopsis, Antirrhinum* (snapdragon), and other plants (a generalized flower is illustrated here), researchers are working with the hypothesis that three classes of organ-identity genes are responsible for the spatial pattern of floral organs. (Chapter 21 presents the experimental support for this hypothesis.) The three classes of organ-identity genes are designated A, B, and C in the upper drawing, a schematic diagram of a floral meristem in transverse view. The products of these genes are transcription factors, which regulate expression of other genes responsible for the development of the specific floral organs: sepals, petals, stamens, and carpels. Sepals develop from the region of the meristem where only the A gene is active. Petals develop from the region where both the A and B genes are expressed. Stamens arise from the meristem region where both the B and C genes are active. And carpels are derived from the region where only the C gene is expressed.

combination of environmental cues, such as daylength, and internal signals, such as hormones, trigger this transition. (You will learn more about the control of flowering in Chapter 39.) The transition from vegetative growth to flowering is associated with the switching on of floral **meristem identity genes**. The protein products of these genes are transcription factors that in turn help activate the genes required for development of the floral meristem.

Once a shoot meristem is induced to flower, what controls the placement of the floral organs? You already know part of the answer. Positional information commits each primordium that arises on the flanks of the shoot tip to develop into an organ of specific structure and function—an anther-bearing stamen, for example. In the past few years, plant biologists have identified some of the genes that are regulated by positional information and that function in this development of floral pattern. Mutations in these genes, called **organ-identity genes**, substitute one type of floral organ where another would normally form. For example, a particular mutation may cause an extra whorl of sepals to develop in a flower where there ought to be petals. The implication is that the wild-type alleles for these organ-identity genes are responsible for the development of normal floral pattern.

Many plant biologists interested in genes that control development have focused on *Arabidopsis thaliana* as their experimental organism. Recall from Chapter 35 that this wild mustard is a small plant with a rapid life cycle, characteristics that make it suitable to culture in the laboratory (see FIGURE 35.1). *Arabidopsis* also has a relatively small genome, a feature that makes the search for specific genes more manageable. Researchers have identified several organ-identity genes that affect flower development in *Arabidopsis* and have used recombinant DNA methods to clone a few of those genes. Very similar organ-identity genes have also been found in snapdragon (*Antirrhinum majus*), which is not closely related to *Arabidopsis*. This suggests that evolution has been very conservative with genes that control the development of an angiosperm's basic body plan.

Organ-identity genes code for transcription factors (see Chapters 19 and 21). Positional information determines which organ-identity genes are expressed in a particular floral-organ primordium. The resulting transcription factors probably induce the expression of those genes responsible for building an organ of specific structure and function. FIGURE 38.21 illustrates a hypothesis for how three regulatory products of organ-identity genes are responsible for normal flower development. (Chapter 21 explores this hypothesis for flower development in much greater detail.) By constructing such hypotheses and designing experiments to test them, researchers are tracing the genetic basis of pattern formation. More generally, with the help of excellent experimental systems such as *Arabidopsis* and the powerful methods of modern cell biology, plant biologists are making rapid progress toward understanding how a single cell becomes a plant.

CHAPTER REVIEW

REVIEW OF KEY CONCEPTS

(with page numbers and key figures)

SEXUAL REPRODUCTION

- **Sporophyte and gametophyte generations alternate in the life**
38.1 **cycles of plants:** ***a review*** **(p. 730, FIGURE 38.1)** The sporophyte, the dominant generation, produces spores that develop within flowers into male gametophytes (pollen grains) and female gametophytes (embryo sacs).
- **Male and female gametophytes develop within a flower's anthers and ovaries, respectively (pp. 730–734, FIGURES 38.2, 38.4)** The four floral organs are sepals, petals, stamens (which form pollen), and carpels (which form embryo sacs within the ovules of their ovaries). Pollen develops from microspores within the sporangia of anthers; embryo sacs develop from megaspores within ovules.
- **Pollination brings male and female gametophytes together (p. 734)** Pollination, which precedes fertilization, is the placing of pollen on the stigma of a carpel.
- **Researchers are unraveling the molecular mechanisms of self-incompatibility (pp. 734–736, FIGURES 38.5, 38.6)** Some plants reject pollen that has an S-allele matching an allele in the stigma cells. Recognition of "self" pollen triggers a signal-transduction pathway leading to a block in growth of a pollen tube.
- **Double fertilization gives rise to the zygote and endosperm (p. 736, FIGURE 38.7)** The pollen tube discharges two sperm into the embryo sac; one fertilizes the egg, while the other fertilizes the polar cell, which gives rise to the food-storing endosperm.
- **The ovule develops into a seed containing an embryo and a sup-**
38.2 **ply of nutrients (pp. 736–738, FIGURES 38.8, 38.9)** The seed coat encloses the embryo, complete with apical meristems and cotyledons, along with a food supply stocked either in the cotyledons or endosperm.
- **The ovary develops into a fruit adapted for seed dispersal (pp. 738–739, FIGURE 38.10 and TABLE 38.1)** The fruit protects the enclosed seeds and aids in wind dispersal or in the attraction of seed-dispersing animals.
- **Evolutionary adaptations of seed germination contribute to seedling survival (pp. 739–740, FIGURES 38.11, 38.12)** Germination, which begins when the seed imbibes water, often requires environmental cues, such as temperature or lighting changes.

ASEXUAL REPRODUCTION

- **Many plants can clone themselves by asexual reproduction (pp. 740–741, FIGURE 38.13)** One important mode of asexual reproduction is the fragmentation of a parent plant into parts that reform whole plants.
- **Vegetative propagation of plants is common in agriculture (pp. 741–743, FIGURE 38.14)** Cloning plants from cuttings is an ancient practice. We can now clone plants from single cells, which can be genetically manipulated before cloning.
- **Sexual and asexual reproduction are complementary in the life histories of many plants:** ***a review*** **(p. 743)** Asexual reproduction enables successful clones to spread; sexual reproduction generates the genetic variation that makes adaptation possible.

CELLULAR MECHANISMS OF PLANT DEVELOPMENT

- **Growth, morphogenesis, and differentiation produce the plant body:** ***an overview of plant development*** **(pp. 743–744, FIGURE 38.16)** You can see how these processes overlap in the development of an organ, such as a leaf or floral part.
- **The cytoskeleton guides the geometry of cell division and expansion (pp. 744–746, FIGURES 38.17, 38.18)** A preprophase band determines where a cell plate will form in a dividing cell. The orientation of the cytoskeleton also affects the direction of cell elongation by controlling the orientation of cellulose microfibrils in the wall.
- **Cellular differentiation depends on the control of gene expression (pp. 746–747, FIGURE 38.20)** The challenge of understanding cellular differentiation is explaining how cells with matching genomes diverge into diverse cell types.
- **Pattern formation determines the location and tissue organization of plant organs (pp. 747–748, FIGURE 38.21)** Development of organs and tissues in specific locations depends on the ability of cells to detect and respond to positional information. Research on organ-identity genes in developing flowers provides an important model system for the study of pattern formation.

SELF-QUIZ

1. Which of the following would definitely be an imperfect flower? A flower that
   a. is also incomplete
   b. lacks sepals
   c. is self-compatible
   d. is staminate
   e. cannot self-pollinate

2. Germinated pollen grain is to ____________ as ____________ is to female gametophyte.
   a. male gametophyte; embryo sac
   b. embryo sac; ovule
   c. ovule; sporophyte
   d. anther; seed
   e. petal; sepal

3. A seed develops from
   a. an ovum
   b. a pollen grain
   c. an ovule
   d. an ovary
   e. an embryo

4. A fruit is a (an)
   a. mature ovary
   b. mature ovule
   c. seed plus its integuments
   d. fused carpel
   e. enlarged embryo sac

5. Which of the following conditions is needed by almost all seeds to break dormancy?
   a. exposure to light
   b. imbibition
   c. abrasion of the seed coat
   d. exposure to cold temperatures
   e. covering of fertile soil

6. Which of the following does *not* appear to have a major effect on plant morphology?
   a. cell division
   b. environmental factors
   c. cell migration
   d. cell expansion
   e. selective expression of genes

7. The direction in which a plant cell will expand seems to a large extent to depend on
   a. the size and position of the central vacuole

b. a ring of actin microfilaments in the cortex of the cell
c. the location of the preprophase band
d. the distribution of solutes in the cytoplasm
e. the orientation of microtubules just inside the plasma membrane

8. The type of mature cell that a particular embryonic plant cell will become appears to be determined mainly by
   a. the selective loss of genes
   b. the cell's final position in a developing organ
   c. the cell's pattern of migration
   d. the cell's age
   e. the cell's particular meristematic lineage

9. Based on the hypothesis presented in FIGURES 21.18 (p. 409) and 38.21, predict the floral morphology of a *double* mutant lacking activity of both gene B and gene C. Explain your answer.
   a. carpel–petal–petal–carpel
   b. petal–petal–petal–petal
   c. sepal–sepal–sepal–sepal
   d. sepal–carpel–carpel–sepal
   e. carpel–carpel–carpel–carpel

10. A plant that is self-incompatible has a genotype of S_5S_9 for the S-locus. It receives pollen from a plant that is S_3S_9. Which of the following is most likely to occur?
   a. All of the pollen will germinate, forming pollen tubes.
   b. None of the pollen will germinate.
   c. About half of the pollen will germinate.
   d. Fertilization will occur in about half of the flowers of the pollinated plant.
   e. Pollen from the S_3S_9 plant will secrete ribonuclease that destroys epidermal cells of the S_5S_9 stigma.

CHALLENGE QUESTIONS

1. Explain some advantages of propagating houseplants by asexual reproduction.
2. Compare and contrast a seed with the amniote egg of a bird as adaptations for reproducing on land.
3. Starting with a zygote in a newly fertilized ovule, trace the developmental pathway leading to a zygote of the next generation.

SCIENCE, TECHNOLOGY, AND SOCIETY

1. As our understanding of how organ-identity genes control the development of floral organs progresses, it will be possible to engineer crop plants to produce a greater or lesser number of specific flower parts. Given that seeds and fruits provide much of our food, discuss how certain floral mutations could possibly increase the food supply.
2. Seed companies gather seeds from all over the world and use them to develop new kinds of food and medicinal plants through crossbreeding and biotechnology. The greatest diversity of potentially useful plants is in the tropics. Some developing tropical countries still lack the resources for conserving plants and developing new crops, but they resent large corporations exploiting their plant resources and selling seeds back to them. What would you suggest to resolve this conflict, allowing both the developing nations and the seed companies to benefit from finding, preserving, and breeding useful plants?

FURTHER READING

Calzada, J-P. V., C. F. Crane, and D. M. Stelly. "Apomixis: The Asexual Revolution." *Science*, November 22, 1996. Describes how genetically engineered plants form seeds without pollination.

Day, S. "Blossoming Talent." *New Scientist*, July 12, 1997. It is now possible to design the timing of flowering and placement of the flowers on certain plants.

Fosket, D. E. *Plant Growth and Development*. San Diego, CA: Academic Press, 1994. A textbook appropriate for undergraduates.

Goldberg, R. B., G. dePaiva, and R. Yadegari. "Plant Embryogenesis: Zygote to Seed." *Science*, October 28, 1994. Describes progress in plant development research.

Grant, M. C. "The Trembling Giant." *Discover*, October 1993. Is an aspen grove the most massive organism?

Mestel, R. "Not Doing It, Plant-Style." *Discover*, January 1995. What prevents certain plants from fertilizing themselves?

Meyerowitz, E. M. "The Genetics of Flower Development." *Scientific American*, November 1994. Relates the progress in our understanding of pattern formation.

Perkins, S. "Transgenic Plants Provoke Petition." *Science News*, September 27, 1997. Some environmental groups are making an issue of engineered gene transfer between species.

WEB LINKS

Visit the special edition of *The Biology Place* for BIOLOGY, Fifth Edition, at http://www.biology.com/campbell. Go to Chapter 38 for online resources, including learning activities, practice exams, and links to the following web sites:

"International Association for Sexual Plant Reproduction Research (IASPRR)"
The IASPRR aims to stimulate scientific research in the field of plant reproduction and related subjects. The site contains links to plant reproduction listservs and to other plant reproduction web sites.

"Welcome to the Amazing Events in Plant Reproduction!"
This interesting site from the Laboratory of Science Education in Hiroshima, Japan, contains numerous images of plant reproductive structures.

"The Missouri Botanical Garden"
Explore the wealth of plant images from one of the world's leading botanical gardens.

"Canola Connection"
Manipulation of plant reproduction to enhance seed yield and nutritive value has been practiced by humans for centuries. This site tells the story of canola, a genetic variation of rapeseed, developed by Canadian plant breeders.

CHAPTER 39

CONTROL SYSTEMS IN PLANTS

Plant Hormones

- Research on how plants grow toward light led to the discovery of plant hormones: *science as a process*
- Plant hormones help coordinate growth, development, and responses to environmental stimuli
- Analysis of mutant plants is extending the list of hormones and their functions
- Signal-transduction pathways link cellular responses to plant hormonal signals and environmental stimuli

Plant Movements as Models for Studying Control Systems

- Tropisms orient the growth of plant organs toward or away from stimuli
- Turgor movements are relatively rapid, reversible plant responses

Control of Daily and Seasonal Responses

- Biological clocks control circadian rhythms in plants and other eukaryotes
- Photoperiodism synchronizes many plant responses to changes of season

Phytochromes

- Phytochromes function as photoreceptors in many plant responses to light and photoperiod
- Phytochromes may entrain the biological clock

Plant Responses to Environmental Stress

- Plants cope with environmental stress through a combination of developmental and physiological responses

Defense Against Pathogens

- Resistance to disease depends on a gene-for-gene recognition between plant and pathogen
- The hypersensitive response (HR) contains an infection
- Systemic acquired resistance (SAR) helps prevent infection throughout the plant

At every stage in the life of a plant, sensitivity to the environment and coordination of responses are evident. One part of a plant can send signals to other parts. For example, the terminal bud at the apex of a shoot is able to suppress the growth of axillary buds that may be many meters away. Plants keep track of the time of day and also the time of year. They can sense gravity and the direction of light and respond to these stimuli in ways that seem to us to be completely appropriate (the grass seedling in the photograph on this page is growing toward light). It is tempting to describe these responses in terms of such human qualities as desire, need, wisdom, and decisiveness. Science, however, searches for a mechanism—a link between cause and effect. A houseplant orients its leaves toward a window not because it is trying to find more light, but because cells on the darker side of stems and petioles grow faster than cells on the brighter side. This mechanism causes the stems and petioles to curve toward the light. Natural selection has favored mechanisms of plant response that enhance reproductive success, but this implies no purposeful planning on the part of the plant.

For the most part, plants and animals respond to environmental stimuli by very different means. Animals, being mobile, respond mainly by behavioral mechanisms, moving toward positive stimuli and away from negative stimuli. Rooted to one location for life, a plant generally responds to environmental cues by adjusting its pattern of growth and development. Plants of the same species vary in body form much more than do animals of the same species. All lions have four limbs and approximately the same body proportions, but oak trees are less regular in their number of limbs and their shapes.

In this chapter you will learn how a plant's morphology and physiology are constantly tuned to the surroundings by complex interactions of environmental factors and internal signals. Some of the most important signals are chemical messengers called hormones.

PLANT HORMONES

The word *hormone* is derived from a Greek verb meaning "to excite." Found in all multicellular organisms, **hormones** are chemical signals that coordinate the parts of the organism. As first defined by animal physiologists, a hormone is a compound that is produced by one part of the body and then transported to other parts of the body, where it triggers responses in target cells and tissues. Another important characteristic of hormones is that only minute concentrations of these chemical messengers are required to induce substantial change in an organism.

Research on how plants grow toward light led to the discovery of plant hormones: *science as a process*

The concept of chemical messengers in plants emerged from a series of classic experiments on how stems respond to light. A houseplant on a windowsill grows toward light. If you rotate the plant, it will soon reorient its growth until its leaves again face the window. The growth of a shoot toward light is called positive **phototropism** (growth away from light is negative phototropism). In a forest or other natural ecosystem where plants may be crowded, phototropism directs growing seedlings toward the sunlight that powers photosynthesis. What is the mechanism for this adaptive response? Much of what is known about phototropism has been learned from studies of grass seedlings, particularly oats. The shoot of a grass seedling is enclosed in a sheath called the coleoptile, which grows straight upward if the seedling is kept in the dark or if it is illuminated uniformly from all sides. If the growing coleoptile is illuminated from one side, it will curve toward the light (see the photograph that opens the chapter). This response results from a differential growth of cells on opposite sides of the coleoptile; the cells on the darker side elongate faster than the cells on the brighter side (FIGURE 39.1).

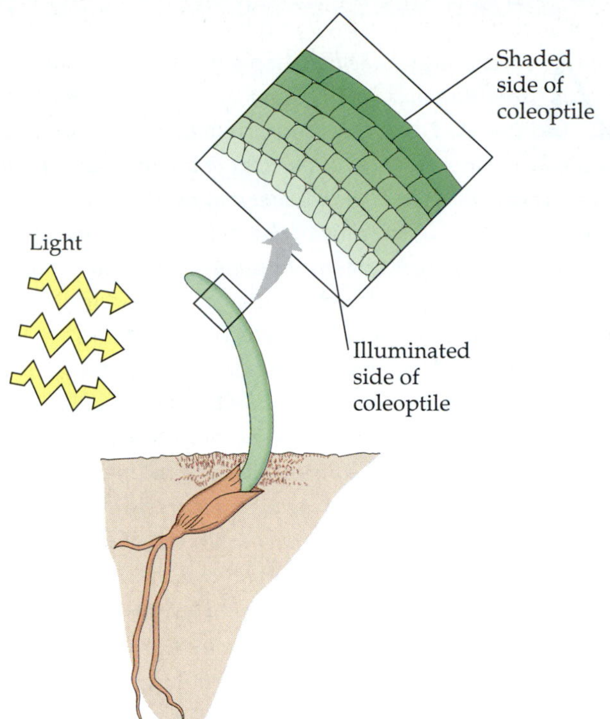

FIGURE 39.1 • Phototropism. The coleoptile of an oat seedling grows toward light because cells on the darker side of the organ elongate faster than cells on the brighter side.

Some of the earliest experiments on phototropism were conducted in the late nineteenth century by Charles Darwin and his son, Francis. They observed that a grass seedling could bend toward light only if the tip of the coleoptile was present (FIGURE 39.2). If the tip was removed, the coleoptile would not curve. The seedling would also fail to grow toward light if the tip was covered with an opaque cap; neither a transparent cap over the tip nor an opaque shield placed farther down the coleoptile prevented the phototropic response. It was the tip of the coleoptile, the Darwins concluded, that was responsible for sensing light. However, the actual growth response, the curvature of the coleoptile, occurred some distance below the

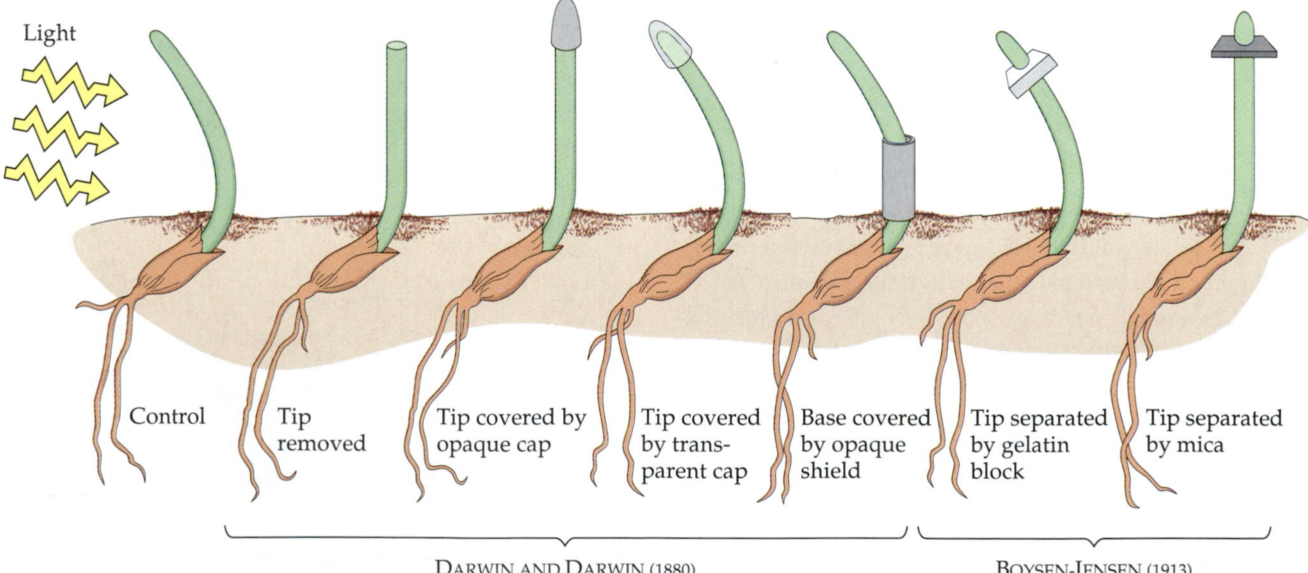

FIGURE 39.2 • Early experiments on phototropism. Only the tip of the coleoptile can sense the direction of light, but the bending response occurs some distance below the tip. A signal of some kind must travel downward from the tip. The signal can pass through a permeable barrier (gelatin block) but not through a solid barrier (a mineral called mica), suggesting that the signal for phototropism is a mobile chemical.

tip. Charles and Francis Darwin proposed the hypothesis that some signal was transmitted downward from the tip to the elongating region of the coleoptile. A few decades later, Peter Boysen-Jensen of Denmark tested this hypothesis and demonstrated that the signal was a mobile substance of some kind. He separated the tip from the remainder of the coleoptile by a block of gelatin, which would prevent cellular contact but allow chemicals to pass. These seedlings behaved normally, bending toward light. However, if the tip was experimentally segregated from the lower coleoptile by an impermeable barrier, no phototropic response occurred.

In 1926, F. W. Went, a young plant physiologist in Holland, extracted the chemical messenger for phototropism by modifying the experiments of Boysen-Jensen (FIGURE 39.3). Went removed the coleoptile tip and placed it on a block of agar, a gelatinous material. The chemical messenger from the tip, Went reasoned, should diffuse into the agar, and the agar block should then be able to substitute for the coleoptile tip. Went placed the agar blocks on decapitated coleoptiles that were kept in the dark. A block that was centered on top of the coleoptile caused the stem to grow straight upward. However, if the block was placed off center, then the coleoptile began to bend away from the side with the agar block, as though growing toward light. Went concluded that the agar block contained a chemical produced in the coleoptile tip, that this chemical stimulated growth as it passed down the coleoptile, and that a coleoptile curved toward light because of a higher concentration of the growth-promoting chemical on the darker side of the coleoptile. For this chemical messenger, or hormone, Went chose the name auxin (Gr. *auxein,* "to increase"). Auxin was later purified and its structure determined by Kenneth Thimann and his colleagues at the California Institute of Technology. Other plant hormones were subsequently discovered.

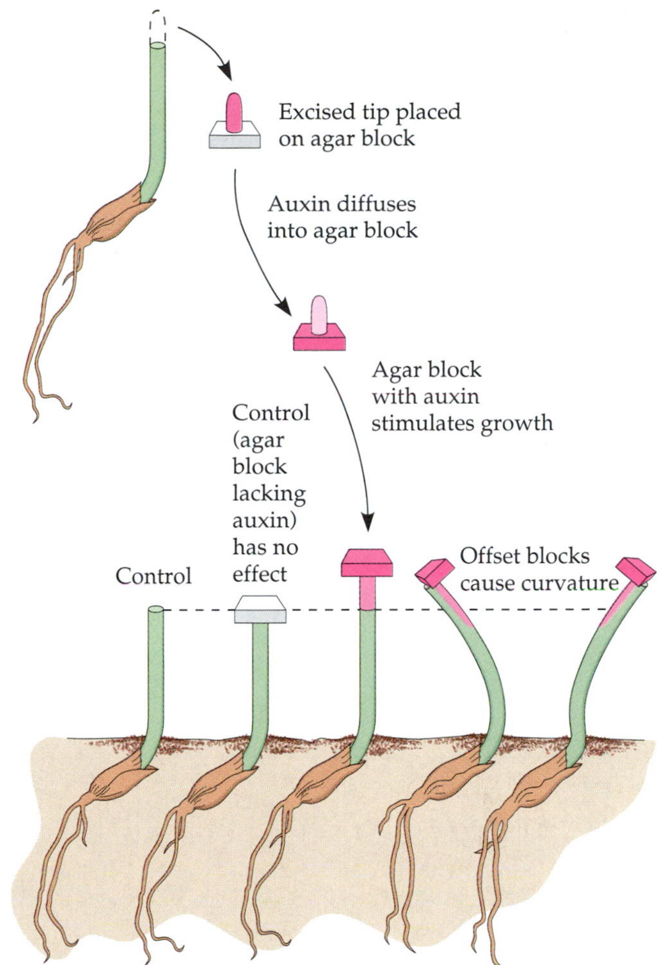

FIGURE 39.3 ▪ The Went experiments. Some chemical (indicated by pink) that can pass into an agar block from a coleoptile tip stimulates elongation of the coleoptile when the block is substituted for a tip. If the block is placed off-center on the top of a decapitated coleoptile kept in the dark, the organ bends as if responding to illumination from one side. The chemical is the hormone auxin, which stimulates elongation of cells in the shoot.

Plant hormones help coordinate growth, development, and responses to environmental stimuli

39.1 TABLE 39.1 on page 754 previews the major classes of plant hormones: auxin, cytokinins (actually a class of related chemicals), gibberellins (also a class of similar chemicals), abscisic acid, and ethylene. The table also includes the oligosaccharins and brassinosteroids as examples of "new" hormone classes that researchers are discovering. In general, hormones control plant growth and development by affecting the division, elongation, and differentiation of cells. Some hormones also mediate shorter-term physiological responses of plants to environmental stimuli.

Each hormone has a multiplicity of effects, depending on its site of action, the developmental stage of the plant, and the concentration of the hormone. Notice in TABLE 39.1 that all the plant hormones are relatively small molecules. Their transport from cell to cell often involves passage across cell walls, a pathway that blocks the movement of large molecules.

Plant hormones are produced in very small concentrations, but a minute amount of hormone can have a profound effect on the growth and development of a plant organ. This implies that the hormonal signal must be amplified in some way. A hormone may act by altering the expression of genes, by affecting the activity of existing enzymes, or by changing properties of membranes. Any of these actions could redirect the metabolism and development of a cell responding to a small number of hormone molecules.

Reaction to a hormone usually depends not so much on the absolute amount of that hormone as on its relative concentration compared with other hormones. It is hormonal balance, rather than hormones acting in isolation, that may control the growth and development of the plant. These interactions will become apparent in the following survey of hormone function.

Auxin

The term **auxin** is actually used to describe any chemical substance that promotes the elongation of coleoptiles, although auxins actually have multiple functions in both monocots and dicots. The natural auxin that has been extracted from plants is a compound named indoleacetic acid, or IAA. In addition to this natural auxin, several other compounds, including some synthetic ones, have auxin activity. Throughout this chapter, however, the name *auxin* is used specifically to refer

39.1

Table 39.1 ▪ An Overview of Plant Hormones

HORMONE	WHERE PRODUCED OR FOUND IN PLANT	MAJOR FUNCTIONS
Auxin (such as IAA)	Embryo of seed, meristems of apical buds, young leaves	Stimulates stem elongation; root growth, differentiation, and branching; development of fruit; apical dominance; phototropism and gravitropism.
Cytokinins (such as zeatin)	Synthesized in roots and transported to other organs	Affect root growth and differentiation; stimulate cell division and growth; stimulate germination; delay senescence.
Gibberellins (such as GA_3)	Meristems of apical buds and roots, young leaves, embryo	Promote seed and bud germination, stem elongation, and leaf growth; stimulate flowering and development of fruit; affect root growth and differentiation.
Abscisic acid	Leaves, stems, roots, green fruit	Inhibits growth; closes stomata during water stress; counteracts breaking of dormancy.
Ethylene	Tissues of ripening fruits, nodes of stems, aging leaves and flowers	Promotes fruit ripening; opposes some auxin effects; promotes or inhibits growth and development of roots, leaves, and flowers, depending on species.
Oligogaccharins (such as oligogalacturonide, a short chain of galacturonic acid, a modified sugar abbreviated here as "Galu") Galu—[Galu]$_n$—Galu	Cell walls	Trigger defense responses against pathogens; regulate growth, cell differentiation, and flowering.
Brassinosteroids (such as brassinolide)	Seeds, fruits, shoots, leaves, and floral buds	Required for normal growth and development.

to IAA. Although auxin affects several aspects of plant development, one of its most important functions is to stimulate the elongation of cells in young developing shoots.

Auxin and Cell Elongation. The apical meristem of a shoot is a major site of auxin synthesis. As auxin from the shoot apex moves down to the region of cell elongation (see Chapter 35), the hormone stimulates growth of the cells. Auxin has this effect only over a certain concentration range, from about 10^{-8} to 10^{-3} *M*. At higher concentrations, auxin may inhibit cell elongation. This is probably due to a high level of auxin inducing the synthesis of another hormone, ethylene, which generally acts as an inhibitor of plant growth due to cell elongation.

The speed at which auxin is transported down the stem from the shoot apex is about 10 mm per hour—much too fast for diffusion, although slower than translocation in phloem. Auxin seems to be transported directly through parenchyma tissue, from one cell to the next. It moves only from shoot tip to base, not in the reverse direction. This unidirectional transport of auxin is called polar transport. Polar transport has nothing to do with gravity, for auxin travels upward in experiments where a stem or coleoptile segment is placed upside down. Polar auxin transport requires energy. FIGURE 39.4 illustrates how proton pumps, driven by ATP, couple metabolic energy to auxin transport. The mechanism of polar auxin transport is another example of cellular work driven by chemiosmosis, the harnessing of H^+ gradients generated by proton pumps.

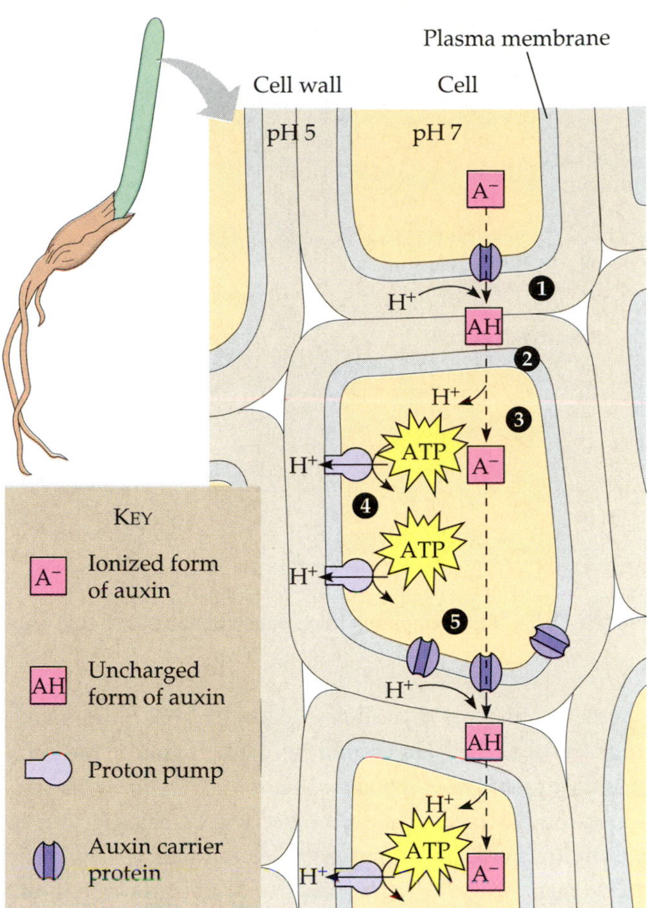

FIGURE 39.4 ▪ Polar auxin transport: a chemiosmotic model. In growing shoots, auxin is transported unidirectionally, from the apex down the shoot. Along this pathway, the hormone enters a cell at the apical end, exits at the basal end, diffuses across the wall, and enters the apical end of the next cell. ① When auxin encounters the acidic environment of the wall, the molecule picks up a hydrogen ion to become electrically neutral. ② As a relatively small, neutral molecule, auxin passes across the plasma membrane. ③ Once inside a cell, the pH 7 environment causes auxin to ionize. This temporarily traps the hormone within the cell, because the plasma membrane is less permeable to ions than to neutral molecules of the same size. ④ ATP-driven proton pumps maintain the pH difference between the inside and outside of the cell. ⑤ Auxin can exit the cell only at the basal end, where specific carrier proteins are built into the membrane. The proton pumps contribute to this auxin efflux by generating a membrane potential (voltage) across the membrane, which favors the transport of anions out of the cell.

According to what is called the acid growth hypothesis, proton pumps located in the plasma membrane also play a role in the growth response of cells to auxin. In a shoot's region of elongation, auxin stimulates the proton pumps, an action that lowers the pH in the wall (FIGURE 39.5, p. 756). This acidification of the wall activates enzymes that break cross-links (hydrogen bonds) between cellulose microfibrils, loosening the fabric of the wall. With its wall now more plastic, the cell is free to take up additional water by osmosis and elongate. For sustained growth after this initial spurt, however, cells must make more cytoplasm and wall material. Auxin also stimulates this sustained growth response.

Later in the chapter we will investigate how cells actually detect the presence of auxin and other signals, and how these stimuli lead to specific responses.

Other Effects of Auxin. In addition to stimulating cell elongation for primary growth, auxin affects secondary growth by inducing cell division in the vascular cambium and by influencing the differentiation of secondary xylem. Auxin also promotes the formation of adventitious roots at the cut base of a stem, an effect employed in horticulture by dipping cuttings in rooting media containing synthetic auxins. Developing seeds also synthesize auxin, which promotes the growth of fruit in many plants. Synthetic auxins sprayed on tomato vines induce fruit development without a need for pollination. This makes it possible to grow seedless tomatoes by substituting for the auxin that would normally be synthesized by seeds.

Cytokinins

Cytokinins were discovered during trial-and-error attempts to find chemical additives that would enhance the growth and development of plant cells in tissue culture. In the 1940s, Johannes van Overbeek, working at the Cold Spring Harbor Laboratory in New York, found he could stimulate the growth

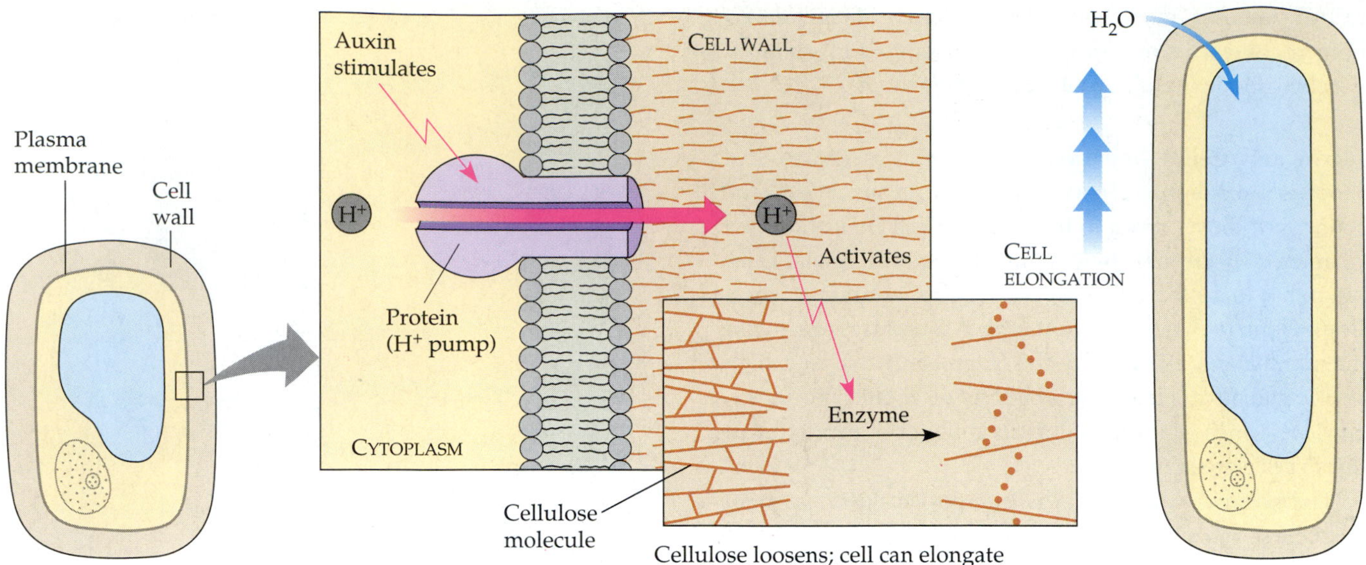

FIGURE 39.5 ▪ Cell elongation in response to auxin: the acid growth hypothesis.

of plant embryos by adding coconut milk, the liquid endosperm of the giant coconut seed, to his culture medium. A decade later, Folke Skoog and Carlos O. Miller, at the University of Wisconsin, induced tobacco cells being grown in culture to divide by adding degraded samples of DNA. The active ingredients of both experimental additives turned out to be modified forms of adenine, one of the components of nucleic acids. These growth regulators were named cytokinins because they stimulate cytokinesis, or cell division. Of the variety of cytokinins that occur naturally in plants, the most common is zeatin, so named because it was first discovered in corn (*Zea mays*). As you read about a few of the functions of cytokinins, notice that these hormones are complemented or countered by other hormones, especially auxin.

Control of Cell Division and Differentiation. Cytokinins are produced in actively growing tissues, particularly in roots, embryos, and fruits. Cytokinins produced in the root reach their target tissues by moving up the plant in the xylem sap. Acting in concert with auxin, cytokinins stimulate cell division and influence the pathway of differentiation.

The effects of cytokinins on cells growing in tissue culture provide clues about how this class of hormones may function in an intact plant. When a piece of parenchyma tissue from a stem is cultured in the absence of cytokinins, the cells grow very large but do not divide. If cytokinins alone are added to the culture, they have no effect. If cytokinins are added along with auxin, however, the cells divide. The ratio of cytokinin to auxin controls the differentiation of the cells. When the concentrations of the two hormones are about equal, the mass of cells continues to grow, but it remains an undifferentiated callus. If there is more cytokinin than auxin, shoot buds develop from the callus. If auxin is more concentrated than cytokinin, roots form. It is remarkable that gene expression can be controlled so extensively by manipulating the concentrations of just two chemical signals.

Control of Apical Dominance. We can see another interaction of cytokinins and auxin in the control of apical dominance, the ability of the terminal bud to suppress the development of axillary buds. In this case, the two hormones are antagonistic. Auxin transported down the shoot from the terminal bud restrains axillary buds from growing, causing a shoot to lengthen at the expense of lateral branching. If the terminal bud is removed, the plant may become bushier (FIGURE 39.6). However, cytokinins entering the shoot system from roots counter the action of auxin by signaling axillary buds to begin growing. Auxin cannot suppress the growth of these buds once it has begun.

The check-and-balance control of lateral branching by auxin and cytokinins may be one way the plant coordinates the growth of its shoot and root systems. As roots become more extensive, the increased level of cytokinins would signal the shoot system to form more branches. The two hormones reverse their roles in the development of lateral roots; auxin stimulates and cytokinins inhibit root branching.

Both auxin and cytokinins may regulate the growth of axillary buds indirectly by changing the concentration of still another hormone, ethylene. The levels of different nutrients in a bud may also affect its response to these hormones.

Cytokinins as Anti-Aging Hormones. Cytokinins can retard the aging of some plant organs, perhaps by inhibiting protein breakdown, by stimulating RNA and protein synthesis, and by mobilizing nutrients from surrounding tissues. If leaves removed from a plant are dipped in a cytokinin solution, they stay green much longer than they otherwise would. It is probable that cytokinins also slow the deterioration of

(a)

(b)

FIGURE 39.6 ▪ Apical dominance. **(a)** Auxin from the apical bud inhibits the growth of axillary buds. This favors elongation of the shoot's main axis over lateral branching. Cytokinins, which are transported upward from roots, counter auxin, stimulating the growth of axillary buds. This explains why, in many plants, axillary buds near the shoot tip are less likely to grow than those closer to the roots. **(b)** Removal of the apical bud from the same plant enabled lateral branches to grow. Unchecked by auxin, cytokinin stimulates the growth of axillary buds, which produce lateral branches of the shoot system.

leaves on intact plants. Because of this anti-aging effect, florists use cytokinin sprays to keep cut flowers fresh.

Gibberellins

A century ago, rice farmers in Asia noticed some exceptionally tall seedlings growing in their paddies. Before these rice seedlings could mature and flower, they grew so tall and spindly that they toppled over. In Japan, this aberration in growth pattern became known as *bakanae*, or "foolish seedling disease." In 1926, E. Kurosawa, a Japanese scientist, discovered that the disease was caused by a fungus of the genus *Gibberella* (FIGURE 39.7). By the 1930s, Japanese scientists had determined that the fungus produced hyper-elongation of rice stems by secreting a chemical, which was given the name **gibberellin**. Western scientists finally learned of gibberellin after World War II. In the past 30 years, scientists have identified more than 80 different gibberellins that occur naturally in plants, although a much smaller number occur in each plant species. Foolish rice seedlings, it seems, suffer from an overdose of growth regulators normally found in plants in lower concentrations. Gibberellins have a variety of effects in plants.

Stem Elongation. Roots and young leaves are major sites of gibberellin production. Gibberellins stimulate growth in both the leaves and the stem, but they have little effect on root growth. In stems, gibberellins stimulate cell elongation *and* cell division. In a growing stem, gibberellins and auxin must be acting simultaneously in some synergistic manner we do not yet understand.

Enhancement of stem elongation by gibberellins can be seen by applying the hormones to certain dwarf varieties of plants. For instance, dwarf pea plants (including the variety Mendel studied; see Chapter 14) grow to normal height if treated with gibberellins. The extent of the growth of dwarf plants is generally correlated with the concentration of the added hormone (see the Methods Box, p. 758). If gibberellins are applied to plants of normal size, there is often no response. Apparently, these plants are already producing their own optimal dose of the hormone.

A specific case in which gibberellins cause rapid elongation of stems is bolting, the growth of a floral stalk. In their nonflowering stage, some plants develop a rosette form; that is, they are low to the ground with very short internodes; a cabbage "head" is an example. As the plant switches to reproductive growth, a surge of gibberellins induces stems to elongate rapidly, which elevates flowers developing from buds at the tips of the stems.

FIGURE 39.7 ▪ "Foolish seedling disease" in rice. The spindly rice plants on the right are infected with the fungus *Gibberella*. The pathogen secretes gibberellins, a growth stimulant that uninfected plants (left) also produce, but in smaller quantity.

METHODS: Bioassay

A bioassay is a technique that determines the concentration of a chemical by measuring the response of living material. In this example, a plant's quantitative response to a gibberellin is used to measure the concentration of the hormone. A sample of unknown concentration is applied to dwarf pea plants like the one on the right, and after a certain amount of growth time their height is compared with dwarfs treated with a range of known gibberellin concentrations. Using the degree of coleoptile curvature in phototropism to determine auxin concentration is another example of a bioassay.

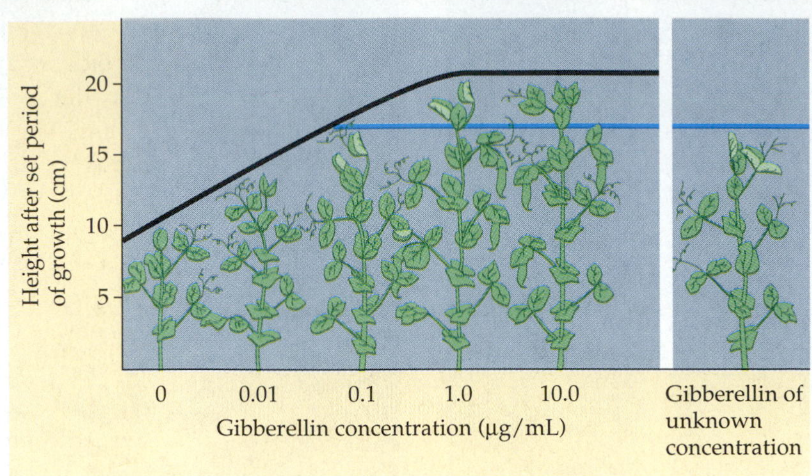

Fruit Growth. Fruit development is another case in which we can observe dual control by auxin and gibberellins. In some plants, both hormones must be present for fruit to set. The most important commercial application of gibberellins is in the spraying of Thompson seedless grapes (FIGURE 39.8). The hormones cause the grapes to grow larger and farther apart.

Germination. Many seeds have a high concentration of gibberellins, particularly in the embryo. After water is imbibed, the release of gibberellins from the embryo signals the seeds to break dormancy and germinate. Some seeds that require special environmental conditions to germinate, such as exposure to light or cold temperatures, will break dormancy if they are treated with a gibberellin solution. In nature, gibberellins in the seed are probably the link between the environmental cue and the metabolic processes that renew the growth of the embryo.

FIGURE 39.8 • The effect of gibberellin treatment on seedless grapes. The grape bunch on the left is an untreated control. The bunch on the right came from a plant that was sprayed with a gibberellin during fruit development.

Gibberellins support the growth of cereal seedlings by stimulating the synthesis of digestive enzymes such as α-amylase that mobilize stored nutrients (see FIGURE 38.11). Even before these enzymes appear, gibberellin stimulates the synthesis of messenger RNA coding for α-amylase. Here is a case in which a hormone controls development by affecting gene expression.

Abscisic Acid

The hormones we have studied so far—auxin, cytokinins, and gibberellins—usually stimulate plant growth. By contrast, there are times in the life of a plant when it is advantageous to slow down growth and assume a dormant ("resting") state. The hormone **abscisic acid (ABA)**, produced in the terminal bud, slows growth and directs leaf primordia to develop into the scales that will protect the dormant bud during winter. The hormone also inhibits cell division in the vascular cambium. Thus, ABA helps prepare the plant for winter by suspending both primary and secondary growth.

Another stage in the life of a plant when it is advantageous to suspend growth is the onset of seed dormancy, and again it may be abscisic acid that acts as the growth inhibitor. The seed will germinate when ABA is overcome by its inactivation or removal or by the increased activity of gibberellins. The seeds of some desert plants break dormancy when a heavy rain washes ABA out of the seed. Other seeds require light or some other stimulus to trigger the degradation of abscisic acid. In most cases the ratio of ABA to gibberellins determines whether the seed will remain dormant or germinate.

In addition to its role as a growth inhibitor, abscisic acid acts as a "stress" hormone, helping the plant cope with adverse conditions. For example, when a plant begins to wilt, ABA accumulates in leaves and causes stomata to close, reducing transpiration and preventing further water loss. This function may depend on ABA that originates in the roots. In some cases, water shortage can stress the root system before the shoot system, and ABA transported from roots to leaves may function as an "early warning system."

Ethylene

Early in the twentieth century, citrus was ripened by "curing" the fruit in sheds equipped with kerosene stoves. Fruit growers believed it was the heat that ripened the fruit, but newer, cleaner-burning stoves did not work. Plant physiologists later learned that ripening in the sheds was actually due to **ethylene**, a gaseous by-product of kerosene combustion. Researchers later demonstrated that plants produce their own ethylene as a hormone, and that this hormone elicits a variety of responses in addition to fruit ripening. Unique among plant hormones because it is a gas, ethylene diffuses through the plant in the air spaces between cells. Dissolved ethylene can also travel from cell to cell in the symplast.

In some cases, ethylene acts to inhibit cell elongation. Many of the inhibitory effects once attributed to auxin are now believed to be the result of ethylene synthesis induced by a high concentration of auxin. It is probably ethylene, for example, that inhibits root elongation and the development of axillary buds in the presence of an excess of auxin. In addition to its role as a growth inhibitor, ethylene is also associated with a variety of aging processes in plants.

Senescence in Plants. Aging, or **senescence**, is a progression of irreversible change that eventually leads to death. A normal part of plant development, senescence may occur at the level of individual cells, entire organs, or the whole plant. Xylem vessel elements and cork cells age and die before assuming their specialized functions. Autumn leaves and withering flower petals are examples of senescent organs. Plants that are annuals age and die soon after flowering. Ethylene probably has important functions in all these cases of senescence, but the aging processes where the effects of the hormone have been studied most extensively are fruit ripening and leaf abscission.

Fruit Ripening. Several changes in structure and metabolism accompany the ripening of an ovary into a fruit. Some of these changes, including the degradation of cell walls, which softens the fruit, and the decrease in chlorophyll content, which causes loss of greenness, can be regarded as aging processes. Ethylene initiates or hastens these deteriorative changes and also causes some ripened fruits to drop from the plant.

A chain reaction occurs during ripening, as ethylene triggers senescence, and the aging cells then release more ethylene. Because ethylene is a gas, the signal to ripen even spreads from fruit to fruit: One bad apple really does spoil the lot. If you pick or buy green fruit, you may be able to speed ripening by storing the fruit in a plastic bag, so that ethylene gas will accumulate. On a commercial scale, many kinds of fruit are ripened in huge storage containers into which ethylene gas is piped—a modern variation on the old curing shed. In other cases, measures are taken to retard ripening caused by natural ethylene. Apples, for instance, are stored in bins flushed with carbon dioxide. Circulating the air prevents ethylene from accumulating, and carbon dioxide inhibits synthesis of new ethylene. Stored in this way, apples picked in autumn can be shipped to grocery stores the following summer. In another application of ethylene research, molecular biologists devised a method to manipulate the expression of one of the genes required for ethylene synthesis (FIGURE 39.9).

Leaf Abscission. The loss of leaves each autumn is an adaptation that keeps deciduous trees from desiccating during winter when the roots cannot absorb water from the frozen ground. Before leaves abscise, many of their essential elements are shunted to storage tissues in the stem. These nutrients are recycled back to developing leaves the following spring. The autumn leaf stops making new chlorophyll and loses its greenness. The fall colors are a combination of new pigments made during autumn and pigments that were already present in the leaf but concealed by the dark green chlorophyll.

FIGURE 39.9 ▪ The control of fruit ripening by antisense RNA. The tomato fruits on the left have ripened naturally as a result of their production of the hormone ethylene. Molecular biologists suppressed ripening of the tomatoes in the center by adding an antisense RNA that blocks transcription of one of the genes required for ethylene synthesis (see Chapter 20). The fruit can then be cued to ripen on demand by adding ethylene gas (tomatoes on the right). As such methods are refined, they may reduce spoilage of fruits and vegetables, a problem that currently ruins almost half the produce harvested in the United States.

When an autumn leaf falls, the breaking point is an abscission layer located near the base of the petiole (FIGURE 39.10). The small parenchyma cells of this layer have very thin walls, and there are no fiber cells around the vascular tissue. The abscission layer is further weakened when enzymes hydrolyze polysaccharides in the cell walls. Finally, the weight of the leaf, with the help of wind, causes a separation within the abscission layer. Even before the leaf falls, a layer of cork forms a protective scar on the twig's side of the abscission layer, preventing pathogens from invading the plant.

A change in the balance of ethylene and auxin controls abscission. An aging leaf produces less and less auxin, and this drop in concentration makes the cells of the abscission layer more sensitive to ethylene. This shift in hormonal balance is self-reinforcing, as cells in the abscission layer begin producing additional ethylene, which inhibits the leaf's synthesis of auxin. As the influence of ethylene on the abscission layer prevails, the cells produce enzymes that digest the cellulose and other components of cell walls.

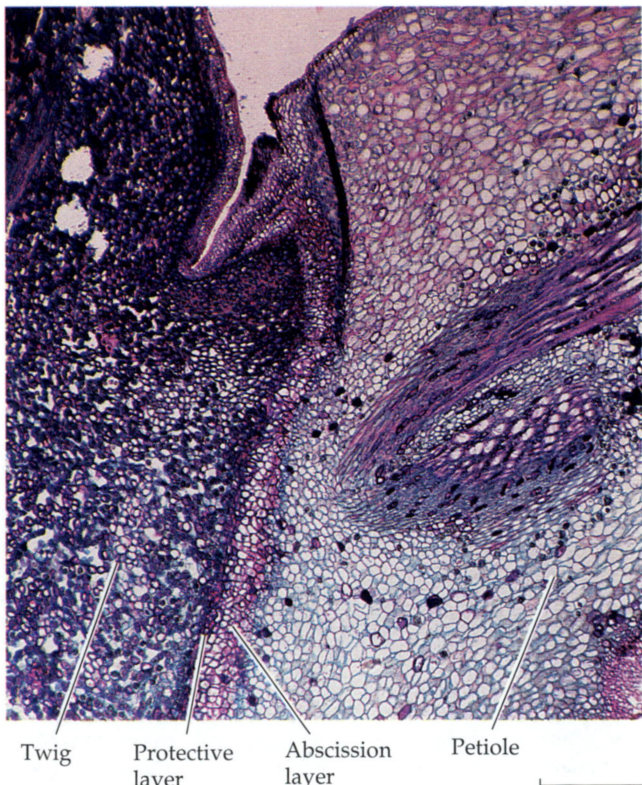

FIGURE 39.10 ▪ Abscission of a maple leaf. Abscission is controlled by a change in the balance of ethylene and auxin. The abscission layer can be seen here as a vertical band at the base of the petiole. After the leaf falls, a protective layer of cork becomes the leaf scar that helps prevent pathogens from invading the plant (LM).

Analysis of mutant plants is extending the list of hormones and their functions

For decades researchers investigated plant hormones mainly by applying the compounds to whole plants or tissue cultures and then measuring the effect on growth and development. In the past few years researchers have started to sort out the molecular biology of hormone synthesis and function by analyzing mutants that grow or develop abnormally. For example, in one mutant variety of tomato, fruit fails to ripen. Researchers have traced this *never-ripe* mutation to a gene that normally codes for a protein that functions as an ethylene receptor. Apparently, *never-ripe* fruit is "deaf" to ethylene. By studying this and other mutations, researchers are beginning to dissect the signal-transduction pathways that link hormonal signals to cellular responses.

In addition to enhancing our understanding of how the well-established plant hormones (auxin, cytokinins, gibberellins, abscissic acid, and ethylene) work, the analysis of mutants has also led to the discovery of new classes of hormones. Two examples are the oligosaccharins and the brassinosteroids.

Oligosaccharins are short chains of sugars released from cell walls by the hydrolytic action of enzymes on cellulose and pectin. Later in the chapter you'll learn that some of these hormones trigger defense responses by plants that have been invaded by pathogens. Other oligosaccharins help regulate growth, cellular differentiation, and flower development.

Brassinosteroids are named for their structure (they are steroids chemically similar to cholesterol and the sex hormones of animals) and the plants in which they were first discovered (members of the mustard family, Brassicaceae). These hormones are now known to occur throughout the plant kingdom and are required for normal growth and development. For example, an *Arabidopsis* mutant with severely stunted growth will grow normally if it is treated with a particular brassinosteroid. Researchers have traced the mutation to a gene that normally codes for one of the enzymes required for synthesis of the steroid hormone.

Signal-transduction pathways link cellular responses to plant hormonal signals and environmental stimuli

Plant hormones activate cellular responses via signal-transduction pathways. You learned in Chapter 11 about the important role that signal transduction plays in cellular regulation. FIGURE 39.11 reviews the basic steps of signal transduction and applies the model to the specific example of auxin-stimulated elongation of cells in a shoot. You will find many more examples of signal-transduction pathways as we continue our study of control systems in plants. For instance, you'll learn how light and other environmental stimuli trigger signal-transduction pathways leading to cellular responses.

■ ■ ■

As we conclude our survey of plant hormones in this first part of the chapter, it is important to remember that these chemical signals are components of control systems that tune a plant's growth, development, reproduction, and physiology to the environment. For example, you learned that auxin functions in the phototropic bending of shoots toward light; that abscissic acid "holds" certain seeds dormant until the environment is suitable for germination; and that ethylene functions in leaf abscission as shorter days and cooler temperatures announce autumn. Such relationships between outside stimuli and the hormonal signals and responses within a plant bring us to an important juncture in our study of control systems. How do plants actually detect changes in their surroundings? Some plant biologists are approaching this problem by studying plant movements.

PLANT MOVEMENTS AS MODELS FOR STUDYING CONTROL SYSTEMS

To the casual observer, most plants do not appear to be very dynamic, but time-lapse photography reveals that they are capable of quite precise movements. The two types of plant movements are tropisms and turgor movements.

Tropisms orient the growth of plant organs toward or away from stimuli

The environment has a great influence on a plant's shape. **Tropisms** are growth responses that result in curvatures of

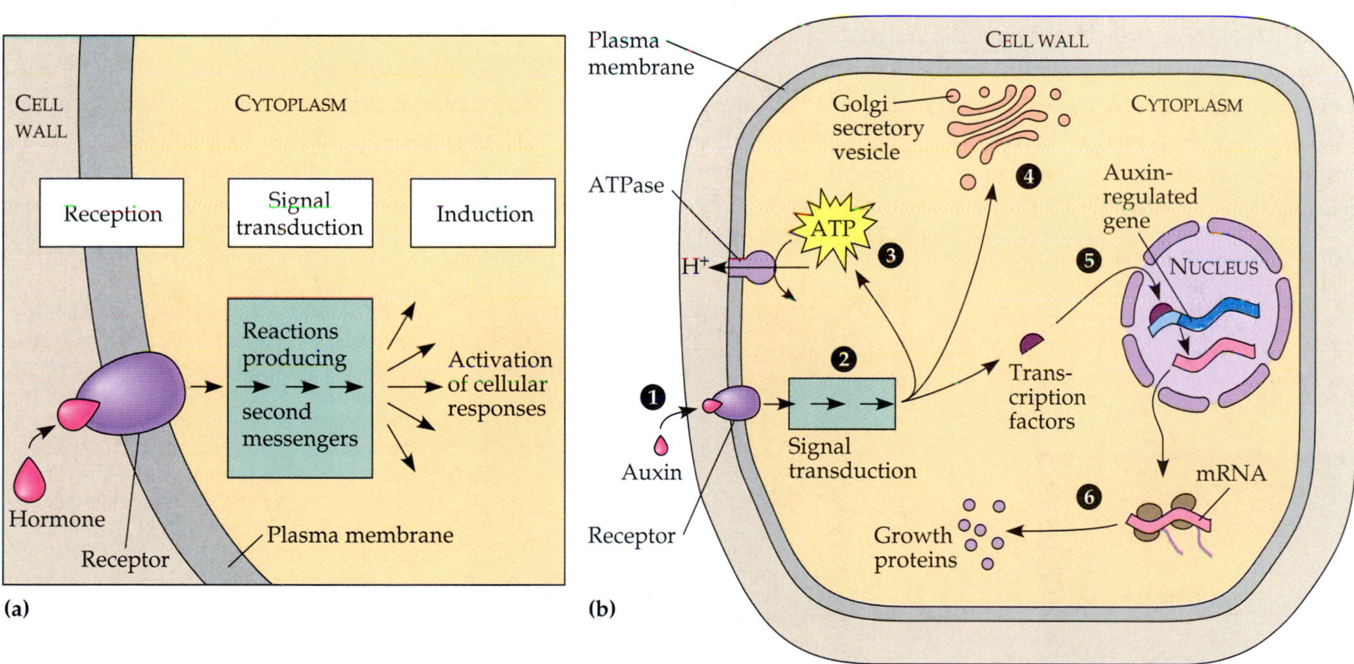

FIGURE 39.11 ▪ Signal-transduction pathways in plant cells. (a) A general model. A hormone binding to a specific receptor stimulates the cell to produce second messengers. A second messenger triggers the cell's various responses to the original signal. In this diagram, the receptor is on the surface of the target cell. In other cases, hormones enter cells and bind to specific receptors inside. Environmental stimuli can also initiate signal-transduction pathways. For example, the absorption of light by a photoreceptor can trigger a plant's response to the light via a signal-transduction pathway. Cyclic AMP and calcium ions (Ca^{2+}) are among the important second messengers in plant cells. Protein kinases, which activate other proteins via phosphorylation, are also common components of signal transduction in plants. (See Chapter 11 to review signal transduction in more detail.) **(b)** A specific example: a hypothetical mechanism for auxin's stimulation of cell elongation. ① The hormone binds to an auxin receptor, and ② this signal is transduced into second messengers within the cell, triggering various responses. ③ Proton pumps are activated, and secretion of acid stimulates enzymes that loosen the wall, enabling the cell to elongate. ④ The Golgi is stimulated to discharge vesicles containing materials that maintain the thickness of the cell wall. ⑤ The signal-transduction pathway also activates transcription factors that induce the transcription of specific genes. ⑥ This leads to the production of proteins required for growth of the cell.

whole plant organs toward or away from stimuli (Gr. *tropos*, "turn"). The mechanism for a tropism is a differential rate of elongation of cells on opposite sides of the organ. Three of the stimuli that induce tropisms, and a consequent change of body shape, are light (phototropism), gravity (gravitropism), and touch (thigmotropism).

Phototropism

As we have seen, the classical hypothesis for what causes grass coleoptiles to grow toward light maintains that cells on the darker side of a stem elongate faster than cells on the brighter side because of an asymmetrical distribution of auxin moving down from the shoot tip. However, studies of phototropism by organs other than grass coleoptiles provide less support for the classical hypothesis. For example, there is no evidence that unilateral light causes an asymmetrical distribution of auxin in the stems of sunflowers, radishes, and other dicots. There *is*, however, an asymmetrical distribution of certain substances that may act as growth inhibitors, with these substances more concentrated on the lighted side of a stem.

Whether phototropism results from auxin stimulating cell elongation on the darker side of a stem or some other chemical messenger inhibiting elongation on the lighter side, most researchers agree that the shoot tip is the site of the photoreception that triggers the growth response. The photoreceptors are pigment molecules called cryptochromes that are most sensitive to blue light. The same blue-light receptor that functions in phototropism may also be involved in stomatal opening.

Gravitropism

Place a seedling on its side, and it will adjust its growth so that the shoot bends upward and the root curves downward. In their responses to gravity, or **gravitropism**, roots display positive gravitropism and shoots exhibit negative gravitropism. Gravitropism functions as soon as a seed germinates, ensuring that the root grows into the soil and the shoot reaches sunlight regardless of how the seed happens to be oriented when it lands.

Plants may tell up from down by the settling of **statoliths**, specialized plastids containing dense starch grains, to the low points of cells (FIGURE 39.12). In roots, statoliths are located in certain cells of the root cap. According to one hypothesis, the aggregation of statoliths at the low points of these cells triggers the redistribution of calcium, which in turn causes lateral transport of auxin within the root. The calcium and auxin accumulate on the lower side of the root's zone of elongation. Because these chemicals are dissolved, they do not respond to gravity but must be actively transported to one side of the root. At high concentration, auxin inhibits cell elongation, an effect that slows growth on the lower side of the root. The more rapid elongation of cells on the upper side causes the root to curve as it grows. This tropism continues until the root is growing straight down.

Plant physiologists are refining the "falling statolith" hypothesis of root gravitropism as they try new experiments. For example, mutants of *Arabidopsis* and tobacco that lack statoliths are still capable of gravitropism, though the response is slower than in wild-type plants. It could be that the entire cell helps the root sense gravity by asymmetric action on proteins

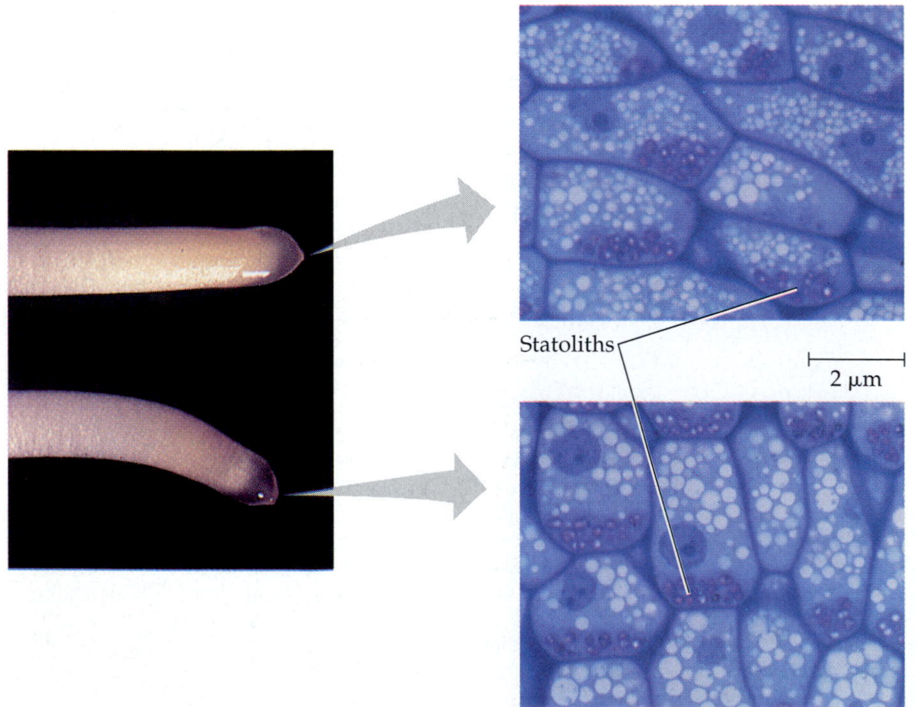

FIGURE 39.12 • The statolith hypothesis for root gravitropism. These corn roots were placed on their sides and photographed before (top) and 1.5 hours after initiation of the gravitropic response. The light micrographs show the locations of statoliths, modified plastids, within cells of the root cap. The settling of statoliths to the low points of these cells may be a gravity-sensing mechanism that leads to the redistribution of auxin and the differential rates of elongation by cells on opposite sides of the root.

that tether the protoplast to the cell wall, stretching the proteins on the "up" side and compressing the proteins on the "down" side of the root cells. Large organelles in addition to starch granules may also contribute by distorting the cytoskeleton as they are pulled by gravity. Statoliths, because of their density, may enhance gravitational sensing by a mechanism that works more slowly in their absence.

Thigmotropism

Most vines and other climbing plants have tendrils that coil around supports (see FIGURE 35.8a). These grasping organs usually grow straight until they touch something; the contact stimulates a coiling response caused by differential growth of cells on opposite sides of the tendril. This directional growth in response to touch is called **thigmotropism** (Gr. *thigma,* "touch").

Mechanical stimulation can also cause a much more general response. One experiment demonstrated that rubbing stems with a stick a few times results in plants that are shorter than controls that are not so manipulated (FIGURE 39.13). In the wild, wind can cause a similar stunting of growth, enabling the plant to hold its ground against strong gusts. A tree growing on a windy mountain ridge, for instance, will usually have a shorter, stockier trunk than a tree of the same species growing in a more sheltered location. This developmental response to mechanical perturbation is referred to as **thigmomorphogenesis**. It usually results from an increased production of ethylene in response to chronic mechanical stimulation.

FIGURE 39.13 ▪ Altering gene expression by touch in *Arabidopsis.* The shorter plant on the left was touched twice a day. Compare it to the unmolested plant on the right, which grew much taller. Researchers have discovered that this developmental response to touch is associated with the induction of five specific genes. A signal-transduction pathway is activated by mechanical stimulation of cells and relays to the genome in the cell nuclei the message to respond.

Turgor movements are relatively rapid, reversible plant responses

In addition to the relatively long-lasting changes in body shape resulting from tropisms, plants are also capable of reversible movements caused by changes in the turgor pressure of specialized cells in response to stimuli.

Rapid Leaf Movements

When the compound leaf of the sensitive plant *Mimosa* is touched, it collapses and its leaflets fold together (FIGURE 39.14, p. 764). This response, which takes only a second or two, results from a rapid loss of turgor by cells within pulvini, specialized motor organs located at the joints of the leaf. The motor cells suddenly become flaccid after stimulation because they lose potassium, which causes water to leave the cells by osmosis. It takes about 10 minutes for the cells to regain their turgor and restore the natural form of the leaf. The function of the sensitive plant's behavior invites speculation. Perhaps by folding its leaves and reducing its surface area when jostled by strong winds, the plant conserves water. Or perhaps because the collapse of the leaves exposes thorns on the stem, the rapid response of the sensitive plant discourages herbivores.

A remarkable feature of rapid leaf movements is the transmission of the stimulus through the plant. If one leaf on a sensitive plant is touched with a hot needle, first that leaf collapses, then the adjacent leaf responds, then the next leaf along the stem, and so on until all the plant's leaves are drooping. From the point of stimulation, the message that produces this response travels wavelike through the plant at a speed of about a centimeter per second. Chemical messengers probably have a role in this transmission, but an electrical impulse can also be detected by attaching electrodes to the plant. These impulses, called **action potentials**, resemble nervous system messages in animals, though the action potentials of plants are thousands of times slower than those of animals. Action potentials, which have been discovered in many species of algae and plants, may be widely used as a form of

FIGURE 39.14 ▪ Rapid turgor movements by the sensitive plant *(Mimosa pudica)*. **(a)** In the unstimulated plant, leaflets are spread apart. **(b)** Within a second or two of being touched, the leaflets have folded together. **(c)** In these light micrographs of a leaflet pair in the closed (stimulated) state, you can see the motor cells in the sectioned pulvini (motor organs). The curvature of a pulvinus is caused by motor cells on one side losing water and becoming flaccid while cells on the opposite side retain their turgor.

(a) Unstimulated

(b) Stimulated

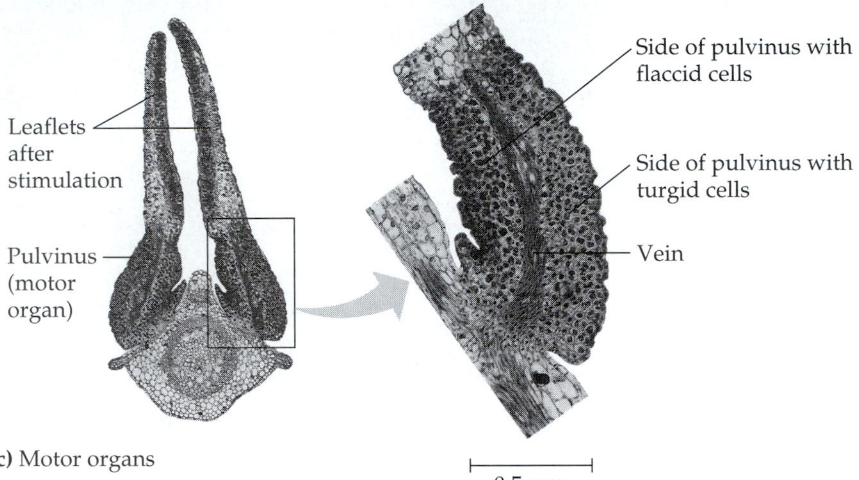

(c) Motor organs

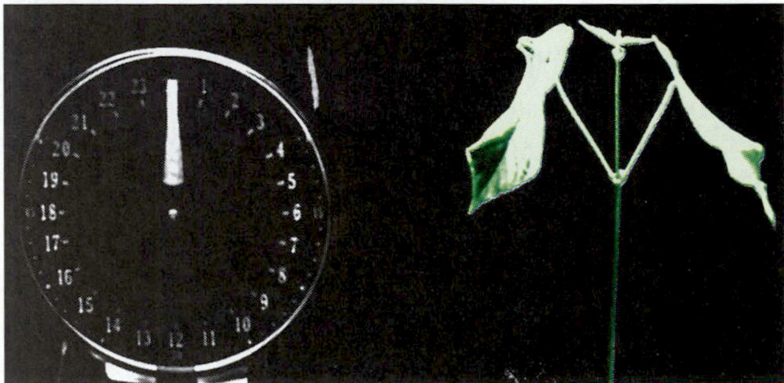

FIGURE 39.15 ▪ Sleep movements of a bean plant. Leaf position at noon (top) and leaf position at midnight. The movements are caused by reversible changes in the turgor pressure of cells on opposing sides of the pulvini, swollen regions of the petiole.

internal communication. Another example is the Venus flytrap, in which action potentials are transmitted from sensory hairs in the trap to the cells that respond by closing the trap (see FIGURE 37.14).

Sleep Movements

Bean plants and many other members of the legume family lower their leaves in the evening and raise them to a horizontal position in the morning (FIGURE 39.15). These **sleep movements** are powered by daily changes in the turgor pressure of motor cells in pulvini similar to those of the sensitive plant. When leaves are horizontal, cells on one side of the pulvinus are turgid, whereas cells on the opposite side are flaccid. This is reversed at night when the leaves close to their "sleeping" position. Paralleling the opposing changes in volume of the motor cells is a massive migration of potassium ions from one side of the pulvinus to the other. Apparently, the potassium is an osmotic agent that leads to the reversible uptake and loss of water by the motor cells. In this respect, the mechanism of sleep movements is similar to stomatal opening and closing. Sleep movements are only one example of the many responses that depend on the ability of plants to keep track of time.

CONTROL OF DAILY AND SEASONAL RESPONSES

Biological clocks control circadian rhythms in plants and other eukaryotes

Your pulse, blood pressure, temperature, rate of cell division, blood cell count, alertness, urine composition, metabolic rate, sex drive, and responsiveness to medications all fluctuate with the time of day. Some insects are more vulnerable to insecticides in the afternoon than they are in the morning. Certain fungi produce spores for several hours the same time each day. Unicellular algae that glow in the dark switch on their bioluminescence like clockwork. Plants also display rhythmic behavior; the sleep movements of legumes and the opening and closing of stomata are examples. All these rhythmic phenomena and many others are controlled by biological clocks, internal oscillators that keep time. Biological clocks seem to be ubiquitous features of eukaryotic organisms, and our first evidence for biological rhythms came from studies of plants.

A physiological cycle with a frequency of about 24 hours is called a **circadian rhythm** (L. *circa,* "approximately," and *dies,* "day"). Are these rhythms truly prompted by an internal clock, or are they merely daily responses to some environmental cycle, such as the rotation of Earth (night and day)? Circadian rhythms persist even when the organism is sheltered from environmental cues. A bean plant, for example, will continue its sleep movements even if kept in constant light or constant darkness; the leaves are not simply responding to sunrise and sunset. Organisms, including humans, continue their rhythms when placed in the deepest mine shafts or when orbited in satellites. All research thus far indicates that the oscillator for circadian rhythms is endogenous (internal). This clock, however, is entrained (set) to a period of precisely 24 hours by daily signals from the environment. If an organism is kept in a constant environment, circadian rhythms deviate from a 24-hour period (a "period" is the duration of one cycle). These free-running periods, as they are called, vary from about 21 to 27 hours, depending on the particular rhythmic response. The sleep movements of bean plants, for instance, have a period of 26 hours when the bean plants are kept under the free-running conditions of constant darkness.

Deviation of the free-running period from exactly 24 hours does not mean that biological clocks drift erratically. The clocks are still keeping perfect time, but they are not synchronized with the outside world. The light-dark cycle resulting from Earth's rotation is the most common factor that entrains biological clocks. If the timing of these cues changes, it takes a few days for the clock to be reset. Thus, a plant kept for several days in the dark will be out of phase with plants growing in a normal environment where the sun rises and sets each day. The same thing happens when we cross several time zones in an airplane; when we reach our destination, the clocks on the wall are not synchronized with our internal clocks. All eukaryotes are probably prone to jet lag.

How do biological clocks work? In attempting to answer this question, we must be careful to differentiate between the oscillator and the rhythmic processes it controls. The sweeping leaves of sleep movements are the "hands" of the biological clock, but these movements are not the essence of the clockwork itself. If the leaves of a bean plant are restrained for several hours so they cannot move, they will, on release, rush to the position appropriate for the time of day. We can interfere with a biological rhythm, but the clock goes right on ticking off the time.

Researchers are tracing the clock to a molecular mechanism that may be common to all eukaryotes. Timekeeping might depend on synthesis of a protein that regulates its own production through feedback control. The protein is a transcription factor that inhibits transcription of the gene that codes for the transcription factor itself. This discovery suggests the hypothesis that the protein undergoes cyclical changes in its concentration as it accumulates and inhibits its own production. Researchers have identified such genes in organisms as diverse as the fruit fly *Drosophila* and the bread mold *Neurospora,* but it is not yet known whether a similar mechanism is associated with the biological clock of plants.

Photoperiodism synchronizes many plant responses to changes of season

Seasonal events are important in the life cycles of most plants. Seed germination, flowering, and the onset and breaking of bud dormancy are examples of stages in plant development that usually occur at specific times of the year. The environmental stimulus plants most often use to detect the time of year is the photoperiod, the relative lengths of night and day. A physiological response to photoperiod, such as flowering, is called **photoperiodism**.

Photoperiodism and the Control of Flowering

39.2 One of the earliest clues to how plants detect the progression of seasons came from a mutant variety of tobacco studied by W. W. Garner and H. A. Allard in 1920. This variety, named Maryland Mammoth, grew exceptionally tall but failed to flower during summer, when normal tobacco plants flowered. Maryland Mammoth finally bloomed in a greenhouse in December. After trying to induce earlier flowering by varying temperature, moisture, and mineral nutrition, Garner and Allard learned that it was the shortening days of winter that stimulated Maryland Mammoth to flower. If the plants were kept in light-tight boxes so that lamps could be used to manipulate durations of "day" and "night," flowering would occur only if the day length were 14 hours or shorter. The

Maryland Mammoths did not flower during summer because, at Maryland's latitude, the days were too long during that season.

Garner and Allard termed Maryland Mammoth a **short-day plant**, because it apparently required a light period *shorter* than a critical length to flower. Chrysanthemums, poinsettias, and some soybean varieties are a few of the other short-day plants, which generally flower in late summer, fall, or winter. Another group of plants dependent on photoperiod will flower only when the light period is *longer* than a certain number of hours. These **long-day plants** generally flower in late spring or early summer. Spinach, for example, flowers when days are 14 hours or longer. Radish, lettuce, iris, and many cereal varieties are also long-day plants. Flowering in a third group, day-neutral plants, is unaffected by photoperiod. Tomatoes, rice, and dandelions are examples of **day-neutral plants** that flower when they reach a certain stage of maturity, regardless of day length at that time.

Critical Night Length. In the 1940s researchers discovered that night length, not day length, actually controls flowering and other responses to photoperiod. Many of these scientists worked with cocklebur, a short-day plant that flowers only when days are 16 hours or shorter (and nights are at least 8 hours long). If the daytime portion of the photoperiod is broken by a brief exposure to darkness, there is no effect on flowering. However, if the nighttime part of the photoperiod is interrupted by even a few minutes of dim light, the plants will not flower (FIGURE 39.16). Cocklebur requires at least 8 hours of *continuous* darkness to flower. Short-day plants are *really* long-night plants, but the older term is embedded firmly in the jargon of plant physiology. Long-day plants are *actually* short-night plants; grown on photoperiods of long nights that would not normally induce flowering, long-day plants will flower if the period of continuous darkness is shortened by a few minutes of light.

Thus, photoperiodic responses depend on a critical night length. Short-day plants will flower if the duration of night is *longer* than the critical length (8 hours for cocklebur); long-day plants will flower when the night is *shorter* than the critical length. The floriculture (flower-growing) industry has applied this knowledge to produce flowers out of season. Chrysanthemums, for instance, are short-day plants that normally bloom in fall, but their blooming can be stalled until Mother's Day in May by punctuating each long night with a flash of light, thus turning one long night into two short nights.

Notice that we distinguish long-day from short-day plants *not* by an absolute night length but by whether the critical night length sets a maximum (long-day plants) or minimum (short-day plants) number of hours of darkness required for flowering.

Some plants bloom after a single exposure to the photoperiod required for flowering. Other species need several successive days of the appropriate photoperiod. Still other plants will respond to photoperiod only if they have been previously exposed to some other environmental stimulus, such as a period of cold temperatures. Winter wheat, for example, will

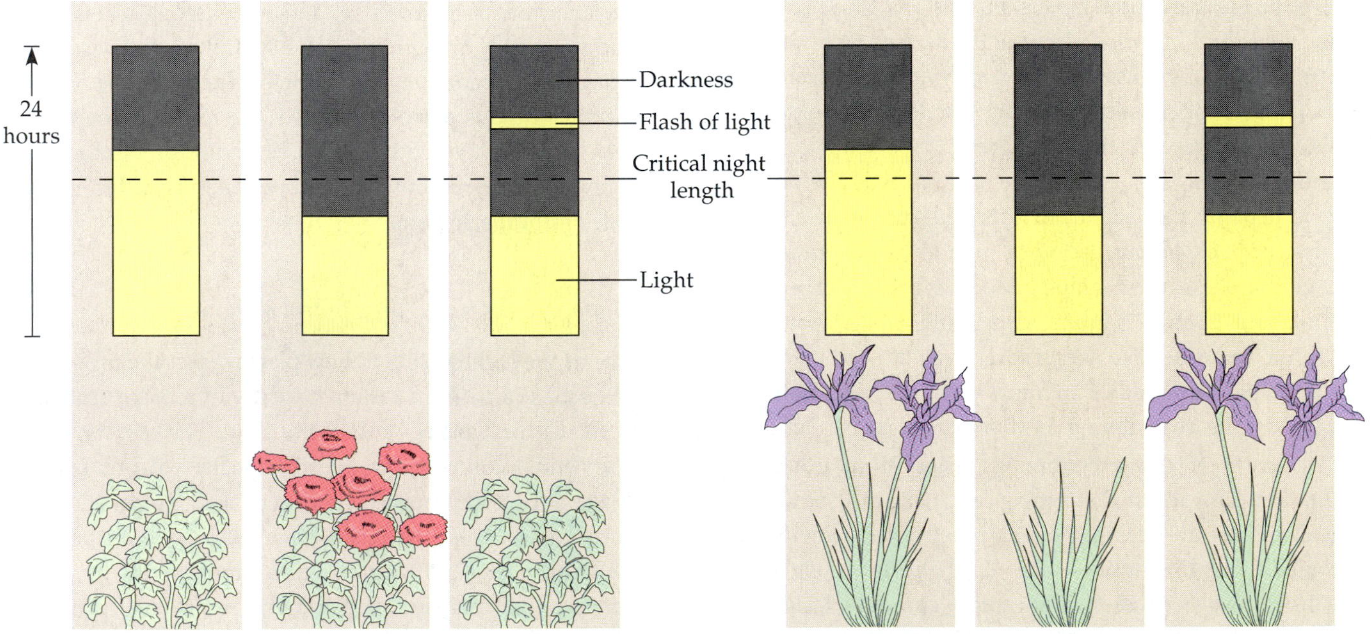

FIGURE 39.16 ▪ Photoperiodic control of flowering. A short-day (long-night) plant flowers when night exceeds a critical dark period. A flash of light interrupting the dark period prevents flowering. A long-day (short-night) plant flowers only if the night is shorter than a critical dark period. The night can be artificially shortened with a flash of light.

39.2

not flower unless it has been exposed to several weeks of temperatures below 10°C. This requirement for pretreatment with cold before flowering is called vernalization. Several weeks after winter wheat is vernalized, a photoperiod with long days (short nights) induces flowering.

Is There a Flowering Hormone? Buds produce flowers, but the photoperiod is detected by leaves. To induce a short-day plant or long-day plant to flower, it is enough in many species to expose a single leaf to the appropriate photoperiod. Indeed, if only one leaf is left attached to the plant, photoperiod is detected and floral buds are induced. If all leaves are removed, however, the plant is blind to photoperiod. Apparently, some message to flower is transported from leaves to buds. Most plant physiologists believe this message is a hormone or some change in the relative concentrations of two or more hormones (FIGURE 39.17). The signal to flower that travels from leaves to buds appears to be the same for short-day and long-day plants, although the two groups of plants differ in the photoperiodic conditions required for leaves to send this signal. The evidence for hormonal signaling in flowering is compelling, but researchers have not yet identified the hormone(s).

Meristem Transition from Vegetative Growth to Flowering. Whatever combination of environmental cues (such as photoperiod) and internal signals (such as hormones) is necessary for flowering to occur, the outcome is the transition of a bud's meristem from a vegetative state to a flowering state. This transition requires changes in the expression of genes that regulate pattern formation. Meristem-identity genes that specify that the bud will now form a flower instead of a vegetative shoot must first be switched on. Then the organ-identity genes that specify the spatial organization of the floral organs—sepals, petals, stamens, and carpels—are activated in the appropriate regions of the meristem (see Chapters 21 and 38). Research on flower development is progressing rapidly, and one goal is to identify the signal-transduction pathways that link such cues as photoperiod and hormonal changes to the gene expression required for flowering.

FIGURE 39.17 ▪ Experimental evidence for some flowering hormone(s). If a plant that has been induced to flower by photoperiod is grafted to a plant that has not been induced, both plants flower, indicating the transmission of a flower-inducing substance. This works in some cases even if one is a short-day plant and the other is a long-day plant.

PHYTOCHROMES

When you consider that plants are solar-powered organisms, it is not surprising that light is an especially important environmental factor in affecting plant growth, development, and reproduction. Plants sense not just the presence of light, but also its direction, intensity, and wavelengths (colors). You have learned, for example, that positive phototropism of a shoot depends on stimulation of a blue-light receptor that is somehow coupled to the growth response that bends the shoot toward light. In this section we'll investigate a family of pigments called phytochromes, which play especially important roles in tuning a plant's form and function to the lighting conditions of its environment.

Phytochromes function as photoreceptors in many plant responses to light and photoperiod

The discovery that night length is a critical factor controlling seasonal responses of plants leads to another question: How does a plant measure the length of darkness in a photoperiod? Pigments named **phytochromes** are part of the answer. They were discovered as a result of studies on how different colors of light affect flowering, seed germination, and other responses to photoperiod.

In experiments on the photoperiodic control of flowering, red light, of wavelength 660 nanometers (nm), is the most effective light in interrupting night length. A short-day plant kept under conditions of critical night length fails to flower if a brief exposure to red light breaks the dark period. Conversely, a flash of red light during the dark period will induce a long-day plant to flower even if the total night length exceeds the critical number of hours (long-day plants, remember,

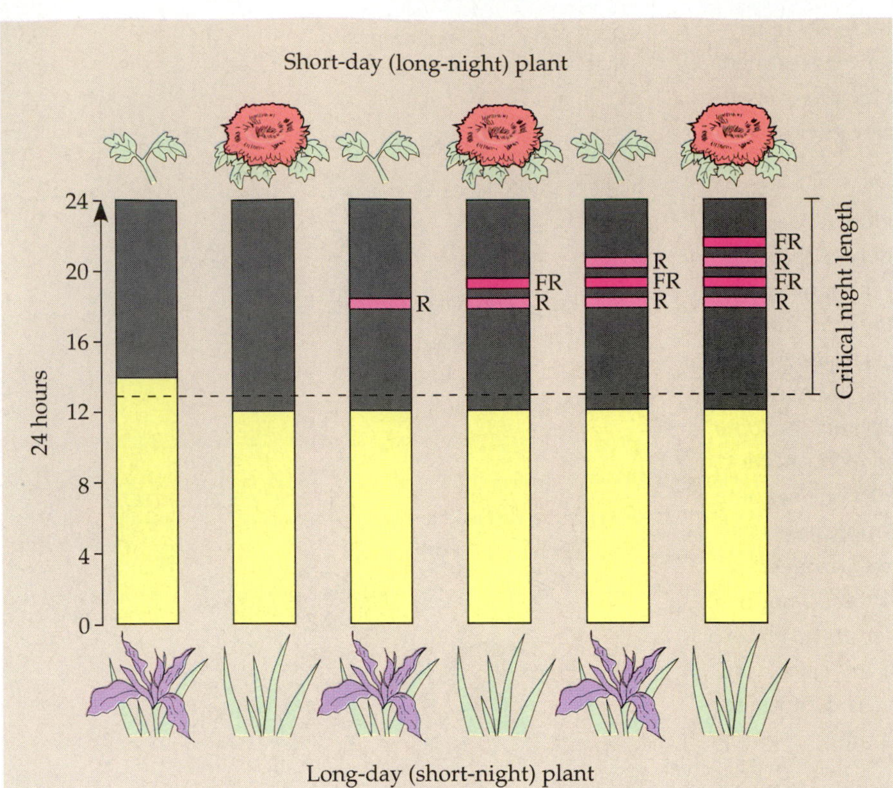

FIGURE 39.18 ▪ Reversible effects of red and far-red light on photoperiodic response. A flash of red light shortens the dark period. A subsequent flash of far-red light cancels the effect of the red flash.

require nights shorter than a critical length). The effect of the red flash is to shorten the plant's perception of night length.

The shortening of night length by red light can be negated by a subsequent flash of light having a wavelength of about 730 nm. This wavelength is in the far-red part of the spectrum, just barely visible to humans. If red light (R) during the dark period is followed by far-red light (FR), the plant perceives no interruption in night length. A short-day plant will not flower if a night of critical length is broken by an R flash, but the plant *will* flower if it receives two flashes, first R and then FR. Reversing this sequence to FR-R prevents flowering. Each wavelength of light cancels the effect of the one that precedes it. No matter how many flashes of light are given, the wavelength of only the last one affects the plant's measurement of night length. Thus, a succession of flashes with the sequence R-FR-R-FR-R prevents short-day plants from flowering, but flowering occurs if the sequence is R-FR-R-FR-R-FR. As expected, the opposite behavior occurs in long-day plants (FIGURE 39.18).

The photoreceptor responsible for the reversible effects of red and far-red light is a phytochrome. It consists of a protein component covalently bonded to a nonprotein part that functions as a chromophore, the light-absorbing part of the molecule (FIGURE 39.19). So far, researchers have identified five phytochromes in *Arabidopsis,* each with a slightly different protein. The different phytochromes may induce different responses of the plant to light, or it may be that two or more phytochromes cooperate in triggering certain responses.

The chromophore of a phytochrome reverts back and forth between two isomeric forms. One isomer absorbs red light, and the other absorbs far-red light. The two variations of phytochrome—P_r (red absorbing) and P_{fr} (far-red absorbing)—are said to be photoreversible:

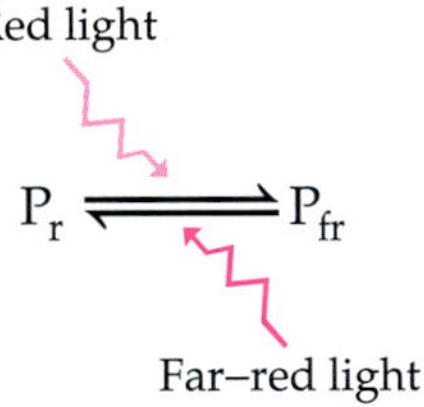

This $P_r \rightleftharpoons P_{fr}$ interconversion acts as a switching mechanism that controls various events in the life of the plant (FIGURE 39.20).

The Ecological Significance of Phytochrome as a Photoreceptor

Light is a key environmental variable in the life of a plant. In the photoperiodic control of flowering and many other plant responses to illumination, phytochrome functions as a photodetector that tells the plant whether light is present. Plants synthesize phytochrome as P_r, and if they are kept in the dark the pigment remains in that form. Then, if the phytochrome is illuminated with sunlight, some of the P_r is converted to P_{fr} because the pigment is exposed to red light (along with all the

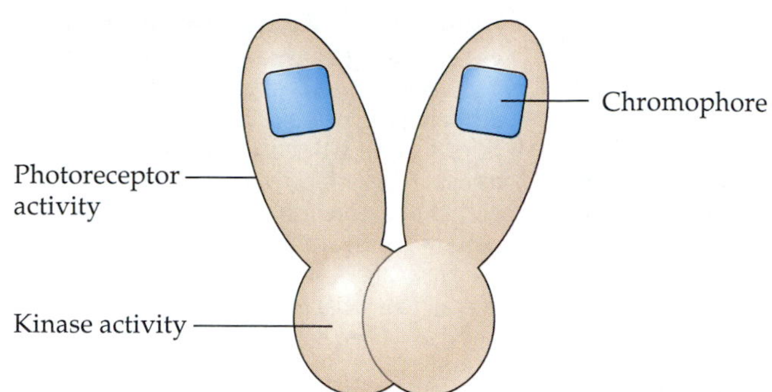

FIGURE 39.19 ▪ Structure of a phytochrome. Phytochromes are homodimers, meaning that each molecule consists of two identical proteins (left and right halves in this simplified diagram) joined to form one functional molecule. Each of these proteins has two domains. One, which functions as the photoreceptor, is covalently bonded to a nonprotein pigment, the chromophore. The other domain joins the protein to its identical partner in the dimer, and this domain also has protein kinase activity. You learned in Chapter 11 that protein kinases are regulatory proteins that activate or inhibit other proteins by phosphorylating them. The structure of a phytochrome molecule suggests that its photoreceptor domains interact with its kinase domains to link light absorption to cellular responses triggered by the kinase.

other wavelengths in sunlight) for the first time. This appearance of P_{fr} is one way plants detect sunlight. P_{fr} is the form of phytochrome that triggers many of a plant's developmental responses to light, such as the germination of seeds that require light to break dormancy.

The phytochrome system also provides the plant with information about the *quality* of light. Sunlight includes both red and far-red radiation. Thus, during the day the $P_r \rightleftharpoons P_{fr}$ photoreversion reaches a dynamic equilibrium, with the ratio of the two phytochrome forms indicating the relative amounts of red and far-red light. This sensing mechanism enables plants to adapt to changes in light conditions. Consider, for example, a "shade-avoiding" tree that requires relatively high light intensity. If this tree becomes shaded by other trees in a forest, the phytochrome ratio shifts in favor of P_r because the forest canopy screens out more red light than far-red light. This cue induces the tree to use most of its resources to grow taller. In contrast, direct sunlight increases the proportion of P_{fr}, which stimulates branching and inhibits vertical growth.

Plant cells actually have very little phytochrome, and yet photoreception by the pigment can have an enormous effect on the whole plant. This implies that the photoconversion of phytochrome from P_r to P_{fr} is a signal that becomes amplified in the signal-transduction pathways that lead to the plant's responses to light. These pathways probably involve the kind of kinase cascades you learned about in Chapter 11.

Phytochromes may entrain the biological clock

In darkness, the phytochrome ratio shifts gradually in favor of the P_r form. This is partly due to turnover in the overall phytochrome pool. The pigment is synthesized in the P_r form, and degradative enzymes destroy more P_{fr} than P_r (see FIGURE 39.20). In addition, in some plant species, P_{fr} present at sundown slowly converts to P_r by a biochemical mechanism. In darkness, there is no means for the accumulating P_r to be reconverted to P_{fr}. Then, at sunrise, the P_{fr} level suddenly increases again by rapid photochemical conversion from P_r.

Thus, phytochrome conversion marks the beginning and end of the dark segment of a photoperiod. However, since the conversion of P_{fr} to P_r is usually completed within a few hours after sunset, P_{fr} disappearance cannot be functioning like an hourglass to measure the length of the dark period.

Night length is measured not by phytochrome but by the biological clock. The role of phytochrome in photoperiodism may be to synchronize the clock to the environment by telling it when the sun sets and rises. If the photoperiodic requirement for flowering is met, the clock triggers some sort of alarm that causes leaves to send a flowering stimulus (perhaps a hormone) to buds. Night length is measured very accurately; some short-day plants will not flower if night is even one minute shorter than the critical length. Some plant species always flower on the same day each year. According to the hypothesis described here, plants tell the season of year by using the clock, apparently entrained with the help of phytochrome, to keep track of photoperiod.

PLANT RESPONSES TO ENVIRONMENTAL STRESS

Environmental fluctuations challenge plants every day of their lives. Occasionally, factors in the environment change

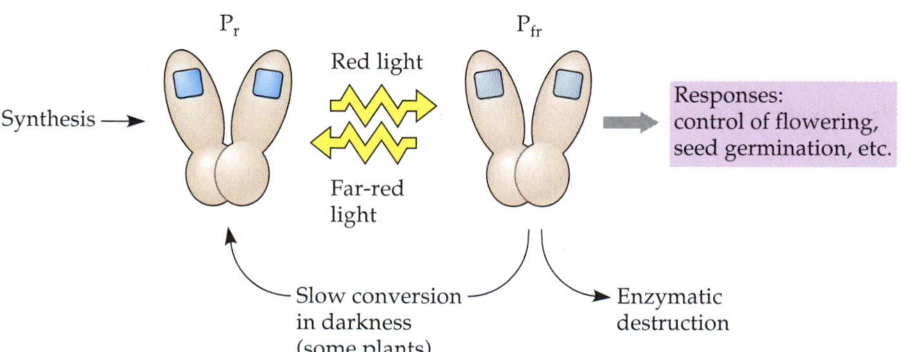

FIGURE 39.20 ▪ Phytochrome: a molecular switching mechanism. Absorption of red light causes the bluish P_r to change to the blue-greenish P_{fr}. Far-red light reverses this conversion. In most cases, it is the P_{fr} form of the pigment that switches on physiological and developmental responses in the plant.

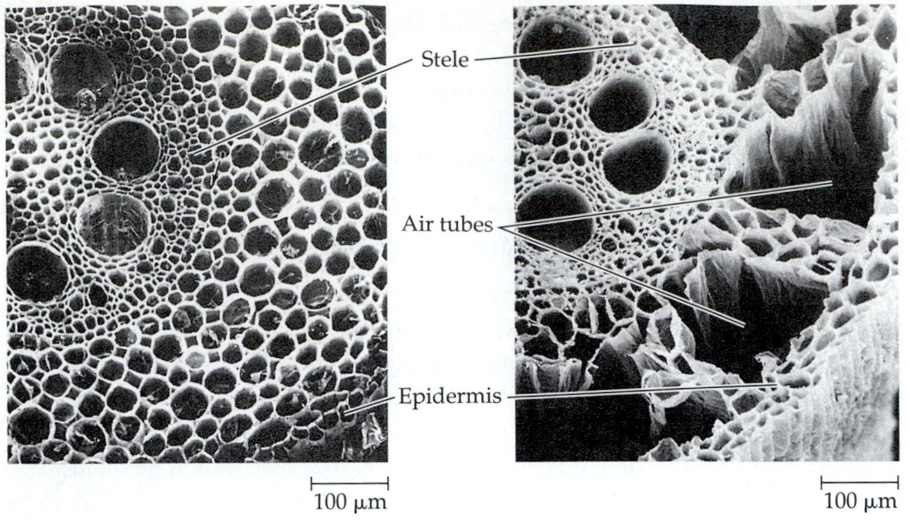

FIGURE 39.21 ▪ A developmental response of corn roots to oxygen deprivation. (a) A transverse section of a control root grown in an aerated hydroponic medium. **(b)** An experimental root grown in a nonaerated hydroponic medium. Oxygen deprivation stimulates the production of the hormone ethylene, which causes some of the cells in the root cortex to age and die. Enzymatic destruction of the cell walls creates air tubes that function as "snorkels," providing oxygen to the submerged roots.

severely enough to put plants under stress. We'll define stress here as an environmental condition that can have an adverse effect on a plant's growth, reproduction, and survival.

Plants cope with environmental stress through a combination of developmental and physiological responses

Responses to Water Deficit

On any bright, warm, dry day, a plant may be stressed by a water deficiency because it is losing water by transpiration faster than the water can be restored by uptake from the soil. Prolonged drought can stress crops and the plants of natural ecosystems for weeks or months. Severe water deficit, of course, will kill a plant, as you may know from experience with neglected houseplants. But plants have control systems that enable them to cope with less extreme water deficits.

Many of a plant's responses to water deficit help the plant conserve water by reducing the rate of transpiration. Water deficit in a leaf causes guard cells to lose turgor, a simple control mechanism that slows transpiration by closing stomata (see Chapter 36). Water deficit also stimulates increased synthesis and release of abscisic acid from mesophyll cells in the leaf, and this hormone helps keep stomata closed by acting on guard cell membranes. Leaves respond to water deficit in several other ways. Because cell expansion is a turgor-dependent process, a water deficit will inhibit the growth (expansion) of young leaves. This response minimizes the transpirational loss of water by slowing the increase in leaf surface. When the leaves of many grasses and other plants wilt from a water deficit, they roll into a shape that reduces transpiration by exposing less leaf surface to the sun. While all of these responses of leaves help the plant conserve water, they also reduce photosynthesis. This is one reason a drought diminishes crop yield.

Root growth also responds to water deficit. During a drought, the soil usually dries from the surface down. This inhibits the growth of shallow roots, partly because cells cannot maintain the turgor required for elongation. Deeper roots surrounded by soil that is still moist continue to grow. Thus, the root system proliferates in a way that maximizes exposure to soil water.

Responses to Oxygen Deprivation

An overwatered houseplant may suffocate because the soil lacks the air spaces that provide oxygen for cellular respiration in the roots. Some plants are structurally adapted to very wet habitats. For example, the submerged roots of trees called mangroves, which inhabit coastal marshes, are continuous with aerial roots that provide access to oxygen. But how do plants less specialized for aquatic environments cope with oxygen deprivation in waterlogged soils? One structural change is the formation of air tubes that provide oxygen to submerged roots (FIGURE 39.21).

Responses to Salt Stress

An excess of sodium chloride or other salts in the soil threatens plants for two reasons. First, by lowering the water potential of the soil solution, salt can cause a water deficit in plants even though the soil has plenty of water. This is because in an environment with a water potential more negative than that of the root tissue, roots will *lose* water rather than absorb it (see Chapter 36). The second problem with saline soil is that sodium and certain other ions are toxic to plants when their concentration is relatively high. The selectively permeable membranes of root cells impede the uptake of most harmful ions, but this only aggravates the problem of acquiring water from soil that is rich in solutes. Many plants can respond to moderate soil salinity by producing *compatible solutes,* organic compounds that keep the water potential of cells more negative than that of the soil solution without admitting toxic quantities of salt. However, most plants cannot survive salt stress for very long. The exceptions are halophytes, salt-tolerant plants with special adaptations such as salt glands, which pump salts out of the plant across the leaf epidermis.

Responses to Heat Stress

Excessive heat can harm and eventually kill a plant by denaturing its enzymes and damaging its metabolism in other ways. One function of transpiration is evaporative cooling. On a warm day, for example, the temperature of a leaf may be 3°–10°C below ambient air temperature. Of course, hot, dry weather also tends to cause water deficiency in many plants; the closing of stomata in response to this stress conserves water but sacrifices evaporative cooling. This dilemma is one of the reasons that very hot, dry days take such a toll on most plants.

Most plants have a backup response that enables them to survive heat stress. Above a certain temperature—about 40°C for most plants that inhabit temperate regions—plant cells begin synthesizing relatively large quantities of special proteins called **heat-shock proteins**. Researchers have also discovered this response in heat-shocked animals and microorganisms. Some heat-shock proteins are identical to chaperone proteins, which function in unstressed cells as temporary scaffolds that help other proteins fold into their functional conformations (see Chapter 5). As heat-shock proteins, perhaps these molecules embrace enzymes and other proteins and help prevent denaturation.

Responses to Cold Stress

One problem plants face when the temperature of the environment falls is a change in the fluidity of cell membranes. Recall from Chapter 8 that a biological membrane is a fluid mosaic, with proteins and lipids moving laterally in the plane of the membrane. When a membrane cools below a critical point, it loses its fluidity as the lipids become locked into crystalline structures. This alters solute transport across the membrane and also adversely affects the functions of membrane proteins. Plants respond to cold stress by altering the lipid composition of their membranes. For example, membrane lipids increase in their proportion of unsaturated fatty acids, which have shapes that help keep membranes fluid at lower temperatures by impeding crystal formation (see FIGURE 8.3). Such molecular modification of the membrane requires from several hours to days, which is one reason rapid chilling is generally more stressful to plants than a more gradual drop in air temperature.

Freezing is a more severe version of cold stress. At subfreezing temperatures, ice crystals begin to form in most plants. If the ice is confined to cell walls and intercellular spaces, the plant will probably survive. However, if ice begins to form within protoplasts, the sharp crystals perforate membranes and organelles, killing the cells. Oaks, maples, roses, rhododendrons, and other woody plants native to regions where winters are cold have special adaptations that enable them to cope with freezing stress. For example, changes in the solute composition of live cells allow the cytosol to cool below 0°C without ice forming, although ice crystals may form in the cell walls.

Responses to Herbivores

Herbivory—animals eating plants—is a stress that plants face in any ecosystem. Plants counter excessive herbivory with both physical defenses, such as thorns, and chemical defenses, such as the production of distasteful or toxic compounds. For example, some plants produce an unusual amino acid called canavanine, named for one of its sources, the jackbean *(Canavalia ensiformis)*. Canavanine resembles arginine, one of the 20 amino acids organisms incorporate into their proteins. If an insect eats a plant containing canavanine, the molecule is incorporated into the insect's proteins in place of arginine. Because canavanine is different enough from arginine to adversely affect the conformation and hence the function of proteins, the insect dies.

At least some plants even recruit predatory animals that help defend the plant against specific herbivores (FIGURE 39.22). For example, insects called parasitoid wasps inject their

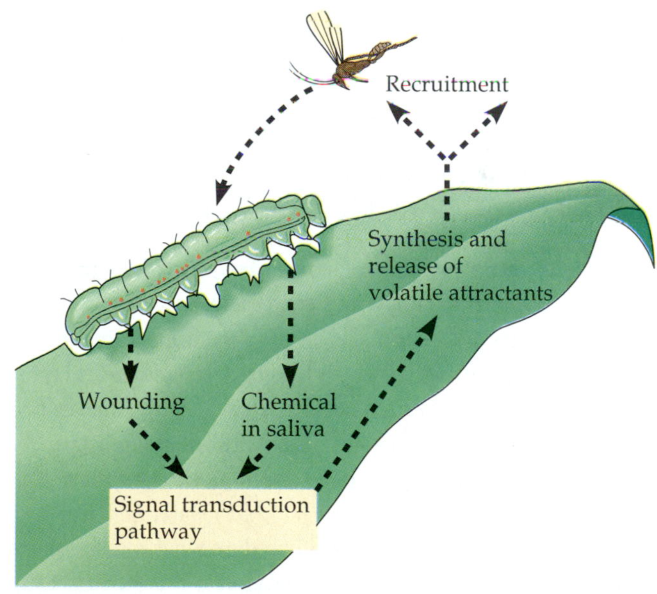

FIGURE 39.22 ▪ Chemical signaling and recruitment of a parasitoid wasp as a response to an herbivore. The physical damage and a salivary chemical from the herbivore, an army-worm caterpillar, triggers a signal-transduction pathway in cells of a corn leaf. The cells respond by synthesizing and releasing volatile attractants that recruit a parasitoid wasp. The wasp will lay its eggs within the caterpillar, and the larvae that hatch from these eggs will destroy the herbivore by consuming its tissues.

eggs into their prey, including caterpillars feeding on plants. The eggs hatch within the caterpillars, and the larvae eat through their organic containers from the inside out. The plant, which benefits from the destruction of the herbivorous caterpillars, has an active role in this ecological drama. In 1997, researchers reported that a leaf damaged by a caterpillar releases volatile compounds that actually attract parasitoid wasps. The stimulus for this response is a combination of the physical damage to the leaf caused by the munching caterpillar and a specific compound present in the caterpillar's saliva.

DEFENSE AGAINST PATHOGENS

A plant, like an animal, is subject to infection by a diversity of pathogenic viruses, bacteria, and fungi that have the potential to damage tissues or even kill the plant. And like an animal, a plant has defense systems that prevent infection and counter pathogens that do manage to infect the plant.

The first line of defense is the physical barrier of the plant's "skin," the epidermis of the primary plant body and the periderm of the secondary plant body (see Chapter 35). This first defense system, however, is not impenetrable. Viruses, bacteria, and the spores and hyphae of fungi can enter the plant through injuries or through natural openings in the epidermis, such as stomata. Once a pathogen invades, the plant mounts a chemical attack as a second line of defense that kills the pathogens and prevents their spread from the site of infection. This second defense system is enhanced by the plant's inherited ability to recognize certain pathogens.

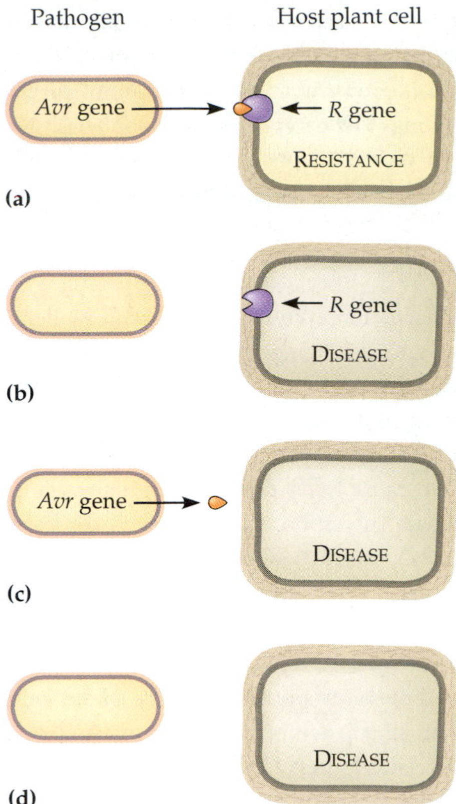

FIGURE 39.23 ▪ Gene-for-gene resistance of plants to pathogens. **(a)** Resistance occurs when the plant has a particular dominant *R* allele that corresponds to a specific dominant *Avr* allele in the pathogen. *R* genes probably code for specific receptors. *Avr* genes produce compounds that function in the pathogen but also act as ligands that bind specifically to the host-plant cell's receptors. Disease occurs if there is no gene-for-gene recognition because **(b)** the pathogen has no dominant *Avr* allele matching an *R* allele in the plant, **(c)** the plant has no dominant *R* allele matching an *Avr* allele in the pathogen, or **(d)** both pathogen and plant lack alleles upon which recognition could be based.

Resistance to disease depends on a gene-for-gene recognition between plant and pathogen

Pathogens against which a plant has little specific defense are said to be virulent. They are the exceptions, for if they were not, then hosts and pathogens would soon perish together. A kind of "compromise" has coevolved between plants and most of their pathogens. In such cases, the pathogen gains enough access to its host to perpetuate itself without severely damaging or killing the plant. The plant is said to be resistant to that particular pathogen.

Specific resistance to disease is based on what is called **gene-for-gene recognition**, because it requires a precise match-up between an allele in the plant and an allele in the pathogen (FIGURE 39.23). A plant has many *R* genes (for resistance), and each pathogen has a set of *Avr* genes (for avirulence, meaning not virulent). The plant is resistant to a pathogen if *any one* of the plant's *R* genes is a dominant allele that corresponds to a dominant *Avr* allele in the pathogen. It is not the genes themselves that interact, of course, but their products. The product of an *R* gene is probably a specific receptor protein inside a plant cell or at its surface. The *Avr* gene probably leads to production of some "signal" molecule from the pathogen, a ligand capable of binding specifically to the plant cell's receptor. The *Avr* product undoubtedly has some function upon which the pathogen itself depends, but the plant is able to "key" on this molecule as an announcement of the pathogen's presence. Binding of the ligand to the receptor triggers a signal-transduction pathway leading to a defense response in the infected plant tissue. This defense includes both an enhancement of the response at the site of infection and a more general response of the whole plant.

The hypersensitive response (HR) contains an infection

Even if a plant is infected by a virulent strain of a pathogen—one for which that particular plant has no genetic resistance—the plant is able to mount a localized chemical attack in response to molecular signals released from cells damaged by

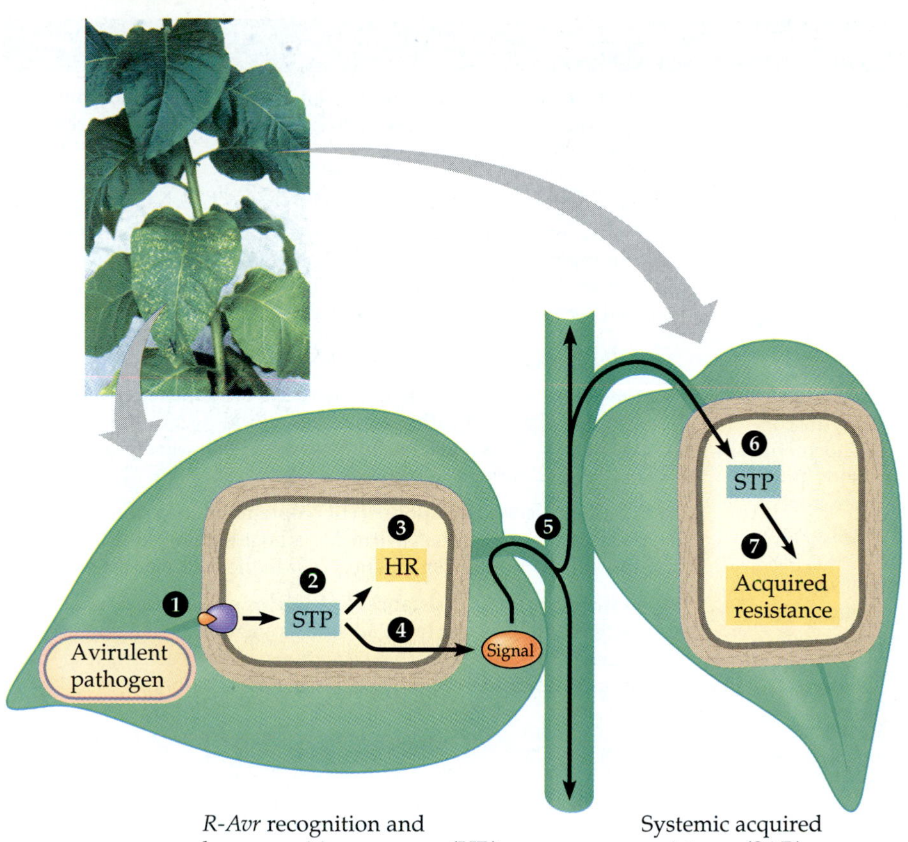

FIGURE 39.24 ▪ Defense responses against an avirulent pathogen. ① Specific resistance is based on the binding of ligands from the pathogen to specific receptors of cells in the infected plant tissue. ② This identification step triggers a signal-transduction pathway (STP) that results in a ③ hypersensitive response (HR). In this HR, the plant cells produce antimicrobial molecules, seal off the infected area by modifying their walls, and then destroy themselves. This localized response produces the lesions in the photograph of an infected leaf. ④ Before they die, the infected cells release a chemical signal, probably salicylic acid, that is ⑤ distributed to the rest of the plant. ⑥ In the cells of leaves and other organs remote from the infection site, the chemical messenger initiates a signal-transduction pathway that ⑦ activates systemic acquired resistance. This SAR includes production of antimicrobial molecules that help protect the cells against a diversity of pathogens for several days.

the infection. The response includes the production of antimicrobial compounds called **phytoalexins**. Infection also activates genes that produce **PR proteins** (for pathogenesis-related). Some of these proteins are antimicrobial, attacking molecules in the cell wall of a bacterium, for example. Others may function as signals that spread "news" of the infection to nearby cells. Infection also stimulates cross-linking of molecules in the cell wall and deposition of lignin, responses that set up a local barricade that slows spread of the pathogen to other parts of the plant.

If the pathogen is avirulent based on an *R-Avr* match, the localized defense response is more vigorous and is called a **hypersensitive response** (abbreviated **HR**). There is an enhanced production of phytoalexins and PR proteins, and the "sealing" response that contains the infection is more effective. After cells at the site of infection mount their chemical defense and seal off the area, they destroy themselves. We can see the result of an HR as lesions on a leaf or other infected organ. As "sick" as such a leaf appears, it will survive, and its defense response will help protect the rest of the plant (FIGURE 39.24).

Systemic acquired resistance (SAR) helps prevent infection throughout the plant

The hypersensitive response, as you have learned, is localized and specific, a containment response based on gene-for-gene (*R* for *Avr*) recognition between host and pathogen. However, this defense response also includes production of chemical signals that "sound the alarm" of infection to the whole plant. Released from the site of infection, the alarm hormones are transported throughout the plant, stimulating production of phytoalexins and PR proteins. This reponse, called **systemic acquired resistance (SAR)**, is nonspecific, providing protection against a diversity of pathogens for days (see FIGURE 39.24).

A good candidate for one of the hormones responsible for activating SAR is salicylic acid. A modified form of this compound, acetylsalicylic acid, is the active ingredient in aspirin. Centuries before aspirin was sold as a pain reliever, some cultures had learned that chewing the bark of a willow tree *(Salix)* would lessen the pain of a toothache or headache. With the discovery of systemic acquired resistance, biologists have finally learned one function of salicylic acid in plants. Aspirin turns out to be a natural medicine in the plants that produce it, but with effects entirely different from the medicinal action in humans who consume the drug.

▪ ▪ ▪

Botanists investigating disease resistance and other control systems are getting to the heart of how a plant adapts to its environment. These scientists, along with thousands of other plant biologists working on other problems and millions of students experimenting with plants in biology courses, are all extending a centuries-old tradition of curiosity about the form and function of the organisms that feed the biosphere.

CHAPTER REVIEW

REVIEW OF KEY CONCEPTS

(with page numbers and key figures)

PLANT HORMONES

- **Research on how plants grow toward light led to the discovery of plant hormones: *science as a process* (pp. 752–753, FIGURES 39.2, 39.3)** Researchers discovered auxin by identifying the compound responsible for transmitting a signal downward through coleoptiles, from the tips to the elongating regions.

- **Plant hormones help coordinate growth, development, and**
39.1 **responses to environmental stimuli (pp. 753–760, TABLE 39.1)** The site of action, the developmental stage of the plant, the concentration of the hormone, and the presence of other hormones all affect reaction to a hormone. This review cites one major function of each hormone. Produced primarily in the apical meristem of the shoot, auxin simulates cell elongation in different target tissues. Cytokinins, produced in actively growing tissues such as roots, embryos, and fruits, stimulate cell division. Gibberellins produced in roots and young leaves stimulate growth in leaves and stems. Abscisic acid slows plant growth and favors the dormant state. Ethylene helps control fruit ripening and senescence of plant cells and organs.

- **Analysis of mutant plants is extending the list of hormones and their functions (p. 760)** Study of *Arabidopsis* mutants with abnormal growth led to the discovery of two new classes of growth regulators: oligosaccharins and brassinosteroids.

- **Signal-transduction pathways link cellular responses to hormonal signals and environmental stimuli (pp. 760–761, FIGURE 39.11)** Binding of hormones to specific receptors activates signal transduction, leading to the cell's responses to that particular hormone.

PLANT MOVEMENTS AS MODELS FOR STUDYING CONTROL SYSTEMS

- **Tropisms orient the growth of plant organs toward or away from stimuli (pp. 760–763, FIGURE 39.12)** These growth responses include phototropism, gravitropism, and thigmotropism (growth responses to touch).

- **Turgor movements are relatively rapid, reversible plant responses (pp. 763–764, FIGURE 9.14)** The motor cells responsible for these movements, including sleep movements, undergo turgor changes caused by potassium transport.

CONTROL OF DAILY AND SEASONAL RESPONSES

- **Biological clocks control circadian rhythms in plants and other eukaryotes (p. 765, FIGURE 39.15)** Free-running circadian cycles are approximately 24 hours long but are entrained to exactly 24 hours by the day/night cycle.

- **Photoperiodism synchronizes many plant responses to changes of**
39.2 **season (pp. 765–767, FIGURE 39.16)** Some developmental processes, including flowering in many plant species, require a certain photoperiod, the relative lengths of night and day. For example, a critical night length sets a minimum (in short-day plants) or maximum (in long-day plants) number of hours of darkness required for flowering.

PHYTOCHROMES

- **Phytochromes function as photoreceptors in many plant responses to light and photoperiod (pp. 767–769, FIGURES 39.18–39.20)** Phytochromes exist in two photoreversible states, with conversion of P_r to P_{fr} triggering many developmental responses.

- **Phytochromes may entrain the biological clock (p. 769)** Phytochrome conversion marks sunrise and sunset, providing the clock with environmental cues.

PLANT RESPONSES TO ENVIRONMENTAL STRESS

- **Plants cope with environmental stress through a combination of developmental and physiological responses (pp. 770–772)** The stresses include water deficit, salinity, oxygen deprivation, heat, cold, herbivory, and pathogens.

DEFENSE AGAINST PATHOGENS

- **Resistance to disease depends on a gene-for-gene recognition between plant and pathogen (p. 772, FIGURE 39.23)** A pathogen is avirulent if it has a specific dominant *Avr* gene corresponding to a particular dominant *R* gene in the host plant.

- **The hypersensitive response (HR) contains an infection (pp. 772–773)** Defense against an avirulent pathogen seals off the infection and kills both pathogen and host cells in the region of the infection.

- **Systemic acquired resistance (SAR) helps prevent infection throughout the plant (p. 773, FIGURE 39.24)** The HR also produces a signal molecule that triggers generalized defense responses in organs distant from the original site of infection.

SELF-QUIZ

1. Which of the following plant hormones is incorrectly paired with its function?
   a. auxin—promotes stem growth through cell elongation
   b. cytokinins—initiate senescence
   c. gibberellins—stimulate seed and bud germination
   d. abscisic acid—promotes seed and bud dormancy
   e. ethylene—inhibits cell elongation

2. Spraying some plants with a combination of auxin and gibberellins
   a. promotes fruit growth
   b. kills broadleaf dicot plants
   c. prevents senescence
   d. promotes fruit ripening
   e. is used to treat dwarfism in plants

3. Buds and sprouts often form on tree stumps. Which of the following hormones would you expect to stimulate their formation?
   a. auxin
   b. cytokinins
   c. abscisic acid
   d. ethylene
   e. gibberellins

4. Which of the following is *not* part of the acid-growth hypothesis?
   a. Auxin stimulates proton pumps in cell membranes.
   b. Lowered pH results in the breakage of cross-links between cellulose microfibrils.
   c. The wall fabric becomes looser (more plastic).
   d. Auxin-activated proton pumps stimulate cell division in meristems.
   e. The turgor pressure of the cell exceeds the restraining pressure of the loosened cell wall, and the cell takes up water and elongates.

5. The signal for flowering could be released earlier than normal in a long-day plant experimentally exposed to flashes of
   a. far-red light during the night
   b. red light during the night

c. red light followed by far-red light during the night
d. far-red light during the day
e. red light during the day

6. The phytochrome system helps set the biological clock by indicating to a plant that light is present when
   a. P_r is rapidly converted to P_{fr}
   b. P_{fr} is slowly converted to P_r
   c. P_r and P_{fr} are equal in concentration
   d. red light is absorbed by P_{fr}
   e. photosynthetic production of ATP powers phytochrome conversion

7. If a long-day plant has a critical night length of 9 hours, which of the following 24-hour cycles would prevent flowering?
   a. 16 hours light/8 hours dark
   b. 14 hours light/10 hours dark
   c. 15.5 hours light/8.5 hours dark
   d. 4 hours light/8 hours dark/4 hours light/8 hours dark
   e. 8 hours light/8 hours dark/light flash/8 hours dark

8. The probable role of salicylic acid in systemic acquired resistance of plants is to
   a. destroy pathogens directly
   b. activate plant defenses throughout the plant before infection spreads
   c. close stomata, thus preventing the entry of pathogens
   d. activate heat-shock proteins
   e. sacrifice infected tissues by hydrolyzing cells

9. Auxin triggers the acidification of cell walls that results in rapid growth, but also stimulates sustained, long-term cell elongation. What best explains how auxin brings about this dual growth response?
   a. Auxin binds to different receptors in different cells.
   b. Different concentrations of auxin have different effects.
   c. Auxin causes a second messenger to activate both proton pumps and certain genes.
   d. The dual effects are due to two different auxins.
   e. Auxin's effects are modified by other antagonistic hormones.

10. The subscripts in the following choices indicate specific *Avr* and *R* genes in pathogens and plant cells. Uppercase letters indicate dominant alleles, while lowercase symbolizes recessive alleles. In which of the situations would the pathogen be avirulent?
   a. Avr_D–R_d
   b. Avr_E–R_G
   c. Avr_M–R_M
   d. Avr_g – R_g
   e. Avr_e – R_E

CHALLENGE QUESTIONS

1. Explain how it is possible that a short-day plant and a long-day plant growing in the same location could flower on the same day of the year.
2. Discuss how day length, phytochrome, the biological clock, gibberellins, and abscisic acid may interrelate in the germination of seeds planted just below the soil surface.
3. Write a short paragraph explaining how pruning results in fruit trees having more branches.
4. Dandelion flowers open each morning and close each evening. Describe an experiment that would determine whether this daily activity is controlled by an internal biological clock or simply by the presence or absence of light.
5. There are some species of bamboo that flower only when they are 120 years old. What mechanisms can you suggest to explain how these plants "count" years? How would you test these hypotheses? What are possible advantages of this reproductive "strategy" relative to seed predation?

SCIENCE, TECHNOLOGY, AND SOCIETY

1. Imagine the following scenario: A plant scientist discovers a synthetic chemical that mimics the effects of a plant hormone. The chemical can be sprayed on apples before harvest to prevent flaking of the natural wax that is formed on the skin. This makes the apples shinier and gives them a deeper red color. What kinds of questions do you think should be answered before farmers start using this chemical on apples?
2. Certain herbicides disrupt growth by mimicking the action of plant hormones. The weed killer 2,4-D, for example, is a synthetic auxin. Recombinant DNA techniques can be used to alter crop plants to make them more resistant to herbicides, so that the herbicides can be used to "weed" crops without harming them. What are some arguments for and against this technology?
3. Based on your study of this chapter, write a short essay explaining at least three examples of how knowledge about the control systems of plants is applied in agriculture or horticulture.

FURTHER READING

Jones, A. M. "Surprising Signals in Plant Cells." *Science,* January 14, 1994. Describes how aspirin and other hormones trigger plant responses.
Mlot, D. "Fine-tuned Plant Response to Insect Attack." *Science News,* December 20, 1997. Explains how plants recruit one animal to kill another.
Page, T. L. "Time Is the Essence: Molecular Analysis of the Biological Clock." *Science,* March 18, 1994. How do cells keep track of time?
Pennisi, E. "Plants Decode a Universal Signal." *Science,* December 19, 1997. Discusses a signal-transduction pathway for stress responses in plants.
Ronald, P. C., "Making Rice Disease-Resistant." *Scientific American,* November 1997. Provides a description of genetic engineering of an important crop.
Vogel, S. "When Leaves Save the Tree." *Natural History,* September 1993. Describes how leaf responses help prevent wind damage to trees.
Wayne, R. "Excitability in Plant Cells." *American Scientist,* March/April 1993. Characterizes nervelike impulses in plants.

WEB LINKS

Visit the special edition of *The Biology Place* for BIOLOGY, Fifth Edition, at http://www.biology.com/campbell. Go to Chapter 39 for online resources, including learning activities, practice exams, and links to the following web sites:

"Plant Hormones"
This site was set up as a freely accessible electronic communication forum for use by plant hormone scientists. It contains a list of plant hormone researchers around the world.

"Everything You Wanted To Know About the Rooting of Plant Cuttings?"
Plant hormones are used extensively for plant propagation in the horticulture and agriculture industries. Find out how it's done from Hortus USA.

"Plant Tissue Culture Information Exchange"
Plant hormones are used extensively in tissue culture. Links are provided to sites dealing with various aspects and uses of plant tissue culture.

"The American Society of Plant Physiologists (ASPP)"
The web site of the ASPP contains useful links to plant science topics.

UNIT SEVEN

ANIMAL FORM AND FUNCTION

Professor Terence Dawson is known throughout the world for his knowledge of kangaroos and other Australian animals. I first met Dr. Dawson when I visited the University of New South Wales in Sydney in 1993. He was explaining that one of the dangers of working on kangaroos at close range is that an animal might grab you by the collar with its forelegs, rear back on its tail, and then fire kicks with its powerful hindlegs at your midsection. "What," I asked him, "is the best way to react in such an emergency?" His advice: "Turn sideways!" I got the picture. The opportunity to learn more about Australian animals from Terry Dawson came in 1997, when I returned to Australia to conduct this interview.

What experiences fueled your interest in Australian animals?

Like so many who grow up in country towns, you just wander around, you fish and hunt. Whenever we were out of school we just wandered around the bush. This was savanna country in western New South Wales, and there were lots of birds and diverse habitats. Probably the first real interest I had in biology, as such, was the tradition of collecting bird eggs, which was one of those things country kids once did in the spring. I got much more interested in it than most of the chaps that I was collecting with, and I actually started to try to find out about the birds. One of the first books my parents gave me was *What Bird Is That?: A Guide to the Birds of Australia.* I got hooked on reading about the birds and trying to understand them. I knew then I wanted to study natural history and be a zoologist, but in those days, one had to have a profession. Being from a farming community (although my parents were not farmers), I decided after I graduated from high school to study agriculture. I went off to a small university, the University of New England, to start an animal science degree. I figured that studying domestic animals and animal science was a good way to operate close to natural history. But I made a decision to move out of agriculture if I got the chance. I got that chance with a postdoctoral fellowship with Knut Schmidt-Nielsen at Duke in North Carolina. I was still interested in large animals, and Schmidt-Nielsen worked on camels. I saw that I could utilize my training on sheep and cattle, which was directed to understanding how animals live in the environment, to understand the animals that I was much more interested in. So I ended up going in the direction of comparative physiology, or ecological physiology. I learned from Schmidt-Nielsen the notion of whole-animal biology—that you can't look at just one system if you're interested in how animals live in their environments.

When did you start to apply these ideas to Australian animals?

My interest in kangaroos developed in an odd way. When I was in the States, people used to ask me about kangaroos, but I really didn't know much about them. I went to the Duke library to find out about kangaroos so I could answer questions people kept asking me. There I realized that basically nothing scientific was known about kangaroos, so I decided

INTERVIEW

TERRY DAWSON

that if I got the opportunity when I got back to Australia I would work on kangaroos. The other motivation stemmed from the idea at the time that kangaroos and other marsupials were primitive and inferior as mammals. That didn't fit very well with my childhood observations of kangaroos. I had seen them out in the open plains where the temperature was about 50° Celsius. Birds were even falling off their perches, but the kangaroos just sat there unperturbed, which didn't seem like primitive thermoregulation. Also, if a kangaroo got moving, hunting dogs couldn't catch it, and that didn't strike me as being inferior. So these notions that marsupials were generally inferior as mammals just didn't sit well. I thought, "Well, we'll go and have a look and see what the real situation is." I returned to Australia when the Zoology Department at the University of New South Wales in Sydney was developing a kangaroo research team and the university had just acquired a 100,000-acre field station in far western NSW. So I joined the Zoology Department here, and it's been too good to move. Our animals are just so fascinating that once you get involved with them, you're hooked. Marsupials and monotremes do things so differently. They break quite a lot of "rules."

But you want to be sure that we understand that "different" doesn't equal "primitive."

The moment you think about how these animals are living in their environments—how they feed, or the stresses they face—you can't possibly think of them as primitive. For example, we always wondered how the platypus, a monotreme, found its food. Platypuses dive and forage on the bottom of streams with their eyes and ears closed. One hypothesis was that platypuses use touch receptors in the bill to feed, but platypuses can locate and feed on guppies in the open water of a large fish tank, and so it's not only touch. They're actively hunting. As it turns out, platypuses also have electroreceptors. They can pick out the twitch of a fish muscle or a worm muscle and home in on their prey. In the 60 or 70 million years that platypuses have been around, there has been a lot of time for such special adaptations to evolve.

Much of your research has focused on marsupials, especially kangaroos. Tell us about some of the reproductive adaptations of kangaroos.

The group consisting of kangaroos and wallabies is actually a fairly modern group. The gray kangaroos go back about 7 million years, but the red kangaroo has a fossil history that goes back only about 100 to 150 thousand years. That is a very modern, recent animal in evolutionary terms. Gray kangaroos and red kangaroos have very different reproductive patterns. Gray kangaroos are seasonal breeders, while red kangaroos breed all the time, unless there is a drought. After a female red kangaroo mates and an egg is fertilized, the embryo doesn't implant in the uterus, but just stops at the 100-cell stage until about 30 days before the previous offspring is expelled from the pouch, and then it starts to develop again. The previous young doesn't just decide to get out; it's expelled when the female contracts the pouch and the young can't get back in. It's controlled by the same maternal hormone that functions in birth in eutherians. A day or so after the big young has been expelled, the next young is born, climbs up, and gets in the pouch. A day or so later the female mates again, and the whole system just keeps going. She's producing a young out of the pouch about every 8 or 9 months. You'd think we'd be up to our ears in kangaroos, but the mortality of young is high.

But during its time in the pouch, the young has good parental care, right?

At that stage, it's much better than human patterns. The kangaroo mother can pull the pouch open and have look inside and scratch its young. She also has to clean the pouch, of course. The young can get in and out. It's like you can have trial births. Once you're a wobbly young, you can get out and have a look around, and if you don't like it you can get straight back in and be carried by mom when she hops off.

Speaking of hopping, tell us more about kangaroos' unique locomotion.

Kangaroos are the only large animals that hop. Most of our work on locomotion is on red kangaroos. These behave differently from other kangaroo species. When they are disturbed, red kangaroos head away from the trees onto the open plain. Off they go, a magnificent flight, and nothing can catch them. Gray kangaroos head into the timber and use the trees as an escape mechanism, but red kangaroos rely entirely on speed. We really don't quite understand the hopping locomotion yet. The energy pattern is unusual. As most animals go faster, from walking to trotting to running, there is a roughly linear increase in energy cost. When moving slowly, kangaroos are rather clumsy, using their tails as fifth legs. But at 6 kilometers an hour (kph), they start to hop, and once they do that, the energy cost doesn't increase as they go faster, at least not up to the speeds we've been able to measure on treadmills. At 30 kph, the energy cost for the hopping is about half that of running for an animal of equivalent size. We initially thought it was related to elasticity in the legs, a bit like the energy saving you get when using a pogo stick. The spring stores the energy when you come down, and you can use that energy for takeoff. This the kangaroos do, and it is probably an important part of the system. But this is not unique to kangaroos; it's an important mechanism in any fast-running animal, including racehorses. There are a variety of other possibilities, but we really don't know the answer. What we need is a much bigger treadmill. You can fool a kangaroo into hopping on a treadmill, but when the length of the hop they would take in the open country becomes considerably longer than the length of the treadmill they start to balk. Their perception of where the end of the treadmill is gets them rather upset, and they start taking smaller steps, which upsets the normal hopping pattern. So we need a very long treadmill or perhaps a nice video in front of them to convince them that they are out there in the country. We're actually designing a bigger treadmill.

How fast can red kangaroos go on the open plain?

They can reach about 45 kph mainly by increasing stride length up to 5–6 meters per hop, and then they can increase to about 60 kph for a short distance by increasing the frequency of the hops. They can only sustain that speed for about 250 meters—to dash across the road in front of a car, for example.

What impact have humans had on the diversity of kangaroos and other marsupials?

The diversity of marsupials, especially large ones, was much greater when the aboriginal people arrived in Australia about 45,000 years ago than it is today. Big animals are easy to hunt, and virtually everywhere in the world where humans came into contact with animals that were naive with respect to humans, there was megafaunal extinction—disappearance of many species of large animals. The next major extinction in Australia occurred with the arrival of Europeans and the grazing industry. Much of Australia is not very fertile, and the climate is dry and erratic. The settlers had European notions of rapid growth rates for the vegetation, and they moved in huge flocks of sheep, and then with the first drought the ecosystem collapsed. The grazing also resulted in severe erosion. A major extinction of native animals occurred within the first 20 years of the arrival of European settlers to the region. The system just collapsed, and the aborigines were also decimated because most of their food was gone. That occurred even before the added problem of introduced feral animals such as cats and foxes, which prey on small marsupials and native rodents, and rabbits, which compete for food with native animals.

Do modern ranchers view kangaroos as pests that compete with sheep for food on the range?

That's changing, to some degree. Certainly when I was a lad, kangaroos were considered to be pests, and the opinion still holds among some graziers or pastoralists. But a lot of research shows that fluctuations in vegetation are due mainly to the erratic nature of our rainfall. Most of the time there's no competition between the kangaroos and the sheep because there is more vegetation than the animals can eat. And even in dry times in arid regions, the kangaroos feed mainly on grasses and don't eat a lot of bushes, which the sheep do eat. So, some of the pastoralists are now accepting that there's not much direct competition. They're also looking toward kangaroos as an extra source of income, both for food and for the tourist industry. People pay big money to come to Australia to see kangaroos. There is also less tension now because most conservationists are convinced that the six extant species of large kangaroos are not becoming extinct. Kangaroos are widespread and abundant. My best guess, based on aerial censuses, is that there are about 50 million kangaroos. We've got lots of kangaroos!

If some ranchers and conservationists are changing their view of kangaroos, is this part of a more general trend in Australian awareness about the native biota?

Yes. When I grew up, the attitude was: If it moves, shoot it; if it doesn't, chop it down. I think the current interest of Australians in the country's natural history is part of a general worldwide increase in awareness of the environment. But, particularly in Australia, we've become aware of the unique nature of our environment and our animals, and scientists have played a major role in uncovering a lot of the interesting things about our animals. Biologists are sharing these interesting things with the public, and as people acquire more knowledge, they become even more interested. Awareness of our unique fauna and understanding that our environment is particularly fragile is now much greater than I would have predicted just 10 years ago.

CHAPTER 40

AN INTRODUCTION TO ANIMAL STRUCTURE AND FUNCTION

The chapters in this unit examine the correlation between structure and function in a variety of animals and selected protists. Our main goal is to see how species of diverse evolutionary history and varying complexity solve problems common to all. For instance, how do organisms as different as an amoeba, a clam, a bumblebee, and a human obtain energy from the environment? And how do they obtain O_2 for cellular respiration while disposing of the waste gas CO_2?

Animals provide vivid examples of biology's core theme: the capacity to adjust to the environment over the long term by adaptation due to natural selection. A bumblebee like the one in the photograph here has a long, thin tongue, an adaptation enabling it to probe deep into flowers for nectar. Further, portions of the bee's hind legs are fringed with long, curved hairs, forming a food basket that collects plant pollen (the yellowish lump on this bee's hind leg).

A foraging bumblebee illustrates another major theme of this unit: the capacity of organisms to adjust to the environment over the short term by physiological responses. Many insects are inactive when it is cold, but bumblebees can forage for nectar and pollen when air temperatures are as low as 5°C. On cool mornings in the early spring, when the first flowers are out, a bumblebee's flight muscles will contract rapidly, much like yours do when you shiver. Heat produced by the contracting muscles warms the bee to over 30°C, the temperature at which its wings can beat fast enough to fly. This unusual ability enables the bumblebee to forage when it is too cold for most other insects to compete with them for resources.

Searching for food, generating body heat, responding to external stimuli, and all other animal activities require fuel in the form of chemical energy. Being warm enough to fly in cold weather depends on a bumblebee's ability to obtain enough fuel to at least match the rate at which its body cells use chemical energy. An appreciation of bioenergetics as it applies to animals—how animals obtain, process, and use chemical energy—is key to understanding animal physiology, and will integrate our comparative study of animals throughout this unit.

This introductory chapter has four main objectives: to illustrate the hierarchy of structural order characterizing animals; to emphasize the importance of energetics in animal life; to examine how the body forms of animals affect interactions with the external environments; and to preview how regulatory systems maintain favorable internal environments.

LEVELS OF STRUCTURAL ORGANIZATION

All life is characterized by hierarchical levels of organization. The cell holds a special place in the hierarchy of life because it is the lowest level of organization that can live independently as an organism. Protists, for example, have specialized organelles that perform particular jobs, enabling them to digest food, sense environmental change, excrete waste products, and reproduce—all within the confines of a single cell. Protists represent the cellular level of organization, the simplest level possible for an organism. Multicellular organisms, including animals, have specialized cells grouped into tissues, the next higher level of structure and function. In most animals, combinations of various tissues make up functional units called organs, and groups of organs that work together form organ systems. For example, the human digestive system consists of a stomach, small intestine, large intestine, gallbladder, and several other organs, each a composite of different types of tissues.

Function correlates with structure in the tissues of animals

40.1 **Tissues** are groups of cells with a common structure and function. Different types of tissues have different structures that are especially suited to their functions. A tissue may be held together by a sticky extracellular matrix that coats the cells (see Chapter 7) or weaves them together in a fabric of fibers. Indeed, the term *tissue* is from a Latin word meaning "weave."

40.1 This symbol links topics in the text to interactive exercises in the CD-ROM that accompanies the book. The number indicates the appropriate activity in the CD.

We can classify tissues into four main categories: epithelial tissue, connective tissue, nervous tissue, and muscle tissue. These are present to some extent in all but the simplest animals; the following survey emphasizes the tissues of vertebrates.

Epithelial Tissue

Occurring in sheets of tightly packed cells, **epithelial tissue** covers the outside of the body and lines organs and cavities within the body. The cells of an epithelium are closely joined, with little material between them. In many epithelia, the cells are riveted together by tight junctions like those described in Chapter 7 (see FIGURE 7.30). This tight packing enables the epithelium to function as a barrier protecting against mechanical injury, invasive microorganisms, and fluid loss. The free surface of the epithelium is exposed to air or fluid, whereas the cells at the base of the barrier are attached to a **basement membrane**, a dense mat of extracellular matrix (FIGURE 40.1). (The term *membrane* in this sense does not refer to a phospholipid

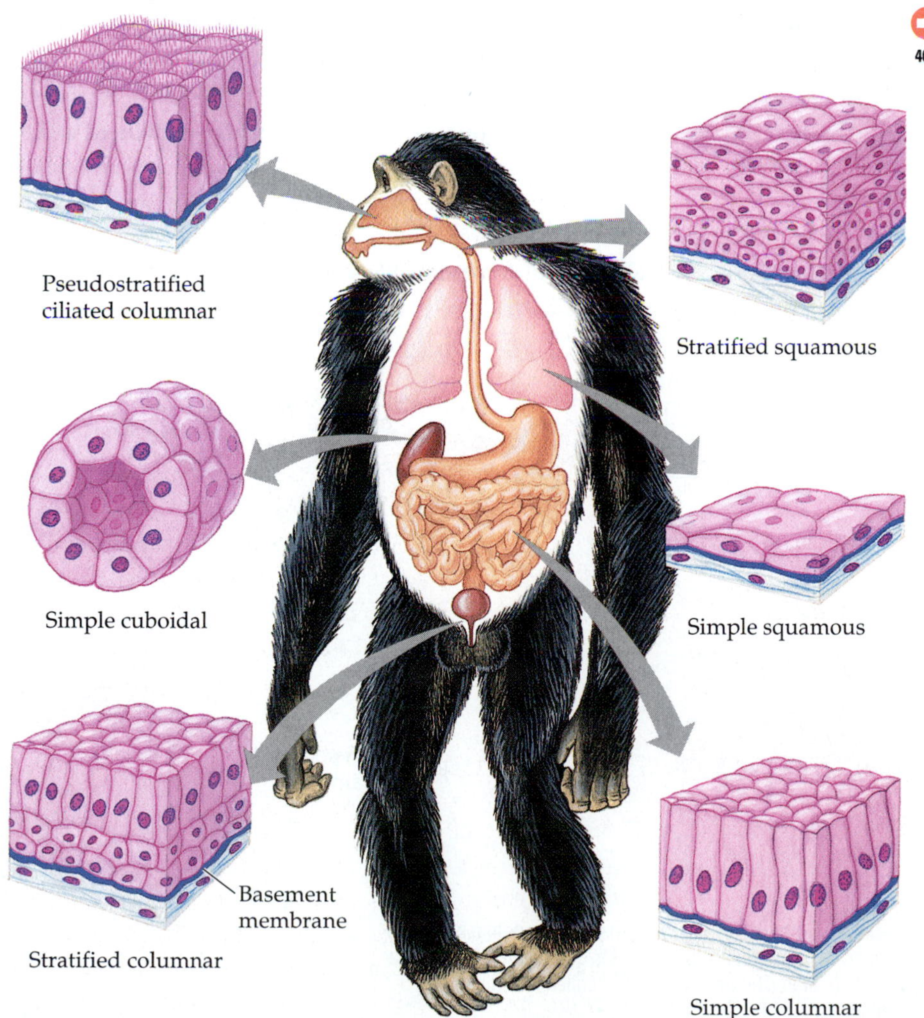

40.1 **FIGURE 40.1 ▪ The structure and function of epithelial tissues.** The structure of an epithelium fits its function. For instance, simple squamous epithelia, which are thin and leaky, function in the exchange of materials by diffusion. These epithelia line blood vessels and the air sacs of the lungs. Stratified squamous epithelia regenerate rapidly by cell division near the basement membrane. The new cells are pushed to the free surface as replacements for cells that are continually sloughed off. This type of epithelium is commonly found on surfaces subject to abrasion, such as the outer skin and linings of the esophagus, anus, and vagina. Columnar epithelia, having cells with relatively large cytoplasmic volumes, are often located where secretion or the active absorption of substances is an important function. For example, the intestines are lined by simple columnar epithelia that secrete digestive juices and absorb nutrients; pseudostratified ciliated columnar epithelia line the nasal passages of many vertebrates; and stratified columnar epithelia line the inner surface of the urinary bladder. Cuboidal cells specialized for secretion make up the epithelia of kidney tubules and many glands, including the thyroid gland and salivary glands.

bilayer, as in a plasma membrane of a cell.) Cell biologists are finding that the basement membrane performs many different functions, such as helping organize sequential events in cellular metabolism, filtering wastes from the blood in the kidney, and providing routes of migration for cells during development.

Two criteria for classifying epithelia are the number of cell layers and the shape of the cells on the free surface. A **simple epithelium** has a single layer of cells, whereas a **stratified epithelium** has multiple tiers of cells. A pseudostratified epithelium is single-layered, but it appears stratified because the cells vary in length. The shape of the cells that are at the free surface of an epithelium may be **cuboidal** (like dice), **columnar** (like bricks on end), or **squamous** (flat like floor tiles). Combining the features of cell shape and number of layers, we get such terms as *simple cuboidal epithelium* and *stratified squamous epithelium* (see FIGURE 40.1).

As well as protecting the organs they line, some epithelia absorb or secrete chemical solutions. For example, the epithelial cells that line the lumen (cavity) of the digestive and respiratory tracts form a **mucous membrane**; they secrete a slimy solution called mucus that lubricates the surface and keeps it moist. The mucous membrane lining the small intestine also releases digestive enzymes and absorbs nutrients. The free epithelial surfaces of some mucous membranes have beating cilia that move the film of mucus along the surface. For example, the ciliated epithelium of our respiratory tubes helps keep our lungs clean by trapping dust and other particles and sweeping them back up the trachea (windpipe).

Connective Tissue

Connective tissue functions mainly to bind and support other tissues. In contrast to epithelia, with their tightly packed cells, connective tissues have a sparse population of cells scattered through an extracellular matrix. The matrix generally consists of a web of fibers embedded in a uniform foundation that may be liquid, jellylike, or solid. In most cases, the substances of the matrix are secreted by the cells of the connective tissue.

Connective tissue fibers, which are made of protein, are of three kinds: collagenous fibers, elastic fibers, and reticular fibers. **Collagenous fibers** are made of collagen, probably the most abundant protein in the animal kingdom (FIGURE 40.2). Collagenous fibers are nonelastic and do not tear easily when pulled lengthwise. If you pinch and pull some skin on the back of your hand, it is mainly collagen that keeps the flesh from tearing away from the bone. **Elastic fibers** are long threads made of a protein called elastin. Elastic fibers provide a rubbery quality that complements the nonelastic strength of collagenous fibers. When you pinch the back of your hand and then let go, elastic fibers quickly restore your skin to its original shape. **Reticular fibers** are very thin and branched. Composed of collagen and continuous with collagenous fibers, they form a tightly woven fabric that joins connective tissue to adjacent tissues.

The major types of connective tissue in vertebrates are loose connective tissue, adipose tissue, fibrous connective tissue, cartilage, bone, and blood (FIGURE 40.3). Each has a structure correlated with its specialized functions.

The most widespread connective tissue in the vertebrate body is **loose connective tissue**. It binds epithelia to underlying tissues and functions as packing material, holding organs in place. This type of connective tissue gets its name from the loose weave of its fibers. Loose connective tissue has all three fiber types: collagenous, elastic, and reticular (see FIGURE 40.3).

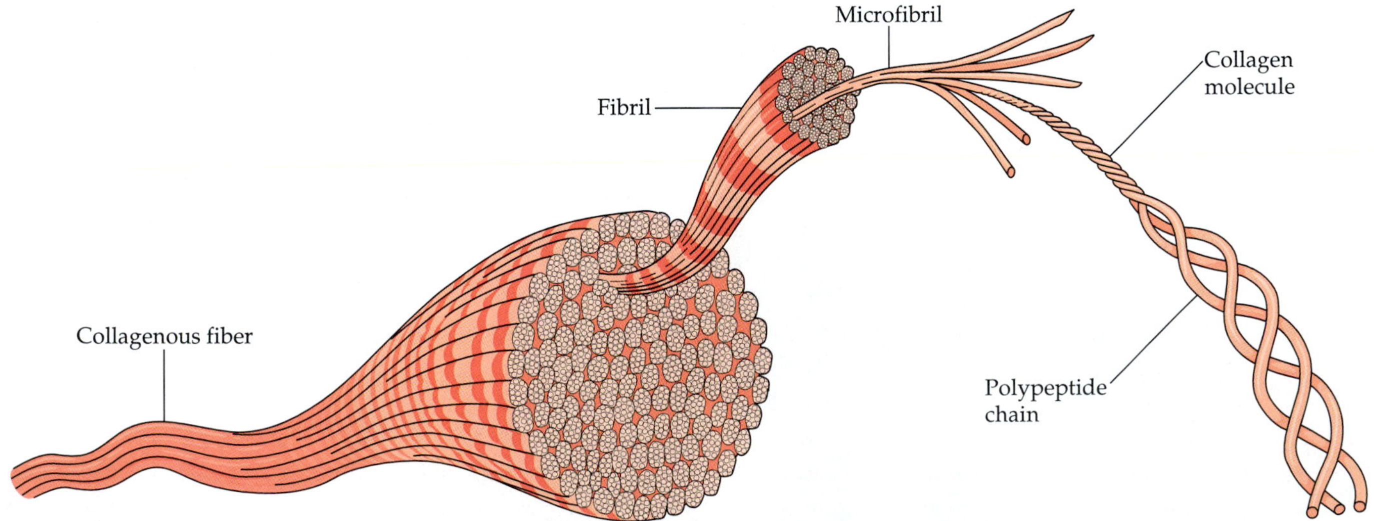

FIGURE 40.2 ▪ The structure of a collagenous fiber. A collagenous fiber is a nonelastic, ropelike bundle of many fibrils, each of which in turn is a bundle of many microfibrils. A microfibril consists of helically coiled collagen molecules, each consisting of three helical polypeptide chains. The arrangement of collagen molecules makes the fibrils appear striped when viewed with an electron microscope.

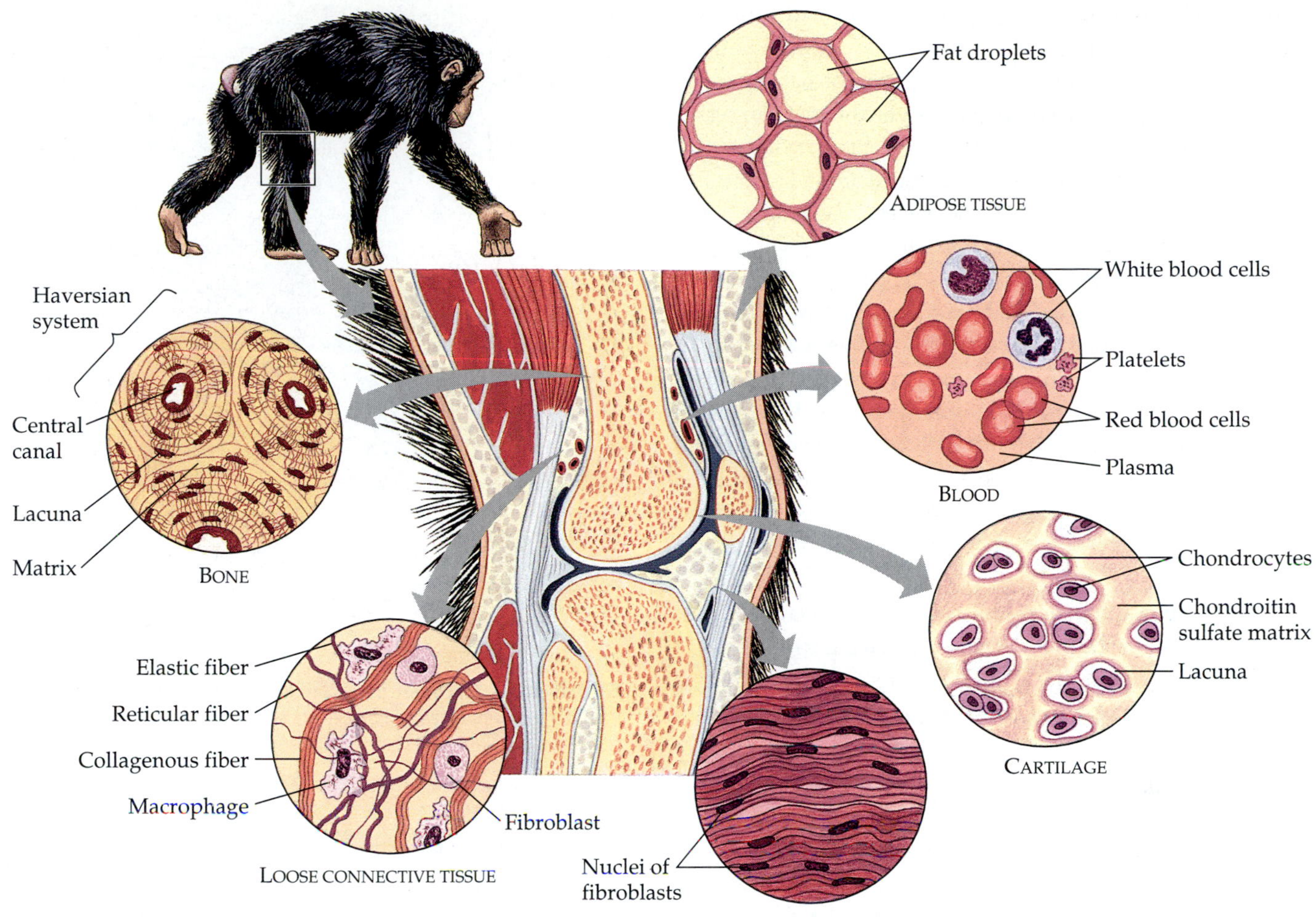

40.1

FIGURE 40.3 ▪ Some representative types of connective tissue. The area shown is the region around the knee joint.

Among the cells scattered in the fibrous mesh of loose connective tissue, two types predominate: fibroblasts and macrophages. **Fibroblasts** secrete the protein ingredients of the extracellular fibers. **Macrophages** are amoeboid cells that roam the maze of fibers, engulfing bacteria and the debris of dead cells by phagocytosis (see Chapter 8). They are weapons in an elaborate arsenal of defense you will learn more about in Chapter 43.

Adipose tissue is a specialized form of loose connective tissue that stores fat in adipose cells distributed throughout its matrix. Adipose tissue pads and insulates the body and stores fuel molecules. Each adipose cell contains a large fat droplet that swells when fat is stored and shrinks when the body uses fat as fuel. Heredity, exercise, and the amount of fat we eat can all affect the amount of fat our adipose cells store. There is also some evidence that the amount of fat we store when we are babies determines in part the number of fat cells in our connective tissues. Fat metabolism is complex, and we will discuss it in more detail in Chapter 41.

Fibrous connective tissue is dense, due to its large numbers of collagenous fibers. The fibers are organized into parallel bundles, an arrangement that maximizes nonelastic strength. We find this type of connective tissue in **tendons**, which attach muscles to bones, and in **ligaments**, which join bones together at joints.

Cartilage has an abundance of collagenous fibers embedded in a rubbery matrix made of a substance called chondroitin sulfate, a protein-carbohydrate complex. Chondroitin sulfate and collagen are secreted by **chondrocytes**, cells confined to scattered spaces in the matrix called lacunae (see FIGURE 40.3). The composite of collagenous fibers and chondroitin sulfate makes cartilage a strong yet somewhat flexible support material. The skeleton of a shark is made of cartilage. Other vertebrates, including humans, have cartilaginous skeletons during the embryo stage, but most of the cartilage is replaced by bone as the embryo matures. We nevertheless retain cartilage as flexible support in certain locations, such as the nose, the ears, the rings that reinforce the windpipe, the discs that act as cushions between our vertebrae, and the caps on the ends of some bones.

The skeleton supporting the body of most vertebrates is made of **bone**, a mineralized connective tissue. Bone-forming cells called **osteoblasts** deposit a matrix of collagen, but they also release calcium, magnesium, and phosphate ions, which chemically combine and harden within the matrix into the mineral hydroxyapatite. The combination of hard mineral and flexible collagen makes bone harder than cartilage without being brittle. The microscopic structure of hard mammalian bone consists of repeating units called **Haversian systems** (see FIGURE 40.3). Each system has concentric layers of the mineralized matrix, which are deposited around a central canal containing blood vessels and nerves that service the bone. Once osteoblasts become trapped in their own secretions, they are called osteocytes. The osteocytes are located in lacunae, spaces surrounded by the hard matrix. Tiny canals in the matrix connect the lacunae and allow nutrients to be supplied to the osteocytes. In long bones, such as the femur in your thigh, only the outer region is hard compact bone built from Haversian systems. The interior is a spongy bone tissue honeycombed with spaces filled with bone marrow. Blood cells are manufactured in red bone marrow located near the ends of long bones. (We will examine bones and skeletons in more detail in Chapter 49.)

Although **blood** functions differently from other connective tissues, it does meet the criterion of having an extensive extracellular matrix. In this case, the matrix is a liquid called plasma, consisting of water, salts, and a variety of dissolved proteins. Suspended in the plasma are two classes of blood cells, erythrocytes (red blood cells) and leukocytes (white blood cells), and cell fragments called platelets. Red cells carry oxygen; white cells function in defense against viruses, bacteria, and other invaders; and platelets aid in blood clotting. Blood will be discussed in detail in Chapters 42 and 43.

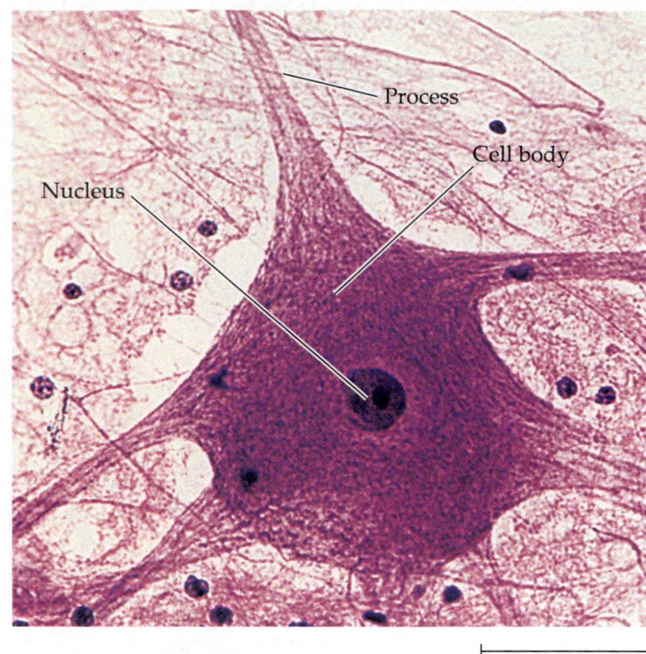

FIGURE 40.4 ▪ The basic structure of a neuron. This nerve cell from the spinal cord has a large cell body with multiple processes that transmit electrical signals called impulses (LM).

Nervous Tissue

Nervous tissue senses stimuli and transmits signals from one part of the animal to another. The functional unit of nervous tissue is the **neuron**, or nerve cell, which is uniquely special-

40.1

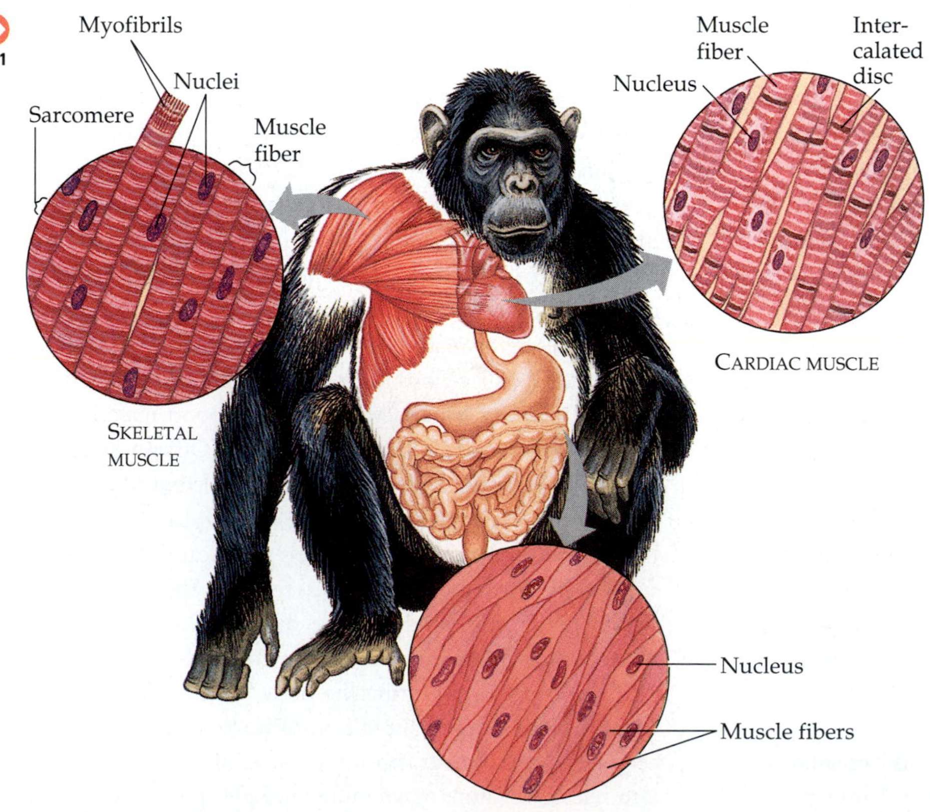

FIGURE 40.5 ▪ Kinds of vertebrate muscle. Skeletal muscle consists of bundles of long cells called fibers; each fiber is a bundle of strands called myofibrils. Each myofibril is a linear array of sarcomeres, the basic contractile units of the muscle. Skeletal muscle is said to be striated because the alignment of sarcomere subunits in adjacent myofibrils forms light and dark bands. Cardiac muscle, also striated, has contractile properties similar to those of skeletal muscle. Unlike skeletal muscle, however, cardiac muscle fibers branch and interconnect via intercalated discs, which help synchronize the heartbeat. Smooth muscle consists of spindle-shaped cells lacking cross-striations.

ized to transmit signals called nerve impulses (FIGURE 40.4). It consists of a cell body and two or more extensions, or processes, called dendrites and axons, which may be as long as a meter in humans. Dendrites transmit impulses from their tips toward the rest of the neuron. Axons transmit impulses toward another neuron or toward an effector, a structure such as a muscle cell that carries out a body response. We postpone a detailed discussion of the structure and function of neurons until Chapter 48.

Muscle Tissue

Muscle tissue is composed of long cells called muscle fibers that are capable of contracting when stimulated by nerve impulses. Arranged in parallel within the cytoplasm of muscle fibers are large numbers of microfilaments made of the contractile proteins actin and myosin. Muscle is the most abundant tissue in most animals, and muscle contraction accounts for much of the energy-consuming cellular work in an active animal.

In the vertebrate body, there are three types of muscle tissue: skeletal muscle, cardiac muscle, and smooth muscle (FIGURE 40.5). Attached to bones by tendons, **skeletal muscle** is responsible for voluntary movements of the body. Adults have a fixed number of muscle cells; weight lifting and other methods of building muscle do not increase the number of cells but simply enlarge those already present. Skeletal muscle is also called **striated muscle** because the arrangement of overlapping filaments gives the cells a striped (striated) appearance under the microscope.

Cardiac muscle forms the contractile wall of the heart. It is striated like skeletal muscle, but cardiac cells are branched, and the ends of the cells are joined by intercalated discs, which relay signals from cell to cell during a heartbeat.

Smooth muscle, so named because it lacks striations, is found in the walls of the digestive tract, bladder, arteries, and other internal organs. The cells are spindle-shaped. They contract more slowly than skeletal muscles but can remain contracted longer. Controlled by different kinds of nerves than those controlling skeletal muscles, smooth muscles are responsible for involuntary body activities, such as churning of the stomach or constriction of arteries. You will learn more about the control and contraction of muscles in Chapter 49.

The organ systems of an animal are interdependent

In all but the simplest animals (sponges and some cnidarians), different tissues are organized into **organs**. In some organs, such as the skin of a vertebrate, the tissues are arranged in layers. The vertebrate stomach has four major tissue layers (FIGURE 40.6). A thick epithelium lines the lumen and secretes mucus and digestive juices into it. Outside this layer is a zone of connective tissue, surrounded by a thick layer of smooth muscle. Yet another layer of connective tissue encapsulates the entire stomach.

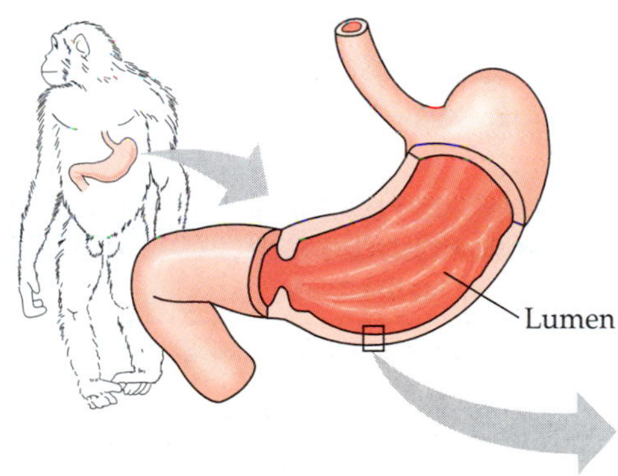

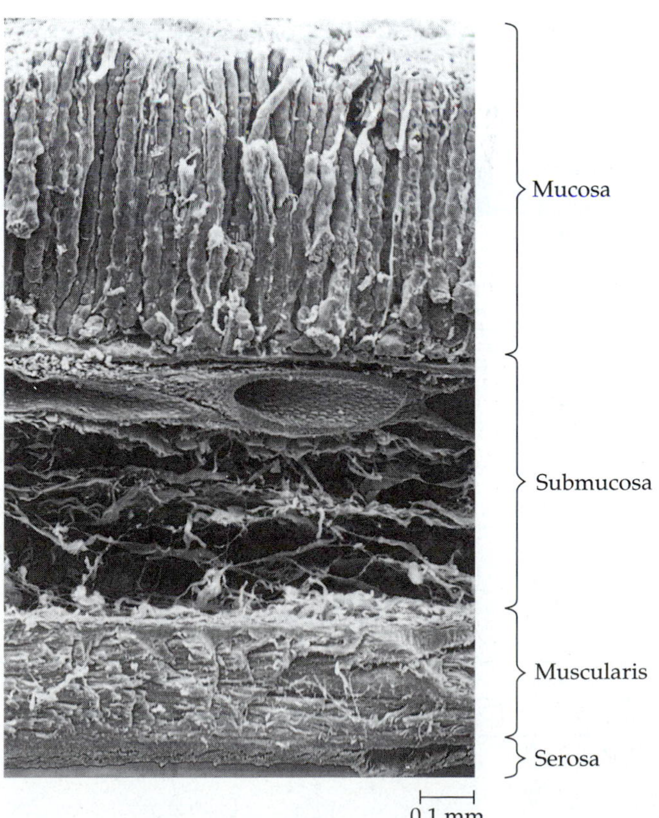

FIGURE 40.6 ▪ Tissue layers of the stomach, a digestive organ. The wall of the stomach and other tubular organs of the digestive system has four main tissue layers: the mucosa, submucosa, muscularis, and serosa. The mucosa is an epithelial layer that lines the lumen. The submucosa is a matrix of connective tissue that contains blood vessels and nerves. The muscularis has an inner layer of circular muscles and an outer layer of longitudinal muscles. External to the muscularis is the serosa, a thin layer of connective tissue and epithelial tissue.

Many of the organs of vertebrates are suspended by sheets of connective tissue called **mesenteries** in body cavities moistened or filled with fluid. Mammals have a **thoracic cavity** housing the lungs and heart that is separated from the lower, **abdominal cavity** by a sheet of muscle called the diaphragm.

A level of organization higher than organs, **organ systems** carry out the major body functions of most animals (TABLE 40.1). Each organ system consists of several organs and has specific functions, but the efforts of all systems must be coordinated for the animal to survive. For instance, nutrients absorbed from the digestive tract are distributed throughout the body by the circulatory system. But the heart that pumps blood through the circulatory system depends on nutrients absorbed by the digestive tract and also on oxygen (O_2) obtained from the air or water by the respiratory system. Any organism, whether a protist or an assembly of organ systems, is a coordinated living whole greater than the sum of its parts.

INTRODUCTION TO THE BIOENERGETICS OF ANIMALS

Animals are heterotrophs that harvest chemical energy from the food they ingest

One of the defining features of life is the exchange of energy with the environment. As heterotrophs, animals obtain chemical energy in food, which contains organic molecules synthesized by other organisms. Food is digested by enzymatic hydrolysis, and energy-containing molecules are absorbed by body cells. The cells harvest chemical energy from some of the molecules, generating ATP by the catabolic processes of cellular respiration and fermentation (see Chapter 9). The chemical energy of ATP powers cellular work, enabling organ systems to perform functions that keep an animal alive. An animal constantly exchanges energy with its environment as cellular work generates heat, which the animal loses to its surroundings. Chemical energy remaining after the needs of staying alive are met can be used in biosynthesis; reproductive products and new tissues (for example, gametes and associated energy reserves such as yolk) develop as cells construct their own macromolecules using chemical energy and carbon skeletons obtained from food molecules (FIGURE 40.7).

Metabolic rate provides clues to an animal's bioenergetic "strategy"

We can learn a lot about an animal's adaptations to its environment by studying bioenergetics. How much of the total energy an animal obtains from food does it need just to stay alive? What are the energy costs for an insect, a bird, or a mammal to walk, run, or swim from one place to another? Physiologists obtain answers to such questions by measuring the rates at which animals use chemical energy. The total amount of energy an animal uses in a unit of time is called its **metabolic rate**. Energy is measured in **calories (cal)** or **kilocalories (kcal)**. (A kilocalorie is 1000 calories. The term *Calorie,* with a capital C, as used by many nutritionists, is actually a kilocalorie.)

Table 40.1 ▪ Organ Systems: Their Main Components and Functions in Mammals

ORGAN SYSTEM	MAIN COMPONENTS	MAIN FUNCTIONS
Digestive	Mouth, pharynx, esophagus, stomach, intestines, liver, pancreas, anus	Food processing (ingestion, digestion, absorption, elimination)
Circulatory	Heart, blood vessels, blood	Internal distribution of materials
Respiratory	Lungs, trachea, other breathing tubes	Gas exchange (uptake of oxygen; disposal of carbon dioxide)
Immune and Lymphatic	Bone marrow, lymph nodes, thymus, spleen, lymph vessels, white blood cells	Body defense (fighting infections and cancer)
Excretory	Kidneys, ureters, urinary bladder, urethra	Disposal of metabolic wastes; regulation of osmotic balance of blood
Endocrine	Pituitary, thyroid, pancreas, other hormone-secreting glands	Coordination of body activities (e.g., digestion, metabolism)
Reproductive	Ovaries, testes, and associated organs	Reproduction
Nervous	Brain, spinal cord, nerves, sensory organs	Coordination of body activities; detection of stimuli and formulation of responses to them
Integumentary	Skin and its derivatives (e.g., hair, claws, skin glands)	Protection against mechanical injury, infection, drying out
Skeletal	Skeleton (bones, tendons, ligaments, cartilage)	Body support, protection of internal organs
Muscular	Skeletal muscles	Movement, locomotion

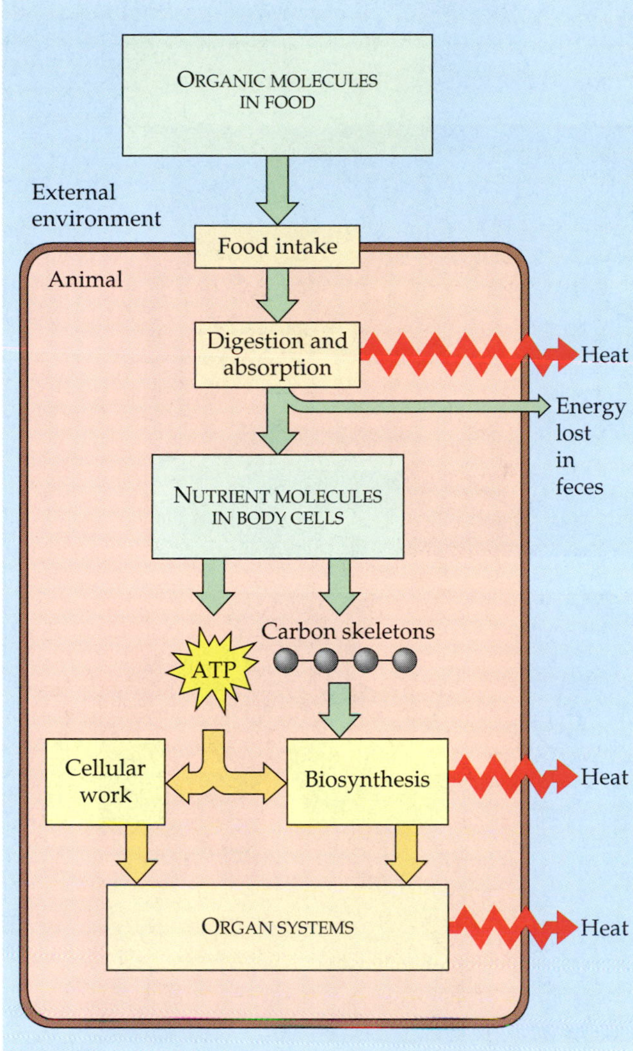

FIGURE 40.7 • Bioenergetics of an animal: an overview. Animals derive chemical energy from the environment in the organic molecules in food. During digestion, hydrolytic enzymes break food down into small nutrient molecules. A portion of the energy in food passes back to the environment in the undigestible material in feces. Body cells absorb the nutrient molecules and oxidize them, producing ATP that fuels the cellular work underlying the life-sustaining activities of organ systems. Body cells also use ATP and carbon skeletons from nutrient molecules for biosynthesis—energy-storing processes that construct macromolecules, forming new tissues for growth and repair of damaged ones and developing the components necessary for reproduction. Biosynthesis thereby keeps organ systems functioning. All cellular work, including work performed in biosynthesis and by the digestive system in processing food, involves a loss of energy to the environment as heat.

Metabolic rate can be determined in several ways. Because heat results from the use of chemical energy, metabolic rate can be measured by monitoring an animal's heat loss per unit of time. An animal is placed in a calorimeter, a closed, insulated chamber equipped with a device that records heat production in calories or kilocalories. Calorimeters are used most often with birds and small mammals, such as mice, which have high metabolic rates. They are cumbersome with large animals and less precise with small animals having low metabolic rates. Another way to measure metabolic rate is to determine the amount of oxygen consumed by an animal's cellular respiration (see the Methods Box, p. 786).

Every animal has a range of metabolic rates. Minimal rates support the basic functions that maintain life, such as breathing and the beating of the heart. Maximal metabolic rates occur during peak activity, such as all-out running or high-speed swimming. Between these extremes, many factors can influence an animal's metabolic rate, including its age, sex, size, body temperature, the temperature of its surroundings, the quality and quantity of its food, its activity level, amount of available oxygen, hormonal balance, and the time of day. Birds, humans, and many insects are usually active and have their highest metabolic rates during daylight hours. By contrast, bats, mice, and many other mammals generally are active and have their highest metabolic rates at night or during the hours of dawn and dusk.

Birds and mammals are mainly endothermic, meaning their bodies are warmed by heat generated by metabolism, and their body temperature must be maintained at a certain level to sustain life. By contrast, most fishes, reptiles, amphibians, and invertebrates are ectothermic, meaning they absorb most of their body heat from the external environment. The number of kilocalories an endotherm requires to sustain minimal life functions is generally higher than that of an ectotherm, because of the energy cost of heating (and cooling) an endothermic body. (We discuss thermoregulation in detail in Chapter 44.)

The metabolic rate of an endotherm at rest, with an empty stomach, and experiencing no stress is called the **basal metabolic rate (BMR)**. The BMR for humans—the number of kilocalories we "burn" lying motionless—averages about 1600 to 1800 kcal per day for adult males and about 1300 to 1500 kcal per day for adult females. These BMRs are about equivalent to the daily energy consumption of a 100-watt light bulb.

Ectotherms are energetically quite different from endotherms. Their body temperature changes with the temperature of their surroundings, and so does their metabolic rate. Unlike BMRs, which can be determined within a range of environmental temperatures, the minimal metabolic rate of an ectotherm must be determined at a specific temperature. The metabolic rate of a resting, fasting, nonstressed ectotherm is called its **standard metabolic rate (SMR)**.

Activity generally has an enormous effect on metabolic rate, and any activity, even a person working quietly at a desk or an insect extending its wings, consumes energy beyond the BMR or SMR. Metabolic rates measured when animals are active give a better idea of the energy costs of everyday life. Metabolic rates are highest during intense physical exercise, with maximal rates about five to ten times greater than the BMR or SMR.

METHODS: Measuring Metabolic Rate

An animal's metabolic rate is the number of kilocalories of energy consumed per unit of time. For an animal that is respiring aerobically, a convenient way to measure metabolic rate is to determine the amount of oxygen (O_2) the animal consumes in a unit of time. The rate of oxygen consumption can be monitored by enclosing the animal in a chamber called a respirometer.

The top photograph shows a ghost crab in a respirometer. Temperature is held constant in the chamber, with air of known O_2 concentration flowing through. The crab's metabolic rate is calculated from the differences between the amounts of O_2 entering and leaving the respirometer. For every liter of O_2 consumed by the crab, respiration liberates about 4.83 kcal of energy from food molecules. If, for example, the crab used 0.05 L of oxygen per hour, then its metabolic rate would be 0.2415 kcal/h (0.05 L/h $\times$ 4.83 kcal/L).

This ghost crab is on a treadmill, running at a constant speed as measurements are made. This experimental design provides estimates of an animal's metabolic rate during activity. Similarly, in the bottom photograph, the metabolic rate of a man fitted with a plastic breathing apparatus is being monitored while he works out on a stationary bike.

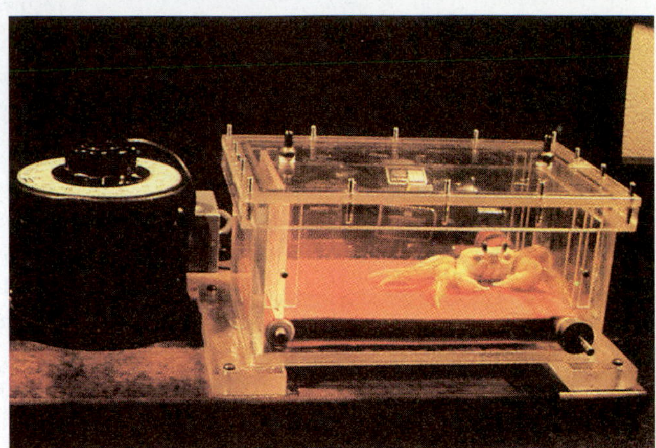

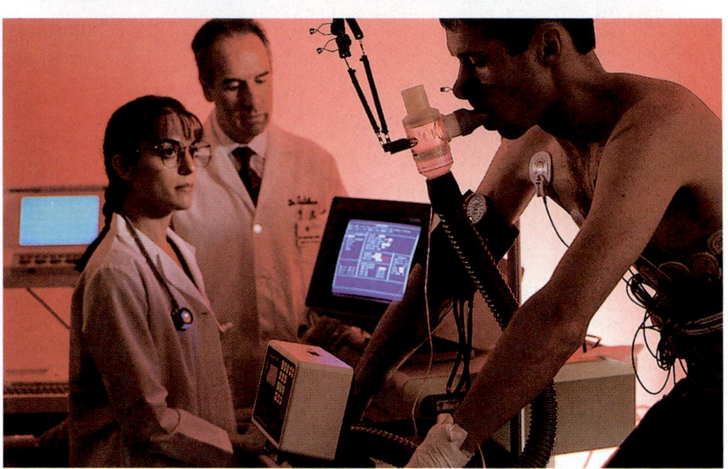

Metabolic rate per gram is inversely related to body size among similar animals

One of animal biology's most intriguing and as yet largely unanswered questions has to do with the relationship between body size and metabolic rate. By monitoring the metabolic rates of many species of vertebrates and invertebrates, physiologists have determined that the amount of energy it takes to maintain each gram of body weight is inversely related to body size. Each gram of a mouse, for instance, consumes about ten times more calories than a gram of an elephant (even though the whole elephant uses far more calories than the whole mouse). The higher metabolic rate of a smaller animal's body tissues demands a proportionately greater rate of delivery of oxygen to the tissues. And correlated with its higher metabolic rate, the smaller mammal also has a higher breathing rate, blood volume (relative to its size), and heart rate (pulse).

What causes the inverse relationship between metabolic rate and size? One hypothesis is that for endotherms, the smaller the animal, the greater the energy cost of maintaining a stable body temperature. This idea stems from the surface area to volume relationship: The smaller an animal, the greater its surface area to volume ratio, and thus the greater the loss of heat to, or gain from, the surroundings. Logical as this hypothesis appears, it is applicable only to endotherms and is not supported by experimental tests. Researchers continue to search for causes underlying the inverse relationship between body size and metabolic rate.

BODY PLANS AND THE EXTERNAL ENVIRONMENT

An animal's size and shape, features that biologists often call body plans or designs, are fundamental aspects of form and function that significantly affect the way an animal interacts with its environment. Use of the terms *plan* and *design* in no way implies that animal body forms are products of a conscious invention. The body plan or design of an animal results from a pattern of development programmed by the genome, itself the product of millions of years of evolution due to natural selection.

Physical support on land depends on adaptations of body proportions and posture

An engineer designing a bridge or tall building must take into account the effects of changes in size, or scale. An increase in size from a small-scale model to the real thing has a significant effect on building design. Physical laws dictate that the strength of a building support depends on its cross-sectional area, which increases as the square of its diameter. In sharp contrast, the strain on the supports depends on the building's weight, which increases as the cube of its height or other linear dimension. In common with a bridge or a building, an animal's body design must account for the greater demand for support that comes with increasing size. Consistent with physical laws, a large animal, such as an elephant, has very different body proportions than a small animal, such as a mouse. Imagine a mouse, with its very slender legs, scaled up to elephant size. If the imaginary animal kept its mouselike body proportions, its legs would collapse under its weight.

In simply applying the building analogy, we might predict the size of an elephant's leg bones to be directly proportional to the strain imposed by its body weight. However, our prediction would be inaccurate; an animal's body is complex and nonrigid, and the building analogy only partly explains the relationship between animal body design and support. The size of an animal's legs relative to its body size is only part of the story. It turns out that body posture—the position of the legs relative to the main body—is a more important design feature in supporting body weight, at least in mammals and birds (FIGURE 40.8). Muscles and tendons, which hold the legs of elephants, coyotes, humans, and other large mammals relatively straight and under the body, bear most of the load. The role of muscle action (powered by chemical energy) in supporting body weight emphasizes the theme of bioenergetics in animal function.

FIGURE 40.8 • The relationship between body weight and posture. A coyote stands, walks, and runs with most of its weight supported by muscles and tendons, which hold its legs directly under the body.

Body size and shape affect interactions with the environment

An animal's size and shape have a direct effect on how the animal exchanges energy and materials with its surroundings. As a requirement for maintaining the fluid integrity of the plasma membranes of its cells, an animal body must be arranged so that all of its living cells are bathed in an aqueous medium. Exchange with the environment occurs as dissolved substances diffuse and are transported across the plasma membranes between the cells and their aqueous surroundings. As shown in FIGURE 40.9a, a single-celled protist living in

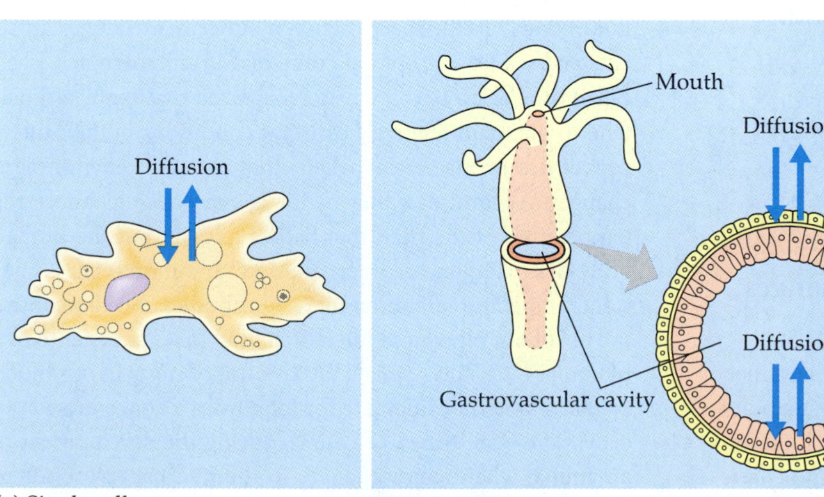

FIGURE 40.9 • Contact with the environment. (a) In a unicellular organism, such as this amoeba, the entire surface area contacts the environment. Because of its small size, the cell has a large surface area relative to its volume through which to exchange materials with the external world. **(b)** A hydra is bilayered. Because the aqueous environment can circulate in and out of its mouth, virtually every one of its cells directly contacts the environment and exchanges materials with it.

water has a sufficient surface area of plasma membrane to service its entire volume of cytoplasm because it is so small. A large cell has less surface area relative to its volume than a smaller cell of the same shape (see FIGURE 7.5). This is one reason nearly all cells are microscopic.

Multicellular animals are composed of microscopic cells, each with its own plasma membrane that functions as a loading and unloading platform for a modest volume of cytoplasm. But this only works if all the cells of the animal have access to a suitable aqueous environment. A hydra, built on the sac plan, has a body wall only two cell layers thick (FIGURE 40.9b). Because its gastrovascular cavity opens to the exterior, both outer and inner layers of cells are bathed in water. A flat body shape is another way to maximize exposure to the surrounding medium. For instance, a tapeworm may be several meters long, but because it is very thin, most of its cells are bathed in the intestinal fluid of the worm's vertebrate host, from which it obtains nutrients.

Two-layered sacs and flat shapes are designs that put a large surface area in contact with the environment, but these simple forms do not allow much complexity in internal organization. Most animals are more complex and made up of compact masses of cells; their outer surfaces are relatively small compared with their volume. As an extreme comparison, the surface-to-volume ratio of a whale is millions of times smaller than that of a protozoan, yet every cell in the whale must be bathed in fluid and have access to oxygen, nutrients, and other resources. Whales and most other animals have extensively folded or branched internal surfaces specialized for exchange with the environment (FIGURE 40.10). The circulatory system shuttles materials among all the exchange surfaces within the animal.

Although exchange with the environment is a problem for animals whose cells are mostly internal, complex body forms have distinct benefits. Because the animal's external surface need not be bathed in water, it is possible for the animal to live on land. Also, because the immediate environment for the cells is the internal body fluid, the animal's organ systems can control the composition of the solution bathing its cells.

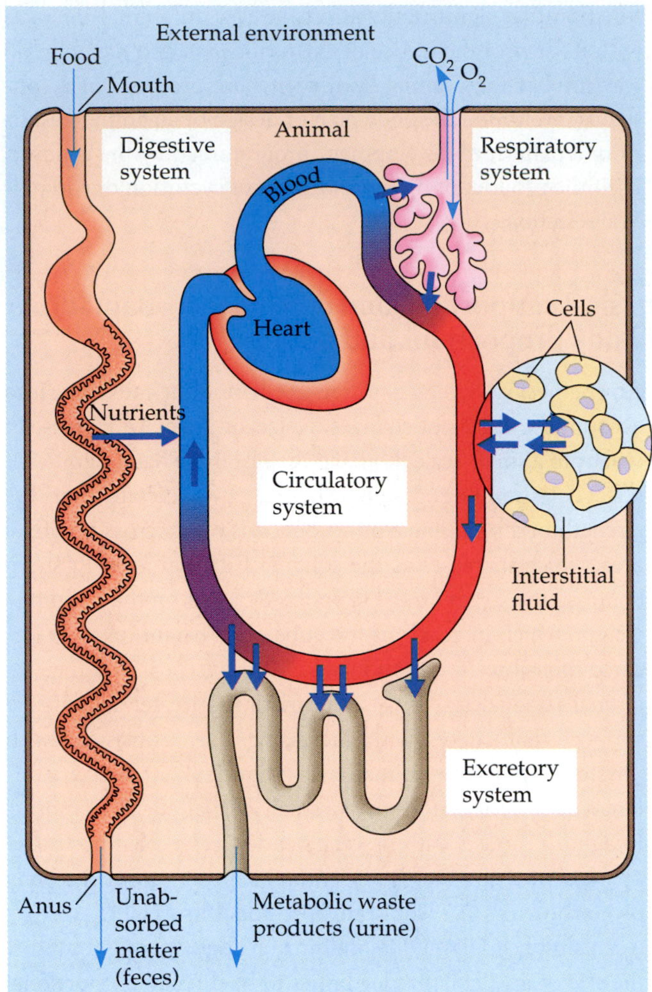

FIGURE 40.10 ▪ Internal surfaces of complex animals. Most animals have specialized surfaces that exchange materials with the environment. Usually internal but connected to the environment via openings on the body surface, these exchange surfaces are branched or folded, and therefore large in surface area. The digestive, respiratory, circulatory, and excretory systems all have specialized exchange surfaces. In addition, the circulatory system transports substances among the various internal surfaces.

REGULATING THE INTERNAL ENVIRONMENT

Mechanisms of homeostasis moderate changes in the internal environment

Over a century ago, French physiologist Claude Bernard made the distinction between the external environment surrounding an animal and the internal environment in which the cells of the animal actually live. The internal environment of vertebrates is called the **interstitial fluid** (see FIGURE 40.10). This fluid, which fills the spaces between our cells, exchanges nutrients and wastes with blood contained in microscopic vessels called capillaries. Bernard also recognized that many animals tend to maintain relatively constant conditions in their internal environment, even when the external environment changes. A pond-dwelling hydra is powerless to affect the temperature of the fluid that bathes its cells, but the human body can maintain its "internal pond" at a more-or-less constant temperature of about 37°C. Our bodies also can control the pH of our blood and interstitial fluid to within a tenth of a pH unit of 7.4, and regulate the amount of sugar in our blood so that it does not fluctuate for long from a concentration of 0.1%. There are times, of course, during the development of an animal when major changes in the internal environment are programmed to occur. For example, the balance of hormones in human blood is altered radically during puberty. Still, the stability of the internal environment is remarkable.

Today, Bernard's "constant internal milieu" is incorporated into the concept of **homeostasis**, which means "steady state." One of the main objectives of modern physiology, and a theme of this unit, is to learn how animals maintain homeostasis. Actually, the internal environment of an animal always fluctuates slightly. Homeostasis is a dynamic state, an interplay between outside forces that tend to change the internal environment and internal control mechanisms that oppose such changes (FIGURE 40.11).

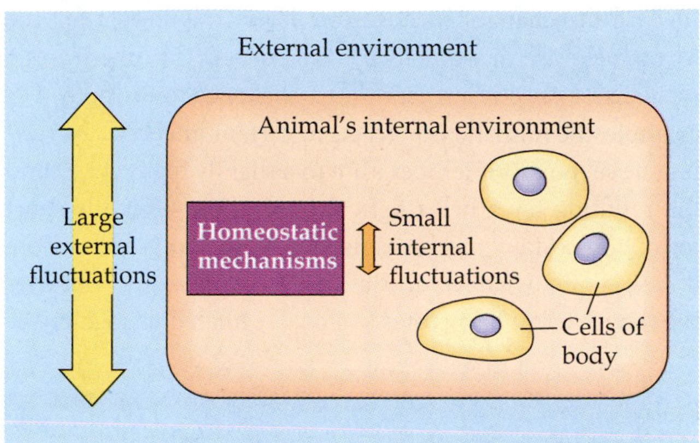

FIGURE 40.11 • Homeostasis. An animal has homeostatic mechanisms that regulate the fluid environment bathing its cells. Conditions of the external environment, such as the salt concentration in the water where the animal lives, may fluctuate widely. By contrast, conditions of the internal environment vary within only a narrow range. Regulating the internal environment as outside forces tend to change it, homeostatic mechanisms maintain conditions within a range in which the animal's metabolic processes can occur.

Homeostasis depends on feedback circuits

Any homeostatic control system has three functional components: a receptor, a control center, and an effector. The *receptor* detects a change in some variable of the animal's internal environment, such as change in body temperature. The *control center* processes information it receives from the receptor and directs an appropriate response by the *effector.* As an example of how these components interact, consider how the temperature of a room is controlled (FIGURE 40.12). In this case, the control center, called a thermostat, also contains the receptor (a thermometer). When room temperature falls below a set point, say 20°C, the thermostat switches on the heater (the effector). When the thermometer detects a temperature above the set point, the thermostat switches the heater off. This type of control circuit is called **negative feedback**, because a change in the variable being monitored triggers the control mechanism to counteract further change in the same direction. Owing to a lag time between reception and response, the variable drifts slightly above and below the set point, but the fluctuations are moderate. Negative-feedback mechanisms prevent small changes from becoming too large. Most homeostatic mechanisms in animals operate on this principle of negative feedback.

Our own body temperature is kept close to a set point of 37°C by the cooperation of several negative-feedback circuits that regulate energy exchange with the environment. One of these involves sweating as a means to dispose of metabolic heat and cool the body. A thermostat in the brain monitors the temperature of the blood. If the thermostat detects a rise in body temperature above the set point, it sends nerve impulses directing sweat glands to increase their production of sweat, thereby lowering body temperature by evaporative cooling (see Chapter 3). When body temperature drops below the set point, the thermostat in the brain stops sending the signals to the glands, and the body retains more of the heat produced by metabolism. We will see several examples of negative feedback in the chapters that follow.

In contrast to negative feedback, **positive feedback** involves a change in some variable that triggers mechanisms that amplify rather than reverse the change. During childbirth, for instance, the pressure of the baby's head against sensors near the opening of the uterus stimulates uterine contractions, which cause greater pressure against the uterine opening, heightening the contractions, which causes still greater pressure. Positive feedback brings childbirth to completion, a very different sort of process from maintaining a steady state.

It is important not to overstate the concept of a constant internal environment. In fact, *regulated change* is essential to normal body functions. In some cases the changes are cyclical,

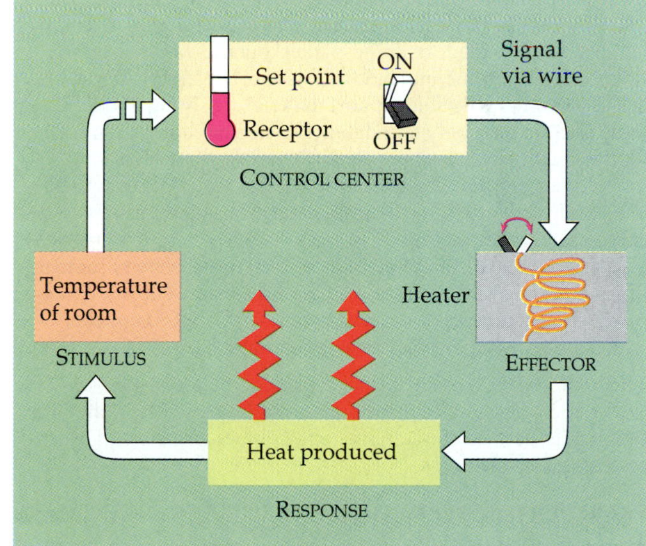

FIGURE 40.12 • Negative feedback as a mechanism of homeostasis. The thermostatic control of room temperature illustrates how negative feedback functions in the maintenance of homeostasis in an animal. A receptor (thermometer), housed with a thermostat in a control center, detects a change in a variable (room temperature). When room temperature falls below a set point, the thermostat switches on an effector (heater). The system is regulated by negative feedback: When the variable exceeds the set point, the control center turns the effector off.

such as the changes in hormone levels responsible for the menstrual cycle in women (see Chapter 46). In other cases a regulated change is a reaction to a challenge to the body. For example, the human body reacts to certain infections by raising the set point for temperature to a slightly higher level, and the resulting fever helps fight the infection. Over the short term, homeostatic mechanisms keep body temperature close to a set point, whatever it is at that particular time. But over the longer term, homeostasis allows regulated change in the body's internal environment, illustrating adaptive response, one of biology's central themes.

■ ■ ■

Having examined some general principles of animal body form and function, we are now ready to compare how diverse animals perform such activities as digestion, circulation, gas exchange, excretion of wastes, reproduction, and coordination—the topics of the next several chapters.

CHAPTER REVIEW

REVIEW OF KEY CONCEPTS

(with page numbers and key figures)

LEVELS OF STRUCTURAL ORGANIZATION

The organization of an animal's body emerges from the grouping of specialized cells into tissues, tissues into functional units called organs, and organs into organ systems.

- **Function correlates with structure in the tissues of animals (pp.
40.1 779–783, FIGURES 40.1, 40.3, 40.5)** Epithelial tissue covers the outside of the body and lines internal organs and cavities. Consistent with their barrier function, epithelial cells are tightly packed. They rest on a dense mat of extracellular matrix, the basement membrane, which may function in metabolism, development, and excretion. Epithelia are described according to the number of cell layers (simple or stratified) and the shape of the surface cells (cuboidal, columnar, or squamous).

  Some epithelia are specialized for absorption and secretion. The mucus secreted by the mucous membranes lining the digestive and respiratory tracts lubricates and moistens these surfaces.

  Connective tissues bind and support other tissues. Loose connective tissue, the body's binding and packing material, consists of fibroblasts and macrophages interspersed among collagenous, elastic, and reticular fibers. Adipose (fat) tissue is a specialized type of loose connective tissue. Fibrous connective tissue, found in tendons and ligaments, is made of dense, parallel bundles of collagenous fibers. Cartilage, bone, and blood are also connective tissues. Cartilage is a strong yet flexible support material consisting of collagenous fibers in a rubbery foundation secreted by chondrocytes. The hard substance of bone is secreted by cells called osteoblasts.

  Neurons, the functional units of nervous tissue, are composed of a cell body with extending dendrites and axons that transmit electrical signals called impulses.

  Muscle tissue is composed of long cells (muscle fibers) containing parallel microfilaments of contractile proteins. The muscle tissue of vertebrates consists of skeletal, cardiac, and smooth muscles, which differ in shape, striation, and nervous control.

- **The organ systems of an animal are interdependent (pp. 783–784, TABLE 40.1)** The body functions as a whole, greater than the sum of its parts, because the activities of all tissues, organs, and organ systems are coordinated.

INTRODUCTION TO THE BIOENERGETICS OF ANIMALS

- **Animals are heterotrophs that harvest chemical energy from the food they ingest (p. 784, FIGURE 40.7)** Animals are heterotrophs that obtain chemical energy by eating and digesting food produced by other organisms. The functions of an animal's cells, tissues, organs, and organ systems depend on cellular work powered by chemical energy in ATP.

- **Metabolic rate provides clues to an animal's bioenergetic "strategy" (pp. 784–785)** An animal's metabolic rate is the total amount of energy used in a unit of time. Measured by the amount of heat an animal gives off or by how much oxygen (O_2) it consumes, metabolic rates range from minimal levels for supporting life to maximal levels for performing peak activities. Minimal metabolic rates for birds and mammals, which maintain a fairly constant body temperature using metabolic heat, are generally higher than those of most fishes, reptiles, amphibians, and invertebrates, whose body temperature changes with that of their surroundings.

- **Metabolic rate per gram is inversely related to body size among similar animals (p. 786)**

BODY PLANS AND THE EXTERNAL ENVIRONMENT

- **Physical support on land depends on adaptations of body proportions and posture (p. 787)** Natural laws govern the relationship between an animal's body weight and the strength of its support structures. Posture often reduces the strain on support structures produced by a larger, heavier body.

- **Body size and shape affect interactions with the environment (pp. 787–788, FIGURES 40.9, 40.10)** Each cell of a multicellular animal must have access to an aqueous environment. Simple two-layered sacs and flat shapes maximize exposure to the surrounding medium. More complex body plans have highly folded internal surfaces specialized for exchanging materials with the environment. A circulatory system conveys these materials throughout an animal's body.

REGULATING THE INTERNAL ENVIRONMENT

- **Mechanisms of homeostasis moderate changes in the internal environment (pp. 788–789, FIGURE 40.11)** The internal environment surrounding the cells making up an animal's body is usually very different from the external environment surrounding the entire animal. In addition, the internal environment is carefully controlled and regulated by the process of homeostasis.

- **Homeostasis depends on feedback circuits (pp. 789–790, FIGURE 40.12)** Homeostatic mechanisms usually involve negative feedback. These mechanisms also enable regulated change as they react to occasional shifts in the body's set points for variables such as temperature.

SELF-QUIZ

1. Simple cuboidal epithelium usually functions in
   a. secretion of substances by glands
   b. lining internal surfaces subject to abrasion
   c. packing and padding body parts
   d. covering body surfaces
   e. attaching muscles to bones

2. Which of the following structures or substances is *incorrectly* paired with a tissue?
   a. Haversian system—bone
   b. platelets—blood

c. fibroblasts—skeletal muscle
d. chondroitin sulfate—cartilage
e. basement membrane—epithelium

3. When an endotherm's body temperature is too low, which of the following occurs?
   a. increased sweating
   b. the brain's thermostat shuts off
   c. increased evaporative cooling
   d. increased rate of adipose tissue formation
   e. the brain's thermostat stops promoting sweating

4. The involuntary muscles that cause the wavelike contractions pushing food along our intestine are
   a. striated muscles
   b. cardiac muscles
   c. skeletal muscles
   d. smooth muscles
   e. intercalated muscles

5. Which of the following is *not* considered to be a tissue?
   a. cartilage
   b. the mucous membrane lining the stomach
   c. blood
   d. brain
   e. cardiac muscle

6. Which of the following statements about bioenergetics is true?
   a. Every animal has a specific metabolic rate that does not change.
   b. A BMR can be determined only at a specific temperature.
   c. Endotherms are warmed by metabolic heat.
   d. An SMR is best measured just after an ectotherm has eaten.
   e. Ectotherms and endotherms are energetically the same.

7. Compared to a smaller cell, a larger cell of the same shape has
   a. less surface area
   b. less surface area per unit of volume
   c. the same surface-to-volume ratio
   d. a smaller average distance between its mitochondria and the external source of oxygen
   e. a smaller cytoplasm-to-nucleus ratio

8. Which of the following vertebrate organ systems does *not* open directly to the external environment?
   a. digestive system
   b. circulatory system
   c. excretory system
   d. respiratory system
   e. reproductive system

9. Most of our cells are surrounded by
   a. blood
   b. connective tissue
   c. interstitial fluid
   d. pure water
   e. air

10. Which of the following physiological responses is an example of *positive* feedback?
    a. An increase in the concentration of glucose in the blood stimulates the pancreas to secrete insulin, a hormone that lowers blood glucose concentration.
    b. A high concentration of carbon dioxide in the blood causes deeper, more rapid breathing, which expels carbon dioxide.
    c. Stimulation of a nerve cell causes sodium ions to leak into the cell, and the sodium influx triggers the inward leaking of even more sodium.
    d. The body's production of red blood cells, which transport oxygen from the lungs to other organs, is stimulated by a low concentration of oxygen.
    e. The pituitary gland secretes a hormone called TSH, which stimulates the thyroid gland to secrete another hormone called thyroxine; a high concentration of thyroxine suppresses the pituitary's secretion of TSH.

CHALLENGE QUESTION

Suggest your own hypothesis to explain the inverse relationship between body size and metabolic rate per gram of tissue. How could you test your hypothesis?

SCIENCE, TECHNOLOGY, AND SOCIETY

Medical researchers are investigating the possibilities of artificial substitutes for various human tissues. Examples are a liquid that could serve as "artificial blood" and a fabric that could temporarily serve as artificial skin for victims of serious burns. In what other situations might artificial blood or skin be useful? What characteristics would these substitutes need in order to function effectively in the body? Why do real tissues work better? Why not use the real things if they work better? Can you think of other artificial tissues that might be useful? What problems do you anticipate in developing and applying them?

FURTHER READING

Adler, T. "Mending Joints." *Science News,* November 12, 1994. Describes progress in treating cartilage injuries with transplanted connective tissue cells.

Ewing, W. A. *Inside Information, Imaging the Human Body.* New York: Simon and Schuster, 1996. Contains striking images created by state-of-the-art techniques of the inner workings of the human body.

Hammond, K. A., and J. Diamond. "Maximal Sustained Energy Budgets in Humans and Animals." *Science,* April 3, 1997. Provides an examination of the upper limits of energy metabolism.

Houston, C. S. "Mountain Sickness." *Scientific American,* October 1992. Discusses how high altitude interferes with homeostasis.

Ingber, D. E. "The Architecture of Life." *Scientific American,* January 1998. Presents an analysis of the universal laws that underlie organic structures.

Knight, J. "Busy Doing Nothing." *New Scientist,* April 25, 1998. Discusses changes in metabolic rates as adaptations.

Marieb, E. *Human Anatomy and Physiology,* 4th ed. Menlo Park, CA: Benjamin/Cummings, 1998. A basic text, beautifully illustrated.

Schmidt-Nielsen, K. *Animal Physiology: Adaptation and Environment,* 5th ed. New York: Cambridge University Press, 1997. A classic text emphasizing form and function, adaptations, and homeostasis.

Stone, G. N. "Hot-Blooded Bees." *Natural History,* July 1993. Discusses temperature regulation by insects as an adaptation to cool northern climates.

Urry, D. W. "Elastic Biomolecular Machines." *Scientific American,* January 1995. An investigation into animal bioenergetics through the design and study of synthetic protein machines that mimic natural functions.

WEB LINKS

Visit the special edition of *The Biology Place* for BIOLOGY, Fifth Edition, at http://www.biology.com/campbell. Go to Chapter 40 for online resources, including learning activities, practice exams, and links to the following web sites:

"JayDoc HistoWeb"
Impressive collection of slides of various tissues and organs.

"PERLjam"
Yet another impressive collection of tissue slides, this time with histopathology emphasized.

"Nutrition on the Web—For Teens"
This interesting site has a number of interactive components including a basal metabolic rate/active metabolic rate calculator.

"A Guided Tour Through the Visible Human Body"
This highly interactive tour is based on the visible human dataset, a collection of over 9000 digital pictures produced by the National Library of Medicine's Visible Human Project.

CHAPTER 41

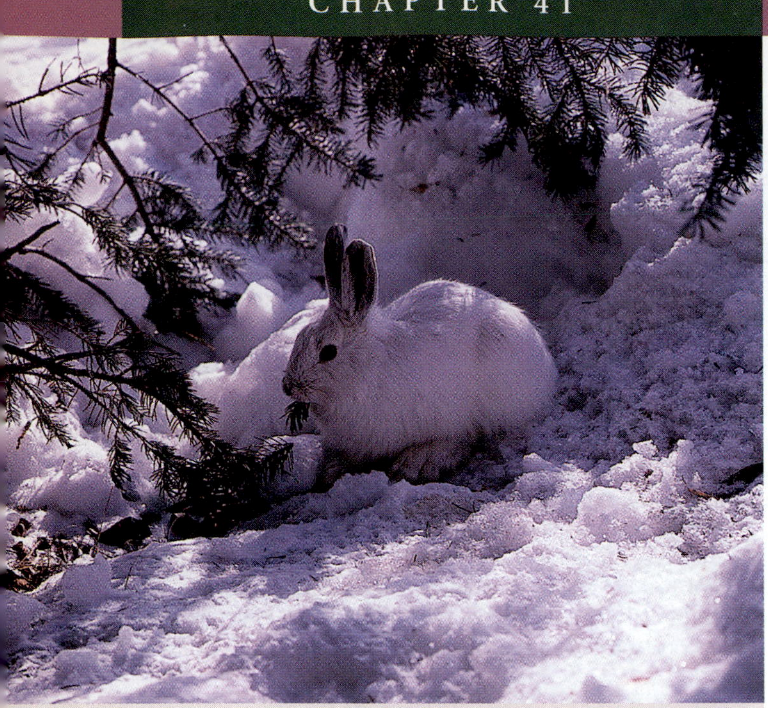

ANIMAL NUTRITION

Every mealtime is a reminder that we are animals, heterotrophs dependent on a regular supply of food derived from other organisms. As a group, animals exhibit a great variety of nutritional adaptations. The snowshoe hare shown in the photograph on this page is adapted for life in northern forests. Able to obtain all their nutritional needs from plants alone, hares and rabbits have a large intestinal pouch housing prokaryotes and protists that digest cellulose. When deep snow covers the ground, a hare can live on pine, fir, or spruce branches—often the only plant materials available.

For any animal, a nutritionally adequate diet is essential for homeostasis, a steady-state balance in body function. A balanced diet provides fuel for cellular work, as well as all the materials the body needs to construct its own organic molecules. In this chapter we examine the nutritional requirements of animals and look at some of the diverse adaptations for obtaining and processing food.

NUTRITIONAL REQUIREMENTS

Animals are heterotrophs that require food for fuel, carbon skeletons, and essential nutrients: *an overview*

A nutritionally adequate diet satisfies three needs: fuel (chemical energy) for all the cellular work of the body; the organic raw materials animals use in biosynthesis (carbon skeletons to make many of their own molecules); and essential nutrients, substances the animal cannot make for itself from *any* raw material and that must be obtained in food in prefabricated form.

Homeostatic mechanisms manage an animal's fuel

The theme of bioenergetics is integral to our study of nutrition. Animals obtain the fuel (chemical energy) that powers the work of their body cells from the oxidation of organic molecules—carbohydrates, proteins, and fats. The monomers of any of these substances can be used as fuel for the generation of ATP by cellular respiration, though priority is usually given to carbohydrates and fats. Fats are especially rich in energy; the oxidation of fat liberates about twice the energy liberated from an equal amount of carbohydrate or protein.

As we saw in Chapter 40, animals have basal energy requirements that must be met to maintain the metabolic functions for sustaining life. When an animal takes in more calories than it consumes to meet its energy requirement, the body tends to store the excess calories. In humans, for example, the liver and muscle cells store energy in the form of glycogen (a polymer made up of many glucose units; see FIGURE 5.6). Glucose is a major fuel molecule for cells, and its metabolism, regulated by hormone action, is an important aspect of homeostasis (FIGURE 41.1). If glycogen stores are full and caloric intake still exceeds caloric expenditure, the excess may be stored as fat.

Between meals or when fewer calories are taken in than are expended (by heavy exercise, for instance), fuel is taken out of storage and oxidized, and weight loss may occur. The human body generally expends liver glycogen first, and then draws on muscle glycogen and fat.

An **undernourished** person or other animal is one whose diet is deficient in calories. When starvation for calories persists, the body begins breaking down its own proteins for fuel, muscles begin to decrease in size, and the brain can become protein-deficient. If an undernourished person survives, some of the damage is irreversible. Even a diet of a single staple such as rice or corn provides calories; thus, undernourishment is generally common only where drought, war, or some other crisis has severely disrupted the food supply. Another cause of undernourishment is anorexia nervosa, an eating disorder associated with a compulsive aversion to body fat.

In the United States and increasingly in other affluent nations, overnourishment, or obesity, is far more common than undernourishment. The human body tends to hoard fat, to immediately store any excess fat molecules obtained from food instead of using them for fuel or biosynthesis. By contrast, when we eat an excess of carbohydrates, the body tends to increase its rate of carbohydrate consumption. Thus, the amount of fat in the diet can have a more direct effect on weight gain than carbohydrates. Even those who are not obese usually have enough fat to sustain them for several weeks. Fat hoarding can be a liability today, but it may have provided a degree of fitness for our hunting/gathering ancestors. Individuals with genes promoting the storage of high-energy molecules during feasts may have been those that survived famines. Despite its tendency to store fat, the human body seems to impose limits on weight gain (and loss). Some people hold a more-or-less constant weight no matter how much they eat. Even obese people usually attain a relatively stable weight, generally unaffected by how much they eat, and most dieters return to their former weight soon after they stop dieting.

Recent discoveries suggest that complex feedback mechanisms regulate fat storage and use in mammals, including humans. A hormone called leptin, produced by adipose cells, is a key player. An increase in adipose tissue increases leptin levels in the blood. A high leptin level cues the brain to depress appetite and increase energy-consuming muscular activity and body-heat production. Conversely, loss of body fat decreases leptin levels in the blood, signaling the brain to increase appetite and weight gain. Apparently these feedback mechanisms regulate body weight around a fairly rigid set point in some individuals and over a relatively wide range in others. Researchers have also identified some of the genes involved in fat homeostasis and several chemical signals that underlie the brain's regulatory roles. Some of the signals and signal antagonists are under development as potential medications for obesity.

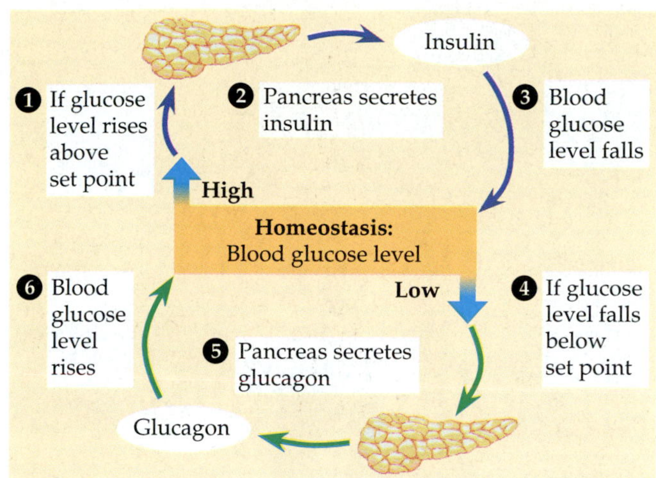

FIGURE 41.1 • Homeostatic regulation of cellular fuel. The body's use and storage of glucose, a major cellular fuel, is closely regulated. After a meal is digested, glucose and other monomers are absorbed into the blood from the digestive tract. ① If the level of glucose in the blood rises above a set point, ② the pancreas secretes insulin, a hormone, into the blood. ③ Insulin enhances the transport of glucose into body cells and stimulates the liver and muscle cells to store glucose as glycogen. As a result, the blood glucose level drops. ④ When the glucose level falls below the set point, ⑤ the pancreas secretes the hormone glucagon, which opposes the effect of insulin. ⑥ Glucagon promotes the breakdown of glycogen and the release of glucose into the blood, increasing the blood glucose level.

An animal's diet must supply essential nutrients and carbon skeletons for biosynthesis

In addition to providing cellular fuel, an animal's diet must also supply all the raw materials needed for biosynthesis. As heterotrophs, animals cannot make organic molecules from raw materials that are entirely inorganic. To synthesize the molecules it needs to grow and replenish itself, an animal must obtain organic precursors (carbon skeletons) from its food. Given a source of organic carbon (such as sugar) and a source of organic nitrogen (such as amino acids from the digestion of protein), the animal can fabricate a great variety of organic molecules. For instance, a single type of amino acid can supply nitrogen for the synthesis of several other types of amino acids that may not be present in the food.

Besides fuel and carbon skeletons, an animal's diet must also supply **essential nutrients**, materials that must be obtained in preassembled form because the animal's cells cannot make them from *any* raw material. An essential nutrient for one animal species is not necessarily required by another. For instance, ascorbic acid (vitamin C) is an essential nutrient for humans, other primates, guinea pigs, and some birds and snakes, but not for most other animals.

FIGURE 41.2 • Obtaining essential nutrients. A giraffe, a plant eater of East Africa, chews on old bones from a dead mammal. Bones contain calcium phosphate, and osteophagia ("bone-eating") is common among herbivores living where soils and plants are deficient in phosphorus, a mineral nutrient required to make ATP, nucleic acids, phospholipids, and bones.

An animal whose diet is missing one or more essential nutrients is said to be **malnourished** (recall that *undernourished* refers to caloric deficiency). For example, cattle and other animals that eat mainly plants may suffer mineral deficiencies if they graze on plants where the soil is deficient in key minerals (FIGURE 41.2). Malnutrition is much more common than undernutrition in human populations, and it is even possible for an overnourished individual also to be malnourished.

There are four classes of essential nutrients: essential amino acids, essential fatty acids, vitamins, and minerals. Animals require 20 amino acids to make proteins, and most animal species can synthesize about half of these, as long as their diet includes organic nitrogen. The remaining ones, the **essential amino acids**, must be obtained from food in prefabricated form. Eight amino acids are essential in the adult human diet (a ninth, histidine, is essential for infants).

A diet that lacks one or more of the essential amino acids results in a form of malnutrition known as protein deficiency. The most common type of malnutrition among humans, protein deficiency is concentrated in geographical regions where there is a great gap between food supply and population size. The victims are usually children, who, if they survive infancy, are likely to be retarded in physical and perhaps mental development.

The most reliable sources of essential amino acids are meat, eggs, cheese, and other animal products. The proteins in animal products are complete, which means that they provide all the essential amino acids in their proper proportions. Most plant proteins, on the other hand, are incomplete, being deficient in one or more essential amino acids. Corn, for example, is deficient in the amino acid lysine. People forced by economic necessity to obtain nearly all their calories from corn would show symptoms of protein deficiency, as would those who eat only rice, wheat, or potatoes. A vegetarian, however, can obtain sufficient quantities of all essential amino acids by eating a combination of plant foods that complement one another. Beans, for example, supply the lysine that is missing in corn; and whereas beans are deficient in methionine, this amino acid is present in corn. Thus, a meal of beans and corn can provide all the essential amino acids (FIGURE 41.3). The combination of vegetables must be consumed during the same day. Because the body cannot store amino acids, a deficiency of a single essential amino acid retards protein synthesis and limits the use of other amino acids. Most cultures have, by trial and error, developed balanced diets that prevent protein deficiency.

Animals can synthesize most of the fatty acids they need. The **essential fatty acids**, the ones they cannot make, are certain unsaturated fatty acids (fatty acids having double bonds; see Chapter 5). In humans, for example, linoleic acid must be present in the diet in preassembled form. This essential fatty acid is required to make some of the phospholipids found in membranes. Most diets furnish ample quantities of essential fatty acids, and thus deficiencies are rare.

Vitamins are organic molecules required in the diet in amounts that are quite small compared with the relatively large quantities of essential amino acids and fatty acids ani-

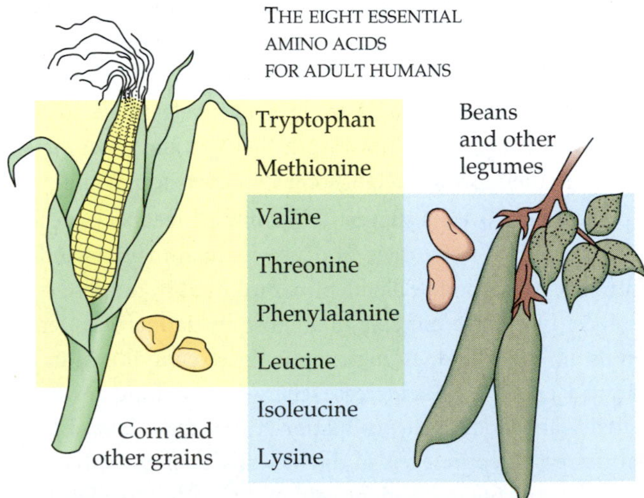

FIGURE 41.3 • Essential amino acids from a vegetarian diet. Corn has little isoleucine and lysine. Beans have ample isoleucine and lysine, but little tryptophan and methionine. An adult human can obtain all eight essential amino acids by eating a meal of corn and beans.

mals need. Tiny amounts of vitamins may suffice, from about 0.01 to 100 mg per day, depending on the vitamin. However, vitamin deficiencies can cause severe problems.

So far, 13 vitamins essential to humans have been identified. The compounds are grouped into two categories: water-soluble vitamins and fat-soluble vitamins (TABLE 41.1). Water-soluble vitamins include the B complex, which consists of several compounds that generally function as coenzymes in key metabolic processes. Vitamin C, also water-soluble, is required for the synthesis of connective tissue. Excesses of water-soluble vitamins are excreted in urine, and moderate overdoses of these vitamins are probably harmless.

The fat-soluble vitamins are A, D, E, and K. Vitamin A is incorporated into visual pigments of the eye. Vitamin D aids in calcium absorption and bone formation. The function of vitamin E is not yet fully understood, but along with vitamin C it seems to protect the phospholipids in membranes from oxidation. Vitamin K is required for blood clotting. Excesses of fat-soluble vitamins are not excreted but are deposited in body fat, so overdoses may result in an accumulation of these compounds to toxic levels.

The subject of vitamin dosage has aroused heated debate. Some believe it is sufficient to meet recommended daily allowances (RDAs), the nutrient intakes proposed by nutritionists to maintain health. Others argue that RDAs are set too low for some vitamins. Research is far from complete, and debate continues, especially over optimal doses of vitamins C and E. All that can be said with any certainty at this time is

Table 41.1 ■ Vitamin Requirements of Humans

VITAMIN	MAJOR DIETARY SOURCES	SOME MAJOR FUNCTIONS IN THE BODY	POSSIBLE SYMPTOMS OF DEFICIENCY OR EXTREME EXCESS
Water-Soluble Vitamins			
Vitamin B_1 (thiamine)	Pork, legumes, peanuts, whole grains	Coenzyme used in removing CO_2 from organic compounds	Beriberi (nerve disorders, emaciation, anemia)
Vitamin B_2 (riboflavin)	Dairy products, meats, enriched grains, vegetables	Component of coenzymes FAD and FMN	Skin lesions such as cracks at corners of mouth
Niacin	Nuts, meats, grains	Component of coenzymes NAD^+ and $NADP^+$	Skin and gastrointestinal lesions, nervous disorders Flushing of face and hands, liver damage
Vitamin B_6 (pyridoxine)	Meats, vegetables, whole grains	Coenzyme used in amino acid metabolism	Irritability, convulsions, muscular twitching, anemia Unstable gait, numb feet, poor coordination
Pantothenic acid	Most foods: meats, dairy products, whole grains, etc.	Component of coenzyme A	Fatigue, numbness, tingling of hands and feet
Folic acid (folacin)	Green vegetables, oranges, nuts, legumes, whole grains (also made by colon bacteria)	Coenzyme in nucleic acid and amino acid metabolism	Anemia, gastrointestinal problems May mask deficiency of vitamin B_{12}
Vitamin B_{12}	Meats, eggs, dairy products	Coenzyme in nucleic acid metabolism; needed for maturation of red blood cells	Anemia, nervous system disorders
Biotin	Legumes, other vegetables, meats	Coenzyme in synthesis of fat, glycogen, and amino acids	Scaly skin inflammation, neuromuscular disorders
Vitamin C (ascorbic acid)	Fruits and vegetables, especially citrus fruits, broccoli, cabbage, tomatoes, green peppers	Used in collagen synthesis (e.g., for bone, cartilage, gums); antioxidant; aids in detoxification; improves iron absorption	Scurvy (degeneration of skin, teeth, blood vessels), weakness, delayed wound healing, impaired immunity Gastrointestinal upset
Fat-Soluble Vitamins			
Vitamin A (retinol)	Provitamin A (beta-carotene) in deep green and orange vegetables and fruits; retinol in dairy products	Component of visual pigments; needed for maintenance of epithelial tissues; antioxidant; helps prevent damage to lipids of cell membranes	Vision problems; dry, scaling skin Headache, irritability, vomiting, hair loss, blurred vision, liver and bone damage
Vitamin D	Dairy products, egg yolk (also made in human skin in presence of sunlight)	Aids in absorption and use of calcium and phosphorus; promotes bone growth	Rickets (bone deformities) in children, bone softening in adults Brain, cardiovascular, and kidney damage
Vitamin E (tocopherol)	Vegetable oils, nuts, seeds	Antioxidant; helps prevent damage to lipids of cell membranes	None well documented in humans; possibly anemia
Vitamin K (phylloquinone)	Green vegetables, tea (also made by colon bacteria)	Important in blood clotting	Defective blood clotting Liver damage and anemia

that people who eat a balanced diet are not likely to develop symptoms of vitamin deficiency.

Minerals are inorganic nutrients, usually required in very small amounts—from less than 1 mg to about 2500 mg per day, depending on the mineral (TABLE 41.2). As with vitamins, mineral requirements vary with animal species. Humans and other vertebrates require relatively large quantities of calcium and phosphorus for the construction and maintenance of bone. Calcium is also necessary for the normal functioning of nerves and muscles, and phosphorus is also an ingredient of ATP and nucleic acid. Iron is a component of the cytochromes that function in cellular respiration (see Chapter 9) and of hemoglobin, the oxygen-binding protein of red blood cells. Magnesium, iron, zinc, copper, manganese, selenium, and molybdenum are cofactors built into the structure of certain enzymes; magnesium, for example, is present in enzymes that split ATP. Vertebrates need iodine to make thyroid hormones, which regulate metabolic rate. Sodium, potassium, and chlorine are important in nerve function and also have a major influence on the osmotic balance between cells and the interstitial fluid. Most people ingest far more salt (sodium chloride) than they need. The average U.S. citizen eats enough salt to provide about 20 times the required amount of sodium. Ingesting an excess of salt and several other minerals can upset homeostatic balance and cause toxic side effects. For example, too much sodium is associated with high blood pressure; excess iron can cause liver damage.

Having surveyed the major nutritional needs, we now shift our attention to the kinds of foods animals eat and the varied mechanisms that allow them to meet their nutritional needs.

Table 41.2 ■ Mineral Requirements of Humans

MINERAL	MAJOR DIETARY SOURCES	SOME MAJOR FUNCTIONS IN THE BODY	POSSIBLE SYMPTOMS OF DEFICIENCY*
Calcium (Ca)	Dairy products, dark green vegetables, legumes	Bone and tooth formation, blood clotting, nerve and muscle function	Retarded growth, possibly loss of bone mass
Phosphorus (P)	Dairy products, meats, grains	Bone and tooth formation, acid-base balance, nucleotide synthesis	Weakness, loss of minerals from bone, calcium loss
Sulfur (S)	Proteins from many sources	Component of certain amino acids	Symptoms of protein deficiency
Potassium (K)	Meats, dairy products, many fruits and vegetables, grains	Acid-base balance, water balance, nerve function	Muscular weakness, paralysis, nausea, heart failure
Chlorine (Cl)	Table salt	Acid-base balance, formation of gastric juice, nerve function, osmotic balance	Muscle cramps, reduced appetite
Sodium (Na)	Table salt	Acid-base balance, water balance, nerve function	Muscle cramps, reduced appetite
Magnesium (Mg)	Whole grains, green leafy vegetables	Cofactor; ATP bioenergetics	Nervous system disturbances
Iron (Fe)	Meats, eggs, legumes, whole grains, green leafy vegetables	Component of hemoglobin and of electron-carriers in energy metabolism; enzyme cofactor	Iron-deficiency anemia, weakness, impaired immunity
Fluorine (F)	Drinking water, tea, seafood	Maintenance of tooth (and probably bone) structure	Higher frequency of tooth decay
Zinc (Zn)	Meats, seafood, grains	Component of certain digestive enzymes and other proteins	Growth failure, scaly skin inflammation, reproductive failure, impaired immunity
Copper (Cu)	Seafood, nuts, legumes, organ meats	Enzyme cofactor in iron metabolism, melanin synthesis, electron transport	Anemia, bone and cardiovascular changes
Manganese (Mn)	Nuts, grains, vegetables, fruits, tea	Enzyme cofactor	Abnormal bone and cartilage
Iodine (I)	Seafood, dairy products, iodized salt	Component of thyroid hormones	Goiter (enlarged thyroid)
Cobalt (Co)	Meats and dairy products	Component of vitamin B_{12}	None, except as B_{12} deficiency
Selenium (Se)	Seafood, meats, whole grains	Enzyme cofactor; antioxidant functioning in close association with vitamin E	Muscle pain, possibly heart muscle deterioration
Chromium (Cr)	Brewer's yeast, liver, seafood, meats, some vegetables	Involved in glucose and energy metabolism	Impaired glucose metabolism
Molybdenum (Mo)	Legumes, grains, some vegetables	Enzyme cofactor	Disorder in excretion of nitrogen-containing compounds

*All of these minerals are also harmful when consumed in excess.

FOOD TYPES AND FEEDING MECHANISMS

Most animals are opportunistic feeders

Most animals eat other organisms, dead or alive, whole or by the piece. (Exceptions are certain parasitic animals, such as tapeworms, which absorb organic molecules across their body surface.) In general, animals fit into one of three dietary categories. **Herbivores**, including gorillas, cows, hares, and many snails, eat mainly autotrophs (plants, algae). **Carnivores**, such as sharks, hawks, spiders, and snakes, eat other animals. **Omnivores** regularly consume animals as well as plant or algal matter. Omnivorous animals include cockroaches, crows, raccoons, and humans, who evolved as hunters, scavengers, and gatherers.

FIGURE 41.4 ▪ Suspension-feeding: a baleen whale. The humpback whale and other baleen whales use comblike plates suspended from the upper jaw to sift small invertebrates from enormous volumes of water. The whale opens its mouth and fills an expandable oral pouch with water, then closes its mouth and contracts the pouch. This forces water out of the mouth through the baleen, leaving a mouthful of trapped food.

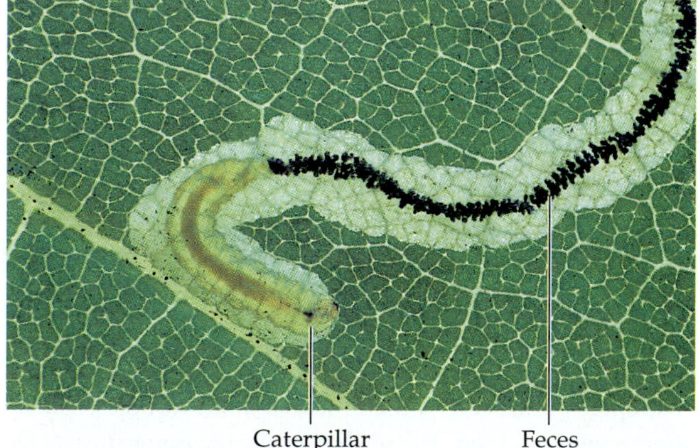

FIGURE 41.5 ▪ Substrate-feeding: a leaf miner. Eating its way through the soft mesophyll of an oak leaf, this caterpillar, the larva of a moth, leaves a trail of dark feces in its wake.

The terms *herbivore, carnivore,* and *omnivore* represent the kinds of food an animal usually eats and the adaptations enabling it to obtain and process that food. However, most animals are opportunistic, eating foods that are outside their main dietary category when these foods are available. For example, cattle and deer, which are herbivores, may occasionally eat small animals or bird eggs, along with grass and other plants. Most carnivores obtain some nutrients from plant materials that remain in the digestive tract of the prey they eat. And all animals consume some bacteria with other types of food.

Diverse feeding adaptations have evolved among animals

The mechanisms by which animals ingest food are highly varied but fall into four main groups. Many aquatic animals are **suspension-feeders** that sift small food particles from the water. Clams and oysters, for example, use their gills to trap tiny morsels, which are then swept along with a film of mucus to the mouth by beating cilia. Baleen whales, the largest animals ever to live, are also suspension-feeders. They swim with their mouths agape, straining millions of small animals from huge volumes of water forced through screenlike plates attached to their jaws (FIGURE 41.4).

Substrate-feeders live in or on their food source, eating their way through the food. Examples are leaf miners, which are larvae of various insects that tunnel through the interior of leaves (FIGURE 41.5). Earthworms are also substrate-feeders or, more specifically, **deposit-feeders**. Eating their way through the dirt, earthworms salvage partially decayed organic material consumed along with soil.

Fluid-feeders make their living by sucking nutrient-rich fluids from a living host (FIGURE 41.6). Mosquitoes and leeches suck blood from animals. Aphids tap the phloem sap of plants. Because these particular fluid-feeders harm their

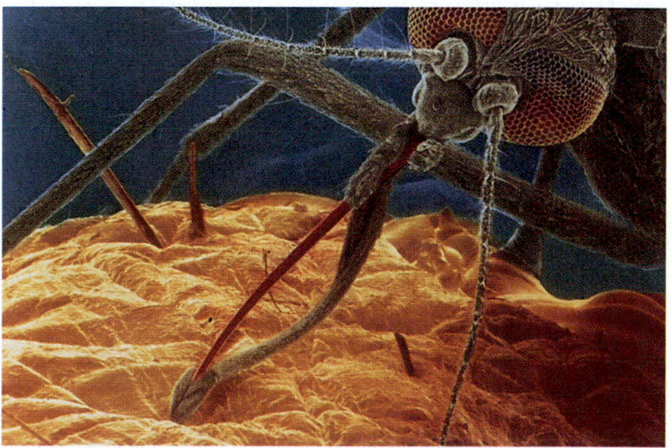

FIGURE 41.6 ▪ Fluid-feeding: a mosquito. This parasite has pierced the skin of its human host with hollow needlelike mouthparts and is filling its digestive tract with a blood meal (colorized SEM).

FIGURE 41.7 ▪ Bulk-feeding: a python. Most animals ingest relatively large pieces of food, though rarely as large as the prey in this case. In this amazing scene, a rock python is beginning to ingest a gazelle. The snake will spend two weeks or more in a quiet place digesting its meal.

hosts, they are considered parasites. Hummingbirds and bees, on the other hand, benefit their host plants, transferring pollen as they move from flower to flower to obtain nectar.

Most animals are **bulk-feeders** that eat relatively large pieces of food (FIGURE 41.7). Their adaptations include such diverse utensils as tentacles, pincers, claws, poisonous fangs, and jaws and teeth that kill their prey or tear off pieces of meat or vegetation.

OVERVIEW OF FOOD PROCESSING

The four main stages of food processing are ingestion, digestion, absorption, and elimination

Ingestion, the act of eating, is the first stage of food processing. **Digestion**, the second stage, is the process of breaking food down into molecules small enough for the body to absorb. The bulk of the organic material in food consists of proteins, fats, and carbohydrates in the form of starch and other polysaccharides. Although these macromolecules are suitable raw materials, animals cannot use them directly, for two reasons. First, macromolecules are too large to pass through membranes and enter the cells of the animal. Second, the macromolecules that make up an animal are not identical to those of its food. In building their macromolecules, however, all organisms use common monomers. For instance, soybeans, cattle, and humans all assemble their proteins from the same 20 amino acids. Digestion cleaves macromolecules into their component monomers, which the animal then uses to make its own molecules. Polysaccharides and disaccharides are split into simple sugars, fats are digested to glycerol and fatty acids, proteins are broken down to amino acids, and nucleic acids are cleaved into nucleotides.

Recall from Chapter 5 that when a cell makes a macromolecule by linking together monomers, it does so by removing a molecule of water for each new covalent bond formed. Digestion reverses this process by breaking bonds with the enzymatic addition of water. This splitting process is called **enzymatic hydrolysis.** Hydrolytic enzymes catalyze the digestion of each of the classes of macromolecules found in food. This chemical digestion is usually preceded by mechanical fragmentation of the food—by chewing, for instance. Breaking food into smaller pieces increases the surface area exposed to digestive juices containing hydrolytic enzymes.

The last two stages of food processing occur after the food is digested. In the third stage, **absorption**, the animal's cells take up (absorb) small molecules such as amino acids and simple sugars from the digestive compartment. Finally, **elimination** occurs, as undigested material passes out of the digestive compartment.

Digestion occurs in specialized compartments

Intracellular Digestion

Food vacuoles, cellular organelles in which hydrolytic enzymes break down food without digesting the cell's own cytoplasm, are the simplest digestive compartments. Heterotrophic protists digest their meals in food vacuoles, usually after engulfing food by phagocytosis or pinocytosis (see Chapter 8). The food vacuoles fuse with lysosomes, which are organelles containing hydrolytic enzymes. This mixes the

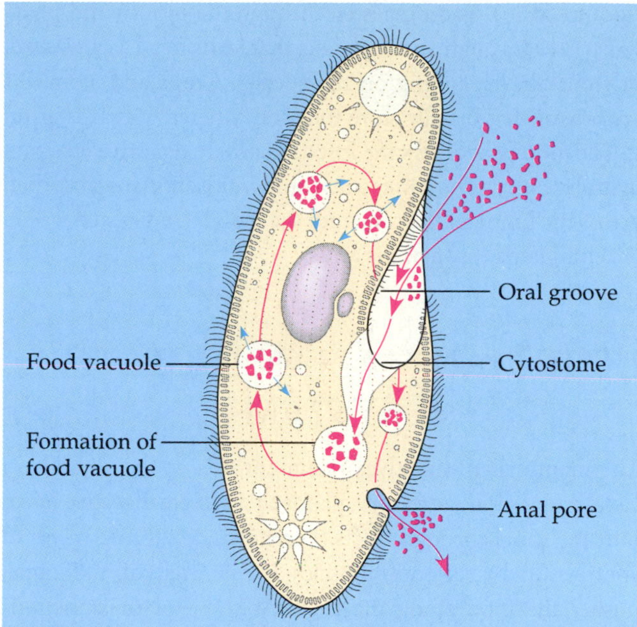

FIGURE 41.8 • Intracellular digestion in *Paramecium.* *Paramecium* has a specialized feeding structure called the oral groove, which leads to the cell's "mouth" (cytostome). Cilia that line the groove draw water and suspended food particles, mostly bacteria, toward the mouth, where the food is packaged into a food vacuole that functions as a miniature digestive compartment. Cytoplasmic streaming carries food vacuoles around the cell, while hydrolytic enzymes are secreted into the vacuoles. As molecules in the food are digested, the nutrients (sugars, amino acids, and other small molecules, represented by blue arrows) are transported across the membrane of the vacuole into the cytoplasm. Later, the vacuole fuses with an anal pore, a specialized region of the plasma membrane where undigested material can be eliminated.

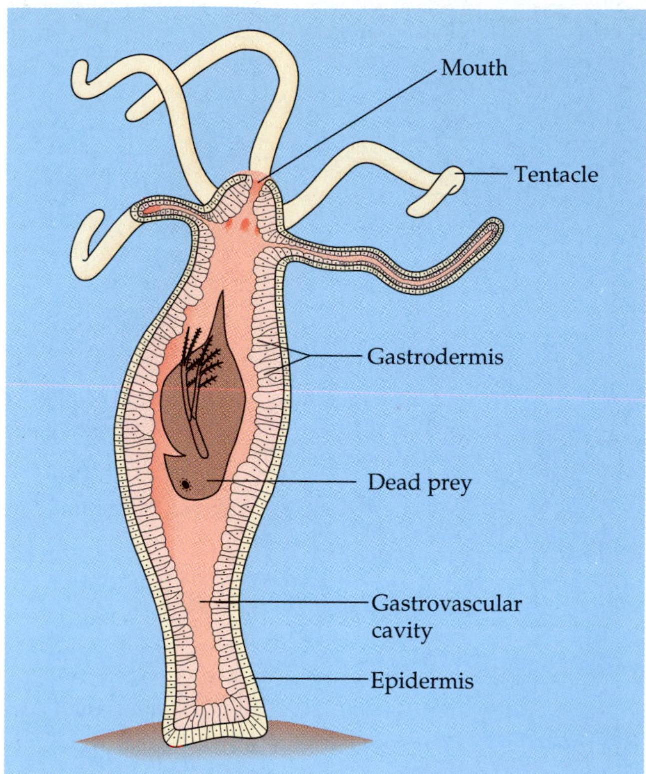

FIGURE 41.9 • Extracellular digestion in a gastrovascular cavity. The outer epidermis of the cnidarian hydra has protective and sensory functions, whereas the inner gastrodermis is specialized for digestion. Digestion begins in the gastrovascular cavity and is completed intracellularly after small food particles are engulfed by the gastrodermal cells.

food with the enzymes, allowing digestion to occur safely within a compartment that is enclosed by a membrane. This digestive mechanism is termed **intracellular digestion** (FIGURE 41.8). Sponges are unusual among animals in that they digest their food (as do protists) entirely by the intracellular mechanism (see FIGURE 33.2).

Extracellular Digestion

In most animals, at least some hydrolysis occurs by **extracellular digestion**, the breakdown of food outside cells. Extracellular digestion occurs within compartments that are continuous, via passages, with the outside of the animal's body.

Many animals with relatively simple body plans have digestive sacs with single openings. These pouches, called **gastrovascular cavities**, function in both digestion and distribution of nutrients throughout the body (hence, the *vascular* part of the term). The cnidarian hydra provides a good example of how a gastrovascular cavity works. Hydras are carnivores that sting prey with specialized organelles called nematocysts and then use tentacles to stuff the food through the mouth into the gastrovascular cavity (FIGURE 41.9). With food in the cavity, specialized cells of the gastrodermis, the tissue layer that lines the cavity, secrete digestive enzymes that break the soft tissues of the prey into tiny pieces. The gastrodermal cells then engulf the food particles, and most of the actual hydrolysis of macromolecules occurs intracellularly as in *Paramecium* and sponges. After a hydra has digested its meal, undigested materials remaining in the gastrovascular cavity, such as the exoskeletons of small crustaceans, are eliminated through the single opening, which functions in the dual role of mouth and anus.

In common with hydras, many flatworms have a gastrovascular cavity with a single opening (see FIGURE 33.9). Also like hydras, digestion begins in the cavity and is completed intracellularly. Having an extracellular cavity for digestion is an adaptation that enables an animal to devour larger prey than can be phagocytosed and digested intracellularly.

In contrast to cnidarians and flatworms, most animals—including nematodes, annelids, mollusks, arthropods, echinoderms, and chordates—have digestive tubes extending between two openings, a mouth and an anus. These tubes are called **complete digestive tracts** or **alimentary canals**. Because food moves along the canal in one direction, the tube can be organized into specialized regions that carry out digestion and nutrient absorption in a stepwise fashion

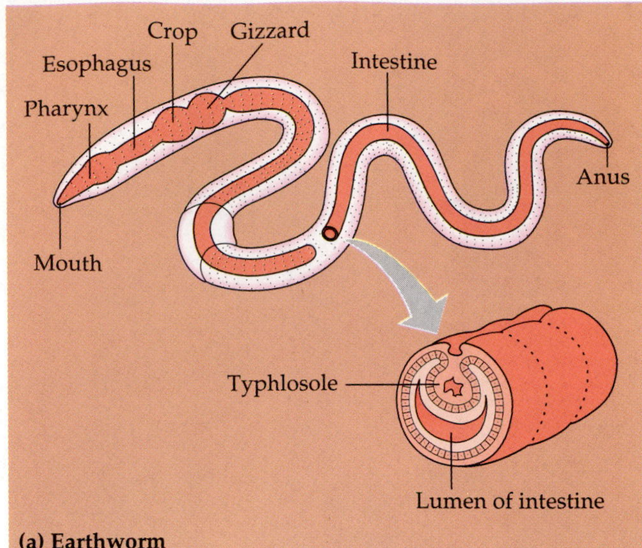

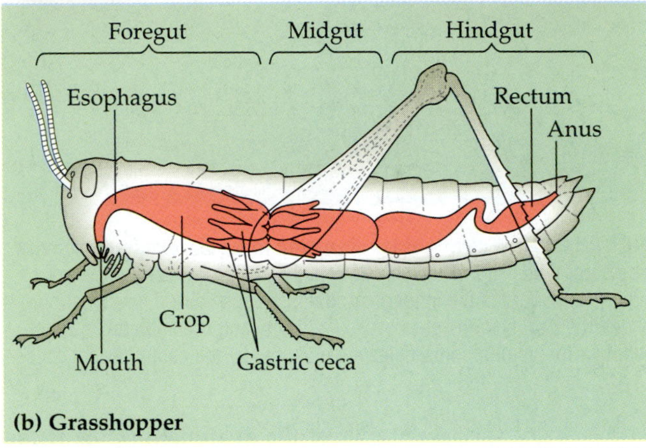

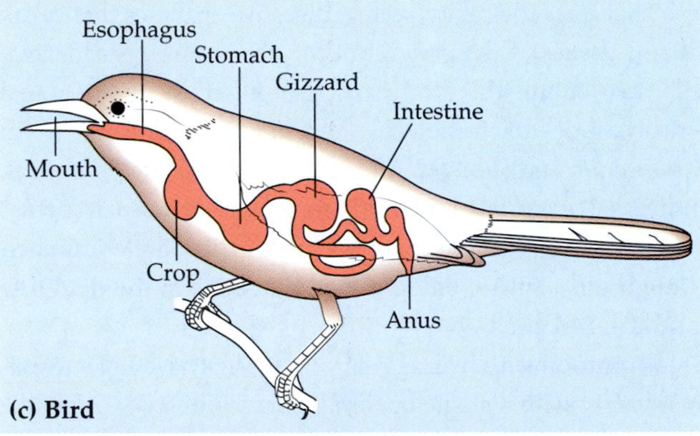

FIGURE 41.10 ▪ Alimentary canals. (a) The digestive tract of an earthworm includes a muscular pharynx that sucks food in through the mouth. Food passes through the esophagus and is stored and moistened in the crop. The muscular gizzard, which contains small bits of sand and gravel, pulverizes the food. Digestion and absorption occur in the intestine, which has a dorsal fold, the typhlosole, that increases the surface area for nutrient absorption. **(b)** A grasshopper has several digestive chambers grouped into three main regions: a foregut, with an esophagus and crop; a midgut; and a hindgut. Food is moistened and stored in the crop, but most digestion occurs in the midgut. Gastric ceca, pouches extending from the midgut, absorb nutrients. **(c)** A bird has three separate chambers—the crop, stomach, and gizzard—where food is pulverized and churned before passing into the intestine. In most birds, chemical digestion and absorption of nutrients occur in the intestine.

(FIGURE 41.10). Food ingested through the mouth and pharynx passes through an esophagus that leads to a crop, gizzard, or stomach, depending on the species. Crops and stomachs are organs that usually store the food, whereas gizzards grind it. The food next enters the intestine, where digestive enzymes hydrolyze the food molecules, and nutrients are absorbed across the lining of the tube into the blood. Undigested wastes are eliminated through the **anus**.

THE MAMMALIAN DIGESTIVE SYSTEM

The mammalian digestive system consists of the alimentary canal and various accessory glands that secrete digestive juices into the canal through ducts. **Peristalsis**, rhythmic waves of contraction by smooth muscles in the wall of the canal, pushes the food along the tract. At some of the junctions between specialized segments of the digestive tube, the muscular layer is modified into ringlike valves called **sphincters**, which close off the tube like drawstrings, regulating the passage of material between chambers of the canal.

The accessory glands of the mammalian digestive system are three pairs of **salivary glands**, the **pancreas**, the **liver**, and its storage organ, the **gallbladder**. Using the human as an example, we now follow a meal through the alimentary canal, seeing in more detail what happens to the food in each of the processing stations along the way (FIGURE 41.11).

The oral cavity, pharynx, and esophagus initiate food processing

The Oral Cavity

Physical and chemical digestion of food both begin in the mouth. During chewing, teeth of various shapes cut, smash, and grind food, making the food easier to swallow and increasing its surface area. The presence of food in the **oral cavity** triggers a nervous reflex that causes the salivary glands to deliver saliva through ducts to the oral cavity. Even before food is actually in the mouth, salivation may occur in anticipation because of learned associations between eating and the time of day, cooking odors, or other stimuli.

In humans, more than a liter of saliva is secreted into the oral cavity each day. Dissolved in the saliva is a slippery glycoprotein (carbohydrate-protein complex) called mucin, which protects the soft lining of the mouth from abrasion and lubricates the food for easier swallowing. Saliva contains buffers that help prevent tooth decay by neutralizing acid in the mouth. Also, antibacterial agents in saliva kill many of the bacteria that enter the mouth with food.

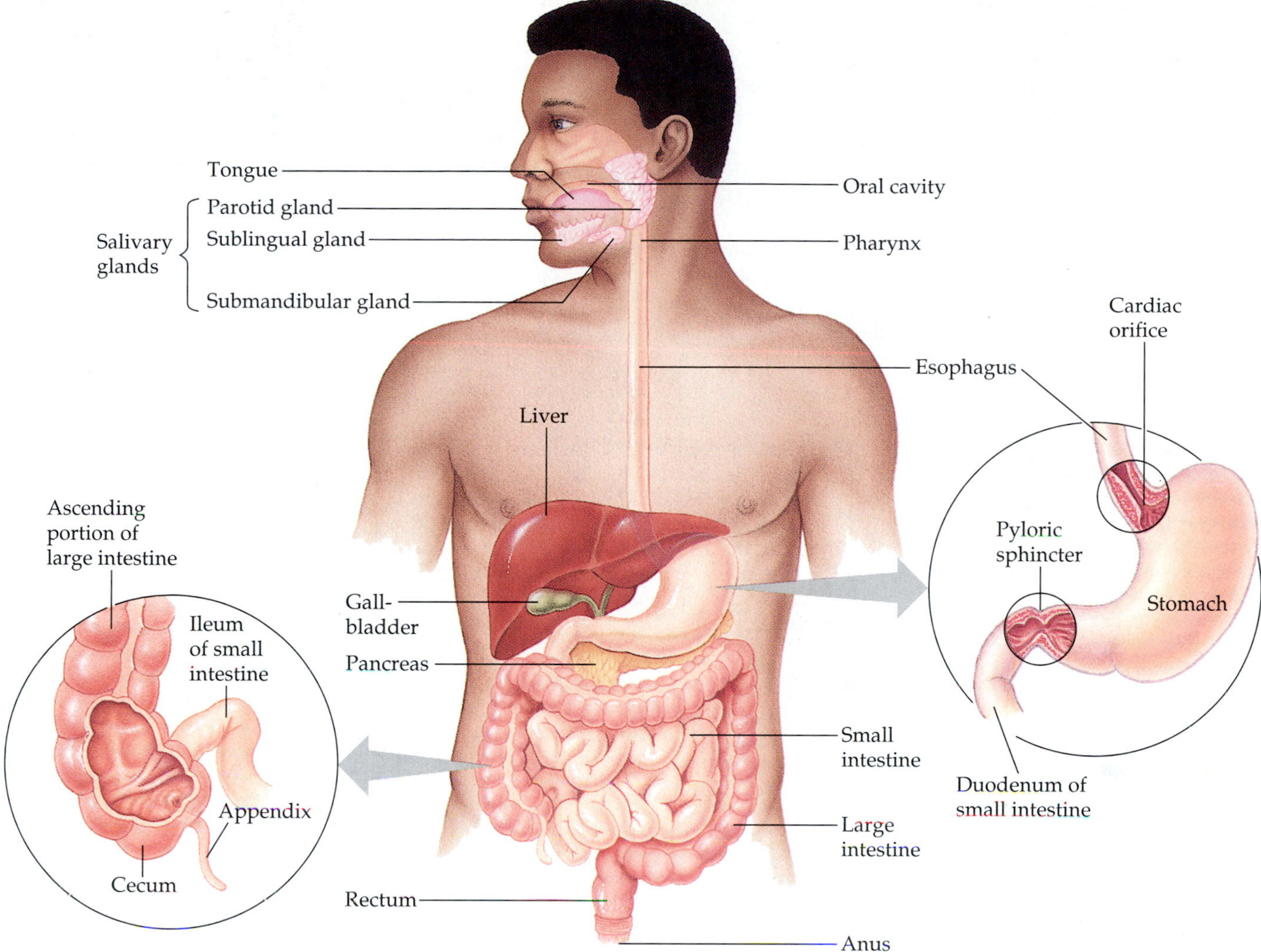

FIGURE 41.11 ▪ The human digestive system. After chewing and swallowing, it takes only 5 to 10 seconds for food to pass down the esophagus and into the stomach, where it spends 2 to 6 hours being partially digested. Final digestion and nutrient absorption occur in the small intestine over a period of 5 to 6 hours. In 12 to 24 hours, any undigested material passes through the large intestine, and feces are expelled through the anus.

Finally, digestion of carbohydrates, the body's main source of chemical energy, begins in the oral cavity. Saliva contains **salivary amylase**, a digestive enzyme that hydrolyzes starch (a glucose polymer from plants) and glycogen (a glucose polymer from animals). The main products of this enzyme's action are smaller polysaccharides and the disaccharide maltose.

The tongue tastes food, manipulates it during chewing, and helps shape the food into a ball called a **bolus**. During swallowing, the tongue pushes a bolus to the very back of the oral cavity and into the pharynx.

The Pharynx

The region we call our throat is the **pharynx**, an intersection that leads to both the esophagus and the windpipe (trachea). When we swallow, the top of the windpipe moves up so that its opening, the glottis, is blocked by a cartilaginous flap, the **epiglottis**. You can see this motion in the bobbing of the "Adam's apple" during swallowing. Closing the opening of the windpipe guards the respiratory system against the entry of food or fluids during swallowing. The swallowing mechanism normally ensures that a bolus will be guided into the entrance of the esophagus (FIGURE 41.12, steps ① and ②, p. 802).

The Esophagus

The **esophagus** conducts food from the pharynx down to the stomach. Peristalsis squeezes a bolus along the narrow esophagus (FIGURE 41.12, step ③). The muscles at the very top of the esophagus are striated (voluntary). Thus, the act of swallowing begins voluntarily, but then the involuntary waves of contraction by smooth muscles in the rest of the esophagus take over. Salivary amylase continues to hydrolyze starch and glycogen as the bolus passes through the esophagus.

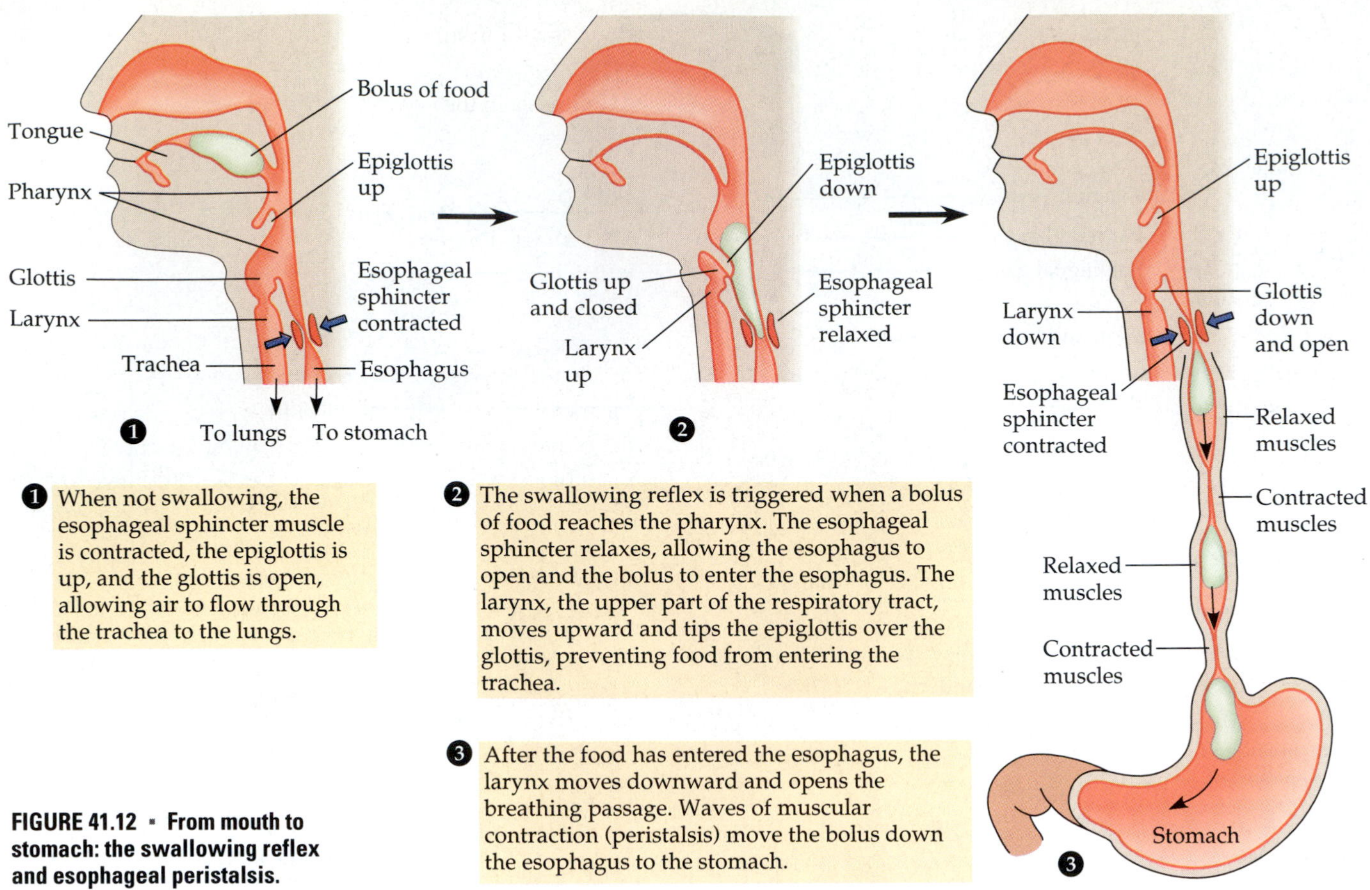

Tongue
Pharynx
Glottis
Larynx
Trachea
Bolus of food
Epiglottis up
Esophageal sphincter contracted
Esophagus
To lungs
To stomach
1
Glottis up and closed
Larynx up
Epiglottis down
Esophageal sphincter relaxed
2
Epiglottis up
Larynx down
Glottis down and open
Esophageal sphincter contracted
Relaxed muscles
Contracted muscles
Relaxed muscles
Contracted muscles
Stomach
3
1 When not swallowing, the esophageal sphincter muscle is contracted, the epiglottis is up, and the glottis is open, allowing air to flow through the trachea to the lungs.
2 The swallowing reflex is triggered when a bolus of food reaches the pharynx. The esophageal sphincter relaxes, allowing the esophagus to open and the bolus to enter the esophagus. The larynx, the upper part of the respiratory tract, moves upward and tips the epiglottis over the glottis, preventing food from entering the trachea.
3 After the food has entered the esophagus, the larynx moves downward and opens the breathing passage. Waves of muscular contraction (peristalsis) move the bolus down the esophagus to the stomach.

FIGURE 41.12 ▪ From mouth to stomach: the swallowing reflex and esophageal peristalsis.

The stomach stores food and performs preliminary digestion

The **stomach** is located on the left side of the abdominal cavity, just below the diaphragm. Because this large organ can store an entire meal, we do not need to eat constantly. With a very elastic wall and accordionlike folds, the stomach can stretch to accommodate about 2 L of food and water.

The epithelium that lines deep pits in the wall of the stomach secretes **gastric juice**, a digestive fluid that mixes with the food. With a high concentration of hydrochloric acid, gastric juice has a pH of about 2—acidic enough to dissolve iron nails. One function of the acid is to disrupt the extracellular matrix that binds cells together in meat and plant material. The acid also kills most bacteria that are swallowed with the food.

Also present in gastric juice is **pepsin**, an enzyme that begins the hydrolysis of proteins. Pepsin breaks peptide bonds adjacent to specific amino acids, cleaving proteins into smaller polypeptides. Pepsin is one of the few enzymes that works best in a strongly acidic environment. Indeed, the low pH of gastric juice denatures the proteins in food, increasing exposure of their peptide bonds to pepsin.

Specialized cells (called chief cells) located in gastric pits synthesize and secrete pepsin in an inactive form called **pepsinogen**. Different cells (parietal cells, also in the pits) secrete hydrochloric acid, which converts pepsinogen to active pepsin by removing a small portion of the molecule and exposing its active site. Activation of pepsinogen in the stomach is an example of positive feedback. Once some pepsinogen is activated by acid, a series of chemical reactions occurs because pepsin itself can activate additional molecules of pepsinogen.

What prevents pepsin from destroying cells of the stomach wall? Because different kinds of cells secrete the acid and pepsinogen, the two ingredients do not mix until their release into the lumen of the stomach. A coating of mucus secreted by the epithelial cells helps protect the stomach lining from being digested. Still, the epithelium is constantly eroded, and mitosis generates enough cells to completely replace the stomach lining every three days. Gastric ulcers, lesions in the stomach lining, are caused mainly by the acid-tolerant bacterium *Helicobacter pylori* and are treated with antibiotics. However, gastric ulcers may worsen if pepsin and acid destroy the lining faster than it can regenerate.

About every 20 seconds, the stomach contents are mixed by the churning action of smooth muscles. You may feel hunger pangs when your empty stomach churns. (Sensations of hunger are also associated with brain centers that monitor the blood's nutritional status.) As a result of mixing and enzyme action, what begins in the stomach as a recently swallowed meal becomes a nutrient broth known as **acid chyme**.

Much of the time, the stomach is closed off at either end (see FIGURE 41.11). The opening from the esophagus to the stomach, the cardiac orifice, normally dilates only when a bolus driven by peristalsis arrives. The occasional backflow of acid chyme from the stomach into the lower end of the esophagus causes heartburn. (If backflow is a persistent problem, an ulcer may develop in the esophagus.) The cardiac orifice opens intermittently with each wave of peristalsis that delivers a bolus. At the opening from the stomach to the small intestine is the **pyloric sphincter**, which helps regulate the passage of chyme into the intestine. A squirt at a time, it takes about 2 to 6 hours after a meal for the stomach to empty.

The small intestine is the major organ of digestion and absorption

With a length of more than 6 m in humans, the **small intestine** is the longest section of the alimentary canal (its name is based on its small diameter, compared with the diameter of the large intestine). The small intestine is the organ in which most enzymatic hydrolysis of the macromolecules in food occurs. It is also responsible for the absorption of most nutrients into the blood.

The pancreas, liver, and gallbladder, as well as the small intestine itself, participate in digestion. The first 25 cm or so of the small intestine is called the **duodenum**. It is here that acid chyme squirting in from the stomach mixes with digestive juices from the pancreas, liver, gallbladder, and gland cells of the intestinal wall itself.

The pancreas produces several hydrolytic enzymes and an alkaline solution rich in bicarbonate. The bicarbonate acts as a buffer, offsetting the acidity of chyme from the stomach.

The liver performs a wide variety of important functions in the body, including the production of **bile**, a mixture of substances that is stored in the gallbladder until needed. Bile contains no digestive enzymes, but it does contain bile salts, which act as detergents and aid in the digestion and absorption of fats. Bile also contains pigments that are by-products of red blood cell destruction in the liver; these bile pigments are eliminated from the body with the feces.

Enzymatic Action in the Small Intestine

41.1 Let's now follow the action of enzymes from the pancreas and intestinal (duodenal) wall in digesting macromolecules.

Carbohydrate Digestion. The digestion of the carbohydrates starch and glycogen begun by salivary amylase in the oral cavity continues in the small intestine (FIGURE 41.13a). Pancreatic amylases hydrolyze starch, glycogen, and smaller

41.1
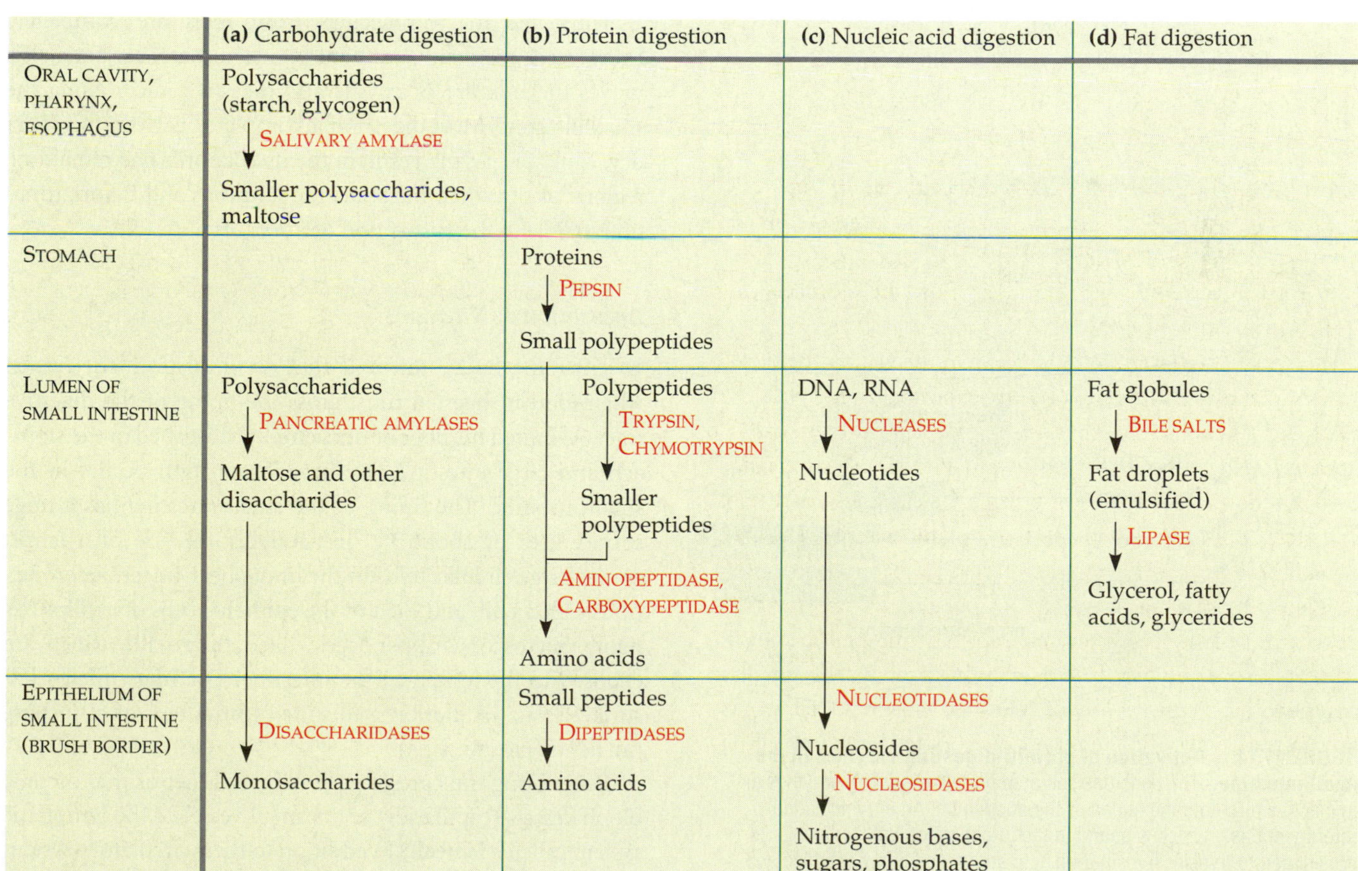

(a) Carbohydrate digestion
(b) Protein digestion
(c) Nucleic acid digestion
(d) Fat digestion
ORAL CAVITY, PHARYNX, ESOPHAGUS
Polysaccharides (starch, glycogen)
SALIVARY AMYLASE
Smaller polysaccharides, maltose
STOMACH
Proteins
PEPSIN
Small polypeptides
LUMEN OF SMALL INTESTINE
Polysaccharides
PANCREATIC AMYLASES
Maltose and other disaccharides
Polypeptides
TRYPSIN, CHYMOTRYPSIN
Smaller polypeptides
AMINOPEPTIDASE, CARBOXYPEPTIDASE
Amino acids
DNA, RNA
NUCLEASES
Nucleotides
Fat globules
BILE SALTS
Fat droplets (emulsified)
LIPASE
Glycerol, fatty acids, glycerides
EPITHELIUM OF SMALL INTESTINE (BRUSH BORDER)
DISACCHARIDASES
Monosaccharides
Small peptides
DIPEPTIDASES
Amino acids
NUCLEOTIDASES
Nucleosides
NUCLEOSIDASES
Nitrogenous bases, sugars, phosphates

FIGURE 41.13 ▪ Enzymatic digestion in the human digestive system.

polysaccharides into disaccharides, including maltose. The enzyme maltase completes the digestion of maltose, splitting it into two molecules of the simple sugar glucose. Maltase is one of a family of disaccharidases, each one specific for the hydrolysis of a different disaccharide. Sucrase, for instance, hydrolyzes table sugar (sucrose), and lactase digests milk sugar (lactose). (In general, adults have much less lactase and much less ability to digest milk sugar than children have.) The disaccharidases are built into the membranes and extracellular matrix covering the intestinal epithelium, which is also the site of sugar absorption. Thus, the terminal steps in carbohydrate digestion—the steps that yield energy-rich monomers—occur where these monomers are actually absorbed into the blood.

Protein Digestion. The digestion of proteins in the small intestine involves completion of the work begun by pepsin in the stomach (FIGURE 41.13b). Enzymes in the duodenum dismantle polypeptides into their component amino acids or into small peptides (fragments only two or three amino acids long). **Trypsin** and **chymotrypsin** are specific for peptide bonds adjacent to certain amino acids, and thus, like pepsin, break large polypeptides into shorter chains. **Carboxypeptidase** splits off one amino acid at a time, beginning at the end of the polypeptide that has a free carboxyl group. **Aminopeptidase** works in the opposite direction. Either aminopeptidase or carboxypeptidase alone could completely digest a protein. But teamwork among these enzymes and the trypsin and chymotrypsin that attack the interior of the protein speeds up hydrolysis tremendously. Other enzymes called **dipeptidases**, attached to the intestinal lining, further hasten digestion by splitting small peptides.

Many protein-digesting enzymes, such as aminopeptidase, are secreted by the intestinal epithelium. In contrast, trypsin, chymotrypsin, and carboxypeptidase are secreted in inactive forms by the pancreas. Another intestinal enzyme called **enteropeptidase** directly or indirectly triggers activation of these enzymes within the lumen of the small intestine (FIGURE 41.14).

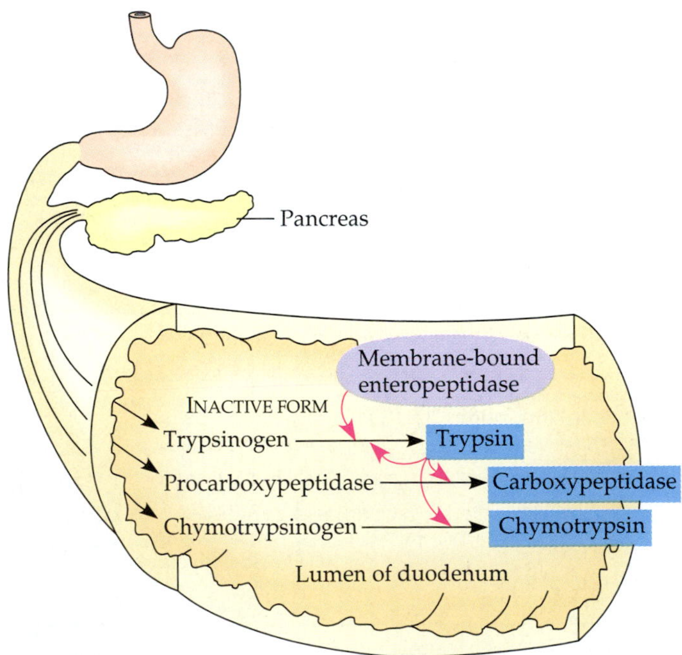

FIGURE 41.14 • Activation of protein-digesting enzymes in the small intestine. The pancreas secretes protein-digesting enzymes in an inactive form into the lumen of the duodenum. An enzyme called enteropeptidase, which is bound to the intestinal epithelium, converts trypsinogen to trypsin. Trypsin then activates procarboxypeptidase and chymotrypsinogen.

Nucleic Acid Digestion. The digestion of nucleic acids involves a hydrolytic assault similar to that mounted on proteins. A team of enzymes called **nucleases** hydrolyzes DNA and RNA in food into their component nucleotides (FIGURE 41.13c). Other hydrolytic enzymes then break nucleotides down further into nucleosides, nitrogenous bases, sugars, and phosphates.

Fat Digestion. Nearly all the fat in a meal reaches the small intestine completely undigested. Hydrolysis of fats is a special problem, because fat molecules are insoluble in water. Bile salts from the gallbladder secreted into the duodenum coat tiny fat droplets and keep them from coalescing, a process called **emulsification**. Because the droplets are small, a large surface area of fat is exposed to **lipase**, an enzyme that hydrolyzes the fat molecules (FIGURE 41.13d).

Thus, the macromolecules from food are completely hydrolyzed to their component monomers as peristalsis moves the mixture of chyme and digestive juices along the small intestine. Most digestion is completed early in this journey, while the chyme is still in the duodenum. The remaining regions of the small intestine, the **jejunum** and **ileum**, function mainly in the absorption of nutrients and water.

Absorption of Nutrients

To enter the body, nutrients that accumulate in the lumen when food is digested must cross the lining of the digestive tract. A limited number of nutrients are absorbed in the stomach and large intestine, but most absorption occurs in the small intestine. The lining of the small intestine has a huge surface area of about 300 m^2, roughly the size of a tennis court. Large circular folds in the lining bear fingerlike projections called **villi**, and each of the epithelial cells of a villus has many microscopic appendages called **microvilli**, which are exposed to the lumen of the intestine. The huge microvillar surface is an adaptation well suited to the task of absorbing nutrients (FIGURE 41.15).

Penetrating the core of each villus is a net of microscopic blood vessels (capillaries) and a small vessel of the lymphatic system called a **lacteal**. (In addition to their circulatory system that carries blood, vertebrates have an auxiliary system of ves-

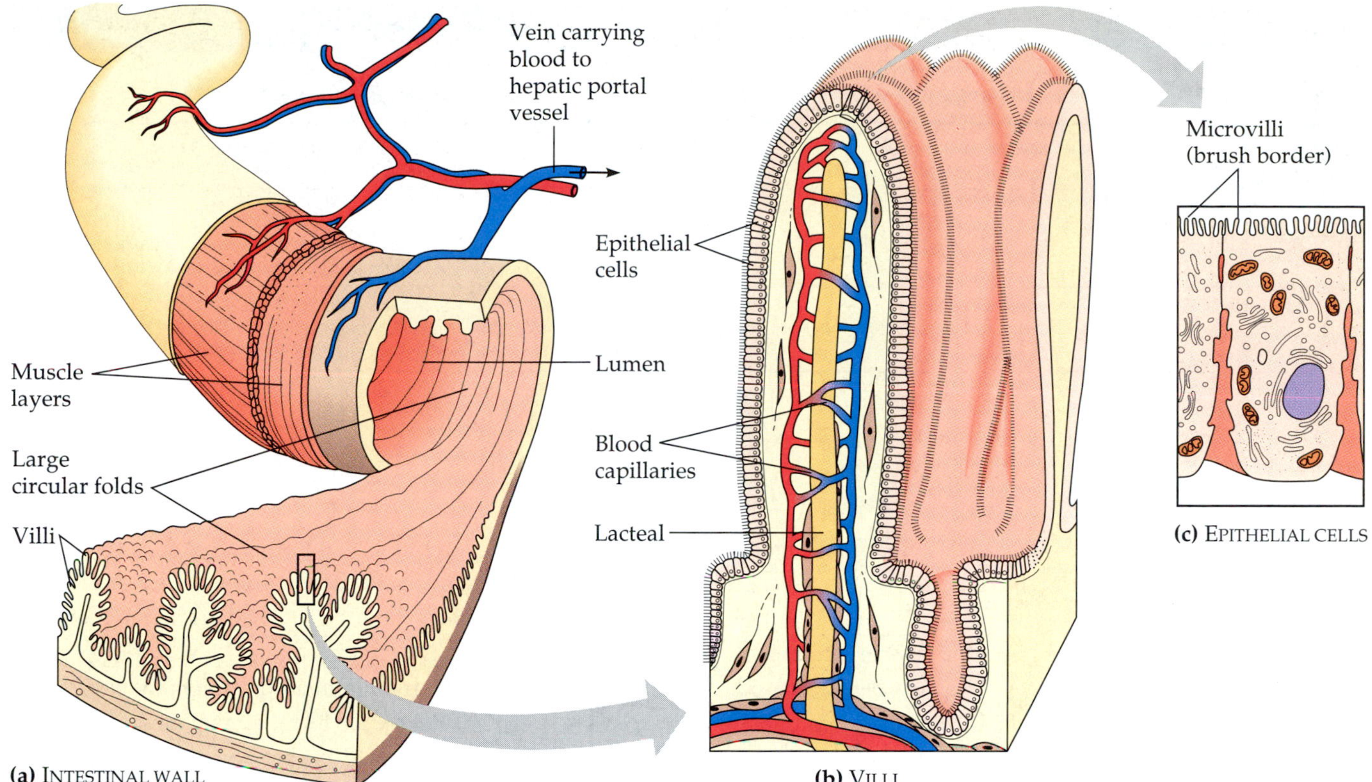

FIGURE 41.15 ▪ The structure of the small intestine. **(a)** Large folds of epithelium line the small intestine. Villi project outward, into the lumen, from the folds. **(b)** Each villus has a small lymphatic vessel called a lacteal and a network of blood capillaries surrounded by a layer of epithelial cells. **(c)** The epithelial cells have microscopic projections, microvilli, that extend into the intestinal lumen. Microfilaments within the microvilli wiggle the tiny projections in the nutrient contents of the intestine's lumen. Nutrients are absorbed across the intestinal epithelium and then across the wall of the capillaries or lacteals. Fats taken up by the lacteals are distributed via the lymphatic system to veins near the heart. Veins distribute blood with absorbed sugars and amino acids to the hepatic portal vessel, which conveys it directly to the liver.

sels—the lymphatic system—that carries a clear fluid called lymph, discussed in Chapter 43.) Nutrients are absorbed across the intestinal epithelium and then across the unicellular epithelium of the capillaries or lacteals. Thus, only two single layers of epithelial cells separate nutrients in the lumen of the intestine from the bloodstream.

In some cases, the transport of nutrients across the epithelial cells is passive. The simple sugar fructose, for example, is apparently absorbed by diffusion down its concentration gradient from the lumen of the intestine into the epithelial cells, then out of the epithelial cells into capillaries. Other nutrients, including amino acids, small peptides, vitamins, glucose, and several other simple sugars, are pumped against gradients by the epithelial membranes.

Amino acids and sugars pass through the epithelium, enter capillaries, and are carried away from the intestine by the bloodstream. After glycerol and fatty acids are absorbed by epithelial cells, they are recombined within those cells to form fats again. The fats are then mixed with cholesterol and coated with special proteins, forming small globules called **chylomicrons**, most of which are transported by exocytosis out of the epithelial cell and into a lacteal. The lacteals converge into larger vessels of the lymphatic system. Lymph, containing chylomicrons, eventually drains from the lymphatic system into large veins that return blood to the heart.

In contrast to the lacteals, the capillaries and veins that drain nutrients away from the villi all converge into a single circulatory channel, the **hepatic portal vessel**, which leads directly to the liver. The rate of flow in this large vessel, about 1 L per minute, ensures that the liver, which has the metabolic versatility to interconvert various organic molecules, has first access to amino acids and sugars absorbed after a meal is digested. The blood that leaves the liver may have a very different balance of these nutrients than the blood that entered via the hepatic portal vessel. For example, the liver plays an important role in regulating the level of glucose fuel molecules in the blood. Blood exiting the liver usually has a glucose concentration of very close to 0.1%, regardless of the carbohydrate content of a meal. From the liver, blood travels to the heart, which pumps the blood and the nutrients it contains to all parts of the body.

Hormones help regulate digestion

41.2 Hormones released by the wall of the stomach and duodenum help ensure that digestive secretions are present only

41.2

Table 41.3 ■ Hormones That Regulate Digestion

DIGESTIVE HORMONE	SITE OF PRODUCTION	STIMULUS FOR PRODUCTION	EFFECT OF HORMONE
Gastrin	Stomach wall	Food in stomach	Stimulates sustained secretion of gastric juice in stomach
Enterogastrones			
Secretin	Duodenum wall	Acidic chyme in duodenum	Stimulates pancreas to release bicarbonate, which neutralizes acid chyme in duodenum
Cholecystokinin (CCK)	Duodenum wall	Amino acids or fatty acids in chyme in duodenum	Stimulates pancreas to release pancreatic enzymes to small intestine; stimulates gallbladder to contract, releasing bile into small intestine
Other enterogastrones	Duodenum wall	Chyme in duodenum	Inhibit peristalsis in stomach (slowing entry of food into small intestine)

when needed (TABLE 41.3). When we see, smell, or taste food, impulses from the brain to the stomach initiate the secretion of gastric juice. Then certain substances in the food stimulate the stomach wall to release the hormone **gastrin** into the circulatory system. As gastrin gradually recirculates in the bloodstream back to the stomach wall, the hormone stimulates further secretion of gastric juice. Thus, an initial burst of gastric secretion at mealtime is followed by a sustained secretion that continues to add gastric juice to the food for some time. If the pH of the stomach contents becomes too low, the acid will inhibit the release of gastrin, decreasing the secretion of gastric juice; this is an example of negative feedback.

Other hormones, collectively called **enterogastrones**, are secreted by the wall of the duodenum. The acidic pH of the chyme that enters the duodenum stimulates cells in the wall to release the hormone **secretin**. This enterogastrone signals the pancreas to release bicarbonate, which neutralizes the acid chyme. A second enterogastrone, **cholecystokinin (CCK)**, is secreted in response to the presence of amino acids or fatty acids. CCK causes the gallbladder to contract and release bile into the small intestine. CCK also triggers the release of pancreatic enzymes. The chyme, particularly if rich in fats, also causes the duodenum to release other enterogastrones that inhibit peristalsis in the stomach, thereby slowing down the entry of food into the small intestine.

Reclaiming water is a major function of the large intestine

The **large intestine**, or **colon**, is connected to the small intestine at a T-shaped junction, where a sphincter (a muscular valve) controls the movement of material. One arm of the T is a pouch called the **cecum** (see FIGURE 41.11). Compared to many other mammals, humans have a relatively small cecum with a fingerlike extension, the **appendix**, which is dispensable. (Lymphoid tissue in the appendix makes a minor contribution to body defense.) The main branch of the human colon is shaped like an upside-down U about 1.5 m long.

A major function of the colon is to reabsorb water that has entered the alimentary canal as the solvent of the various digestive juices. Altogether, about 7 L of fluid are secreted into the lumen of the digestive tract each day. Most reabsorption of water occurs along with nutrient absorption in the small intestine. The colon finishes the job by reclaiming most of the water that remains in the lumen. Together, the small intestine and colon reabsorb about 90% of the water that entered the alimentary canal. The wastes of the digestive tract, the **feces**, become more solid as they are moved along the colon by peristalsis. The movement is sluggish, and it generally takes about 12 to 24 hours for material to travel the length of the organ. If the lining of the colon is irritated—by a viral or bacterial infection, for instance—less water than normal may be reabsorbed, resulting in diarrhea. The opposite problem, constipation, occurs when peristalsis moves the feces along too slowly. An excess of water is reabsorbed, and the feces become compacted.

Living in the large intestine is a rich flora of mostly harmless bacteria. One of the common inhabitants of the human colon is *Escherichia coli*, a favorite research organism of molecular biologists. The presence of *E. coli* in lakes and streams is an indication that the water is contaminated by untreated sewage. Intestinal bacteria live on organic material that would otherwise be eliminated with the feces. As by-products of their metabolism, many colon bacteria generate gases, including methane and hydrogen sulfide, and some produce vitamins (e.g., biotin, folic acid, vitamin K, and several B vitamins), which are absorbed into the blood.

Feces contain masses of bacteria, as well as cellulose and other undigested materials. Although cellulose fibers have no caloric value to humans, their presence in the diet helps move food along the digestive tract. Feces may also contain an abundance of salts. For instance, when iron and calcium concentrations in the blood get too high, the colon lining excretes salts of these elements into the lumen, and they are eliminated in the feces.

The terminal portion of the colon is called the **rectum**, where feces are stored until they can be eliminated. Between the rectum and the anus are two sphincters, one involuntary and the other voluntary. Once or more each day, strong contractions of the colon create an urge to defecate.

EVOLUTIONARY ADAPTATIONS OF VERTEBRATE DIGESTIVE SYSTEMS

Structural adaptations of the digestive system are often associated with diet

The digestive systems of mammals and other vertebrates are variations on a common plan, but there are many intriguing adaptations, often associated with the animal's diet. We will examine just a few.

Dentition, an animal's assortment of teeth, is one example of structural variation reflecting diet. Compare the dentition of carnivorous, herbivorous, and omnivorous mammals in FIGURE 41.16. Nonmammalian vertebrates generally have less specialized dentition, but there are interesting exceptions. For example, poisonous snakes, such as rattlesnakes, have fangs, modified teeth that inject venom into prey. Some fangs are hollow, like syringes, while others drip the poison along grooves on the surfaces of the teeth. Snakes in general have another important anatomical adaptation associated with feeding: The lower jaw is loosely hinged to the skull by an elastic ligament that permits the mouth and throat to open *very* wide for swallowing large prey (once again, witness the astonishing episode recorded in FIGURE 41.7).

The length of the vertebrate digestive system is also correlated with diet. In general, herbivores and omnivores have longer alimentary canals relative to their body size than carnivores (FIGURE 41.17). Vegetation is more difficult to digest

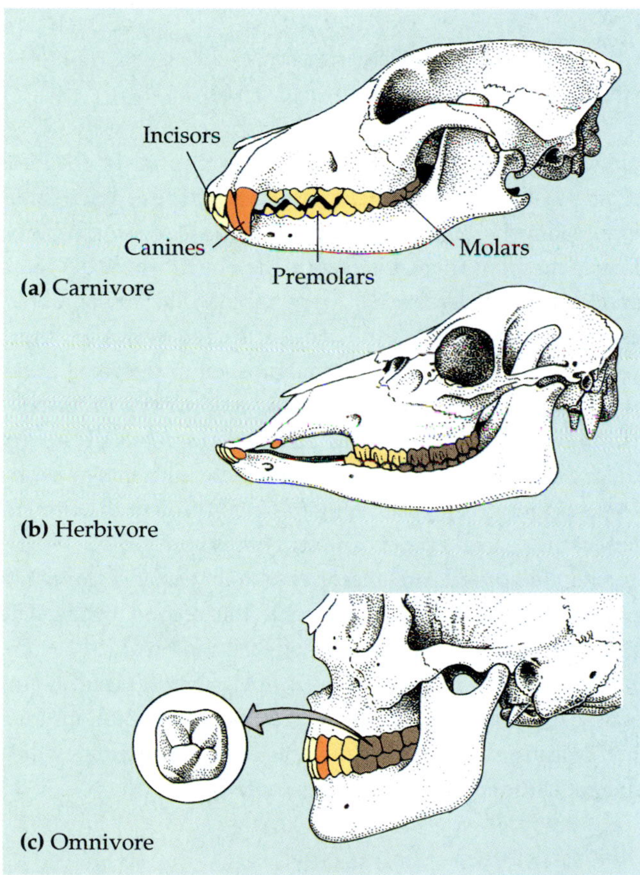

FIGURE 41.16 • Dentition and diet. (a) Carnivores, such as members of the dog and cat families, generally have pointed incisors and canines that can be used to kill prey and rip away pieces of flesh. The jagged premolars and molars crush and shred food. **(b)** In contrast, herbivorous mammals, such as cows, usually have teeth with broad, ridged surfaces that grind tough plant material. The incisors and canines are generally modified for biting off pieces of vegetation. **(c)** Humans, being omnivores adapted for eating both vegetation and meat, have a relatively unspecialized dentition. The permanent (adult) set of teeth is 32 in number. Beginning at the midline of the upper and lower jaw are two blade-like incisors for biting, a pointed canine for tearing, two premolars for grinding, and three molars for crushing.

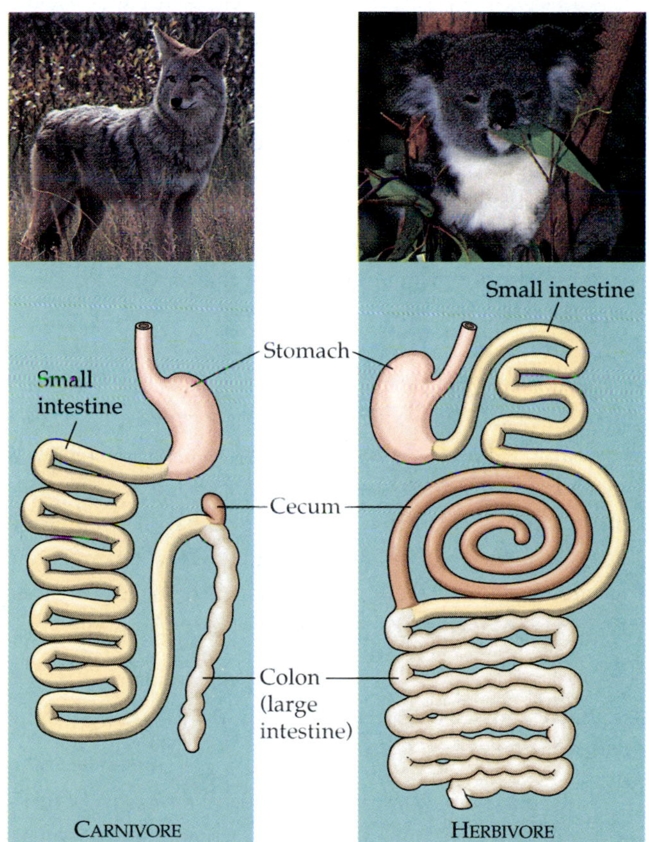

FIGURE 41.17 • The digestive tracts of a carnivore (coyote) and a herbivore (koala) compared. Although these two mammals are about the same size, the koala's intestines are much longer, an adaptation that enhances processing of fibrous, protein-poor eucalyptus leaves from which it obtains virtually all its food and water. Extensive chewing chops the leaves into very small pieces, increasing exposure of the food to digestive juices. The koala's cecum—at 2 m, the longest of any animal of equivalent size—functions as a fermentation chamber where symbiotic bacteria convert the shredded leaves into a more nutritious diet. The shorter length of the coyote's digestive tract is sufficient for digesting meat and absorbing nutrients from this diet.

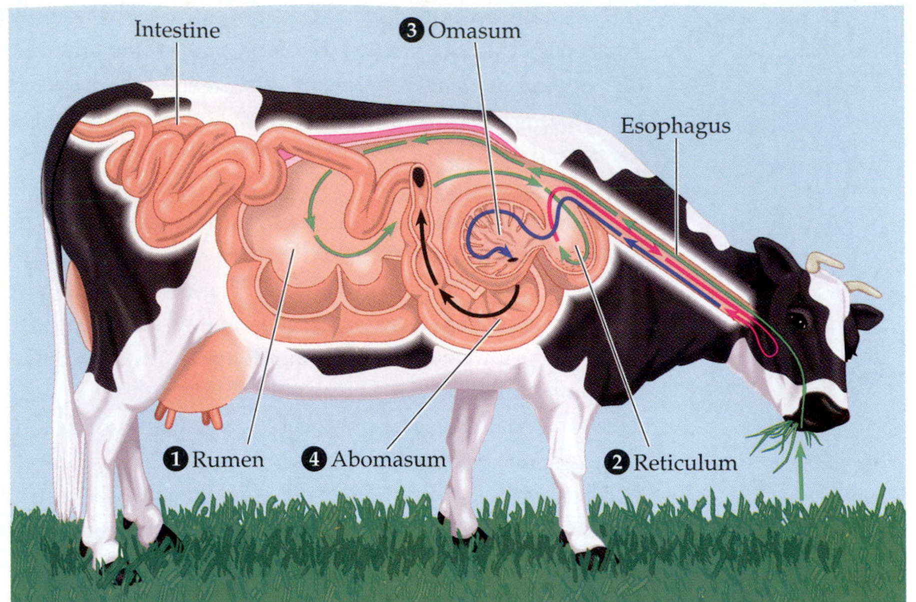

FIGURE 41.18 ▪ Ruminant digestion. The stomach of a ruminant has four chambers. When the cow first chews and swallows a mouthful of grass, boluses (green arrows) enter both ① the rumen and ② the reticulum, where symbiotic prokaryotes and protists (mainly ciliates) go to work on the cellulose-rich meal. As by-products of their metabolism, the microorganisms secrete fatty acids. The cow periodically regurgitates and rechews the cud (red arrows), which further breaks down the fibers, making them more accessible to further microbial action. The cow then reswallows the cud (blue arrows), which moves to ③ the omasum, where water is removed. The cud, containing great numbers of microorganisms, finally passes to ④ the abomasum for digestion by the cow's own enzymes (black arrows). Because of the microbial action, the diet from which a ruminant actually absorbs its nutrients is much richer than the grass the animal originally ate. In fact, a ruminant eating grass or hay obtains many of its nutrients by digesting the symbiotic microorganisms, which reproduce rapidly enough in the rumen to maintain a stable population.

than meat because it contains cell walls. A longer tract furnishes more time for digestion and more surface area for the absorption of nutrients. A model case is the frog, which changes its diet upon metamorphosis. Algae-eating tadpoles (frog larvae) have a coiled intestine that is very long relative to their size. During metamorphosis, the rest of the body grows more than the intestine, leaving the carnivorous adult frogs with a shorter intestine relative to their size.

In some vertebrates, the *functional* length of the alimentary canal is actually longer than superficial appearance reveals. For example, most sharks are carnivores, with a relatively short intestine lacking the "switchbacks" that lengthen the human intestine. However, the lining of a shark's intestine has folds that form a spiral valve, a corkscrewlike apparatus that increases surface area and provides a much longer route than if food were to move through the tube in a straight line.

Symbiotic microorganisms help nourish many vertebrates

In addition to being very long, the alimentary canals of many herbivorous vertebrates have special fermentation chambers where symbiotic bacteria and protists live. These microorganisms have enzymes that can digest cellulose, which the animal itself cannot digest. The microorganisms not only digest the cellulose to simple sugars but also convert the sugars to a variety of nutrients essential to the animal. The hoatzin, a herbivorous bird that lives in South American rain forests, has a large, muscular crop, an esophageal pouch, that houses symbiotic microorganisms. Hard ridges in the wall of the crop grind plant leaves, and the microorganisms break down cellulose. Many herbivores, including horses, house symbiotic microorganisms in a large cecum, the pouch where the small and large intestines connect. The symbiotic bacteria of rabbits and some rodents live in the large intestine as well as in the cecum. Since most nutrients are absorbed in the small intestine, however, potentially nourishing by-products of fermentation by bacteria in the large intestine are initially lost with the feces. Rabbits and rodents obtain these nutrients by eating some of their feces and passing the food through the alimentary canal a second time. (The familiar rabbit "pellets," which are not reingested, are the feces eliminated after food has passed through the digestive tract the second time.) The koala, an Australian marsupial, also has an enlarged cecum, where symbiotic bacteria ferment finely shredded eucalyptus leaves (see FIGURE 41.17). The most elaborate adaptations for a herbivorous diet have evolved in the **ruminants**, which include cattle and sheep (FIGURE 41.18).

▪ ▪ ▪

Our focus in this chapter has been on provisioning the body. We have examined the nutritional requirements of animals and some of the many ways animals obtain nutrients for fuel and building materials. In the next chapter we will see that obtaining food, digesting it, and absorbing nutrients are only parts of a larger story. Indeed, provisioning the body also involves distributing nutrients to cells throughout the body and exchanging respiratory gases with the environment.

CHAPTER REVIEW

REVIEW OF KEY CONCEPTS

(with page numbers and key figures)

NUTRITIONAL REQUIREMENTS

- **Animals are heterotrophs that require food for fuel, carbon skeletons, and essential nutrients: *an overview* (p. 792)**

- **Homeostatic mechanisms manage an animal's fuel (pp. 792–793, FIGURE 41.1)** Carbohydrates and fats are most often used as fuel. Animals store excess calories as glycogen in the liver and muscles and as fat. Undernourished animals have diets deficient in calories.

- **An animal's diet must supply essential nutrients and carbon skeletons for biosynthesis (pp. 793–796, FIGURE 41.3, TABLES 41.1, 41.2)** Carbon skeletons are required in biosynthesis. Essential nutrients must be supplied in preassembled form. Essential amino acids are those an animal cannot make from nitrogen-containing precursors. Essential fatty acids are unsaturated. Vitamins are organic molecules required in small amounts; many serve as coenzymes or parts of coenzymes. Minerals are inorganic nutrients that are required in varying amounts, from about 1 to 2500 mg per day.

FOOD TYPES AND FEEDING MECHANISMS

- **Most animals are opportunistic feeders (p. 797)** All animals are heterotrophs. Herbivores eat mainly plants, and carnivores mainly eat other animals, but most animals in these two categories consume some plant or animal matter. Omnivores regularly eat both plant and animal matter.

- **Diverse feeding adaptations have evolved among animals (pp. 797–798)** Many aquatic animals are suspension-feeders, sifting small particles from the water. Substrate-feeders tunnel through their food, eating as they go. Fluid-feeders suck nutrient-rich fluids from a living host. Most animals are bulk-feeders, eating relatively large pieces of food.

OVERVIEW OF FOOD PROCESSING

- **The four main stages of food processing are ingestion, digestion, absorption, and elimination (p. 798)** Food processing in animals involves ingestion (the act of eating), digestion (enzymatic breakdown of the macromolecules of food into their monomers), absorption (the uptake of nutrients by body cells), and elimination (the passage of undigested materials out of the body in feces).

- **Digestion occurs in specialized compartments (pp. 798–800, FIGURES 41.8–41.10)** In intracellular digestion, food particles are engulfed by endocytosis and digested within food vacuoles. Most animals use extracellular digestion, with enzymatic hydrolysis occurring outside cells in a gastrovascular cavity or alimentary canal. The gastrovascular cavities of cnidarians and flatworms have a single opening through which food enters and undigested wastes pass. Most animals have digestive tracts, or alimentary canals, that move food through a one-way tube with specialized regions for digestion and absorption.

THE MAMMALIAN DIGESTIVE SYSTEM

- Smooth muscles in the wall of the digestive tract propel food along the tract by peristalsis and regulate its passage through strategic points by means of sphincters. The salivary glands, pancreas, and liver add digestive secretions to the tract through ducts (FIGURE 41.11).

- **The oral cavity, pharynx, and esophagus initiate food processing (pp. 800–801, FIGURE 41.12)** Food is lubricated and digestion begins in the oral cavity, where teeth chew food into smaller particles that are exposed to salivary amylase, initiating the breakdown of glucose polymers. The pharynx is the intersection leading to the trachea and the esophagus. The epiglottis usually prevents food from entering the trachea. The esophagus conducts food from the pharynx to the stomach by involuntary peristaltic waves.

- **The stomach stores food and performs preliminary digestion (pp. 802–803)** The stomach stores food and secretes gastric juice, which converts a meal to acid chyme. Gastric juice includes hydrochloric acid and the enzyme pepsin.

- 41.1 **The small intestine is the major organ of digestion and absorption (pp. 803–805, FIGURES 41.13–41.15)** Most digestion and virtually all absorption of nutrients occur in the small intestine. The pancreas and the gallbladder, which stores bile secreted by the liver, empty by ducts into the duodenum, the first part of the small intestine. Carbohydrate digestion continues in the duodenum in the presence of pancreatic amylase in the lumen and disaccharidases built into the plasma membranes of intestinal epithelial cells. The intestinal enzyme enteropeptidase activates the pancreatic enzymes (trypsin, chymotrypsin, carboxypeptidase) and other peptidases, such as aminopeptidase secreted by the intestinal epithelium), which all work together in the small intestine to hydrolyze polypeptides. Dipeptidases on the intestinal cells complete protein digestion. Nucleases, nucleotidases, and nucleosidases in the small intestine digest the nucleic acids DNA and RNA into their component nitrogenous bases, sugars, and phosphates. Fats are broken up into smaller droplets during emulsification by bile salts in the intestinal lumen, thereby allowing maximum exposure to the enzyme lipase. The jejunum and ileum of the small intestine function mainly in absorption. Large folds of the small intestine's lining have fingerlike villi, whose cells have microscopic microvilli, all greatly increasing the surface area. In the core of each villus, a network of capillaries and a lacteal take up and distribute absorbed nutrients. Nutrients are absorbed by diffusion or by active transport. Nutrient-laden blood from the villi is conveyed through the large hepatic portal vessel to the liver, which regulates the nutrient content of the blood.

- **Hormones help regulate digestion (pp. 805–806, TABLE 41.3)**
41.2 Nerve impulses and the hormone gastrin stimulate gastric motility and secretion of gastric juice. A class of intestinal hormones called enterogastrones includes secretin and cholecystokinin, which regulate the activities of the pancreas and gallbladder.

- **Reclaiming water is a major function of the large intestine (pp. 806–807)** The large intestine (colon) aids the small intestine in reabsorbing water and houses bacteria, some of which synthesize vitamins. Feces pass through the rectum and out the anus.

EVOLUTIONARY ADAPTATIONS OF VERTEBRATE DIGESTIVE SYSTEMS

- **Structural adaptations of the digestive system are often associated with diet (pp. 807–808, FIGURES 41.16, 41.17)** A mammal's dentition is generally correlated with its diet. Herbivores generally have longer alimentary canals, reflecting the longer time needed to digest vegetation.

- **Symbiotic microorganisms help nourish many vertebrates (p. 808, FIGURE 41.18)** Many herbivorous vertebrates have special fermentation chambers where symbiotic microorganisms digest cellulose.

SELF-QUIZ

1. Which of the following statements about vitamins is true?
   a. Vitamins are needed in only very small quantities because they are not very important in metabolism.
   b. The energy content of vitamins is so great that we need only small amounts of them.
   c. Excesses of all vitamins are excreted in urine.
   d. Many vitamins are components of nucleic acids and ATP.
   e. Many water-soluble vitamins are coenzymes.

2. The mammalian trachea and esophagus both open into the
   a. larynx
   b. stomach
   c. pharynx
   d. cardiac orifice
   e. epiglottis

3. Our oral cavity, with its dentition, is most functionally analogous to an earthworm's
   a. intestine
   b. pharynx
   c. gizzard
   d. stomach
   e. anus

4. Which of the following enzymes has the lowest pH optimum? (Explain your answer.)
   a. salivary amylase
   b. trypsin
   c. pepsin
   d. pancreatic amylase
   e. pancreatic lipase

5. After surgical removal of an infected gallbladder, a person must be especially careful to restrict his or her dietary intake of
   a. starch
   b. protein
   c. sugar
   d. fat
   e. water

6. Enteropeptidase, a hormone secreted by the small intestine, has which of the following actions?
   a. inhibits bile secretion
   b. inhibits duodenal secretion
   c. activates pancreatic enzymes
   d. inhibits peristalsis in the stomach
   e. increases the pH of chyme

7. Pointed molars and premolars would most likely be part of the dentition of a
   a. human
   b. cow
   c. lion
   d. rabbit
   e. hawk

8. What do the typhlosole of an earthworm, the spiral valve of a shark, and the villi of a mammal all have in common?
   a. All are adaptations for the efficient digestion and absorption of meat.
   b. They are all adaptations of the stomach.
   c. They are all microscopic structures.
   d. They all increase the absorptive surface area of intestinal epithelium.
   e. They are all homologous structures.

9. If you were to sprint 100 m a few hours after lunch, which stored fuel would you probably tap?
   a. muscle proteins
   b. muscle glycogen
   c. fat stored in the liver
   d. fats stored in adipose tissue
   e. blood proteins

10. Chemical digestion is necessary because
    a. food items are typically too large to pass through the digestive system until reduced in size.
    b. an animal can only use the component monomers of the macromolecules it ingests.
    c. without the addition of digestive fluids the bolus would be difficult to move.
    d. animals lack the appropriate enzymes to metabolize food prior to chemical digestion.
    e. the absorption of non-digested materials can disrupt the cell's chemical homeostasis.

CHALLENGE QUESTIONS

1. Trace a bacon, lettuce, and tomato sandwich through the human alimentary canal, describing what happens to the food in each region of the tract.
2. Some essential mineral nutrients, such as selenium and molybdenum, are required in the human diet in minuscule amounts—thousandths of a milligram. What kinds of experiments and observations would help demonstrate that a mineral is required by humans?
3. Compare and contrast the digestive adaptations of a carnivore and an herbivore.

SCIENCE, TECHNOLOGY, AND SOCIETY

1. Famine plagues certain parts of the world today. Some people argue that distributing food more equitably to the various countries would reduce starvation, at least for a while. Others counter that it is erroneous and ultimately harmful to perceive starvation as a global problem, because the causes and long-term solutions are usually regional. Evaluate the biological, political, and ethical aspects of this debate.
2. Contrast undernutrition with malnutrition in humans.

FURTHER READING

Blaser, M. J. "The Bacteria Behind Ulcers." *Scientific American*, February 1996. Recounts the studies demonstrating that acid-tolerant bacteria cause gastric ulcers.

Diamond, J. "Dining with the Snakes." *Discover*, April 1994. Explains how pythons and other large constrictors obtain and digest their food.

Fackelmann, K. "The Fat Fracas." *Science News*, May 2, 1998. A balanced account of the scientific controversy over pros and cons of weight loss.

Friedman, J. M. "The Alphabet of Weight Control." *Nature*, January 9, 1997. Summarizes recent research findings on the homeostatic mechanisms that regulate body weight.

Sanderson, S., and R. Wasserug, "Suspension Feeding Vertebrates." *Scientific American*, March 1990. This article emphasizes suspension-feeding whales.

Seppa, N. "Ulcer Bacterium's Drug Resistance Unmasked." *Science News*, April 25, 1998. A summary of research pointing to specific genetic changes underlying drug resistance and possible links to stomach cancer.

Travis, J. "Gene Heats Up Obesity Research." *Science News*, March 8, 1997. A brief account of a gene that may regulate ATP synthesis and body heat production.

Weindruch, R. "Caloric Restriction and Aging." *Scientific American*, January 1996. Could eating nutritious, low-calorie diets increase longevity in humans, as it appears to in rodents?

Willett, W. C. "Diet and Health: What Should We Eat?" *Science*, April 22, 1994. A critical review of human nutrition and the relationship between diets and major diseases; includes recommendations based on current research.

WEB LINKS

Visit the special edition of *The Biology Place* for BIOLOGY, Fifth Edition, at http://www.biology.com/campbell. Go to Chapter 41 for online resources, including learning activities, practice exams, and links to the following web sites:

"The Food Zone"
Students will find the Food Zone an interesting, colorful, and lively site useful to review the basics of animal nutrition.

"The Center for Food Safety and Applied Nutrition"
Extensive information on all aspects of food, including food safety, food additives, food poisoning, and nutritional information.

"The Atlas of Gastrointestinal Endoscopy"
Interesting site that contains nearly 400 endoscope pictures of normal and diseased parts of the alimentary canal.

"International Food Information Council (IFIC)"
The IFIC is a nonprofit organization based in Washington, D.C., whose mission is to provide sound, scientific information on food safety and nutrition to journalists, health professionals, educators, government officials, and consumers.

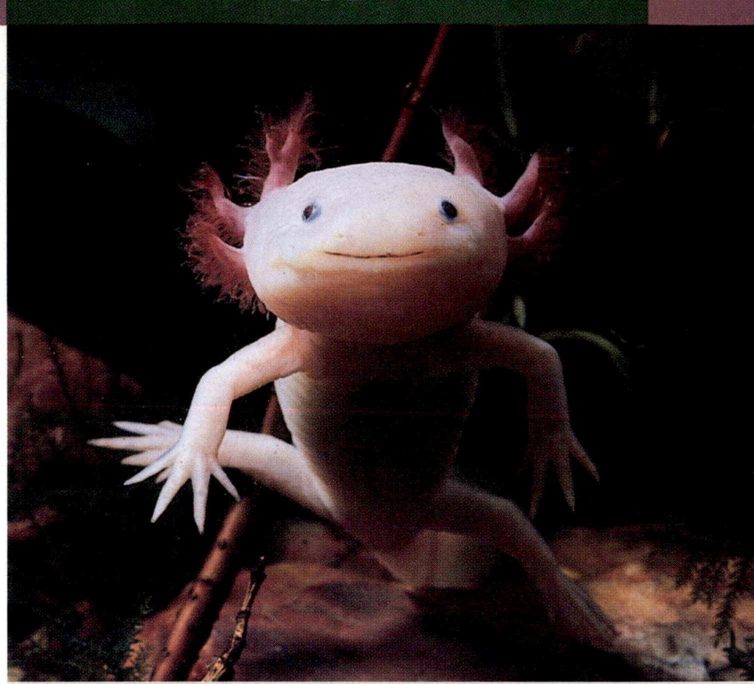

CIRCULATION AND GAS EXCHANGE

Every organism must exchange materials and energy with its environment, and this exchange ultimately occurs at the cellular level. Cells live in aqueous surroundings; the resources they need, such as nutrients and oxygen, pass into the cytoplasm across the plasma membrane, and metabolic wastes, such as carbon dioxide, pass out.

The feathery external gills projecting from the salamander shown here present a sizable surface area to the outside environment. A network of tiny blood vessels (capillaries) lies close to the outside surface of the gills. Oxygen dissolved in the surrounding water diffuses across the thin epithelium covering the gills and into the blood, while carbon dioxide diffuses out into the water.

Salamanders and most other animals have organ systems specialized for exchanging materials with the environment, and many have an internal transport system that conveys fluid (blood or interstitial fluid) throughout the body.

In this chapter you will learn about mechanisms of internal transport in animals. You will also learn about one of the most important cases of chemical transfer between animals and their environment: the exchange of the gases oxygen (O_2) and carbon dioxide (CO_2), which is essential to cellular respiration and bioenergetics.

CIRCULATION IN ANIMALS

Transport systems functionally connect the organs of exchange with the body cells: *an overview*

Diffusion alone is not adequate for transporting chemicals over macroscopic distances in animals—for example, for moving glucose from the digestive tract and oxygen from the lungs to the brain of a mammal. The time it takes for a substance to diffuse from one place to another is proportional to the square of the distance the chemical will travel. For example, if it takes 1 second for a given quantity of glucose to diffuse 100 μm, it will take 100 seconds for the same quantity to diffuse 1 mm. The circulatory system solves this problem by ensuring that no substance must diffuse very far to enter or leave a cell. By transporting fluid throughout the body, it functionally connects the aqueous environment of the body cells to the organs that exchange gases, absorb nutrients, and dispose of wastes. In the lungs of a mammal, for example, oxygen from inhaled air diffuses across a thin epithelium and into the blood, while carbon dioxide diffuses in the opposite direction. The circulatory system then carries the oxygen-rich blood to all parts of the body. As the blood streams through the tissues within microscopic vessels called capillaries, chemicals are transported between the blood and the interstitial fluid that directly bathes the cells.

Internal transport and gas exchange are functionally related; thus, we focus on both the circulatory and respiratory systems in this chapter. We also highlight the role of these two organ systems in homeostasis (see Chapter 40)—for example, in regulating the interstitial fluid's content of nutrients and wastes. Let's turn first to some of the ways fluids are circulated within animals.

Most invertebrates have a gastrovascular cavity or a circulatory system for internal transport

Gastrovascular Cavities

The body plan of a hydra and other cnidarians makes a specialized system for internal transport unnecessary. A body wall only two cells thick encloses a central gastrovascular cavity, which serves the dual functions of digestion and distribution of substances throughout the body (see FIGURE 41.9). The fluid inside the cavity is continuous with the water outside through a single opening; thus, both inner and outer layers of tissue are bathed by fluid. Thin branches of the gastrovascular cavity extend into the tentacles of a hydra, and some cnidarians have even more elaborate gastrovascular cavities (FIGURE 42.1). Since digestion begins in the cavity, only the cells of the inner layer have direct access to nutrients, but the nutrients have only a short distance to diffuse to the cells of the outer layer.

Planarians and other flatworms also have gastrovascular cavities that exchange materials with the environment through a single opening (see FIGURE 33.9). The flat shape of the body and the branching of the gastrovascular cavity throughout the animal ensure that all cells are bathed by a suitable medium.

Open and Closed Circulatory Systems

A gastrovascular cavity is inadequate for internal transport within animals having many layers of cells, especially if the animals live out of water. In insects, other arthropods, and most mollusks, blood bathes the internal organs directly. This arrangement is called an **open circulatory system** (FIGURE 42.2a). There is no distinction between blood and interstitial fluid, and the general body fluid is more correctly termed **hemolymph**. One or more hearts pump the hemolymph into an interconnected system of **sinuses**, which are spaces surrounding the organs. Here, chemical exchange occurs between the hemolymph and body cells. In grasshoppers and other arthropods, the heart is an elongated tube located dorsally. When the heart contracts, it pumps hemolymph through vessels out into sinuses. When the heart relaxes, it draws hemolymph into the circulatory system through pores called ostia. Body movements that squeeze the sinuses help circulate the hemolymph.

In a **closed circulatory system**, blood is confined to vessels and is distinct from the interstitial fluid (FIGURE 42.2b). One or more hearts pump blood into large vessels that branch into smaller ones coursing through the organs. Here, materials are exchanged between the blood and the interstitial fluid bathing the cells. Earthworms, squids, octopuses, and vertebrates have closed circulatory systems.

Vertebrate phylogeny is reflected in adaptations of the cardiovascular system

Internal transport is accomplished in humans and other vertebrates by a closed circulatory system, also called the cardiovascular system. The components of the **cardiovascular system** are the heart, blood vessels, and blood. The heart has one **atrium** or two **atria** (plural), the chambers that receive blood returning to the heart, and one or two **ventricles**, the chambers that pump blood out of the heart. Arteries, veins, and capillaries are the three main kinds of blood vessels, which in the human body have been estimated to extend a total distance of 100,000 km. **Arteries** carry blood away from the heart to organs throughout the body. Within these organs, arteries branch into **arterioles**, small vessels that convey blood

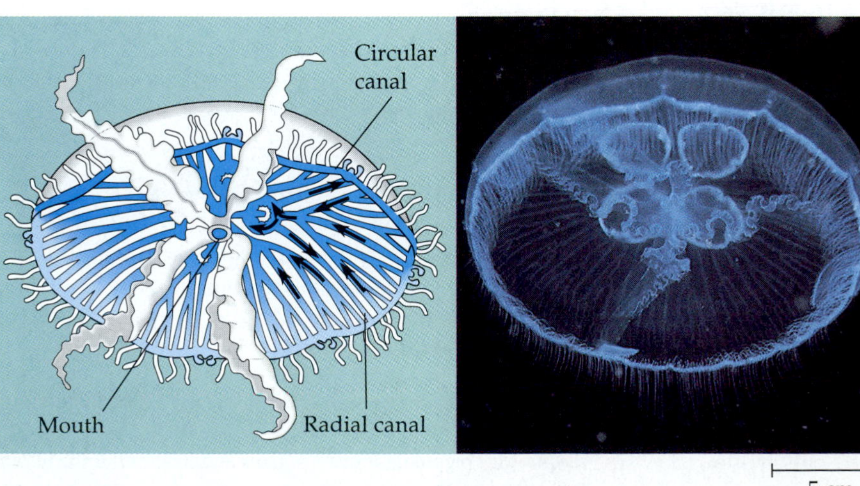

FIGURE 42.1 ▪ Internal transport in the cnidarian *Aurelia*. The mouth leads to an elaborate gastrovascular cavity (shown in blue) that has branches radiating to and from a circular canal. Ciliated cells lining the canals circulate fluid in the directions indicated by the arrows. The animal is viewed here from its underside (oral surface).

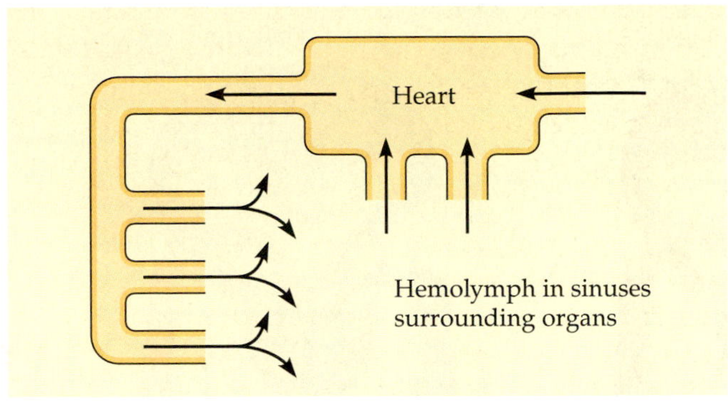

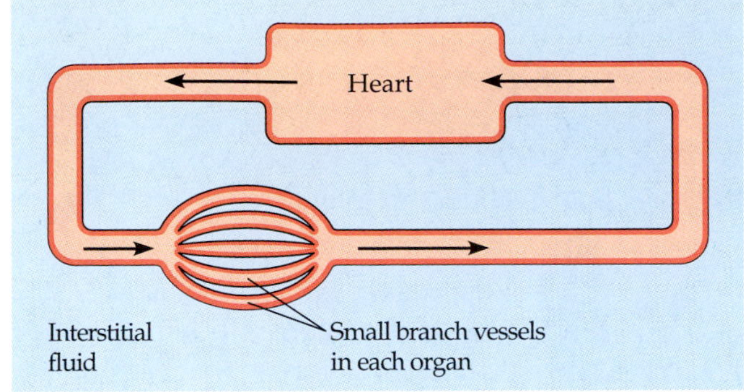

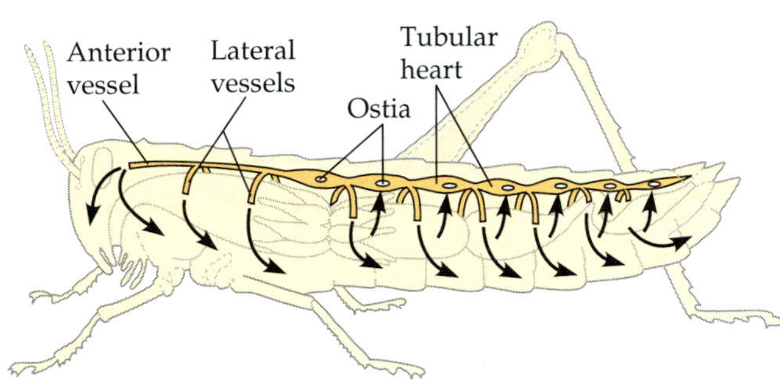

(a) Open circulatory system

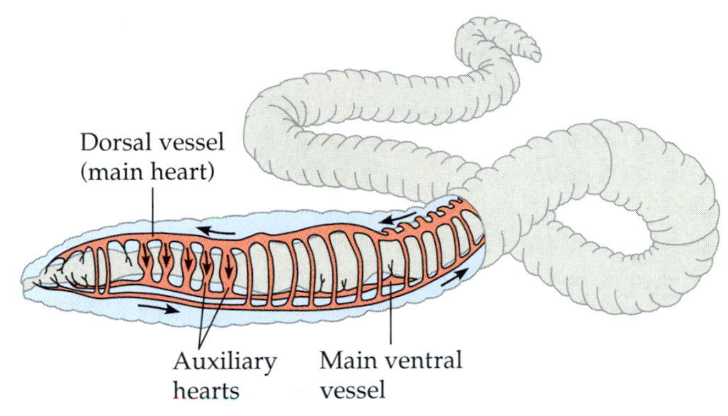

(b) Closed circulatory system

FIGURE 42.2 • Open and closed circulatory systems. (a) In an open circulatory system, blood and interstitial fluid are the same, and this fluid is called hemolymph. The heart pumps hemolymph through vessels into sinuses, where materials are exchanged between the hemolymph and cells. Hemolymph returns to the heart through ostia. In a grasshopper, a dorsal tubular heart pumps hemolymph forward through an anterior vessel and to the sides of the animal through a series of lateral vessels. Squeezed rearward through sinuses by body movements, the hemolymph reenters the heart through ostia, which are equipped with valves that close when the heart contracts. **(b)** A closed circulatory system contains blood within vessels, distinct from the interstitial fluid. As blood passes through small vessels in organs, chemical exchange occurs between the blood and the interstitial fluid, and between the intersititial fluid and body cells. In an earthworm, three major vessels, one dorsal and two ventral, branch into smaller vessels that supply blood to the various organs. The dorsal vessel functions as the main heart, pumping blood forward by peristalsis. Near the worm's anterior end, five pairs of vessels loop around the digestive tract, connecting the dorsal and a ventral vessel. The paired vessels function as auxiliary hearts, propelling blood ventrally. Blood flows rearward in the ventral vessels.

to the capillaries. **Capillaries** are microscopic vessels with very thin, porous walls. Networks of these vessels, called **capillary beds**, infiltrate each tissue. It is across the thin walls of capillaries that chemicals, including gases, are exchanged between the blood and the interstitial fluid surrounding the cells. At their "downstream" end, capillaries converge into **venules**, and venules converge into veins. **Veins** return blood to the heart. Notice that arteries and veins are distinguished by the *direction* in which they carry blood, not by the characteristics of the blood they contain. Not all arteries carry oxygen-rich blood, and not all veins carry blood with a low concentration of oxygen. But all arteries do carry blood from the heart *toward* capillaries, and only veins return blood to the heart *from* capillaries.

The cardiovascular systems of vertebrates are variations of the general circulatory scheme just described. Adaptations of the cardiovascular systems are correlated with gill breathing in aquatic vertebrates and lung breathing in terrestrial vertebrates.

A fish has a heart with two main chambers, one atrium and one ventricle (FIGURE 42.3a, p. 814). Blood pumped from the ventricle travels first to the gills, where the blood picks up oxygen and disposes of carbon dioxide across capillary walls. The gill capillaries converge into a vessel that carries the oxygen-rich blood to capillary beds in all other parts of the body. Blood then returns in veins to the atrium of the heart. Notice that in a fish, blood must pass through *two* capillary beds during each circuit, one in the gills and a second one, called systemic capillaries, in an organ other than the gills. When blood flows through a capillary bed, blood pressure, the hydrostatic pressure that pushes blood through vessels, drops substantially (for reasons that will be explained shortly). Therefore, oxygen-rich blood leaving the gills flows to other organs in the fish quite slowly, but the process is aided by the movements of the body during swimming.

Frogs and other amphibians have a three-chambered heart, with two atria and one ventricle (FIGURE 42.3b). The ventricle pumps blood into a forked artery that directs the blood through two circuits: the pulmocutaneous circuit and the systemic circuit. The **pulmocutaneous circuit** leads to the gas exchange tissues (in the lungs and skin in a frog), where

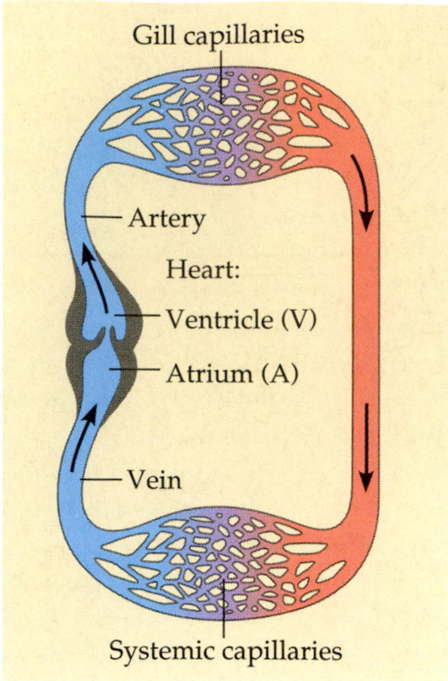

(a) Fish

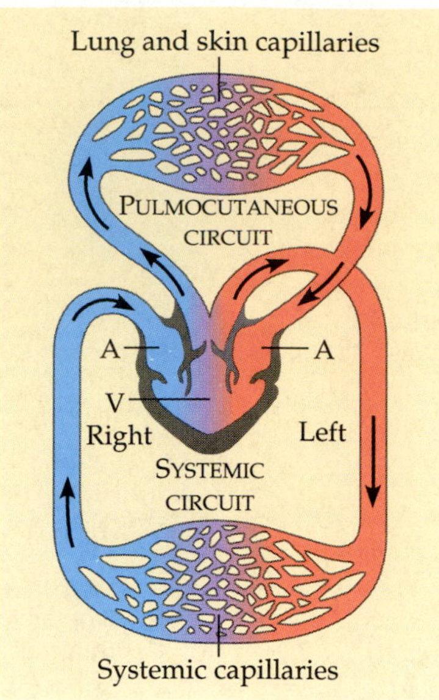

(b) Amphibian

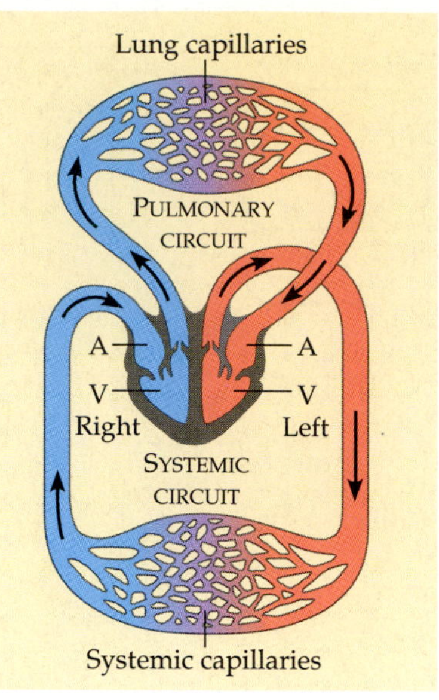

(c) Mammal

FIGURE 42.3 ▪ Generalized circulatory schemes of vertebrates. Red symbolizes oxygen-rich blood, and blue represents oxygen-poor blood. **(a)** Fishes have a two-chambered heart and a single circuit of blood flow, with the atrium of the heart receiving oxygen-poor blood from veins and the ventricle pumping blood to the gills. **(b)** Amphibians have a three-chambered heart and two circuits of blood flow: pulmocutaneous and systemic. This double circulation delivers blood to systemic organs under high pressure. In the single ventricle, there is some mixing of oxygen-rich blood with oxygen-poor blood. (It is customary to illustrate cardiovascular systems with the right side of the heart on the left and the left side on the right, as if the body is facing you from the page.) **(c)** Mammals have a four-chambered heart and double circulation. Within the heart, oxygen-rich blood is kept completely segregated from oxygen-poor blood.

the blood picks up oxygen as it flows through capillaries. The oxygen-rich blood returns to the left atrium of the heart, and then most of it is pumped into the systemic circuit. The **systemic circuit** carries oxygen-rich blood to all body organs and then returns oxygen-poor blood to the right atrium via the veins. This scheme, called **double circulation**, ensures a vigorous flow of blood to the brain, muscles, and other organs because the blood is pumped a second time after it loses pressure in the capillary beds of the lungs or skin. This is distinctly different from the single circulation in the fish, where blood flows directly from the respiratory organs (gills) to other organs under reduced pressure.

In the single ventricle of the frog, there is some mixing of oxygen-rich blood that has returned from the lungs with oxygen-poor blood that has returned from the rest of the body. However, a ridge within the ventricle diverts most of the oxygen-rich blood from the left atrium into the systemic circuit and most of the oxygen-poor blood from the right atrium into the pulmocutaneous circuit.

In reptiles, there is even less mixing of oxygen-rich and oxygen-poor blood. Although the reptilian heart is three-chambered, the single ventricle is partially divided. Reptiles have double circulation, with a systemic circuit and a **pulmonary circuit**, which conveys blood from the heart to the gas-exchange tissues in the lungs and back to the heart. In one order of reptiles, the crocodilians, the ventricle is completely divided into right and left chambers.

The four-chambered heart of a bird or mammal has two
42.1 atria and two completely separated ventricles (FIGURE 42.3C). There is double circulation (systemic and pulmonary circuits); the left side of the heart handles only oxygen-rich blood, while the right side receives and pumps only oxygen-poor blood. Delivery of oxygen to all parts of the body is enhanced because there is no mixing of oxygen-rich blood and oxygen-poor blood, and double circulation restores pressure after blood has passed through the lung capillaries. As endotherms, which use heat released from metabolism to warm the body, birds and mammals require more oxygen per gram of body weight than other vertebrates of equal size. Birds and mammals descended from different reptilian ancestors, and their four-chambered hearts evolved independently—an example of convergent evolution. A more detailed diagram of blood flow through the mammalian cardiovascu-
42.2 lar system is shown in FIGURE 42.4.

Double circulation in mammals depends on the anatomy and pumping cycle of the heart

The Mammalian Heart: A Closer Look

A closer look at the four-chambered mammalian heart provides a more complete understanding of how double circulation works (FIGURE 42.5). Located just beneath the breastbone

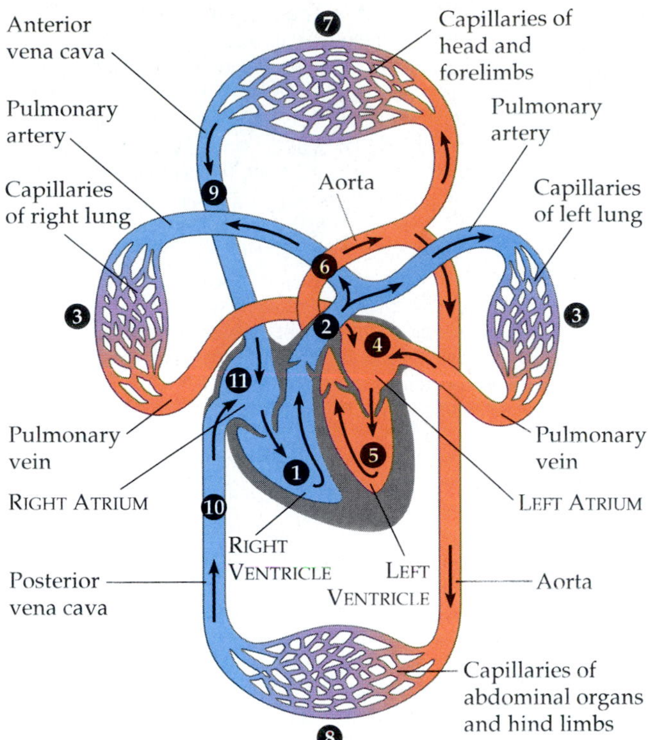

**FIGURE 42.4 ▪ The mammalian cardiovascular system: an
42.1 overview.** The numbers in this diagram trace the flow of blood through
42.2 the heart, main blood vessels, and capillary beds. Beginning with the pulmonary (lung) circuit, ① the right ventricle pumps blood to the lungs via ② the pulmonary arteries. As the blood flows through ③ capillary beds in the right and left lungs, it loads up on oxygen and unloads carbon dioxide. Oxygen-rich blood returns from the lungs via the pulmonary veins to ④ the left atrium of the heart. Next, the oxygen-rich blood flows into ⑤ the left ventricle, as the ventricle opens and the atrium contracts. In turn, the left ventricle pumps the oxygen-rich blood out to body tissues through the systemic circuit. Blood leaves the left ventricle via ⑥ the aorta, which conveys blood to arteries leading throughout the body. The first branches from the aorta are the coronary arteries (not shown), which supply blood to the heart muscle itself. Then come branches leading to capillary beds ⑦ in the head and arms (or forelimbs). The aorta continues in a posterior direction, supplying oxygen-rich blood to arteries leading to ⑧ capillary beds in the abdominal organs and legs (hind limbs). Within each organ, arteries branch into arterioles, which in turn branch into capillaries, where the blood gives up much of its oxygen and picks up the carbon dioxide produced by cellular respiration. Capillaries rejoin to form venules, which convey blood to veins. Oxygen-poor blood from the head, neck, and forelimbs is channeled into a large vein called ⑨ the anterior (or superior) vena cava. Another large vein called ⑩ the posterior (or inferior) vena cava drains blood from the trunk and hind limbs. The two venae cavae empty their blood into ⑪ the right atrium, from which the oxygen-poor blood flows into the right ventricle.

(sternum), the human heart, for example, is about the size of a clenched fist. It is made up mostly of cardiac muscle tissue (see Chapter 40). The two atria have relatively thin walls and function as collection chambers for blood returning to the heart, pumping blood only the short distance to the ventricles. The ventricles have thicker walls and are much more powerful than the atria—especially the left ventricle, which must pump blood out to all body organs through the systemic circuit.

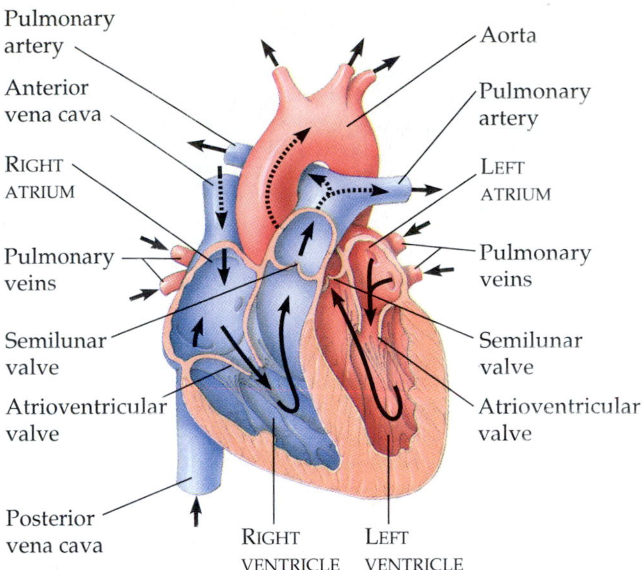

FIGURE 42.5 ▪ The mammalian heart: a closer look. In this detailed view of the structure of the heart, notice the valves, which prevent backflow of blood within the heart, and the relative thickness of the walls of the heart chambers. The atria, which pump blood only into the ventricles, have thinner walls than the ventricles.

Four valves in the heart, each consisting of flaps of connective tissue, prevent backflow of blood. Between each atrium and ventricle is an **atrioventricular (AV) valve.** The AV valves are anchored by strong fibers that prevent them from turning inside out. Pressure generated by the powerful contraction of the ventricles closes the AV valves, keeping blood from flowing back into the atria. **Semilunar valves** are located at the two exits of the heart, where the aorta leaves the left ventricle and the pulmonary artery leaves the right ventricle. The blood is pumped out into the arteries through the semilunar valves, which are forced open by pressure created by ventricular contraction. When the ventricles relax, blood starts to flow back toward the heart, closing the semilunar valves, which prevents blood from flowing back into the ventricles. The elastic walls of the arteries expand when they receive the blood expelled from the ventricles. By taking your **pulse**, which is the rhythmic stretching of the arteries caused by the pressure of blood driven by the powerful contractions of the ventricles, you can measure your **heart rate**, the number of times the heart beats each minute.

The heart sounds we can hear with a stethoscope are caused by the closing of the valves. The sound pattern is "lub-dup, lub-dup, lub-dup." The first heart sound ("lub") is created by the recoil of blood against the closed AV valves. The second sound ("dup") is the recoil of blood against the semilunar valves.

A defect in one or more of the valves causes a condition known as a heart murmur, which may be detectable as a hissing sound when a stream of blood squirts backward through a valve. Some people are born with heart murmurs, while

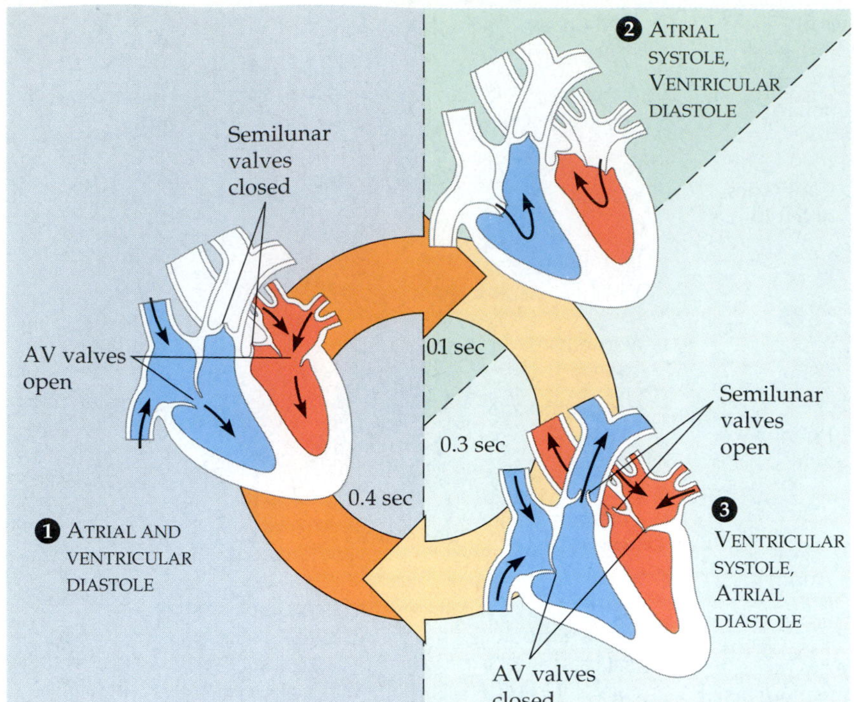

FIGURE 42.6 ▪ The cardiac cycle. The cardiac muscles of the heart contract (systole) and relax (diastole) in a rhythmic cycle. For an adult human at rest with a pulse of about 75 beats per minute, one complete cardiac cycle takes about 0.8 sec. ① During a relaxation phase (atria and ventricles in diastole) lasting about 0.4 sec, blood returning from the large veins flows into the atria and ventricles. ② A brief period (about 0.1 sec) of atrial systole then forces all remaining blood out of the atria into the ventricles. ③ During the remaining 0.3 sec of the cardiac cycle, ventricular systole pumps blood into the large arteries. Note that seven-eighths of the time—all but 0.1 sec of the cardiac cycle—the atria are relaxed and filling with blood returning in the veins.

others have their valves damaged by infection (from rheumatic fever, for instance). Most heart murmurs do not reduce the efficiency of blood flow enough to warrant surgery.

With its right side receiving and pumping only oxygen-poor blood, its left side handling only oxygen-rich blood, and its valves allowing blood to flow in only one direction, the heart is the hub of the cardiovascular system. The heart alternately contracts and relaxes in a rhythmic cycle. When it contracts, it pumps blood; when it relaxes, its chambers fill with blood. One complete sequence of pumping and filling is called the **cardiac cycle**. A contraction phase of the cycle is called **systole**, and a relaxation phase is called **diastole** (FIGURE 42.6).

The volume of blood per minute that the left ventricle pumps into the systemic circuit is called **cardiac output**. This volume depends on two factors: heart rate (pulse) and **stroke volume,** the amount of blood pumped by the left ventricle each time it contracts. The average stroke volume for a human is about 75 mL per beat. A person with this stroke volume and a resting pulse of 70 beats/min has a cardiac output of 5.25 L/min. This is about equivalent to the total volume of blood in the human body. Cardiac output can increase about fivefold during heavy exercise.

Maintaining the Heart's Rhythmic Beat

Certain cells of vertebrate cardiac muscle are self-excitable, meaning they can contract without any signal from the nervous system. Individual cardiac muscle cells removed from a heart and viewed with a microscope can be seen to pulsate. A region of the heart called the **sinoatrial (SA) node**, or **pacemaker**, maintains the heart's pumping rhythm by setting the rate at which all cardiac muscle cells contract. Composed of specialized muscle tissue, the SA node is located in the wall of the right atrium, near the point where the superior vena cava enters the heart (FIGURE 42.7).

The SA node generates electrical impulses much like those produced by nerve cells. Because cardiac muscle cells are electrically coupled (by the intercalated discs between adjacent cells; see FIGURE 40.5), impulses from the SA node spread rapidly through the walls of the atria, making them contract in unison. The impulses also pass to another region of specialized muscle tissue, a relay point called the **atrioventricular (AV) node**, in the wall between the right atrium and right ventricle. Here the impulses are delayed for about 0.1 sec, which ensures that the atria will contract first and empty completely before the ventricles contract. After this delay, specialized muscle fibers conduct the signal to contract throughout the walls of the ventricle.

The impulses that travel through cardiac muscle during the heart cycle produce electrical currents that are conducted through body fluids to the body surface, where the currents can be detected by electrodes placed on the skin and recorded as an **electrocardiogram (ECG or EKG)**.

The SA node sets the tempo for the entire heart, but this pacemaker itself is influenced by a variety of cues. Two sets of nerves oppose each other in adjusting heart rate; one set speeds up the pacemaker, and the other set slows it down. At any given time, heart rate is a compromise regulated by the opposing actions of these two sets of nerves. The pacemaker is

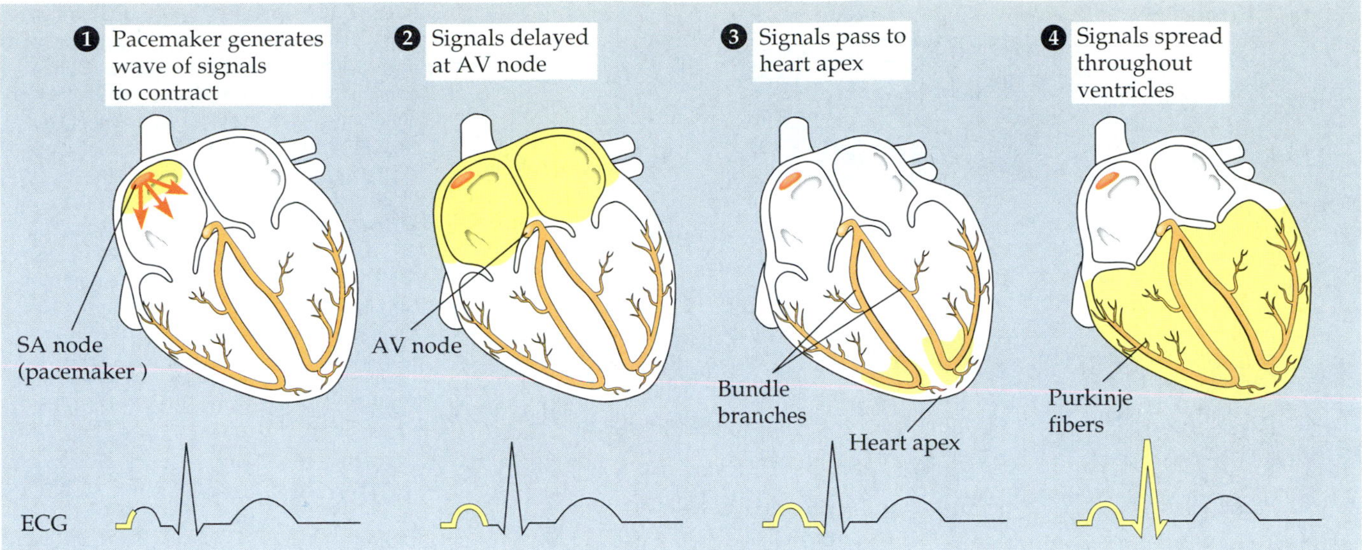

1 Pacemaker generates wave of signals to contract
2 Signals delayed at AV node
3 Signals pass to heart apex
4 Signals spread throughout ventricles
SA node (pacemaker)
AV node
Bundle branches
Heart apex
Purkinje fibers
ECG

FIGURE 42.7 ▪ The control of heart rhythm. ① The SA node, or pacemaker, sets the tempo of the heartbeat by generating electrical signals (yellow) that ② spread through both atria, making them contract simultaneously. The signals to contract are delayed at the AV node for about 0.1 sec, during which blood in the atria empties into the ventricles. ③ Specialized muscle fibers called bundle branches and Purkinje fibers then conduct the signals to the apex of the heart and ④ throughout the ventricular walls. The signals trigger powerful contractions of both ventricles from the apex toward the atria, driving blood into the large arteries. Yellow color in the graphs at the bottom indicates the components of an electrocardiogram (ECG) corresponding to the sequence of electrical events in the heart. The portion of the ECG in step 4 that is not yellow represents electrical activity after the ventricles contract; during this phase of the ECG, the ventricles become reprimed electrically and thus are able to conduct the next round of electrical signals to contract.

also controlled by hormones secreted into the blood by glands. For example, epinephrine, the "fight-or-flight" hormone from the adrenal glands, increases heart rate (see Chapter 45). Body temperature is another factor that affects the pacemaker. A temperature increase of only 1°C raises the heart rate by about 10 beats/min. This is the reason your pulse increases substantially when you have a fever. The heart rate also increases with exercise, an adaptation that enables the circulatory system to provide the additional oxygen needed by muscles hard at work.

Structural differences of arteries, veins, and capillaries vessels correlate with their different functions

Blood vessels are made up of similar tissues. The wall of an artery or vein, for instance, has three similar layers (FIGURE 42.8, p. 818). On the outside, a layer of connective tissue with elastic fibers allows the vessel to stretch and recoil. A middle layer consists of smooth muscle and more elastic fibers. Lining the lumen of all blood vessels, including the capillaries, is an **endothelium**, a single layer of flattened cells (see Chapter 40). The endothelium provides a smooth surface that minimizes resistance to the flow of blood.

Structural differences in the walls of an artery, vein, or capillary correlate with their different functions. For instance, capillaries lack the two outer layers, and their very thin walls consist only of endothelium and its basement membrane. This structure facilitates the exchange of substances between the blood and the interstitial fluid that bathes the cells. Arteries have thicker middle and outer layers than veins. Blood flows through the vessels of the circulatory system at uneven speeds and pressures. The thicker walls of arteries provide strength and elasticity that accommodate the flow of blood pumped rapidly and at high pressure through the arteries by the heart. The thinner-walled veins convey blood back to the heart at low velocity and pressure after the blood has passed through capillary beds. Blood flows through the veins mainly as a result of muscle action; whenever we move, our skeletal muscles pinch our veins and squeeze blood through them. Within our large veins, flaps of tissue that function as one-way valves allow blood to flow only toward the heart (FIGURE 42.9, p. 818).

Physical laws governing the movement of fluids through pipes affect blood flow and blood pressure

Blood Flow Velocity

42.3 Blood travels over a thousand times faster in the aorta (about 30 cm/sec on average) than in the capillaries (about 0.026 cm/sec). To understand why the blood decelerates, we need to consider the *law of continuity*, a rule that governs the flow of

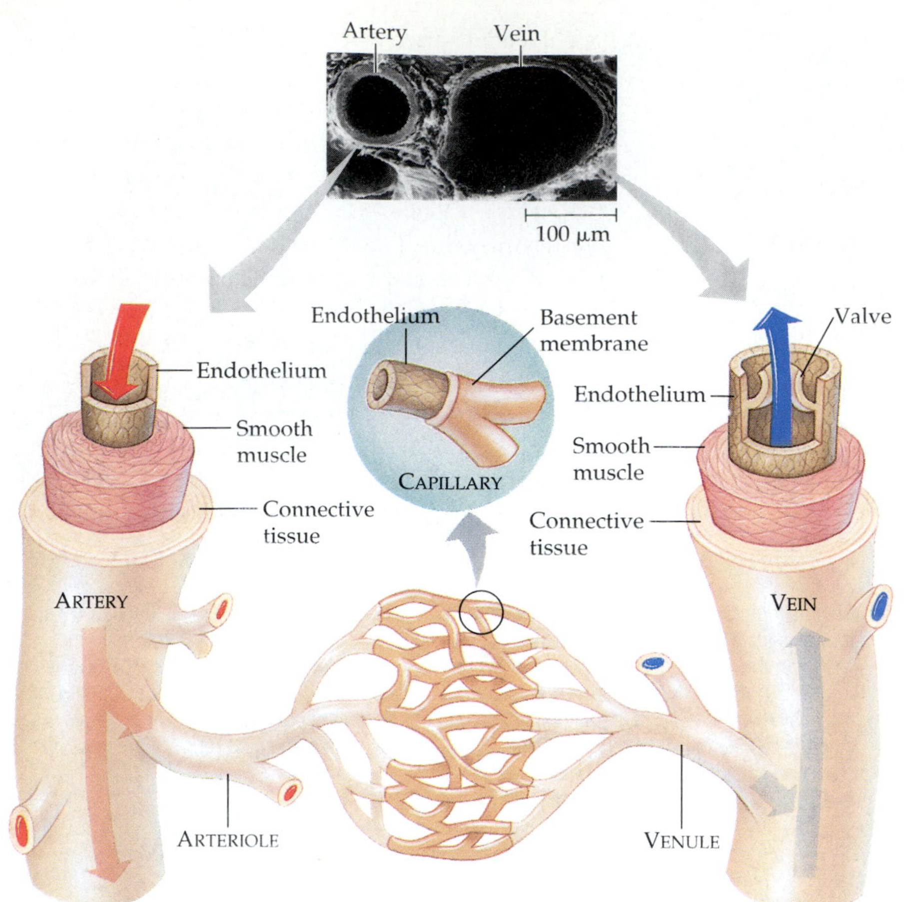

FIGURE 42.8 • The structure of blood vessels. In the micrograph, an artery can be seen next to a thinner-walled vein. The wall of an artery or a vein has three layers: an inner layer of endothelium, a middle layer of smooth muscle with elastic fibers, and an outer layer of connective tissue, also with elastic fibers. Capillaries have much smaller diameters than arteries and veins, and their walls consist of only a single layer of endothelium and a basement membrane. These structural differences correlate with the functional differences among the blood vessels (SEM).

fluids through pipes. If a pipe's diameter changes over its length, a fluid will stream through narrower segments of the pipe faster than it flows through wider segments. The *volume* of flow per second must be constant through the entire pipe, so the fluid must flow faster as the cross-sectional area of the pipe narrows. For instance, compare the velocity of water squirted by a hose with and without a nozzle.

Based on the law of continuity, it may at first seem that blood should travel faster through capillaries than through arteries, because the diameter of capillaries is much smaller.

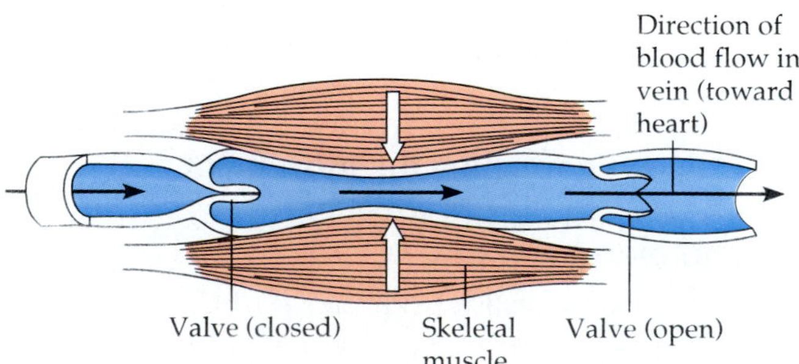

FIGURE 42.9 • Blood flow in veins. Contracting muscles squeeze the veins, in which flaps of tissue act as one-way valves that keep blood moving only toward the heart. If we sit or stand too long, the lack of muscular activity causes our feet to swell with stranded fluid unable to return to the heart.

However, it is the *total* cross-sectional area of the pipes delivering the fluid that determines flow rate. Although an individual capillary is very narrow, each artery conveys blood to such an enormous number of capillaries that the *total* diameter of the vessels is actually much greater in capillary beds than in any other part of the circulatory system. For this reason, the blood slows down substantially as it enters the arterioles from arteries and flows slowest in the capillary beds. Capillaries are the only vessels with walls thin enough to permit the transfer of substances between the blood and interstitial fluid, and the slower flow of blood through these tiny vessels enhances this chemical exchange. As blood leaves the capillary beds and passes to the venules and veins, it speeds up again, a result of the reduction in total cross-sectional area (FIGURE 42.10).

Blood Pressure

Fluids exert a force called hydrostatic pressure against surfaces they contact, and it is that pressure that drives fluids through pipes. The hydrostatic force that blood exerts against the wall of a vessel is called **blood pressure** (see the Methods Box, p. 821). This pressure is much greater in arteries than in veins and is greatest in arteries when the heart contracts during ventricular systole (see FIGURES 42.6 and 42.10). Blood pressure is

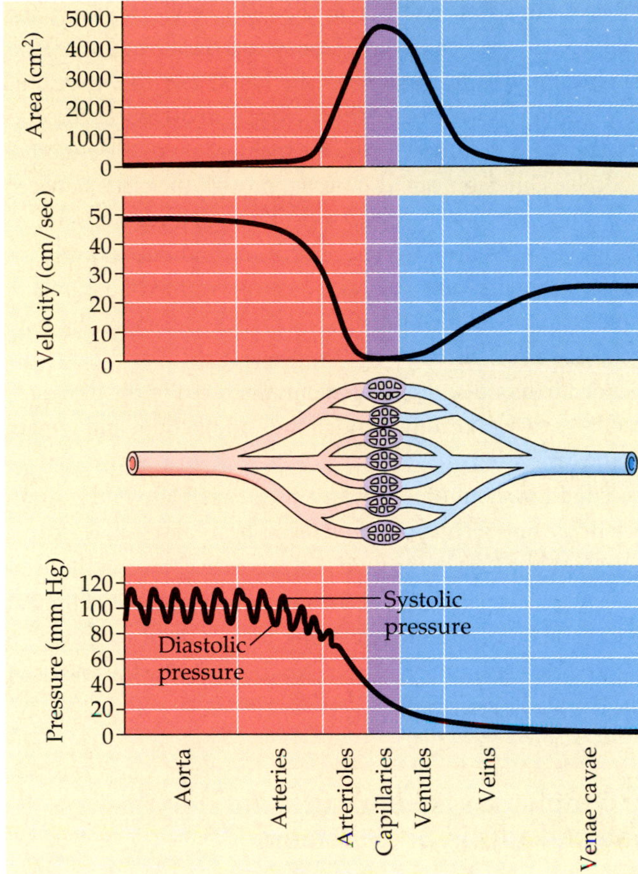

FIGURE 42.10 ▪ The interrelationship of blood flow velocity, cross-sectional area of blood vessels, and blood pressure. Blood flow velocity decreases markedly in the arterioles and is slowest in the capillaries, owing to an increase in total cross-sectional area (smallest in the arteries, greater in the arterioles, and greatest in the capillaries). Blood pressure, the main force driving blood from the heart to the capillaries, is highest in the arteries. Peaks in blood pressure corresponding to ventricular systole alternate with lower pressures corresponding to diastole. Resistance to flow through the arterioles and capillaries, due to contact of the blood with a greater surface area of endothelium, reduces blood pressure and eliminates the pressure peaks.

the main force that propels blood from the heart through the arteries and arterioles to the capillary beds. Fluids always flow from areas of high pressure to areas of lower pressure.

When you take your pulse by placing your fingers on your wrist, you can actually feel an artery bulge with each heartbeat. The surge of pressure is partly due to the narrow openings of arterioles impeding the exit of blood from the arteries. Thus, when the heart contracts, blood enters the arteries faster than it can leave, and the vessels stretch from the pressure. The elastic walls of the arteries snap back during diastole, but the heart contracts again before enough blood has flowed into the arterioles to completely relieve pressure in the arteries. This impedance by the arterioles is called **peripheral resistance**. As a consequence of the elastic arteries working against peripheral resistance, there is a blood pressure even during diastole, driving blood into arterioles and capillaries continuously (see FIGURE 42.10).

Blood pressure is determined partly by cardiac output and partly by the degree of peripheral resistance to blood flow in the arterioles, the bottlenecks of the circulatory system. Contraction of smooth muscles in the walls of the arterioles constricts the tiny vessels, increases resistance, and therefore increases blood pressure upstream in the arteries. When the smooth muscles relax, the arterioles dilate, blood flow through the arterioles increases, and pressure in the arteries falls. Nerve impulses, hormones, and other signals control these arteriole wall muscles. Stress, both physical and emotional, can raise blood pressure by triggering nervous and hormonal responses that constrict blood vessels.

By the time blood reaches the veins, its pressure is not affected much by the heart. This is because the blood encounters so much resistance as it passes through the millions of tiny arterioles and capillaries that the force from the pumping heart can no longer propel the blood in the veins. How, then does blood return to the heart, especially when it must travel from the lower extremities against gravity? Rhythmic contractions of smooth muscles in the walls of venules and veins account for some movement of the blood. But more importantly, the activity of skeletal muscles during exercise squeezes blood through the veins (see FIGURE 42.9). Moreover, when we inhale, the change in pressure within the thoracic (chest) cavity causes the venae cavae and other large veins near the heart to expand and fill with blood.

Transfer of substances between the blood and the interstitial fluid occurs across the thin walls of capillaries

Blood Flow Through Capillary Beds

At any given time, only about 5% to 10% of the body's capillaries have blood flowing through them. Because each tissue has so many capillaries, however, every part of the body is supplied with blood at all times. Capillaries in the brain, heart, kidneys, and liver are usually filled to capacity with blood, but in many other sites, the blood supply varies from time to time.

Two mechanisms, both dependent on smooth muscles controlled by nerve signals and hormones, regulate the distribution of blood in capillary beds. In one mechanism, contraction of the smooth muscle layer in the wall of an arteriole constricts the arteriole, decreasing blood flow through it to a capillary bed. When the muscle layer relaxes, the arteriole dilates, allowing blood to enter the capillaries. In the other mechanism, rings of smooth muscle, called precapillary sphincters because they are located at the entrance to capillary

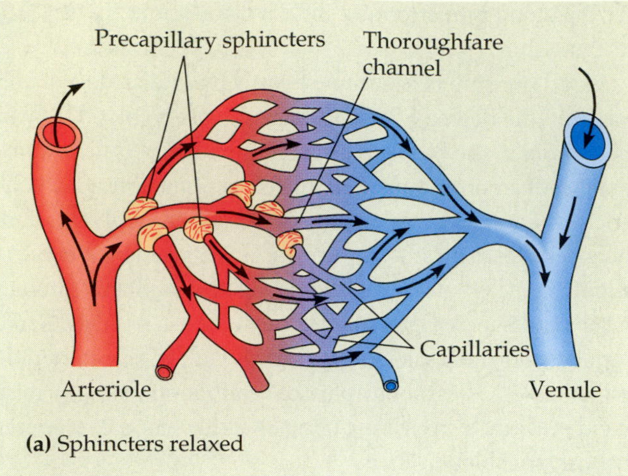

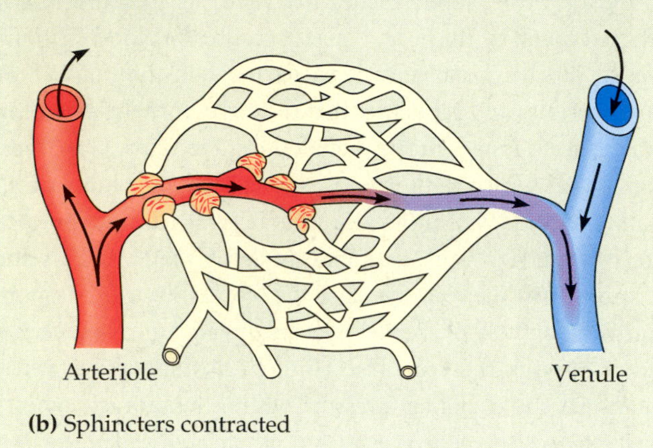

FIGURE 42.11 ▪ Blood flow in capillary beds. Precapillary sphincters regulate the passage of blood into many capillary beds. Some blood flows directly from arterioles to venules through capillaries called thoroughfare channels, which are always open. **(a)** When the precapillary sphincters are relaxed, capillaries branching from the thoroughfare channel are open, and blood flows into the capillary bed. **(b)** When the sphincters are contracted, the capillaries are closed, and blood flow through the capillary bed is reduced.

beds, control the flow of blood between arterioles and venules (FIGURE 42.11).

The blood supply to capillary beds varies locally as blood is diverted from one destination to another. After a meal, for instance, arterioles in the wall of the digestive tract dilate, capillary sphincters relax, and the digestive tract receives a larger share of blood. During strenuous exercise, blood is diverted from the digestive tract and supplied more generously to skeletal muscles and skin. This is one reason heavy exercise immediately after a big meal may cause indigestion.

Capillary Exchange

The all-important transfer of substances between the blood and the interstitial fluid that bathes the cells takes place across the thin endothelial walls of the capillaries. Some substances may be carried across an endothelial cell in vesicles that form by endocytosis on one side of the cell (see FIGURE 8.18) and then release their contents by exocytosis on the opposite side; others simply diffuse between the blood and the interstitial fluid. Small molecules, such as oxygen and carbon dioxide, diffuse down gradients across the membranes of the endothelial cells. Diffusion can also occur through the clefts between adjoining cells. However, transport through these clefts occurs mainly by bulk flow, the movement of fluid due to pressure. Hydrostatic pressure (blood pressure) within the capillary pushes fluid (water and small solutes such as sugars, salts, oxygen, and urea) through the capillary clefts. Blood cells suspended in blood and most proteins dissolved in the blood are too large to pass readily through the endothelium and remain in the capillaries.

Fluid flows out of a capillary at the upstream end near an arteriole, but reenters downstream near a venule (FIGURE 42.12). About 85% of the fluid that leaves the blood at the arterial end of a capillary bed reenters from the interstitial fluid at the venous end, and the remaining 15% of the fluid lost from capillaries is eventually returned to the blood by the vessels of the lymphatic system.

The lymphatic system returns fluid to the blood and aids in body defense

So much blood passes through the capillaries that the cumulative loss of fluid adds up to about 4 L per day. There is also some leakage of blood proteins, even though the capillary wall is not very permeable to these large molecules. The lost fluid and proteins return to the blood via the **lymphatic system**. Fluid enters this system by diffusing into tiny lymph capillaries that are intermingled among capillaries of the cardiovascular system. Once inside the lymphatic system, the fluid is called **lymph**; its composition is about the same as that of interstitial fluid. The lymphatic system drains into the circulatory system near the junction of the venae cavae with the right atrium (see FIGURE 43.4).

Lymph vessels, like veins, have valves that prevent the backflow of fluid toward the capillaries. Rhythmic contractions of the vessel walls help draw fluid into lymphatic capillaries. Also, like veins, lymph vessels depend mainly on the movement of skeletal muscles to squeeze fluid toward the heart.

Along a lymph vessel are organs called **lymph nodes** which filter the lymph. Inside a lymph node is a honeycomb of connective tissue with spaces filled by white blood cells. When the body is fighting an infection, these cells multiply rapidly, and the lymph nodes become swollen and tender (which is why your doctor checks for swollen lymph nodes in your neck). Attacking viruses and bacteria, the white blood cells play an important role in the body's defense.

The lymphatic system, then, helps defend the body against infection and maintains the volume and protein concentration

METHODS: Measurement of Blood Pressure

Blood pressure is recorded as two numbers separated by a slash; the first number is the systolic pressure; the second is the diastolic pressure. ① A typical blood pressure reading for a 20-year-old is 120/70. The units for these numbers are mm of mercury (Hg); a blood pressure of 120 is a force that can support a column of mercury 120 mm high. ② A sphygmomanometer, an inflatable cuff attached to a pressure gauge, measures blood pressure in an artery. The cuff is wrapped around the upper arm and inflated until the pressure closes the artery, so that no blood flows past the cuff. When this occurs, the pressure exerted by the cuff exceeds the blood pressure in the artery. ③ A stethoscope is used to listen for sounds of blood flow below the cuff. If the artery is closed, there is no pulse below the cuff. The cuff is gradually deflated until blood begins to flow into the forearm, and sounds from blood pulsing into the artery below the cuff can be heard with the stethoscope. This occurs when the blood pressure is greater than the pressure exerted by the cuff. The pressure at this point is the systolic pressure, the high pressure exerted by the ventricles contracting. ④ The cuff is loosened further until blood flows freely through the artery, and the sounds below the cuff disappear. The pressure at this point is the diastolic pressure remaining in the artery when the heart is relaxed.

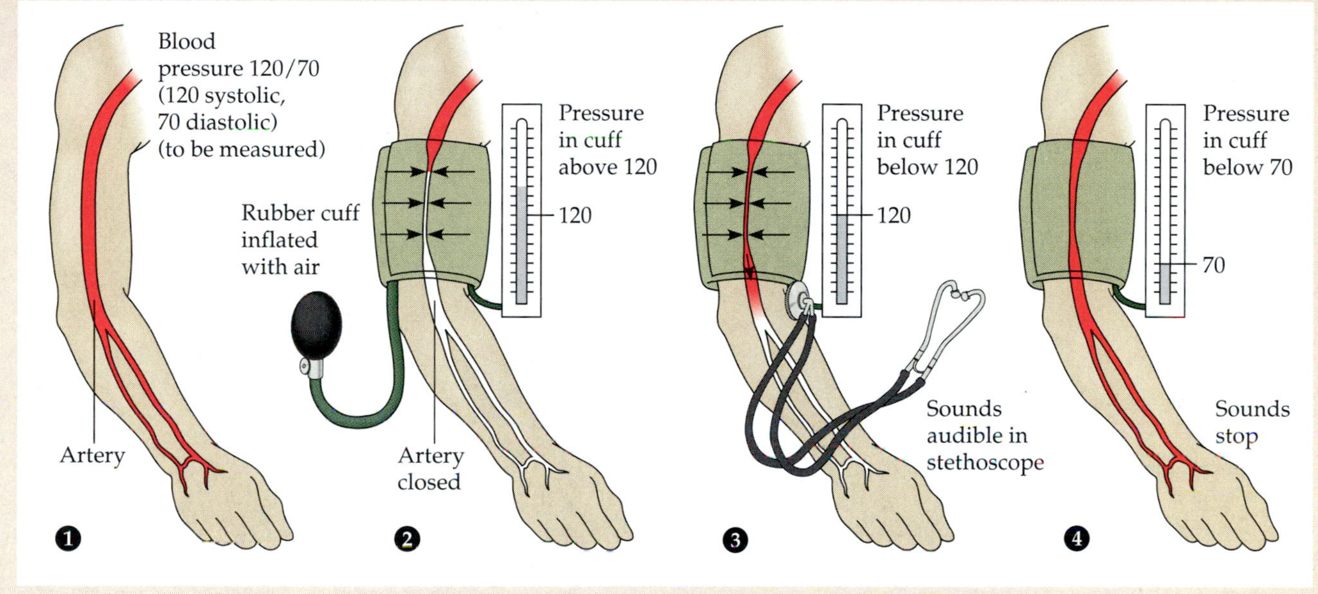

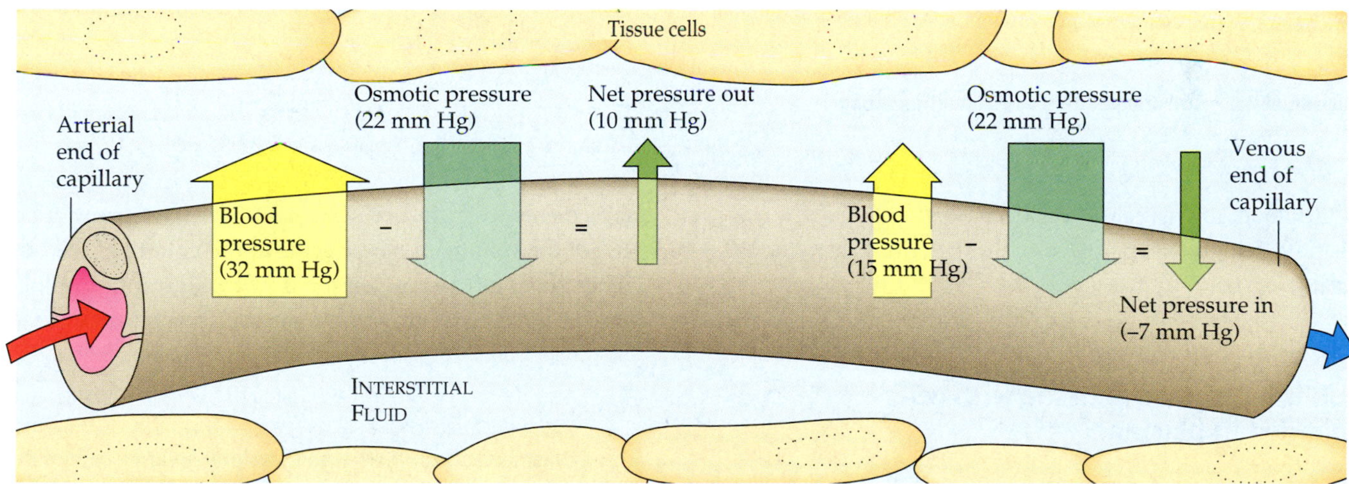

FIGURE 42.12 ▪ The movement of fluid between capillaries and the interstitial fluid. Fluid flows out of a capillary at the upstream end near an arteriole and reenters a capillary downstream near a venule. The direction of fluid movement at any point across the capillary wall depends on the difference between two opposing forces: hydrostatic pressure (blood pressure) and osmotic pressure. Blood pressure tends to force fluid out of the capillary. Osmotic pressure is the tendency for water to enter the capillary because of the relatively high solute concentration of the blood. At the arterial end of the capillary, the blood pressure forcing fluid outward exceeds the osmotic pressure drawing water inward, resulting in a net movement of fluid out of the capillary. The endothelium allows water and small solutes to leave the capillary and retains most proteins in the blood. As blood continues along the capillary, the blood pressure decreases as a result of resistance and loss of fluid volume. At the venous end of the capillary, the tendency for fluid to exit because of blood pressure is overpowered by the tendency for fluid to enter because of osmotic pressure.

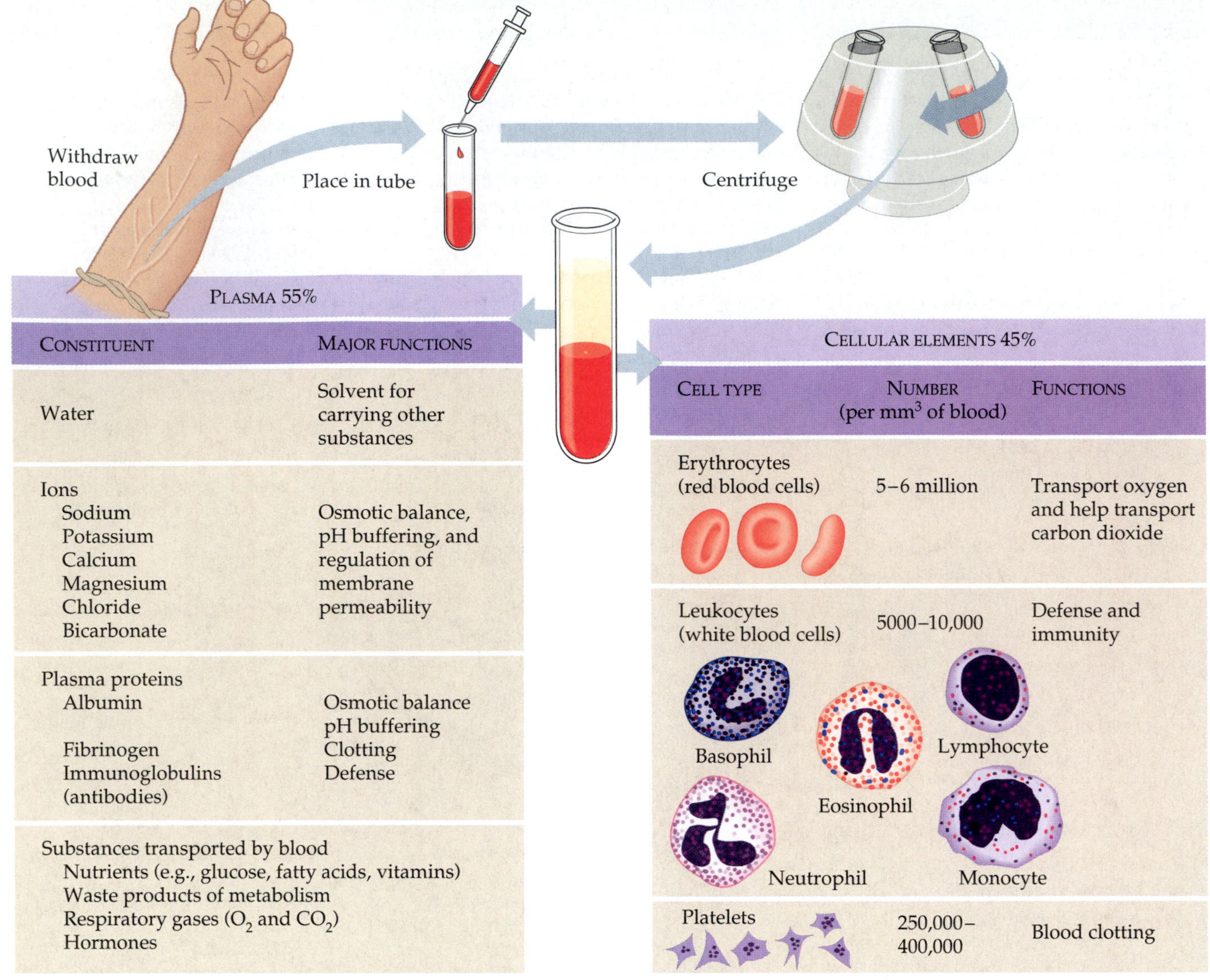

Plasma 55%	
Constituent	**Major functions**
Water	Solvent for carrying other substances
Ions Sodium Potassium Calcium Magnesium Chloride Bicarbonate	Osmotic balance, pH buffering, and regulation of membrane permeability
Plasma proteins Albumin Fibrinogen Immunoglobulins (antibodies)	 Osmotic balance pH buffering Clotting Defense
Substances transported by blood Nutrients (e.g., glucose, fatty acids, vitamins) Waste products of metabolism Respiratory gases (O_2 and CO_2) Hormones	

Cellular elements 45%		
Cell type	**Number (per mm^3 of blood)**	**Functions**
Erythrocytes (red blood cells)	5–6 million	Transport oxygen and help transport carbon dioxide
Leukocytes (white blood cells)	5000–10,000	Defense and immunity
Platelets	250,000–400,000	Blood clotting

FIGURE 42.13 ▪ **The composition of mammalian blood.**

of the blood. You may also recall from Chapter 41 that the lymphatic system transports fats from the digestive tract to the circulatory system.

Blood is a connective tissue with cells suspended in plasma

We now shift our focus from the tubes and pumps of circulatory systems to the fluids being circulated. Vertebrate blood is a type of connective tissue consisting of several kinds of cells suspended in a liquid matrix called **plasma**. The average human body contains about 4 to 6 L of blood. If a blood sample is taken, the cells can be separated from the plasma by spinning the whole blood in a centrifuge. (An anticoagulant must be added to prevent the blood from clotting.) The cellular elements (cells and cell fragments), which occupy about 45% of the volume of blood, settle to the bottom of the centrifuge tube, forming a dense red pellet. Above this cellular pellet is the transparent, straw-colored plasma (FIGURE 42.13).

Plasma

Blood plasma is about 90% water. Among a variety of solutes dissolved in the water are inorganic salts, sometimes referred to as blood electrolytes, present in the plasma in the form of dissolved ions. The combined concentration of these ions is important in maintaining osmotic balance of the blood. Some of the ions also help buffer the blood, which has a pH of 7.4 in humans. And the ability of muscles and nerves to function normally depends on the concentration of key ions in the interstitial fluid, which reflects their concentration in plasma.

The kidney maintains plasma electrolytes at precise concentrations, an example of homeostasis.

Another important class of solutes is the plasma proteins, which have a number of functions. Collectively, they act as buffers against pH changes, help maintain the osmotic balance between the blood and interstitial fluid, and contribute to the blood's viscosity (thickness). The various types of plasma proteins also have specific functions. Some serve as escorts for lipids, which are insoluble in water and can travel in blood only when bound to proteins. Another class of proteins, the immunoglobulins, are the antibodies that help combat viruses and other foreign agents that invade the body (see Chapter 43). And some of the plasma proteins, called fibrinogens, are clotting factors that help plug leaks when blood vessels are injured. Blood plasma that has had these clotting factors removed is called serum.

Plasma also contains various substances in transit from one part of the body to another, including nutrients, metabolic waste products, respiratory gases, and hormones. Blood plasma and interstitial fluid are similar in composition, except that plasma has a much higher protein concentration than interstitial fluid (capillary walls, remember, are not very permeable to proteins).

Cellular Elements

Dispersed throughout blood plasma are two classes of cells: red blood cells, which transport oxygen, and white blood cells, which function in defense. A third cellular element, platelets, are pieces of cells that are involved in blood clotting (see FIGURE 42.13).

Red blood cells, or **erythrocytes**, are by far the most numerous blood cells. Each cubic millimeter of human blood contains 5 to 6 million red cells, and there are about 25 trillion of these cells in the body's 5 L of blood.

The structure of the red blood cell is another excellent example of structure fitting function. A human erythrocyte is a biconcave disk, thinner in the center than at its edges. Mammalian erythrocytes lack nuclei, an unusual characteristic for living cells (the other vertebrate classes have nucleated erythrocytes). Moreover, all red blood cells lack mitochondria and generate their ATP exclusively by anaerobic metabolism. The major function of erythrocytes is to carry oxygen, and they would not be very efficient if their own metabolism were aerobic and consumed some of the oxygen they carry. The small size of erythrocytes (about 12 μm in diameter) also suits their function. For oxygen to be transported, it must diffuse across the plasma membranes of the red blood cells. The smaller the cells, the greater the total area of plasma membrane in a given volume of blood. The biconcave shape of the erythrocyte also adds to its surface area.

As small as a red cell is, it contains about 250 million molecules of **hemoglobin**, an oxygen-carrying protein containing iron (see Chapter 5). Researchers have recently found that hemoglobin binds the gaseous molecule nitric oxide (NO) as well as O_2. As red blood cells pass through the capillary beds of lungs, gills, or other respiratory organs, oxygen diffuses into the erythrocytes and hemoglobin binds O_2 and NO. Hemoglobin unloads its cargo in the capillaries of the systemic circuit. There the O_2 diffuses into body cells. The NO relaxes the walls of the capillaries, allowing them to expand, an effect that probably helps deliver O_2 to the cells.

There are five major types of **white blood cells**, or **leukocytes**: monocytes, neutrophils, basophils, eosinophils, and lymphocytes (see FIGURE 42.13). Their collective function is to fight infections in various ways. For example, monocytes and neutrophils are phagocytes, which engulf and digest bacteria and debris from our own dead cells. As we will see in Chapter 43, lymphocytes become specialized as B cells and T cells, which produce the immune response against foreign substances. White blood cells spend most of their time outside the circulatory system, patrolling through interstitial fluid and the lymphatic system, where most of the battles against pathogens are waged. Normally, a cubic millimeter of human blood has about 5000 to 10,000 leukocytes. The number of these cells increases temporarily whenever the body is fighting an infection.

The third cellular element of blood, **platelets**, are fragments of cells about 2 to 3 μm in diameter. They have no nuclei and originate as pinched-off cytoplasmic fragments of large cells in the bone marrow. Platelets then enter the blood and function in the important process of blood clotting.

Stem Cells and the Replacement of Cellular Elements

The cellular elements of the blood (erythrocytes, leukocytes, and platelets) wear out and are replaced constantly throughout a person's life. Erythrocytes, for example, usually circulate for only about 3 to 4 months and then are destroyed by phagocytic cells located mainly in the liver and spleen. Enzymes digest the old cell's macromolecules, and biosynthetic processes construct new macromolecules using many of the monomers, such as amino acids, obtained from old blood cells, as well as new materials and energy derived from food. Many of the iron atoms derived from the hemoglobin in old red blood cells are built into new hemoglobin molecules.

Erythrocytes, leukocytes, and platelets all develop from a common source, a single population of cells called **pluripotent stem cells** in the red marrow of bones, particularly the ribs, vertebrae, breastbone, and pelvis. ("Pluripotent" means these cells have the potential to differentiate into any type of blood cell or into cells that produce platelets.) Pluripotent

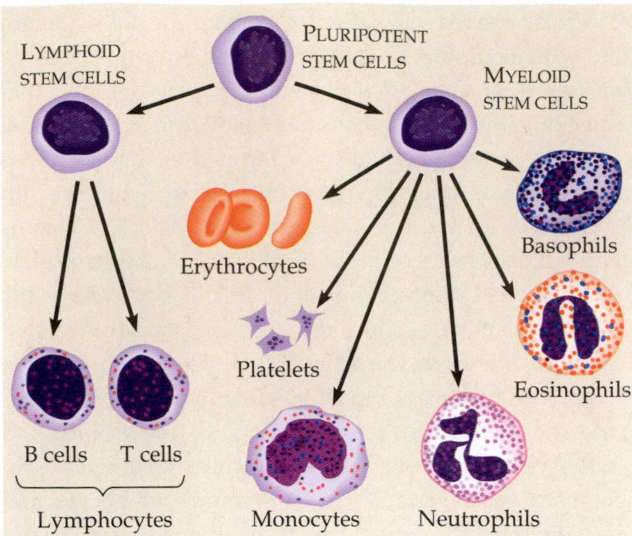

FIGURE 42.14 ▪ Differentiation of blood cells. All blood cells develop from a common source, a population of pluripotent stem cells in red bone marrow. Some of these cells differentiate into lymphoid stem cells, which then develop into B cells and T cells, two types of lymphocytes that function in the immune response (see Chapter 43). All other blood cells differentiate from myeloid stem cells, also derived from the population of pluripotent stem cells.

stem cells arise in the early embryo, and the population then renews itself while replenishing the blood with cellular elements (FIGURE 42.14).

Red blood cell production is controlled by a negative-feedback mechanism that is sensitive to the amount of oxygen reaching the tissues via the blood. If the tissues are not receiving enough oxygen, the kidney converts a plasma protein to a hormone called **erythropoietin**, which stimulates production of erythrocytes in the bone marrow. If blood is delivering more oxygen than the tissues can use, the level of erythropoietin is reduced, and erythrocyte production slows.

Recently, researchers have been able to isolate pluripotent stem cells (by means of monoclonal antibody techniques, discussed in Chapter 43) and grow these cells in laboratory cultures. Purified pluripotent stem cells may soon provide an effective treatment for a number of human diseases, such as leukemia. A person with leukemia has a cancerous line of the stem cells that produce leukocytes. The cancerous stem cells crowd out cells that make red blood cells and produce an unusually high number of leukocytes, many of which are abnormal. One strategy now being used experimentally for treating leukemia is to remove pluripotent stem cells from a patient, destroy the bone marrow, and restock it with noncancerous pluripotent cells. As few as 30 of these cells can completely repopulate the bone marrow.

Blood Clotting

We all get cuts and scrapes from time to time, yet we do not bleed to death because blood contains a self-sealing material that plugs leaks in our vessels. The sealant is always present in our blood in an inactive form called **fibrinogen**. A clot forms only when this plasma protein is converted to its active form, **fibrin**, which aggregates into threads that form the fabric of the clot. The clotting mechanism usually begins with the release of clotting factors from platelets and involves a complex chain of reactions that ultimately transforms fibrinogen to fibrin (FIGURE 42.15). More than a dozen clotting factors have been discovered, and the mechanism is still not fully understood. An inherited defect in any step of the clotting process causes **hemophilia**, a disease characterized by excessive bleeding from even minor cuts and bruises.

Anticlotting factors in the blood normally prevent spontaneous clotting in the absence of injury. Sometimes, however, platelets clump and fibrin coagulates within a blood vessel, blocking the flow of blood. Such a clot is called a **thrombus**. These clots are more likely to form in individuals with **cardiovascular diseases**, diseases of the heart and blood vessels.

Cardiovascular diseases are the leading cause of death in the United States and most other developed nations

More than half of all deaths in the United States are caused by cardiovascular disease; the final blow is usually a heart attack or a stroke. A **heart attack** is the death of cardiac muscle tissue resulting from prolonged blockage of one or more coronary arteries, the vessels that supply oxygen-rich blood to the heart. A **stroke** is the death of nervous tissue in the brain, usually resulting from blockage of arteries in the head. Heart attacks and strokes are frequently associated with a thrombus that clogs an artery. A thrombus that causes a heart attack or stroke may form in a coronary artery or an artery in the brain, or it may develop elsewhere in the circulatory system and reach the heart or brain via the bloodstream. Such a transported clot is called an embolus. The embolus is swept along until it becomes lodged in an artery too small for the clot to pass. Cardiac or brain tissue downstream from the obstruction may die. If the damage in the heart interrupts the conduction of electrical impulses through the cardiac muscle, the heart rate may change drastically or the heart may stop beating altogether. Still, the person may survive if heartbeat is restored by cardiopulmonary resuscitation (CPR) or some other emergency procedure within a few minutes of the attack. The effects of a stroke and the individual's chance of survival depend on the extent and location of the damaged brain tissue.

The suddenness of a heart attack or stroke belies the fact that the arteries of most victims had become gradually impaired by a chronic cardiovascular disease known as **atherosclerosis**. During the course of this disease, growths called plaques develop on the inner walls of the arteries, narrowing their bore (FIGURE 42.16). A plaque forms at a site where the

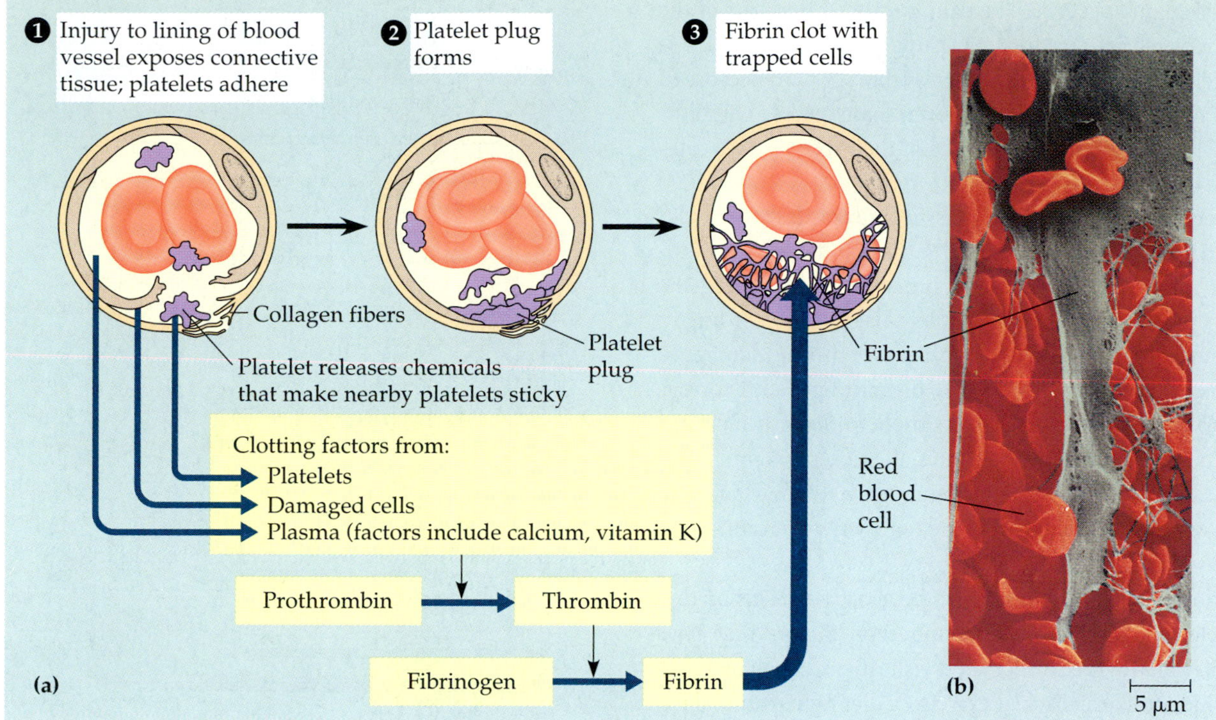

FIGURE 42.15 ▪ **Blood clotting. (a)** ① The clotting process begins when the endothelium of a vessel is damaged and connective tissue in the vessel wall is exposed to blood. Platelets adhere to collagen fibers in the connective tissue and release a substance that makes nearby platelets sticky. ② The platelets form a plug that provides emergency protection against blood loss. ③ This seal is reinforced by a clot of fibrin when vessel damage is more severe. Fibrin is formed via a multistep process: Clotting factors released from the clumped platelets or damaged cells mix with clotting factors in the plasma, forming an activator that converts a plasma protein called prothrombin to its active form, thrombin. Calcium and vitamin K are among the plasma factors required for this step. Thrombin itself is an enzyme that catalyzes the final step of the clotting process, the conversion of fibrinogen to fibrin. Thus, in a cascade of reactions, injury activates prothrombin to thrombin, which then activates fibrinogen to fibrin. The threads of fibrin become interwoven into a patch. **(b)** Red blood cells trapped in a clot of fibrin (colorized SEM).

smooth muscle layer of an artery thickens abnormally and becomes infiltrated with fibrous connective tissue and lipids such as cholesterol. In some cases, the plaques even become hardened by calcium deposits, resulting in a form of atherosclerosis called **arteriosclerosis**, commonly known as hardening of the arteries. An embolus is more likely to become trapped in a vessel that has been narrowed by plaques. Furthermore, plaques are common sites of thrombus formation. Healthy arteries have smooth linings. The rougher lining of an artery affected by atherosclerosis seems to encourage the adhesion of platelets, which triggers the clotting process.

As atherosclerosis progresses, arteries become more and more clogged by plaque, and the threat of heart attack or stroke becomes much greater. Sometimes there are warnings. For example, if a coronary artery is partially blocked by atherosclerosis, a person may feel occasional chest pains, a condition

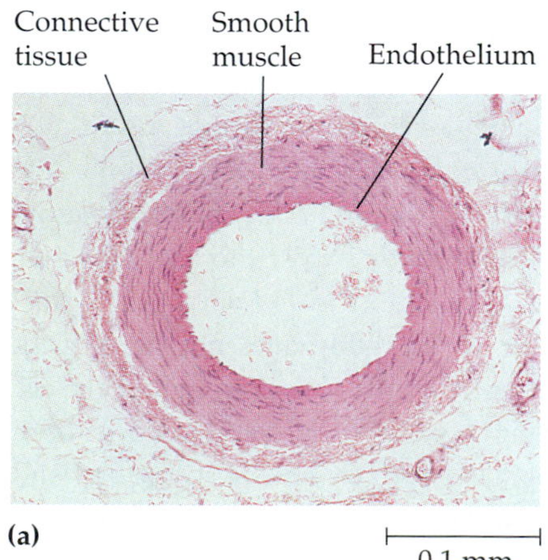

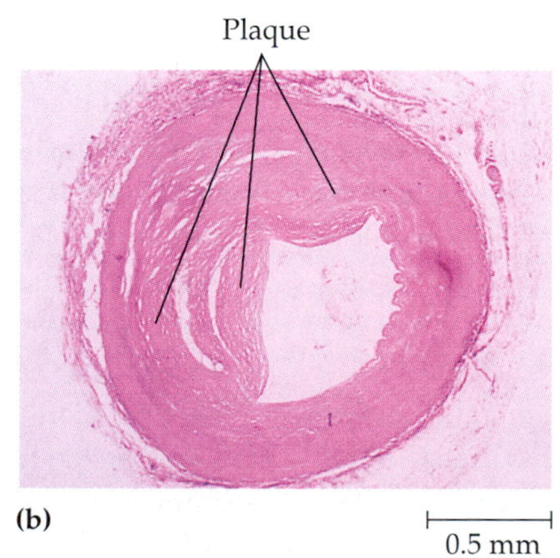

FIGURE 42.16 ▪ **Atherosclerosis.** These light micrographs contrast a cross section of **(a)** a normal artery with **(b)** an artery partially closed by an atherosclerotic plaque. Plaques consist mostly of fibrous connective tissue and smooth muscle cells infiltrated with lipids.

known as angina pectoris. The pain is a signal that part of the heart is not receiving a sufficient supply of oxygen, and it is most likely to occur when the heart is laboring hard because of physical or emotional stress. However, many people with atherosclerosis are completely unaware of their disease until catastrophe strikes.

Hypertension (high blood pressure) promotes atherosclerosis and increases the risk of heart attack and stroke. Atherosclerosis tends to increase blood pressure by narrowing the bore of the vessels and reducing their elasticity. According to one hypothesis, high blood pressure causes chronic damage to the endothelium that lines arteries, promoting plaque formation. Fortunately, hypertension is simple to diagnose and can usually be controlled by medication, diet, exercise, or a combination of these. A diastolic pressure above 90 may be cause for concern, and living with extreme hypertension—say, 200/120—is courting disaster.

To some extent, the tendency to develop hypertension and atherosclerosis is inherited. Noninherited factors that have been correlated with an increased risk of cardiovascular disease include smoking, lack of exercise, a diet rich in animal fat, and an abnormally high concentration of cholesterol in the blood. Cholesterol travels in the blood plasma mainly in the form of particles made up of thousands of cholesterol molecules and other lipids bound to a protein. One form of these particles called **low-density lipoproteins (LDLs)** is associated with the depositing of cholesterol in arterial plaques. Another form, called **high-density lipoproteins (HDLs)**, may reduce cholesterol deposition. Exercise tends to increase HDL concentration, while smoking has the opposite effect on the LDL-to-HDL ratio.

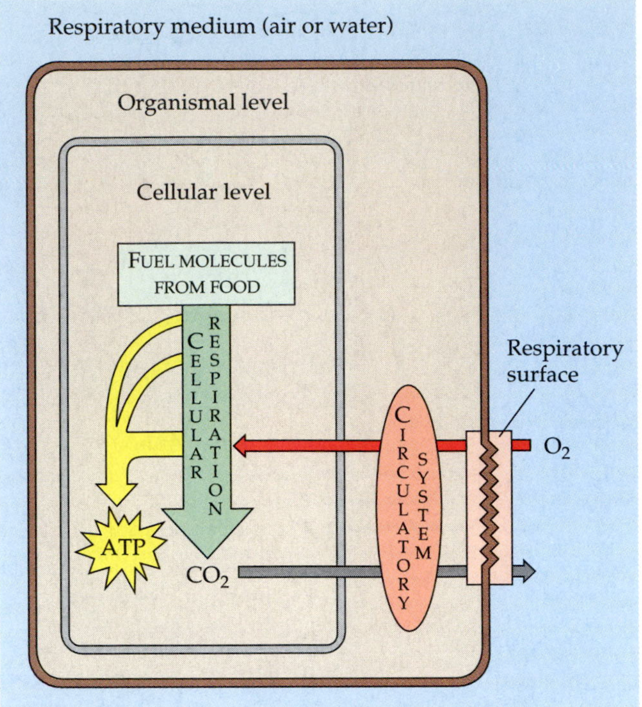

FIGURE 42.17 • The role of gas exchange in bioenergetics. Fueling the body cells requires exchange of gases with the environment: acquiring oxygen (O_2) from the environment and disposing of carbon dioxide (CO_2). In order to make ATP by the process of cellular respiration, an animal's cells must be supplied with O_2 from the environment to break down (oxidize) fuel molecules and release energy from them. Oxygen is the final electron acceptor in the stepwise oxidation of organic fuel molecules (see Chapter 9). Carbon dioxide is the waste product of cellular respiration.

GAS EXCHANGE IN ANIMALS

Gas exchange supplies oxygen for cellular respiration and disposes of carbon dioxide: *an overview*

For the rest of this chapter we focus on **gas exchange** (also called respiration), the uptake of molecular oxygen (O_2) from the environment and the discharge of carbon dioxide (CO_2) to the environment (FIGURE 42.17). Animals require a continuous supply of O_2 for cellular respiration to put the fuel molecules obtained from food to work. They must also dispose of CO_2, the waste product of cellular respiration. As we will see, gas exchange involves both the respiratory system and the circulatory system.

Earth's main reservoir of molecular oxygen is the atmosphere, which is about 21% O_2. Oceans, lakes, and other bodies of water also contain oxygen in the form of dissolved O_2. The source of oxygen, called the **respiratory medium**, is air for terrestrial animals and water for aquatic animals. The part of an animal where oxygen from the environment diffuses into living cells and carbon dioxide diffuses out is called the **respiratory surface**. All living cells must be bathed in water to maintain their plasma membranes. Thus, the respiratory surfaces of terrestrial as well as aquatic animals are moist, and O_2 and CO_2 diffuse across them after first dissolving in water. In addition, an animal's respiratory surfaces must be large enough to provide O_2 and expel CO_2 for the entire body.

A variety of solutions to the problem of providing a large enough respiratory surface have evolved, depending mainly on the size of the organism and whether it lives in water or on land. Gas exchange occurs over the entire surface area of protists and other unicellular organisms. Similarly, for some animals, such as sponges, cnidarians, and flatworms, the plasma membrane of every cell in the body is close enough to the outside environment for gases to diffuse in and out. In many animals, however, the bulk of the body does not have direct access to the respiratory medium, and the respiratory surface is a thin, moist epithelium separating the respiratory medium from the blood or capillaries, which transport gases to and from the rest of the body.

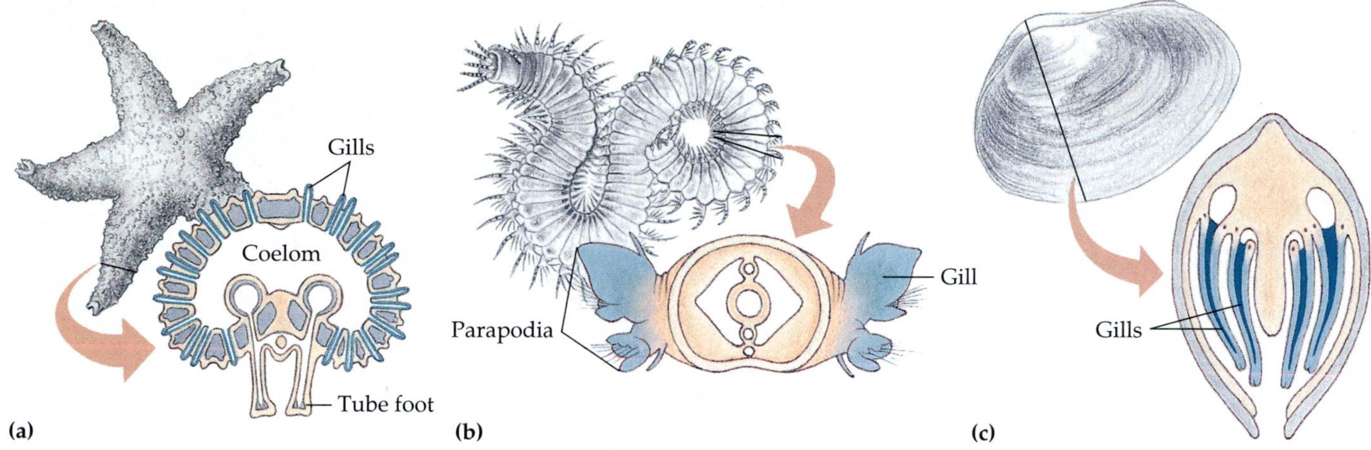

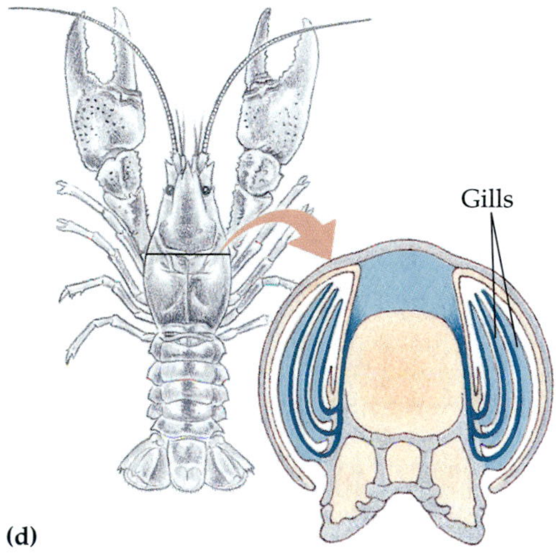

FIGURE 42.18 ▪ Diversity in the structure of gills, external body surfaces functioning in gas exchange. (a) The gills of a sea star are simple tubular projections of the skin. The hollow core of each gill is an extension of the coelom (body cavity). Gas exchange occurs across the moist gill surfaces, and fluid in the coelom circulates in and out of the gills, aiding gas transport. The surfaces of a sea star's tube feet also function in gas exchange. **(b)** Many polychaetes, marine worms of the phylum Annelida, have a pair of flattened appendages called parapodia on each body segment. The parapodia serve as gills and also function in crawling and swimming. **(c)** The gills of a clam are long, flattened plates projecting downward from the main body mass inside the hard shell. Cilia on the gills circulate water around the gill surfaces. **(d)** Crayfish and other crustaceans have long, feathery gills covered by the exoskeleton. Specialized body appendages drive water over the gill surfaces.

Some animals use their entire outer skin as a respiratory organ. An earthworm, for example, has moist skin and exchanges gases by diffusion across its general body surface. Just below the skin is a dense net of capillaries. Because the respiratory surface must be moist, earthworms and many other skin breathers, including some amphibians, must live in water or damp places.

Most animals that use their moist skin as their only respiratory organ are relatively small and are either long and thin or flat in shape, with a high ratio of surface area to volume. For most other animals, the general body surface lacks sufficient area to exchange gases for the whole body. The solution is a region of the body that is extensively folded or branched, thereby enlarging the area of the respiratory surface for gas exchange. Gills, tracheae, and lungs are the three most common respiratory organs.

Gills are respiratory adaptations of most aquatic animals

Gills are variously shaped outfoldings of the body surface specialized for gas exchange. In some invertebrates, such as sea stars, the gills have a simple shape and are distributed over much of the body (FIGURE 42.18a). Many segmented worms have flaplike gills that extend from each segment of the body or long feathery gills clustered at the head or tail (FIGURE 42.18b). The gills of clams, crayfish, and many other animals are restricted to a local body region (FIGURE 42.18c and d). The total surface area of the gills often is much greater than that of the rest of the body.

As a respiratory medium, water has both advantages and disadvantages. There is no problem keeping the cell membranes of the respiratory surface moist, since the gills are completely surrounded by the aqueous environment in which the animal lives. But the O_2 concentration in water is much lower than in air; and the warmer and saltier the water, the less dissolved oxygen it holds. Thus, gills must be very efficient to obtain enough oxygen from water. One process that helps is **ventilation**, increasing the flow of the respiratory medium over the respiratory surface. Crayfish and lobsters have paddle-like appendages that function in ventilation, driving a current of water over the gills. The gills of a bony fish are ventilated continuously by a current of water that enters the mouth, passes through slits in the pharynx, flows over the gills, and

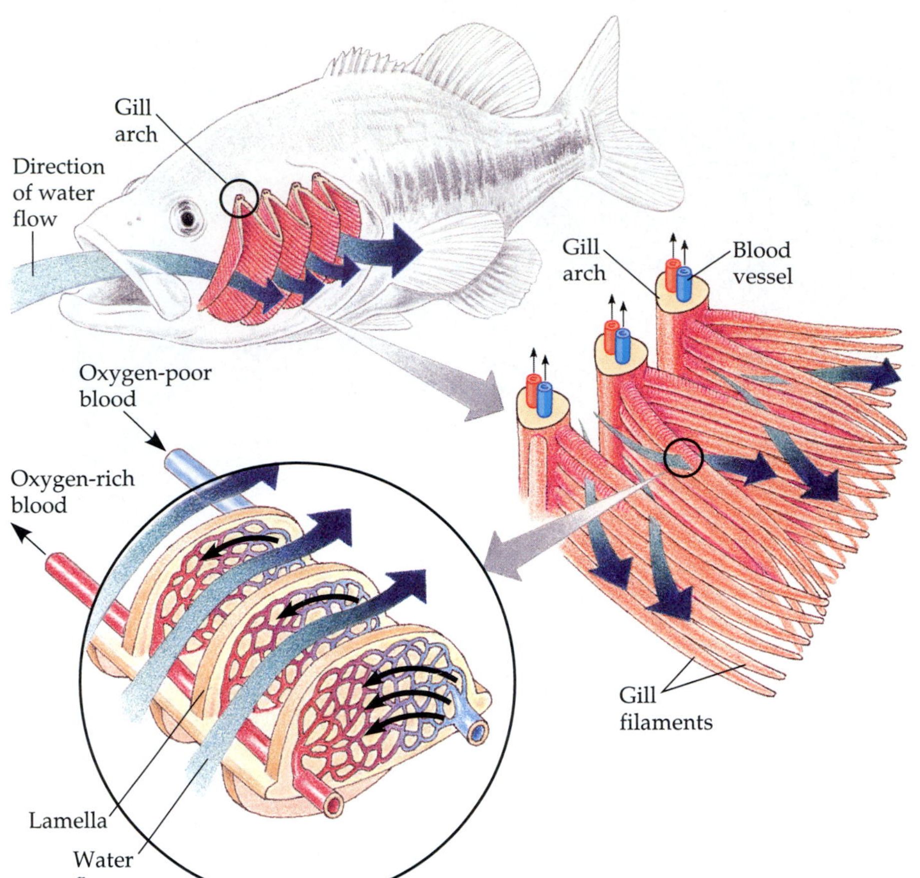

FIGURE 42.19 ▪ The structure and function of fish gills. Fish continuously pump water through the mouth and over the gill arches, using coordinated movements of the jaws and operculum (gill cover) for this ventilation. Each gill arch has two rows of gill filaments, composed of flattened plates called lamellae. Blood flowing through capillaries within the lamellae picks up oxygen from the water. Notice that water flows over the lamellae in a direction opposite to the blood flow, an arrangement called a countercurrent, which enhances the transfer of oxygen from the water to the blood (see FIGURE 42.20).

then out of the body (FIGURE 42.19). Ventilation brings a fresh supply of oxygen and removes carbon dioxide expelled by the gills. Because water is much denser and contains much less oxygen per unit of volume than air, a fish must expend a considerable amount of energy in ventilating its gills.

The arrangement of capillaries in the gills of a fish also enhances gas exchange. Blood flows in the opposite direction to which water passes over the gills. This pattern makes it possible for oxygen to be transferred to the blood by a very efficient process called **countercurrent exchange** (FIGURES 42.19 and 42.20). As blood flows through the capillary, it becomes more and more loaded with oxygen, but it simultaneously encounters water with ever higher oxygen concentrations because the water is just beginning its passage over the gills. This means that along the entire length of the capillary, there is a diffusion gradient favoring the transfer of oxygen from the water to the blood. This countercurrent exchange mechanism is so efficient that the gills can remove more than 80% of the oxygen dissolved in the water passing over the respiratory surface. The mechanism of countercurrent exchange is also important in temperature regulation and several other physiological processes, as we will see in Chapter 44.

Gills are generally unsuitable for an animal living on land, because an expansive surface of wet membrane exposed to air would lose too much water by evaporation, and because the gills would collapse as their fine filaments, no longer supported by water, would cling together. Most terrestrial animals have their respiratory surfaces within the body, opening to the atmosphere only through narrow tubes.

Tracheal systems and lungs are respiratory adaptations of terrestrial animals

As a respiratory medium, air has many advantages, not the least of which is a much higher concentration of oxygen. Also, since O_2 and CO_2 diffuse much faster in air than in water, respiratory surfaces exposed to air do not have to be ventilated as thoroughly as gills. As the respiratory surface removes oxygen from the air and expels carbon dioxide, diffusion rapidly brings more oxygen to the respiratory surface and carries the carbon dioxide away. When a terrestrial animal does ventilate, less energy is expended because air is much easier to move than water. But offsetting these advantages of air as a respiratory medium is a problem: The respiratory surface, which must be large and moist, continuously loses water to the air by evaporation. The problem is solved by a respiratory surface folded into the body.

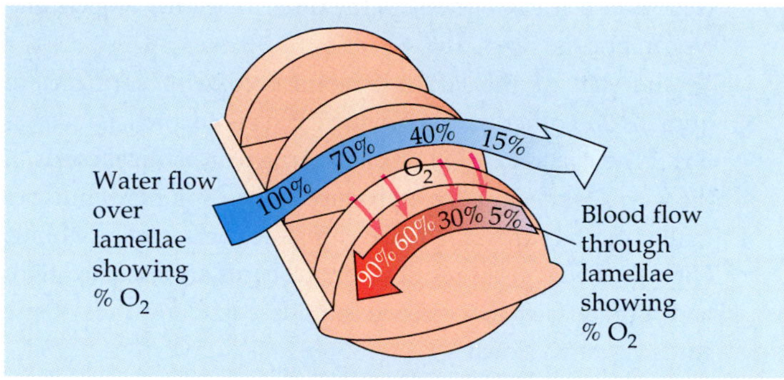

(a) Countercurrent flow (exhibited in fish gills)

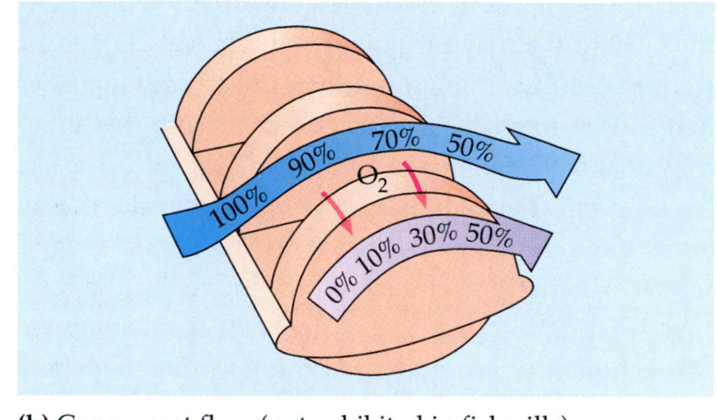

(b) Concurrent flow (not exhibited in fish gills)

FIGURE 42.20 ▪ Countercurrent exchange. The arrangement of blood vessels in fish gills is an adaptation that maximizes oxygen transfer from the water to the blood. **(a)** The direction of blood flow through capillaries in the lamellae of a gill filament is opposite that of water flowing over the lamellae. This countercurrent flow maintains a concentration gradient down which O_2 diffuses from the water into the blood over the whole length of a capillary. As the blood flows through a lamellar capillary, it becomes more and more loaded with oxygen because it continuously passes by water that is more and more concentrated in O_2. Because of countercurrent exchange, a lamella can extract over 80% of the O_2 dissolved in the water passing over it. **(b)** By contrast, if blood flowed through capillaries of the lamellae in the same direction as water flowing over the lamellae (concurrent flow), gills could pick up at most 50% of the oxygen dissolved in the water. As O_2 diffused from the water into the blood, the concentration gradient would become less and less steep and finally cease to exist when the same amount of O_2 was dissolved in both the blood and the water. Fish gills exhibit countercurrent, not concurrent, flow.

Tracheal Systems

The **tracheal system** of insects, made up of air tubes that branch throughout the body, is one variation on the theme of a folded internal respiratory surface. The largest tubes, called tracheae, open to the outside. The finest branches extend to the surface of nearly every cell, where gas is exchanged by diffusion across the moist epithelium that lines the terminal ends of the tracheal system (FIGURE 42.21). With virtually all body cells exposed to the respiratory medium, the open circulatory system of insects is not involved in transporting oxygen and carbon dioxide.

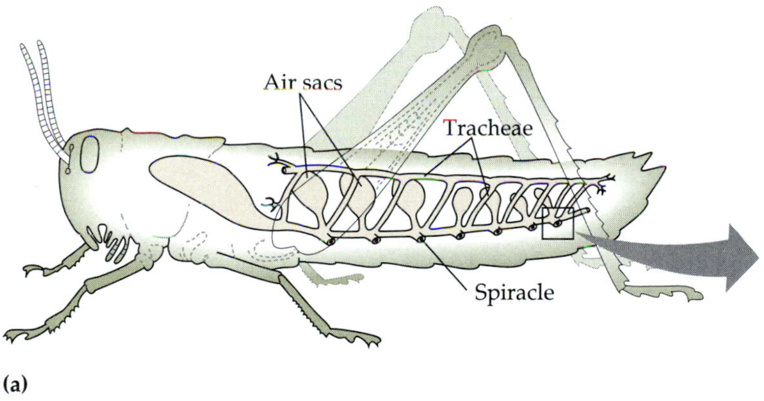

(a)

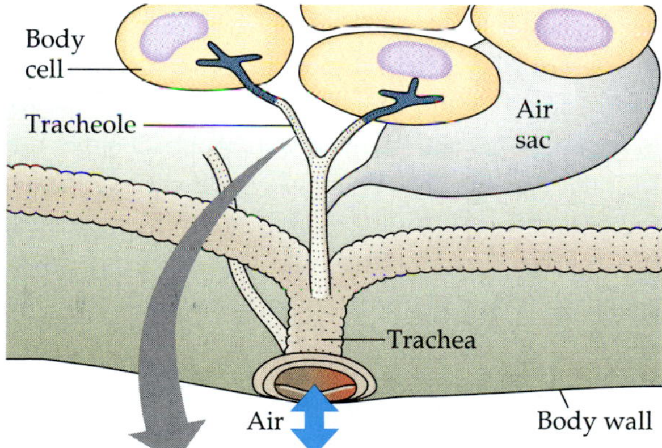

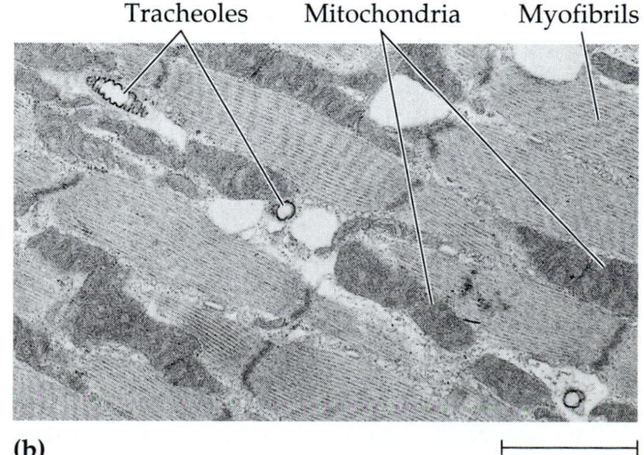

(b)

FIGURE 42.21 ▪ Tracheal systems. (a) The respiratory system of an insect consists of internal tubes that branch repeatedly and deliver air directly to body cells. Rings of chitin reinforce the largest tubes, called tracheae, keeping them from collapsing. Enlarged portions of tracheae form air sacs near organs that require a large supply of oxygen. Air enters the tracheae through openings called spiracles on the insect's body surface and passes into smaller tubes called tracheoles. The tracheoles terminate on the plasma membranes of individual cells. The tiny tips of the tracheoles are closed and contain fluid (dark blue). O_2 diffuses into the body cells from the tracheoles, and CO_2 diffuses from the cells into the tracheoles. When the animal is active and is using more O_2, most of the fluid is withdrawn into the body. This increases the surface area of air in contact with cells. **(b)** This micrograph shows cross sections of tracheoles in a tiny piece of insect flight muscle (TEM). Each of the numerous mitochondria in the muscle cells lies within about 5 μm of one or more tracheoles, and gas exchange occurs by diffusion across this short distance.

For a small insect, diffusion alone brings enough O_2 from the air into the tracheal system and removes enough CO_2 to support cellular respiration. Larger insects with higher energy demands ventilate their tracheal systems with rhythmic body movements that compress and expand the air tubes like bellows. An insect in flight has a high metabolic rate, consuming 10 to 100 times more O_2 than it does at rest. In many flying insects, alternating contraction and relaxation of the flight muscles compress and expand the body, rapidly pumping air through the tracheal system. Also supporting the high metabolic rate, the flight muscle cells are packed with mitochondria, and the tracheal tubes supply each of these ATP-generating organelles with ample oxygen (FIGURE 42.21b). Thus, we see a direct relationship between adaptations of tracheal systems and the theme of bioenergetics.

Lungs

In contrast to the respiratory channels that branch throughout the insect body, **lungs** are restricted to one location. Because the respiratory surface of a lung is not in direct contact with all other parts of the body, the gap must be bridged by the circulatory system, which transports oxygen from the lungs to the rest of the body. Lungs have a dense net of capillaries just under the epithelium, which serves as the respiratory surface. Lungs have evolved in spiders, terrestrial snails, and vertebrates.

Among the vertebrates, amphibians have small lungs that do not provide a large surface (many lack lungs altogether), and they rely heavily on the diffusion of gases across other body surfaces. The skin of frogs, for example, supplements gas exchange in the lungs. In contrast, most reptiles and all birds and mammals rely entirely on their lungs for gas exchange. Turtles are an exception; their rigid shell restricts breathing movements, and they supplement lung breathing with gas exchange across moist epithelial surfaces in their mouth and anus.

Mammalian Respiratory Systems: A Closer Look

Located in the thoracic (chest) cavity, the lungs of mammals have a spongy texture and are honeycombed with a moist epithelium that serves as the respiratory surface. The total surface area of the epithelium (about 100 m^2 in humans) is sufficient to carry out gas exchange for the whole body. A system of branching ducts conveys air to the lungs (FIGURE 42.22). Air enters through the nostrils and is then filtered by hairs, warmed, humidified, and sampled for odors as it flows through a maze of spaces in the nasal cavity. The nasal cavity leads to the pharynx, an intersection where the paths for air and food cross. When food is swallowed, the **larynx** (the upper part of the respiratory tract) moves upward and tips the epiglottis over the glottis (the opening of the windpipe). This allows food to go down the esophagus to the stomach (see FIGURE 41.12). The rest of the time, the glottis is open, and we can breathe.

The wall of the larynx is reinforced with cartilage. In humans and other mammals, the larynx is adapted as a voicebox. When air is exhaled, it rushes by a pair of **vocal cords** in the larynx, and sounds are produced when voluntary muscles in the voicebox are tensed, stretching the cords so they vibrate. High-pitched sounds result when the cords are stretched tight and vibrate fast; low-pitched sounds come from less tense cords vibrating slowly.

From the larynx, air passes into the **trachea**, or windpipe. Rings of cartilage (actually shaped like the letter C) maintain the shape of the trachea. The trachea forks into two **bronchi** (singular, **bronchus**), one leading to each lung. Within the lung, the bronchus branches repeatedly into finer and finer tubes called **bronchioles**. The entire system of air ducts has the appearance of an inverted tree, the trunk being the trachea. The epithelium lining the major branches of this respiratory tree is covered by cilia and a thin film of mucus. The mucus traps dust, pollen, and other particulate contaminants, and the beating cilia move the mucus upward to the pharynx, where it can be swallowed into the esophagus. This process helps cleanse the respiratory system.

At their tips, the tiniest bronchioles dead-end as a cluster of air sacs called **alveoli** (singular, **alveolus**; FIGURE 42.22b). The thin epithelium of the millions of alveoli in the lung serves as the respiratory surface. Oxygen in the air conveyed to the alveoli by the respiratory tree dissolves in the moist film and diffuses across the epithelium and into a web of capillaries that surrounds each alveolus. Carbon dioxide diffuses from the capillaries, across the epithelium of the alveolus, and into the air space (FIGURE 42.22c).

Ventilating the Lungs

Vertebrates ventilate their lungs by **breathing**, the alternate inhalation and exhalation of air. Ventilation maintains a maximal oxygen concentration and minimal carbon dioxide concentration within the alveoli.

A frog ventilates its lungs by **positive pressure breathing**, actually pushing air down its windpipe. Muscles lower the floor of the oral cavity, enlarging it, and air is drawn in through the nostrils. Then, with the nostrils and mouth closed, the floor of the oral cavity rises, and air is forced down the trachea. Elastic recoil of the lungs, together with their compression by the muscular body wall, forces air back out of the lungs during exhalation.

In contrast, mammals ventilate their lungs by **negative pressure breathing**, which works like a suction pump, pulling air, instead of pushing it, down into the lungs.

Negative pressure breathing results from changes in the volume of the lungs instead of the oral cavity. Muscle action changes the volume of the rib cage and the chest cavity, and

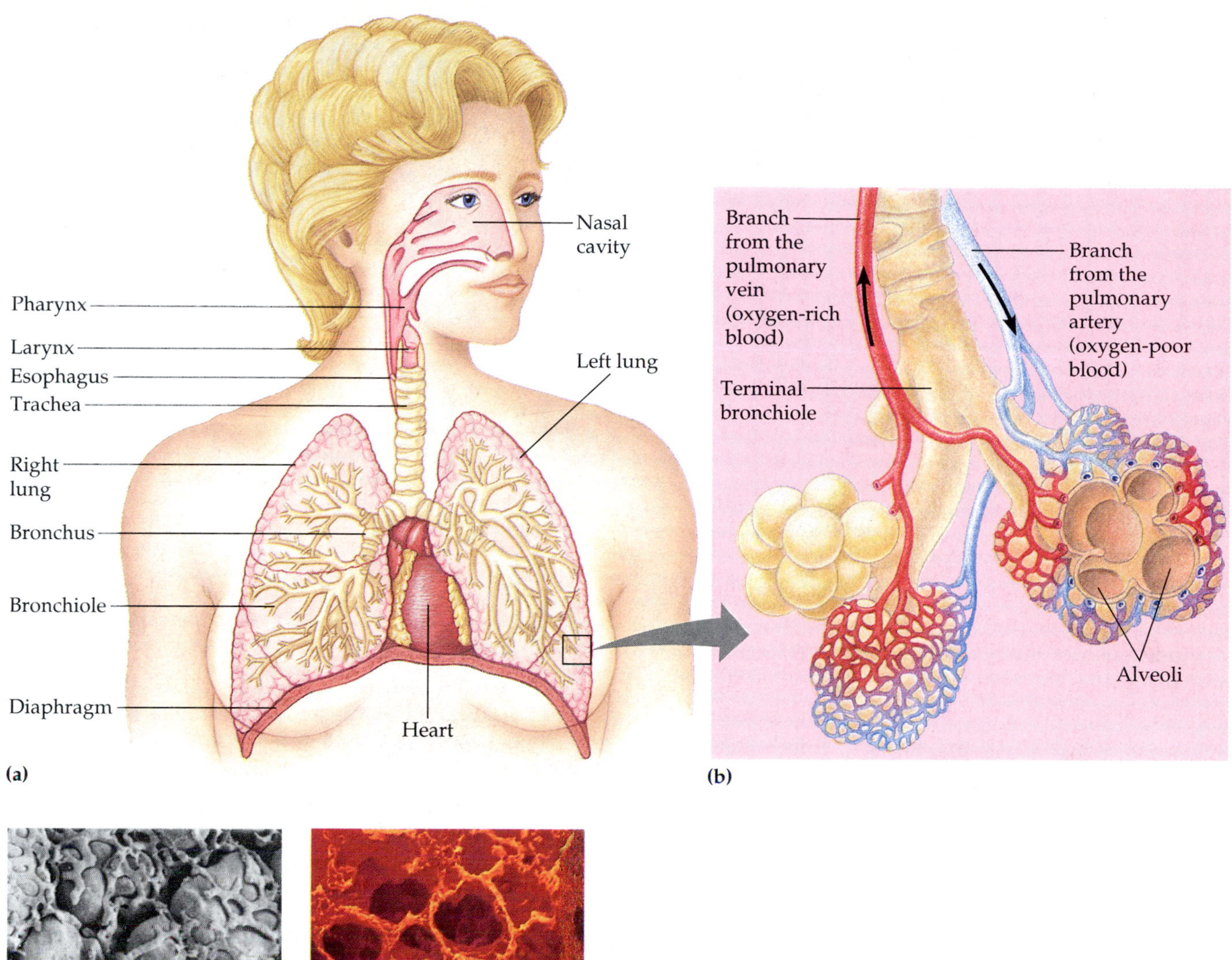

FIGURE 42.22 ▪ The mammalian respiratory system. (a) From the nasal cavity and pharynx, inhaled air passes through the larynx and down the trachea and bronchi to the tiniest bronchioles, which end in lobed, microscopic air sacs, the alveoli. **(b)** A thin, moist epithelium lining the inner surfaces of the alveoli forms the respiratory surface. Branches of the pulmonary artery (see FIGURE 42.4) convey oxygen-poor blood to the alveoli, while branches of the pulmonary vein transport oxygen-rich blood from the alveoli back to the heart. **(c)** The left micrograph (SEM) shows the dense capillary bed that envelops the alveoli. The right micrograph is a cutaway view of alveoli, showing the air spaces.

the lungs follow suit. This can occur because the lungs are enclosed by a double-walled sac. The inner layer of the sac adheres to the outside of the lungs, and the outer layer adheres to the wall of the chest cavity. A thin space filled with fluid separates the two layers. Because of surface tension, the two layers behave like two plates of glass stuck together by a film of water. The layers can slide smoothly past each other, but they cannot be pulled apart easily. Surface tension couples movement of the lungs to movement of the rib cage.

The volume of the lungs increases as a result of contraction of the rib muscles and the **diaphragm**, a sheet of skeletal muscle that forms the bottom wall of the chest cavity. Contraction of the rib muscles expands the rib cage by pulling the ribs upward and the breastbone outward. At the same time, the chest cavity expands as the diaphragm contracts and descends like a piston. All these changes increase the lung volume, and as a result, the air pressure within the alveoli becomes lower than atmospheric pressure. Because air always flows from a region of higher pressure to a region of lower pressure, air rushes through the nostrils and down the breathing tubes to the alveoli. During exhalation, the rib muscles and diaphragm relax, the lung volume is reduced, and the increase in air

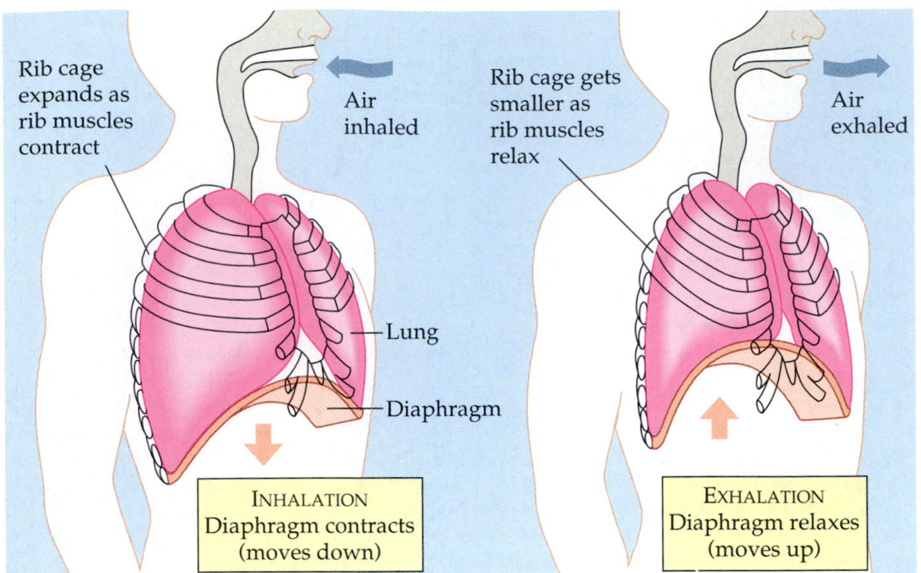

FIGURE 42.23 ▪ Negative pressure breathing. A mammal breathes by changing the air pressure within its lungs relative to the pressure of the outside atmosphere. During inhalation, the rib muscles and diaphragm contract. The volume of the thoracic cavity and lungs increases as the diaphragm moves down and the rib cage expands. Air pressure in the lungs falls below that of the atmosphere, and air rushes into the lungs. Exhalation occurs when the rib muscles and diaphragm relax, restoring the thoracic cavity to its smaller volume.

pressure within the alveoli forces air up the breathing tubes and out through the nostrils (FIGURE 42.23).

Contraction of the rib muscles and diaphragm accounts for the increase in lung volume during shallow breathing, when a mammal is at rest. During vigorous exercise, other muscles of the neck, back, and chest further increase the lung volume by raising the rib cage even more.

The volume of air an animal inhales and exhales with each normal breath is called **tidal volume**. It averages about 500 mL in humans. The maximum volume of air that can be inhaled and exhaled during forced breathing is called **vital capacity**, which averages about 3400 mL and 4800 mL for college-age females and males, respectively. Among other factors, vital capacity depends on the resilience of the lungs. The lungs actually hold more air than the vital capacity, but since it is impossible to completely collapse the alveoli, a **residual volume** of air remains in the lungs even after we forcefully blow out as much air as we can. As lungs lose their resilience as a result of aging or disease (such as emphysema), residual volume increases at the expense of vital capacity.

Ventilation is much more complex in birds than in mammals. Besides lungs, birds have eight or nine air sacs that penetrate the abdomen, neck, and even the wings. The air sacs do not function directly in gas exchange, but act as bellows that keep air flowing through the lungs (FIGURE 42.24). The air sacs also trim the density of the bird, an important adaptation for flight (see Chapter 34). The entire system—lungs and air sacs—is ventilated when the bird inhales and exhales. Air flows through the interconnected system in a circuit that passes through the lungs in one direction only, regardless of whether the bird is inhaling or exhaling. Alveoli, which are dead ends, would not be suitable in such a system. Instead, the lungs of a bird have tiny channels called **parabronchi**, through which air flows continuously in one direction.

Control centers in the brain regulate the rate and depth of breathing

We can hold our breath voluntarily a short while or consciously breathe faster and deeper, but most of the time, auto-

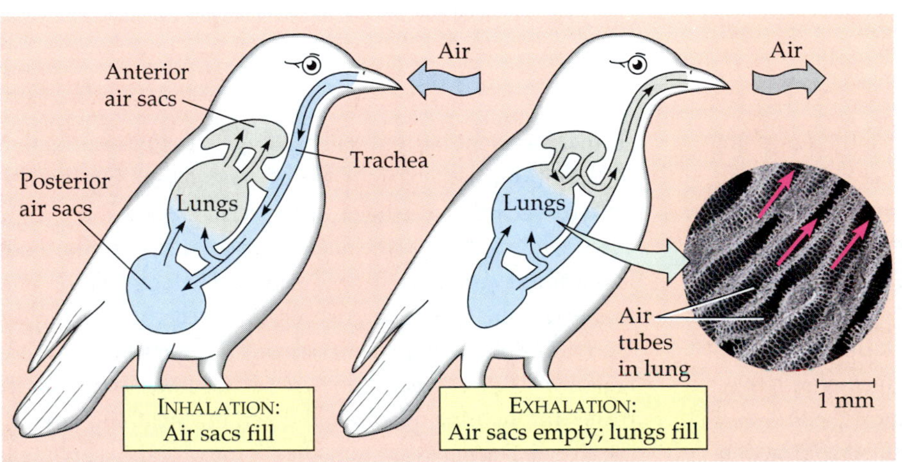

FIGURE 42.24 ▪ The avian respiratory system. A bird has air sacs in addition to lungs. Contraction and relaxation of the air sacs ventilates the lungs, forcing air in one direction (magenta arrows) through tiny parallel tubes in the lungs called parabronchi (inset, SEM). Gas exchange occurs across the walls of the parabronchi. During inhalation, both sets of air sacs expand. The posterior sacs fill with fresh air (blue) from the outside, while the anterior sacs fill with stale air (gray) from the lungs. During exhalation, both sets of air sacs deflate, forcing air from the posterior sacs into the lungs, and air from the anterior sacs out of the system via the trachea. Two cycles of inhalation and exhalation are required for the air to pass all the way through the system and out of the bird.

matic mechanisms regulate our breathing. Automatic control ensures that the work of the respiratory system is coordinated with that of the cardiovascular system.

Our **breathing control centers** are located in two regions of the brain, the medulla oblongata and the pons (FIGURE 42.25). Aided by the control center in the pons, the medulla's center sets the basic breathing rhythm. When we take deep breaths, a negative-feedback mechanism prevents our lungs from overexpanding; stretch sensors in the lung tissue send nerve impulses back to the medulla, inhibiting its breathing control center.

The medulla's control center also helps maintain homeostasis by monitoring the CO_2 level of the blood and regulating the amount of CO_2 our alveoli dispose of when we exhale. Its main cues about CO_2 concentration come from slight changes in the pH of the blood and tissue fluid (cerebrospinal fluid) bathing the brain. Carbon dioxide reacts with water to form carbonic acid, which lowers the pH. When the medulla's control center registers a slight drop in pH (increase in CO_2) of the cerebrospinal fluid or the blood, it increases the depth and rate of breathing, and the excess CO_2 is eliminated in exhaled air. This occurs when we exercise.

The concentration of O_2 in the blood usually has little effect on the breathing control centers. However, when the O_2 level is severely depressed—at very high altitudes, for instance—O_2 sensors in the aorta and carotid arteries in the neck send alarm signals to the breathing control centers, and the centers respond by increasing the breathing rate. A rise in carbon dioxide concentration is usually a good indication of a drop in oxygen concentration, because CO_2 is produced by the same process that consumes O_2—cellular respiration. It is possible, however, to trick the breathing center by hyperventilating. Excessively deep, rapid breathing purges the blood of so much CO_2 that the breathing center temporarily ceases to send impulses to the rib muscles and diaphragm. Breathing stops until the CO_2 level increases enough to switch the breathing center back on.

The breathing center, then, responds to a variety of nervous and chemical signals, adjusting the rate and depth of breathing to meet the changing demands of the body. Control of breathing is only effective, however, if it is coordinated with control of the circulatory system. During exercise, for instance, cardiac output is matched to the increased breathing rate, which enhances O_2 supply and CO_2 removal as blood flows through the lungs.

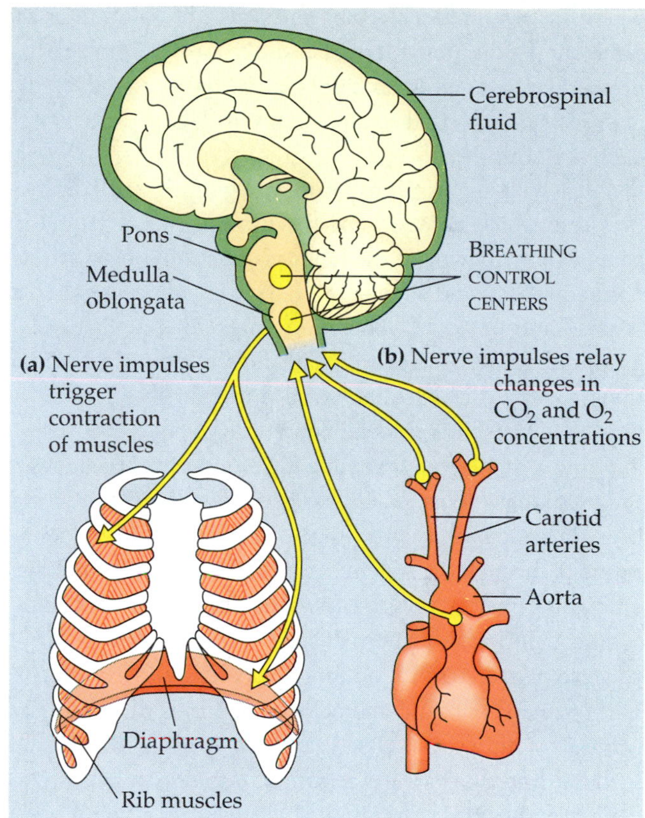

FIGURE 42.25 ▪ Automatic control of breathing. (a) Nerves from a breathing control center in the medulla oblongata of the brain send impulses to the diaphragm and rib muscles, stimulating them to contract and making us inhale. When we are at rest, these nerves send out impulses that result in about 10 to 14 inhalations per minute. Between inhalations, the muscles relax and we exhale. The control center in the medulla sets the basic rhythm, and a control center in the pons moderates it, smoothing out the transitions between inhalations and exhalations. The medulla's control center also helps regulate the CO_2 level of the blood. Sensors in the medulla itself detect changes in the pH of the blood and cerebrospinal fluid bathing the surface of the brain. Changes in pH reflect changes in CO_2 concentration because CO_2 forms carbonic acid in blood; thus, if CO_2 concentration increases, pH decreases (becomes more acidic). **(b)** Other sensors in the walls of the aorta and carotid arteries in the neck detect changes in blood pH and send nerve impulses to the medulla. In response, the medulla's breathing control center alters the rate and depth of breathing, increasing both to dispose of excess CO_2 or decreasing both if CO_2 levels are depressed. The sensors in the aorta and carotid arteries also detect changes in O_2 levels in the blood and signal the medulla to increase the breathing rate when levels become very low.

Gases diffuse down pressure gradients in the lungs and other organs

For a gas, whether present in air or dissolved in water, diffusion depends on differences in a quantity called **partial pressure.** At sea level, the atmosphere exerts a total pressure of 760 mm Hg. This is a downward force equal to that exerted by a column of mercury 760 mm high. Since the atmosphere is 21% oxygen (by volume), the partial pressure of oxygen (abbreviated P_{O_2}) is 0.21 × 760, or about 160 mm Hg. (This is the portion of atmospheric pressure contributed by oxygen; hence the term *partial pressure.*) The partial pressure of carbon dioxide (P_{CO_2}) at sea level is only 0.23 mm Hg. When water is exposed to air, the amount of any gas that dissolves in the water is proportional to its partial pressure in the air and its solubility in water. An equilibrium is eventually reached when the gas molecules enter and leave the solution at the same rate. At this point, the gas is said to have the same partial pressure in the solution as it does in the air. Thus, the P_{O_2} in a glass of water exposed to air is 160 mm Hg, and the P_{CO_2} is 0.23 mm Hg.

A gas will always diffuse from a region of higher partial pressure, which accounts for the movement of respiratory gases. Blood arriving at the lungs via the pulmonary arteries has a lower P_{O_2} and a higher P_{CO_2} than the air in the alveolar spaces (FIGURE 42.26). As blood enters the capillary beds around the alveoli, carbon dioxide diffuses from the blood to the air within the alveoli. Oxygen in the air dissolves in the fluid that coats the epithelium and diffuses across the surface and into a capillary. By the time the blood leaves the lungs in the pulmonary veins, its P_{O_2} has been raised and its P_{CO_2} has been lowered. After returning to the heart, this blood is pumped through the systemic circuit. In the tissue capillaries, gradients of partial pressure favor the diffusion of oxygen out of the blood and carbon dioxide into the blood. This is because cellular respiration rapidly depletes the oxygen content of interstitial fluid and adds carbon dioxide to the fluid (again, by diffusion). After the blood unloads oxygen and loads carbon dioxide, it is returned to the heart by systemic veins. The blood is then pumped to the lungs again, where it exchanges gases with air in the alveoli.

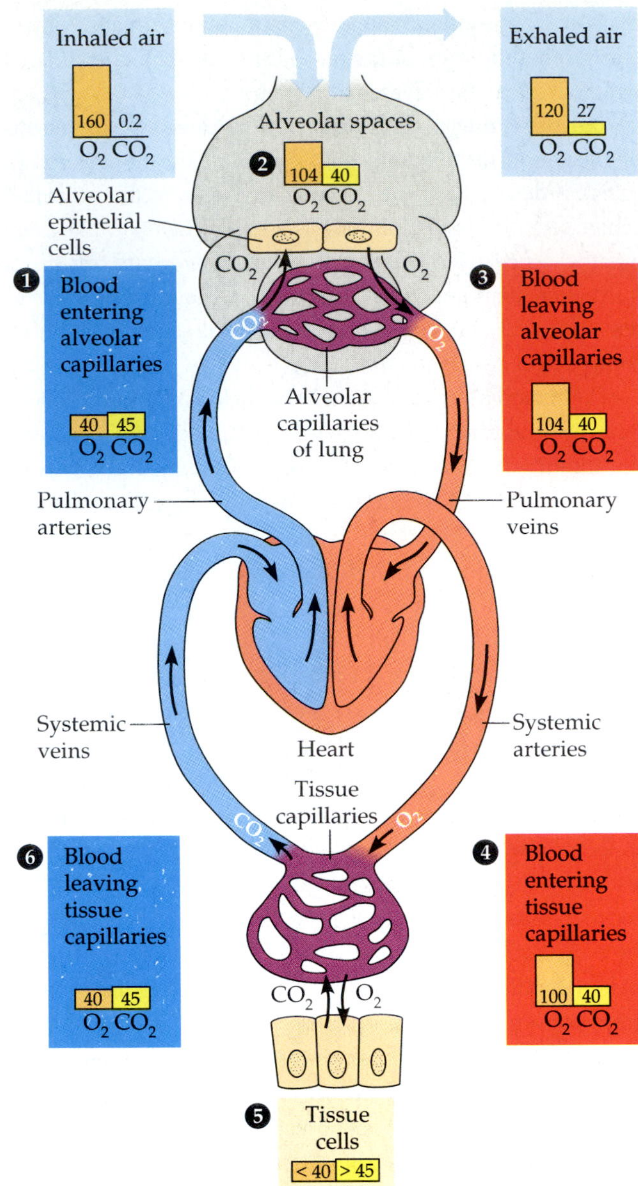

FIGURE 42.26 • Loading and unloading of respiratory gases. The colored bars indicate the partial pressures of O_2 (P_{O_2}) and CO_2 (P_{CO_2}) in mm Hg. The differences between the partial pressure of gases in the alveolar spaces and that in inhaled and exhaled air result from gas exchange in the lungs and from the mixing of alveolar gas that occurs during each breath. For instance, the P_{O_2} in the alveolar spaces is lower than that in the inhaled air (atmosphere) because O_2 diffuses into the alveolar cells and into the blood, and because the inhaled air mixes with the gases that remain in the breathing tubes and alveolar spaces between breaths. ① The blood returning from the body and entering the alveolar capillaries has a lower P_{O_2} than ② the P_{O_2} of the air in the alveolar spaces. Since gases diffuse from a region of higher partial pressure to a region of lower partial pressure, O_2 moves from the alveolar spaces to the blood in the alveolar capillaries of the lung. Conversely, P_{CO_2} is greater in the blood than in the alveolar spaces, so CO_2 moves from the blood to the lungs. ③ The blood leaving the lungs has the same P_{O_2} and P_{CO_2} as the air in the alveolar spaces. ④ When the blood reaches the tissue capillaries, its P_{O_2} is greater than ⑤ the P_{O_2} of the interstitial fluid surrounding the tissue cells, so O_2 moves from the blood to the interstitial fluid, and from there to the tissue cells. Following its pressure gradient, CO_2 moves from the interstitial fluid to the blood. ⑥ After unloading O_2 and loading CO_2, blood leaves the tissue capillaries and returns to the lungs.

Respiratory pigments transport gases and help buffer the blood

Oxygen Transport

Because oxygen is not very soluble in water, very little is transported in blood in the form of dissolved O_2. In most animals, oxygen is carried by **respiratory pigments** in the blood. These pigments are proteins that owe their color to metal atoms built into the molecules. Several oxygen-carrying proteins are found in the blood of various invertebrates. One of these, called **hemocyanin**, has copper as its oxygen-binding component, coloring the blood blueish. Common in arthropods and many mollusks, hemocyanin is dissolved in hemolymph rather than being confined to cells.

The respiratory pigment of almost all vertebrates is hemoglobin, contained in red blood cells. Hemoglobin consists of four subunits, each with a cofactor called a heme group that has an iron atom at its center. It is the iron that actually binds to oxygen; thus, each hemoglobin molecule can carry four molecules of O_2 (see FIGURE 5.23b). To function in O_2 transport, hemoglobin must bind the gas reversibly, loading oxygen in the lungs or gills and unloading it in other parts of the body. Loading and unloading depends on cooperation among the four subunits of the hemoglobin molecule. (See Chapter 6

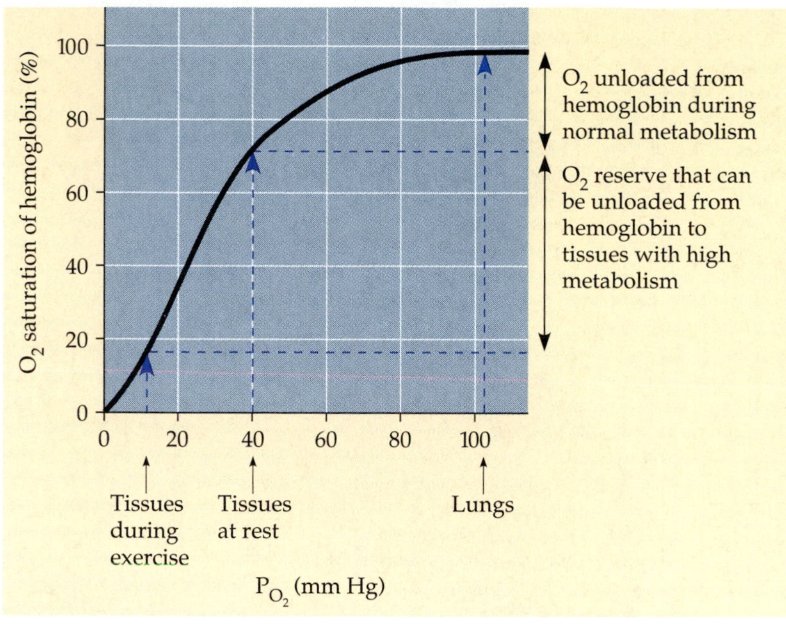

(a)

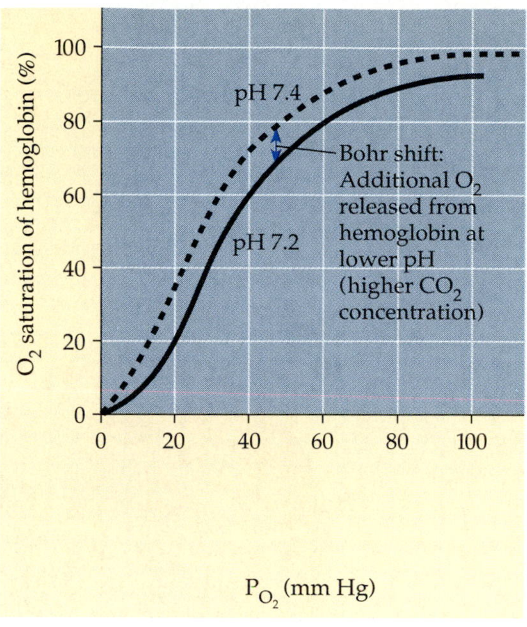

(b)

FIGURE 42.27 ▪ Oxygen dissociation curves for hemoglobin. (a) This is the dissociation curve for hemoglobin at 37°C and pH 7.4. The curve shows the relative amounts of oxygen bound to hemoglobin when the pigment is exposed to solutions varying in their partial pressure of dissolved oxygen. At a P_{O_2} of 104 mm Hg, typical in the lungs, hemoglobin is about 98% saturated with oxygen. At a P_{O_2} of 40 mm Hg, common in the vicinity of tissues, hemoglobin is only about 70% saturated; that is, it gives up about 28% of its oxygen. Hemoglobin can release its reserve of O_2 to tissues that are metabolically very active—muscle tissue during exercise. **(b)** Hydrogen ions affect the conformation of hemoglobin—a drop in pH shifts the oxygen dissociation curve toward the right. Notice that at an equivalent P_{O_2}, say 40 mm Hg, hemoglobin gives up more oxygen at pH 7.2 than it does at pH 7.4, the normal pH of human blood. The pH decreases (becomes more acidic) in very active tissues because the CO_2 produced by respiration reacts with water to form carbonic acid. Hemoglobin then releases more O_2, which supports the higher level of cellular respiration in tissues during exercise.

to review the concept of cooperativity in allosteric proteins.) The binding of oxygen to one subunit induces the remaining subunits to change their shape slightly so that their affinity for oxygen increases. And when one subunit unloads its oxygen, the other three quickly follow suit as a conformational change lowers their affinity for oxygen.

Cooperative oxygen binding and release is evident in the **dissociation curve** for hemoglobin (FIGURE 42.27). Over the range of oxygen partial pressures (P_{O_2}) where the dissociation curve has a steep slope, even a slight change in P_{O_2} causes hemoglobin to load or unload a substantial amount of oxygen. Notice that the steep part of the curve corresponds to the range of oxygen partial pressures found in body tissues. When cells in a particular location begin working harder—during exercise, for instance—P_{O_2} dips in their vicinity as the O_2 is consumed in cellular respiration. Because of the effect of subunit cooperativity, a slight drop in P_{O_2} is enough to cause a relatively large increase in the amount of oxygen the blood unloads.

As with all proteins, hemoglobin's conformation is sensitive to a variety of environmental factors. For example, a drop in pH lowers the affinity of hemoglobin for O_2, an effect called the *Bohr shift* (FIGURE 42.27b). Because CO_2 reacts with water to form carbonic acid, an active tissue will lower the pH of its surroundings and induce hemoglobin to give up more of its oxygen, which can be used for cellular respiration.

Carbon Dioxide Transport

In addition to its role in oxygen transport, hemoglobin also helps the blood transport carbon dioxide and assists in buffering the blood—that is, preventing harmful changes in pH.

FIGURE 42.28, p. 836, illustrates the transport of CO_2 in the blood. Only about 7% of the carbon dioxide released by respiring cells is transported as dissolved CO_2 in blood plasma. Another 23% binds to the multiple amino groups of hemoglobin. Most carbon dioxide, about 70%, is transported in the blood in the form of bicarbonate ions. Carbon dioxide expelled by respiring cells diffuses into the blood plasma and then into the red blood cells, where the CO_2 is converted to bicarbonate. Carbon dioxide first reacts with water to form carbonic acid, which then dissociates into a hydrogen ion and a bicarbonate ion. Most of the hydrogen ions attach to various sites on hemoglobin and other proteins and therefore do not change the pH of blood. The bicarbonate ions diffuse into the plasma. As blood flows through the lungs, the process is reversed. Diffusion of CO_2 out of the blood shifts the chemical

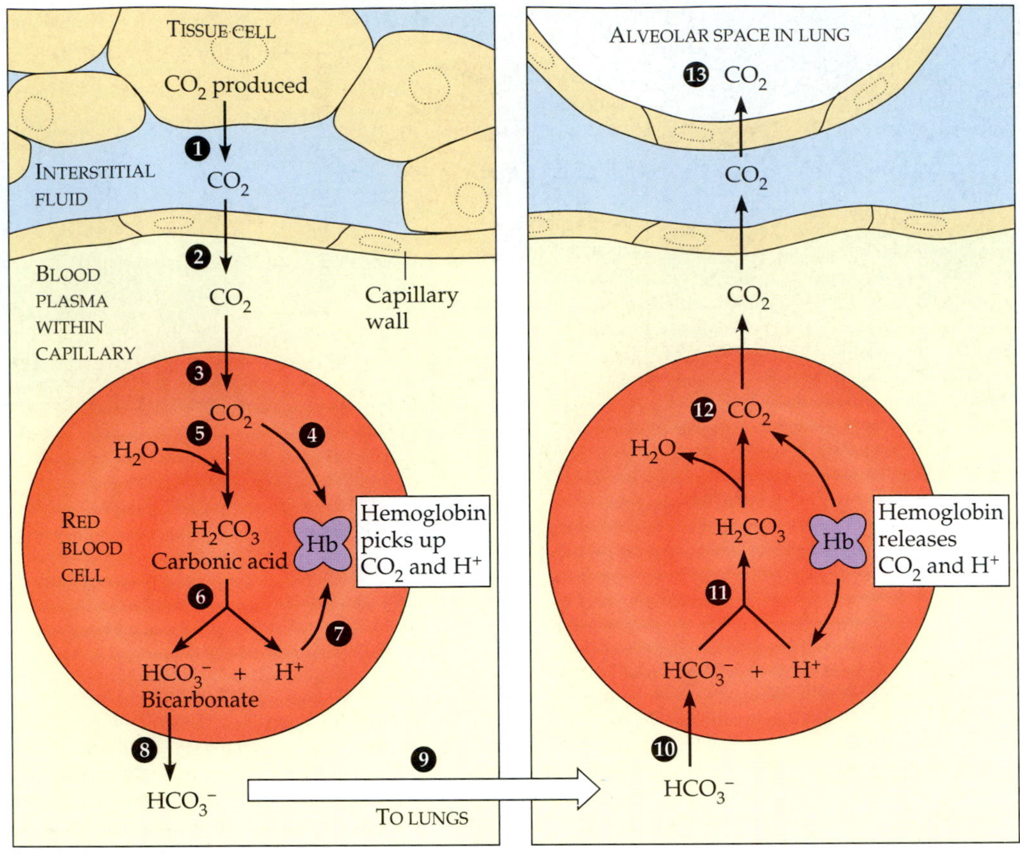

(a) CO_2 transport from tissues

(b) CO_2 transport to lungs

FIGURE 42.28 ▪ Carbon dioxide transport in the blood. (a) ① Carbon dioxide produced by body tissues diffuses into the interstitial fluid and ② into the plasma. Less than 10% remains in the plasma as dissolved CO_2. Over 90% of the CO_2 diffuses into ③ red blood cells. ④ Some is picked up and transported by hemoglobin. ⑤ Most of the CO_2 reacts with H_2O in the erythrocyte to form carbonic acid. (Red blood cells contain the enzyme carbonic anhydrase, which catalyzes this reaction.) ⑥ Carbonic acid dissociates into a bicarbonate ion and a hydrogen ion (H^+). ⑦ Hemoglobin binds most of the hydrogen ions from carbonic acid, preventing them from acidifying the blood. This binding of H^+ causes the Bohr shift shown in FIGURE 42.27b. The reversibility of the carbonic acid–bicarbonate conversion also helps buffer the blood, releasing or removing H^+, depending on pH. ⑧ Most of the bicarbonate ions diffuse into the plasma; ⑨ they are carried in the bloodstream to the lungs. **(b)** The processes that occur in the tissue capillaries are reversed in the lungs. ⑩ Bicarbonate ions diffuse from the plasma into red blood cells. ⑪ H^+, released from hemoglobin, combines with the bicarbonate ions to form carbonic acid. ⑫ CO_2 is formed from carbonic acid and unloaded from hemoglobin. CO_2 diffuses out of the blood, into the interstitial fluid, and into ⑬ the alveolar space, from which it is expelled during exhalation.

equilibrium within red cells in favor of the conversion of bicarbonate to CO_2.

Deep-diving mammals stockpile oxygen and consume it slowly

The Weddell seal shown in FIGURE 42.29 can plunge to depths of 200 to 500 m and remain there for about 20 minutes. It can also occasionally submerge for more than an hour. One of several air-breathing marine mammals adapted to making long underwater dives, the Weddell seal is a predator that eats cod and other deep-sea fishes. What physiological adaptations allow this species and other seals, whales, and dolphins to deep-dive?

One diving adaptation of the Weddell seal is its ability to store oxygen. Compared to humans, the seal contains about twice as much oxygen per kilogram of body weight, mostly in the blood and muscles. About 36% of our total oxygen is in our lungs, and 51% is in our blood. In contrast, the Weddell seal holds only about 5% of its oxygen in its relatively small lungs, stockpiling 70% in the blood. This is possible partly because the seal has about twice the volume of blood per kilogram of body weight as a human. Another adaptation is the seal's huge spleen, which can store about 24 L of blood. The spleen probably contracts after a dive begins, fortifying the blood with red blood cells loaded with oxygen. Diving mammals also have a higher concentration than most other mammals of an oxygen-storing protein called **myoglobin** in their muscles. The Weddell seal can store about 25% of its oxygen in muscle, compared to only 13% in humans.

Diving mammals not only begin an underwater trip with a relatively large O_2 stockpile, they also have adaptations that

FIGURE 42.29 ▪ The Weddell seal, *Leptonychotes weddelli*, a deep-diving mammal. Adaptations of the circulatory and respiratory systems enable this Antarctic seal to travel underwater for more than an hour.

conserve that oxygen. For instance, their heart rate and their O_2 consumption rate decrease. Regulatory mechanisms affecting peripheral resistance route most blood to the brain, spinal cord, eyes, adrenal glands, and placenta (in pregnant seals). Blood supply to the muscles is restricted, and it is shut off altogether during the longest dives. During dives of more than about 20 minutes, the muscles deplete the oxygen stored in their myoglobin and then derive their ATP from anaerobic instead of aerobic respiration (see Chapter 9).

The unusual abilities of the Weddell seal and other diving mammals to power their bodies during long dives showcase a central theme in our study of organisms—the response to environmental pressures over the short term by physiological adjustments and over the long term by natural selection.

▪ ▪ ▪

We have seen throughout this chapter that the circulatory and respiratory systems cooperate extensively. The main function of the respiratory system—to supply O_2 and dispose of CO_2 in support of cellular bioenergetics—depends on the transport work of the circulatory system. Still another activity of the body that depends on the circulatory system is defense, which we will examine in the next chapter.

CHAPTER REVIEW

REVIEW OF KEY CONCEPTS

(with page numbers and key figures)

CIRCULATION IN ANIMALS

- **Transport systems functionally connect the organs of exchange with the body cells: *an overview* (pp. 811–812)** Because most of their cells are too far from the outside environment to be serviced by diffusion or active transport, many animals have an internal transport system that provides a lifeline between the aqueous environment of living cells and the organs, such as the lungs, that exchange chemicals with the outside environment.

- **Most invertebrates have a gastrovascular cavity or a circulatory system for internal transport (p. 812, FIGURES 42.1, 42.2)** Cnidarians and flatworms have gastrovascular cavities that function in circulation as well as digestion. Arthropods and most mollusks have open circulatory systems, in which tissues are bathed directly in hemolymph pumped by a heart into sinuses. Annelids, some mollusks, and vertebrates have closed circulatory systems, with blood confined to vessels, some of which pulsate and function as hearts.

- 42.1 42.2 **Vertebrate phylogeny is reflected in adaptations of the cardiovascular system (pp. 812–814, FIGURES 42.3, 42.4)** In vertebrates, blood flows in a closed cardiovascular system consisting of blood vessels and a two- to four-chambered heart. The heart has one atrium or two atria, which receive blood from veins, and one or two ventricles, which pump blood into arteries. Arteries branch into arterioles, which convey blood to capillaries, the sites of chemical exchange between blood and interstitial fluid. Capillaries rejoin into venules that converge into veins. In fishes, the heart has a single atrium and a single ventricle that pumps blood to gills for oxygenation; the blood then travels to other capillary beds of the body before returning to the heart. Amphibians and most reptiles have a three-chambered heart in which the single ventricle pumps blood via two circuits to gas exchange surfaces and to other parts of the body. The two circuits return blood to separate atria. This double circulation repumps blood returning from the capillary beds of the respiratory organ, thus ensuring a strong flow of blood to the rest of the body. Birds and mammals, both endotherms, have four-chambered hearts that keep oxygen-rich and oxygen-poor blood completely separated.

- **Double circulation in mammals depends on the anatomy and pumping cycle of the heart (pp. 814–817, FIGURES 42.5–42.7)** Heart valves dictate a one-way flow of blood through the heart. The heart rate (pulse) is the number of times the heart beats each minute. The cardiac cycle, one complete sequence of the heart's pumping and filling, consists of periods of contraction, called systole, and periods of relaxation, called diastole. The cardiac output is the volume of blood pumped into the systemic circulation per minute. The intrinsic contraction of the cardiac muscle is coordinated by a conduction system originating in the sinoatrial (SA) node (pacemaker) of the right atrium. The pacemaker initiates a wave of contraction that spreads to both atria, hesitates momentarily at the atrioventricular (AV) node, and then progresses to both ventricles. The pacemaker is influenced by nerves, hormones, body temperature, and exercise.

- **Structural differences of arteries, veins, and capillaries correlate with their different functions (p. 817, FIGURES 42.8, 42.9)** An endothelium lines all blood vessels, including the capillaries. Arteries and veins have two additional outer layers composed of smooth muscle, elastic fibers, and connective tissue. Arteries have the thickest, strongest, and most elastic walls, accommodating high blood pressure and rapid blood flow. Body movements propel blood back to the heart in veins; large veins are equipped with one-way valves.

- 42.3 **Physical laws governing the movement of fluids through pipes affect blood flow and blood pressure (pp. 817–819, FIGURE 42.10)** The velocity of blood flow varies in the circulatory system, being slowest in the capillary beds as a result of the high resistance and large total cross-sectional area of the arterioles and capillaries. This slower flow enhances the exchange of substances between the blood and interstitial fluid. Blood pressure, the hydrostatic force blood exerts against the wall of a vessel, is determined by the cardiac output and peripheral resistance due to variable constriction of the arterioles.

- **Transfer of substances between the blood and the interstitial fluid occurs across the thin walls of capillaries (pp. 819–820, FIGURES 42.11, 42.12)** The steady supply of blood to different organs is determined by variable constriction of arterioles and precapillary sphincters. Substances traverse the endothelium of capillaries in endocytotic-exocytotic vesicles, by diffusion, or dissolved in fluids forced out by blood pressure at the arterial end of the capillary.
- **The lymphatic system returns fluid to the blood and aids in body defense (pp. 820–822)** Fluid reenters the circulation directly at the venous end of the capillary or indirectly through the lymphatic system. White blood cells concentrated in lymph nodes help fight infections.
- **Blood is a connective tissue with cells suspended in plasma (pp. 822–824, FIGURES 42.13–42.15)** Whole blood consists of cellular elements (cells and pieces of cells called platelets) suspended in a liquid matrix called plasma. Plasma is a complex aqueous solution of inorganic electrolytes, proteins, nutrients, metabolic waste products, respiratory gases, and hormones. Plasma proteins influence blood pH, osmotic pressure, and viscosity, and function in lipid transport, immunity (antibodies), and blood clotting (fibrinogens). Red blood cells, or erythrocytes, transport oxygen. Five types of white blood cells, or leukocytes, function in defense by phagocytosing bacteria and debris or by producing antibodies. Pluripotent stem cells in red bone marrow give rise to all types of blood cells. Platelets function in blood clotting, a cascade of complex reactions that converts plasma fibrinogen to fibrin.
- **Cardiovascular diseases are the leading cause of death in the United States and most other developed nations (pp. 824–826)** Cardiovascular disease is a deterioration of the heart and blood vessels. Gradual plaque buildup during atherosclerosis or arteriosclerosis narrows the diameter of blood vessels and may be associated with vessel blockage and consequent heart attack or stroke.

GAS EXCHANGE IN ANIMALS

- **Gas exchange supplies oxygen for cellular respiration and disposes of carbon dioxide: *an overview* (pp. 826–827, FIGURE 42.17)** Cellular respiration demands a constant supply of O_2 and removal of CO_2. Animals require large, moist respiratory surfaces for the adequate diffusion of respiratory gases between their cells and the respiratory medium, either air or water.
- **Gills are respiratory adaptations of most aquatic animals (pp. 827–828, FIGURES 42.18–42.20)** Gills are outfoldings of the body surface specialized for gas exchange. The efficiency of gas exchange in some gills, including those of fishes, is increased by ventilation and countercurrent flow of blood and water.
- **Tracheal systems and lungs are respiratory adaptations of terrestrial animals (pp. 828–832, FIGURES 42.21–42.24)** The tracheae of insects are tiny branching tubes that penetrate the body, bringing O_2 directly to cells. Most terrestrial vertebrates, land snails, and spiders have internal lungs. In mammals, air inhaled through the nostrils passes through the pharynx into the trachea, bronchi, bronchioles, and dead-end alveoli, where gas exchange occurs. Lungs must be ventilated by breathing. Frogs ventilate by positive pressure, pumping air into their lungs. Mammals ventilate with negative pressure by contracting and relaxing rib muscles and the diaphragm, which changes the volume and hence the pressure of the chest cavity and lungs relative to the atmosphere. Birds have one-way ventilation of the lungs, made possible by a system of air sacs and one-way parabronchi in the lungs.
- **Control centers in the brain regulate the rate and depth of breathing (pp. 832–833, FIGURE 42.25)** The control center in the medulla oblongata of the brain sets the basic breathing rhythm. Sensors detect blood pH changes reflecting CO_2 concentrations and O_2 levels in the blood, and the medulla adjusts the rate and depth of breathing to match the metabolic demands of the body. The pons of the brain smooths out the basic rhythm set by the medulla.
- **Gases diffuse down pressure gradients in the lungs and other organs (pp. 833–834, FIGURE 42.26)** O_2 and CO_2 diffuse from where their partial pressures are higher to where they are lower.
- **Respiratory pigments transport gases and help buffer the blood (pp. 834–836, FIGURES 42.27, 42.28)** Respiratory pigments increase the amount of oxygen that blood can carry. Arthropods and many mollusks have copper-containing hemocyanin; vertebrates have hemoglobin. A hemoglobin molecule has four iron-containing subunits, each capable of binding a molecule of oxygen. Most CO_2 generated during metabolism is transported in the form of bicarbonate ions. Hydrogen ions produced when bicarbonate ions form bind to hemoglobin and other proteins, serving to buffer the blood. When blood enters the lungs, CO_2 forms from the bicarbonate and diffuses into the environment.
- **Deep-diving mammals stockpile oxygen and consume it slowly (pp. 836–837, FIGURE 42.29)** Diving mammals can retain an unusually large volume of blood. During a dive, their heart rate slows and blood is shunted away from muscles and noncritical organs, in effect reducing their rate of O_2 consumption.

SELF-QUIZ

1. Which of the following respiratory systems is not closely associated with a blood supply?
   a. vertebrate lungs
   b. fish gills
   c. tracheal systems of insects
   d. the outer skin of an earthworm
   e. the parapodia of a polychaete worm

2. Blood returning to the mammalian heart in a pulmonary vein will drain first into the
   a. vena cava
   b. left atrium
   c. right atrium
   d. left ventricle
   e. right ventricle

3. Pulse is a direct measure of
   a. blood pressure
   b. stroke volume
   c. cardiac output
   d. heart rate
   e. breathing rate

4. When you hold your breath, which of the following blood gas changes first leads to the urge to breath?
   a. rising O_2
   b. falling O_2
   c. rising CO_2
   d. falling CO_2
   e. rising CO_2 and falling O_2

5. In negative pressure breathing, inhalation results from
   a. forcing air from the throat down into the lungs
   b. contracting the diaphragm
   c. relaxing the muscles of the rib cage
   d. using muscles of the lungs to expand the alveoli
   e. contracting the abdominal muscles

6. If you know that an animal lacks a circulatory system, which of the following predictions would you make about the animal?
   a. It has no skeleton.
   b. Its lung capacity is huge.
   c. It has a very thin body wall.
   d. It has hemolymph instead of blood.
   e. It is a marine animal.

7. A decrease in the pH of human blood caused by exercise would
   a. decrease breathing rate
   b. increase heart rate
   c. decrease the amount of O_2 unloaded from hemoglobin
   d. decrease cardiac output
   e. decrease CO_2 binding to hemoglobin

8. Compared to the interstitial fluid that bathes active muscle cells, blood reaching these cells in arteries has a
   a. higher P_{O_2}
   b. higher P_{CO_2}
   c. greater bicarbonate concentration
   d. lower pH
   e. lower osmotic pressure

9. Which of the following reactions prevails in red blood cells traveling through pulmonary capillaries? (Hb = hemoglobin)
   a. $Hb + 4\ O_2 \longrightarrow Hb(O_2)_4$
   b. $Hb(O_2)_4 \longrightarrow Hb + 4\ O_2$
   c. $CO_2 + H_2O \longrightarrow H_2CO_3$
   d. $H_2CO_3 \longrightarrow H^+ + HCO_3^-$
   e. $Hb + 4\ CO_2 \longrightarrow Hb(CO_2)_4$

10. The relationship between blood pressure *(bp)*, cardiac output *(co)*, and peripheral resistance *(pr)* can be expressed as $bp = co \times pr$. All of the following changes would result in an increase in blood pressure *except*
   a. increase in the stroke volume
   b. increase in the heart rate
   c. increase in the duration of ventricular diastole
   d. increase in the tone of the arteriolar smooth muscle
   e. reduction in arteriolar diameter

CHALLENGE QUESTIONS

1. The hemoglobin of a human fetus differs from adult hemoglobin. Compare the dissociation curves of the two hemoglobins in the graph below, and then explain the physiological significance of the difference.

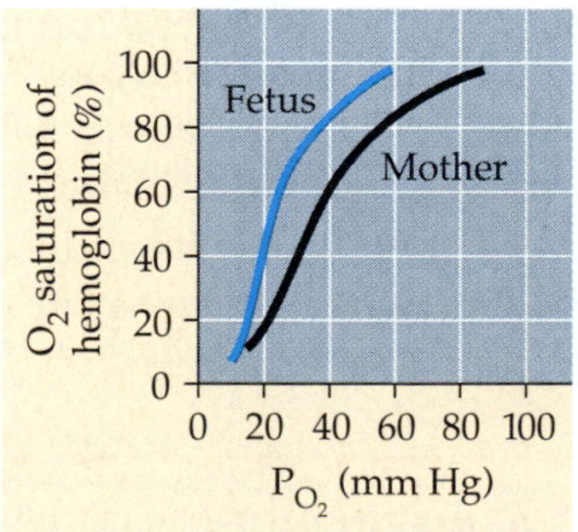

2. Because they support their weight against gravity and repeatedly accelerate limbs from standing starts, terrestrial vertebrates consume more energy in locomotion than do fishes swimming through water. In other words, it takes more calories per gram of animal to move 1 m on land than it does to move 1 m in water (assuming, of course, that the animal is in its natural habitat, either land or water). How does this disparity fit in with the evolution of the vertebrate cardiovascular system?

SCIENCE, TECHNOLOGY, AND SOCIETY

1. The incidence of cardiovascular disease is much lower in the Mediterranean countries (Spain, Italy, Greece, etc.) than in North America. Suggest several alternative hypotheses that might explain this difference. How could you test your hypotheses?

2. Hundreds of studies have linked smoking with cardiovascular and lung disease. According to most health authorities, smoking is the leading cause of preventable, premature death in the United States. Antismoking and health groups have proposed that cigarette advertising be banned entirely. What are some arguments in favor of a total ban on cigarette advertising? What are arguments in opposition? Do you favor or oppose such a ban? Why?

FURTHER READING

Barinaga, M. "How Much Pain for Cardiac Gain?" *Science,* May 30, 1997. A summary of the debate over how much exercise we need.

Buchanan, M. "Fascinating Rhythm." *New Scientist,* January 3, 1998. Discussion of the irregular rhythm of the healthy heart.

Coghlan, A. "Blood-clotting Drug Lays Siege to Tumours." *New Scientist,* October 19, 1996. Cancerous growths may be treated by inducing clots upstream.

Glanz, J. "Hemoglobin Reveals New Role As Blood Pressure Regulator." *Science,* March 22, 1996. Discussion of experiments indicating how hemoglobin modulates blood pressure.

Hademenos, G. J. "The Biophysics of Stroke." *American Scientist,* May–June 1997. A well-illustrated account of blood flow changes in a leading cause of death.

Kooyman, G. L., and P. J. Ponganis. "The Challenges of Diving to Depth." *American Scientist,* November–December 1997. A comparison of the adaptations of deep-diving vertebrates.

Lee, J., and L. Manning. "Environmental Lung Disease." *New Scientist,* September 16, 1995. Describes how airborne pollutants affect our lungs.

Nucci, M. L., and A. Abuchowski. "The Search for Blood Substitutes." *Scientific American,* February 1998. A report on meeting the needs of global blood banks.

Radetsky, P. "The Mother of All Blood Cells." *Discover,* March 1995. Describes the potential application of research on pluripotent stem cells.

Seppa, N. "Secondary Smoke Carries High Price." *Science News,* January 17, 1998. A summary of recent research on the health hazards of tobacco smoke.

WEB LINKS

Visit the special edition of *The Biology Place* for BIOLOGY, Fifth Edition, at http://www.biology.com/campbell. Go to Chapter 42 for online resources, including learning activities, practice exams, and links to the following web sites:

"The Heart: An Online Exploration"
This fascinating site allows you to explore the heart and circulatory system. It is richly illustrated with diagrams and movies.

"The American Heart Association"
Here you will find the most comprehensive information on heart disease and stroke as well as a huge amount of information on keeping your heart healthy.

"The American Lung Association (ALA)"
The mission of the ALA is to prevent lung disease and promote lung health. The site contains a wealth of information on lung diseases.

"The National Heart, Lung, and Blood Institute (NHLBI)"
The NHLBI is part of the National Institutes for Health, and this site provides you with information about their research and educational activities.

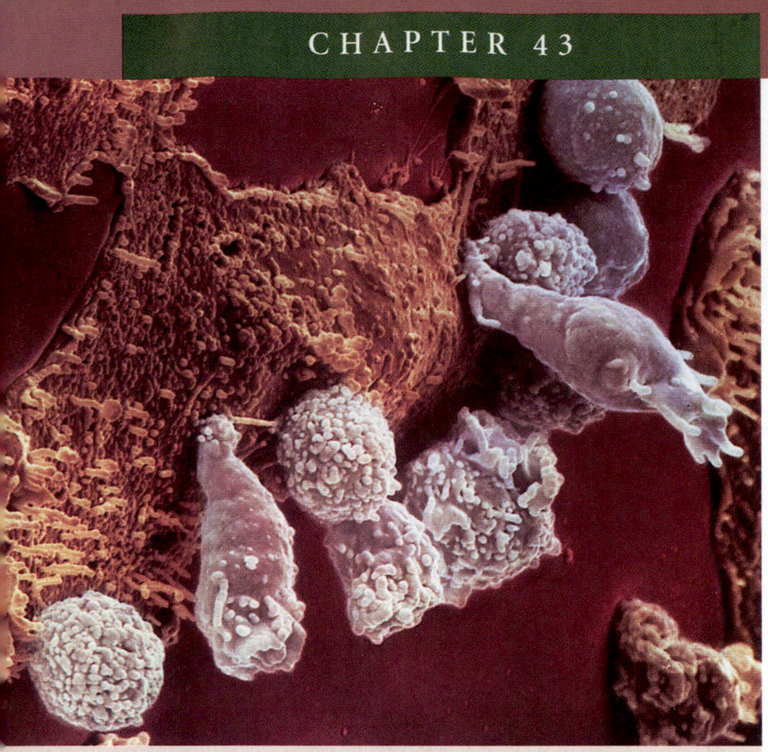

THE BODY'S DEFENSES

An animal must defend itself against unwelcome intruders, the many potentially dangerous viruses, bacteria, and other pathogens it encounters in the air, in food, and in water. It must also deal with abnormal body cells, which, in some cases, may develop into cancer. Three cooperative lines of defense that counter these threats have evolved (FIGURE 43.1). *Two of these are nonspecific—that is, they do not distinguish one infectious agent from another. The first line of nonspecific defense is external, consisting of epithelial tissues that cover and line our bodies (skin and mucous membranes) and the secretions they produce. The second line of nonspecific defense is internal: It is triggered by chemical signals and involves phagocytic cells and antimicrobial proteins that indiscriminately attack invaders that penetrate the body's outer barriers. The appearance of inflammation is a sign that this second line of defense has been deployed.*

The third line of defense is the immune system. The immune system comes into play simultaneously with the second line of defense, but it responds in a specific way to particular microorganisms, aberrant body cells, toxins, and other substances marked by foreign molecules. The body's three lines of defense are somewhat analogous to the defenses of a besieged city: first the city walls, then ordinary soldiers, and finally intelligence officers who identify and track down specific dangerous infiltrators. The immune response, which includes the production of specific defensive proteins called antibodies, involves a diverse group of white blood cells called lymphocytes (see Chapter 42). The photograph on this page (a colorized SEM) shows specialized lymphocytes (light color) attacking a cancer cell (brown). This chapter examines how an animal's nonspecific and specific defenses work together to protect the body. Our main focus is on the defense mechanisms of vertebrates, which have a highly developed immune system.

NONSPECIFIC DEFENSES AGAINST INFECTION

An invading microbe must penetrate the external barrier formed by the skin and mucous membranes, which cover the surface and line the openings of an animal's body (see Chapter 40). If it succeeds in doing so, the pathogen encounters the second line of nonspecific defense, interacting mechanisms that include phagocytosis, the inflammatory response, and antimicrobial proteins.

The skin and mucous membranes provide first-line barriers to infection

Intact skin is a barrier that cannot normally be penetrated by bacteria or viruses, although even minute abrasions may allow their passage. Likewise, the mucous membranes that line the

Nonspecific defense mechanisms		Specific defense mechanisms (immune system)
First line of defense	Second line of defense	Third line of defense
• Skin • Mucous membranes • Secretions of skin and mucous membranes	• Phagocytic white blood cells • Antimicrobial proteins • The inflammatory response	• Lymphocytes • Antibodies

FIGURE 43.1 ▪ An overview of the body's defenses.

digestive, respiratory, and genitourinary tracts bar the entry of potentially harmful microbes. Beyond their role as a physical barrier, the skin and mucous membranes counter pathogens with chemical defenses. In humans, for example, secretions from sebaceous and sweat glands give the skin a pH ranging from 3 to 5, which is acidic enough to prevent colonization by many microbes. (Bacteria that make up the skin's normal flora are adapted to its acidic, relatively dry environment.) Microbial colonization is also inhibited by the washing action of saliva, tears, and mucous secretions that continually bathe the surfaces of exposed epithelia. In addition, all these secretions contain antimicrobial proteins. One of these protective proteins is **lysozyme**, an enzyme that digests the cell walls of many kinds of bacteria and thus destroys many bacteria entering the upper respiratory tract and the openings around the eyes (see FIGURE 5.17).

Mucus, the viscous fluid secreted by cells of mucous membranes, also traps microbes and other particles that contact it. In the trachea, ciliated epithelial cells sweep out mucus with its trapped microbes, preventing them from entering the lungs (FIGURE 43.2). Microbes present in food or water, or those in swallowed mucus, must contend with the highly acidic environment of the stomach. The acid destroys many microbes before they can enter the intestinal tract. There are important exceptions, however: The hepatitis A virus is one of many pathogens that survives gastric acidity and gains access to the body via the digestive tract.

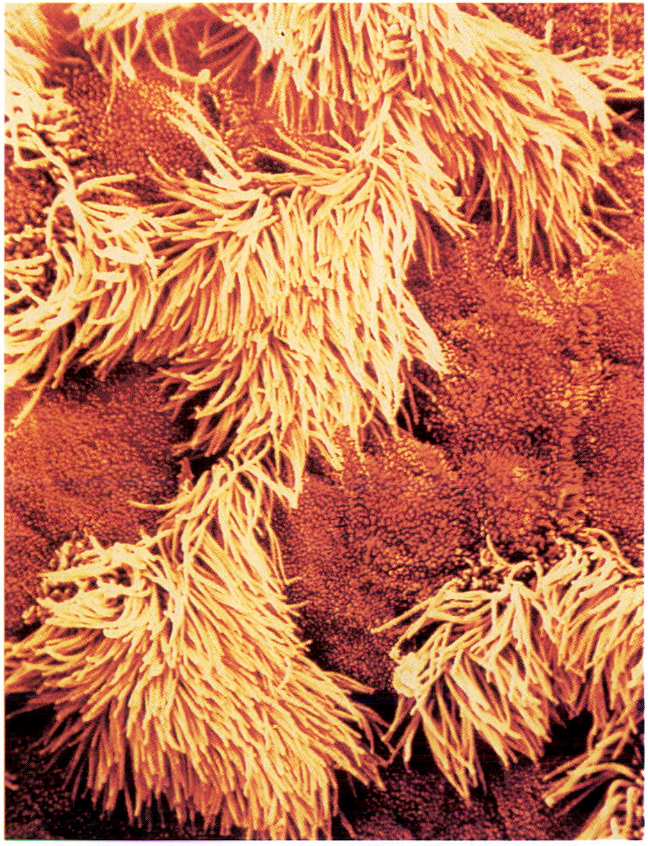

FIGURE 43.2 ▪ First-line respiratory defenses. An important part of the body's first line of defense is the mucous membrane lining the upper part of the respiratory system. In the lining of the trachea, shown here, specialized cells (orange) produce mucus that traps microbes before the invaders can enter the lungs. The mucous membrane is also equipped with ciliated cells (yellow). Synchronized beating of the cilia expels mucus and the trapped microbes upward into the pharynx (colorized SEM).

Phagocytic cells, inflammation, and antimicrobial proteins function early in infection

Microbes that penetrate the first line of defense, such as those that enter through a break in the skin, face the second line of defense. The body's internal mechanisms of nonspecific defense depend mainly on **phagocytosis**, the ingestion of invading organisms by certain types of white cells (see FIGURE 8.18a). As you will see, phagocyte function is intimately associated with an effective inflammatory response and also with certain antimicrobial proteins. These nonspecific mechanisms help limit the spread of microbes in advance of specific immune responses.

Phagocytic and Natural Killer Cells

The phagocytic cells called **neutrophils** constitute about 60% to 70% of all white blood cells (leukocytes). Cells damaged by invading microbes release chemical signals that attract neutrophils from the blood. The neutrophils enter the infected tissue, engulfing and destroying microbes there. (This migration toward the source of a chemical attractant is called *chemotaxis.*) However, neutrophils tend to self-destruct as they destroy foreign invaders, and their average lifespan is only a few days.

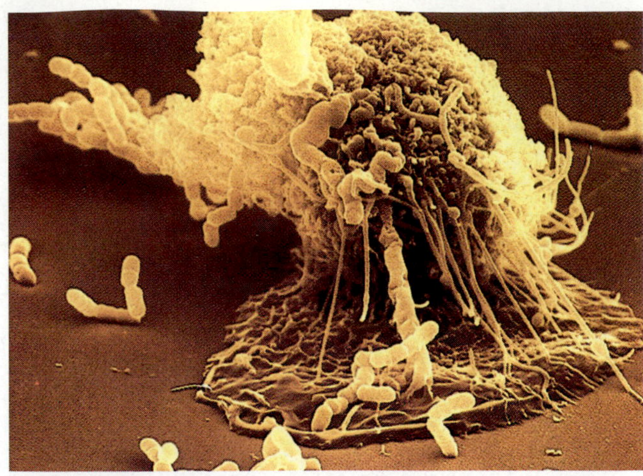

FIGURE 43.3 ▪ Phagocytosis by a macrophage. This micrograph shows fibril-like pseudopodia of a macrophage attaching to rod-shaped bacteria, which will be ingested and destroyed (colorized SEM).

Monocytes, although they constitute only about 5% of leukocytes, provide an even more effective phagocytic defense. New monocytes circulate in the blood for only a few hours, then migrate into tissues, developing into large **macrophages** ("big eaters"). Tissue macrophages, the largest phagocytic cells, are especially effective, long-lived phagocytes. These cells extend long pseudopodia that can attach to polysaccharides on a microbe's surface and engulf the microbe, which is then destroyed by enzymes within macrophage lysosomes (FIGURE 43.3). Interestingly, some microbes have evolved mechanisms for evading phagocytic destruction. Some bacteria have outer capsules to which a macrophage cannot attach. Others, like *Mycobacterium tuberculosis*, are resistant to lysosomal destruction and can even reproduce inside a macrophage. These microorganisms are a particular problem for both nonspecific and specific defenses of the body.

Some macrophages migrate throughout the body, while others reside permanently in certain tissues: in lung (alveolar

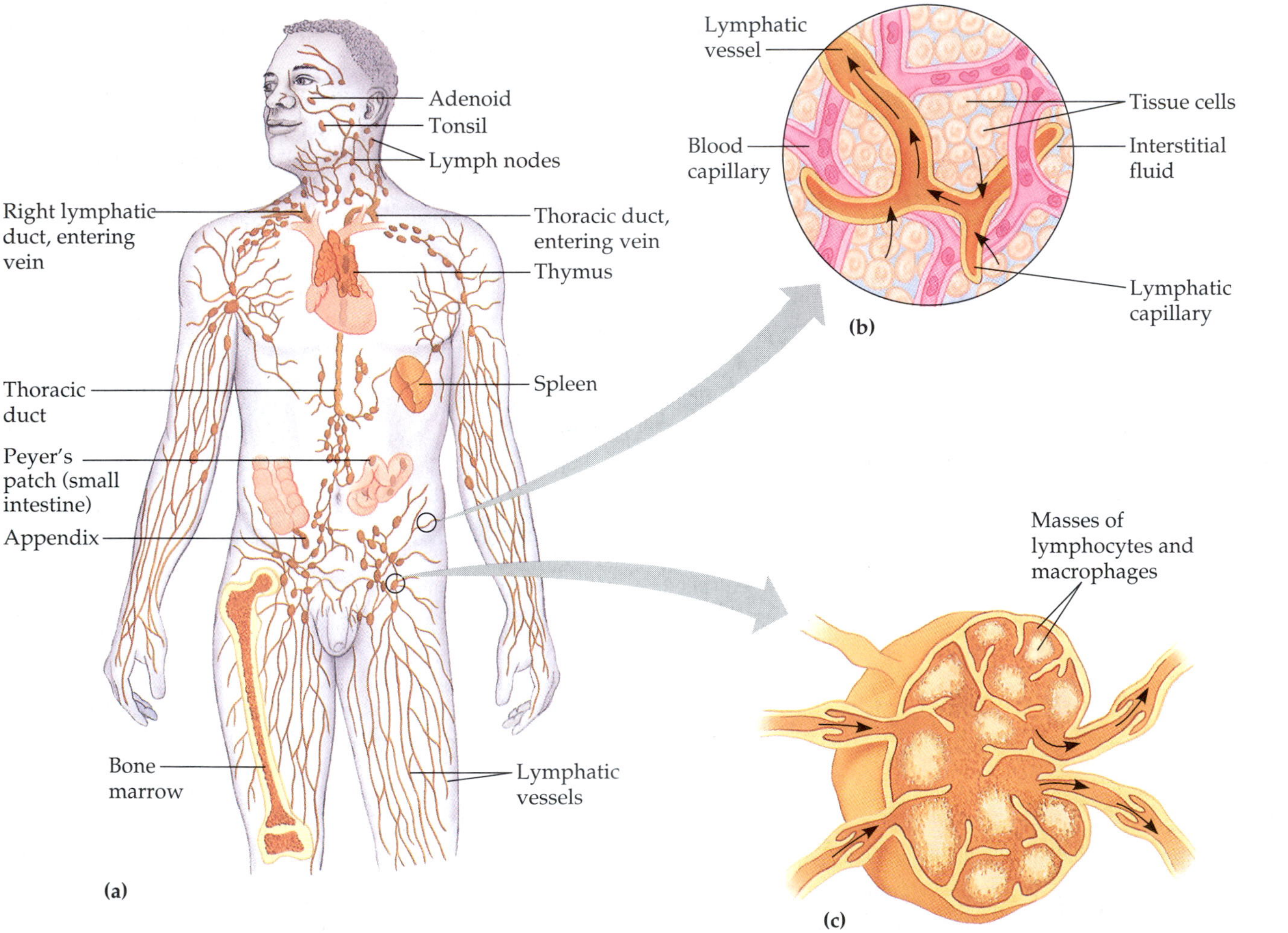

FIGURE 43.4 ▪ The human lymphatic system. The lymphatic system returns fluid from the interstitial spaces to the circulatory system. **(a)** The system includes lymphatic vessels and various satellite organs that are important in the body's defense system, including the spleen, adenoids, tonsils, appendix, Peyer's patches, and numerous lymph nodes. Also depicted in this figure are the bone marrow and the thymus, sites of white blood cell development. **(b)** The interstitial fluid that bathes tissues is continually taken up by the lymphatic capillaries. The fluid, now called lymph, flows through the system of vessels, eventually returning to the blood circulatory system near the shoulders, where the right lymphatic duct and the thoracic duct drain into veins. **(c)** Along the way, lymph must pass through numerous lymph nodes, where any pathogens present in lymph encounter macrophages and lymphocytes, another class of white blood cells with defensive functions.

macrophages), liver (Kupffer's cells), kidney (mesangial cells), brain (microglial cells), connective tissues (histiocytes), and especially in lymph nodes and the spleen, key organs of the lymphatic system (FIGURE 43.4). The fixed macrophages in the spleen, lymph nodes, and other lymphatic tissues are particularly well located to contact infectious agents. Microorganisms, microbial fragments, and foreign molecules that enter the blood encounter macrophages as they become trapped in the net-like architecture of the spleen, while those in tissue fluid flow into lymph and are filtered through the lymph nodes.

About 1.5% of all leukocytes are **eosinophils**. Their main contribution to defense is against larger parasitic invaders, such as the blood fluke *Schistosoma mansoni* (see FIGURE 33.10). Eosinophils position themselves against the external wall of a parasite and discharge destructive enzymes from cytoplasmic granules. These cells have only limited phagocytic activity.

Nonspecific defense also includes **natural killer** (**NK**) **cells.** NK cells do not attack microorganisms directly; instead, they destroy virus-infected body cells as well as abnormal cells that could form tumors. NK cells are not phagocytic; rather, they mount an attack on the cell's membrane, causing it to lyse (burst open).

The Inflammatory Response

Damage to tissue by a physical injury (such as a cut) or by the entry of microorganisms triggers a localized **inflammatory response** (FIGURE 43.5). In the injured area, precapillary arterioles dilate and postcapillary venules constrict, increasing the local blood supply (see FIGURE 42.7). These events are responsible for the characteristic redness and heat of inflammation (L. *inflammo*, "to set on fire"). The blood-engorged capillaries leak fluid into neighboring tissues, causing the edema (swelling) also associated with inflammation.

The inflammatory response is initiated by chemical signals. Some of these signals arise from the invading organism itself. Others, such as **histamine**, are released by cells of the body in response to tissue injury. Histamine is produced by circulating white blood cells called **basophils** and by **mast cells** found in connective tissue. When injured, these cells release histamine, triggering both dilation and increased permeability of nearby capillaries. Leukocytes and damaged tissue cells also discharge **prostaglandins** (see Chapter 45) and other substances that further promote blood flow to the site of injury. Enhanced blood flow and vessel permeability aid in delivering clotting elements to the injured area. Blood clotting marks the beginning of the repair process and helps block the spread of microbes to other parts of the body (see FIGURE 42.15).

Increased local blood flow and capillary permeability also enhance the migration of phagocytic cells from the blood into the injured tissues. Probably the most important element of inflammation—indeed, of nonspecific defense—is phagocytosis. Phagocyte migration usually begins within an hour after injury and is mediated by chemotactic factors called **chemokines**. Neutrophils are the first phagocytes to arrive, followed by blood monocytes, which develop into large, active macrophages. Macrophages not only phagocytose pathogens and their products, but also clean up damaged tissue cells and the remains of neutrophils destroyed in the phagocytic process. The pus that accumulates at the site of some infections consists mostly of dead phagocytic cells and the fluid and proteins that leaked from the capillaries during the inflammatory response. Usually, the pus is absorbed by the body within a few days.

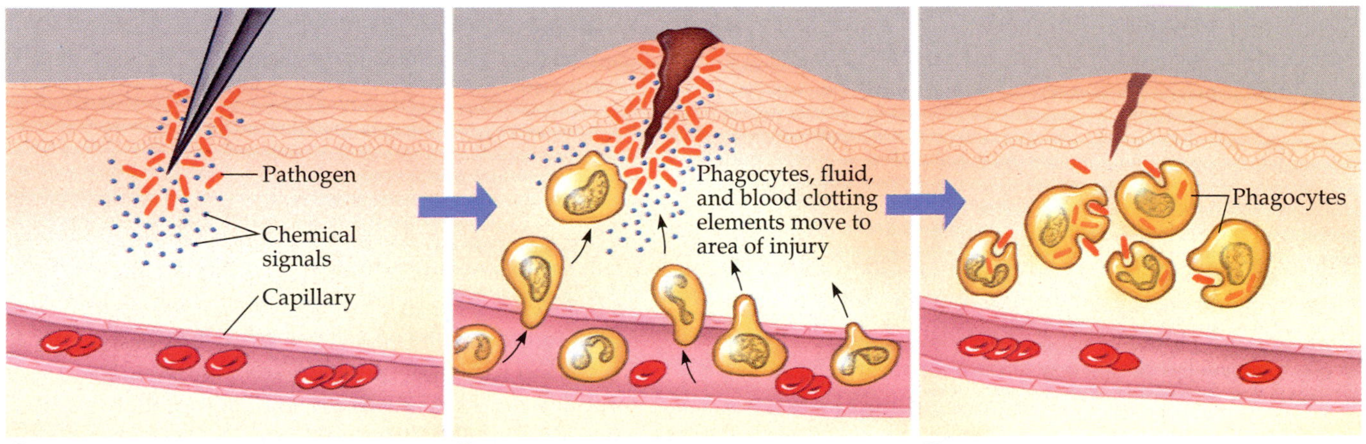

FIGURE 43.5 ▪ A simplified view of the inflammatory response. ① The localized response is triggered when cells of tissue injured by bacteria or physical damage release chemical signals such as histamine and prostaglandins. ② These signals induce capillary dilation (leading to increased blood flow) and increased capillary permeability in the affected area. The tissue cells also release chemicals that attract phagocytic cells and lymphocytes. ③ When phagocytes arrive at the site of injury, they consume pathogens and cell debris, and the tissue heals.

A sliver or other minor injury causes localized inflammation, as you have just seen, but the body may also mount a systemic (widespread) nonspecific response to severe tissue damage or infection. Injured cells often put out a call for reinforcements, emitting chemicals that stimulate the release of more neutrophils from the bone marrow. In a severe infection such as meningitis or appendicitis, the number of leukocytes in the blood may increase several-fold within a few hours of the initial inflammatory events. Another systemic response to infection is fever. Toxins produced by pathogens may trigger the fever, but certain leukocytes also release molecules called **pyrogens**, which set the body's thermostat at a higher temperature. A very high fever may be dangerous, but a moderate fever contributes to defense by inhibiting the growth of some microorganisms. Fever may also facilitate phagocytosis and, by speeding up body reactions, may speed the repair of tissues.

Antimicrobial Proteins

A variety of proteins function in nonspecific defense either by attacking microbes directly or by impeding their reproduction. You have already learned about lysozyme, an antimicrobial enzyme present in tears, saliva, and mucous secretions. Other antimicrobial agents include a set of about 20 serum proteins, known collectively as the **complement system**, that carry out a cascade of steps leading to the lysis of microbes. Some complement components also function along with chemokines in chemotaxis, attracting phagocytic cells to sites of infection. Complement proteins are an essential part of both nonspecific and specific defense. We will learn about them in detail later in this chapter.

Another set of proteins that provide nonspecific defense are the **interferons**, which are secreted by virus-infected cells. While they do not seem to benefit the infected cell, these antiviral proteins diffuse to neighboring cells and induce them to produce other chemicals that inhibit viral reproduction. In this way, interferons limit the cell-to-cell spread of viruses in the body, helping to control viral infections such as colds and influenza. The defense is not virus-specific; interferons produced in response to one virus may confer short-term resistance to unrelated viruses. In addition to its role as an antiviral agent, one type of interferon activates phagocytes, enhancing their ability to ingest and kill microorganisms. Interferons can now be mass-produced by recombinant DNA technology and are being tested clinically for the treatment of viral infections and cancer (see Chapter 20).

Let's review the body's nonspecific forms of defense: The first line of defense, the skin and mucous membranes, prevents most microbes from entering the body; the second line of defense uses phagocytes, natural killer cells, inflammation, and antimicrobial proteins to defend against microbes that have managed to enter the body. These two lines of defense are termed nonspecific because they do not distinguish among specific pathogens.

HOW SPECIFIC IMMUNITY ARISES

While microorganisms are under assault by phagocytic cells, the inflammatory response, and antimicrobial proteins, they inevitably encounter lymphocytes, the key cells of the immune system—the body's third line of defense. Lymphocytes respond to such contacts by generating efficient and selective immune responses that work throughout the body to eliminate the particular invaders. Keep in mind that the cells of the immune system respond similarly to transplanted cells and even cancer cells, which they detect as foreign.

Lymphocytes provide the specificity and diversity of the immune system

The vertebrate body is populated by two main types of lym-
43.1 phocytes: **B lymphocytes** (B cells) and **T lymphocytes** (T cells). Like macrophages, both types of lymphocytes circulate throughout the blood and lymph and are concentrated in the spleen, lymph nodes, and other lymphatic tissues (see FIGURE 43.4). Because lymphocytes recognize and respond to particular microbes and foreign molecules, they are said to display *specificity.* A foreign molecule that elicits a specific response by lymphocytes is called an **antigen**. Antigens include molecules belonging to viruses, bacteria, fungi, protozoa, and parasitic worms. Antigenic molecules are also found on the surfaces of foreign materials such as pollen and transplanted tissue. B cells and T cells specialize in different types of antigen, and they carry out different, but complementary, defensive actions, as we will see later in this chapter. One way that an antigen elicits an immune response is by activating B cells to secrete proteins called **antibodies**. The term *antigen* is a contraction of *anti*body-*gen*erator: Each antigen has a particular molecular shape and stimulates certain B cells to secrete antibodies that interact specifically with it. In fact, B and T lymphocytes even distinguish among antigens with molecular shapes that are only slightly different. So, in contrast to the nonspecific defenses, the immune system targets specific invaders.

The means by which B cells and T cells recognize specific antigens are their plasma membrane-bound **antigen receptors**. Antigen receptors on a B cell are actually transmembrane versions of antibody molecules and are often referred to as *membrane antibodies* (or membrane immunoglobulins). The antigen receptors on a T cell, called **T cell receptors**, are structurally related to membrane antibodies, and they recognize antigens just as specifically. But, unlike antibodies, T cell receptors are never produced in a secreted form. A single T or B lymphocyte bears about 100,000 receptors for antigen, all with exactly the same specificity. The receptors that a single lymphocyte produces are determined by random genetic events that occur in the lymphocyte during its early development (see Chapter 19 and FIGURE 19.6). As an unspecialized

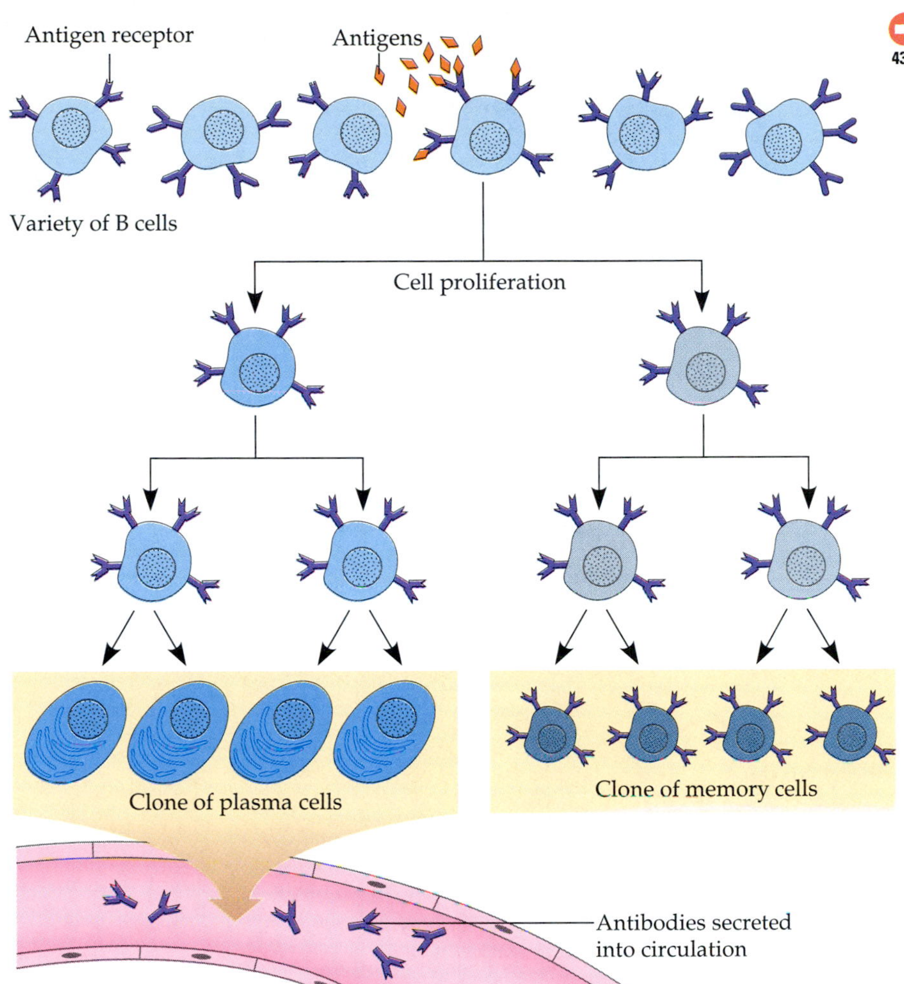

43.1 **FIGURE 43.6 ▪ Clonal selection.** The B cells and T cells of the body collectively recognize an effectively infinite number of antigens, but each individual cell recognizes only one type of antigen. (If you look closely, you will see the variation in the shapes of antigen receptors on the six B cells at the top of this illustration.) When an antigen binds to a B cell or a T cell, that cell proliferates, forming a clone of effector cells with the same specificity. In the diagram here, the antigen "selects" a particular B cell and stimulates it to reproduce and give rise to a population of identical effector cells called plasma cells. The plasma cells secrete antibodies specific for the antigen. Notice that memory cells specific for the antigen are also formed. These relatively long-lived cells can respond rapidly upon subsequent exposure to the same antigen.

cell differentiates into a B or T lymphocyte, segments of antibody genes or receptor genes are linked together by a type of genetic recombination, generating a single functional gene for each polypeptide of an antibody or receptor protein. This process, which occurs before any contact with foreign antigens, creates an enormous variety of B and T lymphocytes in the body, each bearing antigen receptors of particular specificity. With this *diversity* of lymphocytes, the immune system has the capacity to respond to millions of different antigenic molecules (even ones that do not yet exist)—and thus to millions of different potential pathogens.

Antigens interact with specific lymphocytes, inducing immune responses and immunological memory

Although it encounters a large repertoire of B cells and T cells in the body, a microorganism interacts only with lymphocytes bearing receptors specific for its various antigenic molecules. Each selected lymphocyte is activated to divide and to differentiate, eventually forming two clones of cells. One clone consists of a large number of **effector cells**, short-lived cells that combat the same antigen. The other clone consists of **memory cells**, long-lived cells bearing receptors specific for the same antigen. This antigen-driven cloning of lymphocytes is called **clonal selection** (FIGURE 43.6). The concept of clonal selection is so fundamental to understanding immunity that it is worth restating: Each antigen, by binding to specific receptors, selectively activates a tiny fraction of cells from the body's diverse pool of lymphocytes; this relatively small number of selected cells gives rise to clones of thousands of cells, all specific for and dedicated to eliminating that antigen.

The selective proliferation and differentiation of lymphocytes that occurs the first time the body is exposed to an antigen is the **primary immune response.** In the primary response, about 10 to 17 days are required from the initial exposure to antigen for selected lymphocytes to generate the maximum effector cell response. During this period, selected B cells and T cells generate antibody-producing effector B cells, called **plasma cells,** and effector T cells, respectively. While these effector cells are developing, a stricken individual may become ill. Eventually, symptoms of illness diminish and disappear as antibodies and effector T cells clear the antigen from the body. If that individual is exposed to the same antigen at some later time, the response is faster (only 2 to 7 days), of greater magnitude, and more prolonged. This is the

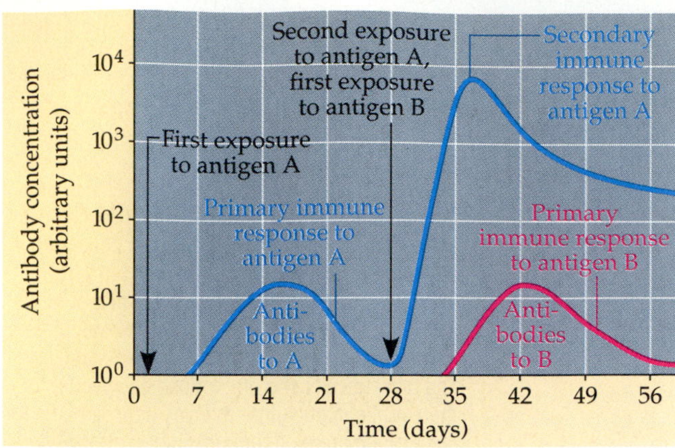

FIGURE 43.7 • Immunological memory. An initial exposure to antigen A stimulates a primary immune response, with the eventual production of antibodies against that antigen. Notice the time lag and relatively small response. A second exposure to antigen A at day 28 produces a faster and larger secondary immune response. If antigen B were also injected at day 28, the reaction to that antigen would be a primary response, not a secondary one. This experiment demonstrates that the secondary response, which is due to the presence of long-lived memory cells, is specific.

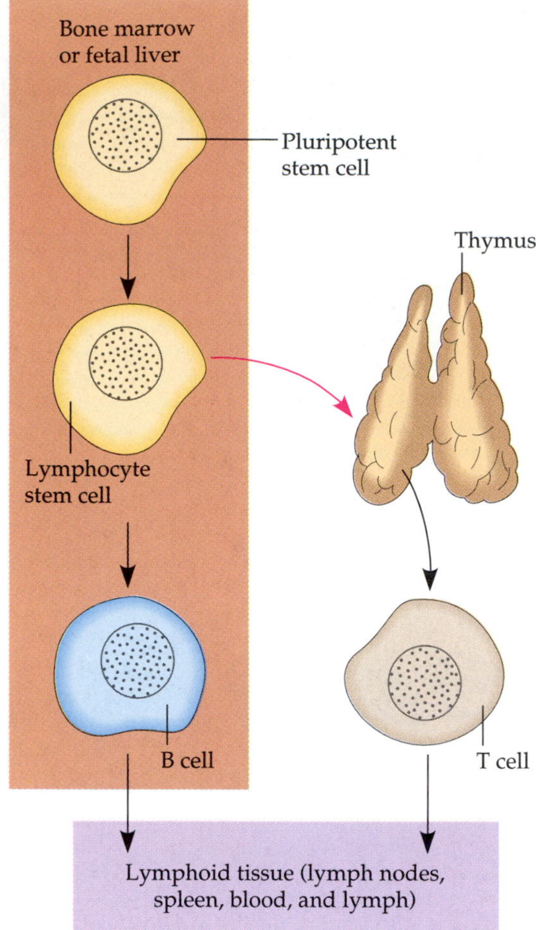

FIGURE 43.8 • The development of lymphocytes. Like other blood cells, lymphocytes differentiate from pluripotent stem cells in bone marrow (see FIGURE 42.14). Lymphocytes that continue their maturation in bone marrow develop into B cells, while lymphocytes that move to the thymus and complete their maturation there differentiate into T cells. Both classes of lymphocytes populate the lymph nodes, spleen, and other lymphatic organs shown in FIGURE 43.4.

secondary immune response. Measures of antibody concentrations in the blood serum over time show clearly the difference between primary and secondary immune responses (FIGURE 43.7). In addition to being more numerous, antibodies produced in the secondary response tend to have greater affinity for the antigen than those secreted in the primary response. The immune system's capacity to generate secondary immune responses is the basis of *immunological memory*. As you have seen, an initial exposure to an antigen gives rise not only to effector cells but to clones of long-lived T and B memory cells (see FIGURE 43.6). The memory cells are poised to proliferate and differentiate rapidly when they later contact the same antigen. The long-term protection developed after exposure to a pathogen was recognized 2400 years ago by Thucydides of Athens, who described how those sick and dying of plague were cared for by others who had recovered, "for no one was ever attacked a second time." This concept of immunity is familiar to all of us: If we had chickenpox as a child, we are unlikely to get it again.

Lymphocyte development gives rise to an immune system that distinguishes self from nonself

Lymphocytes, like all blood cells, originate from pluripotent stem cells in the bone marrow (see FIGURE 42.14) or liver of a developing fetus. Early lymphocytes are all alike, but they later develop into T cells or B cells, depending on where they continue their maturation (FIGURE 43.8). Lymphocytes that migrate from the bone marrow to the thymus, a gland in the thoracic cavity above the heart, develop into T cells ("T" for thymus). Lymphocytes that remain in the bone marrow and continue their maturation there become B cells. The "B" actually stands for the bursa of Fabricius, an organ unique to birds where avian B cells mature, and the place where B lymphocytes were first discovered. But since the B cells of all other vertebrates develop in bone marrow, you can certainly equate "B" with "bone" as well as with "bursa."

Immune Tolerance for Self

While B cells and T cells are maturing in the bone marrow and thymus, their antigen receptors are tested for potential self-reactivity. For the most part, lymphocytes bearing receptors specific for molecules already present in the body are either rendered nonfunctional or destroyed by **programmed cell death** (apoptosis; see FIGURE 21.16), leaving only lymphocytes that react to foreign molecules. This *capacity to distinguish self from nonself* continues to develop even as the cells migrate to lymphatic organs. Thus, the body normally has no mature lymphocytes that react against self components: The immune system exhibits the critical feature of *self-tolerance.* Failure of self-tolerance can lead to autoimmune diseases such as multiple sclerosis. As you will see in the next section, certain body cell surface molecules are essential in the development of T cell self-tolerance as well as in T cell activity.

The Role of Cell Surface Markers in T Cell Function and Development

Lymphocytes do not react to most self antigens, but T cells do have a crucial interaction with one important group of native molecules. These are a collection of cell surface glycoproteins (proteins with attached sugar chains) encoded by a family of genes called the **major histocompatibility complex** (**MHC**). In humans, the MHC glycoproteins are also known as the *HLA* (for *h*uman *l*eukocyte *a*ntigens). Two main classes of MHC molecules mark body cells as "self." **Class I MHC** molecules are found on all nucleated cells—that is, on almost every cell of the body. **Class II MHC** molecules are restricted to a few specialized cell types, including macrophages, B cells, activated T cells, and the cells that make up the interior of the thymus.

Within a single species there are hundreds of different possible alleles for each class I and class II MHC gene. The large number of different MHC molecules means that among humans, for example, it is extremely unlikely that any two people, except identical twins, will have exactly the same set of MHC molecules. Thus, the major histocompatibility complex is a biochemical fingerprint virtually unique to each individual. In fact, the discovery of the MHC occurred in the process of studying the phenomena of skin graft rejection and acceptance; the *histo* in histocompatibility refers to tissues.

The MHC and its role in the body's rejection of tissue grafts initially puzzled scientists: Why would the vertebrate body have evolved markers that prevent members of the same species from sharing tissues? We now know that MHC molecules and their interaction with T cells are critical to a functional immune system. The job of an MHC molecule is **antigen presentation.** Each MHC molecule cradles a fragment of a protein antigen in a hammocklike groove and "presents" it to a T cell. There are two main types of T cells, and each makes specific contacts with MHC molecules on body cell surfaces. **Cytotoxic T cells** (**T_C**) have antigen receptors that bind to fragments of antigens displayed by the body's class I MHC molecules, those that appear on nucleated cells (FIGURE 43.9a). **Helper T cells** (**T_H**) have receptors that bind to fragments of antigens displayed by the body's class II MHC molecules (FIGURE 43.9b). Any MHC molecule can present a variety of antigen fragments. Each MHC-antigen combination forms a unique complex that is recognized by specific antigen receptors on certain T cells.

How does an individual develop T cells bearing receptors that recognize his or her own MHC molecules? Let's return to T cell development in the thymus. Developing T cells interact with thymic cells, which have high levels of both class I MHC molecules (since the cells are nucleated) and class II MHC molecules. Only T cells bearing receptors with affinity for self-MHC reach maturity. Developing T cells having receptors with an affinity for class I MHC become cytotoxic T cells. Those having receptors with moderate affinity for class II MHC become helper T cells.

Now let's review what you've learned so far about the immune system: The immune responses of B and T lymphocytes exhibit four attributes that characterize the immune system as a whole: specificity, diversity, memory, and the capacity to distinguish self from nonself. A critical component of the immune response is the MHC, which displays a combination of self (MHC molecule) and nonself (antigen fragment) that is recognized by specific T lymphocytes. In the following section you will learn more about how lymphocytes recognize and respond to foreign substances, and how they generate immunity.

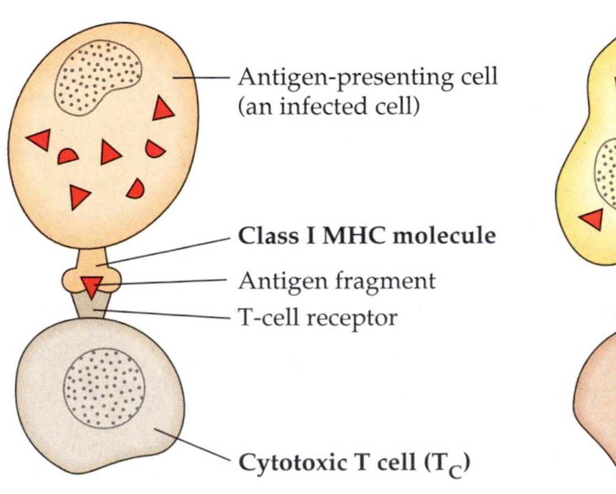

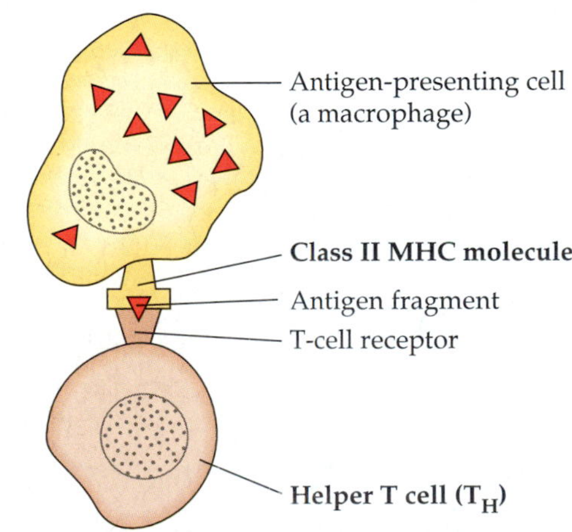

FIGURE 43.9 ▪ The interaction of T cells with MHC molecules. Glycoproteins of the major histocompatibility complex (MHC) on the surfaces of body cells identify "self" to the immune system. Their main role is to "present" fragments of antigens to T cells. There are two main classes of MHC molecules, each of which is recognized by a separate class of T cells. **(a)** Cytotoxic T cells have receptors that bind to antigen-bearing class I MHC molecules. **(b)** Helper T cells have receptors that bind to antigen-bearing class II MHC molecules.

IMMUNE RESPONSES

The immune system can mount two types of responses to antigens: a humoral response and a cell-mediated response. **Humoral immunity** involves B cell activation and results from the production of antibodies that circulate in the blood plasma and lymph, fluids that were long ago called humors. Around the end of the nineteenth century, researchers performed an experiment in which they transferred such fluids from animals that had recovered from an infection to others that had not been exposed to it. For a short time the latter animals were protected from the infection. The investigators had transferred *humor*al immunity (antibodies) from animal to animal. They also found that immunity to some infections could be passed along only if cells, later identified as T lymphocytes, were transferred. This second type of immunity,

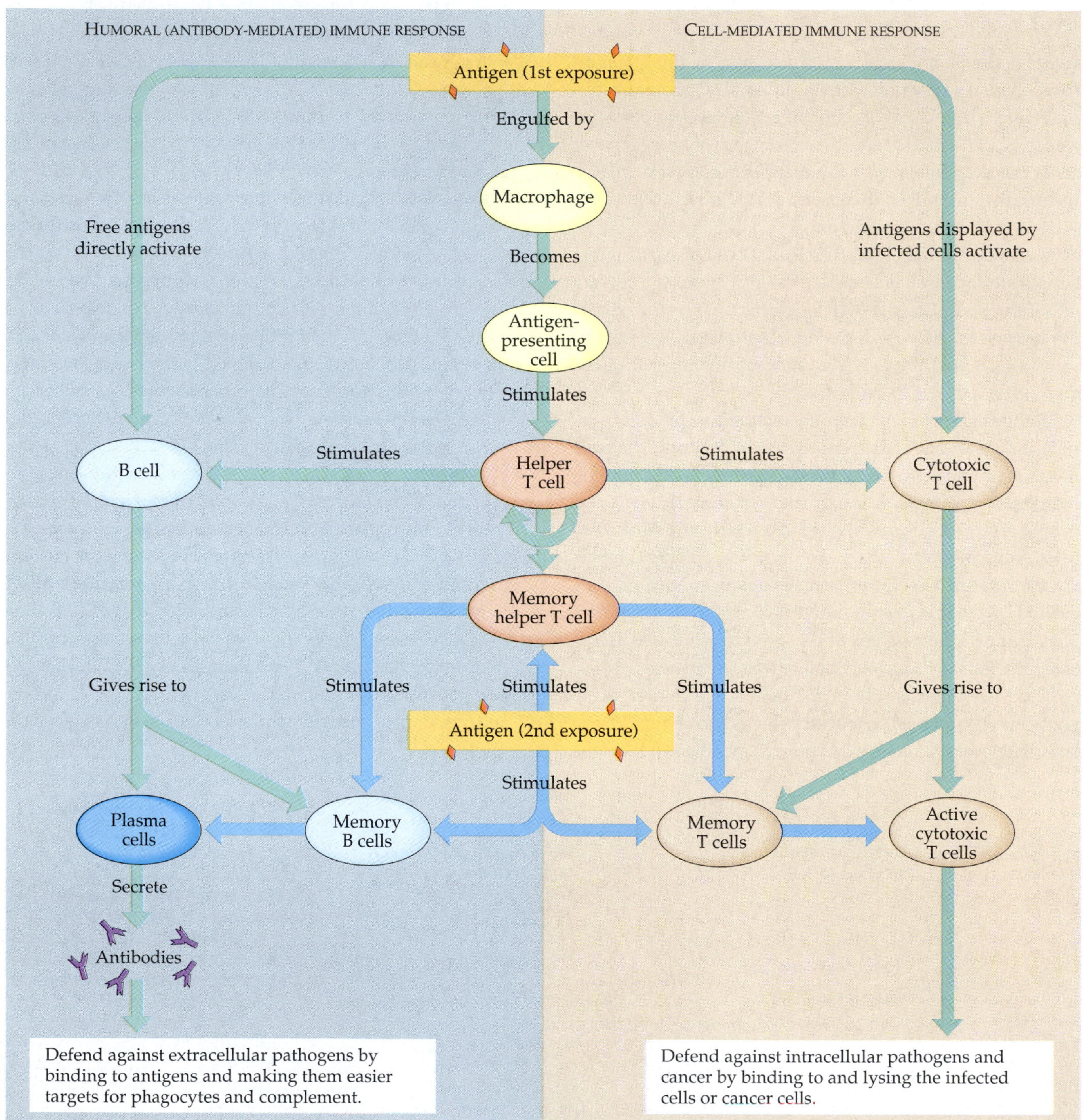

FIGURE 43.10 ▪ An overview of the immune responses. In this simplified flow chart, green arrows track the primary response, and blue arrows track the secondary response. Notice the connections between the humoral response and the cell-mediated response and the central role of the helper T cell.

which depends on the action of T cells, became known as **cell-mediated immunity**.

The circulating antibodies of the humoral response defend mainly against free bacteria, toxins, and viruses present in body fluids. In contrast, the T cells of the cell-mediated response are active against bacteria and viruses within infected body cells and against fungi, protozoa, and parasitic worms. Also, cell-mediated immunity is crucial in the body's response against transplanted tissue and cancer cells, both of which are perceived as "nonself." FIGURE 43.10 provides an overview of the humoral and cell-mediated responses, the two branches of the immune system, which we will explore shortly. In addition to introducing these two branches, the figure shows the connections linking them—cell-signaling interactions among the lymphocytes. Central to this network of cell-signaling is the helper T cell, which responds to antigen presented by a macrophage and stimulates both B cells and other T cells.

Helper T lymphocytes function in both humoral and cell-mediated immunity: *an overview*

Before we look at the function of helper T lymphocytes, we
43.1 must return to the MHC and its role in antigen presentation. Recall that class II MHC molecules, the ones recognized by helper T cells, are found only on certain cell types, mainly those that engulf foreign antigens. The cells that take up antigens include B cells and macrophages. These serve as **antigen-presenting cells** (**APCs**), alerting the immune system, via the helper T cell, that foreign antigen is in the body. For example, a macrophage that has engulfed and broken down a bacterium contains small fragments of bacterial proteins (peptides). As a newly synthesized class II MHC molecule moves toward the macrophage surface, it captures one of these bacterial peptides in its antigen-binding groove and carries it to the surface, revealing the foreign peptide to a helper T cell (FIGURE 43.11). The interaction between an antigen-presenting cell and a helper T cell is greatly enhanced by the presence of a T cell surface protein called **CD4**. Present on most helper T cells, CD4 has an affinity for part of the class II MHC protein. The interaction between CD4 and a class II MHC molecule helps keep the helper T cell and the APC joined while antigen-specific activation occurs.

When a helper T cell is selected by specific contact with the class II MHC–antigen complex on an APC, the T_H cell proliferates and differentiates into a clone of activated helper T cells and memory helper T cells. Activated helper T cells secrete several different **cytokines**, proteins or peptides that serve to stimulate other lymphocytes. For example, the cytokine **interleukin-2** (**IL-2**) helps B cells that have contacted antigen differentiate into antibody-secreting plasma cells. IL-2 also helps cytotoxic T cells to become active killers.

The helper T cell itself is also subject to regulation by cytokines. As a macrophage phagocytoses and presents antigen, the macrophage is stimulated to secrete a cytokine called **interleukin-1** (**IL-1**). IL-1, in combination with the presented antigen, activates the helper T cell to produce IL-2 and other cytokines. Also, in an example of positive feedback, IL-2 secreted by the helper T cell stimulates that same cell to proliferate more rapidly and to become an even more active cytokine producer. In these ways, helper T cells modulate both humoral (B cell) and cell-mediated (cytotoxic T cell) immune responses.

Another type of T lymphocyte, called a **suppressor T cell** (**T_S**), may function in turning off the immune response once

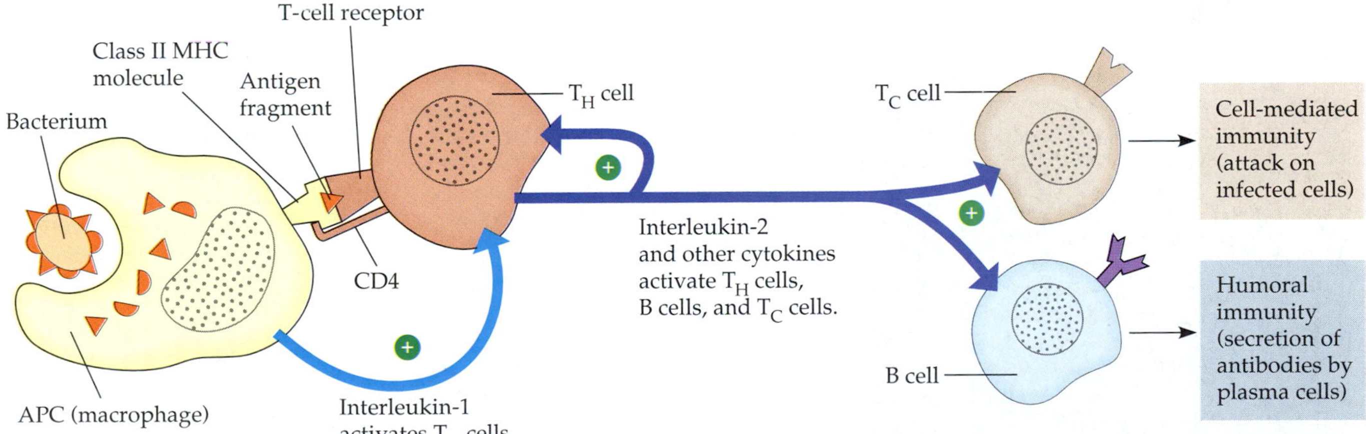

43.1 **FIGURE 43.11 ▪ The central role of helper T cells: a closer look.** Helper T cells mobilize both humoral and cell-mediated branches of the immune response. The receptor of a T_H cell recognizes the class II MHC– antigen complex displayed on the surface of an antigen-presenting cell, usually a macrophage. The interaction of the two cells is enhanced by CD4, a T_H surface protein that binds to the class II MHC protein on the APC. The macrophage secretes interleukin-1, a cytokine that contributes to the activation of the T_H cell. An activated T cell grows and divides, producing a clone of T_H cells, all with receptors keyed to the MHC molecule in combination with the specific antigen that triggered the response. (This figure shows only one T_H cell.) The T_H cells then secrete the cytokine interleukin-2, which amplifies the cell-mediated response by stimulating the proliferation and activity of all the T_H cells in the vicinity. Interleukin-2 and other cytokines also help activate B cells, which function in humoral immunity, and T_C cells, which function in the cell-mediated immune response. The circled plus signs indicate stimulation.

antigen has been eliminated from the body. In fact, it is perhaps as important to end an immune response as it is to induce it in the first place. This cell type, however, is not well understood. In fact, some immunologists believe that T_S cells are actually a variety of helper T cell rather than a separate type of cell.

In the cell-mediated response, cytotoxic T cells counter intracellular pathogens: *a closer look*

43.1 Antigen-activated cytotoxic T lymphocytes kill cancer cells and cells infected by viruses or other intracellular pathogens. Before we look at these events, we must return to class I MHC proteins and their role in presenting antigen to cytotoxic T cells. Recall that all nucleated cells of the body continuously produce class I MHC molecules. As a newly synthesized class I molecule moves toward the cell surface, it captures a small fragment of one of the other proteins synthesized by that cell. If that cell happens to contain a replicating virus, peptide fragments of viral proteins are captured and transported to the cell surface. In this way, class I MHC molecules expose foreign proteins, synthesized in infected or abnormal cells, to cytotoxic T cells. The interaction between the antigen-presenting cell and a cytotoxic T cell is greatly enhanced by the presence of a T cell surface protein called **CD8**. Present on most cytotoxic T cells, CD8 has an affinity for a portion of the class I MHC molecule. The CD8–class I MHC interaction helps keep the two cells in contact while antigen-specific activation is occurring (FIGURE 43.12). Thus, the roles of class I MHC molecules and CD8 are similar to those of class II MHC molecules and CD4, except that different cells are involved.

A cytotoxic T cell, activated by specific contacts with class I MHC–antigen complexes on an infected or tumor cell and further stimulated by IL-2 from a helper T cell, differentiates into an active killer. It kills its so-called **target cell** primarily by releasing **perforin**, a protein that forms pores in the target cell's membrane. As ions and water flow into the target cell, it swells and eventually lyses (see FIGURE 43.12a). The death of

43.1

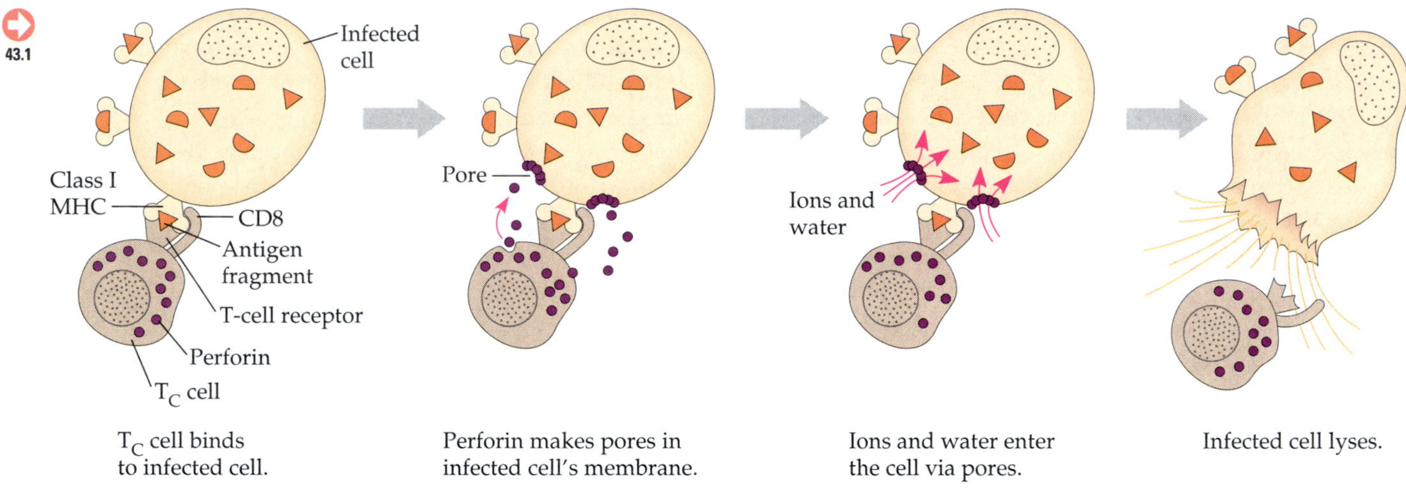

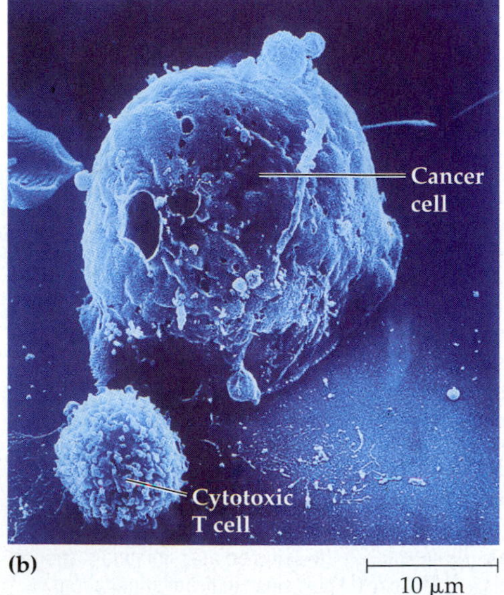

FIGURE 43.12 ▪ The functioning of cytotoxic T cells. (a) The T cell receptor of a cytotoxic T cell recognizes a class I MHC–antigen complex on the surface of an infected cell (or cancer cell). This interaction is enhanced by CD8, which binds to the class I MHC protein, and also by interleukin-2 from helper T cells (not shown). The activated T_C cell discharges the protein perforin, which lyses the antigen-presenting cell. **(b)** The T_C cell (smaller cell) has lysed a cancer cell (SEM).

the infected cell not only deprives the pathogen of a place to reproduce but also exposes it to circulating antibodies, which mark it for disposal. After destroying an infected cell, the T_C cell moves on to kill other cells infected with the same pathogen.

In the same way, cytotoxic T cells defend against malignant tumors. Because tumor cells carry distinctive molecules not found on normal cells, they are identified as foreign by the immune system. Class I MHC molecules on a tumor cell present fragments of **tumor antigen** to cytotoxic T cells. Interestingly, certain cancers, as well as viruses such as Epstein-Barr virus, actively diminish the amounts of class I MHC proteins on affected cells. As a result, these cancers and infections may escape detection by cytotoxic T cells. The body has a backup defense: Natural killer (NK) cells also lyse virus-infected and cancer cells. Although NK cells and T_C cells kill their targets by lysis, NK cells are part of the body's nonspecific defenses—they lack antigen receptors and do not respond to specific antigens.

In the humoral response, B cells make antibodies against extracellular pathogens: *a closer look*

43.1 You have learned that the humoral immune response is initiated when B cells bearing antigen receptors (membrane antibodies) are selected by specific antigens. You also know that B cell activation is aided by IL-2 and other cytokines secreted from helper T cells activated by the same antigen. Stimulated by both antigen and cytokines, the B cell proliferates and differentiates into a clone of antibody-secreting plasma cells and a clone of long-lived memory B cells. Antigens that evoke this type of B cell response are known as **T-dependent antigens** because they can stimulate antibody production only with help from T_H cells (FIGURE 43.13). Most protein antigens are T-dependent.

Other antigens, such as polysaccharides and proteins with many identical polypeptides, function as **T-independent antigens**. Such antigens include the polysaccharides of many bacterial capsules and the proteins that make up bacterial flagella. Apparently, the repeated subunits of these antigens bind simultaneously to a number of membrane antibodies on the B cell surface. This provides enough stimulus to the B cell to generate antibody-secreting plasma cells without benefit of IL-2. The response to T-independent antigens is very important in defending against many bacteria; however, the response is generally weaker than the response to T-dependent antigens, and no memory cells are generated in T-independent responses.

Before we leave our discussion of B cell function, it is important to recall that B cells bear class II MHC molecules: They are antigen-presenting cells. When antigen first binds to membrane antibodies, the B cell takes in a few of the foreign molecules by receptor-mediated endocytosis (see FIGURE 8.18c). In a process very similar to presentation in macro-

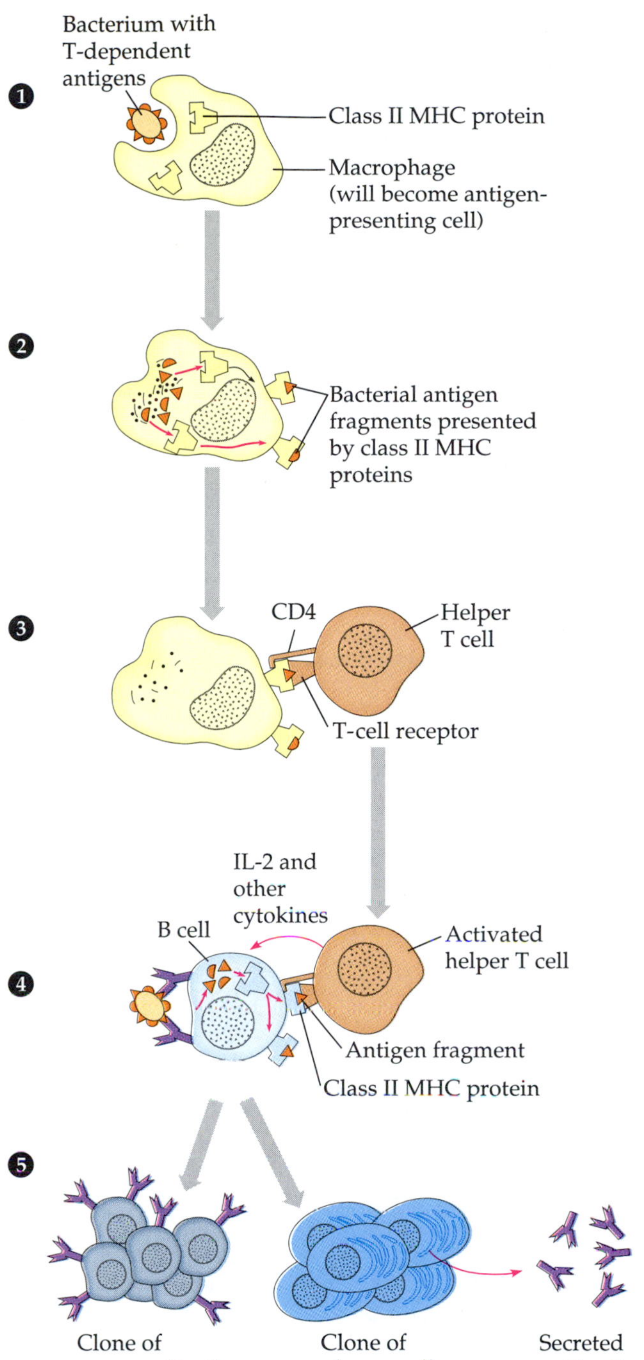

43.1 **FIGURE 43.13 ▪ Humoral response to a T-dependent antigen.** Many antigens can trigger a humoral (antibody-mediated) immune response by B cells only with the participation of helper T cells. Such antigens are called T-dependent antigens, and most protein antigens are of this type. ① A macrophage ingests a pathogen. ② Antigen fragments from the partially digested pathogen form complexes with class II MHC proteins. These self/nonself complexes are then transported to the cell surface, where they are presented to other cells of the immune system. ③ A helper T cell with a receptor specific for the presented antigen interacts with the macrophage by binding to the MHC–antigen complex. ④ The activated helper T cell then interacts with a B cell that has taken in antigens by endocytosis and displays a fragment of the antigen along with class II MHC proteins. The helper T cell secretes IL-2 and other cytokines that activate the B cell. ⑤ The B cell divides repeatedly and differentiates into memory B cells and plasma cells, the antibody-secreting effector cells of humoral immunity.

phages, the B cell presents antigen to a helper T cell. However, although a macrophage can engulf and present peptide fragments from a wide variety of antigens, a B cell internalizes and presents peptides of only the antigen to which it specifically binds. Therefore, immunologists think that macrophages are the main APCs in the primary response (when B cells specific for a particular antigen are rare), whereas B cells, specifically memory B cells, are more important as APCs in secondary responses.

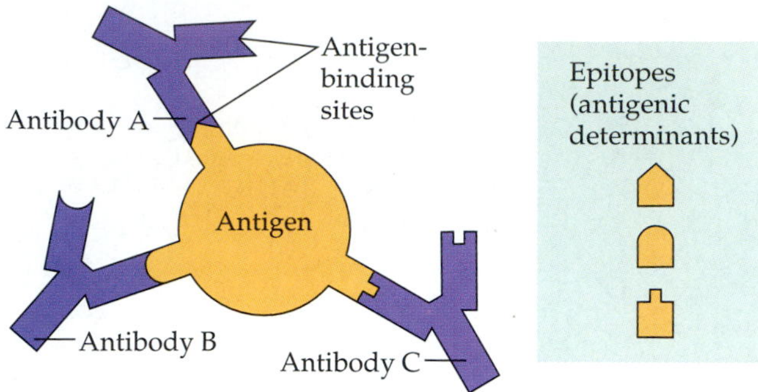

FIGURE 43.14 • Epitopes (antigenic determinants). Antibodies bind to epitopes on the surface of an antigen. In this example, three different antibody molecules react with different epitopes on the same large antigen molecule.

In any given humoral response, the processes just discussed stimulate a variety of different B cells, each giving rise to a clone of thousands of plasma cells. Each plasma cell is estimated to secrete about 2000 antibody molecules per second over the cell's 4- to 5-day lifespan. Next we look more closely at antibodies and how they bind to and mediate the disposal of antigens.

Antibody Structure and Function

Antigens that elicit a humoral immune response are typically the protein and polysaccharide surface components of various microbes, incompatible transplanted tissue, and transfused blood cells. In addition, for some of us, the proteins of foreign substances such as bee venom or pollen act as antigens that induce an allergic, or hypersensitive, humoral response (to be discussed later).

Neither the membrane version of the antibody—that is, the B cell receptor for antigen—nor the secreted antibody actually binds an entire antigen molecule. Rather, an antibody

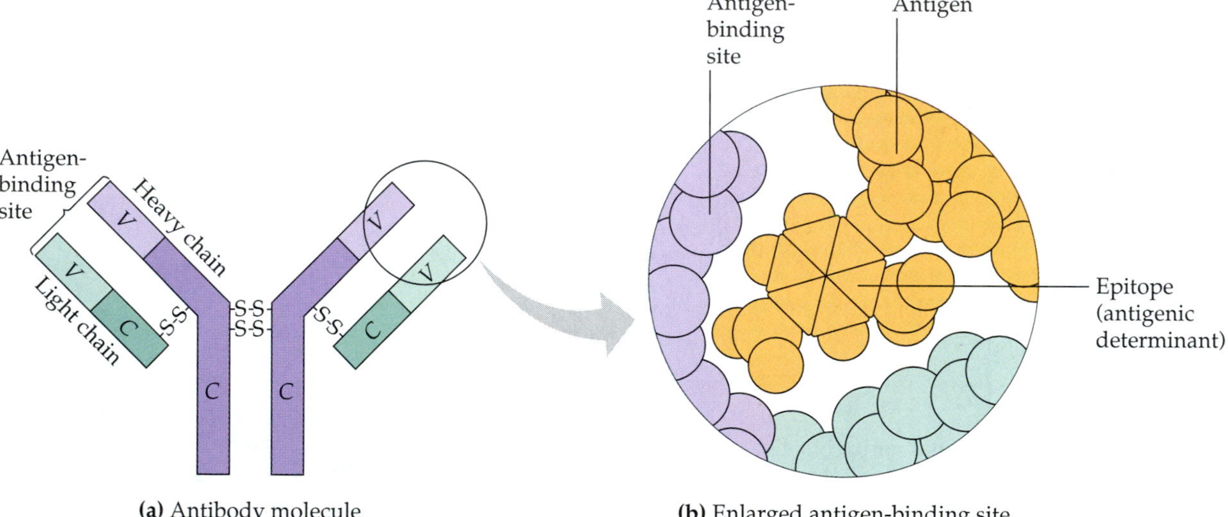

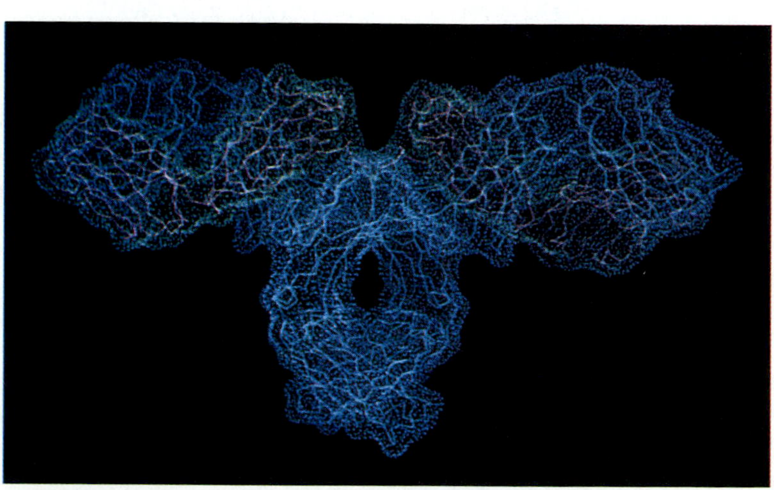

FIGURE 43.15 • The structure of a typical antibody molecule. (a) The Y-shaped molecule is composed of two light and two heavy chains linked together by disulfide bridges (S—S). The amino acid sequences of the variable regions (V), which make up the two identical antigen-binding sites, are different in each specific type of antibody, giving these sites specific shapes that fit certain antigenic epitopes. The remainder of the molecule consists of light and heavy chain constant regions (C), so named because their amino acid sequences vary little from antibody to antibody. **(b)** This enlargement shows an antigenic epitope bound to an antigen-binding site. **(c)** A computer graphic image of an antibody molecule.

interacts with a small, accessible portion of the antigen called an **epitope** or antigenic determinant (FIGURE 43.14). A single antigen such as a bacterial surface protein usually has several effective epitopes, each capable of inducing the production of specific antibody. It is therefore easy to imagine the entire surface of a bacterium coated with many different kinds of antibodies, each specific for a particular protein or polysaccharide epitope. It is estimated that a bacterium can be bound by 4 million antibody molecules!

Antibodies constitute a group of globular serum proteins called **immunoglobulins** (**Igs**). A typical antibody molecule has two identical antigen-binding sites specific for the epitope that provoked its production. Each molecule consists of four polypeptide chains, two identical **heavy chains** and two identical **light chains**, joined by disulfide bridges to form a Y-shaped molecule (FIGURE 43.15a). At the two tips of the Y-shaped molecule are the variable regions (V) of the heavy and light chains, so named because the amino acid sequences in these regions vary extensively from antibody to antibody (see FIGURE 19.6). As shown in FIGURE 43.15b, a heavy-chain V region and a light-chain V region together form the unique contours of an antibody's antigen-binding site. The interaction between an antigen-binding site and its epitope resembles an enzyme-substrate interaction: Multiple noncovalent bonds form between chemical groups on the respective molecules. The power of antibody specificity and of antigen-antibody interactions has been harnessed by the use of antibodies in laboratory research, clinical diagnosis, and the treatment of diseases. In particular, the technology for the production of **monoclonal antibodies** is an extraordinary contribution to biomedical science (see the Methods Box, p. 856).

While the antigen-binding sites are responsible for an antibody's ability to identify a specific epitope as an antigen, the tail of the Y-shaped antibody, formed by the constant regions (C) of the heavy chains (see FIGURE 43.15a), is responsible for its distribution in the body and for the mechanisms by which it mediates antigen disposal. There are five major types of heavy-chain constant regions, and these determine the five major classes of antibodies: IgM, IgG, IgA, IgD, and IgE. The structures and functions of each of the five immunoglobulin classes are summarized in TABLE 43.1. Note that each light chain also has a constant region. It is not a part of the antibody tail and does not contribute to Ig class functions.

Antibody-Mediated Disposal of Antigen

The binding of antibodies to antigens to form antigen-antibody
43.1 complexes is the basis of several antigen disposal mechanisms (FIGURE 43.16, p. 854). The simplest of these is **neutralization**, in which the antibody binds to and blocks the activity of the antigen. For example, antibodies neutralize a virus by attaching to the molecules that the virus must use to infect its host cell. Similarly, antibodies may bind to the surface of a pathogenic bacterium. These microbes, now coated by antibodies, are readily eliminated by phagocytosis. In a process called **opsonization**, the bound antibodies enhance macrophage attachment to, and thus phagocytosis of, the microbes.

Antibody-mediated **agglutination** (clumping) of bacteria or viruses effectively neutralizes and opsonizes the microbes. Agglutination is possible because each antibody molecule has at least two antigen-binding sites. IgG, for example, can bind to identical epitopes on two bacterial cells or viral particles, linking them together. IgM can link together five or more viruses or bacteria (as shown in FIGURE 43.16). These large complexes are readily phagocytosed by macrophages. A similar mechanism is precipitation, the cross-linking of soluble antigen molecules—molecules dissolved in body fluids—to form immobile precipitates that are disposed of by phagocytes.

One of the most important antibody-mediated disposal mechanisms is **complement fixation**, the activation of the complement system by antigen-antibody complexes. Recall that complement consists of about 20 different serum proteins that, in the absence of infection, are inactive. In an infection,

Table 43.1 ▪ The Five Classes of Immunoglobulins

IgM (pentamer) 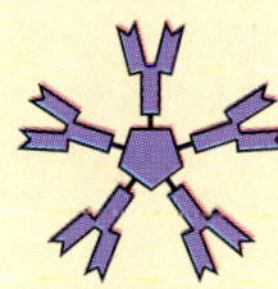	IgMs are the first circulating antibodies to appear in response to an initial exposure to an antigen. Their concentration in the blood declines rapidly. This is diagnostically useful because the presence of IgM usually indicates a current infection by the pathogen causing its formation. IgM consists of five Y-shaped monomers arranged in a pentamer structure. The numerous antigen-binding sites make it very effective in agglutinating antigens and in reactions involving complement. IgM is too large to cross the placenta and does not confer maternal immunity.
IgG (monomer) 	IgG is the most abundant of the circulating antibodies. It readily crosses the walls of blood vessels and enters tissue fluids. IgG also crosses the placenta and confers passive immunity from the mother to the fetus. IgG protects against bacteria, viruses, and toxins circulating in the blood and lymph, and triggers action of the complement system.
IgA (dimer) 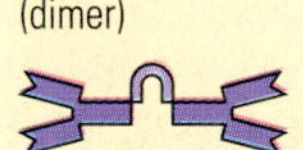	IgA is produced primarily in the form of two Y-shaped monomers (a dimer) by cells abundant in mucous membranes. The main function of IgA is to prevent the attachment of viruses and bacteria to epithelial surfaces. IgA is also found in many body secretions, such as saliva, perspiration, and tears. Its presence in colostrum (the first milk of a nursing mammal) helps protect the infant from gastrointestinal infections.
IgD (monomer)	IgD antibodies do not activate the complement system and cannot cross the placenta. They are mostly found on the surfaces of B cells, probably functioning as an antigen receptor required for initiating the differentiation of B cells into plasma cells and memory B cells.
IgE (monomer)	IgE antibodies are slightly larger than IgG molecules and represent only a very small fraction of the total antibodies in the blood. The tail regions attach to receptors on mast cells and basophils and, when triggered by an antigen, cause the cells to release histamine and other chemicals that cause an allergic reaction.

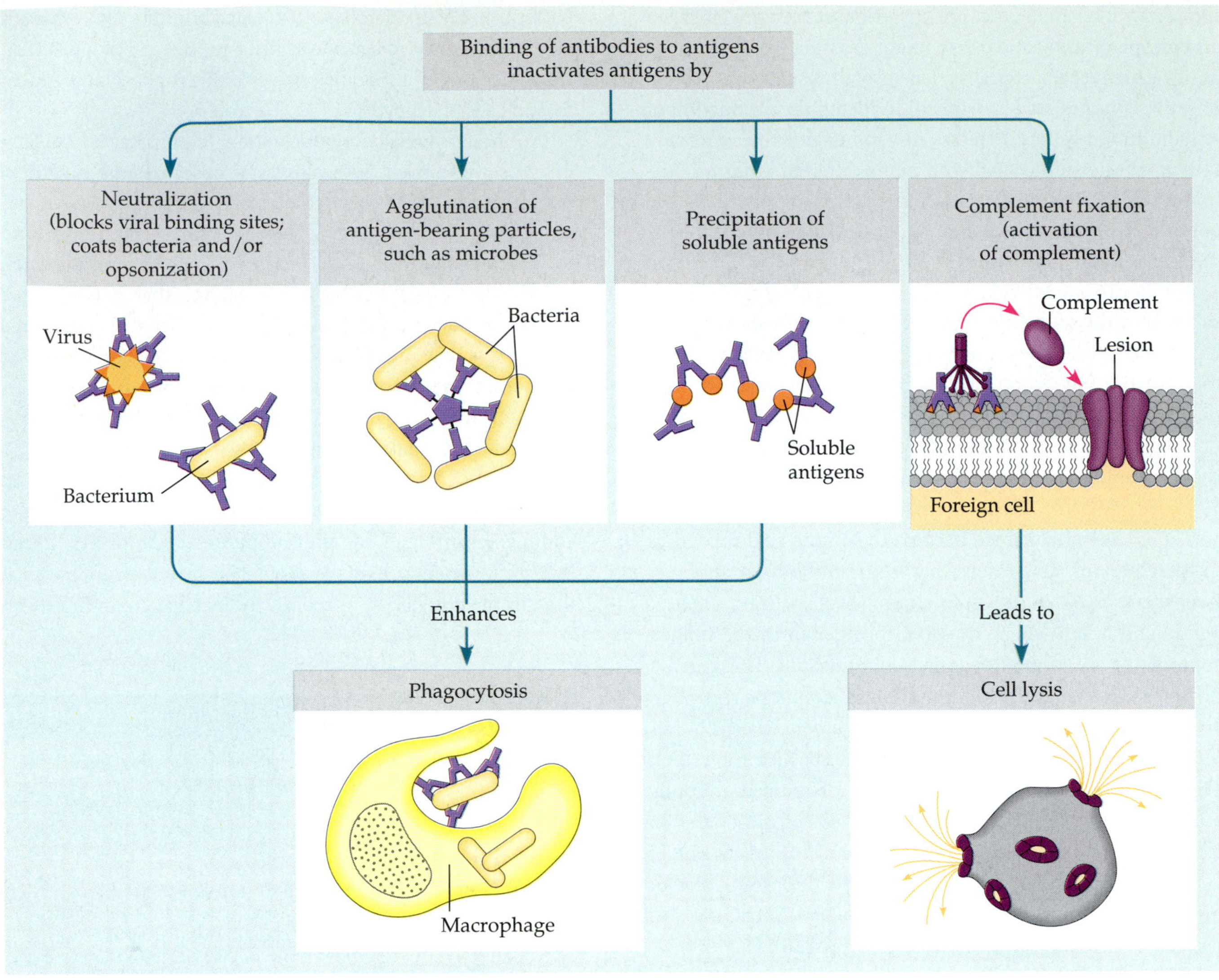

FIGURE 43.16 ▪ **Effector mechanisms of humoral immunity.** The binding of antibodies to antigens tags foreign cells and molecules for destruction by phagocytes or the complement system of proteins.

43.1

however, the first in the series of complement proteins is activated, triggering a cascade of activation steps, each component activating the next in the series. Completion of the complement cascade results in the lysis of many types of viruses and pathogenic cells. Lysis by complement can be achieved in two ways. The *classical pathway* (so called because it was discovered first) is triggered by antibodies bound to antigen and is therefore important in the humoral immune response. The *alternative pathway* is triggered by substances that are naturally present on many bacteria, yeast, viruses, and protozoan parasites; it does not involve antibodies and thus is an important nonspecific defense.

The classical pathway can begin when IgM or IgG antibodies bind to a pathogen, such as a bacterial cell (FIGURE 43.17). The first complement component links two bound antibodies and is thus activated, initiating the cascade. Ultimately, complement proteins generate a **membrane attack complex** (**MAC**), which forms a 7–10 nm-diameter pore in the bacterial membrane. Ions and water rush into the cell, causing it to swell and lyse. The MAC pore is similar to the perforin pore generated by cytotoxic T cells.

In both the classical and alternative pathways, many activated complement proteins contribute to inflammation. By binding to basophils and mast cells, some trigger the release of histamine, the injury-signaling molecule that triggers dilation and increased permeability of blood vessels. Several active complement proteins also attract phagocytes to the site. In addition, one of the activated complement proteins can cause opsonization: Copies of this protein coat bacterial surfaces and, like antibodies, stimulate phagocytosis. In a final example of teamwork in the body's defense systems, antibodies, complement, and phagocytes function together in a phenomenon called **immune adherence**. Microbes coated with antibodies and complement proteins adhere to blood vessel walls, making the pathogens easier prey for phagocytic cells circulating in the blood.

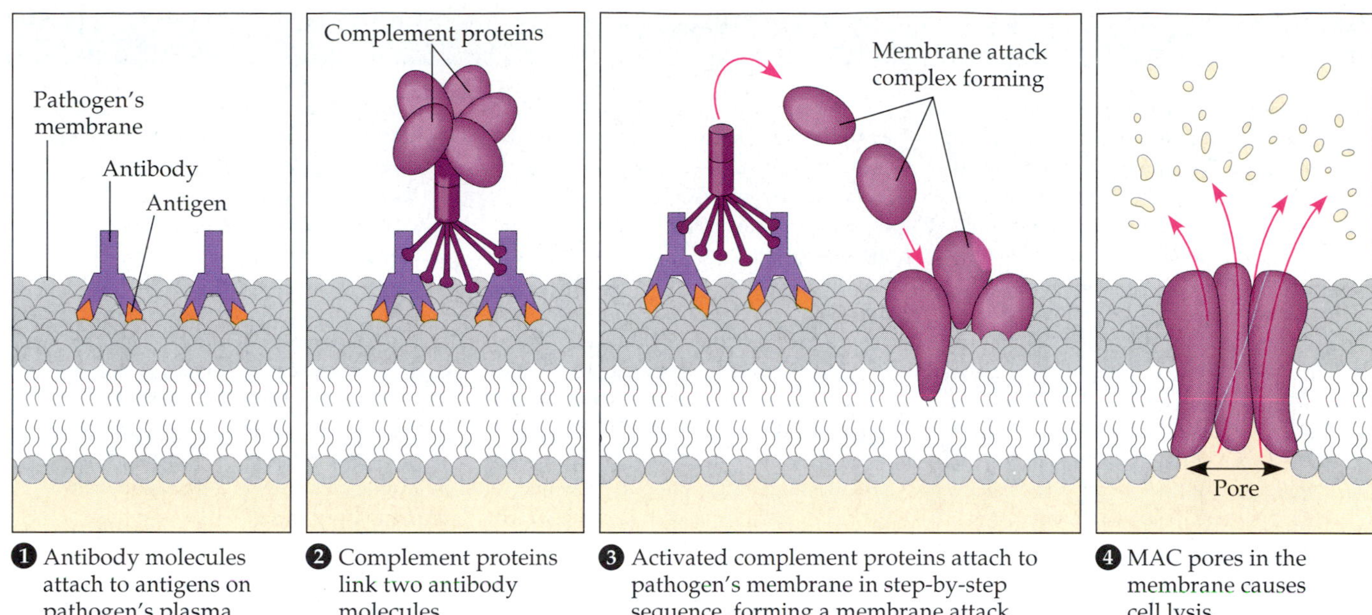

❶ Antibody molecules attach to antigens on pathogen's plasma membrane.

❷ Complement proteins link two antibody molecules.

❸ Activated complement proteins attach to pathogen's membrane in step-by-step sequence, forming a membrane attack complex (MAC).

❹ MAC pores in the membrane causes cell lysis.

FIGURE 43.17 ▪ The classical complement pathway, resulting in lysis of a target cell.

You have seen that antibodies contribute in several ways to phagocytosis by macrophages. Keep in mind that phagocytosis enables macrophages to act as APCs, stimulating helper T cells, which, in turn, stimulate the very B cells whose antibodies contribute to phagocytosis. Thus, positive feedback links together the nonspecific and specific immune systems, resulting in a coordinated and effective response to an active infection.

Invertebrates have a rudimentary immune system

We have emphasized the nonspecific and specific defenses of vertebrates because relatively little is known about how invertebrates react to pathogens that have penetrated the animals' outer barriers. However, experiments reveal that one fundamental facet of defense, the ability to distinguish self from nonself, is well developed in invertebrates. For example, if several cells of two sponges of the same species are mixed, the cells from each sponge sort themselves and reaggregate, excluding cells from the other individual. In many invertebrates, amoeboid cells called coelomocytes identify and destroy foreign materials.

Studies of tissue grafting in earthworms also reveal mechanisms of self/nonself distinction. When a portion of body wall is grafted from one worm to another, the recipient's coelomocytes attack the foreign tissue. If the donor and recipient are from the same population, the grafted tissue survives for about 8 months before it is completely rejected. However, if a worm receives a graft taken from a completely different population, the graft is rejected in just 2 weeks. In addition, the coelomocytes exhibit immunological memory: They reject a second graft from the same distant donor in just a few days. Additional research on the defense systems of invertebrates may help biologists understand how the vertebrate immune system evolved.

IMMUNITY IN HEALTH AND DISEASE

As we continue to learn more about the vertebrate immune system, we expand our understanding of the body's battles against infections and cancer, its establishment of long-term immunity, and its reaction to blood transfusions and tissue transplants. We also gain insight into the pathogenesis, prevention, and treatment of numerous diseases, including disorders of the immune system itself.

Immunity can be achieved naturally or artificially

Immunity conferred by recovering from an infectious disease such as chickenpox is called **active immunity** because it depends on the response of the infected person's own immune system. As is the case with all such infections, this immunity is naturally acquired. Active immunity can also be acquired artificially, by **immunization**, also known as **vaccination**. Vaccines include inactivated bacterial toxins, killed microbes, parts of microbes, and viable but weakened microbes. These agents can no longer cause disease, but they retain the ability to act as antigens, stimulating an immune response, and more importantly, immunological memory. A

METHODS: Monoclonal Antibody Technology

Prior to 1975, the only source of antibodies for research and clinical applications was the blood of immunized animals. Such antibodies are called *polyclonal antibodies* because they arise from many different B cell clones, each specific for a particular epitope on the immunizing antigen. Indeed, all immune responses are polyclonal. While polyclonal antibodies are useful in laboratory applications, the mixture often includes antibodies of undesired specificities. Another limitation is that the source of the antibody, the immunized animal, has a limited lifespan. In 1975, Georges Kohler and Cesar Milstein developed a method that provides an unlimited source of *monoclonal antibodies* (*MAbs*). (In 1984, they were awarded the Nobel Prize for this work.) They are monoclonal because all of the antibodies in a preparation are the product of a single clone of cells, and the secreted monoclonal antibodies are absolutely identical.

The method illustrated here is the preparation of monoclonal antibodies specific for human chorionic gonadotropin (HCG), a protein hormone present in the blood and urine of pregnant women. First, a mouse is immunized with HCG. Later, once an immune response has developed, the mouse's spleen is removed. The spleen cells include many different HCG-reactive B cells as well as B cells specific for other antigens to which the mouse has recently been exposed—a polyclonal population of cells. The spleen cells are then mixed with myeloma cells in conditions that favor fusion. A myeloma is a cancer cell that, by definition, provides each resultant B cell–myeloma hybrid, or *hybridoma*, with the capacity to proliferate indefinitely. The myeloma cells are also special because they harbor a mutation that prevents them from growing in medium containing the drug aminopterin. In the presence of aminopterin, only myelomas fused with lymphocytes survive, having gained that capacity from their normal partners. And only B cells that fuse to myelomas gain the capacity to proliferate in culture indefinitely. The B cells are thus "immortalized"; they become an unlimited source of the antibody they secrete.

Each surviving hybridoma clone is tested for the production of antibody specific for HCG; those producing the desired antibody are cultured for large-scale production of the MAb. This monoclonal antibody now becomes a sensitive diagnostic tool that is used to detect HCG in a blood or urine sample. The binding of the antibody with a test sample (a positive reaction) reveals the presence of the HCG antigen, indicating that the woman who supplied the sample is pregnant.

In late 1998 a monoclonal antibody became available for cancer treatment. Herceptin, a genetically engineered monoclonal antibody, counteracts a common form of aggressive breast cancer by binding to a growth-factor receptor that is present in excess on the cancer cells.

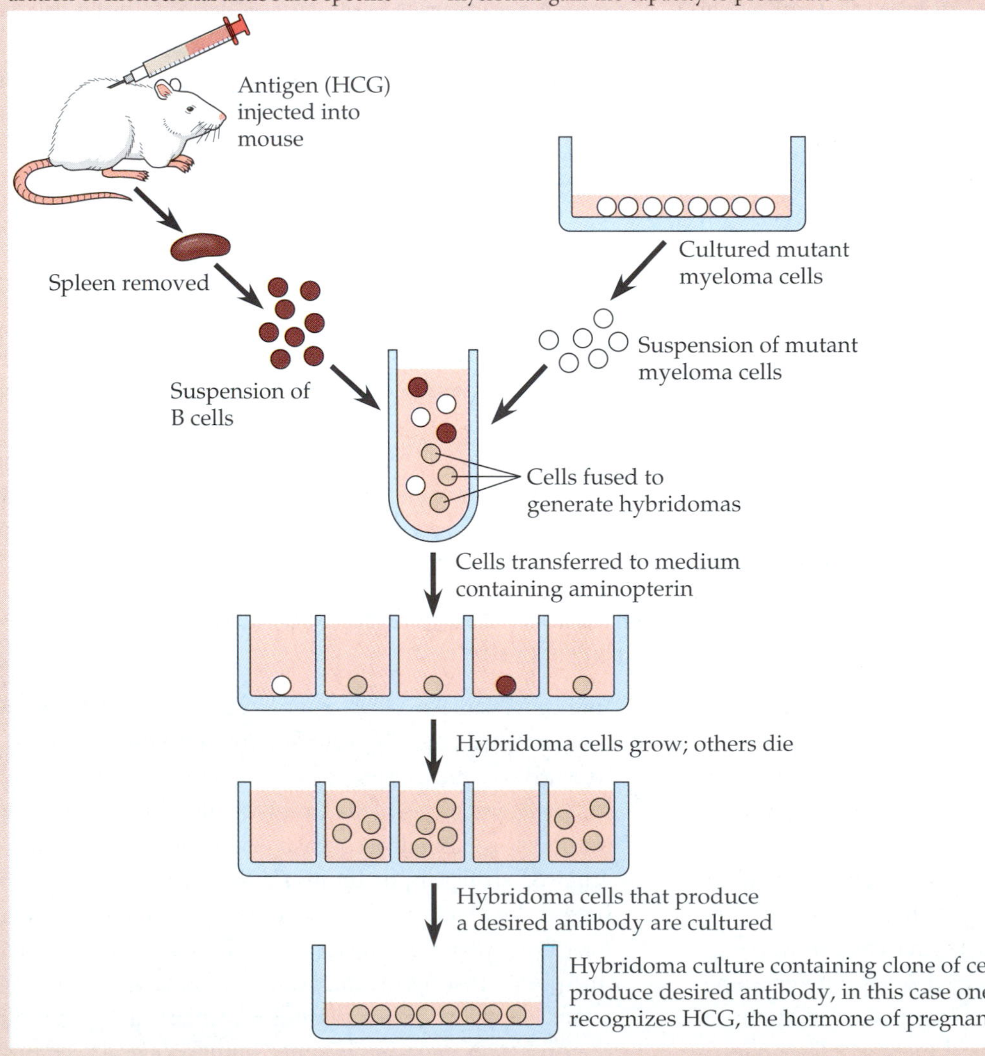

Antigen (HCG) injected into mouse
Spleen removed
Suspension of B cells
Cultured mutant myeloma cells
Suspension of mutant myeloma cells
Cells fused to generate hybridomas
Cells transferred to medium containing aminopterin
Hybridoma cells grow; others die
Hybridoma cells that produce a desired antibody are cultured
Hybridoma culture containing clone of cells that produce desired antibody, in this case one that recognizes HCG, the hormone of pregnancy

vaccinated person who encounters the actual pathogen will have the same quick secondary response based on memory cells as a person who has had the disease. Routine immunization of infants and children has dramatically reduced the incidence of infectious diseases such as measles and whooping cough, and has led to the eradication of smallpox, a disfiguring and often fatal viral disease. Unfortunately, not all infectious agents are easily managed by vaccination. For example, although researchers are working intensively to develop a vaccine for HIV, they face many problems, such as the antigenic variability of the virus.

Antibodies can be transferred from one individual to another, providing **passive immunity**. This occurs naturally when IgG antibodies of a pregnant woman cross the placenta to her fetus. In addition, IgA antibodies are passed from mother to nursing infant in breast milk, especially in the early secretions called colostrum. Passive immunity persists only as long as these antibodies last (a few weeks to a few months), but it provides protection from infections until the baby's own immune system has matured. Passive immunity can also be transferred artificially by injecting antibodies from an animal that is already immune to a disease into another animal, conferring short-term but immediate protection against that disease. Thus, for example, a person bitten by a rabid animal may be injected with antibodies from other people who have been vaccinated against rabies. This measure is important because rabies may progress rapidly, and the response to an active immunization could take too long to save the life of the victim. Actually, most people infected with rabies virus are given both passive and active immunizations: The injected antibodies fight the virus for a few weeks, and then the person's own immune response, induced by the immunization and the infection itself, takes over.

The immune system's capacity to distinguish self from nonself limits blood transfusion and tissue transplantation

In addition to distinguishing between the body's own cells and pathogens such as bacteria, the immune system, given the opportunity, wages war against cells from other individuals. For example, skin transplanted from one person to a nonidentical person will look healthy for a day or two, but it will then be destroyed by immune responses. Interestingly, however, a pregnant woman does not reject the fetus as a foreign body. Apparently, the structure of the placenta is the key to this acceptance (see FIGURE 46.17).

In this section you will learn about the potential problems associated with blood transfusions and tissue transplants. Keep in mind that the body's hostile reaction to an incompatible transfusion or transplant is not a disorder of the immune system, but a normal reaction of a healthy immune system exposed to foreign antigens.

Blood Groups and Blood Transfusion

The genetic basis for the **ABO blood groups** was discussed in Chapter 14. An individual with type A blood has A antigens on the surface of his or her red blood cells. The A molecule is referred to as an antigen because it may be identified as foreign if placed in the body of another person; the A antigen is *not* antigenic to its "owner." Similarly, B antigens are found on type B red blood cells, A and B antigens are found on type AB red blood cells, and neither antigen is found on type O red blood cells.

An individual with type A blood will not, of course, produce antibodies to A antigens. However, a person with type A blood has antibodies to the B antigen, even if the person has never been exposed to type B blood. You may find it odd that antibodies to foreign blood group antigens exist in the body even in the absence of an incompatible blood transfusion. These antibodies arise in response to bacteria (normal flora) that have epitopes very similar to blood group antigens. Thus an individual with type A blood does not make antibodies to A-like bacterial epitopes—the immune system considers these as self—but that person does make antibodies to B-like bacterial epitopes. Therefore, if a person of blood type A receives a transfusion of type B blood, the preexisting anti-B antibodies will induce an immediate and devastating transfusion reaction. The ABO blood types and corresponding antibodies are summarized in FIGURE 14.10.

Because blood group antigens are polysaccharides, they induce T-independent responses, which elicit no memory cells. As a result, each response is like a primary response, and it generates IgM anti-blood-group antibodies, not IgG. This is fortunate; since IgM does not cross the placenta, no harm comes to a developing fetus with a blood type different from its mother's. However, another red blood cell antigen, the **Rh factor**, is able to cause trouble because antibodies produced to it are of the IgG class. A potentially dangerous situation can arise when a mother is Rh-negative (lacks the Rh factor) but has a fetus that is Rh-positive, having inherited the factor from the father. If small amounts of fetal blood cross the placenta, as may happen late in pregnancy or during delivery of the baby, the mother mounts a T-independent humoral response against the Rh factor. The danger occurs in subsequent Rh-positive pregnancies, when the mother's Rh-specific memory B cells are exposed to the Rh factor. These B cells produce IgG antibodies, which can cross the placenta and destroy the red cells of the fetus. To prevent this, the mother is injected with anti-Rh antibodies after delivering her first Rh-positive baby. She is, in effect, passively immunized (artificially) to eliminate the Rh antigen before her own immune system responds and generates immunological memory against the Rh factor, endangering her future Rh-positive babies.

Tissue Grafts and Organ Transplantation

The major histocompatibility complex (MHC), the protein fingerprint unique to each individual, is responsible for stimulating the rejection of tissue grafts and organ transplants. Foreign MHC molecules are antigenic, inducing immune responses against the donated tissue or organ. To minimize rejection, attempts are made to match the MHC of the tissue donor and recipient as closely as possible. In the absence of an identical twin, siblings usually provide the closest tissue-type match. In addition to a close match, various medicines are necessary to suppress the immune response to the transplant. However, this strategy leaves the recipient more susceptible to infection and cancer during the course of treatment. More selective drugs such as cyclosporin A and FK506, which suppress helper T cell activation without crippling nonspecific defense or T-independent humoral responses, have greatly improved the success of organ transplants.

In one type of life-giving transplant, that of bone marrow, the graft itself, rather than the host, is the source of potential immune rejection. Bone marrow transplants are used to treat leukemia and other cancers, as well as various hematological (blood cell) diseases. As in any transplant, the MHC of donor and recipient are matched as closely as possible. Prior to transplant, the recipient is typically treated with irradiation to eliminate his or her own bone marrow cells, including any abnormal cells. This treatment effectively obliterates the recipient's immune system, leaving little chance of graft rejection. However, the great danger in bone marrow transplants is that the donated marrow, containing lymphocytes, will react against the recipient. This **graft versus host reaction** is limited if the MHC molecules of the donor and recipient are well matched. Bone marrow donor programs around the world continually seek volunteers; because of the great variability of the MHC, a diverse pool of potential donors is essential.

Abnormal immune function can lead to disease

The highly regulated interplay of lymphocytes with foreign substances, with each other, and with other body cells provides us with extraordinary protection from many diseases. However, if this delicate balance is disrupted by an immune system malfunction, the effects on the individual can range from the minor inconvenience of some allergies to the serious and often fatal consequences of certain autoimmune and immunodeficiency diseases.

Allergies

Allergies are hypersensitive (exaggerated) responses to certain environmental antigens, called allergens. One hypothesis to explain the origin of allergies is that they are evolutionary remnants of the immune system's response to parasitic worms. The humoral mechanism that combats worms is similar to the allergic response that causes such disorders as hay fever and allergic asthma.

The most common allergies involve antibodies of the IgE class (see TABLE 43.1, p. 853). For example, hay fever occurs when plasma cells secrete IgE specific for pollen allergens. Some of the IgE antibodies attach by their tails to mast cells present in connective tissues, without binding to the pollen. Later, when pollen grains enter the body, they attach to the antigen-binding sites of mast cell–associated IgE, cross-linking adjacent antibody molecules. This event induces the mast cell to degranulate—that is, to release histamine and other inflammatory agents from vesicles called granules (FIGURE 43.18). Recall that histamine causes dilation and increased permeability of small blood vessels. These inflammatory events lead to typical allergy symptoms: sneezing, runny nose, tearing eyes, and smooth muscle contractions that can result in breathing difficulty. Antihistamines diminish allergy symptoms by blocking receptors for histamine.

The most serious consequence of an acute allergic response is **anaphylactic shock**, a life-threatening reaction to injected or ingested allergens. Anaphylactic shock occurs when widespread mast cell degranulation triggers abrupt dilation of peripheral blood vessels, causing a precipitous drop in blood pressure. Death may occur within a few minutes. Allergic responses to bee venom or penicillin can lead to anaphy-

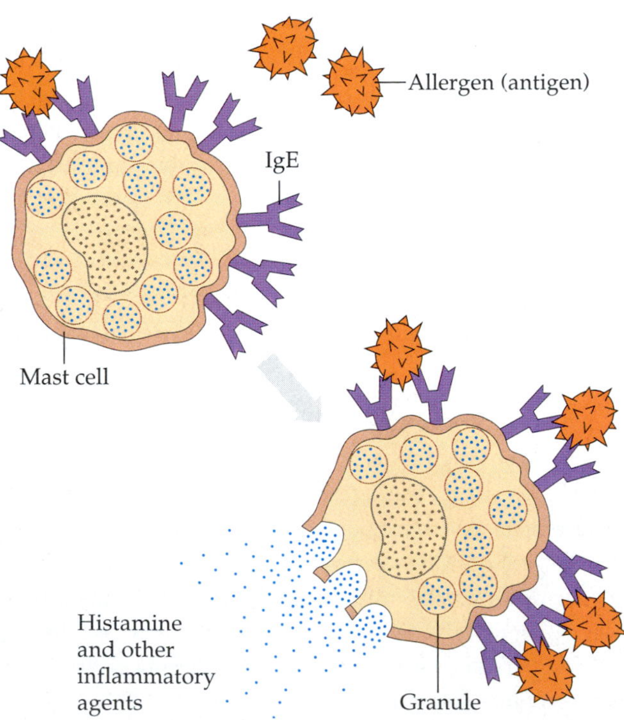

FIGURE 43.18 ▪ Mast cells and the allergic response. Upon first exposure to an allergen, plasma cells secrete IgE antibodies specific for the allergen. Some of these antibodies are attached by their tails to a mast cell. When, upon second exposure, the allergen binds to IgE already on the mast cell, it triggers the degranulation of the cell. Cellular granules (vesicles) release histamine, leading to most of the symptoms of the allergy.

lactic shock in people who are extremely allergic to these substances. Likewise, people very allergic to peanuts, fish, or other foods have died from eating only tiny amounts of these allergens. Some individuals with severe hypersensitivities carry syringes containing the hormone epinephrine, which counteracts this allergic response.

Autoimmune Diseases

Sometimes the immune system loses tolerance for self and turns against certain molecules of the body, causing one of the many autoimmune diseases. In systemic lupus erythematosus (lupus), for example, the immune system generates antibodies (known as auto-antibodies) against all sorts of self molecules, even histones and DNA released by the normal breakdown of body cells. Lupus is characterized by skin rashes, fever, arthritis, and kidney dysfunction. Another antibody-mediated autoimmune disease, rheumatoid arthritis, leads to damage and painful inflammation of the cartilage and bone of joints. In insulin-dependent diabetes mellitus, the insulin-producing beta cells of the pancreas are the targets of autoimmune cell-mediated responses. A final example is multiple sclerosis (MS), the most common chronic neurological disease in developed countries. In MS, T cells reactive against myelin infiltrate the central nervous system and destroy the myelin of neurons (see FIGURE 48.2). People with MS experience a number of serious neurological abnormalities.

The causes of autoimmunity are varied and complex, and there is still much to learn about these diseases. One intriguing finding is that the inheritance of particular MHC alleles is associated with susceptibility to certain autoimmune diseases, such as insulin-dependent diabetes mellitus.

Immunodeficiency Diseases

There are almost as many immunodeficiency diseases as there are components of the immune system. Many inborn deficiencies affect the function of either humoral or cell-mediated immune defenses. In severe combined immunodeficiency (SCID), both branches of the immune system fail to function. For people with this genetic disease, long-term survival requires a bone marrow transplant that will continue to supply functional lymphocytes. For one type of SCID, caused by deficiency of the enzyme adenosine deaminase (ADA), medical scientists have been working to develop a gene therapy in which the individual's own cells are removed, provided with a functional ADA gene, and returned to the body. This treatment would eliminate the danger of graft versus host disease. However, the results to date are equivocal because the patients are also being given additional supplies of the enzyme.

Immunodeficiency is not always an inborn condition; an individual may develop immune dysfunction later in life. For example, certain cancers suppress the immune system, especially Hodgkin's disease, which damages the lymphatic system. Another well-known and devastating acquired immune deficiency is AIDS, described in the next section.

Healthy immune function appears to depend on both the endocrine system and the nervous systems. Nearly 2000 years ago the Greek physician Galen recorded that people suffering from depression were more likely than others to develop cancer. In fact, there is growing evidence that physical and emotional stress can harm immunity. Hormones secreted by the adrenal glands during stress affect the numbers of white blood cells and may suppress the immune system in other ways.

The association between emotional stress and immune function also involves the nervous system. Some neurotransmitters secreted when we are relaxed and happy may enhance immunity. In one study, college students were examined just after a vacation and again during final exams. Their immune systems were impaired in various ways during exam week; for example, interferon levels were lower. These and other observations indicate that general health and state of mind affect immunity. Physiological evidence also points to an immune system–nervous system link: Receptors for neurotransmitters have been discovered on the surfaces of lymphocytes, and a network of nerve fibers penetrates deep into the thymus.

AIDS is an immunodeficiency disease caused by a virus

In 1981, health care workers in the United States noticed an increasing number of cases of Kaposi's sarcoma, a cancer of the skin and blood vessels, and of *Pneumocystis carinii* pneumonia, a protozoan infection. The increased rates were notable because of the rarity of these diseases among the general population; they were known to occur mainly in severely immunosuppressed individuals. These observations led to the recognition of what came to be known as **acquired immunodeficiency syndrome**, or **AIDS**. People with AIDS are highly susceptible to **opportunistic diseases**, infections and cancers that take advantage of an immune system in collapse. The *Pneumocystis* protozoan is a ubiquitous organism, yet it does not cause pneumonia in a person with a healthy immune system. In people with AIDS, opportunistic diseases, neurologic damage, and physiological wasting lead to death.

By 1983, a retrovirus, now called **human immunodeficiency virus** (**HIV**), had been identified as the causative agent of AIDS (FIGURE 43.19, p. 860; also see FIGURE 18.7). With the AIDS mortality rate close to 100%, HIV is the most lethal pathogen ever encountered. The virus probably evolved from another HIV-like virus in central Africa and may have caused unrecognized cases of infection and AIDS there for many years. The virus has been identified in preserved blood samples from as early as 1959 in African nations and in England.

There are two major strains of the virus, HIV-1 and HIV-2. HIV-1 is the more widely distributed and more virulent strain. Both strains infect cells that bear surface CD4 molecules (CD4

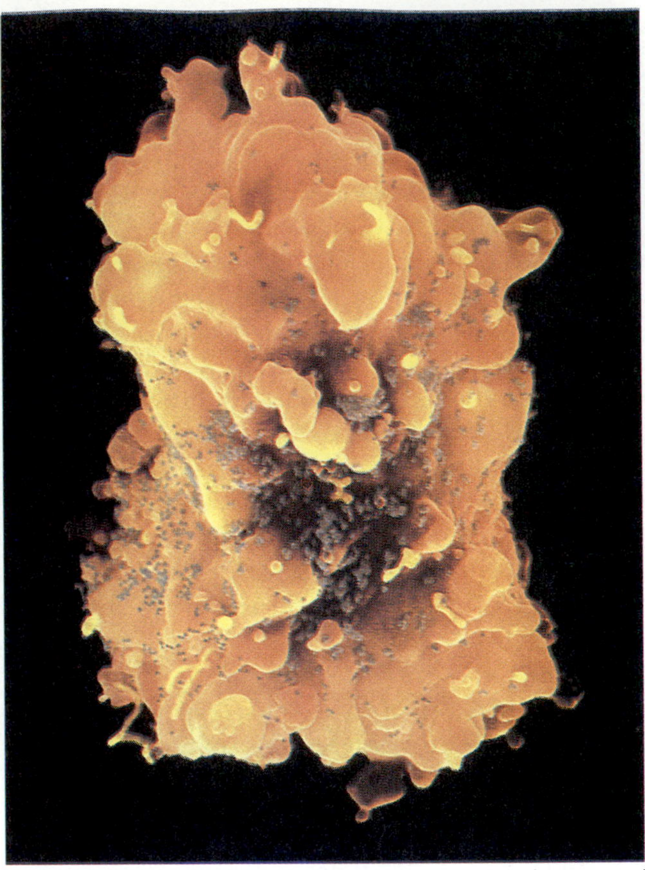

FIGURE 43.19 ▪ A T cell infected with HIV. The viruses (blue) bud continuously from the surface of the T cell (orange; colorized SEM). The cell will die, but only after it produces many copies of its viral killer.

cells). Recall that CD4 molecules on helper T cells assist in the T_H cell–antigen-presenting cell interaction. Because CD4 also functions as the major receptor for the virus, helper T cells are highly susceptible to infection. Other cells that bear fewer molecules of CD4, such as macrophages and some B lymphocytes, are also among the cells infected by HIV.

The cells most critical to HIV pathogenesis are CD4 T cells and macrophages. In both cell types, entry by virus requires not only CD4 but a second protein molecule, a so-called coreceptor. The recently identified coreceptors include fusin, found on helper T cells, and CCR5, found on macrophages. Fusin and CCR5 normally function as receptors for various chemokines. In fact, they were first recognized as HIV coreceptors after it was discovered that some chemokines suppress HIV-1 infection. As it turns out, chemokines inhibit HIV-1 entry because they bind to and block chemokine receptors on potential host cells. In addition, some people who are innately resistant to HIV-1 make defective chemokine receptors. They are not infected because the coreceptor for HIV entry is faulty.

Once inside a cell, HIV RNA is reverse-transcribed, and the
43.2 product DNA is integrated into the host cell genome. In this provirus form, the viral genome directs the production of new virus particles (see FIGURE 18.7). Because a retrovirus exists as a provirus for the life of the infected cell, antibodies fail to eradicate it. Even more challenging to both humoral and cell-mediated responses, however, are the frequent mutational changes that occur in each round of virus replication. Indeed, most HIV particles produced in an infected person differ at least slightly from the original infecting virus.

In spite of these challenges, the immune system responds to the virus. After an initial peak, the number of viruses in the blood falls while anti-HIV antibody rises (FIGURE 43.20). The apparent decrease in the number of viruses in the body represents, in part, the early immune response to HIV. Detection of the anti-HIV antibodies, which appear in the blood about 1 to 12 months after infection, is the most common method for identifying infected individuals. A person who is **HIV-positive** is infected, having tested positive for the presence of antibodies to the virus. The HIV antibody test has also been used to screen all blood supplies in the United States since 1985. Because of the chronic presence of virus, a person continues to have anti-HIV antibodies, perhaps until late in AIDS, when both branches of immunity collapse with the loss of CD4 T cells.

The early fall in the level of HIV in the blood is misleading, however. During this time, HIV replication continues in cells of the lymph nodes and causes structural and functional damage there. In time, the concentration of HIV in the blood (the viral load) increases. Causes of this increase include the breakdown of lymphatic tissue function, the release of virus from these tissues, and the depletion of CD4 and helper T cells (see FIGURE 43.20).

How CD4 T cells become depleted in AIDS is not well understood. The cause is not simply cell death by HIV entry and replication. CD4 T cells that are not infected by virus are also lost. Many different mechanisms for T cell depletion have been proposed and tested. One possibility is that HIV-mediated interactions induce a T cell to undergo inappropriately timed apoptosis, a self-destruction mechanism that is normally highly regulated and crucial to normal immune function.

The time required for an HIV infection to progress to severe CD4 T cell depletion and AIDS varies greatly, but it currently averages about 10 years. During most of this time the individual exhibits only moderate hints of illness, such as swollen lymph nodes (indicating ongoing virus activity) and occasional fever. People with HIV and their physicians follow changes in the level of T cells as an indication of disease progression. However, it has recently been shown that measures of viral load are a better indicator of disease prognosis and of the effectiveness of anti-HIV treatment.

At this time, HIV infection cannot be cured, and the progression to AIDS cannot be prevented. Even though new drug combinations show promise in slowing this progression, the treatments are very expensive and not available to all HIV-positive people. Drugs that seem to slow viral replication when used in various combinations include DNA-synthesis inhibitors, reverse transcriptase inhibitors (such as AZT and

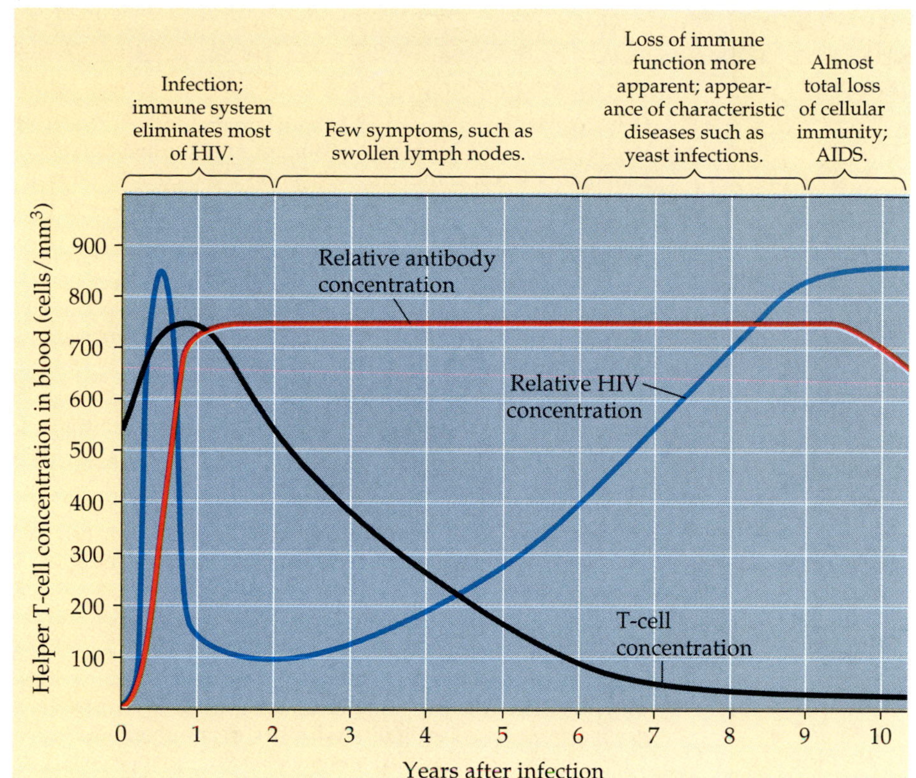

FIGURE 43.20 ▪ The stages of HIV infection. HIV concentration in the blood increases rapidly after the initial infection but falls due to the immune response. However, viruses continue to replicate within lymphatic tissues and cause damage there. The immune system continues to combat the viruses during this period, which may last for years, but in time the virus concentration increases as helper T-cell concentration gradually decreases. AIDS is the last stage of the process.

ddI), and a third, new class of drugs called protease inhibitors. Protease inhibitors prevent a key step in the synthesis of HIV proteins. In addition, many medicines are used to treat the myriad of opportunistic diseases associated with AIDS. These drugs may prolong life, but they do not cure AIDS.

Transmission of HIV requires the transfer of body fluids containing infected cells, such as semen or blood. Unprotected sex (that is, without a condom) among male homosexuals and transmission via nonsterile needles (typically among intravenous drug users) account for most of the AIDS cases reported thus far in the United States and Europe. However, transmission of HIV among heterosexuals is rapidly increasing as a result of unprotected sex with infected partners. In Africa and Asia, transmission has been primarily by heterosexual sex, especially where there is a high incidence of other sexually transmitted diseases that result in genital lesions. These sores facilitate the transmission of HIV, as the skin barrier (first line of defense) is breached and HIV-susceptible cells, mainly macrophages and helper T cells, are attracted to the area by the inflammatory response.

HIV is not transmitted by casual contact. So far, only one case of HIV transmission by kissing has been reported, and both the person who transmitted the virus and the recipient had bleeding gums. The key to recognizing risk is to remember that the virus is best transmitted by *direct* transfer of infected cells, and this is possible when any blood or body secretions are passed from one person to another. Transmission of HIV from mother to child has occurred in two ways: Transmission during fetal development occurs in about 25% of HIV-infected mothers, and the virus can also pass from mother to infant during nursing. HIV antibody screening has virtually eliminated blood transfusions as a route of transmission in developed countries. This test, however, will never completely guarantee a safe blood supply because a person may be infected by the virus for several weeks or months before anti-HIV antibodies become detectable.

In 1997 nearly 6 million people worldwide acquired HIV, and over 2 million perished from AIDS, including about 460,000 children. This number of AIDS cases is expected to grow by nearly 20% per year. The best approach for slowing the spread of HIV is to educate people about the practices that transmit the virus, such as using nonsterile needles and having sex without a condom. Although condoms do not completely eliminate the risk of transmitting HIV (or other similarly transmitted viruses, such as the hepatitis B virus), they do reduce it. Anyone who has sex—vaginal, oral, or anal—with a partner who had unprotected sex with another person during the past two decades risks exposure to HIV.

▪ ▪ ▪

The immune response is one of many adaptations that enable animals to adjust to the adversities of the environment. The next chapter describes several other processes that help maintain favorable conditions within animals as they cope with varying external environments.

CHAPTER REVIEW

REVIEW OF KEY CONCEPTS

(with page numbers and key figures)

NONSPECIFIC DEFENSES AGAINST INFECTION

- **The skin and mucous membranes provide first-line barriers to infection** (pp. 840–841, FIGURE 43.1) The first line of nonspecific defense consists of the intact skin and mucous membranes, mucus, ciliated cells lining the upper respiratory system, lysozyme, and gastric juices.

- **Phagocytic cells, inflammation, and antimicrobial proteins function early in infection** (pp. 841–844, FIGURE 43.5) The second line of nonspecific defense depends primarily upon neutrophils and macrophages, phagocytic white cells in the blood and tissues. Natural killer cells mediate lysis of virus-infected cells and tumor cells. Tissue damage triggers a local inflammatory response. Injured cells release histamine, a chemical signal that causes dilation and increased permeability of blood vessels, allowing fluid and large numbers of phagocytic white blood cells to enter the tissues. The most important antimicrobial proteins in the blood and tissues are the proteins of the complement system, involved in nonspecific and specific defense, and interferons. Secreted by virus-infected cells, interferons inhibit virus production in neighboring cells.

HOW SPECIFIC IMMUNITY ARISES

- 43.1 **Lymphocytes provide the specificity and diversity of the immune system** (pp. 844–845) A substance that elicits an immune response is called an antigen. The immune system recognizes specific antigens (molecules belonging to microbes, toxins, transplanted tissue, or cancer cells) and develops an immune response that inactivates or destroys that substance. B lymphocytes and T lymphocytes recognize antigens via surface antigen receptors: membrane antibodies for B cells, and T cell receptors for T cells. Lymphocytes circulate throughout the blood and lymph and are found in high numbers in lymphatic tissues. The great diversity of lymphocytes, each with receptors of particular specificity, gives the immune system the capacity to respond to virtually any antigen.

- **Antigens interact with specific lymphocytes, inducing immune responses and immunological memory** (pp. 845–846, FIGURE 43.6) Clonal selection occurs when an antigen activates a lymphocyte by binding to specific receptors. In the primary immune response, the body's first exposure to an antigen, the lymphocyte proliferates and differentiates, forming a clone of short-lived, infection-fighting effector cells and a clone of long-lived memory cells, all specific for the antigen. Secondary immune responses to that same antigen, which involve memory cells, are faster and often protective.

- **Lymphocyte development gives rise to an immune system that distinguishes self from nonself** (pp. 846–847, FIGURES 43.8 and 43.9) Lymphocytes develop from pluripotent stem cells in the bone marrow. B cells mature in the marrow, while T cells mature in the thymus. Self-tolerance develops as lymphocytes bearing receptors specific for native molecules are destroyed or rendered nonresponsive. Major histocompatibility complex (MHC) molecules are crucial to T cell function. Class I MHC molecules, located on all nucleated cells of the body, present antigen fragments to cytotoxic T cells. Class II MHC molecules, found mainly on macrophages and B cells, present antigen fragments to helper T cells. Developing T cells are exposed to class I and II MHC molecules on cells of the thymus. Only T cells bearing receptors with affinity for self-MHC reach maturity.

IMMUNE RESPONSES

- 43.1 **Helper T lymphocytes function in both humoral and cell-mediated immunity: *an overview*** (pp. 849–850, FIGURES 43.10 and 43.11) Humoral, or B cell, immunity, based on circulation of antibodies in the blood and lymph, defends against free viruses, bacteria, and other extracellular threats. Cell-mediated, or T cell, immunity defends against intracellular pathogens by destroying infected cells; it also defends against transplanted tissue and cancer cells. A CD4-bearing helper T cell is activated when its receptor binds specifically to a class II MHC–antigen complex on the surface of an antigen-presenting cell (APC). The T cell then secretes interleukin-2 and other cytokines, which activate B cells and cytotoxic T cells.

- 43.1 **In the cell-mediated response, cytotoxic T cells counter intracellular pathogens: *a closer look*** (pp. 850–851, FIGURE 43.12) Most cytotoxic T cells are activated by cytokines and specific binding to class I MHC–antigen complexes on a target (infected, transplanted, or cancerous) cell. The T cell then secretes perforins, which form pores in the target cell membrane, causing the cell to lyse.

- 43.1 **In the humoral response, B cells make antibodies against extracellular pathogens: *a closer look*** (pp. 851–855, FIGURES 43.13–43.16) B cells are activated by cytokines and specific binding of its membrane antibodies to extracellular antigens. Most of these antigens are proteins or large polysaccharides, each with multiple epitopes. Antibodies, also called immunoglobulin (Ig) molecules, are serum proteins. The variable region of an Ig molecule binds to a specific epitope; the constant region determines the antibody's class. The five major immunoglobulin classes are IgG, IgM, IgA, IgD, and IgE. An antibody does not destroy an antigen directly but neutralizes it or targets it for elimination by opsonization, agglutination, precipitation, or complement fixation. Opsonization, agglutination, and precipitation enhance phagocytosis of the antigen-antibody complex; complement fixation leads to lysis of a complement protein–bound bacterium or virus.

- **Invertebrates have a rudimentary immune system** (p. 855) Invertebrates have the ability to distinguish between self and nonself. In many invertebrates, amoeboid cells called coelomocytes can identify and destroy foreign substances. Experiments with earthworms also show that their defense systems form memory against tissue grafts.

IMMUNITY IN HEALTH AND DISEASE

- **Immunity can be achieved naturally or artificially** (pp. 855, 857) Active immunity occurs when the immune system responds to a foreign antigen acquired either by natural infection or artificially, as by immunization. In immunization, a nonpathogenic form of a microbe or part of a microbe generates an immune response to and immunological memory for that microbe. Passive immunity occurs when antibodies are transferred from one individual to another. It occurs naturally, when IgG passes from mother to fetus or when IgA passes from mother to infant in breast milk, or artificially, when antibodies from an animal immune to a disease are injected into another animal, conferring short-term protection.

- **The immune system's capacity to distinguish self from nonself limits blood transfusion and tissue transplantation** (pp. 857–858) Certain antigens on red blood cells determine whether a person has type A, B, AB, or O blood. Antibodies to nonself blood types (generally IgM) already exist in the body. If incompatible blood is transfused, the transfused cells are killed by antibody and complement-mediated lysis. The Rh factor, another red blood cell antigen, creates difficulties when an Rh-negative mother carries successive Rh-positive fetuses. During delivery of the first Rh-positive infant, the mother's immune system develops anti-Rh IgG, which can cross the placenta and may attack the red blood cells of a subsequent Rh-positive fetus. The chances of success in organ or tissue transplantation are improved if the donor and recipient MHC tissue types are well matched. In addition, immunosuppressive drugs help prevent rejection. In bone marrow transplantation, there is danger of a graft versus host reaction.

- **Abnormal immune function can lead to disease (pp. 858–859, FIGURE 43.18)** In allergies such as hay fever, an allergen, such as pollen, triggers histamine release from mast cells, inducing vascular changes and typical symptoms. Sometimes the immune system loses tolerance for self, leading to autoimmune diseases such as rheumatoid arthritis and insulin-dependent diabetes. Some people are naturally deficient in humoral or cell-mediated immune defenses, or both.

- 43.2 **AIDS is an immunodeficiency disease caused by a virus (pp. 859–861, FIGURE 43.20)** Acquired immunodeficiency syndrome (AIDS) is caused by the direct and indirect destruction of CD4-bearing T cells by HIV, the human immunodeficiency virus, over a period of years. AIDS, the final stage of this process, is marked by low helper T cell levels and opportunistic diseases characteristic of a deficient cell-mediated immune response.

SELF-QUIZ

1. An individual with type AB blood
   a. is considered a universal blood donor
   b. is considered a universal blood recipient
   c. produces antibodies to the B antigen
   d. produces antibodies to the A antigen
   e. is Rh-positive

2. Which of the following results in long-term immunity?
   a. the passage of maternal antibodies to her developing fetus
   b. the inflammatory response to a splinter
   c. the administration of serum obtained from people immune to rabies
   d. the administration of the chickenpox vaccine
   e. the passage of maternal antibodies to her nursing infant

3. Foreign antigens synthesized within body cells are presented by
   a. class I MHC molecules to cytotoxic T cells
   b. class II MHC molecules to helper T cells
   c. class I MHC molecules to helper T cells
   d. class II MHC molecules to CD4-bearing cells
   e. class II MHC molecules to cytotoxic T cells

4. Which of the following molecules is *incorrectly* paired with a source?
   a. lysozyme—tears
   b. interferons—virus-infected cells
   c. interleukin-1—macrophages
   d. perforins—cytotoxic T cells
   e. immunoglobulins—helper T cells

5. HIV targets include all of the following *except*
   a. macrophages
   b. cytotoxic T cells
   c. helper T cells
   d. cells bearing CD4 and fusin
   e. cells bearing CD4 and CCR5

6. Which of the following is true about a hybridoma used for the production of monoclonal antibodies?
   a. It is the product of a myeloma cell–B cell fusion.
   b. It will grow in medium containing the drug aminopterin.
   c. It will survive in culture indefinitely.
   d. It produces antibodies of varied specificities.
   e. All of the above except d are true.

7. Which of the following is a characteristic of the early stages of local inflammation?
   a. precapillary arteriole constriction
   b. fever
   c. attack by cytotoxic T cells
   d. release of histamine
   e. antibody-complement–mediated lysis of microbes

8. An epitope associates with which part of an antibody?
   a. the antibody-binding site
   b. the heavy chain constant regions only
   c. the variable regions of a heavy chain and light chain combined
   d. the light chain constant regions only
   e. the antibody tail

9. Which of the following is *not* true about helper T cells?
   a. They function in both cell-mediated and humoral immune responses.
   b. They recognize polysaccharide fragments presented by class II MHC molecules.
   c. They bear surface CD4 molecules.
   d. They are subject to infection by HIV.
   e. When activated, they secrete IL-2 and other cytokines.

10. The body produces antibodies specific to foreign antigens. The process by which the body comes up with the correct antibodies to fight a given pathogen is most like which of the following?
    a. going to a tailor and having a suit made to fit you
    b. ordering Chinese take-out without looking at a menu
    c. testing the fit of a glass slipper on many people, and finding it fits only Cinderella
    d. walking into a video store and choosing the first film you come to
    e. selecting a lottery winner by means of a random drawing

CHALLENGE QUESTIONS

1. Active immunization against hepatitis B virus requires three injections over a period of several months. Explain why multiple injections might be necessary to achieve immunity to this virus. Consider the elements of clonal selection and memory presented in FIGURES 43.6 and 43.7.
2. An Rh-positive baby is born to an Rh-negative mother. After the delivery the mother is passively immunized with antibodies to the Rh factor. Two years later she gives birth to another healthy Rh-positive baby. How did the antibodies administered after the first birth protect the second baby? Should the mother be given antibodies to the Rh factor again? Why or why not?
3. Interferon-gamma has many effects on cells, one of which is to increase the number of class I MHC molecules on the cell surface. Interferon-gamma prepared by recombinant DNA technology is currently being tested clinically for the treatment of some viral infections and cancers. Considering the multiple activities of interferon-gamma, how might it aid in the immune response against these abnormal cells?
4. Draw a diagram depicting the central role of helper T cells in both cell-mediated and humoral responses.

SCIENCE, TECHNOLOGY, AND SOCIETY

1. Concern is increasing over the rising rate of HIV infection among teenagers. Schools in some large cities have instituted programs to make condoms available to students, along with advice and counseling about safer sex. These plans have divided school boards and communities. Some citizens and some church groups are opposed to giving condoms to students, because it might appear to encourage their sexual activity. Many school and public health officials view the situation differently, as a matter of life and death. The heart of the controversy seems to be whether the schools should take such a direct role in this area of student life. What do you think the schools' role should be?

2. Until a vaccine became available, poliomyelitis was one of the most feared diseases, resulting in paralysis when the virus destroyed motor neurons in the brain and spinal cord. Currently, both a live, attenuated oral and an injectable, inactivated (killed) poliovirus vaccine are available. The ease of administration has made the oral vaccine more popular and it is generally credited with controlling the disease worldwide. However, the live virus may mutate to the more virulent form. In any given year approximately 10 people in the United States contract vaccine-associated paralytic polio. Of those cases, the majority are in individuals who were healthy when immunized or who came into contact with someone recently vaccinated. Statistically, the probability of contracting vaccine-associated polio remains low, 1 in 12 million. Do you feel the risk is acceptable when compared to the benefits of oral vaccination? How might public health decisions of this type be made?

FURTHER READING

Beck, G., and G. S. Habicht. "Immunity and the Invertebrates." *Scientific American*, November 1996. A discussion of invertebrate immunity and the evolution of the human immune system.

"Defeating AIDS: What Will It Take?" *Scientific American*, July 1998. A special section on the prevention and treatment of HIV infection and AIDS.

"Elements of Immunity." *Science*, April 5, 1996. A special section on nonspecific and specific immune mechanisms.

"Life, Death, and the Immune System." *Scientific American*, September 1993. A special issue devoted to the immune system.

Nowak, M. A., and A. J. McMichael. "How HIV Defeats the Immune System." *Scientific American*, August 1995. A review of the dynamics of HIV infection and how it ultimately leads to immune system collapse.

Richardson, S. "The End of the Self." *Discover*, April 1996. Polly Matzinger's radical new theory challenges the self/nonself model of immunity.

"Vaccines: Frontiers in Medicine." *Science*, September 2, 1994. A special section devoted to advances in the development of several important vaccines.

WEB LINKS

Visit the special edition of *The Biology Place* for BIOLOGY, Fifth Edition, at http://www.biology.com/campbell. Go to Chapter 43 for online resources, including learning activities, practice exams, and links to the following web sites:

"HIV InSite: Gateway to AIDS Knowledge"
The University of California, San Francisco, has provided this comprehensive and reliable information on HIV/AIDS treatment, policy, research, epidemiology, and prevention.

"Sniffles and Sneezes"
A chatty site that contains numerous articles on two common immunological problems, allergies and asthma.

"The American Association of Immunologists"
At this site you can access abstracts from the *Journal of Immunology.*

"Cell to Cell and Person to Person: Investigating AIDS and HIV"
Visit Campbell's Biology Place to gain access to this investigative project by John Postlethwait, University of Oregon.

CONTROLLING THE INTERNAL ENVIRONMENT

Most animals can survive fluctuations in the external environment that are more extreme than any of their individual cells could tolerate. During the course of a day a human may be exposed to substantial changes in outside temperatures but will die if the internal body temperature goes more than a few degrees above or below an average of about 37°C. The spring peeper, a North American tree frog (shown in the photograph here), often spends an entire winter with much of its body frozen. Most of its blood and interstitial fluid turns to ice, but its cells and a film of water surrounding them are kept from freezing by specialized proteins and a high level of glucose throughout the frog's body. Without these protectants, ice crystals would rupture the cell membranes, and the animal would die.

Wintering spring peepers illustrate several themes central to our study of animals. The theme of bioenergetics is evident in that being frozen and inactive enables a frog to go through the entire winter without eating; its metabolic rate is so low its cells survive on the energy resources they have when the animal freezes. Illustrating the theme of short-term physiological adjustment to environmental change, a spring peeper's organ systems respond quickly to low temperature. For example, the liver secretes freeze-protectant chemicals into the animal's blood as soon as the frog's skin starts freezing. Finally, the mechanisms that maintain homeostasis, keeping the frog's internal environment within ranges that cells can tolerate, are adaptations that have evolved in populations interacting with the particular conditions of their surroundings.

This chapter concentrates on homeostasis, the steady-state balance in the internal environment of an animal's body. We focus on ***thermoregulation****, how animals maintain internal temperature within a tolerable range;* ***osmoregulation****, how they regulate solute balance and the gain and loss of water; and* ***excretion****, how they get rid of the nitrogen-containing waste products of metabolism such as urea.*

REGULATION OF BODY TEMPERATURE

Metabolism is very sensitive to changes in the temperature of an animal's internal environment. For example, the rate of cellular respiration increases with temperature up to a certain point and then declines when temperatures are high enough to begin denaturing enzymes. The properties of membranes also change with temperature.

Although different species of animals are adapted to different temperature ranges, each animal has an optimal temperature range. Within that range, many animals can maintain a constant internal temperature as the external temperature fluctuates. **Thermoregulation** is the maintenance of body temperature within a range that enables cells to function efficiently. To understand the problems and mechanisms of temperature regulation, we first need to consider the exchange of heat between organisms and their environment.

Four physical processes account for heat gain or loss

An organism, like all objects, exchanges heat with its external environment by four physical processes: conduction, convection, radiation, and evaporation (FIGURE 44.1).

Conduction is the direct transfer of thermal motion (heat) between molecules of the environment and those of the body surface, as when an animal sits in a pool of cold water or on a hot rock. Heat will always be conducted from a body of higher temperature to one of lower temperature. Water is 50 to 100 times more effective than air in conducting heat. This is one reason you can rapidly cool your body on a hot day just by standing in cold water.

Convection is the transfer of heat by the movement of air or liquid past the surface of a body, as when a breeze contributes to heat loss from the surface of an animal with dry skin. Convection also contributes to the comfort a fan brings a human on a hot, still day, but most of this effect is due to evaporative cooling. On the other hand, a wind-chill factor compounds the harshness of cold winter temperatures.

Radiation is the emission of electromagnetic waves produced by all objects warmer than absolute zero, including an animal's body and the sun. Radiation can transfer heat between objects that are not in direct contact, as when an animal absorbs heat radiating from the sun. Researchers have discovered a specific adaptation for exploiting solar radiation in polar bears. The fur of these animals is actually clear, not white. Each hair functions somewhat like an optical fiber that transmits ultraviolet radiation to the black skin, where the energy is absorbed and converted to body heat.

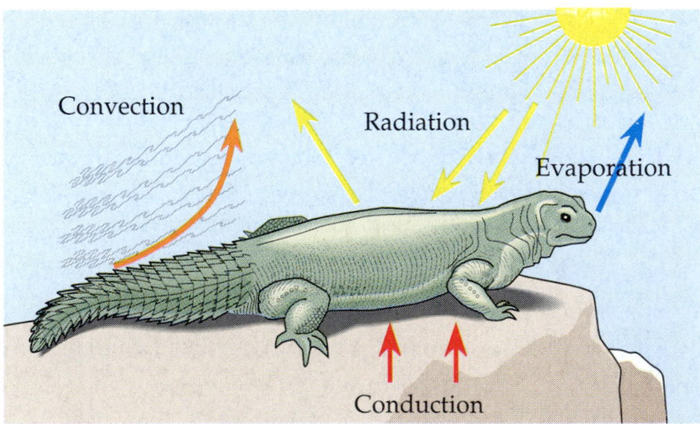

FIGURE 44.1 ▪ Heat exchange between an organism and its environment. Heat is conducted from a body of higher temperature to one of lower temperature. Thus, as indicated here, a lizard can elevate its body temperature using heat conducted from a rock, a common practice among reptiles. At the same time, the animal loses some heat by evaporative cooling from moist surfaces exposed to the environment. Convection may also contribute to its heat loss, when a current of air or water passes over its body. All bodies warmer than absolute zero emit electromagnetic energy. Here, the animal gains more heat via radiation from the sun than its body radiates.

Evaporation is the loss of heat from the surface of a liquid that is losing some of its molecules as gas. Evaporation of water from an animal has a significant cooling effect on the animal's surface.

Convection and evaporation are the most variable causes of heat loss. A breeze of just 15 km/hr will increase total heat loss substantially by increasing convection fivefold. Evaporative cooling is increased greatly by the production of sweat. However, evaporation can only occur if the surrounding air is not saturated with water molecules (that is, if the relative humidity is less than 100%). This is the biological basis for the common complaint, "The heat is not as bad as the humidity."

Ectotherms derive body heat mainly from their surroundings; endotherms derive it mainly from metabolism

One way to classify the thermal characteristics of animals is to emphasize the major source of body heat. An **ectotherm** warms its body mainly by absorbing heat from its surroundings. The amount of heat it derives from its own metabolism is usually negligible. Most invertebrates, fishes, amphibians, and reptiles are ectotherms. In contrast, an **endotherm** derives most or all of its body heat from its own metabolism. Mammals, birds, some fishes, and numerous insects are endotherms. Many endotherms maintain a virtually constant internal temperature even as the temperature of their surroundings fluctuates (FIGURE 44.2). However, a constant body temperature does not distinguish endotherms from ectotherms. For example, many ectothermic marine fishes and invertebrates inhabit water with such stable temperatures that their body temperature varies less than that of humans and other endotherms. Also, the terms *cold-blooded* and *warm-blooded* are misleading. When sitting in the sun, many lizards, which are ectotherms, have higher body temperatures than mammals.

As these examples show, the terms *ectotherm* and *endotherm* are not based on body temperature but rather on the main source of body heat. However, even this distinction of environmental versus metabolic heat sources is not absolute. Many insects and some fishes retain metabolic heat to warm certain parts of their body, such as the thorax, where the muscles used in locomotion are located; and birds and mammals, which are endotherms, may add body heat by basking in the sun.

Endothermy solves certain problems of living on land. It enables terrestrial animals to maintain a constant body temperature in the face of environmental temperature fluctuations that are generally more severe than those an aquatic animal confronts. In general, endothermic vertebrates—birds and mammals—are warmer than their surroundings, but these

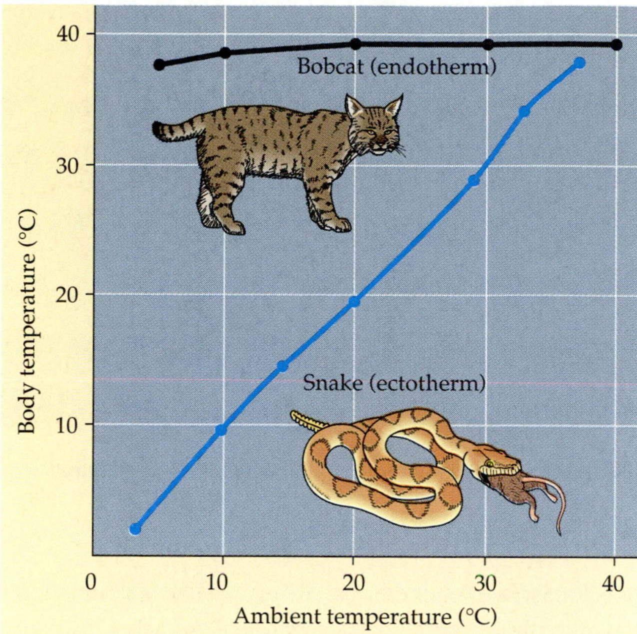

FIGURE 44.2 ▪ The relationship between body temperature and ambient (environmental) temperature in an ectotherm and an endotherm. An ectotherm (such as a snake) obtains most of its body heat from its surroundings. An endotherm (such as a bobcat) derives its body heat mainly from metabolism and uses metabolic energy for heating and cooling mechanisms that keep body temperature relatively constant. (Modified from P. T. Marshall and G. M. Hughes, *Physiology of Mammals and Other Vertebrates,* 2nd ed. Cambridge: Cambridge University Press, 1980.)

animals also have mechanisms for cooling the body in a hot environment. A consistently warm body temperature requires active metabolism, but conversely, a warm body temperature contributes to the high levels of aerobic metabolism (cellular respiration) required to sustain intense physical activity. Warm body temperature is one reason endotherms can generally endure vigorous activity longer than ectotherms.

Being endothermic is liberating, but it is also energetically expensive, especially in a cold environment. For example, at 20°C, a human at rest has a metabolic rate of 1300 to 1800 kcal per day (see Chapter 40). In contrast, a resting ectotherm of similar weight, such as an American alligator, has a metabolic rate of only about 60 kcal per day at 20°C. Thus, endotherms generally consume much more food (as measured in kilocalories) than ectotherms of equivalent size.

The bioenergetic connections between body temperature, active aerobic metabolism, and mobility were important in the evolution of endothermy; moving on land requires considerably more effort than moving in water (see Chapter 49). The efficient circulatory and respiratory systems of birds and mammals can be thought of as adaptations accompanying the evolution of endothermy and a high metabolic rate. This is not to say that ectothermy is incompatible with terrestrial success. Amphibians and reptiles have their own adaptations for coping with the temperature changes of terrestrial environments.

Thermoregulation involves physiological and behavioral adjustments

Both ectothermic and endothermic animals thermoregulate using some combination of up to four general categories of adaptations.

1. *Adjusting the rate of heat exchange between the animal and its surroundings.* Body insulation, such as hair, feathers, and fat located just beneath the skin, reduce an animal's heat loss. Other mechanisms that regulate heat exchange usually involve adaptations of the circulatory system. For example, many endotherms and some ectotherms can alter the amount of blood flowing to their skin. Increased blood flow usually results from **vasodilation**, an increase in the diameter of superficial blood vessels (those near the body surface). Nerve signals usually cause the muscles of the blood vessel walls to relax, and more blood flows through the vessels. When this occurs, more heat is transferred to the environment by conduction, convection, and radiation (see FIGURE 44.1). The reverse adjustment, **vasoconstriction**, reduces blood flow and heat loss by decreasing the diameter of superficial vessels.

Another type of adaptation that alters heat exchange is a special arrangement of arteries and veins called a **countercurrent heat exchanger** (FIGURE 44.3). Countercurrent heat exchange is important in controlling heat loss in many endothermic animals. For example, marine mammals and many birds face the problem of losing large amounts of heat from their extremities. Arteries carrying warm blood down the legs of a bird or the flippers of a dolphin or whale are in close contact with veins conveying blood in the opposite direction, back toward the trunk of the body. This countercurrent arrangement facilitates heat transfer from arteries to veins along the entire length of the blood vessels. Near the end of the leg or flipper, where the blood in an artery has been cooled to a temperature far below the animal's core temperature, the artery can still transfer heat to the even colder blood of an adjacent vein (recall that heat is conducted from a warmer object to a cooler one). The venous blood can continue to absorb heat because it is passing warmer and warmer arterial blood traveling in the opposite direction. As the venous blood approaches the center of body, it is almost as warm as the body core, minimizing the heat lost as a result of supplying blood to body parts immersed in cold water. In some species, blood can enter the limbs either through the heat exchanger or by way of vessels that detour around the exchanger. The relative amount of blood that enters the limbs via the two different paths varies, controlling the rate of heat loss.

2. *Cooling by evaporative heat loss.* Terrestrial endotherms and ectotherms lose water in breathing and across their skin. If the humidity of the air is low enough, the water will evaporate and the animal will lose heat by evaporative cooling. Evaporation from the respiratory system can be increased by panting.

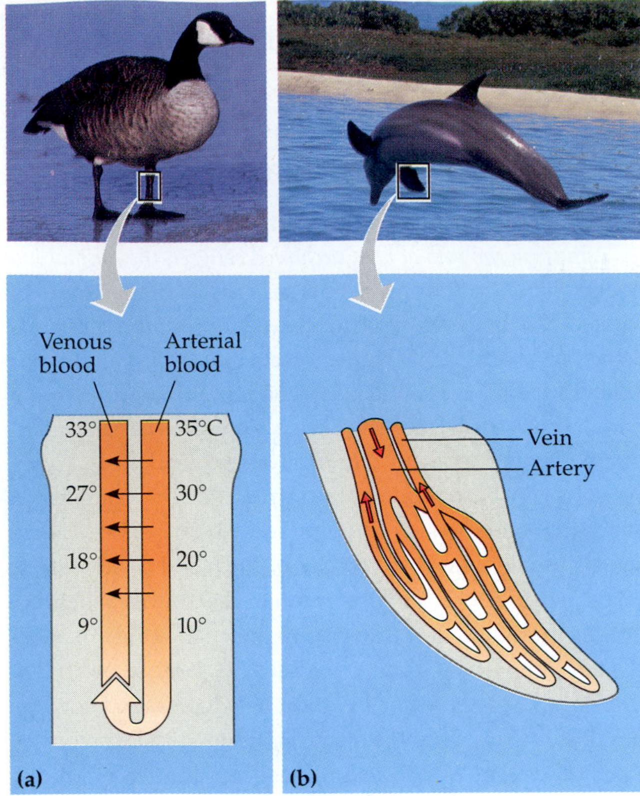

FIGURE 44.3 ▪ Countercurrent heat exchange. (a) Many birds, such as this Canada goose, possess countercurrent systems in their legs that reduce heat loss. The arteries carrying blood down the legs contact the veins that return blood to the body core. In a cold environment, arterial blood transfers heat to venous blood returning to the body (black arrows). The countercurrent flow facilitates heat exchange by setting up a thermal gradient between arterial and venous blood over the entire length of the vessels. The body's core temperature is further protected by the constriction of surface veins. **(b)** In the flippers of marine mammals, such as this Pacific bottlenose dolphin, each artery is surrounded by several veins in a countercurrent arrangement that allows efficient heat exchange between venous and arterial blood.

Evaporative cooling via the skin can be increased by such means as bathing or sweating (FIGURE 44.4).

3. *Behavioral responses.* Many animals can increase or decrease body heat loss by relocating. They will bask in the sun or on warm rocks in winter; find cool, damp areas or burrow in summer; or even migrate to a more suitable climate.

4. *Changing the rate of metabolic heat production.* This fourth category of thermoregulatory adaptation applies only to endotherms, particularly mammals and birds. By means we will discuss in the next section, many species of mammals and birds can double or triple their metabolic heat production when exposed to cold.

Most animals are ectothermic, but endothermy is widespread

In this section we survey several large groups of animals, concentrating on the mechanisms by which they thermoregulate as ectotherms or endotherms. We will see that adaptations for endothermy are found in a wide range of animals.

FIGURE 44.4 ▪ A behavioral adaptation for thermoregulation. Bathing in cool water brings immediate relief from the heat and continues to cool the surface for some time by evaporation.

Invertebrates

Most invertebrates have very little control over their body temperature, but some do adjust temperature by behavioral or physiological mechanisms. The desert locust, for example, must reach a certain temperature to become active. It orients in a direction that maximizes the absorption of sunlight.

Some species of large flying insects, such as bees and large moths, can generate internal heat and are endothermic (see the introduction to Chapter 40). They are able to "warm up" before taking off by contracting all of the flight muscles in synchrony, so that only slight movements of the wings occur but large amounts of heat are produced. This higher temperature of the flight muscles enables the insects to sustain the intense activity required for flight on cold days and at night (FIGURE 44.5a). Endotherms such as bumblebees, honeybees, and certain moths called noctuids that survive and fly during cold winter months have a countercurrent heat exchanger that helps maintain a high temperature in the thorax. For example, the heat exchanger keeps the flight muscles of some noctuid species at about 30°C, even on cold, snowy nights (FIGURE 44.5b).

Honeybees use an additional mechanism that depends on social organization to increase body temperature. In cold weather they increase their movements and huddle together, thereby retaining heat. They maintain a relatively constant temperature by changing the density of the huddling. Individuals move from the cooler outer edges of the cluster to the warmer center and back again, thus circulating and distributing the heat. Honeybees also control the temperature of their hive by transporting water to it in hot weather and fanning with their wings, which promotes evaporation and convection. A honeybee colony uses many of the mechanisms of thermoregulation seen in single organisms.

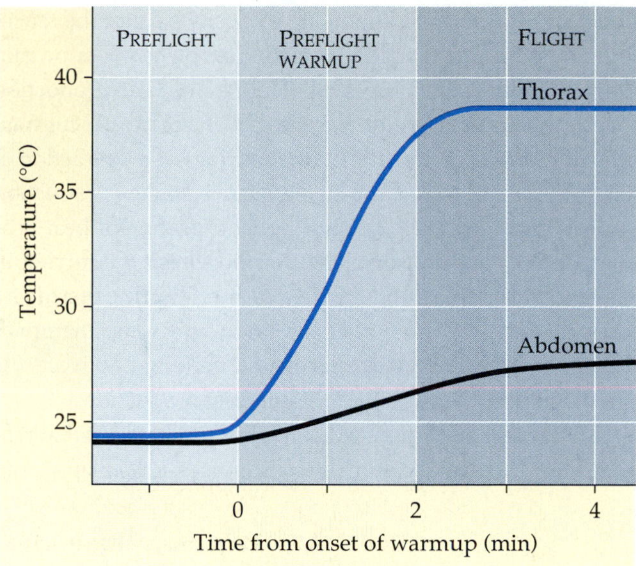

(a) Preflight warmup in the sphinx moth

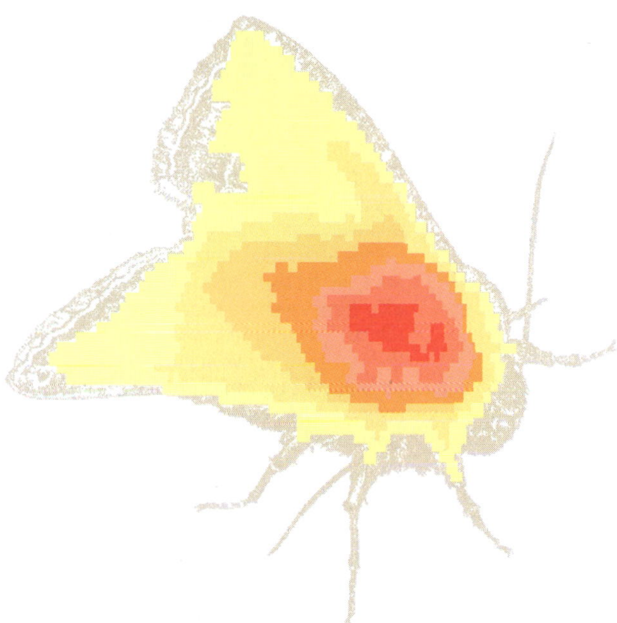

(b) Internal temperature in the winter moth

FIGURE 44.5 ▪ Thermoregulation in moths. (a) The sphinx moth (*Manduca sexta*) is one of the many insect species using a shiveringlike mechanism for preflight warmup of its thoracic flight muscles. Once the moth takes off, metabolic activity of the flight muscles maintains a high thoracic temperature. **(b)** Various endothermic adaptations, including a countercurrent heat exchanger in the thorax, help keep the flight muscles of moths that are active in winter warmed to a temperature of 30°C, even though the external temperature may be subfreezing. An infrared map superimposed over an image of a winter moth depicts the distribution heat immediately after a flight. Red, located here in the thorax region, indicates the highest temperature. Moving outward from the thorax, the variously colored zones correspond to regions of increasingly cooler body temperatures.

Amphibians and Reptiles

Amphibians and reptiles are generally ectotherms with relatively low metabolic rates that contribute little to normal body temperature. The optimal temperature range for amphibians varies substantially with the species. For example, closely related species of salamanders have average body temperatures ranging from 7° to 25°C. Amphibians produce very little heat, and most lose heat rapidly by evaporation from their body surfaces, making it difficult to control body temperature. However, behavioral adaptations enable them to maintain body temperature within a satisfactory range most of the time—by moving to a location where solar heat is available or into water, for instance. When the surroundings are too warm, the animals seek cooler microenvironments, such as shaded areas. Some amphibians, such as bullfrogs, can vary the amount of mucus they secrete from their surface, a physiological response that regulates evaporative cooling.

Reptiles warm themselves mainly by behavioral adaptations. They seek warm places, orienting themselves toward heat sources to increase heat uptake and expanding the body surface exposed to a heat source. Reptiles do not simply maximize heat uptake, however; they may behave in such a way as to truly regulate their temperature within a range. If a sunny spot is too warm, for instance, a lizard may sit alternately in the sun and in the shade or turn in another direction, thereby reducing the surface area exposed to the sun. By seeking favorable sites, many reptiles maintain body temperatures that are quite stable.

Some reptiles also have physiological adaptations that regulate heat loss. For example, in the marine iguana, which inhabits the Galápagos Islands, body heat is conserved by superficial vasoconstriction, routing more blood to the central core of the body when the animal is swimming in the cold ocean. A few reptiles are endothermic for brief periods of time. For instance, when incubating eggs female pythons increase their metabolic rate by shivering, generating enough heat to maintain their body temperature 5°–7°C above the surrounding air. Researchers continue to debate whether certain groups of dinosaurs were endothermic (see Chapter 34).

Fishes

The body temperature of most fishes is usually within 1°–2°C of the surrounding water temperature. Metabolic heat generated by swimming muscles is lost to the surrounding water when blood passes through the gills, and the large dorsal aorta conveys the blood directly inward from the gills, cooling the body core. Endothermic fishes include several large, active species such as the bluefin tuna, swordfish, and the great white shark. Their swimming muscles produce enough metabolic heat to elevate temperatures at the body core, and adaptations of the circulatory system retain the heat. Large arteries convey most of the cold blood from the gills to tissues just under the skin (FIGURE 44.6). Branches deliver blood to the deep muscles, where the small vessels are arranged into a countercurrent heat exchanger. Endothermy enhances vigorous activity by keeping the swimming muscles several degrees warmer than the tissues near the animal's surface, which is about the same temperature as the surrounding water. A current hypothesis is that endothermy in fishes evolved as an adaptation to foraging in cold seas.

Mammals and Birds

Mammals and birds generally maintain high body temperatures within a narrow range of about 36°–38°C for most mammals and about 40°–42°C for most birds. Maintaining these narrow ranges requires the ability to closely balance the rate of metabolic heat production with the rate of heat loss or gain from the outside environment. The rate of heat production can be increased in one of two ways: by the increased contraction of muscles (by moving or by shivering) or by the action of hormones that increase the metabolic rate and the production of heat instead of ATP. The hormonal triggering of heat production is called **nonshivering thermogenesis**. Occurring in numerous mammals and a few birds, nonshivering thermogenesis takes place throughout the body, and some mammals have a tissue called **brown fat** in the neck and between the shoulders that is specialized for rapid heat production.

Mammals and birds have an array of mechanisms that regulate heat exchange with their environment. Vasodilation and

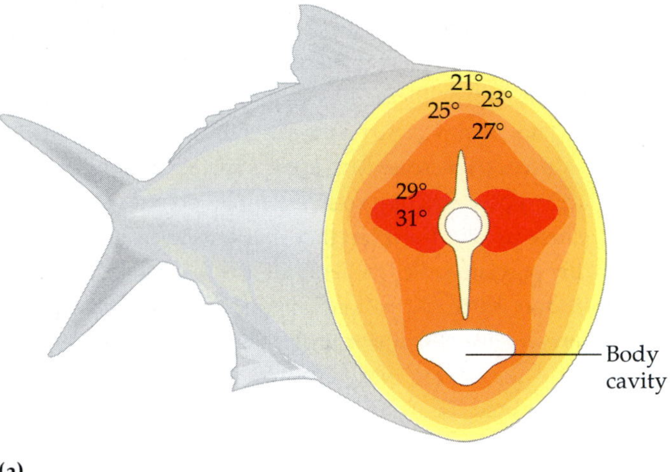

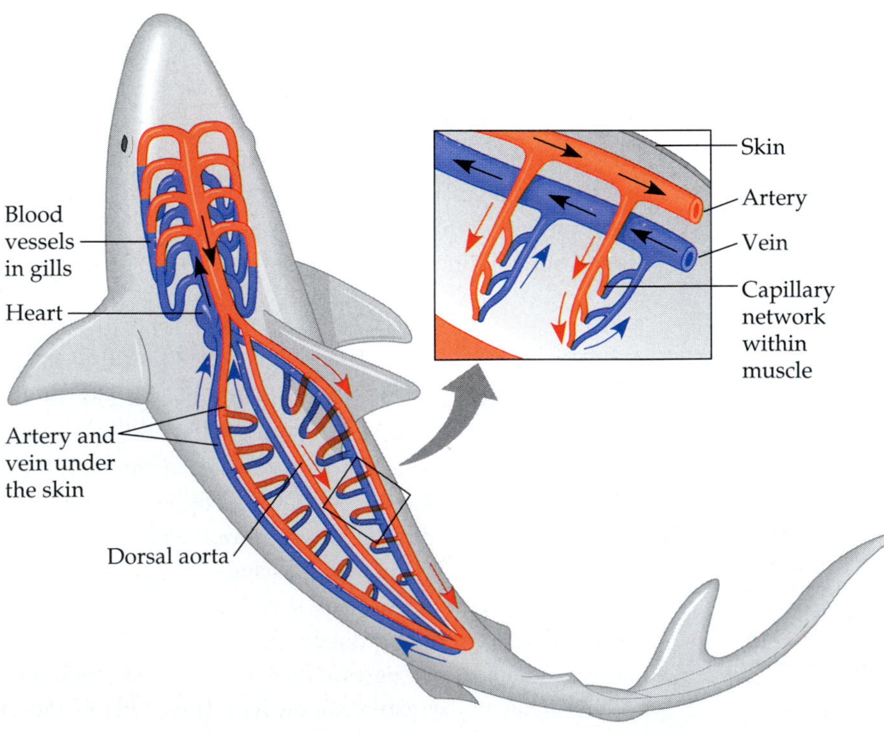

FIGURE 44.6 • Thermoregulation in large, active fishes. (a) The bluefin tuna maintains internal temperatures much higher than the surrounding water. (Colors indicate swimming muscles cut in transverse section.) The temperatures shown were recorded for a tuna in 19°C water. **(b)** Like the bluefin tuna, the great white shark has a countercurrent heat exchanger in its swimming muscles that reduces the loss of metabolic heat. All fishes lose some heat to the surrounding water when their blood passes through the gills. However, in contrast to ectothermic fishes, endothermic species have a relatively small-diameter dorsal aorta, which carries little blood from the gills to the core of the body. Instead, most of the blood from the gills is conveyed via large arteries just under the skin, keeping cool blood away from the body core. As shown in the enlargement, small arteries carrying cool blood inward from the large arteries under the skin are paralleled by small veins carrying warm blood outward from the inner body. This countercurrent flow retains heat in the muscles.

vasoconstriction effect heat exchange and may also contribute to regional temperature differences within the animal. For example, on a cool day a human's temperature may be several degrees lower in the arms and legs than in the trunk, where most vital organs are located. The insulating power of a layer of fur or feathers depends on how much still air the layer traps. Accordingly, most land mammals and birds react to cold by raising their fur or feathers, thereby trapping a thicker layer of still air. Humans rely more on a layer of fat just beneath the skin as insulation against heat loss (FIGURE 44.7). Goose bumps are a vestige of hair-raising left over from our furry ancestors.

Hair loses much of its insulating power in water, and marine mammals such as whales and seals have a very thick layer of insulating fat called blubber, just under the skin. Marine mammals spend time in water colder than their body temperature, and many spend at least part of the year in nearly freezing arctic or antarctic water. Although the loss of heat to water occurs 50 to 100 times more rapidly than heat loss to air, marine mammals maintain body temperatures of about 36°–38°C and have metabolic rates close to those of land mammals of similar size. This suggests that adaptations that conserve heat in marine mammals are more effective than those of land mammals. The flippers or tail of a whale or seal lack insulating blubber, but countercurrent heat exchangers effectively reduce heat loss in these extremities, as they do in the legs of many birds (see FIGURE 44.3).

Many mammals and birds live where endothermy requires both cooling and warming of the body. For example, when a marine mammal moves into warm seas, as do many whales to reproduce, it disposes of excess metabolic heat by vasodilation, enhanced by large numbers of blood vessels in the outer layer of its skin. In hot climates, terrestrial mammals and birds rely heavily on evaporative cooling. Panting is important in most birds and mammals, and some birds have a pouch richly supplied with blood vessels in the floor of the mouth; fluttering the pouch increases evaporation. Many terrestrial mammals have sweat glands, which are controlled by the nervous system (see FIGURE 44.7). Other mechanisms that promote evaporative cooling include spreading saliva on body surfaces, an adaptation of some kangaroos and rodents for combating severe heat stress. Some bats use both saliva and urine to enhance evaporative cooling.

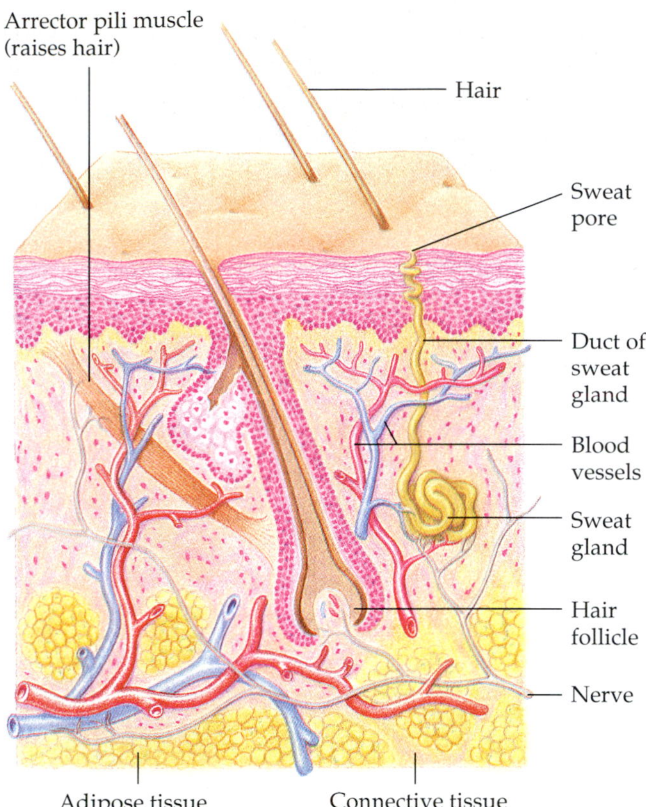

FIGURE 44.7 ▪ Skin as an organ of thermoregulation. Fat (adipose tissue) and hair help insulate mammals. Heat loss to the environment can be regulated by the constriction and dilation of superficial blood vessels and by the erection and compaction of fur. Sweat glands, under nervous control, function in evaporative cooling.

Feedback Mechanisms in Thermoregulation

The regulation of body temperature in humans and other terrestrial mammals is an example of a complex homeostatic system facilitated by feedback mechanisms. Nerve cells that control thermoregulation, as well as those that control many other aspects of homeostasis, are concentrated in the hypothalamus (discussed in detail in Chapter 48). The hypothalamus contains a thermostat that responds to changes in body temperature above and below a set point (actually above or below a normal range) by activating mechanisms that promote heat loss or gain (FIGURE 44.8 on p. 872). Nerve cells that sense body temperatures are located in the skin, the hypothalamus itself, and some other parts of the nervous system. Some of these are warm receptors that signal the thermostat in the hypothalamus when the temperature of the skin or blood increases. Others are cold receptors that signal the thermostat when the temperature decreases. Responding to body temperature below the normal range, the thermostat inhibits heat-loss mechanisms and activates heat-saving ones such as the vasoconstriction of superficial vessels and the erection of fur, while stimulating heat-generating mechanisms (shivering and nonshivering thermogenesis). In response to elevated body temperature, the thermostat shuts down heat-saving mechanisms and promotes body cooling by vasodilation, sweating, or panting.

Temperature Range Adjustments

Many animals can adjust to a new range of environmental temperatures over a period of days or weeks, a physiological response called **acclimatization**. Seasonal change is one

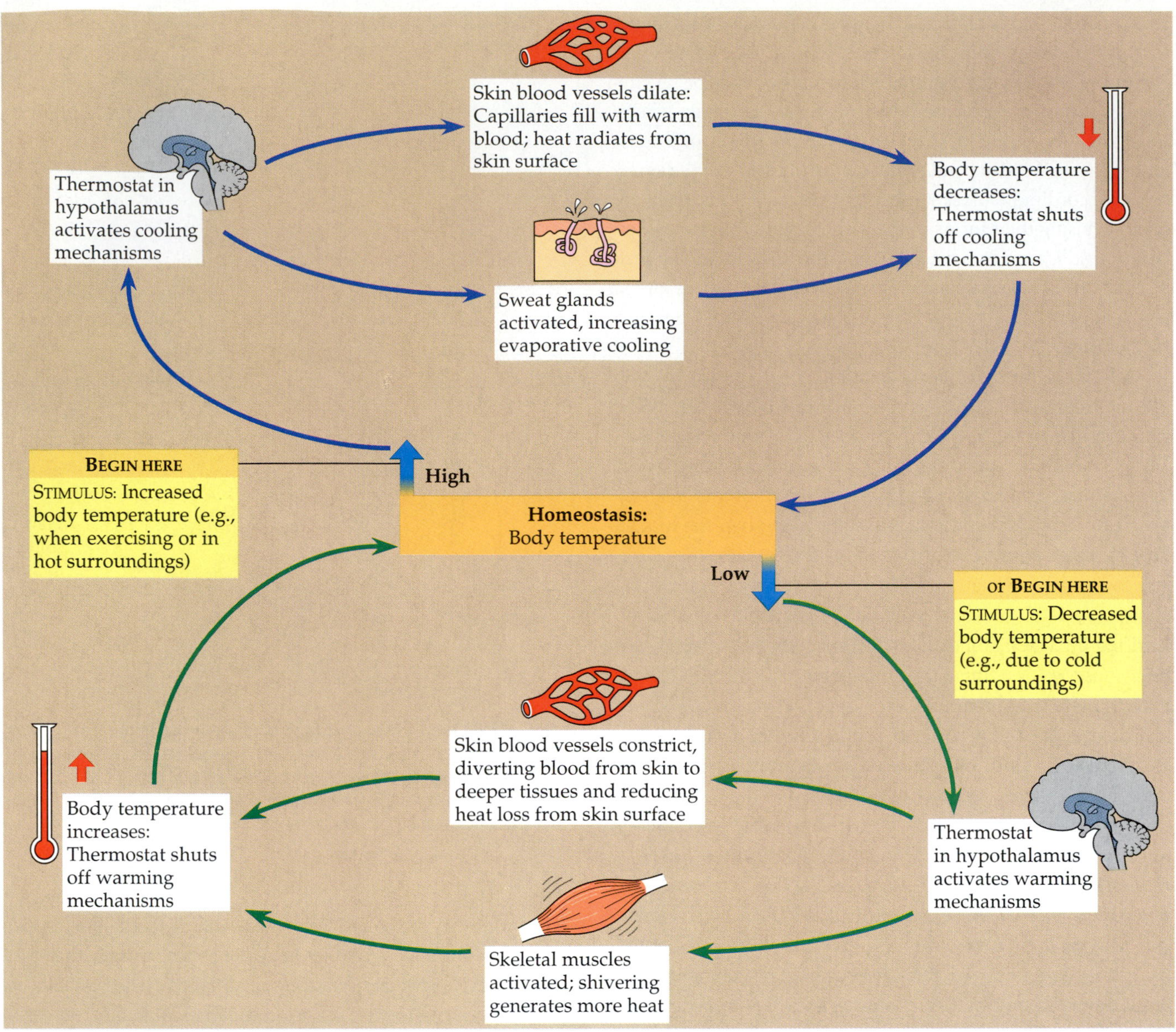

FIGURE 44.8 ▪ The thermostat function of the hypothalamus and feedback mechanisms in human thermoregulation.

context in which physiological adjustments to a new temperature range are important. The bullhead catfish, for instance, can survive water temperatures up to 36°C in summer, but a temperature above 28°C is lethal during winter.

Physiological adjustment to a new external temperature range is multifaceted. It may involve changes in the mechanisms that control an animal's internal temperature. Acclimatization may also involve cellular adjustments. For example, cells may increase the production of certain enzymes, helping compensate for the lowered activity of each enzyme molecule at temperatures that are not optimal. In other cases, cells produce variants of enzymes that have the same function but different temperature optima. Membranes may also change in the proportions of saturated and unsaturated lipids they contain. This response helps keep membranes fluid at different temperatures (see Chapter 8).

Cells can often make rapid adjustments to temperature changes. For example, the cells of tree frogs respond to cold shock within hours. Mammalian cells grown in laboratory cultures respond to a marked increase in temperature and to other forms of severe stress, such as toxins, rapid pH changes, and viral infections, by accumulating special molecules called **stress-induced proteins**, including **heat-shock proteins**. When "shocked" by a rapid change in temperature from 37°C to about 43°C, cultured mammalian cells begin synthesizing heat-shock proteins within minutes. These molecules help maintain the integrity of other proteins that would be denatured by severe heat. Found in bacteria, yeasts, and plant cells as well as in animal cells, stress-induced proteins help prevent cell death when an organism is challenged by severe changes in the external environment.

Torpor conserves energy during environmental extremes

Torpor is an alternative physiological state in which metabolism decreases and the heart and respiratory system slow down. Many endotherms enter a state of torpor in which their body temperature declines. In effect, their body's thermostat is turned down, thereby conserving energy when food supplies are low and environmental temperatures are extreme. **Hibernation** is long-term torpor during which the body temperature is lowered as an adaptation to winter cold and food scarcity. **Estivation**, or summer torpor, is characterized by slow metabolism and inactivity. It enables an animal to survive long periods of high temperatures and scarce water supplies. Hibernation and estivation are often triggered by seasonal changes in the length of daylight. As the days shorten, some animals will eat huge quantities of food before hibernating. Ground squirrels, for instance, will more than double their weight in a month of gorging.

Many small mammals and birds exhibit a daily period of torpor that seems to be adapted to their feeding patterns. For instance, most bats and shrews feed at night and go into torpor when they are inactive during daylight hours. Chickadees and hummingbirds feed during the day and often undergo torpor on cold nights; the body temperature of chickadees that spend winters in cold, northern forests may drop as much as 10°C at night. All endotherms that show a daily torpor are relatively small; when active they have a high metabolic rate and thus a very high rate of energy consumption. During hours when they cannot feed, their daily period of torpor enables them to survive on energy stored in their tissues.

An animal's daily cycle of activity and torpor appears to be a built-in rhythm controlled by the biological clock (see Chapter 48). Even if food is made available to a shrew all day, it still goes through its daily torpor. The need for sleep in humans and the slight drop in body temperature that accompanies it may be a remnant of a more pronounced daily torpor in our early mammalian ancestors.

Torpor is a homeostatic mechanism that involves a relatively broad change in body functions. Many animals can tolerate wide fluctuations in their body temperature, but no animal can withstand high concentrations of its own metabolic wastes or much change in the relative amounts of dissolved solutes and water in its cellular environment.

WATER BALANCE AND WASTE DISPOSAL

In most animals the majority of cells are not in direct contact with the external environment but are bathed by an internal body fluid. Insects and other animals with an open circulatory system have an internal "pond" composed of hemolymph, which bathes all body cells (see Chapter 42). In vertebrates and other animals with a closed circulatory system, the internal pond is interstitial fluid serviced by blood. Maintaining a balance between the uptake of water from the external environment and the loss of water to it is a key aspect of homeostasis. Another is keeping the by-products of metabolism from building up and poisoning the body fluids. Balancing water gain and loss and disposing of metabolic wastes involve movements of solutes (and hence water, which follows solutes by osmosis) between an animal's internal fluids and the external environment.

Water balance and waste disposal depend on transport epithelia

Maintaining water balance and getting rid of metabolic wastes require the passage of certain solutes in specific amounts between an animal and its surroundings. In most cases the solutes pass through a **transport epithelium**—a layer or layers of specialized epithelial cells that regulate solute movements. Some transport epithelia face the outside environment, while others face a tubular channel that leads to the outside through an opening on the animal's body surface. Joined by impermeable tight junctions (see FIGURE 7.30), the cells of the epithelium form a selectively permeable barrier at the tissue-environment boundary. This arrangement, another example of how form fits function, ensures that solute passage is closely regulated; any solute passing between the extracellular fluid and the environment must pass through the selectively permeable membranes of cells.

In most animals, transport epithelia are arranged into tubular networks with extensive surface areas. We find some of the most efficient examples in marine birds that spend months or years at sea and obtain food and water from the ocean (FIGURE 44.9 on p. 874).

The molecular structure of plasma membranes determines the kinds of solutes that move across transport epithelia and the direction they move. For examples, salt-excreting glands remove excess sodium chloride from the blood. By contrast, transport epithelia in the gills of freshwater fishes move salts from the surrounding water into the blood. Transport epithelia in excretory organs often have dual functions of maintaining water balance and disposing of metabolic wastes.

An animal's nitrogenous wastes are correlated with its phylogeny and habitat

Some of the most toxic by-products of metabolism are nitrogen-containing wastes from the breakdown of proteins and nucleic acids. Nitrogen is removed when these

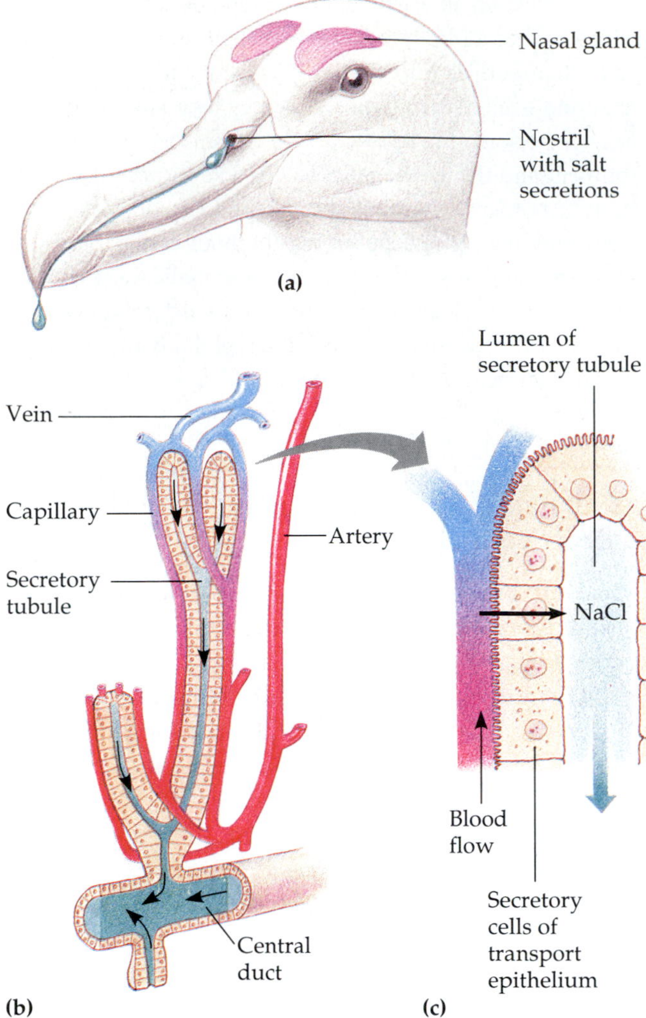

FIGURE 44.9 ▪ Salt-excreting glands in birds. (a) Many marine birds, such as this albatross, can drink seawater because they have a pair of salt-excreting nasal glands. The glands empty via a duct into the nostrils, and the salty solution either runs along a ridge to the tip of the beak or is exhaled in a fine mist from the nostrils. **(b)** This drawing shows one of several thousand secretory tubules in a salt-excreting gland. Each tubule is lined by a transport epithelium surrounded by capillaries, and drains into a central duct. **(c)** The secretory cells of the transport epithelium pump salt from the blood into the tubules. Notice that blood flows counter to the flow of salt secretion. By maintaining a concentration gradient of salt in the tubule (graded blue shading), this countercurrent system enhances salt transfer from the blood to the lumen of the tubule (see Chapter 42).

macromolecules are broken apart for energy, or when they are converted to carbohydrates or fats. The nitrogenous waste product is ammonia, a small and very toxic molecule. Excreting ammonia directly is an efficient way to dispose of waste because virtually no energy is expended. However, many animals first convert ammonia to compounds such as urea or uric acid, which are less toxic but require energy in the form of ATP to produce. The kind of nitrogenous wastes an animal excretes depends on both the animal's evolutionary history and its habitat (FIGURE 44.10).

Ammonia

Most aquatic animals excrete nitrogenous wastes as **ammonia**. Ammonia molecules are very soluble in water and easily pass through membranes. In many invertebrates, ammonia diffuses across the whole body surface into the surrounding water. In fishes, most of the ammonia is lost as ammonium ions (NH_4^+) across the epithelium of the gills, with kidneys excreting only minor amounts of nitrogenous wastes. In freshwater fishes, the epithelium of the gills takes up Na^+ from the water in exchange for NH_4^+, which helps maintain Na^+ concentrations in body fluids much higher than the Na^+ concentration in the surrounding water.

Urea

Although it works in water, ammonia excretion is unsuitable for disposing of nitrogenous wastes on land. Ammonia is so toxic that it can only be transported in an animal and excreted in a very dilute solution, and terrestrial animals simply cannot dispose of it quickly enough. Instead, mammals, most adult amphibians, and many marine fishes and turtles excrete mainly **urea**, a substance that is about 100,000 times less toxic than ammonia. Urea is produced in the vertebrate liver by a metabolic cycle that combines ammonia with carbon dioxide. The circulatory system carries urea to the excretory organs, the kidneys.

Most animals can tolerate high concentrations of urea, and by excreting a concentrated solution of this waste product an animal conserves water, an important adaptation for living on land or in the sea (animals in both environments tend to lose water to their surroundings). As we will see, sharks have a relatively high concentration of urea in the blood, which helps maintain an osmotic balance between their body fluids and seawater.

From a bioenergetic standpoint, we would predict that animals that spend part of their lives in water and part on land would switch between excreting ammonia (thereby saving energy) and urea (expending energy to prevent poisoning). In fact, many amphibians excrete mainly ammonia when they are aquatic tadpoles and mainly urea when they are land-dwelling adults.

Uric Acid

Land snails, insects, birds, and many reptiles excrete **uric acid** as the major nitrogenous waste. Because it is thousands of times less soluble in water than either ammonia or urea, uric acid can be excreted in a pastelike form with very little loss of water.

Uric acid and urea represent two different adaptations that enable terrestrial animals to excrete nitrogenous wastes with a minimal loss of water. One factor that seems to have been important in determining which of these alternatives evolved

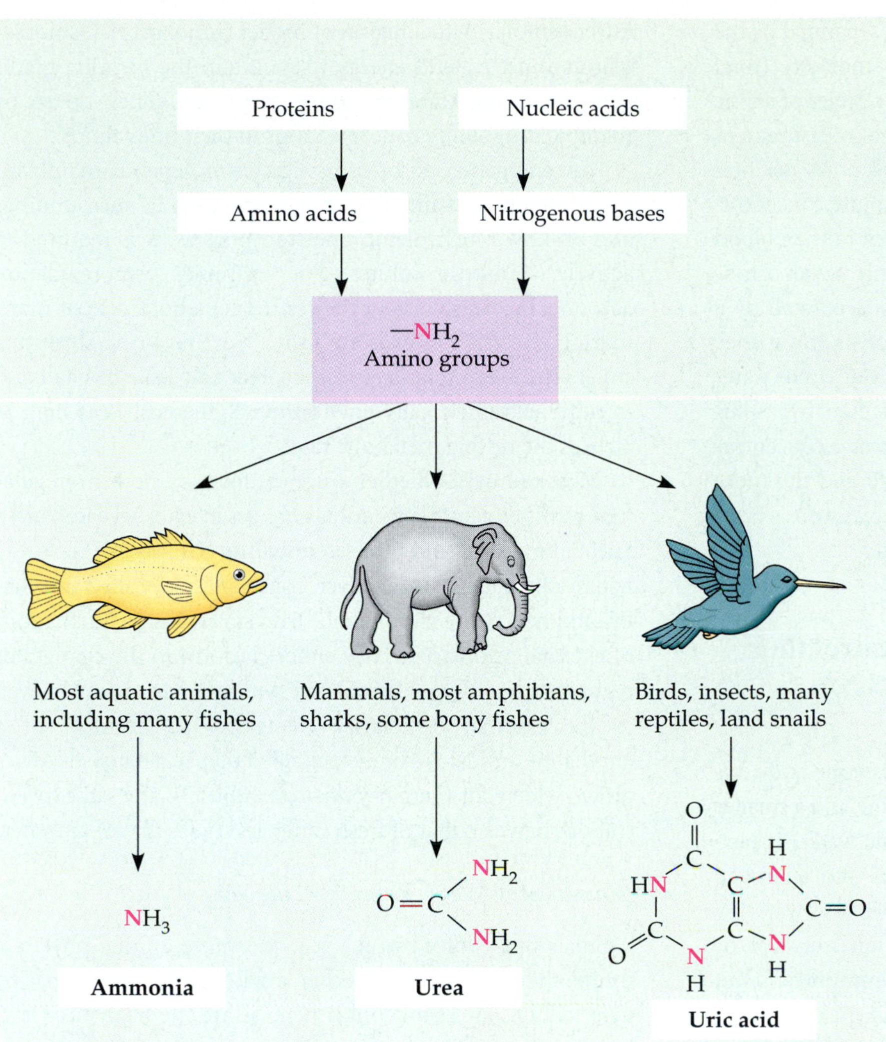

FIGURE 44.10 ▪ Nitrogenous wastes. Ammonia is a toxic by-product of the metabolic removal of nitrogen (deamination) from proteins and nucleic acids. Most aquatic animals excrete ammonia from their body fluids. By contrast, most terrestrial animals first convert the ammonia to urea or uric acid, which costs energy but conserves water because these less-toxic wastes can be transported in the body and hence excreted in more concentrated form.

in a particular group of animals is the mode of reproduction. Soluble wastes can diffuse out of a shell-less amphibian egg or be carried away by the mother's blood in the case of a mammalian embryo. The vertebrates that excrete uric acid, however, produce shelled eggs, which are permeable to gases but not to liquids. If an embryo released mostly ammonia or urea within a shelled egg, the soluble waste would accumulate to toxic concentrations. Uric acid precipitates out of solution and can be stored within the egg as a solid that is left behind when the animal hatches.

In grouping vertebrates according to the nitrogenous wastes they excrete, we find that the boundaries depend on habitat as well as on evolutionary lineage. For example, terrestrial turtles excrete mainly uric acid, whereas aquatic turtles excrete both urea and ammonia. Some turtles can also modify their nitrogenous wastes when their environment changes. For example, a tortoise that usually produces urea can shift to uric acid production when the temperature increases and water becomes less available. This is another example of how response to the environment occurs on two levels: Evolution determines the limits of physiological responses for a species, but individual organisms make adjustments within these evolutionary constraints.

Cells require a balance between osmotic gain and loss of water

Whether an animal inhabits land, fresh water, or salt water, or moves back and forth between these environments, one general problem occurs: The cells of the animal cannot survive a net water gain or loss. Water continuously enters and leaves an animal cell across the plasma membrane; however, uptake and loss must balance. Animal cells swell and burst if there is a net uptake of water or shrivel and die if there is a net loss of water. Even an ice-bound spring peeper, whose blood and body fluids are mostly frozen, has its cells bathed in a film of interstitial fluid that is osmotically balanced with the cytosol. Glucose and proteins in the interstitial fluid ensure the balance.

Recall from Chapter 8 that osmosis, a special case of diffusion, is the movement of water across a selectively permeable membrane. It occurs whenever two solutions separated by the membrane differ in osmotic pressure, or **osmolarity** (total solute concentration expressed as molarity, or moles of solute per liter of solution; see Chapter 3). The unit of measurement for osmolarity used in this chapter is milliosmoles per liter (mosm/L). This unit is equivalent to a total solute concentration of 10^{-3} *M*. For example, the osmolarity of human blood is about 300 mosm/L, while seawater commonly has an osmolarity of about 1000 mosm/L. Two solutions separated by a semipermeable membrane are said to be isoosmotic if they have the same osmolarity. There is no *net* movement of water by osmosis between isoosmotic solutions. When two solutions differ in osmolarity, the one with the greater concentration of solutes is referred to as hyperosmotic and the more dilute solution as hypoosmotic.* Water flows by osmosis from a hypoosmotic solution to a hyperosmotic one.

Osmoregulators expend energy to control their internal osmolarity; osmoconformers are isoosmotic with their surroundings

Think about the ways water enters and leaves your body. We acquire most of our water in our food and drink and a smaller amount by oxidative metabolism. ("Metabolic water" is produced by cellular respiration when electrons and hydrogen ions are added to oxygen; see Chapter 9.) We lose water by urinating and defecating, and by evaporation in sweat or breath. For aquatic animals, evaporation is unimportant, but balancing the osmotic loss and gain of water is vital.

There are two basic solutions to the problem of balancing water gain with water loss. One solution for a marine animal is to be isoosmotic with its saltwater environment. Such an animal, which does not actively adjust its internal osmolarity, is known as an **osmoconformer**. By contrast, an **osmoregulator** is an animal that must adjust its internal osmolarity, since its body fluids are not isoosmotic with the outside environment. An osmoregulator must either discharge excess water if it lives in a hypoosmotic environment or continuously take in water to offset osmotic loss if it inhabits a hyperosmotic environment. The ability to osmoregulate enables animals to live, for example, in fresh water, where the osmolarity is too low to support osmoconformers, and on land, where water is usually in short supply. All freshwater animals and many marine animals are osmoregulators. Humans and other terrestrial animals, also osmoregulators, must compensate for water loss.

Osmoregulation is energetically costly. A net movement of water occurs only in an osmotic gradient (from a region of lower osmolarity to a region of higher osmolarity). Osmoregulators must expend energy to maintain the osmotic gradients that allow water to move in or out. They do so by manipulating solute concentrations in their body fluids.

The energetic cost of osmoregulation depends mainly on how different an animal's osmolarity is from its surroundings and on how much membrane-transport work is required to actively transport solutes. For example, osmoregulation accounts for nearly 5% of the resting metabolic rate of many marine and freshwater bony fishes. For the brine shrimp, a small crustacean that lives in the Great Salt Lake in Utah and in other extremely salty environments, the cost is as high as 30% of the resting metabolic rate.

Most animals, whether osmoconformers or osmoregulators, cannot tolerate substantial changes in external osmolarity. Such animals are said to be **stenohaline** (Gr. *stenos,* "narrow"; haline refers to salt). However, some animals, called **euryhaline** animals (Gr. *eurys,* "broad"), do survive large fluctuations of external osmolarity. They either conform to the changes or regulate their internal osmolarity within a narrow range even as the external osmolarity changes. One example of a euryhaline animal—a bony fish called tilapia, a native of Africa grown widely for human food—can adjust to any salt concentrations between that of fresh water and twice that of seawater.

Maintaining Water Balance in the Sea

Animals first evolved in the sea, and more animal phyla are found there than in any other environment. Most marine invertebrates are osmoconformers, as are the hagfishes (jawless vertebrates); however, even these animals differ from seawater in their concentrations of specific salts. The difference is usually slight, but in some cases it is substantial. Thus, an animal that conforms to the osmolarity of its surroundings may still regulate its internal composition of ions.

Except for the hagfishes, marine vertebrates are osmoregulators. Marine sharks and most other cartilaginous fishes (Class Chondrichthyes) maintain internal salt concentrations lower than that of seawater. Their kidneys excrete some salts, and a salt-excreting organ called the rectal gland excretes sodium chloride out of the body through the anus. However, despite its relatively low salt concentration, a marine shark is slightly hyperosmotic to seawater. If you have ever eaten shark meat, you know it should be soaked in fresh water before cooking. Soaking washes out urea, a waste product of nitrogen metabolism. A large amount of urea dissolved in the body fluids accounts for a shark's being slightly hyperosmotic to seawater. Marine sharks also produce and retain another organic compound, trimethylamine oxide (TMAO), which protects proteins from being damaged by the urea. Water enters the shark's body by osmosis rather than by drinking, and the shark disposes of water in urine, the waste fluid formed by the excretory organs, the kidneys.

* (In this chapter, we use the terms isoosmotic, hypoosmotic, and hyperosmotic, which refer to osmotic pressure, instead of the more familiar terms isotonic, hypotonic, and hypertonic. In a strict sense, the terms for tonicity refer only to the response of animal cells in solutions of known solute concentrations.)

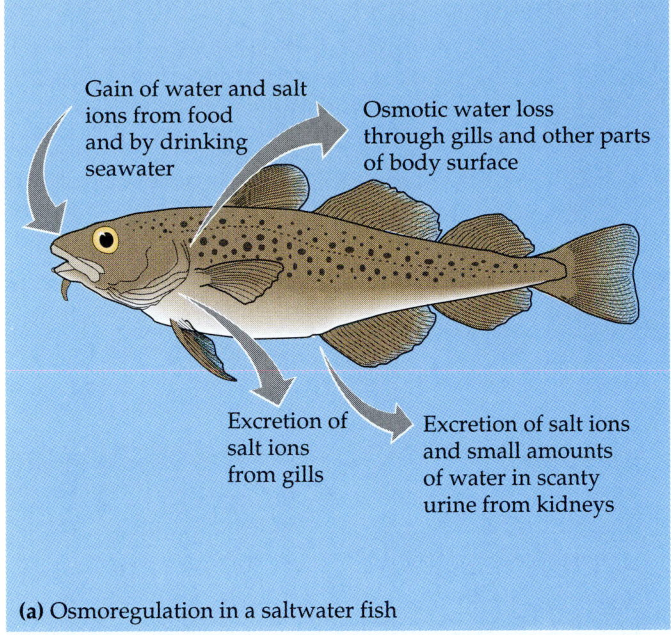

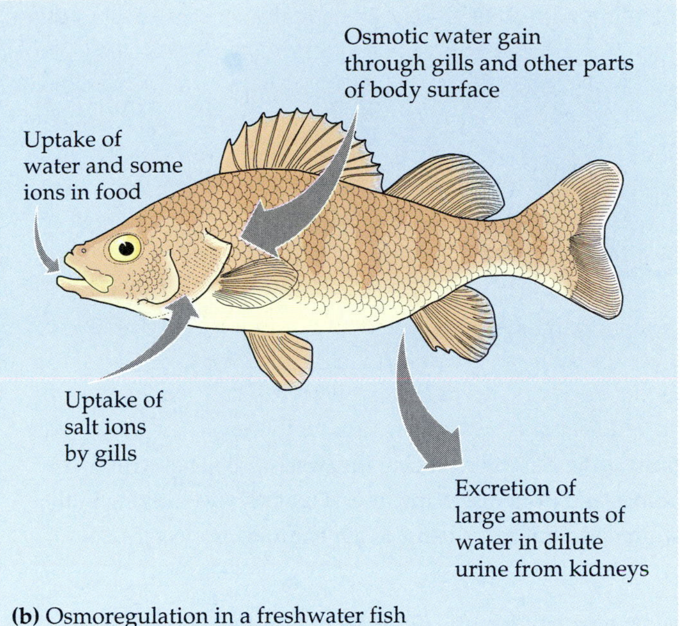

FIGURE 44.11 ▪ Osmoregulation in marine and freshwater bony fishes: a comparison. **(a)** A marine fish, such as this cod, is hypoosmotic to the surrounding seawater and therefore constantly loses water by osmosis. The fish drinks large amounts of seawater; its gills and general body surface dispose of sodium chloride (special cells called chloride cells actively transport Cl^- out, and Na^+ follows passively); and its kidneys dispose of excess calcium (Ca^{2+}), magnesium (Mg^{2+}), and sulfate (SO_4^{2-}) ions while excreting only small amounts of water. **(b)** Facing the opposite situation, a freshwater fish, such as this perch, constantly gains water because it is hyperosmotic to its surroundings. It balances the water gain by excreting large amounts of urine that is hypoosmotic to its body fluids. Salts lost in the urine are replenished by foods and by uptake across the gills; chloride cells in the gills actively transport Cl^- in.

Bony fishes (Class Osteichthyes) evolved from ancestors that entered freshwater habitats. In their subsequent evolution, many groups of bony fishes became marine but internally remained more similar to fresh water in osmolarity. Marine bony fishes constantly lose water by osmosis to their hyperosmotic surroundings (FIGURE 44.11a). They compensate by drinking large amounts of seawater, pumping out excess salts, and excreting urine in relatively small amounts.

Maintaining Osmotic Balance in Fresh Water

The osmoregulatory problems of freshwater animals are opposite those of marine animals. Freshwater animals are constantly taking in water by osmosis because the osmolarity of their internal fluids is much higher than that of their surroundings. Freshwater protists such as *Amoeba* and *Paramecium* have contractile vacuoles that pump out excess water (see FIGURE 28.8). Many freshwater animals, including fishes, bail out water by excreting large amounts of very dilute urine and regaining lost salts in their food or by active uptake from their surroundings (FIGURE 44.11b).

Salmon and other fishes that migrate between seawater and fresh water are euryhaline. While in the ocean, salmon drink seawater and excrete excess salt from the gills, osmoregulating like other marine fishes. With their migration to fresh water, salmon cease drinking, and their gills start taking up salt from the dilute environment, as do the gills of fishes that spend their lives in fresh water.

Special Problems of Living in Temporary Waters

Dehydration dooms most animals, but some aquatic invertebrates living in temporary ponds and films of water around soil particles can lose almost all their body water and survive in a dormant state when their habitats dry up. This remarkable adaptation is called **anhydrobiosis** ("life without water") or cryptobiosis ("hidden life"). Among the most striking examples are the tardigrades, or water bears, tiny invertebrates less than 1 mm long (FIGURE 44.12 on p. 878). In their active, hydrated state, these animals contain about 85% water by weight but can dehydrate to less than 2% water and survive in an inactive state, dry as dust, for a decade or more. Just add water, and within minutes the rehydrated tardigrades are moving about and feeding.

A dehydrated animal and a frozen one (such as the wintering frog discussed earlier) face similar problems. Both must have adaptations that keep cell membranes intact. Researchers are just beginning to learn how tardigrades survive drying out, but studies of roundworms (Class Nematoda) that are anhydrobiotic show that dehydrated individuals contain a large amount of sugars, especially a disaccharide called trehalose. Consisting of two glucose units, trehalose seems to

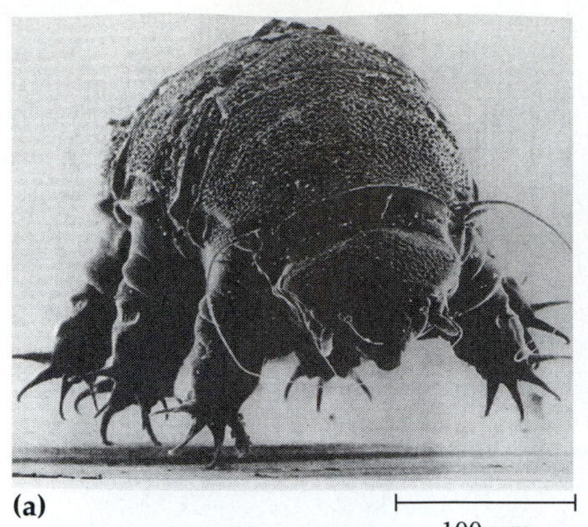

FIGURE 44.12 ▪ Anhydrobiosis. Tardigrades (water bears) inhabit temporary ponds and droplets of water in soil and on moist plants. Even a moss-covered roof provides a suitable habitat. **(a)** When its habitat is wet, a water bear is active and feeds on plant juices. **(b)** When the habitat dries, the animal can lose more than 95% of its body water and survive in a dehydrated, inactive state for decades (SEMs).

protect the cells by replacing the water associated with membranes and proteins. Many insects that survive freezing in the winter also utilize trehalose as a membrane protectant.

Maintaining Osmotic Balance on Land

The threat of desiccation is perhaps the most important problem confronting most terrestrial life, both plants and animals. Humans, for example, die if they lose about 12% of their body water. The severity of this problem may be one reason only two groups of animals, arthropods and vertebrates, have colonized the land with great success. (Although other phyla have some representatives on land, most of their species are aquatic.)

Evolutionary adaptations that conserve water are key to survival on land. Much as a waxy cuticle contributes to the success of plants on land, most terrestrial animals have body coverings that help prevent dehydration. Examples are the waxy layers of the exoskeletons of insects, the shells of land snails, and the multiple layers of dead, keratinized skin cells covering most terrestrial vertebrates. Many terrestrial animals, especially in deserts, are nocturnal, an adaptation that takes advantage of the higher relative humidity of night air, thereby reducing the need to cool the body by evaporation.

Despite these adaptations, many terrestrial animals lose a considerable amount of water, much of it from moist surfaces in their gas exchange organs and in urine. Later in the chapter we discuss the osmoregulatory role of nervous system and hormonal mechanisms in controlling the amount of water in the urine.

Most land animals replenish their water supply by drinking and eating moist foods. However, some are so well adapted to minimizing water loss that they can survive in deserts without drinking. For example, among mammals, kangaroo rats lose so little water that they can recover 90% of the loss by using metabolic water (FIGURE 44.13).

Although the problems of water balance on land or in salt water or fresh water are very different, solutions to the problems have a common theme: the regulation of solute movement (and hence water movement, which follows solutes by

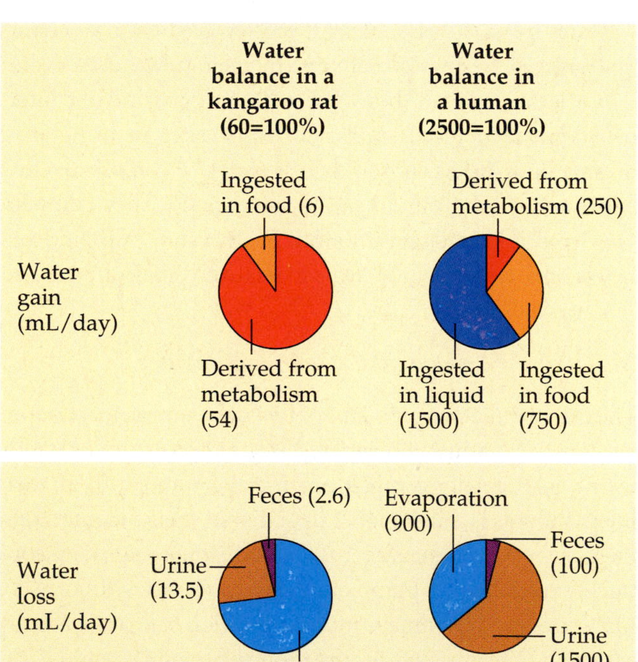

FIGURE 44.13 ▪ Water balance in two terrestrial mammals. **(a)** Kangaroo rats, which live in American southwestern deserts, eat mostly dry seeds and do not drink water. **(b)** A kangaroo rat loses water mainly by evaporation during gas exchange and gains water mainly from cellular metabolism (water produced when hydrogen and oxygen are combined during aerobic respiration). In contrast, a human loses a large amount of water in urine and regains it mostly in food and drink.

osmosis) between the animals' internal fluids and the external environment.

EXCRETORY SYSTEMS

Excretory systems are central to homeostasis because they dispose of metabolic wastes and respond to imbalances in body fluids by excreting more or less of a particular ion.

Most excretory systems produce urine by refining a filtrate derived from body fluids: *an overview*

Excretory systems are diverse, but they have functional similarities. In general, they produce urine by two major processes: filtration of body fluids and refinement of the aqueous solution resulting from filtration (FIGURE 44.14).

First, during **filtration**, blood or other body fluids, depending on the type of excretory system, is exposed to a filtering device made of the selectively permeable membranes of transport epithelia. The membranes retain proteins and other large molecules in the body fluid; hydrostatic pressure (blood pressure in many animals) forces water and small solutes, such as salts, sugars, amino acids, and nitrogenous wastes, through the device into the excretory system. The aqueous solution in the excretory system is called the filtrate.

Excretory systems produce urine from the filtrate by two mechanisms, both involving active transport. Selective transport of water and valuable solutes, such as glucose, salts, and amino acids, from the filtrate back into the body fluids is called **reabsorption**. Because filtration is nonselective, it is important that small molecules essential to the body be returned to the body fluids. In **secretion**, solutes (for example, excess salts and toxins) are removed from the animal's body fluids and added to the filtrate.

The overall effect of filtration, reabsorption, and secretion is analogous to cleaning out a closet (body fluids) by first removing all the small articles, such as shoes and broken coat hangers (filtration), returning the useful items (shoes, perhaps) to the closet (reabsorption), sorting large items like clothes and putting those you don't want into a refuse pile (secretion), and then discarding all the unwanted objects (excretion).

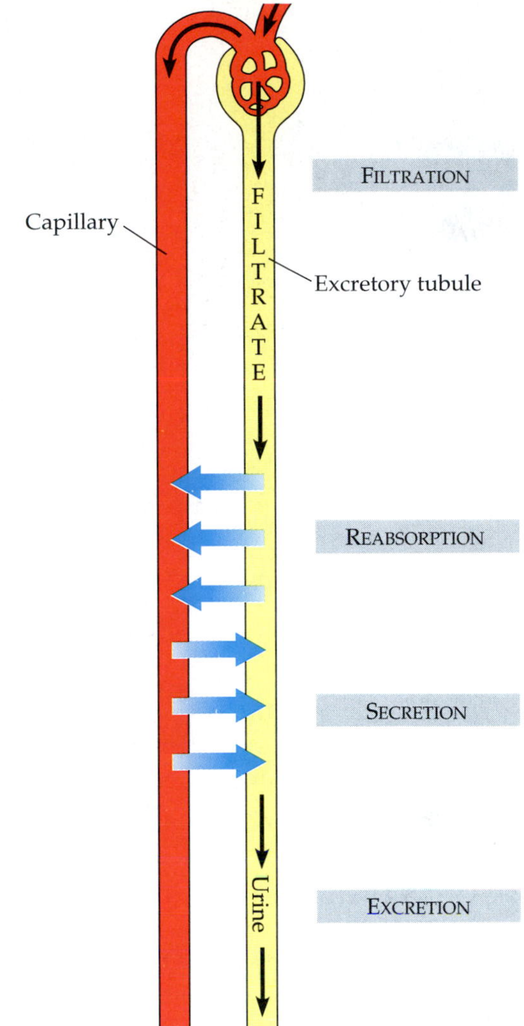

FIGURE 44.14 ▪ Key functions of excretory systems: an overview. Most excretory systems produce a filtrate by pressure-filtering body fluids. In this diagram, modeled after the vertebrate excretory system, the excretory tubule obtains a filtrate from the blood. Water and solutes are forced by blood pressure across the semipermeable membranes of a cluster of capillaries and into the excretory tubule (yellow). The filtrate is then refined by the transport epithelium lining the tubule. In the process of reabsorption, the epithelium reclaims valuable substances from the filtrate; these substances are then returned to the body fluids. In secretion, other substances, such as toxins and excess ions, are extracted from body fluids and added to the contents of the excretory tubule.

Diverse excretory systems are variations on a tubular theme

Protonephridia: Flame-Bulb Systems

Flatworms (Phylum Platyhelminthes) have tubular excretory systems called protonephridia. A **protonephridium** is a network of closed tubules lacking internal openings (FIGURE 44.15 on p. 880). The tubules branch throughout the body, and the smallest branches are capped by a cellular unit called a flame bulb. The flame bulb has a tuft of cilia projecting into the tubule. The beating of the cilia provides the force that draws water and solutes from the interstitial fluid through the flame bulb and into the tubule system. The beating cilia also propel fluid along the tubule, away from the flame bulb. Urine from the tubular system empties into the external environment through openings called nephridiopores. The excreted fluid is very dilute in the case of freshwater flatworms, helping balance the osmotic uptake of water from the environment. Apparently the tubules reabsorb most solutes from the fluid before it exits the body.

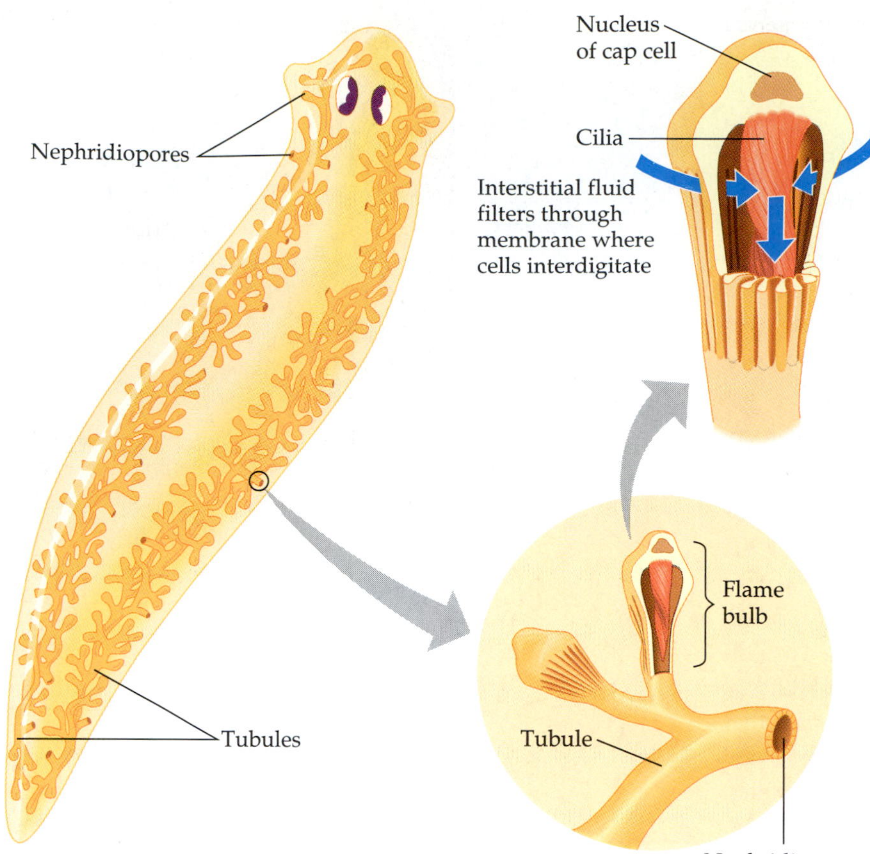

FIGURE 44.15 • Protonephridia: the flame-bulb system of a planarian. Protonephridia are branching internal tubules that function mainly in osmoregulation. A single cell caps the internal end of each tubule and interlocks with a tubule cell, forming a flame bulb. Water and solutes from the interstitial fluid enter the lumen of the tubule across the interdigitating membranes of the cap cells and tubule cells. The beating of cilia on the cap cell keeps the fluid moving into and through the tubules. (The beating cilia resemble a flickering flame, thus the name flame bulb.) The protonephridia of planarians and other freshwater flatworms produce dilute urine, which passes out through small openings called nephridiopores. Thus, freshwater flatworms balance the osmotic uptake of water from their hypotonic environment.

The flame-bulb systems of freshwater flatworms seem to function mainly in osmoregulation; most metabolic wastes diffuse out from the body surface or are excreted into the gastrovascular cavity and eliminated through the mouth (see Chapter 33). However, in some parasitic flatworms, which are isoosmotic to the surrounding fluids of their host organisms, protonephridia function mainly in excretion, disposing of nitrogenous wastes. This difference in function illustrates how structures common to a group of organisms can be adapted in diverse ways by evolution in different environments. Protonephridia are also found in rotifers, some annelids, the larvae of mollusks, and lancelets, which are invertebrate chordates. (See Chapters 33 and 34 to review these animal phyla.)

Metanephridia

Another type of tubular excretory system, the **metanephridium**, has internal openings that collect body fluids (FIGURE 44.16). Metanephridia are found in most annelids, including earthworms. Each segment of a worm has a pair of meta-

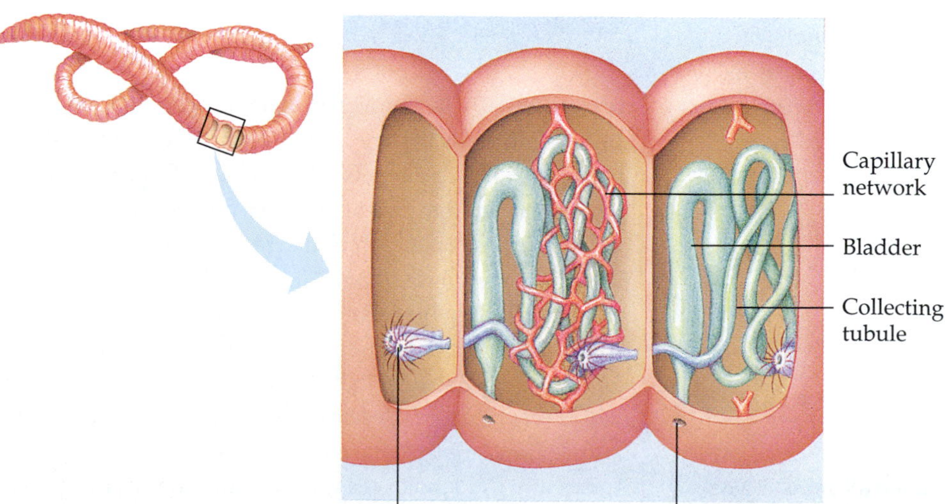

FIGURE 44.16 • Metanephridia of an earthworm. Each segment of the worm contains a pair of metanephridia (shown in green and blue), which collect coelomic fluid from the adjacent anterior segment. Fluid enters the nephrostome and passes through the coiled collecting tubule, which includes a storage bladder that opens to the outside through the nephridiopore. Nitrogenous wastes remain in the fluid, but certain salts are pumped back into the blood. Balancing osmotic uptake of water through the skin, an earthworm's urine is very dilute.

nephridia, which are tubules immersed in coelomic fluid and enveloped by a network of capillaries. The internal opening of a metanephridium is surrounded by a ciliated funnel, the nephrostome, that collects fluid from the coelom.

An earthworm's metanephridia have excretory and osmoregulatory functions. As the fluid moves along the tubule, the transport epithelium bordering the lumen reabsorbs most solutes out of the tubule, and the solutes reenter the blood circulating through the capillaries. Nitrogenous wastes remain in the tubule. Earthworms inhabit damp soil and usually experience a net uptake of water by osmosis. Their metanephridia balance the water influx by producing dilute urine (hypoosmotic to the worm's body fluids). The urine that exits through nephridiopores is mostly water and dissolved nitrogenous wastes.

Malpighian Tubules

The excretory organs of insects and other terrestrial arthropods are called **Malpighian tubules**. They remove nitrogenous wastes from the hemolymph (circulatory fluid) and also function in osmoregulation (FIGURE 44.17). Malpighian tubules open into the digestive tract and dead-end at tips that are immersed in hemolymph. The transport epithelium lining the tubules secretes certain solutes, including nitrogenous wastes, from the hemolymph into the lumen of the tubule. Water follows the solutes into the tubule by osmosis, and the fluid within the tubule then passes into the rectum, where most solutes are pumped back into the hemolymph. Water again follows the solutes, and the nitrogenous wastes are eliminated as nearly dry matter along with the feces. The insect excretory system is one adaptation that has contributed to the tremendous success of these animals on land, where conserving water is essential.

Vertebrate Kidneys

Vertebrates evolved from a group of invertebrate chordates (see Chapter 34). Hagfishes, which are among the most primitive living vertebrates, have kidneys with segmentally arranged excretory tubules, and it is likely that the excretory structures of vertebrate ancestors were also arranged segmentally. By contrast, the kidneys of most vertebrates are compact organs containing numerous tubules that are not segmentally arranged. A dense network of capillaries intimately associated with the tubules is part of the kidney. In vertebrates that osmoregulate, the kidneys function in both excretion and osmoregulation.

The kidneys, the blood vessels that serve them, and the structures that carry urine formed in the kidneys out of the

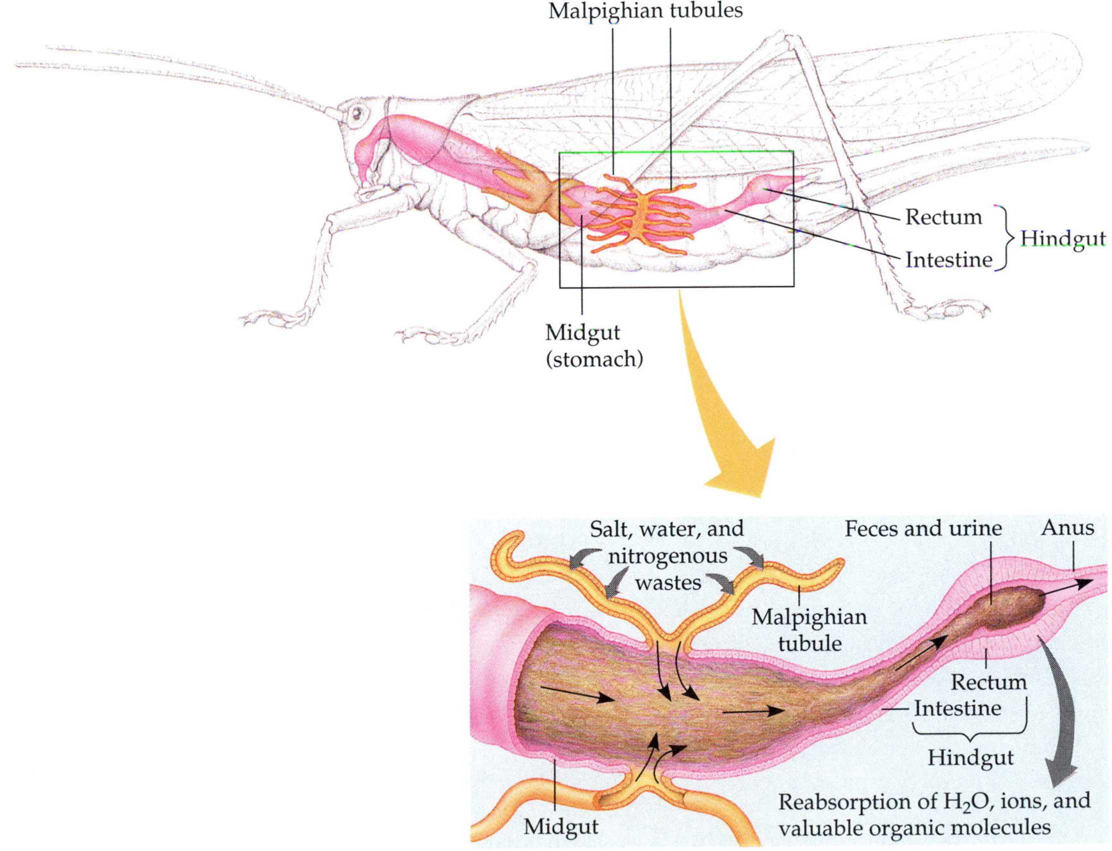

FIGURE 44.17 ▪ Malpighian tubules of insects. Malpighian tubules are outfoldings of the digestive tract. The tubules secrete nitrogenous wastes and salts from the hemolymph, and water follows these solutes by osmosis. Most of the salts and water are reabsorbed across the epithelium of the rectum, and the dry nitrogenous wastes are eliminated with the feces.

body are the components of the vertebrate excretory system (FIGURE 44.18). We will focus first on the mammalian version of the system, using humans as our example, and then compare the excretory systems of the various vertebrate classes.

Nephrons and associated blood vessels are the functional units of the mammalian kidney

In mammals, the kidneys are a pair of bean-shaped organs
44.1 (about 10 cm long in humans, FIGURE 44.18a). Blood enters each kidney via the **renal artery** and leaves each kidney via the **renal vein**. Although the human kidneys account for less than 1% of the weight of the body, they receive about 20% of the blood pumped with each heartbeat. Urine exits the kidney through a duct called the **ureter**. The ureters of both kidneys drain into a common **urinary bladder**. During urination, urine leaves the body from the urinary bladder through a tube called the **urethra**, which empties near the vagina in females or through the penis in males. Sphincter muscles near the junction of the urethra and the bladder control urination.

Structure and Function of the Nephron and Associated Structures

The mammalian kidney has two distinct regions, an outer **renal cortex** and an inner **renal medulla** (FIGURE 44.18b). Packing both regions are microscopic excretory tubules, called nephrons, and collecting ducts, both associated with tiny blood vessels (FIGURE 44.18c). The **nephron**, which is the functional unit of the vertebrate kidney, consists of a single long tubule and a ball of capillaries called the **glomerulus** (FIGURE 44.18d). The blind end of the tubule forms a cup-shaped swelling, called **Bowman's capsule**, which surrounds the glomerulus.

Filtration of the Blood. Filtration occurs as blood pressure forces water, urea, salts, and other small solutes from the blood in the glomerulus into the lumen of Bowman's capsule. The porous capillaries, along with specialized cells of the capsule called **podocytes**, function as a filter, being permeable to water and small solutes but not to blood cells or large molecules such as plasma proteins. Filtration is nonselective with regard to small molecules; any substance small enough to be forced through the capillary wall and between the podocytes by blood pressure enters the lumen of the nephron tubule. The filtrate in the Bowman's capsule contains solutes such as salts, glucose, and vitamins; nitrogenous wastes such as urea; and other small molecules—a mixture that mirrors the concentrations of these substances and the osmolarity of blood plasma.

Pathway of the Filtrate. From Bowman's capsule, the filtrate passes successively through three regions of the nephron: the **proximal tubule**; the **loop of Henle**, a hairpin turn with a descending limb and an ascending limb; and the **distal tubule**. The distal tubule empties into a **collecting duct**, which receives filtrate from many nephrons. The many collecting ducts of the kidney empty into the renal pelvis.

In the human kidney, about 80% of the nephrons, the **cortical nephrons**, have reduced loops of Henle and are almost entirely confined to the renal cortex. The other 20% of the nephrons, the **juxtamedullary nephrons**, have well-developed loops that extend into the renal medulla (see FIGURE 44.18c). Only mammals and birds have juxtamedullary nephrons; the nephrons of other vertebrates lack loops of Henle. As we will soon see, the juxtamedullary nephrons play a key role in the ability of mammals to excrete urine that is hyperosmotic to body fluids, an adaptation that conserves water.

The nephron and the collecting duct are lined by a transport epithelium that processes the filtrate to form the urine. From about 1100 to 2000 L of blood that flows through the human kidneys each day (about 275 times the total volume of blood in the body), the nephrons and collecting ducts process about 180 L of filtrate, but the kidneys excrete only about 1.5 L of urine. The rest of the filtrate, including about 99% of the water, is reabsorbed into the blood.

Blood Vessels Associated with the Nephrons. Each nephron is supplied with blood by an **afferent arteriole**, a branch of the renal artery that subdivides into the capillaries of the glomerulus. The capillaries converge as they leave the glomerulus, forming an **efferent arteriole**. This vessel subdivides again into a second network of capillaries, the **peritubular capillaries**. These capillaries intermingle with the proximal and distal tubules of the nephron. Additional capillaries extend downward to form the **vasa recta**, the capillary system that serves the loop of Henle. The vasa recta is also a loop, with a descending vessel and an ascending vessel conveying blood in opposite directions.

Although the excretory tubules and their surrounding capillaries are closely associated, they do not exchange materials directly. The tubules and capillaries are immersed in interstitial fluid, through which various substances pass back and forth between the plasma within capillaries and the filtrate within the nephron tubule.

Secretion. The proximal and distal tubules are the most common sites of secretion. Secretion is a very selective process involving both passive and active transport. For example, the controlled secretion of hydrogen ions from the interstitial fluid into the nephron tubule is important in maintaining a constant pH for the body fluids.

Reabsorption. The proximal and distal tubules and the loop of Henle all contribute to reabsorption, as does the collecting duct. Nearly all the sugar, vitamins, and other organic nutrients present in the initial filtrate are reabsorbed. Most of the water of the filtrate is also reabsorbed in the kidneys of mammals and birds.

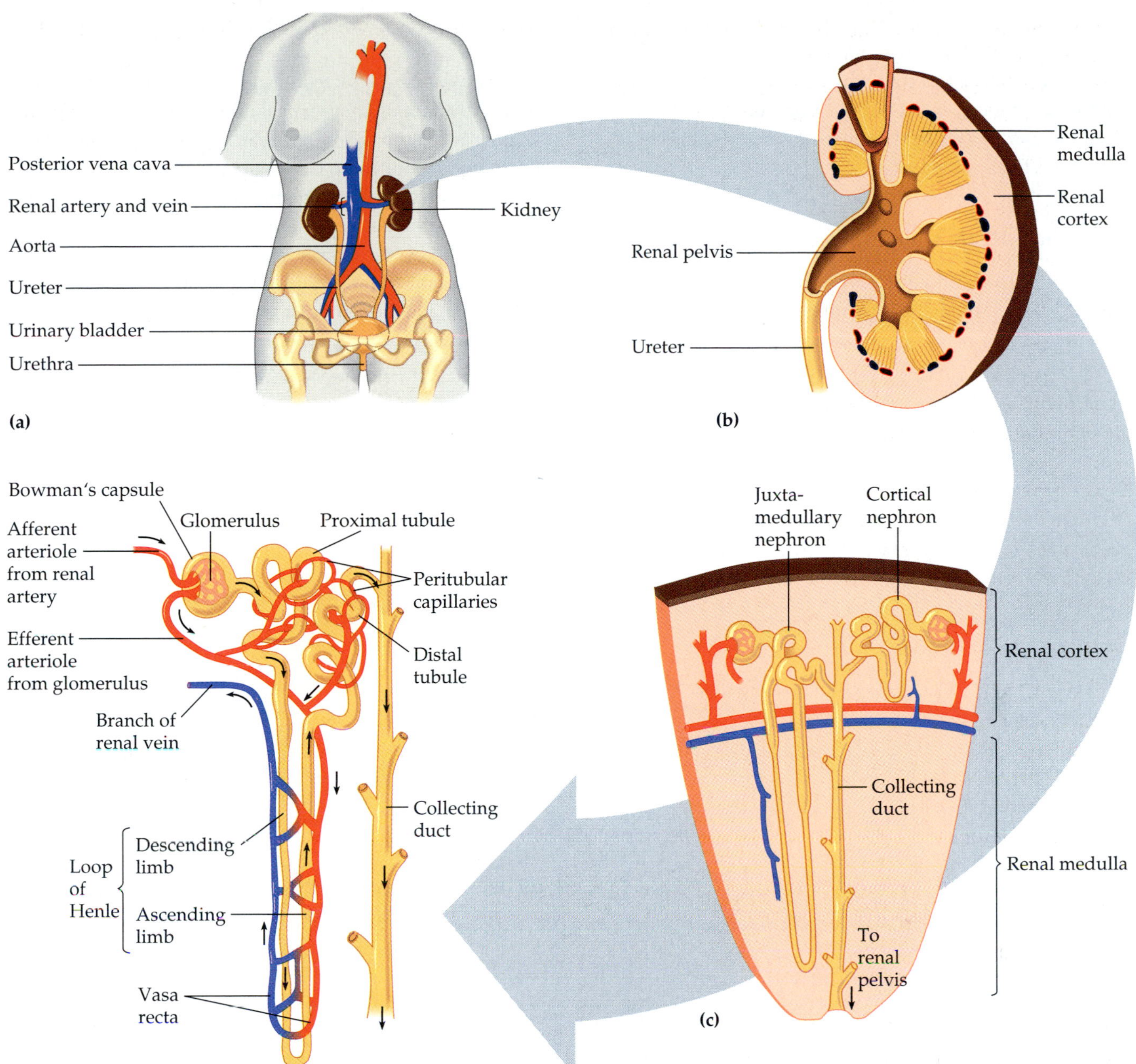

FIGURE 44.18 ▪ The human excretory system at four size scales. 44.1 **(a)** The kidneys produce urine and regulate the composition of the blood. Urine is conveyed to the urinary bladder via the ureter and to the outside via the urethra. Branches of the aorta, the renal arteries, convey blood to the kidneys; renal veins drain blood from the kidneys into the posterior vena cava.
(b) Urine is formed in two distinct regions of the kidney: the renal cortex and renal medulla. It then drains into a central chamber, the renal pelvis, and into the ureters. **(c)** Excretory tubules (nephrons and collecting ducts) and associated blood vessels pack the cortex and medulla. The human kidney has about a million nephrons, representing about 80 km of tubules. Cortical nephrons are restricted mainly to the renal cortex. Juxtamedullary nephrons have a long, hairpinlike portion that extends into the renal medulla. Several nephrons empty into each collecting duct, which drains into the renal pelvis.
(d) Each nephron consists of a glomerulus, or capillary cluster, surrounded by Bowman's capsule; a proximal tubule; a loop of Henle; and a distal tubule. Blood enters the glomerulus via an afferent arteriole and leaves via an efferent arteriole, which conveys it to peritubular capillaries surrounding the proximal and distal tubules, and to the vasa recta, capillaries surrounding the loop of Henle. The nephrons, collecting duct, and associated blood vessels produce urine from a filtrate (water and small solutes) forced by blood pressure into Bowman's capsule from the glomerulus. As the filtrate travels from Bowman's capsule to the collecting duct, its chemical makeup is changed as substances pass via the interstitial fluid between the nephron and the surrounding capillaries. Filtrate processing continues in the collecting duct. The flow of blood in the vasa recta is opposite that of the filtrate in the loop of Henle (arrows).

Together, selective reabsorption and secretion control the concentrations of various salts in body fluids. These key functions of the nephron and collecting ducts modify the composition of the filtrate, increasing the concentrations of some substances and decreasing the concentrations of others in the urine that is finally excreted.

From blood filtrate to urine: *a closer look*

44.2 In this section we concentrate on how the filtrate becomes urine as it flows through the mammalian nephron and collecting duct. The circled numbers correspond to the numbers in FIGURE 44.19.

① *Proximal tubule.* Secretion and reabsorption by the transport epithelium of the proximal tubule substantially alter the volume and composition of filtrate. For example, the cells of the transport epithelium help maintain a constant pH in body fluids by the controlled secretion of hydrogen ions. The cells also synthesize and secrete ammonia, which neutralizes the acid and keeps the filtrate from becoming too acidic. The more acidic the filtrate, the more ammonia the cells produce and secrete, and the urine of a mammal usually contains some

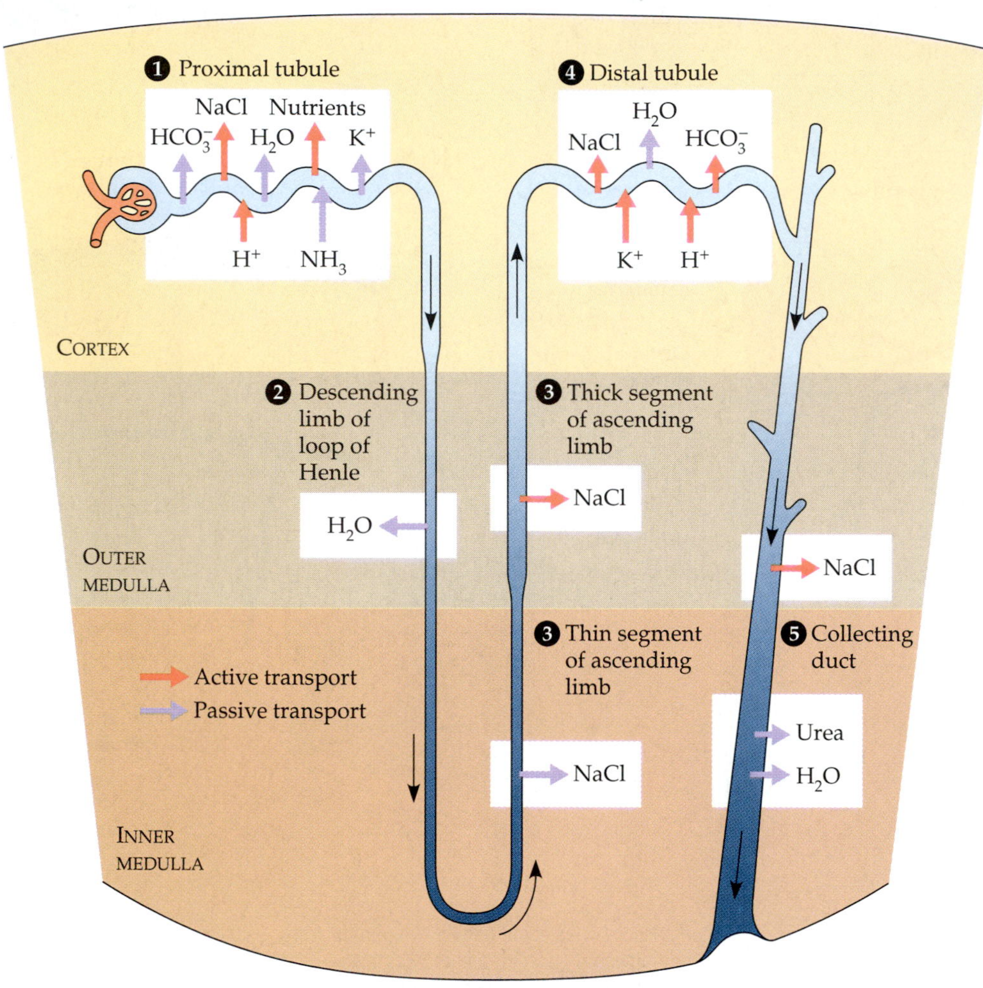

44.2 **FIGURE 44.19 ▪ The nephron and collecting duct: regional functions of the transport epithelium.** In this diagram, purple arrows indicate passive transport and red arrows symbolize active transport. ① The proximal tubule plays an important role in homeostasis by its controlled secretion and reabsorption of several substances. For example, about two-thirds of the NaCl and water filtered from blood into the nephron tubule are reabsorbed across the epithelium of the proximal tubule. This region also functions in the reabsorption of nutrients and in controlling pH by the secretion of H^+ and the reabsorption of HCO_3^-. ② The descending limb of the loop of Henle is permeable to water but not to salt. Osmotic loss of water from the filtrate, as the descending limb penetrates the renal medulla, concentrates NaCl in the filtrate. ③ The ascending limb of the loop of Henle consists of a thin segment and a thick segment. Both have epithelia that are virtually impermeable to water. The thin segment is permeable to NaCl, and the salt that was concentrated in the filtrate within the descending limb now diffuses out of the ascending limb, contributing to a high interstitial osmolarity in the inner medulla of the kidney. The thick segment continues the transfer of salt from the filtrate to the interstitial fluid, but now the transport is active. ④ The distal tubule is another important region of controlled secretion and reabsorption. For example, it helps regulate blood pH by the reabsorption of bicarbonate (HCO_3^-), a buffer. The distal tubule also functions in K^+ and Na^+ homeostasis. ⑤ The specialized epithelium of the collecting duct is permeable to water but not to salt. The duct carries the filtrate toward the renal medulla for a second time, and the filtrate becomes more and more concentrated as water is lost to the interstitial fluid. The bottom portion of the collecting duct is permeable to urea, and leakage of this solute into the interstitial fluid contributes to the high osmolarity of the medulla.

ammonia from this source. The proximal tubule cells also reabsorb about 90% of the important buffer bicarbonate (HCO_3^-) from the filtrate. Drugs and other poisons that have been processed in the liver are secreted into the filtrate by the epithelium of the proximal tubule. They pass from the peritubular capillaries into the interstitial fluid, and then across the epithelium of the tubule into the lumen. On the other hand, nutrients, including glucose and amino acids, are actively transported from the filtrate to the interstitial fluid, and then into the blood within the peritubular capillaries. Without this reabsorption, these nutrients would be lost with the urine. Potassium (K^+) is also reabsorbed.

One of the most important functions of the proximal tubule is the reabsorption of NaCl (salt) and water. Salt in the filtrate diffuses into the cells of the transport epithelium, and the membranes of the cells actively transport Na^+ out of the cells and into the interstitial fluid. This transfer of positive charge is balanced by the passive transport of Cl^- out of the tubule. As salt moves from the filtrate to the interstitial fluid, water follows passively by osmosis. The side of the epithelium facing the exterior of the tubule has a much smaller surface area than the side facing the lumen, which minimizes the leakage of salt and water back into the tubule. Instead, the salt and water now diffuse from the interstitial fluid into the peritubular capillaries.

② *Descending limb of the loop of Henle.* Reabsorption of water continues as the filtrate moves along the tubule to the descending limb of the loop of Henle. Here the transport epithelium is freely permeable to water but not very permeable to salt and other small solutes. For water to move out of the tubule by osmosis, the interstitial fluid bathing the tubule must be hyperosmotic to the filtrate. The osmolarity of the interstitial fluid does in fact increase gradually, becoming progressively greater from the outer cortex to the inner medulla of the kidney. (The mechanism that maintains this gradient will be discussed shortly.) Thus, filtrate moving downward from the cortex to the medulla within the descending limb of the loop of Henle continues to lose water to interstitial fluid of greater and greater osmolarity. At the same time, the NaCl concentration of the filtrate increases as water departs by osmosis.

③ *Ascending limb of the loop of Henle.* The filtrate reaches the tip of the loop, located deep in the renal medulla in the case of juxtamedullary nephrons, then moves to the cortex again within the ascending limb of the loop. In contrast to the descending limb, the transport epithelium of the ascending limb is permeable to salt but not to water. The ascending limb actually has two specialized regions: a thin segment near the loop tip and a thick segment leading to the distal tubule. As filtrate ascends in the thin segment, NaCl, which became concentrated in the descending limb, diffuses out of the tubule into the interstitial fluid. This loss of salt contributes to the high osmolarity of the interstitial fluid in the medulla. The exodus of salt from the filtrate continues in the thick segment of the ascending limb, but here the transport epithelium actively transports NaCl into the interstitial fluid. By losing salt without giving up water, the filtrate becomes progressively more dilute as it moves up to the cortex again in the ascending limb of the loop.

④ *Distal tubule.* The distal tubule is another important site of secretion and reabsorption. For example, the distal tubule plays a key role in regulating the K^+ and NaCl concentration of body fluids by varying the amount of the K^+ that is secreted into the filtrate and the amount of NaCl that is reabsorbed from the filtrate. Like the proximal tubule, the distal tubule also contributes to pH regulation, by the controlled secretion of H^+ and by the reabsorption of bicarbonate (HCO_3^-).

⑤ *Collecting duct.* The collecting duct carries the filtrate back in the direction of the medulla and renal pelvis. The transport epithelium of the collecting duct plays a large role in determining how much salt is actually excreted in the urine by actively reabsorbing NaCl. The epithelium is permeable to water but not to salt. Thus, as the collecting duct traverses the gradient of osmolarity in the interstitial fluid, the filtrate loses more and more water by osmosis to the hyperosmotic fluid outside the duct. Loss of water concentrates the urea in the filtrate, but not all of this urea is immediately passed along to the renal pelvis in the urine. At the bottom of the collecting duct, in the inner medulla, the epithelium of the duct is permeable to urea. Because of the high concentration of urea in the filtrate at this point, some of the urea diffuses out of the duct and into the interstitial fluid, bathing the portions of nephrons in the medulla. This interstitial urea is a major solute contributing, along with the NaCl, to the high osmolarity of the interstitial fluid in the medulla. And it is this high osmolarity of the interstitial fluid that enables the kidney to conserve water by excreting urine that is hyperosmotic to the general body fluids.

The mammalian kidney's ability to conserve water is a key terrestrial adaptation

The osmolarity of human blood is about 300 mosm/L, but the kidney can excrete urine up to four times as concentrated—about 1200 mosm/L. The cooperative action of the loop of Henle and the collecting duct maintain the gradient of osmolarity in the interstitial tissue of the kidney that makes it possible to concentrate the urine. The two solutes responsible for this osmolarity gradient are NaCl, which is deposited in the renal medulla by the loop of Henle, and urea, which leaks across the epithelium of the collecting duct in the inner medulla.

Conservation of Water by Two Solute Gradients

To better understand the physiology of the mammalian kidney as a water-conserving organ, let's retrace the flow of filtrate through the excretory tubule, this time focusing on how the juxtamedullary nephrons maintain an osmolarity gradient in the kidney and use that gradient to excrete a hyperosmotic urine (FIGURE 44.20). Filtrate passing from Bowman's capsule to the proximal tubule has an osmolarity of about 300 mosm/L, the same as blood. As the filtrate flows through the proximal tubule, located in the renal cortex, a large amount of water *and* salt is reabsorbed; thus, the volume of filtrate decreases substantially at this stage, but the osmolarity remains about the same.

As the filtrate flows from cortex to medulla in the descending limb of the loop of Henle, water leaves the tubule by osmosis, and the osmolarity of the filtrate increases as solutes, including NaCl, become more concentrated. Increasing gradually from cortex to medulla, the salt concentration of the filtrate peaks at the elbow of the loop of Henle. This maximizes the diffusion of salt out of the tubule as the filtrate rounds the curve and enters the ascending limb, which, remember, is permeable to salt but not to water. Thus, the two limbs of the loop of Henle cooperate in maintaining the gradient of osmolarity in the interstitial fluid of the kidney. The descending limb produces a progressively saltier filtrate, and the ascending limb exploits this concentration of NaCl to help maintain a high osmolarity in the interstitial fluid of the renal medulla.

Notice that the loop has some of the qualities of a countercurrent system, similar in principle to the countercurrent mechanism that maximizes oxygen absorption by the gills of fishes (see FIGURES 42.19 and 42.20). Although the two limbs of

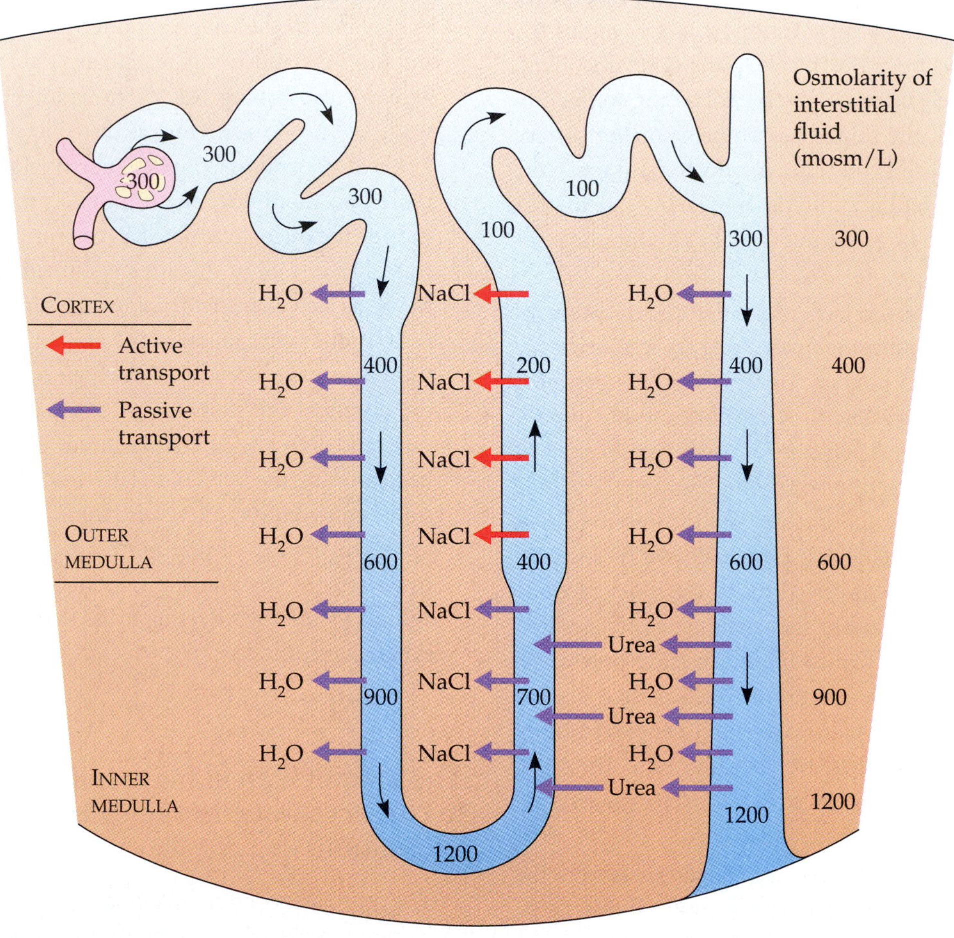

FIGURE 44.20 ▪ How the human kidney concentrates urine: the two-solute model. From the cortex to the inner medulla, the interstitial fluid of the kidney increases in osmolarity from about 300 to 1200 mosm/L. Two solutes contribute to this gradient of osmolarity: NaCl and urea. The loop of Henle maintains the interstitial gradient of NaCl. The filtrate concentration of this salt increases by the loss of water from the descending limb, then the ascending limb leaks the salt into the interstitial fluid. Additional salt is actively transported out of the thick segment of the ascending limb. The second solute, urea, is added to the interstitial fluid of the medulla by diffusion out of the collecting duct (urea remaining in the collecting duct is excreted). Urea reenters the tubule by diffusion into the ascending limb of the loop of Henle. The filtrate makes a total of three trips between the cortex and medulla: first down, then up, and then down one more time in the collecting duct. As the filtrate flows in the collecting duct past interstitial fluid of increasing osmolarity, more and more water moves out of the duct by osmosis, thereby concentrating the solutes, including urea, that are left behind in the filtrate. Under conditions in which the kidney conserves as much water as possible, urine can reach an osmolarity of about 1200 mosm/L, considerably hypertonic to blood (about 300 mosm/L). This ability to excrete nitrogenous wastes with a minimal loss of water is a key terrestrial adaptation of mammals.

the loop of Henle are not in direct physical contact, they are close enough together to affect each other's chemical exchanges with a common interstitial fluid. The loop of Henle can concentrate salt in the inner medulla only because traffic in the descending limb counters the osmolarity gradient produced by the ascending limb in the interstitial fluid.

What prevents the capillaries of the renal medulla from dissipating the osmolarity gradient by carrying away the NaCl that leaks from the ascending limb into the interstitial fluid? Notice in FIGURE 44.18d that the vasa recta is also a countercurrent system, with descending and ascending vessels carrying blood in opposite directions through the kidney's osmolarity gradient. As the descending vessel conveys blood toward the inner medulla, water is lost from the blood and NaCl diffuses into the blood. These fluxes are simply reversed as blood flows back toward the cortex in the ascending vessel, with water reentering the blood and salt diffusing out of the blood. Thus, the vasa recta can supply nutrients and other important substances carried by blood without interfering with the osmolarity gradient that makes it possible for the kidney to excrete a hyperosmotic urine.

By the time the filtrate reaches the distal tubule, it is *not* hyperosmotic to body fluids at all but is actually hypoosmotic. This is because the thick segment of the ascending limb of the loop of Henle actively pumps NaCl out of the tubule, making the filtrate more and more dilute. Now the filtrate descends once again toward the medulla, this time in the collecting duct, which, remember, is permeable to water but not to salt. Flowing from cortex to medulla, the filtrate loses water by osmosis as it encounters interstitial fluid of increasing osmolarity. Urea is thus concentrated in the filtrate. Some of it leaks out of the lower portion of the collecting duct, thereby making a contribution to the high interstitial osmolarity of the inner medulla. (This urea is recycled by its diffusion into the loop of Henle, but continual urea leakage from the collecting duct maintains a high interstitial concentration of this solute.) The urea remaining in the collecting duct is excreted with a minimal loss of water from the body because osmosis causes the filtrate in the collecting duct to equal the osmolarity of the interstitial fluid, which can be as high as 1200 mosm/L in the inner medulla. Notice that urine, at its most concentrated, is actually *isoosmotic* to the interstitial fluid of the inner medulla; however, it is *hyperosmotic* to blood and interstitial fluid elsewhere in the body. The juxtamedullary nephron, with its urine-concentrating features, is a key adaptation to terrestrial life, enabling mammals to get rid of nitrogenous wastes without squandering water.

Nervous system and hormonal feedback circuits regulate kidney functions

Although the kidneys *can* excrete hyperosmotic urine, it is not always beneficial for them to do so. For example, if you have consumed an excessive amount of fluid, the kidneys can actually excrete a large volume of hypoosmotic urine (as dilute as 70 mosm/L, compared to about 300 mosm/L for human blood). This makes it possible to eliminate a lot of water without losing essential salts. The kidney is a versatile osmoregulatory organ, where water and salt reabsorption are subject to a combination of nervous system and hormonal controls.

44.3 One hormone important in osmoregulation is **antidiuretic hormone (ADH)** (FIGURE 44.21a). It is produced in a part of the brain called the hypothalamus and is stored and released from the pituitary gland, which is positioned just below the hypothalamus. Osmoreceptor cells in the hypothalamus monitor the osmolarity of blood, stimulating the release of additional ADH when blood osmolarity rises above a set point of 300 mosm/L.

Excessive water losses due to sweating or diarrhea are examples of crises that could cause an increase in blood osmolarity. More ADH is then discharged into the bloodstream and reaches the kidney. The main targets of ADH are the distal tubules and collecting ducts of the kidney, where the hormone increases the permeability of the epithelium to water. This amplifies water reabsorption, which helps prevent further deviation of blood osmolarity from the set point. By negative feedback, the subsiding osmolarity of the blood reduces the activity of osmoreceptor cells in the hypothalamus, and less ADH is secreted. Only the intake of additional water in food and drink can bring osmolarity all the way back down to 300 mosm/L. When very little ADH is released, as would occur after a large volume of water has lowered the blood osmolarity, the kidneys would absorb little water, resulting in an increased discharge of dilute urine. (Increased urination is called diuresis, and it is because ADH opposes this state that it is called *anti*diuretic hormone.) Alcohol can disturb water balance by inhibiting the release of ADH, causing excessive loss of water in the urine and dehydrating the body. Some of the symptoms of a hangover may be due to this dehydration. Normally, however, blood osmolarity, ADH release, and water reabsorption in the kidney are all linked in a feedback loop that contributes to homeostasis.

A second mechanism that regulates kidney function involves a specialized tissue called the **juxtaglomerular apparatus (JGA)**, located in the vicinity of the afferent arteriole, which supplies blood to the glomerulus (FIGURE 44.21b). When the blood pressure or blood volume in the afferent arteriole drops (sometimes as a result of reduced salt intake), the enzyme renin initiates chemical reactions that convert a plasma protein called angiotensinogen to a peptide called **angiotensin II**. Functioning as a hormone, angiotensin II increases blood pressure and blood volume in several ways. For example, it raises blood pressure by constricting arterioles, decreasing blood flow to many capillaries, including those of the kidney. Angiotensin II also stimulates the proximal tubules of the nephrons to reabsorb more NaCl and

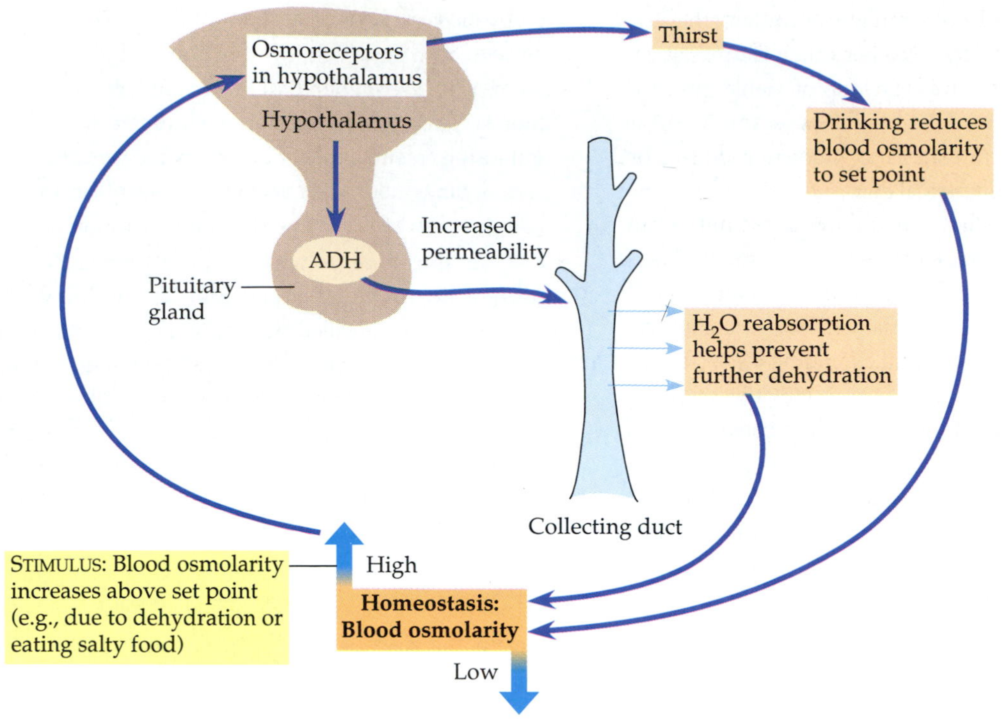

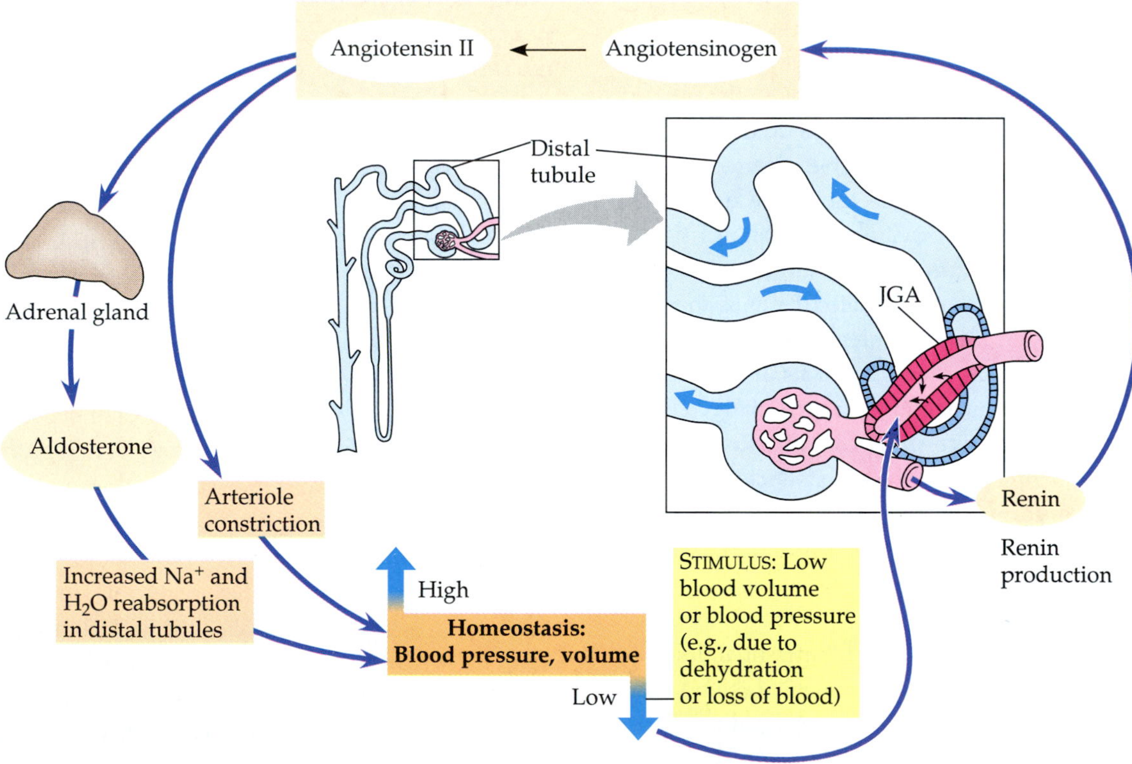

44.3 **FIGURE 44.21 ▪ Hormonal control of the kidney by negative feedback circuits.** **(a)** Antidiuretic hormone (ADH), produced in the hypothalamus of the brain and secreted into the bloodstream from the pituitary gland, enhances fluid retention by making the kidneys reclaim more water. The release of ADH is triggered when osmoreceptor cells in the hypothalamus detect an increase in the osmolarity of the blood. In this situation, the osmoreceptor cells also promote thirst. Drinking reduces the osmolarity of the blood, which inhibits the secretion of ADH, thereby completing the feedback circuit. **(b)** The renin-angiotensin-aldosterone system (RAAS) centers on the juxtaglomerular apparatus (JGA). The JGA responds to a decrease in blood pressure or blood volume by releasing the enzyme renin into the bloodstream (small black arrows). In the blood, renin initiates the conversion of angiotensinogen to angiotensin II. Angiotensin II increases blood pressure by causing arterioles to constrict. It also increases blood volume in two ways: by signaling the proximal tubules of the nephrons to reabsorb more NaCl and water (not illustrated) and by stimulating the adrenal glands to release aldosterone, a hormone that makes the distal tubules reabsorb more Na^+ and water. This leads to an increase in blood volume and pressure, completing the feedback circuit by suppressing the release of renin.

water. This reduces the amount of salt and water excreted in the urine and consequently raises blood volume and pressure. Yet another effect of angiotensin II is stimulation of the adrenal glands, organs located atop the kidneys, to release a hormone called **aldosterone.** This hormone acts on the nephrons' distal tubules, making them reabsorb more sodium (Na^+) and water and increasing blood volume and pressure. In summary, the **renin-angiotensin-aldosterone system (RAAS)** is part of a complex feedback circuit that functions in homeostasis. A drop in blood pressure and blood volume triggers renin release from the JGA. In turn, the rise in blood pressure and volume resulting from the various actions of angiotensin II and aldosterone reduce the release of renin (see FIGURE 44.21b).

It may seem that the functions of ADH and the RAAS are redundant, but this is not the case. It is true that both increase water reabsorption, but they counter different osmoregulatory problems. The release of ADH is a response to an increase in the osmolarity of the blood, as when the body is dehydrated from an inadequate intake of water, for instance. But imagine a situation that causes an excessive loss of salt and body fluids—an injury, for example, or severe diarrhea. This reduces the blood's volume without increasing its osmolarity. The RAAS would save the day by increasing water and Na^+ reabsorption in response to the drop in blood volume caused by fluid loss. Normally, ADH and the RAAS are partners in homeostasis; ADH alone would lower blood Na^+ concentration by stimulating water reabsorption in the kidney, but the RAAS helps maintain balance by stimulating Na^+ reabsorption.

Still another hormone, a peptide called **atrial natriuretic factor (ANF)**, opposes the RAAS. The walls of the atria of the heart release ANF in response to an increase in blood volume and pressure. ANF inhibits the release of renin from the JGA, inhibits NaCl reabsorption by the collecting ducts, and reduces aldosterone release from the adrenal glands. These actions lower blood volume and pressure. Thus, ADH, the RAAS, and ANF provide an elaborate system of checks and balances that regulate the kidney's ability to control the osmolarity, salt concentration, volume, and pressure of blood.

Having considered the mammalian kidney and its regulation in detail, we can now compare the structures and functions of kidneys in other vertebrate classes.

Diverse adaptations of the vertebrate kidney have evolved in different habitats

Variations in nephron structure and function equip the kidneys of different vertebrates for osmoregulation in their various habitats. We have seen, for instance, that nephrons of the mammalian kidney can concentrate urine and conserve water. Mammals that excrete the most hyperosmotic urine, such as kangaroo rats and other mammals adapted to the desert, have exceptionally long loops of Henle. Long loops maintain steep osmotic gradients in the kidney, resulting in urine becoming very concentrated as it passes from cortex to medulla in the collecting ducts. In contrast, beavers, which spend much of their time in fresh water and rarely face problems of dehydration, have nephrons with very short loops, resulting in dilute urine.

Birds, like mammals, have kidneys with juxtamedullary nephrons that specialize in conserving water. However, bird nephrons have much shorter loops of Henle than mammalian nephrons. Bird kidneys, therefore, cannot concentrate urine to the osmolarities achieved by mammalian kidneys.

The kidneys of reptiles, having only cortical nephrons, produce urine that is, at best, isoosmotic to body fluids. However, the epithelium of the cloaca (see Chapter 34) helps conserve fluid by reabsorbing some of the water present in urine and feces. Also, most terrestrial reptiles excrete nitrogenous wastes as uric acid, which, as we discussed earlier, helps conserve water.

In contrast to mammals and birds, a freshwater fish must excrete excess water because the animal is hyperosmotic to its surroundings. Instead of conserving water, the nephrons produce a large volume of very dilute urine. Freshwater fishes conserve salts by efficient reabsorption of ions from the filtrate in the nephrons.

Amphibian kidneys function much like those of freshwater fishes. When in fresh water, the skin of the frog accumulates certain salts from the water by active transport, and the kidneys excrete dilute urine. On land, where dehydration is the most pressing problem of osmoregulation, frogs conserve body fluid by reabsorbing water across the epithelium of the urinary bladder.

Bony fishes that live in seawater, being hypoosmotic to their surroundings, have the opposite problem of their freshwater relatives. In many species, nephrons lack glomeruli and Bowman's capsules, and concentrated urine is formed by secreting ions into the excretory tubules. Thus, as mentioned previously, the kidneys of marine fishes excrete very little urine and function mainly to get rid of divalent ions such as Ca^{2+}, Mg^{2+}, and SO_4^{2-}, which the fish takes in by its incessant drinking of seawater. Its gills excrete mainly monovalent ions such as Na^+ and Cl^- and the bulk of its nitrogenous wastes in the form of NH_4^+ (ammonium ion).

Interacting regulatory systems maintain homeostasis

Numerous regulatory systems are involved in maintaining homeostasis in an animal's internal environment. As we have seen, the mechanisms that rid the body of nitrogenous wastes operate hand in hand with those involved in osmoregulation. Similarly, the regulation of body temperature involves mechanisms that have an impact on metabolic rate, blood pressure, tissue oxygenation, and body weight. Under some conditions, usually at the physical extremes compatible with the organism's life, the demands of one system might come into conflict with those of other systems. For instance, in very warm and dry environments, the conservation of water takes precedence over evaporative heat loss. As a result, many desert animals tolerate occasional abnormally high body temperature. Normally, however, the various regulatory systems act together to maintain homeostasis in the internal environment.

Our discussion of homeostasis would be incomplete without focusing on the liver, the vertebrate body's largest and most functionally diverse organ. Liver functions are pivotal to homeostasis and involve interaction with most of the body's organ systems. For example, liver cells interact with the circulatory system in taking up glucose from the blood. Liver cells store excess glucose as glycogen and, in response to the body's demand for fuel, convert glycogen back to glucose, releasing glucose to the blood. The liver also synthesizes plasma proteins important in blood clotting and in maintaining the osmotic balance of the blood. Assisting the excretory system, liver cells detoxify many chemical poisons and prepare metabolic wastes for disposal. The liver's diverse functions and interactions with other organs accentuate the point that homeostasis requires the concerted action of several body systems.

■ ■ ■

In this chapter we have touched on the complex nervous system and hormonal checks and balances that regulate the activity of organs and organ systems. We concentrate on the hormonal control of homeostasis in the next chapter.

CHAPTER REVIEW

REVIEW OF KEY CONCEPTS

(with page numbers and key figures)

REGULATION OF BODY TEMPERATURE

Thermoregulation, the maintenance of body temperature within a range that enables cells to function efficiently, involves heat transfer between the organism and the external environment.

- **Four physical processes account for heat gain or loss (p. 866, FIGURE 44.1)** They are conduction, convection, radiation, and evaporation.
- **Ectotherms derive body heat mainly from their surroundings; endotherms derive it mainly from metabolism (pp. 866–867, FIGURE 44.2)** Most invertebrates, fishes, amphibians, and reptiles are ectotherms. Endothermy enables animals to maintain a relatively uniform body temperature and a high level of aerobic metabolism.
- **Thermoregulation involves physiological and behavioral adjustments (pp. 867–868, FIGURE 44.3)** Ectotherms and endotherms adjust the rate of heat exchange with their surroundings by evaporative cooling and by behavioral responses. Birds and mammals can also change the rate of metabolic heat production. Insulation, vasodilation, vasoconstriction, and countercurrent heat exchangers alter the rate of heat exchange. Panting, sweating, and bathing increase evaporation.
- **Most animals are ectothermic, but endothermy is widespread (pp. 868–872, FIGURES 44.5–44.7)** Some large active insects and fishes generate metabolic heat by muscle contractions, and many retain it by countercurrent heat exchangers. Some invertebrates, amphibians, and reptiles maintain tolerable internal temperatures by behavioral means. Thermoregulatory mechanisms in mammals and birds include shivering and nonshivering thermogenesis; insulation by fat, hair, or feathers; panting; and countercurrent heat exchange. Physiological adjustment to seasonal temperature changes may involve changes in an animal's thermostatic control mechanisms and various responses at the cellular level.
- **Torpor conserves energy during environmental extremes (p. 873)** Torpor involves a decrease in metabolic, heart, and respiratory rates and enables the animal to temporarily withstand unfavorable temperatures or lack of food and water.

WATER BALANCE AND WASTE DISPOSAL

Disposing of toxic metabolic wastes and balancing the uptake and loss of water are key aspects of homeostasis. Both involve movements of solutes and water (by osmosis) between an animal's internal fluids and the external environment.

- **Water balance and waste disposal depend on transport epithelia (p. 873, FIGURE 44.9)** Specialized cell layers regulate the solute movements required for waste disposal and for tempering changes in body fluids.
- **An animal's nitrogenous wastes are correlated with its phylogeny and habitat (pp. 873–875, FIGURE 44.10)** Protein and nucleic acid metabolism generates ammonia, a toxic waste product excreted in three forms. Most aquatic animals excrete ammonia across the body surface or gill epithelia into the surrounding water. The liver of mammals and most adult amphibians converts ammonia to the less-toxic urea, which is carried to the kidneys, concentrated, and excreted with a minimal loss of water. Uric acid is an insoluble precipitate excreted in the pastelike urine of land snails, insects, birds, and many reptiles.
- **Cells require a balance between osmotic gain and loss of water (pp. 875–876)** Water uptake and loss are balanced by various mechanisms of osmoregulation in different environments. Osmosis occurs whenever two solutions separated by the membrane differ in total solute concentration, or osmolarity (moles of solute per liter of solution).
- **Osmoregulators expend energy to control their internal osmolarity; osmoconformers are isoosmotic with their surroundings (pp. 876–879, FIGURE 44.11)** Osmoconformers, which do not regulate their osmolarity, include most marine invertebrates. Osmoregulators control water uptake and loss in a hyperosmotic or hypoosmotic environment. Stenohaline animals cannot tolerate marked osmotic changes, while euryhaline animals can tolerate such changes, often by altering their osmoregulatory mechanisms. Sharks have an osmolarity slightly higher than seawater because they retain urea. Marine bony fishes lose water to their hyperosmotic environment and drink seawater. Marine vertebrates excrete excess salt through rectal glands, gills, salt-excreting glands, or kidneys. Freshwater organisms constantly take in water from their hypoosmotic environment. Freshwater protists pump out excess water with contractile vacuoles, and freshwater animals excrete dilute urine. Salt loss is replaced by eating or by ion uptake by gills. Terrestrial ani-

mals combat desiccation by the nervous system and hormonal control of thirst, behavioral adaptations, water-conserving excretory organs, and by drinking and eating food with high water content.

EXCRETORY SYSTEMS

Functioning in both waste disposal and water balance, excretory systems are central to homeostasis.

- **Most excretory systems produce urine by refining a filtrate derived from body fluids: *an overview*** **(p. 879, FIGURE 44.14)** Key functions of most excretory systems are filtration (pressure-filtering of body fluids, producing a filtrate) and the production of urine from the filtrate by reabsorption (reclaiming valuable solutes from the filtrate) and secretion (addition of toxins and other solutes from the body fluids to the filtrate).

- **Diverse excretory systems are variations on a tubular theme (pp. 879–882, FIGURES 44.15–44.17)** Extracellular fluid is filtered into the protonephridia of the flame-bulb system in flatworms; these tubules excrete a dilute fluid and also function in osmoregulation. Each segment of an earthworm has a pair of open-ended metanephridia, tubules that collect coelomic fluid and produce dilute urine for excretion. In insects, Malpighian tubules function in osmoregulation and removal of nitrogenous wastes from the hemolymph. Insects produce a relatively dry waste matter, an important adaptation to terrestrial life. Kidneys, the excretory organs of vertebrates, function in both excretion and osmoregulation.

- 44.1 **Nephrons and associated blood vessels are the functional units of the mammalian kidney (pp. 882–884, FIGURE 44.18)** Excretory tubules (consisting of nephrons and collecting ducts) and associated blood vessels pack the kidney. Each nephron consists of a Bowman's capsule, which surrounds a ball of capillaries called the glomerulus; a proximal tubule; a loop of Henle; and a distal tubule. Fluid from several nephrons flows into a collecting duct and passes urine into the kidney's central receptacle, the renal pelvis. A ureter conveys urine from the renal pelvis to the urinary bladder. Blood supply to the nephron is through an afferent arteriole, which divides into the capillaries of the glomerulus. An efferent arteriole carries blood away from Bowman's capsule and subdivides into the peritubular capillaries embracing the proximal and distal tubules. The vasa recta is a system of capillaries that services the loops of Henle. Nephrons control the composition of the blood by filtration, secretion, and reabsorption.

- 44.2 **From blood filtrate to urine: *a closer look*** **(pp. 884–885, FIGURE 44.19)** Substances reabsorbed from the filtrate in the proximal tubule include most of the salt and water filtered from the blood, and glucose and amino acids (by active transport). Drugs, ammonia, and hydrogen ions (for the control of body pH) are selectively secreted into the filtrate. The descending limb of the loop of Henle is permeable to water but not to salt; water moves by osmosis into the hyperosmotic interstitial fluid. Salt diffuses out of the concentrated filtrate as it moves through the salt-permeable ascending limb of the loop of Henle. Secretion and reabsorption by the distal tubule play key roles in regulating potassium concentration and blood pH.

- **The mammalian kidney's ability to conserve water is a key terrestrial adaptation (pp. 885–887, FIGURE 44.20)** The collecting duct, permeable to water but not to salt, carries the filtrate through the kidney's osmolarity gradient, and more water exits by osmosis. Urea also diffuses out of the tubule and, with salt, forms the osmotic gradient that enables the kidney to produce urine that is hyperosmotic to the blood.

- 44.3 **Nervous system and hormonal feedback circuits regulate kidney functions (pp. 887–889, FIGURE 44.21)** The osmolarity of the urine is regulated by nervous system and hormonal control of water and salt reabsorption in the kidneys. Antidiuretic hormone (ADH) increases water reabsorption by the tubule. Renin, released by the juxtaglomerular apparatus (JGA), triggers the formation of angiotensin II, which in turn causes arterioles to constrict and the adrenal glands to release the hormone aldosterone. Aldosterone stimulates the reabsorption of Na^+ and the passive flow of water from the filtrate. Both ADH and the renin-angiotensin-aldosterone system (RAAS) result in the reabsorption of water and more-concentrated urine. Atrial natriuretic factor (ANF), released by the atria of the heart in response to increased blood pressure, inhibits the release of renin and counters the RAAS.

- **Diverse adaptations of the vertebrate kidney have evolved in different habitats (p. 889)** The original function of the vertebrate kidney was osmoregulation. The form and function of nephrons in the various vertebrate classes are related primarily to the requirements for osmoregulation in the animal's habitat. The excretion of nitrogenous wastes became a second function of the kidney in the course of vertebrate evolution.

- **Interacting regulatory systems maintain homeostasis (pp. 889–890)** Homeostasis depends on the interaction of numerous regulatory systems and organ systems. The vertebrate liver performs diverse functions vital to homeostasis. Feedback circuits involving nervous system communication and hormones integrate homeostatic mechanisms.

SELF-QUIZ

1. *Unlike* an earthworm's metanephridia, a mammalian nephron
   a. is intimately associated with a capillary network
   b. forms urine by changing the composition of fluid inside the tubule
   c. functions in both osmoregulation and the excretion of nitrogenous wastes
   d. processes blood instead of coelomic fluid
   e. has a transport epithelium

2. The majority of water and salt filtered into Bowman's capsule is reabsorbed by
   a. the transport epithelia of the proximal tubule
   b. diffusion from the descending limb of the loop of Henle into the hyperosmotic interstitial fluid of the medulla
   c. active transport across the transport epithelium of the thick upper segment of the ascending limb of the loop of Henle
   d. selective secretion and diffusion across the distal tubule
   e. diffusion from the collecting duct into the increasing osmotic gradient of the renal medulla

3. The high osmolarity of the renal medulla is maintained by all the following *except*
   a. diffusion of salt from the ascending limb of the loop of Henle
   b. active transport of salt from the upper region of the ascending limb
   c. the spatial arrangement of juxtamedullary nephrons
   d. diffusion of urea from the collecting duct
   e. diffusion of salt from the descending limb of the loop of Henle

4. Select the pair in which the nitrogenous waste is incorrectly matched with the benefit of its excretion.
   a. urea—low toxicity relative to ammonia
   b. uric acid—can be stored as a precipitate
   c. ammonia—very soluble in water
   d. uric acid—minimal loss of water when excreted
   e. urea—very insoluble in water

5. Nonshivering thermogenesis is
   a. a behavioral adaptation for absorbing heat in ectotherms
   b. a hormone-triggered rise in metabolic rate, often associated with brown fat

c. a countercurrent heat exchange of blood going to the limbs
d. a heat-producing method of large, active fishes
e. muscle contractions that have been identified in winter moths

6. Physiological adjustments, or acclimatization, by an ectotherm to cooler seasonal temperatures may include
   a. an increase in metabolic rate
   b. estivation
   c. changes in the lipid components of cell membranes
   d. increased vasodilation and countercurrent heat exchange
   e. hibernation

7. Which of the following correctly describes a case of osmoregulation?
   a. body fluids that are isoosmotic with the external environment
   b. discharge of excess water in a hypoosmotic environment
   c. expenditure of energy to convert ammonia to less toxic wastes
   d. excretion of salt in a hypoosmotic environment
   e. secretion of drugs and reabsorption of nutrients by the proximal tubule

8. Which process in the nephron is *least* selective?
   a. secretion
   b. reabsorption
   c. transport across the epithelium of a collecting duct
   d. filtration
   e. salt pumping by the loop of Henle

9. The key difference between an ectotherm and an endotherm is that
   a. ectotherms generate energy mainly from fermentation; endotherms mainly from cellular respiration
   b. ectotherms are mostly aquatic animals; endotherms are mostly terrestrial
   c. ectotherms warm their bodies mainly by absorbing environmental heat; endotherms mainly use metabolic heat to warm their bodies
   d. ectotherms are "cold-blooded" animals with body temperatures that cannot reach the high body temperatures of endotherms
   e. ectotherms are all invertebrates; endotherms are all vertebrates

10. The vertebrate liver functions in all of the following regulatory processes *except*
    a. osmoregulation by variable excretion of salts
    b. maintenance of blood sugar concentration
    c. detoxification of harmful substances
    d. production of nitrogenous wastes
    e. caloric storage in the form of glycogen

CHALLENGE QUESTIONS

1. Compare and contrast the osmoregulatory problems and adaptations of (1) a marine bony fish with a freshwater bony fish and (2) an earthworm with an insect.
2. A large part of the terrestrial success of arthropods and vertebrates is attributable to their osmoregulatory capabilities. Compare and contrast the Malpighian tubule with the nephron in regard to anatomy, relationship to circulation, and physiological mechanisms for conserving body water.
3. Write an essay describing the diverse thermoregulatory mechanisms of a human.

SCIENCE, TECHNOLOGY, AND SOCIETY

1. The kidneys remove many drugs from the blood, and these substances show up in the urine. Some employers require a urine drug test at the time of hiring and at intervals during the term of employment. An employee who fails a drug test can lose his or her job. What are some arguments for and against drug testing of individuals in certain occupations?
2. Kidneys were the first organs to be successfully transplanted. A donor can live a normal life with a single kidney, making it possible for individuals to donate a kidney to an ailing relative or even an unrelated individual with a similar tissue type. In some countries, poor people *sell* kidneys to transplant recipients through organ brokers. What are some of the ethical issues associated with this organ commerce?

FURTHER READING

Costanza, J. P., P. A. Callahan, R. E. Lee, Jr., and M. F. Wright. "Frogs Reabsorb Glucose from Urinary Bladder." *Nature*, September 25, 1997. Describes evidence of an unusual mechanism that reclaims glucose, which protects cells from ice crystals, from the urine of freeze-tolerant frogs.

Handrich, Y., R. M. Bevan, J.-B. Charrassin, P. J. Butler, K. Putz, A. J. Woakes, J. Lage, and Y. Le Maho. "Hypothermia in Foraging King Penguins." *Nature*, July 3, 1997. Presents evidence that reduced metabolism may be one reason why penguins can dive for long periods of time.

Heinrich, B. *The Thermal Warriors.* Cambridge, MA: Harvard University Press, 1996. A readable account of thermoregulation in insects.

Heyning, J. E., and J. G. Mead. "Thermoregulation in the Mouths of Feeding Gray Whales." *Science*, November 7, 1997. Countercurrent heat exchangers in the tongue save heat when gray whales feed in cold oceans.

Mares, M. A., R. A Ojeda, C. E. Borghi, S. M. Giannoni, G. B. Diaz, and J. K Braun. "How Desert Rodents Overcome Halophytic Plant Defenses." *BioScience*, November 1997. A discussion of the structural, functional, and behavioral adaptations that allow some mammals to use saline plants as a major food source.

Marieb, E. N. *Human Anatomy and Physiology*, 4th ed. Menlo Park, CA: Benjamin/Cummings, 1998. Chapter 26 describes the human excretory system.

Preest, M. R., and C. A Beuchat. "Ammonia Excretion by Hummingbirds." *Nature*, April 10, 1997. Presents experimental evidence that at least one species of hummingbird can excrete mainly ammonia or mainly uric acid, depending on energy demands at different environmental temperatures.

Schmidt-Nielsen, K. *Animal Physiology, Adaptation and Environment.* Cambridge, UK: Cambridge University Press, 1997. Includes exceptionally lucid units on thermoregulation, water balance, and excretion.

Travis, J. "Chilled Brains." *Science News*, December 6, 1997. Studies of how the brain cells of ground squirrels survive the rigors of torpor may lead to new ways to treat strokes.

Wu, C. "Freeze! Insect Proteins Halt Ice Growth." *Science News*, August 30, 1997. Describes discoveries of antifreeze proteins in insects.

WEB LINKS

Visit the special edition of *The Biology Place* for BIOLOGY, Fifth Edition, at http://www.biology.com/campbell. Go to Chapter 44 for online resources, including learning activities, practice exams, and links to the following web sites:

"Regulatory Physiology"
Explore some of the regulatory physiology experiments carried out by NASA scientists on Spacelab-Mir.

"Butterfly Thermoregulation Page"
A very colorful site!

"Seaworld Animal Resources"
Contains interesting articles on many marine animals, with information on their thermoregulatory and osmoregulatory activities.

"RENALNET Kidney Information Clearinghouse"
RENALNET was created to provide a clearinghouse for information on the cause, treatment, and management of kidney disease. This site contains extensive links to kidney resources.

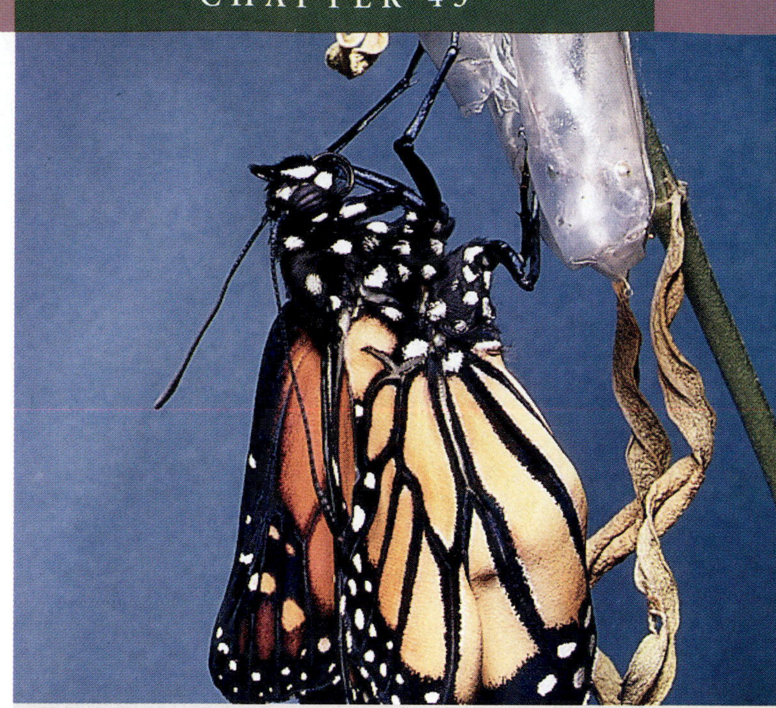

CHEMICAL SIGNALS IN ANIMALS

People offer hormones as an explanation for the howling of alley cats and the moodiness of teenagers. Over a million diabetics in the United States take the hormone insulin, and other hormones are used in cosmetics intended to keep the skin smooth or are added to livestock feed to fatten cattle. The monarch butterfly in the photograph that opens this chapter has just emerged from the silvery cocoon attached to the twig above it. In becoming an adult, the butterfly has undergone a complete change of body form, a metamorphosis regulated by hormones. Internal communication involving those hormones makes it possible for different parts of the insect's body to develop in concert.

An animal **hormone** *(Gr.* hormon, *"excite") is a chemical signal that is secreted into body fluids, most often into the blood, and communicates regulatory messages within the body. Hormones may reach all parts of the body, but only certain types of cells, the* **target cells**, *are equipped to respond. Thus, a given hormone traveling in the bloodstream elicits specific responses—a change in metabolism, for example—from selected target cells, while other cell types ignore that particular hormone.*

This chapter's specific focus is internal chemical signals—what they are and how they function within an animal's body. Our overarching theme is homeostasis—the role of chemical signals in maintaining a dynamic, steady state in body functions.

AN INTRODUCTION TO REGULATORY SYSTEMS

Animals have two systems of internal communication and regulation, the nervous system and the endocrine system. The nervous system, which we will study in Chapters 48 and 49, conveys high-speed signals along specialized cells called neurons. These rapid messages function in such activities as the movement of body parts in response to sudden environmental changes—jerking your hand away from a flame, for example.

Slower means of communication regulate other biological processes, such as the maturation of a butterfly. Different parts of the body must be informed how fast to grow and when to develop the characteristics that distinguish male from female or the juvenile from the adult of a species. This kind of information is often relayed by hormones.

Collectively, all of an animal's hormone-secreting cells constitute the **endocrine system**. Hormone-secreting organs are called **endocrine glands**, also called ductless glands because they secrete their chemical messengers directly into body fluids. In contrast, exocrine glands secrete chemicals, such as sweat, mucus, and digestive enzymes, into ducts that convey the products to the appropriate locations.

Although it is convenient to distinguish between the endocrine and nervous systems, the lines between these two regulatory systems are blurred, and homeostasis depends heavily on overlap between them.

The endocrine system and the nervous system are structurally, chemically, and functionally related

In this chapter we will see many examples of the close relationship between the endocrine system and the nervous system. Many endocrine organs and tissues contain specialized nerve cells, called **neurosecretory cells**, that secrete hormones. Animals as distinct as insects and vertebrates have neurosecretory cells in their brain that secrete hormones into the blood. Several chemicals serve both as hormones of the endocrine system and as signals in the nervous system. Epinephrine, for example, functions in the vertebrate body as the so-called "fight or flight" hormone (produced by the adrenal medulla, an endocrine gland) and as a neurotransmitter that conveys messages between neurons in the nervous system.

The regulation of several physiological processes involves structural and functional overlap between the endocrine and nervous systems. Each system affects the output of the other. For instance, the release of milk by a nursing mother is controlled by a series of interdependent nervous and hormonal signals. Suckling stimulates sensory cells in the nipples, and nervous signals to the part of the brain called the hypothalamus trigger the release of the hormone oxytocin from the pituitary gland. Oxytocin then causes the mammary glands to secrete milk.

Feedback is another feature common to both the endocrine and nervous systems. The chemical and neural events leading to milk release involve positive feedback. The generation of nerve signals (impulses) also involves positive feedback, as we will see in Chapter 48. Negative feedback regulates many endocrine and nervous mechanisms, especially those involved in maintaining homeostasis. FIGURE 45.1 illustrates one example, and we will see several others in this chapter.

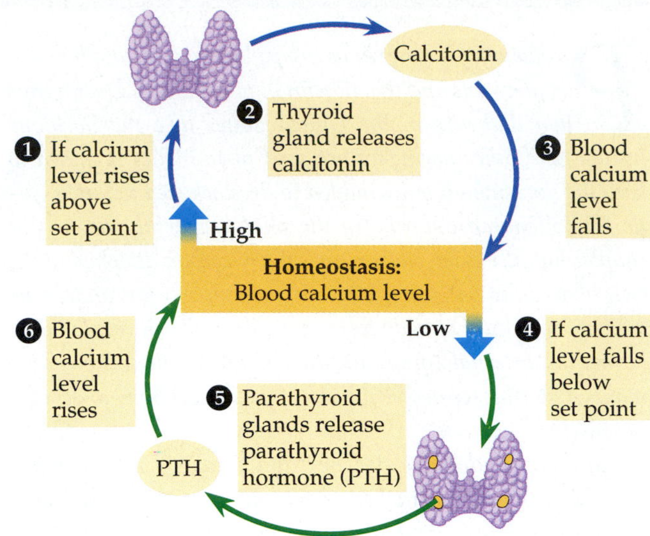

FIGURE 45.1 • Regulation by feedback and homeostasis: an example. Feedback mechanisms in which blood calcium ion concentration controls the secretion of calcitonin and parathyroid hormone (PTH), two hormones with antagonistic action, maintaining calcium levels close to a physiological set point. Later in the chapter you will learn more about this and other cases of antagonistic hormones functioning in homeostasis.

Invertebrate regulatory systems clearly illustrate endocrine and nervous system interactions

Diverse invertebrate hormones function in homeostasis—by regulating water balance, for instance. However, the hormones that have been most extensively studied function in reproduction and development. In a hydra, for example, one hormone stimulates growth and budding (asexual reproduction) but prevents sexual reproduction. In more complex invertebrates the endocrine and nervous systems are generally integrated in the control of reproduction and development. One well-studied example of nerve and hormone interaction in controlling both reproductive functions and behavior involves the hormone that regulates egg laying in the mollusk *Aplysia.* Secreted by specialized neurons, this hormone stimulates the laying of thousands of eggs and also inhibits feeding and locomotion, activities that interfere with reproduction.

All groups of arthropods have extensive endocrine systems. Crustaceans, for example, have hormones for growth and reproduction, water balance, movement of pigments in the integument and in the eyes, and the regulation of metabolism. Having exoskeletons that cannot stretch, crustaceans and insects appear to grow in spurts, shedding the old and secreting a new exoskeleton with each molt. Further, most insects acquire their adult characteristics in a single, terminal molt. In insects and crustaceans (and most likely in all arthropods with exoskeletons), molting is triggered by a hormone called **ecdysone.** In insects, ecdysone is secreted from a pair of endocrine glands, called the prothoracic glands, just behind the head (FIGURE 45.2). Besides stimulating the molt, ecdysone also favors the development of adult characteristics, as in the change from a caterpillar to a butterfly. In insects, ecdysone production is itself controlled by a second hormone, called **brain hormone (BH).** Produced by neurosecretory cells in the brain, this hormone promotes development by stimulating the prothoracic glands to secrete ecdysone.

Brain hormone and ecdysone are balanced by **juvenile hormone (JH)**, the third hormone in this system. JH is secreted by a pair of small glands just behind the brain, the corpora allata. Juvenile hormone promotes the retention of larval characteristics. In the presence of a relatively high concentration of juvenile hormone, ecdysone can still stimulate molting, but the product is a larger larva. Only when the level of juvenile hormone wanes can ecdysone-induced molting produce a developmental stage called a pupa. Within the pupa, metamorphosis replaces larval anatomy with the

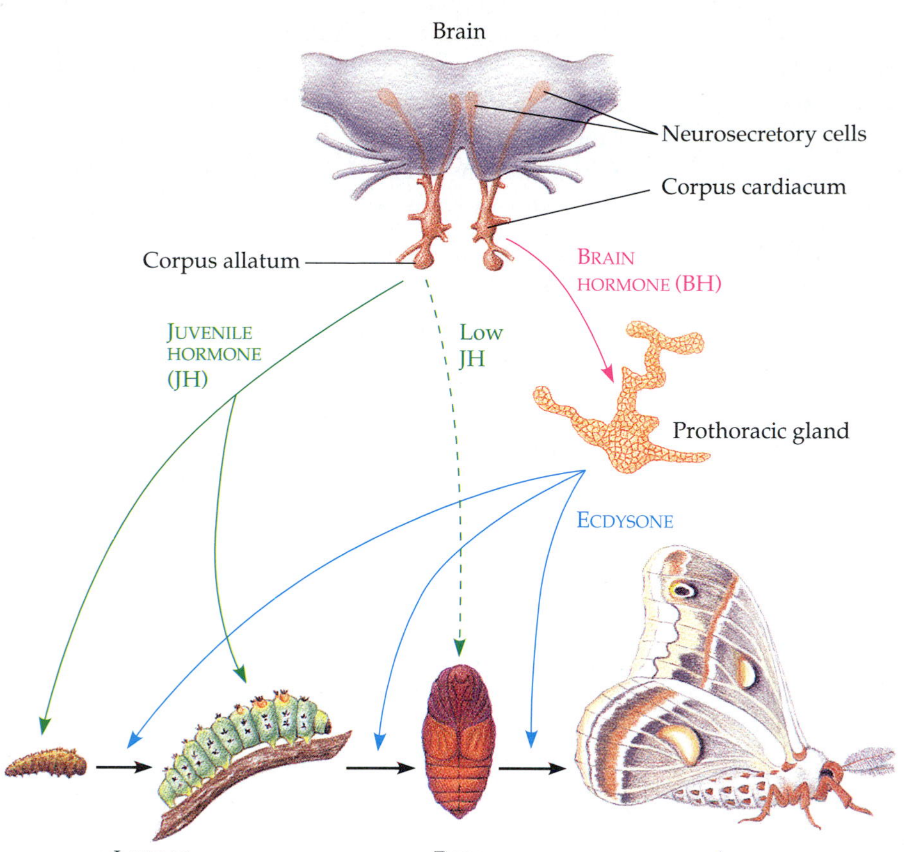

FIGURE 45.2 ▪ Hormonal regulation of insect development. Most insects go through a series of larval stages, with each molt (shedding of the old exoskeleton) leading to a larger larva. Molting of the final larval stage gives rise to a pupa, in which metamorphosis produces the adult form of the insect. Neurosecretory cells in the brain produce brain hormone (BH), but the hormone is stored and released from an organ called the corpus cardiacum. BH signals its main target organ, the prothoracic gland, to produce the hormone ecdysone. Ecdysone secretion is episodic, with each release stimulating a molt. Juvenile hormone (JH), secreted by the corpus allatum, determines the result of the molt. At relatively high concentrations of JH, ecdysone-stimulated molting produces another larval stage. Thus, JH suppresses metamorphosis. But when levels of JH fall below a certain threshold concentration, a pupa forms at the next ecdysone-induced molt. The adult insect emerges from the pupa (see the photo on p. 893).

insect's adult form. (Synthetic versions of JH are now being used as insecticides to prevent insects from maturing into reproducing adults.)

In all these invertebrate examples, we see the importance of the nervous system to hormone activity, and we will see these interactions repeatedly as we survey vertebrate endocrine systems.

CHEMICAL SIGNALS AND THEIR MODES OF ACTION

All animals exhibit coordination by chemical signals. Hormones convey information via the bloodstream to target cells throughout the body, while other types of chemical messengers function in other ways. For example, chemical signals called local regulators act between cells on a localized scale. Once they are outside the cells that produce them, local regulators are taken up by target cells, broken down by enzymes, or held in place by extracellular matrix, all within seconds or milliseconds. Consequently, they can affect only local targets. Still other chemical signals, called pheromones, carry messages between different individuals of a species, as in mate attraction. We will have more to say about pheromones in Chapter 46, which focuses on animal reproduction. Before concentrating on hormones and their modes of action, let's take a brief look at some local regulators and their effects on target cells.

A variety of local regulators affect neighboring target cells

Chemical messengers that affect target cells adjacent to or near their point of secretion function in local regulation (see FIGURE 11.3a). For example, neurotransmitters carry information from one neuron to another or from a neuron to a muscle, gland, or other target. In Chapter 43 we discussed the roles of the local regulators histamine and the interleukins, which coordinate some of the body's defenses. Histamine also promotes the secretion of hydrochloric acid and pepsin by cells of the stomach (see Chapter 41). We will study the role of retinoic acid, another local regulator, in the development of vertebrates in Chapter 47.

Many types of cells produce the gas **nitric oxide (NO)**, which has multiple functions as a local regulator. Highly reactive and potentially toxic, NO usually affects its targets in a few seconds and then breaks down. Secreted by neurons, NO functions as a neurotransmitter (see Chapter 48); secreted by white blood cells, it kills certain cancer cells and bacteria in body fluids. And NO released by endothelial cells in blood vessels makes the adjacent smooth muscles relax, dilating the vessel walls. American pharmacologists Robert Furchgott, Louis Ignarro, and Ferid Murad shared the 1998 Nobel Prize in Medicine for their pioneering work on the physiological effects of NO.

Growth factors are peptides and proteins that function as local regulators. They must be present in the extracellular environment for many types of cells to grow, divide, and develop normally. Growth factors are generally named for the first function discovered for them, and this can be misleading because each growth factor can affect several kinds of cells and have a variety of functions. For instance, a protein called nerve growth factor (NGF), which speeds the rate of development of certain embryonic nerve cells, also affects developing white blood cells and several other kinds of cells such as fibroblasts.

The action of growth factors has been studied mainly in cell cultures, but several experiments show that growth factors work within the animal body as well as in culture. For instance, injecting one called epidermal growth factor (EGF) into fetal mice accelerates epidermal development. And a group of peptides known as insulinlike growth factors (IGFs), produced by the liver, is essential to skeletal development. It is likely that the interaction of numerous growth factors regulates cell behavior in developing tissues and organs of animals. Ongoing research also suggests that growth factors (specifically, one called transforming growth factor, TGF) can enhance the strength of synapses between neurons in the brain of a mature animal.

Prostaglandins (PGs) are modified fatty acids, often derived from lipids of the plasma membrane. They are so named because they were first discovered in the components of semen produced by the human prostate gland. Prostaglandins in semen stimulate contraction of smooth muscles in the wall of the uterus, helping to convey sperm to the egg.

Released from most types of cells into the interstitial fluid, prostaglandins function as local regulators affecting nearby cells in various ways. Some of the best-known actions of prostaglandins are on the female reproductive system. For example, prostaglandins secreted by cells of the placenta cause chemical changes in the nearby muscles of the uterus, making them more excitable and thereby helping to induce labor during childbirth. This is an example of a positive feedback mechanism coordinating a body function (see FIGURE 46.19).

Prostaglandins also function as local regulators in the defense mechanisms of vertebrates. Various prostaglandins help induce fever and inflammation and also intensify the sensation of pain (which can be thought of as contributing to the body's defense by sounding an alarm that something harmful is going on). The anti-inflammatory effects of aspirin and ibuprofen result from these drugs inhibiting enzymes involved in the synthesis of prostaglandins.

Two prostaglandins with very similar molecular structures, prostaglandin E (PGE) and prostaglandin F (PGF), have opposite effects on the smooth muscle cells in the walls of blood vessels serving the lungs. PGE causes the muscles to relax, which dilates the blood vessels and promotes oxygenation of the blood. PGF signals the muscles to contract, which constricts the vessels and reduces blood flow through the lungs. Thus, these two chemical signals are antagonistic, and shifts in their relative concentrations contribute to moment-to-moment adjustments an animal's homeostatic mechanisms make in coping with changing circumstances. The use of antagonistic signals as counterbalances is a common regulatory mechanism in both nervous and chemical coordination of the body (see FIGURE 45.1).

Chemical signals bind to specific receptor proteins within target cells or on their surface

The number of local regulators known to science may be a small fraction of the total number in the animal kingdom. In addition, more than 50 known hormones regulate functions of the human body, and the endocrine systems of other animals are similarly elaborate. A given chemical signal can affect different target cells within an animal differently, or it may affect different species differently. A striking example is thyroxine, a hormone of the thyroid gland. In humans and other vertebrates, thyroxine is responsible for metabolic regulation. But thyroxine also plays diverse roles in animal development. In a specific case, thyroxine triggers the development of an adult frog from the tadpole stage of a frog, stimulating resorption of the tadpole's tail and other morphological changes during frog metamorphosis.

In sharp contrast to the variety of chemical signals and the varied responses they can elicit, the modes of action of all chemical signals are remarkably similar. Each chemical signal has a specific shape that can be recognized by that signal's target cells. A signal's action begins when it binds to a specific receptor. The receptor protein may be built into the plasma membrane of the target cell or may be inside the target cell (FIGURE 45.3). Signal molecules are examples of ligands, small molecules that bind specifically to larger ones, usually proteins. The binding of a chemical signal to a receptor protein triggers chemical events within the target cell that result in a change in its behavior. The diversity of responses of target cells to chemical signals depends on the nature of the target cells and on the number and affinity of receptor proteins on or within the target cells (FIGURE 45.4). Cells are unresponsive to a particular signal if they lack the appropriate receptors. A chemical signal triggers changes in target cells by one of two general mechanisms, depending on whether the signal binds with surface proteins or enters the target cell. We examine these mechanisms in the next two sections.

Most chemical signals bind to plasma-membrane proteins, initiating signal-transduction pathways

A dramatic example of hormone action is illustrated when a frog's skin turns darker or lighter, an adaptation that helps camouflage the frog as lighting changes. This color change is controlled by a peptide hormone called melanocyte-stimulating hormone (MSH), secreted by the pituitary gland at the base of

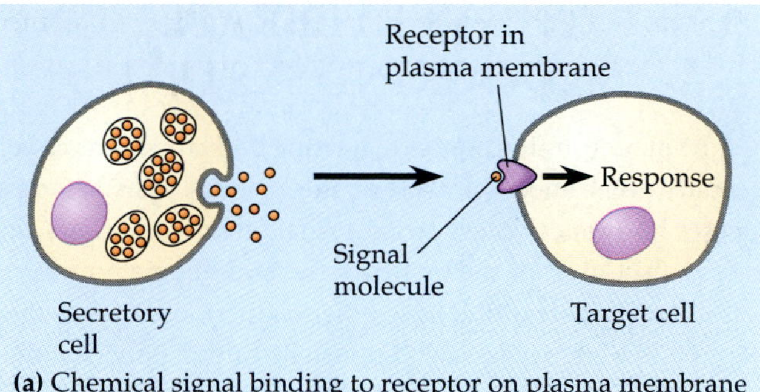

FIGURE 45.3 • How cells communicate. Chemical signals secreted by a sending cell either **(a)** bind to receptor proteins on the surface of a target cell or **(b)** penetrate the target cell's plasma membrane and bind with a receptor protein in the cytoplasm.

the brain. Melanocytes are specialized skin cells that contain the dark brown pigment melanin in cytoplasmic organelles called melanosomes. The frog's skin appears light when the melanosomes are clustered tightly around the cell nucleus and darker when the melanosomes spread throughout the cell. When MSH is added to the interstitial fluid surrounding the pigment-containing cells, the melanosomes disperse. However, direct microinjection of MSH into individual melanocytes does not induce melanosome dispersion.

45.1 Actually, the majority of chemical signals—most neurotransmitters, growth factors, and most hormones, including MSH—are unable to pass through the plasma membrane of their target cells. The receptors of these signals are components of **signal-transduction pathways** that convert the extracellular chemical signal to a specific intracellular response, thereby changing the behavior of the target cells. You may wish to review the basic mechanism of signal transduction, the kinds of receptor molecules and intracellular (second) messengers involved, and the intracellular changes that result, which we discussed in Chapter 11. Hormones are potent regulators, effective in minute amounts, and signal-transduction pathways often trigger elaborate enzyme cascades that amplify the hormonal signal (see FIGURE 11.15).

Steroid hormones, thyroid hormones, and some local regulators enter target cells and bind with intracellular receptors

The general signaling mechanism for steroid hormones was discovered by studying two vertebrate hormones, estrogen and progesterone. In most mammals, including humans, these hormones are necessary for the normal development and function of the female reproductive system. In the early 1960s, researchers demonstrated that cells in the reproductive tract of female rats accumulate estrogen. The hormone was found within the nuclei of these cells, but not in the cells of such tissues as the spleen, which do not respond to estrogen. Progesterone also enters the nuclei of target cells. Such observations led to the hypothesis that cells sensitive to a steroid

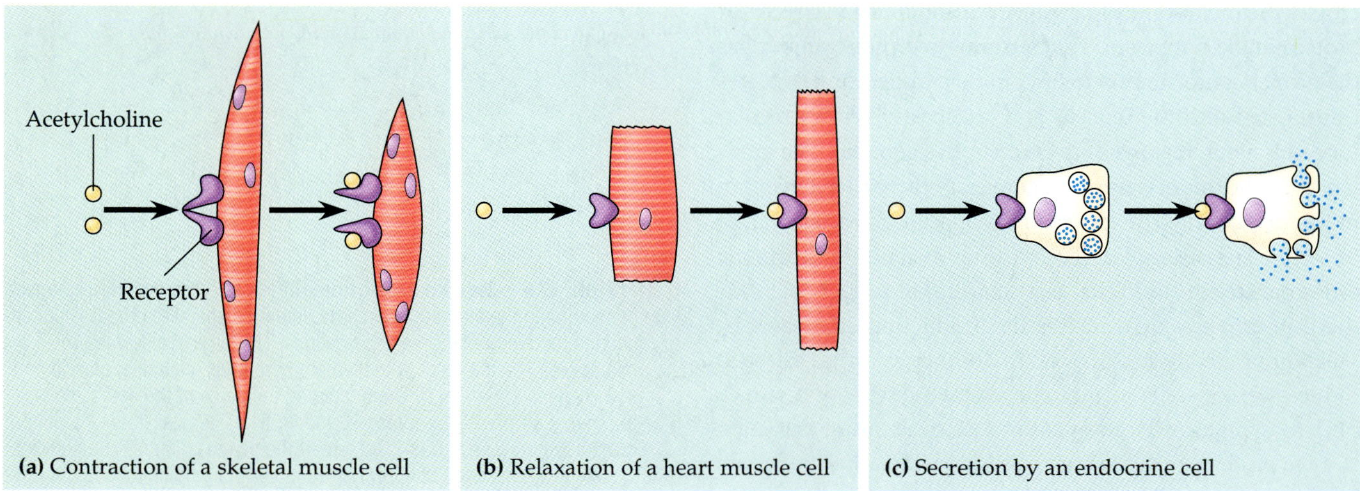

FIGURE 45.4 • One chemical signal, different effects. A single chemical signal, in this case a neurotransmitter called acetylcholine, can trigger different responses in different target cells. As indicated in these three models, different responses may result because the receptors are different [compare (a) with (b) and (c)] or because the target cells are different [compare (b) and (c)].

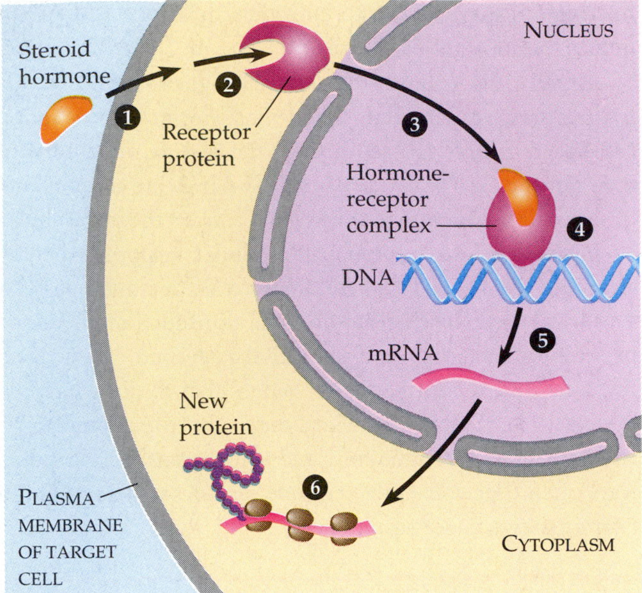

FIGURE 45.5 ▪ Steroid hormone action. In this model, the lipid-soluble steroid hormone (1) passes through the plasma membrane and (2) binds to a receptor protein present only in target cells. The hormone-receptor complex then (3) enters the nucleus and (4) binds to a specific regulatory site, stimulating (5) the transcription of a specific gene into mRNA, which is then (6) translated into a specific protein.

45.2

hormone contain receptor molecules that bind specifically to that hormone. It is now known that most steroid receptors, and the receptors of thyroid hormones and some local regulators such as nitric oxide (NO), are specialized proteins within target cells. The receptor proteins perform the task of transducing the signal within the cell. In effect, the chemical signal activates the receptor, which then triggers a response in the target cell.

45.2 The model in FIGURE 45.5 shows a steroid hormone that crosses the membrane of its target cell and binds to a receptor protein in the cytoplasm. The hormone-receptor complex has the proper conformation to bind to certain sites on DNA. It is transported into the nucleus and attaches to certain regulatory sites along regions of the target cell's genome. The binding of the hormone-receptor complex to these sites can either induce or suppress the expression of specific genes (see Chapter 19). For example, the arthropod molting hormone ecdysone (see FIGURE 45.2) is a steroid that triggers the synthesis of enzymes that catalyze the production of a new exoskeleton of a crab or an insect. In another example, estrogen induces certain cells in the reproductive system of a female bird to synthesize large quantities of ovalbumin, the main storage protein stockpiled in egg white. Different proteins are synthesized in the liver cells of the same animal in response to estrogen. It is likely that the estrogen receptor is the same protein in both target cells, but the hormone-receptor complex probably binds with regulatory sites controlling different genes in the two kinds of cells.

THE VERTEBRATE ENDOCRINE SYSTEM

Of the numerous hormones regulating body functions of vertebrates, some affect only one or a few tissues. Others, such as the sex hormones, which promote male and female characteristics, affect most of the tissues of the body. Some hormones, called **tropic hormones**, have other endocrine glands as their targets and are particularly important to our understanding of chemical coordination. In studying hormonal regulation of the vertebrate body, you may find FIGURE 45.6, showing the locations of the major endocrine glands in the human body, and TABLE 45.1, summarizing the functions of the major vertebrate hormones, especially helpful. In the text, small sketches accompanying each section will help you keep track

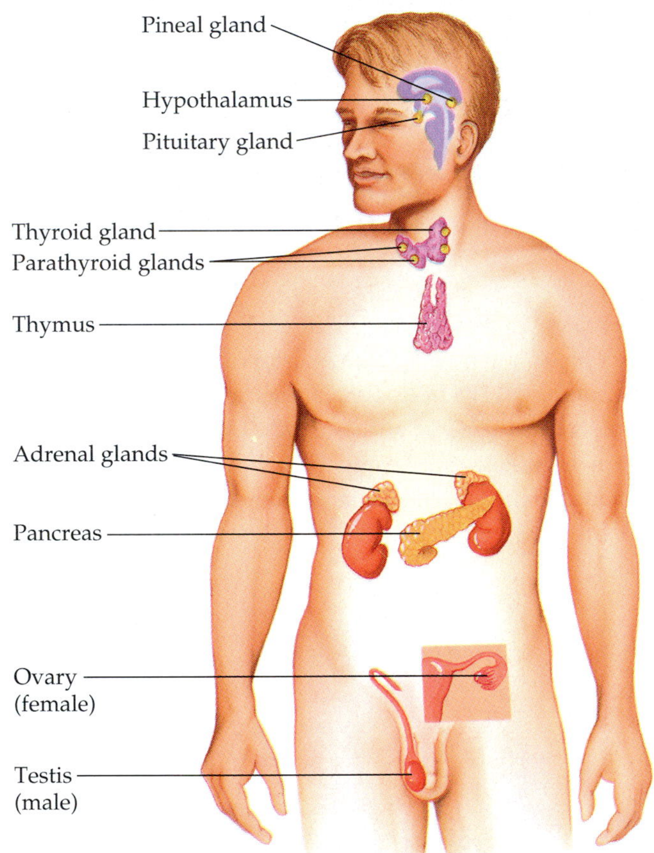

FIGURE 45.6 ▪ Human endocrine glands surveyed in this chapter.
45.3 In addition to the glands shown here, many organs with primarily nonendocrine functions also secrete hormones. The digestive system, for instance, is the source of at least eight hormones, including gastrin and secretin (see Chapter 41). The endocrine functions of the heart and kidneys were covered in Chapters 42 and 44. The liver secretes a number of growth factors (see p. 902). The role of the thymus gland in the immune system is discussed in Chapter 43. Quite large during childhood, the thymus begins to decline at puberty, when the immune system is well established. By adulthood it is largely replaced by adipose and fibrous tissue, but it continues to function throughout life. It secretes several messengers, including thymosin, that stimulate the development and differentiation of T lymphocytes after they leave the thymus. The other endocrine organs in this drawing are discussed in this chapter.

45.3

Table 45.1 ■ Major Vertebrate Endocrine Glands and Some of Their Hormones

GLAND	HORMONE	CHEMICAL CLASS	REPRESENTATIVE ACTIONS	REGULATED BY
Hypothalamus	Hormones released by the posterior pituitary and hormones that regulate the anterior pituitary (see below)			
Pituitary gland				
Posterior pituitary (releases hormones made by hypothalamus)	Oxytocin	Peptide	Stimulates contraction of uterus and mammary gland cells	Nervous system
	Antidiuretic hormone (ADH)	Peptide	Promotes retention of water by kidneys	Water/salt balance
Anterior pituitary	Growth hormone (GH)	Protein	Stimulates growth (especially bones) and metabolic functions	Hypothalamic hormones
	Prolactin (PRL)	Protein	Stimulates milk production and secretion	Hypothalamic hormones
	Follicle-stimulating hormone (FSH)	Glycoprotein	Stimulates production of ova and sperm	Hypothalamic hormones
	Luteinizing hormone (LH)	Glycoprotein	Stimulates ovaries and testes	Hypothalamic hormones
	Thyroid-stimulating hormone (TSH)	Glycoprotein	Stimulates thyroid gland	Thyroxine in blood; hypothalamic hormones
	Adrenocorticotropic hormone (ACTH)	Peptide	Stimulates adrenal cortex to secrete glucocorticoids	Glucocorticoids; hypothalamic hormones
Thyroid gland	Triiodothyronine (T_3) and thyroxine (T_4)	Amine	Stimulate and maintain metabolic processes	TSH
	Calcitonin	Peptide	Lowers blood calcium level	Calcium in blood
Parathyroid glands	Parathyroid hormone (PTH)	Peptide	Raises blood calcium level	Calcium in blood
Pancreas	Insulin	Protein	Lowers blood glucose level	Glucose in blood
	Glucagon	Protein	Raises blood glucose level	Glucose in blood
Adrenal glands				
Adrenal medulla	Epinephrine and norepinephrine	Amine	Raise blood glucose level; increase metabolic activities; constrict certain blood vessels	Nervous system
Adrenal cortex	Glucocorticoids	Steroid	Raise blood glucose level	ACTH
	Mineralocorticoids	Steroid	Promote reabsorption of Na^+ and excretion of K^+ in kidneys	K^+ in blood
Gonads				
Testes	Androgens	Steroid	Support sperm formation; promote development and maintenance of male secondary sex characteristics	FSH and LH
Ovaries	Estrogens	Steroid	Stimulate uterine lining growth; promote development and maintenance of female secondary sex characteristics	FSH and LH
	Progesterone	Steroid	Promotes uterine lining growth	FSH and LH
Pineal gland	Melatonin	Amine	Involved in biological rhythms	Light/dark cycles
Thymus	Thymosin	Peptide	Stimulates T lymphocytes	Not known

of each gland's location. We begin our discussion of the vertebrate endocrine system by looking at the hypothalamus and pituitary gland, which control much of the endocrine system.

The hypothalamus and pituitary integrate many functions of the vertebrate endocrine system

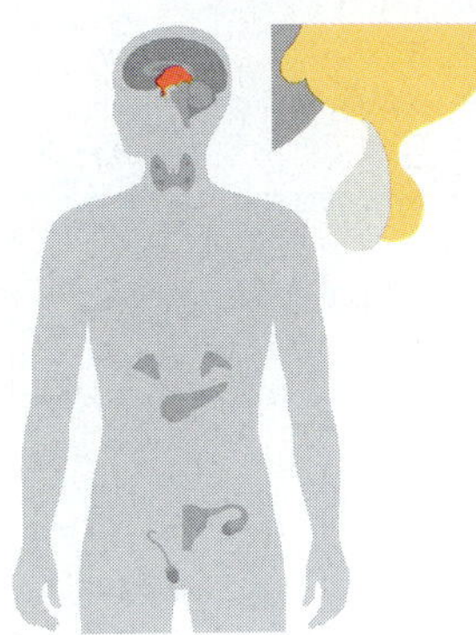

The **hypothalamus** plays an important role in integrating the vertebrate endocrine and nervous systems. This region of the lower brain receives information from nerves throughout the body and from other parts of the brain, then initiates endocrine signals appropriate to environmental conditions. In many vertebrates, for example, the brain passes sensory information about seasonal changes and the availability of a mate to the hypothalamus by means of nerve signals; the hypothalamus then triggers the release of reproductive hormones required for breeding.

The hormone-releasing cells in the hypothalamus are two sets of neurosecretory cells whose secretions are stored in or regulate the activity of the **pituitary gland**, a small organ with multiple endocrine functions. The pituitary gland was formerly called the "master gland" because so many of its hormones regulate other endocrine functions. However, the pituitary itself obeys hormonal orders from the hypothalamus.

No organ illustrates the close structural, functional, and developmental relationships of the endocrine and nervous systems better than the pituitary gland. Located at the base of the hypothalamus, it has two discrete parts that develop from two separate regions of the embryo and have very different functions (FIGURE 45.7). The **anterior pituitary**, or **adenohypophysis**, develops from a fold of the roof of the mouth that grows upward toward the brain and eventually loses its connection to the digestive tract. The anterior pituitary consists of endocrine cells that synthesize and secrete several hormones directly into the blood. A set of neurosecretory cells in the hypothalamus exerts control over the anterior pituitary by secreting two kinds of hormones into the blood. **Releasing hormones** make the anterior pituitary secrete its hormones. **Inhibiting hormones** from the hypothalamus make the anterior pituitary stop secreting hormone.

The hypothalamic releasing and inhibiting hormones are released into capillaries in a region at the base of the hypothalamus (FIGURE 45.7b). The capillaries drain into portal vessels, which are short blood vessels that subdivide into a second capillary bed within the anterior pituitary. In this way, the hypothalamic hormones have direct access to the gland they control. Every anterior pituitary hormone is controlled by at least one releasing hormone, and some have both a releasing hormone and an inhibiting hormone.

Unlike the anterior pituitary, the **posterior pituitary**, or **neurohypophysis**, is an extension of the brain. It develops from a small bulge of the hypothalamus that grows downward toward the mouth fold that forms the anterior pituitary. The posterior pituitary remains attached to the hypothalamus. It stores and secretes two hormones that are made by a set of neurosecretory cells in the hypothalamus.

Posterior Pituitary Hormones

The two hormones released by the posterior pituitary, oxytocin and antidiuretic hormone (ADH), are made by the hypothalamus. Oxytocin acts on muscles of the uterus and ADH acts on the kidneys, rather than affecting other endocrine glands. Oxytocin induces contraction of the uterine muscles during childbirth and causes the mammary glands to eject milk during nursing. ADH acts on the kidneys, increasing water retention and thus decreasing urine volume.

ADH is part of an elaborate feedback scheme that helps regulate the osmolarity of the blood. This mechanism was introduced in Chapter 44, but it is worth reviewing here to illustrate how hormones contribute to homeostasis and how negative feedback controls hormone levels. Blood osmolarity is monitored by a group of nerve cells that function as osmoreceptors in the hypothalamus. When the osmolarity of the plasma increases, these cells shrink slightly (due to osmosis) and transmit nerve impulses to certain neurosecretory cells in the hypothalamus. These cells respond by releasing ADH from their tips, which are located in the posterior pituitary. The posterior pituitary stores ADH and releases it into the general circulation. When ADH reaches the kidneys, it binds to receptors on the surface of the cells lining the collecting ducts. This binding activates a signal-transduction pathway that increases the water permeability of the collecting duct. Water exits the collecting ducts and enters nearby capillaries, helping prevent a further increase in blood osmolarity above the set point. The brain's osmoreceptors also stimulate a thirst drive, and drinking water brings blood osmolarity back down to the set point. Thus, the hormonal reaction to high blood osmolarity is augmented by a behavioral response controlled by the nervous system. As the more dilute blood arrives at the brain, the hypothalamus responds to the reduction in osmolarity by slowing the release of ADH and diminishing the sensation of thirst. The effects of these hormonal and behavioral responses—increased water reabsorption in the kidneys and drinking, respectively—prevent overcompensation by shutting off further secretion of the hormone and quenching thirst. We see once again how negative feedback helps maintain homeostasis. This example also highlights the central role of the hypothalamus as a member of both the endocrine system and the nervous system.

FIGURE 45.7 ▪ Hormones of the hypothalamus and pituitary glands. The pituitary gland, located at the base of the brain and surrounded by bone, consists of the posterior pituitary (neurohypophysis) and the anterior pituitary (adenohypophysis). **(a) The posterior pituitary.** Neurosecretory cells in the hypothalamus synthesize antidiuretic hormone (ADH) and oxytocin, peptide hormones that are transported down the axons to the posterior pituitary, where they are stored. The posterior pituitary releases the hormones into the blood, where they circulate. ADH binds to target cells in the kidneys; oxytocin binds to target cells in the mammary glands and uterus. **(b) The anterior pituitary.** The release of anterior pituitary hormones is controlled by the hypothalamus. Neurosecretory cells in the hypothalamus secrete releasing hormones and inhibiting hormones into a capillary network located above the stalk of the pituitary. Blood containing the hormones travels through short portal vessels and into a second capillary network within the anterior pituitary. In response to specific releasing hormones, endocrine cells in the anterior pituitary secrete certain hormones into the circulation.

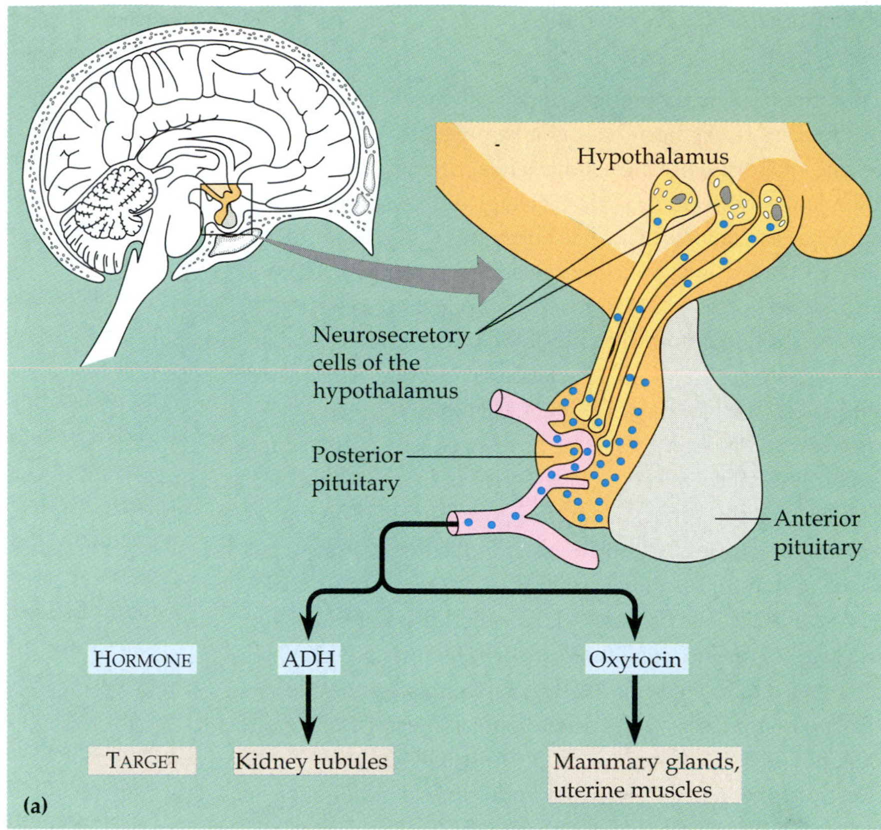

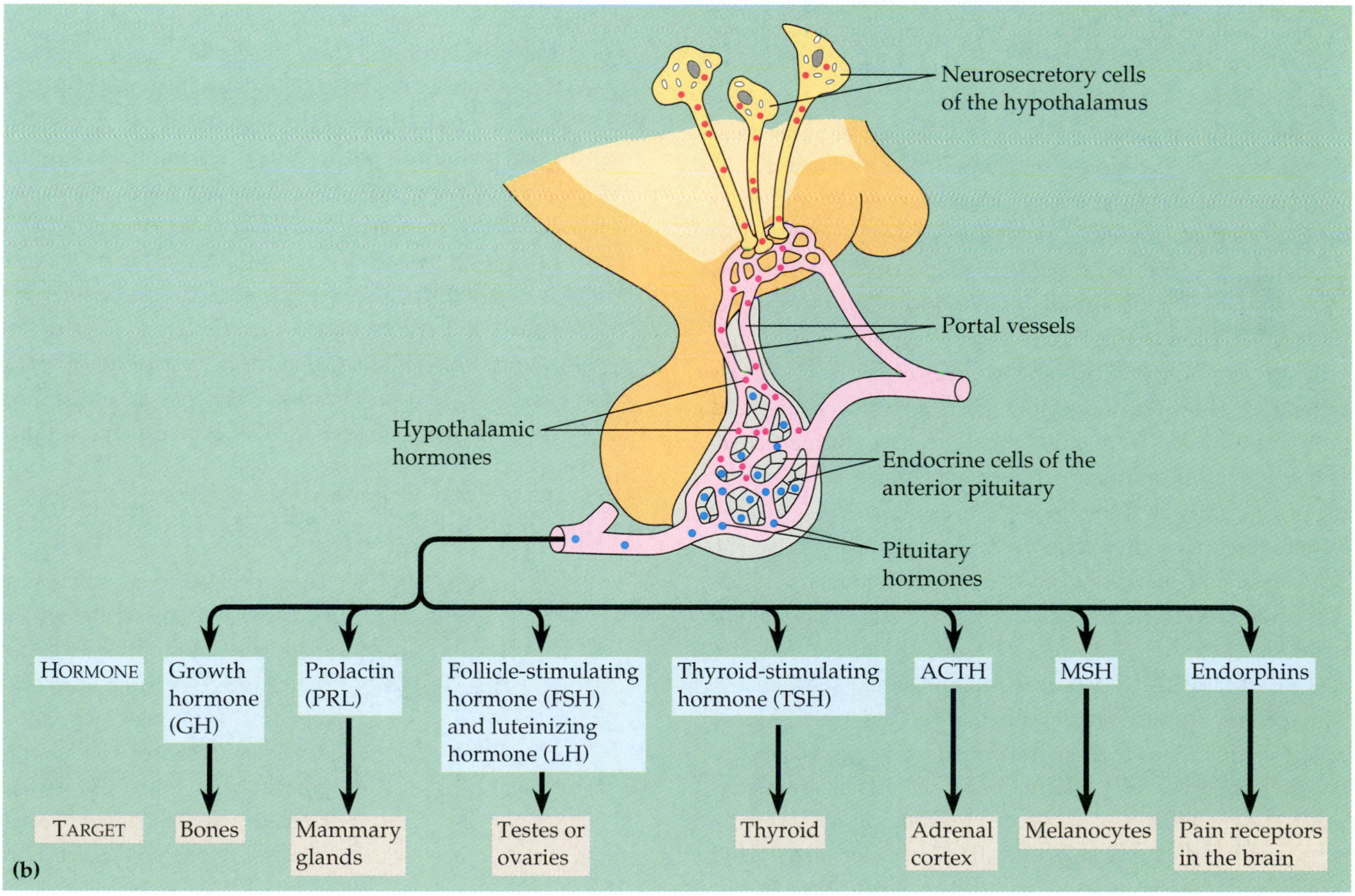

Neurosecretory cells of the hypothalamus

Portal vessels

Hypothalamic hormones

Endocrine cells of the anterior pituitary

Pituitary hormones

HORMONE	Growth hormone (GH)	Prolactin (PRL)	Follicle-stimulating hormone (FSH) and luteinizing hormone (LH)	Thyroid-stimulating hormone (TSH)	ACTH	MSH	Endorphins
TARGET	Bones	Mammary glands	Testes or ovaries	Thyroid	Adrenal cortex	Melanocytes	Pain receptors in the brain

(b)

Anterior Pituitary Hormones

The anterior pituitary produces many different hormones. Four of these are tropic hormones that stimulate the synthesis and release of hormones from other endocrine glands. Thyroid-stimulating hormone (TSH) regulates the release of thyroid hormones; adrenocorticotropic hormone (ACTH) controls the adrenal cortex; and follicle-stimulating hormone (FSH) and luteinizing hormone (LH) govern reproduction by acting on the gonads. Other hormones produced by the anterior pituitary are growth hormone (GH), prolactin (PRL), melanocyte-stimulating hormone (MSH), and the endorphins.

Growth hormone (GH), a protein consisting of almost 200 amino acids, affects a wide variety of target tissues and has both direct effects and tropic effects. GH promotes growth directly and stimulates the production of growth factors. For example, the ability of GH to stimulate the growth of bones and cartilage is partly due to the hormone's signaling the liver to produce **insulinlike growth factors (IGFs)**, which circulate in blood plasma and directly stimulate bone and cartilage growth. (This endocrine response to growth hormone qualifies GH as a tropic hormone. And IGF secretion by the liver qualifies that organ as an endocrine gland, among its many other functions.) In the absence of GH, skeletal growth of an immature animal will stop. If GH is injected into an animal that has been deprived of its own GH, growth will be partially restored.

Several human growth disorders are related to abnormal GH production. Excessive production of GH during development can lead to gigantism, while excessive GH production during adulthood results in the abnormal growth of bones in the hands, feet, and head, a condition known as acromegaly. Deficient GH production in childhood can lead to pituitary dwarfism. Children with GH deficiency have been treated successfully with human growth hormones isolated from cadaver pituitaries. However, the supply falls short of demand, and growth hormones from most other animals are ineffective. One of the most dramatic achievements of genetic engineering has been the production of GH by bacteria with genes for human GH spliced into their genomes (see Chapter 20). The product is now being used to treat children with hypopituitary dwarfism. Some athletes take GH (legally or illegally) to build muscles.

Prolactin (PRL) is a protein so similar to GH that it is believed they are encoded in genes that evolved from the same ancestral gene. The physiological roles of these two hormones, however, are different. Prolactin's most remarkable characteristic is the great diversity of effects it produces in different vertebrate species. For example, PRL stimulates mammary gland growth and milk synthesis in mammals; regulates fat metabolism and reproduction in birds; delays metamorphosis in amphibians, where it may also function as a larval growth hormone; and regulates salt and water balance in freshwater fishes. This list suggests that prolactin is an ancient hormone whose functions have diversified during the evolution of the various vertebrate classes.

Three of the tropic hormones secreted by the anterior pituitary are closely related chemically. **Follicle-stimulating hormone (FSH), luteinizing hormone (LH),** and **thyroid-stimulating hormone (TSH)** are all similar glycoproteins, protein molecules with carbohydrate attached to them. FSH and LH are also called **gonadotropins** because they stimulate the activities of the male and female gonads, the testes and ovaries. TSH stimulates the production of hormones by the thyroid gland.

The remaining hormones from the anterior pituitary all come from a single parent molecule called pro-opiomelanocortin. This large protein is cleaved into several short fragments inside the pituitary cells. At least three of these fragments are active peptide hormones. **Adrenocorticotropic hormone (ACTH)** stimulates the production and secretion of steroid hormones by the adrenal cortex. As described earlier, **melanocyte-stimulating hormone (MSH)** regulates the activity of pigment-containing cells in the skin of some vertebrates. MSH secreted by the mammalian pituitary seems to play a key role in fat metabolism, probably by a feedback mechanism that targets neurons in the hypothalamus (see Chapter 41). The other derivatives of pro-opiomelanocortin are a class of hormones called **endorphins**. These molecules are also produced by certain neurons in the brain. They are sometimes called the body's natural opiates because they inhibit the perception of pain. In fact, heroin and other opiate drugs mimic endorphins and bind to the same receptors in the brain. Some researchers speculate that the so-called runner's high results partly from the release of endorphins when stress and pain in the body reach critical levels. (Endorphins are discussed further in Chapter 48.)

Now that we have examined the regulatory roles of the hypothalamus and pituitary gland, we turn to other vertebrate endocrine glands. The small sketches accompanying each section will help you keep track of each gland's location.

The pineal gland is involved in biorhythms

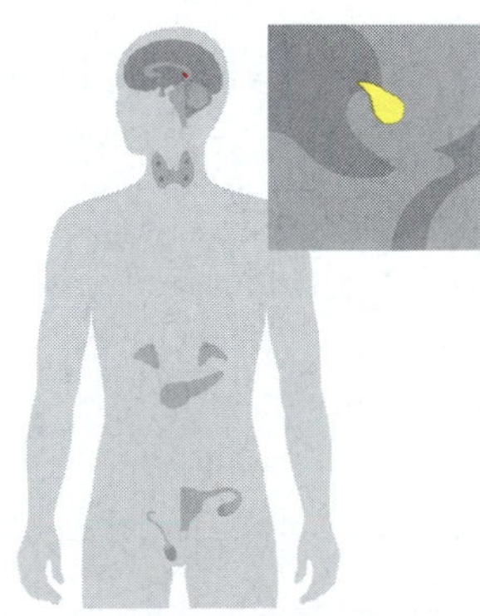

The **pineal gland** is a small mass of tissue near the center of the mammalian brain (closer to the brain surface in some other vertebrates). Although we know considerably more about the pineal than we did when Descartes described it as the seat of the soul, there is much more to learn. The pineal secretes the hormone **melatonin**, a modified amino acid. Depending on the species, the pineal contains light-sensitive cells or has nervous connections from the eyes, and melatonin regulates functions related to light and to seasons marked by changes in day length. For example, melatonin, like melanocyte-stimulating hormone (MSH), affects skin pigmentation in many vertebrates. Most of the pineal's functions, though, are related to biological rhythms associated with reproduction. Since melatonin is secreted at night, the amount secreted depends on the length of the night. In winter, for example, the days are short and the nights are long, so more melatonin is secreted. Thus, melatonin production is a link between a biological clock and daily or seasonal activities, such as reproduction. Recent evidence suggests that the main target cells of melatonin are in the part of the brain called the suprachiasmatic nucleus (SCN), which functions as a biological clock. Melatonin seems to decrease the activity of neurons in the SCN, and this may be related to its role in mediating rhythms. However, much remains to be learned about the precise role of melatonin and about biological clocks in general.

Thyroid hormones function in development, bioenergetics, and homeostasis

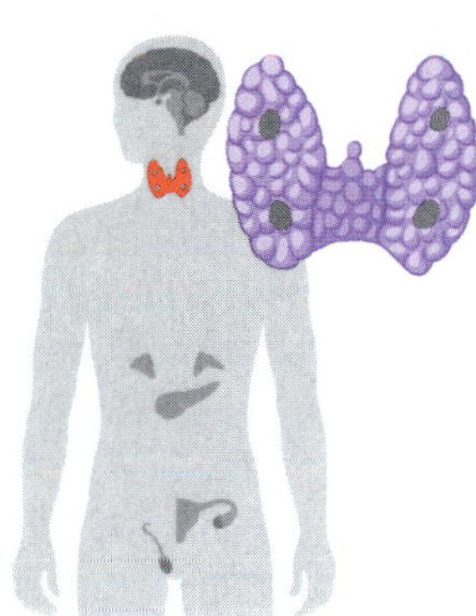

In humans and other mammals, the **thyroid gland** consists of two lobes located on the ventral surface of the trachea (see FIGURE 45.6). In many other vertebrates the two halves of the gland are separated on the two sides of the pharynx. The thyroid gland produces two very similar hormones derived from the amino acid tyrosine: **triiodothyronine** (**T_3**), which contains three iodine atoms, and tetraiodothyronine, or **thyroxine** (**T_4**), which contains four iodine atoms (FIGURE 45.8). In mammals, T_3 is the more active of the two hormones, although both have the same effects on their target cells. The secretion of thyroid hormones is controlled by the hypothalamus and pituitary in a complex negative feedback system (FIGURE 45.9).

The thyroid gland plays a crucial role in vertebrate development and maturation. A striking example is the thyroid's control of metamorphosis of a tadpole into a frog, which involves massive reorganization of many different tissues. The thyroid is equally important in human development. An inherited condition of thyroid deficiency known as cretinism results in markedly retarded skeletal growth and poor mental development. These defects can often be overcome, at least partially, if treatment with thyroid hormones is begun early in

TRIIODOTHYRONINE (T_3)

THYROXINE (T_4)

FIGURE 45.8 ▪ Two thyroid hormones. Structurally identical, except that T_3 has three iodine atoms and T_4 has four, these hormones regulate metabolism in most body cells. The thyroid secretes mainly T_4, most of which is converted to T_3 in target tissues. T_3 binds more quickly to receptors and effects most changes in target cells.

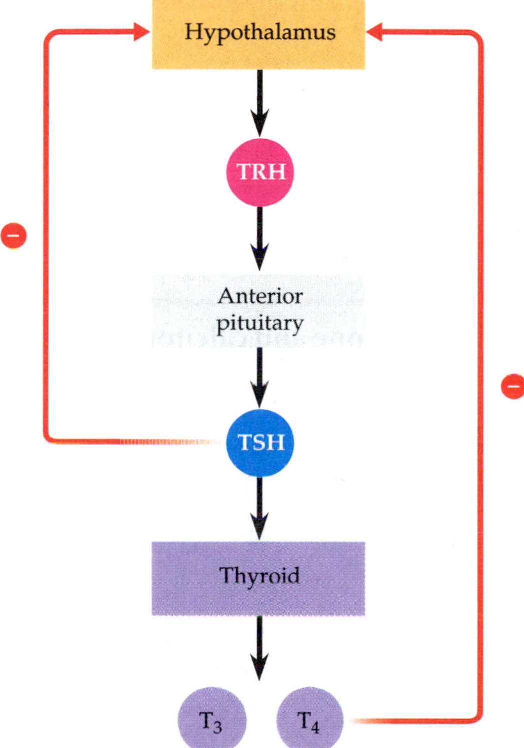

FIGURE 45.9 ▪ Feedback control loops regulating the secretion of thyroid hormones. The hypothalamus secretes TRH (TSH-releasing hormone), which stimulates the anterior pituitary to secrete TSH (thyroid-stimulating hormone). When TSH binds to specific receptors in the thyroid gland, a signal-transduction pathway involving cAMP as a second messenger triggers the synthesis and release of the thyroid hormones T_3 and T_4. The system is balanced by negative feedback loops (red arrows); high levels of T_3, T_4, and TSH in the blood inhibit TRH secretion by the hypothalamus. There is also evidence that additional feedback loops are involved; for example, high levels of thyroid hormones may inhibit the anterior pituitary and the thyroid gland itself, and high levels of TSH may inhibit TSH secretion by the anterior pituitary. The hypothalamus–anterior pituitary–thyroid feedback system explains why iodine deficiencies lead to goiter. In the absence of sufficient iodine, the thyroid gland cannot synthesize adequate amounts of T_3 and T_4. Consequently, the pituitary continues to secrete TSH, leading to an enlargement of the thyroid.

life. Studies with nonhuman animals have shown that thyroid hormones are required for the normal functioning of bone-forming cells and for the branching of nerve cells during embryonic development of the brain.

The thyroid gland also plays a vital role in homeostasis. In adult mammals, for instance, thyroid hormones help maintain normal blood pressure, heart rate, muscle tone, digestion, and reproductive functions. Throughout the body, T_3 and T_4 are important in bioenergetics, generally increasing the rate of oxygen consumption and cellular metabolism. Too much or too little of these hormones in the blood can result in serious metabolic disorders. For example, in humans, an excessive secretion of thyroid hormones, known as hyperthyroidism, produces such symptoms as high body temperature, profuse sweating, weight loss, irritability, and high blood pressure. The opposite condition, hypothyroidism, can cause cretinism in infants and produce symptoms such as weight gain, lethargy, and intolerance to cold in adults. Another condition associated with a shortage of thyroid hormones is an enlargement of the thyroid called goiter, often caused by a deficiency of iodine in the diet.

Besides secreting T_3 and T_4, the mammalian thyroid gland also contains endocrine cells that secrete **calcitonin**. This hormone lowers calcium levels in the blood as part of calcium homeostasis, described in the next section.

Parathyroid hormone and calcitonin balance blood calcium

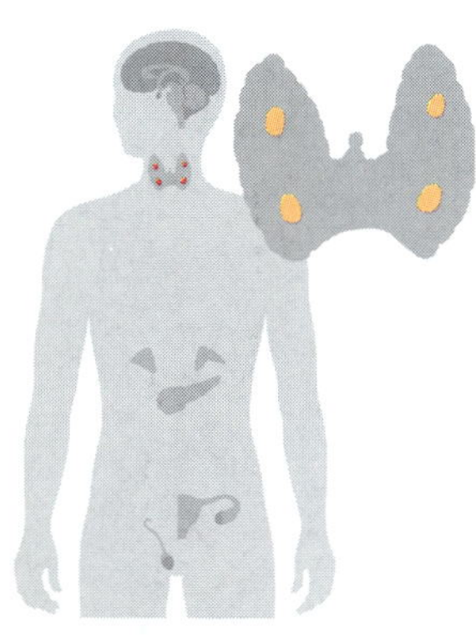

The four **parathyroid glands**, embedded in the surface of the thyroid, function in the homeostasis of calcium ions. They secrete **parathyroid hormone (PTH)**, which raises blood levels of calcium and thus has an effect opposite to that of the thyroid hormone calcitonin. Parathyroid hormone elevates blood Ca^{2+} by stimulating Ca^{2+} reabsorption in the kidneys and by inducing specialized bone cells called osteoclasts to decompose the mineralized matrix of bone and release Ca^{2+} to the blood. Calcitonin has just the opposite effects on the kidneys and bone, thus decreasing blood Ca^{2+}. Vitamin D, synthesized in the skin and converted to its active form in many tissues, is essential to PTH function, so it is also required for complete calcium balance. A lack of PTH causes blood levels of calcium to drop dramatically, leading to convulsive contractions of the skeletal muscles. If unchecked, this condition, known as tetany, is fatal. The control of blood calcium level is an example of how homeostasis is often maintained by the balancing of two antagonistic hormones—in this case, PTH and calcitonin (FIGURE 45.10).

Endocrine tissues of the pancreas secrete insulin and glucagon, antagonistic hormones that regulate blood glucose

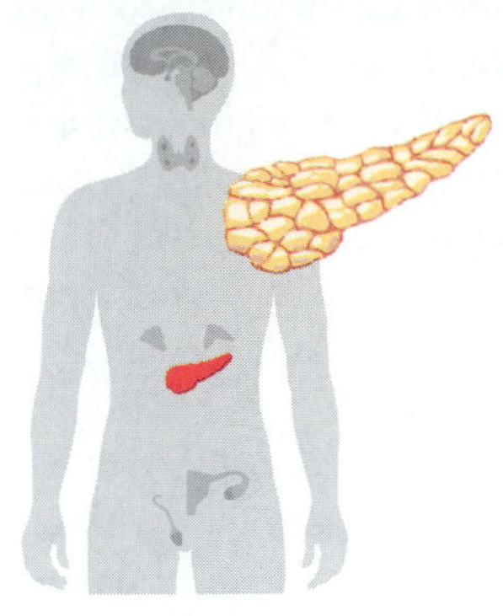

Many organs, such as the **pancreas**, perform both endocrine and exocrine functions. Endocrine cells make up only 1% to 2% of the weight of the pancreas. The rest of the organ is exocrine tissue that produces bicarbonate ions and digestive enzymes that are carried to the small intestine via the pancreatic duct (see Chapter 41). Scattered among this exocrine tissue are the **islets of Langerhans**, clusters of endocrine cells that secrete two hormones directly into the circulatory system. Each islet has a population of **alpha cells**, which secrete the peptide hormone **glucagon**, and a population of **beta cells**, which secrete the hormone **insulin**.

Insulin and glucagon are antagonistic hormones that regulate the concentration of glucose in the blood. This is a critical bioenergetic and homeostatic function, because glucose is a major fuel for cellular respiration and a key source of carbon skeletons for the synthesis of other organic compounds. Metabolic balance depends on the maintenance of blood glucose at a concentration near a set point, which is about 90 mg/100 mL in humans. When blood glucose exceeds this level, insulin is released and acts to lower the glucose concentration. When blood glucose drops below the set point, glucagon increases glucose concentration. By negative feedback, blood glucose concentration determines the relative amounts of insulin and glucagon secreted by the islet cells (FIGURE 45.11, p. 906).

Insulin and glucagon both influence blood glucose concentration by multiple mechanisms. Insulin lowers blood glucose levels by stimulating virtually all body cells except those of the brain to take up glucose from the blood. (Brain cells are unusual in being able to take up glucose without insulin; as a result, the brain has access to circulating fuel molecules almost all the time.) Insulin also decreases blood glucose by slowing glycogen breakdown in the liver and inhibiting the conversion of amino acids and fatty acids to sugar.

The liver, skeletal muscles, and adipose tissues store large amounts of fuel molecules and are especially important in bioenergetics. The liver and muscles store sugar as glycogen, whereas adipose tissue cells convert sugars to fats. The liver is a key fuel-processing center because only liver cells are sensitive to glucagon. Normally, glucagon starts having an effect before blood glucose levels even drop below the set point. In fact, as soon as excess glucose is cleared from the blood, glucagon signals the liver cells to increase glycogen hydrolysis, convert amino acids and fatty acids to glucose, and start slowly releasing glucose back into the circulation.

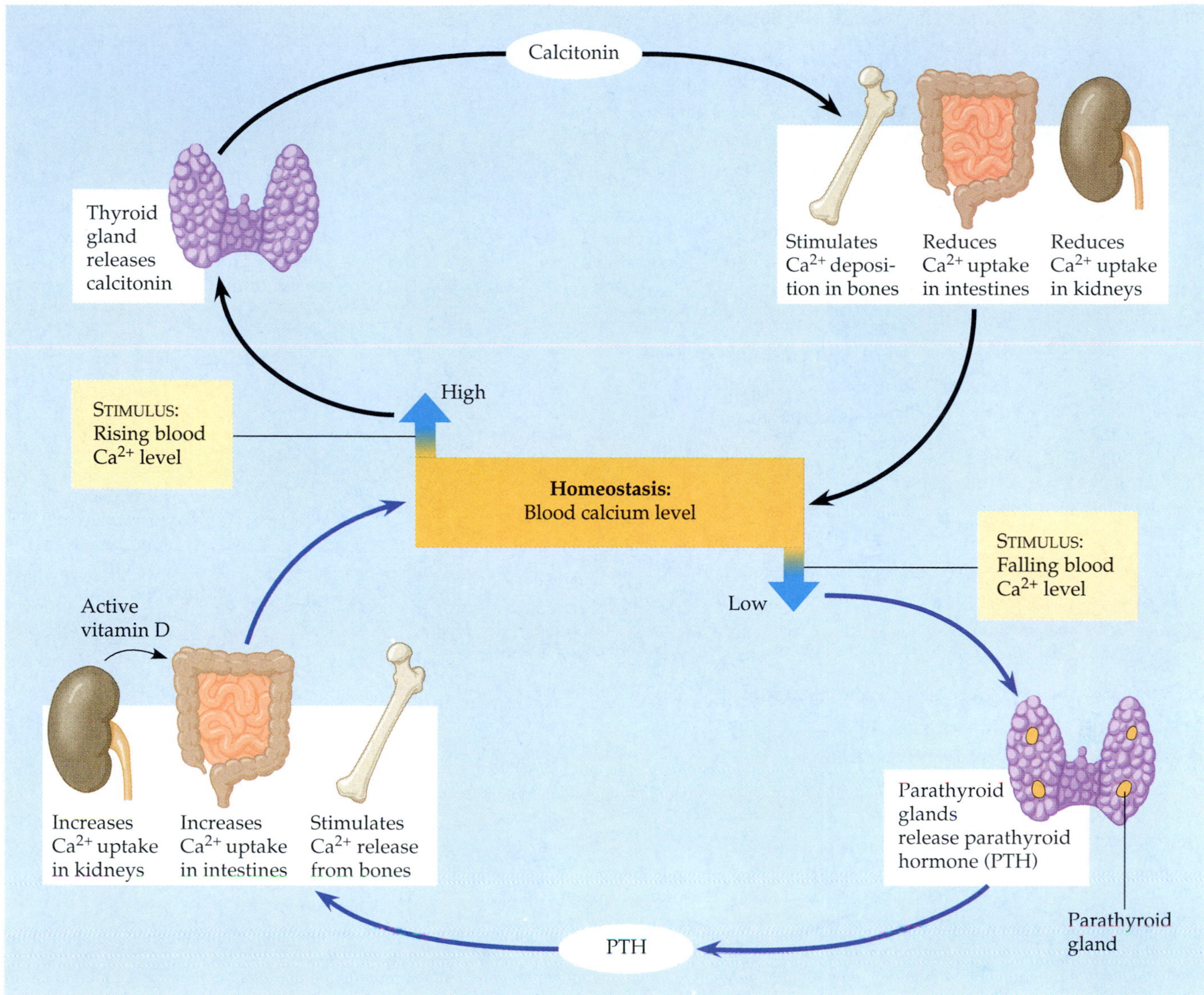

FIGURE 45.10 ▪ Hormonal control of calcium homeostasis in mammals. A negative feedback system involving two antagonistic hormones, calcitonin and parathyroid hormone (PTH), maintains the concentration of calcium in blood within a very narrow range of about 10 mg/100 mL. A rise in blood Ca^{2+} induces the thyroid gland to secrete calcitonin, which lowers the Ca^{2+} concentration by increasing bone deposition, reducing Ca^{2+} uptake in the intestines, and reducing reabsorption in the kidneys. These effects are reversed by PTH, which is secreted from the parathyroid glands when the concentration of blood Ca^{2+} falls below the set point. Blood calcium levels begin to increase as target cells in the kidneys, intestines, and bone respond to PTH. Blood Ca^{2+} will rise only so far before the thyroid counters by secreting more calcitonin. In classic feedback fashion, these two hormones balance each other's effects, thereby minimizing fluctuations in the concentration of blood Ca^{2+}, an ion essential to the normal functioning of all cells. Synthesized in an inactive form by skin exposed to sunlight, vitamin D plays an important role in calcium homeostasis. Vitamin D is carried in the blood and converted to its active form in many tissues such as the liver and kidneys. The active form enables PTH to increase Ca^{2+} uptake by the intestines. (See FIGURE 45.1 for an overview.)

The antagonistic effects of glucagon and insulin are vital to glucose homeostasis, a mechanism that precisely manages both fuel storage and fuel use by body cells. We will revisit the topic of fuel management and see that it involves additional hormones when we study the adrenal glands in the next section. The liver's ability to perform its vital roles in glucose homeostasis results from the metabolic versatility of its cells and its access to absorbed nutrients via the hepatic portal vessel, which carries blood directly from the small intestine to the liver.

When the mechanisms of glucose homeostasis go awry, there are serious consequences. Diabetes mellitus, perhaps the best known endocrine disorder, is caused by a deficiency of insulin or a loss of response to insulin in target tissues. The result is high blood glucose—so high, in fact, that the diabetic's kidneys excrete glucose, which explains why the presence of sugar in urine is one test for diabetes. As more glucose concentrates in the urine, more water is excreted with it, resulting in excessive volumes of urine and persistent thirst.

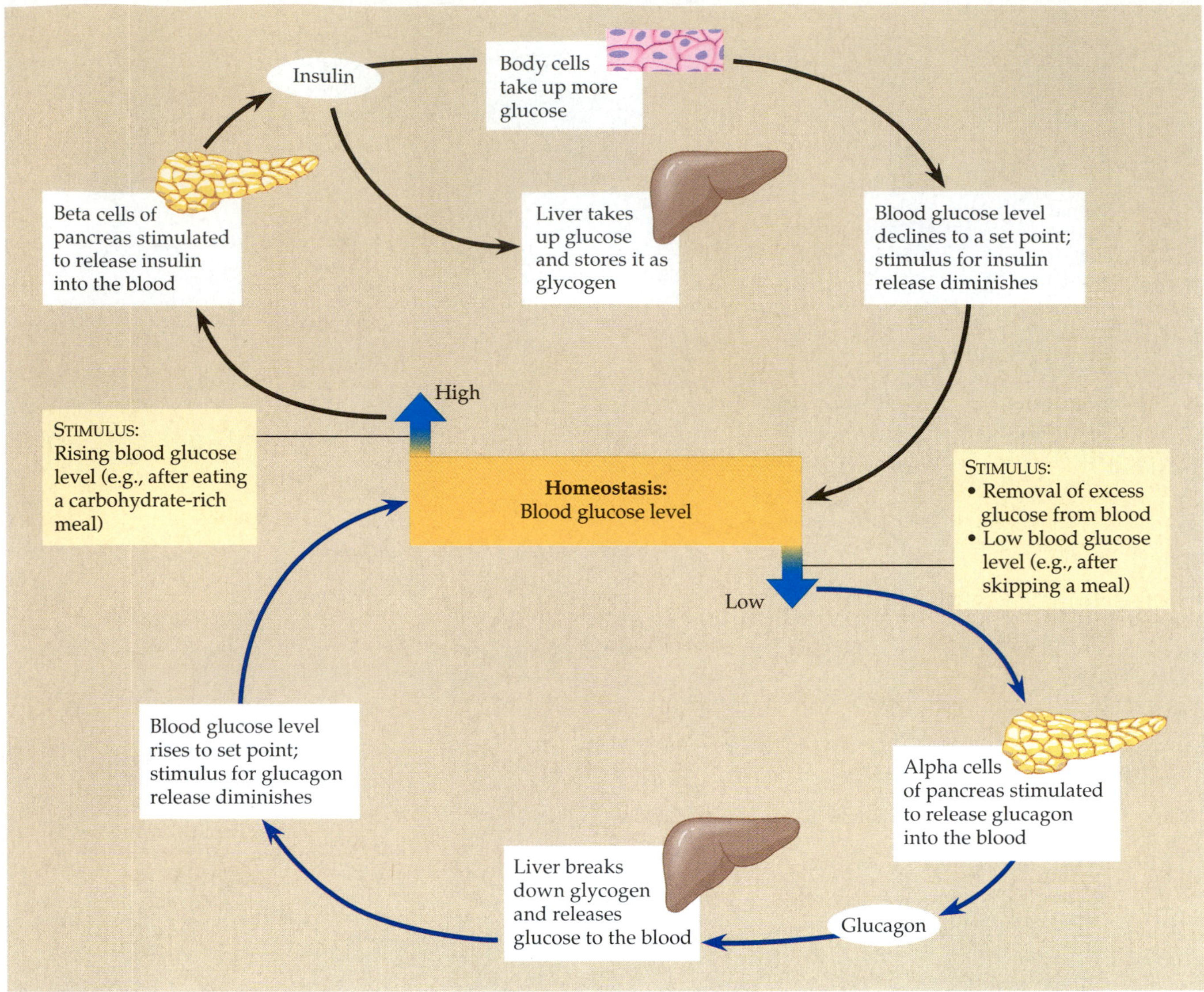

FIGURE 45.11 ▪ Glucose homeostasis maintained by insulin and glucagon. A rise in blood glucose above the set point of about 90 mg/100 mL in humans stimulates the pancreas to secrete insulin, which triggers its target cells to take up the excess glucose from the blood. Once the excess is removed or when blood glucose concentration dips below the set point, the pancreas responds by secreting glucagon, which acts on the liver to raise the blood glucose level.

(*Diabetes,* from the Greek, refers to this copious urination, and *mellitus* is from the Greek word for "honey," referring to the presence of sugar in the urine.) Because glucose is unavailable to most body cells as a major fuel source for diabetics, fat must serve as the main substrate for cellular respiration. In severe cases of diabetes, acidic metabolites formed during fat breakdown accumulate in the blood, threatening life by lowering blood pH.

There are actually two major forms of diabetes with very different causes. **Type I diabetes mellitus** (insulin-dependent diabetes) is an autoimmune disorder, in which the immune system mounts an attack on the cells of the pancreas. (Chapter 43 discusses possible causes of autoimmune reactions.) This disorder usually occurs rather suddenly during childhood, destroying the person's ability to produce insulin. Treatment consists of insulin injections, which are usually taken several times daily. Until recently, insulin for injections was extracted from animal pancreases, but genetic engineering has provided a relatively inexpensive source of human insulin by inserting the gene for the hormone into bacteria (see Chapter 20). **Type II diabetes mellitus** (non-insulin-dependent diabetes) is characterized either by a deficiency of insulin or, more commonly, by reduced responsiveness in target cells due to some change in insulin receptors. Type II diabetes usually occurs after about age 40, becoming more likely with increasing age. More than 90% of diabetics are type II, and many can manage their blood glucose solely by exercise and dietary control. Heredity is a major factor in type II diabetes.

The adrenal medulla and adrenal cortex help the body manage stress

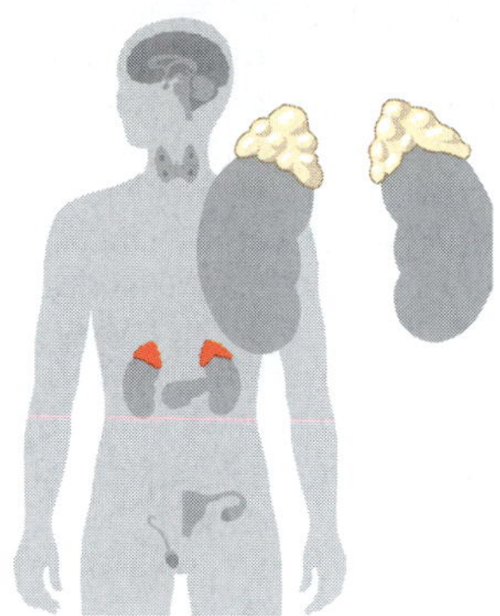

The **adrenal glands** are adjacent to the kidneys. In mammals, each adrenal gland is actually made up of two glands with different cell types, functions, and embryonic origins: the **adrenal cortex**, or outer portion, and the **adrenal medulla**, or central part. Nonmammalian vertebrates have quite different arrangements of the same tissues.

The adrenal medulla has close developmental and functional ties with the nervous system. Secretory cells of the adrenal medulla are derived from cells of the neural crest (see FIGURE 34.5). Some of the neural crest cells in the abdominal region of a vertebrate embryo may differentiate into either endocrine cells of the adrenal medulla or neurons, depending on chemical signals in their immediate surroundings (FIGURE 45.12).

What makes your heart beat faster and your skin develop goose bumps when you sense danger or approach a stressful situation, like speaking in public? These reactions are part of the fight-or-flight response stimulated by two hormones of the adrenal medulla, **epinephrine** (also known as adrenaline) and **norepinephrine** (noradrenaline). These hormones are members of a class of compounds, the **catecholamines**, that are synthesized from the amino acid tyrosine (FIGURE 45.13).

Epinephrine, norepinephrine, and other catecholamines are secreted in response to positive or negative stress—everything from extreme pleasure to increased cold to life-threatening danger. Their release into the blood gives the body a rapid bioenergetic boost, increasing the basal metabolic rate and having dramatic effects on several targets. Epinephrine and norepinephrine increase the rate of glycogen breakdown in the liver and skeletal muscles and glucose release into the blood by liver cells. They also stimulate the release of fatty acids from fat cells. The fatty acids may be used by cells for energy. In addition to increasing the availability of energy sources, epinephrine and norepinephrine have profound effects on the cardiovascular and respiratory systems. For example, they increase both the rate and the stroke volume of

Neural crest cell

Migrates to site of adrenal glands

Gluco-corticoids

Nerve growth factor (NGF)

Endocrine cell of adrenal medulla

Sympathetic neuron

FIGURE 45.12 ▪ Derivation of endocrine cells of the adrenal medulla and neurons from neural crest cells. Studies of neural crest cells in laboratory cultures indicate that the fate of these cells in embryonic vertebrates depends on chemical signals. Embryonic chicken neural crest cells that have migrated to the region of the body where the adrenal glands develop may differentiate into hormone secreting cells of the adrenal medulla if they are given glucocorticoids (hormones secreted by the adrenal cortex). If the same neural crest cells are exposed to nerve growth factor (NGF), they develop into nerve cells called sympathetic neurons (you'll learn about the function of these neurons in Chapter 48).

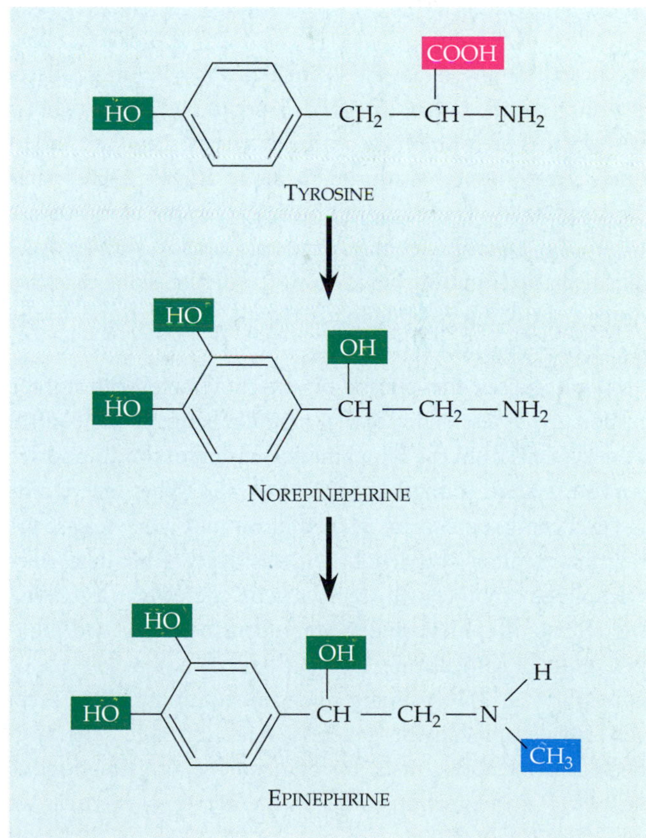

FIGURE 45.13 ▪ The synthesis of catecholamine hormones. Cells in the adrenal medulla synthesize the catecholamines norepinephrine and epinephrine from the amino acid tyrosine. Norepinephrine is made by removing a carboxyl group (red) and adding two hydroxyl groups (green). Epinephrine is made from norepinephrine by adding a methyl group (blue).

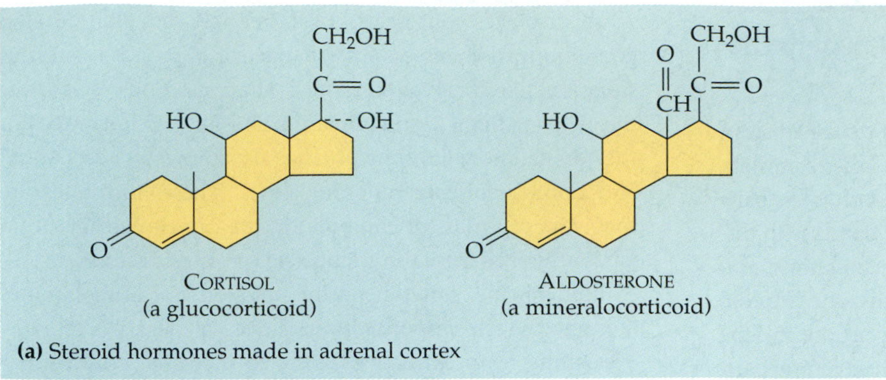

(a) Steroid hormones made in adrenal cortex

CH$_3$
C=O
OH
OH
O
HO
O
TESTOSTERONE
(an androgen)
ESTRADIOL
(an estrogen)
PROGESTERONE
(a progestin)

(b) Steroid hormones made in gonads (and, to some extent, in adrenal cortex)

FIGURE 45.14 ▪ Steroid hormones from the adrenal cortex and gonads. (a) Cortisol (a glucocorticoid) and aldosterone (a mineralocorticoid), both made in the adrenal cortex, are structurally similar to **(b)** the sex hormones testosterone (an androgen), estradiol (an estrogen), and progesterone (a progestin). The precursor for the synthesis of all steroid hormones is cholesterol (see FIGURE 5.14). Most of the androgens (male hormones) that circulate in the blood are made by the testes, and most of the estrogens and progestins (female hormones) are produced by the ovaries; however, small amounts of these sex hormones are also made by the adrenal cortex.

the heartbeat and dilate the bronchioles in the lungs, effects that increase the rate of oxygen delivery to body cells. (This is why doctors prescribe epinephrine as a heart stimulant and to open breathing tubes during asthma attacks.) The catecholamines also cause smooth muscles of some blood vessels to contract and muscles of other vessels to relax, with an overall effect of shunting blood away from the skin, digestive organs, and kidneys while increasing the blood supply to the heart, brain, and skeletal muscles.

What causes the release of catecholamines during the response to stress? The adrenal medulla is under the control of nerve cells from the sympathetic division of the autonomic nervous system (discussed in Chapter 48). When nerve cells are excited by some form of stressful stimulus, they release the neurotransmitter acetylcholine in the adrenal medulla. Acetylcholine combines with receptors on the cells, stimulating the release of epinephrine. Norepinephrine is released independently of epinephrine. Its functions are similar to those of epinephrine, but its primary role is in sustaining blood pressure, while epinephrine generally has a stronger effect on heart and metabolic rates. Norepinephrine also functions as an important neurotransmitter in the nervous system, as we will see in Chapter 48.

The adrenal cortex, like the adrenal medulla, reacts to stress. But it responds to endocrine signals rather than to nervous input. Stressful stimuli cause the hypothalamus to secrete a releasing hormone that stimulates the anterior pituitary to release the tropic hormone ACTH. When it reaches its target via the bloodstream, ACTH stimulates cells of the adrenal cortex to synthesize and secrete a family of steroids called **corticosteroids**. In another case of negative feedback, elevated levels of corticosteroids in the blood suppress the secretion of ACTH.

Many corticosteroids have been isolated from the adrenal cortex. The two main types in humans are the **glucocorticoids**, such as cortisol, and the **mineralocorticoids**, such as aldosterone (FIGURE 45.14).

The primary effect of glucocorticoids is on bioenergetics, specifically on glucose metabolism. Augmenting the fuel-mobilizing effects of glucagon from the pancreas, glucocorticoids promote the synthesis of glucose from noncarbohydrate sources, such as proteins, making more glucose available as fuel. Glucocorticoids act on skeletal muscle, causing a breakdown of muscle proteins. The resulting carbon skeletons are transported to the liver and kidney, where they are converted to glucose and released into the blood. The synthesis of glucose from muscle proteins is a homeostatic mechanism providing circulating fuel when body activities require more than the liver can mobilize from its glycogen stores. It can also be part of a broader role of the glucocorticoids, that of helping the body withstand long-term environmental challenge.

Abnormally high doses of glucocorticoids administered as medication suppress certain components of the body's immune system—for example, the inflammatory reaction that occurs at the site of an infection. Glucocorticoids are used to treat diseases in which excessive inflammation is a problem.

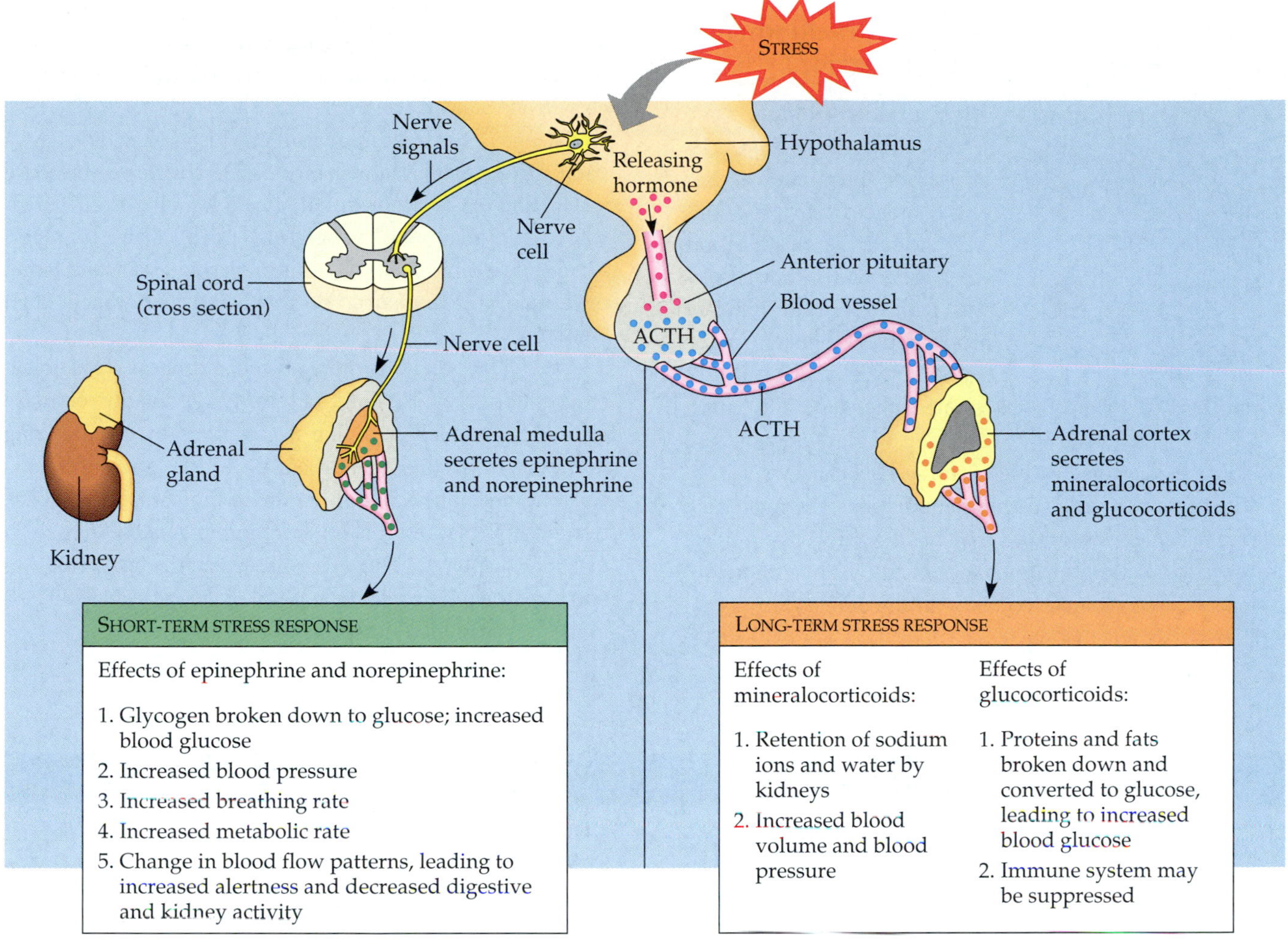

FIGURE 45.15 ▪ Stress and the adrenal gland. Stressful stimuli cause the hypothalamus to activate the adrenal medulla via nerve impulses and the adrenal cortex via hormonal signals. The adrenal medulla mediates short-term responses to stress by secreting the catecholamine hormones epinephrine and norepinephrine. The adrenal cortex controls more prolonged responses by secreting steroid hormones.

Cortisone, for instance, was once thought to be a miracle drug that could cure serious inflammatory conditions such as arthritis. It has become clear, however, that long-term use of corticosteroids can result in increased susceptibility to infection and disease because of the immunosuppressive effects of these powerful drugs.

Mineralocorticoids have their major effects on salt and water balance. Aldosterone, for example, stimulates cells in the kidney to reabsorb sodium ions and water from the filtrate, raising blood pressure and volume. Aldosterone (of the renin-angiotensin-aldosterone system, or RAAS), ADH from the posterior pituitary, and atrial natriuretic factor (ANF) from the heart form a regulatory complex with multiple feedback loops that underlie the kidney's ability to maintain the ion and water homeostasis of the blood (see Chapter 44). Aldosterone secretion is largely regulated by the RAAS itself in response to changes in plasma ion concentration. However, when an individual is under severe stress, the hypothalamus tends to secrete more releasing hormones that increase the rate of secretion of ACTH by the anterior pituitary. The rise in ACTH in the blood then increases the rate of secretion of aldosterone by the adrenal cortex.

Evidence is mounting that both glucocorticoids and mineralocorticoids help maintain homeostasis when the body experiences stress over an extended period of time. FIGURE 45.15 compares the long-term stress-induced actions of corticosteroids with the short-term ones of epinephrine and norepinephrine.

A third group of corticosteroids are sex hormones, mainly androgens (male hormones) similar to testosterone produced by the testes and small amounts of estrogens and progesterone (female hormones). There is evidence that adrenal androgens account for the sex drive in adult females, but otherwise the physiological roles of the adrenal sex hormones are not well understood. We will discuss the sex hormones from the gonads in the next section.

Gonadal steroids regulate growth, development, reproductive cycles, and sexual behavior

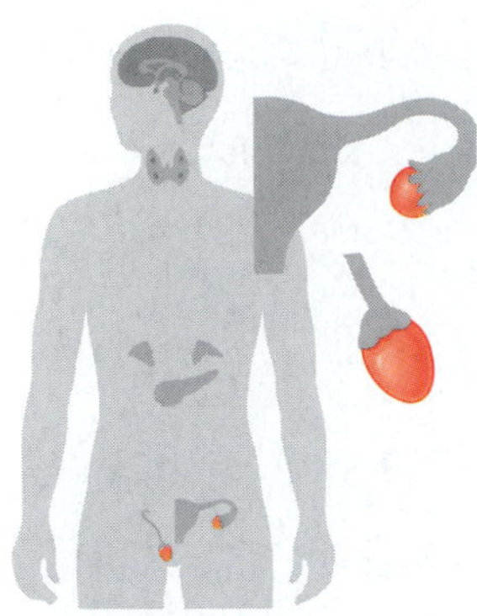

The gonads produce and secrete three major categories of steroid hormones: androgens, estrogens, and progestins (see FIGURE 45.14). All three types are found in both males and females but in different proportions. Produced in the testes of males and ovaries of females, these steroids affect growth and development and also regulate reproductive cycles and sexual behavior.

The testes primarily synthesize **androgens**, the main such hormone being **testosterone**. In general, androgens stimulate the development and maintenance of the male reproductive system. Androgens produced early in the development of an embryo determine that the fetus will develop as a male rather than a female. At puberty, high concentrations of androgens are responsible for the development of human male secondary sex characteristics, such as male patterns of hair growth and a low voice.

Estrogens, the most important of which is estradiol, have a parallel role in the maintenance of the female reproductive system and the development of female secondary sex characteristics. In mammals, **progestins**, which include progesterone, are primarily involved in preparing and maintaining the uterus, which supports the growth and development of an embryo.

The synthesis of both estrogens and androgens is controlled by gonadotropins, FSH and LH, from the anterior pituitary gland. FSH and LH secretion is controlled by one releasing hormone from the hypothalamus, GnRH (gonadotropin-releasing hormone). We will examine the complex feedback relationships that regulate the secretion of gonadal steroids in detail in Chapter 46. There we will see that the endocrine system is central not only to the survival of the individual but also to the propagation of the species.

CHAPTER REVIEW

REVIEW OF KEY CONCEPTS

(with page numbers and key figures)

AN INTRODUCTION TO REGULATORY SYSTEMS

Animals have two regulatory systems that coordinate internal body functions: the nervous system and the endocrine system. Chemical messengers called hormones regulate the activity of target organs at distant sites.

- **The endocrine system and the nervous system are structurally, chemically, and functionally related (p. 894, FIGURE 45.1)** The endocrine and nervous systems often function inseparably in maintaining homeostasis, development, and reproduction. Neurosecretory cells of the nervous system secrete many hormones; several hormones, such as norepinephrine, are used as signals by both the endocrine and the nervous system; and many body functions are regulated by both systems, often by feedback mechanisms.
- **Invertebrate regulatory systems clearly illustrate endocrine and nervous system interactions (pp. 894–895, FIGURE 45.2)** Diverse hormones regulate different aspects of homeostasis in invertebrates. In insects, molting and development are controlled by an interplay between ecdysone, juvenile hormone (JH), and brain hormone secreted by neurosecretory cells.

CHEMICAL SIGNALS AND THEIR MODES OF ACTION

Hormones convey information via the bloodstream. Local regulators function in many cases of homeostasis. Local regulators affect target cells in the immediate vicinity of their secretion. Pheromones act as communication signals between different individuals of the same species.

- **A variety of local regulators affect neighboring target cells (pp. 895–896)** Neurotransmitters are local regulators in the synapses of the nervous system. Growth factors and prostaglandins are secreted into the intercellular fluid between cells.
- **Chemical signals bind to specific receptor proteins within target cells or on their surface (p. 896, FIGURES 45.3, 45.4)** Chemical signals can trigger changes in target cells differently, depending on the type of receptor, the type of receptor cell, and whether they enter cells or attach to surface proteins.

45.1 **Most chemical signals bind to plasma-membrane proteins, initiating signal-transduction pathways (pp. 896–897)** Most chemical signals cannot pass through the cell membrane and bind to specific receptors on the plasma membrane. These signals trigger changes in target cells through signal-transduction pathways.

45.2 **Steroid hormones, thyroid hormones, and some local regulators enter target cells and bind with intracellular receptors (pp. 897–898, FIGURE 45.5)** Chemical signals that actually enter target cells bind to specific protein receptors in the cytoplasm. Hormone-receptor complexes then enter the nucleus, bind to certain sites on DNA, and effect changes in gene expression.

THE VERTEBRATE ENDOCRINE SYSTEM

45.3 FIGURE 45.6 illustrates the major endocrine glands of the human body; TABLE 45.1 lists the major vertebrate hormones. Among the numerous hormones regulating the vertebrate body, tropic hormones have other endocrine glands as their targets. The hormonal functions of the stomach, small intestine, heart, and thymus are discussed in Chapters 41, 42, and 43. The thymus secretes thymosin and other chemical messengers that stimulate the development and differentiation of T lymphocytes.

- **The hypothalamus and pituitary integrate many functions of the vertebrate endocrine system (pp. 900–902, FIGURES 45.6, 45.7)** Neurosecretory cells of the hypothalamus integrate endocrine and neural function by influencing the pituitary gland. The posterior pituitary is an extension of the brain that stores and releases two hormones (oxytocin and antidiuretic hormone, ADH) produced by neurosecretory cells in the hypothalamus. Oxytocin induces uterine contractions and milk ejection, and ADH enhances water reabsorption in the kidneys. Under the direction of releasing and inhibiting hormones conveyed by special portal vessels from the hypothalamus, the anterior pituitary produces an array of hormones, including thyroid-stimulating hormone (TSH), follicle-stimulating hormone (FSH), luteinizing hormone (LH), growth hormone (GH), prolactin (PRL), adrenocorticotropic hormone (ACTH), melanocyte-stimulating hormone (MSH), and the endorphins. The chemically related tropic hormones TSH and the gonadotropins (FSH and LH) stimulate the thyroid gland and the gonads, respectively, to produce their hormones. GH promotes growth directly and stimulates the production of growth factors. Prolactin, named for its stimulation of lactation in mammals, has diverse effects in different vertebrates. ACTH stimulates the adrenal cortex. MSH influences skin pigmentation in some vertebrates and fat metabolism in mammals. Endorphins, the brain's natural opiates, inhibit the perception of pain.

- **The pineal gland is involved in biorhythms (pp. 902–903, FIGURE 45.6, TABLE 45.1)** The pineal gland secretes melatonin, which influences skin pigmentation, biological rhythms, and reproduction in various vertebrates.
- **Thyroid hormones function in development, bioenergetics, and homeostasis (pp. 903–904, FIGURES 45.6, 45.8, 45.9, TABLE 45.1)** The thyroid gland produces iodine-containing hormones (T_3 and T_4) that stimulate metabolism and influence development and maturation. The thyroid also secretes calcitonin, which lowers calcium levels in the blood.
- **Parathyroid hormone and calcitonin balance blood calcium (p. 904, FIGURES 45.6, 45.10, TABLE 45.1)** The parathyroid glands raise plasma calcium levels by secreting parathyroid hormone (PTH). PTH works with calcitonin to affect calcium homeostasis by actions on the intestines, bone, and kidneys.
- **Endocrine tissues of the pancreas secrete insulin and glucagon, antagonistic hormones that regulate blood glucose (pp. 904–906, FIGURES 45.6, 45.11, TABLE 45.1)** The endocrine portion of the pancreas consists of islet cells that secret insulin and glucagon. High blood glucose levels stimulate the release of insulin, which increases the cellular uptake of glucose, promotes the formation and storage of glycogen in the liver, and stimulates protein synthesis and fat storage. Low blood glucose levels trigger glucagon release, which increases blood glucose by stimulating the conversion of glycogen to glucose in the liver and increasing the breakdown of fat and protein. Type I diabetes mellitus is an autoimmune disorder resulting in a lack of insulin. Type II diabetes is usually caused by the loss of responsiveness of target cells to insulin.
- **The adrenal medulla and adrenal cortex help the body manage stress (pp. 907–909, FIGURES 45.6, 45.12–45.15, TABLE 45.1)** The adrenal medulla releases epinephrine and norepinephrine in response to stress-activated impulses from the nervous system. These hormones mediate various fight-or-flight responses. The adrenal cortex releases corticosteroids (including sex hormones), glucocorticoids, and mineralocorticoids. Glucocorticoids influence glucose metabolism and the immune system; mineralocorticoids affect salt and water balance.
- **Gonadal steroids regulate growth, development, reproductive cycles, and sexual behavior (p. 910, FIGURE 45.6, TABLE 45.1)** The gonads—testes and ovaries—produce varying proportions of three kinds of steroid hormones: androgens, estrogens, and progestins.

SELF-QUIZ

1. Which of the following is *not* an accurate statement about hormones?
   a. Hormones are chemical messengers that travel to target cells through the circulatory system.
   b. Hormones often regulate homeostasis through antagonistic functions.
   c. Hormones of the same chemical class usually have the same general function.
   d. Hormones are secreted by specialized cells usually located in endocrine glands.
   e. Hormones are often regulated through feedback loops.

2. A distinctive feature of the mechanism of action of steroid hormones and thyroid hormones is that
   a. these hormones are regulated by feedback loops
   b. target cells react more rapidly to these hormones than to local regulators
   c. these hormones bind with specific receptor proteins on target-cell plasma membranes
   d. these hormones bind to cytoplasmic receptors
   e. these hormones affect metabolism

3. The relationship between ecdysone and BH
   a. is an example of the interaction between the endocrine and nervous systems
   b. illustrates homeostasis achieved by positive feedback
   c. demonstrates that peptide-derived hormones have more widespread effects than steroidal hormones
   d. illustrates homeostasis maintained by antagonistic hormones
   e. demonstrates competitive inhibition for the hormone receptor

4. Growth factors are local regulators that
   a. are produced by the anterior pituitary
   b. are modified fatty acids that stimulate bone and cartilage growth
   c. are found on the surface of cancer cells and stimulate abnormal cell division
   d. are proteins that bind to surface receptors and stimulate target-cell growth and development
   e. include histamines and interleukins and are necessary for cellular differentiation

5. Which of the following hormones is *incorrectly* paired with its action?
   a. oxytocin—stimulates uterine contractions during childbirth
   b. thyroxine—stimulates metabolic processes
   c. insulin—stimulates glycogen breakdown in the liver
   d. ACTH—stimulates the release of glucocorticoids by the adrenal cortex
   e. melatonin—affects biological rhythms, seasonal reproduction

6. An example of antagonistic hormones controlling homeostasis is
   a. thyroxine and parathyroid hormone in calcium balance
   b. insulin and glucagon in glucose metabolism
   c. progestins and estrogens in sexual differentiation
   d. epinephrine and norepinephrine in fight-or-flight responses
   e. oxytocin and prolactin in milk production

7. Which of the following human conditions is *incorrectly* paired with a hormone?
   a. acromegaly—growth hormone
   b. diabetes—insulin
   c. cretinism—thyroid hormone
   d. tetany—PTH
   e. pituitary dwarfism—ACTH

8. A portal vessel carries blood from the hypothalamus directly to the
   a. thyroid
   b. pineal gland
   c. anterior pituitary
   d. posterior pituitary
   e. thymus

9. Which of the following is the most likely explanation for hypothyroidism in a patient whose iodine level is normal?
   a. a disproportionate production of T_3 to T_4
   b. hyposecretion of TSH
   c. hypersecretion of TSH
   d. hypersecretion of MSH
   e. a decrease in the thyroid secretion of calcitonin

10. The main target organs for tropic hormones are
   a. muscles
   b. blood vessels
   c. endocrine glands
   d. kidneys
   e. nerves

CHALLENGE QUESTIONS

1. Describe three specific examples of the interaction between the nervous system and the endocrine system.
2. A woman with a hypothyroid condition is treated with thyroxine. How is this medication likely to affect levels of TSH and TSH-releasing hormone (TRH)?
3. Write a paragraph explaining how the same hormone can have one effect on one type of target cell, a different effect on another type of target cell, and no effect on a third type of target cell.

SCIENCE, TECHNOLOGY, AND SOCIETY

Growth hormone (GH) produced by DNA technology has enabled hundreds of children who suffer from pituitary dwarfism to grow normally and reach a stature within the normal range. Now that the hormone is readily available and relatively inexpensive, many parents who are concerned that their children are not growing fast enough want to use GH to make them grow faster and taller. There can be potentially harmful effects, such as a reduction in body fat and an increase in muscle mass. And no one yet knows if GH injections will have seriously harmful long-term effects in individuals who do not have a hypopituitary condition. What criteria do you think should determine the cases in which GH treatments or other hormone therapies are appropriate?

FURTHER READING

Barinaga, M. "How Jet-Lag Hormone Does Double Duty in the Brain." *Science,* July 25, 1997. Provides a summary of recent research on melatonin and biorhythms.

Lacy, P. E. "Treating Diabetes with Transplanted Cells." *Scientific American,* July 1995. Describes the advancing technology and biology of pancreatic islet transplantation.

Lamberts, S. W. J., A. W. van den Beld, and A. van der Lely. "The Endocrinology of Aging." *Science,* October 17, 1997. Presents evidence that decline in the activity of some hormones is related to aging.

Leutwyler, K. "Don't Stress." *Scientific American,* January 1998. Provides a brief summary of recent research on the long-term effects of stress hormones.

McLachlan, J. A., and S. F. Arnold. "Environmental Estrogens." *American Scientist,* September–October 1996. Presents a discussion of the controversial hypothesis that environmental chemicals function as chemical signals in animals.

Oren, D. A., and M. Terman. "Tweaking the Human Circadian Clock with Light." *Science,* January 16, 1998. Describes experiments on resetting the biological clock.

Schuman, E. "Growth Factors Sculpt the Synapse." *Science,* February 28, 1997. Provides experimental evidence that growth factors can strengthen synaptic signaling.

Snyder, S., and D. Bredt. "Biological Roles of Nitric Oxide." *Scientific American,* May 1992. Describes research on a chemical signal with multiple functions.

WEB LINKS

Visit the special edition of *The Biology Place* for BIOLOGY, Fifth Edition, at http://www.biology.com/campbell. Go to Chapter 45 for online resources, including learning activities, practice exams, and links to the following web sites:

"The American Diabetes Association (ADA)"
Over 1 million diabetics in the United States take the hormone insulin every day. Learn more about this malfunction of the endocrine system from the ADA.

"The International Council on Infertility Information Dissemination"
Malfunctions of the endocrine system often have severe effects. At this site you can explore infertility, one of the effects of malfunctioning sex hormones.

"Endocrine Web"
This site, operated by endocrinologists, contains a vast amount of information about the endocrine system and its disorders.

"Anabolic Steroids: A Threat to Mind and Body"
This site explains the growing evidence that the nonmedical use of anabolic steroids to improve athletic performance and physical appearance poses severe physical and psychological risks to young people.

CHAPTER 46

ANIMAL REPRODUCTION

Overview of Animal Reproduction

- Both asexual and sexual reproduction occur in the animal kingdom
- Diverse mechanisms of asexual reproduction enable animals to produce identical offspring rapidly
- Reproductive cycles and patterns vary extensively among animals

Mechanisms of Sexual Reproduction

- Internal and external fertilization both depend on mechanisms ensuring that mature sperm encounter fertile eggs of the same species
- Species with internal fertilization usually produce fewer zygotes but provide more parental protection than species with external fertilization
- Complex reproductive systems have evolved in many animal phyla

Mammalian Reproduction

- Human reproduction involves intricate anatomy and complex behavior
- Spermatogenesis and oogenesis both involve meiosis but differ in three significant ways
- A complex interplay of hormones regulates reproduction
- Embryonic and fetal development occur during pregnancy in humans and other eutherian (placental) mammals
- Modern technology offers solutions for some reproductive problems

The many aspects of animal form and function we have studied so far can be viewed, in the broadest context, as adaptations contributing to reproductive success. Individuals are transient. A population transcends finite lifespans only by reproduction, the creation of new individuals from existing ones. The two earthworms in the photograph on this page are mating. Unless disturbed, they will remain above the ground and joined like this for several hours. Each worm produces both sperm and eggs, each donates and receives sperm during mating, and each will produce fertilized eggs. In a few weeks, sexual reproduction will be completed when many new individuals hatch from the eggs.

Animal reproduction is the subject of this chapter. We will first compare the diverse reproductive mechanisms that have evolved in the animal kingdom, and then examine the details of mammalian, particularly human, reproduction.

OVERVIEW OF ANIMAL REPRODUCTION

Both asexual and sexual reproduction occur in the animal kingdom

There are two principal modes of animal reproduction. **Asexual** (Greek, "without sex") **reproduction** is the creation of new individuals whose genes all come from one parent without the fusion of egg and sperm. In most cases, asexual reproduction relies entirely on mitotic cell division. **Sexual reproduction** is the creation of offspring by the fusion of haploid **gametes** to form a **zygote** (fertilized egg), which is diploid. Gametes are formed by meiosis (see FIGURE 13.6). The female gamete, the **ovum** (unfertilized egg), is usually a relatively large and nonmotile cell. The male gamete, the **spermatozoon**, is generally a small, motile cell. Sexual reproduction increases genetic variability among offspring by generating unique combinations of genes inherited from two parents. By producing offspring having various phenotypes, sexual reproduction may enhance the reproductive success of parents when pathogens or other environmental factors change relatively rapidly.

Diverse mechanisms of asexual reproduction enable animals to produce identical offspring rapidly

Many invertebrates can reproduce asexually by **fission**, the separation of a parent into two or more individuals of approximately equal size (FIGURE 46.1, p. 914). Also common among invertebrates, **budding** involves new individuals splitting off

FIGURE 46.1 ▪ Two from one: asexual reproduction of a sea anemone *(Anthopleura elegantissima).* The individual in the center of this photograph is undergoing fission, a type of asexual reproduction. Two smaller individuals will soon form as the parent divides approximately in half. The offspring will be genetic copies of the parent.

from existing ones. For example, in certain cnidarians and tunicates, new individuals grow out from the body of a parent (see FIGURE 13.1). The offspring may either detach from the parent or remain joined, eventually forming extensive colonies. Stony corals, which may be more than 1 m across, are cnidarian colonies of several thousand connected individuals. In another form of asexual reproduction, some invertebrates release specialized groups of cells that can grow into new individuals. For example, the **gemmules** of sponges are formed when cells of several types migrate together within the sponge and become surrounded by a protective coat.

Yet another type of asexual reproduction is **fragmentation**, the breaking of the body into several pieces, some or all of which develop into complete adults. For an animal to reproduce this way, fragmentation must be accompanied by **regeneration**, the regrowth of lost body parts. Reproduction by fragmentation and regeneration occurs in many sponges, cnidarians, polychaete annelids, and tunicates. Many animals can also replace lost appendages by regeneration—most sea stars can grow new arms when injured, for example—but this is not reproduction because new individuals are not created. In sea stars of the genus *Linckia,* a whole new individual may develop from an isolated arm. Thus, a single animal with five arms, if broken apart, could asexually give rise to five offspring.

Asexual reproduction has several potential advantages. For instance, it enables animals living in isolation to produce offspring without locating mates. It can also create numerous offspring in a short amount of time, which is ideal for colonizing a habitat rapidly. Theoretically, asexual reproduction is most advantageous in stable, favorable environments because it perpetuates successful genotypes precisely.

Reproductive cycles and patterns vary extensively among animals

Most animals show definite cycles in reproductive activity, often related to changing seasons. The periodic nature of reproduction allows animals to conserve resources and reproduce when more energy is available than is needed for maintenance and when environmental conditions favor the survival of offspring. Ewes (female sheep), for example, have 15-day reproductive cycles and ovulate at the midpoint of each cycle. These cycles generally occur during fall and early winter, so lambs are usually born in the late winter or spring. Even animals that live in apparently stable habitats, such as the tropics or the ocean, generally reproduce only at certain times of the year. Reproductive cycles are controlled by a combination of hormonal and environmental cues, the latter including such factors as seasonal temperature, rainfall, day length, and lunar cycles (see the discussion of the pineal gland in Chapter 45).

Animals may reproduce asexually or sexually exclusively, or they may alternate between the two modes. In aphids, rotifers, and the freshwater crustacean *Daphnia,* each female can produce eggs of two types, depending on environmental conditions such as the time of year. One type of egg is fertilized, but the other type develops by **parthenogenesis**, a process in which the egg develops without being fertilized. The adults produced by parthenogenesis are often haploid, and their cells do not undergo meiosis in forming new eggs. In the case of *Daphnia,* the switch from sexual to asexual reproduction is often related to season. Asexual reproduction occurs under favorable conditions and sexual reproduction during times of environmental stress.

Parthenogenesis has a role in the social organization of certain species of bees, wasps, and ants. Male honeybees, or drones, are produced parthenogenetically, whereas females, both sterile workers and reproductive females (queens), develop from fertilized eggs.

Among vertebrates, several genera of fishes, amphibians, and lizards reproduce exclusively by a complex form of parthenogenesis that involves the doubling of chromosomes after meiosis to create diploid "zygotes." For example, there are about 15 species of whiptail lizards (genus *Cnemidophorus)* that reproduce exclusively by parthenogenesis. There are no males in these species, but the lizards imitate courtship and mating behavior typical of sexual species of the same genus. During the breeding season, one female of each mating pair mimics a male (FIGURE 46.2a). The roles change two or three times during the season, female behavior occurring when the level of the female sex hormone estrogen is high prior to ovulation (the release of eggs), and male behavior occurring after ovulation when the level of estrogen drops (FIGURE 46.2b). In fact, ovulation is more likely to occur if one individual is

(a)

(b)

FIGURE 46.2 ▪ Sexual behavior in parthenogenetic lizards. The desert-grassland whiptail lizard *(Cnemidophorus uniparens)* is an all-female species. These reptiles reproduce by parthenogenesis; eggs undergo a chromosome doubling after meiosis and develop into lizards without being fertilized. However, ovulation is enhanced by courtship and mating rituals that imitate the behavior of closely related species that reproduce sexually. **(a)** Both lizards in this photograph are *C. uniparens* females. The one on top is playing the role of a male. Every two or three weeks during the breeding season, individuals switch sex roles. **(b)** The sexual behavior of *C. uniparens* is correlated with the cycle of ovulation mediated by sex hormones. Before ovulation, an individual behaves like a female. She has relatively large ovaries and her behavior is correlated with high blood levels of estrogen. After ovulation, small ovaries, an abrupt drop in estrogen, and a rapid increase in circulating progesterone are correlated with an individual behaving like a male.

mounted by another during the critical time of the hormone cycle; isolated lizards lay fewer eggs than those that go through the motions of sex. Apparently, these parthenogenetic lizards, which evolved from species having two sexes, still require certain sexual stimuli for maximum reproductive success.

Sexual reproduction presents a special problem for sessile or burrowing animals or for parasites, such as tapeworms, which may have difficulty encountering a member of the opposite sex. One solution to this problem is **hermaphroditism**, in which each individual has both male and female reproductive systems (the term is derived from Hermes and Aphrodite, names of a Greek god and goddess). Although some hermaphrodites fertilize themselves, most must mate with another member of the same species. When this occurs, each animal serves as both male and female, donating and receiving sperm, as we saw for earthworms. Each individual encountered is a potential mate, resulting in twice as many offspring than if only one individual's eggs were fertilized.

Another remarkable reproductive pattern is **sequential hermaphroditism**, in which an individual reverses its sex during its lifetime. In some species, the sequential hermaphrodite is **protogynous** (female first), while other species are **protandrous** (male first). In various species of reef fishes called wrasses, sex reversal is associated with age and size. For example, the Caribbean bluehead wrasse is a protogynous species in which only the largest (usually the oldest) individuals change from female to male (FIGURE 46.3). These fish live in harems consisting of a single male and several females. If the male dies or is removed in experiments, the largest female in the harem changes sex and becomes the new male. Within a week, the transformed individual is producing sperm instead of eggs. In this species, the male defends the harem against intruders, and thus larger size may give a greater reproductive advantage to males than it does to females. In contrast, there are protandrous animals that change from male to female when size increases. In such cases, greater size may increase the reproductive success of females more than it does males. For example, the production of huge numbers of gametes is an important asset for sedentary animals, such as oysters, that release their gametes into the surrounding water. Egg cells are

FIGURE 46.3 ▪ Sex reversal in a sequential hermaphrodite. In many species of reef fishes called wrasses, sex is not fixed but can change at some time in the animal's life. Sex reversal is often correlated with size. In this scene, a male Caribbean bluehead wrasse and two smaller females are feeding on a sea urchin. All wrasses of this species are born females, but the oldest, largest individuals change sex and complete their lives as males.

generally much larger than sperm cells, so females produce fewer gametes than males. Larger females tend to produce more eggs than smaller ones, and species of oysters that are sequential hermaphrodites are generally protandrous.

The diverse reproductive cycles and patterns we observe in the animal kingdom are adaptations that have evolved by natural selection. We will see many other examples as we survey the various mechanisms of sexual reproduction.

MECHANISMS OF SEXUAL REPRODUCTION

The mechanisms of **fertilization**, the union of sperm and egg, play an important part in sexual reproduction. Some species have **external fertilization**; eggs are shed by the female and fertilized by the male in the environment (FIGURE 46.4). Other species have **internal fertilization**; sperm are deposited in or near the female reproductive tract, and fertilization occurs within the tract.

FIGURE 46.4 ▪ The release of eggs and external fertilization. Many amphibians shed gametes into the environment, and fertilization occurs outside the female's body. In most species, behavioral adaptations ensure that a male is present when the female releases eggs. Here, a female frog, clasped by a male (on top), has just released a mass of eggs. The male released sperm (not shown) at the same time, and external fertilization has already occurred in the surrounding water.

Internal and external fertilization both depend on mechanisms ensuring that mature sperm encounter fertile eggs of the same species

Internal fertilization requires cooperative behavior, leading to copulation. In some cases, uncharacteristic sexual behavior is eliminated by natural selection in a direct manner; for example, female spiders will eat males if specific reproductive signals are not followed during mating. (We will discuss other examples of mating behavior in Chapter 51.) Internal fertilization also requires sophisticated reproductive systems, including copulatory organs that deliver sperm and receptacles for its storage and transport to ripe eggs.

Because external fertilization requires an environment where an egg can develop without desiccation or heat stress, it occurs almost exclusively in moist habitats. Many aquatic invertebrates simply shed their eggs and sperm into the surroundings, and fertilization occurs without the parents actually making physical contact. Timing is still crucial to ensure that mature sperm encounter ripe eggs.

Many fishes and amphibians that use external fertilization exhibit specific mating behaviors, resulting in one male fertilizing the eggs of one female. Courtship behavior is a mutual trigger for the release of gametes, with two effects: The probability of successful fertilization is increased, and the choice of mates may be somewhat selective. Environmental cues such as temperature or day length may cause all the individuals of a population to release gametes at once, or chemical signals from one individual releasing gametes may trigger gamete release in others.

Pheromones are chemical signals released by one organism that influence the behavior of other individuals of the same species. Pheromones are small, volatile or water-soluble molecules that can disperse easily into the environment and, like hormones, are active in minute amounts. Many pheromones function as mate attractants. The mate attractants of some female insects can be detected by males as much as a mile away. The pheromone of the female gypsy moth elicits behavioral responses in males at concentrations as low as 1 molecule of pheromone in 10^{17} molecules of other gases in the air. The role of pheromones in social behavior is considered in more detail in Chapter 51.

Species with internal fertilization usually produce fewer zygotes but provide more parental protection than species with external fertilization

All species produce more offspring than survive to reproduce. Species with external fertilization usually produce enormous numbers of zygotes, but the proportion that survive and develop further is often quite small. Internal fertilization usually produces fewer zygotes, but this may be offset by greater protection of the embryos and parental care of the young. Major types of protection include resistant eggshells, development of the embryo within the reproductive tract of the female parent, and parental care of the eggs and offspring.

Many species of terrestrial animals produce eggs that can withstand harsh environments. Birds, reptiles, and monotremes have amniote eggs with calcium and protein shells that resist water loss and physical damage. By contrast, the eggs of fishes and amphibians have only a gelatinous coat.

Rather than secreting a protective shell around the egg, many animals retain the embryo, which develops within the female reproductive tract. Among mammals, marsupials such as kangaroos and opossums retain their embryos for a short period in the uterus; the embryos then crawl out and complete fetal development attached to a mammary gland in the mother's pouch. The embryos of eutherian (placental) mammals develop entirely within the uterus, being nourished by the mother's blood supply through a special organ, the placenta (see Chapter 34).

When a kangaroo crawls out of its mother's pouch for the first time, or when a human is born, it still is not capable of independent existence. We are familiar with adult birds feeding their young and mammals nursing their offspring, but parental care is much more widespread than we might suspect, and it takes a variety of unusual forms. In one species of tropical frog, for instance, the male carries the tadpoles in his stomach until they metamorphose and hop out as young frogs. There are also many cases of parental care among invertebrates (FIGURE 46.5).

FIGURE 46.5 ▪ Parental care in an invertebrate. Compared to many insects and many other arthropods, giant water bugs (*Belostoma*) produce relatively few offspring, but parental protection enhances the survival of those offspring. Fertilization is internal, and the female glues her fertilized eggs to the back of the male (shown here). Whereas the males of most insect species provide no parental care for their offspring, the male giant water bug carries them for days, frequently fanning water over them, which helps keep the eggs moist, aerated, and free of parasites.

Complex reproductive systems have evolved in many animal phyla

To reproduce sexually, animals must have systems that produce and deliver gametes to the gametes of the opposite sex. These reproductive systems are varied. The least complex systems do not even contain distinct **gonads**, the organs that produce gametes in most animals. The most complex reproductive systems contain many sets of accessory tubes and glands that carry and protect the gametes and developing embryos. Many animals whose body plans are otherwise relatively simple possess highly complex reproductive systems. The reproductive systems of parasitic flatworms, for example, are among the most complex in the animal kingdom (FIGURE 46.6).

Among the least complex systems are those of polychaete worms (Phylum Annelida). Most polychaetes have separate sexes but do not have distinct gonads; rather, the eggs and sperm develop from undifferentiated cells lining the coelom. As the gametes mature, they are released from the body wall

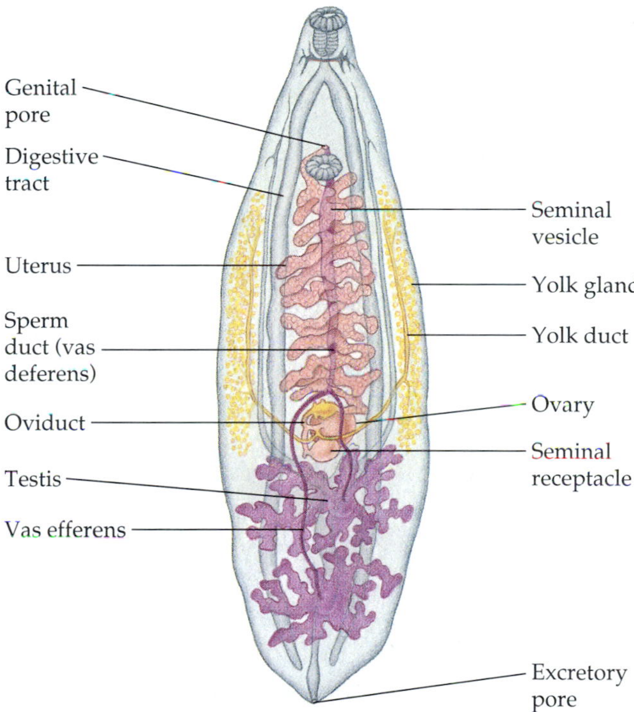

FIGURE 46.6 ▪ Reproductive anatomy of a parasitic flatworm. Most flatworms (Phylum Platyhelminthes) are hermaphroditic, with complex male and female reproductive systems. Both systems open to the outside through the genital pore. Sperm produced in the testes pass through a pair of ducts (vasa efferentia) into a single sperm duct (vas deferens) and are stored in the seminal vesicle. During copulation, sperm are ejaculated into the female system (usually of another individual) and then move through the uterus to the seminal receptacle. (In some flatworms, sperm are injected into the body tissues through the body wall and then migrate to the female reproductive tract.) Eggs from the ovary pass into the oviduct, where they are fertilized by sperm from the seminal receptacle and coated with yolk and tough shell material secreted by the yolk glands. From the oviduct, the fertilized, shelled eggs pass into the uterus from which they are shed through the genital pore. Usually only a small fraction of the eggs will develop into the next generation of adult worms.

and fill the coelom. Depending on the species, mature gametes may be shed through the excretory openings, or the swelling mass of eggs may split the body open, killing the parent and spilling the eggs into the environment.

Most insects have separate sexes with complex reproductive systems (FIGURE 46.7). In the male, sperm develop in a pair of testes and are conveyed along a coiled duct to two seminal vesicles, where they are stored. During mating, sperm are ejaculated into the female reproductive system. In the female, eggs develop in a pair of ovaries and are conveyed through ducts to the vagina, where fertilization occurs. In many species the female reproductive system includes a **spermatheca**, a sac in which sperm may be stored for a year or more.

The basic plan of all vertebrate reproductive systems is quite similar, but there are some important variations. In many nonmammalian vertebrates, the digestive, excretory, and reproductive systems have a common opening to the outside, the **cloaca**, which was probably present in the ancestors of all vertebrates. By contrast, most mammals lack a cloaca and have a separate opening for the digestive tract, and most female mammals have separate openings for the excretory and reproductive systems as well. The uterus of most vertebrates is partly or completely divided into two chambers. However, in humans and other mammals that produce only a few young at a time, as well as in birds and many snakes, the uterus is a single structure. Male reproductive systems differ mainly in the copulatory organs. Many nonmammalian vertebrates do not have a well-developed penis and simply evert the cloaca to ejaculate.

MAMMALIAN REPRODUCTION

Human reproduction involves intricate anatomy and complex behavior

Reproductive Anatomy of the Human Male

46.1 In most mammalian species, including humans, the male's external reproductive organs are the scrotum and penis. The internal reproductive organs consist of gonads that produce gametes (sperm cells) and hormones, accessory glands that secrete products essential to sperm movement, and a set of ducts that carry the sperm and glandular secretions (FIGURE 46.8).

The male gonads, or **testes** (singular, **testis**), consist of many highly coiled tubes surrounded by several layers of connective tissue. These tubes are the **seminiferous tubules**, where sperm form. The **Leydig cells** scattered between the seminiferous tubules produce testosterone and other androgens, the male sex hormones.

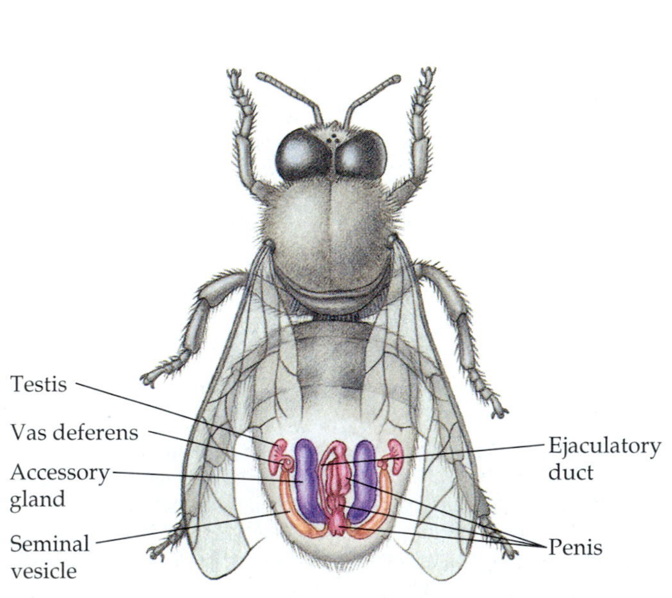

(a) Male honeybee

(b) Female honeybee

FIGURE 46.7 • Insect reproductive anatomy. (a) A male honeybee. Sperm form in the testes, pass through the sperm duct (vas deferens), and are stored in the seminal vesicle. The male ejaculates sperm along with fluid from the accessory glands. Males of some species of insects and other arthropods have appendages called claspers that grasp the female during copulation. **(b)** A female honeybee. Eggs develop in the ovaries, pass through the oviducts, and into the vagina. A pair of accessory glands (only one is shown) add protective secretions to the eggs in the vagina. After mating, sperm are stored in the spermatheca, a sac connected to the vagina by a short duct.

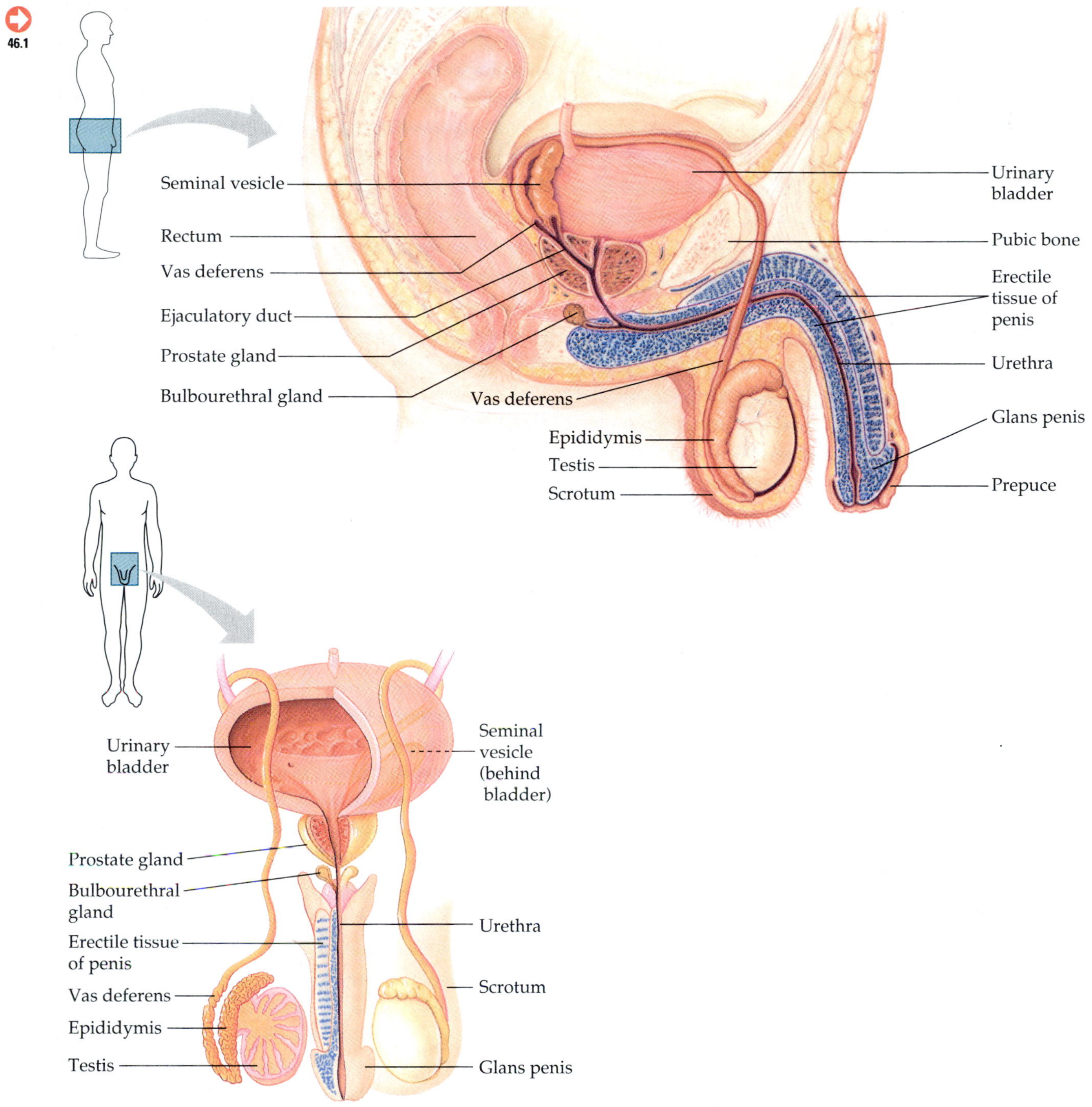

46.1

FIGURE 46.8 ▪ Reproductive anatomy of the human male.

Production of normal sperm cannot occur at the body temperatures of most mammals, and the testes of humans and many other mammals are held outside the abdominal cavity in the **scrotum**, which is a fold of the body wall. The temperature in a scrotum is about 2°C below that in the abdominal cavity. The testes develop high in the abdominal cavity and descend into the scrotum just before birth. In many rodents, the testes are drawn back into the abdominal cavity, and sperm maturation is interrupted between breeding seasons. Some mammals whose body temperature is low enough to allow sperm maturation, such as monotremes, whales, and elephants, retain the testes within the abdominal cavity permanently.

From the seminiferous tubules of a testis, the sperm pass into the coiled tubules of the **epididymis**. It takes about 20 days for sperm to pass through the 6-m-long tubules of each epididymis of a human male. During this passage, the sperm become motile and gain the ability to fertilize. During

ejaculation, the sperm are propelled from the epididymis through the muscular **vas deferens**. These two ducts (one from each epididymis) run from the scrotum around and behind the urinary bladder, where each joins a duct from the seminal vesicle, forming a short **ejaculatory duct**. The ejaculatory ducts open into the **urethra**, the tube that drains both the excretory system and the reproductive system. The urethra runs through the penis and opens to the outside at the tip of the penis.

Three sets of accessory glands (the seminal vesicles, prostate, and bulbourethral glands) add secretions to the **semen**, the fluid that is ejaculated. A pair of **seminal vesicles** contributes about 60% of the total volume of the semen. The fluid from the seminal vesicles is thick, yellowish, and alkaline. It contains mucus, fructose sugar (which provides most of the energy used by the sperm), a coagulating enzyme, ascorbic acid, and prostaglandins (see Chapter 45).

The **prostate gland** is the largest of the semen-secreting glands. It secretes its products directly into the urethra through several small ducts. Prostatic fluid is thin and milky, contains anticoagulant enzymes, citrate (a sperm nutrient), and is slightly acidic. The prostate gland is the source of some of the most common medical problems of men over age 40. Benign (noncancerous) enlargement of the prostate occurs in more than half of all men in this age group and in virtually all men over 70. Prostate cancer is one of the most common cancers in men. It is treated surgically or with drugs that inhibit gonadotropins, resulting in reduced prostate activity and size.

The **bulbourethral glands** are a pair of small glands along the urethra below the prostate. Before ejaculation they secrete a clear mucus that neutralizes any acidic urine remaining in the urethra. Bulbourethral fluid also carries some sperm released before ejaculation, which is one reason for the high failure rate of the withdrawal method of birth control.

A man usually ejaculates about 2 to 5 mL of semen, and each milliliter may contain about 50 to 130 million sperm. Once in the female reproductive tract, prostaglandins in the semen thin the mucus at the opening of the uterus and stimulate contractions of the uterine muscles, which help move the semen up the uterus. The semen is slightly alkaline, and this helps neutralize the acidic environment of the vagina, protecting the sperm and increasing their motility. When first ejaculated, the semen coagulates, making it easier for uterine contractions to move it along; then anticoagulants liquify the semen, and the sperm begin swimming through the female tract.

The human **penis** is composed of three cylinders of spongy erectile tissue derived from modified veins and capillaries. During sexual arousal, the erectile tissue fills with blood from the arteries. As this tissue fills, the increasing pressure seals off the veins that drain the penis, causing it to engorge with blood. The resulting erection is essential to insertion of the penis into the vagina. Rodents, raccoons, walruses, and several other mammals also possess a **baculum**, a bone that is contained in, and helps stiffen, the penis. Temporary impotence, a reversible inability to achieve an erection, can result from alcohol consumption, certain drugs, and emotional problems. Several drugs and penile implant devices are available for men with nonreversible impotence due to nervous system or circulatory problems. A new oral drug called Viagra® promotes the action of the local regulator nitric oxide (NO), enhancing relaxation of smooth muscles in the blood vessels of the penis. This allows blood to enter the erectile tissue and sustain an erection.

The main shaft of the penis is covered by relatively thick skin. The head, or **glans penis**, has a much thinner covering and is consequently more sensitive to stimulation. The human glans is covered by a fold of skin called the foreskin, or **prepuce**, which may be removed by circumcision. Circumcision, which arose from religious traditions, appears to have no verifiable basis in health or hygiene.

Reproductive Anatomy of the Human Female

The female's external reproductive structures are the clitoris
46.2 and two sets of labia surrounding the clitoris and vaginal opening. The internal reproductive organs consist of a pair of gonads and a system of ducts and chambers to conduct the gametes and house the embryo and fetus (FIGURE 46.9).

The female gonads, the **ovaries**, lie in the abdominal cavity, flanking, and attached by a mesentery to, the uterus. Each ovary is enclosed in a tough protective capsule and contains many follicles. A **follicle** consists of one egg cell surrounded by one or more layers of follicle cells, which nourish and protect the developing egg cell. All of the 400,000 follicles a woman will ever have are formed before her birth. Of these, only several hundred will release egg cells during the woman's reproductive years. Starting at puberty and continuing until menopause, usually one follicle matures and releases its egg cell during each menstrual cycle. The cells of the follicle also produce the primary female sex hormones, the estrogens. The egg cell is expelled from the follicle in the process of **ovulation** (FIGURE 46.10). The remaining follicular tissue then grows within the ovary to form a solid mass called the **corpus luteum**. The corpus luteum secretes additional estrogens and progesterone, the hormone that maintains the uterine lining during pregnancy. If the egg cell is not fertilized, the corpus luteum disintegrates, and a new follicle matures during the next cycle.

The female reproductive system is not completely closed, and the egg cell is released into the abdominal cavity near the opening of the **oviduct**, or fallopian tube. The oviduct has a funnel-like opening, and cilia on the inner epithelium lining the duct help collect the egg cell by drawing fluid from the body cavity into the duct. The cilia also convey the egg cell

46.2

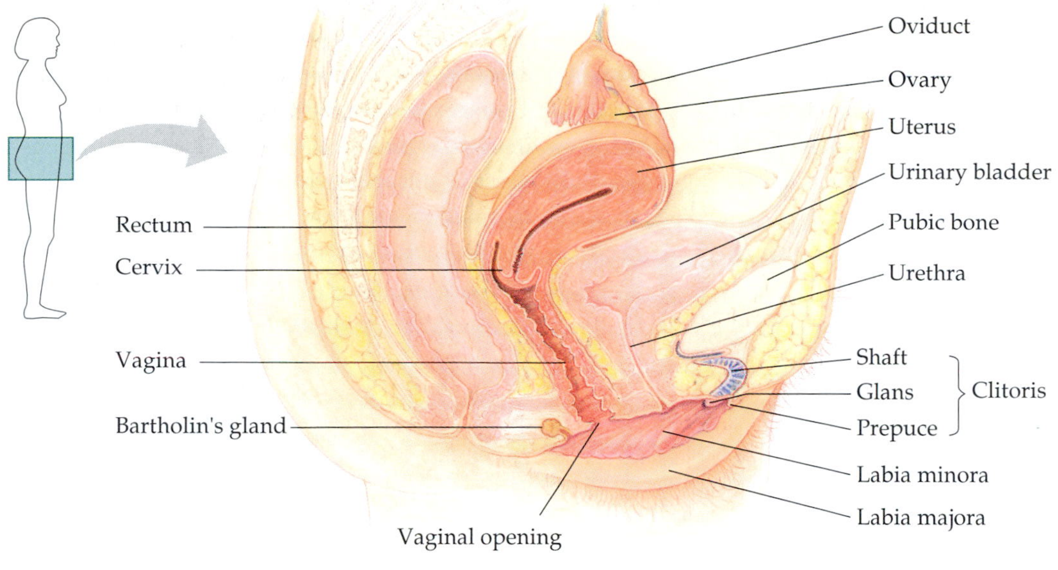

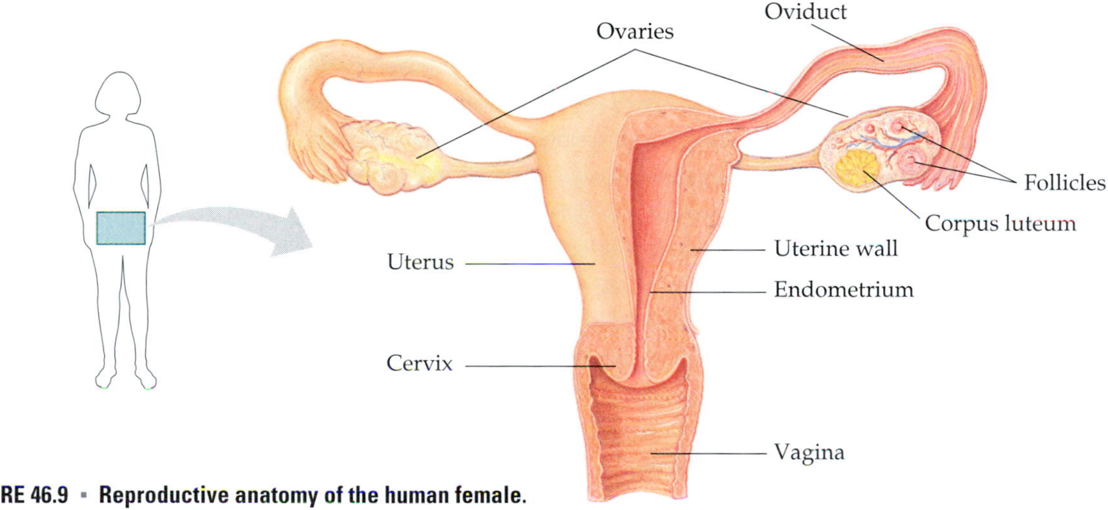

FIGURE 46.9 ▪ Reproductive anatomy of the human female.

down the duct to the **uterus**, also known as the womb. The uterus is a thick, muscular organ that can expand during pregnancy to accommodate a 4-kg fetus. The inner lining of the uterus, the **endometrium**, is richly supplied with blood vessels.

The neck of the uterus is the **cervix**, which opens into the vagina. The **vagina** is a thin-walled chamber that forms the birth canal through which the baby is born; it is also the repository for sperm during copulation.

At birth, and usually until sexual intercourse or vigorous physical activity ruptures it, a vascularized membrane called the **hymen** partly covers the vaginal opening in humans. The vaginal opening and the separate urethral opening are located within a region called the **vestibule**, bordered by a pair of slender skin folds, the **labia minora**. A pair of thick, fatty ridges, the **labia majora**, encloses and protects the labia minora and vestibule. At the front edge of the vestibule, the

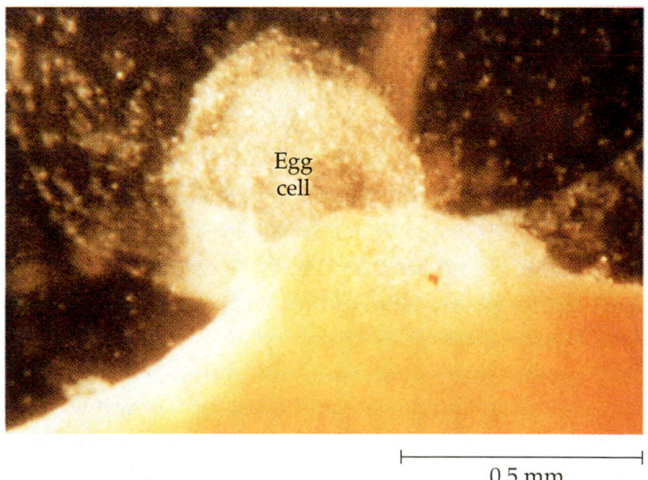

FIGURE 46.10 ▪ Ovulation. An egg cell is released from a follicle at the surface of the ovary. The orangish mass below the ejected egg cell is part of a mammalian ovary.

clitoris consists of a short shaft supporting a rounded glans, or head, covered by a small hood of skin, the prepuce. During sexual arousal, the clitoris, vagina, and labia minora all engorge with blood and enlarge. The clitoris consists largely of erectile tissue. Richly supplied with nerve endings, it is one of the most sensitive points of sexual stimulation. During sexual arousal, **Bartholin's glands**, located near the vaginal opening, secrete mucus into the vestibule, keeping it lubricated and facilitating intercourse.

Mammary glands are present in both sexes but normally function only in women. They are not part of the reproductive system but are important to mammalian reproduction. Within the glands, small sacs of epithelial tissue secrete milk, which drains into a series of ducts opening at the nipple. Fatty (adipose) tissue forms the main mass of the mammary gland of a nonlactating mammal. The low level of estrogen in males prevents the development of both the secretory apparatus and the fat deposits, so male breasts remain small, and the nipples are not connected to the ducts.

Human Sexual Response

Many vertebrates and invertebrates have elaborate and complex mating behavior (see Chapter 51). Human sexuality is characterized by a diversity of stimuli and responses, with a common, underlying physiological pattern.

Two types of physiological reactions predominate in both sexes: **vasocongestion**, the filling of a tissue with blood caused by increased blood flow through the arteries of that tissue, and **myotonia**, increased muscle tension. Both skeletal and smooth muscle may show sustained or rhythmic contractions, including those associated with orgasm.

The sexual response cycle can be divided into four phases: excitement, plateau, orgasm, and resolution. An important function of the excitement phase is preparation of the vagina and penis for **coitus** (sexual intercourse). During this phase, vasocongestion is particularly evident in erection of the penis and clitoris; enlargement of the testes, labia, and breasts; and vaginal lubrication. Myotonia may occur, resulting in nipple erection or tension of the arms and legs.

The plateau phase continues these responses. In females, the outer third of the vagina becomes vasocongested, while the inner two-thirds becomes slightly expanded. This change, coupled with the elevation of the uterus, forms a depression that receives sperm at the back of the vagina. Breathing increases and heart rate rises, sometimes to 150 beats per minute—not in response to the physical effort of sexual activity, but as an involuntary response to stimulation of the autonomic nervous system (see Chapter 48).

Orgasm is characterized by rhythmic, involuntary contractions of the reproductive structures in both sexes. Male orgasm has two stages. Emission is the contraction of the glands and ducts of the reproductive tract, which forces semen into the urethra. Expulsion, or ejaculation, occurs when the urethra contracts and the semen is expelled. During female orgasm, the uterus and outer vagina contract, but the inner two-thirds of the vagina do not. Orgasm is the shortest phase of the sexual response cycle, usually lasting only a few seconds. In members of both sexes, contractions occur at about 0.8-sec intervals and may involve the anal sphincter and several abdominal muscles.

The resolution phase completes the cycle and reverses the responses of the earlier stages. Vasocongested organs return to their normal size and color, and muscles relax. Most of the changes during resolution are completed in 5 minutes. Loss of penile and clitoral erection, however, may take longer. An initial loss of erection is rapid in both sexes, but a return of the organs to their nonaroused size may take as long as an hour.

Spermatogenesis and oogenesis both involve meiosis but differ in three significant ways

Spermatogenesis, the production of mature sperm cells, is a continuous and prolific process in the adult male. Each ejaculation of a human male contains 100 to 650 million sperm cells, and males can ejaculate daily with little loss of fertilizing capacity.

The structure of a sperm cell fits its function (FIGURE 46.11). In most species, a head containing the haploid nucleus is tipped with a special body, the **acrosome**, which contains enzymes that help the sperm penetrate the egg. Behind the head, the sperm cell contains large numbers of mitochondria (or a single large one, in some species) that provide ATP for movement of the tail, which is a flagellum. Mammalian sperm shape varies from species to species, with the head a slender comma shape, an oval form (as in the human sperm), or nearly spherical. Spermatogenesis occurs in the seminiferous tubules of the testes (FIGURE 46.12).

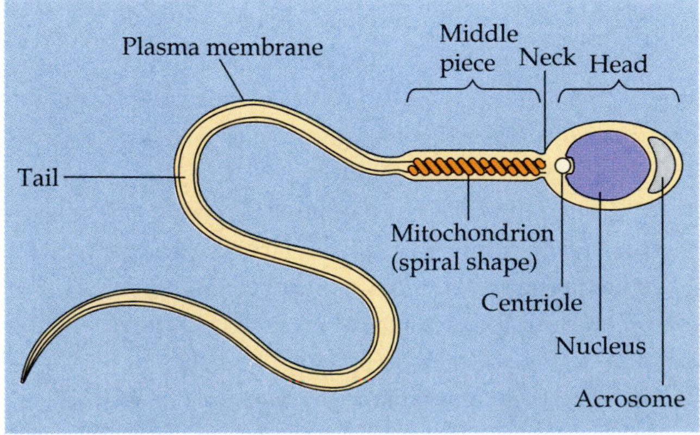

FIGURE 46.11 • Structure of a human sperm cell.

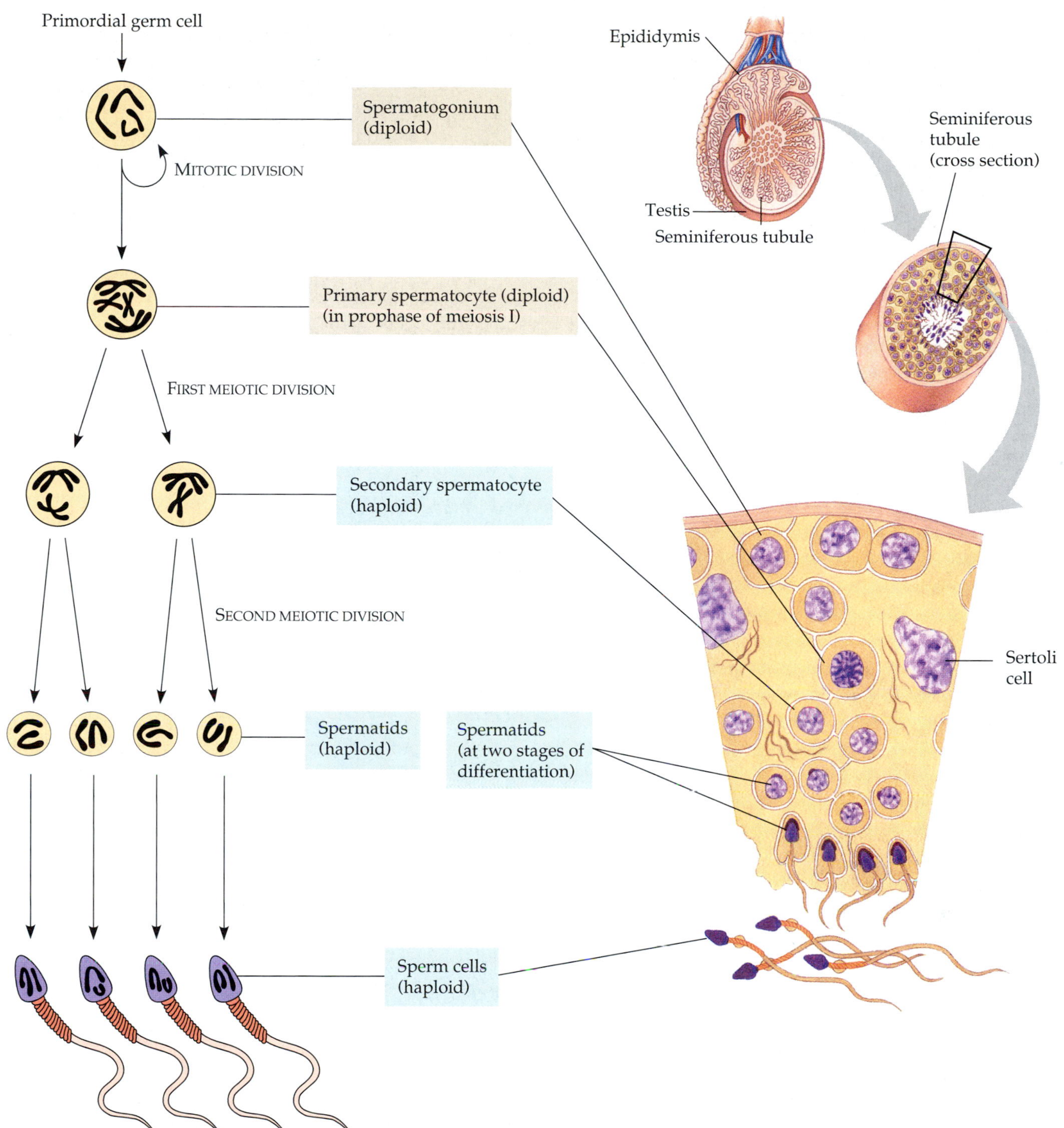

FIGURE 46.12 ▪ Spermatogenesis. These drawings correlate the meiotic stages in sperm development (left) with the structure of seminiferous tubules. Primordial germ cells of the embryonic testes differentiate into spermatogonia, the diploid cells that are the precursors of sperm. Located near the outer wall of the seminiferous tubules, spermatogonia undergo repeated mitoses, which produce large populations of potential sperm. In a mature male, about 3 million spermatogonia per day differentiate into primary spermatocytes. The chromosome number is reduced by half as the primary spermatocytes undergo the first division of meiosis. In this simplified diagram the diploid number ($2n$) is only 4; the actual $2n$ number in humans is 46. Notice that the secondary spermatocytes each have only two chromosomes (the haploid number), and these chromosomes are still duplicated, each consisting of two identical chromatids. The second meiotic division produces four spermatids, each with two single chromosomes. Spermatids then differentiate into mature spermatozoa, or sperm cells. This involves association of the developing sperm with large Sertoli cells, which transfer nutrients to the spermatids. During spermatogenesis, the developing sperm are gradually pushed toward the center of the seminiferous tubule and make their way to the epididymis, where they acquire motility. This process, from spermatogonia to motile sperm, takes 65 to 75 days in the human male.

Oogenesis is the development of ova (mature, unfertilized egg cells) (FIGURE 46.13). Between birth and puberty, the egg cells (primary oocytes) enlarge, and the follicles around them grow. The primary oocytes replicate their DNA and enter prophase I of meiosis but do not change again unless reactivated by hormones. Beginning at puberty, FSH (follicle-stimulating hormone) periodically stimulates a follicle to begin growing again and induces its primary oocyte to complete the first meiotic division. Meiosis then stops again; the secondary oocyte, released during ovulation, does not undergo the second meiotic division right away. In humans, penetration of the egg cell by the sperm triggers the second meiotic division, and only then is oogenesis actually complete.

Oogenesis differs from spermatogenesis in three important ways. First, during the meiotic divisions of oogenesis, cytokinesis is unequal, with almost all the cytoplasm monopolized by a single daughter cell, the secondary oocyte. This large cell can go on to form the ovum; the other products of meiosis, smaller cells called polar bodies, degenerate. This contrasts with spermatogenesis, when all four products of meiosis I and II develop into mature sperm (compare FIGURES 46.12 and 46.13). Second, while the cells from which sperm develop continue to divide by mitosis throughout the male's life, this is not the case for oogenesis in the female. At birth, an ovary already contains all the cells it will ever have that will develop into eggs. Third, oogenesis has long "resting" periods, in con-

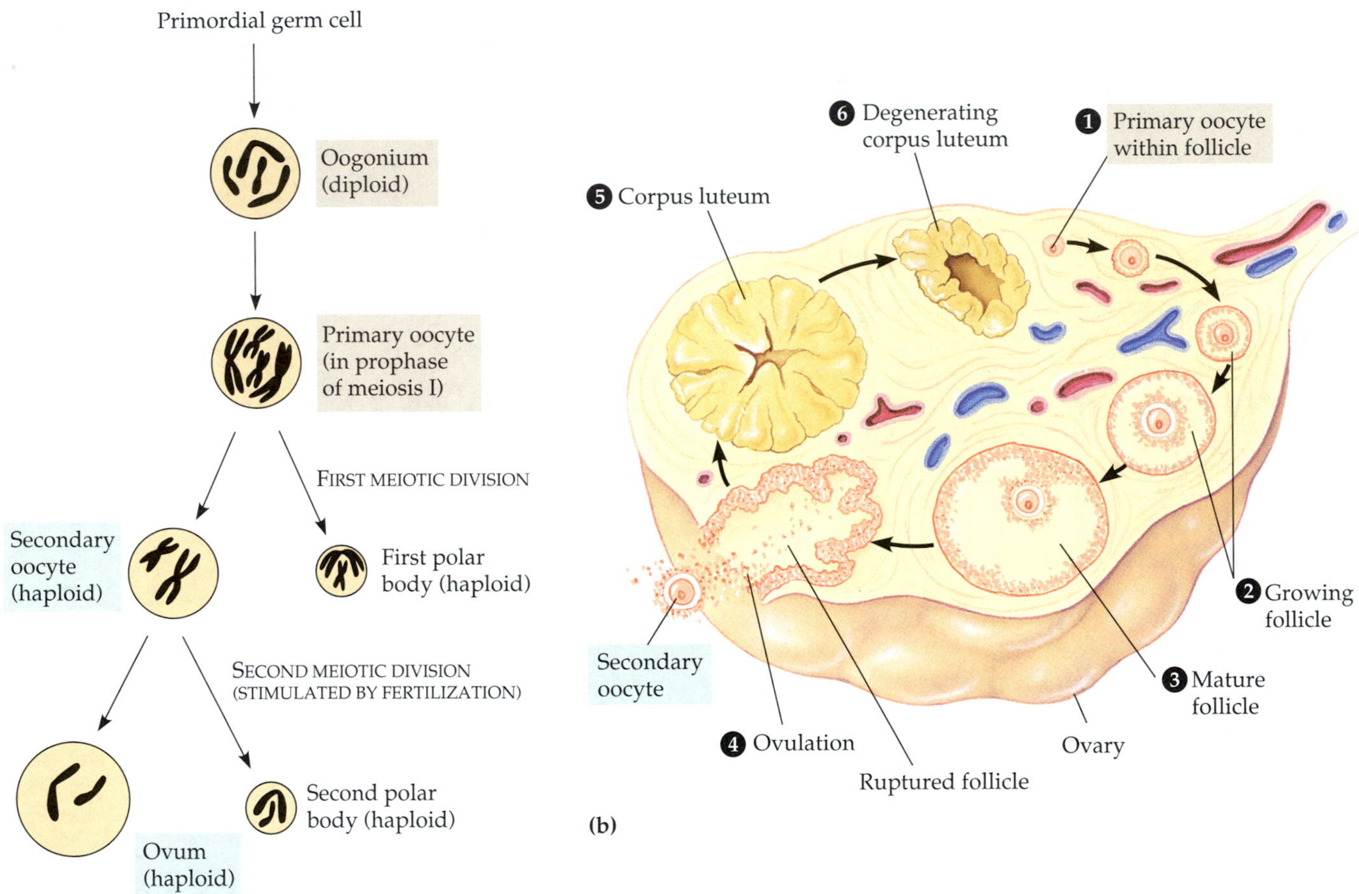

FIGURE 46.13 ▪ **Oogenesis.** **(a)** The production of ova begins with mitosis of the primordial germ cells in the embryo, producing diploid oogonia ($2n = 4$, in this simplified diagram). Each oogonium develops into a primary oocyte, which is also diploid. Starting at puberty, a single primary oocyte usually completes meiosis I each month. The meiotic divisions in oogenesis involve unequal cytokinesis. The first meiotic division produces a large cell, the secondary oocyte, and a much smaller polar body. The second meiotic division, which produces the ovum and another small polar body, occurs only if a sperm cell penetrates the secondary oocyte. After meiosis is completed and the second polar body separates from the ovum, the haploid nuclei of the sperm and the mature ovum fuse in the actual process of fertilization. **(b)** This cutaway view of an ovary illustrates the developmental stages of an ovarian follicle that accompany oogenesis. ① Each primary oocyte develops within a follicle. ② In response to FSH, several follicles grow, but ③ usually only one matures. ④ In the process known as ovulation, the follicle ruptures, releasing a secondary oocyte. ⑤ The remaining follicular tissue develops into the corpus luteum, which ⑥ disintegrates if fertilization does not occur. For convenience, the stages are presented as a cycle (arrows), although they occur at different times and are never actually present simultaneously within the ovary. In a real ovary, each follicle stays in one place as it goes through the sequential stages.

trast to spermatogenesis, which produces mature sperm from precursor cells in an uninterrupted sequence.

A complex interplay of hormones regulates reproduction

The Male Pattern

46.1 In the male, the principal sex hormones are the androgens, of which testosterone is the most important. Androgens, steroid hormones produced mainly by the Leydig cells of the testes, are directly responsible for the primary and secondary sex characteristics of the male. Primary sex characteristics are those associated with the reproductive system: development of the vasa deferentia and other ducts, development of the external reproductive structures, and sperm production. Secondary sex characteristics are features that are not directly related to the reproductive system, including deepening of the voice, distribution of facial and pubic hair, and muscle growth (androgens stimulate protein synthesis). Androgens are also potent determinants of behavior in mammals and other vertebrates. In addition to specific sexual behaviors and sex drive, androgens increase general aggressiveness and are responsible for such actions as singing in birds and calling by frogs. Hormones from the anterior pituitary and hypothalamus control both androgen secretion and sperm production by the testes (FIGURE 46.14).

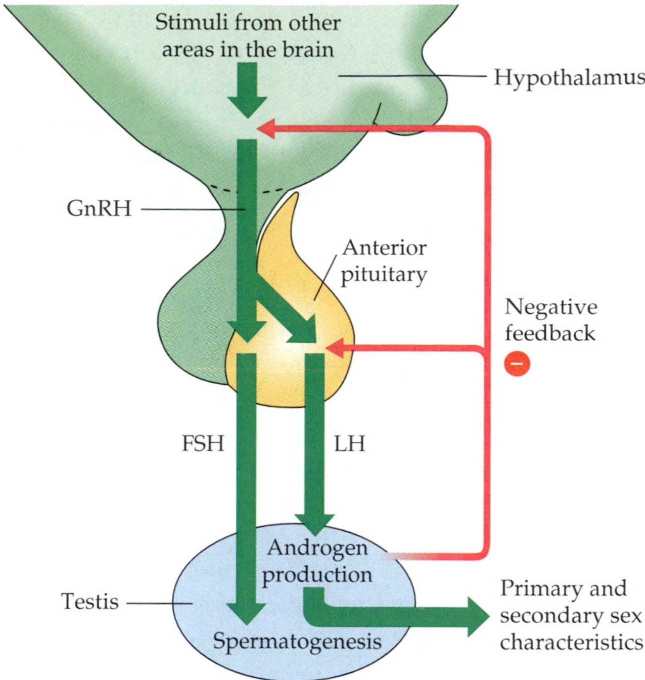

46.1 **FIGURE 46.14 ▪ Hormonal control of the testes.** The pituitary secretes two gonadotropic hormones with different effects on the testes. Luteinizing hormone (LH) stimulates androgen production by the Leydig cells. Follicle-stimulating hormone (FSH) acts on the seminiferous tubules to increase spermatogenesis. Since androgens are also required for sperm production, LH stimulates spermatogenesis indirectly. LH and FSH are in turn regulated by a single hormone from the hypothalamus, gonadotropin-releasing hormone (GnRH). How GnRH controls the release of two different hormones at different times is still unknown. LH, FSH, and GnRH concentrations in the blood are regulated by negative feedback by androgens. GnRH is also controlled by negative feedback from LH and FSH (not shown). In human males, these feedback loops keep the hormones at relatively constant levels, but in many other mammalian species, seasonal cycles in hormone concentration regulate breeding patterns.

The Female Pattern

46.2 In the female, the pattern of hormone secretion and the reproductive events they regulate are cyclic, very different from the male pattern. Whereas males produce sperm continuously, females release only one egg or a few eggs at one time during each cycle. Control of the female cycle is quite complex.

Two different types of cycles occur in female mammals. Humans and many other primates have **menstrual cycles**, whereas other mammals have **estrous cycles**. In both cases, ovulation occurs at a time in the cycle after the endometrium has started to thicken and develop a rich blood supply, which prepares the uterus for the possible implantation of an embryo. One difference between the two types of cycles involves the fate of the uterine lining if pregnancy does not occur. In menstrual cycles, the endometrium is shed from the uterus through the cervix and vagina in a bleeding called **menstruation**. In estrous cycles, the endometrium is reabsorbed by the uterus, and no extensive bleeding occurs.

Other major distinctions include more pronounced behavioral changes during estrous cycles than during menstrual cycles and stronger effects of season and climate on estrous cycles. Whereas human females may be receptive to sexual activity throughout their cycles, most mammals will copulate only during the period surrounding ovulation. This period of sexual activity, called **estrus** (L. *oestrus,* "frenzy," "passion"), is the only time vaginal changes permit mating. Estrus is sometimes called heat, and indeed the female's body temperature increases slightly. The length and frequency of reproductive cycles vary widely among mammals. The human menstrual cycle averages 28 days; the estrous cycle of the rat is only 5 days. Bears and dogs have one cycle per year, but elephants cycle several times per year.

Let's examine the reproductive cycle of the human female in more detail as a case study of how a complex function is coordinated by hormones. Only about 30% of women have cycle lengths within a day or two of the statistical average of 28 days. Cycles vary from one woman to another, ranging from about 20 to 40 days. In some women the cycles are usually very regular, but in other individuals the timing varies from cycle to cycle.

The term *menstrual cycle* refers specifically to the changes that occur in the uterus (FIGURE 46.15). By convention, the first day of a woman's menstrual period, the first day of menstruation, is designated day 1 of the cycle. The **menstrual flow phase** of the cycle, during which menstrual bleeding (loss of most of the functional layer of the endometrium) occurs, usually persists for a few days (FIGURE 46.15d). Then the thin remaining endometrium begins to regenerate and thicken for a week or two, during what is called the **proliferative phase** of the menstrual cycle. During the next phase, the **secretory phase**, usually about two weeks in duration, the endometrium continues to thicken, becomes more vascularized, and develops glands that secrete a fluid rich in glycogen. If an embryo has not implanted in the uterine lining by the end of the secretory phase, a new menstrual flow commences, marking day 1 of the next cycle.

Paralleling the menstrual cycle is an **ovarian cycle** (FIGURE 46.15c). It begins with the **follicular phase**, during which several follicles in the ovary begin to grow. The egg cell enlarges, and the coat of follicle cells becomes multilayered. Of the several follicles that start to grow, only one usually continues to enlarge and mature while the others disintegrate. The maturing follicle develops an internal fluid-filled cavity and grows very large, forming a bulge near the surface of the ovary. The follicular phase ends with **ovulation**, when the follicle and adjacent wall of the ovary rupture, releasing the oocyte. The follicular tissue that remains in the ovary after ovulation develops into the corpus luteum, an endocrine tissue that secretes female hormones during the **luteal phase** of the ovarian cycle. The next cycle begins with a new growth of follicles.

Hormones coordinate the menstrual and ovarian cycles in such a way that growth of the follicle and ovulation are synchronized with preparation of the uterine lining for possible implantation of an embryo. Five hormones participate in an elaborate scheme involving both positive and negative feedback. These hormones are gonadotropin-releasing hormone (GnRH), secreted by the hypothalamus; follicle-stimulating hormone (FSH) and luteinizing hormone (LH), the two gonadotropins secreted by the anterior pituitary; and estrogens (a family of closely related hormones) and progesterone, the female sex hormones secreted by the ovaries. The relative levels of the pituitary and ovarian hormones in blood plasma

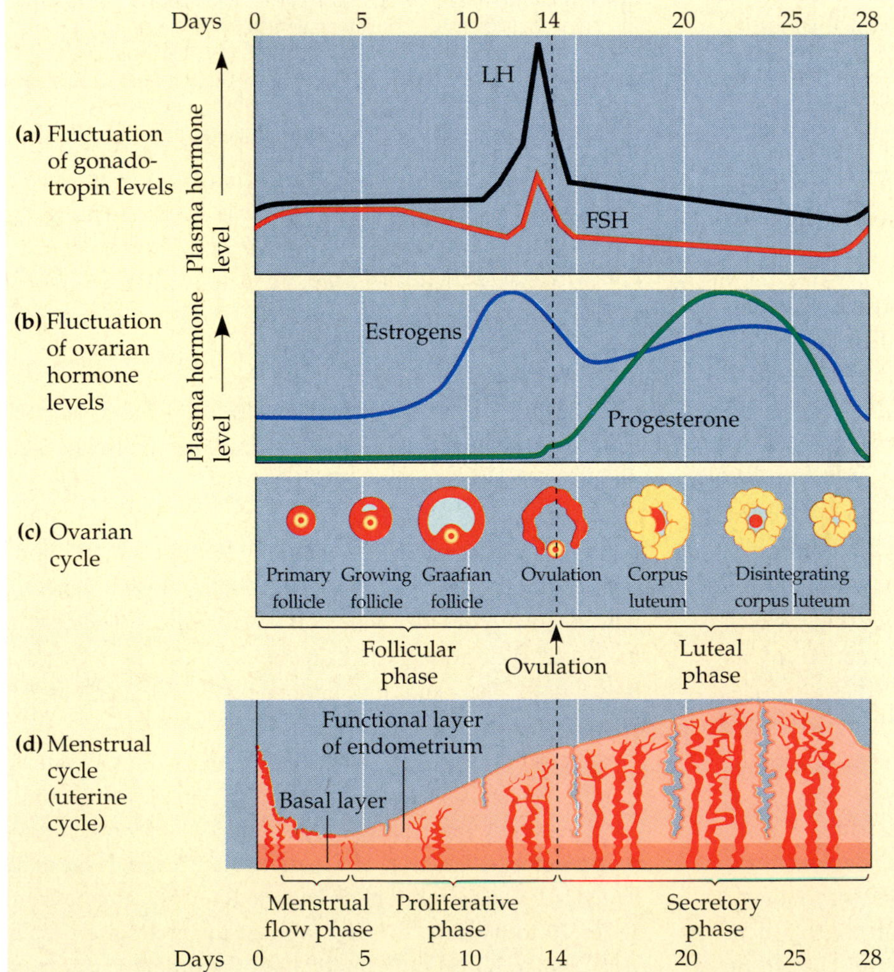

FIGURE 46.15 • The reproductive cycle of the human female. Hormones coordinate the ovarian and menstrual cycles, preparing the uterine lining (endometrium) for implantation of an embryo even before ovulation. **(a)** Changes in LH and FSH levels. **(b)** Changes in the levels of estrogens and progesterone. **(c)** The ovarian cycle consists of a follicular phase, during which follicles grow and secrete increasing amounts of estrogens; ovulation; and a luteal phase, during which the corpus luteum secretes estrogens and progesterone. Length of the follicular phase varies; the luteal phase usually lasts 13 to 15 days. **(d)** The menstrual cycle consists of a menstrual flow phase, a proliferative phase, and a secretory phase. Menstruation, the shedding of the endometrium, occurs during the menstrual flow phase. The first day of flow marks day 1 of the menstrual cycle. During the proliferative phase, estrogens from the growing follicle stimulate the endometrium to thicken and become increasingly vascularized. During the secretory phase, the endometrium continues to thicken, its arteries enlarge, and endometrial glands grow. These endometrial changes require estrogens and progesterone, secreted by the corpus luteum after ovulation. Thus, the secretory phase of the menstrual cycle parallels the luteal phase of the ovarian cycle. Disintegration of the corpus luteum at the end of the luteal phase reduces the amount of estrogens and progesterone available to the endometrium, so it is shed. In the event of pregnancy, additional mechanisms maintain high levels of estrogens and progesterone, preventing loss of the endometrium.

46.2

are graphed in FIGURE 46.15a and b, along with the ovarian and menstrual cycles. As you read the following discussion, refer to the figure as a guide to understanding how the hormones regulate the female reproductive system.

During the follicular phase of the ovarian cycle, the pituitary secretes small amounts of FSH and LH in response to stimulation by GnRH from the hypothalamus. At this time, the cells of immature ovarian follicles have receptors for FSH but not for LH. The FSH stimulates follicle growth, and the cells of these growing follicles secrete estrogens. Notice in FIGURE 46.15b that there is a slow rise in the amount of estrogens secreted during most of the follicular phase. This small increase in estrogens inhibits secretion of the pituitary hormones, keeping the levels of FSH and LH relatively low during most of the follicular phase. These hormonal relationships change radically and rather abruptly when the rate of secretion of estrogens by the growing follicle begins to rise steeply. Whereas a slow rise of estrogens inhibits the secretion of pituitary gonadotropins, a high concentration of estrogens has the opposite effect and *stimulates* the secretion of gonadotropins by acting on the hypothalamus to increase its output of GnRH. You can see this response in FIGURE 46.15a as a steep increase of FSH and LH levels that occurs soon after the increase in the concentration of estrogens. The effect is greater for LH because the high concentration of estrogens, in addition to stimulating GnRH secretion, also increases the sensitivity of LH-releasing mechanisms in the pituitary to the hypothalamic signal (GnRH). By now, the follicles have receptors for LH and can respond to this hormonal cue. In an example of positive feedback, the increase in LH concentration caused by increased secretion of estrogens from the growing follicle induces final maturation of the follicle, and ovulation occurs about a day after the LH surge.

Following ovulation, LH stimulates the transformation of the follicular tissue left behind in the ovary to form the corpus luteum, a glandular structure. (It is for this "luteinizing" function that LH is named.) Under continued stimulation by LH during the luteal phase of the ovarian cycle, the corpus luteum secretes estrogens and a second steroid hormone, progesterone. The corpus luteum usually reaches its maximum development about 8 to 10 days after ovulation. As the levels of progesterone and estrogens rise, the combination of these hormones exerts negative feedback on the hypothalamus and pituitary, inhibiting the secretion of LH and FSH. Near the end of the luteal phase, the corpus luteum disintegrates (possibly a result of prostaglandins secreted by its own cells). Consequently, concentrations of estrogens and progesterone decline sharply. The dropping levels of ovarian hormones liberate the hypothalamus and pituitary from the inhibitory effects of these hormones. The pituitary then begins to secrete enough FSH to stimulate the growth of new follicles in the ovary, initiating the follicular phase of the next ovarian cycle.

How is the ovarian cycle synchronized with the menstrual cycle? Estrogens, secreted in increasing amounts by growing follicles, are a hormonal signal to the uterus, causing the endometrium to thicken. Thus, the follicular phase of the ovarian cycle is coordinated with the proliferative phase of the menstrual cycle. *Before* ovulation, the uterus is already being prepared for a possible embryo. *After* ovulation, estrogens and progesterone secreted by the corpus luteum stimulate continued development and maintenance of the endometrium, including an enlargement of arteries supplying blood to the uterine lining and the growth of endometrial glands that secrete a nutrient fluid that can sustain an early embryo before it actually implants in the uterine lining. Thus, the luteal phase of the ovarian cycle is coordinated with the secretory phase of the menstrual cycle. The rapid drop in the level of ovarian hormones when the corpus luteum disintegrates causes spasms of arteries in the uterine lining that deprive the endometrium of blood. Disintegration of the endometrium results in menstruation and the beginning of a new menstrual cycle. In the meantime, ovarian follicles that will stimulate renewed thickening of the endometrium are just beginning to grow. Cycle after cycle, the maturation and release of egg cells from the ovary is integrated with changes in the uterus, the organ that must accommodate an embryo if the egg cell is fertilized. In the absence of pregnancy, a new cycle begins. We will soon see that there are "override" mechanisms that prevent disintegration of the endometrium in the event of pregnancy.

In addition to their role in coordinating reproductive cycles, estrogens are also responsible for the secondary sex characteristics of the female. The hormones induce deposition of fat in the breasts and hips, increase water retention, affect calcium metabolism, stimulate breast development, and mediate female sexual behavior.

Menopause. On average, human females undergo **menopause**, the cessation of ovulation and menstruation, between the ages of 46 and 54. Apparently, during these years the ovaries lose their responsiveness to gonadotropins (the hormones FSH and LH) from the pituitary, and menopause results from a decline in production of estrogens by the ovary. Menopause is an unusual phenomenon; in most species, females as well as males retain their reproductive capacity throughout life. Is there an evolutionary explanation for menopause? Why might natural selection have favored females who stopped reproducing? One intriguing (and highly controversial) hypothesis proposes that during early human evolution, undergoing menopause after having some children actually increased a woman's fitness; losing the ability to reproduce allowed her to provide better care for her children and grandchildren, thereby increasing the survival of individuals bearing her genes.

Embryonic and fetal development occur during pregnancy in humans and other eutherian (placental) mammals

From Conception to Birth

In placental mammals, **pregnancy**, or **gestation**, is the condition of carrying one or more **embryos**, new developing individuals, in the uterus. Pregnancy is preceded by **conception**, the fertilization of the egg by a sperm cell, and continues until the birth of the offspring. Human pregnancy averages 266 days (38 weeks) from conception, or 40 weeks from the start of the last menstrual cycle. Duration of pregnancy in other species correlates with body size and the extent of development of the young at birth. Many rodents (mice and rats) have gestation periods of about 21 days, whereas those of dogs are closer to 60 days. In cows, gestation averages 270 days (almost the same as humans); in giraffes, it is about 420 days; and in elephants, gestation is more than 600 days.

Human gestation can be divided for convenience of study into three **trimesters** of about 3 months each. The first trimester is the time of most radical change for both the mother and the baby. Fertilization occurs in the oviduct (FIGURE 46.16). About 24 hours later, the resulting zygote begins dividing, a process called **cleavage**. Cleavage continues, with the embryo becoming a ball of cells by the time it reaches the uterus about 3 to 4 days after fertilization. By about 1 week after fertilization, cleavage has produced an embryonic stage called the **blastocyst**, a sphere of cells containing a flattened cavity. In a process that takes about 5 more days, the blastocyst implants into the endometrium. Differentiation of body structures now begins in earnest. (Embryonic development will be described in detail in Chapter 47.) During implantation, the blastocyst becomes embedded in the endometrium, which responds by growing over the blastocyst. The embryo obtains nutrients directly from the endometrium during the first 2 to 4 weeks of development. Meanwhile, tissues grow out from the developing embryo and mingle with the endometrium to form the **placenta**. This disk-shaped organ, containing embryonic and maternal blood vessels, grows to about the size of a dinner plate and weighs somewhat less than 1 kg. Diffusion of material between maternal and embryonic circulations provides nutrients, exchanges respiratory gases, and disposes of metabolic wastes for the embryo. Blood from the embryo travels to the placenta through arteries of the umbilical cord and returns via the umbilical vein, passing through the liver of the embryo (FIGURE 46.17).

The first trimester is also the main period of **organogenesis**, the development of the body organs (FIGURE 46.18). The heart begins beating by the fourth week and can be detected with a stethoscope by the end of the first trimester. By the end of the eighth week, all the major structures of the adult are present in rudimentary form. At this point, the embryo is called a **fetus**. Although well differentiated, the fetus is only 5 cm long by the end of the first trimester. Because of its rapid organogenesis, the embryo is most sensitive during the first trimester to such threats as radiation and drugs that can cause birth defects.

The first trimester is also a time of rapid change for the mother. The embryo secretes hormones that signal its presence and control the mother's reproductive system. One embryonic hormone, **human chorionic gonadotropin (HCG)**, acts like pituitary LH to maintain secretion of progesterone and estrogens by the corpus luteum through the first trimester. In the absence of this hormonal override, the decline in maternal LH due to inhibition of the pituitary by progesterone would result in menstruation and spontaneous abortion of the embryo. Levels of HCG in the maternal blood are so high that some is excreted in the urine, where it can be detected in pregnancy tests (see the Methods Box on p. 856). High levels of progesterone initiate changes in the pregnant woman's reproductive system, including increased mucus in the cervix that forms a protective plug, growth of the maternal part of the placenta, enlargement of the uterus, and (by negative feedback on the

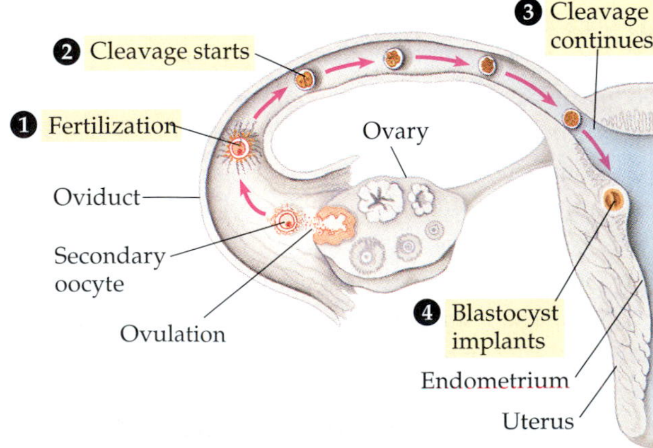

(a) From ovulation to implantation

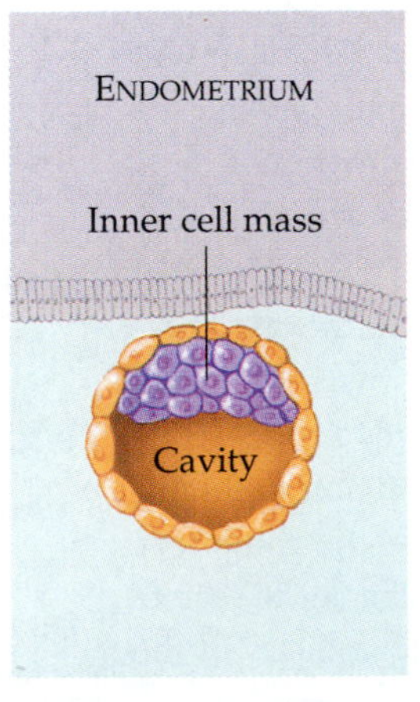

(b) Blastocyst (6 days after conception)

FIGURE 46.16 ▪ Formation of the zygote and early postfertilization events. ① Ovulation releases a secondary oocyte, which enters the oviduct. Fertilization occurs if sperm are present; the union of the ovum and sperm produces a zygote. ② Cleavage, or cell division, begins in the oviduct and continues as the developing embryo moves down the oviduct toward the uterus, propelled by peristalsis and cilia. ③ By the time the embryo reaches the uterus, cleavage has transformed the embryo into a ball of cells. Cleavage continues while the embryo floats freely in the uterus for several days, nourished by fluid secreted by endometrial glands. ④ About 7 days after conception, the embryo is called a blastocyst, which implants into the endometrium.

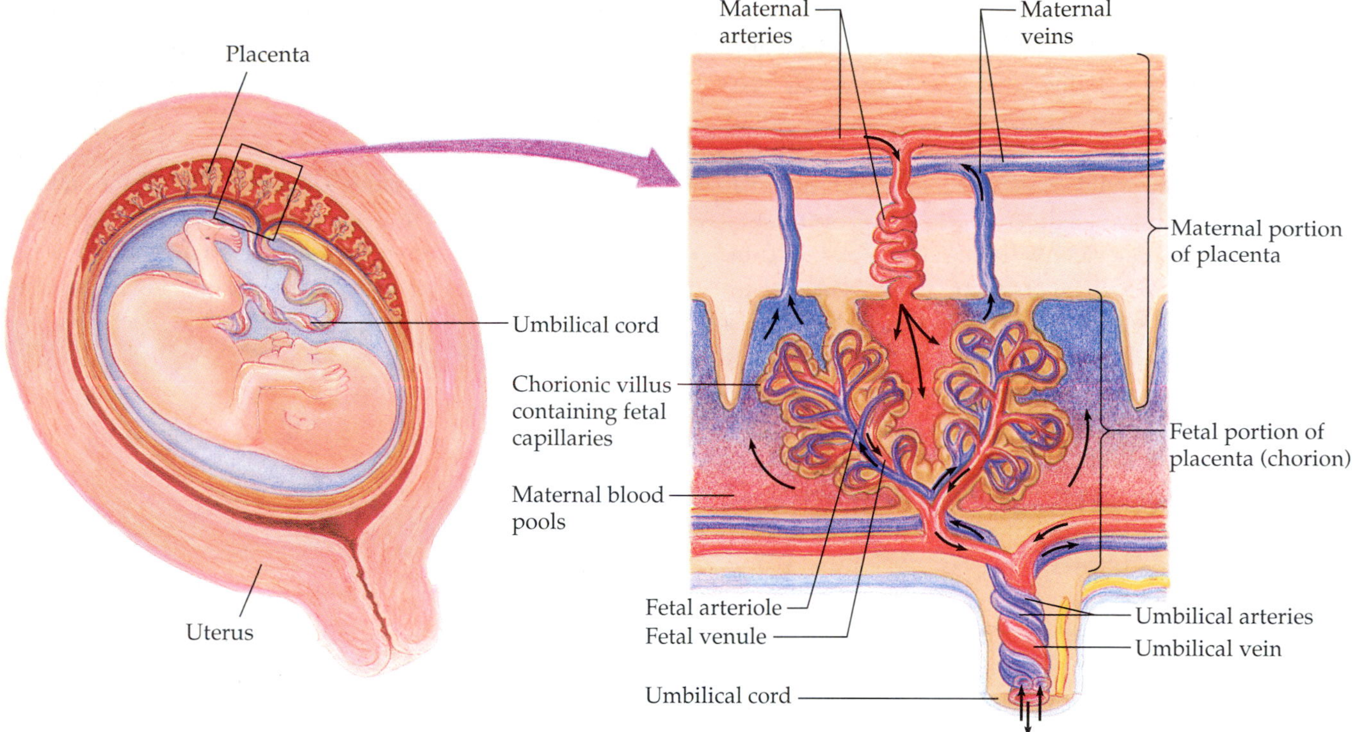

FIGURE 46.17 ▪ Placental circulation. From the fourth week of development until birth, the placenta, a combination of maternal and embryonic tissues, transports nutrients, respiratory gases, and wastes between the embryo or fetus and the mother. Maternal blood enters the placenta in arteries, flows through blood pools in the endometrium, and leaves via veins. Embryonic or fetal blood, which remains in vessels, enters the placenta through arteries and passes through capillaries in fingerlike chorionic villi, where oxygen and nutrients are acquired. As indicated in the drawing, the fetal (or embryonic) capillaries and villi project into the maternal portion of the placenta. Fetal blood leaves the placenta through veins leading back to the fetus. Materials are exchanged by diffusion, active transport, and selective absorption between the fetal capillary bed and the maternal blood pools.

hypothalamus and pituitary) cessation of ovulation and menstrual cycling. The breasts also enlarge rapidly and are often quite tender.

During the second trimester, the fetus grows rapidly to about 30 cm and is very active. The mother may feel movements during the early part of the second trimester, and fetal activity may be visible through the abdominal wall by the middle of this time period. Hormone levels stabilize as HCG declines, the corpus luteum deteriorates, and the placenta secretes its own progesterone, which maintains the pregnancy. During the second trimester, the uterus will grow enough for the pregnancy to become obvious.

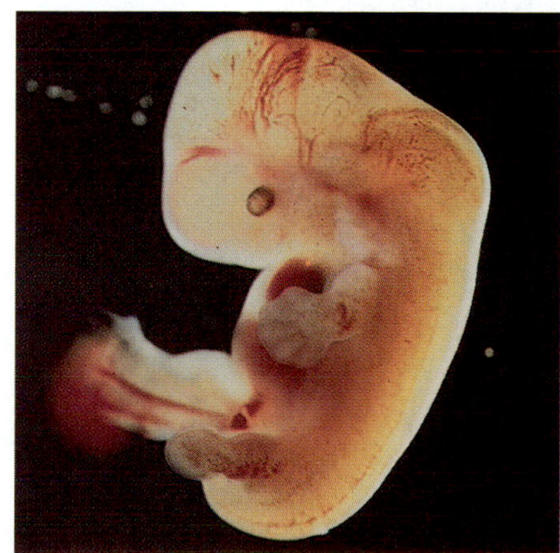

(a) 5 weeks

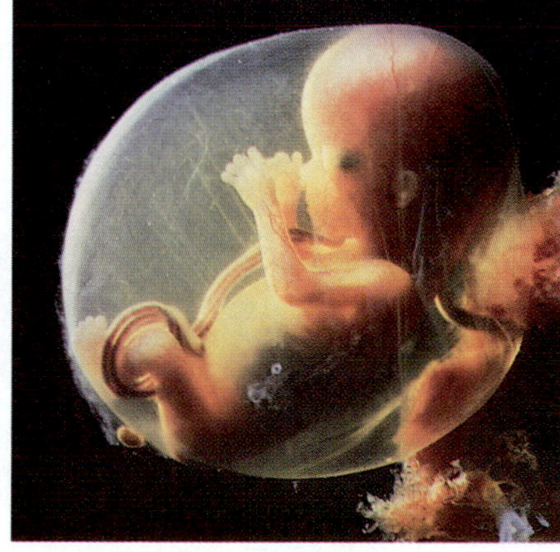

(b) 14 weeks

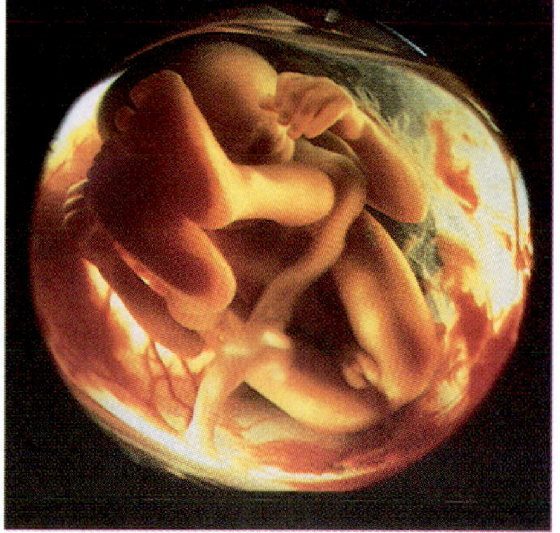

(c) 20 weeks

FIGURE 46.18 ▪ Human fetal development. **(a)** At 5 weeks, limb buds, eyes, the heart, the liver, and rudiments of all other organs have started to develop in the embryo, which is only about 1 cm long. **(b)** Growth and development of the offspring, now called a fetus, continue during the second trimester. This fetus is 14 weeks old and about 6 cm long. **(c)** The fetus in this photograph is 20 weeks old. By the end of the second trimester (at 24 weeks), the fetus grows to about 30 cm in length.

The third and final trimester is one of rapid growth of the fetus to about 3–3.5 kg in weight and 50 cm in length. Fetal activity may decrease as the fetus fills the available space within the embryonic membranes. As the fetus grows and the uterus expands around it, the mother's abdominal organs become compressed and displaced, leading to frequent urination, digestive blockages, and strain in the back muscles. A complex interplay of hormones (estrogens and oxytocin) and local regulators (prostaglandins) induces and regulates labor (FIGURE 46.19). Estrogens, which reach their highest level in the mother's blood during the last weeks of pregnancy, trigger the formation of oxytocin receptors on the uterus. Oxytocin, produced by the fetus and the mother's posterior pituitary, stimulates powerful contractions by the smooth muscles of the uterus. Oxytocin also stimulates the placenta to secrete prostaglandins, which enhance the contractions. In turn, the physical and emotional stresses associated with the contractions stimulate the release of more oxytocin and prostaglandins, a positive feedback system that underlies the three stages of labor.

Birth, or **parturition**, occurs through a series of strong, rhythmic uterine contractions, commonly known as **labor** (FIGURE 46.20). The first stage is the opening up and thinning of the cervix, ending with complete dilation. The second stage is expulsion, or delivery, of the baby. Continuous strong contractions force the fetus down and out of the uterus and vagina. The umbilical cord is cut and clamped at this time. The final stage of labor is delivery of the placenta, which normally follows the baby.

Lactation is an aspect of postnatal care unique to mammals. After birth, decreasing levels of progesterone free the anterior pituitary from negative feedback and allow prolactin secretion. Prolactin stimulates milk production after a delay of 2 or 3 days. The release of milk from the mammary glands is controlled by oxytocin, described in Chapter 45.

FIGURE 46.19 • Hormonal induction of labor.

Reproductive Immunology

Pregnancy is an immunological enigma. Half of the embryo's genes are inherited from the father, and thus many of the chemical markers present on the surface of the embryo will be foreign to the mother. Why, then, does the mother not reject the embryo as a foreign body as she would repel a tissue or

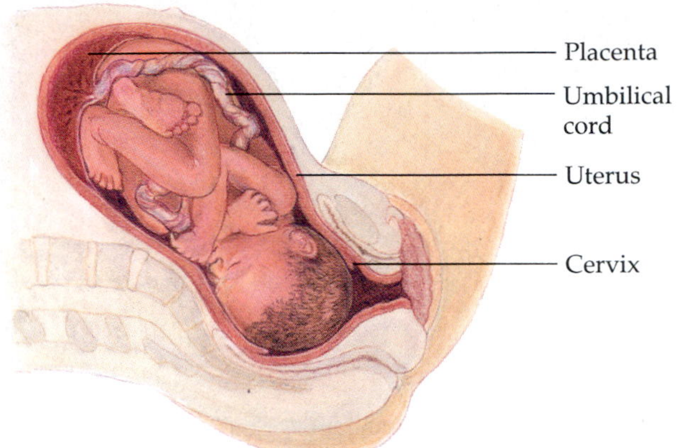

❶ Dilation of the cervix

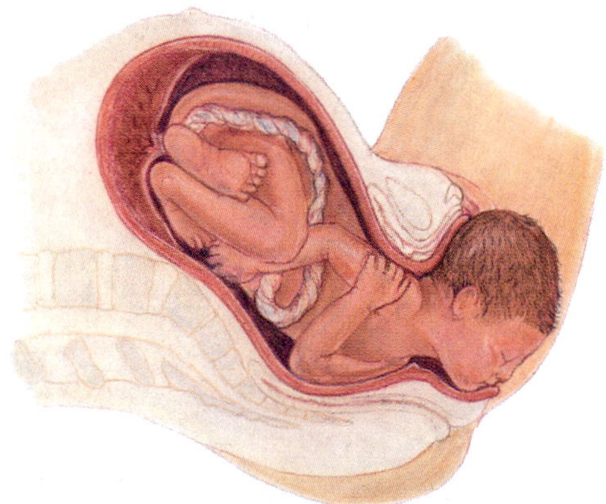

❷ Expulsion: delivery of the infant

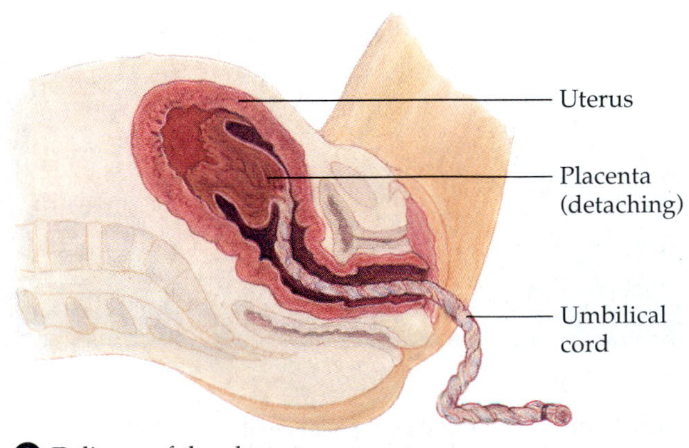

❸ Delivery of the placenta

FIGURE 46.20 • The three stages of labor.

organ graft bearing antigens from another person? Reproductive immunologists are only beginning to solve this puzzle.

Part of the answer is the presence of a physical barrier. A protective layer called the trophoblast prevents the embryo from actually contacting maternal tissue. But the trophoblast develops along with the embryo from the cells of the blastocyst, and this protective barrier, which penetrates the endometrium, may also be foreign to the mother. According to one hypothesis, the trophoblast does *not* develop paternal markers and thus does not trigger an immune response by the mother. However, researchers have discovered paternal antigens on portions of the trophoblast. There is evidence that the trophoblast produces a chemical signal that induces the development of a special type of white blood cell in the uterus that prevents other white cells from mounting an attack on the foreign tissue. This suppressor cell may work by secreting a substance that blocks the action of interleukin-2, the cytokine required for a normal immune response (see Chapter 43). One hypothesis suggests that this local dampening of the immune response occurs, paradoxically, only after nearby white blood cells have first identified the trophoblast as foreign tissue and have taken the first steps of the immune response. If this immunological alarm is not intense enough—that is, if the father's cellular markers are too similar to the mother's—then no suppressor cells are produced.

Some researchers speculate that if the initial immune response is too weak to trigger suppression, the continued attack on the foreign tissue, though weak, may lead to spontaneous abortion of the embryo. According to this view, failure to suppress the immune response in the uterus may account for many cases of women who have multiple miscarriages for no other apparent reason. There has been some success treating frequent miscarriers by sensitizing the woman's immune system to her mate's antigens through immunization—that is, by injecting appropriate chemical markers into the mother prior to pregnancy. The subsequent response to the foreign embryological tissue is intensified to a level that activates the suppressor mechanism. Some critics of this interpretation argue that the psychological support women receive during this experimental treatment counts for more than the immunotherapy itself. Only more research will resolve the interesting and important questions about how a woman's immune system tolerates a 9-month relationship with a large foreign organism.

Contraception

Contraception, the deliberate prevention of fertilization or pregnancy, can be achieved in several ways. Some contraceptive methods prevent the release of mature eggs and sperm from gonads, others prevent fertilization by keeping sperm and eggs apart, and still others prevent implantation of an embryo or abort the embryo (FIGURE 46.21). The following

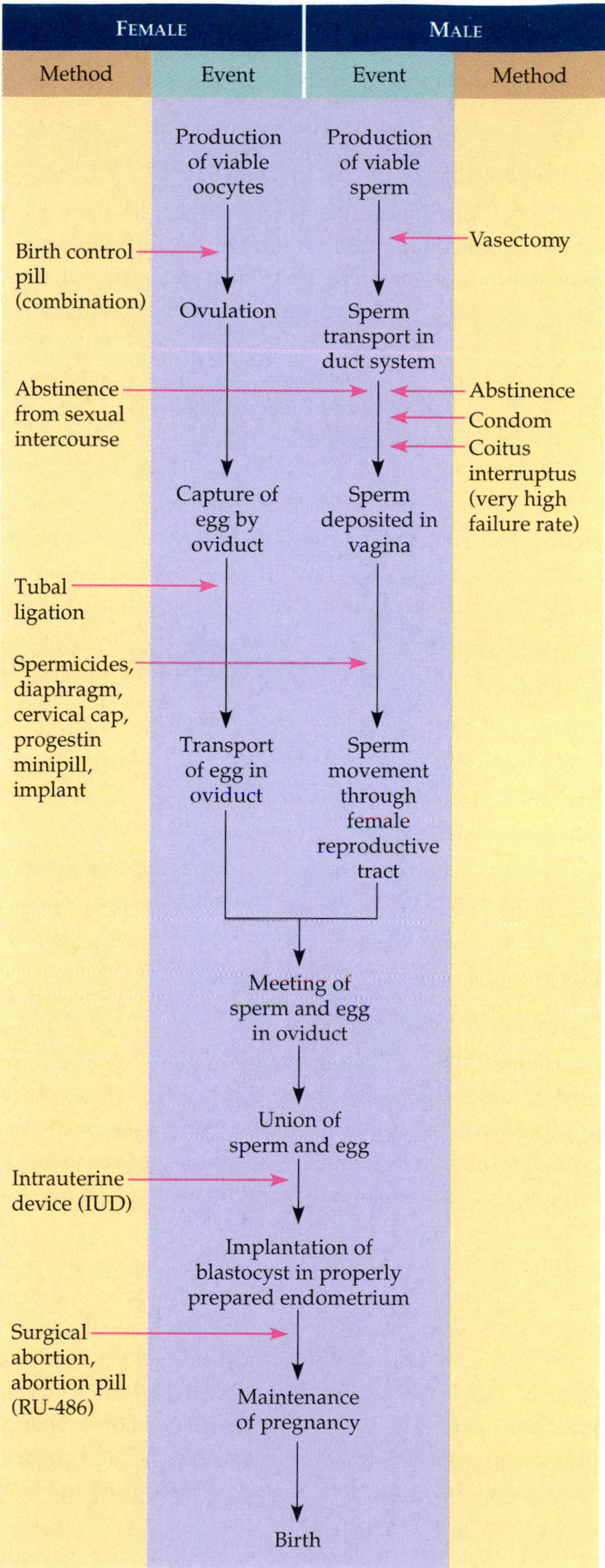

FIGURE 46.21 ▪ Mechanisms of some contraceptive methods. The magenta arrows indicate the stages where contraception works.

brief introduction to the biology of these methods makes no pretense of being a contraception manual. For more complete information, you should consult a physician or health center personnel.

Fertilization can be prevented by abstinence from sexual intercourse or by any of several barriers that keep live sperm from contacting the egg. Temporary abstinence, often called the **rhythm method** of birth control or **natural family planning**, depends on refraining from intercourse when conception is most likely. Because the egg can survive in the oviduct for 24 to 48 hours and sperm for up to 72 hours, a couple practicing temporary abstinence should not engage in intercourse during the few days before and after ovulation. The most effective methods for timing ovulation combine several indicators, including changes in cervical mucus and body temperature during the menstrual cycle. Thus, natural family planning requires that the couple be knowledgeable about these physiological signs. A pregnancy rate of 10% to 20% is typically reported for couples practicing natural family planning. (Pregnancy rate is the number of women who become pregnant during a year out of every 100 women using a particular family-planning method, expressed as a percentage.) Some couples use the natural family planning method to *increase* the probability of conception, so the rate of unplanned pregnancies among couples skilled with this method is actually less than 10% to 20%.

The several **barrier methods** of contraception that block the sperm from meeting the egg have pregnancy rates of less than 10%. For the male, the **condom** is a thin, natural membrane or latex rubber sheath that fits over the penis to collect the semen. For the female, the diaphragm is a dome-shaped rubber cap fitted into the upper portion of the vagina before intercourse. Both of these methods are more effective when used in conjunction with a spermicidal (sperm-killing) foam or jelly. More recently introduced barriers include the cervical cap, which fits tightly around the opening of the cervix and is held in place for a prolonged period by suction, and the new female condom.

Intrauterine devices (IUDs), which are small plastic or metal devices that fit into the uterine cavity, prevent implantation of the blastocyst in the uterus. IUDs have a low pregnancy rate but cause harmful effects in a small percentage of women. Problems of persistent vaginal bleeding, uterine infection, perforation of the uterus, tubal pregnancy (implantation of the embryo in the oviduct), and spontaneous expulsion of the devices have been reported and have led to many lawsuits against IUD manufacturers. New IUD products that prevent implantation by delivering synthetic progesterone locally to the endometrium have recently become available.

As a method of preventing fertilization, coitus interruptus, or withdrawal (removal of the penis from the vagina before ejaculation), is unreliable. Sperm may be present in secretions that precede ejaculation, and a lapse in timing or willpower can result in late withdrawal.

Besides complete abstinence from sexual intercourse, the methods that prevent the release of gametes are the most effective means of birth control. Chemical contraceptives—birth control pills—have pregnancy rates of less than 1%, and sterilization is nearly 100% effective. The most commonly used birth control pills are combinations of synthetic estrogens and a synthetic progestin (progesteronelike hormone). These two hormones act by negative feedback to stop the release of GnRH by the hypothalamus, and of FSH (an estrogen effect) and LH (a progestin effect) by the pituitary. By blocking LH release, the progestin prevents ovulation. As a backup measure, the estrogen inhibits FSH secretion so that no follicles develop. A second type of birth control pill, called the minipill, contains only progestin. The minipill prevents fertilization mainly by altering a woman's cervical mucus so that it blocks sperm from entering the uterus. In 1990, the FDA approved a version of the progestin-only minipill that is implanted under the skin. Steadily releasing a tiny amount of progestin into the blood, the implant produces effective birth control for about 5 years.

Birth control pills have been the center of much debate, particularly because of long-term harmful effects of the estrogens. No solid evidence exists for cancers caused by the pill, but cardiovascular problems are a major concern. Birth control pills have been implicated in abnormal blood clotting, atherosclerosis, and heart attacks. Smoking while using chemical contraception increases the risk of mortality tenfold or more. Although the pill places women at risk for these diseases, it eliminates the dangers of pregnancy; women on birth control pills have mortality rates about one-half those of pregnant women.

Sterilization is the permanent prevention of gamete release. **Tubal ligation** in women usually involves cauterizing or tying off (ligating) a section of the oviducts to prevent eggs from traveling into the uterus. **Vasectomy** in men is the cutting of each vas deferens to prevent sperm from entering the urethra. Both male and female sterilization are relatively safe and free from harmful effects. Both are also difficult to reverse, so the procedures should be considered permanent.

Abortion is the termination of a pregnancy in progress. Spontaneous abortion, or miscarriage, is very common; it occurs in as many as one-third of all pregnancies, often before the woman is even aware she is pregnant. In addition, each year about 1.5 million women in the United States choose abortions performed by physicians. A drug called RU-486, developed in France, enables a woman to terminate pregnancy nonsurgically within the first few weeks. An analog of progesterone, RU-486 blocks progesterone receptors in the uterus, thus preventing progesterone from maintaining pregnancy. The use of RU-486 in the United States has been delayed by the controversy over abortion.

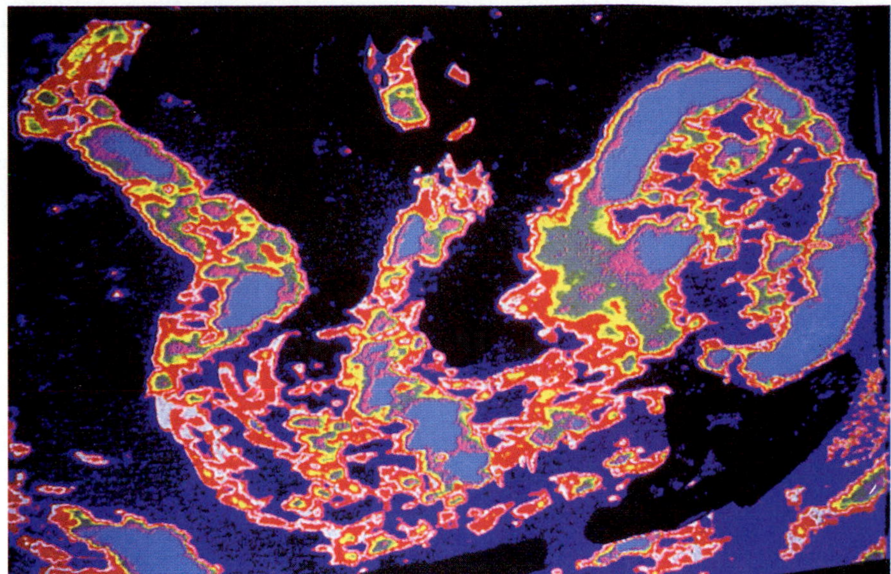

FIGURE 46.22 ▪ Ultrasound imaging. This color-enhanced image shows a fetus in the uterus at about 18 weeks. The image is produced on a computer screen when high-frequency sounds from an ultrasound scanner held against a pregnant woman's abdomen bounce off the fetus.

Of all contraceptives for sexually active individuals, latex condoms are the only ones that offer some protection against sexually transmitted diseases, including AIDS. This protection is, however, not absolute.

Modern technology offers solutions for some reproductive problems

Recent scientific and technological advances have made it possible to deal with problems of reproduction in striking ways. For example, it is now possible to diagnose many genetic diseases and congenital (present at birth) disorders while the fetus is in the uterus. Amniocentesis and chorionic villus sampling are invasive techniques in which amniotic fluid or fetal cells are obtained for genetic analysis (see FIGURE 14.17). Noninvasive procedures usually use high-frequency sound waves, or ultrasound imaging, to detect fetal condition (FIGURE 46.22). An alternate new technique relies on the fact that a few fetal blood cells leak across the placenta into the mother's bloodstream. A blood sample from the mother yields enough fetal cells that can be identified with specific antibodies (which bind to proteins on the surface of fetal cells) and then tested for genetic disorders.

Diagnosing genetic diseases of fetuses poses important ethical questions. To date, essentially all detectable disorders remain untreatable in the uterus, and many cannot be corrected even after birth. Parents may be faced with difficult decisions about whether to terminate a pregnancy or cope with a child who may have profound defects and a short life expectancy. These are complex issues that demand careful, informed thought and competent counseling.

In the past two decades, several techniques have been developed to assist childless couples who want to have offspring. For cases of male infertility, sperm from anonymous donors are widely available from sperm banks. Sperm are deposited in the vagina or cervix when it is possible for the woman to conceive. Another procedure, called **in vitro fertilization**, was developed for women whose oviducts are blocked. Ova are surgically removed following hormonal stimulation of the follicles. The ova are then fertilized in culture dishes in a laboratory. After about 2½ days, when the embryo has reached the eight-cell stage, it is placed in the uterus and allowed to implant. In vitro fertilization is costly but has a success rate similar to the pregnancy rate resulting from insemination by sexual intercourse. Embryos can be frozen for later use if the first attempt is unsuccessful. A couple may choose to have oocytes obtained from another woman fertilized with the male partner's sperm, although more serious ethical issues must be resolved when donors are used. Several thousand children have thus far been conceived by in vitro fertilization, and there is no evidence of any abnormalities associated with this procedure.

One area of reproductive research that has traditionally received little attention is male contraception. Male chemical contraceptives have proved quite elusive. Testosterone will block the release of pituitary gonadotropins, but testosterone itself stimulates spermatogenesis. Estrogens are effective, but they inhibit sex drive and can be feminizing. The best prospects so far are for antagonists of GnRH, which are potent inhibitors of spermatogenesis. Several experimental drugs and birth control techniques for both men and women are currently being tested.

▪ ▪ ▪

In this chapter we have considered the structural and physiological bases of animal reproduction. The next chapter focuses on the mechanics of development that transform a zygote into an animal form and on other topics in animal development.

CHAPTER REVIEW

REVIEW OF KEY CONCEPTS

(with page numbers and key figures)

OVERVIEW OF ANIMAL REPRODUCTION

- **Both asexual and sexual reproduction occur in the animal kingdom (p. 913)** Asexual reproduction produces offspring whose genes all come from a single parent. Sexual reproduction requires the fusion of male and female gametes to form a diploid zygote.
- **Diverse mechanisms of asexual reproduction enable animals to produce identical offspring rapidly (pp. 913–914, FIGURE 46.1)** Fission, budding, and fragmentation with regeneration are mechanisms of asexual reproduction in various invertebrates.
- **Reproductive cycles and patterns vary extensively among animals (pp. 914–916, FIGURE 46.2)** Animals may reproduce exclusively sexually or asexually, or they may alternate between the two, depending on environmental conditions. Variations on these two modes are made possible through parthenogenesis, hermaphroditism, and sequential hermaphroditism. Reproductive cycles are controlled by hormones and environmental cues, such as changes in temperature, rainfall, day length, and seasonal lunar cycles.

MECHANISMS OF SEXUAL REPRODUCTION

In external fertilization, eggs shed by the female are fertilized by sperm in the external environment. In internal fertilization, egg and sperm unite within the female's body.

- **Internal and external fertilization both depend on mechanisms ensuring that mature sperm encounter fertile eggs of the same species (p. 916, FIGURE 46.4)** External and internal fertilization require critical timing, often mediated by environmental cues, pheromones, and/or courtship behavior. Internal fertilization requires important behavioral interactions between male and female animals, as well as compatible copulatory organs.
- **Species with internal fertilization usually produce fewer zygotes but provide more parental protection than species with external fertilization (pp. 916–917, FIGURE 46.5)** Greater protection of embryos and parental care of the young usually follow the production of relatively few offspring by internal fertilization.
- **Complex reproductive systems have evolved in many animal phyla (pp. 917–918, FIGURES 46.6, 46.7)** Reproductive systems range from the production of gametes by undifferentiated cells in the body cavity to complex assemblages of male and female gonads with accessory tubes and glands that carry and protect gametes and developing embryos. The reproductive systems of insects and flatworms are among the most complex in the animal kingdom.

MAMMALIAN REPRODUCTION

- **Human reproduction involves intricate anatomy and complex**
46.1 **behavior (pp. 918–922, FIGURES 46.8, 46.9)** External reproductive
46.2 structures of the human male are the scrotum and penis. The male gonads, or testes, reside in the cool environment of the scrotum. They possess endocrine Leydig cells surrounding sperm-forming seminiferous tubules that successively lead into the epididymis, vas deferens, ejaculatory duct, and urethra, which exits at the tip of the penis. Externally, the human female has a vestibule containing separate openings of the vagina and urethra, the labia minora bordering the vestibule, the labia majora, and the clitoris. Internally, the vagina is connected to the uterus, which connects to two oviducts. Two ovaries (female gonads) are stocked with follicles containing diploid primary oocytes formed before the woman's birth. Beginning at puberty, one or more follicles mature during each menstrual cycle. The oocyte contained in a maturing follicle undergoes the first meiotic division, and a secondary oocyte, which is haploid, is expelled from the surface of the ovary during ovulation. After ovulation, the remaining tissue of the follicle forms a corpus luteum that secretes progesterone and estrogen for a variable duration, depending on whether or not pregnancy occurs. Although separate from the reproductive system, the mammary glands, or breasts, evolved in association with parental care. Both males and females experience the erection of certain body tissues due to vasocongestion and myotonia, which culminate in orgasm.
- **Spermatogenesis and oogenesis both involve meiosis but differ in three significant ways (pp. 922–925, FIGURES 46.12, 46.13)** Cytokinesis is unequal in oogenesis, producing one large ovum. Production of sperm is continuous; in humans, the number of future egg cells is set at birth. Spermatogenesis is an uninterrupted sequence, but there are long delays in oogenesis.
- **A complex interplay of hormones regulates reproduction (pp.**
46.1 **925–927, FIGURES 46.14, 46.15)** Androgens from the testes cause
46.2 the development of primary and secondary sex characteristics in the male. Androgen secretion and sperm production are both controlled by hypothalamic and pituitary hormones. Female hormones are secreted in a rhythmic fashion reflected in the menstrual or estrous cycle. In both types of female cycles, the endometrium thickens in preparation for possible implantation. The menstrual cycle, however, is punctuated by endometrial bleeding and lacks the clear-cut period of sexual receptivity limited to the heat period of the estrous cycle. The human menstrual cycle consists of the menstrual flow phase, proliferative phase, and secretory phase. The ovarian cycle includes the follicular and luteal phases. The female reproductive cycle is orchestrated by cyclic secretion of GnRH from the hypothalamus and of FSH and LH from the anterior pituitary. The developing follicle produces estrogens, and the corpus luteum secretes progesterone and estrogens. Positive and negative feedback produce the changing levels of these five hormones, which coordinate the menstrual and ovarian cycles.
- **Embryonic and fetal development occur during pregnancy in humans and other eutherian (placental) mammals (pp. 928–933, FIGURES 46.16–46.21)** Human pregnancy can be divided into three trimesters. Organogenesis is completed by 8 weeks. Birth, or parturition, results from strong, rhythmic uterine contractions associated with labor. Positive feedback involving the hormones estrogen and oxytocin, and prostaglandins, regulate labor. The ability of a pregnant woman to accept her "foreign" fetus may be due to the suppression of the immune response in her uterus. Contraceptive methods include preventing the release of mature gametes from the gonads, preventing gamete union in the female tract, and preventing implantation of the zygote.
- **Modern technology offers solutions for some reproductive problems (p. 933, FIGURE 46.22)** Current technological methods of detecting fetal condition include ultrasound imaging, amniocentesis, and chorionic villus sampling. Current technology also provides in vitro fertilization.

SELF-QUIZ

1. Which of the following characterizes parthenogenesis?
   a. An individual may change its sex during its lifetime.
   b. Specialized groups of cells may be released and grow into new individuals.
   c. An organism is first a male and then a female.
   d. An egg develops without being fertilized.
   e. Both members of a mating pair have male and female reproductive organs.

2. Which of the following structures is *incorrectly* paired with its function?
   a. gonads—gamete-producing organs
   b. spermatheca—sperm-transferring organ found in male insects
   c. cloaca—common opening for reproductive, excretory, and digestive systems
   d. baculum—bone that stiffens the penis, found in some mammals
   e. endometrium—lining of the uterus, forming the maternal part of the placenta

3. Which of the following male and female structures are *least* alike in function?
   a. seminiferous tubules—vagina
   b. Leydig cells of testes—follicle cells
   c. testes—ovaries
   d. spermatogonia—oogonia
   e. vas deferens—oviduct

4. A difference between estrous and menstrual cycles is that
   a. nonmammalian vertebrates have estrous cycles, whereas mammals have menstrual cycles
   b. the endometrial lining is shed in menstrual cycles but reabsorbed in estrous cycles
   c. estrous cycles occur more frequently than menstrual cycles
   d. estrous cycles are not controlled by hormones
   e. ovulation occurs before the endometrium thickens in estrous cycles

5. Peaks of LH and FSH production occur during
   a. the flow phase of the menstrual cycle
   b. the beginning of the follicular phase of the ovarian cycle
   c. the period surrounding ovulation
   d. the end of the luteal phase of the ovarian cycle
   e. the secretory phase of the menstrual cycle

6. In protandrous hermaphroditism
   a. some individuals may change from male to female
   b. individuals fertilize themselves
   c. males rather than females release pheromones
   d. diploid ova are produced
   e. the adult gonads are undifferentiated

7. During human gestation, organogenesis occurs
   a. in the first trimester
   b. in the second trimester
   c. in the third trimester
   d. while the embryo is in the oviduct
   e. during the blastocyst stage

8. Which pharmacological strategy is most likely to result in a successful male contraceptive?
   a. block FSH receptors on spermatagonia
   b. maintain high circulating concentrations of androgen
   c. block testosterone receptors on Leydig cells
   d. block androgen receptors within the hypothalamus
   e. maintain high circulating concentrations of FSH

9. Fertilization of human eggs most often takes place in the
   a. vagina
   b. ovary
   c. uterus
   d. oviduct (fallopian tube)
   e. vas deferens

10. In mammalian males, the excretory and reproductive systems share the
    a. testes
    b. urethra
    c. ureter
    d. vas deferens
    e. prostate

CHALLENGE QUESTIONS

1. Compare and contrast oogenesis and spermatogenesis.
2. Describe how sexual and asexual reproduction differ in mechanism and result.
3. Explain why menstruation and ovulation do not occur during pregnancy.

SCIENCE, TECHNOLOGY, AND SOCIETY

New techniques for sorting sperm, combined with in vitro fertilization, make it possible for a couple to choose their baby's sex. Would you want to do this if you had the chance? Why or why not? What potential problems can you foresee if this procedure becomes widely available?

FURTHER READING

Butler, V. "Elephants: Trimming the Herd." *BioScience,* February 1998. A discussion of the pros and cons of using humane birth control methods to regulate populations in a national park in South Africa.

Concar, D. "Into the Mind Unborn." *New Scientist,* October 19, 1996. A summary of research and scientific debate about human brain development.

Crews, D. "Animal Sexuality." *Scientific American,* January 1994. A comparison of mechanisms determining gender in several different species, including all-female lizards.

Diamond, J. *Why Is Sex Fun? The Evolution of Human Sexuality.* Weidenfeld and Nicolson: 1997. A brief, thoughtful discussion of fascinating hypotheses.

Loughry, W. M., P. A. Prodöhl, C. M. McDonough, and J. C. Avise. "Polyembryony in Armadillos." *American Scientist,* May–June 1998. A summary of research on a mammal that combines sexual reproduction and cloning.

Rommel, S.A., D.A. Pabst, and W.A. McLellan. "Reproductive Thermoregulation in Marine Mammals." *American Scientist,* September–October, 1998. Explores the circulatory adaptations that cool the testes and fetuses in seals and dolphins.

Sherman, P. "The Evolution of Menopause." *Nature,* April 23, 1998. Provides a discussion of current hypotheses.

Travis, J. "Why Do Women Menstruate?" *Science News,* April 12, 1997. An overview of hypotheses and current debate.

WEB LINKS

Visit the special edition of *The Biology Place* for BIOLOGY, Fifth Edition, at http://www.biology.com/campbell. Go to Chapter 46 for online resources, including learning activities, practice exams, and links to the following web sites:

"The Visible Embryo"
The Visible Embryo looks at the first four weeks of human development, from fertilization to somite development.

"The Embryonic Zoo"
An impressive page of movies and links to other embryological resources.

"Zygote"
This well-designed site has an impressive array of links to scientific, ethical, and philosophical articles on reproduction and development.

"Education Page: Society for Developmental Biology"
Among other things, the education page of the SDB has the "Developmental Cinema," numerous links, and quiz questions. From this page you can submit developmental biology questions to the society.

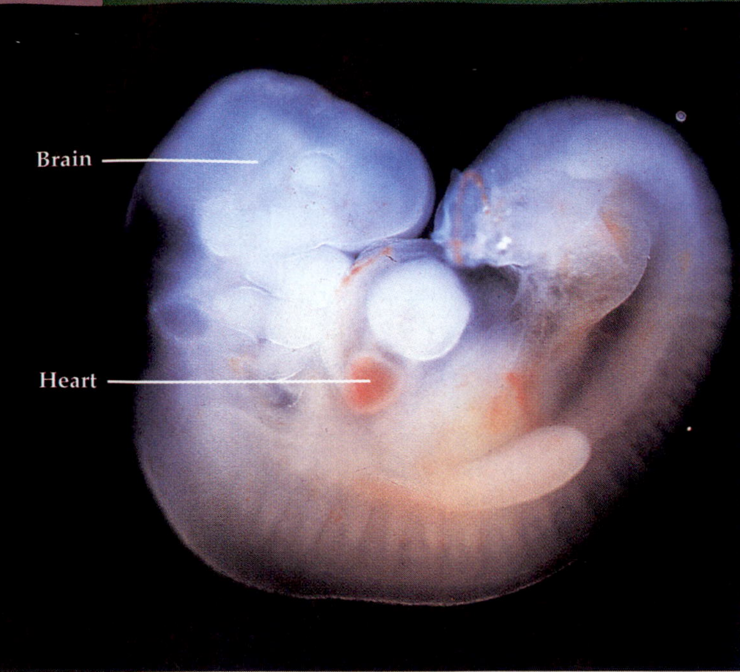

CHAPTER 47

ANIMAL DEVELOPMENT

The Stages of Early Embryonic Development

- From egg to organism, an animal's form develops gradually: *the concept of epigenesis*
- Fertilization activates the egg and brings together the nuclei of sperm and egg
- Cleavage partitions the zygote into many smaller cells
- Gastrulation rearranges the blastula to form a three-layered embryo with a primitive gut
- In organogenesis, the organs of the animal body form from the three embryonic germ layers
- Amniote embryos develop in a fluid-filled sac within a shell or uterus

The Cellular and Molecular Basis of Morphogenesis and Differentiation in Animals

- Morphogenesis in animals involves specific changes in cell shape, position, and adhesion
- The developmental fate of cells depends on cytoplasmic determinants and cell-cell induction: *a review*
- Fate mapping can reveal cell genealogies in chordate embryos
- The eggs of most vertebrates have cytoplasmic determinants that help establish the body axes and differences among cells of the early embryo
- Inductive signals drive differentiation and pattern formation in vertebrates

It is difficult to imagine that each of us began life as a single cell about the size of the period at the end of this sentence. Less than a month after conception, our brains were taking form and our developing hearts had already begun to pulsate (see photograph). It took a total of only about nine months—the length of a school year—to be transfigured from zygote to newborn human, built of billions of differentiated cells organized into specialized tissues and organs.

By combining molecular genetics with classical approaches to embryology, developmental biologists are now beginning to answer many of the questions about how a single fertilized egg cell gives rise to a specific animal. Chapter 21 described some of the recent research, introducing you to a number of basic genetic and cellular mechanisms involved in development. The animals discussed there were mainly invertebrates. In this chapter we concentrate mainly on the embryonic development of animals with backbones.

Molecular biologist Sidney Brenner has wryly commented that developmental biology is about "how to make a mouse," and indeed, mice are favorite experimental models of researchers studying vertebrate development, in particular the development of mammals. Frogs and chickens are also important model organisms for understanding vertebrate development. From the study of these animals and of certain invertebrates (including the ones you met in Chapter 21), scientists have worked out both the stages of embryonic development and many of the underlying cellular and molecular events.

THE STAGES OF EARLY EMBRYONIC DEVELOPMENT

From egg to organism, an animal's form develops gradually: *the concept of epigenesis*

The question of how an egg becomes an animal has been asked for centuries. As recently as the eighteenth century, the prevailing view was that the egg or sperm contains an embryo that is a preformed, miniature infant (FIGURE 47.1). Development was thought to be simply the enlargement of the embryo. This idea of **preformation** came to include the notion that the embryo must contain all its descendants: a series of successively smaller embryos within embryos, like Russian nesting dolls. One theologian proposed that Eve, in the Garden of Eden, stored all future humanity within her.

The competing theory of embryonic development was **epigenesis**, originally proposed 2000 years earlier by Aristotle. According to this idea, the form of an animal emerges gradually from a relatively formless egg. As microscopy improved during the nineteenth century, biologists could see that

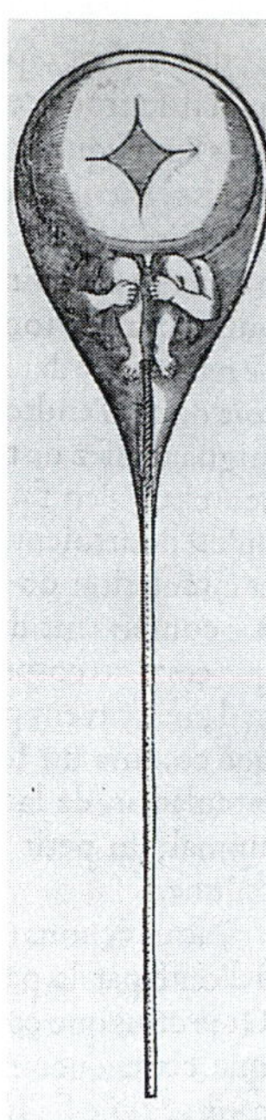

FIGURE 47.1 ▪ A "homunculus" inside the head of a human sperm. According to one version of the preformation idea, a sperm contains a preformed, miniature infant, which simply grows in size during embryonic development. This engraving was made in 1694.

embryos took shape in a series of progressive steps, and epigenesis displaced preformation as the favored explanation among embryologists.

Modern biology, of course, has completely discarded the idea of a tiny person living in an egg or sperm cell. But when interpreted in broader terms, the concept of preformation may have some merit. Although an embryo's form emerges gradually as it develops from a fertilized egg, *something* is preformed in the zygote. As discussed in Chapter 21, an organism's development is largely determined by the genome of the zygote and the organization of the cytoplasm of the egg cell. Messenger RNA, proteins, and other substances made by the mother are heterogeneously distributed in the unfertilized egg, and these substances have a profound effect on the development of the future embryo in most animal species (mammals are an exception, as we will see). After fertilization produces a zygote, cell division partitions the cytoplasm in such a way that nuclei of different embryonic cells are exposed to different cytoplasmic environments. This sets the stage for the expression of different genes in different cells. As cell division continues and the embryo develops, inherited traits emerge by mechanisms that selectively control gene expression, leading to the differentiation (specialization) of cells. The timely communication of instructions, telling cells precisely what to do when, is essential. This information transfer occurs by cell signaling among different embryonic cells. Along with cell division and differentiation, development involves morphogenesis, the process by which an animal takes shape. Thus, the overall process of development is one of epigenesis.

In the first half of the chapter we will survey the early stages of embryonic development, when the body plan of an animal emerges from the fertilized egg. In the second half of the chapter we will look at the cellular and molecular mechanisms that play major roles in the developmental process. We begin with the fertilization of an egg cell by a sperm.

Fertilization activates the egg and brings together the nuclei of sperm and egg

The gametes, sperm and egg that unite during fertilization, are both highly specialized cell types produced by a complex series of developmental events in the testes and ovaries of the parents (see Chapter 46). The main function of fertilization is to combine haploid sets of chromosomes from two individuals into a single diploid cell, the zygote. Another key function is activation of the egg: Contact of the sperm with the egg's surface initiates metabolic reactions within the egg that trigger the onset of embryonic development.

Fertilization has been studied most extensively by combining the gametes of sea urchins in the laboratory. Although the details of fertilization vary with different animal groups, sea urchins (Phylum Echinodermata) provide a good general model for the important events of fertilization. While sea urchins are not vertebrates or even chordates, they are deuterostomes (see Chapter 32), and their early development is similar to that of vertebrates.

The Acrosomal Reaction

The eggs of sea urchins are fertilized externally after the animals release their gametes into the surrounding seawater. When a sperm cell is exposed to molecules from the slowly dissolving jelly coat that surrounds an egg, a vesicle at the tip of the sperm called the acrosome discharges its contents by exocytosis (FIGURE 47.2, p. 938). This **acrosomal reaction** releases hydrolytic enzymes that enable an elongating structure called the *acrosomal process* to penetrate the jelly coat of the egg. The tip of the acrosomal process is coated with a protein that adheres to specific receptor molecules located on the vitelline layer just external to the plasma membrane of the egg. In sea urchins and many other animals, "lock-and-key" recognition of molecules ensures that eggs will be fertilized only by sperm of the same species. This specificity is especially important when fertilization occurs externally in water, where gametes of other species are likely to be present.

The acrosomal reaction leads to the fusion of sperm and egg plasma membranes and the entry of a single sperm nucleus into the cytoplasm of the egg. Fusion of the membranes causes a neuron-like electrical response by the egg's plasma membrane. Ion channels open, allowing sodium ions to flow into the egg cell and change the membrane potential, the voltage across the membrane (see Chapter 8). This membrane depolarization, as the electrical response is called, is

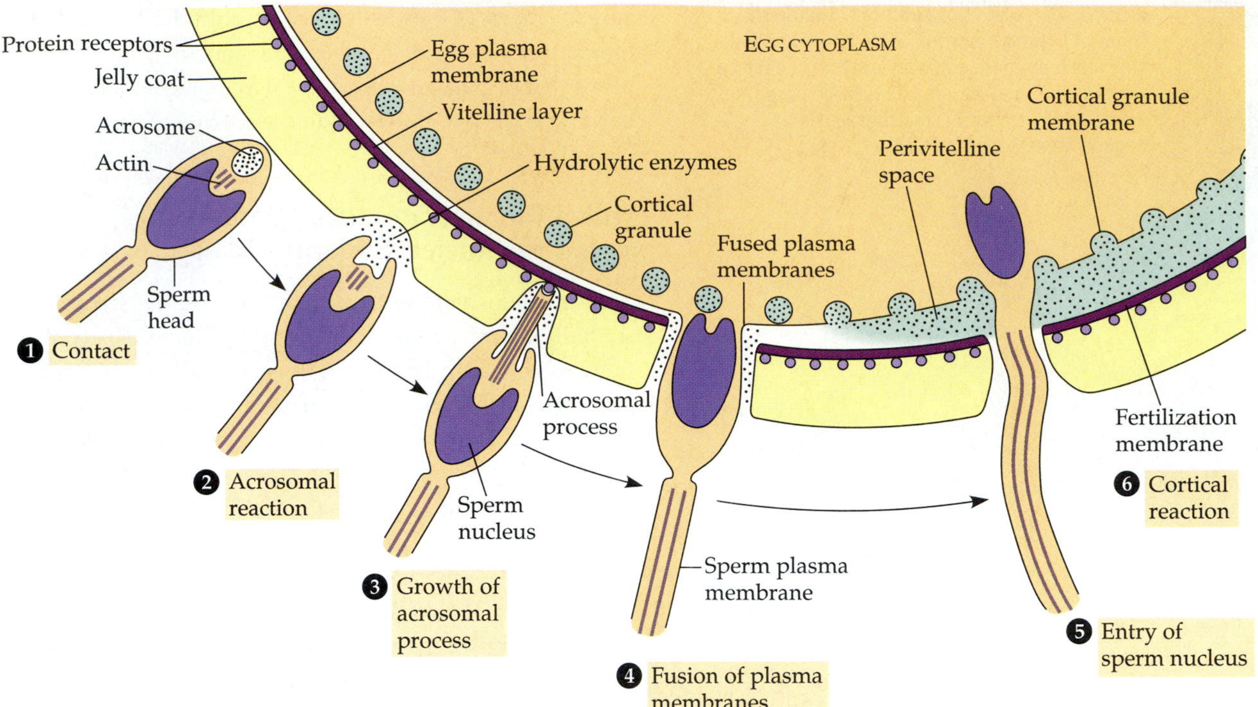

FIGURE 47.2 ▪ The acrosomal and cortical reactions during sea urchin fertilization. The events following contact of a single sperm and egg ensure that only one sperm nucleus enters the cytoplasm of the egg. ① The sperm cell contacts the egg's jelly coat. ② The acrosomal reaction begins with the release of hydrolytic enzymes from the acrosome in the sperm head. The enzymes excavate a hole in the jelly coat, while the growth of actin filaments (see TABLE 7.2) creates a protrusion from the sperm head, the acrosomal process. ③ The acrosomal process extends through the jelly coat and binds to receptors on the egg's vitelline layer. Enzymes on the acrosomal process probably digest a hole in the vitelline layer, and the acrosomal process contacts the egg's plasma membrane. ④ The plasma membranes of the sperm and egg fuse, and ⑤ the sperm nucleus enters the cytoplasm of the egg. Fusion of the gamete membranes triggers both an electrical change in the plasma membrane of the egg and the cortical reaction, blocking entry by other sperm. ⑥ In the cortical reaction, cortical granules in the egg fuse with the plasma membrane and discharge enzymes and other macromolecules that separate the vitelline layer from the plasma membrane and harden it. The vitelline layer becomes a sperm-proof fertilization membrane.

common among animal species. Occurring within about 1 to 3 seconds after a sperm cell binds to the vitelline layer, the depolarization is also called the **fast block to polyspermy** because it prevents more than one sperm cell from fusing with the egg's plasma membrane. Without the block, multiple fertilizations could result in an aberrant chromosome count and abnormal mitosis.

The Cortical Reaction

Another major effect of the fusion of egg and sperm plasma membranes is the **cortical reaction**, a series of changes in the outer zone (cortex) of the egg cytoplasm (see FIGURE 47.2, step 6). The fusion of sperm and egg triggers a signal-transduction pathway that causes the egg's endoplasmic reticulum to release calcium (Ca^{2+}) into the cytosol. The calcium release from the ER begins at the site of sperm entry and then propagates in a wave across the fertilized egg (FIGURE 47.3). Apparently the signaling pathway leads to production of IP_3, which opens ligand-gated calcium channels in the ER membrane (see FIGURE 11.14); the Ca^{2+} released then triggers the opening of other channels, and so on. Within seconds, the high concentration of Ca^{2+} brings about a change in vesicles called **cortical granules**, which lie just under the egg's plasma membrane. Responding to the Ca^{2+} increase, the cortical granules fuse with the plasma membrane and release their contents into the perivitelline space between the plasma membrane and the vitelline layer. Enzymes from the granules separate the vitelline layer from the plasma membrane while mucopolysaccharides produce an osmotic gradient, drawing water into the perivitelline space and swelling it. The swelling pushes the vitelline layer away from the plasma membrane, and other enzymes harden it. The result is that the vitelline layer becomes the **fertilization membrane**, which resists the entry of additional sperm. By this time, usually about a minute after sperm and egg fuse, the voltage across the plasma membrane has returned to normal, and the fast block to polyspermy no longer functions. But the fertilization membrane, along with other changes in the egg's surface, functions as a **slow block to polyspermy**.

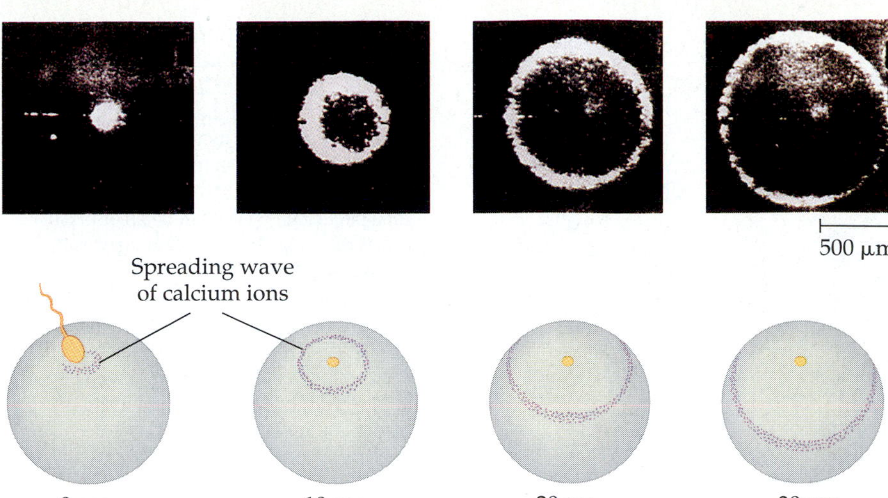

FIGURE 47.3 ▪ A wave of Ca^{2+} release during the cortical reaction. A fluorescent dye that glows when it binds free Ca^{2+} was used in this experiment to track the cortical reaction from the point of sperm contact (0 sec) during the fertilization of a fish egg (LM). The spreading wave of calcium ions, released into the cytosol from the endoplasmic reticulum, triggers the release of more and more Ca^{2+}. The high cytosolic concentration of Ca^{2+} causes the cortical granules to fuse with the plasma membrane, leading to formation of the fertilization membrane. The Ca^{2+} also helps activate metabolic changes inside the fertilized egg.

Activation of the Egg

The sharp rise in the egg's cytosolic concentration of Ca^{2+} not only triggers the cortical reaction but also incites metabolic changes within the egg cell. The unfertilized egg has a very slow metabolism, but within a few minutes of fertilization, the rates of cellular respiration and protein synthesis increase substantially. With these rapid changes, the egg cell is said to be activated. In sea urchins and many other species, the rise in Ca^{2+} triggers a loss of hydrogen ions from the egg, and the egg cytosol becomes slightly alkaline. This pH change seems to be indirectly responsible for the metabolic responses of the egg to fertilization.

Although the binding and fusion of sperm are triggers for egg activation, sperm cells do not contribute any materials required for activation to proceed. Indeed, the unfertilized eggs of many species can be artificially activated by the injection of Ca^{2+} or by a variety of mildly injurious treatments, such as temperature shock. This artificial activation switches on the metabolic responses of the egg and causes it to begin developing by parthenogenesis (without fertilization by a sperm). It is even possible to artificially activate an egg that has had its own nucleus removed. (Of course, embryonic development of such an egg terminates at a very early stage.) The fact that an egg lacking a nucleus can begin making new kinds of proteins upon activation means that mRNA coding for these proteins must be stockpiled in an inactive form in the cytoplasm of the unfertilized egg.

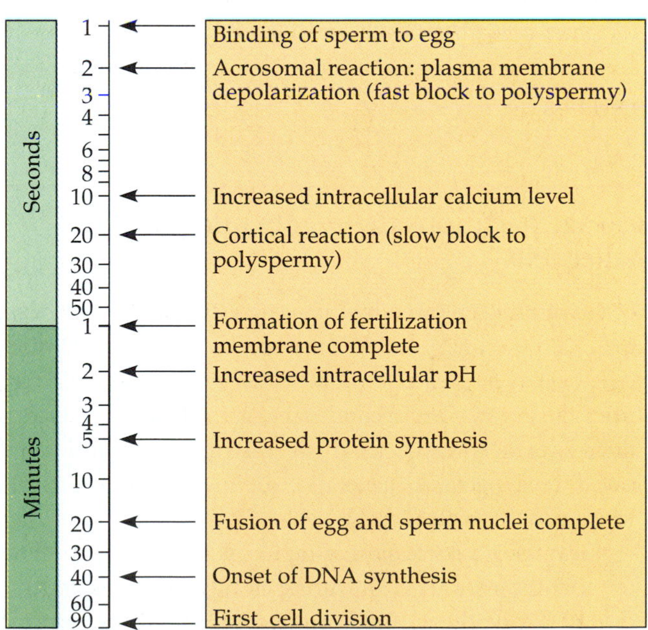

FIGURE 47.4 ▪ Timeline for the fertilization of sea urchin eggs. The process begins when a sperm cell contacts the jelly coat of an egg (top of chart). Notice that the scale is logarithmic.

While the activated egg gears up its metabolism, the nucleus of the sperm cell within the egg starts to swell. After about 20 minutes, the sperm nucleus merges with the egg nucleus, creating the diploid nucleus of the zygote. DNA synthesis begins, and the first cell division occurs in about 90 minutes (in the case of sea urchins and some frogs). The events of fertilization in sea urchins are summarized in FIGURE 47.4.

Fertilization in Mammals

In Chapter 46 you learned how in vitro fertilization of human eggs is enabling some couples with fertility problems to have children. Test-tube fertilization has also made it possible for developmental biologists to study the process of fertilization in mammals. Many of the events turn out to be similar to what has been observed in sea urchins, but there are also important differences.

In contrast to the external fertilization of sea urchins and most other marine invertebrates, fertilization in terrestrial animals, including mammals, is generally internal. Secretions in the mammalian female reproductive tract alter certain molecules on the surface of sperm cells that have been deposited during the male's ejaculation and also increase the motility of the sperm. This enhancement of sperm function

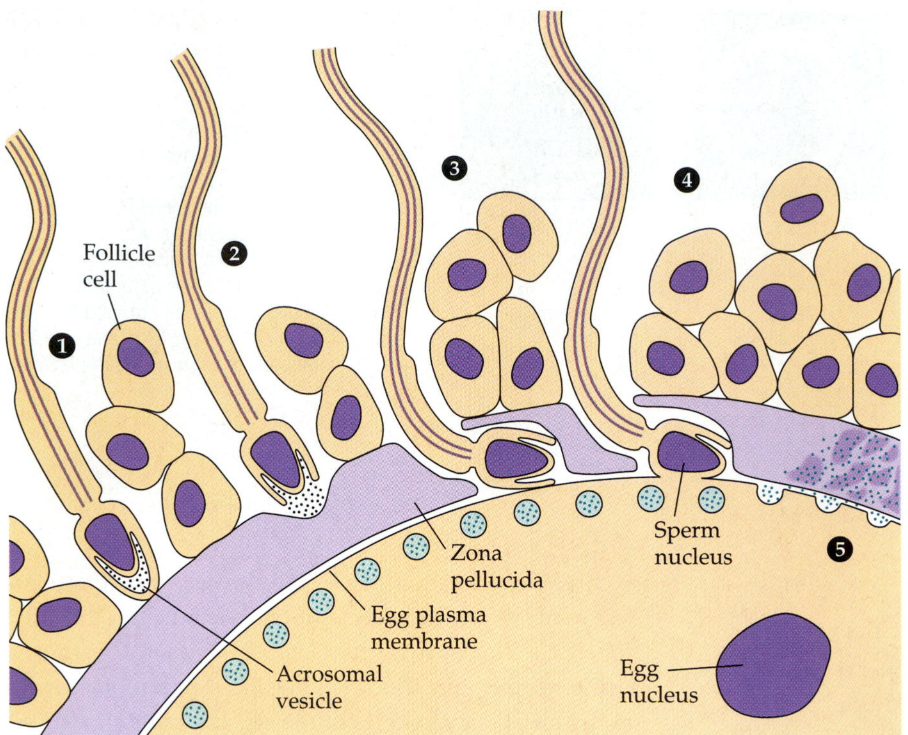

FIGURE 47.5 ▪ Fertilization in mammals. ① The sperm migrates through the coat of follicle cells and binds to receptor molecules in the zona pellucida of the egg. (Receptor molecules are not shown here.) ② This binding induces the acrosomal reaction, in which the sperm releases digestive enzymes into the zona pellucida. ③ With the help of these hydrolytic enzymes, the sperm reaches the plasma membrane of the egg, and membrane proteins of the sperm binds to receptors on the egg membrane. ④ The plasma membranes fuse, making it possible for the contents of the sperm cell to enter the egg. ⑤ Enzymes released during the egg's cortical reaction harden the zona pellucida, which now functions as a block to polyspermy.

in the female reproductive tract, called capacitation, requires about 6 hours in humans (FIGURE 47.5).

The mammalian egg (actually a secondary oocyte at this stage; see Chapter 46) is cloaked by follicle cells that were released with the egg during ovulation. A capacitated sperm cell must migrate through this layer of follicle cells before it reaches the **zona pellucida**, the extracellular matrix of the egg. The zona pellucida consists of three different glycoproteins forming filaments that are cross-linked in a three-dimensional network. One of the glycoproteins, ZP3, also functions as a sperm receptor, binding to a complementary molecule on the surface of the sperm head. (Thus, the zona pellucida is analogous to the vitelline layer of the sea urchin egg, which also has receptors.) The binding of the sperm head to receptor molecules induces the acrosome of the sperm cell to release its contents by exocytosis in an acrosomal reaction similar to that of sea urchin sperm. Protein-digesting enzymes and other hydrolases spilled from the acrosome enable the sperm cell to penetrate the zona pellucida and reach the plasma membrane of the egg. The acrosomal reaction also exposes a protein in the sperm membrane that binds and fuses with the egg membrane.

The binding of a sperm cell to the egg triggers depolarization of the egg membrane, which functions as a fast block to polyspermy, as in sea urchin fertilization. A cortical reaction, in which granules in the cortex of the egg release their contents to the outside of the cell via exocytosis, also occurs. Enzymes released from the cortical granules catalyze alterations of the zona pellucida, which then functions as the slow block to polyspermy.

Fingerlike extensions of the egg cell, called microvilli, take the whole sperm cell, tail and all, into the egg. The basal body of the sperm's flagellum divides and forms two centrosomes (with centrioles) in the zygote. These will generate the mitotic spindle for cell division; unfertilized mammalian eggs have no centrosomes of their own.

In contrast to sea urchin fertilization, the haploid nuclei of sperm and egg do not fuse immediately in mammals. Instead, the envelopes of both nuclei disperse, and the chromosomes from the two gametes share a common spindle apparatus during the first mitotic division of the zygote. Thus, it is not until after this first division, as diploid nuclei form in the two daughter cells, that the chromosomes from the two parents come together in common nuclei to form the genome of the offspring.

Cleavage partitions the zygote into many smaller cells

Fertilization is followed by three successive stages that begin
47.1 to build the animal's body. First, a special type of cell division, called cleavage, creates a multicellular embryo, the blastula, from the zygote. The second stage, gastrulation, produces a three-layered embryo called the gastrula. The third stage, called organogenesis, generates rudimentary organs from which adult structures grow.

Cleavage is a succession of rapid cell divisions that follow fertilization (FIGURE 47.6). During cleavage, the cells undergo the S (DNA synthesis) and M (mitosis) phases of the cell cycle but often virtually skip the G_1 and G_2 phases (see Chapter 12). The embryo does not enlarge during this period of development. Cleavage simply partitions the cytoplasm of one large

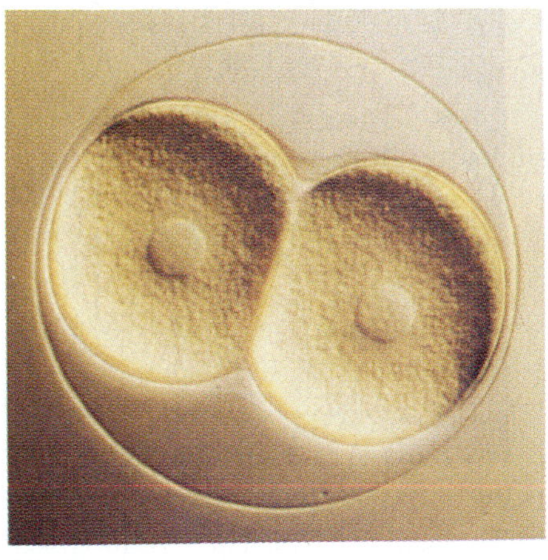

(a)

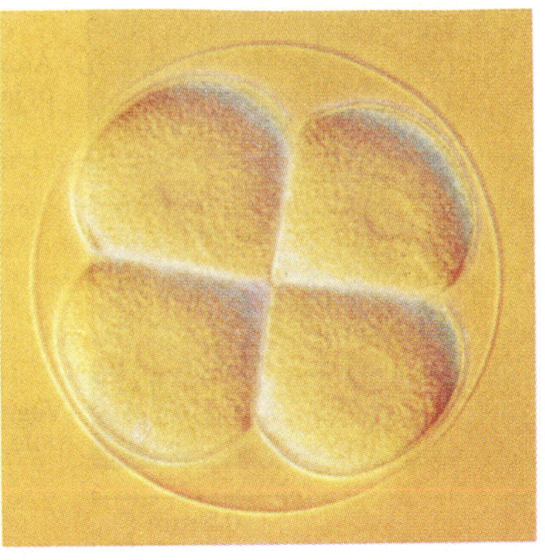

(b)

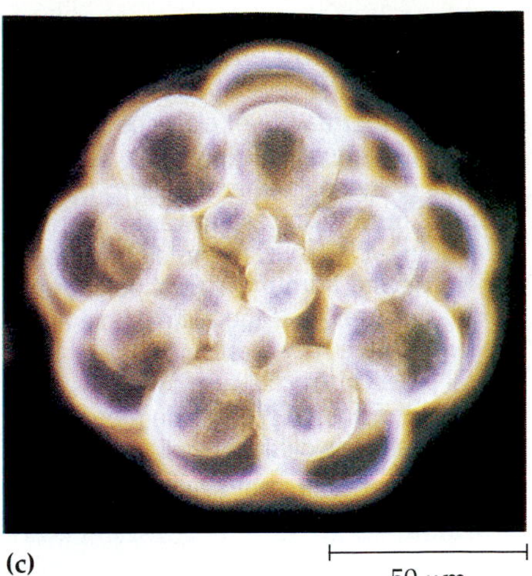

(c)

47.1 **FIGURE 47.6 ▪ Cleavage in an echinoderm (sea urchin) embryo.** Cleavage is a series of mitotic cell divisions that transform the zygote, a single large cell, into a ball of much smaller cells, called blastomeres. These light micrographs of living sea urchin embryos illustrate three stages of cleavage. **(a)** The two-cell stage, following the first cleavage division, occurs about 45–90 minutes after fertilization (notice that the fertilization membrane is still present). **(b)** The second cleavage division produces the four-cell stage. **(c)** In a few hours, repeated cleavage divisions have formed a multicellular ball. The embryo is still retained within the fertilization membrane, from which the larva that develops from the embryo will eventually hatch (LMs).

cell, the zygote, into many smaller cells called **blastomeres**, each with its own nucleus. Thus, different regions of cytoplasm present in the original undivided egg cell end up in separate blastomeres. And because the regions may contain different cytoplasmic components, the partitioning sets the stage for subsequent developmental events.

With the notable exception of mammals, most animals have eggs with a definite polarity. During cleavage in such organisms, the planes of division follow a specific pattern relative to poles of the zygote. The polarity is defined by substances that are heterogeneously distributed in the cytoplasm of the egg, such as mRNA, proteins, and **yolk** (nutrients stored in the egg). In many frogs and other animals, the distribution of yolk is a key factor in influencing the pattern of cleavage. Yolk is most concentrated at one pole of the egg, called the **vegetal pole**, while the opposite pole, the **animal pole**, has the lowest concentration of yolk. The animal pole is also the site where the polar bodies of meiosis are budded from the cell (see Chapter 46), and in some animals it marks the point where the anterior (head) end of the embryo will form.

The hemispheres of the zygote are named for their respective poles. In the eggs of many frogs, the hemispheres have different coloration due to the heterogeneous distribution of cytoplasmic substances. The animal hemisphere has melanin granules embedded in the outer layer of the cytoplasm (the cortex), giving it a deep gray hue, while the vegetal hemisphere contains the yellow yolk (FIGURE 47.7). A rearrangement of the amphibian egg cytoplasm occurs at the time of fertilization. The plasma membrane and cortex rotate toward the point of sperm entry, probably because the centrosome brought into the egg by the sperm cell reorganizes the cytoskeleton. The rotation exposes a light-gray region of cytoplasm, a narrow band called the **gray crescent**. Located near

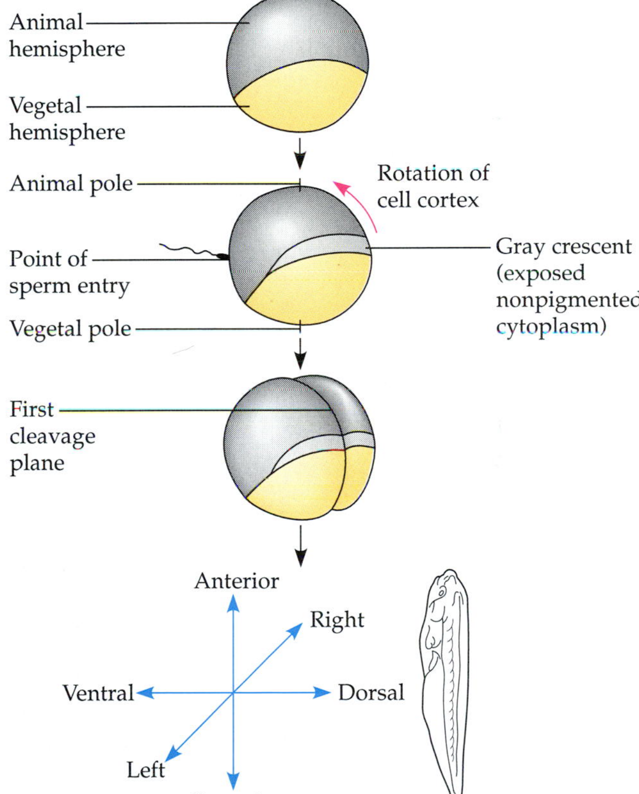

FIGURE 47.7 ▪ The establishment of the body axes and the first cleavage plane in an amphibian. The "animal" hemisphere of the egg, which will become the anterior end of the embryo, is dark gray due to melanin in its outer cytoplasm (cortex); the "vegetal" hemisphere is yellow because it contains the yolk. At fertilization, the pigmented cortex slides over the underlying cytoplasm toward the point of sperm entry, exposing a region of lighter-colored cytoplasm. The exposed region, known as the gray crescent, is opposite the point of sperm entry. The first cleavage division bisects the gray crescent. The crescent marks the location of the dorsal side of the future embryo. Thus, all three axes of the embryo are established before the zygote begins cleavage.

the equator of the egg on the side opposite the point of sperm entry, the gray crescent is an important early marker of the polarity of the amphibian egg.

Because yolk tends to impede cell division, cleavage of the frog zygote occurs more rapidly in the animal hemisphere than in the vegetal hemisphere, resulting in an embryo with different-sized cells. In contrast to frog eggs, those of sea urchins and many other animals have less yolk but still have an animal-vegetal axis, owing to differential distribution of other substances. Without the restraint imposed by yolk, their cleavage divisions all occur at about the same rate, producing blastomeres of virtually equal size.

In both sea urchins and frogs, the first two cleavage divisions are polar (vertical), resulting in four cells that each extend from animal to vegetal pole. The third division is equatorial (horizontal), producing an eight-celled embryo with two tiers of four cells each (FIGURE 47.8a). The general pattern up to this point is the same in both types of embryos. In fact, echinoderms, chordates, and the other animal phyla grouped as deuterostomes share many features of early embryonic development. These similarities distinguish the deuterostomes from the protostomes, the evolutionary branch that includes the mollusks, annelids, and arthropods (see FIGURE 32.6). For example, cleavage in deuterostomes is radial, meaning that the upper (animal) tier of four cells is aligned directly over the lower (vegetal) tier at the eight-cell stage. In contrast, most protostomes exhibit spiral cleavage, in which the cells of the upper tier sit in the grooves between the cells of the lower tier.

Continued cleavage produces a solid ball of cells known as the **morula**, Latin for "mulberry," in reference to the lobed surface of the embryo at this stage (FIGURE 47.8b). A fluid-filled cavity called the **blastocoel** forms within the morula, creating a hollow ball stage of development called the **blastula** (FIGURE 47.8c). In sea urchins, the blastocoel is centrally

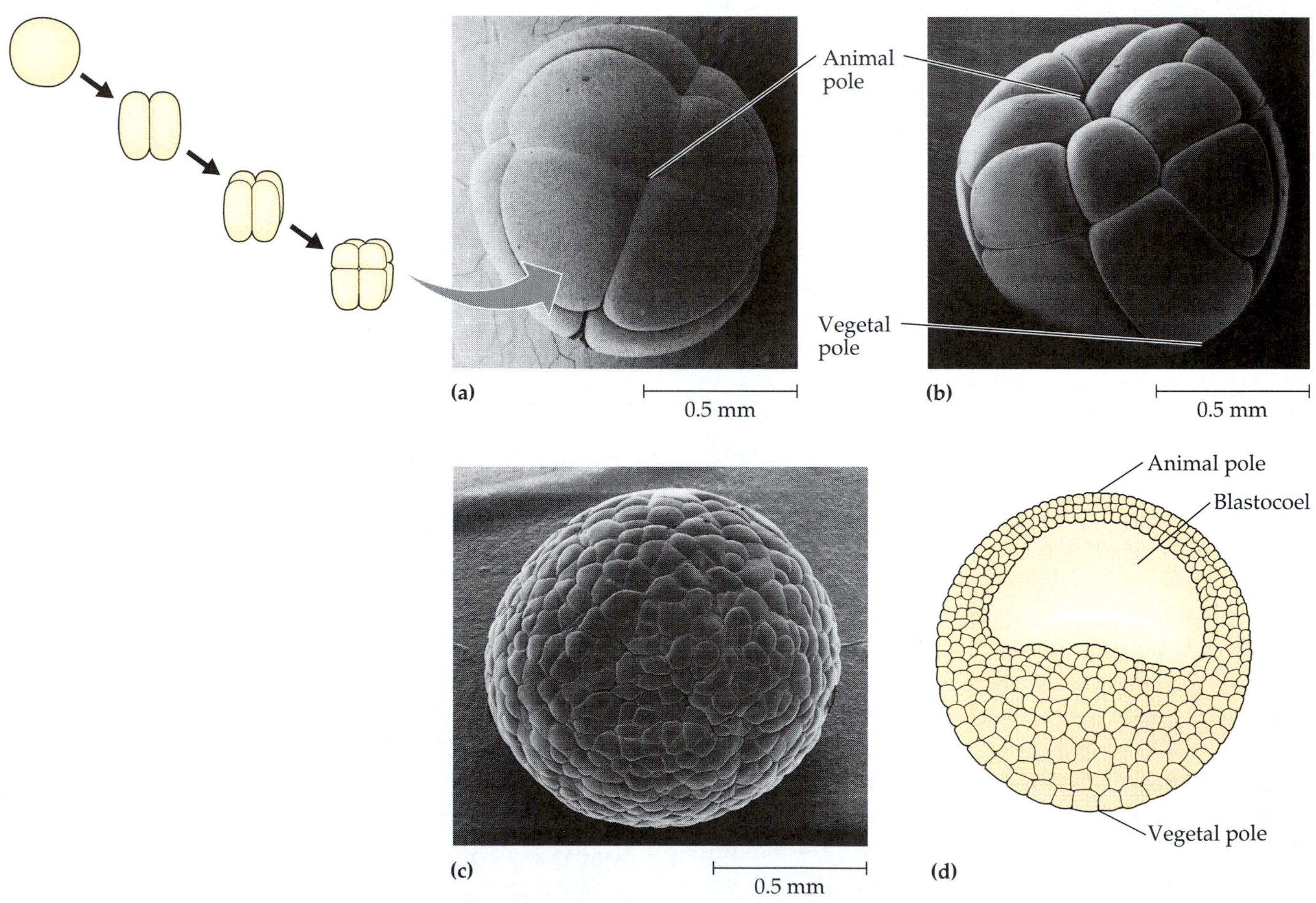

FIGURE 47.8 ▪ Cleavage in a frog embryo.
47.1 Yolk, concentrated near the vegetal pole of the egg, impedes the formation of cleavage furrows. **(a)** After two equal polar divisions, the third cleavage division is perpendicular to the polar axis but is displaced toward the animal pole by yolk. Thus, in this eight-celled embryo viewed from the animal pole (top), the four blastomeres near the animal pole are smaller than the four blastomeres near the vegetal pole. **(b)** As cleavage continues, the cells near the animal pole divide more frequently than the yolk-laden cells near the vegetal pole. A frog embryo consisting of 16–64 cells is called a morula (this is a side view). **(c)** An embryo of 128 cells is a blastula. **(d)** As shown in this diagram of a cross section, the frog blastula contains a blastocoel surrounded by several layers of cells and situated in the animal hemisphere. (a, b, and c are SEMs.)

located in the blastula. However, in frogs, because of unequal cell division the blastocoel is located in the animal hemisphere (FIGURE 47.8d).

Yolk is most plentiful and has its most pronounced effect on cleavage in the eggs of birds, reptiles, many fishes, and insects. In birds, for example, the part of the egg we commonly call the yolk is actually the egg cell (ovum), swollen with yolk nutrients. This enormous cell is surrounded by a protein-rich solution (the egg white) that will provide additional nutrients for the growing embryo. Cleavage of the fertilized egg is restricted to a small disc of yolk-free cytoplasm at the animal pole of the egg cell. This incomplete division of a yolk-rich egg is known as **meroblastic cleavage.** It contrasts with **holoblastic cleavage**, the term for the complete division of eggs having little yolk (as in sea urchins) or a moderate amount of yolk (as in frogs).

The yolk-rich eggs of insects, such as *Drosophila,* undergo a unique type of meroblastic cleavage (see FIGURE 21.9). After fertilization, the zygote's nucleus is situated within a mass of yolk. Cleavage begins with the nucleus undergoing mitotic divisions that are not accompanied by cytokinesis. These mitotic divisions produce several hundred nuclei, which migrate to the outer edge of the egg. After several more rounds of mitosis, plasma membranes form around the nuclei, and the embryo, now a blastula, consists of a single layer of about 6000 cells surrounding a mass of yolk.

Gastrulation rearranges the blastula to form a three-layered embryo with a primitive gut

47.1 The morphogenetic process called **gastrulation** is a dramatic rearrangement of the cells of the blastula. Gastrulation differs in detail from one animal group to another, but a common set of cellular changes drives this spatial rearrangement of an embryo. These general cellular mechanisms are changes in cell motility, changes in cell shape, and changes in cellular adhesion to other cells and to molecules of the extracellular matrix (see Chapter 7). The essential result of gastrulation is that some of the cells at or near the surface of the blastula move to a new, more interior location. This transforms the blastula into a three-layered embryo called the **gastrula**.

The three layers produced by gastrulation are embryonic tissues called **ectoderm**, **endoderm**, and **mesoderm**, also collectively termed the embryonic germ layers. The ectoderm forms the outer layer of the gastrula; the endoderm lines the embryonic digestive tract; and the mesoderm partly fills the space between the ectoderm and the endoderm. Eventually, these three cell layers develop into all the parts of the adult animal. For instance, our nervous system and the outer layer (epidermis) of our skin come from ectoderm; the innermost lining of our digestive tract and associated organs, such as the liver and pancreas, arise from endoderm; and most other organs and tissues, such as the kidney, heart, muscles, and the inner layer of our skin (dermis), develop from mesoderm.

Let's examine gastrulation in a sea urchin embryo (FIGURE 47.9). The wall of the sea urchin blastula consists of a single layer of cells. Gastrulation begins at the vegetal pole where cells detach from the blastula wall and enter the blastocoel as

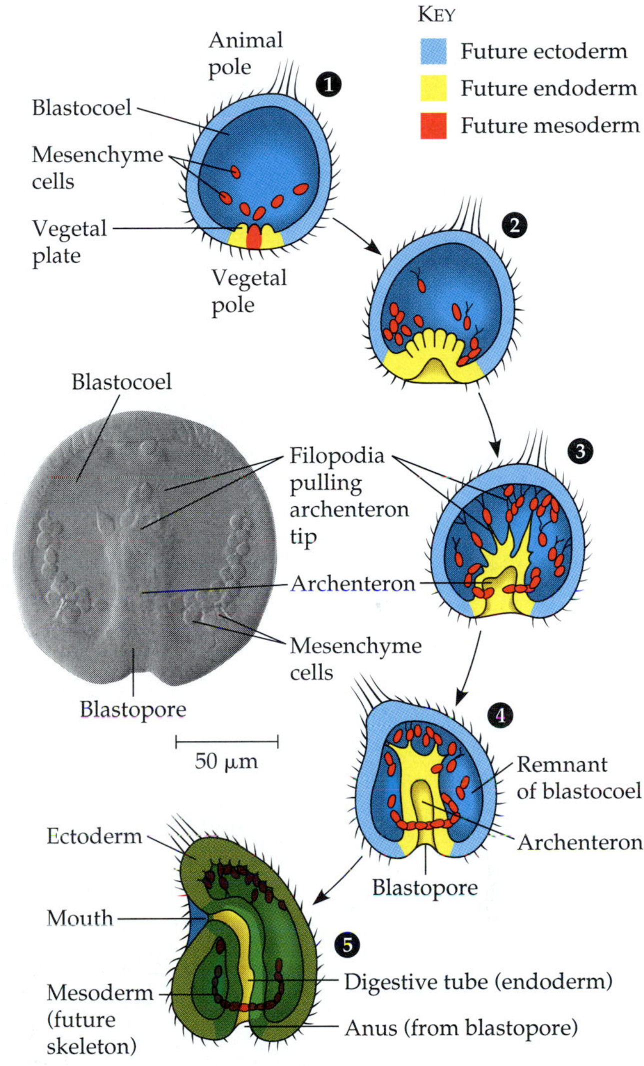

47.1 **FIGURE 47.9 ▪ Sea urchin gastrulation.** ① Formed by cleavage, the blastula consists of a single layer of ciliated cells surrounding the blastocoel. Gastrulation begins with the formation of the vegetal plate at the vegetal pole. Mesenchyme cells (future mesoderm) detach from the vegetal plate and migrate into the blastocoel. ② The vegetal plate in this early gastrula invaginates (buckles inward). Mesenchyme cells begin to form thin extensions (filopodia). ③ Endoderm cells form the archenteron (future digestive tube). Mesenchyme cells form filopodial connections between the tip of the archenteron and the ectoderm cells of the blastocoel wall (inset, LM). ④ Contraction of the filopodia in a late gastrula drags the archenteron the rest of the way across the blastocoel, where the endoderm of the archenteron will fuse with ectoderm of the blastocoel wall. ⑤ Gastrulation is complete. The gastrula has a functional digestive tube formed from the endoderm of the archenteron, with a mouth and an anus. Ectoderm forms the embryo's ciliated outer surface. Some of the mesenchyme cells of the mesoderm have secreted minerals that will form a simple internal skeleton.

migratory cells called mesenchyme cells. The remaining cells flatten slightly to form a vegetal plate that buckles inward, a process called **invagination**. The buckled vegetal plate then undergoes extensive rearrangement of its cells, a process that transforms the shallow invagination into a deeper, narrower pouch called the **archenteron**, or primitive gut. The open end of the archenteron, which will become the anus, is called the **blastopore**. A second opening forms at the other end of the archenteron, forming the mouth end of what is now a rudimentary digestive tube. Gastrulation has produced an embryo with a primitive gut and three germ layers: ectoderm (color-coded blue throughout this chapter), endoderm (yellow), and mesoderm (red). Thus, the triploblastic (three-layered) body plan characteristic of most animal phyla is established very

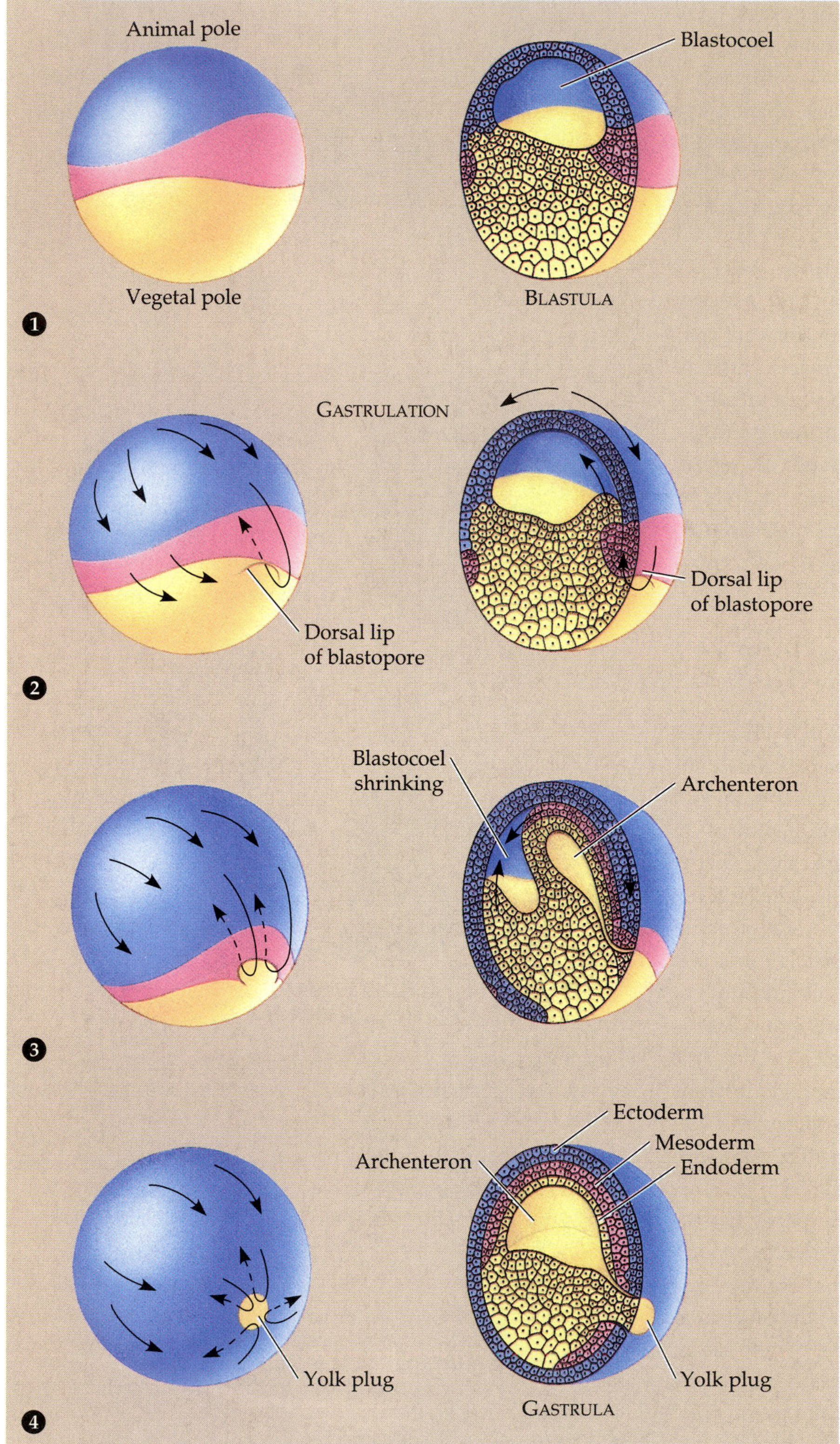

FIGURE 47.10 ▪ Gastrulation in a frog embryo. ① The blastocoel of the frog blastula is off-center and surrounded by a wall that is more than one cell thick. At this stage, the colors indicate regions of the blastula that will become the embryo's three germ layers. ② Gastrulation begins when a small tuck, the dorsal lip of the blastopore, appears on one side of the blastula. The tuck is formed by cells invaginating inward from the surface. Additional cells that will become endoderm and mesoderm then roll inward over the dorsal lip (involution) and move away from the blastopore into the interior of the gastrula. Meanwhile, cells of the animal pole, which will form ectoderm, spread over the embryo's outer surface. ③ Externally, the lip of the blastopore starts becoming circular. Internally, the three germ layers start forming as cells continue migrating inward. The advancing endoderm, mesoderm, and the archenteron, lined by endoderm, are filling the space occupied by the blastocoel. ④ Late in gastrulation, the circular blastopore surrounds a plug of yolk cells (the yolk plug). The three germ layers are now in place, ready for organogenesis.

KEY
- Future ectoderm
- Future endoderm
- Future mesoderm

early in development (see Chapter 32). In the sea urchin, the gastrula eventually develops into a ciliated larva that drifts near the ocean surface as plankton, feeding on bacteria and unicellular algae.

Gastrulation during frog development also produces a three-layered embryo with an archenteron, as shown in FIGURE 47.10. The mechanics of gastrulation are much more complicated in a frog, however, because of the large, yolk-laden cells of the vegetal hemisphere, and because the wall of the blastula is more than one cell thick in most species. The first sign of gastrulation is a small crease on one side of the blastula caused by the invagination of a cluster of cells there. This invagination becomes the blastopore. The cells at the upper edge of the blastopore opening become the **dorsal lip** of the blastopore. The dorsal lip of the blastopore forms where the gray crescent was located in the zygote (see FIGURE 47.7). Gastrulation continues with cells on the surface of the embryo rolling over the edge of the dorsal lip into the interior of the embryo, a process called **involution**. Once inside the embryo, these cells move away from the blastopore along the roof of the blastocoel. Involution continues, with migrating internal cells becoming organized into layered mesoderm and endoderm, and the archenteron forming within the endoderm. Eventually, the complex cell movements of gastrulation produce a three-layered embryo. As the process is completed, the lip of the blastopore encircles a **yolk plug** consisting of large, food-laden cells from the vegetal pole of the embryo. Except for the yolk plug, cells remaining on the surface make up the ectoderm, surrounding the layers of mesoderm and endoderm. With the three germ layers in place, gastrulation is complete, and the embryo's organs begin to form.

Table 47.1 ▪ Derivatives of the Three Embryonic Germ Layers in Vertebrates

GERM LAYER	ORGANS AND TISSUES IN THE ADULT
Ectoderm	Epidermis of skin and its derivatives (e.g., skin glands, nails); epithelial lining of mouth and rectum; sense receptors in epidermis; cornea and lens of eye; nervous system; adrenal medulla; tooth enamel; epithelium of pineal and pituitary glands.
Endoderm	Epithelial lining of digestive tract (except mouth and rectum); epithelial lining of respiratory system; liver; pancreas; thyroid; parathyroids; thymus; lining of urethra, urinary bladder, and reproductive system.
Mesoderm	Notochord; skeletal system; muscular system; circulatory and lymphatic systems; excretory system; reproductive system (except germ cells, which start to differentiate during cleavage); dermis of skin; lining of body cavity; adrenal cortex.

In organogenesis, the organs of the animal body form from the three embryonic germ layers

47.2 Various regions of the three germ layers develop into the rudiments of organs during the process of **organogenesis** (TABLE 47.1). Three kinds of morphogenetic changes—folds, splits, and dense clustering (condensation) of cells—are the first evidence of organ building. The organs that begin to take shape first in the embryos of frogs and other chordates are the neural tube and notochord, the skeletal rod characteristic of all chordate embryos (see Chapter 34).

FIGURE 47.11 on p. 946 shows early organogenesis as it occurs in a frog. The **notochord** is formed from dorsal mesoderm that condenses just above the archenteron, and the neural tube originates as a plate of dorsal ectoderm just above the developing notochord. The neural plate soon folds inward, rolling itself into the **neural tube**, which will become the central nervous system—the brain and spinal cord. These organs are hollow in most chordates because of this mechanism of development. In frogs the notochord elongates and stretches the embryo along its anterior-posterior axis. Later, the notochord will function as a core around which mesodermal cells gather and form the vertebrae.

Other condensations occur in strips of mesoderm lateral to the notochord, which separate into blocks called **somites**. The somites are arranged serially on both sides along the length of the notochord (FIGURE 47.11C). Cells from the somites not only give rise to the vertebrae of the backbone, they also form the muscles associated with the axial skeleton. This serial origin of the axial skeleton and muscles reinforces the point made in Chapter 34 that chordates are basically segmented animals, although the segmentation becomes less obvious later in development. (There are signs of segmentation even in the adult, as in the series of vertebrae in a human or the segments of chevron-shaped muscles in a fish.) Lateral to the somites, the mesoderm splits into two layers that form the lining of the body cavity, or coelom.

As organogenesis progresses, morphogenesis and cellular differentiation continue to refine the organs that arise from the three embryonic germ layers. Review TABLE 47.1, which lists the embryonic sources of the major organs and tissues in frogs and other vertebrates. Unique to vertebrate embryos, a band of cells called the *neural crest* develops along the border where the neural tube pinches off from the ectoderm (see Chapter 34). Cells of the neural crest subsequently migrate to various parts of the embryo, forming pigment cells of the skin, some of the bones and muscles of the skull, the teeth, the medulla of the adrenal glands, and peripheral components of the nervous system, such as sensory and sympathetic ganglia.

Embryonic development of the frog leads to a larval stage, the tadpole, which hatches from the jelly coat that originally cloaked the egg. Later, metamorphosis will transform the frog from the aquatic, herbivorous tadpole to the terrestrial, carnivorous adult.

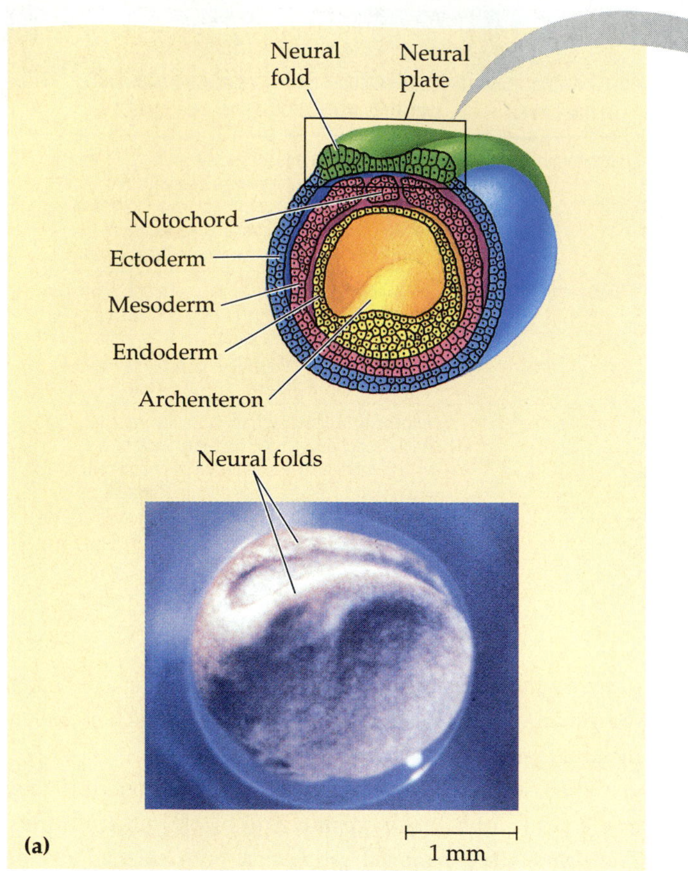

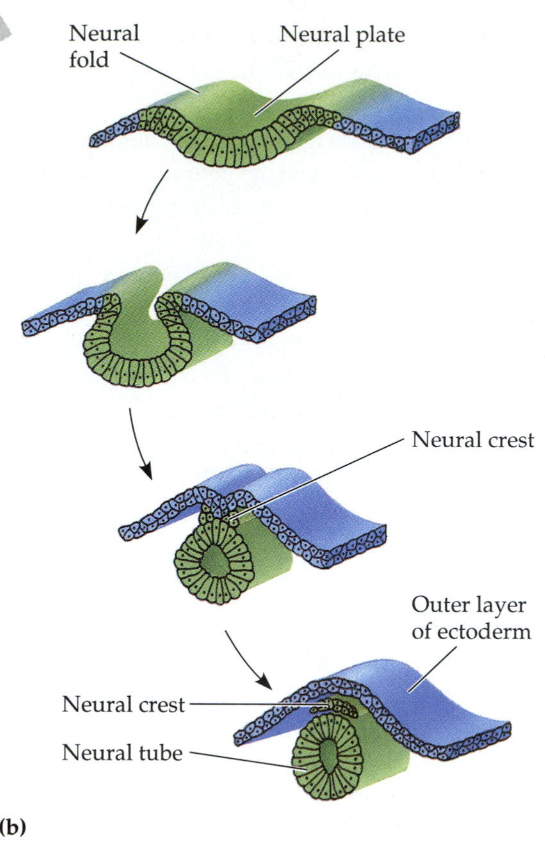

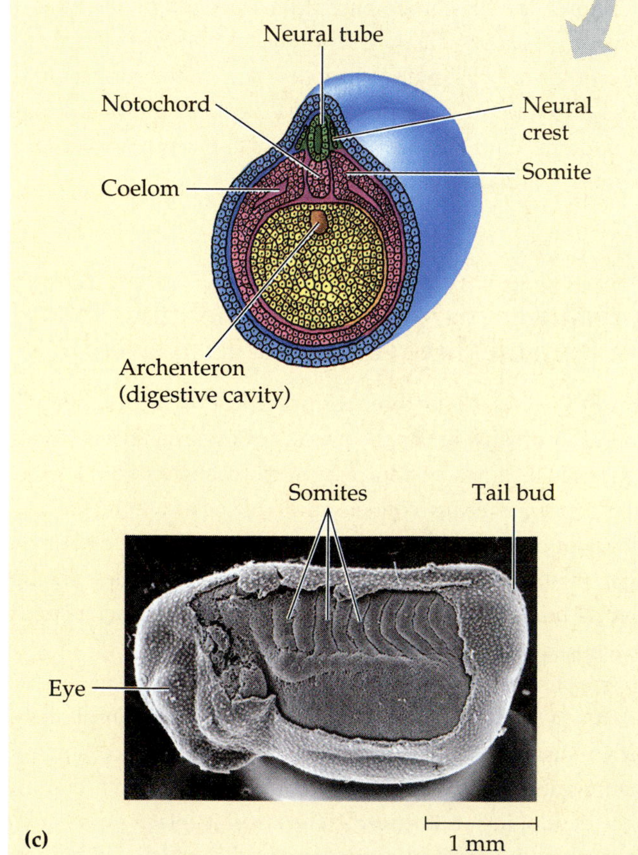

FIGURE 47.11 ▪ Organogenesis in a frog embryo. (a) A cross section of a frog embryo at the beginning of organogenesis shows the three germ layers and the rudiments of the notochord and neural plate. The notochord has developed from dorsal mesoderm, and the dorsal ectoderm has thickened, forming the neural plate. Two pronounced ridges, the neural folds, form the lateral edges of the neural plate. The light micrograph shows a whole frog embryo at the neural plate stage of development. **(b)** These diagrams illustrate how the neural plate rolls into a hollow cylinder, the neural tube. Tissue at the meeting margins of the tube separate from the tube as the neural crest, a source of migrating cells that eventually form many structures, including bones and muscles of the skull, skin pigment cells, adrenal medulla cells, and peripheral ganglia of the nervous system. **(c)** In this cross section, an embryo with a completed neural tube has somites flanking the notochord. Formed from mesoderm, the somites will give rise to segmental structures such as vertebrae and serially arranged skeletal muscles. The lateral mesoderm has begun to separate into the two tissue layers that line the coelom. In the scanning electron micrograph of a whole embryo at the tail-bud stage (side view), part of the ectoderm has been removed to reveal the somites.

47.2

Amniote embryos develop in a fluid-filled sac within a shell or uterus

All vertebrate embryos require an aqueous environment for development. In the case of fish and amphibians, the egg is laid in the surrounding sea or pond and needs no special water-filled enclosure. The vertebrate move onto land required solving the problem of reproduction in dry environments, and two major solutions evolved: the shelled egg of reptiles and birds and the uterus of placental mammals. Within the shell or uterus, the embryos of birds, reptiles, and mammals are surrounded by fluid within a sac formed by a membrane called the amnion. Vertebrates of these three classes are therefore called **amniotes** (see Chapter 34). We have already examined the embryonic development of a vertebrate that lacks an amnion, the frog. For comparison, we now study the early development of two amniotes, a bird and a mammal.

Avian Development

After fertilization, a bird egg undergoes meroblastic cleavage in which cell division occurs only in a small region of yolk-free cytoplasm atop the large mass of yolk. The early cleavage divisions produce a cap of cells called the **blastodisc**, which rests on the undivided yolk. The blastomeres then sort into upper and lower layers, the epiblast and hypoblast (FIGURE 47.12, step ①). The cavity between these two layers is the avian version of the blastocoel, and this embryonic stage is the avian equivalent of the blastula, although its form is different from the hollow ball of an early frog embryo.

Gastrulation, as in the frog embryo, involves cells moving from the surface of the embryo to an interior location. In birds, however, the route of this cell migration is very different (FIGURE 47.12, step ②). Some cells of the upper cell layer (epiblast) move toward the midline of the blastodisc, then detach and move inward toward the yolk. The medial movement on the surface and the inward movement of cells at the blastodisc's midline produce a groove called the **primitive streak**. As the streak lengthens over the surface of the blastodisc, it marks what will become the bird's anterior-posterior axis. The primitive streak is functionally equivalent to the frog blastopore, but it is a linear tuck rather than a ring.

All the cells that will form the embryo come from the epiblast. Some of the epiblast cells that pass through the primitive streak move laterally into the blastocoel, producing the mesoderm. Other epiblast cells, which will produce the endoderm, migrate through the streak and downward, mingling with cells of the hypoblast. The epiblast cells that remain on the surface give rise to the ectoderm. Although the hypoblast contributes no cells to the embryo, it seems to help direct the formation of the primitive streak and is required for normal development. Eventually segregating from the endoderm, the

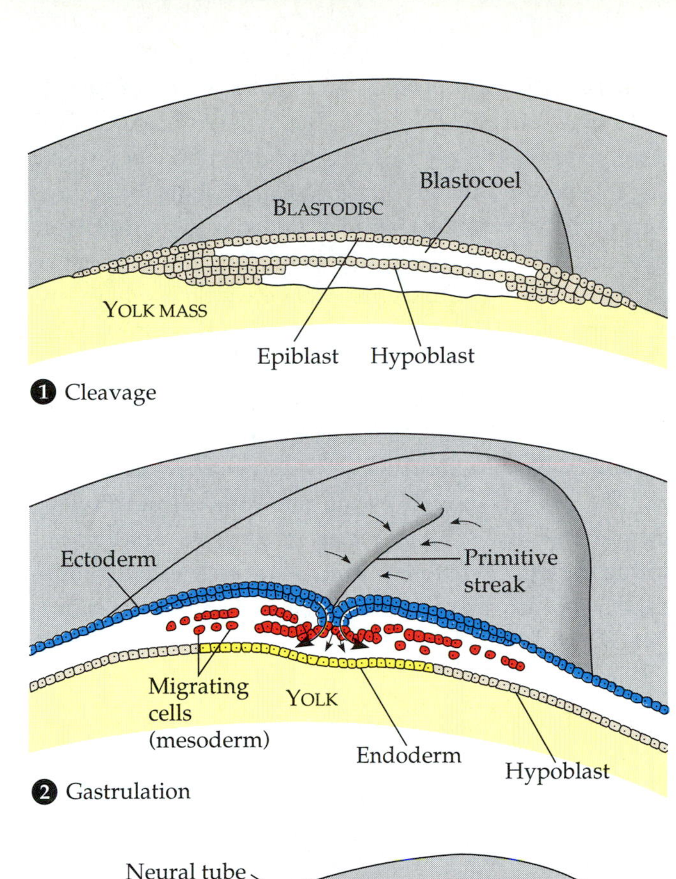

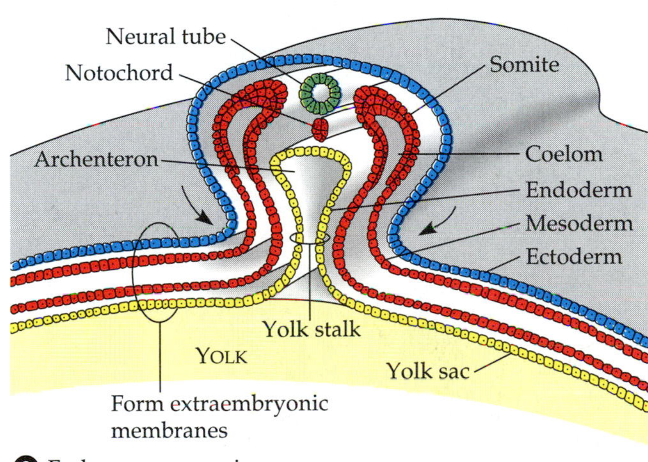

FIGURE 47.12 ▪ Cleavage, gastrulation, and early organogenesis in a chick embryo. ① For the eggs of birds and reptiles, which store a very large amount of yolk, cleavage is meroblastic, or incomplete. Cell division is restricted to a small cap of cytoplasm at the animal pole. Cleavage produces a blastodisc that rests on the large, undivided mass of yolk. The blastodisc becomes arranged into two layers (epiblast and hypoblast) that bound the blastocoel, forming the avian version of a blastula. ② During gastrulation, some cells of the epiblast migrate (arrows) into the interior of the embryo through the primitive streak, shown here in transverse section. Some of these cells move laterally to form mesoderm, while others migrate downward to form the endoderm. ③ The rudimentary gut (archenteron) is formed when lateral folds pinch the embryo away from the yolk. About midway along its length, the embryo will remain attached to the yolk by the yolk stalk formed mainly by hypoblast cells. The notochord, neural tube, and somites form much as they do in the frog. The three germ layers and hypoblast cells also contribute to the system of extraembryonic membranes that support further development.

hypoblast cells form portions of a sac surrounding the yolk and a stalk connecting the yolk mass to the embryo. After the three germ layers are formed, the borders of the embryonic disc fold downward and come together, pinching the embryo into a three-layered tube joined at midbody to the yolk (FIGURE 47.12, step ③). Neural tube formation, development of the notochord and somites, and other events in organogenesis occur much as in the frog embryo. FIGURE 47.13 shows some of the organs in a 2-day-old chick embryo.

Notice in FIGURE 47.12, step 3, that only part of each germ layer contributes to the embryo itself (which is the protrusion that arises from the blastodisc). The tissue layers that are outside the embryo proper develop into four **extraembryonic membranes** that support further embryonic development within the egg. These four "membranes," each a sheet of cells, are the **yolk sac**, the **amnion**, the **chorion**, and the **allantois** (FIGURE 47.14).

Mammalian Development

In most mammalian species, fertilization takes place in the oviduct, and the earliest stages of development occur while the embryo completes its journey down the oviduct to the uterus (see Chapter 46). In contrast to the large, yolky eggs of birds and reptiles, the egg of a placental mammal is quite small, storing little in the way of food reserves. Furthermore, mammalian eggs apparently lack the localized, maternally encoded cytoplasmic determinants that help establish polarity in the eggs of many other animals. As already mentioned the mammalian egg and zygote do not exhibit any obvious polarity, and cleavage of the yolk-lacking zygote is holoblastic. However, mammalian gastrulation and early organogenesis follow a pattern similar to that of birds and reptiles. (Recall from Chapter 34 that mammals descended from reptilian stock during the early Mesozoic era.)

Cleavage is relatively slow in mammals. In the case of humans, the first division is complete about 36 hours after fertilization, the second division at about 60 hours, and the third division at about 72 hours. The cleavage planes seem to be randomly oriented, and the blastomeres are equal in size. An important event during early mammalian development is

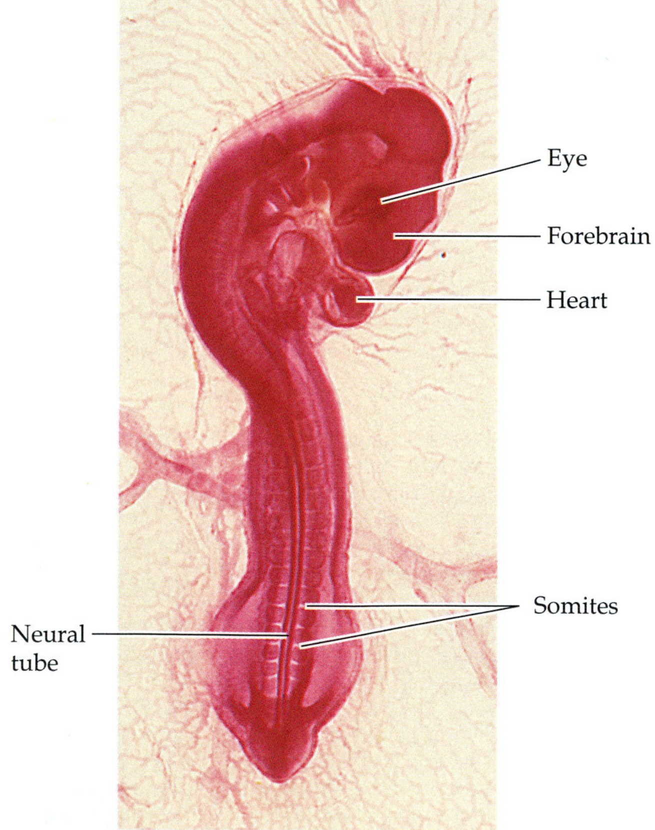

FIGURE 47.13 ▪ Organogenesis in a chick embryo. Rudiments of most major organs have already formed in this chick, which is about 56 hours old (LM).

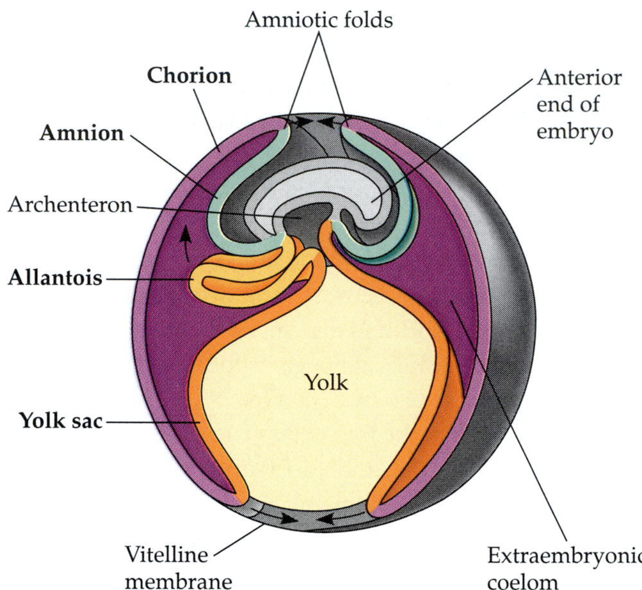

FIGURE 47.14 ▪ The development of extraembryonic membranes in a chick. Each of the four major membranes (labeled in bold type) providing support services for the embryo is a sheet of cells that develops from epithelial sheets external to the embryo proper. The yolk sac expands over the surface of the yolk mass. Cells of the yolk sac will digest yolk, and blood vessels that develop within the membrane will carry nutrients into the embryo. Lateral folds of extraembryonic tissue extend over the top of the embryo and fuse to form two more membranes, the amnion and the chorion, which are separated by an extraembryonic extension of the coelom. The amnion encloses the embryo in a fluid-filled amniotic sac, protecting the embryo from desiccation, and it, along with the chorion, cushions the embryo against mechanical shocks. The fourth membrane, the allantois, originates as an outpocketing of the embryo's hindgut. The allantois is a sac that extends into the extraembryonic coelom. It functions as a disposal sac for uric acid, the insoluble nitrogenous waste of the embryo. As the allantois continues to expand, it presses the chorion against the vitelline membrane, the inner lining of the eggshell. Together, the allantois and chorion form a respiratory organ that serves the embryo. Blood vessels that form in the epithelium of the allantois transport oxygen to the embryonic chick. The extraembryonic membranes of reptiles and birds are adaptations associated with the special problems of development on land.

the process of *compaction,* which occurs at the eight-cell stage. Before compaction, the cells of the early embryo are loosely packed; after compaction the cells tightly adhere to one another. Compaction involves the production of new proteins on the surface of the cells, including proteins called cadherins (discussed later in the chapter).

Further development of the human embryo is shown in FIGURE 47.15. By about 7 days after fertilization, the embryo has over 100 cells arranged around a central cavity. This is the embryonic stage known as the **blastocyst**. Protruding into one end of the blastocyst cavity is a cluster of cells called the **inner cell mass**, which will subsequently develop into the embryo proper and some of the extraembryonic membranes. The outer epithelium surrounding the cavity is the **trophoblast**, which, along with mesodermal tissue, will form the fetal portion of the placenta.

The embryo reaches the uterus by the blastocyst stage and begins to implant soon thereafter. The trophoblast secretes enzymes that enable the blastocyst to penetrate the uterine lining. Bathed in blood spilled from eroded capillaries in the endometrium (uterine lining), the trophoblast thickens and extends fingerlike projections into the surrounding maternal tissue. The placenta eventually forms from this proliferated trophoblast and the region of endometrium it invades (see FIGURE 46.17). About the time the blastocyst implants in the uterus, the inner cell mass forms a flat disc with an upper layer of cells, the epiblast, and a lower layer, the hypoblast. These layers are homologous to those of birds, and, as in birds, the embryo develops entirely from epiblast cells, while the hypoblast cells form the yolk sac. Gastrulation occurs by the inward movement of cells from the upper layer through a primitive streak to form mesoderm and endoderm, just as it does in the chick.

Four extraembryonic membranes homologous to those of reptiles and birds form during mammalian development. The chorion, which develops from the trophoblast, completely surrounds the embryo and the other extraembryonic membranes. The amnion begins as a dome above the proliferating epiblast and eventually encloses the embryo in a fluid-filled amniotic cavity. (The fluid from this cavity is the "water" expelled from the vagina of the mother when the amnion breaks just prior to childbirth.) Below the developing embryo proper, the yolk sac encloses a fluid-filled cavity. Although this cavity contains no yolk, the membrane that surrounds it is given the same name as the homologous membrane in birds and reptiles. The yolk sac membrane of mammals is a site of early formation of blood cells, which later migrate into the embryo proper. The fourth extraembryonic membrane, the allantois, develops as an outpocketing of the embryo's rudimentary gut, as it does in the chick. The allantois is incorporated into the umbilical cord, where it forms blood vessels that transport oxygen and nutrients from the placenta to the embryo and rid the embryo of carbon dioxide and

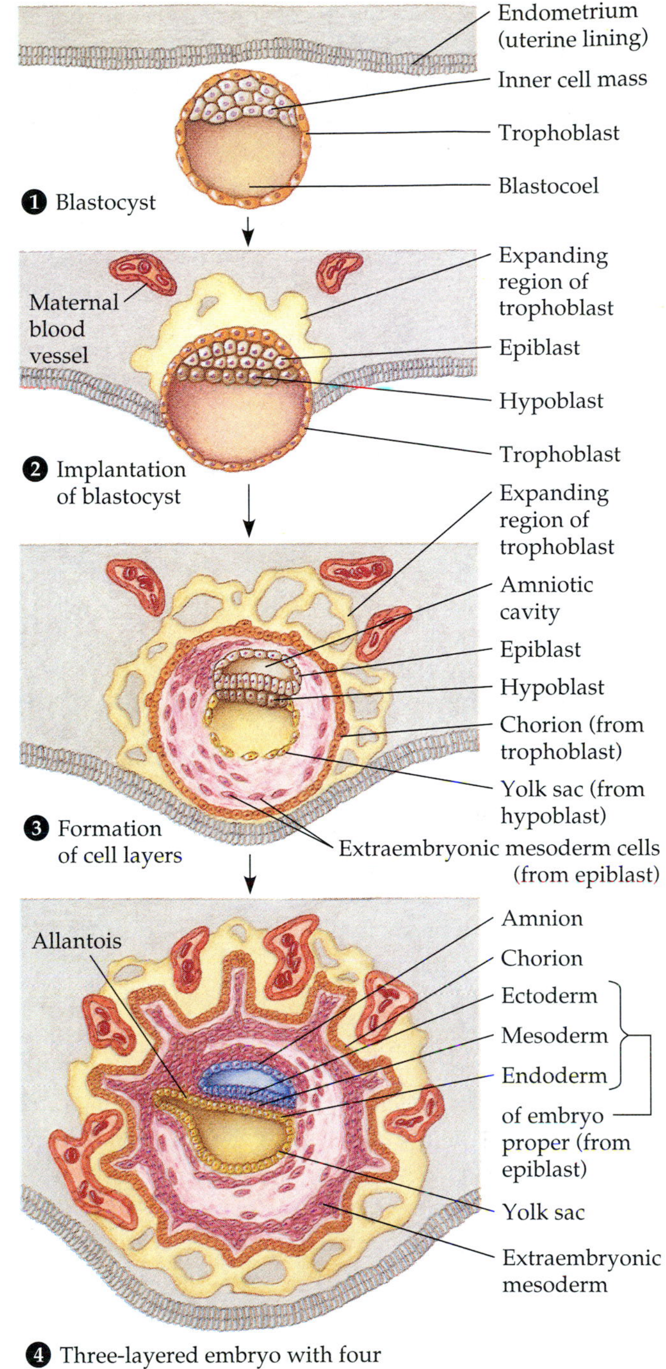

FIGURE 47.15 ▪ Early development of a human embryo and its extraembryonic membranes. This series of drawings illustrates four stages in transverse section. ① Cleavage produces a blastocyst, consisting of a trophoblast surrounding a blastocoel and an inner cell mass. The blastocyst implants in the uterine lining. ② Coincident with implantation, the inner cell mass forms an epiblast cell layer, which will develop into the three germ layers of the embryo, and a hypoblast, which will form the yolk sac. ③ By this stage, the trophoblast has begun to form the chorion and continues to expand into the endometrium. The epiblast has begun to form the amnion, surrounding a fluid-filled cavity. Mesodermal cells that will become part of the placenta are also derived from the epiblast. ④ Gastrulation by the inward movement of epiblast cells has produced a three-layered embryo surrounded by proliferating extraembryonic mesoderm. Note also the four extraembryonic membranes.

nitrogenous wastes. Thus, the extraembryonic membranes of shelled eggs, where embryos are nourished with yolk, were conserved as mammals diverged from reptiles in the course of evolution, but with modifications adapted to development within the reproductive tract of the mother.

Organogenesis begins with the formation of the neural tube, notochord, and somites. By the end of the first trimester of human development, rudiments of all the major organs have developed from the three germ layers, as summarized in TABLE 47.1.

THE CELLULAR AND MOLECULAR BASIS OF MORPHOGENESIS AND DIFFERENTIATION IN ANIMALS

Morphogenesis in animals involves specific changes in cell shape, position, and adhesion

Morphogenesis is a major aspect of development in both animals and plants, but only in animals does it involve the *movement* of cells. Movement of parts of a cell can bring about changes in cell shape or enable a cell to migrate from one place to another within the embryo. Changes in both cell shape and cell position are involved in cleavage, gastrulation, and organogenesis.

Changes in the shape of a cell usually involve reorganization of the cytoskeleton (see Chapter 7). Consider, for example, how the cells of the neural plate form the neural tube (FIGURE 47.16). First, microtubules oriented parallel to the dorsal-ventral axis of the embryo apparently help lengthen the cells in that direction. At the dorsal end of each cell is a parallel array of actin filaments oriented crosswise. These contract, giving the cells a wedge shape that forces the ectoderm layer to bend inward. Similar cell-shape changes are observed for other invaginations (inpocketings) and evaginations (outpocketings) of tissue layers throughout development.

The cytoskeleton also drives the active movement of cells from one place to another in developing animals. Cells "crawl" within the embryo by using cytoskeletal fibers to extend and retract cellular protrusions. This type of motility is akin to the amoeboid movement described in FIGURE 7.27b, but in contrast to the thick pseudopodia of amoeboid cells, the cellular protrusions of migrating embryonic cells are usually flat sheets (lamellipodia) or spikes (filopodia). In the gastrulation of some organisms, invagination is initiated by a wedging of cells on the surface of the blastula, but the penetration of cells deeper into the embryo involves the extension of filopodia by cells on the leading edge of the migrating tissue. By dragging behind them the cells that follow them

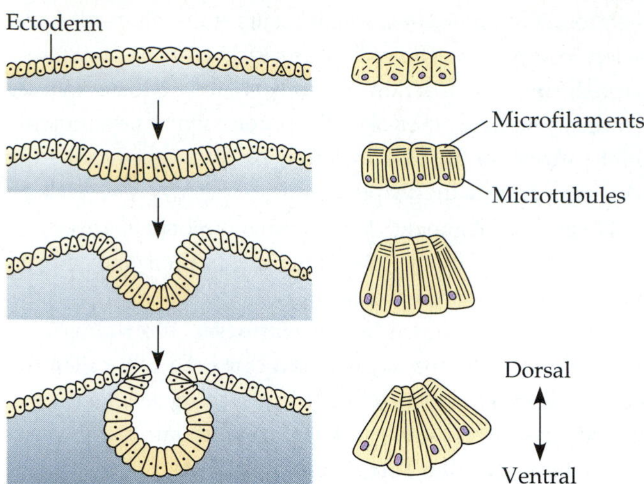

FIGURE 47.16 ▪ Change in cellular shape during morphogenesis. Reorganization of the cytoskeleton is associated with morphogenetic changes in embryonic tissues. As indicated here, in the formation of the neural tube of vertebrates, microtubules help elongate the cells of the neural plate. Microfilaments at the dorsal end of the cells then contract, deforming the cells into wedge shapes and causing the ectoderm to bend inward.

through the blastopore, the leading cells help move a sheet of cells from the embryo's surface into the blastocoel, where it forms the endoderm and mesoderm of the embryo (see FIGURE 47.9). There are also many situations in which cells migrate individually, as when the cells of the neural crest disperse to various parts of the embryo.

Cell crawling is also involved in a type of morphogenetic movement called **convergent extension** (FIGURE 47.17). In convergent extension, the cells of a tissue layer rearrange themselves in such a way that the sheet of cells becomes narrower (converges) while it becomes longer (extends). When many cells wedge between one another, the tissue can extend dramatically. Convergent extension is important in early embryonic development. It occurs, for example, as the archenteron elongates in sea urchin embryos and during involution in the frog gastrula.

Scientists do not yet understand exactly what triggers and directs convergent extension. However, the process probably involves the extracellular matrix (ECM), the mixture of secreted glycoproteins lying outside the plasma membranes of cells (see Chapter 7). The ECM is known to help guide cells in many types of morphogenetic movements. ECM fibers may function as tracks, directing migrating cells along particular routes (FIGURE 47.18). Several kinds of extracellular glycoproteins, including various fibronectins, help cells move by providing anchorage for cell crawling. Other substances in the ECM keep cells on the correct paths by *inhibiting* migration in certain directions. Thus, depending on the substances they secrete, nonmigratory cells situated along migration pathways may promote or inhibit movement of other cells. As migrating cells move along specific paths through the embryo,

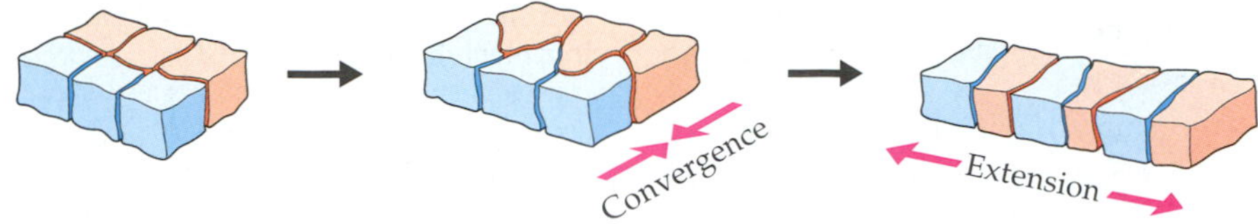

FIGURE 47.17 ▪ Convergent extension of a sheet of cells. In this simplified diagram, changes in cell shape and position cause the layer of cells to become narrower and longer. An unknown signal causes the cells to elongate in a particular directon and to crawl between each other. The result of this convergence is the extension of the cell sheet in a direction perpendicular to the convergence.

receptor proteins on their surfaces pick up directional cues from the immediate environment. Signals from the receptors direct the cytoskeletal elements to propel the cell in the proper direction.

A good example of the role of the ECM in embryonic cell migration is the movement of mesoderm cells along fibronectin during amphibian gastrulation. Fibronectin fibers line the roof of the blastocoel, and as the mesoderm moves into the interior of the embryo, the cells at the free edge of the mesoderm sheet migrate along the fibers. Researchers can disrupt the ability of cells to attach to fibronectin by, for example, injecting embryos with antibodies against fibronectin. Such a disruption can block the inward movement of the mesoderm.

The glycoproteins that attach migrating cells to an underlying ECM also play a role in holding cells together when migrating cells reach their destinations and tissues and organs take shape. Also contributing to cell migration and stable tissue structure are glycoproteins called **cell adhesion molecules (CAMs)**, which are located on the surfaces of cells. CAMs vary in either amount or chemical identity from one type of cell to another, and these differences help regulate morphogenetic movements and tissue building.

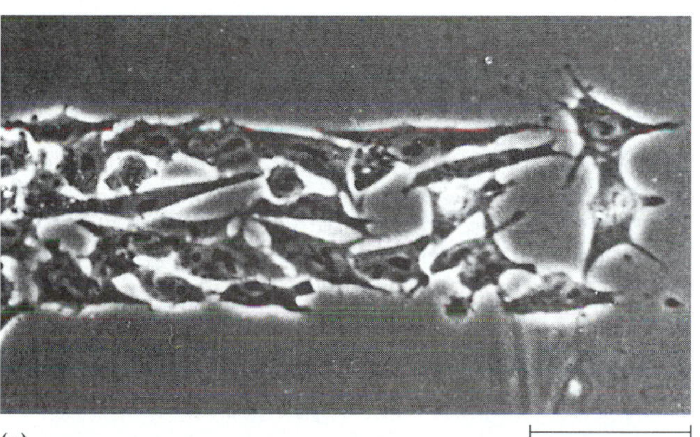

(a) 50 μm

FIGURE 47.18 ▪ The extracellular matrix and cell migration. **(a)** Cells from the neural crest migrate along a strip of fibronectin fibers placed on an artificial underlayer (LM). **(b)** Two different fluorescent dyes demonstrate the close relationship between the orientation of fibronectin fibers in the extracellular matrix (left) and that of contractile microfilaments (right) of the cytoskeleton within two migrating cells (LMs). Notice that the orientations of the intracellular microfilaments and extracellular fibers correspond. The orientation of the extracellular fibers is controlled by the orientation of the cytoskeleton in the cells that secrete the extracellular materials. This is one way that a group of cells can influence the path along which another group of cells migrates during the development of tissues and organs.

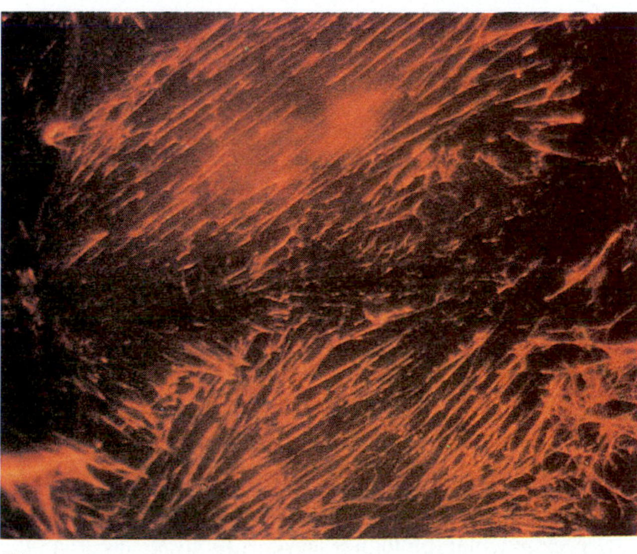

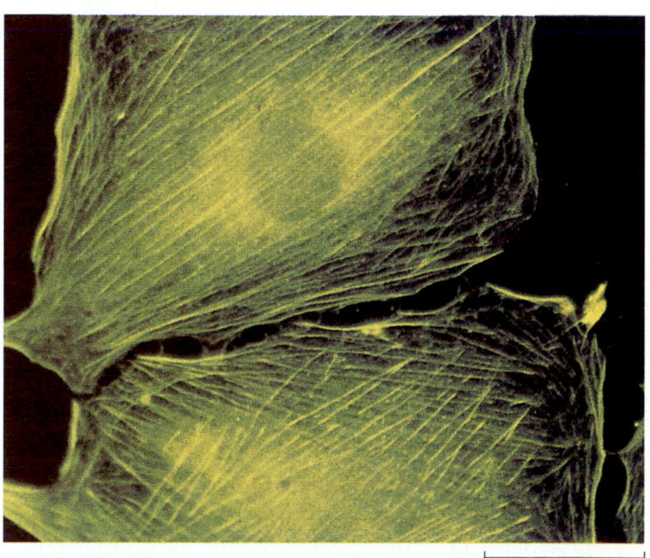

(b) 25 μm

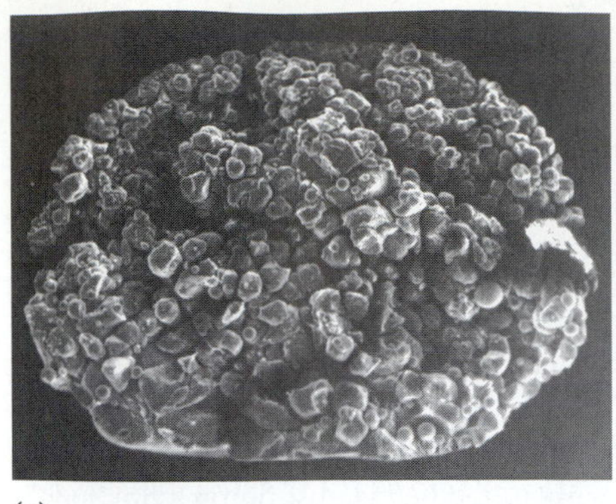
(a)

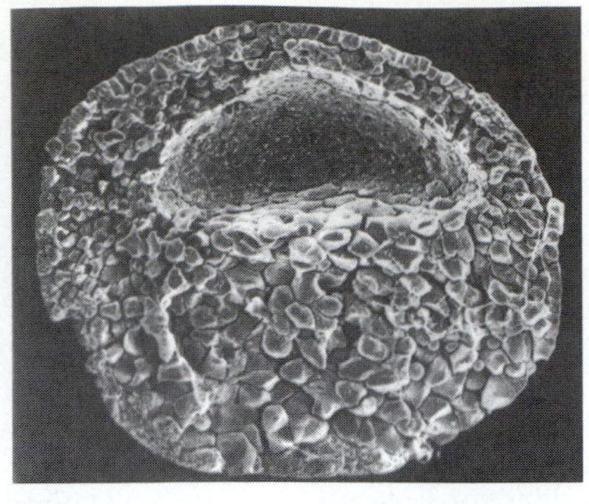
(b)

FIGURE 47.19 ▪ The role of a cadherin in frog blastula formation. (a) Researchers injected *Xenopus* oocytes with single-stranded nucleic acid complementary to the mRNA encoding a cadherin known as EP cadherin. The "antisense" nucleic acid blocked translation of the mRNA after fertilization. The absence of EP cadherin protein interfered with the development of the blastula. The blastocoel did not form properly, and the cells arranged themelves in a disorganized fashion. **(b)** In control embryos, the blastocoel formed normally.

One important class of cell-to-cell adhesion molecules is the **cadherins**, which are so named because they require the presence of calcium ions for proper function. There are many different cadherins, and each cadherin gene is expressed in specific locations at specific times during embryonic development. Researchers have vividly demonstrated the importance of one particular cadherin in the formation of the frog blastula (FIGURE 47.19).

The developmental fate of cells depends on cytoplasmic determinants and cell-cell induction: *a review*

Coupled with the morphogenetic changes that give an animal and its parts their characteristic shapes, development also requires the timely differentiation of many kinds of cells in specific locations. As we discussed in Chapter 21, two general principles integrate our current knowledge of the genetic and cellular mechanisms that underlie differentiation during embryonic development:

1. *In many animal species (mammals being a major exception), the heterogeneous distribution of cytoplasmic determinants in the unfertilized egg leads to regional differences in the early embryo.* By partitioning the heterogeneous cytoplasm of a polarized egg, cleavage parcels out different mRNAs, proteins, and other molecules to different blastomeres. These local differences in cytoplasmic composition help specify the body axes and influence the expression of genes that affect the developmental fate of cells. Thus, cytoplasmic determinants are responsible for the initial differences between cells in the early embryos of many kinds of animals.
2. *Subsequently, in induction, interactions among the embryonic cells themselves induce changes in gene expression. These interactions eventually bring about the differentiation of the many specialized cell types making up a new animal.* Induction may be mediated by diffusible chemical signals, or if the cells are actually in contact, by cell-surface interactions. (In Chapter 21 we saw how both types of cell signaling are involved in the development of the nematode vulva.)

It will help to keep these two principles in mind when we take a closer look at the molecular and cellular mechanisms of differentiation and pattern formation in vertebrate development. First, however, let's look at some historic experiments that provided early researchers with information about cell fates.

Fate mapping can reveal cell genealogies in chordate embryos

You may recall that in the case of the nematode *Caenorhabditis elegans,* it has been possible to map the developmental history of every cell, beginning with the first cleavage division of the zygote (see FIGURE 21.4). This sort of complete cell lineage has not been determined for other animals, but biologists have been making more general territorial diagrams of embryonic development, called **fate maps**, for over 70 years. Classic studies performed in the 1920s by German embryologist W. Vogt established that in embryos whose axes are defined early in development, it is often possible to ascertain which parts of the embryo will be derived from each region of the zygote or blastula. Following the lead of earlier work on marine worms and mollusks, Vogt charted fate maps for amphibian embryos (FIGURE 47.20a). Using nontoxic dyes, he labeled cells of different regions of the surface of amphibian blastulas with different colors and later sectioned the embryos to see where the colors turned up. Vogt's results were among the earliest indications that the lineage (the "genealogy") of cells making up the three germ layers created by gastrulation is traceable to cells in the blastula (compare FIGURES 47.20a and 47.10). Later researchers developed more sophisticated techniques that allowed them to mark an individual blastomere during cleavage and then follow the marker as it was distributed to all the mitotic descendants of that cell (FIGURE 47.20b).

Developmental biologists have combined fate-mapping studies with experimental manipulation of parts of embryos in which they study whether a cell's fate can be changed by mov-

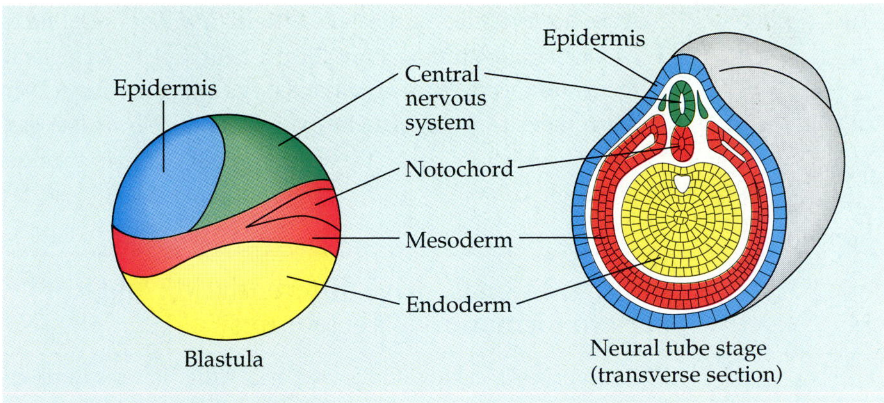

(a) Fate map of a frog embryo

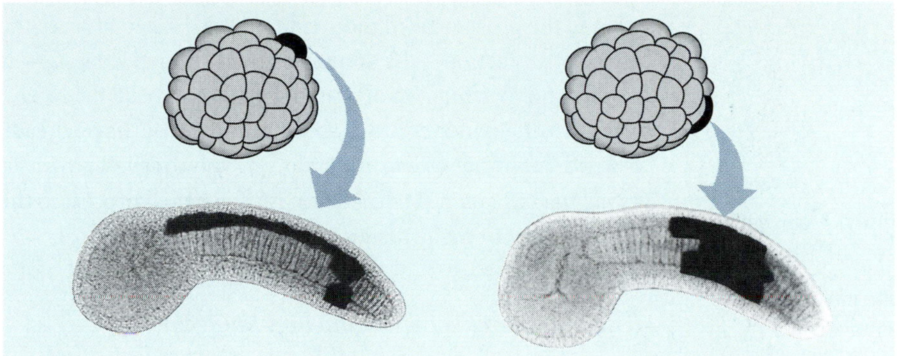

(b) Cell lineage analysis in a tunicate

FIGURE 47.20 • Fate maps for two chordates. (a) The fates of cells of a frog embryo were determined by marking different regions of the blastula surface with dyes of various colors and then determining the locations of dyed cells at later stages of development, such as at this neural tube stage. **(b)** Depicted here are the results of a very specific type of fate map called a cell lineage analysis. The drawings depict 64-cell embryos of a tunicate, an invertebrate chordate. An individual cell can be injected with a dye used as a marker, enabling a researcher to determine which cells in a later embryo are derived from the marked cells. The two light micrographs of larvae contrast the regions that develop from the two different blastomeres indicated in the drawings.

ing it elsewhere in the embryo. Two important conclusions have emerged. First, in most animals, certain early "founder cells" generate specific tissues of the older embryo. Second, as development proceeds, a cell's *developmental potential*—the range of structures that it can give rise to—becomes restricted. Starting with the normal embryo's fate map, researchers can examine how the differentiation of cells is altered in experimental situations or in mutant embryos.

The eggs of most vertebrates have cytoplasmic determinants that help establish the body axes and differences among cells of the early embryo

In order to understand at the molecular level how embryonic cells acquire their fates, we must return to the question of how the basic axes of the embryo are established. It turns out that much of what you learned in Chapter 21 about the roles of cytoplasmic determinants in the development of fruit flies also applies to vertebrate animals.

Polarity and the Basic Body Plan

As you have learned, a bilaterally symmetrical animal has an anterior-posterior axis, a dorsal-ventral axis, and left and right sides. Establishing this basic body plan is a first step in morphogenesis and is prerequisite to the development of tissues and organs. In humans and other mammals, the basic polarities do not seem to be determined until after cleavage, and exactly how the embryo's axes are established is still unknown. However, in most species, fundamental instructions (where the head end will be, and so on) are set down earlier. For example, in many frogs, as we saw in FIGURE 47.7, the locations of yolk and melanin in the unfertilized egg define the animal and vegetal hemispheres, which in turn determine the anterior-posterior body axis. Fertilization then triggers the formation of the gray crescent, whose position determines the dorsal-ventral axis.

Restriction of Cellular Potency

The asymmetric distribution of cytoplasmic determinants in an egg does not necessarily lead to differences among the earliest blastomeres. The first cleavage may occur along an axis that produces two identical blastomeres, which will have equal developmental potential. This is what happens in amphibians, for instance. The first two blastomeres are similar, and if they are experimentally separated, each can develop into a normal tadpole. In other words, both blastomeres are totipotent. However, in experiments in which the entire grey crescent goes to one blastomere, only that blastomere has the capacity to develop into a normal tadpole when the blastomeres are

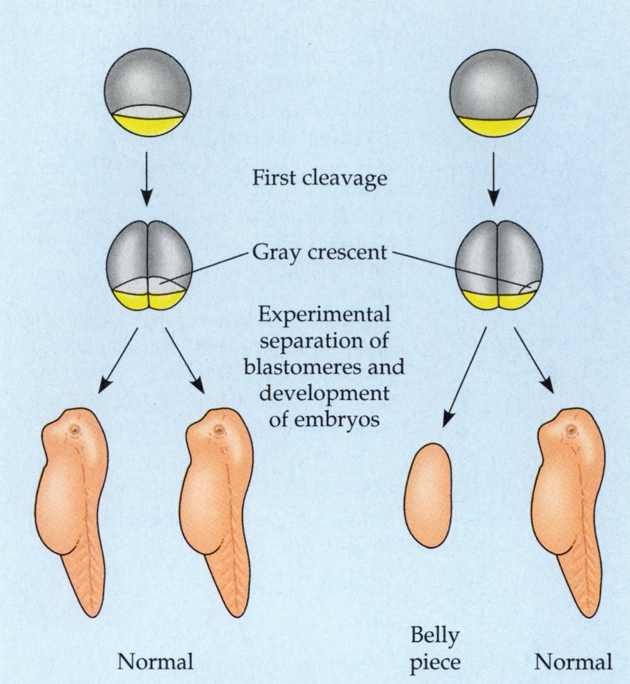

FIGURE 47.21 • Experimental demonstration of the importance of cytoplasmic determinants in amphibians. The first cleavage division of an amphibian zygote normally divides the gray crescent evenly between the two blastomeres, which retain their totipotency after the division (left). However, when one blastomere receives the entire gray crescent (right), only that blastomere develops normally. The other blastomere, lacking cytoplasmic determinants present in the crescent, gives rise to an abnormal embryo without dorsal structures.

separated (FIGURE 47.21). Thus, the fates of embryonic cells can be affected not only by the distribution of cytoplasmic determinants but also by the zygote's characteristic pattern of cleavage.

In many species, only the zygote is totipotent. The first cleavage plane divides cytoplasmic determinants in such a way that each blastomere will give rise only to specific parts of the embryo. In contrast, the cells of mammalian embryos remain totipotent until they become arranged into the trophoblast and inner cell mass of the blastocyst. As mentioned earlier, mammalian eggs appear to lack polarity, and early blastomeres seem to receive equivalent amounts of cytoplasmic components. Indeed, up to the eight-cell stage, the blastomeres of a mammalian embryo all look alike, and each can form a complete embryo if isolated.

However similar or different the cells in an early embryo, the progressive restriction of potency is a general feature of development in all animals. In some species the cells of early gastrulas retain the capacity to give rise to more than one kind of cell, though they have lost their totipotency. If left alone, the dorsal ectoderm of an early amphibian gastrula will develop into a neural plate above the notochord. If the dorsal ectoderm is experimentally replaced with ectoderm from some other location, the transplanted tissue will form a neural plate. If the same experiment is performed on a late-stage gastrula, however, the transplanted ectoderm will not respond to its new location and will not form a neural plate. In general, the tissue-specific fates of cells in late gastrulae are fixed. Even when they are manipulated experimentally, cells of late gastrulae usually give rise to the same types of cell as in the normal embryo.

Inductive signals drive differentiation and pattern formation in vertebrates

As embryonic cell division creates cells with different developmental potential, a new possibility arises. Now one group of cells can influence the development of a neighboring group of cells, in the process called induction. At the molecular level, the effect of induction—the response to an inductive signal—is usually the switching on of a set of genes that make the receiving cells differentiate into a specific tissue. You have already learned about the role of induction in vulval development in *C. elegans* (Chapter 21). Induction is an essential process in the development of many tissues in other animals as well.

The "Organizer" of Spemann and Mangold

The importance of induction in amphibian development was dramatically demonstrated in the 1920s by German zoologist Hans Spemann and his student Hilde Mangold. In a series of transplantation experiments, they discovered that the dorsal lip of the blastopore in the early gastrula plays a key role in embryonic development, initiating a chain of inductions that results in the formation of the neural tube and other organs. In their most famous experiment, they took a piece of dorsal lip from one embryo and grafted it at an abnormal position in a second embryo (FIGURE 47.22). The result was the development of a second notochord and neural tube in the recipient embryo, at the location of the graft. Subsequently, other organs and structures formed, creating a nearly complete second embryo attached to the first. Spemann referred to the dorsal lip of the blastopore as the *primary organizer* of the embryo because of its early and crucial role in development.

Developmental biologists are working intensively to identify the molecular basis of induction by Spemann's organizer. An important clue has come from studies of a growth factor called *bone morphogenetic protein 4* (BMP-4). (Bone morphogenetic proteins, a family of related proteins with a variety of developmental roles, derive their name from members of the family that are important in bone formation.) Amphibian BMP-4 is active exclusively in cells on the *ventral* side of the gastrula. One major function of organizer cells seems to be to inactivate BMP-4 on the dorsal side of the embryo by producing proteins that bind to BMP-4, rendering it unable to signal. Proteins related to BMP-4 and its inhibitors are also found in other animals, including invertebrates such as the fruit fly. The ubiquity of these molecules suggests that they evolved long

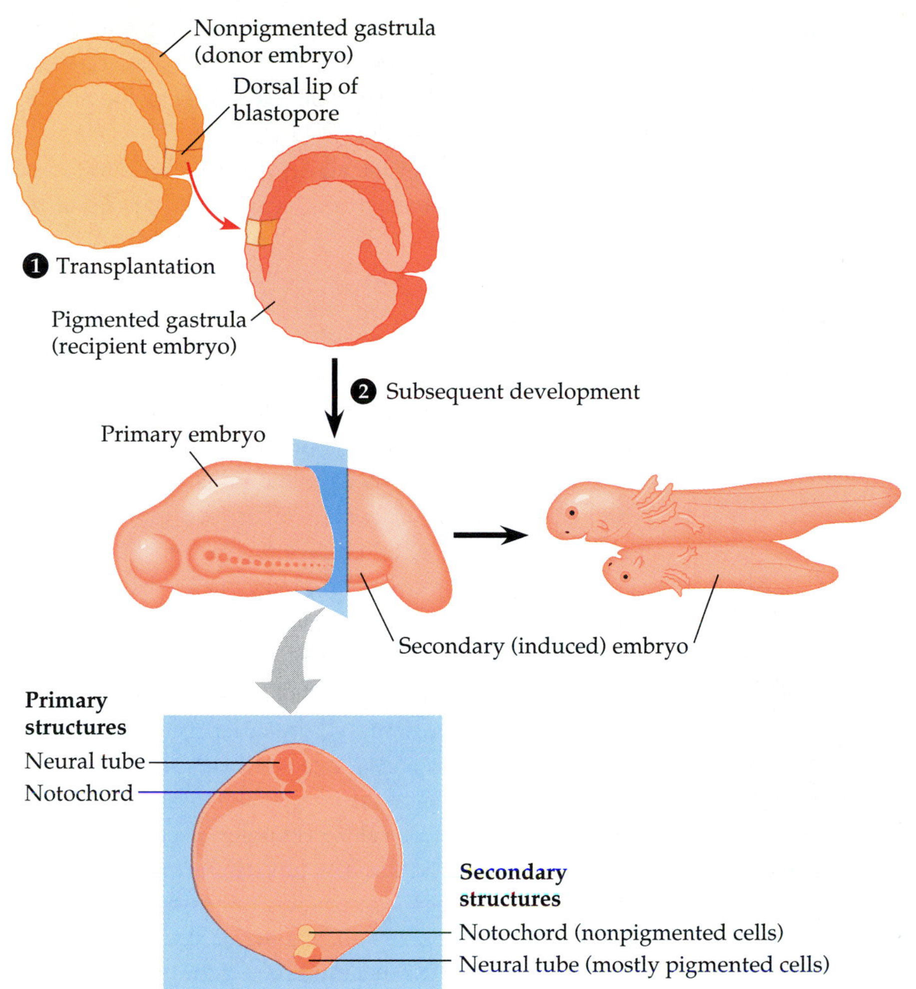

FIGURE 47.22 ▪ The "organizer" of Spemann and Mangold. The dorsal lip of the blastopore in the early gastrula of an amphibian plays a critical role in inducing the development of other parts of the embryo. ① In 1924, Hans Spemann and Hilde Mangold transplanted a piece of the dorsal lip of a nonpigmented newt gastrula (shown in tan) to the ventral side of the early gastrula of a pigmented newt (red). ② The recipient embryo formed a second notochord and neural tube in the region of the transplant, and eventually most of a second embryo. When Mangold and Spemann examined the interior of the double embryo, they found that many cells of the secondary structures came from the (pigmented) recipient, not the (nonpigmented) donor of the graft. This finding indicated that the grafted tissue had "organized" (induced) cells of the recipient to form the extra structures.

ago and may participate in the development of many different organisms.

The induction that causes dorsal ectoderm to develop into the neural tube is only one of many cell-cell interactions that transform the three germ layers into organ systems. Many inductions seem to involve a sequence of inductive steps that progressively determine the fate of cells. For example, in the late gastrula of the frog, ectoderm cells destined to become the lenses of the eyes receive inductive signals from the ectodermal cells that will become the neural plate. Additional inductive signals probably come from endodermal cells and mesodermal cells. Finally, inductive signals from the optic cup, an outgrowth of the developing brain, complete the determination of the lens-forming cells.

Pattern Formation in the Vertebrate Limb

Inductive signals play a major role in **pattern formation**—the development of an animal's spatial organization, the arrangement of organs and tissues in their characteristic places in three-dimensional space. The molecular cues that control pattern formation, called **positional information**, tell a cell where it is with respect to the animal's body axes and help to determine how the cell and its descendants respond to future molecular signals.

In Chapter 21 we discussed pattern formation in the development of the body segments of *Drosophila*. For understanding pattern formation in vertebrates, an important model system has been limb development in the chick. The wings and legs of chicks, like all vertebrate limbs, begin as bumps of tissue called limb buds (FIGURE 47.23, p. 956). Each component of a chick limb, such as a specific bone or muscle, develops with a precise location and orientation relative to three axes: the proximal-distal axis (the "shoulder-to-fingertip" axis), the anterior-posterior axis (the "thumb-to-little finger" axis), and the dorsal-ventral axis (the "knuckle-to-palm" axis). The embryonic cells within a limb bud respond to positional information indicating location along these three axes.

A limb bud consists of a core of mesodermal tissue covered by a layer of ectoderm. By removing or transplanting different pieces of tissue, researchers have discovered two critical organizer regions in the limb bud, regions with profound effects on the limb's development. These two organizer regions are present in all vertebrate limb buds, including those that will develop into forelimbs (such as wings or arms) and those destined to become hindlimbs. In recent years researchers have established that the cells of these regions secrete morphogens that provide key positional information to the other cells of the bud. Recall from Chapter 21 that morphogens are substances that provide positional information in the form of a concentration gradient along an embryonic axis. They are usually proteins that regulate gene expression by functioning as transcription factors.

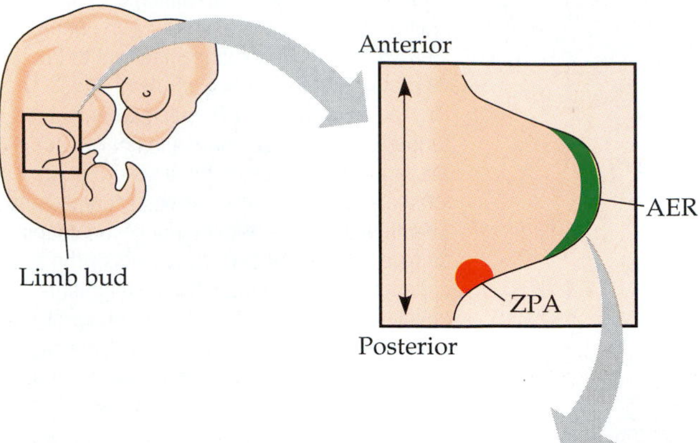

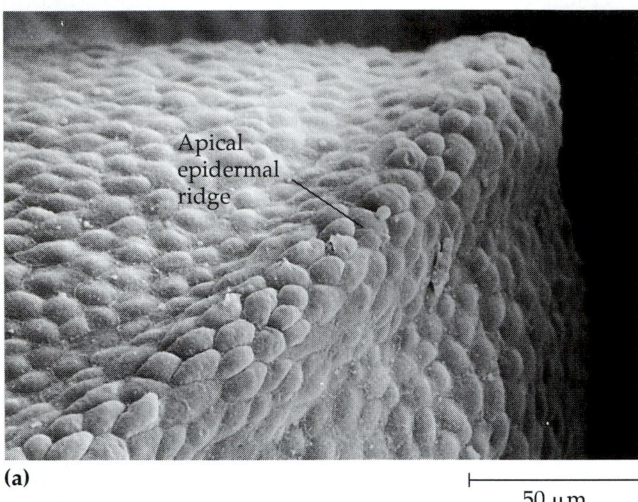

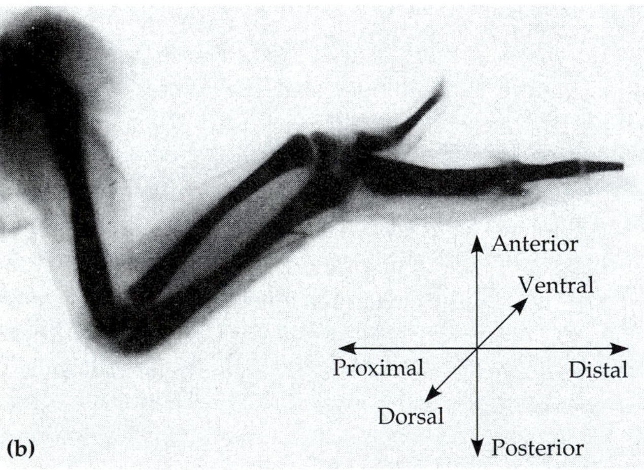

FIGURE 47.23 • Organizer regions in vertebrate limb development. (a) Vertebrate limbs develop from protrusions called limb buds, each consisting of mesoderm cells covered by a layer of ectoderm. Two regions play key "organizer" roles in limb pattern formation: the apical ectodermal ridge (AER) and the zone of polarizing activity (ZPA). (SEM) **(b)** As the bud develops into a limb, such as this wing of a chick embryo, a specific pattern of tissues emerges. Pattern formation requires each embryonic cell to receive some kind of positional information indicating location along the three axes of the limb. The AER and ZPA secrete molecules that help provide this information.

One limb-bud organizer region is the **apical ectodermal ridge (AER)**, a thickened area of ectoderm at the tip of the bud (see FIGURE 47.23a). The AER is required for the outgrowth of the limb along the proximal-distal axis and for patterning along this axis. The cells of the AER produce and secrete several proteins of the fibroblast growth factor (FGF) family. These proteins appear to be the growth signal that promotes limb bud outgrowth. If the AER is surgically removed and beads soaked with FGF are put in its place, a nearly normal limb will develop. The AER and other ectoderm of the limb bud also appear to guide pattern formation along the limb's dorsal-ventral axis. In experiments where the ectoderm of a limb bud is detached from the mesoderm and then replaced with its orientation rotated 180°, the limb elements that form have reversed dorsal-ventral orientation. (This is equivalent to reversing the palm and back of your hand.)

The second major limb-bud organizer region is the **zone of polarizing activity (ZPA)**, located where the posterior side of the bud is attached to the body. The ZPA is necessary for proper pattern formation along the anterior-posterior axis of the limb. Cells nearest the ZPA give rise to posterior structures, such as the digit homologous to our little finger, and cells farthest from the ZPA form anterior structures, such as the avian equivalent of our thumb. A tissue transplantation experiment demonstrating the importance of the ZPA is shown in FIGURE 47.24. Grafting a second ZPA on the anterior side of a limb bud leads to the formation of extra digits, in a mirror-image arrangement.

In the ZPA-transplantation experiment, the extra digits develop from the host limb bud and not from the graft, supporting the hypothesis that the grafted ZPA produces some sort of inductive signal. Indeed, researchers have discovered that the cells of the ZPA secrete morphogens. The most important seems to be a protein growth factor called *Sonic hedgehog*.* If cells genetically engineered to produce large amounts of Sonic hedgehog are implanted in the anterior region of a normal limb bud, a mirror-image limb results—just as if a ZPA had been grafted there. Evidence from studying the mouse version of Sonic hedgehog suggests that extra toes in mice—and perhaps also in humans—can result from the production of the protein in the wrong part of the limb bud. Proteins very similar to Sonic hedgehog have turned out to be important positional cues in a number of developmental situations, including segment formation in the *Drosophila*

* Sonic hedgehog gets its name from two sources, its similarity to a *Drosophila* protein called Hedgehog, which is involved in segmentation of the fly embryo, and a video game character.

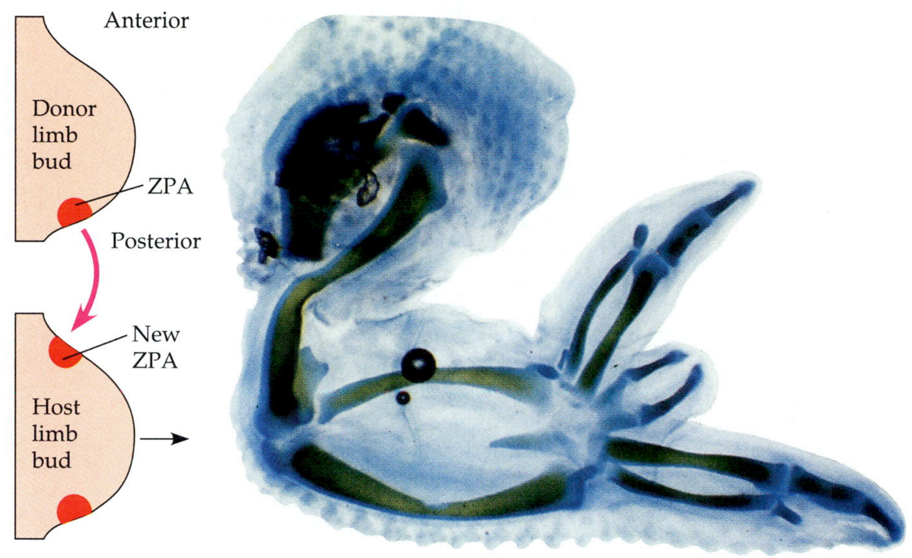

FIGURE 47.24 ▪ The experimental manipulation of positional information. The zone of polarizing activity (ZPA) issues molecular signals that establish the anterior-posterior axis of a vertebrate limb. In one experiment, a second ZPA was added to the anterior margin of a chick limb bud by transplanting ZPA tissue from another limb bud. Cells near the grafted ZPA, like cells near the host bud's own ZPA, apparently receive positional information indicating "posterior." The pattern that emerges in the developing limb is a mirror image, with an arrangement of digits equivalent to two human hands joined together at the thumbs. In a subsequent experiment, researchers showed that a protein called Sonic hedgehog could substitute for the ZPA cells. They genetically engineered cells to produce and secrete Sonic hedgehog and then implanted these cells at an anterior position in a limb bud. As in the ZPA experiment, the result was a mirror-image limb. Thus Sonic hedgehog is a morphogen that functions as a positional cue.

embryo. (The fruit-fly version, called simply Hedgehog, is the product of a segment-polarity gene; see Chapter 21.)

We can conclude from such experiments that pattern formation requires cells to receive and interpret environmental cues that vary from one location to another. These cues, acting together along three axes, tell cells where they are in the three-dimensional realm of a developing organ. In vertebrate limb development, we now know that specific proteins serve as some of these cues. In other words, organizer regions such as the AER and the ZPA function as signaling centers.

What determines whether a limb bud develops into a forelimb or a hindlimb? The cells receiving the signals from the AER and ZPA respond according to their developmental histories. Before the AER or ZPA issues its signals, earlier developmental signals have set up patterns of gene expression distinguishing the future forelimbs from the future hindlimbs. These differences cause the cells of the forelimb and hindlimb limb buds to react differently to the same positional cues.

A hierarchy of gene activations eventually affects the expression of homeobox-containing *(Hox)* genes in cells of the developing limb. *Hox* genes seem to be involved in specifying the identity of various regions of the limbs, as well as of the body as a whole (see Chapter 21). Thus, pattern formation is a chain of events involving many steps of signaling and differentiation. Various pathways of pattern formation occur in all the different parts of the developing embryo, eventually producing the final set of differentiated structures in the fully formed animal.

CHAPTER REVIEW

REVIEW OF KEY CONCEPTS

(with page numbers and key figures)

THE STAGES OF EARLY EMBRYONIC DEVELOPMENT

- **From egg to organism, an animal's form develops gradually: *the concept of epigenesis* (pp. 936–937)** An embryo is not preformed in an egg; it develops by epigenesis, the gradual, gene-directed acquisition of form.
- **Fertilization activates the egg and brings together the nuclei of sperm and egg (pp. 937–940, FIGURE 47.2)** Fertilization both reinstates diploidy and activates the egg to begin a chain of metabolic reactions that triggers the onset of embryonic development. The acrosomal reaction, which occurs when the sperm meets the egg, releases hydrolytic enzymes that digest through material surrounding the egg. Gamete fusion depolarizes the egg cell membrane and sets up a fast block to polyspermy. Sperm-egg fusion also initiates the cortical reaction, involving a signal-transduction pathway in which calcium ions stimulate cortical granules to erect a fertilization membrane that functions as a slow block to polyspermy. In mammalian fertilization, the cortical reaction hardens the zona pellucida as a slow block to polyspermy.
- 47.1 **Cleavage partitions the zygote into many smaller cells (pp. 940–943, FIGURE 47.8)** Fertilization is followed by cleavage, a period of rapid cell division without growth, which results in the production of a large number of cells called blastomeres. Holoblastic cleavage, or division of the entire egg, occurs in species whose eggs have little or moderate amounts of yolk. Meroblastic cleavage, incomplete division of the egg, occurs in species with yolk-rich eggs. Cleavage planes usually follow a specific pattern relative to the animal and vegetal poles of the zygote. In many species, cleavage creates a multicellular ball called the blastula, which contains a fluid-filled cavity, the blastocoel.
- 47.1 **Gastrulation rearranges the blastula to form a three-layered embryo with a primitive gut (pp. 943–945, FIGURE 47.10)** Gastrulation transforms the blastula into an embryo called the gastrula, which has a rudimentary digestive cavity (the archenteron) and three embryonic germ layers: the ectoderm, endoderm, and mesoderm.

47.2 **In organogenesis, the organs of the animal body form from the three embryonic germ layers** (pp. 945–946, FIGURE 47.11) Early events in organogenesis in vertebrates include formation of the notochord by condensation of dorsal mesoderm, development of the neural tube from folding of the ectodermal neural plate, and formation of the coelom from splitting of lateral mesoderm.

- **Amniote embryos develop in a fluid-filled sac within a shell or uterus** (pp. 947–950, FIGURES 47.12, 47.14, 47.15) Meroblastic cleavage in the yolk-rich, shelled eggs of birds and reptiles is restricted to a small disc of cytoplasm at the animal pole. A cap of cells called the blastodisc forms and begins gastrulation with the formation of the primitive streak. In addition to the embryo, the three germ layers give rise to the four extraembryonic membranes: the yolk sac, amnion, chorion, and allantois. The eggs of placental mammals are small and store little food, exhibiting holoblastic cleavage with no apparent polarity. Gastrulation and organogenesis, however, resemble the processes in birds and reptiles. After fertilization and early cleavage in the oviduct, the blastocyst implants in the uterus. The trophoblast initiates formation of the fetal portion of the placenta, and the embryo proper develops from a single layer of cells, the epiblast, within the blastocyst. Extraembryonic membranes homologous to those of birds and reptiles function in intrauterine development.

THE CELLULAR AND MOLECULAR BASIS OF MORPHOGENESIS AND DIFFERENTIATION IN ANIMALS

- **Morphogenesis in animals involves specific changes in cell shape, position, and adhesion** (pp. 950–952, FIGURES 47.16, 47.17) Cytoskeletal rearrangements are responsible for changes in both shape and position of cells. Both kinds of changes are involved in tissue invaginations, as occurs in gastrulation, for example. The extracellular matrix provides anchorage for cells and also helps guide migrating cells toward their destinations. Cell adhesion molecules on cell surfaces are also important for cell migration and for holding cells together in tissues.
- **The developmental fate of cells depends on cytoplasmic determinants and cell-cell induction: *a review*** (p. 952)
- **Fate mapping can reveal cell genealogies in chordate embryos** (pp. 952–953, FIGURE 47.20) Experimentally derived fate maps of embryos have shown that specific regions of the zygote or blastula develop into specific parts of older embryos.
- **The eggs of most vertebrates have cytoplasmic determinants that help establish the body axes and differences among cells of the early embryo** (pp. 953–954, FIGURE 47.21) When cytoplasmic determinants are heterogeneously distributed in an egg, they serve as the basis for setting up differences among parts of the egg and, later, among the blastomeres resulting from cleavage of the zygote. Cells that receive different cytoplasmic determinants acquire different fates.
- **Inductive signals drive differentiation and pattern formation in vertebrates** (pp. 954–957, FIGURES 47.22–47.24) Cells in a developing embryo receive and interpret positional information that varies with location. This information is often in the form of signal molecules secreted by cells in special "organizer" regions of the embryo, such as the dorsal lip of the blastopore in the amphibian gastrula and the apical ectodermal ridge of the vertebrate limb bud. The signal molecules (morphogens) affect gene expression in the cells that receive them, leading to differentiation and the development of particular structures.

SELF-QUIZ

1. The cortical reaction of sea urchin eggs functions directly in the
   a. formation of a fertilization membrane
   b. production of a fast block to polyspermy
   c. release of hydrolytic enzymes from the sperm cell
   d. generation of a nervelike impulse by the egg cell
   e. fusion of egg and sperm nuclei

2. Which of the following is common to both avian and mammalian development?
   a. holoblastic cleavage
   b. primitive streak
   c. trophoblast
   d. yolk plug
   e. gray crescent

3. The archenteron develops into
   a. the mouth in protostomes
   b. the blastocoel
   c. the endoderm
   d. the lumen of the digestive tract
   e. the placenta

4. In a frog embryo, the blastocoel is
   a. completely obliterated by yolk platelets
   b. lined with endoderm during gastrulation
   c. located primarily in the animal hemisphere
   d. the cavity that becomes the coelom
   e. the cavity that later forms the archenteron

5. Amphibians, unlike reptiles, generally lay their eggs in water or moist places. This difference is related to the absence (in amphibians) versus the presence (in reptiles) of
   a. extraembryonic membranes
   b. yolk
   c. cleavage
   d. gastrulation
   e. development of the brain from ectoderm

6. In an amphibian embryo, a band of cells called the neural crest
   a. rolls up to form the neural tube
   b. develops into the main sections of the brain
   c. produces amoeboid cells that migrate to form teeth, skull bones, and other structures in the embryo
   d. has been shown by experiments to be the organizer region of the developing embryo
   e. induces the formation of the notochord

7. Differences in the development of cells in the early embryo (zygote to blastomere) are due to
   a. the differences between meroblastic and holoblastic cleavage
   b. the heterogeneous distribution of cytoplasmic determinants, such as proteins and mRNA
   c. inductive interactions occurring between the developing cells
   d. concentration gradients for various regulatory molecules such as BMP-4
   e. position of the cells relative to the zone of polarizing activity (ZPA)

8. During convergent extension
   a. cells on the opposite side of the embryo follow converging developmental pathways leading to bilateral symmetry
   b. the cells of the neural folds adhere to one another to complete the neural tube
   c. the cells of a tissue layer reorganize forming a narrowed elongated sheet
   d. the dorsal-ventral axis is established
   e. cell adhesion molecules are expressed, causing the 8 blastomeres to tightly adhere to one another

9. In the early development of an amphibian embryo, an important "organizer" is the
   a. neural tube
   b. notochord
   c. archenteron roof
   d. dorsal lip of the blastopore
   e. dorsal ectoderm
10. The occurrence of genetically identical twins indicates that in humans
    a. only the zygote is totipotent
    b. the progressive restriction of potency hypothesis does not apply
    c. the first cleavage event must be transverse to the animal-vegetal axis of the zygote
    d. cell divisions producing the earliest blastomeres do not result in an asymmetric distribution of cytoplasmic determinants
    e. the primary organizer continues to function well past gastrulation

CHALLENGE QUESTIONS

1. Explain how in certain animals (such as *Drosophila*) the organization of the unfertilized egg's cytoplasm affects the form of the adult animal that develops after fertilization.
2. The "snout" of a frog tadpole bears a sucker. A salamander tadpole has a mustache-shaped structure called a balancer in the same area. If ectoderm from the side of a young salamander embryo is transplanted to the snout of a frog embryo, the frog tadpole later has a balancer. If ectoderm is transplanted from the side of a slightly older salamander embryo to the snout of a frog embryo, the frog tadpole ends up with a patch of salamander skin on its snout. Explain the results of this experiment in terms of the mechanisms of development.

SCIENCE, TECHNOLOGY, AND SOCIETY

1. Nerve cells transplanted from aborted fetuses can relieve the symptoms of Parkinson's disease, a brain disorder. Fetal tissue transplants might also be used to treat epilepsy, diabetes, Alzheimer's disease, and spinal cord injuries. Why might tissues from a fetus be particularly useful for replacing diseased or damaged cells? Since 1989, there has been controversy over whether the U.S. government should allow fetal tissues from induced abortions to be used in transplant research. Opponents would allow only tissues from miscarriages to be used. Why would most researchers prefer to use tissues from aborted fetuses? Why do you think some people oppose this? What is your position on this issue, and why?
2. The abortion debate in the United States has provoked thought and created controversy over this question: When does human life begin? What are some of the possible answers? Do you think the continuing study of human embryological development can give a definitive answer and resolve this debate? Why or why not?
3. Spina bifida, a neural tube disorder, is a common and serious birth defect. It results from an error in morphogenesis. The United States Public Health Service recommends that all women of childbearing age in the United States who are capable of becoming pregnant consume 0.4 mg of folic acid per day to reduce the number of cases of neural tube defects. It has also been recommended that the most expeditious way to increase consumption is through fortification of a food staple, such as cereal-grain products. What are your thoughts about this suggestion? Is it reasonable for the entire population to receive an additive that primarily benefits a relatively small number of individuals? Can you identify any other examples of additives in the food supply? If so, is there a difference between additives that improve the safety of the general food supply and those that affect a subset of the population?

FURTHER READING

Gilbert, S. F. *Developmental Biology*, 5th ed. Sunderland, MA: Sinauer Associates, 1997. A widely used upper-division text.

Kessler, D. S., and D. A. Melton. "Vertebrate Embryonic Induction: Mesodermal and Neural Patterning." *Science*, October 28, 1994. A review of progress toward determining the molecular basis of pattern formation in vertebrates.

Newman, R. A. "Adaptive Plasticity in Amphibian Metamorphosis." *BioScience*, October 1992. How does environment affect an individual animal's development?

Vogel, G. "Embryo's Organizational Chart Redrawn." *Science*, June 19, 1998. A recent research report on the roles of the neural plate and notochord in the development of the vertebrate nervous system.

Wolpert, L., R. Beddington, J. Brockes, T. Jessell, P. Lawrence, and E. Meyerowitz. *Principles of Development*. New York: Current Biology Ltd./ Oxford University Press, 1998. A new, concept-oriented upper-division text with excellent full-color illustrations.

WEB LINKS

Visit the special edition of *The Biology Place* for BIOLOGY, Fifth Edition, at http://www.biology.com/campbell. Go to Chapter 47 for online resources, including learning activities, practice exams, and links to the following web sites:

"The Society for Developmental Biology"
The home page of the Society for Developmental Biology provides information on many aspects of developmental biology. The "Interactive Fly" and the "Developmental Biology Cinema" are well worth a visit. The latter requires either the QuickTime or VivoActive plug-in.

"The Edinburgh Digital Atlas of Mouse Development"
At this site you can explore a three-dimensional reconstruction of a developing mouse embryo.

"Morphing Embryos"
Originally shown on the *Nova* TV series, PBS has put on-line these amazing time-lapse sequences of developing embryos from several animal species. Choice of VivoActive, QuickTime, or AVI formats.

"The Amphibian Embryology Tutorial"
Interactive tutorial on embryo formation. For each topic, the amount of detail increases as the pages advance, so you can explore the topic to the depth you find suitable.

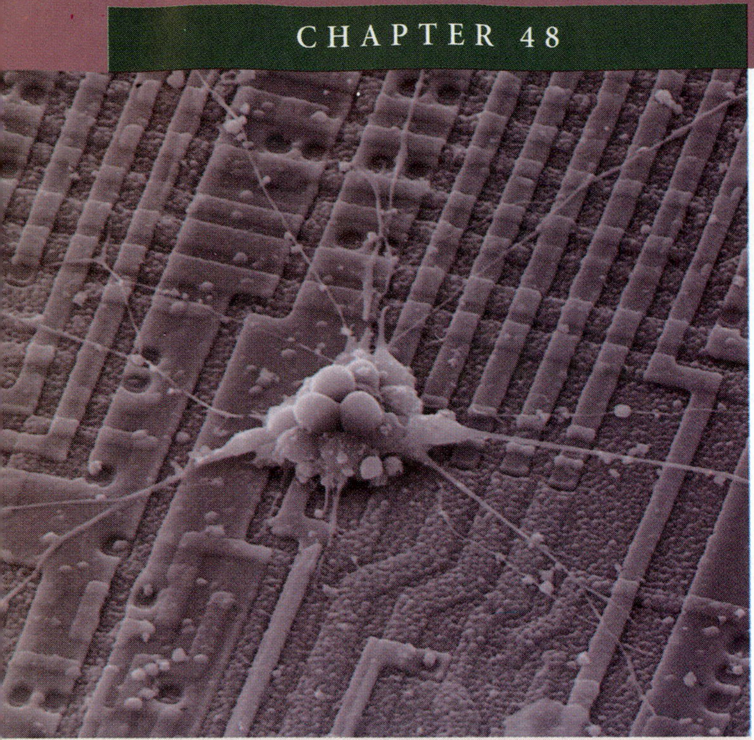

NERVOUS SYSTEMS

The scanning electron micrograph on this page shows an unusual juxtaposition of the basic components of animal nervous systems and computers. A single nerve cell (a neuron) was removed from a nervous system and grown on the surface of a microprocessor. Made of intricate computer chips like this, computers are capable of complex tasks. However, your own nervous system, made of living neurons, is performing vastly more complex tasks as you read and comprehend these words. The human nervous system is probably the most intricately organized aggregate of matter on Earth. A single cubic centimeter of the human brain may contain well over 50 million nerve cells, each of which may communicate with thousands of other neurons in information-processing networks that make the most elaborate computer look primitive. These neural pathways control our every perception and movement and enable us to learn, remember, think, and be conscious of ourselves and our surroundings.

In many cases the nervous system and endocrine system cooperate and interact in regulating internal body functions and behavior. In the maintenance of homeostasis, for example, the hypothalamus and other parts of the brain receive and process data about the body's internal environment and send out commands that correct imbalances to other organs via neurons and neurosecretory cells (see Chapter 45).

Despite their structural and functional linkage, the nervous system and endocrine system play somewhat different roles in body coordination. For example, with its incomparable structural complexity, the nervous system can integrate vast amounts of information, such as that required for human thought and speech. Timing can also be a key difference. The endocrine system may take minutes, hours, or even days to act, partly because of the time it takes for hormones to be made and carried in the blood to their target organs. In contrast, the nervous system is a signaling network with branches carrying information directly to and from specific targets. Neurons are specialized for the fast transmission of impulses—as quickly as 150 m/sec (over 330 mph). As a result, information can go from the brain of a human to the hands (or vice versa) in a few milliseconds.

To survive and reproduce, an animal must respond rapidly and appropriately to its environment. The focus of this chapter is how nervous systems mediate these interactions with the environment while cooperating with the endocrine system in maintaining homeostasis. After studying how nervous systems work, we will examine our own nervous system in some detail.

AN OVERVIEW OF NERVOUS SYSTEMS

Nervous systems are analogous to a telephone network, with complex interconnecting nerves taking the place of wires and a brain functioning as a highly complex central control station. Much can be learned about the way a nervous system works by studying its cellular structure and function. The functional basis of a nervous system is a combination of electrical and chemical signals that enable nerve cells (neurons) to communicate with one another.

Nervous systems perform the three overlapping functions of sensory input, integration, and motor output

In general, a nervous system has three overlapping functions: sensory input, integration, and motor output (FIGURE 48.1). Input is the conduction of signals from sensory receptors, such as the light-detecting cells in the eyes, to integration centers. Integration is the process by which the information from the environmental stimulation of the sensory receptors is interpreted and associated with appropriate responses of the body. For the most part, integration is carried out in the **central nervous system (CNS)**, the brain and spinal cord (in vertebrates). Motor output is the conduction of signals from the integration center, the CNS, to **effector cells**, the muscle cells or gland cells that actually carry out the body's responses to stimuli. The signals are conducted by **nerves**, ropelike bundles of extensions of neurons tightly wrapped in connective tissue. The nerves that communicate motor and sensory signals between the central nervous system and the rest of the body are collectively called the **peripheral nervous system (PNS)**. From receptor to effector, information is communicated in a nerve from one neuron to the next by a combination of electrical and chemical signals. In this chapter we concentrate on communication within the nervous system. Chapter 49 connects the nervous system to its inputs and outputs by discussing sensory receptors and the physiology of movement.

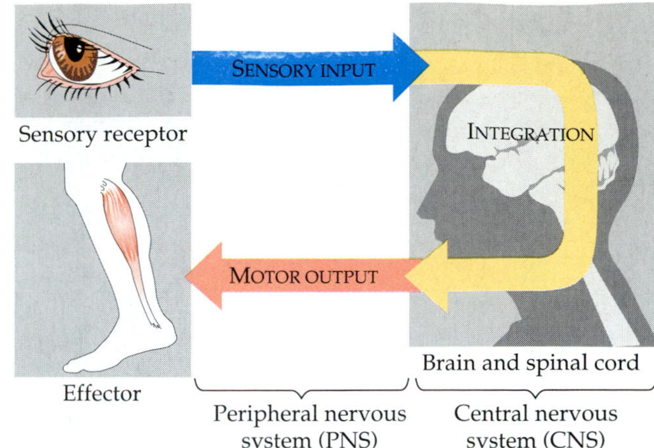

FIGURE 48.1 • Overview of a vertebrate nervous system. The brain and spinal cord together form the central nervous system (CNS), which is responsible for the integration of information. A network of nerves forming the peripheral nervous system (PNS) carries information from sensory receptors (sensory input) to the CNS and motor commands from the CNS (motor output) to various target organs or glands, collectively called effectors.

The nervous system is composed of neurons and supporting cells

Two main classes of cells populate the nervous system: neurons and supporting cells. Neurons are the cells that actually conduct messages along the communication pathways of the nervous system. More numerous are the supporting cells, also called glia, which provide structure in the nervous system and also protect, insulate, and generally assist neurons.

Neurons

48.1 The **neuron** is the functional unit of the nervous system and is specialized for transmitting signals from one location in the body to another. Although there are many different types of neurons, varying in their structural details depending on their function, most neurons share some common features (FIGURE 48.2, p. 962). A neuron has a relatively large **cell body** containing the nucleus and a variety of other cellular organelles. The most striking features of neurons are fiberlike extensions, called processes, that increase the distance over which the cells can conduct messages. Neuronal processes are of two general types: **dendrites**, which convey signals from their tips to the rest of the neuron; and **axons**, which conduct messages toward their tips.

The dendrites of many neurons, such as the one in FIGURE 48.2a, are numerous and extensively branched (the name is derived from the Greek *dendron*, "tree"). Thus, dendrites are structural adaptations that increase the surface area of the neuron where it receives inputs from other neurons or sensory receptors.

Many neurons have a single axon, which may be very long. In fact, the sciatic nerve in your leg contains axons that extend all the way from the lower part of your spinal cord to muscles of the lower leg and foot, a distance of a meter or more. The axon hillock is the region of the cell body where the axon originates. As we will see, this is the region where impulses that pass down the axon are usually generated. Many axons in the vertebrate nervous system are enclosed by an insulating layer called the **myelin sheath**, formed by supporting cells. In the PNS, supporting cells called **Schwann cells** form myelin sheaths; in the CNS, supporting cells called **oligodendrocytes** produce the myelin sheaths. The axon may be branched, and each branch may give rise to hundreds or thousands of specialized endings called **synaptic terminals**, which relay signals to other cells by releasing chemical messengers called neurotransmitters. The site of contact between a synaptic terminal and a target cell (either another neuron or an effector cell, such as a muscle cell) is called a **synapse**. Thus, the synapse is

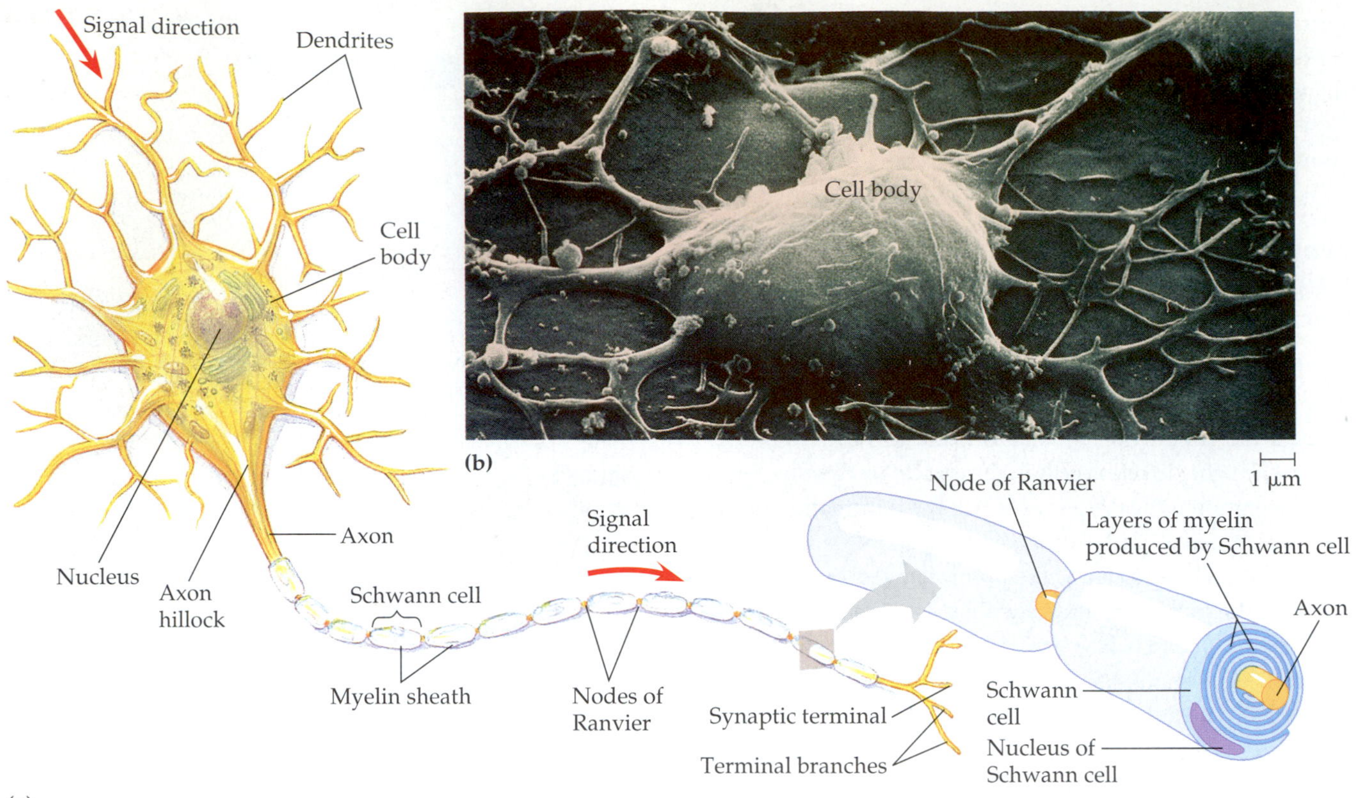

48.1 **FIGURE 48.2 ▪ Structure of a vertebrate neuron.** **(a)** The cell body has two types of processes, or extensions: Dendrites generally receive inputs and conduct signals toward the cell body, whereas axons conduct signals away from the cell body. At the end of the axon, terminal branches each have a synaptic terminal that makes connections with other neurons or target cells. In the peripheral nervous system, supporting cells called Schwann cells wrap many axons with an insulating myelin sheath. Gaps between successive Schwann cells are called nodes of Ranvier. **(b)** A scanning electron micrograph of a neuron.

the junction where one neuron communicates with another neuron in a neural pathway, or where a neuron communicates with a muscle cell or gland cell.

Functional Organization of Neurons

Functionally, there are three classes of neurons, corresponding to the three major functions of the nervous system. **Sensory neurons** communicate information (sensory input; see FIGURE 48.1) about the external and internal environments from sensory receptors to the central nervous system. Most sensory neurons synapse with interneurons, a second class of nerve cells, located within the CNS. **Interneurons** integrate sensory input and motor output; they make synaptic connections only with other neurons. A third class of nerve cells, **motor neurons**, convey impulses (motor output) from the CNS to effector cells. Adapted for different functions, neurons of the three classes differ markedly in shape, and there is also a variety of shapes within each class (FIGURE 48.3). A nervous system involves the coordinated activity of tens of thousands to billions of these diverse nerve cells. We can begin to understand how the whole system works by examining the way groups of neurons are organized.

Neurons are arranged in circuits made up of two or more functional types. The simplest neural circuits involve synapses between only two kinds of neurons, sensory neurons and motor neurons. Each sensory neuron conveys signals from a sensory receptor to a motor neuron, which in turn sends signals to an effector. The result is often a simple, automatic response, called a **reflex**. The knee-jerk reflex is an example of this simple response (FIGURE 48.4). A tap on the knee exerts a pull on the tendon connected to the thigh muscle (quadriceps) that extends the lower leg. Stretch receptors in the muscle detect the pull. (Actually, each stretch receptor is a finely coiled end of a sensory neuron.) When stimulated, the sensory neurons convey signals to synapses with motor neurons in the spinal cord. In turn, the motor neurons send messages back to the quadriceps telling it to contract (extend the leg).

Actually, the knee-jerk reflex involves more than a simple sensory/motor circuit. The contraction of the thigh muscle is accompanied by inhibition of the thigh muscles that flex the lower leg (pull it toward the body). The sensory neurons from the quadriceps synapse not only with motor neurons but also with interneurons in the spinal cord (see FIGURE 48.4). In turn, the interneurons inhibit motor neurons to the flexor muscles, preventing them from contracting.

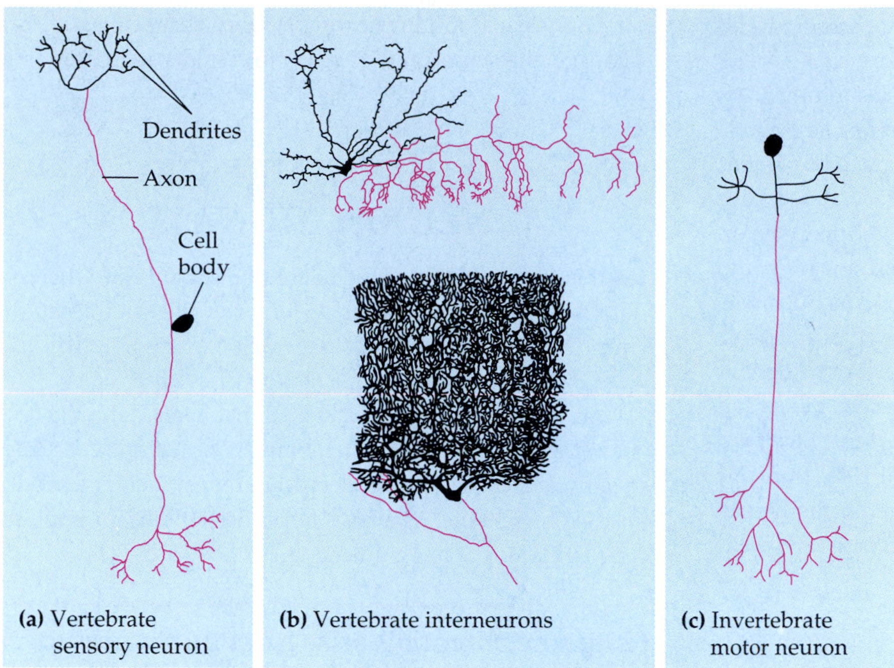

FIGURE 48.3 • Structural diversity of neurons. These examples illustrate some of the variety in neuron shape. Cell bodies and dendrites are black; axons are magenta. **(a)** A vertebrate sensory neuron. The short, multibranched dendrites communicate with sensory receptor cells. A single, long axon, which is usually myelinated, conveys signals from the dendrites to synapses with neurons in the CNS. The cell body is connected only to the axon. This configuration is markedly different from that of the neuron in FIGURE 48.2a, which is a vertebrate motor neuron. **(b)** Two types of interneurons found in the mammalian brain. The top one has multiple dendrites and a branched axon; the bottom one has finely branched, meshlike dendrites. **(c)** An invertebrate motor neuron. In contrast to the vertebrate motor neuron in FIGURE 48.2a, the cell body connects only to the dendrites.

Notice in FIGURE 48.4 that the cell body of the sensory neuron is located outside the spinal cord in a structure called the dorsal root ganglion. A **ganglion** (plural, **ganglia**) is a cluster of nerve cell bodies, often of similar function, in the peripheral nervous system (PNS). Similar functional clusters in the brain are usually called **nuclei** (not to be confused with the nuclei of individual cells). Ganglia and nuclei enable parts of the nervous system to coordinate activities without involving the entire system. For instance, the knee-jerk reflex frees the brain from handling a simple motor task, allowing the brain's greater integrative powers to focus on more complex problems.

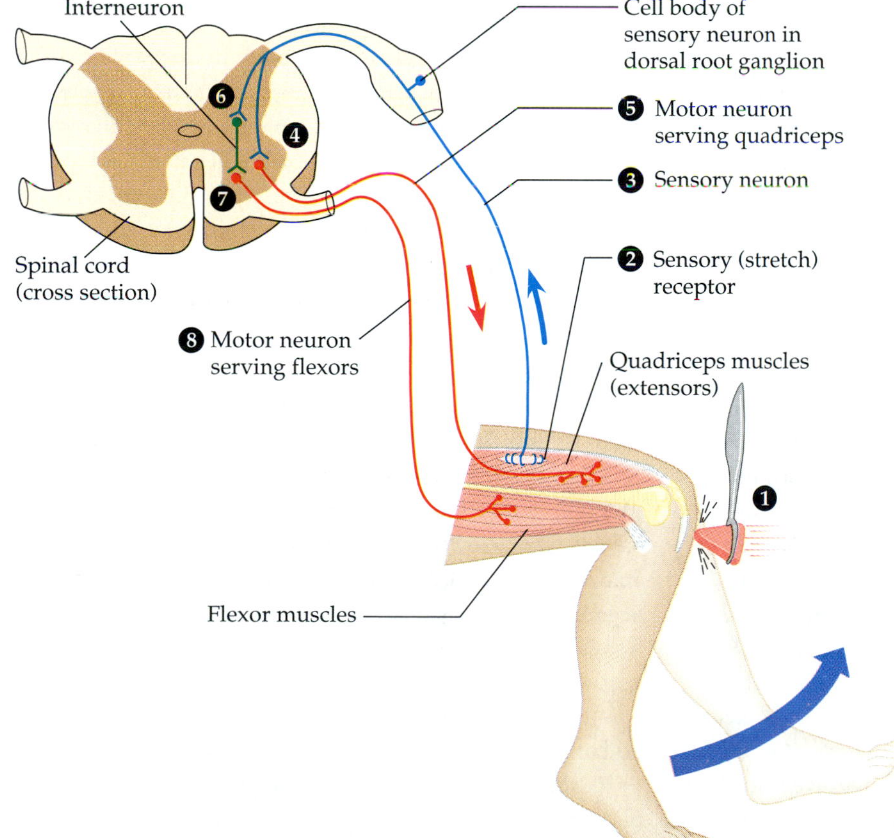

FIGURE 48.4 • The knee-jerk reflex. ① The knee-jerk, or patellar, reflex is caused by tapping the tendon connected to the quadriceps muscle. ② Sensory receptors detect a sudden stretch in the quadriceps (extensor) muscle in the thigh. ③ Sensory neurons convey the information to ④ synapses with motor neurons in the spinal cord. ⑤ The motor neurons convey signals to the quadriceps muscle to contract, jerking the lower leg forward. Only two kinds of neurons (sensory and motor) mediate the actual reflex action, but ⑥ the sensory neurons from the quadriceps also communicate with interneurons in the spinal cord. ⑦ In turn, the interneurons inhibit the motor neurons that ⑧ send signals to different leg muscles (flexors), whose action opposes that of the quadriceps. This inhibition prevents the flexors from contracting, which would resist the action of the quadriceps. For simplicity, only one neuron of each type is shown here, but actually many neurons are involved in the knee-jerk reflex. Ganglia outside the spinal cord (dorsal root ganglia) contain cell bodies of the sensory neurons. The cell bodies of motor neurons and interneurons are contained in gray matter (the butterfly-shaped region) in the spinal cord. The outer white matter in the spinal cord consists of motor and sensory axons.

Most behavioral responses, including reflexes, involve integration by interneurons in the central nervous system. Branches of the interneurons may carry signals to other segments of the spinal cord or to the brain, generating larger-scale or more complex responses. In general, there are three patterns of neural circuits. One kind of circuit takes information from a single source, such as an eye, to several parts of the brain; in this case the information from a single neuron spreads out to several postsynaptic neurons. In a second kind of circuit, information from several presynaptic neurons converges at a single postsynaptic neuron; convergent circuits can bring together information from several sources, such as vision, touch, and hearing, to identify an object in the environment. In a third type of circuit, information flows in a circular path, from one neuron to others and then back to its source. In the human brain, memories may be processed via circular paths and eventually stored at the source.

Supporting Cells (Glia)

Although they do not actually conduct nerve impulses, **supporting cells**, or **glia** ("glue"), are essential for the structural integrity of the nervous system and for the normal functioning of neurons. Glia outnumber neurons by tenfold to fiftyfold.

There are several types of glia in the brain and spinal cord, and as a group these cells do much more than simply glue neurons together. In the developing embryo, supporting cells called *radial glia* form tracks along which neurons migrate or grow out from the neural tube. Other glia called *astrocytes* provide structural and metabolic support for neurons. Astrocytes also induce the formation of tight junctions between cells lining the capillaries in the brain. The result is the **blood-brain barrier**, which restricts the passage of most substances into the brain, allowing the extracellular chemical environment of the CNS to be tightly controlled. Recent evidence also suggests that astrocytes communicate with one another and with other neurons via chemical signals. Still other glia include oligodendrocytes (in the CNS) and Schwann cells (in the PNS) that form insulating myelin sheaths around the axons of many neurons (see FIGURE 48.2a).

Neurons become myelinated in a developing nervous system when Schwann cells or oligodendrocytes grow around axons such that their plasma membranes form concentric layers, somewhat like a jelly roll. The membranes are mostly lipid, which is a poor conductor of electrical currents. Thus, the myelin sheaths provide electrical insulation of the axon, analogous to the rubber insulation covering copper wires. We will see later in this chapter that the myelin sheath also increases the speed of propagation of nerve impulses. In the degenerative disease known as multiple sclerosis, myelin sheaths gradually deteriorate, resulting in a progressive loss of coordination due to the disruption of nerve impulse transmission. Clearly, supporting cells are indispensable partners of neurons in a working nervous system. That being said, let's return to the neuron and its ability to conduct an impulse.

THE NATURE OF NEURAL SIGNALS

In this section we focus on how impulses arise in a neuron and how they are transmitted along the length of a neuron from a dendrite or cell body to the tip of an axon. We will also see how signals pass from one neuron to another.

An impulse is an electrical signal that depends on the flow of ions across the plasma membrane of a neuron. The signal begins as a change in the electrical gradient across the cell's plasma membrane. As a basis for understanding an impulse, let's take a look at this gradient.

Membrane potentials arise from differences in ion concentrations between a cell's contents and the extracellular fluid

48.2 All living cells have an electrical charge difference across their plasma membranes. This difference in charge gives rise to an electrical voltage gradient across the membrane, which can be measured with ultrafine microelectrodes (see the Methods Box, p. 966). The voltage measured across the plasma membrane is called the **membrane potential**, and it is typically in the range of –50 to –100 mV in an animal cell. By convention, the voltage outside the cell is called zero, so the minus sign indicates that the inside of the cell is negative in charge with respect to the outside. For a neuron in its resting state (that is, not transmitting an electrical signal), a membrane potential of –70 mV (about 5% of the voltage in a flashlight battery) is typical.

The membrane potential arises from differences in the ionic composition of the intracellular and extracellular fluids. The selective permeability of the plasma membrane, the barrier between those two fluids, maintains the ionic differences (FIGURE 48.5). Intracellular and extracellular fluids both contain various kinds of dissolved substances, including a variety of electrically charged substances (ions). Inside a cell the principal cation (positively charged ion) is potassium (K^+), although there is also some sodium (Na^+). Outside a cell the situation is reversed, with Na^+ the principal cation and K^+ having a much lower concentration. Inside a cell the principal anions are proteins, amino acids, sulfate, phosphate, and other negatively charged ions that we can group and symbolize by A^-; chloride (Cl^-) is also present but in a relatively low concentration. Outside a cell, Cl^- is the main anion, with other anions present but less important in the context of membrane potentials.

Recall from Chapter 8 that the plasma membrane is a phospholipid bilayer with associated membrane proteins. Ions, being electrically charged, cannot dissolve in lipid and

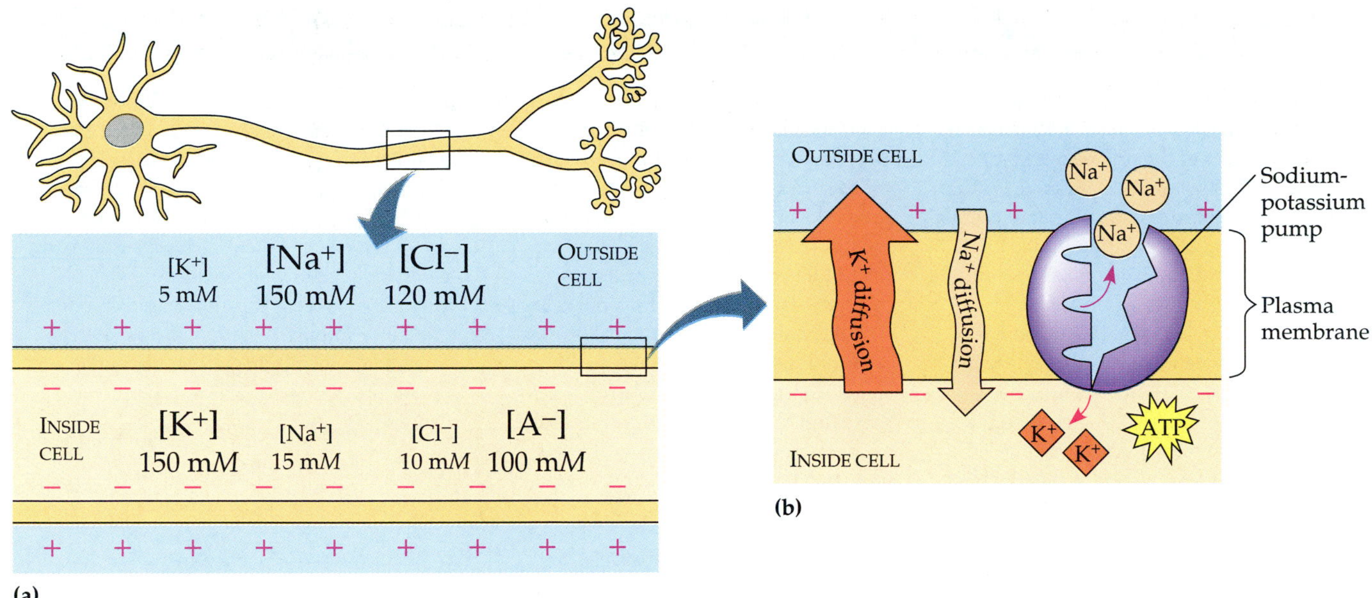

FIGURE 48.5 ▪ The basis of the membrane potential. (a) Intracellular and extracellular fluids have different
48.2 ionic compositions. Shown here are the approximate concentrations for a mammalian cell (in millimoles per liter, abbreviated m*M*) of potassium, $[K^+]$; sodium, $[Na^+]$; chloride, $[Cl^-]$; and anions that remain inside the cell, $[A^-]$. (Brackets indicate concentration of the enclosed substances; e.g., $[K^+]$ stands for potassium ion concentration.) K^+ diffuses out of the cell down its concentration gradient, but the A^- anions cannot follow, so the interior of the cell develops a net negative charge. **(b)** There is a steady diffusion of K^+ out of the cell (orange arrow) and steady diffusion of Na^+ into the cell (beige arrow); the thickness of the arrows indicates the relative permeability of the membrane to K^+ and Na^+. Over time, diffusion would cause the ionic gradients shown in part (a) to dissipate. Dissipation is prevented by the sodium-potassium pump, which uses ATP to actively transport Na^+ out of the cell and K^+ into the cell.

thus cannot directly diffuse across the lipid of the plasma membrane. In order to cross the membrane, ions must either be carried by transport proteins or move through ion channels, which are aqueous pores made up of specific transmembrane protein molecules. There are many different kinds of selective ion channels; some channels allow only Na^+ to cross, others allow only K^+, and still others only Cl^-. Depending on how many ion channels of each kind are present in the plasma membrane of a cell, it is possible for the membrane to have very different permeabilities to each of the different ions. Cells usually have much greater permeability to K^+ than to Na^+, suggesting that the membrane has many more potassium channels than sodium channels; in a resting neuron, for instance, potassium permeability is about fiftyfold higher than sodium permeability. Because the internal anions (A^-) are primarily large organic molecules (proteins and amino acids), they cannot cross the membrane and thus are a pool of internal negative charge that remains in the cell.

How does the distribution of ions shown in FIGURE 48.5 give rise to a membrane potential? Consider the case of potassium ions. There is a large concentration gradient for diffusion of K^+ out of the cell, and the membrane has a high permeability to potassium. Thus, there will be a net flux of K^+ out of the cell (efflux) driven by the concentration gradient. But as K^+ exits, it transfers positive charge from the inside to the outside of the cell. Because A^- stays in the cell, the inside of the cell becomes progressively more negative with respect to the outside. As positive charge is lost while negative charge is trapped within, an electrical gradient builds up across the membrane. In essence this *electrical* gradient competes with the effect of the K^+ *concentration* gradient: The increasing internal negativity attracts positively charged potassium, supporting an influx of K^+ down the electrical gradient. If K^+ were the only ion that could cross the membrane, the voltage across the membrane would continue to build up until the influx of K^+ down the electrical gradient exactly balanced the efflux of K^+ down the concentration gradient. At that point, there would be no further net transfer of charge across the membrane, and the membrane potential would reach a stable, resting value. For the potassium concentration gradient shown in FIGURE 48.5, a stable membrane potential of about −85 mV would be required to exactly counterbalance the concentration gradient in the manner just described. This value of the membrane potential is called the equilibrium potential for potassium ions, because it is the potential at which there will be no net movement of K^+ across the membrane (in other words, potassium is at equilibrium).

Potassium, however, is not the only ion to which the plasma membrane is permeable. Although the membrane is much less permeable to Na^+ than to K^+, the permeability to

METHODS: Measuring Membrane Potentials

Due to a difference in the relative concentrations of cations and anions on opposite sides of the plasma membrane, the cytoplasmic side of the membrane is negative in charge compared to the extracellular side. This charge separation is called a membrane potential, or, for an unstimulated neuron, a resting potential.

Electrophysiologists can measure the membrane potential as a voltage by using microelectrodes (shown in FIGURE a) connected to a sensitive voltmeter or oscilloscope. Precise mechanical devices called micromanipulators (seen next to the microscope in FIGURE b) are used to position one electrode just inside the cell for comparison with a reference electrode located outside the cell. The voltmeter indicates the magnitude of the charge separation across the membrane, typically about −70 mV for an unstimulated neuron. (The minus sign indicates that the inside of the cell is negative in charge with respect to the outside.)

A number of invertebrates, including squids, lobsters, and earthworms, have some unusually large neurons that make excellent model systems for studying nerve impulses. For example, the squid nervous system includes some neurons whose axons have diameters of about 1 mm. These giant axons are relatively easy to impale with microelectrodes. Once the electrodes are in place, they can be used to measure the voltage of the resting potential as well as to record changes in voltage due to ion currents that occur during the transmission of a nerve impulse. Much of the pioneering research on membrane potentials and on the nature of nerve signals was performed using squid giant axons.

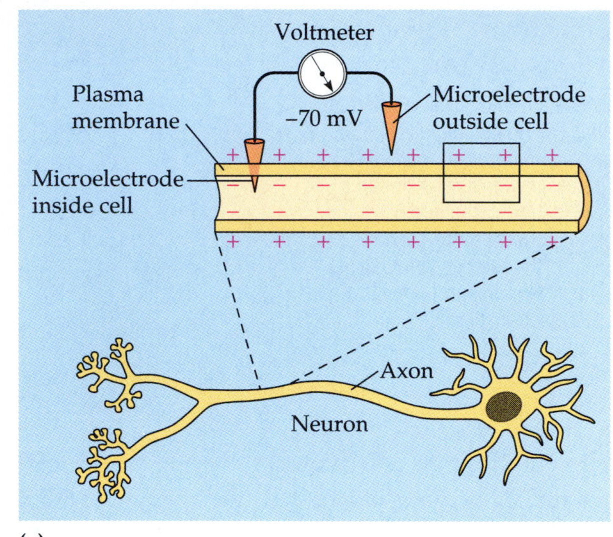

(a)

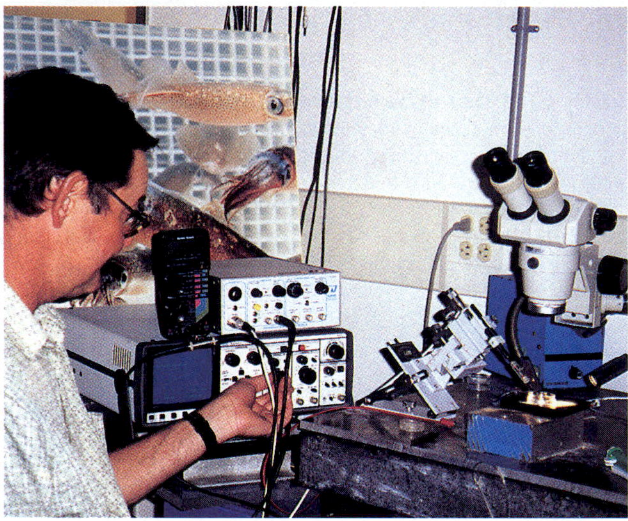

(b)

Na^+ is not zero. For sodium, both the concentration gradient ($[Na^+]$ greater outside the cell) and the electrical gradient (negative inside the cell) tend to move sodium ions into the cell. The resulting trickle of positive charge into the cell, carried by Na^+, makes the actual value of the membrane potential somewhat more positive than the −85 mV that would be expected if the membrane were permeable only to potassium ions. This explains why the membrane potential of a resting neuron is typically about −70 mV rather than about −85 mV.

Over time, the steady influx of sodium would cause a progressive increase in internal sodium concentration. Also, because the influx of Na^+ makes the cell interior less negative than the −85 mV required to balance the potassium concentration gradient, there would be a steady efflux of potassium and a progressive decline in internal K^+ concentration. In other words, if the situation were left unchecked, the concentration gradients for Na^+ and K^+ shown in FIGURE 48.5 would gradually dissipate. This is prevented by a different type of plasma membrane protein also found in abundance in neurons: the sodium-potassium pump (see FIGURE 8.14). This protein uses energy from ATP to drive the active transport of sodium back out of the cell, against both the concentration and electrical gradients for sodium. At the same time, the pump moves potassium into the cell, thus restoring the concentration gradient for this ion as well. In essence, the cell uses metabolic energy, in the form of ATP, to maintain the ionic gradients across the membrane that give rise to the steady-state membrane potential.

An action potential is an all-or-none change in the membrane potential

48.2 All cells have a membrane potential; however, only certain kinds of cells, including neurons and muscle cells, have the ability to generate changes in their membrane potentials. Collectively these cells are called **excitable cells**. The membrane potential of an excitable cell in a resting (unexcited) state is called the **resting potential**. As we are about to see, a change in the membrane potential from its resting state may result in an active electrical impulse.

Neurons have special ion channels, called **gated ion channels**, that allow the cell to change its membrane potential in response to stimuli the cell receives. In the case of a sensory neuron, the stimulus may come from the organism's environment (for example, light in the case of photoreceptors in the eye, or vibrations in the air in the case of receptors in the ear). In the case of an interneuron, the stimulus will ordinarily be produced by the activation of other neurons that provide inputs to the cell. The effect of the stimulus on the neuron depends on the type of gated ion channel that is opened by the stimulus. Some stimuli trigger a **hyperpolarization**, an increase in the electrical gradient across the membrane (FIGURE 48.6a). One of the ways a stimulus can produce a hyperpolarization is by opening a potassium channel, causing an increased outflow of potassium that in turn causes the membrane potential to become more negative. In contrast, a **depolarization** is a reduction in the electrical gradient across the membrane (FIGURE 48.6b). One of the ways this can occur is by a stimulus opening a sodium channel, causing an increased inflow of sodium, which makes the membrane potential less negative. Voltage changes produced by stimulation of this type are called **graded potentials** because the magnitude of change (either hyperpolarization or depolarization) depends on the strength of the stimulus: A larger stimulus will open more channels and produce a larger change in permeability.

In an excitable cell, such as a neuron, the response to a depolarizing stimulus is graded with stimulus intensity only up to a particular level of depolarization, called the **threshold potential**. If a depolarization reaches the threshold, a different type of response, called an **action potential**, will be triggered (FIGURE 48.6c). The threshold potential in a neuron is typically about 15 to 20 mV more positive than the resting potential (that is, at a potential of −50 to −55 mV). Hyperpolarizing stimuli do not produce action potentials; in fact, hyperpolarization makes it less likely that an action potential will be triggered by making it more difficult for a depolarizing stimulus to reach threshold.

The action potential is the nerve impulse. It is a nongraded *all-or-none event,* meaning that the magnitude of the action potential is independent of the strength of the depolarizing stimulus that produced it, provided the depolarization is sufficiently large to reach threshold. Once an action potential is triggered, the membrane potential goes through a stereotypical sequence of changes. During the depolarizing phase, the

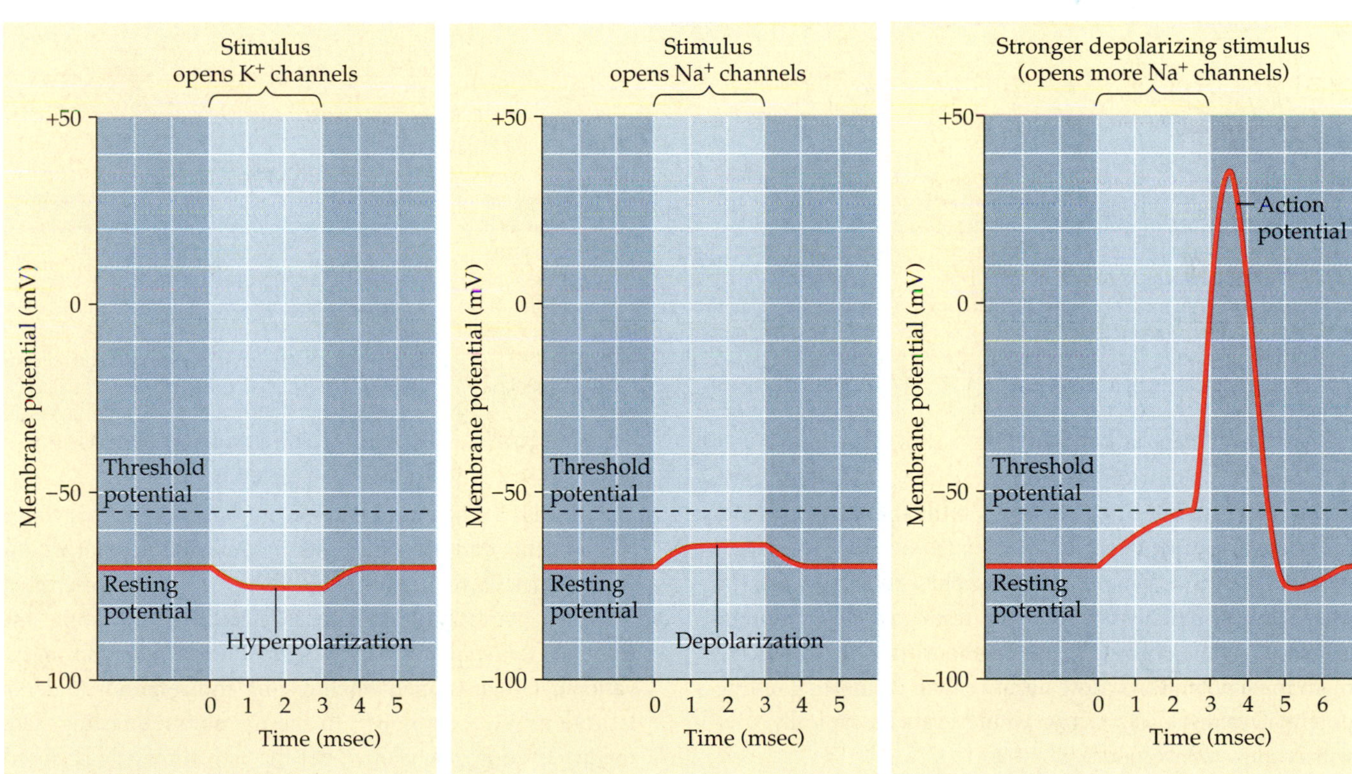

(a) Graded potential: hyperpolarization **(b)** Graded potential: depolarization **(c)** Action potential

FIGURE 48.6 ▪ Graded potentials and the action potential in a neuron. Neurons are stimulated by environmental changes that alter the cell's membrane potential. **(a)** One way a neuron can be hyperpolarized is by stimuli that open potassium channels. **(b)** A neuron can be depolarized by stimuli that open sodium channels. **(c)** A depolarizing stimulus of sufficient strength will change the membrane potential to a critical level called the threshold potential. This triggers an action potential, or nerve impulse, which, unlike a graded potential, is an all-or-none event; the size of the action potential is not affected by the strength of the stimulus that triggered it.

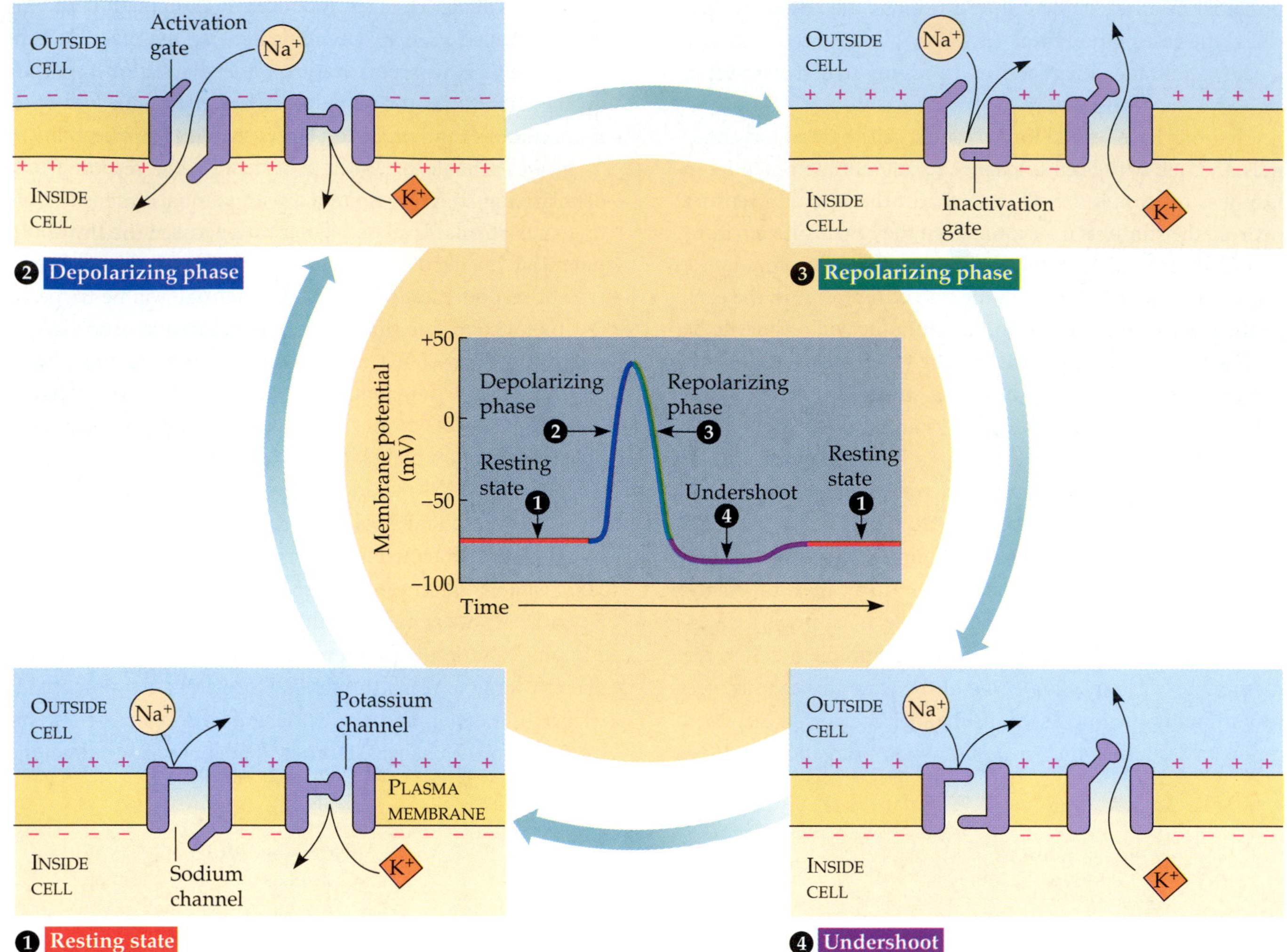

48.2 **FIGURE 48.7 ▪ The role of voltage-gated ion channels in the action potential.** The circled numbers on the action potential graph correspond to the four diagrams of voltage-gated sodium and potassium channels in a neuron's plasma membrane. Opening and closing of the voltage-sensitive gates underlie a sequence of stereotypical changes in membrane potential that occur when a stimulus triggers an action potential. ① In the resting state, both the sodium channel and the potassium channel are closed, and the membrane's resting potential is maintained. ② During the depolarizing phase, the action potential is generated as activation gates of the sodium channels open, and the potassium channels remain closed. Sodium ions rush into the cell, and the interior of the cell becomes more positive. ③ During the repolarizing phase, inactivation gates close the sodium channels, and potassium channels open. Potassium ions leave the cell, and the loss of positive charge causes the inside of the cell to become more negative than the outside of the cell. ④ During the undershoot, both gates of the sodium channels are closed, but potassium channels remain open because their relatively slow gates have not had time to respond to the repolarization of the membrane. Within another millisecond or two, the resting state is restored, and the system is ready to respond to another stimulus.

membrane polarity briefly reverses, with the interior of the cell becoming positive with respect to the outside. This is followed rapidly by a steep repolarizing phase, during which the membrane potential returns to its resting level. There may also be a phase, called the undershoot, during which the membrane potential is more negative than the normal resting potential (FIGURE 48.7). The whole event is typically over within a few milliseconds.

The action potential arises because the plasma membranes of excitable cells have special **voltage-gated ion channels**. These ion channels have gates that open and close in response to changes in membrane potential. Two types of voltage-gated channels contribute to the action potential: potassium channels and sodium channels (see FIGURE 48.7). Each potassium channel has a single gate that is voltage-sensitive; it is closed when resting and *opens slowly* in response to depolarization. By contrast, each sodium channel has two voltage-sensitive gates: an activation gate that is closed when resting and responds to depolarization by *opening rapidly*, and an inactivation gate that is open when resting and responds to depolarization by *closing slowly*. In the membrane's resting state, the inactivation gate is open but the activation gate is closed, so the channel does not allow Na^+ to enter the neuron (see FIGURE 48.7, step ①). Upon depolarization, the activation gate opens quickly, causing an influx of Na^+, which depolarizes the membrane further (see FIGURE 48.7, step ②), opening more voltage-gated sodium channels and causing still more depolarization. This inherently unstable process, an example of positive feedback, continues until all the sodium channels at the stimulated site of the membrane are open. Depolarization to the threshold potential causes further depolarization (the action potential). At this point, the membrane's permeability to Na^+ is about a thousandfold greater than the permeability of the resting membrane. The positive feedback that underlies the rapid depolarizing explains why the action potential, once triggered, is an all-or-none event.

Two factors underlie the rapid repolarizing phase of the action potential as the membrane potential is returned to its resting state. First, the sodium channel inactivation gate, which is slow to respond to changes in voltage, has time to respond to depolarization by closing, returning sodium permeability to its low resting level. Second, potassium channels, whose voltage-sensitive gates respond relatively slowly to depolarization, have had time to open. As a result, during repolarization, K^+ flows rapidly out of the cell, helping restore the internal negativity of the resting neuron (see FIGURE 48.7, step ③). The potassium channel gates are also the main cause of the undershoot, or hyperpolarization, that follows the repolarizing phase. Instead of returning immediately to their resting position, these relatively slow-moving gates remain open during the undershoot, allowing potassium to keep flowing out of the neuron. The continued potassium outflow makes the membrane potential more negative (hyperpolarizes it). Notice that during the undershoot, both the activation gate and the inactivation gate of the sodium channel are closed (see FIGURE 48.7, step ④). If a second depolarizing stimulus arrives during this period, it will be unable to trigger an action potential because the inactivation gates have not had time to reopen after the preceding action potential. This period when the neuron is insensitive to depolarization is called the **refractory period**, and it sets the limit on the maximum rate at which action potentials can be generated.

If the action potential is an all-or-none event with amplitude (size) unaffected by the intensity of the stimulus, how can the nervous system distinguish strong stimuli from weaker ones that are still sufficient to trigger action potentials? Strong stimuli result in a greater *frequency* of action potentials than weaker stimuli; if a stimulus is intense, the neuron will fire repeatedly, producing action potentials as rapidly as the refractory period will allow. Thus, it is the number of action potentials per second, not their amplitude, that codes for stimulus intensity in the nervous system.

Action potentials "travel" along an axon because they are self-propagating

An action potential is a localized electrical event, a membrane depolarization at a specific point of stimulation. A neuron is usually stimulated at its dendrites or cell body; for the resulting action potential to function as a signal, it must somehow "travel" along the axon to the other end of the cell. Actually, the action potential does not travel but is regenerated anew in a sequence along the axon. The effect of an action potential is like tipping the first of a row of standing dominoes. Just as the first domino's fall is relayed to the end of the row, the strong depolarization of one action potential assures that the neighboring region of the neuron will be depolarized above threshold, triggering a new action potential at that position, and so on to the end of the axon (FIGURE 48.8).

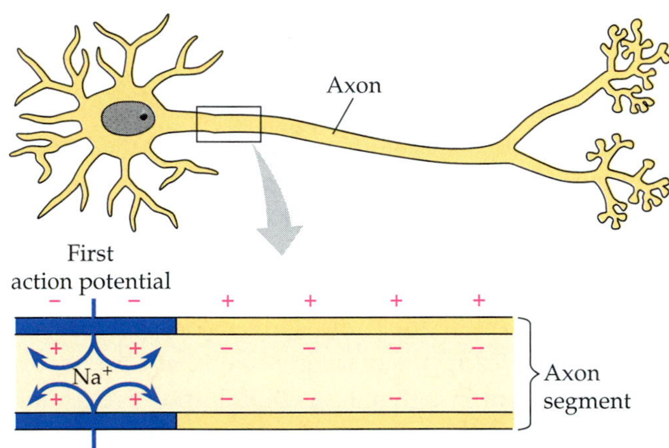

❶ An action potential is generated as sodium ions flow inward across the membrane at one location.

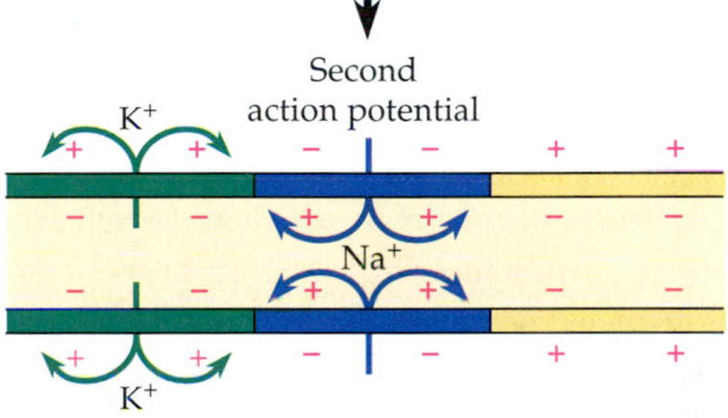

❷ The depolarization of the first action potential has spread to the neighboring region of the membrane, depolarizing it and initiating a second action potential. At the site of the first action potential, the membrane is repolarizing as K^+ flows outward.

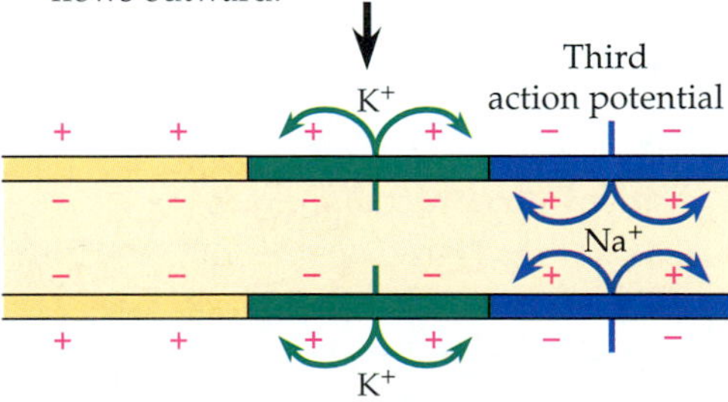

❸ A third action potential follows in sequence, with repolarization in its wake. In this way, local currents of ions across the plasma membrane give rise to a nerve impulse that passes along the axon.

FIGURE 48.8 ▪ Propagation of the action potential. The three parts of this figure show the changes that occur in a portion of an axon at three successive times as a nerve signal passes from left to right. At each point along the axon, the voltage-gated channels go through the sequence described in FIGURE 48.7, reproducing the sequence of voltage changes associated with the action potential.

Once the action potential is propagating along the axon, what prevents Na^+ entry from reexciting the region *behind* the action potential, which would cause depolarization to spread back toward the cell body as well as in the direction of normal propagation? Recall that an action potential is followed by a refractory period, when inactivation gates of sodium channels are closed and an action potential cannot be triggered. A wave of depolarization passing a point along the axon cannot induce another action potential behind it, but only in the forward direction. Thus, the axon is normally a one-way avenue for the conduction of nerve impulses.

Several factors affect the speed at which action potentials propagate along an axon. One factor is the diameter of the axon: The larger the axon's diameter, the faster the speed of transmission. This is because resistance to the flow of electrical current is inversely proportional to the cross-sectional area of the "wire" that conducts the current. In a thick axon, the depolarization associated with an action potential at a particular location can effectively reach farther along the interior of the axon and set up a new action potential at a greater distance away than in a thin axon. Transmission speed varies from several centimeters per second in very thin axons to about 100 m/sec in the giant axons of certain invertebrates, including squids and lobsters (see the Methods Box, p. 966). These giant axons function in behavioral responses requiring great speed, such as the backward tail-flip that enables a threatened lobster or crayfish to escape.

A different means of speeding the propagation of action potentials has evolved in vertebrates. Recall that many axons in vertebrate nervous systems are myelinated, coated with insulating layers of membranes deposited by oligodendrocytes or Schwann cells. The voltage-gated ion channels that produce the action potential are concentrated in the *nodes of Ranvier,* small gaps between successive Schwann cells along the axon. Also, extracellular fluid is in contact with the axon membrane only at the nodes, so that the flow of ions between the inside and outside of the axon can occur only in these regions. For these reasons, the action potential does not propagate over the length of the axon, but rather "jumps" from node to node, skipping the insulated regions of membrane between the nodes (FIGURE 48.9). This mechanism, called **saltatory conduction** (L. *saltare,* "to leap"), results in faster impulse transmission (up to 150 m/sec) in some neurons.

We have seen how stimulating the dendrite of a neuron can produce an action potential that is propagated along the axon to the very tip of the neuron. Our next step is to find out how the impulse is transmitted from one neuron to the next in the nervous system.

Chemical or electrical communication between cells occurs at synapses

Synapses are unique junctions that control communication between a neuron and another cell. Synapses are found between two neurons, between sensory receptors and sensory neurons, between motor neurons and the muscle cells they control, and between neurons and gland cells. Here we focus on synapses between neurons, which usually conduct signals from an axon's synaptic terminals to dendrites or cell bodies of the next cells in a neural pathway. The transmitting cell is called the **presynaptic cell**, and the receiving cell is called the **postsynaptic cell.** Synapses are of two types: electrical synapses and chemical synapses.

Electrical Synapses

An electrical synapse allows action potentials to spread directly from the presynaptic cell to the postsynaptic cell. The cells are connected by gap junctions (see FIGURE 7.30), intercellular channels that allow the local ion currents of an action potential to flow between neurons. The giant axons of lobsters and other crustaceans are connected end to end and coupled by electrical synapses. These make it possible for impulses to travel from neuron to neuron without delay and with no loss of signal strength. Electrical synapses in the central nervous systems of vertebrates synchronize the activity of neurons responsible for some rapid, stereotypical movements. For example, electrical synapses in the brain enable some fishes to flap their tail very rapidly when escaping from predators. However, chemical synapses are much more common than electrical synapses in vertebrates and most invertebrates.

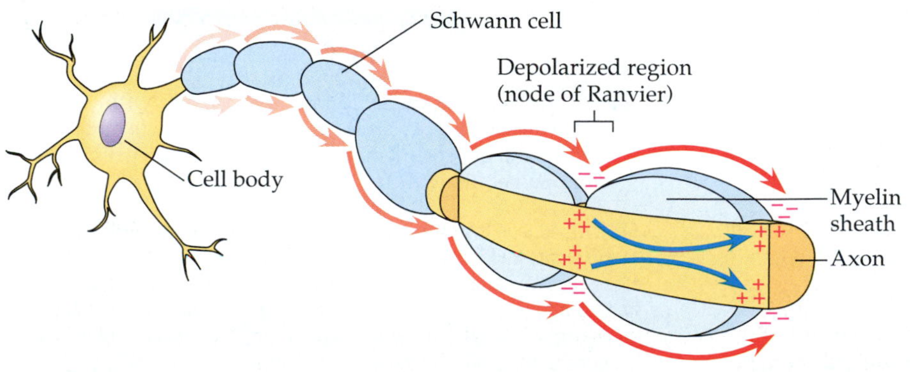

FIGURE 48.9 • Saltatory conduction. In a myelinated axon, the depolarization resulting from an action potential at one node of Ranvier spreads along the interior of the axon to the next node (blue arrows), triggering an action potential there. The action potential thus jumps from node to node as it propagates along the axon (red arrows).

Chemical Synapses

48.3 At a chemical synapse, a narrow gap, or **synaptic cleft**, separates the presynaptic cell from the postsynaptic cell. Because of the cleft, the cells are not electrically coupled, and an action potential occurring in the presynaptic cell cannot be transmitted directly to the membrane of the postsynaptic cell. Instead, a series of events converts the electrical signal of the action potential arriving at the synaptic terminal into a chemical signal that travels across the synapse, where it is converted back into an electrical signal in the postsynaptic cell.

The key to understanding the function of a chemical synapse is to examine its structure. Numerous sacs called **synaptic vesicles** are in the cytoplasm at the tip of the presynaptic axon (FIGURE 48.10). Each vesicle contains thousands of molecules of a **neurotransmitter**, the substance that is released as an intercellular messenger into the synaptic cleft. Many different neurotransmitters have been discovered in the nervous systems of animals. Most neurons secrete only one kind of neurotransmitter. However, a single neuron may receive chemical signals from a variety of neurons that secrete different neurotransmitters from their synaptic terminals.

A neuron dispatches neurotransmitter molecules into the synapse when an action potential arrives at the synaptic terminal and depolarizes the **presynaptic membrane**, the surface of the synaptic terminal that faces the cleft. Calcium ions play a central role in this conversion of the electrical impulse into a chemical signal. Depolarization of the presynaptic membrane causes Ca^{2+} to rush into the neuron through voltage-gated channels. The sudden rise in the cytosolic concentration of Ca^{2+} stimulates the synaptic vesicles to fuse with the presynaptic membrane and spill the neurotransmitter into the synaptic cleft by exocytosis (see Chapter 8). Thousands of vesicles may respond in unison to a single action potential. The neurotransmitter diffuses the short distance from the presynaptic membrane to the **postsynaptic membrane**, the plasma membrane of the cell body or dendrite on the other side of the synapse.

The postsynaptic membrane is specialized to receive the chemical message. Projecting from the extracellular surface of the membrane are proteins that function as specific receptors for neurotransmitters. The receptors are part of selective ion channels that open and close, controlling movements of ions across the postsynaptic membrane. A receptor is keyed to a particular type of neurotransmitter, and when it binds to this

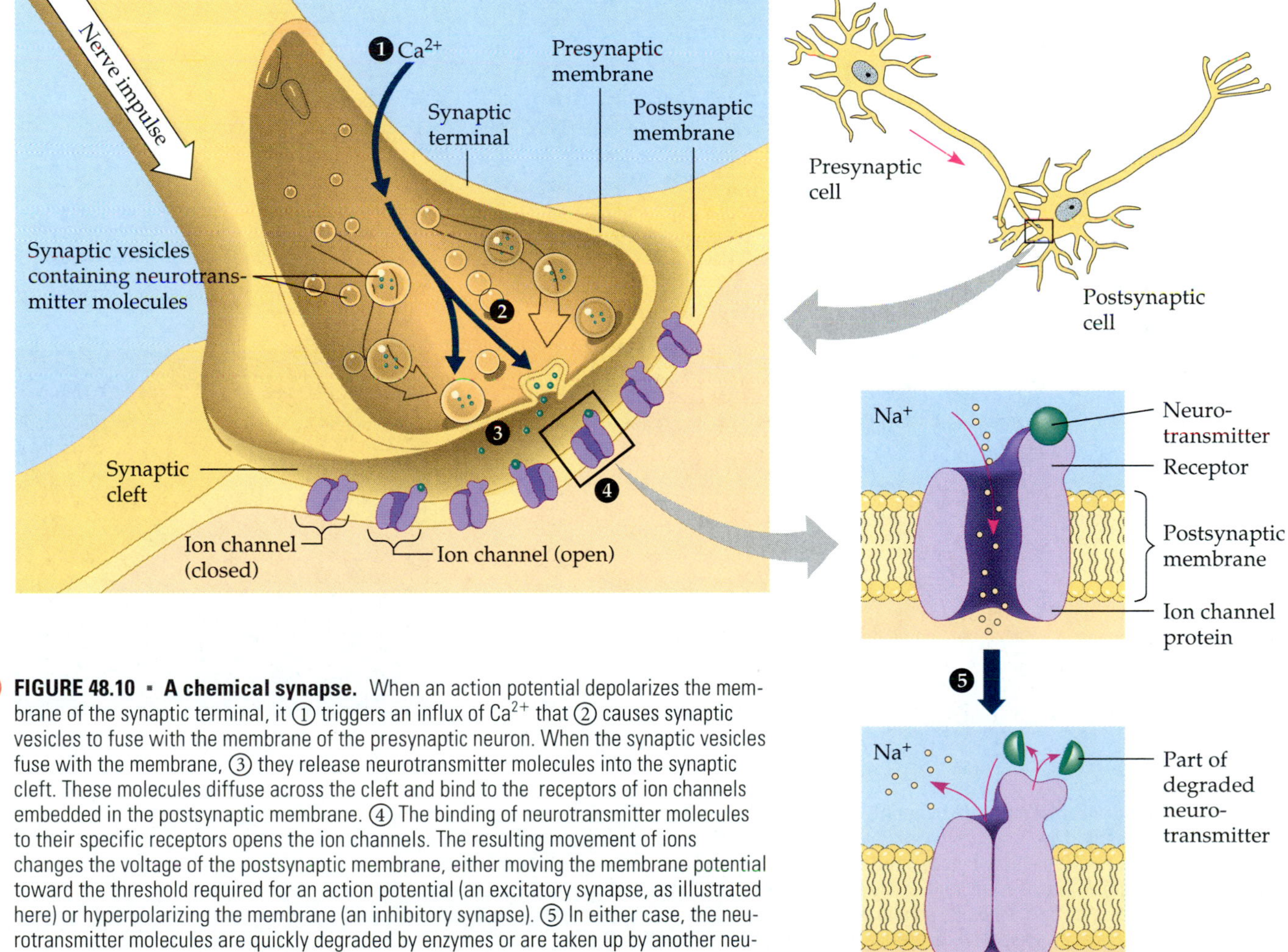

48.3 **FIGURE 48.10 • A chemical synapse.** When an action potential depolarizes the membrane of the synaptic terminal, it ① triggers an influx of Ca^{2+} that ② causes synaptic vesicles to fuse with the membrane of the presynaptic neuron. When the synaptic vesicles fuse with the membrane, ③ they release neurotransmitter molecules into the synaptic cleft. These molecules diffuse across the cleft and bind to the receptors of ion channels embedded in the postsynaptic membrane. ④ The binding of neurotransmitter molecules to their specific receptors opens the ion channels. The resulting movement of ions changes the voltage of the postsynaptic membrane, either moving the membrane potential toward the threshold required for an action potential (an excitatory synapse, as illustrated here) or hyperpolarizing the membrane (an inhibitory synapse). ⑤ In either case, the neurotransmitter molecules are quickly degraded by enzymes or are taken up by another neuron, closing the ion channels and terminating the synaptic response.

chemical, the gate of the ion channel opens, allowing specific ions, such as Na^+, K^+, or Cl^-, to cross the membrane. Thus, the ion channels of the postsynaptic membrane are chemically gated, in contrast to the voltage-gated channels responsible for the action potential.

The ion movements resulting from the binding of neurotransmitter to its receptors alter the membrane potential of the postsynaptic cell. Depending on the type of receptors and the ion channels they control, neurotransmitters binding to the postsynaptic membrane may either excite the membrane by bringing its voltage closer to the threshold potential or inhibit the postsynaptic cell by hyperpolarizing its membrane. In either case, enzymes quickly break down the neurotransmitter, ensuring that its effect on the postsynaptic cell will be brief and precise, and that the next action potential arriving at the synapse will be transmitted. For example, the neurotransmitter known as acetylcholine is rapidly degraded by cholinesterase, an enzyme present in both the synaptic cleft and the postsynaptic membrane.

Note that one important function of the synapse is to allow nerve impulses to be transmitted only in a single direction over a neural pathway. Synaptic vesicles are present only in synaptic terminals, and thus only the presynaptic membrane can discharge neurotransmitters. And receptors are restricted to the postsynaptic membrane, ensuring that only this membrane can receive a chemical signal from another neuron.

A postsynaptic membrane is analogous to a tiny chip in a computer, receiving and processing information in the form of neurotransmitter molecules. At any particular instant, these living computer chips integrate multiple positive and negative inputs and either fire an action potential or not. The ability of the nervous system to perform all of its integrating activities and make precise, appropriate responses to stimuli depends on this complex control of postsynaptic membranes.

Neural integration occurs at the cellular level

A single neuron may receive information from numerous neighboring neurons via thousands of synapses, some of them excitatory and some of them inhibitory (FIGURE 48.11). Excitatory and inhibitory synapses have opposite effects on the membrane potential of the postsynaptic cell.

At an excitatory synapse, neurotransmitter receptors control a type of gated channel that allows Na^+ to enter the cell and K^+ to leave the cell. Because the driving force is greater for Na^+ than for K^+ (remember, voltage and concentration gradients both drive Na^+ into the cell), the effect of opening these channels is a net flow of positive charge into the cell. This depolarizes

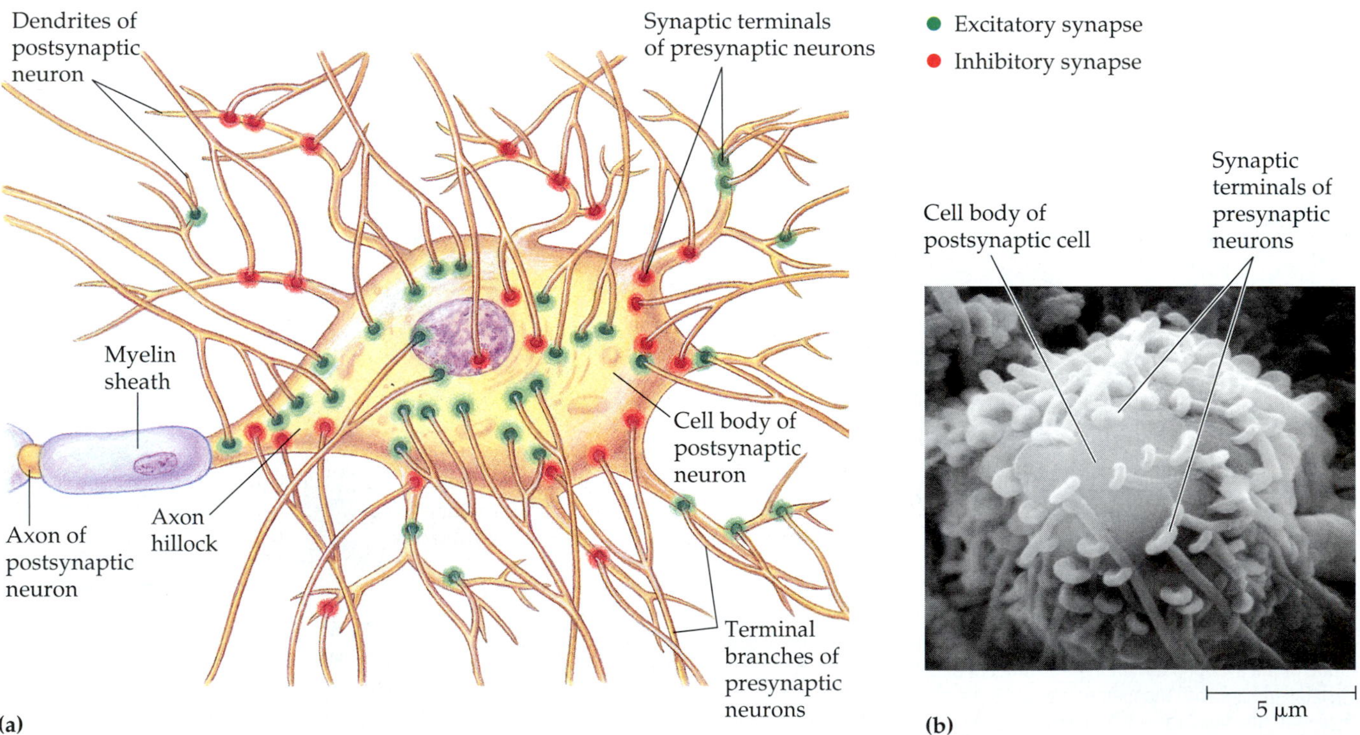

FIGURE 48.11 ▪ Integration of multiple synaptic inputs. (a) Each neuron, especially in the CNS, is on the receiving end of thousands of synapses, some excitatory (green) and others inhibitory (red). At any instant an action potential may be generated at the axon hillock if the combined effect of ion currents induced by excitatory and inhibitory synapses depolarizes the membrane to the threshold potential. Synapses close to the axon hillock generally have a stronger effect on the membrane potential than other synapses. This mechanism of integrating multiple positive and negative inputs by adding their individual effects is called summation. **(b)** This micrograph reveals numerous synaptic terminals of presynaptic neurons that communicate with a single postsynaptic cell (SEM).

the cell, moving the membrane potential closer to the threshold voltage and making it more likely that the postsynaptic cell will generate an action potential. In this case, the electrical change caused by the binding of neurotransmitter to the receptor is called an **excitatory postsynaptic potential**, or **EPSP**.

At an inhibitory synapse, the binding of neurotransmitter molecules to the postsynaptic membrane hyperpolarizes the membrane by opening ion channels that make the membrane more permeable either to K^+, which rushes out of the cell, or to Cl^-, which enters the cell because of a large concentration gradient (see FIGURE 48.5), or to both of these ions. These ion fluxes push the membrane potential to a voltage even more negative than the resting potential, making it more difficult for an action potential to be generated. Therefore, the voltage change associated with chemical signaling at an inhibitory synapse is called an **inhibitory postsynaptic potential**, or **IPSP**. Whether a particular neurotransmitter results in an EPSP or an IPSP depends on the type of receptors and gated ion channels on the postsynaptic membrane responding to that neurotransmitter.

Both EPSPs and IPSPs are graded potentials that vary in magnitude with the number of neurotransmitter molecules binding to receptors on the postsynaptic membrane. The change in voltage, either depolarization or hyperpolarization, lasts only a few milliseconds because the neurotransmitters are inactivated by enzymes soon after their release into the synapse. Also, the electrical impact on the postsynaptic cell decreases with distance away from the synapse. For the postsynaptic cell to fire (generate an action potential), the local ion currents due to EPSPs must be strong enough to depolarize the membrane in the region of the axon hillock to the threshold potential, usually about −50 mV. The axon hillock is the region where voltage-gated sodium channels open and generate an action potential when some stimulus has depolarized the membrane to the threshold.

A single EPSP at one synapse, even one close to the axon hillock, is not usually strong enough to trigger an action potential. However, several synaptic terminals acting simultaneously on the same postsynaptic cell, or a smaller number of synaptic terminals discharging neurotransmitters repeatedly in rapid-fire succession, can have a cumulative impact on the membrane potential at the axon hillock. This additive effect of postsynaptic potentials is called **summation** (FIGURE 48.12).

There are two types of summation: temporal summation and spatial summation. In temporal summation, chemical transmissions from one or more synaptic terminals occur so

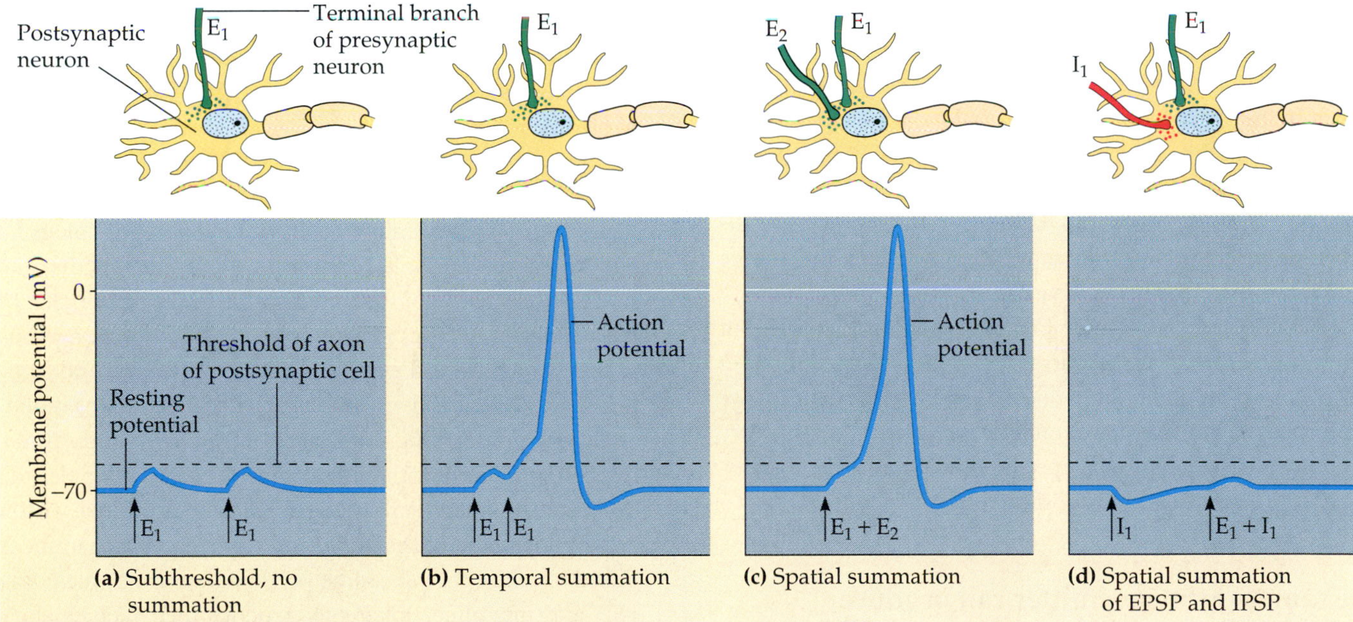

FIGURE 48.12 • Summation of postsynaptic potentials. These graphs trace changes in membrane potentials at a postsynaptic neuron's axon hillock. The arrows indicate times when signals trigger changes in membrane potentials at two excitatory synapses (E_1 and E_2, green) and at one inhibitory synapse (I_1, red). Like most single EPSPs, those shown at E_1 and E_2 cannot depolarize the membrane at the axon hillock all the way to the threshold level, and thus without summation do not trigger an action potential. **(a)** Repeated subthreshold EPSPs that do not overlap in time do not add together and therefore do not depolarize the membrane to threshold. **(b)** Temporal summation occurs when two or more subthreshold EPSPs that overlap in time reinforce each other. Here, a second release of a neurotransmitter at synapse E_1 affects the postsynaptic membrane when it is still partially depolarized from a slightly earlier stimulation. The cumulative effect depolarizes the membrane to threshold, triggering an action potential. **(c)** Spatial summation occurs when two or more presynaptic cells release neurotransmitter at the same time, causing a cumulative voltage change greater than the individual EPSPs. Here, the action potential results from spatial summation of EPSPs at synapses E_1 and E_2, depolarizing the membrane to threshold. **(d)** EPSPs and IPSPs move the membrane potential in opposite directions. An IPSP (I_1) hyperpolarizes the postsynaptic membrane. In an example of spatial summation, if an IPSP (at synapse I_1) releases neurotransmitter at the same time as an EPSP (at synapse E_1), the IPSP subtracts from the depolarization.

close together in time that each postsynaptic potential affects the membrane before the voltage has returned to the resting potential after the previous stimulation (FIGURE 48.12b). In spatial summation, several different synaptic terminals, usually belonging to different presynaptic neurons, stimulate a postsynaptic cell at the same time and have an additive effect on the membrane potential (FIGURE 48.12c).

By reinforcing one another through temporal or spatial summation, the ion currents associated with several EPSPs can depolarize the membrane at the axon hillock to threshold, causing the neuron to fire. Summation also applies to IPSPs; two or more IPSPs can hyperpolarize the membrane to a voltage more negative than any single release of a neurotransmitter at an inhibitory synapse can achieve. Furthermore, IPSPs and EPSPs counter each other's electrical effects (FIGURE 48.12d).

The axon hillock is the neuron's integrating center, the region where the membrane potential represents the effect of all EPSPs and IPSPs. At any instant, the membrane potential at the axon hillock is an average of the depolarization due to summation of all EPSPs and the hyperpolarization due to summation of all IPSPs. (This takes into account the greater impact of synapses at or near the axon hillock.) Whenever EPSPs overpower IPSPs enough for the membrane potential at the axon hillock to reach threshold, an action potential is generated, and the impulse is transmitted along the axon to the next synapse. A few milliseconds later, after the refractory period, the neuron may fire again if the sum of all synaptic inputs at that moment is still sufficient to depolarize the membrane at the axon hillock to threshold level. On the other hand, by that time the sum of all EPSPs and IPSPs may put the membrane potential at the axon hillock at a voltage more negative than the threshold, or may even hyperpolarize the membrane to a potential more negative than the resting potential, thereby desensitizing the neuron for the moment.

Action potentials, remember, are all-or-none events. But now we see that the occurrence of these nerve impulses depends on the ability of the neuron to integrate quantitative information in the form of multiple excitatory and inhibitory inputs, each involving the specific binding of a neurotransmitter to a receptor on the postsynaptic membrane.

The same neurotransmitter can produce different effects on different types of cells

Dozens of different substances, many of them small, nitrogen-containing organic molecules, are known to function as neurotransmitters, and researchers expect to find many more. TABLE 48.1 summarizes the major known neurotransmitters. Notice that a single neurotransmitter can trigger different responses in postsynaptic cells. This versatility depends on the receptors present on different postsynaptic cells and on the receptor's mode of action. In some cases, differences in neurotransmitter effects are due to different protein receptors; in other cases the same receptors trigger different molecular changes in different postsynaptic cells (see FIGURE 45.4). Many neurotransmitters bind with receptors that have a direct effect on ion channel proteins, altering the membrane permeability of the postsynaptic cell. This type of synaptic communication takes only a few milliseconds. Other neurotransmitters take much longer (up to several minutes) because they communicate via complex signal-transduction pathways in the postsynaptic cell.

Acetylcholine is one of the most common neurotransmitters in both invertebrates and vertebrates. In the vertebrate central nervous system, acetylcholine can be inhibitory or excitatory, depending on the type of receptor. At the vertebrate neuromuscular junction, the synapse between a motor neuron and a skeletal muscle cell, acetylcholine is released from the synaptic terminal of the motor neuron. It binds with a receptor that has a direct stimulatory effect on the muscle cell plasma membrane. The effect is excitatory, depolarizing the membrane of the postsynaptic muscle cell. A second type of acetylcholine receptor in heart muscle activates a signal-transduction pathway whose G proteins have two effects: They inhibit adenylyl cyclase and open K^+ channels in the muscle cell membrane, making it less able to generate an action potential. Both effects reduce the strength and rate of cardiac muscle cell contraction.

The **biogenic amines** are neurotransmitters derived from amino acids. One group, known as catecholamines, is produced from the amino acid tyrosine. This group includes **epinephrine** and **norepinephrine**, which also function as hormones (see Chapter 45), and a closely related compound called **dopamine**. Another biogenic amine, **serotonin**, is synthesized from the amino acid tryptophan. The biogenic amines most commonly function as transmitters within the CNS. However, norepinephrine also functions in a branch of the peripheral nervous system called the autonomic nervous system, which we will examine shortly. Dopamine and serotonin are widespread in the brain and affect sleep, mood, attention, and learning. Imbalances of these neurotransmitters are associated with several disorders; for example, the degenerative illness Parkinson's disease is associated with a lack of dopamine in the brain, and an excess of dopamine is linked to schizophrenia. Some psychoactive drugs, including LSD and mescaline, apparently produce their hallucinatory effects by binding to serotonin and dopamine receptors in the brain. The biogenic amines often affect metabolism within the postsynaptic cell. In many cases they bind to specific receptors on the postsynaptic membrane, triggering a signal-transduction pathway that often involves a G protein relay system, an effector enzyme (often adenylyl cyclase), and a second messenger (often cyclic AMP; see FIGURE 11.12).

Four amino acids are known to function as CNS neurotransmitters: **gamma aminobutyric acid (GABA)**, **glycine**, **glutamate**, and **aspartate**. GABA, believed to be the transmit-

Table 48.1 ▪ The Major Known Neurotransmitters

NEUROTRANSMITTER	STRUCTURE	FUNCTIONAL CLASS	SECRETION SITES
Acetylcholine	$H_3C—C(=O)—O—CH_2—CH_2—N^+—(CH_3)_3$	Excitatory to vertebrate skeletal muscles; excitatory or inhibitory at other sites	CNS; PNS; vertebrate neuromuscular junction
Biogenic Amines			
Norepinephrine	(HO)₂-benzene ring—CH(OH)—CH_2—NH_2	Excitatory or inhibitory	CNS; PNS
Dopamine	(HO)₂-benzene ring—CH_2—CH_2—NH_2	Generally excitatory; may be inhibitory at some sites	CNS; PNS
Serotonin	HO-indole ring (N-H, CH)—C—CH_2—CH_2—NH_2	Generally inhibitory	CNS
Amino Acids			
GABA (gamma aminobutyric acid)	$H_2N—CH_2—CH_2—CH_2—COOH$	Inhibitory	CNS; invertebrate neuromuscular junction
Glycine	$H_2N—CH_2—COOH$	Inhibitory	CNS
Glutamate	$H_2N—CH(COOH)—CH_2—CH_2—COOH$	Excitatory	CNS; invertebrate neuromuscular junction
Aspartate	$H_2N—CH(COOH)—CH_2—COOH$	Excitatory	CNS
Neuropeptides			
Substance P	Arg–Pro–Lys–Pro–Gln–Gln–Phe–Phe–Gly–Leu–Met	Excitatory	CNS; PNS
Met-enkephalin (an endorphin)	Tyr–Gly–Gly–Phe–Met	Generally inhibitory	CNS

ter at most inhibitory synapses in the brain, produces IPSPs by increasing the chloride permeability of the postsynaptic membrane.

Several **neuropeptides**, relatively short chains of amino acids, serve as neurotransmitters. In common with the biogenic amines, neuropeptides often operate via signal-transduction pathways. A neuropeptide called **substance P** is a key excitatory signal that mediates our perception of pain. The **endorphins** are neuropeptides that function as natural analgesics, decreasing the perception of pain by the CNS. Endorphins were first discovered in the 1970s by neurochemists studying the mechanism of opium addiction. Candace Pert and Solomon Snyder of Johns Hopkins University found specific receptors for the opiates morphine and heroin on neurons in the brain. It seemed odd that humans would have receptors keyed to chemicals from a plant (the opium poppy). Further research showed that, in fact, the drugs bind to these receptors in the brain by mimicking endorphins, the natural painkillers produced in the brain during times of physical or emotional stress, such as the labor of childbirth. In addition to relieving pain, endorphins also decrease urine output (by affecting ADH secretion; see Chapter 45), depress respiration, produce euphoria, and have other emotional effects through specific pathways in the brain. An endorphin is also released from the anterior pituitary gland as a hormone that affects specific regions of the brain. Once again, we see the overlap between endocrine and nervous system control.

Gaseous Signals of the Nervous System

In common with many other types of cells, some neurons of the vertebrate PNS and CNS utilize gas molecules, notably nitric oxide (NO, see p. 895) and carbon monoxide (CO), as local regulators. For example, during sexual arousal, certain neurons release NO gas into the erectile tissue of the penis. In response, smooth muscle cells in the blood vessel walls of the erectile tissue dilate and fill with blood, producing an erection. The male impotence drug, Viagra, increases the ability to achieve and maintain an erection during sexual arousal by inhibiting an enzyme that slows the muscle-relaxing effects of NO.

Many cells release gas molecules in response to chemical signals. For instance, the neurotransmitter acetylcholine released by neurons into the walls of blood vessels stimulates the endothelial cells of the vessels to synthesize and release NO. In turn, NO signals the neighboring smooth muscle cells to relax, dilating the vessels. The discovery of this mechanism in the late 1980s explained the medicinal action of nitroglycerin, which had been used for a century to treat angina (chest pain associated with reduced blood supply to the heart). Enzymes convert nitroglycerin to NO, which dilates the blood vessels that supply cardiac muscle.

Unlike typical neurotransmitters, NO and other gaseous messengers are not stored in cytoplasmic vesicles; cells synthesize them on demand. They diffuse into neighboring target cells, produce a change, and are broken down—all within a few seconds. In many of its targets, including smooth muscle cells, NO works like many hormones, stimulating a membrane-bound enzyme to synthesize a second messenger that directly affects cellular metabolism.

ORGANIZATION OF NERVOUS SYSTEMS

While there is remarkable uniformity in how nerve cells function throughout the animal kingdom, there is great diversity in how nervous systems as a whole are organized. Some animals even lack a nervous system altogether; sponges, for instance, have no cells specialized for impulse conduction (see FIGURE 33.2). We now survey some of the diversity among animal nervous systems.

Nervous system organization tends to correlate with body symmetry

Most animals have neurons and a nervous system. Structurally, the simplest nervous systems occur in some cnidarians that have a **nerve net**, a system of nerves that branch throughout the body. Hydras, for instance, have a nerve net with no ganglia or distinction between central and peripheral elements (FIGURE 48.13a). Many of the synapses in a hydra's nerve net are electrical, impulses are conducted in both directions, and stimulation at any point on the body of a hydra spreads from that site and may cause movements of the entire body. Certain other cnidarians and ctenophores show evidence of nervous system centralization. In some jellies, for example, clusters of nerve cells at the margin of the bell (associated with simple sensory structures) and pathways around the bell formulate the neural signals that affect activities such as swimming, which require more complex coordination of the entire body.

The nervous system of many echinoderms resembles that of the jellies. In sea stars, for example, radial nerves extend through each arm from a central nerve ring around the oral disk (FIGURE 48.13b). Branches of the radial nerves form an interconnected network similar to the cnidarian nerve net. This system coordinates movement regardless of which arm leads.

Cnidarians, ctenophores, and many echinoderms are radially symmetrical and so are their nervous systems. A radially symmetrical nervous system tends to be uncentralized, but this does not imply structural or functional simplicity, or that radially symmetrical animals are handicapped by their nervous system. The movements of hundreds of tube feet of a sea star during feeding, for example, require complex neural coodination.

Most animals are bilaterally symmetrical and have a bilaterally arranged nervous system with peripheral and central elements. Correlated with bilateral symmetry and centralization, they also exhibit some degree of **cephalization**, a concentration of feeding organs, sensors, and neural structures at the anterior (head) end, the part of the body most likely to make first contact with food or threatening stimuli. In most cases, a brain in the head and one or more nerve cords make up the central nervous system. A **nerve cord** is a thick bundle of nerves usually extending longitudinally through the body from the brain. A main thoroughfare of nerve impulses passing between the brain and the PNS, a nerve cord contains nerve cell bodies that integrate sensory information and formulate command signals to effectors. Brains have evolved from anterior enlargements of nerve cords.

Flatworms represent one of the oldest phylogenetic lineages in which the animals have a clearly defined CNS, composed of a small brain and two or more longitudinal nerve cords (FIGURE 48.13c). Most other invertebrates show a greater degree of centralization and cephalization of the nervous system. For example, annelids and arthropods have a prominent brain and a ventral nerve cord containing segmentally arranged ganglia (FIGURE 48.13d and e).

Mollusks are good examples of how nervous system complexity correlates with an animal's role (niche) in its environment. Sessile or slow-moving mollusks such as chitons and clams have little or no cephalization and relatively simple sense organs (FIGURE 48.13f). In sharp contrast, cephalopods have the most sophisticated nervous systems of any invertebrates, rivaling even those of some vertebrates. The large brain of a squid or octopus accompanied by large, image-forming eyes and rapid conduction along giant axons correlates well with the active predatory life of these animals (FIGURE 48.13g).

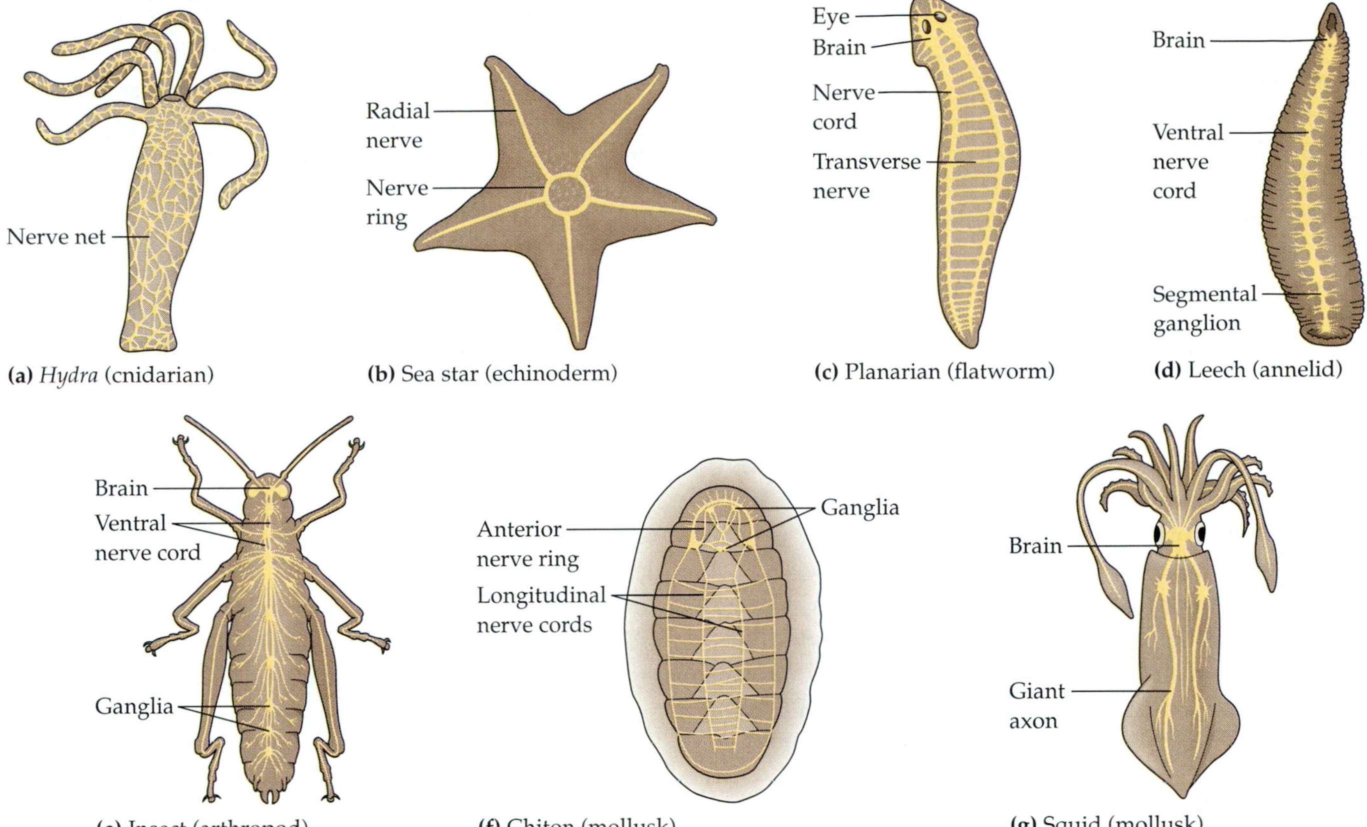

FIGURE 48.13 ▪ Diversity in nervous systems. (a) Hydras have a nerve net with no centralization. **(b)** Echinoderms, such as this sea star, have a central nerve ring with radial nerves connected to a nerve net in each arm. **(c)** The planarian, a flatworm, has a bilaterally symmetrical nervous system. Its CNS is a small brain connected to two longitudinal nerve cords. Its PNS includes a ladderlike system of transverse nerves connecting the nerve cords to smaller nerves that extend throughout the body. **(d)** Annelids, such as this leech, have a brain with a greater concentration of neurons than that of flatworms. Interneurons in ganglia of the ventral nerve cord coordinate many of the actions of the individual body segments. **(e)** The arthropod nervous system, as shown in this insect, probably evolved from an annelidlike system. With extensive fusion of ganglia in the head and ventral nerve cord, it is less uniformly segmented and more centralized than the annelid nervous system.
(f, g) Molluskan nervous systems are highly varied, with the degree of centralization generally correlated with lifestyle. **(f)** At one extreme, the slow-moving chitons and clams have a relatively diffuse system with widely separated ganglia. The central nervous system of the chiton shown here consists of an anterior nerve ring connecting small ganglia and longitudinal nerve cords. **(g)** At the other extreme, cephalopods (squids and octopuses) have a highly centralized and cephalized nervous system that supports their active, predatory lifestyle.

Octopuses can learn to discriminate among visual patterns and perform complex tasks in laboratory experiments.

Mollusks are important research models in neurobiology. Studies of giant squid axons yielded much of our understanding of action potentials, and snails are especially popular among researchers working in such areas as nerve cell regeneration, neurotransmitters, learning, and memory. For example, a marine snail called the sea hare (*Aplysia*) has large neurons and ganglia that are accessible to manipulation. Its entire nervous system consists of only about 20,000 neurons (compared to about 100 billion in the human brain), yet *Aplysia* displays some simple forms of learning. For instance, it learns to ignore mild touch stimuli that are presented repeatedly, a primitive type of learning called habituation. The combination of simplicity and learning ability makes it an ideal model for studying the neuronal basis of learning.

Vertebrate nervous systems are highly centralized and cephalized

Vertebrate nervous systems are structurally and functionally diverse. For instance, the cerebral cortex of a dolphin's brain is more structurally complex and a much more powerful information processor than that of a fish or frog. However, all vertebrate nervous systems have some fundamental similarities, including distinct central and peripheral elements and a high degree of cephalization. In all vertebrates the brain and spinal cord make up the CNS. The brain provides the integrative power that underlies the complex behavior characteristic of all vertebrates. The spinal cord, which runs lengthwise inside the vertebral column, or spine, integrates simple responses to certain kinds of stimuli and conveys information to and from the brain (FIGURE 48.14).

Axons within the CNS are located in well-defined bundles, or tracts, whose myelin sheaths give them a whitish appearance. This **white matter** is clearly distinguishable from **gray matter** (mainly nerve cell bodies, dendrites, and unmyelinated axons) in cross sections of the brain and spinal cord (see FIGURE 48.4).

The vertebrate CNS is derived from the dorsal hollow nerve cord—one of the phylogenetic hallmarks of chordates—and the brain and spinal cord both contain fluid-filled spaces. The narrow **central canal** of the spinal cord is continuous with fluid-filled spaces, called **ventricles**, in the brain. These cavities are filled with **cerebrospinal fluid**, which is formed in the brain by filtration of the blood. Circulating through the central canal and ventricles (and then draining back into the veins), the cerebrospinal fluid conveys nutrients, hormones, and white blood cells across the blood-brain barrier to different parts of the brain. Among the most important functions of cerebrospinal fluid is to act as a shock absorber, cushioning the brain. Also protecting the brain and spinal cord are layers of connective tissue, called **meninges**. In mammals, cerebrospinal fluid circulates between two of the meninges, providing an additional cushion for the brain.

We will return to the vertebrate brain, but first let's take a look at the PNS.

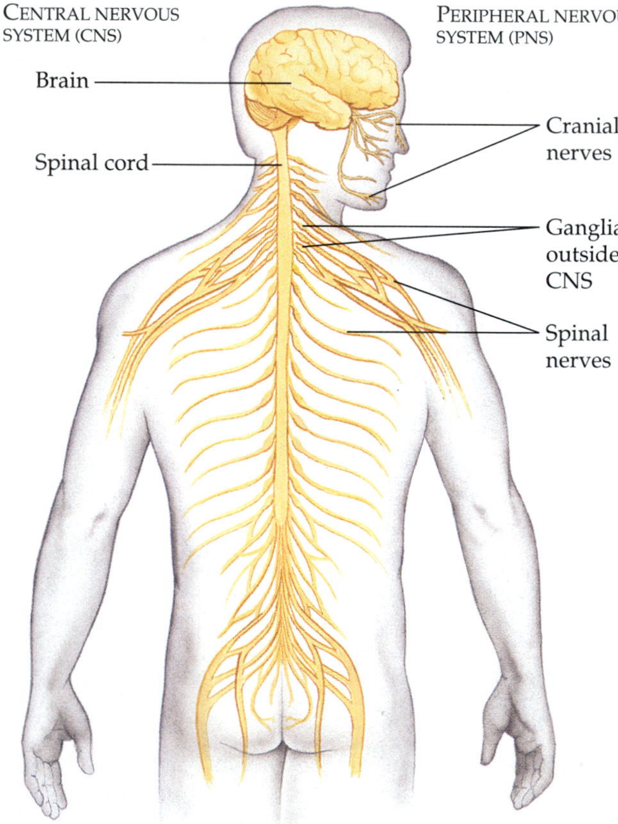

FIGURE 48.14 • The nervous system of a vertebrate. The vertebrate nervous system is highly centralized. The components of the central nervous system (brain and spinal cord) develop from the dorsal, hollow nerve cord, a hallmark of chordates. Cranial nerves (originating in the brain), spinal nerves (originating in the spinal cord), and ganglia outside the central nervous system make up the peripheral nervous system.

The vertebrate PNS has several components differing in organization and function

Structurally, the vertebrate PNS consists of paired cranial and spinal nerves and associated ganglia (see FIGURE 48.14). The **cranial nerves** originate in the brain and innervate organs of the head and upper body. The **spinal nerves** originate in the spinal cord and innervate the entire body. Mammals have 12 pairs of cranial nerves and 31 pairs of spinal nerves. Most of the cranial nerves and all the spinal nerves contain both sensory and motor neurons; a few of the cranial nerves are sensory only (the olfactory and optic nerves, for example).

Because of the complex arrangement of sensory and motor neurons of a vertebrate's cranial and spinal nerves, it is convenient to divide the PNS into a hierarchy of components that differ in function (FIGURE 48.15). The **sensory division** of the PNS is made up of sensory, or afferent, neurons that convey information to the CNS from sensory receptors that monitor the external and internal environment. The **motor division** is composed of efferent neurons that convey signals from the CNS to effector cells. The motor division's **somatic nervous system** carries signals to skeletal muscles mainly in response to *external* stimuli. The somatic nervous system is

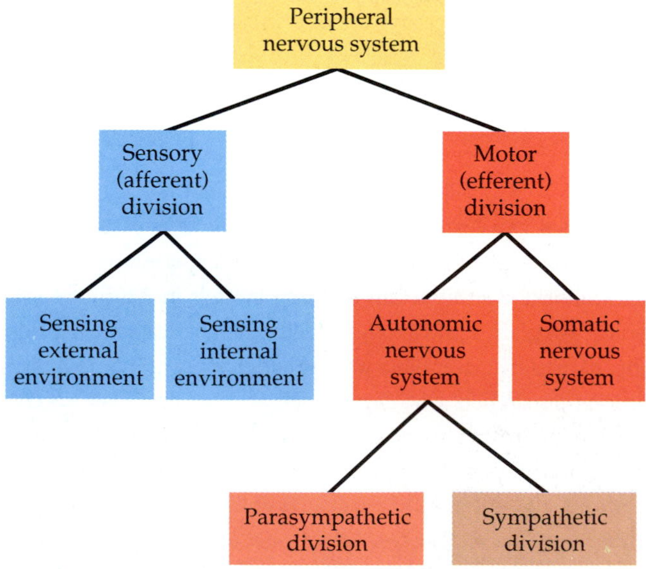

FIGURE 48.15 • Functional hierarchy of the peripheral nervous system. The nerves and ganglia of the vertebrate PNS convey information throughout the body. Sensory and motor divisions of the PNS are organized into a functional hierarchy.

often considered voluntary because it is subject to conscious control, but a substantial proportion of skeletal muscle movement is actually determined by reflexes mediated by the spinal cord or lower brain. The motor division's **autonomic nervous system** conveys signals that regulate the *internal* environment by controlling smooth and cardiac muscles and the organs of the gastrointestinal, cardiovascular, excretory, and endocrine systems. This control is generally involuntary.

The autonomic nervous system consists of two subdivisions that are anatomically, physiologically, and chemically distinguishable: the sympathetic division and the parasympathetic division. When sympathetic and parasympathetic nerves innervate the same organ, they often (but not always) have antagonistic (opposite) effects. In general, signals carried via the **parasympathetic division** enhance activities that gain and conserve energy, such as digestion and slowing the heart rate. In contrast, signals conveyed by the **sympathetic division** generally increase energy consumption and prepare an individual for action by accelerating the heart rate, increasing metabolic rate, and performing related functions (FIGURE 48.16).

The somatic and autonomic nervous systems often cooperate in maintaining the all-important balance among organ system functions essential to homeostasis. In response to a drop in temperature, for example, the hypothalamus signals the autonomic nervous system to constrict surface blood vessels, which reduces heat loss; at the same time the hypothalamus signals the somatic nervous system and causes shivering.

STRUCTURE AND FUNCTION OF THE VERTEBRATE BRAIN

The vertebrate brain develops from three anterior bulges of the spinal cord

We begin studying the vertebrate brain by examining its embryonic development from the neural tube (see FIGURE 47.11). In all vertebrates, three bilaterally symmetrical, anterior bulges of the dorsal hollow nerve cord—the **forebrain**,

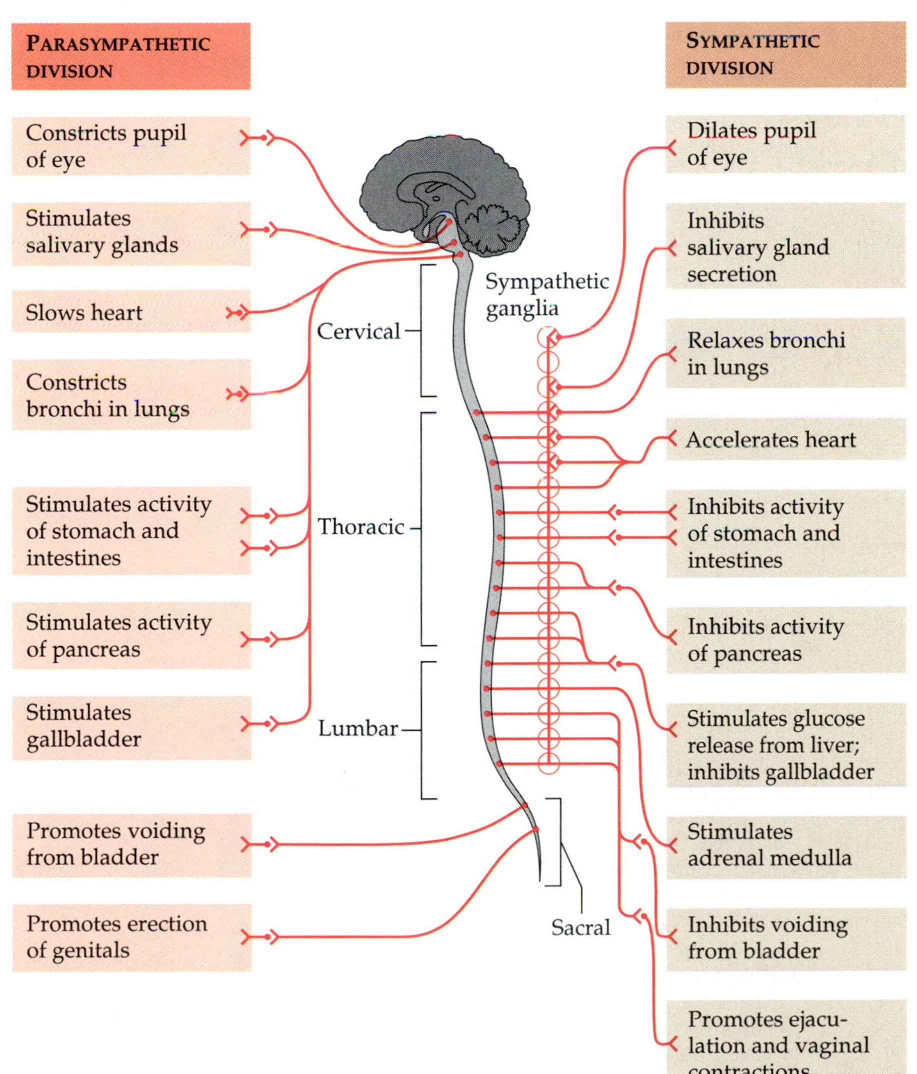

FIGURE 48.16 • The main roles of the parasympathetic and sympathetic components of the motor division's autonomic nervous system in regulating internal body functions. The parasympathetic nerves of the autonomic nervous system originate from the lower brain and the sacral region of the spinal cord. Sympathetic nerves emerge from the middle (thoracic and lumbar) regions of the spinal cord. Most autonomic pathways consist of a chain of two neurons. The synapse between the two neurons is within a ganglion in the PNS. In each pair, the neuron that conveys signals from the CNS to the ganglion is called the preganglionic neuron. Preganglionic neurons release acetylcholine at the synapse. The neuron that conveys signals from the ganglion to the target organ is called the postganglionic neuron. Notice in this diagram that several sympathetic pathways include a synapse in prominent sympathetic ganglia near the spinal cord. Other ganglia are less prominent; those of the parasympathetic neurons reside close to or within the target organs. Most sympathetic axons release the neurotransmitter norepinephrine in their target organs. Parasympathetic neurons release acetylcholine.

midbrain, and **hindbrain**—become evident as the neural tube differentiates (FIGURE 48.17a). These three embryonic structures probably are reminiscent of the brain of an ancestral vertebrate. In the course of vertebrate evolution, the brain became further divided structurally and functionally, providing additional capacity for complex integration.

By the fifth week of development (using the human as our example), five brain regions have formed from the three primary bulges (FIGURE 48.17b). Two of these regions, the **telencephalon** and the **diencephalon**, develop from the forebrain. Of all the brain regions, the telencephalon changes most, developing into the brain's major centers of homeostatic control and integration (FIGURE 48.17c and d). The third brain region, the **mesencephalon**, develops from the midbrain, which remains undivided; the other two regions, the **metencephalon** and **myelencephalon**, develop from the hindbrain.

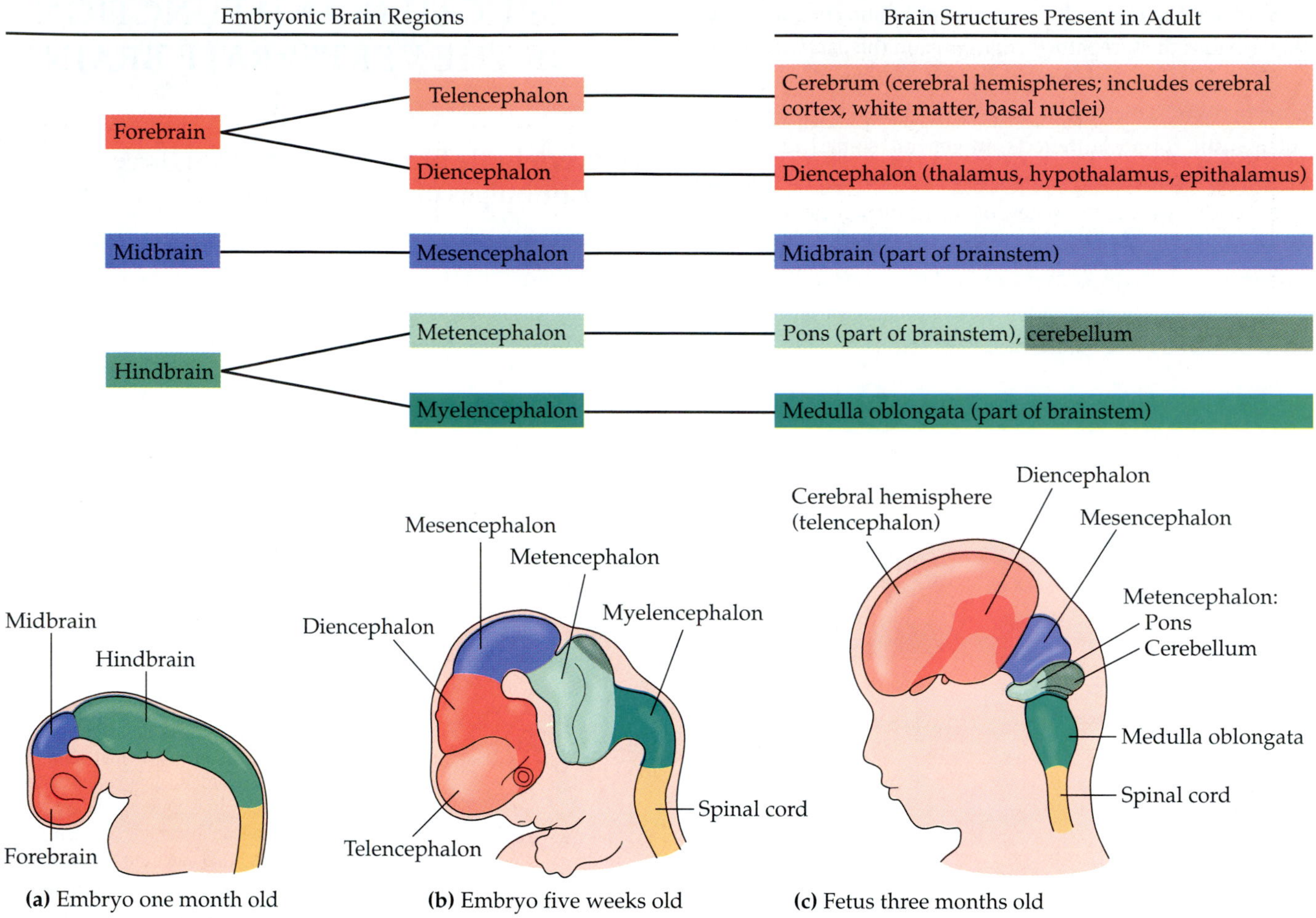

Embryonic Brain Regions		Brain Structures Present in Adult
Forebrain	Telencephalon	Cerebrum (cerebral hemispheres; includes cerebral cortex, white matter, basal nuclei)
	Diencephalon	Diencephalon (thalamus, hypothalamus, epithalamus)
Midbrain	Mesencephalon	Midbrain (part of brainstem)
Hindbrain	Metencephalon	Pons (part of brainstem), cerebellum
	Myelencephalon	Medulla oblongata (part of brainstem)

Midbrain
Hindbrain
Forebrain

(a) Embryo one month old

Mesencephalon
Metencephalon
Diencephalon
Myelencephalon
Spinal cord
Telencephalon

(b) Embryo five weeks old

Cerebral hemisphere (telencephalon)
Diencephalon
Mesencephalon
Metencephalon:
Pons
Cerebellum
Medulla oblongata
Spinal cord

(c) Fetus three months old

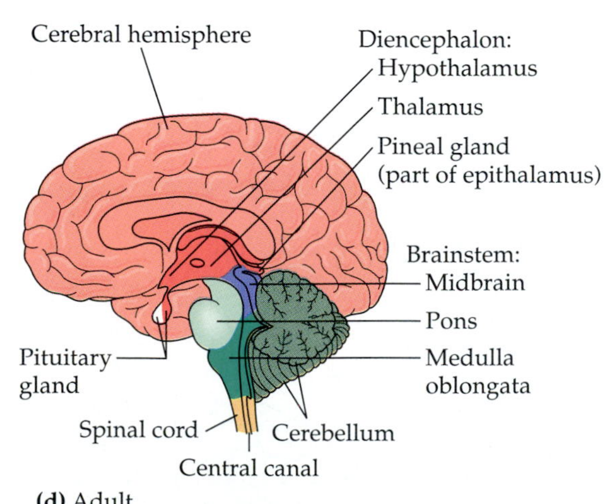

(d) Adult

FIGURE 48.17 ▪ Embryonic development of the brain. Anterior regions of the dorsal nerve cord expand and differentiate extensively during embryonic development. **(a)** The brain in a month-old human embryo. The three primary brain regions—the forebrain, midbrain, and hindbrain (characteristic of all vertebrates)—become evident during the first month of development. **(b)** By five weeks, five regions have differentiated, the telencephalon and diencephalon from the forebrain, the mesencephalon from the midbrain, and the metencephalon and myelencephalon from the hindbrain. **(c)** By three months, the human embryo possesses the major brain centers present in the adult **(d).** Comparing (b), (c), and the sectioned brain in (d), notice that the most profound changes occur in the telencephalon. Rapid, expansive growth of this region during the second and third months creates the cerebrum (the right and left cerebral hemispheres), extending over and around many other brain centers. In the adult, the cerebral hemispheres include a highly convoluted outer region of gray matter called the cerebral cortex, white matter, and internal gray matter, called basal nuclei. Major centers that develop from the diencephalon are the thalamus, epithalamus, and hypothalamus. (The posterior pituitary is an extension of the hypothalamus; see FIGURE 45.7. The pineal gland is a tiny extension of the epithalamus.) In contrast to the telencephalon, the mesencephalon remains undivided and is called the midbrain in the adult. Together, the midbrain, the pons (from the metencephalon), and the medulla oblongata (from the myelencephalon) make up the brainstem in the adult. Another major brain center, the cerebellum, also develops from the metencephalon.

The brainstem conducts data and controls automatic activities essential for survival

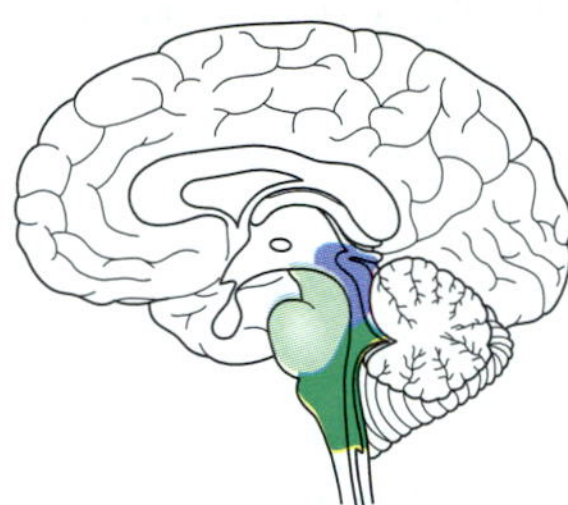

Together, the mesencephalon, part of the metencephalon, and the myelencephalon form the **brainstem**, a stalk and caplike swellings at the anterior end of the spinal cord. The brainstem (sometimes called the "lower brain") has three parts that function in homeostasis, movement coordination, and conduction of information to higher brain centers: the medulla oblongata, the pons, and the midbrain. The **medulla oblongata** (commonly called the medulla) contains centers that control several visceral (autonomic, homeostatic) functions, including breathing, heart and blood vessel activity, swallowing, vomiting, and digestion. The **pons** also participates in some of these activities, having nuclei that regulate the breathing centers in the medulla, for example. All the bundles of axons carrying sensory information to and motor instructions from higher brain regions pass through the brainstem, making data conduction one of the most important functions of the medulla and pons. The brainstem also helps coordinate large-scale body movements such as walking. Most of the descending axons carrying instructions about movement to the spinal cord from the midbrain and forebrain cross from one side of the CNS to the other as they pass through the medulla. As a result, the right side of the brain controls much of the movement of the left side of the body, and vice versa.

The third part of the brainstem, the midbrain, contains centers for the receipt and integration of several types of sensory information. It also serves as a projection center, sending coded sensory information along neurons to specific regions of the forebrain. Prominent nuclei of the midbrain are the **inferior** and **superior colliculi**, which are part of the auditory and visual systems, respectively. All fibers involved in hearing either terminate in or pass through the inferior colliculi. In nonmammalian vertebrates, the superior colliculi take the form of prominent optic lobes and may be the only visual centers. In mammals, vision is integrated in the cerebrum, leaving the superior colliculi to coordinate visual reflexes and perform limited perceptual functions.

The cerebellum controls movement and balance

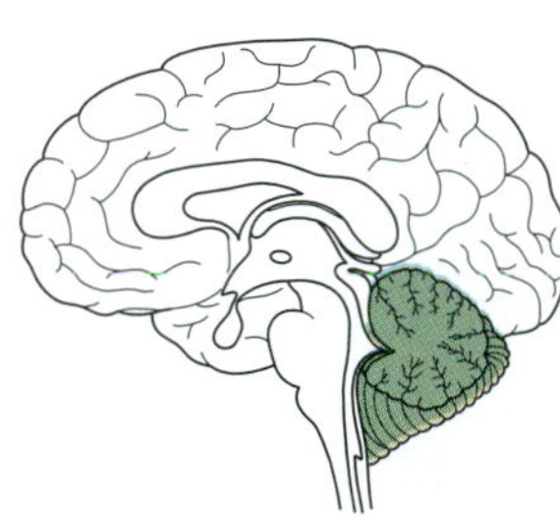

The **cerebellum** develops from part of the metencephalon (see FIGURE 48.17). Its primary function is coordination of movement. The cerebellum receives sensory information about the position of the joints and the length of the muscles, as well as information from the auditory and visual systems. It also receives input from the motor pathways, telling it which actions are being commanded from the cerebrum. The cerebellum uses this information to provide automatic coordination of movements and balance. If one part of the body is moved, the cerebellum will coordinate other parts to ensure smooth action and maintenance of equilibrium. It also plays a role in learning and remembering motor responses. Hand-eye coordination is one example of cerebellar function. If the cerebellum is damaged, the eyes can follow a moving object, but they will not stop at the same place as the object.

The thalamus and hypothalamus are prominent integrating centers of the diencephalon

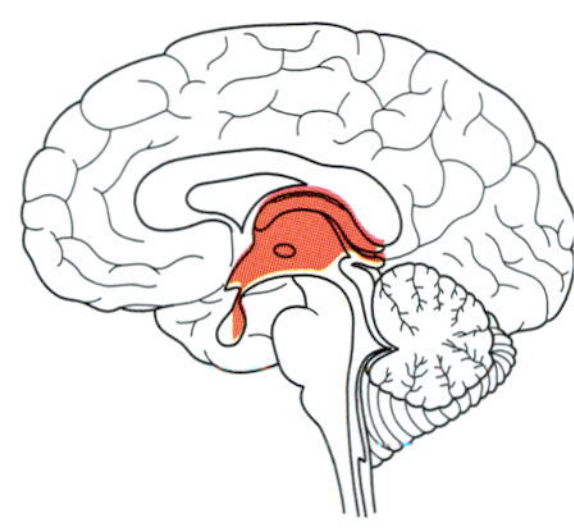

The embryonic diencephalon develops into three adult brain regions: the epithalamus, thalamus, and the hypothalamus (see FIGURE 48.17). The **epithalamus** includes a projection, the pineal gland, and a **choroid plexus**, one of several clusters of capillaries that produce cerebrospinal fluid.

A major integrating center, the **thalamus** is also the main input center for sensory information going to the cerebrum and the main output center for motor information leaving the cerebrum. The thalamus contains many different nuclei, each one dedicated to sensory information of a particular type. Incoming information from all the senses is sorted out in the thalamus and sent on to the appropriate higher brain centers for further interpretation and integration. The thalamus also receives input from the cerebrum and from parts of the brain that regulate emotion and arousal.

Although it weighs only a few grams, the **hypothalamus** is one of the most important brain regions for the homeostatic regulation. We have already seen that the hypothalamus is the source of two sets of hormones, posterior pituitary hormones and releasing hormones that act on the anterior pituitary (see FIGURE 45.7). It contains the body's thermostat, as well as centers for regulating hunger, thirst, and many other basic survival mechanisms. Hypothalamic nuclei also play a role in sexual response and mating behaviors, the fight-or-flight response, and pleasure. The hypothalamic pleasure centers have been given that name because of the responses seen when they are stimulated in experimental animals, although we cannot really know whether a rat experiences what humans interpret as pleasurable sensations.

The Hypothalamus and Circadian Rhythms

Animals, including humans, exhibit all kinds of regularly repeated, rhythmic behaviors. What maintains our daily rhythms when, for example, we sleep, our blood pressure is highest, and our sex drive peaks? We have already discussed circadian (daily) rhythms in plants (see Chapter 39). Many animals also exhibit seasonal rhythms, reproducing or migrating only in the spring or fall, for instance.

Numerous studies have assessed the relative importance of external cues and internal timekeepers in maintaining rhythmic behavior. These studies show that circadian rhythms usually have a strong internal component, referred to as a *biological clock.* Locating the internal mechanisms responsible for behavioral rhythms has been challenging for researchers. An early hypothesis that the location of control mechanisms probably varies across taxonomic boundaries has proved true. For instance, fruit flies (*Drosophila*) appear to have many biological clocks throughout their body and at the outer edges of their wings. In mammals, a pair of structures called the **suprachiasmatic nuclei (SCN)** in the hypothalamus functions as a biological clock. Experiments with rodents have revealed that the cells of the SCN produce specific proteins in response to changing light-dark cycles. The function of this or any other biological clock may be the regulation of a variety of physiological processes, such as hormone release, hunger, and heightened sensitivity to external stimuli that motivate specific rhythmic behaviors.

Researchers have also investigated the role of external cues in circadian rhythms. Usually, a clock's rhythm does not exactly match events in the environment, and external cues are necessary to keep cycles timed to the outside world. Light is a common external cue for circadian rhythms; visual information received by the SCN via sensory neurons from the eyes enables the mammalian clock to remain synchronized with the natural cycles of day length and darkness. For example, activity of the North American flying squirrel normally begins with the onset of darkness and ends at dawn, which suggests that light is an important external cue. If a squirrel is placed in constant light or constant darkness, its rhythmic activity continues, but the duration of its activity cycle (one period of activity plus one period of inactivity) falls a little

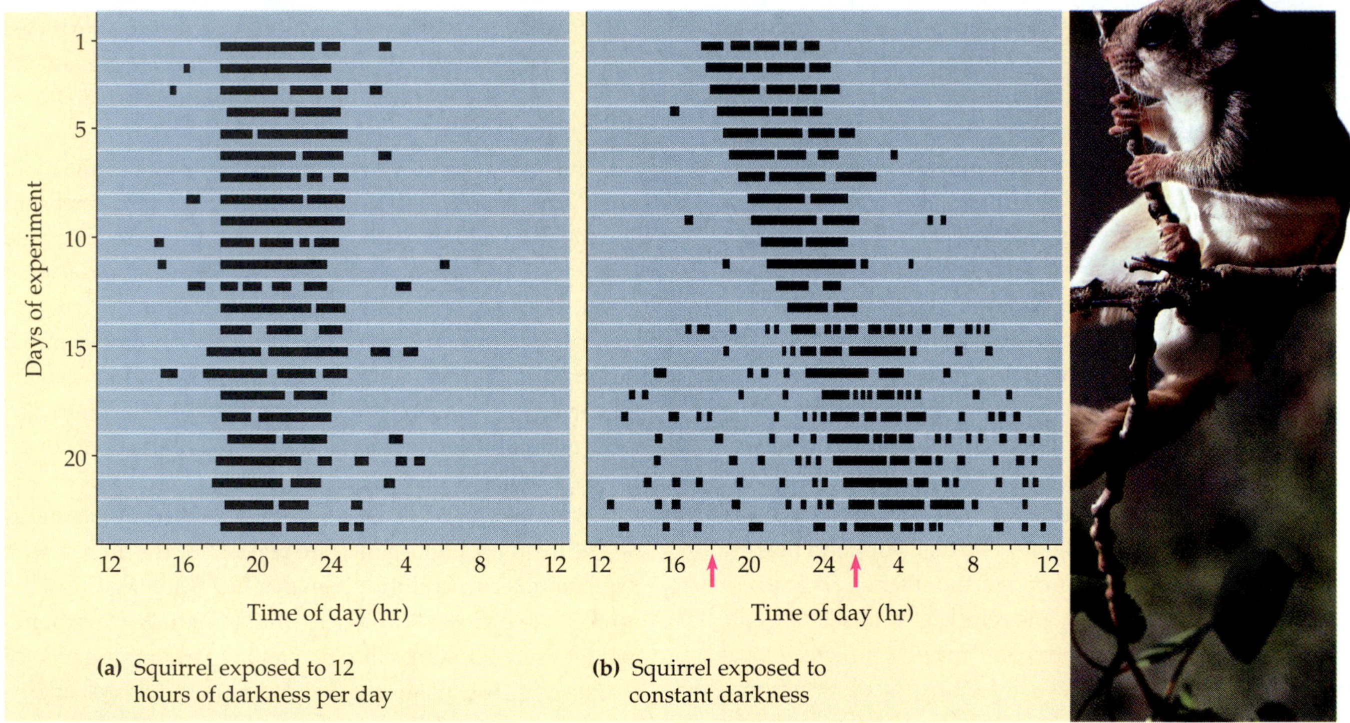

FIGURE 48.18 ▪ Activity rhythms in a nocturnal mammal. An inhabitant of forests in North America, the northern flying squirrel is active at night and usually sleeps in a nest cavity in a hollow tree from dawn to dusk. To study its activity rhythms, researchers have placed flying squirrels in cages containing an exercise wheel in which a squirrel can run. The wheel is connected to a chart recorder, which moves graph paper past a pen at a fixed speed. When a squirrel runs on the wheel, the pen is activated and marks the graph paper. The graphs shown here trace the activity of two flying squirrels held under different light conditions for 23 days. The longer black bars indicate periods of extended activity. **(a)** This graph shows the activity pattern of a squirrel exposed to 12 hours of darkness, simulating natural conditions. **(b)** This graph shows the activity pattern of a squirrel held in constant darkness for 23 days. Although the activity of both squirrels remained rhythmic throughout the recording period, with a distinct period of extended activity every day, the high-activity period of the squirrel in (b) shifted each day (actually by 21 minutes). After 23 days its period of greatest activity was nearly 8 hours out of synchronization with the actual time of day. (The small magenta arrows indicate when the period of greatest activity began on days 1 and 23.) Experiments with flying squirrels, humans, and other animals show that environmental cues are needed to keep biological clocks tuned to external conditions.

more out of sync with the outside world each day (FIGURE 48.18). The internal clock continues to run without external cues, but on its own time. External cues, such as day length and night length, adjust the clock, so that the rhythmic behavior it controls is synchronized with the outside world.

Human circadian rhythms have been studied by placing individuals in comfortable living quarters deep underground, where they could make their own schedules with no external cues of any kind. Under these free-running conditions, the biological clock of humans seems to have a period of about 25 hours, but with much individual variation; like other animals, humans use external cues to adjust their rhythms to 24 hours in the real world.

The cerebrum contains the most sophisticated integrating centers

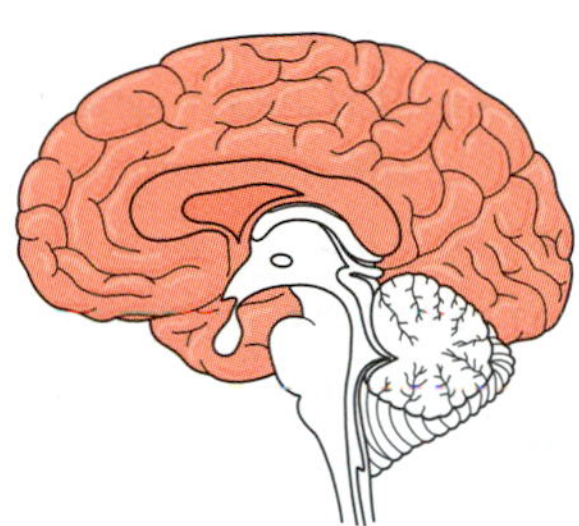

The cerebrum, the most complex integrating center in the CNS, develops from the embryonic telencephalon. The cerebrum is divided into right and left **cerebral hemispheres** (FIGURE 48.19a). Each hemisphere consists of an outer covering of gray matter, the cerebral cortex; internal white matter; and a cluster of nuclei deep within the white matter, the basal nuclei. The **basal nuclei** (also called basal ganglia) are important centers for motor coordination, acting as switches for impulses from other motor systems. If the basal nuclei are damaged, a person may become passive and immobile because the nuclei no longer allow motor impulses to be sent to the muscles. Degeneration of cells entering the basal nuclei occurs in Parkinson's disease.

The **cerebral cortex** is the largest and most complex part of the human brain, and the part that has changed the most during vertebrate evolution. Especially in mammals, sophisticated behavior is associated with the relative size of the cerebral cortex and the presence of convolutions that increase its surface area. Although less than 5 mm thick, the human cerebral cortex has a surface area of about 0.5 m^2 and occupies about 80% of the total brain mass. Primates and cetaceans (whales and porpoises, for example) also have exceptionally large and complex cerebral cortices. In fact, the surface area (relative to body size) of a porpoise's cerebral cortex is second only to that of a human.

Like the rest of the brain, the cerebral cortex is divided into right and left sides. A thick band of fibers (cerebral white matter) known as the **corpus callosum** connects the right and left sides. Each side has four discrete lobes, and researchers have identified a number of functional areas within each lobe (FIGURE 48.19b).

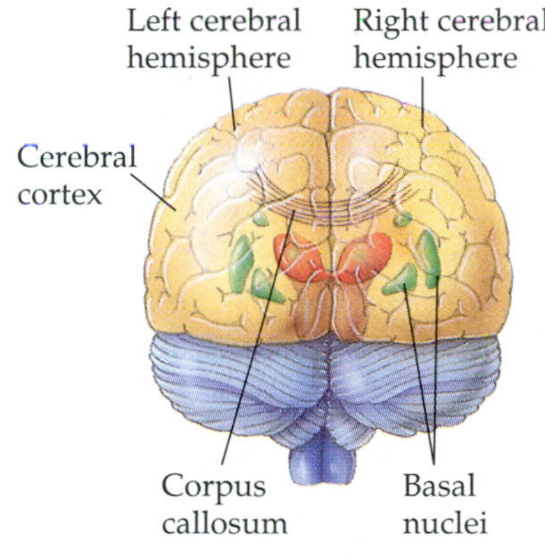

(a) Back of brain

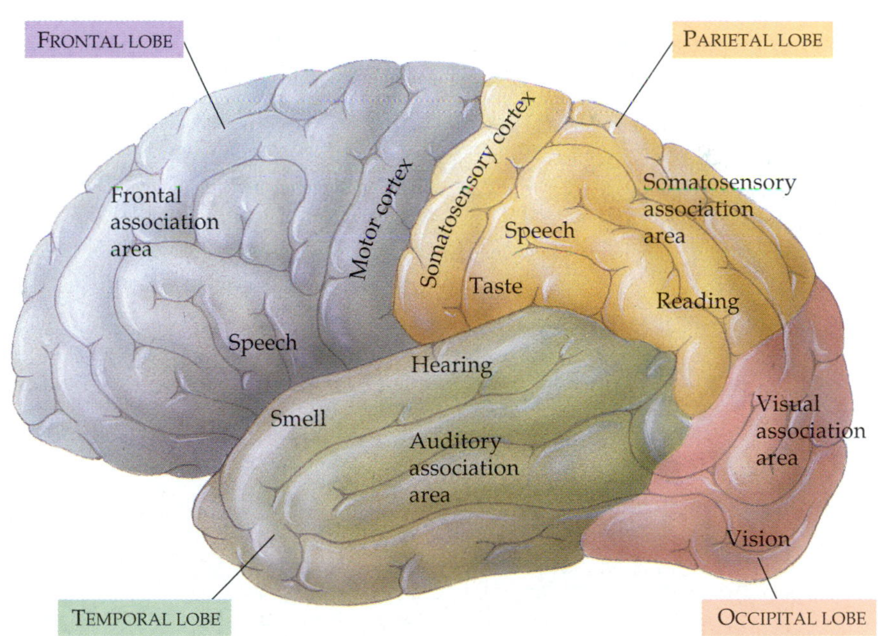

(b) Left side of brain

FIGURE 48.19 • Structure and functional areas of the cerebrum. (a) This rear view of the human brain shows the bilateral nature of the cerebral hemispheres. The corpus collosum (large fiber tracts connecting the hemispheres) and basal nuclei (ganglia) are completely covered (and not visible from the surface) by the convoluted gray matter of the cerebral cortex. **(b)** The surface of each cerebral hemisphere is divided into four lobes. (A fifth lobe, the insula, located within the fold separating the temporal and parietal lobes, is not shown in this drawing of the left hemisphere.) Specialized functions are localized in each lobe. The association areas of the left hemisphere have different functions from those of the right hemisphere.

Two functional cortical areas, the primary motor cortex and the primary somatosensory cortex, form the boundary between the frontal lobe and the parietal lobe. The motor cortex functions mainly in sending commands to skeletal muscles, signaling appropriate responses to sensory stimuli. The somatosensory cortex receives and partially integrates signals from touch, pain, pressure, and temperature receptors throughout the body. The proportion of somatosensory or motor cortex devoted to a particular part of the body is correlated with the importance of sensory or motor information for that part of the body (FIGURE 48.20). Impulses transmitted from receptors to specific areas of somatosensory cortex enable us to associate pain, touch, pressure, heat, or cold with specific parts of the body receiving those stimuli. However, the so-called special senses—vision, hearing, smell, and taste—are integrated by other cortical regions. Each of these functional regions, as well as the somatosensory cortex, cooperates with an adjacent association area (see FIGURE 48.19b). Researchers are beginning to understand how a complicated interchange of signals among receiving centers and association centers produces our sensory perceptions.

The human brain is a major research frontier

Integration of nerve impulses occurs at all levels in the human nervous system. The simplest kinds of integration are the spinal reflexes (see FIGURE 48.4); the most complex integration enables the cerebral cortex to create a work of art or make a scientific discovery. Discovering how our brain works is one of the most interesting and challenging frontiers in modern biology. Some aspects of brain research that are particularly interesting are arousal and sleep; lateralization (hemispheric specialization), language, and speech; emotions; memory and learning; and consciousness.

Arousal and Sleep

As anyone who has sat through a lecture on a warm spring day knows, attentiveness and mental alertness vary from moment to moment. Arousal is a state of awareness of the external world. The counterpart of arousal is sleep, when an individual continues to receive external stimuli but is not conscious of them. The mechanisms of arousal and sleep are fairly well understood, but the question of why we sleep remains a compelling problem. All birds and mammals sleep and show a characteristic sleep-wake cycle, which is probably mediated by the hypothalamus, as it is in humans.

Sleep and wakefulness produce different patterns in the electrical activity of the brain, which can be recorded in an **electroencephalogram**, or **EEG** (FIGURE 48.21). As a general rule, the less mental activity taking place, the more synchronous the brain waves of the EEG. When a healthy person is lying quietly with closed eyes, slow, synchronous *alpha waves* predominate. When the eyes are opened or the person solves a complex problem, faster *beta waves* take over, indicating desynchronization of the parts of the brain.

A sleeping person's EEG reflects the fact that sleep is a dynamic process. In the early stages, *theta waves,* more irregular than beta waves, often predominate. Deeper sleep produces *delta waves,* which are quite slow and highly synchronized. Deep sleep also includes periods when a desynchronized EEG reminiscent of wakefulness occurs. During these periods,

FIGURE 48.20 ▪ Primary motor and somatosensory areas of the human cerebral cortex.

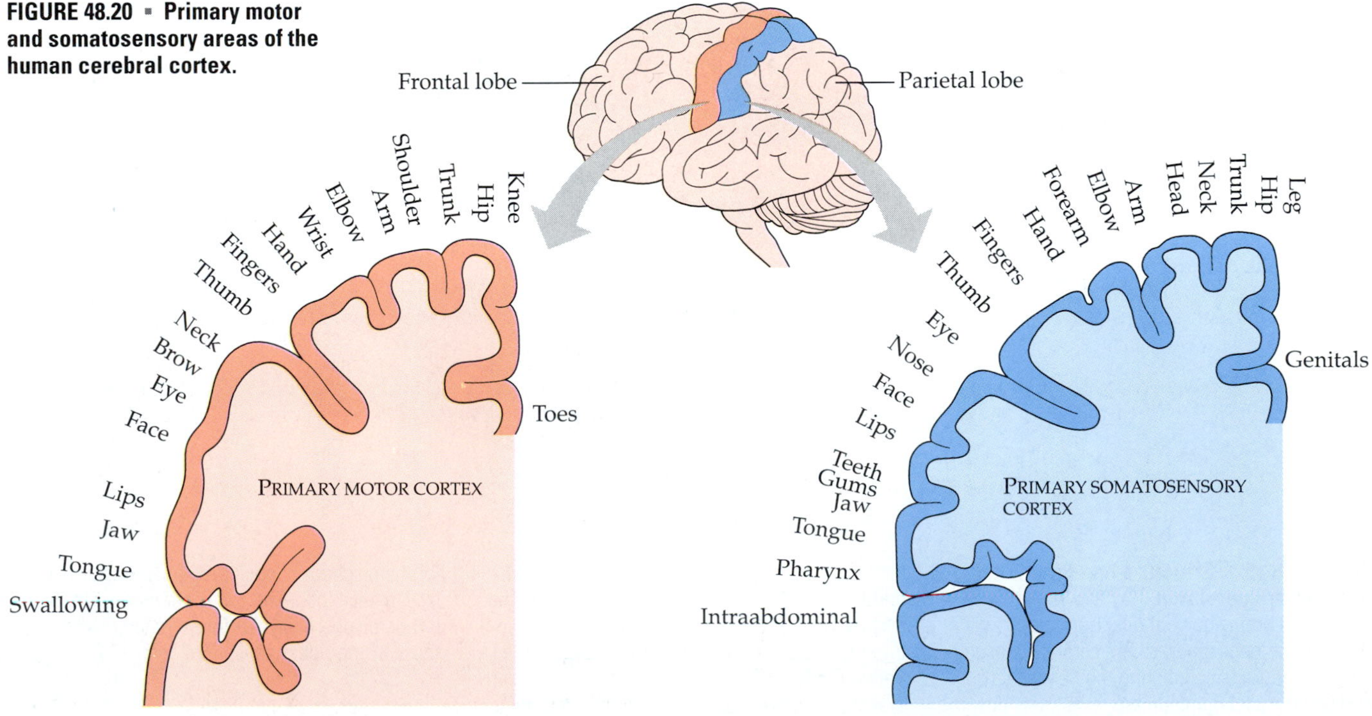

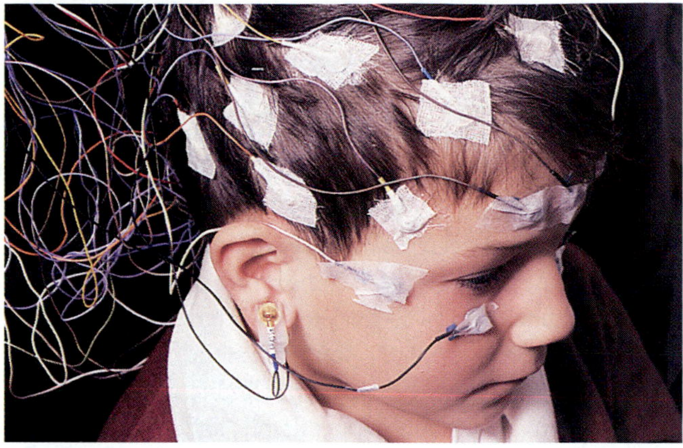

(a) Electrodes on scalp

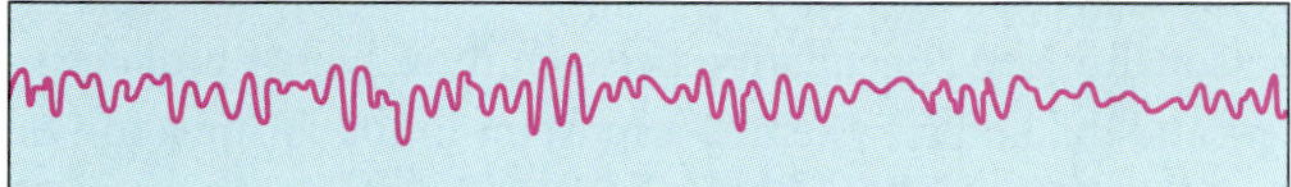

(b) Awake but quiet (alpha waves)

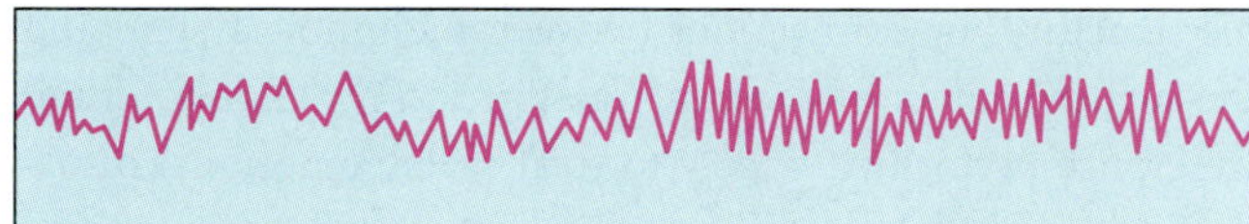

(c) Awake during intense mental activity (beta waves)

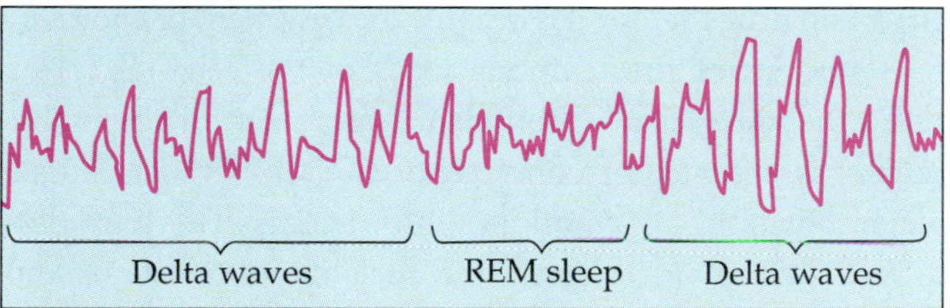

(d) Asleep

FIGURE 48.21 ▪ Brain waves recorded by an electroencephalogram (EEG). (a) Electrical contacts placed on the scalp detect brain waves (electrical activity of the brain). **(b)** When a person is awake but quiet, slow alpha waves dominate the EEG. **(c)** During intense mental activity, rapid, irregular beta waves predominate. **(d)** A sleep cycle usually includes periods of slow, fairly regular delta waves interspersed with periods of REM sleep and dreaming, when brain waves resemble those of the awake state.

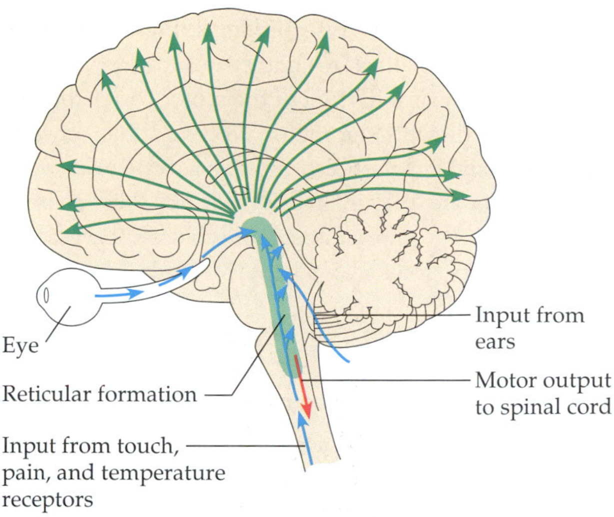

FIGURE 48.22 ▪ The reticular formation. This functional system of over 90 separate nuclei extends through the brainstem (from the medulla to the thalamus). The reticular formation receives input from sensory receptors (blue arrows). Working closely with the cerebral cortex (green arrows), it prevents sensory overload, filtering out familiar and repetitive information that constantly enters the nervous system. Part of this filtering system, the reticular activating system (RAS), is a vital link in determining states of arousal and consciousness.

called REM (rapid eye movement) sleep, the eyes move actively across the visual field behind closed lids. Most dreaming occurs during REM sleep. Like sleep, dreaming has been ascribed magical or prophetic importance, but its true function remains unknown.

Sleep and arousal are controlled by several centers in the cerebrum and the brainstem. A functional system of neurons called the **reticular formation** passes through the core of the brainstem (FIGURE 48.22). Part of the reticular formation, the reticular activating system (RAS), regulates sleep and arousal. Acting as a sensory filter, the RAS selects which information reaches the cortex, and the more input the cortex receives, the more alert and aware a person is. But arousal is not just a generalized phenomenon; certain stimuli can be ignored while the brain is actively processing other input. Also, specific centers regulate sleep and wakefulness. The pons and medulla contain nuclei that cause sleep when stimulated, and the midbrain has a center that causes arousal. Serotonin may be the neurotransmitter of the sleep-producing centers. Drinking milk before bedtime may induce sleep because milk contains large amounts of tryptophan, the amino acid from which serotonin is synthesized.

Lateralization, Language, and Speech

The association areas of the cerebral cortex are lateralized; that is, each side of the brain is specialized for different functions. The processes supporting speech, language, calculation, and the rapid serial processing of detailed information, for example, occur mainly in the left hemisphere. The right hemisphere emphasizes overall context, spatial perception, and creative abilities.

Understanding and generating language requires some very complex interactions between several association areas in the left hemisphere, while the emotional content of language conveyed by tone of voice is processed by association areas of the right hemisphere. Damage to parts of the left hemisphere can cause different sorts of aphasia, the inability to speak coherently. Patients with damage to the association areas of the parietal lobe that translate spoken or visual information (for example, the appearance of words on a page) into meaning can still speak. However, because they cannot understand, such speech frequently consists of long strings of words and nonsense syllables. Damage to another speech association area in the frontal lobe, just in front of the motor cortex that controls the face, disrupts the motor programs that move the tongue, lips, and other speech muscles to articulate words. These patients are able to understand, but their ability to speak language is compromised. You can see these areas "in action" in the PET scans shown in the Methods Box on p. 987.

Emotions

What causes us to laugh, cry, love, envy, worry, and so on has been the subject of much biological and philosophical speculation. Some hypotheses have proposed that emotions result from feedback from the body's organs and muscles to the CNS. Others propose that the brain's perceptions of sensory information first lead to an emotional experience (fear, anger, and so on), and then emotional expression, such as increased heart rate and facial blushing, follow. Emotions are difficult to study experimentally, because even if an experimental animal seems to *show* (express) emotion, we cannot say conclusively that the animal *feels* (experiences) the emotion in the same sense that humans do. Also, experimental animals, such as rats and mice, tend to have relatively stereotyped emotional responses. In contrast, humans tend to have highly individualized emotional experiences and expressions. A stimulus that triggers anger or envy in one person may have no such effect on someone else, or it may elicit different emotional feelings and body responses. Both the autonomic nervous system and somatic motor division mediate the body's expressions of emotion, and the capacity of these PNS components to trigger highly diverse responses in muscles and other organs is a key factor in the varied nature of emotional expression.

Researchers have developed partial maps of some of the brain regions that are involved in emotions. Such maps emphasize that a picture of the neural pathways and functional systems underlying emotions is just starting to emerge, and given the enormous diversity of emotional responses, it is likely that several complex neural pathways and functional systems in the CNS are involved. So far, the maps come mainly from studies of brain-damaged people and other mammals, but the pace of discovery of how the brain integrates all types of information has been greatly speeded by new imaging technology (see the Methods Box). Some human emotions depend on a functional group of nuclei and interconnecting axon tracts in the CNS called the **limbic system** (FIGURE 48.23). Though still only loosely defined, the limbic system includes parts of the thalamus and hypothalamus and portions of the cerebral cortex. The limbic system is linked to areas of the cerebral cortex involved in complex learning, reasoning, and personality, and consultation between these higher brain centers and the limbic system is important in the formulation of emotions. Eliminating this emotional consultation, a surgical procedure called frontal lobotomy involves destruction of limbic structures or connections between the limbic system and the higher centers of the cerebral cortex. Frontal lobotomies were once performed widely to treat severe emotional disorders. However, the resulting docility was usually accompanied by a loss of the ability to concentrate, plan, and work toward goals, and drug therapy has become the treatment of choice.

A nucleus of the temporal lobes of the cerebral cortex called the **amygdala**, a prominent component of the limbic system, is a center of convergence for sensory data and a major organizer of emotional information. The amygdala receives sensory data from the thalamus, brainstem, and olfactory bulb, as well as highly integrated sensory information from association areas of the cerebral cortex. Neural signals pass back and forth between the amygdala and the hypothalamus and brainstem, and a major pathway of signals triggering emotional expression runs from the amygdala to autonomic and somatic motor systems via the hypothalamus and the brainstem's reticular formation.

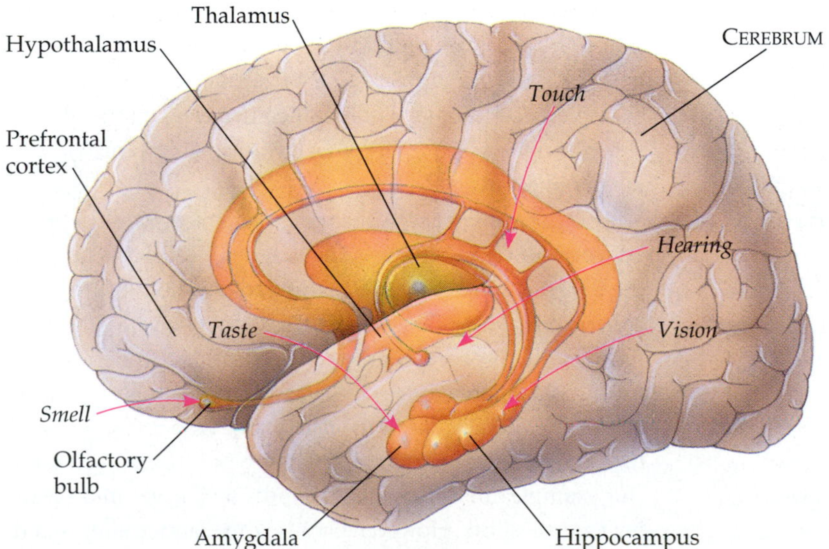

FIGURE 48.23 • The limbic system. Portions of the diencephalon (thalamus and hypothalamus) and inner parts of the cerebral cortex, including the amygdala and hippocampus, make up this functional center of human emotions and memory. Signals from the nose enter the brain through the olfactory bulb, which is part of the limbic system. Other sensory information enters the limbic system via other parts of the cerebral cortex (arrows). A higher center of integration in the cerebrum, the prefrontal cortex, apparently consults the limbic system and other brain centers in processing and retrieving memories and may use memories to modify behavior.

METHODS: Imaging the Nervous System

Several imaging techniques are now available to researchers studying normal body functions and to physicians diagnosing various disorders, including cancer. These techniques are especially useful in studies of the brain.

One widely used method called **magnetic resonance imaging (MRI)** takes advantage of the behavior of hydrogen atoms in water molecules. The nuclei of hydrogen atoms are usually oriented in random directions. MRI uses powerful magnets to align the nuclei, then knocks them out of alignment with a brief pulse of radio waves. Still under the magnets' influence, the hydrogen nuclei spring back into alignment, giving out faint radio signals of their own, signals detected by the MRI scanner and translated by computer into an image. Soft tissues having relatively high water content, including the nervous system, appear more opaque in the images than dense structures, such as bone, which contain relatively little water. For this reason, MRI is useful for detecting problems in the brain and spinal cord, which are surrounded by bone. FIGURE a is an MRI image of a tumor (artificially colored red) in the human spinal cord.

Another imaging technique called **computed tomography (CT)** produces images of a series of thin X-ray sections through the body. A patient is slowly moved through a CT machine as an X-ray source circles around the body, illuminating successive sections from many angles. A computer then produces high-resolution video images of the sections, which can be studied individually or combined into various three-dimensional views. CT is especially useful for detecting ruptured blood vessels on the brain.

A third imaging technology, widely used in brain research, is called **positron-emission tomography (PET)**. PET can reveal the location of a variety of physiological or biochemical processes in the human body. In preparation for a PET scan, water, glucose, or another molecule (depending on the process to be studied) is labeled with a radioactive isotope and injected into a person's bloodstream. (The isotopes are not dangerous because they are used only in small quantities.) The images in FIGURE b are PET scans of the human brain obtained using water labeled with a radioisotope of oxygen that emits particles called positrons. Because radioactive water tends to go to the most active areas of the brain, scientists can use this technique to map those areas used when the brain is performing a particular task. The positrons interact with oppositely charged electrons on the body's atoms, and the resulting radiation is detected by a PET camera. A computer creates the brain maps you see here, which illustrate localized brain activity under four different conditions, all related to language in healthy subjects. Physicians use the PET scanner to evaluate brain and heart disorders and certain types of cancer.

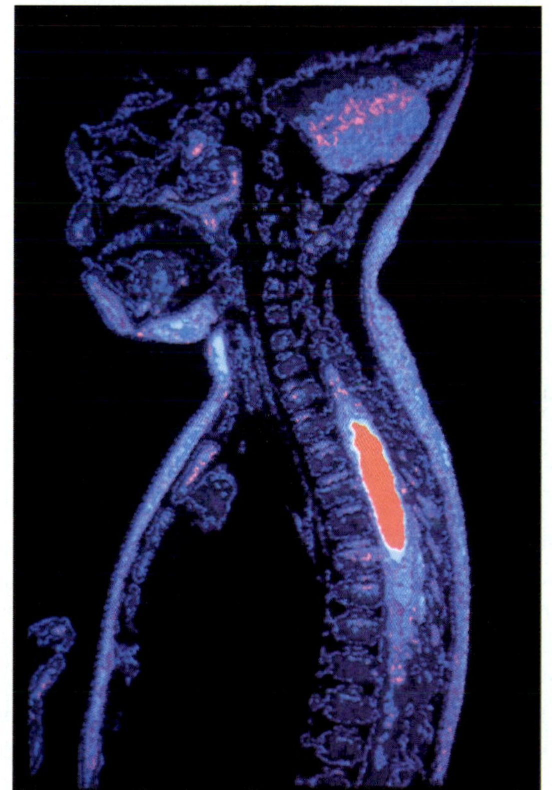

(a)

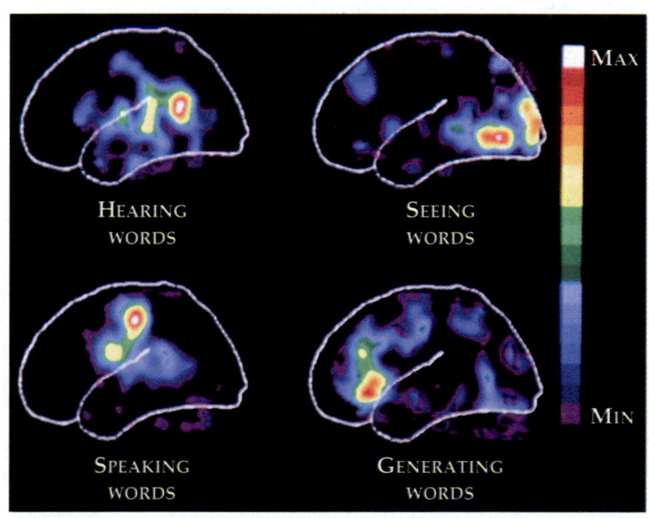

(b)

Memory and Learning

The original manuscript for this chapter was typed on a personal computer that weighs about 11 kg and can retain about 4×10^6 bits (1200 pages) of information. Your brain, a much more powerful information processor, weighs about 1.35 kg and stores billions of bits of information dating back to the beginning of your life.

Memory, which is essential for learning, is the ability to store and retrieve information related to previous experiences. Human memory occurs in two stages. **Short-term memory** reflects an immediate sensory perception of an object or idea and occurs before the image is stored. Short-term memory enables you to dial a phone number after looking it up but without looking at it directly. If you call the number frequently, it becomes stored in **long-term memory** and can be recalled several weeks after you originally looked it up. The transfer of information from short-term to long-term memory is enhanced by rehearsal ("practice makes perfect"), favorable emotional state (we learn best when we are alert and motivated), and the association of new data with data previously learned and stored in long-term memory (it's easier to learn a new card game if you already have "card sense" from playing other games).

In its ability to learn and remember, the human brain apparently distinguishes between facts and skills. When you acquire factual knowledge by memorizing dates, word definitions, the parts of the brain, and other information, this fact memory can be consciously and specifically retrieved from the data bank of your long-term memory. You can even recall visual images, such as the face of a friend. In contrast, skill memory usually involves motor activities that are learned by repetition without consciously remembering specific information. You perform learned motor skills, such as walking, tying your shoes, riding a bicycle, or writing without consciously recalling the individual steps required to do these tasks correctly. Once a skill memory is learned, it is difficult to unlearn. For example, a person who has played tennis for years with a self-taught, awkward backhand has a much tougher time learning the correct form than a beginner just learning the game. Bad habits, as we know, are difficult to break.

By studying amnesia (memory loss) in humans, experimenting with animals, and using brain imaging techniques, neuroscientists are beginning to map the complex brain pathways involved in memory. For instance, in the case of a telephone number and other fact memories, sensory signals from the eyes go to the vision centers in the brain's occipital lobe, where a visual perception is formed (see FIGURES 48.19b and 48.23). They also pass through the brain's sensory filter, the reticular formation (to determine if attention should be paid to the information), to parts of the hypothalamus and limbic system (to determine whether emotion should be involved), and to higher centers of the forebrain, including the prefrontal cortex (for higher-level integration). The pathway is completed when impulses return to the cortical vision centers where the perception first occurred.

Neuroscientists are investigating the cellular changes involved in memory and learning. Two limbic nuclei, the amygdala and **hippocampus**, seem to play key roles (see FIGURE 48.23). The amygdala may act as a type of memory filter, somehow labeling information to be saved by tying it to an event or emotion of the moment. First discovered in the hippocampus, functional changes at certain synapses seem to be directly related to memory storage and learning. One kind of change, called **long-term depression (LTD)**, is decreased responsiveness to an action potential by a postsynaptic cell. LTD is induced by repeated, weak stimulation. A second kind of synaptic change, called **long-term potentiation (LTP)**, is an enhanced responsiveness to action potentials by a postsynaptic cell. It can result when a presynaptic cell bombards a synapse with a series of brief, repeated action potentials that strongly depolarize the postsynaptic membrane. With LTP established, a single action potential from the presynaptic cell has a much greater effect at the synapse than before. Lasting for a matter of hours, days, or weeks, depending on the number and frequency of repeated action potentials, LTP may be what happens when a memory is being stored or when learning takes place.

LTP is associated with the release of the excitatory neurotransmitter glutamate by the presynaptic cell. Glutamate binds with a specific class of receptors in the postsynaptic membrane, opening gated channels that are highly permeable to calcium ions (Ca^{2+}). In turn, the Ca^{2+} triggers a cascade of enzyme activity in the postsynaptic cell, making it more responsive to stimulation. The discovery that both LTP and LTD can occur in several brain centers that were shown by other techniques to be involved in learning and memory supports the hypothesis that these synaptic changes are key elements in learning and memory.

Consciousness

Unraveling more about how networks of neurons in the brain store, retrieve, and use memories, control the internal environment of the body, construct our thoughts and emotions, and make us conscious of ourselves and our surroundings is one of the most challenging and fascinating aspects of modern biology. By **consciousness** we mean subjective awareness, or the ability to be aware of and make conscious judgments about the environment. Neuroscientists and behaviorists are beginning to unify the "mind" and the "brain," but consciousness—exactly what it is and how to study it—remains elusive. Some scientists maintain that consciousness is strictly a human attribute. Others posit that other animals have at least some forms of consciousness and that a continuum of consciousness has evolved in the animal kingdom. For humans, we know that several different stimuli can make us recall a particular memory; for instance, the sight, sound or smell of a sea coast might trigger a memory of childhood experiences. Does it follow that conscious thoughts are triggered by several different cerebral neural pathways?

A central hypothesis about consciousness is that it is an emergent property of the brain, involving simultaneous cooperation of extensive areas of the cerebral cortex. Another idea is that while cerebral neurons and functional groups of neurons are generating conscious thoughts, they are also involved in less complex, more specific tasks. Some neurons, for instance, might be sensing internal or external stimuli, and others might be formulating a command signal to a group of skeletal muscles, while both groups are working together to formulate a conscious awareness of the sensation and the muscular response. Finding ways to test hypotheses about consciousness is part of a challenging and rapidly expanding research effort in neurobiology. So far, researchers have made the most progress using imaging techniques to map the parts of the brain associated with subjective experiences involving the visual system. We return to studies of consciousness in Chapter 51 when we take up the subject of animal cognition.

In the next chapter we examine the relationships of the nervous system to sensory and motor functions: how sense organs gather information about the environment and how skeletal muscles carry out motor commands.

CHAPTER REVIEW

REVIEW OF KEY CONCEPTS

(with page numbers and key figures)

AN OVERVIEW OF NERVOUS SYSTEMS

- **Nervous systems perform the three overlapping functions of sensory input, integration, and motor output (p. 961, FIGURE 48.1)** The nervous system's three main functions are sensory input, integration, and motor output to effector cells. The central nervous system (CNS) integrates information, while the interconnecting nerves of the peripheral nervous system (PNS) communicate sensory and motor signals between the CNS and the rest of the body.

48.1 **The nervous system is composed of neurons and supporting cells (pp. 961–964, FIGURES 48.2–48.4)** Cells of the nervous system include neurons, which transmit the signals, and supporting cells, which support, insulate, and protect the neurons. A neuron's fiber-like dendrites and axons conduct information in the form of nerve impulses. Axons often originate from the axon hillock and terminate in numerous branches. Synaptic terminals at the ends of axons release neurotransmitter molecules into the synapses, thereby relaying neural signals to the dendrites or cell bodies of other neurons or effectors. The CNS consists of the brain and spinal cord. The PNS contains sensory neurons, which transmit information from internal and external environments to the CNS, and motor neurons, which carry information from the CNS to target organs. The spinal cord mediates many reflexes that integrate sensory input with motor output. It also has tracts of neurons that carry information to and from the brain. Interneurons of the CNS integrate sensory input and motor output. Groups of neurons may interact and carry information along specific pathways called circuits. Nerve cell bodies are arranged in functional groups usually called ganglia in the PNS and nuclei in the brain. Glial cells, supporting cells in the CNS, include astrocytes, which line the capillaries in the brain and contribute to the blood-brain barrier, and oligodendrocytes, which wrap and insulate some neurons in a myelin sheath. In the PNS, myelin sheaths are formed by supporting cells called Schwann cells.

THE NATURE OF NEURAL SIGNALS

48.2 **Membrane potentials arise from differences in ion concentrations between a cell's contents and the extracellular fluid (pp. 964–966, FIGURE 48.5)** The membrane potential of a nontransmitting neuron is due to the unequal distribution of ions, particularly sodium and potassium, across the plasma membrane; the cytoplasm is more negatively charged than the extracellular fluid. Membrane potential is maintained by differential ion permeabilities and the sodium-potassium pump.

48.2 **An action potential is an all-or-none change in the membrane potential (pp. 966–969, FIGURES 48.6, 48.7)** A stimulus that affects the membrane's permeability to ions can either depolarize or hyperpolarize the membrane relative to the membrane's resting potential. This local voltage change is called a graded potential, and its magnitude is proportional to the strength of the stimulus. An action potential, or nerve impulse, is a rapid, transient depolarization of the neuron's membrane. A local depolarization to the threshold potential opens voltage-gated sodium channels, and the rapid influx of Na^+ brings the membrane potential to a positive value. The membrane potential is restored to its normal resting value by the closing of the Na^+ channels. A refractory period follows an action potential, corresponding to the period when the voltage-gated Na^+ channels are inactivated. The all-or-none generation of an action potential always creates the same amplitude of voltage change for a given neuron. The frequency of action potentials varies with the intensity of the stimulus.

- **Action potentials "travel" along an axon because they are self-propagating (pp. 969–970, FIGURES 48.8, 48.9)** Once an action potential is initiated in an axon, a wave of depolarization propagates a series of action potentials to the end of the axon. The rate of transmission of a nerve impulse is directly related to the diameter of the axon. Saltatory conduction, a mechanism by which action potentials jump between the nodes of Ranvier of myelinated axons, speeds nervous impulses in vertebrates.

48.3 **Chemical or electrical communication between cells occurs at synapses (pp. 970–972, FIGURE 48.10)** Synapses between neurons conduct signals from the axon of a presynaptic cell to a dendrite or cell body of a postsynaptic cell. Electrical synapses directly pass an action potential between two neurons via gap junctions. In a chemical synapse, a depolarization stimulates the fusion of synaptic vesicles with the presynaptic membrane and the release of neurotransmitter molecules into the synaptic cleft. Neurotransmitters bind to receptor proteins associated with particular ion channels on the postsynaptic membrane. The selective opening of chemically sensitive gates either brings the membrane potential closer to threshold (EPSP) or hyperpolarizes the membrane (IPSP). The neurotransmitter is rapidly broken down by enzymes.

- **Neural integration occurs at the cellular level (pp. 972–974, FIGURES 48.11, 48.12)** Whether or not an action potential is created in the postsynaptic cell depends on temporal or spatial summation of EPSPs and IPSPs at the axon hillock.

- **The same neurotransmitter can produce different effects on different types of cells (pp. 974–976, TABLE 48.1)** One of the most common invertebrate and vertebrate neurotransmitters is acetylcholine. Other transmitters that have been identified include the biogenic amines (epinephrine, norepinephrine, dopamine, and serotonin); several amino acids; and some neuropeptides, such as the analgesic endorphins. Some neurons also release gases, such as nitric oxide, to signal other cells.

ORGANIZATION OF NERVOUS SYSTEMS

- **Nervous system organization tends to correlate with body symmetry (pp. 976–977, FIGURE 48.13)** Invertebrate nervous systems range from the diffuse nerve nets of many cnidarians to the highly centralized nervous systems of the cephalopods, which possess large brains capable of sophisticated learning. The evolution of nervous system centralization was correlated with the evolution of bilateral symmetry.
- **Vertebrate nervous systems are highly centralized and cephalized (pp. 977–978, FIGURE 48.14)** The vertebrate CNS (brain and spinal cord) is the integrating link between the sensory and motor divisions of the PNS. Derived from the embryonic dorsal hollow nerve cord, the CNS contains contiguous spaces filled with cerebrospinal fluid. Gray matter (mainly nerve cell bodies, dendrites, and unmyelinated axons) is distinguishable from white matter (mainly myelinated axons) in the brain and spinal cord.
- **The vertebrate PNS has several components differing in organization and function (pp. 978–979, FIGURES 48.15, 48.16)** The vertebrate PNS consists of paired cranial and spinal nerves and associated ganglia. Functionally, the PNS consists of the sensory, or afferent, division, which brings information from sensory receptors to the CNS, and the motor, or efferent, division, which carries signals away from the CNS to effector cells. The motor division consists of the somatic nervous system, which carries signals to skeletal muscles, and the autonomic nervous system, which regulates the primarily automatic, visceral functions of smooth and cardiac muscles. The autonomic nervous system consists of the parasympathetic and sympathetic divisions, which are anatomically, functionally, and chemically distinct and usually antagonistic in effect on target organs.

STRUCTURE AND FUNCTION OF THE VERTEBRATE BRAIN

- **The vertebrate brain develops from three anterior bulges of the spinal cord (pp. 979–980, FIGURE 48.17)** All vertebrate brains develop and diversify from three embryonic regions: the forebrain, the midbrain, and the hindbrain.
- **The brainstem conducts data and controls automatic activities essential for survival (p. 981)** In the human brain, the midbrain and hindbrain make up the brainstem. The medulla oblongata and pons of the hindbrain work together to control homeostatic functions and conduct sensory and motor signals between the spinal cord and higher brain centers. The midbrain receives, integrates, and projects sensory information to the forebrain.
- **The cerebellum controls movement and balance (p. 981)**
- **The thalamus and hypothalamus are prominent integrating centers of the diencephalon (pp. 981–983)** The forebrain is the site of the most sophisticated neural processing, with major integrating centers in the thalamus, hypothalamus, and cerebrum. The thalamus routes neural input to specific areas of the cerebral cortex, the outer gray matter of the cerebrum. The functions of the hypothalamus range from hormone production to the regulation of body temperature, hunger, thirst, sexual response, the flight-or-fight response, and circadian rhythms.
- **The cerebrum contains the most sophisticated integrating centers (pp. 983–984, FIGURES 48.19, 48.20)** The cerebral cortex contains distinct somatosensory and motor areas, which directly process information, and association areas, which integrate information. Imaging technology enables researchers to identify specific integrating centers in the functioning brain.
- **The human brain is a major research frontier (pp. 984–989, FIGURES 48.21–48.23)** Sleep and arousal are controlled by several areas in the cerebrum and brain stem, the most important being the reticular formation, which filters the sensory input sent to the cortex. The two cerebral hemispheres control different functions. Speech, language, and analytical ability are centered in the left hemisphere, whereas spatial perception and artistic ability predominate in the right. Nerve tracts of the corpus callosum link the two sides and allow the brain to function as an integrated whole. Human emotions are believed to originate from interactions between the limbic system, a group of nuclei in the diencephalon and inner cerebrum. The limbic system interacts closely with the prefrontal cortex, a higher integrating center of the forebrain. Human memory consists of short-term and long-term memories. The process of learning facts appears to differ from that of learning skills. The hippocampus and amygdala, two components of the limbic system, participate in circular brain pathways involved in fact memory. Functional changes at synapses, called LTP (long-term potentiation) and LTD (long-term depression) may underlie memory storage and learning. A heightened sensitivity to a single action potential by a postsynaptic membrane, LTP results from repeated bursts of action potentials. A reduced responsiveness to an action potential by a postsynaptic membrane, LTD results from repeated, weak stimulation.

SELF-QUIZ

1. Which of the following occurs when a stimulus depolarizes a neuron's membrane?
   a. Na^+ diffuses out of the cell.
   b. The action potential approaches zero.
   c. The membrane potential changes from the resting potential to a voltage closer to the threshold potential.
   d. The depolarization is all or none.
   e. The inside of the cell becomes more negative in charge relative to the outside of the cell.
2. Action potentials are usually propagated in only one direction along an axon because
   a. the nodes of Ranvier conduct only in one direction
   b. the brief refractory period prevents opening of voltage-gated Na^+ channels
   c. the axon hillock has a higher membrane potential than the tips of the axon
   d. ions can flow along the axon only in one direction
   e. both sodium and potassium voltage-gated channels open in one direction
3. The depolarization of the presynaptic membrane of an axon *directly* causes
   a. voltage-gated calcium channels in the membrane to open
   b. synaptic vesicles to fuse with the membrane
   c. an action potential in the postsynaptic cell
   d. the opening of chemically sensitive gates that allow neurotransmitter to spill into the synaptic cleft
   e. an EPSP or IPSP in the postsynaptic cell
4. Anesthetics reduce pain by blocking the transmission of nerve impulses. Which of the following might work as an anesthetic?
   a. a chemical that blocks voltage-gated sodium channels in membranes
   b. a chemical that opens voltage-gated potassium channels
   c. a chemical that blocks neurotransmitter receptors

d. a, b, and c
e. only b and c

5. Gray matter is
   a. made up of three protective layers called meninges
   b. located on the outside of the spinal cord
   c. restricted to the brain
   d. populated by cell bodies of neurons
   e. found in the ventricles of the vertebrate brain

6. Which of the following structures or regions is *incorrectly* paired with its function?
   a. limbic system—screening of information between the spinal cord and the brain; regulates arousal and sleep
   b. medulla oblongata—homeostatic control center
   c. cerebellum—coordination of movement and balance
   d. corpus callosum—band of fibers connecting left and right cerebral hemispheres
   e. hypothalamus—production of hormones and regulation of temperature, hunger, and thirst

7. Receptor sites for neurotransmitters are located on the
   a. tips of axons
   b. axon membranes in the regions of the nodes of Ranvier
   c. postsynaptic membrane
   d. membranes of synaptic vesicles
   e. presynaptic membrane

8. All the following electrical changes of neurons are graded events *except*
   a. EPSPs
   b. IPSPs
   c. action potentials
   d. depolarizations caused by stimuli
   e. hyperpolarizations caused by stimuli

9. Of the following components of the nervous system, which is the *most inclusive?*
   a. brain
   b. spinal cord
   c. central nervous system
   d. gray matter
   e. neuron

10. A nerve net is a distinctive feature of which animal phylum?
    a. Chordata
    b. Cnidaria
    c. Annelida
    d. Arthropoda
    e. Platyhelminthes

CHALLENGE QUESTIONS

1. Emergent properties and the hierarchical organization of living matter are two related themes in biology (see Chapter 1). Write an essay discussing some examples of how these themes apply to functions of the nervous system.
2. From what you know about the action potential, propose one feasible mechanism whereby anesthetics might prevent pain.
3. Describe various ways in which drugs that are stimulants could increase the activity of the nervous system by acting at synapses.

SCIENCE, TECHNOLOGY, AND SOCIETY

A woman lay in a coma for nine years after suffering irreversible brain damage in a car accident. Her family asked that she be allowed to die, but legal authorities refused to allow the hospital to withhold life support. A long legal battle went all the way to the U.S. Supreme Court, which ruled in 1990 that a person has a right to refuse medical treatment, but that the state may require clear evidence of the patient's wishes. The woman died 12 days after her feeding tube was removed. Only 10% to 15% of Americans have a living will or other documentation stating their wishes in the event that they become brain-dead and are kept alive by artificial life support. What would you want done if you or a member of your family were in this situation? Why? Who should speak for the patient? Is it a physician's job to ask a patient his or her wishes, if that is possible? What constitutes evidence of a patient's wishes?

FURTHER READING

Gazzaniga, M.S. "The Split Brain Revisited." *Scientific American*, July 1998. Explores the behavioral and evolutionary implications of lateralization.

Gould, J. L., and C. G. Gould. *The Animal Mind.* New York: W. H. Freeman/Scientific American Library, 1994. Provides an introduction to the scientific study of cognition in nonhuman animals.

Marx, J. "Helping Neurons Find Their Way." *Science*, May 19, 1995. Chemotaxis may direct growing axons to specific targets.

Nemeroff, C.B. "The Neurobiology of Depression." *Scientific American*, June 1998. Discusses current research on the role of hormones and the brain's limbic system on a common illness.

Purves, D., et al. (Eds.) *Neuroscience.* Sunderland, MA: Sinauer Associates. 1997. An authoritative, beautifully illustrated text.

Travis, J. "Glia: The Brain's Other Cells." *Science*, November 11, 1994. Presents a concise review of current research on supporting cells of the central nervous system.

Youdim, M. B. H., and P. Riederer. "Understanding Parkinson's Disease." *Scientific American*, January 1997.

WEB LINKS

Visit the special edition of *The Biology Place* for BIOLOGY, Fifth Edition, at http://www.biology.com/campbell. Go to Chapter 48 for online resources, including learning activities, practice exams, and links to the following web sites:

"The Human Brain: Dissections of the Real Brain"
Extensive anatomical images of the brain from the Virtual Hospital Project at the University of Iowa College of Medicine.

"The Digital Rat Atlas"
Navigate your way through the rat brain using this 3-D digital map of rat neuroanatomy.

"Wormland Homepage"
Complex systems are often studied using animals that have the simplest version of a particular system. This page explores various aspects of the neurobiology of several flatworms and nematodes.

"Neuroscience on the Internet"
A comprehensive, searchable site providing links to all aspects of the neurosciences.

CHAPTER 49

SENSORY AND MOTOR MECHANISMS

In the gathering dusk a male moth's antennae detect the chemical attractant of a female moth somewhere upwind. The male takes to the air, following the scent trail toward the female. Suddenly, vibration sensors in the moth's abdomen signal the presence of ultrasonic chirps of a rapidly approaching bat. The bat's sonar enables the mammal to locate moths and other flying insect prey. Reflexively, the moth's nervous system alters the motor output to its wing muscles, sending the insect into an evasive spiral toward the ground. Although it is probably too late for the moth in the photograph on this page, many moths can escape because they can detect a bat's sonar about 30 m away. The bat has to be within 3 m to sense the moth, but since the bat flies faster, it may still have time to detect, home in, and catch its prey.

The outcome of this interaction depends on the abilities of both predator and prey to sense important environmental stimuli and to produce appropriate coordinated movement. Although not all of an animal's moment-by-moment interactions with its environment are as dramatic as such predator-prey struggles, the detection and processing of sensory information and the generation of motor output provide the physiological basis for all animal behavior.

In Chapter 48 we saw how the nervous system transmits and integrates sensory and motor information. We will now examine the input and output of this coordinating system that controls so many aspects of animal behavior. We first consider the sensory receptors that receive information from the environment, and then we examine the structure and function of muscles, the motor effectors that bring about movement in response to that information. We will also study skeletons in the context of body movement.

INTRODUCTION TO SENSORY RECEPTION

Information is transmitted through the nervous system in the form of nerve impulses, or action potentials, which are all-or-none events (see FIGURE 48.6c). An action potential triggered by light striking the eye is the same as an action potential triggered by air vibrating in the ear. The ability to distinguish any type of stimulus, such as sight or sound, depends on the part of the brain that receives the signal. What matters is where impulses go, not what triggers them.

Action potentials that reach the brain via sensory neurons are called **sensations**. Once the brain is aware of sensations, it interprets them, giving us the **perception** of the stimuli. Perceptions, such as colors, smells, sounds, and tastes, are constructions of the brain and do not exist outside it. So, if a tree falls and no one is present to hear it, is there a sound? The fall certainly produces pressure waves in the air, but if sound is defined as a perception, then there is none unless sensory receptors detect the waves and an animal's brain perceives them.

Sensory receptors transduce stimulus energy and transmit signals to the nervous system

Sensations, and the perceptions they evoke in the brain, begin with **sensory reception**, the detection of the energy of a stimulus by sensory cells. Most **sensory receptors** are specialized neurons or epithelial cells that exist singly or in groups with other cell types within sensory organs, such as the eyes and ears. Sensory receptors called **exteroreceptors** detect stimuli outside the body, such as heat, light, pressure, and chemicals. Other sensory receptors called **interoreceptors** detect stimuli within the body, such as blood pressure and body position. All stimuli represent forms of energy, and the general function of receptor cells is to convert the energy of stimuli into changes in membrane potentials and then transmit signals to the nervous system. This task consists of four functions: sensory transduction, amplification, transmission, and integration.

Sensory Transduction

The actual detection of a stimulus involves the conversion of stimulus energy into a change in the membrane potential of a receptor cell, a process called **sensory transduction**. The initial response of the sensory receptor to a stimulus is a change in its membrane permeability, resulting in a graded change in membrane potential called a **receptor potential**. (Recall from Chapter 48 that a graded potential is a change in the voltage across the membrane that is proportional to the strength of the stimulus.) In some cases a stimulus such as pressure can stretch the membrane and increase ion flow. In other cases, specific receptor molecules on the membrane of a receptor cell open or close gates to ion channels when the stimulus is present. We will examine specific examples of sensory transduction later in the chapter.

Amplification

The strengthening of stimulus energy that is otherwise too weak to be carried into the nervous system is called **amplification**. Amplification of the signal may occur in accessory structures of a complex sense organ, as when sound waves are enhanced by a factor of more than 20 before reaching the receptors of the inner ear. Amplification also may be a part of the transduction process itself. An action potential conducted from the eye to the brain has about 100,000 times as much energy as the few photons of light that triggered it.

Transmission

Once the energy in the stimulus has been transduced into a receptor potential, **transmission**, or the conduction of impulses to the CNS, can occur. In some instances, such as in the case of "pain cells," the receptor itself is actually a sensory neuron that conducts action potentials to the CNS. Other receptors are separate cells that must transmit chemical signals (neurotransmitters) across synapses to sensory neurons. If the receptor also functions as the sensory neuron, the intensity of the receptor potential will affect the frequency of action potentials that travel as sensations to the CNS. For separate receptor cells, the strength of the stimulus and receptor potential affect the amount of neurotransmitter released by the receptor at its synapse with a sensory neuron, which in turn determines the frequency of action potentials generated by the sensory neuron. Many sensory neurons spontaneously generate signals at a low rate. Therefore, a stimulus does not really switch the production of action potentials on or off; it modulates their frequency. In this way the CNS is sensitive not only to the presence or absence of a stimulus but also to changes in stimulus intensity.

Integration

The processing of information, or **integration**, begins as soon as information is first received. Signals from receptors are integrated through the summation of graded potentials, as are those within the nervous system. One type of integration by receptor cells is **sensory adaptation**, a decrease in responsiveness during continued stimulation (not to be confused with the term *adaptation* as used in an evolutionary context). Without sensory adaptation, you would feel every beat of your heart and every bit of clothing on your body. Receptors are selective in the information they send to the CNS, and adaptation reduces the likelihood that a continued stimulus will be transmitted.

Another important aspect of sensory integration is the sensitivity of the receptors. The threshold for transduction by receptor cells varies with conditions. For example, the thresholds of glucose receptors in the human mouth can vary over several orders of magnitude of sugar concentration as both the general state of nutrition and the amount of sugar in the diet change.

The integration of sensory information occurs at all levels within the nervous system, and the cellular actions just described are only the first steps. Complex receptors such as the eyes have higher levels of integration as signals converge on sensory nerves, and the CNS further processes all incoming signals.

Sensory receptors are categorized by the type of energy they transduce

Based on the type of energy they detect (transduce), sensory receptors fall into five categories: mechanoreceptors, pain receptors, thermoreceptors, chemoreceptors, and electromagnetic receptors.

Mechanoreceptors are stimulated by physical deformation caused by such stimuli as pressure, touch, stretch, motion, and sound—all forms of mechanical energy. Bending or stretching of the plasma membrane of a mechanoreceptor cell increases its permeability to both sodium and potassium ions, resulting in a depolarization (receptor potential).

The human sense of touch relies on mechanoreceptors that are actually modified dendrites of sensory neurons (FIGURE 49.1). Receptors that detect light touch are close to the surface of the skin; they transduce very slight inputs of mechanical energy into receptor potentials. Receptors responding to strong pressure and vibrations in the body are in deep skin layers. Other touch receptors detect hair movement.

An example of an interoreceptor stimulated by mechanical distortion is the **muscle spindle**, or stretch receptor (see FIGURE 48.4). This mechanoreceptor monitors the length of skeletal muscles. The muscle spindle contains modified muscle fibers attached to sensory neurons and runs parallel to muscle. When the muscle is stretched, the fibers of the spindle are also stretched, depolarizing the sensory neurons and triggering action potentials that are transmitted back to the spinal cord.

The **hair cell** is a common type of mechanoreceptor that detects motion. Hair cells are found in the vertebrate ear and in the lateral line organs of fishes and amphibians, where they detect movement relative to the environment (see Chapter 34). The "hairs" are either specialized cilia or microvilli. They project upward from the surface of the hair cell into either an internal compartment, such as the human inner ear, or an external environment, such as a pond. When the cilia or microvilli bend in one direction, they stretch the hair cell membrane and increase its permeability to sodium and potassium ions, leading to an increase in the rate of impulse production in a sensory neuron. When the cilia bend in the opposite direction, ion permeability decreases, reducing the number of action potentials in the sensory neuron. This specificity allows hair cells to respond to the direction of motion as well as to its strength and speed.

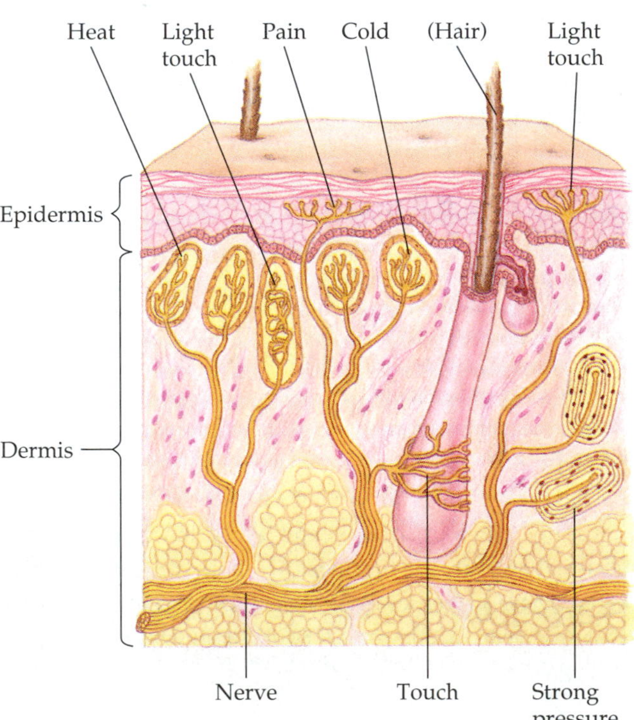

FIGURE 49.1 ▪ Sensory receptors in human skin. Each mechanoreceptor is a modified dendrite of a sensory neuron. Most receptors in the inner layer of the skin (dermis) are encapsulated by one or more layers of connective tissue. Those in the outer skin layers (epidermis) and touch receptors wound around the base of hairs are naked dendrites. Touch receptors at the base of the stout whiskers of mammals such as cats and many rodents are extremely sensitive and enable the animal to detect nearby objects in the dark.

Pain receptors in humans are a class of naked dendrites in the epidermis of the skin called **nociceptors** (see FIGURE 49.1). Most animals probably experience pain, although we cannot say what perceptions other animals actually associate with stimulation of their pain receptors. Pain is one of the most important sensations because the stimulus becomes translated into a negative reaction, such as withdrawal from danger. Rare individuals who are born without any pain sensation may die from such conditions as a ruptured appendix because they cannot feel the associated pain and are unaware of the danger.

Different groups of pain receptors respond to excess heat, pressure, or specific classes of chemicals released from damaged or inflamed tissues. Some of the chemicals that trigger pain include histamine and acids. Prostaglandins increase pain by sensitizing the receptors—that is, lowering their threshold. Aspirin and ibuprofen reduce pain by inhibiting prostaglandin synthesis.

Thermoreceptors, responding to either heat or cold, help regulate body temperature by signaling both surface and body core temperature. There is still debate about the identity of thermoreceptors in the mammalian skin. Possible candidates are two receptors consisting of encapsulated, branched dendrites (see FIGURE 49.1). Many researchers, however, believe that these structures are actually modified pressure receptors and maintain that naked dendrites of certain sensory neurons are the actual thermoreceptors in the skin. There is general agreement that cold and heat receptors in the skin, as well as interothermoreceptors in the anterior hypothalamus of the brain, send information to the body's thermostat, located in the posterior hypothalamus (see Chapter 44).

Chemoreceptors include both general receptors that transmit information about the total solute concentration in a solution and specific receptors that respond to individual kinds of molecules. Osmoreceptors in the mammalian brain, for example, are general receptors that detect changes in the total solute concentration of the blood and stimulate thirst when osmolarity increases (see Chapter 44). Water receptors in the feet of house flies respond to pure water or to a dilute solution of virtually any substance. Most animals also have receptors specific to important molecules, including glucose, oxygen, carbon dioxide, and amino acids. In all these examples, the stimulus molecule binds to a specific site on the membrane of the receptor cell and initiates changes in membrane permeability. Two other groups of chemoreceptors show intermediate specificity. **Gustatory** (taste) and **olfactory** (smell) **receptors** respond to *categories* of related chemicals. Humans often classify such categories as sweet, sour, salty, or bitter. Two of the most sensitive and specific chemoreceptors known are present in the antennae of the male silkworm moth (FIGURE 49.2). They detect the two chemical components of the female moth sex pheromone.

Electromagnetic receptors detect various forms of electromagnetic energy, such as visible light, electricity, and magnetism. **Photoreceptors**, which detect the radiation we know as visible light, are often organized into eyes. Snakes have extremely sensitive infrared receptors that detect the body heat of prey standing out against a colder background (FIGURE 49.3a). Some fishes discharge electric currents and use special electroreceptors to locate objects, such as prey, that disturb the electric currents. The platypus, a monotreme mammal, has electroreceptors on its bill that can probably detect electrical fields generated by the muscles of prey, such as crustaceans, frogs, and small fishes. There is also evidence that many animals that home or migrate use the magnetic field lines of Earth to help orient themselves (FIGURE 49.3b). The iron-containing mineral magnetite is found in the skulls of some birds and mammals (including humans), in the

(a)

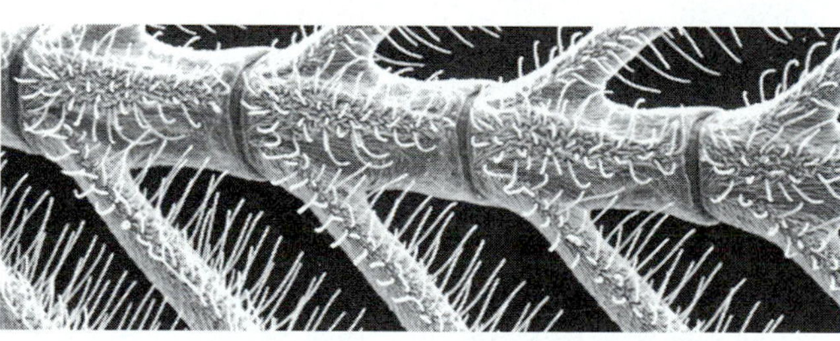

(b)

0.1 mm

FIGURE 49.2 ▪ Chemoreceptors in an insect. (a) The antennae of the male silkworm moth *Bombyx mori* are covered with sensory hairs, visible in **(b)** the SEM enlargement. About 50,000 of these hairs are chemoreceptors that are highly sensitive to the sex pheromone released into the air by the female. Each chemoreceptive hair has thousands of tiny pores that admit air, and each hair contains dendrites of two sensory neurons, one sensitive to bombykol, the other to bombykal, the two components in the pheromone. The male moth begins to respond to the pheromone when as few as 50 of his bombykol receptors come into contact with one bombykol molecule per second.

(a)

(b)

FIGURE 49.3 ▪ Specialized electromagnetic receptors. (a) This rattlesnake and other pit vipers have a pair of infrared receptors, one between each eye and nostril. The organs are sensitive enough to detect the infrared radiation emitted by a warm mouse a meter away. The snake moves its head from side to side until the radiation is detected equally by the two receptors, indicating that the mouse is straight ahead. **(b)** Some migrating animals, such as the beluga whales in this aerial photograph, can apparently sense Earth's magnetic field and use the information, along with other cues, for orientation.

abdomens of bees, in the teeth of some mollusks, and in certain protists and prokaryotes that orient with respect to Earth's magnetic field. Once used by sailors as a primitive compass, magnetite may be part of an important orienting mechanism in many animals. At the forefront of research on magnetoreception, Michael Walker and his colleagues at the University of Auckland have experimentally demonstrated magnetic sensitivity in rainbow trout (*Oncorhynchus mykiss*) and have begun to piece together the sensory pathway underlying the fish's ability to navigate using magnetic fields.

PHOTORECEPTORS

A great variety of light detectors has evolved in the animal kingdom, from simple clusters of cells that detect only the direction and intensity of light to complex organs that form images. Despite their diversity, all photoreceptors contain pigment molecules that absorb light, and molecular evidence indicates that most, if not all, photoreceptors in the animal kingdom may be homologous. Animals as diverse as flatworms, annelids, arthropods, and vertebrates have some of the same, ancient genes associated with photoreceptors. Thus, the genetic underpinnings of all photoreceptors may have evolved in the earliest bilateral animals. The actual eyes that develop in an animal depend on developmental patterns regulated by genetic mechanisms that evolved later in the animal's taxonomic group, and whose effects appear to be superimposed on the ancient, homologous mechanism.

A broad array of photoreceptors has evolved among invertebrates

Most invertebrates have photoreceptors, which range from simple clusters of photoreceptor cells to complex image-forming eyes. One of the simplest is the **eye cup** of planarians, structures that provide information about light intensity and direction without actually forming an image. Photoreceptor cells are located within a cup formed by a layer of cells containing a screening pigment that blocks light. Light can enter the cup and stimulate the photoreceptors only through an opening on one side where there is no screening pigment (FIGURE 49.4). The opening of one eye cup faces left and slightly forward, and the opening of the other cup faces right-forward. Thus, light shining from one side of the planarian can enter only the eye cup on that side. The brain compares the rate of nerve impulses coming from the two eye cups, and the animal turns until the sensations from the two cups are equal and minimal. The result is that the animal moves directly away from the light source and reaches a shaded loca-

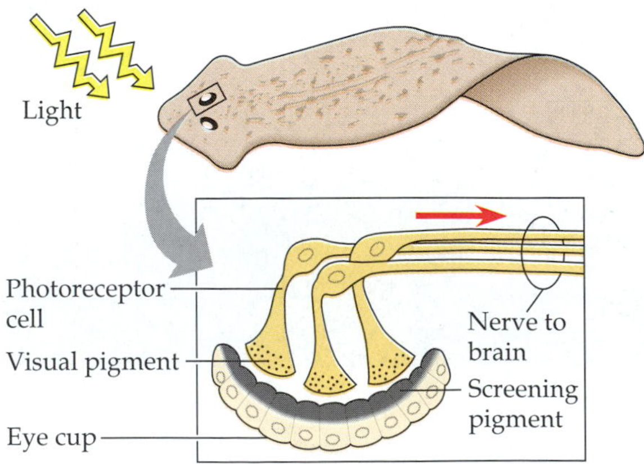

FIGURE 49.4 • Eye cups and orientation behavior of a planarian. The head of the flatworm has two eye cups with photoreceptors that send nerve impulses to the brain. Because each eye cup consists of cells containing screening pigment that shades the photoreceptors, light can reach the photoreceptor cells only through the opening of the cup. The brain directs the body to turn until the sensations from the two cups are equal and minimal, causing the animal to move away from light.

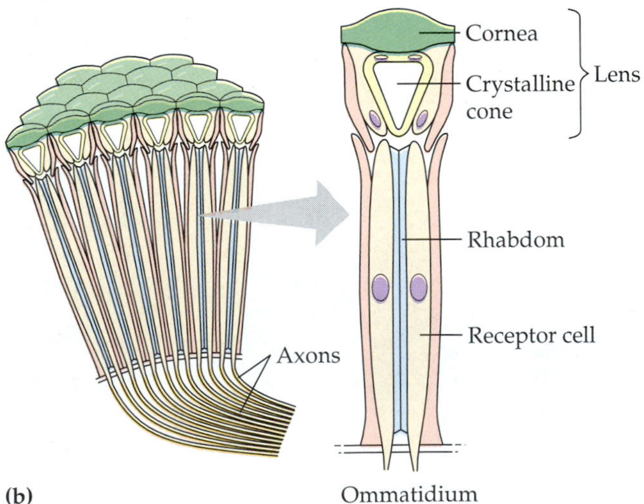

FIGURE 49.5 • Compound eyes. (a) The faceted eyes on the head of a fly, photographed with a stereomicroscope. **(b)** The cornea and crystalline cone of each ommatidium function as a lens that focuses light onto the rhabdom, a stack of pigmented plates on the inside of a circle of receptor cells. The rhabdom traps light and guides it to the receptor cells. The image formed by a compound eye is a mosaic of dots formed by the different intensities of light entering the many ommatidia from different angles.

tion beneath a rock or some other object, a behavioral adaptation that helps hide the planarian from predators.

Image-forming eyes of two major types have evolved in invertebrates: the compound eye and the single-lens eye. **Compound eyes** are found in insects and crustaceans (Phylum Arthropoda) and some polychaete worms (Phylum Annelida). A compound eye consists of up to several thousand light detectors called **ommatidia** (the "facets" of the eye), each with its own light-focusing lens (FIGURE 49.5). Each ommatidium registers light from a tiny portion of the visual field. Differences in the intensity of light entering the many ommatidia result in a mosaic image. The animal's brain may sharpen the image when it integrates the visual information. The compound eye is extremely acute at detecting movement, an important adaptation for flying insects and small animals constantly threatened with predation. This characteristic of the compound eye is partly due to the rapid recovery of the photoreceptors. The human eye can distinguish light flashes up to about 50 flashes per second. For this reason, the individual images of a movie, which flash at a faster rate, fuse together to create the perception of smooth motion. The compound eyes of some insects, however, recover from excitation rapidly enough to detect the flickering of a light flashing 330 times per second. Such an insect viewing a movie could easily resolve each frame of the film as a separate still image. Insects also have excellent color vision, and some (including bees) can see into the ultraviolet range of the spectrum, which is invisible to us. In studying animal behavior, we cannot extrapolate our sensory world to other species; different animals have different sensitivities and different brain organizations.

Among the invertebrates, **single-lens eyes** are found in some jellies, polychaetes, spiders, and many mollusks. A single-lens eye works on a cameralike principle. The eye of an octopus or squid, for example, has a small opening, the pupil, through which light enters. Analogous to a camera's shutter, an adjustable iris changes the diameter of the pupil; behind the pupil, a single lens focuses light onto the retina, which consists of light-transducing receptor cells. Also similar to a camera's action, muscles move the lens forward or backward to focus images on the retina.

Vertebrates have single-lens eyes

Similar to the single-lens eyes of many invertebrates, the eyes of vertebrates are also cameralike, but they evolved in the vertebrate lineage and differ from the single-lens eyes of invertebrates in several details. The human eye, shown in FIGURE 49.6, is capable of detecting an almost countless variety of colors, forming images of objects miles away, and responding to as little as one photon of light. Remember, however, that it is actually the brain that "sees." Thus, to understand vision we must begin by learning how the vertebrate eye generates sensations (action potentials), and then follow these signals to the visual centers of the brain, where images are perceived.

The globe of the vertebrate eye, or eyeball, consists of a tough, white outer layer of connective tissue called the **sclera**

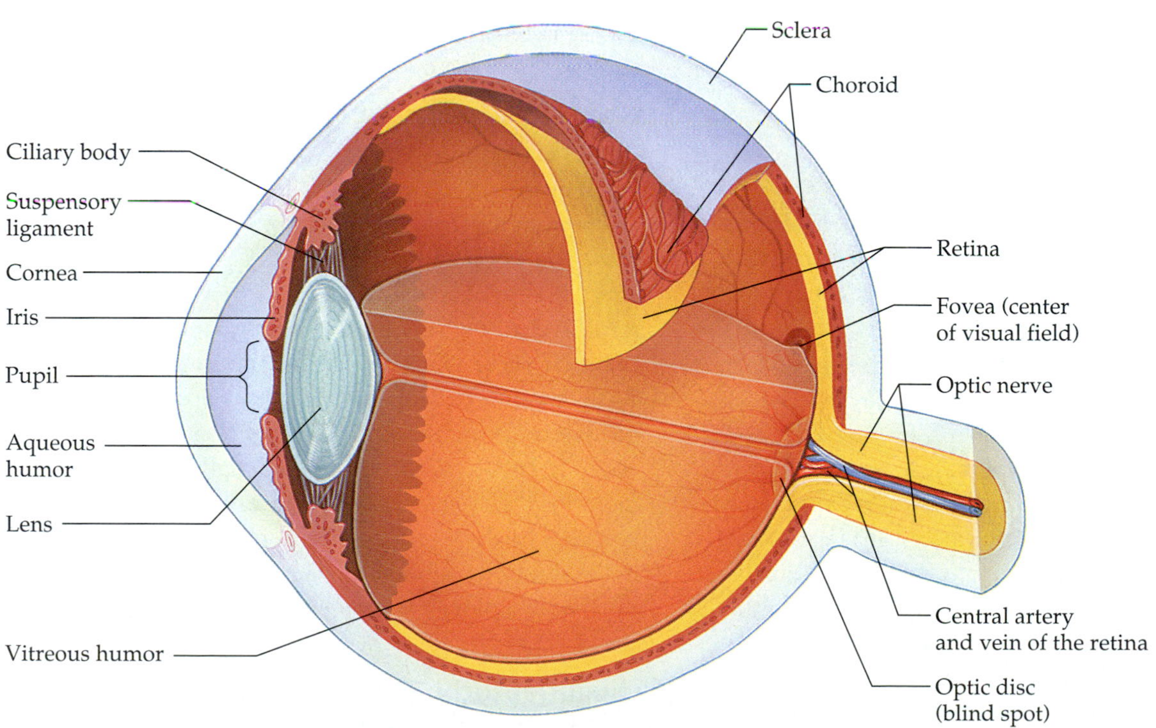

FIGURE 49.6 ▪ Structure of the vertebrate eye. In this longitudinal section of the eye, the jellylike vitreous humor is illustrated only in the lower half of the eyeball. The mucous membrane, or conjunctiva, surrounding the sclera (the white of the eye) is not shown.

and a thin, pigmented inner layer called the **choroid**. A delicate layer of epithelial cells forms a mucous membrane, the **conjunctiva**, that covers the outer surface of the sclera and helps keep the eye moist. At the front of the eye the sclera becomes the transparent **cornea**, which lets light into the eye and acts as a fixed lens. The conjunctiva does not cover the cornea. The anterior choroid forms the donut-shaped **iris**, which gives the eye its color. By changing size, the iris regulates the amount of light entering the **pupil**, the hole in the center of the iris. Just inside the choroid, the **retina** forms the innermost layer of the eyeball and contains the photoreceptor cells. Information from the photoreceptors leaves the eye at the optic disc, where the optic nerve attaches to the eye. Because there are no photoreceptors in the optic disc, this spot on the lower outside of the retina is a blind spot: Light focused onto that part of the retina is not detected.

The **lens** and **ciliary body** divide the eye into two cavities, one between the lens and the cornea, and a much larger cavity behind the lens within the eyeball itself. The ciliary body constantly produces the clear, watery **aqueous humor** that fills the anterior cavity of the eye. Blockage of the ducts that drain the aqueous humor can produce glaucoma, increased pressure that leads to blindness by compressing the retina. The posterior cavity, filled with the jellylike **vitreous humor**, constitutes most of the volume of the eye. The aqueous and vitreous humors function as liquid lenses that help focus light onto the retina. The lens itself is a transparent protein disc that focuses an image onto the retina. Like squids and octopuses, many fishes focus by moving the lens forward or backward, as in a camera. Humans and other mammals, however, focus by changing the *shape* of the lens. When viewing a distant object, the lens is flat. When focusing on a close object, the lens becomes almost spherical, a change called **accommodation** (FIGURE 49.7).

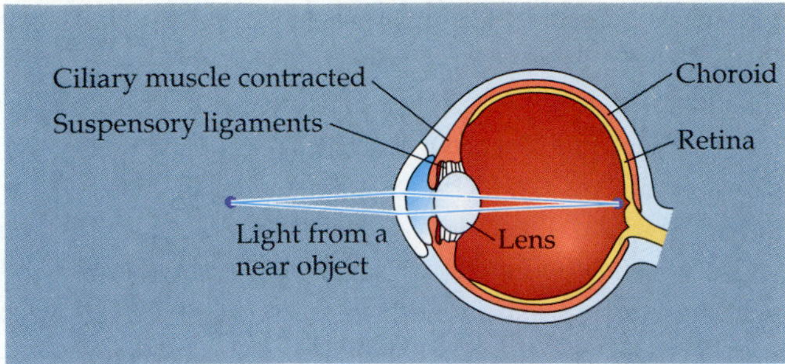

(a) Near vision (accommodation)

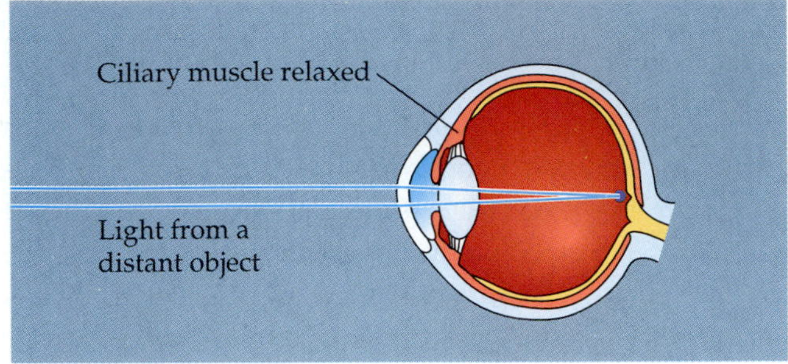

(b) Distance vision

FIGURE 49.7 ▪ Focusing in the mammalian eye. The lens bends light and focuses it onto the retina. The thicker the lens, the more sharply the light is bent. The lens is nearly spherical when focusing on near objects and much flatter when focusing on distant objects. Ciliary muscles control the shape of the lens. **(a)** In near vision, the ciliary muscles contract, pulling the border of the choroid layer of the eye toward the lens and causing the suspensory ligaments to relax. With this reduced tension, the elastic lens becomes thicker and rounder, bending light in such a way that images of near objects can be focused onto the retina. This adjustment of the lens for close vision is known as accommodation. **(b)** In distance vision, the ciliary muscles relax, allowing the choroid to expand and put tension on the suspensory ligaments. The lens is pulled into a flatter shape, and the images of distant objects are focused onto the retina.

The human retina contains about 125 million **rod cells** and 6 million **cone cells**, two types of photoreceptors named for their shapes. They account for 70% of all sensory receptors in the body, a fact that underscores the importance of the eyes and visual information in how humans perceive their environment.

Rods and cones have different functions in vision, and the relative numbers of these two photoreceptors in the retina are partly correlated with whether an animal is most active during the day or at night. Rods are more sensitive to light but do not distinguish colors; they enable us to see at night, but only in black and white. Because it takes more light to stimulate cones, cones do not function in night vision. Cones can distinguish colors in daylight. Color vision is found in all vertebrate classes, though not in all species. Most fishes, amphibians, reptiles, and birds have strong color vision, but humans and other primates are among the minority of mammals with this ability. Most mammals are nocturnal, and a maximum number of rods in the retina is an adaptation that gives these animals keen night vision. Cats, usually most active at night, have limited color vision and probably see a pastel world during the day. In the human eye, rods are found in greatest density at the peripheral regions of the retina and are completely absent from the **fovea**, the center of the visual field (see FIGURE 49.6). You cannot see a dim star at night by looking at it directly; if you view it at an angle, however, focusing the starlight onto the regions of the retinas most populated by rods, you will be able to see the star. You achieve your sharpest daylight vision by looking straight at the object of interest because cones are most dense at the fovea, where there are about 150,000 color receptors per mm^2. Some birds have more than a million cones per mm^2, which enables such species as hawks to spot mice and other small prey from high in the sky. In the retina of the eye, as in all biological structures, variations represent evolutionary adaptations.

The light-absorbing pigment rhodopsin operates via signal transduction

When the vertebrate lens focuses a light image onto the retina, how do the cells of the retina transduce the stimuli into sensations—action potentials that transmit this information about the environment to the brain? Each rod cell or cone cell has an outer segment with a stack of folded membranes, or discs, in which visual pigments are embedded (FIGURE 49.8a). The visual pigments consist of a light-absorbing pigment molecule called **retinal** (a derivative of vitamin A) bonded to a membrane protein called an **opsin**. Opsins vary in structure from one type of photoreceptor to another, and the light-absorbing ability of retinal is affected by the specific identity of its opsin partner.

Rods contain their own type of opsin, which, combined with retinal, makes up the visual pigment **rhodopsin** (FIGURE 49.8b). When rhodopsin absorbs light, its retinal component changes shape, triggering a signal-transduction pathway (see Chapter 11) that ultimately results in a receptor potential in the rod cell membrane. Initially, retinal's shape change causes a conformational change in its opsin partner. The altered opsin molecule then activates a relay molecule in the signal-transduction pathway, a G protein called transducin, which is also in the disc membrane. In turn, transducin activates an effector enzyme that chemically alters the second messenger in the rod cell, a nucleotide called cyclic guanosine monophosphate (cGMP).

In the dark, when rhodopsin is inactive, cGMP is bound to sodium ion channels in the rod cell plasma membrane and keeps those channels open. In this state, the rod cell membrane is actually *depolarized* and releases an inhibitory neurotransmitter at its synapses with other neurons in the retina (FIGURE 49.9, p. 1000). The neurotransmitter inhibits depolarization of the neuron membranes, thus preventing them from developing action potentials. However, when light alters retinal, triggering the rhodopsin signal-transduction pathway, the effector enzyme converts cGMP to GMP, which disengages from the Na^+ channels (FIGURE 49.9b). This closes the channels, decreasing the membrane's permeability to Na^+ and altering the membrane potential. The membrane actually becomes hyperpolarized, and this is the rod cell's receptor potential. In effect, the hyperpolarization slows the rod cell's release of the inhibitory neurotransmitter. Thus, the transduction of light energy into receptor potentials produces a *decrease* in the chemical signal to the cells with which rods synapse, and that decrease is the message that the rods have been stimulated by light.

The light-induced change in retinal, which initiates the light-transducing signal pathway in rod cells, is referred to as "bleaching" of rhodopsin. In the dark, enzymes convert the

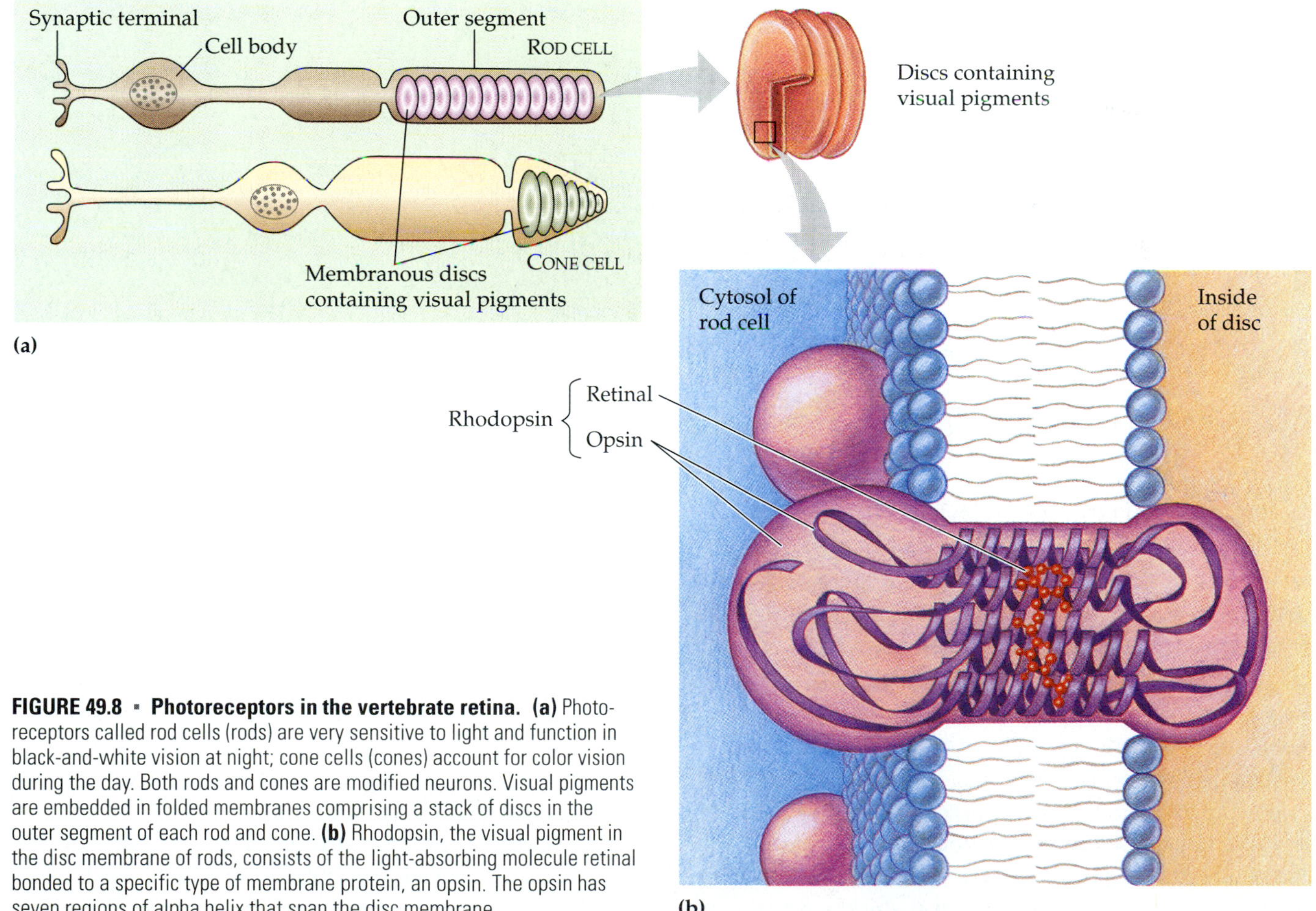

FIGURE 49.8 • Photoreceptors in the vertebrate retina. **(a)** Photoreceptors called rod cells (rods) are very sensitive to light and function in black-and-white vision at night; cone cells (cones) account for color vision during the day. Both rods and cones are modified neurons. Visual pigments are embedded in folded membranes comprising a stack of discs in the outer segment of each rod and cone. **(b)** Rhodopsin, the visual pigment in the disc membrane of rods, consists of the light-absorbing molecule retinal bonded to a specific type of membrane protein, an opsin. The opsin has seven regions of alpha helix that span the disc membrane.

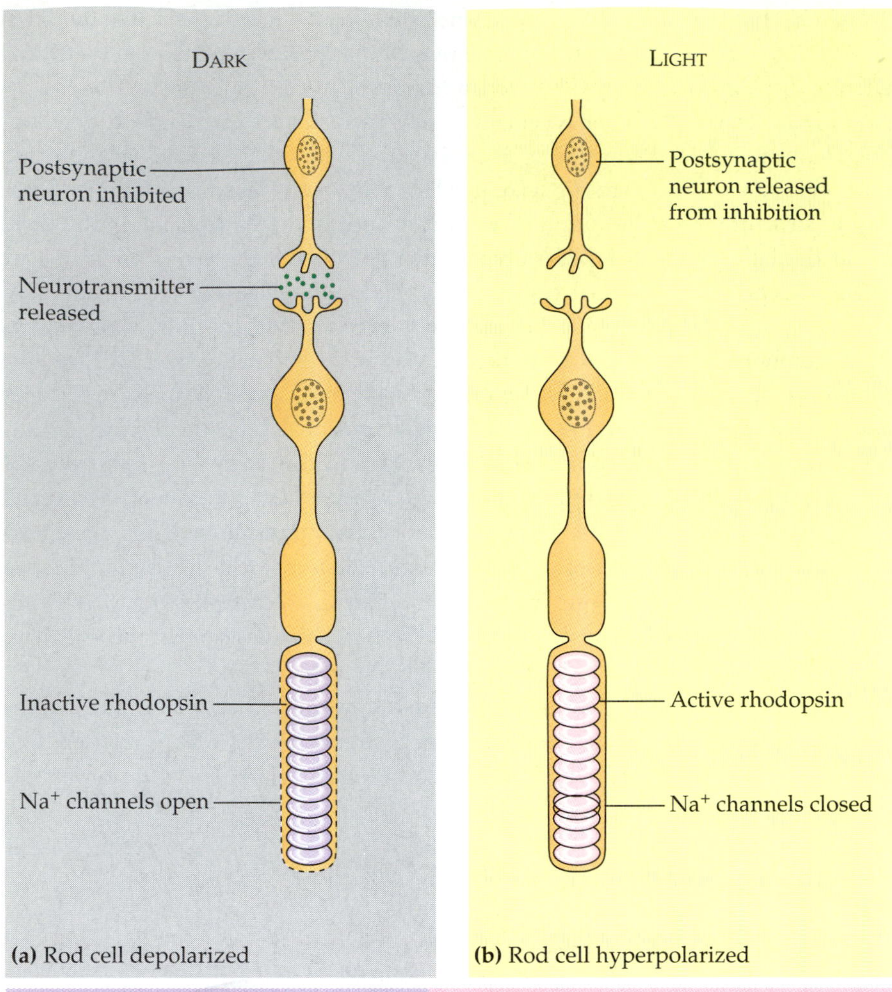

Light causes "bleaching"

Enzyme converts *trans* form to *cis* in the dark

cis form of retinal

trans form of retinal

(c)

FIGURE 49.9 ▪ The effect of light on rod cells and retinal. (a) In the dark, rhodopsin is inactive, and the rod cell membrane is highly permeable to sodium and thus depolarized. In this state, the rod cell releases neurotransmitter molecules that inhibit the firing of postsynaptic neurons in the retina. **(b)** In contrast, when light activates rhodopsin, the rod cell membrane becomes less permeable to sodium, and its membrane potential changes (it develops a receptor potential, a hyperpolarization in this case). The synaptic terminals of the rod cell then slow their release of inhibitory neurotransmitter molecules. Freed from inhibition, the postsynaptic neurons can develop action potentials. **(c)** Retinal exists in two forms, isomers of each other. Absorption of light converts the pigment from the *cis* isomer to the *trans* isomer. This change triggers the signal-transduction pathway that converts the light signal into an electrochemical signal, a receptor potential, in the rod cell membrane. When the photoreceptor is no longer stimulated by light, enzymes convert the retinal back to the *cis* form, which recombines with opsin to form rhodopsin.

retinal back to its original form, and it recombines with opsin to form rhodopsin (FIGURE 49.9C). Bright light keeps the rhodopsin bleached and rods become unresponsive; cones take over. When you walk from a bright environment into a dark place, such as walking into a movie theater in the afternoon, you are initially almost blind to faint light. There is not enough light to stimulate the cones, and it takes at least a few minutes for the bleached rods to become functional again.

Color vision involves more complex signal processing than the rhodopsin mechanism in rods. Color vision results from the presence of three subclasses of cones in the retina, each with its own type of opsin associated with retinal to form visual pigments collectively called **photopsins**. These photoreceptors are known as red cones, green cones, and blue cones, referring to the colors their kind of photopsin is best at absorbing. The absorption spectra for these pigments overlap, and the brain's perception of intermediate hues depends on the differential stimulation of two or more types of cones. For example, when both red and green cones are stimulated, we may see yellow or orange, depending on which of these two populations of cones is most strongly stimulated. Color blindness, more common in males than females because it is gener-

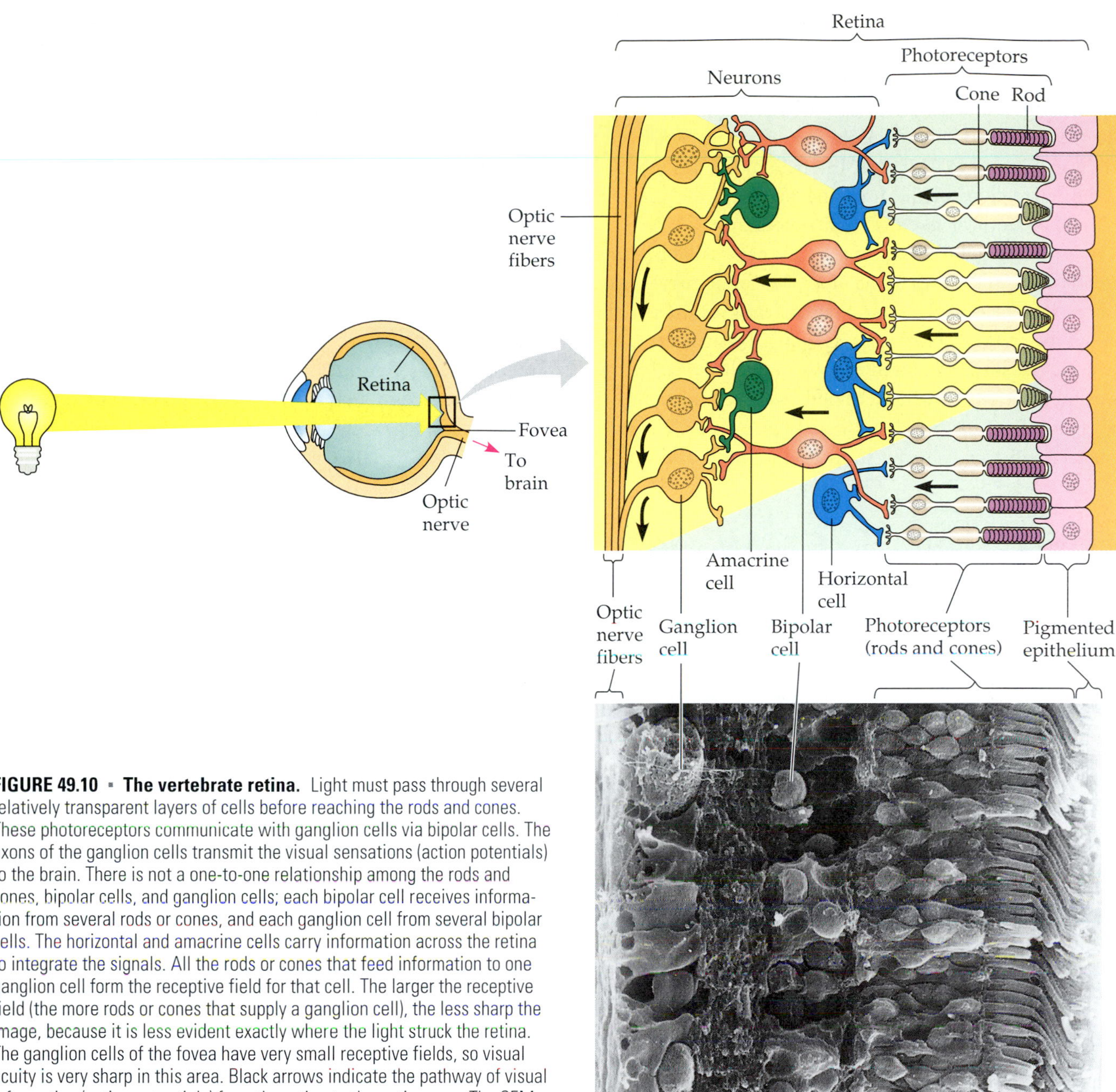

FIGURE 49.10 ▪ The vertebrate retina. Light must pass through several relatively transparent layers of cells before reaching the rods and cones. These photoreceptors communicate with ganglion cells via bipolar cells. The axons of the ganglion cells transmit the visual sensations (action potentials) to the brain. There is not a one-to-one relationship among the rods and cones, bipolar cells, and ganglion cells; each bipolar cell receives information from several rods or cones, and each ganglion cell from several bipolar cells. The horizontal and amacrine cells carry information across the retina to integrate the signals. All the rods or cones that feed information to one ganglion cell form the receptive field for that cell. The larger the receptive field (the more rods or cones that supply a ganglion cell), the less sharp the image, because it is less evident exactly where the light struck the retina. The ganglion cells of the fovea have very small receptive fields, so visual acuity is very sharp in this area. Black arrows indicate the pathway of visual information (action potentials) from the retina to the optic nerve. The SEM shows the retinal cell layers of a rabbit.

ally inherited as a sex-linked trait (see Chapter 15), is due to a deficiency or absence of one or more types of cones.

The retina assists the cerebral cortex in processing visual information

Processing of visual information begins in the retina itself. The axons of rods and cones synapse with neurons called **bipolar cells**, which in turn synapse with **ganglion cells** (FIGURE 49.10). Additional types of neurons in the retina, **horizontal cells** and **amacrine cells**, help integrate the information before it is sent to the brain. The axons of ganglion cells then convey the resulting sensations to the brain as action potentials along the optic nerve.

Signals from the rods and cones may follow either vertical or lateral pathways. In the so-called vertical pathway, information passes directly from the receptor cells to the bipolar cells to the ganglion cells. The horizontal and amacrine cells provide lateral integration of visual signals. Horizontal cells carry signals from one rod or cone to other photoreceptor cells and to several bipolar cells; amacrine cells spread the information from one bipolar cell to several ganglion cells. When a rod or cone stimulates a horizontal cell, the horizontal cell stimulates nearby receptors but inhibits more distant receptors and bipolar cells that are not illuminated, making the light spot appear lighter and the dark surroundings even darker. This integration, called **lateral inhibition**, sharpens edges and enhances contrast in the image. Lateral inhibition is repeated by the interactions of the amacrine cells with the ganglion cells and occurs at all levels of visual processing.

Axons of ganglion cells form the optic nerves that transmit sensations from the eyes to the brain. The optic nerves from the two eyes meet at the **optic chiasm** near the center of the base of the cerebral cortex (FIGURE 49.11). The nerve tracts of the optic chiasm are arranged in such a way that visual sensations from the left visual field of both eyes are transmitted to the right side of the brain, and visual sensations in the right visual field are transmitted to the left side of the brain. Most of the ganglion cell axons lead to the **lateral geniculate nuclei** of the thalamus. Neurons of the lateral geniculate nuclei continue back to the **primary visual cortex** in the occipital lobe of the cerebrum. Additional interneurons carry the information to other, more sophisticated visual processing and integrating centers elsewhere in the cortex.

Point-by-point information in the visual field is projected along neurons onto the visual cortex according to its position in the retina, but the information the brain receives is highly distorted. How does the brain convert a complex set of action potentials representing two-dimensional images projected onto our retinas into three-dimensional perceptions of our surroundings? Researchers estimate that fully 30% of the cerebral cortex—hundreds of millions of interneurons in perhaps dozens of integrating centers—take part in formulating what we actually "see." Determining how these centers integrate such components of our vision as color, motion, depth, shape, and detail is the focus of an exciting, fast-moving research effort.

Left visual field
Right visual field
Left eye
Right eye
Optic nerve
Optic chiasm
Lateral geniculate nucleus
Primary visual cortex

FIGURE 49.11 ▪ Neural pathways for vision. Because of the arrangement of neurons in the retinas, optic nerves, and optic chiasm, the right side of the brain receives sensory information about objects in the left visual field (blue), while the left side of the brain receives information from the right visual field (red). Each optic nerve contains about a million axons that synapse with interneurons in the lateral geniculate nuclei. The nuclei relay sensations to the visual cortex, believed to be the first of many brain centers that cooperate in constructing our visual perceptions.

HEARING AND EQUILIBRIUM

Hearing and the perception of body equilibrium, or balance, are related in most animals. Both involve the formation of sensations by mechanoreceptors containing hair cells that produce receptor potentials when the hairs are bent by settling particles or moving fluid. In mammals and most other terrestrial vertebrates, the sensory organs for hearing and equilibrium are closely associated within fluid-filled canals in the ear.

The mammalian hearing organ is within the inner ear

The mammalian ear can be divided into three regions. The **outer ear** consists of the external pinna and the auditory canal, which collect sound waves and channel them to the **tympanic membrane** (eardrum) separating the outer ear from the middle ear. Within the **middle ear**, vibrations are conducted through three ossicles (small bones)—the **malleus** (hammer), **incus** (anvil), and **stapes** (stirrup)—to the inner ear, passing through the **oval window**, a membrane beneath the stapes (FIGURE 49.12a and b). The middle ear also opens into the **Eustachian tube**, which connects with the pharynx and equalizes pressure between the middle ear and the atmosphere, enabling you to "pop" your ears when changing altitude, for example. The **inner ear** consists of a labyrinth of channels within a skull bone (the temporal bone). These channels are lined by a membrane and contain fluid that moves in response to sound or movement of the head.

The part of the inner ear involved in hearing is a complex coiled organ known as the **cochlea** (L. for "snail"). The cochlea

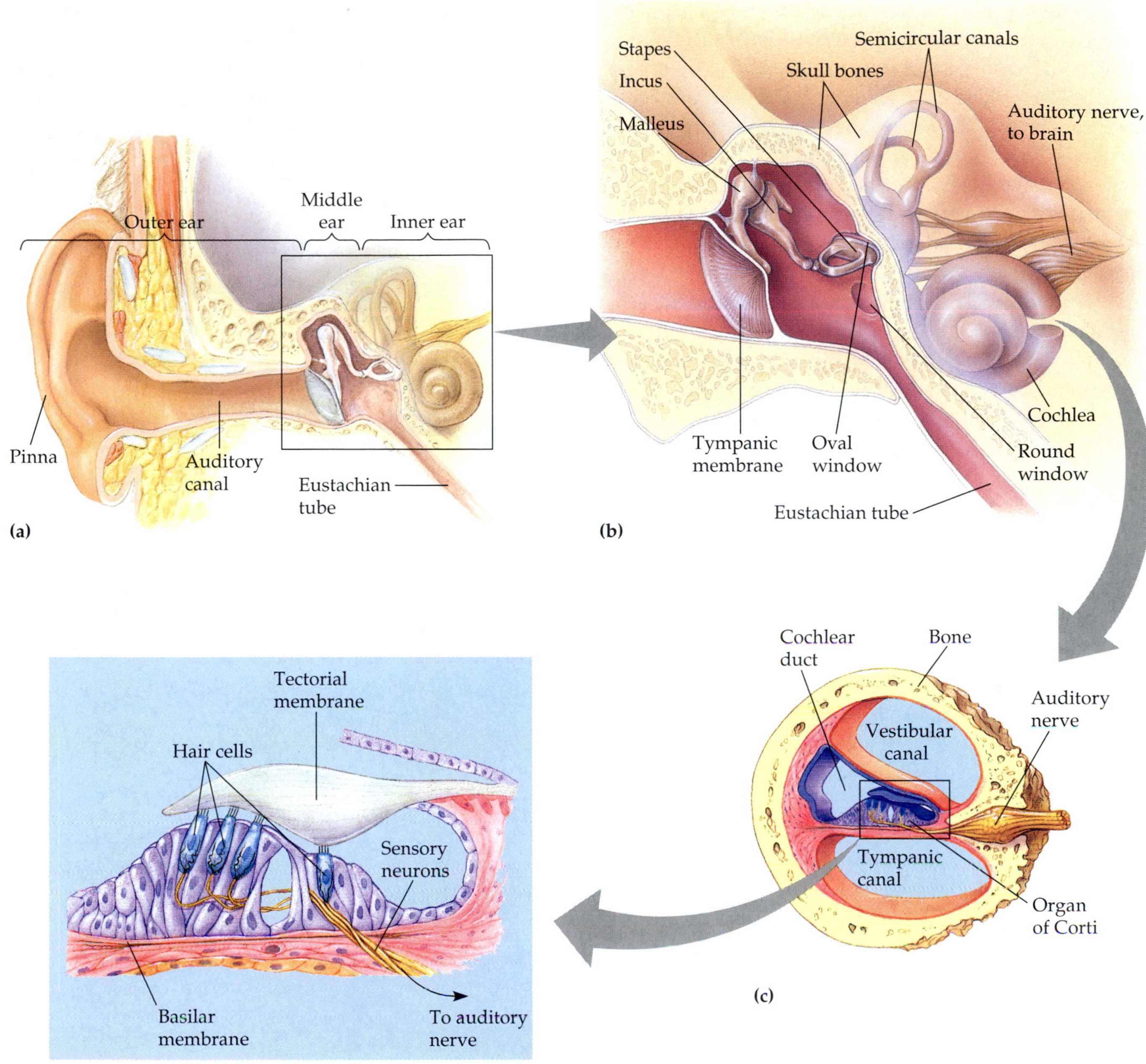

FIGURE 49.12 ▪ Structure and function of the human ear. (a) The pinna and auditory canal of the outer ear collect sound waves. **(b)** The sound waves create vibrations in the tympanic membrane that are conducted via three small bones (malleus, incus, and stapes) in the middle ear to the inner ear. The inner ear contains the cochlea, a long, coiled tube that contains sound detectors, and the semicircular canals, which control balance. **(c)** A cross-sectional view of the cochlea shows three canals. The vestibular canal and tympanic canal contain the fluid perilymph. Between these two canals is a smaller cochlear duct filled with endolymph. The organ of Corti sits on the basilar membrane, which forms the floor of the cochlear duct. **(d)** The receptor cells—hair cells—are part of the organ of Corti. Suspended over the organ of Corti is the tectorial membrane, to which many of the hairs of the receptor cells are attached. Vibrations of the oval window on the surface of the cochlea create pressure waves in the cochlear fluid. As the basilar membrane vibrates, hair cells repeatedly brush against the tectorial membrane. This stimulus causes the hair cells to depolarize and release neurotransmitter, thereby triggering an action potential in a sensory neuron.

has two large chambers, an upper vestibular canal and a lower tympanic canal, separated by a smaller cochlear duct (FIGURE 49.12c). The vestibular and tympanic canals contain a fluid called perilymph, and the cochlear duct is filled with a liquid called endolymph. The floor of the cochlear duct, the basilar membrane, bears the **organ of Corti**, which contains the actual receptor cells of the ear, hair cells with hairs projecting into the cochlear duct (FIGURE 49.12d). Many of the hairs are attached to the tectorial membrane, which hangs over the organ of Corti like a shelf.

How is the complex anatomy of the ear correlated with the function of hearing? The ear converts the energy of pressure waves traveling through air into nerve impulses that the brain perceives as sound. Vibrating objects, such as the reverberating strings of a guitar or the vocal cords of a speaking person, create percussion waves in the surrounding air. These waves cause the tympanic membrane to vibrate with the same frequency as the sound. The three bones of the middle ear amplify and transmit the mechanical movements to the oval window, a membrane on the surface of the cochlea. Vibrations of the oval window produce pressure waves in the fluid within the cochlea.

The cochlea transduces the energy of the vibrating fluid into action potentials. The stapes vibrating against the oval window creates a traveling pressure wave in the fluid of the cochlea that passes into the vestibular canal (FIGURE 49.13a). This wave continues around the tip of the cochlea and through the tympanic canal, dissipating as it strikes the **round window**. The pressure waves in the vestibular canal push downward on the cochlear duct and basilar membrane. The basilar membrane vibrates up and down in response to the pressure waves, and its hair cells alternately brush against and are withdrawn from the tectorial membrane. Deflection of the hairs opens ion channels in the plasma membrane of the hair cells, and positive ions (K^+, in this case) enter. The resulting depolarization increases neurotransmitter release from the hair cell and the frequency of action potentials in the sensory neuron with which the hair cell synapses. This neuron carries the sensations to the brain through the auditory nerve.

Sound is detected by increases in the frequency of impulses in the sensory neuron, but how is the quality of that sound determined? Two important aspects of sound are volume and pitch. Volume (loudness) is determined by the amplitude, or height, of the sound wave. The greater the amplitude of a sound, the more vigorous the vibrations of fluid in the cochlea, the greater the bending of the hair cells, and the more action potentials generated in the sensory neurons. **Pitch** is a function of a sound wave's frequency, or number of vibrations per second, expressed in hertz (Hz). Short, high-frequency waves produce high-pitched sound, while long, low-frequency waves generate low-pitched sound. Healthy young humans can hear sounds in the range of 20 to 20,000 Hz, dogs can hear sounds as high as 40,000 Hz, and bats can emit and hear

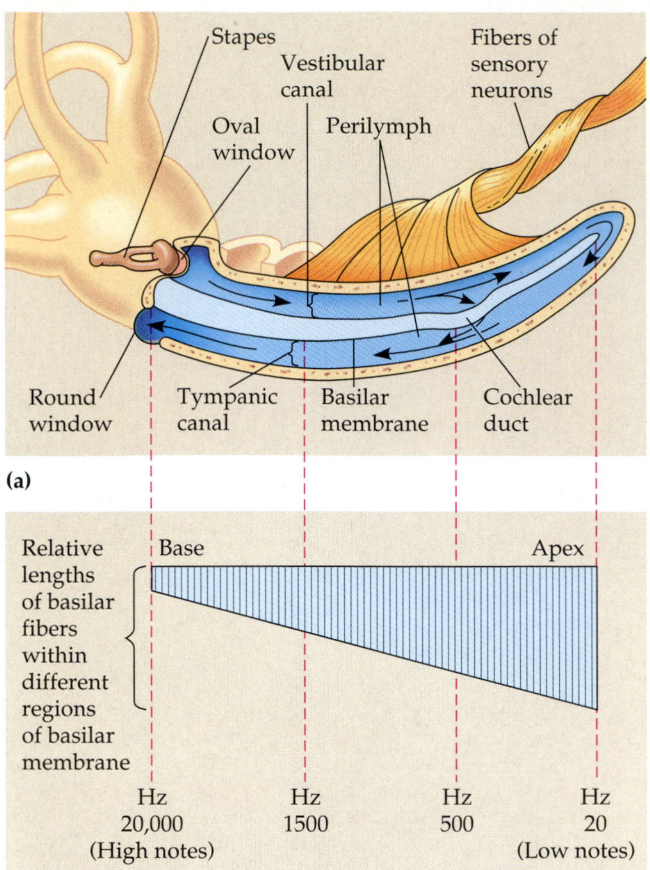

FIGURE 49.13 ▪ How the cochlea distinguishes pitch. (a) Vibrations of the stapes against the oval window agitate the fluid within the cochlea (uncoiled here), causing pressure waves having a frequency equivalent to the sound waves that entered the ear. The waves (black arrows) pass through the vestibular canal to the apex of the cochlea, then back toward the base of the cochlea via the tympanic canal. The energy causes the cochlear duct, with its basilar membrane and organ of Corti, to vibrate up and down. The bouncing of the basilar membrane stimulates the hair cells within the cochlear duct. **(b)** Fibers span the width of the basilar membrane. Like harp strings, these fibers vary in length, being shorter near the base of the membrane and longer near its apex. The length of the fibers "tunes" specific regions of the basilar membrane to vibrate at specific frequencies. **(c)** Different frequencies of pressure waves in the cochlea cause certain places along the basilar membrane to vibrate, stimulating particular hair cells and sensory neurons. The differential stimulation of hair cells is perceived in the brain as sound of a certain pitch.

clicking sounds of even higher frequency, using this ability to locate objects by sonar.

Pitch can be distinguished by the cochlea because the basilar membrane is not uniform along its length (see FIGURE 49.13b and c). The proximal end near the oval window is relatively narrow and stiff, while the distal end near the tip is wider and more flexible. Each region of the basilar membrane is most affected by a particular vibration frequency. The sensory neurons associated with the region vibrating most vigorously at any instant send the most action potentials along the auditory nerve. But the actual perception of pitch depends on neural mapping of the brain. Sensory neurons from the auditory pathway project onto specific auditory areas of the cerebral cortex according to the region of the basilar membrane in which the signal originated. When a particular site of the cortex is stimulated, we perceive a sound of a particular pitch.

The inner ear also contains the organs of equilibrium

Several organs in the inner ear of humans and most other mammals detect body position and balance. Behind the oval window is a vestibule that contains two chambers, the **utricle** and **saccule**. The utricle opens into three **semicircular canals** that complete the apparatus for equilibrium (FIGURE 49.14a).

Sensations related to body position are generated much like sensations of sound in humans and most other mammals. Hair cells in the utricle and saccule respond to changes in head position with respect to gravity and movement in one direction. The hair cells are arranged in clusters, and all the hairs project into a gelatinous material containing many small calcium carbonate particles called otoliths ("ear stones"). Because this material is heavier than the endolymph within the utricle and saccule, gravity is always pulling downward on the hairs of the receptor cells, sending a constant series of action potentials along the sensory neurons of the vestibular branch of the auditory nerve.

Different body angles cause different hair cells and their sensory neurons to be stimulated. When the position of the head changes with respect to gravity (as when the head bends forward), the force on the hair cell changes, and it increases (or decreases) its output of neurotransmitter. The brain interprets the resulting changes in impulse production by the sensory neurons to determine the position of the head. By a similar mechanism, the semicircular canals, arranged in the three spatial planes, detect changes in the rate of rotation or angular movements of the head (FIGURE 49.14b and c).

A lateral line system and inner ear detect pressure waves in most fishes and aquatic amphibians

Like other vertebrates, fishes and aquatic amphibians also have inner ears located near the brain. There is no cochlea, but

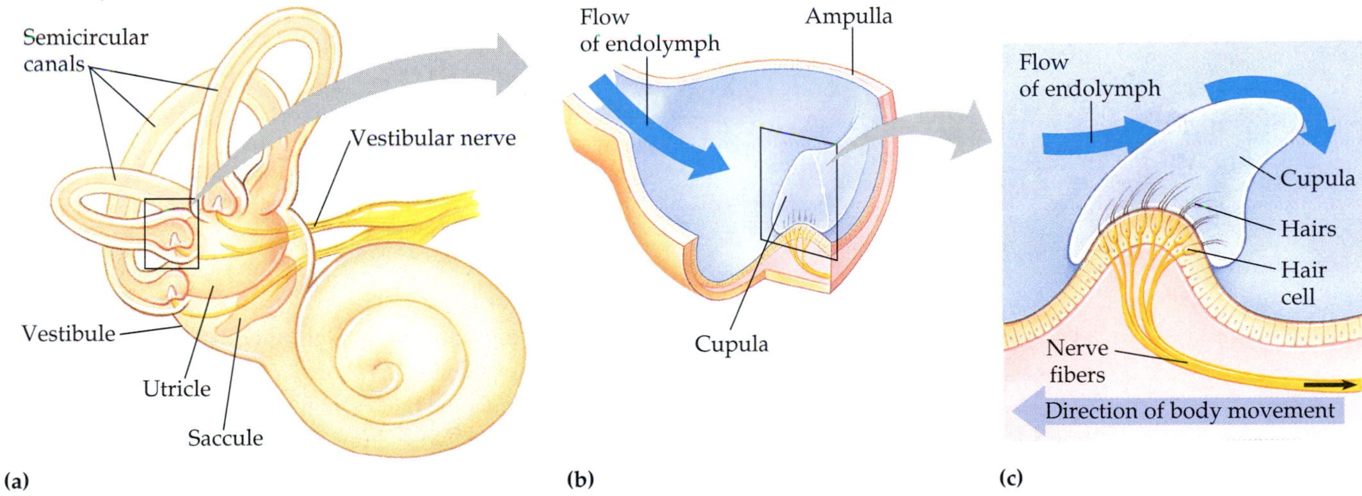

FIGURE 49.14 ▪ Organs of balance in the inner ear. (a) Three structures of the inner ear—the utricle and saccule in the vestibule and the semicircular canals—contain hair cells sensitive to balance and body position. The saccule and utricle tell the brain which way is up and also inform it of the body's fixed position in space or of any linear acceleration associated with movement. The semicircular canals are arranged in the three spatial planes. **(b)** Each canal has at its base a swelling called an ampulla, which contains a cluster of hair cells with hairs projecting into a gelatinous cap called the cupula. **(c)** When the head changes its rate of rotation, inertia prevents the endolymph in the semicircular canals from moving with the head, so the fluid presses against the cupula, bending the hair cells. The bending increases the frequency of action potentials in the sensory neurons in direct proportion to the amount of rotational acceleration. The mechanism adjusts quickly if rotation continues at a constant speed: The endolymph begins moving with the head, and the pressure on the cupula is reduced. If rotation stops suddenly, however, the fluid continues to flow through the semicircular canals and again stimulates the hair cells. This new stimulus can cause dizziness.

there are a saccule, a utricle, and semicircular canals, structures homologous to the equilibrium sensors of our own ears. Within these chambers in the inner ear of a fish, sensory hairs are stimulated by the movement of otoliths. Unlike the mammalian hearing apparatus, the ear of a fish has no eardrum and does not open to the outside of the body. Vibrations of the water caused by sound waves are conducted through the skeleton of the head to the inner ears, setting the otoliths in motion and stimulating the hair cells. The air-filled swim bladder (see Chapter 34) also vibrates in response to sound and may contribute to the transfer of sound to the inner ear. Some fishes, including catfishes and minnows, have a series of bones called the Weberian apparatus, which conducts vibrations from the swim bladder to the inner ear.

Most fishes and aquatic amphibians also have a **lateral line system** along both sides of the body (FIGURE 49.15). The system contains mechanoreceptors that detect low-frequency waves by a mechanism similar to the function of the inner ear. Water from the animal's surroundings enters the lateral line system through numerous pores and flows along a tube past the mechanoreceptors. The receptor units, called **neuromasts**, resemble the ampullae in our semicircular canals. Each neuromast has a cluster of hair cells, with the sensory hairs embedded in a gelatinous cap, the cupula. As the pressure of moving water bends a cupula, the hair cells transduce the energy into receptor potentials and then into action potentials that are transmitted along a nerve to the brain. This information helps the fish perceive its movement through water or the direction and velocity of water currents flowing over its body. The lateral line system also detects water movements or vibrations generated by other moving objects, including prey and predators.

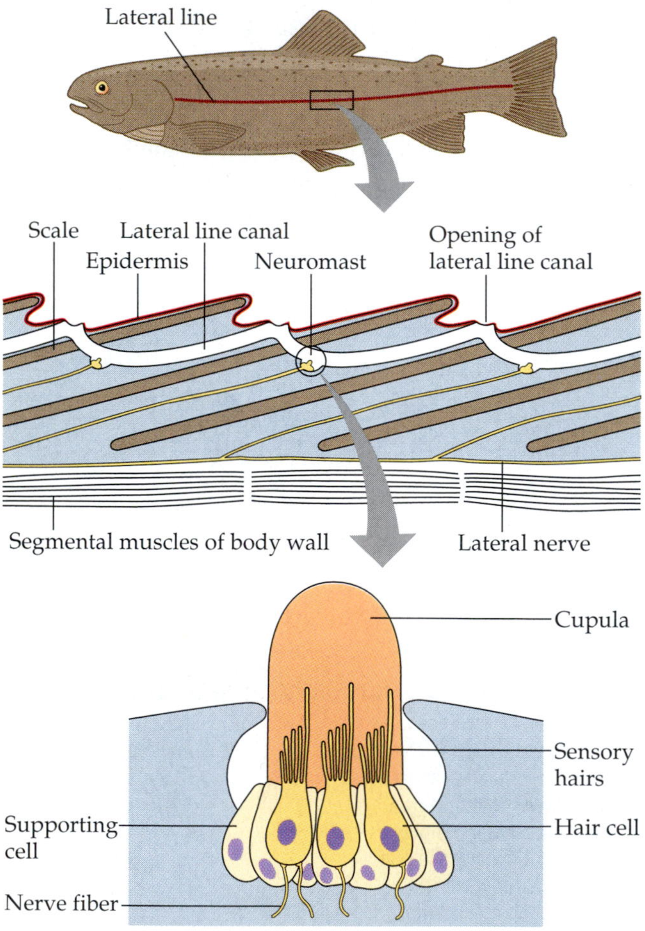

FIGURE 49.15 ▪ The lateral line system in a fish. Water flowing through the system bends hair cells. The hair cells transduce the energy into receptor potentials, triggering action potentials, which are conveyed to the brain. The lateral line system enables a fish to monitor water currents, pressure waves produced by moving objects, and low-frequency sounds conducted through water.

The lateral line system functions only in water. In terrestrial vertebrates, the inner ear has evolved as the main organ of hearing and equilibrium. Some amphibians have a lateral line system as tadpoles, but not as adults living on land. In the ear of a terrestrial frog or toad, sound vibrations traveling in the air are conducted to the inner ear by a tympanic membrane on the body surface and a single middle ear bone. There also is evidence that the lungs of a frog vibrate in response to sound and transmit their vibrations to the eardrum via the auditory tube. A small side pocket of the saccule functions as the main hearing organ of the frog, and it is this outgrowth of the saccule that gave rise to the more elaborate cochlea during the evolution of mammals. Birds also have a cochlea, but like amphibians and reptiles, sound is conducted from the tympanic membrane to the inner ear by a single bone, the stapes.

Many invertebrates have gravity sensors and are sound-sensitive

Most invertebrates have sensory organs called **statocysts** that contain mechanoreceptors and function in their sense of equilibrium (FIGURE 49.16). A common type of statocyst has a layer of hair cells surrounding a chamber containing **statoliths**, which are grains of sand or other dense granules. Gravity causes the statoliths to settle to the low point within the chamber, stimulating hair cells in that location. (This is similar to how the saccule and utricle function in vertebrates, and indeed these structures in the vertebrate inner ear are considered to be specialized types of statocysts.) The statocysts of invertebrates have various locations. For example, many jellies have statocysts at the fringe of the "bell," giving the animals an indication of body position. Lobsters and crayfish have statocysts near the bases of their antennules. Crayfish have been tricked into swimming upside down in experiments in which the statoliths were replaced with metal shavings that could be pulled to the upper end of the statocysts with magnets.

Many invertebrates demonstrate a general sensitivity to sound, although structures specialized for hearing seem to be

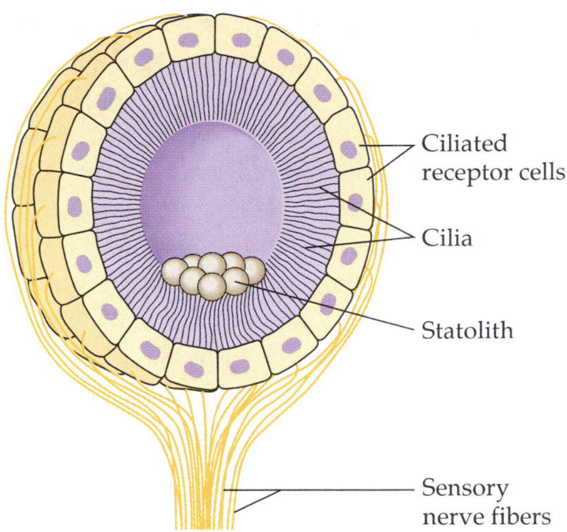

FIGURE 49.16 ▪ The statocyst of an invertebrate. The settling of statoliths to the low point within the chamber bends cilia on receptor cells in that location, providing the brain with information about the position of the body.

less widespread than gravity sensors. Hearing structures have been most extensively studied in terrestrial insects.

Many (perhaps most) insects have body hairs that vibrate in response to sound waves. Hairs of different stiffness and length vibrate at different frequencies. The hairs are commonly tuned to frequencies of sounds produced by other organisms. A male mosquito locates a mate by means of fine hairs on his antennae. The hairs vibrate in a specific way in response to the hum produced by the beating wings of flying females. A tuning fork that vibrates at the same frequency as a female mosquito's wings will also attract males. Some caterpillars (larval moths and butterflies) have vibrating body hairs that detect the buzzing wings of predatory wasps, warning the caterpillars of danger. Many insects also have localized "ears" (FIGURE 49.17). A tympanic membrane (eardrum) is stretched over an internal air chamber. Sound waves vibrate the tympanic membrane, stimulating receptor cells attached to the inside of the membrane and resulting in nerve impulses that are transmitted to the brain. Some moths can hear notes of such high pitch that they detect the sounds bats produce for sonar, and perception of these sounds triggers the moth's escape maneuver, as mentioned at the beginning of this chapter.

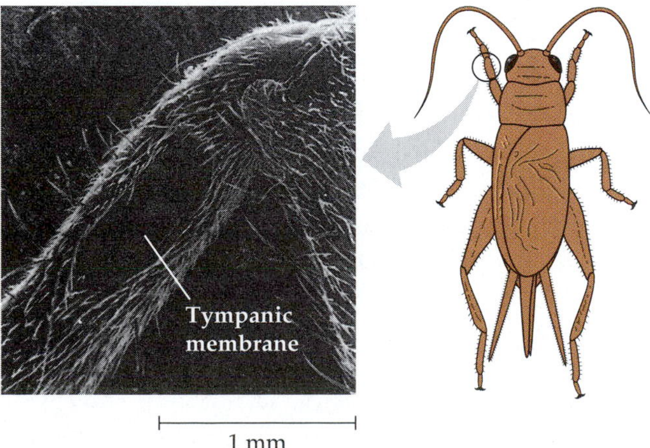

FIGURE 49.17 ▪ An insect ear. The tympanic membrane, this one on the front leg of a cricket, vibrates in response to sound waves (SEM). The vibrations stimulate mechanoreceptor cells attached to the inside of the tympanic membrane.

CHEMORECEPTION—TASTE AND SMELL

Many animals use their chemical senses to find mates, to recognize territory that has been marked with some chemical substance, and to help navigate during migration (FIGURE 49.18). Chemical "conversation" is especially important for

FIGURE 49.18 ▪ Salmon follow their noses home. Various species of Pacific salmon make a one-time round trip from small streams where the fish hatch, to the sea via rivers such as the Columbia, then back to the stream of their origin to spawn. While in the sea, salmon from many river systems school and feed together in the Gulf of Alaska. Sexually mature salmon segregate into groups of common geographic origin and migrate back toward the river from which they emerged as juveniles. During this first stage of the return, they may navigate by the position of the sun. But once a salmon reaches the general region of the river leading to its home stream, a keen sense of smell takes over. The water that flows from each stream into the river carries a unique scent from the types of plants, soil, and other components of that stream. This scent is apparently imprinted in the memory of a young salmon before it migrates to the sea. Years later, on its return journey, a salmon follows these chemical cues at each fork in the river system. Battling white water, sometimes for hundreds of miles, the fish eventually arrives at its place of origin, spawns, and dies.

animals, such as ants and bees, that live in large social groups. In all animals, taste (gustation) and smell (olfaction) are important in feeding behavior. For example, a hydra begins to swallow when chemoreceptors detect the compound glutathione, which is released from prey captured by the hydra's tentacles.

Perceptions of taste and smell are usually interrelated

The perceptions of taste and smell both depend on chemoreceptors that detect specific chemicals in the environment. In the case of terrestrial animals, taste is the detection of certain chemicals that are present in a solution, and smell is the detection of airborne chemicals. However, these chemical senses are usually closely related, and there really is no distinction in aquatic environments.

The taste receptors of insects are located within sensory hairs called sensillae on the feet and mouthparts. The animals use their sense of taste to select food. A tasting hair contains several chemoreceptor cells, each especially responsive to a particular class of chemical stimuli, such as sugar or salt. By integrating sensations (nerve impulses) from these different receptor cells, the insect's brain can apparently distinguish a very large number of tastes (FIGURE 49.19). Insects can also smell airborne chemicals, using olfactory sensillae, usually located on the antennae (see FIGURE 49.2).

In humans and other mammals, the senses of taste and smell are functionally similar and interrelated. In both cases a small molecule must dissolve in liquid to reach the receptor cell and trigger the sensation. That molecule binds to a specific protein in the receptor cell membrane, triggering a depolarization of the membrane and the release of neurotransmitter.

The receptor cells for taste are modified epithelial cells organized into **taste buds** scattered in several areas of the tongue and mouth. Most of the taste buds are on the surface of the tongue or are associated with nipplelike projections called papillae on the tongue. Although we cannot distinguish different types of taste receptors from their structures, we recognize four basic taste perceptions—sweet, sour, salty, and bitter—each detected in a distinct region of the tongue. These basic tastes are associated with specific molecular shapes or charges (the ring structure of glucose for sweetness, for instance, or the positive sodium ion for saltiness) that bind to separate receptor molecules. As with the taste receptors of insects, sensory data transmitted by sensory neurons from taste buds to the mammalian brain represent the differential stimulation of the various classes of receptors. Although each receptor cell is more responsive to a particular type of substance, it can actually be stimulated by a broad range of chemicals. With each taste of food or sip of drink, the brain integrates the differential input from the taste buds, and a complex flavor is perceived.

The olfactory sense of mammals detects certain airborne chemicals. Olfactory receptor cells are neurons that line the

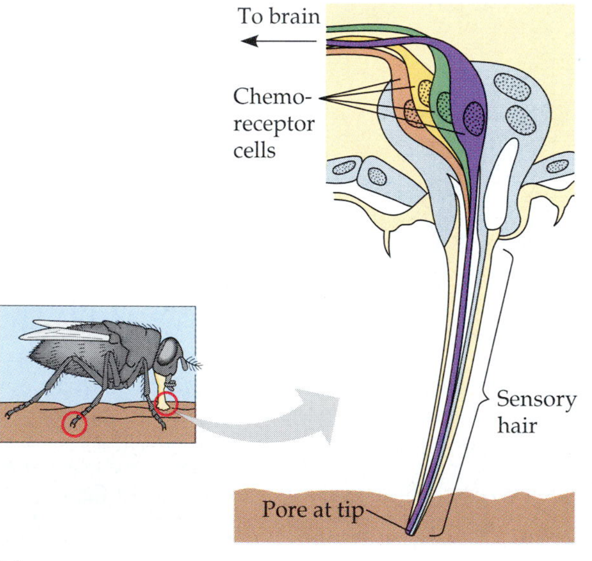

(a)

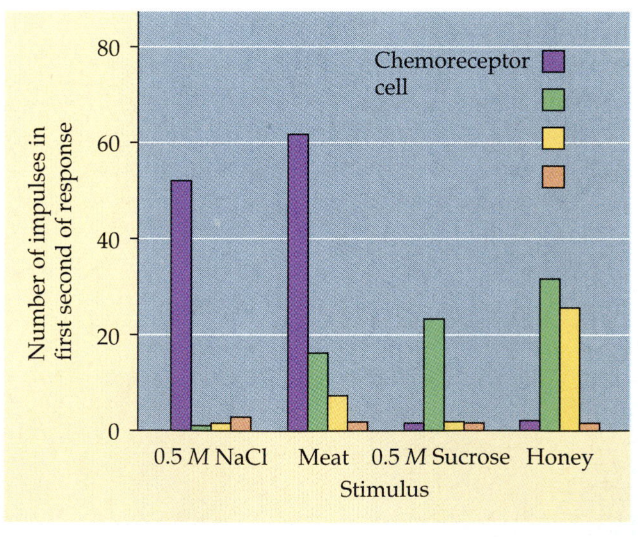

(b)

FIGURE 49.19 • The mechanism of taste in a blowfly. (a) Gustatory sensillae (hairs) on the feet and mouthparts each contain four chemoreceptor cells with dendrites that extend to the pore at the tip of the sensory hair. **(b)** Each chemoreceptor (taste) cell is especially sensitive to a particular class of substance; for example, the receptor colored green here is most responsive to sugars. But this specificity is relative; each cell can respond to some extent to a broad range of chemical stimuli. Thus, any natural food probably stimulates two or more of the receptor cells. The brain apparently integrates the frequencies of impulses arriving along the axons of the four classes of receptor cells and distinguishes a great variety of tastes.

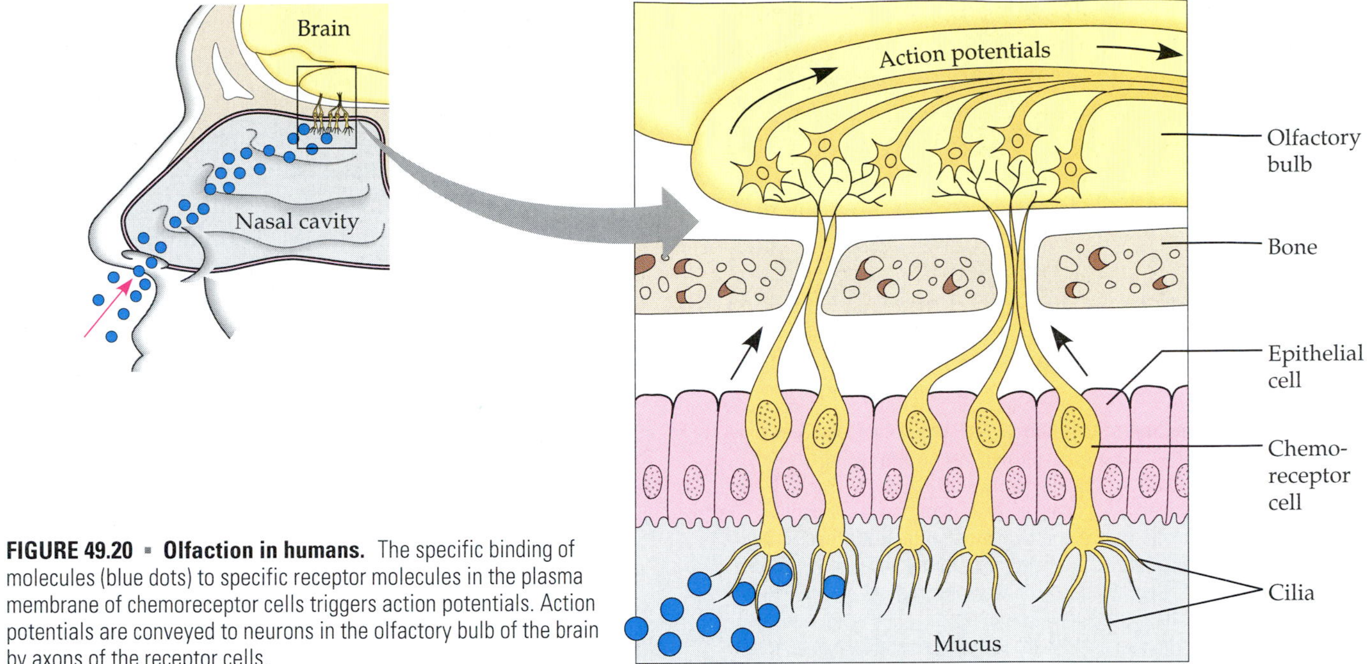

FIGURE 49.20 ■ **Olfaction in humans.** The specific binding of molecules (blue dots) to specific receptor molecules in the plasma membrane of chemoreceptor cells triggers action potentials. Action potentials are conveyed to neurons in the olfactory bulb of the brain by axons of the receptor cells.

upper portion of the nasal cavity and send impulses along their axons directly to the olfactory bulb of the brain (FIGURE 49.20). The receptive ends of the cells contain cilia that extend into the layer of mucus coating the nasal cavity. When an odorous substance diffuses into this region, it binds to specific receptor molecules on the plasma membrane of the olfactory cilia. The binding triggers a signal-transduction pathway involving a G protein relay and, in many cases, the effector enzyme adenylyl cyclase and the second messenger cyclic AMP (see FIGURE 11.12). The second messenger opens Na^+ channels in the olfactory receptor cell membrane, depolarizing it and generating action potentials that go to the brain. Humans can distinguish thousands of different odors, but these are probably based on a few primary odors, analogous to the basic tastes of the gustatory system.

Although the receptors and brain pathways for taste and olfaction are independent, the two senses do interact. Indeed, much of what we call taste is really smell. If the olfactory system is blocked, as by a head cold, the perception of taste is sharply reduced.

■ ■ ■

Throughout our discussions of sensory mechanisms we have seen many examples of how sensory inputs to the nervous system result in the specific body movements that we observe as animal behavior. The swimming of planarians away from light, the escape behavior of a moth that hears bat sonar, the feeding movements of a hydra when it tastes glutathione, and the homing of a salmon that can smell its breeding stream—these are just a few cases that we have mentioned so far. The remainder of the chapter focuses on the motor mechanisms that make these animal responses possible: how animals use their muscles and skeletons to move.

MOVEMENT AND LOCOMOTION

Movement is a hallmark of animals. To catch food, an animal must either move through its environment or move the surrounding water or air past itself. Sessile animals stay put, but they wave tentacles that capture prey or use beating cilia to generate water currents that draw and trap small food particles. Most animals, however, are mobile and spend a considerable portion of their time and energy actively searching for food, as well as escaping from danger and looking for mates. **Locomotion**, or active travel from place to place, is our focus here.

Locomotion requires energy to overcome friction and gravity

The modes of animal locomotion are diverse. Most animal phyla include species that swim. On land and in the sediments on the floor of the sea and lakes, animals crawl, walk, run, or hop. Active flight (in contrast to gliding downward from a tree or elevated ground) has evolved in only a few animal groups: insects, reptiles, birds, and, among the mammals, bats. A large group of flying reptiles died out millions of years ago, leaving birds and bats as the only flying vertebrates.

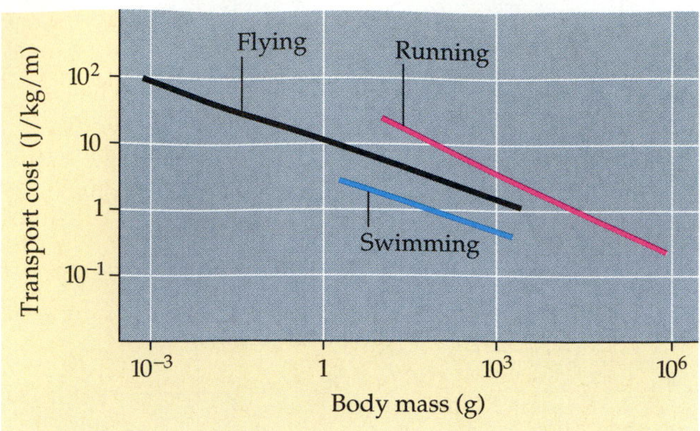

FIGURE 49.21 ▪ The cost of transport. This graph compares the transport cost, in joules per kilogram of body weight per meter traveled, for animals specialized for swimming, flying, and running (1 J = 0.24 cal). Notice that both axes are plotted on logarithmic scales. Running animals generally consume more energy per meter traveled than equivalently sized animals specialized for swimming, partly because to run (or walk), an animal must expend energy to overcome gravity. Swimming is the most efficient mode of transport (assuming, of course, that an animal is specialized for swimming). If we were to compare energy consumption per minute rather than per meter, we would find that flying animals use more energy than animals swimming or walking for the same amount of time. Each line on the graph also shows that a larger animal travels more efficiently than a smaller species specialized for the same mode of transport. For example, a horse consumes less energy per kilogram of body weight than a cat running the same distance. (Of course, total energy consumption is greater for the larger animal.)

In all its forms, locomotion requires that an animal expend energy to overcome two forces that tend to keep it stationary: friction and gravity. Exerting force requires energy-consuming cellular work. Thus, the study of locomotion returns us to the theme of animal bioenergetics. The energetic cost of transport is different for the various modes of locomotion in different environments (FIGURE 49.21).

Swimming

Because most animals are reasonably buoyant in water, overcoming gravity is less of a problem for swimming animals than for species that move on land or through the air. On the other hand, water is a much denser medium than air, and thus the problem of resistance (friction) is a major one for aquatic animals. A sleek, fusiform (torpedolike) shape is a common adaptation of fast swimmers, and swimming tends to be the most energy-efficient means of locomotion.

Animals swim in diverse ways. For instance, many insects and four-legged vertebrates use their legs as oars to push against the water. Squids, scallops, and some cnidarians are jet-propelled, taking in water and squirting it out in bursts. Fishes swim by moving their body and tail from side to side. Whales and other aquatic mammals move by undulating their body and tail up and down.

Locomotion on Land

In general, the problems of locomotion on land are the opposite of those in water. On land, a walking, running, hopping, or crawling animal must be able to support itself and move against gravity, but, at least at moderate speeds, air poses relatively little resistance. When a land animal walks, runs, or hops, its leg muscles expend energy both to propel it and to keep it from falling down. With each step its leg muscles must also overcome inertia by accelerating a leg from a standing start. For moving on land, powerful muscles and strong skeletal support are more important than a streamlined shape.

Traveling mainly by hopping, kangaroos have large muscles that generate a lot of power in their hind legs (FIGURE 49.22). When a kangaroo lands, tendons in its hind legs momentarily store energy. The higher the animal hops, the more energy the tendons store. Analogous to the tension of a spring on a pogo stick, the stored energy is available for the next jump and is a cost-free energy boost that reduces the total amount of energy the animal must expend to travel, as Terry Dawson pointed out in the interview on pp. 776–777. The pogo stick analogy applies to many land animals; the legs of an insect, a dog, or a human, for instance, retain some spring when walking or running, although considerably less than those of a hopping kangaroo.

Maintaining balance is another prerequisite for walking, running, or hopping. A kangaroo's large tail helps balance its body during leaps and also forms a stable tripod with its hind legs when sitting or moving slowly. Illustrating the same principle, a walking cat, dog, or horse keeps three feet on the ground. Bipedal animals, such as humans and birds, keep part of at least one foot on the ground when walking. When running, all four feet (or both feet for bipeds) may be off the ground momentarily, but at running speeds momentum more than foot contact keeps the body upright.

Crawling poses a very different situation. Because much of its body is in contact with the ground, a crawling animal must exert considerable effort to overcome friction. Earthworms crawl by peristalsis, a type of locomotion dependent on a hydrostatic skeleton, which we examine later. Many snakes crawl by undulating the entire body from side to side. Assisted by large, movable scales on the underside, a snake's body pushes against the ground, driving the animal forward. Boa constrictors and pythons creep straight forward, driven by muscles that lift their belly scales off the ground, tilt the scales forward, and then push them backward against the ground.

Flying

Gravity poses a major problem for a flying animal. For an animal to become airborne, its wings must develop enough lift to overcome the downward force of gravity. The key to flight is the shape of the wings. All types of wings, including those of

FIGURE 49.22 ▪ Energy-efficient locomotion on land. Members of the kangaroo family travel from place to place mainly by leaping forward on their large hind legs. Kinetic energy momentarily stored in tendons after each leap provides a cost-free boost for the next leap. The large tail helps balance the animal's body during leaps and while sitting on the ground.

airplanes, are airfoils—structures whose shape alters air currents in a way that creates lift (see FIGURE 34.23).

Cellular and Skeletal Underpinnings of Locomotion

Underlying the diverse forms of locomotion are fundamental mechanisms common to all animals. At the cellular level, all animal movement is based on one of two basic contractile systems, both of which consume energy to move protein strands against one another. These two systems of cell motility—microtubules and microfilaments—were discussed in Chapter 7. Microtubules are responsible for the beating of cilia and the undulations of flagella. Microfilaments play a major role in amoeboid movement, and they are also the contractile elements of muscle cells. It is the contraction of muscles that concerns us in this chapter, but the work of a muscle in itself cannot translate into movement of the animal. Swimming, crawling, running, hopping, and flying all result from muscles working against some type of skeleton.

Skeletons support and protect the animal body and are essential to movement

The three functions of a skeleton are support, protection, and movement. Most land animals would sag from their own weight if they had no skeleton to support them. Even an animal living in water would be a formless mass with no framework to maintain its shape. Many animals have hard skeletons that protect soft tissues. For example, the vertebrate skull protects the brain, and the ribs form a cage around the heart, lungs, and other internal organs. And skeletons aid in movement by giving muscles something firm to work against. There are three main types of skeletons: hydrostatic skeletons, exoskeletons, and endoskeletons.

Hydrostatic Skeletons

A **hydrostatic skeleton** consists of fluid held under pressure in a closed body compartment. This is the main type of skeleton in most cnidarians, flatworms, nematodes, and annelids (see Chapter 33). These animals control their form and movement by using muscles to change the shape of the fluid-filled compartments. Among the cnidarians, for example, a hydra can elongate by closing its mouth and using contractile cells in the body wall to constrict the central gastrovascular cavity. Because water cannot be compressed very much, decreasing the diameter of the cavity forces it to increase in length. In flatworms (planarians), the interstitial fluid is kept under pressure and functions as the main hydrostatic skeleton. Planarian movement results mainly from muscles in the body wall exerting localized forces against the hydrostatic skeleton. Roundworms (nematodes) hold the fluid in the body cavity (a pseudocoelom; see Chapter 32) at a high pressure, and contractions of longitudinal muscles result in thrashing movements. In earthworms and other annelids, the coelomic fluid functions as a hydrostatic skeleton. The coelomic cavity is divided by septa between the segments of the worm, and thus the animal can change the shape of each segment individually, using both circular and longitudinal muscles. The hydrostatic skeleton enables earthworms and most other annelids to move by **peristalsis**, a type of locomotion produced by rhythmic waves of muscle contractions passing from head to tail (FIGURE 49.23).

Hydrostatic skeletons are well suited for life in aquatic environments. They may cushion internal organs from shocks and provide support for crawling and burrowing. However, a hydrostatic skeleton cannot support the forms of terrestrial locomotion in which an animal's body is held off the ground, such as walking or running.

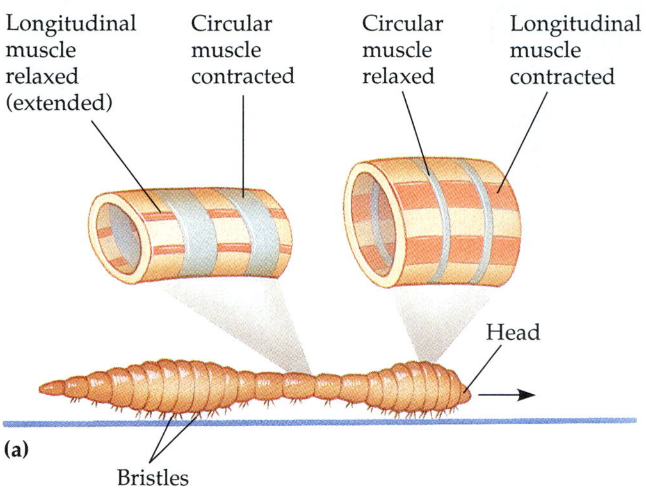

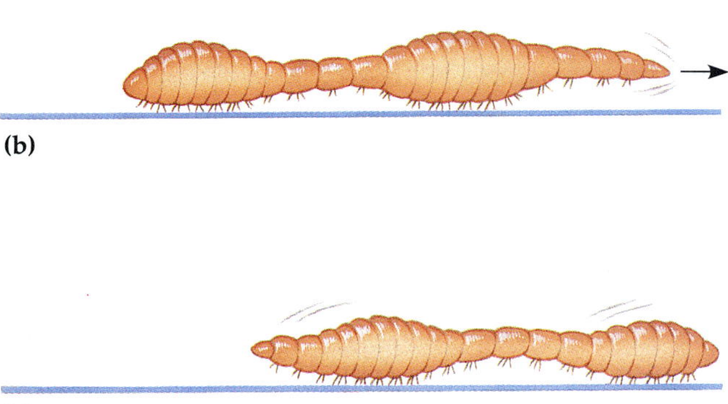

FIGURE 49.23 ▪ Peristaltic locomotion in an earthworm. A hydrostatic skeleton, two sets of muscles (one elongating the body, the other shortening it), and bristles holding to the substrate enable an earthworm to crawl over moist ground or burrow through it. Contraction of longitudinal muscles thickens and shortens the worm, while contraction of circular muscles constricts and elongates it. **(a)** In this example, as the worm crawls forward, body segments at its head and in front of the tail are short and thick (longitudinal muscles contracted; circular muscles relaxed) and anchored to the ground by bristles. Behind the head and at the tail, segments are thin and elongated (circular muscles contracted; longitudinals relaxed). **(b)** The head has moved forward because circular muscles in the head segments have contracted. Segments behind the head and in front of the tail are now thick and anchored, thus preventing the worm from slipping backward. **(c)** The head segments are thick again and anchored in their new position. The rear segments have released their hold on the ground and have been pulled forward.

Exoskeletons

An **exoskeleton** is a hard encasement deposited on the surface of an animal. For example, most mollusks are enclosed in calcareous (calcium carbonate) shells secreted by the mantle, a sheetlike extension of the body wall. As the animal grows, it enlarges the diameter of the shell by adding to its outer edge. Clams and other bivalves close their hinged shells using muscles attached to the inside of this exoskeleton.

The jointed exoskeleton typical of arthropods is a **cuticle**, a nonliving coat secreted by the epidermis. Muscles are attached to knobs and plates of the cuticle that extend into the interior of the body. About 30% to 50% of the cuticle consists of **chitin**, a polysaccharide similar to cellulose. Fibrils of chitin are embedded in a matrix made of protein, forming a composite material that combines strength and flexibility. Where protection is most important, the cuticle is hardened with organic compounds that cross-link the proteins of the exoskeleton. Some crustaceans, such as lobsters, harden portions of their exoskeletons even more by adding calcium salts. In contrast, at the joints of the legs, where the cuticle must be thin and flexible, there is only a small amount of inorganic salts and little cross-linking of proteins. The exoskeleton of an arthropod must periodically be shed (molted) and replaced by a larger case with each spurt of growth by the animal (see FIGURE 5.9).

Endoskeletons

An **endoskeleton** consists of hard supporting elements, such as bones, buried within the soft tissues of an animal. Sponges are reinforced by hard spicules consisting of inorganic material or by softer fibers made of protein (see FIGURE 33.2). Echinoderms have an endoskeleton of hard plates beneath the skin. These ossicles are composed of magnesium carbonate and calcium carbonate crystals, and the separate plates are usually bound together by protein fibers. Sea urchins have a skeleton of tightly bound ossicles, but the ossicles of sea stars are more loosely bound, allowing the animal to change the shape of its arms (see FIGURE 33.37).

Chordates have endoskeletons consisting of cartilage, bone,
49.1 or some combination of these materials (see Chapter 40). The mammalian skeleton is built from more than 200 bones, some fused together and others connected at joints by ligaments that allow freedom of movement (FIGURE 49.24). Anatomists divide the vertebrate frame into an axial skeleton, consisting of the skull, vertebral column (backbone), and rib cage, and an appendicular skeleton, made up of limb bones and the pectoral and pelvic girdles that anchor the appendages to the axial skeleton. In each appendage, several types of joints provide flexibility for body movement and locomotion.

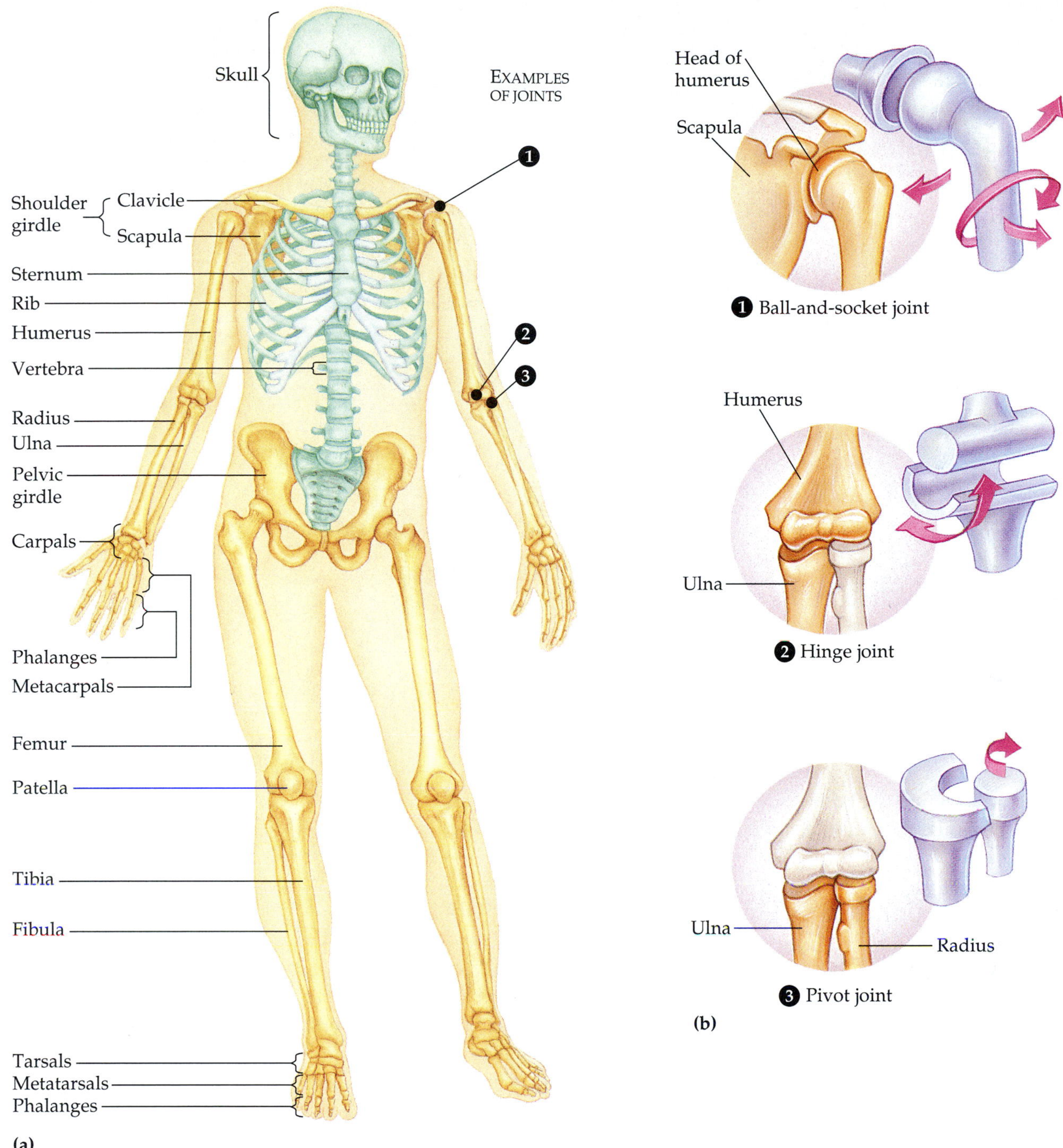

FIGURE 49.24 ▪ The human skeleton. (a) The axial skeleton (green) provides an axis of support for the upright (bipedal) body and surrounds and protects the brain, spinal cord, lungs, and heart. The appendicular skeleton (gold) supports the arms and legs. Evolved from an ancient model, the human skeleton has many features in common with other vertebrate skeletons. For example, most land vertebrates, though quadrupedal, have the same long bones supporting their arms and legs. Circled numbers indicate locations of different types of joints, described in part b. **(b)** Joints allow great flexibility in body movement, indicated by arrows. ① Ball-and-socket joints, where the humerus contacts the shoulder girdle and where the femur contacts the pelvic girdle, enable us to rotate our arms and legs and move them in several planes. ② Hinge joints, such as between the humerus and the head of the ulna, restrict movement to a single plane. ③ A pivot joint allows us to rotate the forearm at the elbow. Hinge and pivot joints between bones in our wrists and hands enable us to make precise manipulations.

49.1

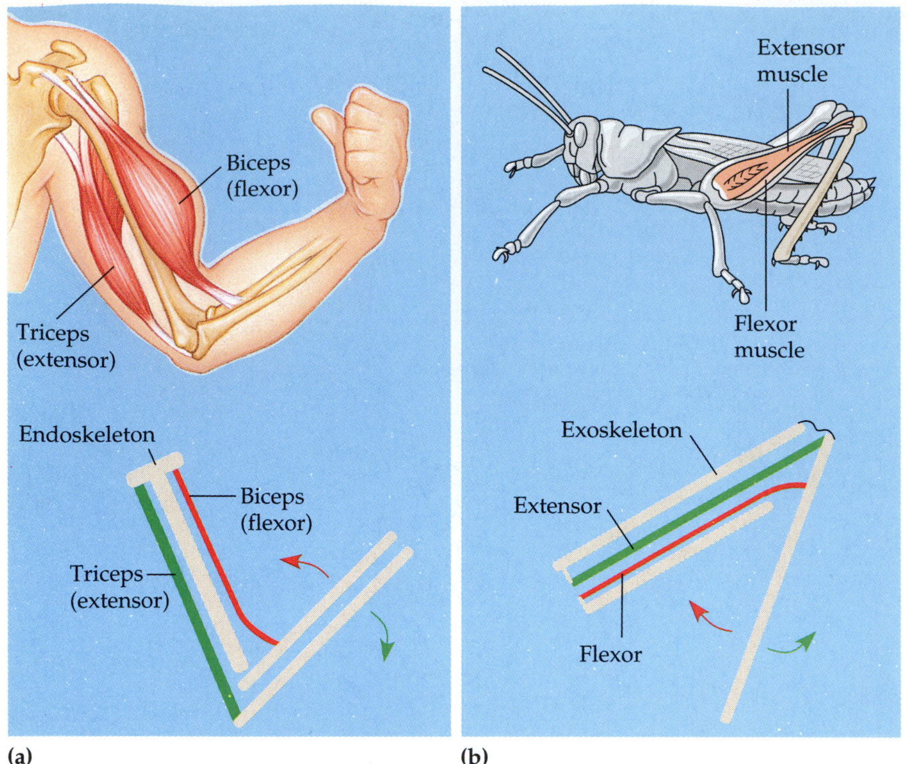

FIGURE 49.25 ▪ The cooperation of muscles and skeletons in movement. Muscles actively contract, but they elongate only when passively stretched. Back-and-forth movement is generally accomplished by antagonistic muscles, each working against the other. This arrangement works with either an endoskeleton or an exoskeleton. **(a)** In humans, contraction of the biceps muscle, represented by red in the bottom diagram, raises (flexes) the forearm. Contraction of the triceps muscle (green) lowers (extends) the forearm. **(b)** Although arthropod muscles are positioned differently and housed within an exoskeleton, the antagonistic action of flexors and extensors is similar to that of a vertebrate. When the flexor muscle (red) in the upper part of a grasshopper's leg contracts, the lower leg is pulled toward the body. In this position the grasshopper is sitting, poised for a jump, as shown here. Alternatively, when the extensor muscle (green) in its upper leg contracts, the leg jerks backward, sending the insect into the air.

Muscles move skeletal parts by contracting

As we mentioned earlier, animal movement is based on the contraction of muscles working against some type of skeleton. The action of a muscle is *always* to contract; muscles can extend only passively.

The ability to move parts of the body in opposite directions requires that muscles be attached to the skeleton in antagonistic pairs, each muscle working against the other (FIGURE 49.25). We flex our arm, for instance, by contracting the biceps, with the hinged joint of the elbow acting as the fulcrum of a lever. To extend the arm, we relax the biceps while the triceps on the opposite side contracts. But how does a muscle actually contract? As always, the key to function is structure. In this section we will examine the structure and mechanism of contraction of vertebrate skeletal muscle and then compare this basic pattern with other types of muscle.

Structure and Function of Vertebrate Skeletal Muscle

49.2 Vertebrate **skeletal muscle**, which is attached to the bones and is responsible for their movement, is characterized by a hierarchy of smaller and smaller parallel units (FIGURE 49.26). A skeletal muscle consists of a bundle of long fibers running the length of the muscle. Each fiber is a single cell with many nuclei, reflecting its formation by the fusion of many embryonic cells. Each fiber is itself a bundle of smaller **myofibrils** arranged longitudinally. The myofibrils, in turn, are composed of two kinds of **myofilaments**. **Thin filaments** consist of two strands of actin and one strand of regulatory protein coiled around one another, while **thick filaments** are staggered arrays of myosin molecules.

Skeletal muscle is also called striated muscle because the regular arrangement of the myofilaments creates a repeating pattern of light and dark bands. Each repeating unit is a **sarcomere**, the basic functional unit of the muscle. The borders of the sarcomere, the **Z lines**, are lined up in adjacent myofibrils and contribute to the striations visible with a light microscope. The thin filaments are attached to the Z lines and project toward the center of the sarcomere, while the thick filaments are centered in the sarcomere. At rest, the thick and thin filaments do not overlap completely, and the area near the edge of the sarcomere where there are only thin filaments is called the **I band.** The **A band** is the broad region that corresponds to the length of the thick filaments. The thin filaments do not extend completely across the sarcomere, so the **H zone** in the center of the A band contains only thick filaments. This arrangement of thick and thin filaments is the key to how the sarcomere, and hence the whole muscle, contracts.

Interactions between myosin and actin underlie muscle contractions

49.3 When a muscle contracts, the length of each sarcomere is reduced; that is, the distance from one Z line to the next becomes shorter. In the contracted sarcomere, the A bands do not change in length, but the I bands shorten and the H zone

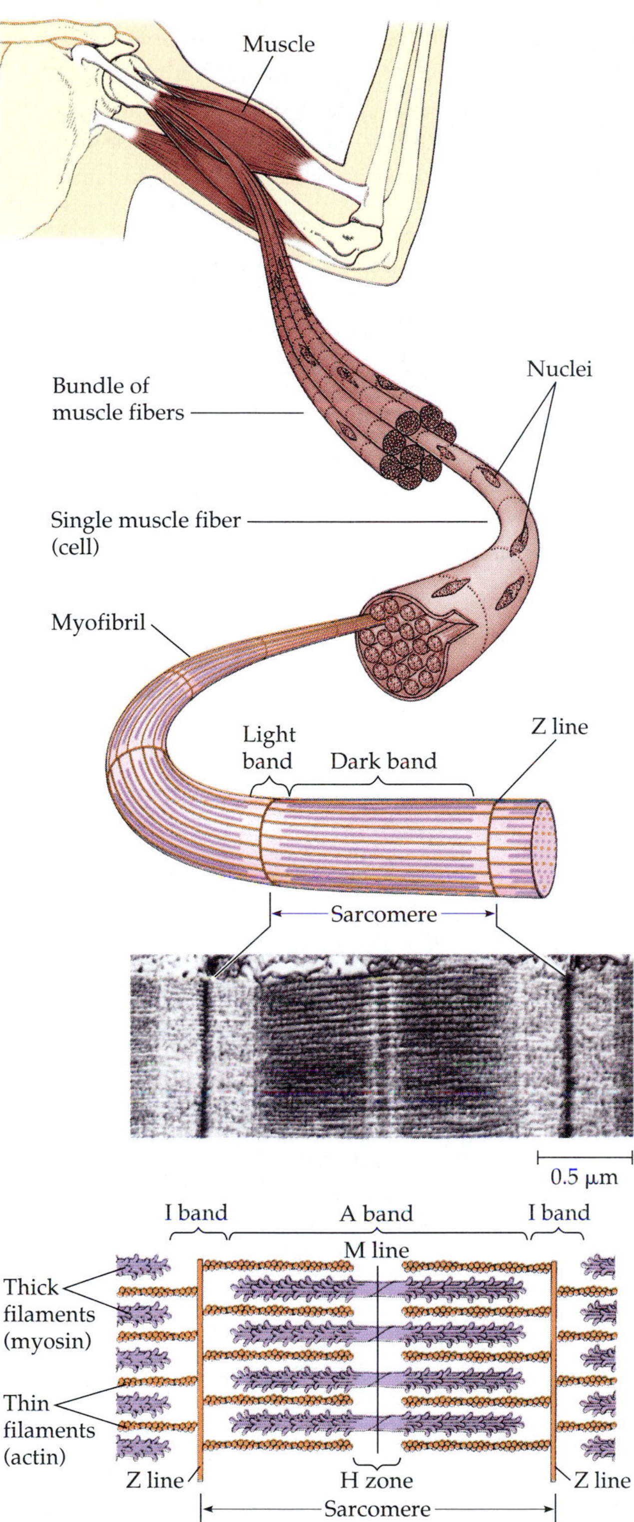

FIGURE 49.26 ▪ The structure of skeletal muscle. A muscle consists of bundles of multinucleated muscle fibers (cells), each of which is a bundle of myofibrils. Each myofibril is made of thick and thin filaments aligned in contractile units called sarcomeres. The arrangement of thick and thin filaments appears as alternating light and dark bands when striated muscle is viewed with a microscope, as in the TEM here. As the bottom diagram indicates, only thin filaments occur in the I bands, and the dark zone in the center of each I band, called the Z line, is attached to the thin filaments. The sarcomere is the entire apparatus between two Z lines. The A band includes regions where thick and thin filaments overlap and a central H zone containing only thick filaments. Connections among the thick filaments form the thin M line within the H zone.

49.2

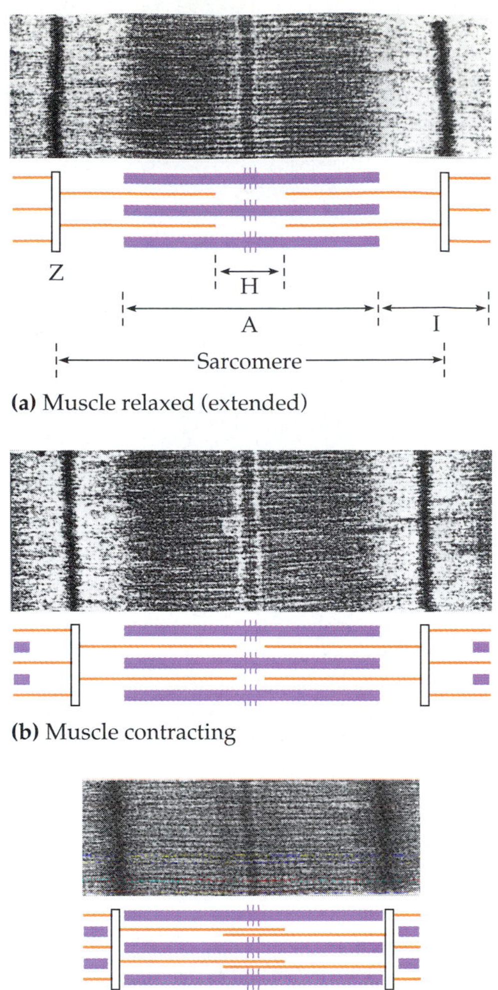

(a) Muscle relaxed (extended)

(b) Muscle contracting

(c) Muscle contracted

FIGURE 49.27 ▪ The sliding-filament model of muscle contraction. As seen in these transmission electron micrographs, the lengths of the thick (myosin) filaments (purple) and thin (actin) filaments (orange) remain the same as contraction occurs. **(a)** In a relaxed muscle, the length of each sarcomere is greater than in a contracting or contracted muscle. **(b)** During contraction, thick and thin filaments slide past each other, shortening the sarcomere. **(c)** When the muscle is fully contracted, the sarcomere is markedly shortened; the thin filaments overlap, and there is little or no space between the ends of the thick filaments and the Z lines.

49.3

disappears (FIGURE 49.27). This behavior can be explained by the **sliding-filament model** of muscle contraction. According to this model, neither the thin filaments nor the thick filaments change in length when the muscle contracts; rather, the filaments slide past each other longitudinally, so that the degree of overlap of the thin and thick filaments increases. If the region of overlap increases, both the length occupied only by thin filaments (the I band) and the length occupied only by thick filaments (the H zone) must decrease.

The sliding of the filaments is based on the interaction of the actin and myosin molecules that make up the thin and thick filaments. Myosin consists of a long, fibrous "tail" region, with a globular "head" region sticking off to the side. The tail is where the individual myosin molecules cohere to form the thick filament. The myosin head is the center of bioenergetic reactions that power muscle contractions. It can bind ATP and hydrolyze it into ADP and inorganic phosphate. Some of the energy released by cleaving the ATP is transferred to the myosin, which changes shape to a high-energy configuration (FIGURE 49.28). This energized myosin binds to a specific site on actin, forming a **cross-bridge**. The stored energy is released, and the myosin head relaxes to its low-energy configuration. This relaxation changes the angle of attachment of the myosin head to the myosin tail; as the

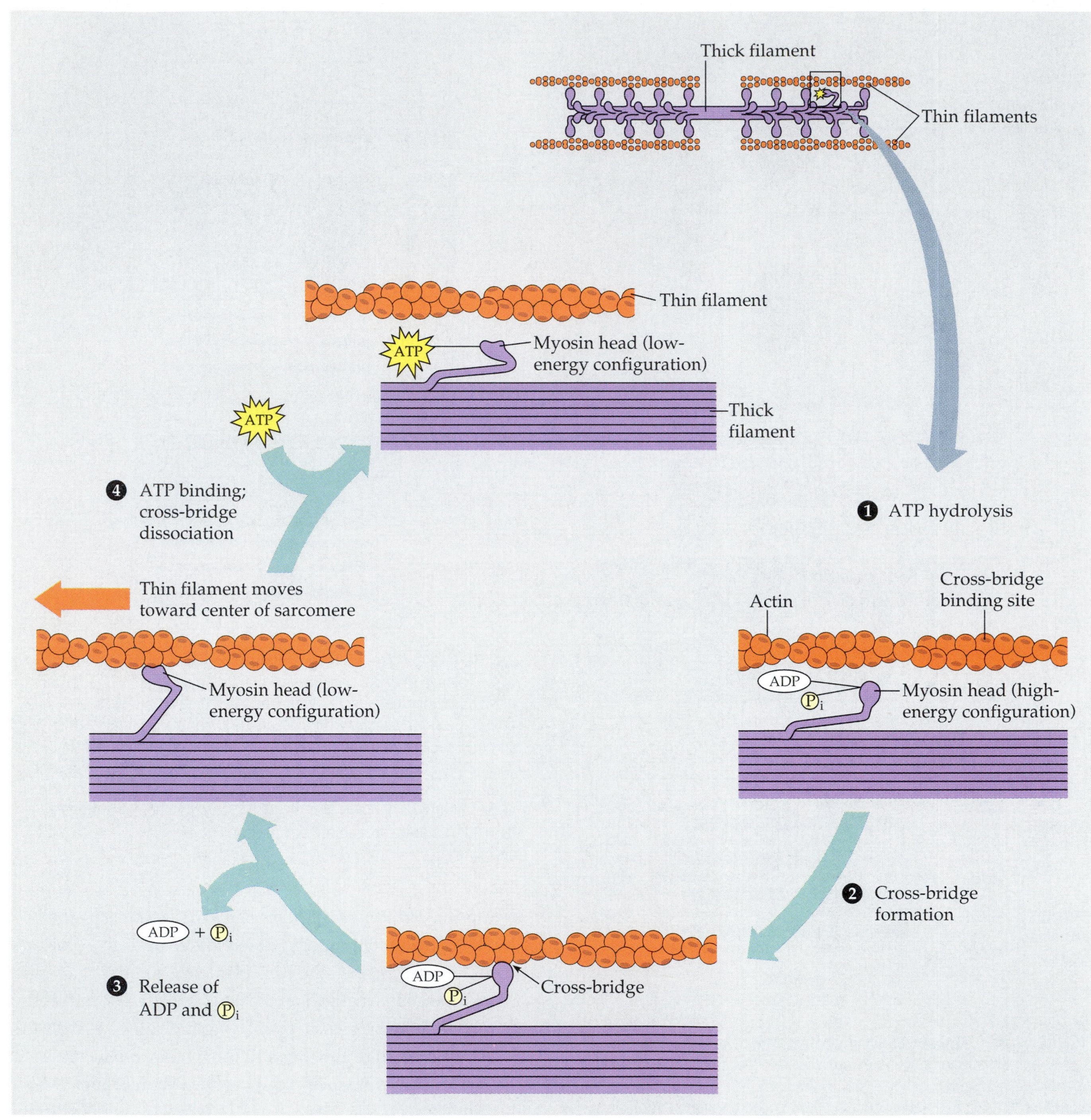

FIGURE 49.28 ▪ The cyclic interaction between myosin and actin in muscle contraction.
49.3 Starting at the top, the myosin head is bound to ATP and is in its low-energy configuration. ① The myosin head hydrolyzes ATP to ADP and inorganic phosphate (Ⓟ$_i$) and is in its high-energy configuration. ② The myosin head binds to actin, forming a cross-bridge. ③ Releasing ADP and Ⓟ$_i$, myosin relaxes to its low-energy state, sliding the thin filament. ④ Binding of a new molecule of ATP releases the myosin head. When the myosin head hydrolyzes the ATP molecule, it returns to the high-energy configuration and begins a new cycle.

myosin bends inward on itself, it exerts tension on the thin filament to which it is bound, pulling the thin filament toward the center of the sarcomere. The bond between the low-energy myosin and actin is broken when a new molecule of ATP binds to the myosin head. In a repeating cycle, the free head can then cleave the new ATP to revert to the high-energy configuration and attach to a new binding site on another actin molecule farther along the thin filament. Each of the approximately 350 heads of a thick filament forms and reforms about five cross-bridges per second, driving filaments past each other.

A muscle cell typically stores only enough ATP for a few contractions. Muscle cells also store glycogen between the myofibrils, but most of the energy needed for repetitive muscle contraction is stored in substances called **phosphagens**. **Creatine phosphate**, the phosphagen of vertebrates, can supply a phosphate group to ADP to make ATP.

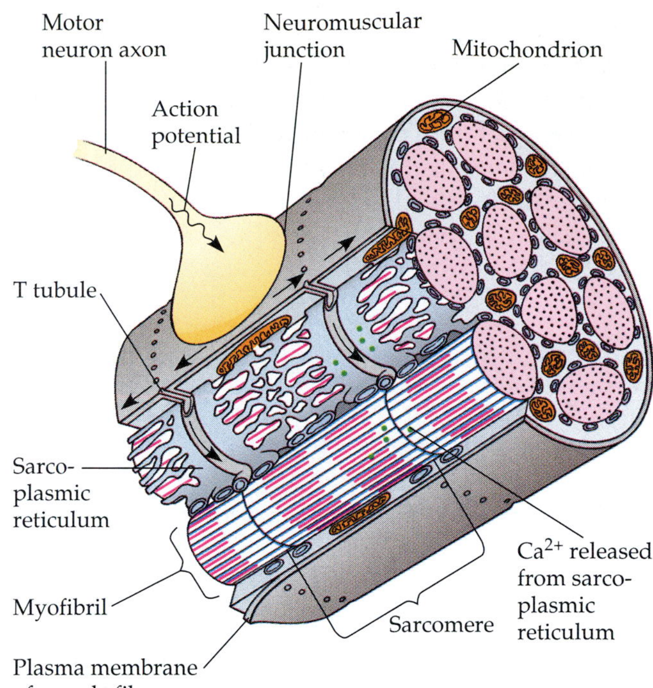

FIGURE 49.30 ▪ The roles of the muscle fiber's sarcoplasmic reticulum and T tubules in contraction. In response to an action potential arriving at the synaptic terminal of a motor neuron, the neuron membrane releases the neurotransmitter acetylcholine. Diffusing across the neuromuscular junction, acetylcholine depolarizes the plasma membrane of the muscle fiber, and action potentials (black arrows) sweep across the fiber and deep into it along T (transverse) tubules. Within the muscle cell the action potentials trigger the release of Ca^{2+} (green dots) from the sarcoplasmic reticulum into the cytoplasm. The Ca^{2+} initiates the sliding of filaments by triggering the binding of myosin to actin.

Calcium ions and regulatory proteins control muscle contraction

A skeletal muscle contracts only when stimulated by a motor neuron. When the muscle is at rest, the myosin binding sites on the actin molecules are blocked by the regulatory protein **tropomyosin**. Another set of regulatory proteins, the **troponin complex**, controls the position of tropomyosin on the thin filament (FIGURE 49.29). For a muscle cell to contract, the myosin binding sites on the actin must be uncovered. This occurs when calcium ions bind to troponin, altering the interaction between troponin and tropomyosin. The Ca^{2+} binding causes the whole tropomyosin-troponin complex to change shape and expose the myosin binding sites on actin. When calcium is present, the sliding of thin and thick filaments can occur, and the muscle contracts. When internal calcium concentration falls, the binding sites of actin are covered, and contraction stops.

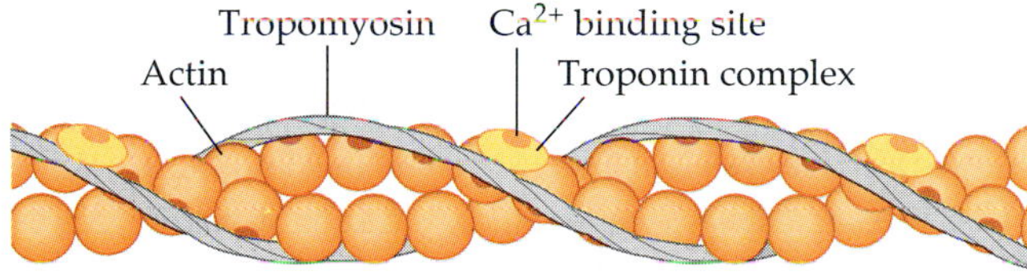

(a) Myosin binding sites blocked; muscle cannot contract

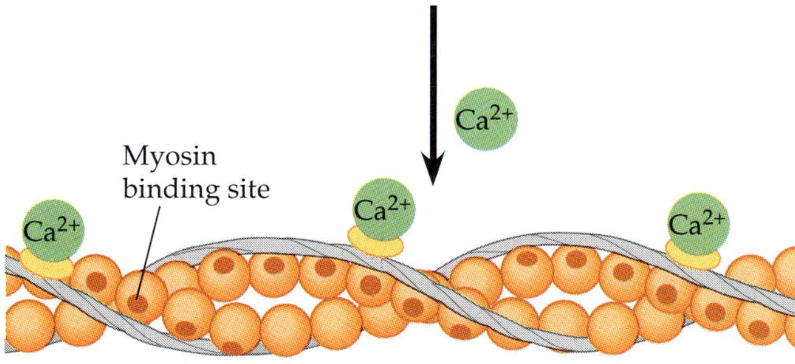

(b) Myosin binding sites exposed; muscle can contract

FIGURE 49.29 ▪ The control of muscle contraction. The thin filament has two strands of actin twisted into a helix. **(a)** When the muscle is at rest, a long, rodlike tropomyosin molecule blocks the myosin binding sites that are instrumental in forming cross-bridges. **(b)** When another protein complex, troponin, binds calcium ions, the binding sites on actin are exposed, cross-bridges with myosin can form, and the muscle contracts.

Calcium concentration in the cytoplasm of the muscle cell is regulated by the **sarcoplasmic reticulum**, a specialized endoplasmic reticulum (FIGURE 49.30). The membrane of the sarcoplasmic reticulum actively transports calcium from the cytoplasm into the interior of the reticulum, which is thus an intracellular storehouse for calcium. The stimulus leading to the contraction of a skeletal muscle cell is an action potential in a motor neuron that makes a synaptic connection with the muscle cell. The synaptic terminal of the motor neuron releases the neurotransmitter acetylcholine at the neuromuscular junction, depolarizing the postsynaptic muscle cell and triggering an action potential in the muscle cell. That action potential is the signal for contraction. The action potential spreads deep into the interior of the muscle cell along infoldings of the plasma membrane called **T (transverse) tubules**. Where the transverse tubules contact the sarcoplasmic reticulum, the action potential changes the permeability of the sarcoplasmic reticulum, causing it to release calcium ions. These calcium ions bind to troponin, allowing the muscle to contract. Muscle contraction stops when the sarcoplasmic reticulum pumps the calcium back out of the cytoplasm, and the tropomyosin-troponin complex again blocks the myosin binding sites as the concentration of calcium falls.

Diverse body movements require variation in muscle activity

Everyday experience suggests that the action of a whole muscle, such as the biceps, is graded; we can voluntarily alter the extent and strength of contraction. Experimental studies also confirm that whole-muscle contractions are graded. However, at the cellular level, any stimulus that depolarizes the plasma membrane of a single muscle fiber triggers an all-or-none contraction, analogous to the response of neurons to depolarizing stimuli. How does the nervous system produce graded contractions of whole muscles? One way is by varying the frequency of action potentials in the motor neurons controlling the muscle. A single action potential will produce an increase in muscle tension lasting about 100 msec or less, a single twitch (FIGURE 49.31). If a second action potential arrives before the response to the first is over, the tension will sum and produce a greater response. If a muscle receives an overlapping series of action potentials, further summation will occur, with the level of tension depending on the rate of stimulation. And if the rate of stimulation is fast enough, the twitches will blur into one smooth and sustained contraction called **tetanus** (not to be confused with the disease of the same name). Motor neurons usually deliver their action potentials in rapid-fire volleys, and the resulting summation of tension results in smooth contraction typical of tetanus rather than the jerky actions of individual muscle twitches.

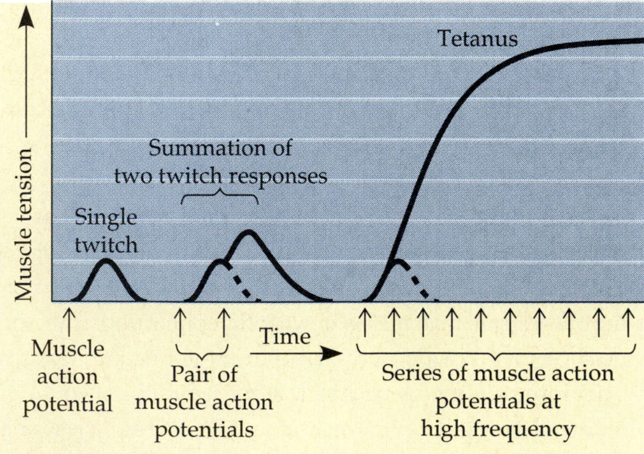

FIGURE 49.31 • Temporal summation of muscle cell contractions. This graph compares the tension developed in a muscle in response to a single action potential, a pair of action potentials, and a series of action potentials. The dashed lines show the response that would have resulted if only the first action potential had occurred.

The nervous system can also produce graded contraction of a whole muscle by taking advantage of the organization of the muscle cells into motor units. In a vertebrate muscle, each muscle cell is innervated by only one motor neuron, but each branched motor neuron may make synaptic connections with many muscle cells (FIGURE 49.32). There may be hundreds of motor neurons controlling an individual muscle, each with its own pool of muscle fibers scattered throughout the muscle. A **motor unit** consists of a single motor neuron and all the muscle fibers it controls. When the motor neuron fires, all the muscle fibers in the motor unit contract as a group. The strength of the resulting contraction will therefore depend on how many muscle fibers the motor neuron controls. In most muscles there is wide variation in the number of muscle fibers among motor units; some motor neurons may control only a few muscle cells, while others may control hundreds. The nervous system can thus regulate the strength of contraction in the whole muscle by both determining how many motor units are activated at a given instant and selecting whether large or small motor units are activated. Tension in a muscle can be progressively increased by activating more and more of the motor neurons controlling the muscle, a process called **recruitment** of motor neurons. Depending on the number and size of motor neurons your brain recruits to the task, you can lift a fork, or something much heavier, like your biology textbook.

Some muscles, especially those that hold the body up and maintain posture, are almost always partially contracted. However, prolonged contraction results in muscle fatigue caused by the depletion of ATP, dissipation of the ion gradients required for normal electrical signaling, and the accumulation of lactate (see Chapter 9). In a mechanism that avoids fatigue in postural muscles, the nervous system alternates activation among the various motor units making up the muscle, so that different motor units take turns maintaining the prolonged contraction.

Fast and Slow Muscle Fibers

We have seen that the action potential in a skeletal muscle fiber is only a trigger for the contraction; the actual duration of the contraction is controlled by how long the calcium concentration in the cytoplasm remains elevated. Not all skeletal muscle fibers are identical in this regard. We can identify fast and slow fibers based on the duration of their twitches. **Fast muscle fibers** are used for rapid, powerful contractions. Some, such as the flight muscles of birds, may be able to sustain long periods of repeated contractions without fatiguing. By contrast, **slow muscle fibers**, which can sustain long contractions, are often found in muscles that maintain posture. A

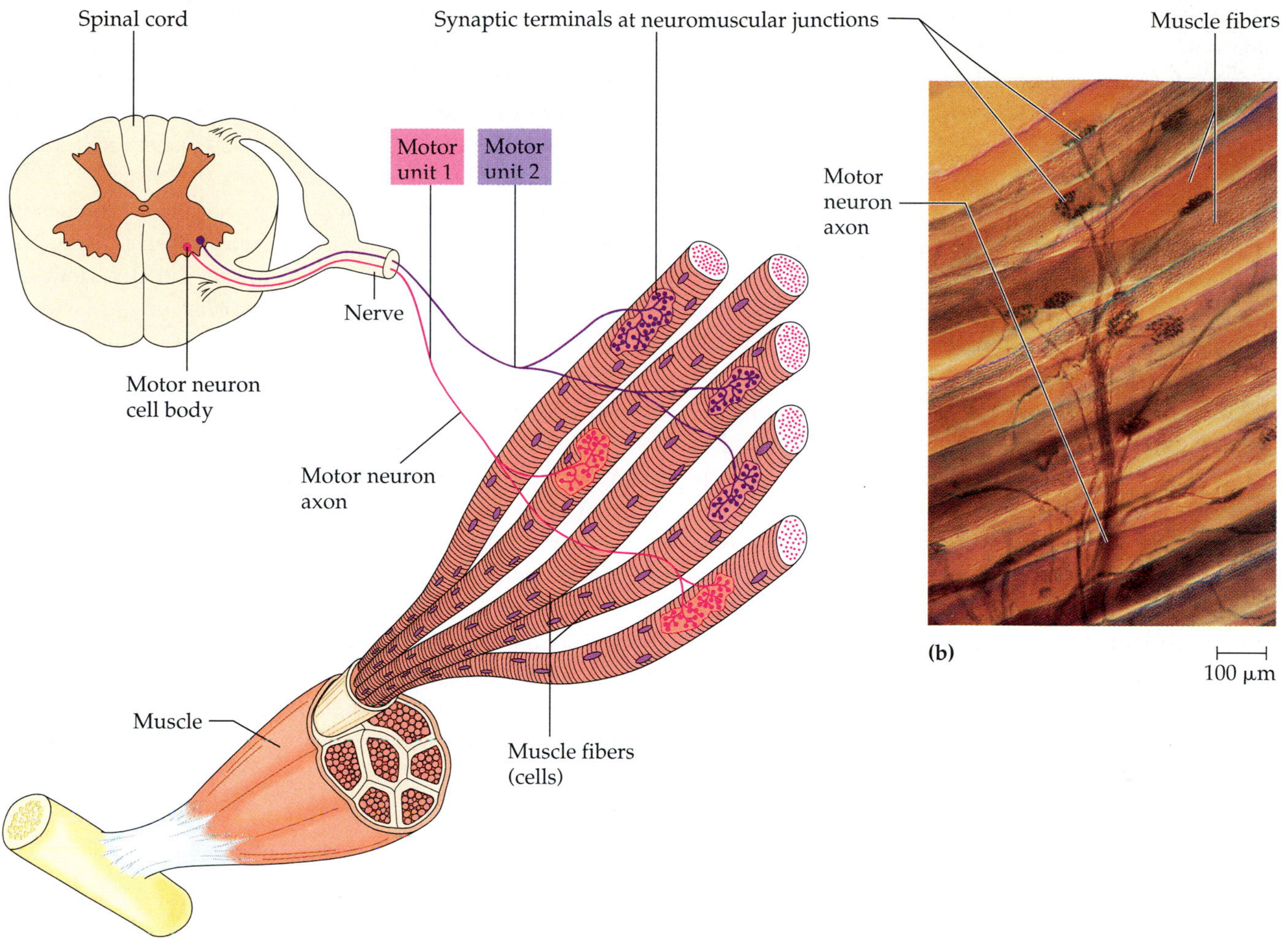

FIGURE 49.32 ▪ Motor units in a vertebrate muscle. (a) Each muscle fiber (cell) has a single neuromuscular junction, or synaptic connection, with the motor neuron that controls it. However, each motor neuron typically branches and controls several or many muscle fibers. A motor neuron and all the fibers it controls constitute a contractile apparatus called a motor unit. **(b)** You can see several synaptic terminals of a branched motor neuron in this light micrograph.

slow fiber has less sarcoplasmic reticulum than a fast fiber, so calcium remains in the cytoplasm longer. This causes a twitch in a slow fiber to last about five times longer than in a fast fiber. Slow fibers are also specialized to make use of a steady supply of energy; they have many mitochondria, a rich blood supply, and an oxygen-storing protein called myoglobin. Myoglobin, the brownish-red pigment in the dark meat of poultry and fish, binds oxygen more tightly than hemoglobin, so it can effectively extract oxygen from the blood.

Other Types of Muscle

There are many types of muscles in the animal kingdom, but as noted before, they all share the same fundamental mechanism of contraction: the sliding of actin filaments and myosin filaments past one another. In addition to skeletal muscle, vertebrates have cardiac muscle and smooth muscle (see FIGURE 40.5).

Vertebrate **cardiac muscle** is found in only one place—the heart. Like skeletal muscle, cardiac muscle is striated. The primary differences between skeletal and cardiac muscle are in their electrical and membrane properties. The junctions between cardiac muscle cells contain specialized regions called **intercalated discs**, where gap junctions provide direct electrical coupling among cells. Thus, an action potential generated in one part of the heart will spread to all the cardiac muscle cells, and the whole heart will contract. Skeletal muscle cells will not contract unless triggered to do so by input from a controlling motor neuron. Cardiac muscle cells, however, can generate action potentials on their own, without any input from the nervous system. The plasma membrane of a cardiac muscle cell has pacemaker properties that cause rhythmic depolarizations, triggering action potentials and causing single cardiac muscle cells to "beat" even when isolated from the heart and placed in cell culture. (But at the whole-organ level, the heart also has a pacemaker, specialized muscle tissue in the wall of the right atrium that coordinates contractions of cardiac muscle cells throughout the heart; see FIGURE 42.7.) The action potentials of cardiac muscle cells are also different from those of skeletal muscle cells, lasting up to twenty times longer. Whereas the action potential of a skeletal muscle cell serves only as a trigger for contraction and does not control the duration of contraction, in a cardiac cell the duration of the action potential plays an important role in controlling the duration of contraction.

Smooth muscle lacks the striations of skeletal and cardiac muscle because the actin and myosin filaments are not all regularly arrayed along the length of the cell. Instead, the filaments may have a spiral arrangement within smooth muscle cells. Smooth muscle also contains less myosin than striated muscle, and the myosin is not associated with specific actin strands. Because of its organization, smooth muscle cannot generate nearly as much tension as striated muscle, but it can contract over a much greater range of lengths. Further, smooth muscle has neither a T tubule system nor a well-developed sarcoplasmic reticulum. Calcium ions must enter the cytoplasm via the plasma membrane during an action potential, and the amount reaching the filaments is rather small. Contractions are relatively slow, but there is a greater range of control than in striated muscle. Smooth muscle is found mainly in the walls of hollow organs such as digestive tract organs and blood vessels. Smooth muscle propels substances through the hollow organ by alternately contracting and relaxing.

Invertebrates possess muscle cells similar to vertebrate skeletal and smooth muscle cells. Arthropod skeletal muscles are nearly identical to vertebrate skeletal muscles. However, the flight muscles of insects are capable of independent, rhythmic contraction, so the wings of some insects can actually beat faster than action potentials can arrive from the central nervous system. Another interesting evolutionary adaptation has been discovered in the muscles that hold clam shells closed. The thick filaments of these muscle fibers contain a unique protein called paramyosin that enables the muscles to remain in a fixed state of contraction with a low rate of energy consumption for as long as a month.

Although we have talked about sensory receptors and muscles separately in this chapter, they are the two ends of a single integrated system. An animal's behavior, so fundamental to how the animal interacts with its environment, is the product of a nervous system connecting sensations to responses. Behavior is discussed in Unit Eight, within the broader context of ecology: the study of interactions between organisms and their environment.

CHAPTER REVIEW

REVIEW OF KEY CONCEPTS

(with page numbers and key figures)

INTRODUCTION TO SENSORY RECEPTION

Animal behavior depends on the ability of the nervous system to integrate the input from sensory receptors about the internal and external environments and translate the information into appropriate responses in muscles and other effectors. Sensations are action potentials traveling along sensory neurons that are interpreted by different parts of the brain as perceptions.

- **Sensory receptors transduce stimulus energy and transmit signals to the nervous system (p. 993)** Sensory receptors are usually modified neurons or epithelial cells that detect environmental stimuli and respond with an electrochemical change in their membrane. Exteroreceptors detect external stimuli; interoreceptors detect internal stimuli. Sensory transduction is the conversion of stimulus energy into a change in a membrane potential called a receptor potential. Stimulus energy may be amplified by accessory structures of sense organs or by transduction. Transmission of the receptor potential to the nervous system occurs either as an action potential (when the receptor is a sensory neuron) or as the release of neurotransmitter, which then initiates action potentials in the sensory neuron with which the receptor cell synapses. Integration of information begins at the receptor level by such processes as sensory adaptation and variation in the sensitivity of the receptor. Integration continues in the CNS.
- **Sensory receptors are categorized by the type of energy they transduce (pp. 993–996, FIGURES 49.1–49.3)** Mechanoreceptors respond to stimuli such as pressure, touch, stretch, motion, and sound. Pain is detected by nociceptors, a group of diverse receptors that respond to excess temperature, pressure, or specific classes of chemicals. Various types of thermoreceptors signal surface and core temperatures of the body. Chemoreceptors respond either to generalized solute concentrations or to specific molecules. Electromagnetic receptors detect energy in the form of different wavelengths of radiation.

PHOTORECEPTORS

- **A broad array of photoreceptors has evolved among invertebrates (pp. 996–997, FIGURES 49.4, 49.5)** The light receptors and visual capabilities of invertebrates vary widely, ranging from the simple light-sensitive eye cup of planarians to the image-forming compound eye of insects and crustaceans and the single-lens eye of some jellies, spiders, and many mollusks.
- **Vertebrates have single-lens eyes (pp. 997–998, FIGURES 49.6, 49.7)** The main parts of the vertebrate eye are an outer layer, the sclera, including the transparent cornea; the conjunctiva, a mucous membrane, surrounding all of the sclera except the cornea; the choroid, a pigmented middle layer, which includes the iris, surrounding the pupil; the retina, an inner layer at the back of the eyeball, containing the photoreceptor cells; and the lens, suspended between two chambers in the eye, which focuses light on the retina.

- **The light-absorbing pigment rhodopsin operates via signal transduction (pp. 998–1001, FIGURES 49.8, 49.9)** Transduction of the light signal occurs in specialized photoreceptors called rods and cones, which contain light-absorbing retinal bonded to specific membrane proteins, collectively called opsins. When rods and cones absorb light, signal-transduction pathways hyperpolarize their membrane, and they release less neurotransmitter.
- **The retina assists the cerebral cortex in processing visual information (pp. 1001–1002, FIGURES 49.10, 49.11)** Chemical messages are transmitted from rods and cones to bipolar cells and then to ganglion cells, whose axons in the optic nerve convey action potentials to the brain. Other neurons in the retina integrate information before it is sent to the brain. Most axons of the optic nerves go to the lateral geniculate nuclei of the thalamus, from which neurons lead to the primary visual cortex. Several integrating centers in the cerebral cortex are active in creating visual perceptions.

HEARING AND EQUILIBRIUM

In mammals and most terrestrial vertebrates, the hair cell mechanoreceptors for hearing and balance are closely associated within fluid-filled canals in the ear.

- **The mammalian hearing organ is within the inner ear (pp. 1002–1005, FIGURES 49.12, 49.13)** The outer ear consists of the external pinna and auditory canal. The tympanic membrane (eardrum) transmits sound waves to three small bones of the middle ear, which amplify and transmit the waves through the oval window to the fluid in the coiled cochlea of the inner ear. Pressure waves vibrate the basilar membrane and the attached organ of Corti, which contains receptor hair cells. The bending of the hairs against the tectorial membrane depolarizes the hair cells, triggering action potentials in the auditory nerve to the brain. Volume (loudness) is a function of the amplitude of the sound wave; increased volume results in greater bending of the hair cells. Pitch is related to frequency of sound waves. Regions of the basilar membrane vibrate more vigorously at different frequencies and transmit to specific auditory areas of the cerebral cortex.
- **The inner ear also contains the organs of equilibrium (p. 1005, FIGURE 49.14)** The utricle, saccule, and three semicircular canals in the inner ear function in balance and equilibrium.
- **A lateral line system and inner ear detect pressure waves in most fishes and aquatic amphibians (pp. 1005–1006, FIGURE 49.15)** The detection of water movement in fishes and aquatic amphibians is accomplished by a lateral line system of clustered hair cell receptors.
- **Many invertebrates have gravity sensors and are sound-sensitive (pp. 1006–1007, FIGURES 49.16, 49.17)** Many arthropods have vibrating exoskeletal hairs and localized "ears," consisting of a tympanic membrane and receptor cells, to sense sounds. Some invertebrates detect position in space by means of statocysts.

CHEMORECEPTION—TASTE AND SMELL

- **Perceptions of taste and smell are usually interrelated (pp. 1008–1009, FIGURES 49.19, 49.20)** Taste and smell both depend on the stimulation of receptor cells by small, dissolved molecules that bind to proteins on a chemoreceptor membrane. Many insects have sensillae, sensory hairs containing taste receptor neurons, on their feet and mouthparts. In mammals, taste receptors are organized into taste buds that respond to distinct shapes of molecules. Olfactory receptor cells line the upper part of the nasal cavity and send their axons to the olfactory bulb of the brain.

MOVEMENT AND LOCOMOTION

Locomotion is active travel from place to place.

- **Locomotion requires energy to overcome friction and gravity (pp. 1009–1011, FIGURE 49.21)** Aquatic, aerial, and terrestrial environments pose different problems for locomotion. Overcoming friction is a major problem for swimmers; gravity is less of a problem for swimming animals than for those that move on land or fly. Walking, running, hopping, or crawling on land requires an animal to support itself and move against gravity. Flight requires that wings develop enough lift to overcome the downward force of gravity. Animal movement uses energy and is a function of protein strands moving past each other, either in the microtubules of cilia and flagella or by microfilaments in amoeboid movement and muscle contraction.
- 49.1 **Skeletons support and protect the animal body and are essential to movement (pp. 1011–1013, FIGURES 49.23, 49.24)** A hydrostatic skeleton, found in most cnidarians, flatworms, nematodes, and annelids, consists of fluid under pressure in a closed body compartment. Hydrostatic skeletons support peristaltic locomotion, produced by rhythmic waves of muscle contractions passing from the head to the tail of many worms. Exoskeletons, found in most mollusks and arthropods, are hard coverings deposited on the surface of an animal. Endoskeletons, found in sponges, echinoderms, and chordates, are hard supporting elements embedded within the animal's body.
- 49.2 **Muscles move skeletal parts by contracting (p. 1014, FIGURES 49.25, 49.26)** Muscles, often present in antagonistic pairs, contract and pull against the skeleton to provide movement. Vertebrate skeletal muscle consists of a bundle of muscle cells, each of which contains myofibrils composed of thin filaments of actin and thick filaments of myosin.
- 49.3 **Interactions between myosin and actin underlie muscle contractions (pp. 1014–1017, FIGURES 49.27, 49.28)** The energy to move myosin heads is provided by ATP. Energized myosin heads bind to actin, forming cross-bridges. Bending of the myosin heads pulls the thin filaments (actin) toward the center of the sarcomere, producing a muscle contraction. When ATP binds to the myosin heads they release, ready to start a new cycle.
- **Calcium ions and regulatory proteins control muscle contraction (pp. 1017–1018, FIGURES 49.29, 49.30)** Contraction begins when impulses from a motor neuron are transmitted to the muscle cell membrane through release of acetylcholine at the neuromuscular junction. Action potentials travel to the interior of the cell along the T tubules, stimulating the release of calcium ions from the sarcoplasmic reticulum. The calcium ions bind to the regulatory troponin-tropomyosin complex on the thin filaments, exposing the myosin binding sites on the actin; muscle contraction proceeds.
- **Diverse body movements require variation in muscle activity (pp. 1018–1020, FIGURES 49.31, 49.32)** A muscle twitch results from a single stimulus. More rapidly delivered signals produce a graded contraction by summation. Tetanus is a state of smooth and sustained contraction obtained when motor neurons deliver a volley of action potentials. A motor unit consists of a branched motor neuron and the muscle fibers it innervates. Multiple motor unit recruitment results in stronger contractions. Cardiac muscle, found only in the heart, consists of striated, branching cells that are electrically connected by intercalated discs. Cardiac muscle cells can generate action potentials without neural input. In smooth muscle, contractions are slow but can be sustained over long periods of time.

SELF-QUIZ

1. Which of the following receptors is *incorrectly* paired with its category?
   a. hair cell—mechanoreceptor
   b. muscle spindle—mechanoreceptor
   c. taste receptor—chemoreceptor
   d. rod—electromagnetic receptor
   e. gustatory receptor—electromagnetic receptor

2. Some sharks close their eyes just before they bite. Although they cannot see their prey, their bites are on target. Researchers have noted that sharks often misdirect their bites at metal objects, and they can find batteries buried under the sand of an aquarium. This evidence suggests that sharks keep track of their prey during the split second before they bite in the same way that
   a. a rattlesnake finds a mouse in its burrow
   b. a male silkworm moth locates a mate
   c. a bat can find moths in the dark
   d. a platypus locates its prey in a muddy river
   e. a migrating bird finds its way on a cloudy night

3. Which of the following is an *incorrect* statement about the vertebrate eye?
   a. The vitreous humor regulates the amount of light entering the pupil.
   b. The transparent cornea is an extension of the sclera.
   c. The fovea is the center of the visual field and contains only cones.
   d. The ciliary muscle functions in accommodation.
   e. The retina lies just inside the choroid and contains the photoreceptor cells.

4. The transduction of sound waves to action potentials takes place
   a. within the tectorial membrane as it is stimulated by the hair cells
   b. when hair cells are bent against the tectorial membrane, causing them to depolarize and release neurotransmitter molecules that stimulate sensory neurons
   c. as the basilar membrane becomes more permeable to sodium ions and depolarizes, initiating an action potential in a sensory neuron
   d. as the basilar membrane vibrates at different frequencies in response to the varying volume of sounds
   e. within the middle ear as the vibrations are amplified by the malleus, incus, and stapes

5. The utricle and saccule are
   a. involved in lateral integration of visual signals
   b. visual centers in the cortex
   c. part of the apparatus responsible for positional information about the head
   d. semicircular canals that respond to head rotation
   e. organs of balance found in insects and crustaceans

6. When you first walk from a brightly lit area into darkness, which of the following occurs?
   a. The photopsins in your cones become bleached.
   b. Your rod cells become hyperpolarized.
   c. The receptor cells release less neurotransmitter.
   d. Lateral inhibition caused by your horizontal cells ceases.
   e. Your rhodopsin is still dissociated into retinal and opsin, and your rods are temporarily nonfunctional.

7. The role of calcium in muscle contraction is
   a. to break the cross-bridges as a cofactor in the hydrolysis of ATP
   b. to bind with troponin, changing its shape so that the myosin binding sites on the actin filament are exposed
   c. to transmit the action potential across the neuromuscular junction
   d. to spread the action potential through the T tubules
   e. to reestablish the polarization of the plasma membrane following an action potential

8. Tetanus refers to
   a. the partial sustained contraction of major supporting muscles
   b. the all-or-none contraction of a single muscle fiber
   c. a stronger contraction resulting from multiple motor unit summation
   d. the result of wave summation, which produces a smooth and sustained contraction of a muscle
   e. the state of muscle fatigue caused by the depletion of ATP and the accumulation of lactate

9. Which of the following is a *true* statement about cardiac muscle cells?
   a. They lack an orderly arrangement of actin and myosin filaments.
   b. They have less extensive sarcoplasmic reticulum and thus contract more slowly than smooth muscle cells.
   c. They are connected by intercalated discs, through which action potentials spread to all cells in the heart.
   d. They have a resting potential more positive than an action potential threshold.
   e. They contract only when stimulated by neurons.

10. Which of the following changes occurs when a skeletal muscle contracts?
   a. The A bands shorten.
   b. The I bands shorten.
   c. The Z lines slide farther apart.
   d. The thin filaments contract.
   e. The thick filaments contract.

CHALLENGE QUESTIONS

1. Compare vision and hearing with respect to receptor type, transduction, amplification, transmission, and perception.
2. Explain the difference between sensation and perception. Give some examples.
3. Although skeletal muscles generally fatigue fairly rapidly, clam shell muscles have a unique protein called paramyosin that allows them to sustain contraction for up to a month. From your knowledge of the cellular mechanism of contraction, propose a hypothesis to explain how paramyosin might work. How would you test your hypothesis experimentally?

SCIENCE, TECHNOLOGY, AND SOCIETY

1. Have you ever felt your ears ringing after listening to loud music from a sound system or at a concert? The sound often tops 90 decibels—loud enough to permanently impair hearing. Do you think people are aware of the possible danger from prolonged exposure to loud music? What, if anything, should be done to warn or protect them?
2. You may know an older person who has broken a bone (often a hip) partly because of osteoporosis, a loss of bone density that affects many women after menopause. Researchers think that prevention is the best way to avoid osteoporosis. They recommend exercise and maximum calcium intake during the teenage years and the twenties. Is it realistic to expect young people to view themselves as future senior citizens? How would you recommend that they be encouraged to develop good health habits that might not pay off for 40 or 50 years?

FURTHER READING

Alexander, R. M. *The Human Machine.* New York: Columbia University Press, 1992. Presents a lucid, well-illustrated account of the biomechanical principles underlying human body movement.

Bennet-Clark, H.C. "How Cicadas Make Their Noise." *Scientific American,* May 1998. Describes the external and internal mechanisms that produce loud, varied sounds of male cicadas.

Deyo, R.A. "Low-Back Pain." *Scientific American,* August 1998. Describes the major problems and compares the effectiveness of treatments.

Freeman, D. H. "In the Realm of the Chemical." *Discover,* June 1993. An essay on the senses of taste and smell and their evolutionary significance in humans.

Hamill, O. P., and D. W. McBride, Jr. "Mechanoreceptive Membrane Channels." *American Scientist,* January/February 1995. Explains research on the cellular mechanisms underlying the senses of touch and hearing: membrane channels with gates sensitive to mechanical stimulation.

Howlett, R. "Flipper's Secret." *New Scientist,* June 28, 1997. Summarizes research on how dolphins obtain information about underwater objects using sonar.

Kirschvink, J. L. "Homing in on Vertebrates." *Nature,* November 27, 1997. Provides a summary of the evidence for magnetic sense in animals.

Lee, J. "Before Your Very Eyes." *New Scientist,* March 15, 1997. Presents an up-to-date summary of progress toward understanding how the brain creates visual perceptions.

Pennisi, E. "The Architecture of Hearing." *Science,* November 14, 1997. Explains how cloning deafness genes aids understanding of ear development and function.

Rome, L. C. "Testing a Muscle's Design." *American Scientist,* July–August 1997. Summarizes research on the molecular basis of functional specificity of muscles.

Thomas, A. L. R. "On the Tails of Birds." *BioScience,* April 1997. Explores the diversity and aerodynamic functions of bird tails.

Travis, J. "Eye-Opening Gene." *Science News,* May 10, 1997. Summarizes recent research on the evolution of animal eyes.

Vogel, S. *Life's Devices.* New York: Prentice-Hall, 1991. An award-winning science writer and researcher explores the biomechanics of locomotion.

White, R.J. "Weightlessness and the Human Body." *Scientific American,* September 1998. Explores the effects of gravity-free environment on the health of astronauts.

WEB LINKS

Visit the special edition of *The Biology Place* for BIOLOGY, Fifth Edition, at **http://www.biology.com/campbell**. Go to Chapter 49 for online resources, including learning activities, practice exams, and links to the following web sites:

"Neuroscience for Kids"

This very richly illustrated site is a great primer for all aspects of neuroscience.

"Cow's Eye Dissection"

The title tells it like it is—a step-by-step guide to dissecting a cow eye.

"Society for Neuroscience"

The Society for Neuroscience is the world's largest organization of scientists and physicians dedicated to understanding the brain, spinal cord, and peripheral nervous system. The section "Briefings" contain many well-written articles on the CNS and the senses.

"The Hosford Muscle Tables"

This web site contains detailed information about the skeletal muscles of the human body.

UNIT EIGHT

ECOLOGY

Michael Dombeck brings unique experience in ecology, academic research, and government to this interview, introducing our unit on Ecology. Formerly head of the U.S. Department of the Interior's Bureau of Land Management, he is currently chief of the U.S. Forest Service, an agency responsible for managing much of the public land in the United States. He is also an adjunct professor of biology at Yale University. Dr. Dombeck received his Ph.D. in fisheries biology at Iowa State University in 1985. His scientific publications and a 1997 book on watershed restoration emphasize the importance of adopting modern ecological principles in managing public lands. Widely respected for his diplomatic skills, Dr. Dombeck helps bridge gaps between the scientists who study the structure and function of biological communities and ecosystems and the lawmakers who may directly affect how ecosystem resources are managed. Larry Mitchell interviewed Dr. Dombeck in Washington, D.C.

Some people say you have one of the toughest jobs in Washington. How do you feel about that?

This job is about managing lands that belong to all U.S. citizens—the public's lands—and this is a time when land and the variety of things it provides people are more highly sought after than ever. We are coming out of a decade and a half when too much of our energy went into what I call red-zone controversies, where you just bring up the topic and we're ready to go at each other's throat: "It's a sin to cut a tree" versus "I want to cut them all."

Besides the red zone of controversy, there is a yellow zone, where we sort of agree on some things and disagree on others. Then there's the green zone, where there is total agreement. The more we can do to encourage people to focus on common goals, on yellow and green zones, the more we can accomplish. The common goals are there, in spite of the many different opinions. We're in a fantastic business, and this is one of the *best* jobs. I have the opportunity to talk about, and help promote, healthy forest ecosystems, clean water, plant and animal diversity, soil stability, hunting, fishing, wildlife watching, and families having open space to recreate on the land.

Tell us what led you to this office.

I grew up 25 miles from a town of 1500 in northern Wisconsin, and like a typical kid growing up in that part of the world, I did a lot of things outside. I know now that I've always been interested in natural history, but I was not sure what I wanted to do until I took some biology courses in college. In fact, *one* professor is the reason I majored in biology; professor George Becker was one of those people with the skill to get you to work at about 120% of your capacity and like it. There are always *some* people around that have that skill, but it's rare. I went through the education program in college, so I have teacher training, along with biology and general science. I also started writing a column over 20 years ago called "Natural History Notes" in a local magazine; I still write for it and am always gleaning the literature for unique little facts about science and animals that stand out and grab people.

INTERVIEW

MICHAEL DOMBECK

Did you always know what you wanted to do?

For awhile, my goal was to be a professor, but like most of us, I didn't have a specific path plotted out. After finishing two master's degrees, I got married, and about that time the Forest Service started hiring fisheries biologists. So I went to work for the Forest Service, thinking, "Well, I like the outdoors, I like the woods, I'll get to see some nice places," and that's one thing that has come true. Eventually, my master's work led to a Ph.D. project at Iowa State on the reproductive ecology of the muskellunge. I had wonderful experiences there, with the students, the faculty, and with our research on the population dynamics during spawning of muskies and northern pike. I was able to do the research as a Forest Service employee. Since then, I like to think I've been at the interface of research and management.

What are main roles of the Forest Service and the Bureau of Land Management (BLM) in the U.S. government?

I like to think of the Forest Service and the BLM, along with the National Park Service, as the people's stewards of the public domain. In the United States, the public domain was at one time about 1.8 billion acres. Some of it was the land resulting from actual purchases, such as the Louisiana Purchase, as well as from various conflicts. Some of these lands were homesteaded; others became Indian reservations, military bases, and other government installations. Other lands were given away to encourage development of the West. If you were a lieutenant colonel in the Revolutionary War, your pension would have been 1100 acres of land; if you were a private in the War of 1812, your pension would have been 80 acres. The railroads got every other section of land, 10 miles on each side of their track as they moved west; in the mountainous areas they got 20 miles on each side. Much of that doesn't belong to the railroads anymore, but much of it remained in private ownership, especially where there is water. Today, if you look at a land tenure map, you'll see a checkerboard pattern that makes absolutely no sense. The Homestead Act of 1862 granted 160 acres of federal land to virtually anyone over 21 years old who would live on it and cultivate it for five years. The U.S. Congress also granted large areas of land to western states to support schools, roads, and other public facilities. Timber resources on some of these lands are still sold to help support many schools.

About 30% of the land in the United States (about 690 million acres) of the initial 1.8 billion acres of public domain lands remains public, owned collectively by its citizens. Millions of Americans use these lands for hunting, fishing, hiking, camping, skiing, and other forms of recreation. Public lands also contain some of the best and last habitats for rare and endangered species. Four federal agencies manage most of these lands—the Bureau of Land Management, the Forest Service, the National Park Service, and the Fish and Wildlife Service. Some 191 million acres are National Forests, congressionally designated tracts of land administered and managed by the U.S. Forest Service, a branch of the Department of Agriculture.

In your book *Watershed Management: Principles and Practices*, you state that federal agencies cannot protect resources such as endangered species or timber without man-

aging them in the context of larger ecosystems. Please elaborate on this.

The BLM and the Forest Service manage large blocks of land, most of it in the western states. In many cases the Forest Service manages lands in the headwaters of watersheds, and the BLM manages much of the land downstream. These agencies are mandated by law to manage their lands to restore and sustain the biological integrity and watershed health. Doing so requires a broad understanding of the watershed as a whole, and a cooperative spirit among all those concerned and charged with managing natural resources. We have to learn to live within the limits of watersheds. An old proverb goes something like: "We haven't inherited the land from our forefathers; we have borrowed it from our children."

What exactly is ecosystem management?

By its very nature, this is not something that can be simply or precisely defined. Ecosystem management is taking a broad view of landscapes and how they are defined by natural processes. By adopting this approach to managing lands, we're moving from a reductionist point of view to a more holistic point of view, taking a "long" view in both space and time and trying to integrate sound ecological science and management principles that are geared to the long-term health of whole ecosystems. We are looking at historic (natural) patterns in ecosystems, or watersheds, and how best to restore and sustain them over the long haul. Instead of looking at single organisms, we're looking at how all the pieces of an ecosystem fit together. Let's work within the limits of the land, and make sure that one use isn't dominant over another. We need to measure our success in management in how we maintain the resilience of an ecosystem.

Ecosystems are not stable entities; they change all the time—some very slowly, others fairly rapidly—and change is part of nature. There are natural disasters and human-inflicted problems; the ability of the ecosystem to heal and continue functioning is the key.

Could you tell us a bit more about ecosystem change and the natural and human forces that can effect it?

Let's try to relate this more to someone's personal experience. If you grow up in an area with a dense forest and learn that forest fire is one of our worst enemies, then in your mind that is the natural state, and in your mind that's the way it should be. But historically this might not have been the case. One thing we know is that forested areas were not all old growth or covered with dense stands of trees. There have always been natural disasters and catastrophes, large and small—fires, earthquakes, hurricanes, tornadoes, and so on—and ecosystems are as diverse as the forces that constantly alter them. People—and government agencies—have not always appreciated the creative side of these forces.

You know, one of the greatest success stories in public education was Smokey Bear. The message was that all fire is bad. The piece we missed is that fire is also an integral part of many ecosystems. Fire can almost be viewed as a cleansing mechanism, like wind and water. If you oversuppress fire, then you get unnaturally high fuel buildup. Then, when you have an extremely dry year, you may end up with such an intense fire that the entire plant community is destroyed. Yellowstone in 1988 was a prime example of that. By being aggressive in fire suppression, we not only allowed fuels to build up, but the species composition of those areas changed. Then we had the extensive fires. Soon after, the Yellowstone ecosystem began showing its resilience, with new plant growth quickly covering the burned areas. Ecosystems have always been varied—mosaics, in a sense—and our challenge today is to maintain those mosaics.

Social debate and education are key components of all this. As we apply the concept of whole-ecosystem management, it's very important to include the socioeconomic aspects. People are an important part of ecosystems, and if you don't take the people with you in decision-making, you don't make things happen on the land. People are the support base and the delivery system for managing healthy, diverse, and productive lands.

Could you discuss the land-management policies of federal agencies in the past, and how they compare with the modern concept of ecosystem management?

When the Forest Service began in the late 1800s, it was driven by conservation ethics and almost single-tree management. This was practiced until shortly after World War II, when the soldiers came home. Then there was prosperity, and there was a tremendous increase in the demand for timber. We were part of turning America into a society of single-family homes. During the 1950s and 1960s, the "can do" culture of the Forest Service led to an increase in timber harvest from about 4 billion board feet in 1950 to 8 to10 billion board feet in the 1960s. Then it oscillated between 9 and 12 billion until the spotted owl injunctions and other controversies of the 1980s.

Starting in the 1960s, a reawakening of the conservation era began with Rachel Carson's book *Silent Spring* and signaled the start of an era of environmental legislation (from the early 1960s to about 1990)—for example, the Wilderness Act, the Clean Water Act, the Clean Air Act, the National Environmental Policy Act, and the Endangered Species Act. With all of these pieces of legislation, society has been pushing us toward the present era of ecosystem management and watershed restoration. But you know, you can't legislate nature. We need to base our decisions on science and be proactive in the political system to provide information to whomever will listen—not the other way around. For example, let's listen to the aquatic ecologists, who know that the streams themselves tell us how we should manage them; the temperature of the water should be determined by such things as groundwater seepage, snow-melt runoff, and rainfall, not politics.

With your background in life science, ecology, and public service, do you have any advice for students who are beginning a career in biology?

The best advice I have for students is that the better grounding you have in basic communication skills—writing, speaking, listening, conversing, persuasion, all those kinds of things—the more successful you will be. For example, in research you can have the best information in the world, but you still have to convince others that your results and conclusions are credible. People skills are more important than ever. In fact, a resource manager of the future is going to be less of a technical expert and more of a facilitator, a catalyst who brings people with diverse points of view together to solve problems. We have a very sophisticated society and lots of technical expertise, but the people who get the most done on the land are those with the best mix of technical and communication skills. I cannot overstate the importance of that.

CHAPTER 50

AN INTRODUCTION TO ECOLOGY AND THE BIOSPHERE

The Scope of Ecology

- Ecology is the scientific study of the interactions between organisms and their environments
- Ecological research ranges from the adaptations of organisms to the dynamics of ecosystems
- Ecology provides a scientific context for evaluating environmental issues

Abiotic Factors of the Biosphere

- Climate and other abiotic factors are important determinants of the biosphere's distribution of organisms

Aquatic and Terrestrial Biomes

- Aquatic biomes occupy the largest part of the biosphere
- The geographical distribution of terrestrial biomes is based mainly on regional variations in climate

Concepts of Organismal Ecology

- The costs and benefits of homeostasis affect an organism's responses to environmental variation
- An organism's short-term responses to environmental variations operate within a long-term evolutionary framework

Organisms as open systems that interact continuously with their environments is a theme that has already surfaced many times in this book. The scientific study of the interactions between organisms and their environments is called **ecology** *(from the Greek* oikos, *"home," and* logos, *"to study"). This straightforward definition masks an enormously complex and exciting area of biology that is also of increasingly critical practical importance. Photographs of the planet taken by Apollo astronauts, such as the one on this page, remind us that Earth is a finite home in the vastness of space, not an unlimited frontier for human activity. The science of ecology can provide the basic understanding of natural processes necessary to manage the planet's limited resources over the long term.*

We begin this unit by scanning the breadth of ecology and with a closer look at three key words in our definition: "scientific," "interactions," and "environment." This chapter also examines the main physical factors of the environment that influence organisms and the ways in which organisms are adapted to those factors. We will also examine some of the main features of Earth's ecosystems.

THE SCOPE OF ECOLOGY

By necessity, humans have always had an absorbing interest in other organisms and their environments. As hunters and gatherers, prehistoric people had to learn where game and edible plants could be found in greatest abundance. Naturalists from Aristotle to Darwin made the process of observing and describing organisms in their natural habitat an end in itself, rather than simply a means of survival. Extraordinary insight can still be gained through this descriptive approach, and thus natural history remains fundamental to ecological science.

Although ecology has a long history as a descriptive science, it is becoming increasingly experimental. In spite of the difficulty of conducting experiments that often involve large amounts of time and space, many ecologists are testing hypotheses in the laboratory and in the field. One early example is a classic 1969 study by Daniel Simberloff and E. O. Wilson that involved eliminating all the insects on small islands in the Florida Keys to study recolonization of the islands from nearby mainland populations. In the example illustrated in FIGURE 50.1, a researcher collects invertebrates in enclosures along the shoreline of a lake as part of an experimental study of the effects of heavy metals on fish behavior.

Ecology is the scientific study of the interactions between organisms and their environments

As an area of *scientific* study, ecology incorporates the hypothetico-deductive approach, using observations and experiments to test hypothetical explanations of ecological phenom-

FIGURE 50.1 ▪ Experimental field research in ecology. Over the past three decades ecology has become more of an experimental science. The plastic-lined enclosures (mesocosms) shown here allowed aquatic ecologist Mark Sandheinrich of the University of Wisconsin-LaCrosse to test the prediction that under field conditions, low levels of heavy metals may alter the feeding behavior of bluegill sunfish on invertebrate prey. Working on his Ph.D. research at Iowa State University at the time, Sandheinrich first confirmed that all of the mesocosms contained virtually the same kinds and numbers of invertebrates, bluegills, and aquatic plants. To test the validity of some preliminary results obtained in the laboratory, he then dosed some mesocosms with copper and allowed others to serve as controls.

ena. Many ecologists devise mathematical models that enable them to simulate large-scale experiments that might be impossible to conduct in the field. In this approach, important variables and their hypothetical relationships are described through mathematical equations. The potential ways in which the variables interact can then be studied. For example, many ecologists, climatologists, paleontologists, and other scientists use sophisticated computer programs to develop models that predict the effects human activities will have on climate, and how climatic changes will affect ecosystems. Of course, such simulations are only as good as the basic information on which the models are based, and obtaining that information still requires extensive fieldwork.

The *environment* includes **abiotic components** (nonliving chemical and physical factors) such as temperature, light, water, and nutrients. Just as important in their effects on organisms are **biotic** (living) **components**—all the other organisms that are part of any individual's environment. Other organisms may compete with an individual for food and other resources, prey upon it, or change its physical and chemical environment. As we will see, questions about the relative importance of various environmental components are frequently at the heart of ecological studies—and the accompanying controversies.

Finally, in this closer look at the definition of ecology, we should highlight the *interactions* among organisms and their environment. Organisms are affected by their environment (both the abiotic and biotic components), but by their very presence and activities they also change it—often dramatically. Possibly the most dramatic effect organisms have had on their environment occurred some three billion years ago, when photosynthetic bacteria first began utilizing sunlight for energy and gave off oxygen (O_2) as a by-product of photosynthesis. The aerobic atmosphere that resulted from this change had a profound effect on the entire planet. On a more localized level, trees reduce light levels on the forest floor as they grow, sometimes making the environment unsuitable for their own offspring.

Many biologists recognize Charles Darwin as one of the earliest ecologists (although he predated the word *ecology*). Indeed, it was the geographical distribution of organisms and their exquisite adaptations to specific environments that provided Darwin with evidence for evolution. An important (although not the only) cause of evolutionary change is the *interaction* of organisms with their environment. Thus, events that occur in the frame of what is sometimes called *ecological time* translate into effects over the longer scale of *evolutionary time*. For instance, hawks feeding on field mice have an impact on the gene pool of the prey population by curtailing the reproductive success of certain individuals. One long-term effect of such a predator-prey interaction may be the prevalence in the mouse population of fur coloration that camouflages the animals.

Throughout this unit on ecology we will see many more examples of how organisms and their environments interact and affect one another, both in ecological time and evolutionary time. We will also relate ecological research to concerns about the impacts of human interactions with the environment.

Ecological research ranges from the adaptations of organisms to the dynamics of ecosystems

Because there are many levels and types of interactions between organisms and their environments, the questions ecologists address are wide-ranging and complex. Ecology can be divided into four increasingly comprehensive levels of study, from the interactions of individual organisms with the abiotic environment to the dynamics of ecosystems.

Organismal ecology is concerned with the behavioral, physiological, and morphological ways in which individual organisms meet the challenges posed by their abiotic environment. The distribution of organisms is limited by the abiotic conditions they can tolerate.

The next level of organization in ecology is the **population**, a group of individuals of the same species living in a particular geographic area. Population ecology concentrates mainly on factors that affect population size and composition.

A **community** consists of all the organisms that inhabit a particular area; it is an assemblage of populations of different species. Questions at this level of analysis involve the ways in which interactions among organisms, such as predation, competition, and disease, affect community structure and organization.

Ecological study of the **ecosystem** includes all the abiotic factors in addition to the community of species that exists in a certain area. Some critical questions at the ecosystem level concern energy flow and the cycling of chemicals among the various biotic and abiotic components.

Ecologists often face extraordinary challenges in their research because of the complexity of their questions, the diversity of their subjects, and the large expanses of time and space over which studies often must be conducted. Ecology is also challenging because of its multidisciplinary nature; ecological questions form a continuum with those from other areas of biology, including genetics, evolution, physiology, and behavior, as well as those from other sciences, such as chemistry, physics, geology, and meteorology. We will see numerous examples of ecology's multidisciplinary nature, especially when we examine the science of conservation biology in the final chapter. Seeking to stem the human-caused loss of species and build a sustainable future, this relatively new science melds ecology and allied sciences with a broad range of fields outside biology, including sociology, economics, and the law.

Ecology provides a scientific context for evaluating environmental issues

The science of ecology should be distinguished from the informal use of the same word by the popular media to refer to environmental concerns. Nevertheless, we need to understand the often complicated and delicate relationships between organisms and their environments in order to address environmental problems.

Much of our current environmental awareness had its beginnings with Rachel Carson's 1962 book *Silent Spring,* which pointed out that the widespread use of pesticides such as DDT was causing population declines in many nontarget organisms. Today, acid precipitation, localized famine aggravated by land misuse and population growth, the poisoning of soil and streams with toxic wastes, and the growing list of species extinct or endangered because of habitat destruction are just a few of the problems that threaten the home we share with millions of other forms of life.

Advances in ecological research have highlighted the complex relationships among organisms and their surroundings. Although it makes ecology a dynamic and exciting science, this complexity can lead to problems. Politicians and lawyers often want specific answers to such questions as: How much forest is needed to save spotted owls? While ecological studies can certainly provide essential information for making policy decisions on habitat preservation, responses to such questions often include further questions: How many owls must be saved? With what certainty must they be saved? How long can they survive in this amount of forest?

Many influential ecologists, including Michael Dombeck, the scientist you met in the preceding interview, recognize their responsibility to educate legislators and the general public about decisions that affect the environment. An important part of this responsibility is communicating the complexity of environmental issues.

ABIOTIC FACTORS OF THE BIOSPHERE

The **biosphere** is the global ecosystem—the sum of all the planet's ecosystems, or all of life and where it lives. The most complex level in ecology, the biosphere includes the atmosphere to an altitude of several kilometers, the land down to and including water-bearing rocks at least 1500 meters belowground, lakes and streams, caves, and the oceans to a depth of several kilometers.

Climate and other abiotic factors are important determinants of the biosphere's distribution of organisms

Ecologists have long recognized striking global and regional patterns in the distribution of organisms within the biosphere (FIGURE 50.2). These patterns mainly reflect regional differences in climate and other abiotic factors in the environment. Most organisms ultimately derive their energy from sunlight, and they must tolerate the ranges of temperature, humidity, salinity, and light in their environment. In this section we will analyze how some important abiotic factors such as temperature influence the distribution of organisms. We also will consider global, regional, and seasonal variations in *climate,* an environmental feature that incorporates several abiotic factors, including temperature and rainfall. Throughout this discussion it is important to remember that the physical environment varies in both space and time. Although two regions of Earth may experience different conditions at any given moment, daily and annual fluctuations of abiotic factors sometimes blur or accentuate the distinctions between those regions.

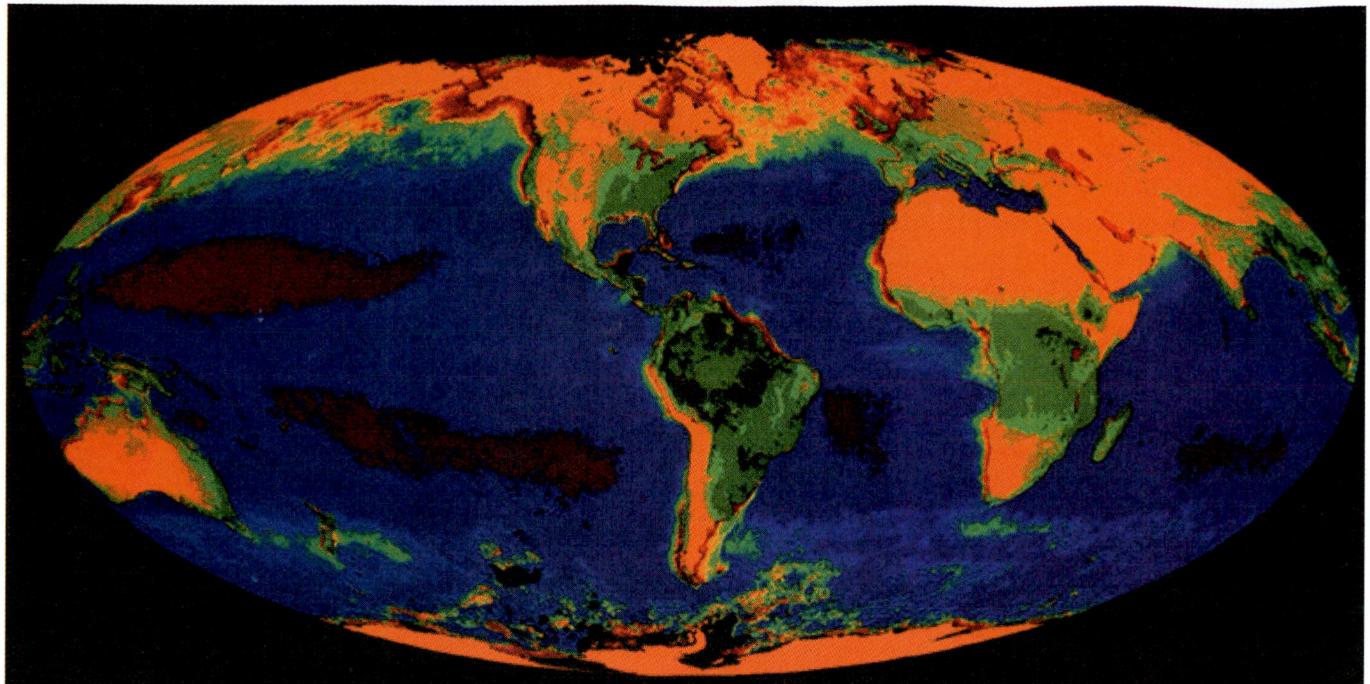

FIGURE 50.2 ▪ Patterns of distribution in the biosphere. The regional distribution of life in the biosphere mainly reflects variations in abiotic factors such as temperature and the availability of water. This global patchiness is highlighted in this image of Earth, which is based on data sent from two satellites over an eight-year period. The colors are keyed to the relative abundance of chlorophyll, an indication of the regional densities of life. Bright red areas along the coasts correspond to nutrient-rich areas where algae abound—some of the most productive oceanic regions. Other productive areas in the oceans—either relatively shallow water or places where upwelling currents bring nutrients to surface waters—are in green. Dark red areas in the oceans are the least productive waters. Green or black areas on land are dense forests, including the tropical forests of South America and Africa. Orange areas on land are relatively barren regions, such as the Sahara.

Major Abiotic Factors

Temperature. Environmental temperature is an important factor in the distribution of organisms because of its effect on biological processes and the inability of most organisms to regulate body temperature precisely. Cells may rupture if the water they contain freezes at temperatures below 0°C, and the proteins of most organisms denature at temperatures above 45°C. In addition, few organisms can maintain a sufficiently active metabolism at very low or very high temperatures. Within this range, most biochemical reactions and physiological processes occur more rapidly at higher temperature. Extraordinary adaptations enable some organisms to live outside this temperature range. The actual internal temperature of an organism is affected by heat exchange with its environment (see Chapter 44), and most organisms cannot maintain body temperatures more than a few degrees above or below the ambient temperature. As endotherms, mammals and birds are the major exceptions, but even endotherms function best within certain environmental temperature ranges that vary with the species.

Water. The unique properties of water have effects on organisms and their environments, as you learned in Chapter 3. Water is essential to life, but its availability varies dramatically among habitats. Freshwater and marine organisms live submerged in an aquatic environment, but they face problems of water balance if their intracellular osmolarity does not match that of the surrounding water. Organisms in terrestrial environments encounter a nearly constant threat of desiccation, and their evolution has been shaped by the requirements for obtaining and conserving adequate supplies of water.

Sunlight. Sunlight provides the energy that drives nearly all ecosystems, although only plants and other photosynthetic organisms use this energy source directly. Light intensity is not the most important factor limiting plant growth in many terrestrial environments, but shading by a forest canopy makes competition for light in the understory intense. In aquatic environments the intensity and quality of light limit the distribution of photosynthetic organisms. Every meter of water depth selectively absorbs about 45% of the red light and about 2% of the blue light that pass through it. As a result, most photosynthesis in aquatic environments occurs relatively near the surface. However, the photosynthetic organisms themselves absorb some of the light that penetrates, further reducing light levels in the waters below.

Light is also important to the development and behavior of the many plants and animals that are sensitive to photoperiod, the relative lengths of daytime and nighttime. Photoperiod is a more reliable indicator than temperature for cueing seasonal events, such as flowering or migration.

Wind. Wind amplifies the effects of environmental temperature on organisms by increasing heat loss due to evaporation and convection (the wind-chill factor). It also contributes to water loss in organisms by increasing the rate of evaporation in animals and transpiration in plants. In addition, wind can have a substantial effect on the growth form of plants by inhibiting the growth of limbs on the windward side of trees; limbs on the leeward side grow normally, resulting in a "flagged" appearance (see FIGURE 35.2).

Rocks and Soil. The physical structure, pH, and mineral composition of rocks and soil limit the distribution of plants and of the animals that feed upon them, thus contributing to the patchiness we see in terrestrial ecosystems. In streams and rivers the composition of the substrate can affect water chemistry, which in turn influences the resident plants and animals. In marine environments the structure of the substrates in the intertidal zone and on seafloors determines the types of organisms that can attach or burrow in those habitats.

Periodic Disturbances. Catastrophic disturbances such as fires, hurricanes, tornadoes, and volcanic eruptions can devastate biological communities. After the disturbance the area is recolonized by organisms or repopulated by survivors, but the structure of the community undergoes a succession of changes during the rebound. Some disturbances, such as volcanic eruptions, are so infrequent and unpredictable over space and time that organisms have not acquired evolutionary adaptations to them. Fire, on the other hand, although unpredictable over the short term, recurs frequently in some communities, and many plants have adapted to this periodic disturbance. In fact, several communities actually depend on periodic fire to maintain them. We will see examples of this when we look at grasslands and other ecosystems later in this chapter, and we will examine the role of disturbance in ecosystems in more detail in Chapters 54 and 55.

Climate and the Distribution of Organisms

The abiotic factors just described have a direct effect on the biology of organisms. The first four factors—temperature, water, light, and wind—are the major components of **climate**, the prevailing weather conditions at a locality. We can see the great impact of climate on the distribution of organisms by constructing a climograph, a plot of the temperature and rainfall in a particular region, often in terms of annual means. For example, FIGURE 50.3 shows a climograph for some of the geographic regions in North America. The term **biome** refers to major types of ecosystems, those that occupy broad geographic regions; coniferous forests, deserts, and grasslands are some examples. Notice that the range of rainfall occurring in northern coniferous forests is similar to that of temperate forests, but that temperatures are different. Grasslands are generally drier than either kind of forest, and deserts are drier still.

Annual means for temperature and rainfall are reasonably well correlated with the biomes we find in different regions. However, we must always be careful to distinguish a *correlation* between variables from *causation*, a cause-and-effect relationship. Although our climograph provides circumstantial evidence that temperature and rainfall are important to the distribution of biomes, it does not prove that these variables govern their geographic location. Only a detailed analysis of the water and temperature tolerances of individual species could establish the controlling effects of these variables.

We can see in our climograph that factors other than mean temperature and precipitation must also play a role in determining which biomes are found where, as there are regions where biomes overlap. For example, there are areas in North America with a certain temperature and precipitation combination that support a temperate forest, but other areas with the same values for these variables support a coniferous forest; still others support a grassland. How do we explain this variation? First, remember that the climograph is based on annual *means*. Often it is not only the mean climate that is important but also the pattern of climatic variation. For example, some areas may get regular precipitation throughout the year, whereas others with the same annual amount have distinct wet and dry seasons. A similar phenomenon may occur with respect to temperature. Other factors, such as the bedrock in

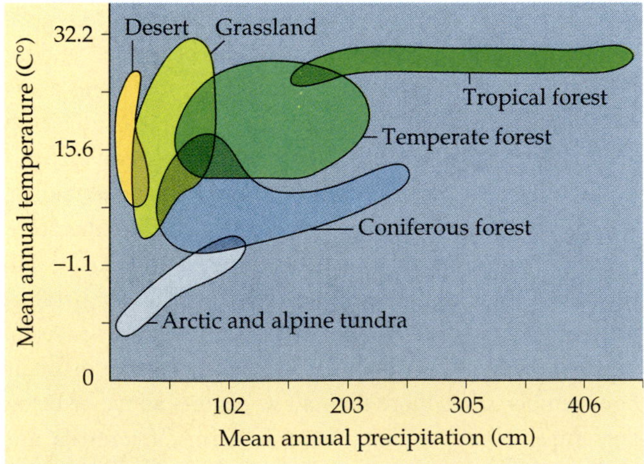

FIGURE 50.3 • A climograph for some major kinds of ecosystems (biomes) in North America. The areas plotted here encompass the ranges of annual mean temperatures and precipitation occurring in the biomes. The climograph provides only circumstantial evidence, however, that these factors are important in explaining the distribution of the biomes. The areas of overlap, for example, show that these variables alone are not sufficient to explain the observed distribution.

an area, may greatly affect mineral nutrient availability and soil structure, which in turn affect the kind of vegetation that will develop.

With these complex considerations in mind, let's take a closer look at global climate patterns, as well as local and seasonal variations in the physical environment, to understand the geographical distribution of organisms.

Global Climate Patterns

Earth's global climate patterns are largely determined by the input of solar energy and the planet's movement in space. About half the solar energy that reaches the upper layers of the atmosphere is absorbed before it reaches Earth; certain wavelengths of light (including the ultraviolet wavelengths that are damaging to biological systems) are more readily absorbed by oxygen molecules and ozone than are others. Much of the energy that strikes Earth itself is absorbed by land and water (and organisms), although some is reflected back into the atmosphere. The sun's warming effect on the atmosphere, land, and water establishes the temperature variations, cycles of air movement, and evaporation of water that are responsible for dramatic latitudinal variations in climate. Because solar radiation is most intense when the sun is directly overhead, the shape of the Earth causes latitudinal variation in the intensity of sunlight (FIGURE 50.4). However, the planet is also tilted on its axis by 23.5° relative to its plane of orbit around the sun, and this tilt causes seasonal variation in the intensity of solar radiation (FIGURE 50.5).

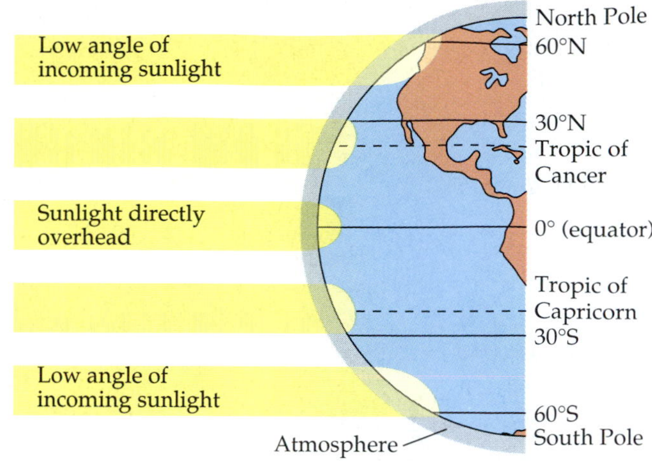

FIGURE 50.4 ▪ Solar radiation and latitude. Because sunlight strikes the equator perpendicularly, more heat and light are delivered there per unit of surface area than at higher northern and southern latitudes, where sunlight has a longer path through the atmosphere and strikes the curved surface of Earth at an oblique angle. This uneven distribution of solar radiation creates latitudinal differences in temperature and light intensity and establishes the vertical air currents illustrated in FIGURE 50.6a and b. In this figure the equator is facing the sun directly, as it does at the equinoxes (see FIGURE 50.5).

The angle of incoming solar radiation changes everywhere from day to day as Earth revolves around the sun, but only the **tropics** (those regions that lie between 23.5° north latitude and 23.5° south latitude) ever receive sunlight from directly overhead. As a result, the tropics experience the greatest annual input and the least seasonal variation in solar radiation of any region on Earth. The seasonality of light and

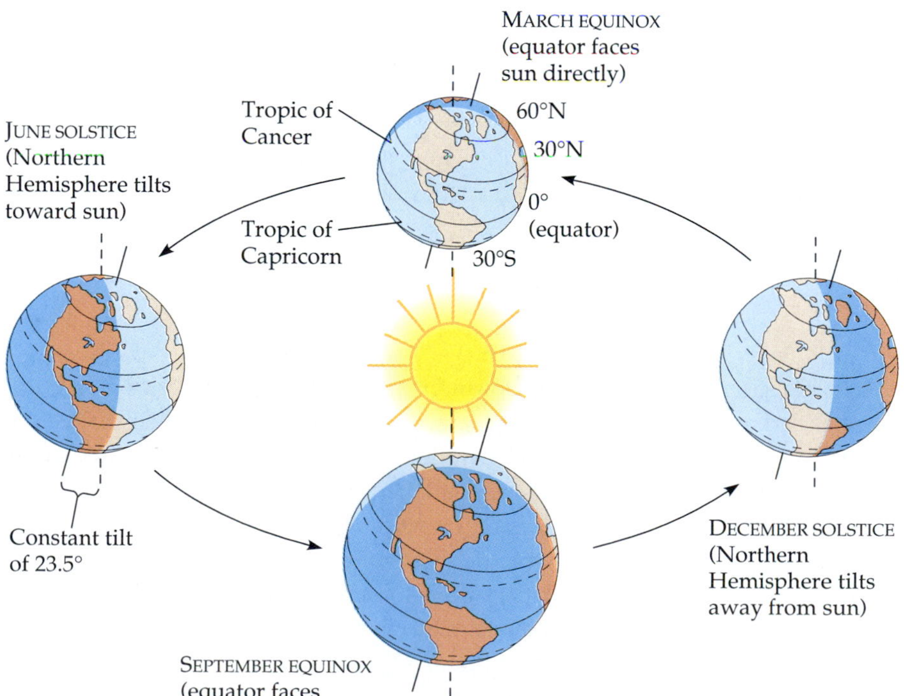

FIGURE 50.5 ▪ The cause of the seasons. The permanent tilt of the Earth on its axis causes seasonal variation in temperature and light intensity as the planet revolves around the sun. The December solstice marks the beginning of winter in the Northern Hemisphere, when the North Pole is maximally tilted *away* from the sun. Day length is reduced and solar radiation arrives at its most oblique angle, producing lower temperatures during short winter days. In contrast, the December solstice marks the beginning of summer in the Southern Hemisphere, when the South Pole is maximally tilted *toward* the sun. Day length increases and the sun is more nearly overhead, increasing temperatures during the long days of summer. Seasons in the two hemispheres are reversed when the North Pole is tilted toward the sun and the South Pole away from the sun on the June solstice. At the March and September equinoxes, neither pole is tilted toward the sun, and all regions on Earth experience 12 hours of daylight and 12 hours of darkness.

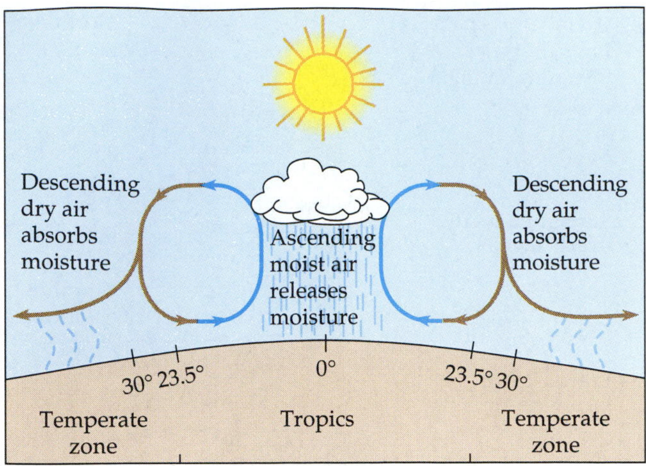

(a) Air circulation and precipitation near the equator

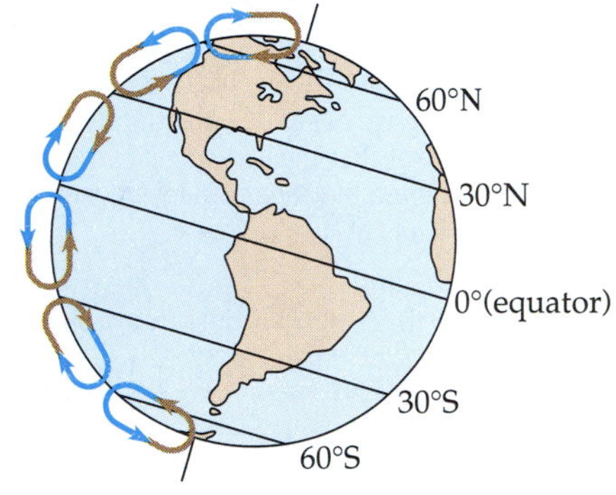

(b) Global air circulation

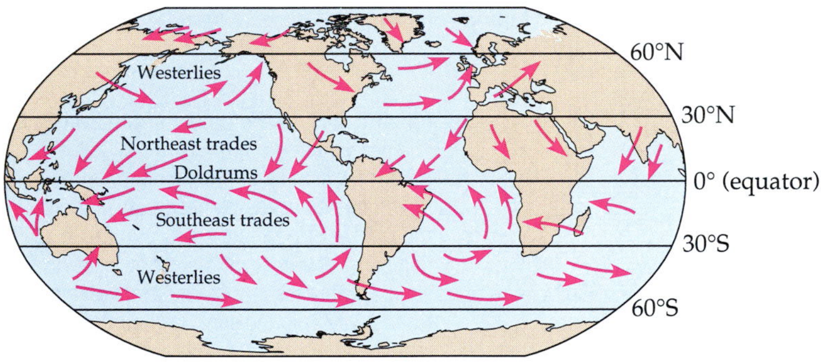

(c) Global wind patterns

FIGURE 50.6 ▪ Global air circulation, precipitation, and winds. (a) Air masses in the lower atmosphere are warmed by solar radiation and by heat radiating from Earth. Air expands, decreases in density, and rises when it is warmed. Thus, warm air at the equator rises, creating an area of light, shifting winds known as the doldrums. The warm air masses expand as they rise, and, because this expansion distributes their internal heat energy over a larger volume, they also cool as they move upward through the atmosphere. Cool air holds less water vapor than warm air, and the rising air masses drop large amounts of rain in the tropics. The now dry air masses flow toward the two poles at high altitude, cooling further as they spread away from the equator. The density of the air masses increases as they cool, and they descend, absorbing water from the land and creating bands of arid climate around 30° latitude. **(b)** The movement of heated air creates three major air circulation cells on either side of the equator. Within each circulation cell, rising air (blue) releases moisture as precipitation, and descending air (brown) absorbs moisture, creating arid conditions. The first circulation cells, described in part (a), are completed when some of the air that descends at 30° latitude flows back toward the equator at low altitude. The remainder of the air flows away from the equator, also at low altitude, initiating midlatitude circulation cells that acquire water and then release it as the air rises around 60° latitude. The third circulation cells carry cool, dry air to the poles, where it descends and flows back toward the equator. Air from the different circulation cells merges where the cells meet, constantly mixing the gases in the lower levels of the atmosphere. **(c)** Air flowing in the lower levels of the circulation cells, close to Earth's surface, creates predictable global wind patterns. However, as Earth rotates on its axis, land near the equator moves faster than that at the poles, deflecting the winds from the vertical paths illustrated in part (b) and creating more easterly and westerly flows. Cooling trade winds blow from east to west in the tropics and subtropics. In contrast, prevailing westerlies blow from west to east in the temperate zones.

temperature increases steadily toward the poles; polar regions have long, cold winters with periods of continual darkness and short summers with periods of continual light.

Intense solar radiation near the equator initiates a global circulation of air, creating precipitation and winds (FIGURE 50.6). High temperatures in the tropics evaporate water from Earth's surface and cause warm, wet air masses to rise and flow toward the poles. The rising air masses release much of their water content, creating abundant precipitation in tropical regions. Thus, high temperatures, intense sunshine, and ample rainfall are all characteristic of a tropical climate, fostering the growth of lush vegetation in some tropical forests and the development of coral reefs. The high-altitude air masses, now dry, descend toward Earth at latitudes around 30° north and south, absorbing moisture from the land and creating an arid climate conducive to the development of the deserts that are common at these latitudes. Some of the descending air flows toward the poles at low altitude, establishing a midlatitude circulation cell that deposits abundant precipitation (though less than in the tropics) where the air masses again rise and release moisture in the vicinity of 60° latitude. Broad expanses of coniferous forest dominate the landscape at these fairly wet, but generally cool, latitudes. A third circulation cell carries some of the cold and dry rising air to the poles, where it descends and flows back toward the equator, absorbing moisture and creating the comparatively rainless and bitterly cold climates of the Arctic and Antarctica.

Local and Seasonal Effects on Climate

Proximity to bodies of water and topographic features such as mountain ranges create a climatic patchiness on a regional scale, and smaller features of the landscape also contribute to local climatic variation. These regional and local variations in climate create a number of ecosystems that are less widely spread than the major biomes.

Ocean currents influence climate along the coasts of continents by heating or cooling overlying air masses, which may then pass across the land. Evaporation from the ocean is also greater than it is over land, and coastal regions are generally moister than inland areas at the same latitude. The cool, misty climate produced by the cold California current that flows southward along the western United States supports a rain forest ecosystem dominated by large coniferous trees in the Pacific Northwest and large redwood groves farther south. Similarly, the warm Gulf Stream flowing north out of the Gulf of Mexico and across the North Atlantic tempers the climate on the west coast of the British Isles, making it warmer than the coast of New England, which is actually farther south but is cooled by a current flowing south from the coast of Greenland.

As every vacationer knows, oceans (and large inland bodies of water) generally moderate the climate of nearby terrestrial environments on a daily cycle. During a warm summer day, when the land is hotter than a large lake or the ocean, air over the land heats and rises, drawing a cool breeze from the water across the land. At night, by contrast, air over the warmer ocean or lake rises, establishing a circulation that draws cooler air from the land out over the water, replacing it with warmer air from offshore. Proximity to water does not always moderate climate, however. Several regions (including the coast of central and southern California) have a Mediterranean-like climate; in summer, cool, dry ocean breezes are warmed when they contact the land, absorbing moisture and creating hot rainless summers just a few miles inland.

Mountains also have a significant effect on solar radiation, local temperature, and rainfall. South-facing slopes in the Northern Hemisphere receive more sunlight than nearby north-facing slopes and are therefore warmer and drier. In many of the mountains of western North America, spruce and other conifers occupy the north-facing slopes, whereas shrubby, drought-resistant vegetation inhabits many south-facing slopes. In addition, at any particular latitude, air temperature declines approximately 6°C with every 1000-m increase in elevation, paralleling the decline of temperature with increasing latitude. In the north temperate zone, for example, a 1000-m increase in elevation produces a temperature change equivalent to that over an 880-km increase in latitude. This is one reason mountain communities are similar to those at lower elevation farther from the equator. When warm, moist air approaches a mountain, it rises and cools, releasing moisture on the windward side of the range. On the leeward side of the mountain, cooler, dry air descends, absorbing moisture and producing a rainshadow (FIGURE 50.7). Deserts commonly occur on the leeward sides of mountain ranges, a phenomenon evident in the Great Basin and Mojave Desert of western North America, the Gobi Desert of Asia, and in the small deserts that characterize the southwest corners of some Caribbean islands.

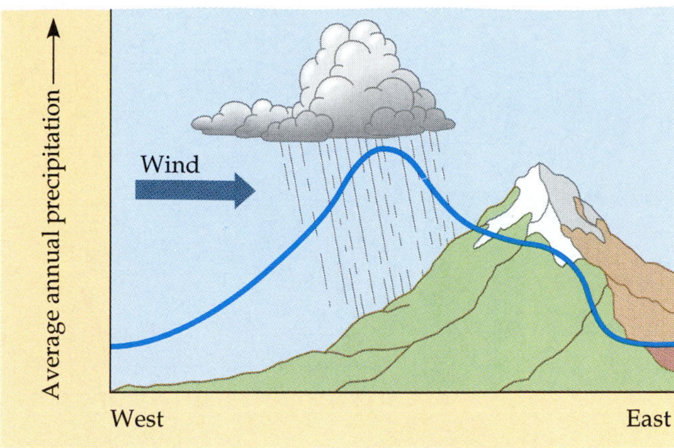

FIGURE 50.7 ▪ Rain shadows. As moisture-laden air is deflected upward by the Sierra Nevada mountain range in California, it cools and the moisture condenses, causing heavy precipitation (peak of the blue line) on the western side of the mountain. When the air follows the eastern slope down, it warms and absorbs moisture from the land, creating dry conditions in the Great Basin of the southwestern United States.

Seasonality generates local environmental variation in addition to the global changes in day length, solar radiation, and temperature described earlier. Because of the changing angle of the sun over the course of the year, the belts of wet and dry air on either side of the equator undergo slight seasonal shifts in latitude that produce marked wet and dry seasons around 20° latitude, where tropical deciduous forests grow. In addition, seasonal changes in wind patterns produce variations in ocean currents, sometimes causing the upwelling of nutrient-rich, cold water from deep ocean layers, thus nourishing organisms that live near the surface.

Ponds and lakes are also sensitive to seasonal temperature changes. During the summer and winter, many lakes in temperate regions are thermally stratified, or layered vertically according to temperature. Such lakes undergo a biannual mixing of their waters as a result of changing water-temperature profiles. This **turnover**, as it is called, brings oxygenated water from the surface of lakes to the bottom and nutrient-rich water from the bottom to the surface in both spring and autumn. These cyclic changes in the abiotic properties of lakes

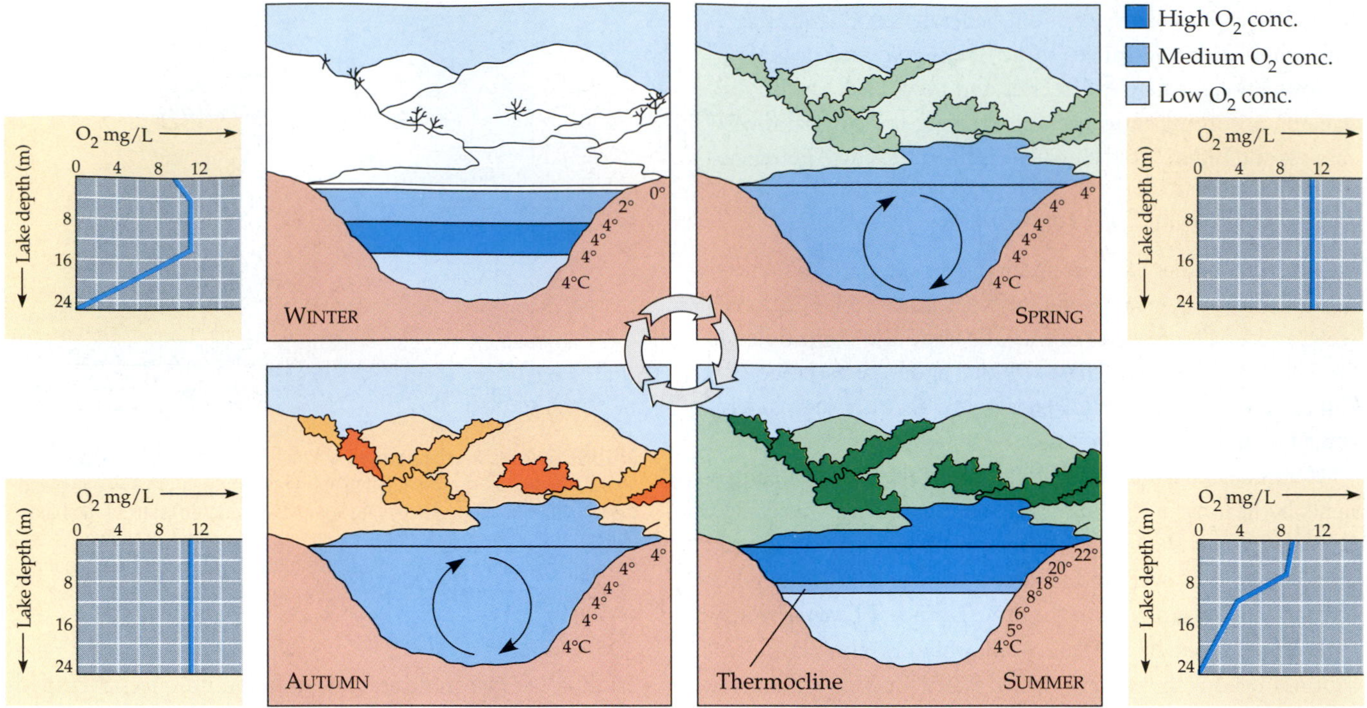

FIGURE 50.8 ▪ Lake stratification and seasonal turnover. Lakes in temperate zones tend to stratify by temperature and density in winter and summer. The biannual mixing of lake waters occurs because water is most dense at 4°C, and water at that temperature sinks below water that is either warmer or colder. In winter, the coldest water in the lake (0°C) lies just below the surface ice; water is progressively warmer at deeper levels of the lake, typically 4°–5°C at the bottom. In spring, as the sun melts ice, the surface water warms to 4°C and sinks below the cooler layers immediately below, eliminating the thermal stratification established in winter. In the absence of thermal layering, spring winds mix the water to great depth, bringing oxygen (O_2) to the bottom waters (see graphs) and nutrients to the surface. In summer, the lake regains a distinctive thermal profile, with warm water at the surface separated from cold bottom water by a narrow vertical zone of rapid temperature change, called a thermocline. In autumn, as surface water cools rapidly it sinks below the underlying layers, remixing the lake water until the surface begins to freeze and the winter temperature profile is reestablished.

are essential for the survival and growth of organisms at all levels within this ecosystem (FIGURE 50.8).

Climate also varies on a very fine scale, called microclimate. For example, ecologists often refer to the microclimate on a forest floor or under a rock. Many features in the environment influence microclimates by casting shade, reducing evaporation from soil, and minimizing the effects of wind. Forest trees frequently moderate the microclimate below. Cleared areas generally experience greater temperature extremes than the forest interior, because of greater solar radiation and wind currents that are established by the rapid heating and cooling of open land; evaporation is generally greater in clearings as well. Low-lying ground is usually wetter than high ground and tends to be occupied by different species of trees within the same forest. If you have ever lifted a log or large stone in the woods, you are well aware that there are organisms (such as salamanders, worms, and some insects) that live in the shelter of this microenvironment, buffered from the extremes of temperature and moisture. Every environment on Earth is similarly characterized by a mosaic of small-scale differences in the abiotic factors that influence the distributions of organisms.

AQUATIC AND TERRESTRIAL BIOMES

Having examined some of the abiotic factors that influence the distribution of organisms, we now turn to a brief survey of the major types of ecosystems, the biomes. The worldwide distribution of aquatic and terrestrial biomes is shown in FIGURES 50.9 and 50.15 (p. 1043).

Aquatic biomes occupy the largest part of the biosphere

Life originated in water and evolved there for almost 3 billion years before plants and animals began moving onto land. Aquatic biomes still account for the largest part of the biosphere.

Ecologists distinguish between freshwater biomes and marine biomes on the basis of physical and chemical differences. For example, marine biomes generally have salt concentrations that average 3%, whereas freshwater biomes are usually characterized by a salt concentration less than 1%. Covering about 75% of Earth's surface, oceans have always

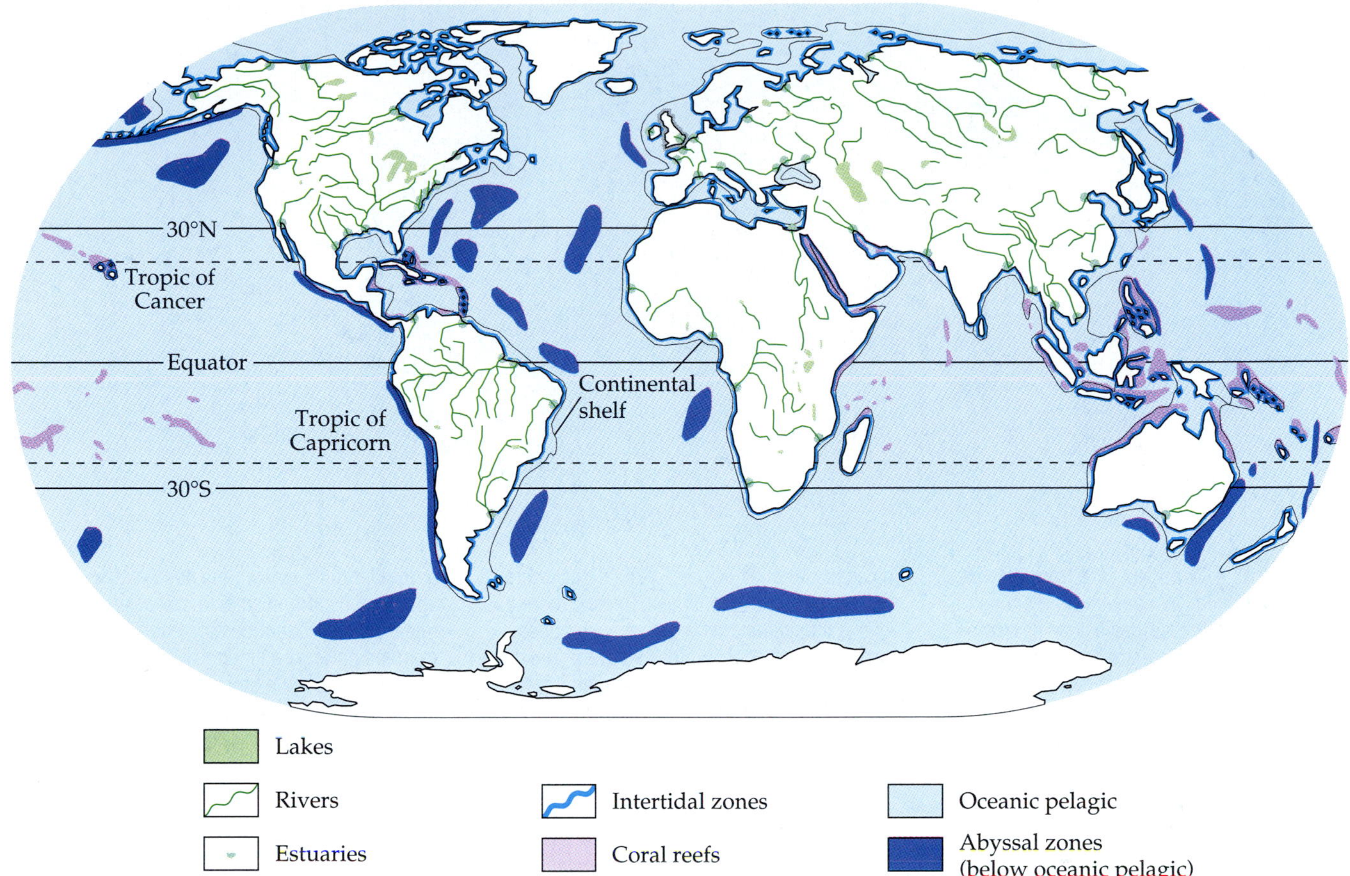

FIGURE 50.9 ▪ The distribution of major aquatic biomes. The characteristics of aquatic biomes are determined by the physical and chemical properties of water in different regions. Rivers and streams are created by runoff from the land that collects in channels and flows into standing bodies of water. Lakes form where water accumulates in inland basins, and estuaries are tidal habitats where rivers flow into the sea. A variety of marine biomes exists in the oceans, where salt concentration is usually high. Intertidal communities flourish along coastlines where rising tides periodically inundate the land. Coral reefs develop in the shallow tropical waters of continental shelves. Pelagic communities exist in the open ocean (over continental shelves and in deeper waters). Benthic communities are found in the substrate below all bodies of water. The map also indicates abyssal zones, the habitat of deep benthic communities.

had an enormous impact on the biosphere. The evaporation of seawater provides most of the planet's rainfall, and ocean temperatures have a major effect on world climate and wind patterns. In addition, marine algae and photosynthetic bacteria supply a substantial portion of the world's oxygen (O_2) and consume huge amounts of atmospheric carbon dioxide. Freshwater biomes are closely linked to the soils and biotic components of the terrestrial biomes through which they pass or in which they are situated. The particular characteristics of a freshwater biome are also influenced by the patterns and speed of water flow and the climate to which the biome is exposed.

Vertical Stratification of Aquatic Biomes

Many aquatic biomes exhibit pronouced vertical stratification of physical and chemical variables. Light is absorbed by both the water itself and the microorganisms in it, so that its intensity decreases rapidly with depth. Ecologists distinguish between the upper **photic zone**, where there is sufficient light for photosynthesis, and the lower **aphotic zone**, where little light penetrates. Water temperature also tends to be stratified, especially during summer and winter (see FIGURE 50.8). Heat energy from sunlight warms the surface waters to whatever depth the sunlight penetrates, but the deeper waters remain quite cold. In the ocean and in many temperate-zone lakes, a narrow stratum of rapid temperature change called a **thermocline** separates a more uniformly warm upper layer from more uniformly cold deeper waters. At the bottom of all aquatic biomes the substrate is called the **benthic zone.** Made up of sand and organic and inorganic sediments ("ooze"), the benthic zone is occupied by communities of organisms collectively called **benthos.** A major source of food for the benthos is dead organic matter called **detritus.** In lakes and oceans, detritus "rains" down from the productive surface waters of the photic zone.

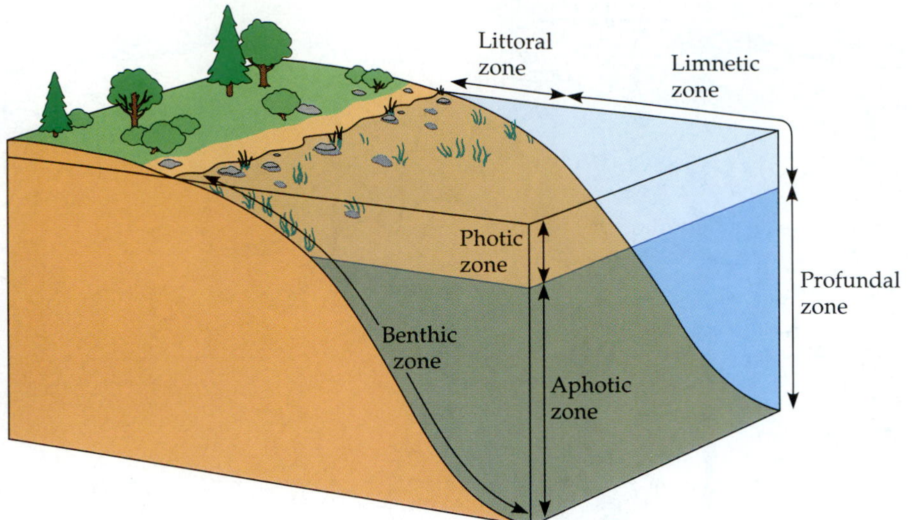

FIGURE 50.10 ▪ Zonation in a lake. Physical criteria define zones in many lakes. If enough light to sustain photosynthesis does not penetrate to the bottom, there are photic and aphotic zones. Rooted and floating plants inhabit the littoral zone, or shallow-water zone; some have stems and leaves that emerge above the water surface. The littoral community in most lakes is diverse, including many species of attached algae, especially diatoms, and a variety of grazing snails, herbivorous and carnivorous arthropods, fishes, and amphibians. Beyond the littoral zone is the limnetic zone, the region of open water in which photosynthesis occurs. The deep, aphotic waters constitute the profundal zone. The entire lake bottom is the benthic zone. When a lake is thermally stratified, its deep waters are colder and thus denser, and do not mix with the surface waters that contact the atmosphere. Decomposers may deplete the O_2 supply in the profundal zone, making the deep waters unsuitable for most organisms. Decomposition also releases large quantities of mineral nutrients from the bottom sediments, but these nutrients are trapped in the waters at the bottom of the lake during stratification. The biannual turnover that results from spring and fall temperature changes in temperate lakes brings O_2 to the profundal zone and nutrients to the limnetic zone (see FIGURE 50.8).

Freshwater Biomes

In this section we examine two general categories of freshwater biomes, standing bodies of water (such as lakes and ponds) and moving types (rivers and streams). Standing bodies of water range from small ponds that are a few square meters to large lakes that are thousands of square kilometers in area. In most lakes, communities of plants and animals are distributed according to the depth of the water and its distance from shore (FIGURE 50.10). Rooted and floating aquatic plants flourish in the **littoral zone**, the shallow, well-lit waters close to shore. In a lake, the well-lit, open surface waters farther from shore, called the **limnetic zone**, are occupied by a variety of phytoplankton consisting of algae and cyanobacteria. These organisms photosynthesize and reproduce at a high rate during spring and summer. Zooplankton, mostly rotifers and small crustaceans, graze on the phytoplankton. The zooplankton are consumed by many small fish, which in turn become food for larger fish, semiaquatic snakes and turtles, and fish-eating birds.

Most of the small organisms of a lake's limnetic zone are short-lived, and their remains continually sink into a deep aphotic region, called the **profundal zone**, and down to the benthic zone. Microbes in the profundal and benthic zones use oxygen for cellular respiration as they decompose this detritus.

Lakes are often classified according to their production of organic matter. **Oligotrophic** lakes are deep and nutrient-poor, and the phytoplankton in the limnetic zone are not very productive (FIGURE 50.11a). **Eutrophic** lakes, in contrast, are usually shallower, and the nutrient content of their water is high. As a result, the phytoplankton are very productive, and the waters are often quite murky (FIGURE 50.11b). Between the oligotrophic and eutrophic extremes are lakes with a moderate amount of nutrients and phytoplankton productivity, called **mesotrophic**.

Over long periods of time, oligotrophic lakes may become mesotrophic and eventually eutrophic as runoff from the surrounding land brings in additional sediments and nutrients. Unfortunately, human activities often speed this natural process dramatically. Runoff from fertilized lawns and agricultural fields and the dumping of municipal wastes enrich the lakes with excessive amounts of nitrogen and phosphorus, mineral nutrients that normally limit the growth of phytoplankton and plants. The result of such pollution is often a population explosion of algae, the production of much detritus, and the eventual depletion of oxygen supplies. Such "cultural eutrophication" makes the water unusable and degrades the lake's aesthetic value (see FIGURE 54.13).

Streams and rivers are bodies of water moving continuously in one direction (FIGURE 50.11c). At the headwaters of a stream (perhaps a spring or snowmelt), the water is often cold and clear, and it carries little sediment and relatively few mineral nutrients. The channel is usually narrow, with a swift current passing over a rocky substrate. Farther downstream, where numerous tributaries may have joined to form a river, the water may be more turbid, carrying substantially more

(a) An oligotrophic lake

(b) A eutrophic lake

(c) A stream flowing into a river

FIGURE 50.11 ▪ Freshwater ecosystems.
(a) Oligotrophic lakes, such as Lake Baikal in Siberia, shown here, are nutrient-poor with a small surface area relative to depth. The bottom sediments are low in decomposable organic matter, limiting bacterial populations in the benthos. The shortage of nutrients in the water limits photosynthesis by plankton in the limnetic zone; as a result, the water is clear and oxygen-rich and usually supports diverse populations of fish and invertebrates.
(b) Eutrophic lakes, such as this one called Peck's Pond in the Pocono mountains of eastern Pennsylvania, are nutrient-rich and generally have a larger surface area relative to depth. The availability of nutrients supports a high rate of photosynthesis, leading to murkier water than that of an oligotrophic lake. High organic content in the benthos leads to high decomposition rates and potentially low oxygen supplies in the profundal and benthic zones.
(c) Rivers and streams support substantially different biological communities than those in ponds and lakes. The photograph shows the mouth of a stream, flowing into the much larger Snake River in Idaho. The nutrient content of flowing water biomes is largely determined by the terrain and vegetation through which they flow. Fallen leaves from dense, overhanging vegetation can add substantial amounts of organic matter, and the weathering of rocks can increase the concentration of inorganic nutrients. The turbulent flow of many streams constantly oxygenates the water, whereas the sometimes murkier, warmer waters of large rivers may contain relatively little oxygen. Stream- and river-dwelling animals exhibit evolutionary adaptations that enable them to resist being swept away. The smaller ones are typically flat in shape and can attach to rocks temporarily. Many arthropods live on the underside or downstream side of rocks, thereby exploiting a microhabitat that is relatively free of turbulent flow.

(a) Wetlands

(b) An estuary

FIGURE 50.12 ▪ Wetlands and estuaries. **(a)** This marsh in Pennsylvania is an example of a basin wetland. Marshes are usually covered with water year-round. Predominant plants are emergent (with stems and leaves extending above the water surface), such as the pond lilies, reeds, and cattails visible in this photograph. Other kinds of wetlands include swamps (dominated by woody plants), bogs (dominated by sphagnum mosses), and seasonal pools (typically flooded during the winter and spring and dry in summer and fall). Wetlands store water, reduce flooding of surrounding lands, filter pollutants, and support a rich assemblage of organisms. **(b)** This view of an estuary that is part of Chesapeake Bay in Maryland shows the intimate association of river mouths and the marine environment into which they carry water. Unfortunately, the land surrounding Chesapeake Bay is heavily populated and industrialized, and pollution that enters the bay through four major rivers has made it unsuitable for many plant and animal species. What was once a bountiful natural source of seafood and other resources has been degraded and rendered less productive by human activity.

sediment (from the erosion of soil) and nutrients. The channel near the mouth of a river is relatively wide, and the substrate is generally silty from the deposition of sediments over long periods of time.

Many streams and rivers have been affected by pollution from human activities, by stream channelization to speed water flow, and by dams that hold water. For centuries humans used streams and rivers as depositories of waste, thinking that these materials would be diluted and carried downstream. While some pollutants are carried far from their source, many settle to the bottom, where they can be taken up by aquatic organisms. Even the pollutants that are carried away contribute to estuary, ocean, and lake pollution. In many cases, dams have completely changed the downstream ecosystems, altering the intensity and volume of water flow and affecting fish and invertebrate populations.

Wetlands

At the simplest level, a **wetland** is an area covered with water that supports aquatic plants (FIGURE 50.12a). In fact, wetlands range from periodically flooded regions to soil that is permanently saturated during the growing season. These conditions favor the growth of specially adapted plants called hydrophytes ("water plants"), which can grow in water or in soil that is periodically anaerobic due to the presence of water. Hydrophytes include floating pond lilies and emergent cattails, many sedges, tamarack, and black spruce. Both the hydrology and the vegetation of an area are important determinants of its classification as a wetland—a classification that can be critical when federal, state, and local governments are making preservation decisions based on rigorous, and often conflicting, definitions.

A wide variety of wetlands types has been recognized, ranging from marshes to swamps to bogs. All these varieties, however, generally form in one of three different topographic situations. Basin wetlands develop in shallow basins, ranging from upland depressions to filled-in lakes and ponds. Riverine wetlands develop along shallow and periodically flooded banks of rivers and streams. Fringe wetlands occur along the coasts of large lakes and seas, where water flows back and forth because of rising lake levels or tidal action. Thus, fringe wetlands include both freshwater and marine biomes. Marine coastal wetlands are closely linked to estuaries, which we examine shortly. The flow of water through a wetland, as well as the duration, frequency, depth, and season of flooding, determine the types of plants that are present.

Ecologically, wetlands are among the richest of biomes. They contain a diverse community of invertebrates, which support a wide variety of birds. Herbivores from crustaceans to muskrats consume algae, detritus, and plants. In addition to the rich diversity of species that is supported by wetlands, the ecological and economic value of wetlands exceeds that expected from their geographic extent alone; they provide water-storage basins that reduce the intensity of flooding, and they improve water quality by filtering pollutants. In the past, humans have often regarded wetlands as wastelands—as sources of mosquitoes, flies, and bad odors—and have destroyed many wetlands, mostly to provide land for agriculture and development. Recently, both governments and private organizations are attempting to protect remaining wetlands through acquisition, economic incentives, and regulation. A great deal of research is underway to determine how wetlands can be created or restored.

Estuaries

The area where a freshwater stream or river merges with the ocean is called an **estuary**; it is often bordered by extensive coastal wetlands called mudflats and saltmarshes (see FIGURE 50.12b). Salinity varies spatially within estuaries, from nearly that of fresh water to that of the ocean; it also varies over the course of a day with the rise and fall of the tides. Nutrients from the river enrich estuarine waters, making estuaries one of the most biologically productive environments on Earth.

Saltmarsh grasses, algae, and phytoplankton are the major producers in estuaries. This environment also supports a variety of worms, oysters, crabs, and many of the fish species that humans consume. Many marine invertebrates and fishes use estuaries as a breeding ground or migrate through them to freshwater habitats upstream. Estuaries are also crucial feeding areas for many semiaquatic vertebrates, particularly waterfowl.

Although estuaries support a wide variety of commercially valuable species, areas around estuaries are also prime locations for commercial and residential developments. In addition, estuaries are unfortunately at the receiving end for pollutants dumped upstream. Very little undisturbed estuarine habitat remains, and a large percentage has been totally eliminated by landfill and development. Many states have now—rather belatedly—taken steps to preserve their remaining estuaries.

Zonation in Marine Communities: An Introduction

Similar to the communities in freshwater lakes, marine communities are distributed according to depth of the water, degree of light penetration, distance from shore, and open water versus bottom (FIGURE 50.13). There is a photic zone where phytoplankton, zooplankton, and many fish species occur, and an aphotic zone below. Because water absorbs light so well and the ocean is so deep, most of the ocean volume is virtually devoid of light, except for tiny amounts produced by

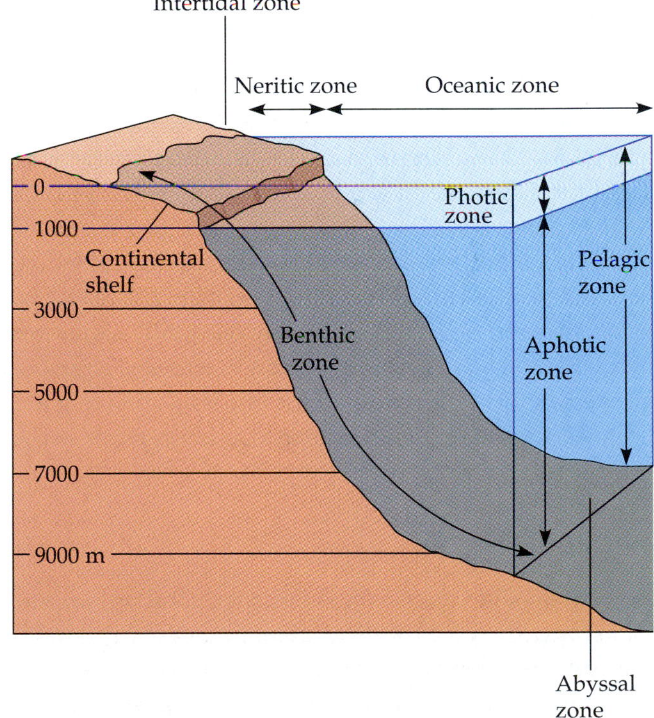

FIGURE 50.13 ▪ Zonation in the marine environment. The marine environment can be classified on the basis of three physical criteria: light penetration (photic and aphotic zones), distance from shore and water depth (intertidal, neritic, and oceanic zones), and whether it is open water or bottom (pelagic and benthic zones). The abyssal zone is the benthic region in deep oceans. Ecologists often combine two designations, such as the oceanic pelagic zone, to identify the location of a biome.

a few luminescent fishes and invertebrates. The zone where land meets water is called the **intertidal zone**; beyond the intertidal zone is the **neritic zone**, the shallow regions over the continental shelves; and past the continental shelf is the **oceanic zone**, reaching very great depths. Finally, open water of any depth is the **pelagic zone**, at the bottom of which is the seafloor, or **benthic zone**.

Intertidal Zones

An intertidal zone is alternately submerged and exposed by the twice-daily cycle of tides. Intertidal communities are therefore subject to huge daily variations in the availability of seawater (and the nutrients it carries) and in temperature. Perhaps most significant of all, intertidal organisms are subject to the mechanical forces of wave action, which can dislodge them from the habitat.

The rocky intertidal zone is vertically stratified (FIGURE 50.14a). Most of the organisms have structural adaptations that enable them to attach to the hard substrate in this physically tumultuous environment. On sandy substrates (beaches) or mudflats, the intertidal zone is not as clearly stratified. Wave action constantly moves the particles of mud and sand, and few large algae or plants occupy these habitats. Many animals, such as suspension-feeding worms and clams and predatory crustaceans, bury themselves in sand or mud, feeding when the tides bring sources of food. Other surface-dwelling organisms, such as crabs and shorebirds, are scavengers or predators on these organisms.

Partly because of our strong attraction to the seashore, humans have had a long-term impact on intertidal ecosystems. The recreational use of ocean shores has caused a severe decline in the numbers of many beach-nesting birds and sea turtles. Incoming tides carry in polluted water and old fishing lines and plastic debris that can harm wildlife. The most dramatic intertidal pollutant is probably oil, which harms not only birds and marine mammals but also intertidal algae and invertebrates. The ultimate outcome of oil pollution in intertidal zones is reduced species diversity, with increases in the populations of a few oil-resistant species.

Coral Reefs

In warm tropical waters in the neritic zone, **coral reefs** constitute a conspicuous and distinctive biome. Currents and waves constantly renew nutrient supplies to the reefs, and sunlight penetrates to the ocean floor, allowing photosynthesis.

Coral reefs are dominated by the structure of the coral itself, formed by a diverse group of cnidarians that secrete hard external skeletons made of calcium carbonate. These skeletons vary in shape, forming a substrate upon which other corals, sponges, and algae grow (FIGURE 50.14b). Multicellular algae that are encrusted with calcium carbonate also add large amounts of limestone to most reefs, as do bryozoans (see Chapter 33). The coral animals themselves feed on microscopic organisms and particles of organic debris. They also obtain organic molecules from the photosynthesis of symbiotic dinoflagellate algae that live in their tissues. Coral animals can survive without the dinoflagellates, but their rate of calcium carbonate deposition is much slower without them; thus, reef formation by corals depends on this symbiotic association.

Some coral reefs cover enormous expanses of shallow ocean, but this delicate biome is easily degraded by pollution and development, as well as by souvenir hunters who gather the coral skeletons. Corals are also subject to damage from native and introduced predators such as the crown-of-thorns sea star, which has undergone a population explosion in many regions and actually destroyed coral reefs in parts of the western Pacific Ocean. Reef communities are very old and grow very slowly, and they may not be able to withstand continued human encroachment.

The Oceanic Pelagic Biome

Most of the ocean's water lies far from shore in the **oceanic pelagic biome**, constantly mixed by ocean currents. Nutrient concentrations are generally lower in the open ocean than in coastal areas because the remains of plankton and other organisms sink below the photic region into the dark, lower benthic zone. In some tropical areas, surface waters are lower in nutrients than the surface waters of temperate oceans because a year-round thermal stratification prevents an exchange of nutrients between the surface and the deep. Temperate oceans generally are more productive, because, like temperate lakes, they experience a nutrient turnover in the spring and, to a limited extent, in the fall. In the spring the recirculation of nutrients from the depths stimulates a surge of photosynthetic plankton growth.

Photosynthetic plankton grow and reproduce rapidly in the photic region of the oceanic biome. Modern sampling methods, which take bacterial photosynthesis into account, show that the plankton's rates of organic food production are higher than formerly thought. Nonetheless, photosynthetic plankton account for less than half the photosynthetic activity on Earth. Zooplankton, including protozoans, worms, copepods, shrimp-like krill, jellies, and the small larvae of invertebrates and fishes, graze on the photosynthetic plankton. Most plankton exhibit morphological structures, such as bubble-trapping spines, lipid droplets, gelatinous capsules, and air bladders, that help them stay afloat within the photic zone.

The oceanic pelagic biome also includes free-swimming animals, called nekton, that can move against the currents to locate food. Large squids, fishes, sea turtles, and marine mammals feed either on plankton or each other. Although many of these animals feed in the photic region of the pelagic zone, others live at great depth where fish may have enlarged eyes, enabling them to see in the very dim light, or luminescent organs that attract mates and prey. Many pelagic birds, such as

(a) Intertidal zone

(c) Benthos

(b) Coral reef

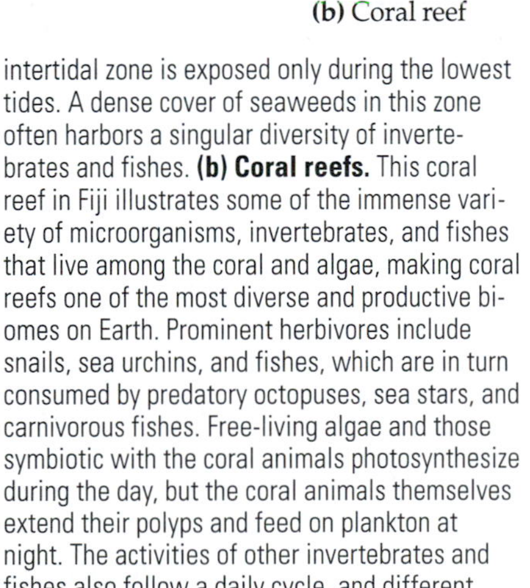

FIGURE 50.14 ▪ Marine biomes. (a) Intertidal zones. This photograph of the rocky intertidal zones at Washington State's Lopez Island was taken at low tide to illustrate the vertical zonation of algae and animals. The density of organisms in each of the three major zones is roughly proportional to the percentage of time the zone is submerged. Organisms in the uppermost zone—grazing mollusks, suspension-feeding barnacles, and a few algae—are submerged only during the highest tides and have numerous adaptations that prevent dehydration and overheating. The middle zone, generally submerged at high tide and exposed at low tide, is inhabited by a diverse array of algae, sponges, sea anemones, mollusks, crustaceans, echinoderms, and small fishes. Tidepool organisms in this zone may experience dramatic increases in salinity as evaporation decreases the volume of water in the pools during low tides. The bottom of the intertidal zone is exposed only during the lowest tides. A dense cover of seaweeds in this zone often harbors a singular diversity of invertebrates and fishes. **(b) Coral reefs.** This coral reef in Fiji illustrates some of the immense variety of microorganisms, invertebrates, and fishes that live among the coral and algae, making coral reefs one of the most diverse and productive biomes on Earth. Prominent herbivores include snails, sea urchins, and fishes, which are in turn consumed by predatory octopuses, sea stars, and carnivorous fishes. Free-living algae and those symbiotic with the coral animals photosynthesize during the day, but the coral animals themselves extend their polyps and feed on plankton at night. The activities of other invertebrates and fishes also follow a daily cycle, and different components of the fauna are active by day and by night. **(c) Benthos: a deep-sea vent community.** Benthic faunas occupy the ocean bottom from the intertidal zone to the abyssal zone. The species composition of benthos varies dramatically with water depth. Pictured here is a vent community, first discovered at a depth of 2500 m in the late 1970s. These communities are found at spreading centers on the seafloor, where hot magma superheats the water. About a dozen species of prokaryotes identified near the vents are chemoautotrophic producers that obtain energy by oxidizing H_2S formed by a reaction of the hot water with dissolved sulfate (SO_4^{2-}). Among the animals in these communities are giant tube-dwelling worms (pictured here), some more than 1 m long. They are apparently nourished by chemosynthetic prokaryotes that live as symbionts within the worms. Many other invertebrates, including arthropods and echinoderms, are also abundant around the vents.

petrels, terns, albatrosses, and boobies, catch fishes in the surface waters.

Benthos

The ocean bottom below the neritic and pelagic zones is the benthic zone, as in other aquatic biomes. Nutrients reach the seafloor from the waters above by "raining down" in the form of detritus. Although the benthic zone in shallow, near-coastal waters may receive substantial sunlight, light and temperature decline dramatically with depth.

Neritic benthic communities are extremely productive, consisting of bacteria, fungi, seaweeds and filamentous algae, numerous invertebrates, and fishes. Species composition of these communities varies with distance from the shore, water depth, and composition of the bottom. Many organisms live buried in soft substrates.

Organisms in deep benthic communities of the **abyssal zone** are adapted to continuous cold (about 3°C), extremely high water pressure, near or total absence of light, and low nutrient concentrations. However, oxygen is usually present in abyssal waters, and a fairly diverse community of invertebrates and fishes occupies this region. Marine scientists have also discovered a unique assemblage of organisms associated with deep-sea hydrothermal vents of volcanic origin in mid-ocean ridges (FIGURE 50.14c). In this dark, hot, oxygen-deficient environment, the food producers are not photosynthesizing organisms but chemoautotrophic prokaryotes (see Chapter 27). The organic molecules they synthesize form the base of a food chain that includes giant polychaete worms, arthropods, echinoderms, and fishes.

The geographical distribution of terrestrial biomes is based mainly on regional variations in climate

All the abiotic factors we covered earlier in the chapter, but
50.1 especially climate, are important in determining why a particular terrestrial biome is found in a particular area. Because there are latitudinal patterns of climate over Earth's surface (see FIGURES 50.4–50.6), there are also latitudinal patterns of biome distribution. For example, coniferous forests extend in a broad band across North America, Europe, and Asia (FIGURE 50.15).

Terrestrial biomes are often named for major physical or climatic features and for their predominant vegetation. For example, temperate grasslands are dominated by various grass species and are generally found in middle latitudes, where the climate is more moderate than in the tropics or polar regions. Each biome is also characterized by microorganisms, fungi, and animals adapted to that particular environment. Temperate grasslands, for example, are more likely than forests to be populated by large grazing mammals.

Vertical stratification is an important feature of terrestrial biomes, and the shapes and sizes of plants largely define the layering. For example, in many forests the layers are the upper **canopy**, then the low-tree stratum, the shrub understory, the ground layer of herbaceous plants, the forest floor (litter layer), and finally the root layer. Other (nonforest) biomes have similar, though usually less pronounced vertical strata. For instance, grasslands have a canopy formed by an herbaceous layer of grass species, a litter layer, and a root layer. The root layer in arctic tundra is shallower than in most other biomes because it is underlain by a permanently frozen stratum called **permafrost**.

Vertical stratification of a biome's vegetation provides many different habitats for animals, which often occupy well-defined feeding groups, from the insectivorous and carnivorous birds and bats that feed above canopies to the small mammals, numerous worms, and arthropods that forage the litter and root layers for food.

Terrestrial biomes usually grade into each other, without sharp boundaries. If the area of intergradation is itself large, it may be recognized as a separate biome, or *ecotone.*

The actual species composition of any one kind of biome varies from one location to another. For instance, in the northern coniferous forest (taiga) of North America, red spruce is common in the east but does not occur in most other areas, where black spruce and white spruce are abundant. Although the vegetation of African deserts superficially resembles that of North American deserts, the plants are in different families. Such "ecological equivalents" can arise because of convergent evolution (see Chapter 25).

Biomes are dynamic, and disturbance rather than stability tends to be the rule. As a result of disturbance, biomes usually exhibit extensive patchiness, with several communities represented. Hurricanes create openings in tropical and temperate forests. In northern coniferous forests, snowfall may break branches and small trees, producing openings that allow deciduous species, such as aspen and birch, to grow. In many biomes even the dominant plants depend on periodic disturbance. For example, fire is an integral component of grasslands, savannas, chaparral, and many coniferous forests. Before agricultural and urban development, much of the southeastern United States was dominated by a single conifer species, the longleaf pine. Without periodic burning, deciduous trees tended to replace the pines. As Michael Dombeck discussed in the interview opening this unit, forest managers now use fire as a tool to help maintain many coniferous forests.

In many biomes today, extensive human activities have radically altered the natural patterns of periodic disturbance. Most of the eastern United States, for example, is classified as

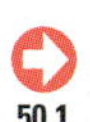
50.1 This symbol links topics in the text to interactive exercises in the CD-ROM that accompanies the book. The number indicates the appropriate activity in the CD.

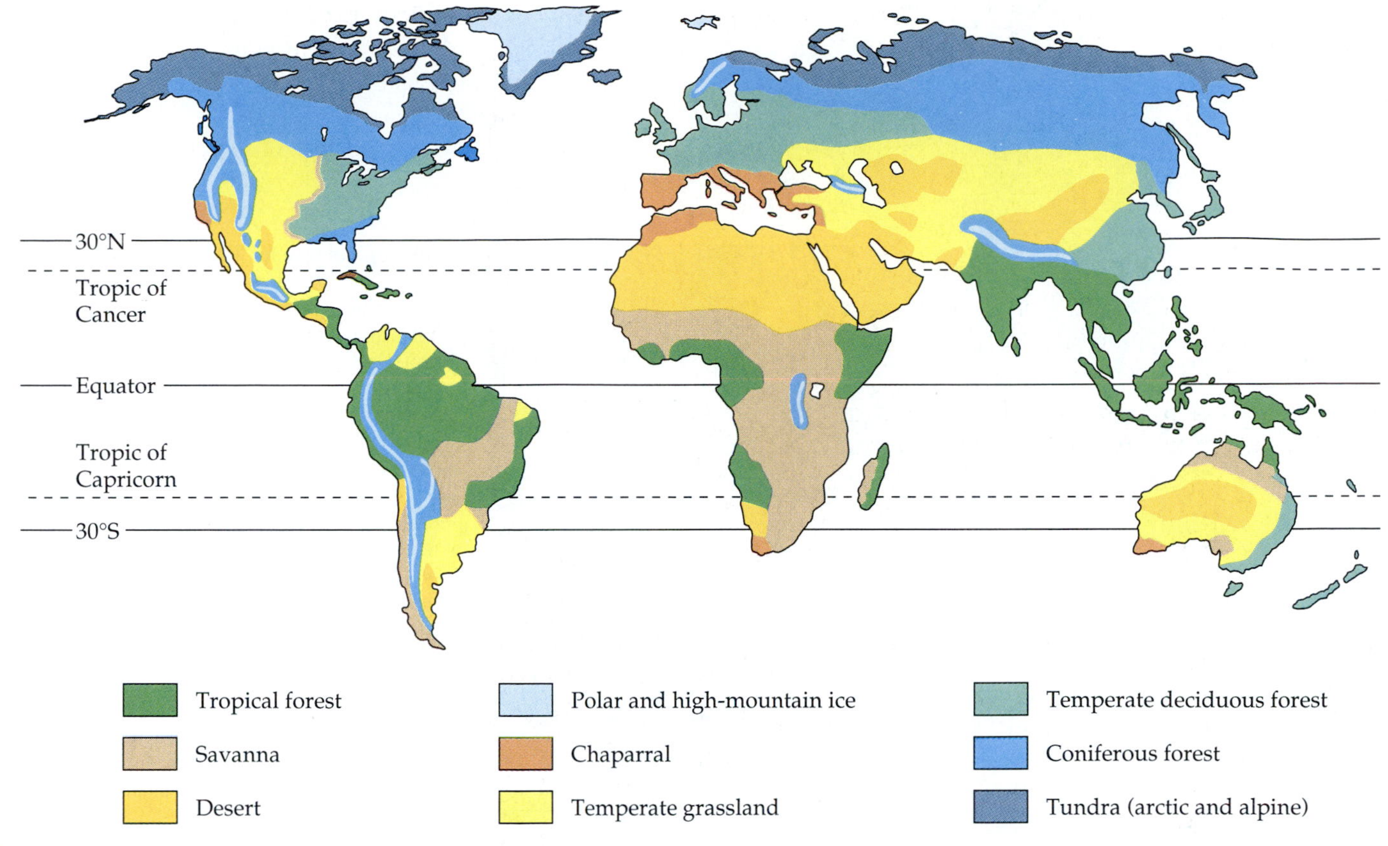

50.1 **FIGURE 50.15 ■ The distribution of major terrestrial biomes.** Although terrestrial biomes are mapped here with sharp boundaries, biomes actually grade into one another, sometimes over relatively large areas. The tropics are the low-latitude regions bordered by the Tropic of Cancer and the Tropic of Capricorn.

temperate deciduous forest, but human activity has eliminated all but a tiny percentage of the original forest. In fact, humans have altered much of Earth's surface, replacing original biomes with urban and agricultural ones.

FIGURE 50.16 on pages 1044–1047 surveys the major terrestrial biomes, beginning near the equator and generally approaching the poles.

CONCEPTS OF ORGANISMAL ECOLOGY

In surveying the biomes we have seen that the abiotic conditions of the biosphere largely determine the distribution of life forms. We now turn to the organisms themselves and look at some of the physiological, morphological, and behavioral mechanisms that allow species to meet the challenges of their environment.

The costs and benefits of homeostasis affect an organism's responses to environmental variation

The success of an organism at survival and reproduction reflects its overall tolerance to the entire set of environmental variables it confronts. In many cases the ability to tolerate a particular factor may depend on another factor. For example, many aquatic ectothermic organisms can survive reduced oxygen levels at low temperatures, but not at high temperatures when their metabolic rates are also high. Coping with a set of environmental problems usually involves imperfect adaptations that represent evolutionary compromises. Panting or sweating, for instance, cools the body on a hot day but can also lead to a water deficit.

Regulators and Conformers

Chapter 40 introduced the term *homeostasis*, the maintenance of a steady-state internal environment in the face of variations in the external environment. Many animals and plants can be described as **regulators** that use behavioral and physiological mechanisms to achieve homeostasis in the face of environmental fluctuations in temperature, moisture, light intensity, and concentrations of a variety of chemical factors. For example, Pacific salmon, which spend part of their lives in salt water and part in fresh water, maintain a constant solute concentration in their blood by osmoregulation. Other organisms, particularly those that live in relatively stable environments,

(a) Tropical forests. The photograph shows a tropical rain forest in Costa Rica. Tropical rain forests have pronounced vertical stratification. Trees in the canopy make up the topmost stratum. The canopy is often closed, so that little light reaches the ground below. When an opening does occur, perhaps because of a fallen tree, other trees and large woody vines grow rapidly, competing for light and space as they fill the gap. Many of the trees are covered with epiphytes (plants that grow on other plants rather than in soil), such as orchids and bromeliads. Rainfall, which is quite variable in the tropics, is the prime determinant of the vegetation growing in an area. In lowland areas that have a prolonged dry season or scarce rainfall at any time, tropical dry forests predominate. The plants found there are a mixture of thorny shrubs and trees and succulents. In regions with distinct wet and dry seasons, tropical deciduous trees are common.

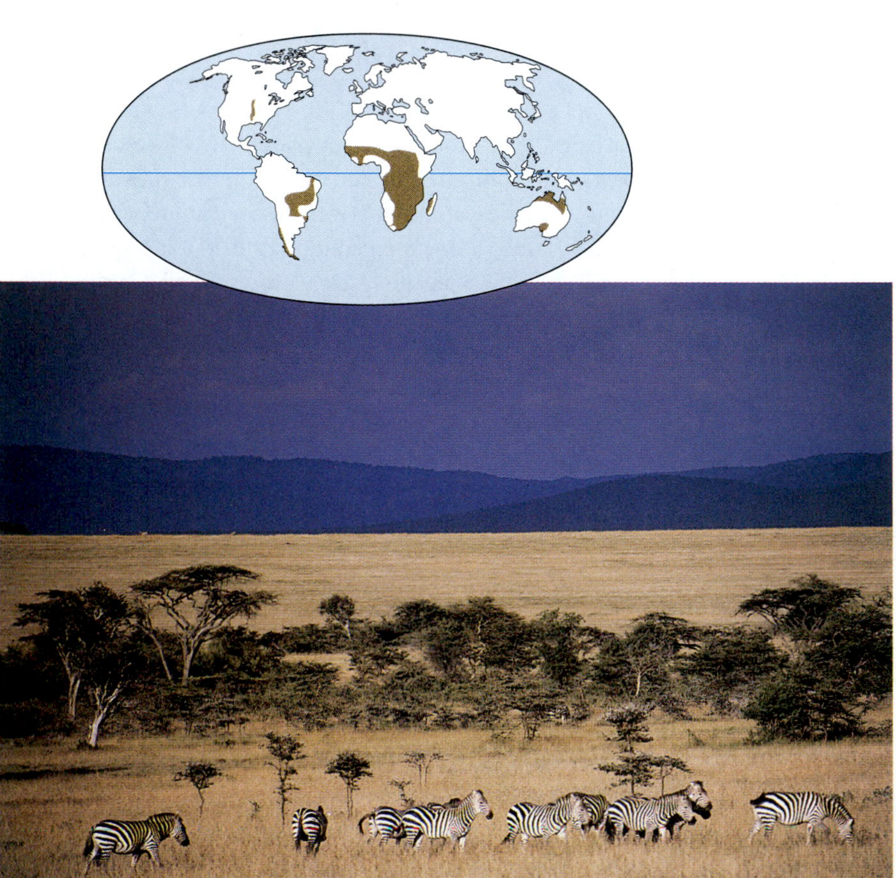

(b) Savanna. This Kenyan savanna is a showcase of large herbivores and their predators. Actually, the dominant herbivores here and in other savannas are insects, especially ants and termites. Grasses and scattered trees are the dominant plants. Fire is an important abiotic component, and the dominant plant species are fire-adapted. The luxuriant growth of grasses and forbs (small broadleaf plants) during the rainy season provides a rich food source for animals. However, large grazing mammals must migrate to greener pastures and scattered watering holes during regular periods of seasonal drought.

FIGURE 50.16 • Examples of terrestrial biomes.

50.1

(c) Desert. Sparse rainfall (less than 30 cm per year) largely determines that an area will be a desert. Some deserts have soil surface temperatures above 60°C during the day. Other deserts, such as those west of the Rocky Mountains and in central Asia, are relatively cold. The Sonoran Desert of southern Arizona (shown here) is characterized by giant saguaro cacti and deeply rooted shrubs. Evolutionary adaptations of desert plants and animals include a remarkable array of mechanisms that store water. The "pleated" structure of saguaro cacti in the photo enables the plants to expand when they absorb water during wet periods. Some desert mice never drink, deriving all their water from the metabolic breakdown of the seeds they eat. Many desert plants also rely on CAM photosynthesis, a metabolic adaptation that conserves water in this arid environment (see Chapter 10). Protective adaptations that deter feeding by mammals and insects, such as spines on cacti and poisons in the leaves of shrubs, are also common in desert plants.

(d) Chaparral. Dense, spiny, evergreen shrubs dominate chaparral biomes, midlatitude coastal areas with mild, rainy winters and long, hot, dry summers. Plants of the chaparral, such as those in this California scrubland, are adapted to and dependent on periodic fires. The dry, woody shrubs are frequently ignited by lightning and by careless human activities, creating summer and autumn brushfires in the densely populated canyons of southern California and elsewhere. Some of the shrubs produce seeds that will germinate only after a hot fire. Food reserves stored in their fire-resistant roots enable them to resprout quickly and use nutrients released by fires.

FIGURE 50.16 ▪ Examples of terrestrial biomes (continued).

(e) Temperate grassland. The veldts of South Africa, the puszta of Hungary, the pampas of Argentina and Uruguay, the steppes of Russia, and the plains and prairies of central North America are all temperate grasslands. The key to the persistence of grasslands is seasonal drought, occasional fires, and grazing by large mammals, all of which prevent establishment of woody shrubs and trees. Temperate grasslands such as the tallgrass prairie in Kansas (shown here) once covered much of central North America. Because grassland soil is both deep and rich in nutrients, these habitats provide fertile land for agriculture. Most grassland in the United States has been converted to farmland, and very little natural prairie exists today.

(f) Temperate deciduous forest. Dense stands of deciduous trees are trademarks of temperate deciduous forests, such as this one in Great Smoky Mountains National Park in North Carolina. Temperate deciduous forests occur throughout midlatitudes where there is sufficient moisture to support the growth of large trees. More open than rain forests and not as tall, a mature temperate deciduous forest has distinct vertical layers, including one or two strata of trees, an understory of shrubs, and an herbaceous stratum. Deciduous forest trees drop their leaves before winter, when temperatures are too low for effective photosynthesis and water lost through transpiration is not easily replaced from frozen soil. Many temperate deciduous forest mammals also enter a dormant winter state called hibernation, and some bird species migrate to warmer climates. Virtually all the original deciduous forests in North America were destroyed by logging and land-clearing for agriculture and urban development. In contrast to drier biomes, these forests tend to recover after disturbance, and today we see deciduous trees dominating undeveloped areas over much of their former range.

FIGURE 50.16 ▪ Examples of terrestrial biomes (continued).

(g) Coniferous forest. Cone-bearing trees such as pine, spruce, fir, and hemlock dominate coniferous forests. Coastal coniferous forests of the U.S. Pacific Northwest, such as the one in Olympic National Park in western Washington shown here, are actually temperate rain forests. Warm, moist air from the Pacific Ocean supports these unique communities, which like most coniferous forests are dominated by one or a few tree species. Extending in a broad band across northern North America and Eurasia to the southern border of the arctic tundra, the northern coniferous forest, or taiga, is the largest terrestrial biome on Earth (see FIGURE 50.15). Taiga receives heavy snowfall during winter. The conical shape of many conifers prevents too much snow from accumulating on and breaking their branches. Coniferous forests are being logged at an alarming rate, and the old-growth stands of these trees may soon disappear.

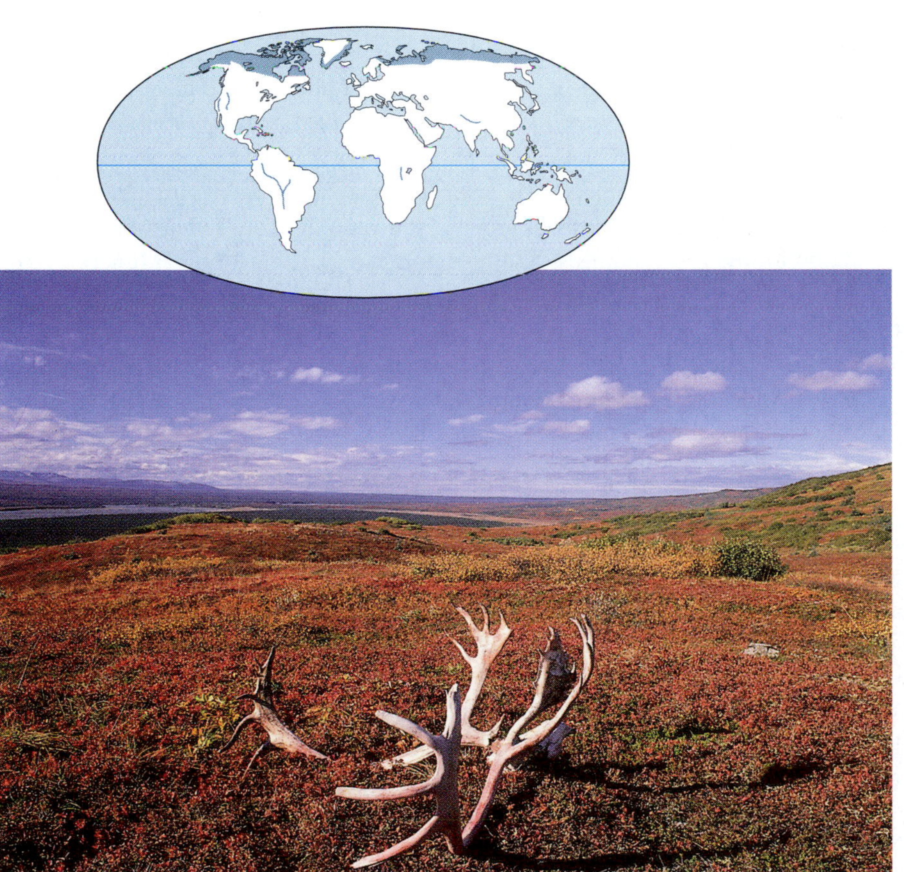

(h) Tundra. Permafrost (permanently frozen subsoil), bitterly cold temperatures, and high winds are responsible for the absence of trees and other tall plants in this arctic tundra in central Alaska (photographed in autumn). Although the arctic tundra receives very little annual rainfall (see FIGURE 50.3), water cannot penetrate the underlying permafrost and accumulates in pools on the shallow topsoil during the short summer. Tundra covers expansive areas of the Arctic, amounting to 20% of Earth's land surface. High winds and cold temperatures create similar plant communities, called alpine tundra, on very high mountaintops at all latitudes, including the tropics.

FIGURE 50.16 ▪ Examples of terrestrial biomes (continued).

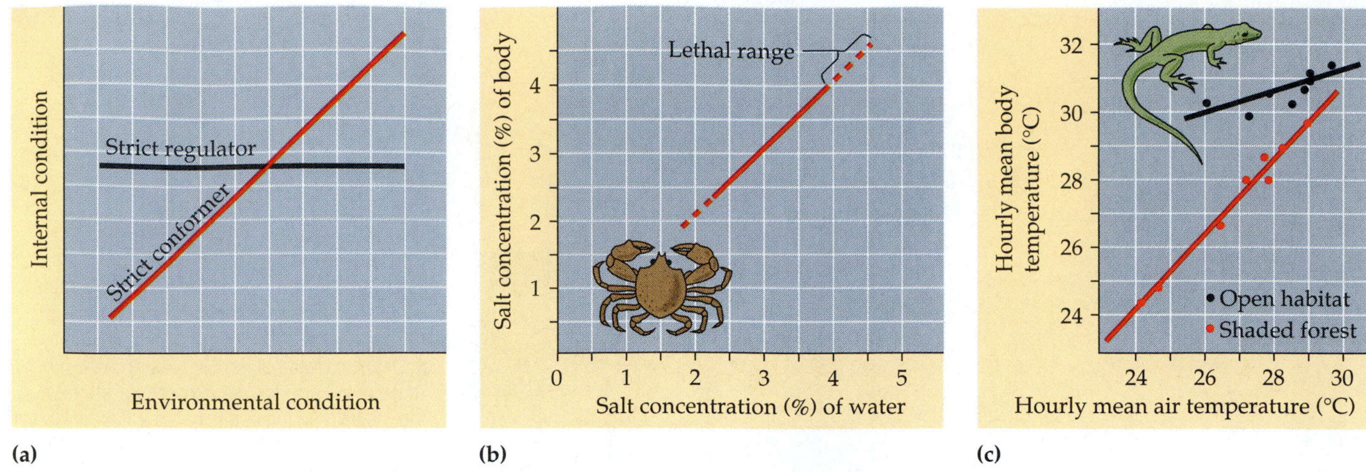

FIGURE 50.17 ▪ Regulators and conformers. (a) With respect to a given environmental variable, organisms can be described as either regulators, which maintain a nearly constant internal environment over a range of external conditions, or conformers, which allow their internal environment to vary. Although the two "strategies" are idealized in the illustration, most organisms are neither strict regulators nor strict conformers. **(b)** Spider crabs (*Libinia*) are osmoconformers with little ability to regulate internal osmolarity in the narrow range of salinity where they live. If exposed to slightly higher or lower salinity in the laboratory, the crabs continue to conform and soon die. **(c)** Some species change their regulatory mode in different environments. *Anolis cristatellus,* a small lizard on Puerto Rico, behaviorally thermoregulates in open habitats where it can bask in the plentiful patches of sun (black dots), but it is a thermoconformer in shaded forests where basking sites are rarer (red dots).

are often **conformers**, allowing some conditions within their bodies to vary with external changes (FIGURE 50.17a). Many marine invertebrates, such as spider crabs of the genus *Libinia,* live in environments where the salinity is very stable. These organisms do not osmoregulate, and if placed in water of varying salinity, they will lose or gain water to conform to the external environment even when this internal adjustment is extreme enough to cause death (see FIGURE 50.17b).

Conforming and regulating represent extremes on a continuum, and few organisms are perfect regulators or conformers. For example, the salmon previously described can osmoregulate, but they conform to external temperatures. Even endotherms such as ourselves are not perfect thermoregulators; anyone living in a cold climate has certainly noticed how cool exposed extremities such as hands, nose, and ears become on a cold day.

Many species are conformers under certain environmental conditions but can regulate to some extent under others. Regulation requires the expenditure of energy, and in some environments the cost of regulation may outweigh the benefits of homeostasis. For example, temperature regulation would require the forest-dwelling lizard *Anolis cristatellus* to travel long distances (and risk capture by a predator) to find an exposed sunny perch. It is therefore likely to survive longer and produce more offspring by allowing its body temperature to conform to that of the forest environment. However, this same species behaviorally thermoregulates in open habitats where it can bask in the plentiful patches of sun (FIGURE 50.17c).

The Principle of Allocation

One concept organismal ecologists have found useful in assessing the responses of organisms to their complex environments is the **principle of allocation**. This principle holds that each organism has a limited amount of energy that can be allocated for obtaining nutrients, escaping from predators, coping with environmental fluctuations (maintaining homeostasis), growth, and reproduction. Energy expended for homeostasis is therefore not available for other functions. For example, in grasshoppers, which are moderately active ectotherms, about 30% of assimilated energy remains after the animal's basic maintenance needs are met. This energy can be channeled into growth or reproduction. In contrast, for a very active endotherm, such as a weasel that uses most of its ingested energy to stay warm and active, only 2.5% of assimilated energy remains, and for a wren, only 0.5% remains. For the latter organisms there must be significant evolutionary advantages to endothermy and higher activity levels that offset the high maintenance costs.

Different priorities in energy allocation are related to the distribution of organisms and their homeostatic mechanisms. Conformers that live in very stable environments, such as the spider crabs in FIGURE 50.17b, might be able to channel more of their energy into growth and reproduction. However, the intolerance of such specialists to environmental change severely restricts their geographical distribution. In contrast, regulators that allocate a larger fraction of their energy to coping with environmental changes may grow and propagate less efficiently, but such organisms are able to survive and reproduce over a wider range of variable environments.

An organism's short-term responses to environmental variations operate within a long-term evolutionary framework

Organisms can react to variations in their environment with a variety of short-term physiological, morphological, and behavioral responses. However, it is important to remember that all such responses occur within a framework of adaptations fashioned by natural selection acting over evolutionary time. For example, all plants are capable of changing the size of the stomata of their leaves, a physiological response that helps prevent desiccation under environmental conditions when transpiration exceeds delivery of water. In plants living in the desert, the ability to adjust the size of the stomatal openings in response to water stress is superimposed on many other anatomical and physiological adaptations that have accumulated over evolutionary time as these plants have evolved in their arid environments. For instance, some desert plants have their stomata in pits, protected from the hot, dry winds that accelerate transpiration. Also common among desert plants is the CAM pathway of photosynthesis (see Chapter 10), which enables the plants to keep their stomata closed during the daytime.

When we observe a certain behavioral, physiological, or morphological response, two general kinds of questions arise. So-called *proximate questions* concern mechanisms operating in ecological time. For instance, what environmental stimuli, if any, trigger the response, and what internal mechanisms of the organism underlie the response? Proximate questions about a biological clock, for example, would be concerned with its structure and function, how it synchronizes an organism's activities with seasonal changes, as well as the genetic mechanisms that specify the clock's development. In contrast, so-called *ultimate questions* are evolutionary ones: Why did natural selection favor a particular clock mechanism and not a different one? Hypotheses that address ultimate questions propose that the response maximized fitness (survival and reproductive success) in some particular way.

The search for answers to ultimate and proximate questions is not an either/or dichotomy. Understanding the relationship between an organism and its environment requires knowledge of both short-term mechanisms and the long-term evolutionary framework in which the mechanisms operate. Indeed, the distinction between short-term mechanisms on the scale of ecological time and adaptation on the scale of evolutionary time begins to blur when we consider that the range of responses of an individual to changes in the environment is itself the product of evolutionary history. For example, when a mammal or bird uses physiological adjustments to maintain constant body temperature in the face of fluctuations in environmental temperature, it is utilizing mechanisms of homeostasis that are adaptations acquired by natural selection.

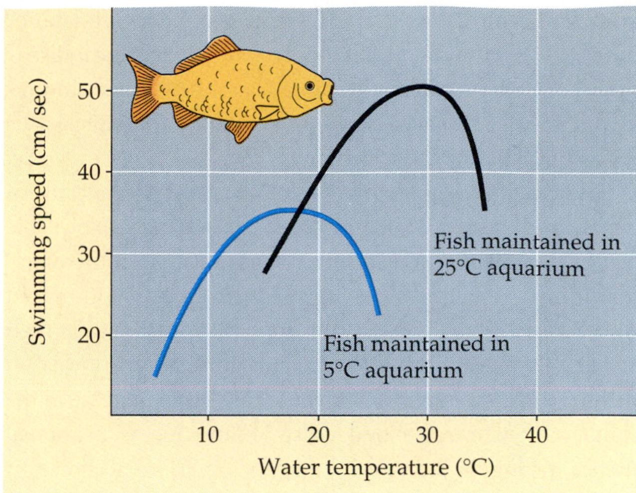

FIGURE 50.18 ▪ Tolerance curves and acclimation. Goldfish, like all ectothermic animals, have an optimal temperature for physiological function as well as a range of temperatures they can tolerate. In this example, swimming speed, which varies with temperature, is used as an index of the fish's general well-being. Goldfish normally live at temperatures between 25° and 30°C, but both the optimal temperature and the tolerance range can be shifted somewhat when the fish is gradually acclimated to a water temperature of 5°C. Acclimation is a slow process, and the fish would not survive a sudden, large change in water temperature. Acclimation also involves trade-offs. Notice that the cold-acclimated fish swims faster than the warm-acclimated fish at 15°C but much more slowly at 25°C.

Physiological Responses

Physiological responses to environmental change may be implemented and changed over time scales ranging from seconds to weeks. Physiological responses that involve relatively small changes in the rates of processes, and that do not require alteration of body structure or biochemical pathways, can occur very rapidly. For example, when you venture outside on a very cold day, blood vessels in your skin may constrict within seconds, a physiological response that minimizes the loss of body heat.

Regulation and homeostasis are the hallmarks of physiological responses. However, all organisms, whether regulators or conformers, function most efficiently under certain environmental conditions. We can study an organism's response to changing environmental conditions in the laboratory by varying a single abiotic factor, such as temperature, and measuring some aspect of the organism's performance. The resulting tolerance (or performance) curves are approximately bell-shaped, with peak performance at some optimal condition and the tails of the curve representing the limits of the organism's tolerance to the particular environmental variable. FIGURE 50.18 shows tolerance curves for the effect of temperature on the swimming speed of goldfish. Tolerance limits are important determinants of the geographical distribution of organisms, though biological interactions can prevent a species from occupying a habitat to which it is physiologically adapted (see Chapter 53).

Physiological responses to environmental variation can also include **acclimation**, which involves substantial but reversible changes that shift an organism's tolerance curve in the direction of the environmental change. For example, if you moved from Boston, which is essentially at sea level, to the mile-high city of Denver, one physiological response to the lower O_2 pressure in your new environment would be an increase in the number of your red blood cells. Acclimation is a gradual process, taking days or weeks, and the ability to acclimate is generally related to the range of environmental conditions the species naturally experiences. Species that live in very hot climates, for example, usually do not acclimate to extreme cold.

Morphological Responses

Morphological responses—those that alter the form or internal anatomy of the body—may develop over the lifetimes of individual organisms or across generations. In some cases these responses are examples of acclimation, since they are reversible. Many mammals and birds, for example, grow a heavier coat of fur or feathers in winter; sometimes coat color changes seasonally as well, camouflaging the animal against winter snow and summer vegetation.

Other morphological changes are irreversible over the lifetime of an individual. In many cases environmental variation can affect growth and differentiation patterns, often leading to remarkable morphological variation within a species. In general, plants are more morphologically plastic than animals; this response helps them compensate for their inability to move from one environmental patch to another. One example is the arrowleaf plant, which can grow on land, rooted in water with its upper leaves emerging above the surface, or completely submerged in water. The leaf structure of this plant varies with the environment in which the leaves grow. Submerged leaves are flexible, bending with the currents, and, lacking a waxy cuticle, are able to absorb mineral nutrients from the surrounding water. Arrowleaf plants growing on land have more extensive root systems, and their leaves are more rigid and covered with a thick cuticle that reduces water loss.

Behavioral Responses

Behavioral responses can be almost instantaneous in their effects and are easily reversed. Behavioral response in the sense of muscular reaction to a stimulus is limited to animals. The quickest response of many animals to an unfavorable change in the environment is to move to a new location. Such movement may be fairly localized. For example, many desert animals escape intense heat by burrowing, and they maintain a reasonably constant body temperature while active by shuttling between sun and shade. Some animals are capable of migrating great distances in response to such environmental cues as changes in temperature or changes in photoperiod associated with seasonal transitions. Many migratory birds overwinter in Central and South America, returning to northern latitudes to breed in the summer.

Some animals are able to modify their immediate environment by cooperative social behavior. Honeybees, for instance, can cool the inside of their hive on hot days by the collective beating of their wings. During cold periods they seal the hive, helping retain the heat generated by their activity inside. Many small mammals huddle within burrows during cold weather, a behavioral mechanism that reduces heat loss by minimizing the total amount of surface the animals expose to the cold air.

■ ■ ■

In Unit Six (for plants) and Unit Seven (for animals) we discussed various structural and physiological mechanisms that have evolved as adaptations to the constraints of specific environments. In the next chapter we focus on behavioral mechanisms and adaptations, which are key aspects of how animals interact with their environments.

CHAPTER REVIEW

REVIEW OF KEY CONCEPTS

(with page numbers and key figures)

THE SCOPE OF ECOLOGY

- **Ecology is the scientific study of the interactions between organisms and their environments (pp. 1026–1027)** Ecology uses observations and experiments to test hypothetical explanations of these interactions. The environment includes both abiotic (nonliving) and biotic (living) components. Ecological interactions affect how organisms evolve, and evolutionary change in turn affects ecological relationships.
- **Ecological research ranges from the adaptations of organisms to the dynamics of ecosystems (pp. 1027–1028)** Ecological research spans increasingly comprehensive levels of organization, from the individual organism through populations, communities, and ecosystems to the biosphere (the global ecosystem).
- **Ecology provides a scientific context for evaluating environmental issues (p. 1028)** Although environmental problems have political, economic, and ethical components, it is ecology that provides the scientific basis for understanding these problems.

ABIOTIC FACTORS OF THE BIOSPHERE

- **Climate and other abiotic factors are important determinants of the biosphere's distribution of organisms (pp. 1028–1034, FIG-**

URES 50.4–50.6) The biosphere is an environmental mosaic in which several abiotic factors affect the distribution and abundance of organisms: temperature, water quality and availability, light intensity, wind, soil characteristics, and less predictable disturbances such as fire. Global climates and seasonality are established by the input of solar energy and Earth's rotation around the sun. Latitudinal variation in temperature and precipitation account for the geographical distribution of major ecosystems. Oceans and lakes moderate the climate in coastal localities, and mountains influence temperature and rainfall.

AQUATIC AND TERRESTRIAL BIOMES

- **Aquatic biomes occupy the largest part of the biosphere (pp. 1034–1042, FIGURE 50.9)** Oceans, covering nearly 75% of Earth's surface, have a major effect on climate, and marine organisms supply a substantial part of the world's oxygen (O_2). Freshwater biomes are closely linked to their surrounding terrestrial biomes. Aquatic biomes are often stratified vertically with regard to light penetration, temperature, and community structure. Eutrophic lakes are high in nutrients and productivity, and oligotrophic lakes are nutrient-poor. Runoff from surrounding land transforms oligotrophic lakes into eutrophic lakes. Rivers and streams contain freshwater communities that change significantly from the source to the final destination in an ocean or lake. Wetlands have soil ranging from periodically flooded to permanently saturated. Wetlands are biologically diverse, containing specially adapted plants called hydrophytes, which can grow in water or soil that is periodically anaerobic. An estuary is the zone of marked fluctuations in salinity where a river or stream enters the ocean.

  Oceanic zones, classified according to depth, light penetration, distance from the shore, and open water versus bottom, include the intertidal zone, the neritic zone, and the oceanic zone. In the tropics, coral reefs are found in the warm, nutrient-rich waters of the neritic zone. The oceanic pelagic biome includes most of the open ocean. Photosynthetic plankton in the photic region of the pelagic zone are the primary food source for the rest of the community. Benthic, or bottom, communities subsist largely on detritus that rains down from the pelagic zone.

- **The geographical distribution of terrestrial biomes is based**
50.1 **mainly on regional variations in climate (pp. 1042–1043, FIGURE 50.15)** Near the equator, where photoperiod and temperature are nearly constant, the amount and pattern of rainfall determines biomes, including tropical rain forest and savanna. Deserts are inhabited by plants and animals adapted to extremely dry conditions. Chaparral is a dry scrubland found where winters are mild and rainy and summers are hot and dry. Temperate grasslands occur on nutrient-rich, deep soils where periodic fires and drought and the grazing of large mammals inhibit the growth of woody plants. Temperate deciduous forests occur in midlatitudes where there is sufficient moisture to support the growth of large, broadleaf deciduous trees. Coniferous forests include coastal temperate rain forests and the northern coniferous forest, or taiga. The largest terrestrial biome, taiga, is characterized by long, cold, snowy winters and short summers. Arctic tundra occurs at the northernmost limits of plant growth, where cold temperatures, wind, and permafrost limit plants to low shrubby or matlike forms. Alpine tundra occurs at high altitudes.

CONCEPTS OF ORGANISMAL ECOLOGY

- **The costs and benefits of homeostasis affect an organism's responses to environmental variation (pp. 1043–1049, FIGURE 50.17)** An organism's adaptations to its environment represent a set of evolutionary compromises. Organisms that are primarily regulators use physiological or behavioral mechanisms to achieve homeostasis, while conformers allow a particular internal condition to vary with external changes. Because the energy an organism allocates to homeostasis is unavailable for its other processes, a homeostatic mechanism must increase its fitness.

- **An organism's short-term responses to environmental variations operate within a long-term evolutionary framework (pp. 1049–1050, FIGURE 50.18)** The relationship between an organism and its environment involves both short-term response mechanisms and the long-term evolutionary framework in which short-term responses operate. Proximate questions concern the mechanisms by which an organism adapts to its environment, while ultimate questions concern the evolutionary explanations of these mechanisms. Short-term mechanisms include physiological, morphological, and behavioral responses to change. Physiological responses can be very rapid and reversible, as occur in acclimation. Many plants exhibit irreversible morphological plasticity to compensate for their immobility. Animals exhibit behavioral responses, including moving or migrating to more favorable locations.

SELF-QUIZ

1. Which statement follows from the principle of allocation?
   a. The number of organisms an area can support is determined by its energy supply.
   b. Physiological adjustments to environmental changes can extend the tolerance limits of organisms.
   c. The total amount of energy available to an organism is partitioned into such processes as reproduction, obtaining nutrients, and coping with the environment.
   d. Organisms that use more energy for growth and reproduction are able to survive in a wider range of variable environments.
   e. Organisms allocate most of their energy for homeostasis.

2. Which statement about tolerance limits is *incorrect*?
   a. They can often be tested experimentally and plotted as a tolerance curve.
   b. They help determine whether organisms can live in particular environments.
   c. They can be extended by acclimation in some cases.
   d. They are likely to be greater in regulators than in conformers.
   e. They are generally greatest for organisms restricted to stable environments.

3. Which of the following biomes is *correctly* paired with the description of its climate?
   a. savanna—cool temperature, precipitation uniform during the year
   b. tundra—long summers, mild winters
   c. temperate deciduous forest—relatively short growing season, mild winters
   d. temperate grasslands—relatively warm winters, most rainfall in summer
   e. tropical forests—nearly constant photoperiod and temperature

4. The oceans affect the biosphere in all of the following ways *except*
   a. producing a substantial amount of the biosphere's oxygen (O_2)
   b. removing carbon dioxide (CO_2) from the atmosphere
   c. moderating the climate of terrestrial biomes
   d. regulating the pH of freshwater biomes and terrestrial groundwater
   e. being the source of most terrestrial rainfall

5. Which of the following is *correctly* paired with its description?
   a. neritic zone—shallow area over continental shelf
   b. benthic zone—surface water of shallow seas
   c. pelagic zone—seafloor
   d. aphotic zone—zone in which light penetrates
   e. intertidal zone—open water at the edge of the continental shelf

6. Which of the following do all terrestrial biomes have in common?
   a. annual average rainfall in excess of 25 centimeters
   b. a distribution predicted almost entirely by rock and soil patterns
   c. clear boundaries between adjacent biomes
   d. vegetation demonstrating vertical stratification
   e. a biodiversity pattern that is directly proportional to latitude

7. Hummingbirds living at elevations above 4500 meters enter torpor when the sun sets and they are unable to forage. This reaction to environmental variation illustrates
   a. a morphological response
   b. ultimate causation
   c. an endothermic response
   d. accommodation
   e. a physiological response

8. The growing season would generally be shortest in which biome?
   a. tropical rain forest
   b. savanna
   c. taiga
   d. temperate deciduous forest
   e. temperate grassland

9. Imagine some cosmic catastrophe that jolts Earth so that it is no longer tilted—instead its axis is perpendicular to the line between the sun and Earth. The most predictable effect of this change would be
   a. no more night and day
   b. a big change in the length of the year
   c. a cooling of the equator
   d. a loss of seasonal variations at northern and southern latitudes
   e. the elimination of ocean currents

10. While climbing in the Rocky Mountains, one observes transitions in biological communities that are analogous to the changes one encounters
    a. in biomes at different latitudes
    b. at different depths in the ocean
    c. in a community through different seasons
    d. in an ecosystem as it evolves over time
    e. traveling across the United States from east to west

CHALLENGE QUESTIONS

1. In which terrestrial biome is your college or university located? Describe how five local features on your campus influence microclimates.
2. Describe five abiotic factors that might be important to the life of a tree in a temperate deciduous forest. How might the tree adjust to day-to-day changes in these factors?

SCIENCE, TECHNOLOGY, AND SOCIETY

1. During the summer of 1988, huge forest fires burned a large portion of Yellowstone National Park. The National Park Service has a natural-burn policy: Fires that start naturally are allowed to burn unless they endanger human settlements. Lightning ignited the Yellowstone fires, so they were allowed to spread and burn themselves out as much as possible; firefighters primarily protected people. This drew a lot of public criticism; the Park Service was accused of letting a national treasure go up in flames. Park Service scientists stuck with the natural-burn policy. Do you think this was the best decision? Support your position.
2. Although chaparral habitats burn regularly, many people in southern California and elsewhere build expensive houses in canyons and on hillsides covered with this vegetation. Should local governments regulate development in such communities? Should the government continue to provide disaster relief to people whose homes burn in these periodic fires so that these people can rebuild in the hills?

FURTHER READING

Adler, T. "The Expiration of Respiration." *Science News,* February 10, 1996. Provides an overview of the significance of oxygen and its paucity in many aquatic ecosystems.

Heinrich, B. "In Plain Sight." *Natural History,* July 1995. Describes how certain bird species hide their nests on the open tundra.

Holland, H. D., and U. Petersen. *Living Dangerously: The Earth, Its Resources, and the Environment.* Princeton, NJ: Princeton University Press, 1995. An introduction to basic ecology and environmental science.

Hunt, L. 1998. "Send in the Clouds." *New Scientist,* May 30, 1998. Introduces a rigorous new look at the controversial Gaia hypothesis—that organisms regulate Earth's ecosystems.

Kleiner, K. "Fanning the Wildfires." *New Scientist,* October 19, 1996. Discusses the biological and political aspects of fire in ecosystems.

Maurer, B. A. "Ecological Science and Statistical Paradigms: At the Threshold." *Science,* January 23, 1998. Examines the scientific training needed to cope with rapid environmental changes.

Nicol, S., and I. Allison. "The Frozen Skin of the Southern Ocean." *American Scientist,* September–October, 1997. Surveys the complex biotic and abiotic factors of Antarctica in a global context.

Reice, S. R. "Nonequilibrium Determinants of Biological Community Structure." *American Scientist,* September–October 1994. An illustrated discussion of the key role of disturbance in ecosystems.

Ricklefs, R. E. *The Economy of Nature,* 3rd ed. New York: W. H. Freeman, 1993.

Rützler, K., and I. C. Feller. "Caribbean Mangrove Swamps." *Scientific American,* March 1996. A look at a highly diverse and delicate tropical ecosystem.

Smith, R. L. *Ecology and Field Biology,* 5th ed. Menlo Park, CA: Benjamin/Cummings, 1996. An introductory text.

"Unknown Oceans." *New Scientist,* supplement, November 2, 1996. Introduces some major research questions.

WEB LINKS

Visit the special edition of *The Biology Place* for BIOLOGY, Fifth Edition, at **http://www.biology.com/campbell**. Go to Chapter 50 for online resources, including learning activities, practice exams, and links to the following web sites:

"The U.S. Environmental Protection Agency"
Explore the EPA, whose mission is to protect human health and to safeguard the natural environment.

"Biosphere 2"
Take a virtual tour through Biosphere 2, a 7,200,000-cubic foot sealed glass and spaceframe structure containing seven wilderness ecosystems.

"A Scientist Alerts the Public to the Hazards of Pesticides"
Find out more about the life and times of Rachel Carson, author of the epic book *Silent Spring.*

"National Oceanic and Atmospheric Administration"
Visit the homesite of NOAA, whose mission is to describe and predict changes in Earth's environment and to conserve and manage wisely the nation's coastal and marine resources.

CHAPTER 51

BEHAVIORIAL BIOLOGY

Introduction to Behavior and Behavioral Ecology

- Behavior results from both genes and environmental factors
- Innate behavior is developmentally fixed
- Classical ethology presaged an evolutionary approach to behavioral biology
- Behavioral ecology emphasizes evolutionary hypotheses: *science as a process*

Learning

- Learning is experience-based modification of behavior
- Imprinting is learning limited to a critical time period
- Many animals can learn to associate one stimulus with another
- Practice and exercise may explain the ultimate bases of play

Animal Cognition

- The study of cognition connects nervous system function with behavior
- Movement from place to place often depends on internal coding of spatial relationships
- The study of consciousness poses a unique challenge for scientists

Social Behavior and Sociobiology

- Sociobiology places social behavior in an evolutionary context
- Competitive social behaviors often represent contests for resources
- Mating behavior relates directly to an animal's fitness
- Social interactions depend on diverse modes of communication
- The concept of inclusive fitness can account for most altruistic behavior
- Sociobiology connects evolutionary theory to human culture

The study of animal behavior is undoubtedly one of the oldest aspects of biology. Tens of thousands of years ago, behavioral knowledge was essential to human survival. By learning the habits of the animals around them, early humans increased their chances of securing a meal and decreased their chances of becoming a meal. Thus, our ancestors' study of animal behavior ultimately enhanced their Darwinian fitness. More generally, our own behavior and that of other animals has its ultimate basis in evolution.

Take the song of the male Magnolia Warbler in the photograph on this page, for instance. To our ears, it is a musical "weete, weete, weetchew," with the last note slurring upward. However, as one experienced birder is fond of telling novices, "It's not music to a bird's ears. Birds sing for practical reasons—to attract mates, to let other males or females know where they are, to hold a territory where they can raise and feed their young. For birds, singing is about survival and passing genes to the next generation."

Bird song provides an excellent introduction to the subject of behavior. For one thing, bird song is amenable to modern experimental methods. As a result, biologists are beginning to understand a great deal about development, functions, and consequences of bird song at virtually all levels of biological organization, from molecules to whole organisms to entire populations. Bird song is also attractive to researchers because of several striking parallels between it and human speech. It has also become a model system for animal behavior research because it demonstrates a very important generalization: Behavior is influenced by both genetic and environmental factors. The study of bird song has provided guideposts for the study of other complex behaviors that are less well understood.

Studying an animal's behavior is essential to understanding the animal's evolution and ecological interactions. This chapter emphasizes the nature of animal behavior, how biologists study it, and the function of behavior in the relationship between an animal and its environment.

INTRODUCTION TO BEHAVIOR AND BEHAVIORAL ECOLOGY

A dictionary definition of behavior may read something like "to act, react, or function in a particular way in response to some stimulus." Much of behavior does indeed consist of externally observable muscular activity, the "act" and "react" components of this definition. But a young bird that hears an adult song may show no obviously related muscular activity. Instead, the memory of the song may be stored in the bird's brain. Although this memory may change the brain's functioning, any observable muscular response will come later, perhaps even months later, when the bird begins to learn to match this stored memory of an adult song. Thus, if we think of **behavior** as what an animal does and *how* it does it, this definition will encompass the nonmotor components of behavior as well as an animal's observable actions.

When we observe a certain behavior, we are apt to ask both *proximate* and *ultimate* questions (see Chapter 50). In the study of animal behavior, proximate questions are mechanistic—concerned with the environmental stimuli, if any, that trigger a behavior, as well as the genetic and physiological mechanisms underlying a behavioral act. Ultimate questions address the evolutionary significance of behavior. To emphasize the distinction (and also the connection) between proximate and ultimate causation, consider the observation that the Magnolia Warbler, like many animals, breeds in spring and early summer. In terms of proximate causation, a reasonable hypothesis is that breeding is triggered by the effect of increased day length on an animal's photoreceptors. Many animals can be stimulated to begin breeding by experimentally lengthening their daily exposure to light. This stimulus results in neural and hormonal changes that induce behavior associated with reproduction, such as singing and nest building in birds.

In contrast to proximate questions, ultimate ones take such forms as, Why did natural selection favor this behavior and not a different one? Hypotheses addressing "why" questions propose that the behavior maximizes fitness in some particular way. A reasonable hypothesis for why many animals reproduce in spring and early summer is that this is when breeding is most productive or adaptive. For warblers and many other birds, an abundant supply of insects in the spring provides ample food for rapid growth of offspring. Individuals that attempt to breed at other times would be at a selective disadvantage. Increased day length itself has little adaptive significance, but since it is the most reliable indicator of time of year, there has been selection for a proximate mechanism that depends on increased day length. In brief, proximate mechanisms produce behaviors that ultimately evolved because they increase fitness in some particular way. Behavioral biologists also use the comparative methods of phylogenetic biology (see Chapter 25) to formulate hypotheses about the evolution of behavior. Phylogenetic trees based on molecular, morphological, or behavioral data and illustrating the most likely evolutionary history of a closely related group of species enable researchers to estimate when a particular behavior arose in a lineage, whether it arose once or repeatedly, and which kinds of behavior occurred in ancestors.

Behavior results from both genes and environmental factors

A myth that is still perpetuated to some extent by popular media is that behavior is due *either* to genes (nature) *or* to environmental influences (nurture). In biology, however, the nature versus nurture debate is not about either/or; it is about the degrees to which genes and the environment influence phenotypic traits, including behavioral ones. As we discussed in Chapter 14, phenotype depends on both genes and the environment; behavioral traits have genetic and environmental components, as do all of an animal's anatomical and physiological features.

Like other phenotypic traits, behaviors exhibit a range of phenotypic variation (a "norm of reaction") depending on the environment in which the genotype is expressed (see Chapter 14). The lovebird case study depicted in FIGURE 51.1 illustrates behavior with a strong genetic influence. Nonetheless there is a norm of reaction: The behavior can be modified by environmental experience. At another extreme, the most advanced forms of problem solving are characterized by very broad norms of reaction. However, they too have a genetic component—they depend on genes whose expression creates a neural system receptive to advanced learning. Most behavioral traits are polygenic, with broad norms of reaction.

So what do the media reports of newly discovered "genes for" complex human behavioral traits, such as depression, violence, or alcoholism, really mean? Robert Plomin, director of the Center for Developmental and Health Genetics at Pennsylvania State University, puts it this way: "Research into heritability [of behavior] is the best demonstration I know of the importance of environment. They [genes and nongenetic environmental factors] build on each other."

The environmental factors that affect behavior are *all* conditions in which the genes underlying behavior are expressed. This includes the chemical environment within cells, as well as all the hormonal and other chemical and physical conditions experienced by a developing animal within an egg or a womb. It also includes the multiple interactions among components of an animal's nervous system and effectors, as well as the varied chemical, visual, auditory, or tactile interactions with other organisms.

Innate behavior is developmentally fixed

If behavior has both genetic and environmental underpinnings, what do we mean when we say a particular behavior is innate? For example, newly hatched, still blind birds of many species beg for food by raising their heads, opening their mouths, and cheeping loudly when a parent lands on the side of the nest. This behavior is often attributed to genetic programming without any environmental influence. However, it is imprecise to say that any behavior is due solely to genes. All genes, including those whose expression underlies innate behavior, require an environment (a physical body) to be expressed. The key point about innate behavior is that the range of environmental differences among individuals does not appear (from all the studies to date) to alter the behavior. Although usage of the term *innate* varies, in behavioral biology it refers to behavior that is *developmentally fixed;* all individuals exhibit virtually the same behavior despite the

① Several species of brightly colored African parrots, commonly known as lovebirds, build cup-shaped nests inside tree cavities. Females typically make nests with thin strips of vegetation (or, in the laboratory, paper) that they cut with their beaks. In one species, Fischer's lovebird *(Agapornis fischeri)*, the bird cuts relatively long strips and carries them back to the nest one at a time in her beak.

❶ Nests made with long strips—no tucking behavior

② In contrast, the peach-faced lovebird *(A. roseicollis)* cuts shorter strips and usually carries several at a time by tucking them into the feathers of the lower back. Tucking is a fairly complex behavior because the strips must be held just right and pushed in firmly, and the feathers then smoothed over.

❷ Nests made with short strips—tucking behavior

③ These two species are closely related and have been experimentally interbred. The resultant hybrid females exhibited an intermediate kind of nest-building behavior. The strips cut by the hybrid birds were intermediate in length; even more interesting was the birds' hybrid manner of handling the strips. They usually made some attempt to tuck them into their rear feathers, but in some cases they did not let go after turning and pushing them a short distance. In other cases the strips were manipulated or inserted improperly or simply dropped. The result was almost a total failure to transport strips by this method. Eventually, the birds learned to transport the strips in their beaks. Even so, they always made at least token tucking attempts.

❸ Hybrid nests made with intermediate-length strips—in first mating season, unsuccessful tucking behavior

④ After several years, the birds still turned their heads to the rear before flying off with a strip. These observations demonstrate that the phenotypic differences in the behavior of the two species are based on different genotypes. We also see that the behavior can be modified by experience; the hybrid birds eventually learned to transport the strips.

❹ In later seasons, only head-turning behavior

FIGURE 51.1 ▪ Genetic and environmental components of behavior: a case study.

inevitable environmental differences within and outside their bodies during development and throughout life.

In a broad sense, the ultimate cause for innate behavior may be that performing some behaviors automatically, without having any specific experience, may have maximized fitness to the point that genes for variant behavior were lost.

Classical ethology presaged an evolutionary approach to behavioral biology

Modern behavioral biology has its roots in a research field known as **ethology**, which originated in the 1930s with naturalists who tried to understand how a variety of animals behave in their natural habitats. Foremost among them were Karl von Frisch, Konrad Lorenz, and Niko Tinbergen, who shared a Nobel Prize in 1973 for their discoveries (FIGURE 51.2, p. 1056). How animals can carry out many behaviors without ever having seen them performed was one of the major subjects of early ethological research. Ethologists focused on proximate mechanisms, but with an eye toward the genetic links to behavior and to the adaptive nature of behavior, an orientation that helped connect behavioral biology with evolution and ecology.

Early ethologists developed the concept of a **fixed action pattern (FAP)**, a sequence of behavioral acts that is essentially unchangeable and usually carried to completion once initiated. A FAP is triggered by an external sensory stimulus known as a **sign stimulus.** In many cases the sign stimulus is some feature of another species. For example, some moths instantly fold their wings and drop to the ground in response

① A female digger wasp excavates and cares for four or five separate underground nests. She will fly to each nest daily, bringing food to the single larva in each nest. Biologist Niko Tinbergen designed field experiments to test his hypothesis that the wasp uses visual landmarks to keep track of where her nests are located. First, Tinbergen marked a wasp's nest with a ring of pinecones.

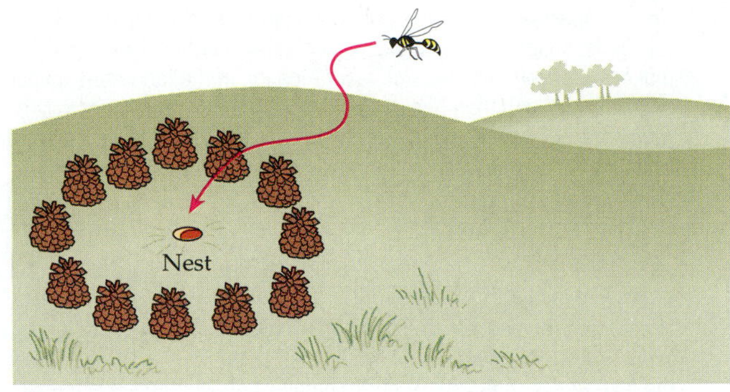

② After the mother wasp visited the nest and flew away, Tinbergen moved the pinecones a few feet to one side of the nest. When the wasp returned, she flew to the center of the pinecone circle instead of the nearby nest. The results of such experiments supported the hypothesis that digger wasps use landmarks to keep track of their nests, and that they can learn new visual cues.

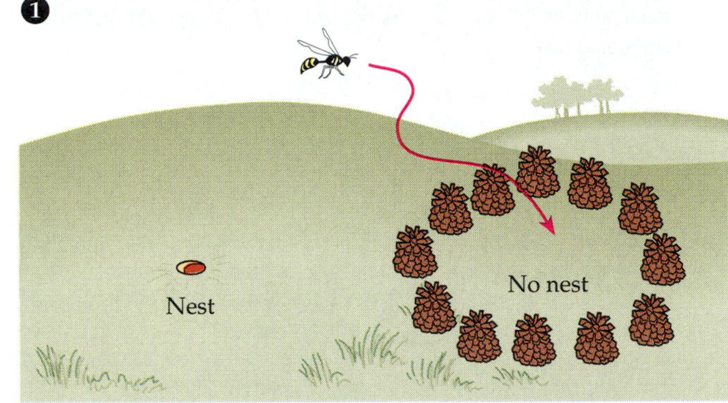

③ In a follow-up experiment, Tinbergen restored the pinecones to the actual nest site but arranged them in a triangle instead of a circle. He placed a circle of stones to one side of the nest. The returning wasp flew to the stone ring, a result supporting the hypothesis that the insect was cued by the arrangement of the landmarks rather than the physical objects themselves. Thus, the proximate cause of the wasp's nest-locating behavior is the environmental cue of the landmark arrangement and the response it elicits in the animal. For ultimate causation, a reasonable hypothesis is that the wasp's fitness is enhanced by the female's ability to store information about nest location and to use that information to find and service her nests.

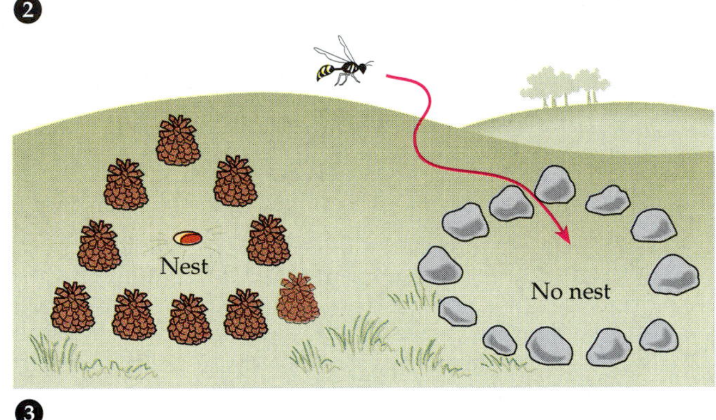

FIGURE 51.2 ▪ Niko Tinbergen's experiments on the digger wasp's nest-locating behavior.

to the ultrasonic signals sent out by predatory bats (see the introduction to Chapter 49, p. 992). The ultrasonic signals are the cue that triggers avoidance behavior in the moths.

A classic case of sign stimuli and FAPs can be observed in the male three-spined stickleback fish, which attacks other males that invade his territory. The stimulus for the attack behavior is the red belly of the intruder. The stickleback will not attack an invading fish lacking a red underside but will readily attack nonfishlike models as long as some red is present (FIGURE 51.3). Tinbergen, who first reported these findings, was inspired to look into the matter by his casual observation that his fish responded aggressively when a red truck passed their tank. As it turns out, red coloration of body parts tends to trigger either aggressive or sexual behavior in many species that have color vision.

Classic experiments on FAPs and sign stimuli indicated that many animals tend to use a relatively limited subset of the sensory information available to them and to behave stereotypically in many situations. In contrast, humans often tend to respond to an entire situation, and we generally base our actions on more diverse information. If a stickleback processed information like a human, it would realize that the models in FIGURE 51.3 are not real fish, despite their red bellies. Actually, relatively simple, stereotypical behaviors seem to occur in all animals, including humans. Human infants grasp strongly with their hands in response to a tactile stimulus. An infant's smile could also be considered a FAP; it is readily induced by simple stimuli such as a sound or a figure consisting of two dark spots on a white circle, a kind of rudimentary representation of a face.

Many researchers now consider the concept of FAPs overly simplistic, and modern behavioral biology is more concerned with understanding the adaptive function of behavior than with defining the precise nature of behavioral sequences. What can we say about the adaptive function of stereotypical behavior? Some ethologists have suggested that FAPs trig-

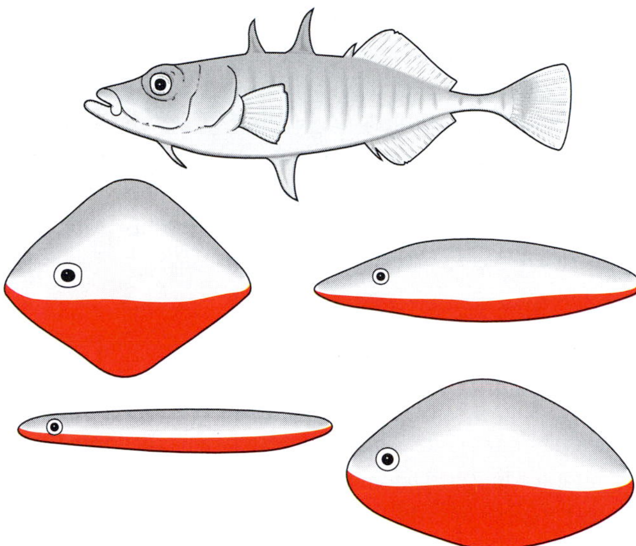

FIGURE 51.3 ▪ Classical demonstration of innate behavior. Aggression in male three-spined stickleback fish is triggered by a simple visual cue. The realistic model at the top without a red underside produces no response. All the others produce strong responses because they have the required red underside.

gered by simple cues prevent an animal from wasting time processing or integrating a wide variety of inputs. Perhaps a better way of interpreting the situation has to do with the limitations of stereotypical behavior and how it evolved. Close correlations exist between an animal's sensitivity to general stimuli and its behavior. For example, frogs have retinal cells that are especially good at detecting movement, and the movement of an object, such as a fly, is the stimulus that triggers a frog's tongue to shoot out and capture prey. A frog will starve if surrounded by dead or motionless flies but will attack one readily if it moves. A frog's sensory-neural network for detecting movement is probably much less complex than the apparatus that would be necessary to rapidly distinguish a fly from another object of similar size. Simple cues and a relatively simple, stereotypical behavior pattern usually work quite well in an animal's normal sensory world. An orientation toward adaptive function and ultimate causes leads us into the modern science of behavioral ecology, which seeks explanations in terms of enhancing fitness.

Behavioral ecology emphasizes evolutionary hypotheses: *science as a process*

Evolution is the core theme of biology, and today the study of behavior in an ecological context emphasizes evolutionary (ultimate) explanations. Because natural selection works on the enormous amount of genetic variation generated by mutation and recombination, we expect organisms to possess features that maximize their genetic representation in the next generation. Applied to behavior, this concept means that we expect animals to behave in ways that maximize their fitness. For example, feeding behavior is likely to optimize net energy gain, and the choice of a healthy mate tends to maximize the number of healthy offspring produced.

The research approach based on the expectation that animals increase their Darwinian fitness by optimal behavior is called **behavioral ecology**. Of course, this expectation is valid only because genes influence behavior; if there were no genetic influence, behavior could not be subjected to natural selection and could not evolve.

In the rest of this section we will explore some examples of research in behavioral ecology.

Songbird Repertoires

Suppose you became interested in the observation that many songbirds have a repertoire of song types. Some of these songs sound identical to us but can be easily distinguished when analyzed with a sound spectrograph (FIGURE 51.4). Why has

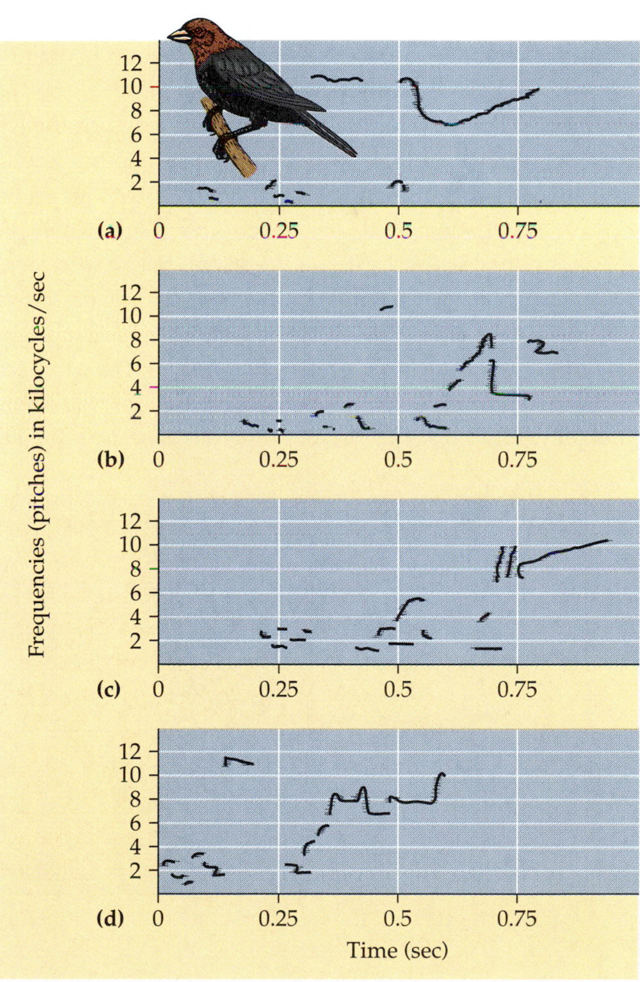

FIGURE 51.4 ▪ The repertoire of a songbird. These sonograms, or "voiceprints," show a graph of a sound's frequency (perceived as pitch) versus time. Shown are four distinct song types of one male brown-headed cowbird. The song in (a) is quite distinct and easily distinguishable from the other three, which would sound similar to us but probably not to a bird. Individual cowbirds generally have three to six song types, but other species have dozens or even hundreds of types.

natural selection favored this multisong behavior over the expert vocalization of a single tune? Following the approach of behavioral ecology, you could formulate several testable hypotheses, all starting with "A repertoire increases fitness because . . ." You might hypothesize that a repertoire increases fitness because it makes an older, more experienced male more attractive to females. For this hypothesis to be true, it must be the case that: (1) males learn more song types as they get older, so that repertoire size is a reliable indicator of age; and (2) females prefer to mate with males having large repertoires. Thus, your hypothesis makes two clearly testable predictions.

To test the first prediction, you can determine whether there is a correlation between a male's age and the size of his song repertoire. If there is not, your hypothesis will be invalidated, which would be informative although perhaps disappointing. As it turns out, some songbird species show this correlation, while others do not. Next, you can determine whether females are more sexually stimulated by a large song repertoire than by a small one. This can be done by playing tape-recorded male songs to females that have been made especially receptive to male song by the temporary administration of a female hormone. Such females indicate their song preferences by assuming a copulatory posture, even though there is no male around. FIGURE 51.5 shows another way to assess female responses. All this work may lead to an evolutionary explanation: Bird-song repertoires result in females mating more often or earlier in the season with experienced males, who will give their offspring a better chance to survive.

Now suppose you did not use evolutionary principles to guide your research on song repertoires. Without an approach that generates testable hypotheses and predictions, you would probably record observations of numerous aspects of singing behavior. Although these efforts may produce interesting data, they would not *explain* the behavior. Alternatively, you might hypothesize that repertoires have nothing to do with fitness, but that male birds simply find variety more pleasurable than singing the same boring song over and over again. The method for testing such a hypothesis is not clear, emphasizing again how much more productive it is to use evolutionary principles as a guide to behavioral research.

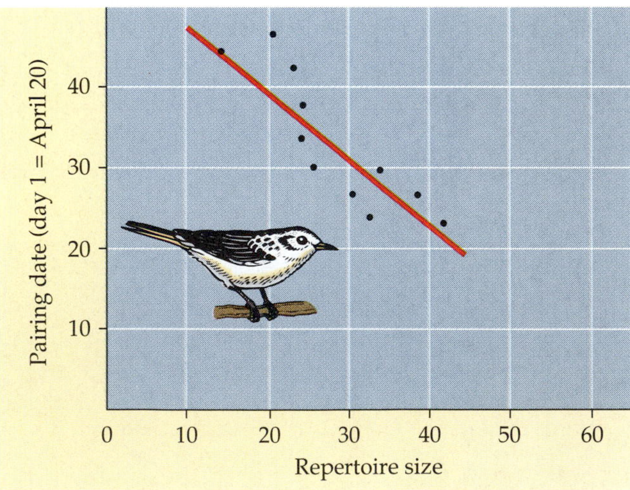

FIGURE 51.5 ▪ Female warblers prefer males with large song repertoires. Male sedge warblers with large repertoires attract females to pair with them earlier in the breeding season, as well as more often, than males with small repertoires. It is therefore likely that females prefer large repertoires. Males with large repertoires benefit from pairing early because breeding early tends to be more successful than breeding late in the season.

Cost/Benefit Analysis of Foraging Behavior

Feeding is obviously essential to survival and reproductive success, but what determines exactly what an animal eats? Because feeding behavior, or foraging, lends itself to experimental analysis at both the proximate and the ultimate level, it has become a favorite research topic among behavioral ecologists.

Animals feed in many ways, using various foraging behaviors that are closely linked to morphological traits. Suspension-feeding, for example, requires behavior different from that required for active predation (see Chapter 41). Ecological and evolutionary considerations are also extremely important; food habits are a fundamental part of an animal's niche and may be shaped in part by competition with other species. In this section, we will examine foraging from a behavioral perspective and see how behavioral ecologists are using cost/benefit analysis to study the proximate and ultimate causes of diverse foraging "strategies."

Hypothetically animals of many species can choose from a large array of potential foods. Some animals tend to be generalists, feeding on a wide variety of items. Gulls feed on material that may be living or dead, aquatic or terrestrial, plant or animal. In contrast, oyster catchers, wading shorebirds that forage in the intertidal zone, often feed on very specific species of shellfish. In fact, oyster catchers are so specialized that they use individual hunting techniques, learned from their parents, that further restrict the size and location of the shellfish they eat. Specialists usually have morphological and behavioral adaptations that are highly specific to their food, and as a result they are extremely efficient at foraging. Generalists cannot be as efficient at securing any one type of food, but they have the advantage of having other options if a preferred food becomes unavailable.

Most generalists do not choose their food randomly. Often an animal will concentrate on a particular item when it is abundant, sometimes to the exclusion of other foods. This behavior seems to depend on the animal's developing a **search image**, a set of key characteristics that will lead it to the desired object. We can understand search images from our own experience. If you were looking for a particular package on a kitchen shelf, you would probably scan rapidly, looking for a package of a particular size and color rather than reading labels. Eventually, if the favored item becomes scarce relative to others, the animal will switch to a new food source and develop a new search image. Search images enable an animal to combine efficient short-term specialization with the flexibility of generalization.

The switching described here, and other choices animals make while foraging, have generated enormous interest among behavioral ecologists. Much recent research has emphasized **optimal foraging**, the concept that natural selection will favor animals that choose foraging strategies that maximize the differential between benefits and costs. Benefits are usually considered in terms of energy (calories) gained. However, other optimization criteria, such as specific nutrients, are sometimes more important than energy. Costs or trade-offs associated with foraging consist of the energy needed to locate, catch, and eat food; the risk of being caught by a predator during feeding; and time taken away from other vital activities, such as searching for a mate.

Many trade-offs must be considered in an optimal foraging study. A given food item may be large and therefore contain considerable energy, but if it is farther away than a small item, moving to it will require more energy. In addition, the larger item may require more time and energy to subdue or manipulate before swallowing, time and energy which could be used to pursue other prey.

Behavioral ecologists use these and many other factors to predict how an animal will forage given a particular set of conditions. Their goal is not to test whether animals in fact forage optimally, but to use the expectation of optimality as a guide to organizing research and generating testable predictions. When their predictions are borne out, researchers come closer to understanding the major factors that shape an animal's foraging behavior. When their predictions are not borne out, they have still made progress because they know they must consider additional factors. Predictions in optimal foraging studies are usually quantitative; they are frequently based on direct measurements of the calories an animal must expend to secure particular food items and the calories it gains by doing so. Numerous studies of many species show that animals tend to modify their behavior in a way that keeps their overall ratio of energy intake to energy expenditure high.

The bluegill sunfish provides an interesting example of how animals maximize the energy intake-to-expenditure ratio. These animals feed on small crustaceans called *Daphnia,* generally selecting larger prey individuals, which supply the most energy. However, smaller prey will be selected if larger prey are too far away (FIGURE 51.6a). The optimal foraging approach predicts that the proportion of small to large prey eaten will also vary with the overall density of prey. At

(a)

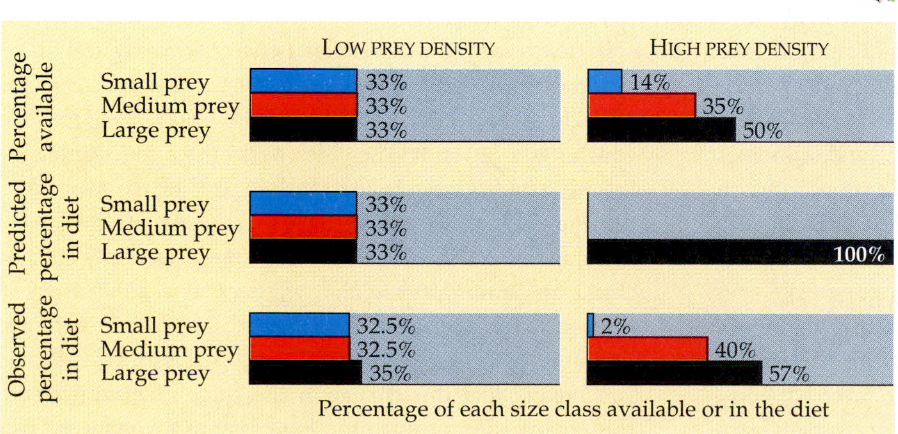

(b)

FIGURE 51.6 ▪ Feeding by young bluegill sunfish. (a) In feeding on *Daphnia* (water fleas), the fish do not feed randomly but tend to select prey based on "apparent size," information about both prey size and distance. Confronted with various potential prey, the animal will pursue the one that looks largest. Small prey (low energy yield) at the middle distance in this example may be ignored. But small prey at the closest distance may be taken with a relatively small energy expenditure. More distant but larger prey will require more energy expenditure but will provide a high yield and thus may be chosen over small prey at middle and even close distances. **(b)** When prey are at low density, calculations based on optimal foraging theory predict that bluegill sunfish will not be selective, but will eat any size prey as it is encountered. At higher prey densities the ratio of energy intake to energy expended can be maximized by concentrating only on larger prey. In the experiments described here, we see that bluegills did forage nonselectively at lower prey density. At higher prey density they favored larger prey, though not to the extent predicted.

very low prey densities, bluegill sunfish should exhibit little size selectivity, because all the prey encountered are needed to meet energy requirements. At higher prey densities it is more efficient to concentrate on larger crustaceans. In actual experiments, bluegill sunfish did become more selective at higher prey densities, though not to the extent that would theoretically maximize efficiency (FIGURE 51.6b). Young bluegill sunfish forage fairly efficiently, but not as close to the optimum as older individuals, who are apparently able to make more complex distinctions. It may be that younger fish judge size and distance less accurately because their vision is not yet completely developed; learning may also be involved.

In another example, the smallmouth bass readily consumes both minnows and crayfish. The fact that it does not show an overall preference suggests that the trade-offs balance, with minnows being the optimal prey in some situations and crayfish in others. Minnows contain more usable energy per unit weight (the crayfish has a lot of hard-to-digest exoskeleton), but they may require a greater energy expenditure to pursue. However, even though crayfish may be easier to catch, their large claws and aggressive resistance make them harder to subdue. Trade-offs also include the relative abundance and size of each type of prey. The brain of the smallmouth bass is somehow able to integrate all the relevant variables and formulate responses resulting in a highly efficient foraging behavior that involves switching between minnows and crayfish as conditions change. The proximate mechanisms responsible for this process are not known, but we might predict that they include specific cues that trigger innate as well as mainly experience-based behavior, to which we turn next.

LEARNING

Learning is experience-based modification of behavior

The fact that behavior is a mix of genetic and environmental influences challenges behavioral biologists to sort out the relative contributions of genes and environment.

Analyzing the genetic and environmental underpinnings of behavior can help scientists understand the extent to which behavior can vary among individuals of a species. In this section we examine various forms of **learning**, which is formally defined as the modification of behavior resulting from specific experiences. As we have discussed, even though an animal need not witness a developmentally fixed (innate) behavior to perform it, experience is still involved. Most innate behaviors improve with performance, as animals learn to carry them out more efficiently. Conversely, it might seem that some things, such as the different human languages, are completely learned. It is true that the ability to speak a specific language, such as English or Spanish, has little or no genetic basis. However, the ability to learn *a* language is a function of a complex brain that develops in a particular environmental context under the guidance of a human genome.

There are many nonhuman examples that parallel the human-language situation. Many songbirds have different songs in different populations. These "dialects" arise because male birds learn to sing like other males, generally early in life, and this sort of learning is prone to occasional variations that will cause populations to differ. Birds reared in one dialect but moved to another population early in life will learn the new dialect, just as a person born in the northern part of the United States who moves to the deep south early in life will speak like a Southerner. Chickadees living in temperate regions learn new song dialects each fall when they shift from living in single-family social groups to large winter flocks. These flock-specific songs help establish a group identity that is useful in defending rare food resources from other conspecific flocks during the winter. As we discuss the various forms of learning we will see additional examples in which it is productive to analyze a behavior and identify those aspects for which genetic or environmental influences are of primary importance.

Learning Versus Maturation

Learning often affects innate (developmentally fixed) behavior, but changes in innate behaviors are not always due to learning. For instance, behavior may change because of ongoing developmental changes in neuromuscular systems, a process called **maturation**. We commonly speak of birds "learning" to fly, and you may have seen fledgling birds awkwardly fluttering about as if they were practicing. However, young birds have been experimentally reared in restrictive devices so that they could never flap their wings until an age when their normal kin were already flying. Such birds flew immediately and normally when released. Thus, the improvement must have resulted from neuromuscular maturation, not from learning.

In many cases the distinction between learning and maturation is less obvious. When an adult herring gull brings food to its chick, the adult bends its head down and moves its beak, which has a red spot. The chick pecks the beak, stimulating the adult to regurgitate the food. Experiments have shown that the red spot swung horizontally at the end of a beak is the signal that elicits the chick's pecking. However, newly hatched chicks are indiscriminate and will peck at a wide variety of objects; chicks a week or two old show their strongest response to accurate models of their parents' beaks. Would you predict that this change in the chick's behavior results from maturation or learning? Experiments involving herring gulls and laughing gulls show that maturation is not involved.

A laughing gull chick that has been reared by a herring gull will respond more strongly to its foster parent's beak type than to the beak of its own species. The reverse is true for a herring gull chick cross-fostered by laughing gulls. This is an example of how learning can modify a behavior that is basically developmentally fixed.

Habituation

Learning ranges from simple behavioral changes to complex problem solving. **Habituation** is a very simple type of learning that involves a loss of responsiveness to stimuli that convey little or no information. Examples are widespread. A hydra contracts when disturbed by a slight touch; it stops responding, however, if disturbed repeatedly by such a stimulus. Many mammals and birds recognize alarm calls of conspecifics threatened by a predator, but they eventually stop responding if these calls are not followed by an actual attack (the "cry-wolf" effect). In terms of ultimate causation, habituation may increase fitness by allowing an animal's nervous system to focus on stimuli that signal food, mates, or real danger instead of wasting time or energy on a vast number of other stimuli that are irrelevant to its survival and reproduction.

FIGURE 51.7 ▪ Imprinting. Konrad Lorenz was "mother" to these imprinted geese.

Imprinting is learning limited to a critical time period

Some of the most interesting cases where learning interacts closely with innate behavior involve the phenomenon known as **imprinting**, learning that is limited to a specific time period in an animal's life and that is generally irreversible. You may have seen young ducks or geese following their mother. Mother-offspring bonding in species with parental care is a critical part of the reproductive cycle. If bonding fails to happen, the parent will not initiate care of the infant. The result is certain death to the offspring and loss of reproductive fitness to the parent. But how do the young know whom—or what—to follow? In his most famous study, Konrad Lorenz divided a clutch of graylag goose eggs, leaving some with the mother and putting the rest in an incubator. The young reared by the mother showed normal behavior, following her about as goslings and eventually growing up to interact and mate with other geese. When the artificially incubated eggs hatched, the geese spent their first few hours with the researcher instead of with their mother. From that day on, they steadfastly followed Lorenz and showed no recognition of their own mother or other adults of their own species (FIGURE 51.7). As adults, the birds continued to prefer the company of Lorenz and other humans over that of their own species, and they sometimes even initiated courtship behavior with humans.

Apparently, graylag geese have no innate sense of "mother" or "I am a goose, you are a goose." Instead, they simply respond to and identify with the first object they encounter that has certain simple characteristics. What is innate in these birds is the ability or tendency to respond; the outside world provides the *imprinting stimulus*, something to which the response will be directed. The most important imprinting stimulus in Lorenz's graylag geese was movement of an object away from the young, although the effect was greater if the object emitted some sound. The sound did not have to be that of a goose, however; Lorenz found that a box with a ticking clock in it was readily and permanently accepted as "mother."

Critical Period

In contrast to other types of learning, imprinting is distinguished by a **critical period**, a limited phase in an individual animal's development when learning of particular behaviors can take place. Lorenz found, for example, that geese totally isolated from any moving objects during the first two days after hatching, which is the critical period for imprinting on parents, failed to imprint on anything afterward. Imprinting has commonly been thought of as involving very young animals and rather short critical periods. But it is now clear that a similar learning process occurs in older animals, and that the critical period may be of various durations. For example, just as a young bird requires imprinting to "know" its parents, the adults must also imprint to recognize their young. For a day or two after their young hatch, adult herring gulls will accept and even defend a strange chick introduced into their nesting territory. After imprinting, which is probably based largely on individually variable cues such as the call notes of chicks, the adults will kill and eat any strange chick.

By imprinting on their parents, young birds first learn who will care for them, and subsequently learn species identity and the kind of bird they should mate with later in life. Not only does this sexual imprinting occur later, but the critical period lasts longer. For example, in one study involving two closely related species of finches, young males of one species were reared first with members of their own species, then with members of the other species during their several-week-long critical period for sexual identity. Later, when exposed to females of their own species, they mated quite reluctantly. They readily mated with females of the other species, however, even when they had not seen any members of that species for as long as eight years. Identification with the second species had been permanently imprinted. While a critical period and irreversibility characterize imprinting, these phenomena are not always rigidly fixed. For example, the cross-fostered finches did eventually mate with females of their own species.

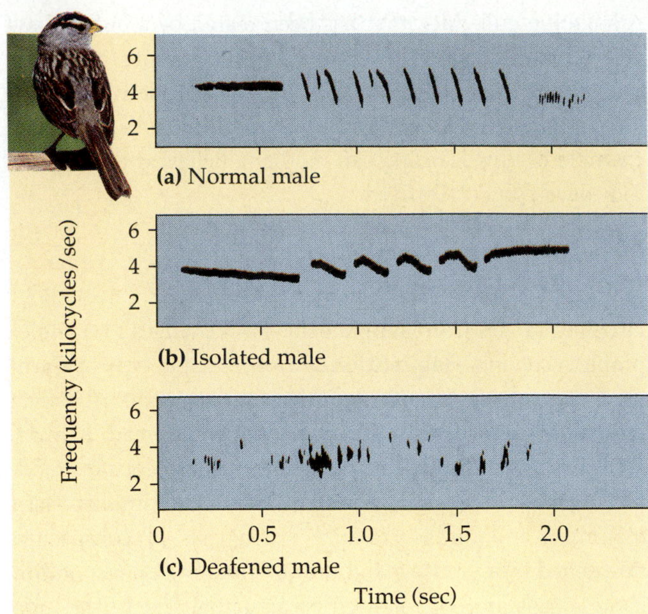

FIGURE 51.8 ▪ Sonograms of songs of white-crowned sparrows reared under three different conditions. (a) Males that heard tape recordings of their species' song before reaching 50 days of age learned to sing a normal song several months later. **(b)** Isolated males were reared in soundproof chambers, which ensured that they never heard their species' song.They sang only the primitive template of the song. **(c)** These males were deafened *after* they had heard recordings of their species' song but *before* they ever started to practice singing.

Song Development in Birds: A Study of Imprinting

Bird song is a rewarding and a perplexing subject of study for biologists who examine not only the proximate mechanisms of song, but the ultimate adaptive nature of song as well. Early experiments on song development showed that male white-crowned sparrows raised in isolation in soundproof chambers developed abnormal songs that had only a slight resemblance to the normal adult song. This crude and undeveloped song is called the template, a neural "scaffolding" upon which the full adult song is developed. However, if tape recordings of typical adult male white-crowned sparrow songs were played to the experimental birds when they were 10 to 50 days old, they developed a normal song several months later. The normal song developed as a result of a bird's comparing its own singing with its memory of the taped song. This was shown by experimentally deafening some birds after exposing them to the taped playbacks: The deaf birds developed even simpler and more abnormal songs than did isolated birds that could hear but were never exposed to songs (FIGURE 51.8).

We see that there are two learning processes in the sparrow's song development: First a bird must acquire a song type by hearing an adult, and then it must learn to match this song by listening to itself. Young males raised in soundproof chambers and exposed to taped songs of white-crowned sparrows after 50 days of age sang like birds that never heard these songs. In other words, the sparrows were receptive to learning the taped songs only during a critical period, which is why song learning can be considered a type of imprinting. Innate (developmentally fixed) influences were also demonstrated: Young white-crowned sparrows that heard taped songs of other sparrow species during the critical period did not adopt these songs.

Additional experiments revealed even more complexity in how birds learn their songs. When isolated white-crowned sparrows more than 50 days old were exposed to live singing adults of another species, they learned the song of the other species. Because a bird can interact socially with it, a live singing adult provides a much stronger and more diverse set of stimuli than a taped song. These strong stimuli can overcome the developmentally fixed tendency to acquire only a white-crowned sparrow song. Again we see that an innate tendency can be modified by experience and is not necessarily inflexible. Furthermore, the critical period proved to be longer when the stimulus was a live bird than when it was a taped playback.

People also have a critical period for learning vocalizations: It is well known that foreign languages are learned most easily up until the teen years. Of course, this critical period is not rigidly fixed. Adults can learn a new language, but they usually require much more time and effort to become fluent than a child does.

Bird species other than the white-crowned sparrow have different programs for song development. The closely related song sparrow develops a nearly normal song even if reared in total isolation. But males have larger repertoires if they can learn song variants from other birds. Mockingbirds have huge repertoires of at least 150 song types, many of which are accurate mimics of sounds made by other kinds of birds, mammals, and even telephone rings. Apparently, the fitness value of a large repertoire is very high for mockingbirds.

Until recently, researchers assumed that all birds had to listen to themselves in order to develop song. But eastern phoebes experimentally deafened early in life, well before they began to sing, developed normal songs of their species. Thus, there are important exceptions to the song-learning scenario based on extensive study of white-crowned sparrows.

Research with canaries has helped elucidate the relationship between the observed behavior of song learning and the neuronal events underlying the process. Canaries are unusual in that they sing variable numbers of songs and develop new songs each year. Thus, they must relearn a new song repertoire before each breeding season. Biologist Fernando Nottebohm identified the region of the forebrain responsible for song learning in canaries. He found that this region in male canaries shows dramatic individual variation in size according to the season and the number of different songs in an individual male's repertoire. Brains are largest during the breeding season and in males with the biggest song repertoire. The shrinking of this brain region at the end of the breeding season may be a mechanism for erasing unneeded songs. Subsequent regeneration of neurons in the brain during courtship provides a way for new song learning to take place. From an adaptive perspective, the ability of male canaries to learn new songs more than once in their lifetime must be critically important to their fitness, as this adaptation requires a considerable investment of energy. Nottebohm's findings illustrate the important results that can be gleaned from studies melding neurobiology and behavioral ecology.

The observation of so much variation in the ways birds acquire songs and then deploy them suggests that this is a fairly flexible behavioral system in many groups of birds. The strong connection between song and reproductive fitness in birds has apparently enabled natural selection to modify nearly all aspects of song and its development into the rich diversity that we find in modern birds.

Many animals can learn to associate one stimulus with another

Numerous classical studies in behavioral science were concerned with the proximate causes of **associative learning**, the ability of many animals to learn to associate one stimulus with another. Well known from the laboratory studies of Russian physiologist Ivan Pavlov in the early 1900s, a type of associative learning called **classical conditioning** is learning to associate an arbitrary stimulus with a reward or punishment. Pavlov sprayed powdered meat into dogs' mouths, causing them to salivate (primarily a physiological rather than a behavioral response). Just before the spraying, however, he exposed the dogs to a sound such as a ringing bell or clicking metronome. Eventually the dogs salivated readily in response to the sound alone, which they had learned to associate with the normal stimulus. Similar types of conditioning experiments have been carried out with other animals.

Another type of associative learning is **operant conditioning**, also called trial-and-error learning. Here, an animal learns to associate one of its own behaviors with a reward or punishment and then tends to repeat or avoid that behavior. For example, predators quickly learn to associate certain kinds of prey with painful experiences (FIGURE 51.9) and modify their behavior accordingly. Because what tastes good or bad to an animal is likely to be affected by natural selection on the basis of the food's value, genes influence the outcome of this operant conditioning.

The best known laboratory studies involving operant conditioning date from the 1930s work of American psychologist

FIGURE 51.9 ▪ Trial-and-error learning. A young coyote has probably learned from receiving a face-full of painful quills to be wary of porcupines.

B. F. Skinner. A rat or other animal placed in a "Skinner box" finds and manipulates a lever in the box, usually by accident, and is rewarded by the release of food. The animal quickly learns to associate manipulation of the lever with a food reward. Such learning is the basis for most of the animal training done by humans, in which the trainer typically encourages a behavior by rewarding the animal. Eventually the animal performs the behavior on command, without always receiving a reward.

Practice and exercise may explain the ultimate bases of play

Many mammals and some birds engage in behavior that can best be described as **play**. Such behavior has no apparent external goal but involves movements closely associated with goal-directed behaviors (FIGURE 51.10). For example, playful stalking and attacking of conspecifics occurs in many predator species, such as those in members of the cat and dog families. Although this behavior does not usually involve painful bites, the animals grab and mouth one another, using movements similar to those used to capture and kill prey. A study of bottlenose dolphins in Australia revealed that young dolphins spend long periods away from their mothers in groups of juveniles engaged in a full range of social and sexual play. African lions and dolphins are social animals, and a social lifestyle is one of the hallmarks of mammalian species that routinely engage in play.

Another common feature of play is that it is potentially dangerous or costly. Baboons sometimes kill and eat juvenile vervet monkeys, and they are most successful in doing so when the monkeys are playing in groups away from adults. In a study of caged ibex, a type of wild goat, play resulted in at least five out of fourteen kids sustaining injuries that produced limps. "Horsing around" often produces similar injuries in human kids.

FIGURE 51.10 ▪ Play behavior. In the view of some behavioral ecologists, the roughhousing behavior of these lion cubs evolved in spite of the energy it consumes and the risks it poses because play improves individual fitness in some way. Practicing survival behavior such as capturing prey, experimenting with social roles, and building a healthy body through exercise are three possible benefits of play.

Play obviously consumes energy, and the risks to life and limb result in significant additional costs. What could be the ultimate adaptive basis for such seemingly pointless behavior? The "practice hypothesis" suggests that play is a type of learning that allows animals to perfect behaviors needed in functional circumstances. This hypothesis is supported by the observation that play is most common in young animals. However, movements used in play show little improvement after their first few practices. An equally likely ultimate explanation for play is the "exercise hypothesis," which suggests that play is adaptive because it keeps the muscular and cardiovascular systems in top condition. The exercise hypothesis also predicts that play should be especially common in young animals, because they typically do not have to exert themselves in useful activities while under the protection and care of their parents. However, recent studies of beluga whales and several species of dolphins indicate that play (creating complex air bubbles) is also common in adults, at least in captivity.

ANIMAL COGNITION

What does an animal's brain do with the information it obtains about the outside world? If a chimpanzee is placed in an area with a banana hung high out of reach and several boxes on the floor, the chimp can "size up" the situation and stack the boxes, enabling it to reach the food. Such novel problem-solving behavior is highly developed in some mammals, especially primates and dolphins, and notable examples have also been observed in some bird species, especially crows, ravens, and jays. Ravens, for example, exhibit marked individual variation in their attempts to solve some of the problems imposed on them by experimenters (FIGURE 51.11). Watching an animal solve a problem makes us aware that its nervous system has a substantial ability to process information. An expanding research effort on animal cognition seeks to understand information processing at all levels, from the nervous system activities that underlie sophisticated behavior, such as problem solving, to the internal representations animals have about physical objects in their surroundings.

The study of cognition connects nervous system function with behavior

The term *cognition* is variously defined. In a narrow sense, it is essentially synonymous with consciousness, or awareness. In a broad sense—the way we use the term in this book—**cognition** is the ability of an animal's nervous system to perceive, store, process, and use information gathered by sensory recep-

FIGURE 51.11 ▪ Problem solving. Behavioral biologist Bernd Heinrich placed ravens in an experimental situation in which they had to solve the problem of retrieving a food reward hanging from a string attached to a wire. The raven shown here solved the problem by using one foot to retrieve the string and the other foot to hold the string as it was being retrieved. Of interest was the tremendous individual variation in behavior that Heinrich observed. Some ravens never learned to get at the food, while others solved the problem in different ways.

tors. The study of animal cognition, called **cognitive ethology**, attempts to illustrate the connection between data processing by nervous systems and animal behavior. Cognitive ethology includes, but is not limited to, the study of animal consciousness, or awareness.

One area of research in cognitive ethology investigates how animal brains represent physical stimuli from the environment. For instance, what kind of calculations, if any, does the brain of a dog make about where a Frisbee it is chasing will land? What is the nature of the dog's internal representation of the spatial relationships between the spinning Frisbee and other objects in the immediate environment? Questions such as these, concerned with internal representations of an animal's physical surroundings, are separate and distinct from questions about consciousness. We look first at some studies of internal representations.

Movement from place to place often depends on internal coding of spatial relationships

A central hypothesis of cognitive ethology is that many animals formulate **cognitive maps**, internal representations, or codes, of the spatial relationships among objects in their surroundings. James Gould of Princeton University has reported that bees can form and use "mental maps" of their foraging areas. In a broad sense, most animals that move through space probably have some internal map, or encoded representation, of spatial relationships of external objects. Exceptions would be animals that exhibit only the simplest kinds of movements.

Two kinds of movement that may occur without any internal representation are kineses and taxes. A **kinesis** involves a simple change in activity rate in response to a stimulus. Sowbugs or woodlice become more active in dry areas and less active in humid ones, a simple behavior that tends to keep these animals in moist environments. The animals do not move toward or away from specific conditions, but since they slow down in a favorable environment, they tend to stay there. In contrast, a **taxis** is a more or less automatic, oriented movement toward or away from some stimulus. For example, housefly larvae are negatively phototactic after feeding, automatically moving away from light; this simple response presumably ensures that the flies remain in an area where they are harder for predators to detect. Trout exhibit positive rheotaxis (Gr. *rheos*, "current"); they automatically swim or orient themselves in an upstream direction, which keeps them from being swept away.

Migration Behavior

The most extensive studies of spatial representations (cognitive maps) have been made for animals that exhibit **migration**, or regular movement over relatively long distances. Migrating animals generally make one round trip between two regions each year, although there is considerable variation among species. The most notable examples are the migrations of birds, whales, a few butterfly species, and certain oceangoing fish. How is it, for example, that golden plovers find their way over 13,000 kilometers from their arctic breeding grounds to southeastern South America? Even more remarkably, some populations of these birds return to the Hawaiian and Marquesas islands, small pieces of land in a vast expanse of ocean (FIGURE 51.12, p. 1066).

Migrating animals use one of three mechanisms, or some combination of these three techniques, to find their way. In *piloting*, an animal moves from one familiar landmark to another until it reaches its destination. Piloting is used mostly for short-distance movements. In *orientation*, an animal can detect compass directions and travels in a particular straight-line path for a certain distance or until it reaches its destination. The most complex process is *navigation*, which involves determining one's present location relative to other locations, in addition to detecting compass direction (orientation). If you were dropped off at an unfamiliar spot and told that your home was directly to the north, you could use a compass and straight-line travel to get home; that is, you could use orientation. But a compass alone would not be adequate if you were not told which way to go; to choose the right direction you would need to determine where you were in relation to home.

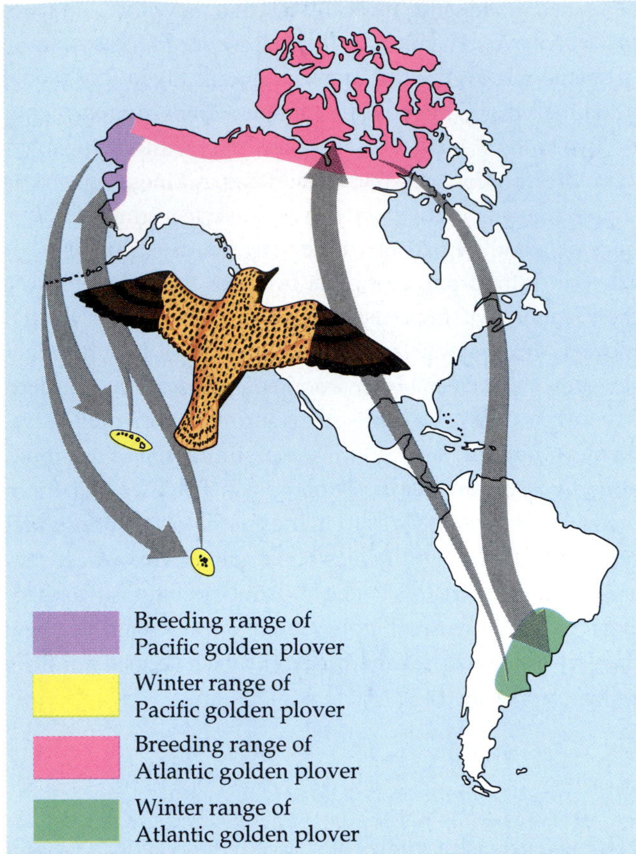

FIGURE 51.12 ▪ Migration routes of the golden plover. These birds are able to navigate across vast expanses of ocean to the relatively small Hawaiian and Marquesas islands (yellow). The ground-nesting plovers migrate from warm winter feeding areas to seasonally rich, essentially predator-free breeding grounds in the Arctic during the short arctic summer.

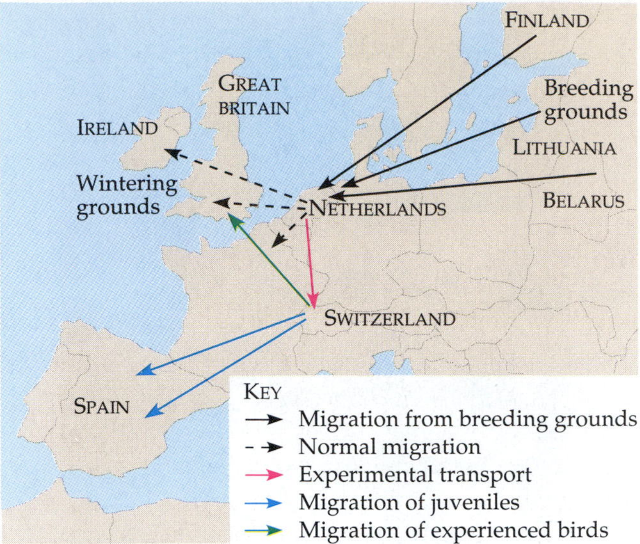

FIGURE 51.13 ▪ Orientation versus navigation in juvenile and adult starlings. About 11,000 starlings were captured in the Netherlands on their migration from their breeding grounds in northeastern Europe to their wintering grounds in Great Britain, Ireland, and northern France. After being transported to Switzerland (red arrow) and released, juvenile starlings, which had never made the journey before, continued to fly west and southwest (blue arrows), which brought them to Spain. Adults, all of whom had made the trip at least once before, flew northwest (green arrow), an atypical direction but one that took them to their usual wintering grounds. Members of both age groups were able to detect direction, but only the adults showed true navigation because they had developed a "map sense" and could determine where their original goal was relative to the site to which they were transported.

You would need a complex mental picture of your surroundings, a so-called map sense. FIGURE 51.13 reinforces this distinction between orientation and navigation.

What sorts of cues do animals use for orientation and navigation? The answer, surprising at first but now firmly established, is that some species of birds and other animals commonly use a combination of compass references: Earth's magnetic field, the sun (for daytime travel), and the stars (celestial rotation, for nighttime). Calibrated against one another, these indicators provide excellent, albeit complex, cues to direction.

Cognitive ethologists are more interested in how animals utilize cues in moving from place to place than in the study of cues themselves. Orienting by the sun or stellar constellations requires an internal timing device to compensate for the continuous daily movement of celestial objects. Consider what would happen if you started walking one day, orienting yourself by keeping the sun on your left. In the morning you would be heading south, but by evening you would be heading back north, having made a circle and gotten nowhere. The stars also shift their apparent position as Earth rotates. One night-migrating bird, the indigo bunting, avoids the need for a timing mechanism by the same means as ancient navigators: fixing on the North Star, which moves little. Many migrants, however, use an internal clock. For example, if an experimental sun is held in a constant position, starlings change their orientation steadily at a rate of about 15° per hour. This normally compensates for the change in the sun's position as Earth rotates on its axis. The calibration problem is very complex because the apparent location of the sun shifts at a variable rate, being fastest at midday. Furthermore, the apparent position of celestial objects changes as the animal moves over its migration route. Recently, Kenneth and Mary Able of the State University of New York at Albany discovered that for one long-distance migrant, the Savannah sparrow, the magnetic and celestial compasses are reset during brief stopovers along the migration route.

The study of consciousness poses a unique challenge for scientists

A simple but profound question is whether nonhuman animals are consciously *aware* of themselves and of the world around them. An equally profound question is whether the study of consciousness (awareness) is within the purview of science.

Many people who have spent a lot of time with pets or wild animals argue that these animals are not behaving like sophisticated robots. But is a dog aware of itself when it is chasing a Frisbee? Do animals "feel" pain or pleasure or sadness as we do? To date we have no way of answering such questions directly, because conscious awareness is known only to the individual that experiences it, and unlike phenomena that can be studied objectively, it is not associated with any observable behavioral or physiological change.

Donald Griffin of Princeton University is a foremost proponent of the view that conscious thinking is an inherent and essential part of the behavior of many nonhuman animals. Griffin argues that if other animals behave in ways we associate with conscious processing in ourselves, perhaps it makes sense to assume that they have the same underlying awareness. In her famous field studies, Jane Goodall has described cognitive decision making in chimpanzees. Griffin suggests that such abilities may extend to many nonprimate branches of the phylogenetic tree as well. He argues, for example, that conscious processes are at the heart of such behaviors as the injury-feigning "strategy" of some species of ground-nesting birds (FIGURE 51.14). This view sees consciousness as arising through the normal process of natural selection, and, like many other major animal functions, as forming a phylogenetic continuum that extends far back in evolutionary history.

Because of the difficulty of pursuing with scientific rigor answers to questions about consciousness, some researchers assume the most conservative position—that most animals are not aware. There are intermediate positions within the argument, and no behavioral biologist would argue that *all* animals behave in ways that indicate consciousness, but these two polar positions help define the debate. Other animals may lack the ability to consciously integrate information (to "think") to the same extent as humans do, but is this a matter of degree—a continuum of abilities—or are humans fundamentally different in some behavioral respect? Ultimately, answers to questions about animal consciousness may profoundly affect how we interact with other animals—and how we view ourselves as well.

FIGURE 51.14 ▪ Injury-feigning display. This killdeer uses deception to defend her nest against predators or human disturbance. When danger threatens she leaves the nest, which is usually concealed, and begins an elaborate display as though she had a broken wing. This behavior has the effect of leading the potential predator away from the nest. As the predator gets close to her, she simply flies away. She returns to the nest only after the danger has passed. Killdeers show extraordinary individual variation in the form and use of this display. In addition, individual killdeers use multiple variations of the display depending on the type of threat to the nest and whether or not they have seen the threat before. Some cognitive ethologists point to versatile behavior such as this to support the hypothesis that nonhuman animals are conscious, thinking beings.

SOCIAL BEHAVIOR AND SOCIOBIOLOGY

Sociobiology places social behavior in an evolutionary context

Social behavior, broadly defined, is any kind of interaction between two or more animals, usually of the same species. Though most sexually reproducing species must be social for part of their life cycle in order to reproduce, some species spend most of their lives in close association with conspecifics. Social interactions have long been a research focus for scientists who study behavior. The complexity of behavior increases dramatically when interactions among individuals are considered. Aggression, courtship, cooperation, and even deception are part of the behavioral landscape of social behavior. Social behavior has both costs and benefits to members of those species that interact extensively.

The relatively new discipline of **sociobiology** applies evolutionary theory as a foundation for the study and interpretation of social behavior. Much of the evolutionary theory underlying the modern study of social behavior stems from the work of British biologist William Hamilton. (Ironically, in the early 1960s, graduate student Hamilton was informed by his superiors that his theories of social evolution were not of Ph.D. quality.) Hamilton used the concepts of fitness and the genetic basis of behavior in analyzing the evolution and maintenance of social behavior in animals. The evolutionary emphasis of Hamilton's work underlies sociobiology and the more general field of behavioral ecology as it is practiced today. The development of sociobiology into a coherent method of analysis and interpretation was catalyzed by E. O. Wilson's watershed book *Sociobiology: The New Synthesis.*

Because members of a population have a common niche, there is a strong potential for conflict, especially among members of species that normally maintain densities near what the environment can sustain. Sometimes social behavior seems to involve cooperation, as when a group carries out behavior

(a)

(b)

FIGURE 51.15 ▪ Cooperative prey capture. (a) This pack of African wild dogs is killing a wildebeest much larger than themselves. **(b)** This line of white pelicans is moving toward a school of fish. This cooperative behavior makes it more difficult for fish to escape by swimming around the birds. Although all participants benefit from these sorts of cooperation, each individual is behaving in a manner that maximizes its own benefits.

more efficiently than is possible for a single individual (FIGURE 51.15). Keep in mind, however, that even when behavior requires some cooperation and seems to be mutually beneficial to interacting individuals, as in mating behavior, each participant usually acts in a way that will maximize its benefits, even if this is at a cost to the other participant. In the next section we examine competitive social interactions, where this "selfish" aspect of behavior is most obvious. Subsequent sections focus on social behaviors involving cooperation.

Competitive social behaviors often represent contests for resources

Agonistic Behavior

In **agonistic behavior**, a contest involving both threatening and submissive behavior determines which competitor gains access to some resource, such as food or a mate. Sometimes the encounter involves tests of strength. More commonly the contestants engage in threat displays that make them look large or fierce, often with exaggerated posturing and vocalizations. Eventually one individual stops threatening and ends with a submissive or appeasement display, in effect surrendering. Much of this behavior includes **ritual**, the use of symbolic activity, so that usually no serious harm is done to either combatant (FIGURE 51.16). Dogs and wolves show aggression by baring their teeth; erecting their ears, tail, and fur; standing upright; and looking directly at their opponent—all of which make the animal appear large and threatening. The eventual loser, on the other hand, sleeks its fur, tucks its tail, and looks away. This appeasement display inhibits any further aggressive activity. The degree to which combat is ritualized depends on the scarcity of the resource and the likelihood that the resource will be available again. For example, male ground squirrels often inflict severe injury upon, or even kill, each other when battling for access to sexually receptive females. In this case the females for which the ground squirrels are competing are in estrus and receptive to male courtship for only a few hours each year, and thus a male's entire reproductive fitness may depend on his ability to compete against other males that one day.

FIGURE 51.16 ▪ Ritual wrestling by rattlesnakes. The rattlesnakes attempt to pin each other to the ground, but they never use their deadly fangs in such combat.

When animals do inflict injury, natural selection favors a strong tendency to end the contest as soon as the winner is established, because violent combat could injure the victor as well as the defeated. Any future interaction between the same two animals is usually settled much more quickly in favor of the original victor.

Dominance Hierarchies

Many animals live in social groups maintained by agonistic behavior. Chickens are an example. If several hens who are unfamiliar with one another are put together, they respond by skirmishing and pecking each other. Eventually the group establishes a clear "pecking order"—a more or less linear **dominance hierarchy**. Within a group, the alpha (top-ranked) hen controls the behavior of all others, often by mere threats rather than actual pecking. The beta (second-ranked) hen similarly subdues all others except the alpha, and so on down the line to the omega, or lowest, animal. The advantage to the top-ranked bird is obvious; it is assured of access to resources such as food. Even for lower-ranked animals the system ensures that they do not waste energy or risk harm in futile combat.

Wolves typically function in packs, cooperation being necessary for killing large prey. Within each pack there is a dominance hierarchy among the females, and the top female controls the mating of the others. When food is generally abundant, the top female mates and allows others to do so also; when food is scarce, she allows less mating by other females, thereby making more food available for her own young. Although these examples illustrate dominance hierarchies among females, male hierarchies are also common.

Territoriality

A **territory** is an area that an individual defends, usually excluding other members of its own species. Territories are typically used for feeding, mating, rearing young, or combinations of these activities. Usually a territory is fixed in location, its size varying with the species, the territory's function, and the amount of resources available. Song sparrow pairs, for example, may have territories of about 3000 m^2, in which they carry out all activities during the several months of their breeding season. Gannets and most other seabirds, in contrast, mate and nest in territories of only a few square meters or less and feed away from their territories (FIGURE 51.17). Bull sea lions defend small territories used only for mating, whereas red squirrels have rather large territories apparently based on feeding patterns. In many species that defend territories only during the breeding season, individuals may form social groups at other times. Chickadees, for instance, form monogamous breeding pairs that defend small territories in the summer. In the winter they form larger flocks, enabling the birds to forage more efficiently and benefit from the increased protection from predators that results from membership in a large group.

Note that there is a distinction between a territory and a home range, which is simply the area in which an animal roams about and which is often not defended. In some species, such as breeding song sparrows, territory and home range are the same; but for other species, such as gannets, a territory is considerably smaller than the home range. The distinction cannot always be made clearly. Gray squirrels, for example, typically have home ranges that overlap extensively, but one individual may defend part of the range from competitors.

FIGURE 51.17 ▪ Territories. Gannets nest virtually only a peck apart and defend their territories by calling and pecking at each other. This is a population of Australian gannets in New Zealand.

Territories are established and defended through agonistic behavior, and an individual that has gained a territory is often difficult to dislodge. Why do owners usually win? One explanation in behavioral ecology is that a territory is worth more to an owner than to an intruder because the owner is already familiar with it. Thus, because it has more to gain from a territory, an owner is more likely to escalate a battle than is an intruder. In addition, established territory holders are likely to be older and more experienced at engaging in agonistic interactions.

Natural selection does not always favor territoriality, and not all species are territorial. However, for those animals that are, the territory can provide exclusive access to food supplies, breeding areas, and places to raise young. Moreover, familiarity with a specific area may help individuals avoid predators. In a territorial species, such benefits outweigh the energy costs of defending territory and thereby increase fitness.

Ownership of territory is usually continually proclaimed; this is a primary function of most familiar bird songs, as well as the noisy bellowing of sea lions and the chatter of the red squirrel. Other animals may use scent marks or frequent patrols that warn potential invaders (FIGURE 51.18, p. 1070). Grey wolves, which live in packs on huge territories (hundreds of square kilometers), use multiple strategies to advertise territorial boundaries, including scent marking and howling. Multiple signals help dispel any ambiguities as to the boundaries of a territory, thereby minimizing the risk that one group will accidentally stray into the territory of a rival pack. This is especially important for wolves because actual face-to-face meetings between groups are often violent.

Defense of territory is usually directed only at conspecifics; a white-crowned sparrow may live within a song sparrow's territory because a different species usually occupies a different niche, or role in the environment, and is less likely to be a

(a)

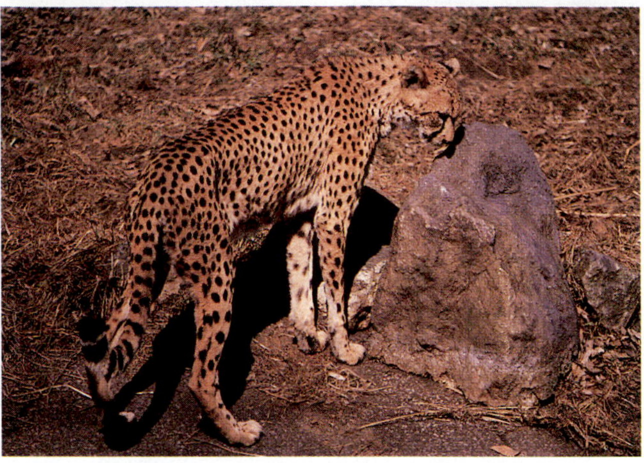

(b)

FIGURE 51.18 • Staking out territory with chemical markers. **(a)** This male cheetah, a resident of Africa's Serengeti National Park, is spray-urinating on a tree. The odor will serve as a chemical "No Trespassing" sign to other males. **(b)** Another male cheetah sniffs a marked rock. With their keen olfactory sense, cheetahs can distinguish their own marks from those left by others. These signals help prevent face-to-face meetings that could escalate into violence.

direct competitor. Another adaptive reason for concentrating defense on conspecifics is that they are likely to mate with a territory holder's mate.

Although dominance hierarchies and territoriality evolved as a result of their advantages to individuals, such systems have important effects at the population level because they tend to stabilize density. If resources were allocated evenly among all members of a population, the "fair share" that each individual received might not be enough to sustain anyone, leading to occasional population crashes. Dominance and territoriality usually mean that at least some individuals receive an adequate amount of a resource. Often, in fact, if a resource such as food becomes scarce, territories expand somewhat. In addition, there are usually individuals low in the hierarchy or lacking territories who are ready to move up or step in if one of the successful individuals dies. The result may be relatively stable populations from year to year.

Mating behavior relates directly to an animal's fitness

All aspects of reproductive behavior receive extensive attention from behavioral ecologists. The reason is that researchers can often determine the number of young an individual produces, which for many species comes close to determining an animal's fitness. The correlation between measurable behaviors and fitness may not be as strong for other aspects of behavioral ecology, such as optimal foraging.

Courtship

Most animals probably do not have any conscious sense of reproduction as an important function in their lives, nor do they have the kind of continuous attraction to members of the opposite sex found in most humans. Often there seems to be a strong tendency for an animal to view any organism of the same species as a threatening competitor to be driven off, if possible. Even within many social species, individuals usually avoid contact with one another. How, then, is mating accomplished? In many animal species, potential partners must go through a complex courtship interaction, unique to the species, before mating. This complex behavior often consists of a sequence of actions that seems to confirm that the animals are of the same species but the opposite sex, that they are in appropriate physiological condition, and that they are not threats to each other.

In some species, courtship also plays an important role in allowing one or both sexes to choose a mate from a number of candidates. When individuals choose mates, females nearly always show greater discrimination than males. We would predict this because in most species females have more parental investment in each offspring. **Parental investment** is defined as the time and resources an individual must expend to produce offspring. Eggs are usually much larger and much more costly to produce than sperm. Placental mammals have less disparity in gamete size than most other animals, but female placental mammals invest considerable time and energy in carrying the young before birth. In most species, females are very choosy; picking a poor-quality male can be a costly error. By contrast, males of most animal species tend to mate with as many females as possible. Males usually compete with one another for mates, sometimes by trying to impress females. Thus, males of most species perform more intense courtship displays than females; in many species, in fact, it is only males who court. Secondary sex characteristics are usually much more pronounced in males; bright plumage in birds and antlers in deer are examples. Differential reproductive success that results from advantages in attracting or competing for mates, called *sexual selection,* was discussed in more detail in Chapter 23.

In some animals, competition among individuals of the same sex (usually males) almost entirely determines which animals of that sex will mate. But in other species, individuals, usually females, actively choose among potential mates based on specific characteristics of the male or resources under his control. This process is called *assessment.* There are two ultimate bases for such choices. First, if the other sex provides parental care, it is advantageous to choose as competent a mate as possible. For example, male common terns (a bird species closely related to gulls) carry fish and display them to potential mates as part of their mating ritual. Eventually a male may begin to feed fish to a female. This behavior may be a good proximate indicator of a male's ability to provide food for tern chicks. In some animal species, females prefer males who are capable of the most extreme and energetic courtship displays or who have the most extreme secondary sex characteristics, such as long tails. Perhaps these characteristics are proximate indicators that the males are vigorous and in good health.

The second ultimate basis for mate choice is genetic quality. This is likely to be most important when males provide no parental care and sperm are their only contribution to offspring. In a number of bird and insect species, males display communally in a small area called a **lek**. Females visit the lek and choose among the displaying males. After mating occurs there is no further contact between females and males. Researchers hypothesize that there is a connection between a dominant male in a lek and his evolutionary fitness. In terms of a female's choice, we would predict that she would tend to mate with a male whose characteristics favor high fitness since the success of her offspring will depend on both her genes and those of her mate. The proximate basis for this sort of choice is also likely to be a preference for males that court the most vigorously and are adorned with the showiest secondary sex characteristics. An especially important factor in male genetic quality may be the ability to counter pathogens and parasites (see FIGURE 23.12).

It is often very difficult to determine the bases for female assessment, and in some cases it is even difficult to determine whether differential mating success among males is due to male-male competition, female assessment, or both. One of the best known examples of ritualized courtship is that of the three-spined stickleback fish previously mentioned (FIGURE 51.19). Although largely innate, the courtship sequence does not necessarily proceed smoothly or quickly. Often a female will begin to follow a male and then hesitate. She may cut off the sequence at any point, sometimes because she is simultaneously being courted by other nearby males. The final result is successful mating by some but not necessarily all males. Males provide all the parental care in this species. Therefore, proximate mechanisms of assessment by the female must gauge such relevant characteristics as nest quality and territorial defense by the male.

Ritualized acts, whether of courtship or of agonistic behavior, probably evolved from actions whose meaning was at one time more direct. We can see an interesting example of this process in a group of insects called dance flies. The males of some species of dance flies spin oval balloons of silk and carry

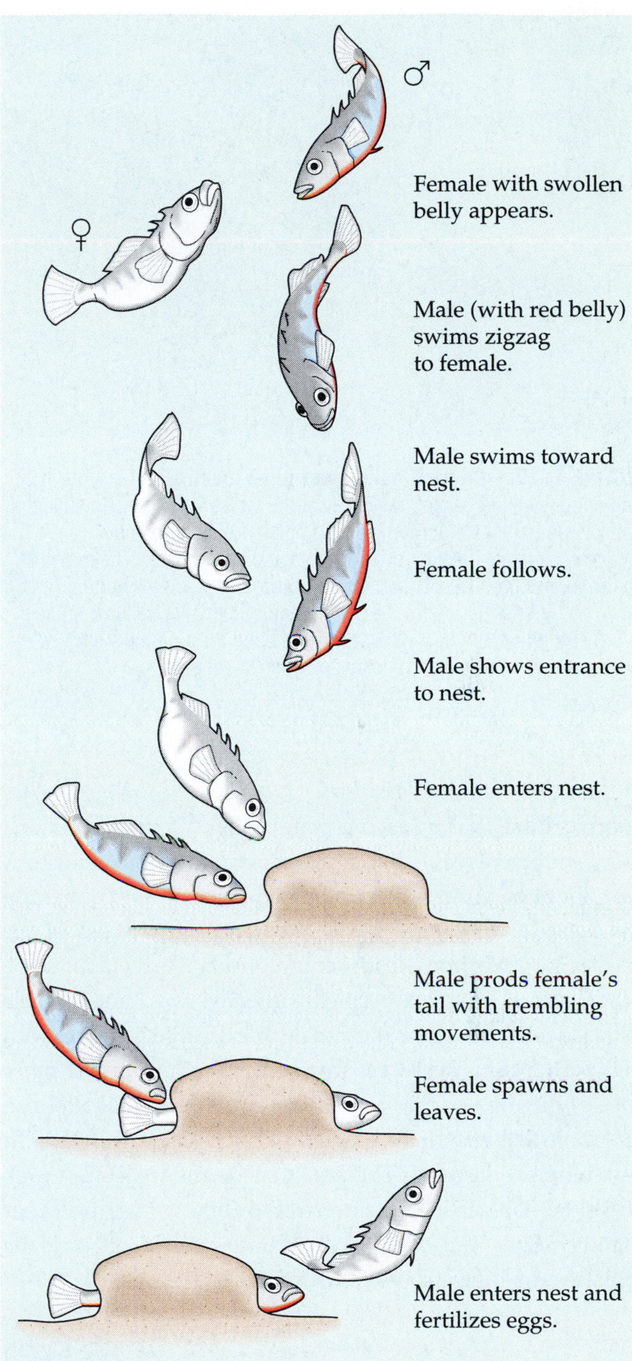

♂
♀
Female with swollen belly appears.
Male (with red belly) swims zigzag to female.
Male swims toward nest.
Female follows.
Male shows entrance to nest.
Female enters nest.
Male prods female's tail with trembling movements.
Female spawns and leaves.
Male enters nest and fertilizes eggs.

FIGURE 51.19 ▪ Courtship behavior in the three-spined stickleback. Males are strongly territorial, defending an area in which they have built a nest. If a gravid (egg-carrying) female approaches, her swollen belly inhibits the male's aggressive behavior and elicits zigzag swimming in the male. This entices the female to swim closer, which in turn stimulates the male to swim to the nest and stick his snout inside. This action stimulates the female to wriggle into the tunnel. The male then nuzzles her tail, which stimulates her to spawn, after which she swims out the other end of the nest. The male then enters and deposits sperm on the eggs, after which he immediately and quite aggressively drives the female out of the area, apparently because she lacks the swollen belly that would inhibit his aggression.

FIGURE 51.20 ▪ Courtship behavior in an insect. The insects in this drawing are called dance flies because the males fly up and down in the air in swarms that attract females. Common around streams and lakeshores, dance flies eat other insects and plant nectar. The males of some species produce silk balloons that they carry while swarming. A female selects a male from the swarm and accepts his balloon, and the pair fly off and mate. In some species the balloons are empty sacs, while in others they contain bits of insect prey or flower petals.

them while flying in a swarm (FIGURE 51.20). The females approach the swarm seeking mates. After a female chooses a male, she accepts his balloon, and the two fly off to copulate. A clue about how this ritual originated comes from comparing the behavior of several species of dance flies. Males of the species just described feed mainly on plant nectar. In some other species males are chiefly insectivorous, and a male brings a dead insect for the female to eat while he mates with her; perhaps this keeps her from attacking him. In yet other species the male carries a dead insect inside a silk balloon, a variation that may have evolved because silk was helpful in subduing the insect or because it made the male's gift look larger. Perhaps in species ancestral to those whose males eat mainly nectar, the ritual evolved into merely offering the female something associated with food.

Mating Systems

The mating relationship between males and females varies a great deal among species. In many species, mating is **promiscuous**, with no strong pair-bonds or lasting relationships. In species where the mates remain together for a longer period, the relationship may be **monogamous** (one male mating with one female) or **polygamous** (an individual of one sex mating with several of the other). Polygamous relationships most often involve a single male and many females, called **polygyny**, which can be explained in terms of a difference in parental investment. However, in some species a single female mates with several males, a relationship called **polyandry**.

The needs of the young are an important ultimate factor in the evolution of mating systems. Most newly hatched birds cannot care for themselves and require a large, continuous food supply that one parent may not be able to provide. In such a system, a male may ultimately leave more viable offspring by helping a single mate than by going off to seek more mates. This may explain why most birds are monogamous. In birds with young that feed and care for themselves almost immediately after hatching, there is less need for parents to stay together. Males of these species can maximize their reproductive success by seeking other mates, and polygyny is relatively common in such birds. In the case of mammals, the lactating female is often the only food source for the young. Males usually play no role, or, if they protect the females and young, they typically take care of many at once in a harem.

Another factor that influences mating systems and parental care is certainty of paternity. Young born or eggs laid by a female definitely contain the female's genes. But even within a normally monogamous relationship, these young may have been fathered by a male other than the female's usual mate. For example, this was the case in a population of red-winged blackbirds studied by researchers who used DNA fingerprinting to determine the paternity of nestlings (see the Methods Box). The certainty of paternity is relatively low in most species with internal fertilization because the acts of mating and birth (or mating and egg laying) are separated over time. This could explain why exclusively male parental care occurs in very few bird or mammal species. However, certainty of paternity is much higher when egg laying and mating occur together, as in external fertilization. This may explain why parental care in fishes and amphibians, when it occurs at all, is at least as likely to be by males as by females. Male parental care occurs in only 2 out of 28 (7%) fish and amphibian families with internal fertilization, but in 61 out of 89 (69%) families with external fertilization. In fish, even when parental care is given exclusively by males, the mating system is often polygynous, with several females laying eggs in a nest tended by one male.

It is important to point out again that when behavioral ecologists use terms such as "certainty of paternity," they do not mean that animals are aware of those factors when they behave a certain way. Parental behavior correlated with certainty of paternity exists because it has been reinforced over generations by natural selection.

Social interactions depend on diverse modes of communication

Much of what we have discussed under the topics of competitive social interactions and mating behaviors involves instances in which animals intentionally, although not neces-

METHODS: Using Molecular Biology to Answer Behavioral Questions

Molecular biology has had an impact on every field of biology, including behavioral ecology. Here we examine the specific case of a research project that applied DNA fingerprinting to determine the paternity of nestlings hatched in a population of red-winged blackbirds (FIGURE a). The DNA fingerprinting was based on restriction fragments (RFLPs), as described in Chapter 20.

In most bird species, mating relationships are fixed, typically involving one male and one to several females that are joint inhabitants of the male's territory. By counting the number of young in the nest or nests within a male's territory, it should be possible to compare the relative reproductive success (fitness) of different males. Behavioral ecologists have sometimes used this approach to compare the effects of behavioral variations on the fitness of different birds in a population. But there is a problem with this method of measuring how many offspring each male sires. Although a male generally guards his mate (or mates) during her fertile period, there is evidence that this vigilance is imperfect at preventing the females from mating with other males, and females may occasionally mate with males from neighboring territories.

In the 1970s, biologists in the U.S. Fish and Wildlife Service tried to control populations of red-winged blackbirds, which sometimes threaten agricultural crops. Rather than kill the birds, the biologists performed vasectomies on the males. The researchers studying paternity found, surprisingly, that 50% or more of the eggs in each vasectomized male's territory had been fertilized, presumably by neighboring males who had not had the procedure. Perhaps vasectomies altered male territorial defense, increasing the frequency of extrapair copulations (EPCs). Thus, there was still the possibility that EPCs were rare for females mated to normal, untreated males.

The problem was not resolved until 1990, when a team of researchers published a DNA fingerprinting analysis of paternity in a population of redwings. Analysis of DNA from blood samples of nestlings and adults revealed that 45% of the nests had at least one offspring fathered by a male other than the one claiming that territory. The DNA fingerprinting even allowed the researchers to identify which males had ventured out of their territories and fathered offspring in the territories of neighbors. In FIGURE b, the green areas represent the territories of individual males. The fractions indicate the number of offspring sired by a male over the total number of young in his territory. The arrows show the direction of EPCs and the number of "extra" young sired by each male. Thus, male X fathered all three of the young in his territory, as well as two offspring in the neighboring territory of male Y and one offspring in the territory of male Z. A relatively large number of EPCs occur in the territories of males that have a larger-than-average number of females. Perhaps this is because a male with multiple mates cannot guard them all effectively.

In the past few years, DNA fingerprinting studies have shown that EPCs are common in many bird species that were previously assumed to be faithful in their pair-bonding. These techniques have reinvigorated the study of monogamous breeding systems by illuminating the variability in reproductive fitness that was previously undetectable.

(b) reprinted with permission from Gibbs, et al. "Realized Reproductive Successes of Polygonous Red-Winged Blackbirds Revealed by Markers," *Science* 250 (1990): 1394–1397. ©1990 American Association for the Advancement of Science.

(a)

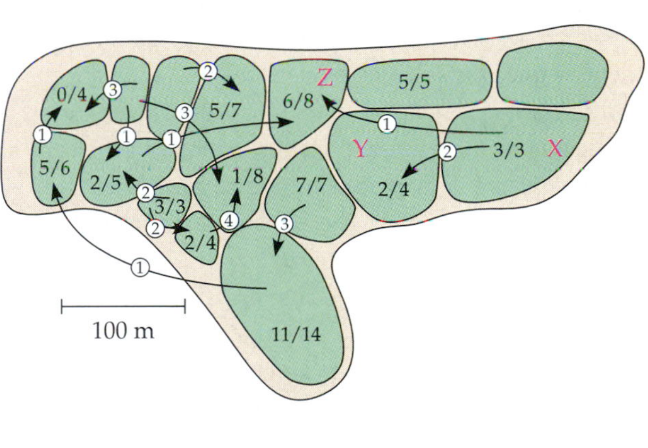

(b)

sarily consciously, transmit information by special behaviors called *displays*. The intentional transmission of information between individuals is the usual definition of communication in behavioral ecology. Singing by male birds is, in effect, a display that transmits the information, "This is my territory. Keep out!" This is almost certainly an important message of singing; if we play the tape-recorded songs of another male in a male bird's territory, he becomes highly agitated, approaching and sometimes even attacking the speaker. Someone else has not only ignored his warning but has claimed the territory for himself. This simple experiment is so infallible that some bird watchers routinely use it to find and see secretive birds that would otherwise stay hidden. The playback procedure makes another important point. We cannot get into an animal's brain to determine whether it has received a message sent by another individual. How, then, do we know when communication has occurred? We usually say that communication has occurred when an act by a "sender" produces a detectable change in the behavior of another individual, the "receiver." Bird song is communication because it produces a response.

Before behavioral ecology was firmly established it was generally assumed that communications systems evolve in ways that maximize the quantity and accuracy of information. Behavioral ecologists take a different evolutionary perspective. They believe that these displays evolve because the fitness of the sender is increased by the effect the message has on the individuals receiving the message. This may explain adaptations in which the function of communication seems to be deception, especially when the message's recipients belong to a different species. For example, male and female *Photinus* fireflies find each other and mate when females give a characteristic pattern of light flashes in response to flashes given by males. However, females of another firefly genus, *Photurus,* respond to these male *Photinus* flashes with the same flashing pattern. A male *Photinus* receives the incorrect message that a female *Photurus* is a suitable mate, and when he finds her she does not mate with him, but kills and eats him instead. Such cases of mimicry, in which communication is adaptive for the sender but maladaptive for the receiver, are widespread.

Deceit also occurs within species. In some mammals, after a dominant male assumes control of a social group, he kills young that are born too soon to be his offspring. Without dependent young, the females ovulate sooner, allowing the new dominant male to father their offspring. In an Asian monkey known as the hanuman langur, females in the early stages of pregnancy actively communicate receptiveness to copulations from new males. These females then give birth only shortly before the young fathered by these males would appear. This may deceive a male into treating the female's young as if they were his own, when in fact they are not. Again, none of this implies that langurs have consciously shaped their behavior to maximize fitness. Instead, it simply implies that females with a tendency to solicit copulations from new males have had a selective advantage over other females without such a tendency.

Another important evolutionary consideration in communication is the sensory mode used to transmit information. Animals transmit information with visual, auditory, chemical (olfactory), tactile, and electrical signals. Which mode is used to transmit information is closely related to an animal's basic lifestyle. Most terrestrial mammals are nocturnal, which makes visual displays relatively ineffective. But olfactory and auditory signals work as well in the dark as in the light, and most mammalian species stress these signals. Birds, by contrast, are mostly diurnal and use mostly visual and auditory signals. They almost never use olfactory signals, probably because they can fly faster than chemical signals can travel. (It is hard to imagine a system in which it would be adaptive for a messenger to arrive before its message.) Unlike most mammals, humans are diurnal and in common with birds use mainly visual and auditory communication. Therefore, we can detect the songs and bright colors that birds use to communicate with each other. This may explain why bird watching is so popular. If humans had the well-developed olfactory abilities of most mammals and could detect their rich world of chemical cues, mammal sniffing might be as popular as bird watching.

Animals that communicate by odors emit chemical signals called **pheromones**. These are especially common among mammals and insects and often relate to reproductive behavior. Female silkworm moths, for example, emit a pheromone that can attract males from several kilometers away. Once the moths are together, pheromones also trigger specific courtship behaviors. Another example is the familiar trailing behavior of ants, in which scouts release scents that guide other ants to the food.

One of the most complex communication systems—certainly among invertebrates—is that of social, or hive, bees. Pheromones produced by a hive's queen and her daughters, the workers, maintain the social order of honeybee colonies. Recent studies indicate that varied blends of two fatty acids, rather than single chemicals, underlie the social behavior and reproduction of honeybees. The context of a chemical signal can be as important as the chemical itself. When male honeybees (drones) are outside the hive (where they can mate with a queen), they are attracted to her pheromone; however, when drones are inside the hive they are unaffected by the queen's pheromone.

For maximum foraging efficiency, worker bees must convey to one another the location of good food sources, which may change frequently as various flowers bloom or new patches are discovered. How do bees communicate? The study of honeybee communication has a long and rich tradition of experimental research that continues to reveal new elements of the bees' language. The problem was first studied in the 1940s by Karl von Frisch, who carefully watched individual European honeybees (*Apis mellifera carnica*) as they returned to special observation hives. A returning bee would quickly become the center of attention by other bees, called followers (FIGURE 51.21a). The returning bee would go through a repetitive behavior that von Frisch called a dance. If the food source was close to the hive (less than 50 m away), the returning bee moved in tight circles while waggling its abdomen from side to side (FIGURE 51.21b). This dance was usually accompanied by the bee's regurgitating some of the acquired nectar. This behavior, which von Frisch called the round dance, had the effect of exciting the follower bees and motivating them to leave the hive and search for food that was nearby.

However, bees often forage at great distances from the hive,
51.1 sometimes in excess of 5 km. In such cases, the round dance is insufficient, lacking both directionality and distance cues necessary for the followers to locate the food source efficiently. A worker returning from a longer distance does a "waggle dance" (FIGURE 51.21C): a half-circle swing in one direction, followed by a straight run and then a half-circle swing in the other direction. This dance seems to indicate both direction and distance. The angle of the straight run in relation to the vertical surface of the open hive is the same as the horizontal angle of the food in relation to the sun. For example, if the bee runs at a 30° angle to the right of vertical, the other workers will fly 30° to the right of the horizontal direction of the sun. Distance to the food is indicated by a variety of elements of the waggle dance. For example, a longer straight run during the dance, and hence an increasing number of abdominal waggles per run, indicates a greater distance to the food source.

During waggle dances, the bee also regurgitates nectar; thus, when bees leave to forage, they already "know" the type of food

51.1

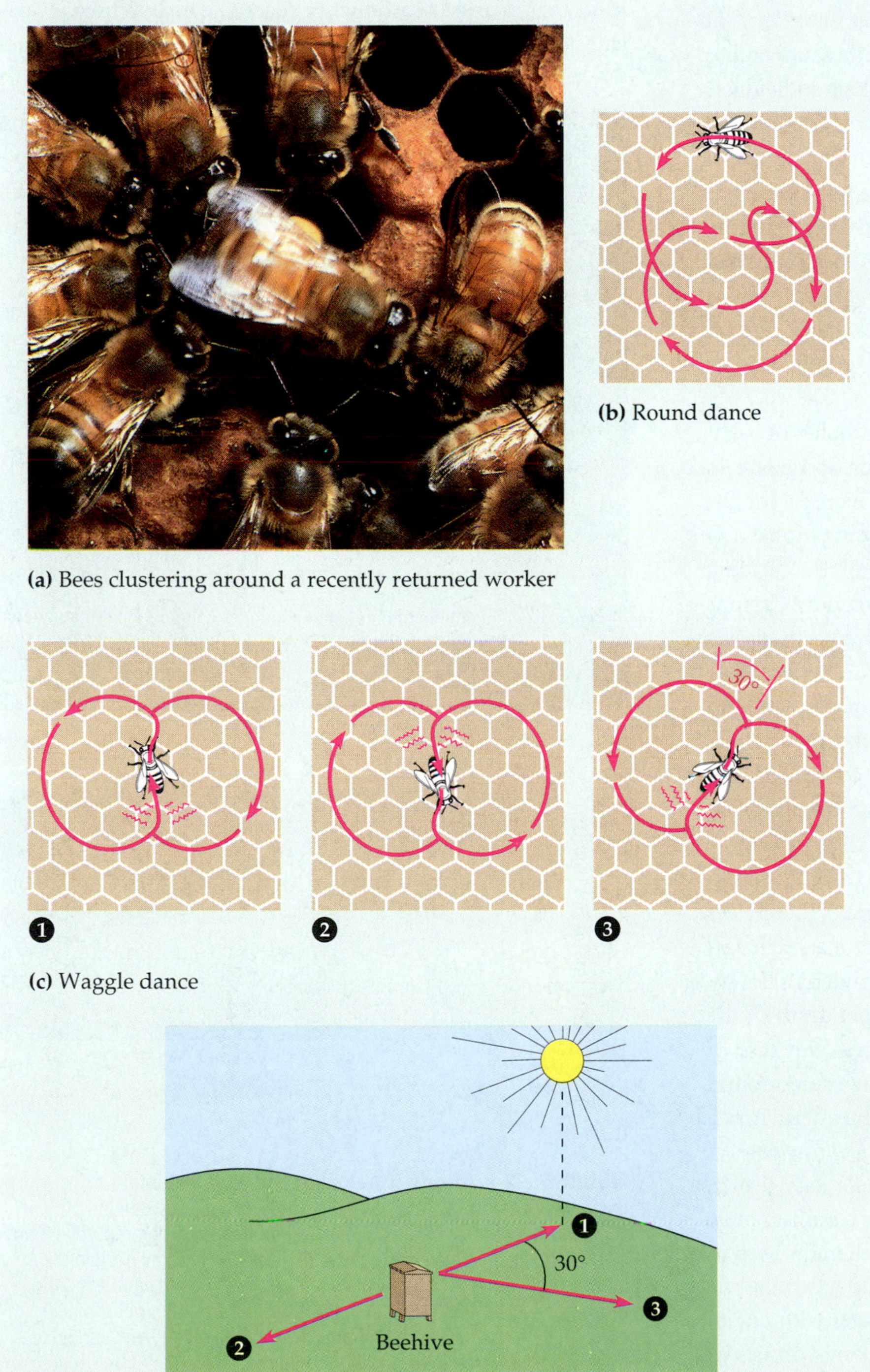

FIGURE 51.21 ▪ Communication in bees: one hypothesis. (a) Worker bees cluster around one of their sisters, recently returned from a foraging trip. **(b)** The round dance indicates that food is near but may provide no information on directionality or specific distance. **(c)** The waggle dance is performed when food is distant. This dance pattern resembles a figure eight, with a straight run between two semicircular movements. According to von Frisch's hypothesis, the waggle dance indicates both distance and direction. Distance is indicated by the duration of each waggle run or dance and the number of abdominal waggles performed per waggle run. Direction is indicated by the angle (in relation to the vertical surface of the hive) of the straight run that forms part of the dance itself. ① For instance, if the straight run is directly upward, this signals that food is in the same direction as the sun. ② If the straight run is directly downward, the food is in the direction opposite the sun. ③ If the angle is 30° to the right of vertical, the food is 30° to the right of the horizontal direction of the sun, and so forth. Odor cues (pheromones) and sound may also convey information about the location and type of food.

to seek, its distance, and its direction. There is also evidence that the sounds and odors emanating from the dancing bee provide information about the food source. In fact, some researchers question the strength of the evidence that the pattern of the waggle dance itself is the key communication and are exploring other hypotheses for how bees report a source of food.

The concept of inclusive fitness can account for most altruistic behavior

Many social behaviors are selfish, meaning that they benefit the individual at the expense of others, especially competitors. A bird that establishes a territory deprives other individuals of one, and if there is not enough habitat, these other individuals may be unable to breed. Even in species in which individuals do not engage in agonistic behavior, most adaptations that benefit one individual will indirectly harm others. For example, superior foraging ability by one individual may leave less food for others. It is easy to understand the pervasive nature of selfishness if natural selection shapes behavior. Behavior that maximizes an individual's reproductive success will be favored by selection, regardless of how much damage such behavior does to another individual, a local population, or even an entire species.

How, then, can we explain observed examples of what appears to be altruism? On occasion, animals do behave in ways that reduce their individual fitness and increase the fitness of the recipient of the behavior; this is our functional definition of **altruism**. Consider the example of the Belding ground squirrel, which lives in some mountainous regions of the western United States and is vulnerable to predators such as coyotes and hawks (FIGURE 51.22a). If a predator approaches, one of the squirrels often gives a high-pitched alarm call. This alerts unaware individuals, who then retreat to their burrows. Careful observations have confirmed that the conspicuous alarm behavior increases the risk of being killed, because it identifies the caller's location.

Another example of altruistic behavior occurs in bee societies, in which the workers are sterile. The workers themselves will never reproduce, but they labor on behalf of a single fertile queen. Furthermore, the workers sting intruders, a behavior that helps defend the hive but results in the death of the individual. Still another example of altruism is the case of mole rats, highly social rodents that live in underground chambers and tunnels in southern and northeastern Africa (FIGURE 51.23). The naked mole rat (*Heterocephalus glaber*) is almost hairless and nearly blind; naked mole rats live in colonies of 75 to 250 or more individuals. The common mole rat (*Cryptomys hottentotus*) is haired and generally lives in smaller colonies. In both species, each colony has only one reproducing female, called the queen, who mates with one to three males, called kings. The rest of the colony consists of nonreproductive females and males who forage for underground roots and tubers and care for the queen, the kings, and new offspring still dependent on the queen. The nonreproductive individuals may sacrifice their own lives in trying to protect the queen or kings from snakes or other predators that invade the colony.

How can a naked mole rat, a worker bee, or a Belding ground squirrel enhance its fitness by aiding other members of the population, which are apt to be its closest competitors? How can altruistic behavior be maintained by evolution if it does not enhance—and, in fact, may even reduce—the reproductive success of the self-sacrificing individuals? Natural selection favors anatomical, physiological, and behavioral traits that increase reproductive success, which in turn propagates the genes responsible for those traits. When parents sac-

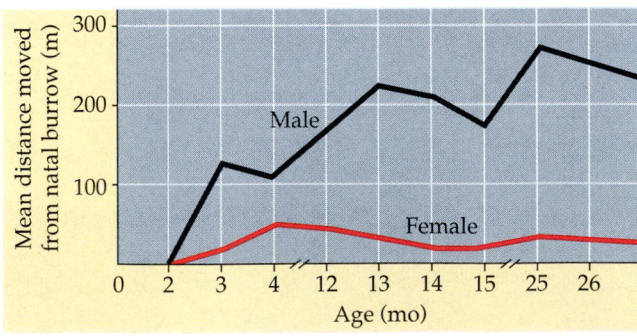

(a)

Mean distance moved from natal burrow (m)
300
200
100
Male
Female
0 2 3 4 12 13 14 15 25 26
Age (mo)

(b)

FIGURE 51.22 • Altruistic behavior and a male-female difference in its expression. **(a)** By sounding an alarm call, this Belding ground squirrel warns others of danger, such as an approaching predator. Nearly all alarm calls are given by females. **(b)** This graph helps explain the male-female difference in altruistic behavior of ground squirrels. After they are weaned, males disperse much farther from their birthplaces than do females. Therefore, females are more likely to live near close relatives, and warning these relatives increases the inclusive fitness of the altruist.

rifice their own personal well-being to produce and aid offspring, this actually increases the fitness of the parents, because it maximizes their genetic representation in the population. But what about helping other close relatives? Like parents and offspring, siblings have half of their genes in common. Therefore, selection might also favor helping one's parents produce more siblings or even helping siblings directly. Behavioral ecologist William Hamilton first realized that selection could result in an animal's increasing its genetic representation in the next generation by "altruistically" helping close relatives other than its own offspring. This realization led to the concept of **inclusive fitness**, which describes the total effect an individual has on proliferating its genes by producing its own offspring *and* by providing aid that enables other close relatives to increase the production of their offspring.

An important quantitative measure of inclusive fitness is the **coefficient of relatedness**, which is the proportion of genes that are identical in two individuals because of common ancestors. For example, the coefficient of relatedness for siblings is 0.5, meaning that they match in 50% of their genes. For cousins, the coefficient of relatedness is 0.125. We expect that the higher the coefficient of relatedness, the more likely an individual is to aid a relative: The individual's altruism can result in more genes identical to its own in the next generation if it aids a sibling rather than a cousin. This mechanism of increasing inclusive fitness is called **kin selection**. The contribution of kin selection to inclusive fitness varies among species. Kin selection may be rare to nonexistent in species that are never social or that disperse so widely that individuals never live near close relatives. In these species, an animal's inclusive fitness is essentially the same as its individual fitness—its fitness based solely on the production of its own offspring.

The British geneticist J. B. S. Haldane anticipated the concepts of inclusive fitness and kin selection by jokingly saying that he would lay down his life for two brothers or eight cousins. In today's terms, we would say that he would do this because either two brothers or eight cousins would result in as much representation of Haldane's genes as would two of his own offspring. But this assumes equal reproductive potential for all individuals. We assume that Haldane was as likely to have as many children as each of his individual brothers and cousins. Often, however, reproductive potentials differ. A sterile individual should theoretically sacrifice its life for one fertile brother or even one cousin; otherwise its inclusive fitness will be zero. Even differences in reproductive potentials that are not as great as between sterile and fertile individuals can still contribute to kin selection. For example, territories might be in limited supply, and one animal might have one while its brother does not. The homeless animal might have a greater increase in its inclusive fitness by helping its brother than by

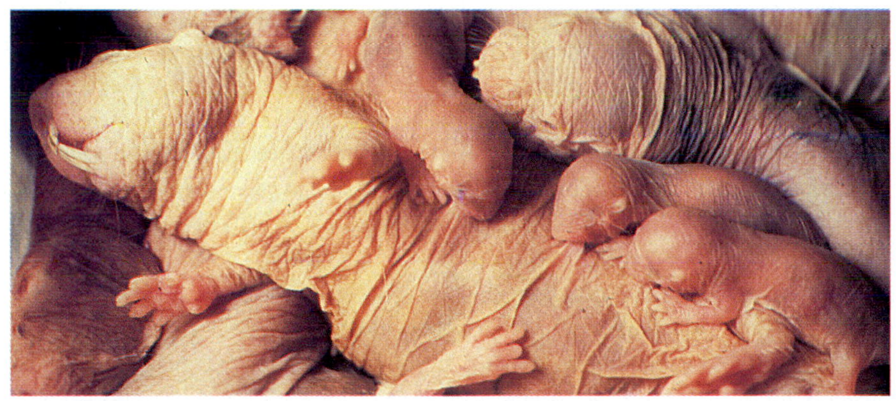

(a)

(b)

FIGURE 51.23 ▪ Two species of colonial mammals. (a) Naked mole rats (*Heterocephalus glaber*) live in underground colonies made up of a single queen, several kings, and often hundreds of nonreproductive individuals. In this photograph, several nonreproductive individuals, who perform all the maintenance activities of the colony, huddle around the queen and her young. Members of each colony are a closely related family unit. **(b)** The common mole rat (*Cryptomys hottentotus*), widespread in southern Africa, is also colonial with only a queen and several kings reproducing. Compared to naked mole rats, however, common mole rat colonies are smaller and more genetically diverse.

trying to win a territory of its own, if the second option has only a small chance of success. In trying to predict whether an individual animal will aid relatives, behavioral ecologists have derived a formula that combines coefficients of relatedness, costs to the altruist, and benefits to the recipient of aid. (Animals, of course, do not work out the kin selection formula for themselves or calculate behavior that increases their inclusive fitness.)

If kin selection explains altruism, then the examples of unselfish behavior we observe should involve close relatives. This expectation is met, but often in complex ways. Like most mammals, female Belding ground squirrels settle close to their site of birth, while males settle at distant sites. Thus, only females are likely to live near close relatives, and nearly all alarm calls are given by females (see FIGURE 51.22b). However, if all of a female's close relatives are dead, she rarely gives alarm calls. In the case of worker bees, the individuals are sterile, and anything they do to help the entire hive benefits the only permanent member who is reproductively active—the queen, who is their mother.

In the case of naked mole rats, DNA analyses have shown that all the individuals in a colony are closely related. Genetically, the queen appears to be a sibling, daughter, or mother of the kings, and the nonreproductive rats are the queen's direct descendants or her siblings (see FIGURE 51.23a). Hence, when a nonreproductive individual enhances a queen's or king's chances of reproducing, it increases the chances that some genes identical to its own will be passed to the next generation.

The common mole rat seems to be different (see FIGURE 51.23b). Some individuals may move from one colony to another, making its colonial groups more genetically diverse, and decreasing the opportunities for kin selection.

Some researchers posit that only by living in cooperative groups can mole rats obtain enough food to survive where resources are in short supply. The cooperative behavior of the common mole rat and the naked mole rat enhances tunneling and underground foraging for roots and tubers, factors that may enable these two species to inhabit arid regions. Three other species of mole rats, which are not colonial, are found only where water and other resources are relatively plentiful. Thus, resource limitation may underlie the evolution of colonial life among these mammals, and perhaps the evolution of the kind of altruistic behavior seen in the naked mole rat.

Some animals occasionally behave altruistically toward others who are not relatives. A baboon may help an unrelated companion in a fight, or a wolf may offer food to another wolf even though they share no kinship. Such behavior can be adaptive if the aided individual returns the favor in the future. This sort of exchange of aid is called **reciprocal altruism** and is commonly invoked to explain altruism in humans. Reciprocal altruism is rare in other animals; it is limited largely to species with social groups stable enough that individuals have many chances to exchange aid. It is likely that all behavior that seems altruistic actually increases fitness in some way. Thus, some behavioral ecologists argue that true altruism never really occurs, except, perhaps, in humans.

Sociobiology connects evolutionary theory to human culture

Recall from earlier in the chapter that the main premise of sociobiology is that behavioral characteristics exist because they are expressions of genes that have been perpetuated by natural selection. In the last chapter of *Sociobiology*, E. O. Wilson speculates about the evolutionary basis of certain kinds of social behavior, including culture, in humans. The debate about the connection between biological evolution and human culture remains heated today.

Let's examine the debate in the context of a specific example: the avoidance of incest. Such avoidance is adaptive because inbreeding may increase the frequency of certain kinds of genetic disorders. Many birds and mammals clearly avoid incest. Similarly, nearly all human cultures have laws or taboos forbidding sexual relations or marriage between brother and sister. Is there an innate aversion to incest that we share with other species, or do we acquire this behavior as part of our socialization? Someone favoring the "nurture" side of the debate might argue that if this behavior were innate, cultural taboos would be superfluous. According to this argument, the avoidance of incest is a learned behavior, and the social stigma attached to incest may be based on experience; people who break the taboo are more likely to have children with congenital disorders. Some sociobiologists, on the other hand, would argue that the occurrence of a specific behavior in diverse cultures is evidence that the behavior has an innate component. According to this argument, incest taboos are simply proximate mechanisms that reinforce a behavior that ultimately evolved because of its effect on fitness.

Some sociobiologists, including Wilson, envision cultural and genetic components of social behavior as linked in a cycle of reinforcement. If most members of a society share an innate avoidance of incest, the aversion is likely to become formalized in laws and taboos. The cultural stigma, in turn, acts as an environmental factor in natural selection, amplifying the evolutionary component of the behavior. In the case of incest taboos, society may shun or imprison offenders, thus lowering their reproductive fitness. According to this view, genes and culture are integrated in human nature. Behavioral biologist Marian Dawkins summarizes it this way: "Culture cannot escape from biology. We are cultural beings because we have been shaped by natural selection to be so."

The spectrum of possible social behaviors may be circumscribed by our genetic potential, but this is very different from saying that genes are rigid determinants of behavior. This is at the core of the debate about sociobiology. Opponents fear that a sociobiological interpretation of human behavior can be used to justify the status quo in human society, thus rationalizing current social injustices. Sociobiologists argue that this is a gross oversimplification and misunderstanding of what the data tell us about human biology. Sociobiology does not reduce us to robots stamped out of rigid genetic molds. Individuals vary extensively in anatomical features, and we should expect inherent variations in behavior as well. Furthermore, though we are locked into our genotypes, our nervous systems are not "hardwired." Environment intervenes in the pathway from genotype to phenotype for physical traits, and even more, in general, for behavioral traits. And because of our capacity for learning and versatility, human behavior is probably more plastic than that of any other animal. Over our recent evolutionary history we have built up structured societies with governments, laws, cultural values, and religions that define what is acceptable behavior and what is not, even when unacceptable behavior might enhance an individual's Darwinian fitness. Perhaps it is our social and cultural institutions that make us truly unique and that provide the only feature in which there is no continuum between humans and other animals (FIGURE 51.24).

■ ■ ■

In this chapter we have examined the role of behavior in the relationship between animals and their environment. We have focused on the proximate mechanisms underlying behavior and on how particular behavior patterns contribute to an animal's survival and reproductive success. In studying an animal's behavior, we are watching it perform on an ecological stage, a setting where evolution by natural selection determines which individuals contribute the greatest number of genes to a population's gene pool. In the next chapter we take a closer look at populations as ecological and evolutionary units.

FIGURE 51.24 ▪ Both genes and culture build human nature. Teaching of the younger generation by the older is one of the basic ways in which all cultures are transmitted. Sociobiologists see tutoring as an innate tendency with adaptive value that has evolved in the human species.

CHAPTER REVIEW

REVIEW OF KEY CONCEPTS

(with page numbers and key figures)

INTRODUCTION TO BEHAVIOR AND BEHAVIORAL ECOLOGY

Behavior is what an animal does and how it does it. It includes both observable motor activity and nonmotor activities such as storing of memories in the animal's brain. Questions about environmental stimuli that provoke an animal's responses and about the response mechanisms themselves address proximate causes of behavior. Questions about why natural selection has favored a particular behavior address ultimate, or evolutionary, causes of behavior. The methods of phylogenetic biology can show when behavioral traits may have arisen in an animal's lineage.

- **Behavior results from both genes and environmental factors (p. 1054, FIGURE 51.1)** Genes and the environment contribute to behavior. Behavioral traits exhibit a range of phenotypic variation, or norm of reaction, depending on the environment in which the genotype is expressed. Environmental factors include all the chemical and physical conditions in the organism in which the gene is expressed.
- **Innate behavior is developmentally fixed (pp. 1054–1055)**
- **Classical ethology presaged an evolutionary approach to behavioral biology (pp. 1055–1057, FIGURES 51.2 AND 51.3)** Lorenz, Tinbergen, von Frisch, and other classical ethologists set the stage for modern research in behavioral ecology. Most of their research focused on proximate causes of behavior. Early ethologists focused on fixed action patterns (FAPs), essentially unchangeable series of acts usually carried to completion once triggered by an external sensory stimulus (sign stimulus).
- **Behavioral ecology emphasizes evolutionary hypotheses: *science as a process* (pp. 1057–1060, FIGURES 51.4–51.6)** Behavioral ecology studies behavior in an ecological context. It is based on the expectation that animals behave in ways that increase their Darwinian fitness (reproductive success). For example, according to the premise of optimal foraging, animals forage efficiently by modifying their behavior in complex ways that favor a high ratio of energy intake to expenditure.

LEARNING

- **Learning is experience-based modification of behavior (pp. 1060–1061)** Learning is modification of behavior resulting from specific experiences. Some apparent learning is due mostly to inherent maturation. Habituation is a simple kind of learning involving loss of sensitivity to unimportant stimuli.

- **Imprinting is learning limited to a critical time period** (pp. 1061–1063, FIGURES 51.7, 51.8) In classical experiments, Lorenz showed that a goose's sense of parental and species identity is not innate, but determined by what it sees during a brief period after hatching. Imprinting occurs in many species, at various ages and for critical periods of varying durations. It is largely irreversible, although modification by later learning can occur. Many birds exhibit a combination of early imprinting and later modification in the learning of specific songs.

- **Many animals can learn to associate one stimulus with another** (pp. 1063–1064, FIGURE 51.9) Associative learning involves linking one stimulus with another. Early experiments demonstrated a type of associative learning called classical conditioning. In operant conditioning, or trial-and-error learning, an animal learns to associate one of its own behaviors with reward or punishment and modifies the behavior accordingly. Operant conditioning is common in nature.

- **Practice and exercise may explain the ultimate bases of play** (p. 1064, FIGURE 51.10) Play, which is especially common among young mammals, is behavior with no apparent goal that resembles some goal-directed behavior. Play is costly in terms of energy and is often dangerous. Its prevalence may be explained by the benefits of practicing functional behavior, by the need for exercise, or both.

ANIMAL COGNITION

- **The study of cognition connects nervous system function with behavior** (pp. 1064–1065) Cognition is the ability of an animal's nervous system to perceive, store, process, and use information gathered by sensory receptors. Cognitive ethology explores the connection between nervous system function and animal behavior.

- **Movement from place to place often depends on internal coding of spatial relationships** (pp. 1065–1066, FIGURES 51.12, 51.13) Many animals construct cognitive maps, or internal representations of the spatial relationships among objects in their surroundings, enabling them to move from place to place. Some migrating birds and other animals navigate by calibrating several cues; the Earth's magnetic field, the sun, and the stars.

- **The study of consciousness poses a unique challenge for scientists** (pp. 1066–1067) A debate centers on whether nonhuman animals are conscious, thinking organisms and whether consciousness is amenable to the scientific process. Consciousness is difficult to study, because it is not associated with observable behavioral or physiological changes.

SOCIAL BEHAVIOR AND SOCIOBIOLOGY

- **Sociobiology places social behavior in an evolutionary context** (pp. 1067–1068) Social behavior encompasses the spectrum of interactions between two or more animals, usually of the same species, and has long been a research focus for behavioral scientists. The relatively new discipline of sociobiology examines the evolution of social behavior.

- **Competitive social behaviors often represent contests for resources** (pp. 1068–1070, FIGURES 51.16, 51.17) Agonistic behavior involves a contest in which one competitor gains an advantage in obtaining access to a limited resource, such as food or a mate. Some animals have dominance hierarchies, with differentially ranked individuals permitted options appropriate to their status. Territoriality is a behavior in which an animal defends a specific, fixed portion of its home range against intrusion by other animals of the same species through agonistic interactions.

- **Mating behavior relates directly to an animal's fitness** (pp. 1070–1072, FIGURES 51.19, 51.20) There is a high correlation between observed mating behaviors and number of offspring, which is often the main determinant of an animal's fitness. Many animals engage in courtship, which announces that the participants are nonthreatening and are potential mates. In most species, females have more parental investment than males and mate more selectively. Males of most species compete for mates; females of many species engage in assessment, or selection of the male with the most favorable characteristics. Mating systems vary widely among different species. They may be promiscuous, monogamous, or polygamous, depending partly on the parental investment of each parent.

- **Social interactions depend on diverse modes of communication**
51.1 (pp. 1072–1076, FIGURE 51.21) Social interactions depend on communication, the transmission of information between individuals. Behavioral ecologists propose that specific communication behaviors evolve because their effect on the recipient maximizes the fitness of the sender. This hypothesis accounts for both transmission of accurate information (usually within a species) and deception (usually between species). Animals communicate with one another by means of visual, auditory, or electrical signals, or by chemical signals called pheromones. Classical studies in ethology showed that bees use complex movements (dances) to convey information about the location of food sources.

- **The concept of inclusive fitness can account for most altruistic behavior** (pp. 1076–1078, FIGURES 51.22, 51.23) Altruistic behavior benefits a conspecific, usually a relative, at the potential expense of the helpful individual. It seems best explained by the theory of inclusive fitness: Genes enhance their proliferation by directing organisms to care for others who share those genes. The coefficient of relatedness is the proportion of genes that are identical in two related individuals. The higher their coefficient of relationship, the more likely two animals are to help each other, a mechanism for promoting inclusive fitness called kin selection. Reciprocal altruism between unrelated individuals may be ultimately advantageous to the altruistic individual when the recipient "returns the favor."

- **Sociobiology connects evolutionary theory to human culture** (pp. 1078–1079) Sociobiology maintains that human social behavior can be understood in evolutionary terms, another source of heated debate. Some sociobiologists envision cultural and genetic components of social behavior as linked in a cycle of reinforcement.

SELF-QUIZ

1. Bees can detect wavelengths of light that we cannot see and sense minute amounts of chemicals we cannot smell. But unlike many insects, bees cannot hear very well. Which of the following statements best fits in the perspective of behavioral ecology?
   a. Bees are too small to have functional ears.
   b. Hearing must not contribute much to a bee's fitness.
   c. If a bee could hear, its tiny brain would be swamped with information.
   d. This is an example of a proximate causation.
   e. If bees could hear, the noise of the hive would distract the bees from their work.

2. The nature-versus-nurture controversy centers on
   a. the distinction between proximate and ultimate causes of behavior
   b. the role of genes in learning
   c. whether animals have conscious feelings or thoughts
   d. the extent to which an animal's behavior is innate or learned
   e. the importance of good parental care

3. Researchers have shown that young female black grouse mate preferentially with males that have mated with other females. Which of the following is *not* a testable hypothesis for the ultimate causation of this behavior?

a. Copying other females speeds mate selection and reduces potential exposure to predators.
b. Imitation helps younger females learn to recognize male traits that may increase fitness.
c. Copying other females speeds mate selection and increases the fitness of young females.
d. Selected males occupy a location near the center of the lek.
e. Mate selection in black grouse is determined by a combination of genetic and environmental factors.

4. Female spotted sandpipers aggressively court males and, after mating, leave the clutch for the male to incubate. This sequence may be repeated several times with different males until no available males remain, forcing the female to incubate her last clutch. All of the following terms describe this behavior *except*
   a. polygamy
   b. polyandry
   c. polygyny
   d. promiscuity
   e. parental investment

5. Which of the following is *least* likely to involve cognition?
   a. navigation of a sparrow during seasonal migration
   b. being aware of your neighbor's lawn care
   c. territoriality
   d. positive rheotaxis of a fish in a current
   e. male birds in a lek

6. Which of the following is *not* true of agonistic behavior?
   a. It is most common among members of the same species.
   b. It may be used to establish and defend territories.
   c. It often involves symbolic conflict and often does not cause serious harm to either the winner or the loser in the encounter.
   d. It is a uniquely male behavior.
   e. It may be used to establish dominance hierarchies.

7. After a jay accidentally found a caterpillar under a walnut shell, researchers observed the bird looking under walnut shells for the next 15 minutes. What term best describes this behavior?
   a. instinct
   b. optimal foraging
   c. developing a search image
   d. innate
   e. agnostic

8. The core idea of sociobiology is that
   a. human behavior is rigidly predetermined by inheritance
   b. humans cannot learn to alter their social behavior
   c. many aspects of social behavior have an evolutionary basis
   d. the social behavior of humans is comparable to that of honeybees
   e. environment outweighs genes in human behavior

9. Which of the following provides an example of habituation?
   a. Humpback whales migrating from Hawaii to Alaska are observed singing songs first identified in humpbacks migrating between Alaska and Baja California.
   b. Male sticklebacks attempt to attack any red-colored object near their tank.
   c. Adult brown pelicans are more successful at capturing fish than are juveniles.
   d. Female warblers incubate the eggs cowbirds deposit in their nests.
   e. Aquarium fish are initially startled by tapping on the aquarium glass but eventually ignore it.

10. A honeybee returning to the hive from a food source performs a waggle dance with the run oriented straight to the left on the vertical surface. Most likely, this means that the food is located
   a. 90° left of the hive
   b. 90° left of the line from the hive to the sun
   c. in the opposite direction, straight to the right of the hive
   d. above the hive and slightly to the left
   e. very close to the hive

CHALLENGE QUESTION

European birds called wagtails eat insects. During winter, when food is scarce, a wagtail may defend a territory where it captures an average of 20 insects per hour, or it may join a flock that ranges widely over the countryside. A bird in a flock averages 25 insects per hour. Discuss the two strategies in terms of optimal foraging. Why do you think the wagtails do not always travel in flocks?

SCIENCE, TECHNOLOGY, AND SOCIETY

Researchers are very interested in studying identical twins who were separated at birth and raised apart. So far, the data suggest that these twins are much more alike than researchers would have predicted; they have similar personalities, mannerisms, habits, and interests. What kind of general question do you think researchers hope to answer by studying twins that have been raised apart? Why do identical twins make good subjects for this kind of research? What do the results suggest to you? What are the potential pitfalls of this research? What abuses might occur if the studies are not evaluated critically and the results are carelessly cited in support of a particular social agenda?

FURTHER READING

Ben-Ari, E.T. "Pheromones: What's in a Name?" *BioScience*, July 1998. Discusses the biological roles of pheromones and evidence for their participation in human behavior.

Buss, D. M. "The Strategies of Human Mating." *American Scientist*, May/June 1994. Presents the perspective of behavioral ecology.

Costa, J. T. "Caterpillars as Social Insects." *American Scientist*, March–April 1997. Provides an interesting challenge to traditional ideas about insect sociality.

Dawkins, M. S. *Unravelling Animal Behavior*. Essex, UK: Longman Scientific, 1995. A thoughtful, succinct analysis of modern behavioral concepts.

Dugatkin, L. A. "The Evolution of Cooperation." *BioScience*, June 1997. Explores the ultimate causes of diverse cooperative behavior.

Lohman, K. "How Sea Turtles Navigate." *Scientific American*, January 1992. Describes one of the most remarkable homing phenomena in nature.

Poulin, R., and A. S. Grutter. "Cleaning Symbiosis: Proximate and Adaptive Explanations." *BioScience*, July/August 1996. Discusses the ecological and evolutionary implications of surface parasite removal from individuals of one species by those of another species.

Queller, D. C., and J. E. Strassmann. "Kin Selection and Social Insects." *BioScience*, March 1998. Illustrates how kin selection explains the evolution of insect societies.

Schoech, S. J. "Physiology of Helping Florida Scrub Jays." *American Scientist*, January–February 1998. Examines proximate causes (hormonal mechanisms) of cooperative breeding.

WEB LINKS

Visit the special edition of *The Biology Place* for BIOLOGY, Fifth Edition, at **http://www.biology.com/campbell**. Go to Chapter 51 for online resources, including learning activities, practice exams, and links to the following web sites:

"The Animal Behavior Society (ABS)"
The web site of the ABS contains a wealth of links to behavioral biology resources around the world.

"Department of Zoological Research at the National Zoological Park"
Visit some of the behavioral ecology research projects being conducted around the world by the National Zoo's Department of Zoological Research.

"The Nobel Prize Internet Archive"
Follow this link to find out more about the three classical ethologists Karl von Frisch, Konrad Lorenz, and Niko Tinbergen.

CHAPTER 52

POPULATION ECOLOGY

Characteristics of Populations

- Two important characteristics of any population are density and the spacing of individuals
- Demography is the study of factors that affect the growth and decline of populations

Life History Traits

- Life histories are highly diverse but exhibit patterns in their variability
- Limited resources mandate trade-offs between investments in reproduction and in survival

Population Growth Models

- An exponential model of population growth describes an idealized population in an unlimited environment
- A logistic model of population growth incorporates the concept of carrying capacity

Population Limiting Factors

- Density-dependent factors regulate population growth by varying with the density
- The occurrence and severity of density-independent factors are unrelated to population density
- A mix of density-dependent and density-independent factors probably limits the growth of most populations
- Some populations have regular boom and bust cycles

Human Population Growth

- The human population has been growing almost exponentially for centuries but cannot do so indefinitely

Every day, we hear about local and global problems that threaten our well-being or provoke disputes between individuals and nations—global warming, toxic waste, conflicts over oil in the Middle East, declining health of the oceans, civil strife aggravated by depressed economics. Contributing to all these apparently unrelated problems is a common factor: the continued increase of the human population in the face of limited resources. The human population explosion is now Earth's most significant biological phenomenon. With a population of approximately 6 billion individuals, our species requires vast amounts of materials and space, including places to live, land to grow our food, and places to dump our waste. (The photo that opens this chapter shows the clearing of land for a shopping center.) Endlessly expanding our presence on Earth, we have devastated the environment for many other species and now threaten to make it unfit for ourselves.

To understand the problem of human population growth on more than a superficial level, we must consider the general principles of population ecology. It is obvious that no population can grow indefinitely. Species other than humans sometimes exhibit population explosions, but their populations inevitably crash. In contrast to these radical booms and busts, many populations are relatively stable over time, with only minor increases or decreases in population size.

In our earlier study of biological populations (Chapter 23) we emphasized the relationship between population genetics—the structure and dynamics of gene pools—and evolution. Evolution remains our central theme as we now take a more ecological focus on populations. Population ecology, the subject of this chapter, is concerned with measuring changes in population size and composition, and with identifying the ecological and evolutionary causes of these fluctuations. Later in this chapter we will return to our discussion of the human population. Let's first examine some of the structural and dynamic aspects of populations as they apply to any species.

CHARACTERISTICS OF POPULATIONS

A population is a more abstract entity than an organism or a cell, yet populations have a set of characteristics that apply only to this level of biological organization. We can think of a **population** as individuals of a single species that simultaneously occupy the same general area; they rely on the same resources, are influenced by similar environmental factors, and have a high likelihood of interacting with one another. The characteristics of a population are shaped by interactions between individuals and their environments on both ecological and evolutionary time scales, and natural selection can modify these characteristics.

Two important characteristics of any population are density and the spacing of individuals

At any given moment, every population has geographical boundaries and a population size (the number of individuals it includes). Ecologists begin studying a population by defining boundaries appropriate to the organisms under study and to the questions being posed. A population's boundaries may be natural ones, such as a specific island in Lake Superior where terns nest, or they may be arbitrarily defined by an investigator, such as the oak trees within a specific county in Minnesota. Regardless of such differences, two important characteristics of any population are its density and its dispersion. Population **density** is the number of individuals per unit area or volume—the number of oak trees per km^2 in the Minnesota county, for example. **Dispersion** is the pattern of spacing among individuals within the geographical boundaries of the population.

Measuring Density

52.1 In rare cases it is possible to determine population size and density by actually counting all individuals within the boundaries of the population. We could count the number of sea stars in a tidepool, for example. Herds of large mammals, such as buffalo or elephants, can sometimes be counted accurately from airplanes. In most cases, however, it is impractical or impossible to count all individuals in a population. Instead, ecologists often use a variety of sampling techniques to estimate densities and total population sizes. For example, they might estimate the number of alligators in the Florida Everglades by counting individuals in a few representative plots of an appropriate size. Such estimates are more accurate when there are more numerous or larger sample plots, and when the habitat is homogeneous.

In some cases, population sizes are estimated not by counts of organisms but by indirect indicators, such as the numbers of nests or burrows (FIGURE 52.1), or signs such as droppings or tracks. Another sampling technique commonly used to estimate wildlife populations is the **mark-recapture method**, described in the Methods Box on p. 1085.

FIGURE 52.1 ▪ An indirect census of a prairie dog population. The number of prairie dogs in this colony in South Dakota can be estimated by counting the number of earthen mounds constructed by these rodents. The estimate is rough because the animals are social, and the number of individuals inhabiting a system of tunnels under each mound is variable.

Patterns of Dispersion

Within a population's geographic range, local densities may vary substantially because the environment is patchy (not all areas provide equally suitable habitat) and because individuals exhibit patterns of spacing in relation to other members of the population.

The most common pattern of dispersion is **clumped**, with the individuals aggregated in patches. Plants may be clumped in certain sites where soil conditions and other environmental factors favor germination and growth. For example, the eastern red cedar is often found clumped on limestone outcrops, where soil is less acidic than in nearby areas. Animals often spend most of their time in a particular microenvironment that satisfies their requirements. For example, many forest insects and salamanders are clumped under logs where the humidity remains high. Herbivorous animals of a particular species are likely to be most abundant where their food plants are concentrated. Clumping of animals may also be associated with mating or other social behavior. For example, mayflies often swarm in great numbers, a behavior that increases mating chances for these insects, which spend only a day or two as reproductive adults. There may also be "safety in numbers"; fish swimming in large schools, for example, are often less likely to be eaten by predators than fish swimming alone or in small groups (FIGURE 52.2a, p. 1084).

The ecological concept of **grain** relates to the spatial variation, or environmental patchiness, of individual organisms. A *coarse-grained environment* is one in which environmental patches are so large (relative to the size and activity of the organism) that an individual organism can distinguish and choose among patches. A field of wildflowers is coarse-grained for a small herbivorous insect because only certain plants are suitable food sources; the insect may spend its entire larval life on one plant, having selected that plant over others.

A *fine-grained environment* is one in which patches are small relative to the size and activities of an organism, and the organism may not even behave as though patches exist. For a large herbivorous mammal, such as a horse, all the plants in a field may be more or less the same. If the horse feeds more or less indiscriminately, the field is a fine-grained environment.

(a) Clumped

(b) Uniform

(c) Random

FIGURE 52.2 ▪ Patterns of dispersion within a population's geographical range. Individuals within a population may be clumped in groups, uniformly (evenly) spaced, or randomly dispersed. In populations with a clumped pattern, the individuals within each clump also show a pattern of dispersion, as do the clumps themselves. **(a)** Butterfly fish, like many fish, are often found clumped in schools. Schooling may increase the hydrodynamic efficiency of swimming, reduce predation risks, and increase feeding efficiency. Within a school, individuals are more or less evenly spaced. **(b)** Birds nesting on small islands, such as these king penguins on South Georgia Island in the South Atlantic Ocean, often exhibit uniform spacing. **(c)** Trees of the same species are often randomly distributed in tropical rain forests, but this pattern of dispersion is relatively rare in nature.

Of course, a large field in which succulent clover grows on one side and spiny thistles on the other represents a coarse-grained environment to the horse, because the animal can select the side of the field it prefers.

Temporal variation in the environment can also be coarse-grained or fine-grained, depending on the periodicity of the variation in relation to the lifespan of the organism. A sudden cold snap might have a dramatic effect on the success of a mayfly, whose entire adult lifespan may be only a few hours, but have little influence on the activities of a long-lived mammal. Daily variations in environmental factors are usually fine-grained for most organisms, whereas seasonal and longer-term shifts in climate are coarse-grained for even the largest organisms.

In contrast to a clumped distribution of individuals within a population, a **uniform**, or evenly spaced, pattern of dispersion may result from direct interactions between individuals in the population. For example, a tendency toward regular spacing of plants may result from shading and competition for water and minerals; some plants also secrete chemicals that inhibit the germination and growth of nearby individuals that could compete for resources. Animals often exhibit uniform dispersion as a result of social interactions (see FIGURE 52.2b).

Random spacing (unpredictable, patternless dispersion) occurs in the absence of strong attractions or repulsions among individuals of a population; the position of each individual is independent of other individuals. For example, forest trees are sometimes randomly distributed (see FIGURE 52.2c). Overall, however, random patterns are not common in nature; most populations show at least a tendency toward either a clumped or uniform distribution.

Estimates of population density and local dispersion patterns within populations are important in analyzing population dynamics. They enable researchers to compare and contrast the growth or stability of populations occupying different geographic areas. On a broader scale, populations within a species also show distributional patterns, often concentrating in clusters within a species' range. For example, populations of cattails are not evenly distributed throughout the range of the species, but are clustered in areas along rivers and lakes and in wetlands. The factors that influence the distribution of a species over its range are the subject of **biogeography**, which we explore in Chapters 53 and 55.

Demography is the study of factors that affect the growth and decline of populations

Changes in population size reflect the relative rates of processes that add individuals to the population and eliminate individuals from it. Additions occur through births (which we will define here to include all forms of reproduction) and

52.1

METHODS: A Mark-Recapture Estimate of Population Size

Traps are placed within the boundaries of the population being studied, and captured animals are marked with tags, collars, bands, or spots of dye and then released. After a few days or a few weeks—enough time for the marked animals to mix randomly with unmarked members of the population—traps are set again. The proportion of marked to unmarked animals that are captured during the second trapping gives an estimate of the size of the entire population. If there have been no births, deaths, immigration, or emigration, the following simple equation can be used to estimate the population size, N:

$$N = \frac{\text{Number marked} \times \text{Total catch second time}}{\text{Number of marked recaptures}}$$

For example, suppose that 50 sanderlings are captured in mist net traps, marked with leg bands (FIGURE a), and released. Two weeks later, 100 sanderlings are captured (FIGURE b). If 10 of this second catch are marked birds that have been recaptured, we would estimate that 10% of the total sanderling population is marked. Since 50 birds were originally marked, we would then estimate that the entire population consists of about 500 birds. This method assumes each marked individual has the same probability of being trapped as each unmarked individual. This is not always a safe assumption, however; an animal that has been trapped once, for instance, may be wary of the traps later.

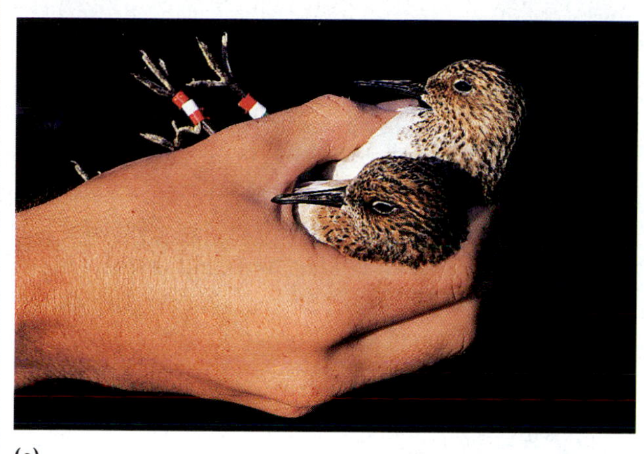

(a)

(b)

immigration, the influx of new individuals from other areas. Opposing these additions are mortality (death) and emigration, the movement of individuals out of a population. Our focus in this chapter is primarily on factors that influence birthrates and death rates.

The study of the vital statistics that affect population size is called **demography**. Birthrates and death rates usually vary among subgroups within a population, depending in particular on age and sex. Thus, a population's age structure and sex ratio are two of its most important demographic features.

Age Structure and Sex Ratio

Many organisms exhibit overlapping generations, or the coexistence of individuals from more than one generation. Only organisms in which adults all reproduce at about the same time and then die, such as annual plants and many insects, do not have overlapping generations. The coexistence of generations gives most populations an **age structure**, which is the relative number of individuals of each age. Demographers often use age pyramids to show the age structure of a population (see FIGURE 52.23). A population's structure is very important in determining the rate at which it is growing.

Every age group has a characteristic birthrate and death rate. **Birthrate**, or **fecundity**, the number of offspring produced during a certain amount of time, is often greatest for individuals of intermediate age. In humans, for example, the birthrate is highest among 20-year-old women. The **death rate** is highest in the first year and, of course, in old age. In many species, juveniles and old individuals are generally more likely to die than individuals of intermediate age, who have the optimum combination of vigor and the ability to find food and avoid predators that comes with maturity. Therefore, a population with a large percentage of individuals of prime reproductive (or slightly younger) age will grow proportionately faster than a population that has an age structure skewed toward older individuals. We will discuss the implications of such a pattern for human population growth later in the chapter.

An important demographic feature related to age structure is **generation time**, the average span between the birth of individuals and the birth of their offspring. In general, generation time is strongly related to body size over a broad range of organisms (FIGURE 52.3). Other factors being equal, a shorter generation time will result in faster population growth (assuming, of course, that the overall birthrate is greater than the death rate). This is simply because the increases in population size attributed to births accumulate more rapidly when individuals reach sexual maturity in a shorter period of time.

The **sex ratio**, the proportion of individuals of each sex, is another important demographic statistic that affects population growth. The number of females is usually directly related to the number of births that can be expected, but the number of males may be less significant because in many species a single male can mate with several females. In herds of elk, for example, there are fewer males of reproductive age than females, but this has no significant effect on the number of births in the overall population, because each male guards a "harem" of females with which he mates. In many bird species, by contrast, individuals form monogamous pair-bonds; any significant reduction in males would be more likely to affect the population's birthrate. Wildlife management often reflects these demographic considerations. For example, deer-hunting regulations are usually more liberal regarding the killing of bucks than of does because each buck typically mates with many does.

Life Tables and Survivorship Curves

About a century ago, when life insurance first became available, insurance companies took notice of some of the early scientific studies of human populations. Needing to determine how long, on average, an individual of a given age could be expected to live, they began using what are called **life tables**. Population ecologists have adapted this approach for nonhuman populations.

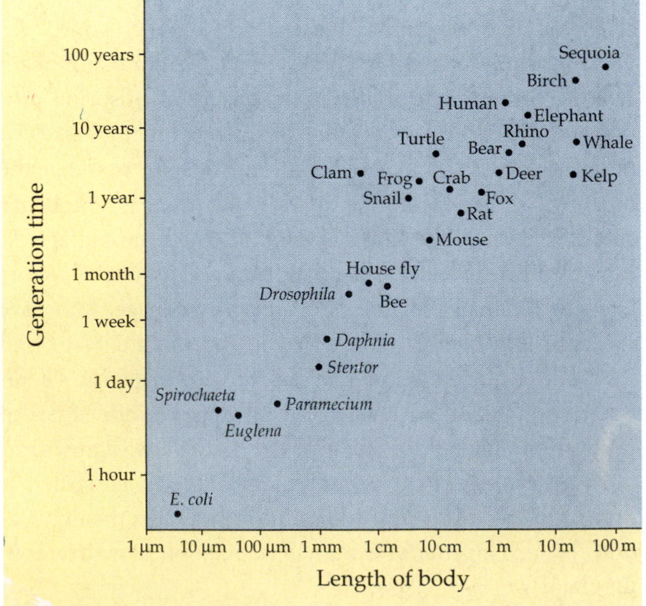

FIGURE 52.3 ▪ Generation time and body size. Small organisms generally have short generation times, achieving reproductive maturity quickly. Generation time increases with body size because larger organisms generally take longer to reach the size at which they can reproduce.

One way to construct a life table is to follow the fate of a **cohort**, a group of individuals of the same age, from birth until all are dead. The table is constructed from the number of individuals that die and the number of offspring born to each member of each age group during the defined time period. This approach can be used for only a very limited number of short-lived species. Fortunately, a cross-sectional life table can also be constructed by knowing age-specific mortality and birthrates in a population during a specified period of time.

TABLE 52.1 is a life table for a cohort of the cactus ground finch (*Geospiza scandens*), one of the Galápagos finches (see FIGURE 25.7). A lot can be learned about a population from age-specific mortalities, death rates, and birthrates (indicated by fledglings, or young birds that have just left the nest, for ground finches). The third column in the table shows the proportion of breeding birds in a cohort that are still alive at a given age. Notice that the death rates are generally highest among the youngest and oldest birds and that if a bird lives past the first year, its life expectancy is relatively high for the next 7+ years. The rate of offspring (fledgling) production is shown in the last two columns; it is highest for 8-year-old birds (both females and males), and lowest for the youngest and oldest birds. If you add all the numbers in each of the last two columns, you will obtain the *net reproductive rates* for the cohort: 2.10 for females and 0.77 for males. When the net reproductive rate equals 1, an organism is replacing itself. Thus, for this cohort, the females slightly more than doubled their numbers, while the number of males decreased.

A graphic way of representing some of the data in a life table is to draw a **survivorship curve**, a plot of the numbers in a cohort still alive at each age (FIGURE 52.4). Survivorship curves can be classified into three general types. A Type I curve is relatively flat at the start, reflecting low death rates during early and middle life, and dropping steeply as death rates increase among older age groups. Humans and many other large mammals that produce relatively few offspring but provide them with good care often exhibit this kind of curve. In contrast, a Type III curve drops sharply at the left of the graph, reflecting very high death rates for the young, but then flattens out as death rates decline for those few individuals that have survived to a certain critical age. This type of curve is usually associated with organisms that produce very large numbers of offspring but provide little or no care, such as many fishes and marine invertebrates. An oyster, for example, may release millions of eggs, but most offspring die as larvae from predation or other causes. Those few that manage to survive long enough to attach to a suitable substrate and begin growing a hard shell, however, will probably survive for a relatively long time. Type II curves are intermediate, with

Table 52.1 ■ Life Table for Cactus Ground Finches (*Geospiza scandens*) on Isla Daphne Major (a Galápagos Island)*

AGE	NUMBER ALIVE AT BEGINNING OF YEAR	PROPORTION OF COHORT SURVIVING TO BEGINNING OF YEAR	NUMBER OF DEATHS DURING YEAR	DEATH RATE	AVERAGE LIFE EXPECTANCY	NUMBER OF SUCCESSFUL FLEDGLINGS PER FEMALE	NUMBER OF SUCCESSFUL FLEDGLINGS PER MALE
0	100	1.00	49	0.49	2.64	0.00	0.00
1	51	0.51	20	0.39	3.70	0.19	0.00
2	31	0.31	10	0.32	4.76	0.05	0.00
3	21	0.21	4	0.19	5.79	0.40	0.03
4	17	0.17	0	0.00	6.02	0.17	0.07
5	17	0.17	1	0.06	5.02	0.10	0.07
6	16	0.16	0	0.00	4.31	0.17	0.12
7	16	0.16	0	0.00	3.31	0.05	0.03
8	16	0.16	7	0.44	2.31	0.70	0.33
9	9	0.09	4	0.44	2.72	0.02	0.07
10	5	0.05	0	0.00	3.50	0.00	0.00
11	5	0.05	0	0.00	2.50	0.00	0.00
12	5	0.05	3	0.60	1.50	0.25	0.05
13	2	0.02	0	0.00	2.00	---	0.00
14	2	0.02	1	0.50	1.00	---	0.00
15	1	0.01	1	1.00	0.50	---	0.00
16	0	0.00	---	---	---	---	---

*The original study included fewer than 100 birds (both sexes included), but we have started with 100 in the cohort to make the table easier to follow.

SOURCE: Adapted from Smith, R. L. *Ecology and Field Biology,* 5th ed. Menlo Park, CA: Benjamin/Cummings, 1996, p. 377; original data from Grant, P. R., and B. R. Grant. "Demography and genetically effective sizes of two populations of Darwin's finches." *Ecology* 73: 766–784, 1992.

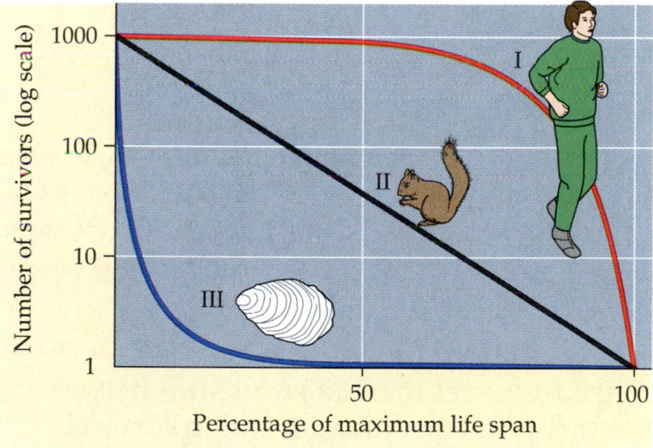

FIGURE 52.4 ▪ Survivorship curves. These three curves illustrate idealized patterns of survivorship in different kinds of organisms. As an example of a Type I curve, humans in developed countries experience high survival rates until old age. At the opposite extreme are Type III curves for organisms such as oysters, which experience very high mortality as larvae but decreased mortality later in life. Type II survivorship curves are intermediate between the other two types, and result when a constant proportion of individuals die at each age. Notice that the *y* axis is logarithmic and that the *x* axis is on a relative scale, so that species with widely varying lifespans can be compared on the same graph.

mortality more constant over the lifespan. This kind of survivorship has been observed in some annual plants, various invertebrates such as *Hydra,* some lizard species, and some rodents, such as the gray squirrel.

Many species, of course, fall somewhere between these basic types of survivorship or show more complex patterns. In birds, for example, mortality is often high among the youngest individuals (as in a Type III curve) but fairly constant among adults (as in a Type II curve); see the death rates in TABLE 52.1. Some invertebrates, such as crabs, may show a "stair-stepped" curve, with brief periods of increased mortality during molts (caused by physiological problems or greater vulnerability to predation), followed by periods of lower mortality (when the exoskeleton is hard).

Survivorship is an important factor in the changes in population size over time. Next we consider some additional characteristics that influence population dynamics.

LIFE HISTORY TRAITS

In Chapter 50 we looked at some of the ways the responses of organisms to environmental variation increase their chances of survival. However, natural selection does not act only on traits that increase survival; organisms that survive a long time but do not reproduce are not at all "fit" in the Darwinian sense. In many cases there are trade-offs between survival and traits such as frequency of reproduction, investment in parental care, and the number of offspring per reproductive episode (often called seed crop for seed plants and litter size or clutch size for animals). The traits that affect an organism's schedule of reproduction and death make up its **life history**. Of course, a particular life history, like most characteristics of an organism, is the result of natural selection operating over evolutionary time. Life history traits help determine how populations grow.

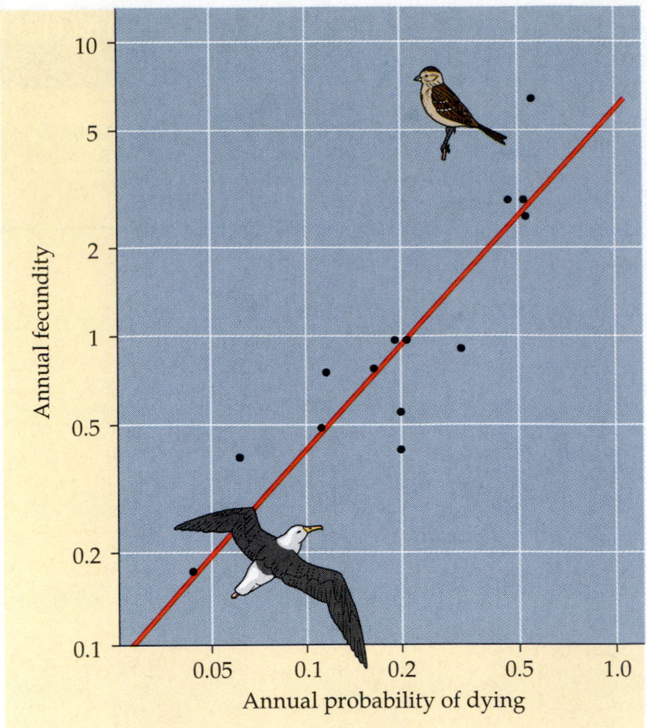

FIGURE 52.5 ▪ The relationship between adult mortality and annual fecundity in fourteen species of birds. Birds that have a high probability of dying during any given year usually raise more offspring each year than those with a low probability of dying. At one end of this scale, the wandering albatross has the lowest fecundity (about 0.2 offspring per year, or only a single surviving offspring every 5 years). The albatross also has the lowest annual mortality. At the other extreme, the tree sparrow has more than a 50% chance of dying from one breeding season to another but produces an average of six fledglings each year.

Life histories are highly diverse but exhibit patterns in their variability

Because of varying pressures of natural selection, life histories are very diverse. Pacific salmon, for example, hatch in the headwaters of a stream, then migrate to open ocean, where they require several years to mature. They eventually return to freshwater streams to spawn, producing millions of small eggs in a single reproductive opportunity, and then they die. In contrast, some lizards produce only a few large eggs during their second year, then repeat the reproductive act annually for several years. The life histories of plants are just as variable. Some species of oaks do not reproduce until the tree is 20 years old, but then produce vast numbers of large seeds each year for a century or more. Annual desert wildflowers generally germinate, grow, produce many small seeds, and then die, all in the span of a month after spring rains. Complicating matters further, important characteristics of life history may vary significantly among populations of a single species, or even among individuals in the same population.

Despite such wide variation in life history traits, there are some patterns in the way in which they vary. Life histories often vary in parallel with environmental factors. Some of the earliest work on life histories resulted from studies in the 1940s by British ornithologist David Lack. Building on previous research, Lack showed that songbirds in the tropics lay fewer eggs than their counterparts at higher latitudes. He suggested that the evolution of clutch size reflected the number of young that parents could successfully feed; in other words, clutch size is an adaptation to food supply. He further suggested that because days are longer during offspring-rearing seasons at high latitudes, temperate birds are able to gather more food than birds in the tropics, where day length is approximately 12 hours all year. Though other researchers have suggested different explanations for this pattern, the number of offspring per reproductive event has been found to vary in parallel with latitude in other taxa as well; for instance, tropical mammals, lizards, and even insects tend to produce fewer eggs than their temperate-region counterparts.

Another pattern in life history traits is that they often vary with respect to each other. For example, among birds, fecundity and mortality tend to vary in close association (FIGURE 52.5). Other traits, such as delayed maturity and high parental investment in each offspring, tend to be correlated with low fecundity and low mortality.

Limited resources mandate trade-offs between investments in reproduction and in survival

Darwinian fitness is measured not by how many offspring are produced but by how many survive to produce their own offspring: Heritable characteristics of life history that result in the most reproductively successful descendants will become more common within the population. If we were to construct a hypothetical life history that would yield the greatest lifetime reproductive output, we might imagine a population of individuals that begin reproducing at an early age, produce many offspring each time they reproduce, and reproduce many times in a lifetime. However, natural selection cannot maximize all these variables simultaneously, because organisms have finite resources that mandate trade-offs. For example, the production of many offspring with little chance of survival may result in fewer descendants than the production of a few well-cared-for offspring that can compete vigorously for limited resources in an already dense population.

The life histories we observe in organisms represent a resolution of several conflicting demands. An important part of the study of life histories has been understanding the relationship between limited resources and competing functions: Time, energy, and nutrients that are used for one thing cannot be used for something else. Several experiments have demonstrated such a trade-off. In one study, the fecundity of female seed beetles was manipulated by depriving them of egg-laying sites or mates. Females that laid fewer eggs lived longer (FIGURE 52.6a), suggesting a trade-off between investing in current reproduction and survival. There can also be trade-offs between current and future reproduction. When researchers experimentally manipulated the number of eggs in collared flycatcher nests, females that reared more chicks one year had smaller clutches the following year (FIGURE 52.6b).

As in our beetle and flycatcher examples, many life history issues involve balancing the profit of immediate investment in offspring against the cost to future prospects of reproduction. These issues can be phrased in terms of three basic questions: How often should an organism breed? When should it begin to reproduce? How many offspring should it produce during each reproductive episode? The way each population resolves these questions results in the integrated life history patterns we see in nature. We will look more closely at each of these choices, but first it is important to clarify our use of the word "choice." Individual organisms rarely choose when to breed and how many offspring to have. (Humans are an important exception we will consider later in the chapter.) Life history traits are evolutionary outcomes that are reflected in the development and physiology of an organism. Organisms generally breed as soon as they reach sexual maturity, and although the environment may influence when this occurs, the organisms do not consciously choose when to become mature. Similarly, the number of offspring produced during a given reproductive episode is associated with the number of gametes fertilized, not the result of a conscious decision.

Number of Reproductive Episodes Per Lifetime

Some plants and animals invest most of their energy in growth and development, expend this energy in a single large reproductive effort, and then die. Most insects have this type of life history, called **semelparity** (L. *semel,* "once," and *parito,* "to beget"), as do some species of salmon, annual plants, and some perennial plants such as bamboos and century plants. Other organisms produce fewer offspring at a time over a span of many seasons, a life history adaptation called **iteroparity** (L. *itero,* "to repeat"). The relative advantage of each "strategy" can be thought of in terms of a trade-off between fecundity and survival probability. Multiple breeding episodes require that an organism allocate some of its resources to survival. For example, perennial plants invest more in their roots, and also in the formation of freeze-resistant or drought-resistant buds, than do plants that live only a single season. Of course, all the resources not allocated to reproduction are wasted if the organism happens to die before reproducing again.

Under what circumstances do we expect semelparity and iteroparity to evolve? Population ecologists have developed mathematical models to determine the relative payoffs of each, and have shown that the relevant considerations are the probability of survival of both the adult and the immature individual. The models predict semelparity when the cost to parents of staying alive between broods is great, or if there is a large trade-off between fecundity and survival. Iteroparity is predicted if individuals survive well once they are established, but immature individuals are unlikely to survive. Thus, in the

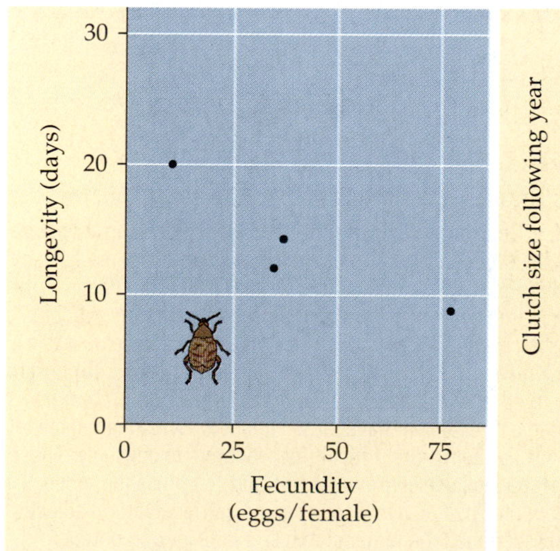

(a) Seed beetles

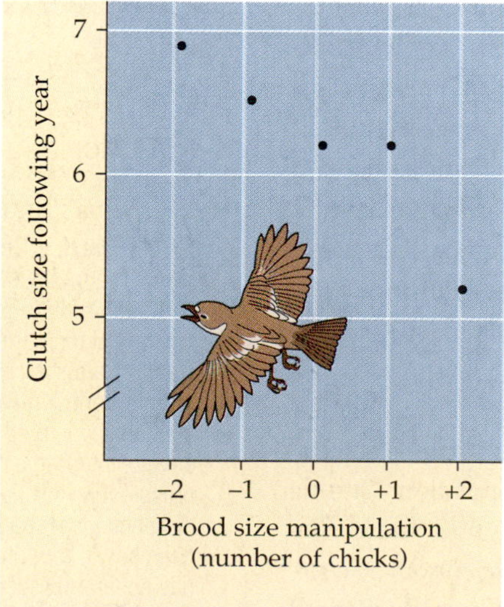

(b) Collared flycatchers

FIGURE 52.6 ▪ Effects of current reproductive effort on future reproductive success. (a) When the fecundity of female seed beetles (Family Bruchidae) is manipulated by denying them access to males or egg-laying sites, there is a trade-off between adult longevity and fecundity. **(b)** When the clutch size of collared flycatchers is manipulated by removing or adding one or two eggs, there is a direct trade-off between the number of chicks raised that year and the next year's fecundity. In this study, there was no effect of current fecundity on adult survival.

harsh climates of the desert, many plants live only a single season and put all their energy into a single reproductive effort. In the moist tropics, where competition and predation make seedling establishment difficult but where plants can usually live a long time once they are established, there are more iteroparous plants.

Semelparity is rarely found in plants and animals that live for longer than one or two years; once organisms have invested the resources necessary for survival between growing seasons, reproduction every year seems most successful. However, some organisms live for several seasons and then invest all their energy into a single immense reproductive effort—often called "big-bang" reproduction. The agave, or century plant, grows in arid climates with sparse and unpredictable rainfall. Agaves grow vegetatively for several years, then send up a large flowering stalk, produce seeds, and die (FIGURE 52.7). The shallow roots of agaves catch water after rain showers but are dry during droughts. This unpredictable water supply may prevent seed production or seedling establishment for several years at a time. By growing and storing nutrients until an unusually wet year and then putting all their resources into reproduction, the agave's big-bang reproduction is a life history adaptation to severe climate. Pacific salmon also exhibit big-bang reproduction. During their single reproductive effort, female salmon convert a large portion of their body tissue to eggs, and males to sperm production and mating. The huge cost of migrating upriver to reach their spawning grounds may make it advantageous to make the trip only a single time, and salmon expend so many resources in this effort that it kills them.

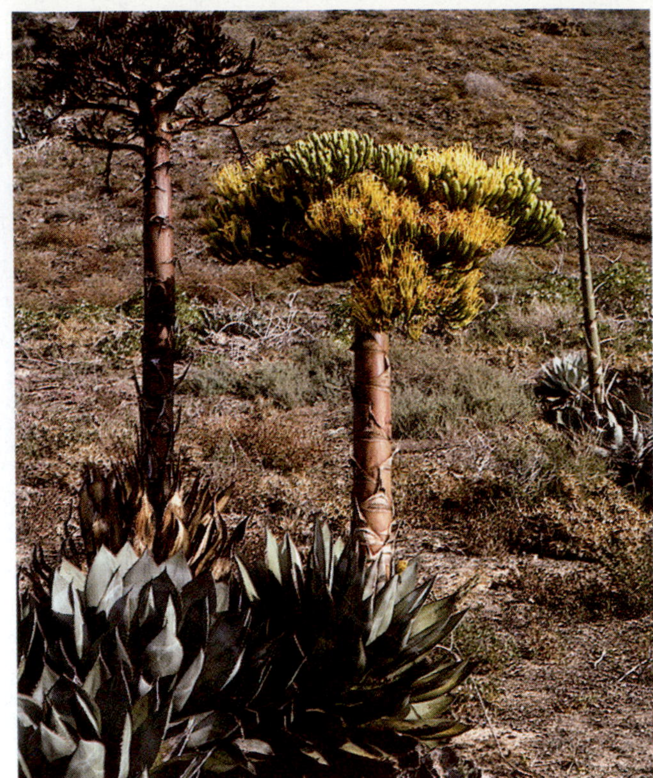

FIGURE 52.7 ▪ An example of big-bang reproduction. Agaves, or century plants, grow without reproducing for several years and then produce a gigantic flowering stalk and many seeds. After this one-time reproductive effort, the plant dies.

Number of Offspring Per Reproductive Episode and Age at First Reproduction

For iteroparous organisms, the total amount of energy invested in a given breeding season could affect an individual's chances of surviving to reproduce again (see FIGURE 52.6a) or the number of offspring it is able to produce during the next season (see FIGURE 52.6b). The trade-off is current fecundity against adult survival and future fecundity. In general, organisms with a lower probability of surviving to another year may maximize reproductive success by investing more in their current effort. Thus, organisms with high losses to predation or high overwintering mortality tend to invest more in a single reproductive episode. Organisms with potentially long lifespans do not generally increase current fecundity enough to jeopardize future reproduction. In some cases, clutch, litter, or seed crop size can vary seasonally within a single population (FIGURE 52.8). There is also a trade-off between the number and quality of offspring produced in a single reproductive episode. Generally, organisms that produce many offspring produce small ones; thus, each offspring is endowed with little energy to start

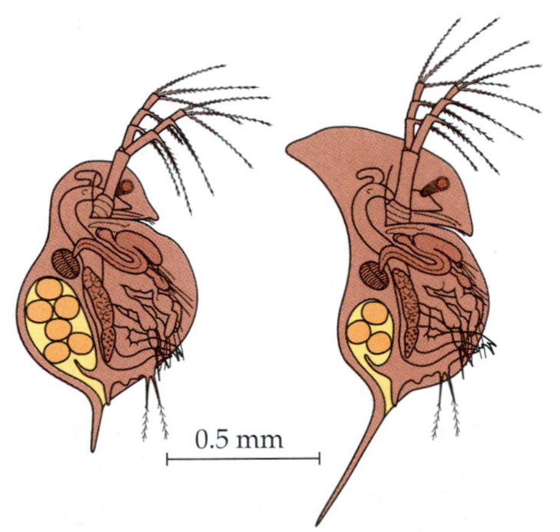

(a) Spring body form **(b)** Summer body form

FIGURE 52.8 ▪ Seasonal variation in life history caused by predation. The freshwater crustacean *Daphnia retrocurva* shows marked seasonal variation in body shape and clutch size. **(a)** In spring, the phytoplankton eaten by *Daphnia* are abundant, and predators that eat *Daphnia* are scarce. Under these conditions, individuals develop into a rounder form with a large brood chamber containing six eggs. **(b)** In summer, however, predators that feed on *Daphnia* are abundant. *Daphnia* developing at this time of year have large "helmets" and long tail spines, presumably making them more difficult for predators to consume. These morphological changes use energy reserves and compress the brood chamber so that only half as many eggs can be carried. Natural selection has apparently sacrificed the larger clutch size in favor of a better chance of producing at least some eggs and surviving to reproduce again.

life on its own. A large number of offspring and small young are typical of organisms with a Type III survivorship pattern (see FIGURE 52.4). In contrast, offspring from small clutches, litters, or seed crops are generally larger, and each stands a better chance of surviving to adulthood, as illustrated by Type I and Type II curves.

As with other life history adaptations, the number and size of offspring depend on the selective pressures under which the organism evolved. Plants and animals whose young are subject to high mortality rates often produce large numbers of relatively small offspring (FIGURE 52.9a). Thus, plants that colonize disturbed environments usually produce many small seeds, many of which will not get to a suitable environment. Small size might actually benefit such seeds if it enables them to be carried long distances. Birds such as quail and mammals such as rabbits and mice that suffer high predation rates also produce large numbers of small offspring. In other organisms, extra investment on the part of the parent greatly increases the offspring's chances of survival. Oak, walnut, and coconut trees all have large seeds with a store of energy that the seedlings can use to get established (FIGURE 52.9b). In animals, parental investment in offspring does not always end with incubation or gestation. Primates generally have only one or two offspring at a time. Parental care and an extended period of learning in the first several years of life are very important to offspring fitness in these mammals.

Age at First Reproduction. For organisms that reproduce repeatedly during their lifespan, the timing of first reproduction can have a large effect on the female's lifetime reproductive output. Again, the balance is between current reproduction and survival plus future reproduction. Organisms that delay first reproduction avoid costs that can include courtship, nest building, gamete production, and migrating to breeding areas. In many animals, life experience might reduce some risks of breeding or increase fecundity. Also, many organisms grow throughout their lives, and these size increases often increase fecundity. On the other hand, a female that delays reproduction must invest more energy in maintenance and growth. Although such an adaptation might increase her potential for future reproduction, she may die before producing any offspring at all, thus reducing her fitness to zero.

Mathematical models of these phenomena suggest that if older females produce many more offspring each time they reproduce than do younger ones, and if there is a good chance of surviving to an advanced age, then female lifetime reproductive output is maximized when first reproduction is delayed. In general, this is what we see in nature. Among many birds, for example, age at sexual maturity varies directly with the annual survival rates of adults. In many cases, birds that live a long time gain experience during a relatively long period as juveniles, which makes them more successful at rearing offspring. Birds that have a low probability of surviving from one

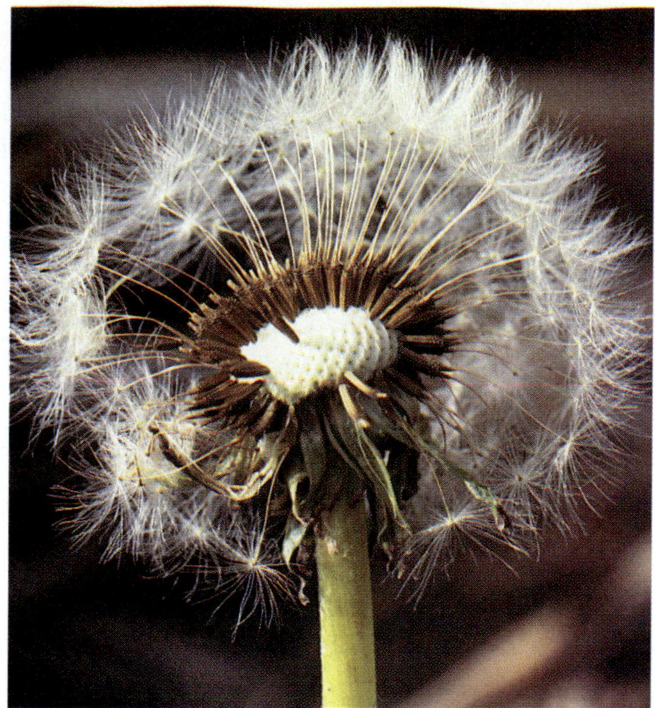

(a) A plant with a large seed crop size

(b) A plant with a moderate seed crop size

FIGURE 52.9 ▪ Variation in seed crop size in plants. (a) Most annual plants, such as this dandelion, grow quickly and produce a large number of seeds. Although most of the seeds will not produce mature plants, their large number and ability to disperse to new habitats ensure that at least some will grow and eventually produce seeds themselves. **(b)** Some plants, such as this coconut palm, produce a moderate number of very large seeds. The large endosperm provides nutrients for the embryo (a plant's version of parental care), an adaptation that helps ensure the success of a relatively large fraction of offspring. Animal species exhibit similar trade-offs between number of offspring and the amount of nutrients provided to each offspring.

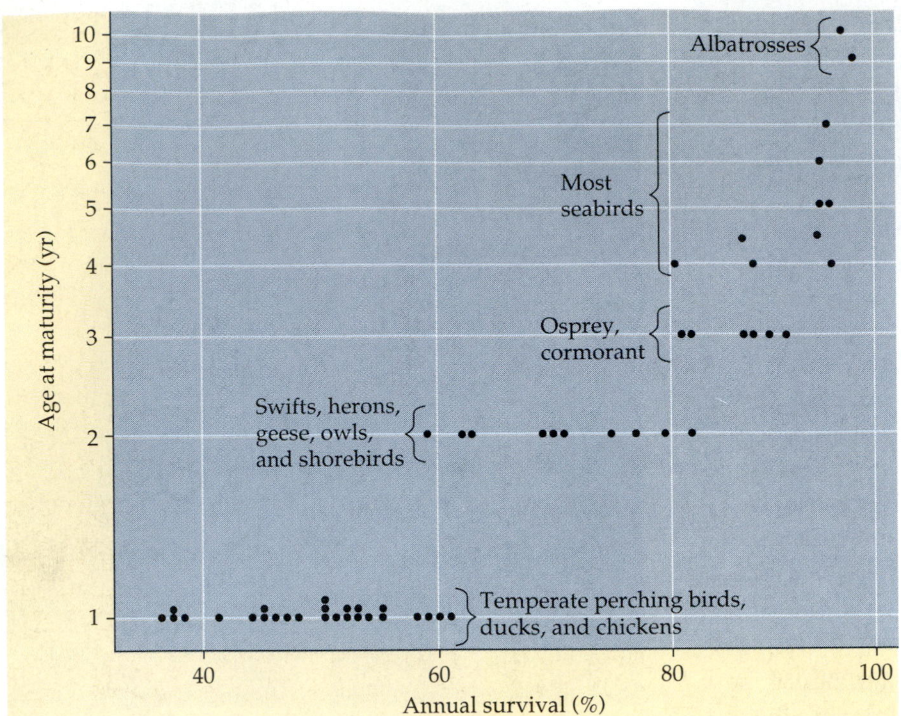

FIGURE 52.10 ▪ The relationship between age at maturity and annual adult survival in birds. Generally, birds that have a good chance of surviving from one year to the next delay reproduction longer than birds with low survival probabilities. The advantage of delayed reproduction in birds may be experience that makes them more successful at rearing offspring. In some organisms, delayed reproduction might allow larger clutches.

year to the next generally start reproducing as soon as possible (FIGURE 52.10).

Now that we have analyzed some patterns that underlie diverse life histories, let's examine the effects of these phenomena on the growth of populations and the factors that regulate population size.

POPULATION GROWTH MODELS

To begin to understand the potential for population increase, consider a single bacterium that can reproduce by fission every 20 minutes under ideal laboratory conditions. At the end of this time, there would be two bacteria, four after 40 minutes, and so on. If this continued for only a day and a half—a mere 36 hours—there would be bacteria enough to form a layer a foot deep over the entire Earth. At the other life history extreme, elephants may produce only six young in a 100-year lifespan. Darwin calculated that it would take only 750 years for a single mating pair of elephants to produce a population of 19 million. Obviously, indefinite increase does not occur either in the laboratory or in nature. A population that begins at a low level in a favorable environment may increase rapidly for a while, but eventually the numbers must, as a result of limited resources and other factors, stop growing.

As we discussed in Chapter 50, finding the answers to many ecological questions depends on a combination of observation, experimentation, and mathematical modeling. The two major forces affecting population growth—birthrates and death rates—can be measured (or observed) in many populations and used to predict how the populations will change in size over time. Small organisms can be studied in the laboratory to determine how various factors affect their population growth rates, and certain natural populations can be experimentally manipulated to answer the same questions. Mathematical models for testing hypotheses about the effects of different factors on population growth can be an alternative to experiments that would be difficult or impossible to conduct.

An exponential model of population growth describes an idealized population in an unlimited environment

Imagine a hypothetical population consisting of a few individuals living in an ideal, unlimited environment. Under these conditions, there are no restrictions on the abilities of individuals to harvest energy, grow, and reproduce, aside from the inherent physiological limitations that are the result of their life histories. The population will increase in size with every birth and with the immigration of individuals from other populations, and decrease in size with every death and with the emigration of individuals out of the population. For simplicity, let's ignore the effects of immigration and emigration (a more complex formulation would certainly include these factors). We can define change in population size during a fixed time interval with the verbal equation shown on the next page.

$$\text{Change in population size during time interval} = \text{Births during time interval} - \text{Deaths during time interval}$$

We use mathematical models as a simple way of generalizing the ideas that are expressed in words. If we let N = population size and t = time, we can also specify that ΔN = change in population size and Δt = the time interval (appropriate to the lifespan and generation time of the species) over which we are evaluating population growth. (The Greek letter delta, Δ, indicates change, such as time elapsing.) We can now rewrite the verbal equation just presented as

$$\Delta N/\Delta t = B - D$$

in which B = the absolute number of births in the population during the time interval and D = the absolute number of deaths.

All populations differ in size, and we want to create a general mathematical model that can be applied to any population. We therefore convert the simple model just presented into one in which births and deaths are expressed as the average number of births and deaths per individual during the specified time interval. Let b = the annual per capita birthrate, or the number of offspring produced per year by an average member of the population. If, for example, a population of 1000 individuals experiences 34 births per year, then the per capita birthrate is $^{34}/_{1000}$, or 0.034. If we know the per capita birthrate and death rate, we can calculate the expected number of births and deaths in a population of any size. For example, if we know that the annual per capita birthrate is 0.034 and the population size is 500, we could use the formula $B = bN$ to calculate the absolute number of births expected in that population: 17 (0.034×500) per year. (To ensure that you are comfortable with this formula, calculate how many births you would expect in a population of 700 and in a population of 1700, for which $b = 0.050$ per year.) Similarly, the per capita death rate, symbolized as d, allows us to calculate the expected number of deaths in a population of any size. If $d = 0.016$ per year, we would expect 16 deaths per year in a population of 1000 individuals. (Using the formula $D = dN$, how many annual deaths would you expect per year if $d = 0.010$ annually in populations numbering 500, 700, and 1700?) For natural populations or those in the laboratory, the per capita birthrates and death rates can be calculated from estimates of population size and data in a life table, such as TABLE 52.1.

We can revise the population growth equation again, this time using per capita birthrates and death rates rather than the absolute numbers of births and deaths:

$$\Delta N/\Delta t = bN - dN$$

One final simplification is in order. Because population ecologists are concerned with overall changes in population size, they use r to identify the difference in the per capita birthrates and death rates ($r = b - d$). This value, the per capita population growth rate, tells whether a population is actually growing (positive value of r) or declining (negative value of r). **Zero population growth (ZPG)** occurs when the per capita birthrates and death rates are equal and r equals 0. Note that births and deaths still occur in the population, but they balance each other exactly. (We will discuss the relevance of ZPG for the human population and the factors preventing the human population from leveling off later in this chapter.)

Using the population growth rate, we rewrite the equation as

$$\Delta N/\Delta t = rN$$

Finally, most ecologists use the notation of differential calculus to express population growth in terms of instantaneous growth rates:

$$dN/dt = rN$$

If you have not yet studied calculus, don't be intimidated by the form of the last equation; it is essentially the same as the previous one, except that the time intervals are very small.

We started this section by describing a population living under ideal conditions. In such a situation, the population grows at the fastest rate possible, because all members have access to abundant food and are free to reproduce at their physiological capacity. This maximum population growth rate, called the **intrinsic rate of increase**, is symbolized as r_{max}. Population increase under these conditions is called **exponential population growth**, and it is described by the equation:

$$dN/dt = r_{max}N$$

The size of a population that is growing exponentially increases rapidly, resulting in a J-shaped growth curve when population size is plotted over time (FIGURE 52.11). Although

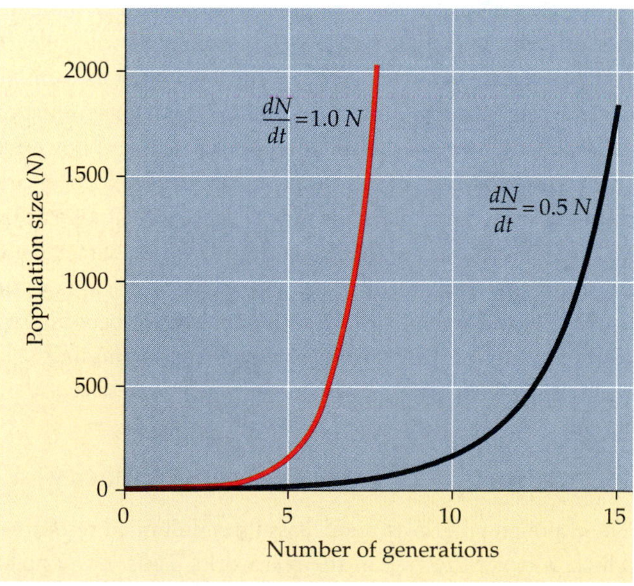

FIGURE 52.11 · Population growth predicted by the exponential model. The exponential growth model predicts unlimited population increase under conditions of unlimited resources. This graph compares growth in populations with two different values of r_{max}: 1.0 and 0.5.

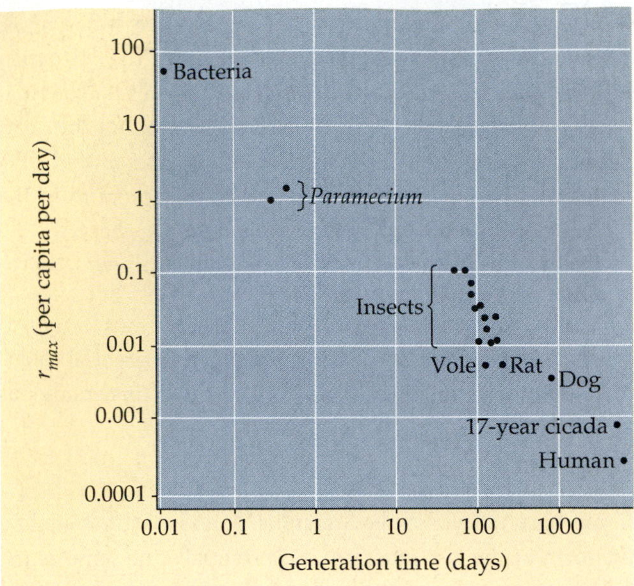

FIGURE 52.12 ▪ Generation time and r_{max} (maximum population growth rate). Small organisms that mature quickly tend to have short generation times and high r_{max}. Larger organisms, which take longer to reach sexual maturity, usually have long generation times and low r_{max}. Notice that for comparison here, r_{max} is presented as per capita per day, even for large animals. Both axes of this graph are logarithmic.

the intrinsic rate of increase is constant as the population grows, the population actually accumulates more new individuals per unit of time when it is large than when it is small—the curves in FIGURE 52.11 get progressively steeper through time. This occurs because population growth is dependent upon N as well as r_{max}, and larger populations experience more births (and deaths) than small ones growing at the same per capita rate.

We can see in FIGURE 52.11 that a population with a higher intrinsic rate of increase will grow faster than one with a lower rate of increase. The value of r_{max} for a population is influenced by life history features, such as age at the beginning of reproduction, the number of young produced, and how well the young survive. Usually, generation time and r_{max} are inversely related over a broad range of species (FIGURE 52.12).

The J-shaped curve of exponential growth is characteristic of some populations that are introduced into a new or unfilled environment, or whose numbers have been drastically reduced by a catastrophic event and are rebounding.

A logistic model of population growth incorporates the concept of carrying capacity

The exponential growth model assumes unlimited resources, which is never the case in the real world. Thus, no population—neither bacteria, elephants, nor any other organisms—can grow exponentially indefinitely. As any population grows larger in size, its increased density may influence the ability of individuals to harvest sufficient resources for maintenance, growth, and reproduction. Populations subsist on a finite amount of available resources, and as the population becomes more crowded, each individual has access to an increasingly smaller share. Ultimately, there is a limit to the number of individuals that can occupy a habitat. Ecologists define **carrying capacity** as the maximum population size that a particular environment can support with no net increase or decrease over a relatively long period of time. Carrying capacity, symbolized as K, is a property of the environment, and it varies over space and time with the abundance of limiting resources. For example, the carrying capacity for bats may be high in a habitat where flying insects are abundant and there are caves for roosting but lower in a habitat where food and suitable shelters are less numerous. Energy limitation is perhaps the most common determinant of K, although other factors, such as shelters or refuges from predators and suitable nesting and roosting sites, are often limiting.

Crowding and resource limitation can have a profound effect on the population growth rate. If individuals cannot obtain sufficient resources to reproduce, per capita birthrates will decline. If they cannot consume enough energy to maintain themselves, per capita death rates may increase. A decrease in b or an increase in d results in a smaller r and a lower overall rate of population growth.

The Logistic Growth Equation

We can modify our mathematical model of population growth to incorporate changes in r as the population size grows toward the carrying capacity (as N grows toward K). A model of **logistic population growth** incorporates the effect of population density on r, allowing it to vary from r_{max} under ideal conditions to zero as carrying capacity is reached. When a population's size is below the carrying capacity, population growth is rapid according to the logistic model, but as N approaches K, population growth is slow.

Mathematically, we construct the logistic model by starting with the model of exponential population growth and creating an expression that reduces the value of r as N increases (FIGURE 52.13). If the maximum sustainable population size is K, then $(K - N)$ tells us how many additional individuals the environment can accommodate, and $(K - N)/K$ tells us what fraction of K is still available for population growth. By multiplying r_{max} by $(K - N)/K$, we reduce the value of r as N increases:

$$\frac{dN}{dt} = r_{max}N\,\frac{K-N}{K}$$

Thus, the actual growth rate of the population at any population size becomes

$$r_{max}N\,\frac{K-N}{K}$$

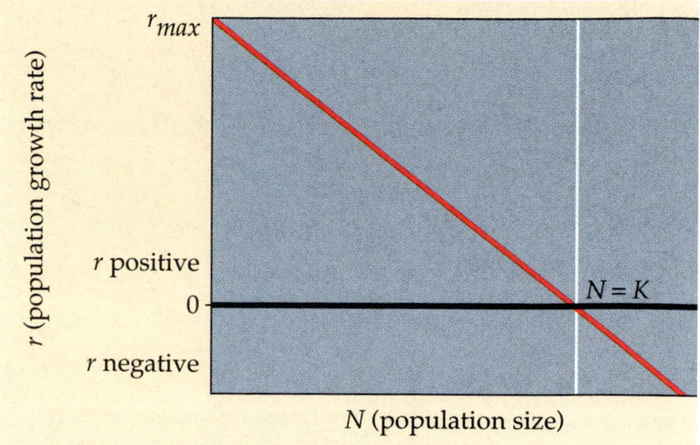

FIGURE 52.13 ▪ Reduction of *r* (population growth rate) with increasing *N* (population size). The logistic model of population growth assumes that *r* decreases as *N* increases. When *N* is close to 0, *r* equals r_{max}, and the population grows rapidly. However, as *N* approaches *K* (the carrying capacity of the environment), *r* approaches 0, and population growth slows. If *N* is greater than *K*, then *r* is negative, and population size decreases.

Table 52.2 ▪ Variations in *r* and Δ*N* in a Hypothetical Population Showing Logistic Growth Where *K* is 1000 and the Intrinsic Rate of Increase is Constant at 0.05 per Capita per Year

N	(*K* − *N*)/*K*	*r*	Δ*N**
20	0.98	0.049	+ 1
100	0.90	0.045	+ 5
250	0.75	0.038	+ 9
500	0.50	0.025	+13
750	0.25	0.013	+ 9
1000	0.00	0.000	0

*Δ*N* is rounded to the nearest whole number.

TABLE 52.2 shows hypothetical calculations of *r* and *N* at various population sizes for a population growing according to the logistic model. Notice that when *N* is low, the term (*K* − *N*)/*K* is large, and *r* is close to r_{max}. But when *N* is large and resources are limiting, the term (*K* − *N*)/*K* will be small, and *r* is substantially reduced below r_{max}. Zero population growth occurs when the numbers of births and deaths are equal and *r* equals 0—in this case, when *N* equals *K*.

The logistic model of population growth produces a sigmoid (S-shaped) growth curve (FIGURE 52.14) when *N* is plotted over time. New individuals are added to the population most rapidly at intermediate population sizes when there is not only a breeding population of substantial size, but also lots of available space and other resources in the environment. The population growth rate slows dramatically as *N* approaches *K*. Because the rate at which a population grows changes with the density of organisms that are currently in the population, the logistic model is said to be density-dependent.

Notice that we haven't said anything about *how* the growth rate of the population changes as *N* approaches *K*. Either the birthrate (*b*) must decrease, the death rate (*d*) must increase, or both. Later in the chapter we will go into some detail about some of the factors that could affect *b* and *d*.

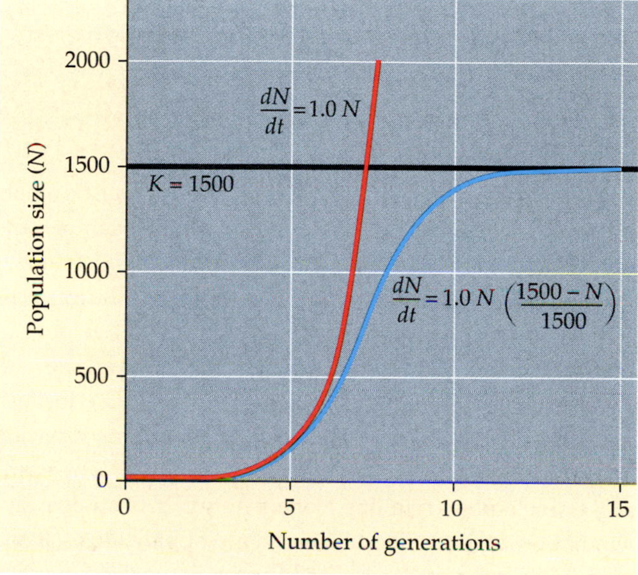

FIGURE 52.14 ▪ Population growth predicted by the logistic model. The logistic growth model assumes that there is a maximum population size that the environment can support—the carrying capacity (*K*). The rate of population growth slows as the population approaches the carrying capacity of the environment. The blue line shows logistic growth in a population with an r_{max} of 1.0 and *K* = 1500 individuals. For comparison, the red line illustrates a population growing exponentially with the same r_{max}.

How Well Does the Logistic Model Fit the Growth of Real Populations?

The growth of laboratory populations of some small animals, such as beetles or crustaceans, and of microorganisms, such as paramecia, yeast, and bacteria, fit S-shaped curves fairly well (FIGURE 52.15a and b, p. 1096). However, these experimental populations are grown in a constant environment lacking predators and other species that may compete for resources, idealized conditions that never occur in nature. Even under these laboratory conditions, not all populations stabilize at a distinct carrying capacity, and most populations show some unpredictable deviations from a smooth sigmoid curve. But studies of some organisms introduced into new habitats or of populations that rebound after being decimated by disease or hunting provide general support for the concept of carrying capacity that underlies logistic population growth (FIGURE 52.15c).

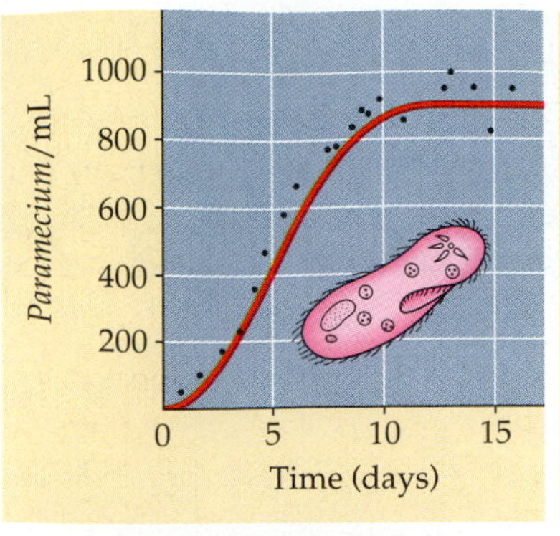

(a) A *Paramecium* population in laboratory culture

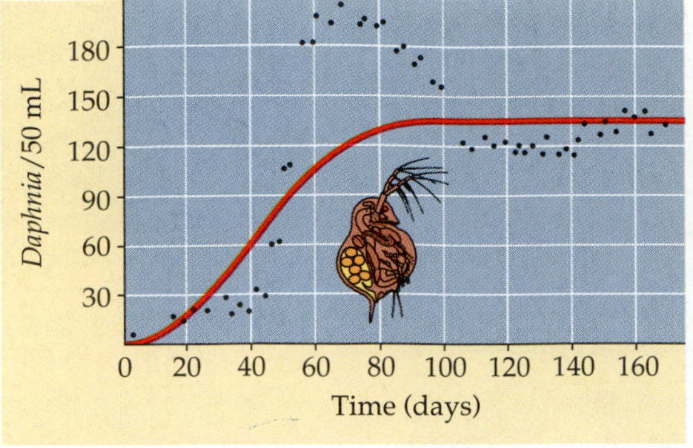

(b) A *Daphnia* population in laboratory culture

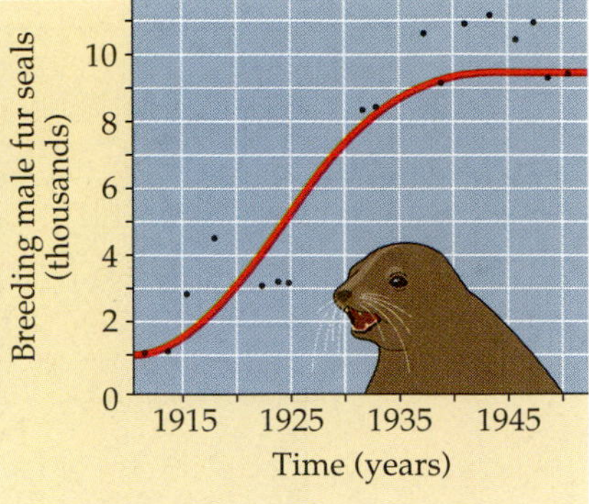

(c) A fur seal (*Callorhinus ursinus*) population on St. Paul Island, Alaska

FIGURE 52.15 ▪ Examples of logistic population growth. In each of these studies, actual data (points on the graph) are compared to an idealized curve (red) based on the logistic model of population growth. **(a)** The growth of *Paramecium aurelia* in small laboratory cultures closely approximates logistic growth if the experimenter maintains a constant environment by regularly adding food and removing toxic wastes. **(b)** Similarly, the growth of a population of *Daphnia* in a small laboratory culture also shows approximate logistic growth. Notice, however, that this population overshot the carrying capacity of its artificial environment and then settled back to a relatively stable population size. **(c)** The numbers of male fur seals with "harems" on St. Paul Island, Alaska, were greatly depressed by hunting until 1911. After hunting was banned, the population increased dramatically and now oscillates around an equilibrium number, presumably the island's carrying capacity for this species.

Some of the basic assumptions built into the logistic model clearly do not apply to all populations. For example, the model incorporates the idea that even at low population levels, each individual added to the population has the same negative effect on population growth rate; that is, *any* increase in N reduces $(K - N)/K$. Some populations, however, show an *Allee effect* (named after the researcher who first described it), in which individuals may have a more difficult time surviving and reproducing if the population size is too small. For example, a single plant standing alone may suffer from excessive wind but would be protected in a clump of individuals. Some oceanic birds require large numbers at their breeding grounds to provide the necessary social stimulation for reproduction. And conservation biologists fear that populations of animals that live solitary lives, such as rhinoceroses, may be so small that individuals will not be able to locate mates in the breeding season. In these cases, a greater number of individuals in the population has an enhancing effect, up to a point, on population growth. In addition, when a population is small, there is a greater possibility that chance events will eliminate all individuals, or that inbreeding will lead to a general reduction in fitness.

The logistic model also makes the assumption that populations approach carrying capacity smoothly. In many populations, however, there is a lag time before the negative effects of an increasing population are realized. For example, as some important resource, such as food, becomes limiting for a population, reproduction will be reduced, but the birthrate may not be affected immediately because the organisms may use their energy reserves to continue producing eggs for a short time. This may cause the population to overshoot the carrying capacity. Eventually, deaths will exceed births, and the population may then drop below carrying capacity; even though reproduction begins again as numbers fall, there is a delay until new individuals actually appear. Many populations oscillate about their carrying capacity (see FIGURE 52.19) or overshoot it at least once before attaining a relatively stable size (see FIGURE 52.15b). We will examine some possible reasons for these oscillations later in the chapter.

As you will see in the next section, some populations do not necessarily remain at, or even reach, levels where population density is an important factor. In many insects and other small, quickly reproducing organisms that are sensitive to environmental fluctuations, physical variables such as temperature or moisture reduce the population well before resources become limiting.

Overall, the logistic model is a useful starting point for thinking about how populations grow and for constructing more complex models. Although it fits few, if any, real populations exactly, the logistic model incorporates basic ideas that, with modification, do apply to many populations. And like any good hypothesis, this model has stimulated many experiments and discussions that, whether they support the model or not, lead to a greater understanding of the field of study.

Population Growth Models and Life Histories

The logistic model predicts different growth rates for populations under conditions of high and low density relative to the carrying capacity of the environment. At high densities, each individual has few resources available, and the population is growing slowly, if at all. At low densities, the opposite is true—resources are abundant and the population is growing rapidly. During the late 1960s, population ecologist Martin Cody introduced the concept that different life history adaptations would be favored under these different conditions. He proposed that at high population density, selection favors adaptations that enable organisms to survive and reproduce with few resources. Thus, competitive ability and maximum efficiency of resource utilization are favored in populations that tend to remain at or near their carrying capacity. At low population density, adaptations that promote rapid reproduction, such as increased fecundity and earlier maturity, are selected. High rates of reproduction, regardless of efficiency, are favored in this case.

These different life history "strategies" are sometimes referred to as *K*-selected and *r*-selected traits, respectively, after the variables of the logistic equation. ***K*-selected populations**, also called **equilibrial populations**, are those that are likely to be living at a density near the limit imposed by their resources (*K*, or carrying capacity). By contrast, ***r*-selected populations**, also called **opportunistic populations**, are likely to be found in variable environments in which population densities fluctuate, or in open habitats where individuals are likely to face little competition. As we have seen, there is a tendency for life history traits to vary, but often they vary in ways represented in TABLE 52.3. Still, it has proven difficult to demonstrate a direct relationship between population growth rate and specific life history traits, and the concepts of *r*-selection and *K*-selection are mainly useful as hypothetical models that help in our thinking about life history patterns. Life history evolves in the context of a complex interplay of factors, and ecologists are increasingly recognizing that populations show a mix of these traits.

Table 52.3 ■ Characteristics of Idealized *r*-Selected (Opportunistic) and *K*-Selected (Equilibrial) Populations

CHARACTERISTIC	*r*-SELECTED POPULATIONS	*K*-SELECTED POPULATIONS
Maturation time	Short	Long
Lifespan	Short	Long
Death rate	Often high	Usually low
Number of offspring produced per reproductive episode	Many	Few
Number of reproductions per lifetime	Usually one	Often several
Timing of first reproduction	Early in life	Late in life
Size of offspring or eggs	Small	Large
Parental care	None	Often extensive

SOURCE: Adapted from Pianka, E. R. *Evolutionary Ecology*, 5th ed. Menlo Park, CA: Benjamin/Cummings, 1994, p. 191.

POPULATION LIMITING FACTORS

The exponential and logistic models predict very different patterns of population growth. The exponential model sets no limit on population increase, whereas the logistic model predicts the regulation of population growth as density increases.

Population ecologists have long debated the most important factors regulating population growth. At one time, one camp emphasized the importance of density-dependent factors in population regulation, and another camp emphasized density-independent factors. The current consensus is that the relative importance of these factors differs among species and their specific circumstances, and that often both density-dependent and density-independent factors interact to affect population densities.

Density-dependent factors regulate population growth by varying with the density

The major biological implication of the logistic model is that increasing population density reduces the resources available for individual organisms and that resource limitation ultimately limits population growth. Indeed, the logistic model is a model of **intraspecific competition:** the reliance of individuals of the same species on the same limited resources. As the population size increases, the competition becomes more intense, and growth rate *(r)* declines in proportion to the intensity of competition; the population growth rate is density-dependent. In restricting population growth, a **density-dependent factor** is one that intensifies as the population increases in size. As noted previously in our description of the logistic model, density-dependent factors reduce the population growth rate by decreasing reproduction or by increasing mortality in a crowded population. In general, the density-dependent factor that limits a population's growth can be said to determine the carrying capacity, or *K*, of the environment. Let's look at some examples of how this occurs.

Resource limitation in crowded populations can influence the future size of a population by profoundly reducing reproduction. Available food supplies often limit the reproductive output of songbirds, for example, and, as bird population density increases in a particular habitat, each female lays fewer eggs. Seed production by plants is similarly affected by crowding (FIGURE 52.16, p. 1098). In both of these cases, increasing population density causes heightened intraspecific competition for declining nutrients, resulting in a lower birthrate. Factors other than intraspecific competition for nutrients can also limit populations. In many vertebrates and some invertebrates, territoriality, the defense of a well-bounded physical space, may reduce intraspecific competition for food and nest sites within the territory, but the space that constitutes the territory becomes the resource for which individuals compete.

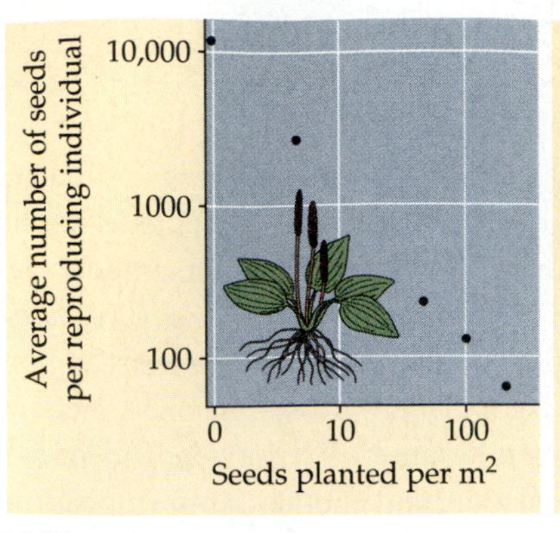

(a) Plantain

(b) Great tit

FIGURE 52.16 • Decreased fecundity at high population densities. (a) The average number of seeds produced by plantain (*Plantago major*), a small herb, decreases with increased sowing density. In this experiment, the germination rate and the proportion of plants that produced any seeds also decrease at higher densities, and mortality rate increases. **(b)** Average clutch size decreases markedly with increasing population density in a woodland population of the great tit (*Parus major*).

For oceanic birds such as gannets, which nest on rocky islands where they are relatively safe from predators, the limited number of suitable nesting sites allows only a certain number of pairs to nest and reproduce. Up to a certain population size, most birds can find a suitable nest site, but few birds beyond that number breed successfully. Thus, the limiting resource that usually determines K for gannets is safe nesting space. As this space is used up, birds that cannot obtain a nesting spot do not reproduce.

Population density also influences the health and survivorship probabilities of plants and animals. Plants grown under crowded conditions tend to be smaller and less robust than those grown at lower densities. Small plants are less likely to survive, and those that do survive produce fewer flowers, fruit, and seeds, a phenomenon well recognized by gardeners who thin their flowers and vegetables to produce the best possible yield. Similarly, animals often experience increased mortality at high population densities. In laboratory studies of flour beetles, for example, the percentage of eggs that hatch and survive to adulthood decreases steadily as density increases from moderate to high levels (FIGURE 52.17).

Predation may also be an important density-dependent regulator for some populations if a predator encounters and captures more prey as the population density of the prey increases. This is only a density-dependent effect, however, if a larger percentage of the prey population is taken as its density increases. Many predators, for example, exhibit switching behavior: They begin to concentrate on a particularly common species of prey because it becomes energetically efficient to do so (see the discussion on optimal foraging in Chapter 51). For example, trout may concentrate for a few days on a particular species of insect that is emerging from its aquatic larval stage, then switch as another insect species becomes more abundant. As a prey population builds up, predators may feed preferentially on that species, consuming a higher percentage of individuals; this can cause density-dependent regulation of the prey population.

The accumulation of toxic wastes is another component of carrying capacity that can contribute to density-dependent regulation of population size. In laboratory cultures of small microorganisms, for example, metabolic by-products accumulate as the population grows, poisoning the population within this limited, artificial environment. Indeed, ethanol accumulates as a waste product when yeast ferments sugar, and the alcohol content of wine is usually less than 13%, the maximum ethanol concentration that wine-producing yeast cells can tolerate. Some scientists think that waste accumulation could eventually limit human populations.

For some animal species, intrinsic factors, rather than the extrinsic factors just discussed, appear to regulate population size. White-footed mice in a small field enclosure will multiply from a few to a colony of 30 to 40 individuals, but eventually reproduction will decline until the population ceases to grow. This change is clearly associated with increasing density, but it occurs even when food and shelter, the main resources needed by the mice, are provided in abundance. Although the exact mechanisms for this phenomenon are not yet understood, we do know that high densities induce a stress syndrome in

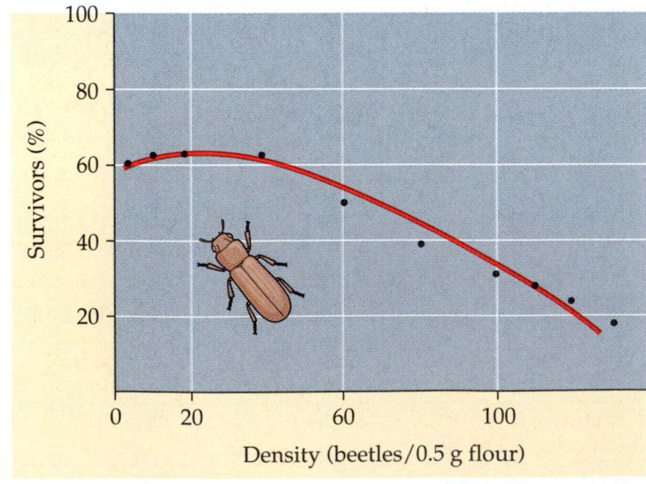

FIGURE 52.17 • Decreased vigor and survivorship at high population densities. The percentage of flour beetles (*Tribolium confusum*) surviving from egg stage to adult in a laboratory culture decreases at moderate to high population densities, reducing the numbers of adults in the next generation.

which hormonal changes delay sexual maturation, cause reproductive organs to shrink, and depress the immune system. In this case, high densities cause both an increase in mortality and a decrease in birthrates. Similar effects of crowding have been observed in wild populations of woodchucks and Old World rabbits.

We have just considered examples where increased densities cause population growth rates to decline by affecting reproduction, growth, and survivorship in the individuals that make up the populations. Other factors can also affect population growth in ways unrelated to density. In some cases, these density-independent factors can be more important than density-dependent factors. Let's see how.

The occurrence and severity of density-independent factors are unrelated to population density

Density-independent factors are unrelated to population size; they affect the same percentage of individuals regardless of population density. The most common and important density-independent factors are related to weather and climate. For example, a freeze in the fall may kill a certain percentage of the insects in a population. Obviously, the timing of the first freeze and just how cold it gets are not affected by the density of the insect population. Some populations grow exponentially, not reaching the carrying capacity of their environment, until their numbers are reduced by weather, predators, or another density-independent component of the environment. While no one argues that this is true of many organisms, ecologists debate the relative importance of density-dependent and density-independent factors in regulating populations.

One of the most frequently cited examples of density-independent population growth is that of a small insect pest of the genus *Thrips* (FIGURE 52.18). These animals feed on the pollen, leaf, and flower tissues of many plant species; those we will discuss here are agricultural pests in Australia, subsisting mainly on the pollen of roses and other cultivated plants. The number of flowers available to thrips is correlated with seasonal weather patterns, but there are always some flowers available, and thrips remain active all year. However, during the winter (which begins in June in Australia), cool temperatures lower the development rate and fecundity of the thrips, and many take so long to mature that the flowers in which they live die and fall to the ground, carrying their resident thrips to their deaths. As warm weather arrives in the spring, rates of development and fecundity increase, and populations rise to high levels. Population growth is not checked until the heat and dryness of summer cause an increase in adult mortality and the population declines. This occurs well before density-dependent factors become important and before the population shows any evidence of reaching carrying capacity. A few individuals remain in the surviving flowers, and these allow population growth to resume again when favorable conditions return. The population size of many other species, particularly insects and other small organisms, is probably limited at some point primarily by density-independent factors.

FIGURE 52.18 ▪ Density-independent regulation of population size. Populations of the insect genus *Thrips* grow rapidly during spring in flowers that provide both food and shelter. This graph shows population growth in Australian thrips feeding on roses. Before the population reaches carrying capacity, the insects' numbers are drastically reduced during the dry Australian summer. (In interpreting this graph, remember that summer in the Southern Hemisphere begins in late December.)

Environmental episodes more sporadic than seasonal changes in weather can also affect populations in a density-independent manner. For example, fires and hurricanes may strike often enough in some areas to have a significant impact on some populations. In contrast, volcanic eruptions, though dramatic and destructive, occur so infrequently that they are not very important in limiting population growth.

A mix of density-dependent and density-independent factors probably limits the growth of most populations

Over the long term, species' populations exhibit varied dynamics. Some remain fairly stable in size most of the time and are presumably close to a carrying capacity that is determined by density-dependent factors. Superimposed on this general stability, however, are short-term fluctuations due to density-independent factors. In one case study, researchers monitored the number of European herons, large wading birds, for 30 years in two different areas of England. The general pattern was the same for both areas: Populations were reasonably stable over the three-decade span, but major declines occurred following unusually cold winters.

In sharp contrast to relatively stable populations, many others are characterized by instability. For example, in very cold, snowy areas, many deer may starve to death during the winter. The severity of this factor is proportional to the season's harshness; colder temperatures increase energy requirements (and therefore the need for food) while deeper snow makes it harder to find food. But the effect is also density-dependent, because each individual in a large population gets a proportionately smaller share of what little food is available.

The relative importance of density-dependent and density-independent controls may also vary seasonally. This complex interaction of factors apparently influences populations of the bobwhite quail. The range of this bird includes southern Wisconsin, where survival is difficult during winter. The number of birds alive at the end of the winter is largely determined by the depth of snow cover, a density-independent factor. If there is little snow, as much as 80% of the population will survive, but with more snow, survivors may decline to only 20%. Whatever happens in winter, however, the population size at the end of summer tends to remain constant from year to year. Apparently, after a harsh winter kills off most of the birds, the surviving adults benefit from the increased per capita resources and have a high reproductive rate.

Among the most erratic populations, those of several species of crabs fluctuate in what appears to be a chaotic manner (FIGURE 52.19). In a recent study, Kevin Higgins and Alan Hastings of the University of California, Davis, found strong evidence that a synergistic combination of density-dependent and density-independent factors accounts for marked fluctuations in some populations of the Dungeness crab. Small changes in environmental variables seemed to be magnified by density-dependent factors. These results support an increasingly popular hypothesis that the dynamics of many (perhaps most) populations result from a complex interaction of density-dependent and density-independent forces.

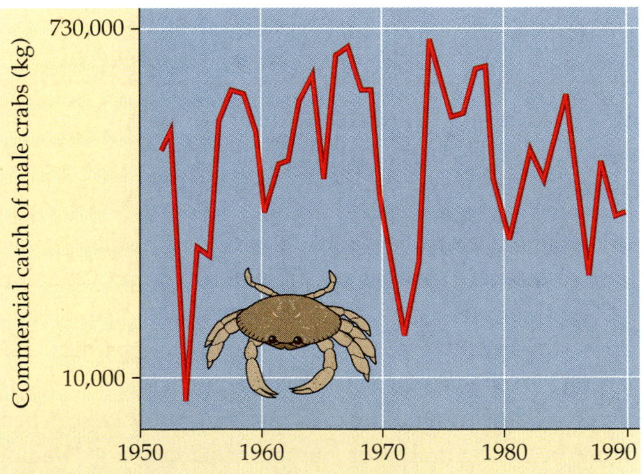

FIGURE 52.19 • Marked population fluctuations. Populations of the Dungeness crab (*Cancer magister*), a commercially important species of the U.S. Pacific Northwest, are well known for their erratic fluctuations. This graph of the commercial catch of male crabs over a 40-year period at Fort Bragg, California, illustrates the general pattern. Researchers developed a mathematical model that simulates a mix of density-dependent factors (for example, intraspecific competition and cannibalism) and density-independent factors (for example, minor changes in water temperatures caused by alterations in ocean currents). The model accurately predicts the fluctuations in the actual catch, indicating that a complex interaction of density-dependent and density-independent forces is responsible for the dynamic changes in the crab populations.

Some populations have regular boom and bust cycles

Some populations of insects, birds, and mammals fluctuate in density with remarkable regularity, more regularity than can be explained by chance alone. Perhaps the most striking population cycles known are those of periodical cicadas, grasshopper-like insects that complete their life cycles every 13 or 17 years, emerging from the ground at phenomenal densities (as high as 600 per m^2). This long life cycle may be an adaptation that reduces predation; few predators can wait 13 or 17 years for their prey to appear. Cicada populations are locally controlled, however, by a fungus whose spores can remain alive underground for the 13 or 17 years between cicada outbreaks.

Some small herbivorous mammals, such as voles and lemmings, tend to have 3- to 5-year cycles; larger mammalian herbivores, such as snowshoe hares and muskrats, as well as some birds such as the ruffed grouse and ptarmigans, have 9- to 11-year cycles. The causes of these cycles undoubtedly vary among species, and perhaps even among populations of the same species. We will explore several hypotheses for these cycles.

One idea is that crowding regulates cyclical populations, perhaps by affecting the organisms' endocrine systems. As described earlier for white-footed mice, stress resulting from a high population density may alter hormonal balance, which in turn reduces fertility and increases aggressiveness. Crowding and hormonal changes may also underlie the mass emigrations of the Norwegian lemming (*Lemmus lemmus*) that are often featured in nature films. However, little is known about how these hormonal changes affect behavior, and whether they are a common occurrence in the many species of animals that have population cycles.

Another hypothesis is that population cycles are caused by a time lag in the response to density-dependent factors, creating large fluctuations of population size above and below carrying capacity. As we noted before, extending the logistic model to incorporate time lags in response can lead to fluctuations similar to those observed in nature. Predation could also cause populations to oscillate if increases in prey populations are followed by increases in the number of predators, which in turn become so numerous that they cause prey densities to decrease. Predator populations would then decrease, allowing prey to increase, and so on. Such a mechanism was once a popular explanation for a correlation in the cycles of the snowshoe hare and one of its main predators, the lynx (FIGURE 52.20).

For the lynx and other predators that depend heavily on a single prey species, the availability of prey can influence popu-

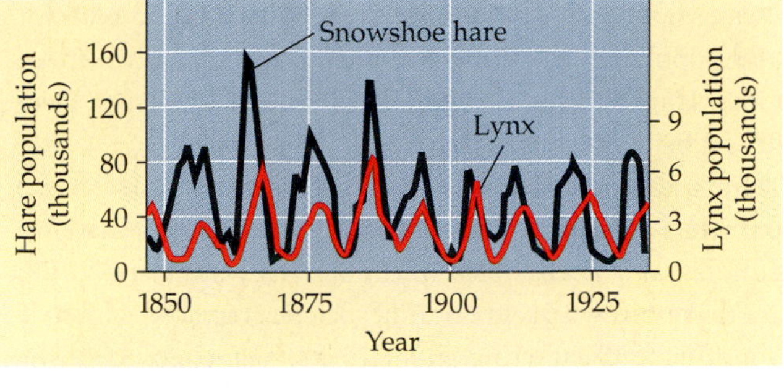

FIGURE 52.20 ▪ Population cycles in snowshoe hare and lynx. Oscillations in the population density of hare followed by corresponding changes in lynx density were once interpreted as strong evidence that these populations of prey and predator regulated each other. (Population counts are based on the number of pelts sold by trappers to the Hudson Bay Company.) However, snowshoe hare populations on islands where lynx are absent show similar cycles. The cycles of lynx populations may indeed be caused by cyclic fluctuations of the hare, a major food source for lynx. Changing ideas about the hare-lynx interaction emphasize the problem of inferring process from pattern. Most patterns of population growth are likely caused by multiple interacting factors that are difficult to untangle without direct experimentation.

lation changes. Thus, 10-year cycles in some lynx populations may result from 10-year cycles in hare populations. However, research shows that snowshoe hare populations cycle whether exposed to lynx predation or not.

An alternative hypothesis for the snowshoe hare cycle is that high hare population density causes a deterioration in the quality of their food. Studies indicate that when certain plants are damaged by herbivores, their nutrient content decreases, and they produce increased amounts of defensive chemicals. As more plant chemicals are studied, some that effectively regulate herbivores may turn up; however, to date no experimental studies have shown a clear relationship between herbivore cycles and food quality.

From 1986 to 1994, a research team led by Charles Krebs of the University of British Columbia experimentally manipulated snowshoe-hare predators (lynx, coyotes, and great horned owls) and hare foods (mainly willow and birch plants) while monitoring snowshoe hare populations in areas of one square kilometer in the Yukon. The results support the hypothesis that the 10-year cycles of snowshoe hares result from the combined effects of predation and fluctuations in the hare's food sources. Krebs and his co-workers call for additional long-term experimental studies to unravel what appear to be complex causes of cycles in population sizes of herbivores.

HUMAN POPULATION GROWTH

The explosive growth of the human population, coupled with massive consumption of the planet's resources by developed nations, is the primary cause of severe environmental degradation. Many of the environmental problems that we now confront cannot be solved without stringent regulation of our numbers.

The human population has been growing almost exponentially for centuries but cannot do so indefinitely

The exponential growth model in FIGURE 52.11 virtually describes the population explosion of humans. Ours is a singular case; it is unlikely that any other population of large animals has ever sustained so much growth for so long. The human population increased relatively slowly until about 1650, when approximately 500 million people inhabited Earth (FIGURE 52.21). The population doubled to 1 billion within the next two centuries, doubled again to 2 billion between 1850 and 1930, and doubled still again by 1975 to more than 4 billion. The population now numbers approximately 6 billion people and

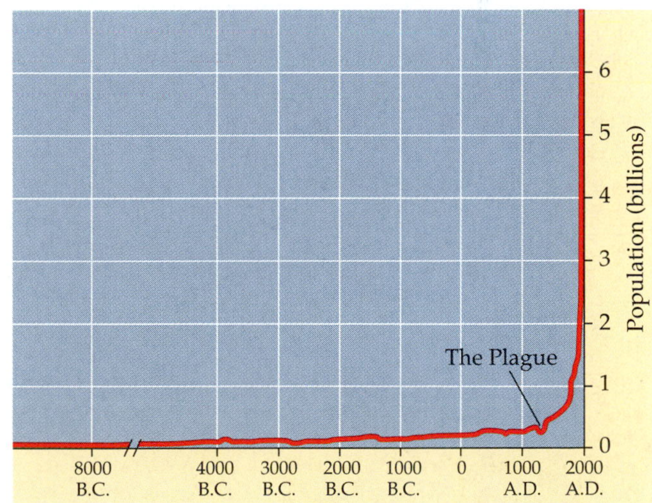

FIGURE 52.21 ▪ Human population growth. The human population has grown almost continuously throughout history, but it has skyrocketed since the Industrial Revolution. No other population of organisms has shown such growth for so long, and the human population must eventually either level off or decline. Whether this reduction in population growth will occur because of decreased birthrates or massive mortality is an open question, one that careful population policies can address.

increases by about 80 million each year. It takes only 3 years for world population growth to add the population equivalent of another United States. If the present growth rate persists, there will be 8 billion people on Earth by the year 2017.

Human population growth is based on the same general parameters that affect other animal and plant populations: birthrates and death rates. Birthrates increased and death rates decreased when agricultural societies replaced a lifestyle of hunting and gathering about 10,000 years ago. Since the Industrial Revolution, virtually exponential growth has resulted mainly from a drop in death rates, especially infant mortality, even in the least developed countries (FIGURE 52.22). Improved nutrition, better medical care, and sanitation have all contributed to an increased percentage of newborns that survive long enough to leave offspring of their own. The effect of decreasing mortality on population growth is coupled with birthrates that are still relatively high in most developing countries, resulting in an actual increase in population growth rates.

Predictions of future trends in global human population growth vary widely. Based heavily on a global decrease in fertility (the number of children per woman) that has occurred in recent years and on an extensive survey of reproductive intentions, one computer model predicts that by about 2080 the human population will peak at about 10.6 billion and then begin a slight decline to about 10.4 billion by the end of the twenty-first century. In sharp contrast are models that predict continuing a high rate of growth through the twenty-first century, with a doubling to about 12 billion perhaps as early as 2050.

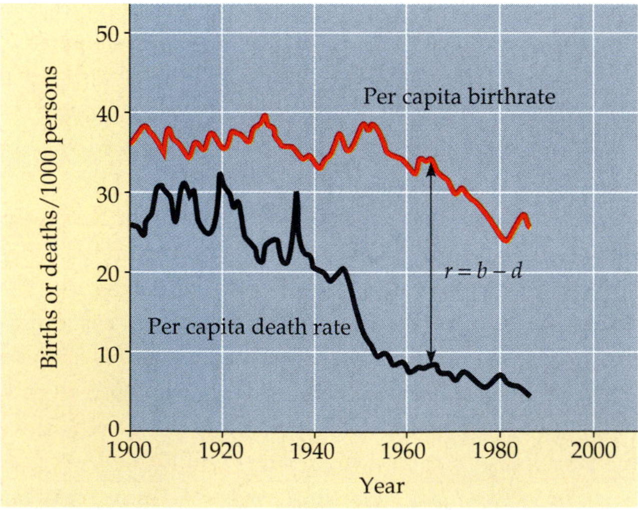

FIGURE 52.22 ▪ Changes in birthrates and death rates in Sri Lanka. Although family planning efforts and better medical care have reduced both per capita birthrates and death rates in Sri Lanka, the population growth rate (r, the difference between the birthrate and death rate at any point in time) increased after 1940. Population control will occur only after a reduction in the birthrate is greater than a reduction in the death rate.

We can only conjecture about Earth's ultimate carrying capacity for the human population, or about what factors will eventually limit our growth. Perhaps food will be the main factor. Malnutrition and famines are common in some countries, but they result mainly from unequal distribution, rather than inadequate production, of food. So far, technological changes in agriculture have allowed food supplies to more or less keep up with global population growth. However, we do know that, for energetic reasons, environments can support a larger number of herbivores than carnivores (see Chapter 54). If everyone ate as much meat as the wealthiest people in the world, less than half of the present world population could be fed on current food harvests. Nevertheless, it seems unlikely that people in wealthier countries will abandon the consumption of meat. Perhaps we will eventually be limited by suitable space, like the gannets on ocean islands. Certainly, as our population grows, the conflict over how space will be utilized will intensify, and agricultural land may be developed for housing. There seem to be few limits, however, on how closely humans can be crowded together.

We could also run out of resources other than nutrients and space. Many people are concerned about supplies of resources such as some metals and fossil fuels, which are nonrenewable. It is also possible that our population will eventually be limited by the capacity of the environment to absorb the wastes and other insults imposed by humans. For instance, heavy use of commercial fertilizers to produce large food crops already threatens the quality of groundwater in some agricultural areas. In cases such as this, current human occupants could lower Earth's long-term carrying capacity for future generations. Because of the interrelatedness of elements such as food production, land use, and energy consumed in producing food, it is likely that the carrying capacity for humans will be determined by several interacting factors.

Current worldwide population growth is actually a mosaic of various rates of growth in different countries. Some developed countries, such as Sweden, are near zero population growth because birthrates and death rates balance. The human population as a whole, however, continues to grow because birthrates exceed death rates in most nations, particularly in developing countries.

A major factor in the variation of growth rates among countries is the variation in their age structures, an important demographic factor in present and future growth trends (FIGURE 52.23). The relatively uniform age distribution in Sweden, for instance, contributes to that country's stable population size; individuals of reproductive age or younger are not disproportionately represented in the population. In contrast, Mexico has an age structure that is bottom-heavy, skewed toward young individuals who will grow up and sustain the explosive growth with their own reproduction. Notice in FIGURE 52.23 that the age structure for the United States is rela-

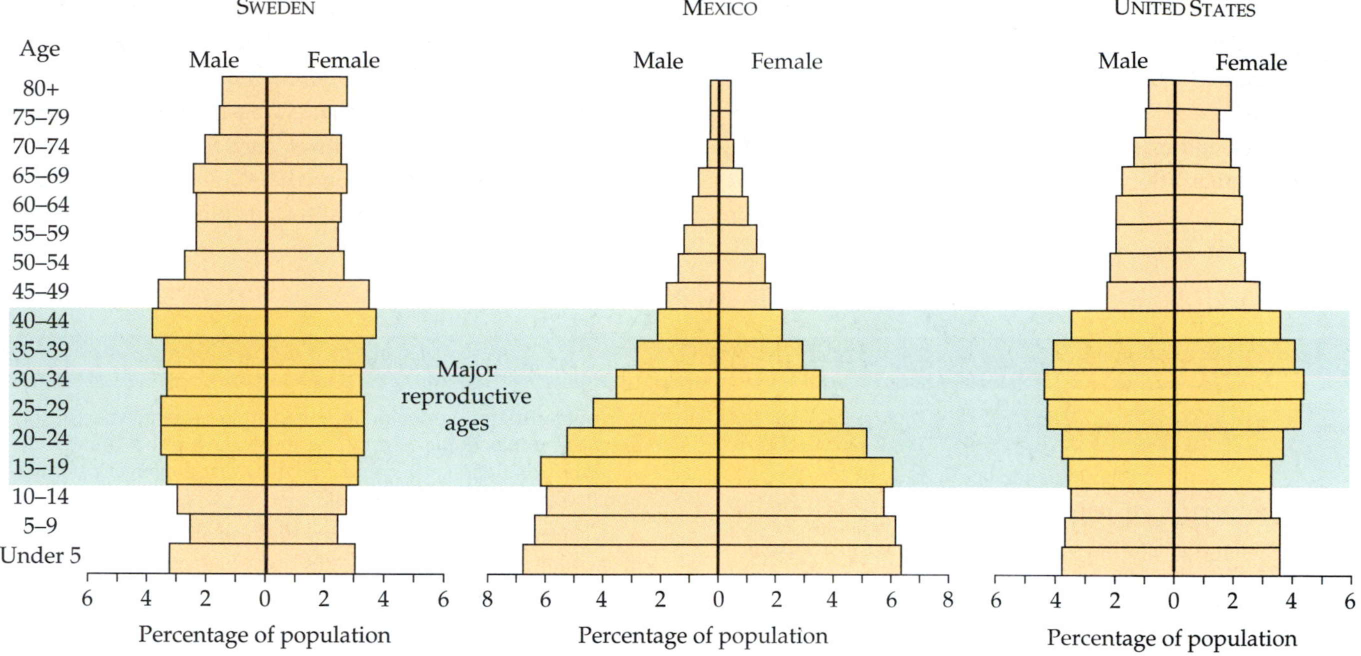

FIGURE 52.23 ▪ Age structures of three nations. The proportion of individuals in different age groups (1990 data) has a significant impact on the potential for future population growth. Mexico, for example, has a large fraction of individuals who are young and likely to reproduce in the near future. In contrast, Sweden's population is distributed more evenly through all age classes, with a high proportion of individuals beyond prime reproductive years. The United States has a fairly even age distribution, except for the bulge corresponding to the post–World War II "baby boom."

tively even except for a bulge that corresponds to the "baby boom" that lasted for about two decades after the end of World War II. Even though couples born during those years have had an average of fewer than two children, the nation's overall birthrate still exceeds the death rate because there are so many "boomers" of reproductive age.

Age-structure diagrams not only reveal a population's growth trends, they also indicate social conditions. Based on the diagrams shown in FIGURE 52.23, we can predict, for instance, that employment for an increasing number of working-age people will continue to be a significant problem for Mexico in the foreseeable future. For both Sweden and the United States, a decreasing number of working-age people will be supporting an increasing number of retired people in the future.

A unique feature of human population growth is our ability to control it with voluntary contraception and government-sponsored family planning. Reduced family size certainly contributes to successful population control. However, there is a great deal of disagreement among world leaders as to how much support should be provided for global family planning efforts. Social change and the rising educational and career aspirations of women in many cultures encourage them to delay marriage and postpone reproduction. Delayed reproduction dramatically decreases population growth rates. You can develop a sense of this phenomenon by imagining two human populations in which women each produce three children but begin reproduction at different ages. In one population, females first give birth at age 15, and in the other, at age 30. If we start with a cohort of newborn girls, then after 30 years, women in the first population will already begin to have grandchildren, whereas women in the second population will be giving birth to their first children. After 60 years, women in the first population will have a large number of great-great grandchildren (who will themselves begin to reproduce 15 years later), but women in the second population will just begin to see their grandchildren being born.

The problem of defining carrying capacity for humans is confounded by the observation that carrying capacity has changed with human cultural evolution (see Chapter 34). The advent of agricultural and industrial technology has significantly increased *K* at least twice during human history, and opponents of population control are counting on some new, as yet unidentified technological breakthrough that will allow our population to grow and plateau at some higher level.

Technology has undoubtedly increased Earth's carrying capacity for humans, but no population can continue to grow indefinitely. Exactly what the world's human carrying capacity is and under what circumstances we will approach it are topics of great concern and debate.

Ideally, human populations would reach carrying capacity smoothly and then level off. This will occur when birthrates and death rates are equal, and a decrease in birthrate is more desirable than an increase in death rate. If, however, the population fluctuated about *K*, we would expect periods of increase followed by mass death, as has occurred during plagues, localized famines, and international military conflicts. In any case, the human population must eventually stop growing. Unlike other organisms, we can decide whether zero population growth will be attained through social changes involving individual choice or government intervention, or through increased mortality due to resource limitation and environmental degradation. For better or worse, we have the unique responsibility to decide the fate of our species and the rest of the biosphere. We will examine these issues in greater depth when we discuss the concept of sustainability in Chapter 55.

CHAPTER REVIEW

REVIEW OF KEY CONCEPTS

(with page numbers and key figures)

CHARACTERISTICS OF POPULATIONS

52.1 **Two important characteristics of any population are density and the spacing of individuals (pp. 1083–1084, FIGURE 52.2)** Density is the number of individuals per unit area or volume, and dispersion is the spacing of individuals. The mark-recapture method is a common technique for estimating population size. Dispersion patterns vary in a population's range due to environmental patchiness. Dispersion may range from clumped (most common) to uniform, to random, as determined by various environmental or social factors.

- **Demography is the study of factors that affect the growth and decline of populations (pp. 1084–1087, FIGURE 52.4)** Age structure and sex ratio are important demographic features. Life tables tabulate mortality rates, survivorship from one age to the next, and average reproductive output for cohorts of populations. Survivorship curves, which plot the numbers in a cohort still alive at each age, can be classified into three general types, depending on the rates of mortality over the entire lifespan.

LIFE HISTORY TRAITS

The traits that affect an organism's schedule of reproduction and death make up its life history.

- **Life histories are highly diverse but exhibit patterns in their variability (p. 1088)** Natural selection has led to the evolution of diverse life history "strategies" that maximize lifetime reproductive success. Life history traits represent trade-offs between conflicting demands for limited time, energy, and nutrients.

- **Limited resources mandate trade-offs between investments in reproduction and in survival (pp. 1088–1092, FIGURES 52.6–52.9)** The number of reproductive episodes per lifetime represents a trade-off between fecundity and the survival probability of the reproductive adult. Semelparous organisms reproduce a single time and then die, while iteroparous organisms reproduce repeatedly over several breeding seasons. When survival between breeding seasons is low or if there is a large trade-off between fecundity and survival, semelparity is favored over iteroparity. Clutch size and age at first reproduction involve trade-offs between current and future fecundity, fecundity and adult survival, or fecundity and survival of the offspring.

POPULATION GROWTH MODELS

The two major forces affecting population growth, birthrates and death rates, can be measured and used to predict how population size will change over time.

- **An exponential model of population growth describes an idealized population in an unlimited environment (pp. 1092–1094, FIGURE 52.11)** Ignoring immigration and emigration, a population's growth rate, *r*, is the birthrate minus the death rate. The exponential growth equation $dN/dt = r_{max}N$ represents a population's potential growth in an unlimited environment, where r_{max} is the maximum possible growth rate and *N* the number of individuals in the population. This model predicts that the larger a population becomes, the faster it grows.

- **A logistic model of population growth incorporates the concept of carrying capacity (pp. 1094–1097, FIGURE 52.14)** Exponential growth cannot be sustained indefinitely in any population. A more realistic model limits growth by incorporating carrying capacity (*K*), the maximum population size that can be sustained by available resources. The logistic equation $dN/dt = r_{max}N(K - N)/K$ describes an S-shaped curve in which population growth levels off as population size approaches carrying capacity. This model predicts different growth rates under different population densities. According to a proposed relationship between population density and life histories, selection should favor traits that allow survival and reproduction with few resources in populations that live at densities near carrying capacity (*K*), while adaptations that promote rapid reproduction (high *r*) should be favored at low densities. Organisms that tend to live at or near their carrying capacity are called *K*-selected; those organisms found in variable environments where their numbers fluctuate, or in open or disturbed habitats, are called *r*-selected. However, life histories evolve in complex contexts and tend to show a mix of *K*- and *r*-selected traits.

POPULATION LIMITING FACTORS

Population growth is limited by both density-dependent and density-independent factors, the relative importance of which varies with the species and the circumstances.

- **Density-dependent factors regulate population growth by varying with the density (pp. 1097–1099, FIGURES 52.16, 52.17)** A density-dependent factor intensifies as population density increases and can eventually stabilize a population near its carrying capacity. Several density-dependent factors—intraspecific competition for limited resources, increased predation, stress due to crowding, or buildup of toxins—can cause population growth rates to decline at high densities.

- **The occurrence and severity of density-independent factors are unrelated to population density (p. 1099, FIGURE 52.18)** Density-independent factors, such as climatic events and fires, reduce population size by a given fraction, regardless of its density. The population size of many species, particularly small organisms such as insects, is limited by seasonally occurring density-independent factors.

- **A mix of density-dependent and density-independent factors probably limits the growth of most populations (pp. 1099–1100, FIGURE 52.19)** Populations that are generally stable are probably close to a carrying capacity determined by density-dependent limits, but their short-term fluctuations are density-independent. Many populations are characterized by instability, such as seasonal variation. For example, deer populations are reduced by starvation in harsh winters; the cold is density-independent, but the effect on the animals is partly density-dependent because of competition for food. Populations of some species fluctuate very erratically. The mix of different types of limiting factors is very complex in the dynamics of many populations.

- **Some populations have regular boom and bust cycles (pp. 1100–1101, FIGURE 52.20)** Some populations have cyclical density fluctuations. Crowding may regulate such populations, or population cycles may be caused by a time lag in response to density-dependent factors, creating large fluctuations above and below carrying capacity. Cyclical variation in the populations of some herbivorous species may cause corresponding fluctuations in their predators' populations. The causes of the herbivore cycles are complex; they include the effects of predation and fluctuation in food sources.

HUMAN POPULATION GROWTH

Explosive human population growth and massive consumption of resources by developed nations are the primary causes of environmental degradation.

- **The human population has been growing almost exponentially for centuries but cannot do so indefinitely (pp. 1101–1104, FIGURES 52.21–52.23)** Since the Industrial Revolution, human population growth has been sustained by such factors as improved nutrition, medical care, and sanitation, which have lowered death rates. We do not know the Earth's carrying capacity for humans or what factors will ultimately limit human growth. The age structure of the population is a major factor in the different growth rates of different countries. The human species is unique in having the ability to consciously control its own population growth.

SELF-QUIZ

1. A uniform dispersion pattern for a population may indicate that
   a. the population is spreading out and increasing its range
   b. resources are heterogeneously distributed
   c. individuals of the population are competing for some resource, such as water and minerals for plants or nesting sites for animals
   d. there is an absence of strong attractions or repulsions among individuals
   e. the density of the population is low

2. A population that has a relatively low r value will most likely
   a. have large clutch sizes with relatively small offspring
   b. be found in environments that are highly variable
   c. have an early age of first reproduction and a short generation time
   d. produce fewer offspring with more competitive capabilities
   e. be regulated by density-independent factors

3. The term $(K - N)/K$ influences dN/dt such that
   a. the increase in actual population numbers is greatest when N is small
   b. as N approaches K, r_{max} (the intrinsic rate of increase) becomes smaller
   c. when N equals K, population growth is zero
   d. when K is small, the influence of density-dependent factors is smaller
   e. as N approaches K, the birthrate approaches zero

4. A population's carrying capacity is
   a. the number of individuals in that population
   b. reached when the number of deaths exceeds the number of births
   c. inversely related to r_{max}
   d. the population size that can be supported by available resources for that species within the habitat
   e. set at 8 billion for the human population

5. A Type III survivorship curve would be expected in a species in which
   a. mortality occurs at a constant rate over the lifespan
   b. parental care is extensive
   c. a large number of offspring are produced but parental care is minimal
   d. mortality rate is quite low for the young
   e. K-selection prevails

6. In a mark-recapture study of a lake trout population, 40 fish were captured, marked, and released. In a second capture, 45 fish were captured; 9 of these were marked. What is the estimated number of individuals in the lake trout population?
   a. 90
   b. 200
   c. 360
   d. 800
   e. 1800

7. The example of the population cycles of the snowshoe hare and its predator, the lynx, illustrates that
   a. predators are the major factor in controlling the size of prey populations and are, in turn, regulated in their numbers by the oscillating supply of prey
   b. the two species must have evolved in close contact with each other because their life histories are intertwined
   c. one should not conclude a cause-and-effect relationship when viewing population patterns without careful observation and experimentation
   d. both populations are controlled by density-independent factors
   e. the hare population is r-selected, whereas the lynx population is K-selected

8. The current size of the human population is closest to
   a. 2 million
   b. 3 billion
   c. 4 billion
   d. 6 billion
   e. 10 billion

9. Consider five human populations that differ demographically *only* in their age structures. The population that will grow the most in the next 30 years is the one with the greatest fraction of people in which age group? Explain your answer.
   a. 10–20
   b. 20–30
   c. 30–40
   d. 40–50
   e. 50–60

10. All these terms are characteristic of human populations in industrialized countries *except*
    a. relatively small family size
    b. several potential reproductions per lifetime
    c. r-selected life history
    d. Type I survivorship curve
    e. relatively even age structure

CHALLENGE QUESTIONS

1. Beavers are released in an effort to repopulate a valley where they were trapped to extinction many years ago. Many of their young are killed by predators, but each original pair produces an average of three offspring that survive to maturity. (This means, in effect, 1.5 offspring per parent per generation.) If this rate of growth continues, roughly how many beavers will there be in the sixth generation for each original pair in the first generation? Graph the number of beavers against generation, and compare to the curves in FIGURES 52.11 and 52.14. What kind of growth is seen in the beaver population? What would the growth curve look like if the beavers averaged 1.1 offspring per parent per generation? Do numbers have to double each generation for exponential growth to occur?

2. The mountain gorilla, spotted owl, giant panda, snow leopard, and grizzly bear are all endangered by human encroachment on their environments. Another thing these animals have in common is that they all have equilibrial life histories. Why might they be more easily endangered than animals with opportunistic life histories? What general type of survivorship curve would you expect these species to exhibit? Explain your answer.

3. A fisheries biologist studied a population of cichlids in a small African lake (120 hectares). He found that the fish lived only in scattered reed beds that accounted for 25% of the area of the lake. He caught 185 fish, tagged them, and released them. Two days later, he netted 208 fish and found that 35 of them were tagged. How many cichlids would you estimate to be in the lake? What is the density of the cichlid population? What is the dispersion pattern of the fish? How might the dispersion pattern affect your interpretation of their density?

SCIENCE, TECHNOLOGY, AND SOCIETY

Many people regard the rapid population growth of developing countries as our most serious environmental problem. Others think that the population growth in developed countries, though smaller, is actually a greater threat to the environment. What kinds of problems result from population growth in (a) developing countries, and (b) the industrialized world? Which do you think is the greater threat, and why?

FURTHER READING

Brown, J. L., and E. Pollitt. "Malnutrition, Poverty, and Intellectual Development." *Scientific American,* February 1996. Describes research on the effects of malnutrition, currently afflicting nearly 200 million children worldwide.

Cohen, J. "Ten Myths of Population." *Discover,* April 1996. A thoughtful analysis of often-stated but unsubstantiated ideas about human population growth.

Dasgupta, P. S. "Population, Poverty, and the Local Environment." *Scientific American,* February 1995. The value of a child's labor to a family is one of the cultural factors behind the population explosion.

Kates, R. W. "Sustaining Life on the Earth." *Scientific American,* October 1994. Explores the cultural context of environmental concerns.

Korpimaki, E., and C. J. Krebs. "Predation and Population Cycles of Small Mammals." *BioScience,* November 1996. Contains thoughtful analysis of research on this topic.

Miller, S. K. "Bats Sow Seeds of Rainforest Recovery." *New Scientist,* June 18, 1994. Discusses hurricanes as density-independent regulators of population size.

Olshansky, S. J., B. A. Carnes, and D. Grahn. "Confronting the Boundaries of Human Longevity." *American Scientist,* January–February 1998. Analyzes the evolution of human life history strategy in light of increasing lifespans.

Tuljapurkar, S. "Taking the Measure of Uncertainty." *Nature,* June 19, 1997. Discusses recent scientific predictions of human population growth through the twenty-first century.

WEB LINKS

Visit the special edition of *The Biology Place* for BIOLOGY, Fifth Edition, at http://www.biology.com/campbell. Go to Chapter 52 for online resources, including learning activities, practice exams, and links to the following web sites:

"Population Ecology"
This site from Virginia Tech provides a good starting point for topics in population ecology.

"NERC Centre for Population Biology"
Check out the Ecotron project funded by the British Natural Environment Research Council at Imperial College, London.

"Center for Computational Ecology Yale Institute for Biospheric Studies"
Visit this Center at Yale University, a leading developer of computational tools for the simulation, analysis, and validation of ecological models.

"US-Demography Home Page"
The mission of the US-Demography Home Page is to identify, document, and provide simple access to demographic information concerning the United States of America.

COMMUNITY ECOLOGY

On your next walk through a field or woodland, or even across campus or through a park, try to observe some of the interactions of the species present. You may see birds using trees as nesting sites, bees pollinating flowers, shelf fungi growing on trees, spiders trapping insects in their webs, ferns growing in shade provided by trees—a sample of the many interactions that exist in any ecological theater. In addition to the physical and chemical factors discussed in Chapter 50, an organism's environment includes other individuals in its population and populations of other species living in the same area. Such an assemblage of species living close enough together for potential interaction is called a ***community****. In the photograph that opens this chapter, the lion, the zebra, the hyena, the vultures, and the grasses and other plants are all members of a community in Kenya.*

This chapter examines the diverse kinds of biotic interactions among organisms and addresses the central issue in community ecology: What factors are most significant in structuring a community—in determining its species composition and both the absolute and the relative abundance of species present? The analysis of communities and the factors that make them more than the sum of their component species' populations is an active area of ecological research.

EARLY HYPOTHESES OF COMMUNITY STRUCTURE

Communities differ dramatically in their **species richness**, the numbers of species they contain. They also differ in their **relative abundance** of species. Some communities consist of a few common species and many rare ones, whereas others contain an equivalent number of species that are all about equally common. The relative abundance of species within a community has an enormous impact on its general character. These concepts are illustrated in TABLE 53.1 (p. 1108), which lists the trees present in a deciduous forest in West Virginia. Two trees, yellow poplar and sassafras, make up almost 84% of the entire stand, and four of the ten tree species are represented by only a single individual each. A different community that had the same species richness, but in which the numbers were more evenly divided among the ten species, would seem more diverse. Indeed, the term **species diversity**, as used by ecologists, considers *both* components of diversity: species richness and relative abundance.

The interactive and individualistic hypotheses pose alternative explanations of community structure: *science as a process*

Why are certain combinations of species found together as members of a community? Two divergent views on this question emerged among ecologists in the 1920s and 1930s, derived primarily from observations of plant distributions.

Table 53.1 ▪ The Trees Present in a Deciduous Forest in West Virginia

SPECIES	NUMBER	PERCENTAGE OF STAND
Yellow poplar (*Liriodendron tulipifera*)	122	44.5
Sassafras (*Sassafras albidum*)	107	39.0
Black cherry (*Prunus serotina*)	12	4.4
Cucumber magnolia (*Magnolia acuminata*)	11	4.0
Red maple (*Acer rubrum*)	10	3.6
Red oak (*Quercus rubra*)	8	2.9
Butternut (*Juglans cinerea*)	1	.4
Shagbark hickory (*Carya ovata*)	1	.4
American beech (*Fagus grandifolia*)	1	.4
Sugar maple (*Acer saccharum*)	1	.4
	274	100.0

SOURCE: Smith, R. L., and T. M. Smith. *Elements of Ecology*, 4th ed. Menlo Park, CA: Benjamin/Cummings, 1998, p. 271.

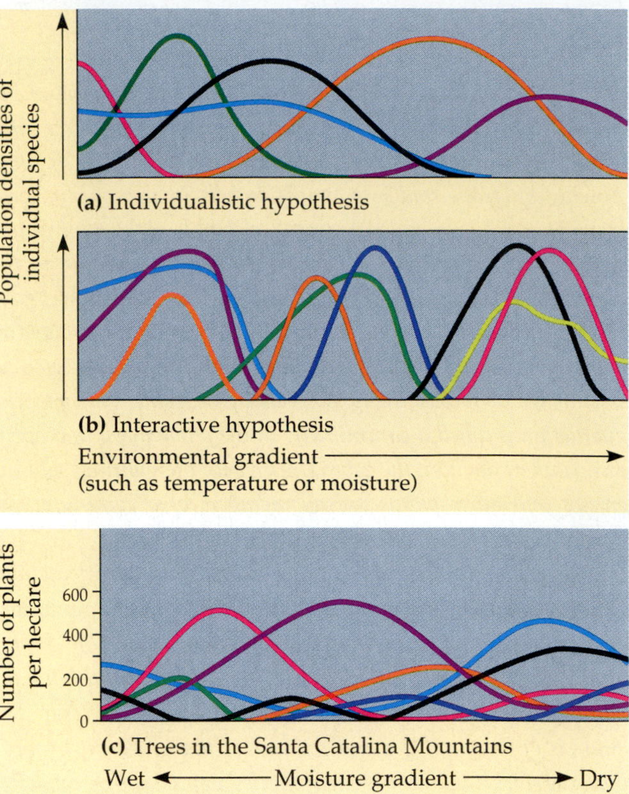

FIGURE 53.1 ▪ Testing the individualistic and interactive hypotheses of communities. Ecologist Robert Whittaker tested these two hypotheses by graphing the abundance of different plant species (*y* axis) along environmental gradients of abiotic factors such as temperature or moisture (*x* axis). Each line on the graphs represents the abundance of one species. **(a)** The individualistic hypothesis proposes that species are independently distributed along gradients and that a community is simply the assemblage of species that happen to occupy the same area. **(b)** The interactive hypothesis suggests that communities are discrete groupings of particular species that are closely interdependent and nearly always occur together. **(c)** The distribution of tree species at one elevation in the Santa Catalina Mountains of Arizona supports the individualistic hypothesis. Each tree species has an independent distribution along the gradient, apparently conforming to its tolerance for moisture, and the species that live together at any point along the gradient have similar physical requirements. Because the vegetation changes continuously along the gradient, it is impossible to delimit sharp boundaries for the communities.

The **individualistic hypothesis**, first enunciated by H. A. Gleason, depicted the community as a chance assemblage of species found in the same area simply because they happen to have similar abiotic requirements—for example, for temperature, rainfall, and soil type. The alternative view, the **interactive hypothesis**, advocated by F. E. Clements, saw the community as an assemblage of closely linked species, locked into association by mandatory biotic interactions that cause the community to function as an integrated unit. Evidence for the interactive view was based on the observation that certain species of plants are consistently found together. For example, deciduous forests in the northeastern United States almost always include certain species of oak, maple, birch, and beech, along with a specific assemblage of shrubs and vines. These two ways of looking at community structure suggest different priorities in studying biological communities. The individualistic hypothesis emphasizes studying single species, while the interactive hypothesis emphasizes entire assemblages of species as the essential units for understanding the interrelationships and distributions of organisms.

We can use the hypothetico-deductive approach to evaluate the individualistic and interactive hypotheses. The two hypotheses of communities make contrasting predictions about how species should be distributed along a gradient of environmental variables, such as moisture or temperature. The individualistic hypothesis predicts that communities should generally lack discrete geographical boundaries because each species has an independent distribution along the environmental gradient. In other words, each species will be distributed according to its tolerance ranges for abiotic factors that vary along the gradient, and communities should change continuously along the gradient with the addition or loss of particular species (FIGURE 53.1a). According to the interactive hypothesis, in contrast, species should be clustered into discrete communities with distinct boundaries, because the presence or absence of a particular species is largely governed by the presence or absence of other species with which it interacts (FIGURE 53.1b).

In most actual cases, especially where there are broad regions characterized by gradients of environmental variation,

FIGURE 53.2 ▪ An example of a distinct boundary between communities. Where environmental factors change abruptly, sharp boundaries do exist between adjacent communities. In some grassland areas of coastal California, weathering of a certain type of rock, called serpentine rock, increases the magnesium content of the soil. Where magnesium-rich soils predominate, native California wildflowers grow and blossom colorfully in the early spring. In nearby areas where the soil is derived from sandstone, however, grasses dominate the community. The sharpness of these local boundaries is apparent in this scene from the Jasper Ridge Biological Preserve near Stanford University.

the composition of plant communities does seem to change on a continuum, with each species more or less independently distributed (FIGURE 53.1C). Such distributions generally support the view of plant communities as relatively loose associations without discrete boundaries. However, where some key factor in the physical environment changes abruptly, adjacent communities are delineated by correspondingly sharp boundaries. In California, for example, many native plants have been replaced by introduced European grasses, but in some strictly delimited areas where serpentine rocks increase the magnesium content of the soil, native wildflowers persist (FIGURE 53.2). The individualistic hypothesis can account for this type of discontinuity without assuming that certain plant species occur in the same area because they are locked into mandatory community relationships.

The individualistic hypothesis may not apply to the animals in a community, as animals are usually linked more intimately to other organisms. For example, limpkins, long-beaked birds of Florida swamps, feed primarily on one species of snail. The birds' evolutionary adaptations make them extremely efficient in their specialized foraging, but their geographical distribution is restricted by the distribution of their prey. On the other hand, gray squirrels of the eastern United States are not as strongly tied to a particular food. They may be found in a variety of habitats, including forests where pines are abundant, although they are most common in areas of mature hardwoods, such as oak and hickory trees.

Thus, simple generalizations can rarely explain why certain species commonly occur together in communities. Just as we saw in Chapter 52 that population sizes are usually affected by a combination of density-dependent and density-independent factors, we will see that the distributions of most populations in communities are affected to some extent by both abiotic factors and interactions with other species. Among the most significant abiotic factors affecting the structure of communities are those that disturb and destabilize existing relationships among organisms, such as fire, floods, and storms. As we discuss later in this chapter, disturbance may be the single most important influence affecting the structure and species composition of many communities. Before turning to disturbance, however, let's take a look at the variety of interactions that occur among the populations that make up communities.

INTERACTIONS BETWEEN POPULATIONS OF DIFFERENT SPECIES

Although the geographical distributions of many species are largely determined by their adaptations to abiotic environmental factors, organisms are also influenced by biotic interactions with other individuals in their immediate vicinity. In Chapter 51 we examined one such interaction, intraspecific competition. In this section we consider **interspecific interactions**, those that occur between populations of different species living together within a community.

Interspecific interactions can be strong selection factors in evolution

Just as the physical and chemical features of the environment are important factors in adaptation by natural selection, so are interactions between species. In some cases, the adaptation of one species to the presence of another has a relatively obvious evolutionary basis. For example, in Great Britain, natural selection has apparently favored peppered moths (*Biston betularia*) of coloration that blends in with the living lichens on which the moths sometimes rest, an adaptation that seems to make it harder for predatory birds to spot the moths.

In a broad sense, **coevolution** describes interactions involving reciprocal evolutionary adaptations in two species: A change in one species acts as a selective force on another species, and counteradaptation by the second species, in turn, is a selective force on individuals in the first species. Coevolution has been studied most extensively in predator-prey relationships, in mutualism, and in parasite-host relationships. One important example of coevolution was discussed in Chapter 30: the interaction between a flowering plant and its exclusive pollinator (see FIGURE 30.9).

Coevolution involves reciprocal genetic change in interacting species. However, the genetic basis of most cases has not yet been determined, and it is often difficult to establish that an evolutionary change in one species is a selective force that induces an evolutionary change in another. As an example, let's examine a specific relationship that probably illustrates coevolution but also demonstrates the complexities that may confound our attempts to unravel evolutionary history. Passionflower vines of the genus *Passiflora* are protected against most herbivorous insects by their production of toxic compounds in young leaves and shoots. However, the larvae of butterflies of the genus *Heliconius* can tolerate these defensive chemicals. This counteradaptation has enabled *Heliconius* larvae to become specialized feeders on plants that few other insects can eat (FIGURE 53.3). Survival of the larvae is further enhanced by a behavioral adaptation of the butterflies. The eggs that female *Heliconius* butterflies lay on the leaves of passionflower vines are bright yellow, and other females generally avoid laying eggs on leaves marked by yellow dots. This behavior presumably reduces intraspecific competition among the larvae for food.

An infestation of *Heliconius* larvae can devastate a passionflower vine, and these poison-resistant insects are likely to be a strong selection force favoring the evolution of more defenses in the plants. In some species of *Passiflora*, the leaves have conspicuous yellow spots that look like *Heliconius* eggs, an adaptation that may divert the butterflies to other plants in their search for egg-laying sites. The story is more complicated, however. The yellow "spots" on the passionflower vine are actually nectaries, which attract ants and wasps that prey on *Heliconius* eggs or larvae. There is evidence, too, that the mere presence of ants on a leaf will discourage a *Heliconius* butterfly from laying eggs there. Thus, adaptations that may seem, on superficial examination, to be coevolutionary responses between just two species may, in fact, result from interactions among many species in the community. It is difficult to sort out the importance of the various selective forces, and the simple idea of coevolution as adaptation-counteradaptation occurring exclusively between two species does not often adequately describe the interactions within communities.

Despite the problems in assessing cause and effect in the evolution of complex ecological relationships, biologists agree that the adaptation of organisms to other species in a community is a fundamental characteristic of life. We will encounter several examples of such adaptations next as we survey the interspecific interactions of greatest importance in community ecology.

Interspecific interactions may have positive, negative, or neutral effects on a population's density: *an overview*

The possible interactions between any two species living together in a community are summarized in TABLE 53.2. For convenience we use a pair of signs, such as +/−, to refer to the effects on population density of the two species involved in

(a)

(b)

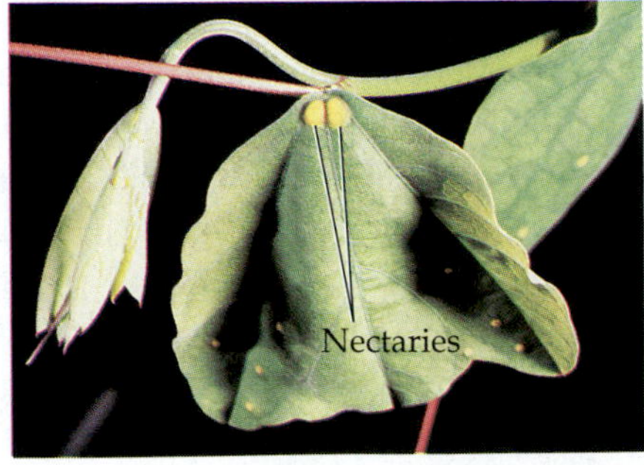

(c)

FIGURE 53.3 • Evolutionary interactions between a plant and an insect. (a) Passionflower vines (*Passiflora*) produce toxic chemicals that help protect leaves from herbivorous insects. A counteradaptation has evolved in the butterfly *Heliconius:* Its larvae can feed on the leaves because they have digestive enzymes that break down the toxic compounds. **(b)** The females of some *Heliconius* species avoid laying eggs on passionflower leaves that already carry bright yellow egg clusters deposited by other females. This behavior reduces intraspecific competition on individual leaves. **(c)** These yellow nectaries that resemble eggs grow on the leaves of some species of passionflowers. Experiments show that *Heliconius* females avoid laying eggs on leaves with these yellow spots. These nectaries, as well as smaller ones scattered over the leaf, also attract ants and other insects that prey on the butterfly eggs and larvae.

53.1

Table 53.2 ■ Interspecific Interactions

INTERACTION	EFFECTS ON POPULATION DENSITY
Predation (+/−) (includes parasitism)	The interaction is beneficial to one species and detrimental to the other.
Competition (−/−)	The interaction is detrimental to both species.
Commensalism (+/0)	One species benefits from the interaction but the other is unaffected.
Mutualism (+/+)	The interaction is beneficial to both species.

each interaction. For example, mutual symbiosis (mutualism) is a +/+ interaction, meaning that the density of each species is increased in the presence of the other. Predation is an example of a +/− interaction, with a positive effect on population density of one species (the predator) and a negative effect on the density of the other population (the prey). The following three sections examine the interspecific interactions of communities in more detail.

Predation and parasitism are +/− interactions: *a closer look*

The most conspicuous population interactions are those
53.1 involving **predation**, in which a **predator** eats its **prey**.

Though most of us associate predation with the kinds of interactions illustrated by the lion eating the zebra in the photograph that opens this chapter, there are other important kinds of +/− interactions. **Parasitism** involves specialized kinds of predators called parasites that live on or in their hosts, seldom killing them outright. In **parasitoidism**, insects, usually small wasps, lay eggs on living hosts. The larvae then feed within the body of the host, eventually causing its death. **Herbivory** occurs when animals eat plants, and we include it here as a form of predation. (Herbivory can kill an entire organism, as when a mouse eats a seed. However, the kind of herbivory called grazing does not usually kill the plant and is actually more similar to parasitism than to predation.) While all of the above interactions involve one kind of organism feeding on another, the negative effects in some +/− interactions are not always due to feeding, as we will see when we discuss brood parasitism.

Predation

Many important feeding adaptations of predators are both obvious and familiar. Most predators have acute senses that enable them to locate and identify potential prey. In addition, many predators have adaptations such as claws, teeth, fangs, stingers, or poison that help catch and subdue, or simply chew, the organisms on which they feed. Rattlesnakes and other pit vipers, for example, locate their prey with special heat-sensing organs located between each eye and nostril, and they kill small birds and mammals by injecting them with toxins through their fangs. Similarly, many herbivorous insects locate appropriate food plants by using chemical sensors on their feet, and their mouthparts are adapted for shredding tough vegetation. Predators that pursue their prey are generally fast and agile, whereas those that lie in ambush are often camouflaged in their environments.

Through repeated encounters with predators over evolutionary time, various defensive adaptations have evolved in prey species, and we will now survey some of these.

Plant Defenses Against Herbivores. In many cases, herbivores do not consume an entire plant, so their impact on prey density is different from that of a predator that consumes whole prey. Nevertheless, the removal of plant tissue—stems, leaves, bark, and sap—usually affects a plant's fitness and ability to survive, and plants exhibit a variety of defensive adaptations against the animals that might eat them. Many of these defenses are mechanical. For example, thorns may discourage large herbivores such as vertebrates, and some plants have microscopic crystals in their tissues, or hooks or spines on their leaves, that make feeding difficult even for small insects.

Many plants produce chemicals that function in defense by making the vegetation distasteful or harmful to an herbivore. Some well-known poisons and drugs probably function in plants as chemical weapons against herbivores: strychnine, produced by plants of the genus *Strychnos;* morphine, from the opium poppy; nicotine, produced by tobacco; and digitoxin, from the common foxglove. Other defensive compounds that are not toxic to humans but may be distasteful to herbivores are responsible for the familiar flavors of cinnamon, cloves, and peppermint. Some plants even produce chemicals that are analogous to insect hormones and cause abnormal development in some insects that eat them.

The specific defenses of a plant population can act as selective agents that provoke the evolution of counteradaptations in herbivore populations. The responses of herbivores may nullify the plant defenses and allow subsequent generations of herbivores to eat the descendants of the original plant population. For example, some insects, such as the *Heliconius* larvae depicted in FIGURE 53.3, can absorb or detoxify certain compounds produced by plants. Some insects even store plant poisons and use them in defense against their own predators; the larvae of monarch butterflies, for instance, store toxins from their host plants (milkweeds) that make them distasteful to some of their predators. Although a plant's defenses may restrict the number of herbivore species that can feed upon it, the herbivores must eat to reproduce, and selection to overcome plant defenses is strong. Hence, no plant defense is likely to provide protection forever.

Animal Defenses Against Predators. Animals can avoid being eaten by using passive defenses, such as hiding, or active ones, such as escaping or defending themselves against predators. Fleeing is a common antipredator response, though it can be very expensive energetically. Many animals flee into a shelter and avoid being caught without expending the energy required for a prolonged flight. Active self-defense is less common, though some large grazing mammals will vigorously defend their young from predators such as lions. Other behavioral defenses include alarm calls, which often bring in many individuals of the prey species that mob the predator. Mobbing can involve either harassment at a safe distance, or direct attack (FIGURE 53.4). Distraction displays direct the attention of the predator away from a vulnerable prey, such as a bird chick, to another potential prey that is more likely to escape, such as the chick's parent.

FIGURE 53.4 ▪ A behavioral defense against predators. Many prey species turn the tables and attack their predator. This is common in bird species that nest in groups. Here, two crows are mobbing a barn owl, a predator that often kills and eats eggs and nestlings. Mobbing behavior has been observed in animals as diverse as baboons and ground squirrels.

Many other defenses rely on adaptive coloration, which has evolved repeatedly among animals. Camouflage, called **cryptic coloration**, is passive defense that makes potential prey difficult to spot against its background (FIGURE 53.5). A camouflaged animal need only remain still on an appropriate substrate to avoid detection. The shape of an animal can also help camouflage it; the cryptic shape of the seaweedlike fish in FIGURE 1.14 is an example. Deceptive markings are another form of adaptive coloration. Large, fake eyes or false heads can apparently deceive predators momentarily, allowing the prey to escape (FIGURE 53.6), or they may induce the predator to strike a nonvital area.

Some animals have mechanical or chemical defenses against would-be predators. Most predators are strongly discouraged by the familiar defenses of skunks and porcupines (see FIGURE 51.9). Some animals, such as poisonous toads and frogs, can synthesize toxins. Others acquire chemical defense passively by accumulating toxins from the plants they eat. Monarch butterfly larvae provide one example of toxin accumulation from plants. Remarkably, the poisons are retained throughout metamorphosis and are present in adult butterflies. Birds that eat monarch butterflies regurgitate their prey and quickly learn to avoid feeding on that species of insect. This is not a foolproof protection, however; monarchs are still subject to predation by many predatory and parasitoid insects, and even by some mammals, such as mice.

Animals with effective chemical defenses are often brightly colored, a warning to predators known as **aposematic colora-**

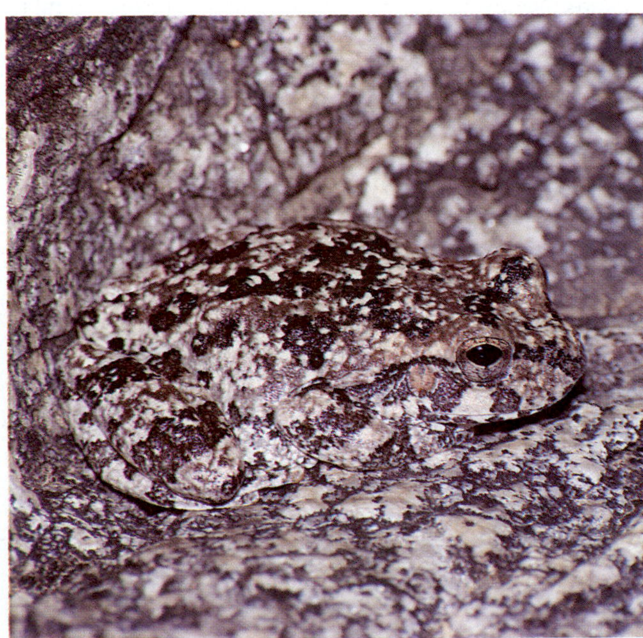

FIGURE 53.5 ▪ Camouflage. Cryptic coloration of the canyon tree frog (*Hyla arenicolor*) allows it to "disappear" on a granite background.

FIGURE 53.6 ▪ Deceptive coloration. The hindwing markings of the io moth (*Automerisio*) resemble the eyes of a much larger animal. When the moth moves its forewings, potential predators may be momentarily startled, enabling the moth to escape.

FIGURE 53.7 ▪ Aposematic (warning) coloration. Many toxic or unpalatable animals are conspicuously colored. The blue-ringed octopus (*Haplochlaena maculosa*), which inhabits Australian coastal waters, is highly venomous. Its iridescent blue rings expand and contract, probably as a warning signal to potential predators. With arms outstretched, the adult octopuses are only about 20 cm long, but a potent neurotoxin in their salivary glands is lethal to many animals, including humans. Symbiotic bacteria living in the salivary glands synthesize the toxin.

tion (FIGURE 53.7). This warning coloration seems to be adaptive; there is evidence that predators are more cautious in dealing with bright color patterns in potential prey, perhaps because so many aposematic animals tend to be dangerous prey. In an example of convergent evolution, unpalatable animals in several different taxa have similar patterns of coloration—black with yellow or red stripes characterize unpalatable animals as diverse as yellow-jacket wasps and coral snakes.

Mimicry. A predator or species of prey may gain a significant advantage through **mimicry**, a phenomenon in which the mimic bears a superficial resemblance to another species, the model. Defensive mimicry in prey often involves aposematic models. While there are many examples of mimicry in which the mimic and the model are related, such as one butterfly mimicking another, mimicry can also cross broad taxonomic lines.

In **Batesian mimicry**, a palatable or harmless species mimics an unpalatable or harmful model. In one intriguing example, the larva of the hawkmoth puffs up its head and thorax when disturbed, looking like the head of a small poisonous snake complete with eyes (FIGURE 53.8). The mimicry even involves behavior; the larva weaves its head back and forth and hisses like a snake. For Batesian mimicry to be effective, models must generally outnumber mimics; otherwise, predators would learn that animals with a particular coloration are good rather than bad to eat. In **Müllerian mimicry**, two or more unpalatable, aposematically colored species resemble each other. Presumably each species gains an additional advantage, because the pooling of numbers causes predators to learn more quickly to avoid any prey with a particular appearance.

Predators also use mimicry in a variety of ways. For example, some snapping turtles have tongues that resemble a wriggling worm, thus luring small fish; any fish that tries to eat the "bait" is itself quickly consumed as the turtle's strong jaws snap closed.

Parasitism

We are classifying predation and parasitism together here because they are both +/− interactions in communities. In parasitism, one organism, the **parasite**, derives its nourishment from another organism, its **host**, which is harmed or at least loses some energy or materials in the process. Organisms that live within their host, such as tapeworms and malarial parasites, are called **endoparasites**; others that feed on the external surface of a host, like mosquitoes and aphids, are called **ectoparasites**.

As in other +/− relationships, natural selection favors parasites that are best able to locate hosts and feed on them.

(a)

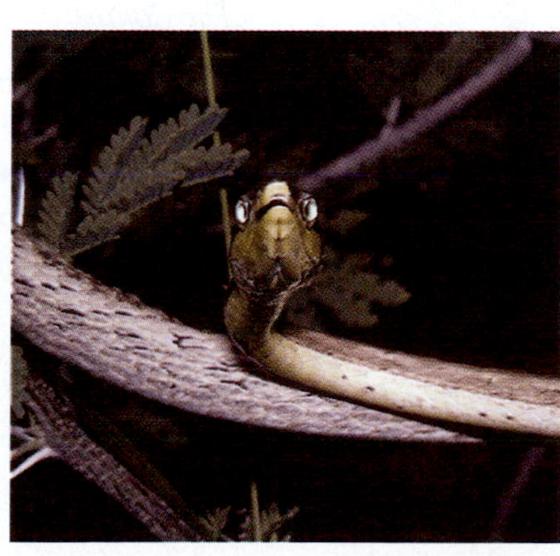

(b)

FIGURE 53.8 ▪ Batesian mimicry. In Batesian mimicry, a palatable species mimics the appearance of a harmful or unpalatable one. When it is disturbed, **(a)** the hawkmoth larva resembles **(b)** a snake.

Many endoparasites acquire hosts by passive mechanisms. For example, *Ascaris*, a nematode endoparasite of the human intestine, produces an abundance of eggs that are passed from a host's digestive tract to the external environment. In places without good sanitation, other humans inadvertently ingest the eggs and acquire the parasite themselves. Ectoparasites and many endoparasites have elaborate host-finding adaptations. Mosquitoes, for example, may first locate a host by detecting its movement and then confirm its identity on the basis of temperature and chemical cues, such as exhaled carbon dioxide and compounds on skin.

Natural selection has also favored the evolution of defensive capabilities in potential hosts. Some plant products that are toxic to herbivores, for instance, are also toxic to such parasites as fungi and bacteria. In vertebrates, the immune system provides defense against internal parasites. Some of the best-documented cases of genetic changes underlying coevolved host-parasite systems involve the vertebrate immune system and trypanosomes, protist parasites that inhabit the blood (see FIGURE 28.5). The host's antibodies and the populations of trypanosomes change in a matter of days, with different trypanosome genes that code for glycoprotein surface coats being switched on and off in tandem with regulatory changes underlying differential antibody production in the vertebrate (FIGURE 53.9).

While we usually think of parasites consuming parts of the host itself, in some cases organisms exploit the behavior of other organisms. The best-known example of this is brood parasitism, which occurs when birds such as cowbirds and European cuckoos lay their eggs in the nests of other species. In some cases the newly hatched brood parasite ejects other eggs from the nest when it hatches, and the host parents often invest more in feeding the imposter than they do their own offspring. The fitness of the foster parents, or hosts, is lowered by the parasite; thus, this is truly a +/− interaction. An evolutionary adaptation exhibited by many potential hosts is the ability to detect parasite eggs in their nests and eject them, or, if this is impossible, to start over in their nesting attempt, often abandoning their own eggs at the same time. As a counteradaptation to this host defense, some brood parasite species have elaborate egg mimicry that tricks the host into accepting the parasite egg as its own.

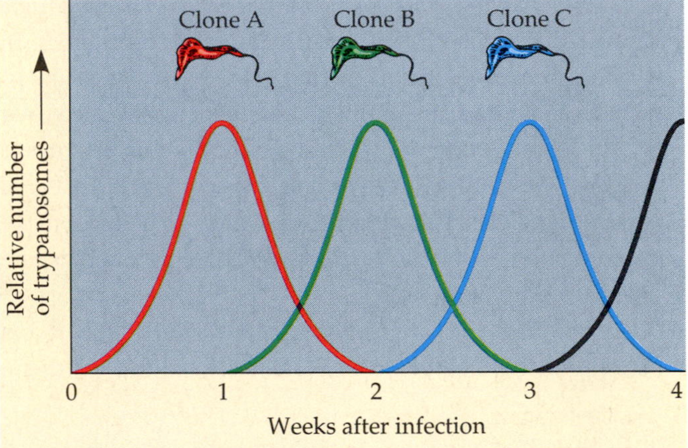

FIGURE 53.9 ▪ Rapid population change in a host-parasite system. Trypanosomes are microscopic parasites that live in the blood of vertebrates, causing diseases such as African sleeping sickness in humans. As this graph indicates, a sequence of trypanosome clones, distinguished by the antigenic nature of their glycoprotein surface coats, appears in the blood in a matter of weeks. (The black line indicates population growth of a fourth clone.) The changes in surface coats result from natural selection by host antibodies against the antigenic glycoproteins of the trypanosomes. Each trypanosome clone multiplies rapidly until host antibodies kill off virtually all of its members. A new clone then replaces the previous one, multiplying until it too is eliminated by specific antibodies. In this coevolutionary host-parasite system, changes in the parasite population parallel changes in the host's lymphocytes that produce specific antibodies. The antigens of the parasite induce the changes in the host's lymphocytes.

Interspecific competitions are −/− interactions: *a closer look*

When populations of two or more species in a community
53.1 rely on similar limiting resources, they may be subject to **interspecific competition**. Competition can occur in different ways. Actual fighting over resources is termed **interference competition**, whereas the consumption or use of similar resources is called **exploitative competition**. The density-dependent effects of interspecific competition are similar to those of intraspecific competition, discussed in Chapter 52. As population densities increase, every individual has access to a smaller share of some limiting resource; as a result, mortality rates increase, birthrates decrease, and population growth is curtailed. In interspecific competition, however, the population growth of a species may be limited by the density of competing species as well as by the density of its own population. For example, if several bird species in a forest feed on a limited population of insects, the density of each species may have a negative impact on population growth in the others. Similarly, species may compete for nesting sites, shelters, or any other resource that is in short supply.

The Competitive Exclusion Principle

In the first quarter of the twentieth century, two mathematician-biologists, A. J. Lotka and V. Volterra, independently modified the logistic model of population growth (see Chapter 52) to incorporate the effects of interspecific competition. They predicted that two species with similar requirements could not coexist in the same community; one species would inevitably harvest resources and reproduce more efficiently, driving the other to local extinction. Even a slight reproductive advantage would eventually lead to the elimination of the inferior competitor and an increase in the density of the superior one.

In 1934, Russian ecologist G. F. Gause tested this hypothesis with laboratory experiments on the effects of interspecific competition between populations of two closely related species of protozoans, *Paramecium aurelia* and *P. caudatum* (FIGURE 53.10). Grown in separate cultures under constant conditions and with a constant amount of bacteria added daily as food, each population of *Paramecium* grew until it leveled off at what was apparently the carrying capacity. When the two species were cultured together on a constant food supply, however, *P. caudatum,* apparently unable to compete with *P. aurelia,* was driven to extinction in the microcosm of the culture dish. Gause's experiments supported the hypothesis that two species with similar needs for the same limiting resources cannot coexist in the same place. This concept was later termed the **competitive exclusion principle**. Subsequent laboratory experiments on several other species of animals and plants reinforced the principle.

Although laboratory studies have generally confirmed the predictions of the competitive exclusion principle, natural communities are infinitely more complex than laboratory conditions. In order to determine how important competition is in structuring natural populations, ecologists have used both field experiments and indirect evidence from observations of how species use resources in the presence and absence of potential competitors. Before describing evidence about the importance of competition, we must consider how ecologists define and visualize resource use.

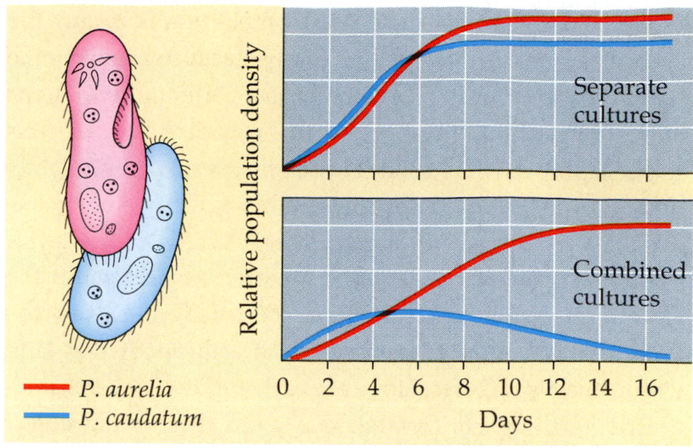

FIGURE 53.10 • Competition in laboratory populations of *Paramecium*. As indicated in the upper graph, in separate laboratory cultures with constant amounts of bacteria added daily for food, populations of the species *Paramecium aurelia* and *P. caudatum* each grow to carrying capacity. In contrast (lower graph), when the two species are grown together, *P. aurelia* has a competitive edge in obtaining food, and *P. caudatum* is driven to extinction in the culture.

We can now restate the competitive exclusion principle to say that two species cannot coexist in a community if their niches are identical. However, ecologically similar species can coexist in a community if there are one or more significant differences in their niches.

Ecological Niches

The concept of the ecological niche is almost inseparable from the concept of interspecific competition, but it is difficult to define rigorously. The **ecological niche** is the sum total of the organism's use of the biotic and abiotic resources in its environment. One way to grasp the concept is through an analogy made by ecologist Eugene Odum: If an organism's habitat is its address, the niche is its occupation. Put another way, an organism's niche is its ecological role—how it "fits into" an ecosystem. The niche of a population of tropical tree lizards, for example, consists of, among many other variables, the temperature range it tolerates, the size of trees upon which it perches, the time of day in which it is active, and the size and type of insects it eats.

The term **fundamental niche** refers to the set of resources a population is theoretically capable of using under ideal circumstances. In reality, each population is embedded in a web of interactions with populations of other species, and biological constraints, such as competition, predation, or the absence of some usable resources, may force the population to use only a subset of its fundamental niche. The resources a population actually uses are collectively called its **realized niche**.

Evidence for Competition in Nature

One problem with evaluating competition in nature is that if competition is as potent a force as the competitive exclusion principle suggests, we would expect it to be quite rare in natural communities. The reason is simple. There are two possible outcomes to competition between species with the same ecological niche: Either the weaker competitor will become extinct, or one of the species will evolve enough to use a different set of resources. In both of these outcomes, there will no longer be competition, which poses an intellectual and practical dilemma for ecologists. It is difficult to demonstrate the existence and importance of a force (competition) that, by its very nature, often cannot operate for long periods of time.

Although we cannot look directly at the evolutionary history of a community, some ecologists emphasize the importance of past competition, citing several lines of circumstantial evidence suggesting that it has been a major factor in shaping some of the ecological relationships we see today. Thus, in many cases we must study what ecologist Joseph H. Connell has called the "ghost of competition past." We'll look at some examples of this, but first we need to review two terms we introduced in Chapter 24. *Sympatric* populations occur in the same geographic area and can thus interact, whereas *allopatric* populations occur in different geographic areas.

One line of evidence for past competition is simply the observation that similar species always seem to exhibit some niche differences when they coexist in a community. Patterns of **resource partitioning**, in which sympatric species consume slightly different foods or use other resources in slightly different ways, are well documented, particularly among animals. Several species of small arboreal lizards of the genus *Anolis*, for example, are often sympatric. At one site in the Dominican Republic, seven *Anolis* species live in close proximity to each other, feeding on small arthropods that land within their territories. However, each species uses a characteristic perching site (FIGURE 53.11), and these perch differences presumably minimize competition among the lizard species. Each species also has structural characteristics, such as body size or leg length, that adapt it to its particular microhabitat, suggesting that natural selection has favored perch site specializations among these sympatric lizards. Similar patterns of resource partitioning are evident in other *Anolis* communities throughout the American tropics. In this case, resource partitioning seems to be the "ghost" that resulted from past competition among these lizard species.

A second line of circumstantial evidence for the importance of competition comes from comparisons of closely related species whose populations are sometimes sympatric and sometimes allopatric. Although allopatric populations of such species are similar in structure and use similar resources, sympatric populations often show differences in body structures and in the resources they use. The tendency for characteristics to be more divergent in sympatric populations of two species than in allopatric populations of the same two species is called **character displacement**. The Galápagos finches described in Chapter 25 provide a good example of character displacement in beak sizes and, presumably, in the seeds that they can eat most efficiently. Allopatric populations of *Geospiza fuliginosa* and *G. fortis* have beaks of similar size, but on an island where both species occur, a significant difference in beak depth has evolved (FIGURE 53.12). This difference presumably enables the two species to avoid competition by feeding on seeds of different sizes and probably represents the "ghost" that resulted from past competition.

Examples of resource partitioning and character displacement are compelling, but controlled field experiments provide stronger evidence that one species can influence the density and distribution of another species when they are in direct competition. In a classic study, Connell manipulated the densities of two barnacle species that ordinarily grow in different strata of the rocky intertidal zone and compete for the limited attachment space on the rocks (FIGURE 53.13). The heavier-shelled *Balanus* grows more rapidly than *Chthamalus*, and *Balanus* shells edge underneath those of *Chthamalus* and literally pry them off the rock. After Connell removed *Balanus* from the lower strata where it was most common, *Chthamalus* was able to grow there. This is an example of interference competition; one species is able to exclude another from the area

(a)

(b)

(c)

FIGURE 53.11 • Resource partitioning in a group of sympatric lizards. (a) Seven species of *Anolis* lizards live in close proximity at La Palma in the Dominican Republic. Each species perches in a characteristic microhabitat, distinguished by the amount of sun it receives and the size of the vegetation. **(b)** *A. distichus*, for example, perches on fenceposts and other sunny surfaces (such as this leaf), whereas **(c)** *A. insolitus* usually perches on shady branches. Such patterns of resource partitioning probably reduce interspecific competition among the members of a community, enabling them to coexist within a small geographic area.

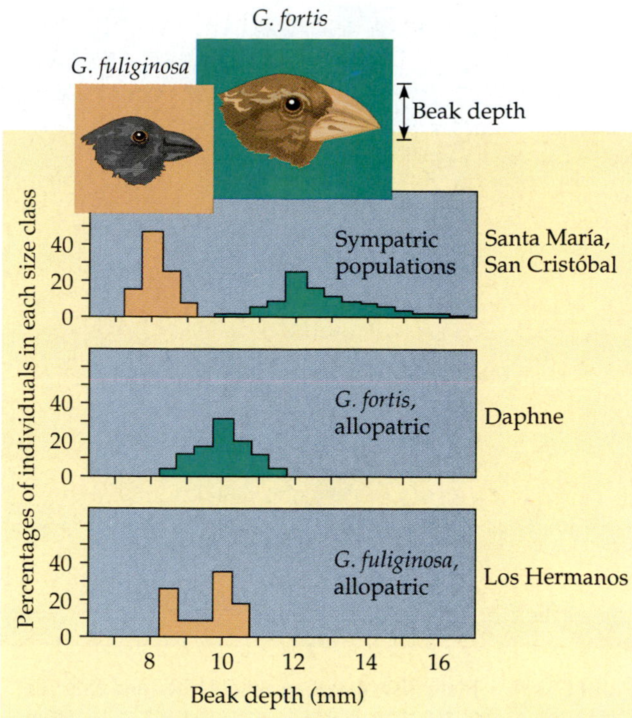

FIGURE 53.12 ▪ Character displacement: circumstantial evidence for competition in nature. Although allopatric populations of potential competitors are often similar in morphology and use equivalent resources, sympatric populations may diverge in both characteristics. In this example, two species of Galápagos finches have similar beak morphologies and presumably eat similarly sized seeds where their populations are allopatric on Daphne and Los Hermanos islands. However, where the two species are sympatric on Santa María and San Cristóbal, *Geospiza fuliginosa* has a shallower, smaller beak and *G. fortis* a deeper, larger one. Such evolutionary changes in body structure are thought to reflect resource partitioning. In this case the two species have adapted to eating different sizes of seeds.

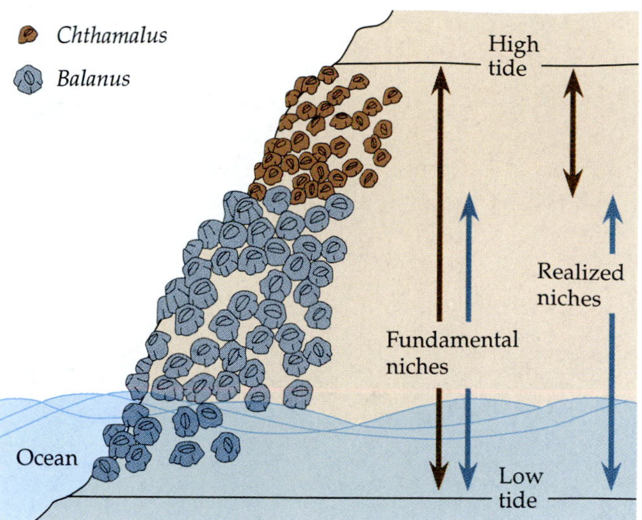

FIGURE 53.13 ▪ Experimental evidence for competition in nature. *Balanus balanoides* and *Chthamalus stellatus* are two species of barnacle that grow on rocks exposed during low tide along the Scottish coast. The barnacles have a stratified distribution, with *Balanus* most concentrated on the lower portions of the rocks and *Chthamalus* on the higher portions. The swimming larvae of the barnacles may settle randomly on the rocks and begin to develop into sessile adults, but *Balanus* fails to survive high on the rocks because it is unable to resist desiccation when these areas are exposed to air for several hours during low tides. Its fundamental niche and its realized niche are similar. Even though *Chthamalus* is concentrated primarily on the upper strata of rocks, when ecologist Joseph H. Connell removed *Balanus* from the lower strata, the *Chthamalus* population spread into that area. Thus, *Chthamalus could* survive lower on the rocks than where it is generally found, were it not for competition from *Balanus*. Its realized niche is only a fraction of its fundamental niche.

where their fundamental niches overlap. Similar field experiments on a variety of plant and animal species suggest that interspecific competition is strong under some circumstances.

Commensalism and mutualism are +/0 and +/+ interactions, respectively: *a closer look*

53.1 As we discussed in Chapter 27, **symbiosis** ("living together") is a term that encompasses a variety of interactions in which two species, a **host** and its **symbiont**, maintain a close association. There are three general types of symbiotic interactions. In **parasitism**, one organism—the parasite—harms the host; in **commensalism**, one partner benefits without significantly affecting the other; and in **mutualism**, both partners benefit from the relationship. In our survey of interspecific interactions in communities, we included parasitism with predation because both are +/− interactions in which one organism generally feeds on another. In this section we complete our survey of community interactions by examining commensal and mutual symbiotic relationships.

Commensalism (+/0)

Few, if any, absolute examples of commensalism exist because it is unlikely that one partner in an ecological interaction will be unaffected by the other. Strict commensalism benefits only one of the species involved, thus any evolutionary change in the relationship is likely to occur only in the beneficiary.

"Hitchhiking" species, such as algae that grow on the shells of aquatic turtles or barnacles that attach to whales, are sometimes considered to be commensal. However, the hitchhikers may actually decrease the reproductive success of their hosts slightly by reducing the efficiency of the hosts' movements in their search for food or escape from predators. Commensal associations sometimes involve one species obtaining food that is inadvertently exposed by another. For instance, cowbirds and cattle egrets feed on insects flushed out of the grass by grazing bison, cattle, horses, and other herbivores. Because the birds increase their feeding rates when following the herbivores, they clearly benefit from the association. Much of the time the herbivores may be unaffected by the relationship. However, in many cases it is likely that they too derive some benefit; the birds tend to be opportunistic feeders and occasionally remove and eat ticks and other ectoparasites from the

FIGURE 53.14 ▪ Possible commensalism between a bird and a mammal. The cattle egret concentrates its feeding where grazing cattle and other large herbivores, such as these cape buffalo in Tanzania, flush insects from the vegetation. This relationship benefits the bird without apparent harm or benefit to the mammal. Of course, the relationship may help or harm the buffalo in ways not obvious here.

FIGURE 53.15 ▪ Mutualism between acacia trees and ants. Certain species of Central and South American acacia trees, called bull's horn acacias, have hollow thorns that house stinging ants of the genus *Pseudomyrmex.* The ants feed on sugar produced by nectaries on the tree and on protein-rich swellings called Beltian bodies (orange in the photograph) that grow at the tips of leaflets. The acacia benefits from housing and feeding a population of pugnacious ants, which attack anything that touches the tree. The ants sting other insects, remove fungal spores and other debris, and clip surrounding vegetation that happens to grow close to the foliage of the acacia. A series of experiments in the 1960s demonstrated the mutualistic nature of the relationship. When the ants on experimental trees were poisoned, the trees died, probably because of damage by herbivores and failure to compete with the surrounding vegetation for light and growing space.

herbivores (FIGURE 53.14). We might also predict a negative effect for the herbivores: The presence of the birds may somehow make an herbivore—especially an immature one—more susceptible to predators.

The difficulty of finding true examples of commensalism illustrates a feature of symbiotic relationships in general: They are dynamic and not as easy to distinguish as their definitions might imply. In many cases, what we mean by calling an association commensalism, parasitism, or mutualism is that for most of the time that two associating populations are together, they exhibit that particular symbiotic relationship.

Mutualism (+/+)

In contrast to commensal relationships, mutualistic relationships require the evolution of adaptations in both participating species, because changes in either species are likely to affect the survival and reproduction of the other. Many coevolved mutualistic adaptations have been described in previous chapters: nitrogen fixation by bacteria in the root nodules of legumes; the digestion of cellulose by microorganisms in the digestive systems of termites and ruminant mammals; photosynthesis by unicellular protists in the tissues of corals; the exchange of nutrients in mycorrhizae, the association of fungi and the roots of plants; and specific interactions of certain pollinators and flowering plants. FIGURE 53.15 illustrates yet another interesting mutualistic interaction: the relationship between certain species of acacia and the ants that protect these trees from herbivores and competitors.

Many mutualistic relationships may have evolved from predator-prey or host-parasite interactions. Certain angiosperm plants, for example, have adaptations that attract animals that function in pollination or seed dispersal; these probably represent counteradaptations to the herbivores' feeding on pollen and seeds. In many cases, pollen is spared when the pollinator is able to consume nectar, and seeds are dispersed by animals that eat the fruit that encloses them. Any plants that could derive some benefit by sacrificing organic materials other than pollen or seeds would increase their reproductive success, and the adaptations for mutualistic interactions would spread through the plant population.

INTERSPECIFIC INTERACTIONS AND COMMUNITY STRUCTURE

Together, a community's predator-prey interactions (including plant-herbivore relationships), host-parasite associations, and mutualistic interactions constitute a complex web of relationships. The web is much more difficult to study than its component parts, but understanding it can provide insights into community structure.

One way to study a community's web of interspecific interactions is to examine feeding (trophic) relationships. One approach is to sort species populations into broad functional groups with similar trophic position. All plants are lumped together as producers, all animals that consume plants—from caterpillars to cattle—as primary consumers (herbivores), and so on. Lumping species into these functional categories provides a broad overview of a community, but it tends to obscure some important aspects of population interactions.

An alternative approach called food web analysis emphasizes the myriad trophic connections among community members. This approach recognizes, for example, that not all plants are consumed by all herbivores. Food webs can be very complex; nevertheless, a reasonably detailed analysis of them can be useful in evaluating the role of interspecific interactions on structure in some communities. We will analyze food webs further in Chapter 54. Let's now reexamine interspecific interactions in light of the influence they might have on community structure and diversity.

Predators can alter community structure by moderating competition among prey species

In simple laboratory experiments where a single predator species is kept with a single prey species having no refuge, the predator may devour all the prey and then perish from starvation. If this scenario were commonly enacted in nature, predators would always reduce the diversity of species in communities. These effects have been documented in some aquatic systems, where the accidental or purposeful introduction of fish and other aquatic predators has drastically reduced diversity in communities.

Probably the most important effect of a predator on community structure is to moderate competition among its prey species. Heavy predation can reduce the density of a strong competitor, thereby allowing weaker competitors to persist in the community. Experiments by ecologist Robert Paine in the 1960s were among the first to provide a clear picture of this complex interaction. Paine removed the dominant predator, a sea star of the genus *Pisaster*, from experimental areas within the intertidal zone on the coast of Washington State. The favorite prey of *Pisaster*, the mussel *Mytilus*, then outnumbered many of the other tidepool organisms that compete for attachment space on the rocks. Because *Mytilus* was such a dominant competitor in the experimentally created predator-free environment, the species richness of the community declined markedly (FIGURE 53.16). This research and numerous other field experiments have given rise to the concept of a **keystone species**, a species (in this case a predator) that makes an unusually strong impact on community structure, an impact that is disproportionately large relative to its own abundance. In some communities, keystone predators maintain higher species diversity by reducing the densities of strong competitors, such that competitive exclusion of other species does not occur.

(a)

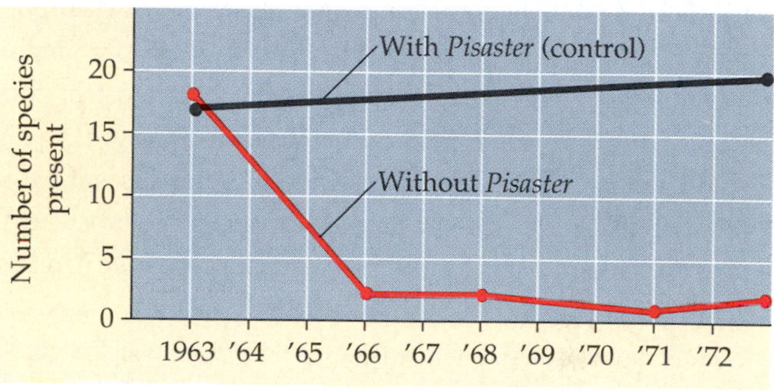

(b)

FIGURE 53.16 ▪ A keystone predator. (a) The rocky intertidal zone on the coast of Washington State contains a variety of algal species and invertebrates, including the sea star *Pisaster ochraceous,* which feeds preferentially on mussels but will also consume other invertebrates. **(b)** When ecologist Robert Paine experimentally removed *Pisaster* from a tidepool in 1963, mussels eventually took over the rock face and eliminated other invertebrates and algae. In a control area from which *Pisaster* was not removed, there was little change in species diversity.

The activities of some keystone species alter the character of whole communities. Beavers, for instance, construct dams that change free-flowing streams into lakes and ponds; terrestrial communities are flooded and replaced by aquatic ones. The foraging activities of African elephants produce truly large-scale changes in community composition and structure. Browsing on trees and shrubs, elephants uproot extensive forested areas, sometimes transforming them into savannahs or grasslands.

Mutualism and parasitism can have community-wide effects

The complex interactions that characterize parasitic and mutualistic associations often play significant roles in communities. Keystone mutualists, such as mycorrhizal fungi and nitrogen-fixing bacteria (see Chapters 31 and 37), help maintain critical ecosystem processes that influence all other species in communities.

Parasitic diseases that reduce populations of one species often affect other species as well. In tropical oceans, bacterial diseases of sea urchins that graze on algae have contributed to recent losses of some coral reefs. Unchecked by grazing when sea urchins die off, algae have overgrown and killed some reefs. In shallow waters off the U.S. and European North Atlantic coasts, dense growths of aquatic flowering plants called sea grass have occasionally been decimated by infections of a protist called *Labyrinthula*. Extensive losses of sea grass destroyed the habitats of virtually all animals in the community.

Another important effect of some parasites is the modification of host behavior. For example, aquatic arthropods called amphipods normally dive when disturbed, but amphipods infected with the larval stages of some tapeworms cling to objects in the water. Laboratory studies indicate that this behavior makes the infected amphipods more susceptible to predatory fishes, which are the next hosts in the tapeworm's life cycle. Thus, a parasite-induced behavioral change in a host may affect populations of the parasite and host.

The trypanosomes that cause African sleeping sickness and related diseases in domestic animals also alter host behavior in a way that seems to enhance the parasites' fitness. African trypanosomes are transmitted to vertebrates by bloodsucking insects called tsetse flies. Tsetses that are infected with trypanosomes tend to probe the skin of their vertebrate hosts for blood more often than uninfected tsetses. Increasing the chance that the trypanosomes will be transmitted to the vertebrate host, the tsetse's altered behavior apparently results from blockage by the trypanosomes of the mechanoreceptors that detect how much blood a fly has ingested. The trypanosome/tsetse relationship has far-reaching effects. Because some of the trypanosomes are deadly to domestic animals, agriculture and urbanization have been stifled in large areas of equatorial Africa; in effect, a host-parasite relationship has prevented the community-wide changes that accompany large-scale human settlement.

More far-reaching yet are the effects of human attempts to control parasites such as the mosquitoes that transmit malaria, yellow fever, and other important diseases. A combination of habitat alteration (draining wetland breeding sites), better sanitation, and chemical pesticides, such as the chlorinated hydrocarbon DDT, has reduced populations of mosquitoes and the incidence of the diseases they carry. Malaria, for instance, was once common in the United States. The disease was virtually eliminated from northern and midwestern regions by wetland drainage for agriculture and increased sanitation during the early twentieth century and from the southern U.S. by the use of DDT in the early 1950s. However, numerous populations of mosquitoes have become resistant to DDT and other chemicals, and widespread habitat change now threatens many species. Chemical pesticides can also have hazardous side effects. Because of extensive spraying before any potential dangers were known, chlorinated hydrocarbons are environmental contaminants throughout the globe. We will address some of the problems posed by persistent pesticides in Chapter 54.

Interspecific competition influences populations of many species and can affect community structure

In the 1960s and 1970s, many ecologists proposed that competition was a major factor limiting the diversity of species that could occupy a community. This hypothesis was largely based on observations of niche differences and resource partitioning among sympatric species. A given quantity of resources, these ecologists argued, could be partitioned only so finely before the effects of competition would inevitably lead to the extinction of poorer competitors, setting a limit on the number of species that could occur together. Numerous studies, such as the barnacle research previously described (see FIGURE 53.13), have documented the effects of competition in nature, suggesting that it may be a potent force in structuring communities.

Still more evidence for the important influence of competition comes from cases where species have been accidentally or purposely introduced into new communities by humans. In many cases, these introduced species, or **exotic species**, have outcompeted native community members and altered community structure. A recent troublesome invader in the United States is the zebra mussel, a bivalve mollusk, which entered the St. Lawrence Seaway in the mid-1980s in ballast water released by a cargo ship that had traveled from the mussels' native Caspian Sea. By the summer of 1993, zebra mussels were found in the Mississippi River as far south as New Orleans. Of the many problems caused by the mussels, those that have provoked the most uproar have resulted from their competition with humans—by clogging reservoir intake pipes, for example. However, zebra mussels also compete with native shellfish for space, and with fish for the plankton used for nourishment. The full extent to which this new competitor is altering community structure is yet to be determined.

Despite the evidence we have just examined, interspecific competition does not always lead to competitive exclusion, and competing species may sometimes coexist, albeit at reduced densities. Whereas competition is probably important in regulating the relative abundance of many species, and perhaps the species richness of many communities, many ecologists are reluctant to embrace it as a significant factor in structuring communities. It is often difficult to demonstrate that two species are currently competing, and it is even more difficult to assess what has happened in their evolutionary past. Furthermore, competition may not be important unless population sizes are near carrying capacity and resources are limiting.

Research on the nesting behavior of two North American bird species that occupy the same trees illustrates the difficulty of identifying competition as a key factor in community structure. MacGillivray's warblers nest low in trees; black-headed grosbeaks nest high. At first this might be interpreted as another example of resource partitioning that reduces competition—in this case, for nesting sites. But ecologist Thomas Martin of the University of Montana has demonstrated that predation may be the key factor in this segregation of nests; predators locate and feed on the young of these birds more successfully when nests are clumped, rather than being more widely distributed in the trees. In other words, the partitioning we see in nest sites reduces predation, and this, rather than a reduction in competition between the bird species, may explain the observed difference.

Determining the effects of interspecific factors (competition, predation, and symbioses) on species diversity and structure in communities is an important research goal in ecology. A related goal is determining the relative significance of environmental patchiness in community structure.

A complex interplay of interspecific interactions and environmental variability characterizes community structure

In general, habitats that are more varied can support a more diverse community for the simple reason that they provide more ecological niches. An important aspect of environmental heterogeneity is the vegetation structure in the community, and, as we saw in our survey of the world's biomes, there is a great deal of diversity in the overall form of vegetation. In turn, the vegetation in a community largely determines the types of animals found there. In general, structurally complex vegetation provides a diversity of microhabitats that can be used by a variety of animal species (see FIGURE 53.11), whereas less complex vegetation does not provide a structural resource that animal populations can partition easily.

Patchiness is a spatial and temporal characteristic of all ecosystems (see Chapter 50). In one example of spatial patchiness, the mineral content of soil varies locally with the chemical composition of the rocks from which it is derived (see FIGURE 53.2). Similarly, soil moisture varies with topography, such that low-lying areas are generally wetter than high ground. If different species are best adapted to different local conditions, environmental patchiness can increase the community's species diversity by facilitating resource partitioning among potential competitors. The distributions of some low-growing plants in the forests of Indiana, for example, are regulated by their seedlings' tolerance of calcium and organic matter. Because these factors vary locally, wild black cherry occurs in some patches and two species of violets are found in others. If the distribution of these abiotic factors were uniform throughout the forest, then one of the species would probably outcompete the others, thereby reducing species diversity in the community.

Temporal heterogeneity of habitats can also affect community diversity. This is especially pronounced with seasonal changes, with spring flowers giving way to summer and then fall-blooming species, and spring migrant birds being replaced by summer residents and then fall migrants. Over the course of a single day, animals that are active during daylight are replaced by nocturnal animals after dark. These temporal changes can greatly increase total species diversity in a community.

Numerous studies lead to the general conclusion that communities are characterized by environmental heterogeneity and a web of interspecific interactions. Which factors are most important to a community's structure can vary from one type of community to another. Because of the variety of communities and the complexity of the interplay of factors, there are few, if any, communities for which we have a good understanding of all the important relationships or how the relationships evolved.

DISTURBANCE AND NONEQUILIBRIUM

A traditional view of biological communities is that they are in a state of equilibrium, a more or less stable balance, unless seriously disturbed by human activities. The focus of this "balance of nature" view is on defining the factors—mainly the interspecific interactions—that seem to maintain stability in communities and that may return stability to disturbed areas. **Stability** in this context is the tendency of a community to reach and maintain an equilibrium, or relatively constant condition, in the face of disturbance. We have seen that interspecific interactions are important *features* of communities, but are they the predominant forces that structure communities and affect their species diversity?

Nonequilibrium resulting from disturbance is a prominent feature of most communities

Disturbances are events, such as storms, fire, floods, droughts, overgrazing, or human activities that damage communities, remove organisms from them, and alter resource availability. Storms disturb virtually all communities, even those in deep oceans. Fire is a significant cause of disturbance in most terrestrial communities; grasslands and chaparral biomes are dependent on regular burning (FIGURE 53.17, p. 1122). Freezing is a frequent occurrence in many rivers, lakes, and ponds, and many streams and ponds are disturbed by seasonal drying. Disturbances often create opportunities for species that have not previously occupied habitats to become established.

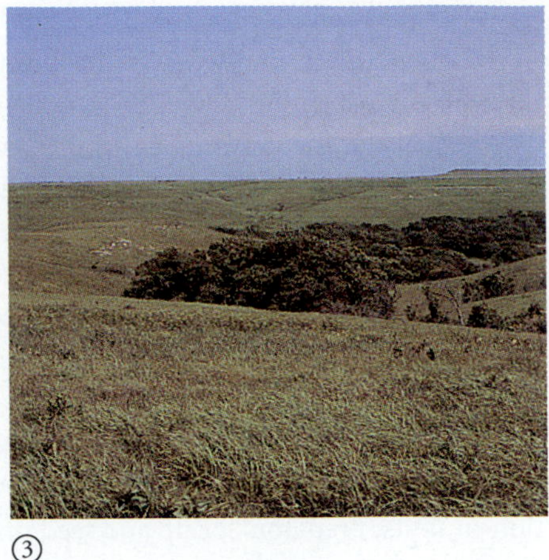

① ② ③

FIGURE 53.17 ▪ Routine disturbance in a grassland community. Historically, prairie grasslands were often swept by fire. These photographs were taken ① before, ② during, and ③ after a controlled burn conducted by ecologists studying the long-term effects of fire on a tallgrass prairie in Kansas. ① A prairie that has not burned for several years has a high proportion of detritus (dead grass) which ② serves as fuel for fires. ③ Prairies regrow rapidly after a fire burns the detritus. Approximately one month after the controlled burn, virtually all of the biomass in this prairie was living. (The trees in the photographs are growing along a stream and were not burned during the study.)

Increasing evidence indicates that some amount of disturbance and nonequilibrium resulting from disturbance is the norm for most communities. One popular hypothesis is that communities are usually in a state of recovery from disturbance. Understanding how communities respond to the spatial and temporal effects of disturbance is an important research goal in community ecology.

Humans are the most widespread agents of disturbance

Many animals are agents of disturbance. Overgrazed grasslands and forests that contain large deer populations both illustrate the impact that herbivores can have on communities. Beavers, elephants, and gypsy moths are other examples of animals that create major disturbances in communities throughout the world.

Of all animals, humans have the greatest impact on communities worldwide. Logging and clearing for urban development, mining, and farming have reduced large tracts of forests to small patches of disconnected woodlots in many parts of the United States and throughout Europe. Similarly, agricultural development has disrupted what were once the vast grasslands of the North American prairie. After forests are clear-cut and left alone, weedy and shrubby vegetation often colonizes the area and dominates it for many years. This type of vegetation is also found extensively in agricultural fields that are no longer under cultivation, and in vacant lots and construction sites that are periodically cleared. Human disturbance of communities is by no means limited to the United States and Europe; nor is it a recent problem. Tropical rain forests are quickly disappearing as a result of clear-cutting for lumber and pastureland. Centuries of overgrazing and agricultural disturbance have contributed to the current famine in parts of Africa by turning seasonal grasslands into great barren areas.

Human disturbance usually reduces species diversity in communities. We currently use about 60% of Earth's land in one way or another, mostly as cropland, forest, and rangeland. Most crops are grown in monocultures, intensive cultivations of a single plant variety over large areas. Even forests that are used to produce pulpwood and lumber are often replanted in single-species stands. And the effects of intensive grazing on rangelands often include the removal of several native plant species and replacement with only a few introduced species.

We tend to think of disturbances as having negative impacts, but this is not always the case. Disturbance can affect community structure on virtually any scale, and small-scale disturbances often enhance environmental patchiness that can be important to the maintenance of species diversity in a community (FIGURE 53.18). Frequent small-scale disturbances can also prevent large-scale disturbances. The fires that occurred in Yellowstone National Park during the summer of 1988 are an example. Much of this park was dominated by lodgepole pine, a tree that requires the rejuvenating influence of periodic fires. Lodgepole cones remain closed until exposed to intense heat. When a forest fire destroys the parent trees, the cones open and the seeds are released. The new generation of lodgepoles can then thrive on nutrients released from the burned trees and on the direct sunlight that was blocked by the old forest. Lodgepole pines that are over 100 years old become increasingly flammable, but fire suppression for decades had prevented small lightning-induced fires that would have resulted in patches of less-flammable trees. As a result, by 1988, about one-third of the Yellowstone forests were 250 to 300 years old. The drought conditions of 1988, combined with the fuel that had accumulated in the forest, resulted in large-scale destruction of the old lodgepole forests (FIGURE 53.19a). Demonstrating that communities can often

(a)

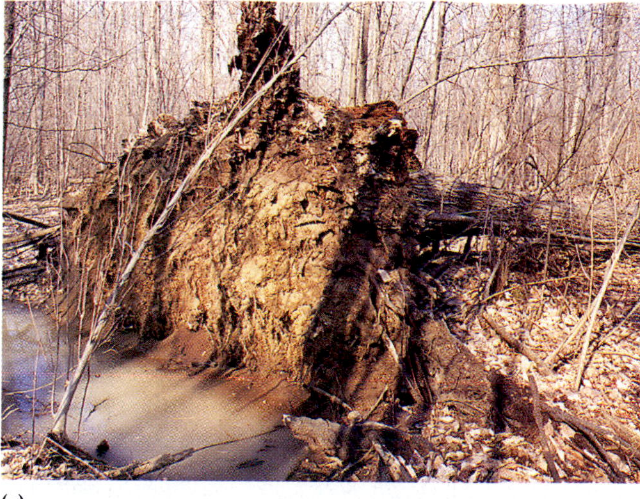

(c)

(b)

FIGURE 53.18 ▪ Environmental patchiness caused by small-scale disturbances. Shown here are three examples of small-scale soil disturbances that may enhance species diversity in communities. **(a)** A large ant colony in a forest in Minnesota. As indicated here, ants often create small open patches that may be colonized by organisms missing from or less abundant in the surrounding forest. **(b)** A mound of soil removed from a burrow by black-tailed prairie dogs at Devil's Tower National Monument in Wyoming. Burrows often create habitats for a variety of organisms in addition to the burrower. **(c)** A root mound and water-filled depression created by a tree felled by a windstorm. The dead tree, the mound, and the depression create new habitats in this forest in Michigan.

(a)

(b)

FIGURE 53.19 ▪ Patchiness and recovery following a large-scale disturbance. The Yellowstone National Park fire of 1988 destroyed large areas of coniferous forests dominated by lodgepole pines. **(a)** As this photograph taken soon after the fire shows, the burn left a patchy landscape; note the unburned trees in the background. **(b)** This photograph of the same general area taken the following year indicates how rapidly the community recovered. A variety of herbaceous plants, different from the species that inhabited the floor of the former forest, virtually cover the ground.

respond very rapidly to massive disturbance, burned areas in Yellowstone were largely covered with new vegetation the following year (FIGURE 53.19b).

Succession is a process of change that results from disturbance in communities

Changes in community composition and structure are most apparent after some disturbance, such as a large fire or a volcanic eruption, strips away the existing vegetation. The disturbed area may be colonized by a variety of species, which are gradually replaced by a succession of other species. Such transitions in species composition over ecological time represent a process called **ecological succession**. The process is called **primary succession** if it begins in a virtually lifeless area where soil has not yet formed, such as on a new volcanic island or on the rubble left behind by a retreating glacier. Often the only life forms initially present are autotrophic bacteria. Lichens and mosses, which grow from windblown spores, are commonly the first macroscopic photosynthesizers to colonize such areas. Soil develops gradually, as rocks weather and organic matter accumulates from the decomposed remains of the early colonizers. Once soil is present, the lichens and mosses are usually overgrown by other plants, such as grasses, shrubs, and trees that sprout from seeds blown in from nearby areas or carried in by animals. Eventually, an area may be colonized by plants that become the community's prevalent form of vegetation. Primary succession from barren soil to such a community may take hundreds or thousands of years.

Secondary succession occurs where an existing community has been cleared by some disturbance that leaves the soil intact. Often the area begins a return to something like its original state. For instance, forested areas that are cleared for farming will, if abandoned, undergo secondary succession and may eventually return to forest. The earliest plants to recolonize such an area are often herbaceous species that grow from windblown or animal-borne seeds. If the area is not burned or heavily grazed, woody shrubs may in time replace most of the herbaceous species, and eventually forest trees may replace most of the shrubs.

What factors determine the course of succession? Competition among individual species for available resources is one possibility. Because resource availability changes over the course of succession, different species compete better at different stages. Early successional stages are often characterized by *r*-selected species that are good colonizers because of their high fecundity and excellent dispersal mechanisms (see Chapter 52). Many of these organisms may be described as "fugitive" or "weedy" species that do not compete well except in newly disturbed areas. Such species maintain themselves by constantly colonizing newly disturbed areas before better competitors can become established in the same places.

Tolerance of the abiotic conditions in a barren area affects species composition during early successional stages. Many *K*-selected species may colonize an area, but they will not grow in abundance if environmental conditions there are at the extremes of their tolerance limits. Variations in the growth rates and maturation times of colonizing species can also be important. For example, if the seeds of two plant species, one an annual herb and the other a tree, colonize a community at the same time, the herbaceous species will have earlier prominence because of its faster growth and shorter generation time. Many of the plants that populated the charred ground in Yellowstone Park the year after the 1988 fire were these kinds of early colonizers (see FIGURE 53.19). The ecological impact of the tree species may not be realized until the trees are relatively large.

Some of the changes in community structure during succession may result from *inhibition* of some species by others through exploitative competition, interference competition, or both. The presence of organisms also affects the abiotic environment by modifying local conditions. This may result in *facilitation,* in which the group of organisms representing one successional stage "paves the way" for species typical of the next stage. For example, alders are abundant in an intermediate stage of succession following glacial removal of soil. Alders lower soil pH as their dropped leaves decompose. The change in pH facilitates the entry of conifer trees, which require acidic soil. Sometimes the changes that facilitate the development of a later stage actually make the environment unsuitable for the very species responsible for the changes.

Both inhibition and facilitation may be involved throughout the successional process. For example, horseweed is one of the earliest colonizers in abandoned farm fields. For a year or two, horseweed may inhibit other species through shading and soil water use. However, as the horseweed and other early species die and decompose, their presence facilitates the entry of later species by adding organic matter to the soil, which aids in holding moisture. Some of the conifer trees that dominate a later successional stage require full sunlight; their growth becomes self-inhibitory as the trees shade the ground and prevent their own offspring from growing. At the same time, the conifers continue to add organic material to the soil, and the shade they cast keeps the forest floor moist, conditions that may facilitate the germination and growth of hardwood trees.

The nonequilibrial model views communities as mosaics of patches at different stages of succession

A focus on interspecific interactions producing a predictable progression of stages is a traditional view of ecological succession. In the early 1900s this general view led some ecologists to hypothesize that succession leads to a stable endpoint, a climax community. A stable climax was predicted to develop when a web of interspecific interactions became so intricate that the community was saturated. No additional species could "fit into" the community unless resources became available through the localized extinction of species that were already present.

Ecologists soon recognized that the notion of stable climax communities was seriously flawed and too simplistic to represent the wealth of variation in nature. Most communities are routinely disturbed by outside factors during the course of succession. Without fire, many grassland areas would develop into forest. We might say that forest is the climax community for such areas, but that would make little sense if the forest community never develops. In this case, periodic fires maintain the community at a stage that simply does not fit the idea of a climax state. Even communities that appear relatively stable change over long periods of time. Studies of pollen preserved in lake sediments provide evidence of community composition over thousands or even millions of years. These studies indicate that common tree species sometimes disappeared from North American forests for hundreds of years, only to eventually return at a later time. The concept of climax communities is no longer considered useful in ecological research, although it remains with us in some popular literature.

In sharp contrast to the concept of succession as an orderly, linear progression driven mainly by interspecific interactions, a more current view emphasizes the inevitability of disturbance and nonequilibrium. Succession is seen as a variable, ongoing process. Most studies indicate that linear progression is not typical of succession, and mounting evidence supports the concept that disturbance is the main force that drives successional changes.

According to this nonequilibrial model, succession can take a variety of pathways, depending on the size, frequency, and severity of disturbance. By destabilizing existing community structures and favoring new ones, disturbances not only initiate succession, they also affect its course. The course of succession may vary, for example, with the identity of the particular species that happen to colonize a disturbed area first, and disturbance may also change the identities and numbers of species during any successional stages. Many ecologists posit that disturbance keeps communities in a continual state of flux, rendering them mosaics of patches at different successional stages and preventing them from ever achieving a state of equilibrium.

What are the actual effects of disturbance in a community? Disturbances result in the loss of individuals, thus creating opportunities in the form of vacated niche spaces that other species may colonize. After a disturbance, some survivors may recolonize an area; grasses may regrow from unburned roots, and trees may sprout from partially burned stumps. Species may also recolonize a disturbed area by migrating from adjacent areas. In general, disturbances followed mainly by regrowth and migration from adjacent areas do not significantly alter community structure because species formerly present in the disturbed areas largely refill it.

Disturbances that trigger major changes in community structure generally are those whose magnitude or frequency results in colonization of disturbed patches by **recruitment**. In this process, species from distant areas that are not directly associated with the disturbed patch or its immediate vicinity are the major colonizers. Examples of recruits include windblown and animal-borne seeds. The usual result of recruitment is a different mix of species than formerly present. Thus, a supply of distant recruits and vacated spaces that they can colonize are prerequisites to significant community alteration by disturbances. Alteration may affect a community's relative abundance of species, or its species richness, or both.

Several hypotheses have been proposed to explain why some communities are more species diverse than others. One hypothesis, the nonequilibrial model, proposes that high species diversity results mainly from environmental patchiness caused by frequent abiotic disturbances. Another idea, the **dynamic equilibrium hypothesis**, proposed in the late 1970s, maintains that species diversity depends mainly on the effect of disturbance on the competitive interactions of populations. According to this model, species diversity is enhanced when the frequency of disturbance prevents competitive exclusion, because poorer competitors persist in the community. The effects of disturbances caused by keystone species, such as elephants and beavers, are consistent with this hypothesis.

Another 1970s proposal, the **intermediate disturbance hypothesis**, posits that species diversity is greatest where disturbances are moderate in both frequency and severity, because organisms typical of different successional stages will be present. When disturbance is severe and frequent, the community may include only good colonizers typical of early stages of succession. If disturbances are mild and rare in a particular location, the late-successional species that are most competitive will make up the community. Studies of species diversity in tropical rain forests provide some evidence for the intermediate disturbance hypothesis. Scattered throughout these forests are gaps where trees and the vines attached to them have fallen. In these disturbed areas, species from various successional stages coexist within a relatively small space.

Clearly, all communities are subject to some disturbance. On a small scale, worms and rodents disturb soil by burrowing through it. Storms create patches of various sizes in virtually all communities. However, communities are also characterized by interspecific interactions that may affect community structure, succession, and species diversity. The challenge of future research is to sort out the relative importance of the disturbances that tend to destabilize communities and the interspecific interactions that tend to stabilize them.

These issues are central to our current dilemma of species conservation, and we will revisit them in Chapter 55.

COMMUNITY ECOLOGY AND BIOGEOGRAPHY

Biogeography is the study of the past and present distribution of individual species and entire communities. The field of biogeography provides a different approach to understanding community properties by analyzing both global and local phenomena, mostly from a historical perspective.

Traditionally, biogeographers have been concerned with the actual identities of the species that make up particular communities. For example, biogeographers show how the present distributions of species reflect their distant evolutionary history, as well as more recent interactions with both biotic and abiotic components of the environment. More than a century ago, Darwin, Wallace, and other naturalists began to recognize biogeographical realms that we now associate with patterns of continental drift that followed the breakup of Pangaea (FIGURE 53.20). We discussed these historical aspects of biogeography in Chapters 22 and 25. Although this is still an active area of research, some biogeographers have more recently applied the principles of community ecology to the analysis of geographical distribution. The intersection of these fields is fertile ground: Biogeographical analyses have contributed as much to our understanding of community patterns and processes as the study of community ecology has contributed to our knowledge of biogeography. Here we consider factors that limit species ranges, and island biogeography, two topics that have captured the attention of ecologists during the last three decades.

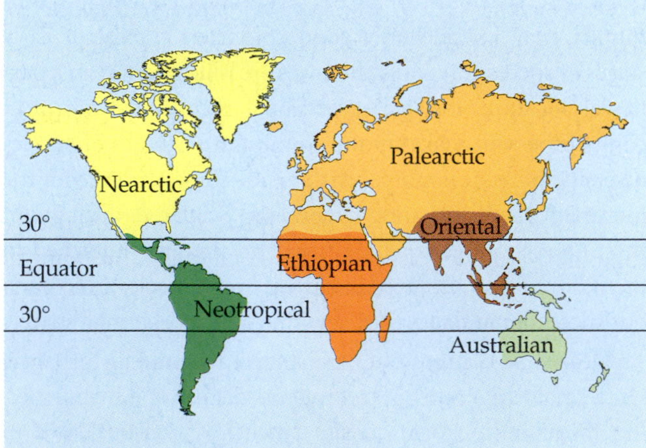

FIGURE 53.20 • Biogeographical realms. Continental drift and barriers such as deserts and mountain ranges all contribute to the distinctive floras and faunas found in Earth's major regions. Except for Australia, the realms are not sharply delineated but grade together in zones where taxa from adjacent realms coexist.

Dispersal and survivability in ecological and evolutionary time account for the geographical ranges of species

Understanding the determinants of a species' geographical range is central to any analysis in community ecology and biogeography. Three general explanations can account for the limitation of a species to a particular present range: (1) The species may never have dispersed beyond its current boundaries; (2) pioneers that did spread beyond the observed range failed to survive; and (3) over evolutionary time the species has retracted from a once-larger range to its present boundaries. Research in paleontology and historical biogeography can identify cases for which the third explanation is valid. For example, we know from fossil evidence that close relatives of living elephants and camels once occupied North America, but local extinctions have caused retractions of the geographical ranges of both groups.

Distinguishing between the first and second explanations is more difficult, because discovering a few colonists that dispersed to a new area but then died is harder than looking for the proverbial needle in a haystack. However, transplant experiments, in which individuals of a plant or animal species are moved to similar environments outside their range, can provide useful information. Survival of the transplants suggests that the species had not dispersed to suitable locations outside the existing range. Failure of the transplants, in contrast, suggests that the species could not expand its range because of its inability to tolerate different abiotic conditions or to compete with resident species. In one simple experiment, researchers transplanted several individuals of the lizard *Anolis cristatellus* from the warm lowlands in Puerto Rico to a cool, forested site at higher elevation. These animals failed to survive for more than a few weeks, presumably because they could not attain sufficiently high body temperatures to capture and digest their food. Therefore, this species does not occupy high-elevation forests because it is not adapted to the physical environment in such habitats.

Species diversity on some islands tends to reach a dynamic equilibrium in ecological time

Because of their limited size and isolation, islands provide excellent opportunities for studying some of the factors that affect the species diversity of communities. By "islands," we mean not only oceanic islands but also habitat islands on land, such as lakes, mountain peaks separated by lowlands, or natural woodland fragments surrounded by areas disturbed by humans—in other words, any areas surrounded by an environment not suitable for the "island" species. In the 1960s, American ecologists Robert MacArthur and E. O. Wil-

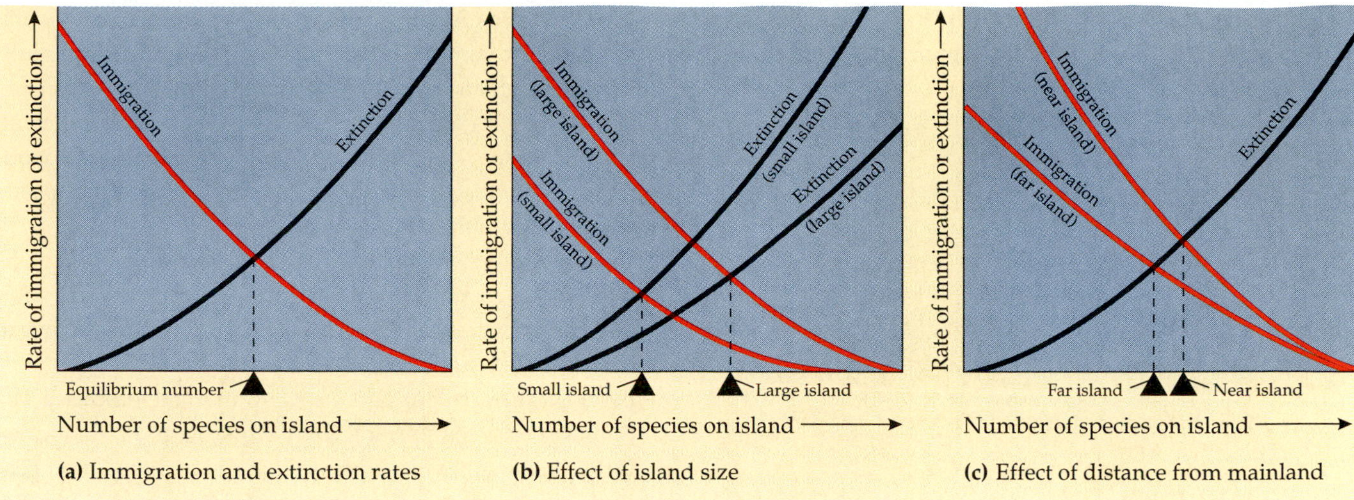

FIGURE 53.21 ▪ The hypothesis of island biogeography. (a) The equilibrium number (black triangle) of species on an island represents a balance between the immigration of new species to the island and the extinction of species that are already there. **(b)** Large islands may ultimately have a larger equilibrium number of species than small islands because immigration rates tend to be higher and extinction rates tend to be lower on large islands. **(c)** Although extinction rates do not differ with an island's distance from a mainland source of species, near islands tend to have larger equilibrium numbers of species than far islands because immigration rates to near islands are higher than those for more distant ones.

son developed a general hypothesis of island biogeography to identify the important determinants of species diversity on an island with a given set of physical characteristics. The study of islands may help us understand some of the interactions in more complex systems.

Imagine a newly formed oceanic island some distance from a mainland that will serve as a source of colonizing species. Two factors will determine the number of species that eventually inhabit the island: the rate at which new species immigrate to the island and the rate at which species become extinct on the island. Immigration and extinction rates are, in turn, affected by two important features of the island: its size and its distance from the mainland. Small islands will generally have lower immigration rates, because potential colonizers are less likely to reach a small island. For example, birds blown out to sea by a storm are more likely to land by chance on a larger island than on a small one. Small islands will also have higher extinction rates. They generally contain fewer resources and less diverse habitats for colonizing species to partition, increasing the likelihood of competitive exclusion. Distance from the mainland is also important; for two islands of equal size, a closer island will have a higher immigration rate than one farther away.

The immigration and extinction rates are also affected at any given time by the number of species already present on the island. As the number of species on the island increases, the immigration rate of new species decreases, because any individual reaching the island is less likely to represent a species that is not yet present. At the same time, as more species inhabit an island, extinction rates on the island increase because of the greater likelihood of competitive exclusion.

These relationships are summarized in FIGURE 53.21, where immigration and extinction rates are plotted as a function of the number of species present on the island. The main prediction of this model is that eventually an equilibrium will be reached where the rate of species immigration matches the rate of species extinction (FIGURE 53.21a). The number of species at this equilibrium point is correlated with the island's size and distance from the mainland. The Wilson/MacArthur hypothesis of island biogeography predicts that the equilibrium species number is greater on a large island than a small island (FIGURE 53.21b), and greater on a near island than a far island (FIGURE 53.21c). Any equilibrium, of course, is dynamic; immigration and extinction continue, and the exact species composition may change over time.

MacArthur and Wilson's studies of the diversity of amphibians and reptiles on many island chains, including the West Indies, support the prediction that species richness increases with island size (FIGURE 53.22, p. 1128). Species counts also fit the prediction that the number of species decreases with increasing remoteness of the island.

In the late 1960s, Wilson and Daniel Simberloff tested the hypothesis of island biogeography in experiments on small islands of mangroves off the southern tip of Florida. Six islands, each about 12 m in diameter, were enclosed in tents and fumigated to kill all resident arthropods. The pesticide used, methyl bromide, decomposes rapidly, and the islands could be recolonized by arthropods from the mainland species pool. Within about a year, species numbers equilibrated on each island, with the fewest species on the island most distant from the coast. Although the equilibrium number for each island was about the same before fumigation and after recolonization, the species composition was different. Chance events—in this case, which species of arthropods happened to disperse to which islands—affected the species composition of the communities.

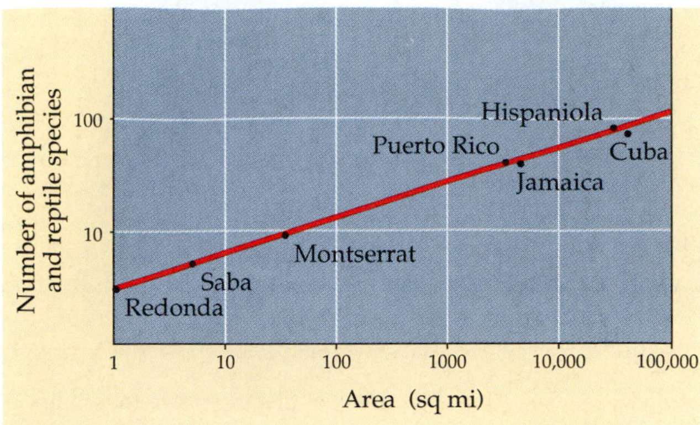

FIGURE 53.22 ▪ Species richness and island size. This species-area graph illustrates that the number of amphibian and reptile species found on West Indian islands is closely related to island size. Large islands harbor more species because greater habitat diversity allows greater resource partitioning among the resident species, reducing the likelihood of competitive exclusion.

In the past several decades the island biogeography hypothesis has come under considerable fire. Its predictions of equilibria in the species composition of communities may apply in only a limited number of cases and over relatively short time periods, where colonization is the important process determining species composition. Over a longer period, abiotic disturbances, adaptive evolutionary changes, and speciation generally alter the species composition and community structure on islands. More important than whether the hypothesis has widespread applicability is that it stimulated discussion and research on the effects of habitat size on species diversity, a topic we will turn to when we discuss habitat fragmentation and biodiversity in Chapter 55.

CHAPTER REVIEW

REVIEW OF KEY CONCEPTS

(with page numbers and key figures)

EARLY HYPOTHESES OF COMMUNITY STRUCTURE

Species diversity within a community, an assemblage of species living close enough together for potential interaction, includes both the species richness (number of species present) and the relative abundance of each species.

- **The interactive and individualistic hypotheses pose alternative explanations of community structure: *science as a process* (pp. 1107–1109, FIGURE 53.1)** The individualistic hypothesis proposes that communities are chance assemblages of independently distributed species with the same abiotic requirements. These assemblages grade into one another, except where abiotic conditions change abruptly. The interactive hypothesis states that the species within a community are locked into biotic interactions and that the community functions as an integrated unit. Observations of plant associations tend to support the individualistic hypothesis, but the associations of animal species are more complex, depending on both abiotic factors and interspecific interactions.

INTERACTIONS BETWEEN POPULATIONS OF DIFFERENT SPECIES

- **Interspecific interactions can be strong selection factors in evolution (pp. 1109–1110)** Coevolution, reciprocal interactions between two species that result in a series of adaptations and counteradaptations, has been studied most extensively in predator-prey relationships, mutualism, and host-parasite relationships.
- **Interspecific interactions may have positive, negative, or neutral effects on a population's density: *an overview* (pp. 1110–1111, TABLE 53.2)** These interactions can be represented by pairs of symbols (+, −, and 0) indicating the idealized effect of the relationship on each of the two interacting species.
- 53.1 **Predation and parasitism are +/− interactions: *a closer look* (pp. 1111–1114, FIGURES 53.4–53.9)** Predation refers to interactions in which animals feed on other organisms. Parasitism is a type of predation in which a parasite lives on or in a host, deriving nourishment from it but usually not killing it outright. Herbivory may be regarded as a form of predation, although grazing is more similar to parasitism. Aposematic coloration warns predators that prey animals have chemical defenses. In mimicry, an animal bears superficial resemblance to an unpalatable or harmful model. Very rapid coevolution occurs in some host-parasite systems in which the vertebrate host response involves the immune system.
- 53.1 **Interspecific competitions are −/− interactions: *a closer look* (pp. 1114–1117, FIGURES 53.10–53.13)** When populations of different species in a community use similar limiting resources, they may fight over resources (interference competition), or each may simply reduce the resources available to the other (exploitative competition). The ecological niche is the sum total of the organism's use of the biotic and abiotic resources in its environment. The competitive exclusion principle states that two species cannot coexist in the same community if their niches are identical. Interspecies competition leads to extinction of the weaker competitor or adaptation of one species to a new niche; thus, it cannot operate for long. Resource partitioning and character displacement provide indirect evidence for the importance of past competition.
- 53.1 **Commensalism and mutualism are +/0 and +/+ interactions, respectively: *a closer look* (pp. 1117–1118, FIGURES 53.14–53.15)** Symbiosis, in which a host and a symbiont maintain a close association, includes parasitism, commensalism, and mutualism. Commensalism refers to symbiotic interactions in which one species benefits and the other is not affected; there are few if any cases of pure commensalism. Mutualism refers to symbiotic interactions in which both species benefit. Many coevolved mutualistic adaptations occur, including nitrogen-fixing bacteria living with legumes and the interactions of pollinating animals with flowering plants.

INTERSPECIFIC INTERACTIONS AND COMMUNITY STRUCTURE

The trophic structure of a community refers to all of the feeding relationships in the community. Food-web analysis emphasizes the complexity of trophic connections in a community.

- **Predators can alter community structure by moderating competition among prey species (p. 1119, FIGURE 53.16)** Predators may reduce or increase diversity in a community. Keystone species have a disproportionately strong impact on community structure relative to their own abundance. Keystone predators increase species diversity by limiting the densities of the most competitive species in a community, allowing more species to coexist.
- **Mutualism and parasitism can have community-wide effects (pp. 1119–1120)** Keystone mutualists such as mycorrhizal fungi and nitrogen-fixing bacteria help maintain key ecosystem processes. A parasitic infection that reduces the population of one species may affect other species, sometimes destroying an entire habitat. A parasite may induce in its host a behavioral change that affects both host and parasite, to the parasite's benefit.
- **Interspecific competition influences populations of many species and can affect community structure (pp. 1120–1121)** Exotic species introduced into a community often outcompete native species, but competing species sometimes coexist at reduced densities. Competition may be important in structuring many communities, serving to regulate relative abundance and perhaps species richness.
- **A complex interplay of interspecific interactions and environmental variability characterizes community structure (p. 1121)** Both environmental heterogeneity (patchiness) and temporal heterogeneity (seasonal and diurnal changes) provide an increased number of habitats and more complex community structure. Communities are generally characterized by both environmental heterogeneity and a complex web of interspecific interactions, but the factors most important to community structure vary from one community to another.

DISTURBANCE AND NONEQUILIBRIUM

- **Nonequilibrium resulting from disturbance is a prominent feature of most communities (pp. 1121–1122, FIGURE 53.17)** Disturbances remove organisms from communities, alter resource availability, and create vacated niches that other species can colonize. Increasingly, evidence suggests that disturbance and nonequilibrium instead of stability and equilibrium are the norm for most communities.
- **Humans are the most widespread agents of disturbance (pp. 1122–1124, FIGURE 53.18)** Among all animals, humans create the greatest disturbances in communities, usually reducing species diversity. Humans also prevent some naturally occurring disturbances such as fire, which can be important to community structure.
- **Succession is a process of change that results from disturbance in communities (p. 1124)** Succession involves changes in species composition of a community over ecological time. Primary succession occurs where no soil previously existed; secondary succession begins in an area where soil remains after a disturbance. Succession may involve competition among species with different abilities to utilize available resources, different tolerances of abiotic conditions, and different growth rates and generation times. Inhibition, in which species inhibit the growth of newcomers, and facilitation, in which species of one stage alter the environment in a way that enables species in the next stage to grow, may be involved throughout succession.
- **The nonequilibrial model views communities as mosaics of patches at different stages of succession (pp. 1124–1126)** Ecologists have generally discarded the early idea that succession leads to stable climax communities, in favor of the nonequilibrial model—that succession is an ongoing process driven by disturbance. Major changes in community structure are caused by large or frequent disturbances that result in colonization of the disturbed area by recruitment from distant areas. The dynamic equilibrium hypothesis proposes that species diversity depends mainly on the effects of disturbance on competitive interactions. The immediate disturbance hypothesis posits that diversity is greatest where disturbances are moderately frequent and severe. The nonequilibrial model proposes that diversity results mainly from environmental patchiness caused by abiotic disturbances.

COMMUNITY ECOLOGY AND BIOGEOGRAPHY

Biogeography, the study of the past and present distribution of species and communities, provides a historical perspective for community ecology. The intersection of community ecology and biogeography is fertile ground for analyzing communities in terms of their evolutionary histories and present interactions (p. 1126, FIGURE 53.20).

- **Dispersal and survivability in ecological and evolutionary time account for the geographical ranges of species (p. 1126)** A species is limited to a given range if it never disperses beyond that range, if it disperses but fails to survive in other locations, or if it has retracted from a larger range. Paleontology and historical biogeography have identified cases of retraction from a larger range. Transplant experiments show that some species fail to survive when dispersed to locations outside their range.
- **Species diversity on some islands tends to reach a dynamic equilibrium in ecological time (pp. 1126–1128, FIGURES 53.21–53.22)** An ecological island is any habitat surrounded by an environment not suitable for the species in that habitat. A hypothesis of island biogeography maintains that species richness on an ecological island levels off at some dynamic equilibrium point, where new immigrations are balanced by extinctions. The hypothesis predicts that species richness is directly proportional to size and inversely proportional to distance of the island from the source of colonizers.

SELF-QUIZ

1. The concept of trophic structure of a community emphasizes the
   a. prevalent form of vegetation
   b. keystone predator
   c. feeding relationships within a community
   d. effects of coevolution
   e. species richness of the community
2. According to the concept of competitive exclusion,
   a. two species cannot coexist in the same habitat
   b. extinction or emigration are the only possible results of competitive interactions
   c. intraspecific competition results in the success of the best-adapted individuals
   d. two species cannot share the same realized niche in a habitat
   e. resource partitioning will allow a species to utilize all the resources of its fundamental niche
3. The effect of a keystone predator within a community may be to
   a. competitively exclude other predators from the community
   b. maintain species diversity by preying on the prey species that is the dominant competitor
   c. increase the relative abundance of the most competitive prey species
   d. encourage the coevolution of predator and prey adaptations
   e. create nonequilibrium in species diversity
4. All the following statements are consistent with the nonequilibrial model of community structure *except*
   a. disturbance plays major roles in determining species diversity
   b. species diversity may be increased by certain types of disturbances
   c. species composition and the number of species may continue to change throughout the course of succession

d. succession reaches a climax when the intricate web of interactions allows for the addition of new species only when resources are made available by extinction
e. the course of succession may vary depending on the chance arrival of early colonizers

5. Transplant experiments provide evidence that
   a. transplants always fail
   b. continental drift accounts for the geographical distribution of species
   c. the hypothesis of island biogeography is valid
   d. keystone species maintain community structure
   e. some species can live outside their normal ranges

6. An example of cryptic coloration is the
   a. green color of a plant
   b. bright markings of a poisonous tropical frog
   c. stripes of a skunk
   d. mottled coloring of moths that rest on lichens
   e. bright colors of an insect-pollinated flower

7. An example of Müllerian mimicry is
   a. a butterfly that resembles a leaf
   b. two poisonous frogs that resemble each other in coloration
   c. a minnow with spots that look like large eyes
   d. a beetle that resembles a scorpion
   e. a carnivorous fish with a wormlike tongue that lures prey

8. Predation and parasitism are similar in that both can be characterized as
   a. +/+ interactions
   b. +/− interactions
   c. +/0 interactions
   d. −/− interactions
   e. symbiotic interactions

9. To be certain that two species had coevolved, one would *ideally* need to establish that
   a. the two species originated about the same time
   b. local extinction of one species dooms the other species
   c. each species affects the population density of the other species
   d. one species has adaptations that *specifically* tracked evolutionary change in the other species, and vice versa
   e. the two species are adapted to a common set of environmental conditions

10. According to the hypothesis of island biogeography, species richness would be greatest on an island that is
    a. small and remote
    b. large and remote
    c. large and close to a mainland
    d. small and close to a mainland
    e. environmentally homogeneous

CHALLENGE QUESTION

An ecologist studying desert plants performed the following experiment. She staked out two identical plots that included a few sagebrush plants and numerous small annual wildflowers. She found the same five wildflower species in similar numbers in both plots. Then she enclosed one of the plots with a fence to keep out kangaroo rats, the most common herbivores in the area. After two years, four species of wildflowers were no longer present in the fenced plot, but one wildflower species had increased dramatically. The control plot had not changed significantly. Using the concepts discussed in the chapter, what do you think happened?

SCIENCE, TECHNOLOGY, AND SOCIETY

1. By 1935, hunting and trapping had eliminated wolves from the United States outside Alaska. Since wolves have been protected as an endangered species, they have moved south from Canada and have become reestablished in the Rocky Mountains and northern Great Lakes. Conservationists who would like to speed up this process have reintroduced wolves into Yellowstone National Park. Local ranchers are opposed to bringing back the wolves because they fear predation on their cattle and sheep. What are some reasons for reestablishing wolves in Yellowstone Park? What effects might the reintroduction of wolves have on the ecological communities in the region? What might be done to mitigate the conflicts between ranchers and wolves?
2. Write a paragraph describing one example of how human activities have reduced species diversity in a biological community within your own local region.

FURTHER READING

American Institute of Biological Sciences. "Special Section on Symbiosis." *BioScience,* April 1998. Contains seven articles about symbiotic associations involving microorganisms.
Anderson, I., and R. Nowak. "Australia's Giant Lab." *New Scientist,* February 22, 1997. Traces the use of viruses to reduce populations of introduced European rabbits.
Bergelson, J. "Competition Between Two Weeds." *American Scientist,* November/December 1996. Contains an experimental analysis of factors influencing interspecific competition between plants.
Holmes, B. "Blazing Chainsaws." *New Scientist,* May 24, 1997. Discusses the advantages of logging operations that mimic natural disturbances.
Power, M.E, et al. "Challenges in the Quest for Keystones." *BioScience,* September 1996. Discusses the importance and difficulty of identifying keystone species in communities.
Reice, S. R. "Nonequilibrium Determinants of Biological Community Structure." *American Scientist,* September/October 1994. Are communities always recovering from disturbance?
Thornton, I. *Krakatau: The Destruction and Reassembly of an Island Ecosystem.* Cambridge, MA: Harvard University Press, 1996.
Tilman, D. "The Benefits of Natural Disasters." *Science,* September 13, 1996. Provides a succinct overview of recent research on the significance of disturbance in communities.
Wills, C. "Safety in Diversity." *New Scientist,* March 23, 1996. Explains the role of parasitism in maintaining species richness.

WEB LINKS

Visit the special edition of *The Biology Place* for BIOLOGY, Fifth Edition, at http://www.biology.com/campbell. Go to Chapter 53 for online resources, including learning activities, practice exams, and links to the following web sites:

"National Estuary Program"
The hardest hit wildlife communities are those in and around river estuaries because of human activity. Find out what the EPA is trying to do to protect these communities.

"The Institute for Terrestrial Ecology"
Explore the diverse ecological research projects being conducted at the six research stations of the United Kingdom's Institute for Terrestrial Ecology.

"On Lotka–Volterra Predator–Prey Equations"
This Swedish site introduces you to the equations that describe the population dynamics in a simple predator–prey system. It explains the equation's components and then allows you to test your own model with an interactive simulator.

"Landscape Ecology and Biogeography"
This site provides links to a list of biogeographic resources.

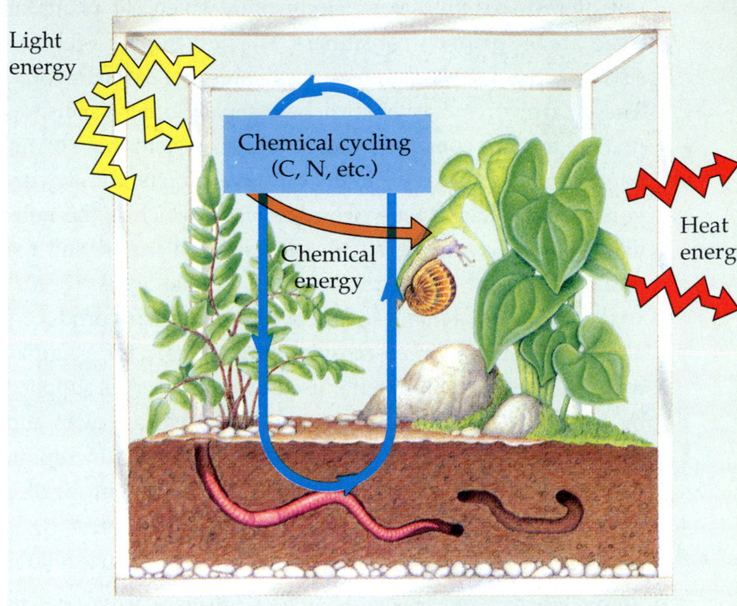

CHAPTER 54

ECOSYSTEMS

Trophic Relationships in Ecosystems

- Trophic relationships determine an ecosystem's routes of energy flow and chemical cycling
- Primary producers include plants, algae, and many species of bacteria
- Many primary and higher-order consumers are opportunistic feeders
- Decomposition interconnects all trophic levels

Energy Flow in Ecosystems

- An ecosystem's energy budget depends on primary productivity
- As energy flows through an ecosystem, much is lost at each trophic level

Cycling of Chemical Elements in Ecosystems

- Biological and geological processes move nutrients among organic and inorganic compartments
- Decomposition rates largely determine the rates of nutrient cycling
- Field experiments reveal how vegetation regulates chemical cycling: *science as a process*

Human Impacts on Ecosystems

- The human population is disrupting chemical cycles throughout the biosphere
- Toxins can become concentrated in successive trophic levels of food webs
- Human activities are causing fundamental changes in the composition of the atmosphere
- The exploding human population is altering habitats and reducing biodiversity worldwide

*An **ecosystem** consists of all the organisms living in a community as well as all the abiotic factors with which they interact. As with populations and communities, the boundaries of ecosystems are usually not discrete. Ecosystems can range from a laboratory microcosm, such as the terrarium depicted here, to lakes and forests. Indeed, many ecologists regard the entire biosphere as a global ecosystem, a composite of all the local ecosystems on Earth.*

The most inclusive level in the hierarchy of biological organization, an ecosystem involves two processes that cannot be fully described at lower levels: energy flow and chemical cycling. Energy enters most ecosystems in the form of sunlight. It is then converted to chemical energy by autotrophic organisms, passed to heterotrophs in the organic compounds of food, and dissipated in the form of heat. Chemical elements such as carbon and nitrogen are cycled between abiotic and biotic components of the ecosystem. Photosynthetic organisms acquire these elements in inorganic form from the air, soil, and water and assimilate them into organic molecules, some of which are consumed by animals. The elements are returned in inorganic form to the air, soil, and water by the metabolism of plants and animals and by other organisms, such as bacteria and fungi, that break down organic wastes and dead organisms. The movements of energy and matter through ecosystems are related because both occur by the transfer of substances through feeding relationships. However, because energy, unlike matter, cannot be recycled, an ecosystem must be powered by a continuous influx of new energy from an external source (the sun). Thus, energy flows through ecosystems, while matter cycles within them.

This chapter describes the dynamics of energy flow and chemical cycling in ecosystems and considers some of the consequences of human intrusions into these processes.

TROPHIC RELATIONSHIPS IN ECOSYSTEMS

Each ecosystem has a **trophic structure** of feeding relationships. Ecologists divide the species in a community or ecosystem into **trophic levels** on the basis of their main source of nutrition.

Trophic relationships determine an ecosystem's routes of energy flow and chemical cycling

The trophic level that ultimately supports all others in an ecosystem consists of autotrophs, or the **primary producers** of the ecosystem. Most primary producers are photosynthetic organisms that use light energy to synthesize sugars and other organic compounds, which they then use as fuel for cellular respiration and as building material for growth. Organisms in trophic levels above the primary producers are heterotrophs that directly or indirectly depend on the photosynthetic output of primary producers. Herbivores, which eat plants or algae, are the **primary consumers**. The next trophic level consists of **secondary consumers**, carnivores that eat herbivores. These carnivores may in turn be eaten by other carnivores that are **tertiary consumers**, and some ecosystems have carnivores of an even higher level. Some consumers, the **detritivores**, derive their energy from **detritus**, which is the nonliving organic material, such as feces, fallen leaves, and the remains of dead organisms, from all trophic levels. Detritivores often form a major link between the primary producers and the consumers in an ecosystem. In streams, for example, much of the organic material used by consumers is supplied by terrestrial plants and enters the ecosystem as leaves and other debris that fall into the water or are washed in by runoff. A crayfish might feed on plant detritus at the bottom of a stream or lake and then be eaten by a fish. In a forest, birds might feed on earthworms that have been feeding on leaf litter in the soil.

An ecosystem's trophic structure determines the routes of energy flow and chemical cycling. The pathway along which food is transferred from trophic level to trophic level, beginning with primary producers, is known as a **food chain** (FIGURE 54.1). As we will discuss later in the chapter, the length of food chains is limited by the amount of energy that gets transferred from one level to the next.

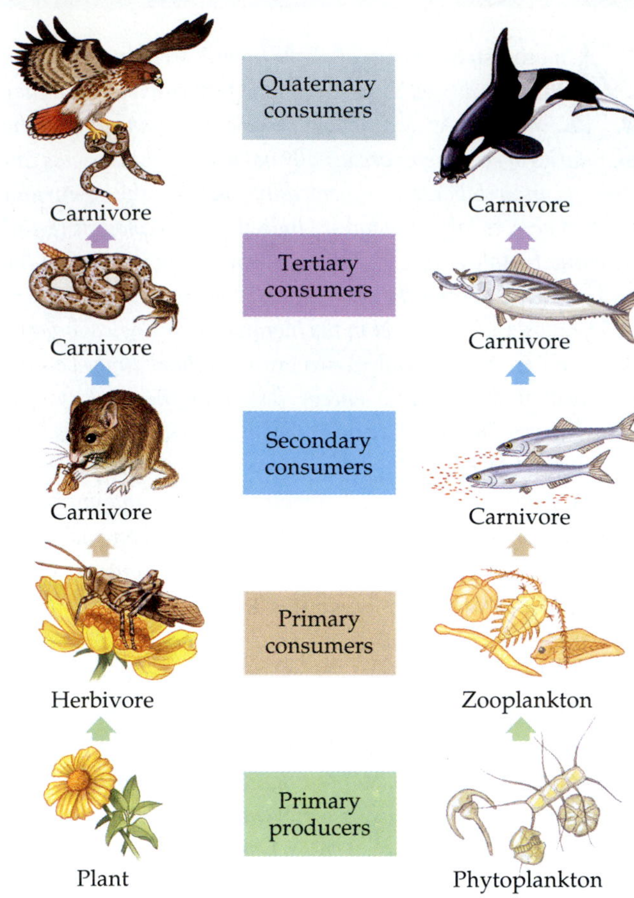

FIGURE 54.1 • Examples of terrestrial and marine food chains. Energy and nutrients pass through the trophic levels of an ecosystem when organisms feed on one another. Decomposers (detritivores), important components of every ecosystem, are not shown here.

54.1 Actually, few ecosystems are so simple that they are characterized by a single, unbranched food chain. Several types of primary consumers usually feed on the same plant species, and one species of primary consumer may eat several different plants. Such branching of food chains occurs at the other trophic levels as well. For example, adult frogs, which are secondary consumers, eat several insect species that may also be eaten by various birds. In addition, some consumers feed at several different trophic levels. An owl, for instance, may eat mice, which are mainly primary consumers, but may also eat some invertebrates; an owl may also feed on snakes, which are strictly carnivorous. Omnivores, including humans, eat producers as well as consumers of different levels. Thus, the feeding relationships in an ecosystem are usually woven into elaborate **food webs** (FIGURE 54.2). The producer ⟶ primary consumer ⟶ secondary consumer chain is thus a simplification of the many permutations that feeding relationships can have.

It is important to make the distinction between ecosystem structure (trophic system) and ecosystem processes, such as production and consumption, that affect energy flow and chemical cycling. In an ecological sense, **production** means the rate of incorporation of energy and materials into the bodies of organisms. Thus, all organisms are producers (although primary producers are sometimes simply called "producers" because their production supports that of all other organisms). **Consumption** is loosely defined but generally refers to the metabolic use for growth and reproduction of assimilated organic materials; in this sense, all organisms, including autotrophs (which metabolize organic compounds they make themselves from materials they assimilate from the environment), are consumers. A third ecosystem process, **decomposition**, is the breakdown of organic materials to inorganic ones. All organisms perform some decomposition; in cellular metabolism, they all break down organic material and release inorganic products, such as carbon dioxide and ammonia, to the environment.

Now let's elaborate on these ecosystem processes by looking at some examples of the main primary producers, consumers, and decomposers in various ecosystems.

Primary producers include plants, algae, and many species of bacteria

The main primary producers in most terrestrial ecosystems are plants. In the limnetic zone of lakes and in the open ocean, phytoplankton (algae and bacteria) are the most important autotrophs, whereas multicellular algae and aquatic plants are

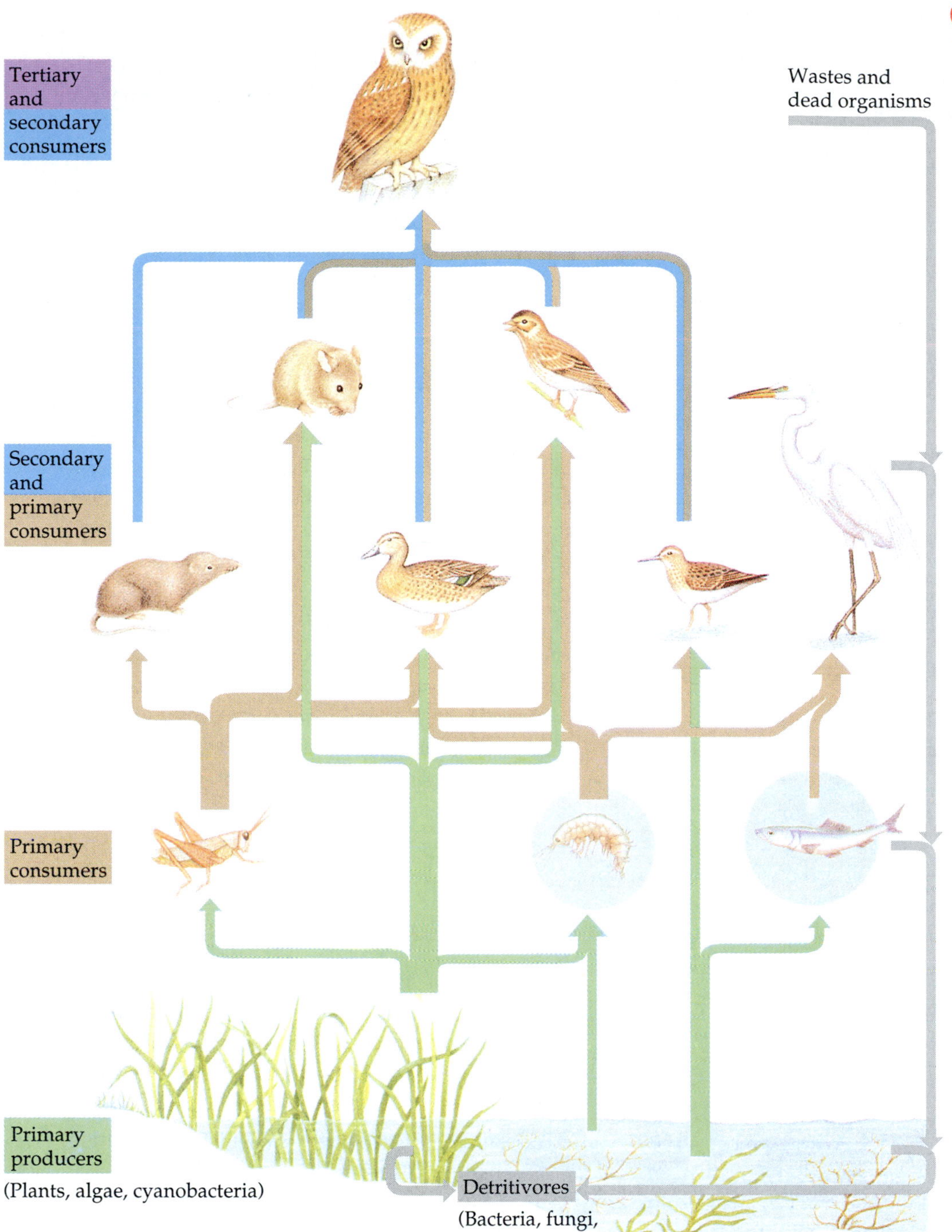

54.1 **FIGURE 54.2 ▪ A food web.** This simplified diagram of feeding relationships does not include all species that feed at each trophic level. Notice that some species feed at more than one trophic level.

often more important primary producers in the littoral (shallow, near-shore) areas of both freshwater and marine ecosystems. In the aphotic zone of the deep sea, however, most life depends on photosynthetic production in the photic zone; energy and nutrients rain down from above in the form of dead plankton and other detritus. One notable exception is the community of organisms that live near hot water vents on the deep-sea floor (see FIGURE 50.14C). Chemoautotrophic bacteria that derive energy from the oxidation of hydrogen sulfide are the main producers in these ecosystems, which are therefore supported by chemical, rather than solar, energy. Because the bacteria need oxygen derived from photosynthesis to oxidize the hydrogen sulfide, these hot water vent ecosystems are not, however, totally independent of solar energy.

Many primary and higher-order consumers are opportunistic feeders

The primary consumers, or herbivores, on land are mostly insects, snails, plant parasites, and certain vertebrates, including grazing mammals and the numerous birds and mammals that eat seeds and fruit. As researchers study the feeding habits of primary consumers, they learn that many of these organisms are opportunistic; they supplement their main diet of autotrophs with some heterotrophic material when it becomes available. Chipmunks and other squirrels, for example, eat mostly seeds and fruit but may also eat bird eggs and nestlings occasionally. Many consumers that mainly eat live organisms also scavenge some dead organic matter.

In aquatic ecosystems, phytoplankton are consumed mainly by zooplankton, which include heterotrophic protists, various small invertebrates (especially crustaceans and, in the ocean, larval stages of many species that live in the benthos as adults), and some fish. In common with terrestrial organisms, many aquatic heterotrophs are opportunistic.

Examples of secondary consumers in terrestrial ecosystems are spiders, frogs, insect-eating birds, carnivorous mammals, and animal parasites. In aquatic habitats, many fish feed on zooplankton and are in turn fed upon by other fish. In the benthic zone of the seas, algae-eating invertebrates are prey to other invertebrates, such as sea stars.

Decomposition interconnects all trophic levels

The organic material that composes the living organisms in an ecosystem is eventually recycled, broken down (decomposed) and returned to the abiotic environment in forms that can be used by autotrophs. Although all organisms carry out decomposition to some degree, an ecosystem's main decomposers are prokaryotes and fungi, which first secrete enzymes that digest organic material and then absorb the breakdown products. Accounting for most of the conversion of organic materials from all trophic levels into inorganic compounds usable by autotrophs, decomposition by prokaryotes and fungi links all trophic levels.

ENERGY FLOW IN ECOSYSTEMS

All organisms require energy for growth, maintenance, reproduction, and, in some species, locomotion. In this section we focus on how energy enters an ecosystem, flows within it, and eventually exits from it.

An ecosystem's energy budget depends on primary productivity

Most primary producers use light energy to synthesize energy-rich organic molecules, which can subsequently be broken down to make ATP. Consumers acquire their organic fuels secondhand (or even thirdhand or fourthhand) through food webs. Therefore, the extent of photosynthetic activity sets the spending limit for the energy budget of the entire ecosystem.

The Global Energy Budget

Every day, Earth is bombarded by about 10^{22} joules (J) of solar radiation (1 J = 0.239 calories). This is the energy equivalent of 100 million atomic bombs the size of the one dropped on Hiroshima. As described in Chapter 50, the intensity of the solar energy striking Earth and its atmosphere varies with latitude, with the tropics receiving the highest input. Most solar radiation is absorbed, scattered, or reflected by the atmosphere in an asymmetrical pattern determined by variations in cloud cover and the quantity of dust in the air over different regions. The amount of solar radiation reaching the globe ultimately limits the photosynthetic output of ecosystems, although photosynthetic productivity is also limited by water, temperature, and nutrient availability.

Much of the solar radiation that reaches the biosphere lands on bare ground and bodies of water that either absorb or reflect the incoming energy. Only a small fraction actually strikes algae, photosynthetic bacteria, and plant leaves, and only some of this is of wavelengths suitable for photosynthesis. Of the visible light that does reach photosynthetic organisms, only about 1% to 2% is converted to chemical energy by photosynthesis, and this efficiency varies with the type of organism, light level, and other factors. Although the fraction of the total incoming solar radiation that is ultimately trapped by photosynthesis is very small, primary producers on Earth collectively create about 170 billion tons of organic material per year—an impressive quantity.

Primary Productivity

The amount of light energy converted to chemical energy (organic compounds) by the autotrophs of an ecosystem during a given time period is called **primary productivity**. Total primary productivity is known as **gross primary productivity (GPP)**. Not all of this productivity is stored as organic material in the growing plants, because the plants use some of the molecules as fuel in their own cellular respiration. Thus, **net primary productivity (NPP)** is equal to the gross primary productivity minus the energy used by the producers for respiration (Rs):

$$\text{NPP} = \text{GPP} - \text{Rs}$$

We may also think of this relationship in terms of the generalized equations for photosynthesis and respiration:

$$6\,CO_2 + 6\,H_2O \underset{\text{Respiration}}{\overset{\text{Photosynthesis}}{\rightleftharpoons}} C_6H_{12}O_6 + 6\,O_2$$

Gross primary productivity results from photosynthesis; net primary productivity is the difference between the yield of photosynthesis and the consumption of organic fuel in respiration.

Net primary productivity is the measurement of interest to us because it represents the storage of chemical energy available to consumers in an ecosystem. Between 50% and 90% of the gross primary productivity of most primary producers remains as net primary productivity after their energetic needs are fulfilled. The NPP-to-GPP ratio is generally smaller for large producers with elaborate nonphotosynthetic structures, such as trees, which support large and metabolically active stem and root systems.

Primary productivity can be expressed in terms of energy per unit area per unit time ($J/m^2/yr$), or as **biomass** (weight) of vegetation added to the ecosystem per unit area per unit time ($g/m^2/yr$). Biomass is usually expressed in terms of the dry weight of organic material because water molecules contain no usable energy, and because the water content of plants varies over short periods of time. An ecosystem's primary productivity should not be confused with the total biomass of photosynthetic autotrophs present at a given time, called the **standing crop biomass**; primary productivity is the *rate* at which the organisms synthesize *new* biomass. Although a forest has a very large standing crop biomass, its primary productivity may actually be less than that of some grasslands, which do not accumulate vegetation because animals consume the plants rapidly and because many of the plants are annuals.

Different ecosystems vary considerably in their productivity as well as in their contribution to the total productivity on Earth (FIGURE 54.3). Tropical rain forests are among the most productive terrestrial ecosystems, and because they cover a large portion of the Earth, they contribute a large proportion of the planet's overall productivity. Estuaries and coral reefs also have very high productivity, but their total contribution to global productivity is relatively small because these ecosystems are not very extensive. The open ocean contributes more primary productivity than any other ecosystem, but this is because of its very large size; productivity per unit area is relatively low. Deserts and tundra also have low productivity.

The factors most important in limiting productivity depend on the type of ecosystem and on seasonal changes in the environment. Productivity in terrestrial ecosystems is generally correlated with precipitation, temperature, and light intensity. Farmers often irrigate their fields, for example, to increase productivity in habitats where the availability of water limits photosynthetic activity; heat and light, as well as water, are provided to plants grown in greenhouses. Productivity generally increases with proximity to the equator because water, heat, and light are more readily available in the tropics.

Inorganic nutrients may also be important in limiting the productivity of many terrestrial ecosystems. Recall from Chapter 37 that plants need a variety of inorganic nutrients, some in relatively large quantities and others only in trace amounts—but all are crucial. Primary productivity removes nutrients from a system, sometimes faster than they are

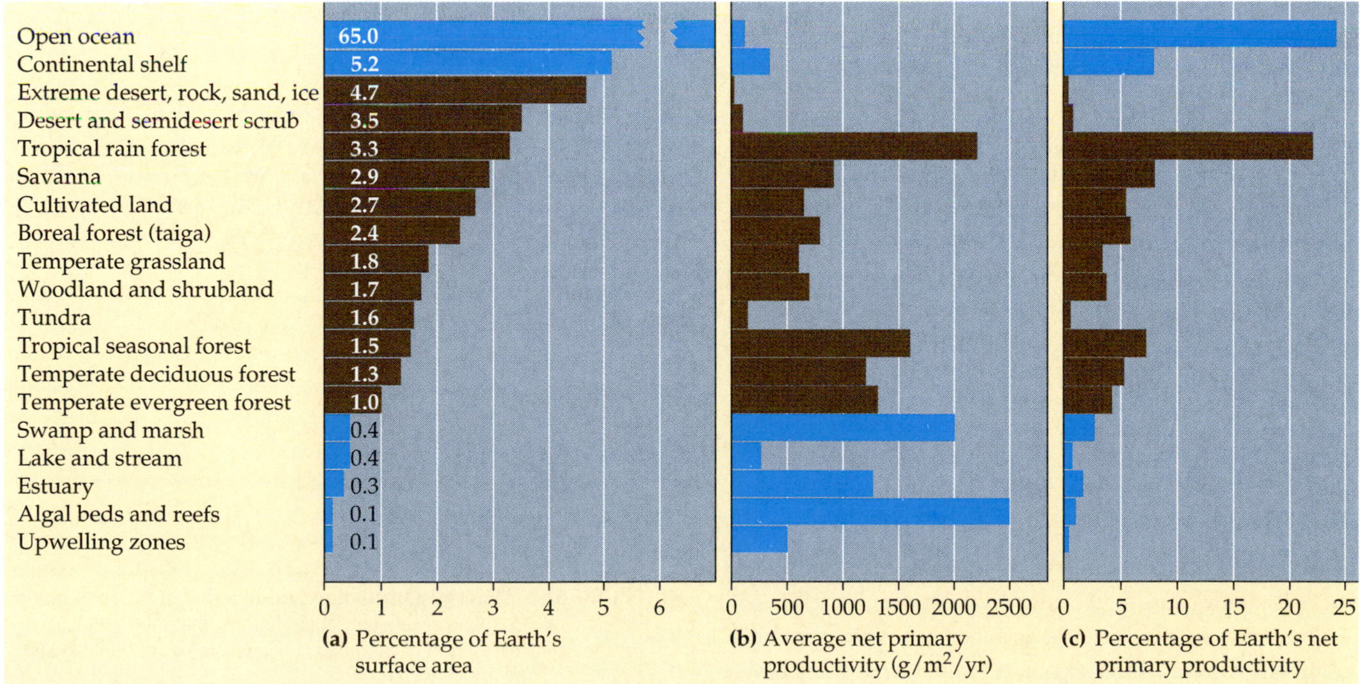

FIGURE 54.3 ▪ Productivity of different ecosystems. (a) The geographical extent and **(b)** the productivity per unit area of different ecosystems determine their total contribution to **(c)** worldwide primary productivity. Open ocean, for example, contributes a lot to the planet's productivity despite its low productivity per unit area because of its large size, whereas tropical rain forest contributes a lot because of its high productivity. (Aquatic ecosystems are color-coded blue in these histograms; terrestrial ecosystems are brown.)

returned. At some point, productivity may slow or cease because a specific nutrient is no longer present in sufficient quantity. It is unlikely that all nutrients will be exhausted simultaneously, so that further productivity is limited by the single nutrient—called the **limiting nutrient**—that is no longer present in adequate supply. Adding other nutrients to the system will not stimulate renewed productivity because they are already present in sufficient quantity. However, the addition of the limiting nutrient will stimulate the system to resume growth until some other nutrient or the same one becomes limiting. In many ecosystems, either nitrogen or phosphorus is a key limiting nutrient. Some evidence also suggests that CO_2 sometimes limits productivity. Later in the chapter we will discuss how human intrusions affecting nutrient balance have disrupted terrestrial and aquatic ecosystems.

Productivity in the seas is generally greatest in the shallow waters near continents and along coral reefs, where light and nutrients are plentiful. In the open oceans, light intensity affects the productivity of phytoplankton communities. Productivity is generally greatest near the surface and declines sharply with depth, as light is rapidly absorbed by water and plankton. The primary productivity per unit area of the open ocean is relatively low because inorganic nutrients, especially nitrogen and phosphorus, are in short supply near the surface; at great depth, where nutrients are abundant, there is insufficient light to support photosynthesis. Phytoplankton communities are most productive where upwelling currents bring nitrogen and phosphorus to the surface. This phenomenon is apparent in antarctic seas, which, in spite of the cold water and low light intensity, are actually more productive than most tropical seas. The chemoautotrophic ecosystems near hot water vents are also very productive, but these communities are not widespread, and their overall contribution to marine productivity is small.

In freshwater ecosystems, as in the open ocean, light intensity and its variation with depth appear to be important determinants of productivity. The availability of inorganic nutrients may also limit productivity in freshwater ecosystems, as it does in the oceans, but the biannual turnover of lakes mixes the waters, carrying nutrients to the well-illuminated surface layers (see FIGURE 50.8).

As energy flows through an ecosystem, much is lost at each trophic level

As energy flows through an ecosystem, much of it is dissipated before it can be consumed by organisms at the next level. If all of the plants in a prairie were piled into a huge mound, another mound of all the herbivores would be dwarfed beside the plants. But the herbivore mound would be much larger than a mound of secondary consumers. The amount of energy available to each trophic level is determined by net primary productivity and the efficiencies with which food energy is converted to biomass in each link of the food chain. As we will see, these efficiencies are never 100%.

Secondary Productivity

The rate at which an ecosystem's consumers convert the chemical energy of the food they eat into their own new biomass is called the **secondary productivity** of the ecosystem. Consider the transfer of organic matter from producers to herbivores, the primary consumers. In most ecosystems, herbivores manage to eat only a small fraction of the plant material produced, and they cannot digest all the organic compounds that they do ingest.

FIGURE 54.4 is a simplified diagram of how the energy a consumer obtains as food might be partitioned. Of the 200 J (48 cal) consumed by a caterpillar, only about 33 J (one-sixth) is used for growth. The rest is passed as feces or used for cellular respiration. Of course, the energy contained in the feces is not lost from the ecosystem; it can still be consumed by detritivores. However, the energy used for respiration is lost from the ecosystem; thus, while solar radiation is the ultimate source of energy for most ecosystems, respiratory heat loss is the ultimate sink. This is why energy is said to flow through, not cycle within, ecosystems. Only the chemical energy stored as growth (or the production of offspring) by herbivores is available as food to secondary consumers. In one sense, our example actually overestimates the conversion of primary productivity into secondary productivity because we did not account for all the plant material that herbivores do not even

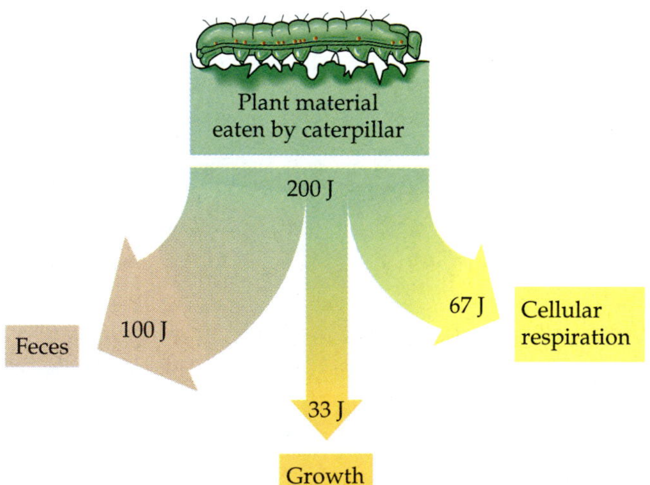

FIGURE 54.4 • Energy partitioning within a link of the food chain. Caterpillars digest and absorb only about half of what they eat, passing the rest as feces. Thus, if a caterpillar consumed leaves containing 200 joules of energy, 100 J would be lost in feces (1 J = 0.239 cal). Approximately two-thirds of the absorbed material, or 67 J, would be used in maintenance—as fuel for cellular respiration, which degrades food molecules to inorganic waste products and heat. The remaining 33 J would be converted into caterpillar biomass and would therefore be available to the next trophic level.

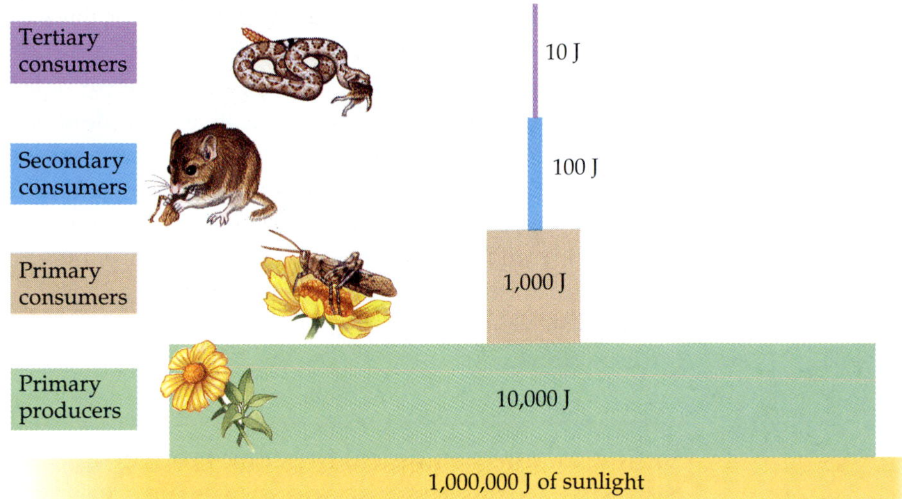

FIGURE 54.5 ▪ An idealized pyramid of net productivity. In the case illustrated here, 10% of the energy available at each trophic level is converted into new biomass in the trophic level above it. Notice that primary producers convert only about 1% of the energy in the sunlight available to them into primary productivity. In actual ecosystems the decline in productivity with the transfer of energy between trophic levels varies with the particular species present; a 10% transfer of energy (ecological efficiency) is a rough average.

consume. The fact that natural ecosystems usually look green—they contain large amounts of plant material—indicates that much net primary productivity is not converted over the short term into secondary productivity.

Carnivores are slightly more efficient at converting food into biomass, mainly because meat is more easily digested than vegetation. But in many cases, secondary consumers use more of the energy they assimilate for cellular respiration, which dramatically decreases the amount of chemical energy available to the next trophic level. Endotherms, in particular, devote a large proportion of their assimilated energy to maintaining a high and constant body temperature.

Ecological Efficiency and Ecological Pyramids

Ecological efficiency is the percentage of energy transferred from one trophic level to the next, or the ratio of net productivity at one trophic level to net productivity at the level below. Ecological efficiencies vary greatly among organisms, usually ranging from 5% to 20%. In other words, 80% to 95% of the energy available at one trophic level never transfers to the next. This multiplicative loss of energy from a food chain can be represented diagrammatically by a **pyramid of productivity**, in which the trophic levels are stacked in blocks, with primary producers forming the foundation of the pyramid. The size of each block is proportional to the productivity of each trophic level (per unit of time). Pyramids of productivity are typically quite bottom-heavy because ecological efficiencies are low (FIGURE 54.5).

One important ecological consequence of decreasing energy transfers through a food web can be represented in a **biomass pyramid**, in which each tier represents the standing crop biomass (the total dry weight of all organisms) in a trophic level. Biomass pyramids generally narrow sharply from producers at the base to top-level carnivores at the apex because energy transfers between trophic levels are so inefficient (FIGURE 54.6a). Some aquatic ecosystems, however, have inverted biomass pyramids, with primary consumers outweighing producers. In the waters of the English Channel, for example, the biomass of zooplankton (consumers) is five times the weight of phytoplankton (producers) (FIGURE 54.6b). Such inverted biomass pyramids occur because the zooplankton consume the phytoplankton so quickly that the producers never develop a large population size or standing crop. Instead, the phytoplankton grow, reproduce, and are consumed rapidly. Phytoplankton have a short **turnover time**, or a low standing crop biomass compared to their productivity:

$$\text{Turnover time} = \frac{\text{Standing crop biomass (mg/m}^2\text{)}}{\text{Productivity (mg/m}^2\text{/day)}}$$

Nevertheless, the *productivity* pyramid for this ecosystem is upright, like the one in FIGURE 54.5, because phytoplankton have a higher productivity than zooplankton.

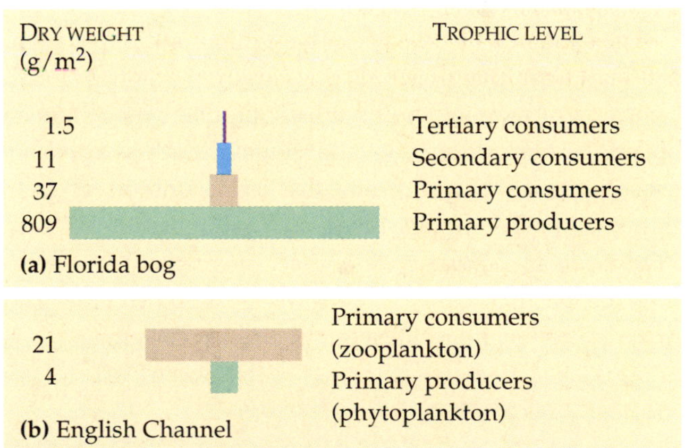

FIGURE 54.6 ▪ Pyramids of standing crop biomass. Numbers denote the dry weight (g/m^2) for all organisms at a trophic level. **(a)** Most biomass pyramids show a sharp decrease in biomass at successively higher trophic levels, as illustrated by data from a bog at Silver Springs, Florida. **(b)** However, in some aquatic ecosystems, such as the English Channel, a small standing crop of primary producers (phytoplankton) supports a larger standing crop of primary consumers (zooplankton). This is because the phytoplankton have a short turnover time: The algae reproduce rapidly and are consumed at a high rate.

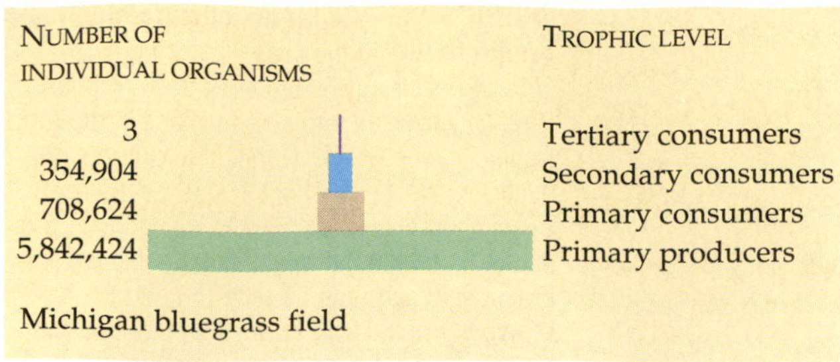

FIGURE 54.7 ▪ A pyramid of numbers. Because top-level predators are generally large animals, the small biomass at the top of a food chain is contained in a relatively small number of individuals. In this pyramid of numbers for a bluegrass field in Michigan, only three top carnivores are supported in an ecosystem based on the productivity of nearly 6 million plants.

The multiplicative loss of energy from food chains severely limits the overall biomass of top-level carnivores that any ecosystem can support. Only about one one-thousandth of the chemical energy fixed by photosynthesis can flow all the way through a food web to a tertiary consumer, such as a hawk or a shark. This explains why food webs usually include only three to five trophic levels; there is simply not enough energy in the webs to support another trophic level.

Because top-level predators tend to be fairly large animals, the limited biomass at the top of an ecological pyramid is concentrated in a relatively small number of individuals. This phenomenon is reflected in a **pyramid of numbers**, in which the size of each block is proportional to the number of individual organisms present in each trophic level (FIGURE 54.7). Populations of top predators are typically very small, and the animals may be widely spaced within their habitats. As a result, many predators are highly susceptible to extinction (as well as to the evolutionary consequences of small population size discussed in Chapter 23).

The dynamics of energy flow have important implications for the human population. Eating meat is a relatively inefficient way of tapping photosynthetic productivity. A human obtains far more calories by eating grains directly as a primary consumer than by processing that same amount of grain through another trophic level and eating grain-fed beef. Worldwide agriculture could, in fact, successfully feed many more people than it does today if we all consumed only plant material, feeding more efficiently as primary consumers.

CYCLING OF CHEMICAL ELEMENTS IN ECOSYSTEMS

Although ecosystems receive an essentially inexhaustible influx of solar energy, chemical elements are available only in limited amounts. (The meteorites that occasionally strike Earth are the only extraterrestrial source of matter.) Life on Earth therefore depends on the recycling of essential chemical elements. Even while an individual organism is alive, much of its chemical stock is rotated continuously, as nutrients are absorbed and waste products released. Atoms present in the complex molecules of an organism at its time of death are returned as simpler compounds to the atmosphere, water, or soil by the action of bacterial and fungal decomposers. This decomposition replenishes the pools of inorganic nutrients that plants and other autotrophs use to build new organic matter. Because nutrient circuits involve both biotic and abiotic components of ecosystems, they are also called **biogeochemical cycles**.

Biological and geological processes move nutrients among organic and inorganic compartments

A chemical's specific route through a biogeochemical cycle varies with the particular element and the trophic structure of the ecosystem. We can, however, recognize two general categories of biogeochemical cycles. Gaseous forms of carbon, oxygen, sulfur, and nitrogen occur in the atmosphere, and cycles of these elements are essentially global. For example, some of the carbon and oxygen atoms a plant acquires from the air as CO_2 may have been released into the atmosphere by the respiration of an animal in some distant locale. Other elements that are less mobile in the environment, including phosphorus, potassium, calcium, and the trace elements, generally cycle on a more localized scale, at least over the short term. Soil is the main abiotic reservoir of these elements, which are absorbed by plant roots and eventually returned to the soil by decomposers, usually in the same general vicinity.

Before examining the details of some individual cycles, let's look at a general model of nutrient cycling that shows the main reservoirs, or compartments, of elements and the processes that transfer elements between reservoirs (FIGURE 54.8). Most nutrients accumulate in four reservoirs, each of which is defined by two characteristics: whether it contains organic or inorganic materials, and whether or not the materials are directly available for use by organisms. One compartment of organic materials is composed of the living organisms themselves and detritus; these nutrients are available to other organisms when consumers feed on one another and when detritivores consume nonliving organic matter. The second organic compartment includes "fossilized" deposits of once-living organisms (coal, oil, and peat), from which nutrients cannot be assimilated directly. Material moved from the living organic compartment to the fossilized organic compartment long ago, when organisms died and were buried by sedimentation over millions of years to become coal or oil.

Nutrients also occur in two inorganic compartments, one in which they are available for use by organisms and one in which they are not. The available inorganic compartment includes matter (elements and compounds) that is dissolved in

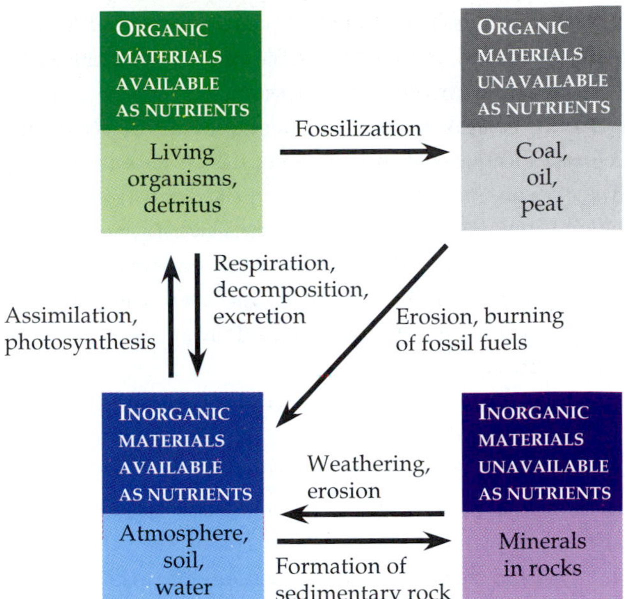

FIGURE 54.8 ▪ A general model of nutrient cycling. Most nutrients cycle within the biosphere among four major compartments, or reservoirs. Each compartment is categorized by type of materials (organic or inorganic) and whether or not the materials are available for use by organisms as a source of nutrients. The biological and geological processes that move nutrients from one compartment to another are indicated on the arrows.

water or present in soil or air. Organisms assimilate materials from this compartment directly and return nutrients to it through the fairly rapid processes of respiration, excretion, and decomposition. Elements in the unavailable inorganic compartment are tied up in rocks. Although organisms cannot tap into this compartment directly, nutrients slowly become available for use through the action of weathering and erosion. Similarly, unavailable organic materials move into the available inorganic nutrients compartment through erosion or when fossil fuels are burned and their elements are vaporized.

Describing biogeochemical cycles in general theory is much simpler than actually tracing elements through these cycles. Not only are ecosystems exceedingly complex, they usually exchange at least some of their materials with other regions. Even in a pond, which has discrete boundaries, several processes add and remove key nutrients. Minerals dissolved in rainwater or runoff from the neighboring land are added to the pond, as are nutrient-rich pollen, fallen leaves, and other airborne material. And, of course, carbon, oxygen, and nitrogen cycle between the pond and the atmosphere. Birds may feed on fish or the aquatic larvae of insects, which derived their store of nutrients from the pond, and some of those nutrients may then be excreted or eliminated on land far from the pond's drainage area. Keeping track of the inflow and outflow is even more challenging in less clearly delineated terrestrial ecosystems. Nevertheless, ecologists have worked out the general schemes for chemical cycling in several ecosystems, often by adding tiny amounts of radioactive tracers that enable the researchers to follow chemical elements through the various biotic and abiotic components of the ecosystems.

The Water Cycle

Although only a very small proportion of Earth's water resides in living material, water is essential to living organisms. In addition to water's direct contributions to the fitness of the environment (see Chapter 3), its movement within and between ecosystems also transfers other materials in several biogeochemical cycles. Driven by solar energy, most of the water cycle occurs between the oceans and the atmosphere through evaporation and precipitation (FIGURE 54.9). The amount of water evaporating from

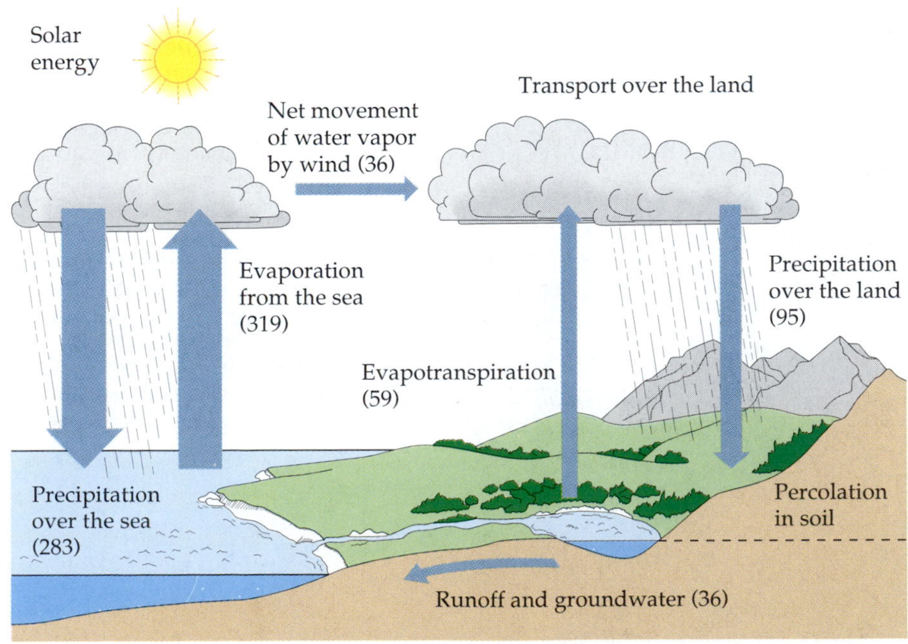

FIGURE 54.9 ▪ The water cycle. On a global scale, evaporation exceeds precipitation over the oceans. The result is a net movement of water vapor, carried by winds, from the ocean to the land. The excess of precipitation over evaporation on land results in the formation of surface and groundwater systems that flow back to the sea, completing the major part of the cycle. Over the sea, evaporation forms most water vapor. On land, however, 90% or more of the vaporization is due to plant transpiration, which together with other types of evaporation is referred to as evapotranspiration. The numbers in this diagram indicate water flow in billion billion (10^{18}) grams per year.

oceans exceeds precipitation over oceans, and the excess water vapor is moved by wind to the land. Over land surfaces, precipitation exceeds evaporation and transpiration, the evaporative loss of water from plants. Runoff and groundwater from the land balance the net flow of water vapor from the ocean to the land. The water cycle differs from the other cycles because most water flux through ecosystems occurs by physical, rather than chemical, processes; during evaporation, transpiration, and precipitation, water maintains its form as H_2O. An ecologically (although not quantitatively) important exception is the chemical transformation of water during photosynthesis.

The Carbon Cycle

54.2 Carbon is a basic constituent of all organic compounds. Its movement through an ecosystem parallels that of energy more closely than do other chemicals; carbohydrates are produced during photosynthesis, and CO_2 is released with energy during respiration. In the carbon cycle, the reciprocal processes of photosynthesis and cellular respiration provide a link between the atmosphere and terrestrial environments (FIGURE 54.10). Plants acquire carbon, in the form of CO_2, from the atmosphere through the stomata of their leaves and incorporate it into the organic matter of their own biomass through the process of photosynthesis. Some of this organic material then becomes the carbon source for consumers. Respiration by all organisms returns CO_2 to the atmosphere.

Although CO_2 is present in the atmosphere at a relatively low concentration (about 0.03%), carbon recycles at a relatively fast rate because plants have a high demand for this gas. Each year, plants remove about one-seventh of the CO_2 in the atmosphere; this is approximately (but not exactly) balanced by respiration. Some carbon may be diverted from the cycle for longer periods. This happens, for example, when it is accumulated in wood and other durable organic material. Metabolic breakdown by detritivores eventually recycles even this carbon to the atmosphere as CO_2, although fires can oxidize such organic material to CO_2 much faster. Some processes, however, can remove carbon from short-term cycling for millions of years; in some environments, detritus accumulates much more quickly than detritivores can break it down. Under certain conditions, these deposits eventually form coal and petroleum that become locked away as unavailable organic nutrients.

The amount of CO_2 in the atmosphere varies slightly with the seasons. Concentrations of CO_2 are lowest during the Northern Hemisphere's summer and highest during winter. This seasonal pulse of CO_2 concentration occurs because there is more land in the Northern Hemisphere than in the Southern, and therefore more vegetation. The vegetation has maximal photosynthetic activity during summer, reducing the global amount of atmospheric CO_2. During winter, the vegetation releases more CO_2 by respiration than it uses for photosynthesis, causing a global increase in the gas (see FIGURE 54.17).

Superimposed on this seasonal fluctuation is a continuing increase in the overall concentration of atmospheric CO_2 caused by the combustion of fossil fuels by humans. From a long-term perspective, this can be viewed as a return to the atmosphere of CO_2 that was removed by photosynthesis long ago. But in the millions of years while this material was effectively out of circulation, a new equilibrium developed in the global carbon cycle. Now that balance is being disrupted, with uncertain consequences that we will consider later.

The cycling of carbon is complicated in aquatic environments by the interaction of CO_2 with water and limestone.

54.2

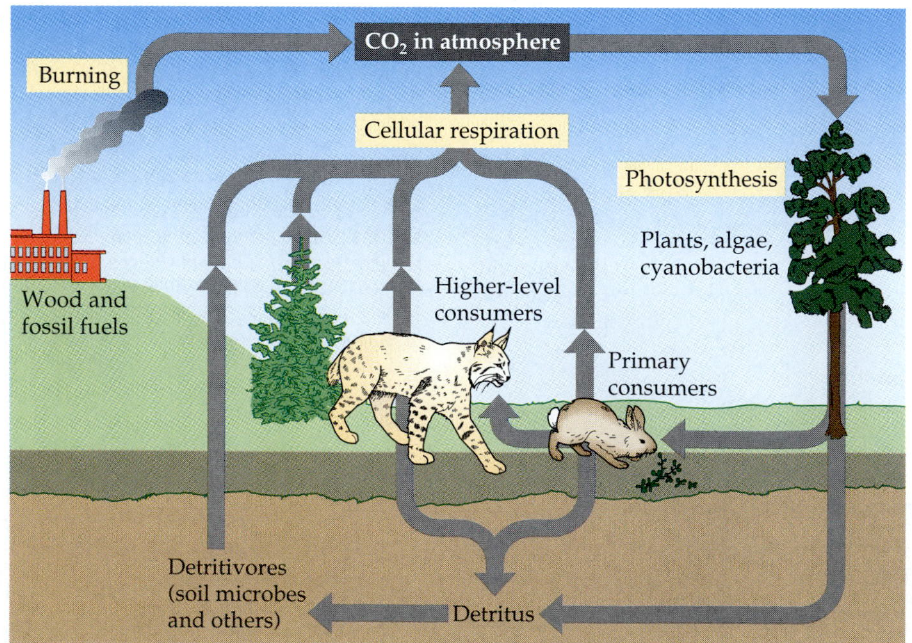

FIGURE 54.10 • The carbon cycle. The reciprocal processes of photosynthesis and cellular respiration are responsible for the major transformations and movements of carbon. A seasonal pulse in atmospheric CO_2 is caused by decreased photosynthetic activity during the Northern Hemisphere's winter. On a global scale, the return of CO_2 to the atmosphere by respiration closely balances its removal by photosynthesis. However, the burning of wood and fossil fuels adds more CO_2 to the atmosphere; as a result, the amount of atmospheric CO_2 is steadily increasing. Atmospheric CO_2 also moves into and out of aquatic systems, where it is involved in a dynamic equilibrium with other inorganic forms, including bicarbonates.

Dissolved carbon dioxide reacts with water to form carbonic acid (H_2CO_3). Carbonic acid in turn reacts with the limestone ($CaCO_3$) that is abundant in many waters, including the ocean, to form bicarbonates and carbonate ions:

$$H_2O + CO_2 \rightleftharpoons H_2CO_3$$

$$H_2CO_3 + CaCO_3 \rightleftharpoons Ca(HCO_3)_2 \rightleftharpoons Ca^{2+} + 2\,HCO_3^-$$

$$2\,HCO_3^- \rightleftharpoons 2\,H^+ + 2\,CO_3^{2-}$$

Bicarbonate Carbonate

As CO_2 is used in photosynthesis in aquatic and marine environments, the equilibrium of this reaction series shifts toward the left, converting bicarbonates back to CO_2. Thus, bicarbonates serve as a CO_2 reservoir. Aquatic autotrophs may also use dissolved bicarbonate directly as their source of carbon. Overall, the amount of carbon present in various inorganic forms in the ocean, not including sediments, is about 50 times that available in the atmosphere. Because of these inorganic reactions of CO_2 in water, as well as its uptake by marine phytoplankton, the ocean may serve as an important "buffer" that can absorb some of the CO_2 being added to the atmosphere by the burning of fossil fuels.

The Nitrogen Cycle

54.3 Nitrogen is another key chemical in ecosystems. It is found in all amino acids, which make up the proteins of organisms. Nitrogen is available to plants only in the form of two minerals: NH_4^+ (ammonium) and NO_3^- (nitrate). Though Earth's atmosphere is almost 80% nitrogen, it is mostly in the form of nitrogen gas (N_2), which is unavailable to plants.

Nitrogen enters ecosystems via two natural pathways, the relative importance of which varies greatly from ecosystem to ecosystem. The first, atmospheric deposition, accounts for approximately 5% to 10% of the usable nitrogen that enters most ecosystems. In this process, NH_4^+ and NO_3^-, the two forms of nitrogen available to plants, are added to soil by being dissolved in rain or by settling as part of fine dust or other particulates. Some plants, such as the epiphytic bromeliads found in the canopy of tropical rain forests, have aerial roots that can take up NH_4^+ and NO_3^- directly from the atmosphere.

The other pathway for nitrogen to enter ecosystems is via **nitrogen fixation**. Only certain prokaryotes can fix nitrogen—that is, convert N_2 into minerals that can be used to synthesize nitrogenous organic compounds such as amino acids. Indeed, prokaryotes are vital links at several points in the nitrogen cycle (FIGURE 54.11). Nitrogen is fixed in terrestrial ecosystems by free-living (nonsymbiotic) soil bacteria as well as by symbiotic bacteria (*Rhizobium*) in the root nodules of legumes and certain other plants (see Chapter 37). Some cyanobacteria fix nitrogen in aquatic ecosystems. Organisms that fix nitrogen, of course, are fulfilling their own metabolic requirements, but the excess ammonia they release becomes available to other organisms. In addition to these natural sources of usable nitrogen, industrial fixation of nitrogen for fertilizer now makes a major contribution to the pool of nitrogenous minerals in terrestrial and aquatic ecosystems.

The direct product of nitrogen fixation is ammonia (NH_3). However, most soils are at least slightly acidic, and NH_3 released into the soil picks up a hydrogen ion (H^+) to form ammonium, NH_4^+, which can be used directly by plants.

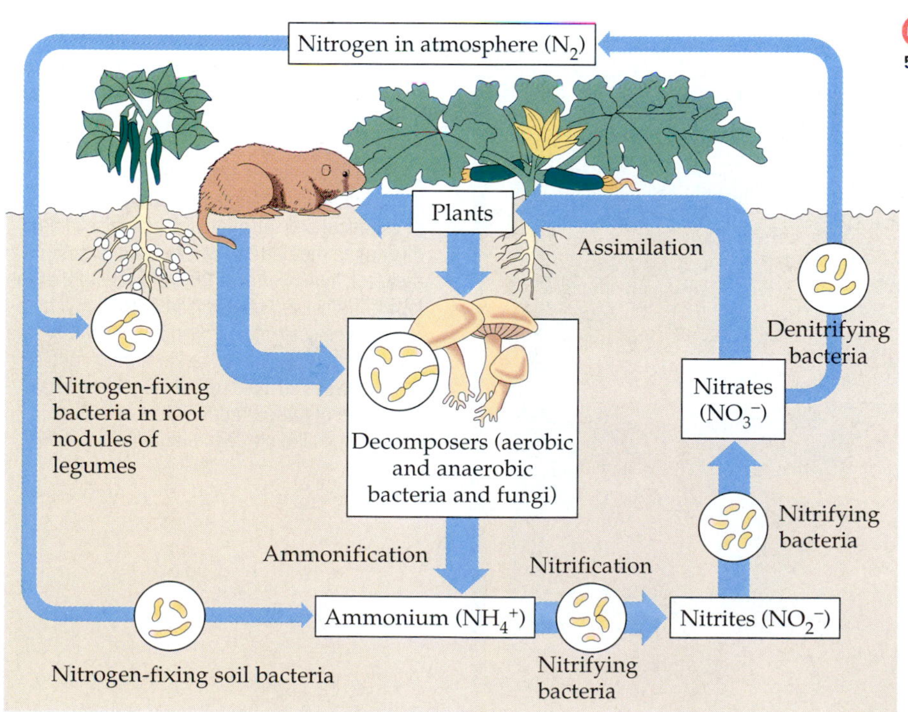

54.3 **FIGURE 54.11 ▪ The nitrogen cycle.** Most of the nitrogen cycling through food webs is taken up by plants in the form of nitrate. Most of this, in turn, comes from the nitrification of ammonium that results from the decomposition of organic material. The addition of nitrogen from the atmosphere and its return via denitrification involve relatively small amounts compared to the local recycling that occurs in the soil or water. The widths of the arrows represent the relative amounts of nitrogen that move between the different sources, but these amounts are extremely variable across ecosystems. Also, in some ecosystems, atmospheric deposition of NH_4^+ and NO_3^- that is dissolved in rain adds nitrogenous minerals to the soil. (The text describes atmospheric deposition, but it is not included in this simplified diagram.)

Because NH_3 is a gas, it can evaporate back to the atmosphere from soils with a pH close to 7 (such as those in the midwestern United States). This NH_3 lost from the soil may then form NH_4^+ in the atmosphere. As a result, NH_4^+ concentrations in rainfall are correlated with soil pH over large regions. This local recycling of nitrogen by atmospheric deposition can be especially pronounced in agricultural areas where both nitrogen fertilizers and lime (a base that decreases soil acidity) are used extensively.

Although plants can use ammonium directly, most of the ammonium in soil is used by certain aerobic bacteria as an energy source; their activity oxidizes ammonium to nitrite (NO_2^-) and then to nitrate (NO_3^-), a process called **nitrification**. Nitrate released from these bacteria can then be assimilated by plants and converted to organic forms, such as amino acids and proteins. Animals can assimilate only organic nitrogen, by eating plants or other animals. Some bacteria can obtain the oxygen they need for metabolism from nitrate rather than from O_2 under anaerobic conditions. As a result of this **denitrification** process, some nitrate is converted back to N_2, returning to the atmosphere. The decomposition of organic nitrogen back to ammonium, a process called **ammonification**, is carried out mainly by bacterial and fungal decomposers. This process recycles large amounts of nitrogen to the soil.

Overall, most of the nitrogen cycling in natural systems involves the nitrogenous compounds in soil and water, not atmospheric N_2. Although nitrogen fixation is important in the buildup of a pool of available nitrogen, it contributes only a tiny fraction of the nitrogen assimilated annually by total vegetation. Nevertheless, many common species of plants depend on their association with nitrogen-fixing bacteria to provide this essential nutrient in a form they can assimilate. The amount of N_2 returned to the atmosphere by denitrification is also relatively small. The important point is that although nitrogen exchanges between soil and atmosphere are significant over the long term, the majority of nitrogen in most ecosystems is recycled locally by decomposition and reassimilation.

The Phosphorus Cycle

Organisms require phosphorus as a major constituent of nucleic acids, phospholipids, and ATP and other energy shuttles, and as a mineral constituent of bones and teeth.

In some respects the phosphorus cycle is simpler than either the carbon or nitrogen cycles. Phosphorus cycling does not include movement through the atmosphere because there are no significant phosphorus-containing gases. In addition, phosphorus occurs in only one important inorganic form, phosphate (PO_4^{3-}), which plants absorb and use for organic synthesis. The weathering of rocks gradually adds phosphate to soil (FIGURE 54.12). After producers incorporate phosphorus into biological molecules, it is transferred to consumers in organic form, and added back to the soil by the excretion of phosphate by animals and by the action of bacterial and fungal decomposers on detritus. Humus and soil particles bind phosphate, so that the recycling of phosphorus tends to be quite localized in ecosystems. However, phosphorus does leach

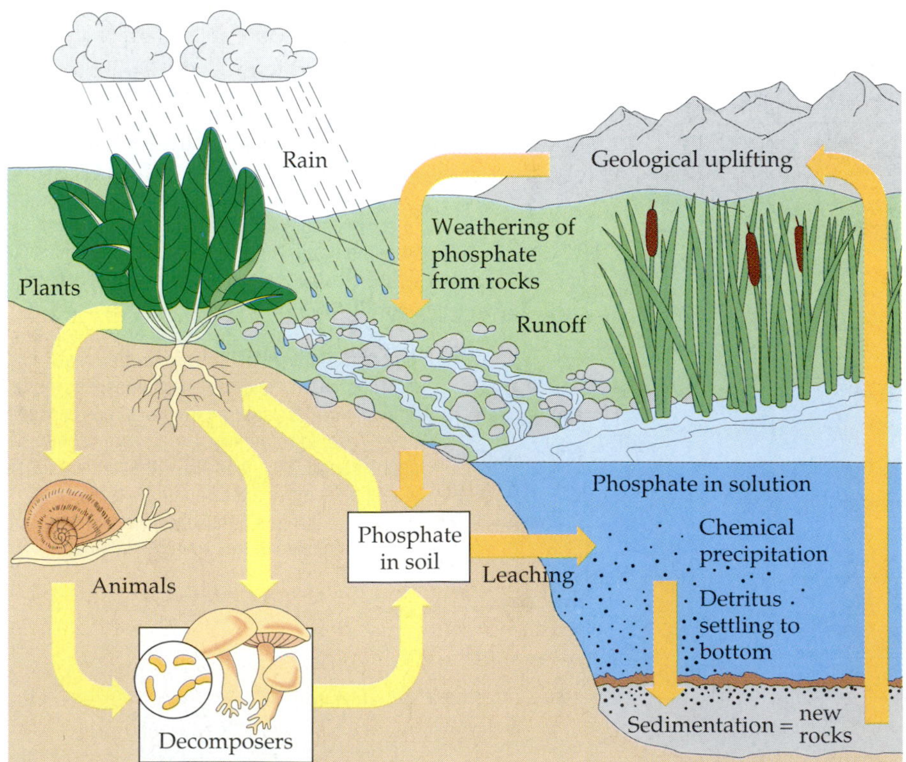

FIGURE 54.12 • The phosphorus cycle. Phosphorus, which does not have an atmospheric component, tends to cycle locally (yellow arrows). Exact rates vary in different systems. Generally, small losses from terrestrial systems caused by leaching are balanced by gains from the weathering of rocks. In aquatic systems, as in terrestrial systems, phosphorus is cycled through food webs. Some phosphorus is lost from the ecosystem because of chemical precipitation or through settling of detritus to the bottom, where sedimentation may lock away some of the nutrient before biological processes can reclaim it. On a much longer time scale, this phosphorus may become available to ecosystems again through geological processes such as uplifting (gold arrows). This general pattern applies to many other nutrients, including trace elements.

FIGURE 54.13 ▪ The experimental eutrophication of a lake. The far basin of this lake was separated from the near basin by a plastic curtain and fertilized with inorganic sources of carbon, nitrogen, and phosphorus. Within two months the fertilized basin was covered with an algal bloom, which appears white in the photograph. The near basin, which was treated with only carbon and nitrogen, remained unchanged. In this case, phosphorus was the key limiting nutrient, and its addition stimulated the explosive growth of algal populations.

into the water table, gradually draining from terrestrial ecosystems to the sea. Severe erosion can hasten this drain, but the weathering of rocks generally keeps pace with the loss of phosphate. Phosphate that reaches the ocean gradually accumulates in sediments and becomes incorporated into rocks that may later be included in terrestrial ecosystems as a result of geological processes that raise the seafloor or lower sea level at a particular location. Thus, most phosphate recycles locally among soil, plants, and consumers on the scale of ecological time, while a parallel sedimentary cycle removes and restores terrestrial phosphorus over geological time. The same general pattern applies to other nutrients that lack atmospheric forms.

In aquatic ecosystems that have not been seriously altered by human activity, a low level of dissolved phosphates often limits primary productivity. In many cases, however, excess (rather than limited) phosphates are a problem. The addition of phosphates in the form of sewage and runoff from fertilized agricultural fields stimulates the growth of algae in aquatic ecosystems, often with negative consequences, such as the eutrophication seen in FIGURE 54.13. For this reason, many states have banned detergents containing phosphates.

A review of FIGURES 54.10–54.12 at this time will reinforce the concept that chemical cycling in ecosystems involves both geological and biological processes.

Decomposition rates largely determine the rates of nutrient cycling

The rates at which nutrients cycle in different ecosystems are extremely variable, mostly due to differences in rates of decomposition. In tropical rain forests, most organic material decomposes in a few months to a few years, while in temperate forests decomposition takes, on average, 4 to 6 years. In the tundra, decomposition can take 50 years, and in aquatic ecosystems, where most decomposition occurs in anaerobic bottom muds, it may occur even more slowly. The temperature and the availability of water and O_2 all affect rates of decomposition, and thus nutrient cycling times. Other factors that can influence nutrient cycling are local soil chemistry and the frequency of fires.

In some parts of a tropical rain forest, key nutrients such as phosphorus occur in the soil at levels far below those typical of a temperate forest. At first this might seem paradoxical because tropical forests generally have very high productivity. The key to this apparent riddle is rapid decomposition in tropical areas because of the warm temperatures and abundant precipitation. In addition, the immense biomass of these forests creates a high demand for nutrients, which are absorbed almost as soon as they become available through the action of decomposers. As a result of rapid decomposition, relatively little organic material accumulates as leaf litter on the floor of tropical rain forests; about 75% of the nutrients in the ecosystem are present in the woody trunks of trees, and about 10% are contained in the soil. The relatively low concentrations of some nutrients in the soil of tropical rain forests result from a fast cycling time, not an overall scantiness of these elements in the ecosystem.

In temperate forests, where decomposition is much slower, the soil may contain 50% of all the organic material in the ecosystem. The nutrients present in temperate forest detritus and soil may remain there for fairly long periods of time before being assimilated by plants.

In aquatic ecosystems, bottom sediments are comparable to the detritus layer in terrestrial ecosystems, except for their very slow decomposition and the fact that algae and aquatic plants usually assimilate nutrients directly from the water. Thus, these sediments often constitute a nutrient sink, and aquatic ecosystems can only be very productive if there is interchange between the bottom layers of water and the surface (see FIGURE 50.8).

Field experiments reveal how vegetation regulates chemical cycling: *science as a process*

Many research groups are conducting **long-term ecological research (LTER)** to track the dynamics of natural ecosystems over relatively long periods of time. In this section we examine an example of LTER as a capstone to our study of chemical cycling in ecosystems.

Since 1963, a team of scientists has been conducting a long-term study of nutrient cycling in a forest ecosystem under both natural conditions and after vegetation is removed. The study site is the Hubbard Brook Experimental Forest in the White Mountains of New Hampshire. It is a deciduous forest with several valleys, each drained by a small creek that is a tributary of Hubbard Brook. Bedrock impenetrable to water is close to the surface of the soil, and each valley constitutes a watershed that can drain only through its creek.

The research team first determined the mineral budget for each of six valleys by measuring the inflow and outflow of several key nutrients. They collected rainfall at several sites to measure the amount of water and dissolved minerals added to the ecosystem. To monitor the loss of water and minerals, they constructed small concrete dams, each with a V-shaped spillway, across the creek at the bottom of each valley (FIGURE 54.14a). About 60% of the water added to the ecosystem as rainfall and snow exits through the stream, and the remaining 40% is lost by transpiration from plants and evaporation from the soil.

Preliminary studies confirmed that internal cycling within a terrestrial ecosystem conserves most of the mineral nutrients. Mineral inflow and outflow balanced and were relatively small compared with the quantity of minerals being recycled within the forest ecosystem. For example, only about 0.3% more Ca^{2+} left a valley via its creek than was added by rainwater, and this small net loss was probably replaced by chemical decomposition of the bedrock. During most years the forest actually registered small net gains of a few mineral nutrients, including nitrogenous ones.

In 1966, one of the valleys, with an area of 15.6 hectares, was completely logged and then sprayed with herbicides for three years to prevent regrowth of plants (FIGURE 54.14b). All the original plant material was left in place to decompose. The inflow and outflow of water and minerals in the experimentally altered watershed were compared with those in a control watershed for 3 years. Water runoff from the altered watershed increased by 30% to 40%, apparently because there were no plants to absorb and transpire water from the soil. Net losses of minerals from the altered watershed were huge. The concentration of Ca^{2+} in the creek increased fourfold, for

(a)

(b)

FIGURE 54.14 • Nutrient recycling in the Hubbard Brook Experimental Forest: an example of long-term ecological research. **(a)** Concrete dams built across streams at the bottom of watersheds enabled researchers to monitor the outflow of water and nutrients from the ecosystem. Because these small dams were anchored to the impervious bedrock, all water draining from the valleys had to pass through the V-shaped spillways at the dams. **(b)** Some watersheds were completely logged to study the effects of the loss of vegetation on drainage and nutrient cycling. All the logs and rubble from the clear-cutting operation were left in place, and care was taken not to disturb the soil any more than necessary.

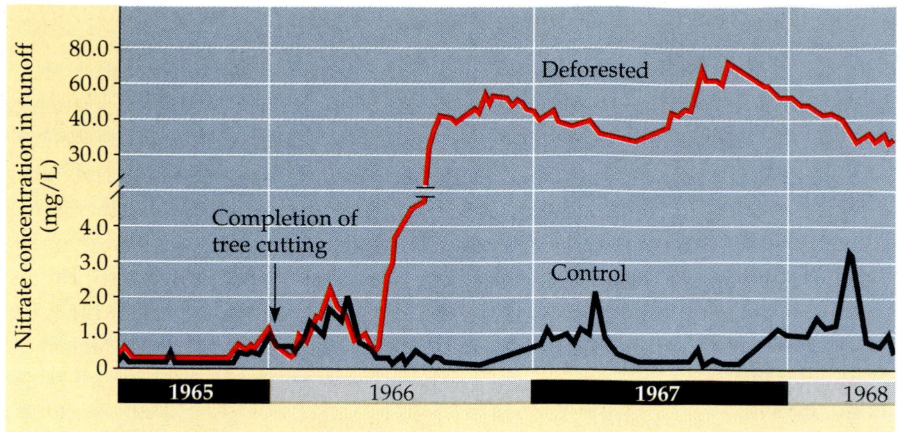

FIGURE 54.15 ▪ The loss of nitrate from a deforested watershed in the Hubbard Brook Experimental Forest. The concentration of nitrate in runoff from the deforested watershed was 60 times greater than in a control unlogged watershed.

example, and the concentration of K^+ increased by a factor of 15. Most remarkable was the loss of nitrate, which increased in concentration in the creek 60-fold (FIGURE 54.15). Not only was this vital mineral nutrient drained from the ecosystem, but nitrate in the creek reached a level considered unsafe for drinking water.

This study demonstrated that the amount of nutrients leaving an intact forest ecosystem is controlled by the plants themselves; when plants are not present to retain them, nutrients are lost from the system. These effects are almost immediate, occurring within a few months of logging, and continue as long as plants are absent.

After 30 years, data from Hubbard Brook point to some other long-term trends. It now appears that none of the watersheds was undisturbed by human activity even when the study began. For instance, before and during the study, acid precipitation seems to have caused an increase in the levels of Ca^{2+} in stream water (see Chapter 3). Apparently, since the 1950s acid rain and snow have dissolved most of the Ca^{2+} from the forest soil, and the streams have carried it away. By the 1990s, forest plants at Hubbard Brook had virtually stopped adding new growth, apparently because of a lack of Ca^{2+}.

While the Hubbard Brook study and several other LTER projects were designed to assess natural ecosystem dynamics, the results continue to provide important insight into the mechanisms by which human activities affect these processes. In the next two sections we will consider other examples of human intrusion.

HUMAN IMPACTS ON ECOSYSTEMS

As the human population has grown to an immense size, our activities and technological capabilities have intruded in one way or another into the dynamics of most ecosystems. Even where we have not completely destroyed a natural system, our actions have disrupted the trophic structure, energy flow, and chemical cycling of ecosystems in most areas of the world. The effects are sometimes local or regional, but the ecological impact of humans can be widespread or even global. For example, acid precipitation may be carried by prevailing winds and fall hundreds or thousands of miles from the smokestacks emitting the chemicals that produce it.

The human population is disrupting chemical cycles throughout the biosphere

Human activity often intrudes in nutrient cycles by removing nutrients from one part of the biosphere and adding them to another. This may result in the depletion of key nutrients in one area, excesses in another place, and the disruption of the natural equilibrium of chemical cycles in both locations. For example, nutrients in the soil of croplands soon appear in the wastes of humans and livestock, and then appear in streams and lakes through runoff from fields and discharge as sewage. Someone eating a piece of broccoli in Washington, D.C., is consuming nutrients that only days before might have been in the soil in California; and a short time later, some of these nutrients will be in the Potomac River on their way to the sea, having passed through an individual's digestive system and the local sewage facilities.

Humans have intruded on nutrient cycles to such an extent that it is no longer possible to understand any cycle without taking human effects into account. In addition to transporting nutrients from one location to another, we have added entirely new materials, many of them toxic, to ecosystems. Next we examine a few examples of the impacts of humans on chemical cycles in ecosystems.

Agricultural Effects on Nutrient Cycling

Food production for Earth's growing human population has many effects on ecosystem dynamics, ranging from near-elimination of many fish species in some areas by overharvesting, to the spread of toxic compounds for controlling pests in croplands, to the depletion of surface and groundwater supplies by irrigation. Let's focus on how agriculture affects nutrient cycling.

After natural vegetation is cleared from an area, crops may be grown for some time without additional supplementation of nutrients because of the existing reserve of nutrients in the soil. However, in agricultural ecosystems a substantial fraction of these nutrients is not recycled but is exported from the area in the form of crop biomass. The "free" period for crop production—when there is no need to add nutrients to the soil—varies greatly. When some of the early North American prairie lands were first tilled, for example, good crops could be produced for many years because the large store of organic materials in the soil continued to decompose and provide nutrients. By contrast, some cleared land in the tropics can be farmed for only one or two years because so few of the ecosystems' nutrients are contained in the soil. Eventually, in any area under intensive agriculture, the natural store of nutrients becomes exhausted. When this happens, fertilizer must be added. The industrially synthesized fertilizers used extensively today are produced at considerable expense of both money and energy.

Agriculture has a great impact on the nitrogen cycle. Cultivation—breaking up and mixing the soil—increases the rate of decomposition of organic matter, releasing usable nitrogen that is then removed from the ecosystem when crops are harvested. As we saw in the case of Hubbard Brook, ecosystems from which living plants are removed lose nitrogen not only because it is removed with the plants themselves, but because without plants to take them up, nitrates continue to be leached from the ecosystem. Industrially synthesized fertilizer is used to make up for the loss of usable nitrogen from agricultural ecosystems. Recent studies indicate that human activities have approximately doubled the globe's supply of fixed nitrogen available to primary producers. The main cause is industrial nitrogen fixation for fertilizers, but increased legume cultivation and burning are also important. (Fire releases nitrogen compounds stored in soils and vegetation, thereby enhancing cycling of nitrogen compounds available to photosynthesizers.) A considerable portion of excess nitrogen in the soil leaches downward, and toxic levels of nitrates in groundwater have shown up in some areas, such as eastern Europe and the midwestern United States. Increased nitrogen fixation is also associated with a greater release of nitrogen compounds (N_2 and nitrogen oxides) into the air by denitrifiers (see FIGURE 54.11). Nitrogen oxides can contribute to atmospheric warming (see greenhouse effect, p. 1148), to the depletion of atmospheric ozone, and in some ecosystems to acid precipitation. Much of the nitrogen in fertilizers also ends up in surface waters where it can stimulate algal and bacterial growth, a problem we examine next.

Accelerated Eutrophication of Lakes

As we discussed in Chapter 50, lakes are classified on a scale of increasing nutrient availability as oligotrophic, mesotrophic, or eutrophic (see FIGURE 50.11). In an oligotrophic lake, primary productivity is relatively low because the mineral nutrients required by phytoplankton are scarce. In other lakes, basin and watershed characteristics cause the addition of more nutrients. These nutrients are captured by the primary producers and then continuously recycled through the lake's food webs. Thus, the overall productivity is higher in mesotrophic lakes and highest in eutrophic (Gr., "well nourished") ones.

Human intrusion has disrupted freshwater ecosystems by what is termed *cultural eutrophication.* Sewage and factory wastes; runoff of animal waste from pastures and stockyards; and the leaching of fertilizer from agricultural, recreational, and urban areas has overloaded many streams, rivers, and lakes with inorganic nutrients. This enrichment often results in an explosive increase in the density of photosynthetic organisms (see FIGURE 54.13). Shallower areas become weed-choked, making boating and fishing impossible. Large "blooms" of algae and cyanobacteria become common, sometimes resulting in increased oxygen production during the day but reduced oxygen levels at night because of respiration by these large populations of organisms. As the photosynthetic organisms die and organic material accumulates at the lake bottom, detritivores use all the oxygen in the deeper waters. All these effects may make it impossible for some organisms to survive. For example, cultural eutrophication of Lake Erie wiped out commercially important fishes such as blue pike, whitefish, and lake trout by the 1960s. Since then, tighter regulations on the dumping of wastes into the lake have enabled some fish populations to rebound, but many of the native species of fishes and invertebrates have not recovered. And the U.S. Congress has passed legislation that loosens the standards for ensuring water quality.

Toxins can become concentrated in successive trophic levels of food webs

Humans produce an immense variety of toxic chemicals, including thousands of synthetics previously unknown in nature, that are dumped into ecosystems with little regard for the ecological consequences. Many of these poisons cannot be degraded by microorganisms and consequently persist in the environment for years or even decades. In other cases, chemicals released into the environment may be relatively harmless but are converted to more toxic products by reaction with other substances or by the metabolism of microorganisms. For example, mercury, a by-product of plastic production, was once routinely expelled into rivers and the sea in an insoluble form. Bacteria in the bottom mud converted the waste to methyl mercury, an extremely toxic soluble compound that

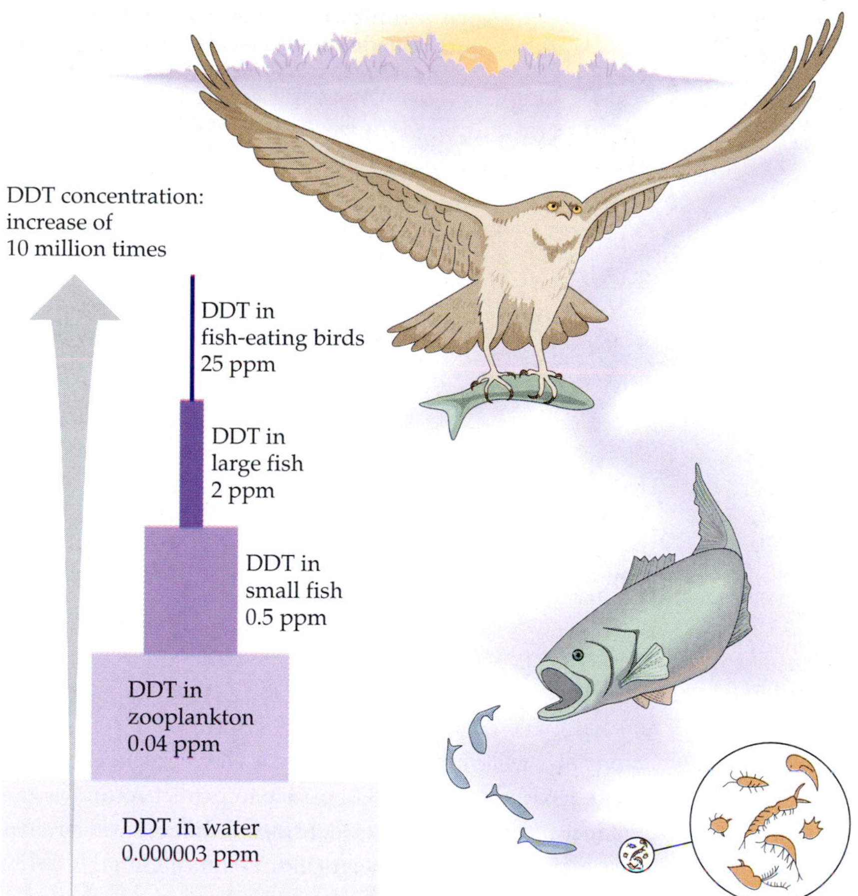

FIGURE 54.16 ▪ Biological magnification of DDT in a food chain. DDT concentration in a Long Island Sound food chain was magnified by a factor of about 10 million, from just 0.000003 parts per million (ppm) as a pollutant in seawater to a concentration of 25 ppm in a bird at the top of this food pyramid, the fish-eating osprey.

then accumulated in the tissues of organisms, including humans who consumed fish from the contaminated waters.

Organisms acquire toxic substances from the environment along with nutrients and water. Some of the poisons are metabolized and excreted, but others accumulate in specific tissues, especially fat. An example of a class of industrially synthesized compounds that act in this manner are the chlorinated hydrocarbons, which include many pesticides, such as DDT, and the industrial chemicals called PCBs (polychlorinated biphenols). Current research is implicating many of these compounds and others in endocrine system disruption in a large number of animal species, including humans. One of the reasons these toxins are so harmful is that they become more concentrated in successive trophic levels of a food web, a process called **biological magnification**. Magnification occurs because the biomass at any given trophic level is produced from a much larger biomass ingested from the level below. Thus, top-level carnivores tend to be the organisms most severely affected by toxic compounds that have been released into the environment.

A well-known example of biological magnification involves DDT, which has been used to control insects such as mosquitoes and agricultural pests. DDT persists in the environment and is transported by water to areas far from where it is applied, and it rapidly became a global problem. Because the compound is soluble in lipids, it collects in the fatty tissues of animals, and its concentration is magnified in higher trophic levels (FIGURE 54.16). Traces of DDT have been found in nearly every organism tested; it has even been found in human breast milk throughout the world. One of the first signs that DDT was a serious environmental problem was a decline in the populations of pelicans, ospreys, and eagles, birds that feed at the top of food chains. The accumulation of DDT (and DDE, a product of its partial breakdown) in the tissues of these birds interfered with the deposition of calcium in their eggshells, a trend that it now seems had already begun, possibly because of other environmental contaminants. When these birds tried to incubate their eggs, the weight of the parents broke the affected eggs, resulting in catastrophic declines in their reproduction rates. DDT was banned in the United States in 1971, and a dramatic recovery in populations of the affected bird species followed. The pesticide is still used in other parts of the world, however.

Human activities are causing fundamental changes in the composition of the atmosphere

Many human activities release a variety of gaseous waste products. We once thought the vastness of the atmosphere could absorb these materials without significant consequences, but we now know that the finite volume of the atmosphere means that human intrusions can cause

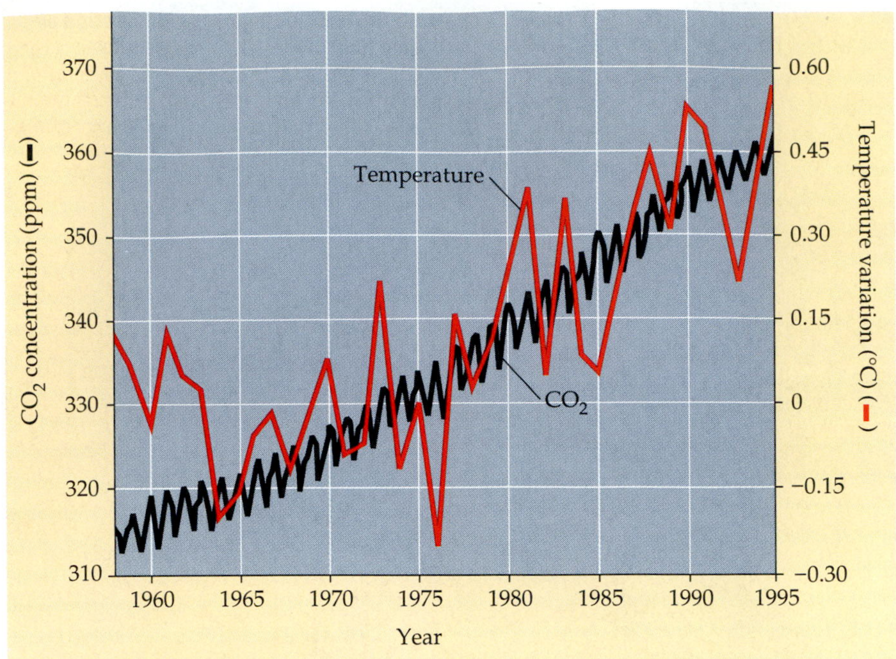

FIGURE 54.17 ▪ The increase in atmospheric carbon dioxide and average temperatures from 1958 to 1995. In addition to normal seasonal fluctuations in CO_2 levels (black), this graph shows a steady increase in the total amount of CO_2 in the atmosphere. These measurements are being taken at a relatively remote site in Hawaii, where the air is free from the variable short-term effects that occur near large urban areas. Though average temperatures over the same time period fluctuate a great deal (red), there is a warming trend. Climatologists predict that temperatures could rise about 2°C over the next 100 years if atmospheric CO_2 levels continue to rise at the current rate.

fundamental changes in its composition and its interactions with the rest of the biosphere. One pressing problem that relates directly to one of the nutrient cycles we examined is the rising level of carbon dioxide in the atmosphere.

Carbon Dioxide Emissions and the Greenhouse Effect

Since the Industrial Revolution, the concentration of CO_2 in the atmosphere has been increasing as a result of the combustion of fossil fuels and the burning of enormous quantities of wood removed by deforestation. Various methods have estimated that the average carbon dioxide concentration in the atmosphere before 1850 was about 274 parts per million (ppm). When a monitoring station on Hawaii's Mauna Loa peak began making very accurate measurements in 1958, the CO_2 concentration was 316 ppm (FIGURE 54.17). Today, the concentration of CO_2 in the atmosphere exceeds 360 ppm, an increase of about 14% since the measurements began. If CO_2 emissions continue to increase at the present rate, by the year 2075 the atmospheric concentration of this gas will be double what it was at the start of the Industrial Revolution.

Increased productivity by vegetation is one predictable consequence of increasing CO_2 levels. In fact, when CO_2 concentrations are raised in experimental chambers such as greenhouses, most plants respond with increased growth. However, because C_3 plants are more limited than C_4 plants by CO_2 availability (see Chapter 10), one effect of increasing CO_2 concentrations on a global scale may be the spread of C_3 species into terrestrial habitats previously favoring C_4 plants. This may have important agricultural implications. For example, corn, a C_4 plant and the most important grain crop in the United States, may be replaced on farms by wheat and soybeans, C_3 crops that will outproduce corn in a CO_2-enriched environment. However, it is not yet possible to predict accurately the gradual and complex effects that rising CO_2 levels will have on species composition in nonagricultural communities.

One factor that complicates predictions about the long-term effects of rising atmospheric CO_2 concentration is its possible influence on Earth's heat budget. Much of the solar radiation that strikes the planet is reflected back into space. Although CO_2 and water vapor in the atmosphere are transparent to visible light, they intercept and absorb much of the reflected infrared radiation, rereflecting it back toward Earth. This process retains some of the solar heat. If it were not for this **greenhouse effect**, the average air temperature at Earth's surface would be −18°C. The marked increase in atmospheric CO_2 concentrations during the last 150 years concerns ecologists and environmentalists because of its potential effect on global temperature.

Scientists continue to construct mathematical models in attempts to predict how increasing levels of atmospheric CO_2 will affect global temperatures. To date, no models are sophisticated enough to include all the biotic and abiotic factors that can influence atmospheric gas concentrations and temperature (for example, cloud cover, CO_2 uptake by photosynthetic organisms, the effects of particles in the air). However, a number of studies predict a doubling of CO_2 concentration by the end of the twenty-first century and an associated average temperature increase of about 2°C. Supporting these models is a correlation between CO_2 levels and temperatures in prehistoric times. Climatologists can actually measure CO_2 levels in bubbles trapped in glacial ice at different times in Earth's history. Past temperatures are inferred by several methods, one of which is described in the Methods Box.

METHODS: Using Fossil Pollen Samples to Deduce Past Climates

To deduce what Earth's climate was like in the past, paleoecologists study the ecology of fossil communities. In this case, based on the work of Margaret Davis, Professor of Ecology at the University of Minnesota, the fossils are pollen grains preserved in the sediments at the bottom of lakes and bogs. Sediments are deposited throughout the life of a lake; thus, the pollen at each level is a record of the surrounding vegetation that was present when that sediment was deposited. Each cubic centimeter of lake mud can hold about 100,000 to 200,000 pollen grains, which retain their original shape and ornamentation (FIGURE a) and can usually be classified, at least at the genus level.

Dr. Davis takes cores of sediments from beneath lakes and bogs (FIGURE b), usually during the winter, when it is possible to bring coring equipment onto the ice. In the laboratory, her research team uses radiometric dating methods to determine the age of different levels of the core (see FIGURE 25.2), removes core samples of different ages, and dissolves the matrix of material surrounding the pollen grains. By analyzing the pollen, paleoecologists can reconstruct the vegetation that was present around the lake at any time during the period represented by the core.

From hundreds of cores taken throughout eastern North America, paleoecologists have reconstructed the historical distributions of populations of many plant species. The maps in FIGURE c were constructed using data from these cores to infer the movement of spruce populations over the past 18,000 years, as the glaciers that had covered much of the continent retreated. These maps show where high, medium, and low amounts of spruce pollen (represented by dark, medium, and light green, respectively) were found in cores. The receding glacier is shown in white. Using knowledge of climatic conditions that spruce can tolerate, it is possible to infer mean temperatures during the time period covered by the cores. And studying how climatic changes in the past affected communities is helping ecologists predict the consequences of the current global warming trend.

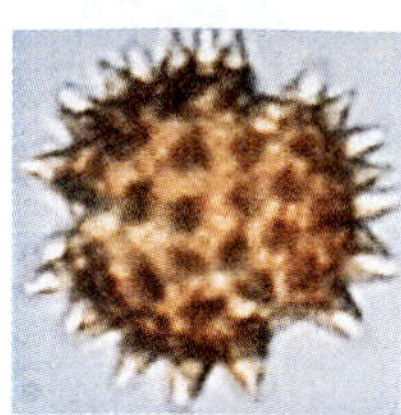

10 μm

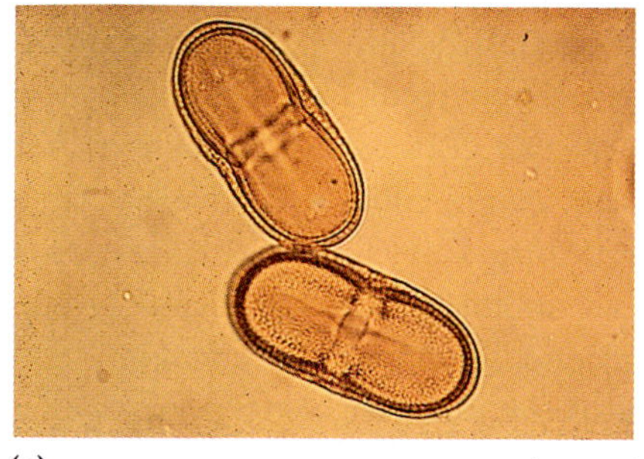

(a) 10 μm

(b)

Spruce (*Picea*)

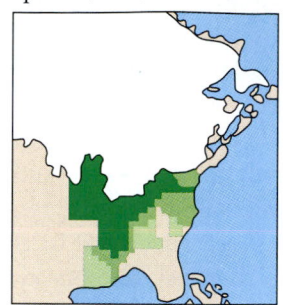

18,000 years ago

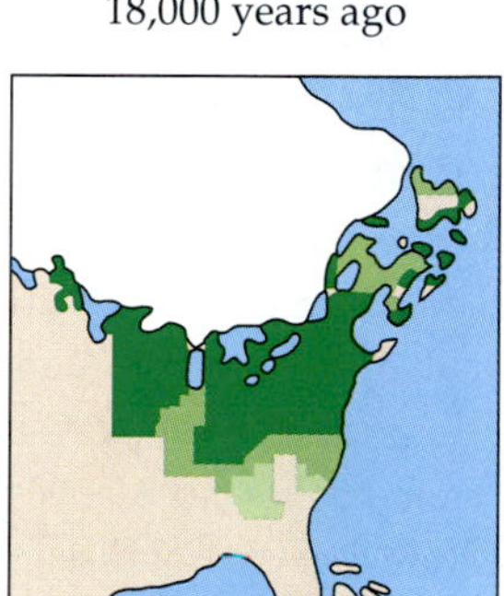

12,000 years ago

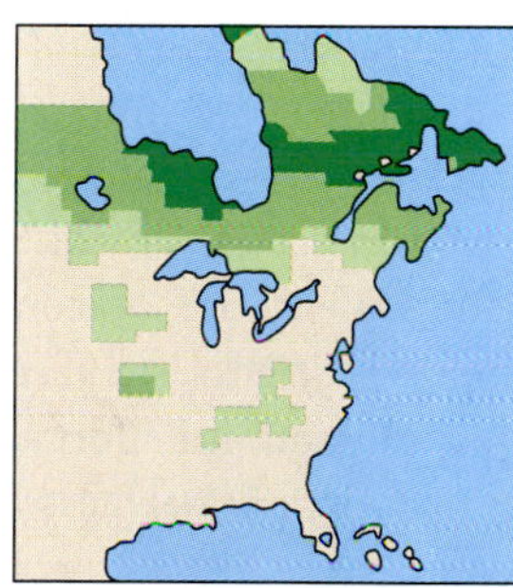

6,000 years ago

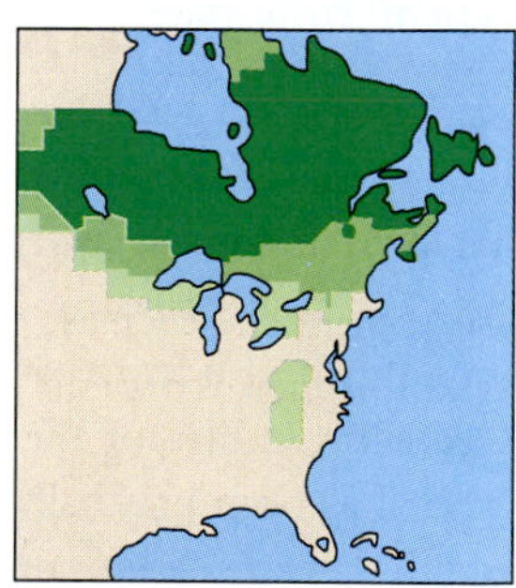

(c) Current

(c) From Thompson Webb III et al., "Vegetation, Lake Levels, and Climate in Eastern North America for the Past 18,000 Years" in H. E. Wright, Jr. et al. (eds), *Global Climates since the Last Glacial Maximum* (Minneapolis, MN: University of Minnesota Press, 1993), part of figure 17.10, p. 449. ©1993 by University of Minnesota Press.

An increase of only 1.3°C would make the world warmer than at any time in the past 100,000 years. A worst-case scenario suggests that the warming would be greatest near the poles; the resultant melting of polar ice might raise sea level by an estimated 100 m, gradually flooding coastal areas 150 or more km inland from the current coastline. New York, Miami, Los Angeles, and many other cities would then be under water. A warming trend would also alter the geographical distribution of precipitation, making major agricultural areas of the central United States much drier. However, the various mathematical models disagree about the details of how climate in each region will be affected. By studying how past periods of global warming and cooling affected plant communities, ecologists are using another strategy to help predict the consequences of future temperature changes. Records from pollen cores provide evidence that plant communities change dramatically with changes in temperature. Past climate changes occurred gradually, and plant and animal populations could migrate into areas where abiotic conditions allowed them to survive. A major concern is the projected high rates of climate change—by some estimates, higher than at any time in the past 10,000 years. Many organisms, especially plants that cannot disperse rapidly over long distances, will probably not be able to survive such rapid changes.

Coal, natural gas, gasoline, wood, and other organic fuels central to modern life cannot be burned without releasing CO_2. The apparent warming of the planet that is now under way as a result of the addition of CO_2 to the atmosphere is a problem of uncertain consequences and no simple solutions. Given the importance of combustion to our increasingly industrialized societies, stabilizing CO_2 emissions will require concerted international effort and the acceptance of dramatic changes in both personal lifestyles and industrial processes.

Depletion of Atmospheric Ozone

Life on Earth is protected from the damaging effects of ultraviolet (UV) radiation by a protective layer of ozone molecules (O_3) that is present in the lower stratosphere between 17 and 25 km above Earth's surface. Ozone absorbs UV radiation, preventing much of it from contacting organisms in the biosphere. Satellite studies of the atmosphere suggest that the ozone layer has been gradually depleted, or "thinned," since 1975, and that the depletion continues at an increasing rate.

The destruction of atmospheric ozone probably results mainly from the accumulation of chlorofluorocarbons, chemicals used for refrigeration, as propellants in aerosol cans, and in certain manufacturing processes. When the breakdown products from these chemicals rise to the stratosphere, the chlorine they contain reacts with ozone, reducing it to molecular O_2. Subsequent chemical reactions liberate the chlorine, allowing it to react with other ozone molecules in a catalytic chain reaction. The effect is most apparent over Antarctica, where cold winter temperatures facilitate these atmospheric reactions. Scientists first described the "ozone hole" over Antarctica in 1985 and have since documented that it is a seasonal phenomenon that grows and shrinks in an annual cycle. However, the magnitude of ozone depletion and the size of the ozone hole have generally increased in recent years, and the hole sometimes extends as far as the southernmost portions of Australia, New Zealand, and South America. At the more heavily populated middle latitudes, ozone levels have decreased 2% to 10% during the past 20 years.

The consequences of ozone depletion for life on Earth may be quite severe. Some scientists expect increases in both lethal and nonlethal forms of skin cancer and in cataracts among humans, as well as unpredictable effects on crops and natural communities, especially the phytoplankton that are responsible for a large proportion of Earth's primary productivity. The danger posed by ozone depletion is so great that many nations have agreed to end the production of chlorofluorocarbons within a decade. Unfortunately, even if all chlorofluorocarbons were banned today, the chlorine molecules that are already in the atmosphere will continue to influence stratospheric ozone levels for at least a century.

The exploding human population is altering habitats and reducing biodiversity worldwide

As we have seen, the activities and technology of the exploding human population have intruded in one way or another into the functioning of many ecosystems. Even where we have not completely destroyed a natural system, our actions have disrupted the trophic structure, energy flow, and chemical cycling of ecosystems in most areas of the world. Here we address ways that humans are directly affecting the biosphere's distribution and diversity of organisms.

Human encroachment on natural ecosystems has reached epidemic proportions. The clearing of natural ecosystems, which is usually necessary for agricultural, industrial, and residential development, undoubtedly causes the greatest local disruption of natural environments. Clear-cut harvesting of timber also destroys vast tracts of forest. Relatively little undisturbed habitat still exists in many countries; in the United States, for example, only 15% of the original primary forest (most of it in Alaska) and less than 1% of the original tallgrass prairie remain. The statistics on forests are even worse in some other regions, such as Europe, China, and Australia. In recent years, environmentalists and conservation biologists have focused attention on the destruction of tropical forests, among the most productive ecosystems on Earth. Some estimates suggest that if tropical forests continue to be cut at the present rate (about 500,000 km^2 worldwide per year), nearly all large tracts of this ecosystem will be eliminated within a decade or two.

(a)

(b)

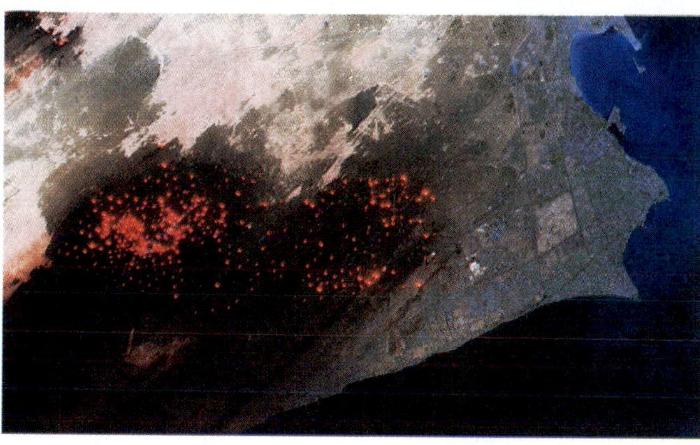

(c)

FIGURE 54.18 ▪ Ecological effects of warfare. Military conflicts devastate entire landscapes as well as particular ecosystems. The use of a defoliant (agent orange) during the Vietnam War quickly destroyed vast areas of tropical forest. **(a)** A Vietnamese tropical rain forest before defoliation. **(b)** The same area afterward. One of the chemicals in agent orange is a carcinogen. **(c)** In this satellite view of Kuwait at the end of the 1991 Persian Gulf War, hundreds of burning oil wells are visible as orange dots. The fires blackened the skies, deposited a thick layer of oily soot on surrounding areas, and squandered precious fossil fuel.

Development and logging are certainly not the only human activities that disrupt entire ecosystems. The ecological consequences of war are devastating. During the Vietnam War, for example, the United States used large quantities of chemical defoliants to kill the vegetation in which Viet Cong soldiers concealed themselves (FIGURE 54.18a and b). More recently, as the Persian Gulf War ended in 1991, Iraqi soldiers created other disasters that ravaged the landscape. Huge oil spills polluted marine ecosystems, and the burning of oil wells in Kuwait blackened the skies and left oily deposits on everything nearby (FIGURE 54.18c).

A major concern about the wholesale destruction of any natural habitat, particularly tropical rain forest, is the loss of biodiversity. We look at Earth's current biodiversity crisis and attempts to curb it in Chapter 55.

CHAPTER REVIEW

REVIEW OF KEY CONCEPTS

(with page numbers and key figures)

TROPHIC RELATIONSHIPS IN ECOSYSTEMS

An ecosystem is the most inclusive level in the hierarchy of biological organization, consisting of the community of organisms in a defined area and the abiotic factors making up the physical environment. Energy flows through ecosystems, and chemicals cycle through them; these interrelated processes occur by the transfer of nutrient substances through feeding relationships.

54.1 **Trophic relationships determine an ecosystem's routes of energy flow and chemical cycling (pp. 1131–1132, FIGURE 54.2)** Species in an ecosystem are divided into different trophic (feeding) levels, depending on their main source of nutrition. Autotrophic organisms are primary producers; heterotrophic organisms are consumers. Herbivores are primary consumers that eat mainly or exclusively autotrophs. Carnivores are secondary and tertiary consumers. Detritivores feed on organic wastes and dead organisms from all trophic levels. The pathway along which food from primary producers is transferred to the various levels of consumers is known as a food chain. However, natural feeding relationships are usually more like webs because consumers feed at more than one trophic level.

- **Primary producers include plants, algae, and many species of bacteria (pp. 1132–1133)**

- **Many primary and higher-order consumers are opportunistic feeders (pp. 1133–1134)**
- **Decomposition interconnects all trophic levels (p. 1134)**

ENERGY FLOW IN ECOSYSTEMS

- **An ecosystem's energy budget depends on primary productivity (pp. 1134–1136, FIGURE 54.3)** The energy assimilated during photosynthesis is a tiny fraction of the solar radiation reaching Earth; it sets the spending limit for the global energy budget. Gross primary productivity is the total energy assimilated by an ecosystem in a given time period. Net primary productivity, the energy accumulated in autotroph biomass, equals gross primary productivity minus the energy used by the primary producers for respiration. Only net primary productivity is available to consumers.
- **As energy flows through an ecosystem, much is lost at each trophic level (pp. 1136–1138, FIGURE 54.5)** The amount of energy available to each trophic level is determined by the net primary productivity and the efficiencies with which food energy is converted into biomass at each link of the food chain. Secondary productivity is the rate at which an ecosystem's consumers convert the chemical energy of food into biomass. The percentage of energy transferred from one trophic level to the next, called ecological efficiency, is generally 5% to 20%.

CYCLING OF CHEMICAL ELEMENTS IN ECOSYSTEMS

Nutrient cycles involve both biotic and abiotic components of ecosystems; they are also called biogeochemical cycles.

- **Biological and geological processes move nutrients among**
54.2 **organic and inorganic compartments (pp. 1138–1143, FIGURES**
54.3 **54.8–54.12)** Water moves in a global cycle driven by solar energy.
The carbon cycle primarily reflects the reciprocal processes of photosynthesis and cellular respiration. Nitrogen enters ecosystems by atmospheric deposition and nitrogen fixation by prokaryotes, but most of the nitrogen cycling in natural ecosystems involves local cycles between organisms and soil or water. The phosphorus cycle does not involve the atmosphere much in the short term, and thus occurs on a more localized scale than the water, carbon, and nitrogen cycles.
- **Decomposition rates largely determine the rates of nutrient cycling (pp. 1143–1144)** The proportion of a nutrient in a particular form and its cycling time in that form vary among ecosystems, mostly because of differences in the rate of decomposition.
- **Field experiments reveal how vegetation regulates chemical cycling: *science as a process* (pp. 1144–1145, FIGURES 54.14, 54.15)** Long-term ecological research projects track ecosystem dynamics over relatively long periods of time. Studies in deciduous forests in New Hampshire have shown that logging increases water runoff and can cause huge losses of minerals.

HUMAN IMPACTS ON ECOSYSTEMS

- **The human population is disrupting chemical cycles throughout the biosphere (pp. 1145–1146)** Agricultural practices result in the constant removal of nutrients from ecosystems, so that large supplements are continually required. Considerable amounts of the nutrients in fertilizer move into aquatic ecosystems, where they can stimulate excess algal growth. The dumping of nutrient-rich wastes into aquatic habitats accelerates eutrophication, and the increased biomass depletes O_2.
- **Toxins can become concentrated in successive trophic levels of food webs (pp. 1146–1147, FIGURE 54.16)** The release of toxic wastes has polluted the environment with harmful substances that often persist for long periods of time. DDT is an example of the many toxic substances that become concentrated along the food chain by biological magnification.
- **Human activities are causing fundamental changes in the composition of the atmosphere (pp. 1147–1150)** Because of the burning of wood and fossil fuels, CO_2 in the atmosphere has been steadily increasing. The ultimate effects may include significant warming and other influences on climate, as well as altered distributions of species and ecosystem productivity. Chlorine-containing pollutants are eroding the ozone layer, which reduces the penetration of UV radiation through the atmosphere.
- **The exploding human population is altering habitats and reducing biodiversity worldwide (pp. 1150–1151)**

SELF-QUIZ

1. Which of the following organisms is *incorrectly* paired with its trophic level?
   a. cyanobacteria—primary producer
   b. grasshopper—primary consumer
   c. zooplankton—secondary consumer
   d. eagle—tertiary consumer
   e. fungi—detritivore

2. One of the lessons from a pyramid of productivity is that
   a. only one-half of the energy in one trophic level is passed on to the next level
   b. most of the energy from one trophic level is incorporated into the biomass of the next level
   c. the energy lost as heat or in cellular respiration is 10% of the available energy of each trophic level
   d. ecological efficiency is highest for primary consumers
   e. eating grain-fed beef is an inefficient means of obtaining the energy trapped by photosynthesis

3. The role of decomposers in the nitrogen cycle is to
   a. fix N_2 into ammonia
   b. release ammonia from organic compounds, thus returning it to the soil
   c. denitrify ammonia, thus returning N_2 to the atmosphere
   d. convert ammonia to nitrate, which can then be absorbed by plants
   e. incorporate nitrogen into amino acids and organic compounds

4. The Hubbard Brook Experimental Forest study demonstrated all of the following *except* that
   a. most minerals were recycled within a forest ecosystem
   b. mineral inflow and outflow within a natural watershed were nearly balanced
   c. deforestation resulted in an increase in water runoff
   d. the nitrate concentration in waters draining the deforested area became dangerously high
   e. deforestation causes a large increase in the density of soil bacteria

5. The recent increase in atmospheric CO_2 concentration is mainly a result of an increase in
   a. primary productivity
   b. the biosphere's biomass
   c. the absorption of infrared radiation escaping from Earth
   d. the burning of fossil fuels and wood
   e. cellular respiration by the exploding human population

6. Which of the following is a result of biological magnification?
   a. Top-level predators may be most harmed by toxic environmental chemicals.
   b. DDT has spread throughout every ecosystem and is found in almost every organism.

c. The greenhouse effect will be most significant at the poles.
d. Energy is lost at each trophic level of a food chain.
e. Many nutrients are being removed from agricultural lands and shunted into aquatic ecosystems.

7. Which of these ecosystems has the *lowest* primary productivity per square meter?
   a. a salt marsh
   b. an open ocean
   c. a coral reef
   d. a grassland
   e. a tropical rain forest

8. Quantities of mineral nutrients in soils of tropical rain forests are relatively low because
   a. the standing crop is small
   b. microorganisms that recycle chemicals are not very abundant in tropical soils
   c. the decomposition of organic refuse and reassimilation of chemicals by plants occur rapidly
   d. nutrient cycles occur at a relatively slow rate in tropical soils
   e. the high temperatures destroy the nutrients

9. Which of the following illustrates secondary productivity?
   a. 170×10^9 tons of organic material are created annually.
   b. $6\,CO_2 + 6\,H_20 \rightleftharpoons C_6H_{12}O_6 + 6\,O_2$
   c. A tropical rain forest adds to the ecosystem more than 2200 g of vegetation per m^2 per yr.
   d. Only 15% of the energy consumed by a caterpillar is converted to biomass.
   e. Carnivores produce proportionately less feces than do herbivores.

10. Which of the following statements concerning the water cycle is *incorrect?*
    a. There is a net movement of water vapor from oceans to terrestrial environments.
    b. Precipitation exceeds evaporation on land.
    c. Most of the water that evaporates from oceans is returned by runoff from land.
    d. Transpiration makes a significant contribution to evaporative water loss from terrestrial ecosystems.
    e. Evaporation exceeds precipitation over the seas.

CHALLENGE QUESTIONS

1. Imagine you have been chosen as the biologist for the design team for a self-contained space station to be assembled in orbit. It will be stocked with organisms you choose to create an ecosystem that will support you and five other people for two years. Describe the main functions you expect the organisms to perform. List the types of organisms you would select, and explain why you chose them.
2. In Southeast Asia there's an old saying, "There is only one tiger to a hill." Explain this saying in terms of the energy flow in ecosystems.
3. A carbon atom in CO_2 might be used by a dandelion to make a sugar molecule. The sugar molecule might then be consumed by a grasshopper when it eats the dandelion, and be deposited in the soil in an organic molecule in the grasshopper's feces. A bacterium could release the carbon atom back into the air as CO_2 as it decomposes the soil litter. Continue the story, describing five more places the carbon atom might go, including your own body.

SCIENCE, TECHNOLOGY, AND SOCIETY

The amount of CO_2 in the atmosphere is increasing, and global temperature has increased over the past century; however, scientists do not agree about the extent to which the two phenomena are related. Most say that greenhouse warming is under way, and we need to take action now to avoid drastic environmental change. Some say it is too soon to tell, and we should gather more data before we act. What are the advantages and disadvantages of doing something now to slow global warming? What are the advantages and disadvantages of waiting until more data are available?

FURTHER READING

Asner, G. P., T. R. Seastedt, and A. R. Townsend. "The Decoupling of Terrestrial Carbon and Nitrogen Cycles." *BioScience,* April 1997. Describes human alteration of key biogeochemical connections.

Kaiser, J. "Acid Rain's Dirty Business: Stealing Minerals From the Soil." *Science,* April 12, 1996. A review of recent research, including that at the Hubbard Brook experimental station.

Karl, T. R., N. Nicholls, and J. Gregory. "The Coming Climate." *Scientific American,* May 1997. Three members of the Intergovernmental Panel on Climate Change discuss possible scenarios.

Pitelka, L. F., and the Plant Migration Workshop Group. "Plant Migration and Climate Change." *American Scientist,* September–October 1997. A summary of a scientific workshop on predicting plant community changes associated with global climate change.

Raloff, J. "Those Old Dioxin Blues." *Science News,* May 17, 1997. Evaluates toxins as the cause of fish kills in the Great Lakes.

Richardson, L. L. "Remote Sensing of Algal Bloom Dynamics." *BioScience,* July/August 1996. Describes the use of algal pigment analyses to track population dynamics in aquatic ecosystems.

Schimel, D. S. "The Carbon Equation." *Nature,* May 21, 1998. Discusses models of the potential effects on global warming of CO_2 uptake and release in oceans and on land.

Smil, V. "Global Population and the Nitrogen Cycle." *Scientific American,* July 1997. Discusses the importance of nitrogen fertilizer in feeding the burgeoning human population, and the potential dangers of its overuse.

Suplee, C. "Unlocking the Climate Puzzle." *National Geographic,* May 1998. Considers the causes and potential consequences of global warming.

WEB LINKS

Visit the special edition of *The Biology Place* for BIOLOGY, Fifth Edition, at http://www.biology.com/campbell. Go to Chapter 54 for online resources, including learning activities, practice exams, and links to the following web sites:

"Critical Ecoregions Program"
Explore the activities the Sierra Club is undertaking to restore the ecological health of 21 different ecoregions.

"Office of Sustainable Ecosystems and Communities"
The mission of this EPA office is to foster the implementation of integrated, geographic approaches to environmental protection with an emphasis on ecological integrity, economic sustainability, and quality of life.

"American Forests: Plant the Future"
American Forests, responsible for the Global ReLeaf Program, are an organization dedicated to forest ecosystem restoration. You can plant a tree on-line if you want!

"Ocean Planet"
A joint venture between NASA and the Smithsonian, this is an online version of a traveling exhibit that looked at marine ecosystems and human interactions with them.

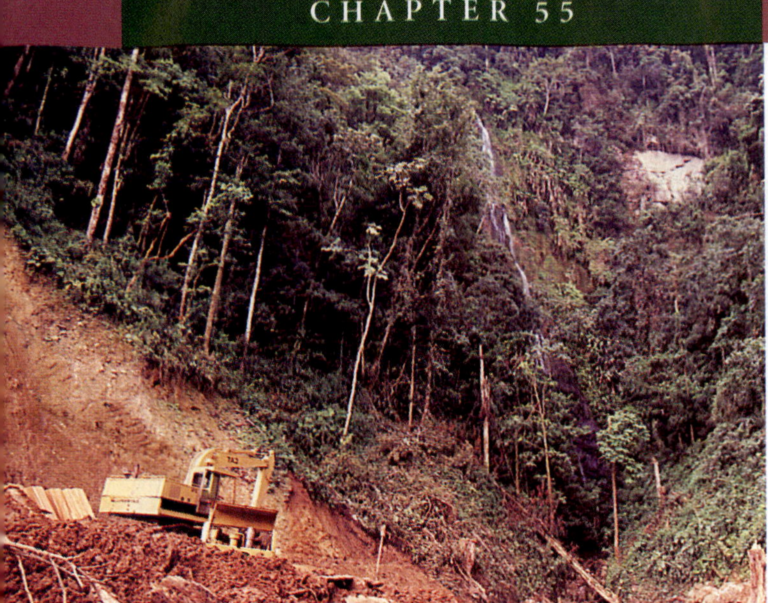

CHAPTER 55

CONSERVATION BIOLOGY

The Biodiversity Crisis: An Overview

- Numerous examples indicate that estimates of extinction rates are on track
- The major threats to biodiversity are habitat destruction, overexploitation, and competition by exotic species
- Biodiversity is vital to human welfare
- Change in ecological and evolutionary time is the focus of conservation biology

The Geographic Distribution of Biodiversity

- Gradual variation in biodiversity correlates with geographical gradients
- Biodiversity hot spots have high concentrations of endemic species
- Migratory species present special problems in conservation

Conservation at the Population and Species Levels

- Sustaining genetic diversity and the environmental arena for evolution is an ultimate goal
- The dynamics of subdivided populations apply to problems caused by habitat fragmentation
- Population viability analyses examine the chances of a species persisting or becoming extinct in the habitats available to it
- Analyzing the viability of selected species may help sustain other species: *science as a process*
- Conserving species involves weighing conflicting demands

Conservation at the Community, Ecosystem, and Landscape Levels

- Edges and corridors can strongly influence landscape biodiversity
- Nature reserves must be functional parts of landscapes
- Restoring degraded areas is an increasingly important conservation effort
- Sustainable development goals are reorienting ecological research and will require changing some human values

Biology is the science of life, thus, it is fitting that our final chapter be about preserving life. ***Conservation biology*** *is a goal-oriented science that seeks to counter the* ***biodiversity crisis****, the current rapid decrease in Earth's great variety of life.*

To date, scientists have described and formally named about 1.5 million species of organisms. We can only estimate how many more currently exist. Some biologists believe that the number is about 10 million, but others estimate it to be between 30 million and 80 million. Some of the greatest concentrations of species are found in the tropics, where the scene in the photograph on this page is commonplace; tropical forests, such as the one in Ecuador shown here, are being destroyed at an alarming rate to make room for, and support, a burgeoning human population.

Throughout the biosphere, human activities are altering trophic structures, energy flow, chemical cycling, and natural disturbance—ecosystem processes upon which we and other species depend. The amount of human-altered land surface is approaching 50%, and we use over half of all accessible surface fresh water. In the oceans, stocks of many fishes are being depleted by overharvest, and some of the most productive and diverse areas, such as coral reefs and estuaries, are being severely stressed. By some estimates we are in the process of doing more damage to the biosphere and pushing more species toward extinction than the large asteroid that seems to have triggered the mass extinctions at the close of the Cretaceous period 65 million years ago (see FIGURE 25.6)*. Globally, the rate of species loss may be as much as 50 times higher than at any time in the past 100,000 years.*

In this chapter we take a closer look at the biodiversity crisis and at the science of conservation biology. We will examine some of the research and conservation strategies biologists are using in attempts to slow the rate of species loss. Along the way we will see that conservation biology relies on research at all levels of ecology, from populations through ecosystems.

THE BIODIVERSITY CRISIS: AN OVERVIEW

Extinction is a natural phenomenon that has been occurring almost since life first evolved; it is the current *rate* of extinction that underlies the biodiversity crisis. Because we can only estimate the number of species currently existing, we cannot determine the actual rate of species loss or the real magnitude of the biodiversity crisis. We do know for certain that we are living in a time of a high rate of species extinction caused by a high rate of ecosystem degradation by a single species, *Homo sapiens*. By some assessments, human activity in tropical areas alone has increased extinction between 1000 and 10,000 times over the normal "background" rate of one out of every million species per year.

Numerous examples indicate that estimates of extinction rates are on track

Rates of extinction are often expressed as the number or percentage of species expected to become extinct in an area in a unit of time. Estimating such rates is difficult at best. The most accurate assessments of extinction rates can be made for taxa that are relatively well known, such as birds and mammals. Birds are among the most thoroughly studied animals, and some analyses suggest that birds are good indicators of the presence of other taxa. Thus, extinction rate estimates for less well-known, nonavian taxa are sometimes based on the rate of loss of bird species.

Most often, extinction rates are estimated from the concept of species-area relations, a generalization based on studies of oceanic islands in the tropics. According to this concept, the number of species in an area is directly related to the size of the area (see FIGURE 53.22). Thus, estimates of the rate of species extinctions usually hinge on assessments of the rate of habitat loss. A rule of thumb based on species-area relations in deforested tropical areas predicts that, on average, about 50% of the total number of species will be lost in an area where 90% of the habitat is lost.

Though widely used, species-area relations can provide only a general idea of species numbers and rates of loss. It is important to keep in mind that any extinction rate is approximate. Nonetheless, many examples indicate that extinction rates *are* unusually high and a cause for concern. For example:

- According to the International Union for Conservation of Nature and Natural Resources (IUCN), 11% of the 9040 known bird species in the world are endangered. In the past 40 years, population densities of migratory songbirds in the mid-Atlantic United States dropped 50%.
- A recent survey conducted by the Center for Plant Conservation showed that of the approximately 20,000 known plant species in the United States, 680 are in danger of becoming extinct by the year 2000.
- About 20% of the known freshwater fishes in the world have either become extinct during historical times or are seriously threatened. One of the largest rapid extinction events yet recorded is the ongoing loss of freshwater fishes in Lake Victoria in east Africa. About 200 of the 300 species of cichlids in the lake have been lost, mainly due to the recent introduction by Europeans of an exotic predator, the Nile perch (FIGURE 55.1).

FIGURE 55.1 • An exotic predator. The Nile perch (*Lates niloticus*), a large predator (up to 2 m long and weighing up to 450 kg), has decimated indigenous fish populations and disrupted community processes in east Africa's Lake Victoria. One of the largest freshwater fishes, it was introduced into the lake in the 1960s by Europeans who wanted to provide additional food and income for the people living in the area. The lake had supported fishing for a variety of fish species found nowhere else—mostly cichlids, which feed on detritus and plants. The introduced perch annihilated the cichlid populations, destroying the native fishery and reducing the perch's own food supply. The perch's numbers are now so diminished that it is not a significant food source for humans.

Extinction is a local, often unseen process. In order to know for certain that a given species is extinct, we must know its exact distribution and its habits. But the fact that millions of the world's species have not even been given names is a clear indication that we do not know most species well at all. Arthropods (especially insects), nematodes, fungi, and prokaryotes head the list of taxa with great numbers of undiscovered species, but even well-studied taxa, such as birds and mammals, are not completely known. In the past decade, scientists have increased the list of known mammals by about 15%. Without a more complete catalogue of biodiversity and knowledge of the geographic distribution and ecological roles of Earth's species, our efforts to understand the structure and function of ecosystems upon which our survival depends will remain incomplete.

The major threats to biodiversity are habitat destruction, overexploitation, and competition by exotic species

Human alteration of habitat is the single greatest threat to biodiversity throughout the biosphere. Massive destruction of habitats throughout the world has been brought about by agriculture, urban development, forestry, mining, and environmental pollution. The IUCN implicates destruction of physical habitat in 73% of its designations of species as extinct, endangered, vulnerable, and rare.

Though most studies have focused on terrestrial ecosystems, habitat loss also appears to be a major threat to marine biodiversity, especially on continental coasts and coral reefs. About 93% of Earth's coral reefs, among the most species-rich aquatic communities, have been damaged by human activities. At the current rate of destruction, 40% to 50% of the reefs could be lost in the next 30 to 40 years. About a third of the planet's marine fish species utilize coral reefs, which occupy only about 0.2% of the ocean floor.

Ranking second behind habitat loss as an important cause of the biodiversity crisis is competition of exotic (nonnative) species with native species. Exotic species are imported in various ways. People inadvertently carry hitchhiking seeds or insects with them when they travel throughout the world, and many foreign plants and animals have been intentionally introduced for agricultural or ornamental purposes. Most transplanted species fail to survive outside their normal range, but there are many examples of viable transplants. If your campus is in an urban area, there is a good chance that the birds you see most frequently as you walk between classes are starlings, house sparrows, and rock doves (commonly called "pigeons")—all exotic species that have displaced native birds in many areas of North America. Many viable exotics have had little effect on existing ecosystems, but some play an important role in their new communities, usually through predation on native species or competition for resources. We have already mentioned the destruction of the local fish population by the introduced Nile perch in Lake Victoria (see FIGURE 55.1). Another striking example is the northward march of fire ants, which were accidentally introduced into the southern United States from Brazil in 1918 (FIGURE 55.2). Displacement by introduced species is considered at least partially responsible for 68% of the listings of extinct, endangered, vulnerable, and rare species published by the IUCN.

Other significant threats to biodiversity, such as overexploitation of wildlife, often compound problems of shrinking habitat and exotic species. Animal species whose numbers have been drastically reduced by excessive commercial harvest or sport hunting include whales, American bison, Galápagos tortoises, and numerous fishes. Little more than a century ago, British biologist T. H. Huxley declared, "Probably all the great sea fisheries are inexhaustible: that is to say that nothing we do seriously affects the number of fish." Today, modern fishing techniques have reduced populations of cod, herring, mackerel, and many other commercially important species to levels that cannot sustain further human exploitation. In addition to the species being sought, many other organisms are often killed by harvesting methods; for example, dolphins, marine turtles, and seabirds are caught in fishing nets, and countless numbers of invertebrates are killed by marine trawls. An expanding, often illegal, world trade in wildlife (such as rare fishes, birds, orchids, and cactuses) and wildlife products (including mammal skins, bird feathers, rhinoceros horns, elephant tusks, and grizzly bear gallbladders) also threatens many species.

Biodiversity is vital to human welfare

Why should we care about the loss of biodiversity? Perhaps the purest reason is what Harvard biologist E. O. Wilson calls *biophilia*, our sense of connection to nature and other forms of life. The concept that other species are important and should be protected is a pervasive theme of many religions. There is also a concern for future generations. Do we have the right to deprive them of Earth's species richness? G. H. Brundtland, former Prime Minister of Norway, put it this way: "We must consider our planet to be on loan from our children, rather than being a gift from our ancestors."

In addition to the aesthetic and ethical reasons for preserving biodiversity, there are practical reasons as well. Biodiversity is a crucial natural resource, and species that are threatened could provide crops, fibers, and medicines. In the United States, 25% of all prescriptions dispensed from pharmacies contain substances derived from plants. In the 1970s, researchers discovered that the rosy periwinkle from Madagascar contains alkaloids that inhibit cancer cell growth (FIGURE 55.3). The result of this discovery is remission for most victims of two potentially deadly forms of cancer, Hodgkin's disease and a childhood leukemia. There are five other species of periwinkles on Madagascar, and one is approaching extinction.

Of utmost importance, the loss of species means the loss of genes. Each species is a unique combination of genetic variability resulting from evolutionary processes, and biodiversity

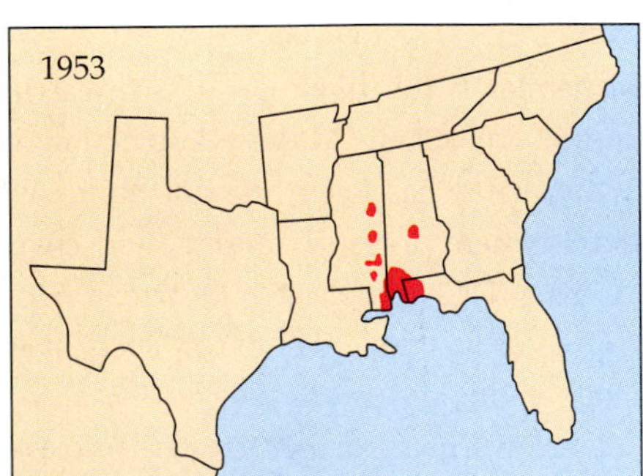

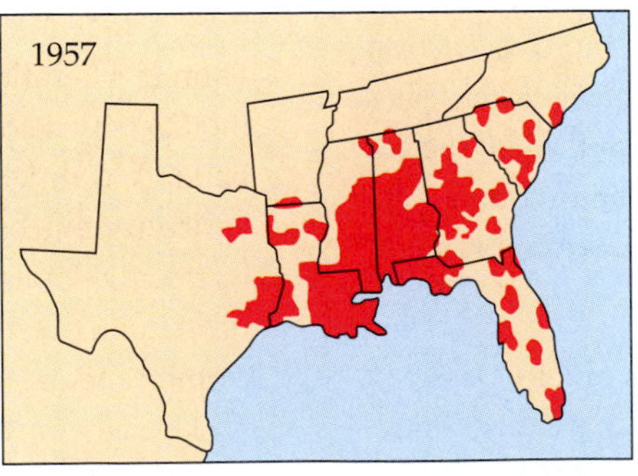

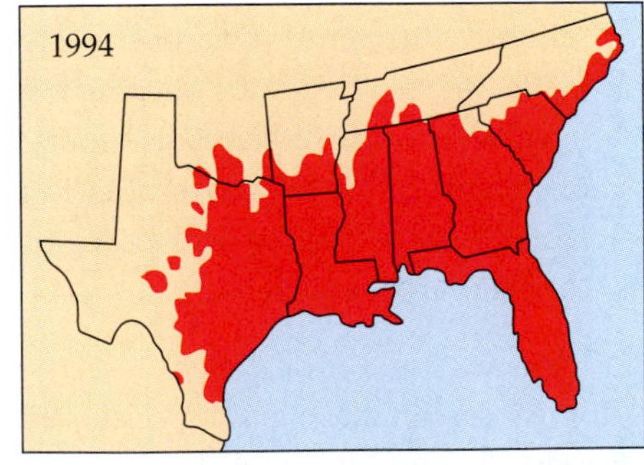

FIGURE 55.2 ▪ March of the fire ants. The fire ant (*Solenopsis invicta*) can inflict a beelike sting and tends to attack in large numbers. In the southern United States, more than 70,000 sting victims seek medical help each year. Fire ants were accidentally imported from South America in 1918, probably in the hold of a ship that docked in Mobile, Alabama. The range of the ant has been spreading ever since, as traced by this series of maps. In their spread across the southeastern United States, they reduced the number of native ant species in one area of Texas from 15 to 5 and killed hatchlings of the then-threatened brown pelican in wildlife refuges.

FIGURE 55.3 ▪ The rosy periwinkle (*Catharanthus roseus*): a plant that saves lives. Before alkaloids that inhibit cancer cell growth were discovered in the rosy periwinkle 20 years ago, Hodgkin's disease and acute lymphocytic leukemia were two of the deadliest cancers. Now most victims are cured. This plant is one of hundreds that are used to treat human diseases.

ultimately is the sum total of all the genomes of all organisms on Earth. Because many millions of species may become extinct before we even know about them, we stand to lose irretrievably all the genetic potential held in their unique libraries of genes.

Recently, U.S. National Park Service officials have been negotiating with private industry to sell samples of extremophilic prokaryotes from the numerous hot springs in Yellowstone National Park. The corporations anticipate using DNA extracted from the prokaryotes to mass-produce industrially useful enzymes. Many researchers and industry officials are enthusiastic about the potential that such "bioprospecting" holds for the future development of new medicines, food, petroleum substitutes, industrial chemicals, and other important products.

The benefits that individual species provide to humans are often substantial, but saving individual species is only part of the rationale for saving habitats. Humans evolved in the living ecosystems of Earth, and our bodies are finely adjusted to these conditions. While it is possible that we could survive in a world with considerably less biodiversity, it is important to realize that humans are dependent on ecosystems and other species. By allowing the extinction of species and the degradation of habitats to continue, we are taking a risk with our own species' survival that seems unwise.

In an attempt to counter what they see as a tendency of policy makers and governments to undervalue the biosphere's life-sustaining features, a team of ecologists and economists recently estimated the cost of replacing ecosystem "services" as a measure of the services' value. For example, part of the value of a wetland was estimated from the increased cost of flood damage that occurred because of the loss of the wetland's ability to hold flood water. The value of recreation in ecosystems was estimated from the funds spent on recreational pursuits. In 1997, the scientists estimated the average annual value of the biosphere at US$33 trillion. In contrast, the global gross national product in 1997 was US$18 trillion. Although rough, these estimates help make the important point that we cannot afford to continue to take ecosystems for granted.

Change in ecological and evolutionary time is the focus of conservation biology

Traditionally, after converting most of their lands to serve human needs, most nations set aside small areas to remain undeveloped. Great Britain, for example, has about 99% of its land altered for human use and less than 1% set aside as nature reserves. Globally, an area about half the size of the United States (about 4.25 million km^2, or about 3% of the planet's land surface) has been protected as natural areas (including national parks, designated wilderness, and nature reserves). For many such areas, however, protection is incomplete and does not preclude continued human disturbance, such as poaching, mineral development, logging, and excessive use by recreationists. *Preservationism*—setting aside select areas to remain natural and undeveloped—is part of conservation biology's history.

Another concept that foreshadowed modern conservation biology, called *resource conservation,* arose from the economic reality that long-term maintenance of natural resources is essential to human welfare. This concept led to the working doctrine called "multiple use" adopted by the U.S. Forest Service and Bureau of Land Management in the early 1900s. Under the multiple use policy, government agencies have tried to make public lands and waters meet the needs of agriculture and extractive industries (for example, logging, geothermal energy production, and mining), as well as nature preservation and recreation. Because of political and economic pressures, multiple use has historically favored industry and agriculture at the expense of environmental concerns. As discussed by current chief of the U.S. Forest Service Dr. Michael Dombeck in the interview on pp. 1024–1025, attempts are being made to modify the multiple use concept with the more scientifically grounded policy of ecosystem management.

Ecosystem management is an outgrowth of a third approach to conservation, an *evolutionary/ecological view.* This view of nature stems from an understanding that natural systems result from millions of years of evolution and that ecosystem processes are necessary to maintain the proper functioning of the biosphere. Evolutionary and ecological principles pertaining to populations, communities, and ecosystems, which we have discussed in the previous three chapters, form the scientific foundation for modern conservation biology.

The science of conservation biology began to emerge in the late 1970s as an interdisciplinary organization of scientists working toward the long-term maintenance of functional ecosystems and a reduction of the rate of species extinction. Fostering research on biodiversity and the means to save it, an international group of scientists and educators founded the Society for Conservation Biology in 1985. Today, with about 5000 members, the society helps unite the conservation efforts of biologists, anthropologists, sociologists, economists, government and industry officials, and private citizens.

The philosophical underpinnings of modern conservation biology are evolution, nonequilibrium ecology, and concern for the human presence. Conservation biologists recognize that biodiversity can be sustained only if the evolutionary mechanisms that have given rise to species and communities of organisms continue to operate. Thus, the goal is not simply to preserve individual species but to sustain ecosystems, where natural selection can continue to function, and to maintain the genetic variability upon which natural selection acts.

As we discussed in Chapter 53, ecologists generally embrace the concept that nonequilibrium and disturbance are significant determinants of the structure and function of biological communities and that disturbance helps maintain biodiversity. Some communities remain relatively stable for long periods of time, but disturbance over ecological time, by living organisms and abiotic forces such as fires, storms, and floods, is inevitable. As functional features of ecosystems, disturbance and nonequilibrium are distinctly different from environmental degradation, which involves structural and functional breakdown.

Consideration of the human presence is vital in conservation biology because no ecosystems are unaffected by humans, and because human diversity—including our global gene pool and diverse cultures—is a significant component of Earth's biodiversity. Conservation biology seeks to foster those human activities that sustain ecosystems and reduce the current rate of environmental degradation.

THE GEOGRAPHIC DISTRIBUTION OF BIODIVERSITY

As we discussed in Chapter 50, habitats and the distribution of organisms are patchy. Biodiversity is not evenly distributed locally or globally; rather, there are discernible patterns in distribution, including clines (gradual variations), hot spots (high diversity areas), and local concentrations of migratory species.

Gradual variation in biodiversity correlates with geographical gradients

Biologists have long recognized the existence of clines in species diversity in the form of major geographical gradients. The number of terrestrial bird species in North and Central America, for example, increases steadily from the Arctic to the tropics (FIGURE 55.4). Similar clines have been observed for most other major groups of organisms, including microbes, flowering plants, reptiles, and mammals. Community ecologists and biogeographers have proposed several hypotheses to explain this pattern (and, more generally, the factors that regulate all patterns of diversity in natural communities). Some of these ideas involve factors that regulate species diversity on more local scales. Four of these hypotheses are:

1. *Energy availability.* Increased solar radiation in the tropics increases the photosynthetic activity of plants, providing an increased resource base for other organisms and thus a capacity to sustain more species.
2. *Habitat heterogeneity.* Compared to other areas, tropical regions commonly experience more local disturbances (such as treefalls, floods, and hurricanes) and have greater environmental patchiness, allowing a greater diversity of plant species to form the resource base for diverse communities of animals.

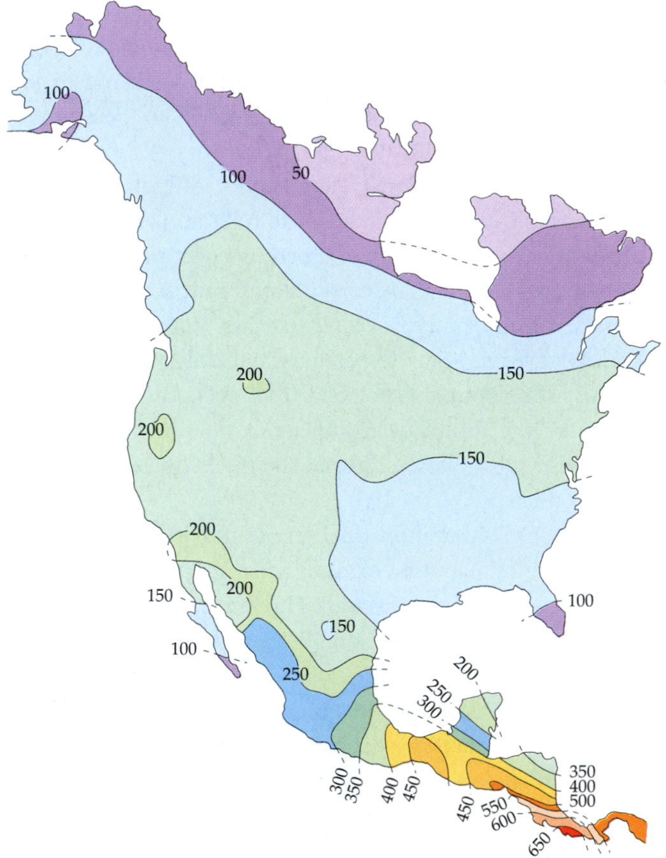

FIGURE 55.4 ▪ Species density of North and Central American birds. Biogeographers often plot latitudinal trends in numbers of species on maps that illustrate how many species occupy different geographic areas. In this species-density map for breeding species of North and Central American birds, we can see that fewer than 100 species are found in arctic areas, whereas more than 600 species occupy some tropical regions.

3. *Niche specialization.* Tropical climates may allow many organisms to specialize on a narrower range of resources. Smaller niches would reduce competition and permit a finer level of resource partitioning among species, which in turn would foster greater species diversity.
4. *Population interactions.* Diversity is, in a sense, self-propagating because complex population interactions coevolve, and the resulting predator-prey and symbiotic interactions in a diverse community prevent any populations from becoming dominant.

The question of what factors determine latitudinal gradients in species diversity is too complex to be addressed with simple experiments either in the laboratory or in the field. Many ecologists consider it likely that a combination of factors underlies most species clines. For instance, research on mammalian clines indicates that energy availability is the most important factor in determining species diversity in high latitudes (where solar energy is most limited), but habitat heterogeneity is the most important factor in more energy-rich areas.

Another clinal diversity pattern is an increase, with depth, in the number of species in some marine benthic faunas. Several ecologists have proposed that the long-term stability of these benthic habitats has fostered many coevolved relationships among the organisms present. However, no research has yet been able to establish a *causal* relationship between environmental stability and species diversity.

Biodiversity hot spots have high concentrations of endemic species

A **biodiversity hot spot** is a relatively small area with an exceptional concentration of species. Many of the organisms in biodiversity hot spots are **endemic species**, meaning they are found nowhere else. For example, nearly 30% of all bird species are endemic and confined to only about 2% of Earth's land area. The International Council for Bird Preservation (ICBP) has identified 221 hot spots where about 70% of all bird species whose long-term existence is in jeopardy are found. Totaling about 5% of Earth's land area, 76% of these bird hot spots are in the tropics. Biogeographers have also identified 18 vascular plant hot spots (FIGURE 55.5). Totaling only about 0.5% of the global land surface, these areas collectively contain the biosphere's greatest concentrations of endemic vascular plants—about 50,000 species, or 20% of all known vascular plant species. Fourteen of the plant hot spots are tropical forest biomes; the other four are dry shrublands (chaparral). In some (but not all) cases, hot spots for one taxonomic group also harbor a large proportion of other endemic taxa. For instance, the 18 vascular-plant hot spots also contain about 7% of all species of land vertebrates.

Because endemic species are limited to specific areas, they are highly sensitive to habitat degradation. To date, six of the 18 hot spot areas shown in FIGURE 55.5 have lost nearly 90% of their original habitats to human development. At the current rate of habitat alteration, the rest could lose similar amounts in the next two decades. Species-area relations indicate that habitat losses of this magnitude could mean the loss of about half of the species in the hot spots. Thus, biodiversity hot spots are also hot spots of extinction and rank high on the list of areas demanding strong global conservation efforts.

Numerous studies indicate that islands are hot spots of bird extinctions. For example, about 90% of the 104 species of birds lost in the past 400 years were endemic on islands. Today, all of the areas where over 10% of the bird species are threatened with extinction are islands: Hawaii currently stands to lose about 33% of its bird species, and the Philippines and New Zealand may lose about 15% of theirs. Although data for nonavian taxa are not as complete, it seems

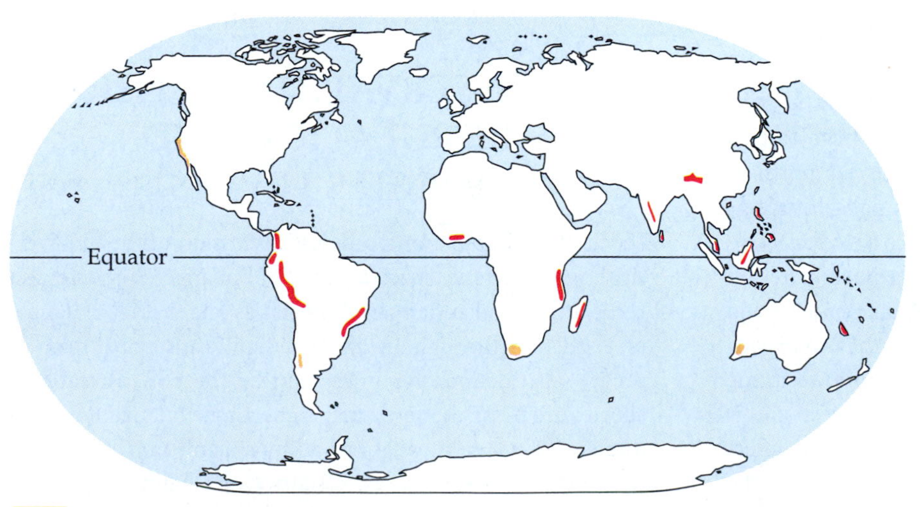

FIGURE 55.5 ▪ Some biodiversity hot spots. Less than 1% of Earth's land surface harbors nearly 20% of the known species of plants and probably a greater percentage of all land animals. The 18 hot spots shown here are limited to only two kinds of biomes, tropical forests and dry shrublands, or chaparral. Biodiversity hot spots are characterized by a large number of endemic species (those found nowhere else) that are highly susceptible to loss of habitat. Islands of the western Pacific are especially rich, and this area is also one of the most diverse marine hot spots.

likely that most species of terrestrial animals currently threatened with extinction are also endemic on islands.

In the United States, the greatest numbers of endangered species (terrestrial and freshwater) occur in Hawaii, southern California, the southern Appalachians, and the southeastern coastal states (especially Florida), areas with the highest numbers of endemic species. Florida's high numbers of endemic species actually illustrates the effect of islands. For most of the last 10 million years, much of the Florida peninsula was submerged beneath the sea. Today's areas of high ground, including a sandy, central area called the Lake Wales Ridge, were formerly islands where many of Florida's endemic species arose. Most of these species are threatened by habitat loss due to explosive human population growth and agriculture (mainly cattle ranches and citrus groves).

Studies of biodiversity and recent extinctions clearly show that many threatened, endangered, and potentially endangered species are concentrated in hot spots. Optimistically, this may indicate that many species can be protected in relatively small areas. Identifying and acting to save the thousands of endemic species in hot spots is an essential part of modern conservation efforts. On the other hand, conservation biologists are quick to point out that hot spots are not the same for every group of organisms, and more research is needed to identify endemic areas for a broad segment of biodiversity. Species endangerment is a truly global problem, and focusing on hot spots should not detract from efforts to conserve habitats and species diversity in other areas.

Migratory species present special problems in conservation

The problem of preserving species is particularly complicated for migratory species that breed in one country and spend some months of the year in another (for instance, to overwinter or to escape severe seasonal drought). Successful conservation efforts for such species usually require international cooperation and the careful preservation of habitat in both parts of the species' range. Monarch butterflies from eastern North America, for example, migrate from as far north as Canada to a few small overwintering sites in high-altitude fir forests of the Transvolcanic Range of Mexico (FIGURE 55.6). Although these migrants have been abundant in some recent autumns, their migration has been termed an "endangered phenomenon" because of the vulnerability of these overwintering sites to human intrusion. In addition, milkweed, the host plant of monarchs, is considered a noxious weed and is destroyed in Canada. Herbicide and insecticide use can affect monarch populations. Habitat preservation in Canada, the United States, or Mexico alone will not remove the threats to this insect, and the situation is similar for many species of migratory songbirds, sea turtles, and marine mammals.

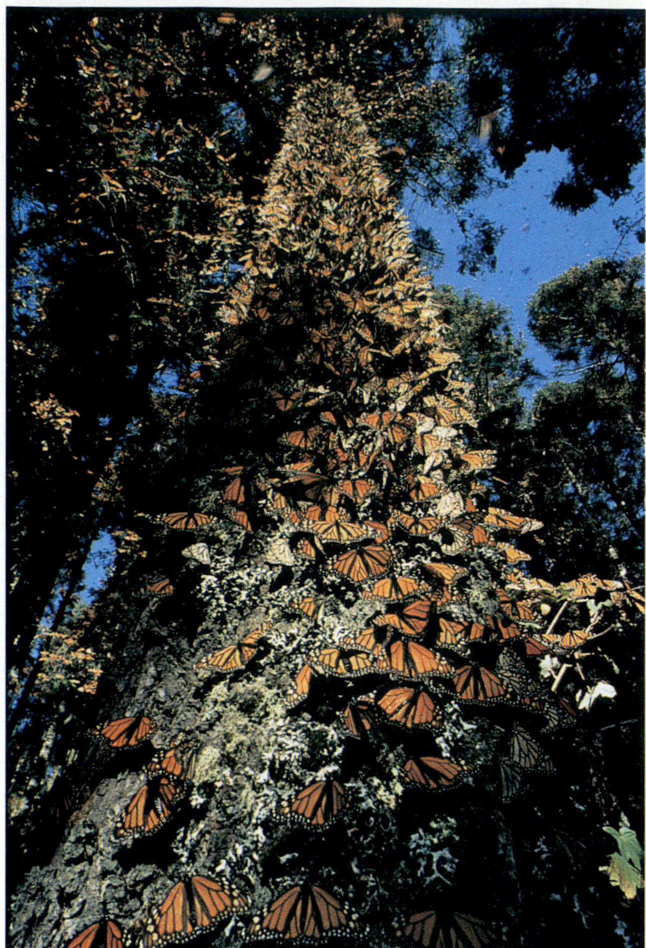

FIGURE 55.6 ▪ A migratory species that overwinters in dense numbers. Monarch butterflies (*Danaus plexippus*) occupy much of the United States and Canada during the summer months. In the autumn they migrate to local sites in Mexico and California where temperatures are above freezing but cold enough to slow the insects' metabolism and allow them to survive on stored food reserves until the following spring. Overwintering populations, such as this one in the highlands of central Mexico, are susceptible to habitat disturbance because they are concentrated in small areas.

CONSERVATION AT THE POPULATION AND SPECIES LEVELS

Much of the popular and political discussion of the biodiversity crisis centers on species. The U.S. Endangered Species Act (ESA) defines an **endangered species** as one that is "in danger of extinction throughout all or a significant portion of its range." Also defined for protection by the ESA, **threatened species** are those that are likely to become endangered in the foreseeable future throughout all or a significant portion of their range. However, as we discussed in Chapter 24, species are often difficult to define, and there is considerable debate over what criteria, such as reproductive isolation, genetic coherency, or morphology, best apply to species definitions.

The U.S. Endangered Species Act also defines units to be conserved as "distinct population segments," and those charged with administering the act have struggled with the meaning of the words "units" and "distinct." So how do we determine priorities for conservation?

Sustaining genetic diversity and the environmental arena for evolution is an ultimate goal

Species, however defined, are actually only part of Earth's biodiversity. In a broad sense, biodiversity and factors that sustain it encompass all the genetic variability within populations of species as well as the myriad ecosystem processes that provide the arena for evolution.

Conservation biology's focus on species and populations has involved understanding the dynamics of small populations, diagnosing declines and assessing the factors responsible for them, and determining how to sustain small, often fragmented populations. Ideally, conservation efforts should begin before serious declines occur, when there is time to save large enough areas of habitat to support natural populations. As the ranks of conservation biologists continue to grow, early detection of declines may become more frequent if funding is available to study populations. At present, however, conservation efforts lag far behind the rate of decline and loss of species. Many species populations have already been reduced to critically low numbers due to habitat alteration by human activities. As the number of these cases mounts, it is becoming less and less practical to focus on conserving genetic diversity. As we will see, the toolbox of conservation biology includes some features of crisis management and the application of some untested hypotheses and concepts. Few members of the conservation community have the luxury of doing "science for science's sake." "Learn as we apply" is a regrettable but necessary part of the effort to slow the biodiversity crisis. And most importantly, conservation science is focusing more on sustaining ecosystem processes and the evolutionary lineages that species represent rather than conserving individual species or populations.

The dynamics of subdivided populations apply to problems caused by habitat fragmentation

Degradation of habitats by human activities often involves reduction in the area of suitable habitat for species populations and fragmentation of the remaining area (FIGURE 55.7). Because habitat fragments resemble islands surrounded by areas variously altered by human activity, the hypothesis of island biogeography (see FIGURE 53.21) has been applied in studying habitat fragmentation and its effects on species populations. However, the island model has proved overly simplistic in many cases. Generally, concepts developed by studying subdivided populations are more useful in conservation work.

FIGURE 55.7 ▪ Fragmentation of a forest ecosystem. This aerial photograph of Mt. Hood National Forest in the western United States shows a common result of ecosystem alteration. In this case, the altered area includes "islands" of coniferous forest created when much of the original forest was cut for timber.

Because of the patchiness of environments, the populations of many species are subdivided even in unaltered areas. A subdivided population (a network of subpopulations) of a single species is called a **metapopulation** (Greek *meta,* "next to"). Metapopulations vary a great deal, depending in part on the size, quality, spatial arrangement, and persistence of habitat patches.

In most cases, the subpopulations of a metapopulation are separated into habitat patches that vary in quality. Other factors being equal, a habitat patch that includes, for example, a source of dissolved calcium ions would be a higher-quality patch for most animals than one where calcium is present but in a chemical form that the animals cannot use. Similarly, in a fragmented forest, the presence of some large, dead trees can make the difference between a high-quality patch and a low-quality one for squirrels, raccoons, and many birds that require rot cavities for their nests. Patches with abundant, high-quality resources tend to have persistent subpopulations that produce more offspring. A persistent subpopulation may serve as a source of new individuals that disperse to low-quality patches—for instance, forest fragments with no suitable nesting sites. In many cases, some subpopulations are subject to temporary extinction, and low-quality patches may be populated only when new individuals reared in high-quality patches disperse to them. Dispersal is also essential to maintain genetic variability within subpopulations. A subpopulation that is cut off from others may eventually become genetically distinct.

Fragmentation of habitats is a common effect of human activities, and many populations that were originally unified have been reduced in numbers and converted to metapopulations. Species that existed as metapopulations before human intervention are often reduced and fragmented even more. For example, the bull trout (*Salvelinus confluentus*), a freshwater

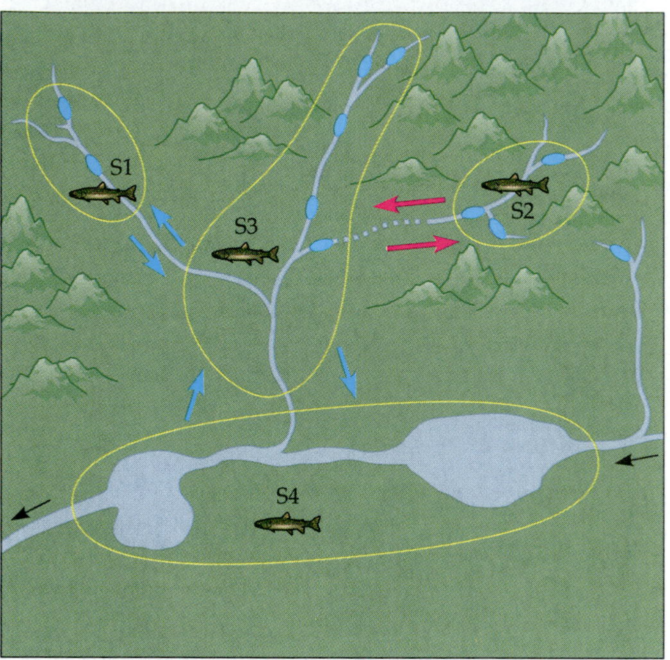

Egg-laying sites in mountain streams

Regular, frequent dispersal and gene flow between subpopulations

Irregular, infrequent dispersal; minimal gene flow between subpopulations

(a)

s1
s2
s3
s4
s5
Mill site for silver mine
Hydroelectric dam
Hydroelectric dam

Egg-laying sites in mountain streams

Clear-cut (logged) areas

Roads

Irregular, infrequent dispersal; minimal gene flow between subpopulations

(b)

FIGURE 55.8 ▪ A metapopulation of a freshwater fish altered by human activity. Many populations are subdivided (exist as metapopulations) in patchy environments. This idealized example, derived from a case study of the bull trout (*Salvelinus confluentus*), an endangered species, illustrates how a metapopulation changed as a result of human activities. Bull trout inhabit lakes, rivers, and mountain streams in the northwestern United States, including Alaska. They require cold, fast-flowing streams with small cobble substrates for breeding and egg-laying. **(a)** Historical records indicate that before human intervention, this bull trout metapopulation consisted of four subpopulations (S1–S4; yellow outlines). S1 and S2 were resident in high-mountain streams. The S3 subpopulation inhabited a larger area of the watershed and used the lakes and river for occasional feeding forays. S4 was mainly resident in the lakes and river but ascended the streams to reproduce. Some S4 fish probably used egg-laying sites in small mountain streams (middle right) with no resident bull trout. Gene flow (indicated by arrows) among the subpopulations was maintained chiefly by movements of individuals from S3 and S4 to and from egg-laying sites in the streams. Located above a portion of the stream that dries up (dashed line) during many summers, S2 was the most isolated subpopulation, but some individuals from other subpopulations used the S2 egg-laying sites in wet years. **(b)** After human alterations (hydroelectric dams, logging, road-building, and mining), the bull trout metapopulation is reduced and more fragmented. The s1 subpopulation (formerly S1) is virtually isolated by mining operations that release toxic metals and nitrates (from dynamite) into the stream. The s2 group is even more isolated than it was historically; logging operations and roads have increased the rate at which water exits the area, thereby reducing the time when the intermittent stream contains water. Clear-cut areas and roads have also increased the amount of silt in neighboring portions of the streams, and increased siltation has destroyed some bull trout egg-laying sites. The s3 subpopulation is reduced in size and distribution by the effects of mining, logging, and roads. Two hydroelectric dams on the river fragmented the original river/lake subpopulation (S4 in part a), creating a fifth subpopulation (s5). The s4 subpopulation has been reduced in numbers because relatively few individuals manage to overcome human-caused obstacles and ascend to egg-laying sites. The s5 fish are completely cut off from the other subpopulations and have lost access to many of their historic breeding sites. Saving the bull trout will require careful scrutiny of their metapopulations and strong measures to prevent further loss of egg-laying sites and access to them. This combination of habitat alterations represents the complex nature of problems many species and conservation biologists now face.

fish native to northwestern North America, is often segregated into three kinds of subpopulations (FIGURE 55.8). One kind of bull trout subpopulation consists mainly of individuals that reside in high-mountain streams throughout life. In a second kind, the majority of individuals spend most of their time in streams but feed intermittently in large rivers and lakes. Bull trout in a third type of subpopulation spend most of their time in lakes and large rivers and migrate into streams only to reproduce. Human activities, especially logging, mining, and dam-building, have reduced the size and number of habitat patches and in turn the size and number of many bull trout subpopulations. Many bull trout metapopulations have suffered drastic declines, and the species was recently declared endangered in the United States. Understanding the dynamics of metapopulations is a key part of efforts to save the bull trout and many other species.

Source and Sink Dynamics in Metapopulations

For many metapopulations, reproductive rates are markedly different from one habitat patch to another. An area of habitat where a population's reproductive success exceeds mortality, and from which excess individuals disperse, is called a **source habitat**. In contrast, a habitat where mortality exceeds reproductive success is a **sink habitat**. In some metapopulations, the percentage of habitat patches that are occupied remains more or less constant because of dispersal from source to sink habitats.

Conservation efforts often require distinguishing sources from sinks. Field studies and computer simulations indicate that simply censusing subpopulations to determine where the greatest numbers of individuals live is potentially misleading. In some cases, source habitats may contain as little as 10% of a metapopulation. Distinguishing sources from sinks requires detailed analyses of birthrates and death rates of individuals in various habitat types. It is also important to identify the conditions that affect dispersal from sources to sinks and the specific combination of habitat factors that is critical for a species. If this critical combination is eliminated, a habitat may become a sink. For instance, the red-cockaded woodpecker of the southeastern and south central United States, which requires open conifer forests or savannas maintained by periodic fires, became endangered when fire suppression resulted in a thicker forest understory (FIGURE 55.9).

Sustaining many of the metapopulations created or altered by human-caused habitat fragmentation requires identifying and protecting source habitats. Until recently, efforts to sustain the peregrine falcon (*Falco peregrinus*) in southern California depended on stocking wild subpopulations with birds raised in captivity. However, when researchers discovered that southern areas were sink habitats and that a peregrine population in northern California was a source of colonizing individuals for the southern subpopulation, they realized that the best way to sustain the southern population was to carefully monitor and bolster the northern population.

Understanding source and sink dynamics is also essential to designing the most effective nature reserves. Working in habitats fragmented by logging activities, Kevin McKelvey and his associates in the U.S. Forest Service found that adding areas that turn out to be sink habitats can hamper attempts to

(a)

(b)

(c)

FIGURE 55.9 ▪ Habitat requirements of the red-cockaded woodpecker (*Picoides borealis*). (a) This endangered species and many others have specific habitat needs essential to their survival. **(b)** Preferred habitats of the red-cockaded woodpeckers are open stands of mature pine forests, preferably longleaf pine (*Pinus palustris*) as shown here. Breeding pairs of woodpeckers excavate nest cavities in mature pines (usually 60+ years old) with heartwood rotted and softened by fungi. **(c)** A habitat becomes a sink, with breeding birds tending to abandon nest cavities, when vegetation among the pines is thick and higher than about 15 feet. Apparently, the birds require a clear flight path between their home trees and neighboring feeding grounds. Historically, forest and savanna communities with a low understory dominated by longleaf pine were common in seasonally dry areas in the southeastern United States. Regular fires destroyed oaks and other hardwoods that tend to fill spaces among the mature pines. With increased human population, logging, and fire suppression, longleaf pine communities and red-cockaded woodpeckers became rare. The bird was listed as an endangered species in 1970. Recently, populations of *P. borealis* have recovered to sustainable levels through intensive management of private and public timberlands and nature preserves. Use of fire to control the underbrush is a key management tool.

sustain metapopulations of the northern spotted owl (*Strix occidentalis caurina*). This species nests in old-growth rain forests of the Pacific Northwest, and its sink habitats include immature forests in various stages of regrowth after logging. If sink habitats surround relatively small source areas, dispersing owls tend to colonize the sinks and fail to reproduce. Considering the shortage of old-growth forests in the area, McKelvey concluded that a key element of an effective spotted owl reserve is a distinct boundary that prevents juvenile birds from dispersing to inhospitable habitats.

Population viability analyses examine the chances of a species persisting or becoming extinct in the habitats available to it

A relatively new and increasingly popular approach to conservation problems, **population viability analysis (PVA)**, is a method of predicting whether or not a species will persist in a particular environment. Population viability analyses incorporate information on a population's genetic variability and life history characteristics, such as sex ratios, age at first reproduction, fecundity, and average birthrates and death rates (see Chapter 52). They also take into account the available data on the population's response to environmental factors such as predation, parasitism, interspecific competition, and the disturbances characteristic of the population's habitat. A PVA is usually generated by computer simulations that incorporate life history data along with mathematical estimates of the population's responses to environmental factors. For instance, useful data for a PVA on a metapopulation of a freshwater fish species such as the bull trout (see FIGURE 55.8) would be the size of the breeding population in each subpopulation, the reproductive rates of each subpopulation, the number of other fish species present (including predators and prey of the bull trout), the availability of other food species (aquatic insects and other invertebrates), the amount of suitable habitat including breeding sites, and the amount of gene flow among the subpopulations. For the bull trout in many areas, a PVA would also factor in the effect of an exotic species. The brook trout, a close relative of the bull trout, has been introduced from the eastern United States throughout much of the bull trout's range (FIGURE 55.10). Brook trout often outcompete bull trout, and anglers may inadvertently mistake a bull trout, which as an endangered species must be released, from a brook trout, which can be legally kept.

A PVA usually predicts viability over the long term, so periodic natural catastrophes, such as 50-year or 100-year floods, are often factored in. However, because threatened populations are usually small, their survival often turns on chance events, and a significant component of population viability analyses is an assessment of the role of chance in causing extinction. A PVA would also consider the potential effects of any human activities, such as logging or mining, that would chemically and physically alter portions of the metapopulation's habitat.

(a)

(b)

FIGURE 55.10 ▪ Two closely related, historically isolated, species of fish whose ranges now overlap. (a) The bull trout (*Salvelinus confluentus*) and **(b)** its close relative, the brook trout (*Salvelinus fontinalis*). Historically, the brook trout was found only in streams and lakes of northeastern North America. Now widely introduced throughout the mountainous west, it can outcompete bull trout and poses a serious threat to them in many areas. (Ironically, many brook trout populations have been seriously reduced in this species' original range.)

The simplest population viability analysis predicts the likelihood that a population will persist for a given number of generations based on current habitat conditions and population numbers. Following an initial prediction, more extensive PVAs may be produced by repeated modeling and research. For example, a field researcher might adjust the number of predators, competitors, or type of habitat features. If multiple populations or subpopulations exist, experiments may be run on the effects of adjustments on some of these units while others serve as controls.

Estimating Minimum Viable Population Size (MVP)

Some of the most immediately pressing conservation work concerns species whose populations are small, fragmented, and threatened with extinction. Thus, population viability analyses are often geared toward predicting a species' **minimum viable population size (MVP)**, the smallest number of individuals needed to perpetuate a population, subpopulation, or species. Such predictions usually specify a percent chance of population survival (for example, 99%) and a time frame (for example, 1000 years). Field studies and computer simulations indicate that MVP sizes vary over a wide range. At the lower end, some populations of rare birds that have been monitored for nearly a century have persisted with as few as ten breeding pairs. MVP estimates provide a rough baseline for planning species recovery programs. Whenever feasible, the goal is to maintain populations well above minimal size.

Estimates of MVP underlie predictions of the **minimum dynamic area**, the amount of suitable habitat needed to sustain a viable population. The minimum dynamic area can be estimated by combining MVP estimates with data on home range (the area over which an individual or group of individuals range in a year). A population's size and the minimum dynamic area must be large enough to accommodate the disturbances that characterize the species' natural habitat.

Estimating Effective Population Size (N_e)

Concerned with genetic diversity and the capacity of populations to function as evolving units, many biologists argue that meaningful estimates of MVP require determination of the **effective population size (N_e)**, which is based on the number of adults that successfully breed (contribute gametes to the next generation). The following formula estimates N_e using the sex ratio of breeding individuals:

$$N_e = \frac{4N_m N_f}{N_m + N_f}$$

where N_f and N_m are, respectively, the numbers of females and males that successfully breed. Applying this formula to an idealized population whose total size is 1000 individuals, N_e will also be 1000 if every individual breeds and the sex ratio is 500 females:500 males. In this case, $N_e = (4 \times 500 \times 500) / (500 + 500) = 1000$. Deviation from these conditions (not all individuals breed and/or there is not a 50:50 sex ratio) reduces N_e. For instance, if the total population size is 1000, but only 400 females breed with 400 males, $N_e = (4 \times 400 \times 400) / (400 + 400) = 800$, or 80% of the total population size.

N_e is always a fraction of the total population, thus simply censusing the population—determining the total number of individuals—does not provide a good measure of the number of individuals needed to prevent extinction. Whenever possible, conservation programs are geared to sustain total population sizes that include, at least, the minimum viable number of reproductively active individuals. Numerous life history traits can influence N_e, and alternate formulas for estimating N_e take into account family size, maturation age, genetic relatedness among population members, the effects of gene flow between geographically separated populations, and population fluctuations.

The Effect of Genetic Diversity on Survivability

The desire to sustain effective population sizes, N_e, stems from the concern that populations retain enough genetic diversity to be evolutionarily adaptable. Populations with low N_e are prone to inbreeding, reduced heterozygosity, and the random effects of genetic drift and bottlenecking (see Chapter 23). However, for many species—especially those that reproduce slowly, such as the cheetah, European brown bear, and grizzly bear—small populations with low genetic variability seem to be the norm. A number of plant species, such as the lousewort *Pedicularis* and several grasses, also seem to have inherently low genetic variability. Furthermore, low genetic variability does not necessarily lead to permanently small populations. For example, after being reduced by hunting to only about 20 individuals, northern elephant seal populations rebounded to a total of about 30,000 individuals today. Among plants, many populations of cord grass (*Spartina anglica*), which thrives in saltmarshes, are genetically uniform at many loci. *S. anglica* arose from a few parent plants only about a century ago by hybridization and allopolyploidy (see Chapter 24). Spreading by cloning, this species now dominates large areas of tidal mudflats in Europe and Asia. Thus, in some cases, low genetic diversity is associated with population expansion rather than decline.

Reduced genetic variability alone may or may not be crucial to the survival of wild populations. A 1998 study of a metapopulation of the fritillary butterfly (*Melitaea cinxia*) in Finland indicated that inbreeding can reduce larval survival, adult longevity, and egg-hatching rate and eventually cause extinction in small, isolated subpopulations. However, for most species, habitat degradation and loss usually take their toll before genetic effects have a chance to cause significant harm. Nonetheless, maintaining genetic diversity is a priority in the long-term management of species whose original populations have been drastically reduced and fragmented by human activities. Genetic considerations are also critical for managing captive populations in sanctuaries and zoos, from which future reintroductions into the wild may be made.

Analyzing the viability of selected species may help sustain other species: *science as a process*

Because modeling requires extensive background research and time, it is likely that populations of relatively few of the many threatened or endangered species will be systematically analyzed. Furthermore, no single mathematical formula or computer model incorporates all factors that affect N_e or

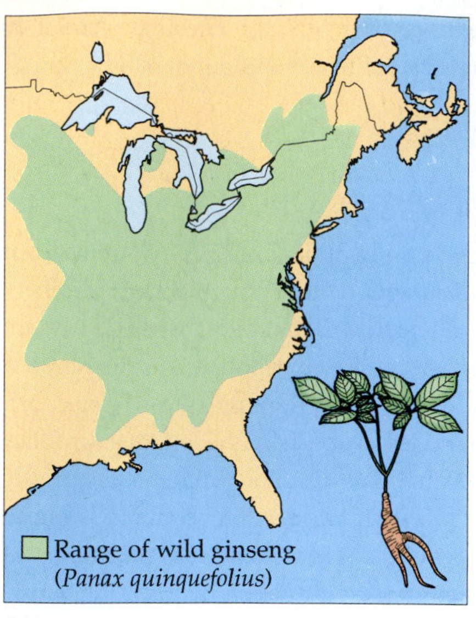

(a)

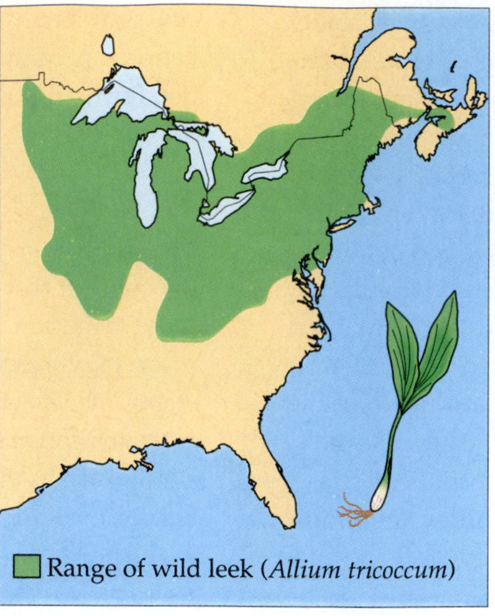

(b)

FIGURE 55.11 ▪ Two species of edible plants whose persistence is threatened by habitat loss and overharvesting. **(a)** Distribution of American ginseng (*Panax quinquefolius*), whose roots bring high prices for their medicinal effects. **(b)** Distribution of wild leek (*Allium tricoccum*), valued for its edible bulb. These perennial herbs are found in deciduous forest communities in eastern North America. Plant ecologists Patrick Nantel, Andrée Nault, and Daniel Gagnon have analyzed the potential viability of the plants in southeastern Canada. Populations of both species are in decline, and many have become extinct because of human activities. Increasing loss of forest habitats to urbanization is a significant factor, and fragmentation of remaining habitats has made the plants more accessible to human harvest. Minimum viable population sizes generated by computer models were about 170 ginseng plants and between 300 and 1030 leek plants. With only about 20 known ginseng populations in Canada having more than 170 individuals, and with leek populations of more than a few hundred being rare, close monitoring of human harvest is essential to the long-term persistence of these forest plants.

MVP. Most often, because of the immediacy of extinction threats, the available data dictate the model used to make estimates. Wildlife biologist Mark Shaffer, a pioneer of population viability analyses, once summarized the status of conservation science this way: "Conservation biology operates in a model-rich but data-poor world, and the tub of species we need to know something about is filling rapidly." However, an increasing number of viability studies are being conducted, and what is learned from a select group of species may help us develop strategies to sustain other species. Let's examine some actual examples of population viability analyses.

For his doctoral dissertation in environmental sciences in 1994 at Université du Québec, Patrick Nantel presented a population viability analysis of two edible herbaceous plants, American ginseng (*Panax quinquefolius*) and wild leek (*Allium tricoccum*). Both of these plants are currently in serious decline in North America (FIGURE 55.11). Nantel's PVA incorporated data on trends in the numbers of individuals capable of reproducing in several populations for two-, three-, and four-year periods. Computer simulations projected the likely effect of environmental influences on these populations. The PVA indicated that most populations of American ginseng and wild leek in Canada are currently too small to persist if harvested at all. Nantel's work is an example of the increasing use of predictive models in planning conservation strategy.

Mark Shaffer, currently with the Wilderness Society, performed one of the first population viability analyses as part of a long-term study of grizzly bears in Yellowstone National Park and its environs (FIGURE 55.12). Grizzly bears require very large areas of habitat. For instance, estimates of the grizzly's minimum habitat needs in western Canada are about 5 million hectares for a population of 50 individuals and about 200 million hectares for 1000 individuals. A threatened species in the United States, the grizzly is currently found in only four of the 48 contiguous states. Its population in those states has been drastically reduced and fragmented: In 1800, an estimated 100,000 individuals ranged over about 500 million hectares of more or less continuous habitat, while today there are six virtually isolated subpopulations totaling about 1000 individuals with a total range of less than 5 million hectares. Most of these subpopulations have fewer than 100 individuals (at least one currently has fewer than 20). The Yellowstone subpopulation is the largest, with about 200 bears in an area of about 1 million hectares.

Attempting to determine viable sizes for the U.S. grizzly subpopulations, Shaffer used life history data obtained for individual Yellowstone bears over a 12-year period and simulated the effects of environmental factors on survival and reproduction. His models predicted that a total grizzly bear population of 70 to 90 individuals in suitable habitat will have about a 95% chance of surviving for 100 years. Achieving a 99% chance of survival for a century or a 95% chance for 200 years requires enough habitat to support at least 100 bears. Because of habitat limitations, however, recovery targets for several of the U.S. subpopulations have been tentatively set at fewer than 100 bears. In such cases, biologists are hopeful that the small populations can be sustained by careful monitoring and special protective measures.

Concerned that policy decisions have been made without information on potential losses of genetic variability in grizzly

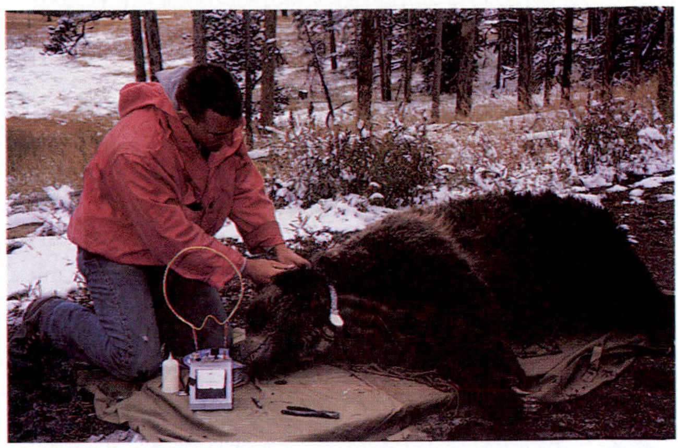

FIGURE 55.12 ▪ Long-term monitoring of a grizzly bear population. Wildlife biologists John and Frank Craighead and graduate student researchers studied population dynamics of the grizzly bear (*Ursus arctos*) in Yellowstone National Park for several decades. The bear in this photograph was being fitted with a radiotransmitter so that its movements could be tracked and compared with other individuals in the population. The Yellowstone bears make up one of a handful of small subpopulations of grizzlies in the contiguous United States. Grizzly bears require extensive areas of remote habitat, and maintaining viable populations involves political as well as ecological decisions.

bear populations, Fred Allendorf and his co-workers at the University of Montana developed a computer model that augmented Shaffer's work. Using detailed life history and kinship data from individual bears in subpopulations in Montana, Wyoming, and British Columbia, Allendorf's model estimated that the effective population size (N_e) of grizzly populations is only about 25% of the total population size. Usually only a few dominant males breed; locating females may be difficult since individuals inhabit extensive areas, and females may reproduce only when there is abundant food. Thus, even the relatively large Yellowstone population of 200 bears has an effective population size of only 50, a level that Allendorf contends would lead to a loss of genetic variability and possibly fitness.

Allendorf also developed a model to predict the effect of introducing bears from distant areas into isolated subpopulations. His results indicate that introducing only two unrelated bears each decade into populations of 100 individuals would reduce the loss of genetic variation by about half. For the grizzly bear, and probably for many other species whose populations are very small, finding ways to promote dispersal among subpopulations may be one of the most urgent conservation needs.

Conserving species involves weighing conflicting demands

Determining viable population numbers and habitat needs is only part of the effort to save species. Most often it is necessary to weigh a species' biological and ecological needs against other conflicting demands. Thus, conservation biology often highlights the relationships between science, technology, and society—one of the themes of this book. For example, an ongoing, sometimes bitter debate in the U.S. Pacific Northwest pits saving habitat for populations of the northern spotted owl, timber wolf, grizzly bear, and bull trout against demands for jobs in the timber, mining, and other resource extraction industries. Programs to restock wolves and to bolster the populations of grizzly bears and other large carnivores are opposed by some recreationists concerned for their safety and by many ranchers concerned with potential losses of livestock.

Large, high-profile vertebrates are not always the focal points in conflicts involving conservation biology, but habitat use is almost always at issue. Should work proceed on a new highway bridge if it destroys the only remaining habitat of a species of freshwater mussel? If you were the owner of a coffee plantation growing varieties that thrive in bright sunlight, do you think you would be willing to change to shade-tolerant coffee varieties that are less productive and less profitable but support large numbers of songbirds?

In addition to questions about human habitat needs, another important factor to weigh is the ecological role of species. Since we will not be able to save every endangered species, we must determine which ones are most important for conserving biodiversity as a whole. Species do not exert equal influence on community and ecosystem processes. Some organisms, called *keystone species,* have disproportionately large impacts relative to their numbers (see Chapter 53). Some keystone species significantly modify habitats, creating diverse patches that support numerous other species. Keystone mutualists provide other species with nutrients, defense against predators and parasites, or, in the case of pollinators, the means to reproduce. Identifying keystone species and finding ways to sustain their populations can ensure the continuance of numerous other species and can be central to the survival of whole communities.

CONSERVATION AT THE COMMUNITY, ECOSYSTEM, AND LANDSCAPE LEVELS

Most preservation efforts in the past have focused on saving endangered species, but today conservation efforts are increasingly aimed at sustaining the biodiversity of entire communities and ecosystems. On a broader scale yet, the principles of community and ecosystem ecology are being brought to bear on studies of the biodiversity of whole landscapes. In an ecological sense, a *landscape* is a regional assemblage of interacting ecosystems, such as a forest or forest patches, adjacent open fields, wetlands, streams, and streamside (riparian) habitats. **Landscape ecology** is the application of ecological principles to the study of land-use patterns. Understanding landscape dynamics is critically important in conservation because many species use more than one kind of ecosystem, and many live on the borders between ecosystems. The goal of landscape ecology, of which ecosystem management is part, is to understand patterns of landscape use in the past, present, and foreseeable future and to make species conservation a functional part of the picture. Such a broad view requires understanding regional community and ecosystem ecology as well as human population dynamics and economics (see the Methods Box).

Edges and corridors can strongly influence landscape biodiversity

The boundaries or *edges* between ecosystems (such as a lake and the surrounding forest, or cropland and suburban housing tracts) and within ecosystems (such as roadsides and rock outcroppings) are defining features of landscapes (FIGURE 55.13). An edge has its own set of physical conditions, such as soil type, topography, and disturbances, that differ from those on either side. For instance, the soil surface of an edge between a forest patch and a burned area receives more sunlight and is usually hotter and drier than the forest interior but cooler and wetter than the soil surface in the burned area. Blown-down trees are a common disturbance feature of forest edges, which are less protected from strong winds than are forest interiors.

Associated with their specific physical features, edges also have their own communities of organisms. Some organisms thrive in edge communities because they require resources of the two adjacent areas. For instance, the ruffed grouse (*Bonasa umbellatus*) needs forest habitat for nesting, winter food, and shelter, as well as forest openings with dense shrubs and herbs for summer food. White-tailed deer also thrive in edge habitats, where they can browse on woody shrubs, and deer populations often expand when forests are logged.

The proliferation of edge species can have positive or negative effects on a community's biodiversity. A recent study of edge communities in a tropical rain forest in Cameroon showed that these areas may be important sites of speciation. On the other hand, communities in which edges have proliferated due to human alterations often have reduced biodiversity due to the preponderance of edge-adapted species. In one example, populations of the brown-headed cowbird (*Molothrus ater*), an edge-adapted species that lays its eggs in the nests of other birds, are currently expanding in many areas of the western United States. Cowbirds forage in open fields on insects disturbed by or attracted to cattle and other large herbivores, but they need forests where they can parasitize the nests of other birds. Cowbird numbers are burgeoning where forests are being heavily cut and fragmented, creating more edge habitat and open land for cattle, horses, and sheep. Increasing cowbird parasitism and loss of habitat are correlated with declining populations of several of the cowbird's host species—migratory songbirds such as the yellow warbler, red-eyed vireo, and American redstart.

Another important landscape feature, especially where habitats have been severely fragmented, is a **movement corridor**, a narrow strip or series of small clumps of quality habitat connecting otherwise isolated patches. Streamside habitats often serve as corridors, and government policy in some nations prohibits destruction of these riparian areas. In areas of heavy human use, artificial corridors are sometimes constructed. For example, highways bisect habitat patches required for survival of the few remaining Florida panthers

(a)

(b)

FIGURE 55.13 ▪ Edges between ecosystems. (a) This photograph illustrates a regional landscape. Edges emphasize the distinctness of dry forest habitat, a rocky area with grassy islands, and a flat, grassy lakeshore in Kakadu National Park in northern Australia. **(b)** Human activities that degrade and fragment habitats often create edges that are more abrupt than those delineating natural landscapes. Pronounced edges surround clear cuts in this photograph of a heavily logged rain forest in Malaysia.

METHODS: Using Computer Mapping Techniques to Inventory and Manage Landscapes: Gap Analysis

Developing a landscape inventory of an area, such as a nation, state, region, or continent, can lead to plans that integrate human needs and biodiversity. Modern mapping and imaging technology now enables ecologists to inventory the major physical features, land ownership, and biodiversity of ecosystems. A method called **gap analysis** uses computerized maps, especially those generated by satellite sensing and aerial photography, along with information on the distribution of species. The most useful gap analyses include confirmation of satellite and aerial data by sampling in the field.

One of the original goals of gap analysis was to discover unprotected habitat areas vital to the survival of rare, endemic populations. Computer software superimposes maps of vegetation and animal distribution on geophysical and land-use maps. Gap analysis often indicates that protected areas do not include some significant habitats of endangered species, especially rare, endemic ones (FIGURE a). The map of a portion of southern California in FIGURE b was created by the National Gap Analysis Program of the U.S. Geological Survey. In this case, the apparent distribution of the orange-throated whiptail, an endemic lizard, is mainly outside protected areas.

Once habitat "gaps" are identified, steps can be taken to help sustain biodiversity. A gap analysis in Hawaii showed that endangered species of forest birds were concentrated outside protected areas. The study led to the establishment of the Hakalau Forest National Wildlife Refuge in an area rich in endemic bird life on the island of Hawaii.

Several nations, including the United States, are using gap analysis to map ecosystems comprising their land areas. Gap analysis also enables graphic correlations of distributional patterns of organisms with abiotic factors, such as temperature, precipitation, and elevation.

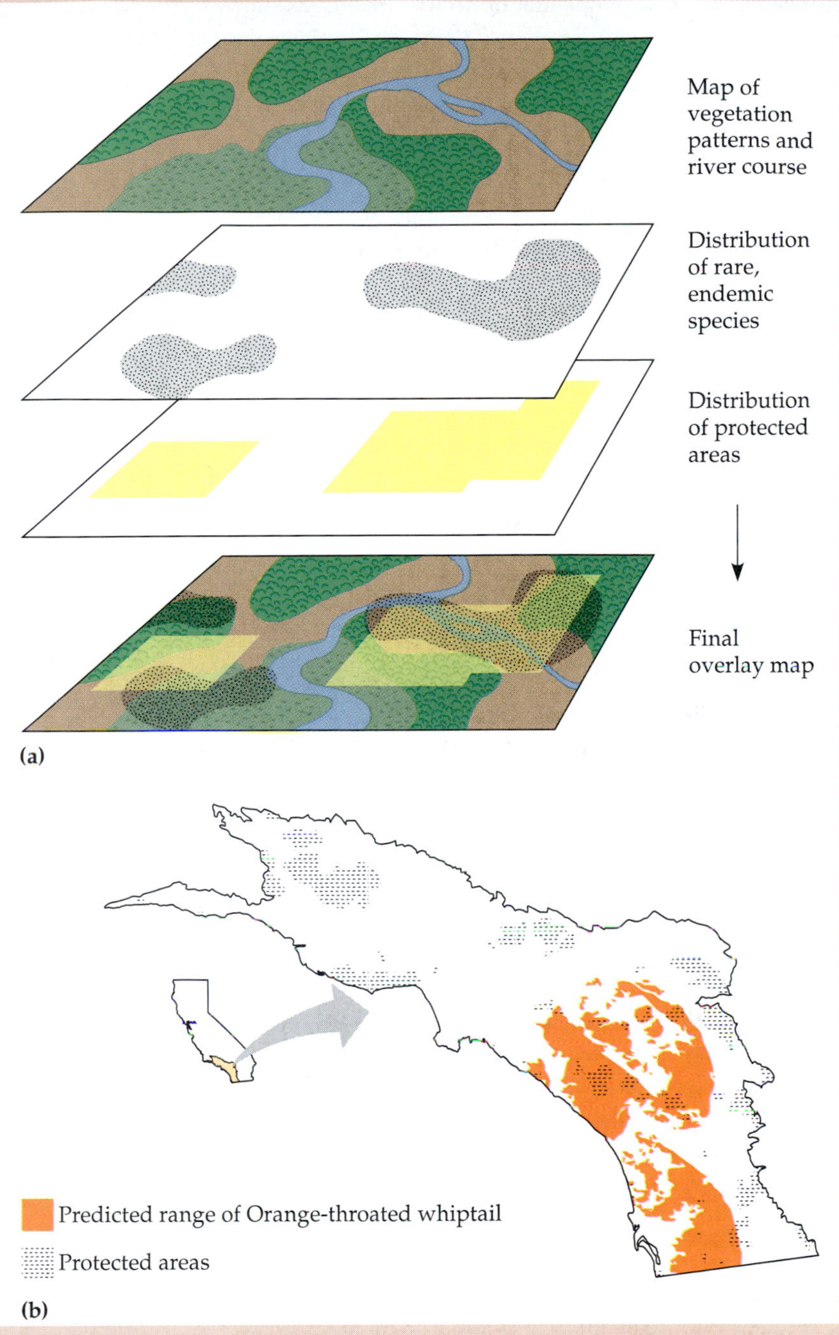

(*Felis concolor coryi*). The state of Florida has erected high fences to reduce road kills and underpasses to allow movements among protected areas for the panthers.

Corridors can promote dispersal and help sustain metapopulations, and they are especially important to species that migrate among different habitats seasonally. On the other hand, a corridor can be harmful—as, for example, in the spread of diseases, especially among small subpopulations in closely situated habitat patches. The effects of corridors have not been thoroughly studied, and researchers tend to evaluate their potential effects on a case-by-case basis.

Nature reserves must be functional parts of landscapes

Terrestrial and aquatic parks, wilderness areas, and other legally protected nature reserves are an important component of efforts to sustain species. In some areas, these refuges are the sole or main sites where rare and endangered species reside. Metaphorically, nature reserves are islands in a sea of habitat degraded to varying degrees by human activity. It is important to realize, however, that protected "islands" are not isolated from their surroundings, and that nonequilibrium ecology applies to nature reserves as well as the landscapes in which they are embedded.

An older policy—that protected areas should be set aside to remain unchanged forever—is consistent with the concept that ecosystems are balanced, self-regulating units. However, as we have discussed, disturbance over ecological time is a functional component of all ecosystems, and management policies that ignore natural disturbances or attempt to prevent them have generally proved to be self-defeating. For instance, setting aside an area of a fire-dependent community (such as a portion of a tallgrass prairie, chaparral, or dry pine forest) with the intention of saving it is unrealistic if periodic burning is excluded. Without the dominant disturbance, the fire-adapted species are usually outcompeted by other species, and biodiversity is reduced.

Because human disturbance and fragmentation are increasingly dominant landscape features, patch dynamics, metapopulation dynamics, edges, and corridor effects are important in the design and management of protected areas. Unfortunately, there are many more questions than answers. For example, is it better to create one large preserve or a group of smaller preserves that collectively include the same total area? One argument for extensive preserves is that large, far-ranging animals with low-density populations, such as the grizzly bear, require extensive habitats. In addition, more extensive areas have proportionately smaller perimeters than smaller areas and are therefore less affected by edge effects. An argument favoring smaller, disjunct preserves is that they may slow the spread of disease throughout a population. Often outweighing all other considerations, recent and ongoing land use by humans may largely dictate the size and shape of protected areas.

Several nations have adopted an approach to landscape management called **zoned reserve systems** (FIGURE 55.14). Protected areas are surrounded by lands that are used and altered more extensively by human activity. The challenge is to develop in the surrounding lands a social and economic climate that sustains itself while remaining compatible with the long-term viability of the protected core area. As ecologist Daniel Janzen, a leader in tropical conservation, has said, "The likelihood of long-term survival of a conserved wildland area is directly proportional to the economic health and stability of the society in which that wildland is embedded."

The continued high rate of human exploitation of ecosystems leads to the prediction that considerably less than 10% of the biosphere will ever be protected as nature reserves. Sustaining biodiversity often involves working in landscapes that are almost entirely human dominated. The Florida scrub jay, an endangered endemic species, inhabits dry scrub oak communities that have nearly been replaced by housing developments and citrus groves (FIGURE 55.15). Attempting to understand if this species could coexist with human development, avian ecologist Reed Bowman at Archbold Biological Station in central Florida examined scrub jay population viability across a gradient of human density. Unfortunately, housing developments, even if they contain some scrub habitats, turned out to be sink habitats for the jay. Bowman is now convinced that longterm survival of this bird depends on reserves of contiguous, intact scrub surrounded by areas where some natural vegetation remains.

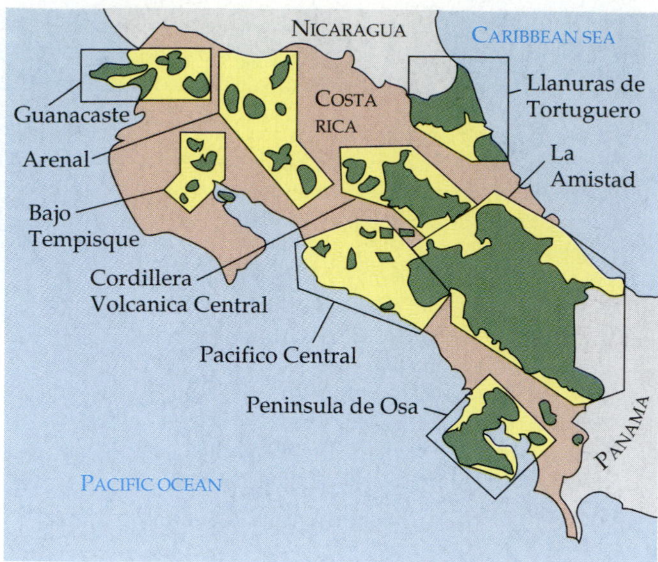

FIGURE 55.14 • The zoned reserve concept of landscape management. This map shows one nation's attempt to accommodate biodiversity. With international financial and scientific help, Costa Rica has established several conservation areas, or zoned reserves. The green areas on the map are national park lands, core areas where human disruption is minimized. Surrounding the cores, yellow areas are buffer zones, or transition areas, mainly privately owned, where most of the human population resides and works. Ideally, the most destructive practices—industries such as mining, large-scale monoculture, and major new urban developments—are confined to the outermost fringes of the transition areas. In the buffer zones, progress has been made in promoting sustainable agriculture and forestry, activities that provide steady economic support for local residents without drastically altering habitats.

Restoring degraded areas is an increasingly important conservation effort

Eventually, some areas that are altered and degraded by human activity are abandoned. For instance, the soils of many tropical areas become unproductive and are abandoned less than 5 years after being cleared for farming. Mining activities may last for several decades, and lands are then abandoned in a degraded state. Many ecosystems are also damaged inadvertently by such mishaps as oil spills. These degraded habitats and ecosystems are increasing in area because the natural rates of recovery by successional processes are slower than the

(a)

(b)

FIGURE 55.15 ▪ An endangered, endemic species in its unique habitat. (a) The Florida scrub jay (*Aphelocoma coerulescens*) inhabits desertlike scrub communities in central Florida. **(b)** New housing developments and expanding citrus groves threaten this bird and the remaining fragments of its unique habitat. Ecologists at the Archbold Biological Station, a small reserve with a stable scrub jay population, have found that housing developments do not provide enough food (arthropods) for the jays and are associated with higher mortality of adults. Even if a housing development contains some scrub habitat, it is a difficult place for these birds to rear enough young to offset the increased death rate. Archbold researchers predict that the Florida scrub jay will survive only if reserves of intact oak scrub habitat are maintained and properly managed with prescribed fire.

rate of degradation by human activities. A new subdiscipline of conservation biology called **restoration ecology** applies ecological principles in developing ways to return degraded ecosystems to conditions as similar as possible to their natural, predegraded state.

Two key strategies in restoration ecology are bioremediation and augmentation of ecosystem processes. **Bioremediation** is the use of living organisms, usually prokaryotes, fungi, or plants, to detoxify polluted ecosystems. Several extremophilic bacteria and archaea thrive in environments similar to industrially polluted sites. Some success has been achieved in using the bacterium *Pseudomonas,* supplied with growth stimulants, to clean up oil spills on beaches. A number of researchers are focusing on the ability of certain prokaryotes, lichens, and plants to concentrate metals. Researchers in the United Kingdom recently discovered a lichen species that grows on soil polluted with uranium dust left over from mining. Useful as a biological monitor of uranium and potentially as a remediator, the lichen concentrates uranium in a dark pigment similar to melanin in human skin. Experiments indicate that some plants not only extract metals from contaminated soils but concentrate them in commercially marketable quantities.

Augmenting ecosystem processes requires determining what factors, such as chemical nutrients, have been removed from an area and are limiting its rate of recovery. Encouraging the growth of plants that thrive in nutrient-poor soils often speeds up the rate of successional changes that can lead to recovery of damaged sites. Ariel Lugo, director of the U.S. Forest Service's Institute of Tropical Forestry in Puerto Rico, has evidence of a positive effect of an exotic plant species on the recovery of native vegetation (FIGURE 55.16). Thriving on nitrogen-poor soils, the leguminous plant *Albizia procera,*

FIGURE 55.16 ▪ Restoration of degraded roadsides in the tropics. Forest ecologist Ariel Lugo has monitored rapid regrowth of indigenous communities along roadsides in Puerto Rico. An exotic plant, *Albizia procera* (shown here), which thrives on nitrogen-poor soils, first colonized these sites after the original forest was removed and soils were depleted of nutrients. Apparently, the rapid buildup of organic material from dense stands of *Albizia* enabled indigenous plants to recolonize the area and overgrow the exotic plant in a relatively brief time.

exotic in Puerto Rico, helps set the stage for recolonization by native tropical forest species.

To date, the most extensive and successful restoration projects have been in marginally disturbed wetlands in landscapes where biodiversity has not been greatly depleted. In these projects, restoring the natural water-flow patterns and replanting indigenous vegetation has led to recolonization by animal populations. Restoring viable populations of highly sensitive wetland species to heavily degraded wetlands is much more challenging, as are similar restoration efforts in most ecosystems.

Because of the novelty of restoration science, the complexity of ecosystems, and the unique features of each situation, restoration ecologists usually must learn as they go. As Princeton University biologist Andrew Dobson puts it, "It is a relatively straightforward exercise to take apart an ecosystem or an automobile engine. In contrast, reassembling the engine (or the ecosystem) will reveal a deeper level of understanding of how each of its components functions." Approximating the original, not duplicating it, is the goal of restoration ecology.

Sustainable development goals are reorienting ecological research and will require changing some human values

Facing increasing loss and fragmentation of habitats, how can we best manage Earth's resources? If we are to conserve most of a nation's species, which habitat patches are most crucial? Among the limited choices, which areas are most practical to protect and manage if we are to save rare species or the greatest number of species?

Clearly, we must understand the complex interconnections
55.1 of the biosphere if we are to make sensible decisions about how to conserve them. To this end, many nations, scientific societies, and private foundations have embraced the concept of **sustainable development**, the long-term prosperity of human societies and the ecosystems that support them. The forward-looking Ecological Society of America, the world's largest organization of professional ecologists, endorses a research agenda termed the **Sustainable Biosphere Initiative.** The goal of this initiative is to define and acquire the basic ecological information necessary for the intelligent and responsible development, management, and conservation of Earth's resources. The research agenda includes studies of global change, including interactions between climate and ecological processes; biological diversity and its role in maintaining ecological processes; and the ways in which the productivity of natural and artificial ecosystems can be sustained. This initiative requires a strong commitment of human and economic resources.

The challenges presented by declining biodiversity, an expanding human population, and environmental degradation are pressing. We must tackle them before we complete all the scientific studies that would allow us to plot the surest path. Despite the uncertainties, now is not a time for gloom and doom, but a time to aggressively pursue more knowledge about life, and to take heart. Quoting Michael Soulé, cofounder of the Society for Conservation Biology, "There are no hopeless cases, only people without hope and expensive cases."

Sustainable development requires a commitment that few people or nations have undertaken to promote ecosystem processes and biodiversity. Those of us living in affluent developed nations are responsible for the greatest amount of environmental degradation. Reality demands that we change some of our values, learn to revere the natural processes that sustain us, and reduce our orientation toward short-term personal gain. The current state of the biosphere demonstrates that we are treading precariously on uncharted ecological ground, and that the importance of our scientific and personal efforts cannot be overstated.

Conservation science is an intersection of numerous facets of biology, including ecology, evolution, physiology, molecular biology, genetics, and behavior. Efforts to sustain ecosystem processes and stem the loss of biodiversity also connect life science with the social sciences, economics, and humanities. By studying life and its diversity and making these connections, we become more aware of ourselves and the roles each of us plays in the biosphere.

CHAPTER REVIEW

REVIEW OF KEY CONCEPTS

(with page numbers and key figures)

THE BIODIVERSITY CRISIS: AN OVERVIEW

Although extinction is a natural process, the current rate of species extinction is alarmingly high.

- **Numerous examples indicate that estimates of extinction rates are on track (p. 1155)**
- **The major threats to biodiversity are habitat destruction, overexploitation, and competition by exotic species (pp. 1155–1156, FIGURE 55.2)** Human alteration of habitat poses the single greatest threat to biodiversity in terrestrial and aquatic ecosystems. Competition by nonnative species and excessive harvesting for commerce and sport are other significant threats.
- **Biodiversity is vital to human welfare (pp. 1156–1157, FIGURE 55.3)** Other species provide humans with food, fiber, medicines, and a sense of connectedness with nature. Estimates by ecologists

and economists indicate the enormous economic value of ecosystem services.

- **Change in ecological and evolutionary time is the focus of conservation biology (pp. 1157–1158)** The modern science of conservation emerged from earlier concepts of preservationism, resource conservation, and multiple use. Emphasizing evolutionary principles, nonequilibrium ecology, and a concern for the human presence, conservation biology's goal is to ensure that the evolutionary processes that have created populations, communities, and ecosystems are perpetuated.

THE GEOGRAPHIC DISTRIBUTION OF BIODIVERSITY

- **Gradual variation in biodiversity correlates with geographical gradients (pp. 1158–1159, FIGURE 55.4)** Clines (gradual variations) in species diversity correlate with latitudinal gradients on land and with depth in marine benthic environments.
- **Biodiversity hot spots have high concentrations of endemic species (pp. 1159–1160, FIGURE 55.5)** Areas with exceptionally high concentrations of endemic species, called biodiversity hot spots, are also hot spots of extinction.
- **Migratory species present special problems in conservation (p. 1160, FIGURE 55.6)** Species that migrate internationally can be especially vulnerable to extinction.

CONSERVATION AT THE POPULATION AND SPECIES LEVELS

The biodiversity crisis extends over a hierarchy from the genetic makeup of populations to communities, ecosystems, and regional assemblages of interacting ecosystems called landscapes. A historical focus on saving individual species and perpetuating genetic diversity continues as a part of modern conservation biology.

- **Sustaining genetic diversity and the environmental arena for evolution is an ultimate goal (p. 1161)**
- **The dynamics of subdivided populations apply to problems caused by habitat fragmentation (pp. 1161–1164, FIGURE 55.8)** Understanding the dynamics and survivability of subdivided populations (metapopulations) is increasingly important as human activities continue to fragment habitats. Reproductive rates are often very different for isolated portions of metapopulations. Protecting source habitats (where a population's reproductive success is greater than its mortality) is a vital aspect of conservation.
- **Population viability analyses examine the chances of a species persisting or becoming extinct in the habitats available to it (pp. 1164–1165)** Conservation biologists use population viability analyses (PVAs) to assess long-term survivability of a species' populations in particular habitats. PVAs are usually generated by computer modeling using life history data, genetic variability, and a population's response to environmental conditions, especially disturbances. PVAs may predict minimum viable population size (MVP), the smallest number of individuals needed to perpetuate a population, subpopulation, or species. Estimates of MVP may be based on determination of the effective population size (N_e), that fraction of the total population that actually contributes to succeeding generations.
- **Analyzing the viability of selected species may help sustain other species: *science as a process* (pp. 1165–1167, FIGURE 55.11, 55.12)** Although time and resources preclude many species and populations from being systematically analyzed, what is learned from the relatively few PVAs that researchers are able to conduct may be applicable to many others.
- **Conserving species involves weighing conflicting demands (p. 1167)** Conservation efforts often involve resolving conflicts between the habitat needs of species and human demands for economic development and living space. Providing habitat that sustains keystone species—those having a large ecological impact relative to their numbers—can be important in sustaining whole communities.

CONSERVATION AT THE COMMUNITY, ECOSYSTEM, AND LANDSCAPE LEVELS

- **Edges and corridors can strongly influence landscape biodiversity (pp. 1168–1169, FIGURE 55.13)** Boundaries (edges) between ecosystems and along prominent features within ecosystems have unique sets of physical conditions and communities of species. Edges become more extensive as habitat fragmentation increases, and edge-adapted species may become more dominant. Movement corridors, strips or clumps of quality habitat connecting habitat patches, may promote dispersal and help sustain metapopulations, or they may promote harmful conditions (such as disease).
- **Nature reserves must be functional parts of landscapes (pp. 1169–1170, FIGURES 55.14, 55.15)** Many questions remain about providing and maintaining nature reserves. Sustaining biodiversity in reserves over the long term requires management to provide adequate disturbance and to ensure that human activities in the surrounding landscape support the protected habitats. Conservation efforts often involve working in landscapes that are largely human dominated.
- **Restoring degraded areas is an increasingly important conservation effort (pp. 1170–1171, FIGURE 55.16)** The new science of restoration ecology seeks to develop ways to restore degraded ecosystems and sustain biodiversity. Restoration efforts often involve bioremediation, the use of organisms to detoxify polluted ecosystems, and augmentation of ecosystem processes, such as successional changes.
- **Sustainable development goals are reorienting ecological research**
55.1 **and will require changing some human values (pp. 1171–1172)** Sustainable development, the long-term prosperity of human societies and the ecosystems supporting them, depends on ecological knowledge and on a commitment to promote ecosystem processes and biodiversity.

SELF-QUIZ

1. All of the following are reasonable hypotheses for the high diversity observed in tropical rain forests *except*
   a. habitat heterogeneity
   b. energy availability and the elevated photosynthetic rate
   c. climatic variability
   d. niche specialization and resource partitioning
   e. population interactions and coevolution

2. Research on the northern spotted owl indicates that
   a. owl populations will stabilize if an adequate number of immature forest source habitats can be preserved
   b. metapopulations must be more closely spaced to compensate for the owl's limited flight capacity
   c. setting aside marginal habitat encourages dispersion of owls into areas where reproductive success is unlikely
   d. rapid ecological succession has produced a patchy metapopulation pattern
   e. the current owl population is too small to disperse into source habitats

3. A population of strictly monogamous swans consists of 40 males and 10 females. The effective population size (N_e) for this population is
   a. 50
   b. 40
   c. 32
   d. 20
   e. 10

4. Which of the following conditions is the most likely indicator of a metapopulation's long-term survival?
   a. The population is not subdivided into subpopulations.
   b. Regular and extensive movement of individuals among patches makes the subpopulations function essentially as a single unit.
   c. Sources and sinks all contain subpopulations.
   d. Sink habitats remain occupied while some source habitats are empty.
   e. All subpopulations are connected by corridors.

5. The application of ecological principles to return a degraded ecosystem to its natural state is characteristic of
   a. bioremediation
   b. landscape ecology
   c. conservation ecology
   d. restoration ecology
   e. resource conservation

6. Not all species contribute equally to the integrity of an ecological community; with limited time and resources, it would be most reasonable to focus conservation efforts on
   a. the large vertebrates
   b. keystone species
   c. primary producers
   d. exotic species
   e. genetic variability of endangered species

7. Which of the following statements most clearly addresses what conservation biologists mean by the "biodiversity crisis"?
   a. Worldwide extinction rates are currently 50 times greater than at any time during the past 100,000 years.
   b. Introduced species, such as house sparrows and starlings, have rapidly expanded their ranges.
   c. Harvests of marine fish, such as cod and menhaden, are declining.
   d. Many pest species have developed resistance and are no longer effectively controlled by insecticide applications.
   e. The rate of patent application for pharmaceuticals developed from living organisms is currently declining.

8. Populations with low effective population sizes are susceptible to all of the following problems *except*
   a. inbreeding
   b. reduced heterozygosity
   c. bottlenecking
   d. genetic drift
   e. adaptive radiation

9. Which of the following errors would result in an underestimate of a species' MVP?
   a. underestimating the maximum age of reproduction
   b. underestimating the average birthrate
   c. overestimating the death rate
   d. overestimating the average fecundity
   e. overestimating the age of sexual maturity

10. European zebra mussels, accidentally released into Lake Erie in 1988, quickly displaced native mussel species. This threat to biodiversity is an example of
   a. exotic introduction
   b. metapopulation expansion
   c. habitat fragmentation
   d. overexploitation
   e. eutrophication

CHALLENGE QUESTION

Suppose that you are in charge of planning a forest reserve, and one of your main goals is to help sustain locally beleaguered populations of woodland birds. Parasitism by the brown-headed cowbird is an escalating problem in the area. Reading research reports, you note that female cowbirds are usually reluctant to penetrate more than about 100 meters into a forest, and that some woodland birds are known to reduce cowbird nest parasitism by restricting their nesting to the denser, more central regions of forests. The forested area you have to work with is about 1000 m by 6000 m. A recent logging operation removed about half of the trees on one of the 6000 m sides; the other three sides are adjacent to deforested pastureland. Your plan must include space for a small maintenance building, which you estimate to take up about 100 square meters. It will also be necessary to build a road, 10 m by 1000 m, across the reserve. Where would you construct the road and the building, and why?

SCIENCE, TECHNOLOGY, AND SOCIETY

Some organizations are starting to envision a sustainable society—one in which each generation inherits sufficient natural and economic resources and a relatively stable environment. The Worldwatch Institute, an environmental policy organization, estimates that we must reach sustainability by the year 2030 to avoid economic and environmental disaster. To get there, we must begin shaping a sustainable society during the next ten years or so. In what ways is our current system not sustainable? What might we do to work toward sustainability, and what are the major roadblocks to achieving it? How would your life be different in a sustainable society?

FURTHER READING

American Association for the Advancement of Science. "Human Dominated Ecosystems." *Science*, July 25, 1997. Contains several news articles and research reports on the human presence in the biosphere.

Amos, A. M. "Turning the Tide." *Global Biodiversity*, Winter 1997. A call for a more scientific study and greater conservation efforts in marine ecosystems.

Baskin, Y. *The Work of Nature: How the Diversity of Life Sustains Us.* Washington, DC: Island Press. 1997. A popular review of research on communities and ecosystems as life-support systems.

Chadwick, D. H. "Blue Refuges, U.S. National Marine Sanctuaries." *National Geographic*, March 1998. A wonderfully illustrated essay on 12 coastal preserves.

Chapin, F. S. III, et al. "Ecosystem Consequences of Changing Biodiversity." *BioScience*, January 1998. Includes documentation of the biodiversity crisis and discussion of research needs.

Dobson, A. P. *Conservation and Biodiversity.* New York: Scientific American. 1996. A brief, elegantly illustrated introduction to conservation biology.

Foin, T. C., S. P. D. Riley, A. L. Pawley, D. R. Ayres, T. M. Carlsen, P. J. Hodum, and P. V. Switzer. "Improving Recovery Planning for Threatened and Endangered Species." *BioScience*, March 1998. A call for more effective planning and the need to identify common objectives in species recovery projects.

Friedrich, R. L., and R. V. Blystone. "Internet Teaching Resources for Remote Sensing and GIS." *BioScience*, March 1998. Discusses the potential use of real data on the world wide web in science education.

Harwell, M. A. "Ecosystem Management of South Florida." *BioScience*, September 1997. A case study of the potential for cooperation between public policy makers and interdisciplinary scientists in achieving sustainability of a threatened ecosystem.

Lubchenco, J. "Entering the Century of the Environment: A New Social Contract for Science." *Science*, January 23, 1998. Describes the challenge to scientists in working toward ecosystem sustainability.

Malakoff, D. "Atlantic Salmon Spawn Fight Over Species Protection." *Science*, February 6, 1998. An example of the many controversies over biodiversity.

Meffe, G. K., and C. R. Carroll. *Principles of Conservation Biology*, 2e. Sunderland, MA: Sinauer Associates. 1997. An authoritative, lucidly written text.

Myers, N. "Mass Extinction and Evolution." *Science*, October 24, 1997. Emphasizes the effect of the biodiversity crisis on the future of evolution.

O'Neill, R. V., et al. "Monitoring Environmental Quality at the Landscape Scale." *BioScience*, September 1997. Presents an overview of landscape ecology in analyzing environmental quality.

Parry-Jones, R., and A. Vincent. "Can We Tame Wild Medicine?" *New Scientist*, January 3, 1998. Provides a discussion of the threat to some species posed by overexploitation for herbal medicines.

Pimentel, D., C. Wilson, C. McCullum, R. Huang, P. Dwen, J. Flack, Q. Tran, F. Saltman, and B. Cliff. "Economic and Environmental Benefits of Biodiversity." *BioScience*, December 1997. A scholarly analysis of the potential dividends of saving biodiversity.

Weeks, W. W. *Beyond the Ark: Tools for an Ecosystem Approach to Conservation*. Washington, DC: Island Press. 1997. A brief introduction to modern concepts in conservation biology.

Wildt, D. E., W. F. Rall, J. K. Critser, S. L. Monfort, and U. S. Seal. "Genome Resource Banks." *BioScience*, November 1997. Discusses the need to institute gene banks for wildlife species.

WEB LINKS

Visit the special edition of *The Biology Place* for BIOLOGY, Fifth Edition, at **http://www.biology.com/campbell**. Go to Chapter 55 for online resources, including learning activities, practice exams, and links to the following web sites:

"The National Wildlife Federation"
The National Wildlife Federation is the largest member-supported conservation group in the U.S., uniting individuals, organizations, businesses, and government to protect wildlife, wild places, and the environment.

"The World Conservation Union"
The mission of this organization is to influence, encourage, and assist societies throughout the world to conserve the integrity and diversity of nature and to ensure that any use of natural resources is equitable and ecologically sustainable.

"Smithsonian Institution's Conservation and Research Center"
Visit one of the world's leading centers for conservation research and training.

"Africa Environment 2000"
This is the story of a conservation organization founded to save rhinoceroses by two young women cycling from Scotland to Zimbabwe.

SELF-QUIZ ANSWERS

CHAPTER 2

1. b
2. a
3. b
4. b
5. c
6. b
7. b
8. a
9. b
10. b

CHAPTER 3

1. d
2. c
3. b
4. c
5. b
6. c
7. c
8. d
9. c
10. c

CHAPTER 4

1. b
2. c
3. d
4. d
5. a
6. b
7. b
8. a
9. d
10. b

CHAPTER 5

1. d
2. c
3. d
4. a
5. b
6. a
7. d

CHAPTER 6

1. b
2. c
3. b
4. b
5. e
6. a
7. c
8. c
9. a
10. c

CHAPTER 7

1. c
2. c
3. b
4. d
5. d
6. b
7. c
8. b
9. e
10. a

CHAPTER 8

1. b
2. c
3. a
4. d
5. b
6. fructose
7. glucose
8. cell contents
9. into cell
10. b

CHAPTER 9

1. d
2. b
3. c
4. c
5. a
6. a
7. b
8. d
9. b
10. b

CHAPTER 10

1. d
2. b
3. d
4. b
5. b
6. c
7. d
8. e
9. d
10. b

CHAPTER 11

1. c
2. d
3. a
4. c
5. c
6. c
7. a
8. b
9. c
10. a

CHAPTER 12

1. b
2. b
3. a
4. a
5. See Fig. 12.5
6. c
7. c
8. c
9. b
10. a

CHAPTER 13

1. d
2. b
3. d
4. d
5. c
6. d
7. a
8. c
9. c
10. d

CHAPTER 14

Genetics Problems

1. Incomplete dominance, with heterozygotes being gray in color. Mating a gray rooster with a black hen should yield approximately equal numbers of gray and black offspring.

2. F_1 cross is $AARR \times aarr$. Genotype of progeny is $AaRr$, phenotype is all axial-pink. F_2 cross is $AaRr \times AaRr$. Genotypes of progeny are 4 $AaRr$: 2 $AaRR$: 2 $AARr$: 2 $aaRr$: 2 $Aarr$: 1 $AARR$: 1 $aaRR$: 1 $AArr$: 1 $aarr$. Phenotypes of progeny are 6 axial-pink : 3 axial-red : 3 axial-white : 2 terminal-pink : 1 terminal-white : 1 terminal-red.

3. a. $^1/_{64}$ b. $^1/_{64}$ c. $^1/_8$ d. $^1/_{32}$

4. Albino is a recessive trait; black is dominant.

Parents	Gametes	Offspring
$BB \times bb$	B and b	All Bb
$bb \times Bb$	b and $^1/_2 B$, $^1/_2 b$	$^1/_2 Bb$, $^1/_2 bb$

5. a. $PPLl \times PPLl$, $PpLl$, or $ppLl$
b. $ppLl \times ppLl$
c. $PPLL \times$ any of the 9 possible genotypes
d. $PpLl \times Ppll$
e. $PpLl \times PpLl$

6. Man I^Ai; woman I^Bi; child ii. Other genotypes for children are $^1/_4$, I^AI^B, $^1/_4$ I^Ai, $^1/_4$ I^Bi.

7. Four

8. a. $^3/_4 \times {}^3/_4 \times {}^3/_4 = {}^{27}/_{64}$
b. $1 - {}^{27}/_{64} = {}^{37}/_{64}$
c. $^1/_4 \times {}^1/_4 \times {}^1/_4 = {}^1/_{64}$
d. $1 - {}^1/_{64} = {}^{63}/_{64}$

9. a. $^1/_{256}$ b. $^1/_{16}$ c. $^1/_{256}$ d. $^1/_{64}$ e. $^1/_{128}$

10. If the "curl" allele is dominant, then the original mutant crossed with noncurl cats will produce both curl and noncurl offspring. If the mutation is recessive, then only curl offspring will result from curl × curl matings. You know that cats are true-breeding when curl × curl matings produce only curl offspring. A pure-bred curl cat is homozygous (as it turns out, for the dominant allele, which causes the curled ears).

11. a. 1 b. $^1/_{32}$ c. $^1/_8$ d. $^1/_2$

12. $^1/_9$ **13.** $^1/_{16}$

14. Twenty-five percent will be cross-eyed; all of the cross-eyed offspring will also be white.

15. The dominant allele I is epistatic to the p locus, and thus the F_1 generation will be:
9 $I\_P\_$: colorless
3 I_pp : colorless
3 $iiP\_$: purple
1 $iipp$: red
Overall, 12 colorless : 3 purple : 1 red.

16. Recessive; George = Aa, Arlene = aa, Sandra = AA or Aa, Tom = aa, Sam = Aa, Wilma = aa, Ann = Aa, Michael = Aa, Daniel = Aa, Alan = Aa, Tina = AA or Aa, Carla = aa, Christopher = AA or Aa

17. $^1/_2$ **18.** $^1/_6$

19. 9 $B\_A\_$: agouti
3 B_aa : black
3 $bbA\_$: white
1 $bbaa$: white
Overall, 9 agouti : 3 black : 4 white.

20. The allele for the disorder is dominant. Individuals 1 and 2 are heterozygous, and individual 3 is homozygous for the recessive (normal) allele.

CHAPTER 15

Genetics Problems

1. 0; $^1/_2$, $^1/_{16}$

2. Recessive. If the disorder were dominant, it would affect at least one parent of a child born with the disorder. For a girl to have the disorder, she would have to inherit recessive alleles from *both* parents. This would be very rare, especially since males with the allele die in their early teens.

3. $^1/_4$ for each daughter ($^1/_2$ chance that child will be female × $^1/_2$ chance of a homozygous recessive genotype); $^1/_2$ for first son.

4. 17%

5. The disorder would always be inherited from the mother.

6. *XXX*

7. *D–A–B–C*

8. In meiosis, the combined 14-21 chromosome will behave as one chromosome. If a gamete receives the combined 14-21 chromosome and a normal copy of chromosome 21, trisomy 21 will result when this gamete combines with a normal gamete.

9. At some point during development, one of the embryo's cells may have failed to carry out mitosis after duplicating its chromosomes. Subsequent normal cell cycles would produce genetic copies of this tetraploid cell.

10. Fifty percent of the offspring would show phenotypes that resulted from crossovers. These results would be the same as those from a cross where *A* and *B* were not linked. Further crosses involving other genes on the same chromosome would reveal the linkage and map distances.

11. One hypothesis is that a translocation has moved one of the genes to a different chromosome.

12. 6%. Wild type (heterozygous for normal wings and red eyes) × recessive homozygote with vestigial wings and purple eyes.

13. Between *T* and *A*, 12%; between *A* and *S*, 5%.

14. Between *T* and *S*, 18%. Sequence of genes is *T-A-S*.

CHAPTER 16

1. c
2. d
3. b
4. c
5. b
6. a
7. c
8. d
9. a
10. b

CHAPTER 17

1. c
2. b
3. d
4. d
5. a
6. a
7. c
8. d
9. e
10. b

CHAPTER 18

1. e
2. d
3. b
4. a
5. e
6. d
7. a
8. c
9. a
10. e

CHAPTER 19

1. c
2. a
3. a
4. a
5. e
6. a
7. e
8. c
9. b
10. b

CHAPTER 20

1. b
2. b
3. c
4. b
5. a
6. c
7. e
8. c
9. d
10. b

CHAPTER 21

1. a
2. e
3. b
4. a
5. a
6. d
7. b
8. a
9. c
10. e

CHAPTER 22

1. a
2. b
3. c
4. d
5. c
6. b
7. b
8. c
9. c
10. d

CHAPTER 23

1. b
2. d
3. c
4. a
5. c
6. a
7. c
8. b
9. e
10. b

CHAPTER 24

1. b
2. b
3. b
4. a
5. b
6. c
7. e
8. a
9. d
10. b

CHAPTER 25

1. d
2. b
3. a
4. e
5. c
6. b
7. a
8. d
9. b
10. e

CHAPTER 26

1. b
2. e
3. d
4. c
5. c
6. b
7. a
8. b
9. d
10. c

CHAPTER 27

1. b
2. a
3. d
4. c
5. c
6. a
7. d
8. c
9. a
10. c

CHAPTER 28

1. a
2. c
3. e
4. a
5. b
6. e
7. b
8. c
9. c
10. b

CHAPTER 29

1. c
2. a
3. a
4. b
5. a
6. a
7. haploid
8. haploid
9. diploid
10. haploid

CHAPTER 30

1. d
2. a
3. b
4. a
5. b
6. d
7. b
8. d
9. c
10. a

CHAPTER 31

1. b
2. c
3. c
4. b
5. e
6. d
7. a
8. e
9. b
10. a

CHAPTER 32

1. c
2. e
3. d
4. c
5. c
6. b
7. b
8. a
9. e
10. b

CHAPTER 33

1. d
2. c
3. a
4. a
5. a
6. a
7. d
8. a
9. e
10. c

CHAPTER 34

1. e
2. c
3. d
4. d
5. c
6. a
7. b
8. e
9. c
10. a

CHAPTER 35

1. d
2. c
3. b
4. c
5. d
6. a
7. d
8. d
9. b
10. c

CHAPTER 36

1. e
2. c
3. d
4. c
5. b
6. c
7. a
8. b
9. c
10. c

CHAPTER 37

1. b
2. b
3. c
4. d
5. b
6. b
7. b
8. c
9. d
10. a

CHAPTER 38

1. d
2. a
3. c
4. a
5. b
6. c
7. e
8. b
9. c
10. c

CHAPTER 39

1. b
2. a
3. b
4. d
5. b
6. a
7. b
8. b
9. c
10. c

CHAPTER 40

1. a
2. c
3. e
4. d
5. d
6. c
7. b
8. b
9. c
10. c

CHAPTER 41

1. e
2. c
3. c
4. c
5. d
6. c
7. c
8. d
9. b
10. b

CHAPTER 42

1. c
2. b
3. d
4. c
5. b
6. c
7. b
8. a
9. a
10. c

CHAPTER 43

1. b
2. d
3. a
4. e
5. b
6. e
7. d
8. c
9. b
10. c

CHAPTER 44

1. d
2. a
3. e
4. e
5. b
6. d
7. b
8. d
9. c
10. a

CHAPTER 45

1. c
2. d
3. a
4. d
5. c
6. b
7. e
8. c
9. b
10. c

CHAPTER 46

1. d
2. b
3. a
4. b
5. c
6. a
7. a
8. a
9. d
10. b

CHAPTER 47

1. a
2. b
3. d
4. c
5. a
6. c
7. b
8. c
9. d
10. d

CHAPTER 48

1. c
2. b
3. a
4. d
5. d
6. a
7. c
8. c
9. c
10. b

CHAPTER 49

1. e
2. d
3. a
4. b
5. c
6. e
7. b
8. d
9. c
10. b

CHAPTER 50

1. c
2. e
3. e
4. d
5. a
6. d
7. e
8. c
9. d
10. a

CHAPTER 51

1. b
2. d
3. d
4. c
5. d
6. d
7. c
8. c
9. e
10. b

CHAPTER 52

1. c
2. d
3. c
4. d
5. c
6. b
7. c
8. d
9. a
10. c

CHAPTER 53

1. c
2. d
3. b
4. d
5. e
6. d
7. b
8. b
9. d
10. c

CHAPTER 54

1. c
2. e
3. b
4. e
5. d
6. a
7. b
8. c
9. d
10. c

CHAPTER 55

1. c
2. c
3. d
4. b
5. d
6. b
7. a
8. e
9. d
10. a

THE METRIC SYSTEM

MEASUREMENT	UNIT AND ABBREVIATION	METRIC EQUIVALENT	METRIC-TO-ENGLISH CONVERSION FACTOR	ENGLISH-TO-METRIC CONVERSION FACTOR
Length	1 kilometer (km)	= 1000 (10^3) meters	1 km = 0.62 mile	1 mile = 1.61 km
	1 meter (m)	= 100 (10^2) centimeters = 1000 millimeters	1 m = 1.09 yards 1 m = 3.28 feet 1 m = 39.37 inches	1 yard = 0.914 m 1 foot = 0.305 m
	1 centimeter (cm)	= 0.01 (10^{-2}) meter	1 cm = 0.394 inch	1 foot = 30.5 cm 1 inch = 2.54 cm
	1 millimeter (mm)	= 0.001 (10^{-3}) meter	1 mm = 0.039 inch	
	1 micrometer (μm) (formerly micron, μ)	= 10^{-6} meter (10^{-3}mm)		
	1 nanometer (nm) (formerly millimicron, mμ)	= 10^{-9} meter (10^{-3}μm)		
	1 angstrom (Å)	= 10^{-10} meter (10^{-4}μm)		
Area	1 hectare (ha)	= 10,000 square meters	1 ha = 2.47 acres	1 acre = 0.0405 ha
	1 square meter (m^2)	= 10,000 square centimeters	1 m^2 = 1.196 square yards 1 m^2 = 10.764 square feet	1 square yard = 0.8361 m^2 1 square foot = 0.0929 m^2
	1 square centimeter (cm^2)	= 100 square millimeters	1 cm^2 = 0.155 square inch	1 square inch = 6.4516 cm^2
Mass	1 metric ton (t)	= 1000 kilograms	1 t = 1.103 tons	1 ton = 0.907 t
	1 kilogram (kg)	= 1000 grams	1 kg = 2.205 pounds	1 pound = 0.4536 kg
	1 gram (g)	= 1000 milligrams	1 g = 0.0353 ounce 1 g = 15.432 grains	1 ounce = 28.35 g
	1 milligram (mg)	= 10^{-3} gram	1 mg = approx. 0.015 grain	
	1 microgram (μg)	= 10^{-6} gram		
Volume (solids)	1 cubic meter (m^3)	= 1,000,000 cubic centimeters	1m^3 = 1.308 cubic yards 1 m^3 = 35.315 cubic feet	1 cubic yard = 0.7646 m^3 1 cubic foot = 0.0283 m^3
	1 cubic centimeter (cm^3 or cc)	= 10^{-6} cubic meter	1 cm^3 = 0.061 cubic inch	1 cubic inch = 16.387 cm^3
	1 cubic millimeter (mm^3)	= 10^{-9} cubic meter (10^{-3} cubic centimeter)		
Volume (Liquids and Gases)	1 kiloliter (kl or kL)	= 1000 liters	1 kL = 264.17 gallons	1 gallon = 3.785 L
	1 liter (L)	= 1000 milliliters	1 L = 0.264 gallons 1 L = 1.057 quarts	1 quart = 0.946 L
	1 milliliter (mL)	= 10^{-3} liter = 1 cubic centimeter	1 mL = 0.034 fluid ounce 1 mL = approx. $\frac{1}{4}$ teaspoon 1 ml = approx. 15–16 drops (gtt.)	1 quart = 946 mL 1 pint = 473 mL 1 fluid ounce = 29.57 mL 1 teaspoon = approx. 5 mL
	1 microliter (μl or μL)	= 10^{-6} liter (10^{-3} milliliters)		
Time	1 second (s)	= $\frac{1}{60}$ minute		
	1 millisecond (ms)	= 10^{-3} second		
Temperature	Degrees Celsius (°C) (Absolute zero, when all molecular motion ceases, is −273 °C. The Kelvin (K) scale, which has the same size degrees as Celsius, has its zero point at absolute zero. Thus, 0° K = −273°C.)		$°F = \frac{9}{5}°C + 32$	$°C = \frac{5}{9}(°F - 32)$

CLASSIFICATION OF LIFE

This appendix presents the taxonomic classification used for the major groups of organisms discussed in this text; not all phyla are included. Plant and fungal divisions are the taxonomic equivalents of phyla. The classification reviewed here is based on the three-domain (superkingdom) system; in the traditional five-kingdom system, all prokaryotes are classified in a single kingdom, the Monera. The rationale for alternative taxonomic systems is discussed in Unit Five of the text.

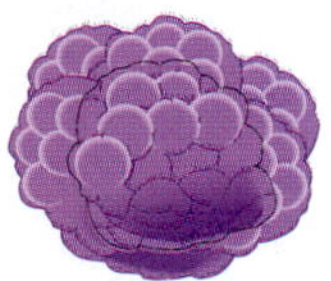

DOMAIN ARCHAEA

Methanogens
Extreme halophiles
Thermoacidophiles

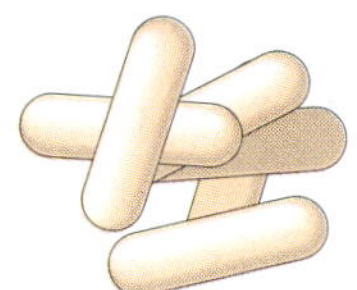

DOMAIN BACTERIA

Proteobacteria
- Purple bacteria
- Chemoautotrophic proteobacteria
- Chemoheterotrophic proteobacteria

Gram-positive eubacteria
Cyanobacteria
Spirochetes
Chlamydias

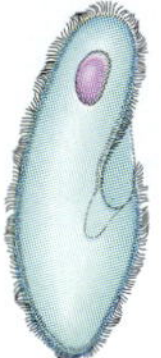

DOMAIN EUKARYA

The five-kingdom classification scheme unites all eukaryotes generally called protists in a single kingdom, the Protista. In this book we follow the lead of current research, classifying some protists in "candidate kingdoms," groups that are phylogenetically unified but not yet recognized as formal taxa. We also discuss some other protists whose affinities are less clear, and these are also listed in this appendix. We use the taxonomic category *phylum* for groups that are chiefly heterotrophic and *division* for those that are mainly autotrophic. Green algae (Division Chlorophyta) are also listed here, although many plant biologists classify them in the plant kingdom; see Chapter 26.

Candidate Kingdom Archaezoa

Phylum Diplomonada (*Giardia* and other diplomonads)
Phylum Trichomonada (trichomonads)
Phylum Microsporida (microsporidians)

Candidate Kingdom Euglenozoa

Division Euglenophyta (euglenoids)
Phylum Kinetoplastida (trypanosomes)

Candidate Kingdom Alveolata

Division Dinoflagellata (dinoflagellates)
Phylum Apicomplexa (*Plasmodium* and other apicomplexans)
Phylum Ciliophora (ciliates)

Candidate Kingdom Stramenopila

Division Phaeophyta (brown algae)
Phylum Oomycota (water molds)
Division Chrysophyta (golden algae)
Division Bacillariophyta (diatoms)

Candidate Kingdom Rhodophyta (red algae)

Groups whose affinities are less certain

Phylum Rhizopoda (some amoebas)
Phylum Actinopoda (heliozoans, radiolarians)
Phylum Foraminifera (forams)
Phylum Myxomycota (plasmodial slime molds)
Phylum Acrasiomycota (cellular slime molds)

Group with close ties to the plant kingdom

Division Chlorophyta (green algae)

Kingdom Plantae

Division Bryophyta (mosses)
Division Hepatophyta (liverworts)
Division Anthocerophyta (hornworts)
Division Lycophyta (club mosses)

Division Sphenophyta (horsetails)

Division Pterophyta (ferns)

Division Coniferophyta (conifers)

Division Cycadophyta (cycads)

Division Ginkgophyta (ginkgos)

Division Gnetophyta (gnetae)

Division Anthophyta (flowering plants)

Class Monocotyledones (monocots)

Class Dicotyledones (dicots)

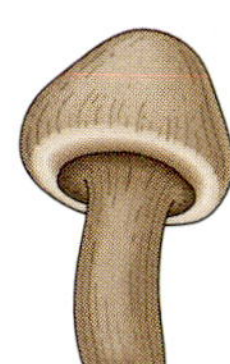

Kingdom Fungi

Division Chytridiomycota (chytrids)

Division Zygomycota (zygomycetes)

Division Ascomycota (sac fungi)

Division Basidiomycota (club fungi)

Division Deuteromycota (imperfect fungi)

Lichens (symbiotic associations of algae and fungi)

Kingdom Animalia

Phylum Porifera (sponges)

Phylum Cnidaria

Class Hydrozoa (hydrozoans)

Class Scyphozoa (jellies)

Class Anthozoa (sea anemones and coral animals)

Phylum Ctenophora (comb jellies)

Phylum Platyhelminthes (flatworms)

Class Turbellaria (free-living flatworms)

Class Trematoda (flukes)

Class Monogenea (flukes)

Class Cestoidea (tapeworms)

Phylum Nemertea (proboscis worms)

Phylum Rotifera (rotifers)

Phylum Nematoda (roundworms)

Phylum Mollusca (mollusks)

Class Polyplacophora (chitons)

Class Gastropoda (gastropods: snails and their relatives)

Class Bivalvia (bivalves)

Class Cephalopoda (cephalopods; squids and octopuses)

Phylum Annelida (segmented worms)

Class Oligochaeta (oligochaetes)

Class Polychaeta (polychaetes)

Class Hirudinea (leeches)

Phylum Onychophora (walking worms; some schemes classify onychophorans as a subphylum or class in the Phylum Arthropoda)

Phylum Arthropoda (arthropods)

Subphylum Trilobitomorpha (trilobites, all extinct)

Subphylum Chelicerata (chelicerates)

Class Eurypterida (water scorpions, all extinct)

Class Arachnida (spiders, ticks, scorpions)

Subphylum Uniramia (uniramians)

Class Diplopoda (millipedes)

Class Chilopoda (centipedes)

Class Insecta (insects)

Subphylum Crustacea (crustaceans)

(*Note:* The validity of subphyla Uniramia and Crustacea is being debated; molecular systematics indicates that insects and crustaceans may constitute a monophyletic group distinct from diplopods and chilopods.)

Phylum Phoronida (phoronids)

Phylum Bryozoa (bryozoans)

Phylum Brachiopoda (brachiopods: lamp shells)

Phylum Echinodermata (echinoderms)

Class Asteroidea (sea stars)

Class Ophiuroidea (brittle stars)

Class Echinoidea (sea urchins and sand dollars)

Class Crinoidea (sea lilies)

Class Concentricycloidea (sea daisies)

Class Holothuroidea (sea cucumbers)

Phylum Chordata (chordates)

Subphylum Urochordata (urochordates: tunicates)

Subphylum Cephalochordata (cephalochordates: lancelets)

Subphylum Vertebrata (vertebrates)

Superclass Agnatha (jawless vertebrates)

Class Myxini (hagfishes)

Class Cephalospidomorphi (lampreys)

Superclass Gnathostomata

Class Placodermi (extinct jawed fishes)

Class Chondrichthyes (cartilaginous fishes)

Class Osteichthyes (bony fishes)

Class Amphibia (amphibians)

Class Reptilia (reptiles)

Class Aves (birds)

Class Mammalia (mammals)

A COMPARISON OF THE LIGHT MICROSCOPE AND THE ELECTRON MICROSCOPE

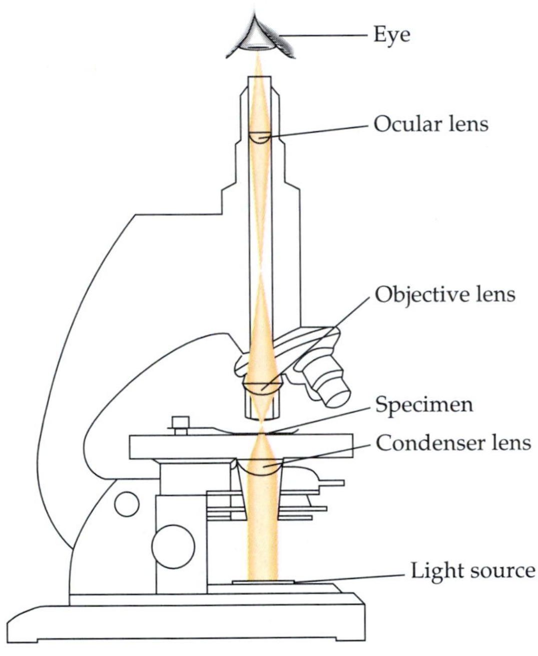

(a) LIGHT MICROSCOPE
In light microscopy, light is focused on a specimen by a glass condenser lens; the image is then magnified by an objective lens and an ocular lens, for projection on the eye or on photographic film.

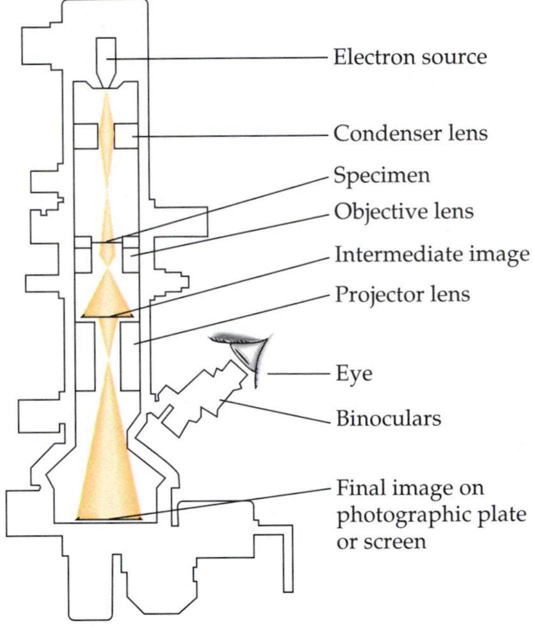

(b) ELECTRON MICROSCOPE
In electron microscopy, a beam of electrons (top of the microscope) is used instead of light, and electromagnets are used instead of glass lenses. The electron beam is focused on the specimen by a condenser lens; the image is magnified by an objective lens and a projector lens, for projection on a screen or on photographic film.

CREDITS

PHOTOGRAPHS

1.CO Courtesy of the Imogen Cunningham Trust. ©1974 Imogen Cunningham.
1.01a Courtesy of NYU/Strongin.
1.01b Photo by Tim White.
1.01c Courtesy of Mary-Claire King, University of Washington.
1.01d ©Paul Anderton/Addison Wesley Longman, Inc.
1.02a ©Robert Fletterick.
1.02b ©Dr. Jeremy Burgess/SPL/Photo Researchers, Inc.
1.02c ©Manfred Kage/Peter Arnold, Inc.
1.02d ©Dr. Jeremy Burgess/SPL/Photo Researchers, Inc.
1.02e ©John Shaw/Tom Stack and Associates.
1.02f ©1997 Photodisc.
1.03a ©Bob Stovall/Bruce Coleman, Inc.
1.03b ©Michio Hoshino/Minden Pictures.
1.03c ©Michael Fogden/Bruce Coleman, Inc.
1.03d ©Wolfgang Bayer/Bruce Coleman, Inc.
1.03e ©Jeff Lepore/Photo Researchers, Inc.
1.03f, g ©Tom and Pat Leeson/Photo Researchers, Inc.
1.05a ©1997 Photodisc.
1.06a ©1998 Photodisc.
1.06b ©Janice Sheldon.
1.06c ©Dr. Alfred Llinas/Peter Arnold, Inc.
1.06d Courtesy of Dr. Nicolae Simionescu.
1.07 ©Ric Ergenbright/Ric Ergenbright Photography.
1.09a ©Manfred Kage/Peter Arnold, Inc.
1.09b Courtesy W. L. Dentler, University of Kansas/BPS.
1.09c ©Omikron/Photo Researchers, Inc.
1.11a ©A.B. Dowsett/Science Source/Photo Researchers, Inc.
1.11b ©Ralph Robinson/Visuals Unlimited.
1.11c ©D.P. Wilson/Science Source/Photo Researchers, Inc.
1.11d, e, f ©1998 Photodisc.
1.12 ©Chip Clark.
1.13 Courtesy of Richard Milner.
1.14 ©Rudi Kuiter. Reprinted with permission of Discover Magazine.
1.15 Courtesy of David Reznick, University of California, Riverside.
1.18a ©Hank Morgan/Photo Researchers Inc.
1.18b ©Peter Menzel.
2.CO ©Stella Johnson.
2.EOC2 ©Phil Degginger/Color-Pic, Inc.
2.MB1a ©Terraphotographics/Biological Photo Service.
2.MB1c From M.C. Ratazzi et al., *Am J Human Genet* 28(1976):143–154.
2.02 ©E. R. Degginger/Color-Pic, Inc.
2.02 Stephen Frisch/Benjamin/Cummings.
2.03 ©Grant Heilman/Grant Heilman Photography, Inc.
2.04 ©Ivan Polunin/Bruce Coleman, Inc.
2.06 ©Igor Kostin/IMAGO/SYGMA.
2.13 ©Stephen Frisch/Benjamin/Cummings.
2.17 ©Annalisa Kraft/Benjamin/Cummings.
2.18 ©Runk Schoenberger/Grant Heilman Photography, Inc.
3.CO Courtesy of NASA.
3.02 ©E.R. Degginger/Color-Pic, Inc.
3.03 ©Stephen Dalton/NHPA.
3.04 ©1996 DUOMO/William Sallaz.
3.06 ©Flip Nicklin/Minden Pictures.
3.10 ©1991 Maresa Pryor/Earth Scenes/Animals Animals.
4.CO Courtesy of P. Shing Ho.
4.01 Roger Ressmeyer/©Corbis.
4.05 ©Manfred Kage/Peter Arnold, Inc.
4.08 (left) ©W.J. Weber/Visuals Unlimited.
4.08 (right) ©Stephen J. Krasemann/Photo Researchers, Inc.
5.CO ©Martin Shields.
5.MB1b, c, d Courtesy of Marie Green, University of California, Riverside.
5.01a Estate of Linus Pauling.
5.01b ©Louise Lockley/SCIRO/Science Photo Library/Photo Researchers, Inc.
5.06a ©Dr. Jeremy Burgess/Photo Researchers, Inc.
5.06b Courtesy of H. Shio and P.B. Lazarow.
5.08 ©J. Litray/Visuals Unlimited.
5.09a ©F. Collet/Photo Researchers, Inc.
5.09b ©George Disario/The Stock Market.
5.11a Courtesy of the American Dairy Association.
5.11b ©Lara Hartley.
5.17 Courtesy of the Graphics Systems Research Group, IBM U.K. Scientific Centre.
5.19a, b ©M. Murayama, Murayama Research Laboratory/BPS.
5.21 (left) ©Martin Shields.
5.21 (right) Vollrath & Edmunds, *Nature* 340: 305–317.
6.CO ©Jean-Marie Bassot/Photo Researchers, Inc.
6.02 ©Eunice Harris 1987/Photo Researchers, Inc.
6.03 ©Manfred Kage/Peter Arnold, Inc.
6.11 Courtesy of Thomas Steitz, Yale University.
6.18 ©R. Rodewald, University of Virginia/Biological Photo Service.
7.CO ©M. Schliwa/Visuals Unlimited.
7.T01.1 ©Biophoto Associates/Photo Researchers, Inc.
7.T01.2 ©David M. Phillips/Visuals Unlimited.
7.T01.3 ©Ed Reschke.
7.T01.4, 5 ©David M. Phillips/Visuals Unlimited.
7.T01.6 Courtesy of Noran Instruments.
7.T02.1 Courtesy of Dr. Mary Osborn, Max Planck Institute.
7.T02.2 Courtesy of Drs. Frank Solomon and J. Dinsmore, MIT.
7.T02.3 Mark S. Ladinsky and J. Richard McIntosh, University of Colorado.
7.02a, b ©William L. Dentler, University of Kansas/Biological Photo Service.
7.04b ©S.C. Holt, University of Texas Health Center/Biological Photo Service.
7.06a Courtesy of J. David Robertson.
7.09e U. Aebi. et al. *Nature* 323(1996):560–564, figure 1a. Used by permission.
7.10 ©D.W. Fawcett/Photo Researchers, Inc.
7.11 ©R. Bolender, D. Fawcett/Photo Researchers, Inc.
7.12 ©G.T. Cole, University of Texas, Austin/Biological Photo Service.
7.13a ©R. Rodewald, University of Virginia/Biological Photo Service.
7.13b Courtesy of Daniel S. Friend, Harvard Medical School.
7.15 Courtesy of E.H. Newcomb, University of Wisconsin.
7.17 Courtesy of Daniel S. Friend, Harvard Medical School.
7.18 ©WP Wergin and E.H. Newcomb, University of Wisconsin, Madison/Biological Photo Service.
7.19 Reproduced from S.E. Frederick and E.H. Newcomb, *The Journal of Cell Biology* 43 (1969):343 by copyright permission of The Rockefeller University Press. Provided by E.H. Newcomb.
7.20 Courtesy of Dr. John E. Heuser, Washington University, St. Louis.
7.22 Courtesy of Kent McDonald, University of California, Berkeley.
7.23a ©Richard Kessel/Visuals Unlimited.
7.23b ©1990 Dennis Kunkel.
7.24a ©OMIKRON/Science Source/Photo Researchers, Inc.
7.24b, c ©W. L. Dentler, University of Kansas/ Biological Photo Service.
7.26 Reproduced from Hirokawa Nobutaka, *The Journal of Cell Biology* 94(1982):425 by copyright permission of The Rockefeller University Press.
7.27 Courtesy of Dr. John E. Heuser, Washington University, St. Louis.
7.28 ©G.F. Leedale/Photo Researchers, Inc.
7.30a Reproduced from Douglas J. Kelly, *The Journal of Cell Biology* 28(1966):51 by permission of The Rockefeller University Press.
7.30c Reproduced from C. Peracchia and A.F. Dulhunty, *The Journal of Cell Biology* 70(1976):419 by permission of The Rockefeller University Press.
7.31 ©Boehringer Ingelheim International GmbH, photo Lennart Nilsson/Albert Bonniers Forlag AB, *The Body Victorious*, Delacorte Press, Dell Publishing Co., Inc.
8.MB1 Courtesy of Philippa Claude.
8.12a, b ©Cabisco/Visuals Unlimited.
8.18a ©R.N. Band and H.S. Pankratz, Michigan State University/Biological Photo Service.
8.18b ©D.W. Fawcett/Photo Researchers, Inc.
8.18c From M.M. Perry and A.B. Gilbert, *J. Cell Sci.* 39(1979):257. ©1979 by The Company of Biologists Ltd.
9.CO ©Gerry Ellis/ENP Images.
10.CO ©E.R. Degginger/Earth Scenes/Animals Animals.
10.01a ©Renee Lynn/Photo Researchers, Inc.
10.01b ©Bob Evans/Peter Arnold, Inc.
10.01c ©Dwight Kuhn.

10.01d ©Tom Adams/Peter Arnold, Inc.
10.01e ©Paul Johnson/Biological Photo Service.
10.02 (bottom left) ©M. Eichelberger/Visuals Unlimited.
10.02 (top right) Courtesy of W.P. Wergin and E.H. Newcomb, University of Wisconsin/ Biological Photo Service.
10.09b ©Christine L. Case, Skyline College.
10.18 (left) ©C.F. Miescke/Biological Photo Service.
10.18 (right) ©Phil Degginger/Color-Pic, Inc.
11.CO ©Eric Schabtach and Ira Herskowitz.
11.02 Courtesy of Dale Kaiser.
12.EOC1 ©Carolina Biological Supply/Phototake.
12.MB1 Courtesy of Dr. Gunter Albrecht-Buehler, Northwestern University.
12.01a ©Biophoto Associates/Photo Researchers, Inc.
12.01b ©C.R. Wyttenbach, University of Kansas/Biological Photo Service.
12.01c ©Biophoto/Science Source/Photo Researchers, Inc.
12.03 ©Biophoto/Photo Researchers, Inc.
12.05 ©Ed Reschke.
12.06a Courtesy of Richard McIntosh, University of Colorado.
12.08a ©David M. Phillips/Visuals Unlimited.
12.08b Micrograph by B.A. Palevitz. Courtesy of E.H. Newcomb, University of Wisconsin.
12.09 ©Carolina Biological Supply/Phototake NYC.
13.CO ©Reuters/Řiha, Jr., Bob/Archive Photos.
13.MB1 ©SIU/Visuals Unlimited.
13.MB1 ©CNRI/SPL/Photo Researchers, Inc.
13.01 ©Roland Birke/OKAPIA/Photo Researchers, Inc.
13.02 Courtesy of the Inouye and Yaneshiro Families.
14.CO The Bettman Archives.
14.EOC1 Breeder/owner: Patricia Speciale; photographer: Norma JubinVille.
14.13 (left) ©Stan Goldblatt/Photo Researchers, Inc.
14.13 (right) ©Gilbert Grant/Photo Researchers, Inc.
14.14a ©Photodisc.
14.14b ©Anthony Loveday/Benjamin/Cummings.
14.15 ©Bill Longcore/Photo Researchers, Inc.
14.16 Courtesy of Dr. Nancy Wexler, Columbia University.
15.02 ©Darwin Dale.
15.10 ©1997 Photodisc.
15.13a ©CNRI/Science Photo Library/Photo Researchers, Inc.
15.13b ©Richard Hutchings/Photo Researchers, Inc.
15.15 From L. P. Hosticka and M. R. Hanson, "Induction of plastid mutations by nitrosomethylurea" *Journal of Heredity* 75(1984):242–246.
16.CO ©Barrington Brown/Photo Researchers, Inc.
16.02a ©Lee D. Simon/Photo Researchers, Inc.
16.03a, b From J.D. Watson, *The Double Helix*, p. 215. (NY: Atheneum Press, 1968). ©1968 by J.D. Watson. Courtesy of Cold Spring Harbor Laboratory Archives.
16.04 Courtesy of Richard Wagner, UCSF Graphics.
16.09b Reproduced from D. J. Burks and P. J. Stambrook, *The Journal of Cell Biology* 77(1978):762 by copyright permission of The Rockefeller University Press. Photos provided by P. J. Stambrook.
16.19a ©Peter Lansdorp.
17.05 Courtesy of Keith V. Wood, University of California, San Diego.
17.18 Courtesy of B. Hamkalo and O.L. Miller, Jr.
18.02a, b, d ©Robley C. Williams, Univ. of California, Berkeley/Biological Photo Service.
18.02c ©John Cardamone Jr., University of Pittsburgh/Biological Photo Service.
18.07 ©C.Dauguet/Institute Pasteur/Photo Researchers, Inc.
18.08a (left) ©Sherman Thomson/Visuals Unlimited.
18.08a (right) ©Norm Thomas/Photo Researchers, Inc.
18.08b Photo provided by N. Obalka, N. Yeager, R. Beachy, and C. Fauquet, The Scripps Research Institute.
18.13 ©Dennis Kunkel/Phototake NYC.
19.CO Courtesy of M. B. Roth and J. Gall.
19.01a1 ©S. C. Holt, University of Texas, Health Science Center, San Antonio/Biological Photo Service.
19.01a2 Courtesy of Victoria Foe.
19.01b Courtesy of Barbara A. Hamkalo.
19.01c From J. R. Paulsen and U. K. Laemmli, *Cell* 12(1977):817–828.
19.01d1, 2 G. F. Bahr/AFIP.
19.02 Courtesy of O. L. Miller Jr., Department of Biology, University of Virginia.
20.CO ©Kyodo News International Inc.
20.MB2 Courtesy of Repligen Corporation.
21.CO ©W. Gehring/Photo Researchers, Inc.
21.MB1a Courtesy of Eric Wieschaus, Princeton University.
21.MB1b Courtesy of Wolfgang Driever.
21.01 (left) ©Dwight Kuhn.
21.01 (right) ©Hans Pfletschinger/Peter Arnold, Inc.
21.03a ©Oliver Meckes/Photo Researchers, Inc.
21.03b ©Dr. Robert Horvitz/Peter Arnold, Inc.
21.03c Courtesy of Stanton Short, Jackson Laboratory.
21.03d Courtesy of Nancy Hopkins, MIT.
21.03e Courtesy of Elliot Meyerowitz, Caltech.
21.04 From J.E. Sulston and H.R. Horvitz, *Developmental Biology* 56(1977):110–156.
21.07 ©Najlah Feanny/SABA.
21.11b Courtesy of Ruth Lahmann, The Whitehead Institution.
21.11c Courtesy of Wolfgang Driever.
21.12a, b, c Drs. Jim Langeland, Steve Paddock, Sean Carroll, University of Wisconsin and The Howard Hughes Medical Institute.
21.13 ©F. R. Turner, Indiana University.
21.17 ©Dwight Kuhn.
21.18b 1, 2, 3 Courtesy of Elliot Meyerowitz, Caltech.
22.CO ©William Ferguson.
22.01 Larry Burrows, Life Magazine ©Time Inc.
22.03a Courtesy of The National Maritime Museum, London.
22.03b ©Joe McDonald/Animals Animals.
22.04a, b, c ©Tui de Roy/Bruce Coleman, Inc.
22.05a ©E.S. Ross, California Academy of Sciences.
22.05b (left) ©K.G. Preston/MAFHA/Animals Animals.
22.05b (right) ©P. and W. Ward/OSF/Animals Animals.
22.06 (center left) ©Jack Wilburn/Earth Scenes/ Animals Animals.
22.06 (upper left) ©Michael P. Gadomski/Photo Researchers, Inc.
22.06 (upper right, center, lower left) ©Dwight R. Kuhn.
22.06 (center right) ©David Cavagnaro/Peter Arnold, Inc.
22.06 (lower right) ©Clyde H. Smith/Peter Arnold, Inc.
22.08 Philip Gingerich ©1991. Reprinted with permission of Discover Magazine.
22.10a ©Dwight Kuhn.
22.10b ©Lennart Nilsson/Albert Bonniers Forlag AB, *A Child Is Born*, Dell Publishing, 1990.
23.CO ©Edward S. Ross/California Academy of Sciences.
23.01 ©J. Antonovics/Visuals Unlimited.
23.02 ©David Cavagnaro.
23.06 ©Ralph Reinhold/Animals Animals.
23.07 Courtesy of Edmund Brodie III, University of Kentucky.
23.09 Courtesy of Thomas B. Smith, San Francisco State University.
23.10 Courtesy of Dr. Michio Hori, Kyoto University.
23.12 ©Jane Burton/Bruce Coleman, Inc.
24.CO ©Frans Lanting/Minden Pictures.
24.02a1 ©Don & Pat Valenti/Tom Stack & Associates.
24.02a2 ©John Shaw/Tom Stack & Associates.
24.02b1, 2, 3, 4 ©Photodisc.
24.03 ©Joe McDonald/Visuals Unlimited.
24.04 ©John Eastcott/The Image Works.
24.07 (bottom left) ©John Shaw/Bruce Coleman, Inc.
24.07 (center) ©Larry Ulrich/DRK Photo.
24.07 (top right) ©Michael Fogden/Bruce Coleman.
24.08 ©David Cavagnaro.
24.12 Courtesy of Kenneth Kaneshiro, University of Hawaii.
24.15 ©Jane Burton/Bruce Coleman, Inc.
25.CO ©Brian Parker/Tom Stack and Associates.
25.01a ©Barbara J. Miller/Biological Photo Service.
25.01b Courtesy of Margo Crabtree.
25.01c ©Tom Till.
25.01d ©Manfred Kage/Peter Arnold, Inc.
25.01e ©Walter H. Hodge/Peter Arnold, Inc.
25.01f Courtesy of Dr. Martin Lockley, University of Colorado.
25.01g Courtesy Dr. David A. Grimaldi. Photo by Jacklyn Beckett/The American Museum of Natural History, N.Y.
25.10 ©Tom McHugh/Photo Researchers, Inc.
25.14 ©Hanny Paul/Gamma Liaison.
26.CO ©Chip Clark NMNH, artist Peter Sawyer.
26.01a Courtesy of John Stolz, University of Massachusetts.
26.01b, c, 26.02 ©S. M. Awramik, University of California/Biological Photo Service.
26.05a ©Sidney Fox, University of Miami/ Biological Photo Service.
26.05b Courtesy of F. M. Menger and Kurt Gabrielson, Emory University.
27.CO Dr. Tony Brain and David Parker/Science Photo Library/Photo Researchers, Inc.
27.MB1 ©Christine Case, Skyline College.
27.T03.1 ©Paul Johnson/Biological Photo Service.
27.T03.2 ©L. Evans Roth/Biological Photo Service.
27.T03.3 ©London School of Hygiene and Tropical Medicine/Science Photo Library/Photo Researchers, Inc.
27.T03.4 H. S. Pankratz, Michigan State University/Biological Photo Service.
27.T03.5 ©Michael Gabridge/Visuals Unlimited.
27.T03.6 ©Biophoto Associates/Science Source/ Photo Researchers, Inc.
27.T03.7 P. W. Johnson and Jon M. Sieburth. University of Rhode Island/Biological Photo Service.
27.T03.8 CNRI/SPL/Photo Researchers, Inc.

27.T03.9 ©Moredon Animal Health/SPL/Photo Researchers, Inc.
27.02a ©Meckes/Ottawa/Photo Researchers, Inc.
27.02b ©Manfred Kage/Peter Arnold, Inc.
27.02c ©David Chase/CNRI/Phototake NYC.
27.03 Courtesy of Esther Angert, Harvard University.
27.04 Fran Heyl Associates, photo by David Hasty, computer enhancement by pixelation.
27.05 Courtesy of J. Adler.
27.06a Courtesy of S. W. Watson.
27.06b N. J. Lang/University of California/ Biological Photo Service.
27.07 ©John Durham/SPL/Photo Researchers, Inc.
27.09a Courtesy of Frederick Atwood, Flint Hill School.
27.09b ©T. E. Adams/Visuals Unlimited.
27.10 ©Helen E. Carr/Biological Photo Service.
27.12 Courtesy of Exxon Corporation.
28.CO ©M. I. Walker/Photo Researchers, Inc.
28.04 ©E. White/Visuals Unlimited.
28.05 Courtesy of John Mansfield, University of Wisconsin, Madison.
28.06a ©Eric Grave/Photo Researchers, Inc.
28.06b Courtesy of JoAnn Burkholder, North Carolina State University.
28.07 Courtesy of Masamichi Aikawa, Case Western Reserve University.
28.08a ©Eric Grave/Photo Researchers, Inc.
28.08b ©Manfred Kage/Peter Arnold, Inc.
28.08c ©M. Abbey/Visuals Unlimited.
28.10 ©Peter Parks/Animals Animals.
28.11a, b ©Eric Grave/Photo Researchers, Inc.
28.12 ©Manfred Kage/Peter Arnold, Inc.
28.12b ©Paola Koch/Photo Researchers, Inc.
28.13 ©Ray Simmons/Photo Researchers, Inc.
28.13 ©R. Calentine/Visuals Unlimited.
28.14 Courtesy of Robert Kay, MRC Cambridge.
28.15a ©Eric Grave/Photo Researchers Inc.
28.15b ©Fred Rhoades/Mycena Consulting.
28.16 ©Biological Photo Service.
28.17 ©Fred Rhoades/Mycena Consulting.
28.18 ©Anne Wertheim/Animals Animals.
28.19 ©W Lewis Trusty/Animals Animals.
28.20 Courtesy of J. R. Waaland, University of Washington/Biological Photo Service.
28.21a Courtesy of J. R. Waaland, University of Washington/Biological Photo Service.
28.21b ©D. P. Wilson and Eric and David Hosking/Photo Researchers, Inc.
28.21c ©Gary Robinson/Visuals Unlimited.
28.22a ©Manfred Kage/Peter Arnold, Inc.
28.22b Estate of Dr. J. Metzner/Peter Arnold, Inc.
28.22c ©Laurie Campbell/NHPA.
28.23 ©M. I. Walker/Science Source/Photo Researchers, Inc.
28.24 Courtesy of W. L. Dentler, University of Kansas.
28.25 ©D. P. Wilson and Eric and David Hosking/Photo Researchers, Inc.
29.CO ©Otto Rogge.
29.EOC1 ©Dale and Marian Zimmerman/Bruce Coleman, Inc.
29.01a (left) ©J. R. Waaland/Biological Photo Service.
29.01a (right) ©Biological Photo Service.
29.01b ©Graham Kent.
29.04a ©Martha Cook and Claudia Lipke.
29.04b ©Linda Graham.
29.06 ©Glenn Oliver/Visuals Unlimited.
29.07 ©Dwight Kuhn.
29.08 ©Science VU.
29.09 From the Botany Department Collection at the University of Wisconsin, Madison. Courtesy of David Hanson, photographed by Claudia Lipke.
29.10 ©Chip Clark, 1988.
29.11 ©Milton Rand/Tom Stack and Associates.
29.12 ©Robert and Linda Mitchell.
29.13 ©E. S. Ross, California Academy of Sciences.
29.14 ©The Field Museum, Neg # 75400c.
30.CO Courtesy of The National Museum of Natural History, Smithsonian Institute.
30.03a ©James L. Castner.
30.03b ©Runk Schoenberger/Grant Heilman Photography, Inc.
30.03c ©Michael Fogden/Bruce Coleman, Inc.
30.03d ©C. P. Hickman/Visuals Unlimited.
30.07 ©Gregory K. Scott/Photo Researchers, Inc.
30.09 ©D. Wilder.
31.01 ©Fred Rhoades/Mycena Consulting.
31.05 J. R. Waaland, University of Washington/ Biological Photo Service.
31.05 Courtesy of William Barstow, University of Georgia, Athens.
31.06 Ed Reschke/Peter Arnold, Inc.
31.07a ©Fred Rhoades/Mycena Consulting.
31.07b ©Jacana/Photo Researchers, Inc.
31.07c ©J. L. Lepore/Photo Researchers, Inc.
31.08 E. R. Degginger/Animals Animals.
31.09a ©Kerry T. Givens/Tom Stack and Associates.
31.09b ©Frans Lanting/Minden Pictures.
31.09c ©Adrian Davies/Bruce Coleman, Inc.
31.10 ©Biophoto Associates/Photo Researchers, Inc.
31.11 (left) ©Jack Bostrack/Visuals Unlimited.
31.11 (right) ©M. F. Brown/Visuals Unlimited.
31.12 ©N. Allin and G. L. Barron, University of Guelph/Biological Photo Service.
31.13 Courtesy of Stephen J. Kron, University of Chicago.
31.14 ©Fred Rhoades/Mycena Consulting.
31.15 ©V. Ahmadijian/Visuals Unlimited.
31.16 ©R.L. Peterson, University of Guelf/ Biological Photo Service.
31.17 ©Robb Walsh.
32.CO ©Fred Bavendam/Minden Pictures.
33.CO ©Gary Braasch.
33.01, 33.05a ©Andrew J. Martinez/Photo Researchers, Inc.
33.05b ©Robert Brons/Biological Photo Service.
33.05c ©Kevin McCarthy/Offshoot Stock.
33.05d ©Chris Huss/The Wildlife Collection.
33.06 ©Robert Brons/Biological Photo Service.
33.07 ©Fred Bavendam/Peter Arnold, Inc.
33.08 ©Bill Wood/Bruce Coleman, Inc.
33.10 Centers for Disease Control.
33.11 ©Stanley Fleger/Visuals Unlimited.
33.12 ©W. I. Walker/Photo Researchers, Inc.
33.13a, b ©L. S. Stepanowicz/Photo Researchers, Inc.
33.14 ©Bill Wodd/NHPA.
33.15a ©Colin Milkins, Oxford Scientific Films/ Animals Animals.
33.15b ©Fred Bavendam/Peter Arnold, Inc.
33.17 ©Jeff Foott/Tom Stack and Associates.
33.19a ©Kevin Schafer.
33.19b ©Chris Huss.
33.20 ©H. W. Pratt/Biological Photo Service.
33.22a ©Tom McHugh/Steinhart Aquarium/Photo Researchers, Inc.
33.22b ©Fred Bavendam/Peter Arnold, Inc.
33.22c ©Douglas Faulkner/Photo Researchers, Inc.
33.24a ©Sea Studio.
33.24b ©Kjell Sandved.
33.24c ©Robert and Linda Mitchell.
33.26 ©Cliff B. Frith/Bruce Coleman, Inc.
33.27 ©Chip Clark.
33.28 ©Milton Tierney, Jr./Visuals Unlimited.
33.29a ©Frans Lantin/Minden Pictures.
33.29b ©David Scharf.
33.29c Courtesy of Diana Sammatoro, Pennsylvania State University.
33.30a ©Paul Skelcher/Rainbow.
33.31a ©Robert and Linda Mitchell.
33.31b ©E. R. Degginger/Color-Pic, Inc.
33.32 ©George Grall/National Geographic Society.
33.34a, b, c, d, e ©John Shaw/Tom Stack and Associates.
33.35a ©Frans Lanting/Minden Pictures.
33.35b ©Tom McHugh/Photo Researchers, Inc.
33.35c ©C. R. Wyttenbach/University of Kansas/ Biological Photo Service.
33.36a ©Jeffrey L. Rotman/Peter Arnold, Inc.
33.36b ©Gary Milburn/Tom Stack and Associates.
33.36c ©Dave Woodward/Tom Stack and Associates.
33.36d ©Marty Snyderman.
33.36e ©Carl Roessler/Animals Animals.
33.36f ©Fred Bavendam/Peter Arnold, Inc.
34.CO ©Biophoto Associates/Photo Researchers, Inc.
34.02a ©Scott Johnson/Animals Animals.
34.04a ©Runk Schoenenberg/Grant Heilman Photography.
34.04c ©Mike Neveux.
34.07 ©Herve Berthoule/Jacana/Photo Researchers, Inc.
34.07 (inset) ©Tom Hugh/Photo Researchers, Inc.
34.09a ©Steinhart Aquarium/Photo Researchers, Inc.
34.09b ©Fred Bavendam/Minden Pictures.
34.10a ©Patrice Ceisel/Visuals Unlimited.
34.10b ©J. M. Labat/Jacana/Photo Researchers, Inc.
34.12 ©Richard Ellis/Photo Researchers, Inc.
34.15a ©Dwight Kuhn.
34.15b, c ©Michael Fogden/Bruce Coleman, Inc.
34.16a, b, c ©Hans Pfletschinger/Peter Arnold, Inc.
34.18 Photograph by Jessie Cohen, National Zoological Park, Smithsonian Institution. ©Smithsonian Institution.
34.21a ©Mark Moffett/Minden Pictures.
34.21b ©A.N.T./NHPA.
34.21c ©Robert and Linda Mitchell.
34.21d ©Medford Taylor/National Geographic Society.
34.22b ©Janice Sheldon.
34.23 ©Stephen J. Kraseman/DRK Photo.
34.25a ©Mitsuaki Iwago/Minden Pictures.
34.25b ©John Henry Dick/Vireo.
34.25c ©Frans Lanting/Minden Pictures.
34.25d ©1997 Photodisc.
34.27a(inset) Courtesy of Mervyn Griffiths/ CSIRO.
34.27a ©D. Parer and E. Parer Cook/Auscape.
34.27b ©Dan Hadden/Ardea Ltd.
34.27c ©Fritz Prenzel/Animals Animals.
34.29 ©Erwin and Peggy Bauer/Bruce Coleman, Inc.
34.31a ©Stephen Dalton/NHPA.
34.31b ©Mickey Gibson/Animals Animals.
34.32a ©E. R. Degginger/Animals Animals.
34.32b ©Frans Lanting/Minden Pictures.
34.32c ©Nancy Adams/Tom Stack and Associates.
34.32d ©Michael K. Nichols/National Geographic Society.
34.34a Cleveland Museum of Natural History.
34.34b ©John Reader/SPL/Photo Researchers, Inc.
34.34c ©Institute of Human Origins, photo by Donald Johanson.
35.01a, c Elliot M. Meyerowitz and John Bowman, *Development* 112(1991):1–2 31.2.
35.01b Courtesy of Elliot Meyerowitz, Caltech.
35.02 ©David Cavagnaro.
35.05, 35.06a, b, c ©Dwight Kuhn.
35.06d ©George Bernard/Animals Animals.

35.08a ©Dwight Kuhn.
35.08b, c ©Kevin Schafer.
35.08d ©Larry Mellichamp/Visuals Unlimited.
35.09a, b ©Micrograph by W. P. Wergin, provided by E. H. Newcomb.
35.09c ©G. F. Leedale/Photo Researchers, Inc.
35.10a ©Dwight Kuhn.
35.10b ©Biophoto/Science Source/Photo Researchers, Inc.
35.10c, d ©Bruce Iverson.
35.10d ©Ed Reschke/Peter Arnold, Inc.
35.10e (left) ©George J. Wilder/Visuals Unlimited.
35.10e (right) ©Randy Moore/BioPhot.
35.15 Carolina Biological Supply/Phototake NYC.
35.16a ©Ed Reschke.
35.16b Carolina Biological Supply/Phototake NYC.
35.17 ©Dwight Kuhn.
35.18, 35.19, 35.20b, 35.20c ©Ed Reschke.
35.22 Courtesy of Dr. Howard F. Towner, Loyola Marymount University.
35.24d ©Ed Reschke/Peter Arnold, Inc.
36.MB1a Courtesy of Gary Tallman, Pepperdine University.
36.07 ©Dana Richter/Visuals Unlimited.
36.12 ©John D. Cunningham/Visuals Unlimited.
36.15a, b, c, Photos by M. H. Zimmerman courtesy of Professor P. B. Tomlinson, Harvard University.
37.CO ©Kim Taylor/Burce Coleman, Inc.
37.03 ©Holt Studios/Nigel Cattlin/Photo Researchers, Inc.
37.04 ©John H. Hoffman/Bruce Coleman.
37.05 ©William E. Ferguson.
37.07 John Reganhold, US Department of Agriculture Soil Conservation Service.
37.09a ©Biophoto Associates/Photo Researchers, Inc.
37.09b E. H. Newcomb/Biological Photo Service.
37.12a ©Gerald van Dyke/Visuals Unlimited.
37.12b ©Carolina Biological Supply/Phototake NYC.
37.13a ©Kevin Schafer.
37.13a (inset) ©Biophoto Associates/Photo Researchers, Inc.
37.13b ©Biological Photo Service.
37.14a ©Jeff Lepore/Photo Researchers, Inc.
37.14b ©John Shaw/Tom Stack and Associates.
38.CO © Stephenie Ferguson.
38.T1.01 ©Harold Taylor/OSF/Earth Scenes.
38.T1.02 ©William J. Weber/Visuals Unlimited.
38.T1.03 ©Bob Gossington/Bruce Coleman, Inc.
38.03a ©Tom Branch, 1978/Photo Researchers, Inc.
38.03b ©Ed Reschke/Peter Arnold, Inc.
38.03c ©Link, 1985/Visuals Unlimited.
38.03d ©Stephen Dalton/Photo Researchers, Inc.
38.03e ©William Ferguson.
38.03f ©Ray Coleman/Photo Researchers, Inc.
38.04a ©Graham Kent.
38.04b ©Ed Reschke.
38.13a ©James L. Castner.
38.13b ©David Cavagnaro/DRK.
38.14a, b ©Bruce Iverson.
38.15a Courtesy of Dr. John C. Sanford/Cornell University.
38.16a ©Jack Bostrack/Visuals Unlimited.
38.16b ©John D. Cunningham/Visuals Unlimited.
38.17 Courtesy of Susan Wick, University of Minnesota.
38.18 ©B. Wells and Kay Roberts.
38.20 ©Bruce Iverson.
39.CO, 39.06a, b ©Malcolm Wilkins, Glasgow University, UK.
39.07 Courtesy of Tugio Sasaki, Institute for Agricultural Research, Japan.
39.08 Courtesy of Fred Jensen, Kearney Agricultural Center.
39.09 Courtesy of Stephen Gladfelter, Stanford University.
39.10 ©Ed Reschke.
39.12 Courtesy of Michael Evans, Ohio State University.
39.13 Courtesy of Janet Braam, from *Cell* 60 (9 February 1990): cover. ©1990 by Cell Press.
39.14a, b ©John Kaprielan/Photo Researchers.
39.14c From K. Esau, *Anatomy of Seed Plants*, 2nd ed. (New York: John Wiley and Sons, 1977), fig. 19.4, p. 358.
39.15 Courtesy of Frank B. Salisbury, Utah State University.
39.21a, b Courtesy of J. L. Basq and M. C. Drew.
39.24 Courtesy of Barbara Baker, University of California, Berkeley.
40.CO ©Dwight R. Kuhn.
40.MB1.1 Courtesy of Robert Full, University of California.
40.MB1.2 ©Yoav Levy/Phototake NYC.
40.04 ©Ed Reschke.
40.08 ©Jeff Foott/Tom Stack & Associates.
41.CO ©Marty Stouffer/Animals Animals.
41.02 ©Carol Hughes/Bruce Coleman, Inc.
41.04 ©Brandon Cole.
41.05 Courtesy of Thomas Eisner, Cornell University.
41.06 ©Lennart Nilsson/Albert Bonniers Forlag AB.
41.07 ©Gunter Ziesler/Peter Arnold, Inc.
41.17 (left) ©Brian Milne/Animals Animals.
41.17 (right) ©Hans and Judy Beste/Animals Animals.
42.CO ©Jane Burton/Bruce Coleman, Inc.
42.01 ©Norbert Wu/Mo Young Productions.
42.15b ©Boehringer Ingelheim International GmbH photo by Lennart Nilsson/Albert Bonniers Verlag AB, *The Body Victorious*, Delacorte Press.
42.16a ©Ed Reschke.
42.16b ©W. Ober/Visuals Unlimited.
42.21 Courtesy of Dr. Peng Chai, University of Texas and Dr. Hong Y. Yan, University of Kentucky.
42.22c (right) ©CNRI/Photo Researchers Inc.
42.24a ©Professor Hans Rainer Dunker, Justus Liebig University, Giessen.
42.29 ©Kevin Schafer.
43.CO ©Lennart Nilsson/Albert Bonniers Forlag AB, *A Child Is Born*, Dell Publishing Company.
43.02 ©Science Photo Library/Photo Researchers, Inc.
43.03, 43.12b ©Lennart Nilsson/Boehringer Ingelheim International GmbH.
43.15c ©Dr. A. J. Olson, The Scripps Research Institute.
43.19 ©Lennart Nilsson/Boehringer Ingelheim International GmbH.
44.CO ©John Shaw/Bruce Coleman, Inc.
44.03a ©Jeff Lepore/Photo Researchers, Inc.
44.03b ©Dave B. Fleetham/Visuals Unlimited.
44.04 ©Daniel Lyons/Bruce Coleman, Inc.
44.12a, b Courtesy of John Crowe, University of California.
44.13a ©Tom McHugh/Photo Researchers, Inc.
45.CO ©William E. Ferguson.
46.CO Hans Pfletschinger/Peter Arnold, Inc.
46.01 ©David Wrobel/Monterey Bay Aquarium.
46.02a Courtesy of David Crews, photo by P. de Vries.
46.03 ©Stephan Myers.
46.04 ©Dwight Kuhn.
46.05 ©William Ferguson.
46.10 ©C. Edelman/La Villette/Photo Researchers, Inc.
46.18a, b, c ©Lennart Nilsson/Albert Bonniers Vorlag AB.
46.22 ©Howard Sochurek.
47.CO ©Lennart Nilsson/Albert Bonniers Vorlag AB.
47.01 Historical Collections, College of Physicians, Philadelphia.
47.03.1, 2, 3, 4 J. Reproduced from C. Rilkey, L. F. Jaffe, E. B. Ridgeway, and G. T. Reynolds, *The Journal of Cell Biology* 76(1978):448–466 by copyright permission of The Rockefeller University Press.
47.06a, b, c, ©George Watchmaker.
47.09 Courtesy of Charles A Ettensohn, Carnegie Mellon University.
47.11a ©CABISCO/Visuals Unlimited.
47.11c Thomas Poole, SUNY Health Science Center.
47.13 Carolina Biological Supply/Phototake NYC.
47.18a Reproduced from Dr. Jean Paul Thiery *The Journal of Cell Biology* 96(1983):462–473 by copyright permission of the Rockefeller University Press.
47.18b1, 2 ©Richard Hynes, from *Scientific American*, June 1986.
47.19a, b Courtesy of Janet Heasman, University of Minnesota.
47.20b1, 2 From Hiroki Nishida, *Dev. Biol.* 121 (1987):526. Reprinted by permission of Academic Press.
47.23a, b Courtesy of Kathryn Tosney, University of Michigan.
47.24 Based on Honig and Summerbell, 1985. Photo courtesy of Dennis Summerbell.
48.CO ©V. I. Lab E. R. I. C./FPG.
48.MB1b Courtesy of W. F. Gilly, Hopkins Marine Station of Stanford University.
48.MB2a ©Howard Sochurek.
48.MB2b Courtesy of Marcus Raichle, MD, Washington University School of Medicine, St Louis, MO.
48.02b ©Manfred Kage/Peter Arnold, Inc.
48.11b Courtesy of E. R. Lewis, University of California, Berkeley.
48.18 ©Stouffer Productions/Animals Animals.
48.21a ©Alexander Tsiarias/Photo Researchers, Inc.
49.CO ©Stephen Dalton/NHPA.
49.02a ©OSF/Animals Animals.
49.02b Courtesy of R. A. Steinbrecht, Max Planck Institute.
49.03a ©Joe McDonald/Animals Animals.
49.03b ©Russ Kinne/Photo Researchers, Inc.
49.05a ©John L. Pontier/Animals Animals.
49.17 From Richard Elzinga, *Fundamentals of Entomology* 3rd ed. ©1987, p. 185. Reprinted by permission of Prentice-Hall, Upper Saddle River, NJ.
49.18 ©Joe Monroe/Photo Researchers, Inc.
49.22 ©John Cancalosi/Tom Stack and Associates.
49.26 Courtesy of Clara Franzini-Armstrong, University of Pennsylvania.
49.27a, b, c, Courtesy of Dr. H. E. Huxley.
49.32b ©Eric Grave/Photo Researchers, Inc.
50.CO Courtesy of NASA - LBJ Space Center.
50.01 Courtesy of Gary Atchison, Iowa State University.
50.02 ©Copyright 1995, Los Angeles Times. Reprinted with permission.
50.11a ©Boyd Norton.
50.11b ©Michael Gadomski/Earth Scenes/Animals Animals.
50.11c ©William H. Mullins/Photo Researchers, Inc.
50.12a ©Steve Solum/ Bruce Coleman, Inc.
50.12b ©M.E. Warre/Photo Researchers, Inc.

50.14a © Chris Huss.
50.14b ©David Hall, 1984/ Photo Researchers, Inc.
50.14c ©Dudley Foster/Woods Hole Oceanographic Institution.
50.16a ©Peter Ward/Bruce Coleman, Inc.
50.16b ©Jonathan Scott/Planet Earth Pictures.
50.16c ©Charlie Ott/Photo Researchers, Inc.
50.16d ©John D. Cunningham/Visuals Unlimited.
50.16e ©Frank Oberle.
50.16f ©Carr Clifton.
50.16g ©Michio Hoshino/Minden Pictures.
50.16h ©J. Warden/Superstock, Inc.
51.CO ©Tom Vezo.
51.MB1a ©E.R. Degginger/Animals Animals.
51.07 Thomas McAvoy, Life Magazine ©Time Inc.
51.08 ©Calvin Larsen/Photo Researchers, Inc.
51.09 (left) ©Harry Engels/Animals Animals.
51.09 (right) ©1992: Stephen J. Krasemann/Photo Researchers, Inc.
51.10 ©Ted Kerasote/Photo Researchers, Inc.
51.11 ©Bernd Heinrich, University of Vermont.
51.14 ©Jeff Foott/Bruce Coleman, Inc.
51.15a ©G.R. Higbee/Photo Researchers, Inc.
51.15b ©Bruce Davidson/Animals Animals.
51.16 ©Visuals Unlimited.
51.17 ©Doug Wechsler/Animals Animals.
51.18a ©Jonathan Scott/Seaphot Ltd./Planet Earth Pictures.
51.18b ©Michael Dick/Animals Animals.
51.21a ©Kenneth Lorenzen, UC Davis.
51.22a ©Stephen Kraseman/Peter Arnold, Inc.
51.23a Courtesy of Dr. Jennifer Jarvis, University of Cape Town.
51.23b ©Tom McHugh/Photo Researchers, Inc.
51.24 ©Ivan Polunin/Bruce Coleman, Inc.
52.CO ©Will and Deni McIntyre/Photo Researchers Inc.
52.MB1a, b ©Frans Lanting/Minden Pictures.
52.01 ©Jim Brandenburg/Minden Pictures.
52.02a ©Sophie de Wilde/Jacana/Photo Researchers, Inc.
52.02b ©Frans Lanting/Minden Pictures.
52.02c ©Will & Deni McIntryre/Photo Researchers, Inc.
52.07 ©E.R. Degginger/Color-Pic, Inc.
52.09a ©R. Calentine/Visuals Unlimited.
52.09b ©Max and Bea Hunn/Visuals Unlimited.
52.20 ©Alan Carey/Photo Researchers, Inc.
53.CO ©Richard Estes/Photo Researchers, Inc.
53.02 Courtesy of Paul Hertz.
53.03a, b, c ©L.E.Gilbert/BPS.
53.04 ©Arthur Morris/VIREO.
53.05 ©C. Allan Morgan/Peter Arnold, Inc.
53.06 ©John L. Tveten.
53.07 ©David B. Fleetham/Tom Stack and Associates.
53.08a ©Adrian Davies/Bruce Coleman, Inc.
53.08b ©Peter J. Mayne.
53.11b ©1985 Joseph T. Collins/Photo Researchers, Inc.
53.11c ©Kevin de Queiroz, National Museum of Natural History.
53.14 ©Stephen Kraseman/Photo Researchers, Inc.
53.15 ©Robert and Linda Mitchell.
53.16a Courtesy of Karen Oberhauser.
53.17a, b, c, Courtesy of Frank Gilliam, Marshall University.
53.18a ©Dominique Braud/Tom Stack and Associates.
53.18b ©Raymond K. Gehman/National Geographic Image Collection.
53.18c ©John Sohlden/Visuals Unlimited.
53.19a ©Grant Heilman/Grant Heilman Photography.
53.19b ©Jeff Foot/Tom Stack and Associates.
54.MB1a Courtesy of Sara Hotchkiss, University of Minnesota.
54.MB1a, b Courtesy of Margaret Davis, University of Minnesota.
54.14a ©John D. Cunningham/Visuals Unlimited.
54.14b Provided by the Northeastern Forest Experiment Station, Forest Service, United States Department of Agriculture.
54.18a, b Matthew Meselson, Harvard University.
54.18c ©Earth Observation Satellite Co., Lanham, MD/SPL/Photo Researchers, Inc.
55.CO ©Juan Manuel Renjifo/Earth Scenes.
55.01 ©Darrel Plowes.
55.03 ©William Ferguson.
55.06 ©Frans Lanting/Minden Pictures.
55.07 ©Gary Braash/Woodfin Camp and Associates.
55.09a, b ©Raymond K. Gehman/National Geographic Society.
55.09c ©Blanche C. Haney.
55.10a ©Jim Yuskavitch.
55.10b ©Robert Maier/Animals Animals.
55.12 ©F. & J. Craighead.
55.13a ©David Hosking/Photo Researchers, Inc.
55.13b ©James P. Blair/National Geographic Image Collection.
55.15a ©Joe McDonald/Bruce Coleman, Inc.
55.15b ©Reed Bowman.
55.16 Courtesy of Ariel E. Lugo/USDA/Forest Service, photo by Mr. Jesus Ayala O'Neill.

Unit openers (interviews):
1.1 ©Stella Johnson.
1.2 ©Stella Johnson.
2.1 ©Linda Bartlett/Addison Wesley Longman.
2.2 ©Linda Bartlett/Addison Wesley Longman.
3.1 ©Mary Levin, University of Washington.
3.2 ©Mary Levin, University of Washington.
4.1 ©Kathleen Cameron/Addison Wesley Longman.
4.2 ©Kathleen Cameron/Addison Wesley Longman.
5.1 ©Sven Martson/Addison Wesley Longman.
5.2 ©Sven Martson/Addison Wesley Longman.
6.1 ©Kathleen Cameron/Addison Wesley Longman.
6.2 ©Kathleen Cameron/Addison Wesley Longman.
7.1 ©Paul Anderton/Addison Wesley Longman.
7.2 ©Paul Anderton/Addison Wesley Longman.
8.1 ©Linda Bartlett/Addison Wesley Longman.
8.2 ©Linda Bartlett/Addison Wesley Longman.

Table of contents:
Unit One: ©Jean-Marie Bassot/Photo Researchers, Inc.
Unit Two: ©Gerry Ellis/ENP Images.
Unit Three: ©Kyodo News International, Inc.
Unit Four: ©Brian Parker/Tom Stack and Associates.
Unit Five: ©Fred Bavendam/Minden Pictures.
Unit Seven: ©Lennart Nilsson/Albert Bonniers Vorlag AB.
Unit Eight: ©Richard Estes/Photo Researchers, Inc.

ILLUSTRATIONS

Contributing Artists to Previous Editions: Nea Bisek, Chris Carothers, Rachel Ciemma, Barbara Cousins, Pamela Drury-Wattenmaker, Cecile Duray-Bito, Janet Hayes, Darwen Hennings, Vally Hennings, Georg Klatt, Sandra McMahon, Linda McVay, Kenneth Miller, Fran Milner, Elizabeth Morales-Denney, Jackie Osborn, Carol Verbeeck, John Waller, and Judy Waller.

The following figures are adapted from Neil A. Campbell, Lawrence G. Mitchell, and Jane B. Reece, *Biology: Concepts and Connections* 2nd Ed. (Menlo Park, CA; Benjamin/Cummings, 1997). ©1997 The Benjamin/Cummings Publishing Company: 2.5, 3.9, 5.UN2, 5.2, 9.16, 10.03, 11.3, 12.03, 12.5, 12.8a, 12.13, 13.MB1, 13.6, 17.6, 18.2, 19.7, 19.14, 22.3a, 22.11, 23.11, 24.11, 25.3, 25.8, 26.7, 27.8, 29.10, 32.1, 32.6, 33.10, 33.11, 34.8, 34.14, 34.23, 35.25, 37.8, 38.10, 38.15, 39.5, 40.11, 41.3, 41.9, 41.10, 41.11, 41.12, 41.15, 42.MB1, 42.3, 42.4, 42.5, 42.6, 42.8, 42.9, 42.11, 42.12, 42.15a, 42.19, 42.20, 42.21, 42.22, 42.23, 42.24, 42.25, 42.26, 43.4, 44.1, 44.6, 44.10, 44.11, 44.14, 44.18, 45.5, 45.6, 46.8, 46.9, 46.11, 46.19, 47.10, 48.UN, 48.1, 48.14, 48.17, 48.18, 48.19, 48.23, 49.4, 49.8, 49.10, 49.23, 52.4, 52.23, 53.10, 54.10.

The following figures are adapted from Christopher K. Mathews and K. E. van Holde, *Biochemistry* 2nd ed. (Menlo Park, CA: Benjamin/Cummings, 1996) ©1996 The Benjamin/Cummings Publishing Company: 4.5, 9.9, 10.10, 17.14, 19.10.

The following figures are adapted from Wayne M. Becker, Jane B. Reece, and Martin F. Poenie, *The World of the Cell* 3rd ed. (Menlo Park, CA: Benjamin/Cummings, 1996) ©1986, 1991, 1996 The Benjamin/Cummings Publishing Company, Inc.: 4.6, 7.T2, 7.5, 8.6, 11.07, 11.12, 12.12, 16.13, 16.16, 17.9, 19.5, 19.14, 20.MB1.

The following figures are adapted from Elaine N. Marieb, *Human Anatomy and Physiology* 4th ed. (Menlo Park, CA: Benjamin/Cummings, 1998). Copyright ©1998 Benjamin/Cummings, an imprint of Addison Wesley Longman, Inc.: 7.11, 7.22, 46.17, 48.02a, 48.16, 48.09, 48.16, 48.22, 49.24a, 49.27.

3.08 Adapted from a figure by Michael Pique, The Scripps Research Institute.
4.07 Adapted from an illustration by Clark Still, Columbia University.
5.12 From Robert Wallace, Gerald P. Sanders, and Robert J. Ferl, *Biology: The Science of Life* 3rd ed. (New York: HarperCollins, 1991). ©1991 HarperCollins Publishers, Inc. Reprinted by permission of Addison-Wesley Educational Publishers.
5.23b ©Irving Geis.
6.01 From Alberts et al., *Molecular Biology of the Cell*, 2nd ed. (New York: Garland Publishing, 1989). ©1989 Garland Publishing Inc. Reprinted by permission.
9.05 Adapted from Alberts et al., *Molecular Biology of the Cell* 2nd ed., fig. 7.17, p. 351 (New York: Garland Publishing, 1989). ©1989 Garland Publishing. Reprinted by permission.
10.12 From Richard and David Walker, *Energy, Plants and Man*, fig. 4.1, p. 69 (Sheffield: University of Sheffield). ©David and Richard Walker. Reprinted by permission.
12.06 Adapted from Neil A. Campbell, Lawrence G. Mitchell, and Jane B. Reece, *Biology: Concepts and Connections* (Menlo Park, CA; Benjamin/Cummings, 1994). Copyright ©1994 The Benjamin/Cummings Publishing Company.
12.11 From Alberts et al., *Molecular Biology of the Cell* 2nd ed. fig. 13.75 (New York: Garland Publishing, 1989). ©1989 Garland Publishing. Reprinted by permission.

18.CO From James D. Watson et al., *Molecular Biology of the Gene* 4th ed., fig. 7.10 (Menlo Park, CA: Benjamin/Cummings, 1987). ©1987 James D. Watson.
19.03 (inset) ©Irving Geis.
21.04 Adapted from Alberts et al., *Molecular Biology of the Cell*, 2nd ed., fig. 16.32, p. 904 (New York: Garland Publishing, 1989). ©1989 Garland Publishing. Reprinted by permission.
21.14 Adapted from an illustration by William McGinnis.
21.18 (upper right illustration only) Adapted from E. Dennis et al., "Manipulating Floral Identity," *Current Biology* 3(1993):90–93. Reprinted by permission.
25.05 Data from M. J. Benton, "Diversification and Extinction in the History of Life" *Science* 268 (1995):55.
25.16 Adapted from Cladogram by Peggy Conversano, illustrations by Ed Heck and Frank Ippolito, courtesy of Natural History.
27.5 From Gerard J. Tortora, Berdell R. Funke, and Christine L. Case, *Microbiology: An Introduction* 6th ed. (Menlo Park, CA: Benjamin/Cummings, 1998). ©1998 Benjamin/Cummings an imprint of Addison Wesley Longman, Inc.
30.10 Data from Berner, Robert A. "The Rise of Plants and Their Effects on Weathering and Atmospheric CO_2" *Science*, 276(1997):545.
34.20 ©Donna Braginetz.
34.26 Stephen J. Gould et al., *The Book of Life* (London: Ebury Press), p. 96. Reprinted by permission of Random House UK Ltd.
34.35 Adapted from an illustration from Laurie Grace, "The Recent African Genesis of Humans" by A. C. Wilson, R. I. Cann, *Scientific American* 1992:73.
43.14, 43.15 From Gerard J. Tortora, Berdell R. Funke, and Christine L. Case, *Microbiology: An Introduction*, 6th ed. (Menlo Park, CA: Benjamin/Cummings, 1998). ©1998 Benjamin/Cummings, an imprint of Addison Wesley Longman, Inc.
43.17 Adapted from Lennart Nilsson and Jan Lindberg, *The Body Victorious* (New York: Delacorte Press, 1987), p. 27. Illustration by Urban Frank, Studio Frank and Co, Illustrator.
44.02 Adapted from P. T. Marshall and G. M. Hughes, *Physiology of Mammals and Other Vertebrates*, 2nd ed. (Cambridge: Cambridge University Press, 1980). Reprinted with the permission of Cambridge University Press.
44.05b Adapted from an illustration by Enid Kotschnig from "Thermoregulation in a Winter Moth" by Bernd Heinrich, *Scientific American* 1987:105.
44.10 Adapted from Lawrence G. Mitchell, John A. Mutchmor, and Warren D. Dolphin, *Zoology* (Menlo Park, CA: Benjamin/Cummings, 1988) ©1988 The Benjamin/Cummings Publishing Company.
44.13b Kangaroo rat data from Schmidt-Nielsen, *Animal Physiology Adaptation and Environment* 4th ed., p. 339 (Cambridge: Cambridge University Press, 1990).
45.04 Adapted from Alberts et al., *Molecular Biology of the Cell*, 2nd ed., fig. 15.9 (New York: Garland Publishing, 1989). ©1989 Garland Publishing. Reprinted by permission.
45.12 Adapted from Gilbert, *Developmental Biology*, 4th ed., fig. 7.39 p. 282. (Sunderland, MA: Sinauer Associates Inc.). Reprinted by permission.
47.17 From Wolpert et al., *Principles of Development*, fig, 8.25, p. 251 right (Oxford: Oxford University Press, 1988). By permission of Oxford University Press.
47.20 Adapted from Alberts et al., *Molecular Biology of the Cell*, 2nd ed., fig. 16.29a, p. 904. (New York: Garland Publishing, 1989). ©1989 Garland Publishing. Reprinted by permission.
47.22 *Left*: From Wolpert et al., *Principles of Development* (Oxford: Oxford University Press. 1988), fig. 1.10 (right). By permission of Oxford University Press. Right: Adapted from Gilbert, *Developmental Biology* 5th Ed., part of fig. 15.12, p. 604 (Sunderland, MA: Sinauer Associates Inc.). ©1997 Sinauer Associates. Reprinted by permission.
48.07 Adapted from G. Matthews, *Cellular Physiology of Nerve and Muscle* (Cambridge, MA: Blackwell Scientific Publications, 1986). Reprinted by permission of Blackwell Science, Inc.
49.25 Adapted from Lawrence G. Mitchell, John A. Mutchmor, and Warren D. Dolphin, *Zoology* (Menlo Park, CA; Benjamin/Cummings, 1988). ©1988 The Benjamin/Cummings Publishing Company.
51.02 Adapted from Lawrence G. Mitchell, John A. Mutchmor, and Warren D. Dolphin, *Zoology* (Menlo Park, CA; Benjamin/Cummings, 1988). ©1988 The Benjamin/Cummings Publishing Company.
51.03 Adapted from N. Tinberger, *The Study of Instinct* (Oxford University Press, Oxford, 1951). By permission of Oxford University Press.
51.08 Courtesy of Masakazu Konishi.
52.10 R. E. Rickleffs, in D. S. Farner, *Breeding Biology of Birds*, pp. 366–435 (National Academy of Sciences, Washington DC, 1973).
52.19 Data from Higgins et al., "Stochastic Dynamics and Deterministic Skeletons: Population Behavior of Dungeness Crab", *Science* May 30, 1997.
53.9 Adapted from Gerard J. Tortora, Berdell R. Funke, and Christine L. Case, *Microbiology: An Introduction*, 6th ed. (Menlo Park, CA: Benjamin/Cummings, 1998). ©1998 Benjamin/Cummings, an imprint of Addison Wesley Longman, Inc.
53.11a A. S. Rand and E. E. Williams, "The Anoles of La Palma: Aspects of Their Ecological Relationships," *Breviora* no. 327, 1969. Museum of Comparative Zoology, Harvard University. ©Presidents and Fellows of Harvard College.
54.02 Adapted from R. Leo Smith, *Ecology and Field Biology*, 4th ed. ©1990 by R. Leo Smith. Reprinted by permission of Addison-Wesley Educational Publishers.
54.09 From R. E. Ricklefs, *The Economy of Nature* 4th ed.©1997 by W. H. Freeman and Company. Used with permission.
54.16 From G. Tyler, *Living in the Environment*, 2nd ed., p. 87 (Belmont, CA: Wadsworth Publishing, 1979). ©1979 Wadsworth Publishing Company.
55.MB1a Adapted from an illustration by Tamara R. Sayre.
55.MB1b USGS-BRD-Gap Analysis.
55.02 Tom Moore ©1986 Discover Magazine. Reprinted by permission of Discover Magazine.
55.14 Adapted from W. Purves and G. Orians, *Life: The Science of Biology*, 5th ed., fig. 55.23, p. 1239. (Sunderland, MA: Sinauer, 1998). Copyright ©1998 by Sinauer and W. H. Freeman and Company. Reprinted by permission.

GLOSSARY

abscisic acid (ABA) *(ab-SIS-ik)* A plant hormone that generally acts to inhibit growth, promote dormancy, and help the plant tolerate stressful conditions.

absorption spectrum The range of a pigment's ability to absorb various wavelengths of light.

abyssal zone *(uh-BIS-ul)* The portion of the ocean floor where light does not penetrate and where temperatures are cold and pressures intense.

acclimatization *(uh-KLY-mih-ty-ZAY-shun)* Physiological adjustment to a change in an environmental factor.

accommodation The automatic adjustment of an eye to focus on near objects.

acetyl CoA The entry compound for the Krebs cycle in cellular respiration; formed from a fragment of pyruvate attached to a coenzyme.

acetylcholine One of the most common neurotransmitters; functions by binding to receptors and altering the permeability of the postsynaptic membrane to specific ions, either depolarizing or hyperpolarizing the membrane.

acid A substance that increases the hydrogen ion concentration in a solution.

acid precipitation Rain, snow, or fog that is more acidic than pH 5.6.

acoelomate *(a-SEEL-oh-mate)* A solid-bodied animal lacking a cavity between the gut and outer body wall.

acrosome *(AK-ruh-some)* An organelle at the tip of a sperm cell that helps the sperm penetrate the egg.

actin *(AK-tin)* A globular protein that links into chains, two of which twist helically about each other, forming microfilaments in muscle and other contractile elements in cells.

action potential A rapid change in the membrane potential of an excitable cell, caused by stimulus-triggered, selective opening and closing of voltage-sensitive gates in sodium and potassium ion channels.

active site The specific portion of an enzyme that attaches to the substrate by means of weak chemical bonds.

active transport The movement of a substance across a biological membrane against its concentration or electrochemical gradient, with the help of energy input and specific transport proteins.

adaptive peak An equilibrium state in a population when the gene pool has allele frequencies that maximize the average fitness of a population's members.

adaptive radiation The emergence of numerous species from a common ancestor introduced into an environment, presenting a diversity of new opportunities and problems.

adenylyl cyclase An enzyme that converts ATP to cyclic AMP in response to a chemical signal.

adrenal gland *(uh-DREE-nul)* An endocrine gland located adjacent to the kidney in mammals; composed of two glandular portions: an outer cortex, which responds to endocrine signals in reacting to stress and effecting salt and water balance, and a central medulla, which responds to nervous inputs resulting from stress.

aerobic *(air-OH-bik)* Containing oxygen; referring to an organism, environment, or cellular process that requires oxygen.

age structure The relative number of individuals of each age in a population.

agnathan *(AG-naa-thun)* A member of a jawless class of vertebrates represented today by the lampreys and hagfishes.

agonistic behavior *(ag-on-IS-tik)* A type of behavior involving a contest of some kind that determines which competitor gains access to some resource, such as food or mates.

AIDS (acquired immunodeficiency syndrome) The name of the late stages of HIV infection; defined by a specified reduction of T cells and the appearance of characteristic secondary infections.

aldehyde *(AL-duh-hyde)* An organic molecule with a carbonyl group located at the end of the carbon skeleton.

aldosterone *(al-DAH-stair-own)* An adrenal hormone that acts on the distal tubules of the kidney to stimulate the reabsorption of sodium (Na^+) and the passive flow of water from the filtrate.

alga (plural, **algae**) A photosynthetic, plantlike protist.

all-or-none event An action that occurs either completely or not at all, such as the generation of an action potential by a neuron.

allantois *(AL-an-TOH-iss)* One of four extraembryonic membranes; serves as a repository for the embryo's nitrogenous waste.

allele *(uh-LEEL)* An alternative form of a gene.

allometric growth *(AL-oh-MET-rik)* The variation in the relative rates of growth of various parts of the body, which helps shape the organism.

allopatric speciation *(AL-oh-PAT-rik)* A mode of speciation induced when the ancestral population becomes segregated by a geographical barrier.

allopolyploid *(AL-oh-POL-ee-ploid)* A common type of polyploid species resulting from two different species interbreeding and combining their chromosomes.

allosteric site *(AL-oh-STEER-ik)* A specific receptor site on an enzyme molecule remote from the active site. Molecules bind to the allosteric site and change the shape of the active site, making it either more or less receptive to the substrate.

alpha (α) **helix** A spiral shape constituting one form of the secondary structure of proteins, arising from a specific hydrogen-bonding structure.

alternation of generations A life cycle in which there is both a multicellular diploid form, the sporophyte, and a multicellular haploid form, the gametophyte; characteristic of plants.

altruistic behavior *(AL-troo-IS-tik)* The aiding of another individual at one's own risk or expense.

alveolus *(al-VEE-oh-lus)* (plural, **alveoli**) (1) One of the deadend, multilobed air sacs that constitute the gas exchange surface of the lungs. (2) One of the milk-secreting sacs of epithelial tissue in the mammary glands.

amino acid *(uh-MEE-noh)* An organic molecule possessing both carboxyl and amino groups. Amino acids serve as the monomers of proteins.

amino group A functional group that consists of a nitrogen atom bonded to two hydrogen atoms; can act as a base in solution, accepting a hydrogen ion and acquiring a charge of +1.

aminoacyl-tRNA synthetases A family of enzymes, at least one for each amino acid, that catalyze the attachment of an amino acid to its specific tRNA molecule.

amniocentesis *(AM-nee-oh-sen-TEE-sis)* A technique for determining genetic abnormalities in a fetus by the presence of certain chemicals or defective fetal cells in the amniotic fluid, obtained by aspiration from a needle inserted into the uterus.

amnion *(AM-nee-on)* The innermost of four extraembryonic membranes; encloses a fluid-filled sac in which the embryo is suspended.

amniote A vertebrate possessing an amnion surrounding the embryo; reptiles, birds, and mammals are amniotes.

amniotic egg *(AM-nee-AH-tik)* A shelled, water-retaining egg that enables reptiles, birds, and egg-laying mammals to complete their life cycles on dry land.

Amphibia The vertebrate class of amphibians, represented by frogs, salamanders, and caecilians.

amphipathic molecule A molecule that has both a hydrophilic region and a hydrophobic region.

anaerobic *(an-air-OH-bik)* Lacking oxygen; referring to an organism, environment, or cellular process that lacks oxygen and may be poisoned by it.

anagenesis *(AN-uh-JEN-eh-sis)* A pattern of evolutionary change involving the transformation of an entire population, sometimes to a state different enough from the ancestral population to justify renaming it as a separate species; also called phyletic evolution.

analogy The similarity of structure between two species that are not closely related; attributable to convergent evolution.

androgens *(AN-droh-jens)* The principal male steroid hormones, such as testosterone, which stimulate the development and maintenance of the male reproductive system and secondary sex characteristics.

aneuploidy *(AN-yoo-ploy-dee)* A chromosomal aberration in which certain chromosomes are present in extra copies or are deficient in number.

angiosperm *(AN-jee-oh-spurm)* A flowering plant, which forms seeds inside a protective chamber called an ovary.

anion *(AN-eye-on)* A negatively charged ion.

annual A plant that completes its entire life cycle in a single year or growing season.

anterior Referring to the head end of a bilaterally symmetrical animal.

anther *(AN-thur)* The terminal pollen sac of a stamen, inside which pollen grains with male gametes form in the flower of an angiosperm.

antheridium *(an-theh-RID-ee-um)* In plants, the male gametangium, a moist chamber in which gametes develop.

antibiotic A chemical that kills bacteria or inhibits their growth.

antibody An antigen-binding immunoglobulin, produced by B cells, that functions as the effector in an immune response.

anticodon *(AN-tee-CO-don)* A specialized base triplet on one end of a tRNA molecule that recognizes a particular complementary codon on an mRNA molecule.

antidiuretic hormone (ADH) A hormone important in osmoregulation.

antigen *(AN-teh-jen)* A foreign macromolecule that does not belong to the host organism and that elicits an immune response.

aphotic zone *(ay-FOE-tik)* The part of the ocean beneath the photic zone, where light does not penetrate sufficiently for photosynthesis to occur.

apical dominance *(AY-pik-ul)* Concentration of growth at the tip of a plant shoot, where a terminal bud partially inhibits axillary bud growth.

apical meristem *(AY-pik-ul MARE-eh-stem)* Embryonic plant tissue in the tips of roots and in the buds of shoots that supplies cells for the plant to grow in length.

apomorphic character *(AP-oh-MORE-fik)* A derived phenotypic character, or homology, that evolved after a branch diverged from a phylogenetic tree.

apoplast *(AP-oh-plast)* In plants, the nonliving continuum formed by the extracellular pathway provided by the continuous matrix of cell walls.

apoptosis Programmed cell death brought about by signals that trigger the activation of a cascade of "suicide" proteins in the cells destined to die.

aposematic coloration *(AP-oh-so-MAT-ik)* The bright coloration of animals with effective physical or chemical defenses that acts as a warning to predators.

aquaporin A transport protein in the plasma membranes of a plant or animal cell that specifically facilitates the diffusion of water across the membrane (osmosis).

aqueous solution *(AY-kwee-us)* A solution in which water is the solvent.

Archaea One of two prokaryotic domains, the other being the Bacteria.

Archaezoa Primitive eukaryotic group that includes diplomonads, such as *Giardia*; some systematists assign kingdom status to archezoans.

archegonium *(ar-kih-GO-nee-um)* In plants, the female gametangium, a moist chamber in which gametes develop.

archenteron *(ark-EN-ter-on)* The endoderm-lined cavity, formed during the gastrulation process, that develops into the digestive tract of an animal.

arteriosclerosis A cardiovascular disease caused by the formation of hard plaques within the arteries.

artery A vessel that carries blood away from the heart to organs throughout the body.

artificial selection The selective breeding of domesticated plants and animals to encourage the occurrence of desirable traits.

ascus (plural, **asci**) A saclike spore capsule located at the tip of the ascocarp in dikaryotic hyphae; defining feature of the Ascomycota division of fungi.

asexual reproduction A type of reproduction involving only one parent that produces genetically identical offspring by budding or by the division of a single cell or the entire organism into two or more parts.

associative learning The acquired ability to associate one stimulus with another; also called classical conditioning.

assortative mating A type of nonrandom mating in which mating partners resemble each other in certain phenotypic characters.

asymmetric carbon A carbon atom covalently bonded to four different atoms or groups of atoms.

atomic number The number of protons in the nucleus of an atom, unique for each element and designated by a subscript to the left of the elemental symbol.

atomic weight The total atomic mass, which is the mass in grams of one mole of the atom.

ATP (adenosine triphosphate) *(uh-DEN-oh-sin try-FOS-fate)* An adenine-containing nucleoside triphosphate that releases free energy when its phosphate bonds are hydrolyzed. This energy is used to drive endergonic reactions in cells.

ATP synthase A cluster of several membrane proteins found in the mitochondrial cristae (and bacterial plasma membrane) that function in chemiosmosis with adjacent electron transport chains, using the energy of a hydrogen-ion concentration gradient to make ATP. ATP synthases provide a port through which hydrogen ions diffuse into the matrix of a mitrochondrion.

atrioventricular valve A valve in the heart between each atrium and ventricle that prevents a backflow of blood when the ventricles contract.

atrium *(AY-tree-um)* (plural, **atria**) A chamber that receives blood returning to the vertebrate heart.

autogenesis model According to this model, eukaryotic cells evolved by the specialization of internal membranes originally derived from prokaryotic plasma membranes.

autoimmune disease An immunological disorder in which the immune system turns against itself.

autonomic nervous system *(AWT-uh-NAHM-ik)* A subdivision of the motor nervous system of vertebrates that regulates the internal environment; consists of the sympathetic and parasympathetic divisions.

autopolyploid *(AW-toe-POL-ee-ploid)* A type of polyploid species resulting from one species doubling its chromosome number to become tetraploid, which may self-fertilize or mate with other tetraploids.

autosome *(AW-tuh-some)* A chromosome that is not directly involved in determining sex, as opposed to the sex chromosomes.

autotroph *(AW-toh-TROHF)* An organism that obtains organic food molecules without eating other organisms. Autotrophs use energy from the sun or from the oxidation of inorganic substances to make organic molecules from inorganic ones.

auxins *(AWK-sins)* A class of plant hormones, including indoleacetic acid (IAA), having a variety of effects, such as phototropic response through the stimulation of cell elongation, stimulation of secondary growth, and the development of leaf traces and fruit.

auxotroph *(AWK-soh-trohf)* A nutritional mutant that is unable to synthesize and that cannot grow on media lacking certain essential molecules normally synthesized by wild-type strains of the same species.

Aves The vertebrate class of birds, characterized by feathers and other flight adaptations.

axillary bud *(AKS-ill-air-ee)* An embryonic shoot present in the angle formed by a leaf and stem.

axon *(AKS-on)* A typically long extension, or process, from a neuron that carries nerve impulses away from the cell body toward target cells.

B cell A type of lymphocyte that develops in the bone marrow and later produces antibodies, which mediate humoral immunity.

Bacteria One of two prokaryotic domains, the other being the Archaea.

bacterium (plural, **bacteria**) A prokaryotic microorganism in Domain Bacteria.

balanced polymorphism A type of polymorphism in which the frequencies of the coexisting forms do not change noticeably over many generations.

bark All tissues external to the vascular cambium in a plant growing in thickness, consisting of phloem, phelloderm, cork cambium, and cork.

Barr body A dense object lying along the inside of the nuclear envelope in female mammalian cells, representing an inactivated X chromosome.

basal body *(BAY-sul)* A eukaryotic cell organelle consisting of a 9 + 0 arrangement of microtubule triplets; may organize the microtubule assembly of a cilium or flagellum; structurally identical to a centriole.

basal metabolic rate (BMR) The minimal number of kilocalories a resting animal requires to fuel itself for a given time.

base A substance that reduces the hydrogen ion concentration in a solution.

basement membrane The floor of an epithelial membrane on which the basal cells rest.

base-pair substitution A point mutation; the replacement of one nucleotide and its partner from the complementary DNA strand by another pair of nucleotides.

basidium (plural, **basidia**) A reproductive appendage that produces sexual spores on the gills of mushrooms. The fungal division Basidiomycota is named for this structure.

Batesian mimicry *(BAYTZ-ee-un MIM-ih-kree)* A type of mimicry in which a harmless species looks like a different species that is poisonous or otherwise harmful to predators.

behavioral ecology A heuristic approach based on the expectation that Darwinian fitness (reproductive success) is improved by optimal behavior.

benthic zone The bottom surfaces of aquatic environments.

biennial *(by-EN-ee-ul)* A plant that requires two years to complete its life cycle.

bilateral symmetry Characterizing a body form with a central longitudinal plane that divides the body into two equal but opposite halves.

bilateria *(BY-leh-TEER-ee-uh)* Members of the branch of eumetazoans possessing bilateral symmetry.

binary fission The type of cell division by which prokaryotes reproduce; each dividing daughter cell receives a copy of the single parental chromosome.

binomial The two-part Latinized name of a species, consisting of genus and specific epithet.

biodiversity hotspot A relatively small area with an exceptional concentration of species.

bioenergetics The study of how organisms manage their energy resources.

biogeochemical cycles The various nutrient circuits, which involve both biotic and abiotic components of ecosystems.

biogeography The study of the past and present distribution of species.

biological magnification A trophic process in which retained substances become more concentrated with each link in the food chain.

biological species A population or group of populations whose members have the potential to interbreed.

biomass The dry weight of organic matter comprising a group of organisms in a particular habitat.

biome *(BY-ome)* One of the world's major communities, classified according to the predominant vegetation and characterized by adaptations of organisms to that particular environment.

biosphere *(BY-oh-sfeer)* The entire portion of Earth that is inhabited by life; the sum of all the planet's communities and ecosystems.

biotechnology The industrial use of living organisms or their components to improve human health and food production.

biotic *(by-OT-ik)* Pertaining to the living organisms in the environment.

blastocoel *(BLAS-toh-seel)* The fluid-filled cavity that forms in the center of the blastula embryo.

blastocyst An embryonic stage in mammals; a hollow ball of cells produced one week after fertilization in humans.

blastopore *(BLAS-toh-por)* The opening of the archenteron in the gastrula that develops into the mouth in protostomes and the anus in deuterostomes.

blastula *(BLAS-tyoo-la)* The hollow ball of cells marking the end stage of cleavage during early embryonic development.

blood-brain barrier A specialized capillary arrangement in the brain that restricts the passage of most substances into the brain, thereby preventing dramatic fluctuations in the brain's environment.

blood pressure The hydrostatic force that blood exerts against the wall of a vessel.

bond energy The quantity of energy that must be absorbed to break a particular kind of chemical bond; equal to the quantity of energy the bond releases when it forms.

book lungs Organs of gas exchange in spiders, consisting of stacked plates contained in an internal chamber.

bottleneck effect Genetic drift resulting from the reduction of a population, typically by a natural disaster, such that the surviving population is no longer genetically representative of the original population.

Bowman's capsule *(BOH-munz)* A cup-shaped receptacle in the vertebrate kidney that is the initial, expanded segment of the nephron where filtrate enters from the blood.

brainstem The hindbrain and midbrain of the vertebrate central nervous system. In humans, it forms a cap on the anterior end of the spinal cord, extending to about the middle of the brain.

bryophytes *(BRY-oh-fites)* The mosses, liverworts, and hornworts; a group of nonvascular plants that inhabit the land but lack many of the terrestrial adaptations of vascular plants.

budding An asexual means of propagation in which outgrowths from the parent form and pinch off to live independently or else remain attached to eventually form extensive colonies.

buffer A substance that consists of acid and base forms in solution and that minimizes changes in pH when extraneous acids or bases are added to the solution.

bulk flow The movement of water due to a difference in pressure between two locations.

C_3 plant A plant that uses the Calvin cycle for the initial steps that incorporate CO_2 into organic material, forming a three-carbon compound as the first stable intermediate.

C_4 plant A plant that prefaces the Calvin cycle with reactions that incorporate CO_2 into four-carbon compounds, the end-product of which supplies CO_2 for the Calvin cycle.

calcitonin *(kal-sih-TOH-nin)* A mammalian thyroid hormone that lowers blood calcium levels.

calmodulin *(kal-MOD-yoo-lin)* An intracellular protein to which calcium binds in its function as a second messenger in hormone action.

calorie (cal) The amount of heat energy required to raise the temperature of 1 g of water 1°C; the amount of heat energy that 1 g of water releases when it cools by 1°C. The Calorie (with a capital C), usually used to indicate the energy content of food, is a kilocalorie.

Calvin cycle The second of two major stages in photosynthesis (following the light reactions), involving atmospheric CO_2 fixation and reduction of the fixed carbon into carbohydrate.

CAM plant A plant that uses crassulacean acid metabolism, an adaptation for photosynthesis in arid conditions, first discovered in the family Crassulaceae. Carbon dioxide entering open stomata during the night is converted into organic acids, which release CO_2 for the Calvin cycle during the day, when stomata are closed.

Cambrian explosion A burst of evolutionary origins when most of the major body plans of animals appeared in a relatively brief time in geological history; recorded in the fossil record about 545 to 525 million years ago.

capillary *(KAP-ill-air-ee)* A microscopic blood vessel that penetrates the tissues and consists of a single layer of endothelial cells that allows exchange between the blood and interstitial fluid.

capsid The protein shell that encloses the viral genome; rod-shaped, polyhedral, or more completely shaped.

carbohydrate *(KAR-bo-HY-drate)* A sugar (monosaccharide) or one of its dimers (disaccharides) or polymers (polysaccharides).

carbon fixation The incorporation of carbon from CO_2 into an organic compound by an autotrophic organism (a plant, another photosynthetic organism, or a chemoautotrophic bacterium).

carbonyl group *(KAR-buh-nil)* A functional group present in aldehydes and ketones, consisting of a carbon atom double-bonded to an oxygen atom.

carboxyl group *(kar-BOX-ul)* A functional group present in organic acids, consisting of a single carbon atom double-bonded to an oxygen atom and also bonded to a hydroxyl group.

carcinogen *(kar-SIN-oh-jen)* A chemical agent that causes cancer.

cardiac muscle *(KAR-dee-ak)* A type of muscle that forms the contractile wall of the heart; its cells are joined by intercalated discs that relay each heartbeat.

cardiac output The volume of blood pumped per minute by the left ventricle of the heart.

carnivore An animal, such as a shark, hawk, or spider, that eats other animals.

carotenoids *(keh-ROT-en-oydz)* Accessory pigments, yellow and orange, in the chloroplasts of plants; by absorbing wavelengths of light that chlorophyll cannot, they broaden the spectrum of colors that can drive photosynthesis.

carpel *(KAR-pel)* The female reproductive organ of a flower, consisting of the stigma, style, and ovary.

carrier In human genetics, an individual who is heterozygous at a given genetic locus, with one normal allele and one potentially harmful recessive allele. The heterozygote is phenotypically normal for the character determined by the gene but can pass on the harmful allele to offspring.

carrying capacity The maximum population size that can be supported by the available resources, symbolized as *K*.

cartilage *(KAR-til-ij)* A type of flexible connective tissue with an abundance of collagenous fibers embedded in chondrin.

Casparian strip *(kas-PAR-ee-un)* A water-impermeable ring of wax around endodermal cells in plants that blocks the passive flow of water and solutes into the stele by way of cell walls.

catabolic pathway *(KAT-uh-BOL-ik)* A metabolic pathway that releases energy by breaking down complex molecules into simpler compounds.

catabolite activator protein (CAP) *(ka-TAB-ul-LITE)* In *E. coli*, a helper protein that stimulates gene expression by binding within the promoter region of an operon and enhancing the promoter's ability to associate with RNA polymerase.

cation *(KAT-eye-on)* An ion with a positive charge, produced by the loss of one or more electrons.

cation exchange A process in which positively charged minerals are made available to a plant when hydrogen ions in the soil displace mineral ions from the clay particles.

cell center A region in the cytoplasm near the nucleus from which microtubules originate and radiate.

cell cycle An ordered sequence of events in the life of a dividing eukaryotic cell, composed of the M, G_1, S, and G_2 phases.

cell fractionation The disruption of a cell and separation of its organelles by centrifugation.

cell-mediated immunity The type of immunity that functions in defense against fungi, protists, bacteria, and viruses inside host cells and against tissue transplants, with highly specialized cells that circulate in the blood and lymphoid tissue.

cell plate A double membrane across the midline of a dividing plant cell, between which the new cell wall forms during cytokinesis.

cell wall A protective layer external to the plasma membrane in plant cells, bacteria, fungi, and some protists. In the case of plant cells, the wall is formed of cellulose fibers embedded in a polysaccharide-protein matrix. The primary cell wall is thin and flexible, whereas the secondary cell wall is stronger and more rigid, and is the primary constituent of wood.

cellular differentiation The structural and functional divergence of cells as they become specialized during a multicellular organism's development; dependent on the control of gene expression.

cellular respiration The most prevalent and efficient catabolic pathway for the production of ATP, in which oxygen is consumed as a reactant along with the organic fuel.

cellulose *(SELL-yoo-lose)* A structural polysaccharide of cell walls, consisting of glucose monomers joined by β-1, 4-glycosidic linkages.

Celsius scale *(SELL-see-us)* A temperature scale (°C) equal to $\frac{5}{9}$ (°F − 32) that measures the freezing point of water at 0°C and the boiling point of water at 100°C.

central nervous system (CNS) In vertebrate animals, the brain and spinal cord.

centriole *(SEN-tree-ole)* A structure in an animal cell, composed of cylinders of microtubule triplets arranged in a 9 + 0 pattern. An animal cell usually has a pair of centrioles, which are involved in cell division.

centromere *(SEN-troh-mere)* The centralized region joining two sister chromatids.

centrosome Material present in the cytoplasm of all eukaryotic cells and important during cell division; also called microtubule-organizing center.

cephalochordate A chordate without a backbone, represented by lancelets, tiny marine animals.

cerebellum *(SEH-reh-BELL-um)* Part of the vertebrate hindbrain (rhombencephalon) located dorsally; functions in unconscious coordination of movement and balance.

cerebral cortex *(seh-REE-brul)* The surface of the cerebrum; the largest and most complex part of the mammalian brain, containing sensory and motor nerve cell bodies of the cerebrum; the part of the vertebrate brain most changed through evolution.

cerebrum *(seh-REE-brum)* The dorsal portion, composed of right and left hemispheres, of the vertebrate forebrain; the integrating center for memory, learning, emotions, and other highly complex functions of the central nervous system.

chaparral *(SHAP-uh-RAL)* A scrubland biome of dense, spiny evergreen shrubs found at mid-latitudes along coasts where cold ocean currents circulate offshore; characterized by mild, rainy winters and long, hot, dry summers.

chemical equilibrium In a reversible chemical reaction, the point at which the rate of the forward reaction equals the rate of the reverse reaction.

chemiosmosis *(KEM-ee-oz-MOH-sis)* The production of ATP using the energy of hydrogen-ion gradients across membranes to phosphorylate ADP; powers most ATP synthesis in cells.

chemoautotroph *(KEE-moh-AW-toh-trohf)* An organism that needs only carbon dioxide as a carbon source but that obtains energy by oxidizing inorganic substances.

chemoheterotroph *(KEE-moh-HET-er-oh-trohf)* An organism that must consume organic molecules for both energy and carbon.

chemoreceptor A receptor that transmits information about the total solute concentration in a solution or about individual kinds of molecules.

chiasma *(KY-as-muh)* (plural, **chiasmata**) The X-shaped, microscopically visible region representing homologous chromatids that have exchanged genetic material through crossing over during meiosis.

chitin *(KY-tin)* A structural polysaccharide of an amino sugar found in many fungi and in the exoskeletons of all arthropods.

chlorophyll A green pigment located within the chloroplasts of plants; chlorophyll *a* can participate directly in the light reactions, which convert solar energy to chemical energy.

chloroplast *(KLOR-oh-plast)* An organelle found only in plants and photosynthetic protists that absorbs sunlight and uses it to drive the synthesis of organic compounds from carbon dioxide and water.

cholesterol *(kol-ESS-teh-rol)* A steroid that forms an essential component of animal cell membranes and acts as a precursor molecule for the synthesis of other biologically important steroids.

Chondrichthyes The vertebrate class of cartilaginous fishes, represented by sharks and their relatives.

chondrin A protein-carbohydrate complex secreted by chondrocytes; chondrin and collagen fibers form cartilage.

chordate *(KOR-date)* A member of a diverse phylum of animals that possess a notochord; a dorsal, hollow nerve cord; pharyngeal gill slits; and a postanal tail as embryos.

chorion *(KOR-ee-on)* The outermost of four extraembryonic membranes; contributes to the formation of the mammalian placenta.

chorionic villus sampling (CVS) *(KOR-ee-on-ik VILL-us)* A technique for diagnosing genetic and congenital defects while the fetus is in the uterus. A small sample of the fetal portion of the placenta is removed and analyzed.

chromatin The complex of DNA and proteins that makes up a eukaryotic chromosome. When the cell is not dividing, chromatin exists as a mass of very long, thin fibers that are not visible with a light microscope.

Chromista In some classification systems, a kingdom consisting of brown algae, golden algae, and diatoms.

chromosome *(KRO-muh-some)* A threadlike, gene-carrying structure found in the nucleus. Each chromosome consists of one very long DNA molecule and associated proteins. *See* chromatin.

chytrid Fungus with flagellated stage; possible evolutionary link between fungi and protists.

cilium *(SILL-ee-um)* (plural, **cilia**) A short cellular appendage specialized for locomotion, formed from a core of nine outer doublet microtubules and two inner single microtubules ensheathed in an extension of plasma membrane.

circadian rhythm *(sur-KAY-dee-un)* A physiological cycle of about 24 hours, present in all eukaryotic organisms, that persists even in the absence of external cues.

cladistics *(kluh-DIS-tiks)* A taxonomic approach that classifies organisms according to the order in time at which branches arise along a phylogenetic tree, without considering the degree of morphological divergence.

cladogenesis *(KLAY-doh-GEN-eh-sis)* A pattern of evolutionary change that produces biological diversity by budding one or more new species from a parent species that continues to exist; also called branching evolution.

cladogram A dichotomous phylogenetic tree that branches repeatedly, suggesting a classification of organisms based on the time sequence in which evolutionary branches arise.

classical conditioning A type of associative learning; the association of a normally irrelevant stimulus with a fixed behavioral response.

cleavage The process of cytokinesis in animal cells, characterized by pinching of the plasma membrane; specifically, the succession of rapid cell divisions without growth during early embryonic development that converts the zygote into a ball of cells.

cleavage furrow The first sign of cleavage in an animal cell; a shallow groove in the cell surface near the old metaphase plate.

cline Variation in features of individuals in a population that parallels a gradient in the environment.

cloaca *(kloh-AY-kuh)* A common opening for the digestive, urinary, and reproductive tracts in all vertebrates except most mammals.

clonal selection *(KLOH-nul)* The mechanism that determines specificity and accounts for antigen memory in the immune system; occurs because an antigen introduced into the body selectively activates only a tiny fraction of inactive lymphocytes, which proliferate to form a clone of effector cells specific for the stimulating antigen.

clone (1) A lineage of genetically identical individuals or cells. (2) In popular usage, a single individual organism that is genetically identical to another individual. (3) As a verb, to make one or more genetic replicas of an individual or cell. Also, *see* gene cloning.

cloning vector An agent used to transfer DNA in genetic engineering, such as a plasmid that moves recombinant DNA from a test tube back into a cell, or a virus that transfers recombinant DNA by infection.

closed circulatory system A type of internal transport in which blood is confined to vessels.

cochlea *(KOH-klee-uh)* The complex, coiled organ of hearing that contains the organ of Corti.

codominance A phenotypic situation in which both alleles are expressed in the heterozygote.

codon *(KOH-don)* A three-nucleotide sequence of DNA or mRNA that specifies a particular amino acid or termination signal; the basic unit of the genetic code.

coelom *(SEE-lome)* A body cavity completely lined with mesoderm.

coelomate *(SEE-loh-mate)* An animal whose body cavity is completely lined by mesoderm, the layers of which connect dorsally and ventrally to form mesenteries.

coenocytic *(SEN-oh-SIT-ik)* Referring to a multinucleated condition resulting from the repeated division of nuclei without cytoplasmic division.

coenzyme *(ko-EN-zyme)* An organic molecule serving as a cofactor. Most vitamins function as coenzymes in important metabolic reactions.

coevolution The mutual influence on the evolution of two different species interacting with each other and reciprocally influencing each other's adaptations.

cofactor Any nonprotein molecule or ion that is required for the proper functioning of an enzyme. Cofactors can be permanently bound to the active site or may bind loosely with the substrate during catalysis.

cohesion The binding together of like molecules, often by hydrogen bonds.

cohesion species concept The idea that specific evolutionary adaptations and discrete complexes of genes define species.

collagen A glycoprotein in the extracellular matrix of animal cells that forms strong fibers, found extensively in connective tissue and bone; the most abundant protein in the animal kingdom.

collecting duct The location in the kidney where filtrate from renal tubules is collected; the filtrate is now called urine.

collenchyma cell *(koal-EN-keh-muh)* A flexible plant cell type that occurs in strands or cylinders that support young parts of the plant without restraining growth.

commensalism *(kuh-MEN-sul-iz-um)* A symbiotic relationship in which the symbiont benefits but the host is neither helped nor harmed.

community All the organisms that inhabit a particular area; an assemblage of populations of different species living close enough together for potential interaction.

companion cell A type of plant cell that is connected to a sieve-tube member by many plasmodesmata and whose nucleus and ribosomes may serve one or more adjacent sieve-tube members.

competitive exclusion principle The concept that when the populations of two species compete for the same limited resources, one population will use the resources more efficiently and have a reproductive advantage that will eventually lead to the elimination of the other population.

competitive inhibitor A substance that reduces the activity of an enzyme by entering the active site in place of the substrate whose structure it mimics.

complement fixation An immune response in which antigen-antibody complexes activate complement proteins.

complement system A group of at least 20 blood proteins that cooperate with other defense mechanisms; may amplify the inflammatory response, enhance phagocytosis, or directly lyse pathogens; activated by the onset of the immune response or by surface antigens on microorganisms or other foreign cells.

complementary DNA (cDNA) A DNA molecule made in vitro using mRNA as a template and the enzyme reverse transcriptase. A cDNA molecule therefore corresponds to a gene, but lacks the introns present in the DNA of the genome.

complete digestive tract A digestive tube that runs between a mouth and an anus; also called alimentary canal. An incomplete digestive tract has only one opening.

complete flower A flower that has sepals, petals, stamens, and carpels.

compound A chemical combination, in a fixed ratio, of two or more elements.

compound eye A type of multifaceted eye in insects and crustaceans consisting of up to several thousand light-detecting, focusing ommatidia; especially good at detecting movement.

condensation reaction A reaction in which two molecules become covalently bonded to each other through the loss of a small molecule, usually water; also called dehydration reaction.

cone cell One of two types of photoreceptors in the vertebrate eye; detects color during the day.

conidium (plural, **conidia**) A naked, asexual spore produced at the ends of hyphae in ascomycetes.

conifer A gymnosperm whose reproductive structure is the cone. Conifers include pines, firs, redwoods, and other large trees.

conjugation *(KON-joo-GAY-shun)* In bacteria, the transfer of DNA between two cells that are temporarily joined.

connective tissue Animal tissue that functions mainly to bind and support other tissues, having a sparse population of cells scattered through an extracellular matrix.

conservation biology A goal-oriented science that seeks to counter the biodiversity crisis, the current rapid decrease in Earth's variety of life.

contraception The prevention of pregnancy.

convection The mass movement of warmed air or liquid to or from the surface of a body or object.

convergent evolution The independent development of similarity between species as a result of their having similar ecological roles and selection pressures.

cooperativity *(koh-OP-ur-uh-TIV-eh-tee)* An interaction of the constituent subunits of a protein causing a conformational change in one subunit to be transmitted to all the others.

cork cambium *(KAM-bee-um)* A cylinder of meristematic tissue in plants that produces cork cells to replace the epidermis during secondary growth.

corpus luteum *(KOR-pus LOO-tee-um)* A secreting tissue in the ovary that forms from the collapsed follicle after ovulation and produces progesterone.

cortex The region of the root between the stele and epidermis filled with ground tissue.

cotransport The coupling of the "downhill" diffusion of one substance to the "uphill" transport of another against its own concentration gradient.

cotyledons *(KOT-eh-LEE-dons)* The one (monocot) or two (dicot) seed leaves of an angiosperm embryo.

countercurrent exchange The opposite flow of adjacent fluids that maximizes transfer rates; for example, blood in the gills flows in the opposite direction in which water passes over the gills, maximizing oxygen uptake and carbon dioxide loss.

covalent bond *(koh-VAY-lent)* A type of strong chemical bond in which two atoms share one pair of electrons in a mutual valence shell.

crista *(KRIS-tuh)* (plural, **cristae**) An infolding of the inner membrane of a mitochondrion that houses the electron transport chain and the enzyme catalyzing the synthesis of ATP.

crossing over The reciprocal exchange of genetic material between nonsister chromatids during synapsis of meiosis I.

cryptic coloration *(KRIP-tik)* A type of camouflage that makes potential prey difficult to spot against its background.

cuticle *(KYOO-teh-kul)* (1) A waxy covering on the surface of stems and leaves that acts as an adaptation to prevent desiccation in terrestrial plants. (2) The exoskeleton of an arthropod, consisting of layers of protein and chitin that are variously modified for different functions.

cyanobacteria *(sy-AN-oh-bak-TEER-ee-uh)* Photosynthetic, oxygen-producing bacteria (formerly known as blue-green algae).

cyclic AMP (cAMP) Cyclic adenosine monophosphate, a ring-shaped molecule made from ATP that is a common intracellular signaling molecule (second messenger) in eukaryotic cells, for example, in vertebrate endocrine cells. It is also a regulator of some bacterial operons.

cyclic electron flow A route of electron flow during the light reactions of photosynthesis that involves only photosystem I and produces ATP but not NADPH or oxygen.

cyclin *(SY-klin)* A regulatory protein whose concentration fluctuates cyclically.

cyclin-dependent kinase (Cdk) A protein kinase that is active only when attached to a particular cyclin.

cytochrome *(SY-toh-krome)* An iron-containing protein, a component of electron transport chains in mitochondria and chloroplasts.

cytokines In the vertebrate immune system, protein factors secreted by macrophages and helper T cells as regulators of neighboring cells.

cytokinesis *(SY-toh-kin-EE-sis)* The division of the cytoplasm to form two separate daughter cells immediately after mitosis.

cytokinins *(SY-toh-KY-nins)* A class of related plant hormones that retard aging and act in concert with auxins to stimulate cell division, influence the pathway of differentiation, and control apical dominance.

cytoplasm *(SY-toh-plaz-um)* The entire contents of the cell, exclusive of the nucleus, and bounded by the plasma membrane.

cytoplasmic determinants In animal development, substances deposited by the mother in the eggs she produces that regulate the expression of genes affecting the early development of the embryo.

cytoplasmic streaming A circular flow of cytoplasm, involving myosin and actin filaments, that speeds the distribution of materials within cells.

cytoskeleton *(SY-toh-SKEL-eh-ton)* A network of microtubules, microfilaments, and intermediate filaments that branch throughout the cytoplasm and serve a variety of mechanical and transport functions.

cytosol *(SY-toh-sol)* The semifluid portion of the cytoplasm.

cytotoxic T cell (T_C) A type of lymphocyte that kills infected cells and cancer cells.

dalton *(DAWL-ton)* The atomic mass unit; a measure of mass for atoms and subatomic particles.

Darwinian fitness A measure of the relative contribution of an individual to the gene pool of the next generation.

day-neutral plant A plant whose flowering is not affected by photoperiod.

decomposers Saprotrophic fungi and bacteria that absorb nutrients from nonliving organic material such as corpses, fallen plant material, and the wastes of living organisms, and convert them into inorganic forms.

dehydration reaction A chemical reaction in which two molecules covalently bond to one another with the removal of a water molecule.

deletion (1) A deficiency in a chromosome resulting from the loss of a fragment through breakage. (2) A mutational loss of a nucleotide from a gene.

demography The study of statistics relating to births and deaths in populations.

denaturation For proteins a process in which a protein unravels and loses its native conformation, thereby becoming biologically inactive. For DNA, the separation of the two strands of the double helix. Denaturation occurs under extreme conditions of pH, salt concentration, and temperature.

dendrite *(DEN-dryt)* One of usually numerous, short, highly branched processes of a neuron that conveys nerve impulses toward the cell body.

density The number of individuals per unit area or volume.

density-dependent factor Any factor influencing population regulation that has a greater impact as population density increases.

density-dependent inhibition The phenomenon observed in normal animal cells that causes them to stop dividing when they come into contact with one another.

density-independent factor Any factor influencing population regulation that acts to reduce population by the same percentage, regardless of size.

deoxyribonucleic acid (DNA) *(DEE-oks-ee-ry-boh-noo-KLAY-ik)* A double-stranded, helical nucleic acid molecule capable of replicating and determining the inherited structure of a cell's proteins.

deoxyribose The sugar component of DNA, having one less hydroxyl group than ribose, the sugar component of RNA.

depolarization An electrical state in an excitable cell whereby the inside of the cell is made less negative relative to the outside than at the resting membrane potential. A neuron membrane is depolarized if a stimulus decreases its voltage from the resting potential of –70 mV in the direction of zero voltage.

deposit-feeder A heterotroph, such as an earthworm, that eats its way through detritus, salvaging bits and pieces of decaying organic matter.

dermal tissue system The protective covering of plants; generally a single layer of tightly packed epidermal cells covering young plant organs formed by primary growth.

desmosome *(DEZ-muh-some)* A type of intercellular junction in animal cells that functions as an anchor.

determinate cleavage A type of embryonic development in protostomes that rigidly casts the developmental fate of each embryonic cell very early.

determinate growth A type of growth characteristic of animals, in which the organism stops growing after it reaches a certain size.

determination The progressive restriction of developmental potential, causing the possible fate of each cell to become more limited as the embryo develops.

detritus *(deh-TRY-tis)* Dead organic matter.

deuterostomes *(DOO-ter-oh-stomes)* One of two distinct evolutionary lines of coelomates, consisting of the echinoderms and chordates and characterized by radial, indeterminate cleavage, enterocoelous formation of the coelom, and development of the anus from the blastopore.

diaphragm A sheet of muscle that forms the bottom wall of the thoracic cavity in mammals; active in ventilating the lungs.

diastole *(dy-ASS-toh-lee)* The stage of the heart cycle in which the heart muscle is relaxed, allowing the chambers to fill with blood.

dicot *(DY-kot)* A subdivision of flowering plants whose members possess two embryonic seed leaves, or cotyledons.

differentiation *See* cellular differentiation.

diffusion The spontaneous tendency of a substance to move down its concentration gradient from a more concentrated to a less concentrated area.

digestion The process of breaking down food into molecules small enough for the body to absorb.

dihybrid cross *(DY-HY-brid)* A breeding experiment in which parental varieties differing in two traits are mated.

dikaryon *(dy-KAH-ree-on)* A mycelium of certain septate fungi that possesses two separate haploid nuclei per cell.

dioecious *(dy-EE-shus)* Referring to a plant species that has staminate and carpellate flowers on separate plants.

diploid cell *(DIP-loyd)* A cell containing two sets of chromosomes ($2n$), one set inherited from each parent.

directional selection Natural selection that favors individuals on one end of the phenotypic range.

disaccharide *(dy-SAK-ur-ide)* A double sugar, consisting of two monosaccharides joined by dehydration synthesis.

dispersion The distribution of individuals within geographical population boundaries.

diversifying selection Natural selection that favors extreme over intermediate phenotypes.

DNA ligase *(LY-gaze)* A linking enzyme essential for DNA replication; catalyzes the covalent bonding of the 3' end of a new DNA fragment to the 5' end of a growing chain.

DNA methylation The addition of methyl groups (—CH_3) to bases of DNA after DNA synthesis; may serve as a long-term control of gene expression.

DNA polymerase An enzyme that catalyzes the elongation of new DNA at a replication fork by the addition of nucleotides to the existing chain.

DNA probe A chemically synthesized, radioactively labeled segment of nucleic acid used to find a gene of interest by hydrogen-bonding to a complementary sequence.

domain A taxonomic category above the kingdom level; the three domains are Archaea, Bacteria, and Eukarya.

dominance hierarchy A linear "pecking order" of animals, where position dictates characteristic social behaviors.

dominant allele In a heterozygote, the allele that is fully expressed in the phenotype.

double circulation A circulation scheme with separate pulmonary and systemic circuits, which ensures vigorous blood flow to all organs.

double fertilization A mechanism of fertilization in angiosperms, in which two sperm cells unite with two cells in the embryo sac to form the zygote and endosperm.

double helix The form of native DNA, referring to its two adjacent polynucleotide strands wound into a spiral shape.

Down syndrome A human genetic disease resulting from having an extra chromosome 21, characterized by mental retardation and heart and respiratory defects.

duodenum *(doo-oh-DEE-num)* The first section of the small intestine, where acid chyme from the stomach mixes with digestive juices from the pancreas, liver, gallbladder, and gland cells of the intestinal wall.

duplication An aberration in chromosome structure resulting from an error in meiosis or mutagens; duplication of a portion of a chromosome resulting from fusion with a fragment from a homologous chromosome.

dynein *(DY-nin)* A large contractile protein forming the sidearms of microtubule doublets in cilia and flagella.

ecdysone *(EK-deh-sone)* A steroid hormone that triggers molting in arthropods.

ecological efficiency The ratio of net productivity at one trophic level to net productivity at the next lower level.

ecological niche *(nich)* The sum total of an organism's utilization of the biotic and abiotic resources of its environment.

ecological species concept The idea that ecological roles (niches) define species.

ecological succession Transition in the species composition of a biological community, often following ecological disturbance of the community; the establishment of a biological community in an area virtually barren of life.

ecology The study of how organisms interact with their environments.

ecosystem A level of ecological study that includes all the organisms in a given area as well as the abiotic factors with which they interact; a community and its physical environment.

ectoderm *(EK-tuh-durm)* The outermost of the three primary germ layers in animal embryos; gives rise to the outer covering and, in some phyla, the nervous system, inner ear, and lens of the eye.

ectotherm *(EK-toh-thurm)* An animal, such as a reptile, fish, or amphibian, that must use environmental energy and behavioral adaptations to regulate its body temperature.

effector cell A muscle cell or gland cell that performs the body's responses to stimuli; responds to signals from the brain or other processing center of the nervous system.

electrochemical gradient The diffusion gradient of an ion, representing a type of potential energy that accounts for both the concentration difference of the ion across a membrane and its tendency to move relative to the membrane potential.

electrogenic pump An ion transport protein generating voltage across the membrane.

electromagnetic spectrum The entire spectrum of radiation; ranges in wavelength from less than a nanometer to more than a kilometer.

electron microscope (EM) A microscope that focuses an electron beam through a specimen, resulting in resolving power a thousandfold greater than that of a light microscope. A transmission electron microscope (TEM) is used to study the internal structure of thin sections of cells. A scanning electron microscope (SEM) is used to study the fine details of cell surfaces.

electron transport chain A sequence of electron-carrier molecules (membrane proteins) that shuttle electrons during the redox reactions that release energy used to make ATP.

electronegativity The tendency for an atom to pull electrons toward itself.

element Any substance that cannot be broken down to any other substance.

embryo sac The female gametophyte of angiosperms, formed from the growth and division of the megaspore into a multicellular structure with eight haploid nuclei.

enantiomer *(eh-NAN-she-uh-mer)* One of a pair of molecules that are mirror-image isomers of each other.

endangered species A species that is in danger of extinction throughout all or a significant portion of its range.

endemic species Species that are confined to a specific, relatively small geographic area.

endergonic reaction *(EN-dur-GON-ik)* A nonspontaneous chemical reaction in which free energy is absorbed from the surroundings.

endocrine gland *(EN-doh-krin)* A ductless gland that secretes hormones directly into the bloodstream.

endocrine system The internal system of chemical communication involving hormones, the ductless glands that secrete hormones, and the molecular receptors on or in target cells that respond to hormones; functions in concert with the nervous system to effect internal regulation and maintain homeostasis.

endocytosis *(EN-doh-sy-TOH-sis)* The cellular uptake of macromolecules and particulate substances by localized regions of the plasma membrane that surround the substance and pinch off to form an intracellular vesicle.

endoderm *(EN-doh-durm)* The innermost of the three primary germ layers in animal embryos; lines the archenteron and gives rise to the liver, pancreas, lungs, and the lining of the digestive tract.

endodermis *(EN-doh-DUR-mis)* The innermost layer of the cortex in plant roots; a cylinder one cell thick that forms the boundary between the cortex and the stele.

endomembrane system The collection of membranes inside and around a eukaryotic cell, related either through direct physical contact or by the transfer of membranous vesicles.

endometrium *(EN-doh-MEE-tree-um)* The inner lining of the uterus, which is richly supplied with blood vessels.

endoplasmic reticulum (ER) *(EN-doh-plaz-mik reh-TIK-yoo-lum)* An extensive membranous network in eukaryotic cells, continuous with the outer nuclear membrane and composed of ribosome-studded (rough) and ribosome-free (smooth) regions.

endorphin *(en-DOR-fin)* A hormone produced in the brain and anterior pituitary that inhibits pain perception.

endoskeleton *(EN-doh-SKEL-eh-ton)* A hard skeleton buried within the soft tissues of an animal, such as the spicules of sponges, the plates of echinoderms, and the bony skeletons of vertebrates.

endosperm *(EN-doh-spurm)* A nutrient-rich tissue formed by the union of a sperm cell with two polar nuclei during double fertilization, which provides nourishment to the developing embryo in angiosperm seeds.

endospore A thick-coated, resistant cell produced within a bacterial cell exposed to harsh conditions.

endosymbiotic theory *(EN-doh-SIM-by-OT-ik)* A hypothesis about the origin of the eukaryotic cell, maintaining that the forerunners of eukaryotic cells were symbiotic associations of prokaryotic cells living inside larger prokaryotes.

endothelium *(EN-doh-THEEL-ee-um)* The innermost, simple squamous layer of cells lining the blood vessels; the only constituent structure of capillaries.

endotherm *(EN-doh-thurm)* An animal that uses metabolic energy to maintain a constant body temperature, such as a bird or mammal.

endotoxin *(EN-doh-TOKS-in)* A component of the outer membranes of certain gram-negative bacteria responsible for generalized symptoms of fever and ache.

energy The capacity to do work by moving matter against an opposing force.

enhancer A DNA sequence that recognizes certain transcription factors that can stimulate transcription of nearby genes.

entropy *(EN-truh-pee)* A quantitative measure of disorder or randomness, symbolized by *S*.

environmental grain An ecological term for the effect of spatial variation, or patchiness, relative to the size and behavior of an organism.

enzymes A class of proteins serving as catalysts, chemical agents that change the rate of a reaction without being consumed by the reaction.

epidermis *(EP-eh-DER-mis)* (1) The dermal tissue system in plants. (2) The outer covering of animals.

epigenesis *(EP-eh-JEN-eh-sis)* The progressive development of form in an embryo.

epiglottis A cartilaginous flap that blocks the top of the windpipe, the glottis, during swallowing, which prevents the entry of food or fluid into the respiratory system.

epinephrine A hormone produced as a response to stress; also called adrenaline.

epiphyte *(EP-eh-fite)* A plant that nourishes itself but grows on the surface of another plant for support, usually on the branches or trunks of tropical trees.

episome *(EP-eh-some)* A plasmid capable of integrating into the bacterial chromosome.

epistasis A phenomenon in which one gene alters the expression of another gene that is independently inherited.

epithelial tissue *(EP-eh-THEEL-ee-ul)* Sheets of tightly packed cells that line organs and body cavities.

epitope A localized region on the surface of an antigen that is chemically recognized by antibodies; also called antigenic determinant.

erythrocyte *(er-RITH-roh-site)* A red blood cell; contains hemoglobin, which functions in transporting oxygen in the circulatory system.

esophagus *(eh-SOF-eh-gus)* A channel that conducts food, by peristalsis, from the pharynx to the stomach.

essential amino acids The amino acids that an animal cannot synthesize itself and must obtain from food. Eight amino acids are essential in the human adult.

estivation *(ES-teh-VAY-shun)* A physiological state characterized by slow metabolism and inactivity, which permits survival during long periods of elevated temperature and diminished water supplies.

estrogens *(ES-troh-jens)* The primary female steroid sex hormones, which are produced in the ovary by the developing follicle during the first half of the cycle and in smaller quantities by the corpus luteum during the second half. Estrogens stimulate the development and maintenance of the female reproductive system and secondary sex characteristics.

estrous cycle *(ES-trus)* A type of reproductive cycle in all female mammals except higher primates, in which the nonpregnant endometrium is reabsorbed rather than shed, and sexual response occurs only during midcycle at estrus.

ethylene *(ETH-ul-een)* The only gaseous plant hormone, responsible for fruit ripening, growth inhibition, leaf abscission, and aging.

euchromatin *(yoo-KROH-muh-tin)* The more open, unraveled form of eukaryotic chromatin, which is available for transcription.

eukaryotic cell *(YOO-kar-ee-OT-ik)* A type of cell with a membrane-enclosed nucleus and membrane-enclosed organelles, present in protists, plants, fungi, and animals; also called eukaryote.

eumetazoa *(YOO-met-uh-ZOH-uh)* Members of the subkingdom that includes all animals except sponges.

eutherian mammals Placental mammals; those whose young complete their embryonic development within the uterus, joined to the mother by the placenta.

eutrophic lake A highly productive lake, having a high rate of biological productivity supported by a high rate of nutrient cycling.

evaporative cooling The property of a liquid whereby the surface becomes cooler during evaporation, owing to a loss of highly kinetic molecules to the gaseous state.

evolution All the changes that have transformed life on Earth from its earliest beginnings to the diversity that characterizes it today.

evolutionary species concept The idea that evolutionary lineages and ecological roles can form the basis of species identification.

exaptation A structure that evolves and functions in one environmental context but that can perform additional functions when placed in some new environment.

excitatory postsynaptic potential (EPSP) *(POST-sin-AP-tik)* An electrical change (depolarization) in the membrane of a postsynaptic neuron caused by the binding of an excitatory neurotransmitter from a presynaptic cell to a postsynaptic receptor; makes it more likely for a postsynaptic neuron to generate an action potential.

excretion The disposal of nitrogen-containing waste products of metabolism.

exergonic reaction *(EKS-ur-GON-ik)* A spontaneous chemical reaction in which there is a net release of free energy.

exocytosis *(EKS-oh-sy-TOH-sis)* The cellular secretion of macromolecules by the fusion of vesicles with the plasma membrane.

exon A coding region of a eukaryotic gene that is expressed. Exons are separated from each other by introns.

exoskeleton A hard encasement on the surface of an animal, such as the shells of mollusks or the cuticles of arthropods, that provides protection and points of attachment for muscles.

exotoxin *(EKS-oh-TOKS-in)* A toxic protein secreted by a bacterial cell that produces specific symptoms even in the absence of the bacterium.

exponential population growth The geometric increase of a population as it grows in an ideal, unlimited environment.

extracellular matrix (ECM) The substance in which animal tissue cells are embedded; consists of protein and polysaccharides.

extraembryonic membranes *(EKS-truh-EM-bree-AHN-ik)* Four membranes (yolk sac, amnion, chorion, allantois) that support the developing embryo in reptiles, birds, and mammals.

F factor A fertility factor in bacteria, a DNA segment that confers the ability to form pili for conjugation and associated functions required for the transfer of DNA from donor to recipient. May exist as a plasmid or integrated into the bacterial chromosome.

F_1 generation The first filial or hybrid offspring in a genetic cross-fertilization.

F_2 generation Offspring resulting from interbreeding of the hybrid F_1 generation.

facilitated diffusion The spontaneous passage of molecules and ions, bound to specific carrier proteins, across a biological membrane down their concentration gradients.

facultative anaerobe *(FAK-ul-tay-tiv AN-uh-robe)* An organism that makes ATP by aerobic respiration if oxygen is present but that switches to fermentation under anaerobic conditions.

fat (triacylglycerol) *(tri-AH-sil-GLIS-er-all)* A biological compound consisting of three fatty acids linked to one glycerol molecule.

fatty acid A long carbon chain carboxylic acid. Fatty acids vary in length and in the number and location of double bonds; three fatty acids linked to a glycerol molecule form fat.

feedback inhibition A method of metabolic control in which the end-product of a metabolic pathway acts as an inhibitor of an enzyme within that pathway.

fermentation A catabolic process that makes a limited amount of ATP from glucose without an electron transport chain and that produces a characteristic end-product, such as ethyl alcohol or lactic acid.

fertilization The union of haploid gametes to produce a diploid zygote.

fiber A lignified cell type that reinforces the xylem of angiosperms and functions in mechanical support; a slender, tapered sclerenchyma cell that usually occurs in bundles.

fibrin *(FY-brin)* The activated form of the blood-clotting protein fibrinogen, which aggregates into threads that form the fabric of the clot.

fibroblast *(FY-broh-blast)* A type of cell in loose connective tissue that secretes the protein ingredients of the extracellular fibers.

first law of thermodynamics *(THUR-moh-dy-NAM-iks)* The principle of conservation of energy. Energy can be transferred and transformed, but it cannot be created or destroyed.

fixed action pattern (FAP) A highly stereotypical behavior that is innate and must be carried to completion once initiated.

flaccid *(FLAS-id)* Limp; walled cells are flaccid in isotonic surroundings, where there is no tendency for water to enter.

flagellum *(fluh-JEL-um)* (plural, **flagella**) A long cellular appendage specialized for locomotion, formed from a core of nine outer doublet microtubules and two inner single microtubules, ensheathed in an extension of plasma membrane.

fluid-feeder An animal that lives by sucking nutrient-rich fluids from another living organism.

fluid mosaic model The currently accepted model of cell membrane structure, which envisions the membrane as a mosaic of individually inserted protein molecules drifting laterally in a fluid bilayer of phospholipids.

follicle *(FOL-eh-kul)* A microscopic structure in the ovary that contains the developing ovum and secretes estrogens.

food chain The pathway along which food is transferred from trophic level to trophic level, beginning with producers.

food web The elaborate, interconnected feeding relationships in an ecosystem.

founder effect A cause of genetic drift attributable to colonization by a limited number of individuals from a parent population.

fragile X syndrome A hereditary mental disorder, partially explained by genomic imprinting and the addition of nucleotides to a triplet repeat near the end of an *X* chromosome.

frameshift mutation A mutation occurring when the number of nucleotides inserted or deleted is not a multiple of 3, thus resulting in improper grouping into codons.

free energy A quantity of energy that interrelates entropy *(S)* and the system's total energy *(H)*; symbolized by *G*. The change in free energy of a system is calculated by the equation $G = \Delta H - T\Delta S$, where *T* is absolute temperature.

free energy of activation The initial investment of energy necessary to start a chemical reaction; also called activation energy.

frequency-dependent selection A decline in the reproductive success of a morph resulting from the morph's phenotype becoming too common in a population; a cause of balanced polymorphism in populations.

fruit A mature ovary of a flower that protects dormant seeds and aids in their dispersal.

functional group A specific configuration of atoms commonly attached to the carbon skeletons of organic molecules and usually involved in chemical reactions.

G protein A GTP-binding protein that relays signals from a plasma-membrane signal receptor, known as a G-protein linked receptor, to other signal-transduction proteins inside the cell. When such a receptor is activated, it in turn activates the G protein, causing it to bind a molecule of GTP in place of GDP. Hydrolysis of the bound GTP to GDP inactivates the G protein.

G-protein linked receptor A signal receptor protein in the plasma membrane that responds to the binding of a signal molecule by activating a G protein.

G_1 phase The first growth phase of the cell cycle, consisting of the portion of interphase before DNA synthesis begins.

G_2 phase The second growth phase of the cell cycle, consisting of the portion of interphase after DNA synthesis occurs.

gametangium *(GAM-eh-TANJ-ee-um)* (plural, **gametangia**) The reproductive organ of bryophytes, consisting of the male antheridium and female archegonium; a multichambered jacket of sterile cells in which gametes are formed.

gamete *(GAM-eet)* A haploid egg or sperm cell; gametes unite during sexual reproduction to produce a diploid zygote.

gametophyte *(guh-MEE-toh-fite)* The multicellular haploid form in organisms undergoing alternation of generations, which mitotically produces haploid gametes that unite and grow into the sporophyte generation.

ganglion *(GANG-lee-un)* (plural, **ganglia**) A cluster (functional group) of nerve cell bodies in a centralized nervous system.

gap junction A type of intercellular junction in animal cells that allows the passage of material or current between cells.

gastrin A digestive hormone, secreted by the stomach, that stimulates the secretion of gastric juice.

gastrovascular cavity The central digestive compartment, usually with a single opening that functions as both mouth and anus.

gastrula *(GAS-troo-la)* The two-layered, cup-shaped embryonic stage.

gastrulation *(GAS-truh-LAY-shun)* The formation of a gastrula from a blastula.

gated ion channel A specific ion channel that opens and closes to allow the cell to alter its membrane potential.

gel electrophoresis *(JELL eh-LEK-troh-for-EE-sis)* The separation of nucleic acids or proteins, on the basis of their size and electrical charge, by measuring their rate of movement through an electrical field in a gel.

gene A discrete unit of hereditary information consisting of a specific nucleotide sequence in DNA (or RNA, in some viruses).

gene amplification The selective synthesis of DNA, which results in multiple copies of a single gene, thereby enhancing expression.

gene cloning The production of multiple copies of a gene.

gene flow The loss or gain of alleles from a population due to the emigration or immigration of fertile individuals, or the transfer of gametes, between populations.

gene pool The total aggregate of genes in a population at any one time.

genetic drift Changes in the gene pool of a small population due to chance.

genetic map An ordered list of genetic loci (genes or other genetic markers) along a chromosome.

genetic recombination The general term for the production of offspring that combine traits of the two parents.

genome *(JEE-nome)* The complete complement of an organism's genes; an organism's genetic material.

genomic imprinting The parental effect on gene expression. Identical alleles may have different effects on offspring, depending on whether they arrive in the zygote via the ovum or via the sperm.

genomic library A set of thousands of DNA segments from a genome, each carried by a plasmid, phage, or other cloning vector.

genotype *(JEE-noh-type)* The genetic makeup of an organism.

genus *(JEE-nus)* (plural, **genera**) A taxonomic category above the species level, designated by the first word of a species' binomial Latin name.

geographical range The geographic area in which a population lives.

geological time scale A time scale established by geologists that reflects a consistent sequence of historical periods, grouped into four eras: Precambrian, Paleozoic, Mesozoic, and Cenozoic.

gibberellins *(JIB-ur-EL-ins)* A class of related plant hormones that stimulate growth in the stem and leaves, trigger the germination of seeds and breaking of bud dormancy, and stimulate fruit development with auxin.

gill A localized extension of the body surface of many aquatic animals, specialized for gas exchange.

glial cell *(GLEE-ul)* A nonconducting cell of the nervous system that provides support, insulation, and protection for the neurons.

glomerulus *(glum-AIR-yoo-lus)* A ball of capillaries surrounded by Bowman's capsule in the nephron and serving as the site of filtration in the vertebrate kidney.

glucagon A peptide hormone secreted by pancreatic endocrine cells that raises blood glucose levels; an antagonistic hormone to insulin.

glucocorticoid A corticosteroid hormone secreted by the adrenal cortex that influences glucose metabolism and immune function.

glycocalyx *(GLY-koh-KAY-liks)* A fuzzy coat on the outside of animal cells, made of sticky oligosaccharides.

glycogen *(GLY-koh-jen)* An extensively branched glucose storage polysaccharide found in the liver and muscle of animals; the animal equivalent of starch.

glycolysis *(gly-KOL-eh-sis)* The splitting of glucose into pyruvate. Glycolysis is the one metabolic pathway that occurs in all living cells, serving as the starting point for fermentation or aerobic respiration.

glycoprotein A protein with covalently attached carbohydrate.

Golgi apparatus *(GOAL-jee)* An organelle in eukaryotic cells consisting of stacks of flat membranous sacs that modify, store, and route products of the endoplasmic reticulum.

gonadotropins *(goh-NAD-oh-TROH-pinz)* Hormones that stimulate the activities of the testes and ovaries; a collective term for follicle-stimulating and luteinizing hormones.

gonads *(GOH-nadz)* The male and female sex organs; the gamete-producing organs in most animals.

graded potential A local voltage change in a neuron membrane induced by stimulation of a neuron, with strength proportional to the strength of the stimulus and lasting about a millisecond.

gradualism A view of Earth's history that attributes profound change to the cumulative product of slow but continuous processes.

Gram stain A staining method that distinguishes between two different kinds of bacterial cell walls.

granum *(GRAN-um)* (plural, **grana**) A stacked portion of the thylakoid membrane in the chloroplast. Grana function in the light reactions of photosynthesis.

gravitropism *(GRAV-eh-TROH-piz-um)* A response of a plant or animal in relation to gravity.

greenhouse effect The warming of planet Earth due to the atmospheric accumulation of carbon dioxide, which absorbs infrared radiation and slows its escape from the irradiated Earth.

gross primary productivity (GPP) The total primary productivity of an ecosystem.

ground meristem A primary meristem that gives rise to ground tissue in plants.

ground tissue system A tissue of mostly parenchyma cells that makes up the bulk of a young plant and fills the space between the dermal and vascular tissue systems.

growth factor A protein that must be present in the extracellular environment (culture medium or animal body) for the growth and normal development of certain types of cells.

guard cell A specialized epidermal plant cell that forms the boundaries of the stomata.

guttation The exudation of water droplets caused by root pressure in certain plants.

gymnosperm *(JIM-noh-spurm)* A vascular plant that bears naked seeds not enclosed in any specialized chambers.

habituation A simple kind of learning involving a loss of sensitivity to unimportant stimuli, allowing an animal to conserve time and energy.

haploid cell *(HAP-loid)* A cell containing only one set of chromosomes *(n)*.

Hardy-Weinberg theorem An axiom maintaining that the sexual shuffling of genes alone cannot alter the overall genetic make-up of a population.

haustorium (plural, **haustoria**) In parasitic fungi, a nutrient-absorbing hyphal tip that penetrates the tissues of the host but remains outside the host cell membranes.

Haversian system *(ha-VER-shun)* One of many structural units of vertebrate bone, consisting of concentric layers of mineralized bone matrix surrounding lacunae, which contain osteocytes, and a central canal, which contains blood vessels and nerves.

heat The total amount of kinetic energy due to molecular motion in a body of matter. Heat is energy in its most random form.

heat-shock protein A protein that helps protect other proteins during heat stress, found in plants, animals, and microorganisms.

helper T cell (T_H) A type of T cell that is required by some B cells to help them make antibodies or that helps other T cells respond to antigens or secrete lymphokines or interleukins.

hemoglobin *(HEE-moh-gloh-bin)* An iron-containing protein in red blood cells that reversibly binds oxygen.

hemolymph In invertebrates with an open circulatory system, the body fluid that bathes tissues.

hepatic portal vessel A large circulatory channel that conveys nutrient-laden blood from the small intestine to the liver, which regulates the blood's nutrient content.

herbivore A heterotrophic animal that eats plants.

hermaphrodite *(her-MAF-roh-dite)* An individual that functions as both male and female in sexual reproduction by producing both sperm and eggs.

heterochromatin *(HET-ur-oh-KROH-muh-tin)* Nontranscribed eukaryotic chromatin that is so highly compacted that it is visible with a light microscope during interphase.

heterochrony Evolutionary changes in the timing or rate of development.

heterocyst *(HET-ur-oh-sist)* A specialized cell that engages in nitrogen fixation on some filamentous cyanobacteria.

heteromorphic *(HET-ur-oh-MOR-fik)* A condition in the life cycle of all modern plants in which the sporophyte and gametophyte generations differ in morphology.

heterosporous *(HET-ur-OS-pur-us)* Referring to plants in which the sporophyte produces two kinds of spores that develop into unisexual gametophytes, either female or male.

heterotroph *(HET-ur-oh-TROHF)* An organism that obtains organic food molecules by eating other organisms or their by-products.

heterozygote advantage *(HET-ur-oh-ZY-gote)* A mechanism that preserves variation in eukaryotic gene pools by conferring greater reproductive success on heterozygotes over individuals homozygous for any one of the associated alleles.

heterozygous *(HET-ur-oh-ZY-gus)* Having two different alleles for a given genetic character.

hibernation A physiological state that allows survival during long periods of cold temperatures and reduced food supplies, in which metabolism decreases, the heart and respiratory system slow down, and body temperature is maintained at a lower level than normal.

histamine *(HISS-tuh-meen)* A substance released by injured cells that causes blood vessels to dilate during an inflammatory response.

histone *(HISS-tone)* A small protein with a high proportion of positively charged amino acids that binds to the negatively charged DNA and plays a key role in its chromatin structure.

HIV (human immunodeficiency virus) The infectious agent that causes AIDS; HIV is an RNA retrovirus.

holoblastic cleavage *(HOH-loh-BLAS-tik)* A type of cleavage in which there is complete division of the egg, as in eggs having little yolk (sea urchin) or a moderate amount of yolk (frog).

homeobox *(HOME-ee-oh-BOX)* A 180-nucleotide sequence within a homeotic gene encoding the part of the protein that binds to the DNA of the genes regulated by the protein.

homeosis Evolutionary alteration in the placement of different body parts.

homeostasis *(HOME-ee-oh-STAY-sis)* The steady-state physiological condition of the body.

homeotic genes *(HOME-ee-OT-ik)* Genes that control the overall body plan of animals by controlling the developmental fate of groups of cells.

homologous chromosomes *(home-OL-uh-gus)* Chromosome pairs of the same length, centromere position, and staining pattern that possess genes for the same characters at corresponding loci. One homologous chromosome is inherited from the organism's father, the other from the mother.

homologous structures Structures in different species that are similar because of common ancestry.

homology Similarity in characteristics resulting from a shared ancestry.

homosporous *(home-OS-pur-us)* Referring to plants in which a single type of spore develops into a bisexual gametophyte having both male and female sex organs.

homozygous *(HOME-oh-ZY-gus)* Having two identical alleles for a given trait.

hormone One of many types of circulating chemical signals in all multicellular organisms that are formed in specialized cells, travel in body fluids, and coordinate the various parts of the organism by interacting with target cells.

Human Genome Project An international collaborative effort to map and sequence the DNA of the entire human genome.

humoral immunity *(HYOO-mur-al)* The type of immunity that fights bacteria and viruses in body fluids with antibodies that circulate in blood plasma and lymph, fluids formerly called humors.

hybrid zone A region where two related populations that diverged after becoming geographically isolated make secondary contact and interbreed where their geographical ranges overlap.

hydrocarbon *(HY-droh-kar-bon)* An organic molecule consisting only of carbon and hydrogen.

hydrogen bond A type of weak chemical bond formed when the slightly positive hydrogen atom of a polar covalent bond in one molecule is attracted to the slightly negative atom of a polar covalent bond in another molecule.

hydrogen ion A single proton with a charge of +1. The dissociation of a water molecule (H_2O)

leads to the generation of a hydroxide ion (OH^-) and a hydrogen ion (H^+).

hydrolysis *(hy-DROL-eh-sis)* A chemical process that lyses or splits molecules by the addition of water; an essential process in digestion.

hydrophilic *(HY-droh-FIL-ik)* Having an affinity for water.

hydrophobic *(HY-droh-FOH-bik)* Having an aversion to water; tending to coalesce and form droplets in water.

hydrophobic interaction A type of weak chemical bond formed when molecules that do not mix with water coalesce to exclude the water.

hydrostatic skeleton *(HY-droh-STAT-ik)* A skeletal system composed of fluid held under pressure in a closed body compartment; the main skeleton of most cnidarians, flatworms, nematodes, and annelids.

hydroxyl group *(hy-DROKS-ul)* A functional group consisting of a hydrogen atom joined to an oxygen atom by a polar covalent bond. Molecules possessing this group are soluble in water and are called alcohols.

hyperpolarization An electrical state whereby the inside of the cell is made more negative relative to the outside than at the resting membrane potential. A neuron membrane is hyperpolarized if a stimulus increases its voltage from the resting potential of -70 mV, reducing the chance that the neuron will transmit a nerve impulse.

hypertonic solution A solution with a greater solute concentration than another, a hypotonic solution.

hypha *(HY-fa)* (plural, **hyphae**) A filament that collectively makes up the body of a fungus.

hypothalmus *(HY-poh-THAL-uh-mus)* The ventral part of the vertebrate forebrain; functions in maintaining homeostasis, especially in coordinating the endocrine and nervous systems; secretes hormones of the posterior pituitary and releasing factors, which regulate the anterior pituitary.

hypotonic solution A solution with a lesser solute concentration than another, a hypertonic solution.

imaginal disk *(i-MAJ-in-ul)* An island of undifferentiated cells in an insect larva, which are committed (determined) to form a particular organ during metamorphosis to the adult.

immunoglobulin (Ig) *(IM-myoo-noh-GLOB-yoo-lin)* One of the class of proteins comprising the antibodies.

imprinting A type of learned behavior with a significant innate component, acquired during a limited critical period.

incomplete dominance A type of inheritance in which F_1 hybrids have an appearance that is intermediate between the phenotypes of the parental varieties.

incomplete flower A flower lacking sepals, petals, stamens, or carpels.

incomplete metamorphosis *(MET-uh-MOR-foh-sis)* A type of development in certain insects, such as grasshoppers, in which the larvae resemble adults but are smaller and have different body proportions. The animal goes through a series of molts, each time looking more like an adult, until it reaches full size.

indeterminate cleavage A type of embryonic development in deuterostomes, in which each cell produced by early cleavage divisions retains the capacity to develop into a complete embryo.

indeterminate growth A type of growth characteristic of plants, in which the organism continues to grow as long as it lives.

induced fit The change in shape of the active site of an enzyme so that it binds more snugly to the substrate, induced by entry of the substrate.

induction The ability of one group of embryonic cells to influence the development of another.

inflammatory response A line of defense triggered by penetration of the skin or mucous membranes, in which small blood vessels in the vicinity of an injury dilate and become leakier, enhancing the infiltration of leukocytes; may also be widespread in the body.

ingestion A heterotrophic mode of nutrition in which other organisms or detritus are eaten whole or in pieces.

inhibitory postsynaptic potential (IPSP) *(POST-sin-AP-tik)* An electrical charge (hyperpolarization) in the membrane of a postsynaptic neuron caused by the binding of an inhibitory neurotransmitter from a presynaptic cell to a postsynaptic receptor; makes it more difficult for a postsynaptic neuron to generate an action potential.

inner cell mass A cluster of cells in a mammalian blastocyst that protrudes into one end of the cavity and subsequently develops into the embryo proper and some of the extraembryonic membranes.

inositol trisphosphate (IP_3) *(in-NOS-i-tahl)* The second messenger, which functions as an intermediate between certain nonsteroid hormones and the third messenger, a rise in cytoplasmic Ca^{2+} concentration.

insertion A mutation involving the addition of one or more nucleotide pairs to a gene.

insertion sequence The simplest kind of a transposon, consisting of inserted repeats of DNA flanking a gene for transposase, the enzyme that catalyzes transposition.

insight learning The ability of an animal to perform a correct or appropriate behavior on the first attempt in a situation with which it has had no prior experience.

insulin *(IN-sul-in)* A vertebrate hormone that lowers blood glucose levels by promoting the uptake of glucose by most body cells and the synthesis and storage of glycogen in the liver; also stimulates protein and fat synthesis; secreted by endocrine cells of the pancreas called islets of Langerhans.

interferon *(IN-tur-FEER-on)* A chemical messenger of the immune system, produced by virus-infected cells and capable of helping other cells resist the virus.

interleukin-1 *(IN-tur-loo-kin)* A chemical regulator (cytokin) secreted by macrophages that have ingested a pathogen or foreign molecule and have bound with a helper T cell; stimulates T cells to grow and divide and elevates body temperature. Interleukin-2, secreted by activated T cells, stimulates helper T cells to proliferate more rapidly.

intermediate filament A component of the cytoskeleton that includes all filaments intermediate in size between microtubules and microfilaments.

interneuron *(IN-tur-NOOR-ahn)* An association neuron; a nerve cell within the central nervous system that forms synapses with sensory and motor neurons and integrates sensory input and motor output.

internode The segment of a plant stem between the points where leaves are attached.

interphase The period in the cell cycle when the cell is not dividing. During interphase, cellular metabolic activity is high, chromosomes and organelles are duplicated, and cell size may increase. Interphase accounts for 90% of the time of each cell cycle.

interstitial cells *(IN-tur-STISH-ul)* Cells scattered among the seminiferous tubules of the vertebrate testis that secrete testosterone and other androgens, the male sex hormones.

interstitial fluid The internal environment of vertebrates, consisting of the fluid filling the spaces between cells.

intertidal zone The shallow zone of the ocean where land meets water.

intrinsic rate of increase The difference between the number of births and the number of deaths, symbolized as r_{max}; the maximum population growth rate.

introgression *(IN-troh-GRES-shun)* The transplantation of genes between species resulting from fertile hybrids mating successfully with one of the parent species.

intron *(IN-tron)* A noncoding, intervening sequence within a eukaryotic gene.

inversion An aberration in chromosome structure resulting from an error in meiosis or from mutagens; reattachment in a reverse orientation of a chromosomal fragment to the chromosome from which the fragment originated.

invertebrate An animal without a backbone; invertebrates make up 95% of animal species.

in vitro fertilization *(VEE-troh)* Fertilization of ova in laboratory containers followed by artificial implantation of the early embryo in the mother's uterus.

ion *(EYE-on)* An atom that has gained or lost electrons, thus acquiring a charge.

ionic bond *(eye-ON-ik)* A chemical bond resulting from the attraction between oppositely charged ions.

isogamy *(eye-SOG-uh-mee)* A condition in which male and female gametes are morphologically indistinguishable.

isomer *(EYE-sum-ur)* One of several organic compounds with the same molecular formula but different structures and therefore different properties. The three types are structural isomers, geometric isomers, and enantiomers.

isomorphic generations Alternating generations in which the sporophytes and gametophytes look alike, although they differ in chromosome number.

isotonic solutions Solutions of equal solute concentration.

isotope *(EYE-so-tope)* One of several atomic forms of an element, each containing a different number of neutrons and thus differing in atomic mass.

joule (J) A unit of energy: $1\ J = 0.239$ cal; $1\ cal = 4.184\ J$.

juvenile hormone (JH) A hormone in arthropods, secreted by the corpora allata glands, that promotes the retention of larval characteristics.

juxtaglomerular apparatus (JGA) Specialized tissue located near the afferent arteriole that supplies blood to the kidney glomerulus; the JGA raises blood pressure by producing renin, which activates angiotensin.

***K*-selection** The concept that in certain (*K*-selected) populations, life history is centered around producing relatively few offspring that have a good chance of survival.

karyogamy The fusion of nuclei of two cells, as part of syngamy.

karyotype *(KAR-ee-oh-type)* A method of organizing the chromosomes of a cell in relation to number, size, and type.

keystone predator A predatory species that helps maintain species richness in a community by reducing the density of populations of the best competitors so that populations of less competitive species are maintained.

kilocalorie (kcal) A thousand calories; the amount of heat energy required to raise the temperature of 1 kg of water 1°C.

kin selection A phenomenon of inclusive fitness, used to explain altruistic behavior between related individuals.

kinesis *(kih-NEE-sis)* A change in activity rate in response to a stimulus.

kinetic energy *(kih-NET-ik)* The energy of motion, which is directly related to the speed of that motion. Moving matter does work by transferring some of its kinetic energy to other matter.

kinetochore *(kih-NET-oh-kor)* A specialized region on the centromere that links each sister chromatid to the mitotic spindle.

kingdom A taxonomic category, the second broadest after domain.

Koch's postulates A set of four criteria for determining whether a specific pathogen is the cause of a disease.

Krebs cycle A chemical cycle involving eight steps that completes the metabolic breakdown of glucose molecules to carbon dioxide; occurs within the mitochondrion; the second major stage in cellular respiration.

lacteal *(lak-TEEL)* A tiny lymph vessel extending into the core of an intestinal villus and serving as the destination for absorbed chylomicrons.

lagging strand A discontinuously synthesized DNA strand that elongates in a direction away from the replication fork.

larva *(LAR-vuh)* (plural, **larvae**) A free-living, sexually immature form in some animal life cycles that may differ from the adult in morphology, nutrition, and habitat.

lateral line system A mechanoreceptor system consisting of a series of pores and receptor units (neuromasts) along the sides of the body of fishes and aquatic amphibians; detects water movements made by an animal itself and by other moving objects.

lateral meristem *(MARE-eh-stem)* The vascular and cork cambium, a cylinder of dividing cells that runs most of the length of stems and roots and is responsible for secondary growth.

law of independent assortment Mendel's second law, stating that each allele pair segregates independently during gamete formation; applies when genes for two traits are located on different pairs of homologous chromosomes.

law of segregation Mendel's first law, stating that allele pairs separate during gamete formation, and then randomly re-form pairs during the fusion of gametes at fertilization.

leading strand The new continuous complementary DNA strand synthesized along the template strand in the mandatory $5' \rightarrow 3'$ direction.

leukocyte *(LOO-koh-site)* A white blood cell; typically functions in immunity, such as phagocytosis or antibody production.

lichen *(LY-ken)* An organism formed by the symbiotic association between a fungus and a photosynthetic alga.

life table A table of data summarizing mortality in a population.

ligament A type of fibrous connective tissue that joins bones together at joints.

ligand *(LIG-und)* A molecule that binds specifically to a receptor site of another molecule.

ligand-gated ion channel receptor A signal receptor protein in a cell membrane that can act as a channel for the passage of a specific ion across the membrane. When activated by a signal molecule, the receptor either allows or blocks passage of the ion, resulting in a change in ion concentration that usually affects cell functioning.

light microscope (LM) An optical instrument with lenses that refract (bend) visible light to magnify images of specimens.

light reactions The steps in photosynthesis that occur on the thylakoid membranes of the chloroplast and convert solar energy to the chemical energy of ATP and NADPH, evolving oxygen in the process.

lignin *(LIG-nin)* A hard material embedded in the cellulose matrix of vascular plant cell walls that functions as an important adaptation for support in terrestrial species.

limbic system *(LIM-bik)* A group of nuclei (clusters of nerve cell bodies) in the lower part of the mammalian forebrain that interact with the cerebral cortex in determining emotions; includes the hippocampus and the amygdala.

linkage map A genetic map *(see)* based on the frequencies of recombination between markers during crossing over of homologous chromosomes. The greater the frequency of recombination between two genetic markers, the farther apart they are assumed to be.

linked genes Genes that are located on the same chromosome.

lipid *(LIH-pid)* One of a family of compounds, including fats, phospholipids, and steroids, that are insoluble in water.

lipoprotein A protein bonded to a lipid; includes the low-density lipoproteins (LDLs) and high-density lipoproteins (HDLs) that transport fats and cholesterol in blood.

locus *(LOH-kus)* (plural, **loci**) A particular place along the length of a certain chromosome where a given gene is located.

logistic population growth A model describing population growth that levels off as population size approaches carrying capacity.

long-day plant A plant that flowers, usually in late spring or early summer, only when the light period is longer than a critical length.

loop of Henle The long hairpin turn, with a descending and ascending limb, of the renal tubule in the vertebrate kidney; functions in water and salt reabsorption.

lungs The invaginated respiratory surfaces of terrestrial vertebrates, land snails, and spiders that connect to the atmosphere by narrow tubes.

lymph *(limf)* The colorless fluid, derived from interstitial fluid, in the lymphatic system of vertebrate animals.

lymphatic system *(lim-FAT-ik)* A system of vessels and lymph nodes, separate from the circulatory system, that returns fluid and protein to the blood.

lymphocyte A white blood cell. The lymphocytes that complete their development in the bone marrow are called B cells, and those that mature in the thymus are called T cells.

lysogenic cycle A type of phage replication cycle in which the viral genome becomes incorporated into the bacterial host chromosome as a prophage.

lysosome *(LY-so-some)* A membrane-enclosed bag of hydrolytic enzymes found in the cytoplasm of eukaryotic cells.

lysozyme *(LY-so-zime)* An enzyme in perspiration, tears, and saliva that attacks bacterial cell walls.

lytic cycle *(LIT-ik)* A type of viral replication cycle resulting in the release of new phages by death or lysis of the host cell.

M phase The mitotic phase of the cell cycle, which includes mitosis and cytokinesis.

macroevolution Evolutionary change on a grand scale, encompassing the origin of novel designs, evolutionary trends, adaptive radiation, and mass extinction.

macromolecule A giant molecule of living matter formed by the joining of smaller molecules, usually by condensation synthesis. Polysaccharides, proteins, and nucleic acids are macromolecules.

macrophage *(MAK-roh-fage)* An amoeboid cell that moves through tissue fibers, engulfing bacteria and dead cells by phagocytosis.

major histocompatibility complex (MHC) A large set of cell surface antigens encoded by a family of genes. Foreign MHC markers trigger T-cell responses that may lead to the rejection of transplanted tissues and organs.

Malpighian tubule *(mal-PIG-ee-un)* A unique excretory organ of insects that empties into the digestive tract, removes nitrogenous wastes from the blood, and functions in osmoregulation.

Mammalia The vertebrate class of mammals, characterized by body hair and mammary glands that produce milk to nourish the young.

mantle A heavy fold of tissue in mollusks that drapes over the visceral mass and may secrete a shell.

marsupial *(mar-SOOP-ee-ul)* A mammal, such as a koala, kangaroo, or opossum, whose young complete their embryonic development inside a maternal pouch called the marsupium.

matrix The nonliving component of connective tissue, consisting of a web of fibers embedded in homogeneous ground substance that may be liquid, jellylike, or solid.

matter Anything that takes up space and has mass.

mechanoreceptor A sensory receptor that detects physical deformations in the body's environment associated with pressure, touch, stretch, motion, and sound.

medulla oblongata *(meh-DOO-luh OBB-long-GAH-tuh)* The lowest part of the vertebrate brain; a swelling of the hindbrain dorsal to the anterior spinal cord that controls autonomic, homeostatic functions, including breathing, heart and blood vessel activity, swallowing, digestion, and vomiting.

medusa *(meh-DOO-suh)* The floating, flattened, mouth-down version of the cnidarian body plan. The alternate form is the polyp.

megapascal (MPa) *(MEG-uh-pass-KAL)* A unit of pressure equivalent to 10 atmospheres of pressure.

meiosis *(my-OH-sis)* A two-stage type of cell division in sexually reproducing organisms that results in gametes with half the chromosome number of the original cell.

membrane potential The charge difference between the cytoplasm and extracellular fluid in all cells, due to the differential distribution of ions. Membrane potential affects the activity of excitable cells and the transmembrane movement of all charged substances.

memory cell A clone of long-lived lymphocytes, formed during the primary immune response, that remains in a lymph node until activated by exposure to the same antigen that triggered its formation. Activated memory cells mount the secondary immune response.

menstrual cycle *(MEN-stroo-ul)* A type of reproductive cycle in higher female primates, in which the nonpregnant endometrium is shed as a bloody discharge through the cervix into the vagina.

meristem *(MARE-eh-stem)* Plant tissue that remains embryonic as long as the plant lives, allowing for indeterminate growth.

meroblastic cleavage *(MARE-oh-BLAS-tik)* A type of cleavage in which there is incomplete division of yolk-rich egg, characteristic of avian development.

mesentery *(MEZ-en-ter-ee)* A membrane that suspends many of the organs of vertebrates inside fluid-filled body cavities.

mesoderm *(MEZ-oh-durm)* The middle primary germ layer of an early embryo that develops into the notochord, the lining of the coelom, muscles, skeleton, gonads, kidneys, and most of the circulatory system.

mesophyll *(MEZ-oh-fil)* The ground tissue of a leaf, sandwiched between the upper and lower epidermis and specialized for photosynthesis.

messenger RNA (mRNA) A type of RNA synthesized from DNA in the genetic material that attaches to ribosomes in the cytoplasm and specifies the primary structure of a protein.

metabolism *(meh-TAB-oh-liz-um)* The totality of an organism's chemical processes, consisting of catabolic and anabolic pathways.

metamorphosis *(MET-uh-MOR-fuh-sis)* The resurgence of development in an animal larva that transforms it into a sexually mature adult.

metanephridium *(MET-uh-neh-FRID-ee-um)* (plural, **metanephridia**) In annelid worms, a type of excretory tubule with internal openings called nephrostomes that collect body fluids and external openings called nephridiopores.

metapopulation A subdivided population of a single species.

metastasis *(meh-TAS-teh-sis)* The spread of cancer cells beyond their original site.

microevolution A change in the gene pool of a population over a succession of generations.

microfilament A solid rod of actin protein in the cytoplasm of almost all eukaryotic cells, making up part of the cytoskeleton and acting alone or with myosin to cause cell contraction.

microtubule A hollow rod of tubulin protein in the cytoplasm of all eukaryotic cells and in cilia, flagella, and the cytoskeleton.

microvillus (plural, **microvilli**) One of many fine, fingerlike projections of the epithelial cells in the lumen of the small intestine that increase its surface area.

middle lamella *(luh-MEL-uh)* A thin layer of adhesive extracellular material, primarily pectins, found between the primary walls of adjacent young plant cells.

mimicry A phenomenon in which one species benefits by a superficial resemblance to an unrelated species. A predator or species of prey may gain a significant advantage through mimicry.

mineralocorticoid A corticosteroid hormone secreted by the adrenal cortex that regulates salt and water homeostasis.

minimum dynamic area The amount of suitable habitat needed to sustain a viable population.

minimum viable population size (MVP) The smallest number of individuals needed to perpetuate a population.

missense mutation The most common type of mutation involving a base-pair substitution within a gene that changes a codon, but the new codon makes sense in that it still codes for an amino acid.

mitochondrial matrix The compartment of the mitochondrion enclosed by the inner membrane and containing enzymes and substrates for the Krebs cycle.

mitochondrion *(MY-toh-KON-dree-un)* (plural, **mitochondria**) An organelle in eukaryotic cells that serves as the site of cellular respiration.

mitosis *(my-TOH-sis)* A process of nuclear division in eukaryotic cells conventionally divided into five stages: prophase, prometaphase, metaphase, anaphase, and telophase. Mitosis conserves chromosome number by equally allocating replicated chromosomes to each of the daughter nuclei.

modern synthesis A comprehensive theory of evolution emphasizing natural selection, gradualism, and populations as the fundamental units of evolutionary change; also called neo-Darwinism.

molarity A common measure of solute concentration, referring to the number of moles of solute in 1 L of solution.

mold A rapidly growing, asexually reproducing fungus.

mole The number of grams of a substance that equals its molecular weight in daltons and contains Avogadro's number of molecules.

molecular formula A type of molecular notation indicating only the quantity of the constituent atoms.

molecule Two or more atoms held together by covalent bonds.

molting A process in arthropods in which the exoskeleton is shed at intervals to allow growth by the secretion of a larger exoskeleton.

monoclonal antibody *(MON-oh-KLONE-ul)* A defensive protein produced by cells descended from a single cell; an antibody that is secreted by a clone of cells and, consequently, is specific for a single antigenic determinant.

monocot *(MON-oh-kot)* A subdivision of flowering plants whose members possess one embryonic seed leaf, or cotyledon.

monoculture Cultivation of large land areas with a single plant variety.

monoecious *(mon-EE-shus)* Referring to a plant species that has both staminate and carpellate flowers on the same individual.

monohybrid cross A breeding experiment that uses parental varieties differing in a single character.

monomer *(MON-uh-mer)* The subunit that serves as the building block of a polymer.

monophyletic *(MON-oh-fy-LEH-tik)* Pertaining to a taxon derived from a single ancestral species that gave rise to no species in any other taxa.

monosaccharide *(MON-oh-SAK-ur-ide)* The simplest carbohydrate, active alone or serving as a monomer for disaccharides and polysaccharides. Also known as simple sugars, the molecular formulas of monosaccharides are generally some multiple of CH_2O.

monotreme *(MON-uh-treem)* An egg-laying mammal, represented by the platypus and echidna.

morphogen A substance, such as bicoid protein, that provides positional information in the form of a concentration gradient along an embryonic axis.

morphogenesis *(MOR-foh-JEN-eh-sis)* The development of body shape and organization during ontogeny.

morphological species concept The idea that species are defined by measurable anatomical criteria.

morphospecies A species defined by its anatomical features.

mosaic development A pattern of development, such as that of a mollusk, in which the early blastomeres each give rise to a specific part of the embryo. In some animals, the fate of the blastomeres is established in the zygote.

mosaic evolution The evolution of different features of an organism at different rates.

motor neuron A nerve cell that transmits signals from the brain or spinal cord to muscles or glands.

motor unit A single motor neuron and all the muscle fibers it controls.

MPF (M-phase promoting factor) A protein complex required for a cell to progress from late interphase to mitosis; the active form consists of cyclin and *cdc2*, a protein kinase.

Müllerian mimicry *(myoo-LER-ee-un)* A mutual mimicry by two unpalatable species.

multigene family A collection of genes with similar or identical sequences, presumably of common origin.

mutagen *(MYOOT-uh-jen)* A chemical or physical agent that interacts with DNA and causes a mutation.

mutagenesis *(MYOOT-uh-JEN-uh-sis)* The creation of mutations.

mutation *(myoo-TAY-shun)* A rare change in the DNA of genes that ultimately creates genetic diversity.

mutualism *(MYOO-choo-ul-iz-um)* A symbiotic relationship in which both the host and the symbiont benefit.

mycelium *(my-SEEL-ee-um)* The densely branched network of hyphae in a fungus.

mycorrhizae *(MY-koh-RY-zee)* Mutualistic associations of plant roots and fungi.

myelin sheath *(MY-eh-lin)* In a neuron, an insulating coat of cell membrane from Schwann cells that is interrupted by nodes of Ranvier where saltatory conduction occurs.

myofibril *(MY-oh-FY-brill)* A fibril collectively arranged in longitudinal bundles in muscle cells (fibers); composed of thin filaments of actin and a regulatory protein and thick filaments of myosin.

myoglobin *(MY-uh-glow-bin)* An oxygen-storing, pigmented protein in muscle cells.

myosin *(MY-uh-sin)* A type of protein filament that interacts with actin filaments to cause cell contraction.

NAD^+ (nicotinamide adenine dinucleotide) A coenzyme present in all cells that helps enzymes transfer electrons during the redox reactions of metabolism.

natural killer cell A nonspecific defensive cell that attacks tumor cells and destroys infected body cells, especially those harboring viruses.

natural selection Differential success in the reproduction of different phenotypes resulting from the interaction of organisms with their environment. Evolution occurs when natural selection causes changes in relative frequencies of alleles in the gene pool.

negative feedback A primary mechanism of homeostasis, whereby a change in a physiological variable that is being monitored triggers a response that counteracts the initial fluctuation.

nephron *(NEF-ron)* The tubular excretory unit of the vertebrate kidney.

neritic zone *(neh-RIT-ik)* The shallow regions of the ocean overlying the continental shelves.

net primary productivity (NPP) The gross primary productivity minus the energy used by the producers for cellular respiration; represents the storage of chemical energy in an ecosystem available to consumers.

neural crest A band of cells along the border where the neural tube pinches off from the ectoderm; the cells migrate to various parts of the embryo and form the pigment cells in the skin, bones of the skull, the teeth, the adrenal glands, and parts of the peripheral nervous system.

neuron *(NOOR-on)* A nerve cell; the fundamental unit of the nervous system, having structure and properties that allow it to conduct signals by taking advantage of the electrical charge across its cell membrane.

neurosecretory cells Hypothalamus cells that receive signals from other nerve cells, but instead of signaling to an adjacent nerve cell or muscle, they release hormones into the bloodstream.

neurotransmitter A chemical messenger released from the synaptic terminal of a neuron at a chemical synapse that diffuses across the synaptic cleft and binds to and stimulates the postsynaptic cell.

neutral variation Genetic diversity that confers no apparent selective advantage.

niche *See* ecological niche.

nitrogen fixation The assimilation of atmospheric nitrogen by certain prokaryotes into nitrogenous compounds that can be directly used by plants.

nitrogenase *(nih-TRAH-juh-nayz)* An enzyme, unique to certain prokaryotes, that reduces N_2 to NH_3.

node A point along the stem of a plant at which leaves are attached.

nodes of Ranvier *(ran-VEER)* The small gaps in the myelin sheath between successive glial cells along the axon of a neuron; also, the site of high concentration of voltage-gated ion channels.

noncompetitive inhibitor A substance that reduces the activity of an enzyme by binding to a location remote from the active site, changing its conformation so that it no longer binds to the substrate.

noncyclic electron flow A route of electron flow during the light reactions of photosynthesis that involves both photosystems and produces ATP, NADPH, and oxygen; the net electron flow is from water to $NADP^+$.

noncyclic photophosphorylation *(FO-toh-fos-FOR-eh-LAY-shun)* The production of ATP by noncyclic electron flow.

nondisjunction An accident of meiosis or mitosis, in which both members of a pair of homologous chromosomes or both sister chromatids fail to move apart properly.

nonpolar covalent bond A type of covalent bond in which electrons are shared equally between two atoms of similar electronegativity.

nonsense mutation A mutation that changes an amino acid codon to one of the three stop codons, resulting in a shorter and usually nonfunctional protein.

norm of reaction The range of phenotypic possibilities for a single genotype, as influenced by the environment.

notochord *(NO-toh-kord)* A longitudinal, flexible rod formed from dorsal mesoderm and located between the gut and the nerve cord in all chordate embryos.

nuclear envelope The membrane in eukaryotes that encloses the nucleus, separating it from the cytoplasm.

nucleic acid (polynucleotide) *(PAHL-ee-NOO-klee-oh-tide)* A polymer consisting of many nucleotide monomers; serves as a blueprint for proteins and, through the actions of proteins, for all cellular activities. The two types are DNA and RNA.

nucleic acid probe In DNA technology, a labeled single-stranded nucleic acid molecule used to tag a specific nucleotide sequence in a nucleic acid sample. Molecules of the probe hydrogen-bond to the complementary sequence wherever it occurs; radioactive or other labeling of the probe allows its location to be detected.

nucleoid *(NOO-klee-oid)* A dense region of DNA in a prokaryotic cell.

nucleoid region The region in a prokaryotic cell consisting of a concentrated mass of DNA.

nucleolus *(noo-KLEE-oh-lus)* (plural, **nucleoli**) A specialized structure in the nucleus, formed from various chromosomes and active in the synthesis of ribosomes.

nucleoside *(NOO-klee-oh-side)* An organic molecule consisting of a nitrogenous base joined to a five-carbon sugar.

nucleosome *(NOO-klee-oh-some)* The basic, beadlike unit of DNA packaging in eukaryotes, consisting of a segment of DNA wound around a protein core composed of two copies of each of four types of histone.

nucleotide *(NOO-klee-oh-tide)* The building block of a nucleic acid, consisting of a five-carbon sugar covalently bonded to a nitrogenous base and a phosphate group.

nucleus (1) An atom's central core, containing protons and neutrons. (2) The chromosome-containing organelle of a eukaryotic cell. (3) A cluster of neurons.

obligate aerobe *(OB-lig-it AIR-obe)* An organism that requires oxygen for cellular respiration and cannot live without it.

obligate anaerobe *(AN-ur-obe)* An organism that cannot use oxygen and is poisoned by it.

oceanic zone The region of water lying over deep areas beyond the continental shelf.

oligotrophic lake A nutrient-poor, clear, deep lake with minimum phytoplankton.

omnivore A heterotrophic animal that consumes both meat and plant material.

oncogene *(ON-koh-jeen)* A gene found in viruses or as part of the normal genome that is involved in triggering cancerous characteristics.

ontogeny *(on-TOJ-en-ee)* The embryonic development of an organism.

oogamy *(oh-OG-um-ee)* A condition in which male and female gametes differ, such that a small, flagellated sperm fertilizes a large, nonmotile egg.

oogenesis *(OO-oh-JEN-eh-sis)* The process in the ovary that results in the production of female gametes.

open circulatory system An arrangement of internal transport in which blood bathes the organs directly and there is no distinction between blood and interstitial fluid.

operant conditioning *(OP-ur-ent)* A type of associative learning that directly affects behavior in a natural context; also called trial-and-error learning.

operon *(OP-ur-on)* A unit of genetic function common in bacteria and phages, consisting of coordinately regulated clusters of genes with related functions.

opsonization An immune response in which the binding of antibodies to the surface of a microbe facilitates phagocytosis of the microbe by a macrophage.

organ A specialized center of body function composed of several different types of tissues.

organ-identity gene A plant gene in which a mutation causes a floral organ to develop in the wrong location.

organ of Corti The actual hearing organ of the vertebrate ear, located in the floor of the cochlear canal in the inner ear; contains the receptor cells (hair cells) of the ear.

organelle *(OR-guh-NEL)* One of several formed bodies with a specialized function, suspended in the cytoplasm and found in eukaryotic cells.

organic chemistry The study of carbon compounds (organic compounds).

organogenesis *(or-GAN-oh-JEN-eh-sis)* An early period of rapid embryonic development in which the organs take form from the primary germ layers.

orgasm Rhythmic, involuntary contractions of certain reproductive structures in both sexes during the human sexual response cycle.

osmoconformer An animal that does not actively adjust its internal osmolarity because it is isotonic with its environment.

osmolarity *(OZ-moh-LAR-eh-tee)* Solute concentration expressed as molarity.

osmoregulation Adaptations to control the water balance in organisms living in hypertonic, hypotonic, or terrestrial environments.

osmoregulator An animal whose body fluids have a different osmolarity than the environment, and that must either discharge excess water if it lives in a hypotonic environment or take in water if it inhabits a hypertonic environment.

osmosis *(oz-MOH-sis)* The diffusion of water across a selectively permeable membrane.

osmotic pressure *(oz-MOT-ik)* A measure of the tendency of a solution to take up water when separated from pure water by a selectively permeable membrane.

Osteichthyes The vertebrate class of bony fishes, characterized by a skeleton reinforced by calcium phosphate; the most abundant and diverse vertebrates.

ostracoderm *(os-TRAK-uh-durm)* An extinct agnathan; a fishlike creature encased in an armor of bony plates.

outgroup A species or group of species that is closely related to the group of species being studied, but clearly not as closely related as any study-group members are to each other.

ovarian cycle *(oh-VAIR-ee-un)* The cyclic recurrence of the follicular phase, ovulation, and the luteal phase in the mammalian ovary, regulated by hormones.

ovary *(OH-vur-ee)* (1) In flowers, the portion of a carpel in which the egg-containing ovules develop. (2) In animals, the structure that produces female gametes and reproductive hormones.

oviduct *(OH-veh-dukt)* A tube passing from the ovary to the vagina in invertebrates or to the uterus in vertebrates.

oviparous *(oh-VIP-ur-us)* Referring to a type of development in which young hatch from eggs laid outside the mother's body.

ovoviviparous *(OH-voh-vy-VIP-ur-us)* Referring to a type of development in which young hatch from eggs that are retained in the mother's uterus.

ovulation The release of an egg from ovaries. In humans, an ovarian follicle releases an egg during each menstrual cycle.

ovule *(OV-yool)* A structure that develops in the plant ovary and contains the female gametophyte.

ovum *(OH-vum)* The female gamete; the haploid, unfertilized egg, which is usually a relatively large, nonmotile cell.

oxidation The loss of electrons from a substance involved in a redox reaction.

oxidative phosphorylation *(FOS-for-eh-LAY-shun)* The production of ATP using energy derived from the redox reactions of an electron transport chain.

oxidizing agent The electron acceptor in a redox reaction.

pacemaker A specialized region of the right atrium of the mammalian heart that sets the rate of contraction; also called the sinoatrial (SA) node.

paedogenesis *(pee-doh-JEN-eh-sis)* The precocious development of sexual maturity in a larva.

paedomorphosis *(PEE-doh-mor-FOH-sis)* The retention in an adult organism of the juvenile features of its evolutionary ancestors.

paleontology *(PAY-lee-un-TOL-uh-jee)* The scientific study of fossils.

Pangaea *(pan-JEE-uh)* The supercontinent formed near the end of the Paleozoic era when plate movements brought all the land masses of Earth together.

paraphyletic *(PAR-uh-FY-leh-tik)* Pertaining to a taxon that excludes some members that share a common ancestor with members included in the taxon.

parasite *(PAR-uh-site)* An organism that absorbs nutrients from the body fluids of living hosts.

parasitism A symbiotic relationship in which the symbiont (parasite) benefits at the expense of the host by living either within the host (endoparasite) or outside the host (ectoparasite).

parasympathetic division One of two divisions of the autonomic nervous system; generally enhances body activities that gain and conserve energy, such as digestion and reduced heart rate.

parathyroid glands Four endocrine glands, embedded in the surface of the thyroid gland, that secrete parathyroid hormone and raise blood calcium levels.

parazoa *(PAR-uh-ZOH-uh)* Members of the subkingdom of animals consisting of the sponges.

parenchyma *(pur-EN-kim-uh)* A relatively unspecialized plant cell type that carries most of the metabolism, synthesizes and stores organic products, and develops into more differentiated cell types.

parthenogenesis *(PAR-then-oh-JEN-eh-sis)* A type of reproduction in which females produce offspring from unfertilized eggs.

partial pressure The concentration of gases; a fraction of total pressure.

passive transport The diffusion of a substance across a biological membrane.

pattern formation The ordering of cells into specific three-dimensional structures, an essential part of shaping an organism and its individual parts during development.

pedigree A family tree describing the occurrence of heritable characters in parents and offspring across as many generations as possible.

pelagic zone *(pel-AY-jik)* The area of the ocean past the continental shelf, with areas of open water often reaching to very great depths.

peptide bond The covalent bond between two amino acid units, formed by condensation synthesis.

peptidoglycan *(PEP-tid-oh-GLY-kan)* A type of polymer in bacterial cell walls consisting of modified sugars cross-linked by short polypeptides.

perception The interpretation of sensations by the brain.

perennial *(pur-EN-ee-ul)* A plant that lives for many years.

pericycle *(PAIR-eh-sy-kul)* A layer of cells just inside the endodermis of a root that may become meristematic and begin dividing again.

periderm *(PAIR-eh-durm)* The protective coat that replaces the epidermis in plants during secondary growth, formed of the cork and cork cambium.

peripheral nervous system The sensory and motor neurons that connect to the central nervous system.

peristalsis *(PAIR-is-TAL-sis)* Rhythmic waves of contraction of smooth muscle that push food along the digestive tract.

peroxisome *(pur-OKS-eh-some)* A microbody containing enzymes that transfer hydrogen from various substrates to oxygen, producing and then degrading hydrogen peroxide.

petiole *(PET-ee-ole)* The stalk of a leaf, which joins the leaf to a node of the stem.

pH scale A measure of hydrogen ion concentration equal to $-\log [H^+]$ and ranging in value from 0 to 14.

phage *(fage)* A virus that infects bacteria; also called a bacteriophage.

phagocytosis *(FAY-goh-sy-TOH-sis)* A type of endocytosis involving large, particulate substances.

pharynx *(FAH-rinks)* An area in the vertebrate throat where air and food passages cross; in flatworms, the muscular tube that protrudes from the ventral side of the worm and ends in the mouth.

phenetics *(feh-NEH-tiks)* An approach to taxonomy based entirely on measurable similarities and differences in phenotypic characters, without consideration of homology, analogy, or phylogeny.

phenotype *(FEE-nuh-type)* The physical and physiological traits of an organism.

pheromone *(FAIR-uh-mone)* A small, volatile chemical signal that functions in communication between animals and acts much like a hormone in influencing physiology and behavior.

phloem *(FLOH-um)* The portion of the vascular system in plants consisting of living cells arranged into elongated tubes that transport sugar and other organic nutrients throughout the plant.

phosphate group *(FOS-fate)* A functional group important in energy transfer.

phospholipids *(FOS-foh-LIP-ids)* Molecules that constitute the inner bilayer of biological membranes, having a polar, hydrophilic head and a nonpolar, hydrophobic tail.

photic zone *(FOH-tik)* The narrow top slice of the ocean, where light permeates sufficiently for photosynthesis to occur.

photoautotroph *(FOH-toh-AW-toh-trohf)* An organism that harnesses light energy to drive the synthesis of organic compounds from carbon dioxide.

photoheterotroph *(FOH-toh-HET-ur-oh-trohf)* An organism that uses light to generate ATP but that must obtain carbon in organic form.

photon *(FOH-tahn)* A quantum, or discrete amount, of light energy.

photoperiodism *(FOH-toh-PEER-ee-od-iz-um)* A physiological response to day length, such as flowering in plants.

photophosphorylation *(FOH-toh-fos-for-uh-LAY-shun)* The process of generating ATP from ADP and phosphate by means of a proton-motive force generated by the thylakoid membrane of the chloroplast during the light reactions of photosynthesis.

photorespiration A metabolic pathway that consumes oxygen, releases carbon dioxide, generates no ATP, and decreases photosynthetic output; generally occurs on hot, dry, bright days, when stomata close and the oxygen concentration in the leaf exceeds that of carbon dioxide.

photosynthesis The conversion of light energy to chemical energy that is stored in glucose or other organic compounds; occurs in plants, algae, and certain prokaryotes.

photosystem The light-harvesting unit in photosynthesis, located on the thylakoid membrane of the chloroplast and consisting of the antenna complex, the reaction-center chlorophyll *a*, and the primary electron acceptor. There are two types of photosystems, I and II; they absorb light best at different wavelengths.

phototropism *(FOH-toh-TROH-piz-um)* Growth of a plant shoot toward or away from light.

phylogeny *(fih-LOJ-en-ee)* The evolutionary history of a species or group of related species.

phylum A taxonomic category; phyla are divided into classes.

phytoalexin *(fy-toh-ah-LEK-sin)* An antibiotic, produced by plants, that destroys microorganisms or inhibits their growth.

phytochrome *(FY-tuh-krome)* A pigment involved in many responses of plants to light.

pilus *(PILL-us)* (plural, **pili**) A surface appendage in certain bacteria that functions in adherence and the transfer of DNA during conjugation.

pineal gland *(PIN-ee-ul)* A small endocrine gland on the dorsal surface of the vertebrate forebrain; secretes the hormone melatonin, which regulates body functions related to seasonal day length.

pinocytosis *(PY-noh-sy-TOH-sis)* A type of endocytosis in which the cell ingests extracellular fluid and its dissolved solutes.

pith The core of the central vascular cylinder of monocot roots, consisting of parenchyma cells, which are ringed by vascular tissue; ground tissue interior to vascular bundles in dicot stems.

pituitary gland *(pih-TOO-ih-tair-ee)* An endocrine gland at the base of the hypothalamus; consists of a posterior lobe (neurohypophysis), which stores and releases two hormones produced by the hypothalamus, and an anterior lobe (adenohypophysis), which produces and secretes many hormones that regulate diverse body functions.

placenta *(pluh-SEN-tuh)* A structure in the pregnant uterus for nourishing a viviparous fetus with the mother's blood supply; formed from the uterine lining and embryonic membranes.

placental mammal A member of a group of mammals, including humans, whose young complete their embryonic development in the uterus, joined to the mother by a placenta.

placoderm *(PLAK-oh-durm)* A member of an extinct class of fishlike vertebrates that had jaws and were enclosed in a tough, outer armor.

plankton Mostly microscopic organisms that drift passively or swim weakly near the surface of oceans, ponds, and lakes.

plasma *(PLAZ-muh)* The liquid matrix of blood in which the cells are suspended.

plasma cell A derivative of B cells that secretes antibodies.

plasma membrane The membrane at the boundary of every cell that acts as a selective barrier, thereby regulating the cell's chemical composition.

plasmid *(PLAZ-mid)* A small ring of DNA that carries accessory genes separate from those of a bacterial chromosome. Also found in some eukaryotes, such as yeast.

plasmodesma *(PLAZ-moh-DEZ-muh)* (plural, **plasmodesmata**) An open channel in the cell wall of plants through which strands of cytosol connect from adjacent cells.

plasmogamy The fusion of the cytoplasm of cells from two individuals; occurs as one stage of syngamy.

plasmolysis *(plaz-MOL-eh-sis)* A phenomenon in walled cells in which the cytoplasm shrivels and the plasma membrane pulls away from the cell wall when the cell loses water to a hypertonic environment.

plastid One of a family of closely related plant organelles, including chloroplasts, chromoplasts, and amyloplasts (leucoplasts).

platelet A small enucleated blood cell important in blood clotting; derived from large cells in the bone marrow.

pleated sheet One form of the secondary structure of proteins in which the polypeptide chain folds back and forth, or where two regions of the chain lie parallel to each other and are held together by hydrogen bonds.

pleiotropy *(PLY-eh-troh-pee)* The ability of a single gene to have multiple effects.

plesiomorphic character *(PLEEZ-ee-oh-MOR-fik)* A primitive phenotypic character possessed by a remote ancestor.

pluripotent stem cell A cell within bone marrow that is a progenitor for any kind of blood cell.

point mutation A change in a gene at a single nucleotide pair.

polar covalent bond A type of covalent bond between atoms that differ in electronegativity. The shared electrons are pulled closer to the more electronegative atom, making it slightly negative and the other atom slightly positive.

polar molecule A molecule (such as water) with opposite charges on opposite sides.

pollen grain An immature male gametophyte that develops within the anthers of stamens in a flower.

pollination *(POL-eh-NAY-shun)* The placement of pollen onto the stigma of a carpel by wind or animal carriers, a prerequisite to fertilization.

polyandry *(POL-ee-AN-dree)* A polygamous mating system involving one female and many males.

polygenic inheritance *(POL-ee-JEN-ik)* An additive effect of two or more gene loci on a single phenotypic character.

polygyny *(pol-IJ-en-ee)* A polygamous mating system involving one male and many females.

polymer *(POL-eh-mur)* A large molecule consisting of many identical or similar monomers linked together.

polymerase chain reaction (PCR) A technique for amplifying DNA in vitro by incubating with special primers, DNA polymerase molecules and nucleotides.

polymorphic *(POL-ee-MOR-fik)* Referring to a population in which two or more physical forms are present in readily noticeable frequencies.

polymorphism *(POL-ee-MOR-fiz-um)* The coexistence of two or more distinct forms of individuals (polymorphic characters) in the same population.

polyp *(POL-ip)* The sessile variant of the cnidarian body plan. The alternate form is the medusa.

polypeptide *(POL-ee-PEP-tide)* A polymer (chain) of many amino acids linked together by peptide bonds.

polyphyletic Pertaining to a taxon whose members were derived from two or more ancestral forms not common to all members.

polyploidy *(POL-ee-ploid-ee)* A chromosomal alteration in which the organism possesses more than two complete chromosome sets.

polyribosome *(POL-ee-RY-boh-some)* An aggregation of several ribosomes attached to one messenger RNA molecule.

polysaccharide *(POL-ee-SAK-ur-ide)* A polymer of up to over a thousand monosaccharides, formed by condensation synthesis.

population A group of individuals of one species that live in a particular geographic area.

population viability analysis (PVA) A method of predicting whether or not a species will persist in a particular environment.

positional information Signals, to which genes regulating development respond, indicating a cell's location relative to other cells in an embryonic structure.

positive feedback A physiological control mechanism in which a change in some variable triggers mechanisms that amplify the change.

postsynaptic membrane *(post-sin-AP-tik)* The surface of the cell on the opposite side of the synapse from the synaptic terminal of the stimulating neuron that contains receptor proteins and degradative enzymes for the neurotransmitter.

postzygotic barrier *(POST-zy-GOT-ik)* Any of several species-isolating mechanisms that prevent hybrids produced by two different species from developing into viable, fertile adults.

potential energy The energy stored by matter as a result of its location or spatial arrangement.

prezygotic barrier *(PREE-zy-GOT-ik)* A reproductive barrier that impedes mating between species or hinders fertilization of ova if interspecific mating is attempted.

primary consumer An herbivore; an organism in the trophic level of an ecosystem that eats plants or algae.

primary germ layers The three layers (ectoderm, mesoderm, endoderm) of the late gastrula, which develop into all parts of an animal.

primary growth Growth initiated by the apical meristems of a plant root or shoot.

primary immune response The initial immune response to an antigen, which appears after a lag of several days.

primary producer An autotroph, which collectively make up the trophic level of an ecosystem that ultimately supports all other levels; usually a photosynthetic organism.

primary productivity The rate at which light energy or inorganic chemical energy is converted to the chemical energy of organic compounds by autotrophs in an ecosystem.

primary structure The level of protein structure referring to the specific sequence of amino acids.

primary succession A type of ecological succession that occurs in an area where there were originally no organisms.

primer An already existing RNA chain bound to template DNA to which DNA nucleotides are added during DNA synthesis.

principle of allocation The concept that each organism has an energy budget, or a limited amount of total energy available for all of its maintenance and reproductive needs.

prion An infectious form of protein that may increase in number by converting related proteins to more prions.

probe *See* nucleic acid probe.

procambium *(pro-KAM-bee-um)* A primary meristem of roots and shoots that forms the vascular tissue.

prokaryotic cell *(pro-KAR-ee-OT-ik)* A type of cell lacking a membrane-enclosed nucleus and membrane-enclosed organelles; found only in the domains Bacteria and Archaea.

promoter A specific nucleotide sequence in DNA that binds RNA polymerase and indicates where to start transcribing RNA.

prophage *(PRO-fage)* A phage genome that has been inserted into a specific site on the bacterial chromosome.

prostaglandin (PG) *(PROS-tuh-GLAN-din)* One of a group of modified fatty acids secreted by virtually all tissues and performing a wide variety of functions as messengers.

protein *(PRO-teen)* A three-dimensional biological polymer constructed from a set of 20 different monomers called amino acids.

protein kinase An enzyme that transfers phosphate groups from ATP to a protein.

protein phosphatase An enzyme that removes phosphate groups from proteins, often functioning to reverse the effect of a protein kinase.

proteoglycans *(pro-tee-oh-GLY-kanz)* A glycoprotein in the extracellular matrix of animal cells, rich in carbohydrate.

proteasome A giant protein complex that recognizes and destroys proteins tagged for elimination by the small protein ubiquitin.

protoderm *(PRO-toh-durm)* The outermost primary meristem, which gives rise to the epidermis of roots and shoots.

proton-motive force The potential energy stored in the form of an electrochemical gradient, generated by the pumping of hydrogen ions across biological membranes during chemiosmosis.

proton pump *(PRO-tahn)* An active transport mechanism in cell membranes that consumes ATP to force hydrogen ions out of a cell and, in the process, generates a membrane potential.

protonephridium *(PRO-toh-nef-RID-ee-um)* An excretory system, such as the flame-cell system of flatworms, consisting of a network of closed tubules having external openings called nephridiopores and lacking internal openings.

proto-oncogene *(PRO-toh-ONK-oh-jeen)* A normal cellular gene corresponding to an oncogene; a gene with a potential to cause cancer, but that requires some alteration to become an oncogene.

protoplast The contents of a plant cell exclusive of the cell wall.

protostome *(PRO-toh-stome)* A member of one of two distinct evolutionary lines of coelomates, consisting of the annelids, mollusks, and arthropods, and characterized by spiral, determinate cleavage, schizocoelous formation of the coelom, and development of the mouth from the blastopore.

protozoan (plural, **protozoa**) A protist that lives primarily by ingesting food, an animal-like mode of nutrition.

provirus Viral DNA that inserts into a host genome.

proximate causation The immediate mechanisms underlying an organism's behavioral, physiological, or morphological response, in contrast to ultimate, or evolutionary causation.

pseudocoelomate *(SOO-doh-SEEL-oh-mate)* An animal whose body cavity is not completely lined by mesoderm.

pseudopodium *(SOO-doh-POH-dee-um)* (plural, **pseudopodia**) A cellular extension of amoeboid cells used in moving and feeding.

punctuated equilibrium A theory of evolution advocating spurts of relatively rapid change followed by long periods of stasis.

quantitative character A heritable feature in a population that varies continuously as a result of environmental influences and the additive effect of two or more genes (polygenic inheritance).

quaternary structure *(KWAT-ur-nair-ee)* The particular shape of a complex, aggregate protein, defined by the characteristic three-dimensional arrangement of its constituent subunits, each a polypeptide.

quiescent center A region located within the zone of cell division in plant roots, containing meristematic cells that divide very slowly.

***r*-selection** The concept that in certain (*r*-selected) populations, a high reproductive rate is the chief determinant of life history.

radial cleavage A type of embryonic development in deuterostomes in which the planes of cell division that transform the zygote into a ball of cells are either parallel or perpendicular to the polar axis, thereby aligning tiers of cells one above the other.

radial symmetry Characterizing a body shaped like a pie or barrel, with many equal parts radiating outward like the spokes of a wheel; present in cnidarians and echinoderms.

radiata Members of the radially symmetrical animal phyla, including cnidarians.

radicle An embryonic root of a plant.

radioactive dating A method of determining the age of fossils and rocks using half-lives of radioactive isotopes.

radioactive isotope An isotope, an atomic form of a chemical element, that is unstable; the nucleus decays spontaneously, giving off detectable particles and energy.

radiometric dating A method paleontologists use for determining the ages of rocks and fossils on a scale of absolute time, based on the half-life of radioactive isotopes.

receptor-mediated endocytosis *(EN-doh-sy-TOH-sis)* The movement of specific molecules into a cell by the inward budding of membranous vesicles containing proteins with receptor sites specific to the molecules being taken in; enables a cell to acquire bulk quantities of specific substances.

receptor potential An initial response of a receptor cell to a stimulus, consisting of a change in voltage across the receptor membrane proportional to the stimulus strength. The intensity of the receptor potential determines the frequency of action potentials traveling to the nervous system.

recessive allele In a heterozygote, the allele that is completely masked in the phenotype.

reciprocal altruism *(AL-troo-IZ-um)* Altruistic behavior between unrelated individuals; believed to produce some benefit to the altruistic individual in the future when the current beneficiary reciprocates.

recognition species concept The idea that specific mating adaptations become fixed in a population and form the basis of species identification.

recombinant An offspring whose phenotype differs from that of the parents.

recombinant DNA A DNA molecule made in vitro with segments from different sources.

redox reaction *(REE-doks)* A chemical reaction involving the transfer of one or more electrons from one reactant to another; also called oxidation-reduction reaction.

reducing agent The electron donor in a redox reaction.

reduction The gaining of electrons by a substance involved in a redox reaction.

reflex An automatic reaction to a stimulus, mediated by the spinal cord or lower brain.

refractory period *(ree-FRAK-tor-ee)* The short time immediately after an action potential in which the neuron cannot respond to another stimulus, owing to an increase in potassium permeability.

regulative development A pattern of development, such as that of a mammal, in which the early blastomeres retain the potential to form the entire animal.

relative fitness The contribution of one genotype to the next generation compared to that of alternative genotypes for the same locus.

releaser A signal stimulus that functions as a communication signal between individuals of the same species.

releasing hormone A hormone produced by neurosecretory cells in the hypothalamus of the vertebrate brain that stimulates or inhibits the secretion of hormones by the anterior pituitary.

repetitive DNA Nucleotide sequences, usually noncoding, that are present in many copies in a eukaryotic genome. The repeated units may be short and arranged tandemly (in series) or long and dispersed in the genome.

replication fork A Y-shaped point on a replicating DNA molecule where new strands are growing.

repressible enzyme An enzyme whose synthesis is inhibited by a specific metabolite.

repressor A protein that suppresses the transcription of a gene.

Reptilia The vertebrate class of reptiles, represented by lizards, snakes, turtles, and crocodilians.

resolving power A measure of the clarity of an image; the minimum distance that two points can be separated and still be distinguished as two separate points.

resource partitioning The division of environmental resources by coexisting species populations such that the niche of each species differs by one or more significant factors from the niches of all coexisting species populations.

resting potential The membrane potential characteristic of a nonconducting, excitable cell, with the inside of the cell more negative than the outside.

restriction enzyme A degradative enzyme that recognizes and cuts up DNA (including that of certain phages) that is foreign to a bacterium.

restriction fragment length polymorphisms (RFLPs) Differences in DNA sequence on homologous chromosomes that result in different patterns of restriction fragment lengths (DNA segments resulting from treatment with restriction enzymes); useful as genetic markers for making linkage maps.

restriction site A specific sequence on a DNA strand that is recognized as a "cut site" by a restriction enzyme.

retina *(REH-tin-uh)* The innermost layer of the vertebrate eye, containing photoreceptor cells (rods and cones) and neurons; transmits images formed by the lens to the brain via the optic nerve.

retinal The light-absorbing pigment in rods and cones of the vertebrate eye.

retrovirus *(REH-troh-VY-rus)* An RNA virus that reproduces by transcribing its RNA into DNA and then inserting the DNA into a cellular chromosome; an important class of cancer-causing viruses.

reverse transcriptase *(trans-KRIP-tase)* An enzyme encoded by some RNA viruses that uses RNA as a template for DNA synthesis.

rhodopsin A visual pigment consisting of retinal and opsin. When rhodopsin absorbs light, the retinal changes shape and dissociates from the opsin, after which it is converted back to its original form.

ribonucleic acid (RNA) *(ry-boh-noo-KLAY-ik)* A type of nucleic acid consisting of nucleotide monomers with a ribose sugar and the nitrogenous bases adenine (A), cytosine (C), guanine (G), and uracil (U); usually single-stranded; functions in protein synthesis and as the genome of some viruses.

ribose The sugar component of RNA.

ribosomal RNA (rRNA) The most abundant type of RNA. Together with proteins, it forms the structure of ribosomes that coordinate the sequential coupling of tRNA molecules to the series of mRNA codons.

ribosome A cell organelle constructed in the nucleolus, functioning as the site of protein synthesis in the cytoplasm. Consists of rRNA and protein molecules, which make up two subunits.

ribozyme An enzymatic RNA molecule that catalyzes reactions during RNA splicing.

RNA polymerase *(pul-IM-ur-ase)* An enzyme that links together the growing chain of ribonucleotides during transcription.

RNA processing Modification of RNA before it leaves the nucleus, a process unique to eukaryotes.

RNA splicing The removal of noncoding portions (introns) of the RNA molecule after initial synthesis.

rod cell One of two kinds of photoreceptors in the vertebrate retina; sensitive to black and white and enables night vision.

root cap A cone of cells at the tip of a plant root that protects the apical meristem.

root hair A tiny projection growing just behind the root tips of plants, increasing surface area for the absorption of water and minerals.

root pressure The upward push of water within the stele of vascular plants, caused by active pumping of minerals into the xylem by root cells.

rough ER That portion of the endoplasmic reticulum studded with ribosomes.

R plasmid A bacterial plasmid carrying genes that confer resistance to certain antibiotics.

rubisco Ribulose carboxylase, the enzyme that catalyzes the first step (the addition of CO_2 to RuBP, or ribulose bisphosphate) of the Calvin cycle.

ruminant An animal, such as a cow or a sheep, with an elaborate, multicompartmentalized stomach specialized for an herbivorous diet.

S phase The synthesis phase of the cell cycle, constituting the portion of interphase during which DNA is replicated.

SA (sinoatrial) node The pacemaker of the heart, located in the wall of the right atrium. At the base of the wall separating the two atria is another patch of nodal tissue called the atrioventricular node (AV).

saltatory conduction *(SAHL-tuh-TOR-ee)* Rapid transmission of a nerve impulse along an axon resulting from the action potential jumping from one node of Ranvier to another, skipping the myelin-sheathed regions of membrane.

saprobe An organism that acts as a decomposer by absorbing nutrients from dead organic matter.

sarcomere *(SAR-koh-meer)* The fundamental, repeating unit of striated muscle, delimited by the Z lines.

sarcoplasmic reticulum *(SAR-koh-PLAZ-mik reh-TIK-yoo-lum)* A modified form of endoplasmic reticulum in striated muscle cells that stores calcium used to trigger contraction during stimulation.

saturated fatty acid A fatty acid in which all carbons in the hydrocarbon tail are connected by single bonds, thus maximizing the number of hydrogen atoms that can attach to the carbon skeleton.

savanna *(suh-VAN-uh)* A tropical grassland biome with scattered individual trees, large herbivores, and three distinct seasons based primarily on rainfall, maintained by occasional fires and drought.

Schwann cells A chain of supporting cells enclosing the axons of many neurons and forming an insulating layer called the myelin sheath.

sclereid *(SKLER-ee-id)* A short, irregular sclerenchyma cell in nutshells and seed coats and scattered through the parenchyma of some plants.

sclerenchyma cell *(skler-EN-kim-uh)* A rigid, supportive plant cell type usually lacking protoplasts and possessing thick secondary walls strengthened by lignin at maturity.

second law of thermodynamics The principle whereby every energy transfer or transformation increases the entropy of the universe. Ordered forms of energy are at least partly converted to heat, and in spontaneous reactions, the free energy of the system also decreases.

second messenger A small, nonprotein, water-soluble molecule or ion, such as calcium ion or cyclic AMP, that relays a signal to a cell's interior in response to a signal received by a signal receptor protein.

secondary compound A chemical compound synthesized through the diversion of products of major metabolic pathways for use in defense by prey species.

secondary consumer A member of the trophic level of an ecosystem consisting of carnivores that eat herbivores.

secondary growth The increase in girth of the stems and roots of many plants, especially woody, perennial dicots.

secondary immune response The immune response elicited when an animal encounters the same antigen at some later time. The secondary immune response is more rapid, of greater magnitude, and of longer duration than the primary immune response.

secondary productivity The rate at which all the heterotrophs in an ecosystem incorporate organic material into new biomass, which can be equated to chemical energy.

secondary structure The localized, repetitive coiling or folding of the polypeptide backbone of a protein due to hydrogen bond formation between peptide linkages.

secondary succession A type of succession that occurs where an existing community has been severely cleared by some disturbance.

sedimentary rock *(SED-eh-MEN-tar-ee)* Rock formed from sand and mud that once settled in layers on the bottom of seas, lakes, and marshes. Sedimentary rocks are often rich in fossils.

seed An adaptation for terrestrial plants consisting of an embryo packaged along with a store of food within a resistant coat.

selection coefficient The difference between two fitness values, representing a relative measure of selection against an inferior genotype.

selective permeability A property of biological membranes that allows some substances to cross more easily than others.

self-incompatibility The capability of certain flowers to block fertilization by pollen from the same or a closely related plant.

semen *(SEE-men)* The fluid that is ejaculated by the male during orgasm; contains sperm and secretions from several glands of the male reproductive tract.

semicircular canals A three-part chamber of the inner ear that functions in maintaining equilibrium.

semilunar valve A valve located at the two exits of the heart, where the aorta leaves the left ventricle and the pulmonary artery leaves the right ventricle.

seminiferous tubules *(SEM-in-IF-er-us)* Highly coiled tubes in the testes in which sperm are produced.

sensation An impulse sent to the brain from activated receptors and sensory neurons.

sensory neuron A nerve cell that receives information from the internal and external environments and transmits the signals to the central nervous system.

sensory receptor A specialized structure that responds to specific stimuli from an animal's external or internal environment; transmits the information of an environmental stimulus to the animal's nervous system by converting stimulus energy to the electrochemical energy of action potentials.

sepal *(SEE-pul)* A whorl of modified leaves in angiosperms that encloses and protects the flower bud before it opens.

sex chromosomes The pair of chromosomes responsible for determining the sex of an individual.

sex-linked gene A gene located on a sex chromosome.

sexual dimorphism *(dy-MOR-fiz-um)* A special case of polymorphism based on the distinction between the secondary sex characteristics of males and females.

sexual reproduction A type of reproduction in which two parents give rise to offspring that have unique combinations of genes inherited from the gametes of the two parents.

sexual selection Selection based on variation in secondary sex characteristics, leading to the enhancement of sexual dimorphism.

shoot system The aerial portion of a plant body, consisting of stems, leaves, and flowers.

short-day plant A plant that flowers, usually in late summer, fall, or winter, only when the light period is shorter than a critical length.

sieve-tube member A chain of living cells that form sieve tubes in phloem.

sign stimulus An external sensory stimulus that triggers a fixed action pattern.

signal peptide A stretch of amino acids on polypeptides that targets proteins to specific destinations in eukaryotic cells.

signal-transduction pathway A mechanism linking a mechanical or chemical stimulus to a cellular response.

sink habitat A habitat where mortality exceeds reproduction.

sister chromatids *(KROH-muh-tidz)* Replicated forms of a chromosome joined together by the centromere and eventually separated during mitosis or meiosis II.

skeletal muscle Striated muscle generally responsible for the voluntary movements of the body.

sliding-filament model The theory explaining how muscle contracts, based on change within a sarcomere, the basic unit of muscle organization, stating that thin (actin) filaments slide across thick (myosin) filaments, shortening the sarcomere; the shortening of all sarcomeres in a myofibril shortens the entire myofibril.

small nuclear ribonucleoprotein (snRNP) *(RY-boh-NOO-klee-oh-pro-teen)* One of a variety of small particles in the cell nucleus, composed of RNA and protein molecules; functions are not fully understood, but some form parts of spliceosomes, active in RNA splicing.

smooth ER That portion of the endoplasmic reticulum that is free of ribosomes.

smooth muscle A type of muscle lacking the striations of skeletal and cardiac muscle because of the uniform distribution of myosin filaments in the cell.

sociobiology The study of social behavior based on evolutionary theory.

sodium-potassium pump A special transport protein in the plasma membrane of animal cells that transports sodium out of and potassium into the cell against their concentration gradients.

solute *(SOL-yoot)* A substance that is dissolved in a solution.

solution A homogeneous, liquid mixture of two or more substances.

solvent The dissolving agent of a solution. Water is the most versatile solvent known.

somatic cell *(soh-MAT-ik)* Any cell in a multicellular organism except a sperm or egg cell.

somatic nervous system The branch of the motor division of the vertebrate peripheral nervous system composed of motor neurons that carry signals to skeletal muscles in response to external stimuli.

source habitat A habitat where reproduction exceeds mortality and from which excess individuals disperse.

Southern blotting A hybridization technique that enables researchers to determine the presence of certain nucleotide sequences in a sample of DNA.

speciation *(SPEE-see-AY-shun)* The origin of new species in evolution.

species A particular kind of organism; members possess similar anatomical characteristics and have the ability to interbreed.

species diversity The number and relative abundance of species in a biological community.

species richness The number of species in a biological community.

species selection A theory maintaining that species living the longest and generating the greatest number of species determine the direction of major evolutionary trends.

specific heat The amount of heat that must be absorbed or lost for 1 g of a substance to change its temperature 1°C.

spectrophotometer An instrument that measures the proportions of light of different wavelengths absorbed and transmitted by a pigment solution.

spermatogenesis The continuous and prolific production of mature sperm cells in the testis.

sphincter *(SFINK-ter)* A ringlike valve, consisting of modified muscles in a muscular tube, such as a digestive tract; closes off the tube like a drawstring.

spindle An assemblage of microtubules that orchestrates chromosome movement during eukaryotic cell division.

spiral cleavage A type of embryonic development in protostomes, in which the planes of cell division that transform the zygote into a ball of cells occur obliquely to the polar axis, resulting in cells of each tier sitting in the grooves between cells of adjacent tiers.

spliceosome *(SPLY-see-oh-some)* A complex assembly that interacts with the ends of an RNA intron in splicing RNA; releases an intron and joins two adjacent exons.

sporangium (plural, **sporangia**) A capsule in fungi and plants in which meiosis occurs and haploid spores develop.

spore In the life cycle of a plant or alga undergoing alternation of generations, a meiotically produced haploid cell that divides mitotically, generating a multicellular individual, the gametophyte, without fusing with another cell.

sporophyte The multicellular diploid form in organisms undergoing alternation of generations that results from a union of gametes and that meiotically produces haploid spores that grow into the gametophyte generation.

sporopollenin A secondary product, a polymer synthesized by a side branch of a major metabolic pathway of plants that is resistant to almost all kinds of environmental damage; especially important in the evolutionary move of plants onto land.

stabilizing selection Natural selection that favors intermediate variants by acting against extreme phenotypes.

stamen The pollen-producing male reproductive organ of a flower, consisting of an anther and filament.

starch A storage polysaccharide in plants consisting entirely of glucose.

statocyst *(STAT-eh-SIST)* A type of mechanoreceptor that functions in equilibrium in invertebrates through the use of statoliths, which stimulate hair cells in relation to gravity.

stele The central vascular cylinder in roots where xylem and phloem are located.

stereoisomer A molecule that is a mirror image of another molecule with the same molecular formula.

steroids A class of lipids characterized by a carbon skeleton consisting of four rings with various functional groups attached.

stoma (plural, **stomata**) A microscopic pore surrounded by guard cells in the epidermis of leaves and stems that allows gas exchange between the environment and the interior of the plant.

strict aerobe *(AIR-obe)* An organism that can survive only in an atmosphere of oxygen, which is used in aerobic respiration.

strict anaerobe An organism that cannot survive in an atmosphere of oxygen. Other substances, such as sulfate or nitrate, are the terminal electron acceptors in the electron transport chains that generate their ATP.

stroma The fluid of the chloroplast surrounding the thylakoid membrane; involved in the synthesis of organic molecules from carbon dioxide and water.

stromatolite Rock made of banded domes of sediment in which are found the most ancient forms of life: prokaryotes dating back as far as 3.5 billion years.

structural formula A type of molecular notation in which the constituent atoms are joined by lines representing covalent bonds.

substrate The substance on which an enzyme works.

substrate-level phosphorylation The formation of ATP by directly transferring a phosphate group to ADP from an intermediate substrate in catabolism.

summation A phenomenon of neural integration in which the membrane potential of the postsynaptic cell in a chemical synapse is determined by the total activity of all excitatory and inhibitory presynaptic impulses acting on it at any one time.

suppressor T cell (T_S) A type of T cell that causes B cells as well as other cells to ignore antigens.

surface tension A measure of how difficult it is to stretch or break the surface of a liquid. Water has a high surface tension because of the hydrogen bonding of surface molecules.

survivorship curve A plot of the number of members of a cohort that are still alive at each age; one way to represent age-specific mortality.

suspension-feeder An aquatic animal, such as a clam or a baleen whale, that sifts small food particles from the water.

sustainable agriculture Long-term productive farming methods that are environmentally safe.

sustainable development The long-term prosperity of human societies and the ecosystems that support them.

swim bladder An adaptation, derived from a lung, that enables bony fishes to adjust their density and thereby control their buoyancy.

symbiont *(SIM-by-ont)* The smaller participant in a symbiotic relationship, living in or on the host.

symbiosis An ecological relationship between organisms of two different species that live together in direct contact.

sympathetic division One of two divisions of the autonomic nervous system of vertebrates; generally increases energy expenditure and prepares the body for action.

sympatric speciation A mode of speciation occurring as a result of a radical change in the genome that produces a reproductively isolated subpopulation in the midst of its parent population.

symplast In plants, the continuum of cytoplasm connected by plasmodesmata between cells.

synapomorphies Shared derived characters; homologies that evolved in an ancestor common to all species on one branch of a fork in a cladogram, but not common to species on the other branch.

synapse *(SIN-aps)* The locus where one neuron communicates with another neuron in a neural pathway; a narrow gap between a synaptic terminal of an axon and a signal-receiving portion (dendrite or cell body) of another neuron or effector cell. Neurotransmitter molecules released by synaptic terminals diffuse across the synapse, relaying messages to the dendrite or effector.

synapsis The pairing of replicated homologous chromosomes during prophase I of meiosis.

synaptic terminal A bulb at the end of an axon in which neurotransmitter molecules are stored and released.

syngamy *(SIN-gam-ee)* The process of cellular union during fertilization.

systematics The branch of biology that studies the diversity of life; encompasses taxonomy and is involved in reconstructing phylogenetic history.

systemic acquired resistance (SAR) A defensive response in infected plants that helps protect healthy tissue from pathogenic invasion.

systole *(SIS-toh-lee)* The stage of the heart cycle in which the heart muscle contracts and the chambers pump blood.

T cell A type of lymphocyte responsible for cell-mediated immunity that differentiates under the influence of the thymus.

taiga *(TY-guh)* The coniferous or boreal forest biome, characterized by considerable snow, harsh winters, short summers, and evergreen trees.

taxis *(TAKS-iss)* A movement toward or away from a stimulus.

taxon (plural, **taxa**) The named taxonomic unit at any given level.

taxonomy The branch of biology concerned with naming and classifying the diverse forms of life.

telomerase An enzyme that catalyzes the lengthening of telomeres; the enzyme includes a molecule of RNA that serves as a template for new telomere segments.

telomere The protective structure at each end of a eukaryotic chromosome. Specifically, the tandemly repetitive DNA (*see* repetitive DNA) at the end of the chromosome's DNA molecule.

temperate deciduous forest A biome located throughout midlatitude regions where there is sufficient moisture to support the growth of large, broad-leaf deciduous trees.

temperate virus A virus that can reproduce without killing the host.

temperature A measure of the intensity of heat in degrees, reflecting the average kinetic energy of the molecules.

tendon A type of fibrous connective tissue that attaches muscle to bone.

tertiary consumer A member of a trophic level of an ecosystem consisting of carnivores that eat mainly other carnivores.

tertiary structure *(TUR-shee-air-ee)* Irregular contortions of a protein molecule due to interactions of side chains involved in hydrophobic interactions, ionic bonds, hydrogen bonds, and disulfide bridges.

testcross Breeding of an organism of unknown genotype with a homozygous recessive individual to determine the unknown genotype. The ratio of phenotypes in the offspring determines the unknown genotype.

testis (plural, **testes**) The male reproductive organ, or gonad, in which sperm and reproductive hormones are produced.

testosterone The most abundant androgen hormone in the male body.

tetanus *(TET-un-us)* The maximal, sustained contraction of a skeletal muscle, caused by a very fast frequency of action potentials elicited by continual stimulation.

tetrapod A vertebrate possessing two pairs of limbs, such as amphibians, reptiles, birds, and mammals.

thalamus *(THAL-uh-mus)* One of two integrating centers of the vertebrate forebrain. Neurons with cell bodies in the thalamus relay neural input to specific areas in the cerebral cortex and regulate what information goes to the cerebral cortex.

thermoregulation The maintenance of internal temperature within a tolerable range.

thick filament A filament composed of staggered arrays of myosin molecules; a component of myofibrils in muscle fibers.

thigmomorphogenesis A response in plants to chronic mechanical stimulation, resulting from increased ethylene production; an example is thickening stems in response to strong winds.

thigmotropism *(THIG-moh-TROH-piz-um)* The directional growth of a plant in relation to touch.

threatened species Species that are likely to become endangered in the foreseeable future throughout all or a significant portion of their range.

threshold potential The potential an excitable cell membrane must reach for an action potential to be initiated.

thylakoid *(THY-luh-koid)* A flattened membrane sac inside the chloroplast, used to convert light energy to chemical energy.

thymus *(THY-mus)* An endocrine gland in the neck region of mammals that is active in establishing the immune system; secretes several messengers, including thymosin, that stimulate T cells.

thyroid gland An endocrine gland that secretes iodine-containing hormones (T_3 and T_4), which stimulate metabolism and influence development and maturation in vertebrates, and cacitonin, which lowers blood calcium levels in mammals.

thyroid-stimulating hormone (TSH) A hormone produced by the anterior pituitary that regulates the release of thyroid hormones.

Ti plasmid A plasmid of a tumor-inducing bacterium that integrates a segment of its DNA into the host chromosome of a plant; frequently used as a carrier for genetic engineering in plants.

tight junction A type of intercellular junction in animal cells that prevents the leakage of material between cells.

tissue An integrated group of cells with a common structure and function.

tonoplast A membrane that encloses the central vacuole in a plant cell, separating the cytosol from the cell sap.

torpor In animals, a physiological state that conserves energy by slowing down the heart and respiratory systems.

totipotency The ability of embryonic cells to retain the potential to form all parts of the animal.

trace element An element indispensable for life but required in extremely minute amounts.

trachea *(TRAY-kee-uh)* The windpipe; that portion of the respiratory tube that has C-shaped cartilagenous rings and passes from the larynx to two bronchi.

tracheae *(TRAY-kee-ee)* Tiny air tubes that branch throughout the insect body for gas exchange.

tracheal system A gas exchange system of branched, chitin-lined tubes that infiltrate the body and carry oxygen directly to cells in insects.

tracheid *(TRAY-kee-id)* A water-conducting and supportive element of xylem composed of long, thin cells with tapered ends and walls hardened with lignin.

transcription The synthesis of RNA on a DNA template.

transcription factor A regulatory protein that binds to DNA and stimulates transcription of specific genes.

transfer RNA (tRNA) An RNA molecule that functions as an interpreter between nucleic acid and protein language by picking up specific amino acids and recognizing the appropriate codons in the mRNA.

transformation (1) The conversion of a normal animal cell to a cancerous cell. (2) A phenomenon in which external DNA is assimilated by a cell.

translation The synthesis of a polypeptide using the genetic information encoded in an mRNA molecule. There is a change of "language" from nucleotides to amino acids.

translocation (1) An aberration in chromosome structure resulting from an error in meiosis or from mutagens; attachment of a chromosomal fragment to a nonhomologous chromosome. (2) During protein synthesis, the third stage in the elongation cycle when the RNA carrying the growing polypeptide moves from the A site to the P site on the ribosome. (3) The transport via phloem of food in a plant.

transpiration The evaporative loss of water from a plant.

transposon *(trans-POH-son)* A transposable genetic element; a mobile segment of DNA that serves as an agent of genetic change.

triplet code A set of three-nucleotide-long words that specify the amino acids for polypeptide chains.

triploblastic Possessing three germ layers: the endoderm, mesoderm, and ectoderm. Most eumetazoa are triploblastic.

trophic level The division of species in an ecosystem on the basis of their main nutritional source. The trophic level that ultimately supports all others consists of autotrophs, or primary producers.

trophic structure The different feeding relationships in an ecosystem that determine the route of energy flow and the pattern of chemical cycling.

trophoblast The outer epithelium of the blastocyst, which forms the fetal part of the placenta.

tropic hormone A hormone that has another endocrine gland as a target.

tropical rain forest The most complex of all communities, located near the equator where rainfall is abundant; harbors more species of plants and animals than all other terrestrial biomes combined.

tropism A growth response that results in the curvature of whole plant organs toward or away from stimuli due to differential rates of cell elongation.

tumor A mass that forms within otherwise normal tissue, caused by the uncontrolled growth of a transformed cell.

tumor-suppressor gene A gene whose protein products inhibit cell division, thereby preventing uncontrolled cell growth (cancer).

tundra A biome at the extreme limits of plant growth; at the northernmost limits, it is called arctic tundra, and at high altitudes, where plant forms are limited to low shrubby or matlike vegetation, it is called alpine tundra.

turgid *(TUR-jid)* Firm; walled cells become turgid as a result of the entry of water from a hypotonic environment.

turgor pressure The force directed against a cell wall after the influx of water and the swelling of a walled cell due to osmosis.

tyrosine kinase An enzyme that catalyzes the transfer of phosphate groups from ATP to the amino acid tyrosine in a substrate protein.

tyrosine kinase receptor A receptor protein in the plasma membrane that responds to the binding of a signal molecule by catalyzing the transfer of phosphate groups from ATP to tyrosines on the cytoplasmic side of the receptor. The phosphorylated tyrosines activate other signal-transduction proteins within the cell.

ultimate causation The evolutionary explanation for a behavioral, physiological, or morphological response, in contrast to proximate causation, the immediate mechanisms that underlie a response.

unsaturated fatty acid A fatty acid possessing one or more double bonds between the carbons in the hydrocarbon tail. Such bonding reduces the number of hydrogen atoms attached to the carbon skeleton.

urea A soluble form of nitrogenous waste excreted by mammals and most adult amphibians.

ureter A duct leading from the kidney to the urinary bladder.

urethra A tube that releases urine from the body near the vagina in females or through the penis in males; also serves in males as the exit tube for the reproductive system.

uric acid An insoluble precipitate of nitrogenous waste excreted by land snails, insects, birds, and some reptiles.

urochordate A chordate without a backbone, commonly called a tunicate, a sessile marine animal.

uterus A female organ where eggs are fertilized and/or development of the young occurs.

vaccine A harmless variant or derivative of a pathogen that stimulates a host's immune system to mount defenses against the pathogen.

vacuole A membrane-enclosed sac taking up most of the interior of a mature plant cell and containing a variety of substances important in plant reproduction, growth, and development.

valence shell The outermost energy shell of an atom, containing the valence electrons involved in the chemical reactions of that atom.

Van der Waals interactions Weak attractions between molecules or parts of molecules that are brought about by localized charge fluctuations.

vascular cambium A continuous cylinder of meristematic cells surrounding the xylem and pith that produces secondary xylem and phloem.

vascular plants Plants with vascular tissue, consisting of all modern species except the mosses and their relatives.

vascular tissue Plant tissue consisting of cells joined into tubes that transport water and nutrients throughout the plant body.

vascular tissue system A system formed by xylem and phloem throughout the plant, serving as a transport system for water and nutrients, respectively.

vas deferens The tube in the male reproductive system in which sperm travel from the epididymis to the urethra.

vegetative reproduction Cloning of plants by asexual means.

vein A vessel that returns blood to the heart.

ventilation Any method of increasing contact between the respiratory medium and the respiratory surface.

vertebrate A chordate animal with a backbone: the mammals, birds, reptiles, amphibians, and various classes of fishes.

vessel element A specialized short, wide cell in angiosperms; arranged end to end, they form continuous tubes for water transport.

vestigial organ A type of homologous structure that is rudimentary and of marginal or no use to the organism.

viroid *(VY-roid)* A plant pathogen composed of molecules of naked RNA only several hundred nucleotides long.

visceral muscle Smooth muscle found in the walls of the digestive tract, bladder, arteries, and other internal organs.

visible light That portion of the electromagnetic spectrum detected as various colors by the human eye, ranging in wavelength from about 400 nm to about 700 nm.

vitalism The belief that natural phenomena are governed by a life force outside the realm of physical and chemical laws.

vitamin An organic molecule required in the diet in very small amounts; vitamins serve primarily as coenzymes or parts of coenzymes.

viviparous *(vy-VIP-er-us)* Referring to a type of development in which the young are born alive after having been nourished in the uterus by blood from the placenta.

voltage-gated channel Ion channel in a membrane that opens and closes in response to changes in membrane potential (voltage); the sodium and potassium channels of neurons are examples.

water potential The physical property predicting the direction in which water will flow, governed by solute concentration and applied pressure.

water vascular system A network of hydraulic canals unique to echinoderms that branches into extensions called tube feet, which function in locomotion, feeding, and gas exchange.

wavelength The distance between crests of waves, such as those of the electromagnetic spectrum.

wild type An individual with the normal phenotype.

wobble A violation of the base-pairing rules in that third nucleotide (5' end) of a tRNA anticodon can form hydrogen bonds with more than one kind of base in the third position (3' end) of a codon.

xylem *(ZY-lum)* The tube-shaped, nonliving portion of the vascular system in plants that carries water and minerals from the roots to the rest of the plant.

yeast A unicellular fungus that lives in liquid or moist habitats, primarily reproducing asexually by simple cell division or by budding of a parent cell.

yolk sac One of four extraembryonic membranes that supports embryonic development; the first site of blood cells and circulatory system function.

zoned reserve systems Habitat areas that are protected from human alteration and surrounded by lands that are used and more extensively altered by human activity.

zygote The diploid product of the union of haploid gametes in conception; a fertilized egg.

INDEX

A *t* following a page number indicates a table; an *f* following a page number indicates a figure; a *b* following a page number indicates a box.